# Stay on top of OSHA standards with expert, easy-to-use information.

No one needs to remind you of the critical necessity to make sure your organization is OSHA compliant. In fact, you probably already take great care to comply… yet OSHA regulations are constantly changing. So, while you no doubt stay on top of things, do you know for a fact that you meet the current standards?

A WoltersKluwer Company

With these best-selling OSHA publications, you can stay in compliance and protect your organization. You will be provided with current OSHA standards for your specific industry, along with the most recent updates on OSHA regulations for inspections, citations, and penalties.

For more information see reverse side.

## ORDER YOUR COPIES TODAY—CALL 888-276-5422, OR FAX THIS ORDER FORM TO 800-224-8299  ATTENTION: BOOKS DEPT

### Order Information:

| Qty | Title | Offer Number | Cost per copy | Total |
|---|---|---|---|---|
| | OSHA Standards for General Industry | 0-5865-401 | $48.95 | |
| | OSHA Standards for General Industry CD | 0-1783-301 | $49.95 | |
| | OSHA Pocket Digest of General Standards | 0-5971-301 | $7.95 | |
| | OSHA Standards for the Construction Industry | 0-5864-401 | $39.95 | |
| | OSHA Standards for the Construction Industry CD | 0-1784-301 | $49.95 | |
| | OSHA Pocket Digest of Construction Standards | 0-4479-301 | $7.95 | |
| | What Every Supervisor Must Know About OSHA - Construction | 0-4410-301 | $38.35 | |
| | What Every Supervisor Must Know About OSHA - Federal | 0-4861-301 | $38.35 | |
| | What Every Supervisor Must Know About OSHA - General Industry | 0-4411-301 | $38.35 | |
| | | | GRAND TOTAL | |

For large orders or custom covers, please call 888-276-5422.

Appropriate sales tax, shipping and handling will be added. Prices quoted apply only in the U.S. and are subject to change.

### Shipping Information:

Name & Title

Company

Address

City    State    Zip

Phone    Fax

### Billing Information:

○ Bill me

○ Charge my CCH Account

Account Number

GCY 5102

# Invaluable desktop references to aid you in making sure your company is protected and compliant.

**CCH's OSHA Standards for General Industry** provides the latest OSHA regulations for general industry. In addition to the text of 29 CFR 1910, it contains rules on inspections, citations and penalties, as well as the requirements for recording and reporting occupational injuries and illness.

Also included in **CCH's OSHA Standards for General Industry** are Part 1903 regulations governing inspections, citations, and proposed penalties; Part 1904 regulations covering occupational injury and illness recordkeeping and reporting; and Part 1990 carcinogen policy and model standards. Additionally, 5(a) of the Occupational Safety and Health Act of 1970 (the "General Duty" clause) is incorporated. Contents: Recent Amendments. Important amendments, additions, and other changes to the OSHA standards have been implemented since the last edition. Continuously updated changes to standards are accessible via the Internet. (Also available as a pocket digest.)

**CCH's OSHA Standards for Construction Industry** covers the full text of OSHA construction industry standards (29 CFR 1926), PLUS cross-referenced general industry standards (29 CFR 1910) wherever applicable, with up-to-date coverage of the complete standards and all amendments. In addition to the text of 29 CFR 1926, the volume contains rules on inspections, citations and penalties, as well as the requirements for recording and reporting occupational injuries and illnesses.

For your convenience, the complete text of each general industry section/appendix immediately follows each cross-reference. Plus, with your purchase you'll gain access—via the Internet—to continuous updates to the standards between publication dates. (Also available as a pocket digest.)

All books are printed in Union shops

**What Every Supervisor Must Know about OSHA Federal** outlines steps for organizations to comply with OSHA 29 CFR 1960 regulation for Federal Supervisors and Safety Committees, all in an easy-to-use, "how-to" format. The Federal Book explains OSHA rules, regulations and penalties, but also provides a basis for accident prevention efforts. While OSHA rules and regulations are explained in this book, the emphasis is on accident prevention using established guidelines. Organizations can use the risk management approach to evaluate their facilities and safeguard their workers. The Book also discusses how to handle an OSHA inspection and the Federal Employees Compensation Act.

**What Every Supervisor Must Know about OSHA—General Industry** was designed for use by the OSHA Training Institute Education Centers as part of the curriculum for the OSHA 501 "Train the Trainer" program. This single source book provides managers, supervisors and safety personnel with the tools and knowledge needed to ensure a safe and healthy workplace. It explains the purpose of key OSHA regulations and what actions OSHA rules require or prohibit. Written with an emphasis on accident prevention using established guidelines, this book provides guidance for organizations to ensure compliance with Federal Regulations. It addresses how accidents affect the costs of lost productivity, workers compensation expense, and legal liabilities for managers. Outlined steps are provided for organizations to comply with OSHA regulations and Safety Committees in an easy-to-use format.

**What Every Supervisor Must Know about OSHA—Construction** provides guidance specifically for construction workers to ensure compliance with mandated Federal regulations, emphasizing accident prevention. The book also outlines steps for organizations to comply with OSHA's regulations in an easy-to-use format, presenting topics in a "how to" format, along with numerous quizzes and a sample fall protection plan. This single desktop reference provides the full text of OSHA construction industry standards (29 CFR 1926), PLUS cross-referenced general industry standards (29 CFR 1910) wherever applicable, with up-to-date coverage of the complete standards and all amendments. For your convenience, the complete text of each general industry section/appendix immediately follows each cross-reference.

With the newest OSHA references at your fingertips, there is no better way to protect your organization and avoid costly fines. To order call **888-276-5422**, fax **800-224-8299**, or visit **www.safety.cch.com/OSHA**.

# CCH® OSHA Standards for General Industry

**(29 CFR PART 1910)**
With Amendments as of August 2006

Promulgated by the Occupational
Safety and Health Administration
United States Department of Labor

**CCH Editorial Staff Publication**

Publisher: Paul Gibson, J.D.

Director, HHR Editorial: Jeff Reinholtz, J.D.

Managing Editors: Joy Waltemath, J.D., David Stephanides, J.D.

Coordinating Editor: Debra Levin, J.D.

Production Coordinators: Theresa J. Jensen, Jennifer Wintczak

Cover Design: Craig Arritola

Interior Design: Erika E. Lenz

Layout: Craig Arritola

This publication is designed to provide accurate and authoritative information in regard to the subject matter covered. It is sold with the understanding that the publisher is not engaged in rendering legal, accounting, or other professional service. If legal advice or other expert assistance is required, the services of a competent professional person should be sought.

ISBN 0-8080-1543-5
©2006 **CCH**. All Rights Reserved.
4025 W. Peterson Ave.
Chicago, IL 60646-6085
1 800 248 3248
www.safety.cch.com

No claim is made to original government works; however, the gathering, compilation, magnetic translation, digital conversion, and arrangement of such materials, the historical, statutory, and other notes and references, as well as commentary and materials in this Product or Publication, are subject to CCH's copyright.

All Rights Reserved

Printed in the United States of America

# About This Book

This book contains the occupational safety and health standards for general industry promulgated by the Occupational Safety and Health Administration (OSHA), effective June 2006. Although this is not an official OSHA publication, it reproduces in full the text of regulations contained in the official OSHA government version at Title 29 of the Code of Federal Regulations (CFR) Part 1910. Also included in this book are Part 1903 regulations on inspections, citations, and proposed penalties; Part 1904 regulations covering occupational injury and illness recordkeeping and reporting; and Part 1990 carcinogen policy and model standards. Additionally, §5(a) of the Occupational Safety and Health Act of 1970 (the "General Duty" clause) is incorporated.

**Hexavalent chromium (Cr(VI)).** An amended OSHA standard reduced the 8-hour time-weighted average exposure for hexavalent chromium form a permissible exposure limit (PEL) of 1 milligram per 10 cubic meters of air to 5 micrograms per cubic meter of air. The rules limiting occupational exposure to Cr(VI) also contained requirements for determining exposure, exposure control methods, respiratory protection, protective clothing and equipment, hygiene areas and practices, medical surveillance and recordkeeping. The start-up date allowed up to four years for implementing engineering controls to achieve the new PEL.

**Technical amendments.** OSHA published amendments making corrections to numerous rules to remove references to effective dates in the past and sections that no longer exist or were cited incorrectly. Sections with changes include §§1910.95; 1910.120; 1910.134; 1910.178; 1910.266; 1910.440; 1910.441; 1910.1000; 1910.1001; 1910.1003; 1910.1017; 1910.1018; 1910.1025; 1910.1027; 1910.1029; 1910.1030; 1910.1043; 1910.1044; 1910.1045; 1910.1047; 1910.1048; 1910.1050; and 1910.1051.

**Interpretation letters.** In this edition, we include some of the more recent guidance OSHA has issued in the form of letters of interpretation. Topics covered include machine guarding based on shaft size or speed; employer's obligation to assure accuracy of sharps injury log; interlock requirements for trash compactor; HAZWOPER standard for conditionally exempt small quantity generators; fire brigade standard for health care facilities; revising baseline audiograms; alternative fall protection when guardrails are infeasible; and determining a case as days away or restricted duty.

**Product enhancements.** As part of CCH's ongoing efforts to improve the quality of our products, this book has been redesigned. In response to customer requests, the editors of this book have chosen a brighter, heavier paper and to improve the brightness and quality of the pages. The green highlights emphasizing standard numbers and other important content remain a continuing feature. Sections have been tabbed throughout the book, each section includes a full citation line for easy reference, the reformatted index has been retained, and includes page numbers, graphics have been separated from text by a color border; and the reproduction of the 2004 recordkeeping forms. Furthermore, "Appendices" contains the following elements: Standards Interpretation Letters; OSHA Posters: State Plan information; CPL-2-0.124 Multi-Employer Citation Policy, and much more. Any questions or suggestions for further enhancements to this standards book may be directed to CCH Health and Human Resources customer service team members at 1-800-449-9525.

**Visit our Safety Website.** To meet our customers' needs, the editors at CCH have created a website for the safety professional. As a CCH subscriber, you are entitled to the most current OSHA information available. CCH delivers the information that you need, quickly, efficiently, and in a format that you want. On the CCH Web site you can keep abreast of updated OSHA regulations as soon as the government publishes them in the *Federal Register*. This will allow you to keep your *OSHA Standards Book* current longer. So check the Web site often and stay up to date with the latest developments; and remember to stay safe!

Your new home for all your professional safety needs can be found at:

**www.safety.cch.com**

**For further information.** The CCH Safety website (www.safety.cch.com), CCH EMPLOYMENT SAFETY AND HEALTH GUIDE (monthly), or CCH HUMAN RESOURCES MANAGEMENT—WORKPLACE SAFETY GUIDE (monthly) should be consulted for further court and administrative actions, as well as amendments. Reports for these publications update the OSHA standards, which are reproduced in the CCH EMPLOYMENT SAFETY AND HEALTH GUIDE in their entirety. The On-Line version of the CCH EMPLOYMENT SAFETY AND HEALTH GUIDE also contains the text of the standards for states that have OSHA-approved state plans, selected environmental safety laws, and the complete library of Occupational Safety and Health Review Commission decisions. The casebook series CCH OCCUPATIONAL SAFETY AND HEALTH DECISIONS is issued as an adjunct to the CCH EMPLOYMENT SAFETY AND HEALTH GUIDE. The CCH SAFETY MANAGEMENT HANDBOOK provides practical information for an effective worker safety and health program, including laws, regulations, OSHA standards, and model programs.

June 2006

CCH, a Wolters Kluwer business

# Table of Contents

## Subpart A—General

| Section | | Page |
|---|---|---|
| 1910.1 | Purpose and scope | 1 |
| 1910.2 | Definitions | 1 |
| 1910.3 | Petitions for the issuance, amendment, or repeal of a standard | 1 |
| 1910.4 | Amendments to this part | 1 |
| 1910.5 | Applicability of standards | 1 |
| 1910.6 | Incorporation by reference | 2 |
| 1910.7 | Definition and requirements for a nationally recognized testing laboratory | 6 |
| 1910.7 | APPENDIX A—OSHA Recognition Process for Nationally Recognized Testing Laboratories | 7 |
| 1910.8 | OMB control numbers under the Paperwork Reduction Act | 9 |

## Subpart B—Adoption and Extension of Established Federal Standards

| Section | | Page |
|---|---|---|
| 1910.11 | Scope and purpose | 11 |
| 1910.12 | Construction work | 11 |
| 1910.15 | Shipyard employment | 11 |
| 1910.16 | Longshoring and marine terminals | 11 |
| 1910.17 | Effective dates | 12 |
| 1910.18 | Changes in established Federal standards | 12 |
| 1910.19 | Special provisions for air contaminants | 12 |

## Subpart C—Reserved

| Section | | Page |
|---|---|---|
| 1910.20 | [Removed and Reserved] | 15 |

## Subpart D—Walking-Working Surfaces

| Section | | Page |
|---|---|---|
| 1910.21 | Definitions | 17 |
| 1910.22 | General requirements | 20 |
| 1910.23 | Guarding floor and wall openings and holes | 20 |
| 1910.24 | Fixed industrial stairs | 22 |
| 1910.25 | Portable wood ladders | 22 |
| 1910.26 | Portable metal ladders | 23 |
| 1910.27 | Fixed ladders | 24 |
| 1910.28 | Safety requirements for scaffolding | 27 |
| 1910.29 | Manually propelled mobile ladder stands and scaffolds (towers) | 35 |
| 1910.30 | Other working surfaces | 36 |

## Subpart E—Exit Routes, Emergency Action Plans, and Fire Prevention Plans

| Section | | Page |
|---|---|---|
| 1910.33 | Table of contents | 37 |
| 1910.34 | Coverage and definitions | 37 |
| 1910.35 | Compliance with NFPA 101-2000, Life Safety Code | 37 |
| 1910.36 | Design and construction requirements for exit routes | 37 |
| 1910.37 | Maintenance, safeguards, and operational features for exit routes | 38 |
| 1910.38 | Emergency action plans | 38 |
| 1910.39 | Fire prevention plans | 39 |
| | *Appendix to Subpart E —* Exit Routes, Emergency Action Plans, and Fire Prevention Plans | 39 |

## Subpart F—Powered Platforms, Manlifts, and Vehicle-Mounted Work Platforms

| Section | | Page |
|---|---|---|
| 1910.66 | Powered platforms for building maintenance | 41 |
| 1910.66 | APPENDIX A—Guidelines (Advisory) | 48 |
| 1910.66 | APPENDIX B—Exhibits (Advisory) | 50 |
| 1910.66 | APPENDIX C—Personal Fall Arrest System (Section I—Mandatory; Section II and III—Non-mandatory) | 51 |
| 1910.66 | APPENDIX D—Existing Installations (Mandatory) | 54 |
| 1910.67 | Vehicle-mounted elevating and rotating work platforms | 58 |
| 1910.68 | Manlifts | 59 |

## Subpart G—Occupational Health and Environmental Control

| Section | | Page |
|---|---|---|
| 1910.94 | Ventilation | 63 |
| 1910.95 | Occupational noise exposure | 73 |
| 1910.95 | APPENDIX A—Noise Exposure Computation | 76 |
| 1910.95 | APPENDIX B—Methods for Estimating the Adequacy of Hearing Protector Attenuation | 78 |
| 1910.95 | APPENDIX C—Audiometric Measuring Instruments | 78 |
| 1910.95 | APPENDIX D—Audiometric Test Rooms | 78 |
| 1910.95 | APPENDIX E—Acoustic Calibration of Audiometers | 78 |

| Section | | Page |
|---|---|---|
| 1910.95 | APPENDIX F—Calculations and Application of Age Corrections to Audiograms | 79 |
| 1910.95 | APPENDIX G—Monitoring Noise Levels Non-mandatory Informational Appendix | 80 |
| 1910.95 | APPENDIX H—Availability of Referenced Documents | 81 |
| 1910.95 | APPENDIX I—Definitions | 81 |
| 1910.97 | Nonionizing radiation | 82 |
| 1910.98 | Effective dates | 82 |

## Subpart H—Hazardous Materials

| Section | | Page |
|---|---|---|
| 1910.101 | Compressed gases (general requirements) | 83 |
| 1910.102 | Acetylene | 83 |
| 1910.103 | Hydrogen | 83 |
| 1910.104 | Oxygen | 88 |
| 1910.105 | Nitrous oxide | 89 |
| 1910.106 | Flammable and combustible liquids | 89 |
| 1910.107 | Spray finishing using flammable and combustible materials | 106 |
| 1910.108 | [Reserved] | 111 |
| 1910.109 | Explosives and blasting agents | 111 |
| 1910.110 | Storage and handling of liquefied petroleum gases | 121 |
| 1910.111 | Storage and handling of anhydrous ammonia | 140 |
| 1910.112 | [Reserved] | 148 |
| 1910.113 | [Reserved] | 148 |
| 1910.119 | Process safety management of highly hazardous chemicals | 148 |
| 1910.119 | APPENDIX A— List of Highly Hazardous Chemicals, Toxics and Reactives (Mandatory) | 152 |
| 1910.119 | APPENDIX B—Block Flow Diagram and Simplified Process Flow Diagram (Non-mandatory) | 153 |
| 1910.119 | APPENDIX C—Compliance Guidelines and Recommendations for Process Safety Management (Non-mandatory) | 154 |
| 1910.119 | APPENDIX D—Sources of Further Information (Non-mandatory) | 159 |
| 1910.120 | Hazardous waste operations and emergency response | 159 |
| 1910.120 | APPENDICES—Hazardous Waste Operations and Emergency Response | 171 |
| 1910.120 | APPENDIX A—Personal Protective Equipment Test Methods | 171 |
| 1910.120 | APPENDIX B—General Description and Discussion of the Levels of Protection and Protective Gear | 173 |
| 1910.120 | APPENDIX C—Compliance Guidelines | 175 |
| 1910.120 | APPENDIX D—References | 177 |
| 1910.120 | APPENDIX E—Training Curriculum Guidelines | 178 |
| 1910.121 | [Reserved] | 183 |
| 1910.122 | Table of contents | 183 |
| 1910.123 | Dipping and coating operations: Coverage and definitions | 184 |
| 1910.124 | General requirements for dipping and coating operations | 184 |
| 1910.125 | Additional requirements for dipping and coating operations that use flammable or combustible liquids | 185 |
| 1910.126 | Additional requirements for special dipping and coating operations | 186 |

## Subpart I—Personal Protective Equipment

| Section | | Page |
|---|---|---|
| 1910.132 | General requirements | 187 |
| 1910.133 | Eye and face protection | 187 |
| 1910.134 | Respiratory protection | 188 |
| 1910.134 | APPENDIX A—Fit Testing Procedures (Mandatory) | 194 |
| 1910.134 | APPENDIX B-1—User Seal Check Procedures (Mandatory) | 200 |
| 1910.134 | APPENDIX B-2—Respirator Cleaning Procedures (Mandatory) | 201 |
| 1910.134 | APPENDIX C—OSHA Respirator Medical Evaluation Questionnaire (Mandatory) | 201 |
| 1910.134 | APPENDIX D—Information for Employees Using Respirators When Not Required Under the Standard (Mandatory) | 204 |
| 1910.135 | Head protection | 204 |
| 1910.136 | Foot protection | 204 |
| 1910.137 | Electrical protective equipment | 204 |
| 1910.138 | Hand protection | 206 |
| | *Appendix A to Subpart I —* References for Further Information (Non-mandatory) | 206 |
| | *Appendix B to Subpart I —* Non-mandatory Compliance Guidelines for Hazard Assessment and Personal Protective Equipment Selection | 207 |

# OSHA Standards for General Industry

vii

| Section | | Page |
|---|---|---|

## Subpart J—General Environmental Controls

| 1910.141 | Sanitation | 211 |
| 1910.142 | Temporary labor camps | 212 |
| 1910.143 | [Reserved] | 213 |
| 1910.144 | Safety color code for marking physical hazards | 213 |
| 1910.145 | Specifications for accident prevention signs and tags | 214 |
| 1910.145(f) | APPENDICES—Accident Prevention Tags | 215 |
| 1910.145(f) | APPENDIX A—Recommended Color Coding | 215 |
| 1910.145(f) | APPENDIX B—References for Further Information | 215 |
| 1910.146 | Permit-required confined spaces | 215 |
| 1910.146 | APPENDICES—Permit-required Confined Spaces | 221 |
| 1910.146 | APPENDIX A—Permit-required Confined Space Decision Flow Chart | 221 |
| 1910.146 | APPENDIX B—Procedures for Atmospheric Testing | 221 |
| 1910.146 | APPENDIX C—Examples of Permit-required Confined Space Programs | 221 |
| 1910.146 | APPENDIX D—Sample Permits | 224 |
| 1910.146 | APPENDIX D-2—Entry Permit | 224 |
| 1910.146 | APPENDIX E—Sewer System Entry | 225 |
| 1910.146 | APPENDIX F—Rescue Team or Rescue Service Evaluation Criteria (Non-mandatory) | 225 |
| 1910.147 | The control of hazardous energy (lockout/tagout) | 226 |
| 1910.147 | APPENDIX A—Typical Minimal Lockout Procedure | 230 |

## Subpart K—Medical and First Aid

| 1910.151 | Medical services and first aid | 231 |
| 1910.151 | APPENDIX A—First Aid Kits (Non-mandatory) | 231 |
| 1910.152 | [Reserved] | 231 |

## Subpart L—Fire Protection

| 1910.155 | Scope, application and definitions applicable to this subpart | 233 |
| 1910.156 | Fire brigades | 234 |
| 1910.157 | Portable fire extinguishers | 236 |
| 1910.158 | Standpipe and hose systems | 237 |
| 1910.159 | Automatic sprinkler systems | 238 |
| 1910.160 | Fixed extinguishing systems, general | 238 |
| 1910.161 | Fixed extinguishing systems, dry chemical | 239 |
| 1910.162 | Fixed extinguishing systems, gaseous agent | 239 |
| 1910.163 | Fixed extinguishing systems, water spray and foam | 240 |
| 1910.164 | Fire detection systems | 240 |
| 1910.165 | Employee alarm systems | 240 |
| | *Appendices to Subpart L—* | 241 |
| | *Appendix A to Subpart L—Fire Protection* | 241 |
| | *Appendix B to Subpart L—National Consensus Standards* | 247 |
| | *Appendix C to Subpart L—Fire Protection References for Further Information* | 248 |
| | *Appendix D to Subpart L—Availability of Publications Incorporated by Reference in Section 1910.156 Fire Brigades* | 249 |
| | *Appendix E to Subpart L—Test Methods for Protective Clothing* | 250 |

## Subpart M—Compressed Gas and Compressed Air Equipment

| 1910.166-1910.168 | [Reserved] | 253 |
| 1910.169 | Air receivers | 253 |

## Subpart N—Materials Handling and Storage

| 1910.176 | Handling materials—general | 255 |
| 1910.177 | Servicing multi-piece and single piece rim wheels | 255 |
| 1910.177 | APPENDIX A—Trajectory | 257 |
| 1910.177 | APPENDIX B—Ordering Information for the OSHA Charts | 257 |
| 1910.178 | Powered industrial trucks | 257 |
| 1910.178 | APPENDIX A—Stability of Powered Industrial Trucks (Non-mandatory Appendix to Paragraph (L) of this Section) | 262 |
| 1910.179 | Overhead and gantry cranes | 263 |
| 1910.180 | Crawler locomotive and truck cranes | 269 |

| Section | | Page |
|---|---|---|
| 1910.181 | Derricks | 272 |
| 1910.183 | Helicopters | 276 |
| 1910.184 | Slings | 277 |

## Subpart O—Machinery and Machine Guarding

| Section | | Page |
|---|---|---|
| 1910.211 | Definitions | 293 |
| 1910.212 | General requirements for all machines | 297 |
| 1910.213 | Woodworking machinery requirements | 297 |
| 1910.214 | [Reserved] | 300 |
| 1910.215 | Abrasive wheel machinery | 300 |
| 1910.216 | Mills and calenders in the rubber and plastics industries | 309 |
| 1910.217 | Mechanical power presses | 309 |
| 1910.217 | APPENDIX A—Mandatory Requirements for Certification/Validation of Safety Systems for Presence Sensing Device Initiation of Mechanical Power Presses | 317 |
| 1910.217 | APPENDIX B—Non-mandatory Guidelines for Certification/Validation of Safety Systems for Presence Sensing Device Initiation of Mechanical Power Presses | 320 |
| 1910.217 | APPENDIX C—Mandatory Requirements for OSHA Recognition of Third-party Validation Organizations for the PSDI Standard | 320 |
| 1910.217 | APPENDIX D—Non-mandatory Supplementary Information | 323 |
| 1910.218 | Forging machines | 324 |
| 1910.219 | Mechanical power-transmission apparatus | 325 |

## Subpart P—Hand and Portable Powered Tools and Other Hand-Held Equipment

| Section | | Page |
|---|---|---|
| 1910.241 | Definitions | 329 |
| 1910.242 | Hand and portable powered tools and equipment, general | 330 |
| 1910.243 | Guarding of portable powered tools | 330 |
| 1910.244 | Other portable tools and equipment | 333 |

## Subpart Q—Welding, Cutting and Brazing

| Section | | Page |
|---|---|---|
| 1910.251 | Definitions | 335 |
| 1910.252 | General requirements | 335 |
| 1910.253 | Oxygen-fuel gas welding and cutting | 339 |
| 1910.254 | Arc welding and cutting | 345 |
| 1910.255 | Resistance welding | 346 |

## Subpart R—Special Industries

| Section | | Page |
|---|---|---|
| 1910.261 | Pulp, paper, and paperboard mills | 349 |
| 1910.262 | Textiles | 355 |
| 1910.263 | Bakery equipment | 358 |
| 1910.264 | Laundry machinery and operations | 362 |
| 1910.265 | Sawmills | 362 |
| 1910.266 | Logging operations | 369 |
| 1910.266 | APPENDIX A—First-aid Kits (Mandatory) | 375 |
| 1910.266 | APPENDIX B—First-aid and CPR Training (Mandatory) | 375 |
| 1910.266 | APPENDIX C—Comparable ISO Standards (Non-mandatory) | 376 |
| 1910.267 | [Reserved] | 376 |
| 1910.268 | Telecommunications | 376 |
| 1910.269 | Electric power generation, transmission, and distribution | 384 |
| 1910.269 | APPENDIX A—Flow Charts | 403 |
| 1910.269 | APPENDIX A-1—Application of §1910.269 and Subpart S of this Part to Electrical Installations | 403 |
| 1910.269 | APPENDIX A-2—Application of §1910.269 and Subpart S of this Part to Electrical Safety-related Work Practices | 404 |
| 1910.269 | APPENDIX A-3—Application of §1910.269 and Subpart S of this Part to Tree-trimming Operations | 404 |
| 1910.269 | APPENDIX A-4—Application of §§1910.147, 1910.269 and 1910.333 to Hazardous Energy Control Procedures (Lockout/Tagout) | 404 |
| 1910.269 | APPENDIX A-5—Application of §§1910.146 and 1910.269 to Permit-required Confined Spaces | 405 |
| 1910.269 | APPENDIX B—Working on Exposed Energized Parts | 405 |
| 1910.269 | APPENDIX C—Protection from Step and Touch Potentials | 410 |
| 1910.269 | APPENDIX D—Methods of Inspecting and Testing Wood Poles | 411 |
| 1910.269 | APPENDIX E—Reference Documents | 412 |
| 1910.272 | Grain handling facilities | 412 |
| 1910.272 | APPENDIX A— Grain Handling Facilities | 415 |

# OSHA Standards for General Industry

| Section | | Page |
|---|---|---|
| 1910.272 | APPENDIX B—Grain Handling Facilities | 418 |
| 1910.272 | APPENDIX C—Grain Handling Facilities | 418 |

## Subpart S—Electrical

| Section | | Page |
|---|---|---|
| 1910.301 | Introduction | 419 |
| 1910.302 | Electric utilization systems | 419 |
| 1910.303 | General requirements | 420 |
| 1910.304 | Wiring design and protection | 421 |
| 1910.305 | Wiring methods, components, and equipment for general use | 424 |
| 1910.306 | Specific purpose equipment and installations | 426 |
| 1910.307 | Hazardous (classified) locations | 429 |
| 1910.308 | Special systems | 429 |
| 1910.309-1910.330 | [Reserved] | 431 |
| 1910.331 | Scope | 431 |
| 1910.332 | Training | 431 |
| 1910.333 | Selection and use of work practices | 432 |
| 1910.334 | Use of equipment | 434 |
| 1910.335 | Safeguards for personnel protection | 435 |
| 1910.336-1910.360 | [Reserved] | 435 |
| 1910.361-1910.380 | [Reserved] | 435 |
| 1910.381-1910.398 | [Reserved] | 435 |
| 1910.399 | Definitions applicable to this subpart | 435 |
| | *Appendix A to Subpart S—Reference Documents* | 440 |
| | *Appendix B to Subpart S—[Reserved]* | 440 |
| | *Appendix C to Subpart S—[Reserved]* | 440 |

## Subpart T—Commercial Diving Operations

| Section | | Page |
|---|---|---|
| 1910.401 | Scope and application | 441 |
| 1910.402 | Definitions | 441 |
| 1910.410 | Qualifications of dive team | 442 |
| 1910.420 | Safe practices manual | 442 |
| 1910.421 | Pre-dive procedures | 442 |
| 1910.422 | Procedures during dive | 443 |
| 1910.423 | Post-dive procedures | 443 |
| 1910.424 | SCUBA diving | 444 |
| 1910.425 | Surface-supplied air diving | 444 |
| 1910.426 | Mixed-gas diving | 444 |
| 1910.427 | Liveboating | 445 |
| 1910.430 | Equipment | 445 |
| 1910.440 | Recordkeeping requirements | 446 |
| | *Appendix A to Subpart T— Examples of Conditions Which May Restrict or Limit Exposure to Hyperbaric Conditions* | 446 |
| | *Appendix B to Subpart T— Guidelines for Scientific Diving* | 447 |
| | *Appendix C to Subpart T— Alternative Conditions Under §1910.401(a)(3) for Recreational Diving Instructors and Diving Guides (Mandatory)* | 447 |

## Subpart U—V [Reserved]

| Section | | Page |
|---|---|---|
| 1910.442-1910.899 | [Reserved] | 451 |

## Subpart W—Program Standards [Reserved]

| Section | | Page |
|---|---|---|
| 1910.900 | [Repealed and Reserved] | 451 |

## Subparts X—Y [Reserved]

| Section | | Page |
|---|---|---|
| 1910.901-1910.999 | [Reserved] | 451 |

## Subpart Z—Toxic and Hazardous Substances

| Section | | Page |
|---|---|---|
| 1910.1000 | Air contaminants | 453 |
| 1910.1001 | Asbestos | 462 |
| 1910.1001 | APPENDIX A—OSHA Reference Method—Mandatory | 468 |
| 1910.1001 | APPENDIX B—Detailed Procedures for Asbestos Sampling and Analysis—Non-mandatory | 469 |
| 1910.1001 | APPENDIX C—[Reserved] | 475 |
| 1910.1001 | APPENDIX D—Medical Questionnaires; Mandatory | 475 |
| 1910.1001 | APPENDIX E—Interpretation and Classification of Chest Roentgenograms—Mandatory | 478 |
| 1910.1001 | APPENDIX F—Work Practices and Engineering Controls for Automotive Brake and Clutch Inspection, Disassembly, Repair and Assembly—Mandatory | 478 |

| Section | | Page |
|---|---|---|
| 1910.1001 | APPENDIX G—Substance Technical Information for Asbestos—Non-mandatory | 478 |
| 1910.1001 | APPENDIX H—Medical Surveillance Guidelines for Asbestos—Non-mandatory | 479 |
| 1910.1001 | APPENDIX I—Smoking Cessation Program Information for Asbestos—Non-mandatory | 480 |
| 1910.1001 | APPENDIX J—Polarized Light Microscopy of Asbestos—Non-mandatory | 480 |
| 1910.1002 | Coal tar pitch volatiles; interpretation of term | 488 |
| 1910.1003 | 13 Carcinogens (4-Nitrobiphenyl, etc ) | 488 |
| 1910.1004 | alpha-Naphthylamine | 490 |
| 1910.1005 | [Reserved] | 490 |
| 1910.1006 | Methyl chloromethyl ether | 490 |
| 1910.1007 | 3,3'-Dichlorobenzidine (and its salts) | 490 |
| 1910.1008 | bis-Chloromethyl ether | 490 |
| 1910.1009 | beta-Naphthylamine | 490 |
| 1910.1010 | Benzidine | 490 |
| 1910.1011 | 4-Aminodiphenyl | 490 |
| 1910.1012 | Ethyleneimine | 490 |
| 1910.1013 | beta-Propiolactone | 490 |
| 1910.1014 | 2-Acetylaminofluorene | 490 |
| 1910.1015 | 4-Dimethylaminoazo-benzene | 490 |
| 1910.1016 | N-Nitrosodimethylamine | 490 |
| 1910.1017 | Vinyl chloride | 490 |
| 1910.1017 | APPENDIX A—Supplementary Medical Information | 493 |
| 1910.1018 | Inorganic arsenic | 493 |
| 1910.1018 | APPENDIX A—Inorganic Arsenic Substance Information Sheet | 498 |
| 1910.1018 | APPENDIX B—Substance Technical Guidelines | 499 |
| 1910.1018 | APPENDIX C—Medical Surveillance Guidelines | 499 |
| 1910.1020 | Access to employee exposure and medical records | 500 |
| 1910.1020 | APPENDIX A—Sample Authorization Letter for the Release of Employee Medical Record Information to a Designated Representative (Non-mandatory) | 504 |
| 1910.1020 | APPENDIX B—Availability of NIOSH Registry of Toxic Effects of Chemical Substances (RTECS) (Non-mandatory) | 504 |
| 1910.1025 | Lead | 505 |
| 1910.1025 | APPENDIX A—Substance Data Sheet for Occupational Exposure to Lead | 511 |
| 1910.1025 | APPENDIX B—Employee Standard Summary | 513 |
| 1910.1025 | APPENDIX C—Medical Surveillance Guidelines | 516 |
| 1910.1026 | Chromium (VI) | 522 |
| 1910.1027 | Cadmium | 525 |
| 1910.1027 | APPENDIX A—Substance Safety Data Sheet | 534 |
| 1910.1027 | APPENDIX B—Substance Technical Guidelines for Cadmium | 539 |
| 1910.1027 | APPENDIX C—[Reserved] | 541 |
| 1910.1027 | APPENDIX D—Occupational Health History Interview with Reference to Cadmium Exposure | 541 |
| 1910.1027 | APPENDIX E—Cadmium in Workplace Atmospheres | 543 |
| 1910.1027 | APPENDIX F—Non-mandatory Protocol for Biological Monitoring | 552 |
| 1910.1028 | Benzene | 581 |
| 1910.1028 | APPENDIX A—Substance Safety Data Sheet, Benzene | 585 |
| 1910.1028 | APPENDIX B—Substance Technical Guidelines, Benzene | 586 |
| 1910.1028 | APPENDIX C—Medical Surveillance Guidelines for Benzene | 587 |
| 1910.1028 | APPENDIX D—Sampling and Analytical Methods for Benzene Monitoring and Measurement Procedures | 588 |
| 1910.1029 | Coke oven emissions | 591 |
| 1910.1029 | APPENDIX A—Coke Oven Emissions Substance Information Sheet | 597 |
| 1910.1029 | APPENDIX B—Industrial Hygiene and Medical Surveillance Guidelines | 598 |
| 1910.1030 | Bloodborne pathogens | 598 |
| 1910.1030 | APPENDIX A—Hepatitis B Vaccine Declination (Mandatory) | 605 |
| 1910.1043 | Cotton dust | 605 |
| 1910.1043 | APPENDIX A—Air Sampling and Analytical Procedures for Determining Concentrations of Cotton Dust | 609 |
| 1910.1043 | APPENDIX B-I—Respiratory Questionnaire | 610 |
| 1910.1043 | APPENDIX B-II—Respiratory Questionnaire for Non-Textile Workers for the Cotton Industry | 612 |

# OSHA Standards for General Industry

| Section | | Page |
|---|---|---|
| 1910.1043 | APPENDIX B-III—Abbreviated Respiratory Questionnaire | **613** |
| 1910.1043 | APPENDIX C—Spirometry Prediction Tables for Normal Males and Females | **615** |
| 1910.1043 | APPENDIX D—Pulmonary Function Standards for Cotton Dust Standard | **622** |
| 1910.1043 | APPENDIX E—Vertical Elutriator Equivalency Protocol | **623** |
| 1910.1044 | 1,2-dibromo-3-chloropropane | **624** |
| 1910.1044 | APPENDIX A—Substance Safety Data Sheet for DBCP | **627** |
| 1910.1044 | APPENDIX B—Substance Technical Guidelines for DBCP | **628** |
| 1910.1044 | APPENDIX C—Medical Surveillance Guidelines for DBCP | **629** |
| 1910.1045 | Acrylonitrile | **630** |
| 1910.1045 | APPENDIX A—Substance Safety Data Sheet for Acrylonitrile | **634** |
| 1910.1045 | APPENDIX B—Substance Technical Guidelines for Acrylonitrile | **635** |
| 1910.1045 | APPENDIX C—Medical Surveillance Guidelines for Acrylonitrile | **637** |
| 1910.1045 | APPENDIX D—Sampling and Analytical Methods for Acrylonitrile | **637** |
| 1910.1047 | Ethylene oxide | **641** |
| 1910.1047 | APPENDIX A—Substance Safety Data Sheet for Ethylene Oxide (Non-mandatory) | **645** |
| 1910.1047 | APPENDIX B—Substance Technical Guidelines for Ethylene Oxide (Non-mandatory) | **647** |
| 1910.1047 | APPENDIX C—Medical Surveillance Guidelines for Ethylene Oxide (Non-Mandatory) | **648** |
| 1910.1047 | APPENDIX D—Sampling and Analytical Methods for Ethylene Oxide (Non-mandatory) | **649** |
| 1910.1048 | Formaldehyde | **654** |
| 1910.1048 | APPENDIX A—Substance Technical Guidelines for Formalin | **658** |
| 1910.1048 | APPENDIX B—Sampling Strategy and Analytical Methods for Formaldehyde | **661** |
| 1910.1048 | APPENDIX C—Medical Surveillance—Formaldehyde | **665** |
| 1910.1048 | APPENDIX D—Non-mandatory Medical Disease Questionnaire | **667** |
| 1910.1050 | Methylenedianiline | **668** |
| 1910.1050 | APPENDIX A—Substance Data Sheet, for 4,4'-methylenedianiline | **674** |
| 1910.1050 | APPENDIX B—Substance Technical Guidelines, MDA | **675** |
| 1910.1050 | APPENDIX C—Medical Surveillance Guidelines for MDA | **676** |
| 1910.1050 | APPENDIX D—Sampling and Analytical Methods For MDA Monitoring and Measurement Procedures | **676** |
| 1910.1051 | 1,3-Butadiene | **678** |
| 1910.1051 | APPENDIX A—Substance Safety Data Sheet for 1,3-Butadiene (Non-mandatory) | **683** |
| 1910.1051 | APPENDIX B—Substance Technical Guidelines for 1,3-Butadiene (Non-mandatory) | **684** |
| 1910.1051 | APPENDIX C—Medical Screening and Surveillance for 1,3-Butadiene (Non-mandatory) | **685** |
| 1910.1051 | APPENDIX D—Sampling and Analytical Method for 1,3-Butadiene (Non-mandatory) | **686** |
| 1910.1051 | APPENDIX E—[Reserved] | **689** |
| 1910.1051 | APPENDIX F—Medical Questionnaires (Non-mandatory) | **689** |
| 1910.1052 | Methylene Chloride | **689** |
| 1910.1052 | APPENDIX A—Substance Safety Data Sheet and Technical Guidelines for Methylene Chloride | **697** |
| 1910.1052 | APPENDIX B—Medical Surveillance for Methylene Chloride | **700** |
| 1910.1052 | APPENDIX C—Questions and Answers—Methylene Chloride Control in Furniture Stripping | **703** |
| 1910.1096 | Ionizing radiation | **706** |
| 1910.1200 | Hazard communication | **710** |
| 1910.1200 | APPENDIX A—Health Hazard Definitions (Mandatory) | **717** |
| 1910.1200 | APPENDIX B—Hazard Determination (Mandatory) | **718** |
| 1910.1200 | APPENDIX C—[Reserved] | **718** |
| 1910.1200 | APPENDIX D—Definition of "Trade Secret" (Mandatory) | **718** |
| 1910.1200 | APPENDIX E—Guidelines for Employer Compliance (Advisory) | **719** |
| 1910.1201 | Retention of DOT markings, placards and labels | **722** |
| 1910.1450 | Occupational exposure to hazardous chemicals in laboratories | **722** |
| 1910.1450 | APPENDIX A—National Research Council Recommendations Concerning Chemical Hygiene in Laboratories (Non-mandatory) | **725** |

| Section | | Page |
|---|---|---|
| 1910.1450 | APPENDIX B—References (Non-mandatory) | 730 |

## Occupational Safety and Health Act, Sec. 5. Duties

| | | |
|---|---|---|
| | "General Duty Clause" | 733 |

## 29 CFR Part 1903—Inspections, Citations, and Proposed Penalties

| Section | | Page |
|---|---|---|
| 1903.1 | Purpose and scope | 735 |
| 1903.2 | Posting of notice; availability of Act, regulations and applicable standards | 735 |
| 1903.3 | Authority for inspection | 735 |
| 1903.4 | Objection to inspection | 735 |
| 1903.5 | Entry not a waiver | 736 |
| 1903.6 | Advance notice of inspections | 736 |
| 1903.7 | Conduct of inspections | 736 |
| 1903.8 | Representatives of employers and employees | 736 |
| 1903.9 | Trade secrets | 737 |
| 1903.10 | Consultation with employees | 737 |
| 1903.11 | Complaints by employees | 737 |
| 1903.12 | Inspection not warranted; informal review | 737 |
| 1903.13 | Imminent danger | 737 |
| 1903.14 | Citations; notices of *de minimis* violations; policy regarding employee rescue activities | 737 |
| 1903.14a | Petitions for modification of abatement date | 738 |
| 1903.15 | Proposed penalties | 738 |
| 1903.16 | Posting of citations | 739 |
| 1903.17 | Employer and employee contests before the Review Commission | 739 |
| 1903.18 | Failure to correct a violation for which a citation has been issued | 739 |
| 1903.19 | Abatement verification | 739 |
| 1903.19 | APPENDICES—Abatement Verification | 740 |
| 1903.19 | APPENDIX A—Sample Abatement-Certification Letter (Non-mandatory) | 741 |
| 1903.19 | APPENDIX B—Sample Abatement Plan or Progress Report (Non-mandatory) | 741 |
| 1903.19 | APPENDIX C—Sample Warning Tag (Non-mandatory) | 741 |

| Section | | Page |
|---|---|---|
| 1903.20 | Informal conferences | 741 |
| 1903.21 | State administration | 741 |
| 1903.22 | Definitions | 741 |

## 29 CFR Part 1904—Recording and Reporting Occupational Injuries and Illnesses

| | | |
|---|---|---|
| 1904.0 | Purpose | 743 |
| 1904.1 | Partial exemption for employers with 10 or fewer employees | 743 |
| 1904.2 | Partial exemption for establishments in certain industries | 743 |
| 1904.3 | Keeping records for more than one agency | 743 |
| 1904.4 | Recording criteria | 744 |
| 1904.5 | Determination of work-relatedness | 744 |
| 1904.6 | Determination of new cases | 745 |
| 1904.7 | General recording criteria | 746 |
| 1904.8 | Recording criteria for needlestick and sharps injuries | 748 |
| 1904.9 | Recording criteria for cases involving medical removal under OSHA standards | 748 |
| 1904.10 | Recording criteria for cases involving occupational hearing loss | 748 |
| 1904.11 | Recording criteria for work-related tuberculosis cases | 749 |
| 1904.13-1904.28 | [Reserved] | 749 |
| 1904.29 | Forms | 749 |
| 1904.30 | Multiple business establishments | 750 |
| 1904.31 | Covered employees | 750 |
| 1904.32 | Annual summary | 750 |
| 1904.33 | Retention and updating | 751 |
| 1904.34 | Change in business ownership | 751 |
| 1904.35 | Employee involvement | 751 |
| 1904.36 | Prohibition against discrimination | 751 |
| 1904.37 | State recordkeeping regulations | 751 |
| 1904.38 | Variances from the recordkeeping rule | 751 |
| 1904.39 | Reporting fatalities and multiple hospitalization incidents to OSHA | 752 |
| 1904.40 | Providing records to government representatives | 752 |
| 1904.41 | Annual OSHA injury and illness survey of ten or more employers | 753 |

# OSHA Standards for General Industry

| Section | | Page |
|---|---|---|
| 1904.42 | Requests from the Bureau of Labor Statistics for data | **753** |
| 1904.43 | Summary and posting of the 2001 data | **753** |
| 1904.44 | Retention and updating of old forms | **753** |
| 1904.45 | OMB control numbers under the Paperwork Reduction Act | **753** |
| 1904.46 | Definitions | **753** |

## 29 CFR Part 1990—Identification, Classification, and Regulation of Potential Occupational Carcinogens

| Section | | Page |
|---|---|---|
| 1990.101 | Scope | **755** |
| 1990.102 | Purpose | **755** |
| 1990.103 | Definitions | **755** |
| 1990.104 | Scientific review panel | **755** |
| 1990.105 | Advisory committees | **755** |
| 1990.106 | Amendments to this policy | **755** |
| 1990.111 | General statement of regulatory policy | **756** |
| 1990.112 | Classification of potential carcinogens | **756** |
| 1990.121 | Candidate list of potential occupational carcinogens | **756** |
| 1990.122 | Response to petitions | **757** |
| 1990.131 | Priority lists for regulating potential occupational carcinogens | **757** |
| 1990.132 | Factors to be considered | **757** |
| 1990.133 | Publication | **757** |
| 1990.141 | Advance notice of proposed rulemaking | **757** |
| 1990.142 | Initiation of a rulemaking | **758** |
| 1990.143 | General provisions for the use of human and animal data | **758** |
| 1990.144 | Criteria for consideration of arguments on certain issues | **759** |
| 1990.145 | Consideration of substantial new issues or substantial new evidence | **759** |
| 1990.146 | Issues to be considered in the rulemaking | **760** |
| 1990.147 | Final action | **760** |
| 1990.151 | Model standard pursuant to section 6(b) of the Act | **761** |
| 1990.152 | Model emergency temporary standard pursuant to section 6(c) of the Act | **764** |

## Appendices

| | Page |
|---|---|
| William Steiger Act | **769** |
| Part 1977 – Discrimination Against Employees Exercising Rights Under The Williams-Steiger Occupational Safety And Health Act of 1970 | **771** |
| Part 1980 – Procedures for the Handling of Discrimination Complaints Under Section 806 of the Corporate and Criminal Fraud Accountability Act of 2002, Title VIII of the Sarbanes-Oxley Act of 2002 | **775** |
| OSHA State, Regional, and National Offices | **780** |
| OSHA Safety Poster (English and Spanish) | **782** |
| OSHA 300 forms | **784** |
| OSHA Training Institute Education Centers | **795** |
| Multi-Employer Citation Policy | **797** |
| Emergency Preparedness and Response Guide | **801** |
| Safety and Health Management Voluntary Guidelines | **807** |
| Local Emphasis Programs | **815** |

## Subject Index for 29 CFR 1910, Sec. 5, 1903, 1904, and 1990—Occupational Safety and Health Standards

| | Page |
|---|---|
| Index | **819** |

## Appendices—Interpretation Letters

§§1910.219(c)(2)(i); 1910.219(c)(5)

01/09/2006 – OSHA's machine guarding requirements provide no exemption from guarding based on shaft size or speed ... **851**

§§1910.1030; 1910.1030(c)(1)(iv); 1910.1030(f)(3); 1910.1030(h)(5)

01/12/2006 – Employer's obligation to assure the accuracy of the sharps injury log ... **851**

§§1910.147; 1910.147(a)(2)(ii); 1910.212

01/18/2006 – Interlock requirements for access door(s) of trash compactor ... **852**

§§1910.120; 1910.120(a)(2)(iii); 1910.120(p)(8)

02/01/2006 – Applicability of HAZWOPER standard to hazardous waste generators that are not conditionally exempt as small quantity generators ... **853**

§§1910.156; 1910.156(a)

02/06/2006 – Applicability of the Fire brigade standard to health care facilities ... **854**

| Section | Page |
|---|---|
| §§1910.95; 1910.95(g)(9) | |
| 02/10/2006 – Clarification on revising a baseline audiogram if a standard threshold shift occurs in only one ear | **854** |

| Section | Page |
|---|---|
| §§1910.23; 1910.23(c) | |
| Alternative fall protection program when guardrails are infeasible to install | **855** |

# Subpart A—General

> **Authority:** Secs. 4, 6, 8, Occupational Safety and Health Act of 1970 (29 U.S.C. 653, 655, 657); Secretary of Labor's Order Numbers 12-71 (36 FR 8754), 8-76 (41 FR 25059), 9-83 (48 FR 35736), 1-90 (55 FR 9033), or 6-96 (62 FR 111), as applicable.
>
> Sections 1910.7 and 1910.8 also issued under 29 CFR Part 1911. Section 1910.7(f) also issued under 31 U.S.C. 9701, 29 U.S.C. 9a, 5 U.S.C. 553; Pub. L. 106-113 (113 Stat. 1501A-222); and OMB Circular A-25 (dated July 8, 1993) (58 FR 38142, July 15, 1993).

## §1910.1 Purpose and scope.

**1910.1(a)** Section 6(a) of the Williams-Steiger Occupational Safety and Health Act of 1970 (84 Stat. 1593) provides that "without regard to chapter 5 of title 5, United States Code, or to the other subsections of this section, the Secretary shall, as soon as practicable during the period beginning with the effective date of this Act and ending 2 years after such date, by rule promulgate as an occupational safety or health standard any national consensus standard, and any established Federal standard, unless he determines that the promulgation of such a standard would not result in improved safety or health for specifically designated employees." The legislative purpose of this provision is to establish, as rapidly as possible and without regard to the rule-making provisions of the Administrative Procedure Act, standards with which industries are generally familiar, and on whose adoption interested and affected persons have already had an opportunity to express their views. Such standards are either (1) national consensus standards on whose adoption affected persons have reached substantial agreement, or (2) Federal standards already established by Federal statutes or regulations.

**1910.1(b)** This part carries out the directive to the Secretary of Labor under section 6(a) of the Act. It contains occupational safety and health standards which have been found to be national consensus standards or established Federal standards.

## §1910.2 Definitions.

As used in this part, unless the context clearly requires otherwise:

**1910.2(a)** *Act* means the Williams-Steiger Occupational Safety and Health Act of 1970 (84 Stat. 1590).

**1910.2(b)** *Assistant Secretary of Labor* means the Assistant Secretary of Labor for Occupational Safety and Health;

**1910.2(c)** *Employer* means a person engaged in a business affecting commerce who has employees, but does not include the United States or any State or political subdivision of a State;

**1910.2(d)** *Employee* means an employee of an employer who is employed in a business of his employer which affects commerce;

**1910.2(e)** *Commerce* means trade, traffic, commerce, transportation, or communication among the several States, or between a State and any place outside thereof, or within the District of Columbia, or a possession of the United States (other than the Trust Territory of the Pacific Islands), or between points in the same State but through a point outside thereof;

**1910.2(f)** *Standard* means a standard which requires conditions, or the adoption or use of one or more practices, means, methods, operations, or processes, reasonably necessary or appropriate to provide safe or healthful employment and places of employment;

**1910.2(g)** *National consensus standard* means any standard or modification thereof which (1) has been adopted and promulgated by a nationally recognized standards-producing organization under procedures whereby it can be determined by the Secretary of Labor or by the Assistant Secretary of Labor that persons interested and affected by the scope or provisions of the standard have reached substantial agreement on its adoption, (2) was formulated in a manner which afforded an opportunity for diverse views to be considered, and (3) has been designated as such a standard by the Secretary or the Assistant Secretary, after consultation with other appropriate Federal agencies; and

**1910.2(h)** *Established Federal standard* means any operative standard established by any agency of the United States and in effect on April 28, 1971, or contained in any Act of Congress in force on the date of enactment of the Williams-Steiger Occupational Safety and Health Act.

## §1910.3 Petitions for the issuance, amendment, or repeal of a standard.

**1910.3(a)** Any interested person may petition in writing the Assistant Secretary of Labor to promulgate, modify, or revoke a standard. The petition should set forth the terms or the substance of the rule desired, the effects thereof if promulgated, and the reasons therefor.

**1910.3(b)(1)** The relevant legislative history of the Act indicates congressional recognition of the American National Standards Institute and the National Fire Protection Association as the major sources of national consensus standards. National consensus standards adopted on May 29, 1971, pursuant to section 6(a) of the Act are from those two sources. However, any organization which deems itself a producer of national consensus standards, within the meaning of section 3(9) of the Act, is invited to submit in writing to the Assistant Secretary of Labor at any time prior to February 1, 1973, all relevant information which may enable the Assistant Secretary to determine whether any of its standards satisfy the requirements of the definition of "national consensus standard" in section 3(9) of the Act.

**1910.3(b)(2)** Within a reasonable time after the receipt of a submission pursuant to paragraph (b)(1) of this section, the Assistant Secretary of Labor shall publish or cause to be published in the **FEDERAL REGISTER** a notice of such submission, and shall afford interested persons a reasonable opportunity to present written data, views, or arguments with regard to the question whether any standards of the organization making the submission are national consensus standards.

## §1910.4 Amendments to this part.

**1910.4(a)** The Assistant Secretary of Labor shall have all of the authority of the Secretary of Labor under sections 3(9) and 6(a) of the Act.

**1910.4(b)** The Assistant Secretary of Labor may at any time before April 28, 1973, on his own motion or upon the written petition of any person, by rule promulgate as a standard any national consensus standard and any established Federal standard, pursuant to and in accordance with section 6(a) of the Act, and, in addition, may modify or revoke any standard in this part 1910. In the event of conflict among any such standards, the Assistant Secretary of Labor shall take the action necessary to eliminate the conflict, including the revocation or modification of a standard in this part, so as to assure the greatest protection of the safety or health of the affected employees.

## §1910.5 Applicability of standards.

**1910.5(a)** Except as provided in paragraph (b) of this section, the standards contained in this part shall apply with respect to employments performed in a workplace in a State, the District of Columbia, the Commonwealth of Puerto Rico, the Virgin Islands, American Samoa, Guam, Trust Territory of the Pacific Islands, Wake Island, Outer Continental Shelf lands defined in the Outer Continental Shelf Lands Act, Johnston Island, and the Canal Zone.

**1910.5(b)** None of the standards in this part shall apply to working conditions of employees with respect to which Federal agencies other than the Department of Labor, or State agencies acting under section 274 of the Atomic Energy Act of 1954, as amended (42 U.S.C. 2021), exercise statutory authority to prescribe or enforce standards or regulations affecting occupational safety or health.

**1910.5(c)(1)** If a particular standard is specifically applicable to a condition, practice, means, method, operation, or process, it shall prevail over any different general standard which might otherwise be applicable to the same condition, practice, means, method, operation, or process. For example, §1915.23(c)(3) of this title prescribes personal protective equipment for certain ship repairmen working in specified areas. Such a standard shall apply, and shall not be deemed modified nor superseded by any different general standard whose provisions might otherwise be applicable, to the ship repairmen working in the areas specified in §1915.23(c)(3).

**1910.5(c)(2)** On the other hand, any standard shall apply according to its terms to any employment and place of employment in any industry, even though particular standards are also prescribed for the industry, as in subpart B or subpart R of this part, to the extent that none of such particular standards applies. To illustrate, the general standard regarding noise exposure in §1910.95 applies to employments and places of employment in pulp, paper, and paperboard mills covered by §1910.261.

**1910.5(d)** In the event a standard protects on its face a class of persons larger than employees, the standard shall be applicable under this part only to employees and their employment and places of employment.

**1910.5(e)** [Reserved]

**1910.5(f)** An employer who is in compliance with any standard in this part shall be deemed to be in compliance with the requirement of section 5(a)(1) of the Act, but only to the extent of the condition, practice, means, method, operation, or process covered by the standard.

[39 FR 23502, June 27, 1974, as amended at 58 FR 35308, June 30, 1993]

### §1910.6 Incorporation by reference.

**1910.6(a)(1)** The standards of agencies of the U.S. Government, and organizations which are not agencies of the U.S. Government which are incorporated by reference in this part, have the same force and effect as other standards in this part. Only the mandatory provisions (i.e., provisions containing the word "shall" or other mandatory language) of standards incorporated by reference are adopted as standards under the Occupational Safety and Health Act.

**1910.6(a)(2)** Any changes in the standards incorporated by reference in this part and an official historic file of such changes are available for inspection at the national office of the Occupational Safety and Health Administration, U.S. Department of Labor, Washington, DC 20210.

**1910.6(a)(3)** The materials listed in paragraphs (b) through (w) of this section are incorporated by reference in the corresponding sections noted as they exist on the date of the approval, and a notice of any change in these materials will be published in the FEDERAL REGISTER. These incorporations by reference were approved by the Director of the Federal Register in accordance with 5 U.S.C. 552(a) and 1 CFR part 51.

**1910.6(a)(4)** Copies of the following standards that are issued by the respective private standards organizations may be obtained from the issuing organizations. The materials are available for purchase at the corresponding addresses of the private standards organizations noted below. In addition, all are available for inspection at the Office of the Federal Register, 800 North Capitol Street, NW., suite 700, Washington DC, and through the OSHA Docket Office, room N2625, U.S. Department of Labor, 200 Constitution Ave., Washington, DC 20210, or any of its regional offices.

**1910.6(b)** The following material is available for purchase from the American Conference of Governmental Industrial Hygienists (ACGIH), 1014 Broadway, Cincinnati OH 45202:

**1910.6(b)(1)** "Industrial Ventilation: A Manual of Recommended Practice" (22nd ed., 1995), incorporation by reference (IBR) approved for §1910.124(b)(4)(iii).

**1910.6(b)(2)** Threshold Limit Values and Biological Exposure Indices for 1986-87 (1986), IBR approved for §1910.120, PEL definition.

**1910.6(c)** The following material is available for purchase from the American Society of Agricultural Engineers (ASAE), 2950 Niles Road, Post Office Box 229, St. Joseph, MI 49085:

**1910.6(c)(1)** ASAE Emblem for Identifying Slow Moving Vehicles, ASAE S276.2 (1968), IBR approved for §1910.145(d)(10).

**1910.6(c)(2)** [Reserved]

**1910.6(d)** The following material is available for purchase from the Agriculture Ammonia Institute-Rubber Manufacturers (AAI-RMA) Association, 1400 K St. NW, Washington DC 20005:

**1910.6(d)(1)** AAI-RMA Specifications for Anhydrous Ammonia Hose, IBR approved for §1910.111(b)(8)(i).

**1910.6(d)(2)** [Reserved]

**1910.6(e)** The following material is available for purchase from the American National Standards Institute (ANSI), 11 West 42nd St., New York, NY 10036:

**1910.6(e)(1)** ANSI A10.2-44 Safety Code for Building Construction, IBR approved for §1910.144(a)(1)(ii).

**1910.6(e)(2)** ANSI A10.3-70 Safety Requirements for Explosive-Actuated Fastening Tools, IBR approved for §1910.243(d)(1)(i).

**1910.6(e)(3)** ANSI A11.1-65 (R 70) Practice for Industrial Lighting, IBR approved for §§1910.219(c)(5)(iii); 1910.261(a)(3)(i), (c)(10), and (k)(21); and 1910.265(c)(2).

**1910.6(e)(4)** ANSI A11.1-65 Practice for Industrial Lighting, IBR approved for §§1910.262(c)(6) and 1910.265(d)(2)(i)(a).

**1910.6(e)(5)** ANSI A12.1-67 Safety Requirements for Floor and Wall Openings, Railings, and Toe Boards, IBR approved for §§1910.66 appendix D, (c)(4); 1910.68(b)(4) and (b)(8)(ii); 1910.261(a)(3)(ii), (b)(3), (c)(3)(i), (c)(15)(ii), (e)(4), (g)(13), (h)(1), (h)(3)(vi), (j)(4) (ii) and (iv), (j)(5)(i), (k)(6), (k)(13)(i), and (k)(15).

**1910.6(e)(6)** ANSI A13.1-56 Scheme for the Identification of Piping Systems, IBR approved for §§1910.253(d)(4)(ii); 1910.261(a)(3)(iii); 1910.262(c)(7).

**1910.6(e)(7)** ANSI A14.1-68 Safety Code for Portable Wood Ladders, Supplemented by ANSI A14.1a-77, IBR approved for §1910.261(a)(3)(iv) and (c)(3)(i).

**1910.6(e)(8)** ANSI A14.2-56 Safety Code for Portable Metal Ladders, Supplemented by ANSI A14.2a-77, IBR approved for §1910.261(a)(3)(v) and (c)(3)(i).

**1910.6(e)(9)** ANSI A14.3-56 Safety Code for Fixed Ladders, IBR approved for §§1910.68(b)(4) and (12); 1910.179(c)(2); and 1910.261(a)(3)(vi) and (c)(3)(i).

**1910.6(e)(10)** ANSI A17.1-65 Safety Code for Elevators, Dumbwaiters and Moving Walks, Including Supplements, A17.1a (1967); A17.1b (1968); A17.1c (1969); A17.1d (1970), IBR approved for §1910.261(a)(3)(vii), (g)(11)(i), and (l)(4).

**1910.6(e)(11)** ANSI A17.2-60 Practice for the Inspection of Elevators, Including Supplements, A17.2a (1965), A17.2b (1967), IBR approved for §1910.261(a)(3)(viii).

**1910.6(e)(12)** ANSI A90.1-69 Safety Standard for Manlifts, IBR approved for §1910.68(b)(3).

**1910.6(e)(13)** ANSI A92.2-69 Standard for Vehicle Mounted Elevating and Rotating Work Platforms, IBR approved for §§1910.67(b)(1), (2), (c)(3), and (4) and 1910.268(s)(1)(v).

**1910.6(e)(14)** ANSI A120.1-70 Safety Code for Powered Platforms for Exterior Building Maintenance, IBR approved for §1910.66 app. D (b) through (d).

**1910.6(e)(15)** ANSI B7.1-70 Safety Code for the Use, Care and Protection of Abrasive Wheels, IBR approved for §§1910.94(b)(5)(i)(a); 1910.215(b)(12); and 1910.218(j)(5).

**1910.6(e)(16)** ANSI B15.1-53 (R 58) Safety Code for Mechanical Power Transmission Apparatus, IBR approved for §§1910.68(b)(4) and 1910.261(a)(3)(ix), (b)(1), (e)(3), (e)(9), (f)(4), (j)(5)(iv), (k)(12), and (l)(3).

**1910.6(e)(17)** ANSI B20.1-57 Safety Code for Conveyors, Cableways, and Related Equipment, IBR approved for §§1910.218(j)(3); 1910.261(a)(3)(x), (b)(1), (c)(15)(iv), (f)(4), and (j)(2); 1910.265(c)(18)(i).

**1910.6(e)(18)** ANSI B30.2-43 (R 52) Safety Code for Cranes, Derricks, and Hoists, IBR approved for §1910.261(a)(3)(xi), (c)(2)(vi), and (c)(8) (i) and (iv).

**1910.6(e)(19)** ANSI B30.2.0-67 Safety Code for Overhead and Gantry Cranes, IBR approved for §§1910.179(b)(2); 1910.261(a)(3)(xii), (c)(2)(v), and (c)(8) (i) and (iv).

**1910.6(e)(20)** ANSI B30.5-68 Safety Code for Crawler, Locomotive, and Truck Cranes, IBR approved for §§1910.180(b)(2) and 1910.261(a)(3)(xiii).

**1910.6(e)(21)** ANSI B30.6-69 Safety Code for Derricks, IBR approved for §§1910.181(b)(2) and 1910.268(j)(4)(iv) (E) and (H).

**1910.6(e)(22)** ANSI B31.1-55 Code for Pressure Piping, IBR approved for §1910.261(g)(18)(iii).

**1910.6(e)(23)** ANSI B31.1-67, IBR approved for §1910.253(d)(1)(i)(A)

**1910.6(e)(24)** ANSI B31.1a-63 Addenda to ANSI B31.1 (1955), IBR approved for §1910.261(g)(18)(iii).

**1910.6(e)(25)** ANSI B31.1-67 and Addenda B31.1 (1969) Code for Pressure Piping, IBR approved for §§1910.103(b)(1)(iii)(b); 1910.104(b)(5)(ii); 1910.218(d)(4) and (e)(1)(iv); and 1910.261(a)(3)(xiv) and (g)(18)(iii).

**1910.6(e)(26)** ANSI B31.2-68 Fuel Gas Piping, IBR approved for §1910.261(g)(18)(iii).

**1910.6(e)(27)** ANSI B31.3-66 Petroleum Refinery Piping, IBR approved for §1910.103(b)(3)(v)(b).

**1910.6(e)(28)** ANSI B31.5-66 Addenda B31.5a (1968) Refrigeration Piping, IB approved for §§1910.103(b)(3)(v)(b) and 1910.111(b)(7)(iii).

**1910.6(e)(29)** ANSI B56.1-69 Safety Standard for Powered Industrial Trucks, IBR approved for §§1910.178(a)(2) and (3) and 1910.261(a)(3)(xv), (b)(6), (m)(2), and (m)(5)(iii).

**1910.6(e)(30)** ANSI B57.1-65 Compressed Gas Cylinder Valve Outlet and Inlet Connections, IBR approved for §1910.253(b)(1)(iii).

**1910.6(e)(31)** [Reserved.]

**1910.6(e)(32)** ANSI B175.1-1991, Safety Requirements for Gasoline-Powered Chain Saws §1910.266(e)(2)(i).

**1910.6(e)(33)** ANSI C1-71 National Electrical Code, IBR approved for §1910.66 appendix D (c)(22) (i) and (vii).

**1910.6(e)(34)** ANSI C33.2-56 Safety Standard for Transformer-Type Arc Welding Machines, IBR approved for §1910.254(b)(1).

**1910.6(e)(35)** [Reserved.]

**1910.6(e)(36)** ANSI H23.1-70 Seamless Copper Water Tube Specification, IBR approved for §1910.110(b)(8)(ii) and (13)(ii)(b)(1).

**1910.6(e)(37)** ANSI H38.7-69 Specification for Aluminum Alloy Seamless Pipe and Seamless Extruded Tube, IBR approved for §1910.110(b)(8)(i).

**1910.6(e)(38)** ANSI J6.4-71 Standard Specification for Rubber Insulating Blankets, IBR approved for §1910.268 (f)(1) and (n)(11)(v).

**1910.6(e)(39)** ANSI J6.6-71 Standard Specification for Rubber Insulating Gloves, IBR approved for §1910.268(f)(1) and (n)(11)(iv).

**1910.6(e)(40)** ANSI K13.1-67 Identification of Gas Mask Canisters, IBR approved for §1910.261(a)(3)(xvi) and (h)(2)(iii).

**1910.6(e)(41)** ANSI K61.1-60 Safety Requirements for the Storage and Handling of Anhydrous Ammonia, IBR approved for §1910.111(b)(11)(i).

**1910.6(e)(42)** ANSI K61.1-66 Safety Requirements for the Storage and Handling of Anhydrous Ammonia, IBR approved for §1910.111(b)(11)(i).

**1910.6(e)(43)** ANSI O1.1-54 (R 61) Safety Code for Woodworking Machinery, IBR approved for §1910.261(a)(3)(xvii), (e)(7), and (i)(2).

**1910.6(e)(44)** ANSI S1.4-71 (R 76) Specification for Sound Level Meters, IBR approved for §1910.95 appendixes D and I.

**1910.6(e)(45)** ANSI S1.11-71 (R 76) Specification for Octave, Half-Octave and Third-Octave Band Filter Sets, IBR approved for §1910.95 appendix D.

**1910.6(e)(46)** ANSI S3.6-69 Specifications for Audiometers, IBR approved for §1910.95(h)(2) and (5)(ii) and appendix D.

**1910.6(e)(47)** ANSI Z4.1-68 Requirements for Sanitation in Places of Employment, IBR approved for §1910.261(a)(3)(xviii) and (g)(15)(vi).

**1910.6(e)(48)** [Reserved.]

**1910.6(e)(49)** ANSI Z9.1-51 Safety Code for Ventilation and Operation of Open Surface Tanks, IBR approved for §§1910.94(c)(5)(iii)(e) and 1910.261 (a)(3)(xix), (g)(18)(v), and (h)(2)(i).

**1910.6(e)(50)** ANSI Z9.1-71 Practices for Ventilation and Operation of Open-Surface Tanks, IBR approved for §1910.124(b)(4)(iv).

**1910.6(e)(51)** ANSI Z9.2-60 Fundamentals Governing the Design and Operation of Local Exhaust Systems, IBR approved for §§1910.94(a)(4)(i) introductory text, (a)(6) introductory text, (b)(3)(ix), (b)(4)(i) and (ii), (c)(3)(i) introductory text, (c)(5)(iii)(b), and (c)(7)(iv)(a); 1910.261(a)(3)(xx), (g)(1)(i) and (iii), and (h)(2)(ii).

**1910.6(e)(52)** ANSI Z9.2-79 Fundamentals Governing the Design and Operation of Local Exhaust Systems, IBR approved for §1910.124(b)(4)(i).

**1910.6(e)(53)** ANSI Z12.12-68 Standard for the Prevention of Sulfur Fires and Explosions, IBR approved for §1910.261 (a)(3)(xxi), (d)(1)(i), (f)(2)(iv), and (g)(1)(i).

**1910.6(e)(54)** ANSI Z12.20-62 (R 69) Code for the Prevention of Dust Explosions in Woodworking and Wood Flour Manufacturing Plants, IBR approved for §1910.265(c)(20)(i).

**1910.6(e)(55)** ANSI Z21.30-64 Requirements for Gas Appliances and Gas Piping Installations, IBR approved for §1910.265(c)(15).

**1910.6(e)(56)** ANSI Z24.22-57 Method of Measurement of Real-Ear Attenuation of Ear Protectors at Threshold, IBR approved for §1910.261(a)(3)(xxii).

**1910.6(e)(57)** ANSI Z33.1-61 Installation of Blower and Exhaust Systems for Dust, Stock, and Vapor Removal or Conveying, IBR approved for §§1910.94(a)(4)(i); 1910.261 (a)(3)(xxiii) and (f)(5); and 1910.265(c)(20)(i).

**1910.6(e)(58)** ANSI Z33.1-66 Installation of Blower and Exhaust Systems for Dust, Stock, and Vapor Removal or Conveying, IBR approved for §1910.94(a)(2)(ii).

**1910.6(e)(59)** ANSI Z35.1-68 Specifications for Accident Prevention Signs, IBR approved for §1910.261(a)(3)(xxiv) and (c)(16).

**1910.6(e)(60)** ANSI Z41.1-67 Men's Safety Toe Footwear, IBR approved for §§1910.94(a)(5)(v); 1910.136(b)(2) and 1910.261(i)(4).

**1910.6(e)(61)** ANSI Z41-91, Personal Protection-Protective Footwear, IBR approved for §1910.136(b)(1).

**1910.6(e)(62)** ANSI Z48.1-54 Method for Marking Portable Compressed Gas Containers to Identify the Material Contained, IBR approved for §§1910.103(b)(1)(i)(*c*); 1910.110(b)(5)(iii); and 1910.253(b)(1)(ii).

**1910.6(e)(63)** ANSI Z48.1-54 (R 70) Method for Marking Portable Compressed Gas Containers To Identify the Material Contained, IBR approved for §§1910.111(e)(1) and 1910.134(d)(4).

**1910.6(e)(64)** ANSI Z49.1-67 Safety in Welding and Cutting, IBR approved for §1910.252(c)(1)(iv) (A) and (B).

**1910.6(e)(65)** ANSI Z53.1-67 Safety Color Code for Marking Physical Hazards and the Identification of Certain Equipment, IBR approved for §§1910.97(a)(3)(ii); 1910.145(d) (2), (4), and (6).

**1910.6(e)(66)** ANSI Z54.1-63 Safety Standard for Non-Medical X-Ray and Sealed Gamma Ray Sources, IBR approved for §1910.252(d)(1)(vii) and (2)(ii).

**1910.6(e)(67)** ANSI Z87.1-68 Practice of Occupational and Educational Eye and Face Protection, IBR approved for §§1910.133(b)(2); 1910.252(b)(2)(ii)(*I*); and 1910.261(a)(3)(xxv), (d)(1)(ii), (f)(5), (g)(10), (g)(15)(v), (g)(18)(ii), and (i)(4).

**1910.6(e)(68)** ANSI Z87.1-89, Practice for Occupational and Educational Eye and Face Protection, IBR approved for §1910.133(b)(1).

**1910.6(e)(69)** ANSI Z88.2-69 Practices for Respiratory Protection, IBR approved for §§1910.94(c)(6)(iii)(*a*); 1910.134(c); and 1910.261(a)(3)(xxvi), (b)(2), (f)(5), (g)(15)(v), (h)(2) (iii) and (iv), and (i)(4).

**1910.6(e)(70)** ANSI Z89.1-69 Safety Requirements for Industrial Head Protection, IBR approved for §§1910.135(b)(2); and 1910.261(a)(3)(xxvii), (b)(2), (g)(15)(v), and (i)(4).

**1910.6(e)(71)** ANSI Z89.1-86, Protective Headwear for Industrial Workers Requirements, IBR approved for §1910.135(b)(1).

**1910.6(e)(72)** ANSI Z89.2-71 Safety Requirements for Industrial Protective Helmets for Electrical Workers, Class B, IBR approved for §1910.268(i)(1).

**1910.6(f)** The following material is available for purchase from the American Petroleum Institute (API), 1220 L Street NW, Washington DC 20005:

**1910.6(f)(1)** [Reserved.]

**1910.6(f)(2)** API 12B (May 1958) Specification for Bolted Production Tanks, 11th Ed., With Supplement No. 1, Mar. 1962, IBR approved for §1910.106(b)(1)(i)(*a*)(*3*).

**1910.6(f)(3)** API 12D (Aug. 1957) Specification for Large Welded Production Tanks, 7th Ed., IBR approved for §1910.106(b)(1)(i)(*a*)(*3*).

**1910.6(f)(4)** API 12F (Mar. 1961) Specification for Small Welded Production Tanks, 5th Ed., IBR approved for §1910.106(b)(1)(i)(*a*)(*3*).

**1910.6(f)(5)** API 620, Fourth Ed. (1970) Including appendix R, Recommended Rules for Design and Construction of Large Welded Low Pressure Storage Tanks, IBR approved for §§1910.103(c)(1)(i)(*a*); 1910.106(b)(1)(iv)(*b*)(*1*); and 1910.111(*d*)(*1*) (ii) and (iii).

**1910.6(f)(6)** API 650 (1966) Welded Steel Tanks for Oil Storage, 3rd Ed., IBR approved for §1910.106(b)(1)(iii)(*a*)(*2*).

**1910.6(f)(7)** API 1104 (1968) Standard for Welding Pipelines and Related Facilities, IBR approved for §1910.252(d)(1)(v).

**1910.6(f)(8)** API 2000 (1968) Venting Atmospheric and Low Pressure Storage Tanks, IBR approved for §1910.106(b)(2)(iv)(*b*)(*1*).

**1910.6(f)(9)** API 2201 (1963) Welding or Hot Tapping on Equipment Containing Flammables, IBR approved for §1910.252(d)(1)(vi).

**1910.6(g)** The following material is available for purchase from the American Society of Mechanical Engineers (ASME), United Engineering Center, 345 East 47th Street, New York, NY 10017:

**1910.6(g)(1)** ASME Boiler and Pressure Vessel Code, Sec. VIII, 1949, 1950, 1952, 1956, 1959, and 1962 Ed., IBR approved for §§1910.110(b)(10)(iii) (Table H-26), (d)(2) (Table H-31); (e)(3)(i) (Table H-32), (h)(2) (Table H-34); and 1910.111(b)(2)(vi);

**1910.6(g)(2)** ASME Code for Pressure Vessels, 1968 Ed., IBR approved for §§1910.106(i)(3)(i); 1910.110(g)(2)(iii)(*b*)(*2*); and 1910.217(b)(12);

**1910.6(g)(3)** ASME Boiler and Pressure Vessel Code, Sec. VIII, 1968, IBR approved for §§1910.103; 1910.104(b)(4)(ii); 1910.106(b)(1)(iv)(*b*)(*2*) and (*i*)(3)(ii); 1910.107; 1910.110(b)(11) (i)(*b*) and (iii)(*a*)(*1*); 1910.111(b)(2) (i), (ii), and (iv); and 1910.169(a)(2)(i) and (ii);

**1910.6(g)(4)** ASME Boiler and Pressure Vessel Code, Sec. VIII, Paragraph UG-84, 1968, IBR approved for §1910.104 (b)(4)(ii) and (b)(5)(iii);

**1910.6(g)(5)** ASME Boiler and Pressure Vessel Code, Sec. VIII, Unfired Pressure Vessels, Including Addenda (1969), IBR approved for §§1910.261; 1910.262; 1910.263(i)(24)(ii);

**1910.6(g)(6)** Code for Unfired Pressure Vessels for Petroleum Liquids and Gases of the API and the ASME, 1951 Ed., IBR approved for §1910.110(b)(3)(iii); and

**1910.6(g)(7)** ASME B56.6-1992 (with addenda), Safety Standard for Rough Terrain Forklift Trucks, IBR approved for §1910.266(f)(4).

**1910.6(h)** The following material is available for purchase from the American Society for Testing and Materials (ASTM), 1916 Race Street, Philadelphia, PA 19103:

**1910.6(h)(1)** ASTM A 47-68 Malleable Iron Castings, IBR approved for §1910.111(b)(7)(vi).

**1910.6(h)(2)** ASTM A 53-69 Welded and Seamless Steel Pipe, IBR approved for §§1910.110(b)(8)(i) (*a*) and (*b*) and 1910.111(b)(7)(iv).

**1910.6(h)(3)** ASTM A 126-66 Gray Iron Casting for Valves, Flanges and Pipe Fitting, IBR approved for §1910.111(b)(7)(vi).

**1910.6(h)(4)** ASTM A 391-65 (ANSI G61.1-1968) Alloy Steel Chain, IBR approved for §1910.184(e)(4).

**1910.6(h)(5)** ASTM A 395-68 Ductile Iron for Use at Elevated Temperatures, IBR approved for §1910.111(b)(7)(vi).

**1910.6(h)(6)** ASTM B 88-69 Seamless Copper Water Tube, IBR approved for §1910.110(b)(8)(i)(*a*) and (13)(ii)(*b*)(*1*).

**1910.6(h)(7)** ASTM B 88-66A Seamless Copper Water Tube, IBR approved for §1910.252(d)(1)(i)(A)(2).

**1910.6(h)(8)** ASTM B 117-64 Salt Spray (Fog) Test, IBR approved for §1910.268(g)(2)(i)(A).

**1910.6(h)(9)** ASTM B 210-68 Aluminum-Alloy Drawn Seamless Tubes, IBR approved for §1910.110(b)(8)(ii).

**1910.6(h)(10)** ASTM B 241-69, IBR approved for §1910.110(b)(8)(i) introductory text.

**1910.6(h)(11)** ASTM D 5-65 Test for Penetration by Bituminous Materials, IBR approved for §1910.106(a)(17).

**1910.6(h)(12)** ASTM D 56-70 Test for Flash Point by Tag Closed Tester, IBR approved for §1910.106(a)(14)(i).

**1910.6(h)(13)** ASTM D 86-62 Test for Distillation of Petroleum Products, IBR approved for §§1910.106(a)(5) and 1910.119(b) "Boiling point."

**1910.6(h)(14)** ASTM D 88-56 Test for Saybolt Viscosity, IBR approved for §1910.106(a)(37).

**1910.6(h)(15)** ASTM D 93-71 Test for Flash Point by Pensky Martens, IBR approved for §1910.106(a)(14)(ii).

**1910.6(h)(16)** ASTM D 323-68, IBR approved for §1910.106(a)(30)

**1910.6(h)(17)** ASTM D 445-65 Test for Viscosity of Transparent and Opaque Liquids, IBR approved for §1910.106(a)(37).

**1910.6(h)(18)** ASTM D 1692-68 Test for Flammability of Plastic Sheeting and Cellular Plastics, IBR approved for §1910.103(c)(1)(v)(*d*).

**1910.6(h)(19)** ASTM D 2161-66 Conversion Tables For SUS, IBR approved for §1910.106(a)(37).

**1910.6(i)** The following material is available for purchase from the American Welding Society (AWS), 550 NW, LeJeune Road, P.O. Box 351040, Miami FL 33135:

**1910.6(i)(1)** AWS A3.0 (1969) Terms and Definitions, IBR approved for §1910.251(c).

**1910.6(i)(2)** [Reserved.]

**1910.6(i)(3)** AWS B3.0-41 Standard Qualification Procedure, IBR approved for §1910.67(c)(5)(i).

**1910.6(i)(4)** AWS D1.0-1966 Code for Welding in Building Construction, IBR approved for §1910.27(b)(6).

**1910.6(i)(5)** AWS D2.0-69 Specifications for Welding Highway and Railway Bridges, IBR approved for §1910.67(c)(5)(iv).

**1910.6(i)(6)** AWS D8.4-61 Recommended Practices for Automotive Welding Design, IBR approved for §1910.67(c)(5)(ii).

**1910.6(i)(7)** AWS D10.9-69 Standard Qualification of Welding Procedures and Welders for Piping and Tubing, IBR approved for §1910.67(c)(5)(iii).

**1910.6(j)** The following material is available for purchase from the Department of Commerce:

**1910.6(j)(1)** Commercial Standard, CS 202-56 (1961) "Industrial Lifts and Hinged Loading Ramps," IBR approved for §1910.30(a)(3).

**1910.6(j)(2)** Publication "Model Performance Criteria for Structural Fire Fighters' Helmets," IBR approved for §1910.156(e)(5)(i).

**1910.6(k)** The following material is available for purchase from the Compressed Gas Association (CGA), 1235 Jefferson Davis Highway, Arlington, VA 22202:

**1910.6(k)(1)** CGA C-6 (1968) Standards for Visual Inspection of Compressed Gas Cylinders, IBR approved for §1910.101(a).

**1910.6(k)(2)** CGA C-8 (1962) Standard for Requalification of ICC-3HT Cylinders, IBR approved for §1910.101(a).

**1910.6(k)(3)** CGA G-1 (1966) Acetylene, IBR approved for §1910.102(a).

**1910.6(k)(4)** CGA G-1.3 (1959) Acetylene Transmission for Chemical Synthesis, IBR approved for §1910.102(b).

**1910.6(k)(5)** CGA G-1.4 (1966) Standard for Acetylene Cylinder Charging Plants, IBR approved for §1910.102(b).

**1910.6(k)(6)** CGA G-7.1 (1966) Commodity Specification, IBR approved for §1910.134(d)(1).

**1910.6(k)(7)** CGA G-8.1 (1964) Standard for the Installation of Nitrous Oxide Systems at Consumer Sites, IBR approved for §1910.105.

**1910.6(k)(8)** CGA P-1 (1965) Safe Handling of Compressed Gases, IBR approved for §1910.101(b).

**1910.6(k)(9)** CGA P-3 (1963) Specifications, Properties, and Recommendations for Packaging, Transportation, Storage and Use of Ammonium Nitrate, IBR approved for §1910.109(i)(1)(ii)(*b*).

**1910.6(k)(10)** CGA S-1.1 (1963) and 1965 Addenda. Safety Release Device Standards—Cylinders for Compressed Gases, IBR approved for §§1910.101(c); 1910.103(c)(1)(iv)(*a*)(*2*).

**1910.6(k)(11)** CGA S-1.2 (1963) Safety Release Device Standards, Cargo and Portable Tanks for Compressed Gases, IBR approved for §§1910.101(c); 1910.103(c)(1)(iv)(*a*)(*2*).

**1910.6(k)(12)** CGA S-1.3 (1959) Safety Release Device Standards-Compressed Gas Storage Containers, IBR approved for §§1910.103(c)(1)(iv)(*a*)(*2*); 1910.104(b)(6)(iii); and 1910.111(d)(4)(ii)(*b*).

**1910.6(k)(13)** CGA 1957 Standard Hose Connection Standard, IBR approved for §1910.253(e)(4)(v) and (5)(iii).

**1910.6(k)(14)** CGA and RMA (Rubber Manufacturer's Association) Specification for Rubber Welding Hose (1958), IBR approved for §1910.253(e)(5)(i).

**1910.6(k)(15)** CGA 1958 Regulator Connection Standard, IBR approved for §1910.253(e)(4)(iv) and (6).

**1910.6(l)** The following material is available for purchase from the Crane Manufacturer's Association of America, Inc. (CMAA), 1 Thomas Circle NW, Washington DC 20005:

**1910.6(l)(1)** CMAA Specification 1B61, Specifications for Electric Overhead Traveling Cranes, IBR approved for §1910.179(b)(6)(i).

**1910.6(l)(2)** [Reserved]

**1910.6(m)** The following material is available for purchase from the General Services Administration:

**1910.6(m)(1)** GSA Pub. GG-B-0067b, Air Compressed for Breathing Purposes, or Interim Federal Specifications, Apr. 1965, IBR approved for §1910.134(d)(4).

**1910.6(m)(2)** [Reserved]

**1910.6(n)** The following material is available for purchase from the Department of Health and Human Services:

# OSHA Standards for General Industry

**1910.6(n)(1)** Publication No. 76-120 (1975), List of Personal Hearing Protectors and Attenuation Data, IBR approved for §1910.95 App. B.

**1910.6(n)(2)** [Reserved]

**1910.6(o)** The following material is available for purchase from the Institute of Makers of Explosives (IME), 420 Lexington Avenue, New York, NY 10017:

**1910.6(o)(1)** IME Pamphlet No. 17, 1960, Safety in the Handling and Use of Explosives, IBR approved for §§1910.261 (a)(4)(iii) and (c)(14)(ii).

**1910.6(o)(2)** [Reserved]

**1910.6(p)** The following material is available for purchase from the National Electrical Manufacturer's Association (NEMA):

**1910.6(p)(1)** NEMA EW-1 (1962) Requirements for Electric Arc Welding Apparatus, IBR approved for §§1910.254(b)(1).

**1910.6(p)(2)** [Reserved]

**1910.6(q)** The following material is available for purchase from the National Fire Protection Association (NFPA), 11 Tracy Drive, Avon, MA 02322:

**1910.6(q)(1)** NFPA 30 (1969) Flammable and Combustible Liquids Code, IBR approved for §1910.178(f)(1).

**1910.6(q)(2)** NFPA 32-1970 Standard for Dry Cleaning Plants, IBR approved for §1910.106(j)(6)(i).

**1910.6(q)(3)** NFPA 33-1969 Standard for Spray Finishing Using Flammable and Combustible Material, IBR approved for §§1910.94(c)(1)(ii), (2), (3) (i) and (iii), and (5).

**1910.6(q)(4)** NFPA 34-1966 Standard for Dip Tanks Containing Flammable or Combustible Liquids, IBR approved for §1910.124(b)(4)(iv).

**1910.6(q)(5)** NFPA 34-1995 Standard for Dip Tanks Containing Flammable or Combustible Liquids, IBR approved for §1910.124(b)(4)(ii).

**1910.6(q)(6)** NFPA 35-1970 Standard for the Manufacture of Organic Coatings, IBR approved for §1910.106(j)(6)(ii).

**1910.6(q)(7)** NFPA 36-1967 Standard for Solvent Extraction Plants, IBR approved for §1910.106(j)(6)(iii).

**1910.6(q)(8)** NFPA 37-1970 Standard for the Installation and Use of Stationary Combustion Engines and Gas Turbines, IBR approved for §§1910.106(j)(6)(iv) and 1910.110 (b)(20)(iv)(c) and (e)(11).

**1910.6(q)(9)** NFPA 51B-1962 Standard for Fire Protection in Use of Cutting and Welding Processes, IBR approved for §1910.252(a)(1) introductory text.

**1910.6(q)(10)** NFPA 54-1969 Standard for the Installation of Gas Appliances and Gas Piping, IBR approved for §1910.110(b)(20)(iv)(a).

**1910.6(q)(11)** NFPA 54A-1969 Standard for the Installation of Gas Piping and Gas Equipment on Industrial Premises and Certain Other Premises, IBR approved for §1910.110(b)(20)(iv)(b).

**1910.6(q)(12)** NFPA 58-1969 Standard for the Storage and Handling of Liquefied Petroleum Gases (ANSI Z106.1-1970), IBR approved for §§1910.110(b)(3)(iv) and (i)(3) (i) and (ii); and 1910.178(f)(2).

**1910.6(q)(13)** NFPA 59-1968 Standard for the Storage and Handling of Liquefied Petroleum Gases at Utility Gas Plants, IBR approved for §§1910.110(b)(3)(iv) and (i)(2)(iv).

**1910.6(q)(14)** NFPA 62-1967 Standard for the Prevention of Dust Explosions in the Production, Packaging, and Handling of Pulverized Sugar and Cocoa, IBR approved for §1910.263(k)(2)(i).

**1910.6(q)(15)** NFPA 68-1954 Guide for Explosion Venting, IBR approved for §1910.94(a)(2)(iii).

**1910.6(q)(16)** NFPA 70-1971 National Electrical Code, IBR approved for §1910.66 App. D(c)(2).

**1910.6(q)(17)** NFPA 78-1968 Lightning Protection Code, IBR approved for §1910.109(i)(6)(ii).

**1910.6(q)(18)** NFPA 80-1968 Standard for Fire Doors and Windows, IBR approved for §1910.106(d)(4)(i).

**1910.6(q)(19)** NFPA 80-1970 Standard for the Installation of Fire Doors and Windows, IBR approved for §1910.253(f)(6)(i)(I).

**1910.6(q)(20)** NFPA 86A-1969 Standard for Oven and Furnaces Design, Location and Equipment, IBR approved for §§1910.107(j)(1) and (l)(3) and 1910.108 (b)(2) and (d)(2).

**1910.6(q)(21)** NFPA 91-1961 Standard for the Installation of Blower and Exhaust Systems for Dust, Stock, and Vapor Removal or Conveying (ANSI Z33.1-61), IBR approved for §1910.107(d)(1).

**1910.6(q)(22)** NFPA 91-1969 Standards for Blower and Exhaust Systems, IBR approved for §1910.108(b)(1).

**1910.6(q)(23)** NFPA 96-1970 Standard for the Installation of Equipment for the Removal of Smoke and Grease Laden Vapors from Commercial Cooking Equipment, IBR approved for §1910.110(b)(20)(iv)(d).

**1910.6(q)(24)** NFPA 101-1970 Code for Life Safety From Fire in Buildings and Structures, IBR approved for §1910.261(a)(4)(ii).

**1910.6(q)(25)** NFPA 203M-1970 Manual on Roof Coverings, IBR approved for §1910.109(i)(1)(iii)(c).

**1910.6(q)(26)** NFPA 251-1969 Standard Methods of Fire Tests of Building Construction and Materials, IBR approved for §§1910.106(d)(3)(ii) introductory text and (d)(4)(i).

**1910.6(q)(27)** NFPA 302-1968 Fire Protection Standard for Motor-Craft (Pleasure and Commercial), IBR approved for §1910.265(d)(2)(iv) introductory text.

**1910.6(q)(28)** NFPA 385-1966 Recommended Regulatory Standard for Tank Vehicles for Flammable and Combustible Liquids, IBR approved for §1910.106(g)(1)(i)(e)(1).

**1910.6(q)(29)** NFPA 496-1967 Standard for Purged Enclosures for Electrical Equipment in Hazardous Locations, IBR approved for §1910.103 (c)(1)(ix)(e)(1).

**1910.6(q)(30)** NFPA 505-1969 Standard for Type Designations, Areas of Use, Maintenance, and Operation of Powered Industrial Trucks, IBR approved for §1910.110(e)(2)(iv).

**1910.6(q)(31)** NFPA 566-1965 Standard for the Installation of Bulk Oxygen Systems at Consumer Sites, IBR approved for §§1910.253(b)(4)(iv) and (c)(2)(v).

**1910.6(q)(32)** NFPA 656-1959 Code for the Prevention of Dust Ignition in Spice Grinding Plants, IBR approved for §1910.263(k)(2)(i).

**1910.6(q)(33)** NFPA 1971-1975 Protective Clothing for Structural Fire Fighting, IBR approved for §1910.156(e)(3)(ii) introductory text.

**1910.6(r)** The following material is available for purchase from the National Food Plant Institute, 1700 K St. NW., Washington, DC 20006:

**1910.6(r)(1)** Definition and Test Procedures for Ammonium Nitrate Fertilizer (Nov. 1964), IBR approved for §1910.109 Table H-22, ftn. 3.

**1910.6(r)(2)** [Reserved]

**1910.6(s)** The following material is available for purchase from the National Institute for Occupational Safety and Health (NIOSH):

**1910.6(s)(1)** Registry of Toxic Effects of Chemical Substances, 1978, IBR approved for §1910.20(c)(13)(i) and appendix B.

**1910.6(s)(2)** Development of Criteria for Fire Fighters Gloves; Vol. II, Part II; Test Methods, 1976, IBR approved for §1910.156(e)(4)(i) introductory text.

**1910.6(s)(3)** NIOSH Recommendations for Occupational Safety and Health Standards (Sept. 1987), IBR approved for §1910.120 PEL definition.

**1910.6(t)** The following material is available for purchase from the Public Health Service:

**1910.6(t)(1)** U.S. Pharmacopeia, IBR approved for §1910.134(d)(1).

**1910.6(t)(2)** Publication No. 934 (1962), Food Service Sanitation Ordinance and Code, Part V of the Food Service Sanitation Manual, IBR approved for §1910.142(i)(1).

**1910.6(u)** The following material is available for purchase from the Society of Automotive Engineers (SAE), 485 Lexington Avenue, New York, NY 10017:

**1910.6(u)(1)** SAE J185, June 1988, Recommended Practice for Access Systems for Off-Road Machines, IBR approved for §1910.266(f)(5)(i).

**1910.6(u)(2)** SAE J231, January 1981, Minimum Performance Criteria for Falling Object Protective Structure (FOPS), IBR approved for §1910.266(f)(3)(ii).

**1910.6(u)(3)** SAE J386, June 1985, Operator Restraint Systems for Off-Road Work Machines, IBR approved for §1910.266(d)(3)(iv).

**1910.6(u)(4)** SAE J397, April 1988, Deflection Limiting Volume-ROPS/FOPS Laboratory Evaluation, IBR approved for §1910.266(f)(3)(iv).

**1910.6(u)(5)** SAE 765 (1961) SAE Recommended Practice: Crane Loading Stability Test Code, IBR approved for §1910.180 (c)(1)(iii) and (e)(2)(iii)(*a*).

**1910.6(u)(6)** SAE J1040, April 1988, Performance Criteria for Rollover Protective Structures (ROPS) for Construction, Earthmoving, Forestry and Mining Machines, IBR approved for §1910.266(f)(3)(ii).

**1910.6(v)** The following material is available for purchase from the Fertilizer Institute, 1015 18th Street NW, Washington, DC 20036:

**1910.6(v)(1)** Standard M-1 (1953, 1955, 1957, 1960, 1961, 1963, 1965, 1966, 1967, 1968), Superseded by ANSI K61.1-1972, IBR approved for §1910.111(b)(1)(i) and (iii).

**1910.6(v)(2)** [Reserved]

**1910.6(w)** The following material is available for purchase from Underwriters Laboratories (UL), 207 East Ohio Street, Chicago, IL 60611:

**1910.6(w)(1)** UL 58-61 Steel Underground Tanks for Flammable and Combustible Liquids, 5th Ed., IBR approved for §1910.106(b)(1)(iii)(*a*)(*1*).

**1910.6(w)(2)** UL 80-63 Steel Inside Tanks for Oil-Burner Fuel, IBR approved for §1910.106(b)(1)(iii)(*a*)(*1*).

**1910.6(w)(3)** UL 142-68 Steel Above Ground Tanks for Flammable and Combustible Liquids, IBR approved for §1910.106(b)(1)(iii)(*a*)(*1*).

[39 FR 23502, June 27, 1974, as amended at 49 FR 5321, Feb. 10, 1984; 61 FR 9231, Mar. 7, 1996; 64 FR 13908, Mar. 23, 1999]

## §1910.7 Definition and requirements for a nationally recognized testing laboratory.

**1910.7(a)** *Application.* This section shall apply only when the term *nationally recognized testing laboratory* is used in other sections of this part.

**1910.7(b)** *Laboratory requirements.* The term *nationally recognized testing laboratory* (NRTL) means an organization which is recognized by OSHA in accordance with appendix A of this section and which tests for safety, and lists or labels or accepts, equipment or materials and which meets all of the following criteria:

**1910.7(b)(1)** For each specified item of equipment or material to be listed, labeled or accepted, the NRTL has the capability (including proper testing equipment and facilities, trained staff, written testing procedures, and calibration and quality control programs) to perform:

**1910.7(b)(1)(i)** Testing and examining of equipment and materials for workplace safety purposes to determine conformance with appropriate test standards; or

**1910.7(b)(1)(ii)** Experimental testing and examining of equipment and materials for workplace safety purposes to determine conformance with appropriate test standards or performance in a specified manner.

**1910.7(b)(2)** The NRTL shall provide, to the extent needed for the particular equipment or materials listed, labeled, or accepted, the following controls or services:

**1910.7(b)(2)(i)** Implements control procedures for identifying the listed and labeled equipment or materials;

**1910.7(b)(2)(ii)** Inspects the run of production of such items at factories for product evaluation purposes to assure conformance with the test standards; and

**1910.7(b)(2)(iii)** Conducts field inspections to monitor and to assure the proper use of its identifying mark or labels on products;

**1910.7(b)(3)** The NRTL is completely independent of employers subject to the tested equipment requirements, and of any manufacturers or vendors of equipment or materials being tested for these purposes; and,

**1910.7(b)(4)** The NRTL maintains effective procedures for:

**1910.7(b)(4)(i)** Producing creditable findings or reports that are objective and without bias; and

**1910.7(b)(4)(ii)** Handling complaints and disputes under a fair and reasonable system.

**1910.7(c)** *Test standards.* An *appropriate test standard* referred to in §1910.7(b)(1) (i) and (ii) is a document which specifies the safety requirements for specific equipment or class of equipment and is:

**1910.7(c)(1)** Recognized in the United States as a safety standard providing an adequate level of safety, and

**1910.7(c)(2)** Compatible with and maintained current with periodic revisions of applicable national codes and installation standards, and

**1910.7(c)(3)** Developed by a standards developing organization under a method providing for input and consideration of views of industry groups, experts, users, consumers, governmental authorities, and others having broad experience in the safety field involved, or

**1910.7(c)(4)** In lieu of paragraphs (c) (1), (2), and (3), the standard is currently designated as an American National Standards Institute (ANSI) safety-designated product standard or an American Society for Testing and Materials (ASTM) test standard used for evaluation of products or materials.

**1910.7(d)** *Alternative test standard.* If a testing laboratory desires to use a test standard other than one allowed under paragraph (c) of this section, then the Assistant Secretary of Labor shall evaluate the proposed standard to determine that it provides an adequate level of safety before it is used.

**1910.7(e)** *Implementation.* A testing organization desiring recognition by OSHA as an NRTL shall request that OSHA evaluate its testing and control programs against the requirements in this section for any equipment or material it may specify. The recognition procedure shall be conducted in accordance with appendix A to this section.

**1910.7(f)** *Fees.* (1) Each applicant for NRTL recognition and each NRTL must pay fees for services provided by OSHA. OSHA will assess fees for the following services:

**1910.7(f)(1)(i)** Processing of applications for initial recognition, expansion of recognition, or renewal of recognition, including on-site reviews; review and evaluation of the applications; and preparation of reports, evaluations and **FEDERAL REGISTER** notices; and

**1910.7(f)(1)(ii)** Audits of sites.

**1910.7(f)(2)** The fee schedule established by OSHA reflects the cost of performing the activities for each service listed in paragraph (f)(1) of this section. OSHA calculates the fees based on either the average or actual time required to perform the work necessary; the staff costs per hour (which include wages, fringe benefits, and expenses other than travel for personnel that perform or administer the activities covered by the fees); and the average or actual costs for travel when on-site reviews are involved. The formula for the fee calculation is as follows:

Activity Fee = [Average (or Actual) Hours to Complete the Activity × Staff Costs per Hour] + Average (or Actual) Travel Costs

**1910.7(f)(3)(i)** OSHA will review costs annually and will propose a revised fee schedule, if warranted. In its review, OSHA will apply the formula established in paragraph (f)(2) of this section to the current estimated costs for the NRTL Program. If a change is warranted, OSHA will follow the implementation table in paragraph (f)(4) of this section.

**1910.7(f)(3)(ii)** OSHA will publish all fee schedules in the FEDERAL REGISTER. Once published, a fee schedule remains in effect until it is superseded by a new fee schedule. Any member of the public may request a change to the fees included in the current fee schedule. Such a request must include appropriate documentation in support of the suggested change. OSHA will consider such requests during its annual review of the fee schedule.

**1910.7(f)(4)** OSHA will implement fee assessment, collection, and payment as follows:

# OSHA Standards for General Industry

| Approximate dates | Action required |
|---|---|
| **I. Annual Review of Fee Schedule** ||
| November 1 | OSHA will publish any proposed new Fee Schedule in the **FEDERAL REGISTER**, if OSHA determines changes in the schedule are warranted. |
| November 16 | Comments due on the proposed new Fee Schedule. |
| December 15 | OSHA will publish the final Fee Schedule in the **FEDERAL REGISTER**, making it effective. |
| **II. Application Processing Fees** ||
| Time of application | Applicant must pay the applicable fees shown in the Fee Schedule when submitting the application; OSHA will not begin processing until fees are received. |
| Publication of preliminary notice | Applicant must pay remainder of fees; OSHA cancels application if fees are not paid when due. |
| **III. Audit Fees** ||
| After audit performed | OSHA will bill each existing NRTL for the audit fees in effect at the time of audit, but will reflect actual travel costs and staff time in the bill. |
| 30 days after bill date | NRTLs must pay audit fees; OSHA will assess late fee if audit fees are not paid. |
| 45 days after bill date | OSHA will send a letter to the NRTL requesting immediate payment of the audit fees and late fee. |
| 60 days after bill date | OSHA will publish a notice in the **FEDERAL REGISTER** announcing its intent to revoke recognition for NRTLs that have not paid these audit fees. |

**1910.7(f)(5)** OSHA will provide details about how to pay the fees through appropriate OSHA Program Directives, which will be available on the OSHA web site.

### APPENDIX A TO §1910.7—
### OSHA RECOGNITION PROCESS FOR NATIONALLY RECOGNIZED TESTING LABORATORIES

#### INTRODUCTION

This appendix provides requirements and criteria which OSHA will use to evaluate and recognize a Nationally Recognized Testing Laboratory (NRTL). This process will include the the evaluation of the product evaluation and control programs being operated by the NRTL, as well as the NRTL's testing facilities being used in its program. In the evaluation of the NRTLs, OSHA will use either consensus-based standards currently in use nationally, or other standards or criteria which may be considered appropriate. This appendix implements the definition of NRTL in 29 CFR 1910.7 which sets out the criteria that a laboratory must meet to be recognized by OSHA (initially and on a continuing basis). The appendix is broader in scope, providing procedures for renewal, expansion and revocation of OSHA recognition. Except as otherwise provided, the burden is on the applicant to establish by a preponderance of the evidence that it is entitled to recognition as an NRTL. If further detailing of these requirements and criteria will assist the NRTLs or OSHA in this activity, this detailing will be done through appropriate OSHA Program Directives.

#### I. Procedures for Initial OSHA Recognition

#### A. Applications.

1. *Eligibility.* a. Any testing agency or organization considering itself to meet the definition of nationally recognized testing laboratory as specified in §1910.7 may apply for OSHA recognition as an NRTL.

b. However, in determining eligibility for a foreign-based testing agency or organization, OSHA shall take into consideration the policy of the foreign government regarding both the acceptance in that country of testing data, equipment acceptances, and listings, and labeling, which are provided through nationally recognized testing laboratories recognized by the Assistant Secretary, and the accessibility to government recognition or a similar system in that country by U.S.-based safety-related testing agencies, whether recognized by the Assistant Secretary or not, if such recognition or a similar system is required by that country.

2. *Content of application.* a. The applicant shall provide sufficient information and detail demonstrating that it meets the requirements set forth in §1910.7, in order for an informed decision concerning recognition to be made by the Assistant Secretary.

b. The applicant also shall identify the scope of the NRTL-related activity for which the applicant wishes to be recognized. This will include identifying the testing methods it will use to test or judge the specific equipment and materials for which recognition is being requested, unless such test methods are already specified in the test standard. If requested to do so by OSHA, the applicant shall provide documentation of the efficacy of these testing methods.

c. The applicant may include whatever enclosures, attachments, or exhibits the applicant deems appropriate. The application need not be submitted on a Federal form.

3. *Filing office location.* The application shall be filed with: NRTL Recognition Program, Occupational Safety and Health Administration, U.S. Department of Labor, 200 Constitution Avenue, NW., Washington, DC 20210.

4. *Amendments and withdrawals.* a. An application may be revised by an applicant at any time prior to the completion of activity under paragraph I.B.4. of this appendix.

b. An application may be withdrawn by an applicant, without prejudice, at any time prior to the final decision by the Assistant Secretary in paragraph I.B.7.c. of this appendix.

#### B. Review and Decision Process; Issuance or Renewal.

1. *Acceptance and on-site review.* a. Applications submitted by eligible testing agencies will be accepted by OSHA, and their receipt acknowledged in writing. After receipt of an application, OSHA may request additional information if it believes information relevant to the requirements for recognition has been omitted.

b. OSHA shall, as necessary, conduct an on-site review of the testing facilities of the applicant, as well as the applicant's administrative and technical practices, and, if necessary, review any additional documentation underlying the application.

c. These on-site reviews will be conducted by qualified individuals technically expert in these matters, including, as appropriate, non-Federal consultants/contractors acceptable to OSHA. The protocol for each review will be based on appropriate national consensus standards or international guides, with such additions, changes, or deletions as may be considered necessary and appropriate in each case by OSHA. A written report shall be made of each on-site review and a copy shall be provided to the applicant.

2. *Positive finding by staff.* If, after review of the application, and additional information, and the on-site review report, the applicant appears to have met the requirements for recognition, a written recommendation shall be submitted by the responsible OSHA personnel to the Assistant Secretary that the application be approved, accompanied by a supporting explanation.

3. *Negative finding by staff.*—a. *Notification to applicant.* If, after review of the application, any additional information and the on-site review report, the applicant does not appear to have met the requirements for recognition, the responsible OSHA personnel shall notify the applicant in writing, listing the specific requirements of §1910.7 and

this appendix which the applicant has not met, and allow a reasonable period for response.

b. *Revision of application.* (i) After receipt of a notification of negative finding (i.e., for intended disapproval of the application), and within the response period provided, the applicant may:

(*a*) Submit a revised application for further review, which could result in a positive finding by the responsible OSHA personnel pursuant to subsection I.B.2. of this appendix; or

(*b*) Request that the original application be submitted to the Assistant Secretary with an attached statement of reasons, supplied by the applicant of why the application should be approved.

(ii) This procedure for applicant notification and potential revision shall be used only once during each recognition process.

4. *Preliminary finding by Assistant Secretary.* a. The Assistant Secretary, or a special designee for this purpose, will make a preliminary finding as to whether the applicant has or has not met the requirements for recognition, based on the completed application file, the written staff recommendation, and the statement of reasons supplied by the applicant if there remains a staff recommendation of disapproval.

b. Notification of this preliminary finding will be sent to the applicant and subsequently published in the FEDERAL REGISTER.

c. This preliminary finding shall not be considered an official decision by the Assistant Secretary or OSHA, and does not confer any change in status or any interim or temporary recognition for the applicant.

5. *Public review and comment period.*—a. The FEDERAL REGISTER notice of preliminary finding will provide a period of not less than 30 calendar days for written comments on the applicant's fulfillment of the requirements for recognition. The application, supporting documents, staff recommendation, statement of applicant's reasons, and any comments received, will be available for public inspection in the OSHA Docket Office.

b. Any member of the public, including the applicant, may supply detailed reasons and evidence supporting or challenging the sufficiency of the applicant's having met the requirements of the definition in 29 CFR §1910.7 and this appendix. Submission of pertinent documents and exhibits shall be made in writing by the close of the comment period.

6. *Action after public comment.*—a. *Final decision by Assistant Secretary.* Where the public review and comment record supports the Assistant Secretary's preliminary finding concerning the application, i.e., absent any serious objections or substantive claims contrary to the preliminary finding having been received in writing from the public during the comment period, the Assistant Secretary will proceed to final written decision on the application. The reasons supporting this decision shall be derived from the evidence available as a result of the full application, the supporting documentation, the staff finding, and the written comments and evidence presented during the public review and comment period.

b. *Public announcement.* A copy of the Assistant Secretary's final decision will be provided to the applicant. Subsequently, a notification of the final decision shall be published in the FEDERAL REGISTER. The publication date will be the effective date of the recognition.

c. *Review of final decision.* There will be no further review activity available within the Department of Labor from the final decision of the Assistant Secretary.

7. *Action after public objection.*—a. *Review of negative information.* At the discretion of the Assistant Secretary or his designee, OSHA may authorize Federal or contract personnel to initiate a special review of any information provided in the public comment record which appears to require resolution, before a final decision can be made.

b. *Supplementation of record.* The contents and results of special reviews will be made part of this record by the Assistant Secretary by either:

(i) Reopening the written comment period for public comments on these reviews; or

(ii) Convening an informal hearing to accept public comments on these reviews, conducted under applicable OSHA procedures for similar hearings.

c. *Final decision by the Assistant Secretary.* The Assistant Secretary shall issue a decision as to whether it has been demonstrated, based on a preponderance of the evidence, that the applicant meets the requirements for recognition. The reasons supporting this decision shall be derived from the evidence available as a result of the full application, the supporting documentation, the staff finding, the comments and evidence presented during the public review and comment period, and written to transcribed evidence received during any subsequent reopening of the written comment period or informal public hearing held.

d. *Public announcement.* A copy of the Assistant Secretary's final decision will be provided to the applicant, and a notification will be published in the **FEDERAL REGISTER** subsequently announcing the decision.

e. *Review of final decision.* There will be no further review activity available within the Department of Labor from the final decision of the Assistant Secretary.

c. *Terms and conditions of recognition.* The following terms and conditions shall be part of every recognition:

1. *Letter of recognition.* The recognition by OSHA of any NRTL will be evidenced by a letter of recognition from OSHA. The letter will provide the specific details of the scope of the OSHA recognition, including the specific equipment or materials for which OSHA recognition has been granted, as well as any specific conditions imposed by OSHA.

2. *Period of recognition.* The recognition by OSHA of each NRTL will be valid for five years, unless terminated before the expiration of the period. The dates of the period of recognition will be stated in the recognition letter.

3. *Constancy in operations.* The recognized NRTL shall continue to satisfy all the requirements or limitations in the letter of recognition during the period of recognition.

4. *Accurate publicity.* The OSHA-recognized NRTL shall not engage in or permit others to engage in misrepresentation of the scope or conditions of its recognition.

5. *Temporary Recognition of Certain NRTLs.* a. Notwithstanding all other requirements and provisions of §1910.7 and this appendix, the following two organizations are recognized temporarily as nationally recognized testing laboratories by the Assistant Secretary for a period of five years beginning June 13, 1988 and ending on July 13, 1993:

(i) Underwriters Laboratories, Inc., 333 Pfingsten Road, Northbrook, Illinois 60062.

(ii) Factory Mutual Research Corporation, 1151 Boston-Providence Turnpike, Norwood, Massachusetts 02062.

b. At the end of the five-year period, the two temporarily recognized laboratories shall apply for renewal of OSHA recognition utilizing the following procedures established for renewal of OSHA recognition.

## II. Supplementary Procedures.
### A. Test standard changes.

A recognized NRTL may change a testing standard or elements incorporated in the standard such as testing methods or pass-fail criteria by notifying the Assistant Secretary of the change, certifying that the revised standard will be at least as effective as the prior standard, and providing the supporting data upon which its conclusions are based. The NRTL need not inform the Assistant Secretary of minor deviations from a test standard such as the use of new instrumentation that is more accurate or sensitive than originally called for in the standard. The NRTL also need not inform the Assistant Secretary of its adoption of revisions to third-party testing standards meeting the requirements of §1910.7(c)(4), if such revisions have been developed by the standards developing organization, or of its adoption of revisions to other third-party test standards which the developing organization has submitted to OSHA. If, upon review, the Assistant Secretary or his designee determines that the proposed revised standard is not "substantially equivalent" to the previous version with regard to the level of safety obtained, OSHA will not accept the proposed testing standard by the recognized NRTL, and will initiate discontinuance of that aspect of OSHA-recognized activity by the NRTL by modification of the official letter of recognition. OSHA will publicly announce this action and the NRTL will be required to communicate this OSHA decision directly to affected manufacturers.

### B. Expansion of current recognition

1. *Eligibility.* A recognized NRTL may apply to OSHA for an expansion of its current recognition to cover other categories of NRTL testing in addition to those included in the current recognition.

2. *Procedure.* a. OSHA will act upon and process the application for expansion in accordance with subsection I.B. of this appendix, except that the period for written comments, specified in paragraph 5.a of subsection I.B. of this appendix, will be not less than 15 calendar days.

b. In that process, OSHA may decide not to conduct an on-site review, where the substantive scope of the request to expand recognition is closely related to the current area of recognition.

c. The expiration date for each expansion of recognition shall coincide with the expiration date of the current basic recognition period.

### C. Renewal of OSHA recognition

1. *Eligibility.* A recognized NRTL may renew its recognition by filing a renewal request at the address in paragraph I.A.3. of this appendix not less than nine months, nor more than one year, before the expiration date of its current recognition.

2. *Procedure.* a. OSHA will process the renewal request in accordance with subsection I.B. of this appendix, except that the period for written comments, specified in paragraph 5.a of subsection I.B. of this appendix, will be not less than 15 calendar days.

b. In that process, OSHA may determine not to conduct the on-site reviews in I.B.1.a. where appropriate.

c. When a recognized NRTL has filed a timely and sufficient renewal request, its current recognition will not expire until a final decision has been made by OSHA on the request.

d. After the first renewal has been granted to the NRTL, the NRTL shall apply for a continuation of its recognition status every five years by submitting a renewal request. In lieu of submitting a renewal request after the initial renewal, the NRTL may certify its continuing compliance with the terms of its letter of recognition and 29 CFR 1910.7.

3. *Alternative procedure.* After the initial recognition and before the expiration thereof, OSHA may (for good cause) determine that there is a sufficient basis to dispense with the renewal requirement for a given laboratory and will so notify the laboratory of such a determination in writing. In lieu of submitting a renewal request, any laboratory so notified shall certify its continuing compliance with the terms of its letter of recognition and 29 CFR 1910.7.

### D. Voluntary termination of recognition.

At any time, a recognized NRTL may voluntarily terminate its recognition, either in its entirety or with respect to any area covered in its recognition, by giving written notice to OSHA. The written notice shall state the date as of which the termination is to take effect. The Assistant Secretary shall inform the public of any voluntary termination by **FEDERAL REGISTER** notice.

### E. Revocation of recognition by OSHA.

1. *Potential causes.* If an NRTL either has failed to continue to substantially satisfy the requirements of §1910.7 or this appendix, or has not been reasonably performing the NRTL testing requirements encompassed within its letter of recognition, or has materially misrepresented itself in its applications or misrepresented the scope or conditions of its recognition, the Assistant Secretary may revoke the recognition of a recognized NRTL, in whole or in part. OSHA may initiate revocation procedures on the basis of information provided by any interested person.

2. *Procedure.* a. Before proposing to revoke recognition, the Agency will notify the recognized NRTL in writing, giving it the opportunity to rebut or correct the alleged deficiencies which would form the basis of the proposed revocation, within a reasonable period.

b. If the alleged deficiencies are not corrected or reconciled within a reasonable period, OSHA will propose, in writing to the recognized NRTL, to revoke recognition. If deemed appropriate, no other announcement need be made by OSHA.

c. The revocation shall be effective in 60 days unless within that period the recognized NRTL corrects the deficiencies or requests a hearing in writing.

d. If a hearing is requested, it shall be held before an administrative law judge of the Department of Labor pursuant to the rules specified in 29 CFR part 1905, subpart C.

e. The parties shall be OSHA and the recognized NRTL. The Assistant Secretary may allow other interested persons to participate in these hearings if such participation would contribute to the resolution of issues germane to the proceeding and not cause undue delay.

f. The burden of proof shall be on OSHA to demonstrate by a preponderance of the evidence that the recognition should be revoked because the NRTL is not meeting the requirements for recognition, has not been reasonably performing the product testing functions as required by §1910.7, this appendix A, or the letter of recognition, or has materially misrepresented itself in its applications or publicity.

3. *Final decision.* a. After the hearing, the Administrative Law Judge shall issue a decision stating the reasons based on the record as to whether it has been demonstrated, based on a preponderance of evidence, that the applicant does not continue to meet the requirements for its current recognition.

b. Upon issuance of the decision, any party to the hearing may file exceptions within 20 days pursuant to 29 CFR 1905.28. If no exceptions are filed, this decision is the final decision of the Assistant Secretary. If objections are filed, the Administrative Law Judge shall forward the decision, exceptions and record to the Assistant Secretary for the final decision on the proposed revocation.

c. The Assistant Secretary will review the record, the decision by the Administrative Law Judge, and the exceptions filed. Based on this, the Assistant Secretary shall issue the final decision as to whether it has been demonstrated, by a preponderance of evidence, that the recognized NRTL has not continued to meet the requirements for OSHA recognition. If the Assistant Secretary finds that the NRTL does not meet the NRTL recognition requirements, the recognition will be revoked.

4. *Public announcement.* A copy of the Assistant Secretary's final decision will be provided to the applicant, and a notification will be published in the FEDERAL REGISTER announcing the decision, and the availability of the complete record of this proceeding at OSHA. The effective date of any revocation will be the date the final decision copy is sent to the NRTL.

5. *Review of final decision.* There will be no further review activity available within the Department of Labor from the final decision of the Assistant Secretary.

[53 FR 12120, Apr. 12, 1988; 53 FR 16838, May 11, 1988, as amended at 54 FR 24333, June 7, 1989; 65 FR 46818, 46819, July 31, 2000]

### §1910.8 OMB control numbers under the Paperwork Reduction Act.

The following sections or paragraphs each contain a collection of information requirement which has been approved by the Office of Management and Budget under the control number listed.

| 29 CFR citation | OMB control No. |
| --- | --- |
| 1910.7 | 1218-0147 |
| 1910.23 | 1218-0199 |
| 1910.66 | 1218-0121 |
| 1910.67(b) | 1218-0230 |
| 1910.68 | 1218-0226 |
| 1910.95 | 1218-0048 |
| 1910.111 | 1218-0208 |
| 1910.119 | 1218-0200 |
| 1910.120 | 1218-0202 |
| 1910.132 | 1218-0205 |
| 1910.134 | 1218-0099 |
| 1910.137 | 1218-0190 |
| 1910.142 | 1218-0096 |
| 1910.145 | 1218-0132 |
| 1910.146 | 1218-0203 |
| 1910.147 | 1218-0150 |
| 1910.156 | 1218-0075 |
| 1910.157(e)(3) | 1218-0210 |
| 1910.157(f)(16) | 1218-0218 |
| 1910.177(d)(3)(iv) | 1218-0219 |
| 1910.179(j)(2)(iii) and (iv) | 1218-0224 |
| 1910.179(m)(1) and (m)(2) | 1218-0224 |
| 1910.180(d)(6) | 1218-0221 |
| 1910.180(g)(1) and (g)(2)(ii) | 1218-0221 |
| 1910.181(g)(1) and (g)(3) | 1218-0222 |
| 1910.184(e)(4), (f)(4) and (i)(8)(ii) | 1218-0223 |
| 1910.217(e)(1)(i) and (ii) | 1218-0229 |
| 1910.217(g) | 1218-0070 |

| 29 CFR citation | OMB control No. |
|---|---|
| 1910.217(h) | 1218-0143 |
| 1910.218(a)(2)(i) and (ii) | 1218-0228 |
| 1910.252(a)(2)(xiii)(c) | 1218-0207 |
| 1910.255(e) | 1218-0207 |
| 1910.266 | 1218-0198 |
| 1910.268 | 1218-0225 |
| 1910.269 | 1218-0190 |
| 1910.272 | 1218-0206 |
| 1910.420 | 1218-0069 |
| 1910.421 | 1218-0069 |
| 1910.423 | 1218-0069 |
| 1910.430 | 1218-0069 |
| 1910.440 | 1218-0069 |
| 1910.1001 | 1218-0133 |
| 1910.1003 | 1218-0085 |
| 1910.1004 | 1218-0084 |
| 1910.1006 | 1218-0086 |
| 1910.1007 | 1218-0083 |
| 1910.1008 | 1218-0087 |
| 1910.1009 | 1218-0089 |
| 1910.1010 | 1218-0082 |
| 1910.1011 | 1218-0090 |
| 1910.1012 | 1218-0080 |
| 1910.1013 | 1218-0079 |
| 1910.1014 | 1218-0088 |
| 1910.1015 | 1218-0044 |
| 1910.1016 | 1218-0081 |
| 1910.1017 | 1218-0010 |
| 1910.1018 | 1218-0104 |
| 1910.1020 | 1218-0065 |
| 1910.1025 | 1218-0092 |
| 1910.1027 | 1218-0185 |
| 1910.1028 | 1218-0129 |
| 1910.1029 | 1218-0128 |
| 1910.1030 | 1218-0180 |
| 1910.1043 | 1218-0061 |
| 1910.1044 | 1218-0101 |
| 1910.1045 | 1218-0126 |
| 1910.1047 | 1218-0108 |
| 1910.1048 | 1218-0145 |
| 1910.1050 | 1218-0184 |
| 1910.1051 | 1218-0170 |
| 1910.1052 | 1218-0179 |
| 1910.1096 | 1218-0103 |
| 1910.1200 | 1218-0072 |
| 1910.1450 | 1218-0131 |

[61 FR 5508, Feb. 13, 1996, as amended at 62 FR 29668, June 2, 1997; 62 FR 42666, Aug. 8, 1997; 62 FR 43581, Aug. 14, 1997; 62 FR 65203, Dec. 11, 1997; 63 FR 13340, Mar. 19, 1998; 63 FR 17093, Apr. 8, 1998]

# Subpart B—
# Adoption and Extension of Established Federal Standards

> **Authority:** Secs. 4, 6, and 8 of the Occupational Safety and Health Act, 29 U.S.C. 653, 655, 657; Walsh-Healey Act, 41 U.S.C. 35 *et seq.*; Service Contract Act of 1965, 41 U.S.C. 351 *et seq.*; Sec.107, Contract Work Hours and Safety Standards Act (Construction Safety Act), 40 U.S.C. 333; Sec. 41, Longshore and Harbor Workers' Compensation Act, 33 U.S.C. 941; National Foundation of Arts and Humanities Act, 20 U.S.C. 951 *et seq.*; Secretary of Labor's Order No. 12-71 (36 FR 8754), 8-76 (41 FR 1911), 9-83 (48 FR 35736), 1-90 (55 FR 9033), or 6-96 (62 FR 111), as applicable.

## §1910.11 Scope and purpose.

**1910.11(a)** The provisions of this subpart B adopt and extend the applicability of, established Federal standards in effect on April 28, 1971, with respect to every employer, employee, and employment covered by the Act.

**1910.11(b)** It bears emphasis that only standards (i.e., substantive rules) relating to safety or health are adopted by any incorporations by reference of standards prescribed elsewhere in this chapter or this title. Other materials contained in the referenced parties are not adopted. Illustrations of the types of materials which are not adopted are these. The incorporations by reference of parts 1915, 1916, 1917, 1918 in §§1910.13, 1910.14, 1910.15, and 1910.16 are not intended to include the discussion in those parts of the coverage of the Longshoremen's and Harbor Workers' Compensation Act or the penalty provisions of the Act. Similarly, the incorporation by reference of part 1926 in §1910.12 is not intended to include references to interpretative rules having relevance to the application of the Construction Safety Act, but having no relevance to the application to the Occupational Safety and Health Act.

## §1910.12 Construction work.

**1910.12(a)** *Standards*. The standards prescribed in part 1926 of this chapter are adopted as occupational safety and health standards under section 6 of the Act and shall apply, according to the provisions thereof, to every employment and place of employment of every employee engaged in construction work. Each employer shall protect the employment and places of employment of each of his employees engaged in construction work by complying with the appropriate standards prescribed in this paragraph.

**1910.12(b)** *Definition*. For purposes of this section, *Construction work* means work for construction, alteration, and/or repair, including painting and decorating. See discussion of these terms in §1926.13 of this title.

**1910.12(c)** *Construction Safety Act distinguished.* This section adopts as occupational safety and health standards under section 6 of the Act the standards which are prescribed in part 1926 of this chapter. Thus, the standards (substantive rules) published in subpart C and the following subparts of part 1926 of this chapter are applied. This section does not incorporate subparts A and B of part 1926 of this chapter. Subparts A and B have pertinence only to the application of section 107 of the Contract Work Hours and Safety Standards Act (the Construction Safety Act). For example, the interpretation of the term "subcontractor" in paragraph (c) of §1926.13 of this chapter is significant in discerning the coverage of the Construction Safety Act and duties thereunder. However, the term "subcontractor" has no significance in the application of the Act, which was enacted under the Commerce Clause and which establishes duties for "employers" which are not dependent for their application upon any contractual relationship with the Federal Government or upon any form of Federal financial assistance.

**1910.12(d)** For the purposes of this part, to the extent that it may not already be included in paragraph (b) of this section, "construction work" includes the erection of new electric transmission and distribution lines and equipment, and the alteration, conversion, and improvement of the existing transmission and distribution lines and equipment.

## §1910.15 Shipyard employment.

**1910.15(a)** *Adoption and extension of established safety and health standards for shipyard employment.* The standards prescribed by part 1915 (formerly parts 1501-1503) of this title and in effect on April 28, 1971 (as revised), are adopted as occupational safety or health standards under section 6(a) of the Act and shall apply, according to the provisions thereof, to every employment and place of employment of every employee engaged in ship repair, shipbreaking, and shipbuilding, or a related employment. Each employer shall protect the employment and places of employment of each of his employees engaged in ship repair, shipbreaking, and shipbuilding, or a related employment, by complying with the appropriate standards prescribed by this paragraph.

**1910.15(b)** *Definitions*. For purposes of this section:

**1910.15(b)(1)** *Ship repair* means any repair of a vessel, including, but not restricted to, alterations, conversions, installations, cleaning, painting, and maintenance work;

**1910.15(b)(2)** *Shipbreaking* means any breaking down of a vessel's structure for the purpose of scrapping the vessel, including the removal of gear, equipment, or any component of a vessel;

**1910.15(b)(3)** *Shipbuilding* means the construction of a vessel, including the installation of machinery and equipment;

**1910.15(b)(4)** *Related employment* means any employment performed as an incident to, or in conjunction with, ship repair, shipbreaking, and shipbuilding work, including, but not restricted to, inspection, testing, and employment as a watchman; and

**1910.15(b)(5)** *Vessel* includes every description of watercraft or other artificial contrivance used, or capable of being used, as a means of transportation on water, including special purpose floating structures not primarily designed for, or used as a means of, transportation on water.

[58 FR 35308, June 30, 1993]

## §1910.16 Longshoring and marine terminals.

**1910.16(a)** *Safety and health standards for longshoring.* **(1)** Part 1918 of this chapter shall apply exclusively, according to the provisions thereof, to all employment of every employee engaged in longshoring operations or related employment aboard any vessel. All cargo transfer accomplished with the use of shore-based material handling devices shall be governed by part 1917 of this chapter.

**1910.16(a)(2)** Part 1910 does not apply to longshoring operations except for the following provisions:

**1910.16(a)(2)(i)** *Access to employee exposure and medical records.* Subpart Z, §1910.1020;

**1910.16(a)(2)(ii)** *Commercial diving operations.* Subpart T;

**1910.16(a)(2)(iii)** *Electrical.* Subpart S when shore-based electrical installations provide power for use aboard vessels;

**1910.16(a)(2)(iv)** *Hazard communication.* Subpart Z, §1910.1200;

**1910.16(a)(2)(v)** *Ionizing radiation.* Subpart Z, §1910.1096;

**1910.16(a)(2)(vi)** *Noise.* Subpart G, §1910.95;

**1910.16(a)(2)(vii)** *Nonionizing radiation.* Subpart G, §1910.97;

**Note to paragraph** (a)(2)(vii): Exposures to nonionizing radiation emissions from commercial vessel transmitters are considered hazardous under the following conditions: (1) where the radar is transmitting, the scanner is stationary, and the exposure distance is 18.7 feet (6 m.) or less; or (2) where the radar is transmitting, the scanner is rotating, and the exposure distance is 5.2 feet (1.8 m.) or less.

**1910.16(a)(2)(viii)** *Respiratory protection.* Subpart I, §1910.134;

**1910.16(a)(2)(ix)** *Toxic and hazardous substances.* Subpart Z applies to marine cargo handling activities except for the following:

**1910.16(a)(2)(ix)(A)** When a substance or cargo is contained within a sealed, intact means of packaging or containment complying with Department of Transportation or International Maritime Organization requirements;[1]

**1910.16(a)(2)(ix)(B)** Bloodborne pathogens, §1910.1030;

---

[1] The International Maritime Organization publishes the International Maritime Dangerous Goods Code to aid compliance with the international legal requirements of the International Convention for the Safety of Life at Sea, 1960.

**1910.16(a)(2)(ix)(C)** Carbon monoxide, §1910.1000 (See §1918.94 (a)); and

**1910.16(a)(2)(ix)(D)** Hydrogen sulfide, §1910.1000 (See §1918.94 (f)).

**1910.16(a)(2)(x)** Powered industrial truck operator training, Subpart N, §1910.178(l).

**1910.16(b)** *Safety and health standards for marine terminals.* Part 1917 of this chapter shall apply exclusively, according to the provisions thereof, to employment within a marine terminal, except as follows:

**1910.16(b)(1)** The provisions of part 1917 of this chapter do not apply to the following:

**1910.16(b)(1)(i)** Facilities used solely for the bulk storage, handling, and transfer of flammable and combustible liquids and gases.

**1910.16(b)(1)(ii)** Facilities subject to the regulations of the Office of Pipeline Safety of the Research and Special Programs Administration, Department of Transportation (49 CFR chapter I, subchapter D), to the extent such regulations apply to specific working conditions.

**1910.16(b)(1)(iii)** Fully automated bulk coal handling facilities contiguous to electrical power generating plants.

**1910.16(b)(2)** Part 1910 does not apply to marine terminals except for the following:

**1910.16(b)(2)(i)** *Abrasive blasting.* Subpart G, §1910.94(a);

**1910.16(b)(2)(ii)** *Access to employee exposure and medical records.* Subpart Z, §1910.1020;

**1910.16(b)(2)(iii)** *Commercial diving operations.* Subpart T;

**1910.16(b)(2)(iv)** *Electrical.* Subpart S;

**1910.16(b)(2)(v)** *Grain handling facilities.* Subpart R, §1910.272;

**1910.16(b)(2)(vi)** *Hazard communication.* Subpart Z, §1910.1200;

**1910.16(b)(2)(vii)** *Ionizing radiation.* Subpart Z, §1910.1096;

**1910.16(b)(2)(viii)** *Noise.* Subpart G, §1910.95;

**1910.16(b)(2)(ix)** *Nonionizing radiation.* Subpart G, §1910.97;

**1910.16(b)(2)(x)** *Respiratory protection.* Subpart I, §1910.134;

**1910.16(b)(2)(xi)** *Safety requirements for scaffolding.* Subpart D, §1910.28;

**1910.16(b)(2)(xii)** *Servicing multi-piece and single piece rim wheels.* Subpart N, §1910.177;

**1910.16(b)(2)(xiii)** *Toxic and hazardous substances.* Subpart Z applies to marine cargo handling activities except for the following:

**1910.16(b)(2)(xiii)(A)** When a substance or cargo is contained within a sealed, intact means of packaging or containment complying with Department of Transportation or International Maritime Organization requirements;[2]

**1910.16(b)(2)(xiii)(B)** Bloodborne pathogens, §1910.1030;

**1910.16(b)(2)(xiii)(C)** Carbon monoxide, §1910.1000 (See §1917.24(a)); and

**1910.16(b)(2)(xiii)(D)** Hydrogen sulfide, §1910.1000 (See §1917.73(a)(2)); and

**1910.16(b)(2)(xiv)** Powered industrial truck operator training, Subpart N, §1910.178(l).

**1910.16(c)** *Definitions.* For purposes of this section:

**1910.16(c)(1)** *Longshoring operation* means the loading, unloading, moving, or handling of, cargo, ship's stores, gear, etc., into, in, on, or out of any vessel;

**1910.16(c)(2)** *Related employment* means any employment performed as an incident to or in conjunction with, longshoring operations including, but not restricted to, securing cargo, rigging, and employment as a porter, checker, or watchman; and

**1910.16(c)(3)** *Vessel* includes every description of watercraft or other artificial contrivance used, or capable of being used, as a means of transportation on water, including special purpose floating structures not primarily designed for, or used as a means of, transportation on water.

**1910.16(c)(4)** *Marine terminal* means wharves, bulkheads, quays, piers, docks and other berthing locations and adjacent storage or adjacent areas and structures associated with the primary movement of cargo or materials from vessel to shore or shore to vessel including structures which are devoted to receiving, handling, holding, consolidation and loading or delivery of waterborne shipments or passengers, including areas devoted to the maintenance of the terminal or equipment. The term does not include production or manufacturing areas having their own docking facilities and located at a marine terminal nor does the term include storage facilities directly associated with those production or manufacturing areas.

[39 FR 23502, June 27, 1974, as amended at 48 FR 30908, July 5, 1983; 52 FR 36026, Sept. 25, 1987; 62 FR 40195, July 25, 1997; 63 FR 66270, Dec. 1, 1998]

### §1910.17 Effective dates.

**1910.17(a)-(b)** [Reserved]

**1910.17(c)** Except whenever any employment or place of employment is, or becomes, subject to any safety and health standard prescribed in part 1915, 1916, 1917, 1918, or 1926 of this title on a date before August 27, 1971, by virtue of the Construction Safety Act or the Longshoremen's and Harbor Workers' Compensation Act, that occupational safety and health standard as incorporated by reference in this subpart shall also become effective under the Williams-Steiger Occupational Safety and Health Act of 1970 on that date.

[39 FR 23502, June 27, 1974, as amended at 61 FR 9235, Mar. 7, 1996]

### §1910.18 Changes in established Federal standards.

Whenever an occupational safety and health standard adopted and incorporated by reference in this subpart B is changed pursuant to section 6(b) of the Act and the statute under which the standard was originally promulgated, and in accordance with part 1911 of this chapter, the standard shall be deemed changed for purposes of that statute and this subpart B, and shall apply under this subpart B. For the purposes of this section, a change in a standard includes any amendment, addition, or repeal, in whole or in part, of any standard.

### §1910.19 Special provisions for air contaminants.

**1910.19(a)** *Asbestos, tremolite, anthophyllite, and actinolite dust.* Section 1910.1001 shall apply to the exposure of every employee to asbestos, tremolite, anthophyllite, and actinolite dust in every employment and place of employment covered by §1910.16, in lieu of any different standard on exposure to asbestos, tremolite, anthophyllite, and actinolite dust which would otherwise be applicable by virtue of any of those sections.

**1910.19(b)** *Vinyl chloride.* Section 1910.1017 shall apply to the exposure of every employee to vinyl chloride in every employment and place of employment covered by §§1910.12, 1910.13, 1910.14, 1910.15, or 1910.16, in lieu of any different standard on exposure to vinyl chloride which would otherwise be applicable by virtue of any of those sections.

**1910.19(c)** *Acrylonitrile.* Section 1910.1045 shall apply to the exposure of every employee to acrylonitrile in every employment and place of employment covered by §§1910.12, 1910.13, 1910.14, 1910.15, or §1910.16, in lieu of any different standard on exposure to acrylonitrile which would otherwise be applicable by virtue of any of those sections.

**1910.19(d)** [Reserved]

**1910.19(e)** *Inorganic arsenic.* Section 1910.1018 shall apply to the exposure of every employee to inorganic arsenic in every employment covered by §§1910.12, 1910.13, 1910.14, 1910.15, or §1910.16, in lieu of any different standard on exposure to inorganic arsenic which would otherwise be applicable by virtue of any of those sections.

**1910.19(f)** [Reserved]

**1910.19(g)** *Lead.* Section 1910.1025 shall apply to the exposure of every employee to lead in every employment and place of employment covered by §§1910.13, 1910.14, 1910.15, and 1910.16, in lieu of any different standard on exposure to lead which would otherwise be applicable by virtue of those sections.

**1910.19(h)** *Ethylene oxide.* Section 1910.1047 shall apply to the exposure of every employee to ethylene oxide in every employment and place of employment covered by §§1910.12, 1910.13, 1910.14, 1910.15, or 1910.16, in

---

[2] The International Maritime Organization publishes the International Maritime Dangerous Goods Code to aid compliance with the international legal requirements of the International Convention of the Safety of Life at Sea, 1960.

lieu of any different standard on exposure to ethylene oxide which would otherwise be applicable by virtue of those sections.

**1910.19(i)** *4,4'-Methylenedianiline (MDA).* Section 1910.1050 shall apply to the exposure of every employee to MDA in every employment and place of employment covered by §§1910.13, 1910.14, 1910.15, or 1910.16, in lieu of any different standard on exposure to MDA which would otherwise be applicable by virtue of those sections.

**1910.19(j)** *Formaldehyde.* Section 1910.1048 shall apply to the exposure of every employee to formaldehyde in every employment and place of employment covered by §1910.12, 1910.13, 1910.14, 1910.15 or 1910.16 in lieu of any different standard on exposure to formaldehyde which would otherwise be applicable by virtue of those sections.

**1910.19(k)** *Cadmium.* Section 1910.1027 shall apply to the exposure of every employee to cadmium in every employment and place of employment covered by 1910.16 in lieu of any different standard on exposures to cadmium that would otherwise be applicable by virtue of those sections.

**1910.19(l)** *1,3-Butadiene (BD).* Section 1910.1051 shall apply to the exposure of every employee to BD in every employment and place of employment covered by §§1910.12, 1910.13, 1910.14, 1910.15, or 1910.16, in lieu of any different standard on exposure to BD which would otherwise be applicable by virtue of those sections.

**1910.19(m)** *Methylene chloride (MC).* Section 1910.1052 shall apply to the exposure of every employee to MC in every employment and place of employment covered by §1910.16 in lieu of any different standard on exposure to MC which would otherwise be applicable by virtue of that section when it is not present in sealed, intact containers.

[43 FR 28473, June 30, 1978, as amended at 43 FR 45809, Oct. 3, 1978; 43 FR 53007, Nov. 14, 1978; 44 FR 5447, Jan. 26, 1979; 46 FR 32022, June 19, 1981; 49 FR 25796, June 22, 1984; 50 FR 51173, Dec. 13, 1985; 52 FR 46291, Dec. 4, 1987; 57 FR 35666, Aug. 10, 1992; 57 FR 42388, Sept. 14, 1992; 59 FR 41057, Aug. 10, 1994; 61 FR 56831, Nov. 4, 1996; 62 FR 1600, Jan. 10, 1997]

# Subpart C—Reserved

### §1910.20 Access to employee exposure and medical records. [Removed.]

[§1910.20 revised at 53 FR 38163, Sept. 29, 1988; removed and redesignated as §1910.1020 at 61 FR 31430, June 20, 1996, effective June 30, 1996]

[Subpart C removed and reserved, at 61 FR 31430, June 20, 1996, effective June 30, 1996]

# Subpart D—Walking-Working Surfaces

> **Authority:** Secs. 4, 6, and 8 of the Occupational Safety and Health Act of 1970 (29 U.S.C. 653, 655, and 657); Secretary of Labor's Order No. 12-71 (36 FR 8754), 8-76 (41 FR 25059), 9-83 (48 FR 35736), or 1-90 (55 FR 9033), as applicable; and 29 CFR part 1911.

## §1910.21 Definitions.

**1910.21(a)** As used in §1910.23, unless the context requires otherwise, floor and wall opening, railing and toe board terms shall have the meanings ascribed in this paragraph.

**1910.21(a)(1)** *Floor hole.* An opening measuring less than 12 inches but more than 1 inch in its least dimension, in any floor, platform, pavement, or yard, through which materials but not persons may fall; such as a belt hole, pipe opening, or slot opening.

**1910.21(a)(2)** *Floor opening.* An opening measuring 12 inches or more in its least dimension, in any floor, platform, pavement, or yard through which persons may fall; such as a hatchway, stair or ladder opening, pit, or large manhole. Floor openings occupied by elevators, dumb waiters, conveyors, machinery, or containers are excluded from this subpart.

**1910.21(a)(3)** *Handrail.* A single bar or pipe supported on brackets from a wall or partition, as on a stairway or ramp, to furnish persons with a handhold in case of tripping.

**1910.21(a)(4)** *Platform.* A working space for persons, elevated above the surrounding floor or ground; such as a balcony or platform for the operation of machinery and equipment.

**1910.21(a)(5)** *Runway.* A passageway for persons, elevated above the surrounding floor or ground level, such as a footwalk along shafting or a walkway between buildings.

**1910.21(a)(6)** *Standard railing.* A vertical barrier erected along exposed edges of a floor opening, wall opening, ramp, platform, or runway to prevent falls of persons.

**1910.21(a)(7)** *Standard strength and construction.* Any construction of railings, covers, or other guards that meets the requirements of §1910.23.

**1910.21(a)(8)** *Stair railing.* A vertical barrier erected along exposed sides of a stairway to prevent falls of persons.

**1910.21(a)(9)** *Toeboard.* A vertical barrier at floor level erected along exposed edges of a floor opening, wall opening, platform, runway, or ramp to prevent falls of materials.

**1910.21(a)(10)** *Wall hole.* An opening less than 30 inches but more than 1 inch high, of unrestricted width, in any wall or partition; such as a ventilation hole or drainage scupper.

**1910.21(a)(11)** *Wall opening.* An opening at least 30 inches high and 18 inches wide, in any wall or partition, through which persons may fall; such as a yard-arm doorway or chute opening.

**1910.21(b)** As used in §1910.24, unless the context requires otherwise, fixed industrial stair terms shall have the meaning ascribed in this paragraph.

**1910.21(b)(1)** *Handrail.* A single bar or pipe supported on brackets from a wall or partition to provide a continuous handhold for persons using a stair.

**1910.21(b)(2)** *Nose, nosing.* That portion of a tread projecting beyond the face of the riser immediately below.

**1910.21(b)(3)** *Open riser.* The air space between the treads of stairways without upright members (risers).

**1910.21(b)(4)** *Platform.* An extended step or landing breaking a continuous run of stairs.

**1910.21(b)(5)** *Railing.* A vertical barrier erected along exposed sides of stairways and platforms to prevent falls of persons. The top member of railing usually serves as a handrail.

**1910.21(b)(6)** *Rise.* The vertical distance from the top of a tread to the top of the next higher tread.

**1910.21(b)(7)** *Riser.* The upright member of a step situated at the back of a lower tread and near the leading edge of the next higher tread.

**1910.21(b)(8)** *Stairs, stairway.* A series of steps leading from one level or floor to another, or leading to platforms, pits, boiler rooms, crossovers, or around machinery, tanks, and other equipment that are used more or less continuously or routinely by employees, or only occasionally by specific individuals. A series of steps and landings having three or more risers constitutes stairs or stairway.

**1910.21(b)(9)** *Tread.* The horizontal member of a step.

**1910.21(b)(10)** *Tread run.* The horizontal distance from the leading edge of a tread to the leading edge of an adjacent tread.

**1910.21(b)(11)** *Tread width.* The horizontal distance from front to back of tread including nosing when used.

**1910.21(c)** As used in §1910.25, unless the context requires otherwise, portable wood ladders terms shall have the meanings ascribed in this paragraph.

**1910.21(c)(1)** *Ladders.* A ladder is an appliance usually consisting of two side rails joined at regular intervals by cross-pieces called steps, rungs, or cleats, on which a person may step in ascending or descending.

**1910.21(c)(2)** *Stepladder.* A stepladder is a self-supporting portable ladder, nonadjustable in length, having flat steps and a hinged back. Its size is designated by the overall length of the ladder measured along the front edge of the side rails.

**1910.21(c)(3)** *Single ladder.* A single ladder is a non-self-supporting portable ladder, nonadjustable in length, consisting of but one section. Its size is designated by the overall length of the side rail.

**1910.21(c)(4)** *Extension ladder.* An extension ladder is a non-self-supporting portable ladder adjustable in length. It consists of two or more sections traveling in guides or brackets so arranged as to permit length adjustment. Its size is designated by the sum of the lengths of the sections measured along the side rails.

**1910.21(c)(5)** *Sectional ladder.* A sectional ladder is a non-self-supporting portable ladder, nonadjustable in length, consisting of two or more sections of ladder so constructed that the sections may be combined to function as a single ladder. Its size is designated by the overall length of the assembled sections.

**1910.21(c)(6)** *Trestle ladder.* A trestle ladder is a self-supporting portable ladder, nonadjustable in length, consisting of two sections hinged at the top to form equal angles with the base. The size is designated by the length of the side rails measured along the front edge.

**1910.21(c)(7)** *Extension trestle ladder.* An extension trestle ladder is a self-supporting portable ladder, adjustable in length, consisting of a trestle ladder base and a vertically adjustable single ladder, with suitable means for locking the ladders together. The size is designated by the length of the trestle ladder base.

**1910.21(c)(8)** *Special-purpose ladder.* A special-purpose ladder is a portable ladder which represents either a modification or a combination of design or construction features in one of the general-purpose types of ladders previously defined, in order to adapt the ladder to special or specific uses.

**1910.21(c)(9)** *Trolley ladder.* A trolley ladder is a semifixed ladder, nonadjustable in length, supported by attachments to an overhead track, the plane of the ladder being at right angles to the plane of motion.

**1910.21(c)(10)** *Side-rolling ladder.* A side-rolling ladder is a semifixed ladder, nonadjustable in length, supported by attachments to a guide rail, which is generally fastened to shelving, the plane of the ladder being also its plane of motion.

**1910.21(c)(11)** *Wood characteristics.* Wood characteristics are distinguishing features which by their extent and number determine the quality of a piece of wood.

**1910.21(c)(12)** *Wood irregularities.* Wood irregularities are natural characteristics in or on wood that may lower its durability, strength, or utility.

**1910.21(c)(13)** *Cross grain.* Cross grain (slope of grain) is a deviation of the fiber direction from a line parallel to the sides of the piece.

**1910.21(c)(14)** *Knot.* A knot is a branch or limb, imbedded in the tree and cut through in the process of lumber manufacture, classified according to size, quality, and occurrence. The size of the knot is determined as the average diameter on the surface of the piece.

**1910.21(c)(15)** *Pitch and bark pockets.* A pitch pocket is an opening extending parallel to the annual growth rings containing, or that has contained, pitch, either solid or liquid. A bark pocket is an opening between annual growth rings that contains bark.

**1910.21(c)(16)** *Shake.* A shake is a separation along the grain, most of which occurs between the rings of annual growth.

**1910.21(c)(17)** *Check.* A check is a lengthwise separation of the wood, most of which occurs across the rings of annual growth.

**1910.21(c)(18)** *Wane.* Wane is bark, or the lack of wood from any cause, on the corner of a piece.

**1910.21(c)(19)** *Decay.* Decay is disintegration of wood substance due to action of wood-destroying fungi. It is also known as dote and rot.

**1910.21(c)(20)** *Compression failure.* A compression failure is a deformation (buckling) of the fibers due to excessive compression along the grain.

**1910.21(c)(21)** *Compression wood.* Compression wood is an aberrant (abnormal) and highly variable type of wood structure occurring in softwood species. The wood commonly has density somewhat higher than does normal wood, but somewhat lower stiffness and tensile strength for its weight in addition to high longitudinal shrinkage.

**1910.21(c)(22)** *Low density.* Low-density wood is that which is exceptionally light in weight and usually deficient in strength properties for the species.

**1910.21(d)** As used in §1910.26, unless the context requires otherwise, portable metal ladder terms shall have the meanings ascribed in this paragraph.

**1910.21(d)(1)** *Ladder.* A ladder is an appliance usually consisting of two side rails joined at regular intervals by cross-pieces called steps, rungs, or cleats, on which a person may step in ascending or descending.

**1910.21(d)(2)** *Step ladder.* A step ladder is a self-supporting portable ladder, nonadjustable in length, having flat steps and a hinged back. Its size is designated by the overall length of the ladder measured along the front edge of the side rails.

**1910.21(d)(3)** *Single ladder.* A single ladder is a non-self-supporting portable ladder, nonadjustable in length, consisting of but one section. Its size is designated by the overall length of the side rail.

**1910.21(d)(4)** *Extension ladder.* An extension ladder is a non-self-supporting portable ladder adjustable in length. It consists of two or more sections traveling in guides or brackets so arranged as to permit length adjustment. Its size is designated by the sum of the lengths of the sections measured along the side rails.

**1910.21(d)(5)** *Platform ladder.* A self-supporting ladder of fixed size with a platform provided at the working level. The size is determined by the distance along the front rail from the platform to the base of the ladder.

**1910.21(d)(6)** *Sectional ladder.* A sectional ladder is a non-self-supporting portable ladder, non-adjustable in length, consisting of two or more sections so constructed that the sections may be combined to function as a single ladder. Its size is designated by the overall length of the assembled sections.

**1910.21(d)(7)** *Trestle ladder.* A trestle ladder is a self-supporting portable ladder, non-adjustable in length, consisting of two sections, hinged at the top to form equal angles with the base. The size is designated by the length of the side rails measured along the front edge.

**1910.21(d)(8)** *Extension trestle ladder.* An extension trestle ladder is a self-supporting portable ladder, adjustable in length, consisting of a trestle ladder base and a vertically adjustable single ladder, with suitable means for locking the ladders together. The size is designated by the length of the trestle ladder base.

**1910.21(d)(9)** *Special-purpose ladder.* A special-purpose ladder is a portable ladder which represents either a modification or a combination of design or construction features in one of the general-purpose types of ladders previously defined, in order to adapt the ladder to special or specific uses.

**1910.21(e)** As used in §1910.27, unless the context requires otherwise, fixed ladder terms shall have the meanings ascribed in this paragraph.

**1910.21(e)(1)** *Ladder.* A ladder is an appliance usually consisting of two side rails joined at regular intervals by cross-pieces called steps, rungs, or cleats, on which a person may step in ascending or descending.

**1910.21(e)(2)** *Fixed ladder.* A fixed ladder is a ladder permanently attached to a structure, building, or equipment.

**1910.21(e)(3)** *Individual-rung ladder.* An individual-rung ladder is a fixed ladder each rung of which is individually attached to a structure, building, or equipment.

**1910.21(e)(4)** *Rail ladder.* A rail ladder is a fixed ladder consisting of side rails joined at regular intervals by rungs or cleats and fastened in full length or in sections to a building, structure, or equipment.

**1910.21(e)(5)** *Railings.* A railing is any one or a combination of those railings constructed in accordance with §1910.23. A standard railing is a vertical barrier erected along exposed edges of floor openings, wall openings, ramps, platforms, and runways to prevent falls of persons.

**1910.21(e)(6)** *Pitch.* Pitch is the included angle between the horizontal and the ladder, measured on the opposite side of the ladder from the climbing side.

**1910.21(e)(7)** *Fastenings.* A fastening is a device to attach a ladder to a structure, building, or equipment.

**1910.21(e)(8)** *Rungs.* Rungs are ladder cross-pieces of circular or oval cross-section on which a person may step in ascending or descending.

**1910.21(e)(9)** *Cleats.* Cleats are ladder cross-pieces of rectangular cross-section placed on edge on which a person may step in ascending or descending.

**1910.21(e)(10)** *Steps.* Steps are the flat cross-pieces of a ladder on which a person may step in ascending or descending.

**1910.21(e)(11)** *Cage.* A cage is a guard that may be referred to as a cage or basket guard which is an enclosure that is fastened to the side rails of the fixed ladder or to the structure to encircle the climbing space of the ladder for the safety of the person who must climb the ladder.

**1910.21(e)(12)** *Well.* A well is a permanent complete enclosure around a fixed ladder, which is attached to the walls of the well. Proper clearances for a well will give the person who must climb the ladder the same protection as a cage.

**1910.21(e)(13)** *Ladder safety device.* A ladder safety device is any device, other than a cage or well, designed to eliminate or reduce the possibility of accidental falls and which may incorporate such features as life belts, friction brakes, and sliding attachments.

**1910.21(e)(14)** *Grab bars.* Grab bars are individual handholds placed adjacent to or as an extension above ladders for the purpose of providing access beyond the limits of the ladder.

**1910.21(e)(15)** *Through ladder.* A through ladder is one from which a man getting off at the top must step through the ladder in order to reach the landing.

**1910.21(e)(16)** *Side-step ladder.* A side-step ladder is one from which a man getting off at the top must step sideways from the ladder in order to reach the landing.

**1910.21(f)** As used in §1910.28, unless the context requires otherwise, scaffolding terms shall have the meaning ascribed in this paragraph.

**1910.21(f)(1)** *Bearer.* A horizontal member of a scaffold upon which the platform rests and which may be supported by ledgers.

**1910.21(f)(2)** *Boatswain's chair.* A seat supported by slings attached to a suspended rope, designed to accommodate one workman in a sitting position.

**1910.21(f)(3)** *Brace.* A tie that holds one scaffold member in a fixed position with respect to another member.

**1910.21(f)(4)** *Bricklayers' square scaffold.* A scaffold composed of framed wood squares which support a platform limited to light and medium duty.

**1910.21(f)(5)** *Carpenters' bracket scaffold.* A scaffold consisting of wood or metal brackets supporting a platform.

**1910.21(f)(6)** *Coupler.* A device for locking together the component parts of a tubular metal scaffold. The material used for the couplers shall be of a structural type, such as a drop-forged steel, malleable iron, or structural grade aluminum. The use of gray cast iron is prohibited.

**1910.21(f)(7)** *Crawling board or chicken ladder.* A plank with cleats spaced and secured at equal intervals, for use by a worker on roofs, not designed to carry any material.

**1910.21(f)(8)** *Double pole or independent pole scaffold.* A scaffold supported from the base by a double row of uprights, independent of support from the walls and constructed of uprights, ledgers, horizontal platform bearers, and diagonal bracing.

**1910.21(f)(9)** *Float or ship scaffold.* A scaffold hung from overhead supports by means of ropes and consisting of a substantial platform having diagonal bracing underneath, resting upon and securely fastened to two parallel plank bearers at right angles to the span.

# OSHA Standards for General Industry

**1910.21(f)(10)** *Guardrail.* A rail secured to uprights and erected along the exposed sides and ends of platforms.

**1910.21(f)(11)** *Heavy duty scaffold.* A scaffold designed and constructed to carry a working load not to exceed 75 pounds per square foot.

**1910.21(f)(12)** *Horse scaffold.* A scaffold for light or medium duty, composed of horses supporting a work platform.

**1910.21(f)(13)** *Interior hung scaffold.* A scaffold suspended from the ceiling or roof structure.

**1910.21(f)(14)** *Ladder jack scaffold.* A light duty scaffold supported by brackets attached to ladders.

**1910.21(f)(15)** *Ledger (stringer).* A horizontal scaffold member which extends from post to post and which supports the putlogs or bearer forming a tie between the posts.

**1910.21(f)(16)** *Light duty scaffold.* A scaffold designed and constructed to carry a working load not to exceed 25 pounds per square foot.

**1910.21(f)(17)** *Manually propelled mobile scaffold.* A portable rolling scaffold supported by casters.

**1910.21(f)(18)** *Masons' adjustable multiple-point suspension scaffold.* A scaffold having a continuous platform supported by bearers suspended by wire rope from overhead supports, so arranged and operated as to permit the raising or lowering of the platform to desired working positions.

**1910.21(f)(19)** *Maximum intended load.* The total of all loads including the working load, the weight of the scaffold, and such other loads as may be reasonably anticipated.

**1910.21(f)(20)** *Medium duty scaffold.* A scaffold designed and constructed to carry a working load not to exceed 50 pounds per square foot.

**1910.21(f)(21)** *Mid-rail.* A rail approximately midway between the guardrail and platform, used when required, and secured to the uprights erected along the exposed sides and ends of platforms.

**1910.21(f)(22)** *Needle beam scaffold.* A light duty scaffold consisting of needle beams supporting a platform.

**1910.21(f)(23)** *Outrigger scaffold.* A scaffold supported by outriggers or thrustouts projecting beyond the wall or face of the building or structure, the inboard ends of which are secured inside of such a building or structure.

**1910.21(f)(24)** *Putlog.* A scaffold member upon which the platform rests.

**1910.21(f)(25)** *Roofing bracket.* A bracket used in sloped roof construction, having provisions for fastening to the roof or supported by ropes fastened over the ridge and secured to some suitable object.

**1910.21(f)(26)** *Runner.* The lengthwise horizontal bracing or bearing members or both.

**1910.21(f)(27)** *Scaffold.* Any temporary elevated platform and its supporting structure used for supporting workmen or materials or both.

**1910.21(f)(28)** *Single-point adjustable suspension scaffold.* A manually or power-operated unit designed for light duty use, supported by a single wire rope from an overhead support so arranged and operated as to permit the raising or lowering of the platform to desired working positions.

**1910.21(f)(29)** *Single pole scaffold.* Platforms resting on putlogs or crossbeams, the outside ends of which are supported on ledgers secured to a single row of posts or uprights and the inner ends of which are supported on or in a wall.

**1910.21(f)(30)** *Stone setters' adjustable multiple-point suspension scaffold.* A swinging-type scaffold having a platform supported by hangers suspended at four points so as to permit the raising or lowering of the platform to the desired working position by the use of hoisting machines.

**1910.21(f)(31)** *Toeboard.* A barrier secured along the sides and ends on a platform, to guard against the falling of material.

**1910.21(f)(32)** *Tube and coupler scaffold.* An assembly consisting of tubing which serves as posts, bearers, braces, ties, and runners, a base supporting the posts, and special couplers which serve to connect the uprights and to join the various members.

**1910.21(f)(33)** *Tubular welded frame scaffold.* A sectional, panel, or frame metal scaffold substantially built up of prefabricated welded sections which consist of posts and horizontal bearer with intermediate members. Panels or frames shall be braced with diagonal or cross braces.

**1910.21(f)(34)** *Two-point suspension scaffold (swinging scaffold).* A scaffold, the platform of which is supported by hangers (stirrups) at two points, suspended from overhead supports so as to permit the raising or lowering of the platform to the desired working position by tackle or hoisting machines.

**1910.21(f)(35)** *Window jack scaffold.* A scaffold, the platform of which is supported by a bracket or jack which projects through a window opening.

**1910.21(f)(36)** *Working load.* Load imposed by men, materials, and equipment.

**1910.21(g)** As used in §1910.29, unless the context requires otherwise, manually propelled mobile ladder stand and scaffold (tower) terms shall have the meaning ascribed in this paragraph.

**1910.21(g)(1)** *Bearer.* A horizontal member of a scaffold upon which the platform rests and which may be supported by ledgers.

**1910.21(g)(2)** *Brace.* A tie that holds one scaffold member in a fixed position with respect to another member.

**1910.21(g)(3)** *Climbing ladder.* A separate ladder with equally spaced rungs usually attached to the scaffold structure for climbing and descending.

**1910.21(g)(4)** *Coupler.* A device for locking together the components of a tubular metal scaffold which shall be designed and used to safely support the maximum intended loads.

**1910.21(g)(5)** *Design working load.* The maximum intended load, being the total of all loads including the weight of the men, materials, equipment, and platform.

**1910.21(g)(6)** *Equivalent.* Alternative design or features, which will provide an equal degree or factor of safety.

**1910.21(g)(7)** *Guardrail.* A barrier secured to uprights and erected along the exposed sides and ends of platforms to prevent falls of persons.

**1910.21(g)(8)** *Handrail.* A rail connected to a ladder stand running parallel to the slope and/or top step.

**1910.21(g)(9)** *Ladder stand.* A mobile fixed size self-supporting ladder consisting of a wide flat tread ladder in the form of stairs. The assembly may include handrails.

**1910.21(g)(10)** *Ledger (stringer).* A horizontal scaffold member which extends from post to post and which supports the bearer forming a tie between the posts.

**1910.21(g)(11)** *Mobile scaffold (tower).* A light, medium, or heavy duty scaffold mounted on casters or wheels.

**1910.21(g)(12)** *Mobile.* "Manually propelled."

**1910.21(g)(13)** *Mobile work platform.* Generally a fixed work level one frame high on casters or wheels, with bracing diagonally from platform to vertical frame.

**1910.21(g)(14)** *Runner.* The lengthwise horizontal bracing and/or bearing members.

**1910.21(g)(15)** *Scaffold.* Any temporary elevated platform and its necessary vertical, diagonal, and horizontal members used for supporting workmen and materials. (Also known as a scaffold tower.)

**1910.21(g)(16)** *Toeboard.* A barrier at platform level erected along the exposed sides and ends of a scaffold platform to prevent falls of materials.

**1910.21(g)(17)** *Tube and coupler scaffold.* An assembly consisting of tubing which serves as posts, bearers, braces, ties, and runners, a base supporting the posts, and uprights, and serves to join the various members, usually used in fixed locations.

**1910.21(g)(18)** *Tubular welded frame scaffold.* A sectional, panel, or frame metal scaffold substantially built up of prefabricated welded sections, which consist of posts and bearers with intermediate connecting members and braced with diagonal or cross braces.

**1910.21(g)(19)** *Tubular welded sectional folding scaffold.* A sectional, folding metal scaffold either of ladder frame or inside stairway design, substantially built of prefabricated welded sections, which consist of end frames, platform frame, inside inclined stairway frame and braces, or hinged connected diagonal and horizontal braces, capable of being folded into a flat package when the scaffold is not in use.

**1910.21(g)(20)** *Work level.* The elevated platform, used for supporting workmen and their materials, comprising the necessary vertical, horizontal, and diagonal braces, guardrails, and ladder for access to the work platform.

## §1910.22 General requirements.

This section applies to all permanent places of employment, except where domestic, mining, or agricultural work only is performed. Measures for the control of toxic materials are considered to be outside the scope of this section.

**1910.22(a)** *Housekeeping.* **(1)** All places of employment, passageways, storerooms, and service rooms shall be kept clean and orderly and in a sanitary condition.

**1910.22(a)(2)** The floor of every workroom shall be maintained in a clean and, so far as possible, a dry condition. Where wet processes are used, drainage shall be maintained, and false floors, platforms, mats, or other dry standing places should be provided where practicable.

**1910.22(a)(3)** To facilitate cleaning, every floor, working place, and passageway shall be kept free from protruding nails, splinters, holes, or loose boards.

**1910.22(b)** *Aisles and passageways.* **(1)** Where mechanical handling equipment is used, sufficient safe clearances shall be allowed for aisles, at loading docks, through doorways and wherever turns or passage must be made. Aisles and passageways shall be kept clear and in good repairs, with no obstruction across or in aisles that could create a hazard.

**1910.22(b)(2)** Permanent aisles and passageways shall be appropriately marked.

**1910.22(c)** *Covers and guardrails.* Covers and/or guardrails shall be provided to protect personnel from the hazards of open pits, tanks, vats, ditches, etc.

**1910.22(d)** *Floor loading protection.* **(1)** In every building or other structure, or part thereof, used for mercantile, business, industrial, or storage purposes, the loads approved by the building official shall be marked on plates of approved design which shall be supplied and securely affixed by the owner of the building, or his duly authorized agent, in a conspicuous place in each space to which they relate. Such plates shall not be removed or defaced but, if lost, removed, or defaced, shall be replaced by the owner or his agent.

**1910.22(d)(2)** It shall be unlawful to place, or cause, or permit to be placed, on any floor or roof of a building or other structure a load greater than that for which such floor or roof is approved by the building official.

## §1910.23 Guarding floor and wall openings and holes.

**1910.23(a)** *Protection for floor openings.* **(1)** Every stairway floor opening shall be guarded by a standard railing constructed in accordance with paragraph (e) of this section. The railing shall be provided on all exposed sides (except at entrance to stairway). For infrequently used stairways where traffic across the opening prevents the use of fixed standard railing (as when located in aisle spaces, etc.), the guard shall consist of a hinged floor opening cover of standard strength and construction and removable standard railings on all exposed sides (except at entrance to stairway).

**1910.23(a)(2)** Every ladderway floor opening or platform shall be guarded by a standard railing with standard toeboard on all exposed sides (except at entrance to opening), with the passage through the railing either provided with a swinging gate or so offset that a person cannot walk directly into the opening.

**1910.23(a)(3)** Every hatchway and chute floor opening shall be guarded by one of the following:

**1910.23(a)(3)(i)** Hinged floor opening cover of standard strength and construction equipped with standard railings or permanently attached thereto so as to leave only one exposed side. When the opening is not in use, the cover shall be closed or the exposed side shall be guarded at both top and intermediate positions by removable standard railings.

**1910.23(a)(3)(ii)** A removable railing with toeboard on not more than two sides of the opening and fixed standard railings with toeboards on all other exposed sides. The removable railings shall be kept in place when the opening is not in use.

Where operating conditions necessitate the feeding of material into any hatchway or chute opening, protection shall be provided to prevent a person from falling through the opening.

**1910.23(a)(4)** Every skylight floor opening and hole shall be guarded by a standard skylight screen or a fixed standard railing on all exposed sides.

**1910.23(a)(5)** Every pit and trapdoor floor opening, infrequently used, shall be guarded by a floor opening cover of standard strength and construction. While the cover is not in place, the pit or trap opening shall be constantly attended by someone or shall be protected on all exposed sides by removable standard railings.

**1910.23(a)(6)** Every manhole floor opening shall be guarded by a standard manhole cover which need not be hinged in place. While the cover is not in place, the manhole opening shall be constantly attended by someone or shall be protected by removable standard railings.

**1910.23(a)(7)** Every temporary floor opening shall have standard railings, or shall be constantly attended by someone.

**1910.23(a)(8)** Every floor hole into which persons can accidentally walk shall be guarded by either:

**1910.23(a)(8)(i)** A standard railing with standard toeboard on all exposed sides, or

**1910.23(a)(8)(ii)** A floor hole cover of standard strength and construction. While the cover is not in place, the floor hole shall be constantly attended by someone or shall be protected by a removable standard railing.

**1910.23(a)(9)** Every floor hole into which persons cannot accidentally walk (on account of fixed machinery, equipment, or walls) shall be protected by a cover that leaves no openings more than 1 inch wide. The cover shall be securely held in place to prevent tools or materials from falling through.

**1910.23(a)(10)** Where doors or gates open directly on a stairway, a platform shall be provided, and the swing of the door shall not reduce the effective width to less than 20 inches.

**1910.23(b)** *Protection for wall openings and holes.* **(1)** Every wall opening from which there is a drop of more than 4 feet shall be guarded by one of the following:

**1910.23(b)(1)(i)** Rail, roller, picket fence, half door, or equivalent barrier. Where there is exposure below to falling materials, a removable toe board or the equivalent shall also be provided. When the opening is not in use for handling materials, the guard shall be kept in position regardless of a door on the opening. In addition, a grab handle shall be provided on each side of the opening with its center approximately 4 feet above floor level and of standard strength and mounting.

**1910.23(b)(1)(ii)** Extension platform onto which materials can be hoisted for handling, and which shall have side rails or equivalent guards of standard specifications.

**1910.23(b)(2)** Every chute wall opening from which there is a drop of more than 4 feet shall be guarded by one or more of the barriers specified in paragraph (b)(1) of this section or as required by the conditions.

**1910.23(b)(3)** Every window wall opening at a stairway landing, floor, platform, or balcony, from which there is a drop of more than 4 feet, and where the bottom of the opening is less than 3 feet above the platform or landing, shall be guarded by standard slats, standard grill work (as specified in paragraph (e)(11) of this section), or standard railing.

Where the window opening is below the landing, or platform, a standard toe board shall be provided.

**1910.23(b)(4)** Every temporary wall opening shall have adequate guards but these need not be of standard construction.

**1910.23(b)(5)** Where there is a hazard of materials falling through a wall hole, and the lower edge of the near side of the hole is less than 4 inches above the floor, and the far side of the hole more than 5 feet above the next lower level, the hole shall be protected by a standard toeboard, or an enclosing screen either of solid construction, or as specified in paragraph (e)(11) of this section.

**1910.23(c)** *Protection of open-sided floors, platforms, and runways.* **(1)** Every open-sided floor or platform 4 feet or more above adjacent floor or ground level shall be guarded by a standard railing (or the equivalent as specified in paragraph (e)(3) of this section) on all open sides except where there is entrance to a ramp, stairway, or fixed ladder. The railing shall be provided with a toeboard wherever, beneath the open sides,

**1910.23(c)(1)(i)** Persons can pass,

**1910.23(c)(1)(ii)** There is moving machinery, or

**1910.23(c)(1)(iii)** There is equipment with which falling materials could create a hazard.

**1910.23(c)(2)** Every runway shall be guarded by a standard railing (or the equivalent as specified in paragraph (e)(3) of this section) on all open sides 4 feet or more above floor or ground level. Wherever tools, machine parts,

# OSHA Standards for General Industry

or materials are likely to be used on the runway, a toeboard shall also be provided on each exposed side.

Runways used exclusively for special purposes (such as oiling, shafting, or filling tank cars) may have the railing on one side omitted where operating conditions necessitate such omission, providing the falling hazard is minimized by using a runway of not less than 18 inches wide. Where persons entering upon runways become thereby exposed to machinery, electrical equipment, or other danger not a falling hazard, additional guarding than is here specified may be essential for protection.

**1910.23(c)(3)** Regardless of height, open-sided floors, walkways, platforms, or runways above or adjacent to dangerous equipment, pickling or galvanizing tanks, degreasing units, and similar hazards shall be guarded with a standard railing and toe board.

**1910.23(d)** *Stairway railings and guards.* **(1)** Every flight of stairs having four or more risers shall be equipped with standard stair railings or standard handrails as specified in paragraphs (d)(1) (i) through (v) of this section, the width of the stair to be measured clear of all obstructions except handrails:

**1910.23(d)(1)(i)** On stairways less than 44 inches wide having both sides enclosed, at least one handrail, preferably on the right side descending.

**1910.23(d)(1)(ii)** On stairways less than 44 inches wide having one side open, at least one stair railing on open side.

**1910.23(d)(1)(iii)** On stairways less than 44 inches wide having both sides open, one stair railing on each side.

**1910.23(d)(1)(iv)** On stairways more than 44 inches wide but less than 88 inches wide, one handrail on each enclosed side and one stair railing on each open side.

**1910.23(d)(1)(v)** On stairways 88 or more inches wide, one handrail on each enclosed side, one stair railing on each open side, and one intermediate stair railing located approximately midway of the width.

**1910.23(d)(2)** Winding stairs shall be equipped with a handrail offset to prevent walking on all portions of the treads having width less than 6 inches.

**1910.23(e)** *Railing, toe boards, and cover specifications.* **(1)** A standard railing shall consist of top rail, intermediate rail, and posts, and shall have a vertical height of 42 inches nominal from upper surface of top rail to floor, platform, runway, or ramp level. The top rail shall be smooth-surfaced throughout the length of the railing. The intermediate rail shall be approximately halfway between the top rail and the floor, platform, runway, or ramp. The ends of the rails shall not overhang the terminal posts except where such overhang does not constitute a projection hazard.

**1910.23(e)(2)** A stair railing shall be of construction similar to a standard railing but the vertical height shall be not more than 34 inches nor less than 30 inches from upper surface of top rail to surface of tread in line with face of riser at forward edge of tread.

**1910.23(e)(3)** [Reserved]

**1910.23(e)(3)(i)** For wood railings, the posts shall be of at least 2-inch by 4-inch stock spaced not to exceed 6 feet; the top and intermediate rails shall be of at least 2-inch by 4-inch stock. If top rail is made of two right-angle pieces of 1-inch by 4-inch stock, posts may be spaced on 8-foot centers, with 2-inch by 4-inch intermediate rail.

**1910.23(e)(3)(ii)** For pipe railings, posts and top and intermediate railings shall be at least 1 1/2 inches nominal diameter with posts spaced not more than 8 feet on centers.

**1910.23(e)(3)(iii)** For structural steel railings, posts and top and intermediate rails shall be of 2-inch by 2-inch by 3/8-inch angles or other metal shapes of equivalent bending strength with posts spaced not more than 8 feet on centers.

**1910.23(e)(3)(iv)** The anchoring of posts and framing of members for railings of all types shall be of such construction that the completed structure shall be capable of withstanding a load of at least 200 pounds applied in any direction at any point on the top rail.

**1910.23(e)(3)(v)** Other types, sizes, and arrangements of railing construction are acceptable provided they meet the following conditions:

**1910.23(e)(3)(v)(*a*)** A smooth-surfaced top rail at a height above floor, platform, runway, or ramp level of 42 inches nominal;

**1910.23(e)(3)(v)(*b*)** A strength to withstand at least the minimum requirement of 200 pounds top rail pressure;

**1910.23(e)(3)(v)(*c*)** Protection between top rail and floor, platform, runway, ramp, or stair treads, equivalent at least to that afforded by a standard intermediate rail;

**1910.23(e)(4)** A standard toeboard shall be 4 inches nominal in vertical height from its top edge to the level of the floor, platform, runway, or ramp. It shall be securely fastened in place and with not more than 1/4-inch clearance above floor level. It may be made of any substantial material either solid or with openings not over 1 inch in greatest dimension.

Where material is piled to such height that a standard toeboard does not provide protection, paneling from floor to intermediate rail, or to top rail shall be provided.

**1910.23(e)(5)(i)** A handrail shall consist of a lengthwise member mounted directly on a wall or partition by means of brackets attached to the lower side of the handrail so as to offer no obstruction to a smooth surface along the top and both sides of the handrail. The handrail shall be of rounded or other section that will furnish an adequate handhold for anyone grasping it to avoid falling. The ends of the handrail should be turned in to the supporting wall or otherwise arranged so as not to constitute a projection hazard.

**1910.23(e)(5)(ii)** The height of handrails shall be not more than 34 inches nor less than 30 inches from upper surface of handrail to surface of tread in line with face of riser or to surface of ramp.

**1910.23(e)(5)(iii)** The size of handrails shall be: When of hardwood, at least 2 inches in diameter; when of metal pipe, at least 1 1/2 inches in diameter. The length of brackets shall be such as will give a clearance between handrail and wall or any projection thereon of at least 3 inches. The spacing of brackets shall not exceed 8 feet.

**1910.23(e)(5)(iv)** The mounting of handrails shall be such that the completed structure is capable of withstanding a load of at least 200 pounds applied in any direction at any point on the rail.

**1910.23(e)(6)** All handrails and railings shall be provided with a clearance of not less than 3 inches between the handrail or railing and any other object.

**1910.23(e)(7)** Floor opening covers may be of any material that meets the following strength requirements:

**1910.23(e)(7)(i)** Trench or conduit covers and their supports, when located in plant roadways, shall be designed to carry a truck rear-axle load of at least 20,000 pounds.

**1910.23(e)(7)(ii)** Manhole covers and their supports, when located in plant roadways, shall comply with local standard highway requirements if any; otherwise, they shall be designed to carry a truck rear-axle load of at least 20,000 pounds.

**1910.23(e)(7)(iii)** The construction of floor opening covers may be of any material that meets the strength requirements. Covers projecting not more than 1 inch above the floor level may be used providing all edges are chamfered to an angle with the horizontal of not over 30 degrees. All hinges, handles, bolts, or other parts shall set flush with the floor or cover surface.

**1910.23(e)(8)** Skylight screens shall be of such construction and mounting that they are capable of withstanding a load of at least 200 pounds applied perpendicularly at any one area on the screen. They shall also be of such construction and mounting that under ordinary loads or impacts, they will not deflect downward sufficiently to break the glass below them. The construction shall be of grillwork with openings not more than 4 inches long or of slatwork with openings not more than 2 inches wide with length unrestricted.

**1910.23(e)(9)** Wall opening barriers (rails, rollers, picket fences, and half doors) shall be of such construction and mounting that, when in place at the opening, the barrier is capable of withstanding a load of at least 200 pounds applied in any direction (except upward) at any point on the top rail or corresponding member.

**1910.23(e)(10)** Wall opening grab handles shall be not less than 12 inches in length and shall be so mounted as to give 3 inches clearance from the side framing of the wall opening. The size, material, and anchoring of the grab handle shall be such that the completed structure is capable of withstanding a load of at least 200 pounds applied in any direction at any point of the handle.

**1910.23(e)(11)** Wall opening screens shall be of such construction and mounting that they are capable of withstanding a load of at least 200 pounds applied horizontally at any point on the near side of the screen. They may be of solid construction, of grillwork with openings not more than 8 inches long, or of slatwork with openings not more than 4 inches wide with length unrestricted.

[39 FR 23502, June 27, 1974, as amended at 43 FR 49744, Oct. 24, 1978; 49 FR 5321, Feb. 10, 1984]

### §1910.24 Fixed industrial stairs.

**1910.24(a)** *Application of requirements.* This section contains specifications for the safe design and construction of fixed general industrial stairs. This classification includes interior and exterior stairs around machinery, tanks, and other equipment, and stairs leading to or from floors, platforms, or pits. This section does not apply to stairs used for fire exit purposes, to construction operations to private residences, or to articulated stairs, such as may be installed on floating roof tanks or on dock facilities, the angle of which changes with the rise and fall of the base support.

**1910.24(b)** *Where fixed stairs are required.* Fixed stairs shall be provided for access from one structure level to another where operations necessitate regular travel between levels, and for access to operating platforms at any equipment which requires attention routinely during operations. Fixed stairs shall also be provided where access to elevations is daily or at each shift for such purposes as gauging, inspection, regular maintenance, etc., where such work may expose employees to acids, caustics, gases, or other harmful substances, or for which purposes the carrying of tools or equipment by hand is normally required. (It is not the intent of this section to preclude the use of fixed ladders for access to elevated tanks, towers, and similar structures, overhead traveling cranes, etc., where the use of fixed ladders is common practice.) Spiral stairways shall not be permitted except for special limited usage and secondary access situations where it is not practical to provide a conventional stairway. Winding stairways may be installed on tanks and similar round structures where the diameter of the structure is not less than five (5) feet.

**1910.24(c)** *Stair strength.* Fixed stairways shall be designed and constructed to carry a load of five times the normal live load anticipated but never of less strength than to carry safely a moving concentrated load of 1,000 pounds.

**1910.24(d)** *Stair width.* Fixed stairways shall have a minimum width of 22 inches.

**1910.24(e)** *Angle of stairway rise.* Fixed stairs shall be installed at angles to the horizontal of between 30° and 50°. Any uniform combination of rise/tread dimensions may be used that will result in a stairway at an angle to the horizontal within the permissible range. Table D-1 gives rise/tread dimensions which will produce a stairway within the permissible range, stating the angle to the horizontal produced by each combination. However, the rise/tread combinations are not limited to those given in Table D-1.

### TABLE D-1

| Angle to horizontal | Rise (in inches) | Tread run (in inches) |
|---|---|---|
| 30° 35' | 6 1/2 | 11 |
| 32° 08' | 6 3/4 | 10 3/4 |
| 33° 41' | 7 | 10 1/2 |
| 35° 16' | 7 1/4 | 10 1/4 |
| 36° 52' | 7 1/2 | 10 |
| 38° 29' | 7 3/4 | 9 3/4 |
| 40° 08' | 8 | 9 1/2 |
| 41° 44' | 8 1/4 | 9 1/4 |
| 43° 22' | 8 1/2 | 9 |
| 45° 00' | 8 3/4 | 8 3/4 |
| 46° 38' | 9 | 8 1/2 |
| 48° 16' | 9 1/4 | 8 1/4 |
| 49° 54' | 9 1/2 | 8 |

**1910.24(f)** *Stair treads.* All treads shall be reasonably slip-resistant and the nosings shall be of nonslip finish. Welded bar grating treads without nosings are acceptable providing the leading edge can be readily identified by personnel descending the stairway and provided the tread is serrated or is of definite nonslip design. Rise height and tread width shall be uniform throughout any flight of stairs including any foundation structure used as one or more treads of the stairs.

**1910.24(g)** *Stairway platforms.* Stairway platforms shall be no less than the width of a stairway and a minimum of 30 inches in length measured in the direction of travel.

**1910.24(h)** *Railings and handrails.* Standard railings shall be provided on the open sides of all exposed stairways and stair platforms. Handrails shall be provided on at least one side of closed stairways preferably on the right side descending. Stair railings and handrails shall be installed in accordance with the provisions of §1910.23.

**1910.24(i)** *Vertical clearance.* Vertical clearance above any stair tread to an overhead obstruction shall be at least 7 feet measured from the leading edge of the tread.

[39 FR 23502, June 27, 1974, as amended at 43 FR 49744, Oct. 24, 1978; 49 FR 5321, Feb. 10, 1984]

### §1910.25 Portable wood ladders.

**1910.25(a)** *Application of requirements.* This section is intended to prescribe rules and establish minimum requirements for the construction, care, and use of the common types of portable wood ladders, in order to insure safety under normal conditions of usage. Other types of special ladders, fruitpicker's ladders, combination step and extension ladders, stockroom step ladders, aisle-way step ladders, shelf ladders, and library ladders are not specifically covered by this section.

**1910.25(b)** *Materials—(1) Requirements applicable to all wood parts.* (i) All wood parts shall be free from sharp edges and splinters; sound and free from accepted visual inspection from shake, wane, compression failures, decay, or other irregularities. Low density wood shall not be used.

**1910.25(b)(1)(ii)** [Reserved]

**1910.25(b)(2)** [Reserved]

**1910.25(c)** *Construction requirements.*

**1910.25(c)(1)** [Reserved]

**1910.25(c)(2)** *Portable stepladders.* Stepladders longer than 20 feet shall not be supplied. Stepladders as hereinafter specified shall be of three types:

Type I—Industrial stepladder, 3 to 20 feet for heavy duty, such as utilities, contractors, and industrial use.

Type II—Commercial stepladder, 3 to 12 feet for medium duty, such as painters, offices, and light industrial use.

Type III—Household stepladder, 3 to 6 feet for light duty, such as light household use.

**1910.25(c)(2)(i)** *General requirements.*

**1910.25(c)(2)(i)(a)** [Reserved]

**1910.25(c)(2)(i)(b)** A uniform step spacing shall be employed which shall be not more than 12 inches. Steps shall be parallel and level when the ladder is in position for use.

**1910.25(c)(2)(i)(c)** The minimum width between side rails at the top, inside to inside, shall be not less than 11 1/2 inches. From top to bottom, the side rails shall spread at least 1 inch for each foot of length of stepladder.

**1910.25(c)(2)(i)(d)-(e)** [Reserved]

**1910.25(c)(2)(i)(f)** A metal spreader or locking device of sufficient size and strength to securely hold the front and back sections in open positions shall be a component of each stepladder. The spreader shall have all sharp points covered or removed to protect the user. For Type III ladder, the pail shelf and spreader may be combined in one unit (the so-called shelf-lock ladder).

**1910.25(c)(3)** *Portable rung ladders.*

**1910.25(c)(3)(i)** [Reserved]

**1910.25(c)(3)(ii)** *Single ladder.* *(a)* Single ladders longer than 30 feet shall not be supplied.

**1910.25(c)(3)(ii)(b)** [Reserved]

**1910.25(c)(3)(iii)** *Two-section ladder.* *(a)* Two-section extension ladders longer than 60 feet shall not be supplied. All ladders of this type shall consist of two sections, one to fit within the side rails of the other, and arranged in such a manner that the upper section can be raised and lowered.

**1910.25(c)(3)(iii)(b)** [Reserved]

**1910.25(c)(3)(iv)** *Sectional ladder.* *(a)* Assembled combinations of sectional ladders longer than lengths specified in this subdivision shall not be used.

**1910.25(c)(3)(iv)(b)** [Reserved]

**1910.25(c)(3)(v)** *Trestle and extension trestle ladder.* **(a)** Trestle ladders, or extension sections or base sections of extension trestle ladders longer than 20 feet shall not be supplied.

**1910.25(c)(3)(v)(b)** [Reserved]

**1910.25(c)(4)** *Special-purpose ladders.*

**1910.25(c)(4)(i)** [Reserved]

**1910.25(c)(4)(ii)** *Painter's stepladder.* **(a)** Painter's stepladders longer than 12 feet shall not be supplied.

**1910.25(c)(4)(ii)(b)** [Reserved]

**1910.25(c)(4)(iii)** *Mason's ladder.* A mason's ladder is a special type of single ladder intended for use in heavy construction work.

**1910.25(c)(4)(iii)(a)** Mason's ladders longer than 40 feet shall not be supplied.

**1910.25(c)(4)(iii)(b)** [Reserved]

**1910.25(c)(5)** *Trolley and side-rolling ladders—(i) Length.* Trolley ladders and side-rolling ladders longer than 20 feet should not be supplied.

**1910.25(c)(5)(ii)** [Reserved]

**1910.25(d)** *Care and use of ladders—(1) Care.* To insure safety and serviceability the following precautions on the care of ladders shall be observed:

**1910.25(d)(1)(i)** Ladders shall be maintained in good condition at all times, the joint between the steps and side rails shall be tight, all hardware and fittings securely attached, and the movable parts shall operate freely without binding or undue play.

**1910.25(d)(1)(ii)** Metal bearings of locks, wheels, pulleys, etc., shall be frequently lubricated.

**1910.25(d)(1)(iii)** Frayed or badly worn rope shall be replaced.

**1910.25(d)(1)(iv)** Safety feet and other auxiliary equipment shall be kept in good condition to insure proper performance.

**1910.25(d)(1)(v)-(ix)** [Reserved]

**1910.25(d)(1)(x)** Ladders shall be inspected frequently and those which have developed defects shall be withdrawn from service for repair or destruction and tagged or marked as "Dangerous, Do Not Use."

**1910.25(d)(1)(xi)** Rungs should be kept free of grease and oil.

**1910.25(d)(2)** *Use.* The following safety precautions shall be observed in connection with the use of ladders:

**1910.25(d)(2)(i)** Portable rung and cleat ladders shall, where possible, be used at such a pitch that the horizontal distance from the top support to the foot of the ladder is one-quarter of the working length of the ladder (the length along the ladder between the foot and the top support). The ladder shall be so placed as to prevent slipping, or it shall be lashed, or held in position. Ladders shall not be used in a horizontal position as platforms, runways, or scaffolds.

**1910.25(d)(2)(ii)** Ladders for which dimensions are specified should not be used by more than one man at a time nor with ladder jacks and scaffold planks where use by more than one man is anticipated. In such cases, specially designed ladders with larger dimensions of the parts should be procured;

**1910.25(d)(2)(iii)** Portable ladders shall be so placed that the side rails have a secure footing. The top rest for portable rung and cleat ladders shall be reasonably rigid and shall have ample strength to support the applied load;

**1910.25(d)(2)(iv)** Ladders shall not be placed in front of doors opening toward the ladder unless the door is blocked upon, locked, or guarded;

**1910.25(d)(2)(v)** Ladders shall not be placed on boxes, barrels, or other unstable bases to obtain additional height;

**1910.25(d)(2)(vi)-(vii)** [Reserved]

**1910.25(d)(2)(viii)** Ladders with broken or missing steps, rungs, or cleats, broken side rails, or other faulty equipment shall not be used; improvised repairs shall not be made;

**1910.25(d)(2)(ix)** Short ladders shall not be spliced together to provide long sections;

**1910.25(d)(2)(x)** Ladders made by fastening cleats across a single rail shall not be used;

**1910.25(d)(2)(xi)** Ladders shall not be used as guys, braces, or skids, or for other than their intended purposes;

**1910.25(d)(2)(xii)** Tops of the ordinary types of stepladders shall not be used as steps;

**1910.25(d)(2)(xiii)** On two-section extension ladders the minimum overlap for the two sections in use shall be as follows:

| Size of ladder (feet) | Overlap (feet) |
| --- | --- |
| Up to and including 36 | 3 |
| Over 36 up to and including 48 | 4 |
| Over 48 up to and including 60 | 5 |

**1910.25(d)(2)(xiv)** Portable rung ladders with reinforced rails (see paragraphs (c)(3) (ii)(c) and (iii)(d) this section) shall be used only with the metal reinforcement on the under side;

**1910.25(d)(2)(xv)** No ladder should be used to gain access to a roof unless the top of the ladder shall extend at least 3 feet above the point of support, at eave, gutter, or roofline;

**1910.25(d)(2)(xvi)** [Reserved]

**1910.25(d)(2)(xvii)** Middle and top sections of sectional or window cleaner's ladders should not be used for bottom section unless the user equips them with safety shoes;

**1910.25(d)(2)(xviii)** [Reserved]

**1910.25(d)(2)(xix)** The user should equip all portable rung ladders with nonslip bases when there is a hazard of slipping. Nonslip bases are not intended as a substitute for care in safely placing, lashing, or holding a ladder that is being used upon oily, metal, concrete, or slippery surfaces;

**1910.25(d)(2)(xx)** The bracing on the back legs of step ladders is designed solely for increasing stability and not for climbing.

[39 FR 23502, June 27, 1974, as amended at 43 FR 49744, Oct. 24, 1978; 49 FR 5321, Feb. 10, 1984]

## §1910.26 Portable metal ladders.

**1910.26(a)** *Requirements—(1) General.* Specific design and construction requirements are not part of this section because of the wide variety of metals and design possibilities. However, the design shall be such as to produce a ladder without structural defects or accident hazards such as sharp edges, burrs, etc. The metal selected shall be of sufficient strength to meet the test requirements, and shall be protected against corrosion unless inherently corrosion-resistant.

**1910.26(a)(1)(i)-(ii)** [Reserved]

**1910.26(a)(1)(iii)** The spacing of rungs or steps shall be on 12-inch centers.

**1910.26(a)(1)(iv)** [Reserved]

**1910.26(a)(1)(v)** Rungs and steps shall be corrugated, knurled, dimpled, coated with skid-resistant material, or otherwise treated to minimize the possibility of slipping.

**1910.26(a)(2)** *General specifications—straight and extension ladders.* **(i)** The minimum width between side rails of a straight ladder or any section of an extension ladder shall be 12 inches.

**1910.26(a)(2)(ii)** The length of single ladders or individual sections of ladders shall not exceed 30 feet. Two-section ladders shall not exceed 48 feet in length and over two-section ladders shall not exceed 60 feet in length.

**1910.26(a)(2)(iii)** Based on the nominal length of the ladder, each section of a multisection ladder shall overlap the adjacent section by at least the number of feet stated in the following:

| Normal length of ladder (feet) | Overlap (feet) |
|---|---|
| Up to and including 36 | 3 |
| Over 36, up to and including 48 | 4 |
| Over 48, up to 60 | 5 |

**1910.26(a)(2)(iv)** Extension ladders shall be equipped with positive stops which will insure the overlap specified in the table above.

**1910.26(a)(3)** *General specifications—step ladders.*

**1910.26(a)(3)(i)-(ii)** [Reserved]

**1910.26(a)(3)(iii)** The length of a stepladder is measured by the length of the front rail. To be classified as a standard length ladder, the measured length shall be within plus or minus one-half inch of the specified length. Stepladders shall not exceed 20 feet in length.

**1910.26(a)(3)(iv)-(vi)** [Reserved]

**1910.26(a)(3)(vii)** The bottoms of the four rails are to be supplied with insulating nonslip material for the safety of the user.

**1910.26(a)(3)(viii)** A metal spreader or locking device of sufficient size and strength to securely hold the front and back sections in the open position shall be a component of each stepladder. The spreader shall have all sharp points or edges covered or removed to protect the user.

**1910.26(a)(4)** *General specifications—trestles and extension trestle ladders.* **(i)** Trestle ladders or extension sections or base sections of extension trestle ladders shall be not more than 20 feet in length.

**1910.26(a)(4)(ii)** [Reserved]

**1910.26(a)(5)** *General specifications—platform ladders.* **(i)** The length of a platform ladder shall not exceed 20 feet. The length of a platform ladder shall be measured along the front rail from the floor to the platform.

**1910.26(a)(5)(ii)** [Reserved]

**1910.26(b)** [Reserved]

**1910.26(c)** *Care and maintenance of ladders—***(1)** *General.* To get maximum serviceability, safety, and to eliminate unnecessary damage of equipment, good safe practices in the use and care of ladder equipment must be employed by the users.

The following rules and regulations are essential to the life of the equipment and the safety of the user.

**1910.26(c)(2)** *Care of ladders.*

**1910.26(c)(2)(i)-(iii)** [Reserved]

**1910.26(c)(2)(iv)** Ladders must be maintained in good usable condition at all times.

**1910.26(c)(2)(v)** [Reserved]

**1910.26(c)(2)(vi)** If a ladder is involved in any of the following, immediate inspection is necessary:

**1910.26(c)(2)(vi)(a)** If ladders tip over, inspect ladder for side rails dents or bends, or excessively dented rungs; check all rung-to- side-rail connections; check hardware connections; check rivets for shear.

**1910.26(c)(2)(vi)(b)-(c)** [Reserved]

**1910.26(c)(2)(vi)(d)** If ladders are exposed to oil and grease, equipment should be cleaned of oil, grease, or slippery materials. This can easily be done with a solvent or steam cleaning.

**1910.26(c)(2)(vii)** Ladders having defects are to be marked and taken out of service until repaired by either maintenance department or the manufacturer.

**1910.26(c)(3)** *Use of ladders.* **(i).** A simple rule for setting up a ladder at the proper angle is to place the base a distance from the vertical wall equal to one-fourth the working length of the ladder.

**1910.26(c)(3)(ii)** Portable ladders are designed as a one-man working ladder based on a 200-pound load.

**1910.26(c)(3)(iii)** The ladder base section must be placed with a secure footing.

**1910.26(c)(3)(iv)** The top of the ladder must be placed with the two rails supported, unless equipped with a single support attachment.

**1910.26(c)(3)(v)** When ascending or descending, the climber must face the ladder.

**1910.26(c)(3)(vi)** Ladders must not be tied or fastened together to provide longer sections. They must be equipped with the hardware fittings necessary if the manufacturer endorses extended uses.

**1910.26(c)(3)(vii)** Ladders should not be used as a brace, skid, guy or gin pole, gangway, or for other uses than that for which they were intended, unless specifically recommended for use by the manufacturer.

**1910.26(c)(3)(viii)** See §1910.333(c) for work practices to be used when work is performed on or near electric circuits.

[39 FR 23502, June 27, 1974, as amended at 43 FR 49745, Oct. 24, 1978; 49 FR 5321, Feb. 10, 1984; 55 FR 32014, Aug. 6, 1990]

### §1910.27 Fixed ladders.

**1910.27(a)** *Design requirements—***(1)** Design considerations. All ladders, appurtenances, and fastenings shall be designed to meet the following load requirements:

**1910.27(a)(1)(i)** The minimum design live load shall be a single concentrated load of 200 pounds.

**1910.27(a)(1)(ii)** The number and position of additional concentrated live-load units of 200 pounds each as determined from anticipated usage of the ladder shall be considered in the design.

**1910.27(a)(1)(iii)** The live loads imposed by persons occupying the ladder shall be considered to be concentrated at such points as will cause the maximum stress in the structural member being considered.

**1910.27(a)(1)(iv)** The weight of the ladder and attached appurtenances together with the live load shall be considered in the design of rails and fastenings.

**1910.27(a)(2)** *Design stresses.* Design stresses for wood components of ladders shall not exceed those specified in §1910.25. All wood parts of fixed ladders shall meet the requirements of §1910.25(b).

For fixed ladders consisting of wood side rails and wood rungs or cleats, used at a pitch in the range 75 degrees to 90 degrees, and intended for use by no more than one person per section, single ladders as described in §1910.25(c)(3)(ii) are acceptable.

**1910.27(b)** *Specific features—***(1)** *Rungs and cleats.* **(i)** All rungs shall have a minimum diameter of three-fourths inch for metal ladders, except as covered in paragraph (b)(7)(i) of this section and a minimum diameter of 1 1/8 inches for wood ladders.

**1910.27(b)(1)(ii)** The distance between rungs, cleats, and steps shall not exceed 12 inches and shall be uniform throughout the length of the ladder.

**1910.27(b)(1)(iii)** The minimum clear length of rungs or cleats shall be 16 inches.

**1910.27(b)(1)(iv)** Rungs, cleats, and steps shall be free of splinters, sharp edges, burrs, or projections which may be a hazard.

**1910.27(b)(1)(v)** The rungs of an individual-rung ladder shall be so designed that the foot cannot slide off the end. A suggested design is shown in figure D-1.

**FIGURE D-1—SUGGESTED DESIGN FOR RUNGS ON INDIVIDUAL-RUNG LADDERS**

## OSHA Standards for General Industry

**1910.27(b)(2)** *Side rails.* Side rails which might be used as a climbing aid shall be of such cross sections as to afford adequate gripping surface without sharp edges, splinters, or burrs.

**1910.27(b)(3)** *Fastenings.* Fastenings shall be an integral part of fixed ladder design.

**1910.27(b)(4)** *Splices.* All splices made by whatever means shall meet design requirements as noted in paragraph (a) of this section. All splices and connections shall have smooth transition with original members and with no sharp or extensive projections.

**1910.27(b)(5)** *Electrolytic action.* Adequate means shall be employed to protect dissimilar metals from electrolytic action when such metals are joined.

**1910.27(b)(6)** *Welding.* All welding shall be in accordance with the "Code for Welding in Building Construction" (AWSD1.0-1966).

**1910.27(b)(7)** *Protection from deterioration.* **(i)** Metal ladders and appurtenances shall be painted or otherwise treated to resist corrosion and rusting when location demands. Ladders formed by individual metal rungs imbedded in concrete, which serve as access to pits and to other areas under floors, are frequently located in an atmosphere that causes corrosion and rusting. To increase rung life in such atmosphere, individual metal rungs shall have a minimum diameter of 1 inch or shall be painted or otherwise treated to resist corrosion and rusting.

**1910.27(b)(7)(ii)** Wood ladders, when used under conditions where decay may occur, shall be treated with a nonirritating preservative, and the details shall be such as to prevent or minimize the accumulation of water on wood parts.

**1910.27(b)(7)(iii)** When different types of materials are used in the construction of a ladder, the materials used shall be so treated as to have no deleterious effect one upon the other.

**FIGURE D-2—RAIL LADDER WITH BAR STEEL RAILS AND ROUND STEEL RUNGS**

**1910.27(c)** *Clearance*—**(1)** *Climbing side.* On fixed ladders, the perpendicular distance from the centerline of the rungs to the nearest permanent object on the climbing side of the ladder shall be 36 inches for a pitch of 76 degrees, and 30 inches for a pitch of 90 degrees (fig. D-2 of this section), with minimum clearances for intermediate pitches varying between these two limits in proportion to the slope, except as provided in subparagraphs (3) and (5) of this paragraph.

**1910.27(c)(2)** *Ladders without cages or wells.* A clear width of at least 15 inches shall be provided each way from the centerline of the ladder in the climbing space, except when cages or wells are necessary.

**1910.27(c)(3)** *Ladders with cages or baskets.* Ladders equipped with cage or basket are excepted from the provisions of subparagraphs (1) and (2) of this paragraph, but shall conform to the provisions of paragraph (d)(1)(v) of this section. Fixed ladders in smooth-walled wells are excepted from the provisions of subparagraph (1) of this paragraph, but shall conform to the provisions of paragraph (d)(1)(vi) of this section.

**1910.27(c)(4)** *Clearance in back of ladder.* The distance from the centerline of rungs, cleats, or steps to the nearest permanent object in back of the ladder shall be not less than 7 inches, except that when unavoidable obstructions are encountered, minimum clearances as shown in figure D-3 shall be provided.

**FIGURE D-3—CLEARANCE FOR UNAVOIDABLE OBSTRUCTION AT REAR OF FIXED LADDER**

**1910.27(c)(5)** *Clearance in back of grab bar.* The distance from the centerline of the grab bar to the nearest permanent object in back of the grab bars shall be not less than 4 inches. Grab bars shall not protrude on the climbing side beyond the rungs of the ladder which they serve.

**1910.27(c)(6)** *Step-across distance.* The step-across distance from the nearest edge of ladder to the nearest edge of equipment or structure shall be not more than 12 inches, or less than 2 1/2 inches (fig. D-4).

**FIGURE D-4—LADDER FAR FROM WALL**

**1910.27(c)(7)** *Hatch cover.* Counterweighted hatch covers shall open a minimum of 60 degrees from the horizontal. The distance from the centerline of rungs or cleats to the edge of the hatch opening on the climbing side shall be not less than 24 inches for offset wells or 30 inches for straight wells. There shall be not protruding potential hazards within 24 inches of the centerline of rungs or cleats; any such hazards within 30 inches of the centerline of the rungs or cleats shall be fitted with deflector plates placed at an angle of 60 degrees from the horizontal as indicated in figure D-5. The relationship of a fixed ladder to an acceptable counterweighted hatch cover is illustrated in figure D-6.

**1910.27(d)** *Special requirements*—(1) *Cages or wells.* (i) Cages or wells (except on chimney ladders) shall be built, as shown on the applicable drawings, covered in detail in figures D-7, D-8, and D-9, or of equivalent construction.

**1910.27(d)(1)(ii)** Cages or wells (except as provided in subparagraph (5) of this paragraph) conforming to the dimensions shown in figures D-7, D-8, and D-9 shall be provided on ladders of more than 20 feet to a maximum unbroken length of 30 feet.

**FIGURE D-5—DEFLECTOR PLATES FOR HEAD HAZARDS**

**FIGURE D-6—RELATIONSHIP OF FIXED LADDER TO A SAFE ACCESS HATCH**

**1910.27(d)(1)(iii)** Cages shall extend a minimum of 42 inches above the top of landing, unless other acceptable protection is provided.

**1910.27(d)(1)(iv)** Cages shall extend down the ladder to a point not less than 7 feet nor more than 8 feet above the base of the ladder, with bottom flared not less than 4 inches, or portion of cage opposite ladder shall be carried to the base.

**1910.27(d)(1)(v)** Cages shall not extend less than 27 nor more than 28 inches from the centerline of the rungs of the ladder. Cage shall not be less than 27 inches in width. The inside shall be clear of projections. Vertical bars shall be located at a maximum spacing of 40 degrees around the circumference of the cage; this will give a maximum spacing of approximately 9 1/2 inches, center to center.

**1910.27(d)(1)(vi)** Ladder wells shall have a clear width of at least 15 inches measured each way from the centerline of the ladder. Smooth-walled wells shall be a minimum of 27 inches from the centerline of rungs to the well wall on the climbing side of the ladder. Where other obstructions on the climbing side of the ladder exist, there shall be a minimum of 30 inches from the centerline of the rungs.

**FIGURE D-7—CAGES FOR LADDERS MORE THAN 20 FEET HIGH**

**FIGURE D-8—CLEARANCE DIAGRAM FOR FIXED LADDER IN WELL**

# OSHA Standards for General Industry

**FIGURE D-9—CAGES—SPECIAL APPLICATIONS**

**1910.27(d)(2)** *Landing platforms.* When ladders are used to ascend to heights exceeding 20 feet (except on chimneys), landing platforms shall be provided for each 30 feet of height or fraction thereof, except that, where no cage, well, or ladder safety device is provided, landing platforms shall be provided for each 20 feet of height or fraction thereof. Each ladder section shall be offset from adjacent sections. Where installation conditions (even for a short, unbroken length) require that adjacent sections be offset, landing platforms shall be provided at each offset.

**1910.27(d)(2)(i)** Where a man has to step a distance greater than 12 inches from the centerline of the rung of a ladder to the nearest edge of structure or equipment, a landing platform shall be provided. The minimum step-across distance shall be 2 1/2 inches.

**1910.27(d)(2)(ii)** All landing platforms shall be equipped with standard railings and toeboards, so arranged as to give safe access to the ladder. Platforms shall be not less than 24 inches in width and 30 inches in length.

**1910.27(d)(2)(iii)** One rung of any section of ladder shall be located at the level of the landing laterally served by the ladder. Where access to the landing is through the ladder, the same rung spacing as used on the ladder shall be used from the landing platform to the first rung below the landing.

**1910.27(d)(3)** *Ladder extensions.* The side rails of through or side-step ladder extensions shall extend 3 1/2 feet above parapets and landings. For through ladder extensions, the rungs shall be omitted from the extension and shall have not less than 18 nor more than 24 inches clearance between rails. For side-step or offset fixed ladder sections, at landings, the side rails and rungs shall be carried to the next regular rung beyond or above the 3 1/2 feet minimum (fig. D-10).

**FIGURE D-10—OFFSET FIXED LADDER SECTIONS**

**1910.27(d)(4)** *Grab bars.* Grab bars shall be spaced by a continuation of the rung spacing when they are located in the horizontal position. Vertical grab bars shall have the same spacing as the ladder side rails. Grab-bar diameters shall be the equivalent of the round-rung diameters.

**1910.27(d)(5)** *Ladder safety devices.* Ladder safety devices may be used on tower, water tank, and chimney ladders over 20 feet in unbroken length in lieu of cage protection. No landing platform is required in these cases. All ladder safety devices such as those that incorporate lifebelts, friction brakes, and sliding attachments shall meet the design requirements of the ladders which they serve.

**1910.27(e)** *Pitch*—**(1)** *Preferred pitch.* The preferred pitch of fixed ladders shall be considered to come in the range of 75 degrees and 90 degrees with the horizontal (fig. D-11).

**FIGURE D-11—PITCH OF FIXED LADDERS**

**1910.27(e)(2)** *Substandard pitch.* Fixed ladders shall be considered as substandard if they are installed within the substandard pitch range of 60 and 75 degrees with the horizontal. Substandard fixed ladders are permitted only where it is found necessary to meet conditions of installation. This substandard pitch range shall be considered as a critical range to be avoided, if possible.

**1910.27(e)(3)** *Scope of coverage in this section.* This section covers only fixed ladders within the pitch range of 60 degrees and 90 degrees with the horizontal.

**1910.27(e)(4)** *Pitch greater than 90 degrees.* Ladders having a pitch in excess of 90 degrees with the horizontal are prohibited.

**1910.27(f)** *Maintenance.* All ladders shall be maintained in a safe condition. All ladders shall be inspected regularly, with the intervals between inspections being determined by use and exposure.

## §1910.28 Safety requirements for scaffolding.

**1910.28(a)** *General requirements for all scaffolds.* **(1)** Scaffolds shall be furnished and erected in accordance with this standard for persons engaged in work that cannot be done safely from the ground or from solid construction, except that ladders used for such work shall conform to §1910.25 and §1910.26.

**1910.28(a)(2)** The footing or anchorage for scaffolds shall be sound, rigid, and capable of carrying the maximum intended load without settling or displacement. Unstable objects such as barrels, boxes, loose brick, or concrete blocks shall not be used to support scaffolds or planks.

**1910.28(a)(3)** [Reserved]

**1910.28(a)(4)** Scaffolds and their components shall be capable of supporting without failure at least four times the maximum intended load.

**1910.28(a)(5)** Scaffolds and other devices mentioned or described in this section shall be maintained in safe condition. Scaffolds shall not be altered or moved horizontally while they are in use or occupied.

**1910.28(a)(6)** Any scaffold damaged or weakened from any cause shall be immediately repaired and shall not be used until repairs have been completed.

**1910.28(a)(7)** Scaffolds shall not be loaded in excess of the working load for which they are intended.

**1910.28(a)(8)** All load-carrying timber members of scaffold framing shall be a minimum of 1,500 $f$. (Stress Grade) construction grade lumber. All dimensions are nominal sizes as provided in the American Lumber Standards, except that where rough sizes are noted, only rough or undressed lumber of the size specified will satisfy minimum requirements. (NOTE: Where nominal sizes of lumber are used in place of rough sizes, the nominal size lumber shall be such as to provide equivalent strength to that specified in tables D-7 through D-12 and D-16.)

**1910.28(a)(9)** All planking shall be Scaffold Grade as recognized by grading rules for the species of wood used. The maximum permissible spans for 2- × 9-inch or wider planks are shown in the following table:

|  | Material |||||
| --- | --- | --- | --- | --- | --- |
|  | Full thickness undressed lumber ||| Nominal thickness lumber ||
| Working load (p.s.f.) | 25 | 50 | 75 | 25 | 50 |
| Permissible span (ft.) | 10 | 8 | 6 | 8 | 9 |

The maximum permissible span for 1 1/4 × 9-inch or wider plank of full thickness is 4 feet with medium loading of 50 p.s.f.

**1910.28(a)(10)** Nails or bolts used in the construction of scaffolds shall be of adequate size and in sufficient numbers at each connection to develop the designed strength of the scaffold. Nails shall not be subjected to a straight pull and shall be driven full length.

**1910.28(a)(11)** All planking or platforms shall be overlapped (minimum 12 inches) or secured from movement.

**1910.28(a)(12)** An access ladder or equivalent safe access shall be provided.

**1910.28(a)(13)** Scaffold planks shall extend over their end supports not less than 6 inches nor more than 18 inches.

**1910.28(a)(14)** The poles, legs, or uprights of scaffolds shall be plumb, and securely and rigidly braced to prevent swaying and displacement.

**1910.28(a)(15)** Materials being hoisted onto a scaffold shall have a tag line.

**1910.28(a)(16)** Overhead protection shall be provided for men on a scaffold exposed to overhead hazards.

**1910.28(a)(17)** Scaffolds shall be provided with a screen between the toeboard and the guardrail, extending along the entire opening, consisting of No. 18 gauge U.S. Standard Wire one-half-inch mesh or the equivalent, where persons are required to work or pass under the scaffolds.

**1910.28(a)(18)** Employees shall not work on scaffolds during storms or high winds.

**1910.28(a)(19)** Employees shall not work on scaffolds which are covered with ice or snow, unless all ice or snow is removed and planking sanded to prevent slipping.

**1910.28(a)(20)** Tools, materials, and debris shall not be allowed to accumulate in quantities to cause a hazard.

**1910.28(a)(21)** Only treated or protected fiber rope shall be used for or near any work involving the use of corrosive substances or chemicals.

**1910.28(a)(22)** Wire or fiber rope used for scaffold suspension shall be capable of supporting at least six times the intended load.

**1910.28(a)(23)** When acid solutions are used for cleaning buildings over 50 feet in height, wire rope supported scaffolds shall be used.

**1910.28(a)(24)** The use of shore scaffolds or lean-to scaffolds is prohibited.

**1910.28(a)(25)** Lumber sizes, when used in this section, refer to nominal sizes except where otherwise stated.

**1910.28(a)(26)** Scaffolds shall be secured to permanent structures, through use of anchor bolts, reveal bolts, or other equivalent means. Window cleaners' anchor bolts shall not be used.

**1910.28(a)(27)** Special precautions shall be taken to protect scaffold members, including any wire or fiber ropes, when using a heat-producing process.

**1910.28(b)** *General requirements for wood pole scaffolds.* **(1)** Scaffold poles shall bear on a foundation of sufficient size and strength to spread the load from the poles over a sufficient area to prevent settlement. All poles shall be set plumb.

**1910.28(b)(2)** Where wood poles are spliced, the ends shall be squared and the upper section shall rest squarely on the lower section. Wood splice plates shall be provided on at least two adjacent sides and shall not be less than 4 feet 0 inches in length, overlapping the abutted ends equally, and have the same width and not less than the cross-sectional area of the pole. Splice plates of other materials of equivalent strength may be used.

**1910.28(b)(3)** Independent pole scaffolds shall be set as near to the wall of the building as practicable.

**1910.28(b)(4)** All pole scaffolds shall be securely guyed or tied to the building or structure. Where the height or length exceeds 25 feet, the scaffold shall be secured at intervals not greater than 25 feet vertically and horizontally.

**1910.28(b)(5)** Putlogs or bearers shall be set with their greater dimensions vertical, long enough to project over the ledgers of the inner and outer rows of poles at least 3 inches for proper support.

**1910.28(b)(6)** Every wooden putlog on single pole scaffolds shall be reinforced with a 3/16 × 2-inch steel strip or equivalent secured to its lower edge throughout its entire length.

**1910.28(b)(7)** Ledgers shall be long enough to extend over two pole spaces. Ledgers shall not be spliced between the poles. Ledgers shall be reinforced by bearing blocks securely nailed to the side of the pole to form a support for the ledger.

**1910.28(b)(8)** Diagonal bracing shall be provided to prevent the poles from moving in a direction parallel with the wall of the building, or from buckling.

**1910.28(b)(9)** Cross bracing shall be provided between the inner and outer sets of poles in independent pole scaffolds. The free ends of pole scaffolds shall be cross braced.

**1910.28(b)(10)** Full diagonal face bracing shall be erected across the entire face of pole scaffolds in both directions. The braces shall be spliced at the poles.

**1910.28(b)(11)** Platform planks shall be laid with their edges close together so the platform will be tight with no spaces through which tools or fragments of material can fall.

**1910.28(b)(12)** Where planking is lapped, each plank shall lap its end supports at least 12 inches. Where the ends of planks abut each other to form a flush floor, the butt joint shall be at the centerline of a pole. The abutted ends shall rest on separate bearers. Intermediate beams shall be provided where necessary to prevent dislodgment of planks due to deflection, and the ends shall be nailed or cleated to prevent their dislodgment.

**1910.28(b)(13)** When a scaffold turns a corner, the platform planks shall be laid to prevent tipping. The planks that meet the corner putlog at an angle shall be laid first, extending over the diagonally placed putlog far enough to have a good safe bearing, but not far enough to involve any danger from tipping. The planking running in the opposite direction at right angles shall be laid so as to extend over and rest on the first layer of planking.

**1910.28(b)(14)** When moving platforms to the next level, the old platform shall be left undisturbed until the new putlogs or bearers have been set in place, ready to receive the platform planks.

**1910.28(b)(15)** Guardrails not less than 2 × 4 inches or the equivalent and not less than 36 inches or more than 42 inches high, with a midrail, when required, of 1 × 4-inch lumber or equivalent, and toeboards, shall be installed at all open sides on all scaffolds more than 10 feet above the ground or floor. Toeboards shall be a minimum of 4 inches in height. Wire mesh shall be installed in accordance with paragraph (a)(17) of this section.

**1910.28(b)(16)** All wood pole scaffolds 60 feet or less in height shall be constructed and erected in accordance with tables D-7 through D-12 of this section. If they are over 60 feet in height they shall be designed by a registered professional engineer and constructed and erected in accordance with such design. A copy of the typical drawings and specifications shall be made available to the employer and for inspection purposes.

**1910.28(b)(17)** Wood-pole scaffolds shall not be erected beyond the reach of effective firefighting apparatus.

### TABLE D-7—MINIMUM NOMINAL SIZE AND MAXIMUM SPACING OF MEMBERS OF SINGLE POLE SCAFFOLDS—LIGHT DUTY

|  | Maximum height of scaffold | |
|---|---|---|
|  | 20 feet | 60 feet |
| Uniformly distributed load | Not to exceed 25 pounds per square foot | |
| Poles or uprights | 2 by 4 in. | 4 by 4 in. |
| Pole spacing (longitudinal) | 6 ft. 0 in. | 10 ft. 0 in. |
| Maximum width of scaffold | 5 ft. 0 in. | 5 ft. 0 in. |
| Bearers or putlogs to 3 ft. 0 in. width | 2 by 4 in. | 2 by 4 in. |
| Bearers or putlogs to 5 ft. 0 in. width | 2 by 6 in. or 3 by 4 in | 2 by 6 in. or 3 by 4 in. (rough). |
| Ledgers | 1 by 4 in. | 1 1/4 by 9 in. |
| Planking | 1 1/4 by 9 in. (rough). | 2 by 9 in. |
| Vertical spacing of horizontal members | 7 ft. 0 in. | 7 ft. 0 in. |
| Bracing, horizontal and diagonal | 1 by 4 in. | 1 by 4 in. |
| Tie-ins | 1 by 4 in. | 1 by 4 in. |
| Toeboards | 4 in. high (minimum). | 4 in. high (minimum). |
| Guardrail | 2 by 4 in. | 2 by 4 in. |

All members except planking are used on edge.

### TABLE D-8—MINIMUM NOMINAL SIZE AND MAXIMUM SPACING OF MEMBERS OF SINGLE POLE SCAFFOLDS—MEDIUM DUTY

| | |
|---|---|
| Uniformly distributed load | Not to exceed 50 pounds per square foot. |
| Maximum height of scaffold | 60 ft. |
| Poles or uprights | 4 by 4 in. |
| Pole spacing (longitudinal) | 8 ft. 0 in. |
| Maximum width of scaffold | 5 ft. 0 in. |
| Bearers or putlogs | 2 by 9 in. or 3 by 4 in. |
| Spacing of bearers or putlogs | 8 ft. 0 in. |
| Ledgers | 2 by 9 in. |
| Vertical spacing of horizontal members | 9 ft. 0 in. |
| Bracing, horizontal | 1 by 6 in. or 1 1/4 by 4 in. |
| Bracing, diagonal | 1 by 4 in. |
| Tie-ins | 1 by 4 in. |
| Planking | 2 by 9 in. |
| Toeboards | 4 in. high (minimum). |
| Guardrail | 2 by 4 in. |

All members except planking are used on edge.

### TABLE D-9—MINIMUM NOMINAL SIZE AND MAXIMUM SPACING OF MEMBERS OF SINGLE POLE SCAFFOLDS—HEAVY DUTY

| | |
|---|---|
| Uniformly distributed load | Not to exceed 75 pounds per square foot |
| Maximum height of scaffold | 60 ft. |
| Poles or uprights | 4 by 4 in. |
| Pole spacing (longitudinal) | 6 ft. 0 in. |
| Maximum width of scaffold | 5 ft. 0 in. |
| Bearers or putlogs | 2 by 9 in. or 3 by 5 in. (rough). |
| Spacing of bearers or putlogs | 6 ft. 0 in. |
| Ledgers | 2 by 9 in. |
| Vertical spacing of horizontal members | 6 ft. 6 in. |
| Bracing horizontal and diagonal | 2 by 4 in. |
| Tie-ins | 1 by 4 in. |
| Planking | 2 by 9 in. |
| Toeboards | 4 in. high (minimum). |
| Guardrail | 2 by 4 in. |

All members except planking are used on edge.

### TABLE D-10—MINIMUM NOMINAL SIZE AND MAXIMUM SPACING OF MEMBERS OF INDEPENDENT POLE SCAFFOLDS—LIGHT DUTY

|  | Maximum height of scaffold | |
|---|---|---|
|  | 20 feet | 60 feet |
| Uniformly distributed load | Not to exceed 25 pounds per square foot | |
| Poles or uprights | 2 by 4 in. | 4 by 4 in. |
| Pole spacing (longitudinal) | 6 ft. 0 in. | 10 ft. 0 in. |
| Pole spacing (transverse) | 6 ft. 0 in. | 10 ft. 0 in. |
| Ledgers | 1 1/4 by 4 in. | 1 1/4 by 9 in. |
| Bearers to 3 ft. 0 in. span | 2 by 4 in. | 2 by 4 in. |
| Bearers to 10 ft. 0 in. span | 2 by 6 in. or 3 by 4 in. | 2 by 9 (rough) or 3 by 8 in. |
| Planking | 1 1/4 by 9 in. | 2 by 9 in. |
| Vertical spacing of horizontal members | 7 ft. 0 in. | 7 ft. 0 in. |
| Bracing, horizontal and diagonal | 1 by 4 in. | 1 by 4 in. |
| Tie-ins | 1 by 4 in. | 1 by 4 in. |
| Toeboards | 4 in. high | 4 in. high (minimum). |
| Guardrail | 2 by 4 in. | 2 by 4 in. |

All members except planking are used on edge.

### TABLE D-11—MINIMUM NOMINAL SIZE AND MAXIMUM SPACING OF MEMBERS OF INDEPENDENT POLE SCAFFOLDS—MEDIUM DUTY

| | |
|---|---|
| Uniformly distributed load | Not to exceed 50 pounds per square foot. |
| Maximum height of scaffold | 60 ft. |
| Poles or uprights | 4 by 4 in. |
| Pole spacing (longitudinal) | 8 ft. 0 in. |
| Pole spacing (transverse) | 8 ft. 0 in. |
| Ledgers | 2 by 9 in. |
| Vertical spacing of horizontal members | 6 ft. 0 in. |
| Spacing of bearers | 8 ft. 0 in. |
| Bearers | 2 by 9 in. (rough) or 2 by 10 in. |
| Bracing, horizontal | 1 by 6 in. or 1 1/4 by 4 in. |
| Bracing, diagonal | 1 by 4 in. |
| Tie-ins | 1 by 4 in. |
| Planking | 2 by 9 in. |
| Toeboards | 4 in. high (minimum). |
| Guardrail | 2 by 4 in. |

All members except planking are used on edge.

### TABLE D-12—MINIMUM NOMINAL SIZE AND MAXIMUM SPACING OF MEMBERS OF INDEPENDENT POLE SCAFFOLDS—HEAVY DUTY

| | |
|---|---|
| Uniformly distributed load | Not to exceed 75 pounds per square foot. |
| Maximum height of scaffold | 60 ft. |
| Poles or uprights | 4 by 4 in. |
| Pole spacing (longitudinal) | 6 ft. 0 in. |
| Pole spacing (transverse) | 8 ft. 0 in. |
| Ledgers | 2 by 9 in. |
| Vertical spacing of horizontal members | 4 ft. 6 in. |
| Bearers | 2 by 9 in. (rough). |
| Bracing, horizontal and diagonal | 2 by 4 in. |
| Tie-ins | 1 by 4 in. |
| Planking | 2 by 9 in. |
| Toeboards | 4 in. high (minimum). |
| Guardrail | 2 by 4 in. |

All members except planking are used on edge.

### TABLE D-13—TUBE AND COUPLER SCAFFOLDS—LIGHT DUTY

| | |
|---|---|
| Uniformly distributed load | Not to exceed 25 p.s.f. |
| Post spacing (longitudinal) | 10 ft. 0 in. |
| Post spacing (transverse) | 6 ft. 0 in. |

| Working levels | Additional planked levels | Maximum height |
|---|---|---|
| 1 | 8 | 125 ft. |
| 2 | 4 | 125 ft. |
| 3 | 0 | 91 ft. 0 in. |

### TABLE D-14—TUBE AND COUPLER SCAFFOLDS—MEDIUM DUTY

| | |
|---|---|
| Uniformly distributed load | Not to exceed 50 p.s.f. |
| Post spacing (longitudinal) | 8 ft. 0 in. |
| Post spacing (transverse) | 6 ft. 0 in. |

| Working levels | Additional planked levels | Maximum height |
|---|---|---|
| 1 | 6 | 125 ft. |
| 2 | 0 | 78 ft. 0 in. |

### TABLE D-15—TUBE AND COUPLER SCAFFOLDS—HEAVY DUTY

| | |
|---|---|
| Uniformly distributed load | Not to exceed 75 p.s.f. |
| Post spacing (longitudinal) | 6 ft. 6 in. |
| Post spacing (transverse) | 6 ft. 0 in. |

| Working levels | Additional planked levels | Maximum height |
|---|---|---|
| 1 | 6 | 125 ft. |

**1910.28(c)** *Tube and coupler scaffolds.* **(1)** A light-duty tube and coupler scaffold shall have all posts, bearers, runners, and bracing of nominal 2-inch O.D. steel tubing. The posts shall be spaced no more than 6 feet apart by 10 feet along the length of the scaffold. Other structural metals when used must be designed to carry an equivalent load.

**1910.28(c)(2)** A medium-duty tube and coupler scaffold shall have all posts, runners, and bracing of nominal 2-inch O.D. steel tubing. Posts spaced not more than 6 feet apart by 8 feet along the length of the scaffold shall have bearers of nominal 2 1/2-inch O.D. steel tubing. Posts spaced not more than 5 feet apart by 8 feet along the length of the scaffold shall have bearers of nominal 2-inch O.D. steel tubing. Other structural metals when used must be designed to carry an equivalent load.

**1910.28(c)(3)** A heavy-duty tube and coupler scaffold shall have all posts, runners, and bracing of nominal 2-inch O.D. steel tubing, with the posts spaced not more than 6 feet apart by 6 feet 6 inches along the length of the scaffold. Other structural metals when used must be designed to carry an equivalent load.

**1910.28(c)(4)** Tube and coupler scaffolds shall be limited in heights and working levels to those permitted in tables D-13, 14, and 15, of this section. Drawings and specifications of all tube and coupler scaffolds above the limitations in tables D-13, 14, and 15 of this section shall be designed by a registered professional engineer and copies made available to the employer and for inspection purposes.

**1910.28(c)(5)** All tube and coupler scaffolds shall be constructed and erected to support four times the maximum intended loads as set forth in tables D-13, 14, and 15 of this section, or as set forth in the specifications by a registered professional engineer, copies which shall be made available to the employer and for inspection purposes.

**1910.28(c)(6)** All tube and coupler scaffolds shall be erected by competent and experienced personnel.

**1910.28(c)(7)** Posts shall be accurately spaced, erected on suitable bases, and maintained plumb.

**1910.28(c)(8)** Runners shall be erected along the length of the scaffold located on both the inside and the outside posts at even height. Runners shall be interlocked to form continuous lengths and coupled to each post. The bottom runners shall be located as close to the base as possible. Runners shall be placed not more than 6 feet 6 inches on centers.

**1910.28(c)(9)** Bearers shall be installed transversely between posts and shall be securely coupled to the posts bearing on the runner coupler. When coupled directly to the runners, the coupler must be kept as close to the posts as possible.

**1910.28(c)(10)** Bearers shall be at least 4 inches but not more than 12 inches longer than the post spacing or runner spacing. Bearers may be cantilevered for use as brackets to carry not more than two planks.

**1910.28(c)(11)** Cross bracing shall be installed across the width of the scaffold at least every third set of posts horizontally and every fourth runner vertically. Such bracing shall extend diagonally from the inner and outer runners upward to the next outer and inner runners.

**1910.28(c)(12)** Longitudinal diagonal bracing shall be installed at approximately a 45-degree angle from near the base of the first outer post upward to the extreme top of the scaffold. Where the longitudinal length of the scaffold permits, such bracing shall be duplicated beginning at every fifth post. In a similar manner, longitudinal diagonal bracing shall also be installed from the last post extending back and upward toward the first post. Where conditions preclude the attachment of this bracing to the posts, it may be attached to the runners.

# OSHA Standards for General Industry

**1910.28(c)(13)** The entire scaffold shall be tied to and securely braced against the building at intervals not to exceed 30 feet horizontally and 26 feet vertically.

**1910.28(c)(14)** Guardrails not less than 2x4 inches or the equivalent and not less than 36 inches or more than 42 inches high, with a mid-rail, when required, of 1x4-inch lumber or equivalent, and toeboards, shall be installed at all open sides on all scaffolds more than 10 feet above the ground or floor. Toeboards shall be a minimum of 4 inches in height. Wire mesh shall be installed in accordance with paragraph (a)(17) of this section.

**1910.28(d)** *Tubular welded frame scaffolds.* **(1)** Metal tubular frame scaffolds, including accessories such as braces, brackets, trusses, screw legs, ladders, etc., shall be designed and proved to safely support four times the maximum intended load.

**1910.28(d)(2)** Spacing of panels or frames shall be consistent with the loads imposed.

**1910.28(d)(3)** Scaffolds shall be properly braced by cross bracing or diagonal braces, or both, for securing vertical members together laterally, and the cross braces shall be of such length as will automatically square and aline vertical members so that the erected scaffold is always plumb, square, and rigid. All brace connections shall be made secure.

**1910.28(d)(4)** Scaffold legs shall be set on adjustable bases or plain bases placed on mud sills or other foundations adequate to support the maximum intended load.

**1910.28(d)(5)** The frames shall be placed one on top of the other with coupling or stacking pins to provide proper vertical alinement of the legs.

**1910.28(d)(6)** Where uplift may occur, panels shall be locked together vertically by pins or other equivalent suitable means.

**1910.28(d)(7)** Guardrails not less than 2 × 4 inches or the equivalent and not less than 36 inches or more than 42 inches high, with a mid-rail, when required, of 1- × 4-inch lumber or equivalent, and toeboards, shall be installed at all open sides on all scaffolds more than 10 feet above the ground or floor. Toeboards shall be a minimum of 4 inches in height. Wire mesh shall be installed in accordance with paragraph (a)(17) of this section.

**1910.28(d)(8)** All tubular metal scaffolds shall be constructed and erected to support four times the maximum intended loads.

**1910.28(d)(9)** To prevent movement, the scaffold shall be secured to the building or structure at intervals not to exceed 30 feet horizontally and 26 feet vertically.

**1910.28(d)(10)** Maximum permissible spans of planking shall be in conformity with paragraph (a)(9) of this section.

**1910.28(d)(11)** Drawings and specifications for all frame scaffolds over 125 feet in height above the base plates shall be designed by a registered professional engineer and copies made available to the employer and for inspection purposes.

**1910.28(d)(12)** All tubular welded frame scaffolds shall be erected by competent and experienced personnel.

**1910.28(d)(13)** Frames and accessories for scaffolds shall be maintained in good repair and every defect, unsafe condition, or noncompliance with this section shall be immediately corrected before further use of the scaffold. Any broken, bent, excessively rusted, altered, or otherwise structurally damaged frames or accessories shall not be used.

**1910.28(d)(14)** Periodic inspections shall be made of all welded frames and accessories, and any maintenance, including painting, or minor corrections authorized by the manufacturer, shall be made before further use.

**1910.28(e)** *Outrigger scaffolds.* **(1)** Outrigger beams shall extend not more than 6 feet beyond the face of the building. The inboard end of outrigger beams, measured from the fulcrum point to the extreme point of support, shall be not less than one and one-half times the outboard end in length. The beams shall rest on edge, the sides shall be plumb, and the edges shall be horizontal. The fulcrum point of the beam shall rest on a secure bearing at least 6 inches in each horizontal dimension. The beam shall be secured in place against movement and shall be securely braced at the fulcrum point against tipping.

**1910.28(e)(2)** The inboard ends of outrigger beams shall be securely supported either by means of struts bearing against sills in contact with the overhead beams or ceiling, or by means of tension members secured to the floor joists underfoot, or by both if necessary. The inboard ends of outrigger beams shall be secured against tipping and the entire supporting structure shall be securely braced in both directions to prevent any horizontal movement.

**1910.28(e)(3)** Unless outrigger scaffolds are designed by a licensed professional engineer, they shall be constructed and erected in accordance with table D-16. Outrigger scaffolds designed by a registered professional engineer shall be constructed and erected in accordance with such design. A copy of the detailed drawings and specifications showing the sizes and spacing of members shall be kept on the job.

**1910.28(e)(4)** Planking shall be laid tight and shall extend to within 3 inches of the building wall. Planking shall be nailed or bolted to outriggers.

**1910.28(e)(5)** Where there is danger of material falling from the scaffold, a wire mesh or other enclosure shall be provided between the guardrail and the toeboard.

**1910.28(e)(6)** Where additional working levels are required to be supported by the outrigger method, the plans and specifications of the outrigger and scaffolding structure shall be designed by a registered professional engineer.

**1910.28(f)** *Masons' adjustable multiple-point suspension scaffolds.* **(1)** The scaffold shall be capable of sustaining a working load of 50 pounds per square foot and shall not be loaded in excess of that figure.

**1910.28(f)(2)** The scaffold shall be provided with hoisting machines that meet the requirements of a nationally recognized testing laboratory. Refer to §1910.7 for definition of nationally recognized testing laboratory.

### TABLE D-16—MINIMUM NOMINAL SIZE AND MAXIMUM SPACING OF MEMBERS OF OUTRIGGER SCAFFOLDS

|  | Light duty | Medium duty |
|---|---|---|
| Maximum scaffold load | 25 p.s.f. | 50 p.s.f. |
| Outrigger size | 2×10 in. | 3×10 in. |
| Maximum outrigger spacing | 10 ft 0 in. | 6 ft 0 in. |
| Planking | 2×9 in. | 2×9 in. |
| Guardrail | 2×4 in. | 2×4 in. |
| Guardrail uprights | 2×4 in. | 2×4 in. |
| Toeboards (minimum) | 4 in. | 4 in. |

**1910.28(f)(3)** The platform shall be supported by wire ropes in conformity with paragraph (a)(22) of this section, suspended from overhead outrigger beams.

**1910.28(f)(4)** The scaffold outrigger beams shall consist of structural metal securely fastened or anchored to the frame or floor system of the building or structure.

**1910.28(f)(5)** Each outrigger beam shall be equivalent in strength to at least a standard 7-inch, 15.3-pound steel I-beam, be at least 15 feet long, and shall not project more than 6 feet 6 inches beyond the bearing point.

**1910.28(f)(6)** Where the overhang exceeds 6 feet 6 inches, outrigger beams shall be composed of stronger beams or multiple beams and be installed in accordance with approved designs and instructions.

**1910.28(f)(7)** If channel iron outrigger beams are used in place of I-beams, they shall be securely fastened together with the flanges turned out.

**1910.28(f)(8)** All outrigger beams shall be set and maintained with their webs into vertical position.

**1910.28(f)(9)** A stop bolt shall be placed at each end of every outrigger beam.

**1910.28(f)(10)** The outrigger beam shall rest on suitable wood-bearing blocks.

**1910.28(f)(11)** All parts of the scaffold such as bolts, nuts, fittings, clamps, wire rope, and outrigger beams and their fastenings, shall be maintained in sound and good working condition and shall be inspected before each installation and periodically thereafter.

**1910.28(f)(12)** The free end of the suspension wire ropes shall be equipped with proper size thimbles and be secured by splicing or other equivalent means. The running ends shall be securely attached to the hoisting drum and at least four turns of rope shall at all times remain on the drum.

**1910.28(f)(13)** Where a single outrigger beam is used, the steel shackles or clevises with which the wire ropes are attached to the outrigger beams shall be placed directly over the hoisting drums.

**1910.28(f)(14)** The scaffold platform shall be equivalent in strength to at least 2-inch planking. (For maximum planking spans see paragraph (a)(9) of this section.)

**1910.28(f)(15)** Guardrails not less than 2 × 4 inches or the equivalent and not less than 36 inches or more than 42 inches high, with a mid-rail, when required, of 1 × 4-inch lumber or equivalent, and toeboards, shall be installed at all open sides on all scaffolds more than 10 feet above the ground or floor. Toeboards shall be a minimum of 4 inches in height. Wire mesh shall be installed in accordance with paragraph (a)(17) of this section.

**1910.28(f)(16)** Overhead protection shall be provided on the scaffold, not more than 9 feet above the platform, consisting of 2-inch planking or material of equivalent strength laid tight, when men are at work on the scaffold and an overhead hazard exists.

**1910.28(f)(17)** Each scaffold shall be installed or relocated in accordance with designs and instructions, of a registered professional engineer, and supervised by a competent, designated person.

**1910.28(g)** *Two-point suspension scaffolds (swinging scaffolds).* **(1)** Two-point suspension scaffold platforms shall be not less than 20 inches no more than 36 inches wide overall. The platform shall be securely fastened to the hangers by U-bolts or by other equivalent means.

**1910.28(g)(2)** The hangers of two-point suspension scaffolds shall be made of wrought iron, mild steel, or other equivalent material having a cross-sectional area capable of sustaining four times the maximum intended load, and shall be designed with a support for guardrail, intermediate rail, and toeboard.

**1910.28(g)(3)** When hoisting machines are used on two-point suspension scaffolds, such machines shall be of a design tested and approved by a nationally recognized testing laboratory. Refer to §1910.7 for definition of nationally recognized testing laboratory.

**1910.28(g)(4)** The roof irons or hooks shall be of wrought iron, mild steel, or other equivalent material of proper size and design, securely installed and anchored. Tie-backs of three-fourth inch manila rope or the equivalent shall serve as a secondary means of anchorage, installed at right angles to the face of the building whenever possible and secured to a structurally sound portion of the building.

**1910.28(g)(5)** Guardrails not less than 2 × 4 inches or the equivalent and not less than 36 inches or more than 42 inches high, with a mid-rail, when required, of 1- × 4-inch lumber or equivalent, and toeboards, shall be installed at all open sides on all scaffolds more than 10 feet above the ground or floor. Toeboards shall be a minimum of 4 inches in height. Wire mesh shall be installed in accordance with paragraph (a)(17) of this section.

**1910.28(g)(6)** Two-point suspension scaffolds shall be suspended by wire or fiber ropes. Wire and fiber ropes shall conform to paragraph (a)(22) of this section.

**1910.28(g)(7)** The blocks for fiber ropes shall be of standard 6-inch size, consisting of at least one double and one single block. The sheaves of all blocks shall fit the size of rope used.

**1910.28(g)(8)** All wire ropes, fiber ropes, slings, hangers, platforms, and other supporting parts shall be inspected before every installation. Periodic inspections shall be made while the scaffold is in use.

**1910.28(g)(9)** On suspension scaffolds designed for a working load of 500 pounds no more than two men shall be permitted to work at one time. On suspension scaffolds with a working load of 750 pounds, no more than three men shall be permitted to work at one time. Each workman shall be protected by a safety lifebelt attached to a lifeline. The lifeline shall be securely attached to substantial members of the structure (not scaffold), or to securely rigged lines, which will safely suspend the workman in case of a fall.

**1910.28(g)(10)** Where acid solutions are used, fiber ropes are not permitted unless acid-proof.

**1910.28(g)(11)** Two-point suspension scaffolds shall be securely lashed to the building or structure to prevent them from swaying. Window cleaners' anchors shall not be used for this purpose.

**1910.28(g)(12)** The platform of every two-point suspension scaffold shall be one of the following types:

**1910.28(g)(12)(i)** The side stringer of ladder-type platforms shall be clear straight-grained spruce or materials of equivalent strength and durability. The rungs shall be of straight-grained oak, ash, or hickory, at least 1 1/8 inch in diameter, with seven-eighth inch tenons mortised into the side stringers at least seven-eighth inch. The stringers shall be tied together with the tie rods not less than one-quarter inch in diameter, passing through the stringers and riveted up tight against washers on both ends. The flooring strips shall be spaced not more than five-eighth inch apart except at the side rails where the space may be 1 inch. Ladder-type platforms shall be constructed in accordance with table D-17.

**1910.28(g)(12)(ii)** Plank-type platforms shall be composed of not less than nominal 2- × 8-inch unspliced planks, properly cleated together on the underside starting 6 inches from each end; intervals in between shall not exceed 4 feet. The plank-type platform shall not extend beyond the hangers more than 18 inches. A bar or other effective means shall be securely fastened to the platform at each end to prevent its slipping off the hanger. The span between hangers for plank-type platforms shall not exceed 10 feet.

**1910.28(g)(12)(iii)** Beam platforms shall have side stringers of lumber not less than 2 × 6 inches set on edge. The span between hangers shall not exceed 12 feet when beam platforms are used. The flooring shall be supported on 2- and 6-inch crossbeams, laid flat and set into the upper edge of the stringers with a snug fit, at intervals of not more than 4 feet, securely nailed in place. The flooring shall be of 1- × 6-inch material properly nailed. Floorboards shall not be spaced more than one-half inch apart.

### TABLE D-17—SCHEDULE FOR LADDER-TYPE PLATFORMS

|  | Length of platform (feet) ||||| 
|---|---|---|---|---|---|
|  | 12 | 14 & 16 | 18 & 20 | 22 & 24 | 28 & 30 |
| Side stringers, minimum cross section (finished sizes): | | | | | |
| At ends (in.) | 1 3/4 × 2 3/4 | 1 3/4 × 2 3/4 | 1 3/4 × 3 | 1 3/4 × 3 | 1 3/4 × 3 1/2 |
| At middle (in.) | 1 3/4 × 3 3/4 | 1 3/4 × 3 3/4 | 1 3/4 × 4 | 1 3/4 × 4 1/4 | 1 3/4 × 5 |
| Reinforcing strip (minimum)[1] | ............ | ............ | ............ | ............ | ............ |
| Rungs[2] | ............ | ............ | ............ | ............ | ............ |
| Tie rods: | | | | | |
| Number (minimum) | 3 | 4 | 4 | 5 | 6 |
| Diameter (minimum) | 1/4 in | 1/4 in | 1/4 in | 1/4 in | 1/4 in. |
| Flooring, minimum finished size (in.) | 1/2 × 2 3/4 | 1/2 × 2 3/4 | 1/2 × 2 3/4 | 1/2 × 3/4 | 1/2 × 2 3/4 |

[1] A 1/8 × 7/8-in. steel reinforcing strip or its equivalent shall be attached to the side or underside full length.
[2] Rungs shall be 1 1/8-in. minimum, diameter with at least 7/8-in. diameter tenons, and the maximum spacing shall be 12 in. center to center.

**1910.28(h)** *Stone setters' adjustable multiple-point suspension scaffolds.* **(1)** The scaffold shall be capable of sustaining a working load of 25 pounds per square foot and shall not be overloaded. Scaffolds shall not be used for storage of stone or other heavy materials.

**1910.28(h)(2)** The hoisting machine and its supports shall be of a type tested and listed by a nationally recognized testing laboratory. Refer to §1910.399(a)(77) for definition of listed, and §1910.7 for nationally recognized testing laboratory.

# OSHA Standards for General Industry

**1910.28(h)(3)** The platform shall be securely fastened to the hangers by U-bolts or other equivalent means.

**1910.28(h)(4)** The scaffold unit shall be suspended from metal outriggers, iron brackets, wire rope slings, or iron hooks which will safely support the maximum intended load.

**1910.28(h)(5)** Outriggers when used shall be set with their webs in a vertical position, securely anchored to the building or structure and provided with stop bolts at each end.

**1910.28(h)(6)** The scaffold shall be supported by wire rope conforming with paragraph (a)(22) of this section, suspended from overhead supports.

**1910.28(h)(7)** The free ends of the suspension wire ropes shall be equipped with proper size thimbles, secured by splicing or other equivalent means. The running ends shall be securely attached to the hoisting drum and at least four turns of rope shall remain on the drum at all times.

**1910.28(h)(8)** Guardrails not less than 2 by 4 inches or the equivalent and not less than 36 inches or more than 42 inches high, with a midrail, when required, of 1- by 4-inch lumber or equivalent, and toeboards, shall be installed at all open sides on all scaffolds more than 10 feet above the ground or floor. Toeboards shall be a minimum of 4 inches in height. Wire mesh shall be installed in accordance with paragraph (a)(17) of this section.

**1910.28(h)(9)** When two or more scaffolds are used on a building or structure they shall not be bridged one to the other but shall be maintained at even height with platforms butting closely.

**1910.28(h)(10)** Each scaffold shall be installed or relocated in accordance with designs and instructions of a registered professional engineer, and such installation or relocation shall be supervised by a competent designated person.

**1910.28(i)** *Single-point adjustable suspension scaffolds.* **(1)** The scaffolding, including power units or manually operated winches, shall be a type tested and listed by a nationally recognized testing laboratory. Refer to §1910.399(a)(77) for definition of listed, and §1910.7 for nationally recognized testing laboratory.

**1910.28(i)(2)** [Reserved]

**1910.28(i)(3)** All power-operated gears and brakes shall be enclosed.

**1910.28(i)(4)** In addition to the normal operating brake, all-power driven units must have an emergency brake which engages automatically when the normal speed of descent is exceeded.

**1910.28(i)(5)** Guards, mid-rails, and toeboards shall completely enclose the cage or basket. Guardrails shall be no less than 2 by 4 inches or the equivalent installed no less than 36 inches nor more than 42 inches above the platform. Mid-rails shall be 1 by 6 inches or the equivalent, installed equidistant between the guardrail and the platform. Toeboards shall be a minimum of 4 inches in height.

**1910.28(i)(6)** The hoisting machines, cables, and equipment shall be regularly serviced and inspected after each installation and every 30 days thereafter.

**1910.28(i)(7)** The units may be combined to form a two-point suspension scaffold. Such scaffold shall comply with paragraph (g) of this section.

**1910.28(i)(8)** The supporting cable shall be straight for its entire length, and the operator shall not sway the basket and fix the cable to any intermediate points to change his original path of travel.

**1910.28(i)(9)** Equipment shall be maintained and used in accordance with the manufacturers' instructions.

**1910.28(i)(10)** Suspension methods shall conform to applicable provisions of paragraphs (f) and (g) of this section.

**1910.28(j)** *Boatswain's chairs.* **(1)** The chair seat shall be not less than 12 by 24 inches, and of 1-inch thickness. The seat shall be reinforced on the underside to prevent the board from splitting.

**1910.28(j)(2)** The two fiber rope seat slings shall be of 5/8-inch diameter, reeved through the four seat holes so as to cross each other on the underside of the seat.

**1910.28(j)(3)** Seat slings shall be of at least 3/8-inch wire rope when a workman is conducting a heat producing process such as gas or arc welding.

**1910.28(j)(4)** The workman shall be protected by a safety life belt attached to a lifeline. The lifeline shall be securely attached to substantial members of the structure (not scaffold), or to securely rigged lines, which will safely suspend the worker in case of a fall.

**1910.28(j)(5)** The tackle shall consist of correct size ball bearing or bushed blocks and properly spliced 5/8-inch diameter first-grade manila rope.

**1910.28(j)(6)** The roof irons, hooks, or the object to which the tackle is anchored shall be securely installed. Tiebacks when used shall be installed at right angles to the face of the building and securely fastened to a chimney.

**1910.28(k)** *Carpenters' bracket scaffolds.* **(1)** The brackets shall consist of a triangular wood frame not less than 2 by 3 inches in cross section, or of metal of equivalent strength. Each member shall be properly fitted and securely joined.

**1910.28(k)(2)** Each bracket shall be attached to the structure by means of one of the following:

**1910.28(k)(2)(i)** A bolt no less than five-eighths inch in diameter which shall extend through the inside of the building wall.

**1910.28(k)(2)(ii)** A metal stud attachment device.

**1910.28(k)(2)(iii)** Welding to steel tanks.

**1910.28(k)(2)(iv)** Hooking over a well-secured and adequately strong supporting member.

The brackets shall be spaced no more than 10 feet apart.

**1910.28(k)(3)** No more than two persons shall occupy any given 10 feet of a bracket scaffold at any one time. Tools and materials shall not exceed 75 pounds in addition to the occupancy.

**1910.28(k)(4)** The platform shall consist of not less than two 2- by 9-inch nominal size planks extending not more than 18 inches or less than 6 inches beyond each end support.

**1910.28(k)(5)** Guardrails not less than 2 by 4 inches or the equivalent and not less than 36 inches or more than 42 inches high, with a midrail, when required, of 1- by 4-inch lumber or equivalent, and toeboards, shall be installed at all open sides on all scaffolds more than 10 feet above the ground or floor. Toeboards shall be a minimum of 4 inches in height. Wire mesh shall be installed in accordance with paragraph (a)(17) of this section.

**1910.28(l)** *Bricklayers' square scaffolds.* **(1)** The squares shall not exceed 5 feet in width and 5 feet in height.

**1910.28(l)(2)** Members shall be not less than those specified in Table D-18.

**1910.28(l)(3)** The squares shall be reinforced on both sides of each corner with 1- by 6-inch gusset pieces. They shall also have braces 1 by 8 inches on both sides running from center to center of each member, or other means to secure equivalent strength and rigidity.

**1910.28(l)(4)** The squares shall be set not more than 5 feet apart for medium duty scaffolds, and not more than 8 feet apart for light duty scaffolds. Bracing 1 × 8 inches, extending from the bottom of each square to the top of the next square, shall be provided on both front and rear sides of the scaffold.

### TABLE D-18—MINIMUM DIMENSIONS FOR BRICKLAYERS' SQUARE SCAFFOLD MEMBERS

| Members | Dimensions (inches) |
|---|---|
| Bearers or horizontal members | 2 by 6. |
| Legs | 2 by 6. |
| Braces at corners | 1 by 6. |
| Braces diagonally from center frame | 1 by 8. |

**1910.28(l)(5)** Platform planks shall be at least 2- by 9-inch nominal size. The ends of the planks shall overlap the bearers of the squares and each plank shall be supported by not less than three squares.

**1910.28(l)(6)** Bricklayers' square scaffolds shall not exceed three tiers in height and shall be so constructed and arranged that one square shall rest directly above the other. The upper tiers shall stand on a continuous row of planks laid across the next lower tier and be nailed down or otherwise secured to prevent displacement.

**1910.28(l)(7)** Scaffolds shall be level and set upon a firm foundation.

**1910.28(m)** *Horse scaffolds.* **(1)** Horse scaffolds shall not be constructed or arranged more than two tiers or 10 feet in height.

**1910.28(m)(2)** The members of the horses shall be not less than those specified in Table D-19.

**1910.28(m)(3)** Horses shall be spaced not more than 5 feet for medium duty and not more than 8 feet for light duty.

**1910.28(m)(4)** When arranged in tiers, each horse shall be placed directly over the horse in the tier below.

**1910.28(m)(5)** On all scaffolds arranged in tiers, the legs shall be nailed down to the planks to prevent displacement or thrust and each tier shall be substantially cross braced.

### TABLE D-19—MINIMUM DIMENSIONS FOR HORSE SCAFFOLD MEMBERS

| Members | Dimensions (inches) |
| --- | --- |
| Horizontal members or bearers | 3 by 4. |
| Legs | 1 1/4 by 4 1/2. |
| Longitudinal brace between legs | 1 by 6. |
| Gusset brace at top of legs | 1 by 8. |
| Half diagonal braces | 1 1/4 by 4 1/2. |

**1910.28(m)(6)** Horses or parts which have become weak or defective shall not be used.

**1910.28(m)(7)** Guardrails not less than 2 by 4 inches or the equivalent and not less than 36 inches or more than 42 inches high with a mid-rail, when required, of 1- by 4-inch lumber or equivalent and toeboards, shall be installed at all open sides on all scaffolds more than 10 feet above the ground or floor. Toeboards shall be a minimum of 4 inches in height. Wire mesh shall be installed in accordance with paragraph (a)(17) of this section.

**1910.28(n)** *Needle beam scaffold.* **(1)** Wood needle beams shall be in accordance with paragraph (a) (5) and (9) of this section, and shall be not less than 4 by 6 inches in size, with the greater dimension placed in a vertical direction. Metal beams or the equivalent conforming to paragraph (a) (4) and (8) of this section may be used.

**1910.28(n)(2)** Ropes or hangers shall be provided for supports. The span between supports on the needle beam shall not exceed 10 feet for 4- by 6-inch timbers. Rope supports shall be equivalent in strength to 1-inch diameter first-grade manila rope.

**1910.28(n)(3)** The ropes shall be attached to the needle beams by a scaffold hitch or a properly made eye splice. The loose end of the rope shall be tied by a bowline knot or by a round turn and one-half hitch.

**1910.28(n)(4)** The platform span between the needle beams shall not exceed 8 feet when using 2-inch scaffold plank. For spans greater than 8 feet, platforms shall be designed based on design requirements for the special span. The overhang of each end of the platform planks shall be not less than 1 foot and not more than 18 inches.

**1910.28(n)(5)** When one needle beam is higher than the other or when the platform is not level the platform shall be secured against slipping.

**1910.28(n)(6)** All unattached tools, bolts, and nuts used on needle beam scaffolds shall be kept in suitable containers.

**1910.28(n)(7)** One end of a needle beam scaffold may be supported by a permanent structural member conforming to paragraphs (a) (4) and (8) of this section.

**1910.28(n)(8)** Each man working on a needle beam scaffold 20 feet or more above the ground or floor and working with both hands, shall be protected by a safety life belt attached to a lifeline. The lifeline shall be securely attached to substantial members of the structure (not scaffold), or to securely rigged lines, which will safely suspend the workman in case of a fall.

**1910.28(o)** *Plasterers', decorators', and large area scaffolds.* **(1)** Plasterers', decorators', lathers', and ceiling workers' inside scaffolds shall be constructed in accordance with the general requirements set forth for independent wood pole scaffolds.

**1910.28(o)(2)** Guardrails not less than 2 by 4 inches or the equivalent and not less than 36 inches or more than 42 inches high, with a mid-rail, when required, of 1- by 4-inch lumber or equivalent, and toeboards, shall be installed at all open sides on all scaffolds more than 10 feet above the ground or floor. Toeboards shall be a minimum of 4 inches in height. Wire mesh shall be installed in accordance with paragraph (a)(17) of this section.

**1910.28(o)(3)** All platform planks shall be laid with the edges close together.

**1910.28(o)(4)** When independent pole scaffold platforms are erected in sections, such sections shall be provided with connecting runways equipped with substantial guardrails.

**1910.28(p)** *Interior hung scaffolds.*

**1910.28(p)(1)** [Reserved]

**1910.28(p)(2)** The suspended steel wire rope shall conform to paragraph (a)(22) of this section. Wire may be used providing the strength requirements of paragraph (a)(22) of this section are met.

**1910.28(p)(3)** For hanging wood scaffolds, the following minimum nominal size material is recommended:

**1910.28(p)(3)(i)** Supporting bearers 2 by 9 inches on edge.

**1910.28(p)(3)(ii)** Planking 2 by 9 inches or 2 by 10 inches, with maximum span 7 feet for heavy duty and 10 feet for light duty or medium duty.

**1910.28(p)(4)** Steel tube and coupler members may be used for hanging scaffolds with both types of scaffold designed to sustain a uniform distributed working load up to heavy duty scaffold loads with a safety factor of four.

**1910.28(p)(5)** When a hanging scaffold is supported by means of wire rope, such wire rope shall be wrapped at least twice around the supporting members and twice around the bearers of the scaffold, with each end of the wire rope secured by at least three standard wire-rope clips.

**1910.28(p)(6)** All overhead supporting members shall be inspected and checked for strength before the scaffold is erected.

**1910.28(p)(7)** Guardrails not less than 2 by 4 inches or the equivalent and not less than 36 inches or more than 42 inches high, with a mid-rail, when required, of 1- by 4-inch lumber or equivalent, and toeboards, shall be installed at all open sides on all scaffolds more than 10 feet above the ground or floor. Toeboards shall be a minimum of 4 inches in height. Wire mesh shall be installed in accordance with paragraph (a)(17) of this section.

**1910.28(q)** *Ladder-jack scaffolds.* **(1)** All ladder-jack scaffolds shall be limited to light duty and shall not exceed a height of 20 feet above the floor or ground.

**1910.28(q)(2)** All ladders used in connection with ladder-jack scaffolds shall be heavy-duty ladders and shall be designed and constructed in accordance with §1910.25 and §1910.26.

**1910.28(q)(3)** The ladder jack shall be so designed and constructed that it will bear on the side rails in addition to the ladder rungs, or if bearing on rungs only, the bearing area shall be at least 10 inches on each rung.

**1910.28(q)(4)** Ladders used in conjunction with ladder jacks shall be so placed, fastened, held, or equipped with devices so as to prevent slipping.

**1910.28(q)(5)** The wood platform planks shall be not less than 2 inches nominal in thickness. Both metal and wood platform planks shall overlap the bearing surface not less than 12 inches. The span between supports for wood shall not exceed 8 feet. Platform width shall be not less than 18 inches.

**1910.28(q)(6)** Not more than two persons shall occupy any given 8 feet of any ladder-jack scaffold at any one time.

**1910.28(r)** *Window-jack scaffolds.* **(1)** Window-jack scaffolds shall be used only for the purpose of working at the window opening through which the jack is placed.

**1910.28(r)(2)** Window jacks shall not be used to support planks placed between one window jack and another or for other elements of scaffolding.

**1910.28(r)(3)** Window-jack scaffolds shall be provided with suitable guardrails unless safety belts with lifelines are attached and provided for the workman. Window-jack scaffolds shall be used by one man only.

**1910.28(s)** *Roofing brackets.* **(1)** Roofing brackets shall be constructed to fit the pitch of the roof.

**1910.28(s)(2)** Brackets shall be secured in place by nailing in addition to the pointed metal projections. The nails shall be driven full length into the roof. When rope supports are used, they shall consist of first-grade manila of at least three-quarter-inch diameter, or equivalent.

**1910.28(s)(3)** A substantial catch platform shall be installed below the working area of roofs more than 20 feet from the ground to eaves with a

# OSHA Standards for General Industry

slope greater than 3 inches in 12 inches without a parapet. In width the platform shall extend 2 feet beyond the projection of the eaves and shall be provided with a safety rail, mid-rail, and toeboard. This provision shall not apply where employees engaged in work upon such roofs are protected by a safety belt attached to a lifeline.

**1910.28(t)** *Crawling boards or chicken ladders.* **(1)** Crawling boards shall be not less than 10 inches wide and 1 inch thick, having cleats 1 × 1 1/2 inches. The cleats shall be equal in length to the width of the board and spaced at equal intervals not to exceed 24 inches. Nails shall be driven through and clinched on the underside. The crawling board shall extend from the ridge pole to the eaves when used in connection with roof construction, repair, or maintenance.

**1910.28(t)(2)** A firmly fastened lifeline of at least three-quarter-inch rope shall be strung beside each crawling board for a handhold.

**1910.28(t)(3)** Crawling boards shall be secured to the roof by means of adequate ridge hooks or equivalent effective means.

**1910.28(u)** *Float or ship scaffolds.* **(1)** Float or ship scaffolds shall support not more than three men and a few light tools, such as those needed for riveting, bolting, and welding. They shall be constructed in accordance with paragraphs (u) (2) through (6) of this section, unless substitute designs and materials provide equivalent strength, stability, and safety.

**1910.28(u)(2)** The platform shall be not less than 3 feet wide and 6 feet long, made of three-quarter-inch plywood, equivalent to American Plywood Association Grade B-B, Group I, Exterior.

**1910.28(u)(3)** Under the platform, there shall be two supporting bearers made from 2- × 4-inch, or 1- × 10-inch rough, selected lumber, or better. They shall be free of knots or other flaws and project 6 inches beyond the platform on both sides. The ends of the platform shall extend about 6 inches beyond the outer edges of the bearers. Each bearer shall be securely fastened to the platform.

**1910.28(u)(4)** An edging of wood not less than 3/4 × 1 1/2 inches, or equivalent, shall be placed around all sides of the platform to prevent tools from rolling off.

**1910.28(u)(5)** Supporting ropes shall be 1-inch diameter manila rope or equivalent, free from deterioration, chemical damage, flaws, or other imperfections. Rope connections shall be such that the platform cannot shift or slip. If two ropes are used with each float, each of the two supporting ropes shall be hitched around one end of a bearer and pass under the platforms to the other end of the bearer where it is hitched again, leaving sufficient rope at each end for the supporting ties.

**1910.28(u)(6)** Each workman shall be protected by a safety lifebelt attached to a lifeline. The lifeline shall be securely attached to substantial members of the structure (not scaffold), or to securely rigged lines, which will safely suspend the workman in case of a fall.

**1910.28(v)** *Scope.* This section establishes safety requirements for the construction, operation, maintenance, and use of scaffolds used in the maintenance of buildings and structures.

[39 FR 23502, June 27, 1974, as amended at 43 FR 49746, Oct. 24, 1978; 49 FR 5321, Feb. 10, 1984; 53 FR 12121, Apr. 12, 1988]

## §1910.29 Manually propelled mobile ladder stands and scaffolds (towers).

**1910.29(a)** *General requirements*—**(1)** *Application.* This section is intended to prescribe rules and requirements for the design, construction, and use of mobile work platforms (including ladder stands but not including aerial ladders) and rolling (mobile) scaffolds (towers). This standard is promulgated to aid in providing for the safety of life, limb, and property, by establishing minimum standards for structural design requirements and for the use of mobile work platforms and towers.

**1910.29(a)(2)** *Working loads.* **(i)** Work platforms and scaffolds shall be capable of carrying the design load under varying circumstances depending upon the conditions of use. Therefore, all parts and appurtenances necessary for their safe and efficient utilization must be integral parts of the design.

**1910.29(a)(2)(ii)** Specific design and construction requirements are not a part of this section because of the wide variety of materials and design possibilities. However, the design shall be such as to produce a mobile ladder stand or scaffold that will safely sustain the specified loads. The material selected shall be of sufficient strength to meet the test requirements and shall be protected against corrosion or deterioration.

**1910.29(a)(2)(ii)(a)** The design working load of ladder stands shall be calculated on the basis of one or more 200-pound persons together with 50 pounds of equipment each.

**1910.29(a)(2)(ii)(b)** The design load of all scaffolds shall be calculated on the basis of:

*Light*—Designed and constructed to carry a working load of 25 pounds per square foot.

*Medium*—Designed and constructed to carry a working load of 50 pounds per square foot.

*Heavy*—Designed and constructed to carry a working load of 75 pounds per square foot.

All ladder stands and scaffolds shall be capable of supporting at least four times the design working load.

**1910.29(a)(2)(iii)** The materials used in mobile ladder stands and scaffolds shall be of standard manufacture and conform to standard specifications of strength, dimensions, and weights, and shall be selected to safely support the design working load.

**1910.29(a)(2)(iv)** Nails, bolts, or other fasteners used in the construction of ladders, scaffolds, and towers shall be of adequate size and in sufficient numbers at each connection to develop the designed strength of the unit. Nails shall be driven full length. (All nails should be immediately withdrawn from dismantled lumber.)

**1910.29(a)(2)(v)** All exposed surfaces shall be free from sharp edges, burrs or other safety hazards.

**1910.29(a)(3)** *Work levels.* **(i)** The maximum work level height shall not exceed four (4) times the minimum or least base dimensions of any mobile ladder stand or scaffold. Where the basic mobile unit does not meet this requirement, suitable outrigger frames shall be employed to achieve this least base dimension, or provisions shall be made to guy or brace the unit against tipping.

**1910.29(a)(3)(ii)** The minimum platform width for any work level shall not be less than 20 inches for mobile scaffolds (towers). Ladder stands shall have a minimum step width of 16 inches.

**1910.29(a)(3)(iii)** The supporting structure for the work level shall be rigidly braced, using adequate cross bracing or diagonal bracing with rigid platforms at each work level.

**1910.29(a)(3)(iv)** The steps of ladder stands shall be fabricated from slip resistant treads.

**1910.29(a)(3)(v)** The work level platform of scaffolds (towers) shall be of wood, aluminum, or plywood planking, steel or expanded metal, for the full width of the scaffold, except for necessary openings. Work platforms shall be secured in place. All planking shall be 2-inch (nominal) scaffold grade minimum 1,500 $f.$ (stress grade) construction grade lumber or equivalent.

**1910.29(a)(3)(vi)** All scaffold work levels 10 feet or higher above the ground or floor shall have a standard (4-inch nominal) toeboard.

**1910.29(a)(3)(vii)** All work levels 10 feet or higher above the ground or floor shall have a guardrail of 2- by 4-inch nominal or the equivalent installed no less than 36 inches or more than 42 inches high, with a mid-rail, when required, of 1- by 4-inch nominal lumber or equivalent.

**1910.29(a)(3)(viii)** A climbing ladder or stairway shall be provided for proper access and egress, and shall be affixed or built into the scaffold and so located that its use will not have a tendency to tip the scaffold. A landing platform shall be provided at intervals not to exceed 30 feet.

**1910.29(a)(4)** *Wheels or casters.* **(i)** Wheels or casters shall be properly designed for strength and dimensions to support four (4) times the design working load.

**1910.29(a)(4)(ii)** All scaffold casters shall be provided with a positive wheel and/or swivel lock to prevent movement. Ladder stands shall have at least two (2) of the four (4) casters and shall be of the swivel type.

**1910.29(a)(4)(iii)** Where leveling of the elevated work platform is required, screw jacks or other suitable means for adjusting the height shall be provided in the base section of each mobile unit.

**1910.29(b)** *Mobile tubular welded frame scaffolds*—**(1)** *General.* Units shall be designed to comply with the requirements of paragraph (a) of this section.

**1910.29(b)(2)** *Bracing.* Scaffolds shall be properly braced by cross braces and/or diagonal braces for securing vertical members together laterally. The cross braces shall be of a length that will automatically square and align vertical members so the erected scaffold is always plumb, square, and rigid.

**1910.29(b)(3)** *Spacing.* Spacing of panels or frames shall be consistent with the loads imposed. The frames shall be placed one on top of the other with coupling or stacking pins to provide proper vertical alignment of the legs.

**1910.29(b)(4)** *Locking.* Where uplift may occur, panels shall be locked together vertically by pins or other equivalent means.

**1910.29(b)(5)** *Erection.* Only the manufacturer of a scaffold or his qualified designated agent shall be permitted to erect or supervise the erection of scaffolds exceeding 50 feet in height above the base, unless such structure is approved in writing by a registered professional engineer, or erected in accordance with instructions furnished by the manufacturer.

**1910.29(c)** *Mobile tubular welded sectional folding scaffolds*—**(1)** *General.* Units including sectional stairway and sectional ladder scaffolds shall be designed to comply with the requirements of paragraph (a) of this section.

**1910.29(c)(2)** *Stairway.* An integral stairway and work platform shall be incorporated into the structure of each sectional folding stairway scaffold.

**1910.29(c)(3)** *Bracing.* An integral set of pivoting and hinged folding diagonal and horizontal braces and a detachable work platform shall be incorporated into the structure of each sectional folding ladder scaffold.

**1910.29(c)(4)** *Sectional folding stairway scaffolds.* Sectional folding stairway scaffolds shall be designed as medium duty scaffolds except for high clearance. These special base sections shall be designed as light duty scaffolds. When upper sectional folding stairway scaffolds are used with a special high clearance base, the load capacity of the entire scaffold shall be reduced accordingly. The width of a sectional folding stairway scaffold shall not exceed 4 1/2 feet. The maximum length of a sectional folding stairway scaffold shall not exceed 6 feet.

**1910.29(c)(5)** *Sectional folding ladder scaffolds.* Sectional folding ladder scaffolds shall be designed as light duty scaffolds including special base (open end) sections which are designed for high clearance. For certain special applications the six-foot (6') folding ladder scaffolds, except for special high clearance base sections, shall be designed for use as medium duty scaffolds. The width of a sectional folding ladder scaffold shall not exceed 4 1/2 feet. The maximum length of a sectional folding ladder scaffold shall not exceed 6 feet 6 inches for a six-foot (6') long unit, 8 feet 6 inches for an eight-foot (8') unit or 10 feet 6 inches for a ten-foot (10') long unit.

**1910.29(c)(6)** *End frames.* The end frames of sectional ladder and stairway scaffolds shall be designed so that the horizontal bearers provide supports for multiple planking levels.

**1910.29(c)(7)** *Erection.* Only the manufacturer of the scaffold or his qualified designated agent shall be permitted to erect or supervise the erection of scaffolds exceeding 50 feet in height above the base, unless such structure is approved in writing by a licensed professional engineer, or erected in accordance with instructions furnished by the manufacturer.

**1910.29(d)** *Mobile tube and coupler scaffolds*—**(1)** *Design.* Units shall be designed to comply with the applicable requirements of paragraph (a) of this section.

**1910.29(d)(2)** *Material.* The material used for the couplers shall be of a structural type, such as a drop-forged steel, malleable iron or structural grade aluminum. The use of gray cast iron is prohibited.

**1910.29(d)(3)** *Erection.* Only the manufacturer of the scaffold or his qualified designated agent shall be permitted to erect or supervise the erection of scaffolds exceeding 50 feet in height above the base, unless such structure is approved in writing by a licensed professional engineer, or erected in accordance with instructions furnished by the manufacturer.

**1910.29(e)** *Mobile work platforms*—**(1)** *Design.* Units shall be designed for the use intended and shall comply with the requirements of paragraph (a) of this section.

**1910.29(e)(2)** *Base width.* The minimum width of the base of mobile work platforms shall not be less than 20 inches.

**1910.29(e)(3)** *Bracing.* Adequate rigid diagonal bracing to vertical members shall be provided.

**1910.29(f)** *Mobile ladder stands*—**(1)** *Design.* Units shall comply with applicable requirements of paragraph (a) of this section.

**1910.29(f)(2)** *Base width.* The minimum base width shall conform to paragraph (a)(3)(i) of this section. The maximum length of the base section shall be the total length of combined steps and top assembly, measured horizontally, plus five-eighths inch per step of rise.

**1910.29(f)(3)** *Steps.* Steps shall be uniformly spaced, and sloped, with a rise of not less than nine (9) inches, nor more than ten (10) inches, and a depth of not less seven (7) inches. The slope of the steps section shall be a minimum of fifty- five (55) degrees and a maximum of sixty (60) degrees measured from the horizontal.

**1910.29(f)(4)** *Handrails.* **(i)** Units having more than five (5) steps or 60 inches vertical height to the top step shall be equipped with handrails.

**1910.29(f)(4)(ii)** Handrails shall be a minimum of 29 inches high. Measurements shall be taken vertically from the center of the step.

**1910.29(f)(5)** *Loading.* The load (see paragraph (a)(2)(ii)(a) of this section) shall be applied uniformly to a 3 1/2 inches wide area front to back at the center of the width span with a safety factor of four (4).

## §1910.30 Other working surfaces.

**1910.30(a)** *Dockboards (bridge plates).* **(1)** Portable and powered dockboards shall be strong enough to carry the load imposed on them.

**1910.30(a)(2)** Portable dockboards shall be secured in position, either by being anchored or equipped with devices which will prevent their slipping.

**1910.30(a)(3)** Powered dockboards shall be designed and constructed in accordance with Commercial Standard CS202-56 (1961) "Industrial Lifts and Hinged Loading Ramps published by the U.S. Department of Commerce, which is incorporated by reference as specified in §1910.6.

**1910.30(a)(4)** Handholds, or other effective means, shall be provided on portable dockboards to permit safe handling.

**1910.30(a)(5)** Positive protection shall be provided to prevent railroad cars from being moved while dockboards or bridge plates are in position.

**1910.30(b)** *Forging machine area.* **(1)** Machines shall be so located as to give **(i)** enough clearance between machines so that the movement of one operator will not interfere with the work of another, **(ii)** ample room for cleaning machines and handling the work, including material and scrap. The arrangement of machines shall be such that operators will not stand in aisles.

**1910.30(b)(2)** Aisles shall be provided of sufficient width to permit the free movement of employees bringing and removing material. This aisle space is to be independent of working and storage space.

**1910.30(b)(3)** Wood platforms used on the floor in front of machines shall be substantially constructed.

**1910.30(c)** *Veneer machinery.* **(1)** Sides of steam vats shall extend to a height of not less than 36 inches above the floor, working platform, or ground.

**1910.30(c)(2)** Large steam vats divided into sections shall be provided with substantial walkways between sections. Each walkway shall be provided with a standard handrail on each exposed side. These handrails may be removable, if necessary.

**1910.30(c)(3)** Covers shall be removed only from that portion of steaming vats on which men are working and a portable railing shall be placed at this point to protect the operators.

**1910.30(c)(4)** Workmen shall not ride or step on logs in steam vats.

[39 FR 23502, June 27, 1974, as amended at 49 FR 5322, Feb. 10, 1984; 61 FR 9235, Mar. 7, 1996]

## §1910.31 Sources of standards [Removed].

[§1910.31 removed at 61 FR 9235, Mar. 7, 1996.]

## §1910.32 Standards organizations [Removed].

[§1910.32 removed at 61 FR 9235, Mar. 7, 1996.]

# Subpart E—Exit Routes, Emergency Action Plans, and Fire Prevention Plans

> **Authority:** Secs. 4, 6, 8, Occupational Safety and Health Act of 1970 (29 U.S.C. 653, 655, 657); Secretary of Labor's Order Nos. 12-71 (36 FR 8754), (8-76 41 FR 25059), 9-83 (48 FR 35736) or 1-90 (55 FR 9033), 6-96 (62 FR 111), or 3-2000 (65 FR 50017), as applicable.

### §1910.33 Table of contents.

This section lists the sections and paragraph headings contained in §§ 1910.34 through 1910.39.

Sec. 1910.34 Coverage and definitions.
 (a) Every employer is covered.
 (b) Exit routes are covered.
 (c) Definitions.

Sec. 1910.35 Compliance with NFPA 101-2000, Life Safety Code.

Sec. 1910.36 Design and construction requirements for exit routes.
 (a) Basic requirements.
 (b) The number of exit routes must be adequate.
 (c) Exit discharge.
 (d) An exit door must be unlocked.
 (e) A side-hinged exit door must be used.
 (f) The capacity of an exit route must be adequate.
 (g) An exit route must meet minimum height and width requirements.
 (h) An outdoor exit route is permitted.

Sec. 1910.37 Maintenance, safeguards, and operational features for exit routes.
 (a) The danger to employees must be minimized.
 (b) Lighting and marking must be adequate and appropriate.
 (c) The fire retardant properties of paints or solutions must be maintained.
 (d) Exit routes must be maintained during construction, repairs, or alterations.
 (e) An employee alarm system must be operable.

Sec. 1910.38 Emergency action plans.
 (a) Application.
 (b) Written and oral emergency action plans.
 (c) Minimum elements of an emergency action plan.
 (d) Employee alarm system.
 (e) Training.
 (f) Review of emergency action plan.

Sec. 1910.39 Fire prevention plans.
 (a) Application.
 (b) Written and oral fire prevention plans.
 (c) Minimum elements of a fire prevention plan.
 (d) Employee information.

### §1910.34 Coverage and definitions.

**1910.34(a)** *Every employer is covered.* Sections 1910.34 through 1910.39 apply to workplaces in general industry except mobile workplaces such as vehicles or vessels.

**1910.34(b)** *Exits routes are covered.* The rules in Sec. Sec. 1910.34 through 1910.39 cover the minimum requirements for exit routes that employers must provide in their workplace so that employees may evacuate the workplace safely during an emergency. Sections 1910.34 through 1910.39 also cover the minimum requirements for emergency action plans and fire prevention plans.

**1910.34(c)** *Definitions.* Electroluminescent means a light-emitting capacitor. Alternating current excites phosphor atoms when placed between the electrically conductive surfaces to produce light. This light source is typically contained inside the device.

*Exit* means that portion of an exit route that is generally separated from other areas to provide a protected way of travel to the exit discharge. An example of an exit is a two-hour fire resistance-rated enclosed stairway that leads from the fifth floor of an office building to the outside of the building.

*Exit access* means that portion of an exit route that leads to an exit. An example of an exit access is a corridor on the fifth floor of an office building that leads to a two-hour fire resistance-rated enclosed stairway (the Exit).

*Exit discharge* means the part of the exit route that leads directly outside or to a street, walkway, refuge area, public way, or open space with access to the outside. An example of an exit discharge is a door at the bottom of a two-hour fire resistance-rated enclosed stairway that discharges to a place of safety outside the building.

*Exit route* means a continuous and unobstructed path of exit travel from any point within a workplace to a place of safety (including refuge areas). An exit route consists of three parts: The exit access; the exit; and, the exit discharge. (An exit route includes all vertical and horizontal areas along the route.)

*High hazard area* means an area inside a workplace in which operations include high hazard materials, processes, or contents.

*Occupant load* means the total number of persons that may occupy a workplace or portion of a workplace at any one time. The occupant load of a workplace is calculated by dividing the gross floor area of the workplace or portion of a workplace by the occupant load factor for that particular type of workplace occupancy. Information regarding "Occupant load" is located in NFPA 101-2000, Life Safety Code.

*Refuge* area means either:

(1) space along an exit route that is protected from the effects of fire by separation from other spaces within the building by a barrier with at least a one-hour fire resistance-rating; or

(2) A floor with at least two spaces, separated from each other by smoke-resistant partitions, in a building protected throughout by an automatic sprinkler system that complies with Sec. 1910.159 of this part.

*Self-luminous* means a light source that is illuminated by a self-contained power source (*e.g.*, tritium) and that operates independently from external power sources. Batteries are not acceptable self-contained power sources. The light source is typically contained inside the device.

### §1910.35 Compliance with NFPA 101-2000, Life Safety Code.

An employer who demonstrates compliance with the exit route provisions of NFPA 101-2000, the Life Safety Code, will be deemed to be in compliance with the corresponding requirements in Sec. Sec. 1910.34, 1910.36, and 1910.37.

### §1910.36 Design and construction requirements for exit routes.

**1910.36(a)** Basic requirements. Exit routes must meet the following design and construction requirements: **(1)** An exit route must be permanent. Each exit route must be a permanent part of the workplace.

**1910.36(a)(2)** An exit must be separated by fire resistant materials. Construction materials used to separate an exit from other parts of the workplace must have a one-hour fire resistance-rating if the exit connects three or fewer stories and a two-hour fire resistance-rating if the exit connects four or more stories.

**1910.36(a)(3)** Openings into an exit must be limited. An exit is permitted to have only those openings necessary to allow access to the exit from occupied areas of the workplace, or to the exit discharge. An opening into an exit must be protected by a self-closing fire door that remains closed or automatically closes in an emergency upon the sounding of a fire alarm or employee alarm system. Each fire door, including its frame and hardware, must be listed or approved by a nationally recognized testing laboratory. Section 1910.155(c)(3)(iv)(A) of this part defines "listed" and Sec. 1910.7 of this part defines a "nationally recognized testing laboratory."

**1910.36(b)** The number of exit routes must be adequate. **(1)** Two exit routes. At least two exit routes must be available in a workplace to permit prompt evacuation of employees and other building occupants during an emergency, except as allowed in paragraph (b)(3) of this section. The exit routes must be located as far away as practical from each other so that if one exit route is blocked by fire or smoke, employees can evacuate using the second exit route.

**1910.36(b)(2)** More than two exit routes. More than two exit routes must be available in a workplace if the number of employees, the size of the building, its occupancy, or the arrangement of the workplace is such that all employees would not be able to evacuate safely during an emergency.

**1910.36(b)(3)** A single exit route. A single exit route is permitted where the number of employees, the size of the building, its occupancy, or the arrangement of the workplace is such that all employees would be able to evacuate safely during an emergency.

**Note to paragraph 1910.36(b):** For assistance in determining the number of exit routes necessary for your workplace, consult NFPA 101-2000, Life Safety Code.

**1910.36(c)** Exit discharge. **(1)** Each exit discharge must lead directly outside or to a street, walkway, refuge area, public way, or open space with access to the outside.

**1910.36(c)(2)** The street, walkway, refuge area, public way, or open space to which an exit discharge leads must be large enough to accommodate the building occupants likely to use the exit route.

**1910.36(c)(3)** Exit stairs that continue beyond the level on which the exit discharge is located must be interrupted at that level by doors, partitions, or other effective means that clearly indicate the direction of travel leading to the exit discharge.

**1910.36(d)** An exit door must be unlocked. **(1)** Employees must be able to open an exit route door from the inside at all times without keys, tools, or special knowledge. A device such as a panic bar that locks only from the outside is permitted on exit discharge doors.

**1910.36(d)(2)** Exit route doors must be free of any device or alarm that could restrict emergency use of the exit route if the device or alarm fails.

**1910.36(d)(3)** An exit route door may be locked from the inside only in mental, penal, or correctional facilities and then only if supervisory personnel are continuously on duty and the employer has a plan to remove occupants from the facility during an emergency.

**1910.36(e)** A side-hinged exit door must be used. **(1)** A side-hinged door must be used to connect any room to an exit route.

**1910.36(e)(2)** The door that connects any room to an exit route must swing out in the direction of exit travel if the room is designed to be occupied by more than 50 people or if the room is a high hazard area (i.e., contains contents that are likely to burn with extreme rapidity or explode).

**1910.36(f)** The capacity of an exit route must be adequate. **(1)** Exit routes must support the maximum permitted occupant load for each floor served.

**1910.36(f)(2)** The capacity of an exit route may not decrease in the direction of exit route travel to the exit discharge.

**Note to paragraph 1910.36(f):** Information regarding "Occupant load" is located in NFPA 101-2000, Life Safety Code.

**1910.36(g)** An exit route must meet minimum height and width requirements.

**1910.36(g)(1)** The ceiling of an exit route must be at least seven feet six inches (2.3 m) high. Any projection from the ceiling must not reach a point less than six feet eight inches (2.0 m) from the floor.

**1910.36(g)(2)** An exit access must be at least 28 inches (71.1 cm) wide at all points. Where there is only one exit access leading to an exit or exit discharge, the width of the exit and exit discharge must be at least equal to the width of the exit access.

**1910.36(g)(3)** The width of an exit route must be sufficient to accommodate the maximum permitted occupant load of each floor served by the exit route.

**1910.36(g)(4)** Objects that project into the exit route must not reduce the width of the exit route to less than the minimum width requirements for exit routes.

**1910.36(h)** An outdoor exit route is permitted. Each outdoor exit route must meet the minimum height and width requirements for indoor exit routes and must also meet the following requirements:

**1910.36(h)(1)** The outdoor exit route must have guardrails to protect unenclosed sides if a fall hazard exists;

**1910.36(h)(2)** The outdoor exit route must be covered if snow or ice is likely to accumulate along the route, unless the employer can demonstrate that any snow or ice accumulation will be removed before it presents a slipping hazard;

**1910.36(h)(3)** The outdoor exit route must be reasonably straight and have smooth, solid, substantially level walkways; and

**1910.36(h)(4)** The outdoor exit route must not have a dead-end that is longer than 20 feet (6.2 m).

### §1910.37 Maintenance, safeguards, and operational features for exit routes.

**1910.37(a)** The danger to employees must be minimized. **(1)** Exit routes must be kept free of explosive or highly flammable furnishings or other decorations.

**1910.37(a)(2)** Exit routes must be arranged so that employees will not have to travel toward a high hazard area, unless the path of travel is effectively shielded from the high hazard area by suitable partitions or other physical barriers.

**1910.37(a)(3)** Exit routes must be free and unobstructed. No materials or equipment may be placed, either permanently or temporarily, within the exit route. The exit access must not go through a room that can be locked, such as a bathroom, to reach an exit or exit discharge, nor may it lead into a dead-end corridor. Stairs or a ramp must be provided where the exit route is not substantially level.

**1910.37(a)(4)** Safeguards designed to protect employees during an emergency (*e.g.*, sprinkler systems, alarm systems, fire doors, exit lighting) must be in proper working order at all times.

**1910.37(b)** Lighting and marking must be adequate and appropriate. **(1)** Each exit route must be adequately lighted so that an employee with normal vision can see along the exit route.

**1910.37(b)(2)** Each exit must be clearly visible and marked by a sign reading "Exit."

**1910.37(b)(3)** Each exit route door must be free of decorations or signs that obscure the visibility of the exit route door.

**1910.37(b)(4)** If the direction of travel to the exit or exit discharge is not immediately apparent, signs must be posted along the exit access indicating the direction of travel to the nearest exit and exit discharge. Additionally, the line-of-sight to an exit sign must clearly be visible at all times.

**1910.37(b)(5)** Each doorway or passage along an exit access that could be mistaken for an exit must be marked "Not an Exit" or similar designation, or be identified by a sign indicating its actual use (*e.g.*, closet).

**1910.37(b)(6)** Each exit sign must be illuminated to a surface value of at least five foot-candles (54 lux) by a reliable light source and be distinctive in color. Self-luminous or electroluminescent signs that have a minimum luminance surface value of at least .06 footlamberts (0.21 cd/m$^2$) are permitted.

**1910.37(b)(7)** Each exit sign must have the word "Exit" in plainly legible letters not less than six inches (15.2 cm) high, with the principal strokes of the letters in the word "Exit" not less than three-fourths of an inch (1.9 cm) wide.

**1910.37(c)** The fire retardant properties of paints or solutions must be maintained. Fire retardant paints or solutions must be renewed as often as necessary to maintain their fire retardant properties.

**1910.37(d)** Exit routes must be maintained during construction, repairs, or alterations. **(1)** During new construction, employees must not occupy a workplace until the exit routes required by this subpart are completed and ready for employee use for the portion of the workplace they occupy.

**1910.37(d)(2)** During repairs or alterations, employees must not occupy a workplace unless the exit routes required by this subpart are available and existing fire protections are maintained, or until alternate fire protection is furnished that provides an equivalent level of safety.

**1910.37(d)(3)** Employees must not be exposed to hazards of flammable or explosive substances or equipment used during construction, repairs, or alterations, that are beyond the normal permissible conditions in the workplace, or that would impede exiting the workplace.

**1910.37(e)** An employee alarm system must be operable. Employers must install and maintain an operable employee alarm system that has a distinctive signal to warn employees of fire or other emergencies, unless employees can promptly see or smell a fire or other hazard in time to provide adequate warning to them. The employee alarm system must comply with Sec. 1910.165.

### §1910.38 Emergency action plans.

**1910.38(a)** Application. An employer must have an emergency action plan whenever an OSHA standard in this part requires one. The requirements in this section apply to each such emergency action plan.

**1910.38(b)** Written and oral emergency action plans. An emergency action plan must be in writing, kept in the workplace, and available to employees for review. However, an employer with 10 or fewer employees may communicate the plan orally to employees.

**1910.38(c)** Minimum elements of an emergency action plan. An emergency action plan must include at a minimum:

**1910.38(c)(1)** Procedures for reporting a fire or other emergency;

**1910.38(c)(2)** Procedures for emergency evacuation, including type of evacuation and exit route assignments;

**1910.38(c)(3)** Procedures to be followed by employees who remain to operate critical plant operations before they evacuate;

**1910.38(c)(4)** Procedures to account for all employees after evacuation;

**1910.38(c)(5)** Procedures to be followed by employees performing rescue or medical duties; and

**1910.38(c)(6)** The name or job title of every employee who may be contacted by employees who need more information about the plan or an explanation of their duties under the plan.

**1910.38(d)** Employee alarm system. An employer must have and maintain an employee alarm system. The employee alarm system must use a distinctive signal for each purpose and comply with the requirements in Sec. 1910.165.

**1910.38(e)** Training. An employer must designate and train employees to assist in a safe and orderly evacuation of other employees.

**1910.38(f)** Review of emergency action plan. An employer must review the emergency action plan with each employee covered by the plan:

**1910.38(f)(1)** When the plan is developed or the employee is assigned initially to a job;

**1910.38(f)(2)** When the employee's responsibilities under the plan change; and

**1910.38(f)(3)** When the plan is changed.

## §1910.39 Fire prevention plans.

**1910.39(a)** Application. An employer must have a fire prevention plan when an OSHA standard in this part requires one. The requirements in this section apply to each such fire prevention plan.

**1910.39(b)** Written and oral fire prevention plans. A fire prevention plan must be in writing, be kept in the workplace, and be made available to employees for review. However, an employer with 10 or fewer employees may communicate the plan orally to employees.

**1910.39(c)** Minimum elements of a fire prevention plan. A fire prevention plan must include:

**1910.39(c)(1)** A list of all major fire hazards, proper handling and storage procedures for hazardous materials, potential ignition sources and their control, and the type of fire protection equipment necessary to control each major hazard;

**1910.39(c)(2)** Procedures to control accumulations of flammable and combustible waste materials;

**1910.39(c)(3)** Procedures for regular maintenance of safeguards installed on heat-producing equipment to prevent the accidental ignition of combustible materials;

**1910.39(c)(4)** The name or job title of employees responsible for maintaining equipment to prevent or control sources of ignition or fires; and

**1910.39(c)(5)** The name or job title of employees responsible for the control of fuel source hazards.

**1910.39(d)** Employee information. An employer must inform employees upon initial assignment to a job of the fire hazards to which they are exposed. An employer must also review with each employee those parts of the fire prevention plan necessary for self-protection.

### APPENDIX TO SUBPART E TO PART 1910— Exit Routes, Emergency Action Plans, and Fire Prevention Plans.

This appendix serves as a nonmandatory guideline to assist employers in complying with the appropriate requirements of subpart E.

### §1910.38 Employee emergency plans.

1. *Emergency action plan elements.* The emergency action plan should address emergencies that the employer may reasonably expect in the workplace. Examples are: fire; toxic chemical releases; hurricanes; tornadoes; blizzards; floods; and others. The elements of the emergency action plan presented in paragraph 1910.38 can be supplemented by the following to more effectively achieve employee safety and health in an emergency. The employer should list in detail the procedures to be taken by those employees who have been selected to remain behind to care for essential plant operations until their evacuation becomes absolutely necessary. Essential plant operations may include the monitoring of plant power supplies, water supplies, and other essential services which cannot be shut down for every emergency alarm. Essential plant operations may also include chemical or manufacturing processes which must be shut down in stages or steps where certain employees must be present to assure that safe shut down procedures are completed.

The use of floor plans or workplace maps which clearly show the emergency escape routes should be included in the emergency action plan. Color coding will aid employees in determining their route assignments. The employer should also develop and explain in detail what rescue and medical first aid duties are to be performed and by whom. All employees are to be told what actions they are to take in these emergency situations that the employer anticipates may occur in the workplace.

2. *Emergency evacuation.* At the time of an emergency, employees should know what type of evacuation is necessary and what their role is in carrying out the plan. In some cases where the emergency is very grave, total and immediate evacuation of all employees is necessary. In other emergencies, a partial evacuation of nonessential employees with a delayed evacuation of others may be necessary for continued plant operation. In some cases, only those employees in the immediate area of the fire may be expected to evacuate or move to a safe area such as when a local application fire suppression system discharge employee alarm is sounded. Employees must be sure that they know what is expected of them in all such emergency possibilities which have been planned in order to provide assurance of their safety from fire or other emergency.

The designation of refuge or safe areas for evacuation should be determined and identified in the plan. In a building divided into fire zones by fire walls, the refuge area may still be within the same building but in a different zone from where the emergency occurs.

Exterior refuge or safe areas may include parking lots, open fields or streets which are located away from the site of the emergency and which provide sufficient space to accommodate the employees. Employees should be instructed to move away from the exit discharge doors of the building, and to avoid congregating close to the building where they may hamper emergency operations.

3. *Emergency action plan training.* The employer should assure that an adequate number of employees are available at all times during working hours to act as evacuation wardens so that employees can be swiftly moved from the danger location to the safe areas. Generally, one warden for each twenty employees in the workplace should be able to provide adequate guidance and instruction at the time of a fire emergency. The employees selected or who volunteer to serve as wardens should be trained in the complete workplace layout and the various alternative escape routes from the workplace. All wardens and fellow employees should be made aware of handicapped employees who may need extra assistance, such as using the buddy system, and of hazardous areas to be avoided during emergencies. Before leaving, wardens should check rooms and other enclosed spaces in the workplace for employees who may be trapped or otherwise unable to evacuate the area.

After the desired degree of evacuation is completed, the wardens should be able to account for or otherwise verify that all employees are in the safe areas.

In buildings with several places of employment, employers are encouraged to coordinate their plans with the other employers in the building. A building-wide or standardized plan for the whole building is acceptable provided that the employers inform their respective employees of their duties and responsibilities under the plan. The standardized plan need not be kept by each employer in the multi-employer building, provided there is an accessible location within the building where the plan can be reviewed by affected employees. When multi-employer building-wide plans are not feasible, employers should coordinate their plans with the other employers within the building to assure that conflicts and confusion are avoided during times of emergencies. In multi-story buildings where more than one employer is on a single floor, it is essential that these employers coordinate their plans with each other to avoid conflicts and confusion.

4. *Fire prevention housekeeping.* The standard calls for the control of accumulations of flammable and combustible waste materials.

It is the intent of this standard to assure that hazardous accumulations of combustible waste materials are controlled so that a fast developing fire, rapid spread of toxic smoke, or an explosion will not occur. This does not necessarily mean that each room has to be swept each day. Employers and employees should be aware of the hazardous properties of materials in their workplaces, and the degree of hazard each poses. Certainly oil soaked rags have to be treated differently than general paper trash in office areas. However, large accumulations of waste paper or corrugated boxes, etc., can pose a significant fire hazard. Accumulations of materials which can cause large fires or generate dense smoke that are easily ignited or may start from spontaneous combustion, are the types of materials with which this standard is concerned. Such combustible materials may be easily ignited by matches, welder's sparks, cigarettes and similar low level energy ignition sources.

5. *Maintenance of equipment under the fire prevention plan.* Certain equipment is often installed in workplaces to control heat sources or to detect fuel leaks. An example is a temperature limit switch often found on deep-fat food fryers found in restaurants. There may be similar switches for high temperature dip tanks, or flame failure and flashback arrester devices on furnaces and similar heat producing equipment. If these devices are not properly maintained or if they become inoperative, a definite fire hazard exists. Again employees and supervisors should be aware of the specific type of control devices on equipment involved with combustible materials in the workplace and should make sure, through periodic inspection or testing, that these controls are operable. Manufacturers' recommendations should be followed to assure proper maintenance procedures.

[45 FR 60714, Sept. 12, 1980]

# Subpart F—Powered Platforms, Manlifts, and Vehicle-Mounted Work Platforms

> **Authority:** Secs. 4, 6, and 8 of the Occupational Safety and Health Act of 1970 (29 U.S.C. 653, 655, and 657); Secretary of Labor's Order No. 12-71 (36 FR 8754), 8-76 (41 FR 25059), 9-83 (48 FR 35736), or 1-90 (55 FR 9033), as applicable; and 29 CFR part 1911.

### §1910.66 Powered platforms for building maintenance.

**1910.66(a)** *Scope.* This section covers powered platform installations permanently dedicated to interior or exterior building maintenance of a specific structure or group of structures. This section does not apply to suspended scaffolds (swinging scaffolds) used to service buildings on a temporary basis and covered under subpart D of this part, nor to suspended scaffolds used for construction work and covered under subpart L of 29 CFR part 1926. Building maintenance includes, but is not limited to, such tasks as window cleaning, caulking, metal polishing and reglazing.

**1910.66(b)** *Application—*(1) *New installations.* This section applies to all permanent installations completed after July 23, 1990. Major modifications to existing installations completed after that date are also considered new installations under this section.

**1910.66(b)(2)** *Existing installations.* (i) Permanent installations in existence and/or completed before July 23, 1990 shall comply with paragraphs (g), (h), (i), (j) and appendix C of this section.

**1910.66(b)(2)(ii)** In addition, permanent installations completed after August 27, 1971, and in existence and/or completed before July 23, 1990, shall comply with appendix D of this section.

**1910.66(c)** *Assurance.* (1) Building owners of new installations shall inform the employer before each use in writing that the installation meets the requirements of paragraphs (e)(1) and (f)(1) of this section and the additional design criteria contained in other provisions of paragraphs (e) and (f) of this section relating to: required load sustaining capabilities of platforms, building components, hoisting and supporting equipment; stability factors for carriages, platforms and supporting equipment; maximum horizontal force for movement of carriages and davits; design of carriages, hoisting machines, wire rope and stabilization systems; and design criteria for electrical wiring and equipment.

**1910.66(c)(2)** Building owners shall base the information required in paragraph (c)(1) of this section on the results of a field test of the installation before being placed into service and following any major alteration to an existing installation, as required in paragraph (g)(1) of this section. The assurance shall also be based on all other relevant available information, including, but not limited to, test data, equipment specifications and verification by a registered professional engineer.

**1910.66(c)(3)** Building owners of all installations, new and existing, shall inform the employer in writing that the installation has been inspected, tested and maintained in compliance with the requirements of paragraphs (g) and (h) of this section and that all protection anchorages meet the requirements of paragraph (l)(c)(10) of appendix C.

**1910.66(c)(4)** The employer shall not permit employees to use the installation prior to receiving assurance from the building owner that the installation meets the requirements contained in paragraphs (c)(1) and (c)(3) of this section.

**1910.66(d)** *Definitions.*

*Anemometer* means an instrument for measuring wind velocity.

*Angulated roping* means a suspension method where the upper point of suspension is inboard from the attachments on the suspended unit, thus causing the suspended unit to bear against the face of the building.

*Building face roller* means a rotating cylindrical member designed to ride on the face of the building wall to prevent the platform from abrading the face of the building and to assist in stabilizing the platform.

*Building maintenance* means operations such as window cleaning, caulking, metal polishing, reglazing, and general maintenance on building surfaces.

*Cable* means a conductor, or group of conductors, enclosed in a weatherproof sheath, that may be used to supply electrical power and/or control current for equipment or to provide voice communication circuits.

*Carriage* means a wheeled vehicle used for the horizontal movement and support of other equipment.

*Certification* means a written, signed and dated statement confirming the performance of a requirement of this section.

*Combination cable* means a cable having both steel structural members capable of supporting the platform, and copper or other electrical conductors insulated from each other and the structural members by nonconductive barriers.

*Competent person* means a person who, because of training and experience, is capable of identifying hazardous or dangerous conditions in powered platform installations and of training employees to identify such conditions.

*Continuous pressure* means the need for constant manual actuation for a control to function.

*Control* means a mechanism used to regulate or guide the operation of the equipment.

*Davit* means a device, used singly or in pairs, for suspending a powered platform from work, storage and rigging locations on the building being serviced. Unlike outriggers, a davit reacts its operating load into a single roof socket or carriage attachment.

*Equivalent* means alternative designs, materials or methods which the employer can demonstrate will provide an equal or greater degree of safety for employees than the methods, materials or designs specified in the standard.

*Ground rigging* means a method of suspending a working platform starting from a safe surface to a point of suspension above the safe surface.

*Ground rigged davit* means a davit which cannot be used to raise a suspended working platform above the building face being serviced.

*Guide button* means a building face anchor designed to engage a guide track mounted on a platform.

*Guide roller* means a rotating cylindrical member, operating separately or as part of a guide assembly, designed to provide continuous engagement between the platform and the building guides or guideways.

*Guide shoe* means a device attached to the platform designed to provide a sliding contact between the platform and the building guides.

*Hoisting machine* means a device intended to raise and lower a suspended or supported unit.

*Hoist rated load* means the hoist manufacturer's maximum allowable operating load.

*Installation* means all the equipment and all affected parts of a building which are associated with the performance of building maintenance using powered platforms.

*Interlock* means a device designed to ensure that operations or motions occur in proper sequence.

*Intermittent stabilization* means a method of platform stabilization in which the angulated suspension wire rope(s) are secured to regularly spaced building anchors.

*Lanyard* means a flexible line of rope, wire rope or strap which is used to secure the body belt or body harness to a deceleration device, lifeline or anchorage.

*Lifeline* means a component consisting of a flexible line for connection to an anchorage at one end to hang vertically (vertical lifeline), or for connection to anchorages at both ends to stretch horizontally (horizontal lifeline), and which serves as a means for connecting other components of a personal fall arrest system to the anchorage.

*Live load* means the total static weight of workers, tools, parts, and supplies that the equipment is designed to support.

*Obstruction detector* means a control that will stop the suspended or supported unit in the direction of travel if an obstruction is encountered, and will allow the unit to move only in a direction away from the obstruction.

*Operating control* means a mechanism regulating or guiding the operation of equipment that ensures a specific operating mode.

*Operating device* means a device actuated manually to activate a control.

*Outrigger* means a device, used singly or in pairs, for suspending a working platform from work, storage, and rigging locations on the building being serviced. Unlike davits, an outrigger reacts its operating moment load as at least two opposing vertical components acting into two or more distinct roof points or attachments.

*Platform rated load* means the combined weight of workers, tools, equipment and other material which is permitted to be carried by the working platform at the installation, as stated on the load rating plate.

*Poured socket* means the method of providing wire rope terminations in which the ends of the rope are held in a tapered socket by means of poured spelter or resins.

*Primary brake* means a brake designed to be applied automatically whenever power to the prime mover is interrupted or discontinued.

*Prime mover* means the source of mechanical power for a machine.

*Rated load* means the manufacturer's recommended maximum load.

*Rated strength* means the strength of wire rope, as designated by its manufacturer or vendor, based on standard testing procedures or acceptable engineering design practices.

*Rated working load* means the combined static weight of men, materials, and suspended or supported equipment.

*Registered professional engineer* means a person who has been duly and currently registered and licensed by an authority within the United States or its territories to practice the profession of engineering.

*Roof powered platform* means a working platform where the hoist(s) used to raise or lower the platform is located on the roof.

*Roof rigged davit* means a davit used to raise the suspended working platform above the building face being serviced. This type of davit can also be used to raise a suspended working platform which has been ground-rigged.

*Rope* means the equipment used to suspend a component of an equipment installation, i.e., wire rope.

*Safe surface* means a horizontal surface intended to be occupied by personnel, which is so protected by a fall protection system that it can be reasonably assured that said occupants will be protected against falls.

*Secondary brake* means a brake designed to arrest the descent of the suspended or supported equipment in the event of an overspeed condition.

*Self powered platform* means a working platform where the hoist(s) used to raise or lower the platform is mounted on the platform.

*Speed reducer* means a positive type speed reducing machine.

*Stability factor* means the ratio of the stabilizing moment to the overturning moment.

*Stabilizer tie* means a flexible line connecting the building anchor and the suspension wire rope supporting the platform.

*Supported equipment* means building maintenance equipment that is held or moved to its working position by means of attachment directly to the building or extensions of the building being maintained.

*Suspended equipment* means building maintenance equipment that is suspended and raised or lowered to its working position by means of ropes or combination cables attached to some anchorage above the equipment.

*Suspended scaffold (swinging scaffold)* means a scaffold supported on wire or other ropes, used for work on, or for providing access to, vertical sides of structures on a temporary basis. Such scaffold is not designed for use on a specific structure or group of structures.

*Tail line* means the nonsupporting end of the wire rope used to suspend the platform.

*Tie-in guides* means the portion of a building that provides continuous positive engagement between the building and a suspended or supported unit during its vertical travel on the face of the building.

*Traction hoist* means a type of hoisting machine that does not accumulate the suspension wire rope on the hoisting drum or sheave, and is designed to raise and lower a suspended load by the application of friction forces between the suspension wire rope and the drum or sheave.

*Transportable outriggers* means outriggers designed to be moved from one work location to another.

*Trolley carriage* means a carriage suspended from an overhead track structure.

*Verified* means accepted by design, evaluation, or inspection by a registered professional engineer.

*Weatherproof* means so constructed that exposure to adverse weather conditions will not affect or interfere with the proper use or functions of the equipment or component.

*Winding drum hoist* means a type of hoisting machine that accumulates the suspension wire rope on the hoisting drum.

*Working platform* means suspended or supported equipment intended to provide access to the face of a building and manned by persons engaged in building maintenance.

*Wrap* means one complete turn of the suspension wire rope around the surface of a hoist drum.

**1910.66(e)** *Powered platform installations—Affected parts of buildings—* **(1)** *General requirements.* The following requirements apply to affected parts of buildings which utilize working platforms for building maintenance.

**1910.66(e)(1)(i)** Structural supports, tie-downs, tie-in guides, anchoring devices and any affected parts of the building included in the installation shall be designed by or under the direction of a registered professional engineer experienced in such design;

**1910.66(e)(1)(ii)** Exterior installations shall be capable of withstanding prevailing climatic conditions;

**1910.66(e)(1)(iii)** The building installation shall provide safe access to, and egress from, the equipment and sufficient space to conduct necessary maintenance of the equipment;

**1910.66(e)(1)(iv)** The affected parts of the building shall have the capability of sustaining all the loads imposed by the equipment; and,

**1910.66(e)(1)(v)** The affected parts of the building shall be designed so as to allow the equipment to be used without exposing employees to a hazardous condition.

**1910.66(e)(2)** *Tie-in guides.* (i) The exterior of each building shall be provided with tie-in guides unless the conditions in paragraph (e)(2)(ii) or (e)(2)(iii) of this section are met.

**NOTE:** See Figure 1 in appendix B of this section for a description of a typical continuous stabilization system utilizing tie-in guides.

**1910.66(e)(2)(ii)** If angulated roping is employed, tie-in guides required in paragraph (e)(2)(i) of this section may be eliminated for not more than 75 feet (22.9 m) of the uppermost elevation of the building, if infeasible due to exterior building design, provided an angulation force of at least 10 pounds (44.4 n) is maintained under all conditions of loading.

**1910.66(e)(2)(iii)** Tie-in guides required in paragraph (e)(2)(i) of this section may be eliminated if one of the guide systems in paragraph (e)(2)(iii)(A), (e)(2)(iii)(B) or (e)(2)(iii)(C) of this section is provided, or an equivalent.

**1910.66(e)(2)(iii)(A)** Intermittent stabilization system. The system shall keep the equipment in continuous contact with the building facade, and shall prevent sudden horizontal movement of the platform. The system may be used together with continuous positive building guide systems using tie-in guides on the same building, provided the requirements for each system are met.

**1910.66(e)(2)(iii)(A)(*1*)** The maximum vertical interval between building anchors shall be three floors or 50 feet (15.3 m), whichever is less.

**1910.66(e)(2)(iii)(A)(*2*)** Building anchors shall be located vertically so that attachment of the stabilizer ties will not cause the platform suspension ropes to angulate the platform horizontally across the face of the building. The anchors shall be positioned horizontally on the building face so as to be symmetrical about the platform suspension ropes.

**1910.66(e)(2)(iii)(A)(*3*)** Building anchors shall be easily visible to employees and shall allow a stabilizer tie attachment for each of the platform suspension ropes at each vertical interval. If more than two suspension ropes are used on a platform, only the two building-side suspension ropes at the platform ends shall require a stabilizer attachment.

**1910.66(e)(2)(iii)(A)(*4*)** Building anchors which extend beyond the face of the building shall be free of sharp edges or points. Where cables, suspension wire ropes and lifelines may be in contact with the building face, external building anchors shall not interfere with their handling or operation.

**1910.66(e)(2)(iii)(A)(*5*)** The intermittent stabilization system building anchors and components shall be capable of sustaining without failure at least four times the maximum anticipated load applied or transmitted to the components and anchors. The minimum design wind load for each anchor shall be 300 (1334 n) pounds, if two anchors share the wind load.

**1910.66(e)(2)(iii)(A)(*6*)** The building anchors and stabilizer ties shall be capable of sustaining anticipated horizontal and vertical loads from winds specified for roof storage design which may act on the platform and wire ropes if the platform is stranded on a building face. If the building anchors have different spacing than the suspension wire rope or if the building requires different suspension spacings on one platform, one building anchor and stabilizer tie shall be capable of sustaining the wind loads.

**NOTE:** See Figure 2 in appendix B of this section for a description of a typical intermittent stabilization system.

**1910.66(e)(2)(iii)(B)** Button guide stabilization system.

**1910.66(e)(2)(iii)(B)(1)** Guide buttons shall be coordinated with platform mounted equipment of paragraph (f)(5)(vi) of this section.

**1910.66(e)(2)(iii)(B)(2)** Guide buttons shall be located horizontally on the building face so as to allow engagement of each of the guide tracks mounted on the platform.

**1910.66(e)(2)(iii)(B)(3)** Guide buttons shall be located in vertical rows on the building face for proper engagement of the guide tracks mounted on the platform.

**1910.66(e)(2)(iii)(B)(4)** Two guide buttons shall engage each guide track at all times except for the initial engagement.

**1910.66(e)(2)(iii)(B)(5)** Guide buttons which extend beyond the face of the building shall be free of sharp edges or points. Where cables, ropes and lifelines may be in contact with the building face, guide buttons shall not interfere with their handling or operation.

**1910.66(e)(2)(iii)(B)(6)** Guide buttons, connections and seals shall be capable of sustaining without damage at least the weight of the platform, or provision shall be made in the guide tracks or guide track connectors to prevent the platform and its attachments from transmitting the weight of the platform to the guide buttons, connections and seals. In either case, the minimum design load shall be 300 pounds (1334 n) per building anchor.

**NOTE:** See paragraph (f)(5)(vi) of this section for relevant equipment provisions.

**NOTE:** See Figure 3 in appendix B of this section for [sic] a description of a typical button guide stabilization system.

**1910.66(e)(2)(iii)(C)** System utilizing angulated roping and building face rollers. The system shall keep the equipment in continuous contact with the building facade, and shall prevent sudden horizontal movement of the platform. This system is acceptable only where the suspended portion of the equipment in use does not exceed 130 feet (39.6 m) above a safe surface or ground level, and where the platform maintains no less than 10 pounds (44.4 n) angulation force on the building facade.

**1910.66(e)(2)(iv)** Tie-in guides for building interiors (atriums) may be eliminated when a registered professional engineer determines that an alternative stabilization system, including systems in paragraphs (e)(2)(iii)(A), (B) and (C), or a platform tie-off at each work station will provide equivalent safety.

**1910.66(e)(3)** *Roof guarding.* **(i)** Employees working on roofs while performing building maintenance shall be protected by a perimeter guarding system which meets the requirements of paragraph (c)(1) of §1910.23 of this part.

**1910.66(e)(3)(ii)** The perimeter guard shall not be more than six inches (152 mm) inboard of the inside face of a barrier, i.e. the parapet wall, or roof edge curb of the building being serviced; however, the perimeter guard location shall not exceed an 18 inch (457 mm) setback from the exterior building face.

**1910.66(e)(4)** *Equipment stops.* Operational areas for trackless type equipment shall be provided with structural stops, such as curbs, to prevent equipment from traveling outside its intended travel areas and to prevent a crushing or shearing hazard.

**1910.66(e)(5)** *Maintenance access.* Means shall be provided to traverse all carriages and their suspended equipment to a safe area for maintenance and storage.

**1910.66(e)(6)** *Elevated track.* **(i)** An elevated track system which is located four feet (1.2 m) or more above a safe surface, and traversed by carriage supported equipment, shall be provided with a walkway and guardrail system; or

**1910.66(e)(6)(ii)** The working platform shall be capable of being lowered, as part of its normal operation, to the lower safe surface for access and egress of the personnel and shall be provided with a safe means of access and egress to the lower safe surface.

**1910.66(e)(7)** *Tie-down anchors.* Imbedded tie-down anchors, fasteners, and affected structures shall be resistant to corrosion.

**1910.66(e)(8)** *Cable stabilization.* **(i)** Hanging lifelines and all cables not in tension shall be stabilized at each 200 foot (61 m) interval of vertical travel of the working platform beyond an initial 200 foot (61 m) distance.

**1910.66(e)(8)(ii)** Hanging cables, other than suspended wire ropes, which are in constant tension shall be stabilized when the vertical travel exceeds an initial 600 foot (183 m) distance, and at further intervals of 600 feet (183 m) or less.

**1910.66(e)(9)** *Emergency planning.* A written emergency action plan shall be developed and implemented for each kind of working platform operation. This plan shall explain the emergency procedures which are to be followed in the event of a power failure, equipment failure or other emergencies which may be encountered. The plan shall also explain that employees inform themselves about the building emergency escape routes, procedures and alarm systems before operating a platform. Upon initial assignment and whenever the plan is changed the employer shall review with each employee those parts of the plan which the employee must know to protect himself or herself in the event of an emergency.

**1910.66(e)(10)** *Building maintenance.* Repairs or major maintenance of those building portions that provide primary support for the suspended equipment shall not affect the capability of the building to meet the requirements of this standard.

**1910.66(e)(11)** *Electrical requirements.* The following electrical requirements apply to buildings which utilize working platforms for building maintenance.

**1910.66(e)(11)(i)** General building electrical installations shall comply with §§1910.302 through 1910.308 of this part, unless otherwise specified in this section;

**1910.66(e)(11)(ii)** Building electrical wiring shall be of such capacity that when full load is applied to the equipment power circuit not more than a five percent drop from building service-vault voltage shall occur at any power circuit outlet used by equipment regulated by this section;

**1910.66(e)(11)(iii)** The equipment power circuit shall be an independent electrical circuit that shall remain separate from all other equipment within or on the building, other than power circuits used for hand tools that will be used in conjunction with the equipment. If the building is provided with an emergency power system, the equipment power circuit may also be connected to this system;

**1910.66(e)(11)(iv)** The power circuit shall be provided with a disconnect switch that can be locked in the "OFF" and "ON" positions. The switch shall be conveniently located with respect to the primary operating area of the equipment to allow the operators of the equipment access to the switch;

**1910.66(e)(11)(v)** The disconnect switch for the power circuit shall be locked in the "ON" position when the equipment is in use; and

**1910.66(e)(11)(vi)** An effective two-way voice communication system shall be provided between the equipment operators and persons stationed within the building being serviced. The communications facility shall be operable and shall be manned at all times by persons stationed within the building whenever the platform is being used.

**1910.66(f)** **Powered platform installations—Equipment—(1)** *General requirements.* The following requirements apply to equipment which are part of a powered platform installation, such as platforms, stabilizing components, carriages, outriggers, davits, hoisting machines, wire ropes and electrical components.

**1910.66(f)(1)(i)** Equipment installations shall be designed by or under the direction of a registered professional engineer experienced in such design;

**1910.66(f)(1)(ii)** The design shall provide for a minimum live load of 250 pounds (113.6 kg) for each occupant of a suspended or supported platform;

**1910.66(f)(1)(iii)** Equipment that is exposed to wind when not in service shall be designed to withstand forces generated by winds of at least 100 miles per hour (44.7 m/s) at 30 feet (9.2 m) above grade; and

**1910.66(f)(1)(iv)** Equipment that is exposed to wind when in service shall be designed to withstand forces generated by winds of at least 50 miles per hour (22.4 m/s) for all elevations.

**1910.66(f)(2)** *Construction requirements.* Bolted connections shall be self-locking or shall otherwise be secured to prevent loss of the connections by vibration.

**1910.66(f)(3)** *Suspension methods.* Elevated building maintenance equipment shall be suspended by a carriage, outriggers, davits or an equivalent method.

**1910.66(f)(3)(i)** *Carriages.* Carriages used for suspension of elevated building maintenance equipment shall comply with the following:

**1910.66(f)(3)(i)(A)** The horizontal movement of a carriage shall be controlled so as to ensure its safe movement and allow accurate positioning of the platform for vertical travel or storage;

**1910.66(f)(3)(i)(B)** Powered carriages shall not exceed a traversing speed of 50 feet per minute (0.3 m/s);

**1910.66(f)(3)(i)(C)** The initiation of a traversing movement for a manually propelled carriage on a smooth level surface shall not require a person to exert a horizontal force greater than 40 pounds (444.8 n);

**1910.66(f)(3)(i)(D)** Structural stops and curbs shall be provided to prevent the traversing of the carriage beyond its designed limits of travel;

**1910.66(f)(3)(i)(E)** Traversing controls for a powered carriage shall be of a continuous pressure weatherproof type. Multiple controls when provided shall be arranged to permit operation from only one control station at a time. An emergency stop device shall be provided on each end of a powered carriage for interrupting power to the carriage drive motors;

**1910.66(f)(3)(i)(F)** The operating control(s) shall be so connected that in the case of suspended equipment, traversing of a carriage is not possible until the suspended portion of the equipment is located at its uppermost designed position for traversing; and is free of contact with the face of the building or building guides. In addition, all protective devices and interlocks are to be in the proper position to allow traversing of the carriage;

**1910.66(f)(3)(i)(G)** Stability for underfoot supported carriages shall be obtained by gravity, by an attachment to a structural support, or by a combination of gravity and a structural support. The use of flowing counterweights to achieve stability is prohibited.

**1910.66(f)(3)(i)(G)(1)** The stability factor against overturning shall not be less than two for horizontal traversing of the carriage, including the effects of impact and wind.

**1910.66(f)(3)(i)(G)(2)** The carriages and their anchorages shall be capable of resisting accidental over-tensioning of the wire ropes suspending the working platform, and this calculated value shall include the effect of one and one-half times the stall capacity of the hoist motor. All parts of the installation shall be capable of withstanding without damage to any part of the installation the forces resulting from the stall load of the hoist and one half the wind load.

**1910.66(f)(3)(i)(G)(3)** Roof carriages which rely on having tie-down devices secured to the building to develop the required stability against overturning shall be provided with an interlock which will prevent vertical platform movement unless the tie-down is engaged;

**1910.66(f)(3)(i)(H)** An automatically applied braking or locking system, or equivalent, shall be provided that will prevent unintentional traversing of power traversed or power assisted carriages;

**1910.66(f)(3)(i)(I)** A manual or automatic braking or locking system or equivalent, shall be provided that will prevent unintentional traversing of manually propelled carriages;

**1910.66(f)(3)(i)(J)** A means to lock out the power supply for the carriage shall be provided;

**1910.66(f)(3)(i)(K)** Safe access to and egress from the carriage shall be provided from a safe surface. If the carriage traverses an elevated area, any operating area on the carriage shall be protected by a guardrail system in compliance with the provisions of paragraph (f)(5)(i)(F) of this section. Any access gate shall be self-closing and self-latching, or provided with an interlock;

**1910.66(f)(3)(i)(L)** Each carriage work station position shall be identified by location markings and/or position indicators; and

**1910.66(f)(3)(i)(M)** The motors shall stall if the load on the hoist motors is at any time in excess of three times that necessary for lifting the working platform with its rated load.

**1910.66(f)(3)(ii)** *Transportable outriggers.* **(A)** Transportable outriggers may be used as a method of suspension for ground rigged working platforms where the point of suspension does not exceed 300 feet (91.5 m) above a safe surface. Tie-in guide system(s) shall be provided which meet the requirements of paragraph (e)(2) of this section.

**1910.66(f)(3)(ii)(B)** Transportable outriggers shall be used only with self-powered, ground rigged working platforms.

**1910.66(f)(3)(ii)(C)** Each transportable outrigger shall be secured with a tie-down to a verified anchorage on the building during the entire period of its use. The anchorage shall be designed to have a stability factor of not less than four against overturning or upsetting of the outrigger.

**1910.66(f)(3)(ii)(D)** Access to and egress from the working platform shall be from and to a safe surface below the point of suspension.

**1910.66(f)(3)(ii)(E)** Each transportable outrigger shall be designed for lateral stability to prevent roll-over in the event an accidental lateral load is applied to the outrigger. The accidental lateral load to be considered in this design shall be not less than 70 percent of the rated load of the hoist.

**1910.66(f)(3)(ii)(F)** Each transportable outrigger shall be designed to support an ultimate load of not less than four times the rated load of the hoist.

**1910.66(f)(3)(ii)(G)** Each transportable outrigger shall be so located that the suspension wire ropes for two point suspended working platforms are hung parallel.

**1910.66(f)(3)(ii)(H)** A transportable outrigger shall be tied-back to a verified anchorage on the building with a rope equivalent in strength to the suspension rope.

**1910.66(f)(3)(ii)(I)** The tie-back rope shall be installed parallel to the centerline of the outrigger.

**1910.66(f)(3)(iii)** *Davits.* **(A)** Every davit installation, fixed or transportable, rotatable or non-rotatable shall be designed and installed to insure that it has a stability factor against overturning of not less than four.

**1910.66(f)(3)(iii)(B)** The following requirements apply to roof rigged davit systems:

**1910.66(f)(3)(iii)(B)(1)** Access to and egress from the working platform shall be from a safe surface. Access or egress shall not require persons to climb over a building's parapet or guard railing; and

**1910.66(f)(3)(iii)(B)(2)** The working platform shall be provided with wheels, casters or a carriage for traversing horizontally.

**1910.66(f)(3)(iii)(C)** The following requirements apply to ground rigged davit systems:

**1910.66(f)(3)(iii)(C)(1)** The point of suspension shall not exceed 300 feet (91.5 m) above a safe surface. Guide system(s) shall be provided which meet the requirements of paragraph (e)(2) of this section;

**1910.66(f)(3)(iii)(C)(2)** Access and egress to and from the working platform shall only be from a safe surface below the point of suspension.

**1910.66(f)(3)(iii)(D)** A rotating davit shall not require a horizontal force in excess of 40 pounds (177.9 n) per person to initiate a rotating movement.

**1910.66(f)(3)(iii)(E)** The following requirements shall apply to transportable davits:

**1910.66(f)(3)(iii)(E)(1)** A davit or part of a davit weighing more than 80 pounds (36 kg) shall be provided with a means for its transport, which shall keep the center of gravity of the davit at or below 36 inches (914 mm) above the safe surface during transport;

**1910.66(f)(3)(iii)(E)(2)** A davit shall be provided with a pivoting socket or with a base that will allow the insertion or removal of a davit at a position of not more than 35 degrees above the horizontal, with the complete davit inboard of the building face being serviced; and

**1910.66(f)(3)(iii)(E)(3)** Means shall be provided to lock the davit to its socket or base before it is used to suspend the platform.

**1910.66(f)(4)** *Hoisting machines.* **(i)** Raising and lowering of suspended or supported equipment shall be performed only by a hoisting machine.

**1910.66(f)(4)(ii)** Each hoisting machine shall be capable of arresting any overspeed descent of the load.

**1910.66(f)(4)(iii)** Each hoisting machine shall be powered only by air, electric or hydraulic sources.

**1910.66(f)(4)(iv)** Flammable liquids shall not be carried on the working platform.

**1910.66(f)(4)(v)** Each hoisting machine shall be capable of raising or lowering 125 percent of the rated load of the hoist.

**1910.66(f)(4)(vi)** Moving parts of a hoisting machine shall be enclosed or guarded in compliance with paragraphs (a)(1) and (2) of §1910.212 of this part.

**1910.66(f)(4)(vii)** Winding drums, traction drums and sheaves and directional sheaves used in conjunction with hoisting machines shall be compatible with, and sized for, the wire rope used.

**1910.66(f)(4)(viii)** Each winding drum shall be provided with a positive means of attaching the wire rope to the drum. The attachment shall be capable of developing at least four times the rated load of the hoist.

**1910.66(f)(4)(ix)** Each hoisting machine shall be provided with a primary brake and at least one independent secondary brake, each capable of stopping and holding not less than 125 percent of the lifting capacity of the hoist.

**1910.66(f)(4)(ix)(A)** The primary brake shall be directly connected to the drive train of the hoisting machine, and shall not be connected through belts, chains, clutches, or set screw type devices. The brake shall automatically set when power to the prime mover is interrupted.

**1910.66(f)(4)(ix)(B)(*1*)** The secondary brake shall be an automatic emergency type of brake that, if actuated during each stopping cycle, shall not engage before the hoist is stopped by the primary brake.

**1910.66(f)(4)(ix)(B)(*2*)** When a secondary brake is actuated, it shall stop and hold the platform within a vertical distance of 24 inches (609.6 mm).

**1910.66(f)(4)(x)** Any component of a hoisting machine which requires lubrication for its protection and proper functioning shall be provided with a means for that lubrication to be applied.

**1910.66(f)(5)** *Suspended equipment—***(i)** *General requirements.* **(A)** Each suspended unit component, except suspension ropes and guardrail systems, shall be capable of supporting, without failure, at least four times the maximum intended live load applied or transmitted to that component.

**1910.66(f)(5)(i)(B)** Each suspended unit component shall be constructed of materials that will withstand anticipated weather conditions.

**1910.66(f)(5)(i)(C)** Each suspended unit shall be provided with a load rating plate, conspicuously located, stating the unit weight and rated load of the suspended unit.

**1910.66(f)(5)(i)(D)** When the suspension points on a suspended unit are not at the unit ends, the unit shall be capable of remaining continuously stable under all conditions of use and position of the live load, and shall maintain at least a 1.5 to 1 stability factor against unit upset.

**1910.66(f)(5)(i)(E)** Guide rollers, guide shoes or building face rollers shall be provided, and shall compensate for variations in building dimensions and for minor horizontal out-of-level variations of each suspended unit.

**1910.66(f)(5)(i)(F)** Each working platform of a suspended unit shall be secured to the building facade by one or more of the following methods, or by an equivalent method:

**1910.66(f)(5)(i)(F)(*1*)** Continuous engagement to building anchors as provided in paragraph (e)(2)(i) of this section;

**1910.66(f)(5)(i)(F)(*2*)** Intermittent engagement to building anchors as provided in paragraph (e)(2)(iii)(A) of this section;

**1910.66(f)(5)(i)(F)(*3*)** Button guide engagement as provided in paragraph (e)(2)(iii)(B) of this section; or

**1910.66(f)(5)(i)(F)(*4*)** Angulated roping and building face rollers as provided in paragraph (e)(2)(iii)(C) of this section.

**1910.66(f)(5)(i)(G)** Each working platform of a suspended unit shall be provided with a guardrail system on all sides which shall meet the following requirements:

**1910.66(f)(5)(i)(G)(*1*)** The system shall consist of a top guardrail, midrail, and a toeboard.

**1910.66(f)(5)(i)(G)(*2*)** The top guardrail shall not be less than 36 inches (914 mm) high and shall be able to withstand at least a 100-pound (444 n) force in any downward or outward direction;

**1910.66(f)(5)(i)(G)(*3*)** The midrail shall be able to withstand at least a 75-pound (333 n) force in any downward or outward direction; and

**1910.66(f)(5)(i)(G)(*4*)** The areas between the guardrail and toeboard on the ends and outboard side, and the area between the midrail and toeboard on the inboard side, shall be closed with a material that is capable of withstanding a load of 100 pounds (45.4 KG.) applied horizontally over any area of one square foot (.09 m$^2$). The material shall have all openings small enough to reject passage of life lines and potential falling objects which may be hazardous to persons below.

**1910.66(f)(5)(i)(G)(*5*)** Toeboards shall be capable of withstanding, without failure, a force of at least 50 pounds (222 n) applied in any downward or horizontal direction at any point along the toeboard.

**1910.66(f)(5)(i)(G)(*6*)** Toeboards shall be three and one-half inches (9 cm) minimum in length from their top edge to the level of the platform floor.

**1910.66(f)(5)(i)(G)(*7*)** Toeboards shall be securely fastened in place at the outermost edge of the platform and have no more than one-half inch (1.3 cm) clearance above the platform floor.

**1910.66(f)(5)(i)(G)(*8*)** Toeboards shall be solid or with an opening not over one inch (2.5 cm) in the greatest dimension.

**1910.66(f)(5)(ii)** *Two and four-point suspended working platforms.* **(A)** The working platform shall be not less than 24 inches (610 mm) wide and shall be provided with a minimum of a 12 inch (305 mm) wide passage at or past any obstruction on the platform.

**1910.66(f)(5)(ii)(B)** The flooring shall be of a slip-resistant type and shall contain no opening that would allow the passage of life lines, cables and other potential falling objects. If a larger opening is provided, it shall be protected by placing a material under the opening which shall prevent the passage of life lines, cables and potential falling objects.

**1910.66(f)(5)(ii)(C)** The working platform shall be provided with a means of suspension that will restrict the platform's inboard to outboard roll about its longitudinal axis to a maximum of 15 degrees from a horizontal plane when moving the live load from the inboard to the outboard side of the platform.

**1910.66(f)(5)(ii)(D)** Any cable suspended from above the platform shall be provided with a means for storage to prevent accumulation of the cable on the floor of the platform.

**1910.66(f)(5)(ii)(E)** All operating controls for the vertical travel of the platform shall be of the continuous-pressure type, and shall be located on the platform.

**1910.66(f)(5)(ii)(F)** Each operating station of every working platform shall be provided with a means of interrupting the power supply to all hoist motors to stop any further powered ascent or descent of the platform.

**1910.66(f)(5)(ii)(G)** The maximum rated speed of the platform shall not exceed 50 feet per minute (0.3 ms) with single speed hoists, nor 75 feet per minute (0.4 ms) with multi-speed hoists.

**1910.66(f)(5)(ii)(H)** Provisions shall be made for securing all tools, water tanks, and other accessories to prevent their movement or accumulation on the floor of the platform.

**1910.66(f)(5)(ii)(I)** Portable fire extinguishers conforming to the provisions of §1910.155 and §1910.157 of this part shall be provided and securely attached on all working platforms.

**1910.66(f)(5)(ii)(J)** Access to and egress from a working platform, except for those that land directly on a safe surface, shall be provided by stairs, ladders, platforms and runways conforming to the provisions of subpart D of this part. Access gates shall be self-closing and self-latching.

**1910.66(f)(5)(ii)(K)** Means of access to or egress from a working platform which is 48 inches (1.2 m) or more above a safe surface shall be provided with a guardrail system or ladder handrails that conform to the provisions of subpart D of this part.

**1910.66(f)(5)(ii)(L)** The platform shall be provided with a secondary wire rope suspension system if the platform contains overhead structures which restrict the emergency egress of employees. A horizontal lifeline or a direct connection anchorage shall be provided, as part of a fall arrest system which meets the requirements of appendix C, for each employee on such a platform.

**1910.66(f)(5)(ii)(M)** A vertical lifeline shall be provided as part of a fall arrest system which meets the requirements of appendix C, for each employee on a working platform suspended by two or more wire ropes, if the failure of one wire rope or suspension attachment will cause the platform to upset. If a secondary wire rope suspension is used, vertical lifelines are not required for the fall arrest system, provided that each employee is attached to a horizontal lifeline anchored to the platform.

**1910.66(f)(5)(ii)(N)** An emergency electric operating device shall be provided on roof powered platforms near the hoisting machine for use in the event of failure of the normal operating device located on the working platform, or failure of the cable connected to the platform. The emergency electric operating device shall be mounted in a secured compartment, and the

**1910.66(f)(5)(iii)** *Single point suspended working platforms.* **(A)** The requirements of paragraphs (f)(5)(ii) (A) through (K) of this section shall also apply to a single point working platform.

**1910.66(f)(5)(iv)** *Ground-rigged working platforms.* **(A)** Groundrigged working platforms shall comply with all the requirements of paragraphs (f)(5)(ii) (A) through (M) of this section.

**1910.66(f)(5)(iv)(B)** After each day's use, the power supply within the building shall be disconnected from a ground-rigged working platform, and the platform shall be either disengaged from its suspension points or secured and stored at grade.

**1910.66(f)(5)(v)** *Intermittently stabilized platforms.* **(A)** The platform shall comply with paragraphs (F)(5)(ii) (A) through (M) of this section.

**1910.66(f)(5)(v)(B)** Each stabilizer tie shall be equipped with a "quick connect-quick disconnect" device which cannot be accidently disengaged, for attachment to the building anchor, and shall be resistant to adverse environmental conditions.

**1910.66(f)(5)(v)(C)** The platform shall be provided with a stopping device that will interrupt the hoist power supply in the event the platform contacts a stabilizer tie during its ascent.

**1910.66(f)(5)(v)(D)** Building face rollers shall not be placed at the anchor setting if exterior anchors are used on the building face.

**1910.66(f)(5)(v)(E)** Stabilizer ties used on intermittently stabilized platforms shall allow for the specific attachment length needed to effect the predetermined angulation of the suspended wire rope. The specific attachment length shall be maintained at all building anchor locations.

**1910.66(f)(5)(v)(F)** The platform shall be in continuous contact with the face of the building during ascent and descent.

**1910.66(f)(5)(v)(G)** The attachment and removal of stabilizer ties shall not require the horizontal movement of the platform.

**1910.66(f)(5)(v)(H)** The platform-mounted equipment and its suspension wire ropes shall not be physically damaged by the loads from the stabilizer tie or its building anchor. The platform, platform mounted equipment and wire ropes shall be able to withstand a load that is at least twice the ultimate strength of the stabilizer tie.

**NOTE:** See Figure II in appendix B of this section for a description of a typical intermittent stabilization system.

**1910.66(f)(5)(vi)** *Button-guide stabilized platforms.* **(A)** The platform shall comply with paragraphs (f)(5)(ii) (A) through (M) of this section.

**1910.66(f)(5)(vi)(B)** Each guide track on the platform shall engage a minimum of two guide buttons during any vertical travel of the platform following the initial button engagement.

**1910.66(f)(5)(vi)(C)** Each guide track on a platform that is part of a roof rigged system shall be provided with a storage position on the platform.

**1910.66(f)(5)(vi)(D)** Each guide track on the platform shall be sufficiently maneuverable by platform occupants to permit easy engagement of the guide buttons, and easy movement into and out of its storage position on the platform.

**1910.66(f)(5)(vi)(E)** Two guide tracks shall be mounted on the platform and shall provide continuous contact with the building face.

**1910.66(f)(5)(vi)(F)** The load carrying components of the button guide stabilization system which transmit the load into the platform shall be capable of supporting the weight of the platform, or provision shall be made in the guide track connectors or platform attachments to prevent the weight of the platform from being transmitted to the platform attachments.

**NOTE:** See Figure III in appendix B of this section for a description of a typical button guide stabilization system.

**1910.66(f)(6)** *Supported equipment.* **(i)** Supported equipment shall maintain a vertical position in respect to the face of the building by means other than friction.

**1910.66(f)(6)(ii)** Cog wheels or equivalent means shall be incorporated to provide climbing traction between the supported equipment and the building guides. Additional guide wheels or shoes shall be incorporated as may be necessary to ensure that the drive wheels are continuously held in positive engagement with the building guides.

**1910.66(f)(6)(iii)** Launch guide mullions indexed to the building guides and retained in alignment with the building guides shall be used to align drive wheels entering the building guides.

**1910.66(f)(6)(iv)** Manned platforms used on supported equipment shall comply with the requirements of paragraphs (f)(5)(ii)(A), (f)(5)(ii)(B), and (f)(5)(ii)(D) through (K) of this section covering suspended equipment.

**1910.66(f)(7)** *Suspension wire ropes and rope connections.* **(i)** Each specific installation shall use suspension wire ropes or combination cable and connections meeting the specification recommended by the manufacturer of the hoisting machine used. Connections shall be capable of developing at least 80 percent of the rated breaking strength of the wire rope.

**1910.66(f)(7)(ii)** Each suspension rope shall have a "Design Factor" of at least 10. The "Design Factor" is the ratio of the rated strength of the suspension wire rope to the rated working load, and shall be calculated using the following formula:

$$F = \frac{S(N)}{W}$$

Where:

F = Design Factor

S = Manufacturer's rated strength of one suspension rope

N = Number of suspension ropes under load

W = Rated working load on all ropes at any point of travel

**1910.66(f)(7)(iii)** Suspension wire rope grade shall be at least improved plow steel or equivalent.

**1910.66(f)(7)(iv)** Suspension wire ropes shall be sized to conform with the required design factor, but shall not be less than 5/16 inch (7.94 mm) in diameter.

**1910.66(f)(7)(v)** No more than one reverse bend in six wire rope lays shall be permitted.

**1910.66(f)(7)(vi)** A corrosion-resistant tag shall be securely attached to one of the wire rope fastenings when a suspension wire rope is to be used at a specific location and will remain in that location. This tag shall bear the following wire rope data:

**1910.66(f)(7)(vi)(A)** The diameter (inches and/or mm);

**1910.66(f)(7)(vi)(B)** Construction classification;

**1910.66(f)(7)(vi)(C)** Whether non-preformed or preformed;

**1910.66(f)(7)(vi)(D)** The grade of material;

**1910.66(f)(7)(vi)(E)** The manufacturer's rated strength;

**1910.66(f)(7)(vi)(F)** The manufacturer's name;

**1910.66(f)(7)(vi)(G)** The month and year the ropes were installed; and

**1910.66(f)(7)(vi)(H)** The name of the person or company which installed the ropes.

**1910.66(f)(7)(vii)** A new tag shall be installed at each rope renewal.

**1910.66(f)(7)(viii)** The original tag shall be stamped with the date of the resocketing, or the original tag shall be retained and a supplemental tag shall be provided when ropes are resocketed. The supplemental tag shall show the date of resocketing and the name of the person or company that resocketed the rope.

**1910.66(f)(7)(ix)** Winding drum type hoists shall contain at least three wraps of the suspension wire rope on the drum when the suspended unit has reached the lowest possible point of its vertical travel.

**1910.66(f)(7)(x)** Traction drum and sheave type hoists shall be provided with a wire rope of sufficient length to reach the lowest possible point of vertical travel of the suspended unit, and an additional length of the wire rope of at least four feet (1.2 m).

**1910.66(f)(7)(xi)** The lengthening or repairing of suspension wire ropes is prohibited.

**1910.66(f)(7)(xii)** Babbitted fastenings for suspension wire rope are prohibited.

**1910.66(f)(8)** *Control circuits, power circuits and their components.* **(i)** Electrical wiring and equipment shall comply with subpart S of this part, except as otherwise required by this section.

**1910.66(f)(8)(ii)** Electrical runway conductor systems shall be of a type designed for use in exterior locations, and shall be located so that they do not come into contact with accumulated snow or water.

**1910.66(f)(8)(iii)** Cables shall be protected against damage resulting from overtensioning or from other causes.

**1910.66(f)(8)(iv)** Devices shall be included in the control system for the equipment which will provide protection against electrical overloads, three phase reversal and phase failure. The control system shall have a separate method, independent of the direction control circuit, for breaking the power circuit in case of an emergency or malfunction.

**1910.66(f)(8)(v)** Suspended or supported equipment shall have a control system which will require the operator of the equipment to follow predetermined procedures.

**1910.66(f)(8)(vi)** The following requirements shall apply to electrical protection devices:

**1910.66(f)(8)(vi)(A)** On installations where the carriage does not have a stability factor of at least four against overturning, electrical contact(s) shall be provided and so connected that the operating devices for the suspended or supported equipment shall be operative only when the carriage is located and mechanically retained at an established operating point.

**1910.66(f)(8)(vi)(B)** Overload protection shall be provided in the hoisting or suspension system to protect against the equipment operating in the "up" direction with a load in excess of 125 percent of the rated load of the platform; and

**1910.66(f)(8)(vi)(C)** An automatic detector shall be provided for each suspension point that will interrupt power to all hoisting motors for travel in the "down" direction, and apply the primary brakes if any suspension wire rope becomes slack. A continuous-pressure rigging-bypass switch designed for use during rigging is permitted. This switch shall only be used during rigging.

**1910.66(f)(8)(vii)** Upper and lower directional switches designed to prevent the travel of suspended units beyond safe upward and downward levels shall be provided.

**1910.66(f)(8)(viii)** Emergency stop switches shall be provided on remote controlled, roof-powered manned platforms adjacent to each control station on the platform.

**1910.66(f)(8)(ix)** Cables which are in constant tension shall have overload devices which will prevent the tension in the cable from interfering with the load limiting device required in paragraph (f)(8)(vi)(B) of this section, or with the platform roll limiting device required in paragraph (f)(5)(ii)(C) of this section. The setting of these devices shall be coordinated with other overload settings at the time of design of the system, and shall be clearly indicated on or near the device. The device shall interrupt the equipment travel in the "down" direction.

**1910.66(g)** *Inspection and tests*—**(1)** *Installations and alterations.* All completed building maintenance equipment installations shall be inspected and tested in the field before being placed in initial service to determine that all parts of the installation conform to applicable requirements of this standard, and that all safety and operating equipment is functioning as required. A similar inspection and test shall be made following any major alteration to an existing installation. No hoist in an installation shall be subjected to a load in excess of 125 percent of its rated load.

**1910.66(g)(2)** *Periodic inspections and tests.* **(i)** Related building supporting structures shall undergo periodic inspection by a competent person at intervals not exceeding 12 months.

**1910.66(g)(2)(ii)** All parts of the equipment including control systems shall be inspected, and, where necessary, tested by a competent person at intervals specified by the manufacturer/supplier, but not to exceed 12 months, to determine that they are in safe operating condition. Parts subject to wear, such as wire ropes, bearings, gears, and governors shall be inspected and/or tested to determine that they have not worn to such an extent as to affect the safe operation of the installation.

**1910.66(g)(2)(iii)** The building owner shall keep a certification record of each inspection and test required under paragraphs (g)(2)(i) and (ii) of this section. The certification record shall include the date of the inspection, the signature of the person who performed the inspection, and the number, or other identifier, of the building support structure and equipment which was inspected. This certification record shall be kept readily available for review by the Assistant Secretary of Labor or the Assistant Secretary's representative and by the employer.

**1910.66(g)(2)(iv)** Working platforms and their components shall be inspected by the employer for visible defects before every use and after each occurrence which could affect the platform's structural integrity.

**1910.66(g)(3)** *Maintenance inspections and tests.* **(i)** A maintenance inspection and, where necessary, a test shall be made of each platform installation every 30 days, or where the work cycle is less than 30 days such inspection and/or test shall be made prior to each work cycle. This inspection and test shall follow procedures recommended by the manufacturer, and shall be made by a competent person.

**1910.66(g)(3)(ii)** The building owner shall keep a certification record of each inspection and test performed under paragraph (g)(3)(i) of this section. The certification record shall include the date of the inspection and test, the signature of the person who performed the inspection and/or test, and an identifier for the platform installation which was inspected. The certification record shall be kept readily available for review by the Assistant Secretary of Labor or the Assistant Secretary's representative and by the employer.

**1910.66(g)(4)** *Special inspection of governors and secondary brakes.* **(i)** Governors and secondary brakes shall be inspected and tested at intervals specified by the manufacturer/supplier but not to exceed every 12 months.

**1910.66(g)(4)(ii)** The results of the inspection and test shall confirm that the initiating device for the secondary braking system operates at the proper overspeed.

**1910.66(g)(4)(iii)** The results of the inspection and test shall confirm that the secondary brake is functioning properly.

**1910.66(g)(4)(iv)** If any hoisting machine or initiating device for the secondary brake system is removed from the equipment for testing, all reinstalled and directly related components shall be reinspected prior to returning the equipment installation to service.

**1910.66(g)(4)(v)** Inspection of governors and secondary brakes shall be performed by a competent person.

**1910.66(g)(4)(vi)** The secondary brake governor and actuation device shall be tested before each day's use. Where testing is not feasible, a visual inspection of the brake shall be made instead to ensure that it is free to operate.

**1910.66(g)(5)** *Suspension wire rope maintenance, inspection and replacement.* **(i)** Suspension wire rope shall be maintained and used in accordance with procedures recommended by the wire rope manufacturer.

**1910.66(g)(5)(ii)** Suspension wire rope shall be inspected by a competent person for visible defects and gross damage to the rope before every use and after each occurrence which might affect the wire rope's integrity.

**1910.66(g)(5)(iii)** A thorough inspection of suspension wire ropes in service shall be made once a month. Suspension wire ropes that have been inactive for 30 days or longer shall have a thorough inspection before they are placed into service. These thorough inspections of suspension wire ropes shall be performed by a competent person.

**1910.66(g)(5)(iv)** The need for replacement of a suspension wire rope shall be determined by inspection and shall be based on the condition of the wire rope. Any of the following conditions or combination of conditions will be cause for removal of the wire rope:

**1910.66(g)(5)(iv)(A)** Broken wires exceeding three wires in one strand or six wires in one rope lay;

**1910.66(g)(5)(iv)(B)** Distortion of rope structure such as would result from crushing or kinking;

**1910.66(g)(5)(iv)(C)** Evidence of heat damage;

**1910.66(g)(5)(iv)(D)** Evidence of rope deterioration from corrosion;

**1910.66(g)(5)(iv)(E)** A broken wire within 18 inches (460.8 mm) of the end attachments;

**1910.66(g)(5)(iv)(F)** Noticeable rusting and pitting;

**1910.66(g)(5)(iv)(G)** Evidence of core failure (a lengthening of rope lay, protrusion of the rope core and a reduction in rope diameter suggests core failure); or

**1910.66(g)(5)(iv)(H)** More than one valley break (broken wire).

**1910.66(g)(5)(iv)(I)** Outer wire wear exceeds one-third of the original outer wire diameter.

**1910.66(g)(5)(iv)(J)** Any other condition which the competent person determines has significantly affected the integrity of the rope.

**1910.66(g)(5)(v)** The building owner shall keep a certification record of each monthly inspection of a suspension wire rope as required in paragraph (g)(5)(iii) of this section. The record shall include the date of the inspection, the signature of the person who performed the inspection, and a number, or other identifier, of the wire rope which was inspected. This record of inspection shall be made available for review by the Assistant Secretary of Labor or the Assistant Secretary's representative and by the employer.

**1910.66(g)(6)** *Hoist inspection.* Before lowering personnel below the top elevation of the building, the hoist shall be tested each day in the lifting direction with the intended load to make certain it has sufficient capacity to raise the personnel back to the boarding level.

**1910.66(h)** *Maintenance*—**(1)** *General maintenance.* All parts of the equipment affecting safe operation shall be maintained in proper working order so that they may perform the functions for which they were intended. The equipment shall be taken out of service when it is not in proper working order.

**1910.66(h)(2)** *Cleaning.* **(i)** Control or power contactors and relays shall be kept clean.

**1910.66(h)(2)(ii)** All other parts shall be kept clean if their proper functioning would be affected by the presence of dirt or other contaminants.

**1910.66(h)(3)** *Periodic resocketing of wire rope fastenings.* **(i)** Hoisting ropes utilizing poured socket fastenings shall be resocketed at the non-drum ends at intervals not exceeding 24 months. In resocketing the ropes, a sufficient length shall be cut from the end of the rope to remove damaged or fatigued portions.

**1910.66(h)(3)(ii)** Resocketed ropes shall conform to the requirements of paragraph (f)(7) of this section.

**1910.66(h)(3)(iii)** Limit switches affected by the resocketed ropes shall be reset, if necessary.

**1910.66(h)(4)** *Periodic reshackling of suspension wire ropes.* The hoisting ropes shall be reshackled at the nondrum ends at intervals not exceeding 24 months. When reshackling the ropes, a sufficient length shall be cut from the end of the rope to remove damaged or fatigued portions.

**1910.66(h)(5)** *Roof systems.* Roof track systems, tie-downs, or similar equipment shall be maintained in proper working order so that they perform the function for which they were intended.

**1910.66(h)(6)** *Building face guiding members.* T-rails, indented mullions, or equivalent guides located in the face of a building shall be maintained in proper working order so that they perform the functions for which they were intended. Brackets for cable stabilizers shall similarly be maintained in proper working order.

**1910.66(h)(7)** *Inoperative safety devices.* No person shall render a required safety device or electrical protective device inoperative, except as necessary for tests, inspections, and maintenance. Immediately upon completion of such tests, inspections and maintenance, the device shall be restored to its normal operating condition.

**1910.66(i)** *Operations*—**(1)** *Training.* **(i)** Working platforms shall be operated only by persons who are proficient in the operation, safe use and inspection of the particular working platform to be operated.

**1910.66(i)(1)(ii)** All employees who operate working platforms shall be trained in the following:

**1910.66(i)(1)(ii)(A)** Recognition of, and preventive measures for, the safety hazards associated with their individual work tasks.

**1910.66(i)(1)(ii)(B)** General recognition and prevention of safety hazards associated with the use of working platforms, including the provisions in the section relating to the particular working platform to be operated.

**1910.66(i)(1)(ii)(C)** Emergency action plan procedures required in paragraph (e)(9) of this section.

**1910.66(i)(1)(ii)(D)** Work procedures required in paragraph (i)(1)(iv) of this section.

**1910.66(i)(1)(ii)(E)** Personal fall arrest system inspection, care, use and system performance.

**1910.66(i)(1)(iii)** Training of employees in the operation and inspection of working platforms shall be done by a competent person.

**1910.66(i)(1)(iv)** Written work procedures for the operation, safe use and inspection of working platforms shall be provided for employee training. Pictorial methods of instruction, may be used, in lieu of written work procedures, if employee communication is improved using this method. The operating manuals supplied by manufacturers for platform system components can serve as the basis for these procedures.

**1910.66(i)(1)(v)** The employer shall certify that employees have been trained in operating and inspecting a working platform by preparing a certification record which includes the identity of the person trained, the signature of the employer or the person who conducted the training and the date that training was completed. The certification record shall be prepared at the completion of the training required in paragraph (i)(1)(ii) of this section, and shall be maintained in a file for the duration of the employee's employment. The certification record shall be kept readily available for review by the Assistant Secretary of Labor or the Assistant Secretary's representative.

**1910.66(i)(2)** *Use.* **(i)** Working platforms shall not be loaded in excess of the rated load, as stated on the platform load rating plate.

**1910.66(i)(2)(ii)** Employees shall be prohibited from working on snow, ice, or other slippery material covering platforms, except for the removal of such materials.

**1910.66(i)(2)(iii)** Adequate precautions shall be taken to protect the platform, wire ropes and life lines from damage due to acids or other corrosive substances, in accordance with the recommendations of the corrosive substance producer, supplier, platform manufacturer or other equivalent information sources. Platform members which have been exposed to acids or other corrosive substances shall be washed down with a neutralizing solution, at a frequency recommended by the corrosive substance producer or supplier.

**1910.66(i)(2)(iv)** Platform members, wire ropes and life lines shall be protected when using a heat producing process. Wire ropes and life lines which have been contacted by the heat producing process shall be considered to be permanently damaged and shall not be used.

**1910.66(i)(2)(v)** The platform shall not be operated in winds in excess of 25 miles per hour (40.2 km/hr) except to move it from an operating to a storage position. Wind speed shall be determined based on the best available information, which includes on-site anemometer readings and local weather forecasts which predict wind velocities for the area.

**1910.66(i)(2)(vi)** On exterior installations, an anemometer shall be mounted on the platform to provide information of on-site wind velocities prior to and during the use of the platform. The anemometer may be a portable (hand held) unit which is temporarily mounted during platform use.

**1910.66(i)(2)(vii)** Tools, materials and debris not related to the work in progress shall not be allowed to accumulate on platforms. Stabilizer ties shall be located so as to allow unencumbered passage along the full length of the platform and shall be of such length so as not to become entangled in rollers, hoists or other machinery.

**1910.66(j)** *Personal fall protection.* Employees on working platforms shall be protected by a personal fall arrest system meeting the requirements of appendix C, section I, of this standard, and as otherwise provided by this standard.

## APPENDIX A TO §1910.66—GUIDELINES (ADVISORY)

1. *Use of the Appendix.* Appendix A provides examples of equipment and methods to assist the employer in meeting the requirements of the indicated provision of the standard. Employers may use other equipment or procedures which conform to the requirements of the standard. This appendix neither adds to nor detracts from the mandatory requirements set forth in §1910.66.

2. *Assurance.* Paragraph (c) of the standard requires the building owner to inform the employer in writing that the powered platform installation complies with certain requirements of the standard, since the employer may not have the necessary information to make these determinations. The employer, however, remains responsible for meeting these requirements which have not been set off in paragraph (c)(1).

3. *Design Requirements.* The design requirements for each installation should be based on the limitations (stresses, deflections, etc.), established by nationally recognized standards as promulgated by the following organizations, or to equivalent standards:

AA—The Aluminum Association, 818 Connecticut Avenue, NW., Washington, DC, 20006

Aluminum Construction Manual

Specifications For Aluminum Structures

Aluminum Standards and Data

AGMA—American Gear Manufacturers Association, 101 North Fort Meyer Dr., Suite 1000, Arlington, VA 22209

AISC—American Institute of Steel Construction, 400 North Michigan Avenue, Chicago, IL 60611

ANSI—American National Standards Institute, Inc., 1430 Broadway, New York, NY 10018

ASCE—American Society of Civil Engineers, 345 East 47th Street, New York, NY 10017

ASME—American Society of Mechanical Engineers, 345 East 47th Street, New York, NY 10017

ASTM—American Society for Testing and Materials, 1916 Race Street, Philadelphia, PA 19103

AWS—American Welding Society, Inc., Box 351040, 550 NW. LeJeunne Road, Miami, FL 33126

JIC—Joint Industrial Council, 2139 Wisconsin Avenue NW., Washington, DC 20007

NEMA—National Electric Manufacturers Association, 2101 L Street, NW., Washington, DC 20037

4. *Tie-in-guides.* Indented mullions, T-rails or other equivalent guides are acceptable as tie-in guides in a building face for a continuous stabilization system. Internal guides are embedded in other building members with only the opening exposed (see Figure 1 of appendix B). External guides, however, are installed external to the other building members and so are fully exposed. The minimum opening for tie-in guides is three-quarters of an inch (19 mm), and the minimum inside dimensions are one-inch (25 mm) deep and two inches (50 mm) wide.

Employers should be aware of the hazards associated with tie-in guides in a continuous stabilization system which was not designed properly. For example, joints in these track systems may become extended or discontinuous due to installation or building settlement. If this alignment problem is not corrected, the system could jam when a guide roller or guide shoe strikes a joint and this would cause a hazardous situation for employees. In another instance, faulty design will result in guide rollers being mounted in a line so they will jam in the track at the slightest misalignment.

5. *Building anchors (intermittent stabilization system).* In the selection of the vertical distance between building anchors, certain factors should be given consideration. These factors include building height and architectural design, platform length and weight, wire rope angulation, and the wind velocities in the building area. Another factor to consider is the material of the building face, since this material may be adversely affected by the building rollers.

External or indented type building anchors are acceptable. Receptacles in the building facade used for the indented type should be kept clear of extraneous materials which will hinder their use. During the inspection of the platform installation, evidence of a failure or abuse of the anchors should be brought to the attention of the employer.

6. *Stabilizer tie length.* A stabilizer tie should be long enough to provide for the planned angulation of the suspension cables. However, the length of the tie should not be excessive and become a problem by possibly becoming entangled in the building face rollers or parts of the platform machinery. The attachment length may vary due to material elongation and this should be considered when selecting the material to be used. Consideration should also be given to the use of ties which are easily installed by employees, since this will encourage their use.

7. *Intermittent stabilization system.* Intermittent stabilization systems may use different equipment, tie-in devices and methods to restrict the horizontal movement of a powered platform with respect to the face of the building. One acceptable method employs corrosion-resistant building anchors secured in the face of the building in vertical rows every third floor or 50 feet (15.3 m), whichever is less. The anchors are spaced horizontally to allow a stabilization attachment (stabilizer tie) for each of the two platform suspension wire ropes. The stabilizer tie consists of two parts. One part is a quick connect-quick disconnect device which utilizes a corrosion-resistant yoke and retainer spring that is designed to fit over the building anchors. The second part of the stabilizer tie is a lanyard which is used to maintain a fixed distance between the suspension wire rope and the face of the building.

In this method, as the suspended powered platform descends past the elevation of each anchor, the descent is halted and each of the platform occupants secures a stabilizer tie between a suspension wire rope and a building anchor. The procedure is repeated as each elevation of a building anchor is reached during the descent of the powered platform.

As the platform ascends, the procedure is reversed; that is, the stabilizer ties are removed as each elevation of a building anchor is reached. The removal of each stabilizer tie is assured since the platform is provided with stopping devices which will interrupt power to its hoist(s) in the event either stopping device contacts a stabilizer during the ascent of the platform.

Figure 2 of appendix B illustrates another type of acceptable intermittent stabilization system which utilizes retaining pins as the quick connect-quick disconnect device in the stabilizer tie.

8. *Wire Rope Inspection.* The inspection of the suspension wire rope is important since the rope gradually loses strength during its useful life. The purpose of the inspection is to determine whether the wire rope has sufficient integrity to support a platform with the required design factor.

If there is any doubt concerning the condition of a wire rope or its ability to perform the required work, the rope should be replaced. The cost of wire rope replacement is quite small if compared to the cost in terms of human injuries, equipment down time and replacement.

No listing of critical inspection factors, which serve as a basis for wire rope replacement in the standard, can be a substitute for an experienced inspector of wire rope. The listing serves as a user's guide to the accepted standards by which ropes must be judged.

Rope life can be prolonged if preventive maintenance is performed regularly. Cutting off an appropriate length of rope at the end termination before the core degrades and valley breaks appear minimizes degradation at these sections.

9. *General Maintenance.* In meeting the general maintenance requirement in paragraph (h)(1) of the standard, the employer should undertake the prompt replacement of broken, worn and damaged parts, switch contacts, brushes, and short flexible conductors of electrical devices. The components of the electrical service system and traveling cables should be replaced when damaged or significantly abraded. In addition, gears, shafts, bearings, brakes and hoisting drums should be kept in proper alignment.

10. *Training.* In meeting the training requirement of paragraph (i)(1) of the standard, employers should use both on the job training and formal classroom training. The written work procedures used for this training should be obtained from the manufacturer, if possible, or prepared as necessary for the employee's information and use.

Employees who will operate powered platforms with intermittent stabilization systems should receive instruction in the specific ascent and descent procedures involving the assembly and disassembly of the stabilizer ties.

An acceptable training program should also include employee instruction in basic inspection procedures for the purpose of determining the need for repair and replacement of platform equipment. In addition, the program should cover the inspection, care and use of the personal fall protection equipment required in paragraph (j)(1) of the standard.

In addition, the training program should also include emergency action plan elements. OSHA brochure #1B3088 (Rev.) 1985, "How to Prepare for Workplace Emergencies," details the basic steps needed to prepare to handle emergencies in the workplace.

Following the completion of a training program, the employee should be required to demonstrate competency in operating the equipment safely. Supplemental training of the employee should be provided by the employer, as necessary, if the equipment used or other working conditions should change.

An employee who is required to work with chemical products on a platform should receive training in proper cleaning procedures, and in the hazards, care and handling of these products. In addition, the employee should be supplied with the appropriate personal protective equipment, such as gloves and eye and face protection.

11. *Suspension and Securing of Powered Platforms (Equivalency).* One acceptable method of demonstrating the equivalency of a method of suspending or securing a powered platform, as required in paragraphs

(e)(2)(iii), (f)(3) and (f)(5)(i)(F), is to provide an engineering analysis by a registered professional engineer. The analysis should demonstrate that the proposed method will provide an equal or greater degree of safety for employees than any one of the methods specified in the standard.

## APPENDIX B TO §1910.66—EXHIBITS (ADVISORY)

The three drawings in appendix B illustrate typical platform stabilization systems which are addressed in the standard. The drawings are to be used for reference purposes only, and do not illustrate all the mandatory requirements for each system.

**FIGURE 1—TYPICAL SELF-POWERED PLATFORM—CONTINUOUS EXTERNAL OR INDENTED MULLION GUIDE SYSTEM**

**FIGURE 2—TYPICAL SELF-POWERED PLATFORM—INTERMITTENT TIE-IN SYSTEM**

**FIGURE 3—TYPICAL SELF-POWERED PLATFORM—BUTTON GUIDE SYSTEM**

# OSHA Standards for General Industry

## APPENDIX C TO §1910.66—PERSONAL FALL ARREST SYSTEM (SECTION I—MANDATORY; SECTION II AND III—NON-MANDATORY)

### Use of the Appendix

Section I of appendix C sets out the mandatory criteria for personal fall arrest systems used by all employees using powered platforms, as required by paragraph (j)(1) of this standard. Section II sets out nonmandatory test procedures which may be used to determine compliance with applicable requirements contained in section I of this appendix. Section III provides nonmandatory guidelines which are intended to assist employers in complying with these provisions.

I. *Personal fall arrest systems*—(a) *Scope and application*. This section establishes the application of and performance criteria for personal fall arrest systems which are required for use by all employees using powered platforms under paragraph 1910.66(j).

(b) *Definitions. Anchorage* means a secure point of attachment for lifelines, lanyards or deceleration devices, and which is independent of the means of supporting or suspending the employee.

*Body belt* means a strap with means both for securing it about the waist and for attaching it to a lanyard, lifeline, or deceleration device.

*Body harness* means a design of straps which may be secured about the employee in a manner to distribute the fall arrest forces over at least the thighs, pelvis, waist, chest and shoulders with means for attaching it to other components of a personal fall arrest system.

*Buckle* means any device for holding the body belt or body harness closed around the employee's body.

*Competent person* means a person who is capable of identifying hazardous or dangerous conditions in the personal fall arrest system or any component thereof, as well as in their application and use with related equipment.

*Connector* means a device which is used to couple (connect) parts of the system together. It may be an independent component of the system (such as a carabiner), or an integral component of part of the system (such as a buckle or dee-ring sewn into a body belt or body harness, or a snap-hook spliced or sewn to a lanyard or self-retracting lanyard).

*Deceleration device* means any mechanism, such as a rope grab, ripstitch lanyard, specially woven lanyard, tearing or deforming lanyard, or automatic self retracting-lifeline/lanyard, which serves to dissipate a substantial amount of energy during a fall arrest, or otherwise limits the energy imposed on an employee during fall arrest.

*Deceleration distance* means the additional vertical distance a falling employee travels, excluding lifeline elongation and free fall distance, before stopping, from the point at which the deceleration device begins to operate. It is measured as the distance between the location of an employee's body belt or body harness attachment point at the moment of activation (at the onset of fall arrest forces) of the deceleration device during a fall, and the location of that attachment point after the employee comes to a full stop.

*Equivalent* means alternative designs, materials or methods which the employer can demonstrate will provide an equal or greater degree of safety for employees than the methods, materials or designs specified in the standard.

*Free fall* means the act of falling before the personal fall arrest system begins to apply force to arrest the fall.

*Free fall distance* means the vertical displacement of the fall arrest attachment point on the employee's body belt or body harness between onset of the fall and just before the system begins to apply force to arrest the fall. This distance excludes deceleration distance, lifeline and lanyard elongation but includes any deceleration device slide distance or self-retracting lifeline/lanyard extension before they operate and fall arrest forces occur.

*Lanyard* means a flexible line of rope, wire rope, or strap which is used to secure the body belt or body harness to a deceleration device, lifeline, or anchorage.

*Lifeline* means a component consisting of a flexible line for connection to an anchorage at one end to hang vertically (vertical lifeline), or for connection to anchorages at both ends to stretch horizontally (horizontal lifeline), and which serves as a means for connecting other components of a personal fall arrest system to the anchorage.

*Personal fall arrest system* means a system used to arrest an employee in a fall from a working level. It consists of an anchorage, connectors, a body belt or body harness and may include a lanyard, deceleration device, lifeline, or suitable combinations of these.

*Qualified person* means one with a recognized degree or professional certificate and extensive knowledge and experience in the subject field who is capable of design, analysis, evaluation and specifications in the subject work, project, or product.

*Rope grab* means a deceleration device which travels on a lifeline and automatically frictionally engages the lifeline and locks so as to arrest the fall of an employee. A rope grab usually employs the principle of inertial locking, cam/lever locking, or both.

*Self-retracting lifeline/lanyard* means a deceleration device which contains a drum-wound line which may be slowly extracted from, or retracted onto, the drum under slight tension during normal employee movement, and which, after onset of a fall, automatically locks the drum and arrests the fall.

*Snap-hook* means a connector comprised of a hookshaped member with a normally closed keeper, or similar arrangement, which may be opened to permit the hook to receive an object and, when released, automatically closes to retain the object. Snap-hooks are generally one of two types:

1. The locking type with a self-closing, self-locking keeper which remains closed and locked until unlocked and pressed open for connection or disconnection, or

2. The non-locking type with a self-closing keeper which remains closed until pressed open for connection or disconnection.

*Tie-off* means the act of an employee, wearing personal fall protection equipment, connecting directly or indirectly to an anchorage. It also means the condition of an employee being connected to an anchorage.

(c) *Design for system components.* (1) Connectors shall be drop forged, pressed or formed steel, or made of equivalent materials.

(2) Connectors shall have a corrosion-resistant finish, and all surfaces and edges shall be smooth to prevent damage to interfacing parts of the system.

(3) Lanyards and vertical lifelines which tie-off one employee shall have a minimum breaking strength of 5,000 pounds (22.2 kN).

(4) Self-retracting lifelines and lanyards which automatically limit free fall distance to two feet (0.61 m) or less shall have components capable of sustaining a minimum static tensile load of 3,000 pounds (13.3 kN) applied to the device with the lifeline or lanyard in the fully extended position.

(5) Self-retracting lifelines and lanyards which do not limit free fall distance to two feet (0.61 m) or less, ripstitch lanyards, and tearing and deforming lanyards shall be capable of sustaining a minimum tensile load of 5,000 pounds (22.2 kN) applied to the device with the lifeline or lanyard in the fully extended position.

(6) Dee-rings and snap-hooks shall be capable of sustaining a minimum tensile load of 5,000 pounds (22.2 kN).

(7) Dee-rings and snap-hooks shall be 100 percent proof-tested to a minimum tensile load of 3,600 pounds (16 kN) without cracking, breaking, or taking permanent deformation.

(8) Snap-hooks shall be sized to be compatible with the member to which they are connected so as to prevent unintentional disengagement of the snap-hook by depression of the snap-hook keeper by the connected member, or shall be a locking type snap-hook designed and used to prevent disengagement of the snap-hook by the contact of the snaphook keeper by the connected member.

(9) Horizontal lifelines, where used, shall be designed, and installed as part of a complete personal fall arrest system, which maintains a safety factor of at least two, under the supervision of a qualified person.

(10) Anchorages to which personal fall arrest equipment is attached shall be capable of supporting at least 5,000 pounds (22.2 kN) per employee attached, or shall be designed, installed, and used as part of a complete personal fall arrest system which maintains a safety factor of at least two, under the supervision of a qualified person.

(11) Ropes and straps (webbing) used in lanyards, lifelines, and strength components of body belts and body harnesses, shall be made from synthetic fibers or wire rope.

(d) *System performance criteria.* (1) Personal fall arrest systems shall, when stopping a fall:

(i) Limit maximum arresting force on an employee to 900 pounds (4 kN) when used with a body belt;

(ii) Limit maximum arresting force on an employee to 1,800 pounds (8 kN) when used with a body harness;

(iii) Bring an employee to a complete stop and limit maximum deceleration distance an employee travels to 3.5 feet (1.07 m); and

(iv) Shall have sufficient strength to withstand twice the potential impact energy of an employee free falling a distance of six feet (1.8 m), or the free fall distance permitted by the system, whichever is less.

(2)(i) When used by employees having a combined person and tool weight of less than 310 pounds (140 kg), personal fall arrest systems which meet the criteria and protocols contained in paragraphs (b), (c) and (d) in section II of this appendix shall be considered as complying with the provisions of paragraphs (d)(1)(i) through (d)(1)(iv) above.

(ii) When used by employees having a combined tool and body weight of 310 pounds (140 kg) or more, personal fall arrest systems which meet the criteria and protocols contained in paragraphs (b), (c) and (d) in section II may be considered as complying with the provisions of paragraphs (d)(1)(i) through (d)(1)(iv) provided that the criteria and protocols are modified appropriately to provide proper protection for such heavier weights.

(e) *Care and use.* (1) Snap-hooks, unless of a locking type designed and used to prevent disengagement from the following connections, shall not be engaged:

(i) Directly to webbing, rope or wire rope;

(ii) To each other;

(iii) To a dee-ring to which another snap-hook or other connector is attached;

(iv) To a horizontal lifeline; or

(v) To any object which is incompatibly shaped or dimensioned in relation to the snap-hook such that the connected object could depress the snap-hook keeper a sufficient amount to release itself.

(2) Devices used to connect to a horizontal lifeline which may become a vertical lifeline shall be capable of locking in either direction on the lifeline.

(3) Personal fall arrest systems shall be rigged such that an employee can neither free fall more than six feet (1.8 m), nor contact any lower level.

(4) The attachment point of the body belt shall be located in the center of the wearer's back. The attachment point of the body harness shall be located in the center of the wearer's back near shoulder level, or above the wearer's head.

(5) When vertical lifelines are used, each employee shall be provided with a separate lifeline.

(6) Personal fall arrest systems or components shall be used only for employee fall protection.

(7) Personal fall arrest systems or components subjected to impact loading shall be immediately removed from service and shall not be used again for employee protection unless inspected and determined by a competent person to be undamaged and suitable for reuse.

(8) The employer shall provide for prompt rescue of employees in the event of a fall or shall assure the self-rescue capability of employees.

(9) Before using a personal fall arrest system, and after any component or system is changed, employees shall be trained in accordance with the requirements of paragraph 1910.66(i)(1), in the safe use of the system.

(f) *Inspections.* Personal fall arrest systems shall be inspected prior to each use for mildew, wear, damage and other deterioration, and defective components shall be removed from service if their strength or function may be adversely affected.

II. *Test methods for personal fall arrest systems (non-mandatory)*—(a) *General.* Paragraphs (b), (c), (d) and (e), of this section II set forth test procedures which may be used to determine compliance with the requirements in paragraph (d)(1)(i) through (d)(1)(iv) of section I of this appendix.

(b) *General conditions for all tests in section II.* (1) Lifelines, lanyards and deceleration devices should be attached to an anchorage and connected to the body-belt or body harness in the same manner as they would be when used to protect employees.

(2) The anchorage should be rigid, and should not have a deflection greater than .04 inches (1 mm) when a force of 2,250 pounds (10 kN) is applied.

(3) The frequency response of the load measuring instrumentation should be 120 Hz.

(4) The test weight used in the strength and force tests should be a rigid, metal, cylindrical or torso-shaped object with a girth of 38 inches plus or minus four inches (96 cm plus or minus 10 cm).

(5) The lanyard or lifeline used to create the free fall distance should be supplied with the system, or in its absence, the least elastic lanyard or lifeline available to be used with the system.

(6) The test weight for each test should be hoisted to the required level and should be quickly released without having any appreciable motion imparted to it.

(7) The system's performance should be evaluated taking into account the range of environmental conditions for which it is designed to be used.

(8) Following the test, the system need not be capable of further operation.

(c) *Strength test.* (1) During the testing of all systems, a test weight of 300 pounds plus or minus five pounds (135 kg plus or minus 2.5 kg) should be used. (See paragraph (b)(4), above.)

(2) The test consists of dropping the test weight once. A new unused system should be used for each test.

(3) For lanyard systems, the lanyard length should be six feet plus or minus two inches (1.83 m plus or minus 5 cm) as measured from the fixed anchorage to the attachment on the body belt or body harness.

(4) For rope-grab-type deceleration systems, the length of the lifeline above the centerline of the grabbing mechanism to the lifeline's anchorage point should not exceed two feet (0.61 m).

(5) For lanyard systems, for systems with deceleration devices which do not automatically limit free fall distance to two feet (0.61 m) or less, and for systems with deceleration devices which have a connection distance in excess of one foot (0.3 m) (measured between the centerline of the lifeline and the attachment point to the body belt or harness), the test weight should be rigged to free fall a distance of 7.5 feet (2.3 m) from a point that is 1.5 feet (46 cm) above the anchorage point, to its hanging location (six feet below the anchorage). The test weight should fall without interference, obstruction, or hitting the floor or ground during the test. In some cases a non-elastic wire lanyard of sufficient length may need to be added to the system (for test purposes) to create the necessary free fall distance.

(6) For deceleration device systems with integral lifelines or lanyards which automatically limit free fall distance to two feet (0.61 m) or less, the test weight should be rigged to free fall a distance of four feet (1.22 m).

(7) Any weight which detaches from the belt or harness should constitute failure for the strength test.

(d) *Force test*—(1) *General.* The test consists of dropping the respective test weight specified in (d)(2)(i) or (d)(3)(i) once. A new, unused system should be used for each test.

(2) *For lanyard systems.* (i) A test weight of 220 pounds plus or minus three pounds (100 kg plus or minus 1.6 kg) should be used. (See paragraph (b)(4), above.)

(ii) Lanyard length should be six feet plus or minus two inches (1.83 m plus or minus 5 cm) as measured from the fixed anchorage to the attachment on the body belt or body harness.

(iii) The test weight should fall free from the anchorage level to its hanging location (a total of six feet (1.83 m) free fall distance) without interference, obstruction, or hitting the floor or ground during the test.

(3) *For all other systems.* (i) A test weight of 220 pounds plus or minus three pounds (100 kg plus or minus 1.6 kg) should be used. (See paragraph (b)(4), above.)

(ii) The free fall distance to be used in the test should be the maximum fall distance physically permitted by the system during normal use conditions, up to a maximum free fall distance for the test weight of six feet (1.83 m), except as follows:

(A) For deceleration systems which have a connection link or lanyard, the test weight should free fall a distance equal to the connection distance (measured between the centerline of the lifeline and the attachment point to the body belt or harness).

(B) For deceleration device systems with integral lifelines or lanyards which automatically limit free fall distance to two feet (0.61 m) or less, the test weight should free fall a distance equal to that permitted by the system in normal use. (For example, to test a system with a self-retracting lifeline or lanyard, the test weight should be supported and the system allowed to retract the lifeline or lanyard as it would in normal use. The test weight would then be released and the force and deceleration distance measured).

(4) A system fails the force test if the recorded maximum arresting force exceeds 1,260 pounds (15.6 kN) when using a body belt, and/or exceeds 2,520 pounds (11.2 kN) when using a body harness.

(5) The maximum elongation and deceleration distance should be recorded during the force test.

(e) *Deceleration device tests*—(1) *General.* The device should be evaluated or tested under the environmental conditions, (such as rain, ice, grease, dirt, type of lifeline, etc.), for which the device is designed.

(2) *Rope-grab-type deceleration devices.* (i) Devices should be moved on a lifeline 1,000 times over the same length of line a distance of not less than one foot (30.5 cm), and the mechanism should lock each time.

(ii) Unless the device is permanently marked to indicate the type(s) of lifeline which must be used, several types (different diameters and different materials), of lifelines should be used to test the device.

(3) *Other self-activating[sic]-type deceleration devices.* The locking mechanisms of other self-activating-type deceleration devices designed for more than one arrest should lock each of 1,000 times as they would in normal service.

III. *Additional non-mandatory guidelines for personal fall arrest systems.* The following information constitutes additional guidelines for use in complying with requirements for a personal fall arrest system.

(a) *Selection and use considerations.* The kind of personal fall arrest system selected should match the particular work situation, and any possible free fall distance should be kept to a minimum. Consideration should be given to the particular work environment. For example, the presence of acids, dirt, moisture, oil, grease, etc., and their effect on the system, should be evaluated. Hot or cold environments may also have an adverse affect on the system. Wire rope should not be used where an electrical hazard is anticipated. As required by the standard, the employer must plan to have means available to promptly rescue an employee should a fall occur, since the suspended employee may not be able to reach a work level independently.

Where lanyards, connectors, and lifelines are subject to damage by work operations such as welding, chemical cleaning, and sandblasting, the component should be protected, or other securing systems should be used. The employer should fully evaluate the work conditions and environment (including seasonal weather changes) before selecting the appropriate personal fall protection system. Once in use, the system's effectiveness should be monitored. In some cases, a program for cleaning and maintenance of the system may be necessary.

(b) *Testing considerations.* Before purchasing or putting into use a personal fall arrest system, an employer should obtain from the supplier information about the system based on its performance during testing so that the employer can know if the system meets this standard. Testing should be done using recognized test methods. Section II of this appendix C contains test methods recognized for evaluating the performance of fall arrest systems. Not all systems may need to be individually tested; the performance of some systems may be based on data and calculations derived from testing of similar systems, provided that enough information is available to demonstrate similarity of function and design.

(c) *Component compatibility considerations.* Ideally, a personal fall arrest system is designed, tested, and supplied as a complete system. However, it is common practice for lanyards, connectors, lifelines, deceleration devices, body belts and body harnesses to be interchanged since some components wear out before others. The employer and employee should realize that not all components are interchangeable. For instance, a lanyard should not be connected between a body belt (or harness) and a deceleration device of the self-retracting type since this can result in additional free fall for which the system was not designed. Any substitution or change to a personal fall arrest system should be fully evaluated or tested by a competent person to determine that it meets the standard, before the modified system is put in use.

(d) *Employee training considerations.* Thorough employee training in the selection and use of personal fall arrest systems is imperative. As stated in the standard, before the equipment is used, employees must be trained in the safe use of the system. This should include the following: Application limits; proper anchoring and tie-off techniques; estimation of free fall distance, including determination of deceleration distance, and total fall distance to prevent striking a lower level; methods of use; and inspection and storage of the system. Careless or improper use of the equipment can result in serious injury or death. Employers and employees should become familiar with the material in this appendix, as well as manufacturer's recommendations, before a system is used. Of uppermost importance is the reduction in strength caused by certain tie-offs (such as using knots, tying around sharp edges, etc.) and maximum permitted free fall distance. Also, to be stressed are the importance of inspections prior to use, the limitations of the equipment, and unique conditions at the worksite which may be important in determining the type of system to use.

(e) *Instruction considerations.* Employers should obtain comprehensive instructions from the supplier as to the system's proper use and application, including, where applicable:

(1) The force measured during the sample force test;

(2) The maximum elongation measured for lanyards during the force test;

(3) The deceleration distance measured for deceleration devices during the force test;

(4) Caution statements on critical use limitations;

(5) Application limits;

(6) Proper hook-up, anchoring and tie-off techniques, including the proper dee-ring or other attachment point to use on the body belt and harness for fall arrest;

(7) Proper climbing techniques;

(8) Methods of inspection, use, cleaning, and storage; and

(9) Specific lifelines which may be used. This information should be provided to employees during training.

(f) *Inspection considerations.* As stated in the standard (section I, Paragraph (f)), personal fall arrest systems must be regularly inspected. Any component with any significant defect, such as cuts, tears, abrasions, mold, or undue stretching; alterations or additions which might affect its efficiency; damage due to deterioration; contact with fire, acids, or other corrosives; distorted hooks or faulty hook springs; tongues unfitted to the shoulder of buckles; loose or damaged mountings; non-functioning parts; or wearing or internal deterioration in the ropes must be withdrawn from service immediately, and should be tagged or marked as unusable, or destroyed.

(g) *Rescue considerations.* As required by the standard (section I, Paragraph (e)(8)), when personal fall arrest systems are used, the employer must assure that employees can be promptly rescued or can rescue themselves should a fall occur. The availability of rescue personnel, ladders or other rescue equipment should be evaluated. In some situations, equipment which allows employees to rescue themselves after the fall has been arrested may be desirable, such as devices which have descent capability.

(h) *Tie-off considerations.* (1) One of the most important aspects of personal fall protection systems is fully planning the system *before* it is put into use. Probably the most overlooked component is planning for suitable anchorage points. Such planning should ideally be done before the structure or building is constructed so that anchorage points can be incorporated during construction for use later for window cleaning or other building maintenance. If properly planned, these anchorage points may be used *during* construction, as well as afterwards.

(2) Employers and employees should at all times be aware that the strength of a personal fall arrest system is based on its being attached to an anchoring system which does not significantly reduce the strength of the system (such as a properly dimensioned eye-bolt/snap-hook anchorage). Therefore, if a means of attachment is used that will reduce the strength of the system, that component should be replaced by a stronger one, but one that will also maintain the appropriate maximum arrest force characteristics.

(3) Tie-off using a knot in a rope lanyard or lifeline (at any location) can reduce the lifeline or lanyard strength by 50 percent or more. Therefore, a stronger lanyard or lifeline should be used to compensate for the weakening effect of the knot, or the lanyard length should be reduced (or the tie-off location raised) to minimize free fall distance, or the lanyard or

lifeline should be replaced by one which has an appropriately incorporated connector to eliminate the need for a knot.

(4) Tie-off of a rope lanyard or lifeline around an "H" or "I" beam or similar support can reduce its strength as much as 70 percent due to the cutting action of the beam edges. Therefore, use should be made of a webbing lanyard or wire core lifeline around the beam; or the lanyard or lifeline should be protected from the edge; or free fall distance should be greatly minimized.

(5) Tie-off where the line passes over or around rough or sharp surfaces reduces strength drastically. Such a tie-off should be avoided or an alternative tie-off rigging should be used. Such alternatives may include use of a snap-hook/dee ring connection, wire rope tie-off, an effective padding of the surfaces, or an abrasion-resistance strap around or over the problem surface.

(6) Horizontal lifelines may, depending on their geometry and angle of sag, be subjected to greater loads than the impact load imposed by an attached component. When the angle of horizontal lifeline sag is less than 30 degrees, the impact force imparted to the lifeline by an attached lanyard is greatly amplified. For example, with a sag angle of 15 degrees, the force amplification is about 2:1 and at 5 degrees sag, it is about 6:1. Depending on the angle of sag, and the line's elasticity, the strength of the horizontal lifeline and the anchorages to which it is attached should be increased a number of times over that of the lanyard. Extreme care should be taken in considering a horizontal lifeline for multiple tie-offs. The reason for this is that in multiple tie-offs to a horizontal lifeline, if one employee falls, the movement of the falling employee and the horizontal lifeline during arrest of the fall may cause other employees to also fall. Horizontal lifeline and anchorage strength should be increased for each additional employee to be tied-off. For these and other reasons, the design of systems using horizontal lifelines must only be done by qualified persons. Testing of installed lifelines and anchors prior to use is recommended.

(7) The strength of an eye-bolt is rated along the axis of the bolt and its strength is greatly reduced if the force is applied at an angle to this axis (in the direction of shear). Also, care should be exercised in selecting the proper diameter of the eye to avoid accidental disengagement of snap-hooks not designed to be compatible for the connection.

(8) Due to the significant reduction in the strength of the lifeline/lanyard (in some cases, as much as a 70 percent reduction), the sliding hitch knot should not be used for lifeline/lanyard connections except in emergency situations where no other available system is practical. The "one-and-one" sliding hitch knot should never be used because it is unreliable in stopping a fall. The "two-and-two," or "three-and-three" knot (preferable), may be used in emergency situations; however, care should be taken to limit free fall distance to a minimum because of reduced lifeline/lanyard strength.

(i) *Vertical lifeline considerations.* As required by the standard, each employee must have a separate lifeline when the lifeline is vertical. The reason for this is that in multiple tie-offs to a single lifeline, if one employee falls, the movement of the lifeline during the arrest of the fall may pull other employees' lanyards, causing them to fall as well.

(j) *Snap-hook considerations.* Although not required by this standard for all connections, locking snap-hooks designed for connection to suitable objects (of sufficient strength) are highly recommended in lieu of the non-locking type. Locking snap-hooks incorporate a positive locking mechanism in addition to the spring loaded keeper, which will not allow the keeper to open under moderate pressure without someone first releasing the mechanism. Such a feature, properly designed, effectively prevents roll-out from occurring.

As required by the standard (section I, paragraph (e)(1)) the following connections must be avoided (unless properly designed locking snap-hooks are used) because they are conditions which can result in roll-out when a nonlocking snap-hook is used:

- Direct connection of a snap-hook to a horizontal lifeline.
- Two (or more) snap-hooks connected to one dee-ring.
- Two snap-hooks connected to each other.
- A snap-hook connected back on its integral lanyard.
- A snap-hook connected to a webbing loop or webbing lanyard.
- Improper dimensions of the dee-ring, rebar, or other connection point in relation to the snap-hook dimensions which would allow the snap-hook keeper to be depressed by a turning motion of the snap-hook.

(k) *Free fall considerations.* The employer and employee should at all times be aware that a system's maximum arresting force is evaluated under normal use conditions established by the manufacturer, and in no case using a free fall distance in excess of six feet (1.8 m). A few extra feet of free fall can significantly increase the arresting force on the employee, possibly to the point of causing injury. Because of this, the free fall distance should be kept at a minimum, and, as required by the standard, in no case greater than six feet (1.8 m). To help assure this, the tie-off attachment point to the lifeline or anchor should be located at or above the connection point of the fall arrest equipment to belt or harness. (Since otherwise additional free fall distance is added to the length of the connecting means (i.e. lanyard)). Attaching to the working surface will often result in a free fall greater than six feet (1.8 m). For instance, if a six foot (1.8 m) lanyard is used, the total free fall distance will be the distance from the working level to the body belt (or harness) attachment point plus the six feet (1.8 m) of lanyard length. Another important consideration is that the arresting force which the fall system must withstand also goes up with greater distances of free fall, possibly exceeding the strength of the system.

(l) *Elongation and deceleration distance considerations.* Other factors involved in a proper tie-off are elongation and deceleration distance. During the arresting of a fall, a lanyard will experience a length of stretching or elongation, whereas activation of a deceleration device will result in a certain stopping distance. These distances should be available with the lanyard or device's instructions and must be added to the free fall distance to arrive at the total fall distance before an employee is fully stopped. The additional stopping distance may be very significant if the lanyard or deceleration device is attached near or at the end of a long lifeline, which may itself add considerable distance due to its own elongation. As required by the standard, sufficient distance to allow for all of these factors must also be maintained between the employee and obstructions below, to prevent an injury due to impact before the system fully arrests the fall. In addition, a minimum of 12 feet (3.7 m) of lifeline should be allowed below the securing point of a rope grab type deceleration device, and the end terminated to prevent the device from sliding off the lifeline. Alternatively, the lifeline should extend to the ground or the next working level below. These measures are suggested to prevent the worker from inadvertently moving past the end of the lifeline and having the rope grab become disengaged from the lifeline.

(m) *Obstruction considerations.* The location of the tie-off should also consider the hazard of obstructions in the potential fall path of the employee. Tie-offs which minimize the possibilities of exaggerated swinging should be considered. In addition, when a body belt is used, the employee's body will go through a horizontal position to a jack-knifed position during the arrest of all falls. Thus, obstructions which might interfere with this motion should be avoided or a severe injury could occur.

(n) *Other considerations.* Because of the design of some personal fall arrest systems, additional considerations may be required for proper tie-off. For example, heavy deceleration devices of the self-retracting type should be secured overhead in order to avoid the weight of the device having to be supported by the employee. Also, if selfretracting equipment is connected to a horizontal lifeline, the sag in the lifeline should be minimized to prevent the device from sliding down the lifeline to a position which creates a swing hazard during fall arrest. In all cases, manufacturer's instructions should be followed.

### APPENDIX D TO § 1910.66— EXISTING INSTALLATIONS (MANDATORY)

#### Use of the Appendix

Appendix D sets out the mandatory building and equipment requirements for applicable permanent installations completed after August 27, 1971, and no later than July 23, 1990 which are exempt from the paragraphs (a), (b)(1), (b)(2), (c), (d), (e), and (f) of this standard. The requirements in appendix D are essentially the same as unrevised building and equipment provisions which previously were designated as 29 CFR 1910.66 (a), (b), (c) and (d) and which were effective on August 27, 1971.

**NOTE:** All existing installations subject to this appendix shall also comply with paragraphs (g), (h), (i), (j) and appendix C of the standard 29 CFR 1910.66.

(a) *Definitions applicable to this appendix*—(1) *Angulated roping.* A system of platform suspension in which the upper wire rope sheaves or suspension points are closer to the plane of the building face than the corresponding attachment points on the platform, thus causing the platform to press against the face of the building during its vertical travel.

(2) *ANSI.* American National Standards Institute.

(3) *Babbitted fastenings.* The method of providing wire rope attachments in which the ends of the wire strands are bent back and are held in a tapered socket by means of poured molten babbitt metal.

(4) *Brake—disc type.* A brake in which the holding effect is obtained by frictional resistance between one or more faces of discs keyed to the rotating member to be held and fixed discs keyed to the stationary or housing member (pressure between the discs being applied axially).

(5) *Brake—self-energizing band type.* An essentially unidirectional brake in which the holding effect is obtained by the snubbing action of a flexible band wrapped about a cylindrical wheel or drum affixed to the rotating member to be held, the connections and linkages being so arranged that the motion of the brake wheel or drum will act to increase the tension or holding force of the band.

(6) *Brake—shoe type.* A brake in which the holding effect is obtained by applying the direct pressure of two or more segmental friction elements held to a stationary member against a cylindrical wheel or drum affixed to the rotating member to be held.

(7) *Building face rollers.* A specialized form of guide roller designed to contact a portion of the outer face or wall structure of the building, and to assist in stabilizing the operators' platform during vertical travel.

(8) *Continuous pressure.* Operation by means of buttons or switches, any one of which may be used to control the movement of the working platform or roof car, only as long as the button or switch is manually maintained in the actuating position.

(9) *Control.* A system governing starting, stopping, direction, acceleration, speed, and retardation of moving members.

(10) *Controller.* A device or group of devices, usually contained in a single enclosure, which serves to control in some predetermined manner the apparatus to which it is connected.

(11) *Electrical ground.* A conducting connection between an electrical circuit or equipment and the earth, or some conducting body which serves in place of the earth.

(12) *Guide roller.* A rotating, bearing-mounted, generally cylindrical member, operating separately or as part of a guide shoe assembly, attached to the platform, and providing rolling contact with building guideways, or other building contact members.

(13) *Guide shoe.* An assembly of rollers, slide members, or the equivalent, attached as a unit to the operators' platform, and designed to engage with the building members provided for the vertical guidance of the operators' platform.

(14) *Interlock.* A device actuated by the operation of some other device with which it is directly associated, to govern succeeding operations of the same or allied devices.

(15) *Operating device.* A pushbutton, lever, or other manual device used to actuate a control.

(16) *Powered platform.* Equipment to provide access to the exterior of a building for maintenance, consisting of a suspended power-operated working platform, a roof car, or other suspension means, and the requisite operating and control devices.

(17) *Rated load.* The combined weight of employees, tools, equipment, and other material which the working platform is designed and installed to lift.

(18) *Relay, direction.* An electrically energized contactor responsive to an initiating control circuit, which in turn causes a moving member to travel in a particular direction.

(19) *Relay, potential for vertical travel.* An electrically energized contactor responsive to initiating control circuit, which in turn controls the operation of a moving member in both directions. This relay usually operates in conjunction with direction relays, as covered under the definition, "relay, direction."

(20) *Roof car.* A structure for the suspension of a working platform, providing for its horizontal movement to working positions.

(21) *Roof-powered platform.* A powered platform having the raising and lowering mechanism located on a roof car.

(22) *Self-powered platform.* A powered platform having the raising and lowering mechanism located on the working platform.

(23) *Traveling cable.* A cable made up of electrical or communication conductors or both, and providing electrical connection between the working platform and the roof car or other fixed point.

(24) *Weatherproof.* Equipment so constructed or protected that exposure to the weather will not interfere with its proper operation.

(25) *Working platform.* The suspended structure arranged for vertical travel which provides access to the exterior of the building or structure.

(26) *Yield point.* The stress at which the material exhibits a permanent set of 0.2 percent.

(27) *Zinced fastenings.* The method of providing wire rope attachments in which the splayed or fanned wire ends are held in a tapered socket by means of poured molten zinc.

(b) *General requirements.* (1) Design requirements. All powered platform installations for exterior building maintenance completed as of August 27, 1971, but no later than [insert date, 180 days after the effective date], shall meet all of the design, construction and installation requirements of Part II and III of the "American National Standard Safety Requirements for Powered Platforms for Exterior Building Maintenance ANSI A120.1-1970" and of this appendix. References shall be made to appropriate parts of ANSI A120.1-1970 for detail specifications for equipment and special installations.

(2) *Limitation.* The requirements of this appendix apply only to electric powered platforms. It is not the intent of this appendix to prohibit the use of other types of power. Installation of powered platforms using other types of power is permitted, provided such platforms have adequate protective devices for the type of power used, and otherwise provide for reasonable safety of life and limb to users of equipment and to others who may be exposed.

(3) *Types of powered platforms.* (i) For the purpose of applying this appendix, powered platforms are divided into two basic types, Type F and Type T.

(ii) Powered platforms designated as Type F shall meet all the requirements in Part II of ANSI A120.1-1970, American National Standard Safety Requirements for Powered Platforms for Exterior Building Maintenance. A basic requirement of Type F equipment is that the work platform is suspended by at least four wire ropes and designed so that failure of any one wire rope will not substantially alter the normal position of the working platform. Another basic requirement of Type F equipment is that only one layer of hoisting rope is permitted on winding drums. Type F powered platforms may be either roof-powered or self-powered.

(iii) Powered platforms designated as Type T shall meet all the requirements in Part III of ANSI A120.1-1970 American National Standard Safety Requirements for Powered Platforms for Exterior Building Maintenance, except for section 28, Safety Belts and Life Lines. A basic requirement of Type T equipment is that the working platform is suspended by at least two wire ropes. Failure of one wire rope would not permit the working platform to fall to the ground, but would upset its normal position. Type T powered platforms may be either roof-powered or self-powered.

(iv) The requirements of this section apply to powered platforms with winding drum type hoisting machines. It is not the intent of this section to prohibit powered platforms using other types of hoisting machines such as, but not limited to, traction drum hoisting machines, air powered machines, hydraulic powered machines, and internal combustion machines. Installation of powered platforms with other types of hoisting machines is permitted, provided adequate protective devices are used, and provided reasonable safety of life and limb to users of the equipment and to others who may be exposed is assured.

(v) Both Type F and Type T powered platforms shall comply with the requirements of appendix C of this standard.

(c) *Type F powered platforms*—(1) *Roof car, general.* (i) A roof car shall be provided whenever it is necessary to move the working platform horizontally to working or storage positions.

(ii) The maximum rated speed at which a power traversed roof car may be moved in a horizontal direction shall be 50 feet per minute.

(2) *Movement and positioning of roof car.* (i) Provision shall be made to protect against having the roof car leave the roof or enter roof areas not designed for travel.

(ii) The horizontal motion of the roof cars shall be positively controlled so as to insure proper movement and positioning of the roof car.

(iii) Roof car positioning devices shall be provided to insure that the working platform is placed and retained in proper position for vertical travel and during storage.

(iv) Mechanical stops shall be provided to prevent the traversing of the roof car beyond its normal limits of travel. Such stops shall be capable of withstanding a force equal to 100 percent of the inertial effect of the roof car in motion with traversing power applied.

(v)(a) The operating device of a power-operated roof car for traversing shall be located on the roof car, the working platform, or both, and shall be of the continuous pressure weather-proof electric type. If more than one operating device is provided, they shall be so arranged that traversing is possible only from one operating device at a time.

(b) The operating device shall be so connected that it is not operable until:

(1) The working platform is located at its uppermost position of travel and is not in contact with the building face or fixed vertical guides in the face of the building; and

(2) All protective devices and interlocks are in a position for traversing.

(3) *Roof car stability.* Roof car stability shall be determined by either paragraph (c)(3) (i) or (ii) of this appendix, whichever is greater.

(i) The roof car shall be continuously stable, considering overturning moment as determined by 125 percent rated load, plus maximum dead load and the prescribed wind loading.

(ii) The roof car and its anchorages shall be capable of resisting accidental over-tensioning of the wire ropes suspending the working platform and this calculated value shall include the effect of one and one-half times the value. For this calculation, the simultaneous effect of one-half wind load shall be included, and the design stresses shall not exceed those referred to in paragraph (b)(1) of this appendix.

(iii) If the load on the motors is at any time in excess of three times that required for lifting the working platform with its rated load the motor shall stall.

(4) *Access to the roof car.* Safe access to the roof car and from the roof car to the working platform shall be provided. If the access to the roof car at any point of its travel is not over the roof area or where otherwise necessary for safety, self-closing, self-locking gates shall be provided. Applicable provisions of the American National Standard Safety Requirements for Floor and Wall Openings, Railings and Toeboard, A12.1-1967, shall apply.

(5) *Means for maintenance, repair, and storage.* Means shall be provided to run the roof car away from the roof perimeter, where necessary, and to provide a safe area for maintenance, repairs, and storage. Provisions shall be made to secure the machine in the stored position. For stored machines subject to wind forces, see special design and anchorage requirements for "wind forces" in Part II, section 10.5.1.1 of ANSI A120.1-1970 American National Standard Safety Requirements for Powered Platforms for Exterior Building Maintenance.

(6) *General requirements for working platforms.* The working platform shall be of girder or truss construction and shall be adequate to support its rated load under any position of loading, and comply with the provisions set forth in section 10 of ANSI A120.1-1970, American National Standard Safety Requirements for Powered Platforms for Exterior Building Maintenance.

(7) *Load rating plate.* Each working platform shall bear a manufacturer's load rating plate, conspicuously posted; stating the maximum permissible rated load. Load rating plates shall be made of noncorrosive material and shall have letters and figures stamped, etched, or cast on the surface. The minimum height of the letters and figures shall be one-fourth inch.

(8) *Minimum size.* The working platform shall have a minimum net width of 24 inches.

(9) *Guardrails.* Working platforms shall be furnished with permanent guard rails not less than 36 inches high, and not more than 42 inches high at the front (building side). At the rear, and on the sides, the rail shall not be less than 42 inches high. An intermediate guardrail shall be provided around the entire platform between the top guardrail and the toeboard.

(10) *Toeboards.* A four-inch toeboard shall be provided along all sides of the working platform.

(11) *Open spaces between guardrails and toeboards.* The spaces between the intermediate guardrail and platform toeboard on the building side of the working platform, and between the top guardrail and the toeboard on other sides of the platform, shall be filled with metallic mesh or similar material that will reject a ball one inch in diameter. The installed mesh shall be capable of withstanding a load of 100 pounds applied horizontally over any area of 144 square inches. If the space between the platform and the building face does not exceed eight inches, and the platform is restrained by guides, the mesh may be omitted on the front side.

(12) *Flooring.* The platform flooring shall be of the nonskid type, and if of open construction, shall reject a 9/16-inch diameter ball, or be provided with a screen below the floor to reject a 9/16-inch diameter ball.

(13) *Access gates.* Where access gates are provided, they shall be self-closing and self-locking.

(14) *Operating device for vertical movement of the working platform.* (i) The normal operating device for the working platform shall be located on the working platform and shall be of the continuous pressure weatherproof electric type.

(ii) The operating device shall be operable only when all electrical protective devices and interlocks on the working platform are in position for normal service and, the roof car, if provided, is at an established operating point.

(15) *Emergency electric operative device.* (i) In addition, on roof-powered platforms, an emergency electric operating device shall be provided near the hoisting machine for use in the event of failure of the normal operating device for the working platform, or failure of the traveling cable system. The emergency operating device shall be mounted in a locked compartment and shall have a legend mounted thereon reading: "For Emergency Operation Only. Establish Communication With Personnel on Working Platform Before Use."

(ii) A key for unlocking the compartment housing the emergency operating device shall be mounted in a break-glass receptacle located near the emergency operating device.

(16) *Manual cranking for emergency operation.* Emergency operation of the main drive machine may be provided to allow manual cranking. This provision for manual operation shall be designed so that not more than two persons will be required to perform this operation. The access to this provision shall include a means to automatically make the machine inoperative electrically while under the emergency manual operation. The design shall be such that the emergency brake is operative at or below governor tripping speed during manual operation.

(17) *Arrangement and guarding of hoisting equipment.*

(i) Hoisting equipment shall consist of a power-driven drum or drum contained in the roof car (roof-powered platforms) or contained on the working platform (self-powered platform).

(ii) The hoisting equipment shall be power-operated in both up and down directions.

(iii) Guard or other protective devices shall be installed wherever rotating shafts or other mechanisms or gears may expose personnel to a hazard.

(iv) Friction devices or clutches shall not be used for connecting the main driving mechanism to the drum or drums. Belt or chain-driven machines are prohibited.

(18) *Hoisting motors.* (i) Hoisting motors shall be electric and of weatherproof construction.

(ii) Hoisting motors shall be in conformance with applicable provisions of paragraph (c)(22) of this appendix, Electric Wiring and Equipment.

(iii) Hoisting motors shall be directly connected to the hoisting machinery. Motor couplings, if used, shall be of steel construction.

(19) *Brakes.* The hoisting machine(s) shall have two independent braking means, each designed to stop and hold the working platform with 125 percent of rated load.

(20) *Hoisting ropes and rope connections.* (i) Working platforms shall be suspended by wire ropes of either 6×19 or 6×37 classification, preformed or nonpreformed.

(ii) [Reserved]

(iii) The minimum factor of safety shall be 10, and shall be calculated by the following formula:

$$F = S \times N/W$$

Where

S = Manufacturer's rated breaking strength of one rope.

N = Number of ropes under load.

W = Maximum static load on all ropes with the platform and its rated load at any point of its travel.

(iv) Hoisting ropes shall be sized to conform with the required factor of safety, but in no case shall the size be less than 5/16 inch diameter.

(v) Winding drums shall have at least three turns of rope remaining when the platform has landed at the lowest possible point of its travel.

(vi) The lengthening or repairing of wire rope by the joining of two or more lengths is prohibited.

(vii) The nondrum ends of the hoisting ropes shall be provided with individual shackle rods which will permit individual adjustment of rope lengths, if required.

(viii) More than two reverse bends in each rope is prohibited.

(21) *Rope tag data.* (i) A metal data tag shall be securely attached to one of the wire rope fastenings. This data tag shall bear the following wire rope data:

(a) The diameter in inches.

(b) Construction classification.

(c) Whether nonpreformed or preformed.

(d) The grade of material used.

(e) The manufacturer's rated breaking strength.

(f) Name of the manufacturer of the rope.

(g) The month and year the ropes were installed.

(22) *Electrical wiring and equipment.* (i) All electrical equipment and wiring shall conform to the requirements of the National Electrical Code, NFPA 70-1971; ANSI C1-1971 (Rev. of C1-1968), except as modified by ANSI A120.1-1970 "American National Standard Safety Requirements for Powered Platforms for Exterior Building Maintenance." For detail design specifications for electrical equipment, see Part 2, ANSI A120.1-1970.

(ii) All motors and operation and control equipment shall be supplied from a single power source.

(iii) The power supply for the powered platform shall be an independent circuit supplied through a fused disconnect switch.

(iv) Electrical conductor parts of the power supply system shall be protected against accidental contact.

(v) Electrical grounding shall be provided.

(a) Provisions for electrical grounding shall be included with the power-supply system.

(b) Controller cabinets, motor frames, hoisting machines, the working platform, roof car and roof car track system, and noncurrent carrying parts of electrical equipment, where provided, shall be grounded.

(c) The controller, where used, shall be so designed and installed that a single ground or short circuit will not prevent both the normal and final stopping device from stopping the working platform.

(d) Means shall be provided on the roof car and working platform for grounding portable electric tools.

(e) The working platform shall be grounded through a grounding connection in a traveling cable. Electrically powered tools utilized on the working platform shall be grounded.

(vi) Electrical receptacles located on the roof or other exterior location shall be of a weatherproof type and shall be located so as not to be subject to contact with water or accumulated snow. The receptacles shall be grounded and the electric cable shall include a grounding conductor. The receptacle and plug shall be a type designed to avoid hazard to persons inserting or withdrawing the plug. Provision shall be made to prevent application of cable strain directly to the plug and receptacle.

(vii) Electric runway conductor systems shall be of the type designed for use in exterior locations and shall be located so as not to be subject to contact with water or accumulated snow. The conductors, collectors, and disconnecting means shall conform to the same requirements as those for cranes and hoists in Article 610 of the National Electrical Code, NFPA 70-1971; ANSI C1-1971 (Rev. of C1-1968). A grounded conductor shall parallel the power conductors and be so connected that it cannot be opened by the disconnecting means. The system shall be designed to avoid hazard to persons in the area.

(viii) Electrical protective devices and interlocks of the weatherproof type shall be provided.

(ix) Where the installation includes a roof car, electric contact(s) shall be provided and so connected that the operating devices for the working platform shall be operative only when the roof car is located and mechanically retained at an established operating point.

(x) Where the powered platform includes a powered-operated roof car, the operating device for the roof car shall be inoperative when the roof car is mechanically retained at an established operating point.

(xi) An electric contact shall be provided and so connected that it will cause the down direction relay for vertical travel to open if the tension in the traveling cable exceeds safe limits.

(xii) An automatic overload device shall be provided to cut off the electrical power to the circuit in all hoisting motors for travel in the up direction, should the load applied to the hoisting ropes at either end of the working platform exceed 125 percent of its normal tension with rated load, as shown on the manufacturer's data plate on the working platform.

(xiii) An automatic device shall be provided for each hoisting rope which will cut off the electrical power to the hoisting motor or motors in the down direction and apply the brakes if any hoisting rope becomes slack.

(xiv) Upper and lower directional limit devices shall be provided to prevent the travel of the working platform beyond the normal upper and lower limits of travel.

(xv) Operation of a directional limit device shall prevent further motion in the appropriate direction, if the normal limit of travel has been reached.

(xvi) Directional limit devices, if driven from the hoisting machine by chains, tapes, or cables, shall incorporate a device to disconnect the electric power from the hoisting machine and apply both the primary and secondary brakes in the event of failure of the driving means.

(xvii) Final terminal stopping devices of the working platform:

(a) Final terminal stopping devices for the working platform shall be provided as a secondary means of preventing the working platform from over-traveling at the terminals.

(b) The device shall be set to function as close to each terminal landing as practical, but in such a way that under normal operating conditions it will not function when the working platform is stopped by the normal terminal stopping device.

(c) Operation of the final terminal stopping device shall open the potential relay for vertical travel, thereby disconnecting the electric power from the hoisting machine, and applying both the primary and secondary brakes.

(d) The final terminal stopping device for the upper limit of travel shall be mounted so that it is operated directly by the motion of the working platform itself.

(xviii) Emergency stop switches shall be provided in or adjacent to each operating device.

(xix) Emergency stop switches shall:

(a) Have red operating buttons or handles.

(b) Be conspicuously and permanently marked "Stop."

(c) Be the manually opened and manually closed type.

(d) Be positively opened with the opening not solely dependent on springs.

(xx) The manual operation of an emergency stop switch associated with an operating device for the working platform shall open the potential relay for vertical travel, thereby disconnecting the electric power from the hoisting machine and applying both the primary and secondary brakes.

(xxi) The manual operation of the emergency stop switch associated with the operating device for a power-driven roof car shall cause the electrical power to the traverse machine to be interrupted, and the traverse machine brake to apply.

(23) *Requirements for emergency communications.* (i) Communication equipment shall be provided for each powered platform for use in an emergency.

(ii) Two-way communication shall be established between personnel on the roof and personnel on the stalled working platform before any emergency operation of the working platform is undertaken by personnel on the roof.

(iii) The equipment shall permit two-way voice communication between the working platform and

(a) Designated personnel continuously available while the powered platform is in use; and

(b) Designated personnel on roof-powered platforms, undertaking emergency operation of the working platform by means of the emergency operating device located near the hoisting machine.

(iv) The emergency communication equipment shall be one of the following types:

(a) Telephone connected to the central telephone exchange system; or

(b) Telephones on a limited system or an approved two-way radio system, provided designated personnel are available to receive a message during the time the powered platform is in use.

(d) *Type T powered platforms*—(1) *Roof car.* The requirements of paragraphs (c)(1) through (c)(5) of this appendix shall apply to Type T powered platforms.

(2) *Working platform.* The requirements of paragraphs (c)(6) through (c)(16) of this appendix apply to Type T powered platforms.

(i) The working platform shall be suspended by at least two wire ropes.

(ii) The maximum rated speed at which the working platform of self-powered platforms may be moved in a vertical direction shall not exceed 35 feet per minute.

(3) *Hoisting equipment.* The requirements of paragraphs (c) (17) and (18) of this appendix shall apply to Type T powered platforms.

(4) *Brakes.* Brakes requirements of paragraph (c)(19) of this appendix shall apply.

(5) *Hoisting ropes and rope connections.* (i) Paragraphs (c)(20) (i) through (vi) and (viii) of this appendix shall apply to Type T powered platforms.

(ii) Adjustable shackle rods in subparagraph (c)(20)(vii) of this appendix shall apply to Type T powered platforms, if the working platform is suspended by more than two wire ropes.

(6) *Electrical wiring and equipment.* (i) The requirements of paragraphs (c)(22) (i) through (vi) of this appendix shall apply to Type T powered platforms. "Circuit protection limitation," "powered platform electrical service system," all operating services and control equipment shall comply with the specifications contained in Part 2, section 26, ANSI A120.1-1970.

(ii) For electrical protective devices the requirements of paragraphs (c)(22) (i) through (viii) of this appendix shall apply to Type T powered platforms. Requirements for the "circuit potential limitation" shall be in accordance with specifications contained in Part 2, section 26, of ANSI A120.1-1970.

(7) *Emergency communications.* All the requirements of paragraph (c)(23) of this appendix shall apply to Type T powered platforms.

[54 FR 31456, July 28, 1989, as amended at 61 FR 9235, Mar. 7, 1996]

## §1910.67 Vehicle-mounted elevating and rotating work platforms.

**1910.67(a)** *Definitions applicable to this section*—(1) *Aerial device.* Any vehicle—mounted device, telescoping or articulating, or both, which is used to position personnel.

**1910.67(a)(2)** *Aerial ladder.* An aerial device consisting of a single- or multiple-section extensible ladder.

**1910.67(a)(3)** *Articulating boom platform.* An aerial device with two or more hinged boom sections.

**1910.67(a)(4)** *Extensible boom platform.* An aerial device (except ladders) with a telescopic or extensible boom. Telescopic derricks with personnel platform attachments shall be considered to be extensible boom platforms when used with a personnel platform.

**1910.67(a)(5)** *Insulated aerial device.* An aerial device designed for work on energized lines and apparatus.

**1910.67(a)(6)** *Mobile unit.* A combination of an aerial device, its vehicle, and related equipment.

**1910.67(a)(7)** *Platform.* Any personnel-carrying device (basket or bucket) which is a component of an aerial device.

**1910.67(a)(8)** *Vehicle.* Any carrier that is not manually propelled.

**1910.67(a)(9)** *Vertical tower.* An aerial device designed to elevate a platform in a substantially vertical axis.

**1910.67(b)** *General requirements.* (1) Unless otherwise provided in this section, aerial devices (aerial lifts) acquired on or after July 1, 1975, shall be designed and constructed in conformance with the applicable requirements of the American National Standard for "Vehicle Mounted Elevating and Rotating Work Platforms," ANSI A92.2—1969, including appendix, which is incorporated by reference as specified in §1910.6. Aerial lifts acquired for use before July 1, 1975 which do not meet the requirements of ANSI A92.2—1969, may not be used after July 1, 1976, unless they shall have been modified so as to conform with the applicable design and construction requirements of ANSI A92.2—1969. Aerial devices include the following types of vehicle-mounted aerial devices used to elevate personnel to jobsites above ground: (i) Extensible boom platforms, (ii) aerial ladders, (iii) articulating boom platforms, (iv) vertical towers, and (v) a combination of any of the above. Aerial equipment may be made of metal, wood, fiberglass reinforced plastic (FRP), or other material; may be powered or manually operated; and are deemed to be aerial lifts whether or not they are capable of rotating about a substantially vertical axis.

**1910.67(b)(2)** Aerial lifts may be "field modified" for uses other than those intended by the manufacturer, provided the modification has been certified in writing by the manufacturer or by any other equivalent entity, such as a nationally recognized testing laboratory, to be in conformity with all applicable provisions of ANSI A92.2—1969 and this section, and to be at least as safe as the equipment was before modification.

**1910.67(b)(3)** The requirements of this section do not apply to firefighting equipment or to the vehicles upon which aerial devices are mounted, except with respect to the requirement that a vehicle be a stable support for the aerial device.

**1910.67(b)(4)** For operations near overhead electric lines, see §1910.333(c)(3).

**1910.67(c)** *Specific requirements*—(1) *Ladder trucks and tower trucks.* Before the truck is moved for highway travel, aerial ladders shall be secured in the lower traveling position by the locking device above the truck cab, and the manually operated device at the base of the ladder, or by other equally effective means (e.g., cradles which prevent rotation of the ladder in combination with positive acting linear actuators).

**1910.67(c)(2)** *Extensible and articulating boom platforms.* (i) Lift controls shall be tested each day prior to use to determine that such controls are in safe working condition.

**1910.67(c)(2)(ii)** Only trained persons shall operate an aerial lift.

**1910.67(c)(2)(iii)** Belting off to an adjacent pole, structure, or equipment while working from an aerial lift shall not be permitted.

**1910.67(c)(2)(iv)** Employees shall always stand firmly on the floor of the basket, and shall not sit or climb on the edge of the basket or use planks, ladders, or other devices for a work position.

**1910.67(c)(2)(v)** A body belt shall be worn and a lanyard attached to the boom or basket when working from an aerial lift.

**1910.67(c)(2)(vi)** Boom and basket load limits specified by the manufacturer shall not be exceeded.

**1910.67(c)(2)(vii)** The brakes shall be set and outriggers, when used, shall be positioned on pads or a solid surface. Wheel chocks shall be installed before using an aerial lift on an incline.

**1910.67(c)(2)(viii)** An aerial lift truck may not be moved when the boom is elevated in a working position with men in the basket, except for equipment which is specifically designed for this type of operation in accordance with the provisions of paragraphs (b)(1) and (b)(2) of this section.

**1910.67(c)(2)(ix)** Articulating boom and extensible boom platforms, primarily designed as personnel carriers, shall have both platform (upper) and lower controls. Upper controls shall be in or beside the platform within easy reach of the operator. Lower controls shall provide for overriding the upper controls. Controls shall be plainly marked as to their function. Lower level controls shall not be operated unless permission has been obtained from the employee in the lift, except in case of emergency.

**1910.67(c)(2)(x)** Climbers shall not be worn while performing work from an aerial lift.

**1910.67(c)(2)(xi)** The insulated portion of an aerial lift shall not be altered in any manner that might reduce its insulating value.

**1910.67(c)(2)(xii)** Before moving an aerial lift for travel, the boom(s) shall be inspected to see that it is properly cradled and outriggers are in stowed position, except as provided in paragraph (c)(2)(viii) of this section.

**1910.67(c)(3)** *Electrical tests.* Electrical tests shall be made in conformance with the requirements of ANSI A92.2—1969, Section 5. However, equivalent DC voltage tests may be used in lieu of the AC voltage test specified in A92.2—1969. DC voltage tests which are approved by the equipment manufacturer or equivalent entity shall be considered an equivalent test for the purpose of this paragraph (c)(3).

**1910.67(c)(4)** *Bursting safety factor.* All critical hydraulic and pneumatic components shall comply with the provisions of the American National Standards Institute standard, ANSI A92.2—1969, Section 4.9 Bursting Safety

# OSHA Standards for General Industry

Factor. Critical components are those in which a failure would result in a free fall or free rotation of the boom. All noncritical components shall have a bursting safety factor of at least two to one.

**1910.67(c)(5)** *Welding standards.* All welding shall conform to the following Automotive Welding Society (AWS) Standards, which are incorporated by reference as specified in §1910.6, as applicable:

**1910.67(c)(5)(i)** Standard Qualification Procedure, AWS B3.0—41.

**1910.67(c)(5)(ii)** Recommended Practices for Automotive Welding Design, AWS D8.4-61.

**1910.67(c)(5)(iii)** Standard Qualification of Welding Procedures and Welders for Piping and Tubing, AWS D10.9-69.

**1910.67(c)(5)(iv)** Specifications for Welding Highway and Railway Bridges, AWS D2.0-69.

[39 FR 23502, June 27, 1974, as amended at 40 FR 13439, Mar. 26, 1975; 55 FR 32014, Aug. 6, 1990; 61 FR 9235, Mar. 7, 1996]

## §1910.68 Manlifts.

**1910.68(a)** *Definitions applicable to this section—***(1)** *Handhold (Handgrip).* A handhold is a device attached to the belt which can be grasped by the passenger to provide a means of maintaining balance.

**1910.68(a)(2)** *Open type.* One which has a handgrip surface fully exposed and capable of being encircled by the passenger's fingers.

**1910.68(a)(3)** *Closed type.* A cup-shaped device, open at the top in the direction of travel of the step for which it is to be used, and closed at the bottom, into which the passenger may place his fingers.

**1910.68(a)(4)** *Limit switch.* A device, the purpose of which is to cut off the power to the motor and apply the brake to stop the carrier in the event that a loaded step passes the terminal landing.

**1910.68(a)(5)** *Manlift.* A device consisting of a power-driven endless belt moving in one direction only, and provided with steps or platforms and handholds attached to it for the transportation of personnel from floor to floor.

**1910.68(a)(6)** *Rated speed.* Rated speed is the speed for which the device is designed and installed.

**1910.68(a)(7)** *Split-rail switch.* An electric limit switch operated mechanically by the rollers on the manlift steps. It consists of an additional hinged or "split" rail, mounted on the regular guide rail, over which the step rollers pass. It is springloaded in the "split" position. If the step supports no load, the rollers will "bump" over the switch; if a loaded step should pass over the section, the split rail will be forced straight, tripping the switch and opening the electrical circuit.

**1910.68(a)(8)** *Step (platform).* A step is a passenger carrying unit.

**1910.68(a)(9)** *Travel.* The travel is the distance between the centers of the top and bottom pulleys.

**1910.68(b)** *General requirements—***(1)** *Application.* This section applies to the construction, maintenance, inspection, and operation of manlifts in relation to accident hazards. Manlifts covered by this section consist of platforms or brackets and accompanying handholds mounted on, or attached to an endless belt, operating vertically in one direction only and being supported by, and driven through pulleys, at the top and bottom. These manlifts are intended for conveyance of persons only. It is not intended that this section cover moving stairways, elevators with enclosed platforms ("Paternoster" elevators), gravity lifts, nor conveyors used only for conveying material. This section applies to manlifts used to carry only personnel trained and authorized by the employer in their use.

**1910.68(b)(2)** *Purpose.* The purpose of this section is to provide reasonable safety for life and limb.

**1910.68(b)(3)** *Design requirements.* All new manlift installations and equipment installed after the effective date of these regulations shall meet the design requirements of the "American National Safety Standard for Manlifts ANSI A90.1-1969", which is incorporated by reference as specified in §1910.6, and the requirements of this section.

**1910.68(b)(4)** *Reference to other codes and subparts.* The following codes, and subparts of this part, are applicable to this section: Safety Code for Mechanical Power Transmission Apparatus, ANSI B15.1-1953 (R 1958), and subpart O; subpart S; Safety Code for Fixed Ladders, ANSI A14.3-1956, and Safety Requirements for Floor and Wall Openings, Railings and Toeboards, ANSI A12.1-1967, and subpart D. The preceding ANSI standards are incorporated by reference as specified in §1910.6.

**1910.68(b)(5)** *Floor openings—***(i)** *Allowable size.* Floor openings for both the "up" and "down" runs shall be not less than 28 inches nor more than 36 inches in width for a 12-inch belt; not less than 34 inches nor more than 38 inches for a 14-inch belt; and not less than 36 inches nor more than 40 inches for a 16-inch belt and shall extend not less than 24 inches, nor more than 28 inches from the face of the belt.

**1910.68(b)(5)(ii)** *Uniformity.* All floor openings for a given manlift shall be uniform in size and shall be approximately circular, and each shall be located vertically above the opening below it.

**1910.68(b)(6)** *Landing—***(i)** *Vertical clearance.* The clearance between the floor or mounting platform and the lower edge for the conical guard above it required by subparagraph (7) of this paragraph shall not be less than 7 feet 6 inches. Where this clearance cannot be obtained no access to the manlift shall be provided and the manlift runway shall be enclosed where it passes through such floor.

**1910.68(b)(6)(ii)** *Clear landing space.* The landing space adjacent to the floor openings shall be free from obstruction and kept clear at all times. This landing space shall be at least 2 feet in width from the edge of the floor opening used for mounting and dismounting.

**1910.68(b)(6)(iii)** *Lighting and landing.* Adequate lighting, not less than 5-foot candles, shall be provided at each floor landing at all times when the lift is in operation.

**1910.68(b)(6)(iv)** *Landing surface.* The landing surfaces at the entrances and exits to the manlift shall be constructed and maintained as to provide safe footing at all times.

**1910.68(b)(6)(v)** *Emergency landings.* Where there is a travel of 50 feet or more between floor landings, one or more emergency landings shall be provided so that there will be a landing (either floor or emergency) for every 25 feet or less of manlift travel.

**1910.68(b)(6)(v)(a)** Emergency landings shall be accessible from both the "up" and "down" rungs of the manlift and shall give access to the ladder required in subparagraph (12) of this paragraph.

**1910.68(b)(6)(v)(b)** Emergency landings shall be completely enclosed with a standard railing and toeboard.

**1910.68(b)(6)(v)(c)** Platforms constructed to give access to bucket elevators or other equipment for the purpose of inspection, lubrication, and repair may also serve as emergency landings under this rule. All such platforms will then be considered part of the emergency landing and shall be provided with standard railings and toeboards.

**1910.68(b)(7)** *Guards on underside of floor openings—***(i)** *Fixed type.* On the ascending side of the manlift floor openings shall be provided with a bevel guard or cone meeting the following requirements:

**1910.68(b)(7)(i)(a)** The cone shall make an angle of not less than 45° with the horizontal. An angle of 60° or greater shall be used where ceiling heights permit.

**1910.68(b)(7)(i)(b)** The lower edge of this guard shall extend at least 42 inches outward from any handhold on the belt. It shall not extend beyond the upper surface of the floor above.

**1910.68(b)(7)(i)(c)** The cone shall be made of not less than No. 18 U.S. gauge sheet steel or material of equivalent strength or stiffness. The lower edge shall be rolled to a minimum diameter of one-half inch and the interior shall be smooth with no rivets, bolts or screws protruding.

**1910.68(b)(7)(ii)** *Floating type.* In lieu of the fixed guards specified in subdivision (i) of this subparagraph a floating type safety cone may be used, such floating cones to be mounted on hinges at least 6 inches below the underside of the floor and so constructed as to actuate a limit switch should a force of 2 pounds be applied on the edge of the cone closest to the hinge. The depth of this floating cone need not exceed 12 inches.

**1910.68(b)(8)** *Protection of entrances and exits—***(i)** *Guard rail requirement.* The entrances and exits at all floor landings affording access to the manlift shall be guarded by a maze (staggered railing) or a handrail equipped with self-closing gates.

**1910.68(b)(8)(ii)** *Construction.* The rails shall be standard guardrails with toeboards meeting the provisions of the Safety Requirements for Floor and Wall Openings, Railings and Toeboards, ANSI A12.1-1967 and §1910.23.

**1910.68(b)(8)(iii)** *Gates.* Gates, if used, shall open outward and shall be self-closing. Corners of gates shall be rounded.

**1910.68(b)(8)(iv)** *Maze.* Maze or staggered openings shall offer no direct passage between enclosure and outer floor space.

**1910.68(b)(8)(v)** Except where building layout prevents, entrances at all landings shall be in the same relative position.

**1910.68(b)(9)** *Guards for openings—(i) Construction.* The floor opening at each landing shall be guarded on sides not used for entrance or exit by a wall, a railing and toeboard or by panels of wire mesh of suitable strength.

**1910.68(b)(9)(ii)** *Height and location.* Such rails or guards shall be at least 42 inches in height on the up-running side and 66 inches on the down-running side.

**1910.68(b)(10)** *Bottom arrangement—(i) Bottom landing.* At the bottom landing the clear area shall be not smaller than the area enclosed by the guardrails on the floors above, and any wall in front of the down-running side of the belt shall be not less than 48 inches from the face of the belt. This space shall not be encroached upon by stairs or ladders.

**1910.68(b)(10)(ii)** *Location of lower pulley.* The lower (boot) pulley shall be installed so that it is supported by the lowest landing served. The sides of the pulley support shall be guarded to prevent contact with the pulley or the steps.

**1910.68(b)(10)(iii)** *Mounting platform.* A mounting platform shall be provided in front or to one side of the uprun at the lowest landing, unless the floor level is such that the following requirement can be met: The floor or platform shall be at or above the point at which the upper surface of the ascending step completes its turn and assumes a horizontal position.

**1910.68(b)(10)(iv)** *Guardrails.* To guard against persons walking under a descending step, the area on the downside of the manlift shall be guarded in accordance with subparagraph (8) of this paragraph. To guard against a person getting between the mounting platform and an ascending step, the area between the belt and the platform shall be protected by a guardrail.

**1910.68(b)(11)** *Top arrangements—(i) Clearance from floor.* A top clearance shall be provided of at least 11 feet above the top terminal landing. This clearance shall be maintained from a plane through each face of the belt to a vertical cylindrical plane having a diameter 2 feet greater than the diameter of the floor opening, extending upward from the top floor to the ceiling on the up-running side of the belt. No encroachment of structural or machine supporting members within this space will be permitted.

**1910.68(b)(11)(ii)** *Pulley clearance.* (*a*) There shall be a clearance of at least 5 feet between the center of the head pulley shaft and any ceiling obstruction.

**1910.68(b)(11)(ii)(*b*)** The center of the head pulley shaft shall be not less than 6 feet above the top terminal landing.

**1910.68(b)(11)(iii)** *Emergency grab rail.* An emergency grab bar or rail and platform shall be provided at the head pulley when the distance to the head pulley is over 6 feet above the top landing, otherwise only a grab bar or rail is to be provided to permit the rider to swing free should the emergency stops become inoperative.

**1910.68(b)(12)** *Emergency exit ladder.* A fixed metal ladder accessible from both the "up" and "down" run of the manlift shall be provided for the entire travel of the manlift. Such ladder shall be in accordance with the existing ANSI A14.3-1956 Safety Code for Fixed Ladders and §1910.27.

**1910.68(b)(13)** *Superstructure bracing.* Manlift rails shall be secured in such a manner as to avoid spreading, vibration, and misalinement.

**1910.68(b)(14)** *Illumination—(i) General.* Both runs of the manlift shall be illuminated at all times when the lift is in operation. An intensity of not less than 1-foot candle shall be maintained at all points. (However, see subparagraph (6)(iii) of this paragraph for illumination requirements at landings.)

**1910.68(b)(14)(ii)** *Control of illumination.* Lighting of manlift runways shall be by means of circuits permanently tied in to the building circuits (no switches), or shall be controlled by switches at each landing. Where separate switches are provided at each landing, any switch shall turn on all lights necessary to illuminate the entire runway.

**1910.68(b)(15)** *Weather protection.* The entire manlift and its driving mechanism shall be protected from the weather at all times.

**1910.68(c)** *Mechanical requirements—(1) Machines, general—(i) Brakes.* Brakes provided for stopping and holding a manlift shall be inherently self-engaging, by requiring power or force from an external source to cause disengagement. The brake shall be electrically released, and shall be applied to the motor shaft for direct-connected units or to the input shaft for belt-driven units. The brake shall be capable of stopping and holding the manlift when the descending side is loaded with 250 lb on each step.

**1910.68(c)(1)(ii)** *Belt.* (*a*) The belts shall be of hard-woven canvas, rubber-coated canvas, leather, or other material meeting the strength requirements of paragraph (b)(3) of this section and having a coefficient of friction such that when used in conjunction with an adequate tension device it will meet the brake test specified in subdivision (i) of this subparagraph.

**1910.68(c)(1)(ii)(*b*)** The width of the belt shall be not less than 12 inches for a travel not exceeding 100 feet, not less than 14 inches for a travel greater than 100 feet but not exceeding 150 feet and 16 inches for a travel exceeding 150 feet.

**1910.68(c)(1)(ii)(*c*)** A belt that has become torn while in use on a manlift shall not be spliced and put back in service.

**1910.68(c)(2)** *Speed—(i) Maximum speed.* No manlift designed for a speed in excess of 80 feet per minute shall be installed.

**1910.68(c)(2)(ii)** [Reserved]

**1910.68(c)(3)** *Platforms or steps—(i) Minimum depth.* Steps or platforms shall be not less than 12 inches nor more than 14 inches deep, measured from the belt to the edge of the step or platform.

**1910.68(c)(3)(ii)** *Width.* The width of the step or platform shall be not less than the width of the belt to which it is attached.

**1910.68(c)(3)(iii)** *Distance between steps.* The distance between steps shall be equally spaced and not less than 16 feet measured from the upper surface of one step to the upper surface of the next step above it.

**1910.68(c)(3)(iv)** *Angle of step.* The surface of the step shall make approximately a right angle with the "up" and "down" run of the belt, and shall travel in the approximate horizontal position with the "up" and "down" run of the belt.

**1910.68(c)(3)(v)** *Surfaces.* The upper or working surfaces of the step shall be of a material having inherent nonslip characteristics (coefficient of friction not less than 0.5) or shall be covered completely by a nonslip tread securely fastened to it.

**1910.68(c)(3)(vi)** *Strength of step supports.* When subjected to a load of 400 pounds applied at the approximate center of the step, step frames, or supports and their guides shall be of adequate strength to:

**1910.68(c)(3)(vi)(*a*)** Prevent the disengagement of any step roller.

**1910.68(c)(3)(vi)(*b*)** Prevent any appreciable misalinement.

**1910.68(c)(3)(vi)(*c*)** Prevent any visible deformation of the steps or its support.

**1910.68(c)(3)(vii)** *Prohibition of steps without handholds.* No steps shall be provided unless there is a corresponding handhold above or below it meeting the requirements of paragraph (c)(4) of this section. If a step is removed for repairs or permanently, the handholds immediately above and below it shall be removed before the lift is again placed in service.

**1910.68(c)(4)** *Handholds—(i) Location.* Handholds attached to the belt shall be provided and installed so that they are not less than 4 feet nor more than 4 feet 8 inches above the step tread. These shall be so located as to be available on the both "up" and "down" run of the belt.

**1910.68(c)(4)(ii)** *Size.* The grab surface of the handhold shall be not less than 4 1/2 inches in width, not less than 3 inches in depth, and shall provide 2 inches of clearance from the belt. Fastenings for handholds shall be located not less than 1 inch from the edge of the belt.

**1910.68(c)(4)(iii)** *Strength.* The handhold shall be capable of withstanding, without damage, a load of 300 pounds applied parallel to the run of the belt.

**1910.68(c)(4)(iv)** *Prohibition of handhold without steps.* No handhold shall be provided without a corresponding step. If a handhold is removed permanently or temporarily, the corresponding step and handhold for the opposite direction of travel shall also be removed before the lift is again placed in service.

**1910.68(c)(4)(v)** *Type.* All handholds shall be of the closed type.

**1910.68(c)(5)** *Up limit stops—(i) Requirements.* Two separate automatic stop devices shall be provided to cut off the power and apply the brake when a loaded step passes the upper terminal landing. One of these shall consist of a split-rail switch mechanically operated by the step roller and located not more than 6 inches above the top terminal landing. The second automatic stop device may consist of any of the following:

**1910.68(c)(5)(i)(a)** Any split-rail switch placed 6 inches above and on the side opposite the first limit switch.

**1910.68(c)(5)(i)(b)** An electronic device.

**1910.68(c)(5)(i)(c)** A switch actuated by a lever, rod, or plate, the latter to be placed on the "up" side of the head pulley so as to just clear a passing step.

**1910.68(c)(5)(ii)** *Manual reset location.* After the manlift has been stopped by a stop device it shall be necessary to reset the automatic stop manually. The device shall be so located that a person resetting it shall have a clear view of both the "up" and "down" runs of the manlift. It shall not be possible to reset the device from any step or platform.

**1910.68(c)(5)(iii)** *Cut-off point.* The initial limit stop device shall function so that the manlift will be stopped before the loaded step has reached a point 24 inches above the top terminal landing.

**1910.68(c)(5)(iv)** *Electrical requirements.* *(a)* Where such switches open the main motor circuit directly they shall be of the multipole type.

**1910.68(c)(5)(iv)(b)** Where electronic devices are used they shall be so designed and installed that failure will result in shutting off the power to the driving motor.

**1910.68(c)(5)(iv)(c)** Where flammable vapors or combustible dusts may be present, electrical installations shall be in accordance with the requirements of subpart S of this part for such locations.

**1910.68(c)(5)(iv)(d)** Unless of the oil-immersed type controller contacts carrying the main motor current shall be copper to carbon or equal, except where the circuit is broken at two or more points simultaneously.

**1910.68(c)(6)** *Emergency stop*—*(i) General.* An emergency stop means shall be provided.

**1910.68(c)(6)(ii)** *Location.* This stop means shall be within easy reach of the ascending and descending runs of the belt.

**1910.68(c)(6)(iii)** *Operation.* This stop means shall be so connected with the control lever or operating mechanism that it will cut off the power and apply the brake when pulled in the direction of travel.

**1910.68(c)(6)(iv)** *Rope.* If rope is used, it shall be not less than three-eights inch in diameter. Wire rope, unless marlin-covered, shall not be used.

**1910.68(c)(7)** *Instruction and warning signs*—*(i) Instruction signs at landings or belts.* Signs of conspicuous and easily read style giving instructions for the use of the manlift shall be posted at each landing or stenciled on the belt.

**1910.68(c)(7)(i)(a)** [Reserved]

**1910.68(c)(7)(i)(b)** The instructions shall read approximately as follows:

Face the Belt.

Use the Handholds.

To Stop—Pull Rope.

**1910.68(c)(7)(ii)** *Top floor warning sign and light.* *(a)* At the top floor an illuminated sign shall be displayed bearing the following wording:

**"TOP FLOOR—GET OFF"**

Signs shall be in block letters not less than 2 inches in height. This sign shall be located within easy view of an ascending passenger and not more than 2 feet above the top terminal landing.

**1910.68(c)(7)(ii)(b)** In addition to the sign required by paragraph (c)(7)(ii)(*a*) of this section, a red warning light of not less than 40- watt rating shall be provided immediately below the upper landing terminal and so located as to shine in the passenger's face.

**1910.68(c)(7)(iii)** *Visitor warning.* A conspicuous sign having the following legend—**AUTHORIZED PERSONNEL ONLY**—shall be displayed at each landing.

**1910.68(d)** *Operating rules*—*(1) Proper use of manlifts.* No freight, packaged goods, pipe, lumber, or construction materials of any kind shall be handled on any manlift.

**1910.68(d)(2)** [Reserved]

**1910.68(e)** *Periodic inspection*—*(1) Frequency.* All manlifts shall be inspected by a competent designated person at intervals of not more than 30 days. Limit switches shall be checked weekly. Manlifts found to be unsafe shall not be operated until properly repaired.

**1910.68(e)(2)** *Items covered.* This periodic inspection shall cover but is not limited to the following items:

Steps.
Step Fastenings.
Rails.
Rail Supports and Fastenings.
Rollers and Slides.
Belt and Belt Tension.
Handholds and Fastenings.
Floor Landings.
Guardrails.
Lubrication.
Limit Switches.
Warning Signs and Lights.
Illumination.
Drive Pulley.
Bottom (boot) Pulley and Clearance.
Pulley Supports.
Motor.
Driving Mechanism.
Brake.
Electrical Switches.
Vibration and Misalignment.
"Skip" on up or down run when mounting step (indicating worn gears).

**1910.68(e)(3)** *Inspection record.* A certification record shall be kept of each inspection which includes the date of the inspection, the signature of the person who performed the inspection and the serial number, or other identifier, of the manlift which was inspected. This record of inspection shall be made available to the Assistant Secretary of Labor or a duly authorized representative.

[39 FR 23502, June 27, 1974, as amended at 43 FR 49746, Oct. 24, 1978; 51 FR 34560, Sept. 29, 1986; 54 FR 24334, June 7, 1989; 55 FR 32014, Aug. 6, 1990; 61 FR 9235, Mar. 7, 1996]

**§1910.69 Source of standards [Removed].**

[§1910.69 amended at 61 FR 9235, Mar. 7, 1996, effective May 6, 1996.]

**§1910.70 Standards organizations [Removed].**

[§1910.70 amended at 61 FR 9235, Mar. 7, 1996, effective May 6, 1996.]

# Subpart G—Occupational Health and Environmental Control

**Authority:** Sections 4, 6, and 8 of the Occupational Safety and Health Act of 1970 (29 U.S.C. 653, 655, 657); Secretary of Labor's Orders Nos. 12-71 (36 FR 8754), 8-76 (41 FR 25059), 9-83 (48 FR 35736), 1-90 (55 FR 9033), or 6-96 (62 FR 111), as applicable; and 29 CFR part 1911.

## §1910.94 Ventilation.

**1910.94(a)** *Abrasive blasting*—**(1)** *Definitions applicable to this paragraph*—(i) *Abrasive.* A solid substance used in an abrasive blasting operation.

**1910.94(a)(1)(ii)** *Abrasive-blasting respirator.* A respirator constructed so that it covers the wearer's head, neck, and shoulders to protect the wearer from rebounding abrasive.

**1910.94(a)(1)(iii)** *Blast cleaning barrel.* A complete enclosure which rotates on an axis, or which has an internal moving tread to tumble the parts, in order to expose various surfaces of the parts to the action of an automatic blast spray.

**1910.94(a)(1)(iv)** *Blast cleaning room.* A complete enclosure in which blasting operations are performed and where the operator works inside of the room to operate the blasting nozzle and direct the flow of the abrasive material.

**1910.94(a)(1)(v)** *Blasting cabinet.* An enclosure where the operator stands outside and operates the blasting nozzle through an opening or openings in the enclosure.

**1910.94(a)(1)(vi)** *Clean air.* Air of such purity that it will not cause harm or discomfort to an individual if it is inhaled for extended periods of time.

**1910.94(a)(1)(vii)** *Dust collector.* A device or combination of devices for separating dust from the air handled by an exhaust ventilation system.

**1910.94(a)(1)(viii)** *Exhaust ventilation system.* A system for removing contaminated air from a space, comprising two or more of the following elements (*a*) enclosure or hood, (*b*) duct work, (*c*) dust collecting equipment, (*d*) exhauster, and (*e*) discharge stack.

**1910.94(a)(1)(ix)** *Particulate-filter respirator.* An air purifying respirator, commonly referred to as a dust or a fume respirator, which removes most of the dust or fume from the air passing through the device.

**1910.94(a)(1)(x)** *Respirable dust.* Airborne dust in sizes capable of passing through the upper respiratory system to reach the lower lung passages.

**1910.94(a)(1)(xi)** *Rotary blast cleaning table.* An enclosure where the pieces to be cleaned are positioned on a rotating table and are passed automatically through a series of blast sprays.

**1910.94(a)(1)(xii)** *Abrasive blasting.* The forcible application of an abrasive to a surface by pneumatic pressure, hydraulic pressure, or centrifugal force.

**1910.94(a)(2)** *Dust hazards from abrasive blasting.* **(i)** Abrasives and the surface coatings on the materials blasted are shattered and pulverized during blasting operations and the dust formed will contain particles of respirable size. The composition and toxicity of the dust from these sources shall be considered in making an evaluation of the potential health hazards.

**1910.94(a)(2)(ii)** The concentration of respirable dust or fume in the breathing zone of the abrasive-blasting operator or any other worker shall be kept below the levels specified in §1910.1000.

**1910.94(a)(2)(iii)** Organic abrasives which are combustible shall be used only in automatic systems. Where flammable or explosive dust mixtures may be present, the construction of the equipment, including the exhaust system and all electric wiring, shall conform to the requirements of American National Standard Installation of Blower and Exhaust Systems for Dust, Stock, and Vapor Removal or Conveying, Z33.1-1961 (NFPA 91-1961), which is incorporated by reference as specified in §1910.6, and subpart S of this part. The blast nozzle shall be bonded and grounded to prevent the build up of static charges. Where flammable or explosive dust mixtures may be present, the abrasive blasting enclosure, the ducts, and the dust collector shall be constructed with loose panels or explosion venting areas, located on sides away from any occupied area, to provide for pressure relief in case of explosion, following the principles set forth in the National Fire Protection Association Explosion Venting Guide, NFPA 68-1954, which is incorporated by reference as specified in §1910.6.

**1910.94(a)(3)** *Blast-cleaning enclosures.* **(i)** Blast-cleaning enclosures shall be exhaust ventilated in such a way that a continuous inward flow of air will be maintained at all openings in the enclosure during the blasting operation.

**1910.94(a)(3)(i)(a)** All air inlets and access openings shall be baffled or so arranged that by the combination of inward air flow and baffling the escape of abrasive or dust particles into an adjacent work area will be minimized and visible spurts of dust will not be observed.

**1910.94(a)(3)(i)(b)** The rate of exhaust shall be sufficient to provide prompt clearance of the dust-laden air within the enclosure after the cessation of blasting.

**1910.94(a)(3)(i)(c)** Before the enclosure is opened, the blast shall be turned off and the exhaust system shall be run for a sufficient period of time to remove the dusty air within the enclosure.

**1910.94(a)(3)(i)(d)** Safety glass protected by screening shall be used in observation windows, where hard deep-cutting abrasives are used.

**1910.94(a)(3)(i)(e)** Slit abrasive-resistant baffles shall be installed in multiple sets at all small access openings where dust might escape, and shall be inspected regularly and replaced when needed.

**1910.94(a)(3)(i)(e)(1)** Doors shall be flanged and tight when closed.

**1910.94(a)(3)(i)(e)(2)** Doors on blast-cleaning rooms shall be operable from both inside and outside, except that where there is a small operator access door, the large work access door may be closed or opened from the outside only.

**1910.94(a)(3)(i)(e)(2)(ii)** [Reserved]

**1910.94(a)(4)** *Exhaust ventilation systems.* **(i)** The construction, installation, inspection, and maintenance of exhaust systems shall conform to the principles and requirements set forth in American National Standard Fundamentals Governing the Design and Operation of Local Exhaust Systems, Z9.2-1960, and ANSI Z33.1-1961, which is incorporated by reference as specified in §1910.6.

**1910.94(a)(4)(i)(a)** When dust leaks are noted, repairs shall be made as soon as possible.

**1910.94(a)(4)(i)(b)** The static pressure drop at the exhaust ducts leading from the equipment shall be checked when the installation is completed and periodically thereafter to assure continued satisfactory operation. Whenever an appreciable change in the pressure drop indicates a partial blockage, the system shall be cleaned and returned to normal operating condition.

**1910.94(a)(4)(i)(b)(ii)** In installations where the abrasive is recirculated, the exhaust ventilation system for the blasting enclosure shall not be relied upon for the removal of fines from the spent abrasive instead of an abrasive separator. An abrasive separator shall be provided for the purpose.

**1910.94(a)(4)(i)(b)(iii)** The air exhausted from blast-cleaning equipment shall be discharged through dust collecting equipment. Dust collectors shall be set up so that the accumulated dust can be emptied and removed without contaminating other working areas.

**1910.94(a)(5)** *Personal protective equipment.* **(i)** Employers must use only respirators approved by the National Institute for Occupational Safety and Health (NIOSH) under 42 CFR part 84 to protect employees from dusts produced during abrasive-blasting operations.

**1910.94(a)(5)(ii)** Abrasive-blasting respirators shall be worn by all abrasive-blasting operators:

**1910.94(a)(5)(ii)(a)** When working inside of blast-cleaning rooms, or

**1910.94(a)(5)(ii)(b)** When using silica sand in manual blasting operations where the nozzle and blast are not physically separated from the operator in an exhaust ventilated enclosure, or

**1910.94(a)(5)(ii)(c)** Where concentrations of toxic dust dispersed by the abrasive blasting may exceed the limits set in §1910.1000 and the nozzle and blast are not physically separated from the operator in an exhaust-ventilated enclosure.

**1910.94(a)(5)(iii)** Properly fitted particulate-filter respirators, commonly referred to as dust-filter respirators, may be used for short, intermittent, or occasional dust exposures such as cleanup, dumping of dust collectors, or unloading shipments of sand at a receiving point when it is not feasible to control the dust by enclosure, exhaust ventilation, or other means. The respirators used must be approved by NIOSH under 42 CFR part 84 for protection against the specific type of dust encountered.

**1910.94(a)(5)(iii)(a)** *Dust-filter respirators* may be used to protect the operator of outside abrasive-blasting operations where nonsilica abrasives are used on materials having low toxicities.

**1910.94(a)(5)(iii)(b)** Dust-filter respirators shall not be used for continuous protection where silica sand is used as the blasting abrasive, or toxic materials are blasted.

**1910.94(a)(5)(iv)** For employees who use respirators required by this section, the employer must implement a respiratory protection program in accordance with 29 CFR 1910.134.

**1910.94(a)(5)(v)** Operators shall be equipped with heavy canvas or leather gloves and aprons or equivalent protection to protect them from the impact of abrasives. Safety shoes shall be worn to protect against foot injury where heavy pieces of work are handled.

**1910.94(a)(5)(v)(a)** Safety shoes shall conform to the requirements of American National Standard for Men's Safety-Toe Footwear, Z41.1-1967, which is incorporated by reference as specified in §1910.6.

**1910.94(a)(5)(v)(b)** Equipment for protection of the eyes and face shall be supplied to the operator when the respirator design does not provide such protection and to any other personnel working in the vicinity of abrasive blasting operations. This equipment shall conform to the requirements of §1910.133.

**1910.94(a)(6)** *Air supply and air compressors.* Air for abrasive-blasting respirators must be free of harmful quantities of dusts, mists, or noxious gases, and must meet the requirements for supplied-air quality and use specified in 29 CFR 1910.134(i).

**1910.94(a)(7)** *Operational procedures and general safety.* Dust shall not be permitted to accumulate on the floor or on ledges outside of an abrasive-blasting enclosure, and dust spills shall be cleaned up promptly. Aisles and walkways shall be kept clear of steel shot or similar abrasive which may create a slipping hazard.

**1910.94(a)(8)** *Scope.* This paragraph (a) applies to all operations where an abrasive is forcibly applied to a surface by pneumatic or hydraulic pressure, or by centrifugal force. It does not apply to steam blasting, or steam cleaning, or hydraulic cleaning methods where work is done without the aid of abrasives.

**1910.94(b)** *Grinding, polishing, and buffing operations—*(1) *Definitions applicable to this paragraph—*(i) *Abrasive cutting-off wheels.* Organic-bonded wheels, the thickness of which is not more than one forty-eighth of their diameter for those up to, and including, 20 inches in diameter, and not more than one-sixtieth of their diameter for those larger than 20 inches in diameter, used for a multitude of operations variously known as cutting, cutting off, grooving, slotting, coping, and jointing, and the like. The wheels may be "solid" consisting of organic-bonded abrasive material throughout, "steel centered" consisting of a steel disc with a rim of organic-bonded material moulded around the periphery, or of the "inserted tooth" type consisting of a steel disc with organic-bonded abrasive teeth or inserts mechanically secured around the periphery.

**1910.94(b)(1)(ii)** *Belts.* All power-driven, flexible, coated bands used for grinding, polishing, or buffing purposes.

**1910.94(b)(1)(iii)** *Branch pipe.* The part of an exhaust system piping that is connected directly to the hood or enclosure.

**1910.94(b)(1)(iv)** *Cradle.* A movable fixture, upon which the part to be ground or polished is placed.

**1910.94(b)(1)(v)** *Disc wheels.* All power-driven rotatable discs faced with abrasive materials, artificial or natural, and used for grinding or polishing on the side of the assembled disc.

**1910.94(b)(1)(vi)** *Entry loss.* The loss in static pressure caused by air flowing into a duct or hood. It is usually expressed in inches of water gauge.

**1910.94(b)(1)(vii)** *Exhaust system.* A system consisting of branch pipes connected to hoods or enclosures, one or more header pipes, an exhaust fan, means for separating solid contaminants from the air flowing in the system, and a discharge stack to outside.

**1910.94(b)(1)(viii)** *Grinding wheels.* All power-driven rotatable grinding or abrasive wheels, except disc wheels as defined in this standard, consisting of abrasive particles held together by artificial or natural bonds and used for peripheral grinding.

**1910.94(b)(1)(ix)** *Header pipe (main pipe).* A pipe into which one or more branch pipes enter and which connects such branch pipes to the remainder of the exhaust system.

**1910.94(b)(1)(x)** *Hoods and enclosures.* The partial or complete enclosure around the wheel or disc through which air enters an exhaust system during operation.

**1910.94(b)(1)(xi)** *Horizontal double-spindle disc grinder.* A grinding machine carrying two power-driven, rotatable, coaxial, horizontal spindles upon the inside ends of which are mounted abrasive disc wheels used for grinding two surfaces simultaneously.

**1910.94(b)(1)(xii)** *Horizontal single-spindle disc grinder.* A grinding machine carrying an abrasive disc wheel upon one or both ends of a power-driven, rotatable single horizontal spindle.

**1910.94(b)(1)(xiii)** *Polishing and buffing wheels.* All power-driven rotatable wheels composed all or in part of textile fabrics, wood, felt, leather, paper, and may be coated with abrasives on the periphery of the wheel for purposes of polishing, buffing, and light grinding.

**1910.94(b)(1)(xiv)** *Portable grinder.* Any power-driven rotatable grinding, polishing, or buffing wheel mounted in such manner that it may be manually manipulated.

**1910.94(b)(1)(xv)** *Scratch brush wheels.* All power-driven rotatable wheels made from wire or bristles, and used for scratch cleaning and brushing purposes.

**1910.94(b)(1)(xvi)** *Swing-frame grinder.* Any power-driven rotatable grinding, polishing, or buffing wheel mounted in such a manner that the wheel with its supporting framework can be manipulated over stationary objects.

**1910.94(b)(1)(xvii)** *Velocity pressure (vp).* The kinetic pressure in the direction of flow necessary to cause a fluid at rest to flow at a given velocity. It is usually expressed in inches of water gauge.

**1910.94(b)(1)(xviii)** *Vertical spindle disc grinder.* A grinding machine having a vertical, rotatable power-driven spindle carrying a horizontal abrasive disc wheel.

**1910.94(b)(2)** *Application.* Wherever dry grinding, dry polishing or buffing is performed, and employee exposure, without regard to the use of respirators, exceeds the permissible exposure limits prescribed in §1910.1000 or other sections of this part, a local exhaust ventilation system shall be provided and used to maintain employee exposures within the prescribed limits.

**1910.94(b)(3)** *Hood and branch pipe requirements.* (i) Hoods connected to exhaust systems shall be used, and such hoods shall be designed, located, and placed so that the dust or dirt particles shall fall or be projected into the hoods in the direction of the air flow. No wheels, discs, straps, or belts shall be operated in such manner and in such direction as to cause the dust and dirt particles to be thrown into the operator's breathing zone.

**1910.94(b)(3)(ii)** Grinding wheels on floor stands, pedestals, benches, and special-purpose grinding machines and abrasive cutting-off wheels shall have not less than the minimum exhaust volumes shown in Table G-4 with a recommended minimum duct velocity of 4,500 feet per minute in the branch and 3,500 feet per minute in the main. The entry losses from all hoods except the vertical-spindle disc grinder hood, shall equal 0.65 velocity pressure for a straight takeoff and 0.45 velocity pressure for a tapered takeoff. The entry loss for the vertical-spindle disc grinder hood is shown in figure G-1 (following §1910.94(b)).

### TABLE G-4—GRINDING AND ABRASIVE CUTTING-OFF WHEELS

| Wheel diameter (inches) | Wheel width (inches) | Minimum exhaust volume (feet$^3$/min.) |
|---|---|---|
| To 9 | 1 1/2 | 220 |
| Over 9 to 16 | 2 | 390 |
| Over 16 to 19 | 3 | 500 |
| Over 19 to 24 | 4 | 610 |
| Over 24 to 30 | 5 | 880 |
| Over 30 to 36 | 6 | 1,200 |

For any wheel wider than wheel diameters shown in Table G-4, increase the exhaust volume by the ratio of the new width to the width shown.

***Example:*** If wheel width = 4 1/2 inches, then

4.5 ÷ 4 × 610 = 686 (rounded to 690).

# OSHA Standards for General Industry

**1910.94(b)(3)(iii)** Scratch-brush wheels and all buffing and polishing wheels mounted on floor stands, pedestals, benches, or special-purpose machines shall have not less than the minimum exhaust volume shown in Table G-5.

### TABLE G-5—BUFFING AND POLISHING WHEELS

| Wheel diameter (inches) | Wheel width (inches) | Minimum exhaust volume (feet$^3$/min.) |
|---|---|---|
| To 9 | 2 | 300 |
| Over 9 to 16 | 3 | 500 |
| Over 16 to 19 | 4 | 610 |
| Over 19 to 24 | 5 | 740 |
| Over 24 to 30 | 6 | 1,040 |
| Over 30 to 36 | 6 | 1,200 |

**1910.94(b)(3)(iv)** Grinding wheels or discs for horizontal single-spindle disc grinders shall be hooded to collect the dust or dirt generated by the grinding operation and the hoods shall be connected to branch pipes having exhaust volumes as shown in Table G-6.

### TABLE G-6—HORIZONTAL SINGLE-SPINDLE DISC GRINDER

| Disc diameter (inches) | Exhaust volume (ft.$^3$/min.) |
|---|---|
| Up to 12 | 220 |
| Over 12 to 19 | 390 |
| Over 19 to 30 | 610 |
| Over 30 to 36 | 880 |

**1910.94(b)(3)(v)** Grinding wheels or discs for horizontal double-spindle disc grinders shall have a hood enclosing the grinding chamber and the hood shall be connected to one or more branch pipes having exhaust volumes as shown in Table G-7.

### TABLE G-7—HORIZONTAL DOUBLE-SPINDLE DISC GRINDER

| Disc diameter (inches) | Exhaust volume (ft.$^3$/min.) |
|---|---|
| Up to 19 | 610 |
| Over 19 to 25 | 880 |
| Over 25 to 30 | 1,200 |
| Over 30 to 53 | 1,770 |
| Over 53 to 72 | 6,280 |

**1910.94(b)(3)(vi)** Grinding wheels or discs for vertical single-spindle disc grinders shall be encircled with hoods to remove the dust generated in the operation. The hoods shall be connected to one or more branch pipes having exhaust volumes as shown in Table G-8.

### TABLE G-8—VERTICAL SPINDLE DISC GRINDER

| Disc diameter (inches) | One-half or more of disc covered Number[1] | Exhaust foot$^3$/min. | Disc not covered Number[1] | Exhaust foot$^3$/min. |
|---|---|---|---|---|
| Up to 20 | 1 | 500 | 2 | 780 |
| Over 20 to 30 | 2 | 780 | 2 | 1,480 |
| Over 30 to 53 | 2 | 1,770 | 4 | 3,530 |
| Over 53 to 72 | 2 | 3,140 | 5 | 6,010 |

[1] Number of exhaust outlets around periphery of hood, or equal distribution provided by other means.

**1910.94(b)(3)(vii)** Grinding and polishing belts shall be provided with hoods to remove dust and dirt generated in the operations and the hoods shall be connected to branch pipes having exhaust volumes as shown in Table G-9.

### TABLE G-9—GRINDING AND POLISHING BELTS

| Belts width (inches) | Exhaust volume (ft.$^3$/min.) |
|---|---|
| Up to 3 | 220 |
| Over 3 to 5 | 300 |
| Over 5 to 7 | 390 |
| Over 7 to 9 | 500 |
| Over 9 to 11 | 610 |
| Over 11 to 13 | 740 |

**1910.94(b)(3)(viii)** *Cradles and swing-frame grinders.* Where cradles are used for handling the parts to be ground, polished, or buffed, requiring large partial enclosures to house the complete operation, a minimum average air velocity of 150 feet per minute shall be maintained over the entire opening of the enclosure. Swing-frame grinders shall also be exhausted in the same manner as provided for cradles. (See fig. G-3)

**1910.94(b)(3)(ix)** Where the work is outside the hood, air volumes must be increased as shown in American Standard Fundamentals Governing the Design and Operation of Local Exhaust Systems, Z9.2-1960 (section 4, exhaust hoods).

**1910.94(b)(4)** *Exhaust systems.* **(i)** Exhaust systems for grinding, polishing, and buffing operations should be designed in accordance with American Standard Fundamentals Governing the Design and Operation of Local Exhaust Systems, Z9.2-1960.

**1910.94(b)(4)(ii)** Exhaust systems for grinding, polishing, and buffing operations shall be tested in the manner described in American Standard Fundamentals Governing the Design and Operation of Local Exhaust Systems, Z9.2-1960.

**1910.94(b)(4)(iii)** All exhaust systems shall be provided with suitable dust collectors.

**1910.94(b)(5)** *Hood and enclosure design.* **(i)(a)** It is the dual function of grinding and abrasive cutting-off wheel hoods to protect the operator from the hazards of bursting wheels as well as to provide a means for the removal of dust and dirt generated. All hoods shall be not less in structural strength than specified in the American National Standard Safety Code for the Use, Care, and Protection of Abrasive Wheels, B7.1-1970, which is incorporated by reference as specified in §1910.6.

**1910.94(b)(5)(i)(b)** Due to the variety of work and types of grinding machines employed, it is necessary to develop hoods adaptable to the particular machine in question, and such hoods shall be located as close as possible to the operation.

**1910.94(b)(5)(ii)** Exhaust hoods for floor stands, pedestals, and bench grinders shall be designed in accordance with figure G-2. The adjustable tongue shown in the figure shall be kept in working order and shall be adjusted within one-fourth inch of the wheel periphery at all times.

**1910.94(b)(5)(iii)** Swing-frame grinders shall be provided with exhaust booths as indicated in figure G-3.

**1910.94(b)(5)(iv)** Portable grinding operations, whenever the nature of the work permits, shall be conducted within a partial enclosure. The opening in the enclosure shall be no larger than is actually required in the operation and an average face air velocity of not less than 200 feet per minute shall be maintained.

**1910.94(b)(5)(v)** Hoods for polishing and buffing and scratch-brush wheels shall be constructed to conform as closely to figure G-4 as the nature of the work will permit.

**1910.94(b)(5)(vi)** Cradle grinding and polishing operations shall be performed within a partial enclosure similar to figure G-5. The operator shall be positioned outside the working face of the opening of the enclosure. The face opening of the enclosure should not be any greater in area than that actually required for the performance of the operation and the average air velocity into the working face of the enclosure shall not be less than 150 feet per minute.

**1910.94(b)(5)(vii)** Hoods for horizontal single-spindle disc grinders shall be constructed to conform as closely as possible to the hood shown in figure G-6. It is essential that there be a space between the back of the wheel and the hood, and a space around the periphery of the wheel of at least 1 inch in order to permit the suction to act around the wheel periphery. The opening on the side of the disc shall be no larger than is required for the grinding operation, but must never be less than twice the area of the branch outlet.

**1910.94(b)(5)(viii)** Horizontal double-spindle disc grinders shall have a hood encircling the wheels and grinding chamber similar to that illustrated in figure G-7. The openings for passing the work into the grinding chamber should be kept as small as possible, but must never be less than twice the area of the branch outlets.

**1910.94(b)(5)(ix)** Vertical-spindle disc grinders shall be encircled with a hood so constructed that the heavy dust is drawn off a surface of the disc and the lighter dust exhausted through a continuous slot at the top of the hood as shown in figure G-1.

**1910.94(b)(5)(x)** Grinding and polishing belt hoods shall be constructed as close to the operation as possible. The hood should extend almost to the belt, and 1-inch wide openings should be provided on either side. Figure G-8 shows a typical hood for a belt operation.

**FIGURE G-1—VERTICAL SPINDLE DISC GRINDER EXHAUST HOOD AND BRANCH PIPE CONNECTIONS**

| Dia. D. inches | | Exhaust E | | Volume Exhausted at 4,500 ft/min ft³/min | Note |
|---|---|---|---|---|---|
| Min. | Max. | No. Pipes | Dia. | | |
| .................. | 20 | 1 | 4 1/4 | 500 | When one-half or more of the disc can be hooded, use exhaust ducts as shown at the left. |
| Over 20 ......... | 30 | 2 | 4 | 780 | |
| Over 30 ......... | 72 | 2 | 6 | 1,770 | |
| Over 53 ......... | 72 | 2 | 8 | 3,140 | |
| .................. | 20 | 2 | 4 | 780 | When no hood can be used over disc, use exhaust ducts as shown at left. |
| Over 20 ......... | 20 | 2 | 4 | 780 | |
| Over 30 ......... | 30 | 2 | 5 1/2 | 1,480 | |
| Over 53 ......... | 53 | 4 | 6 | 3,530 | |
| | 72 | 5 | 7 | 6,010 | |

Entry loss =1.0 slot velocity pressure + 0.5 branch velocity pressure.
Minimum slot velocity = 2,000 ft/min—1/2-inch slot width.

**FIGURE G-2—STANDARD GRINDER HOOD**

| Wheel dimension, inches | | | Exhaust outlet, inches E | Volume of air at 4,500 ft/min |
|---|---|---|---|---|
| Diameter Min = d | Max = D | Width, Max | | |
| ............ | 9 | 1 1/2 | 3 | 220 |
| Over 9 ............ | 16 | 2 | 4 | 390 |
| Over 16 ............ | 19 | 3 | 4 1/2 | 500 |
| Over 19 ............ | 24 | 4 | 5 | 610 |
| Over 24 ............ | 30 | 5 | 6 | 880 |
| Over 30 ............ | 36 | 6 | 7 | 1,200 |

Entry loss = 0.45 velocity pressure for tapered takeoff 0.65 velocity pressure for straight takeoff.

**FIGURE G–3—A METHOD OF APPLYING AN EXHAUST ENCLOSURE TO SWING-FRAME GRINDERS**
**NOTE:** Baffle to reduce front opening as much as possible

**FIGURE G–4—STANDARD BUFFING AND POLISHING HOOD**

| Wheel dimension, inches ||| Exhaust outlet, inches $E$ | Volume of air at 4,500 ft/min |
| Diameter || Width, Max |||
| Min = $d$ | Max = $D$ | | | |
|---|---|---|---|---|
|  | 9 | 2 | 3 1/2 | 300 |
| Over 9 | 16 | 3 | 4 | 500 |
| Over 16 | 19 | 4 | 5 | 610 |
| Over 19 | 24 | 5 | 5 1/2 | 740 |
| Over 24 | 30 | 6 | 6 1/2 | 1,040 |
| Over 30 | 36 | 6 | 7 | 1,200 |

Entry loss = 0.15 velocity pressure for tapered takeoff; 0.65 velocity pressure for straight takeoff.

# OSHA Standards for General Industry

**FIGURE G–5—CRADLE POLISHING OR GRINDING ENCLOSURE**
Entry Loss = 0.45 velocity pressure for tapered takeoff

MINIMUM VELOCITY IN E IS 3,000 FT/MIN
MINIMUM VELOCITY AT ENCLOSURE FACE IS 150 FT/MIN

**FIGURE G–6—HORIZONTAL SINGLE-SPINDLE DISC GRINDER EXHAUST HOOD AND BRANCH PIPE CONNECTIONS**

| Dia D, inches | | Exhaust E, dia. inches | Volume exhausted at 4,500 ft/min. ft³/min. |
|---|---|---|---|
| Min | Max. | | |
| | 12 | 3 | 220 |
| Over 12 | 19 | 4 | 390 |
| Over 19 | 30 | 5 | 610 |
| Over 30 | 36 | 6 | 880 |

**NOTE:** If grinding wheels are used for disc grinding purposes, hoods must conform to structural strength and materials as described in 9.1.

Entry loss = 0.45 velocity pressure for tapered takeoff.

**FIGURE G–7—HORIZONTAL DOUBLE-SPINDLE DISC GRINDER EXHAUST HOOD AND BRANCH PIPE CONNECTIONS**

| Disc dia. inches | | Exhaust E | | Volume exhaust at 4,500 ft/min. ft³/min. | Note |
|---|---|---|---|---|---|
| Min. | Max. | No. Pipes | Dia. | | |
| ............ | 19 | 1 | 5 | 610 | When width "W" permits, exhaust ducts should be as near heaviest grinding as possible. |
| Over 19 ............ | 25 | 1 | 6 | 880 | |
| Over 25 ............ | 30 | 1 | 7 | 1,200 | |
| Over 30 ............ | 53 | 2 | 6 | 1,770 | |
| Over 53 ............ | 72 | 4 | 8 | 6,280 | |

Entry loss = 0.45 velocity pressure for tapered takeoff.

**FIGURE G–8—A TYPICAL HOOD FOR A BELT OPERATION**

| Belt width W. Inches | Exhaust volume. ft.¹/min |
|---|---|
| Up to 3 ............ | 220 |
| 3 to 5 ............ | 300 |
| 5 to 7 ............ | 390 |
| 7 to 9 ............ | 500 |
| 9 to 11 ............ | 610 |
| 11 to 13 ............ | 740 |

Minimum duct velocity = 4,500 ft/min branch, 3,500 ft/min main.

Entry loss = 0.45 velocity pressure for tapered takeoff; 0.65 velocity pressure for straight takeoff.

**1910.94(b)(6)** *Scope.* This paragraph (b), prescribes the use of exhaust hood enclosures and systems in removing dust, dirt, fumes, and gases generated through the grinding, polishing, or buffing of ferrous and nonferrous metals.

**1910.94(c)** *Spray finishing operations*—**(1)** *Definitions applicable to this paragraph*—**(i)** *Spray-finishing operations.* Spray-finishing operations are employment of methods wherein organic or inorganic materials are utilized in dispersed form for deposit on surfaces to be coated, treated, or cleaned. Such methods of deposit may involve either automatic, manual, or electrostatic deposition but do not include metal spraying or metallizing, dipping, flow coating, roller coating, tumbling, centrifuging, or spray washing and degreasing as conducted in self-contained washing and degreasing machines or systems.

**1910.94(c)(1)(ii)** *Spray booth.* Spray booths are defined and described in §1910.107(a). (See sections 103, 104, and 105 of the Standard for Spray Finishing Using Flammable and Combustible Materials, NFPA No. 33-1969, which is incorporated by reference as specified in §1910.6).

**1910.94(c)(1)(iii)** *Spray room.* A spray room is a room in which spray-finishing operations not conducted in a spray booth are performed separately from other areas.

**1910.94(c)(1)(iv)** *Minimum maintained velocity.* Minimum maintained velocity is the velocity of air movement which must be maintained in order to meet minimum specified requirements for health and safety.

**1910.94(c)(2)** *Location and application.* Spray booths or spray rooms are to be used to enclose or confine all operations. Spray-finishing operations shall be located as provided in sections 201 through 206 of the Standard for Spray Finishing Using Flammable and Combustible Materials, NFPA No. 33-1969.

**1910.94(c)(3)** *Design and construction of spray booths.* **(i)** Spray booths shall be designed and constructed in accordance with §1910.107(b) (1) through (4) and (6) through (10) (see sections 301-304 and 306-310 of the Standard for Spray Finishing Using Flammable and Combustible Materials, NFPA No. 33-1969), for general construction specifications. For a more detailed discussion of fundamentals relating to this subject, see ANSI Z9.2-1960

**1910.94(c)(3)(i)(a)** Lights, motors, electrical equipment, and other sources of ignition shall conform to the requirements of §1910.107 (b)(10) and (c). (See section 310 and chapter 4 of the Standard for Spray Finishing Using Flammable and Combustible Materials NFPA No. 33-1969.)

**1910.94(c)(3)(i)(b)** In no case shall combustible material be used in the construction of a spray booth and supply or exhaust duct connected to it.

**1910.94(c)(3)(ii)** Unobstructed walkways shall not be less than 6 1/2 feet high and shall be maintained clear of obstruction from any work location in the booth to a booth exit or open booth front. In booths where the open front is the only exit, such exits shall be not less than 3 feet wide. In booths having multiple exits, such exits shall not be less than 2 feet wide, provided that the maximum distance from the work location to the exit is 25 feet or less. Where booth exits are provided with doors, such doors shall open outward from the booth.

**1910.94(c)(3)(iii)** Baffles, distribution plates, and dry-type overspray collectors shall conform to the requirements of §1910.107(b) (4) and (5). (See sections 304 and 305 of the Standard for Spray Finishing Using Flammable and Combustible Materials, NFPA No. 33-1969.)

**1910.94(c)(3)(iii)(a)** Overspray filters shall be installed and maintained in accordance with the requirements of §1910.107 (b)(5), (see section 305 of the Standard for Spray Finishing Using Flammable and Combustible Materials, NFPA No. 33-1969), and shall only be in a location easily accessible for inspection, cleaning, or replacement.

**1910.94(c)(3)(iii)(b)** Where effective means, independent of the overspray filters, are installed which will result in design air distribution across the booth cross section, it is permissible to operate the booth without the filters in place.

**1910.94(c)(3)(iv)(a)** For wet or water-wash spray booths, the water-chamber enclosure, within which intimate contact of contaminated air and cleaning water or other cleaning medium is maintained, if made of steel, shall be 18 gage or heavier and adequately protected against corrosion.

**1910.94(c)(3)(iv)(b)** Chambers may include scrubber spray nozzles, headers, troughs, or other devices. Chambers shall be provided with adequate means for creating and maintaining scrubbing action for removal of particulate matter from the exhaust air stream.

**1910.94(c)(3)(v)** Collecting tanks shall be of welded steel construction or other suitable non-combustible material. If pits are used as collecting tanks, they shall be concrete, masonry, or other material having similar properties.

**1910.94(c)(3)(v)(a)** Tanks shall be provided with weirs, skimmer plates, or screens to prevent sludge and floating paint from entering the pump suction box. Means for automatically maintaining the proper water level shall also be provided. Fresh water inlets shall not be submerged. They shall terminate at least one pipe diameter above the safety overflow level of the tank.

**1910.94(c)(3)(v)(b)** Tanks shall be so constructed as to discourage accumulation of hazardous deposits.

**1910.94(c)(3)(vi)** Pump manifolds, risers, and headers shall be adequately sized to insure sufficient water flow to provide efficient operation of the water chamber.

**1910.94(c)(4)** *Design and construction of spray rooms.* **(i)** Spray rooms, including floors, shall be constructed of masonry, concrete, or other noncombustible material.

**1910.94(c)(4)(ii)** Spray rooms shall have noncombustible fire doors and shutters.

**1910.94(c)(4)(iii)** Spray rooms shall be adequately ventilated so that the atmosphere in the breathing zone of the operator shall be maintained in accordance with the requirements of paragraph (c)(6)(ii) of this section.

**1910.94(c)(4)(iv)** Spray rooms used for production spray-finishing operations shall conform to the requirements for spray booths.

**1910.94(c)(5)** *Ventilation.* **(i)** Ventilation shall be provided in accordance with provisions of §1910.107(d) (see chapter 5 of the Standard for Spray Finishing Using Flammable or Combustible Materials, NFPA No. 33-1969), and in accordance with the following:

**1910.94(c)(5)(i)(a)** Where a fan plenum is used to equalize or control the distribution of exhaust air movement through the booth, it shall be of sufficient strength or rigidity to withstand the differential air pressure or other superficially imposed loads for which the equipment is designed and also to facilitate cleaning. Construction specifications shall be at least equivalent to those of paragraph (c)(5)(iii) of this section.

**1910.94(c)(5)(i)(b)** [Reserved]

**1910.94(c)(5)(ii)** Inlet or supply ductwork used to transport makeup air to spray booths or surrounding areas shall be constructed of noncombustible materials.

**1910.94(c)(5)(ii)(a)** If negative pressure exists within inlet ductwork, all seams and joints shall be sealed if there is a possibility of infiltration of harmful quantities of noxious gases, fumes, or mists from areas through which ductwork passes.

**1910.94(c)(5)(ii)(b)** Inlet ductwork shall be sized in accordance with volume flow requirements and provide design air requirements at the spray booth.

**1910.94(c)(5)(ii)(c)** Inlet ductwork shall be adequately supported throughout its length to sustain at least its own weight plus any negative pressure which is exerted upon it under normal operating conditions.

**1910.94(c)(5)(iii)(a)** Exhaust ductwork shall be adequately supported throughout its length to sustain its weight plus any normal accumulation in interior during normal operating conditions and any negative pressure exerted upon it.

**1910.94(c)(5)(iii)(b)** Exhaust ductwork shall be sized in accordance with good design practice which shall include consideration of fan capacity, length of duct, number of turns and elbows, variation in size, volume, and character of materials being exhausted. See American National Standard Z9.2-1960 for further details and explanation concerning elements of design.

**1910.94(c)(5)(iii)(c)** Longitudinal joints in sheet steel ductwork shall be either lock-seamed, riveted, or welded. For other than steel construction, equivalent securing of joints shall be provided.

**1910.94(c)(5)(iii)(d)** Circumferential joints in ductwork shall be substantially fastened together and lapped in the direction of airflow. At least every fourth joint shall be provided with connecting flanges, bolted together, or of equivalent fastening security.

**1910.94(c)(5)(iii)(e)** Inspection or clean-out doors shall be provided for every 9 to 12 feet of running length for ducts up to 12 inches in diameter, but the distance between cleanout doors may be greater for larger pipes. (See 8.3.21 of American National Standard Z9.1-1951, which is incorporated by reference as specified in §1910.6.) A clean-out door or doors shall be provided for servicing the fan, and where necessary, a drain shall be provided.

**1910.94(c)(5)(iii)(f)** Where ductwork passes through a combustible roof or wall, the roof or wall shall be protected at the point of penetration by open space or fire-resistive material between the duct and the roof or wall. When ducts pass through firewalls, they shall be provided with automatic fire dampers on both sides of the wall, except that three-eighth-inch steel plates may be used in lieu of automatic fire dampers for ducts not exceeding 18 inches in diameter.

**1910.94(c)(5)(iii)(g)** Ductwork used for ventilating any process covered in this standard shall not be connected to ducts ventilating any other process or any chimney or flue used for conveying any products of combustion.

**1910.94(c)(6)** *Velocity and air flow requirements.* **(i)** Except where a spray booth has an adequate air replacement system, the velocity of air into all openings of a spray booth shall be not less than that specified in Table G-10 for the operating conditions specified. An adequate air replacement system is one which introduces replacement air upstream or above the object being sprayed and is so designed that the velocity of air in the booth cross section is not less than that specified in Table G-10 when measured upstream or above the object being sprayed.

### TABLE G-10—MINIMUM MAINTAINED VELOCITIES INTO SPRAY BOOTHS

| Operating conditions for objects completely inside booth | Crossdraft, f.p.m. | Airflow velocities, f.p.m. Design | Range |
|---|---|---|---|
| Electrostatic and automatic airless operation contained in booth without operator. | Negligible | 50 large booth<br>100 small booth | 50-75<br>75-125 |
| Air-operated guns, manual or automatic | Up to 50 | 100 large booth<br>150 small booth | 75-125<br>125-175 |
| Air-operated guns, manual or automatic | Up to 100 | 150 large booth<br>200 small booth | 125-175<br>150-250 |

**NOTES:**
(1) Attention is invited to the fact that the effectiveness of the spray booth is dependent upon the relationship of the depth of the booth to its height and width.
(2) Crossdrafts can be eliminated through proper design and such design should be sought. Crossdrafts in excess of 100 fpm (feet per minute) should not be permitted.
(3) Excessive air pressures result in loss of both efficiency and material waste in addition to creating a backlash that may carry overspray and fumes into adjacent work areas.
(4) Booths should be designed with velocities shown in the column headed "Design." However, booths operating with velocities shown in the column headed "Range" are in compliance with this standard.

**1910.94(c)(6)(ii)** In addition to the requirements in paragraph (c)(6)(i) of this section the total air volume exhausted through a spray booth shall be such as to dilute solvent vapor to at least 25 percent of the lower explosive limit of the solvent being sprayed. An example of the method of calculating this volume is given below.

**Example:** To determine the lower explosive limits of the most common solvents used in spray finishing, see Table G-11. Column 1 gives the number of cubic feet of vapor per gallon of solvent and column 2 gives the lower explosive limit (LEL) in percentage by volume of air. Note that the quantity of solvent will be diminished by the quantity of solids and nonflammables contained in the finish.

To determine the volume of air in cubic feet necessary to dilute the vapor from 1 gallon of solvent to 25 percent of the lower explosive limit, apply the following formula:

Dilution volume required per gallon of solvent = 4 (100 – LEL) (cubic feet of vapor per gallon) ÷ LEL

Using toluene as the solvent.

(1) LEL of toluene from Table G-11, column 2, is 1.4 percent.
(2) Cubic feet of vapor per gallon from Table G-11, column 1, is 30.4 cubic feet per gallon.
(3) Dilution volume required = 4 (100 – 1.4) 30.4 ÷ 1.4 = 8,564 cubic feet.
(4) To convert to cubic feet per minute of required ventilation, multiply the dilution volume required per gallon of solvent by the number of gallons of solvent evaporated per minute.

### TABLE G-11—LOWER EXPLOSIVE LIMIT OF SOME COMMONLY USED SOLVENTS

| Solvent | Cubic feet per gallon of vapor of liquid at 70 °F. Column 1 | Lower explosive limit in percent by volume of air at 70 °F. Column 2 |
|---|---|---|
| Acetone | 44.0 | 2.6 |
| Amyl Acetate (iso) | 21.6 | [1]1.0 |
| Amyl Alcohol (n) | 29.6 | 1.2 |
| Amyl Alcohol (iso) | 29.6 | 1.2 |
| Benzene | 36.8 | [1]1.4 |
| Butyl Acetate (n) | 24.8 | 1.7 |
| Butyl Alcohol (n) | 35.2 | 1.4 |
| Butyl Cellosolve | 24.8 | 1.1 |
| Cellosolve | 33.6 | 1.8 |
| Cellosolve Acetate | 23.2 | 1.7 |
| Cyclohexanone | 31.2 | [1]1.1 |
| 1,1 Dichloroethylene | 42.4 | 5.9 |
| 1,2 Dichloroethylene | 42.4 | 9.7 |
| Ethyl Acetate | 32.8 | 2.5 |
| Ethyl Alcohol | 55.2 | 4.3 |
| Ethyl Lactate | 28.0 | [1]1.5 |
| Methyl Acetate | 40.0 | 3.1 |
| Methyl Alcohol | 80.8 | 7.3 |
| Methyl Cellosolve | 40.8 | 2.5 |
| Methyl Ethyl Ketone | 36.0 | 1.8 |
| Methyl n-Propyl Ketone | 30.4 | 1.5 |
| Naphtha (VM&P) (76° Naphtha) | 22.4 | 0.9 |
| Naphtha (100° Flash) Safety Solvent—Stoddard Solvent | 23.2 | 1.0 |
| Propyl Acetate (n) | 27.2 | 2.8 |
| Propyl Acetate (iso) | 28.0 | 1.1 |
| Propyl Alcohol (n) | 44.8 | 2.1 |
| Propyl Alcohol (iso) | 44.0 | 2.0 |
| Toluene | 30.4 | 1.4 |
| Turpentine | 20.8 | 0.8 |
| Xylene (o) | 26.4 | 1.0 |

[1] At 212 °F.

**1910.94(c)(6)(iii)(a)** When an operator is in a booth downstream from the object being sprayed, an air-supplied respirator or other type of respirator must be used by employees that has been approved by NIOSH under 42 CFR part 84 for the material being sprayed.

**1910.94(c)(6)(iii)(b)** Where downdraft booths are provided with doors, such doors shall be closed when spray painting.

**1910.94(c)(7)** *Make-up air.* **(i)** Clean fresh air, free of contamination from adjacent industrial exhaust systems, chimneys, stacks, or vents, shall be supplied to a spray booth or room in quantities equal to the volume of air exhausted through the spray booth.

**1910.94(c)(7)(ii)** Where a spray booth or room receives make-up air through self-closing doors, dampers, or louvers, they shall be fully open at all times when the booth or room is in use for spraying. The velocity of air through such doors, dampers, or louvers shall not exceed 200 feet per minute. If the fan characteristics are such that the required air flow through the booth will be provided, higher velocities through the doors, dampers, or louvers may be used.

**1910.94(c)(7)(iii)(a)** Where the air supply to a spray booth or room is filtered, the fan static pressure shall be calculated on the assumption that the filters are dirty to the extent that they require cleaning or replacement.

**1910.94(c)(7)(iii)(b)** The rating of filters shall be governed by test data supplied by the manufacturer of the filter. A pressure gage shall be installed to show the pressure drop across the filters. This gage shall be marked to show the pressure drop at which the filters require cleaning or replacement. Filters shall be replaced or cleaned whenever the pressure drop across them becomes excessive or whenever the air flow through the face of the booth falls below that specified in Table G-10.

**1910.94(c)(7)(iv)(a)** Means for heating make-up air to any spray booth or room, before or at the time spraying is normally performed, shall be provided in all places where the outdoor temperature may be expected to remain below 55 °F for appreciable periods of time during the operation of the booth except where adequate and safe means of radiant heating for all operating personnel affected is provided. The replacement air during the heating seasons shall be maintained at not less than 65 °F. at the point of entry into the spray booth or spray room. When otherwise unheated make-up air would be at a temperature of more than 10 °F below room temperature, its temperature shall be regulated as provided in section 3.6.3 of ANSI Z9.2-1960.

**1910.94(c)(7)(iv)(b)** As an alternative to an air replacement system complying with the preceding section, general heating of the building in which the spray room or booth is located may be employed provided that all occupied parts of the building are maintained at not less than 65 °F when the exhaust system is in operation or the general heating system supplemented by other sources of heat may be employed to meet this requirement.

**1910.94(c)(7)(iv)(c)** No means of heating make-up air shall be located in a spray booth.

**1910.94(c)(7)(iv)(d)** Where make-up air is heated by coal or oil, the products of combustion shall not be allowed to mix with the make-up air, and the products of combustion shall be conducted outside the building through a flue terminating at a point remote from all points where make-up air enters the building.

**1910.94(c)(7)(iv)(e)** Where make-up air is heated by gas, and the products of combustion are not mixed with the make-up air but are conducted through an independent flue to a point outside the building remote from all points where make-up air enters the building, it is not necessary to comply with paragraph (c)(7)(iv)(f) of this section.

**1910.94(c)(7)(iv)(f)** Where make-up air to any manually operated spray booth or room is heated by gas and the products of combustion are allowed to mix with the supply air, the following precautions must be taken:

**1910.94(c)(7)(iv)(f)(1)** The gas must have a distinctive and strong enough odor to warn workmen in a spray booth or room of its presence if in an unburned state in the make-up air.

**1910.94(c)(7)(iv)(f)(2)** The maximum rate of gas supply to the make-up air heater burners must not exceed that which would yield in excess of 200 p.p.m. (parts per million) of carbon monoxide or 2,000 p.p.m. of total combustible gases in the mixture if the unburned gas upon the occurrence of flame failure were mixed with all of the make-up air supplied.

**1910.94(c)(7)(iv)(f)(3)** A fan must be provided to deliver the mixture of heated air and products of combustion from the plenum chamber housing the gas burners to the spray booth or room.

**1910.94(c)(8)** *Scope.* Spray booths or spray rooms are to be used to enclose or confine all spray finishing operations covered by this paragraph (c). This paragraph does not apply to the spraying of the exteriors of buildings, fixed tanks, or similar structures, nor to small portable spraying apparatus not used repeatedly in the same location.

[39 FR 23502, June 27, 1974, as amended at 40 FR 23073, May 28, 1975; 40 FR 24522, June 9, 1975; 43 FR 49746, Oct. 24, 1978; 49 FR 5322, Feb. 10, 1984; 55 FR 32015, Aug. 6, 1990; 58 FR 35308, June 30, 1993; 61 FR 9236, Mar. 7, 1996; 63 FR 1269, Jan. 8, 1998; 64 FR 13909, Mar. 23, 1999]

### §1910.95 Occupational noise exposure.

**1910.95(a)** Protection against the effects of noise exposure shall be provided when the sound levels exceed those shown in Table G-16 when measured on the A scale of a standard sound level meter at slow response. When noise levels are determined by octave band analysis, the equivalent A-weighted sound level may be determined as follows:

### FIGURE G-9

Equivalent sound level contours. Octave band sound pressure levels may be converted to the equivalent A-weighted sound level by plotting them on this graph and noting the A-weighted sound level corresponding to the point of highest penetration into the sound level contours. This equivalent A-weighted sound level, which may differ from the actual A-weighted sound level of the noise, is used to determine exposure limits from Table 1.G-16.

**1910.95(b)(1)** When employees are subjected to sound exceeding those listed in Table G-16, feasible administrative or engineering controls shall be utilized. If such controls fail to reduce sound levels within the levels of Table G-16, personal protective equipment shall be provided and used to reduce sound levels within the levels of the table.

**1910.95(b)(2)** If the variations in noise level involve maxima at intervals of 1 second or less, it is to be considered continuous.

### TABLE G-16—PERMISSIBLE NOISE EXPOSURES[1]

| Duration per day, hours | Sound level dBA slow response |
| --- | --- |
| 8 | 90 |
| 6 | 92 |
| 4 | 95 |
| 3 | 97 |
| 2 | 100 |
| 1 1/2 | 102 |
| 1 | 105 |
| 1/2 | 110 |
| 1/4 or less | 115 |

[1] When the daily noise exposure is composed of two or more periods of noise exposure of different levels, their combined effect should be considered, rather than the individual effect of each. If the sum of the following fractions:

$C_1/T_1 + C_2/T_2 C_n/T_n$ exceeds unity, then, the mixed exposure should be considered to exceed the limit value. Cn indicates the total time of exposure at a specified noise level, and Tn indicates the total time of exposure permitted at that level.

Exposure to impulsive or impact noise should not exceed 140 dB peak sound pressure level.

**1910.95(c)** *Hearing conservation program.* **(1)** The employer shall administer a continuing, effective hearing conservation program, as described in paragraphs (c) through (o) of this section, whenever employee noise exposures equal or exceed an 8-hour time-weighted average sound level (TWA) of 85 decibels measured on the A scale (slow response) or, equivalently, a dose of fifty percent. For purposes of the hearing conservation program, employee noise exposures shall be computed in accordance with appendix A and Table G-16a, and without regard to any attenuation provided by the use of personal protective equipment.

**1910.95(c)(2)** For purposes of paragraphs (c) through (n) of this section, an 8-hour time-weighted average of 85 decibels or a dose of fifty percent shall also be referred to as the action level.

**1910.95(d)** *Monitoring.* **(1)** When information indicates that any employee's exposure may equal or exceed an 8-hour time-weighted average of 85 decibels, the employer shall develop and implement a monitoring program.

**1910.95(d)(1)(i)** The sampling strategy shall be designed to identify employees for inclusion in the hearing conservation program and to enable the proper selection of hearing protectors.

**1910.95(d)(1)(ii)** Where circumstances such as high worker mobility, significant variations in sound level, or a significant component of impulse noise make area monitoring generally inappropriate, the employer shall use representative personal sampling to comply with the monitoring requirements of this paragraph unless the employer can show that area sampling produces equivalent results.

**1910.95(d)(2)(i)** All continuous, intermittent and impulsive sound levels from 80 decibels to 130 decibels shall be integrated into the noise measurements.

**1910.95(d)(2)(ii)** Instruments used to measure employee noise exposure shall be calibrated to ensure measurement accuracy.

**1910.95(d)(3)** Monitoring shall be repeated whenever a change in production, process, equipment or controls increases noise exposures to the extent that:

**1910.95(d)(3)(i)** Additional employees may be exposed at or above the action level; or

**1910.95(d)(3)(ii)** The attenuation provided by hearing protectors being used by employees may be rendered inadequate to meet the requirements of paragraph (j) of this section.

**1910.95(e)** *Employee notification.* The employer shall notify each employee exposed at or above an 8-hour time-weighted average of 85 decibels of the results of the monitoring.

**1910.95(f)** *Observation of monitoring.* The employer shall provide affected employees or their representatives with an opportunity to observe any noise measurements conducted pursuant to this section.

**1910.95(g)** *Audiometric testing program.* **(1)** The employer shall establish and maintain an audiometric testing program as provided in this paragraph by making audiometric testing available to all employees whose exposures equal or exceed an 8-hour time-weighted average of 85 decibels.

**1910.95(g)(2)** The program shall be provided at no cost to employees.

**1910.95(g)(3)** Audiometric tests shall be performed by a licensed or certified audiologist, otolaryngologist, or other physician, or by a technician who is certified by the Council of Accreditation in Occupational Hearing Conservation, or who has satisfactorily demonstrated competence in administering audiometric examinations, obtaining valid audiograms, and properly using, maintaining and checking calibration and proper functioning of the audiometers being used. A technician who operates microprocessor audiometers does not need to be certified. A technician who performs audiometric tests must be responsible to an audiologist, otolaryngologist or physician.

**1910.95(g)(4)** All audiograms obtained pursuant to this section shall meet the requirements of appendix C: *Audiometric Measuring Instruments.*

**1910.95(g)(5)** *Baseline audiogram.* **(i)** Within 6 months of an employee's first exposure at or above the action level, the employer shall establish a valid baseline audiogram against which subsequent audiograms can be compared.

**1910.95(g)(5)(ii)** *Mobile test van exception.* Where mobile test vans are used to meet the audiometric testing obligation, the employer shall obtain a valid baseline audiogram within 1 year of an employee's first exposure at or above the action level. Where baseline audiograms are obtained more than 6 months after the employee's first exposure at or above the action level, employees shall wearing hearing protectors for any period exceeding six months after first exposure until the baseline audiogram is obtained.

**1910.95(g)(5)(iii)** Testing to establish a baseline audiogram shall be preceded by at least 14 hours without exposure to workplace noise. Hearing protectors may be used as a substitute for the requirement that baseline audiograms be preceded by 14 hours without exposure to workplace noise.

**1910.95(g)(5)(iv)** The employer shall notify employees of the need to avoid high levels of non-occupational noise exposure during the 14-hour period immediately preceding the audiometric examination.

**1910.95(g)(6)** *Annual audiogram.* At least annually after obtaining the baseline audiogram, the employer shall obtain a new audiogram for each employee exposed at or above an 8-hour time-weighted average of 85 decibels.

**1910.95(g)(7)** *Evaluation of audiogram.* **(i)** Each employee's annual audiogram shall be compared to that employee's baseline audiogram to determine if the audiogram is valid and if a standard threshold shift as defined in paragraph (g)(10) of this section has occurred. This comparison may be done by a technician.

**1910.95(g)(7)(ii)** If the annual audiogram shows that an employee has suffered a standard threshold shift, the employer may obtain a retest within 30 days and consider the results of the retest as the annual audiogram.

**1910.95(g)(7)(iii)** The audiologist, otolaryngologist, or physician shall review problem audiograms and shall determine whether there is a need for further evaluation. The employer shall provide to the person performing this evaluation the following information:

**1910.95(g)(7)(iii)(A)** A copy of the requirements for hearing conservation as set forth in paragraphs (c) through (n) of this section;

**1910.95(g)(7)(iii)(B)** The baseline audiogram and most recent audiogram of the employee to be evaluated;

**1910.95(g)(7)(iii)(C)** Measurements of background sound pressure levels in the audiometric test room as required in appendix D: Audiometric Test Rooms.

**1910.95(g)(7)(iii)(D)** Records of audiometer calibrations required by paragraph (h)(5) of this section.

**1910.95(g)(8)** *Follow-up procedures.* **(i)** If a comparison of the annual audiogram to the baseline audiogram indicates a standard threshold shift as defined in paragraph (g)(10) of this section has occurred, the employee shall be informed of this fact in writing, within 21 days of the determination.

**1910.95(g)(8)(ii)** Unless a physician determines that the standard threshold shift is not work related or aggravated by occupational noise exposure, the employer shall ensure that the following steps are taken when a standard threshold shift occurs:

**1910.95(g)(8)(ii)(A)** Employees not using hearing protectors shall be fitted with hearing protectors, trained in their use and care, and required to use them.

**1910.95(g)(8)(ii)(B)** Employees already using hearing protectors shall be refitted and retrained in the use of hearing protectors and provided with hearing protectors offering greater attenuation if necessary.

**1910.95(g)(8)(ii)(C)** The employee shall be referred for a clinical audiological evaluation or an otological examination, as appropriate, if additional testing is necessary or if the employer suspects that a medical pathology of the ear is caused or aggravated by the wearing of hearing protectors.

**1910.95(g)(8)(ii)(D)** The employee is informed of the need for an otological examination if a medical pathology of the ear that is unrelated to the use of hearing protectors is suspected.

**1910.95(g)(8)(iii)** If subsequent audiometric testing of an employee whose exposure to noise is less than an 8-hour TWA of 90 decibels indicates that a standard threshold shift is not persistent, the employer:

**1910.95(g)(8)(iii)(A)** Shall inform the employee of the new audiometric interpretation; and

**1910.95(g)(8)(iii)(B)** May discontinue the required use of hearing protectors for that employee.

**1910.95(g)(9)** *Revised baseline.* An annual audiogram may be substituted for the baseline audiogram when, in the judgment of the audiologist, otolaryngologist or physician who is evaluating the audiogram:

**1910.95(g)(9)(i)** The standard threshold shift revealed by the audiogram is persistent; or

**1910.95(g)(9)(ii)** The hearing threshold shown in the annual audiogram indicates significant improvement over the baseline audiogram.

**1910.95(g)(10)** *Standard threshold shift.* **(i)** As used in this section, a standard threshold shift is a change in hearing threshold relative to the baseline audiogram of an average of 10 dB or more at 2000, 3000, and 4000 Hz in either ear.

**1910.95(g)(10)(ii)** In determining whether a standard threshold shift has occurred, allowance may be made for the contribution of aging (presbycusis) to the change in hearing level by correcting the annual audiogram according to the procedure described in appendix F: *Calculation and Application of Age Correction to Audiograms.*

**1910.95(h)** *Audiometric test requirements.* **(1)** Audiometric tests shall be pure tone, air conduction, hearing threshold examinations, with test frequencies including as a minimum 500, 1000, 2000, 3000, 4000, and 6000 Hz. Tests at each frequency shall be taken separately for each ear.

**1910.95(h)(2)** Audiometric tests shall be conducted with audiometers (including microprocessor audiometers) that meet the specifications of, and are maintained and used in accordance with, American National Standard Specification for Audiometers, S3.6-1969, which is incorporated by reference as specified in §1910.6.

**1910.95(h)(3)** Pulsed-tone and self-recording audiometers, if used, shall meet the requirements specified in appendix C: *Audiometric Measuring Instruments.*

**1910.95(h)(4)** Audiometric examinations shall be administered in a room meeting the requirements listed in appendix D: *Audiometric Test Rooms.*

**1910.95(h)(5)** *Audiometer calibration.* **(i)** The functional operation of the audiometer shall be checked before each day's use by testing a person with known, stable hearing thresholds, and by listening to the audiometer's output to make sure that the output is free from distorted or unwanted sounds. Deviations of 10 decibels or greater require an acoustic calibration.

**1910.95(h)(5)(ii)** Audiometer calibration shall be checked acoustically at least annually in accordance with appendix E: *Acoustic Calibration of Audiometers.* Test frequencies below 500 Hz and above 6000 Hz may be omitted from this check. Deviations of 15 decibels or greater require an exhaustive calibration.

**1910.95(h)(5)(iii)** An exhaustive calibration shall be performed at least every two years in accordance with sections 4.1.2; 4.1.3.; 4.1.4.3; 4.2; 4.4.1; 4.4.2; 4.4.3; and 4.5 of the American National Standard Specification for Audiometers, S3.6-1969. Test frequencies below 500 Hz and above 6000 Hz may be omitted from this calibration.

**1910.95(i)** *Hearing protectors.* **(1)** Employers shall make hearing protectors available to all employees exposed to an 8-hour time-weighted average of 85 decibels or greater at no cost to the employees. Hearing protectors shall be replaced as necessary.

**1910.95(i)(2)** Employers shall ensure that hearing protectors are worn:

**1910.95(i)(2)(i)** By an employee who is required by paragraph (b)(1) of this section to wear personal protective equipment; and

**1910.95(i)(2)(ii)** By any employee who is exposed to an 8-hour time-weighted average of 85 decibels or greater, and who:

**1910.95(i)(2)(ii)(A)** Has not yet had a baseline audiogram established pursuant to paragraph (g)(5)(ii); or

**1910.95(i)(2)(ii)(B)** Has experienced a standard threshold shift.

**1910.95(i)(3)** Employees shall be given the opportunity to select their hearing protectors from a variety of suitable hearing protectors provided by the employer.

**1910.95(i)(4)** The employer shall provide training in the use and care of all hearing protectors provided to employees.

**1910.95(i)(5)** The employer shall ensure proper initial fitting and supervise the correct use of all hearing protectors.

**1910.95(j)** *Hearing protector attenuation.* **(1)** The employer shall evaluate hearing protector attenuation for the specific noise environments in which the protector will be used. The employer shall use one of the evaluation methods described in appendix B: *Methods for Estimating the Adequacy of Hearing Protection Attenuation.*

**1910.95(j)(2)** Hearing protectors must attenuate employee exposure at least to an 8-hour time-weighted average of 90 decibels as required by paragraph (b) of this section.

**1910.95(j)(3)** For employees who have experienced a standard threshold shift, hearing protectors must attenuate employee exposure to an 8-hour time-weighted average of 85 decibels or below.

**1910.95(j)(4)** The adequacy of hearing protector attenuation shall be re-evaluated whenever employee noise exposures increase to the extent that the hearing protectors provided may no longer provide adequate attenuation. The employer shall provide more effective hearing protectors where necessary.

**1910.95(k)** *Training program.* **(1)** The employer shall institute a training program for all employees who are exposed to noise at or above an 8-hour time-weighted average of 85 decibels, and shall ensure employee participation in such program.

**1910.95(k)(2)** The training program shall be repeated annually for each employee included in the hearing conservation program. Information provided in the training program shall be updated to be consistent with changes in protective equipment and work processes.

**1910.95(k)(3)** The employer shall ensure that each employee is informed of the following:

**1910.95(k)(3)(i)** The effects of noise on hearing;

**1910.95(k)(3)(ii)** The purpose of hearing protectors, the advantages, disadvantages, and attenuation of various types, and instructions on selection, fitting, use, and care; and

**1910.95(k)(3)(iii)** The purpose of audiometric testing, and an explanation of the test procedures.

**1910.95(l)** *Access to information and training materials.* **(1)** The employer shall make available to affected employees or their representatives copies of this standard and shall also post a copy in the workplace.

**1910.95(l)(2)** The employer shall provide to affected employees any informational materials pertaining to the standard that are supplied to the employer by the Assistant Secretary.

**1910.95(l)(3)** The employer shall provide, upon request, all materials related to the employer's training and education program pertaining to this standard to the Assistant Secretary and the Director.

**1910.95(m)** *Recordkeeping*—**(1)** *Exposure measurements.* The employer shall maintain an accurate record of all employee exposure measurements required by paragraph (d) of this section.

**1910.95(m)(2)** *Audiometric tests.* **(i)** The employer shall retain all employee audiometric test records obtained pursuant to paragraph (g) of this section.

**1910.95(m)(2)(ii)** This record shall include:

**1910.95(m)(2)(ii)(A)** Name and job classification of the employee;

**1910.95(m)(2)(ii)(B)** Date of the audiogram;

**1910.95(m)(2)(ii)(C)** The examiner's name;

**1910.95(m)(2)(ii)(D)** Date of the last acoustic or exhaustive calibration of the audiometer; and

**1910.95(m)(2)(ii)(E)** Employee's most recent noise exposure assessment.

**1910.95(m)(2)(ii)(F)** The employer shall maintain accurate records of the measurements of the background sound pressure levels in audiometric test rooms.

**1910.95(m)(3)** *Record retention.* The employer shall retain records required in this paragraph (m) for at least the following periods.

**1910.95(m)(3)(i)** Noise exposure measurement records shall be retained for two years.

**1910.95(m)(3)(ii)** Audiometric test records shall be retained for the duration of the affected employee's employment.

**1910.95(m)(4)** *Access to records.* All records required by this section shall be provided upon request to employees, former employees, representatives designated by the individual employee, and the Assistant Secretary. The provisions of 29 CFR 1910.1020 (a)-(e) and (g)-(i) apply to access to records under this section.

**1910.95(m)(5)** *Transfer of records.* If the employer ceases to do business, the employer shall transfer to the successor employer all records required to be maintained by this section, and the successor employer shall retain them for the remainder of the period prescribed in paragraph (m)(3) of this section.

**1910.95(n)** *Appendices.* **(1)** Appendices A, B, C, D, and E to this section are incorporated as part of this section and the contents of these appendices are mandatory.

**1910.95(n)(2)** Appendices F and G to this section are informational and are not intended to create any additional obligations not otherwise imposed or to detract from any existing obligations.

**1910.95(o) Exemptions.** Paragraphs (c) through (n) of this section shall not apply to employers engaged in oil and gas well drilling and servicing operations.

## APPENDIX A TO §1910.95— NOISE EXPOSURE COMPUTATION

### This Appendix is Mandatory

### I. Computation of Employee Noise Exposure

(1) Noise dose is computed using Table G-16a as follows:

(i) When the sound level, L, is constant over the entire work shift, the noise dose, D, in percent, is given by: D = 100 C/T where C is the total length of the work day, in hours, and T is the reference duration corresponding to the measured sound level, L, as given in Table G-16a or by the formula shown as a footnote to that table.

(ii) When the workshift noise exposure is composed of two or more periods of noise at different levels, the total noise dose over the work day is given by:

D =100 ($C_1/T_1 + C_2/T_2 + ... + C_n/T_n$), where $C_n$ indicates the total time of exposure at a specific noise level, and $T_n$ indicates the reference duration for that level as given by Table G-16a.

(2) The eight-hour time-weighted average sound level (TWA), in decibels, may be computed from the dose, in percent, by means of the formula: TWA = 16.61 $\log_{10}$ (D/100) + 90. For an eight-hour workshift with the noise level constant over the entire shift, the TWA is equal to the measured sound level.

(3) A table relating dose and TWA is given in Section II.

### TABLE G-16A

| A-weighted sound level, L (decibel) | Reference duration, T (hour) |
|---|---|
| 80 | 32 |
| 81 | 27.9 |
| 82 | 24.3 |
| 83 | 21.1 |
| 84 | 18.4 |
| 85 | 16 |
| 86 | 13.9 |
| 87 | 12.1 |
| 88 | 10.6 |
| 89 | 9.2 |
| 90 | 8 |
| 91 | 7.0 |
| 92 | 6.1 |
| 93 | 5.3 |
| 94 | 4.6 |
| 95 | 4 |
| 96 | 3.5 |
| 97 | 3.0 |
| 98 | 2.6 |
| 99 | 2.3 |
| 100 | 2 |
| 101 | 1.7 |
| 102 | 1.5 |
| 103 | 1.3 |
| 104 | 1.1 |
| 105 | 1 |
| 106 | 0.87 |
| 107 | 0.76 |
| 108 | 0.66 |
| 109 | 0.57 |
| 110 | 0.5 |
| 111 | 0.44 |
| 112 | 0.38 |
| 113 | 0.33 |
| 114 | 0.29 |
| 115 | 0.25 |
| 116 | 0.22 |
| 117 | 0.19 |
| 118 | 0.16 |
| 119 | 0.14 |
| 120 | 0.125 |
| 121 | 0.11 |
| 122 | 0.095 |
| 123 | 0.082 |
| 124 | 0.072 |
| 125 | 0.063 |
| 126 | 0.054 |
| 127 | 0.047 |
| 128 | 0.041 |
| 129 | 0.036 |
| 130 | 0.031 |

In the above table the reference duration, T, is computed by

$$T = \frac{8}{2^{(L-90)/5}}$$

where L is the measured A-weighted sound level.

### II. Conversion Between "Dose" and "8-Hour Time-Weighted Average" Sound Level

Compliance with paragraphs (c)-(r) of this regulation is determined by the amount of exposure to noise in the workplace. The amount of such exposure is usually measured with an audiodosimeter which gives a readout in terms of "dose." In order to better understand the requirements of the amendment, dosimeter readings can be converted to an "8-hour time-weighted average sound level." (TWA).

In order to convert the reading of a dosimeter into TWA, see Table A-1, below. This table applies to dosimeters that are set by the manufacturer to calculate dose or percent exposure according to the relationships in Table G-16a. So, for example, a dose of 91 percent over an eight hour day results in a TWA of 89.3 dB, and, a dose of 50 percent corresponds to a TWA of 85 dB.

If the dose as read on the dosimeter is less than or greater than the values found in Table A-1, the TWA may be calculated by using the formula: TWA=16.61 $\log_{10}$ (D/100) + 90 where TWA=8-hour time-weighted average sound level and D = accumulated dose in percent exposure.

### TABLE A-1—CONVERSION FROM "PERCENT NOISE EXPOSURE" OR "DOSE" TO "8-HOUR TIME-WEIGHTED AVERAGE SOUND LEVEL" (TWA)

| Dose or percent noise exposure | TWA |
|---|---|
| 10 | 73.4 |
| 15 | 76.3 |
| 20 | 78.4 |
| 25 | 80.0 |
| 30 | 81.3 |
| 35 | 82.4 |
| 40 | 83.4 |
| 45 | 84.2 |
| 50 | 85.0 |
| 55 | 85.7 |
| 60 | 86.3 |
| 65 | 86.9 |
| 70 | 87.4 |
| 75 | 87.9 |
| 80 | 88.4 |
| 81 | 88.5 |
| 82 | 88.6 |

# OSHA Standards for General Industry

| Dose or percent noise exposure | TWA |
|---|---|
| 83 | 88.7 |
| 84 | 88.7 |
| 85 | 88.8 |
| 86 | 88.9 |
| 87 | 89.0 |
| 88 | 89.1 |
| 89 | 89.2 |
| 90 | 89.2 |
| 91 | 89.3 |
| 92 | 89.4 |
| 93 | 89.5 |
| 94 | 89.6 |
| 95 | 89.6 |
| 96 | 89.7 |
| 97 | 89.8 |
| 98 | 89.9 |
| 99 | 89.9 |
| 100 | 90.0 |
| 101 | 90.1 |
| 102 | 90.1 |
| 103 | 90.2 |
| 104 | 90.3 |
| 105 | 90.4 |
| 106 | 90.4 |
| 107 | 90.5 |
| 108 | 90.6 |
| 109 | 90.6 |
| 110 | 90.7 |
| 111 | 90.8 |
| 112 | 90.8 |
| 113 | 90.9 |
| 114 | 90.9 |
| 115 | 91.1 |
| 116 | 91.1 |
| 117 | 91.1 |
| 118 | 91.2 |
| 119 | 91.3 |
| 120 | 91.3 |
| 125 | 91.6 |
| 130 | 91.9 |
| 135 | 92.2 |
| 140 | 92.4 |
| 145 | 92.7 |
| 150 | 92.9 |
| 155 | 93.2 |
| 160 | 93.4 |
| 165 | 93.6 |
| 170 | 93.8 |
| 175 | 94.0 |
| 180 | 94.2 |
| 185 | 94.4 |
| 190 | 94.6 |
| 195 | 94.8 |
| 200 | 95.0 |
| 210 | 95.4 |
| 220 | 95.7 |
| 230 | 96.0 |
| 240 | 96.3 |
| 250 | 96.6 |
| 260 | 96.9 |
| 270 | 97.2 |
| 280 | 97.4 |
| 290 | 97.7 |
| 300 | 97.9 |
| 310 | 98.2 |

| Dose or percent noise exposure | TWA |
|---|---|
| 320 | 98.4 |
| 330 | 98.6 |
| 340 | 98.8 |
| 350 | 99.0 |
| 360 | 99.2 |
| 370 | 99.4 |
| 380 | 99.6 |
| 390 | 99.8 |
| 400 | 100.0 |
| 410 | 100.2 |
| 420 | 100.4 |
| 430 | 100.5 |
| 440 | 100.7 |
| 450 | 100.8 |
| 460 | 101.0 |
| 470 | 101.2 |
| 480 | 101.3 |
| 490 | 101.5 |
| 500 | 101.6 |
| 510 | 101.8 |
| 520 | 101.9 |
| 530 | 102.0 |
| 540 | 102.2 |
| 550 | 102.3 |
| 560 | 102.4 |
| 570 | 102.6 |
| 580 | 102.7 |
| 590 | 102.8 |
| 600 | 102.9 |
| 610 | 103.0 |
| 620 | 103.2 |
| 630 | 103.3 |
| 640 | 103.4 |
| 650 | 103.5 |
| 660 | 103.6 |
| 670 | 103.7 |
| 680 | 103.8 |
| 690 | 103.9 |
| 700 | 104.0 |
| 710 | 104.1 |
| 720 | 104.2 |
| 730 | 104.3 |
| 740 | 104.4 |
| 750 | 104.5 |
| 760 | 104.6 |
| 770 | 104.7 |
| 780 | 104.8 |
| 790 | 104.9 |
| 800 | 105.0 |
| 810 | 105.1 |
| 820 | 105.2 |
| 830 | 105.3 |
| 840 | 105.4 |
| 850 | 105.4 |
| 860 | 105.5 |
| 870 | 105.6 |
| 880 | 105.7 |
| 890 | 105.8 |
| 900 | 105.8 |
| 910 | 105.9 |
| 920 | 106.0 |
| 930 | 106.1 |
| 940 | 106.2 |
| 950 | 106.2 |
| 960 | 106.3 |

§1910.95 App. A

Occupational Health and Environmental Control

| Dose or percent noise exposure | TWA |
|---|---|
| 970 | 106.4 |
| 980 | 106.5 |
| 990 | 106.5 |
| 999 | 106.6 |

## APPENDIX B TO §1910.95—
## METHODS FOR ESTIMATING THE ADEQUACY OF HEARING PROTECTOR ATTENUATION

### This Appendix is Mandatory

For employees who have experienced a significant threshold shift, hearing protector attenuation must be sufficient to reduce employee exposure to a TWA of 85 dB. Employers must select one of the following methods by which to estimate the adequacy of hearing protector attenuation.

The most convenient method is the Noise Reduction Rating (NRR) developed by the Environmental Protection Agency (EPA). According to EPA regulation, the NRR must be shown on the hearing protector package. The NRR is then related to an individual worker's noise environment in order to assess the adequacy of the attenuation of a given hearing protector. This appendix describes four methods of using the NRR to determine whether a particular hearing protector provides adequate protection within a given exposure environment. Selection among the four procedures is dependent upon the employer's noise measuring instruments.

Instead of using the NRR, employers may evaluate the adequacy of hearing protector attenuation by using one of the three methods developed by the National Institute for Occupational Safety and Health (NIOSH), which are described in the "List of Personal Hearing Protectors and Attenuation Data," HEW Publication No. 76-120, 1975, pages 21-37. These methods are known as NIOSH methods #1B1, #1B2 and #1B3. The NRR described below is a simplification of NIOSH method #1B2. The most complex method is NIOSH method #1B1, which is probably the most accurate method since it uses the largest amount of spectral information from the individual employee's noise environment. As in the case of the NRR method described below, if one of the NIOSH methods is used, the selected method must be applied to an individual's noise environment to assess the adequacy of the attenuation. Employers should be careful to take a sufficient number of measurements in order to achieve a representative sample for each time segment.

**NOTE:** The employer must remember that calculated attenuation values reflect realistic values only to the extent that the protectors are properly fitted and worn.

When using the NRR to assess hearing protector adequacy, one of the following methods must be used:

(i) When using a dosimeter that is capable of C-weighted measurements:

(A) Obtain the employee's C-weighted dose for the entire workshift, and convert to TWA (see appendix A, II).

(B) Subtract the NRR from the C-weighted TWA to obtain the estimated A-weighted TWA under the ear protector.

(ii) When using a dosimeter that is not capable of C-weighted measurements, the following method may be used:

(A) Convert the A-weighted dose to TWA (see appendix A).

(B) Subtract 7 dB from the NRR.

(C) Subtract the remainder from the A-weighted TWA to obtain the estimated A-weighted TWA under the ear protector.

(iii) When using a sound level meter set to the A-weighting network:

(A) Obtain the employee's A-weighted TWA.

(B) Subtract 7 dB from the NRR, and subtract the remainder from the A-weighted TWA to obtain the estimated A-weighted TWA under the ear protector.

(iv) When using a sound level meter set on the C-weighting network:

(A) Obtain a representative sample of the C-weighted sound levels in the employee's environment.

(B) Subtract the NRR from the C-weighted average sound level to obtain the estimated A-weighted TWA under the ear protector.

(v) When using area monitoring procedures and a sound level meter set to the A-weighing network.

(A) Obtain a representative sound level for the area in question.

(B) Subtract 7 dB from the NRR and subtract the remainder from the A-weighted sound level for that area.

(vi) When using area monitoring procedures and a sound level meter set to the C-weighting network:

(A) Obtain a representative sound level for the area in question.

(B) Subtract the NRR from the C-weighted sound level for that area.

## APPENDIX C TO §1910.95—
## AUDIOMETRIC MEASURING INSTRUMENTS

### This Appendix is Mandatory

1. In the event that pulsed-tone audiometers are used, they shall have a tone on-time of at least 200 milliseconds.

2. Self-recording audiometers shall comply with the following requirements:

(A) The chart upon which the audiogram is traced shall have lines at positions corresponding to all multiples of 10 dB hearing level within the intensity range spanned by the audiometer. The lines shall be equally spaced and shall be separated by at least 1/4 inch. Additional increments are optional. The audiogram pen tracings shall not exceed 2 dB in width.

(B) It shall be possible to set the stylus manually at the 10-dB increment lines for calibration purposes.

(C) The slewing rate for the audiometer attenuator shall not be more than 6 dB/sec except that an initial slewing rate greater than 6 dB/sec is permitted at the beginning of each new test frequency, but only until the second subject response.

(D) The audiometer shall remain at each required test frequency for 30 seconds (±3 seconds). The audiogram shall be clearly marked at each change of frequency and the actual frequency change of the audiometer shall not deviate from the frequency boundaries marked on the audiogram by more than ±3 seconds.

(E) It must be possible at each test frequency to place a horizontal line segment parallel to the time axis on the audiogram, such that the audiometric tracing crosses the line segment at least six times at that test frequency. At each test frequency the threshold shall be the average of the midpoints of the tracing excursions.

## APPENDIX D TO §1910.95—AUDIOMETRIC TEST ROOMS

### This Appendix is Mandatory

Rooms used for audiometric testing shall not have background sound pressure levels exceeding those in Table D-1 when measured by equipment conforming at least to the Type 2 requirements of American National Standard Specification for Sound Level Meters, S1.4-1971 (R1976), and to the Class II requirements of American National Standard Specification for Octave, Half-Octave, and Third-Octave Band Filter Sets, S1.11-1971 (R1976).

### TABLE D-1—MAXIMUM ALLOWABLE OCTAVE-BAND SOUND PRESSURE LEVELS FOR AUDIOMETRIC TEST ROOMS

| Octave-band center frequency (Hz) | 500 | 1000 | 2000 | 4000 | 8000 |
|---|---|---|---|---|---|
| Sound pressure level (dB) | 40 | 40 | 47 | 57 | 62 |

## APPENDIX E TO §1910.95—
## ACOUSTIC CALIBRATION OF AUDIOMETERS

### This Appendix is Mandatory

Audiometer calibration shall be checked acoustically, at least annually, according to the procedures described in this appendix. The equipment necessary to perform these measurements is a sound level meter, octave-band filter set, and a National Bureau of Standards 9A coupler. In making these measurements, the accuracy of the calibrating equipment shall be sufficient to determine that the audiometer is within the tolerances permitted by American Standard Specification for Audiometers, S3.6-1969.

# OSHA Standards for General Industry

### (1) Sound Pressure Output Check

A. Place the earphone coupler over the microphone of the sound level meter and place the earphone on the coupler.

B. Set the audiometer's hearing threshold level (HTL) dial to 70 dB.

C. Measure the sound pressure level of the tones at each test frequency from 500 Hz through 6000 Hz for each earphone.

D. At each frequency the readout on the sound level meter should correspond to the levels in Table E-1 or Table E-2, as appropriate, for the type of earphone, in the column entitled "sound level meter reading."

### (2) Linearity Check

A. With the earphone in place, set the frequency to 1000 Hz and the HTL dial on the audiometer to 70 dB.

B. Measure the sound levels in the coupler at each 10-dB decrement from 70 dB to 10 dB, noting the sound level meter reading at each setting.

C. For each 10-dB decrement on the audiometer the sound level meter should indicate a corresponding 10 dB decrease.

D. This measurement may be made electrically with a voltmeter connected to the earphone terminals.

### (3) Tolerances

When any of the measured sound levels deviate from the levels in Table E-1 or Table E-2 by ±3 dB at any test frequency between 500 and 3000 Hz, 4 dB at 4000 Hz, or 5 dB at 6000 Hz, an exhaustive calibration is advised. An exhaustive calibration is required if the deviations are greater than 15 dB or greater at any test frequency.

#### TABLE E-1—REFERENCE THRESHOLD LEVELS FOR TELEPHONICS—TDH-39 EARPHONES

| Frequency, Hz | Reference threshold level for TDH-39 earphones, dB | Sound level meter reading, dB |
|---|---|---|
| 500 | 11.5 | 81.5 |
| 1000 | 7 | 77 |
| 2000 | 9 | 79 |
| 3000 | 10 | 80 |
| 4000 | 9.5 | 79.5 |
| 6000 | 15.5 | 85.5 |

#### TABLE E-2—REFERENCE THRESHOLD LEVELS FOR TELEPHONICS—TDH-49 EARPHONES

| Frequency, Hz | Reference threshold level for TDH-49 earphones, dB | Sound level meter reading, dB |
|---|---|---|
| 500 | 13.5 | 83.5 |
| 1000 | 7.5 | 77.5 |
| 2000 | 11 | 81.0 |
| 3000 | 9.5 | 79.5 |
| 4000 | 10.5 | 80.5 |
| 6000 | 13.5 | 83.5 |

## APPENDIX F TO §1910.95— CALCULATIONS AND APPLICATION OF AGE CORRECTIONS TO AUDIOGRAMS

### This Appendix Is Non-Mandatory

In determining whether a standard threshold shift has occurred, allowance may be made for the contribution of aging to the change in hearing level by adjusting the most recent audiogram. If the employer chooses to adjust the audiogram, the employer shall follow the procedure described below. This procedure and the age correction tables were developed by the National Institute for Occupational Safety and Health in the criteria document entitled "Criteria for a Recommended Standard ... Occupational Exposure to Noise," ((HSM)-11001).

For each audiometric test frequency:

(i) Determine from Tables F-1 or F-2 the age correction values for the employee by:

(A) Finding the age at which the most recent audiogram was taken and recording the corresponding values of age corrections at 1000 Hz through 6000 Hz;

(B) Finding the age at which the baseline audiogram was taken and recording the corresponding values of age corrections at 1000 Hz through 6000 Hz.

(ii) Subtract the values found in step (i)(B) from the value found in step (i)(A).

(iii) The differences calculated in step (ii) represented that portion of the change in hearing that may be due to aging.

EXAMPLE: Employee is a 32-year-old male. The audiometric history for his right ear is shown in decibels below.

| Employee's age | \multicolumn{5}{c}{Audiometric test frequency (Hz)} |
|---|---|---|---|---|---|
| | 1000 | 2000 | 3000 | 4000 | 6000 |
| 26 | 10 | 5 | 5 | 10 | 5 |
| *27 | 0 | 0 | 0 | 5 | 5 |
| 28 | 0 | 0 | 0 | 10 | 5 |
| 29 | 5 | 0 | 5 | 15 | 5 |
| 30 | 0 | 5 | 10 | 20 | 10 |
| 31 | 5 | 10 | 20 | 15 | 15 |
| *32 | 5 | 10 | 10 | 25 | 20 |

The audiogram at age 27 is considered the baseline since it shows the best hearing threshold levels. Asterisks have been used to identify the baseline and most recent audiogram. A threshold shift of 20 dB exists at 4000 Hz between the audiograms taken at ages 27 and 32.

(The threshold shift is computed by subtracting the hearing threshold at age 27, which was 5, from the hearing threshold at age 32, which is 25). A retest audiogram has confirmed this shift. The contribution of aging to this change in hearing may be estimated in the following manner:

Go to Table F-1 and find the age correction values (in dB) for 4000 Hz at age 27 and age 32.

| | \multicolumn{5}{c}{Frequency (Hz)} |
|---|---|---|---|---|---|
| | 1000 | 2000 | 3000 | 4000 | 6000 |
| Age 32 | 6 | 5 | 7 | 10 | 14 |
| Age 27 | 5 | 4 | 6 | 7 | 11 |
| Difference | 1 | 1 | 1 | 3 | 3 |

The difference represents the amount of hearing loss that may be attributed to aging in the time period between the baseline audiogram and the most recent audiogram. In this example, the difference at 4000 Hz is 3 dB. This value is subtracted from the hearing level at 4000 Hz, which in the most recent audiogram is 25, yielding 22 after adjustment. Then the hearing threshold in the baseline audiogram at 4000 Hz (5) is subtracted from the adjusted annual audiogram hearing threshold at 4000 Hz (22). Thus the age-corrected threshold shift would be 17 dB (as opposed to a threshold shift of 20 dB without age correction).

### TABLE F-1—AGE CORRECTION VALUES IN DECIBELS FOR MALES

| Years | \multicolumn{5}{c|}{Audiometric Test Frequencies (Hz)} |
|---|---|---|---|---|---|
| | 1000 | 2000 | 3000 | 4000 | 6000 |
| 20 or younger | 5 | 3 | 4 | 5 | 8 |
| 21 | 5 | 3 | 4 | 5 | 8 |
| 22 | 5 | 3 | 4 | 5 | 8 |
| 23 | 5 | 3 | 4 | 6 | 9 |
| 24 | 5 | 3 | 5 | 6 | 9 |
| 25 | 5 | 3 | 5 | 7 | 10 |
| 26 | 5 | 4 | 5 | 7 | 10 |
| 27 | 5 | 4 | 6 | 7 | 11 |
| 28 | 6 | 4 | 6 | 8 | 11 |
| 29 | 6 | 4 | 6 | 8 | 12 |
| 30 | 6 | 4 | 6 | 9 | 12 |
| 31 | 6 | 4 | 7 | 9 | 13 |
| 32 | 6 | 5 | 7 | 10 | 14 |
| 33 | 6 | 5 | 7 | 10 | 14 |
| 34 | 6 | 5 | 8 | 11 | 15 |
| 35 | 7 | 5 | 8 | 11 | 15 |
| 36 | 7 | 5 | 9 | 12 | 16 |
| 37 | 7 | 6 | 9 | 12 | 17 |
| 38 | 7 | 6 | 9 | 13 | 17 |
| 39 | 7 | 6 | 10 | 14 | 18 |
| 40 | 7 | 6 | 10 | 14 | 19 |
| 41 | 7 | 6 | 10 | 14 | 20 |
| 42 | 8 | 7 | 11 | 16 | 20 |
| 43 | 8 | 7 | 12 | 16 | 21 |
| 44 | 8 | 7 | 12 | 17 | 22 |
| 45 | 8 | 7 | 13 | 18 | 23 |
| 46 | 8 | 8 | 13 | 19 | 24 |
| 47 | 8 | 8 | 14 | 19 | 24 |
| 48 | 9 | 8 | 14 | 20 | 25 |
| 49 | 9 | 9 | 15 | 21 | 26 |
| 50 | 9 | 9 | 16 | 22 | 27 |
| 51 | 9 | 9 | 16 | 23 | 28 |
| 52 | 9 | 10 | 17 | 24 | 29 |
| 53 | 9 | 10 | 18 | 25 | 30 |
| 54 | 10 | 10 | 18 | 26 | 31 |
| 55 | 10 | 11 | 19 | 27 | 32 |
| 56 | 10 | 11 | 20 | 28 | 34 |
| 57 | 10 | 11 | 21 | 29 | 35 |
| 58 | 10 | 12 | 22 | 31 | 36 |
| 59 | 11 | 12 | 22 | 32 | 37 |
| 60 or older | 11 | 13 | 23 | 33 | 38 |

### TABLE F-2—AGE CORRECTION VALUES IN DECIBELS FOR FEMALES

| Years | \multicolumn{5}{c|}{Audiometric Test Frequencies (Hz)} |
|---|---|---|---|---|---|
| | 1000 | 2000 | 3000 | 4000 | 6000 |
| 20 or younger | 7 | 4 | 3 | 3 | 6 |
| 21 | 7 | 4 | 4 | 3 | 6 |
| 22 | 7 | 4 | 4 | 4 | 6 |
| 23 | 7 | 5 | 4 | 4 | 7 |
| 24 | 7 | 5 | 4 | 4 | 7 |
| 25 | 8 | 5 | 4 | 4 | 7 |
| 26 | 8 | 5 | 5 | 4 | 8 |
| 27 | 8 | 5 | 5 | 5 | 8 |
| 28 | 8 | 5 | 5 | 5 | 8 |
| 29 | 8 | 5 | 5 | 5 | 9 |
| 30 | 8 | 6 | 5 | 5 | 9 |
| 31 | 8 | 6 | 6 | 5 | 9 |
| 32 | 9 | 6 | 6 | 6 | 10 |
| 33 | 9 | 6 | 6 | 6 | 10 |
| 34 | 9 | 6 | 6 | 6 | 10 |
| 35 | 9 | 6 | 7 | 7 | 11 |
| 36 | 9 | 7 | 7 | 7 | 11 |
| 37 | 9 | 7 | 7 | 7 | 12 |
| 38 | 10 | 7 | 7 | 7 | 12 |
| 39 | 10 | 7 | 8 | 8 | 12 |
| 40 | 10 | 7 | 8 | 8 | 13 |
| 41 | 10 | 8 | 8 | 8 | 13 |
| 42 | 10 | 8 | 9 | 9 | 13 |
| 43 | 11 | 8 | 9 | 9 | 14 |
| 44 | 11 | 8 | 9 | 9 | 14 |
| 45 | 11 | 8 | 10 | 10 | 15 |
| 46 | 11 | 9 | 10 | 10 | 15 |
| 47 | 11 | 9 | 10 | 11 | 16 |
| 48 | 12 | 9 | 11 | 11 | 16 |
| 49 | 12 | 9 | 11 | 11 | 16 |
| 50 | 12 | 10 | 11 | 12 | 17 |
| 51 | 12 | 10 | 12 | 12 | 17 |
| 52 | 12 | 10 | 12 | 13 | 18 |
| 53 | 13 | 10 | 13 | 13 | 18 |
| 54 | 13 | 11 | 13 | 14 | 19 |
| 55 | 13 | 11 | 14 | 14 | 19 |
| 56 | 13 | 11 | 14 | 15 | 20 |
| 57 | 13 | 11 | 15 | 15 | 20 |
| 58 | 14 | 12 | 15 | 16 | 21 |
| 59 | 14 | 12 | 16 | 16 | 21 |
| 60 or older | 14 | 12 | 16 | 17 | 22 |

### APPENDIX G TO §1910.95— MONITORING NOISE LEVELS NON-MANDATORY INFORMATIONAL APPENDIX

This appendix provides information to help employers comply with the noise monitoring obligations that are part of the hearing conservation amendment.

#### WHAT IS THE PURPOSE OF NOISE MONITORING?

This revised amendment requires that employees be placed in a hearing conservation program if they are exposed to average noise levels of 85 dB or greater during an 8 hour workday. In order to determine if exposures are at or above this level, it may be necessary to measure or monitor the actual noise levels in the workplace and to estimate the noise exposure or "dose" received by employees during the workday.

#### WHEN IS IT NECESSARY TO IMPLEMENT A NOISE MONITORING PROGRAM?

It is not necessary for every employer to measure workplace noise. Noise monitoring or measuring must be conducted only when exposures are at or above 85 dB. Factors which suggest that noise exposures in the workplace may be at this level include employee complaints about the loudness of noise, indications that employees are losing their hearing, or noisy conditions which make normal conversation difficult. The employer should also consider any information available regarding noise emitted from specific machines. In addition, actual workplace noise measurements can suggest whether or not a monitoring program should be initiated.

#### HOW IS NOISE MEASURED?

Basically, there are two different instruments to measure noise exposures: the sound level meter and the dosimeter. A sound level meter is a device that measures the intensity of sound at a given moment. Since sound level meters provide a measure of sound intensity at only one point in time, it is generally necessary to take a number of measurements at different times during the day to estimate noise exposure over a workday. If noise levels fluctuate, the amount of time noise remains at each of the various measured levels must be determined.

To estimate employee noise exposures with a sound level meter it is also generally necessary to take several measurements at different locations within the workplace. After appropriate sound level meter readings are obtained, people sometimes draw "maps" of the sound levels within different areas of the workplace. By using a sound level "map" and information on employee locations throughout the day, estimates of individual exposure levels can be developed. This measurement method is generally referred to as *area* noise monitoring.

A dosimeter is like a sound level meter except that it stores sound level measurements and integrates these measurements over time, providing an average noise exposure reading for a given period of time, such as an 8-hour workday. With a dosimeter, a microphone is attached to the employee's clothing and the exposure measurement is simply read at the end of the desired time period. A reader may be used to read-out the dosimeter's measurements. Since the dosimeter is worn by the employee, it measures noise levels in those locations in which the employee travels. A sound level meter can also be positioned within the immediate vicinity of the exposed worker to obtain an individual exposure estimate. Such procedures are generally referred to as *personal* noise monitoring.

Area monitoring can be used to estimate noise exposure when the noise levels are relatively constant and employees are not mobile. In workplaces where employees move about in different areas or where the noise intensity tends to fluctuate over time, noise exposure is generally more accurately estimated by the personal monitoring approach.

In situations where personal monitoring is appropriate, proper positioning of the microphone is necessary to obtain accurate measurements. With a dosimeter, the microphone is generally located on the shoulder and remains in that position for the entire workday. With a sound level meter, the microphone is stationed near the employee's head, and the instrument is usually held by an individual who follows the employee as he or she moves about.

Manufacturer's instructions, contained in dosimeter and sound level meter operating manuals, should be followed for calibration and maintenance. To ensure accurate results, it is considered good professional practice to calibrate instruments before and after each use.

### HOW OFTEN IS IT NECESSARY TO MONITOR NOISE LEVELS?

The amendment requires that when there are significant changes in machinery or production processes that may result in increased noise levels, remonitoring must be conducted to determine whether additional employees need to be included in the hearing conservation program. Many companies choose to remonitor periodically (once every year or two) to ensure that all exposed employees are included in their hearing conservation programs.

### WHERE CAN EQUIPMENT AND TECHNICAL ADVICE BE OBTAINED?

Noise monitoring equipment may be either purchased or rented. Sound level meters cost about $500 to $1,000, while dosimeters range in price from about $750 to $1,500. Smaller companies may find it more economical to rent equipment rather than to purchase it. Names of equipment suppliers may be found in the telephone book (Yellow Pages) under headings such as: "Safety Equipment," "Industrial Hygiene," or "Engineers-Acoustical." In addition to providing information on obtaining noise monitoring equipment, many companies and individuals included under such listings can provide professional advice on how to conduct a valid noise monitoring program. Some audiological testing firms and industrial hygiene firms also provide noise monitoring services. Universities with audiology, industrial hygiene, or acoustical engineering departments may also provide information or may be able to help employers meet their obligations under this amendment.

Free, on-site assistance may be obtained from OSHA-supported state and private consultation organizations. These safety and health consultative entities generally give priority to the needs of small businesses.

### APPENDIX H TO §1910.95— AVAILABILITY OF REFERENCED DOCUMENTS

Paragraphs (c) through (o) of 29 CFR 1910.95 and the accompanying appendices contain provisions which incorporate publications by reference. Generally, the publications provide criteria for instruments to be used in monitoring and audiometric testing. These criteria are intended to be mandatory when so indicated in the applicable paragraphs of §1910.95 and appendices.

It should be noted that OSHA does not require that employers purchase a copy of the referenced publications. Employers, however, may desire to obtain a copy of the referenced publications for their own information.

The designation of the paragraph of the standard in which the referenced publications appear, the titles of the publications, and the availability of the publications are as follows:

| Paragraph designation | Referenced publication | Available from— |
|---|---|---|
| Appendix B | "List of Personal Hearing Protectors and Attenuation Data," HEW Pub. No. 76-120, 1975. NTIS-PB267461. | National Technical Information Service, Port Royal Road, Springfield, VA 22161. |
| Appendix D | "Specification for Sound Level Meters," S1.4-1971 (R1976). | American National Standards Institute, Inc., 1430 Broadway, New York, NY 10018. |
| §1910.95(k)(2), appendix E | "Specifications for Audiometers," S3.6-1969. | American National Standards Institute, Inc., 1430 Broadway, New York, NY 10018. |
| Appendix D | "Specification for Octave, Half-Octave and Third-Octave Band Filter Sets," S1.11-1971 (R1976). | Back Numbers Department, Dept. STD, American Institute of Physics, 333 E. 45th St., New York, NY 10017; American National Standards Institute, Inc., 1430 Broadway, New York, NY 10018. |

The referenced publications (or a microfiche of the publications) are available for review at many universities and public libraries throughout the country. These publications may also be examined at the OSHA Technical Data Center, Room N2439, United States Department of Labor, 200 Constitution Avenue, NW., Washington, DC 20210, (202) 219-7500 or at any OSHA Regional Office (see telephone directories under United States Government—Labor Department).

### APPENDIX I TO §1910.95—DEFINITIONS

These definitions apply to the following terms as used in paragraphs (c) through (n) of 29 CFR 1910.95.

Action level—An 8-hour time-weighted average of 85 decibels measured on the A-scale, slow response, or equivalently, a dose of fifty percent.

Audiogram—A chart, graph, or table resulting from an audiometric test showing an individual's hearing threshold levels as a function of frequency.

Audiologist—A professional, specializing in the study and rehabilitation of hearing, who is certified by the American Speech-Language-Hearing Association or licensed by a state board of examiners.

Baseline audiogram—The audiogram against which future audiograms are compared.

Criterion sound level—A sound level of 90 decibels.

Decibel (dB)—Unit of measurement of sound level.

Hertz (Hz)—Unit of measurement of frequency, numerically equal to cycles per second.

Medical pathology—A disorder or disease. For purposes of this regulation, a condition or disease affecting the ear, which should be treated by a physician specialist.

Noise dose—The ratio, expressed as a percentage, of (1) the time integral, over a stated time or event, of the 0.6 power of the measured SLOW exponential time-averaged, squared A-weighted sound pressure and (2) the product of the criterion duration (8 hours) and the 0.6 power of the squared sound pressure corresponding to the criterion sound level (90 dB).

*Noise dosimeter*—An instrument that integrates a function of sound pressure over a period of time in such a manner that it directly indicates a noise dose.

*Otolaryngologist*—A physician specializing in diagnosis and treatment of disorders of the ear, nose and throat.

*Representative exposure*—Measurements of an employee's noise dose or 8-hour time-weighted average sound level that the employers deem to be representative of the exposures of other employees in the workplace.

*Sound level*—Ten times the common logarithm of the ratio of the square of the measured A-weighted sound pressure to the square of the standard reference pressure of 20 microPascals. Unit: decibels (dB). For use with this regulation, SLOW time response, in accordance with ANSI S1.4-1971 (R1976), is required.

*Sound level meter*—An instrument for the measurement of sound level.

*Time-weighted average sound level*—That sound level, which if constant over an 8-hour exposure, would result in the same noise dose as is measured.

[39 FR 23502, June 27, 1974, as amended at 46 FR 4161, Jan. 16, 1981; 46 FR 62845, Dec. 29, 1981; 48 FR 9776, Mar. 8, 1983; 48 FR 29687, June 28, 1983; 54 FR 24333, June 7, 1989; 61 FR 9236, Mar. 7, 1996]

### §1910.96 Ionizing radiation [Removed].

[§1910.96 amended at 61 FR 5507, Feb. 13, 1996; removed and redesignated as §1910.1096 at 61 FR 31430, June 20, 1996, eff. June 30, 1996.]

### §1910.97 Nonionizing radiation.

**1910.97(a)** *Electromagnetic radiation*—**(1)** *Definitions applicable to this paragraph.* **(i)** The term *electromagnetic radiation* is restricted to that portion of the spectrum commonly defined as the radio frequency region, which for the purpose of this specification shall include the microwave frequency region.

**1910.97(a)(1)(ii)** *Partial body irradiation.* Pertains to the case in which part of the body is exposed to the incident electromagnetic energy.

**1910.97(a)(1)(iii)** *Radiation protection guide.* Radiation level which should not be exceeded without careful consideration of the reasons for doing so.

**1910.97(a)(1)(iv)** The word "symbol" as used in this specification refers to the overall design, shape, and coloring of the rf radiation sign shown in figure G-11.

**1910.97(a)(1)(v)** *Whole body irradiation.* Pertains to the case in which the entire body is exposed to the incident electromagnetic energy or in which the cross section of the body is smaller than the cross section of the incident radiation beam.

**1910.97(a)(2)** *Radiation protection guide.* **(i)** For normal environmental conditions and for incident electromagnetic energy of frequencies from 10 MHz to 100 GHz, the radiation protection guide is 10 mW/cm.$^2$ (milliwatt per square centimeter) as averaged over any possible 0.1-hour period. This means the following:

Power density: 10 mW/cm.$^2$ for periods of 0.1-hour or more.

Energy density: 1 mW-hr./cm.$^2$ (milliwatt hour per square centimeter) during any 0.1-hour period.

This guide applies whether the radiation is continuous or intermittent.

**1910.97(a)(2)(ii)** These formulated recommendations pertain to both whole body irradiation and partial body irradiation. Partial body irradiation must be included since it has been shown that some parts of the human body (e.g., eyes, testicles) may be harmed if exposed to incident radiation levels significantly in excess of the recommended levels.

**1910.97(a)(3)** *Warning symbol.* **(i)** The warning symbol for radio frequency radiation hazards shall consist of a red isosceles triangle above an inverted black isosceles triangle, separated and outlined by an aluminum color border. The words "Warning—Radio-Frequency Radiation Hazard" shall appear in the upper triangle. See figure G-11.

**1910.97(a)(3)(ii)** American National Standard Safety Color Code for Marking Physical Hazards and the Identification of Certain Equipment, Z53.1-1953, which is incorporated by reference as specified in §1910.6, shall be used for color specification. All lettering and the border shall be of aluminum color.

**1910.97(a)(3)(iii)** The inclusion and choice of warning information or precautionary instructions is at the discretion of the user. If such information is included it shall appear in the lower triangle of the warning symbol.

**FIGURE G-11—RADIO-FREQUENCY RADIATION HAZARD WARNING SYMBOL**

1. Place handling and mounting instructions on reverse side.
2. D = Scaling unit.
3. Lettering: Ratio of letter height to thickness of letter lines.
   Upper triangle : 5 to 1 Large
   6 to 1 Medium
   Lower triangle : 4 to 1 Small
   6 to 1 Medium
4. Symbol is square, triangles are right-angle isosceles.

**1910.97(a)(4)** *Scope.* This section applies to all radiations originating from radio stations, radar equipment, and other possible sources of electromagnetic radiation such as used for communication, radio navigation, and industrial and scientific purposes. This section does not apply to the deliberate exposure of patients by, or under the direction of, practitioners of the healing arts.

**1910.97(b)** [Reserved]

[39 FR 23502, June 27, 1974, as amended at 61 FR 9236, Mar. 7, 1996]

### §1910.98 Effective dates.

**1910.98(a)** The provisions of this Subpart G shall become effective on August 27, 1971, except as provided in the remaining paragraphs of this section.

**1910.98(b)** The following provisions shall become effective on February 15, 1972:

§1910.94 (a)(2)(iii), (a)(3), (a)(4), (b), (c)(2), (c)(3), (c)(4), (c)(5), (c)(6)(i), (c)(6)(ii), (d)(1)(ii), (d)(3), (d)(4), (d)(5), and (d)(7).

**1910.98(c)** Notwithstanding anything in paragraph (a), (b), or (d) of this section, any provision in any other section of this subpart which contains in itself a specific effective date or time limitation shall become effective on such date or shall apply in accordance with such limitation.

**1910.98(d)** Notwithstanding anything in paragraph (a) of this section, if any standard in 41 CFR part 50-204, other than a national consensus standard incorporated by reference in §50-204.2(a)(1), is or becomes applicable at any time to any employment and place of employment, by virtue of the Walsh-Healey Public Contracts Act, or the Service Contract Act of 1965, or the National Foundation on Arts and Humanities Act of 1965, any corresponding established Federal standard in this Subpart G which is derived from 41 CFR part 50-204 shall also become effective, and shall be applicable to such employment and place of employment, on the same date.

### §1910.99 Sources of standards [Removed].

[§1910.99 removed at 61 FR 9236, Mar. 7, 1996, effective May 6, 1996.]

### §1910.100 Standards organizations [Removed].

[§1910.100 removed at 61 FR 9236, Mar. 7, 1996, effective May 6, 1996.]

# Subpart H—Hazardous Materials

> **Authority:** Sections 4, 6, and 8 of the Occupational Safety and Health Act of 1970 (29 U.S.C. 653, 655, 657); Secretary of Labor's Orders Nos. 12-71 (36 FR 8754), 8-76 (41 FR 25059), 9-83 (48 FR 35736), 1-90 (55 FR 9033), or 6-96 (62 FR 111), 3-2000 (65 FR 50017), as applicable; and 29 CFR Part 1911.
>
> Sections 1910.103, 1910.106 through 1910.111, and 1910.119, 1910.120, and 1910.122 through 1910.126 also issued under 29 CFR part 1911.
>
> Section 1910.119 also issued under Section 304, Clean Air Act Amendments of 1990 (Pub.L. 101-549), reprinted at 29 U.S.C. 655 Note.
>
> Section 1910.120 also issued under Section 126, Superfund Amendments and Reauthorization Act of 1986 as amended (29 U.S.C. 655 Note), and 5 U.S.C. 553.

## §1910.101 Compressed gases (general requirements).

**1910.101(a)** *Inspection of compressed gas cylinders.* Each employer shall determine that compressed gas cylinders under his control are in a safe condition to the extent that this can be determined by visual inspection. Visual and other inspections shall be conducted as prescribed in the Hazardous Materials Regulations of the Department of Transportation (49 CFR parts 171-179 and 14 CFR part 103). Where those regulations are not applicable, visual and other inspections shall be conducted in accordance with Compressed Gas Association Pamphlets C-6-1968 and C-8-1962, which is incorporated by reference as specified in §1910.6.

**1910.101(b)** *Compressed gases.* The in-plant handling, storage, and utilization of all compressed gases in cylinders, portable tanks, rail tankcars, or motor vehicle cargo tanks shall be in accordance with Compressed Gas Association Pamphlet P-1-1965, which is incorporated by reference as specified in §1910.6.

**1910.101(c)** *Safety relief devices for compressed gas containers.* Compressed gas cylinders, portable tanks, and cargo tanks shall have pressure relief devices installed and maintained in accordance with Compressed Gas Association Pamphlets S-1.1-1963 and 1965 addenda and S-1.2-1963, which is incorporated by reference as specified in §1910.6.

[39 FR 23502, June 27, 1974, as amended at 61 FR 9236, Mar. 7, 1996]

## §1910.102 Acetylene.

**1910.102(a)** *Cylinders.* The in-plant transfer, handling, storage, and utilization of acetylene in cylinders shall be in accordance with Compressed Gas Association Pamphlet G-1-1966, which is incorporated by reference as specified in §1910.6.

**1910.102(b)** *Piped systems.* The piped systems for the in-plant transfer and distribution of acetylene shall be designed, installed, maintained, and operated in accordance with Compressed Gas Association Pamphlet G-1.3-1959, which is incorporated by reference as specified in §1910.6.

**1910.102(c)** *Generators and filling cylinders.* Plants for the generation of acetylene and the charging (filling) of acetylene cylinders shall be designed, constructed, and tested in accordance with the standards prescribed in Compressed Gas Association Pamphlet G-1.4-1966, which is incorporated by reference as specified in §1910.6.

[39 FR 23502, June 27, 1974, as amended at 61 FR 9236, Mar. 7, 1996]

## §1910.103 Hydrogen.

**1910.103(a)** *General*—(1) *Definitions.* As used in this section (i) Gaseous hydrogen system is one in which the hydrogen is delivered, stored and discharged in the gaseous form to consumer's piping. The system includes stationary or movable containers, pressure regulators, safety relief devices, manifolds, interconnecting piping and controls. The system terminates at the point where hydrogen at service pressure first enters the consumer's distribution piping.

**1910.103(a)(1)(ii)** Approved—Means, unless otherwise indicated, listed or approved by a nationally recognized testing laboratory. Refer to §1910.7 for definition of nationally recognized testing laboratory.

**1910.103(a)(1)(iii)** Listed—See "approved".

**1910.103(a)(1)(iv)** ASME—American Society of Mechanical Engineers.

**1910.103(a)(1)(v)** DOT Specifications—Regulations of the Department of Transportation published in 49 CFR Chapter I.

**1910.103(a)(1)(vi)** DOT regulations—See §1910.103(a)(1)(v).

**1910.103(a)(2)** *Scope*—(i) *Gaseous hydrogen systems.* (a) Paragraph (b) of this section applies to the installation of gaseous hydrogen systems on consumer premises where the hydrogen supply to the consumer premises originates outside the consumer premises and is delivered by mobile equipment.

**1910.103(a)(2)(i)(b)** Paragraph (b) of this section does not apply to gaseous hydrogen systems having a total hydrogen content of less than 400 cubic feet, nor to hydrogen manufacturing plants or other establishments operated by the hydrogen supplier or his agent for the purpose of storing hydrogen and refilling portable containers, trailers, mobile supply trucks, or tank cars.

**1910.103(a)(2)(ii)** *Liquefied hydrogen systems.* (a) Paragraph (c) of this section applies to the installation of liquefied hydrogen systems on consumer premises.

**1910.103(a)(2)(ii)(b)** Paragraph (c) of this section does not apply to liquefied hydrogen portable containers of less than 150 liters (39.63 gallons) capacity; nor to liquefied hydrogen manufacturing plants or other establishments operated by the hydrogen supplier or his agent for the sole purpose of storing liquefied hydrogen and refilling portable containers, trailers, mobile supply trucks, or tank cars.

**1910.103(b)** *Gaseous hydrogen systems*—(1) *Design*—(i) *Containers.* (a) Hydrogen containers shall comply with one of the following:

**1910.103(b)(1)(i)(a)(1)** Designed, constructed, and tested in accordance with appropriate requirements of ASME Boiler and Pressure Vessel Code, Section VIII—Unfired Pressure Vessels—1968, which is incorporated by reference as specified in §1910.6.

**1910.103(b)(1)(i)(a)(2)** Designed, constructed, tested and maintained in accordance with U.S. Department of Transportation Specifications and Regulations.

**1910.103(b)(1)(i)(b)** Permanently installed containers shall be provided with substantial noncombustible supports on firm noncombustible foundations.

**1910.103(b)(1)(i)(c)** Each portable container shall be legibly marked with the name "Hydrogen" in accordance with "Marking Portable Compressed Gas Containers to Identify the Material Contained" ANSI Z48.1—1954, which is incorporated by reference as specified in §1910.6. Each manifolded hydrogen supply unit shall be legibly marked with the name Hydrogen or a legend such as "This unit contains hydrogen."

**1910.103(b)(1)(ii)** *Safety relief devices.* (a) Hydrogen containers shall be equipped with safety relief devices as required by the ASME Boiler and Pressure Vessel Code, Section VIII Unfired Pressure Vessels, 1968 or the DOT Specifications and Regulations under which the container is fabricated.

**1910.103(b)(1)(ii)(b)** Safety relief devices shall be arranged to discharge upward and unobstructed to the open air in such a manner as to prevent any impingement of escaping gas upon the container, adjacent structure or personnel. This requirement does not apply to DOT Specification containers having an internal volume of 2 cubic feet or less.

**1910.103(b)(1)(ii)(c)** Safety relief devices or vent piping shall be designed or located so that moisture cannot collect and freeze in a manner which would interfere with proper operation of the device.

**1910.103(b)(1)(iii)** *Piping, tubing, and fittings.* (a) Piping, tubing, and fittings shall be suitable for hydrogen service and for the pressures and temperatures involved. Cast iron pipe and fittings shall not be used.

**1910.103(b)(1)(iii)(b)** Piping and tubing shall conform to Section 2—"Industrial Gas and Air Piping"—Code for Pressure Piping, ANSI B31.1-1967 with addenda B31.1-1969, which is incorporated by reference as specified in §1910.6.

**1910.103(b)(1)(iii)(c)** Joints in piping and tubing may be made by welding or brazing or by use of flanged, threaded, socket, or compression fittings. Gaskets and thread sealants shall be suitable for hydrogen service.

**1910.103(b)(1)(iv)** *Equipment assembly.* (a) Valves, gauges, regulators, and other accessories shall be suitable for hydrogen service.

**1910.103(b)(1)(iv)(b)** Installation of hydrogen systems shall be supervised by personnel familiar with proper practices with reference to their construction and use.

**1910.103(b)(1)(iv)(c)** Storage containers, piping, valves, regulating equipment, and other accessories shall be readily accessible, and shall be protected against physical damage and against tampering.

**1910.103(b)(1)(iv)(d)** Cabinets or housings containing hydrogen control or operating equipment shall be adequately ventilated.

**1910.103(b)(1)(iv)(e)** Each mobile hydrogen supply unit used as part of a hydrogen system shall be adequately secured to prevent movement.

**1910.103(b)(1)(iv)(f)** Mobile hydrogen supply units shall be electrically bonded to the system before discharging hydrogen.

**1910.103(b)(1)(v)** *Marking.* The hydrogen storage location shall be permanently placarded as follows: "HYDROGEN—FLAMMABLE GAS —NO SMOKING—NO OPEN FLAMES," or equivalent.

**1910.103(b)(1)(vi)** *Testing.* After installations, all piping, tubing, and fittings shall be tested and proved hydrogen gas tight at maximum operating pressure.

**1910.103(b)(2)** *Location—(i) General. (a)* The system shall be located so that it is readily accessible to delivery equipment and to authorized personnel.

**1910.103(b)(2)(i)(b)** Systems shall be located above ground.

**1910.103(b)(2)(i)(c)** Systems shall not be located beneath electric power lines.

**1910.103(b)(2)(i)(d)** Systems shall not be located close to flammable liquid piping or piping of other flammable gases.

**1910.103(b)(2)(i)(e)** Systems near aboveground flammable liquid storage shall be located on ground higher than the flammable liquid storage except when dikes, diversion curbs, grading, or separating solid walls are used to prevent accumulation of flammable liquids under the system.

**1910.103(b)(2)(ii)** *Specific requirements. (a)* The location of a system, as determined by the maximum total contained volume of hydrogen, shall be in the order of preference as indicated by Roman numerals in Table H-1.

### TABLE H-1

| Nature of location | Less than 3,000 CF | 3,000 CF to 15,000 CF | In excess of 15,000 CF |
|---|---|---|---|
| Outdoors | I | I | I |
| In a separate building | II | II | II. |
| In a special room | III | III | Not permitted. |
| Inside buildings not in a special room and exposed to other occupancies | IV | Not permitted. | Not permitted. |

Size of hydrogen system

**1910.103(b)(2)(ii)(b)** The minimum distance in feet from a hydrogen system of indicated capacity located outdoors, in separate buildings or in special rooms to any specified outdoor exposure shall be in accordance with Table H-2.

**1910.103(b)(2)(ii)(c)** The distances in Table H-2 Items 1 and 3 to 10 inclusive do not apply where protective structures such as adequate fire walls are located between the system and the exposure.

### TABLE H-2

| | Type of outdoor exposure | | Less than 3,000 CF | 3,000 CF to 15,000 CF | In excess of 15,000 CF |
|---|---|---|---|---|---|
| 1. | Building or structure | Wood frame construction[1] | 10 | 25 | 50 |
| | | Heavy timber, noncombustible or ordinary construction[1] | 0 | 10 | [2]25 |
| | | Fire-resistive construction[1] | 0 | 0 | 0 |
| 2. | Wall openings | Not above any part of a system | 10 | 10 | 10 |
| | | Above any part of a system | 25 | 25 | 25 |
| 3. | Flammable liquids above ground | 0 to 1,000 gallons | 10 | 25 | 25 |
| | | In excess of 1,000 gallons | 25 | 50 | 50 |
| 4. | Flammable liquids below ground— 0 to 1,000 gallons. | Tank | 10 | 10 | 10 |
| | | Vent or fill opening of tank | 25 | 25 | 25 |
| 5. | Flammable liquids below ground.— in excess of 1,000 gallons. | Tank | 20 | 20 | 20 |
| | | Vent or fill opening of tank | 25 | 25 | 25 |
| 6. | Flammable gas storage, either high pressure or low pressure. | 0 to 15,000 CF capacity | 10 | 25 | 25 |
| | | In excess of 15,000 CF capacity | 25 | 50 | 50 |
| 7. | Oxygen storage | 12,000 CF or less[4] | | | |
| | | More than 12,000 CF[5] | | | |
| 8. | Fast burning solids such as ordinary lumber, excelsior or paper | | 50 | 50 | 50 |
| 9. | Slow burning solids such as heavy timber or coal | | 25 | 25 | 25 |
| 10. | Open flames and other sources of ignition | | 25 | 25 | 25 |
| 11. | Air compressor intakes or inlets to ventilating or air-conditioning equipment | | 50 | 50 | 50 |
| 12. | Concentration of people[3] | | 25 | 50 | 50 |

Size of hydrogen system

[1] Refer to NFPA No. 220 Standard Types of Building Construction for definitions of various types of construction. (1969 Ed.)
[2] But not less than one-half the height of adjacent side wall of the structure.
[3] In congested areas such as offices, lunchrooms, locker rooms, time-clock areas.
[4] Refer to NFPA No. 51, gas systems for welding and cutting (1969).
[5] Refer to NFPA No. 556, bulk oxygen systems at consumer sites (1969).

# OSHA Standards for General Industry

**1910.103(b)(2)(ii)(d)** Hydrogen systems of less than 3,000 CF when located inside buildings and exposed to other occupancies shall be situated in the building so that the system will be as follows:

**1910.103(b)(2)(ii)(d)(1)** In an adequately ventilated area as in paragraph (b)(3)(ii)(b) of this section.

**1910.103(b)(2)(ii)(d)(2)** Twenty feet from stored flammable materials or oxidizing gases.

**1910.103(b)(2)(ii)(d)(3)** Twenty-five feet from open flames, ordinary electrical equipment or other sources of ignition.

**1910.103(b)(2)(ii)(d)(4)** Twenty-five feet from concentrations of people.

**1910.103(b)(2)(ii)(d)(5)** Fifty feet from intakes of ventilation or air-conditioning equipment and air compressors.

**1910.103(b)(2)(ii)(d)(6)** Fifty feet from other flammable gas storage.

**1910.103(b)(2)(ii)(d)(7)** Protected against damage or injury due to falling objects or working activity in the area.

**1910.103(b)(2)(ii)(d)(8)** More than one system of 3,000 CF or less may be installed in the same room, provided the systems are separated by at least 50 feet. Each such system shall meet all of the requirements of this paragraph.

**1910.103(b)(3)** *Design consideration at specific locations*—**(i)** *Outdoor locations.* **(a)** Where protective walls or roofs are provided, they shall be constructed of noncombustible materials.

**1910.103(b)(3)(i)(b)** Where the enclosing sides adjoin each other, the area shall be properly ventilated.

**1910.103(b)(3)(i)(c)** Electrical equipment within 15 feet shall be in accordance with subpart S of this part.

**1910.103(b)(3)(ii)** *Separate buildings.* **(a)** Separate buildings shall be built of at least noncombustible construction. Windows and doors shall be located so as to be readily accessible in case of emergency. Windows shall be of glass or plastic in metal frames.

**1910.103(b)(3)(ii)(b)** Adequate ventilation to the outdoors shall be provided. Inlet openings shall be located near the floor in exterior walls only. Outlet openings shall be located at the high point of the room in exterior walls or roof. Inlet and outlet openings shall each have minimum total area of one (1) square foot per 1,000 cubic feet of room volume. Discharge from outlet openings shall be directed or conducted to a safe location.

**1910.103(b)(3)(ii)(c)** Explosion venting shall be provided in exterior walls or roof only. The venting area shall be equal to not less than 1 square foot per 30 cubic feet of room volume and may consist of any one or any combination of the following: Walls of light, noncombustible material, preferably single thickness, single strength glass; lightly fastened hatch covers; lightly fastened swinging doors in exterior walls opening outward; lightly fastened walls or roof designed to relieve at a maximum pressure of 25 pounds per square foot.

**1910.103(b)(3)(ii)(d)** There shall be no sources of ignition from open flames, electrical equipment, or heating equipment.

**1910.103(b)(3)(ii)(e)** Electrical equipment shall be in accordance with subpart S of this part for Class I, Division 2 locations.

**1910.103(b)(3)(ii)(f)** Heating, if provided, shall be by steam, hot water, or other indirect means.

**1910.103(b)(3)(iii)** *Special rooms.* **(a)** Floor, walls, and ceiling shall have a fire-resistance rating of at least 2 hours. Walls or partitions shall be continuous from floor to ceiling and shall be securely anchored. At least one wall shall be an exterior wall. Openings to other parts of the building shall not be permitted. Windows and doors shall be in exterior walls and shall be located so as to be readily accessible in case of emergency. Windows shall be of glass or plastic in metal frames.

**1910.103(b)(3)(iii)(b)** Ventilation shall be as provided in paragraph (b)(3)(ii)(b) of this section.

**1910.103(b)(3)(iii)(c)** Explosion venting shall be as provided in paragraph (b)(3)(ii)(c) of this section.

**1910.103(b)(3)(iii)(d)** There shall be no sources of ignition from open flames, electrical equipment, or heating equipment.

**1910.103(b)(3)(iii)(e)** Electric equipment shall be in accordance with the requirements of subpart S of this part for Class I, Division 2 locations.

**1910.103(b)(3)(iii)(f)** Heating, if provided, shall be by steam, hot water, or indirect means.

**1910.103(b)(4)** *Operating instructions.* For installations which require any operation of equipment by the user, legible instructions shall be maintained at operating locations.

**1910.103(b)(5)** *Maintenance.* The equipment and functioning of each charged gaseous hydrogen system shall be maintained in a safe operating condition in accordance with the requirements of this section. The area within 15 feet of any hydrogen container shall be kept free of dry vegetation and combustible material.

**1910.103(c)** *Liquefied hydrogen systems*—**(1)** *Design*—**(i)** *Containers.* **(a)** Hydrogen containers shall comply with the following: Storage containers shall be designed, constructed, and tested in accordance with appropriate requirements of the ASME Boiler and Pressure Vessel Code, Section VIII—Unfired Pressure Vessels (1968) or applicable provisions of API Standard 620, Recommended Rules for Design and Construction of Large, Welded, Low-Pressure Storage Tanks, Second Edition (June 1963) and appendix R (April 1965), which is incorporated by reference as specified in §1910.6.

**1910.103(c)(1)(i)(b)** Portable containers shall be designed, constructed and tested in accordance with DOT Specifications and Regulations.

**1910.103(c)(1)(ii)** *Supports.* Permanently installed containers shall be provided with substantial noncombustible supports securely anchored on firm noncombustible foundations. Steel supports in excess of 18 inches in height shall be protected with a protective coating having a 2-hour fire-resistance rating.

**1910.103(c)(1)(iii)** *Marking.* Each container shall be legibly marked to indicate "LIQUEFIED HYDROGEN—FLAMMABLE GAS."

**1910.103(c)(1)(iv)** *Safety relief devices.* **(a)(1)** Stationary liquefied hydrogen containers shall be equipped with safety relief devices sized in accordance with CGA Pamphlet S-1, Part 3, Safety Relief Device Standards for Compressed Gas Storage Containers, which is incorporated by reference as specified in §1910.6.

**1910.103(c)(1)(iv)(a)(2)** Portable liquefied hydrogen containers complying with the U.S. Department of Transportation Regulations shall be equipped with safety relief devices as required in the U.S. Department of Transportation Specifications and Regulations. Safety relief devices shall be sized in accordance with the requirements of CGA Pamphlet S-1, Safety Relief Device Standards, Part 1, Compressed Gas Cylinders and Part 2, Cargo and Portable Tank Containers.

**1910.103(c)(1)(iv)(b)** Safety relief devices shall be arranged to discharge unobstructed to the outdoors and in such a manner as to prevent impingement of escaping liquid or gas upon the container, adjacent structures or personnel. See paragraph (c)(2)(i)(f) of this section for venting of safety relief devices in special locations.

**1910.103(c)(1)(iv)(c)** Safety relief devices or vent piping shall be designed or located so that moisture cannot collect and freeze in a manner which would interfere with proper operation of the device.

**1910.103(c)(1)(iv)(d)** Safety relief devices shall be provided in piping wherever liquefied hydrogen could be trapped between closures.

**1910.103(c)(1)(v)** *Piping, tubing, and fittings.* **(a)** Piping, tubing, and fittings and gasket and thread sealants shall be suitable for hydrogen service at the pressures and temperatures involved. Consideration shall be given to the thermal expansion and contraction of piping systems when exposed to temperature fluctuations of ambient to liquefied hydrogen temperatures.

**1910.103(c)(1)(v)(b)** Gaseous hydrogen piping and tubing (above -20 °F.) shall conform to the applicable sections of Pressure Piping Section 2—Industrial Gas and Air Piping, ANSI B31.1-1967 with addenda B31.1-1969. Design of liquefied hydrogen or cold (-20 °F. or below) gas piping shall use Petroleum Refinery Piping ANSI B31.3-1966 or Refrigeration Piping ANSI B31.5-1966 with addenda B31.5a-1968 as a guide, which are incorporated by reference as specified in §1910.6.

**1910.103(c)(1)(v)(c)** Joints in piping and tubing shall preferably be made by welding or brazing; flanged, threaded, socket, or suitable compression fittings may be used.

**1910.103(c)(1)(v)(d)** Means shall be provided to minimize exposure of personnel to piping operating at low temperatures and to prevent air condensate from contacting piping, structural members, and surfaces not suitable for cryogenic temperatures. Only those insulating materials which are rated nonburning in accordance with ASTM Procedures D1692-68, which is incorporated by reference as specified in §1910.6, may be used. Other protective means may be used to protect personnel. The insulation shall be designed to have a vapor-tight seal in the outer covering to prevent the condensation of air and subsequent oxygen enrichment within the insulation. The insulation material and outside shield shall also be of adequate design to prevent attrition of the insulation due to normal operating conditions.

**1910.103(c)(1)(v)(e)** Uninsulated piping and equipment which operate at liquefied-hydrogen temperature shall not be installed above asphalt surfaces or other combustible materials in order to prevent contact of liquid air with such materials. Drip pans may be installed under uninsulated piping and equipment to retain and vaporize condensed liquid air.

**1910.103(c)(1)(vi)** *Equipment assembly. (a)* Valves, gauges, regulators, and other accessories shall be suitable for liquefied hydrogen service and for the pressures and temperatures involved.

**1910.103(c)(1)(vi)(b)** Installation of liquefied hydrogen systems shall be supervised by personnel familiar with proper practices and with reference to their construction and use.

**1910.103(c)(1)(vi)(c)** Storage containers, piping, valves, regulating equipment, and other accessories shall be readily accessible and shall be protected against physical damage and against tampering. A shutoff valve shall be located in liquid product withdrawal lines as close to the container as practical. On containers of over 2,000 gallons capacity, this shutoff valve shall be of the remote control type with no connections, flanges, or other appurtenances (other than a welded manual shutoff valve) allowed in the piping between the shutoff valve and its connection to the inner container.

**1910.103(c)(1)(vi)(d)** Cabinets or housings containing hydrogen control equipment shall be ventilated to prevent any accumulation of hydrogen gas.

**1910.103(c)(1)(vii)** *Testing. (a)* After installation, all field-erected piping shall be tested and proved hydrogen gas-tight at operating pressure and temperature.

**1910.103(c)(1)(vii)(b)** Containers if out of service in excess of 1 year shall be inspected and tested as outlined in (a) of this subdivision. The safety relief devices shall be checked to determine if they are operable and properly set.

**1910.103(c)(1)(viii)** *Liquefied hydrogen vaporizers. (a)* The vaporizer shall be anchored and its connecting piping shall be sufficiently flexible to provide for the effect of expansion and contraction due to temperature changes.

**1910.103(c)(1)(viii)(b)** The vaporizer and its piping shall be adequately protected on the hydrogen and heating media sections with safety relief devices.

**1910.103(c)(1)(viii)(c)** Heat used in a liquefied hydrogen vaporizer shall be indirectly supplied utilizing media such as air, steam, water, or water solutions.

**1910.103(c)(1)(viii)(d)** A low temperature shutoff switch shall be provided in the vaporizer discharge piping to prevent flow of liquefied hydrogen in the event of the loss of the heat source.

**1910.103(c)(1)(ix)** *Electrical systems. (a)* Electrical wiring and equipment located within 3 feet of a point where connections are regularly made and disconnected, shall be in accordance with subpart S of this part, for Class I, Group B, Division 1 locations.

**1910.103(c)(1)(ix)(b)** Except as provided in (a) of this subdivision, electrical wiring, and equipment located within 25 feet of a point where connections are regularly made and disconnected or within 25 feet of a liquid hydrogen storage container, shall be in accordance with subpart S of this part, for Class I, Group B, Division 2 locations. When equipment approved for class I, group B atmospheres is not commercially available, the equipment may be—

**1910.103(c)(1)(ix)(b)(1)** Purged or ventilated in accordance with NFPA No. 496-1967, Standard for Purged Enclosures for Electrical Equipment in Hazardous Locations,

**1910.103(c)(1)(ix)(b)(2)** Intrinsically safe, or

**1910.103(c)(1)(ix)(b)(3)** Approved for Class I, Group C atmospheres. This requirement does not apply to electrical equipment which is installed on mobile supply trucks or tank cars from which the storage container is filled.

**1910.103(c)(1)(x)** *Bonding and grounding.* The liquefied hydrogen container and associated piping shall be electrically bonded and grounded.

**1910.103(c)(2)** *Location of liquefied hydrogen storage—(i) General requirements.* (a) The storage containers shall be located so that they are readily accessible to mobile supply equipment at ground level and to authorized personnel.

**1910.103(c)(2)(i)(b)** The containers shall not be exposed by electric power lines, flammable liquid lines, flammable gas lines, or lines carrying oxidizing materials.

**1910.103(c)(2)(i)(c)** When locating liquefied hydrogen storage containers near above-ground flammable liquid storage or liquid oxygen storage, it is advisable to locate the liquefied hydrogen container on ground higher than flammable liquid storage or liquid oxygen storage.

**1910.103(c)(2)(i)(d)** Where it is necessary to locate the liquefied hydrogen container on ground that is level with or lower than adjacent flammable liquid storage or liquid oxygen storage, suitable protective means shall be taken (such as by diking, diversion curbs, grading), with respect to the adjacent flammable liquid storage or liquid oxygen storage, to prevent accumulation of liquids within 50 feet of the liquefied hydrogen container.

**1910.103(c)(2)(i)(e)** Storage sites shall be fenced and posted to prevent entrance by unauthorized personnel. Sites shall also be placarded as follows: "Liquefied Hydrogen—Flammable Gas—No Smoking—No Open Flames."

**1910.103(c)(2)(i)(f)** If liquefied hydrogen is located in (as specified in Table H-3) a separate building, in a special room, or inside buildings when not in a special room and exposed to other occupancies, containers shall have the safety relief devices vented unobstructed to the outdoors at a minimum elevation of 25 feet above grade to a safe location as required in paragraph (c)(1)(iv)(b) of this section.

**1910.103(c)(2)(ii)** *Specific requirements. (a)* The location of liquefied hydrogen storage, as determined by the maximum total quantity of liquefied hydrogen, shall be in the order of preference as indicated by Roman numerals in the following Table H-3.

### TABLE H-3—MAXIMUM TOTAL QUANTITY OF LIQUEFIED HYDROGEN STORAGE PERMITTED

| Nature of location | 39.63 (150 liters) to 50 | 51 to 300 | 301 to 600 | In excess of 600 |
|---|---|---|---|---|
| Outdoors | I | I | I | I |
| In a separate building | II | II | II | Not permitted |
| In a special room | III | III | Not permitted | Do |
| Inside buildings not in a special room and exposed to other occupancies | IV | Not permitted | Do | Do |

*Size of hydrogen storage (capacity of gallons)*

NOTE: This table does not apply to the storage in dewars of the type generally used in laboratories for experimental purposes.

**1910.103(c)(2)(ii)(b)** The minimum distance in feet from liquefied hydrogen systems of indicated storage capacity located outdoors, in a separate building, or in a special room to any specified exposure shall be in accordance with Table H-4.

## TABLE H-4—MINIMUM DISTANCE (FEET) FROM LIQUEFIED HYDROGEN SYSTEMS TO EXPOSURE[1][2]

| Type of Exposure | 39.63 (150 liters) to 3,500 | 3,501 to 15,000 | 15,001 to 30,000 |
|---|---|---|---|
| 1. Fire-resistive building and fire walls[3] | 5 | 5 | 5 |
| 2. Noncombustible building[3] | 25 | 50 | 75 |
| 3. Other buildings[3] | 50 | 75 | 100 |
| 4. Wall openings, air-compressor intakes, inlets for air-conditioning or ventilating equipment | 75 | 75 | 75 |
| 5. Flammable liquids (above ground and vent or fill openings if below ground) (see 513 and 514) | 50 | 75 | 100 |
| 6. Between stationary liquefied hydrogen containers | 5 | 5 | 5 |
| 7. Flammable gas storage | 50 | 75 | 100 |
| 8. Liquid oxygen storage and other oxidizers (see 513 and 514) | 100 | 100 | 100 |
| 9. Combustible solids | 50 | 75 | 100 |
| 10. Open flames, smoking and welding | 50 | 50 | 50 |
| 11. Concentrations of people | 75 | 75 | 75 |

[1] The distance in Nos. 2, 3, 5, 7, 9, and 12 in Table H-4 may be reduced where protective structures, such as firewalls equal to height of top of the container, to safeguard the liquefied hydrogen storage system, are located between the liquefied hydrogen storage installation and the exposure.

[2] Where protective structures are provided, ventilation and confinement of product should be considered. The 5-foot distance in Nos. 1 and 6 facilitates maintenance and enhances ventilation.

[3] Refer to Standard Types of Building Construction, NFPA No. 220-1969 for definitions of various types of construction. In congested areas such as offices, lunchrooms, locker rooms, time-clock areas.

**1910.103(c)(2)(iii)** *Handling of liquefied hydrogen inside buildings other than separate buildings and special rooms.* Portable liquefied hydrogen containers of 50 gallons or less capacity as permitted in Table H-3 and in compliance with subdivision (i)(f) of this subparagraph when housed inside buildings not located in a special room and exposed to other occupancies shall comply with the following minimum requirements:

**1910.103(c)(2)(iii)(a)** Be located 20 feet from flammable liquids and readily combustible materials such as excelsior or paper.

**1910.103(c)(2)(iii)(b)** Be located 25 feet from ordinary electrical equipment and other sources of ignition including process or analytical equipment.

**1910.103(c)(2)(iii)(c)** Be located 25 feet from concentrations of people.

**1910.103(c)(2)(iii)(d)** Be located 50 feet from intakes of ventilation and air-conditioning equipment or intakes of compressors.

**1910.103(c)(2)(iii)(e)** Be located 50 feet from storage of other flammable-gases or storage of oxidizing gases.

**1910.103(c)(2)(iii)(f)** Containers shall be protected against damage or injury due to falling objects or work activity in the area.

**1910.103(c)(2)(iii)(g)** Containers shall be firmly secured and stored in an upright position.

**1910.103(c)(2)(iii)(h)** Welding or cutting operations, and smoking shall be prohibited while hydrogen is in the room.

**1910.103(c)(2)(iii)(i)** The area shall be adequately ventilated. Safety relief devices on the containers shall be vented directly outdoors or to a suitable hood. See paragraphs (c)(1)(iv)(b) and (c)(2)(i)(f) of this section.

**1910.103(c)(3)** *Design considerations at specific locations—(i) Outdoor locations. (a)* Outdoor location shall mean outside of any building or structure, and includes locations under a weather shelter or canopy provided such locations are not enclosed by more than two walls set at right angles and are provided with vent-space between the walls and vented roof or canopy.

**1910.103(c)(3)(i)(b)** Roadways and yard surfaces located below liquefied hydrogen piping, from which liquid air may drip, shall be constructed of noncombustible materials.

**1910.103(c)(3)(i)(c)** If protective walls are provided, they shall be constructed of noncombustible materials and in accordance with the provisions of paragraph (c)(3)(i)(a) of this section.

**1910.103(c)(3)(i)(d)** Electrical wiring and equipment shall comply with paragraph (c)(1)(ix) (a) and (b) of this section.

**1910.103(c)(3)(i)(e)** Adequate lighting shall be provided for nighttime transfer operation.

**1910.103(c)(3)(ii)** *Separate buildings. (a)* Separate buildings shall be of light noncombustible construction on a substantial frame. Walls and roofs shall be lightly fastened and designed to relieve at a maximum internal pressure of 25 pounds per square foot. Windows shall be of shatterproof glass or plastic in metal frames. Doors shall be located in such a manner that they will be readily accessible to personnel in an emergency.

**1910.103(c)(3)(ii)(b)** Adequate ventilation to the outdoors shall be provided. Inlet openings shall be located near the floor level in exterior walls only. Outlet openings shall be located at the high point of the room in exterior walls or roof. Both the inlet and outlet vent openings shall have a minimum total area of 1 square foot per 1,000 cubic feet of room volume. Discharge from outlet openings shall be directed or conducted to a safe location.

**1910.103(c)(3)(ii)(c)** There shall be no sources of ignition.

**1910.103(c)(3)(ii)(d)** Electrical wiring and equipment shall comply with paragraphs (c)(1)(ix) (a) and (b) of this section except that the provisions of paragraph (c)(1)(ix)(b) of this section shall apply to all electrical wiring and equipment in the separate building.

**1910.103(c)(3)(ii)(e)** Heating, if provided, shall be by steam, hot water, or other indirect means.

**1910.103(c)(3)(iii)** *Special rooms. (a)* Floors, walls, and ceilings shall have a fire resistance rating of at least 2 hours. Walls or partitions shall be continuous from floor to ceiling and shall be securely anchored. At least one wall shall be an exterior wall. Openings to other parts of the building shall not be permitted. Windows and doors shall be in exterior walls and doors shall be located in such a manner that they will be accessible in an emergency. Windows shall be of shatterproof glass or plastic in metal frames.

**1910.103(c)(3)(iii)(b)** Ventilation shall be as provided in paragraph (c)(3)(ii)(b) of this section.

**1910.103(c)(3)(iii)(c)** Explosion venting shall be provided in exterior walls or roof only. The venting area shall be equal to not less than 1 square foot per 30 cubic feet of room volume and may consist of any one or any combination of the following: Walls of light noncombustible material; lightly fastened hatch covers; lightly fastened swinging doors opening outward in exterior walls; lightly fastened walls or roofs designed to relieve at a maximum pressure of 25 pounds per square foot.

**1910.103(c)(3)(iii)(d)** There shall be no sources of ignition.

**1910.103(c)(3)(iii)(e)** Electrical wiring and equipment shall comply with paragraph (c)(1)(ix) (a) and (b) of this section except that the provision of paragraph (c)(1)(ix)(b) of this section shall apply to all electrical wiring and equipment in the special room.

**1910.103(c)(3)(iii)(f)** Heating, if provided, shall be steam, hot water, or by other indirect means.

**1910.103(c)(4)** *Operating instructions—(i) Written instructions.* For installation which require any operation of equipment by the user, legible instructions shall be maintained at operating locations.

**1910.103(c)(4)(ii)** *Attendant.* A qualified person shall be in attendance at all times while the mobile hydrogen supply unit is being unloaded.

**1910.103(c)(4)(iii)** *Security.* Each mobile liquefied hydrogen supply unit used as part of a hydrogen system shall be adequately secured to prevent movement.

**1910.103(c)(4)(iv)** *Grounding.* The mobile liquefied hydrogen supply unit shall be grounded for static electricity.

**1910.103(c)(5)** *Maintenance.* The equipment and functioning of each charged liquefied hydrogen system shall be maintained in a safe operating condition in accordance with the requirements of this section. Weeds or similar combustibles shall not be permitted within 25 feet of any liquefied hydrogen equipment.

[39 FR 23502, June 27, 1974, as amended at 43 FR 49746, Oct. 24, 1978; 53 FR 12121, Apr. 12, 1988; 55 FR 32015, Aug. 6, 1990; 58 FR 35309, June 30, 1993; 61 FR 9236, 9237, Mar. 7, 1996]

## §1910.104 Oxygen.

**1910.104(a)** *Scope.* This section applies to the installation of bulk oxygen systems on industrial and institutional consumer premises. This section does not apply to oxygen manufacturing plants or other establishments operated by the oxygen supplier or his agent for the purpose of storing oxygen and refilling portable containers, trailers, mobile supply trucks, or tank cars, nor to systems having capacities less than those stated in paragraph (b)(1) of this section.

**1910.104(b)** *Bulk oxygen systems*—**(1)** *Definition.* As used in this section: A bulk oxygen system is an assembly of equipment, such as oxygen storage containers, pressure regulators, safety devices, vaporizers, manifolds, and interconnecting piping, which has storage capacity of more than 13,000 cubic feet of oxygen, Normal Temperature and Pressure (NTP), connected in service or ready for service, or more than 25,000 cubic feet of oxygen (NTP) including unconnected reserves on hand at the site. The bulk oxygen system terminates at the point where oxygen at service pressure first enters the supply line. The oxygen containers may be stationary or movable, and the oxygen may be stored as gas or liquid.

**1910.104(b)(2)** *Location*—**(i)** *General.* Bulk oxygen storage systems shall be located above ground out of doors, or shall be installed in a building of noncombustible construction, adequately vented, and used for that purpose exclusively. The location selected shall be such that containers and associated equipment shall not be exposed by electric power lines, flammable or combustible liquid lines, or flammable gas lines.

**1910.104(b)(2)(ii)** *Accessibility.* The system shall be located so that it is readily accessible to mobile supply equipment at ground level and to authorized personnel.

**1910.104(b)(2)(iii)** *Leakage.* Where oxygen is stored as a liquid, noncombustible surfacing shall be provided in an area in which any leakage of liquid oxygen might fall during operation of the system and filling of a storage container. For purposes of this paragraph, asphaltic or bituminous paving is considered to be combustible.

**1910.104(b)(2)(iv)** *Elevation.* When locating bulk oxygen systems near above-ground flammable or combustible liquid storage which may be either indoors or outdoors, it is advisable to locate the system on ground higher than the flammable or combustible liquid storage.

**1910.104(b)(2)(v)** *Dikes.* Where it is necessary to locate a bulk oxygen system on ground lower than adjacent flammable or combustible liquid storage suitable means shall be taken (such as by diking, diversion curbs, or grading) with respect to the adjacent flammable or combustible liquid storage to prevent accumulation of liquids under the bulk oxygen system.

**1910.104(b)(3)** *Distance between systems and exposures*—**(i)** *General.* The minimum distance from any bulk oxygen storage container to exposures, measured in the most direct line except as indicated in paragraphs (b)(3)(vi) and (viii) of this section, shall be as indicated in paragraphs (b)(3)(ii) to (xviii) of this section inclusive.

**1910.104(b)(3)(ii)** *Combustible structures.* Fifty feet from any combustible structures.

**1910.104(b)(3)(iii)** *Fire resistive structures.* Twenty-five feet from any structures with fire-resistive exterior walls or sprinklered buildings of other construction, but not less than one-half the height of adjacent side wall of the structure.

**1910.104(b)(3)(iv)** *Openings.* At least 10 feet from any opening in adjacent walls of fire resistive structures. Spacing from such structures shall be adequate to permit maintenance, but shall not be less than 1 foot.

**1910.104(b)(3)(v)** *Flammable liquid storage above-ground.*

| Distance (feet) | Capacity (gallons) |
| --- | --- |
| 50 | 0 to 1000 |
| 90 | 1001 or more |

**1910.104(b)(3)(vi)** *Flammable liquid storage below-ground.*

| Distance measured horizontally from oxygen storage container to flammable liquid tank (feet) | Distance from oxygen storage container to filling and vent connections or openings to flammable liquid tank (feet) | Capacity gallons |
| --- | --- | --- |
| 15 | 50 | 0 to 1000 |
| 30 | 50 | 1001 or more |

**1910.104(b)(3)(vii)** *Combustible liquid storage above-ground.*

| Distance (feet) | Capacity (gallons) |
| --- | --- |
| 25 | 0 to 1000 |
| 50 | 1001 or more |

**1910.104(b)(3)(viii)** *Combustible liquid storage below-ground.*

| Distance measured horizontally from oxygen storage container to combustible liquid tank (feet) | Distance from oxygen storage container to filling and vent connections or openings to combustible liquid tank (feet) |
| --- | --- |
| 15 | 40 |

**1910.104(b)(3)(ix)** *Flammable gas storage.* (Such as compressed flammable gases, liquefied flammable gases and flammable gases in low pressure gas holders):

| Distance (feet) | Capacity (cu. ft. NTP) |
| --- | --- |
| 50 | Less than 5000 |
| 90 | 5000 or more |

**1910.104(b)(3)(x)** *Highly combustible materials.* Fifty feet from solid materials which burn rapidly, such as excelsior or paper.

**1910.104(b)(3)(xi)** *Slow-burning materials.* Twenty-five feet from solid materials which burn slowly, such as coal and heavy timber.

**1910.104(b)(3)(xii)** *Ventilation.* Seventy-five feet in one direction and 35 feet in approximately 90° direction from confining walls (not including firewalls less than 20 feet high) to provide adequate ventilation in courtyards and similar confining areas.

**1910.104(b)(3)(xiii)** *Congested areas.* Twenty-five feet from congested areas such as offices, lunchrooms, locker rooms, time clock areas, and similar locations where people may congregate.

**1910.104(b)(3)(xiv)-(xvii)** [Reserved]

**1910.104(b)(3)(xviii)** *Exceptions.* The distances in paragraphs (b)(3)(ii), (iii), (v) to (xi) inclusive, of this section do not apply where protective structures such as firewalls of adequate height to safeguard the oxygen storage systems are located between the bulk oxygen storage installation and the exposure. In such cases, the bulk oxygen storage installation may be a minimum distance of 1 foot from the firewall.

**1910.104(b)(4)** *Storage containers*—**(i)** *Foundations and supports.* Permanently installed containers shall be provided with substantial noncombustible supports on firm noncombustible foundations.

# OSHA Standards for General Industry

**1910.104(b)(4)(ii)** *Construction—liquid.* Liquid oxygen storage containers shall be fabricated from materials meeting the impact test requirements of paragraph UG-84 of ASME Boiler and Pressure Vessel Code, Section VIII—Unfired Pressure Vessels—1968, which is incorporated by reference as specified in §1910.6. Containers operating at pressures above 15 pounds per square inch gage (p.s.i.g.) shall be designed, constructed, and tested in accordance with appropriate requirements of ASME Boiler and Pressure Vessel Code, Section VII—Unfired Pressure Vessels—1968. Insulation surrounding the liquid oxygen container shall be noncombustible.

**1910.104(b)(4)(iii)** *Construction—gaseous.* High-pressure gaseous oxygen containers shall comply with one of the following:

**1910.104(b)(4)(iii)(a)** Designed, constructed, and tested in accordance with appropriate requirements of ASME Boiler and Pressure Vessel Code, Section VIII—Unfired Pressure Vessels—1968.

**1910.104(b)(4)(iii)(b)** Designed, constructed, tested, and maintained in accordance with DOT Specifications and Regulations.

**1910.104(b)(5)** *Piping, tubing, and fittings—(i) Selection.* Piping, tubing, and fittings shall be suitable for oxygen service and for the pressures and temperatures involved.

**1910.104(b)(5)(ii)** *Specification.* Piping and tubing shall conform to Section 2—Gas and Air Piping Systems of Code for Pressure Piping, ANSI, B31.1-1967 with addenda B31.10a-1969, which is incorporated by reference as specified in §1910.6.

**1910.104(b)(5)(iii)** *Fabrication.* Piping or tubing for operating temperatures below -20 °F. shall be fabricated from materials meeting the impact test requirements of paragraph UG-84 of ASME Boiler and Pressure Vessel Code, Section VIII—Unfired Pressure Vessels—1968, when tested at the minimum operating temperature to which the piping may be subjected in service.

**1910.104(b)(6)** *Safety relief devices—(i) General.* Bulk oxygen storage containers, regardless of design pressure shall be equipped with safety relief devices as required by the ASME code or the DOT specifications and regulations.

**1910.104(b)(6)(ii)** *DOT containers.* Bulk oxygen storage containers designed and constructed in accordance with DOT specification shall be equipped with safety relief devices as required thereby.

**1910.104(b)(6)(iii)** *ASME containers.* Bulk oxygen storage containers designed and constructed in accordance with the ASME Boiler and Pressure Vessel Code, Section VIII—Unfired Pressure Vessel—1968 shall be equipped with safety relief devices meeting the provisions of the Compressed Gas Association Pamphlet "Safety Relief Device Standards for Compressed Gas Storage Containers," S-1, Part 3, which is incorporated by reference as specified in §1910.6.

**1910.104(b)(6)(iv)** *Insulation.* Insulation casings on liquid oxygen containers shall be equipped with suitable safety relief devices.

**1910.104(b)(6)(v)** *Reliability.* All safety relief devices shall be so designed or located that moisture cannot collect and freeze in a manner which would interfere with proper operation of the device.

**1910.104(b)(7)** *Liquid oxygen vaporizers—(i) Mounts and couplings.* The vaporizer shall be anchored and its connecting piping be sufficiently flexible to provide for the effect of expansion and contraction due to temperature changes.

**1910.104(b)(7)(ii)** *Relief devices.* The vaporizer and its piping shall be adequately protected on the oxygen and heating medium sections with safety relief devices.

**1910.104(b)(7)(iii)** *Heating.* Heat used in an oxygen vaporizer shall be indirectly supplied only through media such as steam, air, water, or water solutions which do not react with oxygen.

**1910.104(b)(7)(iv)** *Grounding.* If electric heaters are used to provide the primary source of heat, the vaporizing system shall be electrically grounded.

**1910.104(b)(8)** *Equipment assembly and installation—(i) Cleaning.* Equipment making up a bulk oxygen system shall be cleaned in order to remove oil, grease or other readily oxidizable materials before placing the system in service.

**1910.104(b)(8)(ii)** *Joints.* Joints in piping and tubing may be made by welding or by use of flanged, threaded, slip, or compression fittings. Gaskets or thread sealants shall be suitable for oxygen service.

**1910.104(b)(8)(iii)** *Accessories.* Valves, gages, regulators, and other accessories shall be suitable for oxygen service.

**1910.104(b)(8)(iv)** *Installation.* Installation of bulk oxygen systems shall be supervised by personnel familiar with proper practices with reference to their construction and use.

**1910.104(b)(8)(v)** *Testing.* After installation all field erected piping shall be tested and proved gas tight at maximum operating pressure. Any medium used for testing shall be oil free and nonflammable.

**1910.104(b)(8)(vi)** *Security.* Storage containers, piping, valves, regulating equipment, and other accessories shall be protected against physical damage and against tampering.

**1910.104(b)(8)(vii)** *Venting.* Any enclosure containing oxygen control or operating equipment shall be adequately vented.

**1910.104(b)(8)(viii)** *Placarding.* The bulk oxygen storage location shall be permanently placarded to indicate: "OXYGEN—NO SMOKING—NO OPEN FLAMES", or an equivalent warning.

**1910.104(b)(8)(ix)** *Electrical wiring.* Bulk oxygen installations are not hazardous locations as defined and covered in subpart S of this part. Therefore, general purpose or weatherproof types of electrical wiring and equipment are acceptable depending upon whether the installation is indoors or outdoors. Such equipment shall be installed in accordance with the applicable provisions of subpart S of this part.

**1910.104(b)(9)** *Operating instructions.* For installations which require any operation of equipment by the user, legible instructions shall be maintained at operating locations.

**1910.104(b)(10)** *Maintenance.* The equipment and functioning of each charged bulk oxygen system shall be maintained in a safe operating condition in accordance with the requirements of this section. Wood and long dry grass shall be cut back within 15 feet of any bulk oxygen storage container.

[39 FR 23502, June 27, 1974, as amended at 43 FR 49746, Oct. 24, 1978; 61 FR 9237, Mar. 7, 1996]

## §1910.105 Nitrous oxide.

The piped systems for the in-plant transfer and distribution of nitrous oxide shall be designed, installed, maintained, and operated in accordance with Compressed Gas Association Pamphlet G-8.1-1964, which is incorporated by reference as specified in §1910.6.

[39 FR 23502, June 27, 1974, as amended at 61 FR 9237, Mar. 7, 1996]

## §1910.106 Flammable and combustible liquids.

**1910.106(a)** *Definitions.* As used in this section:

**1910.106(a)(1)** Aerosol shall mean a material which is dispensed from its container as a mist, spray, or foam by a propellant under pressure.

**1910.106(a)(2)** Atmospheric tank shall mean a storage tank which has been designed to operate at pressures from atmospheric through 0.5 p.s.i.g.

**1910.106(a)(3)** Automotive service station shall mean that portion of property where flammable or combustible liquids used as motor fuels are stored and dispensed from fixed equipment into the fuel tanks of motor vehicles and shall include any facilities available for the sale and service of tires, batteries, and accessories, and for minor automotive maintenance work. Major automotive repairs, painting, body and fender work are excluded.

**1910.106(a)(4)** Basement shall mean a story of a building or structure having one-half or more of its height below ground level and to which access for fire fighting purposes is unduly restricted.

**1910.106(a)(5)** Boiling point shall mean the boiling point of a liquid at a pressure of 14.7 pounds per square inch absolute (p.s.i.a.) (760 mm.). Where an accurate boiling point is unavailable for the material in question, or for mixtures which do not have a constant boiling point, for purposes of this section the 10 percent point of a distillation performed in accordance with the Standard Method of Test for Distillation of Petroleum Products, ASTM D-86-62, which is incorporated by reference as specified in §1910.6, may be used as the boiling point of the liquid.

**1910.106(a)(6)** Boilover shall mean the expulsion of crude oil (or certain other liquids) from a burning tank. The light fractions of the crude oil burnoff producing a heat wave in the residue, which on reaching a water strata may result in the expulsion of a portion of the contents of the tank in the form of froth.

**1910.106(a)(7)** Bulk plant shall mean that portion of a property where flammable or combustible liquids are received by tank vessel, pipelines, tank car, or tank vehicle, and are stored or blended in bulk for the purpose of distributing such liquids by tank vessel, pipeline, tank car, tank vehicle, or container.

**1910.106(a)(8)** Chemical plant shall mean a large integrated plant or that portion of such a plant other than a refinery or distillery where flammable or combustible liquids are produced by chemical reactions or used in chemical reactions.

**1910.106(a)(9)** Closed container shall mean a container as herein defined, so sealed by means of a lid or other device that neither liquid nor vapor will escape from it at ordinary temperatures.

**1910.106(a)(10)** Crude petroleum shall mean hydrocarbon mixtures that have a flash point below 150 °F. and which have not been processed in a refinery.

**1910.106(a)(11)** Distillery shall mean a plant or that portion of a plant where flammable or combustible liquids produced by fermentation are concentrated, and where the concentrated products may also be mixed, stored, or packaged.

**1910.106(a)(12)** Fire area shall mean an area of a building separated from the remainder of the building by construction having a fire resistance of at least 1 hour and having all communicating openings properly protected by an assembly having a fire resistance rating of at least 1 hour.

**1910.106(a)(13)** Flammable aerosol shall mean an aerosol which is required to be labeled "Flammable" under the Federal Hazardous Substances Labeling Act (15 U.S.C. 1261). For the purposes of paragraph (d) of this section, such aerosols are considered Class IA liquids.

**1910.106(a)(14)** *Flashpoint* means the minimum temperature at which a liquid gives off vapor within a test vessel in sufficient concentration to form an ignitable mixture with air near the surface of the liquid, and shall be determined as follows:

**1910.106(a)(14)(i)** For a liquid which has a viscosity of less than 45 SUS at 100 °F. (37.8 °C.), does not contain suspended solids, and does not have a tendency to form a surface film while under test, the procedure specified in the Standard Method of Test for Flashpoint by Tag Closed Tester (ASTM D-56-70), which is incorporated by reference as specified in §1910.6, shall be used.

**1910.106(a)(14)(ii)** For a liquid which has a viscosity of 45 SUS or more at 100 °F. (37.8 °C.), or contains suspended solids, or has a tendency to form a surface film while under test, the Standard Method of Test for Flashpoint by Pensky-Martens Closed Tester (ASTM D-93-71) shall be used, except that the methods specified in Note 1 to section 1.1 of ASTM D-93-71 may be used for the respective materials specified in the Note. The preceding ASTM standards are incorporated by reference as specified in §1910.6.

**1910.106(a)(14)(iii)** For a liquid that is a mixture of compounds that have different volatilities and flashpoints, its flashpoint shall be determined by using the procedure specified in paragraph (a)(14) (i) or (ii) of this section on the liquid in the form it is shipped. If the flashpoint, as determined by this test, is 100 °F. (37.8 °C.) or higher, an additional flashpoint determination shall be run on a sample of the liquid evaporated to 90 percent of its original volume, and the lower value of the two tests shall be considered the flashpoint of the material.

**1910.106(a)(14)(iv)** Organic peroxides, which undergo autoaccelerating thermal decomposition, are excluded from any of the flashpoint determination methods specified in this subparagraph.

**1910.106(a)(15)** Hotel shall mean buildings or groups of buildings under the same management in which there are sleeping accommodations for hire, primarily used by transients who are lodged with or without meals including but not limited to inns, clubs, motels, and apartment hotels.

**1910.106(a)(16)** Institutional occupancy shall mean the occupancy or use of a building or structure or any portion thereof by persons harbored or detained to receive medical, charitable or other care or treatment, or by persons involuntarily detained.

**1910.106(a)(17)** Liquid shall mean, for the purpose of this section, any material which has a fluidity greater than that of 300 penetration asphalt when tested in accordance with ASTM Test for Penetration for Bituminous Materials, D-5-65, which is incorporated by reference as specified in §1910.6. When not otherwise identified, the term liquid shall include both flammable and combustible liquids.

**1910.106(a)(18)** *Combustible liquid* means any liquid having a flashpoint at or above 100 °F. (37.8 °C.) Combustible liquids shall be divided into two classes as follows:

**1910.106(a)(18)(i)** *Class II liquids* shall include those with flashpoints at or above 100 °F. (37.8 °C.) and below 140 °F. (60 °C.), except any mixture having components with flashpoints of 200 °F. (93.3 °C.) or higher, the volume of which make up 99 percent or more of the total volume of the mixture.

**1910.106(a)(18)(ii)** *Class III liquids* shall include those with flashpoints at or above 140 °F. (60 °C.) Class III liquids are subdivided into two subclasses:

**1910.106(a)(18)(ii)(a)** *Class IIIA liquids* shall include those with flashpoints at or above 140 °F. (60 °C.) and below 200 °F. (93.3 °C.), except any mixture having components with flashpoints of 200 °F. (93.3 °C.), or higher, the total volume of which make up 99 percent or more of the total volume of the mixture.

**1910.106(a)(18)(ii)(b)** *Class IIIB liquids* shall include those with flashpoints at or above 200 °F. (93.3 °C.). This section does not cover Class IIIB liquids. Where the term "Class III liquids is used in this section, it shall mean only Class IIIA liquids.

**1910.106(a)(18)(iii)** When a combustible liquid is heated for use to within 30 °F. (16.7 °C.) of its flashpoint, it shall be handled in accordance with the requirements for the next lower class of liquids.

**1910.106(a)(19)** *Flammable liquid* means any liquid having a flashpoint below 100 °F. (37.8 °C.), except any mixture having components with flashpoints of 100 °F. (37.8 °C.) or higher, the total of which make up 99 percent or more of the total volume of the mixture. Flammable liquids shall be known as Class I liquids. Class I liquids are divided into three classes as follows:

**1910.106(a)(19)(i)** Class IA shall include liquids having flashpoints below 73 °F. (22.8 °C.) and having a boiling point below 100 °F. (37.8 °C.).

**1910.106(a)(19)(ii)** Class IB shall include liquids having flashpoints below 73 °F. (22.8 °C.) and having a boiling point at or above 100 °F. (37.8 °C.).

**1910.106(a)(19)(iii)** Class IC shall include liquids having flashpoints at or above 73 °F. (22.8 °C.) and below 100 °F. (37.8 °C.).

**1910.106(a)(20)** Unstable (reactive) liquid shall mean a liquid which in the pure state or as commercially produced or transported will vigorously polymerize, decompose, condense, or will become self-reactive under conditions of shocks, pressure, or temperature.

**1910.106(a)(21)** Low-pressure tank shall mean a storage tank which has been designed to operate at pressures above 0.5 p.s.i.g. but not more than 15 p.s.i.g.

**1910.106(a)(22)** Marine service station shall mean that portion of a property where flammable or combustible liquids used as fuels are stored and dispensed from fixed equipment on shore, piers, wharves, or floating docks into the fuel tanks of self-propelled craft, and shall include all facilities used in connection therewith.

**1910.106(a)(23)** Mercantile occupancy shall mean the occupancy or use of a building or structure or any portion thereof for the displaying, selling, or buying of goods, wares, or merchandise.

**1910.106(a)(24)** Office occupancy shall mean the occupancy or use of a building or structure or any portion thereof for the transaction of business, or the rendering or receiving of professional services.

**1910.106(a)(25)** Portable tank shall mean a closed container having a liquid capacity over 60 U.S. gallons and not intended for fixed installation.

**1910.106(a)(26)** Pressure vessel shall mean a storage tank or vessel which has been designed to operate at pressures above 15 p.s.i.g.

**1910.106(a)(27)** Protection for exposure shall mean adequate fire protection for structures on property adjacent to tanks, where there are employees of the establishment.

**1910.106(a)(28)** Refinery shall mean a plant in which flammable or combustible liquids are produced on a commercial scale from crude petroleum, natural gasoline, or other hydrocarbon sources.

**1910.106(a)(29)** Safety can shall mean an approved container, of not more than 5 gallons capacity, having a spring-closing lid and spout cover and so designed that it will safely relieve internal pressure when subjected to fire exposure.

**1910.106(a)(30)** Vapor pressure shall mean the pressure, measured in pounds per square inch (absolute) exerted by a volatile liquid as determined by the "Standard Method of Test for Vapor Pressure of Petroleum Products (Reid Method)," American Society for Testing and Materials ASTM D323-68, which is incorporated by reference as specified in §1910.6.

**1910.106(a)(31)** Ventilation as specified in this section is for the prevention of fire and explosion. It is considered adequate if it is sufficient to prevent accumulation of significant quantities of vapor-air mixtures in concentration over one-fourth of the lower flammable limit.

**1910.106(a)(32)** Storage: Flammable or combustible liquids shall be stored in a tank or in a container that complies with paragraph (d)(2) of this section.

**1910.106(a)(33)** Barrel shall mean a volume of 42 U.S. gallons.

**1910.106(a)(34)** Container shall mean any can, barrel, or drum.

**1910.106(a)(35)** Approved unless otherwise indicated, approved, or listed by a nationally recognized testing laboratory. Refer to §1910.7 for definition of nationally recognized testing laboratory.

**1910.106(a)(36)** Listed see "approved" in §1910.106(a)(35).

**1910.106(a)(37)** *SUS* means Saybolt Universal Seconds as determined by the Standard Method of Test for Saybolt Viscosity (ASTM D-88-56), and may be determined by use of the SUS conversion tables specified in ASTM Method D2161-66 following determination of viscosity in accordance with the procedures specified in the Standard Method of Test for Viscosity of Transparent and Opaque Liquids (ASTM D445-65).

**1910.106(a)(38)** *Viscous* means a viscosity of 45 SUS or more.

**1910.106(b) Tank storage**—(1) *Design and construction of tanks*—(i) *Materials.* (a) Tanks shall be built of steel except as provided in paragraphs (b)(1)(i) (b) through (e) of this section.

**1910.106(b)(1)(i)(b)** Tanks may be built of materials other than steel for installation underground or if required by the properties of the liquid stored. Tanks located above ground or inside buildings shall be of noncombustible construction.

**1910.106(b)(1)(i)(c)** Tanks built of materials other than steel shall be designed to specifications embodying principles recognized as good engineering design for the material used.

**1910.106(b)(1)(i)(d)** Unlined concrete tanks may be used for storing flammable or combustible liquids having a gravity of 40° API or heavier. Concrete tanks with special lining may be used for other services provided the design is in accordance with sound engineering practice.

**1910.106(b)(1)(i)(e)** [Reserved]

**1910.106(b)(1)(i)(f)** Special engineering consideration shall be required if the specific gravity of the liquid to be stored exceeds that of water or if the tanks are designed to contain flammable or combustible liquids at a liquid temperature below 0 °F.

**1910.106(b)(1)(ii)** *Fabrication.* (a) [Reserved]

**1910.106(b)(1)(ii)(b)** Metal tanks shall be welded, riveted, and caulked, brazed, or bolted, or constructed by use of a combination of these methods. Filler metal used in brazing shall be nonferrous metal or an alloy having a melting point above 1000 °F. and below that of the metal joined.

**1910.106(b)(1)(iii)** *Atmospheric tanks.* (a) Atmospheric tanks shall be built in accordance with acceptable good standards of design. Atmospheric tanks may be built in accordance with the following consensus standards that are incorporated by reference as specified in §1910.6:

**1910.106(b)(1)(iii)(a)(1)** Underwriters' Laboratories, Inc., Subjects No. 142, Standard for Steel Aboveground Tanks for Flammable and Combustible Liquids, 1968; No. 58, Standard for Steel Underground Tanks for Flammable and Combustible Liquids, Fifth Edition, December 1961; or No. 80, Standard for Steel Inside Tanks for Oil-Burner Fuel, September 1963.

**1910.106(b)(1)(iii)(a)(2)** American Petroleum Institute Standards No. 650, Welded Steel Tanks for Oil Storage, Third Edition, 1966.

**1910.106(b)(1)(iii)(a)(3)** American Petroleum Institute Standards No. 12B, Specification for Bolted Production Tanks, Eleventh Edition, May 1958, and Supplement 1, March 1962; No. 12D, Specification for Large Welded Production Tanks, Seventh Edition, August 1957; or No. 12F, Specification for Small Welded Production Tanks, Fifth Edition, March 1961. Tanks built in accordance with these standards shall be used only as production tanks for storage of crude petroleum in oil-producing areas.

**1910.106(b)(1)(iii)(b)** Tanks designed for underground service not exceeding 2,500 gallons capacity may be used aboveground.

**1910.106(b)(1)(iii)(c)** Low-pressure tanks and pressure vessels may be used as atmospheric tanks.

**1910.106(b)(1)(iii)(d)** Atmospheric tanks shall not be used for the storage of a flammable or combustible liquid at a temperature at or above its boiling point.

**1910.106(b)(1)(iv)** *Low pressure tanks.* (a) The normal operating pressure of the tank shall not exceed the design pressure of the tank.

**1910.106(b)(1)(iv)(b)** Low-pressure tanks shall be built in accordance with acceptable standards of design. Low-pressure tanks may be built in accordance with the following consensus standards that are incorporated by reference as specified in §1910.6:

**1910.106(b)(1)(iv)(b)(1)** American Petroleum Institute Standard No. 620. Recommended Rules for the Design and Construction of Large, Welded, Low-Pressure Storage Tanks, Third Edition, 1966.

**1910.106(b)(1)(iv)(b)(2)** The principles of the Code for Unfired Pressure Vessels, Section VIII of the ASME Boiler and Pressure Vessels Code, 1968.

**1910.106(b)(1)(iv)(c)** Atmospheric tanks built according to Underwriters' Laboratories, Inc., requirements in subdivision (iii)(a) of and shall be limited to 2.5 p.s.i.g. under emergency venting conditions.

This paragraph may be used for operating pressures not exceeding 1 p.s.i.g.

**1910.106(b)(1)(iv)(d)** Pressure vessels may be used as low-pressure tanks.

**1910.106(b)(1)(v)** *Pressure vessels.* (a) The normal operating pressure of the vessel shall not exceed the design pressure of the vessel.

**1910.106(b)(1)(v)(b)** Pressure vessels shall be built in accordance with the Code for Unfired Pressure Vessels, Section VIII of the ASME Boiler and Pressure Vessel Code 1968.

**1910.106(b)(1)(vi)** *Provisions for internal corrosion.* When tanks are not designed in accordance with the American Petroleum Institute, American Society of Mechanical Engineers, or the Underwriters' Laboratories, Inc.'s, standards, or if corrosion is anticipated beyond that provided for in the design formulas used, additional metal thickness or suitable protective coatings or linings shall be provided to compensate for the corrosion loss expected during the design life of the tank.

**1910.106(b)(2)** *Installation of outside aboveground tanks.*

**1910.106(b)(2)(i)** [Reserved]

**1910.106(b)(2)(ii)** *Spacing (shell-to-shell) between aboveground tanks.* (a) The distance between any two flammable or combustible liquid storage tanks shall not be less than 3 feet.

**1910.106(b)(2)(ii)(b)** Except as provided in paragraph (b)(2)(ii)(c) of this section, the distance between any two adjacent tanks shall not be less than one-sixth the sum of their diameters. When the diameter of one tank is less than one-half the diameter of the adjacent tank, the distance between the two tanks shall not be less than one-half the diameter of the smaller tank.

**1910.106(b)(2)(ii)(c)** Where crude petroleum in conjunction with production facilities are located in noncongested areas and have capacities not exceeding 126,000 gallons (3,000 barrels), the distance between such tanks shall not be less than 3 feet.

**1910.106(b)(2)(ii)(d)** Where unstable flammable or combustible liquids are stored, the distance between such tanks shall not be less than one-half the sum of their diameters.

**1910.106(b)(2)(ii)(e)** When tanks are compacted in three or more rows or in an irregular pattern, greater spacing or other means shall be provided so that inside tanks are accessible for firefighting purposes.

**1910.106(b)(2)(ii)(f)** The minimum separation between a liquefied petroleum gas container and a flammable or combustible liquid storage tank shall be 20 feet, except in the case of flammable or combustible liquid tanks operating at pressures exceeding 2.5 p.s.i.g. or equipped with emergency venting which will permit pressures to exceed 2.5 p.s.i.g. in which case the provisions of subdivisions (a) and (b) of this subdivision shall apply. Suitable means shall be taken to prevent the accumulation of flammable or combustible liquids under adjacent liquefied petroleum gas containers such as by diversion curbs or grading. When flammable or combustible liquid storage tanks are within a diked area, the liquefied petroleum gas containers shall be outside the diked area and at least 10 feet away from the centerline of the wall of the diked area. The foregoing provisions shall not apply when liquefied petroleum gas containers of 125 gallons or less capacity are installed adjacent to fuel oil supply tanks of 550 gallons or less capacity.

**1910.106(b)(2)(iii)** [Reserved]

**1910.106(b)(2)(iv)** *Normal venting for aboveground tanks.* *(a)* Atmospheric storage tanks shall be adequately vented to prevent the development of vacuum or pressure sufficient to distort the roof of a cone roof tank or exceeding the design pressure in the case of other atmospheric tanks, as a result of filling or emptying, and atmospheric temperature changes.

**1910.106(b)(2)(iv)***(b)* Normal vents shall be sized either in accordance with: (*1*) The American Petroleum Institute Standard 2000 (1968), Venting Atmospheric and Low-Pressure Storage Tanks, which is incorporated by reference as specified in §1910.6; or (*2*) other accepted standard; or (*3*) shall be at least as large as the filling or withdrawal connection, whichever is larger but in no case less than 1 1/4 inch nominal inside diameter.

**1910.106(b)(2)(iv)***(c)* Low-pressure tanks and pressure vessels shall be adequately vented to prevent development of pressure or vacuum, as a result of filling or emptying and atmospheric temperature changes, from exceeding the design pressure of the tank or vessel. Protection shall also be provided to prevent overpressure from any pump discharging into the tank or vessel when the pump discharge pressure can exceed the design pressure of the tank or vessel.

**1910.106(b)(2)(iv)***(d)* If any tank or pressure vessel has more than one fill or withdrawal connection and simultaneous filling or withdrawal can be made, the vent size shall be based on the maximum anticipated simultaneous flow.

**1910.106(b)(2)(iv)***(e)* Unless the vent is designed to limit the internal pressure 2.5 p.s.i. or less, the outlet of vents and vent drains shall be arranged to discharge in such a manner as to prevent localized overheating of any part of the tank in the event vapors from such vents are ignited.

**1910.106(b)(2)(iv)***(f)* Tanks and pressure vessels storing Class IA liquids shall be equipped with venting devices which shall be normally closed except when venting to pressure or vacuum conditions. Tanks and pressure vessels storing Class IB and IC liquids shall be equipped with venting devices which shall be normally closed except when venting under pressure or vacuum conditions, or with approved flame arresters.

Exemption: Tanks of 3,000 bbls. capacity or less containing crude petroleum in crude-producing areas; and, outside aboveground atmospheric tanks under 1,000 gallons capacity containing other than Class IA flammable liquids may have open vents. (See subdivision (vi)(*b*) of this subparagraph.)

**1910.106(b)(2)(iv)***(g)* Flame arresters or venting devices required in subdivision (*f*) of this subdivision may be omitted for Class IB and IC liquids where conditions are such that their use may, in case of obstruction, result in tank damage.

**1910.106(b)(2)(v)** *Emergency relief venting for fire exposure for aboveground tanks.* *(a)* Every aboveground storage tank shall have some form of construction or device that will relieve excessive internal pressure caused by exposure fires.

**1910.106(b)(2)(v)***(b)* In a vertical tank the construction referred to in subdivision (*a*) of this subdivision may take the form of a floating roof, lifter roof, a weak roof-to-shell seam, or other approved pressure relieving construction. The weak roof-to-shell seam shall be constructed to fail preferential to any other seam.

**1910.106(b)(2)(v)***(c)* Where entire dependence for emergency relief is placed upon pressure relieving devices, the total venting capacity of both normal and emergency vents shall be enough to prevent rupture of the shell or bottom of the tank if vertical, or of the shell or heads if horizontal. If unstable liquids are stored, the effects of heat or gas resulting from polymerization, decomposition, condensation, or self-reactivity shall be taken into account. The total capacity of both normal and emergency venting devices shall be not less than that derived from Table H-10 except as provided in subdivision (*e*) or (*f*) of this subdivision. Such device may be a self-closing manhole cover, or one using long bolts that permit the cover to lift under internal pressure, or an additional or larger relief valve or valves. The wetted area of the tank shall be calculated on the basis of 55 percent of the total exposed area of a sphere or spheroid, 75 percent of the total exposed area of a horizontal tank and the first 30 feet above grade of the exposed shell area of a vertical tank.

### TABLE H-10—WETTED AREA VERSUS CUBIC FEET FREE AIR PER HOUR
(14.7 p.s.i.a. and 60 °F.)

| Square feet | CFH | Squarefeet | CFH | Square feet | CFH |
|---|---|---|---|---|---|
| 20 | 21,100 | 200 | 211,000 | 1,000 | 524,000 |
| 30 | 31,600 | 250 | 239,000 | 1,200 | 557,000 |
| 40 | 42,100 | 300 | 265,000 | 1,400 | 587,000 |
| 50 | 52,700 | 350 | 288,000 | 1,600 | 614,000 |
| 60 | 63,200 | 400 | 312,000 | 1,800 | 639,000 |
| 70 | 73,700 | 500 | 354,000 | 2,000 | 662,000 |
| 80 | 84,200 | 600 | 392,000 | 2,400 | 704,000 |
| 90 | 94,800 | 700 | 428,000 | 2,800 | 742,000 |
| 100 | 105,000 | 800 | 462,000 | and | |
| 120 | 126,000 | 900 | 493,000 | over | |
| 140 | 147,000 | 1,000 | 524,000 | | |
| 160 | 168,000 | | | | |
| 180 | 190,000 | | | | |
| 200 | 211,000 | | | | |

**1910.106(b)(2)(v)***(d)* For tanks and storage vessels designed for pressure over 1 p.s.i.g., the total rate of venting shall be determined in accordance with Table H-10, except that when the exposed wetted area of the surface is greater than 2,800 square feet, the total rate of venting shall be calculated by the following formula:

$$CFH = 1{,}107 A^{0.82}$$

Where;

$CFH$ = Venting requirement, in cubic feet of free air per hour.

$A$ = Exposed wetted surface, in square feet.

**NOTE:** The foregoing formula is based on $Q = 21{,}000 A^{0.82}$.

**1910.106(b)(2)(v)***(e)* The total emergency relief venting capacity for any specific stable liquid may be determined by the following formula:

$$V = 1337 \div L\sqrt{M}$$

$V$ = Cubic feet of free air per hour from Table H-10.

$L$ = Latent heat of vaporization of specific liquid in B.t.u. per pound.

$M$ = Molecular weight of specific liquids.

**1910.106(b)(2)(v)***(f)* The required airflow rate of subdivision (*c*) or (*e*) of this subdivision may be multiplied by the appropriate factor listed in the following schedule when protection is provided as indicated. Only one factor may be used for any one tank.

0.5 for drainage in accordance with subdivision (vii)(*b*) of this subparagraph for tanks over 200 square feet of wetted area.
0.3 for approved water spray.
0.3 for approved insulation.
0.15 for approved water spray with approved insulation.

**1910.106(b)(2)(v)***(g)* The outlet of all vents and vent drains on tanks equipped with emergency venting to permit pressures exceeding 2.5 p.s.i.g. shall be arranged to discharge in such a way as to prevent localized overheating of any part of the tank, in the event vapors from such vents are ignited.

**1910.106(b)(2)(v)***(h)* Each commercial tank venting device shall have stamped on it the opening pressure, the pressure at which the valve reaches the full open position, and the flow capacity at the latter pressure, expressed in cubic feet per hour of air at 60 °F. and at a pressure of 14.7 p.s.i.a.

**1910.106(b)(2)(v)***(i)* The flow capacity of tank venting devices 12 inches and smaller in nominal pipe size shall be determined by actual test of each type and size of vent. These flow tests may be conducted by the manufacturer if certified by a qualified impartial observer, or may be conducted by an outside agency. The flow capacity of tank venting devices larger than 12 inches nominal pipe size, including manhole covers with long bolts or equivalent, may be calculated provided that the opening pressure is actually measured, the rating pressure and corresponding free orifice area are stated, the word "calculated" appears on the nameplate, and the computation is based on a flow coefficient of 0.5 applied to the rated orifice area.

**1910.106(b)(2)(vi)** *Vent piping for aboveground tanks.* **(a)** Vent piping shall be constructed in accordance with paragraph (c) of this section.

**1910.106(b)(2)(vi)(b)** Where vent pipe outlets for tanks storing Class I liquids are adjacent to buildings or public ways, they shall be located so that the vapors are released at a safe point outside of buildings and not less than 12 feet above the adjacent ground level. In order to aid their dispersion, vapors shall be discharged upward or horizontally away from closely adjacent walls. Vent outlets shall be located so that flammable vapors will not be trapped by eaves or other obstructions and shall be at least five feet from building openings.

**1910.106(b)(2)(vi)(c)** When tank vent piping is manifolded, pipe sizes shall be such as to discharge, within the pressure limitations of the system, the vapors they may be required to handle when manifolded tanks are subject to the same fire exposure.

**1910.106(b)(2)(vii)** *Drainage, dikes, and walls for aboveground tanks—(a) Drainage and diked areas.* The area surrounding a tank or a group of tanks shall be provided with drainage as in subdivision (b) of this subdivision, or shall be diked as provided in subdivision (c) of this subdivision, to prevent accidental discharge of liquid from endangering adjoining property or reaching waterways.

**1910.106(b)(2)(vii)(b)** *Drainage.* Where protection of adjoining property or waterways is by means of a natural or manmade drainage system, such systems shall comply with the following:

**1910.106(b)(2)(vii)(b)(1)** [Reserved]

**1910.106(b)(2)(vii)(b)(2)** The drainage system shall terminate in vacant land or other area or in an impounding basin having a capacity not smaller than that of the largest tank served. This termination area and the route of the drainage system shall be so located that, if the flammable or combustible liquids in the drainage system are ignited, the fire will not seriously expose tanks or adjoining property.

**1910.106(b)(2)(vii)(c)** **Diked areas.** Where protection of adjoining property or waterways is accomplished by retaining the liquid around the tank by means of a dike, the volume of the diked area shall comply with the following requirements:

**1910.106(b)(2)(vii)(c)(1)** Except as provided in subdivision (2) of this subdivision, the volumetric capacity of the diked area shall not be less than the greatest amount of liquid that can be released from the largest tank within the diked area, assuming a full tank. The capacity of the diked area enclosing more than one tank shall be calculated by deducting the volume of the tanks other than the largest tank below the height of the dike.

**1910.106(b)(2)(vii)(c)(2)** For a tank or group of tanks with fixed roofs containing crude petroleum with boilover characteristics, the volumetric capacity of the diked area shall be not less than the capacity of the largest tank served by the enclosure, assuming a full tank. The capacity of the diked enclosure shall be calculated by deducting the volume below the height of the dike of all tanks within the enclosure.

**1910.106(b)(2)(vii)(c)(3)** Walls of the diked area shall be of earth, steel, concrete or solid masonry designed to be liquidtight and to withstand a full hydrostatic head. Earthen walls 3 feet or more in height shall have a flat section at the top not less than 2 feet wide. The slope of an earthen wall shall be consistent with the angle of repose of the material of which the wall is constructed.

**1910.106(b)(2)(vii)(c)(4)** The walls of the diked area shall be restricted to an average height of 6 feet above interior grade.

**1910.106(b)(2)(vii)(c)(5)** [Reserved]

**1910.106(b)(2)(vii)(c)(6)** No loose combustible material, empty or full drum or barrel, shall be permitted within the diked area.

**1910.106(b)(2)(viii)** *Tank openings other than vents for aboveground tanks.*

**1910.106(b)(2)(viii)(a)-(c)** [Reserved]

**1910.106(b)(2)(viii)(d)** Openings for gaging shall be provided with a vaportight cap or cover.

**1910.106(b)(2)(viii)(e)** For Class IB and Class IC liquids other than crude oils, gasolines, and asphalts, the fill pipe shall be so designed and installed as to minimize the possibility of generating static electricity. A fill pipe entering the top of a tank shall terminate within 6 inches of the bottom of the tank and shall be installed to avoid excessive vibration.

**1910.106(b)(6)(viii)(f)** Filling and emptying connections which are made and broken shall be located outside of buildings at a location free from any source of ignition and not less than 5 feet away from any building opening. Such connection shall be closed and liquidtight when not in use. The connection shall be properly identified.

**1910.106(b)(3)** *Installation of underground tanks—(i) Location.* Excavation for underground storage tanks shall be made with due care to avoid undermining of foundations of existing structures. Underground tanks or tanks under buildings shall be so located with respect to existing building foundations and supports that the loads carried by the latter cannot be transmitted to the tank. The distance from any part of a tank storing Class I liquids to the nearest wall of any basement or pit shall be not less than 1 foot, and to any property line that may be built upon, not less than 3 feet. The distance from any part of a tank storing Class II or Class III liquids to the nearest wall of any basement, pit or property line shall be not less than 1 foot.

**1910.106(b)(3)(ii)** *Depth and cover.* Underground tanks shall be set on firm foundations and surrounded with at least 6 inches of noncorrosive, inert materials such as clean sand, earth, or gravel well tamped in place. The tank shall be placed in the hole with care since dropping or rolling the tank into the hole can break a weld, puncture or damage the tank, or scrape off the protective coating of coated tanks. Tanks shall be covered with a minimum of 2 feet of earth, or shall be covered with not less than 1 foot of earth, on top of which shall be placed a slab of reinforced concrete not less than 4 inches thick. When underground tanks are, or are likely to be, subject to traffic, they shall be protected against damage from vehicles passing over them by at least 3 feet of earth cover, or 18 inches of well-tamped earth, plus 6 inches of reinforced concrete or 8 inches of asphaltic concrete. When asphaltic or reinforced concrete paving is used as part of the protection, it shall extend at least 1 foot horizontally beyond the outline of the tank in all directions.

**1910.106(b)(3)(iii)** *Corrosion protection.* Corrosion protection for the tank and its piping shall be provided by one or more of the following methods:

**1910.106(b)(3)(iii)(a)** Use of protective coatings or wrappings;

**1910.106(b)(3)(iii)(b)** Cathodic protection; or,

**1910.106(b)(3)(iii)(c)** Corrosion resistant materials of construction.

**1910.106(b)(3)(iv)** *Vents.* **(a)** Location and arrangement of vents for Class I liquids. Vent pipes from tanks storing Class I liquids shall be so located that the discharge point is outside of buildings, higher than the fill pipe opening, and not less than 12 feet above the adjacent ground level. Vent pipes shall discharge only upward in order to disperse vapors. Vent pipes 2 inches or less in nominal inside diameter shall not be obstructed by devices that will cause excessive back pressure. Vent pipe outlets shall be so located that flammable vapors will not enter building openings, or be trapped under eaves or other obstructions. If the vent pipe is less than 10 feet in length, or greater than 2 inches in nominal inside diameter, the outlet shall be provided with a vacuum and pressure relief device or there shall be an approved flame arrester located in the vent line at the outlet or within the approved distance from the outlet.

**1910.106(b)(3)(iv)(b)** *Size of vents.* Each tank shall be vented through piping adequate in size to prevent blow-back of vapor or liquid at the fill opening while the tank is being filled. Vent pipes shall be not less than 1 1/4 inch nominal inside diameter.

### TABLE H-11—VENT LINE DIAMETERS

| Maximum flow GPM | Pipe length[1] 50 feet | 100 feet | 200 feet |
|---|---|---|---|
|  | Inches | Inches | Inches |
| 100 | 1 1/4 | 1 1/4 | 1 1/4 |
| 200 | 1 1/4 | 1 1/4 | 1 1/4 |
| 300 | 1 1/4 | 1 1/4 | 1 1/2 |
| 400 | 1 1/4 | 1 1/2 | 2 |
| 500 | 1 1/2 | 1 1/2 | 2 |
| 600 | 1 1/2 | 2 | 2 |
| 700 | 2 | 2 | 2 |
| 800 | 2 | 2 | 3 |
| 900 | 2 | 2 | 3 |
| 1,000 | 2 | 2 | 3 |

[1] Vent lines of 50 ft., 100 ft., and 200 ft. of pipe plus 7 ells.

**1910.106(b)(3)(iv)(c)** Location and arrangement of vents for Class II or Class III liquids. Vent pipes from tanks storing Class II or Class III flammable liquids shall terminate outside of the building and higher than the fill pipe opening. Vent outlets shall be above normal snow level. They may be fitted with return bends, coarse screens or other devices to minimize ingress of foreign material.

**1910.106(b)(3)(iv)(d)** Vent piping shall be constructed in accordance with paragraph (c) of this section. Vent pipes shall be so laid as to drain toward the tank without sags or traps in which liquid can collect. They shall be located so that they will not be subjected to physical damage. The tank end of the vent pipe shall enter the tank through the top.

**1910.106(b)(3)(iv)(e)** When tank vent piping is manifolded, pipe sizes shall be such as to discharge, within the pressure limitations of the system, the vapors they may be required to handle when manifolded tanks are filled simultaneously.

**1910.106(b)(3)(v)** *Tank openings other than vents.* (a) Connections for all tank openings shall be vapor or liquid tight.

**1910.106(b)(3)(v)(b)** Openings for manual gaging, if independent of the fill pipe, shall be provided with a liquid-tight cap or cover. If inside a building, each such opening shall be protected against liquid overflow and possible vapor release by means of a spring loaded check valve or other approved device.

**1910.106(b)(3)(v)(c)** Fill and discharge lines shall enter tanks only through the top. Fill lines shall be sloped toward the tank.

**1910.106(b)(3)(v)(d)** For Class IB and Class IC liquids other than crude oils, gasolines, and asphalts, the fill pipe shall be so designed and installed as to minimize the possibility of generating static electricity by terminating within 6 inches of the bottom of the tank.

**1910.106(b)(3)(v)(e)** Filling and emptying connections which are made and broken shall be located outside of buildings at a location free from any source of ignition and not less than 5 feet away from any building opening. Such connection shall be closed and liquidtight when not in use. The connection shall be properly identified.

**1910.106(b)(4)** *Installation of tanks inside of buildings*—(i) *Location.* Tanks shall not be permitted inside of buildings except as provided in paragraphs (e), (g), (h), or (i) of this section.

**1910.106(b)(4)(ii)** *Vents.* Vents for tanks inside of buildings shall be as provided in subparagraphs (2) (iv), (v), (vi)(b), and (3)(iv) of this paragraph, except that emergency venting by the use of weak roof seams on tanks shall not be permitted. Vents shall discharge vapors outside the buildings.

**1910.106(b)(4)(iii)** *Vent piping.* Vent piping shall be constructed in accordance with paragraph (c) of this section.

**1910.106(b)(4)(iv)** *Tank openings other than vents.* (a) Connections for all tank openings shall be vapor or liquidtight. Vents are covered in subdivision (ii) of this subparagraph.

**1910.106(b)(4)(iv)(b)** Each connection to a tank inside of buildings through which liquid can normally flow shall be provided with an internal or an external valve located as close as practical to the shell of the tank. Such valves, when external, and their connections to the tank shall be of steel except when the chemical characteristics of the liquid stored are incompatible with steel. When materials other than steel are necessary, they shall be suitable for the pressures, structural stresses, and temperatures involved, including fire exposures.

**1910.106(b)(4)(iv)(c)** Flammable or combustible liquid tanks located inside of buildings, except in one-story buildings designed and protected for flammable or combustible liquid storage, shall be provided with an automatic-closing heat-actuated valve on each withdrawal connection below the liquid level, except for connections used for emergency disposal, to prevent continued flow in the event of fire in the vicinity of the tank. This function may be incorporated in the valve required in (b) of this subdivision, and if a separate valve, shall be located adjacent to the valve required in (b) of this subdivision.

**1910.106(b)(4)(iv)(d)** Openings for manual gaging, if independent of the fill pipe (see (f) of this subdivision), shall be provided with a vaportight cap or cover. Each such opening shall be protected against liquid overflow and possible vapor release by means of a spring loaded check valve or other approved device.

**1910.106(b)(4)(iv)(e)** For Class IB and Class IC liquids other than crude oils, gasolines, and asphalts, the fill pipe shall be so designed and installed as to minimize the possibility of generating static electricity by terminating within 6 inches of the bottom of the tank.

**1910.106(b)(4)(iv)(f)** The fill pipe inside of the tank shall be installed to avoid excessive vibration of the pipe.

**1910.106(b)(4)(iv)(g)** The inlet of the fill pipe shall be located outside of buildings at a location free from any source of ignition and not less than 5 feet away from any building opening. The inlet of the fill pipe shall be closed and liquidtight when not in use. The fill connection shall be properly identified.

**1910.106(b)(4)(iv)(h)** Tanks inside buildings shall be equipped with a device, or other means shall be provided, to prevent overflow into the building.

**1910.106(b)(5)** *Supports, foundations, and anchorage for all tank locations*—(i) *General.* Tank supports shall be installed on firm foundations. Tank supports shall be of concrete, masonry, or protected steel. Single wood timber supports (not cribbing) laid horizontally may be used for outside aboveground tanks if not more than 12 inches high at their lowest point.

**1910.106(b)(5)(ii)** *Fire resistance.* Steel supports or exposed piling shall be protected by materials having a fire resistance rating of not less than 2 hours, except that steel saddles need not be protected if less than 12 inches high at their lowest point. Water spray protection or its equivalent may be used in lieu of fire-resistive materials to protect supports.

**1910.106(b)(5)(iii)** *Spheres.* The design of the supporting structure for tanks such as spheres shall receive special engineering consideration.

**1910.106(b)(5)(iv)** *Load distribution.* Every tank shall be so supported as to prevent the excessive concentration of loads on the supporting portion of the shell.

**1910.106(b)(5)(v)** *Foundations.* Tanks shall rest on the ground or on foundations made of concrete, masonry, piling, or steel. Tank foundations shall be designed to minimize the possibility of uneven settling of the tank and to minimize corrosion in any part of the tank resting on the foundation.

**1910.106(b)(5)(vi)** *Flood areas.* Where a tank is located in an area that may be subjected to flooding, the applicable precautions outlined in this subdivision shall be observed.

**1910.106(b)(5)(vi)(a)** No aboveground vertical storage tank containing a flammable or combustible liquid shall be located so that the allowable liquid level within the tank is below the established maximum flood stage, unless the tank is provided with a guiding structure such as described in (m), (n), and (o) of this subdivision.

**1910.106(b)(5)(vi)(b)** Independent water supply facilities shall be provided at locations where there is no ample and dependable public water supply available for loading partially empty tanks with water.

**1910.106(b)(5)(vi)(c)** In addition to the preceding requirements, each tank so located that more than 70 percent, but less than 100 percent, of its allowable liquid storage capacity will be submerged at the established maximum flood stage, shall be safeguarded by one of the following methods: Tank shall be raised, or its height shall be increased, until its top extends above the maximum flood stage a distance equivalent to 30 percent or more of its allowable liquid storage capacity: *Provided, however,* That the submerged part of the tank shall not exceed two and one-half times the diameter. Or, as an alternative to the foregoing, adequate noncombustible structural guides, designed to permit the tank to float vertically without loss of product, shall be provided.

**1910.106(b)(5)(vi)(d)** Each horizontal tank so located that more than 70 percent of its storage capacity will be submerged at the established flood stage, shall be anchored, attached to a foundation of concrete or of steel and concrete, of sufficient weight to provide adequate load for the tank when filled with flammable or combustible liquid and submerged by flood waters to the established flood stage, or adequately secured by other means.

**1910.106(b)(5)(vi)(e)** [Reserved]

**1910.106(b)(5)(vi)(f)** At locations where there is no ample and dependable water supply, or where filling of underground tanks with liquids is impracticable because of the character of their contents, their use, or for other reasons, each tank shall be safeguarded against movement when empty and submerged by high ground water or flood waters by anchoring, weighting with concrete or other approved solid loading material, or securing by other means. Each such tank shall be so constructed and installed that it will safely resist external pressures due to high ground water or flood waters.

**1910.106(b)(5)(vi)(g)** At locations where there is an ample and dependable water supply available, underground tanks containing flammable or combustible liquids, so installed that more than 70 percent of their storage capacity will be submerged at the maximum flood stage, shall be so anchored, weighted, or secured by other means, as to prevent movement

**1910.106(b)(5)(vi)(h)** Pipe connections below the allowable liquid level in a tank shall be provided with valves or cocks located as closely as practicable to the tank shell. Such valves and their connections to tanks shall be of steel or other material suitable for use with the liquid being stored. Cast iron shall not be permitted.

**1910.106(b)(5)(vi)(i)** At locations where an independent water supply is required, it shall be entirely independent of public power and water supply. Independent source of water shall be available when flood waters reach a level not less than 10 feet below the bottom of the lowest tank on a property.

**1910.106(b)(5)(vi)(j)** The self-contained power and pumping unit shall be so located or so designed that pumping into tanks may be carried on continuously throughout the rise in flood waters from a level 10 feet below the lowest tank to the level of the potential flood stage.

**1910.106(b)(5)(vi)(k)** Capacity of the pumping unit shall be such that the rate of rise of water in all tanks shall be equivalent to the established potential average rate of rise of flood waters at any stage.

**1910.106(b)(5)(vi)(l)** Each independent pumping unit shall be tested periodically to insure that it is in satisfactory operating condition.

**1910.106(b)(5)(vi)(m)** Structural guides for holding floating tanks above their foundations shall be so designed that there will be no resistance to the free rise of a tank, and shall be constructed of noncombustible material.

**1910.106(b)(5)(vi)(n)** The strength of the structure shall be adequate to resist lateral movement of a tank subject to a horizontal force in any direction equivalent to not less than 25 pounds per square foot acting on the projected vertical cross-sectional area of the tank.

**1910.106(b)(5)(vi)(o)** Where tanks are situated on exposed points or bends in a shoreline where swift currents in flood waters will be present, the structures shall be designed to withstand a unit force of not less than 50 pounds per square foot.

**1910.106(b)(5)(vi)(p)** The filling of a tank to be protected by water loading shall be started as soon as flood waters reach a dangerous flood stage. The rate of filling shall be at least equal to the rate of rise of the floodwaters (or the established average potential rate of rise).

**1910.106(b)(5)(vi)(q)** Sufficient fuel to operate the water pumps shall be available at all times to insure adequate power to fill all tankage with water.

**1910.106(b)(5)(vi)(r)** All valves on connecting pipelines shall be closed and locked in closed position when water loading has been completed.

**1910.106(b)(5)(vi)(s)** Where structural guides are provided for the protection of floating tanks, all rigid connections between tanks and pipelines shall be disconnected and blanked off or blinded before the floodwaters reach the bottom of the tank, unless control valves and their connections to the tank are of a type designed to prevent breakage between the valve and the tank shell.

**1910.106(b)(5)(vi)(t)** All valves attached to tanks other than those used in connection with water loading operations shall be closed and locked.

**1910.106(b)(5)(vi)(u)** If a tank is equipped with a swing line, the swing pipe shall be raised to and secured at its highest position.

**1910.106(b)(5)(vi)(v)** *Inspections.* The Assistant Secretary or his designated representative shall make periodic inspections of all plants where the storage of flammable or combustible liquids is such as to require compliance with the foregoing requirements, in order to assure the following:

**1910.106(b)(5)(vi)(v)(1)** That all flammable or combustible liquid storage tanks are in compliance with these requirements and so maintained.

**1910.106(b)(5)(vi)(v)(2)** That detailed printed instructions of what to do in flood emergencies are properly posted.

**1910.106(b)(5)(vi)(v)(3)** That station operators and other employees depended upon to carry out such instructions are thoroughly informed as to the location and operation of such valves and other equipment necessary to effect these requirements.

**1910.106(b)(5)(vii)** *Earthquake areas.* In areas subject to earthquakes, the tank supports and connections shall be designed to resist damage as a result of such shocks.

**1910.106(b)(6)** *Sources of ignition.* In locations where flammable vapors may be present, precautions shall be taken to prevent ignition by eliminating or controlling sources of ignition. Sources of ignition may include open flames, lightning, smoking, cutting and welding, hot surfaces, frictional heat, sparks (static, electrical, and mechanical), spontaneous ignition, chemical and physical-chemical reactions, and radiant heat.

**1910.106(b)(7)** *Testing*—**(i)** *General.* All tanks, whether shop built or field erected, shall be strength tested before they are placed in service in accordance with the applicable paragraphs of the code under which they were built. The American Society of Mechanical Engineers (ASME) code stamp, American Petroleum Institute (API) monogram, or the label of the Underwriters' Laboratories, Inc., on a tank shall be evidence of compliance with this strength test. Tanks not marked in accordance with the above codes shall be strength tested before they are placed in service in accordance with good engineering principles and reference shall be made to the sections on testing in the codes listed in subparagraphs (1) (iii)(a), (iv)(b), or (v)(b) of this paragraph.

**1910.106(b)(7)(ii)** *Strength.* When the vertical length of the fill and vent pipes is such that when filled with liquid the static head imposed upon the bottom of the tank exceeds 10 pounds per square inch, the tank and related piping shall be tested hydrostatically to a pressure equal to the static head thus imposed.

**1910.106(b)(7)(iii)** *Tightness.* In addition to the strength test called for in subdivisions (i) and (ii) of this subparagraph, all tanks and connections shall be tested for tightness. Except for underground tanks, this tightness test shall be made at operating pressure with air, inert gas, or water prior to placing the tank in service. In the case of field-erected tanks the strength test may be considered to be the test for tank tightness. Underground tanks and piping, before being covered, enclosed, or placed in use, shall be tested for tightness hydrostatically, or with air pressure at not less than 3 pounds per square inch and not more than 5 pounds per square inch.

**1910.106(b)(7)(iv)** *Repairs.* All leaks or deformations shall be corrected in an acceptable manner before the tank is placed in service. Mechanical caulking is not permitted for correcting leaks in welded tanks except pinhole leaks in the roof.

**1910.106(b)(7)(v)** *Derated operations.* Tanks to be operated at pressures below their design pressure may be tested by the applicable provisions of subdivision (i) or (ii) of this subparagraph, based upon the pressure developed under full emergency venting of the tank.

**1910.106(c)** *Piping, valves, and fittings*—**(1)** *General*—**(i)** *Design.* The design (including selection of materials) fabrication, assembly, test, and inspection of piping systems containing flammable or combustible liquids shall be suitable for the expected working pressures and structural stresses. Conformity with the applicable provisions of Pressure Piping, ANSI B31 series and the provisions of this paragraph, shall be considered prima facie evidence of compliance with the foregoing provisions.

**1910.106(c)(1)(ii)** *Exceptions.* This paragraph does not apply to any of the following:

**1910.106(c)(1)(ii)(a)** Tubing or casing on any oil or gas wells and any piping connected directly thereto.

**1910.106(c)(1)(ii)(b)** Motor vehicle, aircraft, boat, or portable or stationary engines.

**1910.106(c)(1)(ii)(c)** Piping within the scope of any applicable boiler and pressures vessel code.

**1910.106(c)(1)(iii)** *Definitions.* As used in this paragraph, piping systems consist of pipe, tubing, flanges, bolting, gaskets, valves, fittings, the pressure containing parts of other components such as expansion joints and strainers, and devices which serve such purposes as mixing, separating, snubbing, distributing, metering, or controlling flow.

**1910.106(c)(2)** *Materials for piping, valves, and fittings*—**(i)** *Required materials.* Materials for piping, valves, or fittings shall be steel, nodular iron, or malleable iron, except as provided in paragraph (c)(2) (ii), (iii) and (iv) of this section.

**1910.106(c)(2)(ii)** *Exceptions.* Materials other than steel, nodular iron, or malleable iron may be used underground, or if required by the properties of the flammable or combustible liquid handled. Material other than steel, nodular iron, or malleable iron shall be designed to specifications embodying principles recognized as good engineering practices for the material used.

**1910.106(c)(2)(iii)** *Linings.* Piping, valves, and fittings may have combustible or noncombustible linings.

**1910.106(c)(2)(iv)** *Low-melting materials.* When low-melting point materials such as aluminum and brass or materials that soften on fire exposure such as plastics, or non-ductile materials such as cast iron, are necessary, special consideration shall be given to their behavior on fire exposure. If such materials are used in above ground piping systems or inside buildings, they shall be suitably protected against fire exposure or so located that any spill resulting from the failure of these materials could not unduly expose persons, important buildings or structures or can be readily controlled by remote valves.

**1910.106(c)(3)** *Pipe joints.* Joints shall be made liquid tight. Welded or screwed joints or approved connectors shall be used. Threaded joints and connections shall be made up tight with a suitable lubricant or piping compound. Pipe joints dependent upon the friction characteristics of combustible materials for mechanical continuity of piping shall not be used inside buildings. They may be used outside of buildings above or below ground. If used above ground, the piping shall either be secured to prevent disengagement at the fitting or the piping system shall be so designed that any spill resulting from such disengagement could not unduly expose persons, important buildings or structures, and could be readily controlled by remote valves.

**1910.106(c)(4)** *Supports.* Piping systems shall be substantially supported and protected against physical damage and excessive stresses arising from settlement, vibration, expansion, or contraction.

**1910.106(c)(5)** *Protection against corrosion.* All piping for flammable or combustible liquids, both aboveground and underground, where subject to external corrosion, shall be painted or otherwise protected.

**1910.106(c)(6)** *Valves.* Piping systems shall contain a sufficient number of valves to operate the system properly and to protect the plant. Piping systems in connection with pumps shall contain a sufficient number of valves to control properly the flow of liquid in normal operation and in the event of physical damage. Each connection to pipelines, by which equipments such as tankcars or tank vehicles discharge liquids by means of pumps into storage tanks, shall be provided with a check valve for automatic protection against backflow if the piping arrangement is such that backflow from the system is possible.

**1910.106(c)(7)** *Testing.* All piping before being covered, enclosed, or placed in use shall be hydrostatically tested to 150 percent of the maximum anticipated pressure of the system, or pneumatically tested to 110 percent of the maximum anticipated pressure of the system, but not less than 5 pounds per square inch gage at the highest point of the system. This test shall be maintained for a sufficient time to complete visual inspection of all joints and connections, but for at least 10 minutes.

**1910.106(d)** *Container and portable tank storage*—**(1)** *Scope*—**(i)** *General.* This paragraph shall apply only to the storage of flammable or combustible liquids in drums or other containers (including flammable aerosols) not exceeding 60 gallons individual capacity and those portable tanks not exceeding 660 gallons individual capacity.

**1910.106(d)(1)(ii)** *Exceptions.* This paragraph shall not apply to the following:

**1910.106(d)(1)(ii)(a)** Storage of containers in bulk plants, service stations, refineries, chemical plants, and distilleries;

**1910.106(d)(1)(ii)(b)** Class I or Class II liquids in the fuel tanks of a motor vehicle, aircraft, boat, or portable or stationary engine;

**1910.106(d)(1)(ii)(c)** Flammable or combustible paints, oils, varnishes, and similar mixtures used for painting or maintenance when not kept for a period in excess of 30 days;

**1910.106(d)(1)(ii)(d)** Beverages when packaged in individual containers not exceeding 1 gallon in size.

**1910.106(d)(2)** *Design, construction, and capacity of containers*—**(i)** *General.* Only approved containers and portable tanks shall be used. Metal containers and portable tanks meeting the requirements of and containing products authorized by chapter I, title 49 of the Code of Federal Regulations (regulations issued by the Hazardous Materials Regulations Board, Department of Transportation), shall be deemed to be acceptable.

**1910.106(d)(2)(ii)** *Emergency venting.* Each portable tank shall be provided with one or more devices installed in the top with sufficient emergency venting capacity to limit internal pressure under fire exposure conditions to 10 p.s.i.g., or 30 percent of the bursting pressure of the tank, whichever is greater. The total venting capacity shall be not less than that specified in paragraphs (b)(2)(v) (c) or (e) of this section. At least one pressure-activated vent having a minimum capacity of 6,000 cubic feet of free air (14.7 p.s.i.a. and 60 °F.) shall be used. It shall be set to open at not less than 5 p.s.i.g. If fusible vents are used, they shall be actuated by elements that operate at a temperature not exceeding 300 °F.

**1910.106(d)(2)(iii)** *Size.* Flammable and combustible liquid containers shall be in accordance with Table H-12, except that glass or plastic containers of no more than 1-gallon capacity may be used for a Class IA or IB flammable liquid if:

**1910.106(d)(2)(iii)(a)(1)** Such liquid either would be rendered unfit for its intended use by contact with metal or would excessively corrode a metal container so as to create a leakage hazard; and

**1910.106(d)(2)(iii)(a)(2)** The user's process either would require more than 1 pint of a Class IA liquid or more than 1 quart of a Class IB liquid of a single assay lot to be used at one time, or would require the maintenance of an analytical standard liquid of a quality which is not met by the specified standards of liquids available, and the quantity of the analytical standard liquid required to be used in any one control process exceeds one-sixteenth the capacity of the container allowed under Table H-12 for the class of liquid; or

**1910.106(d)(2)(iii)(b)** The containers are intended for direct export outside the United States.

### TABLE H-12—MAXIMUM ALLOWABLE SIZE OF CONTAINERS AND PORTABLE TANKS

| Container type | Flammable liquids ||| Combustible liquids ||
|---|---|---|---|---|---|
| | Class IA | Class IB | Class IC | Class II | Class III |
| Glass or approved plastic | 1 pt | 1 qt | 1 gal | 1 gal | 1 gal |
| Metal (other than DOT drums) | 1 gal | 5 gal | 5 gal | 5 gal | 5 gal |
| Safety cans | 2 gal | 5 gal | 5 gal | 5 gal | 5 gal |
| Metal drums (DOT specifications) | 60 gal | 60 gal | 60 gal | 60 gal | 60 gal |
| Approved portable tanks | 660 gal | 660 gal | 660 gal | 660 gal | 660 gal |

NOTE: Container exemptions: (a) Medicines, beverages, foodstuffs, cosmetics, and other common consumer items, when packaged according to commonly accepted practices, shall be exempt from the requirements of §1910.106(d)(2)(i) and (ii).

**1910.106(d)(3)** *Design, construction, and capacity of storage cabinets*—**(i)** *Maximum capacity.* Not more than 60 gallons of Class I or Class II liquids, nor more than 120 gallons of Class III liquids may be stored in a storage cabinet.

**1910.106(d)(3)(ii)** *Fire resistance.* Storage cabinets shall be designed and constructed to limit the internal temperature to not more than 325 °F. when subjected to a 10-minute fire test using the standard time-temperature curve as set forth in Standard Methods of Fire Tests of Building Construction and Materials, NFPA 251-1969, which is incorporated by reference as specified in §1910.6. All joints and seams shall remain tight and the door shall remain securely closed during the fire test. Cabinets shall be labeled in conspicuous lettering, "Flammable—Keep Fire Away."

**1910.106(d)(3)(ii)(a)** Metal cabinets constructed in the following manner shall be deemed to be in compliance. The bottom, top, door, and sides of cabinet shall be at least No. 18 gage sheet iron and double walled with 1 1/2-inch air space. Joints shall be riveted, welded or made tight by some equally effective means. The door shall be provided with a three-point lock, and the door sill shall be raised at least 2 inches above the bottom of the cabinet.

**1910.106(d)(3)(ii)(b)** Wooden cabinets constructed in the following manner shall be deemed in compliance. The bottom, sides, and top shall be constructed of an approved grade of plywood at least 1 inch in thickness, which shall not break down or delaminate under fire conditions. All joints shall be rabbetted and shall be fastened in two directions with flathead woodscrews. When more than one door is used, there shall be a rabbetted overlap of not less than 1 inch. Hinges shall be mounted in such a manner as not to lose their holding capacity due to loosening or burning out of the screws when subjected to the fire test.

# OSHA Standards for General Industry

**1910.106(d)(4)** *Design and construction of inside storage rooms*—**(i)** *Construction.* Inside storage rooms shall be constructed to meet the required fire-resistive rating for their use. Such construction shall comply with the test specifications set forth in Standard Methods of Fire Tests of Building Construction and Materials, NFPA 251-1969. Where an automatic sprinkler system is provided, the system shall be designed and installed in an acceptable manner. Openings to other rooms or buildings shall be provided with noncombustible liquid-tight raised sills or ramps at least 4 inches in height, or the floor in the storage area shall be at least 4 inches below the surrounding floor. Openings shall be provided with approved self-closing fire doors. The room shall be liquid-tight where the walls join the floor. A permissible alternate to the sill or ramp is an open-grated trench inside of the room which drains to a safe location. Where other portions of the building or other properties are exposed, windows shall be protected as set forth in the Standard for Fire Doors and Windows, NFPA No. 80-1968, which is incorporated by reference as specified in §1910.6, for Class E or F openings. Wood at least 1 inch nominal thickness may be used for shelving, racks, dunnage, scuffboards, floor overlay, and similar installations.

**1910.106(d)(4)(ii)** *Rating and capacity.* Storage in inside storage rooms shall comply with Table H-13.

### TABLE H-13—STORAGE IN INSIDE ROOMS

| Fire protection[1] provided | Fire resistance | Maximum size | Total allowable quantities (gals./ sq ft./ floor area) |
|---|---|---|---|
| Yes | 2 hours | 500 sq. ft. | 10 |
| No | 2 hours | 500 sq. ft. | 5 |
| Yes | 1 hour | 150 sq. ft. | 4 |
| No | 1 hour | 150 sq. ft. | 2 |

[1] Fire protection system shall be sprinkler, water spray, carbon dioxide, or other system.

**1910.106(d)(4)(iii)** *Wiring.* Electrical wiring and equipment located in inside storage rooms used for Class I liquids shall be approved under Subpart S of this part for Class I, Division 2 Hazardous Locations; for Class II and Class III liquids, shall be approved for general use.

**1910.106(d)(4)(iv)** *Ventilation.* Every inside storage room shall be provided with either a gravity or a mechanical exhaust ventilation system. Such system shall be designed to provide for a complete change of air within the room at least six times per hour. If a mechanical exhaust system is used, it shall be controlled by a switch located outside of the door. The ventilating equipment and any lighting fixtures shall be operated by the same switch. A pilot light shall be installed adjacent to the switch if Class I flammable liquids are dispensed within the room. Where gravity ventilation is provided, the fresh air intake, as well as the exhaust outlet from the room, shall be on the exterior of the building in which the room is located.

**1910.106(d)(4)(v)** *Storage in inside storage rooms.* In every inside storage room there shall be maintained one clear aisle at least 3 feet wide. Containers over 30 gallons capacity shall not be stacked one upon the other. Dispensing shall be by approved pump or self-closing faucet only.

**1910.106(d)(5)** *Storage inside building*—**(i)** *Egress.* Flammable or combustible liquids, including stock for sale, shall not be stored so as to limit use of exits, stairways, or areas normally used for the safe egress of people.

**1910.106(d)(5)(ii)** *Containers.* The storage of flammable or combustible liquids in containers or portable tanks shall comply with subdivisions (iii) through (v) of this subparagraph.

**1910.106(d)(5)(iii)** *Office occupancies.* Storage shall be prohibited except that which is required for maintenance and operation of building and operation of equipment. Such storage shall be kept in closed metal containers stored in a storage cabinet or in safety cans or in an inside storage room not having a door that opens into that portion of the building used by the public.

**1910.106(d)(5)(iv)** *Mercantile occupancies and other retail stores.*

**1910.106(d)(5)(iv)(a)-(d)** [Reserved]

**1910.106(d)(5)(iv)(e)** Leaking containers shall be removed to a storage room or taken to a safe location outside the building and the contents transferred to an undamaged container.

**1910.106(d)(5)(v)** *General purpose public warehouses.* Storage shall be in accordance with Table H-14 or H-15 and in buildings or in portions of such buildings cut off by standard firewalls. Material creating no fire exposure hazard to the flammable or combustible liquids may be stored in the same area.

### TABLE H-14—INDOOR CONTAINER STORAGE

| Class liquid | Storage level | Protected storage maximum per pile Gallons | Unprotected storage maximum per pile Gallons |
|---|---|---|---|
| A | Ground and upper floors | 2,750 | 660 |
|   | Basement | Not permitted | Not permitted |
| B | Ground and upper floors | 5,500 | 1,375 |
|   | Basement | Not permitted | Not permitted |
| C | Ground and upper floors | 16,500 | 4,125 |
|   | Basement | Not permitted | Not permitted |
| II | Ground and upper floors | 16,500 | 4,125 |
|   | Basement | 5,500 | Not permitted |
| III | Ground and upper floors | 55,000 | 13,750 |
|   | Basement | 8,250 | Not permitted |

NOTE 1: When 2 or more classes of materials are stored in a single pile, the maximum gallonage permitted in that pile shall be the smallest of the 2 or more separate maximum gallonages.

NOTE 2: Aisles shall be provided so that no container is more than 12 ft. from an aisle. Main aisles shall be at least 8 ft. wide and side aisles at least 4 ft. wide.

NOTE 3: Each pile shall be separated from each other by at least 4 ft.

(Numbers in parentheses indicate corresponding number of 55-gal. drums.)

### TABLE H-15—INDOOR PORTABLE TANK STORAGE

| Class liquid | Storage level | Protected storage maximum per pile Gallons | Unprotected storage maximum per pile Gallons |
|---|---|---|---|
| IA | Ground and upper floors | Not permitted | Not permitted |
|    | Basement | Not permitted | Not permitted |
| IB | Ground and upper floors | 20,000 | 2,000 |
|    | Basement | Not permitted | Not permitted |
| IC | Ground and upper floors | 40,000 | 5,500 |
|    | Basement | Not permitted | Not permitted |
| II | Ground and upper floors | 40,000 | 5,500 |
|    | Basement | 20,000 | Not permitted |
| III | Ground and upper floors | 60,000 | 22,000 |
|     | Basement | 20,000 | Not permitted |

NOTE 1: When 2 or more classes of materials are stored in a single pile, the maximum gallonage permitted in that pile shall be the smallest of the 2 or more separate maximum gallonages.

NOTE 2: Aisles shall be provided so that no container is more than 12 ft. from an aisle. Main aisles shall be at least 8 ft. wide and side aisles at least 4 ft. wide.

NOTE 3: Each pile shall be separated from each other by at least 4 ft.

**1910.106(d)(5)(vi)** *Flammable and combustible liquid warehouses or storage buildings.* **(a)** If the storage building is located 50 feet or less from a building or line of adjoining property that may be built upon, the exposing wall shall be a blank wall having a fire-resistance rating of at least 2 hours.

**1910.106(d)(5)(vi)(b)** The total quantity of liquids within a building shall not be restricted, but the arrangement of storage shall comply with Table H-14 or H-15.

**1910.106(d)(5)(vi)(c)** Containers in piles shall be separated by pallets or dunnage where necessary to provide stability and to prevent excessive stress on container walls.

**1910.106(d)(5)(vi)(d)** Portable tanks stored over one tier high shall be designed to nest securely, without dunnage, and adequate materials handing equipment shall be available to handle tanks safely at the upper tier level.

**1910.106(d)(5)(vi)(e)** No pile shall be closer than 3 feet to the nearest beam, chord, girder, or other obstruction, and shall be 3 feet below sprinkler deflectors or discharge orifices of water spray, or other overhead fire protection systems.

**1910.106(d)(5)(vi)(f)** Aisles of at least 3 feet wide shall be provided where necessary for reasons of access to doors, windows or standpipe connections.

**1910.106(d)(6)** *Storage outside buildings*—(i) *General.* Storage outside buildings shall be in accordance with Table H–16 or H–17, and subdivisions (ii) and (iv) of this subparagraph.

### TABLE H-16—OUTDOOR CONTAINER STORAGE

| 1-Class | 2-Maximum per pile | 3-Distance between piles | 4-Distance to property line that can be built upon | 5-Distance to street, alley, public way |
|---|---|---|---|---|
| | gal. | ft. | ft. | ft. |
| IA | 1,100 | 5 | 20 | 10 |
| IB | 2,200 | 5 | 20 | 10 |
| IC | 4,400 | 5 | 20 | 10 |
| II | 8,800 | 5 | 10 | 5 |
| III | 22,000 | 5 | 10 | 5 |

NOTE 1: When 2 or more classes of materials are stored in a single pile, the maximum gallonage in that pile shall be the smallest of the 2 or more separate gallonages.

NOTE 2: Within 200 ft. of each container, there shall be a 12-ft. wide access way to permit approach of fire control apparatus.

NOTE 3: The distances listed apply to properties that have protection for exposures as defined. If there are exposures, and such protection for exposures does not exist, the distances in column 4 shall be doubled.

NOTE 4: When total quantity stored does not exceed 50 percent of maximum per pile, the distances in columns 4 and 5 may be reduced 50 percent, but not less than 3 ft.

**1910.106(d)(6)(ii)** *Maximum storage.* A maximum of 1,100 gallons of flammable or combustible liquids may be located adjacent to buildings located on the same premises and under the same management provided the provisions of subdivisions (a) and (b) of this subdivision are complied with.

**1910.106(d)(6)(ii)(a)** [Reserved]

**1910.106(d)(6)(ii)(b)** Where quantity stored exceeds 1,100 gallons, or provisions of subdivision (a) of this subdivision cannot be met, a minimum distance of 10 feet between buildings and nearest container of flammable or combustible liquid shall be maintained.

**1910.106(d)(6)(iii)** *Spill containment.* The storage area shall be graded in a manner to divert possible spills away from buildings or other exposures or shall be surrounded by a curb at least 6 inches high. When curbs are used, provisions shall be made for draining of accumulations of ground or rain water or spills of flammable or combustible liquids. Drains shall terminate at a safe location and shall be accessible to operation under fire conditions.

**1910.106(d)(6)(iv)** *Security.* The storage area shall be protected against tampering or trespassers where necessary and shall be kept free of weeds, debris and other combustible material not necessary to the storage.

**1910.106(d)(7)** *Fire control*—(i) *Extinguishers.* Suitable fire control devices, such as small hose or portable fire extinguishers, shall be available at locations where flammable or combustible liquids are stored.

### TABLE H-17—OUTDOOR PORTABLE TANK STORAGE

| 1-Class | 2-Maximum per pile | 3-Distance between piles | 4-Distance to property line that can be built upon | 5-Distance to street, alley, public way |
|---|---|---|---|---|
| | gal. | ft. | ft. | ft. |
| IA | 2,200 | 5 | 20 | 10 |
| IB | 4,400 | 5 | 20 | 10 |
| IC | 8,800 | 5 | 10 | 10 |
| II | 17,600 | 5 | 10 | 5 |
| III | 44,000 | 5 | 10 | 5 |

NOTE 1: When 2 or more classes of materials are stored in a single pile, the maximum gallonage in that pile shall be the smallest of the 2 or more separate gallonages.

NOTE 2: Within 200 ft. of each container, there shall be a 12-ft. wide access way to permit approach of fire control apparatus.

NOTE 3: The distances listed apply to properties that have protection for exposures as defined. If there are exposures, and such protection for exposures does not exist, the distances in column 4 shall be doubled.

NOTE 4: When total quantity stored does not exceed 50 percent of maximum per pile, the distances in columns 4 and 5 may be reduced 50 percent, but not less than 3 ft.

**1910.106(d)(7)(i)(a)** At least one portable fire extinguisher having a rating of not less than 12-B units shall be located outside of, but not more than 10 feet from, the door opening into any room used for storage.

**1910.106(d)(7)(i)(b)** At least one portable fire extinguisher having a rating of not less than 12-B units must be located not less than 10 feet, nor more than 25 feet, from any Class I or Class II liquid storage area located outside of a storage room but inside a building.

**1910.106(d)(7)(ii)** *Sprinklers.* When sprinklers are provided, they shall be installed in accordance with §1910.159.

**1910.106(d)(7)(iii)** *Open flames and smoking.* Open flames and smoking shall not be permitted in flammable or combustible liquid storage areas.

**1910.106(d)(7)(iv)** *Water reactive materials.* Materials which will react with water shall not be stored in the same room with flammable or combustible liquids.

**1910.106(e)** **Industrial plants**—(1) *Scope*—(i) *Application.* This paragraph shall apply to those industrial plants where:

**1910.106(e)(1)(i)(a)** The use of flammable or combustible liquids is incidental to the principal business, or

**1910.106(e)(1)(i)(b)** Where flammable or combustible liquids are handled or used only in unit physical operations such as mixing, drying, evaporating, filtering, distillation, and similar operations which do not involve chemical reaction. This paragraph shall not apply to chemical plants, refineries or distilleries.

**1910.106(e)(1)(ii)** *Exceptions.* Where portions of such plants involve chemical reactions such as oxidation, reduction, halogenation, hydrogenation, alkylation, polymerization, and other chemical processes, those portions of the plant shall be in accordance with paragraph (h) of this section.

**1910.106(e)(2)** *Incidental storage or use of flammable and combustible liquids*—(i) *Application.* This subparagraph shall be applicable to those portions of an industrial plant where the use and handling of flammable or combustible liquids is only incidental to the principal business, such as automobile assembly, construction of electronic equipment, furniture manufacturing, or other similar activities.

**1910.106(e)(2)(ii)** *Containers.* Flammable or combustible liquids shall be stored in tanks or closed containers.

**1910.106(e)(2)(ii)(a)** Except as provided in subdivisions (*b*) and (*c*) of this subdivision, all storage shall comply with paragraph (d) (3) or (4) of this section.

**1910.106(e)(2)(ii)(b)** The quantity of liquid that may be located outside of an inside storage room or storage cabinet in a building or in any one fire area of a building shall not exceed:

**1910.106(e)(2)(ii)(b)(1)** 25 gallons of Class IA liquids in containers

**1910.106(e)(2)(ii)(b)(2)** 120 gallons of Class IB, IC, II, or III liquids in containers

**1910.106(e)(2)(ii)(b)(3)** 660 gallons of Class IB, IC, II, or III liquids in a single portable tank.

**1910.106(e)(2)(ii)(c)** Where large quantities of flammable or combustible liquids are necessary, storage may be in tanks which shall comply with the applicable requirements of paragraph (b) of this section.

**1910.106(e)(2)(iii)** *Separation and protection.* Areas in which flammable or combustible liquids are transferred from one tank or container to another container shall be separated from other operations in the building by adequate distance or by construction having adequate fire resistance. Drainage or other means shall be provided to control spills. Adequate natural or mechanical ventilation shall be provided.

**1910.106(e)(2)(iv)** *Handling liquids at point of final use.* (*a*) Flammable liquids shall be kept in covered containers when not actually in use.

**1910.106(e)(2)(iv)(b)** Where flammable or combustible liquids are used or handled, except in closed containers, means shall be provided to dispose promptly and safely of leakage or spills.

**1910.106(e)(2)(iv)(c)** Class I liquids may be used only where there are no open flames or other sources of ignition within the possible path of vapor travel.

**1910.106(e)(2)(iv)(d)** Flammable or combustible liquids shall be drawn from or transferred into vessels, containers, or portable tanks within a building only through a closed piping system, from safety cans, by means of a device drawing through the top, or from a container or portable tanks by gravity through an approved self-closing valve. Transferring by means of air pressure on the container or portable tanks shall be prohibited.

**1910.106(e)(3)** *Unit physical operations*—(i) *Application.* This subparagraph shall be applicable in those portions of industrial plants where flammable or combustible liquids are handled or used in unit physical operations such as mixing, drying, evaporating, filtering, distillation, and similar operations which do not involve chemical change. Examples are plants compounding cosmetics, pharmaceuticals, solvents, cleaning fluids, insecticides, and similar types of activities.

**1910.106(e)(3)(ii)** *Location.* Industrial plants shall be located so that each building or unit of equipment is accessible from at least one side for firefighting and fire control purposes. Buildings shall be located with respect to lines of adjoining property which may be built upon as set forth in paragraph (h)(2) (i) and (ii) of this section except that the blank wall referred to in paragraph (h)(2)(ii) of this section shall have a fire resistance rating of at least 2 hours.

**1910.106(e)(3)(iii)** *Chemical processes.* Areas where unstable liquids are handled or small scale unit chemical processes are carried on shall be separated from the remainder of the plant by a fire wall of 2-hour minimum fire resistance rating.

**1910.106(e)(3)(iv)** *Drainage.* (*a*) Emergency drainage systems shall be provided to direct flammable or combustible liquid leakage and fire protection water to a safe location. This may require curbs, scuppers, or special drainage systems to control the spread of fire; see paragraph (b)(2)(vii)(*b*) of this section.

**1910.106(e)(3)(iv)(b)** Emergency drainage systems, if connected to public sewers or discharged into public waterways, shall be equipped with traps or separator.

**1910.106(e)(3)(v)** *Ventilation.* (*a*) Areas as defined in subdivision (i) of this subparagraph using Class I liquids shall be ventilated at a rate of not less than 1 cubic foot per minute per square foot of solid floor area. This shall be accomplished by natural or mechanical ventilation with discharge or exhaust to a safe location outside of the building. Provision shall be made for introduction of makeup air in such a manner as not to short circuit the ventilation. Ventilation shall be arranged to include all floor areas or pits where flammable vapors may collect.

**1910.106(e)(3)(v)(b)** Equipment used in a building and the ventilation of the building shall be designed so as to limit flammable vapor-air mixtures under normal operating conditions to the interior of equipment, and to not more than 5 feet from equipment which exposes Class I liquids to the air. Examples of such equipment are dispensing stations, open centrifuges, plate and frame filters, open vacuum filters, and surfaces of open equipment.

**1910.106(e)(3)(vi)** *Storage and handling.* The storage, transfer, and handling of liquid shall comply with paragraph (h)(4) of this section.

**1910.106(e)(4)** *Tank vehicle and tank car loading and unloading.* (i) Tank vehicle and tank car loading or unloading facilities shall be separated from aboveground tanks, warehouses, other plant buildings or nearest line of adjoining property which may be built upon by a distance of 25 feet for Class I liquids and 15 feet for Class II and Class III liquids measured from the nearest position of any fill stem. Buildings for pumps or shelters for personnel may be a part of the facility. Operations of the facility shall comply with the appropriate portions of paragraph (f)(3) of this section.

**1910.106(e)(4)(ii)** [Reserved]

**1910.106(e)(5)** *Fire control*—(i) *Portable and special equipment.* Portable fire extinguishment and control equipment shall be provided in such quantities and types as are needed for the special hazards of operation and storage.

**1910.106(e)(5)(ii)** *Water supply.* Water shall be available in volume and at adequate pressure to supply water hose streams, foam-producing equipment, automatic sprinklers, or water spray systems as the need is indicated by the special hazards of operation, dispensing and storage.

**1910.106(e)(5)(iii)** *Special extinguishers.* Special extinguishing equipment such as that utilizing foam, inert gas, or dry chemical shall be provided as the need is indicated by the special hazards of operation dispensing and storage.

**1910.106(e)(5)(iv)** *Special hazards.* Where the need is indicated by special hazards of operation, flammable or combustible liquid processing equipment, major piping, and supporting steel shall be protected by approved water spray systems, deluge systems, approved fire-resistant coatings, insulation, or any combination of these.

**1910.106(e)(5)(v)** *Maintenance.* All plant fire protection facilities shall be adequately maintained and periodically inspected and tested to make sure they are always in satisfactory operating condition, and they will serve their purpose in time of emergency.

**1910.106(e)(6)** *Sources of ignition*—(i) *General.* Adequate precautions shall be taken to prevent the ignition of flammable vapors. Sources of ignition include but are not limited to open flames; lightning; smoking; cutting and welding; hot surfaces; frictional heat; static, electrical, and mechanical sparks; spontaneous ignition, including heat-producing chemical reactions; and radiant heat.

**1910.106(e)(6)(ii)** *Grounding.* Class I liquids shall not be dispensed into containers unless the nozzle and container are electrically interconnected. Where the metallic floorplate on which the container stands while filling is electrically connected to the fill stem or where the fill stem is bonded to the container during filling operations by means of a bond wire, the provisions of this section shall be deemed to have been complied with.

**1910.106(e)(7)** *Electrical*—(i) *Equipment.* (*a*) All electrical wiring and equipment shall be installed according to the requirements of Subpart S of this part.

**1910.106(e)(7)(i)(b)** Locations where flammable vapor-air mixtures may exist under normal operations shall be classified Class I, Division 1 according to the requirements of Subpart S of this part. For those pieces of equipment installed in accordance with subparagraph (3)(v)(*b*) of this paragraph, the Division 1 area shall extend 5 feet in all directions from all points of vapor liberation. All areas within pits shall be classified Division 1 if any part of the pit is within a Division 1 or 2 classified area, unless the pit is provided with mechanical ventilation.

**1910.106(e)(7)(i)(c)** Locations where flammable vapor-air mixtures may exist under abnormal conditions and for a distance beyond Division 1 locations shall be classified Division 2 according to the requirements of Subpart S of this part. These locations include an area within 20 feet horizontally, 3 feet vertically beyond a Division 1 area, and up to 3 feet above floor or grade level within 25 feet, if indoors, or 10 feet if outdoors, from any pump, bleeder, withdrawal fitting, meter, or similar device handling Class I liquids. Pits provided with adequate mechanical ventilation within a Division 1 or

2 area shall be classified Division 2. If Class II or Class III liquids only are handled, then ordinary electrical equipment is satisfactory though care shall be used in locating electrical apparatus to prevent hot metal from falling into open equipment.

**1910.106(e)(7)(i)(d)** Where the provisions of subdivisions (*a*), (*b*), and (*c*), of this subdivision require the installation of electrical equipment suitable for Class I, Division 1 or Division 2 locations, ordinary electrical equipment including switchgear may be used if installed in a room or enclosure which is maintained under positive pressure with respect to the hazardous area. Ventilation makeup air shall be uncontaminated by flammable vapors.

**1910.106(e)(8)** *Repairs to equipment.* Hot work, such as welding or cutting operations, use of spark-producing power tools, and chipping operations shall be permitted only under supervision of an individual in responsible charge. The individual in responsible charge shall make an inspection of the area to be sure that it is safe for the work to be done and that safe procedures will be followed for the work specified.

**1910.106(e)(9)** *Housekeeping—(i) General.* Maintenance and operating practices shall be in accordance with established procedures which will tend to control leakage and prevent the accidental escape of flammable or combustible liquids. Spills shall be cleaned up promptly.

**1910.106(e)(9)(ii)** *Access.* Adequate aisles shall be maintained for unobstructed movement of personnel and so that fire protection equipment can be brought to bear on any part of flammable or combustible liquid storage, use, or any unit physical operation.

**1910.106(e)(9)(iii)** *Waste and residue.* Combustible waste material and residues in a building or unit operating area shall be kept to a minimum, stored in covered metal receptacles and disposed of daily.

**1910.106(e)(9)(iv)** *Clear zone.* Ground area around buildings and unit operating areas shall be kept free of weeds, trash, or other unnecessary combustible materials.

**1910.106(f) Bulk plants—(1)** *Storage—(i) Class I liquids.* Class I liquids shall be stored in closed containers, or in storage tanks above ground outside of buildings, or underground in accordance with paragraph (b) of this section.

**1910.106(f)(1)(ii)** *Class II and III liquids.* Class II and Class III liquids shall be stored in containers, or in tanks within buildings or above ground outside of buildings, or underground in accordance with paragraph (b) of this section.

**1910.106(f)(1)(iii)** *Piling containers.* Containers of flammable or combustible liquids when piled one upon the other shall be separated by dunnage sufficient to provide stability and to prevent excessive stress on container walls. The height of the pile shall be consistent with the stability and strength of containers.

**1910.106(f)(2)** *Buildings—(i) Exits.* Rooms in which flammable or combustible liquids are stored or handled by pumps shall have exit facilities arranged to prevent occupants from being trapped in the event of fire.

**1910.106(f)(2)(ii)** *Heating.* Rooms in which Class I liquids are stored or handled shall be heated only by means not constituting a source of ignition, such as steam or hot water. Rooms containing heating appliances involving sources of ignition shall be located and arranged to prevent entry of flammable vapors.

**1910.106(f)(2)(iii)** *Ventilation.* (*a*) Ventilation shall be provided for all rooms, buildings, or enclosures in which Class I liquids are pumped or dispensed. Design of ventilation systems shall take into account the relatively high specific gravity of the vapors. Ventilation may be provided by adequate openings in outside walls at floor level unobstructed except by louvers or coarse screens. Where natural ventilation is inadequate, mechanical ventilation shall be provided.

**1910.106(f)(2)(iii)(b)** Class I liquids shall not be stored or handled within a building having a basement or pit into which flammable vapors may travel, unless such area is provided with ventilation designed to prevent the accumulation of flammable vapors therein.

**1910.106(f)(2)(iii)(c)** Containers of Class I liquids shall not be drawn from or filled within buildings unless provision is made to prevent the accumulation of flammable vapors in hazardous concentrations. Where mechanical ventilation is required, it shall be kept in operation while flammable liquids are being handled.

**1910.106(f)(3)** *Loading and unloading facilities—(i) Separation.* Tank vehicle and tank car loading or unloading facilities shall be separated from aboveground tanks, warehouses, other plant buildings or nearest line of adjoining property that may be built upon by a distance of 25 feet for Class I liquids and 15 feet for Class II and Class III liquids measured from the nearest position of any fill spout. Buildings for pumps or shelters for personnel may be a part of the facility.

**1910.106(f)(3)(ii)** *Class restriction.* Equipment such as piping, pumps, and meters used for the transfer of Class I liquids between storage tanks and the fill stem of the loading rack shall not be used for the transfer of Class II or Class III liquids.

**1910.106(f)(3)(iii)** *Valves.* Valves used for the final control for filling tank vehicles shall be of the self-closing type and manually held open except where automatic means are provided for shutting off the flow when the vehicle is full or after filling of a preset amount.

**1910.106(f)(3)(iv)** *Static protection.* (*a*) Bonding facilities for protection against static sparks during the loading of tank vehicles through open domes shall be provided:

**1910.106(f)(3)(iv)(a)(1)** Where Class I liquids are loaded, or

**1910.106(f)(3)(iv)(a)(2)** Where Class II or Class III liquids are loaded into vehicles which may contain vapors from previous cargoes of Class I liquids.

**1910.106(f)(3)(iv)(b)** Protection as required in (*a*) of this subdivision (iv) shall consist of a metallic bond wire permanently electrically connected to the fill stem or to some part of the rack structure in electrical contact with the fill stem. The free end of such wire shall be provided with a clamp or equivalent device for convenient attachment to some metallic part in electrical contact with the cargo tank of the tank vehicle.

**1910.106(f)(3)(iv)(c)** Such bonding connection shall be made fast to the vehicle or tank before dome covers are raised and shall remain in place until filling is completed and all dome covers have been closed and secured.

**1910.106(f)(3)(iv)(d)** Bonding as specified in (*a*), (*b*), and (*c*) of this subdivision is not required:

**1910.106(f)(3)(iv)(d)(1)** Where vehicles are loaded exclusively with products not having a static accumulating tendency, such as asphalt, most crude oils, residual oils, and water soluble liquids;

**1910.106(f)(3)(iv)(d)(2)** Where no Class I liquids are handled at the loading facility and the tank vehicles loaded are used exclusively for Class II and Class III liquids; and

**1910.106(f)(3)(iv)(d)(3)** Where vehicles are loaded or unloaded through closed bottom or top connections.

**1910.106(f)(3)(iv)(e)** Filling through open domes into the tanks of tank vehicles or tank cars, that contain vapor-air mixtures within the flammable range or where the liquid being filled can form such a mixture, shall be by means of a downspout which extends near the bottom of the tank. This precaution is not required when loading liquids which are nonaccumulators of static charges.

**1910.106(f)(3)(v)** *Stray currents.* Tank car loading facilities where Class I liquids are loaded through open domes shall be protected against stray currents by bonding the pipe to at least one rail and to the rack structure if of metal. Multiple lines entering the rack area shall be electrically bonded together. In addition, in areas where excessive stray currents are known to exist, all pipe entering the rack area shall be provided with insulating sections to electrically isolate the rack piping from the pipelines. No bonding between the tank car and the rack or piping is required during either loading or unloading of Class II or III liquids.

**1910.106(f)(3)(vi)** *Container filling facilities.* Class I liquids shall not be dispensed into containers unless the nozzle and container are electrically interconnected. Where the metallic floorplate on which the container stands while filling is electrically connected to the fill stem or where the fill stem is bonded to the container during filling operations by means of a bond wire, the provisions of this section shall be deemed to have been complied with.

**1910.106(f)(4)** *Wharves—(i) Definition, application.* The term wharf shall mean any wharf, pier, bulkhead, or other structure over or contiguous to navigable water used in conjunction with a bulk plant, the primary function of which is the transfer of flammable or combustible liquid cargo in bulk between the bulk plant and any tank vessel, ship, barge, lighter boat, or other mobile floating craft; and this subparagraph shall apply to all such installations except Marine Service Stations as covered in paragraph (g) of this section.

**1910.106(f)(4)(ii)-(iii)** [Reserved]

**1910.106(f)(4)(iv)** *Design and construction.* Substructure and deck shall be substantially designed for the use intended. Deck may employ any material which will afford the desired combination of flexibility, resistance to shock, durability, strength, and fire resistance. Heavy timber construction is acceptable.

**1910.106(f)(4)(v)** [Reserved]

**1910.106(f)(4)(vi)** *Pumps.* Loading pumps capable of building up pressures in excess of the safe working pressure of cargo hose or loading arms shall be provided with bypasses, relief valves, or other arrangement to protect the loading facilities against excessive pressure. Relief devices shall be tested at not more than yearly intervals to determine that they function satisfactorily at the pressure at which they are set.

**1910.106(f)(4)(vii)** *Hoses and couplings.* All pressure hoses and couplings shall be inspected at intervals appropriate to the service. The hose and couplings shall be tested with the hose extended and using the "inservice maximum operating pressures." Any hose showing material deteriorations, signs of leakage, or weakness in its carcass or at the couplings shall be withdrawn from service and repaired or discarded.

**1910.106(f)(4)(viii)** *Piping and fittings.* Piping, valves, and fittings shall be in accordance with paragraph (c) of this section, with the following exceptions and additions:

**1910.106(f)(4)(viii)(a)** Flexibility of piping shall be assured by appropriate layout and arrangement of piping supports so that motion of the wharf structure resulting from wave action, currents, tides, or the mooring of vessels will not subject the pipe to repeated strain beyond the elastic limit.

**1910.106(f)(4)(viii)(b)** Pipe joints depending upon the friction characteristics of combustible materials or grooving of pipe ends for mechanical continuity of piping shall not be used.

**1910.106(f)(4)(viii)(c)** Swivel joints may be used in piping to which hoses are connected, and for articulated swivel-joint transfer systems, provided that the design is such that the mechanical strength of the joint will not be impaired if the packing material should fail, as by exposure to fire.

**1910.106(f)(4)(viii)(d)** Piping systems shall contain a sufficient number of valves to operate the system properly and to control the flow of liquid in normal operation and in the event of physical damage.

**1910.106(f)(4)(viii)(e)** In addition to the requirements of subdivision (d) of this subdivision, each line conveying flammable liquids leading to a wharf shall be provided with a readily accessible block valve located on shore near the approach to the wharf and outside of any diked area. Where more than one line is involved, the valves shall be grouped in one location.

**1910.106(f)(4)(viii)(f)** Means of easy access shall be provided for cargo line valves located below the wharf deck.

**1910.106(f)(4)(viii)(g)** Pipelines on flammable or combustible liquids wharves shall be adequately bonded and grounded. If excessive stray currents are encountered, insulating joints shall be installed. Bonding and grounding connections on all pipelines shall be located on wharfside of hose-riser insulating flanges, if used, and shall be accessible for inspection.

**1910.106(f)(4)(viii)(h)** Hose or articulated swivel-joint pipe connections used for cargo transfer shall be capable of accommodating the combined effects of change in draft and maximum tidal range, and mooring lines shall be kept adjusted to prevent the surge of the vessel from placing stress on the cargo transfer system.

**1910.106(f)(4)(viii)(i)** Hose shall be supported so as to avoid kinking and damage from chafing.

**1910.106(f)(4)(ix)** *Fire protection.* Suitable portable fire extinguishers with a rating of not less than 12-BC shall be located within 75 feet of those portions of the facility where fires are likely to occur, such as hose connections, pumps, and separator tanks.

**1910.106(f)(4)(ix)(a)** Where piped water is available, ready-connected fire hose in size appropriate for the water supply shall be provided so that manifolds where connections are made and broken can be reached by at least one hose stream.

**1910.106(f)(4)(ix)(b)** Material shall not be placed on wharves in such a manner as to obstruct access to firefighting equipment, or important pipeline control valves.

**1910.106(f)(4)(ix)(c)** Where the wharf is accessible to vehicle traffic, an unobstructed roadway to the shore end of the wharf shall be maintained for access of firefighting apparatus.

**1910.106(f)(4)(x)** *Operations control.* Loading or discharging shall not commence until the wharf superintendent and officer in charge of the tank vessel agree that the tank vessel is properly moored and all connections are properly made. Mechanical work shall not be performed on the wharf during cargo transfer, except under special authorization based on a review of the area involved, methods to be employed, and precautions necessary.

**1910.106(f)(5)** *Electrical equipment*—**(i)** *Application.* This subparagraph shall apply to areas where Class I liquids are stored or handled. For areas where Class II or Class III liquids only are stored or handled, the electrical equipment may be installed in accordance with the provisions of Subpart S of this part, for ordinary locations.

**1910.106(f)(5)(ii)** *Conformance.* All electrical equipment and wiring shall be of a type specified by and shall be installed in accordance with Subpart S of this part.

**1910.106(f)(5)(iii)** *Classification.* So far as it applies Table H-18 shall be used to delineate and classify hazardous areas for the purpose of installation of electrical equipment under normal circumstances. In Table H-18 a classified area shall not extend beyond an unpierced wall, roof, or other solid partition. The area classifications listed shall be based on the premise that the installation meets the applicable requirements of this section in all respects.

**1910.106(f)(6)** *Sources of ignition.* Class I liquids shall not be handled, drawn, or dispensed where flammable vapors may reach a source of ignition. Smoking shall be prohibited except in designated localities. "No Smoking" signs shall be conspicuously posted where hazard from flammable liquid vapors is normally present.

**1910.106(f)(7)** *Drainage and waste disposal.* Provision shall be made to prevent flammable or combustible liquids which may be spilled at loading or unloading points from entering public sewers and drainage systems, or natural waterways. Connection to such sewers, drains, or waterways by which flammable or combustible liquids might enter shall be provided with separator boxes or other approved means whereby such entry is precluded. Crankcase drainings and flammable or combustible liquids shall not be dumped into sewers, but shall be stored in tanks or tight drums outside of any building until removed from the premises.

**1910.106(f)(8)** *Fire control.* Suitable fire-control devices, such as small hose or portable fire extinguishers, shall be available to locations where fires are likely to occur. Additional fire-control equipment may be required where a tank of more than 50,000 gallons individual capacity contains Class I liquids and where an unusual exposure hazard exists from surrounding property. Such additional fire-control equipment shall be sufficient to extinguish a fire in the largest tank. The design and amount of such equipment shall be in accordance with approved engineering standards.

**1910.106(g)** **Service stations**—**(1)** *Storage and handling*—**(i)** *General provisions.* **(a)** Liquids shall be stored in approved closed containers not exceeding 60 gallons capacity, in tanks located underground, in tanks in special enclosures as described in paragraph (g)(i) of this section, or in aboveground tanks as provided for in paragraphs (g)(4)(ii), (b), (c) and (d) of this section.

**1910.106(g)(1)(i)(b)** Aboveground tanks, located in an adjoining bulk plant, may be connected by piping to service station underground tanks if, in addition to valves at aboveground tanks, a valve is also installed within control of service station personnel.

**1910.106(g)(1)(i)(c)** Apparatus dispensing Class I liquids into the fuel tanks of motor vehicles of the public shall not be located at a bulk plant unless separated by a fence or similar barrier from the area in which bulk operations are conducted.

**1910.106(g)(1)(i)(d)** [Reserved]

**1910.106(g)(1)(i)(e)** The provisions of paragraph (g)(1)(i)(a) of this section shall not prohibit the dispensing of flammable liquids in the open from a tank vehicle to a motor vehicle. Such dispensing shall be permitted provided:

**1910.106(g)(1)(i)(e)(1)** The tank vehicle complies with the requirements covered in the Standard on Tank Vehicles for Flammable Liquids, NFPA 385-1966.

**1910.106(g)(1)(i)(e)(2)** The dispensing is done on premises not open to the public.

**1910.106(g)(1)(i)(e)(3)** [Reserved]

**1910.106(g)(1)(i)(e)(4)** The dispensing hose does not exceed 50 feet in length.

**1910.106(g)(1)(i)(e)(5)** The dispensing nozzle is a listed automatic-closing type without a latchopen device.

**1910.106(g)(1)(i)(f)** Class I liquids shall not be stored or handled within a building having a basement or pit into which flammable vapors may travel, unless such area is provided with ventilation designed to prevent the accumulation of flammable vapors therein.

**1910.106(g)(1)(i)(g)** [Reserved]

## TABLE H-18—ELECTRICAL EQUIPMENT HAZARDOUS AREAS—BULK PLANTS

| Location | Class I Group D division | Extent of classified area |
|---|---|---|
| Tank vehicle and tank car:[1] | | |
|    Loading through open dome | 1 | Within 3 feet of edge of dome, extending in all directions. |
| | 2 | Area within 3 feet and 5 feet from edge of dome, extending in all directions. |
|    Loading through bottom connections with atmospheric venting | 1 | Within 3 feet of point of venting to atmosphere extending in all directions. |
| | 2 | Area between 3 feet and 5 feet from point of venting to atmosphere, extending in all directions. Also up to 18 inches above grade within a horizontal radius of 10 feet from point of loading connection. |
|    Loading through closed dome with atmospheric venting | 1 | Within 3 feet of open end of vent, extending in all directions. |
| | 2 | Area between 3 feet and 5 feet from open end of vent, extending in all directions. Also within 3 feet of edge of dome, extending in all directions. |
|    Loading through closed dome with vapor recovery | 2 | Within 3 feet of point of connection of both fill and vapor lines, extending in all directions. |
|    Bottom loading with vapor recovery or any bottom unloading | 2 | Within 3 feet of point of connections extending in all directions. Also up to 18 inches above grade within a horizontal radius of 10 feet from point of connection. |
| Drum and container filling: | | |
|    Outdoors, or indoors with adequate ventilation | 1 | Within 3 feet of vent and fill opening, extending in all directions. |
| | 2 | Area between 3 feet and 5 feet from vent or fill opening, extending in all directions. Also up to 18 inches above floor or grade level within a horizontal radius of 10 feet from vent or fill opening. |
|    Outdoors, or indoors with adequate ventilation | 1 | Within 3 feet of vent and fill opening, extending in all directions. |
| | 2 | Area between 3 feet and 5 feet from vent or fill opening, extending in all directions. Also up to 18 inches above floor or grade level within a horizontal radius of 10 feet from vent or fill opening. |
| Tank—Aboveground: | | |
|    Shell, ends, or roof and dike area | 2 | Within 10 feet from shell, ends, or roof of tank, Area inside dikes to level of top of dike. |
|    Vent | 1 | Within 5 feet of open end of vent, extending in all directions. |
| | 2 | Area between 5 feet and 10 feet from open end of vent, extending in all directions. |
|    Floating roof | 1 | Area above the roof and within the shell. |
| Pits: | | |
|    Without mechanical ventilation | 1 | Entire area within pit if any part is within a Division 1 or 2 classified area. |
|    With mechanical ventilation | 2 | Entire area within pit if any part is within a Division 1 or 2 classified area. |
|    Containing valves, fittings or piping, and not within a division 1 or 2 classified area | 2 | Entire pit. |
| Pumps, bleeders, withdrawal fittings, meters and similar devices: | | |
|    Indoors | 2 | Within 5 feet of any edge of such devices, extending in all directions. Also up to 3 feet above floor or grade level within 25 feet horizontally from any edge of such devices. |
|    Outdoors | 2 | Within 3 feet of any edge of such devices, extending in all directions. Also up to 18 inches above grade level within 10 feet horizontally from any edge of such devices. |
| Storage and repair garage for tank vehicles | 1 | All pits or spaces below floor level. |
| | 2 | Area up to 18 inches above floor or grade level for entire storage or repair garage. |
| Drainage ditches, separators, impounding basins | 2 | Area up to 18 inches above ditch, separator or basin. Also up to 18 inches above grade within 15 feet horizontally from any edge. |
| Garages for other than tank vehicles | ([2]) | If there is any opening to these rooms within the extent of an outdoor classified area, the entire room shall be classified the same as the area classification at the point of the opening. |
| Outdoor drum storage | ([2]) | |
| Indoor warehousing where there is no flammable liquid transfer | ([2]) | If there is any opening to these rooms within the extent of an indoor classified area, the room shall be classified the same as if the wall, curb or partition did not exist. |
| Office and rest rooms | ([2]) | |

[1] When classifying the extent of the area, consideration shall be given to the fact that tank cars or tank vehicles may be spotted at varying points. Therefore, the extremities of the loading or unloading positions shall be used.

[2] Ordinary.

**1910.106(g)(1)(ii)** *Special enclosures.* **(a)** When installation of tanks in accordance with paragraph (b)(3) of this section is impractical because of property or building limitations, tanks for flammable or combustible liquids may be installed in buildings if properly enclosed.

**1910.106(g)(1)(ii)(b)** The enclosure shall be substantially liquid and vaportight without backfill. Sides, top, and bottom of the enclosure shall be of reinforced concrete at least 6 inches thick, with openings for inspection through the top only. Tank connections shall be so piped or closed that neither vapors nor liquid can escape into the enclosed space.

Means shall be provided whereby portable equipment may be employed to discharge to the outside any liquid or vapors which might accumulate should leakage occur.

**1910.106(g)(1)(iii)** *Inside buildings.* **(a)** Except where stored in tanks as provided in subdivision (ii) of this subparagraph, no Class I liquids shall be stored within any service station building except in closed containers of aggregate capacity not exceeding 60 gallons. One container not exceeding 60 gallons capacity equipped with an approved pump is permitted.

**1910.106(g)(1)(iii)(b)** Class I liquids may be transferred from one container to another in lubrication or service rooms of a service station building provided the electrical installation complies with Table H-19 and provided that any heating equipment complies with subparagraph (6) of this paragraph.

**1910.106(g)(1)(iii)(c)** Class II and Class III liquids may be stored and dispensed inside service station buildings from tanks of not more than 120 gallons capacity each.

**1910.106(g)(1)(iv)** [Reserved]

**1910.106(g)(1)(v)** *Dispensing into portable containers.* No delivery of any Class I liquids shall be made into portable containers unless the container is constructed of metal, has a tight closure with screwed or spring cover, and is fitted with a spout or so designed that the contents can be poured without spilling.

**1910.106(g)(2)** [Reserved]

**1910.106(g)(3)** *Dispensing systems—(i) Location.* Dispensing devices at automotive service stations shall be so located that all parts of the vehicle being served will be on the premises of the service station.

**1910.106(g)(3)(ii)** *Inside location.* Approved dispensing units may be located inside of buildings. The dispensing area shall be separated from other areas in an approved manner. The dispensing unit and its piping shall be mounted either on a concrete island or protected against collision damage by suitable means and shall be located in a position where it cannot be struck by a vehicle descending a ramp or other slope out of control. The dispensing area shall be provided with an approved mechanical or gravity ventilation system. When dispensing units are located below grade, only approved mechanical ventilation shall be used and the entire dispensing area shall be protected by an approved automatic sprinkler system. Ventilating systems shall be electrically interlocked with gasoline dispensing units so that the dispensing units cannot be operated unless the ventilating fan motors are energized.

**1910.106(g)(3)(iii)** *Emergency power cutoff.* A clearly identified and easily accessible switch(es) or a circuit breaker(s) shall be provided at a location remote from dispensing devices, including remote pumping systems, to shut off the power to all dispensing devices in the event of an emergency.

**1910.106(g)(3)(iv)** *Dispensing units. (a)* Class I liquids shall be transferred from tanks by means of fixed pumps so designed and equipped as to allow control of the flow and to prevent leakage or accidental discharge.

**1910.106(g)(3)(iv)(b)(1)** Only listed devices may be used for dispensing Class I liquids. No such device may be used if it shows evidence of having been dismantled.

**1910.106(g)(3)(iv)(b)(2)** Every dispensing device for Class I liquids installed after December 31, 1978, shall contain evidence of listing so placed that any attempt to dismantle the device will result in damage to such evidence, visible without disassembly or dismounting of the nozzle.

**1910.106(g)(3)(iv)(c)** Class I liquids shall not be dispensed by pressure from drums, barrels, and similar containers. Approved pumps taking suction through the top of the container or approved self-closing faucets shall be used.

**1910.106(g)(3)(iv)(d)** The dispensing units, except those attached to containers, shall be mounted either on a concrete island or protected against collision damage by suitable means.

**1910.106(g)(3)(v)** *Remote pumping systems.*

**1910.106(g)(3)(v)(a)** This subdivision shall apply to systems for dispensing Class I liquids where such liquids are transferred from storage to individual or multiple dispensing units by pumps located elsewhere than at the dispensing units.

**1910.106(g)(3)(v)(b)** Pumps shall be designed or equipped so that no part of the system will be subjected to pressures above its allowable working pressure. Pumps installed above grade, outside of buildings, shall be located not less than 10 feet from lines of adjoining property which may be built upon, and not less than 5 feet from any building opening. When an outside pump location is impractical, pumps may be installed inside of buildings, as provided for dispensers in subdivision (ii) of this subparagraph, or in pits as provided in subdivision (c) of this subdivision. Pumps shall be substantially anchored and protected against physical damage by vehicles.

**1910.106(g)(3)(v)(c)** Pits for subsurface pumps or piping manifolds of submersible pumps shall withstand the external forces to which they may be subjected without damage to the pump, tank, or piping. The pit shall be no larger than necessary for inspection and maintenance and shall be provided with a fitted cover.

**1910.106(g)(3)(v)(d)** A control shall be provided that will permit the pump to operate only when a dispensing nozzle is removed from its bracket on the dispensing unit and the switch on this dispensing unit is manually actuated. This control shall also stop the pump when all nozzles have been returned to their brackets.

**1910.106(g)(3)(v)(e)** An approved impact valve, incorporating a fusible link, designed to close automatically in the event of severe impact or fire exposure shall be properly installed in the dispensing supply line at the base of each individual dispensing device.

**1910.106(g)(3)(v)(f)** *Testing.* After the completion of the installation, including any paving, that section of the pressure piping system between the pump discharge and the connection for the dispensing facility shall be tested for at least 30 minutes at the maximum operating pressure of the system. Such tests shall be repeated at 5-year intervals thereafter.

**1910.106(g)(3)(vi)** *Delivery nozzles. (a)* A listed manual or automatic-closing type hose nozzle valve shall be provided on dispensers used for the dispensing of Class I liquids.

**1910.106(g)(3)(vi)(b)** Manual-closing type valves shall be held open manually during dispensing. Automatic-closing type valves may be used in conjunction with an approved latch-open device.

**1910.106(g)(4)** *Marine service stations—(i) Dispensing. (a)* The dispensing area shall be located away from other structures so as to provide room for safe ingress and egress of craft to be fueled. Dispensing units shall in all cases be at least 20 feet from any activity involving fixed sources of ignition.

**1910.106(g)(4)(i)(b)** Dispensing shall be by approved dispensing units with or without integral pumps and may be located on open piers, wharves, or floating docks or on shore or on piers of the solid fill type.

**1910.106(g)(4)(i)(c)** Dispensing nozzles shall be automatic-closing without a hold-open latch.

**1910.106(g)(4)(ii)** *Tanks and pumps. (a)* Tanks, and pumps not integral with the dispensing unit, shall be on shore or on a pier of the solid fill type, except as provided in paragraphs (g)(4)(ii) (b) and (c) of this section.

**1910.106(g)(4)(ii)(b)** Where shore location would require excessively long supply lines to dispensers, tanks may be installed on a pier provided that applicable portions of paragraph (b) of this section relative to spacing, diking, and piping are complied with and the quantity so stored does not exceed 1,100 gallons aggregate capacity.

**1910.106(g)(4)(ii)(c)** Shore tanks supplying marine service stations may be located above ground, where rock ledges or high water table make underground tanks impractical.

**1910.106(g)(4)(ii)(d)** Where tanks are at an elevation which would produce gravity head on the dispensing unit, the tank outlet shall be equipped with a pressure control valve positioned adjacent to and outside the tank block valve specified in paragraph (b)(2)(ix)(b) of this section, so adjusted that liquid cannot flow by gravity from the tank in case of piping or hose failure.

**1910.106(g)(4)(iii)** *Piping. (a)* Piping between shore tanks and dispensing units shall be as described in paragraph (c) of this section, except that, where dispensing is from a floating structure, suitable lengths of oil-resistant flexible hose may be employed between the shore piping and the piping on the floating structure as made necessary by change in water level or shoreline.

## TABLE H-19—ELECTRICAL EQUIPMENT HAZARDOUS AREAS—SERVICE STATIONS

| Location | Class I Group D division | Extent of classified area |
|---|---|---|
| Underground tank | | |
| Fill opening | 1 | Any pit, box or space below grade level, any part of which is within the Division 1 or 2 classified area. |
| | 2 | Up to 18 inches above grade level within a horizontal radius of 10 feet from a loose fill connection and within a horizontal radius of 5 feet from a tight fill connection. |
| Vent—Discharging upward | 1 | Within 3 feet of open end of vent, extending in all directions. |
| | 2 | Area between 3 feet and 5 feet of open end of vent, extending in all directions. |
| Dispenser: | | |
| Pits | 1 | Any pit, box or space below grade level, any part of which is within the Division 1 or 2 |
| Dispenser enclosure | 1 | The area 4 feet vertically above base within the enclosure and 18 inches horizontally in all directions. |
| Outdoor | 2 | Up to 18 inches above grade level within 20 feet horizontally of any edge of enclosure. |
| Indoor: | | |
| With mechanical ventilation | 2 | Up to 18 inches above grade or floor level within 20 feet horizontally of any edge of enclosure. |
| With gravity ventilation | 2 | Up to 18 inches above grade or floor level within 20 feet horizontally of any edge of enclosure. |
| Remote pump—Outdoor | 1 | Any pit, box or space below grade level if any part is within a horizontal distance of 10 feet from any edge of pump. |
| | 2 | Within 3 feet of any edge of pump, extending in all directions. Also up to 18 inches above grade level within 10 feet horizontally from any edge of pump. |
| Remote pump—Indoor | 1 | Entire area within any pit. |
| | 2 | Within 5 feet of any edge of pump, extending in all directions. Also up to 3 feet above floor or grade level within 25 feet horizontally from any edge of pump. |
| Lubrication or service room | 1 | Entire area within any pit. |
| | 2 | Area up to 18 inches above floor or grade level within entire lubrication room. |
| Dispenser for Class I Liquids | 2 | Within 3 feet of any fill or dispensing point, extending in all directions. |
| Special enclosure inside building per §1910.106(f)(1)(ii). | 1 | Entire enclosure. |
| Sales, storage and rest rooms | (1) | If there is any opening to these rooms within the extent of a Division 1 area, the entire room shall be classified as Division 1. |

[1] Ordinary.

**1910.106(g)(4)(iii)***(b)* A readily accessible valve to shut off the supply from shore shall be provided in each pipeline at or near the approach to the pier and at the shore end of each pipeline adjacent to the point where flexible hose is attached.

**1910.106(g)(4)(iii)***(c)* Piping shall be located so as to be protected from physical damage.

**1910.106(g)(4)(iii)***(d)* Piping handling Class I liquids shall be grounded to control stray currents.

**1910.106(g)(5)** *Electrical equipment—(i) Application.* This subparagraph shall apply to areas where Class I liquids are stored or handled. For areas where Class II or Class III liquids are stored or handled the electrical equipment may be installed in accordance with the provisions of subpart S of this part, for ordinary locations.

**1910.106(g)(5)(ii)** All electrical equipment and wiring shall be of a type specified by and shall be installed in accordance with subpart S of this part.

**1910.106(g)(5)(iii)** So far as it applies, Table H-19 shall be used to delineate and classify hazardous areas for the purpose of installation of electrical equipment under normal circumstances. A classified area shall not extend beyond an unpierced wall, roof, or other solid partition.

**1910.106(g)(5)(iv)** The area classifications listed shall be based on the assumption that the installation meets the applicable requirements of this section in all respects.

**1910.106(g)(6)** *Heating equipment—(i) Conformance.* Heating equipment shall be installed as provided in paragraphs (g)(6) (ii) through (v) of this section.

**1910.106(g)(6)(ii)** *Application.* Heating equipment may be installed in the conventional manner in an area except as provided in paragraph (g)(6) (iii), (iv), or (v) of this section.

**1910.106(g)(6)(iii)** *Special room.* Heating equipment may be installed in a special room separated from an area classified by Table H-19 by walls having a fire resistance rating of at least 1 hour and without any openings in the walls within 8 feet of the floor into an area classified in Table H-19. This room shall not be used for combustible storage and all air for combustion purposes shall come from outside the building.

**1910.106(g)(6)(iv)** *Work areas.* Heating equipment using gas or oil fuel may be installed in the lubrication, sales, or service room where there is no dispensing or transferring of Class I liquids provided the bottom of the combustion chamber is at least 18 inches above the floor and the heating equipment is protected from physical damage by vehicles. Heating equipment using gas or oil fuel listed for use in garages may be installed in the lubrication or service room where Class I liquids are dispensed provided the equipment is installed at least 8 feet above the floor.

**1910.106(g)(6)(v)** *Electric heat.* Electrical heating equipment shall conform to paragraph (g)(5) of this section.

**1910.106(g)(7)** *Drainage and waste disposal.* Provision shall be made in the area where Class I liquids are dispensed to prevent spilled liquids from flowing into the interior of service station buildings. Such provision may be by grading driveways, raising door sills, or other equally effective means. Crankcase drainings and flammable or combustible liquids shall not be dumped into sewers but shall be stored in tanks or drums outside of any building until removed from the premises.

**1910.106(g)(8)** *Sources of ignition.* In addition to the previous restrictions of this paragraph, the following shall apply: There shall be no smoking or open flames in the areas used for fueling, servicing fuel systems for internal combustion engines, receiving or dispensing of flammable or combustible liquids. Conspicuous and legible signs prohibiting smoking shall be posted within sight of the customer being served. The motors of all equipment being fueled shall be shut off during the fueling operation.

**1910.106(g)(9)** *Fire control.* Each service station shall be provided with at least one fire extinguisher having a minimum approved classification of 6 B,

C, located so that an extinguisher, will be within 75 feet of each pump, dispenser, underground fill pipe opening, and lubrication or service room.

**1910.106(h)** *Processing plants*—**(1)** *Scope.* This paragraph shall apply to those plants or buildings which contain chemical operations such as oxidation, reduction, halogenation, hydrogenation, alkylation, polymerization, and other chemical processes but shall not apply to chemical plants, refineries or distilleries.

**1910.106(h)(2)** *Location*—**(i)** *Classification.* The location of each processing vessel shall be based upon its flammable or combustible liquid capacity.

**1910.106(h)(2)(ii)** [Reserved]

**1910.106(h)(3)** *Processing building*—**(i)** *Construction.* **(a)** Processing buildings shall be of fire-resistance or noncombustible construction, except heavy timber construction with load-bearing walls may be permitted for plants utilizing only stable Class II or Class III liquids. Except as provided in paragraph (h)(2)(ii) of this section or in the case of explosion resistant walls used in conjunction with explosion relieving facilities, see paragraph (h)(3)(iv) of this section, load-bearing walls are prohibited. Buildings shall be without basements or covered pits.

**1910.106(h)(3)(i)(b)** Areas shall have adequate exit facilities arranged to prevent occupants from being trapped in the event of fire. Exits shall not be exposed by the drainage facilities described in paragraph (h)(ii) of this section.

**1910.106(h)(3)(ii)** *Drainage.* **(a)** Emergency drainage systems shall be provided to direct flammable or combustible liquid leakage and fire protection water to a safe location. This may require curbs, scuppers, or special drainage systems to control the spread of fire, see paragraph (b)(2)(vii)(b) of this section.

**1910.106(h)(3)(ii)(b)** Emergency drainage systems, if connected to public sewers or discharged into public waterways, shall be equipped with traps or separators.

**1910.106(h)(3)(iii)** *Ventilation.* **(a)** Enclosed processing buildings shall be ventilated at a rate of not less than 1 cubic foot per minute per square foot of solid floor area. This shall be accomplished by natural or mechanical ventilation with discharge or exhaust to a safe location outside of the building. Provisions shall be made for introduction of makeup air in such a manner as not to short circuit the ventilation. Ventilation shall be arranged to include all floor areas or pits where flammable vapors may collect.

**1910.106(h)(3)(iii)(b)** Equipment used in a building and the ventilation of the building shall be designed so as to limit flammable vapor-air mixtures under normal operating conditions to the interior of equipment, and to not more than 5 feet from equipment which exposes Class I liquids to the air. Examples of such equipment are dispensing stations, open centrifuges, plate and frame filters, open vacuum filters, and surfaces of open equipment.

**1910.106(h)(3)(iv)** *Explosion relief.* Areas where Class IA or unstable liquids are processed shall have explosion venting through one or more of the following methods:

**1910.106(h)(3)(iv)(a)** Open air construction.

**1910.106(h)(3)(iv)(b)** Lightweight walls and roof.

**1910.106(h)(3)(iv)(c)** Lightweight wall panels and roof hatches.

**1910.106(h)(3)(iv)(d)** Windows of explosion venting type.

**1910.106(h)(4)** *Liquid handling*—**(i)** *Storage.* **(a)** The storage of flammable or combustible liquids in tanks shall be in accordance with the applicable provisions of paragraph (b) of this section.

**1910.106(h)(4)(i)(b)** If the storage of flammable or combustible liquids in outside aboveground or underground tanks is not practical because of temperature or production considerations, tanks may be permitted inside of buildings or structures in accordance with the applicable provisions of paragraph (b) of this section.

**1910.106(h)(4)(i)(c)** Storage tanks inside of buildings shall be permitted only in areas at or above grade which have adequate drainage and are separated from the processing area by construction having a fire resistance rating of at least 2 hours.

**1910.106(h)(4)(i)(d)** The storage of flammable or combustible liquids in containers shall be in accordance with the applicable provisions of paragraph (d) of this section.

**1910.106(h)(4)(ii)** *Piping, valves, and fittings.* **(a)** Piping, valves, and fittings shall be in accordance with paragraph (c) of this section.

**1910.106(h)(4)(ii)(b)** Approved flexible connectors may be used where vibration exists or where frequent movement is necessary. Approved hose may be used at transfer stations.

**1910.106(h)(4)(ii)(c)** Piping containing flammable or combustible liquids shall be identified.

**1910.106(h)(4)(iii)** *Transfer.* **(a)** The transfer of large quantities of flammable or combustible liquids shall be through piping by means of pumps or water displacement. Except as required in process equipment, gravity flow shall not be used. The use of compressed air as a transferring medium is prohibited.

**1910.106(h)(4)(iii)(b)** Positive displacement pumps shall be provided with pressure relief discharging back to the tank or to pump suction.

**1910.106(h)(4)(iv)** *Equipment.* **(a)** Equipment shall be designed and arranged to prevent the unintentional escape of liquids and vapors and to minimize the quantity escaping in the event of accidental release.

**1910.106(h)(4)(iv)(b)** Where the vapor space of equipment is usually within the flammable range, the probability of explosion damage to the equipment can be limited by inerting, by providing an explosion suppression system, or by designing the equipment to contain the peak explosion pressure which may be modified by explosion relief. Where the special hazards of operation, sources of ignition, or exposures indicate a need, consideration shall be given to providing protection by one or more of the above means.

**1910.106(h)(5)** *Tank vehicle and tank car loading and unloading.* Tank vehicle and tank car loading or unloading facilities shall be separated from aboveground tanks, warehouses, other plant buildings, or nearest line of adjoining property which may be built upon by a distance of 25 feet for Class I liquids and 15 feet for Class II and Class III liquids measured from the nearest position of any fill stem. Buildings for pumps or shelters for personnel may be a part of the facility. Operations of the facility shall comply with the appropriate portions of paragraph (f)(3) of this section.

**1910.106(h)(6)** *Fire control*—**(i)** *Portable extinguishers.* Approved portable fire extinguishers of appropriate size, type, and number shall be provided.

**1910.106(h)(6)(ii)** *Other controls.* Where the special hazards of operation or exposure indicate a need, the following fire control provision shall be provided.

**1910.106(h)(6)(ii)(a)** A reliable water supply shall be available in pressure and quantity adequate to meet the probable fire demands.

**1910.106(h)(6)(ii)(b)** Hydrants shall be provided in accordance with accepted good practice.

**1910.106(h)(6)(ii)(c)** Hose connected to a source of water shall be installed so that all vessels, pumps, and other equipment containing flammable or combustible liquids can be reached with at least one hose stream. Nozzles that are capable of discharging a water spray shall be provided.

**1910.106(h)(6)(ii)(d)** Processing plants shall be protected by an approved automatic sprinkler system or equivalent extinguishing system. If special extinguishing systems including but not limited to those employing foam, carbon dioxide, or dry chemical are provided, approved equipment shall be used and installed in an approved manner.

**1910.106(h)(6)(iii)** *Alarm systems.* An approved means for prompt notification of fire to those within the plant and any public fire department available shall be provided. It may be advisable to connect the plant system with the public system where public fire alarm system is available.

**1910.106(h)(6)(iv)** *Maintenance.* All plant fire protection facilities shall be adequately maintained and periodically inspected and tested to make sure they are always in satisfactory operating condition and that they will serve their purpose in time of emergency.

**1910.106(h)(7)** *Sources of ignition*—**(i)** *General.* **(a)** Precautions shall be taken to prevent the ignition of flammable vapors. Sources of ignition include but are not limited to open flames; lightning; smoking; cutting and welding; hot surfaces; frictional heat; static, electrical, and mechanical sparks; spontaneous ignition, including heat-producing chemical reactions; and radiant heat.

**1910.106(h)(7)(i)(b)** Class I liquids shall not be dispensed into containers unless the nozzle and container are electrically interconnected. Where the metallic floorplate on which the container stands while filling is electrically connected to the fill stem or where the fill stem is bonded to the container during filling operations by means of a bond wire, the provisions of this section shall be deemed to have been complied with.

**1910.106(h)(7)(ii)** *Maintenance and repair.* **(a)** When necessary to do maintenance work in a flammable or combustible liquid processing area, the work shall be authorized by a responsible representative of the employer.

**1910.106(h)(7)(ii)(b)** Hot work, such as welding or cutting operations, use of spark-producing power tools, and chipping operations shall be permitted only under supervision of an individual in responsible charge who shall make an inspection of the area to be sure that it is safe for the work to be done and that safe procedures will be followed for the work specified.

**1910.106(h)(7)(iii)** *Electrical.* **(a)** All electric wiring and equipment shall be installed in accordance with subpart S of this part.

**1910.106(h)(7)(iii)(b)** Locations where flammable vapor-air mixtures may exist under normal operations shall be classified Class I, Division 1 according to the requirements of subpart S of this part. For those pieces of equipment installed in accordance with paragraph (h)(3)(iii)(b) of this section, the Division 1 area shall extend 5 feet in all directions from all points of vapor liberation. All areas within pits shall be classified Division 1 if any part of the pit is within a Division 1 or 2 classified area, unless the pit is provided with mechanical ventilation.

**1910.106(h)(7)(iii)(c)** Locations where flammable vapor-air mixtures may exist under abnormal conditions and for a distance beyond Division 1 locations shall be classified Division 2 according to the requirements of subpart S of this part. These locations include an area within 20 feet horizontally, 3 feet vertically beyond a Division 1 area, and up to 3 feet above floor or grade level within 25 feet, if indoors, or 10 feet if outdoors, from any pump, bleeder, withdrawal fitting, meter, or similar device handling Class I liquids. Pits provided with adequate mechanical ventilation within a Division 1 or 2 area shall be classified Division 2. If Class II or Class III liquids only are handled, then ordinary electrical equipment is satisfactory though care shall be used in locating electrical apparatus to prevent hot metal from falling into open equipment.

**1910.106(h)(7)(iii)(d)** Where the provisions of paragraphs (h)(7)(iii) (a), (b), and (c) of this section require the installation of explosion-proof equipment, ordinary electrical equipment including switchgear may be used if installed in a room or enclosure which is maintained under positive pressure with respect to the hazardous area. Ventilation makeup air shall be uncontaminated by flammable vapors.

**1910.106(h)(8)** *Housekeeping*—**(i)** *General.* Maintenance and operating practices shall be in accordance with established procedures which will tend to control leakage and prevent the accidental escape of flammable or combustible liquids. Spills shall be cleaned up promptly.

**1910.106(h)(8)(ii)** *Access.* Adequate aisles shall be maintained for unobstructed movement of personnel and so that fire protection equipment can be brought to bear on any part of the processing equipment.

**1910.106(h)(8)(iii)** *Waste and residues.* Combustible waste material and residues in a building or operating area shall be kept to a minimum, stored in closed metal waste cans, and disposed of daily.

**1910.106(h)(8)(iv)** *Clear zone.* Ground area around buildings and operating areas shall be kept free of tall grass, weeds, trash, or other combustible materials.

**1910.106(i)** *Refineries, chemical plants, and distilleries*—**(1)** *Storage tanks.* Flammable or combustible liquids shall be stored in tanks, in containers, or in portable tanks. Tanks shall be installed in accordance with paragraph (b) of this section. Tanks for the storage of flammable or combustible liquids in tank farms and in locations other than process areas shall be located in accordance with paragraph (b)(2) (i) and (ii) of this section.

**1910.106(i)(2)** *Wharves.* Wharves handling flammable or combustible liquids shall be in accordance with paragraph (f)(4) of this section.

**1910.106(i)(3)** *Fired and unfired pressure vessels*—**(i)** *Fired vessels.* Fired pressure vessels shall be constructed in accordance with the Code for Fired Pressure Vessels, Section I of the ASME Boiler and Pressure Vessel Code—1968.

**1910.106(i)(3)(ii)** Unfired vessels shall be constructed in accordance with the Code for Unfired Pressure Vessels, Section VIII of the ASME Boiler and Pressure Vessel Code—1968.

**1910.106(i)(4)** *Location of process units.* Process units shall be located so that they are accessible from at least one side for the purpose of fire control.

**1910.106(i)(5)** *Fire control*—**(i)** *Portable equipment.* Portable fire extinguishment and control equipment shall be provided in such quantities and types as are needed for the special hazards of operation and storage.

**1910.106(i)(5)(ii)** *Water supply.* Water shall be available in volume and at adequate pressure to supply water hose streams, foam producing equipment, automatic sprinklers, or water spray systems as the need is indicated by the special hazards of operation and storage.

**1910.106(i)(5)(iii)** *Special equipment.* Special extinguishing equipment such as that utilizing foam, inert gas, or dry chemical shall be provided as the need is indicated by the special hazards of operation and storage.

**1910.106(j)** *Scope.* This section applies to the handling, storage, and use of flammable and combustible liquids with a flashpoint below 200 °F. This section does not apply to:

**1910.106(j)(1)** Bulk transportation of flammable and combustible liquids;

**1910.106(j)(2)** Storage, handling, and use of fuel oil tanks and containers connected with oil burning equipment;

**1910.106(j)(3)** Storage of flammable and combustible liquids on farms;

**1910.106(j)(4)** Liquids without flashpoints that may be flammable under some conditions, such as certain halogenated hydrocarbons and mixtures containing halogenated hydrocarbons;

**1910.106(j)(5)** Mists, sprays, or foams, except flammable aerosols covered in paragraph (d) of this section; or

**1910.106(j)(6)** Installations made in accordance with requirements of the following standards that are incorporated by reference as specified in §1910.6:

**1910.106(j)(6)(i)** National Fire Protection Association Standard for Drycleaning Plants, NFPA No. 32-1970;

**1910.106(j)(6)(ii)** National Fire Protection Association Standard for the Manufacture of Organic Coatings, NFPA No. 35-1970;

**1910.106(j)(6)(iii)** National Fire Protection Association Standard for Solvent Extraction Plants, NFPA No. 36-1967; or

**1910.106(j)(6)(iv)** National Fire Protection Association Standard for the Installation and Use of Stationary Combustion Engines and Gas Turbines, NFPA No. 37-1970.

[39 FR 23502, June 27, 1974, as amended at 40 FR 3982, Jan. 27, 1975; 40 FR 23743, June 2, 1975; 43 FR 49746, Oct. 24, 1978; 43 FR 51759, Nov. 7, 1978; 47 FR 39164, Sept. 7, 1982; 51 FR 34560, Sept. 29, 1986; 53 FR 12121, Apr. 12, 1988; 55 FR 32015, Aug. 6, 1990; 61 FR 9237, Mar. 7, 1996]

## §1910.107 Spray finishing using flammable and combustible materials.

**1910.107(a)** *Definitions applicable to this section*—**(1)** *Aerated solid powders.* Aerated powders shall mean any powdered material used as a coating material which shall be fluidized within a container by passing air uniformly from below. It is common practice to fluidize such materials to form a fluidized powder bed and then dip the part to be coated into the bed in a manner similar to that used in liquid dipping. Such beds are also used as sources for powder spray operations.

**1910.107(a)(2)** *Spraying area.* Any area in which dangerous quantities of flammable vapors or mists, or combustible residues, dusts, or deposits are present due to the operation of spraying processes.

**1910.107(a)(3)** *Spray booth.* A power-ventilated structure provided to enclose or accommodate a spraying operation to confine and limit the escape of spray, vapor, and residue, and to safely conduct or direct them to an exhaust system.

**1910.107(a)(4)** *Waterwash spray booth.* A spray booth equipped with a water washing system designed to minimize dusts or residues entering exhaust ducts and to permit the recovery of overspray finishing material.

**1910.107(a)(5)** *Dry spray booth.* A spray booth not equipped with a water washing system as described in subparagraph (4) of this paragraph. A dry spray booth may be equipped with (i) distribution or baffle plates to promote an even flow of air through the booth or cause the deposit of overspray before it enters the exhaust duct; or (ii) overspray dry filters to minimize dusts; or (iii) overspray dry filters to minimize dusts or residues entering exhaust ducts; or (iv) overspray dry filter rolls designed to minimize dusts or residues entering exhaust ducts; or (v) where dry powders are being sprayed, with powder collection systems so arranged in the exhaust to capture oversprayed material.

**1910.107(a)(6)** *Fluidized bed.* A container holding powder coating material which is aerated from below so as to form an air-supported expanded cloud of such material through which the preheated object to be coated is immersed and transported.

**1910.107(a)(7)** *Electrostatic fluidized bed.* A container holding powder coating material which is aerated from below so as to form an air-supported expanded cloud of such material which is electrically charged with a charge opposite to the charge of the object to be coated; such object is transported, through the container immediately above the charged and aerated materials in order to be coated.

**1910.107(a)(8)** *Approved.* Shall mean approved and listed by a nationally recognized testing laboratory. Refer to §1910.7 for definition of nationally recognized testing laboratory.

**1910.107(a)(9)** *Listed.* See "approved" in §1910.107(a)(8).

**1910.107(b)** **Spray booths—(1)** *Construction.* Spray booths shall be substantially constructed of steel, securely and rigidly supported, or of concrete or masonry except that aluminum or other substantial noncombustible material may be used for intermittent or low volume spraying. Spray booths shall be designed to sweep air currents toward the exhaust outlet.

**1910.107(b)(2)** *Interiors.* The interior surfaces of spray booths shall be smooth and continuous without edges and otherwise designed to prevent pocketing of residues and facilitate cleaning and washing without injury.

**1910.107(b)(3)** *Floors.* The floor surface of a spray booth and operator's working area, if combustible, shall be covered with noncombustible material of such character as to facilitate the safe cleaning and removal of residues.

**1910.107(b)(4)** *Distribution or baffle plates.* Distribution or baffle plates, if installed to promote an even flow of air through the booth or cause the deposit of overspray before it enters the exhaust duct, shall be of non-combustible material and readily removable or accessible on both sides for cleaning. Such plates shall not be located in exhaust ducts.

**1910.107(b)(5)** *Dry type overspray collectors—(exhaust air filters).* In conventional dry type spray booths, overspray dry filters or filter rolls, if installed, shall conform to the following:

**1910.107(b)(5)(i)** The spraying operations except electrostatic spraying operations shall be so designed, installed and maintained that the average air velocity over the open face of the booth (or booth cross section during spraying operations) shall be not less than 100 linear feet per minute. Electrostatic spraying operations may be conducted with an air velocity over the open face of the booth of not less than 60 linear feet per minute, or more, depending on the volume of the finishing material being applied and its flammability and explosion characteristics. Visible gauges or audible alarm or pressure activated devices shall be installed to indicate or insure that the required air velocity is maintained. Filter rolls shall be inspected to insure proper replacement of filter media.

**1910.107(b)(5)(ii)** All discarded filter pads and filter rolls shall be immediately removed to a safe, well-detached location or placed in a water-filled metal container and disposed of at the close of the day's operation unless maintained completely in water.

**1910.107(b)(5)(iii)** The location of filters in a spray booth shall be so as to not reduce the effective booth enclosure of the articles being sprayed.

**1910.107(b)(5)(iv)** Space within the spray booth on the downstream and upstream sides of filters shall be protected with approved automatic sprinklers.

**1910.107(b)(5)(v)** Filters or filter rolls shall not be used when applying a spray material known to be highly susceptible to spontaneous heating and ignition.

**1910.107(b)(5)(vi)** Clean filters or filter rolls shall be noncombustible or of a type having a combustibility not in excess of class 2 filters as listed by Underwriters' Laboratories, Inc. Filters and filter rolls shall not be alternately used for different types of coating materials, where the combination of materials may be conducive to spontaneous ignition. See also paragraph (g)(6) of this section.

**1910.107(b)(6)** *Frontal area.* Each spray booth having a frontal area larger than 9 square feet shall have a metal deflector or curtain not less than 2 1/2 inches deep installed at the upper outer edge of the booth over the opening.

**1910.107(b)(7)** *Conveyors.* Where conveyors are arranged to carry work into or out of spray booths, the openings therefor shall be as small as practical.

**1910.107(b)(8)** *Separation of operations.* Each spray booth shall be separated from other operations by not less than 3 feet, or by a greater distance, or by such partition or wall as to reduce the danger from juxtaposition of hazardous operations. See also paragraph (c)(1) of this section.

**1910.107(b)(9)** *Cleaning.* Spray booths shall be so installed that all portions are readily accessible for cleaning. A clear space of not less than 3 feet on all sides shall be kept free from storage or combustible construction.

**1910.107(b)(10)** *Illumination.* When spraying areas are illuminated through glass panels or other transparent materials, only fixed lighting units shall be used as a source of illumination. Panels shall effectively isolate the spraying area from the area in which the lighting unit is located, and shall be of a noncombustible material of such a nature or so protected that breakage will be unlikely. Panels shall be so arranged that normal accumulations of residue on the exposed surface of the panel will not be raised to a dangerous temperature by radiation or conduction from the source of illumination.

**1910.107(c)** *Electrical and other sources of ignition—(1)* *Conformance.* All electrical equipment, open flames and other sources of ignition shall conform to the requirements of this paragraph, except as follows:

**1910.107(c)(1)(i)** Electrostatic apparatus shall conform to the requirements of paragraphs (h) and (i) of this section;

**1910.107(c)(1)(ii)** Drying, curing, and fusion apparatus shall conform to the requirements of paragraph (j) of this section;

**1910.107(c)(1)(iii)** Automobile undercoating spray operations in garages shall conform to the requirements of paragraph (k) of this section;

**1910.107(c)(1)(iv)** Powder coating equipment shall conform to the requirements of paragraph (c)(1) of this section.

**1910.107(c)(2)** *Minimum separation.* There shall be no open flame or spark producing equipment in any spraying area nor within 20 feet thereof, unless separated by a partition.

**1910.107(c)(3)** *Hot surfaces.* Space-heating appliances, steampipes, or hot surfaces shall not be located in a spraying area where deposits of combustible residues may readily accumulate.

**1910.107(c)(4)** *Wiring conformance.* Electrical wiring and equipment shall conform to the provisions of this paragraph and shall otherwise be in accordance with subpart S of this part.

**1910.107(c)(5)** *Combustible residues, areas.* Unless specifically approved for locations containing both deposits of readily ignitable residue and explosive vapors, there shall be no electrical equipment in any spraying area, whereon deposits of combustible residues may readily accumulate, except wiring in rigid conduit or in boxes or fittings containing no taps, splices, or terminal connections.

**1910.107(c)(6)** *Wiring type approved.* Electrical wiring and equipment not subject to deposits of combustible residues but located in a spraying area as herein defined shall be of explosion-proof type approved for Class I, group D locations and shall otherwise conform to the provisions of subpart S of this part, for Class I, Division 1, Hazardous Locations. Electrical wiring, motors, and other equipment outside of but within twenty (20) feet of any spraying area, and not separated therefrom by partitions, shall not produce sparks under normal operating conditions and shall otherwise conform to the provisions of subpart S of this part for Class I, Division 2 Hazardous Locations.

**1910.107(c)(7)** *Lamps.* Electric lamps outside of, but within twenty (20) feet of any spraying area, and not separated therefrom by a partition, shall be totally enclosed to prevent the falling of hot particles and shall be protected from mechanical injury by suitable guards or by location.

**1910.107(c)(8)** *Portable lamps.* Portable electric lamps shall not be used in any spraying area during spraying operations. Portable electric lamps, if used during cleaning or repairing operations, shall be of the type approved for hazardous Class I locations.

**1910.107(c)(9)** *Grounding.* (i) All metal parts of spray booths, exhaust ducts, and piping systems conveying flammable or combustible liquids or aerated solids shall be properly electrically grounded in an effective and permanent manner.

**1910.107(c)(9)(ii)** [Reserved]

**1910.107(d)** *Ventilation*—**(1)** *Conformance.* Ventilating and exhaust systems shall be in accordance with the Standard for Blower and Exhaust Systems for Vapor Removal, NFPA No. 91-1961, which is incorporated by reference as specified in §1910.6, where applicable and shall also conform to the provisions of this section.

**1910.107(d)(2)** *General.* All spraying areas shall be provided with mechanical ventilation adequate to remove flammable vapors, mists, or powders to a safe location and to confine and control combustible residues so that life is not endangered. Mechanical ventilation shall be kept in operation at all times while spraying operations are being conducted and for a sufficient time thereafter to allow vapors from drying coated articles and drying finishing material residue to be exhausted.

**1910.107(d)(3)** *Independent exhaust.* Each spray booth shall have an independent exhaust duct system discharging to the exterior of the building, except that multiple cabinet spray booths in which identical spray finishing material is used with a combined frontal area of not more than 18 square feet may have a common exhaust. If more than one fan serves one booth, all fans shall be so interconnected that one fan cannot operate without all fans being operated.

**1910.107(d)(4)** *Fan-rotating element.* The fan-rotating element shall be nonferrous or nonsparking or the casing shall consist of or be lined with such material. There shall be ample clearance between the fan-rotating element and the fan casing to avoid a fire by friction, necessary allowance being made for ordinary expansion and loading to prevent contact between moving parts and the duct or fan housing. Fan blades shall be mounted on a shaft sufficiently heavy to maintain perfect alignment even when the blades of the fan are heavily loaded, the shaft preferably to have bearings outside the duct and booth. All bearings shall be of the self-lubricating type, or lubricated from the outside duct.

**1910.107(d)(5)** *Electric motors.* Electric motors driving exhaust fans shall not be placed inside booths or ducts. See also paragraph (c) of this section.

**1910.107(d)(6)** *Belts.* Belts shall not enter the duct or booth unless the belt and pulley within the duct or booth are thoroughly enclosed.

**1910.107(d)(7)** *Exhaust ducts.* Exhaust ducts shall be constructed of steel and shall be substantially supported. Exhaust ducts without dampers are preferred; however, if dampers are installed, they shall be maintained so that they will be in a full open position at all times the ventilating system is in operation.

**1910.107(d)(7)(i)** Exhaust ducts shall be protected against mechanical damage and have a clearance from unprotected combustible construction or other combustible material of not less than 18 inches.

**1910.107(d)(7)(ii)** If combustible construction is provided with the following protection applied to all surfaces within 18 inches, clearances may be reduced to the distances indicated:

| | |
|---|---|
| (a) 28-gage sheet metal on 1/4-inch asbestos mill board. | 12 inches. |
| (b) 28-gage sheet metal on 1/8-inch asbestos mill board spaced out 1 inch on noncombustible spacers. | 9 inches. |
| (c) 22-gage sheet metal on 1-inch rockwool batts reinforced with wire mesh or the equivalent. | 3 inches. |

(d) Where ducts are protected with an approved automatic sprinkler system, properly maintained, the clearance required in subdivision (i) of this subparagraph may be reduced to 6 inches.

**1910.107(d)(8)** *Discharge clearance.* Unless the spray booth exhaust duct terminal is from a water-wash spray booth, the terminal discharge point shall be not less than 6 feet from any combustible exterior wall or roof nor discharge in the direction of any combustible construction or unprotected opening in any noncombustible exterior wall within 25 feet.

**1910.107(d)(9)** *Air exhaust.* Air exhaust from spray operations shall not be directed so that it will contaminate makeup air being introduced into the spraying area or other ventilating intakes, nor directed so as to create a nuisance. Air exhausted from spray operations shall not be recirculated.

**1910.107(d)(10)** *Access doors.* When necessary to facilitate cleaning, exhaust ducts shall be provided with an ample number of access doors.

**1910.107(d)(11)** *Room intakes.* Air intake openings to rooms containing spray finishing operations shall be adequate for the efficient operation of exhaust fans and shall be so located as to minimize the creation of dead air pockets.

**1910.107(d)(12)** *Drying spaces.* Freshly sprayed articles shall be dried only in spaces provided with adequate ventilation to prevent the formation of explosive vapors. In the event adequate and reliable ventilation is not provided such drying spaces shall be considered a spraying area. See also paragraph (j) of this section.

**1910.107(e)** *Flammable and combustible liquids—storage and handling*—**(1)** *Conformance.* The storage of flammable or combustible liquids in connection with spraying operations shall conform to the requirements of §1910.106, where applicable.

**1910.107(e)(2)** *Quantity.* The quantity of flammable or combustible liquids kept in the vicinity of spraying operations shall be the minimum required for operations and should ordinarily not exceed a supply for 1 day or one shift. Bulk storage of portable containers of flammable or combustible liquids shall be in a separate, constructed building detached from other important buildings or cut off in a standard manner.

**1910.107(e)(3)** *Containers.* Original closed containers, approved portable tanks, approved safety cans or a properly arranged system of piping shall be used for bringing flammable or combustible liquids into spray finishing room. Open or glass containers shall not be used.

**1910.107(e)(4)** *Transferring liquids.* Except as provided in paragraph (e)(5) of this section the withdrawal of flammable and combustible liquids from containers having a capacity of greater than 60 gallons shall be by approved pumps. The withdrawal of flammable or combustible liquids from containers and the filling of containers, including portable mixing tanks, shall be done only in a suitable mixing room or in a spraying area when the ventilating system is in operation. Adequate precautions shall be taken to protect against liquid spillage and sources of ignition.

**1910.107(e)(5)** *Spraying containers.* Containers supplying spray nozzles shall be of closed type or provided with metal covers kept closed. Containers not resting on floors shall be on metal supports or suspended by wire cables. Containers supplying spray nozzles by gravity flow shall not exceed 10 gallons capacity. Original shipping containers shall not be subject to air pressure for supplying spray nozzles. Containers under air pressure supplying spray nozzles shall be of limited capacity, not exceeding that necessary for 1 day's operation; shall be designed and approved for such use; shall be provided with a visible pressure gage; and shall be provided with a relief valve set to operate in conformance with the requirements of the Code for Unfired Pressure Vessels, Section VIII of the ASME Boiler and Pressure Vessel Code—1968, which is incorporated by reference as specified in §1910.6. Containers under air pressure supplying spray nozzles, air-storage tanks and coolers shall conform to the standards of the Code for Unfired Pressure Vessels, Section VIII of the ASME Boiler and Pressure Vessel Code —1968 for construction, tests, and maintenance.

**1910.107(e)(6)** *Pipes and hoses.* **(i)** All containers or piping to which is attached a hose or flexible connection shall be provided with a shutoff valve at the connection. Such valves shall be kept shut when spraying operations are not being conducted.

**1910.107(e)(6)(ii)** When a pump is used to deliver products, automatic means shall be provided to prevent pressure in excess of the design working pressure of accessories, piping, and hose.

**1910.107(e)(6)(iii)** All pressure hose and couplings shall be inspected at regular intervals appropriate to this service. The hose and couplings shall be tested with the hose extended, and using the "inservice maximum operating pressures." Any hose showing material deteriorations, signs of leakage, or weakness in its carcass or at the couplings, shall be withdrawn from service and repaired or discarded.

**1910.107(e)(6)(iv)** Piping systems conveying flammable or combustible liquids shall be of steel or other material having comparable properties of resistance to heat and physical damage. Piping systems shall be properly bonded and grounded.

**1910.107(e)(7)** *Spray liquid heaters.* Electrically powered spray liquid heaters shall be approved and listed for the specific location in which used (see paragraph (c) of this section). Heaters shall not be located in spray booths nor other locations subject to the accumulation of deposits or combustible residue. If an electric motor is used, see paragraph (c) of this section.

**1910.107(e)(8)** *Pump relief.* If flammable or combustible liquids are supplied to spray nozzles by positive displacement pumps, the pump discharge line shall be provided with an approved relief valve discharging to a pump suction or a

**1910.107(e)(9)** *Grounding.* Whenever flammable or combustible liquids are transferred from one container to another, both containers shall be effectively bonded and grounded to prevent discharge sparks of static electricity.

**1910.107(f)** **Protection—(1)** *Conformance.* In sprinklered buildings, the automatic sprinkler system in rooms containing spray finishing operations shall conform to the requirements of §1910.159. In unsprinklered buildings where sprinklers are installed only to protect spraying areas, the installation shall conform to such standards insofar as they are applicable. Sprinkler heads shall be located so as to provide water distribution throughout the entire booth.

**1910.107(f)(2)** *Valve access.* Automatic sprinklers protecting each spray booth (together with its connecting exhaust) shall be under an accessibly located separate outside stem and yoke (OS&Y) subcontrol valve.

**1910.107(f)(3)** *Cleaning of heads.* Sprinklers protecting spraying areas shall be kept as free from deposits as practical by cleaning daily if necessary. (See also paragraph (g) of this section.)

**1910.107(f)(4)** *Portable extinguishers.* An adequate supply of suitable portable fire extinguishers shall be installed near all spraying areas.

**1910.107(g)** **Operations and maintenance—(1)** *Spraying.* Spraying shall not be conducted outside of predetermined spraying areas.

**1910.107(g)(2)** *Cleaning.* All spraying areas shall be kept as free from the accumulation of deposits of combustible residues as practical, with cleaning conducted daily if necessary. Scrapers, spuds, or other such tools used for cleaning purposes shall be of nonsparking material.

**1910.107(g)(3)** *Residue disposal.* Residue scrapings and debris contaminated with residue shall be immediately removed from the premises and properly disposed of. Approved metal waste cans shall be provided wherever rags or waste are impregnated with finishing material and all such rags or waste deposited therein immediately after use. The contents of waste cans shall be properly disposed of at least once daily or at the end of each shift.

**1910.107(g)(4)** *Clothing storage.* Spray finishing employees' clothing shall not be left on the premises overnight unless kept in metal lockers.

**1910.107(g)(5)** *Cleaning solvents.* The use of solvents for cleaning operations shall be restricted to those having flashpoints not less than 100 °F.; however, for cleaning spray nozzles and auxiliary equipment, solvents having flashpoints not less than those normally used in spray operations may be used. Such cleaning shall be conducted inside spray booths and ventilating equipment operated during cleaning.

**1910.107(g)(6)** *Hazardous materials combinations.* Spray booths shall not be alternately used for different types of coating materials, where the combination of the materials may be conducive to spontaneous ignition, unless all deposits of the first used material are removed from the booth and exhaust ducts prior to spraying with the second used material.

**1910.107(g)(7)** *"No Smoking" signs.* "No smoking" signs in large letters on contrasting color background shall be conspicuously posted at all spraying areas and paint storage rooms.

**1910.107(h)** **Fixed electrostatic apparatus—(1)** *Conformance.* Where installation and use of electrostatic spraying equipment is used, such installation and use shall conform to all other paragraphs of this section, and shall also conform to the requirements of this paragraph.

**1910.107(h)(2)** *Type approval.* Electrostatic apparatus and devices used in connection with coating operations shall be of approved types.

**1910.107(h)(3)** *Location.* Transformers, power packs, control apparatus, and all other electrical portions of the equipment, with the exception of high-voltage grids, electrodes, and electrostatic atomizing heads and their connections, shall be located outside of the spraying area, or shall otherwise conform to the requirements of paragraph (c) of this section.

**1910.107(h)(4)** *Support.* Electrodes and electrostatic atomizing heads shall be adequately supported in permanent locations and shall be effectively insulated from the ground. Electrodes and electrostatic atomizing heads which are permanently attached to their bases, supports, or reciprocators, shall be deemed to comply with this section. Insulators shall be nonporous and noncombustible.

**1910.107(h)(5)** *Insulators, grounding.* High-voltage leads to electrodes shall be properly insulated and protected from mechanical injury or exposure to destructive chemicals. Electrostatic atomizing heads shall be effectively and permanently supported on suitable insulators and shall be effectively guarded against accidental contact or grounding. An automatic means shall be provided for grounding the electrode system when it is electrically deenergized for any reason. All insulators shall be kept clean and dry.

**1910.107(h)(6)** *Safe distance.* A safe distance shall be maintained between goods being painted and electrodes or electrostatic atomizing heads or conductors of at least twice the sparking distance. A suitable sign indicating this safe distance shall be conspicuously posted near the assembly.

**1910.107(h)(7)** *Conveyors required.* Goods being painted using this process are to be supported on conveyors. The conveyors shall be so arranged as to maintain safe distances between the goods and the electrodes or electrostatic atomizing heads at all times. Any irregularly shaped or other goods subject to possible swinging or movement shall be rigidly supported to prevent such swinging or movement which would reduce the clearance to less than that specified in paragraph (h)(6) of this section.

**1910.107(h)(8)** *Prohibition.* This process is not acceptable where goods being coated are manipulated by hand. When finishing materials are applied by electrostatic equipment which is manipulated by hand, see paragraph (i) of this section for applicable requirements.

**1910.107(h)(9)** *Fail-safe controls.* Electrostatic apparatus shall be equipped with automatic controls which will operate without time delay to disconnect the power supply to the high voltage transformer and to signal the operator under any of the following conditions:

**1910.107(h)(9)(i)** Stoppage of ventilating fans or failure of ventilating equipment from any cause.

**1910.107(h)(9)(ii)** Stoppage of the conveyor carrying goods through the high voltage field.

**1910.107(h)(9)(iii)** Occurrence of a ground or of an imminent ground at any point on the high voltage system.

**1910.107(h)(9)(iv)** Reduction of clearance below that specified in paragraph (h)(6) of this section.

**1910.107(h)(10)** *Guarding.* Adequate booths, fencing, railings, or guards shall be so placed about the equipment that they, either by their location or character or both, assure that a safe isolation of the process is maintained from plant storage or personnel. Such railings, fencing, and guards shall be of conducting material, adequately grounded.

**1910.107(h)(11)** *Ventilation.* Where electrostatic atomization is used the spraying area shall be so ventilated as to insure safe conditions from a fire and health standpoint.

**1910.107(h)(12)** *Fire protection.* All areas used for spraying, including the interior of the booth, shall be protected by automatic sprinklers where this protection is available. Where this protection is not available, other approved automatic extinguishing equipment shall be provided.

**1910.107(i)** **Electrostatic hand spraying equipment—(1)** *Application.* This paragraph shall apply to any equipment using electrostatically charged elements for the atomization and/or, precipitation of materials for coatings on articles, or for other similar purposes in which the atomizing device is hand held and manipulated during the spraying operation.

**1910.107(i)(2)** *Conformance.* Electrostatic hand spraying equipment shall conform with the other provisions of this section.

**1910.107(i)(3)** *Equipment approval and specifications.* Electrostatic hand spray apparatus and devices used in connection with coating operations shall be of approved types. The high voltage circuits shall be designed so as to not produce a spark of sufficient intensity to ignite any vapor-air mixtures nor result in appreciable shock hazard upon coming in contact with a grounded object under all normal operating conditions. The electrostatically charged exposed elements of the handgun shall be capable of being energized only by a switch which also controls the coating material supply.

**1910.107(i)(4)** *Electrical support equipment.* Transformers, powerpacks, control apparatus, and all other electrical portions of the equipment, with the exception of the handgun itself and its connections to the power supply shall be located outside of the spraying area or shall otherwise conform to the requirements of paragraph (c) of this section.

**1910.107(i)(5)** *Spray gun ground.* The handle of the spraying gun shall be electrically connected to ground by a metallic connection and to be so constructed that the operator in normal operating position is in intimate electrical contact with the grounded handle.

**1910.107(i)(6)** *Grounding—general.* All electrically conductive objects in the spraying area shall be adequately grounded. This requirement shall apply to paint containers, wash cans, and any other objects or devices in the area. The equipment shall carry a prominent permanently installed warning regarding the necessity for this grounding feature.

**1910.107(i)(7)** *Maintenance of grounds.* Objects being painted or coated shall be maintained in metallic contact with the conveyor or other grounded support. Hooks shall be regularly cleaned to insure this contact and areas of contact shall be sharp points or knife edges where possible. Points of support of the object shall be concealed from random spray where feasible and where the objects being sprayed are supported from a conveyor, the point of attachment to the conveyor shall be so located as to not collect spray material during normal operation.

**1910.107(i)(8)** *Interlocks.* The electrical equipment shall be so interlocked with the ventilation of the spraying area that the equipment cannot be operated unless the ventilation fans are in operation.

**1910.107(i)(9)** *Ventilation.* The spraying operation shall take place within a spray area which is adequately ventilated to remove solvent vapors released from the operation.

**1910.107(j)** *Drying, curing, or fusion apparatus—*(1) *Conformance.* Drying, curing, or fusion apparatus in connection with spray application of flammable and combustible finishes shall conform to the Standard for Ovens and Furnaces, NFPA 86A-1969, which is incorporated by reference as specified in §1910.6, where applicable and shall also conform with the following requirements of this paragraph.

**1910.107(j)(2)** *Alternate use prohibited.* Spray booths, rooms, or other enclosures used for spraying operations shall not alternately be used for the purpose of drying by any arrangement which will cause a material increase in the surface temperature of the spray booth, room, or enclosure.

**1910.107(j)(3)** *Adjacent system interlocked.* Except as specifically provided in paragraph (j)(4) of this section, drying, curing, or fusion units utilizing a heating system having open flames or which may produce sparks shall not be installed in a spraying area, but may be installed adjacent thereto when equipped with an interlocked ventilating system arranged to:

**1910.107(j)(3)(i)** Thoroughly ventilate the drying space before the heating system can be started;

**1910.107(j)(3)(ii)** Maintain a safe atmosphere at any source of ignition;

**1910.107(j)(3)(iii)** Automatically shut down the heating system in the event of failure of the ventilating system.

**1910.107(j)(4)** *Alternate use permitted.* Automobile refinishing spray booths or enclosures, otherwise installed and maintained in full conformity with this section, may alternately be used for drying with portable electrical infrared drying apparatus when conforming with the following:

**1910.107(j)(4)(i)** Interior (especially floors) of spray enclosures shall be kept free of overspray deposits.

**1910.107(j)(4)(ii)** During spray operations, the drying apparatus and electrical connections and wiring thereto shall not be located within spray enclosure nor in any other location where spray residues may be deposited thereon.

**1910.107(j)(4)(iii)** The spraying apparatus, the drying apparatus, and the ventilating system of the spray enclosure shall be equipped with suitable interlocks so arranged that:

**1910.107(j)(4)(iii)***(a)* The spraying apparatus cannot be operated while the drying apparatus is inside the spray enclosure.

**1910.107(j)(4)(iii)***(b)* The spray enclosure will be purged of spray vapors for a period of not less than 3 minutes before the drying apparatus can be energized.

**1910.107(j)(4)(iii)***(c)* The ventilating system will maintain a safe atmosphere within the enclosure during the drying process and the drying apparatus will automatically shut off in the event of failure of the ventilating system.

**1910.107(j)(4)(iv)** All electrical wiring and equipment of the drying apparatus shall conform with the applicable sections of subpart S of this part. Only equipment of a type approved for Class I, Division 2 hazardous locations shall be located within 18 inches of floor level. All metallic parts of the drying apparatus shall be properly electrically bonded and grounded.

**1910.107(j)(4)(v)** The drying apparatus shall contain a prominently located, permanently attached warning sign indicating that ventilation should be maintained during the drying period and that spraying should not be conducted in the vicinity that spray will deposit on apparatus.

**1910.107(k)** *Automobile undercoating in garages.* Automobile undercoating spray operations in garages, conducted in areas having adequate natural or mechanical ventilation, are exempt from the requirements pertaining to spray finishing operations, when using undercoating materials not more hazardous than kerosene (as listed by Underwriters' Laboratories in respect to fire hazard rating 30-40) or undercoating materials using only solvents listed as having a flash point in excess of 100 °F. Undercoating spray operations not conforming to these provisions are subject to all requirements of this section pertaining to spray finishing operations.

**1910.107(l)** *Powder coating—*(1) *Electrical and other sources of ignition.* Electrical equipment and other sources of ignition shall conform to the requirements of paragraphs (c)(1) (i)-(iv), (8) and (9)(i) of this section and subpart S of this part.

**1910.107(l)(2)** *Ventilation.* (i) In addition to the provisions of paragraph (d) of this section, where applicable, exhaust ventilation shall be sufficient to maintain the atmosphere below the lowest explosive limits for the materials being applied. All nondeposited air-suspended powders shall be safely removed via exhaust ducts to the powder recovery cyclone or receptacle. Each installation shall be designed and operated to meet the foregoing performance specification.

**1910.107(l)(2)(ii)** Powders shall not be released to the outside atmosphere.

**1910.107(l)(3)** *Drying, curing, or fusion equipment.* The provisions of the Standard for ovens and furnaces, NFPA No. 86A-969 shall apply where applicable.

**1910.107(l)(4)** *Operation and maintenance.* **(i)** All areas shall be kept free of the accumulation of powder coating dusts, particularly such horizontal surfaces as ledges, beams, pipes, hoods, booths, and floors.

**1910.107(l)(4)(ii)** Surfaces shall be cleaned in such manner as to avoid scattering dust to other places or creating dust clouds.

**1910.107(l)(4)(iii)** "No Smoking" signs in large letters on contrasting color background shall be conspicuously posted at all powder coating areas and powder storage rooms.

**1910.107(l)(5)** *Fixed electrostatic spraying equipment.* The provisions of paragraph (h) of this section and other subparagraphs of this paragraph shall apply to fixed electrostatic equipment, except that electrical equipment not covered therein shall conform to paragraph (l)(1) of this section.

**1910.107(l)(6)** *Electrostatic hand spraying equipment.* The provisions of paragraph (i) of this section and other subparagraphs of this paragraph, shall apply to electrostatic handguns when used in powder coating, except that electrical equipment not covered therein shall conform to paragraph (l)(1) of this section.

**1910.107(l)(7)** *Electrostatic fluidized beds.* **(i)** Electrostatic fluidized beds and associated equipment shall be of approved types. The maximum surface temperature of this equipment in the coating area shall not exceed 150 °F. The high voltage circuits shall be so designed as to not produce a spark of sufficient intensity to ignite any powder-air mixtures nor result in appreciable shock hazard upon coming in contact with a grounded object under normal operating conditions.

**1910.107(l)(7)(ii)** Transformers, powerpacks, control apparatus, and all other electrical portions of the equipment, with the exception of the charging electrodes and their connections to the power supply shall be located outside of the powder coating area or shall otherwise conform to the requirements of paragraph (l)(1) of this section.

**1910.107(l)(7)(iii)** All electrically conductive objects within the charging influence of the electrodes shall be adequately grounded. The powder coating equipment shall carry a prominent, permanently installed warning regarding the necessity for grounding these objects.

**1910.107(l)(7)(iv)** Objects being coated shall be maintained in contact with the conveyor or other support in order to insure proper grounding. Hangers shall be regularly cleaned to insure effective contact and areas of contact shall be sharp points or knife edges where possible.

**1910.107(l)(7)(v)** The electrical equipment shall be so interlocked with the ventilation system that the equipment cannot be operated unless the ventilation fans are in operation.

**1910.107(m)** *Organic peroxides and dual component coatings—*(1) *Conformance.* All spraying operations involving the use of organic peroxides and other dual component coatings shall be conducted in approved sprinklered spray booths meeting the requirements of this section.

**1910.107(m)(2)** *Smoking.* Smoking shall be prohibited and "No Smoking" signs shall be prominently displayed and only nonsparking tools shall be used in any area where organic peroxides are stored, mixed or applied.

**1910.107(n)** *Scope.* This section applies to flammable and combustible finishing materials when applied as a spray by compressed air, "airless" or "hydraulic atomization," steam, electrostatic methods, or by any other means in continuous or intermittent processes. The section also covers the application of combustible powders by powder spray guns, electrostatic powder spray guns, fluidized beds, or electrostatic fluidized beds. The section does not apply to outdoor spray application of buildings, tanks, or other similar structures, nor to small portable spraying apparatus not used repeatedly in the same location.

[39 FR 23502, June 27, 1974, as amended at 45 FR 60704, Sept. 12, 1980; 49 FR 5322, Feb. 10, 1984; 53 FR 12121, Apr. 12, 1988; 61 FR 9237, Mar. 7, 1996]

### §1910.108 [Reserved]

[39 FR 23502, June 27, 1974, as amended at 45 FR 60704, Sept. 12, 1980; 49 FR 5322, Feb. 10, 1984; 53 FR 12121, Apr. 12, 1988; 61 FR 9237, Mar. 7, 1996]

### §1910.109 Explosives and blasting agents.

**1910.109(a)** *Definitions applicable to this section—***(1)** *Blasting agent.* Blasting agent—any material or mixture, consisting of a fuel and oxidizer, intended for blasting, not otherwise classified as an explosive and in which none of the ingredients are classified as an explosive, provided that the finished product, as mixed and packaged for use or shipment, cannot be detonated by means of a No. 8 test blasting cap when unconfined.

**1910.109(a)(2)** *Explosive-actuated power devices.* Explosive-actuated power device—any tool or special mechanized device which is actuated by explosives, but not including propellant-actuated power devices. Examples of explosive-actuated power devices are jet tappers and jet perforators.

**1910.109(a)(3)** *Explosive.* Explosive—any chemical compound, mixture, or device, the primary or common purpose of which is to function by explosion, i.e., with substantially instantaneous release of gas and heat, unless such compound, mixture, or device is otherwise specifically classified by the U.S. Department of Transportation; see 49 CFR chapter I. The term "explosives" shall include all material which is classified as Class A, Class B, and Class C explosives by the U.S. Department of Transportation, and includes, but is not limited to dynamite, black powder, pellet powders, initiating explosives, blasting caps, electric blasting caps, safety fuse, fuse lighters, fuse igniters, squibs, cordeau detonant fuse, instantaneous fuse, igniter cord, igniters, small arms ammunition, small arms ammunition primers, smokeless propellant, cartridges for propellant-actuated power devices, and cartridges for industrial guns. Commercial explosives are those explosives which are intended to be used in commercial or industrial operations.

**NOTE 1:** Classification of explosives is described by the U.S. Department of Transportation as follows (see 49 CFR chapter I):

**1910.109(a)(3)(i)** *Class A explosives.* Possessing, detonating, or otherwise maximum hazard; such as dynamite, nitroglycerin, picric acid, lead azide, fulminate of mercury, black powder, blasting caps, and detonating primers.

**1910.109(a)(3)(ii)** *Class B explosives.* Possessing flammable hazard, such as propellant explosives (including some smokeless propellants), photographic flash powders, and some special fireworks.

**1910.109(a)(3)(iii)** *Class C explosives.* Includes certain types of manufactured articles which contain Class A or Class B explosives, or both, as components but in restricted quantities.

**1910.109(a)(3)(iv)** *Forbidden or not acceptable explosives.* Explosives which are forbidden or not acceptable for transportation by common carriers by rail freight, rail express, highway, or water in accordance with the regulations of the U.S. Department of Transportation, 49 CFR chapter I.

**1910.109(a)(4)** *Highway.* Highway—any public street, public alley, or public road.

**1910.109(a)(5)** [Reserved]

**1910.109(a)(6)** *Magazine.* Magazine—any building or structure, other than an explosives manufacturing building, used for the storage of explosives.

**1910.109(a)(7)** *Motor vehicle.* Motor vehicle—any self-propelled vehicle, truck, tractor, semitrailer, or truck-full trailers used for the transportation of freight over public highways.

**1910.109(a)(8)** *Propellant-actuated power devices.* Propellant-actuated power devices—any tool or special mechanized device or gas generator system which is actuated by a smokeless propellant or which releases and directs work through a smokeless propellant charge.

**1910.109(a)(9)** [Reserved]

**1910.109(a)(10)** *Pyrotechnics.* Pyrotechnics—any combustible or explosive compositions or manufactured articles designed and prepared for the purpose of producing audible or visible effects which are commonly referred to as fireworks.

**1910.109(a)(11)** [Reserved]

**1910.109(a)(12)** *Semiconductive hose.* Semiconductive hose—a hose with an electrical resistance high enough to limit flow of stray electric currents to safe levels, yet not so high as to prevent drainage of static electric charges to ground; hose of not more than 2 megohms resistance over its entire length and of not less than 5,000 ohms per foot meets the requirement.

**1910.109(a)(13)** *Small arms ammunition.* Small arms ammunition—any shotgun, rifle, pistol, or revolver cartridge, and cartridges for propellant-actuated power devices and industrial guns. Military-type ammunition containing explosive-bursting charges, incendiary, tracer, spotting, or pyrotechnic projectiles is excluded from this definition.

**1910.109(a)(14)** *Small arms ammunition primers.* Small arms ammunition primers—small percussion-sensitive explosive charges, encased in a cup, used to ignite propellant powder.

**1910.109(a)(15)** *Smokeless propellants.* Smokeless propellants—solid propellants, commonly called smokeless powders in the trade, used in small arms ammunition, cannon, rockets, propellant-actuated power devices, etc.

**1910.109(a)(16)** *Special industrial explosives devices.* Special industrial explosives devices—explosive-actuated power devices and propellant-actuated power devices.

**1910.109(a)(17)** *Special industrial explosives materials.* Special industrial explosives materials—shaped materials and sheet forms and various other extrusions, pellets, and packages of high explosives, which include dynamite, trinitrotoluene (TNT), pentaerythritol tetranitrate (PETN), hexahydro-1,3,5-trinitro-s-triazine (RDX), and other similar compounds used for high-energy-rate forming, expanding, and shaping in metal fabrication, and for dismemberment and quick reduction of scrap metal.

**1910.109(a)(18)** *Water gels or slurry explosives.* These comprise a wide variety of materials used for blasting. They all contain substantial proportions of water and high proportions of ammonium nitrate, some of which is in solution in the water. Two broad classes of water gels are (i) those which are sensitized by a material classed as an explosive, such as TNT or smokeless powder, (ii) those which contain no ingredient classified as an explosive; these are sensitized with metals such as aluminum or with other fuels. Water gels may be premixed at an explosives plant or mixed at the site immediately before delivery into the borehole.

**1910.109(a)(19)** *DOT specifications.* Regulations of the Department of Transportation published in 49 CFR chapter I.

**1910.109(b)** *Miscellaneous provisions—***(1)** *General hazard.* No person shall store, handle, or transport explosives or blasting agents when such storage, handling, and transportation of explosives or blasting agents constitutes an undue hazard to life.

**1910.109(b)(2)** [Reserved]

**1910.109(c)** *Storage of explosives—***(1)** *General provisions.* **(i)** All Class A, Class B, Class C explosives, and special industrial explosives, and any newly developed and unclassified explosives, shall be kept in magazines which meet the requirements of this paragraph.

**1910.109(c)(1)(ii)** Blasting caps, electric blasting caps, detonating primers, and primed cartridges shall not be stored in the same magazine with other explosives.

**1910.109(c)(1)(iii)** Ground around magazines shall slope away for drainage. The land surrounding magazines shall be kept clear of brush, dried grass, leaves, and other materials for a distance of at least 25 feet.

**1910.109(c)(1)(iv)** Magazines as required by this paragraph shall be of two classes; namely, Class I magazines, and Class II magazines.

**1910.109(c)(1)(v)** Class I magazines shall be required where the quantity of explosives stored is more than 50 pounds. Class II magazines may be used where the quantity of explosives stored is 50 pounds or less.

**1910.109(c)(1)(vi)** Class I magazines shall be located away from other magazines in conformity with Table H-21.

### TABLE H-21—AMERICAN TABLE OF DISTANCES FOR STORAGE OF EXPLOSIVES[1-5]

[As revised and approved by the Institute of Makers of Explosives, June 5, 1964]

| Explosives Pounds over | Pounds not over | Distance in feet when storage is barricaded: Separation of magazines |
|---|---|---|
| 2 | 5 | 6 |
| 5 | 10 | 8 |
| 10 | 20 | 10 |
| 20 | 30 | 11 |
| 30 | 40 | 12 |
| 40 | 50 | 14 |
| 50 | 75 | 15 |
| 75 | 100 | 16 |
| 100 | 125 | 18 |
| 125 | 150 | 19 |
| 150 | 200 | 21 |
| 200 | 250 | 23 |
| 250 | 300 | 24 |
| 300 | 400 | 27 |
| 400 | 500 | 29 |
| 500 | 600 | 31 |
| 600 | 700 | 32 |
| 700 | 800 | 33 |
| 800 | 900 | 35 |
| 900 | 1,000 | 36 |
| 1,000 | 1,200 | 39 |
| 1,200 | 1,400 | 41 |
| 1,400 | 1,600 | 43 |
| 1,600 | 1,800 | 44 |
| 1,800 | 2,000 | 45 |
| 2,000 | 2,500 | 49 |
| 2,500 | 3,000 | 52 |
| 3,000 | 4,000 | 58 |
| 4,000 | 5,000 | 61 |
| 5,000 | 6,000 | 65 |
| 6,000 | 7,000 | 68 |
| 7,000 | 8,000 | 72 |
| 8,000 | 9,000 | 75 |
| 9,000 | 10,000 | 78 |
| 10,000 | 12,000 | 82 |
| 12,000 | 14,000 | 87 |
| 14,000 | 16,000 | 90 |
| 16,000 | 18,000 | 94 |
| 18,000 | 20,000 | 98 |
| 20,000 | 25,000 | 105 |
| 25,000 | 30,000 | 112 |
| 30,000 | 35,000 | 119 |
| 35,000 | 40,000 | 124 |
| 40,000 | 45,000 | 129 |
| 45,000 | 50,000 | 135 |
| 50,000 | 55,000 | 140 |
| 55,000 | 60,000 | 145 |
| 60,000 | 65,000 | 150 |
| 65,000 | 70,000 | 155 |
| 70,000 | 75,000 | 160 |
| 75,000 | 80,000 | 165 |
| 80,000 | 85,000 | 170 |
| 85,000 | 90,000 | 175 |
| 90,000 | 95,000 | 180 |
| 95,000 | 100,000 | 185 |
| 100,000 | 110,000 | 195 |
| 110,000 | 120,000 | 205 |
| 120,000 | 130,000 | 215 |
| 130,000 | 140,000 | 225 |
| 140,000 | 150,000 | 235 |
| 150,000 | 160,000 | 245 |
| 160,000 | 170,000 | 255 |
| 170,000 | 180,000 | 265 |
| 180,000 | 190,000 | 275 |
| 190,000 | 200,000 | 285 |
| 200,000 | 210,000 | 295 |
| 210,000 | 230,000 | 315 |
| 230,000 | 250,000 | 335 |
| 250,000 | 275,000 | 360 |
| 275,000 | 300,000 | 385 |

[1] "Natural barricade" means natural features of the ground, such as hills, or timber of sufficient density that the surrounding exposures which require protection cannot be seen from the magazine when the trees are bare of leaves.

[2] "Artificial barricade" means an artificial mound or revetted wall of earth of a minimum thickness of three feet.

[3] "Barricaded" means that a building containing explosives is effectually screened from a magazine, building, railway, or highway, either by a natural barricade, or by an artificial barricade of such height that a straight line from the top of any sidewall of the building containing explosives to the eave line of any magazine, or building, or to a point 12 feet above the center of a railway or highway, will pass through such intervening natural or artificial barricade.

[4] When two or more storage magazines are located on the same property, each magazine must comply with the minimum distances specified from inhabited buildings, railways, and highways, and in addition, they should be separated from each other by not less than the distances shown for "Separation of Magazines," except that the quantity of explosives contained in cap magazines shall govern in regard to the spacing of said cap magazines from magazines containing other explosives. If any two or more magazines are separated from each other by less than the specified "Separation of Magazines" distances, then such two or more magazines, as a group, must be considered as one magazine, and the total quantity of explosives stored in such group must be treated as if stored in a single magazine located on the site of any magazine of the group, and must comply with the minimum of distances specified from other magazines, inhabited buildings, railways, and highways.

[5] This table applies only to the permanent storage of commercial explosives. It is not applicable to transportation of explosives, or any handling or temporary storage necessary or incident thereto. It is not intended to apply to bombs, projectiles, or other heavily encased explosives.

**1910.109(c)(1)(vii)** Except as provided in subdivision (viii) of this subparagraph, class II magazines shall be located in conformity with Table H-21, but may be permitted in warehouses and in wholesale and retail establishments when located on a floor which has an entrance at outside grade level and the magazine is located not more than 10 feet from such an entrance. Two class II magazines may be located in the same building when one is used only for blasting caps in quantities not in excess of 5,000 caps and a distance of 10 feet is maintained between magazines.

**1910.109(c)(1)(viii)** When used for temporary storage at a site for blasting operations, class II magazines shall be located away from other magazines. A distance of at least one hundred and fifty (150) feet shall be maintained between class II magazines and the work in progress when the quantity of explosives kept therein is in excess of 25 pounds, and at least 50 feet when the quantity of explosives is 25 pounds, or less.

**1910.109(c)(1)(ix)** This paragraph (c) does not apply to:

**1910.109(c)(1)(ix)(a)** Stocks of small arms ammunition, propellant-actuated power cartridges, small arms ammunition primers in quantities of less than 750,000, or of smokeless propellants in quantities less than 750 pounds;

**1910.109(c)(1)(ix)(b)** Explosive-actuated power devices when in quantities less than 50 pounds net weight of explosives;

**1910.109(c)(1)(ix)(c)** Fuse lighters and fuse igniters;

**1910.109(c)(1)(ix)(d)** Safety fuses other than cordeau detonant fuses.

**1910.109(c)(2)** *Construction of magazines—general.* **(i)** Magazines shall be constructed in conformity with the provisions of this paragraph.

**1910.109(c)(2)(ii)** Magazines for the storage of explosives, other than black powder, Class B and Class C explosives shall be bullet resistant, weather resistant, fire resistant, and ventilated sufficiently to protect the explosive

# OSHA Standards for General Industry

in the specific locality. Magazines used only for storage of black powder, Class B and Class C explosives shall be weather resistant, fire-resistant, and have ventilation. Magazines for storage of blasting and electric blasting caps shall be weather resistant, fire-resistant, and ventilated.

**1910.109(c)(2)(iii)** Property upon which Class I magazines are located and property where Class II magazines are located outside of buildings shall be posted with signs reading "Explosives—Keep Off."

**1910.109(c)(2)(iv)** Magazines requiring heat shall be heated by either hot-water radiant heating with the magazine building; or air directed into the magazine building over either hot water or low pressure steam (15 p.s.i.g.) coils located outside the magazine building.

**1910.109(c)(2)(v)** The magazine heating systems shall meet the following requirements:

**1910.109(c)(2)(v)(a)** The radiant heating coils within the building shall be installed in such a manner that the explosives or explosives containers cannot contact the coils and air is free to circulate between the coils and the explosives or explosives containers.

**1910.109(c)(2)(v)(b)** The heating ducts shall be installed in such a manner that the hot-air discharge from the duct is not directed against the explosives or explosives containers.

**1910.109(c)(2)(v)(c)** The heating device used in connection with a magazine shall have controls which prevent the ambient building temperature from exceeding 130 °F.

**1910.109(c)(2)(v)(d)** The electric fan or pump used in the heating system for a magazine shall be mounted outside and separate from the wall of the magazine and shall be grounded.

**1910.109(c)(2)(v)(e)** The electric fan motor and the controls for electrical heating devices used in heating water or steam shall have overloads and disconnects, which comply with subpart S of this part. All electrical switch gear shall be located a minimum distance of 25 feet from the magazine.

**1910.109(c)(2)(v)(f)** The heating source for water or steam shall be separated from the magazine by a distance of not less than 25 feet when electrical and 50 feet when fuel fired. The area between the heating unit and the magazine shall be cleared of all combustible materials.

**1910.109(c)(2)(v)(g)** The storage of explosives and explosives containers in the magazine shall allow uniform air circulation so product temperature uniformity can be maintained.

**1910.109(c)(2)(vi)** When lights are necessary inside the magazine, electric safety flashlight, or electric safety lanterns shall be used.

**1910.109(c)(3)** *Construction of Class I magazines.* **(i)** Class I magazines shall be of masonry construction or of wood or of metal construction, or a combination of these types. Thickness of masonry units shall not be less than 8 inches. Hollow masonry units used in construction required to be bullet resistant shall have all hollow spaces filled with weak cement or well-tamped sand. Wood constructed walls, required to be bullet resistant, shall have at least a 6- inch space between interior and exterior sheathing and the space between sheathing shall be filled with well-tamped sand. Metal wall construction, when required to be bullet resistant, shall be lined with brick at least 4 inches in thickness or shall have at least a 6-inch sandfill between interior and exterior walls.

**1910.109(c)(3)(ii)** Floors and roofs of masonry magazines may be of wood construction. Wood floors shall be tongue and grooved lumber having a nominal thickness of 1 inch.

**1910.109(c)(3)(iii)** Roofs required to be bullet resistant shall be protected by a sand tray located at the line of eaves and covering the entire area except that necessary for ventilation. Sand in the sand tray shall be maintained at a depth of not less than 4 inches.

**1910.109(c)(3)(iv)** All wood at the exterior of magazines, including eaves, shall be protected by being covered with black or galvanized steel or aluminum metal of thickness of not less than No. 26 gage. All nails exposed to the interior of magazines shall be well countersunk.

**1910.109(c)(3)(v)** Foundations for magazines shall be of substantial construction and arranged to provide good cross ventilation.

**1910.109(c)(3)(vi)** Magazines shall be ventilated sufficiently to prevent dampness and heating of stored explosives. Ventilating openings shall be screened to prevent the entrance of sparks.

**1910.109(c)(3)(vii)** Openings to magazines shall be restricted to that necessary for the placement and removal of stocks of explosives. Doors for openings in magazines for Class A explosives shall be bullet resistant. Doors for magazines not required to be bullet resistant shall be designed to prevent unauthorized entrance to the magazine.

**1910.109(c)(3)(viii)** [Reserved]

**1910.109(c)(3)(ix)** Provisions shall be made to prevent the piling of stocks of explosives directly against masonry walls, brick-lined or sand-filled metal walls and single-thickness metal walls; such protection, however, shall not interfere with proper ventilation at the interior of side and end walls.

**1910.109(c)(4)** *Construction of Class II magazines.* **(i)** Class II magazines shall be of wood or metal construction, or a combination thereof.

**1910.109(c)(4)(ii)** Wood magazines of this class shall have sides, bottom, and cover constructed of 2-inch hardwood boards well braced at corners and protected by being entirely covered with sheet metal of not less than No. 20 gage. All nails exposed to the interior of the magazine shall be well countersunk. All metal magazines of this class shall have sides, bottom, and cover constructed of sheet metal, and shall be lined with three-eighths-inch plywood or equivalent. Edges of metal covers shall overlap sides at least 1 inch.

**1910.109(c)(4)(iii)** Covers for both wood- and metal-constructed magazines of this class shall be provided with substantial strap hinges and shall be provided with substantial means for locking.

**1910.109(c)(4)(iv)** Magazines of this class shall be painted red and shall bear lettering in white, on all sides and top, at least 3 inches high, "Explosives—Keep Fire Away." Class II magazines when located in warehouses, and in wholesale and retail establishments shall be provided with substantial wheels or casters to facilitate easy removal in the case of fire. Where necessary due to climatic conditions, Class II magazines shall be ventilated.

**1910.109(c)(5)** *Storage within magazines.* **(i)** Packages of explosives shall be laid flat with top side up. Black powder when stored in magazines with other explosives shall be stored separately. Black powder stored in kegs shall be stored on ends, bungs down, or on side, seams down. Corresponding grades and brands shall be stored together in such a manner that brands and grade marks show. All stocks shall be stored so as to be easily counted and checked. Packages of explosives shall be piled in a stable manner. When any kind of explosive is removed from a magazine for use, the oldest explosive of that particular kind shall always be taken first.

**1910.109(c)(5)(ii)** Packages of explosives shall not be unpacked or repacked in a magazine nor within 50 feet of a magazine or in close proximity to other explosives. Tools used for opening packages of explosives shall be constructed of nonsparking materials, except that metal slitters may be used for opening fiberboard boxes. A wood wedge and a fiber, rubber, or wood mallet shall be used for opening or closing wood packages of explosives. Opened packages of explosives shall be securely closed before being returned to a magazine.

**1910.109(c)(5)(iii)** Magazines shall not be used for the storage of any metal tools nor any commodity except explosives, but this restriction shall not apply to the storage of blasting agents and blasting supplies.

**1910.109(c)(5)(iv)** Magazine floors shall be regularly swept, kept clean, dry, free of grit, paper, empty used packages, and rubbish. Brooms and other cleaning utensils shall not have any spark-producing metal parts. Sweepings from floors of magazines shall be properly disposed of. Magazine floors stained with nitroglycerin shall be cleaned according to instructions by the manufacturer.

**1910.109(c)(5)(v)** When any explosive has deteriorated to an extent that it is in an unstable or dangerous condition, or if nitroglycerin leaks from any explosives, then the person in possession of such explosive shall immediately proceed to destroy such explosive in accordance with the instructions of the manufacturer. Only experienced persons shall be allowed to do the work of destroying explosives.

**1910.109(c)(5)(vi)** When magazines need inside repairs, all explosives shall be removed therefrom and the floors cleaned. In making outside repairs, if there is a possibility of causing sparks or fire the explosives shall be removed from the magazine. Explosives removed from a magazine under repair shall either be placed in another magazine or placed a safe distance from the magazine where they shall be properly guarded and protected until repairs have been completed, when they shall be returned to the magazine.

**1910.109(c)(5)(vii)** Smoking, matches, open flames, spark-producing devices, and firearms (except firearms carried by guards) shall not be permitted inside of or within 50 feet of magazines. The land surrounding a magazine shall be kept clear of all combustible materials for a distance of at least 25 feet. Combustible materials shall not be stored within 50 feet of magazines.

**1910.109(c)(5)(viii)** Magazines shall be in the charge of a competent person at all times and who shall be held responsible for the enforcement of all safety precautions.

**1910.109(c)(5)(ix)** Explosives recovered from blasting misfires shall be placed in a separate magazine until competent personnel has determined from the manufacturer the method of disposal. Caps recovered from blasting misfires shall not be reused. Such explosives and caps shall then be disposed of in the manner recommended by the manufacturer.

**1910.109(d)** *Transportation of explosives*—**(1)** *General provisions.* **(i)** No employee shall be allowed to smoke, carry matches or any other flame-producing device, or carry any firearms or loaded cartridges while in or near a motor vehicle transporting explosives; or drive, load, or unload such vehicle in a careless or reckless manner.

**1910.109(d)(1)(ii)** [Reserved]

**1910.109(d)(1)(iii)** Explosives shall not be transferred from one vehicle to another within the confines of any jurisdiction (city, county, State, or other area) without informing the fire and police departments thereof. In the event of breakdown or collision the local fire and police departments shall be promptly notified to help safeguard such emergencies. Explosives shall be transferred from the disabled vehicle to another only, when proper and qualified supervision is provided.

**1910.109(d)(1)(iv)** Blasting caps or electric blasting caps shall not be transported over the highways on the same vehicles with other explosives, unless packaged, segregated, and transported in accordance with the Department of Transportation's Hazardous Materials Regulations (49 CFR parts 177-180).

**1910.109(d)(2)** *Transportation vehicles.* **(i)** Vehicles used for transporting explosives shall be strong enough to carry the load without difficulty and be in good mechanical condition. If vehicles do not have a closed body, the body shall be covered with a flameproof and moistureproof tarpaulin or other effective protection against moisture and sparks. All vehicles used for the transportation of explosives shall have tight floors and any exposed spark-producing metal on the inside of the body shall be covered with wood or other nonsparking materials to prevent contact with packages of explosives. Packages of explosives shall not be loaded above the sides of an open-body vehicle.

**1910.109(d)(2)(ii)** Every vehicle used for transporting explosives and oxidizing materials listed in paragraph (d)(2)(ii)(a) of this section shall be marked as follows:

**1910.109(d)(2)(ii)(a)** Exterior markings or placards required on applicable vehicles shall be as follows for the various classes of commodities:

| Commodity | Type of marking or placard |
| --- | --- |
| Explosives, Class A, any quantity or a combination of Class A and Class B explosives. | Explosives A (Red letters on white background). |
| Explosives, Class B, any quantity. | Explosives B (Red letters on white background). |
| Oxidizing material (blasting agents, ammonium nitrate, etc.), 1,000 pounds or more gross weight. | Oxidizers (Yellow letters on black background). |

**1910.109(d)(2)(ii)(b)** [Reserved]

**1910.109(d)(2)(ii)(c)** Such markings or placards shall be displayed at the front, rear, and on each side of the motor vehicle or trailer, or other cargo carrying body while it contains explosives or other dangerous articles of such type and in such quantity as specified in paragraph (d)(1)(ii)(a) of this subdivision. The front marking or placard may be displayed on the front of either the truck, truck body, truck tractor or the trailer.

**1910.109(d)(2)(ii)(d)** Any motor vehicle, trailer, or other cargo-carrying body containing more than one kind of explosive as well as an oxidizing material requiring a placard under the provisions of paragraph (d)(2)(ii)(a), the aggregate gross weight of which totals 1,000 pounds or more, shall be marked or placarded "Dangerous" as well as "Explosive A" or "Explosive B" as appropriate. If explosives Class A and explosives Class B are loaded on the same vehicle, the "Explosives B" marking need not be displayed.

**1910.109(d)(2)(ii)(e)** In any combination of two or more vehicles containing explosives or other dangerous articles each vehicle shall be marked or placarded as to its contents and in accordance with paragraphs (d)(2)(ii)(a) and (c) of this subdivision.

**1910.109(d)(2)(iii)** Each motor vehicle used for transporting explosives shall be equipped with a minimum of two extinguishers, each having a rating of at least 10-BC.

**1910.109(d)(2)(iii)(a)** Only extinguishers listed or approved by a nationally recognized testing laboratory shall be deemed suitable for use on explosives-carrying vehicles. Refer to §1910.155(c)(3)(iv)(A) for definition of listed, and §1910.7 for nationally recognized testing laboratory.

**1910.109(d)(2)(iii)(b)** Extinguishers shall be filled and ready for immediate use and located near the driver's seat. Extinguishers shall be examined periodically by a competent person.

**1910.109(d)(2)(iv)** A motor vehicle used for transporting explosives shall be given the following inspection to determine that it is in proper condition for safe transportation of explosives:

**1910.109(d)(2)(iv)(a)** Fire extinguishers shall be filled and in working order.

**1910.109(d)(2)(iv)(b)** All electrical wiring shall be completely protected and securely fastened to prevent short-circuiting.

**1910.109(d)(2)(iv)(c)** Chassis, motor, pan, and underside of body shall be reasonably clean and free of excess oil and grease.

**1910.109(d)(2)(iv)(d)** Fuel tank and feedline shall be secure and have no leaks.

**1910.109(d)(2)(iv)(e)** Brakes, lights, horn, windshield wipers, and steering apparatus shall function properly.

**1910.109(d)(2)(iv)(f)** Tires shall be checked for proper inflation and defects.

**1910.109(d)(2)(iv)(g)** The vehicle shall be in proper condition in every other respect and acceptable for handling explosives.

**1910.109(d)(3)** *Operation of transportation vehicles.* **(i)** Vehicles transporting explosives shall only be driven by and be in the charge of a driver who is familiar with the traffic regulations, State laws, and the provisions of this section.

**1910.109(d)(3)(ii)** Except under emergency conditions, no vehicle transporting explosives shall be parked before reaching its destination, even though attended, on any public street adjacent to or in proximity to any place where people work.

**1910.109(d)(3)(iii)** Every motor vehicle transporting any quantity of Class A or Class B explosives shall, at all times, be attended by a driver or other attendant of the motor carrier. This attendant shall have been made aware of the class of the explosive material in the vehicle and of its inherent dangers, and shall have been instructed in the measures and procedures to be followed in order to protect the public from those dangers. He shall have been made familiar with the vehicle he is assigned, and shall be trained, supplied with the necessary means, and authorized to move the vehicle when required.

**1910.109(d)(3)(iii)(a)** For the purpose of this subdivision, a motor vehicle shall be deemed "attended" only when the driver or other attendant is physically on or in the vehicle, or has the vehicle within his field of vision and can reach it quickly and without any kind of interference "attended" also means that the driver or attendant is awake, alert, and not engaged in other duties or activities which may divert his attention from the vehicle, except for necessary communication with public officers, or representatives of the carrier shipper, or consignee, or except for necessary absence from the vehicle to obtain food or to provide for his physical comfort.

**1910.109(d)(3)(iii)(b)** However, an explosive-laden vehicle may be left unattended if parked within a securely fenced or walled area with all gates or entrances locked where parking of such vehicle is otherwise permissible, or at a magazine site established solely for the purpose of storing explosives.

**1910.109(d)(3)(iv)** No spark-producing metal, spark-producing metal tools, oils, matches, firearms, electric storage batteries, flammable substances, acids, oxidizing materials, or corrosive compounds shall be carried in the body of any motor truck and/or vehicle transporting explosives, unless the loading of such dangerous articles and the explosives comply with U.S. Department of Transportation regulations.

**1910.109(d)(3)(v)** Vehicles transporting explosives shall avoid congested areas and heavy traffic. Where routes through congested areas have been designated by local authorities such routes shall be followed.

**1910.109(d)(3)(vi)** Delivery shall only be made to authorized persons and into authorized magazines or authorized temporary storage or handling areas.

**1910.109(e)** *Use of explosives and blasting agents*—**(1)** *General provisions.* **(i)** While explosives are being handled or used, smoking shall not be permitted and no one near the explosives shall possess matches, open light or other fire or flame. No person shall be allowed to handle explosives while under the influence of intoxicating liquors, narcotics, or other dangerous drugs.

**1910.109(e)(1)(ii)** Original containers or Class II magazines shall be used for taking detonators and other explosives from storage magazines to the blasting area.

**1910.109(e)(1)(iii)** When blasting is done in congested areas or in close proximity to a structure, or any other installation that may be damaged, the blast shall be covered before firing with a mat constructed so that it is capable of preventing fragments from being thrown.

**1910.109(e)(1)(iv)** Persons authorized to prepare explosive charges or conduct blasting operations shall use every reasonable precaution, including but not limited to warning signals, flags, barricades, or woven wire mats to insure the safety of the general public and workmen.

**1910.109(e)(1)(v)** Blasting operations shall be conducted during daylight hours.

**1910.109(e)(1)(vi)** Whenever blasting is being conducted in the vicinity of gas, electric, water, fire alarm, telephone, telegraph, and steam utilities, the blaster shall notify the appropriate representatives of such utilities at least 24 hours in advance of blasting, specifying the location and intended time of such blasting. Verbal notice shall be confirmed with written notice.

**1910.109(e)(1)(vii)** Due precautions shall be taken to prevent accidental discharge of electric blasting caps from current induced by radar, radio transmitters, lightning, adjacent powerlines, dust storms, or other sources of extraneous electricity. These precautions shall include:

**1910.109(e)(1)(vii)***(a)* The suspension of all blasting operations and removal of persons from the blasting area during the approach and progress of an electric storm.

**1910.109(e)(1)(vii)***(b)* The posting of signs warning against the use of mobile radio transmitters on all roads within 350 feet of the blasting operations.

**1910.109(e)(2)** *Storage at use sites.* **(i)** Empty containers and paper and fiber packing materials which have previously contained explosive materials shall be disposed of in a safe manner, or reused in accordance with the Department of Transportation's Hazardous Materials Regulations (49 CFR parts 177-180).

**1910.109(e)(2)(ii)** Containers of explosives shall not be opened in any magazine or within 50 feet of any magazine. In opening kegs or wooden cases, no sparking metal tools shall be used; wooden wedges and either wood, fiber or rubber mallets shall be used. Nonsparking metallic slitters may be used for opening fiberboard cases.

**1910.109(e)(2)(iii)** Explosives or blasting equipment that are obviously deteriorated or damaged shall not be used.

**1910.109(e)(2)(iv)** No explosives shall be abandoned.

**1910.109(e)(3)** *Loading of explosives in blast holes.* **(i)** All drill holes shall be sufficiently large to admit freely the insertion of the cartridges of explosives.

**1910.109(e)(3)(ii)** Tamping shall be done only with wood rods without exposed metal parts, but nonsparking metal connectors may be used for jointed poles. Violent tamping shall be avoided. Primed cartridges shall not be tamped.

**1910.109(e)(3)(iii)** When loading blasting agents pneumatically over electric blasting caps, semiconductive delivery hose shall be used and the equipment shall be bonded and grounded.

**1910.109(e)(3)(iv)** No holes shall be loaded except those to be fired in the next round of blasting. After loading, all remaining explosives shall be immediately returned to an authorized magazine.

**1910.109(e)(3)(v)** Drilling shall not be started until all remaining butts of old holes are examined with a wooden stick for unexploded charges, and if any are found, they shall be refired before work proceeds.

**1910.109(e)(3)(vi)** No person shall be allowed to deepen drill holes which have contained explosives.

**1910.109(e)(3)(vii)** After loading for a blast is completed, all excess blasting caps or electric blasting caps and other explosives shall immediately be returned to their separate storage magazines.

**1910.109(e)(4)** *Initiation of explosive charges.*

**1910.109(e)(4)(i)** [Reserved]

**1910.109(e)(4)(ii)** When fuse is used, the blasting cap shall be securely attached to the safety fuse with a standard-ring type cap crimper. All primers shall be assembled at least 50 feet from any magazine.

**1910.109(e)(4)(iii)** Primers shall be made up only as required for each round of blasting.

**1910.109(e)(4)(iv)** No blasting cap shall be inserted in the explosives without first making a hole in the cartridge for the cap with a wooden punch of proper size or standard cap crimper.

**1910.109(e)(4)(v)** Explosives shall not be extracted from a hole that has once been charged or has misfired unless it is impossible to detonate the unexploded charge by insertion of a fresh additional primer.

**1910.109(e)(4)(vi)** If there are any misfires while using cap and fuse, all persons shall be required to remain away from the charge for at least 1 hour. If electric blasting caps are used and a misfire occurs, this waiting period may be reduced to 30 minutes. Misfires shall be handled under the direction of the person in charge of the blasting and all wires shall be carefully traced and search made for unexploded charges.

**1910.109(e)(4)(vii)** Blasters, when testing circuits to charged holes, shall use only blasting galvanometers designed for this purpose.

**1910.109(e)(4)(viii)** Only the employee making leading wire connections in electrical firing shall be allowed to fire the shot. Leading wires shall remain shorted and not be connected to the blasting machine or other source of current until the charge is to be fired.

**1910.109(e)(5)** *Warning required.* Before a blast is fired, the employer shall require that a loud warning signal be given by the person in charge, who has made certain that all surplus explosives are in a safe place, all persons and vehicles are at a safe distance or under sufficient cover, and that an adequate warning has been given.

**1910.109(f)** *Explosives at piers, railway stations, and cars or vessels not otherwise specified in this standard*—**(1)** *Railway cars.* Except in an emergency and with permission of the local authority, no person shall have or keep explosives in a railway car unless said car and contents and methods of loading are in accordance with the U.S. Department of Transportation Regulations for the Transportation of Explosives, 49 CFR chapter I.

**1910.109(f)(2)** *Packing and marking.* No person shall deliver any explosive to any carrier unless such explosive conforms in all respects, including marking and packing, to the U.S. Department of Transportation Regulations for the Transportation of Explosives.

**1910.109(f)(3)** *Marking cars.* Every railway car containing explosives which has reached its designation, or is stopped in transit so as no longer to be in interstate commerce, shall have attached to both sides and ends of the car, cards with the words "Explosives—Handle Carefully—Keep Fire Away" in red letters at least 1 1/2 inches high on a white background.

**1910.109(f)(4)** *Storage.* Any explosives at a railway facility, truck terminal, pier, wharf harbor facility, or airport terminal whether for delivery to a consignee, or forwarded to some other destination shall be kept in a safe place, isolated as far as practicable and in such manner that they can be easily and quickly removed.

**1910.109(f)(5)** *Hours of transfer.* Explosives shall not be delivered to or received from any railway station, truck terminal, pier, wharf, harbor facility, or airport terminal between the hours of sunset and sunrise.

**1910.109(g)** *Blasting agents*—**(1)** *General.* Unless otherwise set forth in this paragraph, blasting agents, excluding water gels, shall be transported, stored, and used in the same manner as explosives. Water gels are covered in paragraph (h) of this section.

**1910.109(g)(2)** *Fixed location mixing.* **(i)** [Reserved]

**1910.109(g)(2)(ii)** Buildings used for the mixing of blasting agents shall conform to the requirements of this section.

**1910.109(g)(2)(ii)(a)** Buildings shall be of noncombustible construction or sheet metal on wood studs.

**1910.109(g)(2)(ii)(b)** Floors in a mixing plant shall be of concrete or of other nonabsorbent materials.

**1910.109(g)(2)(ii)(c)** All fuel oil storage facilities shall be separated from the mixing plant and located in such a manner that in case of tank rupture, the oil will drain away from the mixing plant building.

**1910.109(g)(2)(ii)(d)** The building shall be well ventilated.

**1910.109(g)(2)(ii)(e)** Heating units which do not depend on combustion processes, when properly designed and located, may be used in the building. All direct sources of heat shall be provided exclusively from units located outside the mixing building.

**1910.109(g)(2)(ii)(f)** All internal-combustion engines used for electric power generation shall be located outside the mixing plant building, or shall be properly ventilated and isolated by a firewall. The exhaust systems on all such engines shall be located so any spark emission cannot be a hazard to any materials in or adjacent to the plant.

**1910.109(g)(2)(iii)** Equipment used for mixing blasting agents shall conform to the requirements of this subdivision.

**1910.109(g)(2)(iii)(a)** The design of the mixer shall minimize the possibility of frictional heating, compaction, and especially confinement. All bearings and drive assemblies shall be mounted outside the mixer and protected against the accumulation of dust. All surfaces shall be accessible for cleaning.

**1910.109(g)(2)(iii)(b)** Mixing and packaging equipment shall be constructed of materials compatible with the fuel-ammonium nitrate composition.

**1910.109(g)(2)(iii)(c)** Suitable means shall be provided to prevent the flow of fuel oil to the mixer in case of fire. In gravity flow systems an automatic spring-loaded shutoff valve with fusible link shall be installed.

**1910.109(g)(2)(iv)** The provisions of this subdivision shall be considered when determining blasting agent compositions.

**1910.109(g)(2)(iv)(a)** The sensitivity of the blasting agent shall be determined by means of a No. 8 test blasting cap at regular intervals and after every change in formulation.

**1910.109(g)(2)(iv)(b)** Oxidizers of small particle size, such as crushed ammonium nitrate prills or fines, may be more sensitive than coarser products and shall, therefore, be handled with greater care.

**1910.109(g)(2)(iv)(c)** No hydrocarbon liquid fuel with flashpoint lower than that of No. 2 diesel fuel oil 125 °F. minimum shall be used.

**1910.109(g)(2)(iv)(d)** Crude oil and crankcase oil shall not be used.

**1910.109(g)(2)(iv)(e)** Metal powders such as aluminum shall be kept dry and shall be stored in containers or bins which are moisture-resistant or weathertight. Solid fuels shall be used in such manner as to minimize dust explosion hazards.

**1910.109(g)(2)(iv)(f)** Peroxides and chlorates shall not be used.

**1910.109(g)(2)(v)** All electrical switches, controls, motors, and lights located in the mixing room shall conform to the requirements in subpart S of this part for Class II, Division 2 locations; otherwise they shall be located outside the mixing room. The frame of the mixer and all other equipment that may be used shall be electrically bonded and be provided with a continuous path to the ground.

**1910.109(g)(2)(vi)** Safety precautions at mixing plants shall include the requirements of this subdivision.

**1910.109(g)(2)(vi)(a)** Floors shall be constructed so as to eliminate floor drains and piping into which molten materials could flow and be confined in case of fire.

**1910.109(g)(2)(vi)(b)** The floors and equipment of the mixing and packaging room shall be cleaned regularly and thoroughly to prevent accumulation of oxidizers or fuels and other sensitizers.

**1910.109(g)(2)(vi)(c)** The entire mixing and packaging plant shall be cleaned regularly and thoroughly to prevent excessive accumulation of dust.

**1910.109(g)(2)(vi)(d)** Smoking, matches, open flames, spark-producing devices, and firearms (except firearms carried by guards) shall not be permitted inside of or within 50 feet of any building or facility used for the mixing of blasting agents.

**1910.109(g)(2)(vi)(e)** The land surrounding the mixing plant shall be kept clear of brush, dried grass, leaves, and other materials for a distance of at least 25 feet.

**1910.109(g)(2)(vi)(f)** Empty ammonium nitrate bags shall be disposed of daily in a safe manner.

**1910.109(g)(2)(vi)(g)** No welding shall be permitted or open flames used in or around the mixing or storage area of the plant unless the equipment or area has been completely washed down and all oxidizer material removed.

**1910.109(g)(2)(vi)(h)** Before welding or repairs to hollow shafts, all oxidizer material shall be removed from the outside and inside of the shaft and the shaft vented with a minimum one-half inch diameter opening.

**1910.109(g)(2)(vi)(i)** Explosives shall not be permitted inside of or within 50 feet of any building or facility used for the mixing of blasting agents.

**1910.109(g)(3)** *Bulk delivery and mixing vehicles.* **(i)** The provisions of this paragraph shall apply to off-highway private operations as well as to all public highway movements.

**1910.109(g)(3)(ii)** A bulk vehicle body for delivering and mixing blasting agents shall conform with the requirements of this paragraph (ii).

**1910.109(g)(3)(ii)(a)** The body shall be constructed of noncombustible materials.

**1910.109(g)(3)(ii)(b)** Vehicles used to transport bulk premixed blasting agents on public highways shall have closed bodies.

**1910.109(g)(3)(ii)(c)** All moving parts of the mixing system shall be designed as to prevent a heat buildup. Shafts or axles which contact the product shall have outboard bearings with 1-inch minimum clearance between the bearings and the outside of the product container. Particular attention shall be given to the clearances on all moving parts.

**1910.109(g)(3)(ii)(d)** A bulk delivery vehicle shall be strong enough to carry the load without difficulty and be in good mechanical condition.

**1910.109(g)(3)(iii)** Operation of bulk delivery vehicles shall conform to the requirements of this subdivision. These include the placarding requirements as specified by Department of Transportation.

**1910.109(g)(3)(iii)(a)** The operator shall be trained in the safe operation of the vehicle together with its mixing, conveying, and related equipment. The employer shall assure that the operator is familiar with the commodities being delivered and the general procedure for handling emergency situations.

**1910.109(g)(3)(iii)(b)** The hauling of either blasting caps or other explosives but not both, shall be permitted on bulk trucks provided that a special wood or nonferrous-lined container is installed for the explosives. Such blasting caps or other explosives shall be in DOT-specified shipping containers: see 49 CFR chapter I.

**1910.109(g)(3)(iii)(c)** No person shall smoke, carry matches or any flame-producing device, or carry any firearms while in or about bulk vehicles effecting the mixing transfer or down-the-hole loading of blasting agents at or near the blasting site.

**1910.109(g)(3)(iii)(d)** Caution shall be exercised in the movement of the vehicle in the blasting area to avoid driving the vehicle over or dragging hoses over firing lines, cap wires, or explosive materials. The employer shall assure that the driver, in moving the vehicle, has assistance of a second person to guide his movements.

**1910.109(g)(3)(iii)(e)** No intransit mixing of materials shall be performed.

**1910.109(g)(3)(iv)** Pneumatic loading from bulk delivery vehicles into blast-holes primed with electric blasting caps or other static-sensitive systems shall conform to the requirements of this subdivision.

**1910.109(g)(3)(iv)(a)** A positive grounding device shall be used to prevent the accumulation of static electricity.

**1910.109(g)(3)(iv)(b)** A discharge hose shall be used that has a resistance range that will prevent conducting stray currents, but that is conductive enough to bleed off static buildup.

**1910.109(g)(3)(iv)(c)** A qualified person shall evaluate all systems to determine if they will adequately dissipate static under potential field conditions.

**1910.109(g)(3)(v)** Repairs to bulk delivery vehicles shall conform to the requirements of this section.

**1910.109(g)(3)(v)(a)** No welding or open flames shall be used on or around any part of the delivery equipment unless it has been completely washed down and all oxidizer material removed.

# OSHA Standards for General Industry

**1910.109(g)(3)(v)(b)** Before welding or making repairs to hollow shafts, the shaft shall be thoroughly cleaned inside and out and vented with a minimum one-half-inch diameter opening.

**1910.109(g)(4)** *Bulk storage bins.* **(i)** The bin, including supports, shall be constructed of compatible materials, waterproof, and adequately supported and braced to withstand the combination of all loads including impact forces arising from product movement within the bin and accidental vehicle contact with the support legs.

**1910.109(g)(4)(ii)** The bin discharge gate shall be designed to provide a closure tight enough to prevent leakage of the stored product. Provision shall also be made so that the gate can be locked.

**1910.109(g)(4)(iii)** Bin loading manways or access hatches shall be hinged or otherwise attached to the bin and be designed to permit locking.

**1910.109(g)(4)(iv)** Any electrically driven conveyors for loading or unloading bins shall conform to the requirements of subpart S of this part. They shall be designed to minimize damage from corrosion.

**1910.109(g)(4)(v)** Bins containing blasting agent shall be located, with respect to inhabited buildings, passenger railroads, and public highways, in accordance with Table-21 and separation from other blasting agent storage and explosives storage shall be in conformity with Table H-22.

**1910.109(g)(4)(vi)** Bins containing ammonium nitrate shall be separated from blasting agent storage and explosives storage in conformity with Table H-22.

### TABLE H-22—TABLE OF RECOMMENDED SEPARATION DISTANCES OF AMMONIUM NITRATE AND BLASTING AGENTS FROM EXPLOSIVES OR BLASTING AGENTS[1-6]

| Donor weight Pounds over | Pounds not over | Ammonium nitrate[3] | Blasting agent[4] | Minimum thickness of artificial barricades[5] (in) |
|---|---|---|---|---|
|  | 100 | 3 | 11 | 12 |
| 100 | 300 | 4 | 14 | 12 |
| 300 | 600 | 5 | 18 | 12 |
| 600 | 1,000 | 6 | 22 | 12 |
| 1,000 | 1,600 | 7 | 25 | 12 |
| 1,600 | 2,000 | 8 | 29 | 12 |
| 2,000 | 3,000 | 9 | 32 | 15 |
| 3,000 | 4,000 | 10 | 36 | 15 |
| 4,000 | 6,000 | 11 | 40 | 15 |
| 6,000 | 8,000 | 12 | 43 | 20 |
| 8,000 | 10,000 | 13 | 47 | 20 |
| 10,000 | 12,000 | 14 | 50 | 20 |
| 12,000 | 16,000 | 15 | 54 | 25 |
| 16,000 | 20,000 | 16 | 58 | 25 |
| 20,000 | 25,000 | 18 | 65 | 25 |
| 25,000 | 30,000 | 19 | 68 | 30 |
| 30,000 | 35,000 | 20 | 72 | 30 |
| 35,000 | 40,000 | 21 | 76 | 30 |
| 40,000 | 45,000 | 22 | 79 | 35 |
| 45,000 | 50,000 | 23 | 83 | 35 |
| 50,000 | 55,000 | 24 | 86 | 35 |
| 55,000 | 60,000 | 25 | 90 | 35 |
| 60,000 | 70,000 | 26 | 94 | 40 |
| 70,000 | 80,000 | 28 | 101 | 40 |
| 80,000 | 90,000 | 30 | 108 | 40 |
| 90,000 | 100,000 | 32 | 115 | 40 |
| 100,000 | 120,000 | 34 | 122 | 50 |
| 120,000 | 140,000 | 37 | 133 | 50 |
| 140,000 | 160,000 | 40 | 144 | 50 |
| 160,000 | 180,000 | 44 | 158 | 50 |
| 180,000 | 200,000 | 48 | 173 | 50 |
| 200,000 | 220,000 | 52 | 187 | 60 |
| 220,000 | 250,000 | 56 | 202 | 60 |
| 250,000 | 275,000 | 60 | 216 | 60 |
| 275,000 | 300,000 | 64 | 230 | 60 |

Column header: Minimum separation distance of receptor when barricaded[2] (ft)

[1] These distances apply to the separation of stores only. Table H-21 shall be used in determining separation distances from inhabited buildings, passenger railways, and public highways.

[2] When the ammonium nitrate and/or blasting agent is not barricaded, the distances shown in the table shall be multiplied by six. These distances allow for the possibility of high velocity metal fragments from mixers, hoppers, truck bodies, sheet metal structures, metal container, and the like which may enclose the "donor". Where storage is in bullet-resistant magazines recommended for explosives or where the storage is protected by a bullet-resistant wall, distances, and barricade thicknesses in excess of those prescribed in Table H-21 are not required.

[3] The distances in the table apply to ammonium nitrate that passes the insensitivity test prescribed in the definition of ammonium nitrate fertilizer promulgated by the National Plant Food Institute*; and ammonium nitrate failing to pass said test shall be stored at separation distances determined by competent persons. (*Definition and Test Procedures for Ammonium Nitrate Fertilizer, National Plant Food Institute, November 1964.)

[4] These distances apply to nitro-carbo-nitrates and blasting agents which pass the insensitivity test prescribed in the U.S. Department of Transportation (DOT) regulations.

[5] Earth, or sand dikes, or enclosures filled with the prescribed minimum thickness of earth or sand are acceptable artificial barricades. Natural barricades, such as hills or timber of sufficient density that the surrounding exposures which require protection cannot be seen from the "donor" when the trees are bare of leaves, are also acceptable.

[6] When the ammonium nitrate must be counted in determining the distances to be maintained from inhabited buildings, passenger railways and public highways, it may be counted at one-half its actual weight because its blast effect is lower.

Note 7: Guide to use of table of recommended separation distances of ammonium nitrate and blasting agents from explosives or blasting agents.

(a) Sketch location of all potential donor and acceptor materials together with the maximum mass of material to be allowed in that vicinity. (Potential donors are high explosives, blasting agents, and combination of masses of detonating materials. Potential acceptors are high explosives, blasting agents, and ammonium nitrate.)

(b) Consider separately each donor mass in combination with each acceptor mass. If the masses are closer than table allowance (distances measured between nearest edges), the combination of masses becomes a new potential donor of weight equal to the total mass. When individual masses are considered as donors, distances to potential acceptors shall be measured between edges. When combined masses within propagating distance of each other are considered as a donor, the appropriate distance to the edge of potential acceptors shall be computed as a weighted distance from the combined masses.

Calculation of weighted distance from combined masses:

Let $M_2, M_3 ... M_n$ be donor masses to be combined.

$M_1$ is a potential acceptor mass.

$D_{12}$ is Distance from $M_1$ to $M_2$ (edge to dege).

$D_{13}$ is distance from $M_1$ to $M_2$ (edge to edge), etc.

To find weighted distance $[D_{1(2,3...n)}]$ from combined masses to $M_1$, add the products of the individual masses and distances and divide the total by the sum of the masses thus:

$$D_{1(2,3...n)} = M_2 \times D_{12} + M_3 \times D_{12}... + M_n \times D_{12} M_2 + M_3... = M_n$$

Propagation is possible if either an individual donor mass is less than the tabulated distance from an acceptor or a combined mass is less than the weighted distance from an acceptor.

(c) In determining the distances separating highways, railroads, and inhabited buildings from potential explosions (as prescribed in Table H-21), the sum of all masses which may propagate (i.e., lie at distances less than prescribed in the Table) from *either* individual *or* combined donor masses are included. However, when the ammonium nitrate must be included, only 50 percent of its weight shall be used because of its reduced blast effects. In applying Table H-21 to distances from highways, railroads, and inhabited buildings, distances are measured from the nearest edge of potentially explodable material as prescribed in Table H-21, Note 5.

(d) When all or part of a potential acceptor comprises Explosives Class A as defined in DOT regulations, storage in bullet-resistant magazines is required. Safe distances to stores in bullet-resistant magazines may be obtained from the intermagazine distances prescribed in Table H-21.

(e) Barricades must not have line-of-sight openings between potential donors and acceptors which permit blast or missiles to move directly between masses.

(f) Good housekeeping practices shall be maintained around any bin containing ammonium nitrate or blasting agent. This includes keeping weeds and other combustible materials cleared within 25 feet of such bin. Accumulation of spilled product on the ground shall be prevented.

**1910.109(g)(5)** *Storage of blasting agents and supplies.* **(i)** Blasting agents and oxidizers used for mixing of blasting agents shall be stored in the manner set forth in this subdivision.

**1910.109(g)(5)(i)(a)** Blasting agents or ammonium nitrate, when stored in conjunction with explosives, shall be stored in the manner set forth in paragraph (c) of this section for explosives. The mass of blasting agents and one-half the mass of ammonium nitrate shall be included when computing the total quantity of explosives for determining distance requirements.

**1910.109(g)(5)(i)(b)** Blasting agents, when stored entirely separate from explosives, may be stored in the manner set forth in paragraph (c) of this section or in one-story warehouses (without basements) which shall be:

**1910.109(g)(5)(i)(b)(1)** Noncombustible or fire resistive;

**1910.109(g)(5)(i)(b)(2)** Constructed so as to eliminate open floor drains and piping into which molten materials could flow and be confined in case of fire;

**1910.109(g)(5)(i)(b)(3)** Weather resistant;

**1910.109(g)(5)(i)(b)(4)** Well ventilated; and

**1910.109(g)(5)(i)(b)(5)** Equipped with a strong door kept securely locked except when open for business.

**1910.109(g)(5)(i)(c)** Semitrailer or full-trailer vans used for highway or onsite transportation of the blasting agents are satisfactory for temporarily storing these materials, provided they are located in accordance with Table H-22 with respect to one another. Trailers shall be provided with substantial means for locking, and the trailer doors shall be kept locked, except during the time of placement and removal of stocks of blasting agents.

**1910.109(g)(5)(ii)** Warehouses used for the storage of blasting agents separate from explosives shall be located as set forth in this subdivision.

**1910.109(g)(5)(ii)(a)** Warehouses used for the storage of blasting agents shall be located in Table H-22 with respect to one another.

**1910.109(g)(5)(ii)(b)** If both blasting agents and ammonium nitrate are handled or stored within the distance limitations prescribed through paragraph (g)(2) of this section, one-half the mass of the ammonium nitrate shall be added to the mass of the blasting agent when computing the total quantity of explosives for determining the proper distance for compliance with Table H-21.

**1910.109(g)(5)(iii)** Smoking, matches, open flames, spark producing devices, and firearms are prohibited inside of or within 50 feet of any warehouse used for the storage of blasting agents. Combustible materials shall not be stored within 50 feet of warehouses used for the storage of blasting agents.

**1910.109(g)(5)(iv)** The interior of warehouses used for the storage of blasting agents shall be kept clean and free from debris and empty containers. Spilled materials shall be cleaned up promptly and safely removed. Combustible materials, flammable liquids, corrosive acids, chlorates, or nitrates shall not be stored in any warehouse used for blasting agents unless separated therefrom by a fire resistive separation of not less than 1 hour resistance. The provisions of this subdivision shall not prohibit the storage of blasting agents together with nonexplosive blasting supplies.

**1910.109(g)(5)(v)** Piles of ammonium nitrate and warehouses containing ammonium nitrate shall be adequately separated from readily combustible fuels.

**1910.109(g)(5)(vi)** Caked oxidizers, either in bags or in bulk, shall not be loosened by blasting.

**1910.109(g)(5)(vii)** Every warehouse used for the storage of blasting agents shall be under the supervision of a competent person.

**1910.109(g)(6)** *Transportation of packaged blasting agents.* **(i)** When blasting agents are transported in the same vehicle with explosives, all of the requirements of paragraph (d) of this section shall be complied with.

**1910.109(g)(6)(ii)** Vehicles transporting blasting agents shall only be driven by and be in charge of a driver in possession of a valid motor vehicle operator's license. Such a person shall also be familiar with the State's vehicle and traffic laws.

**1910.109(g)(6)(iii)** No matches, firearms, acids, or other corrosive liquids shall be carried in the bed or body of any vehicle containing blasting agents.

**1910.109(g)(6)(iv)** No person shall be permitted to ride upon, drive, load, or unload a vehicle containing blasting agents while smoking or under the influence of intoxicants, narcotics, or other dangerous drugs.

**1910.109(g)(6)(v)** [Reserved]

**1910.109(g)(6)(vi)** Vehicles transporting blasting agents shall be in safe operating condition at all times.

**1910.109(g)(7)** *Use of blasting agents.* Persons using blasting agents shall comply with all of the applicable provisions of paragraph (e) of this section.

**1910.109(h)** *Water gel (Slurry) explosives and blasting agents*—**(1)** *General provisions.* Unless otherwise set forth in this paragraph, water gels shall be transported, stored and used in the same manner as explosives or blasting agents in accordance with the classification of the product.

**1910.109(h)(2)** *Types and classifications.* **(i)** Water gels containing a substance in itself classified as an explosive shall be classified as an explosive and manufactured, transported, stored, and used as specified for "explosives" in this section, except as noted in subdivision (iv) of this subparagraph.

**1910.109(h)(2)(ii)** Water gels containing no substance in itself classified as an explosive and which are cap-sensitive as defined in paragraph (a) of this section under Blasting Agent shall be classified as an explosive and manufactured, transported, stored and used as specified for "explosives" in this section.

**1910.109(h)(2)(iii)** Water gels containing no substance in itself classified as an explosive and which are not cap-sensitive as defined in paragraph (a) of this section under Blasting Agent shall be classified as blasting agents and manufactured, transported, stored, and used as specified for "blasting agents" in this section.

**1910.109(h)(2)(iv)** When tests on specific formulations of water gels result in Department of Transportation classification as a Class B explosive, bullet-resistant magazines are not required, see paragraph (c)(2)(ii) of this section.

**1910.109(h)(3)** *Fixed location mixing.*

**1910.109(h)(3)(i)** [Reserved]

**1910.109(h)(3)(ii)** Buildings used for the mixing of water gels shall conform to the requirements of this subdivision.

**1910.109(h)(3)(ii)(a)** Buildings shall be of noncombustible construction or sheet metal on wood studs.

**1910.109(h)(3)(ii)(b)** Floors in a mixing plant shall be of concrete or of other nonabsorbent materials.

**1910.109(h)(3)(ii)(c)** Where fuel oil is used all fuel oil storage facilities shall be separated from the mixing plant and located in such a manner that in case of tank rupture, the oil will drain away from the mixing plant building.

**1910.109(h)(3)(ii)(d)** The building shall be well ventilated.

**1910.109(h)(3)(ii)(e)** Heating units that do not depend on combustion processes, when properly designed and located, may be used in the building. All direct sources of heat shall be provided exclusively from units located outside of the mixing building.

**1910.109(h)(3)(ii)(f)** All internal-combustion engines used for electric power generation shall be located outside the mixing plant building, or shall be properly ventilated and isolated by a firewall. The exhaust systems on all such engines shall be located so any spark emission cannot be a hazard to any materials in or adjacent to the plant.

**1910.109(h)(3)(iii)** Ingredients of water gels shall conform to the requirements of this subdivision.

**1910.109(h)(3)(iii)(a)** Ingredients in themselves classified as Class A or Class B explosives shall be stored in conformity with paragraph (c) of this section.

**1910.109(h)(3)(iii)(b)** Nitrate-water solutions may be stored in tank cars, tank trucks, or fixed tanks without quantity or distance limitations. Spills or leaks which may contaminate combustible materials shall be cleaned up immediately.

**1910.109(h)(3)(iii)(c)** Metal powders such as aluminum shall be kept dry and shall be stored in containers or bins which are moisture-resistant or weathertight. Solid fuels shall be used in such manner as to minimize dust explosion hazards.

**1910.109(h)(3)(iii)(d)** Ingredients shall not be stored with incompatible materials.

**1910.109(h)(3)(iii)(e)** Peroxides and chlorates shall not be used.

**1910.109(h)(3)(iv)** Mixing equipment shall comply with the requirements of this subdivision.

**1910.109(h)(3)(iv)(a)** The design of the processing equipment, including mixing and conveying equipment, shall be compatible with the relative sensitivity of the materials being handled. Equipment shall be designed to minimize the possibility of frictional heating, compaction, overloading, and confinement.

# OSHA Standards for General Industry

**1910.109(h)(3)(iv)(b)** Both equipment and handling procedures shall be designed to prevent the introduction of foreign objects or materials.

**1910.109(h)(3)(iv)(c)** Mixers, pumps, valves, and related equipment shall be designed to permit regular and periodic flushing, cleaning, dismantling, and inspection.

**1910.109(h)(3)(iv)(d)** All electrical equipment including wiring, switches, controls, motors, and lights, shall conform to the requirements of subpart S of this part.

**1910.109(h)(3)(iv)(e)** All electric motors and generators shall be provided with suitable overload protection devices. Electrical generators, motors, proportioning devices, and all other electrical enclosures shall be electrically bonded. The grounding conductor to all such electrical equipment shall be effectively bonded to the service-entrance ground connection and to all equipment ground connections in a manner so as to provide a continuous path to ground.

**1910.109(h)(3)(v)** Mixing facilities shall comply with the fire prevention requirements of this subdivision.

**1910.109(h)(3)(v)(a)** The mixing, loading, and ingredient transfer areas where residues or spilled materials may accumulate shall be cleaned periodically. A cleaning and collection system for dangerous residues shall be provided.

**1910.109(h)(3)(v)(b)** A daily visual inspection shall be made of mixing, conveying, and electrical equipment to establish that such equipment is in good operating condition. A program of systematic maintenance shall be conducted on regular schedule.

**1910.109(h)(3)(v)(c)** Heaters which are not dependent on the combustion process within the heating unit may be used within the confines of processing buildings, or compartments, if provided with temperature and safety controls and located away from combustible materials and the finished product.

**1910.109(h)(4)** *Bulk delivery and mixing vehicles.* **(i)** The design of vehicles shall comply with the requirements of this subdivision.

**1910.109(h)(4)(i)(a)** Vehicles used over public highways for the bulk transportation of water gels or of ingredients classified as dangerous commodities, shall meet the requirements of the Department of Transportation and shall meet the requirements of paragraphs (d) and (g)(6) of this section.

**1910.109(h)(4)(i)(b)** When electric power is supplied by a self-contained motor generator located on the vehicle the generator shall be at a point separate from where the water gel is discharged.

**1910.109(h)(4)(i)(c)** The design of processing equipment and general requirements shall conform to subparagraphs (3) (iii) and (iv) of this paragraph.

**1910.109(h)(4)(i)(d)** A positive action parking brake, which will set the wheel brakes on at least one axle shall be provided on vehicles when equipped with air brakes and shall be used during bulk delivery operations. Wheel chocks shall supplement parking brakes whenever conditions may require.

**1910.109(h)(4)(ii)** Operation of bulk delivery and mixing vehicles shall comply with the requirements of this subdivision.

**1910.109(h)(4)(ii)(a)** The placarding requirements contained in DOT regulations apply to vehicles carrying water gel explosives or blasting agents.

**1910.109(h)(4)(ii)(b)** The operator shall be trained in the safe operation of the vehicle together with its mixing, conveying, and related equipment. He shall be familiar with the commodities being delivered and the general procedure for handling emergency situations.

**1910.109(h)(4)(ii)(c)** The hauling of either blasting caps or other explosives, but not both, shall be permitted on bulk trucks provided that a special wood or nonferrous-lined container is installed for the explosives. Such blasting caps or other explosives shall be in DOT-specified shipping containers; see 49 CFR chapter I.

**1910.109(h)(4)(ii)(d)** No person shall be allowed to smoke, carry matches or any flame-producing device, or carry any firearms while in or about bulk vehicles effecting the mixing, transfer, or down-the-hole loading of water gels at or near the blasting site.

**1910.109(h)(4)(ii)(e)** Caution shall be exercised in the movement of the vehicle in the blasting area to avoid driving the vehicle over or dragging hoses over firing lines, cap wires, or explosive materials. The employer shall furnish the driver the assistance of a second person to guide the driver's movements.

**1910.109(h)(4)(ii)(f)** No intransit mixing of materials shall be performed.

**1910.109(h)(4)(ii)(g)** The location chosen for water gel or ingredient transfer from a support vehicle into the borehole loading vehicle shall be away from the blasthole site when the boreholes are loaded or in the process of being loaded.

**1910.109(i)** *Storage of ammonium nitrate—***(1)** *Scope and definitions.* **(i)(a)** Except as provided in paragraph (i)(1)(i)(d) of this paragraph applies to the storage of ammonium nitrate in the form of crystals, flakes, grains, or prills including fertilizer grade, dynamite grade, nitrous oxide grade, technical grade, and other mixtures containing 60 percent or more ammonium nitrate by weight but does not apply to blasting agents.

**1910.109(i)(1)(i)(b)** This paragraph does not apply to the transportation of ammonium nitrate.

**1910.109(i)(1)(i)(c)** This paragraph does not apply to storage under the jurisdiction of and in compliance with the regulations of the U.S. Coast Guard (see 46 CFR parts 146-149).

**1910.109(i)(1)(i)(d)** The storage of ammonium nitrate and ammonium nitrate mixtures that are more sensitive than allowed by the "Definition of Test Procedures for Ammonium Nitrate Fertilizer" is prohibited.

**1910.109(i)(1)(ii)(a)** [Reserved]

**1910.109(i)(1)(ii)(b)** The standards for ammonium nitrate (nitrous oxide grade) are those found in the "Specifications, Properties, and Recommendations for Packaging, Transportation, Storage, and Use of Ammonium Nitrate", available from the Compressed Gas Association, Inc., which is incorporated by reference as specified in §1910.6.

**1910.109(i)(2)** *General provisions.* **(i)** This paragraph applies to all persons storing, having, or keeping ammonium nitrate, and to the owner or lessee of any building, premises, or structure in which ammonium nitrate is stored in quantities of 1,000 pounds or more.

**1910.109(i)(2)(ii)** Approval of large quantity storage shall be subject to due consideration of the fire and explosion hazards, including exposure to toxic vapors from burning or decomposing ammonium nitrate.

**1910.109(i)(2)(iii)(a)** Storage buildings shall not have basements unless the basements are open on at least one side. Storage buildings shall not be over one story in height.

**1910.109(i)(2)(iii)(b)** Storage buildings shall have adequate ventilation or be of a construction that will be self-ventilating in the event of fire.

**1910.109(i)(2)(iii)(c)** The wall on the exposed side of a storage building within 50 feet of a combustible building, forest, piles of combustible materials and similar exposure hazards shall be of fire-resistive construction. In lieu of the fire-resistive wall, other suitable means of exposure protection such as a free standing wall may be used. The roof coverings shall be Class C or better, as defined in the Manual on Roof Coverings, NFPA 203M-1970, which is incorporated by reference as specified in §1910.6.

**1910.109(i)(2)(iii)(d)** All flooring in storage and handling areas, shall be of noncombustible material or protected against impregnation by ammonium nitrate and shall be without open drains, traps, tunnels, pits, or pockets into which any molten ammonium nitrate could flow and be confined in the event of fire.

**1910.109(i)(2)(iii)(e)** The continued use of an existing storage building or structure not in strict conformity with this paragraph may be approved in cases where such continued use will not constitute a hazard to life.

**1910.109(i)(2)(iii)(f)** Buildings and structures shall be dry and free from water seepage through the roof, walls, and floors.

**1910.109(i)(3)** *Storage of ammonium nitrate in bags, drums, or other containers.* **(i)(a)** Bags and containers used for ammonium nitrate must comply with specifications and standards required for use in interstate commerce (see 49 CFR chapter I).

**1910.109(i)(3)(i)(b)** Containers used on the premises in the actual manufacturing or processing need not comply with provisions of paragraph (i)(3)(i)(a) of this paragraph.

**1910.109(i)(3)(ii)(a)** Containers of ammonium nitrate shall not be accepted for storage when the temperature of the ammonium nitrate exceeds 130 °F.

**1910.109(i)(3)(ii)(b)** Bags of ammonium nitrate shall not be stored within 30 inches of the storage building walls and partitions.

**1910.109(i)(3)(ii)(c)** The height of piles shall not exceed 20 feet. The width of piles shall not exceed 20 feet and the length 50 feet except that where the building is of noncombustible construction or is protected by automatic sprinklers the length of piles shall not be limited. In no case shall the ammonium nitrate be stacked closer than 36 inches below the roof or supporting and spreader beams overhead.

**1910.109(i)(3)(ii)(d)** Aisles shall be provided to separate piles by a clear space of not less than 3 feet in width. At least one service or main aisle in the storage area shall be not less than 4 feet in width.

**1910.109(i)(4)** *Storage of bulk ammonium nitrate.* **(i)***(a)* Warehouses shall have adequate ventilation or be capable of adequate ventilation in case of fire.

**1910.109(i)(4)(i)(b)** Unless constructed of noncombustible material or unless adequate facilities for fighting a roof fire are available, bulk storage structures shall not exceed a height of 40 feet.

**1910.109(i)(4)(ii)(a)** Bins shall be clean and free of materials which may contaminate ammonium nitrate.

**1910.109(i)(4)(ii)(b)** Due to the corrosive and reactive properties of ammonium nitrate, and to avoid contamination, galvanized iron, copper, lead, and zinc shall not be used in a bin construction unless suitably protected. Aluminum bins and wooden bins protected against impregnation by ammonium nitrate are permissible. The partitions dividing the ammonium nitrate storage from other products which would contaminate the ammonium nitrate shall be of tight construction.

**1910.109(i)(4)(ii)(c)** The ammonium nitrate storage bins or piles shall be clearly identified by signs reading "Ammonium Nitrate" with letters at least 2 inches high.

**1910.109(i)(4)(iii)(a)** Piles or bins shall be so sized and arranged that all material in the pile is moved out periodically in order to minimize possible caking of the stored ammonium nitrate.

**1910.109(i)(4)(iii)(b)** Height or depth of piles shall be limited by the pressure-setting tendency of the product. However, in no case shall the ammonium nitrate be piled higher at any point than 36 inches below the roof or supporting and spreader beams overhead.

**1910.109(i)(4)(iii)(c)** Ammonium nitrate shall not be accepted for storage when the temperature of the product exceeds 130 °F.

**1910.109(i)(4)(iii)(d)** Dynamite, other explosives, and blasting agents shall not be used to break up or loosen caked ammonium nitrate.

**1910.109(i)(5)** *Contaminants.* **(i)***(a)* Ammonium nitrate shall be in a separate building or shall be separated by approved type firewalls of not less than 1 hour fire-resistance rating from storage of organic chemicals, acids, or other corrosive materials, materials that may require blasting during processing or handling, compressed flammable gases, flammable and combustible materials or other contaminating substances, including but not limited to animal fats, baled cotton, baled rags, baled scrap paper, bleaching powder, burlap or cotton bags, caustic soda, coal, coke, charcoal, cork, camphor, excelsior, fibers of any kind, fish oils, fish meal, foam rubber, hay, lubricating oil, linseed oil, or other oxidizable or drying oils, naphthalene, oakum, oiled clothing, oiled paper, oiled textiles, paint, straw, sawdust, wood shavings, or vegetable oils. Walls referred to in this subdivision need extend only to the underside of the roof.

**1910.109(i)(5)(i)(b)** In lieu of separation walls, ammonium nitrate may be separated from the materials referred to in paragraph (*a*) of this section by a space of at least 30 feet.

**1910.109(i)(5)(i)(c)** Flammable liquids such as gasoline, kerosene, solvents, and light fuel oils shall not be stored on the premises except when such storage conforms to §1910.106, and when walls and sills or curbs are provided in accordance with paragraphs (i)(5)(i) (*a*) or (*b*) of this section.

**1910.109(i)(5)(i)(d)** LP-Gas shall not be stored on the premises except when such storage conforms to §1910.110.

**1910.109(i)(5)(ii)(a)** Sulfur and finely divided metals shall not be stored in the same building with ammonium nitrate except when such storage conforms to paragraphs (a) through (h) of this section.

**1910.109(i)(5)(ii)(b)** Explosives and blasting agents shall not be stored in the same building with ammonium nitrate except on the premises of makers, distributors, and user-compounders of explosives or blasting agents.

**1910.109(i)(5)(ii)(c)** Where explosives or blasting agents are stored in separate buildings, other than on the premises of makers, distributors, and user-compounders of explosives or blasting agents, they shall be separated from the ammonium nitrate by the distances and/or barricades specified in Table H-22 of this subpart, but by not less than 50 feet.

**1910.109(i)(5)(ii)(d)** Storage and/or operations on the premises of makers, distributors, and user-compounders of explosives or blasting agents shall be in conformity with paragraphs (a) through (h) of this section.

**1910.109(i)(6)** *General precautions.* (i) Electrical installations shall conform to the requirements of subpart S of this part, for ordinary locations. They shall be designed to minimize damage from corrosion.

**1910.109(i)(6)(ii)** In areas where lightning storms are prevalent, lightning protection shall be provided. (See the Lightning Protection Code, NFPA 78-1968, which is incorporated by reference as specified in §1910.6.)

**1910.109(i)(6)(iii)** Provisions shall be made to prevent unauthorized personnel from entering the ammonium nitrate storage area.

**1910.109(i)(7)** *Fire protection.* **(i)** Not more than 2,500 tons (2270 tonnes) of bagged ammonium nitrate shall be stored in a building or structure not equipped with an automatic sprinkler system. Sprinkler systems shall be of the approved type and installed in accordance with §1910.159.

**1910.109(i)(7)(ii)(a)** Suitable fire control devices such as small hose or portable fire extinguishers shall be provided throughout the warehouse and in the loading and unloading areas. Suitable fire control devices shall comply with the requirements of §§1910.157 and 1910.158.

**1910.109(i)(7)(ii)(b)** Water supplies and fire hydrants shall be available in accordance with recognized good practices.

**1910.109(j)** *Small arms ammunition, small arms primers, and small arms propellants*—**(1)** *Scope.* This paragraph does not apply to in-process storage and intraplant transportation during manufacture of small arms ammunition, small arms primers, and smokeless propellants.

**1910.109(j)(2)** *Small arms ammunition.* **(i)** No quantity limitations are imposed on the storage of small arms ammunition in warehouses, retail stores, and other general occupancy facilities, except those imposed by limitations of storage facilities.

**1910.109(j)(2)(ii)** Small arms ammunition shall be separated from flammable liquids, flammable solids as classified in 49 CFR part 172, and from oxidizing materials, by a fire-resistive wall of 1-hour rating or by a distance of 25 feet.

**1910.109(j)(2)(iii)** Small arms ammunition shall not be stored together with Class A or Class B explosives unless the storage facility is adequate for this latter storage.

**1910.109(j)(3)** *Smokeless propellants.* **(i)** All smokeless propellants shall be stored in shipping containers specified in 49 CFR 173.93 for smokeless propellants.

**1910.109(j)(3)(ii)** [Reserved]

**1910.109(j)(3)(iii)** Commercial stocks of smokeless propellants over 20 pounds and not more than 100 pounds shall be stored in portable wooden boxes having walls of at least 1 inch nominal thickness.

**1910.109(j)(3)(iv)** Commercial stocks in quantities not to exceed 750 pounds shall be stored in nonportable storage cabinets having wooden walls of at least 1 inch nominal thickness. Not more than 400 pounds shall be permitted in any one cabinet.

**1910.109(j)(3)(v)** Quantities in excess of 750 pounds shall be stored in magazines in accordance with paragraph (c) of this section.

**1910.109(j)(4)** *Small arms ammunition primers.* **(i)** Small arms ammunition primers shall not be stored except in the original shipping container in accordance with the requirements of 49 CFR 173.107 for small arms ammunition primers.

**1910.109(j)(4)(ii)** [Reserved]

**1910.109(j)(4)(iii)** Small arms ammunition primers shall be separated from flammable liquids, flammable solids as classified in 49 CFR part 172, and oxidizing materials by a fire-resistive wall of 1-hour rating or by a distance of 25 feet.

**1910.109(j)(4)(iv)** Not more than 750,000 small arms ammunition primers shall be stored in any one building, except as provided in paragraph (j)(4)(v) of this paragraph. Not more than 100,000 shall be stored in any one pile. Piles shall be at least 15 feet apart.

# OSHA Standards for General Industry

**1910.109(j)(4)(v)** Quantities of small arms ammunition primers in excess of 750,000 shall be stored in magazines in accordance with paragraph (c) of this section.

**1910.109(k)** *Scope.* **(1)** This section applies to the manufacture, keeping, having, storage, sale, transportation, and use of explosives, blasting agents, and pyrotechnics. The section does not apply to the sale and use (public display) of pyrotechnics, commonly known as fireworks, nor the use of explosives in the form prescribed by the official U.S. Pharmacopeia.

**1910.109(k)(2)** The manufacture of explosives as defined in paragraph (a)(3) of this section shall also meet the requirements contained in §1910.119.

**1910.109(k)(3)** The manufacture of pyrotechnics as defined in paragraph (a)(10) of this section shall also meet the requirements contained in §1910.119.

[39 FR 23502, June 27, 1974, as amended at 43 FR 49747, Oct. 24, 1978; 45 FR 60704, Sept. 12, 1980; 53 FR 12122, Apr. 12, 1988; 57 FR 6403, Feb. 24, 1992; 58 FR 35309, June 30, 1993; 61 FR 9237, Mar. 7, 1996; 63 FR 33466, June 18, 1998]

## §1910.110 Storage and handling of liquefied petroleum gases.

**1910.110(a)** *Definitions applicable to this section.* As used in this section:

**1910.110(a)(1)** API-ASME container—A container constructed in accordance with the requirements of paragraph (b)(3)(iii) of this section.

**1910.110(a)(2)** ASME container—A container constructed in accordance with the requirements of paragraph (b)(3)(i) of this section.

**1910.110(a)(3)** Container assembly—An assembly consisting essentially of the container and fittings for all container openings, including shutoff valves, excess flow valves, liquid-level gaging devices, safety relief devices, and protective housing.

**1910.110(a)(4)** Containers—All vessels, such as tanks, cylinders, or drums, used for transportation or storing liquefied petroleum gases.

**1910.110(a)(5)** DOT—Department of Transportation.

**1910.110(a)(6)** DOT container—A container constructed in accordance with the applicable requirements of 49 CFR chapter 1.

**1910.110(a)(7)** "Liquified petroleum gases"—"LPG" and "LP-Gas"—Any material which is composed predominantly of any of the following hydrocarbons, or mixtures of them; propane, propylene, butanes (normal butane or iso-butane), and butylenes.

**1910.110(a)(8)** Movable fuel storage tenders or farm carts—Containers not in excess of 1,200 gallons water capacity, equipped with wheels to be towed from one location of usage to another. They are basically non-highway vehicles, but may occasionally be moved over public roads or highways. They are used as a fuel supply for farm tractors, construction machinery and similar equipment.

**1910.110(a)(9)** P.S.I.G.—pounds per square inch gauge.

**1910.110(a)(10)** P.S.I.A.—pounds per square inch absolute.

**1910.110(a)(11)** Systems—an assembly of equipment consisting essentially of the container or containers, major devices such as vaporizers, safety relief valves, excess flow valves, regulators, and piping connecting such parts.

**1910.110(a)(12)** Vaporizer-burner—an integral vaporizer-burner unit, dependent upon the heat generated by the burner as the source of heat to vaporize the liquid used for dehydrators or dryers.

**1910.110(a)(13)** Ventilation, adequate—when specified for the prevention of fire during normal operation, ventilation shall be considered adequate when the concentration of the gas in a gas-air mixture does not exceed 25 percent of the lower flammable limit.

**1910.110(a)(14)** Approved—unless otherwise indicated, listing or approval by a nationally recognized testing laboratory. Refer to §1910.7 for definition of nationally recognized testing laboratory.

**1910.110(a)(15)** Listed—see "approved" in §1910.110(14).

**1910.110(a)(16)** DOT Specifications—regulations of the Department of Transportation published in 49 CFR chapter I.

**1910.110(a)(17)-(18)** [Reserved]

**1910.110(a)(19)** DOT cylinders—cylinders meeting the requirements of 49 CFR chapter I.

**1910.110(b)** *Basic rules*—**(1)** *Odorizing gases.* **(i)** All liquefied petroleum gases shall be effectively odorized by an approved agent of such character as to indicate positively, by distinct odor, the presence of gas down to concentration in air of not over one-fifth the lower limit of flammability. Odorization, however, is not required if harmful in the use of further processing of the liquefied petroleum gas, or if odorization will serve no useful purpose as a warning agent in such use or further processing.

**1910.110(b)(1)(ii)** The odorization requirement of paragraph (b)(1)(i) of this section shall be considered to be met by the use of 1.0 pounds of ethyl mercaptan, 1.0 pounds of thiophane or 1.4 pounds of amyl mercaptan per 10,000 gallons of LP-Gas. However, this listing of odorants and quantities shall not exclude the use of other odorants that meet the odorization requirements of paragraph (b)(1)(i) of this section.

**1910.110(b)(2)** *Approval of equipment and systems.* **(i)** Each system utilizing DOT containers in accordance with 49 CFR part 178 shall have its container valves, connectors, manifold valve assemblies, and regulators approved.

**1910.110(b)(2)(ii)** Each system for domestic or commercial use utilizing containers of 2,000 gallons or less water capacity, other than those constructed in accordance with 49 CFR part 178, shall consist of a container assembly and one or more regulators, and may include other parts. The system as a unit or the container assembly as a unit, and the regulator or regulators, shall be individually listed.

**1910.110(b)(2)(iii)** In systems utilizing containers of over 2,000 gallons water capacity, each regulator, container valve, excess flow valve, gaging device, and relief valve installed on or at the container, shall have its correctness as to design, construction, and performance determined by listing by a nationally recognized testing laboratory. Refer to §1910.7 for definition of nationally recognized testing laboratory.

**1910.110(b)(3)** *Requirements for construction and original test of containers.* **(i)** Containers used with systems embodied in paragraphs (d), (e), (g), and (h) of this section, except as provided in paragraphs (e)(3)(iii) and (g)(2)(i) of this section, shall be designed, constructed, and tested in accordance with the Rules for Construction of Unfired Pressure Vessels, section VIII, Division 1, American Society of Mechanical Engineers (ASME) Boiler and Pressure Vessel Code, 1968 edition, which is incorporated by reference as specified in §1910.6.

**1910.110(b)(3)(ii)** Containers constructed according to the 1949 and earlier editions of the ASME Code do not have to comply with paragraphs U-2 through U-10 and U-19 thereof. Containers constructed according to paragraph U-70 in the 1949 and earlier editions are not authorized.

**1910.110(b)(3)(iii)** Containers designed, constructed, and tested prior to July 1, 1961, according to the Code for Unfired Pressure Vessels for Petroleum Liquids and Gases, 1951 edition with 1954 Addenda, of the American Petroleum Institute and the American Society of Mechanical Engineers, which is incorporated by reference as specified in §1910.6, shall be considered in conformance. Containers constructed according to API-ASME Code do not have to comply with section I or with appendix to section I. Paragraphs W-601 to W-606 inclusive in the 1943 and earlier editions do not apply.

**1910.110(b)(3)(iv)** The provisions of paragraph (b)(3)(i) of this section shall not be construed as prohibiting the continued use or reinstallation of containers constructed and maintained in accordance with the standard for the Storage and Handling of Liquefied Petroleum Gases NFPA No. 58 in effect at the time of fabrication.

**1910.110(b)(3)(v)** Containers used with systems embodied in paragraph (b), (d)(3)(iii), and (f) of this section, shall be constructed, tested, and stamped in accordance with DOT specifications effective at the date of their manufacture.

**1910.110(b)(4)** *Welding of containers.* Welding to the shell, head, or any other part of the container subject to internal pressure, shall be done in compliance with the code under which the tank was fabricated. Other welding is permitted only on saddle plates, lugs, or brackets attached to the container by the tank manufacturer.

**1910.110(b)(4)(ii)** Where repair or modification involving welding of DOT containers is required, the container shall be returned to a qualified manufacturer making containers of the same type, and the repair or modification made in compliance with DOT regulations.

**1910.110(b)(5)** *Markings on containers.* **(i)** Each container covered in paragraph (b)(3)(i) of this section, except as provided in paragraph (b)(3)(iv) of this section shall be marked as specified in the following:

**1910.110(b)(5)(i)(a)** With a marking identifying compliance with, and other markings required by, the rules of the reference under which the container is constructed; or with the stamp and other markings required by the National Board of Boiler and Pressure Vessel Inspectors.

**1910.110(b)(5)(i)(b)** With notation as to whether the container is designed for underground or aboveground installation or both. If intended for both and different style hoods are provided, the marking shall indicate the proper hood for each type of installation.

**1910.110(b)(5)(i)(c)** With the name and address of the supplier of the container, or with the trade name of the container.

**1910.110(b)(5)(i)(d)** With the water capacity of the container in pounds or gallons, U.S. Standard.

**1910.110(b)(5)(i)(e)** With the pressure in p.s.i.g., for which the container is designed.

**1910.110(b)(5)(i)(f)** With the wording "This container shall not contain a product having a vapor pressure in excess of—p.s.i.g. at 100 °F.," see subparagraph (14)(viii) of this paragraph.

**1910.110(b)(5)(i)(g)** With the tare weight in pounds or other identified unit of weight for containers with a water capacity of 300 pounds or less.

**1910.110(b)(5)(i)(h)** With marking indicating the maximum level to which the container may be filled with liquid at temperatures between 20 °F. and 130 °F., except on containers provided with fixed maximum level indicators or which are filled by weighing. Markings shall be increments of not more than 20 °F. This marking may be located on the liquid level gaging device.

**1910.110(b)(5)(i)(i)** With the outside surface area in square feet.

**1910.110(b)(5)(ii)** Markings specified shall be on a metal nameplate attached to the container and located in such a manner as to remain visible after the container is installed.

**1910.110(b)(5)(iii)** When LP-Gas and one or more other gases are stored or used in the same area, the containers shall be marked to identify their content. Marking shall be in compliance with American National Standard Z48.1-1954, "Method of Marking Portable Compressed Gas Containers To Identify the Material Contained", which is incorporated by reference as specified in §1910.6.

**1910.110(b)(6)** *Location of containers and regulating equipment.* **(i)** Containers, and first stage regulating equipment if used, shall be located outside of buildings, except under one or more of the following:

**1910.110(b)(6)(i)(a)** In buildings used exclusively for container charging, vaporization pressure reduction, gas mixing, gas manufacturing, or distribution.

**1910.110(b)(6)(i)(b)** When portable use is necessary and in accordance with paragraph (c)(5) of this section.

**1910.110(b)(6)(i)(c)** LP-Gas fueled stationary or portable engines in accordance with paragraph (e)(11) or (12) of this section.

**1910.110(b)(6)(i)(d)** LP-Gas fueled industrial trucks used in accordance with paragraph (e)(13) of this section.

**1910.110(b)(6)(i)(e)** LP-Gas fueled vehicles garaged in accordance with paragraph (e)(14) of this section.

**1910.110(b)(6)(i)(f)** Containers awaiting use or resale when stored in accordance with paragraph (f) of this section.

**1910.110(b)(6)(ii)** Each individual container shall be located with respect to the nearest important building or group of buildings in accordance with Table H-23.

### TABLE H-23

| Water capacity per container | Minimum distances — Containers Underground | Minimum distances — Containers Aboveground | Between aboveground containers |
|---|---|---|---|
| Less than 125 gals[1] | 10 feet | None | None. |
| 125 to 250 gallons | 10 feet | 10 | None. |
| 251 to 500 gallons | 10 feet | 10 feet | 3 feet. |
| 501 to 2,000 gallons | 25 feet[2] | 25 feet[2] | 3 feet. |
| 2,001 to 30,000 gallons | 50 feet | 50 feet | 5 feet. |
| 30,001 to 70,000 gallons | 50 feet | 75 feet[3] | |
| 70,001 to 90,000 gallons | 50 feet | 100 feet[3] | |

[1] If the aggregate water capacity of a multi-container installation at a consumer site is 501 gallons or greater, the minimum distance shall comply with the appropriate portion of this table, applying the aggregate capacity rather than the capacity per container. If more than one installation is made, each installation shall be separated from another installation by at least 25 feet. Do not apply the MINIMUM DISTANCES BETWEEN ABOVE-GROUND CONTAINERS to such installations.

[2] The above distance requirements may be reduced to not less than 10 feet for a single container of 1,200 gallons water capacity or less, providing such a container is at least 25 feet from any other LP-Gas container of more than 125 gallons water capacity.

[3] 1/4 of sum of diameters of adjacent containers.

**1910.110(b)(6)(iii)** Containers installed for use shall not be stacked one above the other.

**1910.110(b)(6)(iv)** [Reserved]

**1910.110(b)(6)(v)** In the case of buildings devoted exclusively to gas manufacturing and distributing operations, the distances required by Table H-23 may be reduced provided that in no case shall containers of water capacity exceeding 500 gallons be located closer than 10 feet to such gas manufacturing and distributing buildings.

**1910.110(b)(6)(vi)** Readily ignitible material such as weeds and long dry grass shall be removed within 10 feet of any container.

**1910.110(b)(6)(vii)** The minimum separation between liquefied petroleum gas containers and flammable liquid tanks shall be 20 feet, and the minimum separation between a container and the centerline of the dike shall be 10 feet. The foregoing provision shall not apply when LP-Gas containers of 125 gallons or less capacity are installed adjacent to Class III flammable liquid tanks of 275 gallons or less capacity.

**1910.110(b)(6)(viii)** Suitable means shall be taken to prevent the accumulation of flammable liquids under adjacent liquefied petroleum gas containers, such as by diking, diversion curbs, or grading.

**1910.110(b)(6)(ix)** When dikes are used with flammable liquid tanks, no liquefied petroleum gas containers shall be located within the diked area.

**1910.110(b)(7)** *Container valves and container accessories.* **(i)** Valves, fittings, and accessories connected directly to the container including primary shutoff valves, shall have a rated working pressure of at least 250 p.s.i.g. and shall be of material and design suitable for LP-Gas service. Cast iron shall not be used for container valves, fittings, and accessories. This does not prohibit the use of container valves made of malleable or nodular iron.

**1910.110(b)(7)(ii)** Connections to containers, except safety relief connections, liquid level gaging devices, and plugged openings, shall have shutoff valves located as close to the container as practicable.

**1910.110(b)(7)(iii)** Excess flow valves, where required shall close automatically at the rated flows of vapor or liquid as specified by the manufacturer. The connections or line including valves, fittings, etc., being protected by

## OSHA Standards for General Industry

an excess flow valve shall have a greater capacity than the rated flow of the excess flow valve.

**1910.110(b)(7)(iv)** Liquid level gaging devices which are so constructed that outward flow of container contents shall not exceed that passed by a No. 54 drill size opening, need not be equipped with excess flow valves.

**1910.110(b)(7)(v)** Openings from container or through fittings attached directly on container to which pressure gage connection is made, need not be equipped with shutoff or excess flow valves if such openings are restricted to not larger than No. 54 drill size opening.

**1910.110(b)(7)(vi)** Except as provided in paragraph (c)(5)(i)(b) of this section, excess flow and back pressure check valves where required by this section shall be located inside of the container or at a point outside where the line enters the container; in the latter case, installation shall be made in such manner that any undue strain beyond the excess flow or back pressure check valve will not cause breakage between the container and such valve.

**1910.110(b)(7)(vii)** Excess flow valves shall be designed with a bypass, not to exceed a No. 60 drill size opening to allow equalization of pressures.

**1910.110(b)(7)(viii)** Containers of more than 30 gallons water capacity and less than 2,000 gallons water capacity, filled on a volumetric basis, and manufactured after December 1, 1963, shall be equipped for filling into the vapor space.

**1910.110(b)(8)** *Piping—including pipe, tubing, and fittings.* **(i)** Pipe, except as provided in paragraphs (e)(6)(i) and (g)(10)(iii), of this section shall be wrought iron or steel (black or galvanized), brass, copper, or aluminum alloy. Aluminum alloy pipe shall be at least Schedule 40 in accordance with the specifications for Aluminum Alloy Pipe, American National Standards Institute (ANSI) H38.7-1969 (ASTM, B241-69), which is incorporated by reference as specified in §1910.6, except that the use of alloy 5456 is prohibited and shall be suitably marked at each end of each length indicating compliance with American National Standard Institute Specifications. Aluminum Alloy pipe shall be protected against external corrosion when it is in contact with dissimilar metals other than galvanized steel, or its location is subject to repeated wetting by such liquids as water (except rain water), detergents, sewage, or leaking from other piping, or it passes through flooring, plaster, masonry, or insulation. Galvanized sheet steel or pipe, galvanized inside and out, may be considered suitable protection. The maximum nominal pipe size for aluminum pipe shall be three-fourths inch and shall not be used for pressures exceeding 20 p.s.i.g. Aluminum alloy pipe shall not be installed within 6 inches of the ground.

**1910.110(b)(8)(i)(a)** Vapor piping with operating pressures not exceeding 125 p.s.i.g. shall be suitable for a working pressure of at least 125 p.s.i.g. Pipe shall be at least Schedule 40 (ASTM A-53-69, Grade B Electric Resistance Welded and Electric Flash Welded Pipe, which is incorporated by reference as specified in §1910.6, or equal).

**1910.110(b)(8)(i)(b)** Vapor piping with operating pressures over 125 p.s.i.g. and all liquid piping shall be suitable for a working pressure of at least 250 p.s.i.g. Pipe shall be at least Schedule 80 if joints are threaded or threaded and back welded. At least Schedule 40 (ASTM A-53-69 Grade B Electric Resistance Welded and Electric Flash Welded Pipe or equal) shall be used if joints are welded, or welded and flanged.

**1910.110(b)(8)(ii)** Tubing shall be seamless and of copper, brass, steel, or aluminum alloy. Copper tubing shall be of type K or L or equivalent as covered in the Specification for Seamless Copper Water Tube, ANSI H23.1-1970 (ASTM B88-69), which is incorporated by reference as specified in §1910.6. Aluminum alloy tubing shall be of Type A or B or equivalent as covered in Specification ASTM B210-68 (which is incorporated by reference as specified in §1910.6) and shall be suitably marked every 18 inches indicating compliance with ASTM Specifications. The minimum nominal wall thickness of copper tubing and aluminum alloy tubing shall be as specified in Table H-24 and Table H-25.

Aluminum alloy tubing shall be protected against external corrosion when it is in contact with dissimilar metals other than galvanized steel, or its location is subject to repeated wetting by liquids such as water (except rainwater), detergents, sewage, or leakage from other piping, or it passes through flooring, plaster, masonry, or insulation. Galvanized sheet steel or pipe, galvanized inside and out, may be considered suitable protection. The maximum outside diameter for aluminum alloy tubing shall be three-fourths inch and shall not be used for pressures exceeding 20 p.s.i.g. Aluminum alloy tubing shall not be installed within 6 inches of the ground.

### TABLE H-24—WALL THICKNESS OF COPPER TUBING[1]

| Standard size (inches) | Nominal outside diameter (inches) | Nominal wall thickness (inches) Type K | Nominal wall thickness (inches) Type L |
|---|---|---|---|
| 1/4 | 0.375 | 0.035 | 0.030 |
| 3/8 | 0.500 | 0.049 | 0.035 |
| 1/2 | 0.625 | 0.049 | 0.040 |
| 5/8 | 0.750 | 0.049 | 0.042 |
| 3/4 | 0.875 | 0.065 | 0.045 |
| 1 | 1.125 | 0.065 | 0.050 |
| 1 1/4 | 1.375 | 0.065 | 0.055 |
| 1 1/2 | 1.625 | 0.072 | 0.060 |
| 2 | 2.125 | 0.083 | 0.070 |

[1] Based on data in Specification for Seamless Copper Water Tube, ANSI H23.1-1970 (ASTM B-88-69).

NOTE: The standard size by which tube is designated is 1/8-inch smaller than its nominal outside diameter.

### TABLE H-25—WALL THICKNESS OF ALUMINUM ALLOY TUBING[1]

| Outside diameter (inches) | Nominal wall thickness (inches) Type A | Nominal wall thickness (inches) Type B |
|---|---|---|
| 3/8 | 0.035 | 0.049 |
| 1/2 | 0.035 | 0.049 |
| 5/8 | 0.042 | 0.049 |
| 3/4 | 0.049 | 0.058 |

[1] Based on data in Standard Specification for Aluminum-Alloy Drawn Seamless Coiled Tubes for Special Purpose Applications, ASTM B210-68.

**1910.110(b)(8)(iii)** In systems where the gas in liquid form without pressure reduction enters the building, only heavy walled seamless brass or copper tubing with an internal diameter not greater than three thirty-seconds inch, and a wall thickness of not less than three sixty-fourths inch shall be used. This requirement shall not apply to research and experimental laboratories, buildings, or separate fire divisions of buildings used exclusively for housing internal combustion engines, and to commercial gas plants or bulk stations where containers are charged, nor to industrial vaporizer buildings, nor to buildings, structures, or equipment under construction or undergoing major renovation.

**1910.110(b)(8)(iv)** Pipe joints may be screwed, flanged, welded, soldered, or brazed with a material having a melting point exceeding 1,000 °F. Joints on seamless copper, brass, steel, or aluminum alloy gas tubing shall be made by means of approved gas tubing fittings, or soldered or brazed with a material having a melting point exceeding 1,000 °F.

**1910.110(b)(8)(v)** For operating pressures of 125 p.s.i.g. or less, fittings shall be designed for a pressure of at least 125 p.s.i.g. For operating pressures above 125 p.s.i.g., fittings shall be designed for a minimum of 250 p.s.i.g.

**1910.110(b)(8)(vi)** The use of threaded cast iron pipe fittings such as ells, tees, crosses, couplings, and unions is prohibited. Aluminum alloy fittings shall be used with aluminum alloy pipe and tubing. Insulated fittings shall be used where aluminum alloy pipe or tubing connects with a dissimilar metal.

**1910.110(b)(8)(vii)** Strainers, regulators, meters, compressors, pumps, etc., are not to be considered as pipe fittings. This does not prohibit the use of malleable, nodular, or higher strength gray iron for such equipment.

**1910.110(b)(8)(viii)** All materials such as valve seats, packing, gaskets, diaphragms, etc., shall be of such quality as to be resistant to the action of liquefied petroleum gas under the service conditions to which they are subjected.

**1910.110(b)(8)(ix)** All piping, tubing, or hose shall be tested after assembly and proved free from leaks at not less than normal operating pressures. After installation, piping and tubing of all domestic and commercial systems shall be tested and proved free of leaks using a manometer or equivalent device that will indicate a drop in pressure. Test shall not be made with a flame.

**1910.110(b)(8)(x)** Provision shall be made to compensate for expansion, contraction, jarring, and vibration, and for settling. This may be accomplished by flexible connections.

**1910.110(b)(8)(xi)** Piping outside buildings may be buried, above ground, or both, but shall be well supported and protected against physical damage. Where soil conditions warrant, all piping shall be protected against corrosion. Where condensation may occur, the piping shall be pitched back to the container, or suitable means shall be provided for revaporization of the condensate.

**1910.110(b)(9)** *Hose specifications.* **(i)** Hose shall be fabricated of materials that are resistant to the action of LP-Gas in the liquid and vapor phases. If wire braid is used for reinforcing the hose, it shall be of corrosion-resistant material such as stainless steel.

**1910.110(b)(9)(ii)** Hose subject to container pressure shall be marked "LP-Gas" or "LPG" at not greater than 10-foot intervals.

**1910.110(b)(9)(iii)** Hose subject to container pressure shall be designed for a bursting pressure of not less than 1,250 p.s.i.g.

**1910.110(b)(9)(iv)** Hose subject to container pressure shall have its correctness as to design construction and performance determined by being listed (see §1910.110(a)(15)).

**1910.110(b)(9)(v)** Hose connections subject to container pressure shall be capable of withstanding, without leakage, a test pressure of not less than 500 p.s.i.g.

**1910.110(b)(9)(vi)** Hose and hose connections on the low-pressure side of the regulator or reducing valve shall be designed for a bursting pressure of not less than 125 p.s.i.g. or five times the set pressure of the relief devices protecting that portion of the system, whichever is higher.

**1910.110(b)(9)(vii)** Hose may be used on the low-pressure side of regulators to connect to other than domestic and commercial gas appliances under the following conditions:

**1910.110(b)(9)(vii)(a)** The appliances connected with hose shall be portable and need a flexible connection.

**1910.110(b)(9)(vii)(b)** For use inside buildings the hose shall be of minimum practical length, but shall not exceed 6 feet except as provided in paragraph (c)(5)(i)(g) of this section and shall not extend from one room to another, nor pass through any walls, partitions, ceilings, or floors. Such hose shall not be concealed from view or used in a concealed location. For use outside of buildings, the hose may exceed this length but shall be kept as short as practical.

**1910.110(b)(9)(vii)(c)** The hose shall be approved and shall not be used where it is likely to be subjected to temperatures above 125 °F. The hose shall be securely connected to the appliance and the use of rubber slip ends shall not be permitted.

**1910.110(b)(9)(vii)(d)** The shutoff valve for an appliance connected by hose shall be in the metal pipe or tubing and not at the appliance end of the hose. When shutoff valves are installed close to each other, precautions shall be taken to prevent operation of the wrong valve.

**1910.110(b)(9)(vii)(e)** Hose used for connecting to wall outlets shall be protected from physical damage.

**1910.110(b)(10)** *Safety devices.* **(i)** Every container except those constructed in accordance with DOT specifications and every vaporizer (except motor fuel vaporizers and except vaporizers described in paragraph (b)(11)(ii)(c) of this section and paragraph (d)(4)(v)(a) of this section) whether heated by artificial means or not, shall be provided with one or more safety relief valves of spring-loaded or equivalent type. These valves shall be arranged to afford free vent to the outer air with discharge not less than 5 feet horizontally away from any opening into the building which is below such discharge. The rate of discharge shall be in accordance with the requirements of paragraph (b)(10)(ii) or (b)(10)(iii) of this section in the case of vaporizers.

**1910.110(b)(10)(ii)** Minimum required rate of discharge in cubic feet per minute of air at 120 percent of the maximum permitted start to discharge pressure for safety relief valves to be used on containers other than those constructed in accordance with DOT specification shall be as follows:

| Surface area (sq. ft.) | Flow rate CFM air |
|---|---|
| 20 or less | 626 |
| 25 | 751 |
| 30 | 872 |
| 35 | 990 |
| 40 | 1,100 |
| 45 | 1,220 |
| 50 | 1,330 |
| 55 | 1,430 |
| 60 | 1,540 |
| 65 | 1,640 |
| 70 | 1,750 |
| 75 | 1,850 |
| 80 | 1,950 |
| 85 | 2,050 |
| 90 | 2,150 |
| 95 | 2,240 |
| 100 | 2,340 |
| 105 | 2,440 |
| 110 | 2,530 |
| 115 | 2,630 |
| 120 | 2,720 |
| 125 | 2,810 |
| 130 | 2,900 |
| 135 | 2,990 |
| 140 | 3,080 |
| 145 | 3,170 |
| 150 | 3,260 |
| 155 | 3,350 |
| 160 | 3,440 |
| 165 | 3,530 |
| 170 | 3,620 |
| 175 | 3,700 |
| 180 | 3,790 |
| 185 | 3,880 |
| 190 | 3,960 |
| 195 | 4,050 |
| 200 | 4,130 |
| 210 | 4,300 |
| 220 | 4,470 |
| 230 | 4,630 |
| 240 | 4,800 |
| 250 | 4,960 |
| 260 | 5,130 |
| 270 | 5,290 |
| 280 | 5,450 |
| 290 | 5,610 |
| 300 | 5,760 |
| 310 | 5,920 |
| 320 | 6,080 |
| 330 | 6,230 |
| 340 | 6,390 |
| 350 | 6,540 |
| 360 | 6,690 |
| 370 | 6,840 |
| 380 | 7,000 |
| 390 | 7,150 |
| 400 | 7,300 |
| 450 | 8,040 |
| 500 | 8,760 |
| 550 | 9,470 |
| 600 | 10,170 |
| 650 | 10,860 |

# OSHA Standards for General Industry

| Surface area (sq. ft.) | Flow rate CFM air |
|---|---|
| 700 | 11,550 |
| 750 | 12,220 |
| 800 | 12,880 |
| 850 | 13,540 |
| 900 | 14,190 |
| 950 | 14,830 |
| 1,000 | 15,470 |
| 1,050 | 16,100 |
| 1,100 | 16,720 |
| 1,150 | 17,350 |
| 1,200 | 17,960 |
| 1,250 | 18,570 |
| 1,300 | 19,180 |
| 1,350 | 19,780 |
| 1,400 | 20,380 |
| 1,450 | 20,980 |
| 1,500 | 21,570 |
| 1,550 | 22,160 |
| 1,600 | 22,740 |
| 1,650 | 23,320 |
| 1,700 | 23,900 |
| 1,750 | 24,470 |
| 1,800 | 25,050 |
| 1,850 | 25,620 |
| 1,900 | 26,180 |
| 1,950 | 26,750 |
| 2,000 | 27,310 |

Surface area = total outside surface area of container in square feet.

When the surface area is not stamped on the nameplate or when the marking is not legible, the area can be calculated by using one of the following formulas:

(1) Cylindrical container with hemispherical heads:

**Area = Overall length × outside diameter × 3.1416.**

(2) Cylindrical container with other than hemispherical heads:

**Area = (Overall length + 0.3 outside diameter) × outside diameter × 3.1416.**

**NOTE:** This formula is not exact, but will give results within the limits of practical accuracy for the sole purpose of sizing relief valves.

(3) Spherical container:

**Area = Outside diameter squared × 3.1416.**

Flow Rate-CFM Air = Required flow capacity in cubic feet per minute of air at standard conditions, 60 F. and atmospheric pressure (14.7 p.s.i.a.).

The rate of discharge may be interpolated for intermediate values of surface area. For containers with total outside surface area greater than 2,000 square feet, the required flow rate can be calculated using the formula, Flow Rate-CFM Air=53.632 A0.82.

**A = total outside surface area of the container in square feet.**

Valves not marked "Air" have flow rate marking in cubic feet per minute of liquefied petroleum gas. These can be converted to ratings in cubic feet per minute of air by multiplying the liquefied petroleum gas ratings by factors listed below. Air flow ratings can be converted to ratings in cubic feet per minute of liquefied petroleum gas by dividing the air ratings by the factors listed below.

### AIR CONVERSION FACTORS

| Container type | 100 | 125 | 150 | 175 | 200 |
|---|---|---|---|---|---|
| Air conversion factor | 1.162 | 1.142 | 1.113 | 1.078 | 1.010 |

**1910.110(b)(10)(iii)** Minimum Required Rate of Discharge for Safety Relief Valves for Liquefied Petroleum Gas Vaporizers (Steam Heated, Water Heated, and Direct Fired).

The minimum required rate of discharge for safety relief valves shall be determined as follows:

**1910.110(b)(10)(iii)***(a)* Obtain the total surface area by adding the surface area of vaporizer shell in square feet directly in contact with LP-Gas and the heat exchanged surface area in square feet directly in contact with LP-Gas.

**1910.110(b)(10)(iii)***(b)* Obtain the minimum required rate of discharge in cubic feet of air per minute, at 60 °F. and 14.7 p.s.i.a. from paragraph (b)(10)(ii) of this section, for this total surface area.

**1910.110(b)(10)(iv)** Container and vaporizer safety relief valves shall be set to start-to-discharge, with relation to the design pressure of the container, in accordance with Table H-26.

**1910.110(b)(10)(v)** Safety relief devices used with systems employing containers other than those constructed according to DOT specifications shall be so constructed as to discharge at not less than the rates shown in paragraph (b)(10)(ii) of this section, before the pressure is in excess of 120 percent of the maximum (not including the 10 percent referred to in paragraph (b)(10)(iv) of this section) permitted start to discharge pressure setting of the device.

### TABLE H-26

| Containers | Minimum (percent) | Maximum (percent) |
|---|---|---|
| ASME Code; Par. U-68, U-69—1949 and earlier editions | 110 | [1]25 |
| ASME Code; Par. U-200, U-201—1949 edition | 88 | [1]100 |
| ASME Code—1950, 1952, 1956, 1959, 1962, 1965 and 1968 (Division I) editions | 88 | [1]100 |
| API—ASME Code—all editions | 88 | [1]100 |
| DOT—As prescribed in 49 CFR Chapter I | | |

[1] Manufacturers of safety relief valves are allowed a plus tolerance not exceeding 10 percent of the set pressure marked on the valve.

**1910.110(b)(10)(vi)** In certain locations sufficiently sustained high temperatures prevail which require the use of a lower vapor pressure product to be stored or the use of a higher designed pressure vessel in order to prevent the safety valves opening as the result of these temperatures. As an alternative the tanks may be protected by cooling devices such as by spraying, by shading, or other effective means.

**1910.110(b)(10)(vii)** Safety relief valves shall be arranged so that the possibility of tampering will be minimized. If pressure setting or adjustment is external, the relief valves shall be provided with approved means for sealing adjustment.

**1910.110(b)(10)(viii)** Shutoff valves shall not be installed between the safety relief devices and the container, or the equipment or piping to which the safety relief device is connected except that a shutoff valve may be used where the arrangement of this valve is such that full required capacity flow through the safety relief device is always afforded.

**1910.110(b)(10)(ix)** Safety relief valves shall have direct communication with the vapor space of the container at all times.

**1910.110(b)(10)(x)** Each container safety relief valve used with systems covered by paragraphs (d), (e), (g), and (h) of this section, except as provided in paragraph (e)(3)(iii) of this section shall be plainly and permanently marked with the following: "Container Type" of the pressure vessel on which the valve is designed to be installed; the pressure in p.s.i.g. at which the valve is set to discharge; the actual rate of discharge of the valve in cubic feet per minute of air at 60 °F. and 14.7 p.s.i.a.; and the manufacturer's name and catalog number, for example: T200-250-4050 AIR—indicating that the valve is suitable for use on a Type 200 container, that it is set to start to discharge at 250 p.s.i.g.; and that its rate of discharge is 4,050 cubic feet per minute of air as determined in subdivision (ii) of this subparagraph.

**1910.110(b)(10)(xi)** Safety relief valve assemblies, including their connections, shall be of sufficient size so as to provide the rate of flow required for the container on which they are installed.

**1910.110(b)(10)(xii)** A hydrostatic relief valve shall be installed between each pair of shut-off valves on liquefied petroleum gas liquid piping so as to relieve into a safe atmosphere. The start-to-discharge pressure setting of such relief valves shall not be in excess of 500 p.s.i.g. The minimum setting on relief valves installed in piping connected to other than DOT containers shall not be lower than 140 percent of the container relief valve setting and in piping connected to DOT containers not lower than

400 p.s.i.g. The start-to-discharge pressure setting of such a relief valve, if installed on the discharge side of a pump, shall be greater than the maximum pressure permitted by the recirculation device in the system.

**1910.110(b)(10)(xiii)** The discharge from any safety relief device shall not terminate in or beneath any building, except relief devices covered by paragraphs (b)(6)(i) (*a*) through (*e*) of this section, or paragraphs (c) (4)(i) or (5) of this section.

**1910.110(b)(10)(xiv)** Container safety relief devices and regulator relief vents shall be located not less than five (5) feet in any direction from air openings into sealed combustion system appliances or mechanical ventilation air intakes.

**1910.110(b)(11)** *Vaporizer and housing.* **(i)** Indirect fired vaporizers utilizing steam, water, or other heating medium shall be constructed and installed as follows:

**1910.110(b)(11)(i)***(a)* Vaporizers shall be constructed in accordance with the requirements of paragraph (b)(3) (i)-(iii) of this section and shall be permanently marked as follows:

**1910.110(b)(11)(i)***(a)(1)* With the code marking signifying the specifications to which the vaporizer is constructed.

**1910.110(b)(11)(i)***(a)(2)* With the allowable working pressure and temperature for which the vaporizer is designed.

**1910.110(b)(11)(i)***(a)(3)* With the sum of the outside surface area and the inside heat exchange surface area expressed in square feet.

**1910.110(b)(11)(i)***(a)(4)* With the name or symbol of the manufacturer.

**1910.110(b)(11)(i)***(b)* Vaporizers having an inside diameter of 6 inches or less exempted by the ASME Unfired Pressure Vessel Code, Section VIII of the ASME Boiler and Pressure Vessel Code—1968 shall have a design pressure not less than 250 p.s.i.g. and need not be permanently marked.

**1910.110(b)(11)(i)***(c)* Heating or cooling coils shall not be installed inside a storage container.

**1910.110(b)(11)(i)***(d)* Vaporizers may be installed in buildings, rooms, sheds, or lean-tos used exclusively for gas manufacturing or distribution, or in other structures of light, noncombustible construction or equivalent, well ventilated near the floor line and roof.

When vaporizing and/or mixing equipment is located in a structure or building not used exclusively for gas manufacturing or distribution, either attached to or within a building, such structure or room shall be separated from the remainder of the building by a wall designed to withstand a static pressure of at least 100 pounds per square foot. This wall shall have no openings or pipe or conduit passing through it. Such structure or room shall be provided with adequate ventilation and shall have a roof or at least one exterior wall of lightweight construction.

**1910.110(b)(11)(i)***(e)* Vaporizers shall have, at or near the discharge, a safety relief valve providing an effective rate of discharge in accordance with paragraph (b)(10)(iii) of this section, except as provided in paragraph (d)(4)(v)(*a*), of this section.

**1910.110(b)(11)(i)***(f)* The heating medium lines into and leaving the vaporizer shall be provided with suitable means for preventing the flow of gas into the heat systems in the event of tube rupture in the vaporizer. Vaporizers shall be provided with suitable automatic means to prevent liquid passing through the vaporizers to the gas discharge piping.

**1910.110(b)(11)(i)***(g)* The device that supplies the necessary heat for producing steam, hot water, or other heating medium may be installed in a building, compartment, room, or lean-to which shall be ventilated near the floorline and roof to the outside. The device location shall be separated from all compartments or rooms containing liquefied petroleum gas vaporizers, pumps, and central gas mixing devices by a wall designed to withstand a static pressure of at least 100 pounds per square foot. This wall shall have no openings or pipes or conduit passing through it. This requirement does not apply to the domestic water heaters which may supply heat for a vaporizer in a domestic system.

**1910.110(b)(11)(i)***(h)* Gas-fired heating systems supplying heat exclusively for vaporization purposes shall be equipped with automatic safety devices to shut off the flow of gas to main burners, if the pilot light should fail.

**1910.110(b)(11)(i)***(i)* Vaporizers may be an integral part of a fuel storage container directly connected to the liquid section or gas section or both.

**1910.110(b)(11)(i)***(j)* Vaporizers shall not be equipped with fusible plugs.

**1910.110(b)(11)(i)***(k)* Vaporizer houses shall not have unprotected drains to sewers or sump pits.

**1910.110(b)(11)(ii)** Atmospheric vaporizers employing heat from the ground or surrounding air shall be installed as follows:

**1910.110(b)(11)(ii)***(a)* Buried underground, or

**1910.110(b)(11)(ii)***(b)* Located inside the building close to a point at which pipe enters the building provided the capacity of the unit does not exceed 1 quart.

**1910.110(b)(11)(ii)***(c)* Vaporizers of less than 1 quart capacity heated by the ground or surrounding air, need not be equipped with safety relief valves provided that adequate tests demonstrate that the assembly is safe without safety relief valves.

**1910.110(b)(11)(iii)** Direct gas-fired vaporizers shall be constructed, marked, and installed as follows:

**1910.110(b)(11)(iii)***(a)(1)* In accordance with the requirements of the American Society of Mechanical Engineers Boiler and Pressure Vessel Code—1968 that are applicable to the maximum working conditions for which the vaporizer is designed.

**1910.110(b)(11)(iii)***(a)(2)* With the name of the manufacturer; rated BTU input to the burner; the area of the heat exchange surface in square feet; the outside surface of the vaporizer in square feet; and the maximum vaporizing capacity in gallons per hour.

**1910.110(b)(11)(iii)***(b)(1)* Vaporizers may be connected to the liquid section or the gas section of the storage container, or both; but in any case there shall be at the container a manually operated valve in each connection to permit completely shutting off when desired, of all flow of gas or liquid from container to vaporizer.

**1910.110(b)(11)(iii)***(b)(2)* Vaporizers with capacity not exceeding 35 gallons per hour shall be located at least 5 feet from container shutoff valves. Vaporizers having capacity of more than 35 gallons but not exceeding 100 gallons per hour shall be located at least 10 feet from the container shutoff valves. Vaporizers having a capacity greater than 100 gallons per hour shall be located at least 15 feet from container shutoff valves.

**1910.110(b)(11)(iii)***(c)* Vaporizers may be installed in buildings, rooms, housings, sheds, or lean-tos used exclusively for vaporizing or mixing of liquefied petroleum gas. Vaporizing housing structures shall be of noncombustible construction, well ventilated near the floorline and the highest point of the roof. When vaporizer and/or mixing equipment is located in a structure or room attached to or within a building, such structure or room shall be separated from the remainder of the building by a wall designed to withstand a static pressure of at least 100 pounds per square foot. This wall shall have no openings or pipes or conduit passing through it. Such structure or room shall be provided with adequate ventilation, and shall have a roof or at least one exterior wall of lightweight construction.

**1910.110(b)(11)(iii)***(d)* Vaporizers shall have at or near the discharge, a safety relief valve providing an effective rate of discharge in accordance with paragraph (b)(10)(iii) of this section. The relief valve shall be so located as not to be subjected to temperatures in excess of 140 °F.

**1910.110(b)(11)(iii)***(e)* Vaporizers shall be provided with suitable automatic means to prevent liquid passing from the vaporizer to the gas discharge piping of the vaporizer.

**1910.110(b)(11)(iii)***(f)* Vaporizers shall be provided with means for manually turning off the gas to the main burner and pilot.

**1910.110(b)(11)(iii)***(g)* Vaporizers shall be equipped with automatic safety devices to shut off the flow of gas to main burners if the pilot light should fail. When the flow through the pilot exceeds 2,000 B.t.u. per hour, the pilot also shall be equipped with an automatic safety device to shut off the flow of gas to the pilot should the pilot flame be extinguished.

**1910.110(b)(11)(iii)***(h)* Pressure regulating and pressure reducing equipment if located within 10 feet of a direct fire vaporizer shall be separated from the open flame by a substantially airtight noncombustible partition or partitions.

**1910.110(b)(11)(iii)***(i)* Except as provided in (*c*) of this subdivision, the following minimum distances shall be maintained between direct fired vaporizers and the nearest important building or group of buildings:

Ten feet for vaporizers having a capacity of 15 gallons per hour or less vaporizing capacity.

Twenty-five feet for vaporizers having a vaporizing capacity of 16 to 100 gallons per hour.

Fifty feet for vaporizers having a vaporizing capacity exceeding 100 gallons per hour.

**1910.110(b)(11)(iii)(j)** Direct fired vaporizers shall not raise the product pressure above the design pressure of the vaporizer equipment nor shall they raise the product pressure within the storage container above the pressure shown in the second column of Table H-31.

**1910.110(b)(11)(iii)(k)** Vaporizers shall not be provided with fusible plugs.

**1910.110(b)(11)(iii)(l)** Vaporizers shall not have unprotected drains to sewers or sump pits.

**1910.110(b)(11)(iv)** Direct gas-fired tank heaters shall be constructed and installed as follows:

**1910.110(b)(11)(iv)(a)** Direct gas-fired tank heaters, and tanks to which they are applied, shall only be installed above ground.

**1910.110(b)(11)(iv)(b)** Tank heaters shall be permanently marked with the name of the manufacturer, the rated B.t.u. input to the burner, and the maximum vaporizing capacity in gallons per hour.

**1910.110(b)(11)(iv)(c)** Tank heaters may be an integral part of a fuel storage container directly connected to the container liquid section, or vapor section, or both.

**1910.110(b)(11)(iv)(d)** Tank heaters shall be provided with a means for manually turning off the gas to the main burner and pilot.

**1910.110(b)(11)(iv)(e)** Tank heaters shall be equipped with an automatic safety device to shut off the flow of gas to main burners, if the pilot light should fail. When flow through pilot exceeds 2,000 B.t.u. per hour, the pilot also shall be equipped with an automatic safety device to shut off the flow of gas to the pilot should the pilot flame be extinguished.

**1910.110(b)(11)(iv)(f)** Pressure regulating and pressure reducing equipment if located within 10 feet of a direct fired tank heater shall be separated from the open flame by a substantially airtight noncombustible partition.

**1910.110(b)(11)(iv)(g)** The following minimum distances shall be maintained between a storage tank heated by a direct fired tank heater and the nearest important building or group of buildings:

Ten feet for storage containers of less than 500 gallons water capacity.

Twenty-five feet for storage containers of 500 to 1,200 gallons water capacity.

Fifty feet for storage containers of over 1,200 gallons water capacity.

**1910.110(b)(11)(iv)(h)** No direct fired tank heater shall raise the product pressure within the storage container over 75 percent of the pressure set out in the second column of Table H-31.

**1910.110(b)(11)(v)** The vaporizer section of vaporizer-burners used for dehydrators or dryers shall be located outside of buildings; they shall be constructed and installed as follows:

**1910.110(b)(11)(v)(a)** Vaporizer-burners shall have a minimum design pressure of 250 p.s.i.g. with a factor of safety of five.

**1910.110(b)(11)(v)(b)** Manually operated positive shut-off valves shall be located at the containers to shut off all flow to the vaporizer-burners.

**1910.110(b)(11)(v)(c)** Minimum distances between storage containers and vaporizer-burners shall be as follows:

| Water capacity per container (gallons) | Minimum distances (feet) |
|---|---|
| Less than 501 | 10 |
| 501 to 2,000 | 25 |
| Over 2,000 | 50 |

**1910.110(b)(11)(v)(d)** The vaporizer section of vaporizer-burners shall be protected by a hydrostatic relief valve. The relief valve shall be located so as not to be subjected to temperatures in excess of 140 °F. The start-to-discharge pressure setting shall be such as to protect the components involved, but not less than 250 p.s.i.g. The discharge shall be directed upward and away from component parts of the equipment and away from operating personnel.

**1910.110(b)(11)(v)(e)** Vaporizer-burners shall be provided with means for manually turning off the gas to the main burner and pilot.

**1910.110(b)(11)(v)(f)** Vaporizer-burners shall be equipped with automatic safety devices to shut off the flow of gas to the main burner and pilot in the event the pilot is extinguished.

**1910.110(b)(11)(v)(g)** Pressure regulating and control equipment shall be located or protected so that the temperatures surrounding this equipment shall not exceed 140 °F. except that equipment components may be used at higher temperatures if designed to withstand such temperatures.

**1910.110(b)(11)(v)(h)** Pressure regulating and control equipment when located downstream of the vaporizer shall be designed to withstand the maximum discharge temperature of the vapor.

**1910.110(b)(11)(v)(i)** The vaporizer section of vaporizer-burners shall not be provided with fusible plugs.

**1910.110(b)(11)(v)(j)** Vaporizer coils or jackets shall be made of ferrous metal or high temperature alloys.

**1910.110(b)(11)(v)(k)** Equipment utilizing vaporizer-burners shall be equipped with automatic shutoff devices upstream and downstream of the vaporizer section connected so as to operate in the event of excessive temperature, flame failure, and, if applicable, insufficient airflow.

**1910.110(b)(12)** *Filling densities.* **(i)** The "filling density" is defined as the percent ratio of the weight of the gas in a container to the weight of water the container will hold at 60 °F. All containers shall be filled according to the filling densities shown in Table H-27.

### TABLE H-27—MAXIMUM PERMITTED FILLING DENSITY

| Specific gravity at 60 °F (15.6 °C) | Above ground containers 0 to 1,200 U.S.Gals. (1,000 imp. gal., 4,550 liters) total water cap. | Above ground containers Over 1,200 U.S. Gals. (1,000 imp. gal., 4,550 liters) total water cap. | Underground containers, all capacities |
|---|---|---|---|
| | Percent | Percent | Percent |
| 0.496-0.503 | 41 | 44 | 45 |
| .504-.510 | 42 | 45 | 46 |
| .511-.519 | 43 | 46 | 47 |
| .520-.527 | 44 | 47 | 48 |
| .528-.536 | 45 | 48 | 49 |
| .537-.544 | 46 | 49 | 50 |
| .545-.552 | 47 | 50 | 51 |
| .553-.560 | 48 | 51 | 52 |
| .561-.568 | 49 | 52 | 53 |
| .569-.576 | 50 | 53 | 54 |
| .577-.584 | 51 | 54 | 55 |
| .585-.592 | 52 | 55 | 56 |
| .593-.600 | 53 | 56 | 57 |

**1910.110(b)(12)(ii)** Except as provided in paragraph (b)(12)(iii) of this section, any container including mobile cargo tanks and portable tank containers regardless of size or construction, shipped under DOT jurisdiction or constructed in accordance with 49 CFR chapter I Specifications shall be charged according to 49 CFR chapter I requirements.

**1910.110(b)(12)(iii)** Portable containers not subject to DOT jurisdiction (such as, but not limited to, motor fuel containers on industrial and lift trucks, and farm tractors covered in paragraph (e) of this section, or containers recharged at the installation) may be filled either by weight, or by volume using a fixed length dip tube gaging device.

**1910.110(b)(13)** *LP-Gas in buildings.* **(i)** Vapor shall be piped into buildings at pressures in excess of 20 p.s.i.g. only if the buildings or separate areas thereof, (*a*) are constructed in accordance with this section; (*b*) are used excusively to house equipment for vaporization, pressure reduction, gas mixing, gas manufacturing, or distribution, or to house internal combustion engines, industrial processes, research and experimental laboratories, or equipment and processes using such gas and having similar hazard; (*c*) buildings, structures, or equipment under construction or undergoing major renovation.

**1910.110(b)(13)(ii)** Liquid may be permitted in buildings as follows:

**1910.110(b)(13)(ii)(a)** Buildings, or separate areas of buildings, used exclusively to house equipment for vaporization, pressure reduction, gas mixing, gas manufacturing, or distribution, or to house internal combustion engines, industrial processes, research and experimental laboratories, or equipment and processes using such gas and having similar hazard; and when such buildings, or separate areas thereof are constructed in accordance with this section.

**1910.110(b)(13)(ii)(b)** Buildings, structures, or equipment under construction or undergoing major renovation provided the temporary piping meets the following conditions:

**1910.110(b)(13)(ii)(b)(1)** Liquid piping inside the building shall conform to the requirements of paragraph (b)(8) of this section, and shall not exceed three-fourths iron pipe size. Copper tubing with an outside diameter of three-fourths inch or less may be used provided it conforms to Type K of Specifications for Seamless Water Tube, ANSI H23.1-1970 (ASTM B88-69) (see Table H-24). All such piping shall be protected against construction hazards. Liquid piping inside buildings shall be kept to a minimum. Such piping shall be securely fastened to walls or other surfaces so as to provide adequate protection from breakage and so located as to subject the liquid line to lowest ambient temperatures.

**1910.110(b)(13)(ii)(b)(2)** A shutoff valve shall be installed in each intermediate branch line where it takes off the main line and shall be readily accessible. A shutoff valve shall also be placed at the appliance end of the intermediate branch line. Such shutoff valve shall be upstream of any flexible connector used with the appliance.

**1910.110(b)(13)(ii)(b)(3)** Suitable excess flow valves shall be installed in the container outlet line supplying liquid LP-Gas to the building. A suitable excess flow valve shall be installed immediately downstream of each shutoff valve. Suitable excess flow valves shall be installed where piping size is reduced and shall be sized for the reduced size piping.

**1910.110(b)(13)(ii)(b)(4)** Hydrostatic relief valves shall be installed in accordance with paragraph (b)(10)(xii) of this section.

**1910.110(b)(13)(ii)(b)(5)** The use of hose to carry liquid between the container and the building or at any point in the liquid line, except at the appliance connector, shall be prohibited.

**1910.110(b)(13)(ii)(b)(6)** Where flexible connectors are necessary for appliance installation, such connectors shall be as short as practicable and shall comply with paragraph (b)(8)(ii) or (9) of this section.

**1910.110(b)(13)(ii)(b)(7)** Release of fuel when any section of piping or appliances is disconnected shall be minimized by either of the following methods:

**1910.110(b)(13)(ii)(b)(7)(i)** Using an approved automatic quick-closing coupling (a type closing in both directions when coupled in the fuel line), or

**1910.110(b)(13)(ii)(b)(7)(ii)** Closing the valve nearest to the appliance and allowing the appliance to operate until the fuel in the line is consumed.

**1910.110(b)(13)(ii)(b)(7)(iii)** Portable containers shall not be taken into buildings except as provided in paragraph (b)(6)(i) of this section.

**1910.110(b)(14)** *Transfer of liquids.* The employer shall assure that (i) at least one attendant shall remain close to the transfer connection from the time the connections are first made until they are finally disconnected, during the transfer of the product.

**1910.110(b)(14)(ii)** Containers shall be filled or used only upon authorization of the owner.

**1910.110(b)(14)(iii)** Containers manufactured in accordance with specifications of 49 CFR part 178 and authorized by 49 CFR chapter 1 as a "single trip" or "nonrefillable container" shall not be refilled or reused in LP-Gas service.

**1910.110(b)(14)(iv)** Gas or liquid shall not be vented to the atmosphere to assist in transferring contents of one container to another, except as provided in paragraph (e)(5)(iv) of this section and except that this shall not preclude the use of listed pump utilizing LP-Gas in the vapor phase as a source of energy and venting such gas to the atmosphere at a rate not to exceed that from a No. 31 drill size opening and provided that such venting and liquid transfer shall be located not less than 50 feet from the nearest important building.

**1910.110(b)(14)(v)** Filling of fuel containers for industrial trucks or motor vehicles from industrial bulk storage containers shall be performed not less than 10 feet from the nearest important masonry-walled building or not less than 25 feet from the nearest important building or other construction and, in any event, not less than 25 feet from any building opening.

**1910.110(b)(14)(vi)** Filling of portable containers, containers mounted on skids, fuel containers on farm tractors, or similar applications, from storage containers used in domestic or commercial service, shall be performed not less than 50 feet from the nearest important building.

**1910.110(b)(14)(vii)** The filling connection and the vent from the liquid level gages in containers, filled at point of installation, shall not be less than 10 feet in any direction from air openings into sealed combustion system appliances or mechanical ventilation air intakes.

**1910.110(b)(14)(viii)** Fuel supply containers shall be gaged and charged only in the open air or in buildings especially provided for that purpose.

**1910.110(b)(14)(ix)** The maximum vapor pressure of the product at 100 °F. which may be transferred into a container shall be in accordance with paragraphs (d)(2) and (e)(3) of this section. (For DOT containers use DOT requirements.)

**1910.110(b)(14)(x)** Marketers and users shall exercise precaution to assure that only those gases for which the system is designed, examined, and listed, are employed in its operation, particularly with regard to pressures.

**1910.110(b)(14)(xi)** Pumps or compressors shall be designed for use with LP-Gas. When compressors are used they shall normally take suction from the vapor space of the container being filled and discharge to the vapor space of the container being emptied.

**1910.110(b)(14)(xii)** Pumping systems, when equipped with a positive displacement pump, shall include a recirculating device which shall limit the differential pressure on the pump under normal operating conditions to the maximum differential pressure rating of the pump. The discharge of the pumping system shall be protected so that pressure does not exceed 350 p.s.i.g. If a recirculation system discharges into the supply tank and contains a manual shutoff valve, an adequate secondary safety recirculation system shall be incorporated which shall have no means of rendering it inoperative. Manual shutoff valves in recirculation systems shall be kept open except during an emergency or when repairs are being made to the system.

**1910.110(b)(14)(xiii)** When necessary, unloading piping or hoses shall be provided with suitable bleeder valves for relieving pressure before disconnection.

**1910.110(b)(14)(xiv)** Agricultural air moving equipment, including crop dryers, shall be shut down when supply containers are being filled unless the air intakes and sources of ignition on the equipment are located 50 feet or more from the container.

**1910.110(b)(14)(xv)** Agricultural equipment employing open flames or equipment with integral containers, such as flame cultivators, weed burners, and, in addition, tractors, shall be shut down during refueling.

**1910.110(b)(15)** *Tank car or transport truck loading or unloading points and operations.* (i) The track of tank car siding shall be relatively level.

**1910.110(b)(15)(ii)** A "Tank Car Connected" sign, as covered by DOT rules, shall be installed at the active end or ends of the siding while the tank car is connected.

**1910.110(b)(15)(iii)** While cars are on sidetrack for loading or unloading, the wheels at both ends shall be blocked on the rails.

**1910.110(b)(15)(iv)** The employer shall insure that an employee is in attendance at all times while the tank car, cars, or trucks are being loaded or unloaded.

**1910.110(b)(15)(v)** A backflow check valve, excess-flow valve, or a shutoff valve with means of remote closing, to protect against uncontrolled discharge of LP-Gas from storage tank piping shall be installed close to the point where the liquid piping and hose or swing joint pipe is connected.

**1910.110(b)(15)(vi)** Where practical, the distance of the unloading or loading point shall conform to the distances in subparagraph (6)(ii) of this paragraph.

**1910.110(b)(16)** *Instructions.* Personnel performing installation, removal, operation, and maintenance work shall be properly trained in such function.

**1910.110(b)(17)** *Electrical equipment and other sources of ignition.* (i) Electrical equipment and wiring shall be of a type specified by and shall be installed in accordance with subpart S of this part, for ordinary locations except that fixed electrical equipment in classified areas shall comply with subparagraph (18) of this paragraph.

**1910.110(b)(17)(ii)** Open flames or other sources of ignition shall not be permitted in vaporizer rooms (except those housing direct-fired vaporizers), pumphouses, container charging rooms or other similar locations. Direct-fired vaporizers shall not be permitted in pumphouses or container charging rooms.

**1910.110(b)(17)(iii)** Liquefied petroleum gas storage containers do not require lightning protection.

# OSHA Standards for General Industry

**1910.110(b)(17)(iv)** Since liquefied petroleum gas is contained in a closed system of piping and equipment, the system need not be electrically conductive or electrically bonded for protection against static electricity.

**1910.110(b)(17)(v)** Open flames (except as provided for in paragraph (b)(11) of this section), cutting or welding, portable electric tools, and extension lights capable of igniting LP-Gas, shall not be permitted within classified areas specified in Table H-28 unless the LP-Gas facilities have been freed of all liquid and vapor, or special precautions observed under carefully controlled conditions.

## TABLE H-28

| Part | Location | Extent of classified area[1] | Equipment shall be suitable for Class 1, Group D[2] |
|------|----------|------------------------------|------------------------------------------------------|
| A | Storage containers other than DOT cylinders. | Within 15 feet in all directions from connections, except connections otherwise covered in Table H-28 | Division 2. |
| B | Tank vehicle and tank car loading and unloading.[3] | Within 5 feet in all directions from connections regularly made or disconnected for product transfer. | Division 1. |
|   |   | Beyond 5 feet but within 15 feet in all directions from a point where connections are regularly made or disconnected and within the cylindrical volume between the horizontal equator of the sphere and grade. (See Figure H-1). | Division 2. |
| C | Gage vent openings other than those on DOT cylinders | Within 5 feet in all directions from point of discharge | Division 1. |
| D | Relief valve discharge other than those on DOT cylinders. | Within direct path of discharge. | Division 1. |
|   |   | Within 5 feet in all directions from point of discharge. | Division 1. |
|   |   | Beyond 5 feet but within 15 feet in all directions from point of discharge except within the direct path of discharge. | Division 2. |
| E | Pumps, compressors, gas-air mixers and vaporizers other than direct fired. |   |   |
|   | Indoors without ventilation | Entire room and any adjacent room not separated by a gastight partition. | Division 1. |
|   |   | Within 15 feet of the exterior side of any exterior wall or roof that is not vaportight or within 15 feet of any exterior opening. | Division 2. |
|   | Indoors with adequate ventilation.[4] | Entire room and any adjacent room not separated by a gastight partition. | Division 2. |
|   | Outdoors in open air at or abovegrade. | Within 15 feet in all directions from this equipment and within the cylindrical volume between the horizontal equator of the sphere and grade. See Figure H-1. | Division 2. |
| F | Service Station Dispensing Units. | Entire space within dispenser enclosure, and 18 inches horizontally from enclosure exterior up to an elevation 4 ft. above dispenser base. Entire pit or open space beneath dispenser. | Division 1. |
|   |   | Up to 18 inches abovegrade within 20 ft. horizontally from any edge of enclosure. Note: For pits within this area, see Part F of this table. | Division 2. |
| G | Pits or trenches containing or located beneath LP-Gas valves, pumps, compressors, regulators, and similar equipment. |   |   |
|   | Without mechanical ventilation. | Entire pit or trench | Division 1. |
|   |   | Entire room and any adjacent room not separated by a gastight partition. | Division 2. |
|   |   | Within 15 feet in all directions from pit or trench when located outdoors. | Division 2. |
|   | With adequate mechanical ventilation. | Entire pit or trench | Division 2. |
|   |   | Entire room and any adjacent room not separated by a gastight partition. | Division 2. |
|   |   | Within 15 feet in all directions from pit or trench when located outdoors. | Division 2. |
| H | Special buildings or rooms for storage of portable containers. | Entire room | Division 2. |
| I | Pipelines and connections containing operational bleeds, drips, vents or drains. | Within 5 ft. in all directions from point of discharge | Division 1. |
|   |   | Beyond 5 ft. from point of discharge, same as Part E of this table. |   |
| J | Container filling: |   |   |
|   | Indoors without ventilation. | Entire room | Division 1. |
|   | Indoors with adequate ventilation.[4] | Within 5 feet in all directions from connections regularly made or disconnected for product transfer. | Division 1. |
|   |   | Beyond 5 feet and entire room | Division 2. |
|   | Outdoors in open air | Within 5 feet in all directions from connections regularly made or disconnected for product transfer | Division 1. |
|   |   | Beyond 5 feet but within 15 feet in all directions from a point where connections are regularly made or disconnected and within the cylindrical volume between the horizontal equator of the sphere and grade. (See Figure H-1) | Division 2. |

[1] The classified area shall not extend beyond an unpierced wall, roof, or solid vaportight partition.

[2] See Subpart S of this part.

[3] When classifying extent of hazardous area, consideration shall be given to possible variations in the spotting of tank cars and tank vehicles at the unloading points and the effect these variations of actual spotting point may have on the point of connection.

[4] Ventilation, either natural or mechanical, is considered adequate when the concentration of the gas in a gas-air mixture does not exceed 25 percent of the lower flammable limit under normal operating conditions.

**FIGURE H–1**

Labels on figure:
- SPHERE HAVING RADIUS OF 15 FT. - DIVISION 2
- SPHERE HAVING RADIUS OF 5 FT. - DIVISION 1
- HORIZONTAL EQUATOR OF SPHERE
- POINT OF CONNECTION OR SOURCE OF EXCESSIVE RELEASE OF LIQUID OR GAS
- GRADE LEVEL
- CYLINDRICAL VOLUME - DIVISION 2 (INCLUDES BOTTOM HALF OF 15 FT. RADIUS SPHERE)

**1910.110(b)(18)** *Fixed electrical equipment in classified areas.* Fixed electrical equipment and wiring installed within classified areas specified in Table H-28 shall comply with Table H-28 and shall be installed in accordance with Subpart S of this part. This provision does not apply to fixed electrical equipment at residential or commercial installations of LP-Gas systems or to systems covered by paragraph (e) or (g) of this section.

**1910.110(b)(19)** *Liquid-level gaging device.* Each container manufactured after December 31, 1965, and filled on a volumetric basis shall be equipped with a fixed liquid-level gage to indicate the maximum permitted filling level as provided in paragraph (b)(19)(v) of this section. Each container manufactured after December 31, 1969, shall have permanently attached to the container adjacent to the fixed level gage a marking showing the percentage full that will be shown by that gage. When a variable liquid-level gage is also provided, the fixed liquid-level gage will also serve as a means for checking the variable gage. These gages shall be used in charging containers as required in paragraph (b)(12) of this section.

**1910.110(b)(19)(ii)** All variable gaging devices shall be arranged so that the maximum liquid level for butane, for a 50-50 mixture of butane and propane, and for propane, to which the container may be charged is readily determinable. The markings indicating the various liquid levels from empty to full shall be on the system nameplate or gaging device or part may be on the system nameplate and part on the gaging device. Dials of magnetic or rotary gages shall show whether they are for cylindrical or spherical containers and whether for aboveground or underground service. The dials of gages intended for use only on aboveground containers of over 1,200 gallons water capacity shall be so marked.

**1910.110(b)(19)(iii)** Gaging devices that require bleeding of the product to the atmosphere, such as the rotary tube, fixed tube, and slip tube, shall be designed so that the bleed valve maximum opening is not larger than a No. 54 drill size, unless provided with excess flow valve.

**1910.110(b)(19)(iv)** Gaging devices shall have a design working pressure of at least 250 p.s.i.g.

**1910.110(b)(19)(v)** Length of tube or position of fixed liquid-level gage shall be designed to indicate the maximum level to which the container may be filled for the product contained. This level shall be based on the volume of the product at 40 °F. at its maximum permitted filling density for aboveground containers and at 50 °F. for underground containers. The employer shall calculate the filling point for which the fixed liquid level gage shall be designed according to the method in this subdivision.

**1910.110(b)(19)(v)***(a)* It is impossible to set out in a table the length of a fixed dip tube for various capacity tanks because of the varying tank diameters and lengths and because the tank may be installed either in a vertical or horizontal position. Knowing the maximum permitted filling volume in gallons, however, the length of the fixed tube can be determined by the use of a strapping table obtained from the container manufacturer. The length of the fixed tube should be such that when its lower end touches the surface of the liquid in the container, the contents of the container will be the maximum permitted volume as determined by the following formula:

$$\frac{\text{Water capacity (gals) of container* × filling density**}}{\text{Specific gravity of LP-Gas* × volume correction factor × 100}} = \text{Maximum volume of LP-Gas}$$

\* Measured at 60 °F.
\*\* From subparagraph (12) of this paragraph "Filling Densities."

For aboveground containers the liquid temperature is assumed to be 40 °F, and for underground containers the liquid temperature is assumed to be 50 °F. To correct the liquid volumes at these temperatures to 60 °F. the following factors shall be used.

**1910.110(b)(19)(v)***(b)* Formula for determining maximum volume of liquefied petroleum gas for which a fixed length of dip tube shall be set:

**TABLE H-29—VOLUME CORRECTION FACTORS**

| Specific gravity | Aboveground | Underground |
|---|---|---|
| 0.500 | 1.033 | 1.017 |
| .510 | 1.031 | 1.016 |
| .520 | 1.029 | 1.015 |
| .530 | 1.028 | 1.014 |
| .540 | 1.026 | 1.013 |
| .550 | 1.025 | 1.013 |
| .560 | 1.024 | 1.012 |
| .570 | 1.023 | 1.011 |
| .580 | 1.021 | 1.011 |
| .590 | 1.020 | 1.010 |

**1910.110(b)(19)(v)***(c)* The maximum volume of LP-Gas which can be placed in a container when determining the length of the dip tube expressed as a percentage of total water content of the container is calculated by the following formula.

**1910.110(b)(19)(v)***(d)* The maximum weight of LP-Gas which may be placed in a container *for determining the length of a fixed dip tube* is determined by multiplying the maximum volume of liquefied petroleum gas obtained by the formula in paragraph (b)(19)(b) of this section by the pounds of liquefied petroleum gas in a gallon at 40 °F. for aboveground and at 50 °F. for underground containers. For example, typical pounds per gallon are specified below:

**Example:** Assume a 100-gallon total water capacity tank for aboveground storage of propane having a specific gravity of 0.510 of 60 °F.

[(100 (gals.) × 42 (filling density from subparagraph (12) of this paragraph)) ÷ (0.510 × 1.031 (correction factor from Table H-29 × 100)] = (4200 ÷ 52.6)

(4200 ÷ 52.6) = 79.8 gallons propane, the maximum amount permitted to be placed in a 100-gallon total water capacity aboveground container equipped with a fixed dip tube.

[(Maximum volume of LP-Gas (from formula in subdivision (*b*) of this subdivision) × 100) ÷ Total water content of container in gallons] = Maximum percent of LP-Gas

|  | Aboveground, pounds per gallon | Underground, pounds per gallon |
|---|---|---|
| Propane | 4.37 | 4.31 |
| N Butane | 4.97 | 4.92 |

**1910.110(b)(19)(vi)** Fixed liquid-level gages used on containers other than DOT containers shall be stamped on the exterior of the gage with the letters "DT" followed by the vertical distance (expressed in inches and carried out to one decimal place) from the top of container to the end of the dip tube or to the centerline of the gage when it is located at the maximum permitted filling level. For portable containers that may be filled in the horizontal and/or vertical position the letters "DT" shall be followed by "V" with the vertical distance from the top of the container to the end of the dip tube for vertical filling and with "H" followed by the proper distance for horizontal filling. For DOT containers the stamping shall be placed both on the exterior of the gage and on the container. On above-ground or cargo containers where the gages are positioned at specific levels, the marking may be specified in percent of total tank contents and the marking shall be stamped on the container.

**1910.110(b)(19)(vii)** Gage glasses of the columnar type shall be restricted to charging plants where the fuel is withdrawn in the liquid phase only. They shall be equipped with valves having metallic handwheels, with excess flow valves, and with extra-heavy glass adequately protected with a metal housing applied by the gage manufacturer. They shall be shielded against the direct rays of the sun. Gage glasses of the columnar type are prohibited on tank trucks, and on motor fuel tanks, and on containers used in domestic, commercial, and industrial installations.

**1910.110(b)(19)(viii)** Gaging devices of the float, or equivalent type which do not require flow for their operation and having connections extending to a point outside the container do not have to be equipped with excess flow valves provided the piping and fittings are adequately designed to withstand the container pressure and are properly protected against physical damage and breakage.

**1910.110(b)(20)** *Requirements for appliances.* **(i)** Except as provided in paragraph (b)(20)(ii) of this section, new commercial and industrial gas consuming appliances shall be approved.

**1910.110(b)(20)(ii)** Any appliance that was originally manufactured for operation with a gaseous fuel other than LP-Gas and is in good condition may be used with LP-Gas only after it is properly converted, adapted, and tested for performance with LP-Gas before the appliance is placed in use.

**1910.110(b)(20)(iii)** Unattended heaters used inside buildings for the purpose of animal or poultry production or care shall be equipped with an approved automatic device designed to shut off the flow of gas to the main burners, and pilot if used, in the event of flame extinguishment.

**1910.110(b)(20)(iv)** All commercial, industrial, and agricultural appliances or equipment shall be installed in accordance with the requirements of this section and in accordance with the following NFPA consensus standards, which are incorporated by reference as specified in §1910.6:

**1910.110(b)(20)(iv)(a)** Domestic and commercial appliances—NFPA 54-1969, Standard for the Installation of Gas Appliances and Gas Piping.

**1910.110(b)(20)(iv)(b)** Industrial appliances—NFPA 54A-1969, Standard for the Installation of Gas Piping and Gas Equipment on Industrial Premises and Certain Other Premises.

**1910.110(b)(20)(iv)(c)** Standard for the Installation and Use of Stationary Combustion Engines and Gas Turbines—NFPA 37-1970.

**1910.110(b)(20)(iv)(d)** Standard for the Installation of Equipment for the Removal of Smoke and Grease-Laden Vapors from Commercial Cooking Equipment, NFPA 96-1970.

**1910.110(c)** *Cylinder systems*—**(1)** *Application.* This paragraph applies specifically to systems utilizing containers constructed in accordance with DOT Specifications. All requirements of paragraph (b) of this section apply to this paragraph unless otherwise noted in paragraph (b) of this section.

**1910.110(c)(2)** *Marking of containers.* Containers shall be marked in accordance with DOT regulations. Additional markings not in conflict with DOT regulations may be used.

**1910.110(c)(3)** *Description of a system.* A system shall include the container base or bracket, containers, container valves, connectors, manifold valve assembly, regulators, and relief valves.

**1910.110(c)(4)** *Containers and regulating equipment installed outside of buildings or structures.* **(i)** Containers shall not be buried below ground. However, this shall not prohibit the installation in a compartment or recess below grade level such as a niche in a slope or terrace wall which is used for no other purpose, providing that the container and regulating equipment are not in contact with the ground and the compartment or recess is drained and ventilated horizontally to the outside air from its lowest level, with the outlet at least 3 feet away from any building opening which is below the level of such outlet.

Except as provided in paragraph (b)(10)(xiii) of this section, the discharge from safety relief devices shall be located not less than 3 feet horizontally away from any building opening which is below the level of such discharge and shall not terminate beneath any building unless such space is well ventilated to the outside and is not enclosed on more than two sides.

**1910.110(c)(4)(ii)** Containers shall be set upon firm foundation or otherwise firmly secured; the possible effect on the outlet piping of settling shall be guarded against by a flexible connection or special fitting.

**1910.110(c)(5)** *Containers and equipment used inside of buildings or structures.* **(i)** When operational requirements make portable use of containers necessary and their location outside of buildings or structure is impracticable, containers and equipment are permitted to be used inside of buildings or structures in accordance with (*a*) through (*l*) of this subdivision, and, in addition, such other provisions of this subparagraph as are applicable to the particular use or occupancy.

**1910.110(c)(5)(i)(a)** Containers in use shall mean connected for use.

**1910.110(c)(5)(i)(b)** Systems utilizing containers having a water capacity greater than 2 1/2 pounds (nominal 1 pound LP-Gas capacity) shall be equipped with excess flow valves. Such excess flow valves shall be either integral with the container valves or in the connections to the container valve outlets. In either case, an excess flow valve shall be installed in such a manner that any undue strain beyond the excess flow valve will not cause breakage between the container and the excess flow valve. The installation of excess flow valves shall take into account the type of valve protection provided.

**1910.110(c)(5)(i)(c)** Regulators, if used, shall be either directly connected to the container valves or to manifolds connected to the container values. The regulator shall be suitable for use with LP-Gas. Manifolds and fittings connecting containers to pressure regulator inlets shall be designed for at least 250 p.s.i.g. service pressure.

**1910.110(c)(5)(i)(d)** Valves on containers having a water capacity greater than 50 pounds (nominal 20 pounds LP-Gas capacity) shall be protected while in use.

**1910.110(c)(5)(i)(e)** Containers shall be marked in accordance with paragraph (b)(5)(iii) of this section and paragraph (c)(2) of this section.

**1910.110(c)(5)(i)(f)** Pipe or tubing shall conform to paragraph (b)(8) of this section except that aluminum pipe or tubing shall not be used.

**1910.110(c)(5)(i)(g)(1)** Hose shall be designed for a working pressure of at least 250 p.s.i.g. Hose and hose connections shall have their correctness as to design, construction and performance determined by listing by a nationally recognized testing laboratory. The hose length may exceed the length specified in paragraph (b)(9)(vii)(*b*) of this section, but shall be as short as practicable. Refer to §1910.7 for definition of nationally recognized testing laboratory.

**1910.110(c)(5)(i)(g)(2)** Hose shall be long enough to permit compliance with spacing provisions of this subparagraph without kinking or straining or causing hose to be so close to a burner as to be damaged by heat.

**1910.110(c)(5)(i)(h)** Portable heaters, including salamanders, shall be equipped with an approved automatic device to shut off the flow of gas to the main burner, and pilot if used, in the event of flame extinguishment. Such heaters having inputs above 50,000 B.t.u. manufactured on or after May 17, 1967, and such heaters having inputs above 100,000 B.t.u. manufactured before May 17, 1967, shall be equipped with either.

**1910.110(c)(5)(i)(h)(1)** A pilot which must be lighted and proved before the main burner can be turned on; or

**1910.110(c)(5)(i)(h)(2)** An electric ignition system.

The provisions of this paragraph (*h*) do not apply to tar kettle burners, torches, melting pots, nor do they apply to portable heaters under 7,500 B.t.u.h. input when used with containers having a maximum water capacity of 2 1/2 pounds. Container valves, connectors, regulators, manifolds, piping, and tubing shall not be used as structural supports for heaters.

**1910.110(c)(5)(i)(i)** Containers, regulating equipment, manifolds, pipe, tubing, and hose shall be located so as to minimize exposure to abnormally high temperatures (such as may result from exposure to convection or radiation from heating equipment or installation in confined spaces), physical damage, or tampering by unauthorized persons.

**1910.110(c)(5)(i)(j)** Heat producing equipment shall be located and used so as to minimize the possibility of ignition of combustibles.

**1910.110(c)(5)(i)(k)** Containers having a water capacity greater than 2 1/2 pounds (nominal 1 pound LP-Gas capacity) connected for use, shall stand on a firm and substantially level surface and, when necessary, shall be secured in an upright position.

**1910.110(c)(5)(i)(l)** Containers, including the valve protective devices, shall be installed so as to minimize the probability of impingement of discharge of safety relief devices upon containers.

**1910.110(c)(5)(ii)** Containers having a maximum water capacity of 2 1/2 pounds (nominal 1 pound LP-Gas capacity) are permitted to be used inside of buildings as part of approved self-contained hand torch assemblies or similar appliances.

**1910.110(c)(5)(iii)** Containers having a maximum water capacity of 12 pounds (nominal 5 pounds LP-Gas capacity) are permitted to be used temporarily inside of buildings for public exhibition or demonstration purposes, including use for classroom demonstrations.

**1910.110(c)(5)(iv)** [Reserved]

**1910.110(c)(5)(v)** Containers are permitted to be used in buildings or structures under construction or undergoing major renovation when such buildings or structures are not occupied by the public, as follows:

**1910.110(c)(5)(v)(a)** The maximum water capacity of individual containers shall be 245 pounds (nominal 100 pounds LP-Gas capacity).

**1910.110(c)(5)(v)(b)** For temporary heating such as curing concrete, drying plaster and similar applications, heaters (other than integral heater-container units) shall be located at least 6 feet from any LP-Gas container. This shall not prohibit the use of heaters specifically designed for attachment to the container or to a supporting standard, provided they are designed and installed so as to prevent direct or radiant heat application from the heater onto the container. Blower and radiant type heaters shall not be directed toward any LP-Gas container within 20 feet.

**1910.110(c)(5)(v)(c)** If two or more heater-container units, of either the integral or nonintegral type, are located in an unpartitioned area on the same floor, the container or containers of each unit shall be separated from the container or containers of any other unit by at least 20 feet.

**1910.110(c)(5)(v)(d)** When heaters are connected to containers for use in an unpartitioned area on the same floor, the total water capacity of containers manifolded together for connection to a heater or heaters shall not be greater than 735 pounds (nominal 300 pounds LP-Gas capacity). Such manifolds shall be separated by at least 20 feet.

**1910.110(c)(5)(v)(e)** On floors on which heaters are not connected for use, containers are permitted to be manifolded together for connection to a heater or heaters on another floor, *Provided:*

**1910.110(c)(5)(v)(e)(1)** The total water capacity of containers connected to any one manifold is not greater than 2,450 pounds (nominal 1,000 pounds LP-Gas capacity) and;

**1910.110(c)(5)(v)(e)(2)** Where more than one manifold having a total water capacity greater than 735 pounds (nominal 300 pounds LP-Gas capacity) are located in the same unpartitioned area, they shall be separated by at least 50 feet.

**1910.110(c)(5)(v)(f)** Storage of containers awaiting use shall be in accordance with paragraph (f) of this section.

**1910.110(c)(5)(vi)** Containers are permitted to be used in industrial occupancies for processing, research, or experimental purposes as follows:

**1910.110(c)(5)(vi)(a)** The maximum water capacity of individual containers shall be 245 pounds (nominal 100 pounds LP-Gas capacity).

**1910.110(c)(5)(vi)(b)** Containers connected to a manifold shall have a total water capacity not greater than 735 pounds (nominal 300 pounds LP-Gas capacity) and not more than one such manifold may be located in the same room unless separated at least 20 feet from a similar unit.

**1910.110(c)(5)(vi)(c)** The amount of LP-Gas in containers for research and experimental use shall be limited to the smallest practical quantity.

**1910.110(c)(5)(vii)(a)** Containers are permitted to be used in industrial occupancies with essentially noncombustible contents where portable equipment for space heating is essential and where a permanent heating installation is not practical, as follows:

**1910.110(c)(5)(vii)(b)** Containers and heaters shall comply with and be used in accordance with paragraph (c)(5)(v) of this section.

**1910.110(c)(5)(viii)** Containers are permitted to be used in buildings for temporary emergency heating purposes, if necessary to prevent damage to the buildings or contents, when the permanent heating system is temporarily out of service, as follows:

**1910.110(c)(5)(viii)(a)** Containers and heaters shall comply with and be used in accordance with paragraph (c)(5)(v) of this section.

**1910.110(c)(5)(viii)(b)** The temporary heating equipment shall not be left unattended.

**1910.110(c)(5)(ix)** Containers are permitted to be used temporarily in buildings for training purposes related in installation and use of LP-Gas systems, as follows:

**1910.110(c)(5)(ix)(a)** The maximum water capacity of individual containers shall be 245 pounds (nominal 100 pounds LP-Gas capacity), but the maximum quantity of LP-Gas that may be placed in each container shall be 20 pounds.

**1910.110(c)(5)(ix)(b)** If more than one such container is located in the same room, the containers shall be separated by at least 20 feet.

**1910.110(c)(6)** *Container valves and accessories.*

**1910.110(c)(6)(i)** Valves in the assembly of multiple container systems shall be arranged so that replacement of containers can be made without shutting off the flow of gas in the system.

**NOTE:** This provision is not to be construed as requiring an automatic changeover device.

**1910.110(c)(6)(ii)** Regulators and low-pressure relief devices shall be rigidly attached to the cylinder valves, cylinders, supporting standards, the building walls or otherwise rigidly secured and shall be so installed or protected that the elements (sleet, snow, or ice) will not affect their operation.

**1910.110(c)(6)(iii)** Valves and connections to the containers shall be protected while in transit, in storage, and while being moved into final utilization, as follows:

**1910.110(c)(6)(iii)(a)** By setting into the recess of the container to prevent the possibility of their being struck if the container is dropped upon a flat surface, or

**1910.110(c)(6)(iii)(b)** By ventilated cap or collar, fastened to the container capable of withstanding a blow from any direction equivalent to that of a 30-pound weight dropped 4 feet. Construction must be such that a blow will not be transmitted to the valve or other connection.

**1910.110(c)(6)(iv)** When containers are not connected to the system, the outlet valves shall be kept tightly closed or plugged, even though containers are considered empty.

**1910.110(c)(6)(v)** Containers having a water capacity in excess of 50 pounds (approximately 21 pounds LP-Gas capacity), recharged at the installation, shall be provided with excess flow or backflow check valves to prevent the discharge of container contents in case of failure of the filling or equalizing connection.

**1910.110(c)(7)** *Safety devices.* **(i)** Containers shall be provided with safety devices as required by DOT regulations.

**1910.110(c)(7)(ii)** A final stage regulator of an LP-Gas system (excluding any appliance regulator) shall be equipped on the low-pressure side with a relief valve which is set to start to discharge within the limits specified in Table H-30.

### TABLE H-30

| Regulator delivery pressure | Relief valve start-to-discharge pressure setting (percent of regulator delivery pressure) | |
|---|---|---|
| | Minimum | Maximum |
| 1 p.s.i.g. or less | 200 | 300 |
| Above 1 p.s.i.g. but not over 3 p.s.i.g. | 140 | 200 |
| Above 3 p.s.i.g | 125 | 200 |

**1910.110(c)(7)(iii)** When a regulator or pressure relief valve is used inside a building for other than purposes specified in paragraphs (b)(6)(i) (*a*)-(*g*) of this section, the relief valve and the space above the regulator and relief valve diaphragms shall be vented to the outside air with the discharge outlet located not less than 3 feet horizontally away from any building opening which is below such discharge. These provisions do not apply to individual appliance regulators when protection is otherwise provided nor to paragraph (c)(5) of this section and paragraph (b)(10)(xiii) of this section. In buildings devoted exclusively to gas distribution purposes, the space above the diaphragm need not be vented to the outside.

**1910.110(c)(8)** *Reinstallation of containers.* Containers shall not be reinstalled unless they are requalified in accordance with DOT regulations.

**1910.110(c)(9)** *Permissible product.* A product shall not be placed in a container marked with a service pressure less than four-fifths of the maximum vapor pressure of product at 130 °F.

**1910.110(d)** *Systems utilizing containers other than DOT containers*—**(1)** *Application.* This paragraph applies specifically to systems utilizing storage containers other than those constructed in accordance with DOT specifications. Paragraph (b) of this section applies to this paragraph unless otherwise noted in paragraph (b) of this section.

**1910.110(d)(2)** *Design pressure and classification of storage containers.* Storage containers shall be designed and classified in accordance with Table H-31.

### TABLE H-31

| Container type | For gases with vapor press Not to exceed lb. gage at 100 °F. (37.8 °C) | 1949 and earlier editions of ASME Code (Par. U-68, U-69) | Minimum design pressure of container, lb. per sq. in. gage — 1949 edition of ASME Code (Par. U-200, U-201); 1950, 1952, 1956, 1959, 1962, 1965, and 1968 (Division I) editions of ASME Code; All editions of API-ASME Code[3] |
|---|---|---|---|
| [1]80 | [1]80 | [1]80 | [1]100 |
| 100 | 100 | 100 | 125 |
| 125 | 125 | 125 | 156 |
| 150 | 150 | 150 | 187 |
| 175 | 175 | 175 | 219 |
| [2]200 | 215 | 200 | 250 |

[1] New storage containers of the 80 type have not been authorized since Dec. 31, 1947.

[2] Container type may be increased by increments of 25. The minimum design pressure of containers shall be 100% of the container type designation when constructed under 1949 or earlier editions of the ASME Code (Par. U-68 and U-69). The minimum design pressure of containers shall be 125% of the container type designation when constructed under: (1) The 1949 ASME Code (¶¶U-200 and U-201), (2) 1950, 1952, 1956, 1959, 1962, 1965, and 1968 (Division I) editions of the ASME Code, and (3) all editions of the API-ASME Code.

[3] Construction of containers under the API-ASME Code is not authorized after July 1, 1961.

**1910.110(d)(3)** *Container valves and accessories, filler pipes, and discharge pipes.* **(i)** The filling pipe inlet terminal shall not be located inside a building. For containers with a water capacity of 125 gallons or more, such terminals shall be located not less than 10 feet from any building (see paragraph (b)(6)(ii) of this section), and preferably not less than 5 feet from any driveway, and shall be located in a protective housing built for the purpose.

**1910.110(d)(3)(ii)** The filling connection shall be fitted with one of the following:

**1910.110(d)(3)(ii)***(a)* Combination back-pressure check valve and excess flow valve.

**1910.110(d)(3)(ii)***(b)* One double or two single back-pressure check valves.

**1910.110(d)(3)(ii)***(c)* A positive shutoff valve, in conjunction with either:

**1910.110(d)(3)(ii)***(c)(1)* An internal back-pressure valve, or

**1910.110(d)(3)(ii)***(c)(2)* An internal excess flow valve.

**1910.110(d)(3)(iii)** All openings in a container shall be equipped with approved automatic excess flow valves except in the following: Filling connections as provided in paragraph (d)(3)(ii) of this section; safety relief connections, liquid-level gaging devices as provided in paragraphs (b)(7)(iv), (19)(iii), and (19)(viii) of this section; pressure gage connections as provided in paragraph (b)(7)(v) of this section, as provided in paragraphs (d) (iv), (vi), and (vii) of this section.

**1910.110(d)(3)(iv)** An excess flow valve is not required in the withdrawal service line providing the following are complied with:

**1910.110(d)(3)(iv)***(a)* Such systems' total water capacity does not exceed 2,000 U.S. gallons.

**1910.110(d)(3)(iv)***(b)* The discharge from the service outlet is controlled by a suitable manually operated shutoff valve which is:

**1910.110(d)(3)(iv)***(b)(1)* Threaded directly into the service outlet of the container; or

**1910.110(d)(3)(iv)***(b)(2)* Is an integral part of a substantial fitting threaded into or on the service outlet of the container; or

**1910.110(d)(3)(iv)***(b)(3)* Threaded directly into a substantial fitting threaded into or on the service outlet of the container.

**1910.110(d)(3)(iv)***(c)* The shutoff valve is equipped with an attached handwheel or the equivalent.

**1910.110(d)(3)(iv)***(d)* The controlling orifice between the contents of the container and the outlet of the shutoff valve does not exceed five-sixteenths inch in diameter for vapor withdrawal systems and one-eighth inch in diameter for liquid withdrawal systems.

**1910.110(d)(3)(iv)***(e)* An approved pressure-reducing regulator is directly attached to the outlet of the shutoff valve and is rigidly supported, or that an approved pressure-reducing regulator is attached to the outlet of the shutoff valve by means of a suitable flexible connection, provided the regulator is adequately supported and properly protected on or at the tank.

**1910.110(d)(3)(v)** All inlet and outlet connections except safety relief valves, liquid level gaging devices and pressure gages on containers of 2,000 gallons water capacity, or more, and on any container used to supply fuel directly to an internal combustion engine, shall be labeled to designate whether they communicate with vapor or liquid space. Labels may be on valves.

**1910.110(d)(3)(vi)** In lieu of an excess flow valve openings may be fitted with a quick-closing internal valve which, except during operating periods shall remain closed. The internal mechanism for such valves may be provided with a secondary control which shall be equipped with a fusible plug (not over 220 °F. melting point) which will cause the internal valve to close automatically in case of fire.

**1910.110(d)(3)(vii)** Not more than two plugged openings shall be permitted on a container of 2,000 gallons or less water capacity.

**1910.110(d)(3)(viii)** Containers of 125 gallons water capacity or more manufactured after July 1, 1961, shall be provided with an approved device for liquid evacuation, the size of which shall be three-fourths inch National Pipe Thread minimum. A plugged opening will not satisfy this requirement.

**1910.110(d)(4)** *Safety devices.* **(i)** All safety devices shall comply with the following:

**1910.110(d)(4)(i)***(a)* All container safety relief devices shall be located on the containers and shall have direct communication with the vapor of space of the container.

**1910.110(d)(4)(i)***(b)* In industrial and gas manufacturing plants, discharge pipe from safety relief valves on pipe lines within a building shall discharge vertically upward and shall be piped to a point outside a building.

**1910.110(d)(4)(i)***(c)* Safety relief device discharge terminals shall be so located as to provide protection against physical damage and such discharge pipes shall be fitted with loose raincaps. Return bends and restrictive pipefittings shall not be permitted.

**1910.110(d)(4)(i)***(d)* If desired, discharge lines from two or more safety relief devices located on the same unit, or similar lines from two or more different units, may be run into a common discharge header, provided that the cross-sectional area of such header be at least equal to the sum of the cross-sectional area of the individual discharge lines, and that the setting of safety relief valves are the same.

**1910.110(d)(4)(i)***(e)* Each storage container of over 2,000 gallons water capacity shall be provided with a suitable pressure gage.

**1910.110(d)(4)(i)***(f)* A final stage regulator of an LP-Gas system (excluding any appliance regulator) shall be equipped on the low-pressure side with a relief valve which is set to start to discharge within the limits specified in Table H-30.

**1910.110(d)(4)(i)***(g)* When a regulator or pressure relief valve is installed inside a building, the relief valve and the space above the regulator and relief valve diaphragms shall be vented to the outside air with the discharge outlet located not less than 3 feet horizontally away from any opening into the building which is below such discharge. (These provisions do not apply to individual appliance regulators when protection is otherwise provided. In buildings devoted exclusively to gas distribution purposes, the space above the diaphragm need not be vented to the outside.)

**1910.110(d)(4)(ii)** Safety devices for aboveground containers shall be provided as follows:

**1910.110(d)(4)(ii)(a)** Containers of 1,200 gallons water capacity or less which may contain liquid fuel when installed above ground shall have the rate of discharge required by paragraph (b)(10)(ii) of this section provided by a spring-loaded relief valve or valves. In addition to the required spring-loaded relief valve(s), suitable fuse plug(s) may be used provided the total discharge area of the fuse plug(s) for each container does not exceed 0.25 square inch.

**1910.110(d)(4)(ii)(b)** The fusible metal of the fuse plugs shall have a yield temperature of 208 °F. minimum and 220 °F. maximum. Relief valves and fuse plugs shall have direct communication with the vapor space of the container.

**1910.110(d)(4)(ii)(c)** On a container having a water capacity greater than 125 gallons, but not over 2,000 gallons, the discharge from the safety relief valves shall be vented away from the container vertically upwards and unobstructed to the open air in such a manner as to prevent any impingement of escaping gas upon the container; loose-fitting rain caps shall be used. Suitable provision shall be made for draining condensate which may accumulate in the relief valve or its discharge pipe.

**1910.110(d)(4)(ii)(d)** On containers of 125 gallons water capacity or less, the discharge from safety relief devices shall be located not less than 5 feet horizontally away from any opening into the building below the level of such discharge.

**1910.110(d)(4)(ii)(e)** On a container having a water capacity greater than 2,000 gallons, the discharge from the safety relief valves shall be vented away from the container vertically upwards to a point at least 7 feet above the container, and unobstructed to the open air in such a manner as to prevent any impingement of escaping gas upon the container; loose-fitting rain caps shall be used. Suitable provision shall be made so that any liquid or condensate that may accumulate inside of the safety relief valve or its discharge pipe will not render the valve inoperative. If a drain is used, a means shall be provided to protect the container, adjacent containers, piping, or equipment against impingement of flame resulting from ignition of product escaping from the drain.

**1910.110(d)(4)(iii)** On all containers which are installed underground and which contain no liquid fuel until buried and covered, the rate of discharge of the spring-loaded relief valve installed thereon may be reduced to a minimum of 30 percent of the rate of discharge specified in paragraph (b)(10)(ii) of this section. Containers so protected shall not be uncovered after installation until the liquid fuel has been removed therefrom. Containers which may contain liquid fuel before being installed under ground and before being completely covered with earth are to be considered aboveground containers when determining the rate of discharge requirement of the relief valves.

**1910.110(d)(4)(iv)** On underground containers of more than 2,000 gallons water capacity, the discharge from safety relief devices shall be piped vertically and directly upward to a point at least 7 feet above the ground.

Where there is a probability of the manhole or housing becoming flooded, the discharge from regulator vent lines shall be above the highest probable water level. All manholes or housings shall be provided with ventilated louvers or their equivalent, the area of such openings equaling or exceeding the combined discharge areas of the safety relief valves and other vent lines which discharge their content into the manhole housing.

**1910.110(d)(4)(v)** Safety devices for vaporizers shall be provided as follows:

**1910.110(d)(4)(v)(a)** Vaporizers of less than 1 quart total capacity, heated by the ground or the surrounding air, need not be equipped with safety relief valves provided that adequate tests certified by any of the authorities referred to in paragraph (b)(2) of this section, demonstrate that the assembly is safe without safety relief valves.

**1910.110(d)(4)(v)(b)** No vaporizer shall be equipped with fusible plugs.

**1910.110(d)(4)(v)(c)** In industrial and gas manufacturing plants, safety relief valves on vaporizers within a building shall be piped to a point outside the building and be discharged upward.

**1910.110(d)(5)** *Reinstallation of containers.* Containers may be reinstalled if they do not show any evidence of harmful external corrosion or other damage. Where containers are reinstalled underground, the corrosion resistant coating shall be put in good condition (see paragraph (c)(7)(vi) of this section).

Where containers are reinstalled above ground, the safety devices and gaging devices shall comply with paragraph (c)(4) of this section and paragraph (b)(19) of this section respectively for aboveground containers.

**1910.110(d)(6)** *Capacity of containers.* A storage container shall not exceed 90,000 gallons water capacity.

**1910.110(d)(7)** *Installation of storage containers.* **(i)** Containers installed above ground, except as provided in paragraph (c)(7)(vii) of this section, shall be provided with substantial masonry or noncombustible structural supports on firm masonry foundation.

**1910.110(d)(7)(ii)** Aboveground containers shall be supported as follows:

**1910.110(d)(7)(ii)(a)** Horizontal containers shall be mounted on saddles in such a manner as to permit expansion and contraction. Structural metal supports may be employed when they are protected against fire in an approved manner. Suitable means of preventing corrosion shall be provided on that portion of the container in contact with the foundations or saddles.

**1910.110(d)(7)(ii)(b)** Containers of 2,000 gallons water capacity or less may be installed with nonfireproofed ferrous metal supports if mounted on concrete pads or footings, and if the distance from the outside bottom of the container shell to the concrete pad, footing, or the ground does not exceed 24 inches.

**1910.110(d)(7)(iii)** Any container may be installed with nonfireproofed ferrous metal supports if mounted on concrete pads or footings, and if the distance from the outside bottom of the container to the ground does not exceed 5 feet, provided the container is in an isolated location.

**1910.110(d)(7)(iv)** Containers may be partially buried providing the following requirements are met:

**1910.110(d)(7)(iv)(a)** The portion of the container below the surface and for a vertical distance not less than 3 inches above the surface of the ground is protected to resist corrosion, and the container is protected against settling and corrosion as required for fully buried containers.

**1910.110(d)(7)(iv)(b)** Spacing requirements shall be as specified for underground tanks in paragraph (b)(6)(ii) of this section.

**1910.110(d)(7)(iv)(c)** Relief valve capacity shall be as required for aboveground containers.

**1910.110(d)(7)(iv)(d)** Container is located so as not to be subject to vehicular damage, or is adequately protected against such damage.

**1910.110(d)(7)(iv)(e)** Filling densities shall be as required for aboveground containers.

**1910.110(d)(7)(v)** Containers buried underground shall be placed so that the top of the container is not less than 6 inches below grade. Where an underground container might be subject to abrasive action or physical damage due to vehicular traffic or other causes, then it shall be:

**1910.110(d)(7)(v)(a)** Placed not less than 2 feet below grade, or

**1910.110(d)(7)(v)(b)** Otherwise protected against such physical damage.

It will not be necessary to cover the portion of the container to which manhole and other connections are affixed; however, where necessary, protection shall be provided against vehicular damage. When necessary to prevent floating, containers shall be securely anchored or weighted.

**1910.110(d)(7)(vi)(a)** Containers shall be given a protective coating before being placed under ground. This coating shall be equivalent to hot-dip galvanizing or to two coatings of red lead followed by a heavy coating of coal tar or asphalt. In lowering the container into place, care shall be exercised to prevent damage to the coating. Any damage to the coating shall be repaired before backfilling.

**1910.110(d)(7)(vi)(b)** Containers shall be set on a firm foundation (firm earth may be used) and surrounded with earth or sand firmly tamped in place.

**1910.110(d)(7)(vii)** Containers with foundations attached (portable or semi-portable containers with suitable steel "runners" or "skids" and popularly known in the industry as "skid tanks") shall be designed, installed, and used in accordance with these rules subject to the following provisions:

**1910.110(d)(7)(vii)(a)** If they are to be used at a given general location for a temporary period not to exceed 6 months they need not have fire-resisting foundations or saddles but shall have adequate ferrous metal supports.

**1910.110(d)(7)(vii)(b)** They shall not be located with the outside bottom of the container shell more than 5 feet above the surface of the ground unless fire-resisting supports are provided.

**1910.110(d)(7)(vii)(c)** The bottom of the skids shall not be less than 2 inches or more than 12 inches below the outside bottom of the container shell.

**1910.110(d)(7)(vii)(d)** Flanges, nozzles, valves, fittings, and the like, having communication with the interior of the container, shall be protected against physical damage.

**1910.110(d)(7)(vii)(e)** When not permanently located on fire-resisting foundations, piping connections shall be sufficiently flexible to minimize the possibility of breakage or leakage of connections if the container settles, moves, or is otherwise displaced.

**1910.110(d)(7)(vii)(f)** Skids, or lugs for attachment of skids, shall be secured to the container in accordance with the code or rules under which the container is designed and built (with a minimum factor of safety of four) to withstand loading in any direction equal to four times the weight of the container and attachments when filled to the maximum permissible loaded weight.

**1910.110(d)(7)(viii)** Field welding where necessary shall be made only on saddle plates or brackets which were applied by the manufacturer of the tank.

**1910.110(d)(7)(ix)** For aboveground containers, secure anchorage or adequate pier height shall be provided against possible container flotation wherever sufficiently high floodwater might occur.

**1910.110(d)(7)(x)** When permanently installed containers are interconnected, provision shall be made to compensate for expansion, contraction, vibration, and settling of containers, and interconnecting piping. Where flexible connections are used, they shall be of an approved type and shall be designed for a bursting pressure of not less than five times the vapor pressure of the product at 100 °F. The use of nonmetallic hose is prohibited for permanently interconnecting such containers.

**1910.110(d)(7)(xi)** Container assemblies listed for interchangeable installation above ground or under ground shall conform to the requirements for aboveground installations with respect to safety relief capacity and filling density. For installation above ground all other requirements for aboveground installations shall apply. For installation under ground all other requirements for underground installations shall apply.

**1910.110(d)(8)** *Protection of container accessories.* **(i)** Valves, regulating, gaging, and other container accessory equipment shall be protected against tampering and physical damage. Such accessories shall also be so protected during the transit of containers intended for installation underground.

**1910.110(d)(8)(ii)** On underground or combination aboveground-underground containers, the service valve handwheel, the terminal for connecting the hose, and the opening through which there can be a flow from safety relief valves shall be at least 4 inches above the container and this opening shall be located in the dome or housing. Underground systems shall be so installed that all the above openings, including the regulator vent, are located above the normal maximum water table.

**1910.110(d)(8)(iii)** All connections to underground containers shall be located within a substantial dome, housing, or manhole and with access thereto protected by a substantial cover.

**1910.110(d)(9)** *Drips for condensed gas.* Where vaporized gas on the low-pressure side of the system may condense to a liquid at normal operating temperatures and pressures, suitable means shall be provided for revaporization of the condensate.

**1910.110(d)(10)** *Damage from vehicles.* When damage to LP-Gas systems from vehicular traffic is a possibility, precautions against such damage shall be taken.

**1910.110(d)(11)** *Drains.* No drains or blowoff lines shall be directed into or in proximity to sewer systems used for other purposes.

**1910.110(d)(12)** *General provisions applicable to systems in industrial plants (of 2,000 gallons water capacity and more) and to bulk filling plants.* **(i)** When standard watch service is provided, it shall be extended to the LP-Gas installation and personnel properly trained.

**1910.110(d)(12)(ii)** If loading and unloading are normally done during other than daylight hours, adequate lights shall be provided to illuminate storage containers, control valves, and other equipment.

**1910.110(d)(12)(iii)** Suitable roadways or means of access for extinguishing equipment such as wheeled extinguishers or fire department apparatus shall be provided.

**1910.110(d)(12)(iv)** To minimize trespassing or tampering, the area which includes container appurtenances, pumping equipment, loading and unloading facilities, and cylinder-filling facilities shall be enclosed with at least a 6-foot-high industrial type fence unless otherwise adequately protected. There shall be at least two means of emergency access.

**1910.110(d)(13)** *Container-charging plants.* **(i)** The container-charging room shall be located not less than:

**1910.110(d)(13)(i)(a)** Ten feet from bulk storage containers.

**1910.110(d)(13)(i)(b)** [Reserved]

**1910.110(d)(13)(ii)** Tank truck filling station outlets shall be located not less than:

**1910.110(d)(13)(ii)(a)** [Reserved]

**1910.110(d)(13)(ii)(b)** Ten feet from pumps and compressors if housed in one or more separate buildings.

**1910.110(d)(13)(iii)** The pumps or compressors may be located in the container-charging room or building, in a separate building, or outside of buildings. When housed in a separate building, such building (a small noncombustible weather cover is not to be construed as a building) shall be located not less than:

**1910.110(d)(13)(iii)(a)** Ten feet from bulk storage tanks.

**1910.110(d)(13)(iii)(b)** [Reserved]

**1910.110(d)(13)(iii)(c)** Twenty-five feet from sources of ignition.

**1910.110(d)(13)(iv)** When a part of the container-charging building is to be used for a boiler room or where open flames or similar sources of ignition exist or are employed, the space to be so occupied shall be separated from container charging room by a partition wall or walls of fire-resistant construction continuous from floor to roof or ceiling. Such separation walls shall be without openings and shall be joined to the floor, other walls, and ceiling or roof in a manner to effect a permanent gas-tight joint.

**1910.110(d)(13)(v)** Electrical equipment and installations shall conform with paragraphs (b) (17) and (18) of this section.

**1910.110(d)(14)** *Fire protection.* **(i)** Each bulk plant shall be provided with at least one approved portable fire extinguisher having a minimum rating of 12-B, C.

**1910.110(d)(14)(ii)** In industrial installations involving containers of 150,000 gallons aggregate water capacity or more, provision shall be made for an adequate supply of water at the container site for fire protection in the container area, unless other adequate means for fire control are provided. Water hydrants shall be readily accessible and so spaced as to provide water protection for all containers. Sufficient lengths of firehose shall be provided at each hydrant location on a hose cart, or other means provided to facilitate easy movement of the hose in the container area. It is desirable to equip the outlet of each hose line with a combination fog nozzle. A shelter shall be provided to protect the hose and its conveyor from the weather.

**1910.110(d)(15)** [Reserved]

**1910.110(d)(16)** *Lighting.* Electrical equipment and installations shall conform to paragraphs (b) (17) and (18) of this section.

**1910.110(d)(17)** *Vaporizers for internal combustion engines.* The provisions of paragraph (e)(8) of this section shall apply.

**1910.110(d)(18)** *Gas regulating and mixing equipment for internal combustion engines.* The provisions of paragraph (e)(9) of this section shall apply.

**1910.110(e)** *Liquefied petroleum gas as a motor fuel*—**(1)** *Application.* **(i)** This paragraph applies to internal combustion engines, fuel containers, and pertinent equipment for the use of liquefied petroleum gases as a motor fuel on easily movable, readily portable units including self-propelled vehicles.

**1910.110(e)(1)(ii)** Fuel containers and pertinent equipment for internal combustion engines using liquefied petroleum gas where installation is of the stationary type are covered by paragraph (d) of this section. This paragraph does not apply to containers for transportation of liquefied petroleum gases nor to marine fuel use. All requirements of paragraph (b) of this section apply to this paragraph, unless otherwise noted in paragraph (b) of this section.

**1910.110(e)(2)** *General.* **(i)** Fuel may be used from the cargo tank of a truck while in transit, but not from cargo tanks on trailers or semitrailers. The use of fuel from the cargo tanks to operate stationary engines is permitted providing wheels are securely blocked.

**1910.110(e)(2)(ii)** Passenger-carrying vehicles shall not be fueled while passengers are on board.

**1910.110(e)(2)(iii)** Industrial trucks (including lift trucks) equipped with permanently mounted fuel containers shall be charged outdoors.

Charging equipment shall comply with the provisions of paragraph (h) of this section.

**1910.110(e)(2)(iv)** LP-Gas fueled industrial trucks shall comply with the Standard for Type Designations, Areas of Use, Maintenance and Operation of Powered Industrial Trucks, NFPA 505-1969, which is incorporated by reference as specified in §1910.6.

**1910.110(e)(2)(v)** Engines on vehicles shall be shut down while fueling if the fueling operation involves venting to the atmosphere.

**1910.110(e)(3)** *Design pressure and classification of fuel containers.* (i) Except as covered in paragraphs (e)(3) (ii) and (iii) of this section, containers shall be in accordance with Table H-32.

**1910.110(e)(3)(ii)** Fuel containers for use in industrial trucks (including lift trucks) shall be either DOT containers authorized for LP-Gas service having a minimum service pressure of 240 p.s.i.g. or minimum Container Type 250. Under 1950 and later ASME codes, this means a 312.5-p.s.i.g. design pressure container.

### TABLE H-32

| Container type | For gases with vapor press Not to exceed lb. gage at 100 °F. (37.8 °C) | 1949 and earlier editions of ASME Code (Par. U-68, U-69) | Minimum design pressure of container, lb. per sq. in. gage — 1949 edition of ASME Code (Par. U-200, U-201); 1950, 1952, 1956, 1959, 1962, 1965, and 1968 (Division I) editions of ASME Code; All editions of API-ASME Code[2] |
|---|---|---|---|
| [1]200 | 215Z | 200 | 250 |

[1] Container type may be increased by increments of 25. The minimum design pressure of containers shall be 100% of the container type designation when constructed under 1949 or earlier editions of the ASME Code (Par. U-68 and U-69). The minimum design pressure of containers shall be 125% of the container type designation when constructed under: (1) The 1949 ASME Code (par. U-200 and U-201), (2) 1950, 1952, 1956, 1959, 1962, 1965, and 1968 (Division 1) editions of the ASME Code, and (3) all editions of the API-ASME Code.

[2] Construction of containers under the API-ASME Code is not authorized after July 1, 1961.

**1910.110(e)(3)(iii)** Containers manufactured and maintained under DOT specifications and regulations may be used as fuel containers. When so used they shall conform to all requirements of this paragraph.

**1910.110(e)(3)(iv)** All container inlets and outlets except safety relief valves and gaging devices shall be labeled to designate whether they communicate with vapor or liquid space. Labels may be on valves.

**1910.110(e)(4)** *Installation of fuel containers.* (i) Containers shall be located in a place and in a manner to minimize the possibility of damage to the container. Containers located in the rear of trucks and buses, when protected by substantial bumpers, will be considered in conformance with this requirement. Fuel containers on passenger-carrying vehicles shall be installed as far from the engine as is practicable, and the passenger space and any space containing radio equipment shall be sealed from the container space to prevent direct seepage of gas to these spaces. The container compartment shall be vented to the outside. In case the fuel container is mounted near the engine or the exhaust system, the container shall be shielded against direct heat radiation.

**1910.110(e)(4)(ii)** Containers shall be installed with as much clearance as practicable but never less than the minimum road clearance of the vehicle under maximum spring deflection. This minimum clearance shall be to the bottom of the container or to the lowest fitting on the container or housing, whichever is lower.

**1910.110(e)(4)(iii)** Permanent and removable fuel containers shall be securely mounted to prevent jarring loose, slipping, or rotating, and the fastenings shall be designed and constructed to withstand static loading in any direction equal to twice the weight of the tank and attachments when filled with fuel using a safety factor of not less than four based on the ultimate strength of the material to be used. Field welding, when necessary, shall be made only on saddle plates, lugs or brackets, originally attached to the container by the tank manufacturer.

**1910.110(e)(4)(iv)** Fuel containers on buses shall be permanently installed.

**1910.110(e)(4)(v)** Containers from which vapor only is to be withdrawn shall be installed and equipped with suitable connections to minimize the accidental withdrawal of liquid.

**1910.110(e)(5)** *Valves and accessories.* (i) Container valves and accessories shall have a rated working pressure of at least 250 p.s.i.g., and shall be of a type suitable for liquefied petroleum gas service.

**1910.110(e)(5)(ii)** The filling connection shall be fitted with an approved double back-pressure check valve, or a positive shutoff in conjunction with an internal back-pressure check valve. On a removable container the filler valve may be a hand operated shutoff valve with an internal excess flow valve. Main shutoff valves on the container on liquid and vapor lines must be readily accessible.

**1910.110(e)(5)(iii)** With the exceptions of paragraph (e)(5)(iv)(c) of this section, filling connections equipped with approved automatic back-pressure check valves, and safety relief valves, all connections to containers having openings for the flow of gas in excess of a No. 54 drill size shall be equipped with approved automatic excess flow valves to prevent discharge of content in case connections are broken.

**1910.110(e)(5)(iv)** Liquid-level gaging devices:

**1910.110(e)(5)(iv)(a)** Variable liquid-level gages which require the venting of fuel to the atmosphere shall not be used on fuel containers of industrial trucks (including lift trucks).

**1910.110(e)(5)(iv)(b)** On portable containers that may be filled in the vertical and/or horizontal position, the fixed liquid-level gage must indicate maximum permitted filling level for both vertical and horizontal filling with the container oriented to place the safety relief valve in communication with the vapor space.

**1910.110(e)(5)(iv)(c)** In the case of containers used solely in farm tractor service, and charged at a point at least 50 feet from any important building, the fixed liquid-level gaging device may be so constructed that the outward flow of container content exceeds that passed by a No. 54 drill size opening, but in no case shall the flow exceed that passed by a No. 31 drill-size opening.

An excess flow valve is not required. Fittings equipped with such restricted drill size opening and container on which they are used shall be marked to indicate the size of the opening.

**1910.110(e)(5)(iv)(d)** All valves and connections on containers shall be adequately protected to prevent damage due to accidental contact with stationary objects or from loose objects thrown up from the road, and all valves shall be safeguarded against damage due to collision, overturning or other accident. For farm tractors where parts of the vehicle provide such protection to valves and fittings, the foregoing requirements shall be considered fulfilled. However, on removable type containers the protection for the fittings shall be permanently attached to the container.

**1910.110(e)(5)(iv)(e)** When removable fuel containers are used, means shall be provided in the fuel system to minimize the escape of fuel when the containers are exchanged. This may be accomplished by either of the following methods:

**1910.110(e)(5)(iv)(e)(1)** Using an approved automatic quick-closing coupling (a type closing in both directions when uncoupled) in the fuel line, or

**1910.110(e)(5)(iv)(e)(2)** Closing the valve at the fuel container and allowing the engine to run until the fuel in the line is consumed.

**1910.110(e)(6)** *Piping—including pipe, tubing, and fittings.* (i) Pipe from fuel container to first-stage regulator shall be not less than schedule 80 wrought iron or steel (black or galvanized), brass or copper; or seamless copper, brass, or steel tubing. Steel tubing shall have a minimum wall thickness of 0.049 inch. Steel pipe or tubing shall be adequately protected against exterior corrosion. Copper tubing shall be types K or L or equivalent having a minimum wall thickness of 0.032 inch. Approved flexible connections may be used between container and regulator or between regulator and gas-air mixer within the limits of approval. The use of aluminum pipe or tubing is prohibited. In the case of removable containers an approved flexible connection shall be used between the container and the fuel line.

**1910.110(e)(6)(ii)** All piping shall be installed, braced, and supported so as to reduce to a minimum the possibility of vibration strains or wear.

**1910.110(e)(7)** *Safety devices.* **(i)** Spring-loaded internal type safety relief valves shall be used on all motor fuel containers.

**1910.110(e)(7)(ii)** The discharge outlet from safety relief valves shall be located on the outside of enclosed spaces and as far as practicable from possible sources of ignition, and vented upward within 45 degrees of the vertical in such a manner as to prevent impingement of escaping gas upon containers, or parts of vehicles, or on vehicles in adjacent lines of traffic. A rain cap or other protector shall be used to keep water and dirt from collecting in the valve.

**1910.110(e)(7)(iii)** When a discharge line from the container safety relief valve is used, the line shall be metallic, other than aluminum, and shall be sized, located, and maintained so as not to restrict the required flow of gas from the safety relief valve. Such discharge line shall be able to withstand the pressure resulting from the discharge of vapor when the safety relief valve is in the full open position. When flexibility is necessary, flexible metal hose or tubing shall be used.

**1910.110(e)(7)(iv)** Portable containers equipped for volumetric filling may be filled in either the vertical or horizontal position only when oriented to place the safety relief valve in communication with the vapor space.

**1910.110(e)(7)(v)** Paragraph (b)(10)(xii) of this section for hydrostatic relief valves shall apply.

**1910.110(e)(8)** *Vaporizers.* **(i)** Vaporizers and any part thereof and other devices that may be subjected to container pressure shall have a design pressure of at least 250 p.s.i.g.

**1910.110(e)(8)(ii)** Each vaporizer shall have a valve or suitable plug which will permit substantially complete draining of the vaporizer. It shall be located at or near the lowest portion of the section occupied by the water or other heating medium.

**1910.110(e)(8)(iii)** Vaporizers shall be securely fastened so as to minimize the possibility of becoming loosened.

**1910.110(e)(8)(iv)** Each vaporizer shall be permanently marked at a visible point as follows:

**1910.110(e)(8)(iv)***(a)* With the design pressure of the fuel-containing portion in p.s.i.g.

**1910.110(e)(8)(iv)***(b)* With the water capacity of the fuel-containing portion of the vaporizer in pounds.

**1910.110(e)(8)(v)** Devices to supply heat directly to a fuel container shall be equipped with an automatic device to cut off the supply of heat before the pressure inside the fuel container reaches 80 percent of the start to discharge pressure setting of the safety relief device on the fuel container.

**1910.110(e)(8)(vi)** Engine exhaust gases may be used as a direct source of heat supply for the vaporization of fuel if the materials of construction of those parts of the vaporizer in contact with exhaust gases are resistant to the corrosive action of exhaust gases and the vaporizer system is designed to prevent excessive pressures.

**1910.110(e)(8)(vii)** Vaporizers shall not be equipped with fusible plugs.

**1910.110(e)(9)** *Gas regulating and mixing equipment.* **(i)** Approved automatic pressure reducing equipment shall be installed in a secure manner between the fuel supply container and gas-air mixer for the purpose of reducing the pressure of the fuel delivered to the gas-air mixer.

**1910.110(e)(9)(ii)** An approved automatic shutoff valve shall be provided in the fuel system at some point ahead of the inlet of the gas-air mixer, designed to prevent flow of fuel to the mixer when the ignition is off and the engine is not running. In the case of industrial trucks and engines operating in buildings other than those used exclusively to house engines, the automatic shutoff valve shall be designed to operate if the engine should stop. Atmospheric type regulators (zero governors) shall be considered adequate as an automatic shutoff valve only in cases of outdoor operation such as farm tractors, construction equipment, irrigation pump engines, and other outdoor stationary engine installations.

**1910.110(e)(9)(iii)** The source of the air for combustion shall be completely isolated from the passenger compartment, ventilating system, or air-conditioning system.

**1910.110(e)(10)** [Reserved]

**1910.110(e)(11)** *Stationary engines in buildings.* Stationary engines and gas turbines installed in buildings, including portable engines used instead of or to supplement stationary engines, shall comply with the Standard for the Institution and Use of Stationary Combustion Engines and Gas Turbines, NFPA 37-1970, and the appropriate provisions of paragraphs (b), (c), and (d) of this section.

**1910.110(e)(12)** *Portable engines in buildings.* **(i)** Portable engines may be used in buildings only for emergency use, except as provided by subparagraph (11) of this paragraph.

**1910.110(e)(12)(ii)** Exhaust gases shall be discharged to outside the building or to an area where they will not constitute a hazard.

**1910.110(e)(12)(iii)** Provision shall be made to supply sufficient air for combustion and cooling.

**1910.110(e)(12)(iv)** An approved automatic shutoff valve shall be provided in the fuel system ahead of the engine, designed to prevent flow of fuel to the engine when the ignition is off or if the engine should stop.

**1910.110(e)(12)(v)** The capacity of LP-Gas containers used with such engines shall comply with the applicable occupancy provision of paragraph (c)(5) of this section.

**1910.110(e)(13)** *Industrial trucks inside buildings.* **(i)** LP-Gas-fueled industrial trucks are permitted to be used in buildings and structures.

**1910.110(e)(13)(ii)** No more than two LP-Gas containers shall be used on an industrial truck for motor fuel purposes.

**1910.110(e)(13)(iii)-(iv)** [Reserved]

**1910.110(e)(13)(v)** Industrial trucks shall not be parked and left unattended in areas of possible excessive heat or sources of ignition.

**1910.110(e)(14)** *Garaging LP-Gas-fueled vehicles.* **(i)** LP-Gas-fueled vehicles may be stored or serviced inside garages provided there are no leaks in the fuel system and the fuel tanks are not filled beyond the maximum filling capacity specified in paragraph (b)(12)(i) of this section.

**1910.110(e)(14)(ii)** LP-Gas-fueled vehicles being repaired in garages shall have the container shutoff valve closed except when fuel is required for engine operation.

**1910.110(e)(14)(iii)** Such vehicles shall not be parked near sources of heat, open flames, or similar sources of ignition or near open pits unless such pits are adequately ventilated.

**1910.110(f)** *Storage of containers awaiting use or resale*—**(1)** *Application.* This paragraph shall apply to the storage of portable containers not in excess of 1,000 pounds water capacity, filled or partially filled, at user location but not connected for use, or in storage for resale by dealers or resellers. This paragraph shall not apply to containers stored at charging plants or at plants devoted primarily to the storage and distribution of LP-Gas or other petroleum products.

**1910.110(f)(2)** *General.* **(i)** Containers in storage shall be located so as to minimize exposure to excessive temperature rise, physical damage, or tampering by unauthorized persons.

**1910.110(f)(2)(ii)** Containers when stored inside shall not be located near exits, stairways, or in areas normally used or intended for the safe exit of people.

**1910.110(f)(2)(iii)** Container valves shall be protected while in storage as follows:

**1910.110(f)(2)(iii)***(a)* By setting into recess of container to prevent the possibility of their being struck if the container is dropped upon a flat surface, or

**1910.110(f)(2)(iii)***(b)* By ventilated cap or collar, fastened to container capable of withstanding blow from any direction equivalent to that of a 30-pound weight dropped 4 feet. Construction must be such that a blow will not be transmitted to a valve or other connection.

**1910.110(f)(2)(iv)** The outlet valves of containers in storage shall be closed.

**1910.110(f)(2)(v)** Empty containers which have been in LP-Gas service when stored inside, shall be considered as full containers for the purpose of determining the maximum quantity of LP-Gas permitted by this paragraph.

**1910.110(f)(3)** [Reserved]

**1910.110(f)(4)** *Storage within buildings not frequented by the public (such as industrial buildings).* **(i)** The quantity of LP-Gas stored shall not exceed

300 pounds (approximately 2,550 cubic feet in vapor form) except as provided in subparagraph (5) of this paragraph.

**1910.110(f)(4)(ii)** Containers carried as a part of service equipment on highway mobile vehicles are not to be considered in the total storage capacity in subdivision (i) of this subparagraph provided such vehicles are stored in private garages, and are limited to one container per vehicle with an LP-Gas capacity of not more than 100 pounds. All container valves shall be closed.

**1910.110(f)(5)** *Storage within special buildings or rooms.* **(i)** The quantity of LP-Gas stored in special buildings or rooms shall not exceed 10,000 pounds.

**1910.110(f)(5)(ii)** The walls, floors, and ceilings of container storage rooms that are within or adjacent to other parts of the building shall be constructed of material having at least a 2-hour fire resistance rating.

**1910.110(f)(5)(iii)** A portion of the exterior walls or roof having an area not less than 10 percent of that of the combined area of the enclosing walls and roof shall be of explosion relieving construction.

**1910.110(f)(5)(iv)** Each opening from such storage rooms to other parts of the building shall be protected by a 1 1/2 hour (B) fire door listed by a nationally recognized testing laboratory. Refer to §1910.7 for definition of nationally recognized testing laboratory.

**1910.110(f)(5)(v)** Such rooms shall have no open flames for heating or lighting.

**1910.110(f)(5)(vi)** Such rooms shall be adequately ventilated both top and bottom to the outside only. The openings from such vents shall be at least 5 feet away from any other opening into any building.

**1910.110(f)(5)(vii)** The floors of such rooms shall not be below ground level. Any space below the floor shall be of solid fill or properly ventilated to the open air.

**1910.110(f)(5)(viii)** Such storage rooms shall not be located adjoining the line of property occupied by schools, churches, hospitals, athletic fields or other points of public gathering.

**1910.110(f)(5)(ix)** Fixed electrical equipment shall be installed in accordance with paragraph (b)(18) of this section.

**1910.110(f)(6)** *Storage outside of buildings.* **(i)** Storage outside of buildings, for containers awaiting use or resale, shall be located in accordance with Table H-33 with respect to:

**1910.110(f)(6)(i)(a)** The nearest important building or group of buildings;

**1910.110(f)(6)(i)(b)** [Reserved]

**1910.110(f)(6)(i)(c)** Busy thoroughfares;

### TABLE H-33

| Quantity of LP-Gas Stored | Distance |
|---|---|
| 500 pounds or less | 0 |
| 501 to 2,500 pounds | [1]0 |
| 2,501 to 6,000 pounds | 10 feet |
| 6,001 to 10,000 pounds | 20 feet |
| Over 10,000 pounds | 25 feet |

[1] Container or containers shall be at least 10 feet from any building on adjoining property, any sidewalk, or any of the exposures described in §1910.110(f)(6)(i)(c) or (d) of this paragraph.

**1910.110(f)(6)(ii)** Containers shall be in a suitable enclosure or otherwise protected against tampering.

**1910.110(f)(7)** *Fire protection.* Storage locations other than supply depots separated and located apart from dealer, reseller, or user establishments shall be provided with at least one approved portable fire extinguisher having a minimum rating of 8-B, C.

**1910.110(g)** [Reserved]

**1910.110(h)** *Liquefied petroleum gas service stations*—**(1)** *Application.* This paragraph applies to storage containers, and dispensing devices, and pertinent equipment in service stations where LP-Gas is stored and is dispensed into fuel tanks of motor vehicles. See paragraph (e) of this section for requirements covering use of LP-Gas as a motor fuel. All requirements of paragraph (b) of this section apply to this paragraph unless otherwise noted.

**1910.110(h)(2)** *Design pressure and classification of storage containers.* Storage containers shall be designed and classified in accordance with Table H-34.

### TABLE H-34

| Container type | For gases with vapor press Not to exceed lb. gage at 100 °F. (37.8 °C) | 1949 and earlier editions of ASME Code (Par. U-68, U-69) | 1949 edition of ASME Code (Par. U-200, U-201); 1950, 1952, 1956, 1959, 1962, 1965, and 1968 (Division I) editions of ASME Code; All editions of API-ASME Code[2] |
|---|---|---|---|
| [1]200 | 215 | 200 | 250 |

[1] Container type may be increased by increments of 25. The minimum design pressure of containers shall be 100% of the container type designation when constructed under 1949 or earlier editions of the ASME Code (Par. U-68 and U-69). The minimum design pressure of containers shall be 125% of the container type designation when constructed under: (1) The 1949 ASME Code (par. U-200 and U-201), (2) 1950, 1952, 1956, 1959, 1962, 1965, and 1968 (Division 1) editions of the ASME Code, and (3) all editions of the API-ASME Code.

[2] Construction of containers under the API-ASME Code is not authorized after July 1, 1961.

**1910.110(h)(3)** *Container valves and accessories.* **(i)** A filling connection on the container shall be fitted with one of the following:

**1910.110(h)(3)(i)(a)** A combination back-pressure check and excess flow valve.

**1910.110(h)(3)(i)(b)** One double or two single back-pressure valves.

**1910.110(h)(3)(i)(c)** A positive shutoff valve, in conjunction with either,

**1910.110(h)(3)(i)(c)(1)** An internal back-pressure valve, or

**1910.110(h)(3)(i)(c)(2)** On internal excess flow valve.

In lieu of an excess flow valve, filling connections may be fitted with a quick-closing internal valve, which shall remain closed except during operating periods. The mechanism for such valves may be provided with a secondary control which will cause it to close automatically in case of fire. When a fusible plug is used its melting point shall not exceed 220 °F.

**1910.110(h)(3)(ii)** A filling pipe inlet terminal not on the container shall be fitted with a positive shutoff valve in conjunction with either;

**1910.110(h)(3)(ii)(a)** A black pressure check valve, or

**1910.110(h)(3)(ii)(b)** An excess flow check valve.

**1910.110(h)(3)(iii)** All openings in the container except those listed below shall be equipped with approved excess flow check valves:

**1910.110(h)(3)(iii)(a)** Filling connections as provided in subdivision (i) of this subparagraph.

**1910.110(h)(3)(iii)(b)** Safety relief connections as provided in paragraph (b)(7)(ii) of this section.

**1910.110(h)(3)(iii)(c)** Liquid-level gaging devices as provided in paragraphs (b)(7)(iv) and (19)(iv) of this section.

**1910.110(h)(3)(iii)(d)** Pressure gage connections as provided in paragraph (b)(7)(v) of this section.

**1910.110(h)(3)(iv)** All container inlets and outlets except those listed below shall be labeled to designate whether they connect with vapor or liquid (labels may be on valves):

**1910.110(h)(3)(iv)(a)** Safety relief valves.

**1910.110(h)(3)(iv)(b)** Liquid-level gaging devices.

**1910.110(h)(3)(iv)(c)** Pressure gages.

**1910.110(h)(3)(v)** Each storage container shall be provided with a suitable pressure gage.

**1910.110(h)(4)** *Safety-relief valves.* **(i)** All safety-relief devices shall be installed as follows:

**1910.110(h)(4)(i)(a)** On the container and directly connected with the vapor space.

**1910.110(h)(4)(i)(b)** Safety-relief valves and discharge piping shall be protected against physical damage. The outlet shall be provided with loose-fitting rain caps. There shall be no return bends or restrictions in the discharge piping.

**1910.110(h)(4)(i)(c)** The discharge from two or more safety relief valves having the same pressure settings may be run into a common discharge header. The cross-sectional area of such header shall be at least equal to the sum of the cross-sectional areas of the individual discharges.

**1910.110(h)(4)(i)(d)** Discharge from any safety relief device shall not terminate in any building nor beneath any building.

**1910.110(h)(4)(ii)** Aboveground containers shall be provided with safety relief valves as follows:

**1910.110(h)(4)(ii)(a)** The rate of discharge, which may be provided by one or more valves, shall be not less than that specified in paragraph (b)(10)(ii) of this section.

**1910.110(h)(4)(ii)(b)** The discharge from safety relief valves shall be vented to the open air unobstructed and vertically upwards in such a manner as to prevent any impingement of escaping gas upon the container; loose-fitting rain caps shall be used. On a container having a water capacity greater than 2,000 gallons, the discharge from the safety relief valves shall be vented away from the container vertically upwards to a point at least 7 feet above the container. Suitable provisions shall be made so that any liquid or condensate that may accumulate inside of the relief valve or its discharge pipe will not render the valve inoperative. If a drain is used, a means shall be provided to protect the container, adjacent containers, piping, or equipment against impingement of flame resulting from ignition of the product escaping from the drain.

**1910.110(h)(4)(iii)** Underground containers shall be provided with safety relief valves as follows:

**1910.110(h)(4)(iii)(a)** The discharge from safety-relief valves shall be piped vertically upward to a point at least 10 feet above the ground. The discharge lines or pipes shall be adequately supported and protected against physical damage.

**1910.110(h)(4)(iii)(b)** [Reserved]

**1910.110(h)(4)(iii)(c)** If no liquid is put into a container until after it is buried and covered, the rate of discharge of the relief valves may be reduced to not less than 30 percent of the rate shown in paragraph (b)(10)(ii) of this section. If liquid fuel is present during installation of containers, the rate of discharge shall be the same as for aboveground containers. Such containers shall not be uncovered until emptied of liquid fuel.

**1910.110(h)(5)** *Capacity of liquid containers.* Individual liquid storage containers shall not exceed 30,000 gallons water capacity.

**1910.110(h)(6)** *Installation of storage containers.* **(i)(a)** Each storage container used exclusively in service station operation shall comply with the following table which specifies minimum distances to a building and groups of buildings.

| Water capacity per container (gallons) | Minimum distances Above ground and underground (feet) | Between aboveground containers (feet) |
|---|---|---|
| Up to 2,000 | 25 | 3 |
| Over 2,000 | 50 | 5 |

Note: The above distances may be reduced to not less than 10 feet for service station buildings of other than wood frame construction.

**1910.110(h)(6)(i)(b)** Readily ignitible material including weeds and long dry grass, shall be removed within 10 feet of containers.

**1910.110(h)(6)(i)(c)** The minimum separation between LP-Gas containers and flammable liquid tanks shall be 20 feet and the minimum separation between a container and the centerline of the dike shall be 10 feet.

**1910.110(h)(6)(i)(d)** LP-Gas containers located near flammable liquid containers shall be protected against the flow or accumulation of flammable liquids by diking, diversion curbs, or grading.

**1910.110(h)(6)(i)(e)** LP-Gas containers shall not be located within diked areas for flammable liquid containers.

**1910.110(h)(6)(i)(f)** Field welding is permitted only on saddle plates or brackets which were applied by the container manufacturer.

**1910.110(h)(6)(i)(g)** When permanently installed containers are interconnected, provision shall be made to compensate for expansion, contraction, vibration, and settling of containers and interconnecting piping. Where flexible connections are used, they shall be of an approved type and shall be designed for a bursting pressure of not less than five times the vapor pressure of the product at 100 °F. The use of nonmetallic hose is prohibited for interconnecting such containers.

**1910.110(h)(6)(i)(h)** Where high water table or flood conditions may be encountered protection against container flotation shall be provided.

**1910.110(h)(6)(ii)** Aboveground containers shall be installed in accordance with this subdivision.

**1910.110(h)(6)(ii)(a)** Containers may be installed horizontally or vertically.

**1910.110(h)(6)(ii)(b)** Containers shall be protected by crash rails or guards to prevent physical damage unless they are so protected by virtue of their location. Vehicles shall not be serviced within 10 feet of containers.

**1910.110(h)(6)(ii)(c)** Container foundations shall be of substantial masonry or other noncombustible material. Containers shall be mounted on saddles which shall permit expansion and contraction, and shall provide against the excessive concentration of stresses. Corrosion protection shall be provided for tank-mounting areas. Structural metal container supports shall be protected against fire. This protection is not required on prefabricated storage and pump assemblies, mounted on a common base, with container bottom not more than 24 inches above ground and whose water capacity is 2,000 gallons or less if the piping connected to the storage and pump assembly is sufficiently flexible to minimize the possibility of breakage or leakage in the event of failure of the container supports.

**1910.110(h)(6)(iii)** Underground containers shall be installed in accordance with this subdivision.

**1910.110(h)(6)(iii)(a)** Containers shall be given a protective coating before being placed under ground. This coating shall be equivalent to hot-dip galvanizing or to two coatings of red lead followed by a heavy coating of coal tar or asphalt. In lowering the container into place, care shall be exercised to minimize abrasion or other damage to the coating. Damage to the coating shall be repaired before back-filling.

**1910.110(h)(6)(iii)(b)** Containers shall be set on a firm foundation (firm earth may be used) and surrounded with earth or sand firmly tamped in place. Backfill should be free of rocks or other abrasive materials.

**1910.110(h)(6)(iii)(c)** A minimum of 2 feet of earth cover shall be provided. Where ground conditions make compliance with this requirement impractical, equivalent protection against physical damage shall be provided. The portion of the container to which manhole and other connections are attached need not be covered. If the location is subjected to vehicular traffic, containers shall be protected by a concrete slab or other cover adequate to prevent the weight of a loaded vehicle imposing concentrated direct loads on the container shell.

**1910.110(h)(7)** *Protection of container fittings.* Valves, regulators, gages, and other container fittings shall be protected against tampering and physical damage.

**1910.110(h)(8)** *Transport truck unloading point.* **(i)** During unloading, the transport truck shall not be parked on public thoroughfares and shall be at least 5 feet from storage containers, and shall be positioned so that shutoff valves are readily accessible.

**1910.110(h)(8)(ii)** The filling pipe inlet terminal shall not be located within a building nor within 10 feet of any building or driveway. It shall be protected against physical damage.

**1910.110(h)(9)** *Piping, valves, and fittings.* **(i)** Piping may be underground, above ground, or a combination of both. It shall be well supported and protected against physical damage and corrosion.

**1910.110(h)(9)(ii)** Piping laid beneath driveways shall be installed to prevent physical damage by vehicles.

**1910.110(h)(9)(iii)** Piping shall be wrought iron or steel (black or galvanized), brass or copper pipe; or seamless copper, brass, or steel tubing and shall be suitable for a minimum pressure of 250 p.s.i.g. Pipe joints may be screwed, flanged, brazed, or welded. The use of aluminum alloy piping or tubing is prohibited.

**1910.110(h)(9)(iv)** All shutoff valves (liquid or gas) shall be suitable for liquefied petroleum gas service and designed for not less than the maximum pressure to which they may be subjected. Valves which may be subjected to container pressure shall have a rated working pressure of at least 250 p.s.i.g.

**1910.110(h)(9)(v)** All materials used for valve seats, packing, gaskets, diaphragms, etc., shall be resistant to the action of LP-Gas.

**1910.110(h)(9)(vi)** Fittings shall be steel, malleable iron, or brass having a minimum working pressure of 250 p.s.i.g. Cast iron pipe fittings, such as ells, tees, and unions shall not be used.

**1910.110(h)(9)(vii)** All piping shall be tested after assembly and proved free from leaks at not less than normal operating pressures.

**1910.110(h)(9)(viii)** Provision shall be made for expansion, contraction, jarring, and vibration, and for settling. This may be accomplished by flexible connections.

**1910.110(h)(10)** *Pumps and accessories.* All pumps and accessory equipment shall be suitable for LP-Gas service, and designed for not less than the maximum pressure to which they may be subjected. Accessories shall have a minimum rated working pressure of 250 p.s.i.g. Positive displacement pumps shall be equipped with suitable pressure actuated bypass valves permitting flow from pump discharge to storage container or pump suction.

**1910.110(h)(11)** *Dispensing devices.* **(i)** Meters, vapor separators, valves, and fittings in the dispenser shall be suitable for LP-Gas service and shall be designed for a minimum working pressure of 250 p.s.i.g.

**1910.110(h)(11)(ii)** Provisions shall be made for venting LP-Gas contained in a dispensing device to a safe location.

**1910.110(h)(11)(iii)** Pumps used to transfer LP-Gas shall be equipped to allow control of the flow and to prevent leakage or accidental discharge. Means shall be provided outside the dispensing device to readily shut off the power in the event of fire or accident.

**1910.110(h)(11)(iv)** A manual shutoff valve and an excess flow check valve shall be installed downstream of the pump and ahead of the dispenser inlet.

**1910.110(h)(11)(v)***(a)* Dispensing hose shall be resistant to the action of LP-Gas in the liquid phase and designed for a minimum bursting pressure of 1,250 p.s.i.g.

**1910.110(h)(11)(v)***(b)* An excess flow check valve or automatic shutoff valve shall be installed at the terminus of the liquid line at the point of attachment of the dispensing hose.

**1910.110(h)(11)(vi)***(a)* LP-Gas dispensing devices shall be located not less than 10 feet from aboveground storage containers greater than 2,000 gallons water capacity. The dispensing devices shall not be less than 20 feet from any building (not including canopies), basement, cellar, pit, or line of adjoining property which may be built upon and not less than 10 feet from sidewalks, streets, or thoroughfares. No drains or blowoff lines shall be directed into or in proximity to the sewer systems used for other purposes.

**1910.110(h)(11)(vi)***(b)* LP-Gas dispensing devices shall be installed on a concrete foundation or as part of a complete storage and dispensing assembly mounted on a common base, and shall be adequately protected from physical damage.

**1910.110(h)(11)(vi)***(c)* LP-Gas dispensing devices shall not be installed within a building except that they may be located under a weather shelter or canopy provided this area is not enclosed on more than two sides. If the enclosing sides are adjacent to each other, the area shall be properly ventilated.

**1910.110(h)(11)(vii)** The dispensing of LP-Gas into the fuel container of a vehicle shall be performed by a competent attendant who shall remain at the LP-Gas dispenser during the entire transfer operation.

**1910.110(h)(12)** *Additional rules.* There shall be no smoking on the driveway of service stations in the dispensing areas or transport truck unloading areas. Conspicuous signs prohibiting smoking shall be posted within sight of the customer being served. Letters on such signs shall be not less than 4 inches high. The motors of all vehicles being fueled shall be shut off during the fueling operations.

**1910.110(h)(13)** *Electrical.* Electrical equipment and installations shall conform to paragraphs (b) (17) and (18) of this section.

**1910.110(h)(14)** *Fire protection.* Each service station shall be provided with at least one approved portable fire extinguisher having at least an 8-B, C, rating.

**1910.110(i)** *Scope*—**(1)** *Application.* **(i)** Paragraph (b) of this section applies to installations made in accordance with the requirements of paragraphs (c), (d), (e), (g), and (h) of this section, except as noted in each of those paragraphs.

**1910.110(i)(1)(ii)** Paragraphs (c) through (h) of this section apply as provided in each of those paragraphs.

**1910.110(i)(2)** *Inapplicability.* This section does not apply to:

**1910.110(i)(2)(i)** Marine and pipeline terminals, natural gas processing plants, refineries, or tank farms other than those at industrial sites.

**1910.110(i)(2)(ii)** LP-Gas refrigerated storage systems;

**1910.110(i)(2)(iii)** LP-Gas when used with oxygen. The requirements of §1910.253 shall apply to such use;

**1910.110(i)(2)(iv)** LP-Gas when used in utility gas plants. The National Fire Protection Association Standard for the Storage and Handling of Liquefied Petroleum Gases at Utility Gas Plants, NFPA No. 59-1968, shall apply to such use;

**1910.110(i)(2)(v)** Low-pressure (not in excess of one-half pound per square inch or 14 inches water column) LP-Gas piping systems, and the installation and operation of residential and commercial appliances including their inlet connections, supplied through such systems. For such systems, the National Fire Protection Association Standard for the Installation of Gas Appliances and Gas Piping, NFPA 54-1969 shall apply.

**1910.110(i)(3)** *Retroactivity.* Unless otherwise stated, it is not intended that the provisions of this section be retroactive.

**1910.110(i)(3)(i)** Existing plants, appliances, equipment, buildings, structures, and installations for the storage, handling or use of LP-Gas, which were in compliance with the current provisions of the National Fire Protection Association Standard for the Storage and Handling of Liquefied Petroleum Gases NFPA No. 58, at the time of manufacture or installation may be continued in use, if such continued use does not constitute a recognized hazard that is causing or is likely to cause death or serious physical harm to employees.

**1910.110(i)(3)(ii)** Stocks of equipment and appliances on hand in such locations as manufacturers' storage, distribution warehouses, and dealers' storage and showrooms, which were in compliance with the current provisions of the National Fire Protection Association Standard for the Storage and Handling of Liquefied Petroleum Gases, NFPA No. 58, at the time of manufacture, may be placed in service, if such use does not constitute a recognized hazard that is causing or is likely to cause death or serious physical harm to employees.

[39 FR 23502, June 27, 1974, as amended at 43 FR 49747, Oct. 24, 1978; 49 FR 5322, Feb. 10, 1984; 53 FR 12122, Apr. 12, 1988; 55 FR 25094, June 20, 1990; 55 FR 32015, Aug. 6, 1990; 58 FR 35309, June 30, 1993; 61 FR 9237, 9238, Mar. 7, 1996; 63 FR 33466, June 18, 1998]

## §1910.111 Storage and handling of anhydrous ammonia.

**1910.111(a)** *General*—**(1)** *Scope.* **(i)** This standard is intended to apply to the design, construction, location, installation, and operation of anhydrous ammonia systems including refrigerated ammonia storage systems.

**1910.111(a)(1)(ii)** This standard does not apply to:

**1910.111(a)(1)(ii)***(a)* Ammonia manufacturing plants.

**1910.111(a)(1)(ii)***(b)* Refrigeration plants where ammonia is used solely as a refrigerant.

**1910.111(a)(2)** *Definitions.* As used in this section.

**1910.111(a)(2)(i)** *Appurtenances.* All devices such as pumps, compressors, safety relief devices, liquid-level gaging devices, valves and pressure gages.

**1910.111(a)(2)(ii)** *Cylinder.* A container of 1,000 pounds of water capacity or less constructed in accordance with Department of Transportation specifications.

**1910.111(a)(2)(iii)** *Code.* The Boiler and Pressure Vessel Code, Section VIII, Unfired Pressure Vessels of the American Society of Mechanical Engineers (ASME)—1968.

**1910.111(a)(2)(iv)** *Container.* Includes all vessels, tanks, cylinders, or spheres used for transportation, storage, or application of anhydrous ammonia.

**1910.111(a)(2)(v)** *DOT.* U.S. Department of Transportation.

**1910.111(a)(2)(vi)** *Design pressure* is identical to the term *Maximum Allowable Working Pressure* used in the Code.

# OSHA Standards for General Industry

**1910.111(a)(2)(vii)** *Farm vehicle* (implement of husbandry). A vehicle for use on a farm on which is mounted a container of not over 1,200 gallons water capacity.

**1910.111(a)(2)(viii)** *Filling density.* the percent ratio of the weight of the gas in a container to the weight of water at 60 °F. that the container will hold.

**1910.111(a)(2)(ix)** *Gas.* Anhydrous ammonia in either the gaseous or liquefied state.

**1910.111(a)(2)(x)** *Gas masks.* Gas masks must be approved by the National Institute for Occupational Safety and Health (NIOSH) under 42 CFR part 84 for use with anhydrous ammonia.

**1910.111(a)(2)(xi)** *Capacity.* Total volume of the container in standard U.S. gallons.

**1910.111(a)(2)(xii)** DOT specifications—Regulations of the Department of Transportation published in 49 CFR Chapter I.

**1910.111(b)** **Basic rules.** This paragraph applies to all paragraphs of this section unless otherwise noted.

**1910.111(b)(1)** *Approval of equipment and systems.* Each appurtenance shall be approved in accordance with paragraph (b)(1) (i), (ii), (iii), or (iv) of this section.

**1910.111(b)(1)(i)** It was installed before February 8, 1973, and was approved, tested, and installed in accordance with either the provisions of the American National Standard for the Storage and Handling of Anhydrous Ammonia, K61.1, or the Fertilizer Institute Standards for the Storage and Handling of Agricultural Anhydrous Ammonia, M-1, (both of which are incorporated by reference as specified in §1910.6) in effect at the time of installation; or

**1910.111(b)(1)(ii)** It is accepted, or certified, or listed, or labeled, or otherwise determined to be safe by a nationally recognized testing laboratory; or

**1910.111(b)(1)(iii)** It is a type which no nationally recognized testing laboratory does, or will undertake to, accept, certify, list, label, or determine to be safe; and such equipment is inspected or tested by any Federal, State, municipal, or other local authority responsible for enforcing occupational safety provisions of a Federal, State, municipal or other local law, code, or regulation pertaining to the storage, handling, transport, and use of anhydrous ammonia, and found to be in compliance with either the provisions of the American National Standard for the Storage and Handling of Anhydrous Ammonia, K61.1, or the Fertilizer Institute Standards for the Storage and Handling of Agricultural Anhydrous Ammonia, M-1, in effect at the time of installation; or

**1910.111(b)(1)(iv)** It is a custom-designed and custom-built unit, which no nationally recognized testing laboratory, or Federal, State, municipal or local authority responsible for the enforcement of a Federal, State, municipal, or local law, code or regulation pertaining to the storage, transportation and use of anhydrous ammonia is willing to undertake to accept, certify, list, label or determine to be safe, and the employer has on file a document attesting to its safe condition following the conduct of appropriate tests. The document shall be signed by a registered professional engineer or other person having special training or experience sufficient to permit him to form an opinion as to safety of the unit involved. The document shall set forth the test bases, test data and results, and also the qualifications of the certifying person.

**1910.111(b)(1)(v)** For the purposes of this paragraph (b)(1), the word *listed* means that equipment is of a kind mentioned in a list which is published by a nationally recognized laboratory which makes periodic inspection of the production of such equipment, and states such equipment meets nationally recognized standards or has been tested and found safe for use in a specified manner. *Labeled* means there is attached to it a label, symbol, or other identifying mark of a nationally recognized testing laboratory which, makes periodic inspections of the production of such equipment, and whose labeling indicates compliance with nationally recognized standards or tests to determine safe use in a specified manner. *Certified* means it has been tested and found by a nationally recognized testing laboratory to meet nationally recognized standards or to be safe for use in a specified manner, or is of a kind whose production is periodically inspected by a nationally recognized testing laboratory, and it bears a label, tag, or other record of certification.

**1910.111(b)(1)(vi)** For the purposes of this paragraph (b)(1), refer to §1910.7 for definition of nationally recognized testing laboratory.

**1910.111(b)(2)** *Requirements for construction, original test and requalification of nonrefrigerated containers.* **(i)** Containers used with systems covered in paragraphs (c), (f), (g), and (h) of this section shall be constructed and tested in accordance with the Code except that construction under Table UW12 at a basic joint efficiency of under 80 percent is not authorized.

**1910.111(b)(2)(ii)** Containers built according to the Code do not have to comply with Paragraphs UG125 to UG128 inclusive, and Paragraphs UG132 and UG133 of the Code.

**1910.111(b)(2)(iii)** Containers exceeding 36 inches in diameter or 250 gallons water capacity shall be constructed to comply with one or more of the following:

**1910.111(b)(2)(iii)(a)** Containers shall be stress relieved after fabrication in accordance with the Code, or

**1910.111(b)(2)(iii)(b)** Cold-form heads when used, shall be stress relieved, or

**1910.111(b)(2)(iii)(c)** Hot-formed heads shall be used.

**1910.111(b)(2)(iv)** Welding to the shell, head, or any other part of the container subject to internal pressure shall be done in compliance with the Code. Other welding is permitted only on saddle plates, lugs, or brackets attached to the container by the container manufacturer.

**1910.111(b)(2)(v)** Containers used with systems covered in paragraph (e) of this section shall be constructed and tested in accordance with the DOT specifications.

**1910.111(b)(2)(vi)** The provisions of subdivision **(i)** of this subparagraph shall not be construed as prohibiting the continued use or reinstallation of containers constructed and maintained in accordance with the 1949, 1950, 1952, 1956, 1959, and 1962 editions of the Code or any revisions thereof in effect at the time of fabrication.

**1910.111(b)(3)** *Marking nonrefrigerated containers.* **(i)** System nameplates, when required, shall be permanently attached to the system so as to be readily accessible for inspection and shall include markings as prescribed in subdivision (ii) of this subparagraph.

**1910.111(b)(3)(ii)** Each container or system covered in paragraphs (c), (f), (g), and (h) of this section shall be marked as specified in the following:

**1910.111(b)(3)(ii)(a)** With a notation "Anhydrous Ammonia."

**1910.111(b)(3)(ii)(b)** With a marking identifying compliance with the rules of the Code under which the container is constructed.

Under ground: Container and system nameplate.

Above ground: Container.

**1910.111(b)(3)(ii)(c)** With a notation whether the system is designed for underground or aboveground installation or both.

**1910.111(b)(3)(ii)(d)** With the name and address of the supplier of the system or the trade name of the system and with the date of fabrication.

Under ground and above ground: System nameplate.

**1910.111(b)(3)(ii)(e)** With the water capacity of the container in pounds at 60 °F. or gallons, U.S. Standard.

Under ground: Container and system nameplate.

Above ground: Container.

**1910.111(b)(3)(ii)(f)** With the design pressure in pounds per square inch.

Under ground: Container and system nameplate.

Above ground: Container.

**1910.111(b)(3)(ii)(g)** With the wall thickness of the shell and heads.

Under ground: Container and system nameplate.

Above ground: Container.

**1910.111(b)(3)(ii)(h)** With marking indicating the maximum level to which the container may be filled with liquid anhydrous ammonia at temperatures between 20 °F. and 130 °F. except on containers provided with fixed level indicators, such as fixed length dip tubes, or containers that are filled with weight. Markings shall be in increments of not more than 20 °F.

Above ground and under ground: System nameplate or on liquid-level gaging device.

**1910.111(b)(3)(ii)(i)** With the total outside surface area of the container in square feet.

Under ground: System nameplate.

Above ground: No requirement.

**1910.111(b)(3)(ii)(***j***)** Marking specified on the container shall be on the container itself or on a nameplate permanently attached to it.

**1910.111(b)(4)** *Marking refrigerated containers.* Each refrigerated container shall be marked with nameplate on the outer covering in an accessible place as specified in the following:

**1910.111(b)(4)(i)** With the notation, "Anhydrous Ammonia."

**1910.111(b)(4)(ii)** With the name and address of the builder and the date of fabrication.

**1910.111(b)(4)(iii)** With the water capacity of the container in gallons, U.S. Standard.

**1910.111(b)(4)(iv)** With the design pressure.

**1910.111(b)(4)(v)** With the minimum temperature in degrees Fahrenheit for which the container was designed.

**1910.111(b)(4)(vi)** The maximum allowable water level to which the container may be filled for test purposes.

**1910.111(b)(4)(vii)** With the density of the product in pounds per cubic foot for which the container was designed.

**1910.111(b)(4)(viii)** With the maximum level to which the container may be filled with liquid anhydrous ammonia.

**1910.111(b)(5)** *Location of containers.* **(i)** Consideration shall be given to the physiological effects of ammonia as well as to adjacent fire hazards in selecting the location for a storage container. Containers shall be located outside of buildings or in buildings or sections thereof especially provided for this purpose.

**1910.111(b)(5)(ii)** Permanent storage containers shall be located at least 50 feet from a dug well or other sources of potable water supply, unless the container is a part of a water-treatment installation.

**1910.111(b)(5)(iii)-(iv)** [Reserved]

**1910.111(b)(5)(v)** Storage areas shall be kept free of readily ignitible materials such as waste, weeds, and long dry grass.

**1910.111(b)(6)** *Container appurtenances.* **(i)** All appurtenances shall be designed for not less than the maximum working pressure of that portion of the system on which they are installed. All appurtenances shall be fabricated from materials proved suitable for anhydrous ammonia service.

**1910.111(b)(6)(ii)** All connections to containers except safety relief devices, gaging devices, or those fitted with No. 54 drill-size orifice shall have shutoff valves located as close to the container as practicable.

**1910.111(b)(6)(iii)** Excess flow valves where required by these standards shall close automatically at the rated flows of vapor or liquid as specified by the manufacturer. The connections and line including valves and fittings being protected by an excess flow valve shall have a greater capacity than the rated flow of the excess flow valve so that the valve will close in case of failure of the line or fittings.

**1910.111(b)(6)(iv)** Liquid-level gaging devices that require bleeding of the product to the atmosphere and which are so constructed that outward flow will not exceed that passed by a No. 54 drill-size opening need not be equipped with excess flow valves.

**1910.111(b)(6)(v)** Openings from the container or through fittings attached directly on the container to which pressure gage connections are made need not be equipped with excess flow valves if such openings are not larger than No. 54 drill size.

**1910.111(b)(6)(vi)** Excess flow and back pressure check valves where required by the standards in this section shall be located inside of the container or at a point outside as close as practicable to where the line enters the container. In the latter case installation shall be made in such manner that any undue strain beyond the excess flow or back pressure check valve will not cause breakage between the container and the valve.

**1910.111(b)(6)(vii)** Excess flow valves shall be designed with a bypass, not to exceed a No. 60 drill-size opening to allow equalization of pressures.

**1910.111(b)(6)(viii)** All excess flow valves shall be plainly and permanently marked with the name or trademark of the manufacturer, the catalog number, and the rated capacity.

**1910.111(b)(7)** *Piping, tubing, and fittings.* **(i)** All piping, tubing, and fittings shall be made of material suitable for anhydrous ammonia service.

**1910.111(b)(7)(ii)** All piping, tubing, and fittings shall be designed for a pressure not less than the maximum pressure to which they may be subjected in service.

**1910.111(b)(7)(iii)** All refrigerated piping shall conform to the Refrigeration Piping Code, American National Standards Institute, B31.5-1966 with addenda B31.1a-1968, which is incorporated by reference as specified in §1910.6, as it applies to ammonia.

**1910.111(b)(7)(iv)** Piping used on non-refrigerated systems shall be at least American Society for Testing and Materials (ASTM) A-53-69 Grade B Electric Resistance Welded and Electric Flash Welded Pipe, which is incorporated by reference as specified in §1910.6, or equal. Such pipe shall be at least schedule 40 when joints are welded, or welded and flanged. Such pipe shall be at least schedule 80 when joints are threaded. Threaded connections shall not be back-welded. Brass, copper, or galvanized steel pipe shall not be used.

**1910.111(b)(7)(v)** Tubing made of brass, copper, or other material subject to attack by ammonia shall not be used.

**1910.111(b)(7)(vi)** Cast iron fittings shall not be used but this shall not prohibit the use of fittings made specifically for ammonia service of malleable, nodular, or high strength gray iron meeting American Society for Testing and Materials (ASTM) A47-68, ASTM 395-68, or ASTM A126-66 Class B or C all of which are incorporated by reference as specified in §1910.6.

**1910.111(b)(7)(vii)** Joint compounds shall be resistant to ammonia.

**1910.111(b)(8)** *Hose specifications.* **(i)** Hose used in ammonia service shall conform to the joint Agricultural Ammonia Institute —Rubber Manufacturers Association Specifications for Anhydrous Ammonia Hose.

**1910.111(b)(8)(ii)** Hose subject to container pressure shall be designed for a minimum working pressure of 350 p.s.i.g. and a minimum burst pressure of 1,750 p.s.i.g. Hose assemblies, when made up, shall be capable of withstanding a test pressure of 500 p.s.i.g.

**1910.111(b)(8)(iii)** Hose and hose connections located on the low-pressure side of flow control of pressure-reducing valves shall be designed for a bursting pressure of not less than 5 times the pressure setting of the safety relief devices protecting that portion of the system but not less than 125 p.s.i.g. All connections shall be so designed and constructed that there will be no leakage when connected.

**1910.111(b)(8)(iv)** Where hose is to be used for transferring liquid from one container to another, "wet" hose is recommended. Such hose shall be equipped with approved shutoff valves at the discharge end. Provision shall be made to prevent excessive pressure in the hose.

**1910.111(b)(8)(v)** On all hose one-half inch outside diameter and larger, used for the transfer of anhydrous ammonia liquid or vapor, there shall be etched, cast, or impressed at 5-foot intervals the following information.

"Anhydrous Ammonia" xxx p.s.i.g. (maximum working pressure), manufacturer's name or trademark, year of manufacture.

In lieu of this requirement the same information may be contained on a nameplate permanently attached to the hose.

### TABLE H-36
[Minimum required rate of discharge in cubic feet per minute of air at 120 percent of the maximum permitted start to discharge pressure for safety relief valves.]

| Surface area (sq. ft.) | Flow rate CFM air |
|---|---|
| 20 | 258 |
| 25 | 310 |
| 30 | 360 |
| 35 | 408 |
| 40 | 455 |
| 45 | 501 |
| 50 | 547 |
| 55 | 591 |
| 60 | 635 |
| 65 | 678 |
| 70 | 720 |

# OSHA Standards for General Industry

| Surface area (sq. ft.) | Flow rate CFM air |
|---|---|
| 75 | 762 |
| 80 | 804 |
| 85 | 845 |
| 90 | 885 |
| 95 | 925 |
| 100 | 965 |
| 105 | 1,010 |
| 110 | 1,050 |
| 115 | 1,090 |
| 120 | 1,120 |
| 125 | 1,160 |
| 130 | 1,200 |
| 135 | 1,240 |
| 140 | 1,280 |
| 145 | 1,310 |
| 150 | 1,350 |
| 155 | 1,390 |
| 160 | 1,420 |
| 165 | 1,460 |
| 170 | 1,500 |
| 175 | 1,530 |
| 180 | 1,570 |
| 185 | 1,600 |
| 190 | 1,640 |
| 195 | 1,670 |
| 200 | 1,710 |
| 210 | 1,780 |
| 220 | 1,850 |
| 230 | 1,920 |
| 240 | 1,980 |
| 250 | 2,050 |
| 260 | 2,120 |
| 270 | 2,180 |
| 280 | 2,250 |
| 290 | 2,320 |
| 300 | 2,380 |
| 310 | 2,450 |
| 320 | 2,510 |
| 330 | 2,570 |
| 340 | 2,640 |
| 350 | 2,700 |
| 360 | 2,760 |
| 370 | 2,830 |
| 380 | 2,890 |
| 390 | 2,950 |
| 400 | 3,010 |
| 450 | 3,320 |
| 500 | 3,620 |
| 550 | 3,910 |
| 600 | 4,200 |
| 650 | 4,480 |
| 700 | 4,760 |
| 750 | 5,040 |
| 800 | 5,300 |
| 850 | 5,590 |
| 900 | 5,850 |
| 950 | 6,120 |
| 1,000 | 6,380 |
| 1,050 | 6,640 |
| 1,100 | 6,900 |
| 1,150 | 7,160 |
| 1,200 | 7,410 |
| 1,250 | 7,660 |
| 1,300 | 7,910 |
| 1,350 | 8,160 |
| 1,400 | 8,410 |

| Surface area (sq. ft.) | Flow rate CFM air |
|---|---|
| 1,450 | 8,650 |
| 1,500 | 8,900 |
| 1,550 | 9,140 |
| 1,600 | 9,380 |
| 1,650 | 9,620 |
| 1,700 | 9,860 |
| 1,750 | 10,090 |
| 1,800 | 10,330 |
| 1,850 | 10,560 |
| 1,900 | 10,800 |
| 1,950 | 11,030 |
| 2,000 | 11,260 |
| 2,050 | 11,490 |
| 2,100 | 11,720 |
| 2,150 | 11,950 |
| 2,200 | 12,180 |
| 2,250 | 12,400 |
| 2,300 | 12,630 |
| 2,350 | 12,850 |
| 2,400 | 13,080 |
| 2,450 | 13,300 |
| 2,500 | 13,520 |

Surface Area = total outside surface area of container in square feet. When the surface area is not stamped on the nameplate or when the marking is not legible the area can be calculated by using one of the following formulas:

(1) Cylindrical container with hemispherical heads:

Area = overall length in feet times outside diameter in feet times 3.1416.

(2) Cylindrical container with other than hemispherical heads:

Area = (overall length in feet plus 0.3 outside diameter in feet) times outside diameter in feet times 3.1416.

(3) Spherical container:

Area = outside diameter in feet squared times 3.1416.

Flow Rate—CFM Air = cubic feet per minute of air required at standard conditions, 60 °F. and atmospheric pressure (14.7 p.s.i.a.).

The rate of discharge may be interpolated for intermediate values of surface area. For containers with total outside surface area greater than 2,500 square feet, the required flow rate can be calculated using the formula: Flow Rate CFM Air = 22.11 $A^{0.82}$, where A = outside surface area of the container in square feet.

**1910.111(b)(9)** *Safety relief devices.* **(i)** Every container used in systems covered by paragraphs (c), (f), (g), and (h) of this section shall be provided with one or more safety relief valves of the spring-loaded or equivalent type. The discharge from safety-relief valves shall be vented away from the container upward and unobstructed to the atmosphere. All relief-valve discharge openings shall have suitable rain caps that will allow free discharge of the vapor and prevent entrance of water. Provision shall be made for draining condensate which may accumulate. The rate of the discharge shall be in accordance with the provisions of Table H-36.

**1910.111(b)(9)(ii)** Container safety-relief valves shall be set to start-to-discharge as follows, with relation to the design pressure of the container:

| Containers | Minimum (percent) | Maximum (percent) |
|---|---|---|
| ASME-U-68, U-69 | 110 | 125 |
| ASME-U-200, U-201 | 95 | 100 |
| ASME 1959, 1956, 1952, or 1962 | 95 | 100 |
| API-ASME | 95 | 100 |
| U.S. Coast Guard | 95 | 100 |

As required by DOT Regulations.

**1910.111(b)(9)(iii)** Safety relief devices used in systems covered by paragraphs (c), (f), (g), and (h) of this section shall be constructed to discharge at not less than the rates required in paragraph (b)(9)(i) of this section before the pressure is in excess of 120 percent (not including the 10 percent tolerance referred to in paragraph (b)(9)(ii) of this section) of the maximum permitted start-to-discharge pressure setting of the device.

**1910.111(b)(9)(iv)** Safety-relief valves shall be so arranged that the possibility of tampering will be minimized. If the pressure setting adjustment is external, the relief valves shall be provided with means for sealing the adjustment.

**1910.111(b)(9)(v)** Shutoff valves shall not be installed between the safety-relief valves and the container; except, that a shutoff valve may be used where the arrangement of this valve is such as always to afford full required capacity flow through the relief valves.

**1910.111(b)(9)(vi)** Safety-relief valves shall have direct communication with the vapor space of the container.

**1910.111(b)(9)(vii)** Each container safety-relief valve used with systems covered by paragraphs (c), (f), (g), and (h) of this section shall be plainly and permanently marked with the symbol "NH$_3$" or "AA"; with the pressure in pounds-per-square-inch gage at which the valve is set to start-to-discharge; with the actual rate of discharge of the valve at its full open position in cubic feet per minute of air at 60 °F. and atmospheric pressure; and with the manufacturer's name and catalog number. Example: "NH$_3$ 250-4050 Air" indicates that the valve is suitable for use on an anhydrous ammonia container, is set to start-to-discharge at a pressure of 250 p.s.i.g., and that its rate of discharge at full open position (subdivisions (ii) and (iii) of this subparagraph) is 4,050 cubic feet per minute of air.

**1910.111(b)(9)(viii)** The flow capacity of the relief valve shall not be restricted by any connection to it on either the upstream or downstream side.

**1910.111(b)(9)(ix)** A hydrostatic relief valve shall be installed between each pair of valves in the liquid ammonia piping or hose where liquid may be trapped so as to relieve into the atmosphere at a safe location.

**1910.111(b)(10)** *General.*

**1910.111(b)(10)(i)** [Reserved]

**1910.111(b)(10)(ii)** Stationary storage installations must have at least two suitable gas masks in readily-accessible locations. Full-face masks with ammonia canisters that have been approved by NIOSH under 42 CFR part 84 are suitable for emergency action involving most anhydrous ammonia leaks, particularly leaks that occur outdoors. For respiratory protection in concentrated ammonia atmospheres, a self-contained breathing apparatus is required.

**1910.111(b)(10)(iii)** Stationary storage installations shall have an easily accessible shower or a 50-gallon drum of water.

**1910.111(b)(10)(iv)** Each vehicle transporting ammonia in bulk except farm applicator vehicles shall carry a container of at least 5 gallons of water and shall be equipped with a full face mask.

**1910.111(b)(11)** *Charging of containers.* **(i)** The filling densities for containers that are not refrigerated shall not exceed the following:

| Type of container | Percent by weight | Percent by volume |
|---|---|---|
| Aboveground—Uninsulated | 56 | 82 |
| Aboveground—Uninsulated | | 87.5 |
| Aboveground—Insulated | 57 | 83.5 |
| Underground—Uninsulated | 58 | 85 |

DOT—According to CFR in accord with DOT regulations.

**1910.111(b)(11)(ii)** Aboveground uninsulated containers may be charged 87.5 percent by volume provided the temperature of the anhydrous ammonia being charged is determined to be not lower than 30 °F. or provided the charging of the container is stopped at the first indication of frost or ice formation on its outside surface and is not resumed until such frost or ice has disappeared.

**1910.111(b)(12)** *Transfer of liquids.* **(i)** Anhydrous ammonia shall always be at a temperature suitable for the material of construction and the design of the receiving container.

**1910.111(b)(12)(ii)** The employer shall require the continuous presence of an attendant in the vicinity of the operation during such time as ammonia is being transferred.

**1910.111(b)(12)(iii)** Containers shall be charged or used only upon authorization of the owner.

**1910.111(b)(12)(iv)** Containers shall be gaged and charged only in the open atmosphere or in buildings or areas thereof provided for that purpose.

**1910.111(b)(12)(v)** Pumps used for transferring ammonia shall be those manufactured for that purpose.

**1910.111(b)(12)(v)(a)** Pumps shall be designed for at least 250 p.s.i.g. working pressure.

**1910.111(b)(12)(v)(b)** Positive displacement pumps shall have, installed off the discharged port, a constant differential relief valve discharging into the suction port of the pump through a line of sufficient size to carry the full capacity of the pump at relief valve setting, which setting and installation shall be according to the pump manufacturer's recommendations.

**1910.111(b)(12)(v)(c)** On the discharge side of the pump, before the relief valve line, there shall be installed a pressure gage graduated from 0 to 400 p.s.i.

**1910.111(b)(12)(v)(d)** Plant piping shall contain shutoff valves located as close as practical to pump connections.

**1910.111(b)(12)(vi)** Compressors used for transferring or refrigerating ammonia shall be recommended for ammonia service by the manufacturer.

**1910.111(b)(12)(vi)(a)** Compressors shall be designed for at least 250 p.s.i.g. working pressure.

**1910.111(b)(12)(vi)(b)** Plant piping shall contain shutoff valves located as close as practical to compressor connections.

**1910.111(b)(12)(vi)(c)** A relief valve large enough to discharge the full capacity of the compressor shall be connected to the discharge before any shutoff valve.

**1910.111(b)(12)(vi)(d)** Compressors shall have pressure gages at suction and discharge graduated to at least one and one-half times the maximum pressure that can be developed.

**1910.111(b)(12)(vi)(e)** Adequate means, such as drainable liquid trap, shall be provided on the compressor suction to minimize the entry of liquid into the compressor.

**1910.111(b)(12)(vii)** Loading and unloading systems shall be protected by suitable devices to prevent emptying of the storage container or the container being loaded or unloaded in the event of severance of the hose. Backflow check valves or properly sized excess flow valves shall be installed where necessary to provide such protection. In the event that such valves are not practical, remotely operated shutoff valves may be installed.

**1910.111(b)(13)** *Tank car unloading points and operations.* **(i)** Provisions for unloading tank cars shall conform to the applicable recommendations contained in the DOT regulations.

**1910.111(b)(13)(ii)** The employer shall insure that unloading operations are performed by reliable persons properly instructed and given the authority to monitor careful compliance with all applicable procedures.

**1910.111(b)(13)(iii)** Caution signs shall be so placed on the track or car as to give necessary warning to persons approaching the car from open end or ends of siding and shall be left up until after the car is unloaded and disconnected from discharge connections. Signs shall be of metal or other suitable material, at least 12 by 15 inches in size and bear the words "STOP—Tank Car Connected" or "STOP—Men at Work" the word, "STOP," being in letters at least 4 inches high and the other words in letters at least 2 inches high.

**1910.111(b)(13)(iv)** The track of a tank car siding shall be substantially level.

**1910.111(b)(13)(v)** Brakes shall be set and wheels blocked on all cars being unloaded.

**1910.111(b)(14)** *Liquid-level gaging device.* **(i)** Each container except those filled by weight shall be equipped with an approved liquid-level gaging device. A thermometer well shall be provided in all containers not utilizing a fixed liquid-level gaging device.

**1910.111(b)(14)(ii)** All gaging devices shall be arranged so that the maximum liquid level to which the container is filled is readily determined.

**1910.111(b)(14)(iii)** Gaging devices that require bleeding of the product to the atmosphere such as the rotary tube, fixed tube, and slip tube devices shall be designed so that the maximum opening of the bleed valve is not larger than No. 54 drill size unless provided with an excess flow valve. (This requirement does not apply to farm vehicles used for the application of ammonia as covered in paragraph (h) of this section.)

**1910.111(b)(14)(iv)** Gaging devices shall have a design pressure equal to or greater than the design pressure of the container on which they are installed.

**1910.111(b)(14)(v)** Fixed tube liquid-level gages shall be designed and installed to indicate that level at which the container is filled to 85 percent of its water capacity in gallons.

**1910.111(b)(14)(vi)** Gage glasses of the columnar type shall be restricted to stationary storage installations. They shall be equipped with shutoff valves having metallic handwheels, with excess-flow valves, and with extra heavy glass adequately protected with a metal housing applied by the gage manufacturer. They shall be shielded against the direct rays of the sun.

**1910.111(b)(15)** [Reserved]

**1910.111(b)(16)** *Electrical equipment and wiring.* **(i)** Electrical equipment and wiring for use in ammonia installations shall be general purpose or weather resistant as appropriate.

**1910.111(b)(16)(ii)** Electrical systems shall be installed and maintained in accordance with subpart S of this part.

**1910.111(c)** **Systems utilizing stationary, nonrefrigerated storage containers.** This paragraph applies to stationary, nonrefrigerated storage installations utilizing containers other than those covered in paragraph (e) of this section. Paragraph (b) of this section applies to this paragraph unless otherwise noted.

**1910.111(c)(1)** *Design pressure and construction of containers.* The minimum design pressure for nonrefrigerated containers shall be 250 p.s.i.g.

**1910.111(c)(2)** *Container valves and accessories, filling and discharge connections.* (i) Each filling connection shall be provided with combination back-pressure check valve and excess-flow valve; one double or two single back-pressure check valves; or a positive shutoff valve in conjunction with either an internal back-pressure check valve or an internal excess flow valve.

**1910.111(c)(2)(ii)** All liquid and vapor connections to containers except filling pipes, safety relief connections, and liquid-level gaging and pressure gage connections provided with orifices not larger than No. 54 drill size as required in paragraphs (b)(6) (iv) and (v) of this section shall be equipped with excess-flow valves.

**1910.111(c)(2)(iii)** Each storage container shall be provided with a pressure gage graduated from 0 to 400 p.s.i. Gages shall be designated for use in ammonia service.

**1910.111(c)(2)(iv)** All containers shall be equipped with vapor return valves.

**1910.111(c)(3)** *Safety-relief devices.* **(i)** Every container shall be provided with one or more safety-relief valves of the spring-loaded or equivalent type in accordance with paragraph (b)(9) of this section.

**1910.111(c)(3)(ii)** The rate of discharge of spring-loaded safety relief valves installed on underground containers may be reduced to a minimum of 30 percent of the rate of discharge specified in Table H-36. Containers so protected shall not be uncovered after installation until the liquid ammonia has been removed. Containers which may contain liquid ammonia before being installed underground and before being completely covered with earth are to be considered aboveground containers when determining the rate of discharge requirements of the safety-relief valves.

**1910.111(c)(3)(iii)** On underground installations where there is a probability of the manhole or housing becoming flooded, the discharge from vent lines shall be located above the high water level. All manholes or housings shall be provided with ventilated louvers or their equivalent, the area of such openings equalling or exceeding combined discharge areas of safety-relief valves and vent lines which discharge their content into the manhole housing.

**1910.111(c)(3)(iv)** Vent pipes, when used, shall not be restricted or of smaller diameter than the relief-valve outlet connection.

**1910.111(c)(3)(v)** If desired, vent pipes from two or more safety-relief devices located on the same unit, or similar lines from two or more different units may be run into a common discharge header, provided the capacity of such header is at least equal to the sum of the capacities of the individual discharge lines.

**1910.111(c)(4)** *Reinstallation of containers.* **(i)** Containers once installed under ground shall not later be reinstalled above ground or under ground, unless they successfully withstand hydrostatic pressure retests at the pressure specified for the original hydrostatic test as required by the code under which constructed and show no evidence of serious corrosion.

**1910.111(c)(4)(ii)** Where containers are reinstalled above ground, safety devices or gaging devices shall comply with paragraph (b)(9) of this section and this paragraph respectively for aboveground containers.

**1910.111(c)(5)** *Installation of storage containers.* **(i)** Containers installed above ground, except as provided in paragraph (c)(5)(v) of this section shall be provided with substantial concrete or masonry supports, or structural steel supports on firm concrete or masonry foundations. All foundations shall extend below the frost line.

**1910.111(c)(5)(ii)** Horizontal aboveground containers shall be so mounted on foundations as to permit expansion and contraction. Every container shall be supported to prevent the concentration of excessive loads on the supporting portion of the shell. That portion of the container in contact with foundations or saddles shall be protected against corrosion.

**1910.111(c)(5)(iii)** Containers installed under ground shall be so placed that the top of the container is below the frost line and in no case less than 2 feet below the surface of the ground. Should ground conditions make compliance with these requirements impracticable, installation shall be made otherwise to prevent physical damage. It will not be necessary to cover the portion of the container to which manhole and other connections are affixed. When necessary to prevent floating, containers shall be securely anchored or weighted.

**1910.111(c)(5)(iv)** Underground containers shall be set on a firm foundation (firm earth may be used) and surrounded with earth or sand well tamped in place. The container, prior to being placed under ground, shall be given a corrosion resisting protective coating. The container thus coated shall be so lowered into place as to prevent abrasion or other damage to the coating.

**1910.111(c)(5)(v)** Containers with foundations attached (portable or semiportable tank containers with suitable steel "runners" or "skids" and commonly known in the industry as "skid tanks") shall be designed and constructed in accordance with paragraph (c)(1) of this section.

**1910.111(c)(5)(vi)** Secure anchorage or adequate pier height shall be provided against container flotation wherever sufficiently high flood water might occur.

**1910.111(c)(5)(vii)** The distance between underground containers of over 2,000 gallons capacity shall be at least 5 feet.

**1910.111(c)(6)** *Protection of appurtenances.* **(i)** Valves, regulating, gaging, and other appurtenances shall be protected against tampering and physical damage. Such appurtenances shall also be protected during transit of containers.

**1910.111(c)(6)(ii)** All connections to underground containers shall be located within a dome, housing, or manhole and with access thereto by means of a substantial cover.

**1910.111(c)(7)** *Damage from vehicles.* Precaution shall be taken against damage to ammonia systems from vehicles.

**1910.111(d)** **Refrigerated storage systems.** This paragraph applies to systems utilizing containers with the storage of anhydrous ammonia under refrigerated conditions. All applicable rules of paragraph (b) of this section apply to this paragraph unless otherwise noted.

**1910.111(d)(1)** *Design of containers.* **(i)** The design temperature shall be the minimum temperature to which the container will be refrigerated.

**1910.111(d)(1)(ii)** Containers with a design pressure exceeding 15 p.s.i.g. shall be constructed in accordance with paragraph (b)(2) of this section, and the materials shall be selected from those listed in API Standard 620, Recommended Rules for Design and Construction of Large, Welded, Low-Pressure Storage Tanks, Fourth Edition, 1970, Tables 2.02, R2.2, R2.2(A), R2.2.1, or R2.3 which are incorporated by reference as specified in §1910.6.

**1910.111(d)(1)(iii)** Containers with a design pressure of 15 p.s.i.g. and less shall be constructed in accordance with the applicable requirements of API Standard 620 including its appendix R.

**1910.111(d)(1)(iv)** When austenitic steels or nonferrous materials are used, the Code shall be used as a guide in the selection of materials for use at the design temperature.

**1910.111(d)(1)(v)** The filling density for refrigerated storage containers shall be such that the container will not be liquid full at a liquid temperature corresponding to the vapor pressure at the start-to-discharge pressure setting of the safety-relief valve.

**1910.111(d)(2)** *Installation of refrigerated storage containers.* **(i)** Containers shall be supported on suitable noncombustible foundations designed to accommodate the type of container being used.

**1910.111(d)(2)(ii)** Adequate protection against flotation or other water damage shall be provided wherever high flood water might occur.

**1910.111(d)(2)(iii)** Containers for product storage at less than 32 °F. shall be supported in such a way, or heat shall be supplied, to prevent the effects of freezing and consequent frost heaving.

**1910.111(d)(3)** *Shutoff valves.* When operating conditions make it advisable, a check valve shall be installed on the fill connection and a remotely operated shutoff valve on other connections located below the maximum liquid level.

**1910.111(d)(4)** *Safety relief devices.* **(i)** Safety relief valves shall be set to start-to-discharge at a pressure not in excess of the design pressure of the container and shall have a total relieving capacity sufficient to prevent a maximum pressure in the container of more than 120 percent of the design pressure. Relief valves for refrigerated storage containers shall be self-contained spring-loaded, weight-loaded, or self-contained pilot-operated type.

**1910.111(d)(4)(ii)** The total relieving capacity shall be the larger of:

**1910.111(d)(4)(ii)**(*a*) Possible refrigeration system upset such as (*1*) cooling water failure, (*2*) power failure, (*3*) instrument air or instrument failure, (*4*) mechanical failure of any equipment, (*5*) excessive pumping rates.

**1910.111(d)(4)(ii)**(*b*) Fire exposure determined in accordance with Compressed Gas Association (CGA) S-1, Part 3, Safety Relief Device Standards for Compressed Gas Storage Containers, 1959, which is incorporated by reference as specified in §1910.6, except that "A" shall be the total exposed surface area in square feet up to 25 foot above grade or to the equator of the storage container if it is a sphere, whichever is greater. If the relieving capacity required for fire exposure is greater than that required by (*a*) of this subdivision, the additional capacity may be provided by weak roof to shell seams in containers operating at essentially atmospheric pressure and having an inherently weak roof-to-shell seam. The weak roof-to-shell seam is not to be considered as providing any of the capacity required in (*a*) of this subdivision.

**1910.111(d)(4)(iii)** If vent lines are installed to conduct the vapors from the relief valve, the back pressure under full relieving conditions shall not exceed 50 percent of the start-to-discharge pressure for pressure balanced valves or 10 percent of the start-to-discharge pressure for conventional valves. The vent lines shall be installed to prevent accumulation of liquid in the lines.

**1910.111(d)(4)(iv)** The valve or valve installation shall provide weather protection.

**1910.111(d)(4)(v)** Atmospheric storage shall be provided with vacuum breakers. Ammonia gas, nitrogen, methane, or other inert gases can be used to provide a pad.

**1910.111(d)(5)** *Protection of container appurtenances.* Appurtenances shall be protected against tampering and physical damage.

**1910.111(d)(6)** *Reinstallation of refrigerated storage containers.* Containers of such size as to require field fabrication shall, when moved and reinstalled, be reconstructed and reinspected in complete accordance with the requirements under which they were constructed. The containers shall be subjected to a pressure retest and if rerating is necessary, rerating shall be in accordance with applicable requirements.

**1910.111(d)(7)** *Damage from vehicles.* Precaution shall be taken against damage from vehicles.

**1910.111(d)(8)** *Refrigeration load and equipment.* **(i)** The total refrigeration load shall be computed as the sum of the following:

**1910.111(d)(8)(i)**(*a*) Load imposed by heat flow into the container caused by the temperature differential between design ambient temperature and storage temperature.

**1910.111(d)(8)(i)**(*b*) Load imposed by heat flow into the container caused by maximum sun radiation.

**1910.111(d)(8)(i)**(*c*) Maximum load imposed by filling the container with ammonia warmer than the design storage temperature.

**1910.111(d)(8)(ii)** More than one storage container may be handled by the same refrigeration system.

**1910.111(d)(9)** *Compressors.* **(i)** A minimum of two compressors shall be provided either of which shall be of sufficient size to handle the loads listed in paragraphs (d)(8)(i) (*a*) and (*b*) of this section. Where more than two compressors are provided minimum standby equipment equal to the largest normally operating equipment shall be installed. Filling compressors may be used as standby equipment for holding compressors.

**1910.111(d)(9)(ii)** Compressors shall be sized to operate with a suction pressure at least 10 percent below the minimum setting of the safety valve(s) on the storage container and shall withstand a suction pressure at least equal to 120 percent of the design pressure of the container.

**1910.111(d)(10)** *Compressor drives.* **(i)** Each compressor shall have its individual driving unit.

**1910.111(d)(10)(ii)** An emergency source of power of sufficient capacity to handle the loads listed in paragraphs (d)(8)(i) (*a*) and (*b*) of this section shall be provided unless facilities are available to safely dispose of vented vapors while the refrigeration system is not operating.

**1910.111(d)(11)** *Automatic control equipment.* **(i)** The refrigeration system shall be arranged with suitable controls to govern the compressor operation in accordance with the load as evidenced by the pressure in the container(s).

**1910.111(d)(11)(ii)** An emergency alarm system shall be installed to function in the event the pressure in the container(s) rises to the maximum allowable operating pressure.

**1910.111(d)(11)(iii)** An emergency alarm and shut-off shall be located in the condenser system to respond to excess discharge pressure caused by failure of the cooling medium.

**1910.111(d)(11)(iv)** All automatic controls shall be installed in a manner to preclude operation of alternate compressors unless the controls will function with the alternate compressors.

**1910.111(d)(12)** *Separators for compressors.* **(i)** An entrainment separator of suitable size and design pressure shall be installed in the compressor suction line of lubricated compression. The separator shall be equipped with a drain and gaging device.

**1910.111(d)(12)(ii)** [Reserved]

**1910.111(d)(13)** *Condensers.* The condenser system may be cooled by air or water or both. The condenser shall be designed for at least 250 p.s.i.g. Provision shall be made for purging noncondensibles either manually or automatically.

**1910.111(d)(14)** *Receiver and liquid drain.* A receiver shall be provided with a liquid-level control to discharge the liquid ammonia to storage. The receiver shall be designed for at least 250 p.s.i.g. and be equipped with the necessary connections, safety valves, and gaging device.

**1910.111(d)(15)** *Insulation.* Refrigerated containers and pipelines which are insulated shall be covered with a material of suitable quality and thickness for the temperatures encountered. Insulation shall be suitably supported and protected against the weather. Weatherproofing shall be of a type which will not support flame propagation.

**1910.111(e)** **Systems utilizing portable DOT containers—(1)** *Conformance.* Cylinders shall comply with DOT specifications and shall be maintained, filled, packaged, marked, labeled, and shipped to comply with 49 CFR chapter I and Marking Portable Compressed Gas Containers to Identify the Material Contained, ANSI Z48.1-1954 (R1970), which is incorporated by reference as specified in §1910.6.

**1910.111(e)(2)** *Storage.* Cylinders shall be stored in an area free from ignitable debris and in such manner as to prevent external corrosion. Storage may be indoors or outdoors.

**1910.111(e)(3)** *Heat protection.* Cylinders filled in accordance with DOT regulations will become liquid full at 145 °F. Cylinders shall be protected from heat sources such as radiant flame and steampipes. Heat shall not be applied directly to cylinders to raise the pressure.

**1910.111(e)(4)** *Protection.* Cylinders shall be stored in such manner as to protect them from moving vehicles or external damage.

**1910.111(e)(5)** *Valve cap.* Any cylinder which is designed to have a valve protection cap shall have the cap securely in place when the cylinder is not in service.

**1910.111(f)** *Tank motor vehicles for the transportation of ammonia.* **(1)** This paragraph applies to containers and pertinent equipment mounted on tank motor vehicles including semitrailers and full trailers used for the transportation of ammonia. This paragraph does not apply to farm vehicles. For requirements covering farm vehicles, refer to paragraphs (g) and (h) of this section.

Paragraph (b) of this section applies to this paragraph unless otherwise noted. Containers and pertinent equipment for tank motor vehicles for the transportation of anhydrous ammonia, in addition to complying with the requirements of this section, shall also comply with the requirements of DOT.

**1910.111(f)(2)** *Design pressure and construction of containers.* **(i)** The minimum design pressure for containers shall be that specified in the regulations of the DOT.

**1910.111(f)(2)(ii)** The shell or head thickness of any container shall not be less than three-sixteenth inch.

**1910.111(f)(2)(iii)** All container openings, except safety relief valves, liquid-level gaging devices, and pressure gages, shall be labeled to designate whether they communicate with liquid or vapor space.

**1910.111(f)(3)** *Container appurtenances.* **(i)** All appurtenances shall be protected against physical damage.

**1910.111(f)(3)(ii)** All connections to containers, except filling connections, safety relief devices, and liquid-level and pressure gage connections, shall be provided with suitable automatic excess flow valves, or in lieu thereof, may be fitted with quick-closing internal valves, which shall remain closed except during delivery operations. The control mechanism for such valves may be provided with a secondary control remote from the delivery connections and such control mechanism shall be provided with a fusible section (melting point 208 °F. to 220 °F.) which will permit the internal valve to close automatically in case of fire.

**1910.111(f)(3)(iii)** Filling connections shall be provided with automatic back-pressure check valves, excess-flow valves, or quick-closing internal valves, to prevent back-flow in case the filling connection is broken. Where the filling and discharge connect to a common opening in the container shell and that opening is fitted with a quick-closing internal valve as specified in paragraph (f)(3)(ii) of this section, the automatic valve shall not be required.

**1910.111(f)(3)(iv)** All containers shall be equipped for spray loading (filling in the vapor space) or with an approved vapor return valve of adequate capacity.

**1910.111(f)(4)** *Piping and fittings.* **(i)** All piping, tubing, and fittings shall be securely mounted and protected against damage. Means shall be provided to protect hoses while the vehicle is in motion.

**1910.111(f)(4)(ii)** Fittings shall comply with paragraph (b)(6) of this section. Pipe shall be Schedule 80.

**1910.111(f)(5)** *Safety relief devices.* **(i)** The discharge from safety relief valves shall be vented away from the container upward and unobstructed to the open air in such a manner as to prevent any impingement of escaping gas upon the container; loose-fitting rain caps shall be used. Size of discharge lines from safety valves shall not be smaller than the nominal size of the safety-relief valve outlet connection. Suitable provision shall be made for draining condensate which may accumulate in the discharge pipe.

**1910.111(f)(5)(ii)** Any portion of liquid ammonia piping which at any time may be closed at both ends shall be provided with a hydrostatic relief valve.

**1910.111(f)(6)** *Transfer of liquids.* **(i)** The content of tank motor vehicle containers shall be determined by weight, by a suitable liquid-level gaging device, or other approved methods. If the content of a container is to be determined by liquid-level measurement, the container shall have a thermometer well so that the internal liquid temperature can be easily determined. This volume when converted to weight shall not exceed the filling density specified by the DOT.

**1910.111(f)(6)(ii)** Any pump, except a constant speed centrifugal pump, shall be equipped with a suitable pressure actuated bypass valve permitting flow from discharge to suction when the discharge pressure rises above a predetermined point. Pump discharge shall also be equipped with a spring-loaded safety relief valve set at a pressure not more than 135 percent of the setting of the bypass valve or more than 400 p.s.i.g., whichever is larger.

**1910.111(f)(6)(iii)** Compressors shall be equipped with manually operated shutoff valves on both suction and discharge connections. Pressure gages of bourdon-tube type shall be installed on the suction and discharge of the compressor before the shutoff valves. The compressor shall not be operated if either pressure gage is removed or is inoperative. A spring-loaded, safety-relief valve capable of discharging to atmosphere the full flow of gas from the compressor at a pressure not exceeding 300 p.s.i.g. shall be connected between the compressor discharge and the discharge shutoff valve.

**1910.111(f)(6)(iv)** Valve functions shall be clearly and legibly identified by metal tags or nameplates permanently affixed to each valve.

**1910.111(f)(7)-(8)** [Reserved]

**1910.111(f)(9)** *Chock blocks.* At least two chock blocks shall be provided. These blocks shall be placed to prevent rolling of the vehicle whenever it is parked during loading and unloading operations.

**1910.111(f)(10)** *Portable tank containers (skid tanks).* Where portable tank containers are used for farm storage they shall comply with paragraph (c)(1) of this section. When portable tank containers are used in lieu of cargo tanks and are permanently mounted on tank motor vehicles for the transportation of ammonia, they shall comply with the requirements of this paragraph.

**1910.111(g)** *Systems mounted on farm vehicles other than for the application of ammonia*—**(1)** *Application.* This paragraph applies to containers of 1,200 gallons capacity or less and pertinent equipment mounted on farm vehicles (implements of husbandry) and used other than for the application of ammonia to the soil. Paragraph (b) of this section applies to this paragraph unless otherwise noted.

**1910.111(g)(2)** *Design pressure and classification of containers.* **(i)** The minimum design pressure for containers shall be 250 p.s.i.g.

**1910.111(g)(2)(ii)** The shell or head thickness of any container shall be not less than three-sixteenths of an inch.

**1910.111(g)(3)** *Mounting containers.* **(i)** A suitable "stop" or "stops" shall be mounted on the vehicle or on the container in such a way that the container shall not be dislodged from its mounting due to the vehicle coming to a sudden stop. Back slippage shall also be prevented by proper methods.

**1910.111(g)(3)(ii)** A suitable "hold down" device shall be provided which will anchor the container to the vehicle at one or more places on each side of the container.

**1910.111(g)(3)(iii)** When containers are mounted on four-wheel trailers, care shall be taken to insure that the weight is distributed evenly over both axles.

**1910.111(g)(3)(iv)** When the cradle and the tank are not welded together suitable material shall be used between them to eliminate metal-to-metal friction.

**1910.111(g)(4)** *Container appurtenances.* **(i)** All containers shall be equipped with a fixed liquid-level gage.

**1910.111(g)(4)(ii)** All containers with a capacity exceeding 250 gallons shall be equipped with a pressure gage having a dial graduated from 0-400 p.s.i.

**1910.111(g)(4)(iii)** The filling connection shall be fitted with combination back-pressure check valve and excess-flow valve; one double or two single back-pressure check valves; or a positive shutoff valve in conjunction with either an internal back-pressure check valve or an internal excess flow valve.

**1910.111(g)(4)(iv)** All containers with a capacity exceeding 250 gallons shall be equipped for spray loading or with an approved vapor return valve.

**1910.111(g)(4)(v)** All vapor and liquid connections except safety-relief valves and those specifically exempted by paragraph (b)(6)(v) of this section shall be equipped with approved excess-flow valves or may be fitted with quick-closing internal valves which, except during operating periods, shall remain closed.

**1910.111(g)(4)(vi)** Fittings shall be adequately protected from damage by a metal box or cylinder with open top securely fastened to the container or by rigid guards, well braced, welded to the container on both sides of the fittings or by a metal dome. If a metal dome is used, the relief valve shall be properly vented through the dome.

**1910.111(g)(4)(vii)** If a liquid withdrawal line is installed in the bottom of a container, the connections thereto, including hose, shall not be lower than the lowest horizontal edge of the vehicle axle.

**1910.111(g)(4)(viii)** Provision shall be made to secure both ends of the hose while in transit.

**1910.111(g)(5)** *Marking the container.* There shall appear on each side and on the rear end of the container in letters at least 4 inches high, the words, "Caution—Ammonia" or the container shall be marked in accordance with DOT regulations.

**1910.111(g)(6)** *Farm vehicles.* **(i)** Farm vehicles shall conform with State regulations.

**1910.111(g)(6)(ii)** All trailers shall be securely attached to the vehicle drawing them by means of drawbars supplemented by suitable safety chains.

**1910.111(g)(6)(iii)** A trailer shall be constructed so that it will follow substantially in the path of the towing vehicle and will not whip or swerve dangerously from side to side.

**1910.111(g)(6)(iv)** All vehicles shall carry a can containing 5 gallons or more of water.

**1910.111(h)** *Systems mounted on farm vehicles for the application of ammonia.* **(1)** This paragraph applies to systems utilizing containers of 250 gallons capacity or less which are mounted on farm vehicles (implement of husbandry) and used for the application of ammonia to the soil. Paragraph (b) of this section applies to this paragraph unless otherwise noted. Where larger containers are used, they shall comply with paragraph (g) of this section.

**1910.111(h)(2)** *Design pressure and classification of containers.* **(i)** The minimum design pressure for containers shall be 250 p.s.i.g.

**1910.111(h)(2)(ii)** The shell or head thickness of any container shall not be less than three-sixteenths inch.

**1910.111(h)(3)** *Mounting of containers.* All containers and flow-control devices shall be securely mounted.

**1910.111(h)(4)** *Container valves and accessories.* **(i)** Each container shall have a fixed liquid-level gage.

**1910.111(h)(4)(ii)** The filling connection shall be fitted with a combination back-pressure check valve and an excess-flow valve; one double or two single back-pressure check valves: or a positive shutoff valve in conjunction with an internal back-pressure check valve or an internal excess-flow valve.

**1910.111(h)(4)(iii)** The applicator tank may be filled by venting to open air provided the bleeder valve orifice does not exceed seven-sixteenths inch in diameter.

**1910.111(h)(4)(iv)** Regulation equipment may be connected directly to the tank coupling or flange, in which case a flexible connection shall be used between such regulating equipment and the remainder of the liquid withdrawal system. Regulating equipment not so installed shall be flexibly connected to the container shutoff valve.

**1910.111(h)(4)(v)** No excess flow valve is required in the liquid withdrawal line provided the controlling orifice between the contents of the container and the outlet of the shutoff valve does not exceed seven-sixteenths inch in diameter.

[39 FR 23502, June 27, 1974, as amended at 43 FR 49748, Oct. 24, 1978; 49 FR 5322, Feb. 10, 1984; 53 FR 12122, Apr. 12, 1988; 61 FR 9238, Mar. 7, 1996; 63 FR 1269, Jan. 8, 1998; 63 FR 33466, June 18, 1998]

## §1910.112 [Reserved].

## §1910.113 [Reserved].

## §1910.114 Effective dates [Removed].

[§1910.114 removed at 61 FR 9238, Mar. 7, 1996, effective May 6, 1996.]

## §1910.115 Sources of standards [Removed].

[§1910.115 removed at 61 FR 9238, Mar. 7, 1996, effective May 6, 1996.]

## §1910.116 Standards organizations [Removed].

[§1910.116 removed at 61 FR 9238, Mar. 7, 1996, effective May 6, 1996.]

## §1910.119 Process safety management of highly hazardous chemicals.

*Purpose.* This section contains requirements for preventing or minimizing the consequences of catastrophic releases of toxic, reactive, flammable, or explosive chemicals. These releases may result in toxic, fire or explosion hazards.

**1910.119(a)** *Application.* **(1)** This section applies to the following:

**1910.119(a)(1)(i)** A process which involves a chemical at or above the specified threshold quantities listed in appendix A to this section;

**1910.119(a)(1)(ii)** A process which involves a flammable liquid or gas (as defined in 1910.1200(c) of this part) on site in one location, in a quantity of 10,000 pounds (4535.9 kg) or more except for:

**1910.119(a)(1)(ii)(A)** Hydrocarbon fuels used solely for workplace consumption as a fuel (e.g., propane used for comfort heating, gasoline for vehicle refueling), if such fuels are not a part of a process containing another highly hazardous chemical covered by this standard;

**1910.119(a)(1)(ii)(B)** Flammable liquids stored in atmospheric tanks or transferred which are kept below their normal boiling point without benefit of chilling or refrigeration.

**1910.119(a)(2)** This section does not apply to:

**1910.119(a)(2)(i)** Retail facilities;

**1910.119(a)(2)(ii)** Oil or gas well drilling or servicing operations; or,

**1910.119(a)(2)(iii)** Normally unoccupied remote facilities.

**1910.119(b)** *Definitions.* *Atmospheric tank* means a storage tank which has been designed to operate at pressures from atmospheric through 0.5 p.s.i.g. (pounds per square inch gauge, 3.45 Kpa).

*Boiling point* means the boiling point of a liquid at a pressure of 14.7 pounds per square inch absolute (p.s.i.a.) (760 mm.). For the purposes of this section, where an accurate boiling point is unavailable for the material in question, or for mixtures which do not have a constant boiling point, the 10 percent point of a distillation performed in accordance with the Standard Method of Test for Distillation of Petroleum Products, ASTM D-86-62, which is incorporated by reference as specified in §1910.6, may be used as the boiling point of the liquid.

*Catastrophic release* means a major uncontrolled emission, fire, or explosion, involving one or more highly hazardous chemicals, that presents serious danger to employees in the workplace.

*Facility* means the buildings, containers or equipment which contain a process.

*Highly hazardous chemical* means a substance possessing toxic, reactive, flammable, or explosive properties and specified by paragraph (a)(1) of this section.

*Hot work* means work involving electric or gas welding, cutting, brazing, or similar flame or spark-producing operations.

*Normally unoccupied remote facility* means a facility which is operated, maintained or serviced by employees who visit the facility only periodically to check its operation and to perform necessary operating or maintenance tasks. No employees are permanently stationed at the facility. Facilities meeting this definition are not contiguous with, and must be geographically remote from all other buildings, processes or persons.

*Process* means any activity involving a highly hazardous chemical including any use, storage, manufacturing, handling, or the on-site movement of such chemicals, or combination of these activities. For purposes of this definition, any group of vessels which are interconnected and separate vessels which are located such that a highly hazardous chemical could be involved in a potential release shall be considered a single process.

*Replacement in kind* means a replacement which satisfies the design specification.

*Trade secret* means any confidential formula, pattern, process, device, information or compilation of information that is used in an employer's business, and that gives the employer an opportunity to obtain an advantage over competitors who do not know or use it. Appendix D contained in §1910.1200 sets out the criteria to be used in evaluating trade secrets.

**1910.119(c)** *Employee participation.* **(1)** Employers shall develop a written plan of action regarding the implementation of the employee participation required by this paragraph.

**1910.119(c)(2)** Employers shall consult with employees and their representatives on the conduct and development of process hazards analyses and on the development of the other elements of process safety management in this standard.

**1910.119(c)(3)** Employers shall provide to employees and their representatives access to process hazard analyses and to all other information required to be developed under this standard.

# OSHA Standards for General Industry

**1910.119(d)** *Process safety information.* In accordance with the schedule set forth in paragraph (e)(1) of this section, the employer shall complete a compilation of written process safety information before conducting any process hazard analysis required by the standard. The compilation of written process safety information is to enable the employer and the employees involved in operating the process to identify and understand the hazards posed by those processes involving highly hazardous chemicals. This process safety information shall include information pertaining to the hazards of the highly hazardous chemicals used or produced by the process, information pertaining to the technology of the process, and information pertaining to the equipment in the process.

**1910.119(d)(1)** *Information pertaining to the hazards of the highly hazardous chemicals in the process.* This information shall consist of at least the following:

**1910.119(d)(1)(i)** Toxicity information;

**1910.119(d)(1)(ii)** Permissible exposure limits;

**1910.119(d)(1)(iii)** Physical data;

**1910.119(d)(1)(iv)** Reactivity data;

**1910.119(d)(1)(v)** Corrosivity data;

**1910.119(d)(1)(vi)** Thermal and chemical stability data; and

**1910.119(d)(1)(vii)** Hazardous effects of inadvertent mixing of different materials that could foreseeably occur.

**Note:** Material Safety Data Sheets meeting the requirements of 29 CFR 1910.1200(g) may be used to comply with this requirement to the extent they contain the information required by this subparagraph.

**1910.119(d)(2)** *Information pertaining to the technology of the process.* (i) Information concerning the technology of the process shall include at least the following:

**1910.119(d)(2)(i)(A)** A block flow diagram or simplified process flow diagram (see appendix B to this section);

**1910.119(d)(2)(i)(B)** Process chemistry;

**1910.119(d)(2)(i)(C)** Maximum intended inventory;

**1910.119(d)(2)(i)(D)** Safe upper and lower limits for such items as temperatures, pressures, flows or compositions; and,

**1910.119(d)(2)(i)(E)** An evaluation of the consequences of deviations, including those affecting the safety and health of employees.

**1910.119(d)(2)(ii)** Where the original technical information no longer exists, such information may be developed in conjunction with the process hazard analysis in sufficient detail to support the analysis.

**1910.119(d)(3)** *Information pertaining to the equipment in the process.* (i) Information pertaining to the equipment in the process shall include:

**1910.119(d)(3)(i)(A)** Materials of construction;

**1910.119(d)(3)(i)(B)** Piping and instrument diagrams (P&ID's);

**1910.119(d)(3)(i)(C)** Electrical classification;

**1910.119(d)(3)(i)(D)** Relief system design and design basis;

**1910.119(d)(3)(i)(E)** Ventilation system design;

**1910.119(d)(3)(i)(F)** Design codes and standards employed;

**1910.119(d)(3)(i)(G)** Material and energy balances for processes built after May 26, 1992; and,

**1910.119(d)(3)(i)(H)** Safety systems (e.g., interlocks, detection or suppression systems).

**1910.119(d)(3)(ii)** The employer shall document that equipment complies with recognized and generally accepted good engineering practices.

**1910.119(d)(3)(iii)** For existing equipment designed and constructed in accordance with codes, standards, or practices that are no longer in general use, the employer shall determine and document that the equipment is designed, maintained, inspected, tested, and operating in a safe manner.

**1910.119(e)** *Process hazard analysis.* **(1)** The employer shall perform an initial process hazard analysis (hazard evaluation) on processes covered by this standard. The process hazard analysis shall be appropriate to the complexity of the process and shall identify, evaluate, and control the hazards involved in the process. Employers shall determine and document the priority order for conducting process hazard analyses based on a rationale which includes such considerations as extent of the process hazards, number of potentially affected employees, age of the process, and operating history of the process. The process hazard analysis shall be conducted as soon as possible, but not later than the following schedule:

**1910.119(e)(1)(i)** No less than 25 percent of the initial process hazards analyses shall be completed by May 26, 1994;

**1910.119(e)(1)(ii)** No less than 50 percent of the initial process hazards analyses shall be completed by May 26, 1995;

**1910.119(e)(1)(iii)** No less than 75 percent of the initial process hazards analyses shall be completed by May 26, 1996;

**1910.119(e)(1)(iv)** All initial process hazards analyses shall be completed by May 26, 1997.

**1910.119(e)(1)(v)** Process hazards analyses completed after May 26, 1987 which meet the requirements of this paragraph are acceptable as initial process hazards analyses. These process hazard analyses shall be updated and revalidated, based on their completion date, in accordance with paragraph (e)(6) of this section.

**1910.119(e)(2)** The employer shall use one or more of the following methodologies that are appropriate to determine and evaluate the hazards of the process being analyzed.

**1910.119(e)(2)(i)** What-If;

**1910.119(e)(2)(ii)** Checklist;

**1910.119(e)(2)(iii)** What-If/Checklist;

**1910.119(e)(2)(iv)** Hazard and Operability Study (HAZOP);

**1910.119(e)(2)(v)** Failure Mode and Effects Analysis (FMEA);

**1910.119(e)(2)(vi)** Fault Tree Analysis; or

**1910.119(e)(2)(vii)** An appropriate equivalent methodology.

**1910.119(e)(3)** The process hazard analysis shall address:

**1910.119(e)(3)(i)** The hazards of the process;

**1910.119(e)(3)(ii)** The identification of any previous incident which had a likely potential for catastrophic consequences in the workplace;

**1910.119(e)(3)(iii)** Engineering and administrative controls applicable to the hazards and their interrelationships such as appropriate application of detection methodologies to provide early warning of releases. (Acceptable detection methods might include process monitoring and control instrumentation with alarms, and detection hardware such as hydrocarbon sensors.);

**1910.119(e)(3)(iv)** Consequences of failure of engineering and administrative controls;

**1910.119(e)(3)(v)** Facility siting;

**1910.119(e)(3)(vi)** Human factors; and

**1910.119(e)(3)(vii)** A qualitative evaluation of a range of the possible safety and health effects of failure of controls on employees in the workplace.

**1910.119(e)(4)** The process hazard analysis shall be performed by a team with expertise in engineering and process operations, and the team shall include at least one employee who has experience and knowledge specific to the process being evaluated. Also, one member of the team must be knowledgeable in the specific process hazard analysis methodology being used.

**1910.119(e)(5)** The employer shall establish a system to promptly address the team's findings and recommendations; assure that the recommendations are resolved in a timely manner and that the resolution is documented; document what actions are to be taken; complete actions as soon as possible; develop a written schedule of when these actions are to be completed; communicate the actions to operating, maintenance and other employees whose work assignments are in the process and who may be affected by the recommendations or actions.

**1910.119(e)(6)** At least every five (5) years after the completion of the initial process hazard analysis, the process hazard analysis shall be updated and revalidated by a team meeting the requirements in paragraph (e)(4) of this section, to assure that the process hazard analysis is consistent with the current process.

**1910.119(e)(7)** Employers shall retain process hazards analyses and updates or revalidations for each process covered by this section, as well as the documented resolution of recommendations described in paragraph (e)(5) of this section for the life of the process.

**1910.119(f)** *Operating procedures*—**(1)** The employer shall develop and implement written operating procedures that provide clear instructions for safely conducting activities involved in each covered process consistent with the process safety information and shall address at least the following elements.

**1910.119(f)(1)(i)** *Steps for each operating phase:*

**1910.119(f)(1)(i)(A)** Initial startup;

**1910.119(f)(1)(i)(B)** Normal operations;

**1910.119(f)(1)(i)(C)** Temporary operations;

**1910.119(f)(1)(i)(D)** Emergency shutdown including the conditions under which emergency shutdown is required, and the assignment of shutdown responsibility to qualified operators to ensure that emergency shutdown is executed in a safe and timely manner.

**1910.119(f)(1)(i)(E)** Emergency Operations;

**1910.119(f)(1)(i)(F)** Normal shutdown; and,

**1910.119(f)(1)(i)(G)** Startup following a turnaround, or after an emergency shutdown.

**1910.119(f)(1)(ii)** *Operating limits:*

**1910.119(f)(1)(ii)(A)** Consequences of deviation; and

**1910.119(f)(1)(ii)(B)** Steps required to correct or avoid deviation.

**1910.119(f)(1)(iii)** *Safety and health considerations:*

**1910.119(f)(1)(iii)(A)** Properties of, and hazards presented by, the chemicals used in the process;

**1910.119(f)(1)(iii)(B)** Precautions necessary to prevent exposure, including engineering controls, administrative controls, and personal protective equipment;

**1910.119(f)(1)(iii)(C)** Control measures to be taken if physical contact or airborne exposure occurs;

**1910.119(f)(1)(iii)(D)** Quality control for raw materials and control of hazardous chemical inventory levels; and,

**1910.119(f)(1)(iii)(E)** Any special or unique hazards.

**1910.119(f)(1)(iv)** *Safety systems and their functions.*

**1910.119(f)(2)** Operating procedures shall be readily accessible to employees who work in or maintain a process.

**1910.119(f)(3)** The operating procedures shall be reviewed as often as necessary to assure that they reflect current operating practice, including changes that result from changes in process chemicals, technology, and equipment, and changes to facilities. The employer shall certify annually that these operating procedures are current and accurate.

**1910.119(f)(4)** The employer shall develop and implement safe work practices to provide for the control of hazards during operations such as lockout/tagout; confined space entry; opening process equipment or piping; and control over entrance into a facility by maintenance, contractor, laboratory, or other support personnel. These safe work practices shall apply to employees and contractor employees.

**1910.119(g)** *Training*—**(1)** *Initial training.* **(i)** Each employee presently involved in operating a process, and each employee before being involved in operating a newly assigned process, shall be trained in an overview of the process and in the operating procedures as specified in paragraph (f) of this section. The training shall include emphasis on the specific safety and health hazards, emergency operations including shutdown, and safe work practices applicable to the employee's job tasks.

**1910.119(g)(1)(ii)** In lieu of initial training for those employees already involved in operating a process on May 26, 1992, an employer may certify in writing that the employee has the required knowledge, skills, and abilities to safely carry out the duties and responsibilities as specified in the operating procedures.

**1910.119(g)(2)** *Refresher training.* Refresher training shall be provided at least every three years, and more often if necessary, to each employee involved in operating a process to assure that the employee understands and adheres to the current operating procedures of the process. The employer, in consultation with the employees involved in operating the process, shall determine the appropriate frequency of refresher training.

**1910.119(g)(3)** *Training documentation.* The employer shall ascertain that each employee involved in operating a process has received and understood the training required by this paragraph. The employer shall prepare a record which contains the identity of the employee, the date of training, and the means used to verify that the employee understood the training.

**1910.119(h)** *Contractors*—**(1)** *Application.* This paragraph applies to contractors performing maintenance or repair, turnaround, major renovation, or specialty work on or adjacent to a covered process. It does not apply to contractors providing incidental services which do not influence process safety, such as janitorial work, food and drink services, laundry, delivery or other supply services.

**1910.119(h)(2)** *Employer responsibilities.* **(i)** The employer, when selecting a contractor, shall obtain and evaluate information regarding the contract employer's safety performance and programs.

**1910.119(h)(2)(ii)** The employer shall inform contract employers of the known potential fire, explosion, or toxic release hazards related to the contractor's work and the process.

**1910.119(h)(2)(iii)** The employer shall explain to contract employers the applicable provisions of the emergency action plan required by paragraph (n) of this section.

**1910.119(h)(2)(iv)** The employer shall develop and implement safe work practices consistent with paragraph (f)(4) of this section, to control the entrance, presence and exit of contract employers and contract employees in covered process areas.

**1910.119(h)(2)(v)** The employer shall periodically evaluate the performance of contract employers in fulfilling their obligations as specified in paragraph (h)(3) of this section.

**1910.119(h)(2)(vi)** The employer shall maintain a contract employee injury and illness log related to the contractor's work in process areas.

**1910.119(h)(3)** *Contract employer responsibilities.* **(i)** The contract employer shall assure that each contract employee is trained in the work practices necessary to safely perform his/her job.

**1910.119(h)(3)(ii)** The contract employer shall assure that each contract employee is instructed in the known potential fire, explosion, or toxic release hazards related to his/her job and the process, and the applicable provisions of the emergency action plan.

**1910.119(h)(3)(iii)** The contract employer shall document that each contract employee has received and understood the training required by this paragraph. The contract employer shall prepare a record which contains the identity of the contract employee, the date of training, and the means used to verify that the employee understood the training.

**1910.119(h)(3)(iv)** The contract employer shall assure that each contract employee follows the safety rules of the facility including the safe work practices required by paragraph (f)(4) of this section.

**1910.119(h)(3)(v)** The contract employer shall advise the employer of any unique hazards presented by the contract employer's work, or of any hazards found by the contract employer's work.

**1910.119(i)** *Pre-startup safety review.* **(1)** The employer shall perform a pre-startup safety review for new facilities and for modified facilities when the modification is significant enough to require a change in the process safety information.

**1910.119(i)(2)** The pre-startup safety review shall confirm that prior to the introduction of highly hazardous chemicals to a process:

**1910.119(i)(2)(i)** Construction and equipment is in accordance with design specifications;

**1910.119(i)(2)(ii)** Safety, operating, maintenance, and emergency procedures are in place and are adequate;

**1910.119(i)(2)(iii)** For new facilities, a process hazard analysis has been performed and recommendations have been resolved or implemented before startup; and modified facilities meet the requirements contained in management of change, paragraph (l).

**1910.119(i)(2)(iv)** Training of each employee involved in operating a process has been completed.

**1910.119(j)** *Mechanical integrity*—**(1)** *Application.* Paragraphs (j)(2) through (j)(6) of this section apply to the following process equipment:

**1910.119(j)(1)(i)** Pressure vessels and storage tanks;

# OSHA Standards for General Industry

**1910.119(j)(1)(ii)** Piping systems (including piping components such as valves);

**1910.119(j)(1)(iii)** Relief and vent systems and devices;

**1910.119(j)(1)(iv)** Emergency shutdown systems;

**1910.119(j)(1)(v)** Controls (including monitoring devices and sensors, alarms, and interlocks) and,

**1910.119(j)(1)(vi)** Pumps.

**1910.119(j)(2)** *Written procedures.* The employer shall establish and implement written procedures to maintain the on-going integrity of process equipment.

**1910.119(j)(3)** *Training for process maintenance activities.* The employer shall train each employee involved in maintaining the on-going integrity of process equipment in an overview of that process and its hazards and in the procedures applicable to the employee's job tasks to assure that the employee can perform the job tasks in a safe manner.

**1910.119(j)(4)** *Inspection and testing.* **(i)** Inspections and tests shall be performed on process equipment.

**1910.119(j)(4)(ii)** Inspection and testing procedures shall follow recognized and generally accepted good engineering practices.

**1910.119(j)(4)(iii)** The frequency of inspections and tests of process equipment shall be consistent with applicable manufacturers' recommendations and good engineering practices, and more frequently if determined to be necessary by prior operating experience.

**1910.119(j)(4)(iv)** The employer shall document each inspection and test that has been performed on process equipment. The documentation shall identify the date of the inspection or test, the name of the person who performed the inspection or test, the serial number or other identifier of the equipment on which the inspection or test was performed, a description of the inspection or test performed, and the results of the inspection or test.

**1910.119(j)(5)** *Equipment deficiencies.* The employer shall correct deficiencies in equipment that are outside acceptable limits (defined by the process safety information in paragraph (d) of this section) before further use or in a safe and timely manner when necessary means are taken to assure safe operation.

**1910.119(j)(6)** *Quality assurance.* **(i)** In the construction of new plants and equipment, the employer shall assure that equipment as it is fabricated is suitable for the process application for which they will be used.

**1910.119(j)(6)(ii)** Appropriate checks and inspections shall be performed to assure that equipment is installed properly and consistent with design specifications and the manufacturer's instructions.

**1910.119(j)(6)(iii)** The employer shall assure that maintenance materials, spare parts and equipment are suitable for the process application for which they will be used.

**1910.119(k)** *Hot work permit.* **(1)** The employer shall issue a hot work permit for hot work operations conducted on or near a covered process.

**1910.119(k)(2)** The permit shall document that the fire prevention and protection requirements in 29 CFR 1910.252(a) have been implemented prior to beginning the hot work operations; it shall indicate the date(s) authorized for hot work; and identify the object on which hot work is to be performed. The permit shall be kept on file until completion of the hot work operations.

**1910.119(l)** *Management of change.* **(1)** The employer shall establish and implement written procedures to manage changes (except for "replacements in kind") to process chemicals, technology, equipment, and procedures; and, changes to facilities that affect a covered process.

**1910.119(l)(2)** The procedures shall assure that the following considerations are addressed prior to any change:

**1910.119(l)(2)(i)** The technical basis for the proposed change;

**1910.119(l)(2)(ii)** Impact of change on safety and health;

**1910.119(l)(2)(iii)** Modifications to operating procedures;

**1910.119(l)(2)(iv)** Necessary time period for the change; and,

**1910.119(l)(2)(v)** Authorization requirements for the proposed change.

**1910.119(l)(3)** Employees involved in operating a process and maintenance and contract employees whose job tasks will be affected by a change in the process shall be informed of, and trained in, the change prior to start-up of the process or affected part of the process.

**1910.119(l)(4)** If a change covered by this paragraph results in a change in the process safety information required by paragraph (d) of this section, such information shall be updated accordingly.

**1910.119(l)(5)** If a change covered by this paragraph results in a change in the operating procedures or practices required by paragraph (f) of this section, such procedures or practices shall be updated accordingly.

**1910.119(m)** *Incident investigation.* **(1)** The employer shall investigate each incident which resulted in, or could reasonably have resulted in a catastrophic release of highly hazardous chemical in the workplace.

**1910.119(m)(2)** An incident investigation shall be initiated as promptly as possible, but not later than 48 hours following the incident.

**1910.119(m)(3)** An incident investigation team shall be established and consist of at least one person knowledgeable in the process involved, including a contract employee if the incident involved work of the contractor, and other persons with appropriate knowledge and experience to thoroughly investigate and analyze the incident.

**1910.119(m)(4)** A report shall be prepared at the conclusion of the investigation which includes at a minimum:

**1910.119(m)(4)(i)** Date of incident;

**1910.119(m)(4)(ii)** Date investigation began;

**1910.119(m)(4)(iii)** A description of the incident;

**1910.119(m)(4)(iv)** The factors that contributed to the incident; and,

**1910.119(m)(4)(v)** Any recommendations resulting from the investigation.111(5) The employer shall establish a system to promptly address and resolve the incident report findings and recommendations. Resolutions and corrective actions shall be documented.

**1910.119(m)(6)** The report shall be reviewed with all affected personnel whose job tasks are relevant to the incident findings including contract employees where applicable.

**1910.119(m)(7)** Incident investigation reports shall be retained for five years.

**1910.119(n)** *Emergency planning and response.* The employer shall establish and implement an emergency action plan for the entire plant in accordance with the provisions of 29 CFR 1910.38. In addition, the emergency action plan shall include procedures for handling small releases. Employers covered under this standard may also be subject to the hazardous waste and emergency response provisions contained in 29 CFR 1910.120(a), (p) and (q).

**1910.119(o)** *Compliance Audits.* **(1)** Employers shall certify that they have evaluated compliance with the provisions of this section at least every three years to verify that the procedures and practices developed under the standard are adequate and are being followed.

**1910.119(o)(2)** The compliance audit shall be conducted by at least one person knowledgeable in the process.

**1910.119(o)(3)** A report of the findings of the audit shall be developed.

**1910.119(o)(4)** The employer shall promptly determine and document an appropriate response to each of the findings of the compliance audit, and document that deficiencies have been corrected.

**1910.119(o)(5)** Employers shall retain the two (2) most recent compliance audit reports.

**1910.119(p)** *Trade secrets.* **(1)** Employers shall make all information necessary to comply with the section available to those persons responsible for compiling the process safety information (required by paragraph (d) of this section), those assisting in the development of the process hazard analysis (required by paragraph (e) of this section), those responsible for developing the operating procedures (required by paragraph (f) of this section), and those involved in incident investigations (required by paragraph (m) of this section), emergency planning and response (paragraph (n) of this section) and compliance audits (paragraph (o) of this section) without regard to possible trade secret status of such information.

**1910.119(p)(2)** Nothing in this paragraph shall preclude the employer from requiring the persons to whom the information is made available under paragraph (p)(1) of this section to enter into confidentiality agreements not to disclose the information as set forth in 29 CFR 1910.1200.

**1910.119(p)(3)** Subject to the rules and procedures set forth in 29 CFR 1910.1200(i)(1) through 1910.1200(i)(12), employees and their designated representatives shall have access to trade secret information contained within the process hazard analysis and other documents required to be developed by this standard.

## APPENDIX A TO §1910.119—LIST OF HIGHLY HAZARDOUS CHEMICALS, TOXICS AND REACTIVES (MANDATORY)

This appendix contains a listing of toxic and reactive highly hazardous chemicals which present a potential for a catastrophic event at or above the threshold quantity.

| CHEMICAL NAME | CAS* | TQ** |
|---|---|---|
| Acetaldehyde | 75-07-0 | 2500 |
| Acrolein (2-Propenal) | 107-02-8 | 150 |
| Acrylyl Chloride | 814-68-6 | 250 |
| Allyl Chloride | 107-05-1 | 1000 |
| Allylamine | 107-11-9 | 1000 |
| Alkylaluminums | Varies | 5000 |
| Ammonia, Anhydrous | 7664-41-7 | 10000 |
| Ammonia solutions (>44% ammonia by weight) | 7664-41-7 | 15000 |
| Ammonium Perchlorate | 7790-98-9 | 7500 |
| Ammonium Permanganate | 7787-36-2 | 7500 |
| Arsine (also called Arsenic Hydride) | 7784-42-1 | 100 |
| Bis(Chloromethyl) Ether | 542-88-1 | 100 |
| Boron Trichloride | 10294-34-5 | 2500 |
| Boron Trifluoride | 7637-07-2 | 250 |
| Bromine | 7726-95-6 | 1500 |
| Bromine Chloride | 13863-41-7 | 1500 |
| Bromine Pentafluoride | 7789-30-2 | 2500 |
| Bromine Trifluoride | 7787-71-5 | 15000 |
| 3-Bromopropyne (also called Propargyl Bromide) | 106-96-7 | 100 |
| Butyl Hydroperoxide (Tertiary) | 75-91-2 | 5000 |
| Butyl Perbenzoate (Tertiary) | 614-45-9 | 7500 |
| Carbonyl Chloride (see Phosgene) | 75-44-5 | 100 |
| Carbonyl Fluoride | 353-50-4 | 2500 |
| Cellulose Nitrate (concentration >12.6% nitrogen) | 9004-70-0 | 2500 |
| Chlorine | 7782-50-5 | 1500 |
| Chlorine Dioxide | 10049-04-4 | 1000 |
| Chlorine Pentrafluoride | 13637-63-3 | 1000 |
| Chlorine Trifluoride | 7790-91-2 | 1000 |
| Chlorodiethylaluminum (also called Diethylaluminum Chloride) | 96-10-6 | 5000 |
| 1-Chloro-2,4-Dinitrobenzene | 97-00-7 | 5000 |
| Chloromethyl Methyl Ether | 107-30-2 | 500 |
| Chloropicrin | 76-06-2 | 500 |
| Chloropicrin and Methyl Bromide mixture | None | 1500 |
| Chloropicrin and Methyl Chloride mixture | None | 1500 |
| Cumene Hydroperoxide | 80-15-9 | 5000 |
| Cyanogen | 460-19-5 | 2500 |
| Cyanogen Chloride | 506-77-4 | 500 |
| Cyanuric Fluoride | 675-14-9 | 100 |
| Diacetyl Peroxide (Concentration >70%) | 110-22-5 | 5000 |
| Diazomethane | 334-88-3 | 500 |
| Dibenzoyl Peroxide | 94-36-0 | 7500 |
| Diborane | 19287-45-7 | 100 |
| Dibutyl Peroxide (Tertiary) | 110-05-4 | 5000 |
| Dichloro Acetylene | 7572-29-4 | 250 |
| Dichlorosilane | 4109-96-0 | 2500 |
| Diethylzinc | 557-20-0 | 10000 |
| Diisopropyl Peroxydicarbonate | 105-64-6 | 7500 |
| Dilauroyl Peroxide | 105-74-8 | 7500 |

| CHEMICAL NAME | CAS* | TQ** |
|---|---|---|
| Dimethyldichlorosilane | 75-78-5 | 1000 |
| Dimethylhydrazine, 1,1- | 57-14-7 | 1000 |
| Dimethylamine, Anhydrous | 124-40-3 | 2500 |
| 2,4-Dinitroaniline | 97-02-9 | 5000 |
| Ethyl Methyl Ketone Peroxide (also Methyl Ethyl Ketone Peroxide; concentration >60%) | 1338-23-4 | 5000 |
| Ethyl Nitrite | 109-95-5 | 5000 |
| Ethylamine | 75-04-7 | 7500 |
| Ethylene Fluorohydrin | 371-62-0 | 100 |
| Ethylene Oxide | 75-21-8 | 5000 |
| Ethyleneimine | 151-56-4 | 1000 |
| Fluorine | 7782-41-4 | 1000 |
| Formaldehyde (Formalin) | 50-00-0 | 1000 |
| Furan | 110-00-9 | 500 |
| Hexafluoroacetone | 684-16-2 | 5000 |
| Hydrochloric Acid, Anhydrous | 7647-01-0 | 5000 |
| Hydrofluoric Acid, Anhydrous | 7664-39-3 | 1000 |
| Hydrogen Bromide | 10035-10-6 | 5000 |
| Hydrogen Chloride | 7647-01-0 | 5000 |
| Hydrogen Cyanide, Anhydrous | 74-90-8 | 1000 |
| Hydrogen Fluoride | 7664-39-3 | 1000 |
| Hydrogen Peroxide (52% by weight or greater) | 7722-84-1 | 7500 |
| Hydrogen Selenide | 7783-07-5 | 150 |
| Hydrogen Sulfide | 7783-06-4 | 1500 |
| Hydroxylamine | 7803-49-8 | 2500 |
| Iron, Pentacarbonyl | 13463-40-6 | 250 |
| Isopropylamine | 75-31-0 | 5000 |
| Ketene | 463-51-4 | 100 |
| Methacrylaldehyde | 78-85-3 | 1000 |
| Methacryloyl Chloride | 920-46-7 | 150 |
| Methacryloyloxyethyl Isocyanate | 30674-80-7 | 100 |
| Methyl Acrylonitrile | 126-98-7 | 250 |
| Methylamine, Anhydrous | 74-89-5 | 1000 |
| Methyl Bromide | 74-83-9 | 2500 |
| Methyl Chloride | 74-87-3 | 15000 |
| Methyl Chloroformate | 79-22-1 | 500 |
| Methyl Ethyl Ketone Peroxide (concentration >60%) | 1338-23-4 | 5000 |
| Methyl Fluoroacetate | 453-18-9 | 100 |
| Methyl Fluorosulfate | 421-20-5 | 100 |
| Methyl Hydrazine | 60-34-4 | 100 |
| Methyl Iodide | 74-88-4 | 7500 |
| Methyl Isocyanate | 624-83-9 | 250 |
| Methyl Mercaptan | 74-93-1 | 5000 |
| Methyl Vinyl Ketone | 79-84-4 | 100 |
| Methyltrichlorosilane | 75-79-6 | 500 |
| Nickel Carbonyl (Nickel Tetracarbonyl) | 13463-39-3 | 150 |
| Nitric Acid (94.5% by weight or greater) | 7697-37-2 | 500 |
| Nitric Oxide | 10102-43-9 | 250 |
| Nitroaniline (para Nitroaniline) | 100-01-6 | 5000 |
| Nitromethane | 75-52-5 | 2500 |
| Nitrogen Dioxide | 10102-44-0 | 250 |
| Nitrogen Oxides (NO; $NO_2$; $N_2O_4$; $N_2O_3$) | 10102-44-0 | 250 |
| Nitrogen Tetroxide (also called Nitrogen Peroxide) | 10544-72-6 | 250 |
| Nitrogen Trifluoride | 7783-54-2 | 5000 |
| Nitrogen Trioxide | 10544-73-7 | 250 |
| Oleum (65% to 80% by weight; also called Fuming Sulfuric Acid) | 8014-94-7 | 1000 |
| Osmium Tetroxide | 20816-12-0 | 100 |

# OSHA Standards for General Industry

| CHEMICAL NAME | CAS* | TQ** |
|---|---|---|
| Oxygen Difluoride (Fluorine Monoxide) | 7783-41-7 | 100 |
| Ozone | 10028-15-6 | 100 |
| Pentaborane | 19624-22-7 | 100 |
| Peracetic Acid (concentration >60% Acetic Acid; also called Peroxyacetic Acid) | 79-21-0 | 1000 |
| Perchloric Acid (concentration >60% by weight) | 7601-90-3 | 5000 |
| Perchloromethyl Mercaptan | 594-42-3 | 150 |
| Perchloryl Fluoride | 7616-94-6 | 5000 |
| Peroxyacetic Acid (concentration >60% Acetic Acid; also called Peracetic Acid) | 79-21-0 | 1000 |
| Phosgene (also called Carbonyl Chloride) | 75-44-5 | 100 |
| Phosphine (Hydrogen Phosphide) | 7803-51-2 | 100 |
| Phosphorus Oxychloride (also called Phosphoryl Chloride) | 10025-87-3 | 1000 |
| Phosphorus Trichloride | 7719-12-2 | 1000 |
| Phosphoryl Chloride (also called Phosphorus Oxychloride) | 10025-87-3 | 1000 |
| Propargyl Bromide | 106-96-7 | 100 |
| Propyl Nitrate | 627-3-4 | 2500 |
| Sarin | 107-44-8 | 100 |
| Selenium Hexafluoride | 7783-79-1 | 1000 |
| Stibine (Antimony Hydride) | 7803-52-3 | 500 |
| Sulfur Dioxide (liquid) | 7446-09-5 | 1000 |
| Sulfur Pentafluoride | 5714-22-7 | 250 |
| Sulfur Tetrafluoride | 7783-60-0 | 250 |
| Sulfur Trioxide (also called Sulfuric Anhydride) | 7446-11-9 | 1000 |
| Sulfuric Anhydride (also called Sulfur Trioxide) | 7446-11-9 | 1000 |
| Tellurium Hexafluoride | 7783-80-4 | 250 |
| Tetrafluoroethylene | 116-14-3 | 5000 |
| Tetrafluorohydrazine | 10036-47-2 | 5000 |
| Tetramethyl Lead | 75-74-1 | 1000 |
| Thionyl Chloride | 7719-09-7 | 250 |
| Trichloro (chloromethyl) Silane | 1558-25-4 | 100 |
| Trichloro (dichlorophenyl) Silane | 27137-85-5 | 2500 |
| Trichlorosilane | 10025-78-2 | 5000 |
| Trifluorochloroethylene | 79-38-9 | 10000 |
| Trimethoxysilane | 2487-90-3 | 1500 |

\* Chemical Abstract Service Number.

\*\* Threshold Quantity in Pounds (Amount necessary to be covered by this standard).

## APPENDIX B TO §1910.119—BLOCK FLOW DIAGRAM AND SIMPLIFIED PROCESS FLOW DIAGRAM (NON–MANDATORY)

### EXAMPLE OF A BLOCK FLOW DIAGRAM

### EXAMPLE OF A PROCESS FLOW DIAGRAM

## APPENDIX C TO §1910.119—COMPLIANCE GUIDELINES AND RECOMMENDATIONS FOR PROCESS SAFETY MANAGEMENT (NON-MANDATORY)

This appendix serves as a nonmandatory guideline to assist employers and employees in complying with the requirements of this section, as well as provides other helpful recommendations and information. Examples presented in this appendix are not the only means of achieving the performance goals in the standard. This appendix neither adds nor detracts from the requirements of the standard.

1. *Introduction to Process Safety Management.* The major objective of process safety management of highly hazardous chemicals is to prevent unwanted releases of hazardous chemicals especially into locations which could expose employees and others to serious hazards. An effective process safety management program requires a systematic approach to evaluating the whole process. Using this approach the process design, process technology, operational and maintenance activities and procedures, nonroutine activities and procedures, emergency preparedness plans and procedures, training programs, and other elements which impact the process are all considered in the evaluation. The various lines of defense that have been incorporated into the design and operation of the process to prevent or mitigate the release of hazardous chemicals need to be evaluated and strengthened to assure their effectiveness at each level. Process safety management is the proactive identification, evaluation and mitigation or prevention of chemical releases that could occur as a result of failures in process, procedures or equipment.

The process safety management standard targets highly hazardous chemicals that have the potential to cause a catastrophic incident. This standard as a whole is to aid employers in their efforts to prevent or mitigate episodic chemical releases that could lead to a catastrophe in the workplace and possibly to the surrounding community. To control these types of hazards, employers need to develop the necessary expertise, experiences, judgement and proactive initiative within their workforce to properly implement and maintain an effective process safety management program as envisioned in the OSHA standard. This OSHA standard is required by the Clean Air Act Amendments as is the Environmental Protection Agency's Risk Management Plan. Employers, who merge the two sets of requirements into their process safety management program, will better assure full compliance with each as well as enhancing their relationship with the local community.

While OSHA believes process safety management will have a positive effect on the safety of employees in workplaces and also offers other potential benefits to employers (increased productivity), smaller businesses which may have limited resources available to them at this time, might consider alternative avenues of decreasing the risks associated with highly hazardous chemicals at their workplaces. One method which might be considered is the reduction in the inventory of the highly hazardous chemical. This reduction in inventory will result in a reduction of the risk or potential for a catastrophic incident. Also, employers including small employers may be able to establish more efficient inventory control by reducing the quantities of highly hazardous chemicals on site below the established threshold quantities. This reduction can be accomplished by ordering smaller shipments and maintaining the minimum inventory necessary for efficient and safe operation. When reduced inventory is not feasible, then the employer might consider dispersing inventory to several locations on site. Dispersing storage into locations where a release in one location will not cause a release in another location is a practical method to also reduce the risk or portential for catastrophic incidents.

2. *Employee Involvement in Process Safety Management.* Section 304 of the Clean Air Act Amendments states that employers are to consult with their employees and their representatives regarding the employers efforts in the development and implementation of the process safety management program elements and hazard assessments. Section 304 also requires employers to train and educate their employees and to inform affected employees of the findings from incident investigations required by the process safety management program. Many employers, under their safety and health programs, have already established means and methods to keep employees and their representatives informed about relevant safety and health issues and employers may be able to adapt these practices and procedures to meet their obligations under this standard. Employers who have not implemented an occupational safety and health program may wish to form a safety and health committee of employees and management representatives to help the employer meet the obligations specified by this standard. These committees can become a significant ally in helping the employer to implement and maintain an effective process safety managment program for all employees.

3. *Process Safety Information.* Complete and accurate written information concerning process chemicals, process technology, and process equipment is essential to an effective process safety management program and to a process hazards analysis. The compiled information will be a necessary resource to a variety of users including the team that will perform the process hazards analysis as required under paragraph (e); those developing the training programs and the operating procedures; contractors whose employees will be working with the process; those conducting the pre-startup reviews; local emergency preparedness planners; and insurance and enforcement officials.

The information to be compiled about the chemicals, including process intermediates, needs to be comprehensive enough for an accurate assessment of the fire and explosion characteristics, reactivity hazards, the safety and health hazards to workers, and the corrosion and erosion effects on the process equipment and monitoring tools. Current material safety data sheet (MSDS) information can be used to help meet this requirement which must be supplemented with process chemistry information including runaway reaction and over pressure hazards if applicable.

Process technology information will be a part of the process safety information package and it is expected that it will include diagrams of the type shown in appendix B of this section as well as employer established criteria for maximum inventory levels for process chemicals; limits beyond which would be considered upset conditions; and a qualitative estimate of the consequences or results of deviation that could occur if operating beyond the established process limits. Employers are encouraged to use diagrams which will help users understand the process.

A block flow diagram is used to show the major process equipment and interconnecting process flow lines and show flow rates, stream composition, temperatures, and pressures when necessary for clarity. The block flow diagram is a simplified diagram.

Process flow diagrams are more complex and will show all main flow streams including valves to enhance the understanding of the process, as well as pressures and temperatures on all feed and product lines within all major vessels, in and out of headers and heat exchangers, and points of pressure and temperature control. Also, materials of construction information, pump capacities and pressure heads, compressor horsepower and vessel design pressures and temperatures are shown when necessary for clarity. In addition, major components of control loops are usually shown along with key utilities on process flow diagrams.

Piping and instrument diagrams (P&IDs) may be the more appropriate type of diagrams to show some of the above details and to display the information for the piping designer and engineering staff. The P&IDs are to be used to describe the relationships between equipment and instrumentation as well as other relevant information that will enhance clarity. Computer software programs which do P&IDs or other diagrams useful to the information package, may be used to help meet this requirement.

The information pertaining to process equipment design must be documented. In other words, what were the codes and standards relied on to establish good engineering practice. These codes and standards are published by such organizations as the American Society of Mechanical Engineers, American Petroleum Institute, American National Standards Institute, National Fire Protection Association, American Society for Testing and Materials, National Board of Boiler and Pressure Vessel Inspectors, National Association of Corrosion Engineers, American Society of Exchange Manufacturers Association, and model building code groups.

In addition, various engineering societies issue technical reports which impact process design. For example, the American Institute of Chemical Engineers has published technical reports on topics such as two phase flow for venting devices. This type of technically recognized report would constitute good engineering practice.

For existing equipment designed and constructed many years ago in accordance with the codes and standards available at that time and no longer in general use today, the employer must document which codes and standards were used and that the design and construction along with the testing, inspection and operation are still suitable for the intended use. Where the process technology requires a design which departs from the applicable codes and standards, the employer must document that the design and construction is suitable for the intended purpose.

4. *Process Hazard Analysis.* A process hazard analysis (PHA), sometimes called a process hazard evaluation, is one of the most important elements of the process safety management program. A PHA is an organized and systematic effort to identify and analyze the significance of potential hazards associated with the processing or handling of highly hazardous chemicals. A PHA provides information which will assist employers and employees in making decisions for improving safety and reducing the consequences of unwanted or unplanned releases of hazardous chemicals. A PHA is directed toward analyzing potential causes and consequences of fires, explosions, releases of toxic or flammable chemicals and major spills of hazardous chemicals. The PHA focuses on equipment, instrumentation, utilities, human actions (routine and nonroutine), and external factors that might impact the process. These considerations assist in determining the hazards and potential failure points or failure modes in a process.

The selection of a PHA methodology or technique will be influenced by many factors including the amount of existing knowledge about the process. Is it a process that has been operated for a long period of time with little or no innovation and extensive experience has been generated with its use? Or, is it a new process or one which has been changed frequently by the inclusion of innovative features? Also, the size and complexity of the process will influence the decision as to the appropriate PHA methodology to use. All PHA methodologies are subject to certain limitations. For example, the checklist methodology works well when the process is very stable and no changes are made, but it is not as effective when the process has undergone extensive change. The checklist may miss the most recent changes and consequently the changes would not be evaluated. Another limitation to be considered concerns the assumptions made by the team or analyst. The PHA is dependent on good judgement and the assumptions made during the study need to be documented and understood by the team and reviewer and kept for a future PHA.

The team conducting the PHA need to understand the methodology that is going to be used. A PHA team can vary in size from two people to a number of people with varied operational and technical backgrounds. Some team members may only be a part of the team for a limited time. The team leader needs to be fully knowledgeable in the proper implementation of the PHA methodology that is to be used and should be impartial in the evaluation. The other full or part time team members need to provide the team with expertise in areas such as process technology, process design, operating procedures and practices, including how the work is actually performed, alarms, emergency procedures, instrumentation, maintenance procedures, both routine and nonroutine tasks, including how the tasks are authorized, procurement of parts and supplies, safety and health, and any other relevant subject as the need dictates. At least one team member must be familiar with the process.

The ideal team will have an intimate knowledge of the standards, codes, specifications and regulations applicable to the process being studied. The selected team members need to be compatible and the team leader needs to be able to manage the team, and the PHA study. The team needs to be able to work together while benefiting from the expertise of others on the team or outside the team, to resolve issues, and to forge a consensus on the findings of the study and recommendations.

The application of a PHA to a process may involve the use of different methodologies for various parts of the process. For example, a process involving a series of unit operation of varying sizes, complexities, and ages may use different methodologies and team members for each operation. Then the conclusions can be integrated into one final study and evaluation. A more specific example is the use of a checklist PHA for a standard boiler or heat exchanger and the use of a Hazard and Operability PHA for the overall process. Also, for batch type processes like custom batch operations, a generic PHA of a representative batch may be used where there are only small changes of monomer or other ingredient ratios and the chemistry is documented for the full range and ratio of batch ingredients. Another process that might consider using a generic type of PHA is a gas plant. Often these plants are simply moved from site to site and therefore, a generic PHA may be used for these movable plants. Also, when an employer has several similar size gas plants and no sour gas is being processed at the site, then a generic PHA is feasible as long as the variations of the individual sites are accounted for in the PHA. Finally, when an employer has a large continuous process which has several control rooms for different portions of the process such as for a distillation tower and a blending operation, the employer may wish to do each segment separately and then integrate the final results.

Additionally, small businesses which are covered by this rule, will often have processes that have less storage volume, less capacity, and less complicated than processes at a large facility. Therefore, OSHA would anticipate that the less complex methodologies would be used to meet the process hazard analysis criteria in the standard. These process hazard analyses can be done in less time and with a few people being involved. A less complex process generally means that less data, P&IDs, and process information is needed to perform a process hazard analysis.

Many small businesses have processes that are not unique, such as cold storage lockers or water treatment facilities. Where employer associations have a number of members with such facilities, a generic PHA, evolved from a checklist or what-if questions, could be developed and used by each employer effectively to reflect his/her particular process; this would simplify compliance for them.

When the employer has a number of processes which require a PHA, the employer must set up a priority system of which PHAs to conduct first. A preliminary or gross hazard analysis may be useful in prioritizing the processes that the employer has determined are subject to coverage by the process safety management standard. Consideration should first be given to those processes with the potential of adversely affecting the largest number of employees. This prioritizing should consider the potential severity of a chemical release, the number of potentially affected employees, the operating history of the process such as the frequency of chemical releases, the age of the process and any other relevant factors. These factors would suggest a ranking order and would suggest either using a weighing factor system or a systematic ranking method. The use of a preliminary hazard analysis would assist an employer in determining which process should be of the highest priority and thereby the employer would obtain the greatest improvement in safety at the facility.

Detailed guidance on the content and application of process hazard analysis methodologies is available from the American Institute of Chemical Engineers' Center for Chemical Process Safety (see appendix D).

5. *Operating Procedures and Practices.* Operating procedures describe tasks to be performed, data to be recorded, operating conditions to be maintained, samples to be collected, and safety and health precautions to be taken. The procedures need to be technically accurate, understandable to employees, and revised periodically to ensure that they reflect current operations. The process safety information package is to be used as a resource to better assure that the operating procedures and practices are consistent with the known hazards of the chemicals in the process and that the operating parameters are accurate. Operating procedures should be reviewed by engineering staff and operating personnel to ensure that they are accurate and provide practical instructions on how to actually carry out job duties safely.

Operating procedures will include specific instructions or details on what steps are to be taken or followed in carrying out the stated procedures. These operating instructions for each procedure should include the applicable safety precautions and should contain appropriate information on safety implications. For example, the operating procedures addressing operating parameters will contain operating instructions about pressure limits, temperature ranges, flow rates, what to do when an upset condition occurs, what alarms and instruments are pertinent if an upset condition occurs, and other subjects. Another example of using operating instructions to properly implement operating procedures is in starting up or shutting down the process. In these cases, different parameters will be required from those of normal operation. These operating instructions need to clearly indicate the distinctions between startup and normal operations such as the appropriate allowances for heating up a unit to reach the normal operating parameters. Also the operating instructions need to describe the proper method for increasing the temperature of the unit until the normal operating temperature parameters are achieved.

Computerized process control systems add complexity to operating instructions. These operating instructions need to describe the logic of the software as well as the relationship between the equipment and the control system; otherwise, it may not be apparent to the operator.

Operating procedures and instructions are important for training operating personnel. The operating procedures are often viewed as the standard operating practices (SOPs) for operations. Control room personnel and operating staff, in general, need to have a full understanding of operating procedures. If workers are not fluent in English then procedures and instructions need to be prepared in a second language understood by the workers. In addition, operating procedures need to be changed when there is a change in the process as a result of the management of change procedures. The consequences of operating procedure changes need to be fully evaluated and the information conveyed to the personnel. For example, mechanical changes to the process made by the maintenance department (like changing a valve from steel to brass or other subtle changes) need to be evaluated to determine if operating procedures and practices also need to be changed. All management of change actions must be coordinated and integrated with current operating procedures and operating personnel must be oriented to the changes in procedures before the change is made. When the process is shut down in order to make a change, then the operating procedures must be updated before startup of the process.

Training in how to handle upset conditions must be accomplished as well as what operating personnel are to do in emergencies such as when a pump seal fails or a pipeline ruptures. Communication between operating personnel and workers performing work within the process area, such as nonroutine tasks, also must be maintained. The hazards of the tasks are to be conveyed to operating personnel in accordance with established procedures and to those performing the actual tasks. When the work is completed, operating personnel should be informed to provide closure on the job.

6. *Employee Training.* All employees, including maintenance and contractor employees, involved with highly hazardous chemicals need to fully understand the safety and health hazards of the chemicals and processes they work with for the protection of themselves, their fellow employees and the citizens of nearby communities. Training conducted in compliance with §1910.1200, the Hazard Communication standard, will help employees to be more knowledgeable about the chemicals they work with as well as familiarize them with reading and understanding MSDS. However, additional training in subjects such as operating procedures and safety work practices, emergency evacuation and response, safety procedures, routine and nonroutine work authorization activities, and other areas pertinent to process safety and health will need to be covered by an employer's training program.

In establishing their training programs, employers must clearly define the employees to be trained and what subjects are to be covered in their training. Employers in setting up their training program will need to clearly establish the goals and objectives they wish to achieve with the training that they provide to their employees. The learning goals or objectives should be written in clear measurable terms before the training begins. These goals and objectives need to be tailored to each of the specific training modules or segments. Employers should describe the important actions and conditions under which the employee will demonstrate competence or knowledge as well as what is acceptable performance.

Hands-on-training where employees are able to use their senses beyond listening, will enhance learning. For example, operating personnel, who will work in a control room or at control panels, would benefit by being trained at a simulated control panel or panels. Upset conditions of various types could be displayed on the simulator, and then the employee could go through the proper operating procedures to bring the simulator panel back to the normal operating parameters. A training environment could be created to help the trainee feel the full reality of the situation but, of course, under controlled conditions. This realistic type of training can be very effective in teaching employees correct procedures while allowing them to also see the consequences of what might happen if they do not follow established operating procedures. Other training techniques using videos or on-the-job training can also be very effective for teaching other job tasks, duties, or other important information. An effective training program will allow the employee to fully participate in the training process and to practice their skill or knowledge.

Employers need to periodically evaluate their training programs to see if the necessary skills, knowledge, and routines are being properly understood and implemented by their trained employees. The means or methods for evaluating the training should be developed along with the training program goals and objectives. Training program evaluation will help employers to determine the amount of training their employees understood, and whether the desired results were obtained. If, after the evaluation, it appears that the trained employees are not at the level of knowledge and skill that was expected, the employer will need to revise the training program, provide retraining, or provide more frequent refresher training sessions until the deficiency is resolved. Those who conducted the training and those who received the training should also be consulted as to how best to improve the training process. If there is a language barrier, the language known to the trainees should be used to reinforce the training messages and information.

Careful consideration must be given to assure that employees including maintenance and contract employees receive current and updated training. For example, if changes are made to a process, impacted employees must be trained in the changes and understand the effects of the changes on their job tasks (e.g., any new operating procedures pertinent to their tasks). Additionally, as already discussed the evaluation of the employee's absorption of training will certainly influence the need for training.

7. *Contractors.* Employers who use contractors to perform work in and around processes that involve highly hazardous chemicals, will need to establish a screening process so that they hire and use contractors who accomplish the desired job tasks without compromising the safety and health of employees at a facility. For contractors, whose safety performance on the job is not known to the hiring employer, the employer will need to obtain information on injury and illness rates and experience and should obtain contractor references. Additionally, the employer must assure that the contractor has the appropriate job skills, knowledge and certifications (such as for pressure vessel welders). Contractor work methods and experiences should be evaluated. For example, does the contractor conducting demolition work swing loads over operating processes or does the contractor avoid such hazards?

Maintaining a site injury and illness log for contractors is another method employers must use to track and maintain current knowledge of work activities involving contract employees working on or adjacent to covered processes. Injury and illness logs of both the employer's employees and contract employees allow an employer to have full knowledge of process injury and illness experience. This log will also contain information which will be of use to those auditing process safety management compliance and those involved in incident investigations.

Contract employees must perform their work safely. Considering that contractors often perform very specialized and potentially hazardous tasks such as confined space entry activities and nonroutine repair activities it is quite important that their activities be controlled while they are working on or near a covered process. A permit system or work authorization system for these activities would also be helpful to all affected employers. The use of a work authorization system keeps an employer informed of contract employee activities, and as a benefit the employer will have better coordination and more management control over the work being performed in the process area. A well run and well maintained process where employee safety is fully recognized will benefit all of those who work in the facility whether they be contract employees or employees of the owner.

8. *Pre-Startup Safety.* For new processes, the employer will find a PHA helpful in improving the design and construction of the process from a reliability and quality point of view. The safe operation of the new process will be enhanced by making use of the PHA recommendations before final installations are completed. P&IDs are to be completed along with having the operating procedures in place and the operating staff trained to run the process before startup. The initial startup procedures and normal operating procedures need to be fully evaluated as part of the pre-startup review to assure a safe transfer into the normal operating mode for meeting the process parameters.

For existing processes that have been shutdown for turnaround, or modification, etc., the employer must assure that any changes other than "replacement in kind" made to the process during shutdown go through the management of change procedures. P&IDs will need to be updated as necessary, as well as operating procedures and instructions. If the changes made to the process during shutdown are significant and impact the training program, then operating personnel as well as employees engaged in routine and nonroutine work in the process area may need some refresher or additional training in light of the changes. Any incident investigation recommendations, compliance audits or PHA recommendations need to be reviewed as well to see what impacts they may have on the process before beginning the startup.

9. *Mechanical Integrity.* Employers will need to review their maintenance programs and schedules to see if there are areas where "breakdown" maintenance is used rather than an on-going mechanical integrity program. Equipment used to process, store, or handle highly hazardous chemicals needs to be designed, constructed, installed and maintained to minimize the risk of releases of such chemicals. This requires that a mechanical integrity program be in place to assure the continued integrity of process equipment. Elements of a mechanical integrity program include the identification and categorization of equipment and instrumentation, inspections and tests, testing and inspection frequencies, development of maintenance procedures, training of maintenance personnel, the establishment of criteria for acceptable test results, documentation of test and inspection results, and documentation of manufacturer recommendations as to meantime to failure for equipment and instrumentation.

# OSHA Standards for General Industry

The first line of defense an employer has available is to operate and maintain the process as designed, and to keep the chemicals contained. This line of defense is backed up by the next line of defense which is the controlled release of chemicals through venting to scrubbers or flares, or to surge or overflow tanks which are designed to receive such chemicals, etc. These lines of defense are the primary lines of defense or means to prevent unwanted releases. The secondary lines of defense would include fixed fire protection systems like sprinklers, water spray, or deluge systems, monitor guns, etc., dikes, designed drainage systems, and other systems which would control or mitigate hazardous chemicals once an unwanted release occurs. These primary and secondary lines of defense are what the mechanical integrity program needs to protect and strengthen these primary and secondary lines of defenses where appropriate.

The first step of an effective mechanical integrity program is to compile and categorize a list of process equipment and instrumentation for inclusion in the program. This list would include pressure vessels, storage tanks, process piping, relief and vent systems, fire protection system components, emergency shutdown systems and alarms and interlocks and pumps. For the categorization of instrumentation and the listed equipment the employer would prioritize which pieces of equipment require closer scrutiny than others. Meantime to failure of various instrumentation and equipment parts would be known from the manufacturers data or the employer's experience with the parts, which would then influence the inspection and testing frequency and associated procedures. Also, applicable codes and standards such as the National Board Inspection Code, or those from the American Society for Testing and Material, American Petroleum Institute, National Fire Protection Association, American National Standards Institute, American Society of Mechanical Engineers, and other groups, provide information to help establish an effective testing and inspection frequency, as well as appropriate methodologies.

The applicable codes and standards provide criteria for external inspections for such items as foundation and supports, anchor bolts, concrete or steel supports, guy wires, nozzles and sprinklers, pipe hangers, grounding connections, protective coatings and insulation, and external metal surfaces of piping and vessels, etc. These codes and standards also provide information on methodologies for internal inspection, and a frequency formula based on the corrosion rate of the materials of construction. Also, erosion both internal and external needs to be considered along with corrosion effects for piping and valves. Where the corrosion rate is not known, a maximum inspection frequency is recommended, and methods of developing the corrosion rate are available in the codes. Internal inspections need to cover items such as vessel shell, bottom and head; metallic linings; nonmetallic linings; thickness measurements for vessels and piping; inspection for erosion, corrosion, cracking and bulges; internal equipment like trays, baffles, sensors and screens for erosion, corrosion or cracking and other deficiencies.

Some of these inspections may be performed by state of local government inspectors under state and local statutes. However, each employer needs to develop procedures to ensure that tests and inspections are conducted properly and that consistency is maintained even where different employees may be involved. Appropriate training is to be provided to maintenance personnel to ensure that they understand the preventive maintenance program procedures, safe practices, and the proper use amd application of special equipment or unique tools that may be required. This training is part of the overall training program called for in the standard.

A quality assurance system is needed to help ensure that the proper materials of construction are used, that fabrication and inspection procedures are proper, and that installation procedures recognize field installation concerns. The quality assurance program is an essential part of the mechanical integrity program and will help to maintain the primary and secondary lines of defense that have been designed into the process to prevent unwanted chemical releases or those which control or mitigate a release. "As built" drawings, together with certifications of coded vessels and other equipment, and materials of construction need to be verified and retained in the quality assurance documentation. Equipment installation jobs need to be properly inspected in the field for use of proper materials and procedures and to assure that qualified craftsmen are used to do the job. The use of appropriate gaskets, packing, bolts, valves, lubricants and welding rods need to be verified in the field. Also procedures for installation of safety devices need to be verified, such as the torque on the bolts on ruptured disc installations, uniform torque on flange bolts, proper installation of pump seals, etc. If the quality of parts is a problem, it may be appropriate to conduct audits of the equipment supplier's facilities to better assure proper purchases of required equipment which is suitable for its intended service. Any changes in equipment that may become necessary will need to go through the management of change procedures.

10. *Nonroutine Work Authorizations.* Nonroutine work which is conducted in process areas needs to be controlled by the employer in a consistent manner. The hazards identified involving the work that is to be accomplished must be communicated to those doing the work, but also to those operating personnel whose work could affect the safety of the process. A work authorization notice or permit must have a procedure that describes the steps the maintenance supervisor, contractor representative or other person needs to follow to obtain the necessary clearance to get the job started. The work authorization procedures need to reference and coordinate, as applicable, lockout/tagout procedures, line breaking procedures, confined space entry procedures and hot work authorizations. This procedure also needs to provide clear steps to follow once the job is completed in order to provide closure for those that need to know the job is now completed and equipment can be returned to normal.

11. *Managing Change.* To properly manage changes to process chemicals, technology, equipment and facilities, one must define what is meant by change. In this process safety management standard, change includes all modifications to equipment, procedures, raw materials and processing conditions other than "replacement in kind". These changes need to be properly managed by identifying and reviewing them prior to implementation of the change. For example, the operating procedures contain the operating parameters (pressure limits, temperature ranges, flow rates, etc.) and the importance of operating within these limits. While the operator must have the flexibility to maintain safe operation within the established parameters, any operation outside of these parameters requires review and approval by a written management of change procedure.

Management of change covers such as changes in process technology and changes to equipment and instrumentation. Changes in process technology can result from changes in production rates, raw materials, experimentation, equipment unavailability, new equipment, new product development, change in catalyst and changes in operating conditions to improve yield or quality. Equipment changes include among others change in materials of construction, equipment specifications, piping pre-arrangements, experimental equipment, computer program revisions and changes in alarms and interlocks. Employers need to establish means and methods to detect both technical changes and mechanical changes.

Temporary changes have caused a number of catastrophes over the years, and employers need to establish ways to detect temporary changes as well as those that are permanent. It is important that a time limit for temporary changes be established and monitored since, without control, these changes may tend to become permanent. Temporary changes are subject to the management of change provisions. In addition, the management of change procedures are used to insure that the equipment and procedures are returned to their original or designed conditions at the end of the temporary change. Proper documentation and review of these changes is invaluable in assuring that the safety and health considerations are being incorporated into the operating procedures and the process.

Employers may wish to develop a form or clearance sheet to facilitate the processing of changes through the management of change procedures. A typical change form may include a description and the purpose of the change, the technical basis for the change, safety and health considerations, documentation of changes for the operating procedures, maintenance procedures, inspection and testing, P&IDs, electrical classification, training and communications, pre-startup inspection, duration if a temporary change, approvals and authorization. Where the impact of the change is minor and well understood, a check list reviewed by an authorized person with proper communication to others who are affected may be sufficient. However, for a more complex or significant design change, a hazard evaluation procedure with approvals by operations, maintenance, and safety departments may be appropriate. Changes in documents such as P&IDs, raw materials, operating procedures, mechanical integrity programs, electrical classifications, etc., need to be noted so that these revisions can be made permanent when the drawings and procedure manuals are updated. Copies of process changes need to be kept in an accessible location to ensure that design changes are available to operating personnel as well as to PHA team members when a PHA is being done or one is being updated.

12. *Investigation of Incidents.* Incident investigation is the process of identifying the underlying causes of incidents and implementing steps to prevent similar events from occurring. The intent of an incident investigation is for employers to learn from past experiences and thus avoid repeating past mistakes. The incidents for whicn OSHA expects employers to become aware and to investigate are the types of events which result in or could reasonably have resulted in a catastrophic release. Some of the events are sometimes referred to as "near misses," meaning that a serious consequence did not occur, but could have.

Employers need to develop in-house capability to investigate incidents that occur in their facilities. A team needs to be assembled by the employer and trained in the techniques of investigation including how to conduct interviews of witnesses, needed documentation and report writing. A multi-disciplinary team is better able to gather the facts of the event and to analyze them and develop plausible scenarios as to what happened, and why. Team members should be selected on the basis of their training, knowledge and ability to contribute to a team effort to fully investigate the incident. Employees in the process area where the incident occurred should be consulted, interviewed or made a member of the team. Their knowledge of the events form a significant set of facts about the incident which occurred. The report, its findings and recommendations are to be shared with those who can benefit from the information. The cooperation of employees is essential to an effective incident investigation. The focus of the investigation should be to obtain facts, and not to place blame. The team and the investigation process should clearly deal with all involved individuals in a fair, open and consistent manner.

13. *Emergency Preparedness.* Each employer must address what actions employees are to take when there is an unwanted release of highly hazardous chemicals. Emergency preparedness or the employer's tertiary (third) lines of defense are those that will be relied on along with the secondary lines of defense when the primary lines of defense which are used to prevent an unwanted release fail to stop the release. Employers will need to decide if they want employees to handle and stop small or minor incidental releases. Whether they wish to mobilize the available resources at the plant and have them brought to bear on a more significant release. Or whether employers want their employees to evacuate the danger area and promptly escape to a preplanned safe zone area, and allow the local community emergency response organizations to handle the release. Or whether the employer wants to use some combination of these actions. Employers will need to select how many different emergency preparedness or tertiary lines of defense they plan to have and then develop the necessary plans and procedures, and appropriately train employees in their emergency duties and responsibilities and then implement these lines of defense.

Employers at a minimum must have an emergency action plan which will facilitate the prompt evacuation of employees due to an unwanted release of a highly hazardous chemical. This means that the employer will have a plan that will be activated by an alarm system to alert employees when to evacuate and, that employees who are physically impaired, will have the necessary support and assistance to get them to the safe zone as well. The intent of these requirements is to alert and move employees to a safe zone quickly. Delaying alarms or confusing alarms are to be avoided. The use of process control centers or similar process buildings in the process area as safe areas is discouraged. Recent catastrophes have shown that a large life loss has occurred in these structures because of where they have been sited and because they are not necessarily designed to withstand over-pressures from shockwaves resulting from explosions in the process area.

Unwanted incidental releases of highly hazardous chemicals in the process area must be addressed by the employer as to what actions employees are to take. If the employer wants employees to evacuate the area, then the emergency action plan will be activated. For outdoor processes where wind direction is important for selecting the safe route to a refuge area, the employer should place a wind direction indicator such as a wind sock or pennant at the highest point that can be seen throughout the process area. Employees can move in the direction of cross wind to upwind to gain safe access to the refuge area by knowing the wind direction.

If the employer wants specific employees in the release area to control or stop the minor emergency or incidental release, these actions must be planned for in advance and procedures developed and implemented. Preplanning for handling incidental releases for minor emergencies in the process area needs to be done, appropriate equipment for the hazards must be provided, and training conducted for those employees who will perform the emergency work before they respond to handle an actual release. The employer's training program, including the Hazard Communication standard training is to address the training needs for employees who are expected to handle incidental or minor releases.

Preplanning for releases that are more serious than incidental releases is another important line of defense to be used by the employer. When a serious release of a highly hazardous chemical occurs, the employer through preplanning will have determined in advance what actions employees are to take. The evacuation of the immediate release area and other areas as necessary would be accomplished under the emergency action plan. If the employer wishes to use plant personnel such as a fire brigade, spill control team, a hazardous materials team, or use employees to render aid to those in the immediate release area and control or mitigate the incident, these actions are covered by §1910.120, the Hazardous Waste Operations and Emergency Response (HAZWOPER) standard. If outside assistance is necessary, such as through mutual aid agreements between employers or local government emergency response organizations, these emergency responders are also covered by HAZWOPER. The safety and health protections required for emergency responders are the responsibility of their employers and of the on-scene incident commander.

Responders may be working under very hazardous conditions and therefore the objective is to have them competently led by an on-scene incident commander and the commander's staff, properly equipped to do their assigned work safely, and fully trained to carry out their duties safely before they respond to an emergency. Drills, training exercises, or simulations with the local community emergency response planners and responder organizations is one means to obtain better preparedness. This close cooperation and coordination between plant and local community emergency preparedness managers will also aid the employer in complying with the Environmental Protection Agency's Risk Management Plan criteria.

One effective way for medium to large facilities to enhance coordination and communication during emergencies for on plant operations and with local community organizations is for employers to establish and equip an emergency control center. The emergency control center would be sited in a safe zone area so that it could be occupied throughout the duration of an emergency. The center would serve as the major ccommunication link between the on-scene incident commander and plant or corporate management as well as with the local community officials. The communication equipment in the emergency control center should include a network to receive and transmit information by telephone, radio or other means. It is important to have a backup communication network in case of power failure or one communication means fails. The center should also be equipped with the plant layout and community maps, utility drawings including fire water, emergency lighting, appropriate reference materials such as a government agency notification list, company personnel phone list, SARA Title III reports and material safety data sheets, emergency plans and procedures manual, a listing with the location of emergency response equipment, mutual aid information, and access to meteorological or weather condition data and any dispersion modeling data.

14. *Compliance Audits.* Employers need to select a trained individual or assemble a trained team of people to audit the process safety management system and program. A small process or plant may need only one knowledgeable person to conduct an audit. The audit is to include an evaluation of the design and effectiveness of the process safety management system and a field inspection of the safety and health conditions and practices to verify that the employer's systems are effectively implemented. The audit should be conducted or lead by a person knowledgeable in audit techniques and who is impartial towards the facility or area being audited. The essential elements of an audit program include planning, staffing, conduting the audit, evaluation and corrective action, follow-up and documentation.

Planning in advance is essential to the success of the auditing process. Each employer needs to establish the format, staffing, scheduling and verification methods prior to conducting the audit. The format should be designed to provide the lead auditor with a procedure or checklist which details the requirements of each section of the standard. The names of the audit team members should be listed as part of the format as well. The checklist, if properly designed, could serve as the verification sheet which provides the auditor with the necessary information to expedite the review and assure that no requirements of the standard are omitted. This verification sheet format could also identify those elements that will require evaluation or a response to correct deficiencies. This sheet could also be used for developing the follow-up and documentation requirements.

The selection of effective audit team members is critical to the success of the program. Team members should be chosen for their experience, knowledge, and training and should be familiar with the processes and with auditing techniques, practices and procedures. The size of the team will vary depending on the size and complexity of the process under consideration. For a large, complex, highly instrumented plant, it may be desirable to have team members with expertise in process engineering and design, process chemistry, instrumentation and computer controls, electrical hazards and classifications, safety and health disciplines, maintenance, emergency preparedness, warehousing or shipping, and process safety auditing. The team may use part-time members to provide for the depth of expertise required as well as for what is actually done or followed, compared to what is written.

An effective audit includes a review of the relevant documentation and process safety information, inspection of the physical facilities, and interviews with all levels of plant personnel. Utilizing the audit procedure and checklist developed in the preplanning stage, the audit team can systematically analyze compliance with the provisions of the standard and any other corporate policies that are relevant. For example, the audit team will review all aspects of the training program as part of the overall audit. The team will review the written training program for adequacy of content, frequency of training, effectiveness of training in terms of its goals and objectives as well as to how it fits into meeting the standard's requirements, documentation, etc. Through interviews, the team can determine the employee's knowledge and awareness of the safety procedures, duties, rules, emergency response assignments, etc. During the inspection, the team can observe actual practices such as safety and health policies, procedures, and work authorization practices. This approach enables the team to identify deficiencies and determine where corrective actions or improvements are necessary.

An audit is a technique used to gather sufficient facts and information, including statistical information, to verify compliance with standards. Auditors should select as part of their preplanning a sample size sufficient to give a degree of confidence that the audit reflects the level of compliance with the standard. The audit team, through this systematic analysis, should document areas which require corrective action as well as those areas where the process safety management system is effective and working in an effective manner. This provides a record of the audit procedures and findings, and serves as a baseline of operation data for future audits. It will assist future auditors in determining changes or trends from previous audits.

Corrective action is one of the most important parts of the audit. It includes not only addressing the identified deficiencies, but also planning, followup, and documentation. The corrective action process normally begins with a management review of the audit findings. The purpose of this review is to determine what actions are appropriate, and to establish priorities, timetables, resource allocations and requirements and responsibilities. In some cases, corrective action may involve a simple change in procedure or minor maintenance effort to remedy the concern. Management of change procedures need to be used, as appropriate, even for what may seem to be a minor change. Many of the deficiencies can be acted on promptly, while some may require engineering studies or indepth review of actual procedures and practices. There may be instances where no action is necessary and this is a valid response to an audit finding. All actions taken, including an explanation where no action is taken on a finding, needs to be documented as to what was done and why.

It is important to assure that each deficiency identified is addressed, the corrective action to be taken noted, and the audit person or team responsible be properly documented by the employer. To control the corrective action process, the employer should consider the use of a tracking system. This tracking system might include periodic status reports shared with affected levels of management, specific reports such as completion of an engineering study, and a final implementation report to provide closure for audit findings that have been through management of change, if appropriate, and then shared with affected employees and management. This type of tracking system provides the employer with the status of the corrective action. It also provides the documentation required to verify that appropriate corrective actions were taken on deficiencies identified in the audit.

## APPENDIX D TO §1910.119—SOURCES OF FURTHER INFORMATION (NON–MANDATORY)

1. Center for Chemical Process Safety, American Institute of Chemical Engineers, 345 East 47th Street, New York, NY 10017, (212) 705-7319.

2. "Guidelines for Hazard Evaluation Procedures," American Institute of Chemical Engineers; 345 East 47th Street, New York, NY 10017.

3. "Guidelines for Technical Management of Chemical Process Safety," Center for Chemical Process Safety of the American Institute of Chemical Engineers; 345 East 47th Street, New York, NY 10017.

4. "Evaluating Process Safety in the Chemical Industry," Chemical Manufacturers Association; 2501 M Street NW, Washington, DC 20037.

5. "Safe Warehousing of Chemicals," Chemical Manufacturers Association; 2501 M Street NW, Washington, DC 20037.

6. "Management of Process Hazards," American Petroleum Institute (API Recommended Practice 750); 1220 L Street, N.W., Washington, D.C. 20005.

7. "Improving Owner and Contractor Safety Performance," American Petroleum Institute (API Recommended Practice 2220); API, 1220 L Street N.W., Washington, D.C. 20005.

8. Chemical Manufacturers Association (CMA's Manager Guide), First Edition, September 1991; CMA, 2501 M Street, N.W., Washington, D.C. 20037.

9. "Improving Construction Safety Performance," Report A-3, The Business Roundtable; The Business Roundtable, 200 Park Avenue, New York, NY 10166. (Report includes criteria to evaluate contractor safety performance and criteria to enhance contractor safety performance).

10. "Recommended Guidelines for Contractor Safety and Health," Texas Chemical Council; Texas Chemical Council, 1402 Nueces Street, Austin, TX 78701-1534.

11. "Loss Prevention in the Process Industries," Volumes I and II; Frank P. Lees, Butterworth; London 1983.

12. "Safety and Health Program Management Guidelines," 1989; U.S. Department of Labor, Occupational Safety and Health Administration.

13. "Safety and Health Guide for the Chemical Industry," 1986, (OSHA 3091); U.S. Department of Labor, Occupational Safety and Health Administration; 200 Constitution Avenue, N.W., Washington, D.C. 20210.

14. "Review of Emergency Systems," June 1988; U.S. Environmental Protection Agency (EPA), Office of Solid Waste and Emergency Response, Washington, DC 20460.

15. "Technical Guidance for Hazards Analysis, Emergency Planning for Extremely Hazardous Substances," December 1987; U.S. Environmental Protection Agency (EPA), Federal Emergency Management Administration (FEMA) and U.S. Department of Transportation (DOT), Washington, DC 20460.

16. "Accident Investigation * * * A New Approach," 1983, National Safety Council; 444 North Michigan Avenue, Chicago, IL 60611-3991.

17. "Fire & Explosion Index Hazard Classification Guide," 6th Edition, May 1987, Dow Chemical Company; Midland, Michigan 48674.

18. "Chemical Exposure Index," May 1988, Dow Chemical Company; Midland, Michigan 48674.

[57 FR 6403, Feb. 24, 1992; 57 FR 7847, Mar. 4, 1992, as amended at 61 FR 9238, Mar. 7, 1996]

## §1910.120 Hazardous waste operations and emergency response.

**1910.120(a)** *Scope, application, and definitions*—**(1)** *Scope.* This section covers the following operations, unless the employer can demonstrate that the operation does not involve employee exposure or the reasonable possibility for employee exposure to safety or health hazards:

**1910.120(a)(1)(i)** Clean-up operations required by a governmental body, whether Federal, state, local or other involving hazardous substances that are conducted at uncontrolled hazardous waste sites (including, but not limited to, the EPA's National Priority Site List (NPL), state priority site lists, sites recommended for the EPA NPL, and initial investigations of government identified sites which are conducted before the presence or absence of hazardous substances has been ascertained);

**1910.120(a)(1)(ii)** Corrective actions involving clean-up operations at sites covered by the Resource Conservation and Recovery Act of 1976 (RCRA) as amended (42 U.S.C. 6901 *et seq.*);

**1910.120(a)(1)(iii)** Voluntary clean-up operations at sites recognized by Federal, state, local or other governmental bodies as uncontrolled hazardous waste sites;

**1910.120(a)(1)(iv)** Operations involving hazardous wastes that are conducted at treatment, storage, and disposal (TSD) facilities regulated by 40

CFR parts 264 and 265 pursuant to RCRA; or by agencies under agreement with U.S.E.P.A. to implement RCRA regulations; and

**1910.120(a)(1)(v)** Emergency response operations for releases of, or substantial threats of releases of, hazardous substances without regard to the location of the hazard.

**1910.120(a)(2)** *Application.* **(i)** All requirements of part 1910 and part 1926 of title 29 of the Code of Federal Regulations apply pursuant to their terms to hazardous waste and emergency response operations whether covered by this section or not. If there is a conflict or overlap, the provision more protective of employee safety and health shall apply without regard to 29 CFR 1910.5(c)(1).

**1910.120(a)(2)(ii)** Hazardous substance clean-up operations within the scope of paragraphs (a)(1)(i) through (a)(1)(iii) of this section must comply with all paragraphs of this section except paragraphs (p) and (q).

**1910.120(a)(2)(iii)** Operations within the scope of paragraph (a)(1)(iv) of this section must comply only with the requirements of paragraph (p) of this section.

*Notes and Exceptions:* (A) All provisions of paragraph (p) of this section cover any treatment, storage or disposal (TSD) operation regulated by 40 CFR parts 264 and 265 or by state law authorized under RCRA, and required to have a permit or interim status from EPA pursuant to 40 CFR 270.1 or from a state agency pursuant to RCRA.

(B) Employers who are not required to have a permit or interim status because they are conditionally exempt small quantity generators under 40 CFR 261.5 or are generators who qualify under 40 CFR 262.34 for exemptions from regulation under 40 CFR parts 264, 265 and 270 ("excepted employers") are not covered by paragraphs (p)(1) through (p)(7) of this section. Excepted employers who are required by the EPA or state agency to have their employees engage in emergency response or who direct their employees to engage in emergency response are covered by paragraph (p)(8) of this section, and cannot be exempted by (p)(8)(i) of this section. Excepted employers who are not required to have employees engage in emergency response, who direct their employees to evacuate in the case of such emergencies and who meet the requirements of paragraph (p)(8)(i) of this section are exempt from the balance of paragraph (p)(8) of this section.

(C) If an area is used primarily for treatment, storage or disposal, any emergency response operations in that area shall comply with paragraph (p)(8) of this section. In other areas not used primarily for treatment, storage, or disposal, any emergency response operations shall comply with paragraph (q) of this section. Compliance with the requirements of paragraph (q) of this section shall be deemed to be in compliance with the requirements of paragraph (p)(8) of this section.

**1910.120(a)(2)(iv)** Emergency response operations for releases of, or substantial threats of releases of, hazardous substances which are not covered by paragraphs (a)(1)(i) through (a)(1)(iv) of this section must only comply with the requirements of paragraph (q) of this section.

**1910.120(a)(3)** *Definitions—Buddy system* means a system of organizing employees into work groups in such a manner that each employee of the work group is designated to be observed by at least one other employee in the work group. The purpose of the buddy system is to provide rapid assistance to employees in the event of an emergency.

*Clean-up operation* means an operation where hazardous substances are removed, contained, incinerated, neutralized, stabilized, cleared-up, or in any other manner processed or handled with the ultimate goal of making the site safer for people or the environment.

*Decontamination* means the removal of hazardous substances from employees and their equipment to the extent necessary to preclude the occurrence of foreseeable adverse health affects.

*Emergency response* or *responding to emergencies* means a response effort by employees from outside the immediate release area or by other designated responders (i.e., mutual-aid groups, local fire departments, etc.) to an occurrence which results, or is likely to result, in an uncontrolled release of a hazardous substance. Responses to incidental releases of hazardous substances where the substance can be absorbed, neutralized, or otherwise controlled at the time of release by employees in the immediate release area, or by maintenance personnel are not considered to be emergency responses within the scope of this standard. Responses to releases of hazardous substances where there is no potential safety or health hazard (i.e., fire, explosion, or chemical exposure) are not considered to be emergency responses.

*Facility* means (A) any building, structure, installation, equipment, pipe or pipeline (including any pipe into a sewer or publicly owned treatment works), well, pit, pond, lagoon, impoundment, ditch, storage container, motor vehicle, rolling stock, or aircraft, or (B) any site or area where a hazardous substance has been deposited, stored, disposed of, or placed, or otherwise come to be located; but does not include any consumer product in consumer use or any water-borne vessel.

*Hazardous materials response (HAZMAT) team* means an organized group of employees, designated by the employer, who are expected to perform work to handle and control actual or potential leaks or spills of hazardous substances requiring possible close approach to the substance. The team members perform responses to releases or potential releases of hazardous substances for the purpose of control or stabilization of the incident. A HAZMAT team is not a fire brigade nor is a typical fire brigade a HAZMAT team. A HAZMAT team, however, may be a separate component of a fire brigade or fire department.

*Hazardous substance* means any substance designated or listed under paragraphs (A) through (D) of this definition, exposure to which results or may result in adverse affects on the health or safety of employees:

**1910.120(a)(3)(A)** Any substance defined under section 101(14) of CERCLA;

**1910.120(a)(3)(B)** Any biological agent and other disease-causing agent which after release into the environment and upon exposure, ingestion, inhalation, or assimilation into any person, either directly from the environment or indirectly by ingestion through food chains, will or may reasonably be anticipated to cause death, disease, behavioral abnormalities, cancer, genetic mutation, physiological malfunctions (including malfunctions in reproduction) or physical deformations in such persons or their offspring;

**1910.120(a)(3)(C)** Any substance listed by the U.S. Department of Transportation as hazardous materials under 49 CFR 172.101 and appendices; and

**1910.120(a)(3)(D)** Hazardous waste as herein defined.

*Hazardous waste* means—

(A) A waste or combination of wastes as defined in 40 CFR 261.3, or

(B) Those substances defined as hazardous wastes in 49 CFR 171.8.

*Hazardous waste operation* means any operation conducted within the scope of this standard.

*Hazardous waste site* or *Site* means any facility or location within the scope of this standard at which hazardous waste operations take place.

*Health hazard* means a chemical, mixture of chemicals or a pathogen for which there is statistically significant evidence based on at least one study conducted in accordance with established scientific principles that acute or chronic health effects may occur in exposed employees. The term "health hazard" includes chemicals which are carcinogens, toxic or highly toxic agents, reproductive toxins, irritants, corrosives, sensitizers, heptaotoxins, nephrotoxins, neurotoxins, agents which act on the hematopoietic system, and agents which damage the lungs, skin, eyes, or mucous membranes. It also includes stress due to temperature extremes. Further definition of the terms used above can be found in appendix A to 29 CFR 1910.1200.

*IDLH* or [sic]*Immediately dangerous to life or health* means an atmospheric concentration of any toxic, corrosive or asphyxiant substance that poses an immediate threat to life or would cause irreversible or delayed adverse health effects or would interfere with an individual's ability to escape from a dangerous atmosphere.

*Oxygen deficiency* means that concentration of oxygen by volume below which atmosphere supplying respiratory protection must be provided. It exists in atmospheres where the percentage of oxygen by volume is less than 19.5 percent oxygen.

*Permissible exposure limit* means the exposure, inhalation or dermal permissible exposure limit specified in 29 CFR part 1910, subparts G and Z.

*Published exposure level* means the exposure limits published in "NIOSH Recommendations for Occupational Health Standards" dated 1986, which is incorporated by reference as specified in §1910.6 or if none is specified, the exposure limits published in the standards specified by the American Conference of Governmental Industrial Hygienists in their publication "Threshold Limit Values and Biological Exposure Indices for 1987-88" dated 1987, which is incorporated by reference as specified in §1910.6.

*Post emergency response* means that portion of an emergency response performed after the immediate threat of a release has been stabilized or eliminated and clean-up of the site has begun. If post emergency response is performed by an employer's own employees who were part of the initial emergency response, it is considered to be part of the initial response and not post emergency response. However, if a group of an employer's own employees, separate from the group providing initial response, performs the clean-up operation, then the separate group of employees would be considered to be performing post-emergency response and subject to paragraph (q)(11) of this section.

*Qualified person* means a person with specific training, knowledge and experience in the area for which the person has the responsibility and the authority to control.

*Site safety and health supervisor (or official)* means the individual located on a hazardous waste site who is responsible to the employer and has the authority and knowledge necessary to implement the site safety and health plan and verify compliance with applicable safety and health requirements.

*Small quantity generator* means a generator of hazardous wastes who in any calendar month generates no more than 1,000 kilograms (2,205 pounds) of hazardous waste in that month.

*Uncontrolled hazardous waste site,* means an area identified as an uncontrolled hazardous waste site by a governmental body, whether Federal, state, local or other where an accumulation of hazardous substances creates a threat to the health and safety of individuals or the environment or both. Some sites are found on public lands such as those created by former municipal, county or state landfills where illegal or poorly managed waste disposal has taken place. Other sites are found on private property, often belonging to generators or former generators of hazardous substance wastes. Examples of such sites include, but are not limited to, surface impoundments, landfills, dumps, and tank or drum farms. Normal operations at TSD sites are not covered by this definition.

**1910.120(b)** *Safety and health program.*

**Note to (b):** Safety and health programs developed and implemented to meet other Federal, state, or local regulations are considered acceptable in meeting this requirement if they cover or are modified to cover the topics required in this paragraph. An additional or separate safety and health program is not required by this paragraph.

**1910.120(b)(1)** *General.* **(i)** Employers shall develop and implement a written safety and health program for their employees involved in hazardous waste operations. The program shall be designed to identify, evaluate, and control safety and health hazards, and provide for emergency response for hazardous waste operations.

**1910.120(b)(1)(ii)** The written safety and health program shall incorporate the following:

**1910.120(b)(1)(ii)(A)** An organizational structure;

**1910.120(b)(1)(ii)(B)** A comprehensive workplan;

**1910.120(b)(1)(ii)(C)** A site-specific safety and health plan which need not repeat the employer's standard operating procedures required in paragraph (b)(1)(ii)(F) of this section;

**1910.120(b)(1)(ii)(D)** The safety and health training program;

**1910.120(b)(1)(ii)(E)** The medical surveillance program;

**1910.120(b)(1)(ii)(F)** The employer's standard operating procedures for safety and health; and

**1910.120(b)(1)(ii)(G)** Any necessary interface between general program and site specific activities.

**1910.120(b)(1)(iii)** *Site excavation.* Site excavations created during initial site preparation or during hazardous waste operations shall be shored or sloped as appropriate to prevent accidental collapse in accordance with subpart P of 29 CFR part 1926.

**1910.120(b)(1)(iv)** *Contractors and sub-contractors.* An employer who retains contractor or sub-contractor services for work in hazardous waste operations shall inform those contractors, sub-contractors, or their representatives of the site emergency response procedures and any potential fire, explosion, health, safety or other hazards of the hazardous waste operation that have been identified by the employer, including those identified in the employer's information program.

**1910.120(b)(1)(v)** *Program availability.* The written safety and health program shall be made available to any contractor or subcontractor or their representative who will be involved with the hazardous waste operation; to employees; to employee designated representatives; to OSHA personnel, and to personnel of other Federal, state, or local agencies with regulatory authority over the site.

**1910.120(b)(2)** *Organizational structure part of the site program*—**(i)** The organizational structure part of the program shall establish the specific chain of command and specify the overall responsibilities of supervisors and employees. It shall include, at a minimum, the following elements:

**1910.120(b)(2)(i)(A)** A general supervisor who has the responsibility and authority to direct all hazardous waste operations.

**1910.120(b)(2)(i)(B)** A site safety and health supervisor who has the responsibility and authority to develop and implement the site safety and health plan and verify compliance.

**1910.120(b)(2)(i)(C)** All other personnel needed for hazardous waste site operations and emergency response and their general functions and responsibilities.

**1910.120(b)(2)(i)(D)** The lines of authority, responsibility, and communication.

**1910.120(b)(2)(ii)** The organizational structure shall be reviewed and updated as necessary to reflect the current status of waste site operations.

**1910.120(b)(3)** *Comprehensive workplan part of the site program.* The comprehensive workplan part of the program shall address the tasks and objectives of the site operations and the logistics and resources required to reach those tasks and objectives.

**1910.120(b)(3)(i)** The comprehensive workplan shall address anticipated clean-up activities as well as normal operating procedures which need not repeat the employer's procedures available elsewhere.

**1910.120(b)(3)(ii)** The comprehensive workplan shall define work tasks and objectives and identify the methods for accomplishing those tasks and objectives.

**1910.120(b)(3)(iii)** The comprehensive workplan shall establish personnel requirements for implementing the plan.

**1910.120(b)(3)(iv)** The comprehensive workplan shall provide for the implementation of the training required in paragraph (e) of this section.

**1910.120(b)(3)(v)** The comprehensive workplan shall provide for the implementation of the required informational programs required in paragraph (i) of this section.

**1910.120(b)(3)(vi)** The comprehensive workplan shall provide for the implementation of the medical surveillance program described in paragraph (f) of this section.

**1910.120(b)(4)** *Site-specific safety and health plan part of the program*—**(i)** *General.* The site safety and health plan, which must be kept on site, shall address the safety and health hazards of each phase of site operation and include the requirements and procedures for employee protection.

**1910.120(b)(4)(ii)** *Elements.* The site safety and health plan, as a minimum, shall address the following:

**1910.120(b)(4)(ii)(A)** A safety and health risk or hazard analysis for each site task and operation found in the workplan.

**1910.120(b)(4)(ii)(B)** Employee training assignments to assure compliance with paragraph (e) of this section.

**1910.120(b)(4)(ii)(C)** Personal protective equipment to be used by employees for each of the site tasks and operations being conducted as required by the personal protective equipment program in paragraph (g)(5) of this section.

**1910.120(b)(4)(ii)(D)** Medical surveillance requirements in accordance with the program in paragraph (f) of this section.

**1910.120(b)(4)(ii)(E)** Frequency and types of air monitoring, personnel monitoring, and environmental sampling techniques and instrumentation to be used, including methods of maintenance and calibration of monitoring and sampling equipment to be used.

**1910.120(b)(4)(ii)(F)** Site control measures in accordance with the site control program required in paragraph (d) of this section.

**1910.120(b)(4)(ii)(G)** Decontamination procedures in accordance with paragraph (k) of this section.

**1910.120(b)(4)(ii)(H)** An emergency response plan meeting the requirements of paragraph (l) of this section for safe and effective responses to emergencies, including the necessary PPE and other equipment.

**1910.120(b)(4)(ii)(I)** Confined space entry procedures.

**1910.120(b)(4)(ii)(J)** A spill containment program meeting the requirements of paragraph (j) of this section.

**1910.120(b)(4)(iii)** *Pre-entry briefing.* The site specific safety and health plan shall provide for pre-entry briefings to be held prior to initiating any site activity, and at such other times as necessary to ensure that employees are apprised of the site safety and health plan and that this plan is being followed. The information and data obtained from site characterization and analysis work required in paragraph (c) of this section shall be used to prepare and update the site safety and health plan.

**1910.120(b)(4)(iv)** *Effectiveness of site safety and health plan.* Inspections shall be conducted by the site safety and health supervisor or, in the absence of that individual, another individual who is knowledgeable in occupational safety and health, acting on behalf of the employer as necessary to determine the effectiveness of the site safety and health plan. Any deficiencies in the effectiveness of the site safety and health plan shall be corrected by the employer.

**1910.120(c)** *Site characterization and analysis*—**(1)** *General.* Hazardous waste sites shall be evaluated in accordance with this paragraph to identify specific site hazards and to determine the appropriate safety and health control procedures needed to protect employees from the identified hazards.

**1910.120(c)(2)** *Preliminary evaluation.* A preliminary evaluation of a site's characteristics shall be performed prior to site entry by a qualified person in order to aid in the selection of appropriate employee protection methods prior to site entry. Immediately after initial site entry, a more detailed evaluation of the site's specific characteristics shall be performed by a qualified person in order to further identify existing site hazards and to further aid in the selection of the appropriate engineering controls and personal protective equipment for the tasks to be performed.

**1910.120(c)(3)** *Hazard identification.* All suspected conditions that may pose inhalation or skin absorption hazards that are immediately dangerous to life or health (IDLH), or other conditions that may cause death or serious harm, shall be identified during the preliminary survey and evaluated during the detailed survey. Examples of such hazards include, but are not limited to, confined space entry, potentially explosive or flammable situations, visible vapor clouds, or areas where biological indicators such as dead animals or vegetation are located.

**1910.120(c)(4)** *Required information.* The following information to the extent available shall be obtained by the employer prior to allowing employees to enter a site:

**1910.120(c)(4)(i)** Location and approximate size of the site.

**1910.120(c)(4)(ii)** Description of the response activity and/or the job task to be performed.

**1910.120(c)(4)(iii)** Duration of the planned employee activity.

**1910.120(c)(4)(iv)** Site topography and accessibility by air and roads.

**1910.120(c)(4)(v)** Safety and health hazards expected at the site.

**1910.120(c)(4)(vi)** Pathways for hazardous substance dispersion.

**1910.120(c)(4)(vii)** Present status and capabilities of emergency response teams that would provide assistance to hazardous waste clean-up site employees at the time of an emergency.

**1910.120(c)(4)(viii)** Hazardous substances and health hazards involved or expected at the site, and their chemical and physical properties.

**1910.120(c)(5)** *Personal protective equipment.* Personal protective equipment (PPE) shall be provided and used during initial site entry in accordance with the following requirements:

**1910.120(c)(5)(i)** Based upon the results of the preliminary site evaluation, an ensemble of PPE shall be selected and used during initial site entry which will provide protection to a level of exposure below permissible exposure limits and published exposure levels for known or suspected hazardous substances and health hazards, and which will provide protection against other known and suspected hazards identified during the preliminary site evaluation. If there is no permissible exposure limit or published exposure level, the employer may use other published studies and information as a guide to appropriate personal protective equipment.

**1910.120(c)(5)(ii)** If positive-pressure self-contained breathing apparatus is not used as part of the entry ensemble, and if respiratory protection is warranted by the potential hazards identified during the preliminary site evaluation, an escape self-contained breathing apparatus of at least five minute's duration shall be carried by employees during initial site entry.

**1910.120(c)(5)(iii)** If the preliminary site evaluation does not produce sufficient information to identify the hazards or suspected hazards of the site, an ensemble providing protection equivalent to Level B PPE shall be provided as minimum protection, and direct reading instruments shall be used as appropriate for identifying IDLH conditions. (See appendix B for a description of Level B hazards and the recommendations for Level B protective equipment.)

**1910.120(c)(5)(iv)** Once the hazards of the site have been identified, the appropriate PPE shall be selected and used in accordance with paragraph (g) of this section.

**1910.120(c)(6)** *Monitoring.* The following monitoring shall be conducted during initial site entry when the site evaluation produces information that shows the potential for ionizing radiation or IDLH conditions, or when the site information is not sufficient reasonably to eliminate these possible conditions:

**1910.120(c)(6)(i)** Monitoring with direct reading instruments for hazardous levels of ionizing radiation.

**1910.120(c)(6)(ii)** Monitoring the air with appropriate direct reading test equipment (i.e., combustible gas meters, detector tubes) for IDLH and other conditions that may cause death or serious harm (combustible or explosive atmospheres, oxygen deficiency, toxic substances).

**1910.120(c)(6)(iii)** Visually observing for signs of actual or potential IDLH or other dangerous conditions.

**1910.120(c)(6)(iv)** An ongoing air monitoring program in accordance with paragraph (h) of this section shall be implemented after site characterization has determined the site is safe for the start-up of operations.

**1910.120(c)(7)** *Risk identification.* Once the presence and concentrations of specific hazardous substances and health hazards have been established, the risks associated with these substances shall be identified. Employees who will be working on the site shall be informed of any risks that have been identified. In situations covered by the Hazard Communication Standard, 29 CFR 1910.1200, training required by that standard need not be duplicated.

**Note to paragraph (c)(7):** Risks to consider include, but are not limited to:

**1910.120(c)(7)(a)** Exposures exceeding the permissible exposure limits and published exposure levels.

**1910.120(c)(7)(b)** IDLH concentrations.

**1910.120(c)(7)(c)** Potential skin absorption and irritation sources.

**1910.120(c)(7)(d)** Potential eye irritation sources.

**1910.120(c)(7)(e)** Explosion sensitivity and flammability ranges.

**1910.120(c)(7)(f)** Oxygen deficiency.

**1910.120(c)(8)** *Employee notification.* Any information concerning the chemical, physical, and toxicologic properties of each substance known or expected to be present on site that is available to the employer and relevant to the duties an employee is expected to perform shall be made available to the affected employees prior to the commencement of their work activities. The employer may utilize information developed for the hazard communication standard for this purpose.

**1910.120(d)** *Site control*—**(1)** *General.* Appropriate site control procedures shall be implemented to control employee exposure to hazardous substances before clean-up work begins.

**1910.120(d)(2)** *Site control program.* A site control program for protecting employees which is part of the employer's site safety and health program required in paragraph (b) of this section shall be developed during the planning stages of a hazardous waste clean-up operation and modified as necessary as new information becomes available.

**1910.120(d)(3)** *Elements of the site control program.* The site control program shall, as a minimum, include: A site map; site work zones; the use of a "buddy system"; site communications including alerting means for emergencies; the standard operating procedures or safe work practices; and, identification of the nearest medical assistance. Where these requirements are covered elsewhere they need not be repeated.

**1910.120(e)** *Training*—**(1)** *General.* **(i)** All employees working on site (such as but not limited to equipment operators, general laborers and others) exposed to hazardous substances, health hazards, or safety hazards and their supervisors and management responsible for the site shall receive training meeting the requirements of this paragraph before they are permitted to engage in hazardous waste operations that could expose them to hazardous substances, safety, or health hazards, and they shall receive review training as specified in this paragraph.

**1910.120(e)(1)(ii)** Employees shall not be permitted to participate in or supervise field activities until they have been trained to a level required by their job function and responsibility.

**1910.120(e)(2)** *Elements to be covered.* The training shall thoroughly cover the following:

**1910.120(e)(2)(i)** Names of personnel and alternates responsible for site safety and health;

**1910.120(e)(2)(ii)** Safety, health and other hazards present on the site;

**1910.120(e)(2)(iii)** Use of personal protective equipment;

**1910.120(e)(2)(iv)** Work practices by which the employee can minimize risks from hazards;

**1910.120(e)(2)(v)** Safe use of engineering controls and equipment on the site;

**1910.120(e)(2)(vi)** Medical surveillance requirements, including recognition of symptoms and signs which might indicate overexposure to hazards; and

**1910.120(e)(2)(vii)** The contents of paragraphs (G) through (J) of the site safety and health plan set forth in paragraph (b)(4)(ii) of this section.

**1910.120(e)(3)** *Initial training.* **(i)** General site workers (such as equipment operators, general laborers and supervisory personnel) engaged in hazardous substance removal or other activities which expose or potentially expose workers to hazardous substances and health hazards shall receive a minimum of 40 hours of instruction off the site, and a minimum of three days actual field experience under the direct supervision of a trained, experienced supervisor.

**1910.120(e)(3)(ii)** Workers on site only occasionally for a specific limited task (such as, but not limited to, ground water monitoring, land surveying, or geo-physical surveying) and who are unlikely to be exposed over permissible exposure limits and published exposure limits shall receive a minimum of 24 hours of instruction off the site, and the minimum of one day actual field experience under the direct supervision of a trained, experienced supervisor.

**1910.120(e)(3)(iii)** Workers regularly on site who work in areas which have been monitored and fully characterized indicating that exposures are under permissible exposure limits and published exposure limits where respirators are not necessary, and the characterization indicates that there are no health hazards or the possibility of an emergency developing, shall receive a minimum of 24 hours of instruction off the site and the minimum of one day actual field experience under the direct supervision of a trained, experienced supervisor.

**1910.120(e)(3)(iv)** Workers with 24 hours of training who are covered by paragraphs (e)(3)(ii) and (e)(3)(iii) of this section, and who become general site workers or who are required to wear respirators, shall have the additional 16 hours and two days of training necessary to total the training specified in paragraph (e)(3)(i).

**1910.120(e)(4)** *Management and supervisor training.* On-site management and supervisors directly responsible for, or who supervise employees engaged in, hazardous waste operations shall receive 40 hours initial training, and three days of supervised field experience (the training may be reduced to 24 hours and one day if the only area of their responsibility is employees covered by paragraphs (e)(3)(ii) and (e)(3)(iii)) and at least eight additional hours of specialized training at the time of job assignment on such topics as, but not limited to, the employer's safety and health program and the associated employee training program, personal protective equipment program, spill containment program, and health hazard monitoring procedure and techniques.

**1910.120(e)(5)** *Qualifications for trainers.* Trainers shall be qualified to instruct employees about the subject matter that is being presented in training. Such trainers shall have satisfactorily completed a training program for teaching the subjects they are expected to teach, or they shall have the academic credentials and instructional experience necessary for teaching the subjects. Instructors shall demonstrate competent instructional skills and knowledge of the applicable subject matter.

**1910.120(e)(6)** *Training certification.* Employees and supervisors that have received and successfully completed the training and field experience specified in paragraphs (e)(1) through (e)(4) of this section shall be certified by their instructor or the head instructor and trained supervisor as having successfully completed the necessary training. A written certificate shall be given to each person so certified. Any person who has not been so certified or who does not meet the requirements of paragraph (e)(9) of this section shall be prohibited from engaging in hazardous waste operations.

**1910.120(e)(7)** *Emergency response.* Employees who are engaged in responding to hazardous emergency situations at hazardous waste clean-up sites that may expose them to hazardous substances shall be trained in how to respond to such expected emergencies.

**1910.120(e)(8)** *Refresher training.* Employees specified in paragraph (e)(1) of this section, and managers and supervisors specified in paragraph (e)(4) of this section, shall receive eight hours of refresher training annually on the items specified in paragraph (e)(2) and/or (e)(4) of this section, any critique of incidents that have occurred in the past year that can serve as training examples of related work, and other relevant topics.

**1910.120(e)(9)** *Equivalent training.* Employers who can show by documentation or certification that an employee's work experience and/or training has resulted in training equivalent to that training required in paragraphs (e)(1) through (e)(4) of this section shall not be required to provide the initial training requirements of those paragraphs to such employees and shall provide a copy of the certification or documentation to the employee upon request. However, certified employees or employees with equivalent training new to a site shall receive appropriate, site specific training before site entry and have appropriate supervised field experience at the new site. Equivalent training includes any academic training or the training that existing employees might have already received from actual hazardous waste site work experience.

**1910.120(f) Medical surveillance—(1)** *General.* Employers engaged in operations specified in paragraphs (a)(1)(i) through (a)(1)(iv) of this section and not covered by (a)(2)(iii) exceptions and employers of employees specified in paragraph (q)(9) shall institute a medical surveillance program in accordance with this paragraph.

**1910.120(f)(2)** *Employees covered.* The medical surveillance program shall be instituted by the employer for the following employees:

**1910.120(f)(2)(i)** All employees who are or may be exposed to hazardous substances or health hazards at or above the permissible exposure limits or, if there is no permissible exposure limit, above the published exposure levels for these substances, without regard to the use of respirators, for 30 days or more a year;

**1910.120(f)(2)(ii)** All employees who wear a respirator for 30 days or more a year or as required by §1910.134;

**1910.120(f)(2)(iii)** All employees who are injured, become ill or develop signs or symptoms due to possible overexposure involving hazardous substances or health hazards from an emergency response or hazardous waste operation; and

**1910.120(f)(2)(iv)** Members of HAZMAT teams.

**1910.120(f)(3)** *Frequency of medical examinations and consultations.* Medical examinations and consultations shall be made available by the employer to each employee covered under paragraph (f)(2) of this section on the following schedules:

**1910.120(f)(3)(i)** For employees covered under paragraphs (f)(2)(i), (f)(2)(ii), and (f)(2)(iv):

**1910.120(f)(3)(i)(A)** Prior to assignment;

**1910.120(f)(3)(i)(B)** At least once every twelve months for each employee covered unless the attending physician believes a longer interval (not greater than biennially) is appropriate;

**1910.120(f)(3)(i)(C)** At termination of employment or reassignment to an area where the employee would not be covered if the employee has not had an examination within the last six months;

**1910.120(f)(3)(i)(D)** As soon as possible upon notification by an employee that the employee has developed signs or symptoms indicating possible overexposure to hazardous substances or health hazards, or that the employee has been injured or exposed above the permissible exposure limits or published exposure levels in an emergency situation;

**1910.120(f)(3)(i)(E)** At more frequent times, if the examining physician determines that an increased frequency of examination is medically necessary.

**1910.120(f)(3)(ii)** For employees covered under paragraph (f)(2)(iii) and for all employees including those of employers covered by paragraph (a)(1)(v) who may have been injured, received a health impairment, developed signs or symptoms which may have resulted from exposure to hazardous substances resulting from an emergency incident, or exposed during an emergency incident to hazardous substances at concentrations above the permissible exposure limits or the published exposure levels without the necessary personal protective equipment being used:

**1910.120(f)(3)(ii)(A)** As soon as possible following the emergency incident or development of signs or symptoms;

**1910.120(f)(3)(ii)(B)** At additional times, if the examining physician determines that follow-up examinations or consultations are medically necessary.

**1910.120(f)(4)** *Content of medical examinations and consultations.* **(i)** Medical examinations required by paragraph (f)(3) of this section shall include a medical and work history (or updated history if one is in the employee's file) with special emphasis on symptoms related to the handling of hazardous substances and health hazards, and to fitness for duty including the ability to wear any required PPE under conditions (i.e., temperature extremes) that may be expected at the work site.

**1910.120(f)(4)(ii)** The content of medical examinations or consultations made available to employees pursuant to paragraph (f) shall be determined by the attending physician. The guidelines in the *Occupational Safety and Health Guidance Manual for Hazardous Waste Site Activities* (See appendix D, Reference #10) should be consulted.

**1910.120(f)(5)** *Examination by a physician and costs.* All medical examinations and procedures shall be performed by or under the supervision of a licensed physician, preferably one knowledgeable in occupational medicine, and shall be provided without cost to the employee, without loss of pay, and at a reasonable time and place.

**1910.120(f)(6)** *Information provided to the physician.* The employer shall provide one copy of this standard and its appendices to the attending physician, and in addition the following for each employee:

**1910.120(f)(6)(i)** A description of the employee's duties as they relate to the employee's exposures.

**1910.120(f)(6)(ii)** The employee's exposure levels or anticipated exposure levels.

**1910.120(f)(6)(iii)** A description of any personal protective equipment used or to be used.

**1910.120(f)(6)(iv)** Information from previous medical examinations of the employee which is not readily available to the examining physician.

**1910.120(f)(6)(v)** Information required by §1910.134.

**1910.120(f)(7)** *Physician's written opinion.* **(i)** The employer shall obtain and furnish the employee with a copy of a written opinion from the attending physician containing the following:

**1910.120(f)(7)(i)(A)** The physician's opinion as to whether the employee has any detected medical conditions which would place the employee at increased risk of material impairment of the employee's health from work in hazardous waste operations or emergency response, or from respirator use.

**1910.120(f)(7)(i)(B)** The physician's recommended limitations upon the employee's assigned work.

**1910.120(f)(7)(i)(C)** The results of the medical examination and tests if requested by the employee.

**1910.120(f)(7)(i)(D)** A statement that the employee has been informed by the physician of the results of the medical examination and any medical conditions which require further examination or treatment.

**1910.120(f)(7)(ii)** The written opinion obtained by the employer shall not reveal specific findings or diagnoses unrelated to occupational exposures.

**1910.120(f)(8)** *Recordkeeping.* **(i)** An accurate record of the medical surveillance required by paragraph (f) of this section shall be retained. This record shall be retained for the period specified and meet the criteria of 29 CFR 1910.1020.

**1910.120(f)(8)(ii)** The record required in paragraph (f)(8)(i) of this section shall include at least the following information:

**1910.120(f)(8)(ii)(A)** The name and social security number of the employee;

**1910.120(f)(8)(ii)(B)** Physician's written opinions, recommended limitations, and results of examinations and tests;

**1910.120(f)(8)(ii)(C)** Any employee medical complaints related to exposure to hazardous substances;

**1910.120(f)(8)(ii)(D)** A copy of the information provided to the examining physician by the employer, with the exception of the standard and its appendices.

**1910.120(g)** *Engineering controls, work practices, and personal protective equipment for employee protection.* Engineering controls, work practices, personal protective equipment, or a combination of these shall be implemented in accordance with this paragraph to protect employees from exposure to hazardous substances and safety and health hazards.

**1910.120(g)(1)** *Engineering controls, work practices and PPE for substances regulated in subparts G and Z.* **(i)** Engineering controls and work practices shall be instituted to reduce and maintain employee exposure to or below the permissible exposure limits for substances regulated by 29 CFR part 1910, to the extent required by subpart Z, except to the extent that such controls and practices are not feasible.

**Note to paragraph (g)(1)(i):** Engineering controls which may be feasible include the use of pressurized cabs or control booths on equipment, and/or the use of remotely operated material handling equipment. Work practices which may be feasible are removing all non-essential employees from potential exposure during opening of drums, wetting down dusty operations and locating employees upwind of possible hazards.

**1910.120(g)(1)(ii)** Whenever engineering controls and work practices are not feasible or not required, any reasonable combination of engineering controls, work practices and PPE shall be used to reduce and maintain employee exposures to or below the permissible exposure limits or dose limits for substances regulated by 29 CFR part 1910, subpart Z.

**1910.120(g)(1)(iii)** The employer shall not implement a schedule of employee rotation as a means of compliance with permissible exposure limits or dose limits except when there is no other feasible way of complying with the airborne or dermal dose limits for ionizing radiation.

**1910.120(g)(1)(iv)** The provisions of 29 CFR, subpart G, shall be followed.

**1910.120(g)(2)** *Engineering controls, work practices, and PPE for substances not regulated in subparts G and Z.* An appropriate combination of engineering controls, work practices and personal protective equipment shall be used to reduce and maintain employee exposure to or below published exposure levels for hazardous substances and health hazards not regulated by 29 CFR part 1910, subparts G and Z. The employer may use the published literature and MSDS as a guide in making the employer's determination as to what level of protection the employer believes is appropriate for hazardous substances and health hazards for which there is no permissible exposure limit or published exposure limit.

**1910.120(g)(3)** *Personal protective equipment selection.* **(i)** Personal protective equipment (PPE) shall be selected and used which will protect employees from the hazards and potential hazards they are likely to encounter as identified during the site characterization and analysis.

**1910.120(g)(3)(ii)** Personal protective equipment selection shall be based on an evaluation of the performance characteristics of the PPE relative to the requirements and limitations of the site, the task-specific conditions and duration, and the hazards and potential hazards identified at the site.

**1910.120(g)(3)(iii)** Positive pressure self-contained breathing apparatus, or positive pressure air-line respirators equipped with an escape air supply, shall be used when chemical exposure levels present will create a substantial possibility of immediate death, immediate serious illness or injury, or impair the ability to escape.

**1910.120(g)(3)(iv)** Totally-encapsulating chemical protective suits (protection equivalent to Level A protection as recommended in appendix B) shall be used in conditions where skin absorption of a hazardous substance may result in a substantial possibility of immediate death, immediate serious illness or injury, or impair the ability to escape.

**1910.120(g)(3)(v)** The level of protection provided by PPE selection shall be increased when additional information on site conditions indicates that increased protection is necessary to reduce employee exposures below permissible exposure limits and published exposure levels for hazardous substances and health hazards. (See appendix B for guidance on selecting PPE ensembles.)

**Note to paragraph (g)(3):** The level of employee protection provided may be decreased when additional information or site conditions show that decreased protection will not result in hazardous exposures to employees.

**1910.120(g)(3)(vi)** Personal protective equipment shall be selected and used to meet the requirements of 29 CFR part 1910, subpart I, and additional requirements specified in this section.

# OSHA Standards for General Industry

**1910.120(g)(4)** *Totally-encapsulating chemical protective suits.* **(i)** Totally-encapsulating suits shall protect employees from the particular hazards which are identified during site characterization and analysis.

**1910.120(g)(4)(ii)** Totally-encapsulating suits shall be capable of maintaining positive air pressure. (See appendix A for a test method which may be used to evaluate this requirement.)

**1910.120(g)(4)(iii)** Totally-encapsulating suits shall be capable of preventing inward test gas leakage of more than 0.5 percent. (See appendix A for a test method which may be used to evaluate this requirement.)

**1910.120(g)(5)** *Personal protective equipment (PPE) program.* A written personal protective equipment program, which is part of the employer's safety and health program required in paragraph (b) of this section or required in paragraph (p)(1) of this section and which is also a part of the site-specific safety and health plan shall be established. The PPE program shall address the elements listed below. When elements, such as donning and doffing procedures, are provided by the manufacturer of a piece of equipment and are attached to the plan, they need not be rewritten into the plan as long as they adequately address the procedure or element.

**1910.120(g)(5)(i)** PPE selection based upon site hazards,

**1910.120(g)(5)(ii)** PPE use and limitations of the equipment,

**1910.120(g)(5)(iii)** Work mission duration,

**1910.120(g)(5)(iv)** PPE maintenance and storage,

**1910.120(g)(5)(v)** PPE decontamination and disposal,

**1910.120(g)(5)(vi)** PPE training and proper fitting,

**1910.120(g)(5)(vii)** PPE donning and doffing procedures,

**1910.120(g)(5)(viii)** PPE inspection procedures prior to, during, and after use,

**1910.120(g)(5)(ix)** Evaluation of the effectiveness of the PPE program, and

**1910.120(g)(5)(x)** Limitations during temperature extremes, heat stress, and other appropriate medical considerations.

**1910.120(h)** *Monitoring*—**(1)** *General.* **(i)** Monitoring shall be performed in accordance with this paragraph where there may be a question of employee exposure to hazardous concentrations of hazardous substances in order to assure proper selection of engineering controls, work practices and personal protective equipment so that employees are not exposed to levels which exceed permissible exposure limits, or published exposure levels if there are no permissible exposure limits, for hazardous substances.

**1910.120(h)(1)(ii)** Air monitoring shall be used to identify and quantify airborne levels of hazardous substances and safety and health hazards in order to determine the appropriate level of employee protection needed on site.

**1910.120(h)(2)** *Initial entry.* Upon initial entry, representative air monitoring shall be conducted to identify any IDLH condition, exposure over permissible exposure limits or published exposure levels, exposure over a radioactive material's dose limits or other dangerous condition such as the presence of flammable atmospheres or oxygen-deficient environments.

**1910.120(h)(3)** *Periodic monitoring.* Periodic monitoring shall be conducted when the possibility of an IDLH condition or flammable atmosphere has developed or when there is indication that exposures may have risen over permissible exposure limits or published exposure levels since prior monitoring. Situations where it shall be considered whether the possibility that exposures have risen are as follows:

**1910.120(h)(3)(i)** When work begins on a different portion of the site.

**1910.120(h)(3)(ii)** When contaminants other than those previously identified are being handled.

**1910.120(h)(3)(iii)** When a different type of operation is initiated (e.g., drum opening as opposed to exploratory well drilling).

**1910.120(h)(3)(iv)** When employees are handling leaking drums or containers or working in areas with obvious liquid contamination (e.g., a spill or lagoon).

**1910.120(h)(4)** *Monitoring of high-risk employees.* After the actual clean-up phase of any hazardous waste operation commences; for example, when soil, surface water or containers are moved or disturbed; the employer shall monitor those employees likely to have the highest exposures to hazardous substances and health hazards likely to be present above permissible exposure limits or published exposure levels by using personal sampling frequently enough to characterize employee exposures. If the employees likely to have the highest exposure are over permissible exposure limits or published exposure limits, then monitoring shall continue to determine all employees likely to be above those limits. The employer may utilize a representative sampling approach by documenting that the employees and chemicals chosen for monitoring are based on the criteria stated above.

**Note to paragraph (h):** It is not required to monitor employees engaged in site characterization operations covered by paragraph (c) of this section.

**1910.120(i)** *Informational programs.* Employers shall develop and implement a program, which is part of the employer's safety and health program required in paragraph (b) of this section, to inform employees, contractors, and subcontractors (or their representative) actually engaged in hazardous waste operations of the nature, level and degree of exposure likely as a result of participation in such hazardous waste operations. Employees, contractors and subcontractors working outside of the operations part of a site are not covered by this standard.

**1910.120(j)** *Handling drums and containers*—**(1)** *General.* **(i)** Hazardous substances and contaminated soils, liquids, and other residues shall be handled, transported, labeled, and disposed of in accordance with this paragraph.

**1910.120(j)(1)(ii)** Drums and containers used during the clean-up shall meet the appropriate DOT, OSHA, and EPA regulations for the wastes that they contain.

**1910.120(j)(1)(iii)** When practical, drums and containers shall be inspected and their integrity shall be assured prior to being moved. Drums or containers that cannot be inspected before being moved because of storage conditions (i.e., buried beneath the earth, stacked behind other drums, stacked several tiers high in a pile, etc.) shall be moved to an accessible location and inspected prior to further handling.

**1910.120(j)(1)(iv)** Unlabelled drums and containers shall be considered to contain hazardous substances and handled accordingly until the contents are positively identified and labeled.

**1910.120(j)(1)(v)** Site operations shall be organized to minimize the amount of drum or container movement.

**1910.120(j)(1)(vi)** Prior to movement of drums or containers, all employees exposed to the transfer operation shall be warned of the potential hazards associated with the contents of the drums or containers.

**1910.120(j)(1)(vii)** U.S. Department of Transportation specified salvage drums or containers and suitable quantities of proper absorbent shall be kept available and used in areas where spills, leaks, or ruptures may occur.

**1910.120(j)(1)(viii)** Where major spills may occur, a spill containment program, which is part of the employer's safety and health program required in paragraph (b) of this section, shall be implemented to contain and isolate the entire volume of the hazardous substance being transferred.

**1910.120(j)(1)(ix)** Drums and containers that cannot be moved without rupture, leakage, or spillage shall be emptied into a sound container using a device classified for the material being transferred.

**1910.120(j)(1)(x)** A ground-penetrating system or other type of detection system or device shall be used to estimate the location and depth of buried drums or containers.

**1910.120(j)(1)(xi)** Soil or covering material shall be removed with caution to prevent drum or container rupture.

**1910.120(j)(1)(xii)** Fire extinguishing equipment meeting the requirements of 29 CFR part 1910, subpart L, shall be on hand and ready for use to control incipient fires.

**1910.120(j)(2)** *Opening drums and containers.* The following procedures shall be followed in areas where drums or containers are being opened:

**1910.120(j)(2)(i)** Where an airline respirator system is used, connections to the source of air supply shall be protected from contamination and the entire system shall be protected from physical damage.

**1910.120(j)(2)(ii)** Employees not actually involved in opening drums or containers shall be kept a safe distance from the drums or containers being opened.

**1910.120(j)(2)(iii)** If employees must work near or adjacent to drums or containers being opened, a suitable shield that does not interfere with the work operation shall be placed between the employee and the drums or containers being opened to protect the employee in case of accidental explosion.

**1910.120(j)(2)(iv)** Controls for drum or container opening equipment, monitoring equipment, and fire suppression equipment shall be located behind the explosion-resistant barrier.

**1910.120(j)(2)(v)** When there is a reasonable possibility of flammable atmospheres being present, material handling equipment and hand tools shall be of the type to prevent sources of ignition.

**1910.120(j)(2)(vi)** Drums and containers shall be opened in such a manner that excess interior pressure will be safely relieved. If pressure can not be relieved from a remote location, appropriate shielding shall be placed between the employee and the drums or containers to reduce the risk of employee injury.

**1910.120(j)(2)(vii)** Employees shall not stand upon or work from drums or containers.

**1910.120(j)(3)** *Material handling equipment.* Material handling equipment used to transfer drums and containers shall be selected, positioned and operated to minimize sources of ignition related to the equipment from igniting vapors released from ruptured drums or containers.

**1910.120(j)(4)** *Radioactive wastes.* Drums and containers containing radioactive wastes shall not be handled until such time as their hazard to employees is properly assessed.

**1910.120(j)(5)** *Shock sensitive wastes.* As a minimum, the following special precautions shall be taken when drums and containers containing or suspected of containing shock-sensitive wastes are handled:

**1910.120(j)(5)(i)** All non-essential employees shall be evacuated from the area of transfer.

**1910.120(j)(5)(ii)** Material handling equipment shall be provided with explosive containment devices or protective shields to protect equipment operators from exploding containers.

**1910.120(j)(5)(iii)** An employee alarm system capable of being perceived above surrounding light and noise conditions shall be used to signal the commencement and completion of explosive waste handling activities.

**1910.120(j)(5)(iv)** Continuous communications (i.e., portable radios, hand signals, telephones, as appropriate) shall be maintained between the employee-in-charge of the immediate handling area and both the site safety and health supervisor and the command post until such time as the handling operation is completed. Communication equipment or methods that could cause shock sensitive materials to explode shall not be used.

**1910.120(j)(5)(v)** Drums and containers under pressure, as evidenced by bulging or swelling, shall not be moved until such time as the cause for excess pressure is determined and appropriate containment procedures have been implemented to protect employees from explosive relief of the drum.

**1910.120(j)(5)(vi)** Drums and containers containing packaged laboratory wastes shall be considered to contain shock-sensitive or explosive materials until they have been characterized.

CAUTION: Shipping of shock sensitive wastes may be prohibited under U.S. Department of Transportation regulations. Employers and their shippers should refer to 49 CFR 173.21 and 173.50.

**1910.120(j)(6)** *Laboratory waste packs.* In addition to the requirements of paragraph (j)(5) of this section, the following precautions shall be taken, as a minimum, in handling laboratory waste packs (lab packs):

**1910.120(j)(6)(i)** Lab packs shall be opened only when necessary and then only by an individual knowledgeable in the inspection, classification, and segregation of the containers within the pack according to the hazards of the wastes.

**1910.120(j)(6)(ii)** If crystalline material is noted on any container, the contents shall be handled as a shock-sensitive waste until the contents are identified.

**1910.120(j)(7)** *Sampling of drum and container contents.* Sampling of containers and drums shall be done in accordance with a sampling procedure which is part of the site safety and health plan developed for and available to employees and others at the specific worksite.

**1910.120(j)(8)** *Shipping and transport.* **(i)** Drums and containers shall be identified and classified prior to packaging for shipment.

**1910.120(j)(8)(ii)** Drum or container staging areas shall be kept to the minimum number necessary to identify and classify materials safely and prepare them for transport.

**1910.120(j)(8)(iii)** Staging areas shall be provided with adequate access and egress routes.

**1910.120(j)(8)(iv)** Bulking of hazardous wastes shall be permitted only after a thorough characterization of the materials has been completed.

**1910.120(j)(9)** *Tank and vault procedures.* **(i)** Tanks and vaults containing hazardous substances shall be handled in a manner similar to that for drums and containers, taking into consideration the size of the tank or vault.

**1910.120(j)(9)(ii)** Appropriate tank or vault entry procedures as described in the employer's safety and health plan shall be followed whenever employees must enter a tank or vault.

**1910.120(k)** **Decontamination**—**(1)** *General.* Procedures for all phases of decontamination shall be developed and implemented in accordance with this paragraph.

**1910.120(k)(2)** *Decontamination procedures.* **(i)** A decontamination procedure shall be developed, communicated to employees and implemented before any employees or equipment may enter areas on site where potential for exposure to hazardous substances exists.

**1910.120(k)(2)(ii)** Standard operating procedures shall be developed to minimize employee contact with hazardous substances or with equipment that has contacted hazardous substances.

**1910.120(k)(2)(iii)** All employees leaving a contaminated area shall be appropriately decontaminated; all contaminated clothing and equipment leaving a contaminated area shall be appropriately disposed of or decontaminated.

**1910.120(k)(2)(iv)** Decontamination procedures shall be monitored by the site safety and health supervisor to determine their effectiveness. When such procedures are found to be ineffective, appropriate steps shall be taken to correct any deficiencies.

**1910.120(k)(3)** *Location.* Decontamination shall be performed in geographical areas that will minimize the exposure of uncontaminated employees or equipment to contaminated employees or equipment.

**1910.120(k)(4)** *Equipment and solvents.* All equipment and solvents used for decontamination shall be decontaminated or disposed of properly.

**1910.120(k)(5)** *Personal protective clothing and equipment.* **(i)** Protective clothing and equipment shall be decontaminated, cleaned, laundered, maintained or replaced as needed to maintain their effectiveness.

**1910.120(k)(5)(ii)** Employees whose non-impermeable clothing becomes wetted with hazardous substances shall immediately remove that clothing and proceed to shower. The clothing shall be disposed of or decontaminated before it is removed from the work zone.

**1910.120(k)(6)** *Unauthorized employees.* Unauthorized employees shall not remove protective clothing or equipment from change rooms.

**1910.120(k)(7)** *Commercial laundries or cleaning establishments.* Commercial laundries or cleaning establishments that decontaminate protective clothing or equipment shall be informed of the potentially harmful effects of exposures to hazardous substances.

**1910.120(k)(8)** *Showers and change rooms.* Where the decontamination procedure indicates a need for regular showers and change rooms outside of a contaminated area, they shall be provided and meet the requirements of 29 CFR 1910.141. If temperature conditions prevent the effective use of water, then other effective means for cleansing shall be provided and used.

**1910.120(l)** **Emergency response by employees at uncontrolled hazardous waste sites**—**(1)** *Emergency response plan.* **(i)** An emergency response plan shall be developed and implemented by all employers within the scope of paragraphs (a)(1) (i)-(ii) of this section to handle anticipated emergencies prior to the commencement of hazardous waste operations. The plan shall be in writing and available for inspection and copying by employees, their representatives, OSHA personnel and other governmental agencies with relevant responsibilities.

**1910.120(l)(1)(ii)** Employers who will evacuate their employees from the danger area when an emergency occurs, and who do not permit any of their employees to assist in handling the emergency, are exempt from the requirements of this paragraph if they provide an emergency action plan complying with §1910.38 of this part.

**1910.120(l)(2)** *Elements of an emergency response plan.* The employer shall develop an emergency response plan for emergencies which shall address, as a minimum, the following:

**1910.120(l)(2)(i)** Pre-emergency planning.

**1910.120(l)(2)(ii)** Personnel roles, lines of authority, and communication.

# OSHA Standards for General Industry

**1910.120(l)(2)(iii)** Emergency recognition and prevention.

**1910.120(l)(2)(iv)** Safe distances and places of refuge.

**1910.120(l)(2)(v)** Site security and control.

**1910.120(l)(2)(vi)** Evacuation routes and procedures.

**1910.120(l)(2)(vii)** Decontamination procedures which are not covered by the site safety and health plan.

**1910.120(l)(2)(viii)** Emergency medical treatment and first aid.

**1910.120(l)(2)(ix)** Emergency alerting and response procedures.

**1910.120(l)(2)(x)** Critique of response and follow-up.

**1910.120(l)(2)(xi)** PPE and emergency equipment.

**1910.120(l)(3)** *Procedures for handling emergency incidents.* **(i)** In addition to the elements for the emergency response plan required in paragraph (l)(2) of this section, the following elements shall be included for emergency response plans:

**1910.120(l)(3)(i)(A)** Site topography, layout, and prevailing weather conditions.

**1910.120(l)(3)(i)(B)** Procedures for reporting incidents to local, state, and federal governmental agencies.

**1910.120(l)(3)(ii)** The emergency response plan shall be a separate section of the Site Safety and Health Plan.

**1910.120(l)(3)(iii)** The emergency response plan shall be compatible and integrated with the disaster, fire and/or emergency response plans of local, state, and federal agencies.

**1910.120(l)(3)(iv)** The emergency response plan shall be rehearsed regularly as part of the overall training program for site operations.

**1910.120(l)(3)(v)** The site emergency response plan shall be reviewed periodically and, as necessary, be amended to keep it current with new or changing site conditions or information.

**1910.120(l)(3)(vi)** An employee alarm system shall be installed in accordance with 29 CFR 1910.165 to notify employees of an emergency situation; to stop work activities if necessary; to lower background noise in order to speed communication; and to begin emergency procedures.

**1910.120(l)(3)(vii)** Based upon the information available at time of the emergency, the employer shall evaluate the incident and the site response capabilities and proceed with the appropriate steps to implement the site emergency response plan.

**1910.120(m)** *Illumination.* Areas accessible to employees shall be lighted to not less than the minimum illumination intensities listed in the following Table H-120.1 while any work is in progress:

### TABLE H-120.1.—MINIMUM ILLUMINATION INTENSITIES IN FOOT-CANDLES

| Foot-candles | Area or operations |
| --- | --- |
| 5 | General site areas. |
| 3 | Excavation and waste areas, accessways, active storage areas, loading platforms, refueling, and field maintenance areas. |
| 5 | Indoors: Warehouses, corridors, hallways, and exitways. |
| 5 | Tunnels, shafts, and general underground work areas. (Exception: minimum of 10 foot-candles is required at tunnel and shaft heading during drilling, mucking, and scaling. Mine Safety and Health Administration approved cap lights shall be acceptable for use in the tunnel heading.) |
| 10 | General shops (e.g., mechanical and electrical equipment rooms, active storerooms, barracks or living quarters, locker or dressing rooms, dining areas, and indoor toilets and workrooms.) |
| 30 | First aid stations, infirmaries, and offices. |

**1910.120(n)** *Sanitation at temporary workplaces*—**(1)** *Potable water.* **(i)** An adequate supply of potable water shall be provided on the site.

**1910.120(n)(1)(ii)** Portable containers used to dispense drinking water shall be capable of being tightly closed, and equipped with a tap. Water shall not be dipped from containers.

**1910.120(n)(1)(iii)** Any container used to distribute drinking water shall be clearly marked as to the nature of its contents and not used for any other purpose.

**1910.120(n)(1)(iv)** Where single service cups (to be used but once) are supplied, both a sanitary container for the unused cups and a receptacle for disposing of the used cups shall be provided.

**1910.120(n)(2)** *Nonpotable water.* **(i)** Outlets for nonpotable water, such as water for firefighting purposes, shall be identified to indicate clearly that the water is unsafe and is not to be used for drinking, washing, or cooking purposes.

**1910.120(n)(2)(ii)** There shall be no cross-connection, open or potential, between a system furnishing potable water and a system furnishing nonpotable water.

**1910.120(n)(3)** *Toilet facilities.* **(i)** Toilets shall be provided for employees according to the following Table H-120.2.

### TABLE H-120.2.—TOILET FACILITIES

| Number of employees | Minimum number of facilities |
| --- | --- |
| 20 or fewer | One. |
| More than 20, fewer than 200 | One toilet seat and one urinal per 40 employees. |
| More than 200 | One toilet seat and one urinal per 50 employees. |

**1910.120(n)(3)(ii)** Under temporary field conditions, provisions shall be made to assure that at least one toilet facility is available.

**1910.120(n)(3)(iii)** Hazardous waste sites not provided with a sanitary sewer shall be provided with the following toilet facilities unless prohibited by local codes:

**1910.120(n)(3)(iii)(A)** Chemical toilets;

**1910.120(n)(3)(iii)(B)** Recirculating toilets;

**1910.120(n)(3)(iii)(C)** Combustion toilets; or

**1910.120(n)(3)(iii)(D)** Flush toilets.

**1910.120(n)(3)(iv)** The requirements of this paragraph for sanitation facilities shall not apply to mobile crews having transportation readily available to nearby toilet facilities.

**1910.120(n)(3)(v)** Doors entering toilet facilities shall be provided with entrance locks controlled from inside the facility.

**1910.120(n)(4)** *Food handling.* All food service facilities and operations for employees shall meet the applicable laws, ordinances, and regulations of the jurisdictions in which they are located.

**1910.120(n)(5)** *Temporary sleeping quarters.* When temporary sleeping quarters are provided, they shall be heated, ventilated, and lighted.

**1910.120(n)(6)** *Washing facilities.* The employer shall provide adequate washing facilities for employees engaged in operations where hazardous substances may be harmful to employees. Such facilities shall be in near proximity to the worksite; in areas where exposures are below permissible exposure limits and published exposure levels and which are under the controls of the employer; and shall be so equipped as to enable employees to remove hazardous substances from themselves.

**1910.120(n)(7)** *Showers and change rooms.* When hazardous waste clean-up or removal operations commence on a site and the duration of the work will require six months or greater time to complete, the employer shall provide showers and change rooms for all employees exposed to hazardous substances and health hazards involved in hazardous waste clean-up or removal operations.

**1910.120(n)(7)(i)** Showers shall be provided and shall meet the requirements of 29 CFR 1910.141(d)(3).

**1910.120(n)(7)(ii)** Change rooms shall be provided and shall meet the requirements of 29 CFR 1910.141(e). Change rooms shall consist of two separate change areas separated by the shower area required in paragraph (n)(7)(i) of this section. One change area, with an exit leading off the worksite, shall provide employees with a clean area where they can remove, store, and put on street clothing. The second area, with an exit to the worksite, shall provide employees with an area where they can put on, remove and store work clothing and personal protective equipment.

**1910.120(n)(7)(iii)** Showers and change rooms shall be located in areas where exposures are below the permissible exposure limits and published exposure levels. If this cannot be accomplished, then a ventilation system shall be provided that will supply air that is below the permissible exposure limits and published exposure levels.

**1910.120(n)(7)(iv)** Employers shall assure that employees shower at the end of their work shift and when leaving the hazardous waste site.

**1910.120(o)** *New technology programs.* **(1)** The employer shall develop and implement procedures for the introduction of effective new technologies and equipment developed for the improved protection of employees working with hazardous waste clean-up operations, and the same shall be implemented as part of the site safety and health program to assure that employee protection is being maintained.

**1910.120(o)(2)** New technologies, equipment or control measures available to the industry, such as the use of foams, absorbents, adsorbents, neutralizers, or other means to suppress the level of air contaminates while excavating the site or for spill control, shall be evaluated by employers or their representatives. Such an evaluation shall be done to determine the effectiveness of the new methods, materials, or equipment before implementing their use on a large scale for enhancing employee protection. Information and data from manufacturers or suppliers may be used as part of the employer's evaluation effort. Such evaluations shall be made available to OSHA upon request.

**1910.120(p)** *Certain Operations Conducted Under the Resource Conservation and Recovery Act of 1976 (RCRA).* Employers conducting operations at treatment, storage and disposal (TSD) facilities specified in paragraph (a)(1)(iv) of this section shall provide and implement the programs specified in this paragraph. See the "Notes and Exceptions" to paragraph (a)(2)(iii) of this section for employers not covered.

**1910.120(p)(1)** *Safety and health program.* The employer shall develop and implement a written safety and health program for employees involved in hazardous waste operations that shall be available for inspection by employees, their representatives and OSHA personnel. The program shall be designed to identify, evaluate and control safety and health hazards in their facilities for the purpose of employee protection, to provide for emergency response meeting the requirements of paragraph (p)(8) of this section and to address as appropriate site analysis, engineering controls, maximum exposure limits, hazardous waste handling procedures and uses of new technologies.

**1910.120(p)(2)** *Hazard communication program.* The employer shall implement a hazard communication program meeting the requirements of 29 CFR 1910.1200 as part of the employer's safety and program.

**Note to §1910.120:** The exemption for hazardous waste provided in §1910.1200 is applicable to this section.

**1910.120(p)(3)** *Medical surveillance program.* The employer shall develop and implement a medical surveillance program meeting the requirements of paragraph (f) of this section.

**1910.120(p)(4)** *Decontamination program.* The employer shall develop and implement a decontamination procedure meeting the requirements of paragraph (k) of this section.

**1910.120(p)(5)** *New technology program.* The employer shall develop and implement procedures meeting the requirements of paragraph (o) of this section for introducing new and innovative equipment into the workplace.

**1910.120(p)(6)** *Material handling program.* Where employees will be handling drums or containers, the employer shall develop and implement procedures meeting the requirements of paragraphs (j)(1) (ii) through (viii) and (xi) of this section, as well as (j)(3) and (j)(8) of this section prior to starting such work.

**1910.120(p)(7)** *Training program*—**(i)** *New employees.* The employer shall develop and implement a training program, which is part of the employer's safety and health program, for employees exposed to health hazards or hazardous substances at TSD operations to enable the employees to perform their assigned duties and functions in a safe and healthful manner so as not endanger themselves or other employees. The initial training shall be for 24 hours and refresher training shall be for eight hours annually. Employees who have received the initial training required by this paragraph shall be given a written certificate attesting that they have successfully completed the necessary training.

**1910.120(p)(7)(ii)** *Current employees.* Employers who can show by an employee's previous work experience and/or training that the employee has had training equivalent to the initial training required by this paragraph, shall be considered as meeting the initial training requirements of this paragraph as to that employee. Equivalent training includes the training that existing employees might have already received from actual site work experience. Current employees shall receive eight hours of refresher training annually.

**1910.120(p)(7)(iii)** *Trainers.* Trainers who teach initial training shall have satisfactorily completed a training course for teaching the subjects they are expected to teach or they shall have the academic credentials and instruction experience necessary to demonstrate a good command of the subject matter of the courses and competent instructional skills.

**1910.120(p)(8)** *Emergency response program*—**(i)** *Emergency response plan.* An emergency response plan shall be developed and implemented by all employers. Such plans need not duplicate any of the subjects fully addressed in the employer's contingency planning required by permits, such as those issued by the U.S. Environmental Protection Agency, provided that the contingency plan is made part of the emergency response plan. The emergency response plan shall be a written portion of the employers safety and health program required in paragraph (p)(1) of this section. Employers who will evacuate their employees from the worksite location when an emergency occurs and who do not permit any of their employees to assist in handling the emergency are exempt from the requirements of paragraph (p)(8) if they provide an emergency action plan complying with §1910.38 of this part.

**1910.120(p)(8)(ii)** *Elements of an emergency response plan.* The employer shall develop an emergency response plan for emergencies which shall address, as a minimum, the following areas to the extent that they are not addressed in any specific program required in this paragraph:

**1910.120(p)(8)(ii)(A)** Pre-emergency planning and coordination with outside parties.

**1910.120(p)(8)(ii)(B)** Personnel roles, lines of authority, and communication.

**1910.120(p)(8)(ii)(C)** Emergency recognition and prevention.

**1910.120(p)(8)(ii)(D)** Safe distances and places of refuge.

**1910.120(p)(8)(ii)(E)** Site security and control.

**1910.120(p)(8)(ii)(F)** Evacuation routes and procedures.

**1910.120(p)(8)(ii)(G)** Decontamination procedures.

**1910.120(p)(8)(ii)(H)** Emergency medical treatment and first aid.

**1910.120(p)(8)(ii)(I)** Emergency alerting and response procedures.

**1910.120(p)(8)(ii)(J)** Critique of response and follow-up.

**1910.120(p)(8)(ii)(K)** PPE and emergency equipment.

**1910.120(p)(8)(iii)** *Training.* **(A)** Training for emergency response employees shall be completed before they are called upon to perform in real emergencies. Such training shall include the elements of the emergency response plan, standard operating procedures the employer has established for the job, the personal protective equipment to be worn and procedures for handling emergency incidents.

*Exception #1:* An employer need not train all employees to the degree specified if the employer divides the work force in a manner such that a sufficient number of employees who have responsibility to control emergencies have the training specified, and all other employees, who may first respond to an emergency incident, have sufficient awareness training to recognize that an emergency response situation exists and that they are instructed in that case to summon the fully trained employees and not attempt control activities for which they are not trained.

*Exception #2:* An employer need not train all employees to the degree specified if arrangements have been made in advance for an outside fully-trained emergency response team to respond in a reasonable period and all employees, who may come to the incident first, have sufficient awareness training to recognize that an emergency response situation exists and they have been instructed to call the designated outside fully-trained emergency response team for assistance.

**1910.120(p)(8)(iii)(B)** Employee members of TSD facility emergency response organizations shall be trained to a level of competence in the recognition of health and safety hazards to protect themselves and other employees. This would include training in the methods used to minimize the risk from safety and health hazards; in the safe use of control equipment; in the selection and use of appropriate personal protective equipment; in the safe operating procedures to be used at the incident scene; in the techniques of coordination with other employees to minimize risks; in the appropriate response to over exposure from health hazards or injury to themselves and other employees; and in the recognition of subsequent symptoms which may result from over exposures.

**1910.120(p)(8)(iii)(C)** The employer shall certify that each covered employee has attended and successfully completed the training required in paragraph (p)(8)(iii) of this section, or shall certify the employee's competency at least yearly. The method used to demonstrate competency for certification of training shall be recorded and maintained by the employer.

**1910.120(p)(8)(iv)** *Procedures for handling emergency incidents.* **(A)** In addition to the elements for the emergency response plan required in paragraph (p)(8)(ii) of this section, the following elements shall be included for emergency response plans to the extent that they do not repeat any information already contained in the emergency response plan:

**1910.120(p)(8)(iv)(A)***(1)* Site topography, layout, and prevailing weather conditions.

**1910.120(p)(8)(iv)(A)***(2)* Procedures for reporting incidents to local, state, and federal governmental agencies.

**1910.120(p)(8)(iv)(B)** The emergency response plan shall be compatible and integrated with the disaster, fire and/or emergency response plans of local, state, and federal agencies.

**1910.120(p)(8)(iv)(C)** The emergency response plan shall be rehearsed regularly as part of the overall training program for site operations.

**1910.120(p)(8)(iv)(D)** The site emergency response plan shall be reviewed periodically and, as necessary, be amended to keep it current with new or changing site conditions or information.

**1910.120(p)(8)(iv)(E)** An employee alarm system shall be installed in accordance with 29 CFR 1910.165 to notify employees of an emergency situation; to stop work activities if necessary; to lower background noise in order to speed communication; and to begin emergency procedures.

**1910.120(p)(8)(iv)(F)** Based upon the information available at time of the emergency, the employer shall evaluate the incident and the site response capabilities and proceed with the appropriate steps to implement the site emergency response plan.

**1910.120(q)** *Emergency response to hazardous substance releases.* This paragraph covers employers whose employees are engaged in emergency response no matter where it occurs except that it does not cover employees engaged in operations specified in paragraphs (a)(1)(i) through (a)(1)(iv) of this section. Those emergency response organizations who have developed and implemented programs equivalent to this paragraph for handling releases of hazardous substances pursuant to section 303 of the Superfund Amendments and Reauthorization Act of 1986 (Emergency Planning and Community Right-to-Know Act of 1986, 42 U.S.C. 11003) shall be deemed to have met the requirements of this paragraph.

**1910.120(q)(1)** *Emergency response plan.* An emergency response plan shall be developed and implemented to handle anticipated emergencies prior to the commencement of emergency response operations. The plan shall be in writing and available for inspection and copying by employees, their representatives and OSHA personnel. Employers who will evacuate their employees from the danger area when an emergency occurs, and who do not permit any of their employees to assist in handling the emergency, are exempt from the requirements of this paragraph if they provide an emergency action plan in accordance with §1910.38 of this part.

**1910.120(q)(2)** *Elements of an emergency response plan.* The employer shall develop an emergency response plan for emergencies which shall address, as a minimum, the following to the extent that they are not addressed elsewhere:

**1910.120(q)(2)(i)** Pre-emergency planning and coordination with outside parties.

**1910.120(q)(2)(ii)** Personnel roles, lines of authority, training, and communication.

**1910.120(q)(2)(iii)** Emergency recognition and prevention.

**1910.120(q)(2)(iv)** Safe distances and places of refuge.

**1910.120(q)(2)(v)** Site security and control.

**1910.120(q)(2)(vi)** Evacuation routes and procedures.

**1910.120(q)(2)(vii)** Decontamination.

**1910.120(q)(2)(viii)** Emergency medical treatment and first aid.

**1910.120(q)(2)(ix)** Emergency alerting and response procedures.

**1910.120(q)(2)(x)** Critique of response and follow-up.

**1910.120(q)(2)(xi)** PPE and emergency equipment.

**1910.120(q)(2)(xii)** Emergency response organizations may use the local emergency response plan or the state emergency response plan or both, as part of their emergency response plan to avoid duplication. Those items of the emergency response plan that are being properly addressed by the SARA Title III plans may be substituted into their emergency plan or otherwise kept together for the employer and employee's use.

**1910.120(q)(3)** *Procedures for handling emergency response.* **(i)** The senior emergency response official responding to an emergency shall become the individual in charge of a site-specific Incident Command System (ICS). All emergency responders and their communications shall be coordinated and controlled through the individual in charge of the ICS assisted by the senior official present for each employer.

**Note to paragraph (q)(3)(i):** The "senior official" at an emergency response is the most senior official on the site who has the responsibility for controlling the operations at the site. Initially it is the senior officer on the first-due piece of responding emergency apparatus to arrive on the incident scene. As more senior officers arrive (i.e., battalion chief, fire chief, state law enforcement official, site coordinator, etc.) the position is passed up the line of authority which has been previously established.

**1910.120(q)(3)(ii)** The individual in charge of the ICS shall identify, to the extent possible, all hazardous substances or conditions present and shall address as appropriate site analysis, use of engineering controls, maximum exposure limits, hazardous substance handling procedures, and use of any new technologies.

**1910.120(q)(3)(iii)** Based on the hazardous substances and/or conditions present, the individual in charge of the ICS shall implement appropriate emergency operations, and assure that the personal protective equipment worn is appropriate for the hazards to be encountered. However, personal protective equipment shall meet, at a minimum, the criteria contained in 29 CFR 1910.156(e) when worn while performing fire fighting operations beyond the incipient stage for any incident.

**1910.120(q)(3)(iv)** Employees engaged in emergency response and exposed to hazardous substances presenting an inhalation hazard or potential inhalation hazard shall wear positive pressure self-contained breathing apparatus while engaged in emergency response, until such time that the individual in charge of the ICS determines through the use of air monitoring that a decreased level of respiratory protection will not result in hazardous exposures to employees.

**1910.120(q)(3)(v)** The individual in charge of the ICS shall limit the number of emergency response personnel at the emergency site, in those areas of potential or actual exposure to incident or site hazards, to those who are actively performing emergency operations. However, operations in hazardous areas shall be performed using the buddy system in groups of two or more.

**1910.120(q)(3)(vi)** Back-up personnel shall stand by with equipment ready to provide assistance or rescue. Advance first aid support personnel, as a minimum, shall also stand by with medical equipment and transportation capability.

**1910.120(q)(3)(vii)** The individual in charge of the ICS shall designate a safety official, who is knowledgable in the operations being implemented at the emergency response site, with specific responsibility to identify and evaluate hazards and to provide direction with respect to the safety of operations for the emergency at hand.

**1910.120(q)(3)(viii)** When activities are judged by the safety official to be an IDLH condition and/or to involve an imminent danger condition, the safety official shall have the authority to alter, suspend, or terminate those activities. The safety official shall immediately inform the individual in charge of the ICS of any actions needed to be taken to correct these hazards at the emergency scene.

**1910.120(q)(3)(ix)** After emergency operations have terminated, the individual in charge of the ICS shall implement appropriate decontamination procedures.

**1910.120(q)(3)(x)** When deemed necessary for meeting the tasks at hand, approved self-contained compressed air breathing apparatus may be used with approved cylinders from other approved self-contained compressed air breathing apparatus provided that such cylinders are of the same capacity and pressure rating. All compressed air cylinders used with self-contained breathing apparatus shall meet U.S. Department of Transportation and National Institute for Occupational Safety and Health criteria.

**1910.120(q)(4)** *Skilled support personnel.* Personnel, not necessarily an employer's own employees, who are skilled in the operation of certain equipment, such as mechanized earth moving or digging equipment or crane and hoisting equipment, and who are needed temporarily to perform immediate emergency support work that cannot reasonably be performed in a timely fashion by an employer's own employees, and who will be or may be exposed to the hazards at an emergency response scene, are not required to meet the training required in this paragraph for the employer's regular employees. However, these personnel shall be given an initial briefing at the site prior to their participation in any emergency response. The initial briefing shall include instruction in the wearing of appropriate personal protective equipment, what chemical hazards are involved, and what duties are to be performed. All other appropriate safety and health precautions provided to the employer's own employees shall be used to assure the safety and health of these personnel.

**1910.120(q)(5)** *Specialist employees.* Employees who, in the course of their regular job duties, work with and are trained in the hazards of specific hazardous substances, and who will be called upon to provide technical advice or assistance at a hazardous substance release incident to the individual in charge, shall receive training or demonstrate competency in the area of their specialization annually.

**1910.120(q)(6)** *Training.* Training shall be based on the duties and function to be performed by each responder of an emergency response organization. The skill and knowledge levels required for all new responders, those hired after the effective date of this standard, shall be conveyed to them through training before they are permitted to take part in actual emergency operations on an incident. Employees who participate, or are expected to participate, in emergency response, shall be given training in accordance with the following paragraphs:

**1910.120(q)(6)(i)** *First responder awareness level.* First responders at the awareness level are individuals who are likely to witness or discover a hazardous substance release and who have been trained to initiate an emergency response sequence by notifying the proper authorities of the release. They would take no further action beyond notifying the authorities of the release. First responders at the awareness level shall have sufficient training or have had sufficient experience to objectively demonstrate competency in the following areas:

**1910.120(q)(6)(i)(A)** An understanding of what hazardous substances are, and the risks associated with them in an incident.

**1910.120(q)(6)(i)(B)** An understanding of the potential outcomes associated with an emergency created when hazardous substances are present.

**1910.120(q)(6)(i)(C)** The ability to recognize the presence of hazardous substances in an emergency.

**1910.120(q)(6)(i)(D)** The ability to identify the hazardous substances, if possible.

**1910.120(q)(6)(i)(E)** An understanding of the role of the first responder awareness individual in the employer's emergency response plan including site security and control and the U.S. Department of Transportation's Emergency Response Guidebook.

**1910.120(q)(6)(i)(F)** The ability to realize the need for additional resources, and to make appropriate notifications to the communication center.

**1910.120(q)(6)(ii)** *First responder operations level.* First responders at the operations level are individuals who respond to releases or potential releases of hazardous substances as part of the initial response to the site for the purpose of protecting nearby persons, property, or the environment from the effects of the release. They are trained to respond in a defensive fashion without actually trying to stop the release. Their function is to contain the release from a safe distance, keep it from spreading, and prevent exposures. First responders at the operational level shall have received at least eight hours of training or have had sufficient experience to objectively demonstrate competency in the following areas in addition to those listed for the awareness level and the employer shall so certify:

**1910.120(q)(6)(ii)(A)** Knowledge of the basic hazard and risk assessment techniques.

**1910.120(q)(6)(ii)(B)** Know how to select and use proper personal protective equipment provided to the first responder operational level.

**1910.120(q)(6)(ii)(C)** An understanding of basic hazardous materials terms.

**1910.120(q)(6)(ii)(D)** Know how to perform basic control, containment and/or confinement operations within the capabilities of the resources and personal protective equipment available with their unit.

**1910.120(q)(6)(ii)(E)** Know how to implement basic decontamination procedures.

**1910.120(q)(6)(ii)(F)** An understanding of the relevant standard operating procedures and termination procedures.

**1910.120(q)(6)(iii)** *Hazardous materials technician.* Hazardous materials technicians are individuals who respond to releases or potential releases for the purpose of stopping the release. They assume a more aggressive role than a first responder at the operations level in that they will approach the point of release in order to plug, patch or otherwise stop the release of a hazardous substance. Hazardous materials technicians shall have received at least 24 hours of training equal to the first responder operations level and in addition have competency in the following areas and the employer shall so certify:

**1910.120(q)(6)(iii)(A)** Know how to implement the employer's emergency response plan.

**1910.120(q)(6)(iii)(B)** Know the classification, identification and verification of known and unknown materials by using field survey instruments and equipment.

**1910.120(q)(6)(iii)(C)** Be able to function within an assigned role in the Incident Command System.

**1910.120(q)(6)(iii)(D)** Know how to select and use proper specialized chemical personal protective equipment provided to the hazardous materials technician.

**1910.120(q)(6)(iii)(E)** Understand hazard and risk assessment techniques.

**1910.120(q)(6)(iii)(F)** Be able to perform advance control, containment, and/or confinement operations within the capabilities of the resources and personal protective equipment available with the unit.

**1910.120(q)(6)(iii)(G)** Understand and implement decontamination procedures.

**1910.120(q)(6)(iii)(H)** Understand termination procedures.

**1910.120(q)(6)(iii)(I)** Understand basic chemical and toxicological terminology and behavior.

**1910.120(q)(6)(iv)** *Hazardous materials specialist.* Hazardous materials specialists are individuals who respond with and provide support to hazardous materials technicians. Their duties parallel those of the hazardous materials technician, however, those duties require a more directed or specific knowledge of the various substances they may be called upon to contain. The hazardous materials specialist would also act as the site liaison with Federal, state, local and other government authorities in regards to site activities. Hazardous materials specialists shall have received at least 24 hours of training equal to the technician level and in addition have competency in the following areas and the employer shall so certify:

**1910.120(q)(6)(iv)(A)** Know how to implement the local emergency response plan.

**1910.120(q)(6)(iv)(B)** Understand classification, identification and verification of known and unknown materials by using advanced survey instruments and equipment.

**1910.120(q)(6)(iv)(C)** Know of the state emergency response plan.

**1910.120(q)(6)(iv)(D)** Be able to select and use proper specialized chemical personal protective equipment provided to the hazardous materials specialist.

**1910.120(q)(6)(iv)(E)** Understand in-depth hazard and risk techniques.

**1910.120(q)(6)(iv)(F)** Be able to perform specialized control, containment, and/or confinement operations within the capabilities of the resources and personal protective equipment available.

**1910.120(q)(6)(iv)(G)** Be able to determine and implement decontamination procedures.

**1910.120(q)(6)(iv)(H)** Have the ability to develop a site safety and control plan.

**1910.120(q)(6)(iv)(I)** Understand chemical, radiological and toxicological terminology and behavior.

**1910.120(q)(6)(v)** *On scene incident commander.* Incident commanders, who will assume control of the incident scene beyond the first responder awareness level, shall receive at least 24 hours of training equal to the first responder operations level and in addition have competency in the following areas and the employer shall so certify:

**1910.120(q)(6)(v)(A)** Know and be able to implement the employer's incident command system.

**1910.120(q)(6)(v)(B)** Know how to implement the employer's emergency response plan.

**1910.120(q)(6)(v)(C)** Know and understand the hazards and risks associated with employees working in chemical protective clothing.

**1910.120(q)(6)(v)(D)** Know how to implement the local emergency response plan.

**1910.120(q)(6)(v)(E)** Know of the state emergency response plan and of the Federal Regional Response Team.

**1910.120(q)(6)(v)(F)** Know and understand the importance of decontamination procedures.

**1910.120(q)(7)** *Trainers.* Trainers who teach any of the above training subjects shall have satisfactorily completed a training course for teaching the subjects they are expected to teach, such as the courses offered by the U.S. National Fire Academy, or they shall have the training and/or academic credentials and instructional experience necessary to demonstrate competent instructional skills and a good command of the subject matter of the courses they are to teach.

**1910.120(q)(8)** *Refresher training.* **(i)** Those employees who are trained in accordance with paragraph (q)(6) of this section shall receive annual refresher training of sufficient content and duration to maintain their competencies, or shall demonstrate competency in those areas at least yearly.

**1910.120(q)(8)(ii)** A statement shall be made of the training or competency, and if a statement of competency is made, the employer shall keep a record of the methodology used to demonstrate competency.

**1910.120(q)(9)** *Medical surveillance and consultation.* **(i)** Members of an organized and designated HAZMAT team and hazardous materials specialists shall receive a baseline physical examination and be provided with medical surveillance as required in paragraph (f) of this section.

**1910.120(q)(9)(ii)** Any emergency response employees who exhibits signs or symptoms which may have resulted from exposure to hazardous substances during the course of an emergency incident, either immediately or subsequently, shall be provided with medical consultation as required in paragraph (f)(3)(ii) of this section.

**1910.120(q)(10)** *Chemical protective clothing.* Chemical protective clothing and equipment to be used by organized and designated HAZMAT team members, or to be used by hazardous materials specialists, shall meet the requirements of paragraphs (g) (3) through (5) of this section.

**1910.120(q)(11)** *Post-emergency response operations.* Upon completion of the emergency response, if it is determined that it is necessary to remove hazardous substances, health hazards, and materials contaminated with them (such as contaminated soil or other elements of the natural environment) from the site of the incident, the employer conducting the clean-up shall comply with one of the following:

**1910.120(q)(11)(i)** Meet all of the requirements of paragraphs (b) through (o) of this section; or

**1910.120(q)(11)(ii)** Where the clean-up is done on plant property using plant or workplace employees, such employees shall have completed the training requirements of the following: 29 CFR 1910.38; 1910.134; 1910.1200, and other appropriate safety and health training made necessary by the tasks that they are expected to be performed such as personal protective equipment and decontamination procedures. All equipment to be used in the performance of the clean-up work shall be in serviceable condition and shall have been inspected prior to use.

### APPENDICES TO §1910.120—HAZARDOUS WASTE OPERATIONS AND EMERGENCY RESPONSE

**Note:** The following appendices serve as non-mandatory guidelines to assist employees and employers in complying with the appropriate requirements of this section. However paragraph 1910.120(g) makes mandatory in certain circumstances the use of Level A and Level B PPE protection.

### APPENDIX A TO §1910.120— PERSONAL PROTECTIVE EQUIPMENT TEST METHODS

This appendix sets forth the non-mandatory examples of tests which may be used to evaluate compliance with §1910.120 (g)(4) (ii) and (iii). Other tests and other challenge agents may be used to evaluate compliance.

### A. Totally-encapsulating chemical protective suit pressure test

1.0—Scope

1.1 This practice measures the ability of a gas tight totally-encapsulating chemical protective suit material, seams, and closures to maintain a fixed positive pressure. The results of this practice allow the gas tight integrity of a totally-encapsulating chemical protective suit to be evaluated.

1.2 Resistance of the suit materials to permeation, penetration, and degradation by specific hazardous substances is not determined by this test method.

2.0—Definition of terms

2.1 *Totally-encapsulated chemical protective suit (TECP suit)* means a full body garment which is constructed of protective clothing materials; covers the wearer's torso, head, arms, legs and respirator; may cover the wearer's hands and feet with tightly attached gloves and boots; completely encloses the wearer and respirator by itself or in combination with the wearer's gloves and boots.

2.2 *Protective clothing material* means any material or combination of materials used in an item of clothing for the purpose of isolating parts of the body from direct contact with a potentially hazardous liquid or gaseous chemicals.

2.3 *Gas tight* means, for the purpose of this test method, the limited flow of a gas under pressure from the inside of a TECP suit to atmosphere at a prescribed pressure and time interval.

3.0—Summary of test method

3.1 The TECP suit is visually inspected and modified for the test. The test apparatus is attached to the suit to permit inflation to the pre-test suit expansion pressure for removal of suit wrinkles and creases. The pressure is lowered to the test pressure and monitored for three minutes. If the pressure drop is excessive, the TECP suit fails the test and is removed from service. The test is repeated after leak location and repair.

4.0—Required supplies

4.1 Source of compressed air.

4.2 Test apparatus for suit testing, including a pressure measurement device with a sensitivity of at least 1/4 inch water gauge.

4.3 Vent valve closure plugs or sealing tape.

4.4 Soapy water solution and soft brush.

4.5 Stop watch or appropriate timing device.

5.0—Safety Precautions

5.1 Care shall be taken to provide the correct pressure safety devices required for the source of compressed air used.

6.0—Test Procedure

6.1 Prior to each test, the tester shall perform a visual inspection of the suit. Check the suit for seam integrity by visually examining the seams and gently pulling on the seams. Ensure that all air supply lines, fittings, visor, zippers, and valves are secure and show no signs of deterioration.

6.1.1 Seal off the vent valves along with any other normal inlet or exhaust points (such as umbilical air line fittings or face piece opening) with tape or other appropriate means (caps, plugs, fixture, etc.). Care should be exercised in the sealing process not to damage any of the suit components.

6.1.2 Close all closure assemblies.

6.1.3 Prepare the suit for inflation by providing an improvised connection point on the suit for connecting an airline. Attach the pressure test apparatus to the suit to permit suit inflation from a compressed air source equipped with a pressure indicating regulator. The leak tightness of the pressure test apparatus should be tested before and after each test by closing off the end of the tubing attached to the suit and assuring a pressure of three inches water gauge for three minutes can be maintained. If a component is removed for the test, that component shall be replaced and a second test conducted with another component removed to permit a complete test of the ensemble.

6.1.4 The pre-test expansion pressure (A) and the suit test pressure (B) shall be supplied by the suit manufacturer, but in no case shall they be less than: (A)=three inches water gauge; and (B)=two inches water gauge. The ending suit pressure (C) shall be no less than 80 percent of the test pressure (B); i.e., the pressure drop shall not exceed 20 percent of the test pressure (B).

6.1.5 Inflate the suit until the pressure inside is equal to pressure (A), the pre-test expansion suit pressure. Allow at least one minute to fill out the wrinkles in the suit. Release sufficient air to reduce the suit pressure to pressure (B), the suit test pressure. Begin timing. At the end of three minutes, record the suit pressure as pressure (C), the ending suit pressure. The difference between the suit test pressure and the ending suit test pressure (B-C) shall be defined as the suit pressure drop.

6.1.6 If the suit pressure drop is more than 20 percent of the suit test pressure (B) during the three-minute test period, the suit fails the test and shall be removed from service.

7.0—Retest Procedure

7.1 If the suit fails the test check for leaks by inflating the suit to pressure (A) and brushing or wiping the entire suit (including seams, closures, lens gaskets, glove-to-sleeve joints, etc.) with a mild soap and water solution. Observe the suit for the formation of soap bubbles, which is an indication of a leak. Repair all identified leaks.

7.2 Retest the TECP suit as outlined in Test procedure 6.0.

8.0—Report

8.1 Each TECP suit tested by this practice shall have the following information recorded:

8.1.1 Unique identification number, identifying brand name, date of purchase, material of construction, and unique fit features, e.g., special breathing apparatus.

8.1.2 The actual values for test pressures (A), (B), and (C) shall be recorded along with the specific observation times. If the ending pressure (C) is less than 80 percent of the test pressure (B), the suit shall be identified as failing the test. When possible, the specific leak location shall be identified in the test records. Retest pressure data shall be recorded as an additional test.

8.1.3 The source of the test apparatus used shall be identified and the sensitivity of the pressure gauge shall be recorded.

8.1.4 Records shall be kept for each pressure test even if repairs are being made at the test location.

### CAUTION

Visually inspect all parts of the suit to be sure they are positioned correctly and secured tightly before putting the suit back into service. Special care should be taken to examine each exhaust valve to make sure it is not blocked.

Care should also be exercised to assure that the inside and outside of the suit is completely dry before it is put into storage.

### B. Totally-encapsulating chemical protective suit qualitative leak test

1.0—Scope

1.1 This practice semi-qualitatively tests gas tight totally-encapsulating chemical protective suit integrity by detecting inward leakage of ammonia vapor. Since no modifications are made to the suit to carry out this test, the results from this practice provide a realistic test for the integrity of the entire suit.

1.2 Resistance of the suit materials to permeation, penetration, and degradation is not determined by this test method. ASTM test methods are available to test suit materials for these characteristics and the tests are usually conducted by the manufacturers of the suits.

2.0—Definition of terms

2.1 *Totally-encapsulated chemical protective suit (TECP suit)* means a full body garment which is constructed of protective clothing materials; covers the wearer's torso, head, arms, legs and respirator; may cover the wearer's hands and feet with tightly attached gloves and boots; completely encloses the wearer and respirator by itself or in combination with the wearer's gloves, and boots.

2.2 *Protective clothing material* means any material or combination of materials used in an item of clothing for the purpose of isolating parts of the body from direct contact with a potentially hazardous liquid or gaseous chemicals.

2.3 *Gas tight* means, for the purpose of this test method, the limited flow of a gas under pressure from the inside of a TECP suit to atmosphere at a prescribed pressure and time interval.

2.4 *Intrusion Coefficient* means a number expressing the level of protection provided by a gas tight totally-encapsulating chemical protective suit. The intrusion coefficient is calculated by dividing the test room challenge agent concentration by the concentration of challenge agent found inside the suit. The accuracy of the intrusion coefficient is dependent on the challenge agent monitoring methods. The larger the intrusion coefficient the greater the protection provided by the TECP suit.

3.0—Summary of recommended practice

3.1 The volume of concentrated aqueous ammonia solution (ammonia hydroxide NH4 OH) required to generate the test atmosphere is determined using the directions outlined in 6.1. The suit is donned by a person wearing the appropriate respiratory equipment (either a positive pressure self-contained breathing apparatus or a positive pressure supplied air respirator) and worn inside the enclosed test room. The concentrated aqueous ammonia solution is taken by the suited individual into the test room and poured into an open plastic pan. A two-minute evaporation period is observed before the test room concentration is measured, using a high range ammonia length of stain detector tube. When the ammonia vapor reaches a concentration of between 1000 and 1200 ppm, the suited individual starts a standardized exercise protocol to stress and flex the suit. After this protocol is completed, the test room concentration is measured again. The suited individual exits the test room and his stand-by person measures the ammonia concentration inside the suit using a low range ammonia length of stain detector tube or other more sensitive ammonia detector. A stand-by person is required to observe the test individual during the test procedure; aid the person in donning and doffing the TECP suit; and monitor the suit interior. The intrusion coefficient of the suit can be calculated by dividing the average test area concentration by the interior suit concentration. A colorimetric ammonia indicator strip of bromophenol blue or equivalent is placed on the inside of the suit face piece lens so that the suited individual is able to detect a color change and know if the suit has a significant leak. If a color change is observed the individual shall leave the test room immediately.

4.0—Required supplies

4.1 A supply of concentrated aqueous ammonium hydroxide (58% by weight).

4.2 A supply of bromophenol/blue indicating paper or equivalent, sensitive to 5-10 ppm ammonia or greater over a two-minute period of exposure. [pH 3.0 (yellow) to pH 4.6 (blue)]

4.3 A supply of high range (0.5-10 volume percent) and low range (5-700 ppm) detector tubes for ammonia and the corresponding sampling pump. More sensitive ammonia detectors can be substituted for the low range detector tubes to improve the sensitivity of this practice.

4.4 A shallow plastic pan (PVC) at least 12":14":1" and a half pint plastic container (PVC) with tightly closing lid.

4.5 A graduated cylinder or other volumetric measuring device of at least 50 milliliters in volume with an accuracy of at least ±1 milliliters.

5.0—Safety precautions

5.1 Concentrated aqueous ammonium hydroxide, $NH_4 OH$, is a corrosive volatile liquid requiring eye, skin, and respiratory protection. The person conducting the test shall review the MSDS for aqueous ammonia.

5.2 Since the established permissible exposure limit for ammonia is 35 ppm as a 15 minute STEL, only persons wearing a positive pressure self-contained breathing apparatus or a positive pressure supplied air respirator shall be in the chamber. Normally only the person wearing the totally-encapsulating suit will be inside the chamber. A stand-by person shall have a positive pressure self-contained breathing apparatus, or a positive pressure supplied air respirator available to enter the test area should the suited individual need assistance.

5.3 A method to monitor the suited individual must be used during this test. Visual contact is the simplest but other methods using communication devices are acceptable.

5.4 The test room shall be large enough to allow the exercise protocol to be carried out and then to be ventilated to allow for easy exhaust of the ammonia test atmosphere after the test(s) are completed.

5.5 Individuals shall be medically screened for the use of respiratory protection and checked for allergies to ammonia before participating in this test procedure.

6.0—Test procedure

6.1.1 Measure the test area to the nearest foot and calculate its volume in cubic feet. Multiply the test area volume by 0.2 milliliters of concentrated aqueous ammonia solution per cubic foot of test area volume to determine the approximate volume of concentrated aqueous ammonia required to generate 1000 ppm in the test area.

6.1.2 Measure this volume from the supply of concentrated aqueous ammonia and place it into a closed plastic container.

6.1.3 Place the container, several high range ammonia detector tubes, and the pump in the clean test pan and locate it near the test area entry door so that the suited individual has easy access to these supplies.

6.2.1 In a non-contaminated atmosphere, open a pre-sealed ammonia indicator strip and fasten one end of the strip to the inside of the suit face shield lens where it can be seen by the wearer. Moisten the indicator strip with distilled water. Care shall be taken not to contaminate the detector part of the indicator paper by touching it. A small piece of masking tape or equivalent should be used to attach the indicator strip to the interior of the suit face shield.

6.2.2 If problems are encountered with this method of attachment, the indicator strip can be attached to the outside of the respirator face piece lens being used during the test.

6.3 Don the respiratory protective device normally used with the suit, and then don the TECP suit to be tested. Check to be sure all openings which are intended to be sealed (zippers, gloves, etc.) are completely sealed. DO NOT, however, plug off any venting valves.

6.4 Step into the enclosed test room such as a closet, bathroom, or test booth, equipped with an exhaust fan. No air should be exhausted from the chamber during the test because this will dilute the ammonia challenge concentrations.

6.5 Open the container with the pre-measured volume of concentrated aqueous ammonia within the enclosed test room, and pour the liquid into the empty plastic test pan. Wait two minutes to allow for adequate volatilization of the concentrated aqueous ammonia. A small mixing fan can be used near the evaporation pan to increase the evaporation rate of the ammonia solution.

6.6 After two minutes a determination of the ammonia concentration within the chamber should be made using the high range colorimetric detector tube. A concentration of 1000 ppm ammonia or greater shall be generated before the exercises are started.

6.7 To test the integrity of the suit the following four minute exercise protocol should be followed:

6.7.1 Raising the arms above the head with at least 15 raising motions completed in one minute.

6.7.2 Walking in place for one minute with at least 15 raising motions of each leg in a one-minute period.

6.7.3 Touching the toes with a least 10 complete motions of the arms from above the head to touching of the toes in a one-minute period.

6.7.4 Knee bends with at least 10 complete standing and squatting motions in a one-minute period.

6.8 If at any time during the test the colorimetric indicating paper should change colors, the test should be stopped and section 6.10 and 6.12 initiated (See ¶4.2).

6.9 After completion of the test exercise, the test area concentration should be measured again using the high range colorimetric detector tube.

6.10 Exit the test area.

6.11 The opening created by the suit zipper or other appropriate suit penetration should be used to determine the ammonia concentration in the suit with the low range length of stain detector tube or other ammonia monitor. The internal TECP suit air should be sampled far enough from the enclosed test area to prevent a false ammonia reading.

6.12 After completion of the measurement of the suit interior ammonia concentration the test is concluded and the suit is doffed and the respirator removed.

6.13 The ventilating fan for the test room should be turned on and allowed to run for enough time to remove the ammonia gas. The fan shall be vented to the outside of the building.

6.14 Any detectable ammonia in the suit interior (five ppm ammonia ($NH_3$) or more for the length of stain detector tube) indicates that the suit has failed the test. When other ammonia detectors are used a lower level of detection is possible and it should be specified as the pass/fail criteria.

6.15 By following this test method, an intrusion coefficient of approximately 200 or more can be measured with the suit in a completely operational condition. If the intrusion coefficient is 200 or more, then the suit is suitable for emergency response and field use.

7.0—Retest procedures

7.1 If the suit fails this test, check for leaks by following the pressure test in test A above.

7.2 Retest the TECP suit as outlined in the test procedure 6.0.

8.0—Report

8.1 Each gas tight totally-encapsulating chemical protective suit tested by this practice shall have the following information recorded.

8.1.1 Unique identification number, identifying brand name, date of purchase, material of construction, and unique suit features; e.g., special breathing apparatus.

8.1.2 General description of test room used for test.

8.1.3 Brand name and purchase date of ammonia detector strips and color change data.

8.1.4 Brand name, sampling range, and expiration date of the length of stain ammonia detector tubes. The brand name and model of the sampling pump should also be recorded. If another type of ammonia detector is used, it should be identified along with its minimum detection limit for ammonia.

8.1.5 Actual test results shall list the two test area concentrations, their average, the interior suit concentration, and the calculated intrusion coefficient. Retest data shall be recorded as an additional test.

8.2 The evaluation of the data shall be specified as "suit passed" or "suit failed," and the date of the test. Any detectable ammonia (five ppm or greater for the length of stain detector tube) in the suit interior indicates the suit has failed this test. When other ammonia detectors are used, a lower level of detection is possible and it should be specified as the pass fail criteria.

### CAUTION

Visually inspect all parts of the suit to be sure they are positioned correctly and secured tightly before putting the suit back into service. Special care should be taken to examine each exhaust valve to make sure it is not blocked.

Care should also be exercised to assure that the inside and outside of the suit is completely dry before it is put into storage.

## APPENDIX B TO §1910.120— GENERAL DESCRIPTION AND DISCUSSION OF THE LEVELS OF PROTECTION AND PROTECTIVE GEAR

This appendix sets forth information about personal protective equipment (PPE) protection levels which may be used to assist employers in complying with the PPE requirements of this section.

As required by the standard, PPE must be selected which will protect employees from the specific hazards which they are likely to encounter during their work on-site.

Selection of the appropriate PPE is a complex process which should take into consideration a variety of factors. Key factors involved in this process are identification of the hazards, or suspected hazards; their routes of potential hazard to employees (inhalation, skin absorption, ingestion, and eye or skin contact); and the performance of the PPE *materials* (and seams) in providing a barrier to these hazards. The amount of protection provided by PPE is material-hazard specific. That is, protective equipment materials will protect well against some hazardous substances and poorly, or not at all, against others. In many instances, protective equipment materials cannot be found which will provide continuous protection from the particular hazardous substance. In these cases the breakthrough time of the protective material should exceed the work durations.

Other factors in this selection process to be considered are matching the PPE to the employee's work requirements and task-specific conditions. The durability of PPE materials, such as tear strength and seam strength, should be considered in relation to the employee's tasks. The effects of PPE in relation to heat stress and task duration are a factor in selecting and using PPE. In some cases layers of PPE may be necessary to provide sufficient protection, or to protect expensive PPE inner garments, suits or equipment.

The more that is known about the hazards at the site, the easier the job of PPE selection becomes. As more information about the hazards and conditions at the site becomes available, the site supervisor can make decisions to up-grade or down-grade the level of PPE protection to match the tasks at hand.

The following are guidelines which an employer can use to begin the selection of the appropriate PPE. As noted above, the site information may suggest the use of combinations of PPE selected from the different protection levels (i.e., A, B, C, or D) as being more suitable to the hazards of the work. It should be cautioned that the listing below does not fully address the performance of the specific PPE material in relation to the specific hazards at the job site, and that PPE selection, evaluation and re-selection is an ongoing process until sufficient information about the hazards and PPE performance is obtained.

*Part A.* Personal protective equipment is divided into four categories based on the degree of protection afforded. (See Part B of this appendix for further explanation of Levels A, B, C, and D hazards.)

I. *Level A*—To be selected when the greatest level of skin, respiratory, and eye protection is required.

The following constitute Level A equipment; it may be used as appropriate;

1. Positive pressure, full face-piece self-contained breathing apparatus (SCBA), or positive pressure supplied air respirator with escape SCBA, approved by the National Institute for Occupational Safety and Health (NIOSH).

2. Totally-encapsulating chemical-protective suit.

3. Coveralls.[1]

4. Long underwear.[1]

5. Gloves, outer, chemical-resistant.

6. Gloves, inner, chemical-resistant.

7. Boots, chemical-resistant, steel toe and shank.

8. Hard hat (under suit).[1]

9. Disposable protective suit, gloves and boots (depending on suit construction, may be worn over totally-encapsulating suit).

II. *Level B*—The highest level of respiratory protection is necessary but a lesser level of skin protection is needed.

The following constitute Level B equipment; it may be used as appropriate.

1. Positive pressure, full-facepiece self-contained breathing apparatus (SCBA), or positive pressure supplied air respirator with escape SCBA (NIOSH approved).

2. Hooded chemical-resistant clothing (overalls and long-sleeved jacket; coveralls; one or two-piece chemical-splash suit; disposable chemical-resistant overalls).

3. Coveralls.[1]

4. Gloves, outer, chemical-resistant.

5. Gloves, inner, chemical-resistant.

6. Boots, outer, chemical-resistant steel toe and shank.

7. Boot-covers, outer, chemical-resistant (disposable).[1]

8. Hard hat.

9. [Reserved]

10. Face shield.[1]

III. *Level C*—The concentration(s) and type(s) of airborne substance(s) is known and the criteria for using air purifying respirators are met.

The following constitute Level C equipment; it may be used as appropriate.

1. Full-face or half-mask, air purifying respirators (NIOSH approved).

2. Hooded chemical-resistant clothing (overalls; two-piece chemical-splash suit; disposable chemical-resistant overalls).

3. Coveralls.[1]

4. Gloves, outer, chemical-resistant.

5. Gloves, inner, chemical-resistant.

6. Boots (outer), chemical-resistant steel toe and shank.[1]

7. Boot-covers, outer, chemical-resistant (disposable).[1]

8. Hard hat.[1]

9. Escape mask.[1]

10. Face shield.[1]

IV. *Level D*—A work uniform affording minimal protection, used for nuisance contamination only.

The following constitute Level D equipment; it may be used as appropriate:

1. Coveralls.

2. Gloves.[1]

3. Boots/shoes, chemical-resistant steel toe and shank.

4. Boots, outer, chemical-resistant (disposable).[1]

5. Safety glasses or chemical splash goggles*.

6. Hard hat.[1]

7. Escape mask.[1]

8. Face shield.[1]

*Part B.* The types of hazards for which levels A, B, C, and D protection are appropriate are described below:

I. *Level A*—Level A protection should be used when:

1. The hazardous substance has been identified and requires the highest level of protection for skin, eyes, and the respiratory system based on either the measured (or potential for) high concentration of atmospheric vapors, gases, or particulates; or the site operations and work functions involve a high potential for splash, immersion, or exposure to unexpected vapors, gases, or particulates of materials that are harmful to skin or capable of being absorbed through the skin;

2. Substances with a high degree of hazard to the skin are known or suspected to be present, and skin contact is possible; or

3. Operations are being conducted in confined, poorly ventilated areas, and the absence of conditions requiring Level A have not yet been determined.

II. *Level B*—Level B protection should be used when:

1. The type and atmospheric concentration of substances have been identified and require a high level of respiratory protection, but less skin protection;

2. The atmosphere contains less than 19.5 percent oxygen; or

3. The presence of incompletely identified vapors or gases is indicated by a direct-reading organic vapor detection instrument, but vapors and gases are not suspected of containing high levels of chemicals harmful to skin or capable of being absorbed through the skin.

**NOTE:** This involves atmospheres with IDLH concentrations of specific substances that present severe inhalation hazards and that do not represent a severe skin hazard; or that do not meet the criteria for use of air-purifying respirators.

III. *Level C*—Level C protection should be used when:

1. The atmospheric contaminants, liquid splashes, or other direct contact will not adversely affect or be absorbed through any exposed skin;

2. The types of air contaminants have been identified, concentrations measured, and an air-purifying respirator is available that can remove the contaminants; and

3. All criteria for the use of air-purifying respirators are met.

IV. *Level D*—Level D protection should be used when:

1. The atmosphere contains no known hazard; and

2. Work functions preclude splashes, immersion, or the potential for unexpected inhalation of or contact with hazardous levels of any chemicals.

**NOTE:** As stated before, combinations of personal protective equipment other than those described for Levels A, B, C, and D protection may be more appropriate and may be used to provide the proper level of protection.

As an aid in selecting suitable chemical protective clothing, it should be noted that the National Fire Protection Association (NFPA) has developed standards on chemical protective clothing. The standards that have been adopted by include:

---

[1] Optional, as applicable.

NFPA 1991—Standard on Vapor-Protective Suits for Hazardous Chemical Emergencies (EPA Level A Protective Clothing).

NFPA 1992—Standard on Liquid Splash-Protective Suits for Hazardous Chemical Emergencies (EPA Level B Protective Clothing).

NFPA 1993—Standard on Liquid Splash-Protective Suits for Non-emergency, Non-flammable Hazardous Chemical Situations (EPA Level B Protective Clothing).

These standards apply documentation and performance requirements to the manufacture of chemical protective suits. Chemical protective suits meeting these requirements are labelled as compliant with the appropriate standard. It is recommended that chemical protective suits that meet these standards be used.

### APPENDIX C TO §1910.120— COMPLIANCE GUIDELINES

1. *Occupational Safety and Health Program.* Each hazardous waste site clean-up effort will require an occupational safety and health program headed by the site coordinator or the employer's representative. The purpose of the program will be the protection of employees at the site and will be an extension of the employer's overall safety and health program. The program will need to be developed before work begins on the site and implemented as work proceeds as stated in paragraph (b). The program is to facilitate coordination and communication of safety and health issues among personnel responsible for the various activities which will take place at the site. It will provide the overall means for planning and implementing the needed safety and health training and job orientation of employees who will be working at the site. The program will provide the means for identifying and controlling worksite hazards and the means for monitoring program effectiveness. The program will need to cover the responsibilities and authority of the site coordinator or the employer's manager on the site for the safety and health of employees at the site, and the relationships with contractors or support services as to what each employer's safety and health responsibilities are for their employees on the site. Each contractor on the site needs to have its own safety and health program so structured that it will smoothly interface with the program of the site coordinator or principal contractor.

Also those employers involved with treating, storing or disposal of hazardous waste as covered in paragraph (p) must have implemented a safety and health program for their employees. This program is to include the hazard communication program required in paragraph (p)(1) and the training required in paragraphs (p)(7) and (p)(8) as parts of the employers comprehensive overall safety and health program. This program is to be in writing.

Each site or workplace safety and health program will need to include the following: (1) Policy statements of the line of authority and accountability for implementing the program, the objectives of the program and the role of the site safety and health supervisor or manager and staff; (2) means or methods for the development of procedures for identifying and controlling workplace hazards at the site; (3) means or methods for the development and communication to employees of the various plans, work rules, standard operating procedures and practices that pertain to individual employees and supervisors; (4) means for the training of supervisors and employees to develop the needed skills and knowledge to perform their work in a safe and healthful manner; (5) means to anticipate and prepare for emergency situations; and (6) means for obtaining information feedback to aid in evaluating the program and for improving the effectiveness of the program. The management and employees should be trying continually to improve the effectiveness of the program thereby enhancing the protection being afforded those working on the site.

Accidents on the site or workplace should be investigated to provide information on how such occurrences can be avoided in the future. When injuries or illnesses occur on the site or workplace, they will need to be investigated to determine what needs to be done to prevent this incident from occurring again. Such information will need to be used as feedback on the effectiveness of the program and the information turned into positive steps to prevent any reoccurrence. Receipt of employee suggestions or complaints relating to safety and health issues involved with site or workplace activities is also a feedback mechanism that can be used effectively to improve the program and may serve in part as an evaluative tool(s).

For the development and implementation of the program to be the most effective, professional safety and health personnel should be used. Certified Safety Professionals, Board Certified Industrial Hygienists or Registered Professional Safety Engineers are good examples of professional stature for safety and health managers who will administer the employer's program.

2. *Training.* The training programs for employees subject to the requirements of paragraph (e) of this standard should address: the safety and health hazards employees should expect to find on hazardous waste clean-up sites; what control measures or techniques are effective for those hazards; what monitoring procedures are effective in characterizing exposure levels; what makes an effective employer's safety and health program; what a site safety and health plan should include; hands on training with personal protective equipment and clothing they may be expected to use; the contents of the OSHA standard relevant to the employee's duties and function; and, employee's responsibilities under OSHA and other regulations. Supervisors will need training in their responsibilities under the safety and health program and its subject areas such as the spill containment program, the personal protective equipment program, the medical surveillance program, the emergency response plan and other areas.

The training programs for employees subject to the requirements of paragraph (p) of this standard should address: the employers safety and health program elements impacting employees; the hazard communication program; the medical surveillance program; the hazards and the controls for such hazards that employees need to know for their job duties and functions. All require annual refresher training.

The training programs for employees covered by the requirements of paragraph (q) of this standard should address those competencies required for the various levels of response such as: the hazards associated with hazardous substances; hazard identification and awareness; notification of appropriate persons; the need for and use of personal protective equipment including respirators; the decontamination procedures to be used; preplanning activities for hazardous substance incidents including the emergency reponse plan; company standard operating procedures for hazardous substance emergency responses; the use of the incident command system and other subjects. Hands-on training should be stressed whenever possible. Critiques done after an incident which include an evaluation of what worked and what did not and how could the incident be better handled the next time may be counted as training time.

For hazardous materials specialists (usually members of hazardous materials teams), the training should address the care, use and/or testing of chemical protective clothing including totally encapsulating suits, the medical surveillance program, the standard operating procedures for the hazardous materials team including the use of plugging and patching equipment and other subject areas.

Officers and leaders who may be expected to be in charge at an incident should be fully knowledgeable of their company's incident command system. They should know where and how to obtain additional assistance and be familiar with the local district's emergency response plan and the state emergency response plan.

Specialist employees such as technical experts, medical experts or environmental experts that work with hazardous materials in their regular jobs, who may be sent to the incident scene by the shipper, manufacturer or governmental agency to advise and assist the person in charge of the incident should have training on an annual basis. Their training should include the care and use of personal protective equipment including respirators; knowledge of the incident command system and how they are to relate to it; and those areas needed to keep them current in their respective field as it relates to safety and health involving specific hazardous substances.

Those skilled support personnel, such as employees who work for public works departments or equipment operators who operate bulldozers, sand trucks, backhoes, etc., who may be called to the incident scene to provide emergency support assistance, should have at least a safety and health briefing before entering the area of potential or actual exposure. These skilled support personnel, who have not been a part of the emergency response plan and do not meet the training requirements, should be made aware of the hazards they face and should be provided all necessary protective clothing and equipment required for their tasks.

There are two National Fire Protection Association standards, NFPA 472—"Standard for Professional Competence of Responders to Hazardous Material Incidents" and NFPA 471—"Recommended Practice for Responding to Hazardous Material Incidents", which are excellent resource documents to aid fire departments and other emergency response organizations in developing their training program materials. NFPA 472 provides guidance on the skills and knowledge needed for first responder awareness level, first responder operations level, hazmat technicians, and hazmat specialist. It also offers guidance for the officer corp who will be in charge of hazardous substance incidents.

3. *Decontamination.* Decontamination procedures should be tailored to the specific hazards of the site, and may vary in complexity and number of steps, depending on the level of hazard and the employee's exposure to the hazard. Decontamination procedures and PPE decontamination methods will vary depending upon the specific substance, since one procedure or method may not work for all substances. Evaluation of decontamination methods and procedures should be performed, as necessary, to assure that employees are not exposed to hazards by re-using PPE. References in appendix D may be used for guidance in establishing an effective decontamination program. In addition, the U.S. Coast Guard's Manual, "Policy Guidance for Response to Hazardous Chemical Releases," U.S. Department of Transportation, Washington, DC (COMDTINST M16465.30) is a good reference for establishing an effective decontamination program.

4. *Emergency response plans.* States, along with designated districts within the states, will be developing or have developed local emergency response plans. These state and district plans should be utilized in the emergency response plans called for in the standard. Each employer should assure that its emergency response plan is compatible with the local plan. The major reference being used to aid in developing the state and local district plans is the *Hazardous Materials Emergency Planning Guide*, NRT-1. The current Emergency Response Guidebook from the U.S. Department of Transportation, CMA's CHEMTREC and the Fire Service Emergency Management Handbook may also be used as resources.

Employers involved with treatment, storage, and disposal facilities for hazardous waste, which have the required contingency plan called for by their permit, would not need to duplicate the same planning elements. Those items of the emergency response plan that are properly addressed in the contingency plan may be substituted into the emergency response plan required in 1910.120 or otherwise kept together for employer and employee use.

5. *Personal protective equipment programs.* The purpose of personal protective clothing and equipment (PPE) is to shield or isolate individuals from the chemical, physical, and biologic hazards that may be encountered at a hazardous substance site.

As discussed in appendix B, no single combination of protective equipment and clothing is capable of protecting against all hazards. Thus PPE should be used in conjunction with other protective methods and its effectiveness evaluated periodically.

The use of PPE can itself create significant worker hazards, such as heat stress, physical and psychological stress, and impaired vision, mobility, and communication. For any given situation, equipment and clothing should be selected that provide an adequate level of protection. However, overprotection, as well as under-protection, can be hazardous and should be avoided where possible.

Two basic objectives of any PPE program should be to protect the wearer from safety and health hazards, and to prevent injury to the wearer from incorrect use and/or malfunction of the PPE. To accomplish these goals, a comprehensive PPE program should include hazard identification, medical monitoring, environmental surveillance, selection, use, maintenance, and decontamination of PPE and its associated training.

The written PPE program should include policy statements, procedures, and guidelines. Copies should be made available to all employees, and a reference copy should be made available at the worksite. Technical data on equipment, maintenance manuals, relevant regulations, and other essential information should also be collected and maintained.

6. *Incident command system (ICS).* Paragraph 1910.120(q)(3)(ii) requires the implementation of an ICS. The ICS is an organized approach to effectively control and *manage* operations at an emergency incident. The individual in charge of the ICS is the senior official responding to the incident. The ICS is not much different than the "command post" approach used for many years by the fire service. During large complex fires involving several companies and many pieces of apparatus, a command post would be established. This enabled *one* individual to be in charge of managing the incident, rather than having several officers from different companies making separate, and sometimes conflicting, decisions. The individual in charge of the command post would delegate responsibility for performing various tasks to subordinate officers. Additionally, all communications were routed through the command post to reduce the number of radio transmissions and eliminate confusion. However, strategy, tactics, and all decisions were made by one individual.

The ICS is a very similar system, except it is implemented for emergency response to all incidents, both large and small, that involve hazardous substances.

For a small incident, the individual in charge of the ICS may perform many tasks of the ICS. There may not be any, or little, delegation of tasks to subordinates. For example, in response to a small incident, the individual in charge of the ICS, in addition to normal command activities, may become the safety officer and may designate only one employee (with proper equipment) as a back-up to provide assistance if needed. OSHA does recommend, however, that at least two employees be designated as back-up personnel since the assistance needed may include rescue.

To illustrate the operation of the ICS, the following scenario might develop during a small incident, such as an overturned tank truck with a small leak of flammable liquid.

The first responding senior officer would implement and take command of the ICS. That person would size-up the incident and determine if additional personnel and apparatus were necessary; would determine what actions to take to control the leak; and, determine the proper level of personal protective equipment. If additional assistance is not needed, the individual in charge of the ICS would implement actions to stop and control the leak using the fewest number of personnel that can effectively accomplish the tasks. The individual in charge of the ICS then would designate himself as the safety officer and two other employees as a back-up in case rescue may become necessary. In this scenario, decontamination procedures would not be necessary.

A large complex incident may require many employees and difficult, time-consuming efforts to control. In these situations, the individual in charge of the ICS will want to delegate different tasks to subordinates in order to maintain a span of control that will keep the number of subordinates, that are reporting, to a manageable level.

Delegation of task at large incidents may be by location, where the incident scene is divided into sectors, and subordinate officers coordinate activities within the sector that they have been assigned.

Delegation of tasks can also be by function. Some of the functions that the individual in charge of the ICS may want to delegate at a large incident are: medical services; evacuation; water supply; resources (equipment, apparatus); media relations; safety; and, site control (integrate activities with police for crowd and traffic control). Also for a large incident, the individual in charge of the ICS will designate several employees as back-up personnel; and a number of safety officers to monitor conditions and recommend safety precautions.

Therefore, no matter what size or complexity an incident may be, by implementing an ICS there will be *one individual in charge* who makes the decisions and gives directions; and, all actions, and communications are coordinated through one central point of command. Such a system should reduce confusion, improve safety, organize and coordinate actions, and should facilitate effective management of the incident.

7. *Site Safety and Control Plans.* The safety and security of response personnel and others in the area of an emergeny response incident site should be of primary concern to the incident commander. The use of a site safety and control plan could greatly assist those in charge of assuring the safety and health of employees on the site.

A comprehensive site safety and control plan should include the following: summary analysis of hazards on the site and a risk analysis of those hazards; site map or sketch; site work zones (clean zone, transition or decontamination zone, work or hot zone); use of the buddy system; site communications; command post or command center; standard operating procedures and safe work practices; medical assistance and triage area; hazard monitoring plan (air contaminate monitoring, etc.); decontamination procedures and area; and other relevant areas. This plan should be a part of the employer's emergency response plan or an extension of it to the specific site.

8. *Medical surveillance programs.* Workers handling hazardous substances may be exposed to toxic chemicals, safety hazards, biologic hazards, and radiation. Therefore, a medical surveillance program is essential to assess and monitor workers' health and fitness for employment in hazardous waste operations and during the course of work; to provide emergency and other treatment as needed; and to keep accurate records for future reference.

The *Occupational Safety and Health Guidance Manual for Hazardous Waste Site Activities* developed by the National Institute for Occupational Safety and Health (NIOSH), the Occupational Safety and Health Administration (OSHA), the U.S. Coast Guard (USCG), and the Environmental Protection Agency (EPA); October 1985 provides an excellent example of the types of medical testing that should be done as part of a medical surveillance program.

9. *New Technology and Spill Containment Programs.* Where hazardous substances may be released by spilling from a container that will expose employees to the hazards of the materials, the employer will need to implement a program to contain and control the spilled material. Diking and ditching, as well as use of absorbents like diatomaceous earth, are traditional techniques which have proven to be effective over the years. However, in recent years new products have come into the marketplace, the use of which complement and increase the effectiveness of these traditional methods. These new products also provide emergency responders and others with additional tools or agents to use to reduce the hazards of spilled materials.

These agents can be rapidly applied over a large area and can be uniformly applied or otherwise can be used to build a small dam, thus improving the workers' ability to control spilled material. These application techniques enhance the intimate contact between the agent and the spilled material allowing for the quickest effect by the agent or quickest control of the spilled material. Agents are available to solidify liquid spilled materials, to suppress vapor generation from spilled materials, and to do both. Some special agents, which when applied as recommended by the manufacturer, will react in a controlled manner with the spilled material to neutralize acids or caustics, or greatly reduce the level of hazard of the spilled material.

There are several modern methods and devices for use by emergency response personnel or others involved with spill control efforts to safely apply spill control agents to control spilled material hazards. These include portable pressurized applicators similar to hand-held portable fire extinguishing devices, and nozzle and hose systems similar to portable fire fighting foam systems which allow the operator to apply the agent without having to come into contact with the spilled material. The operator is able to apply the agent to the spilled material from a remote position.

The solidification of liquids provides for rapid containment and isolation of hazardous substance spills. By directing the agent at run-off points or at the edges of the spill, the reactant solid will automatically create a barrier to slow or stop the spread of the material. Clean-up of hazardous substances is greatly improved when solidifying agents, acid or caustic neutralizers, or activated carbon adsorbents are used. Properly applied, these agents can totally solidify liquid hazardous substances or neutralize or absorb them, which results in materials which are less hazardous and easier to handle, transport, and dispose of. The concept of spill treatment, to create less hazardous substances, will improve the safety and level of protection of employees working at spill clean-up operations or emergency response operations to spills of hazardous substances.

The use of vapor suppression agents for volatile hazardous substances, such as flammable liquids and those substances which present an inhalation hazard, is important for protecting workers. The rapid and uniform distribution of the agent over the surface of the spilled material can provide quick vapor knockdown. There are temporary and long-term foam-type agents which are effective on vapors and dusts, and activated carbon adsorption agents which are effective for vapor control and soaking-up of the liquid. The proper use of hose lines or hand-held portable pressurized applicators provides good mobility and permits the worker to deliver the agent from a safe distance without having to step into the untreated spilled material. Some of these systems can be recharged in the field to provide coverage of larger spill areas than the design limits of a single charged applicator unit. Some of the more effective agents can solidify the liquid flammable hazardous substances and at the same time elevate the flashpoint above 140 °F so the resulting substance may be handled as a nonhazardous waste material if it meets the U.S. Environmental Protection Agency's 40 CFR part 261 requirements (See particularly §261.21).

All workers performing hazardous substance spill control work are expected to wear the proper protective clothing and equipment for the materials present and to follow the employer's established standard operating procedures for spill control. All involved workers need to be trained in the established operating procedures; in the use and care of spill control equipment; and in the associated hazards and control of such hazards of spill containment work.

These new tools and agents are the things that employers will want to evaluate as part of their new technology program. The treatment of spills of hazardous substances or wastes at an emergency incident as part of the immediate spill containment and control efforts is sometimes acceptable to EPA and a permit exception is described in 40 CFR 264.1(g)(8) and 265.1(c)(11).

## APPENDIX D TO §1910.120—REFERENCES

The following references may be consulted for further information on the subject of this standard:

1. OSHA Instruction DFO CPL 2.70—January 29, 1986, *Special Emphasis Program: Hazardous Waste Sites.*

2. OSHA Instruction DFO CPL 2-2.37A—January 29, 1986, *Technical Assistance and Guidelines for Superfund and Other Hazardous Waste Site Activities.*

3. OSHA Instruction DTS CPL 2.74—January 29, 1986, *Hazardous Waste Activity Form, OSHA 175.*

4. *Hazardous Waste Inspections Reference Manual,* U.S. Department of Labor, Occupational Safety and Health Administration, 1986.

5. Memorandum of Understanding Among the National Institute for Occupational Safety and Health, the Occupational Safety and Health Administration, the United States Coast Guard, and the United States Environmental Protection Agency, *Guidance for Worker Protection During Hazardous Waste Site Investigations and Clean-up and Hazardous Substance Emergencies.* December 18, 1980.

6. *National Priorities List,* 1st Edition, October 1984; U.S. Environmental Protection Agency, Revised periodically.

7. *The Decontamination of Response Personnel,* Field Standard Operating Procedures (F.S.O.P.) 7; U.S. Environmental Protection Agency, Office of Emergency and Remedial Response, Hazardous Response Support Division, December 1984.

8. *Preparation of a Site Safety Plan,* Field Standard Operating Procedures (F.S.O.P.) 9; U.S. Environmental Protection Agency, Office of Emergency and Remedial Response, Hazardous Response Support Division, April 1985.

9. *Standard Operating Safety Guidelines;* U.S. Environmental Protection Agency, Office of Emergency and Remedial Response, Hazardous Response Support Division, Environmental Response Team; November 1984.

10. *Occupational Safety and Health Guidance Manual for Hazardous Waste Site Activities,* National Institute for Occupational Safety and Health (NIOSH), Occupational Safety and Health Administration (OSHA), U.S. Coast Guard (USCG), and Environmental Protection Agency (EPA); October 1985.

11. *Protecting Health and Safety at Hazardous Waste Sites: An Overview,* U.S. Environmental Protection Agency, EPA/625/9-85/006; September 1985.

12. *Hazardous Waste Sites and Hazardous Substance Emergencies,* NIOSH Worker Bulletin, U.S. Department of Health and Human Services, Public Health Service, Centers for Disease Control, National Institute for Occupational Safety and Health; December 1982.

13. *Personal Protective Equipment for Hazardous Materials Incidents: A Selection Guide;* U.S. Department of Health and Human Services, Public Health Service, Centers for Disease Control, National Institute for Occupational Safety and Health; October 1984.

14. *Fire Service Emergency Management Handbook,* International Association of Fire Chiefs Foundation, 101 East Holly Avenue, Unit 10B, Sterling, VA 22170, January 1985.

15. *Emergency Response Guidebook,* U.S Department of Transportation, Washington, DC, 1987.

16. *Report to the Congress on Hazardous Materials Training, Planning and Preparedness,* Federal Emergency Management Agency, Washington, DC, July 1986.

17. *Workbook for Fire Command,* Alan V. Brunacini and J. David Beageron, National Fire Protection Association, Batterymarch Park, Quincy, MA 02269, 1985.

18. *Fire Command,* Alan V. Brunacini, National Fire Protection Association, Batterymarch Park,, Quincy, MA 02269, 1985.

19. *Incident Command System,* Fire Protection Publications, Oklahoma State University, Stillwater, OK 74078, 1983.

20. *Site Emergency Response Planning,* Chemical Manufacturers Association, Washington, DC 20037, 1986.

21. *Hazardous Materials Emergency Planning Guide,* NRT-1, Environmental Protection Agency, Washington, DC, March 1987.

22. *Community Teamwork: Working Together to Promote Hazardous Materials Transportation Safety.* U.S. Department of Transportation, Washington, DC, May 1983.

23. *Disaster Planning Guide for Business and Industry,* Federal Emergency Management Agency, Publication No. FEMA 141, August 1987.

(The Office of Management and Budget has approved the information collection requirements in this section under control number 1218-0139)

### APPENDIX E TO §1910.120—TRAINING CURRICULUM GUIDELINES

The following non-mandatory general criteria may be used for assistance in developing site-specific training curriculum used to meet the training requirements of 29 CFR 1910.120(e); 29 CFR 1910.120(p)(7), (p)(8)(iii); and 29 CFR 1910.120(q)(6), (q)(7), and (q)(8). These are generic guidelines and they are not presented as a complete training curriculum for any specific employer. Site-specific training programs must be developed on the basis of a needs assessment of the hazardous waste site, RCRA/TSDF, or emergency response operation in accordance with 29 CFR 1910.120.

It is noted that the legal requirements are set forth in the regulatory text of §1910.120. The guidance set forth here presents a highly effective program that in the areas covered would meet or exceed the regulatory requirements. In addition, other approaches could meet the regulatory requirements.

#### *Suggested General Criteria*

*Definitions:*

"Competent" means possessing the skills, knowledge, experience, and judgment to perform assigned tasks or activities satisfactorily as determined by the employer.

"Demonstration" means the showing by actual use of equipment or procedures.

"Hands-on training" means training in a simulated work environment that permits each student to have experience performing tasks, making decisions, or using equipment appropriate to the job assignment for which the training is being conducted.

"Initial training" means training required prior to beginning work.

"Lecture" means an interactive discourse with a class lead by an instructor.

"Proficient" means meeting a stated level of achievement.

"Site-specific" means individual training directed to the operations of a specific job site.

"Training hours" means the number of hours devoted to lecture, learning activities, small group work sessions, demonstration, evaluations, or hands-on experience.

*Suggested core criteria:*

1. *Training facility.* The training facility should have available sufficient resources, equipment, and site locations to perform didactic and hands-on training when appropriate. Training facilities should have sufficient organization, support staff, and services to conduct training in each of the courses offered.

2. *Training Director.* Each training program should be under the direction of a training director who is responsible for the program. The Training Director should have a minimum of two years of employee education experience.

3. *Instructors.* Instructors should be deem competent on the basis of previous documented experience in their area of instruction, successful completion of a "train-the-trainer" program specific to the topics they will teach, and an evaluation of instructional competence by the Training Director.

Instructors should be required to maintain professional competency by participating in continuing education or professional development programs or by completing successfully an annual refresher course and having an annual review by the Training Director.

The annual review by the Training Director should include observation of an instructor's delivery, a review of those observations with the trainer, and an analysis of any instructor or class evaluations completed by the students during the previous year.

4. *Course materials.* The Training Director should approve all course materials to be used by the training provider. Course materials should be reviewed and updated at least annually. Materials and equipment should be in good working order and maintained properly.

All written and audio-visual materials in training curricula should be peer reviewed by technically competent outside reviewers or by a standing advisory committee.

Reviews should possess expertise in the following disciplines were applicable: occupational health, industrial hygiene and safety, chemical/environmental engineering, employee education, or emergency response. One or more of the peer reviewers should be a employee experienced in the work activities to which the training is directed.

5. *Students.* The program for accepting students should include:

a. Assurance that the student is or will be involved in work where chemical exposures are likely and that the student possesses the skills necessary to perform the work.

b. A policy on the necessary medical clearance.

6. *Ratios.* Student-instructor ratios should not exceed 30 students per instructor. Hands-on activity requiring the use of personal protective equipment should have the following student-instructor ratios. For Level C or Level D personal protective equipment the ratio should be 10 students per instructor. For Level A or Level B personal protective equipment the ratio should be 5 students per instructor.

7. *Proficiency assessment.* Proficiency should be evaluated and documented by the use of a written assessment and a skill demonstration selected and developed by the Training Director and training staff. The assessment and demonstration should evaluate the knowledge and individual skills developed in the course of training. The level of minimum achievement necessary for proficiency shall be specified in writing by the Training Director.

If a written test is used, there should be a minimum of 50 questions. If a written test is used in combination with a skills demonstration, a minimum of 25 questions should be used. If a skills demonstration is used, the tasks chosen and the means to rate successful completion should be fully documented by the Training Director.

The content of the written test or of the skill demonstration shall be relevant to the objectives of the course. The written test and skill demonstration should be updated as necessary to reflect changes in the curriculum and any update should be approved by the Training Director.

The proficiency assessment methods, regardless of the approach or combination of approaches used, should be justified, documented and approved by the Training Director.

The proficiency of those taking the additional courses for supervisors should be evaluated and documented by using proficiency assessment methods acceptable to the Training Director. These proficiency assessment methods must reflect the additional responsibilities borne by supervisory personnel in hazardous waste operations or emergency response.

8. *Course certificate.* Written documentation should be provided to each student who satisfactorily completes the training course. The documentation should include:

a. Student's name.

b. Course title.

c. Course date.

d. Statement that the student has successfully completed the course.

e. Name and address of the training provider.

f. An individual identification number for the certificate.

g. List of the levels of personal protective equipment used by the student to complete the course.

This documentation may include a certificate and an appropriate wallet-sized laminated card with a photograph of the student and the above information. When such course certificate cards are used, the individual identification number for the training certificate should be shown on the card.

9. *Recordkeeping.* Training providers should maintain records listing the dates courses were presented, the names of the individual course attenders, the names of those students successfully completing each course, and the number of training certificates issued to each successful student. These records should be maintained for a minimum of five years after the date an individual participated

in a training program offered by the training provider. These records should be available and provided upon the student's request or as mandated by law.

10. *Program quality control.* The Training Director should conduct or direct an annual written audit of the training program. Program modifications to address deficiencies, if any, should be documented, approved, and implemented by the training provider. The audit and the program modification documents should be maintained at the training facility.

### Suggested Program Quality Control Criteria

Factors listed here are suggested criteria for determining the quality and appropriateness of employee health and safety training for hazardous waste operations and emergency response.

*A. Training Plan.*

Adequacy and appropriateness of the training program's curriculum development, instructor training, distribution of course materials, and direct student training should be considered, including

1. The duration of training, course content, and course schedules/agendas;

2. The different training requirements of the various target populations, as specified in the appropriate generic training curriculum;

3. The process for the development of curriculum, which includes appropriate technical input, outside review, evaluation, program pretesting.

4. The adequate and appropriate inclusion of hands-on, demonstration, and instruction methods;

5. Adequate monitoring of student safety, progress, and performance during the training.

*B. Program management, Training Director, staff, and consultants.*

Adequacy and appropriateness of staff performance and delivering an effective training program should be considered, including

1. Demonstration of the training director's leadership in assuring quality of health and safety training.

2. Demonstration of the competency of the staff to meet the demands of delivering high quality hazardous waste employee health and safety training.

3. Organization charts establishing clear lines of authority.

4. Clearly defined staff duties including the relationship of the training staff to the overall program.

5. Evidence that the training organizational structure suits the needs of the training program.

6. Appropriateness and adequacy of the training methods used by the instructors.

7. Sufficiency of the time committed by the training director and staff to the training program.

8. Adequacy of the ratio of training staff to students.

9. Availability and commitment of the training program of adequate human and equipment resources in the areas of

a. Health effects,

b. Safety,

c. Personal protective equipment (PPE),

d. Operational procedures,

e. Employee protection practices/procedures.

10. Appropriateness of management controls.

11. Adequacy of the organization and appropriate resources assigned to assure appropriate training.

12. In the case of multiple-site training programs, adequacy of satellite centers management.

*C. Training facilities and resources.*

Adequacy and appropriateness of the facilities and resources for supporting the training program should be considered, including,

1. Space and equipment to conduct the training.

2. Facilities for representative hands-on training.

3. In the case of multiple-site programs, equipment and facilities at the satellite centers.

4. Adequacy and appropriateness of the quality control and evaluations program to account for instructor performance.

5. Adequacy and appropriateness of the quality control and evaluation program to ensure appropriate course evaluation, feedback, updating, and corrective action.

6. Adequacy and appropriateness of disciplines and expertise being used within the quality control and evaluation program.

7. Adequacy and appropriateness of the role of student evaluations to provide feedback for training program improvement.

*D. Quality control and evaluation.*

Adequacy and appropriateness of quality control and evaluation plans for training programs should be considered, including:

1. A balanced advisory committee and/or competent outside reviewers to give overall policy guidance;

2. Clear and adequate definition of the composition and active programmatic role of the advisory committee or outside reviewers.

3. Adequacy of the minutes or reports of the advisory committee or outside reviewers' meetings or written communication.

4. Adequacy and appropriateness of the quality control and evaluations program to account for instructor performance.

5. Adequacy and appropriateness of the quality control and evaluation program to ensure appropriate course evaluation, feedback, updating, and corrective action.

6. Adequacy and appropriateness of disciplines and expertise being used within the quality control and evaluation program.

7. Adequacy and appropriateness of the role of student evaluations to provide feedback for training program improvement.

*E. Students*

Adequacy and appropriateness of the program for accepting students should be considered, including

1. Assurance that the student already possess the necessary skills for their job, including necessary documentation.

2. Appropriateness of methods the program uses to ensure that recruits are capable of satisfactorily completing training.

3. Review and compliance with any medical clearance policy.

*F. Institutional Environment and Administrative Support*

The adequacy and appropriateness of the institutional environment and administrative support system for the training program should be considered, including

1. Adequacy of the institutional commitment to the employee training program.

2. Adequacy and appropriateness of the administrative structure and administrative support.

*G. Summary of Evaluation Questions*

Key questions for evaluating the quality and appropriateness of an overall training program should include the following:

1. Are the program objectives clearly stated?

2. Is the program accomplishing its objectives?

3. Are appropriate facilities and staff available?

4. Is there an appropriate mix of classroom, demonstration, and hands-on training?

5. Is the program providing quality employee health and safety training that fully meets the intent of regulatory requirements?

6. What are the program's main strengths?

7. What are the program's main weaknesses?

8. What is recommended to improve the program?

9. Are instructors instructing according to their training outlines?

10. Is the evaluation tool current and appropriate for the program content?

11. Is the course material current and relevant to the target group?

## Suggested Training Curriculum Guidelines

The following training curriculum guidelines are for those operations specifically identified in 29 CFR 1910.120 as requiring training. Issues such as qualifications of instructors, training certification, and similar criteria appropriate to all categories of operations addressed in 1910.120 have been covered in the preceding section and are not re-addressed in each of the generic guidelines. Basic core requirements for training programs that are addressed include

1. General Hazardous Waste Operations
2. RCRA operations—Treatment, storage, and disposal facilities.
3. Emergency Response.

A. *General Hazardous Waste Operations and Site-specific Training*

1. *Off-site training.* Training course content for hazardous waste operations, required by 29 CFR 1910.120(e), should include the following topics or procedures:

a. *Regulatory knowledge.*

(1) An review of 29 CFR 1910.120 and the core elements of an occupational safety and health program.

(2) The content of a medical surveillance program as outlined in 29 CFR 1910.120(f).

(3) The content of an effective site safety and health plan consistent with the requirements of 29 CFR 1910.120(b)(4)(ii).

(4) Emergency response plan and procedures as outlined in 29 CFR 1910.38 and 29 CFR 1910.120(l).

(5) Adequate illumination.

(6) Sanitation recommendation and equipment.

(7) Review and explanation of OSHA's hazard-communication standard (29 CFR 1910.1200) and lock-out-tag-out standard (29 CFR 1910.147).

(8) Review of other applicable standards including but not limited to those in the construction standards (29 CFR Part 1926).

(9) Rights and responsibilities of employers and employees under applicable OSHA and EPA laws.

b. *Technical knowledge.*

(1) Type of potential exposures to chemical, biological, and radiological hazards; types of human responses to these hazards and recognition of those responses; principles of toxicology and information about acute and chronic hazards; health and safety considerations of new technology.

(2) Fundamentals of chemical hazards including but not limited to vapor pressure, boiling points, flash points, ph, other physical and chemical properties.

(3) Fire and explosion hazards of chemicals.

(4) General safety hazards such as but not limited to electrical hazards, powered equipment hazards, motor vehicle hazards, walking-working surface hazards, excavation hazards, and hazards associated with working in hot and cold temperature extremes.

(5) Review and knowledge of confined space entry procedures in 29 CFR 1910.146.

(6) Work practices to minimize employee risk from site hazards.

(7) Safe use of engineering controls, equipment, and any new relevant safety technology or safety procedures.

(8) Review and demonstration of competency with air sampling and monitoring equipment that may be used in a site monitoring program.

(9) Container sampling procedures and safeguarding; general drum and container handling procedures including special requirement for laboratory waste packs, shock-sensitive wastes, and radioactive wastes.

(10) The elements of a spill control program.

(11) Proper use and limitations of material handling equipment.

(12) Procedures for safe and healthful preparation of containers for shipping and transport.

(13) Methods of communication including those used while wearing respiratory protection.

c. *Technical skills.*

(1) Selection, use maintenance, and limitations of personal protective equipment including the components and procedures for carrying out a respirator program to comply with 29 CFR 1910.134.

(2) Instruction in decontamination programs including personnel, equipment, and hardware; hands-on training including level A, B, and C ensembles and appropriate decontamination lines; field activities including the donning and doffing of protective equipment to a level commensurate with the employee's anticipated job function and responsibility and to the degree required by potential hazards.

(3) Sources for additional hazard information; exercises using relevant manuals and hazard coding systems.

d. *Additional suggested items.*

(1) A laminated, dated card or certificate with photo, denoting limitations and level of protection for which the employee is trained should be issued to those students successfully completing a course.

(2) Attendance should be required at all training modules, with successful completion of exercises and a final written or oral examination with at least 50 questions.

(3) A minimum of one-third of the program should be devoted to hands-on exercises.

(4) A curriculum should be established for the 8-hour refresher training required by 29 CFR 1910.120(e)(8), with delivery of such courses directed toward those areas of previous training that need improvement or reemphasis.

(5) A curriculum should be established for the required 8-hour training for supervisors. Demonstrated competency in the skills and knowledge provided in a 40-hour course should be a prerequisite for supervisor training.

2. *Refresher training.*

The 8-hour annual refresher training required in 29 CFR 1910.120(e)(8) should be conducted by qualified training providers. Refresher training should include at a minimum the following topics and procedures:

(a) Review of and retraining on relevant topics covered in the 40-hour program, as appropriate, using reports by the students on their work experiences.

(b) Update on developments with respect to material covered in the 40-hour course.

(c) Review of changes to pertinent provisions of EPA or OSHA standards or laws.

(d) Introduction of additional subject areas as appropriate.

(e) Hands-on review of new or altered PPE or decontamination equipment or procedures. Review of new developments in personal protective equipment.

(f) Review of newly developed air and contaminant monitoring equipment.

3. *On-site training.*

a. The employer should provide employees engaged in hazardous waste site activities with information and training prior to initial assignment into their work area, as follows:

(1) The requirements of the hazard communication program including the location and availability of the written program, required lists of hazardous chemicals, and material safety data sheets.

(2) Activities and locations in their work area where hazardous substance may be present.

(3) Methods and observations that may be used to detect the present or release of a hazardous chemical in the work area (such as monitoring conducted by the employer, continuous monitoring devices, visual appearances, or other evidence (sight, sound or smell) of hazardous chemicals being released, and applicable alarms from monitoring devices that record chemical releases.

(4) The physical and health hazards of substances known or potentially present in the work area.

(5) The measures employees can take to help protect themselves from worksite hazards, including specific procedures the employer has implemented.

(6) An explanation of the labeling system and material safety data sheets and how employees can obtain and use appropriate hazard information.

(7) The elements of the confined space program including special PPE, permits, monitoring requirements, communication procedures, emergency response, and applicable lock-out procedures.

# OSHA Standards for General Industry

b. The employer should provide hazardous waste employees information and training and should provide a review and access to the site safety and plan as follows:

(1) Names of personnel and alternate responsible for site safety and health.

(2) Safety and health hazards present on the site.

(3) Selection, use, maintenance, and limitations of personal protective equipment specific to the site.

(4) Work practices by which the employee can minimize risks from hazards.

(5) Safe use of engineering controls and equipment available on site.

(6) Safe decontamination procedures established to minimize employee contact with hazardous substances, including:

(A) Employee decontamination,

(B) Clothing decontamination, and

(C) Equipment decontamination.

(7) Elements of the site emergency response plan, including:

(A) Pre-emergency planning.

(B) Personnel roles and lines of authority and communication.

(C) Emergency recognition and prevention.

(D) Safe distances and places of refuge.

(E) Site security and control.

(F) Evacuation routes and procedures.

(G) Decontamination procedures not covered by the site safety and health plan.

(H) Emergency medical treatment and first aid.

(I) Emergency equipment and procedures for handling emergency incidents.

c. The employer should provide hazardous waste employees information and training on personal protective equipment used at the site, such as the following:

(1) PPE to be used based upon known or anticipated site hazards.

(2) PPE limitations of materials and construction; limitations during temperature extremes, heat stress, and other appropriate medical considerations; use and limitations of respirator equipment as well as documentation procedures as outlined in 29 CFR 1910.134.

(3) PPE inspection procedures prior to, during, and after use.

(4) PPE donning and doffing procedures.

(5) PPE decontamination and disposal procedures.

(6) PPE maintenance and storage.

(7) Task duration as related to PPE limitations.

d. The employer should instruct the employee about the site medical surveillance program relative to the particular site, including

(1) Specific medical surveillance programs that have been adapted for the site.

(2) Specific signs and symptoms related to exposure to hazardous materials on the site.

(3) The frequency and extent of periodic medical examinations that will be used on the site.

(4) Maintenance and availability of records.

(5) Personnel to be contacted and procedures to be followed when signs and symptoms of exposures are recognized.

e. The employees will review and discuss the site safety plan as part of the training program. The location of the site safety plan and all written programs should be discussed with employees including a discussion of the mechanisms for access, review, and references described.

B. *RCRA Operations Training for Treatment, Storage and Disposal Facilities.*

1. As a minimum, the training course required in 29 CFR 1910.120(p) should include the following topics:

(a) Review of the applicable paragraphs of 29 CFR 1910.120 and the elements of the employer's occupational safety and health plan.

(b) Review of relevant hazards such as, but not limited to, chemical, biological, and radiological exposures; fire and explosion hazards; thermal extremes; and physical hazards.

(c) General safety hazards including those associated with electrical hazards, powered equipment hazards, lock-out-tag-out procedures, motor vehicle hazards and walking-working surface hazards.

(d) Confined-space hazards and procedures.

(e) Work practices to minimize employee risk from workplace hazards.

(f) Emergency response plan and procedures including first aid meeting the requirements of paragraph (p)(8).

(g) A review of procedures to minimize exposure to hazardous waste and various type of waste streams, including the materials handling program and spill containment program.

(h) A review of hazard communication programs meeting the requirements of 29 CFR 1910.1200.

(i) A review of medical surveillance programs meeting the requirements of 29 CFR 1910.120(p)(3) including the recognition of signs and symptoms of overexposure to hazardous substance including known synergistic interactions.

(j) A review of decontamination programs and procedures meeting the requirements of 29 CFR 1910.120(p)(4).

(k) A review of an employer's requirements to implement a training program and its elements.

(l) A review of the criteria and programs for proper selection and use of personal protective equipment, including respirators.

(m) A review of the applicable appendices to 29 CFR 1910.120.

(n) Principles of toxicology and biological monitoring as they pertain to occupational health.

(o) Rights and responsibilities of employees and employers under applicable OSHA and EPA laws.

(p) Hands-on exercises and demonstrations of competency with equipment to illustrate the basic equipment principles that may be used during the performance of work duties, including the donning and doffing of PPE.

(q) Sources of reference, efficient use of relevant manuals, and knowledge of hazard coding systems to include information contained in hazardous waste manifests.

(r) At least 8 hours of hands-on training.

(s) Training in the job skills required for an employee's job function and responsibility before they are permitted to participate in or supervise field activities.

2. The individual employer should provide hazardous waste employees with information and training prior to an employee's initial assignment into a work area. The training and information should cover the following topics:

(a) The Emergency response plan and procedures including first aid.

(b) A review of the employer's hazardous waste handling procedures including the materials handling program and elements of the spill containment program, location of spill response kits or equipment, and the names of those trained to respond to releases.

(c) The hazardous communication program meeting the requirements of 29 CFR 1910.1200.

(d) A review of the employer's medical surveillance program including the recognition of signs and symptoms of exposure to relevant hazardous substance including known synergistic interactions.

(e) A review of the employer's decontamination program and procedures.

(f) An review of the employer's training program and the parties responsible for that program.

(g) A review of the employer's personal protective equipment program including the proper selection and use of PPE based upon specific site hazards.

(h) All relevant site-specific procedures addressing potential safety and health hazards. This may include, as appropriate, biological and radiological exposures, fire and explosion hazards, thermal hazards, and physical hazards such as electrical hazards, powered equipment hazards, lock-out-tag-out hazards, motor vehicle hazards, and walking-working surface hazards.

(i) Safe use engineering controls and equipment on site.

(j) Names of personnel and alternates responsible for safety and health.

C. *Emergency response training.*

Federal OSHA standards in 29 CFR 1910.120(q) are directed toward private sector emergency responders. Therefore, the guidelines provided in this portion of the appendix are directed toward that employee population. However, they also impact indirectly through State OSHA or USEPA regulations some public sector emergency responders. Therefore, the guidelines provided in this portion of the appendix may be applied to both employee populations.

States with OSHA state plans must cover their employees with regulations at least as effective as the Federal OSHA standards. Public employees in states without approved state OSHA programs covering hazardous waste operations and emergency response are covered by the U.S. EPA under 40 CFR 311, a regulation virtually identical to §1910.120.

Since this is a non-mandatory appendix and therefore not an enforceable standard, OSHA recommends that those employers, employees or volunteers in public sector emergency response organizations outside Federal OSHA jurisdiction consider the following criteria in developing their own training programs. A unified approach to training at the community level between emergency response organizations covered by Federal OSHA and those not covered directly by Federal OSHA can help ensure an effective community response to the release or potential release of hazardous substances in the community.

a. *General considerations.*

Emergency response organizations are required to consider the topics listed in §1910.120(q)(6). Emergency response organizations may use some or all of the following topics to supplement those mandatory topics when developing their response training programs. Many of the topics would require an interaction between the response provider and the individuals responsible for the site where the response would be expected.

(1) Hazard recognition, including:

(A) Nature of hazardous substances present,

(B) Practical applications of hazard recognition, including presentations on biology, chemistry, and physics.

(2) Principles of toxicology, biological monitoring, and risk assessment.

(3) Safe work practices and general site safety.

(4) Engineering controls and hazardous waste operations.

(5) Site safety plans and standard operating procedures.

(6) Decontamination procedures and practices.

(7) Emergency procedures, first aid, and self-rescue.

(8) Safe use of field equipment.

(9) Storage, handling, use and transportation of hazardous substances.

(10) Use, care, and limitations of personal protective equipment.

(11) Safe sampling techniques.

(12) Rights and responsibilities of employees under OSHA and other related laws concerning right-to-know, safety and health, compensations and liability.

(13) Medical monitoring requirements.

(14) Community relations.

b. *Suggested criteria for specific courses.*

(1) *First responder awareness level.*

(A) Review of and demonstration of competency in performing the applicable skills of 29 CFR 1910.120(q).

(B) Hands-on experience with the U.S. Department of Transportation's *Emergency Response Guidebook* (ERG) and familiarization with OSHA standard 29 CFR 1910.1201.

(C) Review of the principles and practices for analyzing an incident to determine both the hazardous substances present and the basic hazard and response information for each hazardous substance present.

(D) Review of procedures for implementing actions consistent with the local emergency response plan, the organization's standard operating procedures, and the current edition of DOT's ERG including emergency notification procedures and follow-up communications.

(E) Review of the expected hazards including fire and explosions hazards, confined space hazards, electrical hazards, powered equipment hazards, motor vehicle hazards, and walking-working surface hazards.

(F) Awareness and knowledge of the competencies for the First Responder at the Awareness Level covered in the National Fire Protection Association's Standard No. 472, *Professional Competence of Responders to Hazardous Materials Incidents.*

(2) *First responder operations level.*

(A) Review of and demonstration of competency in performing the applicable skills of 29 CFR 1910.120(q).

(B) Hands-on experience with the U.S. Department of Transportation's *Emergency Response Guidebook* (ERG), manufacturer material safety data sheets, CHEMTREC/CANUTEC, shipper or manufacturer contacts, and other relevant sources of information addressing hazardous substance releases. Familiarization with OSHA standard 29 CFR 1910.1201.

(C) Review of the principles and practices for analyzing an incident to determine the hazardous substances present, the likely behavior of the hazardous substance and its container, the types of hazardous substance transportation containers and vehicles, the types and selection of the appropriate defensive strategy for containing the release.

(D) Review of procedures for implementing continuing response actions consistent with the local emergency response plan, the organization's standard operating procedures, and the current edition of DOT's ERG including extended emergency notification procedures and follow-up communications.

(E) Review of the principles and practice for proper selection and use of personal protective equipment.

(F) Review of the principles and practice of personnel and equipment decontamination.

(G) Review of the expected hazards including fire and explosions hazards, confined space hazards, electrical hazards, powered equipment hazards, motor vehicle hazards, and walking-working surface hazards.

(H) Awareness and knowledge of the competencies for the First Responder at the Operations Level covered in the National Fire Protection Association's Standard No. 472, *Professional Competence of Responders to Hazardous Materials Incidents.*

(3) *Hazardous materials technician.*

(A) Review of and demonstration of competency in performing the applicable skills of 29 CFR 1910.120(q).

(B) Hands-on experience with written and electronic information relative to response decision making including but not limited to the U.S. Department of Transportation's *Emergency Response Guidebook* (ERG), manufacturer material safety data sheets, CHEMTREC/CANUTEC, shipper or manufacturer contacts, computer data bases and response models, and other relevant sources of information addressing hazardous substance releases. Familiarization with OSHA standard 29 CFR 1910.1201.

(C) Review of the principles and practices for analyzing an incident to determine the hazardous substances present, their physical and chemical properties, the likely behavior of the hazardous substance and its container, the types of hazardous substance transportation containers and vehicles involved in the release, the appropriate strategy for approaching release sites and containing the release.

(D) Review of procedures for implementing continuing response actions consistent with the local emergency response plan, the organization's standard operating procedures, and the current edition of DOT's ERG including extended emergency notification procedures and follow-up communications.

(E) Review of the principles and practice for proper selection and use of personal protective equipment.

(F) Review of the principles and practices of establishing exposure zones, proper decontamination and medical surveillance stations and procedures.

(G) Review of the expected hazards including fire and explosions hazards, confined space hazards, electrical hazards, powered equipment hazards, motor vehicle hazards, and walking-working surface hazards.

(H) Awareness and knowledge of the competencies for the Hazardous Materials Technician covered in the National Fire Protection Association's Standard No. 472, *Professional Competence of Responders to Hazardous Materials Incidents.*

(4) *Hazardous materials specialist.*

(A) Review of and demonstration of competency in performing the applicable skills of 29 CFR 1910.120(q).

(B) Hands-on experience with retrieval and use of written and electronic information relative to response decision making including but not limited to the U.S. Department of Transportation's *Emergency Response Guidebook* (ERG), manufacturer material safety data sheets, CHEMTREC/CANUTEC, shipper or manufacturer contacts, computer data bases and response models, and other relevant sources of information addressing hazardous substance releases. Familiarization with OSHA standard 29 CFR 1910.1201.

(C) Review of the principles and practices for analyzing an incident to determine the hazardous substances present, their physical and chemical properties, and the likely behavior of the hazardous substance and its container, vessel, or vehicle.

(D) Review of the principles and practices for identification of the types of hazardous substance transportation containers, vessels and vehicles involved in the release; selecting and using the various types of equipment available for plugging or patching transportation containers, vessels or vehicles; organizing and directing the use of multiple teams of hazardous material technicians and selecting the appropriate strategy for approaching release sites and containing or stopping the release.

(E) Review of procedures for implementing continuing response actions consistent with the local emergency response plan, the organization's standard operating procedures, including knowledge of the available public and private response resources, establishment of an incident command post, direction of hazardous material technician teams, and extended emergency notification procedures and follow-up communications.

(F) Review of the principles and practice for proper selection and use of personal protective equipment.

(G) Review of the principles and practices of establishing exposure zones and proper decontamination, monitoring and medical surveillance stations and procedures.

(H) Review of the expected hazards including fire and explosions hazards, confined space hazards, electrical hazards, powered equipment hazards, motor vehicle hazards, and walking-working surface hazards.

(I) Awareness and knowledge of the competencies for the Off-site Specialist Employee covered in the National Fire Protection Association's Standard No. 472, *Professional Competence of Responders to Hazardous Materials Incidents.*

(5) *Incident commander.*

The incident commander is the individual who, at any one time, is responsible for and in control of the response effort. This individual is the person responsible for the direction and coordination of the response effort. An incident commander's position should be occupied by the most senior, appropriately trained individual present at the response site. Yet, as necessary and appropriate by the level of response provided, the position may be occupied by many individuals during a particular response as the need for greater authority, responsibility, or training increases. It is possible for the first responder at the awareness level to assume the duties of incident commander until a more senior and appropriately trained individual arrives at the response site.

Therefore, any emergency responder expected to perform as an incident commander should be trained to fulfill the obligations of the position at the level of response they will be providing including the following:

(A) Ability to analyze a hazardous substance incident to determine the magnitude of the response problem.

(B) Ability to plan and implement an appropriate response plan within the capabilities of available personnel and equipment.

(C) Ability to implement a response to favorably change the outcome of the incident in a manner consistent with the local emergency response plan and the organization's standard operating procedures.

(D) Ability to evaluate the progress of the emergency response to ensure that the response objectives are being met safely, effectively, and efficiently.

(E) Ability to adjust the response plan to the conditions of the response and to notify higher levels of response when required by the changes to the response plan.

[54 FR 9317, Mar. 6, 1989, as amended at 55 FR 14073, Apr. 13, 1990; 56 FR 15832, Apr. 18, 1991; 59 FR 43270, Aug. 22, 1994; 61 FR 9238, Mar. 7, 1996]

### §1910.121 [Reserved].

Dipping and Coating Operations

**Source:** 64 FR 13909, Mar. 23, 1999, unless otherwise noted.

### §1910.122 Table of contents.

This section lists the paragraph headings contained in §§1910.123 through 1910.126.

§1910.123 Dipping and coating operations: Coverage and definitions.

| | |
|---|---|
| **1910.123(a)** | Does this rule apply to me? |
| **1910.123(b)** | What operations are covered? |
| **1910.123(c)** | What operations are not covered? |
| **1910.123(d)** | How are terms used in §§1910.123 through 1910.126 defined? |

§1910.124 General requirements for dipping and coating operations.

| | |
|---|---|
| **1910.124(a)** | What construction requirements apply to dip tanks? |
| **1910.124(b)** | What ventilation requirements apply to vapor areas? |
| **1910.124(c)** | What requirements must I follow to recirculate exhaust air into the workplace? |
| **1910.124(d)** | What must I do when I use an exhaust hood? |
| **1910.124(e)** | What requirements must I follow when an employee enters a dip tank? |
| **1910.124(f)** | What first-aid procedures must my employees know? |
| **1910.124(g)** | What hygiene facilities must I provide? |
| **1910.124(h)** | What treatment and first aid must I provide? |
| **1910.124(i)** | What must I do before an employee cleans a dip tank? |
| **1910.124(j)** | What must I do to inspect and maintain my dipping or coating operation? |

§1910.125 Additional requirements for dipping and coating operations that use flammable or combustible liquids.

| | |
|---|---|
| **1910.125(a)** | What type of construction material must be used in making my dip tank? |
| **1910.125(b)** | When must I provide overflow piping? |
| **1910.125(c)** | When must I provide a bottom drain? |
| **1910.125(d)** | When must my conveyer system shut down automatically? |
| **1910.125(e)** | What ignition and fuel sources must be controlled? |
| **1910.125(f)** | What fire protection must I provide? |
| **1910.125(g)** | To what temperature may I heat a liquid in a dip tank? |

§1910.126 Additional requirements for special dipping and coating operations.

| | |
|---|---|
| **1910.126(a)** | What additional requirements apply to hardening or tempering tanks? |
| **1910.126(b)** | What additional requirements apply to flow coating? |
| **1910.126(c)** | What additional requirements apply to roll coating, roll spreading, or roll impregnating? |
| **1910.126(d)** | What additional requirements apply to vapor degreasing tanks? |
| **1910.126(e)** | What additional requirements apply to cyanide tanks? |
| **1910.126(f)** | What additional requirements apply to spray cleaning tanks and spray degreasing tanks? |
| **1910.126(g)** | What additional requirements apply to electrostatic paint detearing? |

## §1910.123 Dipping and coating operations: Coverage and definitions.

**1910.123(a)** Does this rule apply to me? **(1)** This rule (§§1910.123 through 1910.126) applies when you use a dip tank containing a liquid other than water. It applies when you use the liquid in the tank or its vapor to:

**1910.123(a)(1)(i)** Clean an object;

**1910.123(a)(1)(ii)** Coat an object;

**1910.123(a)(1)(iii)** Alter the surface of an object; or

**1910.123(a)(1)(iv)** Change the character of an object.

**1910.123(a)(2)** This rule also applies to the draining or drying of an object you have dipped or coated.

**1910.123(b)** What operations are covered? Examples of covered operations are paint dipping, electroplating, pickling, quenching, tanning, degreasing, stripping, cleaning, roll coating, flow coating, and curtain coating.

**1910.123(c)** What operations are not covered? You are not covered by this rule if your dip-tank operation only uses a molten material (a molten metal, alloy, or salt, for example).

**1910.123(d)** How are terms used in §§1910.123 through 1910.126 defined?

*Adjacent area* means any area within 20 feet (6.1 m) of a vapor area that is not separated from the vapor area by tight partitions.

*Approved* means that the equipment so designated is listed or approved by a nationally recognized testing laboratory, as defined by §1910.7.

*Autoignition temperature* means the minimum temperature required to cause self-sustained combustion, independent of any other source of heat.

*Combustible liquid* means a liquid having a flash point of 100 °F (37.8 °C) or above.

*Dip tank* means a container holding a liquid other than water and that is used for dipping or coating. An object may be immersed (or partially immersed) in a dip tank or it may be suspended in a vapor coming from the tank.

*Flammable liquid* means a liquid having a flashpoint below 100 °F (37.8 °C).

*Flashpoint* means the minimum temperature at which a liquid gives off a vapor in sufficient concentration to ignite if tested in accordance with the definition of "flashpoint" in §1910.1200(c).

*Lower flammable limit (LFL)* means the lowest concentration of a material that will propagate a flame. The LFL is usually expressed as a percent by volume of the material in air (or other oxidant).

*Vapor area* means any space containing a dip tank, including its drain boards, associated drying or conveying equipment, and any surrounding area where the vapor concentration exceeds 25% of the LFL of the liquid in the tank.

*You* means the employer, as defined by the Occupational Safety and Health Act of 1970 (29 U.S.C. 651 *et seq.*).

## §1910.124 General requirements for dipping and coating operations.

**1910.124(a)** What construction requirements apply to dip tanks? Any container that you use as a dip tank must be strong enough to withstand any expected load.

**1910.124(b)** What ventilation requirements apply to vapor areas? **(1)** The ventilation that you provide to a vapor area must keep the airborne concentration of any substance below 25% of its LFL.

**1910.124(b)(2)** When a liquid in a dip tank creates an exposure hazard covered by a standard listed in subpart Z of this part, you must control worker exposure as required by that standard.

**1910.124(b)(3)** You may use a tank cover or material that floats on the surface of the liquid in a dip tank to replace or supplement ventilation. The method or combination of methods you choose must maintain the airborne concentration of the hazardous material and the worker's exposure within the limits specified in paragraphs (b)(1) and (b)(2) of this section.

**1910.124(b)(4)** When you use mechanical ventilation, it must conform to the following standards that are incorporated by reference as specified in §1910.6:

**1910.124(b)(4)(i)** ANSI Z9.2-1979, Fundamentals Governing the Design and Operation of Local Exhaust Systems;

**1910.124(b)(4)(ii)** NFPA 34-1995, Standard for Dip Tanks Containing Flammable or Combustible Liquids;

**1910.124(b)(4)(iii)** ACGIH's "Industrial Ventilation: A Manual of Recommended Practice" (22nd ed., 1995); or

**1910.124(b)(4)(iv)** ANSI Z9.1-1971, Practices for Ventilation and Operation of Open-Surface Tanks, and NFPA 34-1966, Standard for Dip Tanks Containing Flammable or Combustible Liquids.

**1910.124(b)(5)** When you use mechanical ventilation, it must draw the flow of air into a hood or exhaust duct.

**1910.124(b)(6)** When you use mechanical ventilation, each dip tank must have an independent exhaust system unless the combination of substances being removed will not cause a:

**1910.124(b)(6)(i)** Fire;

**1910.124(b)(6)(ii)** Explosion; or

**1910.124(b)(6)(iii)** Chemical reaction.

**1910.124(c)** What requirements must I follow to recirculate exhaust air into the workplace? **(1)** You may not recirculate exhaust air when any substance in that air poses a health hazard to employees or exceeds 25% of its LFL.

**1910.124(c)(2)** You must ensure that any exhaust air recirculated from a dipping or coating operation using flammable or combustible liquids is:

**1910.124(c)(2)(i)** Free of any solid particulate that poses a health or safety hazard for employees; and

**1910.124(c)(2)(ii)** Monitored by approved equipment.

**1910.124(c)(3)** You must have a system that sounds an alarm and automatically shuts down the operation when the vapor concentration for any substance in the exhaust airstream exceeds 25% of its LFL.

**1910.124(d)** What must I do when I use an exhaust hood? You must:

**1910.124(d)(1)** Provide each room having exhaust hoods with a volume of outside air that is at least 90 percent of the volume of the exhaust air; and

**1910.124(d)(2)** Ensure that the outside air supply does not damage exhaust hoods.

**1910.124(e)** What requirements must I follow when an employee enters a dip tank? When an employee enters a dip tank, you must meet the entry requirements of §1910.146, OSHA's standard for Permit-Required Confined Spaces, as applicable.

**1910.124(f)** What first-aid procedures must my employees know? Your employees must know the first-aid procedures that are appropriate to the dipping or coating hazards to which they are exposed.

**1910.124(g)** What hygiene facilities must I provide? When your employees work with liquids that may burn, irritate, or otherwise harm their skin, you must provide:

**1910.124(g)(1)** Locker space or other storage space to prevent contamination of the employee's street clothes;

**1910.124(g)(2)** An emergency shower and eye-wash station close to the dipping or coating operation. In place of this equipment, you may use a water hose that is at least 4 feet (1.22 m) long and at least 3/4 of an inch (18 mm) thick with a quick-opening valve and carrying a pressure of 25 pounds per square inch (1.62 k/cm$^2$) or less; and

**1910.124(g)(3)** At least one basin with a hot-water faucet for every 10 employees who work with such liquids. (See paragraph (d) of §1910.141.)

**1910.124(h)** What treatment and first aid must I provide? When your employees work with liquids that may burn, irritate, or otherwise harm their skin, you must provide:

**1910.124(h)(1)** A physician's approval before an employee with a sore, burn, or other skin lesion that requires medical treatment works in a vapor area;

**1910.124(h)(2)** Treatment by a properly designated person of any small skin abrasion, cut, rash, or open sore;

**1910.124(h)(3)** Appropriate first-aid supplies that are located near the dipping or coating operation; and

**1910.124(h)(4)** For employees who work with chromic acid, periodic examinations of their exposed body parts, especially their nostrils.

**1910.124(i)** What must I do before an employee cleans a dip tank? Before permitting an employee to clean the interior of a dip tank, you must:

**1910.124(i)(1)** Drain the contents of the tank and open the cleanout doors; and

**1910.124(i)(2)** Ventilate and clear any pockets where hazardous vapors may have accumulated.

**1910.124(j)** What must I do to inspect and maintain my dipping or coating operation? You must:

**1910.124(j)(1)** Inspect the hoods and ductwork of the ventilation system for corrosion or damage:

**1910.124(j)(1)(i)** At least quarterly during operation; and

**1910.124(j)(1)(ii)** Prior to operation after a prolonged shutdown.

**1910.124(j)(2)** Ensure that the airflow is adequate:

**1910.124(j)(2)(i)** At least quarterly during operation; and

**1910.124(j)(2)(ii)** Prior to operation after a prolonged shutdown.

**1910.124(j)(3)** Periodically inspect all dipping and coating equipment, including covers, drains, overflow piping, and electrical and fire-extinguishing systems, and promptly correct any deficiencies;

**1910.124(j)(4)** Provide mechanical ventilation or respirators (selected and used as specified in §1910.134, OSHA's Respiratory Protection standard) to protect employees in the vapor area from exposure to toxic substances released during welding, burning, or open-flame work; and

**1910.124(j)(5)** Have dip tanks thoroughly cleaned of solvents and vapors before permitting welding, burning, or open-flame work on them.

### §1910.125 Additional requirements for dipping and coating operations that use flammable or combustible liquids.

If you use flammable or combustible liquids, you must comply with the requirements of this section as well as the requirements of sections 1910.123, 1910.124, and 1910.126, as applicable.

| You must comply with this section if: | And: |
|---|---|
| • The flashpoint of the flammable or combustible liquid is 200 °F (93.3 °C) or above | • The liquid is heated as part of the operation; or<br>• A heated object is placed in the liquid. |

**1910.125(a)** What type of construction material must be used in making my dip tank? Your dip tank must be made of noncombustible material.

**1910.125(b)** When must I provide overflow piping? **(1)** You must provide properly trapped overflow piping that discharges to a safe location for any dip tank having:

**1910.125(b)(i)** A capacity greater than 150 gallons (568 L); or

**1910.125(b)(ii)** A liquid surface area greater than 10 feet$^2$ (0.95 m$^2$).

**1910.125(b)(2)** You must also ensure that:

**1910.125(b)(2)(i)** Any overflow piping is at least 3 inches (7.6 cm) in diameter and has sufficient capacity to prevent the dip tank from overflowing;

**1910.125(b)(2)(ii)** Piping connections on drains and overflow pipes allow ready access to the interior of the pipe for inspection and cleaning; and

**1910.125(b)(2)(iii)** The bottom of the overflow connection is at least 6 inches (15.2 cm) below the top of the dip tank.

**1910.125(c)** When must I provide a bottom drain? **(1)** You must provide a bottom drain for dip tanks that contain more than 500 gallons (1893 L) of liquid, unless:

**1910.125(c)(1)(i)** The dip tank is equipped with an automatic closing cover meeting the requirements of paragraph (f)(3) of this section; or

**1910.125(c)(1)(ii)** The viscosity of the liquid at normal atmospheric temperature does not allow the liquid to flow or be pumped easily.

**1910.125(c)(2)** You must ensure that the bottom drain required by this section:

**1910.125(c)(2)(i)** Will empty the dip tank during a fire;

**1910.125(c)(2)(ii)** Is properly trapped;

**1910.125(c)(2)(iii)** Has pipes that permit the dip tank's contents to be removed within five minutes after a fire begins; and

**1910.125(c)(2)(iv)** Discharges to a safe location.

**1910.125(c)(3)** Any bottom drain you provide must be capable of manual and automatic operation, and manual operation must be from a safe and accessible location.

**1910.125(c)(4)** You must ensure that automatic pumps are used when gravity flow from the bottom drain is impractical.

**1910.125(d)** When must my conveyor system shut down automatically? If your conveyor system is used with a dip tank, the system must shut down automatically:

**1910.125(d)(1)** If there is a fire; or

**1910.125(d)(2)** If the ventilation rate drops below what is required by paragraph (b) of §1910.124.

**1910.125(e)** What ignition and fuel sources must be controlled? **(1)** In each vapor area and any adjacent area, you must ensure that:

**1910.125(e)(1)(i)** All electrical wiring and equipment conform to the applicable hazardous (classified)-area requirements of subpart S of this part (except as specifically permitted in paragraph (g) of §1910.126); and

**1910.125(e)(1)(ii)** There are no flames, spark-producing devices, or other surfaces that are hot enough to ignite vapors.

**1910.125(e)(2)** You must ensure that any portable container used to add liquid to the tank is electrically bonded to the dip tank and positively grounded to prevent static electrical sparks or arcs.

**1910.125(e)(3)** You must ensure that a heating system that is used in a drying operation and could cause ignition:

**1910.125(e)(3)(i)** Is installed in accordance with NFPA 86A-1969, Standard for Ovens and Furnaces (which is incorporated by reference in §1910.6 of this part);

**1910.125(e)(3)(ii)** Has adequate mechanical ventilation that operates before and during the drying operation; and

**1910.125(e)(3)(iii)** Shuts down automatically if any ventilating fan fails to maintain adequate ventilation.

**1910.125(e)(4)** You also must ensure that:

**1910.125(e)(4)(i)** All vapor areas are free of combustible debris and as free as practicable of combustible stock;

**1910.125(e)(4)(ii)** Rags and other material contaminated with liquids from dipping or coating operations are placed in approved waste cans immediately after use; and

**1910.125(e)(4)(iii)** Waste can contents are properly disposed of at the end of each shift.

**1910.125(e)(5)** You must prohibit smoking in a vapor area and must post a readily visible "No Smoking" sign near each dip tank.

**1910.125(f)** What fire protection must I provide? **(1)** You must provide the fire protection required by this paragraph (f) for:

**1910.125(f)(1)(i)** Any dip tank having a capacity of at least 150 gallons (568 L) or a liquid surface area of at least 4 feet$^2$ (0.38 m$^1$); and

**1910.125(f)(1)(ii)** Any hardening or tempering tank having a capacity of at least 500 gallons (1893 L) or a liquid surface area of at least 25 feet$^2$ (2.37 m$^2$).

**1910.125(f)(2)** For every vapor area, you must provide:

**1910.125(f)(2)(i)** Manual fire extinguishers that are suitable for flammable and combustible liquid fires and that conform to the requirements of §1910.157; and

**1910.125(f)(2)(ii)** An automatic fire-extinguishing system that conforms to the requirements of subpart L of this part.

**1910.125(f)(3)** You may substitute a cover that is closed by an approved automatic device for the automatic fire-extinguishing system if the cover:

**1910.125(f)(3)(i)** Can also be activated manually;

**1910.125(f)(3)(ii)** Is noncombustible or tin-clad, with the enclosing metal applied with locked joints; and

**1910.125(f)(3)(iii)** Is kept closed when the dip tank is not in use.

**1910.125(g)** To what temperature may I heat a liquid in a dip tank? You must maintain the temperature of the liquid in a dip tank:

**1910.125(g)(1)** Below the liquid's boiling point; and

**1910.125(g)(2)** At least 100 °F (37.8 °C) below the liquid's autoignition temperature.

## §1910.126 Additional requirements for special dipping and coating operations.

In addition to the requirements in §§1910.123 through 1910.125, you must comply with any requirement in this section that applies to your operation.

**1910.126(a)** What additional requirements apply to hardening or tempering tanks?

**1910.126(a)(1)** You must ensure that hardening or tempering tanks:

**1910.126(a)(1)(i)** Are located as far as practicable from furnaces;

**1910.126(a)(1)(ii)** Are on noncombustible flooring; and

**1910.126(a)(1)(iii)** Have noncombustible hoods and vents (or equivalent devices) for venting to the outside. For this purpose, vent ducts must be treated as flues and kept away from combustible materials, particularly roofs.

**1910.126(a)(2)** You must equip each tank with an alarm that will sound if the temperature of the liquid comes within 50 °F (10 °C) of its flashpoint (the alarm set point).

**1910.126(a)(3)** When practicable, you must also provide each tank with a limit switch to shut down the conveyor supplying work to the tank.

**1910.126(a)(4)** If the temperature of the liquid can exceed the alarm set point, you must equip the tank with a circulating cooling system.

**1910.126(a)(5)** If the tank has a bottom drain, the bottom drain may be combined with the oil-circulating system.

**1910.126(a)(6)** You must not use air under pressure when you fill the dip tank or agitate the liquid in the dip tank.

**1910.126(b)** What additional requirements apply to flow coating? **(1)** You must use a direct low-pressure pumping system or a 10-gallon (38 L) or smaller gravity tank to supply the paint for flow coating. In case of fire, an approved heat-actuated device must shut down the pumping system.

**1910.126(b)(2)** You must ensure that the piping is substantial and rigidly supported.

**1910.126(c)** What additional requirements apply to roll coating, roll spreading, or roll impregnating? When these operations use a flammable or combustible liquid that has a flashpoint below 140 °F (60 °C), you must prevent sparking of static electricity by:

**1910.126(c)(1)** Bonding and grounding all metallic parts (including rotating parts) and installing static collectors; or

**1910.126(c)(2)** Maintaining a conductive atmosphere (for example, one with a high relative humidity) in the vapor area.

**1910.126(d)** What additional requirements apply to vapor degreasing tanks? **(1)** You must ensure that the condenser or vapor-level thermostat keeps the vapor level at least 36 inches (91 cm) or one-half the tank width, whichever is less, below the top of the vapor degreasing tank.

**1910.126(d)(2)** When you use gas as a fuel to heat the tank liquid, you must prevent solvent vapors from entering the air-fuel mixture. To do this, you must make the combustion chamber airtight (except for the flue opening).

**1910.126(d)(3)** The flue must be made of corrosion-resistant material, and it must extend to the outside. You must install a draft diverter if mechanical exhaust is used on the flue.

**1910.126(d)(4)** You must not allow the temperature of the heating element to cause a solvent or mixture to decompose or to generate an excessive amount of vapor.

**1910.126(e)** What additional requirements apply to cyanide tanks? You must ensure that cyanide tanks have a dike or other safeguard to prevent cyanide from mixing with an acid if a dip tank fails.

**1910.126(f)** What additional requirements apply to spray cleaning tanks and spray degreasing tanks? If you spray a liquid in the air over an open-surface cleaning or degreasing tank, you must control the spraying to the extent feasible by:

**1910.126(f)(1)** Enclosing the spraying operation; and

**1910.126(f)(2)** Using mechanical ventilation to provide enough inward air velocity to prevent the spray from leaving the vapor area.

**1910.126(g)** What additional requirements apply to electrostatic paint detearing? **(1)** You must use only approved electrostatic equipment in paint-detearing operations. Electrodes in such equipment must be substantial, rigidly supported, permanently located, and effectively insulated from ground by nonporous, noncombustible, clean, dry insulators.

**1910.126(g)(2)** You must use conveyors to support any goods being paint deteared.

**1910.126(g)(3)** You must ensure that goods being electrostatically deteared are not manually handled.

**1910.126(g)(4)** Between goods being electrostatically deteared and the electrodes or conductors of the electrostatic equipment, you must maintain a minimum distance of twice the sparking distance. This minimum distance must be displayed conspicuously on a sign located near the equipment.

**1910.126(g)(5)** You must ensure that the electrostatic equipment has automatic controls that immediately disconnect the power supply to the high-voltage transformer and signal the operator if:

**1910.126(g)(5)(i)** Ventilation or the conveyors fail to operate;

**1910.126(g)(5)(ii)** A ground (or imminent ground) occurs anywhere in the high-voltage system; or

**1910.126(g)(5)(iii)** Goods being electrostatically deteared come within twice the sparking distance of the electrodes or conductors of the equipment.

**1910.126(g)(6)** You must use fences, rails, or guards, made of conducting material and adequately grounded, to separate paint-detearing operations from storage areas and from personnel.

**1910.126(g)(7)** To protect paint-detearing operations from fire, you must have in place:

**1910.126(g)(7)(i)** Automatic sprinklers; or

**1910.126(g)(7)(ii)** An automatic fire-extinguishing system conforming to the requirements of subpart L of this part.

**1910.126(g)(8)** To collect paint deposits, you must:

**1910.126(g)(8)(i)** Provide drip plates and screens; and

**1910.126(g)(8)(ii)** Clean these plates and screens in a safe location.

# Subpart I—Personal Protective Equipment

> **Authority:** Sections 4, 6, and 8, Occupational Safety and Health Act of 1970 (29 U.S.C. 653, 655, 657); Secretary of Labor's Order No. 12-71 (36 FR 8754), 8-76 (41 FR 25059), 9-83 (48 FR 35736), 1-90 (55 FR 9033), or 6-96 (62 FR 111), as applicable.
>
> Sections 1910.132, 1910.134, and 1910.138 also issued under 29 CFR part 1911.
>
> Sections 1910.133, 1910.135, and 1910.136 also issued under 29 CFR part 1911 and 5 U.S.C. 553.

## §1910.132 General requirements.

**1910.132(a)** *Application.* Protective equipment, including personal protective equipment for eyes, face, head, and extremities, protective clothing, respiratory devices, and protective shields and barriers, shall be provided, used, and maintained in a sanitary and reliable condition wherever it is necessary by reason of hazards of processes or environment, chemical hazards, radiological hazards, or mechanical irritants encountered in a manner capable of causing injury or impairment in the function of any part of the body through absorption, inhalation or physical contact.

**1910.132(b)** *Employee-owned equipment.* Where employees provide their own protective equipment, the employer shall be responsible to assure its adequacy, including proper maintenance, and sanitation of such equipment.

**1910.132(c)** *Design.* All personal protective equipment shall be of safe design and construction for the work to be performed.

**1910.132(d)** *Hazard assessment and equipment selection.* **(1)** The employer shall assess the workplace to determine if hazards are present, or are likely to be present, which necessitate the use of personal protective equipment (PPE). If such hazards are present, or likely to be present, the employer shall:

**1910.132(d)(1)(i)** Select, and have each affected employee use, the types of PPE that will protect the affected employee from the hazards identified in the hazard assessment;

**1910.132(d)(1)(ii)** Communicate selection decisions to each affected employee; and,

**1910.132(d)(1)(iii)** Select PPE that properly fits each affected employee.

**NOTE:** Non-mandatory Appendix B contains an example of procedures that would comply with the requirement for a hazard assessment.

**1910.132(d)(2)** The employer shall verify that the required workplace hazard assessment has been performed through a written certification that identifies the workplace evaluated; the person certifying that the evaluation has been performed; the date(s) of the hazard assessment; and, which identifies the document as a certification of hazard assessment.

**1910.132(e)** *Defective and damaged equipment.* Defective or damaged personal protective equipment shall not be used.

**1910.132(f)** *Training.* **(1)** The employer shall provide training to each employee who is required by this section to use PPE. Each such employee shall be trained to know at least the following:

**1910.132(f)(1)(i)** When PPE is necessary;

**1910.132(f)(1)(ii)** What PPE is necessary;

**1910.132(f)(1)(iii)** How to properly don, doff, adjust, and wear PPE;

**1910.132(f)(1)(iv)** The limitations of the PPE; and,

**1910.132(f)(1)(v)** The proper care, maintenance, useful life and disposal of the PPE.

**1910.132(f)(2)** Each affected employee shall demonstrate an understanding of the training specified in paragraph (f)(1) of this section, and the ability to use PPE properly, before being allowed to perform work requiring the use of PPE.

**1910.132(f)(3)** When the employer has reason to believe that any affected employee who has already been trained does not have the understanding and skill required by paragraph (f)(2) of this section, the employer shall retrain each such employee. Circumstances where retraining is required include, but are not limited to, situations where:

**1910.132(f)(3)(i)** Changes in the workplace render previous training obsolete; or

**1910.132(f)(3)(ii)** Changes in the types of PPE to be used render previous training obsolete; or

**1910.132(f)(3)(iii)** Inadequacies in an affected employee's knowledge or use of assigned PPE indicate that the employee has not retained the requisite understanding or skill.

**1910.132(f)(4)** The employer shall verify that each affected employee has received and understood the required training through a written certification that contains the name of each employee trained, the date(s) of training, and that identifies the subject of the certification.

**1910.132(g)** Paragraphs (d) and (f) of this section apply only to §§1910.133, 1910.135, 1919.136, and 1910.138. Paragraphs (d) and (f) of this section do not apply to §§1910.134 and 1910.137.

[39 FR 23502, June 27, 1974, as amended at 59 FR 16334, Apr. 6, 1994; 59 FR 33910, July 1, 1994]; 59 FR 34580, July 6, 1994].

## §1910.133 Eye and face protection.

**1910.133(a)** *General requirements.* **(1)** The employer shall ensure that each affected employee uses appropriate eye or face protection when exposed to eye or face hazards from flying particles, molten metal, liquid chemicals, acids or caustic liquids, chemical gases or vapors, or potentially injurious light radiation.

**1910.133(a)(2)** The employer shall ensure that each affected employee uses eye protection that provides side protection when there is a hazard from flying objects. Detachable side protectors (e.g. clip-on or slide-on side shields) meeting the pertinent requirements of this section are acceptable.

**1910.133(a)(3)** The employer shall ensure that each affected employee who wears prescription lenses while engaged in operations that involve eye hazards wears eye protection that incorporates the prescription in its design, or wears eye protection that can be worn over the prescription lenses without disturbing the proper position of the prescription lenses or the protective lenses.

**1910.133(a)(4)** Eye and face PPE shall be distinctly marked to facilitate identification of the manufacturer.

**1910.133(a)(5)** The employer shall ensure that each affected employee uses equipment with filter lenses that have a shade number appropriate for the work being performed for protection from injurious light radiation. The following is a listing of appropriate shade numbers for various operations.

## FILTER LENSES FOR PROTECTION AGAINST RADIANT ENERGY

| Operations | Electrode Size 1/32 in. | Arc Current | Minimum* Protective Shade |
|---|---|---|---|
| Shielded metal arc welding | Less than 3 | Less than 60 | 7 |
|  | 3-5 | 60-160 | 8 |
|  | 5-8 | 160-250 | 10 |
|  | More than 8 | 250-550 | 11 |
| Gas metal arc welding and flux cored arc welding |  | less than 60 | 7 |
|  |  | 60-160 | 10 |
|  |  | 60-250 | 10 |
|  |  | 250-500 | 10 |
| Gas Tungsten arc welding |  | less than 50 | 8 |
|  |  | 50-150 | 8 |
|  |  | 150-500 | 10 |
| Air carbon | (Light) | less than 500 | 10 |
| Arc cutting | (Heavy) | 500-1000 | 11 |
| Plasma arc welding |  | less than 20 | 6 |
|  |  | 20-100 | 8 |
|  |  | 100-400 | 10 |
|  |  | 400-800 | 11 |
| Plasma arc cutting | (light) ** | less than 300 | 8 |
|  | (medium) ** | 300-400 | 9 |
|  | (heavy) ** | 400-800 | 10 |
| Torch brazing |  |  | 3 |
| Torch soldering |  |  | 2 |
| Carbon arc welding |  |  | 14 |

## FILTER LENSES FOR PROTECTION AGAINST RADIANT ENERGY

| Operations | Plate thickness—inches | Plate thickness—mm | Minimum* Protective Shade |
|---|---|---|---|
| Gas Welding: |  |  |  |
| Light | Under 1/8 | Under 3.2 | 4 |
| Medium | 1/8 to 1/2 | 3.2 to 12.7 | 5 |
| Heavy | Over 1/2 | Over 12.7 | 6 |
| Oxygen cutting: |  |  |  |
| Light | Under 1 | Under 25 | 3 |
| Medium | 1 to 6 | 25 to 150 | 4 |
| Heavy | Over 6 | Over 150 | 5 |

\* As a rule of thumb, start with a shade that is too dark to see the weld zone. Then go to a lighter shade which gives sufficient view of the weld zone without going below the minimum. In oxyfuel gas welding or cutting where the torch produces a high yellow light, it is desirable to use a filter lens that absorbs the bellow or sodium line in the visible light of the (spectrum) operation.

\*\* These values apply where the actual arc is clearly seen. Experience has shown that lighter filters may be used when the arc is hidden by the workpiece.

**1910.133(b) Criteria for protective eye and face devices. (1)** Protective eye and face devices purchased after July 5, 1994 shall comply with ANSI Z87.1-1989, "American National Standard Practice for Occupational and Educational Eye and Face Protection," which is incorporated by reference as specified in §1910.6, or shall be demonstrated by the employer to be equally effective.

**1910.133(b)(2)** Eye and face protective devices purchased before July 5, 1994 shall comply with the ANSI "USA standard for Occupational and Educational Eye and Face Protection," Z87.1-1968, which is incorporated by reference as specified in §1910.6, or shall be demonstrated by the employer to be equally effective.

[59 FR 16360, Apr. 6, 1994; 59 FR 33910, July 1, 1994; 61 FR 9227, Mar. 7, 1996; 61 FR 19547, May 2, 1996]

### §1910.134 Respiratory protection.

This section applies to General Industry (part 1910), Shipyards (part 1915), Marine Terminals (part 1917), Longshoring (part 1918), and Construction (part 1926).

**1910.134(a) Permissible practice. (1)** In the control of those occupational diseases caused by breathing air contaminated with harmful dusts, fogs, fumes, mists, gases, smokes, sprays, or vapors, the primary objective shall be to prevent atmospheric contamination. This shall be accomplished as far as feasible by accepted engineering control measures (for example, enclosure or confinement of the operation, general and local ventilation, and substitution of less toxic materials). When effective engineering controls are not feasible, or while they are being instituted, appropriate respirators shall be used pursuant to this section.

**1910.134(a)(2)** Respirators shall be provided by the employer when such equipment is necessary to protect the health of the employee. The employer shall provide the respirators which are applicable and suitable for the purpose intended. The employer shall be responsible for the establishment and maintenance of a respiratory protection program which shall include the requirements outlined in paragraph (c) of this section.

**1910.134(b) Definitions.** The following definitions are important terms used in the respiratory protection standard in this section.

*Air-purifying respirator* means a respirator with an air-purifying filter, cartridge, or canister that removes specific air contaminants by passing ambient air through the air-purifying element.

*Assigned protection factor (APF)* [Reserved]

# OSHA Standards for General Industry

*Atmosphere-supplying respirator* means a respirator that supplies the respirator user with breathing air from a source independent of the ambient atmosphere, and includes supplied-air respirators (SARs) and self-contained breathing apparatus (SCBA) units.

*Canister or cartridge* means a container with a filter, sorbent, or catalyst, or combination of these items, which removes specific contaminants from the air passed through the container.

*Demand respirator* means an atmosphere-supplying respirator that admits breathing air to the facepiece only when a negative pressure is created inside the facepiece by inhalation.

*Emergency situation* means any occurrence such as, but not limited to, equipment failure, rupture of containers, or failure of control equipment that may or does result in an uncontrolled significant release of an airborne contaminant.

*Employee exposure* means exposure to a concentration of an airborne contaminant that would occur if the employee were not using respiratory protection.

*End-of-service-life indicator (ESLI)* means a system that warns the respirator user of the approach of the end of adequate respiratory protection, for example, that the sorbent is approaching saturation or is no longer effective.

*Escape-only respirator* means a respirator intended to be used only for emergency exit.

*Filter or air purifying element* means a component used in respirators to remove solid or liquid aerosols from the inspired air.

*Filtering facepiece* (dust mask) means a negative pressure particulate respirator with a filter as an integral part of the facepiece or with the entire facepiece composed of the filtering medium.

*Fit factor* means a quantitative estimate of the fit of a particular respirator to a specific individual, and typically estimates the ratio of the concentration of a substance in ambient air to its concentration inside the respirator when worn.

*Fit test* means the use of a protocol to qualitatively or quantitatively evaluate the fit of a respirator on an individual. (See also Qualitative fit test QLFT and Quantitative fit test QNFT.)

*Helmet* means a rigid respiratory inlet covering that also provides head protection against impact and penetration.

*High efficiency particulate air (HEPA) filter* means a filter that is at least 99.97% efficient in removing monodisperse particles of 0.3 micrometers in diameter. The equivalent NIOSH 42 CFR 84 particulate filters are the N100, R100, and P100 filters.

*Hood* means a respiratory inlet covering that completely covers the head and neck and may also cover portions of the shoulders and torso.

*Immediately dangerous to life or health (IDLH)* means an atmosphere that poses an immediate threat to life, would cause irreversible adverse health effects, or would impair an individual's ability to escape from a dangerous atmosphere.

*Interior structural firefighting* means the physical activity of fire suppression, rescue or both, inside of buildings or enclosed structures which are involved in a fire situation beyond the incipient stage. (See 29 CFR 1910.155)

*Loose-fitting facepiece* means a respiratory inlet covering that is designed to form a partial seal with the face.

*Maximum use concentration (MUC)* [Reserved].

*Negative pressure respirator (tight fitting)* means a respirator in which the air pressure inside the facepiece is negative during inhalation with respect to the ambient air pressure outside the respirator.

*Oxygen deficient atmosphere* means an atmosphere with an oxygen content below 19.5% by volume.

*Physician or other licensed health care professional (PLHCP)* means an individual whose legally permitted scope of practice (*i.e.*, license, registration, or certification) allows him or her to independently provide, or be delegated the responsibility to provide, some or all of the health care services required by paragraph (e) of this section.

*Positive pressure respirator* means a respirator in which the pressure inside the respiratory inlet covering exceeds the ambient air pressure outside the respirator.

*Powered air-purifying respirator (PAPR)* means an air-purifying respirator that uses a blower to force the ambient air through air-purifying elements to the inlet covering.

*Pressure demand respirator* means a positive pressure atmosphere-supplying respirator that admits breathing air to the facepiece when the positive pressure is reduced inside the facepiece by inhalation.

*Qualitative fit test (QLFT)* means a pass/fail fit test to assess the adequacy of respirator fit that relies on the individual's response to the test agent.

*Quantitative fit test (QNFT)* means an assessment of the adequacy of respirator fit by numerically measuring the amount of leakage into the respirator.

*Respiratory inlet covering* means that portion of a respirator that forms the protective barrier between the user's respiratory tract and an air-purifying device or breathing air source, or both. It may be a facepiece, helmet, hood, suit, or a mouthpiece respirator with nose clamp.

*Self-contained breathing apparatus (SCBA)* means an atmosphere-supplying respirator for which the breathing air source is designed to be carried by the user.

*Service life* means the period of time that a respirator, filter or sorbent, or other respiratory equipment provides adequate protection to the wearer.

*Supplied-air respirator (SAR) or airline respirator* means an atmosphere-supplying respirator for which the source of breathing air is not designed to be carried by the user.

*This section* means this respiratory protection standard.

*Tight-fitting facepiece* means a respiratory inlet covering that forms a complete seal with the face.

*User seal check* means an action conducted by the respirator user to determine if the respirator is properly seated to the face.

**1910.134(c) Respiratory protection program.** This paragraph requires the employer to develop and implement a written respiratory protection program with required worksite-specific procedures and elements for required respirator use. The program must be administered by a suitably trained program administrator. In addition, certain program elements may be required for voluntary use to prevent potential hazards associated with the use of the respirator. The Small Entity Compliance Guide contains criteria for the selection of a program administrator and a sample program that meets the requirements of this paragraph. Copies of the Small Entity Compliance Guide will be available on or about April 8, 1998 from the Occupational Safety and Health Administration's Office of Publications, Room N 3101, 200 Constitution Avenue, NW, Washington, DC, 20210 (202-219-4667).

**1910.134(c)(1)** In any workplace where respirators are necessary to protect the health of the employee or whenever respirators are required by the employer, the employer shall establish and implement a written respiratory protection program with worksite-specific procedures. The program shall be updated as necessary to reflect those changes in workplace conditions that affect respirator use. The employer shall include in the program the following provisions of this section, as applicable:

**1910.134(c)(1)(i)** Procedures for selecting respirators for use in the workplace;

**1910.134(c)(1)(ii)** Medical evaluations of employees required to use respirators;

**1910.134(c)(1)(iii)** Fit testing procedures for tight-fitting respirators;

**1910.134(c)(1)(iv)** Procedures for proper use of respirators in routine and reasonably foreseeable emergency situations;

**1910.134(c)(1)(v)** Procedures and schedules for cleaning, disinfecting, storing, inspecting, repairing, discarding, and otherwise maintaining respirators;

**1910.134(c)(1)(vi)** Procedures to ensure adequate air quality, quantity, and flow of breathing air for atmosphere-supplying respirators;

**1910.134(c)(1)(vii)** Training of employees in the respiratory hazards to which they are potentially exposed during routine and emergency situations;

**1910.134(c)(1)(viii)** Training of employees in the proper use of respirators, including putting on and removing them, any limitations on their use, and their maintenance; and

**1910.134(c)(1)(ix)** Procedures for regularly evaluating the effectiveness of the program.

**1910.134(c)(2)** Where respirator use is not required:

**1910.134(c)(2)(i)** An employer may provide respirators at the request of employees or permit employees to use their own respirators, if the employer determines that such respirator use will not in itself create a hazard. If the employer determines that any voluntary respirator use is permissible, the employer shall provide the respirator users with the information contained in Appendix D to this section ("Information for Employees Using Respirators When Not Required Under the Standard"); and

**1910.134(c)(2)(ii)** In addition, the employer must establish and implement those elements of a written respiratory protection program necessary to ensure that any employee using a respirator voluntarily is medically able to use that respirator, and that the respirator is cleaned, stored, and maintained so that its use does not present a health hazard to the user. Exception: Employers are not required to include in a written respiratory protection program those employees whose only use of respirators involves the voluntary use of filtering facepieces (dust masks).

**1910.134(c)(3)** The employer shall designate a program administrator who is qualified by appropriate training or experience that is commensurate with the complexity of the program to administer or oversee the respiratory protection program and conduct the required evaluations of program effectiveness.

**1910.134(c)(4)** The employer shall provide respirators, training, and medical evaluations at no cost to the employee.

**1910.134(d)** *Selection of respirators.* This paragraph requires the employer to evaluate respiratory hazard(s) in the workplace, identify relevant workplace and user factors, and base respirator selection on these factors. The paragraph also specifies appropriately protective respirators for use in IDLH atmospheres, and limits the selection and use of air-purifying respirators.

**1910.134(d)(1)** *General requirements.* **(i)** The employer shall select and provide an appropriate respirator based on the respiratory hazard(s) to which the worker is exposed and workplace and user factors that affect respirator performance and reliability.

**1910.134(d)(1)(ii)** The employer shall select a NIOSH-certified respirator. The respirator shall be used in compliance with the conditions of its certification.

**1910.134(d)(1)(iii)** The employer shall identify and evaluate the respiratory hazard(s) in the workplace; this evaluation shall include a reasonable estimate of employee exposures to respiratory hazard(s) and an identification of the contaminant's chemical state and physical form. Where the employer cannot identify or reasonably estimate the employee exposure, the employer shall consider the atmosphere to be IDLH.

**1910.134(d)(1)(iv)** The employer shall select respirators from a sufficient number of respirator models and sizes so that the respirator is acceptable to, and correctly fits, the user.

**1910.134(d)(2)** *Respirators for IDLH atmospheres.* **(i)** The employer shall provide the following respirators for employee use in IDLH atmospheres:

**1910.134(d)(2)(i)(A)** A full facepiece pressure demand SCBA certified by NIOSH for a minimum service life of thirty minutes, or

**1910.134(d)(2)(i)(B)** A combination full facepiece pressure demand supplied-air respirator (SAR) with auxiliary self-contained air supply.

**1910.134(d)(2)(ii)** Respirators provided only for escape from IDLH atmospheres shall be NIOSH-certified for escape from the atmosphere in which they will be used.

**1910.134(d)(2)(iii)** All oxygen-deficient atmospheres shall be considered IDLH. Exception: If the employer demonstrates that, under all foreseeable conditions, the oxygen concentration can be maintained within the ranges specified in Table II of this section (i.e., for the altitudes set out in the table), then any atmosphere-supplying respirator may be used.

**1910.134(d)(3)** *Respirators for atmospheres that are not IDLH.* **(i)** The employer shall provide a respirator that is adequate to protect the health of the employee and ensure compliance with all other OSHA statutory and regulatory requirements, under routine and reasonably foreseeable emergency situations.

**1910.134(d)(3)(i)(A)** *Assigned Protection Factors (APFs)* [Reserved]

**1910.134(d)(3)(i)(B)** *Maximum Use Concentration (MUC)* [Reserved]

**1910.134(d)(3)(ii)** The respirator selected shall be appropriate for the chemical state and physical form of the contaminant.

**1910.134(d)(3)(iii)** For protection against gases and vapors, the employer shall provide:

**1910.134(d)(3)(iii)(A)** An atmosphere-supplying respirator, or

**1910.134(d)(3)(iii)(B)** An air-purifying respirator, provided that:

**1910.134(d)(3)(iii)(B)(1)** The respirator is equipped with an end-of-service-life indicator (ESLI) certified by NIOSH for the contaminant; or

**1910.134(d)(3)(iii)(B)(2)** If there is no ESLI appropriate for conditions in the employer's workplace, the employer implements a change schedule for canisters and cartridges that is based on objective information or data that will ensure that canisters and cartridges are changed before the end of their service life. The employer shall describe in the respirator program the information and data relied upon and the basis for the canister and cartridge change schedule and the basis for reliance on the data.

**1910.134(d)(3)(iv)** For protection against particulates, the employer shall provide:

**1910.134(d)(3)(iv)(A)** An atmosphere-supplying respirator; or

**1910.134(d)(3)(iv)(B)** An air-purifying respirator equipped with a filter certified by NIOSH under 30 CFR part 11 as a high efficiency particulate air (HEPA) filter, or an air-purifying respirator equipped with a filter certified for particulates by NIOSH under 42 CFR part 84; or

**1910.134(d)(3)(iv)(C)** For contaminants consisting primarily of particles with mass median aerodynamic diameters (MMAD) of at least 2 micrometers, an air-purifying respirator equipped with any filter certified for particulates by NIOSH.

### TABLE I—ASSIGNED PROTECTION FACTORS [RESERVED]

### TABLE II

| Altitude (ft.) | Oxygen deficient Atmospheres (% $O_2$) for which the employer may rely on atmosphere-supplying respirators |
|---|---|
| Less than 3,001 | 16.0-19.5 |
| 3,001-4,000 | 16.4-19.5 |
| 4,001-5,000 | 17.1-19.5 |
| 5,001-6,000 | 17.8-19.5 |
| 6,001-7,000 | 18.5-19.5 |
| 7,001-8,000[1] | 19.3-19.5 |

[1] Above 8,000 feet the exception does not apply. Oxygen-enriched breathing air must be supplied above 14,000 feet.

**1910.134(e)** *Medical evaluation.* Using a respirator may place a physiological burden on employees that varies with the type of respirator worn, the job and workplace conditions in which the respirator is used, and the medical status of the employee. Accordingly, this paragraph specifies the minimum requirements for medical evaluation that employers must implement to determine the employee's ability to use a respirator.

**1910.134(e)(1)** *General.* The employer shall provide a medical evaluation to determine the employee's ability to use a respirator, before the employee is fit tested or required to use the respirator in the workplace. The employer may discontinue an employee's medical evaluations when the employee is no longer required to use a respirator.

**1910.134(e)(2)** *Medical evaluation procedures.* **(i)** The employer shall identify a physician or other licensed health care professional (PLHCP) to perform medical evaluations using a medical questionnaire or an initial medical examination that obtains the same information as the medical questionnaire.

**1910.134(e)(2)(ii)** The medical evaluation shall obtain the information requested by the questionnaire in Sections 1 and 2, Part A of Appendix C of this section.

**1910.134(e)(3)** *Follow-up medical examination.* **(i)** The employer shall ensure that a follow-up medical examination is provided for an employee who gives a positive response to any question among questions 1 through 8 in Section 2, Part A of Appendix C or whose initial medical examination demonstrates the need for a follow-up medical examination.

**1910.134(e)(3)(ii)** The follow-up medical examination shall include any medical tests, consultations, or diagnostic procedures that the PLHCP deems necessary to make a final determination.

# OSHA Standards for General Industry

**1910.134(e)(4)** *Administration of the medical questionnaire and examinations.* **(i)** The medical questionnaire and examinations shall be administered confidentially during the employee's normal working hours or at a time and place convenient to the employee. The medical questionnaire shall be administered in a manner that ensures that the employee understands its content.

**1910.134(e)(4)(ii)** The employer shall provide the employee with an opportunity to discuss the questionnaire and examination results with the PLHCP.

**1910.134(e)(5)** *Supplemental information for the PLHCP.* **(i)** The following information must be provided to the PLHCP before the PLHCP makes a recommendation concerning an employee's ability to use a respirator:

**1910.134(e)(5)(i)(A)** The type and weight of the respirator to be used by the employee;

**1910.134(e)(5)(i)(B)** The duration and frequency of respirator use (including use for rescue and escape);

**1910.134(e)(5)(i)(C)** The expected physical work effort;

**1910.134(e)(5)(i)(D)** Additional protective clothing and equipment to be worn; and

**1910.134(e)(5)(i)(E)** Temperature and humidity extremes that may be encountered.

**1910.134(e)(5)(ii)** Any supplemental information provided previously to the PLHCP regarding an employee need not be provided for a subsequent medical evaluation if the information and the PLHCP remain the same.

**1910.134(e)(5)(iii)** The employer shall provide the PLHCP with a copy of the written respiratory protection program and a copy of this section.

**Note to paragraph** (e)(5)(iii): When the employer replaces a PLHCP, the employer must ensure that the new PLHCP obtains this information, either by providing the documents directly to the PLHCP or having the documents transferred from the former PLHCP to the new PLHCP. However, OSHA does not expect employers to have employees medically reevaluated solely because a new PLHCP has been selected.

**1910.134(e)(6)** *Medical determination.* In determining the employee's ability to use a respirator, the employer shall:

**1910.134(e)(6)(i)** Obtain a written recommendation regarding the employee's ability to use the respirator from the PLHCP. The recommendation shall provide only the following information:

**1910.134(e)(6)(i)(A)** Any limitations on respirator use related to the medical condition of the employee, or relating to the workplace conditions in which the respirator will be used, including whether or not the employee is medically able to use the respirator;

**1910.134(e)(6)(i)(B)** The need, if any, for follow-up medical evaluations; and

**1910.134(e)(6)(i)(C)** A statement that the PLHCP has provided the employee with a copy of the PLHCP's written recommendation.

**1910.134(e)(6)(ii)** If the respirator is a negative pressure respirator and the PLHCP finds a medical condition that may place the employee's health at increased risk if the respirator is used, the employer shall provide a PAPR if the PLHCP's medical evaluation finds that the employee can use such a respirator; if a subsequent medical evaluation finds that the employee is medically able to use a negative pressure respirator, then the employer is no longer required to provide a PAPR.

**1910.134(e)(7)** *Additional medical evaluations.* At a minimum, the employer shall provide additional medical evaluations that comply with the requirements of this section if:

**1910.134(e)(7)(i)** An employee reports medical signs or symptoms that are related to ability to use a respirator;

**1910.134(e)(7)(ii)** A PLHCP, supervisor, or the respirator program administrator informs the employer that an employee needs to be reevaluated;

**1910.134(e)(7)(iii)** Information from the respiratory protection program, including observations made during fit testing and program evaluation, indicates a need for employee reevaluation; or

**1910.134(e)(7)(iv)** A change occurs in workplace conditions (e.g., physical work effort, protective clothing, temperature) that may result in a substantial increase in the physiological burden placed on an employee.

**1910.134(f)** *Fit testing.* This paragraph requires that, before an employee may be required to use any respirator with a negative or positive pressure tight-fitting facepiece, the employee must be fit tested with the same make, model, style, and size of respirator that will be used. This paragraph specifies the kinds of fit tests allowed, the procedures for conducting them, and how the results of the fit tests must be used.

**1910.134(f)(1)** The employer shall ensure that employees using a tight-fitting facepiece respirator pass an appropriate qualitative fit test (QLFT) or quantitative fit test (QNFT) as stated in this paragraph.

**1910.134(f)(2)** The employer shall ensure that an employee using a tight-fitting facepiece respirator is fit tested prior to initial use of the respirator, whenever a different respirator facepiece (size, style, model or make) is used, and at least annually thereafter.

**1910.134(f)(3)** The employer shall conduct an additional fit test whenever the employee reports, or the employer, PLHCP, supervisor, or program administrator makes visual observations of, changes in the employee's physical condition that could affect respirator fit. Such conditions include, but are not limited to, facial scarring, dental changes, cosmetic surgery, or an obvious change in body weight.

**1910.134(f)(4)** If after passing a QLFT or QNFT, the employee subsequently notifies the employer, program administrator, supervisor, or PLHCP that the fit of the respirator is unacceptable, the employee shall be given a reasonable opportunity to select a different respirator facepiece and to be retested.

**1910.134(f)(5)** The fit test shall be administered using an OSHA-accepted QLFT or QNFT protocol. The OSHA-accepted QLFT and QNFT protocols and procedures are contained in Appendix A of this section.

**1910.134(f)(6)** QLFT may only be used to fit test negative pressure air-purifying respirators that must achieve a fit factor of 100 or less.

**1910.134(f)(7)** If the fit factor, as determined through an OSHA-accepted QNFT protocol, is equal to or greater than 100 for tight-fitting half facepieces, or equal to or greater than 500 for tight-fitting full facepieces, the QNFT has been passed with that respirator.

**1910.134(f)(8)** Fit testing of tight-fitting atmosphere-supplying respirators and tight-fitting powered air-purifying respirators shall be accomplished by performing quantitative or qualitative fit testing in the negative pressure mode, regardless of the mode of operation (negative or positive pressure) that is used for respiratory protection.

**1910.134(f)(8)(i)** Qualitative fit testing of these respirators shall be accomplished by temporarily converting the respirator user's actual facepiece into a negative pressure respirator with appropriate filters, or by using an identical negative pressure air-purifying respirator facepiece with the same sealing surfaces as a surrogate for the atmosphere-supplying or powered air-purifying respirator facepiece.

**1910.134(f)(8)(ii)** Quantitative fit testing of these respirators shall be accomplished by modifying the facepiece to allow sampling inside the facepiece in the breathing zone of the user, midway between the nose and mouth. This requirement shall be accomplished by installing a permanent sampling probe onto a surrogate facepiece, or by using a sampling adapter designed to temporarily provide a means of sampling air from inside the facepiece.

**1910.134(f)(8)(iii)** Any modifications to the respirator facepiece for fit testing shall be completely removed, and the facepiece restored to NIOSH-approved configuration, before that facepiece can be used in the workplace.

**1910.134(g)** *Use of respirators.* This paragraph requires employers to establish and implement procedures for the proper use of respirators. These requirements include prohibiting conditions that may result in facepiece seal leakage, preventing employees from removing respirators in hazardous environments, taking actions to ensure continued effective respirator operation throughout the work shift, and establishing procedures for the use of respirators in IDLH atmospheres or in interior structural firefighting situations.

**1910.134(g)(1)** *Facepiece seal protection.* **(i)** The employer shall not permit respirators with tight-fitting facepieces to be worn by employees who have:

**1910.134(g)(1)(i)(A)** Facial hair that comes between the sealing surface of the facepiece and the face or that interferes with valve function; or

**1910.134(g)(1)(i)(B)** Any condition that interferes with the face-to-facepiece seal or valve function.

**1910.134(g)(1)(ii)** If an employee wears corrective glasses or goggles or other personal protective equipment, the employer shall ensure that such equipment is worn in a manner that does not interfere with the seal of the facepiece to the face of the user.

**1910.134(g)(1)(iii)** For all tight-fitting respirators, the employer shall ensure that employees perform a user seal check each time they put on the respirator using the procedures in Appendix B-1 or procedures recommended by the respirator manufacturer that the employer demonstrates are as effective as those in Appendix B-1 of this section.

**1910.134(g)(2)** *Continuing respirator effectiveness.* **(i)** Appropriate surveillance shall be maintained of work area conditions and degree of employee exposure or stress. When there is a change in work area conditions or degree of employee exposure or stress that may affect respirator effectiveness, the employer shall reevaluate the continued effectiveness of the respirator.

**1910.134(g)(2)(ii)** The employer shall ensure that employees leave the respirator use area:

**1910.134(g)(2)(ii)(A)** To wash their faces and respirator facepieces as necessary to prevent eye or skin irritation associated with respirator use; or

**1910.134(g)(2)(ii)(B)** If they detect vapor or gas breakthrough, changes in breathing resistance, or leakage of the facepiece; or

**1910.134(g)(2)(ii)(C)** To replace the respirator or the filter, cartridge, or canister elements.

**1910.134(g)(2)(iii)** If the employee detects vapor or gas breakthrough, changes in breathing resistance, or leakage of the facepiece, the employer must replace or repair the respirator before allowing the employee to return to the work area.

**1910.134(g)(3)** *Procedures for IDLH atmospheres.* For all IDLH atmospheres, the employer shall ensure that:

**1910.134(g)(3)(i)** One employee or, when needed, more than one employee is located outside the IDLH atmosphere;

**1910.134(g)(3)(ii)** Visual, voice, or signal line communication is maintained between the employee(s) in the IDLH atmosphere and the employee(s) located outside the IDLH atmosphere;

**1910.134(g)(3)(iii)** The employee(s) located outside the IDLH atmosphere are trained and equipped to provide effective emergency rescue;

**1910.134(g)(3)(iv)** The employer or designee is notified before the employee(s) located outside the IDLH atmosphere enter the IDLH atmosphere to provide emergency rescue;

**1910.134(g)(3)(v)** The employer or designee authorized to do so by the employer, once notified, provides necessary assistance appropriate to the situation;

**1910.134(g)(3)(vi)** Employee(s) located outside the IDLH atmospheres are equipped with:

**1910.134(g)(3)(vi)(A)** Pressure demand or other positive pressure SCBAs, or a pressure demand or other positive pressure supplied-air respirator with auxiliary SCBA; and either

**1910.134(g)(3)(vi)(B)** Appropriate retrieval equipment for removing the employee(s) who enter(s) these hazardous atmospheres where retrieval equipment would contribute to the rescue of the employee(s) and would not increase the overall risk resulting from entry; or

**1910.134(g)(3)(vi)(C)** Equivalent means for rescue where retrieval equipment is not required under paragraph (g)(3)(vi)(B).

**1910.134(g)(4)** *Procedures for interior structural firefighting.* In addition to the requirements set forth under paragraph (g)(3), in interior structural fires, the employer shall ensure that:

**1910.134(g)(4)(i)** At least two employees enter the IDLH atmosphere and remain in visual or voice contact with one another at all times;

**1910.134(g)(4)(ii)** At least two employees are located outside the IDLH atmosphere; and

**1910.134(g)(4)(iii)** All employees engaged in interior structural firefighting use SCBAs.

**Note 1 to paragraph** (g): One of the two individuals located outside the IDLH atmosphere may be assigned to an additional role, such as incident commander in charge of the emergency or safety officer, so long as this individual is able to perform assistance or rescue activities without jeopardizing the safety or health of any firefighter working at the incident.

**Note 2 to paragraph** (g): Nothing in this section is meant to preclude firefighters from performing emergency rescue activities before an entire team has assembled.

**1910.134(h)** *Maintenance and care of respirators.* This paragraph requires the employer to provide for the cleaning and disinfecting, storage, inspection, and repair of respirators used by employees.

**1910.134(h)(1)** *Cleaning and disinfecting.* The employer shall provide each respirator user with a respirator that is clean, sanitary, and in good working order. The employer shall ensure that respirators are cleaned and disinfected using the procedures in Appendix B-2 of this section, or procedures recommended by the respirator manufacturer, provided that such procedures are of equivalent effectiveness. The respirators shall be cleaned and disinfected at the following intervals:

**1910.134(h)(1)(i)** Respirators issued for the exclusive use of an employee shall be cleaned and disinfected as often as necessary to be maintained in a sanitary condition;

**1910.134(h)(1)(ii)** Respirators issued to more than one employee shall be cleaned and disinfected before being worn by different individuals;

**1910.134(h)(1)(iii)** Respirators maintained for emergency use shall be cleaned and disinfected after each use; and

**1910.134(h)(1)(iv)** Respirators used in fit testing and training shall be cleaned and disinfected after each use.

**1910.134(h)(2)** *Storage.* The employer shall ensure that respirators are stored as follows:

**1910.134(h)(2)(i)** All respirators shall be stored to protect them from damage, contamination, dust, sunlight, extreme temperatures, excessive moisture, and damaging chemicals, and they shall be packed or stored to prevent deformation of the facepiece and exhalation valve.

**1910.134(h)(2)(ii)** In addition to the requirements of paragraph (h)(2)(i) of this section, emergency respirators shall be:

**1910.134(h)(2)(ii)(A)** Kept accessible to the work area;

**1910.134(h)(2)(ii)(B)** Stored in compartments or in covers that are clearly marked as containing emergency respirators; and

**1910.134(h)(2)(ii)(C)** Stored in accordance with any applicable manufacturer instructions.

**1910.134(h)(3)** *Inspection.* **(i)** The employer shall ensure that respirators are inspected as follows:

**1910.134(h)(3)(i)(A)** All respirators used in routine situations shall be inspected before each use and during cleaning;

**1910.134(h)(3)(i)(B)** All respirators maintained for use in emergency situations shall be inspected at least monthly and in accordance with the manufacturer's recommendations, and shall be checked for proper function before and after each use; and

**1910.134(h)(3)(ii)(C)** Emergency escape-only respirators shall be inspected before being carried into the workplace for use.

**1910.134(h)(3)(ii)** The employer shall ensure that respirator inspections include the following:

**1910.134(h)(3)(ii)(A)** A check of respirator function, tightness of connections, and the condition of the various parts including, but not limited to, the facepiece, head straps, valves, connecting tube, and cartridges, canisters or filters; and

**1910.134(h)(3)(ii)(B)** A check of elastomeric parts for pliability and signs of deterioration.

**1910.134(h)(3)(iii)** In addition to the requirements of paragraphs (h)(3)(i) and (ii) of this section, self-contained breathing apparatus shall be inspected monthly. Air and oxygen cylinders shall be maintained in a fully charged state and shall be recharged when the pressure falls to 90% of the manufacturer's recommended pressure level. The employer shall determine that the regulator and warning devices function properly.

**1910.134(h)(3)(iv)** For respirators maintained for emergency use, the employer shall:

**1910.134(h)(3)(iv)(A)** Certify the respirator by documenting the date the inspection was performed, the name (or signature) of the person who made the inspection, the findings, required remedial action, and a serial number or other means of identifying the inspected respirator; and

**1910.134(h)(3)(iv)(B)** Provide this information on a tag or label that is attached to the storage compartment for the respirator, is kept with the respirator, or is included in inspection reports stored as paper or electronic files. This information shall be maintained until replaced following a subsequent certification.

**1910.134(h)(4)** *Repairs.* The employer shall ensure that respirators that fail an inspection or are otherwise found to be defective are removed from service, and are discarded or repaired or adjusted in accordance with the following procedures:

**1910.134(h)(4)(i)** Repairs or adjustments to respirators are to be made only by persons appropriately trained to perform such operations and shall use only the respirator manufacturer's NIOSH-approved parts designed for the respirator;

**1910.134(h)(4)(ii)** Repairs shall be made according to the manufacturer's recommendations and specifications for the type and extent of repairs to be performed; and

**1910.134(h)(4)(iii)** Reducing and admission valves, regulators, and alarms shall be adjusted or repaired only by the manufacturer or a technician trained by the manufacturer.

**1910.134(i)** *Breathing air quality and use.* This paragraph requires the employer to provide employees using atmosphere-supplying respirators (supplied-air and SCBA) with breathing gases of high purity.

**1910.134(i)(1)** The employer shall ensure that compressed air, compressed oxygen, liquid air, and liquid oxygen used for respiration accords with the following specifications:

**1910.134(i)(1)(i)** Compressed and liquid oxygen shall meet the United States Pharmacopoeia requirements for medical or breathing oxygen; and

**1910.134(i)(1)(ii)** Compressed breathing air shall meet at least the requirements for Grade D breathing air described in ANSI/Compressed Gas Association Commodity Specification for Air, G-7.1-1989, to include:

**1910.134(i)(1)(ii)(A)** Oxygen content (v/v) of 19.5-23.5%;

**1910.134(i)(1)(ii)(B)** Hydrocarbon (condensed) content of 5 milligrams per cubic meter of air or less;

**1910.134(i)(1)(ii)(C)** Carbon monoxide (CO) content of 10 ppm or less;

**1910.134(i)(1)(ii)(D)** Carbon dioxide content of 1,000 ppm or less; and

**1910.134(i)(1)(ii)(E)** Lack of noticeable odor.

**1910.134(i)(2)** The employer shall ensure that compressed oxygen is not used in atmosphere-supplying respirators that have previously used compressed air.

**1910.134(i)(3)** The employer shall ensure that oxygen concentrations greater than 23.5% are used only in equipment designed for oxygen service or distribution.

**1910.134(i)(4)** The employer shall ensure that cylinders used to supply breathing air to respirators meet the following requirements:

**1910.134(i)(4)(i)** Cylinders are tested and maintained as prescribed in the Shipping Container Specification Regulations of the Department of Transportation (49 CFR part 173 and part 178);

**1910.134(i)(4)(ii)** Cylinders of purchased breathing air have a certificate of analysis from the supplier that the breathing air meets the requirements for Grade D breathing air; and

**1910.134(i)(4)(iii)** The moisture content in the cylinder does not exceed a dew point of -50 °F (-45.6 °C) at 1 atmosphere pressure.

**1910.134(i)(5)** The employer shall ensure that compressors used to supply breathing air to respirators are constructed and situated so as to:

**1910.134(i)(5)(i)** Prevent entry of contaminated air into the air-supply system;

**1910.134(i)(5)(ii)** Minimize moisture content so that the dew point at 1 atmosphere pressure is 10 degrees F (5.56 °C) below the ambient temperature;

**1910.134(i)(5)(iii)** Have suitable in-line air-purifying sorbent beds and filters to further ensure breathing air quality. Sorbent beds and filters shall be maintained and replaced or refurbished periodically following the manufacturer's instructions.

**1910.134(i)(5)(iv)** Have a tag containing the most recent change date and the signature of the person authorized by the employer to perform the change. The tag shall be maintained at the compressor.

**1910.134(i)(6)** For compressors that are not oil-lubricated, the employer shall ensure that carbon monoxide levels in the breathing air do not exceed 10 ppm.

**1910.134(i)(7)** For oil-lubricated compressors, the employer shall use a high-temperature or carbon monoxide alarm, or both, to monitor carbon monoxide levels. If only high-temperature alarms are used, the air supply shall be monitored at intervals sufficient to prevent carbon monoxide in the breathing air from exceeding 10 ppm.

**1910.134(i)(8)** The employer shall ensure that breathing air couplings are incompatible with outlets for nonrespirable worksite air or other gas systems. No asphyxiating substance shall be introduced into breathing air lines.

**1910.134(i)(9)** The employer shall use breathing gas containers marked in accordance with the NIOSH respirator certification standard, 42 CFR part 84.

**1910.134(j)** *Identification of filters, cartridges, and canisters.* The employer shall ensure that all filters, cartridges and canisters used in the workplace are labeled and color coded with the NIOSH approval label and that the label is not removed and remains legible.

**1910.134(k)** *Training and information.* This paragraph requires the employer to provide effective training to employees who are required to use respirators. The training must be comprehensive, understandable, and recur annually, and more often if necessary. This paragraph also requires the employer to provide the basic information on respirators in Appendix D of this section to employees who wear respirators when not required by this section or by the employer to do so.

**1910.134(k)(1)** The employer shall ensure that each employee can demonstrate knowledge of at least the following:

**1910.134(k)(1)(i)** Why the respirator is necessary and how improper fit, usage, or maintenance can compromise the protective effect of the respirator;

**1910.134(k)(1)(ii)** What the limitations and capabilities of the respirator are;

**1910.134(k)(1)(iii)** How to use the respirator effectively in emergency situations, including situations in which the respirator malfunctions;

**1910.134(k)(1)(iv)** How to inspect, put on and remove, use, and check the seals of the respirator;

**1910.134(k)(1)(v)** What the procedures are for maintenance and storage of the respirator;

**1910.134(k)(1)(vi)** How to recognize medical signs and symptoms that may limit or prevent the effective use of respirators; and

**1910.134(k)(1)(vii)** The general requirements of this section.

**1910.134(k)(2)** The training shall be conducted in a manner that is understandable to the employee.

**1910.134(k)(3)** The employer shall provide the training prior to requiring the employee to use a respirator in the workplace.

**1910.134(k)(4)** An employer who is able to demonstrate that a new employee has received training within the last 12 months that addresses the elements specified in paragraph (k)(1)(i) through (vii) is not required to repeat such training provided that, as required by paragraph (k)(1), the employee can demonstrate knowledge of those element(s). Previous training not repeated initially by the employer must be provided no later than 12 months from the date of the previous training.

**1910.134(k)(5)** Retraining shall be administered annually, and when the following situations occur:

**1910.134(k)(5)(i)** Changes in the workplace or the type of respirator render previous training obsolete;

**1910.134(k)(5)(ii)** Inadequacies in the employee's knowledge or use of the respirator indicate that the employee has not retained the requisite understanding or skill; or

**1910.134(k)(5)(iii)** Any other situation arises in which retraining appears necessary to ensure safe respirator use.

**1910.134(k)(6)** The basic advisory information on respirators, as presented in Appendix D of this section, shall be provided by the employer in any written or oral format, to employees who wear respirators when such use is not required by this section or by the employer.

**1910.134(l)** *Program evaluation.* This section requires the employer to conduct evaluations of the workplace to ensure that the written respiratory protection program is being properly implemented, and to consult employees to ensure that they are using the respirators properly.

**1910.134(l)(1)** The employer shall conduct evaluations of the workplace as necessary to ensure that the provisions of the current written program are being effectively implemented and that it continues to be effective.

**1910.134(l)(2)** The employer shall regularly consult employees required to use respirators to assess the employees' views on program effectiveness and to identify any problems. Any problems that are identified during this assessment shall be corrected. Factors to be assessed include, but are not limited to:

**1910.134(l)(2)(i)** Respirator fit (including the ability to use the respirator without interfering with effective workplace performance);

**1910.134(l)(2)(ii)** Appropriate respirator selection for the hazards to which the employee is exposed;

**1910.134(l)(2)(iii)** Proper respirator use under the workplace conditions the employee encounters; and

**1910.134(l)(2)(iv)** Proper respirator maintenance.

**1910.134(m)** *Recordkeeping.* This section requires the employer to establish and retain written information regarding medical evaluations, fit testing, and the respirator program. This information will facilitate employee involvement in the respirator program, assist the employer in auditing the adequacy of the program, and provide a record for compliance determinations by OSHA.

**1910.134(m)(1)** *Medical evaluation.* Records of medical evaluations required by this section must be retained and made available in accordance with 29 CFR 1910.1020.

**1910.134(m)(2)** *Fit testing.* **(i)** The employer shall establish a record of the qualitative and quantitative fit tests administered to an employee including:

**1910.134(m)(2)(i)(A)** The name or identification of the employee tested;

**1910.134(m)(2)(i)(B)** Type of fit test performed;

**1910.134(m)(2)(i)(C)** Specific make, model, style, and size of respirator tested;

**1910.134(m)(2)(i)(D)** Date of test; and

**1910.134(m)(2)(i)(E)** The pass/fail results for QLFTs or the fit factor and strip chart recording or other recording of the test results for QNFTs.

**1910.134(m)(2)(ii)** Fit test records shall be retained for respirator users until the next fit test is administered.

**1910.134(m)(3)** A written copy of the current respirator program shall be retained by the employer.

**1910.134(m)(4)** Written materials required to be retained under this paragraph shall be made available upon request to affected employees and to the Assistant Secretary or designee for examination and copying.

**1910.134(n)** [Reserved.]

**1910.134(o)** *Appendices.* **(1)** Compliance with Appendix A, Appendix B-1, Appendix B-2, and Appendix C of this section is mandatory.

**1910.134(o)(2)** Appendix D of this section is non-mandatory and is not intended to create any additional obligations not otherwise imposed or to detract from any existing obligations.

## APPENDIX A TO §1910.134— FIT TESTING PROCEDURES (MANDATORY)

### Part I. OSHA-Accepted Fit Test Protocols

#### A. Fit Testing Procedures—General Requirements

The employer shall conduct fit testing using the following procedures. The requirements in this appendix apply to all OSHA-accepted fit test methods, both QLFT and QNFT.

1. The test subject shall be allowed to pick the most acceptable respirator from a sufficient number of respirator models and sizes so that the respirator is acceptable to, and correctly fits, the user.

2. Prior to the selection process, the test subject shall be shown how to put on a respirator, how it should be positioned on the face, how to set strap tension and how to determine an acceptable fit. A mirror shall be available to assist the subject in evaluating the fit and positioning of the respirator. This instruction may not constitute the subject's formal training on respirator use, because it is only a review.

3. The test subject shall be informed that he/she is being asked to select the respirator that provides the most acceptable fit. Each respirator represents a different size and shape, and if fitted and used properly, will provide adequate protection.

4. The test subject shall be instructed to hold each chosen facepiece up to the face and eliminate those that obviously do not give an acceptable fit.

5. The more acceptable facepieces are noted in case the one selected proves unacceptable; the most comfortable mask is donned and worn at least five minutes to assess comfort. Assistance in assessing comfort can be given by discussing the points in the following item A.6. If the test subject is not familiar with using a particular respirator, the test subject shall be directed to don the mask several times and to adjust the straps each time to become adept at setting proper tension on the straps.

6. Assessment of comfort shall include a review of the following points with the test subject and allowing the test subject adequate time to determine the comfort of the respirator:

(a) Position of the mask on the nose;

(b) Room for eye protection;

(c) Room to talk;

(d) Position of mask on face and cheeks;

7. The following criteria shall be used to help determine the adequacy of the respirator fit:

(a) Chin properly placed;

(b) Adequate strap tension, not overly tightened;

(c) Fit across nose bridge;

(d) Respirator of proper size to span distance from nose to chin;

(e) Tendency of respirator to slip;

(f) Self-observation in mirror to evaluate fit and respirator position.

8. The test subject shall conduct a user seal check, either the negative and positive pressure seal checks described in Appendix B-1 of this section or those recommended by the respirator manufacturer which provide equivalent protection to the procedures in Appendix B-1. Before conducting the negative and positive pressure checks, the subject shall be told to seat the mask on the face by moving the head from side-to-side and up and down slowly while taking in a few slow deep breaths. Another facepiece shall be selected and retested if the test subject fails the user seal check tests.

9. The test shall not be conducted if there is any hair growth between the skin and the facepiece sealing surface, such as stubble beard growth, beard, mustache or sideburns which cross the respirator sealing surface. Any type of apparel which interferes with a satisfactory fit shall be altered or removed.

10. If a test subject exhibits difficulty in breathing during the tests, she or he shall be referred to a physician or other licensed health care professional, as appropriate, to determine whether the test subject can wear a respirator while performing her or his duties.

11. If the employee finds the fit of the respirator unacceptable, the test subject shall be given the opportunity to select a different respirator and to be retested.

12. Exercise regimen. Prior to the commencement of the fit test, the test subject shall be given a description of the fit test and the test subject's responsibilities during the test procedure. The description of the process shall include a description of the test exercises that the subject will be performing. The respirator to be tested shall be worn for at least 5 minutes before the start of the fit test.

13. The fit test shall be performed while the test subject is wearing any applicable safety equipment that may be worn during actual respirator use which could interfere with respirator fit.

14. Test Exercises. (a) Employers must perform the following test exercises for all fit testing methods prescribed in this appendix, except for the CNP quantitative fit testing protocol and the CNP REDON quantitative fit testing

# OSHA Standards for General Industry

protocol. For these two protocols, employers must ensure that the test subjects (i.e., employees) perform the exercise procedure specified in Part I.C.4(b) of this appendix for the CNP quantitative fit testing protocol, or the exercise procedure described in Part I.C.5(b) of this appendix for the CNP REDON quantitative fit-testing protocol. For the remaining fit testing methods, employers must ensure that employees perform the test exercises in the appropriate test environment in the following manner:

(1) Normal breathing. In a normal standing position, without talking, the subject shall breathe normally.

(2) Deep breathing. In a normal standing position, the subject shall breathe slowly and deeply, taking caution so as not to hyperventilate.

(3) Turning head side to side. Standing in place, the subject shall slowly turn his/her head from side to side between the extreme positions on each side. The head shall be held at each extreme momentarily so the subject can inhale at each side.

(4) Moving head up and down. Standing in place, the subject shall slowly move his/her head up and down. The subject shall be instructed to inhale in the up position (i.e., when looking toward the ceiling).

(5) Talking. The subject shall talk out loud slowly and loud enough so as to be heard clearly by the test conductor. The subject can read from a prepared text such as the Rainbow Passage, count backward from 100, or recite a memorized poem or song.

### Rainbow Passage

When the sunlight strikes raindrops in the air, they act like a prism and form a rainbow. The rainbow is a division of white light into many beautiful colors. These take the shape of a long round arch, with its path high above, and its two ends apparently beyond the horizon. There is, according to legend, a boiling pot of gold at one end. People look, but no one ever finds it. When a man looks for something beyond reach, his friends say he is looking for the pot of gold at the end of the rainbow.

(6) Grimace. The test subject shall grimace by smiling or frowning. (This applies only to QNFT testing; it is not performed for QLFT)

(7) Bending over. The test subject shall bend at the waist as if he/she were to touch his/her toes. Jogging in place shall be substituted for this exercise in those test environments such as shroud type QNFT or QLFT units that do not permit bending over at the waist.

(8) Normal breathing. Same as exercise (1).

(b) Each test exercise shall be performed for one minute except for the grimace exercise which shall be performed for 15 seconds. The test subject shall be questioned by the test conductor regarding the comfort of the respirator upon completion of the protocol. If it has become unacceptable, another model of respirator shall be tried. The respirator shall not be adjusted once the fit test exercises begin. Any adjustment voids the test, and the fit test must be repeated.

## B. Qualitative Fit Test (QLFT) Protocols

### 1. General

(a) The employer shall ensure that persons administering QLFT are able to prepare test solutions, calibrate equipment and perform tests properly, recognize invalid tests, and ensure that test equipment is in proper working order.

(b) The employer shall ensure that QLFT equipment is kept clean and well maintained so as to operate within the parameters for which it was designed.

### 2. Isoamyl Acetate Protocol

**NOTE:** This protocol is not appropriate to use for the fit testing of particulate respirators. If used to fit test particulate respirators, the respirator must be equipped with an organic vapor filter.

### (a) Odor Threshold Screening

Odor threshold screening, performed without wearing a respirator, is intended to determine if the individual tested can detect the odor of isoamyl acetate at low levels.

(1) Three 1 liter glass jars with metal lids are required.

(2) Odor-free water (e.g., distilled or spring water) at approximately 25 °C (77 °F) shall be used for the solutions.

(3) The isoamyl acetate (IAA) (also known at isopentyl acetate) stock solution is prepared by adding 1 ml of pure IAA to 800 ml of odor-free water in a 1 liter jar, closing the lid and shaking for 30 seconds. A new solution shall be prepared at least weekly.

(4) The screening test shall be conducted in a room separate from the room used for actual fit testing. The two rooms shall be well-ventilated to prevent the odor of IAA from becoming evident in the general room air where testing takes place.

(5) The odor test solution is prepared in a second jar by placing 0.4 ml of the stock solution into 500 ml of odor-free water using a clean dropper or pipette. The solution shall be shaken for 30 seconds and allowed to stand for two to three minutes so that the IAA concentration above the liquid may reach equilibrium. This solution shall be used for only one day.

(6) A test blank shall be prepared in a third jar by adding 500 cc of odor-free water.

(7) The odor test and test blank jar lids shall be labeled (e.g., 1 and 2) for jar identification. Labels shall be placed on the lids so that they can be peeled off periodically and switched to maintain the integrity of the test.

(8) The following instruction shall be typed on a card and placed on the table in front of the two test jars (i.e., 1 and 2): "The purpose of this test is to determine if you can smell banana oil at a low concentration. The two bottles in front of you contain water. One of these bottles also contains a small amount of banana oil. Be sure the covers are on tight, then shake each bottle for two seconds. Unscrew the lid of each bottle, one at a time, and sniff at the mouth of the bottle. Indicate to the test conductor which bottle contains banana oil."

(9) The mixtures used in the IAA odor detection test shall be prepared in an area separate from where the test is performed, in order to prevent olfactory fatigue in the subject.

(10) If the test subject is unable to correctly identify the jar containing the odor test solution, the IAA qualitative fit test shall not be performed.

(11) If the test subject correctly identifies the jar containing the odor test solution, the test subject may proceed to respirator selection and fit testing.

### (b) Isoamyl Acetate Fit Test

(1) The fit test chamber shall be a clear 55-gallon drum liner suspended inverted over a 2-foot diameter frame so that the top of the chamber is about 6 inches above the test subject's head. If no drum liner is available, a similar chamber shall be constructed using plastic sheeting. The inside top center of the chamber shall have a small hook attached.

(2) Each respirator used for the fitting and fit testing shall be equipped with organic vapor cartridges or offer protection against organic vapors.

(3) After selecting, donning, and properly adjusting a respirator, the test subject shall wear it to the fit testing room. This room shall be separate from the room used for odor threshold screening and respirator selection, and shall be well-ventilated, as by an exhaust fan or lab hood, to prevent general room contamination.

(4) A copy of the test exercises and any prepared text from which the subject is to read shall be taped to the inside of the test chamber.

(5) Upon entering the test chamber, the test subject shall be given a 6-inch by 5-inch piece of paper towel, or other porous, absorbent, single-ply material, folded in half and wetted with 0.75 ml of pure IAA. The test subject shall hang the wet towel on the hook at the top of the chamber. An IAA test swab or ampule may be substituted for the IAA wetted paper towel provided it has been demonstrated that the alternative IAA source will generate an IAA test atmosphere with a concentration equivalent to that generated by the paper towel method.

(6) Allow two minutes for the IAA test concentration to stabilize before starting the fit test exercises. This would be an appropriate time to talk with the test subject; to explain the fit test, the importance of his/her cooperation, and the purpose for the test exercises; or to demonstrate some of the exercises.

(7) If at any time during the test, the subject detects the banana-like odor of IAA, the test is failed. The subject shall quickly exit from the test chamber and leave the test area to avoid olfactory fatigue.

(8) If the test is failed, the subject shall return to the selection room and remove the respirator. The test subject shall repeat the odor sensitivity test, select and put on another respirator, return to the test area and again begin the fit test procedure described in (b) (1) through (7) above. The process continues until a respirator that fits well has been found. Should the odor sensitivity test be failed, the subject shall wait at least

5 minutes before retesting. Odor sensitivity will usually have returned by this time.

(9) If the subject passes the test, the efficiency of the test procedure shall be demonstrated by having the subject break the respirator face seal and take a breath before exiting the chamber.

(10) When the test subject leaves the chamber, the subject shall remove the saturated towel and return it to the person conducting the test, so that there is no significant IAA concentration buildup in the chamber during subsequent tests. The used towels shall be kept in a self-sealing plastic bag to keep the test area from being contaminated.

### 3. Saccharin Solution Aerosol Protocol

The entire screening and testing procedure shall be explained to the test subject prior to the conduct of the screening test.

(a) Taste threshold screening. The saccharin taste threshold screening, performed without wearing a respirator, is intended to determine whether the individual being tested can detect the taste of saccharin.

(1) During threshold screening as well as during fit testing, subjects shall wear an enclosure about the head and shoulders that is approximately 12 inches in diameter by 14 inches tall with at least the front portion clear and that allows free movements of the head when a respirator is worn. An enclosure substantially similar to the 3M hood assembly, parts # FT 14 and # FT 15 combined, is adequate.

(2) The test enclosure shall have a 3/4-inch (1.9 cm) hole in front of the test subject's nose and mouth area to accommodate the nebulizer nozzle.

(3) The test subject shall don the test enclosure. Throughout the threshold screening test, the test subject shall breathe through his/her slightly open mouth with tongue extended. The subject is instructed to report when he/she detects a sweet taste.

(4) Using a DeVilbiss Model 40 Inhalation Medication Nebulizer or equivalent, the test conductor shall spray the threshold check solution into the enclosure. The nozzle is directed away from the nose and mouth of the person. This nebulizer shall be clearly marked to distinguish it from the fit test solution nebulizer.

(5) The threshold check solution is prepared by dissolving 0.83 gram of sodium saccharin USP in 100 ml of warm water. It can be prepared by putting 1 ml of the fit test solution (see (b)(5) below) in 100 ml of distilled water.

(6) To produce the aerosol, the nebulizer bulb is firmly squeezed so that it collapses completely, then released and allowed to fully expand.

(7) Ten squeezes are repeated rapidly and then the test subject is asked whether the saccharin can be tasted. If the test subject reports tasting the sweet taste during the ten squeezes, the screening test is completed. The taste threshold is noted as ten regardless of the number of squeezes actually completed.

(8) If the first response is negative, ten more squeezes are repeated rapidly and the test subject is again asked whether the saccharin is tasted. If the test subject reports tasting the sweet taste during the second ten squeezes, the screening test is completed. The taste threshold is noted as twenty regardless of the number of squeezes actually completed.

(9) If the second response is negative, ten more squeezes are repeated rapidly and the test subject is again asked whether the saccharin is tasted. If the test subject reports tasting the sweet taste during the third set of ten squeezes, the screening test is completed. The taste threshold is noted as thirty regardless of the number of squeezes actually completed.

(10) The test conductor will take note of the number of squeezes required to solicit a taste response.

(11) If the saccharin is not tasted after 30 squeezes (step 10), the test subject is unable to taste saccharin and may not perform the saccharin fit test.

**Note to paragraph 3.** (a): If the test subject eats or drinks something sweet before the screening test, he/she may be unable to taste the weak saccharin solution.

(12) If a taste response is elicited, the test subject shall be asked to take note of the taste for reference in the fit test.

(13) Correct use of the nebulizer means that approximately 1 ml of liquid is used at a time in the nebulizer body.

(14) The nebulizer shall be thoroughly rinsed in water, shaken dry, and refilled at least each morning and afternoon or at least every four hours.

(b) Saccharin solution aerosol fit test procedure.

(1) The test subject may not eat, drink (except plain water), smoke, or chew gum for 15 minutes before the test.

(2) The fit test uses the same enclosure described in 3. (a) above.

(3) The test subject shall don the enclosure while wearing the respirator selected in section I. A. of this appendix. The respirator shall be properly adjusted and equipped with a particulate filter(s).

(4) A second DeVilbiss Model 40 Inhalation Medication Nebulizer or equivalent is used to spray the fit test solution into the enclosure. This nebulizer shall be clearly marked to distinguish it from the screening test solution nebulizer.

(5) The fit test solution is prepared by adding 83 grams of sodium saccharin to 100 ml of warm water.

(6) As before, the test subject shall breathe through the slightly open mouth with tongue extended, and report if he/she tastes the sweet taste of saccharin.

(7) The nebulizer is inserted into the hole in the front of the enclosure and an initial concentration of saccharin fit test solution is sprayed into the enclosure using the same number of squeezes (either 10, 20 or 30 squeezes) based on the number of squeezes required to elicit a taste response as noted during the screening test. A minimum of 10 squeezes is required.

(8) After generating the aerosol, the test subject shall be instructed to perform the exercises in section I. A. 14. of this appendix.

(9) Every 30 seconds the aerosol concentration shall be replenished using one half the original number of squeezes used initially (e.g., 5, 10 or 15).

(10) The test subject shall indicate to the test conductor if at any time during the fit test the taste of saccharin is detected. If the test subject does not report tasting the saccharin, the test is passed.

(11) If the taste of saccharin is detected, the fit is deemed unsatisfactory and the test is failed. A different respirator shall be tried and the entire test procedure is repeated (taste threshold screening and fit testing).

(12) Since the nebulizer has a tendency to clog during use, the test operator must make periodic checks of the nebulizer to ensure that it is not clogged. If clogging is found at the end of the test session, the test is invalid.

### 4. Bitrex™ (Denatonium Benzoate) Solution Aerosol Qualitative Fit Test Protocol

The Bitrex™ (Denatonium benzoate) solution aerosol QLFT protocol uses the published saccharin test protocol because that protocol is widely accepted. Bitrex is routinely used as a taste aversion agent in household liquids which children should not be drinking and is endorsed by the American Medical Association, the National Safety Council, and the American Association of Poison Control Centers. The entire screening and testing procedure shall be explained to the test subject prior to the conduct of the screening test.

(a) Taste Threshold Screening.

The Bitrex taste threshold screening, performed without wearing a respirator, is intended to determine whether the individual being tested can detect the taste of Bitrex.

(1) During threshold screening as well as during fit testing, subjects shall wear an enclosure about the head and shoulders that is approximately 12 inches (30.5 cm) in diameter by 14 inches (35.6 cm) tall. The front portion of the enclosure shall be clear from the respirator and allow free movement of the head when a respirator is worn. An enclosure substantially similar to the 3M hood assembly, parts # FT 14 and # FT 15 combined, is adequate.

(2) The test enclosure shall have a 3/4-inch (1.9 cm) hole in front of the test subject's nose and mouth area to accommodate the nebulizer nozzle.

(3) The test subject shall don the test enclosure. Throughout the threshold screening test, the test subject shall breathe through his or her slightly open mouth with tongue extended. The subject is instructed to report when he/she detects a bitter taste.

(4) Using a DeVilbiss Model 40 Inhalation Medication Nebulizer or equivalent, the test conductor shall spray the Threshold Check Solution into the enclosure. This Nebulizer shall be clearly marked to distinguish it from the fit test solution nebulizer.

(5) The Threshold Check Solution is prepared by adding 13.5 milligrams of Bitrex to 100 ml of 5% salt (NaCl) solution in distilled water.

# OSHA Standards for General Industry

(6) To produce the aerosol, the nebulizer bulb is firmly squeezed so that the bulb collapses completely, and is then released and allowed to fully expand.

(7) An initial ten squeezes are repeated rapidly and then the test subject is asked whether the Bitrex can be tasted. If the test subject reports tasting the bitter taste during the ten squeezes, the screening test is completed. The taste threshold is noted as ten regardless of the number of squeezes actually completed.

(8) If the first response is negative, ten more squeezes are repeated rapidly and the test subject is again asked whether the Bitrex is tasted. If the test subject reports tasting the bitter taste during the second ten squeezes, the screening test is completed. The taste threshold is noted as twenty regardless of the number of squeezes actually completed.

(9) If the second response is negative, ten more squeezes are repeated rapidly and the test subject is again asked whether the Bitrex is tasted. If the test subject reports tasting the bitter taste during the third set of ten squeezes, the screening test is completed. The taste threshold is noted as thirty regardless of the number of squeezes actually completed.

(10) The test conductor will take note of the number of squeezes required to solicit a taste response.

(11) If the Bitrex is not tasted after 30 squeezes (step 10), the test subject is unable to taste Bitrex and may not perform the Bitrex fit test.

(12) If a taste response is elicited, the test subject shall be asked to take note of the taste for reference in the fit test.

(13) Correct use of the nebulizer means that approximately 1 ml of liquid is used at a time in the nebulizer body.

(14) The nebulizer shall be thoroughly rinsed in water, shaken to dry, and refilled at least each morning and afternoon or at least every four hours.

(b) Bitrex Solution Aerosol Fit Test Procedure.

(1) The test subject may not eat, drink (except plain water), smoke, or chew gum for 15 minutes before the test.

(2) The fit test uses the same enclosure as that described in 4. (a) above.

(3) The test subject shall don the enclosure while wearing the respirator selected according to section I. A. of this appendix. The respirator shall be properly adjusted and equipped with any type particulate filter(s).

(4) A second DeVilbiss Model 40 Inhalation Medication Nebulizer or equivalent is used to spray the fit test solution into the enclosure. This nebulizer shall be clearly marked to distinguish it from the screening test solution nebulizer.

(5) The fit test solution is prepared by adding 337.5 mg of Bitrex to 200 ml of a 5% salt (NaCl) solution in warm water.

(6) As before, the test subject shall breathe through his or her slightly open mouth with tongue extended, and be instructed to report if he/she tastes the bitter taste of Bitrex.

(7) The nebulizer is inserted into the hole in the front of the enclosure and an initial concentration of the fit test solution is sprayed into the enclosure using the same number of squeezes (either 10, 20 or 30 squeezes) based on the number of squeezes required to elicit a taste response as noted during the screening test.

(8) After generating the aerosol, the test subject shall be instructed to perform the exercises in section I. A. 14. of this appendix.

(9) Every 30 seconds the aerosol concentration shall be replenished using one half the number of squeezes used initially (e.g., 5, 10 or 15).

(10) The test subject shall indicate to the test conductor if at any time during the fit test the taste of Bitrex is detected. If the test subject does not report tasting the Bitrex, the test is passed.

(11) If the taste of Bitrex is detected, the fit is deemed unsatisfactory and the test is failed. A different respirator shall be tried and the entire test procedure is repeated (taste threshold screening and fit testing).

## 5. Irritant Smoke (Stannic Chloride) Protocol

This qualitative fit test uses a person's response to the irritating chemicals released in the "smoke" produced by a stannic chloride ventilation smoke tube to detect leakage into the respirator.

### (a) General Requirements and Precautions

(1) The respirator to be tested shall be equipped with high efficiency particulate air (HEPA) or P100 series filter(s).

(2) Only stannic chloride smoke tubes shall be used for this protocol.

(3) No form of test enclosure or hood for the test subject shall be used.

(4) The smoke can be irritating to the eyes, lungs, and nasal passages. The test conductor shall take precautions to minimize the test subject's exposure to irritant smoke. Sensitivity varies, and certain individuals may respond to a greater degree to irritant smoke. Care shall be taken when performing the sensitivity screening checks that determine whether the test subject can detect irritant smoke to use only the minimum amount of smoke necessary to elicit a response from the test subject.

(5) The fit test shall be performed in an area with adequate ventilation to prevent exposure of the person conducting the fit test or the build-up of irritant smoke in the general atmosphere.

### (b) Sensitivity Screening Check

The person to be tested must demonstrate his or her ability to detect a weak concentration of the irritant smoke.

(1) The test operator shall break both ends of a ventilation smoke tube containing stannic chloride, and attach one end of the smoke tube to a low flow air pump set to deliver 200 milliliters per minute, or an aspirator squeeze bulb. The test operator shall cover the other end of the smoke tube with a short piece of tubing to prevent potential injury from the jagged end of the smoke tube.

(2) The test operator shall advise the test subject that the smoke can be irritating to the eyes, lungs, and nasal passages and instruct the subject to keep his/her eyes closed while the test is performed.

(3) The test subject shall be allowed to smell a weak concentration of the irritant smoke before the respirator is donned to become familiar with its irritating properties and to determine if he/she can detect the irritating properties of the smoke. The test operator shall carefully direct a small amount of the irritant smoke in the test subject's direction to determine that he/she can detect it.

### (c) Irritant Smoke Fit Test Procedure

(1) The person being fit tested shall don the respirator without assistance, and perform the required user seal check(s).

(2) The test subject shall be instructed to keep his/her eyes closed.

(3) The test operator shall direct the stream of irritant smoke from the smoke tube toward the faceseal area of the test subject, using the low flow pump or the squeeze bulb. The test operator shall begin at least 12 inches from the facepiece and move the smoke stream around the whole perimeter of the mask. The operator shall gradually make two more passes around the perimeter of the mask, moving to within six inches of the respirator.

(4) If the person being tested has not had an involuntary response and/or detected the irritant smoke, proceed with the test exercises.

(5) The exercises identified in section I.A. 14. of this appendix shall be performed by the test subject while the respirator seal is being continually challenged by the smoke, directed around the perimeter of the respirator at a distance of six inches.

(6) If the person being fit tested reports detecting the irritant smoke at any time, the test is failed. The person being retested must repeat the entire sensitivity check and fit test procedure.

(7) Each test subject passing the irritant smoke test without evidence of a response (involuntary cough, irritation) shall be given a second sensitivity screening check, with the smoke from the same smoke tube used during the fit test, once the respirator has been removed, to determine whether he/she still reacts to the smoke. Failure to evoke a response shall void the fit test.

(8) If a response is produced during this second sensitivity check, then the fit test is passed.

### C. Quantitative Fit Test (QNFT) Protocols

The following quantitative fit testing procedures have been demonstrated to be acceptable: Quantitative fit testing using a non-hazardous test aerosol (such as corn oil, polyethylene glycol 400 [PEG 400], di-2-ethyl hexyl sebacate

[DEHS], or sodium chloride) generated in a test chamber, and employing instrumentation to quantify the fit of the respirator; Quantitative fit testing using ambient aerosol as the test agent and appropriate instrumentation (condensation nuclei counter) to quantify the respirator fit; Quantitative fit testing using controlled negative pressure and appropriate instrumentation to measure the volumetric leak rate of a facepiece to quantify the respirator fit.

## 1. General

(a) The employer shall ensure that persons administering QNFT are able to calibrate equipment and perform tests properly, recognize invalid tests, calculate fit factors properly and ensure that test equipment is in proper working order.

(b) The employer shall ensure that QNFT equipment is kept clean, and is maintained and calibrated according to the manufacturer's instructions so as to operate at the parameters for which it was designed.

## 2. Generated Aerosol Quantitative Fit Testing Protocol

(a) Apparatus.

(1) Instrumentation. Aerosol generation, dilution, and measurement systems using particulates (corn oil, polyethylene glycol 400 [PEG 400], di-2-ethyl hexyl sebacate [DEHS] or sodium chloride) as test aerosols shall be used for quantitative fit testing.

(2) Test chamber. The test chamber shall be large enough to permit all test subjects to perform freely all required exercises without disturbing the test agent concentration or the measurement apparatus. The test chamber shall be equipped and constructed so that the test agent is effectively isolated from the ambient air, yet uniform in concentration throughout the chamber.

(3) When testing air-purifying respirators, the normal filter or cartridge element shall be replaced with a high efficiency particulate air (HEPA) or P100 series filter supplied by the same manufacturer.

(4) The sampling instrument shall be selected so that a computer record or strip chart record may be made of the test showing the rise and fall of the test agent concentration with each inspiration and expiration at fit factors of at least 2,000. Integrators or computers that integrate the amount of test agent penetration leakage into the respirator for each exercise may be used provided a record of the readings is made.

(5) The combination of substitute air-purifying elements, test agent and test agent concentration shall be such that the test subject is not exposed in excess of an established exposure limit for the test agent at any time during the testing process, based upon the length of the exposure and the exposure limit duration.

(6) The sampling port on the test specimen respirator shall be placed and constructed so that no leakage occurs around the port (e.g., where the respirator is probed), a free air flow is allowed into the sampling line at all times, and there is no interference with the fit or performance of the respirator. The in-mask sampling device (probe) shall be designed and used so that the air sample is drawn from the breathing zone of the test subject, midway between the nose and mouth and with the probe extending into the facepiece cavity at least 1/4 inch.

(7) The test setup shall permit the person administering the test to observe the test subject inside the chamber during the test.

(8) The equipment generating the test atmosphere shall maintain the concentration of test agent constant to within a 10 percent variation for the duration of the test.

(9) The time lag (interval between an event and the recording of the event on the strip chart or computer or integrator) shall be kept to a minimum. There shall be a clear association between the occurrence of an event and its being recorded.

(10) The sampling line tubing for the test chamber atmosphere and for the respirator sampling port shall be of equal diameter and of the same material. The length of the two lines shall be equal.

(11) The exhaust flow from the test chamber shall pass through an appropriate filter (i.e., high efficiency particulate filter) before release.

(12) When sodium chloride aerosol is used, the relative humidity inside the test chamber shall not exceed 50 percent.

(13) The limitations of instrument detection shall be taken into account when determining the fit factor.

(14) Test respirators shall be maintained in proper working order and be inspected regularly for deficiencies such as cracks or missing valves and gaskets.

(b) Procedural Requirements.

(1) When performing the initial user seal check using a positive or negative pressure check, the sampling line shall be crimped closed in order to avoid air pressure leakage during either of these pressure checks.

(2) The use of an abbreviated screening QLFT test is optional. Such a test may be utilized in order to quickly identify poor fitting respirators that passed the positive and/or negative pressure test and reduce the amount of QNFT time. The use of the CNC QNFT instrument in the count mode is another optional method to obtain a quick estimate of fit and eliminate poor fitting respirators before going on to perform a full QNFT.

(3) A reasonably stable test agent concentration shall be measured in the test chamber prior to testing. For canopy or shower curtain types of test units, the determination of the test agent's stability may be established after the test subject has entered the test environment.

(4) Immediately after the subject enters the test chamber, the test agent concentration inside the respirator shall be measured to ensure that the peak penetration does not exceed 5 percent for a half mask or 1 percent for a full facepiece respirator.

(5) A stable test agent concentration shall be obtained prior to the actual start of testing.

(6) Respirator restraining straps shall not be over-tightened for testing. The straps shall be adjusted by the wearer without assistance from other persons to give a reasonably comfortable fit typical of normal use. The respirator shall not be adjusted once the fit test exercises begin.

(7) The test shall be terminated whenever any single peak penetration exceeds 5 percent for half masks and 1 percent for full facepiece respirators. The test subject shall be refitted and retested.

(8) Calculation of fit factors.

(i) The fit factor shall be determined for the quantitative fit test by taking the ratio of the average chamber concentration to the concentration measured inside the respirator for each test exercise except the grimace exercise.

(ii) The average test chamber concentration shall be calculated as the arithmetic average of the concentration measured before and after each test (i.e., 7 exercises) or the arithmetic average of the concentration measured before and after each exercise or the true average measured continuously during the respirator sample.

(iii) The concentration of the challenge agent inside the respirator shall be determined by one of the following methods:

(A) Average peak penetration method means the method of determining test agent penetration into the respirator utilizing a strip chart recorder, integrator, or computer. The agent penetration is determined by an average of the peak heights on the graph or by computer integration, for each exercise except the grimace exercise. Integrators or computers that calculate the actual test agent penetration into the respirator for each exercise will also be considered to meet the requirements of the average peak penetration method.

(B) Maximum peak penetration method means the method of determining test agent penetration in the respirator as determined by strip chart recordings of the test. The highest peak penetration for a given exercise is taken to be representative of average penetration into the respirator for that exercise.

(C) Integration by calculation of the area under the individual peak for each exercise except the grimace exercise. This includes computerized integration.

(D) The calculation of the overall fit factor using individual exercise fit factors involves first converting the exercise fit factors to penetration values, determining the average, and then converting that result back to a fit factor. This procedure is described in the following equation:

**Overall Fit Factor =**

$$\frac{\text{Number of exercises}}{1/ff_1 + 1/ff_2 + 1/ff_3 + 1/ff_4 + 1/ff_5 + 1/ff6 + 1/ff_7 + 1/ff_8}$$

Where $ff_1$, $ff_2$, $ff_3$, etc. are the fit factors for exercises 1, 2, 3, etc.

(9) The test subject shall not be permitted to wear a half mask or quarter facepiece respirator unless a minimum fit factor of 100 is obtained, or a full facepiece respirator unless a minimum fit factor of 500 is obtained.

(10) Filters used for quantitative fit testing shall be replaced whenever increased breathing resistance is encountered, or when the test agent has altered the integrity of the filter media.

### 3. Ambient aerosol condensation nuclei counter (CNC) quantitative fit testing protocol.

The ambient aerosol condensation nuclei counter (CNC) quantitative fit testing (Portacount™) protocol quantitatively fit tests respirators with the use of a probe. The probed respirator is only used for quantitative fit tests. A probed respirator has a special sampling device, installed on the respirator, that allows the probe to sample the air from inside the mask. A probed respirator is required for each make, style, model, and size that the employer uses and can be obtained from the respirator manufacturer or distributor. The CNC instrument manufacturer, TSI Inc., also provides probe attachments (TSI sampling adapters) that permit fit testing in an employee's own respirator. A minimum fit factor pass level of at least 100 is necessary for a half-mask respirator and a minimum fit factor pass level of at least 500 is required for a full facepiece negative pressure respirator. The entire screening and testing procedure shall be explained to the test subject prior to the conduct of the screening test.

(a) Portacount Fit Test Requirements. (1) Check the respirator to make sure the sampling probe and line are properly attached to the facepiece and that the respirator is fitted with a particulate filter capable of preventing significant penetration by the ambient particles used for the fit test (e.g., NIOSH 42 CFR 84 series 100, series 99, or series 95 particulate filter) per manufacturer's instruction.

(2) Instruct the person to be tested to don the respirator for five minutes before the fit test starts. This purges the ambient particles trapped inside the respirator and permits the wearer to make certain the respirator is comfortable. This individual shall already have been trained on how to wear the respirator properly.

(3) Check the following conditions for the adequacy of the respirator fit: Chin properly placed; Adequate strap tension, not overly tightened; Fit across nose bridge; Respirator of proper size to span distance from nose to chin; Tendency of the respirator to slip; Self-observation in a mirror to evaluate fit and respirator position.

(4) Have the person wearing the respirator do a user seal check. If leakage is detected, determine the cause. If leakage is from a poorly fitting facepiece, try another size of the same model respirator, or another model of respirator.

(5) Follow the manufacturer's instructions for operating the Portacount and proceed with the test.

(6) The test subject shall be instructed to perform the exercises in section I. A. 14. of this appendix.

(7) After the test exercises, the test subject shall be questioned by the test conductor regarding the comfort of the respirator upon completion of the protocol. If it has become unacceptable, another model of respirator shall be tried.

(b) Portacount Test Instrument.

(1) The Portacount will automatically stop and calculate the overall fit factor for the entire set of exercises. The overall fit factor is what counts. The Pass or Fail message will indicate whether or not the test was successful. If the test was a Pass, the fit test is over.

(2) Since the pass or fail criterion of the Portacount is user programmable, the test operator shall ensure that the pass or fail criterion meet the requirements for minimum respirator performance in this Appendix.

(3) A record of the test needs to be kept on file, assuming the fit test was successful. The record must contain the test subject's name; overall fit factor; make, model, style, and size of respirator used; and date tested.

### 4. Controlled negative pressure (CNP) quantitative fit testing protocol.

The CNP protocol provides an alternative to aerosol fit test methods. The CNP fit test method technology is based on exhausting air from a temporarily sealed respirator facepiece to generate and then maintain a constant negative pressure inside the facepiece. The rate of air exhaust is controlled so that a constant negative pressure is maintained in the respirator during the fit test. The level of pressure is selected to replicate the mean inspiratory pressure that causes leakage into the respirator under normal use conditions. With pressure held constant, air flow out of the respirator is equal to air flow into the respirator. Therefore, measurement of the exhaust stream that is required to hold the pressure in the temporarily sealed respirator constant yields a direct measure of leakage air flow into the respirator. The CNP fit test method measures leak rates through the facepiece as a method for determining the facepiece fit for negative pressure respirators. The CNP instrument manufacturer Occupational Health Dynamics of Birmingham, Alabama also provides attachments (sampling manifolds) that replace the filter cartridges to permit fit testing in an employee's own respirator. To perform the test, the test subject closes his or her mouth and holds his/her breath, after which an air pump removes air from the respirator facepiece at a pre-selected constant pressure. The facepiece fit is expressed as the leak rate through the facepiece, expressed as milliliters per minute. The quality and validity of the CNP fit tests are determined by the degree to which the in-mask pressure tracks the test pressure during the system measurement time of approximately five seconds. Instantaneous feedback in the form of a real-time pressure trace of the in-mask pressure is provided and used to determine test validity and quality. A minimum fit factor pass level of 100 is necessary for a half-mask respirator and a minimum fit factor of at least 500 is required for a full facepiece respirator. The entire screening and testing procedure shall be explained to the test subject prior to the conduct of the screening test.

(a) CNP Fit Test Requirements.

(1) The instrument shall have a non-adjustable test pressure of 15.0 mm water pressure.

(2) The CNP system defaults selected for test pressure shall be set at -15 mm of water (-0.58 inches of water) and the modeled inspiratory flow rate shall be 53.8 liters per minute for performing fit tests.

**NOTE:** CNP systems have built-in capability to conduct fit testing that is specific to unique work rate, mask, and gender situations that might apply in a specific workplace. Use of system default values, which were selected to represent respirator wear with medium cartridge resistance at a low-moderate work rate, will allow inter-test comparison of the respirator fit.)

(3) The individual who conducts the CNP fit testing shall be thoroughly trained to perform the test.

(4) The respirator filter or cartridge needs to be replaced with the CNP test manifold. The inhalation valve downstream from the manifold either needs to be temporarily removed or propped open.

(5) The employer must train the test subject to hold his or her breath for at least 10 seconds.

(6) The test subject must don the test respirator without any assistance from the test administrator who is conducting the CNP fit test. The respirator must not be adjusted once the fit-test exercises begin. Any adjustment voids the test, and the test subject must repeat the fit test.

(7) The QNFT protocol shall be followed according to section I. C. 1. of this appendix with an exception for the CNP test exercises.

(b) CNP Test Exercises.

(1) Normal breathing. In a normal standing position, without talking, the subject shall breathe normally for 1 minute. After the normal breathing exercise, the subject needs to hold head straight ahead and hold his or her breath for 10 seconds during the test measurement.

(2) Deep breathing. In a normal standing position, the subject shall breathe slowly and deeply for 1 minute, being careful not to hyperventilate. After the deep breathing exercise, the subject shall hold his or her head straight ahead and hold his or her breath for 10 seconds during test measurement.

(3) Turning head side to side. Standing in place, the subject shall slowly turn his or her head from side to side between the extreme positions on each side for 1 minute. The head shall be held at each extreme momentarily so the subject can inhale at each side. After the turning head side to side exercise, the subject needs to hold head full left and hold his or her breath for 10 seconds during test measurement. Next, the subject needs to hold head full right and hold his or her breath for 10 seconds during test measurement.

(4) Moving head up and down. Standing in place, the subject shall slowly move his or her head up and down for 1 minute. The subject shall be instructed to inhale in the up position (i.e., when looking toward the ceiling). After the moving head up and down exercise, the subject shall hold his or her head full up and hold his or her breath for 10 seconds during test measurement. Next, the subject shall hold his or her head full down and hold his or her breath for 10 seconds during test measurement.

(5) Talking. The subject shall talk out loud slowly and loud enough so as to be heard clearly by the test conductor. The subject can read from a prepared text such as the Rainbow Passage, count backward from 100,

or recite a memorized poem or song for 1 minute. After the talking exercise, the subject shall hold his or her head straight ahead and hold his or her breath for 10 seconds during the test measurement.

(6) Grimace. The test subject shall grimace by smiling or frowning for 15 seconds.

(7) Bending Over. The test subject shall bend at the waist as if he or she were to touch his or her toes for 1 minute. Jogging in place shall be substituted for this exercise in those test environments such as shroud-type QNFT units that prohibit bending at the waist. After the bending over exercise, the subject shall hold his or her head straight ahead and hold his or her breath for 10 seconds during the test measurement.

(8) Normal Breathing. The test subject shall remove and re-don the respirator within a one-minute period. Then, in a normal standing position, without talking, the subject shall breathe normally for 1 minute. After the normal breathing exercise, the subject shall hold his or her head straight ahead and hold his or her breath for 10 seconds during the test measurement. After the test exercises, the test subject shall be questioned by the test conductor regarding the comfort of the respirator upon completion of the protocol. If it has become unacceptable, another model of a respirator shall be tried.

(c) CNP Test Instrument.

(1) The test instrument must have an effective audio-warning device, or a visual-warning device in the form of a screen tracing, that indicates when the test subject fails to hold his or her breath during the test. The test must be terminated and restarted from the beginning when the test subject fails to hold his or her breath during the test. The test subject then may be refitted and retested.

(2) A record of the test shall be kept on file, assuming the fit test was successful. The record must contain the test subject's name; overall fit factor; make, model, style and size of respirator used; and date tested.

5. Controlled negative pressure (CNP) REDON quantitative fit testing protocol.

(a) When administering this protocol to test subjects, employers must comply with the requirements specified in paragraphs (a) and (c) of Part I.C.4 of this appendix ("Controlled negative pressure (CNP) quantitative fit testing protocol"), as well as use the test exercises described below in paragraph (b) of this protocol instead of the test exercises specified in paragraph (b) of Part I.C.4 of this appendix.

(b) Employers must ensure that each test subject being fit tested using this protocol follows the exercise and measurement procedures, including the order of administration, described below in Table A-1 of this appendix.

(c) After completing the test exercises, the test administrator must question each test subject regarding the comfort of the respirator. When a test subject states that the respirator is unacceptable, the employer must ensure that the test administrator repeats the protocol using another respirator model.

(d) Employers must determine the overall fit factor for each test subject by calculating the harmonic mean of the fit testing exercises as follows:

$$\text{Overall Fit Factor} = \frac{N}{[1/FF_1 + 1/FF_2 + \ldots 1/FF_N]}$$

Where:
N = The number of exercises;
FF1 = The fit factor for the first exercise;
FF2 = The fit factor for the second exercise; and
FFN = The fit factor for the nth exercise

### TABLE A-1.—CNP REDON QUANTITATIVE FIT TESTING PROTOCOL

| Exercises[1] | Exercise procedure | Measurement procedure |
|---|---|---|
| Facing Forward | Stand and breathe normally, without talking, for 30 seconds. | Face forward, while holding breath for 10 seconds. |
| Bending Over | Bend at the waist, as if going to touch his or her toes, for 30 seconds. | Face parallel to the floor, while holding breath for 10 seconds |
| Head Shaking | For about three seconds, shake head back and forth vigorously several times while shouting. | Face forward, while holding breath for 10 seconds. |
| REDON 1 | Remove the respirator mask, loosen all facepiece straps, and then redon the respirator mask. | Face forward, while holding breath for 10 seconds. |
| REDON 2 | Remove the respirator mask, loosen all facepiece straps, and then redon the respirator mask again. | Face forward, while holding breath for 10 seconds. |

[1] Exercises are listed in the order in which they are to be administered.

### Part II. New Fit Test Protocols

A. Any person may submit to OSHA an application for approval of a new fit test protocol. If the application meets the following criteria, OSHA will initiate a rulemaking proceeding under section 6(b)(7) of the OSH Act to determine whether to list the new protocol as an approved protocol in this Appendix A.

B. The application must include a detailed description of the proposed new fit test protocol. This application must be supported by either:

1. A test report prepared by an independent government research laboratory (e.g., Lawrence Livermore National Laboratory, Los Alamos National Laboratory, the National Institute for Standards and Technology) stating that the laboratory has tested the protocol and had found it to be accurate and reliable; or

2. An article that has been published in a peer-reviewed industrial hygiene journal describing the protocol and explaining how test data support the protocol's accuracy and reliability.

C. If OSHA determines that additional information is required before the Agency commences a rulemaking proceeding under this section, OSHA will so notify the applicant and afford the applicant the opportunity to submit the supplemental information. Initiation of a rulemaking proceeding will be deferred until OSHA has received and evaluated the supplemental information.

### APPENDIX B-1 TO §1910.134— USER SEAL CHECK PROCEDURES (MANDATORY)

The individual who uses a tight-fitting respirator is to perform a user seal check to ensure that an adequate seal is achieved each time the respirator is put on. Either the positive and negative pressure checks listed in this appendix, or the respirator manufacturer's recommended user seal check method shall be used. User seal checks are not substitutes for qualitative or quantitative fit tests.

### I. Facepiece Positive and/or Negative Pressure Checks

A. *Positive pressure check.* Close off the exhalation valve and exhale gently into the facepiece. The face fit is considered satisfactory if a slight positive pressure can be built up inside the facepiece without any evidence of outward leakage of air at the seal. For most respirators this method of leak testing requires the wearer to first remove the exhalation valve cover before closing off the exhalation valve and then carefully replacing it after the test.

B. *Negative pressure check.* Close off the inlet opening of the canister or cartridge(s) by covering with the palm of the hand(s) or by replacing the filter seal(s), inhale gently so that the facepiece collapses slightly, and hold the breath for ten seconds. The design of the inlet opening of some cartridges cannot be effectively covered with the palm of the hand. The test can be performed by covering the inlet opening of the cartridge with a thin latex or nitrile glove. If the facepiece remains in its slightly collapsed condition and no inward leakage of air is detected, the tightness of the respirator is considered satisfactory.

# OSHA Standards for General Industry

## II. Manufacturer's Recommended User Seal Check Procedures

The respirator manufacturer's recommended procedures for performing a user seal check may be used instead of the positive and/or negative pressure check procedures provided that the employer demonstrates that the manufacturer's procedures are equally effective.

## APPENDIX B-2 TO §1910.134— RESPIRATOR CLEANING PROCEDURES (MANDATORY)

These procedures are provided for employer use when cleaning respirators. They are general in nature, and the employer as an alternative may use the cleaning recommendations provided by the manufacturer of the respirators used by their employees, provided such procedures are as effective as those listed here in Appendix B-2. Equivalent effectiveness simply means that the procedures used must accomplish the objectives set forth in Appendix B-2, i.e., must ensure that the respirator is properly cleaned and disinfected in a manner that prevents damage to the respirator and does not cause harm to the user.

### I. Procedures for Cleaning Respirators

A. Remove filters, cartridges, or canisters. Disassemble facepieces by removing speaking diaphragms, demand and pressure-demand valve assemblies, hoses, or any components recommended by the manufacturer. Discard or repair any defective parts.

B. Wash components in warm (43 °C [110 °F] maximum) water with a mild detergent or with a cleaner recommended by the manufacturer. A stiff bristle (not wire) brush may be used to facilitate the removal of dirt.

C. Rinse components thoroughly in clean, warm (43 °C [110 °F] maximum), preferably running water. Drain.

D. When the cleaner used does not contain a disinfecting agent, respirator components should be immersed for two minutes in one of the following:

1. Hypochlorite solution (50 ppm of chlorine) made by adding approximately one milliliter of laundry bleach to one liter of water at 43 °C (110 °F); or,

2. Aqueous solution of iodine (50 ppm iodine) made by adding approximately 0.8 milliliters of tincture of iodine (6-8 grams ammonium and/or potassium iodide/100 cc of 45% alcohol) to one liter of water at 43 °C (110 °F); or,

3. Other commercially available cleansers of equivalent disinfectant quality when used as directed, if their use is recommended or approved by the respirator manufacturer.

E. Rinse components thoroughly in clean, warm (43 °C [110 °F] maximum), preferably running water. Drain. The importance of thorough rinsing cannot be overemphasized. Detergents or disinfectants that dry on facepieces may result in dermatitis. In addition, some disinfectants may cause deterioration of rubber or corrosion of metal parts if not completely removed.

F. Components should be hand-dried with a clean lint-free cloth or air-dried.

G. Reassemble facepiece, replacing filters, cartridges, and canisters where necessary.

H. Test the respirator to ensure that all components work properly.

## APPENDIX C TO §1910.134—OSHA RESPIRATOR MEDICAL EVALUATION QUESTIONNAIRE (MANDATORY)

To the employer: Answers to questions in Section 1, and to question 9 in Section 2 of Part A, do not require a medical examination.

To the employee:

Can you read (circle one): Yes/No

Your employer must allow you to answer this questionnaire during normal working hours, or at a time and place that is convenient to you. To maintain your confidentiality, your employer or supervisor must not look at or review your answers, and your employer must tell you how to deliver or send this questionnaire to the health care professional who will review it.

Part A. Section 1. (Mandatory) The following information must be provided by every employee who has been selected to use any type of respirator (please print).

1. Today's date: _____

2. Your name: _____

3. Your age (to nearest year): ____

4. Sex (circle one): Male/Female

5. Your height: ____ ft. ____ in.

6. Your weight: ____ lbs.

7. Your job title: _____

8. A phone number where you can be reached by the health care professional who reviews this questionnaire (include the Area Code): _____

9. The best time to phone you at this number: ____

10. Has your employer told you how to contact the health care professional who will review this questionnaire (circle one): Yes/No

11. Check the type of respirator you will use (you can check more than one category):

a. ____ N, R, or P disposable respirator (filter-mask, non-cartridge type only).

b. ____ Other type (for example, half- or full-facepiece type, powered-air purifying, supplied-air, self-contained breathing apparatus).

12. Have you worn a respirator (circle one): Yes/No

7 If "yes," what type(s): _____

Part A. Section 2. (Mandatory) Questions 1 through 9 below must be answered by every employee who has been selected to use any type of respirator (please circle "yes" or "no").

1. Do you *currently* smoke tobacco, or have you smoked tobacco in the last month: Yes/No

2. Have you *ever had* any of the following conditions?

  a. Seizures (fits): Yes/No

  b. Diabetes (sugar disease): Yes/No

  c. Allergic reactions that interfere with your breathing: Yes/No

  d. Claustrophobia (fear of closed-in places): Yes/No

  e. Trouble smelling odors: Yes/No

3. Have you *ever had* any of the following pulmonary or lung problems?

  a. Asbestosis: Yes/No

  b. Asthma: Yes/No

  c. Chronic bronchitis: Yes/No

  d. Emphysema: Yes/No

  e. Pneumonia: Yes/No

  f. Tuberculosis: Yes/No

  g. Silicosis: Yes/No

  h. Pneumothorax (collapsed lung): Yes/No

  i. Lung cancer: Yes/No

  j. Broken ribs: Yes/No

  k. Any chest injuries or surgeries: Yes/No

l. Any other lung problem that you've been told about: Yes/No

4. Do you *currently* have any of the following symptoms of pulmonary or lung illness?

   a. Shortness of breath: Yes/No

   b. Shortness of breath when walking fast on level ground or walking up a slight hill or incline: Yes/No

   c. Shortness of breath when walking with other people at an ordinary pace on level ground: Yes/No

   d. Have to stop for breath when walking at your own pace on level ground: Yes/No

   e. Shortness of breath when washing or dressing yourself: Yes/No

   f. Shortness of breath that interferes with your job: Yes/No

   g. Coughing that produces phlegm (thick sputum): Yes/No

   h. Coughing that wakes you early in the morning: Yes/No

   i. Coughing that occurs mostly when you are lying down: Yes/No

   j. Coughing up blood in the last month: Yes/No

   k. Wheezing: Yes/No

   l. Wheezing that interferes with your job: Yes/No

   m. Chest pain when you breathe deeply: Yes/No

   n. Any other symptoms that you think may be related to lung problems: Yes/No

5. Have you *ever had* any of the following cardiovascular or heart problems?

   a. Heart attack: Yes/No

   b. Stroke: Yes/No

   c. Angina: Yes/No

   d. Heart failure: Yes/No

   e. Swelling in your legs or feet (not caused by walking): Yes/No

   f. Heart arrhythmia (heart beating irregularly): Yes/No

   g. High blood pressure: Yes/No

   h. Any other heart problem that you've been told about: Yes/No

6. Have you *ever had* any of the following cardiovascular or heart symptoms?

   a. Frequent pain or tightness in your chest: Yes/No

   b. Pain or tightness in your chest during physical activity: Yes/No

   c. Pain or tightness in your chest that interferes with your job: Yes/No

   d. In the past two years, have you noticed your heart skipping or missing a beat: Yes/No

   e. Heartburn or indigestion that is not related to eating: Yes/No

   f. Any other symptoms that you think may be related to heart or circulation problems: Yes/No

7. Do you *currently* take medication for any of the following problems?

   a. Breathing or lung problems: Yes/No

   b. Heart trouble: Yes/No

   c. Blood pressure: Yes/No

   d. Seizures (fits): Yes/No

8. If you've used a respirator, have you *ever had* any of the following problems? (If you've never used a respirator, check the following space and go to question 9:)

   a. Eye irritation: Yes/No

   b. Skin allergies or rashes: Yes/No

   c. Anxiety: Yes/No

   d. General weakness or fatigue: Yes/No

   e. Any other problem that interferes with your use of a respirator: Yes/No

9. Would you like to talk to the health care professional who will review this questionnaire about your answers to this questionnaire: Yes/No

Questions 10 to 15 below must be answered by every employee who has been selected to use either a full-facepiece respirator or a self-contained breathing apparatus (SCBA). For employees who have been selected to use other types of respirators, answering these questions is voluntary.

10. Have you *ever lost* vision in either eye (temporarily or permanently): Yes/No

11. Do you *currently* have any of the following vision problems?

    a. Wear contact lenses: Yes/No

    b. Wear glasses: Yes/No

    c. Color blind: Yes/No

    d. Any other eye or vision problem: Yes/No

12. Have you *ever had* an injury to your ears, including a broken ear drum: Yes/No

13. Do you *currently* have any of the following hearing problems?

    a. Difficulty hearing: Yes/No

    b. Wear a hearing aid: Yes/No

    c. Any other hearing or ear problem: Yes/No

14. Have you *ever had* a back injury: Yes/No

15. Do you *currently* have any of the following musculoskeletal problems?

    a. Weakness in any of your arms, hands, legs, or feet: Yes/No

    b. Back pain: Yes/No

    c. Difficulty fully moving your arms and legs: Yes/No

    d. Pain or stiffness when you lean forward or backward at the waist: Yes/No

    e. Difficulty fully moving your head up or down: Yes/No

    f. Difficulty fully moving your head side to side: Yes/No

    g. Difficulty bending at your knees: Yes/No

    h. Difficulty squatting to the ground: Yes/No

    i. Climbing a flight of stairs or a ladder carrying more than 25 lbs: Yes/No

    j. Any other muscle or skeletal problem that interferes with using a respirator: Yes/No

Part B. Any of the following questions, and other questions not listed, may be added to the questionnaire at the discretion of the health care professional who will review the questionnaire.

1. In your present job, are you working at high altitudes (over 5,000 feet) or in a place that has lower than normal amounts of oxygen: Yes/No

If "yes," do you have feelings of dizziness, shortness of breath, pounding in your chest, or other symptoms when you're working under these conditions:

Yes/No

2. At work or at home, have you ever been exposed to hazardous solvents, hazardous airborne chemicals (*e.g.*, gases, fumes, or dust), or have you come into skin contact with hazardous chemicals: Yes/No

If "yes," name the chemicals if you know them:
_____

3. Have you ever worked with any of the materials, or under any of the conditions, listed below:

   a. Asbestos: Yes/No

   b. Silica (*e.g.*, in sandblasting): Yes/No

   c. Tungsten/cobalt (*e.g.*, grinding or welding this material): Yes/No

   d. Beryllium: Yes/No

   e. Aluminum: Yes/No

   f. Coal (for example, mining): Yes/No

   g. Iron: Yes/No

   h. Tin: Yes/No

   i. Dusty environments: Yes/No

   j. Any other hazardous exposures: Yes/No

If "yes," describe these exposures:
_____

4. List any second jobs or side businesses you have:
_____

5. List your previous occupations:
_____

6. List your current and previous hobbies:
_____

7. Have you been in the military services? Yes/No

If "yes," were you exposed to biological or chemical agents (either in training or combat): Yes/No

8. Have you ever worked on a HAZMAT team? Yes/No

9. Other than medications for breathing and lung problems, heart trouble, blood pressure, and seizures mentioned earlier in this questionnaire, are you taking any other medications for any reason (including over-the-counter medications): Yes/No

If "yes," name the medications if you know them:
_____

10. Will you be using any of the following items with your respirator(s)?

    a. HEPA Filters: Yes/No

    b. Canisters (for example, gas masks): Yes/No

    c. Cartridges: Yes/No

11. How often are you expected to use the respirator(s) (circle "yes" or "no" for all answers that apply to you)?:

    a. Escape only (no rescue): Yes/No

    b. Emergency rescue only: Yes/No

    c. Less than 5 hours *per week*: Yes/No

    d. Less than 2 hours *per day*: Yes/No

    e. 2 to 4 hours *per day*: Yes/No

    f. Over 4 hours *per day*: Yes/No

12. During the period you are using the respirator(s), is your work effort:

   a. *Light* (less than 200 kcal per hour): Yes/No

If "yes," how long does this period last during the average shift: ____ hrs. ____ mins.

Examples of a light work effort are *sitting* while writing, typing, drafting, or performing light assembly work; or *standing* while operating a drill press (1-3 lbs.) or controlling machines.

   b. *Moderate* (200 to 350 kcal per hour): Yes/No

If "yes," how long does this period last during the average shift: ____ hrs. ____ mins.

Examples of moderate work effort are *sitting* while nailing or filing; *driving* a truck or bus in urban traffic; *standing* while drilling, nailing, performing assembly work, or transferring a moderate load (about 35 lbs.) at trunk level; *walking* on a level surface about 2 mph or down a 5-degree grade about 3 mph; or *pushing* a wheelbarrow with a heavy load (about 100 lbs.) on a level surface.

   c. *Heavy* (above 350 kcal per hour): Yes/No

If "yes," how long does this period last during the average shift: ____ hrs. ____ mins.

Examples of heavy work are *lifting* a heavy load (about 50 lbs.) from the floor to your waist or shoulder; *working* on a loading dock; *shoveling*; *standing* while bricklaying or chipping castings; *walking* up an 8-degree grade about 2 mph; *climbing* stairs with a heavy load (about 50 lbs.).

13. Will you be wearing protective clothing and/or equipment (other than the respirator) when you're using your respirator: Yes/No

If "yes," describe this protective clothing and/or equipment:
_____

14. Will you be working under hot conditions (temperature exceeding 77 °F): Yes/No

15. Will you be working under humid conditions: Yes/No

16. Describe the work you'll be doing while you're using your respirator(s):
_____

17. Describe any special or hazardous conditions you might encounter when you're using your respirator(s) (for example, confined spaces, life-threatening gases):
_____

18. Provide the following information, if you know it, for each toxic substance that you'll be exposed to when you're using your respirator(s):

Name of the first toxic substance:

Estimated maximum exposure level per shift: ____

Duration of exposure per shift ____

Name of the second toxic substance:
_____

Estimated maximum exposure level per shift: ____

Duration of exposure per shift: ____

Name of the third toxic substance:
_____

Estimated maximum exposure level per shift: ____

Duration of exposure per shift: ____

The name of any other toxic substances that you'll be exposed to while using your respirator:
_____

19. Describe any special responsibilities you'll have while using your respirator(s) that may affect the safety and well-being of others (for example, rescue, security):
_____

## APPENDIX D TO §1910.134 (MANDATORY) INFORMATION FOR EMPLOYEES USING RESPIRATORS WHEN NOT REQUIRED UNDER THE STANDARD

Respirators are an effective method of protection against designated hazards when properly selected and worn. Respirator use is encouraged, even when exposures are below the exposure limit, to provide an additional level of comfort and protection for workers. However, if a respirator is used improperly or not kept clean, the respirator itself can become a hazard to the worker. Sometimes, workers may wear respirators to avoid exposures to hazards, even if the amount of hazardous substance does not exceed the limits set by OSHA standards. If your employer provides respirators for your voluntary use, of if you provide your own respirator, you need to take certain precautions to be sure that the respirator itself does not present a hazard.

You should do the following:

1. Read and heed all instructions provided by the manufacturer on use, maintenance, cleaning and care, and warnings regarding the respirators limitations.

2. Choose respirators certified for use to protect against the contaminant of concern. NIOSH, the National Institute for Occupational Safety and Health of the U.S. Department of Health and Human Services, certifies respirators. A label or statement of certification should appear on the respirator or respirator packaging. It will tell you what the respirator is designed for and how much it will protect you.

3. Do not wear your respirator into atmospheres containing contaminants for which your respirator is not designed to protect against. For example, a respirator designed to filter dust particles will not protect you against gases, vapors, or very small solid particles of fumes or smoke.

4. Keep track of your respirator so that you do not mistakenly use someone else's respirator.

[63 FR 1270, Jan. 8, 1998; 63 FR 20098, 20099, Apr. 23, 1998]

## §1910.135 Head protection.

**1910.135(a)** *General requirements.* **(1)** The employer shall ensure that each affected employee wears a protective helmet when working in areas where there is a potential for injury to the head from falling objects.

**1910.135(a)(2)** The employer shall ensure that a protective helmet designed to reduce electrical shock hazard is worn by each such affected employee when near exposed electrical conductors which could contact the head.

**1910.135(b)** *Criteria for protective helmets.* **(1)** Protective helmets purchased after July 5, 1994 shall comply with ANSI Z89.1-1986, "American National Standard for Personnel Protection—Protective Headwear for Industrial Workers-Requirements," which is incorporated by reference as specified in §1910.6, or shall be demonstrated to be equally effective.

**1910.135(b)(2)** Protective helmets purchased before July 5, 1994 shall comply with the ANSI standard "American National Standard Safety Requirements for Industrial Head Protection," ANSI Z89.1-1969, which is incorporated by reference as specified in §1910.6, or shall be demonstrated by the employer to be equally effective.

[59 FR 16362, Apr. 6, 1994; 61 FR 9227, Mar. 7, 1996; 61 FR 19547, May 1, 1996]

## §1910.136 Foot protection.

**1910.136(a)** *General requirements.* The employer shall ensure that each affected employee uses protective footwear when working in areas where there is a danger of foot injuries due to falling or rolling objects, or objects piercing the sole, and where such employee's feet are exposed to electrical hazards.

**1910.136(b)** *Criteria for protective footwear.* **(1)** Protective footwear purchased after July 5, 1994 shall comply with ANSI Z41-1991, "American National Standard for Personal Protection—Protective Footwear," which is incorporated by reference as specified in §1910.6, or shall be demonstrated by the employer to be equally effective.

**1910.136(b)(2)** Protective footwear purchased before July 5, 1994 shall comply with the ANSI standard "USA Standard for Men's Safety-Toe Footwear," Z41.1-1967, which is incorporated by reference as specified in §1910.6, or shall be demonstrated by the employer to be equally effective.

[59 FR 16360, Apr. 6, 1994; 59 FR 33910, July 1, 1994; 61 FR 9227, Mar. 7, 1996; 61 FR 19547, May 2, 1996; 61 FR 21228, May 9, 1996]

## §1910.137 Electrical protective equipment.

**1910.137(a)** *Design requirements.* Insulating blankets, matting, covers, line hose, gloves, and sleeves made of rubber shall meet the following requirements:

**1910.137(a)(1)** *Manufacture and marking.* **(i)** Blankets, gloves, and sleeves shall be produced by a seamless process.

**1910.137(a)(1)(ii)** Each item shall be clearly marked as follows:

**1910.137(a)(1)(ii)(A)** Class 0 equipment shall be marked Class 0.

**1910.137(a)(1)(ii)(B)** Class 1 equipment shall be marked Class 1.

**1910.137(a)(1)(ii)(C)** Class 2 equipment shall be marked Class 2.

**1910.137(a)(1)(ii)(D)** Class 3 equipment shall be marked Class 3.

**1910.137(a)(1)(ii)(E)** Class 4 equipment shall be marked Class 4.

**1910.137(a)(1)(ii)(F)** Non-ozone-resistant equipment other than matting shall be marked Type I.

**1910.137(a)(1)(ii)(G)** Ozone-resistant equipment other than matting shall be marked Type II.

**1910.137(a)(1)(ii)(H)** Other relevant markings, such as the manufacturer's identification and the size of the equipment, may also be provided.

**1910.137(a)(1)(iii)** Markings shall be nonconducting and shall be applied in such a manner as not to impair the insulating qualities of the equipment.

**1910.137(a)(1)(iv)** Markings on gloves shall be confined to the cuff portion of the glove.

**1910.137(a)(2)** *Electrical requirements.* **(i)** Equipment shall be capable of withstanding the a-c proof-test voltage specified in Table I-2 or the d-c proof-test voltage specified in Table I-3.

**1910.137(a)(2)(i)(A)** The proof test shall reliably indicate that the equipment can withstand the voltage involved.

**1910.137(a)(2)(i)(B)** The test voltage shall be applied continuously for 3 minutes for equipment other than matting and shall be applied continuously for 1 minute for matting.

**1910.137(a)(2)(i)(C)** Gloves shall also be capable of withstanding the a-c proof-test voltage specified in Table I-2 after a 16-hour water soak. (See the note following paragraph (a)(3)(ii)(B) of this section.)

**1910.137(a)(2)(ii)** When the a-c proof test is used on gloves, the 60-hertz proof-test current may not exceed the values specified in Table I-2 at any time during the test period.

**1910.137(a)(2)(ii)(A)** If the a-c proof test is made at a frequency other than 60 Hertz, the permissible proof-test current shall be computed from the direct ratio of the frequencies.

**1910.137(a)(2)(ii)(B)** For the test, gloves (right side out) shall be filled with tap water and immersed in water to a depth that is in accordance with Table I-4. Water shall be added to or removed from the glove, as necessary, so that the water level is the same inside and outside the glove.

**1910.137(a)(2)(ii)(C)** After the 16-hour water soak specified in paragraph (a)(2)(i)(C) of this section, the 60-Hertz proof-test current may exceed the values given in Table I-2 by not more than 2 milliamperes.

**1910.137(a)(2)(iii)** Equipment that has been subjected to a minimum breakdown voltage test may not be used for electrical protection. (See the note following paragraph (a)(3)(ii)(B) of this section.)

**1910.137(a)(2)(iv)** Material used for Type II insulating equipment shall be capable of withstanding an ozone test, with no visible effects. The ozone test shall reliably indicate that the material will resist ozone exposure in actual use. Any visible signs of ozone deterioration of the material, such as checking, cracking, breaks, or pitting, is evidence of failure to meet the requirements for ozone-resistant material. (See the note following paragraph (a)(3)(ii)(B) of this section.)

**1910.137(a)(3)** *Workmanship and finish.* **(i)** Equipment shall be free of harmful physical irregularities that can be detected by the tests or inspections required under this section.

**1910.137(a)(3)(ii)** Surface irregularities that may be present on all rubber goods because of imperfections on forms or molds or because of inherent difficulties in the manufacturing process and that may appear as indentations, protuberances, or imbedded foreign material are acceptable under the following conditions:

**1910.137(a)(3)(ii)(A)** The indentation or protuberance blends into a smooth slope when the material is stretched.

**1910.137(a)(3)(ii)(B)** Foreign material remains in place when the insulating material is folded and stretches with the insulating material surrounding it.

**NOTE:** Rubber insulating equipment meeting the following national consensus standards is deemed to be in compliance with paragraph (a) of this section:

American Society for Testing and Materials (ASTM) D 120-87, Specification for Rubber Insulating Gloves.

ASTM D 178-93 (or D 178-88) Specification for Rubber Insulating Matting.

ASTM D 1048-93 (or D 1048-88a) Specification for Rubber Insulating Blankets.

ASTM D 1049-93 (or D 1049-88) Specification for Rubber Insulating Covers.

ASTM D 1050-90, Specification for Rubber Insulating Line Hose.

ASTM D 1051-87, Specification for Rubber Insulating Sleeves.

These standards contain specifications for conducting the various tests required in paragraph (a) of this section. For example, the a-c and d-c proof tests, the breakdown test, the water soak procedure, and the ozone test mentioned in this paragraph are described in detail in the ASTM standards.

**1910.137(b)** *In-service care and use.* **(1)** Electrical protective equipment shall be maintained in a safe, reliable condition.

**1910.137(b)(2)** The following specific requirements apply to insulating blankets, covers, line hose, gloves, and sleeves made of rubber:

**1910.137(b)(2)(i)** Maximum use voltages shall conform to those listed in Table I-5.

**1910.137(b)(2)(ii)** Insulating equipment shall be inspected for damage before each day's use and immediately following any incident that can reasonably be suspected of having caused damage. Insulating gloves shall be given an air test, along with the inspection.

**1910.137(b)(2)(iii)** Insulating equipment with any of the following defects may not be used:

**1910.137(b)(2)(iii)(A)** A hole, tear, puncture, or cut;

**1910.137(b)(2)(iii)(B)** Ozone cutting or ozone checking (the cutting action produced by ozone on rubber under mechanical stress into a series of interlacing cracks);

**1910.137(b)(2)(iii)(C)** An embedded foreign object;

**1910.137(b)(2)(iii)(D)** Any of the following texture changes: swelling, softening, hardening, or becoming sticky or inelastic.

**1910.137(b)(2)(iii)(E)** Any other defect that damages the insulating properties.

**1910.137(b)(2)(iv)** Insulating equipment found to have other defects that might affect its insulating properties shall be removed from service and returned for testing under paragraphs (b)(2)(viii) and (b)(2)(ix) of this section.

**1910.137(b)(2)(v)** Insulating equipment shall be cleaned as needed to remove foreign substances.

**1910.137(b)(2)(vi)** Insulating equipment shall be stored in such a location and in such a manner as to protect it from light, temperature extremes, excessive humidity, ozone, and other injurious substances and conditions.

**1910.137(b)(2)(vii)** Protector gloves shall be worn over insulating gloves, except as follows:

**1910.137(b)(2)(vii)(A)** Protector gloves need not be used with Class 0 gloves, under limited-use conditions, where small equipment and parts manipulation necessitate unusually high finger dexterity.

**NOTE:** Extra care is needed in the visual examination of the glove and in the avoidance of handling sharp objects.

**1910.137(b)(2)(vii)(B)** Any other class of glove may be used for similar work without protector gloves if the employer can demonstrate that the possibility of physical damage to the gloves is small and if the class of glove is one class higher than that required for the voltage involved. Insulating gloves that have been used without protector gloves may not be used at a higher voltage until they have been tested under the provisions of paragraphs (b)(2)(viii) and (b)(2)(ix) of this section.

**1910.137(b)(2)(viii)** Electrical protective equipment shall be subjected to periodic electrical tests. Test voltages and the maximum intervals between tests shall be in accordance with Table I-5 and Table I-6.

**1910.137(b)(2)(ix)** The test method used under paragraphs (b)(2)(viii) and (b)(2)(ix) of this section shall reliably indicate whether the insulating equipment can withstand the voltages involved.

**NOTE:** Standard electrical test methods considered as meeting this requirement are given in the following national consensus standards:

American Society for Testing and Materials (ASTM) D 120-87, Specification for Rubber Insulating Gloves.

ASTM D 1048-93, Specification for Rubber Insulating Blankets.

ASTM D 1049-93, Specification for Rubber Insulating Covers.

ASTM D 1050-90, Specification for Rubber Insulating Line Hose.

ASTM D 1051-87, Specification for Rubber Insulating Sleeves.

ASTM F 478-92, Specification for In-Service Care of Insulating Line Hose and Covers.

ASTM F 479-93, Specification for In-Service Care of Insulating Blankets.

ASTM F 496-93b Specification for In-Service Care of Insulating Gloves and Sleeves.

**1910.137(b)(2)(x)** Insulating equipment failing to pass inspections or electrical tests may not be used by employees, except as follows:

**1910.137(b)(2)(x)(A)** Rubber insulating line hose may be used in shorter lengths with the defective portion cut off.

**1910.137(b)(2)(x)(B)** Rubber insulating blankets may be repaired using a compatible patch that results in physical and electrical properties equal to those of the blanket.

**1910.137(b)(2)(x)(C)** Rubber insulating blankets may be salvaged by severing the defective area from the undamaged portion of the blanket. The resulting undamaged area may not be smaller than 22 inches by 22 inches (560 mm by 560 mm) for Class 1, 2, 3, and 4 blankets.

**1910.137(b)(2)(x)(D)** Rubber insulating gloves and sleeves with minor physical defects, such as small cuts, tears, or punctures, may be repaired by the application of a compatible patch. Also, rubber insulating gloves and sleeves with minor surface blemishes may be repaired with a compatible liquid compound. The patched area shall have electrical and physical properties equal to those of the surrounding material. Repairs to gloves are permitted only in the area between the wrist and the reinforced edge of the opening.

**1910.137(b)(2)(xi)** Repaired insulating equipment shall be retested before it may be used by employees.

**1910.137(b)(2)(xii)** The employer shall certify that equipment has been tested in accordance with the requirements of paragraphs (b)(2)(viii), (b)(2)(ix), and (b)(2)(xi) of this section. The certification shall identify the equipment that passed the test and the date it was tested.

**NOTE:** Marking of equipment and entering the results of the tests and the dates of testing onto logs are two acceptable means of meeting this requirement.

### TABLE I-2—A-C PROOF-TEST REQUIREMENTS

| Class of equipment | Proof-test voltage rms V | Maximum proof-test current, mA (gloves only) ||||
|---|---|---|---|---|---|
| | | 267-mm (10.5-in) glove | 356-mm (14-in) glove | 406-mm (16-in) glove | 457-mm (18-in) glove |
| 0 | 5,000 | 8 | 12 | 14 | 16 |
| 1 | 10,000 | ........ | 14 | 16 | 18 |
| 2 | 20,000 | ........ | 16 | 18 | 20 |
| 3 | 30,000 | ........ | 18 | 20 | 22 |
| 4 | 40,000 | ........ | ........ | 22 | 24 |

## TABLE I-3—D-C PROOF-TEST REQUIREMENTS

| Class of equipment | Proof-test voltage |
|---|---|
| 0 | 20,000 |
| 1 | 40,000 |
| 2 | 50,000 |
| 3 | 60,000 |
| 4 | 70,000 |

NOTE: The d-c voltages listed in this table are not appropriate for proof testing rubber insulating line hose or covers. For this equipment, d-c proof tests shall use a voltage high enough to indicate that the equipment can be safely used at the voltages listed in Table I-4. See ASTM D 1050-90 and ASTM D 1049-88 for further information on proof tests for rubber insulating line hose and covers.

## TABLE I-4—GLOVE TESTS—WATER LEVEL [1,2]

| Class of glove | AC proof test mm. | AC proof test in. | DC proof test mm. | DC proof test in. |
|---|---|---|---|---|
| 0 | 38 | 1.5 | 38 | 1.5 |
| 1 | 38 | 1.5 | 51 | 2.0 |
| 2 | 64 | 2.5 | 76 | 3.0 |
| 3 | 89 | 3.5 | 102 | 4.0 |
| 4 | 127 | 5.0 | 153 | 6.0 |

[1] The water level is given as the clearance from the cuff of the glove to the water line, with a tolerance of ±13 mm. (±0.5 in.).

[2] If atmospheric conditions make the specified clearances impractical, the clearances may be increased by a maximum of 25 mm. (1 in.).

## TABLE I-5—RUBBER INSULATING EQUIPMENT VOLTAGE REQUIREMENTS

| Class of equipment | Maximum use voltage[1] a-c—rms | Retest voltage[2] a-c—rms | Retest voltage[2] d-c—avg |
|---|---|---|---|
| 0 | 1,000 | 5,000 | 20,000 |
| 1 | 7,500 | 10,000 | 40,000 |
| 2 | 17,000 | 20,000 | 50,000 |
| 3 | 26,500 | 30,000 | 60,000 |
| 4 | 36,000 | 40,000 | 70,000 |

[1] The maximum use voltage is the a-c voltage (rms) classification of the protective equipment that designates the maximum nominal design voltage of the energized system that may be safely worked. The nominal design voltage is equal to the phase-to-phase voltage on multiphase circuits. However, the phase-to-ground potential is considered to be the nominal design voltage:

(1) If there is no multiphase exposure in a system area and if the voltage exposure is limited to the phase-to- ground potential, or

(2) If the electrical equipment and devices are insulated or isolated or both so that the multiphase exposure on a grounded wye circuit is removed.

[2] The proof-test voltage shall be applied continuously for at least 1 minute, but no more than 3 minutes.

## TABLE I-6—RUBBER INSULATING EQUIPMENT TEST INTERVALS

| Type of equipment | When to test |
|---|---|
| Rubber insulating line hose | Upon indication that insulating value is suspect. |
| Rubber insulating covers | Upon indication that insulating value is suspect. |
| Rubber insulating blankets | Before first issue and every 12 months thereafter.[1] |
| Rubber insulating gloves | Before first issue and every 6 months thereafter.[1] |
| Rubber insulating sleeves | Before first issue and every 12 months thereafter.[1] |

[1] If the insulating equipment has been electrically tested but not issued for service, it may not be placed into service unless it has been electrically tested within the previous 12 months.

[59 FR 4435, Jan. 31, 1994; 59 FR 33658, June 30, 1994]

### §1910.138 Hand protection.

**1910.138(a)** *General requirements.* Employers shall select and require employees to use appropriate hand protection when employees' hands are exposed to hazards such as those from skin absorption of harmful substances; severe cuts or lacerations; severe abrasions; punctures; chemical burns; thermal burns; and harmful temperature extremes.

**1910.138(b)** *Selection.* Employers shall base the selection of the appropriate hand protection on an evaluation of the performance characteristics of the hand protection relative to the task(s) to be performed, conditions present, duration of use, and the hazards and potential hazards identified.

[59 FR 16362, Apr. 6, 1994; 59 FR 33911, July 1, 1994]

### §1910.139 Respiratory protection for M. tuberculosis. [Removed.]

[§1910.139 removed at 68 FR 75776-75780, Dec. 31, 2003, effective Dec. 31, 2003.]

### §1910.140 Standards organizations [Removed.]

[§1910.140 removed at 61 FR 9238, Mar. 7, 1996, effective May 6, 1996.]

## APPENDIX A TO SUBPART I TO PART 1910—REFERENCES FOR FURTHER INFORMATION (NON-MANDATORY)

The documents in appendix A provide information which may be helpful in understanding and implementing the standards in Subpart I.

1. Bureau of Labor Statistics (BLS). "Accidents Involving Eye Injuries." Report 597, Washington, D.C.: BLS, 1980.

2. Bureau of Labor Statistics (BLS). "Accidents Involving Face Injuries." Report 604, Washington, D.C.: BLS, 1980.

3. Bureau of Labor Statistics (BLS). "Accidents Involving Head Injuries." Report 605, Washington, D.C.: BLS, 1980.

4. Bureau of Labor Statistics (BLS). "Accidents Involving Foot Injuries." Report 626, Washington, D.C.: BLS, 1981.

5. National Safety Council. "Accident Facts", Annual edition, Chicago, IL: 1981.

6. Bureau of Labor Statistics (BLS). "Occupational Injuries and Illnesses in the United States by Industry," Annual edition, Washington, D.C.: BLS.

7. National Society to Prevent Blindness. "A Guide for Controlling Eye Injuries in Industry," Chicago, Il: 1982.

[59 FR 16362, Apr. 6, 1994]

## APPENDIX B TO SUBPART I TO PART 1910— NON-MANDATORY COMPLIANCE GUIDELINES FOR HAZARD ASSESSMENT AND PERSONAL PROTECTIVE EQUIPMENT SELECTION

This appendix is intended to provide compliance assistance for employers and employees in implementing requirements for a hazard assessment and the selection of personal protective equipment.

1. *Controlling hazards.* PPE devices alone should not be relied on to provide protection against hazards, but should be used in conjunction with guards, engineering controls, and sound manufacturing practices.

2. *Assessment and selection.* It is necessary to consider certain general guidelines for assessing the foot, head, eye and face, and hand hazard situations that exist in an occupational or educational operation or process, and to match the protective devices to the particular hazard. It should be the responsibility of the safety officer to exercise common sense and appropriate expertise to accomplish these tasks.

3. *Assessment guidelines.* In order to assess the need for PPE the following steps should be taken:

a. *Survey.* Conduct a walk-through survey of the areas in question. The purpose of the survey is to identify sources of hazards to workers and co-workers. Consideration should be given to the basic hazard categories:

(a) Impact

(b) Penetration

(c) Compression (roll-over)

(d) Chemical

(e) Heat

(f) Harmful dust

(g) Light (optical) radiation

b. *Sources.* During the walk-through survey the safety officer should observe: (a) sources of motion; i.e., machinery or processes where any movement of tools, machine elements or particles could exist, or movement of personnel that could result in collision with stationary objects; (b) sources of high temperatures that could result in burns, eye injury or ignition of protective equipment, etc.; (c) types of chemical exposures; (d) sources of harmful dust; (e) sources of light radiation, i.e., welding, brazing, cutting, furnaces, heat treating, high intensity lights, etc.; (f) sources of falling objects or potential for dropping objects; (g) sources of sharp objects which might pierce the feet or cut the hands; (h) sources of rolling or pinching objects which could crush the feet; (i) layout of workplace and location of co-workers; and (j) any electrical hazards. In addition, injury/accident data should be reviewed to help identify problem areas.

c. *Organize data.* Following the walk-through survey, it is necessary to organize the data and information for use in the assessment of hazards. The objective is to prepare for an analysis of the hazards in the environment to enable proper selection of protective equipment.

d. *Analyze data.* Having gathered and organized data on a workplace, an estimate of the potential for injuries should be made. Each of the basic hazards (paragraph 3.a.) should be reviewed and a determination made as to the type, level of risk, and seriousness of potential injury from each of the hazards found in the area. The possibility of exposure to several hazards simultaneously should be considered.

4. *Selection guidelines.* After completion of the procedures in paragraph 3, the general procedure for selection of protective equipment is to: a) Become familiar with the potential hazards and the type of protective equipment that is available, and what it can do; i.e., splash protection, impact protection, etc.; b) compare the hazards associated with the environment; i.e., impact velocities, masses, projectile shape, radiation intensities, with the capabilities of the available protective equipment; c) select the protective equipment which ensures a level of protection greater than the minimum required to protect employees from the hazards; and d) fit the user with the protective device and give instructions on care and use of the PPE. It is very important that end users be made aware of all warning labels for and limitations of their PPE.

5. *Fitting the device.* Careful consideration must be given to comfort and fit. PPE that fits poorly will not afford the necessary protection. Continued wearing of the device is more likely if it fits the wearer comfortably. Protective devices are generally available in a variety of sizes. Care should be taken to ensure that the right size is selected.

6. *Devices with adjustable features.* Adjustments should be made on an individual basis for a comfortable fit that will maintain the protective device in the proper position. Particular care should be taken in fitting devices for eye protection against dust and chemical splash to ensure that the devices are sealed to the face. In addition, proper fitting of helmets is important to ensure that it will not fall off during work operations. In some cases a chin strap may be necessary to keep the helmet on an employee's head. (Chin straps should break at a reasonably low force, however, so as to prevent a strangulation hazard). Where manufacturer's instructions are available, they should be followed carefully.

7. *Reassessment of hazards.* It is the responsibility of the safety officer to reassess the workplace hazard situation as necessary, by identifying and evaluating new equipment and processes, reviewing accident records, and reevaluating the suitability of previously selected PPE.

8. *Selection chart guidelines for eye and face protection.* Some occupations (not a complete list) for which eye protection should be routinely considered are: carpenters, electricians, machinists, mechanics and repairers, millwrights, plumbers and pipe fitters, sheet metal workers and tinsmiths, assemblers, sanders, grinding machine operators, lathe and milling machine operators, sawyers, welders, laborers, chemical process operators and handlers, and timber cutting and logging workers. The following chart provides general guidance for the proper selection of eye and face protection to protect against hazards associated with the listed hazard "source" operations.

## EYE AND FACE PROTECTION SELECTION CHART

| Source | Assessment of Hazard | Protection |
|---|---|---|
| IMPACT—Chipping, grinding machining, masonry work, woodworking, sawing, drilling, chiseling, powered fastening, riveting and | Flying fragments, objects, large chips, particles, sand, dirt, etc. | Spectacles with side protection, goggles, face sanding. shields. See notes [1, 3, 5, 6, 10]. For severe exposure, use faceshield. |
| HEAT—Furnace operations, pouring, casting, hot dipping, and welding. | Hot sparks | Faceshields, goggles, spectacles with side protection. For severe exposure use faceshield. See notes [1, 2, 3] |
|  | Splash from molten metals. | Faceshields worn over goggles. See notes [1, 2, 3]. |
|  | High temperature exposure. | Screen face shields, reflective face shields. See notes [1, 2, 3]. |
| CHEMICALS—Acid and chemicals handling, degreasing plating. | Splash | Goggles, eyecup and cover types. For severe exposure, use face shield. See notes [3, 11]. |
|  | Irritating mists. | Special-purpose goggles. |
| DUST—Woodworking, buffing, general dusty conditions. | Nuisance dust | Goggles, eyecup and cover types. See note [8]. |
| LIGHT and/or RADIATION— |  |  |
| Welding: Electric arc | Optical radiation | Welding helmets or welding shields. Typical shades: 10-14. See notes [9, 12] |
| Welding: Gas | Optical radiation | Welding goggles or welding face shield. Typical shades: gas welding 4-8, cutting 3-6, brazing 3-4. See note [9] |
| Cutting, Torch brazing, Torch soldering | Optical radiation | Spectacles or welding faceshield. Typical shades, 1.5-3. See notes [3, 9] |
| Glare | Poor vision | Spectacles with shaded or special-purpose lenses, as suitable. See notes [9, 10]. |

Notes to Eye and Face Protection Selection Chart:

[1] Care should be taken to recognize the possibility of multiple and simultaneous exposure to a variety of hazards. Adequate protection against the highest level of each of the hazards should be provided. Protective devices do not provide unlimited protection.

[2] Operations involving heat may also involve light radiation. As required by the standard, protection from both hazards must be provided.

[3] Faceshields should only be worn over primary eye protection (spectacles or goggles).

[4] As required by the standard, filter lenses must meet the requirements for shade designations in Sec. 1910.133(a)(5). Tinted and shaded lenses are not filter lenses unless they are marked or identified as such.

[5] As required by the standard, persons whose vision requires the use of prescription (Rx) lenses must wear either protective devices fitted with prescription (Rx) lenses or protective devices designed to be worn over regular prescription (Rx) eyewear.

[6] Wearers of contact lenses must also wear appropriate eye and face protection devices in a hazardous environment. It should be recognized that dusty and/or chemical environments may represent an additional hazard to contact lens wearers.

[7] Caution should be exercised in the use of metal frame protective devices in electrical hazard areas.

[8] Atmospheric conditions and the restricted ventilation of the protector can cause lenses to fog. Frequent cleansing may be necessary.

[9] Welding helmets or faceshields should be used only over primary eye protection (spectacles or goggles).

[10] Non-sideshield spectacles are available for frontal protection only, but are not acceptable eye protection for the sources and operations listed for "impact."

[11] Ventilation should be adequate, but well protected from splash entry. Eye and face protection should be designed and used so that it provides both adequate ventilation and protects the wearer from splash entry.

[12] Protection from light radiation is directly related to filter lens density. See note (4). Select the darkest shade that allows task performance.

9. *Selection guidelines for head protection.* All head protection (helmets) is designed to provide protection from impact and penetration hazards caused by falling objects. Head protection is also available which provides protection from electric shock and burn. When selecting head protection, knowledge of potential electrical hazards is important. Class A helmets, in addition to impact and penetration resistance, provide electrical protection from low-voltage conductors (they are proof tested to 2,200 volts). Class B helmets, in addition to impact and penetration resistance, provide electrical protection from high-voltage conductors (they are proof tested to 20,000 volts). Class C helmets provide impact and penetration resistance (they are usually made of aluminum which conducts electricity), and should not be used around electrical hazards.

Where falling object hazards are present, helmets must be worn. Some examples include: working below other workers who are using tools and materials which could fall; working around or under conveyor belts which are carrying parts or materials; working below machinery or processes which might cause material or objects to fall; and working on exposed energized conductors.

Some examples of occupations for which head protection should be routinely considered are: carpenters, electricians, linemen, mechanics and repairers, plumbers and pipe fitters, assemblers, packers, wrappers, sawyers, welders, laborers, freight handlers, timber cutting and logging, stock handlers, and warehouse laborers.

10. *Selection guidelines for foot protection.* Safety shoes and boots which meet the ANSI Z41-1991 Standard provide both impact and compression protection. Where necessary, safety shoes can be obtained which provide puncture protection. In some work situations, metatarsal protection should be provided, and in other special situations electrical conductive or insulating safety shoes would be appropriate.

Safety shoes or boots with impact protection would be required for carrying or handling materials such as packages, objects, parts or heavy tools, which could be dropped; and, for other activities where objects might fall onto the feet. Safety shoes or boots with compression protection would be required for work activities involving skid trucks (manual material handling carts) around bulk rolls (such as paper rolls) and around heavy pipes, all of which could potentially roll over an employee's feet. Safety shoes or boots with puncture protection would be required where sharp objects such as nails, wire, tacks, screws, large staples, scrap metal etc., could be stepped on by employees causing a foot injury.

Some occupations (not a complete list) for which foot protection should be routinely considered are: shipping and receiving clerks, stock clerks, carpenters, electricians, machinists, mechanics and repairers, plumbers and pipe fitters, structural metal workers, assemblers, drywall installers and lathers, packers, wrappers, craters, punch and stamping press operators, sawyers, welders, laborers, freight handlers, gardeners and grounds-keepers, timber cutting and logging workers, stock handlers and warehouse laborers.

11. *Selection guidelines for hand protection.* Gloves are often relied upon to prevent cuts, abrasions, burns, and skin contact with chemicals that are capable of causing local or systemic effects following dermal exposure. OSHA is unaware of any gloves that provide protection against *all* potential hand hazards, and commonly available glove materials provide only limited protection against many chemicals. Therefore, it is important to select the

most appropriate glove for a particular application and to determine how long it can be worn, and whether it can be reused.

It is also important to know the performance characteristics of gloves relative to the specific hazard anticipated; e.g., chemical hazards, cut hazards, flame hazards, etc. These performance characteristics should be assessed by using standard test procedures. Before purchasing gloves, the employer should request documentation from the manufacturer that the gloves meet the appropriate test standard(s) for the hazard(s) anticipated.

Other factors to be considered for glove selection in general include:

(A) As long as the performance characteristics are acceptable, in certain circumstances, it may be more cost effective to regularly change cheaper gloves than to reuse more expensive types; and,

(B) The work activities of the employee should be studied to determine the degree of dexterity required, the duration, frequency, and degree of exposure of the hazard, and the physical stresses that will be applied.

With respect to selection of gloves for protection against chemical hazards:

(A) The toxic properties of the chemical(s) must be determined; in particular, the ability of the chemical to cause local effects on the skin and/or to pass through the skin and cause systemic effects;

(B) Generally, any "chemical resistant" glove can be used for dry powders;

(C) For mixtures and formulated products (unless specific test data are available), a glove should be selected on the basis of the chemical component with the shortest breakthrough time, since it is possible for solvents to carry active ingredients through polymeric materials; and,

(D) Employees must be able to remove the gloves in such a manner as to prevent skin contamination.

12. *Cleaning and maintenance.* It is important that all PPE be kept clean and properly maintained. Cleaning is particularly important for eye and face protection where dirty or fogged lenses could impair vision.

For the purposes of compliance with §1910.132(a) and (b), PPE should be inspected, cleaned, and maintained at regular intervals so that the PPE provides the requisite protection.

It is also important to ensure that contaminated PPE which cannot be decontaminated is disposed of in a manner that protects employees from exposure to hazards.

[59 FR 16362, Apr. 6, 1994]

# Subpart J—General Environmental Controls

**Authority:** Secs. 4, 6, and 8, Occupational Safety and Health Act of 1970, 29 U.S.C. 653, 655, 657; Secretary of Labor's Order No. 12-71 (36 FR 8754), 8-76 (41 FR 25059), 9-83 (48 FR 35736), 1-90 (55 FR 9033), or 6-96 (62 FR 111), as applicable. Sections 1910.141, 1910.142, 1910.145, 1910.146, and 1910.147 also issued under 29 CFR part 1911.

## §1910.141 Sanitation.

**1910.141(a)** *General*—**(1)** *Scope.* This section applies to permanent places of employment.

**1910.141(a)(2)** *Definitions applicable to this section.*

*Nonwater carriage toilet facility,* means a toilet facility not connected to a sewer.

*Number of employees* means, unless otherwise specified, the maximum number of employees present at any one time on a regular shift.

*Personal service room,* means a room used for activities not directly connected with the production or service function performed by the establishment. Such activities include, but are not limited to, first-aid, medical services, dressing, showering, toilet use, washing, and eating.

*Potable water* means water which meets the quality standards prescribed in the U.S. Public Health Service Drinking Water Standards, published in 42 CFR part 72, or water which is approved for drinking purposes by the State or local authority having jurisdiction.

*Toilet facility,* means a fixture maintained within a toilet room for the purpose of defecation or urination, or both.

*Toilet room,* means a room maintained within or on the premises of any place of employment, containing toilet facilities for use by employees.

*Toxic material* means a material in concentration or amount which exceeds the applicable limit established by a standard, such as §§1910.1000 and 1910.1001 or, in the absence of an applicable standard, which is of such toxicity so as to constitute a recognized hazard that is causing or is likely to cause death or serious physical harm.

*Urinal* means a toilet facility maintained within a toilet room for the sole purpose of urination.

*Water closet* means a toilet facility maintained within a toilet room for the purpose of both defecation and urination and which is flushed with water.

*Wet process* means any process or operation in a workroom which normally results in surfaces upon which employees may walk or stand becoming wet.

**1910.141(a)(3)** *Housekeeping.* **(i)** All places of employment shall be kept clean to the extent that the nature of the work allows.

**1910.141(a)(3)(ii)** The floor of every workroom shall be maintained, so far as practicable, in a dry condition. Where wet processes are used, drainage shall be maintained and false floors, platforms, mats, or other dry standing places shall be provided, where practicable, or appropriate waterproof footgear shall be provided.

**1910.141(a)(3)(iii)** To facilitate cleaning, every floor, working place, and passageway shall be kept free from protruding nails, splinters, loose boards, and unnecessary holes and openings.

**1910.141(a)(4)** *Waste disposal.* **(i)** Any receptacle used for putrescible solid or liquid waste or refuse shall be so constructed that it does not leak and may be thoroughly cleaned and maintained in a sanitary condition. Such a receptacle shall be equipped with a solid tight-fitting cover, unless it can be maintained in a sanitary condition without a cover. This requirement does not prohibit the use of receptacles which are designed to permit the maintenance of a sanitary condition without regard to the aforementioned requirements.

**1910.141(a)(4)(ii)** All sweepings, solid or liquid wastes, refuse, and garbage shall be removed in such a manner as to avoid creating a menace to health and as often as necessary or appropriate to maintain the place of employment in a sanitary condition.

**1910.141(a)(5)** *Vermin control.* Every enclosed workplace shall be so constructed, equipped, and maintained, so far as reasonably practicable, as to prevent the entrance or harborage of rodents, insects, and other vermin. A continuing and effective extermination program shall be instituted where their presence is detected.

**1910.141(b)** *Water supply*—**(1)** *Potable water.* **(i)** Potable water shall be provided in all places of employment, for drinking, washing of the person, cooking, washing of foods, washing of cooking or eating utensils, washing of food preparation or processing premises, and personal service rooms.

**1910.141(b)(1)(ii)** [Reserved]

**1910.141(b)(1)(iii)** Portable drinking water dispensers shall be designed, constructed, and serviced so that sanitary conditions are maintained, shall be capable of being closed, and shall be equipped with a tap.

**1910.141(b)(1)(iv)** [Reserved]

**1910.141(b)(1)(v)** Open containers such as barrels, pails, or tanks for drinking water from which the water must be dipped or poured, whether or not they are fitted with a cover, are prohibited.

**1910.141(b)(1)(vi)** A common drinking cup and other common utensils are prohibited.

**1910.141(b)(2)** *Nonpotable water.* **(i)** Outlets for nonpotable water, such as water for industrial or firefighting purposes, shall be posted or otherwise marked in a manner that will indicate clearly that the water is unsafe and is not to be used for drinking, washing of the person, cooking, washing of food, washing of cooking or eating utensils, washing of food preparation or processing premises, or personal service rooms, or for washing clothes.

**1910.141(b)(2)(ii)** Construction of nonpotable water systems or systems carrying any other nonpotable substance shall be such as to prevent backflow or backsiphonage into a potable water system.

**1910.141(b)(2)(iii)** Nonpotable water shall not be used for washing any portion of the person, cooking or eating utensils, or clothing. Nonpotable water may be used for cleaning work premises, other than food processing and preparation premises and personal service rooms: *Provided,* That this nonpotable water does not contain concentrations of chemicals, fecal coliform, or other substances which could create insanitary conditions or be harmful to employees.

**1910.141(c)** *Toilet facilities*—**(1)** *General.* **(i)** Except as otherwise indicated in this paragraph (c)(1)(i), toilet facilities, in toilet rooms separate for each sex, shall be provided in all places of employment in accordance with table J-1 of this section. The number of facilities to be provided for each sex shall be based on the number of employees of that sex for whom the facilities are furnished. Where toilet rooms will be occupied by no more than one person at a time, can be locked from the inside, and contain at least one water closet, separate toilet rooms for each sex need not be provided. Where such single-occupancy rooms have more than one toilet facility, only one such facility in each toilet room shall be counted for the purpose of table J-1.

### TABLE J-1

| Number of employees | Minimum number of water closets[1] |
|---|---|
| 1 to 15 | 1 |
| 16 to 35 | 2 |
| 36 to 55 | 3 |
| 56 to 80 | 4 |
| 81 to 110 | 5 |
| 111 to 150 | 6 |
| Over 150 | ([2]) |

[1] Where toilet facilities will not be used by women, urinals may be provided instead of water closets, except that the number of water closets in such cases shall not be reduced to less than 2/3 of the minimum specified.

[2] 1 additional fixture for each additional 40 employees.

**1910.141(c)(1)(ii)** The requirements of paragraph (c)(1)(i) of this section do not apply to mobile crews or to normally unattended work locations so long as employees working at these locations have transportation immediately available to nearby toilet facilities which meet the other requirements of this subparagraph.

**1910.141(c)(1)(iii)** The sewage disposal method shall not endanger the health of employees.

**1910.141(c)(2)** *Construction of toilet rooms.* **(i)** Each water closet shall occupy a separate compartment with a door and walls or partitions between fixtures sufficiently high to assure privacy.

**1910.141(c)(2)(ii)** [Reserved]

**1910.141(d)** *Washing facilities—* **(1)** *General.* Washing facilities shall be maintained in a sanitary condition.

**1910.141(d)(2)** *Lavatories.* **(i)** Lavatories shall be made available in all places of employment. The requirements of this subdivision do not apply to mobile crews or to normally unattended work locations if employees working at these locations have transportation readily available to nearby washing facilities which meet the other requirements of this paragraph.

**1910.141(d)(2)(ii)** Each lavatory shall be provided with hot and cold running water, or tepid running water.

**1910.141(d)(2)(iii)** Hand soap or similar cleansing agents shall be provided.

**1910.141(d)(2)(iv)** Individual hand towels or sections thereof, of cloth or paper, warm air blowers or clean individual sections of continuous cloth toweling, convenient to the lavatories, shall be provided.

**1910.141(d)(3)** *Showers.* **(i)** Whenever showers are required by a particular standard, the showers shall be provided in accordance with paragraphs (d)(3) (ii) through (v) of this section.

**1910.141(d)(3)(ii)** One shower shall be provided for each 10 employees of each sex, or numerical fraction thereof, who are required to shower during the same shift.

**1910.141(d)(3)(iii)** Body soap or other appropriate cleansing agents convenient to the showers shall be provided as specified in paragraph (d)(2)(iii) of this section.

**1910.141(d)(3)(iv)** Showers shall be provided with hot and cold water feeding a common discharge line.

**1910.141(d)(3)(v)** Employees who use showers shall be provided with individual clean towels.

**1910.141(e)** *Change rooms.* Whenever employees are required by a particular standard to wear protective clothing because of the possibility of contamination with toxic materials, change rooms equipped with storage facilities for street clothes and separate storage facilities for the protective clothing shall be provided.

**1910.141(f)** *Clothes drying facilities.* Where working clothes are provided by the employer and become wet or are washed between shifts, provision shall be made to insure that such clothing is dry before reuse.

**1910.141(g)** *Consumption of food and beverages on the premises—* **(1)** *Application.* This paragraph shall apply only where employees are permitted to consume food or beverages, or both, on the premises.

**1910.141(g)(2)** *Eating and drinking areas.* No employee shall be allowed to consume food or beverages in a toilet room nor in any area exposed to a toxic material.

**1910.141(g)(3)** *Waste disposal containers.* Receptacles constructed of smooth, corrosion resistant, easily cleanable, or disposable materials, shall be provided and used for the disposal of waste food. The number, size, and location of such receptacles shall encourage their use and not result in overfilling. They shall be emptied not less frequently than once each working day, unless unused, and shall be maintained in a clean and sanitary condition. Receptacles shall be provided with a solid tight-fitting cover unless sanitary conditions can be maintained without use of a cover.

**1910.141(g)(4)** *Sanitary storage.* No food or beverages shall be stored in toilet rooms or in an area exposed to a toxic material.

**1910.141(h)** *Food handling.* All employee food service facilities and operations shall be carried out in accordance with sound hygienic principles. In all places of employment where all or part of the food service is provided, the food dispensed shall be wholesome, free from spoilage, and shall be processed, prepared, handled, and stored in such a manner as to be protected against contamination.

[39 FR 23502, June 27, 1974, as amended at 40 FR 18446, April 28, 1975; 40 FR 23073, May 28, 1975; 43 FR 49748, Oct. 24, 1978; 63 FR 33466, June 18, 1998]

## §1910.142 Temporary labor camps.

**1910.142(a)** *Site.* **(1)** All sites used for camps shall be adequately drained. They shall not be subject to periodic flooding, nor located within 200 feet of swamps, pools, sink holes, or other surface collections of water unless such quiescent water surfaces can be subjected to mosquito control measures. The camp shall be located so the drainage from and through the camp will not endanger any domestic or public water supply. All sites shall be graded, ditched, and rendered free from depressions in which water may become a nuisance.

**1910.142(a)(2)** All sites shall be adequate in size to prevent overcrowding of necessary structures. The principal camp area in which food is prepared and served and where sleeping quarters are located shall be at least 500 feet from any area in which livestock is kept.

**1910.142(a)(3)** The grounds and open areas surrounding the shelters shall be maintained in a clean and sanitary condition free from rubbish, debris, waste paper, garbage, or other refuse.

**1910.142(b)** *Shelter.* **(1)** Every shelter in the camp shall be constructed in a manner which will provide protection against the elements.

**1910.142(b)(2)** Each room used for sleeping purposes shall contain at least 50 square feet of floor space for each occupant. At least a 7-foot ceiling shall be provided.

**1910.142(b)(3)** Beds, cots, or bunks, and suitable storage facilities such as wall lockers for clothing and personal articles shall be provided in every room used for sleeping purposes. Such beds or similar facilities shall be spaced not closer than 36 inches both laterally and end to end, and shall be elevated at least 12 inches from the floor. If double-deck bunks are used, they shall be spaced not less than 48 inches both laterally and end to end. The minimum clear space between the lower and upper bunk shall be not less than 27 inches. Triple-deck bunks are prohibited.

**1910.142(b)(4)** The floors of each shelter shall be constructed of wood, asphalt, or concrete. Wooden floors shall be of smooth and tight construction. The floors shall be kept in good repair.

**1910.142(b)(5)** All wooden floors shall be elevated not less than 1 foot above the ground level at all points to prevent dampness and to permit free circulation of air beneath.

**1910.142(b)(6)** Nothing in this section shall be construed to prohibit "banking" with earth or other suitable material around the outside walls in areas subject to extreme low temperatures.

**1910.142(b)(7)** All living quarters shall be provided with windows the total of which shall be not less than one-tenth of the floor area. At least one-half of each window shall be so constructed that it can be opened for purposes of ventilation.

**1910.142(b)(8)** All exterior openings shall be effectively screened with 16-mesh material. All screen doors shall be equipped with self-closing devices.

**1910.142(b)(9)** In a room where workers cook, live, and sleep a minimum of 100 square feet per person shall be provided. Sanitary facilities shall be provided for storing and preparing food.

**1910.142(b)(10)** In camps where cooking facilities are used in common, stoves (in ratio of one stove to 10 persons or one stove to two families) shall be provided in an enclosed and screened shelter. Sanitary facilities shall be provided for storing and preparing food.

**1910.142(b)(11)** All heating, cooking, and water heating equipment shall be installed in accordance with State and local ordinances, codes, and regulations governing such installations. If a camp is used during cold weather, adequate heating equipment shall be provided.

**1910.142(c)** *Water supply.* **(1)** An adequate and convenient water supply, approved by the appropriate health authority, shall be provided in each camp for drinking, cooking, bathing, and laundry purposes.

**1910.142(c)(2)** A water supply shall be deemed adequate if it is capable of delivering 35 gallons per person per day to the campsite at a peak rate of 2 1/2 times the average hourly demand.

**1910.142(c)(3)** The distribution lines shall be capable of supplying water at normal operating pressures to all fixtures for simultaneous operation. Water outlets shall be distributed throughout the camp in such a manner that no shelter is more than 100 feet from a yard hydrant if water is not piped to the shelters.

**1910.142(c)(4)** Where water under pressure is available, one or more drinking fountains shall be provided for each 100 occupants or fraction thereof.

**1910.142(d)** *Toilet facilities.* **(1)** Toilet facilities adequate for the capacity of the camp shall be provided.

**1910.142(d)(2)** Each toilet room shall be located so as to be accessible without any individual passing through any sleeping room. Toilet rooms shall have a window not less than 6 square feet in area opening directly to the outside area or otherwise be satisfactorily ventilated. All outside openings shall be screened with 16-mesh material. No fixture, water closet, chemical toilet, or urinal shall be located in a room used for other than toilet purposes.

**1910.142(d)(3)** A toilet room shall be located within 200 feet of the door of each sleeping room. No privy shall be closer than 100 feet to any sleeping room, dining room, lunch area, or kitchen.

**1910.142(d)(4)** Where the toilet rooms are shared, such as in multifamily shelters and in barracks type facilities, separate toilet rooms shall be provided for each sex. These rooms shall be distinctly marked "for men" and "for women" by signs printed in English and in the native language of the persons occupying the camp, or marked with easily understood pictures or symbols. If the facilities for each sex are in the same building, they shall be separated by solid walls or partitions extending from the floor to the roof or ceiling.

**1910.142(d)(5)** Where toilet facilities are shared, the number of water closets or privy seats provided for each sex shall be based on the maximum number of persons of that sex which the camp is designed to house at any one time, in the ratio of one such unit to each 15 persons, with a minimum of two units for any shared facility.

**1910.142(d)(6)** Urinals shall be provided on the basis of one unit or 2 linear feet of urinal trough for each 25 men. The floor from the wall and for a distance not less than 15 inches measured from the outward edge of the urinals shall be constructed of materials impervious to moisture. Where water under pressure is available, urinals shall be provided with an adequate water flush. Urinal troughs in privies shall drain freely into the pit or vault and the construction of this drain shall be such as to exclude flies and rodents from the pit.

**1910.142(d)(7)** Every water closet installed on or after August 31, 1971, shall be located in a toilet room.

**1910.142(d)(8)** Each toilet room shall be lighted naturally, or artificially by a safe type of lighting at all hours of the day and night.

**1910.142(d)(9)** An adequate supply of toilet paper shall be provided in each privy, water closet, or chemical toilet compartment.

**1910.142(d)(10)** Privies and toilet rooms shall be kept in a sanitary condition. They shall be cleaned at least daily.

**1910.142(e)** *Sewage disposal facilities.* In camps where public sewers are available, all sewer lines and floor drains from buildings shall be connected thereto.

**1910.142(f)** *Laundry, handwashing, and bathing facilities.* **(1)** Laundry, handwashing, and bathing facilities shall be provided in the following ratio:

**1910.142(f)(1)(i)** Handwash basin per family shelter or per six persons in shared facilities.

**1910.142(f)(1)(ii)** Shower head for every 10 persons.

**1910.142(f)(1)(iii)** Laundry tray or tub for every 30 persons.

**1910.142(f)(1)(iv)** Slop sink in each building used for laundry, hand washing, and bathing.

**1910.142(f)(2)** Floors shall be of smooth finish but not slippery materials; they shall be impervious to moisture. Floor drains shall be provided in all shower baths, shower rooms, or laundry rooms to remove waste water and facilitate cleaning. All junctions of the curbing and the floor shall be coved. The walls and partitions of shower rooms shall be smooth and impervious to the height of splash.

**1910.142(f)(3)** An adequate supply of hot and cold running water shall be provided for bathing and laundry purposes. Facilities for heating water shall be provided.

**1910.142(f)(4)** Every service building shall be provided with equipment capable of maintaining a temperature of at least 70 °F. during cold weather.

**1910.142(f)(5)** Facilities for drying clothes shall be provided.

**1910.142(f)(6)** All service buildings shall be kept clean.

**1910.142(g)** *Lighting.* Where electric service is available, each habitable room in a camp shall be provided with at least one ceiling-type light fixture and at least one separate floor- or wall-type convenience outlet. Laundry and toilet rooms and rooms where people congregate shall contain at least one ceiling- or wall-type fixture. Light levels in toilet and storage rooms shall be at least 20 foot-candles 30 inches from the floor. Other rooms, including kitchens and living quarters, shall be at least 30 foot-candles 30 inches from the floor.

**1910.142(h)** *Refuse disposal.* **(1)** Fly-tight, rodent-tight, impervious, cleanable or single service containers, approved by the appropriate health authority shall be provided for the storage of garbage. At least one such container shall be provided for each family shelter and shall be located within 100 feet of each shelter on a wooden, metal, or concrete stand.

**1910.142(h)(2)** Garbage containers shall be kept clean.

**1910.142(h)(3)** Garbage containers shall be emptied when full, but not less than twice a week.

**1910.142(i)** *Construction and operation of kitchens, dining hall, and feeding facilities.* **(1)** In all camps where central dining or multiple family feeding operations are permitted or provided, the food handling facilities shall comply with the requirements of the "Food Service Sanitation Ordinance and Code," Part V of the "Food Service Sanitation Manual," U.S. Public Health Service Publication 934 (1965), which is incorporated by reference as specified in §1910.6.

**1910.142(i)(2)** A properly constructed kitchen and dining hall adequate in size, separate from the sleeping quarters of any of the workers or their families, shall be provided in connection with all food handling facilities. There shall be no direct opening from living or sleeping quarters into a kitchen or dining hall.

**1910.142(i)(3)** No person with any communicable disease shall be employed or permitted to work in the preparation, cooking, serving, or other handling of food, foodstuffs, or materials used therein, in any kitchen or dining room operated in connection with a camp or regularly used by persons living in a camp.

**1910.142(j)** *Insect and rodent control.* Effective measures shall be taken to prevent infestation by and harborage of animal or insect vectors or pests.

**1910.142(k)** *First aid.* **(1)** Adequate first aid facilities approved by a health authority shall be maintained and made available in every labor camp for the emergency treatment of injured persons.

**1910.142(k)(2)** Such facilities shall be in charge of a person trained to administer first aid and shall be readily accessible for use at all times.

**1910.142(l)** *Reporting communicable disease.* **(1)** It shall be the duty of the camp superintendent to report immediately to the local health officer the name and address of any individual in the camp known to have or suspected of having a communicable disease.

**1910.142(l)(2)** Whenever there shall occur in any camp a case of suspected food poisoning or an unusual prevalence of any illness in which fever, diarrhea, sore throat, vomiting, or jaundice is a prominent symptom, it shall be the duty of the camp superintendent to report immediately the existence of the outbreak to the health authority " by telegram, telephone, electronic mail or any method that is equally fast.".

[39 FR 23502, June 27, 1974, as amended at 47 FR 14696, Apr. 6, 1982; 49 FR 18295, Apr. 30, 1984; 61 FR 9238, Mar. 7, 1996; 63 FR 33466, June 18, 1998]

## §1910.143 Nonwater carriage disposal systems [Reserved].

## §1910.144 Safety color code for marking physical hazards.

**1910.144(a)** *Color identification*—**(1)** *Red.* Red shall be the basic color for the identification of:

**1910.144(a)(1)(i)** *Fire protection equipment and apparatus.* [Reserved]

**1910.144(a)(1)(ii)** *Danger.* Safety cans or other portable containers of flammable liquids having a flash point at or below 80 °F, table containers of flammable liquids (open cup tester), excluding shipping containers, shall be painted red with some additional clearly visible identification either in the form of a yellow band around the can or the name of the contents conspicuously stenciled or painted on the can in yellow. Red lights shall be provided at barricades and at temporary obstructions, as specified in ANSI Safety Code for Building Construction, A10.2-1944, which is incorporated by reference as specified in §1910.6. Danger signs shall be painted red.

**1910.144(a)(1)(iii)** *Stop.* Emergency stop bars on hazardous machines such as rubber mills, wire blocks, flat work ironers, etc., shall be red. Stop buttons or electrical switches which letters or other markings appear, used for emergency stopping of machinery shall be red.

**1910.144(a)(2)** [Reserved]

**1910.144(a)(3)** *Yellow.* Yellow shall be the basic color for designating caution and for marking physical hazards such as: Striking against, stumbling, falling, tripping, and "caught in between."

**1910.144(b)** [Reserved]

[39 FR 23502, June 27, 1974, as amended at 43 FR 49748, Oct. 24, 1978; 49 FR 5322, Feb. 10, 1984; 61 FR 9239, Mar. 7, 1996]

### §1910.145 Specifications for accident prevention signs and tags.

**1910.145(a)** *Scope.* **(1)** These specifications apply to the design, application, and use of signs or symbols (as included in paragraphs (c) through (e) of this section) intended to indicate and, insofar as possible, to define specific hazards of a nature such that failure to designate them may lead to accidental injury to workers or the public, or both, or to property damage. These specifications are intended to cover all safety signs except those designed for streets, highways, railroads, and marine regulations. These specifications do not apply to plant bulletin boards or to safety posters.

**1910.145(a)(2)** All new signs and replacements of old signs shall be in accordance with these specifications.

**1910.145(b)** *Definitions.* As used in this section, the word *sign* refers to a surface on prepared for the warning of, or safety instructions of, industrial workers or members of the public who may be exposed to hazards. Excluded from this definition, however, are news releases, displays commonly known as safety posters, and bulletins used for employee education.

**1910.145(c)** *Classification of signs according to use*—**(1)** *Danger signs.* (i) There shall be no variation in the type of design of signs posted to warn of specific dangers and radiation hazards.

**1910.145(c)(1)(ii)** All employees shall be instructed that danger signs indicate immediate danger and that special precautions are necessary.

**1910.145(c)(2)** *Caution signs.* (i) Caution signs shall be used only to warn against potential hazards or to caution against unsafe practices.

**1910.145(c)(2)(ii)** All employees shall be instructed that caution signs indicate a possible hazard against which proper precaution should be taken.

**1910.145(c)(3)** *Safety instruction signs.* Safety instruction signs shall be used where there is a need for general instructions and suggestions relative to safety measures.

**1910.145(d)** *Sign design*—**(1)** *Design features.* All signs shall be furnished with rounded or blunt corners and shall be free from sharp edges, burrs, splinters, or other sharp projections. The ends or heads of bolts or other fastening devices shall be located in such a way that they do not constitute a hazard.

**1910.145(d)(2)** *Danger signs.* The colors red, black, and white shall be those of opaque glossy samples as specified in Table 1 of Fundamental Specification of Safety Colors for CIE Standard Source "C", American National Standard Z53.1-1967, which is incorporated by reference as specified in §1910.6.

**1910.145(d)(3)** [Reserved]

**1910.145(d)(4)** *Caution signs.* Standard color of the background shall be yellow; and the panel, black with yellow letters. Any letters used against the yellow background shall be black. The colors shall be those of opaque glossy samples as specified in Table 1 of American National Standard Z53.1-1967.

**1910.145(d)(5)** [Reserved]

**1910.145(d)(6)** *Safety instruction signs.* Standard color of the background shall be white; and the panel, green with white letters. Any letters used against the white background shall be black. The colors shall be those of opaque glossy samples as specified in Table 1 of American National Standard Z53.1-1967.

**1910.145(d)(7)-(9)** [Reserved]

**1910.145(d)(10)** *Slow-moving vehicle emblem.* This emblem (see fig. J-7) consists of a fluorescent yellow-orange triangle with a dark red reflective border. The yellow-orange fluorescent triangle is a highly visible color for daylight exposure. The reflective border defines the shape of the fluorescent color in daylight and creates a hollow red triangle in the path of motor vehicle headlights at night. The emblem is intended as a unique identification for, and it shall be used only on, vehicles which by design move slowly (25 m.p.h. or less) on the public roads. The emblem is not a clearance marker for wide machinery nor is it intended to replace required lighting or marking of slow-moving vehicles. Neither the color film pattern and its dimensions nor the backing shall be altered to permit use of advertising or other markings. The material, location, mounting, etc., of the emblem shall be in accordance with the American Society of Agricultural Engineers Emblem for Identifying Slow-Moving Vehicles, ASAE R276, 1967, or ASAE S276.2 (ANSI B114.1-1971), which are incorporated by reference as specified in §1910.6.

**FIGURE J-7—SLOW-MOVING VEHICLE EMBLEM**
**Note:** All dimensions are in inches.

**1910.145(e)** *Sign wordings.*

**1910.145(e)(1)** [Reserved]

**1910.145(e)(2)** *Nature of wording.* The wording of any sign should be easily read and concise. The sign should contain sufficient information to be easily understood. The wording should make a positive, rather than negative suggestion and should be accurate in fact.

**1910.145(e)(3)** [Reserved]

**1910.145(e)(4)** *Biological hazard signs.* The biological hazard warning shall be used to signify the actual or potential presence of a biohazard and to identify equipment, containers, rooms, materials, experimental animals, or combinations thereof, which contain, or are contaminated with, viable hazardous agents. For the purpose of this subparagraph the term "biological hazard," or "biohazard," shall include only those infectious agents presenting a risk or potential risk to the well-being of man.

**1910.145(f)** *Accident prevention tags*—**(1)** *Scope and application.* (i) This paragraph (f) applies to all accident prevention tags used to identify hazardous conditions and provide a message to employees with respect to hazardous conditions as set forth in paragraph (f)(3) of this section, or to meet the specific tagging requirements of other OSHA standards.

**1910.145(f)(1)(ii)** This paragraph (f) does not apply to construction, maritime or agriculture.

**1910.145(f)(2)** *Definitions.* Biological hazard or BIOHAZARD means those infectious agents presenting a risk of death, injury or illness to employees.

*Major message* means that portion of a tag's inscription that is more specific than the signal word and that indicates the specific hazardous condition or the instruction to be communicated to the employee. Examples include: "High Voltage," "Close Clearance," "Do Not Start," or "Do Not Use" or a corresponding pictograph used with a written text or alone.

*Pictograph* means a pictorial representation used to identify a hazardous condition or to convey a safety instruction.

*Signal word* means that portion of a tag's inscription that contains the word or words that are intended to capture the employee's immediate attention.

*Tag* means a device usually made of card, paper, pasteboard, plastic or other material used to identify a hazardous condition.

**1910.145(f)(3)** *Use.* Tags shall be used as a means to prevent accidental injury or illness to employees who are exposed to hazardous or potentially

# OSHA Standards for General Industry

hazardous conditions, equipment or operations which are out of the ordinary, unexpected or not readily apparent. Tags shall be used until such time as the identified hazard is eliminated or the hazardous operation is completed. Tags need not be used where signs, guarding or other positive means of protection are being used.

**1910.145(f)(4)** *General tag criteria.* All required tags shall meet the following criteria:

**1910.145(f)(4)(i)** Tags shall contain a signal word and a major message.

**1910.145(f)(4)(i)(A)** The signal word shall be either "Danger," "Caution," or "Biological Hazard," "BIOHAZARD," or the biological hazard symbol.

**1910.145(f)(4)(i)(B)** The major message shall indicate the specific hazardous condition or the instruction to be communicated to the employee.

**1910.145(f)(4)(ii)** The signal word shall be readable at a minimum distance of five feet (1.52 m) or such greater distance as warranted by the hazard.

**1910.145(f)(4)(iii)** The tag's major message shall be presented in either pictographs, written text or both.

**1910.145(f)(4)(iv)** The signal word and the major message shall be understandable to all employees who may be exposed to the identified hazard.

**1910.145(f)(4)(v)** All employees shall be informed as to the meaning of the various tags used throughout the workplace and what special precautions are necessary.

**1910.145(f)(4)(vi)** Tags shall be affixed as close as safely possible to their respective hazards by a positive means such as string, wire, or adhesive that prevents their loss or unintentional removal.

**1910.145(f)(5)** *Danger tags.* Danger tags shall be used in major hazard situations where an immediate hazard presents a threat of death or serious injury to employees. Danger tags shall be used only in these situations.

**1910.145(f)(6)** *Caution tags.* Caution tags shall be used in minor hazard situations where a non-immediate or potential hazard or unsafe practice presents a lesser threat of employee injury. Caution tags shall be used only in these situations.

**1910.145(f)(7)** *Warning tags.* Warning tags may be used to represent a hazard level between "Caution" and "Danger," instead of the required "Caution" tag, provided that they have a signal word of "Warning," an appropriate major message, and otherwise meet the general tag criteria of paragraph (f)(4) of this section.

**1910.145(f)(8)** *Biological hazard tags.* **(i)** Biological hazard tags shall be used to identify the actual or potential presence of a biological hazard and to identify equipment, containers, rooms, experimental animals, or combinations thereof, that contain or are contaminated with hazardous biological agents.

**1910.145(f)(8)(ii)** The symbol design for biological hazard tags shall conform to the design shown below:

**BIOLOGICAL HAZARD SYMBOL CONFIGURATION**

**1910.145(f)(9)** *Other tags.* Other tags may be used in addition to those required by this paragraph (f), or in other situations where this paragraph (f) does not require tags, provided that they do not detract from the impact or visibility of the signal word and major message of any required tag.

## APPENDICES TO §1910.145(f)—ACCIDENT PREVENTION TAGS

### APPENDIX A TO §1910.145(f)—RECOMMENDED COLOR CODING

While the standard does not specifically mandate colors to be used on accident prevention tags, the following color scheme is recommended by OSHA for meeting the requirements of this section:

"DANGER"—Red, or predominantly red, with lettering or symbols in a contrasting color.

"CAUTION"—Yellow, or predominantly yellow, with lettering or symbols in a contrasting color.

"WARNING"—Orange, or predominantly orange, with lettering or symbols in a contrasting color.

"BIOLOGICAL HAZARD"—Fluorescent orange or orange-red, or predominantly so, with lettering or symbols in a contrasting color.

### APPENDIX B TO §1910.145(f)—REFERENCES FOR FURTHER INFORMATION

The following references provide information which can be helpful in understanding the requirements contained in various sections of the standard:

1. Bresnahan, Thomas F., and Bryk, Joseph, "The Hazard Association Values of Accident Prevention Signs", *Journal of American Society of Safety Engineers*; January 1975.

2. Dreyfuss, H., *Symbol Sourcebook*, McGraw Hill; New York, NY, 1972.

3. Glass, R.A. and others, *Some Criteria for Colors and Signs in Workplaces*, National Bureau of Standards, Washington DC, 1983.

4. *Graphic Symbols for Public Areas and Occupational Environments*, Treasury Board of Canada, Ottawa, Canada, July 1980.

5. Howett, G.L., *Size of Letters Required for Visibility as a Function of Viewing Distance and Observer Acuity*, National Bureau of Standards, Washington DC, July 1983.

6. Lerner, N.D. and Collins, B.L., *The Assessment of Safety Symbol Understandability by Different Testing Methods*, National Bureau of Standards, Washington DC, 1980.

7. Lerner, N.D. and Collins, B.L., *Workplace Safety Symbols*, National Bureau of Standards, Washington DC, 1980.

8. Modley, R. and Meyers, W.R., *Handbook of Pictorial Symbols*, Dover Publication, New York, NY, 1976.

9. *Product Safety Signs and Labels*, FMC Corporation, Santa Clara, CA, 1978.

10. *Safety Color Coding for Marking Physical Hazards*, Z53.1, American National Standards Institute, New York, NY, 1979.

11. *Signs and Symbols for the Occupational Environment*, Can. 3-Z-321-77, Canadian Standards Association, Ottawa, September 1977.

12. *Symbols for Industrial Safety*, National Bureau of Standards, Washington DC, April 1982.

13. *Symbol Signs*, U.S. Department of Transportation, Washington DC, November 1974.

[39 FR 23502, June 27, 1974, as amended at 43 FR 49749, Oct. 24, 1978; 43 FR 51759, Nov. 7, 1978; 49 FR 5322, Feb. 10, 1984; 51 FR 33260, Sept. 19, 1986; 61 FR 9239, Mar. 7, 1996]

## §1910.146 Permit-required confined spaces.

**1910.146(a)** *Scope and application.* This section contains requirements for practices and procedures to protect employees in general industry from the hazards of entry into permit-required confined spaces. This section does not apply to agriculture, to construction, or to shipyard employment (Parts 1928, 1926, and 1915 of this chapter, respectively).

**1910.146(b)** *Definitions.*

*Acceptable entry conditions* means the conditions that must exist in a permit space to allow entry and to ensure that employees involved with a permit-required confined space entry can safely enter into and work within the space.

*Attendant* means an individual stationed outside one or more permit spaces who monitors the authorized entrants and who performs all attendant's duties assigned in the employer's permit space program.

*Authorized entrant* means an employee who is authorized by the employer to enter a permit space.

*Blanking or blinding* means the absolute closure of a pipe, line, or duct by the fastening of a solid plate (such as a spectacle blind or a skillet blind) that completely covers the bore and that is capable of withstanding the maximum pressure of the pipe, line, or duct with no leakage beyond the plate.

*Confined space* means a space that:

(1) Is large enough and so configured that an employee can bodily enter and perform assigned work; and

(2) Has limited or restricted means for entry or exit (for example, tanks, vessels, silos, storage bins, hoppers, vaults, and pits are spaces that may have limited means of entry.); and

(3) Is not designed for continuous employee occupancy.

*Double block and bleed* means the closure of a line, duct, or pipe by closing and locking or tagging two in-line valves and by opening and locking or tagging a drain or vent valve in the line between the two closed valves.

*Emergency* means any occurrence (including any failure of hazard control or monitoring equipment) or event internal or external to the permit space that could endanger entrants.

*Engulfment* means the surrounding and effective capture of a person by a liquid or finely divided (flowable) solid substance that can be aspirated to cause death by filling or plugging the respiratory system or that can exert enough force on the body to cause death by strangulation, constriction, or crushing.

*Entry* means the action by which a person passes through an opening into a permit-required confined space. Entry includes ensuing work activities in that space and is considered to have occurred as soon as any part of the entrant's body breaks the plane of an opening into the space.

*Entry permit* (permit) means the written or printed document that is provided by the employer to allow and control entry into a permit space and that contains the information specified in paragraph (f) of this section.

*Entry supervisor* means the person (such as the employer, foreman, or crew chief) responsible for determining if acceptable entry conditions are present at a permit space where entry is planned, for authorizing entry and overseeing entry operations, and for terminating entry as required by this section.

**NOTE:** An entry supervisor also may serve as an attendant or as an authorized entrant, as long as that person is trained and equipped as required by this section for each role he or she fills. Also, the duties of entry supervisor may be passed from one individual to another during the course of an entry operation.

*Hazardous atmosphere* means an atmosphere that may expose employees to the risk of death, incapacitation, impairment of ability to self-rescue (that is, escape unaided from a permit space), injury, or acute illness from one or more of the following causes:

(1) Flammable gas, vapor, or mist in excess of 10 percent of its lower flammable limit (LFL);

(2) Airborne combustible dust at a concentration that meets or exceeds its LFL;

**NOTE:** This concentration may be approximated as a condition in which the dust obscures vision at a distance of 5 feet (1.52 m) or less.

(3) Atmospheric oxygen concentration below 19.5 percent or above 23.5 percent;

(4) Atmospheric concentration of any substance for which a dose or a permissible exposure limit is published in Subpart G, *Occupational Health and Environmental Control*, or in Subpart Z, *Toxic and Hazardous Substances*, of this part and which could result in employee exposure in excess of its dose or permissible exposure limit;

**NOTE:** An atmospheric concentration of any substance that is not capable of causing death, incapacitation, impairment of ability to self-rescue, injury, or acute illness due to its health effects is not covered by this provision.

(5) Any other atmospheric condition that is immediately dangerous to life or health.

**NOTE:** For air contaminants for which OSHA has not determined a dose or permissible exposure limit, other sources of information, such as Material Safety Data Sheets that comply with the Hazard Communication Standard, §1910.1200 of this part, published information, and internal documents can provide guidance in establishing acceptable atmospheric conditions.

*Hot work permit* means the employer's written authorization to perform operations (for example, riveting, welding, cutting, burning, and heating) capable of providing a source of ignition.

*Immediately dangerous to life or health* (IDLH) means any condition that poses an immediate or delayed threat to life or that would cause irreversible adverse health effects or that would interfere with an individual's ability to escape unaided from a permit space.

**NOTE:** Some materials—hydrogen fluoride gas and cadmium vapor, for example—may produce immediate transient effects that, even if severe, may pass without medical attention, but are followed by sudden, possibly fatal collapse 12–72 hours after exposure. The victim "feels normal" from recovery from transient effects until collapse. Such materials in hazardous quantities are considered to be "immediately" dangerous to life or health.

*Inerting* means the displacement of the atmosphere in a permit space by a noncombustible gas (such as nitrogen) to such an extent that the resulting atmosphere is noncombustible.

**NOTE:** This procedure produces an IDLH oxygen-deficient atmosphere.

*Isolation* means the process by which a permit space is removed from service and completely protected against the release of energy and material into the space by such means as: blanking or blinding; misaligning or removing sections of lines, pipes, or ducts; a double block and bleed system; lockout or tagout of all sources of energy; or blocking or disconnecting all mechanical linkages.

*Line breaking* means the intentional opening of a pipe, line, or duct that is or has been carrying flammable, corrosive, or toxic material, an inert gas, or any fluid at a volume, pressure, or temperature capable of causing injury.

*Non-permit confined space* means a confined space that does not contain or, with respect to atmospheric hazards, have the potential to contain any hazard capable of causing death or serious physical harm.

*Oxygen deficient atmosphere* means an atmosphere containing less than 19.5 percent oxygen by volume.

*Oxygen enriched atmosphere* means an atmosphere containing more than 23.5 percent oxygen by volume.

*Permit-required confined space* (permit space) means a confined space that has one or more of the following characteristics:

(1) Contains or has a potential to contain a hazardous atmosphere;

(2) Contains a material that has the potential for engulfing an entrant;

(3) Has an internal configuration such that an entrant could be trapped or asphyxiated by inwardly converging walls or by a floor which slopes downward and tapers to a smaller cross-section; or

(4) Contains any other recognized serious safety or health hazard.

*Permit-required confined space program* (permit space program) means the employer's overall program for controlling, and, where appropriate, for protecting employees from, permit space hazards and for regulating employee entry into permit spaces.

*Permit system* means the employer's written procedure for preparing and issuing permits for entry and for returning the permit space to service following termination of entry.

*Prohibited condition* means any condition in a permit space that is not allowed by the permit during the period when entry is authorized.

*Rescue service* means the personnel designated to rescue employees from permit spaces.

*Retrieval system* means the equipment (including a retrieval line, chest or full-body harness, wristlets, if appropriate, and a lifting device or anchor) used for non-entry rescue of persons from permit spaces.

*Testing* means the process by which the hazards that may confront entrants of a permit space are identified and evaluated. Testing includes specifying the tests that are to be performed in the permit space.

**NOTE:** Testing enables employers both to devise and implement adequate control measures for the protection of authorized entrants and to determine if acceptable entry conditions are present immediately prior to, and during, entry.

**1910.146(c)** *General requirements.* **(1)** The employer shall evaluate the workplace to determine if any spaces are permit-required confined spaces.

**NOTE:** Proper application of the decision flow chart in appendix A to §1910.146 would facilitate compliance with this requirement.

**1910.146(c)(2)** If the workplace contains permit spaces, the employer shall inform exposed employees, by posting danger signs or by any other equally effective means, of the existence and location of and the danger posed by the permit spaces.

**NOTE:** A sign reading "DANGER—PERMIT-REQUIRED CONFINED SPACE, DO NOT ENTER" or using other similar language would satisfy the requirement for a sign.

**1910.146(c)(3)** If the employer decides that its employees will not enter permit spaces, the employer shall take effective measures to prevent its employees from entering the permit spaces and shall comply with paragraphs (c)(1), (c)(2), (c)(6), and (c)(8) of this section.

**1910.146(c)(4)** If the employer decides that its employees will enter permit spaces, the employer shall develop and implement a written permit space program that complies with this section. The written program shall be available for inspection by employees and their authorized representatives.

**1910.146(c)(5)** An employer may use the alternate procedures specified in paragraph (c)(5)(ii) of this section for entering a permit space under the conditions set forth in paragraph (c)(5)(i) of this section.

**1910.146(c)(5)(i)** An employer whose employees enter a permit space need not comply with paragraphs (d) through (f) and (h) through (k) of this section, provided that:

**1910.146(c)(5)(i)(A)** The employer can demonstrate that the only hazard posed by the permit space is an actual or potential hazardous atmosphere;

**1910.146(c)(5)(i)(B)** The employer can demonstrate that continuous forced air ventilation alone is sufficient to maintain that permit space safe for entry;

**1910.146(c)(5)(i)(C)** The employer develops monitoring and inspection data that supports the demonstrations required by paragraphs (c)(5)(i)(A) and (c)(5)(i)(B) of this section;

**1910.146(c)(5)(i)(D)** If an initial entry of the permit space is necessary to obtain the data required by paragraph (c)(5)(i)(C) of this section, the entry is performed in compliance with paragraphs (d) through (k) of this section;

**1910.146(c)(5)(i)(E)** The determinations and supporting data required by paragraphs (c)(5)(i)(A), (c)(5)(i)(B), and (c)(5)(i)(C) of this section are documented by the employer and are made available to each employee who enters the permit space under the terms of paragraph (c)(5) of this section or to that employee's authorized representative; and

**1910.146(c)(5)(i)(F)** Entry into the permit space under the terms of paragraph (c)(5)(i) of this section is performed in accordance with the requirements of paragraph (c)(5)(ii) of this section.

**NOTE:** See paragraph (c)(7) of this section for reclassification of a permit space after all hazards within the space have been eliminated.

**1910.146(c)(5)(ii)** The following requirements apply to entry into permit spaces that meet the conditions set forth in paragraph (c)(5)(i) of this section.

**1910.146(c)(5)(ii)(A)** Any conditions making it unsafe to remove an entrance cover shall be eliminated before the cover is removed.

**1910.146(c)(5)(ii)(B)** When entrance covers are removed, the opening shall be promptly guarded by a railing, temporary cover, or other temporary barrier that will prevent an accidental fall through the opening and that will protect each employee working in the space from foreign objects entering the space.

**1910.146(c)(5)(ii)(C)** Before an employee enters the space, the internal atmosphere shall be tested, with a calibrated direct-reading instrument, for oxygen content, for flammable gases and vapors, and for potential toxic air contaminants, in that order. Any employee who enters the space, or that employee's authorized representative, shall be provided an opportunity to observe the pre-entry testing required by this paragraph.

**1910.146(c)(5)(ii)(D)** There may be no hazardous atmosphere within the space whenever any employee is inside the space.

**1910.146(c)(5)(ii)(E)** Continuous forced air ventilation shall be used, as follows:

**1910.146(c)(5)(ii)(E)(1)** An employee may not enter the space until the forced air ventilation has eliminated any hazardous atmosphere;

**1910.146(c)(5)(ii)(E)(2)** The forced air ventilation shall be so directed as to ventilate the immediate areas where an employee is or will be present within the space and shall continue until all employees have left the space;

**1910.146(c)(5)(ii)(E)(3)** The air supply for the forced air ventilation shall be from a clean source and may not increase the hazards in the space.

**1910.146(c)(5)(ii)(F)** The atmosphere within the space shall be periodically tested as necessary to ensure that the continuous forced air ventilation is preventing the accumulation of a hazardous atmosphere. Any employee who enters the space, or that employee's authorized representative, shall be provided with an opportunity to observe the periodic testing required by this paragraph.

**1910.146(c)(5)(ii)(G)** If a hazardous atmosphere is detected during entry:

**1910.146(c)(5)(ii)(G)(1)** Each employee shall leave the space immediately;

**1910.146(c)(5)(ii)(G)(2)** The space shall be evaluated to determine how the hazardous atmosphere developed; and

**1910.146(c)(5)(ii)(G)(3)** Measures shall be implemented to protect employees from the hazardous atmosphere before any subsequent entry takes place.

**1910.146(c)(5)(ii)(H)** The employer shall verify that the space is safe for entry and that the pre-entry measures required by paragraph (c)(5)(ii) of this section have been taken, through a written certification that contains the date, the location of the space, and the signature of the person providing the certification. The certification shall be made before entry and shall be made available to each employee entering the space or to that employee's authorized representative.

**1910.146(c)(6)** When there are changes in the use or configuration of a non-permit confined space that might increase the hazards to entrants, the employer shall reevaluate that space and, if necessary, reclassify it as a permit-required confined space.

**1910.146(c)(7)** A space classified by the employer as a permit-required confined space may be reclassified as a non-permit confined space under the following procedures:

**1910.146(c)(7)(i)** If the permit space poses no actual or potential atmospheric hazards and if all hazards within the space are eliminated without entry into the space, the permit space may be reclassified as a non-permit confined space for as long as the non-atmospheric hazards remain eliminated.

**1910.146(c)(7)(ii)** If it is necessary to enter the permit space to eliminate hazards, such entry shall be performed under paragraphs (d) through (k) of this section. If testing and inspection during that entry demonstrate that the hazards within the permit space have been eliminated, the permit space may be reclassified as a non-permit confined space for as long as the hazards remain eliminated.

**NOTE:** Control of atmospheric hazards through forced air ventilation does not constitute elimination of the hazards. Paragraph (c)(5) covers permit space entry where the employer can demonstrate that forced air ventilation alone will control all hazards in the space.

**1910.146(c)(7)(iii)** The employer shall document the basis for determining that all hazards in a permit space have been eliminated, through a certification that contains the date, the location of the space, and the signature of the person making the determination. The certification shall be made available to each employee entering the space or to that employee's authorized representative.

**1910.146(c)(7)(iv)** If hazards arise within a permit space that has been declassified to a non-permit space under paragraph (c)(7) of this section, each employee in the space shall exit the space. The employer shall then reevaluate the space and determine whether it must be reclassified as a permit space, in accordance with other applicable provisions of this section.

**1910.146(c)(8)** When an employer (host employer) arranges to have employees of another employer (contractor) perform work that involves permit space entry, the host employer shall:

**1910.146(c)(8)(i)** Inform the contractor that the workplace contains permit spaces and that permit space entry is allowed only through compliance with a permit space program meeting the requirements of this section;

**1910.146(c)(8)(ii)** Apprise the contractor of the elements, including the hazards identified and the host employer's experience with the space, that make the space in question a permit space;

**1910.146(c)(8)(iii)** Apprise the contractor of any precautions or procedures that the host employer has implemented for the protection of employees in or near permit spaces where contractor personnel will be working;

**1910.146(c)(8)(iv)** Coordinate entry operations with the contractor, when both host employer personnel and contractor personnel will be working in or near permit spaces, as required by paragraph (d)(11) of this section; and

**1910.146(c)(8)(v)** Debrief the contractor at the conclusion of the entry operations regarding the permit space program followed and regarding any hazards confronted or created in permit spaces during entry operations.

**1910.146(c)(9)** In addition to complying with the permit space requirements that apply to all employers, each contractor who is retained to perform permit space entry operations shall:

**1910.146(c)(9)(i)** Obtain any available information regarding permit space hazards and entry operations from the host employer;

**1910.146(c)(9)(ii)** Coordinate entry operations with the host employer, when both host employer personnel and contractor personnel will be working in or near permit spaces, as required by paragraph (d)(11) of this section; and

**1910.146(c)(9)(iii)** Inform the host employer of the permit space program that the contractor will follow and of any hazards confronted or created in permit spaces, either through a debriefing or during the entry operation.

**1910.146(d)** *Permit-required confined space program (permit space program).* Under the permit space program required by paragraph (c)(4) of this section, the employer shall:

**1910.146(d)(1)** Implement the measures necessary to prevent unauthorized entry;

**1910.146(d)(2)** Identify and evaluate the hazards of permit spaces before employees enter them;

**1910.146(d)(3)** Develop and implement the means, procedures, and practices necessary for safe permit space entry operations, including, but not limited to, the following:

**1910.146(d)(3)(i)** Specifying acceptable entry conditions;

**1910.146(d)(3)(ii)** Providing each authorized entrant or that employee's authorized representative with the opportunity to observe any monitoring or testing of permit spaces;

**1910.146(d)(3)(iii)** Isolating the permit space;

**1910.146(d)(3)(iv)** Purging, inerting, flushing, or ventilating the permit space as necessary to eliminate or control atmospheric hazards;

**1910.146(d)(3)(v)** Providing pedestrian, vehicle, or other barriers as necessary to protect entrants from external hazards; and

**1910.146(d)(3)(vi)** Verifying that conditions in the permit space are acceptable for entry throughout the duration of an authorized entry.

**1910.146(d)(4)** Provide the following equipment (specified in paragraphs (d)(4)(i) through (d)(4)(ix) of this section) at no cost to employees, maintain that equipment properly, and ensure that employees use that equipment properly:

**1910.146(d)(4)(i)** Testing and monitoring equipment needed to comply with paragraph (d)(5) of this section;

**1910.146(d)(4)(ii)** Ventilating equipment needed to obtain acceptable entry conditions;

**1910.146(d)(4)(iii)** Communications equipment necessary for compliance with paragraphs (h)(3) and (i)(5) of this section;

**1910.146(d)(4)(iv)** Personal protective equipment insofar as feasible engineering and work practice controls do not adequately protect employees;

**1910.146(d)(4)(v)** Lighting equipment needed to enable employees to see well enough to work safely and to exit the space quickly in an emergency;

**1910.146(d)(4)(vi)** Barriers and shields as required by paragraph (d)(3)(iv) of this section;

**1910.146(d)(4)(vii)** Equipment, such as ladders, needed for safe ingress and egress by authorized entrants;

**1910.146(d)(4)(viii)** Rescue and emergency equipment needed to comply with paragraph (d)(9) of this section, except to the extent that the equipment is provided by rescue services; and

**1910.146(d)(4)(ix)** Any other equipment necessary for safe entry into and rescue from permit spaces.

**1910.146(d)(5)** Evaluate permit space conditions as follows when entry operations are conducted:

**1910.146(d)(5)(i)** Test conditions in the permit space to determine if acceptable entry conditions exist before entry is authorized to begin, except that, if isolation of the space is infeasible because the space is large or is part of a continuous system (such as a sewer), pre-entry testing shall be performed to the extent feasible before entry is authorized and, if entry is authorized, entry conditions shall be continuously monitored in the areas where authorized entrants are working;

**1910.146(d)(5)(ii)** Test or monitor the permit space as necessary to determine if acceptable entry conditions are being maintained during the course of entry operations; and

**1910.146(d)(5)(iii)** When testing for atmospheric hazards, test first for oxygen, then for combustible gases and vapors, and then for toxic gases and vapors.

**1910.146(d)(5)(iv)** Provide each authorized entrant or that employee's authorized representative an opportunity to observe the pre-entry and any subsequent testing or monitoring of permit spaces;

**1910.146(d)(5)(v)** Reevaluate the permit space in the presence of any authorized entrant or that employee's authorized representative who requests that the employer conduct such reevaluation because the entrant or representative has reason to believe that the evaluation of that space may not have been adequate;

**1910.146(d)(5)(vi)** Immediately provide each authorized entrant or that employee's authorized representative with the results of any testing conducted in accord with paragraph (d) of this section.

**NOTE:** Atmospheric testing conducted in accordance with appendix B to §1910.146 would be considered as satisfying the requirements of this paragraph. For permit space operations in sewers, atmospheric testing conducted in accordance with appendix B, as supplemented by appendix E to §1910.146, would be considered as satisfying the requirements of this paragraph.

**1910.146(d)(6)** Provide at least one attendant outside the permit space into which entry is authorized for the duration of entry operations;

**NOTE:** Attendants may be assigned to monitor more than one permit space provided the duties described in paragraph (i) of this section can be effectively performed for each permit space that is monitored. Likewise, attendants may be stationed at any location outside the permit space to be monitored as long as the duties described in paragraph (i) of this section can be effectively performed for each permit space that is monitored.

**1910.146(d)(7)** If multiple spaces are to be monitored by a single attendant, include in the permit program the means and procedures to enable the attendant to respond to an emergency affecting one or more of the permit spaces being monitored without distraction from the attendant's responsibilities under paragraph (i) of this section;

**1910.146(d)(8)** Designate the persons who are to have active roles (as, for example, authorized entrants, attendants, entry supervisors, or persons who test or monitor the atmosphere in a permit space) in entry operations, identify the duties of each such employee, and provide each such employee with the training required by paragraph (g) of this section;

**1910.146(d)(9)** Develop and implement procedures for summoning rescue and emergency services, for rescuing entrants from permit spaces, for providing necessary emergency services to rescued employees, and for preventing unauthorized personnel from attempting a rescue;

**1910.146(d)(10)** Develop and implement a system for the preparation, issuance, use, and cancellation of entry permits as required by this section;

**1910.146(d)(11)** Develop and implement procedures to coordinate entry operations when employees of more than one employer are working simultaneously as authorized entrants in a permit space, so that employees of one employer do not endanger the employees of any other employer;

**1910.146(d)(12)** Develop and implement procedures (such as closing off a permit space and canceling the permit) necessary for concluding the entry after entry operations have been completed;

**1910.146(d)(13)** Review entry operations when the employer has reason to believe that the measures taken under the permit space program may not protect employees and revise the program to correct deficiencies found to exist before subsequent entries are authorized; and

**NOTE:** Examples of circumstances requiring the review of the permit space program are: any unauthorized entry of a permit space, the detection of a permit space hazard not covered by the permit, the detection of a condition prohibited by the permit, the occurrence of an injury or near-miss during entry, a change in the use or configuration of a permit space, and employee complaints about the effectiveness of the program.

**1910.146(d)(14)** Review the permit space program, using the canceled permits retained under paragraph (e)(6) of this section within 1 year after each entry and revise the program as necessary, to ensure that employees participating in entry operations are protected from permit space hazards.

**NOTE:** Employers may perform a single annual review covering all entries performed during a 12-month period. If no entry is performed during a 12-month period, no review is necessary.

Appendix C to §1910.146 presents examples of permit space programs that are considered to comply with the requirements of paragraph (d) of this section.

**1910.146(e)** *Permit system.* **(1)** Before entry is authorized, the employer shall document the completion of measures required by paragraph (d)(3) of this section by preparing an entry permit.

**NOTE:** Appendix D to §1910.146 presents examples of permits whose elements are considered to comply with the requirements of this section.

**1910.146(e)(2)** Before entry begins, the entry supervisor identified on the permit shall sign the entry permit to authorize entry.

**1910.146(e)(3)** The completed permit shall be made available at the time of entry to all authorized entrants or their authorized representatives, by posting it at the entry portal or by any other equally effective means, so that the entrants can confirm that pre-entry preparations have been completed.

**1910.146(e)(4)** The duration of the permit may not exceed the time required to complete the assigned task or job identified on the permit in accordance with paragraph (f)(2) of this section.

**1910.146(e)(5)** The entry supervisor shall terminate entry and cancel the entry permit when:

**1910.146(e)(5)(i)** The entry operations covered by the entry permit have been completed; or

**1910.146(e)(5)(ii)** A condition that is not allowed under the entry permit arises in or near the permit space.

**1910.146(e)(6)** The employer shall retain each canceled entry permit for at least 1 year to facilitate the review of the permit-required confined space program required by paragraph (d)(14) of this section. Any problems encountered during an entry operation shall be noted on the pertinent permit so that appropriate revisions to the permit space program can be made.

**1910.146(f)** *Entry permit.* The entry permit that documents compliance with this section and authorizes entry to a permit space shall identify:

**1910.146(f)(1)** The permit space to be entered;

**1910.146(f)(2)** The purpose of the entry;

**1910.146(f)(3)** The date and the authorized duration of the entry permit;

**1910.146(f)(4)** The authorized entrants within the permit space, by name or by such other means (for example, through the use of rosters or tracking systems) as will enable the attendant to determine quickly and accurately, for the duration of the permit, which authorized entrants are inside the permit space;

**NOTE:** This requirement may be met by inserting a reference on the entry permit as to the means used, such as a roster or tracking system, to keep track of the authorized entrants within the permit space.

**1910.146(f)(5)** The personnel, by name, currently serving as attendants;

**1910.146(f)(6)** The individual, by name, currently serving as entry supervisor, with a space for the signature or initials of the entry supervisor who originally authorized entry;

**1910.146(f)(7)** The hazards of the permit space to be entered;

**1910.146(f)(8)** The measures used to isolate the permit space and to eliminate or control permit space hazards before entry;

**NOTE:** Those measures can include the lockout or tagging of equipment and procedures for purging, inerting, ventilating, and flushing permit spaces.

**1910.146(f)(9)** The acceptable entry conditions;

**1910.146(f)(10)** The results of initial and periodic tests performed under paragraph (d)(5) of this section, accompanied by the names or initials of the testers and by an indication of when the tests were performed;

**1910.146(f)(11)** The rescue and emergency services that can be summoned and the means (such as the equipment to use and the numbers to call) for summoning those services;

**1910.146(f)(12)** The communication procedures used by authorized entrants and attendants to maintain contact during the entry;

**1910.146(f)(13)** Equipment, such as personal protective equipment, testing equipment, communications equipment, alarm systems, and rescue equipment, to be provided for compliance with this section;

**1910.146(f)(14)** Any other information whose inclusion is necessary, given the circumstances of the particular confined space, in order to ensure employee safety; and

**1910.146(f)(15)** Any additional permits, such as for hot work, that have been issued to authorize work in the permit space.

**1910.146(g)** *Training.* **(1)** The employer shall provide training so that all employees whose work is regulated by this section acquire the understanding, knowledge, and skills necessary for the safe performance of the duties assigned under this section.

**1910.146(g)(2)** Training shall be provided to each affected employee:

**1910.146(g)(2)(i)** Before the employee is first assigned duties under this section;

**1910.146(g)(2)(ii)** Before there is a change in assigned duties;

**1910.146(g)(2)(iii)** Whenever there is a change in permit space operations that presents a hazard about which an employee has not previously been trained;

**1910.146(g)(2)(iv)** Whenever the employer has reason to believe either that there are deviations from the permit space entry procedures required by paragraph (d)(3) of this section or that there are inadequacies in the employee's knowledge or use of these procedures.

**1910.146(g)(3)** The training shall establish employee proficiency in the duties required by this section and shall introduce new or revised procedures, as necessary, for compliance with this section.

**1910.146(g)(4)** The employer shall certify that the training required by paragraphs (g)(1) through (g)(3) of this section has been accomplished. The certification shall contain each employee's name, the signatures or initials of the trainers, and the dates of training. The certification shall be available for inspection by employees and their authorized representatives.

**1910.146(h)** *Duties of authorized entrants.* The employer shall ensure that all authorized entrants:

**1910.146(h)(1)** Know the hazards that may be faced during entry, including information on the mode, signs or symptoms, and consequences of the exposure;

**1910.146(h)(2)** Properly use equipment as required by paragraph (d)(4) of this section;

**1910.146(h)(3)** Communicate with the attendant as necessary to enable the attendant to monitor entrant status and to enable the attendant to alert entrants of the need to evacuate the space as required by paragraph (i)(6) of this section;

**1910.146(h)(4)** Alert the attendant whenever:

**1910.146(h)(4)(i)** The entrant recognizes any warning sign or symptom of exposure to a dangerous situation, or

**1910.146(h)(4)(ii)** The entrant detects a prohibited condition; and

**1910.146(h)(5)** Exit from the permit space as quickly as possible whenever:

**1910.146(h)(5)(i)** An order to evacuate is given by the attendant or the entry supervisor,

**1910.146(h)(5)(ii)** The entrant recognizes any warning sign or symptom of exposure to a dangerous situation,

**1910.146(h)(5)(iii)** The entrant detects a prohibited condition, or

**1910.146(h)(5)(iv)** An evacuation alarm is activated.

**1910.146(i)** *Duties of attendants.* The employer shall ensure that each attendant:

**1910.146(i)(1)** Knows the hazards that may be faced during entry, including information on the mode, signs or symptoms, and consequences of the exposure;

**1910.146(i)(2)** Is aware of possible behavioral effects of hazard exposure in authorized entrants;

**1910.146(i)(3)** Continuously maintains an accurate count of authorized entrants in the permit space and ensures that the means used to identify authorized entrants under paragraph (f)(4) of this section accurately identifies who is in the permit space;

**1910.146(i)(4)** Remains outside the permit space during entry operations until relieved by another attendant;

**NOTE:** When the employer's permit entry program allows attendant entry for rescue, attendants may enter a permit space to attempt a rescue if they have been trained and equipped for rescue operations as required by paragraph (k)(1) of this section and if they have been relieved as required by paragraph (i)(4) of this section.

**1910.146(i)(5)** Communicates with authorized entrants as necessary to monitor entrant status and to alert entrants of the need to evacuate the space under paragraph (i)(6) of this section;

**1910.146(i)(6)** Monitors activities inside and outside the space to determine if it is safe for entrants to remain in the space and orders the authorized entrants to evacuate the permit space immediately under any of the following conditions;

**1910.146(i)(6)(i)** If the attendant detects a prohibited condition;

**1910.146(i)(6)(ii)** If the attendant detects the behavioral effects of hazard exposure in an authorized entrant;

**1910.146(i)(6)(iii)** If the attendant detects a situation outside the space that could endanger the authorized entrants; or

**1910.146(i)(6)(iv)** If the attendant cannot effectively and safely perform all the duties required under paragraph (i) of this section;

**1910.146(i)(7)** Summon rescue and other emergency services as soon as the attendant determines that authorized entrants may need assistance to escape from permit space hazards;

**1910.146(i)(8)** Takes the following actions when unauthorized persons approach or enter a permit space while entry is underway:

**1910.146(i)(8)(i)** Warn the unauthorized persons that they must stay away from the permit space;

**1910.146(i)(8)(ii)** Advise the unauthorized persons that they must exit immediately if they have entered the permit space; and

**1910.146(i)(8)(iii)** Inform the authorized entrants and the entry supervisor if unauthorized persons have entered the permit space;

**1910.146(i)(9)** Performs non-entry rescues as specified by the employer's rescue procedure; and

**1910.146(i)(10)** Performs no duties that might interfere with the attendant's primary duty to monitor and protect the authorized entrants.

**1910.146(j)** *Duties of entry supervisors.* The employer shall ensure that each entry supervisor:

**1910.146(j)(1)** Knows the hazards that may be faced during entry, including information on the mode, signs or symptoms, and consequences of the exposure;

**1910.146(j)(2)** Verifies, by checking that the appropriate entries have been made on the permit, that all tests specified by the permit have been conducted and that all procedures and equipment specified by the permit are in place before endorsing the permit and allowing entry to begin;

**1910.146(j)(3)** Terminates the entry and cancels the permit as required by paragraph (e)(5) of this section;

**1910.146(j)(4)** Verifies that rescue services are available and that the means for summoning them are operable;

**1910.146(j)(5)** Removes unauthorized individuals who enter or who attempt to enter the permit space during entry operations; and

**1910.146(j)(6)** Determines, whenever responsibility for a permit space entry operation is transferred and at intervals dictated by the hazards and operations performed within the space, that entry operations remain consistent with terms of the entry permit and that acceptable entry conditions are maintained.

**1910.146(k)** *Rescue and emergency services.*

**1910.146(k)(1)** An employer who designates rescue and emergency services, pursuant to paragraph (d)(9) of this section, shall:

**1910.146(k)(1)(i)** Evaluate a prospective rescuer's ability to respond to a rescue summons in a timely manner, considering the hazard(s) identified;

**Note** to paragraph (k)(1)(i): What will be considered timely will vary according to the specific hazards involved in each entry. For example, §1910.134, Respiratory Protection, requires that employers provide a standby person or persons capable of immediate action to rescue employee(s) wearing respiratory protection while in work areas defined as IDLH atmospheres.

**1910.146(k)(1)(ii)** Evaluate a prospective rescue service's ability, in terms of proficiency with rescue-related tasks and equipment, to function appropriately while rescuing entrants from the particular permit space or types of permit spaces identified;

**1910.146(k)(1)(iii)** Select a rescue team or service from those evaluated that:

**1910.146(k)(1)(iii)(A)** Has the capability to reach the victim(s) within a time frame that is appropriate for the permit space hazard(s) identified;

**1910.146(k)(1)(iii)(B)** Is equipped for and proficient in performing the needed rescue services;

**1910.146(k)(1)(iv)** Inform each rescue team or service of the hazards they may confront when called on to perform rescue at the site; and

**1910.146(k)(1)(v)** Provide the rescue team or service selected with access to all permit spaces from which rescue may be necessary so that the rescue service can develop appropriate rescue plans and practice rescue operations.

**Note** to paragraph (k)(1): Non-mandatory appendix F contains examples of criteria which employers can use in evaluating prospective rescuers as required by paragraph (k)(1) of this section.

**1910.146(k)(2)** An employer whose employees have been designated to provide permit space rescue and emergency services shall take the following measures:

**1910.146(k)(2)(i)** Provide affected employees with the personal protective equipment (PPE) needed to conduct permit space rescues safely and train affected employees so they are proficient in the use of that PPE, at no cost to those employees;

**1910.146(k)(2)(ii)** Train affected employees to perform assigned rescue duties. The employer must ensure that such employees successfully complete the training required to establish proficiency as an authorized entrant, as provided by paragraphs (g) and (h) of this section;

**1910.146(k)(2)(iii)** Train affected employees in basic first-aid and cardiopulmonary resuscitation (CPR). The employer shall ensure that at least one member of the rescue team or service holding a current certification in first aid and CPR is available; and

**1910.146(k)(2)(iv)** Ensure that affected employees practice making permit space rescues at least once every 12 months, by means of simulated rescue operations in which they remove dummies, manikins, or actual persons from the actual permit spaces or from representative permit spaces. Representative permit spaces shall, with respect to opening size, configuration, and accessibility, simulate the types of permit spaces from which rescue is to be performed.

**1910.146(k)(3)** To facilitate non-entry rescue, retrieval systems or methods shall be used whenever an authorized entrant enters a permit space, unless the retrieval equipment would increase the overall risk of entry or would not contribute to the rescue of the entrant. Retrieval systems shall meet the following requirements.

**1910.146(k)(3)(i)** Each authorized entrant shall use a chest or full body harness, with a retrieval line attached at the center of the entrant's back near shoulder level, above the entrant's head, or at another point which the employer can establish presents a profile small enough for the successful removal of the entrant. Wristlets may be used in lieu of the chest or full body harness if the employer can demonstrate that the use of a chest or full body harness is infeasible or creates a greater hazard and that the use of wristlets is the safest and most effective alternative.

**1910.146(k)(3)(ii)** The other end of the retrieval line shall be attached to a mechanical device or fixed point outside the permit space in such a manner that rescue can begin as soon as the rescuer becomes aware that rescue is necessary. A mechanical device shall be available to retrieve personnel from vertical type permit spaces more than 5 feet (1.52 m) deep.

**1910.146(k)(4)** If an injured entrant is exposed to a substance for which a Material Safety Data Sheet (MSDS) or other similar written information is required to be kept at the worksite, that MSDS or written information shall be made available to the medical facility treating the exposed entrant.

**1910.146(l)** *Employee participation.* **(1)** Employers shall consult with affected employees and their authorized representatives on the development and implementation of all aspects of the permit space program required by paragraph (c) of this section.

**1910.146(l)(2)** Employers shall make available to affected employees and their authorized representatives all information required to be developed by this section.

# OSHA Standards for General Industry

## APPENDICES TO §1910.146—PERMIT-REQUIRED CONFINED SPACES

**NOTE:** Appendices A through F serve to provide information and non-mandatory guidelines to assist employers and employees in complying with the appropriate requirements of this section.

[58 FR 4549, Jan. 14, 1993; 58 FR 34845, June 29, 1993; 59 FR 2665, May 19, 1994; 63 FR 66038, Dec. 1, 1998]

## APPENDIX A TO §1910.146—PERMIT-REQUIRED CONFINED SPACE DECISION FLOW CHART

[Flow chart showing decision process for permit-required confined spaces, with decision nodes including: Does the workplace contain Confined Spaces as defined by §1910.146(b)?; Does the workplace contain Permit-required Confined Spaces as defined by §1910.146(b)?; Inform employees as required by §1910.146(c)(2); Will permit spaces be entered?; Will contractors enter?; Both contractors and host employees will enter the space?; Will host employees enter to perform entry tasks?; Does space have known or potential hazards?; Can the hazards be eliminated?; Can the space be maintained in a condition safe to enter by continuous forced air ventilation only?; Prepare for entry via permit procedures; Verify acceptable entry conditions; Permit issued by authorizing signature; Entry tasks completed. Permit returned and canceled; Audit permit program and permit based on evaluation of entry by entrants, attendants, testers and preparers, etc.]

[1] Spaces may have to be evacuated and re-evaluated if hazards arise during entry.

[58 FR 4549, Jan. 14, 1993; 58 FR 34846, June 29, 1993; 63 FR 66039, Dec. 1, 1998]

## APPENDIX B TO §1910.146—PROCEDURES FOR ATMOSPHERIC TESTING

Atmospheric testing is required for two distinct purposes: evaluation of the hazards of the permit space and verification that acceptable entry conditions for entry into that space exist.

(1) *Evaluation testing.* The atmosphere of a confined space should be analyzed using equipment of sufficient sensitivity and specificity to identify and evaluate any hazardous atmospheres that may exist or arise, so that appropriate permit entry procedures can be developed and acceptable entry conditions stipulated for that space. Evaluation and interpretation of these data, and development of the entry procedure, should be done by, or reviewed by, a technically qualified professional (e.g., OSHA consultation service, or certified industrial hygienist, registered safety engineer, certified safety professional, certified marine chemist, etc.) based on evaluation of all serious hazards.

(2) *Verification testing.* The atmosphere of a permit space which may contain a hazardous atmosphere should be tested for residues of all contaminants identified by evaluation testing using permit specified equipment to determine that residual concentrations at the time of testing and entry are within the range of acceptable entry conditions. Results of testing (i.e., actual concentration, etc.) should be recorded on the permit in the space provided adjacent to the stipulated acceptable entry condition.

(3) *Duration of testing.* Measurement of values for each atmospheric parameter should be made for at least the minimum response time of the test instrument specified by the manufacturer.

(4) *Testing stratified atmospheres.* When monitoring for entries involving a descent into atmospheres that may be stratified, the atmospheric envelope should be tested a distance of approximately 4 feet (1.22 m) in the direction of travel and to each side. If a sampling probe is used, the entrant's rate of progress should be slowed to accommodate the sampling speed and detector response.

(5) *Order of testing.* A test for oxygen is performed first because most combustible gas meters are oxygen dependent and will not provide reliable readings in an oxygen deficient atmosphere. Combustible gasses are tested for next because the threat of fire or explosion is both more immediate and more life threatening, in most cases, than exposure to toxic gasses and vapors. If tests for toxic gasses and vapors are necessary, they are performed last.

[58 FR 4549, Jan. 14, 1993; 58 FR 34846, June 29, 1993]

## APPENDIX C TO §1910.146—EXAMPLES OF PERMIT-REQUIRED CONFINED SPACE PROGRAMS

### Example 1.

*Workplace.* Sewer entry.

*Potential hazards.* The employees could be exposed to the following:

*Engulfment.*

*Presence of toxic gases.* Equal to or more than 10 ppm hydrogen sulfide measured as an 8-hour time-weighted average. If the presence of other toxic contaminants is suspected, specific monitoring programs will be developed.

*Presence of explosive/flammable gases.* Equal to or greater than 10% of the lower flammable limit (LFL).

*Oxygen Deficiency.* A concentration of oxygen in the atmosphere equal to or less than 19.5% by volume.

A. Entry Without Permit/Attendant

*Certification.* Confined spaces may be entered without the need for a written permit or attendant provided that the space can be maintained in a safe condition for entry by mechanical ventilation alone, as provided in §1910.146(c)(5). All spaces shall be considered permit-required confined spaces until the pre-entry procedures demonstrate otherwise. Any employee required or permitted to pre-check or enter an enclosed/confined space shall have successfully completed, -as a minimum, the training as required by the following sections of these procedures. *A written copy of operating and rescue procedures as required by these procedures shall be at the work site for the duration of the job.* The Confined Space Pre-Entry Check List must be completed by the LEAD WORKER before entry into a confined space. This list verifies completion of items listed below. This check list shall be kept at the job site for duration of the job. If circumstances dictate an interruption in the work, the permit space must be re-evaluated and a new check list must be completed.

*Control of atmospheric and engulfment hazards.*

*Pumps and Lines.* All pumps and lines which may reasonably cause contaminants to flow into the space shall be disconnected, blinded and locked out, or effectively isolated by other means to prevent development of dangerous air contamination or engulfment. Not all laterals to sewers or storm drains require blocking. However, where experience or knowledge of industrial use indicates there is a reasonable potential for contamination of air or engulfment into an occupied sewer, then all affected laterals shall be blocked. If blocking and/or isolation requires entry into the space the provisions for entry into a permit-required confined space must be implemented.

*Surveillance.* The surrounding area shall be surveyed to avoid hazards such as drifting vapors from the tanks, piping, or sewers.

*Testing.* The atmosphere within the space will be tested to determine whether dangerous air contamination and/or oxygen deficiency exists. Detector tubes, alarm only gas monitors and explosion meters are examples of monitoring equipment that may be used to test permit space atmospheres. Testing shall be performed by the LEAD WORKER who has successfully completed the Gas Detector training for the monitor he will use. The minimum parameters to be monitored are oxygen deficiency, LFL, and hydrogen sulfide concentration. A written record of the pre-entry test results shall be made and kept at the work site for the duration of the job. The supervisor will certify in writing, based upon the results of the pre-entry testing, that all hazards have been eliminated. Affected employees shall be able to review the testing results. The most hazardous conditions shall govern when work is being performed in two adjoining, connecting spaces.

*Entry Procedures.* If there are no non-atmospheric hazards present and if the pre-entry tests show there is no dangerous air contamination and/or oxygen deficiency within the space and there is no reason to believe that any is likely to develop, entry into and work within may proceed. Continuous testing of the atmosphere in the immediate vicinity of the workers within the space shall be accomplished. The workers will immediately leave the permit space when any of the gas monitor alarm set points are reached as defined. Workers will not return to the area until a SUPERVISOR who has completed the gas detector training has used a direct reading gas detector to evaluate the situation and has determined that it is safe to enter.

*Rescue.* Arrangements for rescue services are not required where there is no attendant. See the rescue portion of section B., below, for instructions regarding rescue planning where an entry permit is required.

### B. Entry Permit Required

*Permits.* Confined Space Entry Permit. All spaces shall be considered permit-required confined spaces until the pre-entry procedures demonstrate otherwise. Any employee required or permitted to pre-check or enter a permit-required confined space shall have successfully completed, as a minimum, the training as required by the following sections of these procedures. *A written copy of operating and rescue procedures as required by these procedures shall be at the work site for the duration of the job.* The Confined Space Entry Permit must be completed before approval can be given to enter a permit-required confined space. This permit verifies completion of items listed below. This permit shall be kept at the job site for the duration of the job. If circumstances cause an interruption in the work or a change in the alarm conditions for which entry was approved, a new Confined Space Entry Permit must be completed.

*Control of atmospheric and engulfment hazards.*

*Surveillance.* The surrounding area shall be surveyed to avoid hazards such as drifting vapors from tanks, piping or sewers.

*Testing.* The confined space atmosphere shall be tested to determine whether dangerous air contamination and/or oxygen deficiency exists. A direct reading gas monitor shall be used. Testing shall be performed by the SUPERVISOR who has successfully completed the gas detector training for the monitor he will use. The minimum parameters to be monitored are oxygen deficiency, LFL and hydrogen sulfide concentration. A written record of the pre-entry test results shall be made and kept at the work site for the duration of the job. Affected employees shall be able to review the testing results. The most hazardous conditions shall govern when work is being performed in two adjoining, connected spaces.

*Space Ventilation.* Mechanical ventilation systems, where applicable, shall be set at 100% outside air. Where possible, open additional manholes to increase air circulation. Use portable blowers to augment natural circulation if needed. After a suitable ventilating period, repeat the testing. Entry may not begin until testing has demonstrated that the hazardous atmosphere has been eliminated.

*Entry Procedures.* The following procedure shall be observed under any of the following conditions: 1.) Testing demonstrates the existence of dangerous or deficient conditions and additional ventilation cannot reduce concentrations to safe levels; 2.) The atmosphere tests as safe but unsafe conditions can reasonably be expected to develop; 3.) It is not feasible to provide for ready exit from spaces equipped with automatic fire suppression systems and it is not practical or safe to deactivate such systems; or 4.) An emergency exists and it is not feasible to wait for pre-entry procedures to take effect.

All personnel must be trained. A self contained breathing apparatus shall be worn by any person entering the space. At least one worker shall stand by the outside of the space ready to give assistance in case of emergency. The standby worker shall have a self contained breathing apparatus available for immediate use. There shall be at least one additional worker within sight or call of the standby worker. Continuous powered communications shall be maintained between the worker within the confined space and standby personnel.

If at any time there is any questionable action or non-movement by the worker inside, a verbal check will be made. If there is no response, the worker will be moved immediately. *Exception:* If the worker is disabled due to falling or impact, he/she shall not be removed from the confined space unless there is immediate danger to his/her life. Local fire department rescue personnel shall be notified immediately. The standby worker may only enter the confined space in case of an emergency (wearing the self contained breathing apparatus) and only after being relieved by another worker. Safety belt or harness with attached lifeline shall be used by all workers entering the space with the free end of the line secured outside the entry opening. The standby worker shall attempt to remove a disabled worker via his lifeline before entering the space.

When practical, these spaces shall be entered through side openings—those within 3 1/2 feet (1.07 m) of the bottom. When entry must be through a top opening, the safety belt shall be of the harness type that suspends a person upright and a hoisting device or similar apparatus shall be available for lifting workers out of the space.

In any situation where their use may endanger the worker, use of a hoisting device or safety belt and attached lifeline may be discontinued.

When dangerous air contamination is attributable to flammable and/or explosive substances, lighting and electrical equipment shall be Class 1, Division 1 rated per National Electrical Code and no ignition sources shall be introduced into the area.

Continuous gas monitoring shall be performed during all confined space operations. If alarm conditions change adversely, entry personnel shall exit the confined space and a new confined space permit issued.

*Rescue.* Call the fire department services for rescue. Where immediate hazards to injured personnel are present, workers at the site shall implement emergency procedures to fit the situation.

### Example 2.

*Workplace.* Meat and poultry rendering plants.

Cookers and dryers are either batch or continuous in their operation. Multiple batch cookers are operated in parallel. When one unit of a multiple set is shut down for repairs, means are available to isolate that unit from the others which remain in operation.

Cookers and dryers are horizontal, cylindrical vessels equipped with a center, rotating shaft and agitator paddles or discs. If the inner shell is jacketed, it is usually heated with steam at pressures up to 150 psig (1034.25 kPa). The rotating shaft assembly of the continuous cooker or dryer is also steam heated.

*Potential Hazards.* The recognized hazards associated with cookers and dryers are the risk that employees could be:

1. Struck or caught by rotating agitator;

2. Engulfed in raw material or hot, recycled fat;

3. Burned by steam from leaks into the cooker/dryer steam jacket or the condenser duct system if steam valves are not properly closed and locked out;

4. Burned by contact with hot metal surfaces, such as the agitator shaft assembly, or inner shell of the cooker/dryer;

5. Heat stress caused by warm atmosphere inside cooker/dryer;

6. Slipping and falling on grease in the cooker/dryer;

7. Electrically shocked by faulty equipment taken into the cooker/dryer;

8. Burned or overcome by fire or products of combustion; or

9. Overcome by fumes generated by welding or cutting done on grease covered surfaces.

*Permits.* The supervisor in this case is always present at the cooker/dryer or other permit entry confined space when entry is made. The supervisor must follow the pre-entry isolation procedures described in the entry permit in preparing for entry, and ensure that the protective clothing, ventilating equipment and any other equipment required by the permit are at the entry site.

*Control of hazards.* Mechanical. Lock out main power switch to agitator motor at main power panel. Affix tag to the lock to inform others that a permit entry confined space entry is in progress.

*Engulfment.* Close all valves in the raw material blow line. Secure each valve in its closed position using chain and lock. Attach a tag to the valve and chain warning that a permit entry confined space entry is in progress. The same procedure shall be used for securing the fat recycle valve.

*Burns and heat stress.* Close steam supply valves to jacket and secure with chains and tags. Insert solid blank at flange in cooker vent line to condenser manifold duct system. Vent cooker/dryer by opening access door at discharge end and top center door to allow natural ventilation throughout the entry. If faster cooling is needed, use an portable ventilation fan to increase ventilation. Cooling water may be circulated through the jacket to reduce both outer and inner surface temperatures of cooker/dryers faster. Check air and inner surface temperatures in cooker/dryer to assure they are within acceptable limits before entering, or use proper protective clothing.

*Fire and fume hazards.* Careful site preparation, such as cleaning the area within 4 inches (10.16 cm) of all welding or torch cutting operations, and proper ventilation are the preferred controls. All welding and cutting operations shall be done in accordance with the requirements of 29 CFR Part 1910, Subpart Q, OSHA's welding standard. Proper ventilation may be achieved by local exhaust ventilation, or the use of portable ventilation fans, or a combination of the two practices.

*Electrical shock.* Electrical equipment used in cooker/dryers shall be in serviceable condition.

*Slips and falls.* Remove residual grease before entering cooker/dryer.

*Attendant.* The supervisor shall be the attendant for employees entering cooker/dryers.

*Permit.* The permit shall specify how isolation shall be done and any other preparations needed before making entry. This is especially important in parallel arrangements of cooker/dryers so that the entire operation need not be shut down to allow safe entry into one unit.

*Rescue.* When necessary, the attendant shall call the fire department as previously arranged.

### Example 3.

*Workplace.* Workplaces where tank cars, trucks, and trailers, dry bulk tanks and trailers, railroad tank cars, and similar portable tanks are fabricated or serviced.

A. *During fabrication.* These tanks and dry-bulk carriers are entered repeatedly throughout the fabrication process. These products are not configured identically, but the manufacturing processes by which they are made are very similar.

*Sources of hazards.* In addition to the mechanical hazards arising from the risks that an entrant would be injured due to contact with components of the tank or the tools being used, there is also the risk that a worker could be injured by breathing fumes from welding materials or mists or vapors from materials used to coat the tank interior. In addition, many of these vapors and mists are flammable, so the failure to properly ventilate a tank could lead to a fire or explosion.

*Control of hazards.*

*Welding.* Local exhaust ventilation shall be used to remove welding fumes once the tank or carrier is completed to the point that workers may enter and exit only through a manhole. (Follow the requirements of 29 CFR 1910, Subpart Q, OSHA's welding standard, at all times.) Welding gas tanks may never be brought into a tank or carrier that is a permit entry confined space.

*Application of interior coatings/linings.* Atmospheric hazards shall be controlled by forced air ventilation sufficient to keep the atmospheric concentration of flammable materials below 10% of the lower flammable limit (LFL) (or lower explosive limit (LEL), whichever term is used locally). The appropriate respirators are provided and shall be used in addition to providing forced ventilation if the forced ventilation does not maintain acceptable respiratory conditions.

*Permits.* Because of the repetitive nature of the entries in these operations, an "Area Entry Permit" will be issued for a 1 month period to cover those production areas where tanks are fabricated to the point that entry and exit are made using manholes.

*Authorization.* Only the area supervisor may authorize an employee to enter a tank within the permit area. The area supervisor must determine that conditions in the tank trailer, dry bulk trailer or truck, etc. meet permit requirements before authorizing entry.

*Attendant.* The area supervisor shall designate an employee to maintain communication by employer specified means with employees working in tanks to ensure their safety. The attendant may not enter any permit entry confined space to rescue an entrant or for any other reason, unless authorized by the rescue procedure and, and even then, only after calling the rescue team and being relieved by as attendant by another worker.

*Communications and observation.* Communications between attendant and entrant(s) shall be maintained throughout entry. Methods of communication that may be specified by the permit include voice, voice powered radio, tapping or rapping codes on tank walls, signalling tugs on a rope, and the attendant's observation that work activities such as chipping, grinding, welding, spraying, etc., which require deliberate operator control continue normally. These activities often generate so much noise that the necessary hearing protection makes communication by voice difficult.

*Rescue procedures.* Acceptable rescue procedures include entry by a team of employee-rescuers, use of public emergency services, and procedures for breaching the tank. The area permit specifies which procedures are available, but the area supervisor makes the final decision based on circumstances. (Certain injuries may make it necessary to breach the tank to remove a person rather than risk additional injury by removal through an existing manhole. However, the supervisor must ensure that no breaching procedure used for rescue would violate terms of the entry permit. For instance, if the tank must be breached by cutting with a torch, the tank surfaces to be cut must be free of volatile or combustible coatings within 4 inches (10.16 cm) of the cutting line and the atmosphere within the tank must be below the LFL.

*Retrieval line and harnesses.* The retrieval lines and harnesses generally required under this standard are usually impractical for use in tanks because the internal configuration of the tanks and their interior baffles and other structures would prevent rescuers from hauling out injured entrants. However, unless the rescue procedure calls for breaching the tank for rescue, the rescue team shall be trained in the use of retrieval lines and harnesses for removing injured employees through manholes.

B. *Repair or service of "used" tanks and bulk trailers.*

*Sources of hazards.* In addition to facing the potential hazards encountered in fabrication or manufacturing, tanks or trailers which have been in service may contain residues of dangerous materials, whether left over from the transportation of hazardous cargoes or generated by chemical or bacterial action on residues of non-hazardous cargoes.

*Control of atmospheric hazards.* A "used" tank shall be brought into areas where tank entry is authorized only after the tank has been emptied, cleansed (without employee entry) of any residues, and purged of any potential atmospheric hazards.

*Welding.* In addition to tank cleaning for control of atmospheric hazards, coating and surface materials shall be removed 4 inches (10.16 cm) or more from any surface area where welding or other torch work will be done and care taken that the atmosphere within the tank remains well below the LFL. (Follow the requirements of 29 CFR 1910, Subpart Q, OSHA's welding standard, at all times.)

*Permits.* An entry permit valid for up to 1 year shall be issued prior to authorization of entry into used tank trailers, dry bulk trailers or trucks. In addition to the pre-entry cleaning requirement, this permit shall require the employee safeguards specified for new tank fabrication or construction permit areas.

*Authorization.* Only the area supervisor may authorize an employee to enter a tank trailer, dry bulk trailer or truck within the permit area. The area supervisor must determine that the entry permit requirements have been met before authorizing entry.

[58 FR 4549, Jan. 14, 1993; 58 FR 34846, June 29, 1993.]

## APPENDIX D TO §1910.146—SAMPLE PERMITS

Appendix D—1
Confined Space Entry Permit
Date and Time Issued: _____
Job site/Space I.D.: _____
Equipment to be worked on: _____
Stand-by personnel _____

Date and Time Expires: _____
Job Supervisor: _____
Work to be performed: _____

1. Atmospheric Checks: Time _____
   Oxygen _____ %
   Explosive _____ % L.F.L.
   Toxic _____ PPM
2. Tester's signature _____
3. Source isolation (No Entry):      N/A    Yes    No
   Pumps or lines blinded,            ( )    ( )   ( )
   disconnected, or blocked           ( )    ( )   ( )
4. Ventilation Modification:         N/A    Yes    No
   Mechanical                         ( )    ( )   ( )
   Natural Ventilation only           ( )    ( )   ( )
5. Atmospheric check after isolation and Ventilation:
   Oxygen _____ %         > 19.5 %
   Explosive _____ %      L.F.L. < 10 %
   Toxic _____            PPM < 10   PPM $H_2S$
   Tester's signature _____
6. Communication procedures: _____
7. Rescue procedures: _____

8. Entry, standby, and back up persons:   Yes    No
   Successfully completed required training? ( )  ( )

Is it current?                                    ( )      ( )
9. Equipment:                              N/A    Yes      No
   Direct reading gas monitor—tested       ( )    ( )      ( )
   Safety harnesses and lifelines for
   entry and standby persons               ( )    ( )      ( )
   Hoisting equipment                      ( )    ( )      ( )
   Powered communications                  ( )    ( )      ( )
   SCBA's for entry and standby persons    ( )    ( )      ( )
   Protective Clothing                     ( )    ( )      ( )
   All electric equipment listed Class I,
   Vivision I, Group D and Non-sparking tools ( ) ( )      ( )
10. Periodic atmospheric tests:
    Oxygen _____ % Time _____    Oxygen _____ % Time _____
    Oxygen _____ % Time _____    Oxygen _____ % Time _____
    Explosive _____ % Time _____  Explosive _____ % Time _____
    Explosive _____ % Time _____  Explosive _____ % Time _____
    Toxic _____ % Time _____      Toxic _____ % Time _____
    Toxic _____ % Time _____      Toxic _____ % Time _____

We have reviewed the work authorized by this permit and the information contained here-in. Written instructions and safety procedures have been received and are understood. Entry cannot be approved if any squares are marked in the "No" column. This permit is not valid unless all appropriate items are completed.

Permit Prepared By: (Supervisor) _____
Approved By: (Unit Supervisor) _____
Reviewed By (Cs Operations Personnel): _____
                                        printed name                          (signature)

This permit to be kept at job site. Return job site copy to Safety Office following job completion.
Copies: White Original (Safety Office)   Yellow (Unit Supervisor)    Hard (Job site)

## APPENDIX D-2 ENTRY PERMIT

**PERMIT VALID FOR 8 HOURS ONLY. ALL PERMIT COPIES REMAIN AT SITE UNTIL JOB COMPLETED**

DATE: _____ SITE LOCATION/DESCRIPTION _____
PURPOSE OF ENTRY _____
SUPERVISOR(S) in charge of crews _____ Type of Crew _____ Phone # _____

COMMUNICATION PROCEDURES _____
RESCUE PROCEDURES (PHONE NUMBERS AT BOTTOM)
* BOLD DENOTES MINIMUM REQUIREMENTS TO BE COMPLETED AND REVIEWED PRIOR TO ENTRY *

| REQUIREMENTS COMPLETED | DATE | TIME | REQUIREMENTS COMPLETED | DATE | TIME |
|---|---|---|---|---|---|
| **Lock Out/De-energize/Try-out** | | | **Full Body Harness w/"D" ring** | | |
| **Line(s) Broken-Capped-Blank** | | | **Emergency Escape Retrieval Equip** | | |
| **Purge-Flush and Vent** | | | **Lifelines** | | |
| **Ventilation** | | | Fire Extinguishers | | |
| **Secure Area (Post and Flag)** | | | Lighting (Explosive Proof) | | |
| **Breathing Apparatus** | | | Protective Clothing | | |
| **ResuscitatorInhalator** | | | Respirator(s) (Air Purifying) | | |
| **Standby Safety Personnel** | | | Burning and Welding Permit | | |

Note: Items that do not apply enter N/A in the blank.
** RECORD CONTINUOUS MONITORING RESULTS EVERY 2 HOURS **

| CONTINUOUS MONITORING** TEST(S) TO BE TAKEN | Permissible Entry Level | | | | | | | |
|---|---|---|---|---|---|---|---|---|
| **PERCENT OF OXYGEN** | **19.5% to 23.5%** | | | | | | | |
| **LOWER FLAMMABLE LIMIT** | **Under 10%** | | | | | | | |
| **CARBON MONOXIDE** | **+35 PPM** | | | | | | | |
| Aromatic Hydrocarbon | +1 PPM  *5PPM | | | | | | | |
| Hydrogen Cyanide | (Skin)  *4PPM | | | | | | | |
| Hydrogen Sulfide | +10 PPM  *15PPM | | | | | | | |
| Sulfur Dioxide | +2 PPM  *5PPM | | | | | | | |
| Ammonia | *35PPM | | | | | | | |

* Short-term exposure limit: Employee can work in the area up to 15 minutes.
+ 8 hr. Time Weighted Avg.: Employee can work in area 8 hrs (longer with appropriate respiratory protection).
REMARKS: _____

GAS TESTER NAME & CHECK # _____ INSTRUMENT(S) USED _____ MODEL &OR TYPE _____ SERIAL &/OR UNIT # _____

SAFETY STANDBY PERSON IS REQUIRED FOR ALL CONFINED SPACE WORK
SAFETY STANDBY PERSON(S) _____ CHECK # _____ CONFINED SPACE ENTRANT(S) _____ CHECK # _____ CONFINED SPACE ENTRANT(S) _____ CHECK # _____

SUPERVISOR AUTHORIZATION - ALL CONDITIONS SATISFIED _____ DEPARTMENT/PHONE _____
AMBULANCE 2800     FIRE 2900     Safety 4901     Gas Coordinator 4529/5387

# OSHA Standards for General Industry

## APPENDIX E TO §1910.146—SEWER SYSTEM ENTRY

Sewer entry differs in three vital respects from other permit entries: first, there rarely exists any way to completely isolate the space (a section of a continuous system) to be entered; second, because isolation is not complete, the atmosphere may suddenly and unpredictably become lethally hazardous (toxic, flammable or explosive) from causes beyond the control of the entrant or employer, and third, experienced sewer workers are especially knowledgeable in entry and work in their permit spaces because of their frequent entries. Unlike other employments where permit space entry is a rare and exceptional event, sewer workers' usual work environment is a permit space.

(1) *Adherence to procedure.* The employer should designate as entrants only employees who are thoroughly trained in the employer's sewer entry procedures and who demonstrate that they follow these entry procedures exactly as prescribed when performing sewer entries.

(2) *Atmospheric monitoring.* Entrants should be trained in the use of, and be equipped with, atmospheric monitoring equipment which sounds an audible alarm, in addition to its visual readout, whenever one of the following conditions are encountered: Oxygen concentration less than 19.5 percent; flammable gas or vapor at 10 percent or more of the lower flammable limit (LFL); or hydrogen sulfide or carbon monoxide at or above 10 ppm or 35 ppm, respectively, measured as an 8-hour time-weighted average. Atmospheric monitoring equipment needs to be calibrated according to the manufacturer's instructions. The oxygen sensor/broad range sensor is best suited for initial use in situations where the actual or potential contaminants have not been identified, because broad range sensors, unlike substance-specific sensors, enable employers to obtain an overall reading of the hydrocarbons (flammables) present in the space. However, such sensors only indicate that a hazardous threshold of a class of chemicals has been exceeded. They do not measure the levels of contamination of specific substances. Therefore, substance-specific devices, which measure the actual levels of specific substances, are best suited for use where actual and potential contaminants have been identified. The measurements obtained with substance-specific devices are of vital importance to the employer when decisions are made concerning the measures necessary to protect entrants (such as ventilation or personal protective equipment) and the setting and attainment of appropriate entry conditions. However, the sewer environment may suddenly and unpredictably change, and the substance-specific devices may not detect the potentially lethal atmospheric hazards which may enter the sewer environment.

Although OSHA considers the information and guidance provided above to be appropriate and useful in most sewer entry situations, the Agency emphasizes that each employer must consider the unique circumstances, including the predictability of the atmosphere, of the sewer permit spaces in the employer's workplace in preparing for entry. Only the employer can decide, based upon his or her knowledge of, and experience with permit spaces in sewer systems, what the best type of testing instrument may be for any specific entry operation.

The selected testing instrument should be carried and used by the entrant in sewer line work to monitor the atmosphere in the entrant's environment, and in advance of the entrant's direction of movement, to warn the entrant of any deterioration in atmospheric conditions. Where several entrants are working together in the same immediate location, one instrument, used by the lead entrant, is acceptable.

(3) *Surge flow and flooding.* Sewer crews should develop and maintain liaison, to the extent possible, with the local weather bureau and fire and emergency services in their area so that sewer work may be delayed or interrupted and entrants withdrawn whenever sewer lines might be suddenly flooded by rain or fire suppression activities, or whenever flammable or other hazardous materials are released into sewers during emergencies by industrial or transportation accidents.

(4) *Special Equipment.* Entry into large bore sewers may require the use of special equipment. Such equipment might include such items as atmosphere monitoring devices with automatic audible alarms, escape self-contained breathing apparatus (ESCBA) with at least 10 minute air supply (or other NIOSH approved self-rescuer), and waterproof flashlights, and may also include boats and rafts, radios and rope stand-offs for pulling around bends and corners as needed.

[58 FR 4549, Jan. 14, 1993; 58 FR 34845, June 29, 1993; 63 FR 26115, May 19, 1994]

## APPENDIX F TO §1910.146—RESCUE TEAM OR RESCUE SERVICE EVALUATION CRITERIA (NON-MANDATORY)

(1) This appendix provides guidance to employers in choosing an appropriate rescue service. It contains criteria that may be used to evaluate the capabilities both of prospective and current rescue teams. Before a rescue team can be trained or chosen, however, a satisfactory permit program, including an analysis of all permit-required confined spaces to identify all potential hazards in those spaces, must be completed. OSHA believes that compliance with all the provisions of §1910.146 will enable employers to conduct permit space operations without recourse to rescue services in nearly all cases. However, experience indicates that circumstances will arise where entrants will need to be rescued from permit spaces. It is therefore important for employers to select rescue services or teams, either on-site or off-site, that are equipped and capable of minimizing harm to both entrants and rescuers if the need arises.

(2) For all rescue teams or services, the employer's evaluation should consist of two components: an initial evaluation, in which employers decide whether a potential rescue service or team is adequately trained and equipped to perform permit space rescues of the kind needed at the facility and whether such rescuers can respond in a timely manner, and a performance evaluation, in which employers measure the performance of the team or service during an actual or practice rescue. For example, based on the initial evaluation, an employer may determine that maintaining an on-site rescue team will be more expensive than obtaining the services of an off-site team, without being significantly more effective, and decide to hire a rescue service. During a performance evaluation, the employer could decide, after observing the rescue service perform a practice rescue, that the service's training or preparedness was not adequate to effect a timely or effective rescue at his or her facility and decide to select another rescue service, or to form an internal rescue team.

### A. Initial Evaluation

I. The employer should meet with the prospective rescue service to facilitate the evaluations required by §1910.146(k)(1)(i) and §1910.146(k)(1)(ii). At a minimum, if an off-site rescue service is being considered, the employer must contact the service to plan and coordinate the evaluations required by the standard. Merely posting the service's number or planning to rely on the 911 emergency phone number to obtain these services at the time of a permit space emergency would not comply with paragraph (k)(1) of the standard.

II. The capabilities required of a rescue service vary with the type of permit spaces from which rescue may be necessary and the hazards likely to be encountered in those spaces. Answering the questions below will assist employers in determining whether the rescue service is capable of performing rescues in the permit spaces present at the employer's workplace.

1. What are the needs of the employer with regard to response time (time for the rescue service to receive notification, arrive at the scene, and set up and be ready for entry)? For example, if entry is to be made into an IDLH atmosphere, or into a space that can quickly develop an IDLH atmosphere (if ventilation fails or for other reasons), the rescue team or service would need to be standing by at the permit space. On the other hand, if the danger to entrants is restricted to mechanical hazards that would cause injuries (e.g., broken bones, abrasions) a response time of 10 or 15 minutes might be adequate.

2. How quickly can the rescue team or service get from its location to the permit spaces from which rescue may be necessary? Relevant factors to consider would include: the location of the rescue team or service relative to the employer's workplace, the quality of roads and highways to be traveled, potential bottlenecks or traffic congestion that might be encountered in transit, the reliability of the rescuer's vehicles, and the training and skill of its drivers.

3. What is the availability of the rescue service? Is it unavailable at certain times of the day or in certain situations? What is the likelihood that key personnel of the rescue service might be unavailable at times? If the rescue service becomes unavailable while an entry is underway, does it have the capability of notifying the employer so that the employer can instruct the attendant to abort the entry immediately?

4. Does the rescue service meet all the requirements of paragraph (k)(2) of the standard? If not, has it developed a plan that will enable it to meet those requirements in the future? If so, how soon can the plan be implemented?

5. For off-site services, is the service willing to perform rescues at the employer's workplace? (An employer may not rely on a rescuer who declines, for whatever reason, to provide rescue services.)

6. Is an adequate method for communications between the attendant, employer and prospective rescuer available so that a rescue request can be transmitted to the rescuer without delay? How soon after notification can a prospective rescuer dispatch a rescue team to the entry site?

7. For rescues into spaces that may pose significant atmospheric hazards and from which rescue entry, patient packaging and retrieval cannot be safely accomplished in a relatively short time (15-20 minutes), employers should consider using airline respirators (with escape bottles) for the rescuers and to supply rescue air to the patient. If the employer decides to use SCBA, does the prospective rescue service have an ample supply of replacement cylinders and procedures for rescuers to enter and exit (or be retrieved) well within the SCBA's air supply limits?

8. If the space has a vertical entry over 5 feet in depth, can the prospective rescue service properly perform entry rescues? Does the service have the technical knowledge and equipment to perform rope work or elevated rescue, if needed?

9. Does the rescue service have the necessary skills in medical evaluation, patient packaging and emergency response?

10. Does the rescue service have the necessary equipment to perform rescues, or must the equipment be provided by the employer or another source?

### B. Performance Evaluation

Rescue services are required by paragraph (k)(2)(iv) of the standard to practice rescues at least once every 12 months, provided that the team or service has not successfully performed a permit space rescue within that time. As part of each practice session, the service should perform a critique of the practice rescue, or have another qualified party perform the critique, so that deficiencies in procedures, equipment, training, or number of personnel can be identified and corrected. The results of the critique, and the corrections made to respond to the deficiencies identified, should be given to the employer to enable it to determine whether the rescue service can quickly be upgraded to meet the employer's rescue needs or whether another service must be selected. The following questions will assist employers and rescue teams and services evaluate their performance.

1. Have all members of the service been trained as permit space entrants, at a minimum, including training in the potential hazards of all permit spaces, or of representative permit spaces, from which rescue may be needed? Can team members recognize the signs, symptoms, and consequences of exposure to any hazardous atmospheres that may be present in those permit spaces?

2. Is every team member provided with, and properly trained in, the use and need for PPE, such as SCBA or fall arrest equipment, which may be required to perform permit space rescues in the facility? Is every team member properly trained to perform his or her functions and make rescues, and to use any rescue equipment, such as ropes and backboards, that may be needed in a rescue attempt?

3. Are team members trained in the first aid and medical skills needed to treat victims overcome or injured by the types of hazards that may be encountered in the permit spaces at the facility?

4. Do all team members perform their functions safely and efficiently? Do rescue service personnel focus on their own safety before considering the safety of the victim?

5. If necessary, can the rescue service properly test the atmosphere to determine if it is IDLH?

6. Can the rescue personnel identify information pertinent to the rescue from entry permits, hot work permits, and MSDSs?

7. Has the rescue service been informed of any hazards to personnel that may arise from outside the space, such as those that may be caused by future work near the space?

8. If necessary, can the rescue service properly package and retrieve victims from a permit space that has a limited size opening (less than 24 inches (60.9 cm) in diameter), limited internal space, or internal obstacles or hazards?

9. If necessary, can the rescue service safely perform an elevated (high angle) rescue?

10. Does the rescue service have a plan for each of the kinds of permit space rescue operations at the facility? Is the plan adequate for all types of rescue operations that may be needed at the facility? Teams may practice in representative spaces, or in spaces that are "worst-case" or most restrictive with respect to internal configuration, elevation, and portal size. The following characteristics of a practice space should be considered when deciding whether a space is truly representative of an actual permit space:

(1) Internal configuration.

(a) Open—there are no obstacles, barriers, or obstructions within the space. One example is a water tank.

(b) Obstructed—the permit space contains some type of obstruction that a rescuer would need to maneuver around. An example would be a baffle or mixing blade. Large equipment, such as a ladder or scaffold, brought into a space for work purposes would be considered an obstruction if the positioning or size of the equipment would make rescue more difficult.

(2) Elevation.

(a) Elevated—a permit space where the entrance portal or opening is above grade by 4 feet or more. This type of space usually requires knowledge of high angle rescue procedures because of the difficulty in packaging and transporting a patient to the ground from the portal.

(b) Non-elevated—a permit space with the entrance portal located less than 4 feet above grade. This type of space will allow the rescue team to transport an injured employee normally.

(3) Portal size.

(a) Restricted—A portal of 24 inches or less in the least dimension. Portals of this size are too small to allow a rescuer to simply enter the space while using SCBA. The portal size is also too small to allow normal spinal immobilization of an injured employee.

(b) Unrestricted—A portal of greater than 24 inches in the least dimension. These portals allow relatively free movement into and out of the permit space.

(4) Space access.

(a) Horizontal—The portal is located on the side of the permit space. Use of retrieval lines could be difficult.

(b) Vertical—The portal is located on the top of the permit space, so that rescuers must climb down, or the bottom of the permit space, so that rescuers must climb up to enter the space. Vertical portals may require knowledge of rope techniques, or special patient packaging to safely retrieve a downed entrant.

### §1910.147 The control of hazardous energy (lockout/tagout).

**1910.147(a)** *Scope, application and purpose*—**(1)** *Scope.* **(i)** This standard covers the servicing and maintenance of machines and equipment in which the *unexpected* energization or start up of the machines or equipment, or release of stored energy could cause injury to employees. This standard establishes minimum performance requirements for the control of such hazardous energy.

**1910.147(a)(1)(ii)** This standard does not cover the following:

**1910.147(a)(1)(ii)(A)** Construction, agriculture and maritime employment;

**1910.147(a)(1)(ii)(B)** Installations under the exclusive control of electric utilities for the purpose of power generation, transmission and distribution, including related equipment for communication or metering; and

**1910.147(a)(1)(ii)(C)** Exposure to electrical hazards from work on, near, or with conductors or equipment in electric utilization installations, which is covered by subpart S of this part; and

**1910.147(a)(1)(ii)(D)** Oil and gas well drilling and servicing.

**1910.147(a)(2)** *Application.* **(i)** This standard applies to the control of energy during servicing and/or maintenance of machines and equipment.

**1910.147(a)(2)(ii)** Normal production operations are not covered by this standard (See subpart O of this part). Servicing and/or maintenance which takes place during normal production operations is covered by this standard only if;:

**1910.147(a)(2)(ii)(A)** An employee is required to remove or bypass a guard or other safety device; or

**1910.147(a)(2)(ii)(B)** An employee is required to place any part of his or her body into an area on a machine or piece of equipment where work is actually performed upon the material being processed (point of operation) or where an associated danger zone exists during a machine operating cycle.

**NOTE:** *Exception to paragraph (a)(2)(ii):* Minor tool changes and adjustments, and other minor servicing activities, which take place during normal production operations, are not covered by this standard if they are routine, repetitive, and integral to the use of the equipment for production, provided that the work is performed using alternative measures which provide effective protection (See subpart 0 of this part).

**1910.147(a)(2)(iii)** This standard does not apply to the following.

**1910.147(a)(2)(iii)(A)** Work on cord and plug connected electric equipment for which exposure to the hazards of unexpected energization or start up of the equipment is controlled by the unplugging of the equipment from the energy source and by the plug being under the exclusive control of the employee performing the servicing or maintenance.

**1910.147(a)(2)(iii)(B)** Hot tap operations involving transmission and distribution systems for substances such as gas, steam, water or petroleum products when they are performed on pressurized pipelines, provided that the employer demonstrates that (*1*) continuity of service is essential; (*2*) shutdown of the system is impractical; and (*3*) documented procedures are followed, and special equipment is used which will provide proven effective protection for employees.

**1910.147(a)(3)** *Purpose.* **(i)** This section requires employers to establish a program and utilize procedures for affixing appropriate lockout devices or tagout devices to energy isolating devices, and to otherwise disable machines or equipment to prevent unexpected energization, start-up or release of stored energy in order to prevent injury to employees.

**1910.147(a)(3)(ii)** When other standards in this part require the use of lockout or tagout, they shall be used and supplemented by the procedural and training requirements of this section.

**1910.147(b)** *Definitions applicable to this section.*

*Affected employee.* An employee whose job requires him/her to operate or use a machine or equipment on which servicing or maintenance is being performed under lockout or tagout, or whose job requires him/her to work in an area in which such servicing or maintenance is being performed.

*Authorized employee.* A person who locks out or tags out machines or equipment in order to perform servicing or maintenance on that machine or equipment. An affected employee becomes an authorized employee when that employee's duties include performing servicing or maintenance covered under this section.

*Capable of being locked out.* An energy isolating device is capable of being locked out if it has a hasp or other means of attachment to which, or through which, a lock can be affixed, or it has a locking mechanism built into it. Other energy isolating devices are capable of being locked out, if lockout can be achieved without the need to dismantle, rebuild, or replace the energy isolating device or permanently alter its energy control capability.

*Energized.* Connected to an energy source or containing residual or stored energy.

*Energy isolating device.* A mechanical device that physically prevents the transmission or release of energy, including but not limited to the following: A manually operated electrical circuit breaker; a disconnect switch; a manually operated switch by which the conductors of a circuit can be disconnected from all ungrounded supply conductors, and, in addition, no pole can be operated independently; a line valve; a block; and any similar device used to block or isolate energy. Push buttons, selector switches and other control circuit type devices are not energy isolating devices.

*Energy source.* Any source of electrical, mechanical, hydraulic, pneumatic, chemical, thermal, or other energy.

*Hot tap.* A procedure used in the repair, maintenance and services activities which involves welding on a piece of equipment (pipelines, vessels or tanks) under pressure, in order to install connections or appurtenances. It is commonly used to replace or add sections of pipeline without the interruption of service for air, gas, water, steam, and petrochemical distribution systems.

*Lockout.* The placement of a lockout device on an energy isolating device, in accordance with an established procedure, ensuring that the energy isolating device and the equipment being controlled cannot be operated until the lockout device is removed.

*Lockout device.* A device that utilizes a positive means such as a lock, either key or combination type, to hold an energy isolating device in a safe position and prevent the energizing of a machine or equipment. Included are blank flanges and bolted slip blinds.

*Normal production operations.* The utilization of a machine or equipment to perform its intended production function.

*Servicing and/or maintenance.* Workplace activities such as constructing, installing, setting up, adjusting, inspecting, modifying, and maintaining and/or servicing machines or equipment. These activities include lubrication, cleaning or unjamming of machines or equipment and making adjustments or tool changes, where the employee may be exposed to the *unexpected* energization or startup of the equipment or release of hazardous energy.

*Setting up.* Any work performed to prepare a machine or equipment to perform its normal production operation.

*Tagout.* The placement of a tagout device on an energy isolating device, in accordance with an established procedure, to indicate that the energy isolating device and the equipment being controlled may not be operated until the tagout device is removed.

*Tagout device.* A prominent warning device, such as a tag and a means of attachment, which can be securely fastened to an energy isolating device in accordance with an established procedure, to indicate that the energy isolating device and the equipment being controlled may not be operated until the tagout device is removed.

**1910.147(c)** *General*—**(1)** *Energy control program.* The employer shall establish a program consisting of energy control procedures, employee training and periodic inspections to ensure that before any employee performs any servicing or maintenance on a machine or equipment where the unexpected energizing, start up or release of stored energy could occur and cause injury, the machine or equipment shall be isolated from the energy source, and rendered inoperative.

**1910.147(c)(2)** *Lockout/tagout.* **(i)** If an energy isolating device is not capable of being locked out, the employer's energy control program under paragraph (c)(1) of this section shall utilize a tagout system.

**1910.147(c)(2)(ii)** If an energy isolating device is capable of being locked out, the employer's energy control program under paragraph (c)(1) of this section shall utilize lockout, unless the employer can demonstrate that the utilization of a tagout system will provide full employee protection as set forth in paragraph (c)(3) of this section.

**1910.147(c)(2)(iii)** After January 2, 1990, whenever replacement or major repair, renovation or modification of a machine or equipment is performed, and whenever new machines or equipment are installed, energy isolating devices for such machine or equipment shall be designed to accept a lockout device.

**1910.147(c)(3)** *Full employee protection.* **(i)** When a tagout device is used on an energy isolating device which is capable of being locked out, the tagout device shall be attached at the same location that the lockout device would have been attached, and the employer shall demonstrate that the tagout program will provide a level of safety equivalent to that obtained by using a lockout program.

**1910.147(c)(3)(ii)** In demonstrating that a level of safety is achieved in the tagout program which is equivalent to the level of safety obtained by using a lockout program, the employer shall demonstrate full compliance with all tagout-related provisions of this standard together with such additional elements as are necessary to provide the equivalent safety available from the use of a lockout device. Additional means to be considered as part of the demonstration of full employee protection shall include the implementation of additional safety measures such as the removal of an isolating circuit element, blocking of a controlling switch, opening of an extra disconnecting device, or the removal of a valve handle to reduce the likelihood of inadvertent energization.

**1910.147(c)(4)** *Energy control procedure.* **(i)** Procedures shall be developed, documented and utilized for the control of potentially hazardous energy when employees are engaged in the activities covered by this section.

**NOTE:** *Exception:* The employer need not document the required procedure for a particular machine or equipment, when all of the following elements exist: (1) The machine or equipment has no potential for stored or residual energy or reaccumulation of stored energy after shut down which could endanger employees; (2) the machine or equipment has a single energy source which can be readily identified and isolated; (3) the isolation and locking out of that energy source will completely deenergize and deactivate the machine or equipment; (4) the machine or equipment is isolated from that energy source and locked out during servicing or maintenance; (5) a single lockout device will achieve a locked-out condition; (6) the lockout device is under the exclusive control of the authorized employee performing the servicing or maintenance; (7) the servicing or maintenance does not create hazards for other employees; and (8) the employer, in utilizing this exception, has had no accidents involving the unexpected activation or reenergization of the machine or equipment during servicing or maintenance.

**1910.147(c)(4)(ii)** The procedures shall clearly and specifically outline the scope, purpose, authorization, rules, and techniques to be utilized for the control of hazardous energy, and the means to enforce compliance including, but not limited to, the following:

**1910.147(c)(4)(ii)(A)** A specific statement of the intended use of the procedure;

**1910.147(c)(4)(ii)(B)** Specific procedural steps for shutting down, isolating, blocking and securing machines or equipment to control hazardous energy;

**1910.147(c)(4)(ii)(C)** Specific procedural steps for the placement, removal and transfer of lockout devices or tagout devices and the responsibility for them; and

**1910.147(c)(4)(ii)(D)** Specific requirements for testing a machine or equipment to determine and verify the effectiveness of lockout devices, tagout devices, and other energy control measures.

**1910.147(c)(5)** *Protective materials and hardware.* **(i)** Locks, tags, chains, wedges, key blocks, adapter pins, self-locking fasteners, or other hardware shall be provided by the employer for isolating, securing or blocking of machines or equipment from energy sources.

**1910.147(c)(5)(ii)** Lockout devices and tagout devices shall be singularly identified; shall be the only device(s) used for controlling energy; shall not be used for other purposes; and shall meet the following requirements:

**1910.147(c)(5)(ii)(A)** *Durable.* **(1)** Lockout and tagout devices shall be capable of withstanding the environment to which they are exposed for the maximum period of time that exposure is expected.

**1910.147(c)(5)(ii)(A)(2)** Tagout devices shall be constructed and printed so that exposure to weather conditions or wet and damp locations will not cause the tag to deteriorate or the message on the tag to become illegible.

**1910.147(c)(5)(ii)(A)(3)** Tags shall not deteriorate when used in corrosive environments such as areas where acid and alkali chemicals are handled and stored.

**1910.147(c)(5)(ii)(B)** *Standardized.* Lockout and tagout devices shall be standardized within the facility in at least one of the following criteria: Color; shape; or size; and additionally, in the case of tagout devices, print and format shall be standardized.

**1910.147(c)(5)(ii)(C)** *Substantial—(1) Lockout devices.* Lockout devices shall be substantial enough to prevent removal without the use of excessive force or unusual techniques, such as with the use of bolt cutters or other metal cutting tools.

**1910.147(c)(5)(ii)(C)(2)** *Tagout devices.* Tagout devices, including and their means of attachment, shall be substantial enough to prevent inadvertent or accidental removal. Tagout device attachment means shall be of a non-reusable type, attachable by hand, self-locking, and non-releasable with a minimum unlocking strength of no less than 50 pounds and having the general design and basic characteristics of being at least equivalent to a one-piece, all-environment-tolerant nylon cable tie.

**1910.147(c)(5)(ii)(D)** *Identifiable.* Lockout devices and tagout devices shall indicate the identity of the employee applying the device(s).

**1910.147(c)(5)(iii)** Tagout devices shall warn against hazardous conditions if the machine or equipment is energized and shall include a legend such as the following: *Do Not Start, Do Not Open, Do Not Close, Do Not Energize, Do Not Operate.*

**1910.147(c)(6)** *Periodic inspection.* **(i)** The employer shall conduct a periodic inspection of the energy control procedure at least annually to ensure that the procedure and the requirements of this standard are being followed.

**1910.147(c)(6)(i)(A)** The periodic inspection shall be performed by an authorized employee other than the one(s) utilizing the energy control procedure being inspected.

**1910.147(c)(6)(i)(B)** The periodic inspection shall be conducted to correct any deviations or inadequacies identified.

**1910.147(c)(6)(i)(C)** Where lockout is used for energy control, the periodic inspection shall include a review, between the inspector and each authorized employee, of that employee's responsibilities under the energy control procedure being inspected.

**1910.147(c)(6)(i)(D)** Where tagout is used for energy control, the periodic inspection shall include a review, between the inspector and each authorized and affected employee, of that employee's responsibilities under the energy control procedure being inspected, and the elements set forth in paragraph (c)(7)(ii) of this section.

**1910.147(c)(6)(ii)** The employer shall certify that the periodic inspections have been performed. The certification shall identify the machine or equipment on which the energy control procedure was being utilized, the date of the inspection, the employees included in the inspection, and the person performing the inspection.

**1910.147(c)(7)** *Training and communication.* **(i)** The employer shall provide training to ensure that the purpose and function of the energy control program are understood by employees and that the knowledge and skills required for the safe application, usage, and removal of the energy controls are acquired by employees. The training shall include the following:

**1910.147(c)(7)(i)(A)** Each authorized employee shall receive training in the recognition of applicable hazardous energy sources, the type and magnitude of the energy available in the workplace, and the methods and means necessary for energy isolation and control.

**1910.147(c)(7)(i)(B)** Each affected employee shall be instructed in the purpose and use of the energy control procedure.

**1910.147(c)(7)(i)(C)** All other employees whose work operations are or may be in an area where energy control procedures may be utilized, shall be instructed about the procedure, and about the prohibition relating to attempts to restart or reenergize machines or equipment which are locked out or tagged out.

**1910.147(c)(7)(ii)** When tagout systems are used, employees shall also be trained in the following limitations of tags:

**1910.147(c)(7)(ii)(A)** Tags are essentially warning devices affixed to energy isolating devices, and do not provide the physical restraint on those devices that is provided by a lock.

**1910.147(c)(7)(ii)(B)** When a tag is attached to an energy isolating means, it is not to be removed without authorization of the authorized person responsible for it, and it is never to be bypassed, ignored, or otherwise defeated.

**1910.147(c)(7)(ii)(C)** Tags must be legible and understandable by all authorized employees, affected employees, and all other employees whose work operations are or may be in the area, in order to be effective.

**1910.147(c)(7)(ii)(D)** Tags and their means of attachment must be made of materials which will withstand the environmental conditions encountered in the workplace.

**1910.147(c)(7)(ii)(E)** Tags may evoke a false sense of security, and their meaning needs to be understood as part of the overall energy control program.

**1910.147(c)(7)(ii)(F)** Tags must be securely attached to energy isolating devices so that they cannot be inadvertently or accidentally detached during use.

**1910.147(c)(7)(iii)** *Employee retraining.*

**1910.147(c)(7)(iii)(A)** Retraining shall be provided for all authorized and affected employees whenever there is a change in their job assignments, a change in machines, equipment or processes that present a new hazard, or when there is a change in the energy control procedures.

**1910.147(c)(7)(iii)(B)** Additional retraining shall also be conducted whenever a periodic inspection under paragraph (c)(6) of this section reveals, or whenever the employer has reason to believe, that there are deviations from or inadequacies in the employee's knowledge or use of the energy control procedures.

**1910.147(c)(7)(iii)(C)** The retraining shall reestablish employee proficiency and introduce new or revised control methods and procedures, as necessary.

**1910.147(c)(7)(iv)** The employer shall certify that employee training has been accomplished and is being kept up to date. The certification shall contain each employee's name and dates of training.

**1910.147(c)(8)** *Energy isolation.* Lockout or tagout shall be performed only by the authorized employees who are performing the servicing or maintenance.

**1910.147(c)(9)** *Notification of employees.* Affected employees shall be notified by the employer or authorized employee of the application and removal of lockout devices or tagout devices. Notification shall be given before the controls are applied, and after they are removed from the machine or equipment.

**1910.147(d)** *Application of control.* The established procedures for the application of energy control (the lockout or tagout procedures) shall cover the following elements and actions and shall be done in the following sequence:

**1910.147(d)(1)** *Preparation for shutdown.* Before an authorized or affected employee turns off a machine or equipment, the authorized employee shall have knowledge of the type and magnitude of the energy, the hazards of the energy to be controlled, and the method or means to control the energy.

**1910.147(d)(2)** *Machine or equipment shutdown.* The machine or equipment shall be turned off or shut down using the procedures established for the machine or equipment. An orderly shutdown must be utilized to avoid any additional or increased hazard(s) to employees as a result of the equipment stoppage.

**1910.147(d)(3)** *Machine or equipment isolation.* All energy isolating devices that are needed to control the energy to the machine or equipment shall be physically located and operated in such a manner as to isolate the machine or equipment from the energy source(s).

**1910.147(d)(4)** *Lockout or tagout device application.* **(i)** Lockout or tagout devices shall be affixed to each energy isolating device by authorized employees.

**1910.147(d)(4)(ii)** Lockout devices, where used, shall be affixed in a manner to that will hold the energy isolating devices in a "safe" or "off" position.

**1910.147(d)(4)(iii)** Tagout devices, where used, shall be affixed in such a manner as will clearly indicate that the operation or movement of energy isolating devices from the "safe" or "off" position is prohibited.

**1910.147(d)(4)(iii)(A)** Where tagout devices are used with energy isolating devices designed with the capability of being locked, the tag attachment shall be fastened at the same point at which the lock would have been attached.

**1910.147(d)(4)(iii)(B)** Where a tag cannot be affixed directly to the energy isolating device, the tag shall be located as close as safely possible to the device, in a position that will be immediately obvious to anyone attempting to operate the device.

**1910.147(d)(5)** *Stored energy.* **(i)** Following the application of lockout or tagout devices to energy isolating devices, all potentially hazardous stored or residual energy shall be relieved, disconnected, restrained, and otherwise rendered safe.

**1910.147(d)(5)(ii)** If there is a possibility of reaccumulation of stored energy to a hazardous level, verification of isolation shall be continued until the servicing or maintenance is completed, or until the possibility of such accumulation no longer exists.

**1910.147(d)(6)** *Verification of isolation.* Prior to starting work on machines or equipment that have been locked out or tagged out, the authorized employee shall verify that isolation and deenergization of the machine or equipment have been accomplished.

**1910.147(e)** *Release from lockout or tagout.* Before lockout or tagout devices are removed and energy is restored to the machine or equipment, procedures shall be followed and actions taken by the authorized employee(s) to ensure the following:

**1910.147(e)(1)** *The machine or equipment.* The work area shall be inspected to ensure that nonessential items have been removed and to ensure that machine or equipment components are operationally intact.

**1910.147(e)(2)** *Employees.* **(i)** The work area shall be checked to ensure that all employees have been safely positioned or removed.

**1910.147(e)(2)(ii)** After lockout or tagout devices have been removed and before a machine or equipment is started, affected employees shall be notified that the lockout or tagout device(s) have been removed.

**1910.147(e)(3)** *Lockout or tagout devices removal.* Each lockout or tagout device shall be removed from each energy isolating device by the employee who applied the device. *Exception to paragraph (e)(3):* When the authorized employee who applied the lockout or tagout device is not available to remove it, that device may be removed under the direction of the employer, provided that specific procedures and training for such removal have been developed, documented and incorporated into the employer's energy control program. The employer shall demonstrate that the specific procedure provides equivalent safety to the removal of the device by the authorized employee who applied it. The specific procedure shall include at least the following elements:

**1910.147(e)(3)(i)** Verification by the employer that the authorized employee who applied the device is not at the facility;

**1910.147(e)(3)(ii)** Making all reasonable efforts to contact the authorized employee to inform him/her that his/her lockout or tagout device has been removed; and

**1910.147(e)(3)(iii)** Ensuring that the authorized employee has this knowledge before he/she resumes work at that facility.

**1910.147(f)** *Additional requirements*—**(1)** *Testing or positioning of machines, equipment or components thereof.* In situations in which lockout or tagout devices must be temporarily removed from the energy isolating device and the machine or equipment energized to test or position the machine, equipment or component thereof, the following sequence of actions shall be followed:

**1910.147(f)(1)(i)** Clear the machine or equipment of tools and materials in accordance with paragraph (e)(1) of this section;

**1910.147(f)(1)(ii)** Remove employees from the machine or equipment area in accordance with paragraph (e)(2) of this section;

**1910.147(f)(1)(iii)** Remove the lockout or tagout devices as specified in paragraph (e)(3) of this section;

**1910.147(f)(1)(iv)** Energize and proceed with testing or positioning;

**1910.147(f)(1)(v)** Deenergize all systems and reapply energy control measures in accordance with paragraph (d) of this section to continue the servicing and/or maintenance.

**1910.147(f)(2)** *Outside personnel (contractors, etc.).* **(i)** Whenever outside servicing personnel are to be engaged in activities covered by the scope and application of this standard, the on-site employer and the outside employer shall inform each other of their respective lockout or tagout procedures.

**1910.147(f)(2)(ii)** The on-site employer shall ensure that his/her employees understand and comply with the restrictions and prohibitions of the outside employer's energy control program.

**1910.147(f)(3)** *Group lockout or tagout.* **(i)** When servicing and/or maintenance is performed by a crew, craft, department or other group, they shall utilize a procedure which affords the employees a level of protection equivalent to that provided by the implementation of a personal lockout or tagout device.

**1910.147(f)(3)(ii)** Group lockout or tagout devices shall be used in accordance with the procedures required by paragraph (c)(4) of this section including, but not necessarily limited to, the following specific requirements:

**1910.147(f)(3)(ii)(A)** Primary responsibility is vested in an authorized employee for a set number of employees working under the protection of a group lockout or tagout device (such as an operations lock);

**1910.147(f)(3)(ii)(B)** Provision for the authorized employee to ascertain the exposure status of individual group members with regard to the lockout or tagout of the machine or equipment and

**1910.147(f)(3)(ii)(C)** When more than one crew, craft, department, etc. is involved, assignment of overall job-associated lockout or tagout control responsibility to an authorized employee designated to coordinate affected work forces and ensure continuity of protection; and

**1910.147(f)(3)(ii)(D)** Each authorized employee shall affix a personal lockout or tagout device to the group lockout device, group lockbox, or comparable mechanism when he or she begins work, and shall remove those devices when he or she stops working on the machine or equipment being serviced or maintained.

**1910.147(f)(4)** *Shift or personnel changes.* Specific procedures shall be utilized during shift or personnel changes to ensure the continuity of lockout or tagout protection, including provision for the orderly transfer of lockout or tagout device protection between off-going and oncoming employees, to minimize exposure to hazards from the unexpected energization or start-up of the machine or equipment, or the release of stored energy.

**NOTE:** The following appendix to §1910.147 services as a non-mandatory guideline to assist employers and employees in complying with the requirements of this section, as well as to provide other helpful information. Nothing in the appendix adds to or detracts from any of the requirements of this section.

[58 FR 36687 Sept. 1, 1989; as amended at 51 FR 42498, Oct. 17, 1989; 55 FR 38685, 38686, Sept. 20, 1990; 61 FR 5507, Feb. 13, 1996]

## APPENDIX A TO §1910.147—TYPICAL MINIMAL LOCKOUT PROCEDURE

### General

The following simple lockout procedure is provided to assist employers in developing their procedures so they meet the requirements of this standard. When the energy isolating devices are not lockable, tagout may be used, provided the employer complies with the provisions of the standard which require additional training and more rigorous periodic inspections. When tagout is used and the energy isolating devices are lockable, the employer must provide full employee protection (see paragraph (c)(3)) and additional training and more rigorous periodic inspections are required. For more complex systems, more comprehensive procedures may need to be developed, documented and utilized.

### Lockout Procedure

Lockout procedure for

_____

(Name of Company for single procedure or identification of equipment if multiple procedures are used)

### Purpose

This procedure establishes the minimum requirements for the lockout of energy isolating devices whenever maintenance or servicing is done on machines or equipment. It shall be used to ensure that the machine or equipment is stopped, isolated from all potentially hazardous energy sources and locked out before employees perform any servicing or maintenance where the unexpected energization or start-up of the machine or equipment or release of stored energy could cause injury.

### Compliance With This Program

All employees are required to comply with the restrictions and limitations imposed upon them during the use of lockout. The authorized employees are required to perform the lockout in accordance with this procedure. All employees, upon observing a machine or piece of equipment which is locked out to perform servicing or maintenance shall not attempt to start, energize or use that machine or equipment.

_____

Type of compliance enforcement to be taken for violation of the above.

### Sequence of Lockout

(1) Notify all affected employees that servicing or maintenance is required on a machine or equipment and that the machine or equipment must be shut down and locked out to perform the servicing or maintenance.

_____

Name(s)/Job Title(s) of affected employees and how to notify.

(2) The authorized employee shall refer to the company procedure to identify the type and magnitude of the energy that the machine or equipment utilizes, shall understand the hazards of the energy, and shall know the methods to control the energy.

_____

Type(s) and magnitude(s) of energy, its hazards and the methods to control the energy.

(3) If the machine or equipment is operating, shut it down by the normal stopping procedure (depress stop button, open switch, close valve, etc.).

_____

Type(s) and location(s) of machine or equipment operating controls.

(4) De-activate the energy isolating device(s) so that the machine or equipment is isolated from the energy source(s).

_____

Type(s) and location(s) of energy isolating devices.

(5) Lock out the energy isolating device(s) with assigned individual lock(s).

(6) Stored or residual energy (such as that in capacitors, springs, elevated machine members, rotating flywheels, hydraulic systems, and air, gas, steam, or water pressure, etc.) must be dissipated or restrained by methods such as grounding, repositioning, blocking, bleeding down, etc.

_____

Type(s) of stored energy—methods to dissipate or restrain.

(7) Ensure that the equipment is disconnected from the energy source(s) by first checking that no personnel are exposed, then verify the isolation of the equipment by operating the push button or other normal operating control(s) or by testing to make certain the equipment will not operate.

CAUTION: Return operating control(s) to neutral or "off" position after verifying the isolation of the equipment.

_____

Method of verifying the isolation of the equipment.

(8) The machine or equipment is now locked out.

*Restoring Equipment to Service.* When the servicing or maintenance is completed and the machine or equipment is ready to return to normal operating condition, the following steps shall be taken.

(1) Check the machine or equipment and the immediate area around the machine or equipment to ensure that nonessential items have been removed and that the machine or equipment components are operationally intact.

(2) Check the work area to ensure that all employees have been safely positioned or removed from the area.

(3) Verify that the controls are in neutral.

(4) Remove the lockout devices and reenergize the machine or equipment.

**NOTE:** The removal of some forms of blocking may require reenergization of the machine before safe removal.

(5) Notify affected employees that the servicing or maintenance is completed and the machine or equipment is ready for use.

[54 FR 36687, Sept. 1, 1989, as amended at 54 FR 42498, Oct. 17, 1989; 55 FR 38685, 38686, Sept. 20, 1990]

### §1910.148 Standards organizations [Removed].

[§1910.148 removed at 61 FR 9239, Mar. 7, 1996, effective May 6, 1996.]

### §1910.149 Effective dates [Removed].

[§1910.149 removed at 61 FR 9239, Mar. 7, 1996, effective May 6, 1996.]

### §1910.150 Sources of standards [Removed].

[§1910.147 redesignated as §1910.150 at 54 FR 36687, Sept. 1, 1989; removed at 61 FR 9239, Mar. 7, 1996, effective May 6, 1996.]

# Subpart K—Medical and First Aid

> **Authority:** Sections 4, 6, and 8 of the Occupational Safety and Health Act of 1970 (29 U.S.C. 653, 655, 657); Secretary of Labor's Order No. 12-71 (36 FR 8754), 8-76 (41 FR 25059), 9-83 (48 FR 35736), or 6-96 (62 FR 111), as applicable, 29 CFR part 1911.

### §1910.151 Medical services and first aid.

**1910.151(a)** The employer shall ensure the ready availability of medical personnel for advice and consultation on matters of plant health.

**1910.151(b)** In the absence of an infirmary, clinic, or hospital in near proximity to the workplace which is used for the treatment of all injured employees, a person or persons shall be adequately trained to render first aid. Adequate first aid supplies shall be readily available.

**1910.151(c)** Where the eyes or body of any person may be exposed to injurious corrosive materials, suitable facilities for quick drenching or flushing of the eyes and body shall be provided within the work area for immediate emergency use.

#### APPENDIX A TO §1910.151—
#### FIRST AID KITS (NON-MANDATORY)

First aid supplies are required to be readily available under paragraph §1910.151(b). An example of the minimal contents of a generic first aid kit is described in American National Standard (ANSI) Z308.1–1998 "Minimum Requirements for Workplace First-aid Kits." The contents of the kit listed in the ANSI standard should be adequate for small worksites.

When larger operations or multiple operations are being conducted at the same location, employers should determine the need for additional first aid kits at the worksite, additional types of first aid equipment and supplies and additional quantities and types of supplies and equipment in the first aid kits.

In a similar fashion, employers who have unique or changing first-aid needs in their workplace may need to enhance their first-aid kits. The employer can use the OSHA 200 log, OSHA 101's or other reports to identify these unique problems. Consultation from the local fire/rescue department, appropriate medical professional, or local emergency room may be helpful to employers in these circumstances. By assessing the specific needs of their workplace, employers can ensure that reasonably anticipated supplies are available. Employers should assess the specific needs of their worksite periodically and augment the first aid kit appropriately.

If it is reasonably anticipated that employees will be exposed to blood or other potentially infectious materials while using first aid supplies, employers are required to provide appropriate personal protective equipment (PPE) in compliance with the provisions of the Occupational Exposure to Blood borne Pathogens standard, §1910.1030(d)(3) (56 FR 64175). This standard lists appropriate PPE for this type of exposure, such as gloves, gowns, face shields, masks, and eye protection.

[39 FR 23502, June 27, 1974, as amended at 63 FR 33466, June 18, 1998]

### §1910.152 [Reserved].

### §1910.153 Sources of standards [Removed].

[§1910.153 removed at 61 FR 9239, Mar. 7, 1996, effective May 6, 1996.]

# Subpart L—Fire Protection

**Authority:** Sections 4, 6, and 8 of the Occupational Safety and Health Act of 1970 (29 U.S.C. 653, 655, 657); Secretary of Labor's Order No. 12-71 (36 FR 8754), 8-76 (41 FR 25059), 9-83 (48 FR 35736), or 6-96 (62 FR 111) or 3-2000 (65 FR 50017), as applicable; and 29 CFR part 1911.

## §1910.155 Scope, application and definitions applicable to this subpart.

**1910.155(a)** *Scope.* This subpart contains requirements for fire brigades, and all portable and fixed fire suppression equipment, fire detection systems, and fire or employee alarm systems installed to meet the fire protection requirements of 29 CFR part 1910.

**1910.155(b)** *Application.* This subpart applies to all employments except for maritime, construction, and agriculture.

**1910.155(c)** *Definitions applicable to this subpart.* **(1)** *After-flame* means the time a test specimen continues to flame after the flame source has been removed.

**1910.155(c)(2)** *Aqueous film forming foam (AFFF)* means a fluorinated surfactant with a foam stabilizer which is diluted with water to act as a temporary barrier to exclude air from mixing with the fuel vapor by developing an aqueous film on the fuel surface of some hydrocarbons which is capable of suppressing the generation of fuel vapors.

**1910.155(c)(3)** *Approved* means acceptable to the Assistant Secretary under the following criteria:

**1910.155(c)(3)(i)** If it is accepted, or certified, or listed, or labeled or otherwise determined to be safe by a nationally recognized testing laboratory; or

**1910.155(c)(3)(ii)** With respect to an installation or equipment of a kind which no nationally recognized testing laboratory accepts, certifies, lists, labels, or determines to be safe, if it is inspected or tested by another Federal agency and found in compliance with the provisions of the applicable National Fire Protection Association Fire Code; or

**1910.155(c)(3)(iii)** With respect to custom-made equipment or related installations which are designed, fabricated for, and intended for use by its manufacturer on the basis of test data which the employer keeps and makes available for inspection to the Assistant Secretary.

**1910.155(c)(3)(iv)** For the purposes of paragraph (c)(3) of this section:

**1910.155(c)(3)(iv)(A)** Equipment is listed if it is of a kind mentioned in a list which is published by a nationally recognized testing laboratory which makes periodic inspections of the production of such equipment and which states that such equipment meets nationally recognized standards or has been tested and found safe for use in a specified manner;

**1910.155(c)(3)(iv)(B)** Equipment is labeled if there is attached to it a label, symbol, or other identifying mark of a nationally recognized testing laboratory which makes periodic inspections of the production of such equipment, and whose labeling indicates compliance with nationally recognized standards or tests to determine safe use in a specified manner;

**1910.155(c)(3)(iv)(C)** Equipment is accepted if it has been inspected and found by a nationally recognized testing laboratory to conform to specified plans or to procedures of applicable codes; and

**1910.155(c)(3)(iv)(D)** Equipment is certified if it has been tested and found by a nationally recognized testing laboratory to meet nationally recognized standards or to be safe for use in a specified manner or is of a kind whose production is periodically inspected by a nationally recognized testing laboratory, and if it bears a label, tag, or other record of certification.

**1910.155(c)(3)(iv)(E)** Refer to §1910.7 for definition of nationally recognized testing laboratory.

**1910.155(c)(4)** *Assistant Secretary* means the Assistant Secretary of Labor for Occupational Safety and Health or designee.

**1910.155(c)(5)** *Automatic fire detection device* means a device designed to automatically detect the presence of fire by heat, flame, light, smoke or other products of combustion.

**1910.155(c)(6)** *Buddy-breathing device* means an accessory to self-contained breathing apparatus which permits a second person to share the same air supply as that of the wearer of the apparatus.

**1910.155(c)(7)** *Carbon dioxide* means a colorless, odorless, electrically nonconductive inert gas (chemical formula $CO_2$) that is a medium for extinguishing fires by reducing the concentration of oxygen or fuel vapor in the air to the point where combustion is impossible.

**1910.155(c)(8)** *Class A fire* means a fire involving ordinary combustible materials such as paper, wood, cloth, and some rubber and plastic materials.

**1910.155(c)(9)** *Class B fire* means a fire involving flammable or combustible liquids, flammable gases, greases and similar materials, and some rubber and plastic materials.

**1910.155(c)(10)** *Class C fire* means a fire involving energized electrical equipment where safety to the employee requires the use of electrically nonconductive extinguishing media.

**1910.155(c)(11)** *Class D fire* means a fire involving combustible metals such as magnesium, titanium, zirconium, sodium, lithium and potassium.

**1910.155(c)(12)** *Dry chemical* means an extinguishing agent composed of very small particles of chemicals such as, but not limited to, sodium bicarbonate, potassium bicarbonate, urea-based potassium bicarbonate, potassium chloride, or monoammonium phosphate supplemented by special treatment to provide resistance to packing and moisture absorption (caking) as well as to provide proper flow capabilities. Dry chemical does not include dry powders.

**1910.155(c)(13)** *Dry powder* means an compound used to extinguish or control Class D fires.

**1910.155(c)(14)** *Education* means the process of imparting knowledge or skill through systematic instruction. It does not require formal classroom instruction.

**1910.155(c)(15)** *Enclosed structure* means a structure with a roof or ceiling and at least two walls which may present fire hazards to employees, such as accumulations of smoke, toxic gases and heat, similar to those found in buildings.

**1910.155(c)(16)** *Extinguisher classification* means the letter classification given an extinguisher to designate the class or classes of fire on which an extinguisher will be effective.

**1910.155(c)(17)** *Extinguisher rating* means the numerical rating given to an extinguisher which indicates the extinguishing potential of the unit based on standardized tests developed by Underwriters' Laboratories, Inc.

**1910.155(c)(18)** *Fire brigade* (private fire department, industrial fire department) means an organized group of employees who are knowledgeable, trained, and skilled in at least basic fire fighting operations.

**1910.155(c)(19)** *Fixed extinguishing system* means a permanently installed system that either extinguishes or controls a fire at the location of the system.

**1910.155(c)(20)** *Flame resistance* is the property of materials, or combinations of component materials, to retard ignition and restrict the spread of flame.

**1910.155(c)(21)** *Foam* means a stable aggregation of small bubbles which flow freely over a burning liquid surface and form a coherent blanket which seals combustible vapors and thereby extinguishes the fire.

**1910.155(c)(22)** *Gaseous agent* is a fire extinguishing agent which is in the gaseous state at normal room temperature and pressure. It has low viscosity, can expand or contract with changes in pressure and temperature, and has the ability to diffuse readily and to distribute itself uniformly throughout an enclosure.

**1910.155(c)(23)** *Halon 1211* means a colorless, faintly sweet smelling, electrically nonconductive liquefied gas (chemical formula $CBrClF_2$) which is a medium for extinguishing fires by inhibiting the chemical chain reaction of fuel and oxygen. It is also known as bromochlorodifluoromethane.

**1910.155(c)(24)** *Halon 1301* means a colorless, odorless, electrically nonconductive gas (chemical formula $CBrF_3$) which is a medium for extinguishing fires by inhibiting the chemical chain reaction of fuel and oxygen. It is also known as bromotrifluoromethane.

**1910.155(c)(25)** *Helmet* is a head protective device consisting of a rigid shell, energy absorption system, and chin strap intended to be worn to provide protection for the head or portions thereof, against impact, flying or falling objects, electric shock, penetration, heat and flame.

**1910.155(c)(26)** *Incipient stage fire* means a fire which is in the initial or beginning stage and which can be controlled or extinguished by portable fire extinguishers, Class II standpipe or small hose systems without the need for protective clothing or breathing apparatus.

**1910.155(c)(27)** *Inspection* means a visual check of fire protection systems and equipment to ensure that they are in place, charged, and ready for use in the event of a fire.

**1910.155(c)(28)** *Interior structural fire fighting* means the physical activity of fire suppression, rescue or both, inside of buildings or enclosed structures which are involved in a fire situation beyond the incipient stage.

**1910.155(c)(29)** *Lining* means a material permanently attached to the inside of the outer shell of a garment for the purpose of thermal protection and padding.

**1910.155(c)(30)** *Local application system* means a fixed fire suppression system which has a supply of extinguishing agent, with nozzles arranged to automatically discharge extinguishing agent directly on the burning material to extinguish or control a fire.

**1910.155(c)(31)** *Maintenance* means the performance of services on fire protection equipment and systems to assure that they will perform as expected in the event of a fire. Maintenance differs from inspection in that maintenance requires the checking of internal fittings, devices and agent supplies.

**1910.155(c)(32)** *Multipurpose dry chemical* means a dry chemical which is approved for use on Class A, Class B and Class C fires.

**1910.155(c)(33)** *Outer shell* is the exterior layer of material on the fire coat and protective trousers which forms the outermost barrier between the fire fighter and the environment. It is attached to the vapor barrier and liner and is usually constructed with a storm flap, suitable closures, and pockets.

**1910.155(c)(34)** *Positive-pressure breathing apparatus* means self-contained breathing apparatus in which the pressure in the breathing zone is positive in relation to the immediate environment during inhalation and exhalation.

**1910.155(c)(35)** *Pre-discharge employee alarm* means an alarm which will sound at a set time prior to actual discharge of an extinguishing system so that employees may evacuate the discharge area prior to system discharge.

**1910.155(c)(36)** *Quick disconnect valve* means a device which starts the flow of air by inserting of the hose (which leads from the facepiece) into the regulator of self-contained breathing apparatus, and stops the flow of air by disconnection of the hose from the regulator.

**1910.155(c)(37)** *Sprinkler alarm* means an approved device installed so that any waterflow from a sprinkler system equal to or greater than that from single automatic sprinkler will result in an audible alarm signal on the premises.

**1910.155(c)(38)** *Sprinkler system* means a system of piping designed in accordance with fire protection engineering standards and installed to control or extinguish fires. The system includes an adequate and reliable water supply, and a network of specially sized piping and sprinklers which are interconnected. The system also includes a control valve and a device for actuating an alarm when the system is in operation.

**1910.155(c)(39)** *Standpipe systems.* **(i)** *Class I standpipe system* means a 2 1/2" (6.3 cm) hose connection for use by fire departments and those trained in handling heavy fire streams.

**1910.155(c)(39)(ii)** *Class II standpipe system* means a 1 1/2" (3.8 cm) hose system which provides a means for the control or extinguishment of incipient stage fires.

**1910.155(c)(39)(iii)** *Class III standpipe system* means a combined system of hose which is for the use of employees trained in the use of hose operations and which is capable of furnishing effective water discharge during the more advanced stages of fire (beyond the incipient stage) in the interior of workplaces. Hose outlets are available for both 1 1/2" (3.8 cm) and 2 1/2" (6.3 cm) hose.

**1910.155(c)(39)(iv)** *Small hose system* means a system of hose ranging in diameter from 5/8" (1.6 cm up to 1 1/2" (3.8 cm) which is for the use of employees and which provides a means for the control and extinguishment of incipient stage fires.

**1910.155(c)(40)** *Total flooding system* means a fixed suppression system which is arranged to automatically discharge a predetermined concentration of agent into an enclosed space for the purpose of fire extinguishment or control.

**1910.155(c)(41)** *Training* means the process of making proficient through instruction and hands-on practice in the operation of equipment, including respiratory protection equipment, that is expected to be used and in the performance of assigned duties.

**1910.155(c)(42)** *Vapor barrier* means that material used to prevent or substantially inhibit the transfer of water, corrosive liquids and steam or other hot vapors from the outside of a garment to the wearer's body.

[45 FR 60704, Sept. 12, 1980, as amended at 53 FR 12122, Apr. 12, 1988]

### §1910.156 Fire brigades.

**1910.156(a)** *Scope and application—***(1)** *Scope.* This section contains requirements for the organization, training, and personal protective equipment of fire brigades whenever they are established by an employer.

**1910.156(a)(2)** *Application.* The requirements of this section apply to fire brigades, industrial fire departments and private or contractual type fire departments. Personal protective equipment requirements apply only to members of fire brigades performing interior structural fire fighting. The requirements of this section do not apply to airport crash rescue or forest fire fighting operations.

**1910.156(b)** *Organization—***(1)** *Organizational statement.* The employer shall prepare and maintain a statement or written policy which establishes the existence of a fire brigade; the basic organizational structure; the type, amount, and frequency of training to be provided to fire brigade members; the expected number of members in the fire brigade; and the functions that the fire brigade is to perform at the workplace. The organizational statement shall be available for inspection by the Assistant Secretary and by employees or their designated representatives.

**1910.156(b)(2)** *Personnel.* The employer shall assure that employees who are expected to do interior structural fire fighting are physically capable of performing duties which may be assigned to them during emergencies. The employer shall not permit employees with known heart disease, epilepsy, or emphysema, to participate in fire brigade emergency activities unless a physician's certificate of the employees' fitness to participate in such activities is provided. For employees assigned to fire brigades before September 15, 1980, this paragraph is effective on September 15, 1990. For employees assigned to fire brigades on or after September 15, 1980, this paragraph is effective December 15, 1980.

**1910.156(c)** *Training and education.* **(1)** The employer shall provide training and education for all fire brigade members commensurate with those duties and functions that fire brigade members are expected to perform. Such training and education shall be provided to fire brigade members before they perform fire brigade emergency activities. Fire brigade leaders and training instructors shall be provided with training and education which is more comprehensive than that provided to the general membership of the fire brigade.

**1910.156(c)(2)** The employer shall assure that training and education is conducted frequently enough to assure that each member of the fire brigade is able to perform the member's assigned duties and functions satisfactorily and in a safe manner so as not to endanger fire brigade members or other employees. All fire brigade members shall be provided with training at least annually. In addition, fire brigade members who are expected to perform interior structural fire fighting shall be provided with an education session or training at least quarterly.

**1910.156(c)(3)** The quality of the training and education program for fire brigade members shall be similar to those conducted by such fire training schools as the Maryland Fire and Rescue Institute; Iowa Fire Service Extension; West Virginia Fire Service Extension; Georgia Fire Academy, New York State Department, Fire Prevention and Control; Louisiana State University Firemen Training Program, or Washington State's Fire Service Training Commission for Vocational Education. (For example, for the oil refinery industry, with its unique hazards, the training and education program for those fire brigade members shall be similar to those conducted by Texas A & M University, Lamar University, Reno Fire School, or the Delaware State Fire School.)

**1910.156(c)(4)** The employer shall inform fire brigade members about special hazards such as storage and use of flammable liquids and gases, toxic chemicals, radioactive sources, and water reactive substances, to which they may be exposed during fire and other emergencies. The fire brigade members shall also be advised of any changes that occur in relation to the special hazards. The employer shall develop and make available for inspection by fire brigade members, written procedures that describe the actions to be taken in situations involving the special hazards and shall include these in the training and education program.

**1910.156(d)** *Fire fighting equipment.* The employer shall maintain and inspect, at least annually, fire fighting equipment to assure the safe operational condition of the equipment. Portable fire extinguishers and respirators shall be inspected at least monthly. Fire fighting equipment that is in damaged or unserviceable condition shall be removed from service and replaced.

**1910.156(e)** *Protective clothing.* The following requirements apply to those employees who perform interior structural fire fighting. The requirements do not apply to employees who use fire extinguishers or standpipe systems to control or extinguish fires only in the incipient stage.

**1910.156(e)(1)** *General.* **(i)** The employer shall provide at no cost to the employee and assure the use of protective clothing which complies with the requirements of this paragraph. The employer shall assure that protective clothing ordered or purchased after July 1, 1981, meets the requirements contained in this paragraph. As the new equipment is provided, the employer shall assure that all fire brigade members wear the equipment when performing interior structural fire fighting. After July 1, 1985, the employer shall assure that all fire brigade members wear protective clothing meeting the requirements of this paragraph when performing interior structural fire fighting.

**1910.156(e)(1)(ii)** The employer shall assure that protective clothing protects the head, body, and extremities, and consists of at least the following components: foot and leg protection; hand protection; body protection; eye, face and head protection.

**1910.156(e)(2)** *Foot and leg protection.* **(i)** Foot and leg protection shall meet the requirements of paragraphs (e)(2)(ii) and (e)(2)(iii) of this section, and may be achieved by either of the following methods:

**1910.156(e)(2)(i)(A)** Fully extended boots which provide protection for the legs; or

**1910.156(e)(2)(i)(B)** Protective shoes or boots worn in combination with protective trousers that meet the requirements of paragraph (e)(3) of this section.

**1910.156(e)(2)(ii)** Protective footwear shall meet the requirements of §1910.136 for Class 75 footwear. In addition, protective footwear shall be water-resistant for at least 5 inches (12.7 cm) above the bottom of the heel and shall be equipped with slip-resistant outer soles.

**1910.156(e)(2)(iii)** Protective footwear shall be tested in accordance with paragraph (1) of appendix E, and shall provide protection against penetration of the midsole by a size 8D common nail when at least 300 pounds (1330 N) of static force is applied to the nail.

**1910.156(e)(3)** *Body protection.* **(i)** Body protection shall be coordinated with foot and leg protection to ensure full body protection for the wearer. This shall be achieved by one of the following methods:

**1910.156(e)(3)(i)(A)** Wearing of a fire-resistive coat meeting the requirements of paragraph (e)(3)(ii) of this section in combination with fully extended boots meeting the requirements of paragraphs (e)(2)(ii) and (e)(2)(iii) of this section; or

**1910.156(e)(3)(i)(B)** Wearing of a fire-resistive coat in combination with protective trousers both of which meet the requirements of paragraph (e)(3)(ii) of this section.

**1910.156(e)(3)(ii)** The performance, construction, and testing of fire-resistive coats and protective trousers shall be at least equivalent to the requirements of the National Fire Protection Association (NFPA) standard NFPA No. 1971-1975, "Protective Clothing for Structural Fire Fighting," which is incorporated by reference as specified in §1910.6, (See appendix D to subpart L) with the following permissible variations from those requirements:

**1910.156(e)(3)(ii)(A)** Tearing strength of the outer shell shall be a minimum of 8 pounds (35.6 N) in any direction when tested in accordance with paragraph (2) of appendix E; and

**1910.156(e)(3)(ii)(B)** The outer shell may discolor but shall not separate or melt when placed in a forced air laboratory oven at a temperature of 500 °F (260 °C) for a period of five minutes. After cooling to ambient temperature and using the test method specified in paragraph (3) of appendix E, char length shall not exceed 4.0 inches (10.2 cm) and after-flame shall not exceed 2.0 seconds.

**1910.156(e)(4)** *Hand protection.* **(i)** Hand protection shall consist of protective gloves or glove system which will provide protection against cut, puncture, and heat penetration. Gloves or glove system shall be tested in accordance with the test methods contained in the National Institute for Occupational Safety and Health (NIOSH) 1976 publication, "The Development of Criteria for Fire Fighter's Gloves; Vol. II, Part II: Test Methods," which is incorporated by reference as specified in §1910.6, (See appendix D to subpart L) and shall meet the following criteria for cut, puncture, and heat penetration:

**1910.156(e)(4)(i)(A)** Materials used for gloves shall resist surface cut by a blade with an edge having a 60° included angle and a .001 inch (.0025 m.) radius, under an applied force of 16 lbf (72N), and at a slicing velocity of greater or equal to 60 in/min (2.5 cm./sec);

**1910.156(e)(4)(i)(B)** Materials used for the palm and palm side of the fingers shall resist puncture by a penetrometer (simulating a 4d lath nail), under an applied force of 13.2 lbf (60N), and at a velocity greater or equal to 20 in/min (.85 cm./sec); and

**1910.156(e)(4)(i)(C)** The temperature inside the palm and gripping surface of the fingers of gloves shall not exceed 135 °F (57 °C) when gloves or glove system are exposed to 932 °F (500 °C) for five seconds at 4 psi (28 kPa) pressure.

**1910.156(e)(4)(ii)** Exterior materials of gloves shall be flame resistant and shall be tested in accordance with paragraph (3) of appendix E. Maximum allowable afterflame shall be 2.0 seconds, and the maximum char length shall be 4.0 inches (10.2 cm).

**1910.156(e)(4)(iii)** When design of the fire-resistive coat does not otherwise provide protection for the wrists, protective gloves shall have wristlets of at least 4.0 inches (10.2 cm) in length to protect the wrist area when the arms are extended upward and outward from the body.

**1910.156(e)(5)** *Head, eye and face protection.* **(i)** Head protection shall consist of a protective head device with ear flaps and chin strap which meet the performance, construction, and testing requirements of the National Fire Safety and Research Office of the National Fire Prevention and Control Administration, U.S. Department of Commerce (now known as the U.S. Fire Administration), which are contained in "Model Performance Criteria for Structural Firefighters' Helmets" (August 1977) which is incorporated by reference as specified in §1910.6, (See appendix D to subpart L).

**1910.156(e)(5)(ii)** Protective eye and face devices which comply with §1910.133 shall be used by fire brigade members when performing operations where the hazards of flying or falling materials which may cause eye and face injuries are present. Protective eye and face devices provided as accessories to protective head devices (face shields) are permitted when such devices meet the requirements of §1910.133.

**1910.156(e)(5)(iii)** Full facepieces, helmets, or hoods of breathing apparatus which meet the requirements of §1910.134 and paragraph (f) of this section, shall be acceptable as meeting the eye and face protection requirements of paragraph (e)(5)(ii) of this section.

**1910.156(f)** *Respiratory protection devices.* **(1)** *General requirements.* **(i)** The employer must ensure that respirators are provided to, and used by, fire brigade members, and that the respirators meet the requirements of 29 CFR 1910.134 and this paragraph.

**1910.156(f)(1)(ii)** Approved self-contained breathing apparatus with full-facepiece, or with approved helmet or hood configuration, shall be provided to and worn by fire brigade members while working inside buildings or confined spaces where toxic products of combustion or an oxygen deficiency may be present.

Such apparatus shall also be worn during emergency situations involving toxic substances.

**1910.156(f)(1)(iii)** Approved self-contained breathing apparatus may be equipped with either a "buddy-breathing" device or a quick disconnect valve, even if these devices are not certified by NIOSH. If these accessories are used, they shall not cause damage to the apparatus, or restrict the air flow of the apparatus, or obstruct the normal operation of the apparatus.

**1910.156(f)(1)(iv)** Approved self-contained compressed air breathing apparatus may be used with approved cylinders from other approved self-contained compressed air breathing apparatus provided that such cylinders are of the same capacity and pressure rating. All compressed air cylinders used with self-contained breathing apparatus shall meet DOT and NIOSH criteria.

**1910.156(f)(1)(v)** Self-contained breathing apparatuses must have a minimum service-life rating of 30 minutes in accordance with the methods and requirements specified by NIOSH under 42 CFR part 84, except for escape self-contained breathing apparatus (ESCBAs) used only for emergency escape purposes.

**1910.156(f)(1)(vi)** Self-contained breathing apparatus shall be provided with an indicator which automatically sounds an audible alarm when the remaining service life of the apparatus is reduced to within a range of 20 to 25 percent of its rated service time.

**1910.156(f)(2)** *Positive-pressure breathing apparatus.* **(i)** The employer shall assure that self-contained breathing apparatus ordered or purchased after July 1, 1981, for use by fire brigade members performing interior structural fire fighting operations, are of the pressure-demand or other positive-pressure type. Effective July 1, 1983, only pressure-demand or other positive-pressure self-contained breathing apparatus shall be worn by fire brigade members performing interior structural fire fighting.

**1910.156(f)(2)(ii)** This paragraph does not prohibit the use of a self-contained breathing apparatus where the apparatus can be switched from a demand to a positive-pressure mode. However, such apparatus shall be in the positive-pressure mode when fire brigade members are performing interior structural fire fighting operations.

[45 FR 60706, Sept. 12, 1980; 46 FR 24557, May 1, 1981; 49 FR 18295, Apr. 30, 1984; 61 FR 9239, Mar. 7, 1996; 63 FR 1284, Jan. 8, 1998; 63 FR 33467, June 18, 1998]

## Portable Fire Suppression Equipment

### §1910.157 Portable fire extinguishers.

**1910.157(a)** *Scope and application.* The requirements of this section apply to the placement, use, maintenance, and testing of portable fire extinguishers provided for the use of employees. Paragraph (d) of this section does not apply to extinguishers provided for employee use on the outside of workplace buildings or structures. Where extinguishers are provided but are not intended for employee use and the employer has an emergency action plan and a fire prevention plan which meet the requirements of 29 CFR 1910.38 and 29 CFR 1910.39, respectively, then only the requirements of paragraphs (e) and (f) of this section apply.

**1910.157(b)** *Exemptions.* **(1)** Where the employer has established and implemented a written fire safety policy which requires the immediate and total evacuation of employees from the workplace upon the sounding of a fire alarm signal and which includes an emergency action plan and a fire prevention plan which meet the requirements of 29 CFR 1910.38 and 29 CFR 1910.39, respectively, and when extinguishers are not available in the workplace, the employer is exempt from all requirements of this section unless a specific standard in part 1910 requires that a portable fire extinguisher be provided.

**1910.157(b)(2)** Where the employer has an emergency action plan meeting the requirements of §1910.38 which designates certain employees to be the only employees authorized to use the available portable fire extinguishers, and which requires all other employees in the fire area to immediately evacuate the affected work area upon the sounding of the fire alarm, the employer is exempt from the distribution requirements in paragraph (d) of this section.

**1910.157(c)** *General requirements.* **(1)** The employer shall provide portable fire extinguishers and shall mount, locate and identify them so that they are readily accessible to employees without subjecting the employees to possible injury.

**1910.157(c)(2)** Only approved portable fire extinguishers shall be used to meet the requirements of this section.

**1910.157(c)(3)** The employer shall not provide or make available in the workplace portable fire extinguishers using carbon tetrachloride or chlorobromomethane extinguishing agents.

**1910.157(c)(4)** The employer shall assure that portable fire extinguishers are maintained in a fully charged and operable condition and kept in their designated places at all times except during use.

**1910.157(c)(5)** The employer shall remove from service all soldered or riveted shell self-generating soda acid or self-generating foam or gas cartridge water type portable fire extinguishers which are operated by inverting the extinguisher to rupture the cartridge or to initiate an uncontrollable pressure generating chemical reaction to expel the agent.

**1910.157(d)** *Selection and distribution.* **(1)** Portable fire extinguishers shall be provided for employee use and selected and distributed based on the classes of anticipated workplace fires and on the size and degree of hazard which would affect their use.

**1910.157(d)(2)** The employer shall distribute portable fire extinguishers for use by employees on Class A fires so that the travel distance for employees to any extinguisher is 75 feet (22.9 m) or less.

**1910.157(d)(3)** The employer may use uniformly spaced standpipe systems or hose stations connected to a sprinkler system installed for emergency use by employees instead of Class A portable fire extinguishers, provided that such systems meet the respective requirements of §1910.158 or §1910.159, that they provide total coverage of the area to be protected, and that employees are trained at least annually in their use.

**1910.157(d)(4)** The employer shall distribute portable fire extinguishers for use by employees on Class B fires so that the travel distance from the Class B hazard area to any extinguisher is 50 feet (15.2 m) or less.

**1910.157(d)(5)** The employer shall distribute portable fire extinguishers used for Class C hazards on the basis of the appropriate pattern for the existing Class A or Class B hazards.

**1910.157(d)(6)** The employer shall distribute portable fire extinguishers or other containers of Class D extinguishing agent for use by employees so that the travel distance from the combustible metal working area to any extinguishing agent is 75 feet (22.9 m) or less. Portable fire extinguishers for Class D hazards are required in those combustible metal working areas where combustible metal powders, flakes, shavings, or similarly sized products are generated at least once every two weeks.

**1910.157(e)** *Inspection, maintenance and testing.* **(1)** The employer shall be responsible for the inspection, maintenance and testing of all portable fire extinguishers in the workplace.

**1910.157(e)(2)** Portable extinguishers or hose used in lieu thereof under paragraph (d)(3) of this section shall be visually inspected monthly.

**1910.157(e)(3)** The employer shall assure that portable fire extinguishers are subjected to an annual maintenance check. Stored pressure extinguishers do not require an internal examination. The employer shall record the annual maintenance date and retain this record for one year after the last entry or the life of the shell, whichever is less. The record shall be available to the Assistant Secretary upon request.

**1910.157(e)(4)** The employer shall assure that stored pressure dry chemical extinguishers that require a 12-year hydrostatic test are emptied and subjected to applicable maintenance procedures every 6 years. Dry chemical extinguishers having non-refillable disposable containers are exempt from this requirement. When recharging or hydrostatic testing is performed, the 6-year requirement begins from that date.

**1910.157(e)(5)** The employer shall assure that alternate equivalent protection is provided when portable fire extinguishers are removed from service for maintenance and recharging.

**1910.157(f)** *Hydrostatic testing.* **(1)** The employer shall assure that hydrostatic testing is performed by trained persons with suitable testing equipment and facilities.

**1910.157(f)(2)** The employer shall assure that portable extinguishers are hydrostatically tested at the intervals listed in Table L-1 of this section, except under any of the following conditions:

**1910.157(f)(2)(i)** When the unit has been repaired by soldering, welding, brazing, or use of patching compounds;

**1910.157(f)(2)(ii)** When the cylinder or shell threads are damaged;

**1910.157(f)(2)(iii)** When there is corrosion that has caused pitting, including corrosion under removable name plate assemblies;

**1910.157(f)(2)(iv)** When the extinguisher has been burned in a fire; or

**1910.157(f)(2)(v)** When a calcium chloride extinguishing agent has been used in a stainless steel shell.

**1910.157(f)(3)** In addition to an external visual examination, the employer shall assure that an internal examination of cylinders and shells to be tested is made prior to the hydrostatic tests.

# OSHA Standards for General Industry

### TABLE L-1

| Type of extinguishers | Test interval (years) |
|---|---|
| Soda acid (soldered brass shells) (until 1/1/82) | (1) |
| Soda acid (stainless steel shell) | 5 |
| Cartridge operated water and/or antifreeze | 5 |
| Stored pressure water and/or antifreeze | 5 |
| Wetting agent | 5 |
| Foam (soldered brass shells) (until 1/1/82) | (1) |
| Foam (stainless steel shell) | 5 |
| Aqueous Film Forming foam (AFFF) | 5 |
| Loaded stream | 5 |
| Dry chemical with stainless steel | 5 |
| Carbon dioxide | 5 |
| Dry chemical, stored pressure, with mild steel, brazed brass or aluminum shells | 12 |
| Dry chemical, cartridge or cylinder operated, with mild steel shells | 12 |
| Halon 1211 | 12 |
| Halon 1301 | 12 |
| Dry powder, cartridge or cylinder operated with mild steel shells | 12 |

[1] Extinguishers having shells constructed of copper or brass joined by soft solder or rivets shall not be hydrostatically tested and shall be removed from service by January 1, 1982. (Not permitted)

**1910.157(f)(4)** The employer shall assure that portable fire extinguishers are hydrostatically tested whenever they show new evidence of corrosion or mechanical injury, except under the conditions listed in paragraphs (f)(2)(i)-(v) of this section.

**1910.157(f)(5)** The employer shall assure that hydrostatic tests are performed on extinguisher hose assemblies which are equipped with a shut-off nozzle at the discharge end of the hose. The test interval shall be the same as specified for the extinguisher on which the hose is installed.

**1910.157(f)(6)** The employer shall assure that carbon dioxide hose assemblies with a shut-off nozzle are hydrostatically tested at 1,250 psi (8,620 kPa).

**1910.157(f)(7)** The employer shall assure that dry chemical and dry powder hose assemblies with a shut-off nozzle are hydrostatically tested at 300 psi (2,070 kPa).

**1910.157(f)(8)** Hose assemblies passing a hydrostatic test do not require any type of recording or stamping.

**1910.157(f)(9)** The employer shall assure that hose assemblies for carbon dioxide extinguishers that require a hydrostatic test are tested within a protective cage device.

**1910.157(f)(10)** The employer shall assure that carbon dioxide extinguishers and nitrogen or carbon dioxide cylinders used with wheeled extinguishers are tested every 5 years at 5/3 of the service pressure as stamped into the cylinder. Nitrogen cylinders which comply with 49 CFR 173.34(e)(15) may be hydrostatically tested every 10 years.

**1910.157(f)(11)** The employer shall assure that all stored pressure and Halon 1211 types of extinguishers are hydrostatically tested at the factory test pressure not to exceed two times the service pressure.

**1910.157(f)(12)** The employer shall assure that acceptable self-generating type soda acid and foam extinguishers are tested at 350 psi (2,410 kPa).

**1910.157(f)(13)** Air or gas pressure may not be used for hydrostatic testing.

**1910.157(f)(14)** Extinguisher shells, cylinders, or cartridges which fail a hydrostatic pressure test, or which are not fit for testing shall be removed from service and from the workplace.

**1910.157(f)(15)(i)** The equipment for testing compressed gas type cylinders shall be of the water jacket type. The equipment shall be provided with an expansion indicator which operates with an accuracy within one percent of the total expansion or .1 cc (.1 mL) of liquid.

**1910.157(f)(15)(ii)** The equipment for testing non-compressed gas type cylinders shall consist of the following:

**1910.157(f)(15)(ii)(A)** A hydrostatic test pump, hand or power operated, capable of producing not less than 150 percent of the test pressure, which shall include appropriate check valves and fittings;

**1910.157(f)(15)(ii)(B)** A flexible connection for attachment to fittings to test through the extinguisher nozzle, test bonnet, or hose outlet, as is applicable; and

**1910.157(f)(15)(ii)(C)** A protective cage or barrier for personal protection of the tester, designed to provide visual observation of the extinguisher under test.

**1910.157(f)(16)** The employer shall maintain and provide upon request to the Assistant Secretary evidence that the required hydrostatic testing of fire extinguishers has been performed at the time intervals shown in Table L-1. Such evidence shall be in the form of a certification record which includes the date of the test, the signature of the person who performed the test and the serial number, or other identifier, of the fire extinguisher that was tested. Such records shall be kept until the extinguisher is hydrostatically retested at the time interval specified in Table L-1 or until the extinguisher is taken out of service, whichever comes first.

**1910.157(g)** *Training and education.* **(1)** Where the employer has provided portable fire extinguishers for employee use in the workplace, the employer shall also provide an educational program to familiarize employees with the general principles of fire extinguisher use and the hazards involved with incipient stage fire fighting.

**1910.157(g)(2)** The employer shall provide the education required in paragraph (g)(1) of this section upon initial employment and at least annually thereafter.

**1910.157(g)(3)** The employer shall provide employees who have been designated to use fire fighting equipment as part of an emergency action plan with training in the use of the appropriate equipment.

**1910.157(g)(4)** The employer shall provide the training required in paragraph (g)(3) of this section upon initial assignment to the designated group of employees and at least annually thereafter.

[45 FR 60708, Sept. 12, 1980; 46 FR 24557, May 1, 1981, as amended at 51 FR 34560, Sept. 29, 1986; 61 FR 9239, Mar. 7, 1996]

## §1910.158 Standpipe and hose systems.

**1910.158(a)** *Scope and application*—**(1)** *Scope.* This section applies to all small hose, Class II, and Class III standpipe systems installed to meet the requirements of a particular OSHA standard.

**1910.158(a)(2)** *Exception.* This section does not apply to Class I standpipe systems.

**1910.158(b)** *Protection of standpipes.* The employer shall assure that standpipes are located or otherwise protected against mechanical damage. Damaged standpipes shall be repaired promptly.

**1910.158(c)** *Equipment*—**(1)** *Reels and cabinets.* Where reels or cabinets are provided to contain fire hose, the employer shall assure that they are designed to facilitate prompt use of the hose valves, the hose, and other equipment at the time of a fire or other emergency. The employer shall assure that the reels and cabinets are conspicuously identified and used only for fire equipment.

**1910.158(c)(2)** *Hose outlets and connections.* **(i)** The employer shall assure that hose outlets and connections are located high enough above the floor to avoid being obstructed and to be accessible to employees.

**1910.158(c)(2)(ii)** The employer shall standardize screw threads or provide appropriate adapters throughout the system and assure that the hose connections are compatible with those used on the supporting fire equipment.

**1910.158(c)(3)** *Hose.* **(i)** The employer shall assure that every 1 1/2" (3.8 cm) or smaller hose outlet used to meet this standard is equipped with hose connected and ready for use. In extremely cold climates where such installation may result in damaged equipment, the hose may be stored in another location provided it is readily available and can be connected when needed.

**1910.158(c)(3)(ii)** Standpipe systems installed after January 1, 1981, for use by employees, shall be equipped with lined hose. Unlined hose may remain in use on existing systems. However, after the effective date of this standard, unlined hose which becomes unserviceable shall be replaced with lined hose.

**1910.158(c)(3)(iii)** The employer shall provide hose of such length that friction loss resulting from water flowing through the hose will not decrease the pressure at the nozzle below 30 psi (210 kPa). The dynamic pressure at the nozzle shall be within the range of 30 psi (210 kPa) to 125 psi (860 kPa).

**1910.158(c)(4)** *Nozzles.* The employer shall assure that standpipe hose is equipped with shut-off type nozzles.

**1910.158(d)** *Water supply.* The minimum water supply for standpipe and hose systems, which are provided for the use of employees, shall be sufficient to provide 100 gallons per minute (6.3 l/s) for a period of at least thirty minutes.

**1910.158(e)** *Tests and maintenance*—**(1)** *Acceptance tests.* **(i)** The employer shall assure that the piping of Class II and Class III systems installed after January 1, 1981, including yard piping, is hydrostatically tested for a period of at least 2 hours at not less than 200 psi (1380 kPa), or at least 50 psi (340 kPa) in excess of normal pressure when such pressure is greater than 150 psi (1030 kPa).

**1910.158(e)(1)(ii)** The employer shall assure that hose on all standpipe systems installed after January 1, 1981, is hydrostatically tested with couplings in place, at a pressure of not less than 200 psi (1380 kPa), before it is placed in service. This pressure shall be maintained for at least 15 seconds and not more than one minute during which time the hose shall not leak nor shall any jacket thread break during the test.

**1910.158(e)(2)** *Maintenance.* **(i)** The employer shall assure that water supply tanks are kept filled to the proper level except during repairs. When pressure tanks are used, the employer shall assure that proper pressure is maintained at all times except during repairs.

**1910.158(e)(2)(ii)** The employer shall assure that valves in the main piping connections to the automatic sources of water supply are kept fully open at all times except during repair.

**1910.158(e)(2)(iii)** The employer shall assure that hose systems are inspected at least annually and after each use to assure that all of the equipment and hose are in place, available for use, and in serviceable condition.

**1910.158(e)(2)(iv)** When the system or any portion thereof is found not to be serviceable, the employer shall remove it from service immediately and replace it with equivalent protection such as extinguishers and fire watches.

**1910.158(e)(2)(v)** The employer shall assure that hemp or linen hose on existing systems is unracked, physically inspected for deterioration, and reracked using a different fold pattern at least annually. The employer shall assure that defective hose is replaced in accordance with paragraph (c)(3)(ii) of this section.

**1910.158(e)(2)(vi)** The employer shall designate trained persons to conduct all inspections required under this section.

[45 FR 60710, Sept. 12, 1980, as amended at 61 FR 9239, Mar. 7, 1996]

### Fixed Fire Suppression Equipment
### §1910.159 Automatic sprinkler systems.

**1910.159(a)** *Scope and application.* **(1)** The requirements of this section apply to all automatic sprinkler systems installed to meet a particular OSHA standard.

**1910.159(a)(2)** For automatic sprinkler systems used to meet OSHA requirements and installed prior to the effective date of this standard, compliance with the National Fire Protection Association (NFPA) or the National Board of Fire Underwriters (NBFU) standard in effect at the time of the system's installation will be acceptable as compliance with this section.

**1910.159(b)** *Exemptions.* Automatic sprinkler systems installed in workplaces, but not required by OSHA, are exempt from the requirements of this section.

**1910.159(c)** *General requirements*—**(1)** *Design.* **(i)** All automatic sprinkler designs used to comply with this standard shall provide the necessary discharge patterns, densities, and water flow characteristics for complete coverage in a particular workplace or zoned subdivision of the workplace.

**1910.159(c)(1)(ii)** The employer shall assure that only approved equipment and devices are used in the design and installation of automatic sprinkler systems used to comply with this standard.

**1910.159(c)(2)** *Maintenance.* The employer shall properly maintain an automatic sprinkler system installed to comply with this section. The employer shall assure that a main drain flow test is performed on each system annually. The inspector's test valve shall be opened at least every two years to assure that the sprinkler system operates properly.

**1910.159(c)(3)** *Acceptance tests.* The employer shall conduct proper acceptance tests on sprinkler systems installed for employee protection after January 1, 1981, and record the dates of such tests. Proper acceptance tests include the following:

**1910.159(c)(3)(i)** Flushing of underground connections;

**1910.159(c)(3)(ii)** Hydrostatic tests of piping in system;

**1910.159(c)(3)(iii)** Air tests in dry-pipe systems;

**1910.159(c)(3)(iv)** Dry-pipe valve operation; and

**1910.159(c)(3)(v)** Test of drainage facilities.

**1910.159(c)(4)** *Water supplies.* The employer shall assure that every automatic sprinkler system is provided with at least one automatic water supply capable of providing design water flow for at least 30 minutes. An auxiliary water supply or equivalent protection shall be provided when the automatic water supply is out of service, except for systems of 20 or fewer sprinklers.

**1910.159(c)(5)** *Hose connections for fire fighting use.* The employer may attach hose connections for fire fighting use to wet pipe sprinkler systems provided that the water supply satisfies the combined design demand for sprinklers and standpipes.

**1910.159(c)(6)** *Protection of piping.* The employer shall assure that automatic sprinkler system piping is protected against freezing and exterior surface corrosion.

**1910.159(c)(7)** *Drainage.* The employer shall assure that all dry sprinkler pipes and fittings are installed so that the system may be totally drained.

**1910.159(c)(8)** *Sprinklers.* **(i)** The employer shall assure that only approved sprinklers are used on systems.

**1910.159(c)(8)(ii)** The employer may not use older style sprinklers to replace standard sprinklers without a complete engineering review of the altered part of the system.

**1910.159(c)(8)(iii)** The employer shall assure that sprinklers are protected from mechanical damage.

**1910.159(c)(9)** *Sprinkler alarms.* On all sprinkler systems having more than twenty (20) sprinklers, the employer shall assure that a local waterflow alarm is provided which sounds an audible signal on the premises upon water flow through the system equal to the flow from a single sprinkler.

**1910.159(c)(10)** *Sprinkler spacing.* The employer shall assure that sprinklers are spaced to provide a maximum protection area per sprinkler, a minimum of interference to the discharge pattern by building or structural members or building contents and suitable sensitivity to possible fire hazards. The minimum vertical clearance between sprinklers and material below shall be 18 inches (45.7 cm).

**1910.159(c)(11)** *Hydraulically designed systems.* The employer shall assure that hydraulically designed automatic sprinkler systems or portions thereof are identified and that the location, number of sprinklers in the hydraulically designed section, and the basis of the design is indicated. Central records may be used in lieu of signs at sprinkler valves provided the records are available for inspection and copying by the Assistant Secretary.

[45 FR 60710, Sept. 12, 1980; 46 FR 24557, May 1, 1981]

### §1910.160 Fixed extinguishing systems, general.

**1910.160(a)** *Scope and application.* **(1)** This section applies to all fixed extinguishing systems installed to meet a particular OSHA standard except for automatic sprinkler systems which are covered by §1910.159.

**1910.160(a)(2)** This section also applies to fixed systems not installed to meet a particular OSHA standard, but which, by means of their operation, may expose employees to possible injury, death, or adverse health consequences caused by the extinguishing agent. Such systems are only subject to the requirements of paragraphs (b)(4) through (b)(7) and (c) of this section.

**1910.160(a)(3)** Systems otherwise covered in paragraph (a)(2) of this section which are installed in areas with no employee exposure are exempted from the requirements of this section.

**1910.160(b)** *General requirements.* **(1)** Fixed extinguishing system components and agents shall be designed and approved for use on the specific fire hazards they are expected to control or extinguish.

**1910.160(b)(2)** If for any reason a fixed extinguishing system becomes inoperable, the employer shall notify employees and take the necessary temporary precautions to assure their safety until the system is restored to operating order. Any defects or impairments shall be properly corrected by trained personnel.

**1910.160(b)(3)** The employer shall provide a distinctive alarm or signaling system which complies with §1910.165 and is capable of being perceived

**1910.160(b)(4)** The employer shall provide effective safeguards to warn employees against entry into discharge areas where the atmosphere remains hazardous to employee safety or health.

**1910.160(b)(5)** The employer shall post hazard warning or caution signs at the entrance to, and inside of, areas protected by fixed extinguishing systems which use agents in concentrations known to be hazardous to employee safety and health.

**1910.160(b)(6)** The employer shall assure that fixed systems are inspected annually by a person knowledgeable in the design and function of the system to assure that the system is maintained in good operating condition.

**1910.160(b)(7)** The employer shall assure that the weight and pressure of refillable containers is checked at least semi-annually. If the container shows a loss in net content or weight of more than 5 percent, or a loss in pressure of more than 10 percent, it shall be subjected to maintenance.

**1910.160(b)(8)** The employer shall assure that factory charged nonrefillable containers which have no means of pressure indication are weighed at least semi-annually. If a container shows a loss in net weight or more than 5 percent it shall be replaced.

**1910.160(b)(9)** The employer shall assure that inspection and maintenance dates are recorded on the container, on a tag attached to the container, or in a central location. A record of the last semi-annual check shall be maintained until the container is checked again or for the life of the container, whichever is less.

**1910.160(b)(10)** The employer shall train employees designated to inspect, maintain, operate, or repair fixed extinguishing systems and annually review their training to keep them up-to-date in the functions they are to perform.

**1910.160(b)(11)** The employer shall not use chlorobromomethane or carbon tetrachloride as an extinguishing agent where employees may be exposed.

**1910.160(b)(12)** The employer shall assure that systems installed in the presence of corrosive atmospheres are constructed of non-corrosive material or otherwise protected against corrosion.

**1910.160(b)(13)** Automatic detection equipment shall be approved, installed and maintained in accordance with §1910.164.

**1910.160(b)(14)** The employer shall assure that all systems designed for and installed in areas with climatic extremes shall operate effectively at the expected extreme temperatures.

**1910.160(b)(15)** The employer shall assure that at least one manual station is provided for discharge activation of each fixed extinguishing system.

**1910.160(b)(16)** The employer shall assure that manual operating devices are identified as to the hazard against which they will provide protection.

**1910.160(b)(17)** The employer shall provide and assure the use of the personal protective equipment needed for immediate rescue of employees trapped in hazardous atmospheres created by an agent discharge.

**1910.160(c)** *Total flooding systems with potential health and safety hazards to employees.* **(1)** The employer shall provide an emergency action plan in accordance with §1910.38 for each area within a workplace that is protected by a total flooding system which provides agent concentrations exceeding the maximum safe levels set forth in paragraphs (b)(5) and (b)(6) of §1910.162.

**1910.160(c)(2)** Systems installed in areas where employees cannot enter during or after the system's operation are exempt from the requirements of paragraph (c) of this section.

**1910.160(c)(3)** On all total flooding systems the employer shall provide a pre-discharge employee alarm which complies with §1910.165, and is capable of being perceived above ambient light or noise levels before the system discharges, which will give employees time to safely exit from the discharge area prior to system discharge.

**1910.160(c)(4)** The employer shall provide automatic actuation of total flooding systems by means of an approved fire detection device installed and interconnected with a pre-discharge employee alarm system to give employees time to safely exit from the discharge area prior to system discharge.

[45 FR 60711, Sept. 12, 1980]

## §1910.161 Fixed extinguishing systems, dry chemical.

**1910.161(a)** *Scope and application.* This section applies to all fixed extinguishing systems, using dry chemical as the extinguishing agent, installed to meet a particular OSHA standard. These systems shall also comply with §1910.160.

**1910.161(b)** *Specific requirements.* **(1)** The employer shall assure that dry chemical agents are compatible with any foams or wetting agents with which they are used.

**1910.161(b)(2)** The employer may not mix together dry chemical extinguishing agents of different compositions. The employer shall assure that dry chemical systems are refilled with the chemical stated on the approval nameplate or an equivalent compatible material.

**1910.161(b)(3)** When dry chemical discharge may obscure vision, the employer shall provide a pre-discharge employee alarm which complies with §1910.165 and which will give employees time to safely exit from the discharge area prior to system discharge.

**1910.161(b)(4)** The employer shall sample the dry chemical supply of all but stored pressure systems at least annually to assure that the dry chemical supply is free of moisture which may cause the supply to cake or form lumps.

**1910.161(b)(5)** The employer shall assure that the rate of application of dry chemicals is such that the designed concentration of the system will be reached within 30 seconds of initial discharge.

[45 FR 60712, Sept. 12, 1980]

## §1910.162 Fixed extinguishing systems, gaseous agent.

**1910.162(a)** *Scope and application*—**(1)** *Scope.* This section applies to all fixed extinguishing systems, using a gas as the extinguishing agent, installed to meet a particular OSHA standard. These systems shall also comply with §1910.160. In some cases, the gas may be in a liquid state during storage.

**1910.162(a)(2)** *Application.* The requirements of paragraphs (b)(2) and (b)(4) through (b)(6) shall apply only to total flooding systems.

**1910.162(b)** *Specific requirements.* **(1)** Agents used for initial supply and replenishment shall be of the type approved for the system's application. Carbon dioxide obtained by dry ice conversion to liquid is not acceptable unless it is processed to remove excess water and oil.

**1910.162(b)(2)** Except during overhaul, the employer shall assure that the designed concentration of gaseous agents is maintained until the fire has been extinguished or is under control.

**1910.162(b)(3)** The employer shall assure that employees are not exposed to toxic levels of gaseous agent or its decomposition products.

**1910.162(b)(4)** The employer shall assure that the designed extinguishing concentration is reached within 30 seconds of initial discharge except for Halon systems which must achieve design concentration within 10 seconds.

**1910.162(b)(5)** The employer shall provide a distinctive pre-discharge employee alarm capable of being perceived above ambient light or noise levels when agent design concentrations exceed the maximum safe level for employee exposure. A pre-discharge employee alarm for alerting employees before system discharge shall be provided on Halon 1211 and carbon dioxide systems with a design concentration of 4 percent or greater and for Halon 1301 systems with a design concentration of 10 percent or greater. The pre-discharge employee alarm shall provide employees time to safely exit the discharge area prior to system discharge.

**1910.162(b)(6)(i)** Where egress from an area cannot be accomplished within one minute, the employer shall not use Halon 1301 in concentrations greater than 7 percent.

**1910.162(b)(6)(ii)** Where egress takes greater than 30 seconds but less than one minute, the employer shall not use Halon 1301 in a concentration greater than 10 percent.

**1910.162(b)(6)(iii)** Halon 1301 concentrations greater than 10 percent are only permitted in areas not normally occupied by employees provided that any employee in the area can escape within 30 seconds. The employer shall assure that no unprotected employees enter the area during agent discharge.

[45 FR 60712, Sept. 12, 1980; 46 FR 24557, May 1, 1981]

## §1910.163 Fixed extinguishing systems, water spray and foam.

**1910.163(a)** *Scope and application.* This section applies to all fixed extinguishing systems, using water or foam solution as the extinguishing agent, installed to meet a particular OSHA standard. These systems shall also comply with §1910.160. This section does not apply to automatic sprinkler systems which are covered under §1910.159.

**1910.163(b)** *Specific requirements.* **(1)** The employer shall assure that foam and water spray systems are designed to be effective in at least controlling fire in the protected area or on protected equipment.

**1910.163(b)(2)** The employer shall assure that drainage of water spray systems is directed away from areas where employees are working and that no emergency egress is permitted through the drainage path.

[45 FR 60712, Sept. 12, 1980]

### Other Fire Protection Systems

## §1910.164 Fire detection systems.

**1910.164(a)** *Scope and application.* This section applies to all automatic fire detection systems installed to meet the requirements of a particular OSHA standard.

**1910.164(b)** *Installation and restoration.* **(1)** The employer shall assure that all devices and equipment constructed and installed to comply with this standard are approved for the purpose for which they are intended.

**1910.164(b)(2)** The employer shall restore all fire detection systems and components to normal operating condition as promptly as possible after each test or alarm. Spare detection devices and components which are normally destroyed in the process of detecting fires shall be available on the premises or from a local supplier in sufficient quantities and locations for prompt restoration of the system.

**1910.164(c)** *Maintenance and testing.* **(1)** The employer shall maintain all systems in an operable condition except during repairs or maintenance.

**1910.164(c)(2)** The employer shall assure that fire detectors and fire detection systems are tested and adjusted as often as needed to maintain proper reliability and operating condition except that factory calibrated detectors need not be adjusted after installation.

**1910.164(c)(3)** The employer shall assure that pneumatic and hydraulic operated detection systems installed after January 1, 1981, are equipped with supervised systems.

**1910.164(c)(4)** The employer shall assure that the servicing, maintenance and testing of fire detection systems, including cleaning and necessary sensitivity adjustments are performed by a trained person knowledgeable in the operations and functions of the system.

**1910.164(c)(5)** The employer shall also assure that fire detectors that need to be cleaned of dirt, dust, or other particulates in order to be fully operational are cleaned at regular periodic intervals.

**1910.164(d)** *Protection of fire detectors.* **(1)** The employer shall assure that fire detection equipment installed outdoors or in the presence of corrosive atmospheres be protected from corrosion. The employer shall provide a canopy, hood, or other suitable protection for detection equipment requiring protection from the weather.

**1910.164(d)(2)** The employer shall locate or otherwise protect detection equipment so that it is protected from mechanical or physical impact which might render it inoperable.

**1910.164(d)(3)** The employer shall assure that detectors are supported independently of their attachment to wires or tubing.

**1910.164(e)** *Response time.* **(1)** The employer shall assure that fire detection systems installed for the purpose of actuating fire extinguishment or suppression systems shall be designed to operate in time to control or extinguish a fire.

**1910.164(e)(2)** The employer shall assure that fire detection systems installed for the purpose of employee alarm and evacuation be designed and installed to provide a warning for emergency action and safe escape of employees.

**1910.164(e)(3)** The employer shall not delay alarms or devices initiated by fire detector actuation for more than 30 seconds unless such delay is necessary for the immediate safety of employees. When such delay is necessary, it shall be addressed in an emergency action plan meeting the requirements of §1910.38.

**1910.164(f)** *Number, location and spacing of detecting devices.* The employer shall assure that the number, spacing and location of fire detectors is based upon design data obtained from field experience, or tests, engineering surveys, the manufacturer's recommendations, or a recognized testing laboratory listing.

[45 FR 60713, Sept. 12, 1980]

## §1910.165 Employee alarm systems.

**1910.165(a)** *Scope and application.* **(1)** This section applies to all emergency employee alarms installed to meet a particular OSHA standard. This section does not apply to those discharge or supervisory alarms required on various fixed extinguishing systems or to supervisory alarms on fire suppression, alarm or detection systems unless they are intended to be employee alarm systems.

**1910.165(a)(2)** The requirements in this section that pertain to maintenance, testing and inspection shall apply to all local fire alarm signaling systems used for alerting employees regardless of the other functions of the system.

**1910.165(a)(3)** All pre-discharge employee alarms installed to meet a particular OSHA standard shall meet the requirements of paragraphs (b)(1) through (4), (c), and (d)(1) of this section.

**1910.165(b)** *General requirements.* **(1)** The employee alarm system shall provide warning for necessary emergency action as called for in the emergency action plan, or for reaction time for safe escape of employees from the workplace or the immediate work area, or both.

**1910.165(b)(2)** The employee alarm shall be capable of being perceived above ambient noise or light levels by all employees in the affected portions of the workplace. Tactile devices may be used to alert those employees who would not otherwise be able to recognize the audible or visual alarm.

**1910.165(b)(3)** The employee alarm shall be distinctive and recognizable as a signal to evacuate the work area or to perform actions designated under the emergency action plan.

**1910.165(b)(4)** The employer shall explain to each employee the preferred means of reporting emergencies, such as manual pull box alarms, public address systems, radio or telephones. The employer shall post emergency telephone numbers near telephones, or employee notice boards, and other conspicuous locations when telephones serve as a means of reporting emergencies. Where a communication system also serves as the employee alarm system, all emergency messages shall have priority over all non-emergency messages.

**1910.165(b)(5)** The employer shall establish procedures for sounding emergency alarms in the workplace. For those employers with 10 or fewer employees in a particular workplace, direct voice communication is an acceptable procedure for sounding the alarm provided all employees can hear the alarm. Such workplaces need not have a back-up system.

**1910.165(c)** *Installation and restoration.* **(1)** The employer shall assure that all devices, components, combinations of devices or systems constructed and installed to comply with this standard are approved. Steam whistles, air horns, strobe lights or similar lighting devices, or tactile devices meeting the requirements of this section are considered to meet this requirement for approval.

**1910.165(c)(2)** The employer shall assure that all employee alarm systems are restored to normal operating condition as promptly as possible after each test or alarm. Spare alarm devices and components subject to wear or destruction shall be available in sufficient quantities and locations for prompt restoration of the system.

**1910.165(d)** *Maintenance and testing.* **(1)** The employer shall assure that all employee alarm systems are maintained in operating condition except when undergoing repairs or maintenance.

**1910.165(d)(2)** The employer shall assure that a test of the reliability and adequacy of non-supervised employee alarm systems is made every two months. A different actuation device shall be used in each test of a multi-actuation device system so that no individual device is used for two consecutive tests.

**1910.165(d)(3)** The employer shall maintain or replace power supplies as often as is necessary to assure a fully operational condition. Back-up means of alarm, such as employee runners or telephones, shall be provided when systems are out of service.

**1910.165(d)(4)** The employer shall assure that employee alarm circuitry installed after January 1, 1981, which is capable of being supervised is supervised and that it will provide positive notification to assigned personnel whenever a deficiency exists in the system. The employer shall assure that all supervised employee alarm systems are tested at least annually for reliability and adequacy.

**1910.165(d)(5)** The employer shall assure that the servicing, maintenance and testing of employee alarms are done by persons trained in the designed operation and functions necessary for reliable and safe operation of the system.

**1910.165(e)** *Manual operation.* The employer shall assure that manually operated actuation devices for use in conjunction with employee alarms are unobstructed, conspicuous and readily accessible.

[45 FR 60713, Sept. 12, 1980]

### APPENDICES TO SUBPART L TO PART 1910

**Note:** The following appendices to subpart L, except appendix E, serve as nonmandatory guidelines to assist employers in complying with the appropriate requirements of subpart L.

### Appendix A to Subpart L to Part 1910—Fire Protection

*§1910.156 Fire brigades.*

1. *Scope.* This section does not require an employer to organize a fire brigade. However, if an employer does decide to organize a fire brigade, the requirements of this section apply.

2. *Pre-fire planning.* It is suggested that pre-fire planning be conducted by the local fire department and/or the workplace fire brigade in order for them to be familiar with the workplace and process hazards. Involvement with the local fire department or fire prevention bureau is encouraged to facilitate coordination and cooperation between members of the fire brigade and those who might be called upon for assistance during a fire emergency.

3. *Organizational statement.* In addition to the information required in the organizational statement, paragraph 1910.156(b)(1), it is suggested that the organizational statement also contain the following information: a description of the duties that the fire brigade members are expected to perform; the line authority of each fire brigade officer; the number of the fire brigade officers and number of training instructors; and a list and description of the types of awards or recognition that brigade members may be eligible to receive.

4. *Physical capability.* The physical capability requirement applies only to those fire brigade members who perform interior structural fire fighting. Employees who cannot meet the physical capability requirement may still be members of the fire brigade as long as such employees do not perform interior structural fire fighting. It is suggested that fire brigade members who are unable to perform interior structural fire fighting be assigned less stressful and physically demanding fire brigade duties, e.g., certain types of training, recordkeeping, fire prevention inspection and maintenance, and fire pump operations.

Physically capable can be defined as being able to perform those duties specified in the training requirements of section 1910.156(c). Physically capable can also be determined by physical performance tests or by a physical examination when the examining physician is aware of the duties that the fire brigade member is expected to perform.

It is also recommended that fire brigade members participate in a physical fitness program. There are many benefits which can be attributed to being physically fit. It is believed that physical fitness may help to reduce the number of sprain and strain injuries as well as contributing to the improvement of the cardiovascular system.

5. *Training and education.* The paragraph on training and education does not contain specific training and education requirements because the type, amount, and frequency of training and education will be as varied as are the purposes for which fire brigades are organized. However, the paragraph does require that training and education be commensurate with those functions that the fire brigade is expected to perform; i.e., those functions specified in the organizational statement. Such a performance requirement provides the necessary flexibility to design a training program which meets the needs of individual fire brigades.

At a minimum, hands-on training is required to be conducted annually for all fire brigade members. However, for those fire brigade members who are expected to perform interior structural fire fighting, some type of training or education session must be provided at least quarterly.

In addition to the required hands-on training, it is strongly recommended that fire brigade members receive other types of training and education such as: classroom instruction, review of emergency action procedures, pre-fire planning, review of special hazards in the workplace, and practice in the use of self-contained breathing apparatus.

It is not necessary for the employer to duplicate the same training or education that a fire brigade member receives as a member of a community volunteer fire department, rescue squad, or similar organization. However, such training or education must have been provided to the fire brigade member within the past year and it must be documented that the fire brigade member has received the training or education. For example: there is no need for a fire brigade member to receive another training class in the use of positive-pressure self-contained breathing apparatus if the fire brigade member has recently completed such training as a member of a community fire department. Instead, the fire brigade member should receive training or education covering other important equipment or duties of the fire brigade as they relate to the workplace hazards, facilities and processes.

It is generally recognized that the effectiveness of fire brigade training and education depends upon the expertise of those providing the training and education as well as the motivation of the fire brigade members. Fire brigade training instructors must receive a higher level of training and education than the fire brigade members they will be teaching. This includes being more knowledgeable about the functions to be performed by the fire brigade and the hazards involved. The instructors should be qualified to train fire brigade members and demonstrate skills in communication, methods of teaching, and motivation. It is important for instructors and fire brigade members alike to be motivated toward the goals of the fire brigade and be aware of the importance of the service that they are providing for the protection of other employees and the workplace.

It is suggested that publications from the International Fire Service Training Association, the National Fire Protection Association (NFPA-1041), the International Society of Fire Service Instructors and other fire training sources be consulted for recommended qualifications of fire brigade training instructors.

In order to be effective, fire brigades must have competent leadership and supervision. It is important for those who supervise the fire brigade during emergency situations, e.g., fire brigade chiefs, leaders, etc., to receive the necessary training and education for supervising fire brigade activities during these hazardous and stressful situations. These fire brigade members with leadership responsibilities should demonstrate skills in strategy and tactics, fire suppression and prevention techniques, leadership principles, pre-fire planning, and safety practices. It is again suggested that fire service training sources be consulted for determining the kinds of training and education which are necessary for those with fire brigade leadership responsibilities.

It is further suggested that fire brigade leaders and fire brigade instructors receive more formalized training and education on a continuing basis by attending classes provided by such training sources as universities and university fire extension services.

The following recommendations should not be considered to be all of the necessary elements of a complete comprehensive training program, but the information may be helpful as a guide in developing a fire brigade training program.

All fire brigade members should be familiar with exit facilities and their location, emergency escape routes for handicapped workers, and the workplace "emergency action plan."

In addition, fire brigade members who are expected to control and extinguish fires in the incipient stage should, at a minimum, be trained in the use of fire extinguishers, standpipes, and other fire equipment they are assigned to use. They should also be aware of first aid medical procedures and procedures for dealing with special hazards to which they may be exposed. Training and education should include both classroom instruction and actual operation of the equipment under simulated emergency conditions. Hands-on type training must be conducted at least annually but some functions should be reviewed more often.

In addition to the above training, fire brigade members who are expected to perform emergency rescue and interior structural fire fighting should, at a minimum, be familiar with the proper techniques in rescue and fire suppression procedures. Training and education should include fire protection courses, classroom training, simulated fire situations including "wet drills" and, when feasible, extinguishment of actual mock fires.

Frequency of training or education must be at least quarterly, but some drills or classroom training should be conducted as often as monthly or even weekly to maintain the proficiency of fire brigade members.

There are many excellent sources of training and education that the employer may want to use in developing a training program for the workplace fire brigade. These sources include publications, seminars, and courses offered by universities.

There are also excellent fire school courses by such facilities as Texas A and M University, Delaware State Fire School, Lamar University, and Reno Fire School, that deal with those unique hazards which may be encountered by fire brigades in the oil and chemical industry. These schools, and others, also offer excellent training courses which would be beneficial to fire brigades in other types of industries. These courses should be a continuing part of the training program, and employers are strongly encouraged to take advantage of these excellent resources.

It is also important that fire brigade members be informed about special hazards to which they may be exposed during fire and other emergencies. Such hazards as storage and use areas of flammable liquids and gases, toxic chemicals, water-reactive substances, etc., can pose difficult problems. There must be written procedures developed that describe the actions to be taken in situations involving special hazards. Fire brigade members must be trained in handling these special hazards as well as keeping abreast of any changes that occur in relation to these special hazards.

6. *Fire fighting equipment.* It is important that fire fighting equipment that is in damaged or unserviceable condition be removed from service and replaced. This will prevent fire brigade members from using unsafe equipment by mistake.

Fire fighting equipment, except portable fire extinguishers and respirators, must be inspected at least annually. Portable fire extinguishers and respirators are required to be inspected at least monthly.

7. *Protective clothing.* (A) *General.* Paragraph (e) of §1910.156 does not require all fire brigade members to wear protective clothing. It is not the intention of these standards to require employers to provide a full ensemble of protective clothing for every fire brigade member without consideration given to the types of hazardous environments to which the fire brigade member might be exposed. It is the intention of these standards to require adequate protection for those fire brigade members who might be exposed to fires in an advanced stage, smoke, toxic gases, and high temperatures. Therefore, the protective clothing requirements only apply to those fire brigade members who perform interior structural fire fighting operations.

Additionally, the protective clothing requirements do not apply to the protective clothing worn during outside fire fighting operations (brush and forest fires, crash crew operations) or other special fire fighting activities. It is important that the protective clothing to be worn during these types of fire fighting operations reflect the hazards which are expected to be encountered by fire brigade members.

(B) *Foot and leg protection.* Section 1910.156 permits an option to achieve foot and leg protection.

The section recognizes the interdependence of protective clothing to cover one or more parts of the body. Therefore, an option is given so that fire brigade members may meet the foot and leg requirements by either wearing long fire-resistive coats in combination with fully extended boots, or by wearing shorter fire-resistive costs in combination with protective trousers and protective shoes or shorter boots.

(C) *Body protection.* Paragraph (e)(3) of §1910.156 provides an option for fire brigade members to achieve body protection. Fire brigade members may wear a fire-resistive coat in combination with fully extended boots, or they may wear a fire-resistive coat in combination with protective trousers.

Fire-resistive coats and protective trousers meeting all of the requirements contained in NFPA 1971-1975 "Protective Clothing for Structural Fire Fighters," are acceptable as meeting the requirements of this standard.

The lining is required to be permanently attached to the outer shell. However, it is permissible to attach the lining to the outer shell material by stitching in one area such as at the neck. Fastener tape or snap fasteners may be used to secure the rest of the lining to the outer shell to facilitate cleaning. Reference to permanent lining does not refer to a winter liner which is a detachable extra lining used to give added protection to the wearer against the effects of cold weather and wind.

(D) *Hand protection.* The requirements of the paragraph on hand protection may be met by protective gloves or a glove system. A glove system consists of a combination of different gloves. The usual components of a glove system consist of a pair of gloves, which provide thermal insulation to the hands, worn in combination with a second pair of gloves which provide protection against flame, cut, and puncture.

It is suggested that protective gloves provide dexterity and a sense of feel for objects. Criteria and test methods for dexterity are contained in the NIOSH publications, "The Development of Criteria for Firefighters' Gloves; Vol. I: Glove Requirements" and "Vol. II: Glove Criteria and Test Methods." These NIOSH publications also contain a permissible modified version of Federal Test Method 191, Method 5903, (paragraph (3) of appendix E) for flame resistance when gloves, rather than glove material, are tested for flame resistance.

(E) *Head, eye, and face protection.* Head protective devices which meet the requirements contained in NFPA No. 1972 are acceptable as meeting the requirements of this standard for head protection.

Head protective devices are required to be provided with ear flaps so that the ear flaps will be available if needed. It is recommended that ear protection always be used while fighting interior structural fires.

Many head protective devices are equipped with face shields to protect the eyes and face. These face shields are permissible as meeting the eye and face protection requirements of this paragraph as long as such face shields meet the requirements of §1910.133 of the General Industry Standards.

Additionally, full facepieces, helmets or hoods of approved breathing apparatus which meet the requirements of §1910.134 and paragraph (f) of §1910.156 are also acceptable as meeting the eye and face protection requirements.

It is recommended that a flame resistant protective head covering such as a hood or snood, which will not adversely affect the seal of a respirator facepiece, be worn during interior structural fire fighting operations to protect the sides of the face and hair.

8. *Respiratory protective devices.* Respiratory protection is required to be worn by fire brigade members while working inside buildings or confined spaces where toxic products of combustion or an oxygen deficiency is likely to be present; respirators are also to be worn during emergency situations involving toxic substances. When fire brigade members respond to emergency situations, they may be exposed to unknown contaminants in unknown concentrations. Therefore, it is imperative that fire brigade members wear proper respiratory protective devices during these situations. Additionally, there are many instances where toxic products of combustion are still present during mop-up and overhaul operations. Therefore, fire brigade members should continue to wear respirators during these types of operations.

Self-contained breathing apparatus are not required to be equipped with either a buddy-breathing device or a quick-disconnect valve. However, these accessories may be very useful and are acceptable as long as such accessories do not cause damage to the apparatus, restrict the air flow of the apparatus, or obstruct the normal operation of the apparatus.

Buddy-breathing devices are useful for emergency situations where a victim or another fire brigade member can share the same air supply with the wearer of the apparatus for emergency escape purposes.

The employer is encouraged to provide fire brigade members with an alternative means of respiratory protection to be used only for emergency escape purposes if the self-contained breathing apparatus becomes inoperative. Such alternative means of respiratory protection may be either a buddy-breathing device or an escape self-contained breathing apparatus (ESCBA). The ESCBA is a short-duration respiratory protective device which is approved for only emergency escape purposes. It is suggested that if ESCBA units are used, that they be of at least 5 minutes service life.

Quick-disconnect valves are devices which start the flow of air by insertion of the hose (which leads to the facepiece) into the regulator of self-contained breathing apparatus, and stop the flow of air by disconnecting the hose from the regulator. These devices are particularly useful for those positive-pressure self-contained breathing apparatus which do not have the capability of being switched from the demand to the positive-pressure mode.

The use of a self-contained breathing apparatus where the apparatus can be switched from a demand to a positive-pressure mode is acceptable as long as the apparatus is in the positive-pressure mode when performing interior structural fire fighting operations. Also acceptable are approved respiratory protective devices which have been converted to the positive-pressure type when such modification is accomplished by trained and experienced persons using kits or parts approved by NIOSH and provided by the manufacturer and by following the manufacturer's instructions.

# OSHA Standards for General Industry

There are situations which require the use of respirators which have a duration of 2 hours or more. Presently, there are no approved positive-pressure apparatus with a rated service life of more than 2 hours. Consequently, negative-pressure self-contained breathing apparatus with a rated service life of more than 2 hours and which have a minimum protection factor of 5,000 as determined by an acceptable quantitative fit test performed on each individual, will be acceptable for use during situations which require long duration apparatus. Long duration apparatus may be needed in such instances as working in tunnels, subway systems, etc. Such negative-pressure breathing apparatus will continue to be acceptable for a maximum of 18 months after a positive-pressure apparatus with the same or longer rated service life of more than 2 hours is certified by NIOSH/MSHA. After this 18 month phase-in period, all self-contained breathing apparatus used for these long duration situations will have to be of the positive-pressure type.

Protection factor (sometimes called fit factor) is defined as the ratio of the contaminant concentrations outside of the respirator to the contaminant concentrations inside the facepiece of the respirator.

$$PF = \frac{\text{Concentration outside respirator}}{\text{Concentration inside facepiece}}$$

Protection factors are determined by quantitative fit tests. An acceptable quantitative fit test should include the following elements:

1. A fire brigade member who is physically and medically capable of wearing respirators, and who is trained in the use of respirators, dons a self-contained breathing apparatus equipped with a device that will monitor the concentration of a contaminant inside the facepiece.

2. The fire brigade member then performs a qualitative fit test to assure the best face to facepiece seal as possible. A qualitative fit test can consist of a negative-pressure test, positive-pressure test, isoamyl acetate vapor (banana oil) test, or an irritant smoke test. For more details on respirator fitting see the NIOSH booklet entitled "A Guide to Industrial Respiratory Protection" June, 1976, and HEW publication No. (NIOSH) 76-189.

3. The wearer should then perform physical activity which reflects the level of work activity which would be expected during fire fighting activities. The physical activity should include simulated fire-ground work activity or physical exercise such as running-in-place, a step test, etc.

4. Without readjusting the apparatus, the wearer is placed in a test atmosphere containing a non-toxic contaminant with a known, constant, concentration.

The protection factor is then determined by dividing the known concentration of the contaminant in the test atmosphere by the concentration of the contaminant inside the facepiece when the following exercises are performed:

(a) Normal breathing with head motionless for one minute;

(b) Deep breathing with head motionless for 30 seconds;

(c) Turning head slowly from side to side while breathing normally, pausing for at least two breaths before changing direction. Continue for at least one minute;

(d) Moving head slowly up and down while breathing normally, pausing for at least two breaths before changing direction. Continue for at least two minutes;

(e) Reading from a prepared text, slowly and clearly, and loudly enough to be heard and understood. Continue for one minute; and

(f) Normal breathing with head motionless for at least one minute.

The protection factor which is determined must be at least 5,000. The quantitative fit test should be conducted at least three times. It is acceptable to conduct all three tests on the same day. However, there should be at least one hour between tests to reflect the protection afforded by the apparatus during different times of the day.

The above elements are not meant to be a comprehensive, technical description of a quantitative fit test protocol. However, quantitative fit test procedures which include these elements are acceptable for determining protection factors. Procedures for a quantitative fit test are required to be available for inspection by the Assistant Secretary or authorized representative.

Organizations such as Los Alamos Scientific Laboratory, Lawrence Livermore Laboratory, NIOSH, and American National Standards Institute (ANSI) are excellent sources for additional information concerning qualitative and quantitative fit testing.

§1910.157 Portable fire extinguishers.

1. *Scope and application.* The scope and application of this section is written to apply to three basic types of workplaces. First, there are those workplaces where the employer has chosen to evacuate all employees from the workplace at the time of a fire emergency. Second, there are those workplaces where the employer has chosen to permit certain employees to fight fires and to evacuate all other non-essential employees at the time of a fire emergency. Third, there are those workplaces where the employer has chosen to permit all employees in the workplace to use portable fire extinguishers to fight fires.

The section also addresses two kinds of work areas. The entire workplace can be divided into outside (exterior) work areas and inside (interior) work areas. This division of the workplace into two areas is done in recognition of the different types of hazards employees may be exposed to during fire fighting operations. Fires in interior workplaces, pose a greater hazard to employees; they can produce greater exposure to quantities of smoke, toxic gases, and heat because of the capability of a building or structure to contain or entrap these products of combustion until the building can be ventilated. Exterior work areas, normally open to the environment, are somewhat less hazardous, because the products of combustion are generally carried away by the thermal column of the fire. Employees also have a greater selection of evacuation routes if it is necessary to abandon fire fighting efforts.

In recognition of the degree of hazard present in the two types of work areas, the standards for exterior work areas are somewhat less restrictive in regards to extinguisher distribution. Paragraph (a) explains this by specifying which paragraphs in the section apply.

2. *Portable fire extinguisher exemptions.* In recognition of the three options given to employers in regard to the amount of employee evacuation to be carried out, the standards permit certain exemptions based on the number of employees expected to use fire extinguishers.

Where the employer has chosen to totally evacuate the workplace at the time of a fire emergency and when fire extinguishers are not provided, the requirements of this section do not apply to that workplace.

Where the employer has chosen to partially evacuate the workplace or the effected area at the time of a fire emergency and has permitted certain designated employees to remain behind to operate critical plant operations or to fight fires with extinguishers, then the employer is exempt from the distribution requirements of this section. Employees who will be remaining behind to perform incipient fire fighting or members of a fire brigade must be trained in their duties. The training must result in the employees becoming familiar with the locations of fire extinguishers. Therefore, the employer must locate the extinguishers in convenient locations where the employees know they can be found. For example, they could be mounted in the fire truck or cart that the fire brigade uses when it responds to a fire emergency. They can also be distributed as set forth in the National Fire Protection Association's Standard No. 10, "Portable Fire Extinguishers."

Where the employer has decided to permit all employees in the workforce to use fire extinguishers, then the entire OSHA section applies.

3. *Portable fire extinguisher mounting.* Previous standards for mounting fire extinguishers have been criticized for requiring specific mounting locations. In recognition of this criticism, the standard has been rewritten to permit as much flexibility in extinguisher mounting as is acceptable to assure that fire extinguishers are available when needed and that employees are not subjected to injury hazards when they try to obtain an extinguisher.

It is the intent of OSHA to permit the mounting of extinguishers in any location that is accessible to employees without the use of portable devices such as a ladder. This limitation is necessary because portable devices can be moved or taken from the place where they are needed and, therefore, might not be available at the time of an emergency.

Employers are given as much flexibility as possible to assure that employees can obtain extinguishers as fast as possible. For example, an acceptable method of mounting extinguishers in areas where fork lift trucks or tow-motors are used is to mount the units on retractable boards which, by means of counterweighting, can be raised above the level where they could be struck by vehicular traffic. When needed, they can be lowered quickly for use. This method of mounting can also reduce vandalism and unauthorized use of extinguishers. The extinguishers may also be mounted as outlined in the National Fire Protection Association's Standard No. 10, "Portable Fire Extinguishers."

4. *Selection and distribution.* The employer is responsible for the proper selection and distribution of fire extinguishers and the determination of the necessary degree of protection. The selection and distribution of fire extinguishers must reflect the type and class of fire hazards associated with a particular workplace.

Extinguishers for protecting Class A hazards may be selected from the following types: water, foam, loaded stream, or multipurpose dry chemical. Extinguishers for protecting Class B hazards may be selected from the following types: Halon 1301, Halon 1211, carbon dioxide, dry chemicals, foam, or loaded stream. Extinguishers for Class C hazards may be selected from the following types: Halon 1301, Halon 1211, carbon dioxide, or dry chemical.

Combustible metal (Class D hazards) fires pose a different type of fire problem in the workplace. Extinguishers using water, gas, or certain dry chemicals cannot extinguish or control this type of fire. Therefore, certain metals have specific dry powder extinguishing agents which can extinguish or control this type of fire. Those agents which have been specifically approved for use on certain metal fires provide the best protection; however, there are also some "universal" type agents which can be used effectively on a variety of combustible metal fires if necessary. The "universal" type agents include: Foundry flux, Lith-X powder, TMB liquid, pyromet powder, TEC powder, dry talc, dry graphite powder, dry sand, dry sodium chloride, dry soda ash, lithium chloride, zirconium silicate, and dry dolomite.

Water is not generally accepted as an effective extinguishing agent for metal fires. When applied to hot burning metal, water will break down into its basic atoms of oxygen and hydrogen.

This chemical breakdown contributes to the combustion of the metal. However, water is also a good universal coolant and can be used on some combustible metals, but only under proper conditions and application, to reduce the temperature of the burning metal below the ignition point. For example, automatic deluge systems in magnesium plants can discharge such large quantities of water on burning magnesium that the fire will be extinguished. The National Fire Protection Association has specific standards for this type of automatic sprinkler system. Further information on the control of metal fires with water can be found in the National Fire Protection Association's *Fire Protection Handbook*.

An excellent source of selection and distribution criteria is found in the National Fire Protection Association's Standard No. 10. Other sources of information include the National Safety Council and the employer's fire insurance carrier.

5. *Substitution of standpipe systems for portable fire extinguishers.* The employer is permitted to substitute acceptable standpipe systems for portable fire extinguishers under certain circumstances. It is necessary to assure that any substitution will provide the same coverage that portable units provide. This means that fire hoses, because of their limited portability, must be spaced throughout the protected area so that they can reach around obstructions such as columns, machinery, etc. and so that they can reach into closets and other enclosed areas.

6. *Inspection, maintenance and testing.* The ultimate responsibility for the inspection, maintenance and testing of portable fire extinguishers lies with the employer. The actual inspection, maintenance, and testing may, however, be conducted by outside contractors with whom the employer has arranged to do the work. When contracting for such work, the employer should assure that the contractor is capable of performing the work that is needed to comply with this standard.

If the employer should elect to perform the inspection, maintenance, and testing requirements of this section in-house, then the employer must make sure that those persons doing the work have been trained to do the work and to recognize problem areas which could cause an extinguisher to be inoperable. The National Fire Protection Association provides excellent guidelines in its standard for portable fire extinguishers. The employer may also check with the manufacturer of the unit that has been purchased and obtain guidelines on inspection, maintenance, and testing. Hydrostatic testing is a process that should be left to contractors or individuals using suitable facilities and having the training necessary to perform the work.

Anytime the employer has removed an extinguisher from service to be checked or repaired, alternate equivalent protection must be provided. Alternate equivalent protection could include replacing the extinguisher with one or more units having equivalent or equal ratings, posting a fire watch, restricting the unprotected area from employee exposure, or providing a hose system ready to operate.

7. *Hydrostatic testing.* As stated before, the employer may contract for hydrostatic testing. However, if the employer wishes to provide the testing service, certain equipment and facilities must be available. Employees should be made aware of the hazards associated with hydrostatic testing and the importance of using proper guards and water pressures. Severe injury can result if extinguisher shells fail violently under hydrostatic pressure.

Employers are encouraged to use contractors who can perform adequate and reliable service. Firms which have been certified by the Materials Transportation Board (MTB) of the U.S. Department of Transportation (DOT) or State licensed extinguisher servicing firms or recognized by the National Association of Fire Equipment Distributors in Chicago, Illinois, are generally acceptable for performing this service.

8. *Training and education.* This part of the standard is of the utmost importance to employers and employees if the risk of injury or death due to extinguisher use is to be reduced. If an employer is going to permit an employee to fight a workplace fire of any size, the employer must make sure that the employee knows everything necessary to assure the employee's safety.

Training and education can be obtained through many channels. Often, local fire departments in larger cities have fire prevention bureaus or similar organizations which can provide basic fire prevention training programs. Fire insurance companies will have data and information available. The National Fire Protection Association and the National Safety Council will provide, at a small cost, publications that can be used in a fire prevention program.

Actual fire fighting training can be obtained from various sources in the country. The Texas A & M University, the University of Maryland's Fire and Rescue Institute, West Virginia University's Fire Service Extension, Iowa State University's Fire Service Extension and other State training schools and land grant colleges have fire fighting programs directed to industrial applications. Some manufacturers of extinguishers, such as the Ansul Company and Safety First, conduct fire schools for customers in the proper use of extinguishers. Several large corporations have taken time to develop their own on-site training programs which expose employees to the actual "feeling" of fire fighting. Simulated fires for training of employees in the proper use of extinguishers are also an acceptable part of a training program.

In meeting the requirements of this section, the employer may also provide educational materials, without classroom instruction, through the use of employee notice campaigns using instruction sheets or flyers or similar types of informal programs. The employer must make sure that employees are trained and educated to recognize not only what type of fire is being fought and how to fight it, but also when it is time to get away from it and leave fire suppression to more experienced fire fighters.

§1910.158 Standpipe and hose systems.

1. *Scope and application.* This section has been written to provide adequate coverage of those standpipe and hose systems that an employer may install in the workplace to meet the requirements of a particular OSHA standard. For example, OSHA permits the substitution of hose systems for portable fire extinguishers in §1910.157. If an employer chooses to provide hose systems instead of portable Class A fire extinguishers, then those hose systems used for substitution would have to meet the applicable requirements of §1910.157. All other standpipe and hose systems not used as a substitute would be exempt from these requirements.

The section specifically exempts Class I large hose systems. By large hose systems, OSHA means those 2 1/2" (6.3 cm) hose lines that are usually associated with fire departments of the size that provide their own water supply through fire apparatus. When the fire gets to the size that outside protection of that degree is necessary, OSHA believes that in most industries employees will have been evacuated from the fire area and the "professional" fire fighters will take control.

2. *Protection of standpipes.* Employers must make sure that standpipes are protected so that they can be relied upon during a fire emergency. This means protecting the pipes from mechanical and physical damage. There are various means for protecting the equipment such as, but not limited to, enclosing the supply piping in the construction of the building, locating the standpipe in an area which is inaccessible to vehicles, or locating the standpipe in a stairwell.

3. *Hose covers and cabinets.* The employer should keep fire protection hose equipment in cabinets or inside protective covers which will protect it from the weather elements, dirt or other damaging sources. The use of protective covers must be easily removed or opened to assure that hose and nozzle are accessible. When the employer places hose in a cabinet, the employer must make sure that the hose and nozzle are accessible to

# OSHA Standards for General Industry

employees without subjecting them to injury. In order to make sure that the equipment is readily accessible, the employer must also make sure that the cabinets used to store equipment are kept free of obstructions and other equipment which may interfere with the fast distribution of the fire hose stored in the cabinet.

4. *Hose outlets and connections.* The employer must assure that employees who use standpipe and hose systems can reach the hose rack and hose valve without the use of portable equipment such as ladders. Hose reels are encouraged for use because one employee can retrieve the hose, charge it, and place it into service without much difficulty.

5. *Hose.* When the employer elects to provide small hose in lieu of portable fire extinguishers, those hose stations being used for the substitution must have hose attached and ready for service. However, if more than the necessary amount of small hose outlets are provided, hose does not have to be attached to those outlets that would provide redundant coverage. Further, where the installation of hose on outlets may expose the hose to extremely cold climates, the employer may store the hose in houses or similar protective areas and connect it to the outlet when needed.

There is approved lined hose available that can be used to replace unlined hose which is stored on racks in cabinets. The lined hose is constructed so that it can be folded and placed in cabinets in the same manner as unlined hose.

Hose is considered to be unserviceable when it deteriorates to the extent that it can no longer carry water at the required pressure and flow rates. Dry rotted linen or hemp hose, cross threaded couplings, and punctured hose are examples of unserviceable hose.

6. *Nozzles.* Variable stream nozzles can provide useful variations in water flow and spray patterns during fire fighting operations and they are recommended for employee use. It is recommended that 100 psi (700 kPa) nozzle pressure be used to provide good flow patterns for variable stream nozzles. The most desirable attribute for nozzles is the ability of the nozzle person to shut off the water flow at the nozzle when it is necessary. This can be accomplished in many ways. For example, a shut-off nozzle with a lever or rotation of the nozzle to stop flow would be effective, but in other cases a simple globe valve placed between a straight stream nozzle and the hose could serve the same purpose. For straight stream nozzles 50 psi nozzle pressure is recommended. The intent of this standard is to protect the employee from "run-away" hoses if it becomes necessary to drop a pressurized hose line and retreat from the fire front and other related hazards.

7. *Design and installation.* Standpipe and hose systems designed and installed in accordance with NFPA Standard No. 14, "Standpipe and Hose Systems," are considered to be in compliance with this standard.

§1910.159 *Automatic sprinkler systems.*

1. *Scope and application.* This section contains the minimum requirements for design, installation and maintenance of sprinkler systems that are needed for employee safety. The Occupational Safety and Health Administration is aware of the fact that the National Board of Fire Underwriters is no longer an active organization, however, sprinkler systems still exist that were designed and installed in accordance with that organization's standards. Therefore, OSHA will recognize sprinkler systems designed to, and maintained in accordance with, NBFU and earlier NFPA standards.

2. *Exemptions.* In an effort to assure that employers will continue to use automatic sprinkler systems as the primary fire protection system in workplaces, OSHA is exempting from coverage those systems not required by a particular OSHA standard and which have been installed in workplaces solely for the purpose of protecting property. Many of these types of systems are installed in areas or buildings with little or no employee exposure. An example is those warehouses where employees may enter occasionally to take inventory or move stock. Some employers may choose to shut down those systems which are not specifically required by OSHA rather than upgrade them to comply with the standards. OSHA does not intend to regulate such systems. OSHA only intends to regulate those systems which are installed to comply with a particular OSHA standard.

3. *Design.* There are two basic types of sprinkler system design. Pipe schedule designed systems are based on pipe schedule tables developed to protect hazards with standard sized pipe, number of sprinklers, and pipe lengths. Hydraulic designed systems are based on an engineered design of pipe size which will produce a given water density or flow rate at any particular point in the system. Either design can be used to comply with this standard.

The National Fire Protection Association's Standard No. 13, "Automatic Sprinkler Systems," contains the tables needed to design and install either type of system. Minimum water supplies, densities, and pipe sizes are given for all types of occupancies.

The employer may check with a reputable fire protection engineering consultant or sprinkler design company when evaluating existing systems or designing a new installation.

With the advent of new construction materials for the manufacture of sprinkler pipe, materials, other than steel have been approved for use as sprinkler pipe. Selection of pipe material should be made on the basis of the type of installation and the acceptability of the material to local fire and building officials where such systems may serve more than one purpose.

Before new sprinkler systems are placed into service, an acceptance test is to be conducted. The employer should invite the installer, designer, insurance representative, and a local fire official to witness the test. Problems found during the test are to be corrected before the system is placed into service.

4. *Maintenance.* It is important that any sprinkler system maintenance be done only when there is minimal employee exposure to the fire hazard. For example, if repairs or changes to the system are to be made, they should be made during those hours when employees are not working or are not occupying that portion of the workplace protected by the portion of the system which has been shut down.

The procedures for performing a flow test via a main drain test or by the use of an inspector's test valve can be obtained from the employer's fire insurance company or from the National Fire Protection Association's Standard No. 13A, "Sprinkler System, Maintenance."

5. *Water supplies.* The water supply to a sprinkler system is one of the most important factors an employer should consider when evaluating a system. Obviously, if there is no water supply, the system is useless. Water supplies can be lost for various reasons such as improperly closed valves, excessive demand, broken water mains, and broken fire pumps. The employer must be able to determine if or when this type of condition exists either by performing a main drain test or visual inspection. Another problem may be an inadequate water supply. For example, a light hazard occupancy may, through rehabilitation or change in tenants, become an ordinary or high hazard occupancy. In such cases, the existing water supply may not be able to provide the pressure or duration necessary for proper protection. Employers must assure that proper design and tests have been made to assure an adequate water supply. These tests can be arranged through the employer's fire insurance carrier or through a local sprinkler maintenance company or through the local fire prevention organization.

Anytime the employer must shut down the primary water supply for a sprinkler system, the standard requires that equivalent protection be provided. Equivalent protection may include a fire watch with extinguishers or hose lines in place and manned, or a secondary water supply such as a tank truck and pump, or a tank or fire pond with fire pumps, to protect the areas where the primary water supply is limited or shut down. The employer may also require evacuation of the workplace and have an emergency action plan which specifies such action.

6. *Protection of piping.* Piping which is exposed to corrosive atmospheres, either chemical or natural, can become defective to the extent that it is useless. Employers must assure that piping is protected from corrosion by its material of construction, e.g., stainless steel, or by a protective coating, e.g., paint.

7. *Sprinklers.* When an employer finds it necessary to replace sprinkler system components or otherwise change a sprinkler's design, employer should make a complete fire protection engineering survey of that part of the system being changed. This review should assure that the changes to the system will not alter the effectiveness of the system as it is presently designed. Water supplies, densities and flow characteristics should be maintained.

8. *Protection of sprinklers.* All components of the system must be protected from mechanical impact damage. This can be achieved with the use of mechanical guards or screens or by locating components in areas where physical contact is impossible or limited.

9. *Sprinkler alarms.* The most recognized sprinkler alarm is the water motor gong or bell that sounds when water begins to flow through the system. This is not however, the only type of acceptable water flow alarm. Any alarm that gives an indication that water is flowing through the system is acceptable. For example, a siren, a whistle, a flashing light, or similar

alerting device which can transmit a signal to the necessary persons would be acceptable. The purpose of the alarm is to alert persons that the system is operating, and that some type of planned action is necessary.

10. *Sprinkler spacing.* For a sprinkler system to be effective there must be an adequate discharge of water spray from the sprinkler head. Any obstructions which hinder the designed density or spray pattern of the water may create unprotected areas which can cause fire to spread. There are some sprinklers that, because of the system's design, are deflected to specific areas. This type of obstruction is acceptable if the system's design takes it into consideration in providing adequate coverage.

### §1910.160 Fixed extinguishing systems, general.

1. *Scope and application.* This section contains the general requirements that are applicable to all fixed extinguishing systems installed to meet OSHA standards. It also applies to those fixed extinguishing systems, generally total flooding, which are not required by OSHA, but which, because of the agent's discharge, may expose employees to hazardous concentrations of extinguishing agents or combustion by-products. Employees who work around fixed extinguishing systems must be warned of the possible hazards associated with the system and its agent. For example, fixed dry chemical extinguishing systems may generate a large enough cloud of dry chemical particles that employees may become visually disoriented. Certain gaseous agents can expose employees to hazardous by-products of combustion when the agent comes into contact with hot metal or other hot surface. Some gaseous agents may be present in hazardous concentrations when the system has totally discharged because an extra rich concentration is necessary to extinguish deep-seated fires. Certain local application systems may be designed to discharge onto the flaming surface of a liquid, and it is possible that the liquid can splatter when hit with the discharging agent. All of these hazards must be determined before the system is placed into operation, and must be discussed with employees.

Based on the known toxicological effects of agents such as carbon tetrachloride and chlorobromomethane, OSHA is not permitting the use of these agents in areas where employees can be exposed to the agent or its side effects. However, chlorobromomethane has been accepted and may be used as an explosion suppression agent in unoccupied spaces. OSHA is permitting the use of this agent only in areas where employees will not be exposed.

2. *Distinctive alarm signals.* A distinctive alarm signal is required to indicate that a fixed system is discharging. Such a signal is necessary on those systems where it is not immediately apparent that the system is discharging. For example, certain gaseous agents make a loud noise when they discharge. In this case no alarm signal is necessary. However, where systems are located in remote locations or away from the general work area and where it is possible that a system could discharge without anyone knowing that it is doing so, then a distinctive alarm is necessary to warn employees of the hazards that may exist. The alarm can be a bell, gong, whistle, horn, flashing light, or any combination of signals as long as it is identifiable as a discharge alarm.

3. *Maintenance.* The employer is responsible for the maintenance of all fixed systems, but this responsibility does not preclude the use of outside contractors to do such work. New systems should be subjected to an acceptance test before placed in service. The employer should invite the installer, designer, insurance representative and others to witness the test. Problems found during the test need to be corrected before the system is considered operational.

4. *Manual discharge stations.* There are instances, such as for mechanical reasons and others, where the standards call for a manual back-up activation device. While the location of this device is not specified in the standard, the employer should assume that the device should be located where employees can easily reach it. It could, for example, be located along the main means of egress from the protected area so that employees could activate the system as they evacuate the work area.

5. *Personal protective equipment.* The employer is required to provide the necessary personal protective equipment to rescue employees who may be trapped in a totally flooded environment which may be hazardous to their health. This equipment would normally include a positive-pressure self-contained breathing apparatus and any necessary first aid equipment. In cases where the employer can assure the prompt arrival of the local fire department or plant emergency personnel which can provide the equipment, this can be considered as complying with the standards.

### §1910.161 Fixed extinguishing systems, dry chemical.

1. *Scope and application.* The requirements of this section apply only to dry chemical systems. These requirements are to be used in conjunction with the requirements of §1910.160.

2. *Maintenance.* The employer is responsible for assuring that dry chemical systems will operate effectively. To do this, periodic maintenance is necessary. One test that must be conducted during the maintenance check is one which will determine if the agent has remained free of moisture. If an agent absorbs any moisture, it may tend to cake and thereby clog the system. An easy test for acceptable moisture content is to take a lump of dry chemical from the container and drop it from a height of four inches. If the lump crumbles into fine particles, the agent is acceptable.

### §1910.162 Fixed extinguishing systems, gaseous agent.

1. *Scope and application.* This section applies only to those systems which use gaseous agents. The requirements of §1910.160 also apply to the gaseous agent systems covered in this section.

2. *Design concentrations.* Total flooding gaseous systems are based on the volume of gas which must be discharged in order to produce a certain designed concentration of gas in an enclosed area. The concentration needed to extinguish a fire depends on several factors including the type of fire hazard and the amount of gas expected to leak away from the area during discharge. At times it is necessary to "super-saturate" a work area to provide for expected leakage from the enclosed area. In such cases, employers must assure that the flooded area has been ventilated before employees are permitted to reenter the work area without protective clothing and respirators.

3. *Toxic decomposition.* Certain halogenated hydrocarbons will break down or decompose when they are combined with high temperatures found in the fire environment. The products of the decomposition can include toxic elements or compounds. For example, when Halon 1211 is placed into contact with hot metal it will break down and form bromide or fluoride fumes. The employer must find out which toxic products may result from decomposition of a particular agent from the manufacturer, and take the necessary precautions to prevent employee exposure to the hazard.

### §1910.163 Fixed extinguishing systems, water spray and foam.

1. *Scope and application.* This section applies to those systems that use water spray or foam. The requirements of §1910.160 also apply to this type of system.

2. *Characteristics of foams.* When selecting the type of foam for a specific hazard, the employer should consider the following limitations of some foams.

a. Some foams are not acceptable for use on fires involving flammable gases and liquefied gases with boiling points below ambient workplace temperatures. Other foams are not effective when used on fires involving polar solvent liquids.

b. Any agent using water as part of the mixture should not be used on fire involving combustible metals unless it is applied under proper conditions to reduce the temperature of burning metal below the ignition temperature. The employer should use only those foams that have been tested and accepted for this application by a recognized independent testing laboratory.

c. Certain types of foams may be incompatible and break down when they are mixed together.

d. For fires involving water miscible solvents, employers should use only those foams tested and approved for such use. Regular protein foams may not be effective on such solvents.

Whenever employers provide a foam or water spray system, drainage facilities must be provided to carry contaminated water or foam overflow away from the employee work areas and egress routes. This drainage system should drain to a central impounding area where it can be collected and disposed of properly. Other government agencies may have regulations concerning environmental considerations.

### §1910.164 Fire detection systems.

1. *Installation and restoration.* Fire detection systems must be designed by knowledgeable engineers or other professionals, with expertise in fire detection systems and when the systems are installed, there should be an acceptance test performed on the system to insure it operates properly. The manufacturer's recommendations for system design should be consulted. While entire systems may not be approved, each component used in the system is required to be approved. Custom fire detection systems should be designed by knowledgeable fire protection or electrical engineers who are familiar with the workplace hazards and conditions.

# OSHA Standards for General Industry

Some systems may only have one or two individual detectors for a small workplace, but good design and installation is still important. An acceptance test should be performed on all systems, including these smaller systems.

OSHA has a requirement that spare components used to replace those which may be destroyed during an alarm situation be available in sufficient quantities and locations for prompt restoration of the system. This does not mean that the parts or components have to be stored at the workplace. If the employer can assure that the supply of parts is available in the local community or the general metropolitan area of the workplace, then the requirements for storage and availability have been met. The intent is to make sure that the alarm system is fully operational when employees are occupying the workplace, and that when the system operates it can be returned to full service the next day or sooner.

2. *Supervision.* Fire detection systems should be supervised. The object of supervision is detection of any failure of the circuitry, and the employer should use any method that will assure that the system's circuits are operational. Electrically operated sensors for air pressure, fluid pressure, or electrical circuits, can provide effective monitoring and are the typical types of supervision.

3. *Protection of fire detectors.* Fire detectors must be protected from corrosion either by protective coatings, by being manufactured from non-corrosive materials or by location. Detectors must also be protected from mechanical impact damage, either by suitable cages or metal guards where such hazards are present, or by locating them above or out of contact with materials or equipment which may cause damage.

4. *Number, location, and spacing of detectors.* This information can be obtained from the approval listing for detectors or NFPA standards. It can also be obtained from fire protection engineers or consultants or manufacturers of equipment who have access to approval listings and design methods.

§1910.165 Employee alarm systems.

1. *Scope and application.* This section is intended to apply to employee alarm systems used for all types of employee emergencies except those which occur so quickly and at such a rapid rate (e.g., explosions) that any action by the employee is extremely limited following detection.

In small workplaces with 10 or less employees the alarm system can be by direct voice communication (shouting) where any one individual can quickly alert all other employees. Radio may be used to transmit alarms from remote workplaces where telephone service is not available, provided that radio messages will be monitored by emergency services, such as fire, police or others, to insure alarms are transmitted and received.

2. *Alarm signal alternatives.* In recognition of physically impaired individuals, OSHA is accepting various methods of giving alarm signals. For example, visual, tactile or audible alarm signals are acceptable methods for giving alarms to employees. Flashing lights or vibrating devices can be used in areas where the employer has hired employees with hearing or vision impairments. Vibrating devices, air fans, or other tactile devices can be used where visually and hearing impaired employees work. Employers are cautioned that certain frequencies of flashing lights have been claimed to initiate epileptic seizures in some employees and that this fact should be considered when selecting an alarm device. Two way radio communications would be most appropriate for transmitting emergency alarms in such workplaces which may be remote or where telephones may not be available.

3. *Reporting alarms.* Employee alarms may require different means of reporting, depending on the workplace involved. For example, in small workplaces, a simple shout throughout the workplace may be sufficient to warn employees of a fire or other emergency. In larger workplaces, more sophisticated equipment is necessary so that entire plants or high-rise buildings are not evacuated for one small emergency. In remote areas, such as pumping plants, radio communication with a central base station may be necessary. The goal of this standard is to assure that all employees who need to know that an emergency exists can be notified of the emergency. The method of transmitting the alarm should reflect the situation found at the workplace.

Personal radio transmitters, worn by an individual, can be used where the individual may be working such as in a remote location. Such personal radio transmitters shall send a distinct signal and should clearly indicate who is having an emergency, the location, and the nature of the emergency. All radio transmitters need a feedback system to assure that the emergency alarm is sent to the people who can provide assistance.

For multi-story buildings or single story buildings with interior walls for subdivisions, the more traditional alarm systems are recommended for these types of workplaces. Supervised telephone or manual fire alarm or pull box stations with paging systems to transmit messages throughout the building is the recommended alarm system. The alarm box stations should be available within a travel distance of 200 feet. Water flow detection on a sprinkler system, fire detection systems (guard's supervisory station) or tour signal (watchman's service), or other related systems may be part of the overall system. The paging system may be used for nonemergency operations provided the emergency messages and uses will have precedence over all other uses of the system.

4. *Supervision.* The requirements for supervising the employee alarm system circuitry and power supply may be accomplished in a variety of ways. Typically, electrically operated sensors for air pressure, fluid pressure, steam pressure, or electrical continuity of circuitry may be used to continuously monitor the system to assure it is operational and to identify trouble in the system and give a warning signal.

[45 FR 60715, Sept. 12, 1980; 46 FR 24557, May 1, 1981]

### Appendix B to Subpart L to Part 1910—National Consensus Standards

The following table contains a cross-reference listing of those current national consensus standards which contains information and guidelines that would be considered acceptable in complying with requirements in the specific sections of subpart L.

| Subpart L section | National consensus standard |
|---|---|
| 1910.156 | ANSI/NFPA No. 1972; Structural Fire Fighter's Helmets. ANSI Z88.5 American National Standard, Practice for Respirator Protection for the Fire Service. ANSI/NFPA No. 1971, Protective Clothing for Structural Fire Fighters. NFPA No. 1041, Fire Service Instructor Professional Qualifications. |
| 1910.157 | ANSI/NFPA No. 10, Portable Fire Extinguishers. |
| 1910.158 | NSI/NFPA No. 18, Wetting Agents. ANSI/NFPA No. 20, Centrifugal Fire Pumps. NFPA No. 21, Steam Fire Pumps. ANSI/NFPA No. 22, Water Tanks. NFPA No. 24, Outside Protection. NFPA No. 26, Supervision of Valves. NFPA No. 13E, Fire Department Operations in Properties Protected by Sprinkler, Standpipe Systems. ANSI/NFPA No. 194, Fire Hose Connections. NFPA No. 197, Initial Fire Attack, Training for. NFPA No. 1231, Water Supplies for Suburban and Rural Fire Fighting. |
| 1910.159 | ANSI-NFPA No. 13, Sprinkler Systems. NFPA No. 13A, Sprinkler Systems, Maintenance. ANSI/NFPA No. 18, Wetting Agents. ANSI/NFPA No. 20, Centrifugal Fire Pumps. ANSI/NFPA No. 22, Water Tanks. NFPA No. 24, Outside Protection. NFPA No. 26, Supervision of Valves. ANSI/NFPA No. 72B, Auxiliary Signaling Systems. NFPA No. 1231, Water Supplies for ]Suburban and Rural Fire Fighting. |
| 1910.160 | ANSI/NFPA No. 11, Foam Systems. ANSI/NFPA 11A, High Expansion Foam Extinguishing Systems. ANSI/NFPA No. 11B, Synthetic Foam and Combined Agent Systems. ANSI/NFPA No. 12, Carbon Dioxide Systems. ANSI/NFPA No. 12A, Halon 1301 Systems. ANSI/NFPA No. 12B, Halon 1211 Systems. ANSI/NFPA No. 15, Water Spray Systems. ANSI/NFPA 16 Foam-Water Spray Systems. ANSI/NFPA No. 17, Dry Chemical Systems. ANSI/NFPA 69, Explosion Suppression Systems. |
| 1910.161 | ANSI/NFPA No. 11B, Synthetic Foam and Combined Agent Systems. ANSI/NFPA No. 17, Dry Chemical Systems. |
| 1910.162 | ANSI/NFPA No. 12, Carbon Dioxide Systems. ANSI/NFPA No. 12A, Halon 1211 Systems. ANSI/NFPA No. 12B, Halon 1301 Systems. ANSI/NFPA No. 69, Explosion Suppression Systems. |

| Subpart L section | National consensus standard |
|---|---|
| 1910.163 | ANSI/NFPA No. 11, Foam Extinguishing Systems. |
| | ANSI/NFPA No. 11A, High Expansion Foam Extinguishing Systems. |
| | ANSI/NFPA No. 11B, Synthetic Foam and Combined Agent Systems. |
| | ANSI/NFPA No. 15, Water Spray Fixed Systems. |
| | ANSI/NFPA No. 16, Foam-Water Spray Systems. |
| | ANSI/NFPA No. 18, Wetting Agents. |
| | NFPA No. 26, Supervision of Valves. |
| 1910.164 | ANSI/NFPA No. 71, Central Station Signaling Systems. |
| | ANSI/NFPA No. 72A, Local Protective Signaling Systems. |
| | ANSI/NFPA No. 72B, Auxiliary Signaling Systems. |
| | ANSI/NFPA No. 72D, Proprietary Protective Signaling Systems. |
| | ANSI/NFPA No. 72E, Automatic Fire Detectors. |
| | ANSI/NFPA No. 101, Life Safety Code. |
| 1910.165 | ANSI/NFPA No. 71, Central Station Signaling Systems. |
| | ANSI/NFPA No. 72A, Local Protective Signaling Systems. |
| | ANSI/NFPA No. 72B, Auxiliary Protective Signaling Systems. |
| | ANSI/NFPA No. 72C, Remote Station Protective Signaling Systems. |
| | ANSI/NFPA No. 72D, Proprietary Protective Signaling Systems. |
| | ANSI/NFPA No. 101, Life Safety Code. |
| Metric Conversion | ANSI/ASTM No. E380, American National Standard for Metric Practice. |

NFPA standards are available from the National Fire Protection Association, Batterymarch Park, Quincy, MA 02269.

ANSI Standards are available from the American National Standards Institute, 1430 Broadway, New York, NY 10018.

[45 FR 60715, Sept. 12, 1980, as amended at 58 FR 35309, June 30, 1993]

## Appendix C to Subpart L to Part 1910—Fire Protection References For Further Information

I. *Appendix general references.* The following references provide information which can be helpful in understanding the requirements contained in all of the sections of subpart L:

A. *Fire Protection Handbook,* National Fire Protection Association, Batterymarch Park, Quincy, MA 02269.

B. *Accident Prevention Manual for Industrial Operations,* National Safety Council; 425 North Michigan Avenue, Chicago, IL 60611.

C. Various associations also publish information which may be useful in understanding these standards. Examples of these associations are: Fire Equipment Manufacturers Association (FEMA) of Arlington, VA 22204 and the National Association of Fire Equipment Distributors (NAFED) of Chicago, IL 60601.

II. *Appendix references applicable to individual sections.* The following references are grouped according to individual sections contained in subpart L. These references provide information which may be helpful in understanding and implementing the standards of each section of subpart L.

A. *§1910.156. Fire brigades:*

1. *Private Fire Brigades,* NFPA 27; National Fire Protection Association, Batterymarch Park, Quincy, MA 02269.

2. *Initial Fire Attack, Training Standard On,* NFPA 197; National Fire Protection Association, Batterymarch Park, Quincy, MA 02269.

3. *Fire Fighter Professional Qualifications,* NFPA 1001; National Fire Protection Association, Batterymarch Park, Quincy, MA 02269.

4. *Organization for Fire Services,* NFPA 1201; National Fire Protection Association, Batterymarch Park, Quincy, MA 02269.

5. *Organization of a Fire Department,* NFPA 1202; National Fire Protection Association, Batterymarch Park, Quincy, MA 02269.

6. *Protective Clothing for Structural Fire Fighting,* ANSI/NFPA 1971;.

7. *American National Standard for Men's Safety-Toe Footwear,* ANSI Z41.1; American National Standards Institute, New York, NY 10018.

8. *American National Standard for Occupational and Educational Eye and Face Protection,* ANSI Z87.1; American National Standards Institute, New York, NY 10018.

9. *American National Standard, Safety Requirements for Industrial Head Protection,* ANSI Z89.1; American National Standards Institute, New York, NY 10018.

10. *Specifications for Protective Headgear for Vehicular Users,* ANSI Z90.1; American National Standards Institute, New York, NY 10018.

11. *Testing Physical Fitness;* Davis and Santa Maria. Fire Command. April 1975.

12. *Development of a Job-Related Physical Performance Examination for Fire Fighters;* Dotson and Others. A summary report for the National Fire Prevention and Control Administration. Washington, DC. March 1977.

13. *Proposed Sample Standards for Fire Fighters' Protective Clothing and Equipment;* International Association of Fire Fighters, Washington, DC.

14. *A Study of Facepiece Leakage of Self-Contained Breathing Apparatus by DOP Man Tests;* Los Alamos Scientific Laboratory, Los Alamos, NM.

15. *The Development of Criteria for Fire Fighters' Gloves; Vol. II: Glove Criteria and Test Methods;* National Institute for Occupational Safety and Health, Cincinnati, OH. 1976.

16. *Model Performance Criteria for Structural Fire Fighters' Helmets;* National Fire Prevention and Control Administration, Washington, DC. 1977.

17. *Firefighters;* Job Safety and Health Magazine, Occupational Safety and Health Administration, Washington, DC. June 1978.

18. *Eating Smoke—The Dispensable Diet;* Utech, H.P. The Fire Independent, 1975.

19. *Project Monoxide—A Medical Study of an Occupational Hazard of Fire Fighters;* International Association of Fire Fighters, Washington, DC.

20. *Occupational Exposures to Carbon Monoxide in Baltimore Firefighters;* Radford and Levine. Johns Hopkins University, Baltimore, MD. Journal of Occupational Medicine, September, 1976.

21. *Fire Brigades;* National Safety Council, Chicago, IL. 1966.

22. *American National Standard, Practice for Respiratory Protection for the Fire Service;* ANSI Z88.5; American National Standards Institute, New York, NY 10018.

23. *Respirator Studies for the Nuclear Regulatory Commission; October 1, 1977—September 30, 1978. Evaluation and Performance of Open Circuit Breathing Apparatus.* NU REG/CR-1235. Los Alamos Scientific Laboratory; Los Alamos, NM. 87545, January, 1980.

B. *§1910.157. Portable fire extinguishers:*

1. *Standard for Portable Fire Extinguishers,* ANSI/NFPA 10; National Fire Protection Association, Batterymarch Park, Quincy, MA 02269.

2. *Methods for Hydrostatic Testing of Compressed Gas Cylinders,* C-1; Compressed Gas Association, 1235 Jefferson Davis Highway, Arlington, VA 22202.

3. *Recommendations for the Disposition of Unserviceable Compressed Gas Cylinders,* C-2; Compressed Gas Association, 1235 Jefferson Davis Highway, Arlington, VA 22202.

4. *Standard for Visual Inspection of Compressed Gas Cylinders,* C-6; Compressed Gas Association, 1235 Jefferson Davis Highway, Arlington, VA 22202.

5. *Portable Fire Extinguisher Selection Guide,* National Association of Fire Equipment Distributors; 111 East Wacker Drive, Chicago, IL 60601.

C. *§1910.158. Standpipe and hose systems:*

1. *Standard for the Installation of Sprinkler Systems,* ANSI/NFPA 13; National Fire Protection Association, Batterymarch Park, Quincy, MA 02269.

2. *Standard of the Installation of Standpipe and Hose Systems,* ANSI/NFPA 14; National Fire Protection Association, Batterymarch Park, Quincy, MA 02269.

3. *Standard for the Installation of Centrifugal Fire Pumps,* ANSI/NFPA 20; National Fire Protection Association, Batterymarch Park, Quincy, MA 02269.

4. *Standard for Water Tanks for Private Fire Protection,* ANSI/NFPA 22; National Fire Protection Association, Batterymarch Park, Quincy, MA 02269.

5. *Standard for Screw Threads and Gaskets for Fire Hose Connections,* ANSI/NFPA 194; National Fire Protection Association, Batterymarch Park, Quincy, MA 02269.

6. *Standard for Fire Hose,* NFPA 196; National Fire Protection Association, Batterymarch Park, Quincy, MA 02269.

7. *Standard for the Care of Fire Hose*, NFPA 198; National Fire Protection Association, Batterymarch Park, Quincy, MA 02269.

D. §1910.159. Automatic sprinkler systems:

1. *Standard of the Installation of Sprinkler Systems*, ANSI-NFPA 13; National Fire Protection Association, Batterymarch Park, Quincy, MA 02269.

2. *Standard for the Care and Maintenance of Sprinkler Systems*, ANSI/NFPA 13A; National Fire Protection Association, Batterymarch Park, Quincy, MA 02269.

3. *Standard for the Installation of Standpipe and Hose Systems*, ANSI/NFPA 14; National Fire Protection Association, Batterymarch Park, Quincy, MA 02269.

4. *Standard for the Installation of Centrifugal Fire Pumps*, ANSI/NFPA 20; National Fire Protection Association, Batterymarch Park, Quincy, MA 02269.

5. *Standard for Water Tanks for Private Fire Protection*, ANSI-NFPA 22; National Fire Protection Association, Batterymarch Park, Quincy, MA 02269.

6. *Standard for Indoor General Storage*, ANSI/NFPA 231; National Fire Protection Association, Batterymarch Park, Quincy, MA 02269.

7. *Standard for Rack Storage of Materials*, ANSI/NFPA 231C; National Fire Protection Association, Batterymarch Park, Quincy, MA 02269.

E. §1910.160. Fixed extinguishing systems—general information:

1. *Standard for Foam Extinguishing Systems*, ANSI-NFPA 11; National Fire Protection Association, Batterymarch Park, Quincy, MA 02269.

2. *Standard for Hi-Expansion Foam Systems*, ANSI/NFPA 11A; National Fire Protection Association, Batterymarch Park, Quincy, MA 02269.

3. *Standard on Synthetic Foam and Combined Agent Systems*, ANSI/NFPA 11B; National Fire Protection Association, Batterymarch Park, Quincy, MA 02269.

4. *Standard on Carbon Dioxide Extinguishing Systems*, ANSI/NFPA 12; National Fire Protection Association, Batterymarch Park, Quincy, MA 02269.

5. *Standard on Halon 1301*, ANSI/NFPA 12A; National Fire Protection Association, Batterymarch Park, Quincy, MA 02269.

6. *Standard on Halon 1211*, ANSI/NFPA 12B; National Fire Protection Association, Batterymarch Park, Quincy, MA 02269.

7. *Standard for Water Spray Systems*, ANSI/NFPA 15; National Fire Protection Association, Batterymarch Park, Quincy, MA 02269.

8. *Standard for Foam-Water Sprinkler Systems and Foam-Water Spray Systems*, ANSI/NFPA 16; National Fire Protection Association, National Fire Protection Association, Batterymarch Park, Quincy, MA 02269.

9. *Standard for Dry Chemical Extinguishing Systems*, ANSI/NFPA 17; National Fire Protection Association, Batterymarch Park, Quincy, MA 02269.

F. §1910.161. Fixed extinguishing systems—dry chemical:

1. *Standard for Dry Chemical Extinguishing Systems*, ANSI/NFPA 17; National Fire Protection Association, Batterymarch Park, Quincy, MA 02269.

2. *National Electrical Code*, ANSI/NFPA 70; National Fire Protection Association, Batterymarch Park, Quincy, MA 02269.

3. *Standard for the Installation of Equipment for the Removal of Smoke and Grease-Laden Vapor from Commercial Cooking Equipment*, NFPA 96; National Fire Protection Association, Batterymarch Park, Quincy, MA 02269.

G. §1910.162. Fixed extinguishing systems—gaseous agents:

1. *Standard on Carbon Dioxide Extinguishing Systems*, ANSI/NFPA 12; National Fire Protection Association, Batterymarch Park, Quincy, MA 02269.

2. *Standard on Halon 1301*, ANSI/NFPA 12B; National Fire Protection Association, Batterymarch Park, Quincy, MA 02269.

3. *Standard on Halon 1211*, ANSI/NFPA 12B; National Fire Protection Association, Batterymarch Park, Quincy, MA 02269.

4. *Standard on Explosion Prevention Systems*, ANSI/NFPA 69; National Fire Protection Association, Batterymarch Park, Quincy, MA 02269.

5. *National Electrical Code*, ANSI/NFPA 70; National Fire Protection Association, Batterymarch Park, Quincy, MA 02269.

6. *Standard on Automatic Fire Detectors*, ANSI/NFPA 72E; National Fire Protection Association, Batterymarch Park, Quincy, MA 02269.

7. *Determination of Halon 1301/1211 Threshold Extinguishing Concentrations Using the Cup Burner Method*; Riley and Olson, Ansul Report AL-530-A.

H. §1910.163. Fixed extinguishing systems—water spray and foam agents:

1. *Standard for Foam Extinguisher Systems*, ANSI/NFPA 11; National Fire Protection Association, Batterymarch Park, Quincy, MA 02269.

2. *Standard for High Expansion Foam Systems*, ANSI/NFPA 11A; National Fire Protection Association, Batterymarch Park, Quincy, MA 02269.

3. *Standard for Water Spray Fixed Systems for Fire Protection*, ANSI/NFPA 15; National Fire Protection Association, Batterymarch Park, Quincy, MA 02269.

4. *Standard for the Installation of Foam-Water Sprinkler Systems and Foam-Water Spray Systems*, ANSI/NFPA 16; National Fire Protection Association, Batterymarch Park, Quincy, MA 02269.

I. §1910.164. Fire Detection systems:

1. *National Electrical Code*, ANSI/NFPA 70; National Fire Protection Association, Batterymarch Park, Quincy, MA 02269.

2. *Standard for Central Station Signaling Systems*, ANSI/NFPA 71; National Fire Protection Association, Batterymarch Park, Quincy, MA 02269.

3. *Standard on Automatic Fire Detectors*, ANSI/NFPA 72E; National Fire Protection Association, Batterymarch Park, Quincy, MA 02269.

J. §1910.165. Employee alarm systems:

1. *National Electrical Code*, ANSI/NFPA 70; National Fire Protection Association, Batterymarch Park, Quincy, MA 02269.

2. *Standard for Central Station Signaling systems*, ANSI/NFPA 71; National Fire Protection Association, Batterymarch Park, Quincy, MA 02269.

3. *Standard for Local Protective Signaling Systems*, ANSI/NFPA 72A; National Fire Protection Association, Batterymarch Park, Quincy, MA 02269.

4. *Standard for Auxiliary Protective Signaling Systems*, ANSI/NFPA 72B; National Fire Protection Association, Batterymarch Park, Quincy, MA 02269.

5. *Standard for Remote Station Protective Signaling Systems*, ANSI/NFPA 72C; National Fire Protection Association, Batterymarch Park, Quincy, MA 02269.

6. *Standard for Proprietary Protective Signaling Systems*, ANSI/NFPA 72D; National Fire Protection Association, Batterymarch Park, Quincy, MA 02269.

7. *Vocal Emergency Alarms in Hospitals and Nursing Facilities: Practice and Potential*. National Bureau of Standards. Washington, D.C., July 1977.

8. *Fire Alarm and Communication Systems*. National Bureau of Standards. Washington, D.C., April 1978.

[45 FR 60715, Sept. 12, 1980, as amended at 58 FR 35309, June 30, 1993]

### Appendix D to Subpart L to Part 1910— Availability of Publications Incorporated by Reference in Section 1910.156 Fire Brigades

The final standard for fire brigades, Section 1910.156, contains provisions which incorporate certain publications by reference. The publications provide criteria and test methods for protective clothing worn by those fire brigade members who are expected to perform interior structural fire fighting. The standard references the publications as the chief sources of information for determining if the protective clothing affords the required level of protection.

It is appropriate to note that the final standard does not require employers to purchase a copy of the referenced publications. Instead, employers can specify (in purchase orders to the manufacturers) that the protective clothing meet the criteria and test methods contained in the referenced publications and can rely on the manufacturers' assurances of compliance. Employers, however, may desire to obtain a copy of the referenced publications for their own information.

The paragraph designation of the standard where the referenced publications appear, the title of the publications, and the availability of the publications are as follows:

| Paragraph designation | Referenced Publication | Available From |
|---|---|---|
| 1910.156(e)(3)(ii) | "Protective Clothing for Structural Fire Fighting," NFPA No. 1971 (1975). | National Fire Protection Association, Batterymarch Park, Quincy, MA 02269. |
| 1910.156(e)(4)(i) | "Development of Criteria for Fire Fighter's Gloves; Vol. II, Part II: Test Methods" (1976). | U.S. Government Printing Office, Washington, DC 20402. Stock No for Vol. II is: 071-033-0201-1. |
| 1910.156(e)(5)(i) | "Model Performance Criteria for Structural Firefighter's Helmets" (1977). | U.S. Fire Administration, National Fire Safety and Research Office, Washington, D.C. 20230. |

The referenced publications (or a microfiche of the publications) are available for review at many universities and public libraries throughout the country. These publications may also be examined at the OSHA Technical Data Center, Room N2439-Rear, United States Department of Labor, 200 Constitution Ave., N.W., Washington, D.C. 20210 (202-219-7500), or at any OSHA Regional Office (see telephone directories under United States Government-Labor Department).

[45 FR 60715, Sept. 12, 1980, as amended at 58 FR 33509, June 30, 1993; 61 FR 9239, Mar. 7, 1996]

### Appendix E to Subpart L to Part 1910— Test Methods for Protective Clothing

This appendix contains test methods which must be used to determine if protective clothing affords the required level of protection as specified in §1910.156, fire brigades.

(1) *Puncture resistance test method for foot protection.*

A. *Apparatus.* The puncture resistance test shall be performed on a testing machine having a movable platform adjusted to travel at 1/4-inch/min (0.1 cm/sec). Two blocks of hardwood, metal, or plastic shall be prepared as follows: the blocks shall be of such size and thickness as to insure a suitable rigid test ensemble and allow for at least one-inch of the pointed end of an 8D nail to be exposed for the penetration. One block shall have a hole drilled to hold an 8D common nail firmly at an angle of 98°. The second block shall have a maximum 1/2-inch (1.3 cm) diameter hole drilled through it so that the hole will allow free passage of the nail after it penetrates the insole during the test.

B. *Procedure.* The test ensemble consisting of the sample unit, the two prepared blocks, a piece of leather outsole 10 to 11 irons thick, and a new 8D nail, shall be placed as follows: the 8D nail in the hole, the sample of outsole stock superimposed above the nail, the area of the sole plate to be tested placed on the outsole, and the second block with hole so placed as to allow for free passage of the nail after it passes through the outsole stock and sole plate in that order. The machine shall be started and the pressure, in pounds required for the nail to completely penetrate the outsole and sole plate, recorded to the nearest five pounds. Two determinations shall be made on each sole plate and the results averaged. A new nail shall be used for each determination.

C. *Source.* These test requirements are contained in "Military Specification For Fireman's Boots," MIL-B-2885D (1973 and amendment dated 1975) and are reproduced for your convenience.

(2) *Test method for determining the strength of cloth by tearing: Trapezoid Method.*

A. *Test specimen.* The specimen shall be a rectangle of cloth 3-inches by 6-inches (7.6 cm by 15.2 cm). The long dimension shall be parallel to the warp for warp tests and parallel to the filling for filling tests. No two specimens for warp tests shall contain the same warp yarns, nor shall any two specimens for filling tests contain the same filling yarns. The specimen shall be taken no nearer the selvage than 1/10 the width of the cloth. An isosceles trapezoid having an altitude of 3-inches (7.6 cm) and bases of 1 inch (2.5cm) and 4 inches (10.2 cm) in length, respectively, shall be marked on each specimen, preferably with the aid of a template. A cut approximately 3/8-inch (1 cm) in length shall then be made in the center of a perpendicular to the 1-inch (2.5 cm) edge.

B. *Apparatus.* (i) Six-ounce (.17 kg) weight tension clamps shall be used so designed that the six ounces (.17 kg) of weight are distributed evenly across the complete width of the sample.

(ii) The machine shall consist of three main parts: Straining mechanism, clamps for holding specimen, and load and elongation recording mechanisms.

(iii) A machine wherein the specimen is held between two clamps and strained by a uniform movement of the pulling clamp shall be used.

(iv) The machine shall be adjusted so that the pulling clamp shall have a uniform speed of 12 ±10.5 inches per minute (0.5 ±.02 cm/sec).

(v) The machine shall have two clamps with two jaws on each clamp. The design of the two clamps shall be such that one gripping surface or jaw may be an integral part of the rigid frame of the clamp or be fastened to allow a slight vertical movement, while the other gripping surface or jaw shall be completely moveable. The dimension of the immovable jaw of each clamp parallel to the application of the load shall measure one-inch, and the dimension of the jaw perpendicular to this direction shall measure three inches or more. The face of the movable jaw of each clamp shall measure one-inch by three inches.

Each jaw face shall have a flat smooth, gripping surface. All edges which might cause a cutting action shall be rounded to a radius of not over 1/64-inch (.04 cm). In cases where a cloth tends to slip when being tested, the jaws may be faced with rubber or other material to prevent slippage. The distance between the jaws (gage length) shall be one-inch at the start of the test.

(vi) Calibrated dial; scale or chart shall be used to indicate applied load and elongation. The machine shall be adjusted or set, so that the maximum load required to break the specimen will remain indicated on the calibrated dial or scale after the test specimen has ruptured.

(vii) The machine shall be of such capacity that the maximum load required to break the specimen shall be not greater than 85 percent or less than 15 percent of the rated capacity.

(viii) The error of the machine shall not exceed 2 percent up to and including a 50-pound load (22.6 kg) and 1 percent over a 50-pound load (22.6 kg) at any reading within its loading range.

(ix) All machine attachments for determining maximum loads shall be disengaged during this test.

C. *Procedure.* (i) The specimen shall be clamped in the machine along the nonparallel sides of the trapezoid so that these sides lie along the lower edge of the upper clamp and the upper edge of the lower clamp with the cut halfway between the clamps. The short trapezoid base shall be held taut and the long trapezoid base shall lie in the folds.

(ii) The machine shall be started and the force necessary to tear the cloth shall be observed by means of an autographic recording device. The speed of the pulling clamp shall be 12 inches ±0.5 inch per minute (±0.5 + .02 cm/sec).

(iii) If a specimen slips between the jaws, breaks in or at the edges of the jaws, or if for any reason attributable to faulty technique, an individual measurement falls markedly below the average test results for the sample unit, such result shall be discarded and another specimen shall be tested.

(iv) The tearing strength of the specimen shall be the average of the five highest peak loads of resistance registered for 3 inches (7.6 cm) of separation of the tear.

D. *Report.* (i) Five specimens in each of the warp and filling directions shall be tested from each sample unit.

(ii) The tearing strength of the sample unit shall be the average of the results obtained from the specimens tested in each of the warp and filling directions and shall be reported separately to the nearest 0.1-pound (.05 kg).

E. *Source.* These test requirements are contained in "Federal Test Method Standard 191, Method 5136" and are reproduced for your convenience.

(3) *Test method for determining flame resistance of cloth; vertical.*

A. *Test specimen.* The specimen shall be a rectangle of cloth 2 3/4 inches (7.0 cm) by 12 inches (30.5 cm) with the long dimension parallel to either the warp or filling direction of the cloth. No two warp specimens shall contain the same warp yarns, and no two filling specimens shall contain the same filling yarn.

B. *Number of determinations.* Five specimens from each of the warp and filling directions shall be tested from each sample unit.

C. *Apparatus.* (i) Cabinet. A cabinet and accessories shall be fabricated in accordance with the requirements specified in Figures L-1, L-2, and L-3. Galvanized sheet metal or other suitable metal shall be used. The entire inside back wall of the cabinet shall be painted black to facilitate the viewing of the test specimen and pilot flame.

(ii) Burner. The burner shall be equipped with a variable orifice to adjust the flame height, a barrel having a 3/8-inch (1 cm) inside diameter and a pilot light.

(a) The burner may be constructed by combining a 3/8-inch (1 cm) inside diameter barrel 3 ±1/4 inches (7.6 ±.6 cm) long from a fixed orifice burner with a base from a variable orifice burner.

(b) The pilot light tube shall have a diameter of approximately 1/16-inch (.2 cm) and shall be spaced 1/8-inch (.3 cm) away from the burner edge with a pilot flame 1/8-inch (.3 cm) long.

(c) The necessary gas connections and the applicable plumbing shall be as specified in Figure L-4 except that a solenoid valve may be used in lieu of the stopcock valve to which the burner is attached. The stopcock valve or solenoid valve, whichever is used, shall be capable of being fully opened or fully closed in 0.1-second.

(d) On the side of the barrel of the burner, opposite the pilot light there shall be a metal rod of approximately 1/8-inch (.3 cm) diameter spaced 1/2-inch (1.3 cm) from the barrel and extending above the burner. The rod shall have two 5/16-inch (.8 cm) prongs marking the distances of 3/4-inch (1.9 cm) and 1 1/2 inches (3.8 cm) above the top of the burner.

(e) The burner shall be fixed in a position so that the center of the barrel of the burner is directly below the center of the specimen.

(iii) There shall be a control valve system with a delivery rate designed to furnish gas to the burner under a pressure of 2 1/2 ±1/4 (p.s.i.) (17.5 ±1.8 kPa) per square inch at the burner inlet (see (g)(3)(vi)(A)). The manufacturer's recommended delivery rate for the valve system shall be included in the required pressure.

(iv) A synthetic gas mixture shall be of the following composition within the following limits (analyzed at standard conditions): 55 ±3 percent hydrogen, 24 ±1 percent methane, 3 ±1 percent ethane, and 18 ±1 percent carbon monoxide which will give a specific gravity of 0.365 ± 0.018 (air = 1) and a B.T.U. content of 540 ±20 per cubic foot (20.1 ±3.7 kJ/L) (dry basis) at 69.8 °F (21 °C).

(v) There shall be metal hooks and weights to produce a series of total loads to determine length of char. The metal hooks shall consist of No. 19 gage steel wire or equivalent and shall be made from 3-inch (7.6 cm) lengths of wire and bent 1/2-inch (1.3 cm) from one end to a 45 degree hook. One end of the hook shall be fastened around the neck of the weight to be used.

(vi) There shall be a stop watch or other device to measure the burning time to 0.2-second.

(vii) There shall be a scale, graduated in 0.1 inch (.3 cm) to measure the length of char.

D. *Procedure.* (i) The material undergoing test shall be evaluated for the characteristics of after-flame time and char length on each specimen.

(ii) All specimens to be tested shall be at moisture equilibrium under standard atmospheric conditions in accordance with paragraph (3)C of this appendix. Each specimen to be tested shall be exposed to the test flame within 20 seconds after removal from the standard atmosphere. In case of dispute, all testing will be conducted under Standard Atmospheric Conditions in accordance with paragraph (3)C of this appendix.

(iii) The specimen in its holder shall be suspended vertically in the cabinet in such a manner that the entire length of the specimen is exposed and the lower end is 3/4-inch (1.9 cm) above the top of the gas burner. The apparatus shall be set up in a draft free area.

(iv) Prior to inserting the specimen, the pilot flame shall be adjusted to approximately 1/8-inch (.3 cm) in height measured from its lowest point to the tip.

The burner flame shall be adjusted by means of the needle valve in the base of the burner to give a flame height of 1 1/2 inches (3.8 cm) with the stopcock fully open and the air supply to the burner shut off and taped. The 1 1/2-inch (3.8 cm) flame height is obtained by adjusting the valve so that the uppermost portion (tip) of the flame is level with the tip of the metal prong (see Figure L-2) specified for adjustment of flame height. It is an important aspect of the evaluation that the flame height be adjusted with the tip of the flame level with the tip of the metal prong. After inserting the specimen, the stopcock shall be fully opened, and the burner flame applied vertically at the middle of the lower edge of the specimen for 12 seconds and the burner turned off. The cabinet door shall remain shut during testing.

(v) The after-flame shall be the time the specimen continues to flame after the burner flame is shut off.

(vi) After each specimen is removed, the test cabinet shall be cleared of fumes and smoke prior to testing the next specimen.

(vii) After both flaming and glowing have ceased, the char length shall be measured. The char length shall be the distance from the end of the specimen, which was exposed to the flame, to the end of a tear (made lengthwise) of the specimen through the center of the charred area as follows: The specimen shall be folded lengthwise and creased by hand along a line through the highest peak of the charred area. The hook shall be inserted in the specimen (or a hole, 1/4-inch (.6 cm) diameter or less, punched out for the hook) at one side of the charred area 1/4-inch (.6 cm) from the adjacent outside edge and 1/4-inch (.6 cm) in from the lower end. A weight of sufficient size such that the weight and hook together shall equal the total tearing load required in Table L-2 of this section shall be attached to the hook.

(viii) A tearing force shall be applied gently to the specimen by grasping the corner of the cloth at the opposite edge of the char from the load and raising the specimen and weight clear of the supporting surface. The end of the tear shall be marked off on the edge and the char length measurement made along the undamaged edge.

Loads for determining char length applicable to the weight of the test cloth shall be as shown in Table L-2.

### TABLE L-21

| Specified weight per square yard of cloth before any fire retardant treatment or coating—ounces | Total tearing weight for determining the charred length—pound |
| --- | --- |
| 2.0 to 6.0 | 0.25 |
| Over 6.0 to 15.0 | 0.50 |
| Over 15.0 to 23.0 | 0.75 |
| Over 23.0 | 1.0 |

[1] To change into S.I. (System International) units, 1 ounce = 28.35 grams, 1 pound = 453 grams, 1 yard = .91 meter.

(ix) The after-flame time of the specimen shall be recorded to the nearest 0.2-second and the char length to the nearest 0.1-inch (.3 cm).

E. *Report.* (i) The after-flame time and char length of the sample unit shall be the average of the results obtained from the individual specimens tested. All values obtained from the individual specimens shall be recorded.

(ii) The after-flame time shall be reported to the nearest 0.2-second and the char length to the nearest 0.1-inch (.3 cm).

F. *Source.* These test requirements are contained in "Federal Test Method Standard 191, Method 5903 (1971)" and are reproduced for your convenience.

**FIGURE L-1—VERTICAL FLAME RESISTANCE TEXTILE APPARATUS**

All given System International (S.I.) Unit: 1-inch = 2/54 cm. [2.54 cm. sic] dimensions are in inches.

**FIGURE L-2—VERTICAL FLAME RESISTANCE TEXTILE APPARATUS, DOOR AND TOP VIEW W/BAFFLE**

All given dimensions are in inches. System International (S.I.) Unit: 1-inch = 2.54 cm.

**FIGURE L-3—VERTICAL FLAME RESISTANCE TEXTILE APPARATUS, VIEW AND DETAILS**

All given dimensions are in inches. System International (S.I.) Unit: 1-inch = 2.54 cm.

[45 FR 60715, Sept. 12, 1980; 46 FR 24557, May 1, 1981]

**FIGURE L-4—VERTICAL FLAME RESISTANCE TEXTILE APPARATUS**

All given dimensions are in inches. System International (S.I.) Unit: 1 inch = 2.54 cm.

# Subpart M—Compressed Gas and Compressed Air Equipment

**Authority:** Sections 4, 6, and 8 of the Occupational Safety and Health Act of 1970 (29 U.S.C. 653, 655, 657); Secretary of Labor's Order No. 12-71 (36 FR 8754), 8-76 (41 FR 25059), 9-83 (48 FR 35736), or 1-90 (55 FR 9033), as applicable.

### §§1910.166-1910.168 [Reserved].

### §1910.169 Air receivers.

**1910.169(a)** *General requirements*—**(1)** *Application.* This section applies to compressed air receivers, and other equipment used in providing and utilizing compressed air for performing operations such as cleaning, drilling, hoisting, and chipping. On the other hand, however, this section does not deal with the special problems created by using compressed air to convey materials nor the problems created when men work in compressed air as in tunnels and caissons. This section is not intended to apply to compressed air machinery and equipment used on transportation vehicles such as steam railroad cars, electric railway cars, and automotive equipment.

**1910.169(a)(2)** *New and existing equipment.* **(i)** All new air receivers installed after the effective date of these regulations shall be constructed in accordance with the 1968 edition of the A.S.M.E. Boiler and Pressure Vessel Code Section VIII, which is incorporated by reference as specified in §1910.6.

**1910.169(a)(2)(ii)** All safety valves used shall be constructed, installed, and maintained in accordance with the A.S.M.E. Boiler and Pressure Vessel Code, Section VIII Edition 1968.

**1910.169(b)** *Installation and equipment requirements*—**(1)** *Installation.* Air receivers shall be so installed that all drains, handholes, and manholes therein are easily accessible. Under no circumstances shall an air receiver be buried underground or located in an inaccessible place.

**1910.169(b)(2)** *Drains and traps.* A drain pipe and valve shall be installed at the lowest point of every air receiver to provide for the removal of accumulated oil and water. Adequate automatic traps may be installed in addition to drain valves. The drain valve on the air receiver shall be opened and the receiver completely drained frequently and at such intervals as to prevent the accumulation of excessive amounts of liquid in the receiver.

**1910.169(b)(3)** *Gages and valves.* **(i)** Every air receiver shall be equipped with an indicating pressure gage (so located as to be readily visible) and with one or more spring-loaded safety valves. The total relieving capacity of such safety valves shall be such as to prevent pressure in the receiver from exceeding the maximum allowable working pressure of the receiver by more than 10 percent.

**1910.169(b)(3)(ii)** No valve of any type shall be placed between the air receiver and its safety valve or valves.

**1910.169(b)(3)(iii)** Safety appliances, such as safety valves, indicating devices and controlling devices, shall be constructed, located, and installed so that they cannot be readily rendered inoperative by any means, including the elements.

**1910.169(b)(3)(iv)** All safety valves shall be tested frequently and at regular intervals to determine whether they are in good operating condition.

[39 FR 23502, June 27, 1974, as amended at 49 FR 5322, Feb. 10, 1984; 61 FR 9239, Mar. 7, 1996]

### §1910.170 Sources of standards [Removed].

[§1910.170 removed at 61 FR 9239, Mar. 7, 1996, effective May 6, 1996.]

### §1910.171 Standards organizations [Removed].

[§1910.171 removed at 61 FR 9239, Mar. 7, 1996, effective May 6, 1996.]

# Subpart N—Materials Handling and Storage

**Authority:** Sec. 4, 6, 8, Occupational Safety and Health Act of 1970 (29 U.S.C. 653, 655, 657); Secretary of Labor's Order No. 12-71 (36 FR 8754), 8-76 (41 FR 25059), 9-83 (48 FR 35736), 1-90 (55 FR 9033) or 6-96 (62 FR 111), 3-2000 (65 FR 50017) or 5-2002 (67 FR 65008 as applicable. Section 1910.178 also amended under Section 4 of the Administrative Procedure Act (5 U.S.C. 653).

Sections 1910.176, 1910.178, 1910.179, 1910.180, 1910.181, and 1910.184 also issued under 29 CFR part 1911.

## §1910.176 Handling materials—general.

**1910.176(a)** *Use of mechanical equipment.* Where mechanical handling equipment is used, sufficient safe clearances shall be allowed for aisles, at loading docks, through doorways and wherever turns or passage must be made. Aisles and passageways shall be kept clear and in good repair, with no obstruction across or in aisles that could create a hazard. Permanent aisles and passageways shall be appropriately marked.

**1910.176(b)** *Secure storage.* Storage of material shall not create a hazard. Bags, containers, bundles, etc., stored in tiers shall be stacked, blocked, interlocked and limited in height so that they are stable and secure against sliding or collapse.

**1910.176(c)** *Housekeeping.* Storage areas shall be kept free from accumulation of materials that constitute hazards from tripping, fire, explosion, or pest harborage. Vegetation control will be exercised when necessary.

**1910.176(d)** [Reserved]

**1910.176(e)** *Clearance limits.* Clearance signs to warn of clearance limits shall be provided.

**1910.176(f)** *Rolling railroad cars.* Derail and/or bumper blocks shall be provided on spur railroad tracks where a rolling car could contact other cars being worked, enter a building, work or traffic area.

**1910.176(g)** *Guarding.* Covers and/or guard-rails shall be provided to protect personnel from the hazards of open pits, tanks, vats, ditches, etc.

[39 FR 23052, June 27, 1974, as amended at 43 FR 49749, Oct. 24, 1978]

## §1910.177 Servicing multi-piece and single piece rim wheels.

**1910.177(a)** *Scope.* **(1)** This section applies to the servicing of multi-piece and single piece rim wheels used on large vehicles such as trucks, tractors, trailers, buses and off-road machines. It does not apply to the servicing of rim wheels used on automobiles, or on pickup trucks and vans utilizing automobile tires or truck tires designated "LT".

**1910.177(a)(2)** This section does not apply to employers and places of employment regulated under the Construction Safety Standards, 29 CFR part 1926; the Agriculture Standards, 29 CFR part 1928; the Shipyard Standards, 29 CFR part 1915; or the Longshoring Standards, 29 CFR part 1918.

**1910.177(a)(3)** All provisions of this section apply to the servicing of both single piece rim wheels and multi-piece rim wheels unless designated otherwise.

**1910.177(b)** *Definitions. Barrier* means a fence, wall or other structure or object placed between a single piece rim wheel and an employee during tire inflation, to contain the rim wheel components in the event of the sudden release of the contained air of the single piece rim wheel.

*Charts* means the U.S. Department of Labor, Occupational Safety and Health Administration publications entitled "Demounting and Mounting Procedures for Truck/Bus Tires" and "Multi-piece Rim Matching Chart," the National Highway Traffic Safety Administration (NHTSA) publications entitled "Demounting and Mounting Procedures Truck/Bus Tires" and "Multi-piece Rim Matching Chart," or any other poster which contains at least the same instructions, safety precautions and other information contained in the charts that is applicable to the types of wheels being serviced.

*Installing a rim wheel* means the transfer and attachment of an assembled rim wheel onto a vehicle axle hub. *Removing* means the opposite of installing.

*Mounting a tire* means the assembly or putting together of the wheel and tire components to form a rim wheel, including inflation. *Demounting* means the opposite of mounting.

*Multi-piece rim wheel* means the assemblage of a multi-piece wheel with the tire tube and other components.

*Multi-piece wheel* means a vehicle wheel consisting of two or more parts, one of which is a side or locking ring designed to hold the tire on the wheel by interlocking components when the tire is inflated.

*Restraining device* means an apparatus such as a cage, rack, assemblage of bars and other components that will constrain all rim wheel components during an explosive separation of a multi-piece rim wheel, or during the sudden release of the contained air of a single piece rim wheel.

*Rim manual* means a publication containing instructions from the manufacturer or other qualified organization for correct mounting, demounting, maintenance, and safety precautions peculiar to the type of wheel being serviced.

*Rim wheel* means an assemblage of tire, tube and liner (where appropriate), and wheel components.

*Service* or *servicing* means the mounting and demounting of rim wheels, and related activities such as inflating, deflating, installing, removing, and handling.

*Service area* means that part of an employer's premises used for the servicing of rim wheels, or any other place where an employee services rim wheels.

*Single piece rim wheel* means the assemblage of single piece rim wheel with the tire and other components.

*Single piece wheel* means a vehicle wheel consisting of one part, designed to hold the tire on the wheel when the tire is inflated.

*Trajectory* means any potential path or route that a rim wheel component may travel during an explosive separation, or the sudden release of the pressurized air, or an area at which an airblast from a single piece rim wheel may be released. The trajectory may deviate from paths which are perpendicular to the assembled position of the rim wheel at the time of separation or explosion. (See appendix A for examples of trajectories.)

*Wheel* means that portion of a rim wheel which provides the method of attachment of the assembly to the axle of a vehicle and also provides the means to contain the inflated portion of the assembly (i.e., the tire and/or tube).

**1910.177(c)** *Employee training.* **(1)** The employer shall provide a program to train all employees who service rim wheels in the hazards involved in servicing those rim wheels and the safety procedures to be followed.

**1910.177(c)(1)(i)** The employer shall assure that no employee services any rim wheel unless the employee has been trained and instructed in correct procedures of servicing the type of wheel being serviced, and in the safe operating procedures described in paragraphs (f) and (g) of this section.

**1910.177(c)(1)(ii)** Information to be used in the training program shall include, at a minimum, the applicable data contained in the charts (rim manuals) and the contents of this standard.

**1910.177(c)(1)(iii)** Where an employer knows or has reason to believe that any of his employees is unable to read and understand the charts or rim manual, the employer shall assure that the employee is instructed concerning the contents of the charts and rim manual in a manner which the employee is able to understand.

**1910.177(c)(2)** The employer shall assure that each employee demonstrates and maintains the ability to service rim wheels safely, including performance of the following tasks:

**1910.177(c)(2)(i)** Demounting of tires (including deflation);

**1910.177(c)(2)(ii)** Inspection and identification of the rim wheel components;

**1910.177(c)(2)(iii)** Mounting of tires (including inflation with a restraining device or other safeguard required by this section);

**1910.177(c)(2)(iv)** Use of the restraining device or barrier, and other equipment required by this section;

**1910.177(c)(2)(v)** Handling of rim wheels;

**1910.177(c)(2)(vi)** Inflation of the tire when a single piece rim wheel is mounted on a vehicle;

**1910.177(c)(2)(vii)** An understanding of the necessity of standing outside the trajectory both during inflation of the tire and during inspection of the rim wheel following inflation; and

**1910.177(c)(2)(viii)** Installation and removal of rim wheels.

**1910.177(c)(3)** The employer shall evaluate each employee's ability to perform these tasks and to service rim wheels safely, and shall provide additional training as necessary to assure that each employee maintains his or her proficiency.

**1910.177(d)** *Tire servicing equipment.* **(1)** The employer shall furnish a restraining device for inflating tires on multi-piece wheels.

**1910.177(d)(2)** The employer shall provide a restraining device or barrier for inflating tires on single piece wheels unless the rim wheel will be bolted onto a vehicle during inflation.

**1910.177(d)(3)** Restraining devices and barriers shall comply with the following requirements:

**1910.177(d)(3)(i)** Each restraining device or barrier shall have the capacity to withstand the maximum force that would be transferred to it during a rim wheel separation occurring at 150 percent of the maximum tire specification pressure for the type of rim wheel being serviced.

**1910.177(d)(3)(ii)** Restraining devices and barriers shall be capable of preventing the rim wheel components from being thrown outside or beyond the device or barrier for any rim wheel positioned within or behind the device;

**1910.177(d)(3)(iii)** Restraining devices and barriers shall be visually inspected prior to each day's use and after any separation of the rim wheel components or sudden release of contained air. Any restraining device or barrier exhibiting damage such as the following defects shall be immediately removed from service:

**1910.177(d)(3)(A)** Cracks at welds;

**1910.177(d)(3)(B)** Cracked or broken components;

**1910.177(d)(3)(C)** Bent or sprung components caused by mishandling, abuse, tire explosion or rim wheel separation;

**1910.177(d)(3)(D)** Pitting of components due to corrosion; or

**1910.177(d)(3)(E)** Other structural damage which would decrease its effectiveness.

**1910.177(d)(3)(iv)** Restraining devices or barriers removed from service shall not be returned to service until they are repaired and reinspected. Restraining devices or barriers requiring structural repair such as component replacement or rewelding shall not be returned to service until they are certified by either the manufacturer or a Registered Professional Engineer as meeting the strength requirements of paragraph (d)(3)(i) of this section.

**1910.177(d)(4)** The employer shall furnish and assure that an air line assembly consisting of the following components be used for inflating tires:

**1910.177(d)(4)(i)** A clip-on chuck;

**1910.177(d)(4)(ii)** An in-line valve with a pressure gauge or a presettable regulator; and

**1910.177(d)(4)(iii)** A sufficient length of hose between the clip-on chuck and the in-line valve (if one is used) to allow the employee to stand outside the trajectory.

**1910.177(d)(5)** Current charts or rim manuals containing instructions for the type of wheels being serviced shall be available in the service area.

**1910.177(d)(6)** The employer shall furnish and assure that only tools recommended in the rim manual for the type of wheel being serviced are used to service rim wheels.

**1910.177(e)** *Wheel component acceptability.* **(1)** Multi-piece wheel components shall not be interchanged except as provided in the charts or in the applicable rim manual.

**1910.177(e)(2)** Multi-piece wheel components and single piece wheels shall be inspected prior to assembly. Any wheel or wheel component which is bent out of shape, pitted from corrosion, broken, or cracked shall not be used and shall be marked or tagged unserviceable and removed from the service area. Damaged or leaky valves shall be replaced.

**1910.177(e)(3)** Rim flanges, rim gutters, rings, bead seating surfaces and the bead areas of tires shall be free of any dirt, surface rust, scale or loose or flaked rubber build-up prior to mounting and inflation.

**1910.177(e)(4)** The size (bead diameter and tire/wheel widths) and type of both the tire and the wheel shall be checked for compatibility prior to assembly of the rim wheel.

**1910.177(f)** *Safe operating procedure—multi-piece rim wheels.* The employer shall establish a safe operating procedure for servicing multi-piece rim wheels and shall assure that employees are instructed in and follow that procedure. The procedure shall include at least the following elements:

**1910.177(f)(1)** Tires shall be completely deflated before demounting by removal of the valve core.

**1910.177(f)(2)** Tires shall be completely deflated by removing the valve core before a rim wheel is removed from the axle in either of the following situations:

**1910.177(f)(2)(i)** When the tire has been driven underinflated at 80% or less of its recommended pressure, or

**1910.177(f)(2)(ii)** When there is obvious or suspected damage to the tire or wheel components.

**1910.177(f)(3)** Rubber lubricant shall be applied to bead and rim mating surfaces during assembly of the wheel and inflation of the tire, unless the tire or wheel manufacturer recommends against it.

**1910.177(f)(4)** If a tire on a vehicle is underinflated but has more than 80% of the recommended pressure, the tire may be inflated while the rim wheel is on the vehicle provided remote control inflation equipment is used, and no employees remain in the trajectory during inflation.

**1910.177(f)(5)** Tires shall be inflated outside a restraining device only to a pressure sufficient to force the tire bead onto the rim ledge and create an airtight seal with the tire and bead.

**1910.177(f)(6)** Whenever a rim wheel is in a restraining device the employee shall not rest or lean any part of his body or equipment on or against the restraining device.

**1910.177(f)(7)** After tire inflation, the tire and wheel components shall be inspected while still within the restraining device to make sure that they are properly seated and locked. If further adjustment to the tire or wheel components is necessary, the tire shall be deflated by removal of the valve core before the adjustment is made.

**1910.177(f)(8)** No attempt shall be made to correct the seating of side and lock rings by hammering, striking or forcing the components while the tire is pressurized.

**1910.177(f)(9)** Cracked, broken, bent or otherwise damaged rim components shall not be reworked, welded, brazed, or otherwise heated.

**1910.177(f)(10)** Whenever multi-piece rim wheels are being handled, employees shall stay out of the trajectory unless the employer can demonstrate that performance of the servicing makes the employee's presence in the trajectory necessary.

**1910.177(f)(11)** No heat shall be applied to a multi-piece wheel or wheel component.

**1910.177(g)** *Safe operating procedure—single piece rim wheels.* The employer shall establish a safe operating procedure for servicing single piece rim wheels and shall assure that employees are instructed in and follow that procedure. The procedure shall include at least the following elements:

**1910.177(g)(1)** Tires shall be completely deflated by removal of the valve core before demounting.

**1910.177(g)(2)** Mounting and demounting of the tire shall be done only from the narrow ledge side of the wheel. Care shall be taken to avoid damaging the tire beads while mounting tires on wheels. Tires shall be mounted only on compatible wheels of matching bead diameter and width.

**1910.177(g)(3)** Nonflammable rubber lubricant shall be applied to bead and wheel mating surfaces before assembly of the rim wheel, unless the tire or wheel manufacturer recommends against the use of any rubber lubricant.

**1910.177(g)(4)** If a tire changing machine is used, the tire shall be inflated only to the minimum pressure necessary to force the tire bead onto the rim ledge while on the tire changing machine.

**1910.177(g)(5)** If a bead expander is used, it shall be removed before the valve core is installed and as soon as the rim wheel becomes airtight (the tire bead slips onto the bead seat).

**1910.177(g)(6)** Tires may be inflated only when contained within a restraining device, positioned behind a barrier or bolted on the vehicle with the lug nuts fully tightened.

**1910.177(g)(7)** Tires shall not be inflated when any flat, solid surface is in the trajectory and within one foot of the sidewall.

# OSHA Standards for General Industry

**1910.177(g)(8)** Employees shall stay out of the trajectory when inflating a tire.

**1910.177(g)(9)** Tires shall not be inflated to more than the inflation pressure stamped in the sidewall unless a higher pressure is recommended by the manufacturer.

**1910.177(g)(10)** Tires shall not be inflated above the maximum pressure recommended by the manufacturer to seat the tire bead firmly against the rim flange.

**1910.177(g)(11)** No heat shall be applied to a single piece wheel.

**1910.177(g)(12)** Cracked, broken, bent, or otherwise damaged wheels shall not be reworked, welded, brazed, or otherwise heated.

### APPENDIX A — TRAJECTORY

**WARNING**
STAY OUT OF THE TRAJECTORY AS INDICATED BY SHADED AREA
Note: Under some circumstances, the trajectory may deviate from its expected path

### APPENDIX B TO §1910.177— ORDERING INFORMATION FOR THE OSHA CHARTS

OSHA has printed two charts entitled "Demounting and Mounting Procedures for Truck/Bus Tires" and "Multi-piece Rim Matching Chart," as part of a continuing campaign to reduce accidents among employees who service large vehicle rim wheels.

Reprints of the charts are available through the Occupational Safety and Health Administration (OSHA) Area and Regional Offices. The address and telephone number of the nearest OSHA office can be obtained by looking in the local telephone directory under U.S. Government, U.S. Department of Labor, Occupational Safety and Health Administration. Single copies are available without charge.

Individuals, establishments and other organizations desiring single or multiple copies of these charts may order them from the OSHA Publications Office, U.S. Department of Labor, Room N-3101, Washington, DC 20210, Telephone (202) 219-4667.

[49 FR 4350, Feb. 3, 1984, as amended at 52 FR 36026, Sept. 25, 1987; 53 FR 34737, Sept. 8, 1988; 61 FR 9239, Mar. 7, 1996]

## §1910.178 Powered industrial trucks.

**1910.178(a)** *General requirements.* **(1)** This section contains safety requirements relating to fire protection, design, maintenance, and use of fork trucks, tractors, platform lift trucks, motorized hand trucks, and other specialized industrial trucks powered by electric motors or internal combustion engines. This section does not apply to compressed air or nonflammable compressed gas-operated industrial trucks, nor to farm vehicles, nor to vehicles intended primarily for earth moving or over-the-road hauling.

**1910.178(a)(2)** All new powered industrial trucks acquired and used by an employer shall meet the design and construction requirements for powered industrial trucks established in the "American National Standard for Powered Industrial Trucks, Part II, ANSI B56.1-1969", which is incorporated by reference as specified in §1910.6, except for vehicles intended primarily for earth moving or over-the-road hauling.

**1910.178(a)(3)** Approved trucks shall bear a label or some other identifying mark indicating approval by the testing laboratory. See paragraph (a)(7) of this section and paragraph 405 of "American National Standard for Powered Industrial Trucks, Part II, ANSI B56.1-1969", which is incorporated by reference in paragraph (a)(2) of this section and which provides that if the powered industrial truck is accepted by a nationally recognized testing laboratory it should be so marked.

**1910.178(a)(4)** Modifications and additions which affect capacity and safe operation shall not be performed by the customer or user without manufacturers prior written approval. Capacity, operation, and maintenance instruction plates, tags, or decals shall be changed accordingly.

**1910.178(a)(5)** If the truck is equipped with front-end attachments other than factory installed attachments, the user shall request that the truck be marked to identify the attachments and show the approximate weight of the truck and attachment combination at maximum elevation with load laterally centered.

**1910.178(a)(6)** The user shall see that all nameplates and markings are in place and are maintained in a legible condition.

**1910.178(a)(7)** As used in this section, the term, *approved truck* or *approved industrial truck* means a truck that is listed or approved for fire safety purposes for the intended use by a nationally recognized testing laboratory, using nationally recognized testing standards. Refer to §1910.155(c)(3)(iv)(A) for definition of listed, and to §1910.7 for definition of nationally recognized testing laboratory.

**1910.178(b)** *Designations.* For the purpose of this standard there are eleven different designations of industrial trucks or tractors as follows: D, DS, DY, E, ES, EE, EX, G, GS, LP, and LPS.

**1910.178(b)(1)** The D designated units are units similar to the G units except that they are diesel engine powered instead of gasoline engine powered.

**1910.178(b)(2)** The DS designated units are diesel powered units that are provided with additional safeguards to the exhaust, fuel and electrical systems. They may be used in some locations where a D unit may not be considered suitable.

**1910.178(b)(3)** The DY designated units are diesel powered units that have all the safeguards of the DS units and in addition do not have any electrical equipment including the ignition and are equipped with temperature limitation features.

**1910.178(b)(4)** The E designated units are electrically powered units that have minimum acceptable safeguards against inherent fire hazards.

**1910.178(b)(5)** The ES designated units are electrically powered units that, in addition to all of the requirements for the E units, are provided with additional safeguards to the electrical system to prevent emission of hazardous sparks and to limit surface temperatures. They may be used in some locations where the use of an E unit may not be considered suitable.

**1910.178(b)(6)** The EE designated units are electrically powered units that have, in addition to all of the requirements for the E and ES units, the electric motors and all other electrical equipment completely enclosed. In certain locations the EE unit may be used where the use of an E and ES unit may not be considered suitable.

**1910.178(b)(7)** The EX designated units are electrically powered units that differ from the E, ES, or EE units in that the electrical fittings and equipment are so designed, constructed and assembled that the units may be used in certain atmospheres containing flammable vapors or dusts.

**1910.178(b)(8)** The G designated units are gasoline powered units having minimum acceptable safeguards against inherent fire hazards.

**1910.178(b)(9)** The GS designated units are gasoline powered units that are provided with additional safeguards to the exhaust, fuel, and electrical systems. They may be used in some locations where the use of a G unit may not be considered suitable.

**1910.178(b)(10)** The LP designated unit is similar to the G unit except that liquefied petroleum gas is used for fuel instead of gasoline.

**1910.178(b)(11)** The LPS designated units are liquefied petroleum gas powered units that are provided with additional safeguards to the exhaust, fuel, and electrical systems. They may be used in some locations where the use of an LP unit may not be considered suitable.

**1910.178(b)(12)** The atmosphere or location shall have been classified as to whether it is hazardous or nonhazardous prior to the consideration of industrial trucks being used therein and the type of industrial truck required shall be as provided in paragraph (d) of this section for such location.

**1910.178(c)** *Designated locations.* **(1)** The industrial trucks specified under subparagraph (2) of this paragraph are the minimum types required but industrial trucks having greater safeguards may be used if desired.

**1910.178(c)(2)** For specific areas of use, see Table N-1 which tabulates the information contained in this section. References are to the corresponding classification as used in subpart S of this part.

**1910.178(c)(2)(i)** Power-operated industrial trucks shall not be used in atmospheres containing hazardous concentration of acetylene, butadiene, ethylene oxide, hydrogen (or gases or vapors equivalent in hazard to hydrogen, such as manufactured gas), propylene oxide, acetaldehyde, cyclopropane, diethyl ether, ethylene, isoprene, or unsymmetrical dimethyl hydrazine (UDMH).

**1910.178(c)(2)(ii)(a)** Power-operated industrial trucks shall not be used in atmospheres containing hazardous concentrations of metal dust, including aluminum, magnesium, and their commercial alloys, other metals of similarly hazardous characteristics, or in atmospheres containing carbon black, coal or coke dust except approved power-operated industrial trucks designated as EX may be used in such atmospheres.

**1910.178(c)(2)(ii)(b)** In atmospheres where dust of magnesium, aluminum or aluminum bronze may be present, fuses, switches, motor controllers, and circuit breakers of trucks shall have enclosures specifically approved for such locations.

**1910.178(c)(2)(iii)** Only approved power-operated industrial trucks designated as EX may be used in atmospheres containing acetone, acrylonitrile, alcohol, ammonia, benzine, benzol, butane, ethylene dichloride, gasoline, hexane, lacquer solvent vapors, naphtha, natural gas, propane, propylene, styrene, vinyl acetate, vinyl chloride, or xylenes in quantities sufficient to produce explosive or ignitable mixtures and where such concentrations of these gases or vapors exist continuously, intermittently or periodically under normal operating conditions or may exist frequently because of repair, maintenance operations, leakage, breakdown or faulty operation of equipment.

**1910.178(c)(2)(iv)** Power-operated industrial trucks designated as DY, EE, or EX may be used in locations where volatile flammable liquids or flammable gases are handled, processed or used, but in which the hazardous liquids, vapors or gases will normally be confined within closed containers or closed systems from which they can escape only in case of accidental rupture or breakdown of such containers or systems, or in the case of abnormal operation of equipment; also in locations in which hazardous concentrations of gases or vapors are normally prevented by positive mechanical ventilation but which might become hazardous through failure or abnormal operation of the ventilating equipment; or in locations which are adjacent to Class I, Division 1 locations, and to which hazardous concentrations of gases or vapors might occasionally be communicated unless such communication is prevented by adequate positive-pressure ventilation from a source of clear air, and effective safeguards against ventilation failure are provided.

### TABLE N-1—SUMMARY TABLE ON USE OF INDUSTRIAL TRUCKS IN VARIOUS LOCATIONS

| Classes | Unclassified | Class I locations | | | | Class II locations | | | Class III locations |
|---|---|---|---|---|---|---|---|---|---|
| Description of classes | Locations not possessing atmospheres as described in other columns. | Locations in which flammable gases or vapors are, or may be, present in the air in quantities sufficient to produce explosive or ignitable mixtures. | | | | Locations which are hazardous because of the presence of combustible dust. | | | Locations where easily ignitable fibers or flyings are present but not likely to be in suspension in quantities sufficient to produce ignitable mixtures. |
| Groups in classes | None | A | B | C | D | E | F | G | None |
| Examples of locations or atmospheres in classes and groups. | Piers and wharves inside and outside general storage, general industrial or commercial properties. | Acetylene | Hydrogen | Ethyl ether | Gasoline Naphtha Alcohols Acetone Lacquer solvent Benzene | Metal dust | Carbon black coal dust, cokedust | Grain dust, flour dust, starch dust, organic dust | Baled waste, cocoa fiber, cotton, excelsior, hemp, istle, jute, kapok, oakum, sisal, Spanish moss, synthetic fibers, tow. |
| | | 1 | 2 | 1 | | 2 | 1 | | 2 |
| Divisions (nature of hazardous conditions) | None | Above condition exists continuously, intermittently, or periodically under normal operating conditions. | Above condition may occur accidentally as due to a puncture of a storage drum. | Explosive mixture may be present under normal operating conditions, or where failure of equipment may cause the condition to exist simultaneously with arcing or sparking of electrical equipment, or where dusts of an electrically conducting nature may be present. | | Explosive mixture not normally present, but where deposits of dust may cause heat rise in electrical equipment, or where such deposits may be ignited by arcs or sparks from electrical equipment. | Locations in which easily ignitable fibers or materials producing combustible flyings are handled, manufactured, or used. | | Locations in which easily ignitable fibers are stored or handled (except in the process of manufacture). |

## Authorized uses of trucks by types in groups of classes and divisions

| Groups in classes | None | A | B | C | D | A | B | C | D | E | F | G | E | F | G | None | None |
|---|---|---|---|---|---|---|---|---|---|---|---|---|---|---|---|---|---|
| **Type of truck authorized:** | | | | | | | | | | | | | | | | | |
| *Diesel:* | | | | | | | | | | | | | | | | | |
| Type D | D** | | | | | | | | | | | | | | | | |
| Type DS | | | | | | | | | DS | | | | | | DS | | DS |
| Type DY | | | | | | | | | DY | | | | | | DY | DY | DY |
| *Electric:* | | | | | | | | | | | | | | | | | |
| Type E | E** | | | | | | | | | | | | | | | | E |
| Type ES | | | | | | | | | ES | | | | | | ES | | ES |
| Type EE | | | | | | | | | EE | | | | | | EE | EE | EE |
| Type EX | | | | | EX | | | | EX | | EX | EX | | | EX | EX | EX |
| *Gasoline:* | | | | | | | | | | | | | | | | | |
| Type G | G** | | | | | | | | | | | | | | | | |
| Type GS | | | | | | | | | GS | | | | | | GS | | GS |
| *LP-Gas:* | | | | | | | | | | | | | | | | | |
| Type LP | LP** | | | | | | | | | | | | | | | | |
| Type LPS | | | | | | | | | LPS | | | | | | LPS | | LPS |
| Paragraph Ref. in No. 505. | 210.211 | 201(a) | | 203(a) | | 209(a) | | | 204(a), (b) | 202(a) | 205(a) | | 209(a) | | 206(a), (b) | 207(a) | 208(a) |

** Trucks conforming to these types may also be used·see subdivision (c)(2)(x) and (c)(2)(xii) of this section.

**1910.178(c)(2)(v)** In locations used for the storage of hazardous liquids in sealed containers or liquefied or compressed gases in containers, approved power-operated industrial trucks designated as DS, ES, GS, or LPS may be used. This classification includes locations where volatile flammable liquids or flammable gases or vapors are used, but which, would become hazardous only in case of an accident or of some unusual operating condition. The quantity of hazardous material that might escape in case of accident, the adequacy of ventilating equipment, the total area involved, and the record of the industry or business with respect to explosions or fires are all factors that should receive consideration in determining whether or not the DS or DY, ES, EE, GS, LPS designated truck possesses sufficient safeguards for the location. Piping without valves, checks, meters and similar devices would not ordinarily be deemed to introduce a hazardous condition even though used for hazardous liquids or gases. Locations used for the storage of hazardous liquids or of liquified or compressed gases in sealed containers would not normally be considered hazardous unless subject to other hazardous conditions also.

**1910.178(c)(2)(vi)(a)** Only approved power operated industrial trucks designated as EX shall be used in atmospheres in which combustible dust is or may be in suspension continuously, intermittently, or periodically under normal operating conditions, in quantities sufficient to produce explosive or ignitable mixtures, or where mechanical failure or abnormal operation of machinery or equipment might cause such mixtures to be produced.

**1910.178(c)(2)(vi)(b)** The EX classification usually includes the working areas of grain handling and storage plants, room containing grinders or pulverizers, cleaners, graders, scalpers, open conveyors or spouts, open bins or hoppers, mixers, or blenders, automatic or hopper scales, packing machinery, elevator heads and boots, stock distributors, dust and stock collectors (except all-metal collectors vented to the outside), and all similar dust producing machinery and equipment in grain processing plants, starch plants, sugar pulverizing plants, malting plants, hay grinding plants, and other occupancies of similar nature; coal pulverizing plants (except where the pulverizing equipment is essentially dust tight); all working areas where metal dusts and powders are produced, processed, handled, packed, or stored (except in tight containers); and other similar locations where combustible dust may, under normal operating conditions, be present in the air in quantities sufficient to produce explosive or ignitable mixtures.

**1910.178(c)(2)(vii)** Only approved power-operated industrial trucks designated as DY, EE, or EX shall be used in atmospheres in which combustible dust will not normally be in suspension in the air or will not be likely to be thrown into suspension by the normal operation of equipment or apparatus in quantities sufficient to produce explosive or ignitable mixtures but where deposits or accumulations of such dust may be ignited by arcs or sparks originating in the truck.

**1910.178(c)(2)(viii)** Only approved power-operated industrial trucks designated as DY, EE, or EX shall be used in locations which are hazardous because of the presence of easily ignitable fibers or flyings but in which such fibers or flyings are not likely to be in suspension in the air in quantities sufficient to produce ignitable mixtures.

**1910.178(c)(2)(ix)** Only approved power-operated industrial trucks designated as DS, DY, ES, EE, EX, GS, or LPS shall be used in locations where easily ignitable fibers are stored or handled, including outside storage, but are not being processed or manufactured. Industrial trucks designated as E, which have been previously used in these locations may be continued in use.

**1910.178(c)(2)(x)** On piers and wharves handling general cargo, any approved power-operated industrial truck designated as Type D, E, G, or LP may be used, or trucks which conform to the requirements for these types may be used.

**1910.178(c)(2)(xi)** If storage warehouses and outside storage locations are hazardous only the approved power-operated industrial truck specified for such locations in this paragraph (c)(2) shall be used. If not classified as hazardous, any approved power-operated industrial truck designated as Type D, E, G, or LP may be used, or trucks which conform to the requirements for these types may be used.

**1910.178(c)(2)(xii)** If general industrial or commercial properties are hazardous, only approved power-operated industrial trucks specified for such locations in this paragraph (c)(2) shall be used. If not classified as hazardous, any approved power-operated industrial truck designated as Type D, E, G, or LP may be used, or trucks which conform to the requirements of these types may be used.

**1910.178(d)** *Converted industrial trucks.* Power-operated industrial trucks that have been originally approved for the use of gasoline for fuel, when converted to the use of liquefied petroleum gas fuel in accordance with paragraph (q) of this section, may be used in those locations where G, GS or LP, and LPS designated trucks have been specified in the preceding paragraphs.

**1910.178(e)** *Safety guards.* **(1)** High Lift Rider trucks shall be fitted with an overhead guard manufactured in accordance with paragraph (a)(2) of this section, unless operating conditions do not permit.

**1910.178(e)(2)** If the type of load presents a hazard, the user shall equip fork trucks with a vertical load backrest extension manufactured in accordance with paragraph (a)(2) of this section.

**1910.178(f)** *Fuel handling and storage.* **(1)** The storage and handling of liquid fuels such as gasoline and diesel fuel shall be in accordance with NFPA Flammable and Combustible Liquids Code (NFPA No. 30-1969), which is incorporated by reference as specified in §1910.6.

**1910.178(f)(2)** The storage and handling of liquefied petroleum gas fuel shall be in accordance with NFPA Storage and Handling of Liquefied Petroleum Gases (NFPA No. 58-1969), which is incorporated by reference as specified in §1910.6.

**1910.178(g)** *Changing and charging storage batteries.* **(1)** Battery charging installations shall be located in areas designated for that purpose.

**1910.178(g)(2)** Facilities shall be provided for flushing and neutralizing spilled electrolyte, for fire protection, for protecting charging apparatus from damage by trucks, and for adequate ventilation for dispersal of fumes from gassing batteries.

**1910.178(g)(3)** [Reserved]

**1910.178(g)(4)** A conveyor, overhead hoist, or equivalent material handling equipment shall be provided for handling batteries.

**1910.178(g)(5)** Reinstalled batteries shall be properly positioned and secured in the truck.

**1910.178(g)(6)** A carboy tilter or siphon shall be provided for handling electrolyte.

**1910.178(g)(7)** When charging batteries, acid shall be poured into water; water shall not be poured into acid.

**1910.178(g)(8)** Trucks shall be properly positioned and brake applied before attempting to change or charge batteries.

**1910.178(g)(9)** Care shall be taken to assure that vent caps are functioning. The battery (or compartment) cover(s) shall be open to dissipate heat.

**1910.178(g)(10)** Smoking shall be prohibited in the charging area.

**1910.178(g)(11)** Precautions shall be taken to prevent open flames, sparks, or electric arcs in battery charging areas.

**1910.178(g)(12)** Tools and other metallic objects shall be kept away from the top of uncovered batteries.

**1910.178(h)** *Lighting for operating areas.*

**1910.178(h)(1)** [Reserved]

**1910.178(h)(2)** Where general lighting is less than 2 lumens per square foot, auxiliary directional lighting shall be provided on the truck.

**1910.178(i)** *Control of noxious gases and fumes.* **(1)** Concentration levels of carbon monoxide gas created by powered industrial truck operations shall not exceed the levels specified in §1910.1000.

**1910.178(j)** *Dockboards (bridge plates).* See §1910.30(a).

**1910.178(k)** *Trucks and railroad cars.* **(1)** The brakes of highway trucks shall be set and wheel chocks placed under the rear wheels to prevent the trucks from rolling while they are boarded with powered industrial trucks.

**1910.178(k)(2)** Wheel stops or other recognized positive protection shall be provided to prevent railroad cars from moving during loading or unloading operations.

**1910.178(k)(3)** Fixed jacks may be necessary to support a semitrailer and prevent upending during the loading or unloading when the trailer is not coupled to a tractor.

**1910.178(k)(4)** Positive protection shall be provided to prevent railroad cars from being moved while dockboards or bridge plates are in position.

**1910.178(l)** *Operator training.* **(1)** *Safe operation.* **(i)** The employer shall ensure that each powered industrial truck operator is competent to operate a powered industrial truck safely, as demonstrated by the successful completion of the training and evaluation specified in this paragraph (l).

**1910.178(l)(1)(ii)** Prior to permitting an employee to operate a powered industrial truck (except for training purposes), the employer shall ensure that each operator has successfully completed the training required by this paragraph (l), except as permitted by paragraph (l)(5).

**1910.178(l)(2)** *Training program implementation.* **(i)** Trainees may operate a powered industrial truck only:

**1910.178(l)(2)(i)(A)** Under the direct supervision of persons who have the knowledge, training, and experience to train operators and evaluate their competence; and

**1910.178(l)(2)(i)(B)** Where such operation does not endanger the trainee or other employees.

**1910.178(l)(2)(ii)** Training shall consist of a combination of formal instruction (e.g., lecture, discussion, interactive computer learning, video tape, written material), practical training (demonstrations performed by the trainer and practical exercises performed by the trainee), and evaluation of the operator's performance in the workplace.

**1910.178(l)(2)(iii)** All operator training and evaluation shall be conducted by persons who have the knowledge, training, and experience to train powered industrial truck operators and evaluate their competence.

**1910.178(l)(3)** *Training program content.* Powered industrial truck operators shall receive initial training in the following topics, except in topics which the employer can demonstrate are not applicable to safe operation of the truck in the employer's workplace.

**1910.178(l)(3)(i)** Truck-related topics:

**1910.178(l)(3)(i)(A)** Operating instructions, warnings, and precautions for the types of truck the operator will be authorized to operate;

**1910.178(l)(3)(i)(B)** Differences between the truck and the automobile;

**1910.178(l)(3)(i)(C)** Truck controls and instrumentation: where they are located, what they do, and how they work;

**1910.178(l)(3)(i)(D)** Engine or motor operation;

**1910.178(l)(3)(i)(E)** Steering and maneuvering;

**1910.178(l)(3)(i)(F)** Visibility (including restrictions due to loading);

**1910.178(l)(3)(i)(G)** Fork and attachment adaptation, operation, and use limitations;

**1910.178(l)(3)(i)(H)** Vehicle capacity;

**1910.178(l)(3)(i)(I)** Vehicle stability;

**1910.178(l)(3)(i)(J)** Any vehicle inspection and maintenance that the operator will be required to perform;

**1910.178(l)(3)(i)(K)** Refueling and/or charging and recharging of batteries;

**1910.178(l)(3)(i)(L)** Operating limitations;

**1910.178(l)(3)(i)(M)** Any other operating instructions, warnings, or precautions listed in the operator's manual for the types of vehicle that the employee is being trained to operate.

**1910.178(l)(3)(ii)** Workplace-related topics:

**1910.178(l)(3)(ii)(A)** Surface conditions where the vehicle will be operated;

**1910.178(l)(3)(ii)(B)** Composition of loads to be carried and load stability;

**1910.178(l)(3)(ii)(C)** Load manipulation, stacking, and unstacking;

**1910.178(l)(3)(ii)(D)** Pedestrian traffic in areas where the vehicle will be operated;

**1910.178(l)(3)(ii)(E)** Narrow aisles and other restricted places where the vehicle will be operated;

**1910.178(l)(3)(ii)(F)** Hazardous (classified) locations where the vehicle will be operated;

**1910.178(l)(3)(ii)(G)** Ramps and other sloped surfaces that could affect the vehicle's stability;

**1910.178(l)(3)(ii)(H)** Closed environments and other areas where insufficient ventilation or poor vehicle maintenance could cause a buildup of carbon monoxide or diesel exhaust;

**1910.178(l)(3)(ii)(I)** Other unique or potentially hazardous environmental conditions in the workplace that could affect safe operation.

**1910.178(l)(3)(iii)** The requirements of this section.

**1910.178(l)(4)** *Refresher training and evaluation.* **(i)** Refresher training, including an evaluation of the effectiveness of that training, shall be conducted as required by paragraph (l)(4)(ii) to ensure that the operator has the knowledge and skills needed to operate the powered industrial truck safely.

**1910.178(l)(4)(ii)** Refresher training in relevant topics shall be provided to the operator when:

**1910.178(l)(4)(ii)(A)** The operator has been observed to operate the vehicle in an unsafe manner;

**1910.178(l)(4)(ii)(B)** The operator has been involved in an accident or near-miss incident;

**1910.178(l)(4)(ii)(C)** The operator has received an evaluation that reveals that the operator is not operating the truck safely;

**1910.178(l)(4)(ii)(D)** The operator is assigned to drive a different type of truck; or

**1910.178(l)(4)(ii)(E)** A condition in the workplace changes in a manner that could affect safe operation of the truck.

**1910.178(l)(4)(iii)** An evaluation of each powered industrial truck operator's performance shall be conducted at least once every three years.

**1910.178(l)(5)** *Avoidance of duplicative training.* If an operator has previously received training in a topic specified in paragraph (l)(3) of this section, and such training is appropriate to the truck and working conditions encountered, additional training in that topic is not required if the operator has been evaluated and found competent to operate the truck safely.

**1910.178(l)(6)** *Certification.* The employer shall certify that each operator has been trained and evaluated as required by this paragraph (l). The certification shall include the name of the operator, the date of the training, the date of the evaluation, and the identity of the person(s) performing the training or evaluation.

**1910.178(l)(7)** *Dates.* The employer shall ensure that operators of powered industrial trucks are trained, as appropriate, by the dates shown in the following table.

| If the employee was hired: | The initial training and evaluation of that employee must be completed: |
| --- | --- |
| Before December 1, 1999 | By December 1, 1999. |
| After December 1, 1999 | Before the employee is assigned to operate a powered industrial truck. |

**1910.178(l)(8)** Appendix A to this section provides non-mandatory guidance to assist employers in implementing this paragraph (l). This appendix does not add to, alter, or reduce the requirements of this section.

**1910.178(m)** *Truck operations.* **(1)** Trucks shall not be driven up to anyone standing in front of a bench or other fixed object.

**1910.178(m)(2)** No person shall be allowed to stand or pass under the elevated portion of any truck, whether loaded or empty.

**1910.178(m)(3)** Unauthorized personnel shall not be permitted to ride on powered industrial trucks. A safe place to ride shall be provided where riding of trucks is authorized.

**1910.178(m)(4)** The employer shall prohibit arms or legs from being placed between the uprights of the mast or outside the running lines of the truck.

**1910.178(m)(5)(i)** When a powered industrial truck is left unattended, load engaging means shall be fully lowered, controls shall be neutralized, power shall be shut off, and brakes set. Wheels shall be blocked if the truck is parked on an incline.

**1910.178(m)(5)(ii)** A powered industrial truck is unattended when the operator is 25 ft. or more away from the vehicle which remains in his view, or whenever the operator leaves the vehicle and it is not in his view.

**1910.178(m)(5)(iii)** When the operator of an industrial truck is dismounted and within 25 ft. of the truck still in his view, the load engaging means shall be fully lowered, controls neutralized, and the brakes set to prevent movement.

**1910.178(m)(6)** A safe distance shall be maintained from the edge of ramps or platforms while on any elevated dock, or platform or freight car. Trucks shall not be used for opening or closing freight doors.

**1910.178(m)(7)** Brakes shall be set and wheel blocks shall be in place to prevent movement of trucks, trailers, or railroad cars while loading or unloading. Fixed jacks may be necessary to support a semitrailer during loading or unloading when the trailer is not coupled to a tractor. The flooring of trucks, trailers, and railroad cars shall be checked for breaks and weakness before they are driven onto.

**1910.178(m)(8)** There shall be sufficient headroom under overhead installations, lights, pipes, sprinkler system, etc.

**1910.178(m)(9)** An overhead guard shall be used as protection against falling objects. It should be noted that an overhead guard is intended to offer protection from the impact of small packages, boxes, bagged material, etc., representative of the job application, but not to withstand the impact of a falling capacity load.

**1910.178(m)(10)** A load backrest extension shall be used whenever necessary to minimize the possibility of the load or part of it from falling rearward.

**1910.178(m)(11)** Only approved industrial trucks shall be used in hazardous locations.

**1910.178(m)(12)** [Removed and Reserved]

**1910.178(m)(13)** [Reserved]

**1910.178(m)(14)** Fire aisles, access to stairways, and fire equipment shall be kept clear.

**1910.178(n)** *Traveling.* **(1)** All traffic regulations shall be observed, including authorized plant speed limits. A safe distance shall be maintained approximately three truck lengths from the truck ahead, and the truck shall be kept under control at all times.

**1910.178(n)(2)** The right of way shall be yielded to ambulances, fire trucks, or other vehicles in emergency situations.

**1910.178(n)(3)** Other trucks traveling in the same direction at intersections, blind spots, or other dangerous locations shall not be passed.

**1910.178(n)(4)** The driver shall be required to slow down and sound the horn at cross aisles and other locations where vision is obstructed. If the load being carried obstructs forward view, the driver shall be required to travel with the load trailing.

**1910.178(n)(5)** Railroad tracks shall be crossed diagonally wherever possible. Parking closer than 8 feet from the center of railroad tracks is prohibited.

**1910.178(n)(6)** The driver shall be required to look in the direction of, and keep a clear view of the path of travel.

**1910.178(n)(7)** Grades shall be ascended or descended slowly.

**1910.178(n)(7)(i)** When ascending or descending grades in excess of 10 percent, loaded trucks shall be driven with the load upgrade.

**1910.178(n)(7)(ii)** [Reserved]

**1910.178(n)(7)(iii)** On all grades the load and load engaging means shall be tilted back if applicable, and raised only as far as necessary to clear the road surface.

**1910.178(n)(8)** Under all travel conditions the truck shall be operated at a speed that will permit it to be brought to a stop in a safe manner.

**1910.178(n)(9)** Stunt driving and horseplay shall not be permitted.

**1910.178(n)(10)** The driver shall be required to slow down for wet and slippery floors.

**1910.178(n)(11)** Dockboard or bridgeplates, shall be properly secured before they are driven over. Dockboard or bridgeplates shall be driven over carefully and slowly and their rated capacity never exceeded.

**1910.178(n)(12)** Elevators shall be approached slowly, and then entered squarely after the elevator car is properly leveled. Once on the elevator, the controls shall be neutralized, power shut off, and the brakes set.

**1910.178(n)(13)** Motorized hand trucks must enter elevator or other confined areas with load end forward.

**1910.178(n)(14)** Running over loose objects on the roadway surface shall be avoided.

**1910.178(n)(15)** While negotiating turns, speed shall be reduced to a safe level by means of turning the hand steering wheel in a smooth, sweeping motion. Except when maneuvering at a very low speed, the hand steering wheel shall be turned at a moderate, even rate.

**1910.178(o)** *Loading.* **(1)** Only stable or safely arranged loads shall be handled. Caution shall be exercised when handling off-center loads which cannot be centered.

**1910.178(o)(2)** Only loads within the rated capacity of the truck shall be handled.

**1910.178(o)(3)** The long or high (including multiple-tiered) loads which may affect capacity shall be adjusted.

**1910.178(o)(4)** Trucks equipped with attachments shall be operated as partially loaded trucks when not handling a load.

**1910.178(o)(5)** A load engaging means shall be placed under the load as far as possible; the mast shall be carefully tilted backward to stabilize the load.

**1910.178(o)(6)** Extreme care shall be used when tilting the load forward or backward, particularly when high tiering. Tilting forward with load engaging means elevated shall be prohibited except to pick up a load. An elevated load shall not be tilted forward except when the load is in a deposit position over a rack or stack. When stacking or tiering, only enough backward tilt to stabilize the load shall be used.

**1910.178(p)** *Operation of the truck.* **(1)** If at any time a powered industrial truck is found to be in need of repair, defective, or in any way unsafe, the truck shall be taken out of service until it has been restored to safe operating condition.

**1910.178(p)(2)** Fuel tanks shall not be filled while the engine is running. Spillage shall be avoided.

**1910.178(p)(3)** Spillage of oil or fuel shall be carefully washed away or completely evaporated and the fuel tank cap replaced before restarting engine.

**1910.178(p)(4)** No truck shall be operated with a leak in the fuel system until the leak has been corrected.

**1910.178(p)(5)** Open flames shall not be used for checking electrolyte level in storage batteries or gasoline level in fuel tanks.

**1910.178(q)** *Maintenance of industrial trucks.* **(1)** Any power-operated industrial truck not in safe operating condition shall be removed from service. All repairs shall be made by authorized personnel.

**1910.178(q)(2)** No repairs shall be made in Class I, II, and III locations.

**1910.178(q)(3)** Those repairs to the fuel and ignition systems of industrial trucks which involve fire hazards shall be conducted only in locations designated for such repairs.

**1910.178(q)(4)** Trucks in need of repairs to the electrical system shall have the battery disconnected prior to such repairs.

**1910.178(q)(5)** All parts of any such industrial truck requiring replacement shall be replaced only by parts equivalent as to safety with those used in the original design.

**1910.178(q)(6)** Industrial trucks shall not be altered so that the relative positions of the various parts are different from what they were when originally received from the manufacturer, nor shall they be altered either by the addition of extra parts not provided by the manufacturer or by the elimination of any parts, except as provided in paragraph (q)(12) of this section. Additional counterweighting of fork trucks shall not be done unless approved by the truck manufacturer.

**1910.178(q)(7)** Industrial trucks shall be examined before being placed in service, and shall not be placed in service if the examination shows any condition adversely affecting the safety of the vehicle. Such examination shall be made at least daily.

Where industrial trucks are used on a round-the-clock basis, they shall be examined after each shift. Defects when found shall be immediately reported and corrected.

**1910.178(q)(8)** Water mufflers shall be filled daily or as frequently as is necessary to prevent depletion of the supply of water below 75 percent of the filled capacity. Vehicles with mufflers having screens or other parts that may become clogged shall not be operated while such screens or parts are clogged. Any vehicle that emits hazardous sparks or flames from the exhaust system shall immediately be removed from service, and not returned to service until the cause for the emission of such sparks and flames has been eliminated.

**1910.178(q)(9)** When the temperature of any part of any truck is found to be in excess of its normal operating temperature, thus creating a hazardous condition, the vehicle shall be removed from service and not returned to service until the cause for such overheating has been eliminated.

**1910.178(q)(10)** Industrial trucks shall be kept in a clean condition, free of lint, excess oil, and grease. Noncombustible agents should be used for cleaning trucks. Low flash point (below 100 °F.) solvents shall not be used. High flash point (at or above 100 °F.) solvents may be used. Precautions regarding toxicity, ventilation, and fire hazard shall be consonant with the agent or solvent used.

**1910.178(q)(11)** [Reserved]

**1910.178(q)(12)** Industrial trucks originally approved for the use of gasoline for fuel may be converted to liquefied petroleum gas fuel provided the complete conversion results in a truck which embodies the features specified for LP or LPS designated trucks. Such conversion equipment shall be approved. The description of the component parts of this conversion system and the recommended method of installation on specific trucks are contained in the "Listed by Report."

[39 FR 23502, June 27, 1974, as amended at 40 FR 23073, May 28, 1975; 43 FR 49749, Oct. 24, 1978; 49 FR 5322, Feb. 10, 1984; 53 FR 12122, Apr. 12, 1988; 55 FR 32015, Aug. 6 1990; 61 FR 9227, March 7, 1996; 63 FR 66270, Dec. 1, 1998; 68 FR 32368, June 2, 2003]

### APPENDIX A TO §1910.178—STABILITY OF POWERED INDUSTRIAL TRUCKS (NON-MANDATORY APPENDIX TO PARAGRAPH (L) OF THIS SECTION)

A-1. *Definitions.*

The following definitions help to explain the principle of stability:

*Center of gravity* is the point on an object at which all of the object's weight is concentrated. For symmetrical loads, the center of gravity is at the middle of the load.

*Counterweight* is the weight that is built into the truck's basic structure and is used to offset the load's weight and to maximize the vehicle's resistance to tipping over.

*Fulcrum* is the truck's axis of rotation when it tips over.

*Grade* is the slope of a surface, which is usually measured as the number of feet of rise or fall over a hundred foot horizontal distance (the slope is expressed as a percent).

*Lateral stability* is a truck's resistance to overturning sideways.

*Line of action* is an imaginary vertical line through an object's center of gravity.

*Load center* is the horizontal distance from the load's edge (or the fork's or other attachment's vertical face) to the line of action through the load's center of gravity.

*Longitudinal stability* is the truck's resistance to overturning forward or rearward.

*Moment* is the product of the object's weight times the distance from a fixed point (usually the fulcrum). In the case of a powered industrial truck, the distance is measured from the point at which the truck will tip over to the object's line of action. The distance is always measured perpendicular to the line of action.

*Track* is the distance between the wheels on the same axle of the truck.

*Wheelbase* is the distance between the centerline of the vehicle's front and rear wheels.

A-2. *General.*

A-2.1. Determining the stability of a powered industrial truck is simple once a few basic principles are understood. There are many factors that contribute to a vehicle's stability: the vehicle's wheelbase, track, and height; the load's weight distribution; and the vehicle's counterweight location (if the vehicle is so equipped).

A-2.2. The "stability triangle," used in most stability discussions, demonstrates stability simply.

A-3. *Basic Principles.*

A-3.1. Whether an object is stable depends on the object's moment at one end of a system being greater than, equal to, or smaller than the object's moment at the system's other end. This principle can be seen in the way a see-saw or teeter-totter works: that is, if the product of the load and distance from the fulcrum (moment) is equal to the moment at the device's other end, the device is balanced and it will not move. However, if there is a greater moment at one end of the device, the device will try to move downward at the end with the greater moment.

A-3.2. The longitudinal stability of a counterbalanced powered industrial truck depends on the vehicle's moment and the load's moment. In other words, if the mathematic product of the load moment (the distance from the front

# OSHA Standards for General Industry

wheels, the approximate point at which the vehicle would tip forward) to the load's center of gravity times the load's weight is less than the vehicle's moment, the system is balanced and will not tip forward. However, if the load's moment is greater than the vehicle's moment, the greater load-moment will force the truck to tip forward.

A-4. *The Stability Triangle.*

A-4.1. Almost all counterbalanced powered industrial trucks have a three-point suspension system, that is, the vehicle is supported at three points. This is true even if the vehicle has four wheels. The truck's steer axle is attached to the truck by a pivot pin in the axle's center. When the points are connected with imaginary lines, this three-point support forms a triangle called the stability triangle. Figure 1 depicts the stability triangle.

**FIGURE 1**

NOTES:
1. When the vehicle is loaded, the combined center of gravity (CG) shifts toward line B-C. Theoretically the maximum load will result in the CG at the line B-C. In actual practice, the combined CG should never be at line B-C.
2. The addition of additional counterweight will cause the truck CG to shift toward point A and result in a truck that is less stable laterally.

A-4.2. When the vehicle's line of action, or load center, falls within the stability triangle, the vehicle is stable and will not tip over. However, when the vehicle's line of action or the vehicle/load combination falls outside the stability triangle, the vehicle is unstable and may tip over. (See Figure 2.)

**FIGURE 2**

THE VEHICLE IS STABLE

THIS VEHICLE IS UNSTABLE AND WILL CONTINUE TO TIP OVER

A-5. *Longitudinal Stability.*

A-5.1. The axis of rotation when a truck tips forward is the front wheels' points of contact with the pavement. When a powered industrial truck tips forward, the truck will rotate about this line. When a truck is stable, the vehicle-moment must exceed the load-moment. As long as the vehicle-moment is equal to or exceeds the load-moment, the vehicle will not tip over. On the other hand, if the load moment slightly exceeds the vehicle-moment, the truck will begin to tip forward, thereby causing the rear to lose contact with the floor or ground and resulting in loss of steering control. If the load-moment greatly exceeds the vehicle moment, the truck will tip forward.

A-5.2. To determine the maximum safe load-moment, the truck manufacturer normally rates the truck at a maximum load at a given distance from the front face of the forks. The specified distance from the front face of the forks to the line of action of the load is commonly called the load center. Because larger trucks normally handle loads that are physically larger, these vehicles have greater load centers. Trucks with a capacity of 30,000 pounds or less are normally rated at a given load weight at a 24-inch load center. Trucks with a capacity greater than 30,000 pounds are normally rated at a given load weight at a 36-or 48-inch load center. To safely operate the vehicle, the operator should always check the data plate to determine the maximum allowable weight at the rated load center.

A-5.3. Although the true load-moment distance is measured from the front wheels, this distance is greater than the distance from the front face of the forks. Calculating the maximum allowable load-moment using the load-center distance always provides a lower load-moment than the truck was designed to handle. When handling unusual loads, such as those that are larger than 48 inches long (the center of gravity is greater than 24 inches) or that have an offset center of gravity, etc., a maximum allowable load-moment should be calculated and used to determine whether a load can be safely handled. For example, if an operator is operating a 3000 pound capacity truck (with a 24-inch load center), the maximum allowable load-moment is 72,000 inch-pounds (3,000 times 24). If a load is 60 inches long (30-inch load center), then the maximum that this load can weigh is 2,400 pounds (72,000 divided by 30).

A-6. *Lateral Stability.*

A-6.1. The vehicle's lateral stability is determined by the line of action's position (a vertical line that passes through the combined vehicle's and load's center of gravity) relative to the stability triangle. When the vehicle is not loaded, the truck's center of gravity location is the only factor to be considered in determining the truck's stability. As long as the line of action of the combined vehicle's and load's center of gravity falls within the stability triangle, the truck is stable and will not tip over. However, if the line of action falls outside the stability triangle, the truck is not stable and may tip over. Refer to Figure 2.

A-6.2. Factors that affect the vehicle's lateral stability include the load's placement on the truck, the height of the load above the surface on which the vehicle is operating, and the vehicle's degree of lean.

A-7. *Dynamic Stability.*

A-7.1. Up to this point, the stability of a powered industrial truck has been discussed without considering the dynamic forces that result when the vehicle and load are put into motion. The weight's transfer and the resultant shift in the center of gravity due to the dynamic forces created when the machine is moving, braking, cornering, lifting, tilting, and lowering loads, etc., are important stability considerations.

A-7.2. When determining whether a load can be safely handled, the operator should exercise extra caution when handling loads that cause the vehicle to approach its maximum design characteristics. For example, if an operator must handle a maximum load, the load should be carried at the lowest position possible, the truck should be accelerated slowly and evenly, and the forks should be tilted forward cautiously. However, no precise rules can be formulated to cover all of these eventualities.

### §1910.179 Overhead and gantry cranes.

**1910.179(a)** *Definitions applicable to this section.* **(1)** A *crane* is a machine for lifting and lowering a load and moving it horizontally, with the hoisting mechanism an integral part of the machine. Cranes whether fixed or mobile are driven manually or by power.

**1910.179(a)(2)** An *automatic crane* is a crane which when activated operates through a preset cycle or cycles.

**1910.179(a)(3)** A *cab-operated crane* is a crane controlled by an operator in a cab located on the bridge or trolley.

**1910.179(a)(4)** *Cantilever gantry crane* means a gantry or semigantry crane in which the bridge girders or trusses extend transversely beyond the crane runway on one or both sides.

**1910.179(a)(5)** *Floor-operated crane* means a crane which is pendant or nonconductive rope controlled by an operator on the floor or an independent platform.

**1910.179(a)(6)** *Gantry crane* means a crane similar to an overhead crane except that the bridge for carrying the trolley or trolleys is rigidly supported on two or more legs running on fixed rails or other runway.

**1910.179(a)(7)** *Hot metal handling crane* means an overhead crane used for transporting or pouring molten material.

**1910.179(a)(8)** *Overhead crane* means a crane with a movable bridge carrying a movable or fixed hoisting mechanism and traveling on an overhead fixed runway structure.

**1910.179(a)(9)** *Power-operated crane* means a crane whose mechanism is driven by electric, air, hydraulic, or internal combustion means.

**1910.179(a)(10)** A *pulpit-operated crane* is a crane operated from a fixed operator station not attached to the crane.

**1910.179(a)(11)** A *remote-operated crane* is a crane controlled by an operator not in a pulpit or in the cab attached to the crane, by any method other than pendant or rope control.

**1910.179(a)(12)** A *semigantry crane* is a gantry crane with one end of the bridge rigidly supported on one or more legs that run on a fixed rail or runway, the other end of the bridge being supported by a truck running on an elevated rail or runway.

**1910.179(a)(13)** *Storage bridge crane* means a gantry type crane of long span usually used for bulk storage of material; the bridge girders or trusses are rigidly or nonrigidly supported on one or more legs. It may have one or more fixed or hinged cantilever ends.

**1910.179(a)(14)** *Wall crane* means a crane having a jib with or without trolley and supported from a side wall or line of columns of a building. It is a traveling type and operates on a runway attached to the side wall or columns.

**1910.179(a)(15)** *Appointed* means assigned specific responsibilities by the employer or the employer's representative.

**1910.179(a)(16)** *ANSI* means the American National Standards Institute.

**1910.179(a)(17)** An *auxiliary hoist* is a supplemental hoisting unit of lighter capacity and usually higher speed than provided for the main hoist.

**1910.179(a)(18)** A *brake* is a device used for retarding or stopping motion by friction or power means.

**1910.179(a)(19)** A *drag brake* is a brake which provides retarding force without external control.

**1910.179(a)(20)** A *holding brake* is a brake that automatically prevents motion when power is off.

**1910.179(a)(21)** *Bridge* means that part of a crane consisting of girders, trucks, end ties, footwalks, and drive mechanism which carries the trolley or trolleys.

**1910.179(a)(22)** *Bridge travel* means the crane movement in a direction parallel to the crane runway.

**1910.179(a)(23)** A *bumper* (buffer) is an energy absorbing device for reducing impact when a moving crane or trolley reaches the end of its permitted travel; or when two moving cranes or trolleys come in contact.

**1910.179(a)(24)** The *cab* is the operator's compartment on a crane.

**1910.179(a)(25)** *Clearance* means the distance from any part of the crane to a point of the nearest obstruction.

**1910.179(a)(26)** *Collectors current* are contacting devices for collecting current from runway or bridge conductors.

**1910.179(a)(27)** *Conductors, bridge* are the electrical conductors located along the bridge structure of a crane to provide power to the trolley.

**1910.179(a)(28)** *Conductors, runway* (main) are the electrical conductors located along a crane runway to provide power to the crane.

**1910.179(a)(29)** The *control braking means* is a method of controlling crane motor speed when in an overhauling condition.

**1910.179(a)(30)** *Countertorque* means a method of control by which the power to the motor is reversed to develop torque in the opposite direction.

**1910.179(a)(31)** *Dynamic* means a method of controlling crane motor speeds when in the overhauling condition to provide a retarding force.

**1910.179(a)(32)** *Regenerative* means a form of dynamic braking in which the electrical energy generated is fed back into the power system.

**1910.179(a)(33)** *Mechanical* means a method of control by friction.

**1910.179(a)(34)** *Controller, spring return* means a controller which when released will return automatically to a neutral position.

**1910.179(a)(35)** *Designated* means selected or assigned by the employer or the employer's representative as being qualified to perform specific duties.

**1910.179(a)(36)** A *drift point* means a point on a travel motion controller which releases the brake while the motor is not energized. This allows for coasting before the brake is set.

**1910.179(a)(37)** The *drum* is the cylindrical member around which the ropes are wound for raising or lowering the load.

**1910.179(a)(38)** An *equalizer* is a device which compensates for unequal length or stretch of a rope.

**1910.179(a)(39)** *Exposed* means capable of being contacted inadvertently. Applied to hazardous objects not adequately guarded or isolated.

**1910.179(a)(40)** *Fail-safe* means a provision designed to automatically stop or safely control any motion in which a malfunction occurs.

**1910.179(a)(41)** *Footwalk* means the walkway with handrail, attached to the bridge or trolley for access purposes.

**1910.179(a)(42)** A *hoist* is an apparatus which may be a part of a crane, exerting a force for lifting or lowering.

**1910.179(a)(43)** *Hoist chain* means the load bearing chain in a hoist.

**NOTE:** Chain properties do not conform to those shown in ANSI B30.9-1971, Safety Code for Slings.

**1910.179(a)(44)** *Hoist motion* means that motion of a crane which raises and lowers a load.

**1910.179(a)(45)** *Load* means the total superimposed weight on the load block or hook.

**1910.179(a)(46)** The *load block* is the assembly of hook or shackle, swivel, bearing, sheaves, pins, and frame suspended by the hoisting rope.

**1910.179(a)(47)** *Magnet* means an electromagnetic device carried on a crane hook to pick up loads magnetically.

**1910.179(a)(48)** *Main hoist* means the hoist mechanism provided for lifting the maximum rated load.

**1910.179(a)(49)** A *man trolley* is a trolley having an operator's cab attached thereto.

**1910.179(a)(50)** *Rated load* means the maximum load for which a crane or individual hoist is designed and built by the manufacturer and shown on the equipment nameplate(s).

**1910.179(a)(51)** *Rope* refers to wire rope, unless otherwise specified.

**1910.179(a)(52)** *Running sheave* means a sheave which rotates as the load block is raised or lowered.

**1910.179(a)(53)** *Runway* means an assembly of rails, beams, girders, brackets, and framework on which the crane or trolley travels.

**1910.179(a)(54)** *Side pull* means that portion of the hoist pull acting horizontally when the hoist lines are not operated vertically.

**1910.179(a)(55)** *Span* means the horizontal distance center to center of runway rails.

**1910.179(a)(56)** *Standby crane* means a crane which is not in regular service but which is used occasionally or intermittently as required.

**1910.179(a)(57)** A *stop* is a device to limit travel of a trolley or crane bridge. This device normally is attached to a fixed structure and normally does not have energy absorbing ability.

# OSHA Standards for General Industry

**1910.179(a)(58)** A *switch* is a device for making, breaking, or for changing the connections in an electric circuit.

**1910.179(a)(59)** An *emergency stop switch* is a manually or automatically operated electric switch to cut off electric power independently of the regular operating controls.

**1910.179(a)(60)** A *limit switch* is a switch which is operated by some part or motion of a power-driven machine or equipment to alter the electric circuit associated with the machine or equipment.

**1910.179(a)(61)** A *main switch* is a switch controlling the entire power supply to the crane.

**1910.179(a)(62)** A *master switch* is a switch which dominates the operation of contactors, relays, or other remotely operated devices.

**1910.179(a)(63)** The *trolley* is the unit which travels on the bridge rails and carries the hoisting mechanism.

**1910.179(a)(64)** *Trolley travel* means the trolley movement at right angles to the crane runway.

**1910.179(a)(65)** *Truck* means the unit consisting of a frame, wheels, bearings, and axles which supports the bridge girders or trolleys.

**1910.179(b) General requirements—(1)** *Application.* This section applies to overhead and gantry cranes, including semigantry, cantilever gantry, wall cranes, storage bridge cranes, and others having the same fundamental characteristics. These cranes are grouped because they all have trolleys and similar travel characteristics.

**1910.179(b)(2)** *New and existing equipment.* All new overhead and gantry cranes constructed and installed on or after August 31, 1971, shall meet the design specifications of the American National Standard Safety Code for Overhead and Gantry Cranes, ANSI B30.2.0-1967, which is incorporated by reference as specified in §1910.6.

**1910.179(b)(3)** *Modifications.* Cranes may be modified and rerated provided such modifications and the supporting structure are checked thoroughly for the new rated load by a qualified engineer or the equipment manufacturer. The crane shall be tested in accordance with paragraph (k)(2) of this section. New rated load shall be displayed in accordance with subparagraph (5) of this paragraph.

**1910.179(b)(4)** *Wind indicators and rail clamps.* Outdoor storage bridges shall be provided with automatic rail clamps. A wind-indicating device shall be provided which will give a visible or audible alarm to the bridge operator at a predetermined wind velocity. If the clamps act on the rail heads, any beads or weld flash on the rail heads shall be ground off.

**1910.179(b)(5)** *Rated load marking.* The rated load of the crane shall be plainly marked on each side of the crane, and if the crane has more than one hoisting unit, each hoist shall have its rated load marked on it or its load block and this marking shall be clearly legible from the ground or floor.

**1910.179(b)(6)** *Clearance from obstruction.* **(i)** Minimum clearance of 3 inches overhead and 2 inches laterally shall be provided and maintained between crane and obstructions in conformity with Crane Manufacturers Association of America, Inc., Specification No. 61, which is incorporated by reference as specified in §1910.6 (formerly the Electric Overhead Crane Institute, Inc).

**1910.179(b)(6)(ii)** Where passageways or walkways are provided obstructions shall not be placed so that safety of personnel will be jeopardized by movements of the crane.

**1910.179(b)(7)** *Clearance between parallel cranes.* If the runways of two cranes are parallel, and there are no intervening walls or structure, there shall be adequate clearance provided and maintained between the two bridges.

**1910.179(b)(8)** *Designated personnel*—Only designated personnel shall be permitted to operate a crane covered by this section.

**1910.179(c) Cabs—(1)** *Cab location.* **(i)** The general arrangement of the cab and the location of control and protective equipment shall be such that all operating handles are within convenient reach of the operator when facing the area to be served by the load hook, or while facing the direction of travel of the cab. The arrangement shall allow the operator a full view of the load hook in all positions.

**1910.179(c)(1)(ii)** The cab shall be located to afford a minimum of 3 inches clearance from all fixed structures within its area of possible movement.

**1910.179(c)(2)** *Access to crane.* Access to the cab and/or bridge walkway shall be by a conveniently placed fixed ladder, stairs, or platform requiring no step over any gap exceeding 12 inches. Fixed ladders shall be in conformance with the American National Standard Safety Code for Fixed Ladders, ANSI A14.3-1956, which is incorporated by reference as specified in §1910.6.

**1910.179(c)(3)** *Fire extinguisher.* Carbon tetrachloride extinguishers shall not be used.

**1910.179(c)(4)** *Lighting.* Light in the cab shall be sufficient to enable the operator to see clearly enough to perform his work.

**1910.179(d) Footwalks and ladders—(1)** *Location of footwalks.* **(i)** If sufficient headroom is available on cab-operated cranes, a footwalk shall be provided on the drive side along the entire length of the bridge of all cranes having the trolley running on the top of the girders.

**1910.179(d)(1)(ii)** Where footwalks are located in no case shall less than 48 inches of headroom be provided.

**1910.179(d)(2)** *Construction of footwalks.* **(i)** Footwalks shall be of rigid construction and designed to sustain a distributed load of at least 50 pounds per square foot.

**1910.179(d)(2)(ii)** Footwalks shall have a walking surface of antislip type.

**NOTE:** Wood will meet this requirement.

**1910.179(d)(2)(iii)** [Reserved]

**1910.179(d)(2)(iv)** The inner edge shall extend at least to the line of the outside edge of the lower cover plate or flange of the girder.

**1910.179(d)(3)** *Toeboards and handrails for footwalks.* Toeboards and handrails shall be in compliance with section 1910.23 of this part.

**1910.179(d)(4)** *Ladders and stairways.* **(i)** Gantry cranes shall be provided with ladders or stairways extending from the ground to the footwalk or cab platform.

**1910.179(d)(4)(ii)** Stairways shall be equipped with rigid and substantial metal handrails. Walking surfaces shall be of an antislip type.

**1910.179(d)(4)(iii)** Ladders shall be permanently and securely fastened in place and shall be constructed in compliance with §1910.27.

**1910.179(e) Stops, bumpers, rail sweeps, and guards—(1)** *Trolley stops.* **(i)** Stops shall be provided at the limits of travel of the trolley.

**1910.179(e)(1)(ii)** Stops shall be fastened to resist forces applied when contacted.

**1910.179(e)(1)(iii)** A stop engaging the tread of the wheel shall be of a height at least equal to the radius of the wheel.

**1910.179(e)(2)** *Bridge bumpers*—**(i)** A crane shall be provided with bumpers or other automatic means providing equivalent effect, unless the crane travels at a slow rate of speed and has a faster deceleration rate due to the use of sleeve bearings, or is not operated near the ends of bridge and trolley travel, or is restricted to a limited distance by the nature of the crane operation and there is no hazard of striking any object in this limited distance, or is used in similar operating conditions. The bumpers shall be capable of stopping the crane (not including the lifted load) at an average rate of deceleration not to exceed 3 ft/s/s when traveling in either direction at 20 percent of the rated load speed.

**1910.179(e)(2)(i)(*a*)** The bumpers shall have sufficient energy absorbing capacity to stop the crane when traveling at a speed of at least 40 percent of rated load speed.

**1910.179(e)(2)(i)(*b*)** The bumper shall be so mounted that there is no direct shear on bolts.

**1910.179(e)(2)(ii)** Bumpers shall be so designed and installed as to minimize parts falling from the crane in case of breakage.

**1910.179(e)(3)** *Trolley bumpers*—**(i)** A trolley shall be provided with bumpers or other automatic means of equivalent effect, unless the trolley travels at a slow rate of speed, or is not operated near the ends of bridge and trolley travel, or is restricted to a limited distance of the runway and there is no hazard of striking any object in this limited distance, or is used in similar operating conditions. The bumpers shall be capable of stopping the trolley (not including the lifted load) at an average rate of deceleration not to exceed 4.7 ft/s/s when traveling in either direction at one-third of the rated load speed.

**1910.179(e)(3)(ii)** When more than one trolley is operated on the same bridge, each shall be equipped with bumpers or equivalent on their adjacent ends.

**1910.179(e)(3)(iii)** Bumpers or equivalent shall be designed and installed to minimize parts falling from the trolley in case of age.

**1910.179(e)(4)** *Rail sweeps.* Bridge trucks shall be equipped with sweeps which extend below the top of the rail and project in front of the truck wheels.

**1910.179(e)(5)** *Guards for hoisting ropes.* **(i)** If hoisting ropes run near enough to other parts to make fouling or chafing possible, guards shall be installed to prevent this condition.

**1910.179(e)(5)(ii)** A guard shall be provided to prevent contact between bridge conductors and hoisting ropes if they could come into contact.

**1910.179(e)(6)** *Guards for moving parts.* **(i)** Exposed moving parts such as gears, set screws, projecting keys, chains, chain sprockets, and reciprocating components which might constitute a hazard under normal operating conditions shall be guarded.

**1910.179(e)(6)(ii)** Guards shall be securely fastened.

**1910.179(e)(6)(iii)** Each guard shall be capable of supporting without permanent distortion the weight of a 200-pound person unless the guard is located where it is impossible for a person to step on it.

**1910.179(f)** *Brakes*—**(1)** *Brakes for hoists.* **(i)** Each independent hoisting unit of a crane shall be equipped with at least one self-setting brake, hereafter referred to as a holding brake, applied directly to the motor shaft or some part of the gear train.

**1910.179(f)(1)(ii)** Each independent hoisting unit of a crane, except worm-geared hoists, the angle of whose worm is such as to prevent the load from accelerating in the lowering direction shall, in addition to a holding brake, be equipped with control braking means to prevent overspeeding.

**1910.179(f)(2)** *Holding brakes.* **(i)** Holding brakes for hoist motors shall have not less than the following percentage of the full load hoisting torque at the point where the brake is applied.

**1910.179(f)(2)(i)(a)** 125 percent when used with a control braking means other than mechanical.

**1910.179(f)(2)(i)(b)** 100 percent when used in conjunction with a mechanical control braking means.

**1910.179(f)(2)(i)(c)** 100 percent each if two holding brakes are provided.

**1910.179(f)(2)(ii)** Holding brakes on hoists shall have ample thermal capacity for the frequency of operation required by the service.

**1910.179(f)(2)(iii)** Holding brakes on hoists shall be applied automatically when power is removed.

**1910.179(f)(2)(iv)** Where necessary holding brakes shall be provided with adjustment means to compensate for wear.

**1910.179(f)(2)(v)** The wearing surface of all holding-brake drums or discs shall be smooth.

**1910.179(f)(2)(vi)** Each independent hoisting unit of a crane handling hot metal and having power control braking means shall be equipped with at least two holding brakes.

**1910.179(f)(3)** *Control braking means.* **(i)** A power control braking means such as regenerative, dynamic or countertorque braking, or a mechanically controlled braking means shall be capable of maintaining safe lowering speeds of rated loads.

**1910.179(f)(3)(ii)** The control braking means shall have ample thermal capacity for the frequency of operation required by service.

**1910.179(f)(4)** *Brakes for trolleys and bridges.* **(i)** Foot-operated brakes shall not require an applied force of more than 70 pounds to develop manufacturer's rated brake torque.

**1910.179(f)(4)(ii)** Brakes may be applied by mechanical, electrical, pneumatic, hydraulic, or gravity means.

**1910.179(f)(4)(iii)** Where necessary brakes shall be provided with adjustment means to compensate for wear.

**1910.179(f)(4)(iv)** The wearing surface of all brakedrums or discs shall be smooth.

**1910.179(f)(4)(v)** All foot-brake pedals shall be constructed so that the operator's foot will not easily slip off the pedal.

**1910.179(f)(4)(vi)** Foot-operated brakes shall be equipped with automatic means for positive release when pressure is released from the pedal.

**1910.179(f)(4)(vii)** Brakes for stopping the motion of the trolley or bridge shall be of sufficient size to stop the trolley or bridge within a distance in feet equal to 10 percent of full load speed in feet per minute when traveling at full speed with full load.

**1910.179(f)(4)(viii)** If holding brakes are provided on the bridge or trolleys, they shall not prohibit the use of a drift point in the control circuit.

**1910.179(f)(4)(ix)** Brakes on trolleys and bridges shall have ample thermal capacity for the frequency of operation required by the service to prevent impairment of functions from overheating.

**1910.179(f)(5)** *Application of trolley brakes.* **(i)** On cab-operated cranes with cab on trolley, a trolley brake shall be required as specified under paragraph (f)(4) of this section.

**1910.179(f)(5)(ii)** A drag brake may be applied to hold the trolley in a desired position on the bridge and to eliminate creep with the power off.

**1910.179(f)(6)** *Application of bridge brakes.* **(i)** On cab-operated cranes with cab on bridge, a bridge brake is required as specified under paragraph (f)(4) of this section.

**1910.179(f)(6)(ii)** On cab-operated cranes with cab on trolley, a bridge brake of the holding type shall be required.

**1910.179(f)(6)(iii)** On all floor, remote and pulpit-operated crane bridge drives, a brake of noncoasting mechanical drive shall be provided.

**1910.179(g)** *Electric equipment*—**(1)** *General.* **(i)** Wiring and equipment shall comply with subpart S of this part.

**1910.179(g)(1)(ii)** The control circuit voltage shall not exceed 600 volts for a.c. or d.c. current.

**1910.179(g)(1)(iii)** The voltage at pendant push-buttons shall not exceed 150 volts for a.c. and 300 volts for d.c.

**1910.179(g)(1)(iv)** Where multiple conductor cable is used with a suspended pushbutton station, the station must be supported in some satisfactory manner that will protect the electrical conductors against strain.

**1910.179(g)(1)(v)** Pendant control boxes shall be constructed to prevent electrical shock and shall be clearly marked for identification of functions.

**1910.179(g)(2)** *Equipment.* **(i)** Electrical equipment shall be so located or enclosed that live parts will not be exposed to accidental contact under normal operating conditions.

**1910.179(g)(2)(ii)** Electric equipment shall be protected from dirt, grease, oil, and moisture.

**1910.179(g)(2)(iii)** Guards for live parts shall be substantial and so located that they cannot be accidently deformed so as to make contact with the live parts.

**1910.179(g)(3)** *Controllers.* **(i)** Cranes not equipped with spring-return controllers or momentary contact pushbuttons shall be provided with a device which will disconnect all motors from the line on failure of power and will not permit any motor to be restarted until the controller handle is brought to the "off" position, or a reset switch or button is operated.

**1910.179(g)(3)(ii)** Lever operated controllers shall be provided with a notch or latch which in the "off" position prevents the handle from being inadvertently moved to the "on" position. An "off" detent or spring return arrangement is acceptable.

**1910.179(g)(3)(iii)** The controller operating handle shall be located within convenient reach of the operator.

**1910.179(g)(3)(iv)** As far as practicable, the movement of each controller handle shall be in the same general directions as the resultant movements of the load.

**1910.179(g)(3)(v)** The control for the bridge and trolley travel shall be so located that the operator can readily face the direction of travel.

**1910.179(g)(3)(vi)** For floor-operated cranes, the controller or controllers if rope operated, shall automatically return to the "off" position when released by the operator.

**1910.179(g)(3)(vii)** Pushbuttons in pendant stations shall return to the "off" position when pressure is released by the crane operator.

**1910.179(g)(3)(viii)** Automatic cranes shall be so designed that all motions shall fail-safe if any malfunction of operation occurs.

# OSHA Standards for General Industry

**1910.179(g)(3)(ix)** Remote-operated cranes shall function so that if the control signal for any crane motion becomes ineffective the crane motion shall stop.

**1910.179(g)(4)** *Resistors.* **(i)** Enclosures for resistors shall have openings to provide adequate ventilation, and shall be installed to prevent the accumulation of combustible matter too near to hot parts.

**1910.179(g)(4)(ii)** Resistor units shall be supported so as to be as free as possible from vibration.

**1910.179(g)(4)(iii)** Provision shall be made to prevent broken parts or molten metal falling upon the operator or from the crane.

**1910.179(g)(5)** *Switches.* **(i)** The power supply to the runway conductors shall be controlled by a switch or circuit breaker located on a fixed structure, accessible from the floor, and arranged to be locked in the open position.

**1910.179(g)(5)(ii)** On cab-operated cranes a switch or circuit breaker of the enclosed type, with provision for locking in the open position, shall be provided in the leads from the runway conductors. A means of opening this switch or circuit breaker shall be located within easy reach of the operator.

**1910.179(g)(5)(iii)** On floor-operated cranes, a switch or circuit breaker of the enclosed type, with provision for locking in the open position, shall be provided in the leads from the runway conductors. This disconnect shall be mounted on the bridge or footwalk near the runway collectors. One of the following types of floor-operated disconnects shall be provided:

**1910.179(g)(5)(iii)(*a*)** Nonconductive rope attached to the main disconnect switch.

**1910.179(g)(5)(iii)(*b*)** An undervoltage trip for the main circuit breaker operated by an emergency stop button in the pendant pushbutton in the pendant pushbutton station.

**1910.179(g)(5)(iii)(*c*)** A main line contactor operated by a switch or pushbutton in the pendant pushbutton station.

**1910.179(g)(5)(iv)** The hoisting motion of all electric traveling cranes shall be provided with an overtravel limit switch in the hoisting direction.

**1910.179(g)(5)(v)** All cranes using a lifting magnet shall have a magnet circuit switch of the enclosed type with provision for locking in the open position. Means for discharging the inductive load of the magnet shall be provided.

**1910.179(g)(6)** *Runway conductors.* Conductors of the open type mounted on the crane runway beams or overhead shall be so located or so guarded that persons entering or leaving the cab or crane footwalk normally could not come into contact with them.

**1910.179(g)(7)** *Extension lamps.* If a service receptacle is provided in the cab or on the bridge of cab-operated cranes, it shall be a grounded three-prong type permanent receptacle, not exceeding 300 volts.

**1910.179(h)** *Hoisting equipment*—**(1)** *Sheaves.* **(i)** Sheave grooves shall be smooth and free from surface defects which could cause rope damage.

**1910.179(h)(1)(ii)** Sheaves carrying ropes which can be momentarily unloaded shall be provided with close-fitting guards or other suitable devices to guide the rope back into the groove when the load is applied again.

**1910.179(h)(1)(iii)** The sheaves in the bottom block shall be equipped with close-fitting guards that will prevent ropes from becoming fouled when the block is lying on the ground with ropes loose.

**1910.179(h)(1)(iv)** Pockets and flanges of sheaves used with hoist chains shall be of such dimensions that the chain does not catch or bind during operation.

**1910.179(h)(1)(v)** All running sheaves shall be equipped with means for lubrication. Permanently lubricated, sealed and/or shielded bearings meet this requirement.

**1910.179(h)(2)** *Ropes.* **(i)** In using hoisting ropes, the crane manufacturer's recommendation shall be followed. The rated load divided by the number of parts of rope shall not exceed 20 percent of the nominal breaking strength of the rope.

**1910.179(h)(2)(ii)** Socketing shall be done in the manner specified by the manufacturer of the assembly.

**1910.179(h)(2)(iii)** Rope shall be secured to the drum as follows:

**1910.179(h)(2)(iii)(*a*)** No less than two wraps of rope shall remain on the drum when the hook is in its extreme low position.

**1910.179(h)(2)(iii)(*b*)** Rope end shall be anchored by a clamp securely attached to the drum, or by a socket arrangement approved by the crane or rope manufacturer.

**1910.179(h)(2)(iv)** Eye splices. [Reserved]

**1910.179(h)(2)(v)** Rope clips attached with U-bolts shall have the U-bolts on the dead or short end of the rope. Spacing and number of all types of clips shall be in accordance with the clip manufacturer's recommendation. Clips shall be drop-forged steel in all sizes manufactured commercially. When a newly installed rope has been in operation for an hour, all nuts on the clip bolts shall be retightened.

**1910.179(h)(2)(vi)** Swaged or compressed fittings shall be applied as recommended by the rope or crane manufacturer.

**1910.179(h)(2)(vii)** Wherever exposed to temperatures, at which fiber cores would be damaged, rope having an independent wirerope or wire-strand core, or other temperature-damage resistant core shall be used.

**1910.179(h)(2)(viii)** Replacement rope shall be the same size, grade, and construction as the original rope furnished by the crane manufacturer, unless otherwise recommended by a wire rope manufacturer due to actual working condition requirements.

**1910.179(h)(3)** *Equalizers.* If a load is supported by more than one part of rope, the tension in the parts shall be equalized.

**1910.179(h)(4)** *Hooks.* Hooks shall meet the manufacturer's recommendations and shall not be overloaded.

**1910.179(i)** **Warning device.** Except for floor-operated cranes a gong or other effective warning signal shall be provided for each crane equipped with a power traveling mechanism.

**1910.179(j)** **Inspection**—**(1)** *Inspection classification.* **(i)** Initial inspection. Prior to initial use all new and altered cranes shall be inspected to insure compliance with the provisions of this section.

**1910.179(j)(1)(ii)** Inspection procedure for cranes in regular service is divided into two general classifications based upon the intervals at which inspection should be performed. The intervals in turn are dependent upon the nature of the critical components of the crane and the degree of their exposure to wear, deterioration, or malfunction. The two general classifications are herein designated as "frequent" and "periodic" with respective intervals between inspections as defined below:

**1910.179(j)(1)(ii)(*a*)** Frequent inspection—Daily to monthly intervals.

**1910.179(j)(1)(ii)(*b*)** Periodic inspection—1 to 12- month intervals.

**1910.179(j)(2)** *Frequent inspection.* The following items shall be inspected for defects at intervals as defined in paragraph (j)(1)(ii) of this section or as specifically indicated, including observation during operation for any defects which might appear between regular inspections. All deficiencies such as listed shall be carefully examined and determination made as to whether they constitute a safety hazard:

**1910.179(j)(2)(i)** All functional operating mechanisms for maladjustment interfering with proper operation. Daily.

**1910.179(j)(2)(ii)** Deterioration or leakage in lines, tanks, valves, drain pumps, and other parts of air or hydraulic systems. Daily.

**1910.179(j)(2)(iii)** Hooks with deformation or cracks. Visual inspection daily; monthly inspection with a certification record which includes the date of inspection, the signature of the person who performed the inspection and the serial number, or other identifier, of the hook inspected. For hooks with cracks or having more than 15 percent in excess of normal throat opening or more than 10° twist from the plane of the unbent hook refer to paragraph (l)(3)(iii)(*a*) of this section.

**1910.179(j)(2)(iv)** Hoist chains, including end connections, for excessive wear, twist, distorted links interfering with proper function, or stretch beyond manufacturer's recommendations. Visual inspection daily; monthly inspection with a certification record which includes the date of inspection, the signature of the person who performed the inspection and an identifier of the chain which was inspected.

**1910.179(j)(2)(v)** [Reserved]

**1910.179(j)(2)(vi)** All functional operating mechanisms for excessive wear of components.

**1910.179(j)(2)(vii)** Rope reeving for noncompliance with manufacturer's recommendations.

**1910.179(j)(3)** *Periodic inspection.* Complete inspections of the crane shall be performed at intervals as generally defined in paragraph (j)(1)(ii)(b) of this section, depending upon its activity, severity of service, and environment, or as specifically indicated below. These inspections shall include the requirements of paragraph (j)(2) of this section and in addition, the following items. Any deficiencies such as listed shall be carefully examined and determination made as to whether they constitute a safety hazard:

**1910.179(j)(3)(i)** Deformed, cracked, or corroded members.

**1910.179(j)(3)(ii)** Loose bolts or rivets.

**1910.179(j)(3)(iii)** Cracked or worn sheaves and drums.

**1910.179(j)(3)(iv)** Worn, cracked or distorted parts such as pins, bearings, shafts, gears, rollers, locking and clamping devices.

**1910.179(j)(3)(v)** Excessive wear on brake system parts, linings, pawls, and ratchets.

**1910.179(j)(3)(vi)** Load, wind, and other indicators over their full range, for any significant inaccuracies.

**1910.179(j)(3)(vii)** Gasoline, diesel, electric, or other powerplants for improper performance or noncompliance with applicable safety requirements.

**1910.179(j)(3)(viii)** Excessive wear of chain drive sprockets and excessive chain stretch.

**1910.179(j)(3)(ix)** [Reserved]

**1910.179(j)(3)(x)** Electrical apparatus, for signs of pitting or any deterioration of controller contactors, limit switches and pushbutton stations.

**1910.179(j)(4)** *Cranes not in regular use.* **(i)** A crane which has been idle for a period of 1 month or more, but less than 6 months, shall be given an inspection conforming with requirements of paragraph (j)(2) of this section and paragraph (m)(2) of this section before placing in service.

**1910.179(j)(4)(ii)** A crane which has been idle for a period of over 6 months shall be given a complete inspection conforming with requirements of paragraphs (j) (2) and (3) of this section and paragraph (m)(2) of this section before placing in service.

**1910.179(j)(4)(iii)** Standby cranes shall be inspected at least semi-annually in accordance with requirements of paragraph (j)(2) of this section and paragraph (m)(2) of this section.

**1910.179(k)** *Testing*—**(1)** *Operational tests.* **(i)** Prior to initial use all new and altered cranes shall be tested to insure compliance with this section including the following functions:

**1910.179(k)(1)(i)(a)** Hoisting and lowering.

**1910.179(k)(1)(i)(b)** Trolley travel.

**1910.179(k)(1)(i)(c)** Bridge travel.

**1910.179(k)(1)(i)(d)** Limit switches, locking and safety devices.

**1910.179(k)(1)(ii)** The trip setting of hoist limit switches shall be determined by tests with an empty hook traveling in increasing speeds up to the maximum speed. The actuating mechanism of the limit switch shall be located so that it will trip the switch, under all conditions, in sufficient time to prevent contact of the hook or hook block with any part of the trolley.

**1910.179(k)(2)** *Rated load test.* Test loads shall not be more than 125 percent of the rated load unless otherwise recommended by the manufacturer. The test reports shall be placed on file where readily available to appointed personnel.

**1910.179(l)** *Maintenance*—**(1)** *Preventive maintenance.* A preventive maintenance program based on the crane manufacturer's recommendations shall be established.

**1910.179(l)(2)** *Maintenance procedure.* **(i)** Before adjustments and repairs are started on a crane the following precautions shall be taken:

**1910.179(l)(2)(i)(a)** The crane to be repaired shall be run to a location where it will cause the least interference with other cranes and operations in the area.

**1910.179(l)(2)(i)(b)** All controllers shall be at the off position.

**1910.179(l)(2)(i)(c)** The main or emergency switch shall be open and locked in the open position.

**1910.179(l)(2)(i)(d)** Warning or "out of order" signs shall be placed on the crane, also on the floor beneath or on the hook where visible from the floor.

**1910.179(l)(2)(i)(e)** Where other cranes are in operation on the same runway, rail stops or other suitable means shall be provided to prevent interference with the idle crane.

**1910.179(l)(2)(ii)** After adjustments and repairs have been made the crane shall not be operated until all guards have been reinstalled, safety devices reactivated and maintenance equipment removed.

**1910.179(l)(3)** *Adjustments and repairs.* **(i)** Any unsafe conditions disclosed by the inspection requirements of paragraph (j) of this section shall be corrected before operation of the crane is resumed. Adjustments and repairs shall be done only by designated personnel.

**1910.179(l)(3)(ii)** Adjustments shall be maintained to assure correct functioning of components. The following are examples:

**1910.179(l)(3)(ii)(a)** All functional operating mechanisms.

**1910.179(l)(3)(ii)(b)** Limit switches.

**1910.179(l)(3)(ii)(c)** Control systems.

**1910.179(l)(3)(ii)(d)** Brakes.

**1910.179(l)(3)(ii)(e)** Power plants.

**1910.179(l)(3)(iii)** Repairs or replacements shall be provided promptly as needed for safe operation. The following are examples:

**1910.179(l)(3)(iii)(a)** Crane hooks showing defects described in paragraph (j)(2)(iii) of this section shall be discarded. Repairs by welding or reshaping are not generally recommended. If such repairs are attempted they shall only be done under competent supervision and the hook shall be tested to the load requirements of paragraph (k)(2) of this section before further use.

**1910.179(l)(3)(iii)(b)** Load attachment chains and rope slings showing defects described in paragraph (j)(2) (iv) and (v) of this section respectively.

**1910.179(l)(3)(iii)(c)** All critical parts which are cracked, broken, bent, or excessively worn.

**1910.179(l)(3)(iii)(d)** Pendant control stations shall be kept clean and function labels kept legible.

**1910.179(m)** *Rope inspection*—**(1)** *Running ropes.* A thorough inspection of all ropes shall be made at least once a month and a certification record which includes the date of inspection, the signature of the person who performed the inspection and an identifier for the ropes which were inspected shall be kept on file where readily available to appointed personnel. Any deterioration, resulting in appreciable loss of original strength, shall be carefully observed and determination made as to whether further use of the rope would constitute a safety hazard. Some of the conditions that could result in an appreciable loss of strength are the following:

**1910.179(m)(1)(i)** Reduction of rope diameter below nominal diameter due to loss of core support, internal or external corrosion, or wear of outside wires.

**1910.179(m)(1)(ii)** A number of broken outside wires and the degree of distribution or concentration of such broken wires.

**1910.179(m)(1)(iii)** Worn outside wires.

**1910.179(m)(1)(iv)** Corroded or broken wires at end connections.

**1910.179(m)(1)(v)** Corroded, cracked, bent, worn, or improperly applied end connections.

**1910.179(m)(1)(vi)** Severe kinking, crushing, cutting, or unstranding.

**1910.179(m)(2)** *Other ropes.* All rope which has been idle for a period of a month or more due to shutdown or storage of a crane on which it is installed shall be given a thorough inspection before it is used. This inspection shall be for all types of deterioration and shall be performed by an appointed person whose approval shall be required for further use of the rope. A certification record shall be available for inspection which includes the date of inspection, the signature of the person who performed the inspection and an identifier for the rope which was inspected.

**1910.179(n)** *Handling the load*—**(1)** *Size of load.* The crane shall not be loaded beyond its rated load except for test purposes as provided in paragraph (k) of this section.

**1910.179(n)(2)** *Attaching the load.* **(i)** The hoist chain or hoist rope shall be free from kinks or twists and shall not be wrapped around the load.

**1910.179(n)(2)(ii)** The load shall be attached to the load block hook by means of slings or other approved devices.

**1910.179(n)(2)(iii)** Care shall be taken to make certain that the sling clears all obstacles.

**1910.179(n)(3)** *Moving the load.* **(i)** The load shall be well secured and properly balanced in the sling or lifting device before it is lifted more than a few inches.

**1910.179(n)(3)(ii)** Before starting to hoist the following conditions shall be noted:

**1910.179(n)(3)(ii)(***a***)** Hoist rope shall not be kinked.

**1910.179(n)(3)(ii)(***b***)** Multiple part lines shall not be twisted around each other.

**1910.179(n)(3)(ii)(***c***)** The hook shall be brought over the load in such a manner as to prevent swinging.

**1910.179(n)(3)(iii)** During hoisting care shall be taken that:

**1910.179(n)(3)(iii)(***a***)** There is no sudden acceleration or deceleration of the moving load.

**1910.179(n)(3)(iii)(***b***)** The load does not contact any obstructions.

**1910.179(n)(3)(iv)** Cranes shall not be used for side pulls except when specifically authorized by a responsible person who has determined that the stability of the crane is not thereby endangered and that various parts of the crane will not be overstressed.

**1910.179(n)(3)(v)** While any employee is on the load or hook, there shall be no hoisting, lowering, or traveling.

**1910.179(n)(3)(vi)** The employer shall require that the operator avoid carrying loads over people.

**1910.179(n)(3)(vii)** The operator shall test the brakes each time a load approaching the rated load is handled. The brakes shall be tested by raising the load a few inches and applying the brakes.

**1910.179(n)(3)(viii)** The load shall not be lowered below the point where less than two full wraps of rope remain on the hoisting drum.

**1910.179(n)(3)(ix)** When two or more cranes are used to lift a load one qualified responsible person shall be in charge of the operation. He shall analyze the operation and instruct all personnel involved in the proper positioning, rigging of the load, and the movements to be made.

**1910.179(n)(3)(x)** The employer shall insure that the operator does not leave his position at the controls while the load is suspended.

**1910.179(n)(3)(xi)** When starting the bridge and when the load or hook approaches near or over personnel, the warning signal shall be sounded.

**1910.179(n)(4)** *Hoist limit switch.* **(i)** At the beginning of each operator's shift, the upper limit switch of each hoist shall be tried out under no load. Extreme care shall be exercised; the block shall be "inched" into the limit or run in at slow speed. If the switch does not operate properly, the appointed person shall be immediately notified.

**1910.179(n)(4)(ii)** The hoist limit switch which controls the upper limit of travel of the load block shall never be used as an operating control.

**1910.179(o)** *Other requirements, general*—**(1)** *Ladders.* **(i)** The employer shall insure that hands are free from encumbrances while personnel are using ladders.

**1910.179(o)(1)(ii)** Articles which are too large to be carried in pockets or belts shall be lifted and lowered by hand line.

**1910.179(o)(2)** *Cabs.* **(i)** Necessary clothing and personal belongings shall be stored in such a manner as not to interfere with access or operation.

**1910.179(o)(2)(ii)** Tools, oil cans, waste, extra fuses, and other necessary articles shall be stored in the tool box, and shall not be permitted to lie loose in or about the cab.

**1910.179(o)(3)** *Fire extinguishers.* The employer shall insure that operators are familiar with the operation and care of fire extinguishers provided.

[39 FR 23502, June 27, 1974, as amended at 40 FR 27400, June 27, 1975; 49 FR 5322, Feb. 10, 1984; 51 FR 34560, Sept. 29, 1986; 55 FR 32015, Aug. 6, 1990; 61 FR 9239, Mar. 7, 1996]

## §1910.180 Crawler locomotive and truck cranes.

**1910.180(a)** *Definitions applicable to this section.* **(1)** A *crawler crane* consists of a rotating superstructure with power plant, operating machinery, and boom, mounted on a base, equipped with crawler treads for travel. Its function is to hoist and swing loads at various radii.

**1910.180(a)(2)** A *locomotive crane* consists of a rotating superstructure with power-plant, operating machinery and boom, mounted on a base or car equipped for travel on railroad track. It may be self-propelled or propelled by an outside source. Its function is to hoist and swing loads at various radii.

**1910.180(a)(3)** A *truck crane* consists of a rotating superstructure with powerplant, operating machinery and boom, mounted on an automotive truck equipped with a powerplant for travel. Its function is to hoist and swing loads at various radii.

**1910.180(a)(4)** A *wheel mounted crane* (wagon crane) consists of a rotating superstructure with powerplant, operating machinery and boom, mounted on a base or platform equipped with axles and rubber-tired wheels for travel. The base is usually propelled by the engine in the superstructure, but it may be equipped with a separate engine controlled from the superstructure. Its function is to hoist and swing loads at various radii.

**1910.180(a)(5)** An *accessory* is a secondary part or assembly of parts which contributes to the overall function and usefulness of a machine.

**1910.180(a)(6)** *Appointed* means assigned specific responsibilities by the employer or the employer's representative.

**1910.180(a)(7)** *ANSI* means the American National Standards Institute.

**1910.180(a)(8)** An *angle indicator* (boom) is an accessory which measures the angle of the boom to the horizontal.

**1910.180(a)(9)** The *axis of rotation* is the vertical axis around which the crane superstructure rotates.

**1910.180(a)(10)** *Axle* means the shaft or spindle with which or about which a wheel rotates. On truck- and wheel-mounted cranes it refers to an automotive type of axle assembly including housings, gearing, differential, bearings, and mounting appurtenances.

**1910.180(a)(11)** *Axle* (bogie) means two or more automotive-type axles mounted in tandem in a frame so as to divide the load between the axles and permit vertical oscillation of the wheels.

**1910.180(a)(12)** The *base* (mounting) is the traveling base or carrier on which the rotating superstructure is mounted such as a car, truck, crawlers, or wheel platform.

**1910.180(a)(13)** The *boom* (crane) is a member hinged to the front of the rotating superstructure with the outer end supported by ropes leading to a gantry or A-frame and used for supporting the hoisting tackle.

**1910.180(a)(14)** The *boom angle* is the angle between the longitudinal centerline of the boom and the horizontal. The boom longitudinal centerline is a straight line between the boom foot pin (heel pin) centerline and boom point sheave pin centerline.

**1910.180(a)(15)** The *boom hoist* is a hoist drum and rope reeving system used to raise and lower the boom. The rope system may be all live reeving or a combination of live reeving and pendants.

**1910.180(a)(16)** The *boom stop* is a device used to limit the angle of the boom at the highest position.

**1910.180(a)(17)** A *brake* is a device used for retarding or stopping motion by friction or power means.

**1910.180(a)(18)** A *cab* is a housing which covers the rotating superstructure machinery and/or operator's station. On truck-crane trucks a separate cab covers the driver's station.

**1910.180(a)(19)** The *clutch* is a friction, electromagnetic, hydraulic, pneumatic, or positive mechanical device for engagement or disengagement of power.

**1910.180(a)(20)** The *counterweight* is a weight used to supplement the weight of the machine in providing stability for lifting working loads.

**1910.180(a)(21)** *Designated* means selected or assigned by the employer or the employer's representative as being qualified to perform specific duties.

**1910.180(a)(22)** The *drum* is the cylindrical members around which ropes are wound for raising and lowering the load or boom.

**1910.180(a)(23)** *Dynamic* (loading) means loads introduced into the machine or its components by forces in motion.

**1910.180(a)(24)** The *gantry* (A-frame) is a structural frame, extending above the superstructure, to which the boom support ropes are reeved.

**1910.180(a)(25)** A *jib* is an extension attached to the boom point to provide added boom length for lifting specified loads. The jib may be in line with the boom or offset to various angles.

**1910.180(a)(26)** *Load* (working) means the external load, in pounds, applied to the crane, including the weight of load-attaching equipment such as load blocks, shackles, and slings.

**1910.180(a)(27)** *Load block* (upper) means the assembly of hook or shackle, swivel, sheaves, pins, and frame suspended from the boom point.

**1910.180(a)(28)** *Load block* (lower) means the assembly of hook or shackle, swivel, sheaves, pins, and frame suspended by the hoisting ropes.

**1910.180(a)(29)** A *load hoist* is a hoist drum and rope reeving system used for hoisting and lowering loads.

**1910.180(a)(30)** *Load ratings* are crane ratings in pounds established by the manufacturer in accordance with paragraph (c) of this section.

**1910.180(a)(31)** *Outriggers* are extendable or fixed metal arms, attached to the mounting base, which rest on supports at the outer ends.

**1910.180(a)(32)** *Rail clamp* means a tong-like metal device, mounted on a locomotive crane car, which can be connected to the track.

**1910.180(a)(33)** *Reeving* means a rope system in which the rope travels around drums and sheaves.

**1910.180(a)(34)** *Rope* refers to a wire rope unless otherwise specified.

**1910.180(a)(35)** *Side loading* means a load applied at an angle to the vertical plane of the boom.

**1910.180(a)(36)** A *standby crane* is a crane which is not in regular service but which is used occasionally or intermittently as required.

**1910.180(a)(37)** A *standing (guy) rope* is a supporting rope which maintains a constant distance between the points of attachment to the two components connected by the rope.

**1910.180(a)(38)** *Structural competence* means the ability of the machine and its components to withstand the stresses imposed by applied loads.

**1910.180(a)(39)** *Superstructure* means the rotating upper frame structure of the machine and the operating machinery mounted thereon.

**1910.180(a)(40)** *Swing* means the rotation of the superstructure for movement of loads in a horizontal direction about the axis of rotation.

**1910.180(a)(41)** *Swing mechanism* means the machinery involved in providing rotation of the superstructure.

**1910.180(a)(42)** *Tackle* is an assembly of ropes and sheaves arranged for hoisting and pulling.

**1910.180(a)(43)** *Transit* means the moving or transporting of a crane from one jobsite to another.

**1910.180(a)(44)** *Travel* means the function of the machine moving from one location to another, on a jobsite.

**1910.180(a)(45)** The *travel mechanism* is the machinery involved in providing travel.

**1910.180(a)(46)** *Wheelbase* means the distance between centers of front and rear axles. For a multiple axle assembly the axle center for wheelbase measurement is taken as the midpoint of the assembly.

**1910.180(a)(47)** The *whipline* (auxiliary hoist) is a separate hoist rope system of lighter load capacity and higher speed than provided by the main hoist.

**1910.180(a)(48)** A *winch head* is a power driven spool for handling of loads by means of friction between fiber or wire rope and spool.

**1910.180(b)** *General requirements*—**(1)** *Application.* This section applies to crawler cranes, locomotive cranes, wheel mounted cranes of both truck and self-propelled wheel type, and any variations thereof which retain the same fundamental characteristics. This section includes only cranes of the above types, which are basically powered by internal combustion engines or electric motors and which utilize drums and ropes. Cranes designed for railway and automobile wreck clearances are excepted. The requirements of this section are applicable only to machines when used as lifting cranes.

**1910.180(b)(2)** *New and existing equipment.* All new crawler, locomotive, and truck cranes constructed and utilized on or after August 31, 1971, shall meet the design specifications of the American National Standard Safety Code for Crawler, Locomotive, and Truck Cranes, ANSI B30.5-1968, which is incorporated by reference as specified in §1910.6. Crawler, locomotive, and truck cranes constructed prior to August 31, 1971, should be modified to conform to those design specifications by February 15, 1972, unless it can be shown that the crane cannot feasibly or economically be altered and that the crane substantially complies with the requirements of this section.

**1910.180(b)(3)** *Designated personnel.* Only designated personnel shall be permitted to operate a crane covered by this section.

**1910.180(c)** *Load ratings*—**(1)** *Load ratings—where stability governs lifting performance.* **(i)** The margin of stability for determination of load ratings, with booms of stipulated lengths at stipulated working radii for the various types of crane mountings, is established by taking a percentage of the loads which will produce a condition of tipping or balance with the boom in the least stable direction, relative to the mounting. The load ratings shall not exceed the following percentages for cranes, with the indicated types of mounting under conditions stipulated in paragraphs (c)(1) (ii) and (iii) of this section.

| Type of crane mounting | Maximum load ratings (percent of tipping loads) |
|---|---|
| Locomotive, without outriggers: | |
|   Booms 60 feet or less............................................. | [1]85 |
|   Booms over 60 feet .................................................. | [1]85 |
| Locomotive, using outriggers fully extended .............. | 80 |
| Crawler, without outriggers........................................... | 75 |
| Crawler, using outriggers fully extended ................... | 85 |
| Truck and wheel mounted without outriggers or using outriggers fully extended ............................. | 85 |

[1] Unless this results in less than 30,000 pound-feet net stabilizing moment about the rail, which shall be minimum with such booms.

**1910.180(c)(1)(ii)** The following stipulations shall govern the application of the values in paragraph (c)(1)(i) of this section for locomotive cranes:

**1910.180(c)(1)(ii)(a)** Tipping with or without the use of outriggers occurs when half of the wheels farthest from the load leave the rail.

**1910.180(c)(1)(ii)(b)** The crane shall be standing on track which is level within 1 percent grade.

**1910.180(c)(1)(ii)(c)** Radius of the load is the horizontal distance from a projection of the axis of rotation to the rail support surface, before loading, to the center of vertical hoist line or tackle with load applied.

**1910.180(c)(1)(ii)(d)** Tipping loads from which ratings are determined shall be applied under static conditions only, i.e., without dynamic effect of hoisting, lowering, or swinging.

**1910.180(c)(1)(ii)(e)** The weight of all auxiliary handling devices such as hoist blocks, hooks, and slings shall be considered a part of the load rating.

**1910.180(c)(1)(iii)** Stipulations governing the application of the values in paragraph (c)(1)(i) of this section for crawler, truck, and wheel-mounted cranes shall be in accordance with Crane Load-Stability Test Code, Society of Automotive Engineers (SAE) J765, which is incorporated by reference as specified in §1910.6.

**1910.180(c)(1)(iv)** The effectiveness of these preceding stability factors will be influenced by such additional factors as freely suspended loads, track, wind, or ground conditions, condition and inflation of rubber tires, boom lengths, proper operating speeds for existing conditions, and, in general, careful and competent operation. All of these shall be taken into account by the user.

**1910.180(c)(2)** *Load rating chart.* A substantial and durable rating chart with clearly legible letters and figures shall be provided with each crane and securely fixed to the crane cab in a location easily visible to the operator while seated at his control station.

**1910.180(d)** *Inspection classification*—**(1)** *Initial inspection.* Prior to initial use all new and altered cranes shall be inspected to insure compliance with provisions of this section.

**1910.180(d)(2)** *Regular inspection.* Inspection procedure for cranes in regular service is divided into two general classifications based upon the intervals at which inspection should be performed. The intervals in turn are dependent upon the nature of the critical components of the crane and the degree of their exposure to wear, deterioration, or malfunction. The two general classifications are herein designated as "frequent" and "periodic", with respective intervals between inspections as defined below:

# OSHA Standards for General Industry

**1910.180(d)(2)(i)** Frequent inspection: Daily to monthly intervals.

**1910.180(d)(2)(ii)** Periodic inspection: 1- to 12-month intervals, or as specifically recommended by the manufacturer.

**1910.180(d)(3)** *Frequent inspection.* Items such as the following shall be inspected for defects at intervals as defined in paragraph (d)(2)(i) of this section or as specifically indicated including observation during operation for any defects which might appear between regular inspections. Any deficiencies such as listed shall be carefully examined and determination made as to whether they constitute a safety hazard:

**1910.180(d)(3)(i)** All control mechanisms for maladjustment interfering with proper operation: Daily.

**1910.180(d)(3)(ii)** All control mechanisms for excessive wear of components and contamination by lubricants or other foreign matter.

**1910.180(d)(3)(iii)** All safety devices for malfunction.

**1910.180(d)(3)(iv)** Deterioration or leakage in air or hydraulic systems: Daily.

**1910.180(d)(3)(v)** Crane hooks with deformations or cracks. For hooks with cracks or having more than 15 percent in excess of normal throat opening or more than 10° twist from the plane of the unbent hook.

**1910.180(d)(3)(vi)** Rope reeving for noncompliance with manufacturer's recommendations.

**1910.180(d)(3)(vii)** Electrical apparatus for malfunctioning, signs of excessive deterioration, dirt, and moisture accumulation.

**1910.180(d)(4)** *Periodic inspection.* Complete inspections of the crane shall be performed at intervals as generally defined in paragraph (d)(2)(ii) of this section depending upon its activity, severity of service, and environment, or as specifically indicated below. These inspections shall include the requirements of paragraph (d)(3) of this section and in addition, items such as the following. Any deficiencies such as listed shall be carefully examined and determination made as to whether they constitute a safety hazard:

**1910.180(d)(4)(i)** Deformed, cracked, or corroded members in the crane structure and boom.

**1910.180(d)(4)(ii)** Loose bolts or rivets.

**1910.180(d)(4)(iii)** Cracked or worn sheaves and drums.

**1910.180(d)(4)(iv)** Worn, cracked, or distorted parts such as pins, bearings, shafts, gears, rollers and locking devices.

**1910.180(d)(4)(v)** Excessive wear on brake and clutch system parts, linings, pawls, and ratchets.

**1910.180(d)(4)(vi)** Load, boom angle, and other indicators over their full range, for any significant inaccuracies.

**1910.180(d)(4)(vii)** Gasoline, diesel, electric, or other power plants for improper performance or noncompliance with safety requirements.

**1910.180(d)(4)(viii)** Excessive wear of chain-drive sprockets and excessive chain stretch.

**1910.180(d)(4)(ix)** Travel steering, braking, and locking devices, for malfunction.

**1910.180(d)(4)(x)** Excessively worn or damaged tires.

**1910.180(d)(5)** *Cranes not in regular use.* **(i)** A crane which has been idle for a period of one month or more, but less than 6 months, shall be given an inspection conforming with requirements of paragraph (d)(3) of this section and paragraph (g)(2)(ii) of this section before placing in service.

**1910.180(d)(5)(ii)** A crane which has been idle for a period of six months shall be given a complete inspection conforming with requirements of paragraphs (d) (3) and (4) of this section and paragraph (g)(2)(ii) of this section before placing in service.

**1910.180(d)(5)(iii)** Standby cranes shall be inspected at least semiannually in accordance with requirements of paragraph (d)(3) of this section and paragraph (g)(2)(ii) of this section. Such cranes which are exposed to adverse environment should be inspected more frequently.

**1910.180(d)(6)** *Inspection records.* Certification records which include the date of inspection, the signature of the person who performed the inspection and the serial number, or other identifier, of the crane which was inspected shall be made monthly on critical items in use such as brakes, crane hooks, and ropes. This certification record shall be kept readily available.

**1910.180(e)** *Testing*—**(1)** *Operational tests.* **(i)** In addition to prototype tests and quality-control measures, each new production crane shall be tested by the manufacturer to the extent necessary to insure compliance with the operational requirements of this paragraph including functions such as the following:

**1910.180(e)(1)(i)(a)** Load hoisting and lowering mechanisms.

**1910.180(e)(1)(i)(b)** Boom hoisting and lower mechanisms.

**1910.180(e)(1)(i)(c)** Swinging mechanism.

**1910.180(e)(1)(i)(d)** Travel mechanism.

**1910.180(e)(1)(i)(e)** Safety devices.

**1910.180(e)(1)(ii)** Where the complete production crane is not supplied by one manufacturer such tests shall be conducted at final assembly.

**1910.180(e)(1)(iii)** Certified production-crane test results shall be made available.

**1910.180(e)(2)** *Rated load test.* **(i)** Written reports shall be available showing test procedures and confirming the adequacy of repairs or alterations.

**1910.180(e)(2)(ii)** Test loads shall not exceed 110 percent of the rated load at any selected working radius.

**1910.180(e)(2)(iii)** Where rerating is necessary:

**1910.180(e)(2)(iii)(a)** Crawler, truck, and wheel-mounted cranes shall be tested in accordance with SAE Recommended Practice, Crane Load Stability Test Code J765 (April 1961).

**1910.180(e)(2)(iii)(b)** Locomotive cranes shall be tested in accordance with paragraph (c)(1)(i) and (ii) of this section.

**1910.180(e)(2)(iii)(c)** Rerating test report shall be readily available.

**1910.180(e)(2)(iv)** No cranes shall be rerated in excess of the original load ratings unless such rating changes are approved by the crane manufacturer or final assembler.

**1910.180(f)** *Maintenance procedure—General.* After adjustments and repairs have been made the crane shall not be operated until all guards have been reinstalled, safety devices reactivated, and maintenance equipment removed.

**1910.180(g)** *Rope inspection*—**(1)** *Running ropes.* A thorough inspection of all ropes in use shall be made at least once a month and a certification record which includes the date of inspection, the signature of the person who performed the inspection and an identifier for the ropes shall be prepared and kept on file where readily available. All inspections shall be performed by an appointed or authorized person. Any deterioration, resulting in appreciable loss of original strength shall be carefully observed and determination made as to whether further use of the rope would constitute a safety hazard. Some of the conditions that could result in an appreciable loss of strength are the following:

**1910.180(g)(1)(i)** Reduction of rope diameter below nominal diameter due to loss of core support, internal or external corrosion, or wear of outside wires.

**1910.180(g)(1)(ii)** A number of broken outside wires and the degree of distribution of concentration of such broken wires.

**1910.180(g)(1)(iii)** Worn outside wires.

**1910.180(g)(1)(iv)** Corroded or broken wires at end connections.

**1910.180(g)(1)(v)** Corroded, cracked, bent, worn, or improperly applied end connections.

**1910.180(g)(1)(vi)** Severe kinking, crushing, cutting, or unstranding.

**1910.180(g)(2)** *Other ropes.* **(i)** Heavy wear and/or broken wires may occur in sections in contact with equalizer sheaves or other sheaves where rope travel is limited, or with saddles. Particular care shall be taken to inspect ropes at these locations.

**1910.180(g)(2)(ii)** All rope which has been idle for a period of a month or more due to shutdown or storage of a crane on which it is installed shall be given a thorough inspection before it is used. This inspection shall be for all types of deterioration and shall be performed by an appointed or authorized person whose approval shall be required for further use of the rope. A certification record which includes the date of inspection, the signature of the person who performed the inspection, and an identifier for the rope which was inspected shall be prepared and kept readily available.

**1910.180(g)(2)(iii)** Particular care shall be taken in the inspection of non-rotating rope.

**1910.180(h)** *Handling the load*—**(1)** *Size of load.* **(i)** No crane shall be loaded beyond the rated load, except for test purposes as provided in paragraph (e) of this section.

**1910.180(h)(1)(ii)** When loads which are limited by structural competence rather than by stability are to be handled, it shall be ascertained that the weight of the load has been determined within plus or minus 10 percent before it is lifted.

**1910.180(h)(2)** *Attaching the load.* **(i)** The hoist rope shall not be wrapped around the load.

**1910.180(h)(2)(ii)** The load shall be attached to the hook by means of slings or other approved devices.

**1910.180(h)(3)** *Moving the load.* **(i)** The employer shall assure that:

**1910.180(h)(3)(i)***a* The crane is level and where necessary blocked properly.

**1910.180(h)(3)(i)***b* The load is well secured and properly balanced in the sling or lifting device before it is lifted more than a few inches.

**1910.180(h)(3)(ii)** Before starting to hoist, the following conditions shall be noted:

**1910.180(h)(3)(ii)***a* Hoist rope shall not be kinked.

**1910.180(h)(3)(ii)***b* Multiple part lines shall not be twisted around each other.

**1910.180(h)(3)(ii)***c* The hook shall be brought over the load in such a manner as to prevent swinging.

**1910.180(h)(3)(iii)** During hoisting care shall be taken that:

**1910.180(h)(3)(iii)***a* There is no sudden acceleration or deceleration of the moving load.

**1910.180(h)(3)(iii)***b* The load does not contact any obstructions.

**1910.180(h)(3)(iv)** Side loading of booms shall be limited to freely suspended loads. Cranes shall not be used for dragging loads sideways.

**1910.180(h)(3)(v)** No hoisting, lowering, swinging, or traveling shall be done while anyone is on the load or hook.

**1910.180(h)(3)(vi)** The operator should avoid carrying loads over people.

**1910.180(h)(3)(vii)** On truck-mounted cranes, no loads shall be lifted over the front area except as approved by the crane manufacturer.

**1910.180(h)(3)(viii)** The operator shall test the brakes each time a load approaching the rated load is handled by raising it a few inches and applying the brakes.

**1910.180(h)(3)(ix)** Outriggers shall be used when the load to be handled at that particular radius exceeds the rated load without outriggers as given by the manufacturer for that crane. Where floats are used they shall be securely attached to the outriggers. Wood blocks used to support outriggers shall:

**1910.180(h)(3)(ix)***a* Be strong enough to prevent crushing.

**1910.180(h)(3)(ix)***b* Be free from defects.

**1910.180(h)(3)(ix)***c* Be of sufficient width and length to prevent shifting or toppling under load.

**1910.180(h)(3)(x)** Neither the load nor the boom shall be lowered below the point where less than two full wraps of rope remain on their respective drums.

**1910.180(h)(3)(xi)** Before lifting loads with locomotive cranes without using outriggers, means shall be applied to prevent the load from being carried by the truck springs.

**1910.180(h)(3)(xii)** When two or more cranes are used to lift one load, one designated person shall be responsible for the operation. He shall be required to analyze the operation and instruct all personnel involved in the proper positioning, rigging of the load, and the movements to be made.

**1910.180(h)(3)(xiii)** In transit the following additional precautions shall be exercised:

**1910.180(h)(3)(xiii)***a* The boom shall be carried in line with the direction of motion.

**1910.180(h)(3)(xiii)***b* The superstructure shall be secured against rotation, except when negotiating turns when there is an operator in the cab or the boom is supported on a dolly.

**1910.180(h)(3)(xiii)***c* The empty hook shall be lashed or otherwise restrained so that it cannot swing freely.

**1910.180(h)(3)(xiv)** Before traveling a crane with load, a designated person shall be responsible for determining and controlling safety. Decisions such as position of load, boom location, ground support, travel route, and speed of movement shall be in accord with his determinations.

**1910.180(h)(3)(xv)** A crane with or without load shall not be traveled with the boom so high that it may bounce back over the cab.

**1910.180(h)(3)(xvi)** When rotating the crane, sudden starts and stops shall be avoided. Rotational speed shall be such that the load does not swing out beyond the radii at which it can be controlled. A tag or restraint line shall be used when rotation of the load is hazardous.

**1910.180(h)(3)(xvii)** When a crane is to be operated at a fixed radius, the boom-hoist pawl or other positive locking device shall be engaged.

**1910.180(h)(3)(xviii)** Ropes shall not be handled on a winch head without the knowledge of the operator.

**1910.180(h)(3)(xix)** While a winch head is being used, the operator shall be within convenient reach of the power unit control lever.

**1910.180(h)(4)** *Holding the load.* **(i)** The operator shall not be permitted to leave his position at the controls while the load is suspended.

**1910.180(h)(4)(ii)** No person should be permitted to stand or pass under a load on the hook.

**1910.180(h)(4)(iii)** If the load must remain suspended for any considerable length of time, the operator shall hold the drum from rotating in the lowering direction by activating the positive controllable means of the operator's station.

**1910.180(i)** *Other requirements*—**(1)** *Rail clamps.* Rail clamps shall not be used as a means of restraining tipping of a locomotive crane.

**1910.180(i)(2)** *Ballast or counterweight.* Cranes shall not be operated without the full amount of any ballast or counterweight in place as specified by the maker, but truck cranes that have dropped the ballast or counterweight may be operated temporarily with special care and only for light loads without full ballast or counterweight in place. The ballast or counterweight in place specified by the manufacturer shall not be exceeded.

**1910.180(i)(3)** *Cabs.* **(i)** Necessary clothing and personal belongings shall be stored in such a manner as to not interfere with access or operation.

**1910.180(i)(3)(ii)** Tools, oil cans, waste, extra fuses, and other necessary articles shall be stored in the tool box, and shall not be permitted to lie loose in or about the cab.

**1910.180(i)(4)** *Refueling.* **(i)** Refueling with small portable containers shall be done with an approved safety type can equipped with an automatic closing cap and flame arrester. Refer to §1910.155(c)(3) for definition of approved.

**1910.180(i)(4)(ii)** Machines shall not be refueled with the engine running.

**1910.180(i)(5)** *Fire extinguishers.* **(i)** A carbon dioxide, dry chemical, or equivalent fire extinguisher shall be kept in the cab or vicinity of the crane.

**1910.180(i)(5)(ii)** Operating and maintenance personnel shall be made familiar with the use and care of the fire extinguishers provided.

**1910.180(i)(6)** *Swinging locomotive cranes.* A locomotive crane shall not be swung into a position where railway cars on an adjacent track might strike it, until it has been ascertained that cars are not being moved on the adjacent track and proper flag protection has been established.

**1910.180(j)** *Operations near overhead lines.* For operations near overhead electric lines, see §1910.333(c)(3).

[39 FR 23502, June 27, 1974, as amended at 49 FR 5323, Feb. 10, 1984; 51 FR 34561, Sept. 29, 1986; 53 FR 12122, Apr. 12, 1988; 55 FR 32015, Aug 6, 1990; 61 FR 9239, Mar. 7, 1996]

### §1910.181 Derricks.

**1910.181(a)** *Definitions applicable to this section.* **(1)** A *derrick* is an apparatus consisting of a mast or equivalent member held at the head by guys or braces, with or without a boom, for use with a hoisting mechanism and operating ropes.

**1910.181(a)(2)** *A-frame derrick* means a derrick in which the boom is hinged from a cross member between the bottom ends of two upright members spread apart at the lower ends and joined at the top; the boom point secured to the junction of the side members, and the side members are braced or guyed from this junction point.

## OSHA Standards for General Industry

**1910.181(a)(3)** A *basket derrick* is a derrick without a boom, similar to a gin pole, with its base supported by ropes attached to corner posts or other parts of the structure. The base is at a lower elevation than its supports. The location of the base of a basket derrick can be changed by varying the length of the rope supports. The top of the pole is secured with multiple reeved guys to position the top of the pole to the desired location by varying the length of the upper guy lines. The load is raised and lowered by ropes through a sheave or block secured to the top of the pole.

**1910.181(a)(4)** *Breast derrick* means a derrick without boom. The mast consists of two side members spread farther apart at the base than at

A-FRAME

the top and tied together at top and bottom by rigid members. The mast is prevented from tipping forward by guys connected to its top. The load is raised and lowered by ropes through a sheave or block secured to the top crosspiece.

**1910.181(a)(5)** *Chicago boom derrick* means a boom which is attached to a structure, an outside upright member of the structure serving as the mast, and the boom being stepped in a fixed socket clamped to the upright. The derrick is complete with load, boom, and boom point swing line falls.

BASKET

**1910.181(a)(6)** A *gin pole derrick* is a derrick without a boom. Its guys are so arranged from its top as to permit leaning the mast in any direction. The load is raised and lowered by ropes reeved through sheaves or blocks at the top of the mast.

BREAST

**1910.181(a)(7)** *Guy derrick* means a fixed derrick consisting of a mast capable of being rotated, supported in a vertical position by guys, and a boom whose bottom end is hinged or pivoted to move in a vertical plane with a reeved rope between the head of the mast and the boom point for raising and lowering the boom, and a reeved rope from the boom point for raising and lowering the load.

CHICAGO BOOM

**1910.181(a)(8)** *Shearleg derrick* means a derrick without a boom and similar to a breast derrick. The mast, wide at the bottom and narrow at the top, is hinged at the bottom and has its top secured by a multiple

GIN POLE

reeved guy to permit handling loads at various radii by means of load tackle suspended from the mast top.

**1910.181(a)(9)** A *stiffleg derrick* is a derrick similar to a guy derrick except that the mast is supported or held in place by two or more stiff members, called stifflegs, which are capable of resisting either tensile or

GUY

compressive forces. Sills are generally provided to connect the lower ends of the stifflegs to the foot of the mast.

**1910.181(a)(10)** *Appointed* means assigned specific responsibilities by the employer or the employer's representative.

**1910.181(a)(11)** *ANSI* means the American National Standards Institute.

**1910.181(a)(12)** A *boom* is a timber or metal section or strut, pivoted or hinged at the heel (lower end) at a location fixed in height on a frame or mast or vertical member, and with its point (upper end) supported by chains, ropes, or rods to the upper end of the frame, mast, or vertical

STIFF LEG

member. A rope for raising and lowering the load is reeved through sheaves or a block at the boom point. The length of the boom shall be taken as the straight line distance between the axis of the foot pin and the axis of the boom point sheave pin, or where used, the axis of the upper load block attachment pin.

**1910.181(a)(13)** *Boom harness* means the block and sheave arrangement on the boom point to which the topping lift cable is reeved for lowering and raising the boom.

**1910.181(a)(14)** The *boom point* is the outward end of the top section of the boom.

**1910.181(a)(15)** *Derrick bullwheel* means a horizontal ring or wheel, fastened to the foot of a derrick, for the purpose of turning the derrick by means of ropes leading from this wheel to a powered drum.

**1910.181(a)(16)** *Designated* means selected or assigned by the employer or employer's representative as being qualified to perform specific duties.

**1910.181(a)(17)** *Eye* means a loop formed at the end of a rope by securing the dead end to the live end at the base of the loop.

**1910.181(a)(18)** A *fiddle block* is a block consisting of two sheaves in the same plane held in place by the same cheek plates.

**1910.181(a)(19)** The *foot bearing* or *foot block* (sill block) is the lower support on which the mast rotates.

**1910.181(a)(20)** A *gudgeon pin* is a pin connecting the mast cap to the mast allowing rotation of the mast.

**1910.181(a)(21)** A *guy* is a rope used to steady or secure the mast or other member in the desired position.

**1910.181(a)(22)** *Load, working* means the external load, in pounds, applied to the derrick, including the weight of load attaching equipment such as load blocks, shackles, and slings.

**1910.181(a)(23)** *Load block, lower* means the assembly of sheaves, pins, and frame suspended by the hoisting rope.

**1910.181(a)(24)** *Load block, upper* means the assembly of sheaves, pins, and frame suspended from the boom.

**1910.181(a)(25)** *Mast* means the upright member of the derrick.

**1910.181(a)(26)** *Mast cap (spider)* means the fitting at the top of the mast to which the guys are connected.

**1910.181(a)(27)** *Reeving* means a rope system in which the rope travels around drums and sheaves.

**1910.181(a)(28)** *Rope* refers to wire rope unless otherwise specified.

**1910.181(a)(29)** *Safety Hook* means a hook with a latch to prevent slings or load from accidentally slipping off the hook.

**1910.181(a)(30)** *Side loading* is a load applied at an angle to the vertical plane of the boom.

**1910.181(a)(31)** The *sill* is a member connecting the foot block and stiffleg or a member connecting the lower ends of a double member mast.

**1910.181(a)(32)** A *standby derrick* is a derrick not in regular service which is used occasionally or intermittently as required.

**1910.181(a)(33)** *Stiffleg* means a rigid member supporting the mast at the head.

**1910.181(a)(34)** *Swing* means rotation of the mast and/or boom for movements of loads in a horizontal direction about the axis of rotation.

**1910.181(b) General requirements—(1)** *Application*. This section applies to guy, stiffleg, basket, breast, gin pole, Chicago boom and A-frame derricks of the stationary type, capable of handling loads at variable reaches and powered by hoists through systems of rope reeving, used to perform lifting hook work, single or multiple line bucket work, grab, grapple, and magnet work. Derricks may be permanently installed for temporary use as in construction work. The requirements of this section also apply to any modification of these types which retain their fundamental features, except for floating derricks.

**1910.181(b)(2)** *New and existing equipment*. All new derricks constructed and installed on or after August 31, 1971, shall meet the design specifications of the American National Standard Safety Code for Derricks, ANSI B30.6-1969, which is incorporated by reference as specified in §1910.6.

**1910.181(b)(3)** *Designated personnel*. Only designated personnel shall be permitted to operate a derrick covered by this section.

**1910.181(c) Load ratings—(1)** *Rated load marking*. For permanently installed derricks with fixed lengths of boom, guy, and mast, a substantial, durable, and clearly legible rating chart shall be provided with each derrick and securely affixed where it is visible to personnel responsible for the safe operation of the equipment. The chart shall include the following data:

**1910.181(c)(1)(i)** Manufacturer's approved load ratings at corresponding ranges of boom angle or operating radii.

**1910.181(c)(1)(ii)** Specific lengths of components on which the load ratings are based.

**1910.181(c)(1)(iii)** Required parts for hoist reeving. Size and construction of rope may be shown either on the rating chart or in the operating manual.

**1910.181(c)(2)** *Nonpermanent installations*. For nonpermanent installations, the manufacturer shall provide sufficient information from which capacity charts can be prepared for the particular installation. The capacity charts shall be located at the derricks or the jobsite office.

**1910.181(d) Inspection—(1)** *Inspection classification*. **(i)** Prior to initial use all new and altered derricks shall be inspected to insure compliance with the provisions of this section.

**1910.181(d)(1)(ii)** Inspection procedure for derricks in regular service is divided into two general classifications based upon the intervals at which inspection should be performed. The intervals in turn are dependent upon the nature of the critical components of the derrick and the degree of their exposure to wear, deterioration, or malfunction. The two general classifications are herein designated as frequent and periodic with respective intervals between inspections as defined below:

**1910.181(d)(1)(ii)(a)** Frequent inspection—Daily to monthly intervals.

**1910.181(d)(1)(ii)(b)** Periodic inspection—1- to 12-month intervals, or as specified by the manufacturer.

**1910.181(d)(2)** *Frequent inspection*. Items such as the following shall be inspected for defects at intervals as defined in paragraph (d)(1)(ii)(a) of this section or as specifically indicated, including observation during operation for any defects which might appear between regular inspections. Deficiencies shall be carefully examined for any safety hazard:

**1910.181(d)(2)(i)** All control mechanisms: Inspect daily for adjustment, wear, and lubrication.

**1910.181(d)(2)(ii)** All chords and lacing: Inspect daily, visually.

**1910.181(d)(2)(iii)** Tension in guys: Daily.

**1910.181(d)(2)(iv)** Plumb of the mast.

# OSHA Standards for General Industry

**1910.181(d)(2)(v)** Deterioration or leakage in air or hydraulic systems: Daily.

**1910.181(d)(2)(vi)** Derrick hooks for deformations or cracks; for hooks with cracks or having more than 15 percent in excess of normal throat opening or more than 10° twist from the plane of the unbent hook, refer to paragraph (e)(3)(iii) of this section.

**1910.181(d)(2)(vii)** Rope reeving; visual inspection for noncompliance with derrick manufacturer's recommendations.

**1910.181(d)(2)(viii)** Hoist brakes, clutches, and operating levers: check daily for proper functioning before beginning operations.

**1910.181(d)(2)(ix)** Electrical apparatus for malfunctioning, signs of excessive deterioration, dirt, and moisture accumulation.

**1910.181(d)(3)** *Periodic inspection.* **(i)** Complete inspections of the derrick shall be performed at intervals as generally defined in paragraph (d)(1)(ii)(b) of this section depending upon its activity, severity of service, and environment, or as specifically indicated below. These inspections shall include the requirements of paragraph (d)(2) of this section and in addition, items such as the following. Deficiencies shall be carefully examined and a determination made as to whether they constitute a safety hazard:

**1910.181(d)(3)(i)(a)** Structural members for deformations, cracks, and corrosion.

**1910.181(d)(3)(i)(b)** Bolts or rivets for tightness.

**1910.181(d)(3)(i)(c)** Parts such as pins, bearings, shafts, gears, sheaves, drums, rollers, locking and clamping devices, for wear, cracks, and distortion.

**1910.181(d)(3)(i)(d)** Gudgeon pin for cracks, wear, and distortion each time the derrick is to be erected.

**1910.181(d)(3)(i)(e)** Powerplants for proper performance and compliance with applicable safety requirements.

**1910.181(d)(3)(i)(f)** Hooks.

**1910.181(d)(3)(ii)** Foundation or supports shall be inspected for continued ability to sustain the imposed loads.

**1910.181(d)(4)** *Derricks not in regular use.* **(i)** A derrick which has been idle for a period of 1 month or more, but less than 6 months, shall be given an inspection conforming with requirements of paragraph (d)(2) of this section and paragraph (g)(3) of this section before placing in service.

**1910.181(d)(4)(ii)** A derrick which has been idle for a period of over 6 months shall be given a complete inspection conforming with requirements of paragraphs (d) (2) and (3) of this section and paragraph (g)(3) of this section before placing in service.

**1910.181(d)(4)(iii)** Standby derricks shall be inspected at least semiannually in accordance with requirements of paragraph (d)(2) of this section and paragraph (g)(3) of this section.

**1910.181(e)** *Testing*—**(1)** *Operational tests.* Prior to initial use all new and altered derricks shall be tested to insure compliance with this section including the following functions:

**1910.181(e)(1)(i)** Load hoisting and lowering.

**1910.181(e)(1)(ii)** Boom up and down.

**1910.181(e)(1)(iii)** Swing.

**1910.181(e)(1)(iv)** Operation of clutches and brakes of hoist.

**1910.181(e)(2)** *Anchorages.* All anchorages shall be approved by the appointed person. Rock and hairpin anchorages may require special testing.

**1910.181(f)** *Maintenance*—**(1)** *Preventive maintenance.* A preventive maintenance program based on the derrick manufacturer's recommendations shall be established.

**1910.181(f)(2)** *Maintenance procedure.* **(i)** Before adjustments and repairs are started on a derrick the following precautions shall be taken:

**1910.181(f)(2)(i)(a)** The derrick to be repaired shall be arranged so it will cause the least interference with other equipment and operations in the area.

**1910.181(f)(2)(i)(b)** All hoist drum dogs shall be engaged.

**1910.181(f)(2)(i)(c)** The main or emergency switch shall be locked in the open position, if an electric hoist is used.

**1910.181(f)(2)(i)(d)** Warning or out of order signs shall be placed on the derrick and hoist.

**1910.181(f)(2)(i)(e)** The repairs of booms of derricks shall either be made when the booms are lowered and adequately supported or safely tied off.

**1910.181(f)(2)(i)(f)** A good communication system shall be set up between the hoist operator and the appointed individual in charge of derrick operations before any work on the equipment is started.

**1910.181(f)(2)(ii)** After adjustments and repairs have been made the derrick shall not be operated until all guards have been reinstalled, safety devices reactivated, and maintenance equipment removed.

**1910.181(f)(3)** *Adjustments and repairs.* **(i)** Any unsafe conditions disclosed by inspection shall be corrected before operation of the derrick is resumed.

**1910.181(f)(3)(ii)** Adjustments shall be maintained to assure correct functioning of components.

**1910.181(f)(3)(iii)** Repairs or replacements shall be provided promptly as needed for safe operation. The following are examples of conditions requiring prompt repair or replacement:

**1910.181(f)(3)(iii)(a)** Hooks showing defects described in paragraph (d)(2)(vi) of this section shall be discarded.

**1910.181(f)(3)(iii)(b)** All critical parts which are cracked, broken, bent, or excessively worn.

**1910.181(f)(3)(iii)(c)** [Reserved]

**1910.181(f)(3)(iii)(d)** All replacement and repaired parts shall have at least the original safety factor.

**1910.181(g)** *Rope inspection*—**(1)** *Running ropes.* A thorough inspection of all ropes in use shall be made at least once a month and a certification record which includes the date of inspection, the signature of the person who performed the inspection, and an identifier for the ropes which were inspected shall be prepared and kept on file where readily available. Any deterioration, resulting in appreciable loss of original strength shall be carefully observed and determination made as to whether further use of the rope would constitute a safety hazard. Some of the conditions that could result in an appreciable loss of strength are the following:

**1910.181(g)(1)(i)** Reduction of rope diameter below nominal diameter due to loss of core support, internal or external corrosion, or wear of outside wires.

**1910.181(g)(1)(ii)** A number of broken outside wires and the degree of distribution or concentration of such broken wires.

**1910.181(g)(1)(iii)** Worn outside wires.

**1910.181(g)(1)(iv)** Corroded or broken wires at end connections.

**1910.181(g)(1)(v)** Corroded, cracked, bent, worn, or improperly applied end connections.

**1910.181(g)(1)(vi)** Severe kinking, crushing, cutting, or unstranding.

**1910.181(g)(2)** *Limited travel ropes.* Heavy wear and/or broken wires may occur in sections in contact with equalizer sheaves or other sheaves where rope travel is limited, or with saddles. Particular care shall be taken to inspect ropes at these locations.

**1910.181(g)(3)** *Idle ropes.* All rope which has been idle for a period of a month or more due to shutdown or storage of a derrick on which it is installed shall be given a thorough inspection before it is used. This inspection shall be for all types of deterioration. A certification record shall be prepared and kept readily available which includes the date of inspection, the signature of the person who performed the inspection, and an identifier for the ropes which were inspected.

**1910.181(g)(4)** *Nonrotating ropes.* Particular care shall be taken in the inspection of nonrotating rope.

**1910.181(h)** *Operations of derricks.* Derrick operations shall be directed only by the individual specifically designated for that purpose.

**1910.181(i)** *Handling the load*—**(1)** *Size of load.* **(i)** No derrick shall be loaded beyond the rated load.

**1910.181(i)(1)(ii)** When loads approach the maximum rating of the derrick, it shall be ascertained that the weight of the load has been determined within plus or minus 10 percent before it is lifted.

**1910.181(i)(2)** *Attaching the load.* **(i)** The hoist rope shall not be wrapped around the load.

**1910.181(i)(2)(ii)** The load shall be attached to the hook by means of slings or other suitable devices.

**1910.181(i)(3)** *Moving the load.* **(i)** The load shall be well secured and properly balanced in the sling or lifting device before it is lifted more than a few inches.

**1910.181(i)(3)(ii)** Before starting to hoist, the following conditions shall be noted:

**1910.181(i)(3)(ii)(a)** Hoist rope shall not be kinked.

**1910.181(i)(3)(ii)(b)** Multiple part lines shall not be twisted around each other.

**1910.181(i)(3)(ii)(c)** The hook shall be brought over the load in such a manner as to prevent swinging.

**1910.181(i)(3)(iii)** During hoisting, care shall be taken that:

**1910.181(i)(3)(iii)(a)** There is no sudden acceleration or deceleration of the moving load.

**1910.181(i)(3)(iii)(b)** Load does not contact any obstructions.

**1910.181(i)(3)(iv)** A derrick shall not be used for side loading except when specifically authorized by a responsible person who has determined that the various structural components will not be overstressed.

**1910.181(i)(3)(v)** No hoisting, lowering, or swinging shall be done while anyone is on the load or hook.

**1910.181(i)(3)(vi)** The operator should avoid carrying loads over people.

**1910.181(i)(3)(vii)** The operator shall test the brakes each time a load approaching the rated load is handled by raising it a few inches and applying the brakes.

**1910.181(i)(3)(viii)** Neither the load nor boom shall be lowered below the point where less than two full wraps of rope remain on their respective drums.

**1910.181(i)(3)(ix)** When rotating a derrick, sudden starts and stops shall be avoided. Rotational speed shall be such that the load does not swing out beyond the radius at which it can be controlled.

**1910.181(i)(3)(x)** Boom and hoisting rope systems shall not be twisted.

**1910.181(i)(4)** *Holding the load.* **(i)** The operator shall not be allowed to leave his position at the controls while the load is suspended.

**1910.181(i)(4)(ii)** People should not be permitted to stand or pass under a load on the hook.

**1910.181(i)(4)(iii)** If the load must remain suspended for any considerable length of time, a dog, or pawl and ratchet, or other equivalent means, rather than the brake alone, shall be used to hold the load.

**1910.181(i)(5)** *Use of winch heads.* **(i)** Ropes shall not be handled on a winch head without the knowledge of the operator.

**1910.181(i)(5)(ii)** While a winch head is being used, the operator shall be within convenient reach of the power unit control lever.

**1910.181(i)(6)** *Securing boom.* Dogs, pawls, or other positive holding mechanism on the hoist shall be engaged. When not in use, the derrick boom shall:

**1910.181(i)(6)(i)** Be laid down;

**1910.181(i)(6)(ii)** Be secured to a stationary member, as nearly under the head as possible, by attachment of a sling to the load block; or

**1910.181(i)(6)(iii)** Be hoisted to a vertical position and secured to the mast.

**1910.181(j)** *Other requirements*—**(1)** *Guards.* **(i)** Exposed moving parts, such as gears, ropes, setscrews, projecting keys, chains, chain sprockets, and reciprocating components, which constitute a hazard under normal operating conditions shall be guarded.

**1910.181(j)(1)(ii)** Guards shall be securely fastened.

**1910.181(j)(1)(iii)** Each guard shall be capable of supporting without permanent distortion, the weight of a 200-pound person unless the guard is located where it is impossible for a person to step on it.

**1910.181(j)(2)** *Hooks.* **(i)** Hooks shall meet the manufacturer's recommendations and shall not be overloaded.

**1910.181(j)(2)(ii)** Safety latch type hooks shall be used wherever possible.

**1910.181(j)(3)** *Fire extinguishers.* **(i)** A carbon dioxide, dry chemical, or equivalent fire extinguisher shall be kept in the immediate vicinity of the derrick.

**1910.181(j)(3)(ii)** Operating and maintenance personnel shall be familiar with the use and care of the fire extinguishers provided.

**1910.181(j)(4)** *Refueling.* **(i)** Refueling with portable containers shall be done with approved safety type containers equipped with automatic closing cap and flame arrester. Refer to §1910.155(c)(3) for definition of Approved.

**1910.181(j)(4)(ii)** Machines shall not be refueled with the engine running.

**1910.181(j)(5)** *Operations near overhead lines.* For operations near overhead electric lines, see §1910.333(c)(3).

**1910.181(j)(6)** *Cab or operating enclosure.* **(i)** Necessary clothing and personal belongings shall be stored in such a manner as to not interfere with access or operation.

**1910.181(j)(6)(ii)** Tools, oilcans, waste, extra fuses, and other necessary articles shall be stored in the toolbox, and shall not be permitted to lie loose in or about the cab or operating enclosure.

[37 FR 22102, Oct. 18, 1972, as amended at 38 FR 14373, June 1, 1973; 43 FR 49750, Oct. 24, 1978; 49 FR 5323, Feb. 10, 1984; 51 FR 34561, Sept. 29, 1986; 53 FR 12122, Apr. 12, 1988; 55 FR 32015, Aug. 6, 1990; 61 FR 9240, Mar. 7, 1996]

### §1910.182 Effective dates [Removed].

[§1910.182 removed at 61 FR 9240, Mar. 7, 1996, effective May 6, 1996.]

### §1910.183 Helicopters.

**1910.183(a)** [Reserved]

**1910.183(b)** *Briefing.* Prior to each day's operation a briefing shall be conducted. This briefing shall set forth the plan of operation for the pilot and ground personnel.

**1910.183(c)** *Slings and tag lines.* Loads shall be properly slung. Tag lines shall be of a length that will not permit their being drawn up into the rotors. Pressed sleeve, swedged eyes, or equivalent means shall be used for all freely suspended loads to prevent hand splices from spinning open or cable clamps from loosening.

**1910.183(d)** *Cargo hooks.* All electrically operated cargo hooks shall have the electrical activating device so designed and installed as to prevent inadvertent operation. In addition, these cargo hooks shall be equipped with an emergency mechanical control for releasing the load. The employer shall ensure that the hooks are tested prior to each day's operation by a competent person to determine that the release functions properly, both electrically and mechanically.

**1910.183(e)** *Personal protective equipment.* **(1)** Personal protective equipment shall be provided and the employer shall ensure its use by employees receiving the load. Personal protective equipment shall consist of complete eye protection and hardhats secured by chinstraps.

**1910.183(e)(2)** Loose-fitting clothing likely to flap in rotor downwash, and thus be snagged on the hoist line, may not be worn.

**1910.183(f)** *Loose gear and objects.* The employer shall take all necessary precautions to protect employees from flying objects in the rotor downwash. All loose gear within 100 feet of the place of lifting the load or depositing the load, or within all other areas susceptible to rotor downwash, shall be secured or removed.

**1910.183(g)** *Housekeeping.* Good housekeeping shall be maintained in all helicopter loading and unloading areas.

**1910.183(h)** *Load safety.* The size and weight of loads, and the manner in which loads are connected to the helicopter shall be checked. A lift may not be made if the helicopter operator believes the lift cannot be made safely.

**1910.183(i)** *Hooking and unhooking loads.* When employees perform work under hovering craft, a safe means of access shall be provided for employees to reach the hoist line hook and engage or disengage cargo slings. Employees may not be permitted to perform work under hovering craft except when necessary to hook or unhook loads.

**1910.183(j)** *Static charge.* Static charge on the suspended load shall be dissipated with a grounding device before ground personnel touch the suspended load, unless protective rubber gloves are being worn by all ground personnel who may be required to touch the suspended load.

**1910.183(k)** *Weight limitation.* The weight of an external load shall not exceed the helicopter manufacturer's rating.

**1910.183(l)** *Ground lines.* Hoist wires or other gear, except for pulling lines

# OSHA Standards for General Industry

or conductors that are allowed to "pay out" from a container or roll off a reel, shall not be attached to any fixed ground structure, or allowed to foul on any fixed structure.

**1910.183(m) Visibility.** Ground personnel shall be instructed and the employer shall ensure that when visibility is reduced by dust or other conditions, they shall exercise special caution to keep clear of main and stabilizing rotors. Precautions shall also be taken by the employer to eliminate, as far as practical, the dust or other conditions reducing the visibility.

**1910.183(n) Signal systems.** The employer shall instruct the aircrew and ground personnel on the signal systems to be used and shall review the system with the employees in advance of hoisting the load. This applies to both radio and hand signal systems. Hand signals, where used, shall be as shown in Figure N-1.

**1910.183(o) Approach distance.** No employee shall be permitted to approach within 50 feet of the helicopter when the rotor blades are turning, unless his work duties require his presence in that area.

**1910.183(p) Approaching helicopter.** The employer shall instruct employees, and shall ensure, that whenever approaching or leaving a helicopter which has its blades rotating, all employees shall remain in full view of the pilot and keep in a crouched position. No employee shall be permitted to work in the area from the cockpit or cabin rearward while blades are rotating, unless authorized by the helicopter operator to work there.

**1910.183(q) Personnel.** Sufficient ground personnel shall be provided to ensure that helicopter loading and unloading operations can be performed safely.

**1910.183(r) Communications.** There shall be constant reliable communication between the pilot and a designated employee of the ground crew who acts as a signalman during the period of loading and unloading. The signalman shall be clearly distinguishable from other ground personnel.

**1910.183(s) Fires.** Open fires shall not be permitted in areas where they could be spread by the rotor downwash.

[40 FR 13440, Mar. 26, 1975, as amended at 63 FR 33467, June 18, 1998]

## §1910.184 Slings.

**1910.184(a) Scope.** This section applies to slings used in conjunction with other material handling equipment for the movement of material by hoisting, in employments covered by this part. The types of slings covered are those made from alloy steel chain, wire rope, metal mesh, natural or

### FIGURE N-1—HELICOPTER HAND SIGNAL

**MOVE RIGHT** — Left arm extended horizontally; right arm sweeps upward to position over head.

**HOLD-HOVER** — The signal "Hold" is executed by placing arms over head with clenched fists.

**MOVE LEFT** — Right arm extended horizontally; left arm sweeps upward to position over head.

**TAKEOFF** — Right hand behind back; left hand pointing up.

**MOVE FOWARD** — Combination of arm and hand movement in a collecting motion pulling toward body.

**LAND** — Arms crossed in front of body and pointing downward.

**MOVE REARWARD** — Hands above arm, palms out using a noticeable shoving motion.

**MOVE UPWARD** — Arms extended, palms up; arms sweeping up.

**RELEASE SLING LOAD** — Left arm held down away from body. Right arm cuts across left arm in a slashing movement from above.

**MOVE DOWNWARD** — Arms extended, palms down; arms sweeping down.

synthetic fiber rope (conventional three strand construction), and synthetic web (nylon, polyester, and polypropylene).

**1910.184(b) Definitions.** *Angle of loading* is the inclination of a leg or branch of a sling measured from the horizontal or vertical plane as shown in Fig. N-184-5; provided that an angle of loading of five degrees or less from the vertical may be considered a vertical angle of loading.

*Basket hitch* is a sling configuration whereby the sling is passed under the load and has both ends, end attachments, eyes or handles on the hook or a single master link.

*Braided wire rope* is a wire rope formed by plaiting component wire ropes.

*Bridle wire rope sling* is a sling composed of multiple wire rope legs with the top ends gathered in a fitting that goes over the lifting hook.

*Cable laid endless sling-mechanical joint* is a wire rope sling made endless by joining the ends of a single length of cable laid rope with one or more metallic fittings.

*Cable laid grommet-hand tucked* is an endless wire rope sling made from one length of rope wrapped six times around a core formed by hand tucking the ends of the rope inside the six wraps.

*Cable laid rope* is a wire rope composed of six wire ropes wrapped around a fiber or wire rope core.

*Cable laid rope sling-mechanical joint* is a wire rope sling made from a cable laid rope with eyes fabricated by pressing or swaging one or more metal sleeves over the rope junction.

*Choker hitch* is a sling configuration with one end of the sling passing under the load and through an end attachment, handle or eye on the other end of the sling.

*Coating* is an elastomer or other suitable material applied to a sling or to a sling component to impart desirable properties.

*Cross rod* is a wire used to join spirals of metal mesh to form a complete fabric. (See Fig. N-184-2.)

*Designated* means selected or assigned by the employer or the employer's representative as being qualified to perform specific duties.

*Equivalent entity* is a person or organization (including an employer) which, by possession of equipment, technical knowledge and skills, can perform with equal competence the same repairs and tests as the person or organization with which it is equated.

*Fabric (metal mesh)* is the flexible portion of a metal mesh sling consisting of a series of transverse coils and cross rods.

*Female handle (choker)* is a handle with a handle eye and a slot of such dimension as to permit passage of a male handle thereby allowing the use of a metal mesh sling in a choker hitch. (See Fig. N-184-1.)

*Handle* is a terminal fitting to which metal mesh fabric is attached. (See Fig. N-184-1.)

*Handle eye* is an opening in a handle of a metal mesh sling shaped to accept a hook, shackle or other lifting device. (See Fig. N-184-1.)

*Hitch* is a sling configuration whereby the sling is fastened to an object or load, either directly to it or around it.

*Link* is a single ring of a chain.

*Male handle (triangle)* is a handle with a handle eye.

*Master coupling link* is an alloy steel welded coupling link used as an intermediate link to join alloy steel chain to master links. (See Fig. N-184-3.)

*Master link* or *gathering ring* is a forged or welded steel link used to support all members (legs) of an alloy steel chain sling or wire rope sling. (See Fig. N-184-3.)

*Mechanical coupling link* is a nonwelded, mechanically closed steel link used to attach master links, hooks, etc., to alloy steel chain.

*Proof load* is the load applied in performance of a proof test.

*Proof test* is a nondestructive tension test performed by the sling manufacturer or an equivalent entity to verify construction and workmanship of a sling.

**FIG. N-184-1—METAL MESH SLING (TYPICAL)**

**FIG. N-184-2—METAL MESH CONSTRUCTION**

**FIG. N-184-3—MAJOR COMPONENTS OF A QUADRUPLE SLING**

# OSHA Standards for General Industry

*Rated capacity* or *working load limit* is the maximum working load permitted by the provisions of this section.

*Reach* is the effective length of an alloy steel chain sling measured from the top bearing surface of the upper terminal component to the bottom bearing surface of the lower terminal component.

*Selvage edge* is the finished edge of synthetic webbing designed to prevent unraveling.

*Sling* is an assembly which connects the load to the material handling equipment.

*Sling manufacturer* is a person or organization that assembles sling components into their final form for sale to users.

*Spiral* is a single transverse coil that is the basic element from which metal mesh is fabricated. (See Fig. N-184-2.)

*Strand laid endless sling-mechanical joint* is a wire rope sling made endless from one length of rope with the ends joined by one or more metallic fittings.

*Strand laid grommet-hand tucked* is an endless wire rope sling made from one length of strand wrapped six times around a core formed by hand tucking the ends of the strand inside the six wraps.

*Strand laid rope* is a wire rope made with strands (usually six or eight) wrapped around a fiber core, wire strand core, or independent wire rope core (IWRC).

*Vertical hitch* is a method of supporting a load by a single, vertical part or leg of the sling. (See Fig. N-184-4.)

**1910.184(c) Safe operating practices.** Whenever any sling is used, the following practices shall be observed:

**1910.184(c)(1)** Slings that are damaged or defective shall not be used.

**1910.184(c)(2)** Slings shall not be shortened with knots or bolts or other makeshift devices.

**1910.184(c)(3)** Sling legs shall not be kinked.

**1910.184(c)(4)** Slings shall not be loaded in excess of their rated capacities.

**1910.184(c)(5)** Slings used in a basket hitch shall have the loads balanced to prevent slippage.

**1910.184(c)(6)** Slings shall be securely attached to their loads.

**1910.184(c)(7)** Slings shall be padded or protected from the sharp edges of their loads.

**1910.184(c)(8)** Suspended loads shall be kept clear of all obstructions.

**1910.184(c)(9)** All employees shall be kept clear of loads about to be lifted and of suspended loads.

**1910.184(c)(10)** Hands or fingers shall not be placed between the sling and its load while the sling is being tightened around the load.

**1910.184(c)(11)** Shock loading is prohibited.

**1910.184(c)(12)** A sling shall not be pulled from under a load when the load is resting on the sling.

**1910.184(d) Inspections.** Each day before being used, the sling and all fastenings and attachments shall be inspected for damage or defects by a competent person designated by the employer. Additional inspections shall be performed during sling use, where service conditions warrant. Damaged or defective slings shall be immediately removed from service.

**1910.184(e) Alloy steel chain slings. (1)** *Sling identification.* Alloy steel chain slings shall have permanently affixed durable identification stating size, grade, rated capacity, and reach.

**1910.184(e)(2)** *Attachments.* **(i)** Hooks, rings, oblong links, pear shaped links, welded or mechanical coupling links or other attachments shall have a rated capacity at least equal to that of the alloy steel chain with which they are used or the sling shall not be used in excess of the rated capacity of the weakest component.

**1910.184(e)(2)(ii)** Makeshift links or fasteners formed from bolts or rods, or other such attachments, shall not be used.

**1910.184(e)(3)** *Inspections.* **(i)** In addition to the inspection required by paragraph (d) of this section, a thorough periodic inspection of alloy steel chain slings in use shall be made on a regular basis, to be determined on the basis of (A) frequency of sling use; (B) severity of service conditions; (C) nature of lifts being made; and (D) experience gained on the service life of slings used in similar circumstances. Such inspections shall in no event be at intervals greater than once every 12 months.

**1910.184(e)(3)(ii)** The employer shall make and maintain a record of the most recent month in which each alloy steel chain sling was thoroughly inspected, and shall make such record available for examination.

**1910.184(e)(3)(iii)** The thorough inspection of alloy steel chain slings shall be performed by a competent person designated by the employer, and shall include a thorough inspection for wear, defective welds, deformation and increase in length. Where such defects or deterioration are present, the sling shall be immediately removed from service.

**1910.184(e)(4)** *Proof testing.* The employer shall ensure that before use, each new, repaired, or reconditioned alloy steel chain sling, including all welded components in the sling assembly, shall be proof tested by the sling manufacturer or equivalent entity, in accordance with paragraph 5.2 of the American Society of Testing and Materials Specification A391-65, which is incorporated by reference as specified in §1910.6 (ANSI G61.1-1968). The employer shall retain a certificate of the proof test and shall make it available for examination.

**1910.184(e)(5)** *Sling use.* Alloy steel chain slings shall not be used with loads in excess of the rated capacities prescribed in Table N-184-1. Slings not included in this table shall be used only in accordance with the manufacturer's recommendations.

**1910.184(e)(6)** *Safe operating temperatures.* Alloy steel chain slings shall be permanently removed from service if they are heated above 1000 °F. When exposed to service temperatures in excess of 600 °F, maximum working load limits permitted in Table N-184-1 shall be reduced in accordance with the chain or sling manufacturer's recommendations.

**1910.184(e)(7)** *Repairing and reconditioning alloy steel chain slings.* **(i)** Worn or damaged alloy steel chain slings or attachments shall not be used until repaired. When welding or heat testing is performed, slings shall not be used unless repaired, reconditioned and proof tested by the sling manufacturer or an equivalent entity.

**1910.184(e)(7)(ii)** Mechanical coupling links or low carbon steel repair links shall not be used to repair broken lengths of chain.

**1910.184(e)(8)** *Effects of wear.* If the chain size at any point of any link is less than that stated in Table N-184-2, the sling shall be removed from service.

**1910.184(e)(9)** *Deformed attachments.* **(i)** Alloy steel chain slings with cracked or deformed master links, coupling links or other components shall be removed from service.

## TABLE N-184-1—RATED CAPACITY (WORKING LOAD LIMIT), FOR ALLOY STEEL CHAIN SLINGS

Rated Capacity (Working Load Limit), Pounds
[Horizontal angles shown in parentheses]

| Chain size, inches | Single branch sling—90° loading | Double sling vertical angle (1) ||| Triple and quadruple sling (3) vertical angle (1) |||
|---|---|---|---|---|---|---|---|
| | | 30° (60°) | 45° (45°) | 60° (30°) | 30° (60°) | 45° (45°) | 60° (30°) |
| 1/4 | 3,250 | 5,650 | 4,550 | 3,250 | 8,400 | 6,800 | 4,900 |
| 3/8 | 6,600 | 11,400 | 9,300 | 6,600 | 17,000 | 14,000 | 9,900 |
| 1/2 | 11,250 | 19,500 | 15,900 | 11,250 | 29,000 | 24,000 | 17,000 |
| 5/8 | 16,500 | 28,500 | 23,300 | 16,500 | 43,000 | 35,000 | 24,500 |
| 3/4 | 23,000 | 39,800 | 32,500 | 23,000 | 59,500 | 48,500 | 34,500 |
| 7/8 | 28,750 | 49,800 | 40,600 | 28,750 | 74,500 | 61,000 | 43,000 |
| 1 | 38,750 | 67,100 | 5,800 | 38,750 | 101,000 | 82,000 | 58,000 |
| 1 1/8 | 44,500 | 77,000 | 63,000 | 44,500 | 115,500 | 94,500 | 66,500 |
| 1 1/4 | 57,500 | 99,500 | 61,000 | 57,500 | 149,000 | 121,500 | 86,000 |
| 1 3/8 | 67,000 | 116,000 | 94,000 | 67,000 | 174,000 | 141,000 | 100,500 |
| 1 1/2 | 80,000 | 138,000 | 112,900 | 80,000 | 207,000 | 169,000 | 119,500 |
| 1 3/4 | 100,000 | 172,000 | 140,000 | 100,000 | 258,000 | 210,000 | 150,000 |

(1) Rating of multileg slings adjusted for angle of loading measured as the included angle between the inclined leg and the vertical as shown in Figure N-184-5.
(2) Rating of multileg slings adjusted for angle of loading between the inclined leg and the horizontal plane of the load, as shown in Figure N-184-5.
(3) Quadruple sling rating is same as triple sling because normal lifting practice may not distribute load uniformly to all 4 legs.

### TABLE N-184-2—MINIMUM ALLOWABLE CHAIN SIZE AT ANY POINT OF LINK

| Chain size, inches | Minimum allowable chain size, inches |
|---|---|
| 1/4 | 13/64 |
| 3/8 | 19/64 |
| 1/2 | 25/64 |
| 5/8 | 31/64 |
| 3/4 | 19/32 |
| 7/8 | 45/64 |
| 1 | 13/16 |
| 1 1/8 | 29/32 |
| 1 1/4 | 1 |
| 1 3/8 | 1 3/32 |
| 1 1/2 | 1 3/16 |
| 1 3/4 | 1 13/32 |

**1910.184(e)(9)(ii)** Slings shall be removed from service if hooks are cracked, have been opened more than 15 percent of the normal throat opening measured at the narrowest point or twisted more than 10 degrees from the plane of the unbent hook.

**1910.184(f)** *Wire rope slings.* **(1)** *Sling use.* Wire rope slings shall not be used with loads in excess of the rated capacities shown in Tables N-184-3 through N-184-14. Slings not included in these tables shall be used only in accordance with the manufacturer's recommendations.

**1910.184(f)(2)** *Minimum sling lengths.* **(i)** Cable laid and 6×19 and 6×37 slings shall have a minimum clear length of wire rope 10 times the component rope diameter between splices, sleeves or end fittings.

**1910.184(f)(2)(ii)** Braided slings shall have a minimum clear length of wire rope 40 times the component rope diameter between the loops or end fittings.

**1910.184(f)(2)(iii)** Cable laid grommets, strand laid grommets and endless slings shall have a minimum circumferential length of 96 times their body diameter.

**1910.184(f)(3)** *Safe operating temperatures.* Fiber core wire rope slings of all grades shall be permanently removed from service if they are exposed to temperatures in excess of 200 °F. When nonfiber core wire rope slings of any grade are used at temperatures above 400 °F. or below minus 60 °F, recommendations of the sling manufacturer regarding use at that temperature shall be followed.

**1910.184(f)(4)** *End attachments.* **(i)** Welding of end attachments, except covers to thimbles, shall be performed prior to the assembly of the sling.

**1910.184(f)(4)(ii)** All welded end attachments shall not be used unless proof tested by the manufacturer or equivalent entity at twice their rated capacity prior to initial use. The employer shall retain a certificate of the proof test, and make it available for examination.

### TABLE N-184-3—RATED CAPACITIES FOR SINGLE LEG SLINGS
6 × 19 and 6 × 37 Classification Improved Plow Steel Grade Rope With Fiber Core (FC)

| Rope | | Rated capacities, tons (2,000 lb) | | | | | | | | |
|---|---|---|---|---|---|---|---|---|---|---|
| | | Vertical | | | Choker | | | Vertical basket[1] | | |
| Dia (inches) | Constr | HT | MS | S | HT | MS | S | HT | MS | S |
| 1/4 | 6 × 19 | 0.49 | 0.51 | 0.55 | 0.37 | 0.38 | 0.41 | 0.99 | 1.0 | 1.1 |
| 5/16 | 6 × 19 | 0.76 | 0.79 | 0.85 | 0.57 | 0.59 | 0.64 | 1.5 | 1.6 | 1.7 |
| 3/8 | 6 × 19 | 1.1 | 1.1 | 1.2 | 0.80 | 0.85 | 0.91 | 2.1 | 2.2 | 2.4 |
| 7/16 | 6 × 19 | 1.4 | 1.5 | 1.6 | 1.1 | 1.1 | 1.2 | 2.9 | 3.0 | 3.3 |
| 1/2 | 6 × 19 | 1.8 | 2.0 | 2.1 | 1.4 | 1.5 | 12.6 | 3.7 | 3.9 | 4.3 |
| 9/16 | 6 × 19 | 2.3 | 2.5 | 2.7 | 1.7 | 1.9 | 2.0 | 4.6 | 5.0 | 5.4 |
| 5/8 | 6 × 19 | 2.8 | 3.1 | 3.3 | 2.1 | 2.3 | 2.5 | 5.6 | 6.2 | 6.7 |
| 3/4 | 6 × 19 | 3.9 | 4.4 | 4.8 | 2.9 | 3.3 | 3.6 | 7.8 | 8.8 | 9.5 |
| 7/8 | 6 × 19 | 5.1 | 5.9 | 6.4 | 3.9 | 4.5 | 4.8 | 10.0 | 12.0 | 13.0 |
| 1 | 6 × 19 | 6.7 | 7.7 | 8.4 | 5.0 | 5.8 | 6.3 | 13.0 | 15.0 | 17.0 |
| 1 1/8 | 6 × 19 | 8.4 | 9.5 | 10.0 | 6.3 | 7.1 | 7.9 | 17.0 | 19.0 | 21.0 |
| 1 1/4 | 6 × 19 | 9.8 | 11.0 | 12.0 | 7.4 | 8.3 | 9.2 | 20.0 | 22.0 | 25.0 |
| 1 3/8 | 6 × 37 | 12.0 | 13.0 | 15.0 | 8.9 | 10.0 | 11.0 | 24.0 | 27.0 | 30.0 |
| 1 1/2 | 6 × 37 | 14.0 | 16.0 | 15.0 | 10.0 | 12.0 | 13.0 | 28.0 | 32.0 | 35.0 |
| 1 5/8 | 6 × 37 | 16.0 | 18.0 | 21.0 | 12.0 | 14.0 | 15.0 | 33.0 | 27.0 | 41.0 |
| 1 3/4 | 6 × 37 | 19.0 | 21.0 | 24.0 | 14.0 | 16.0 | 18.0 | 38.0 | 43.0 | 48.0 |
| 2 | 6 × 37 | 25.0 | 28.0 | 31.0 | 18.0 | 21.0 | 23.0 | 49.0 | 55.0 | 62.0 |

HT = Hand Tucked Splice and Hidden Tuck Splice. For hidden tuck splice (IWRC) use values in HT columns.

MS = Mechanical Splice.

S = Swaged or Zinc Poured Socket.

[1] These values only apply when the D/d ratio for HT slings is 10 or greater, and for MS and S slings is 20 or greater where: D = Diameter of curvature around which the body of the sling is bent; d = Diameter of rope.

### TABLE N-184-4—RATED CAPACITIES FOR SINGLE LEG SLINGS
6 × 19 and 6 × 37 Classification Improved Plow Steel Grade Rope With Independent Wire Rope Core (IWRC)

| Rope | | Rated capacities, tons (2,000 lb) | | | | | | | | |
|---|---|---|---|---|---|---|---|---|---|---|
| | | Vertical | | | Choker | | | Vertical basket[1] | | |
| Dia (in.) | Constr | HT | MS | S | HT | MS | S | HT | MS | S |
| 1/4 | 6 × 19 | 0.53 | 0.56 | 0.59 | 0.40 | 0.42 | 0.44 | 1.0 | 1.1 | 1.2 |
| 5/16 | 6 × 19 | 0.81 | 0.87 | 0.92 | 0.61 | 0.65 | 0.69 | 1.6 | 1.7 | 1.8 |
| 3/8 | 6 × 19 | 1.1 | 1.2 | 1.3 | 0.86 | 0.93 | 0.98 | 2.3 | 2.5 | 2.6 |
| 7/16 | 6 × 19 | 1.5 | 1.7 | 1.8 | 1.2 | 1.3 | 1.3 | 3.1 | 3.4 | 3.5 |
| 1/2 | 6 × 19 | 2.0 | 2.2 | 2.3 | 1.5 | 1.6 | 1.7 | 3.9 | 4.4 | 4.6 |
| 9/16 | 6 × 19 | 2.5 | 2.7 | 2.9 | 1.8 | 2.1 | 2.2 | 4.9 | 5.5 | 5.8 |
| 5/8 | 6 × 19 | 3.0 | 3.4 | 3.6 | 2.2 | 2.5 | 2.7 | 6.0 | 6.8 | 7.2 |
| 3/4 | 6 × 19 | 4.2 | 4.9 | 5.1 | 3.1 | 3.6 | 3.8 | 8.4 | 9.7 | 10.0 |
| 7/8 | 6 × 19 | 5.5 | 6.6 | 6.9 | 4.1 | 4.9 | 5.2 | 11.0 | 13.0 | 14.0 |
| 1 | 6 × 19 | 7.2 | 8.5 | 9.0 | 5.4 | 6.4 | 6.7 | 14.0 | 17.0 | 18.0 |
| 1 1/8 | 6 × 19 | 9.0 | 10.0 | 11.0 | 6.8 | 7.8 | 8.5 | 18.0 | 21.0 | 23.0 |
| 1 1/4 | 6 × 37 | 10.0 | 12.0 | 13.0 | 7.9 | 9.2 | 9.9 | 21.0 | 24.0 | 26.0 |
| 1 3/8 | 6 × 37 | 13.0 | 15.0 | 16.0 | 9.6 | 11.0 | 12.0 | 25.0 | 29.0 | 32.0 |
| 1 1/2 | 6 × 37 | 15.0 | 17.0 | 19.0 | 11.0 | 13.0 | 14.0 | 30.0 | 35.0 | 38.0 |
| 1 5/8 | 6 × 37 | 18.0 | 20.0 | 22.0 | 13.0 | 15.0 | 17.0 | 35.0 | 41.0 | 44.0 |
| 1 3/4 | 6 × 37 | 20.0 | 24.0 | 26.0 | 15.0 | 18.0 | 19.0 | 41.0 | 47.0 | 51.0 |
| 2 | 6 × 37 | 26.0 | 30.0 | 33.0 | 20.0 | 23.0 | 25.0 | 53.0 | 61.0 | 66.0 |

HT = Hand Tucked Splice. For hidden tuck splice (IWRC) use Table I values in HT column.

MS = Mechanical Splice.

S = Swaged or Zinc Poured Socket.

[1] These values only apply when the D/d ratio for HT slings is 10 or greater, and for MS and S Slings is 20 or greater where: D = Diameter of curvature around which the body of the sling is bent; d = Diameter of rope.

## TABLE N-184-5—RATED CAPACITIES FOR SINGLE LEG SLINGS
Cable Laid Rope—Mechanical Splice Only
7 × 7 × 7 & 7 × 19 Constructions Galvanized Aircraft Grade Rope
7 × 6 × 19 IWRC Construction Improved Plow Steel Grade Rope

| Rope Diameter (inches) | Constr | Vertical | Choker | Vertical basket[1] |
|---|---|---|---|---|
| 1/4 | 7 × 7 × 7 | 0.50 | 0.38 | 1.0 |
| 3/8 | 7 × 7 × 7 | 1.1 | 0.81 | 2.0 |
| 1/2 | 7 × 7 × 7 | 1.8 | 1.4 | 3.7 |
| 5/8 | 7 × 7 × 7 | 2.8 | 2.1 | 5.5 |
| 3/4 | 7 × 7 × 7 | 3.8 | 2.9 | 7.6 |
| 5/8 | 7 × 7 × 19 | 2.9 | 2.2 | 5.8 |
| 3/4 | 7 × 7 × 19 | 4.1 | 3.0 | 8.1 |
| 7/8 | 7 × 7 × 19 | 5.4 | 4.0 | 11.0 |
| 1 | 7 × 7 × 19 | 6.9 | 5.1 | 14.0 |
| 1 1/8 | 7 × 7 × 19 | 8.2 | 6.2 | 16.0 |
| 1 1/4 | 7 × 7 × 19 | 9.9 | 7.4 | 20.0 |
| 3/4 | 7 × 6 × 19 IWRC | 3.8 | 2.8 | 7.6 |
| 7/8 | 7 × 6 × 19 IWRC | 5.0 | 3.8 | 10.0 |
| 1 | 7 × 6 × 19 IWRC | 6.4 | 4.8 | 13.0 |
| 1 1/8 | 7 × 6 × 19 IWRC | 7.7 | 5.8 | 15.0 |
| 1 1/4 | 7 × 6 × 19 IWRC | 9.2 | 6.9 | 18.0 |
| 1 5/16 | 7 × 6 × 19 IWRC | 10.0 | 7.5 | 20.0 |
| 1 3/8 | 7 × 6 × 19 IWRC | 11.0 | 8.2 | 22.0 |
| 1 1/2 | 7 × 6 × 19 IWRC | 13.0 | 9.6 | 26.0 |

[1] These values only apply when the D/d ratio is 10 or greater where: D = Diameter of curvature around which the body of the sling is bent; d = Diameter of rope.

## TABLE N-184-6—RATED CAPACITIES FOR SINGLE LEG SLINGS
8-Part and 6-Part Braided Rope
6 × 7 and 6 × 19 Construction Improved Plow Steel Grade Rope
7 × 7 Construction Galvanized Aircraft Grade Rope

| Component ropes Diameter (inches) | Constr | Vertical 8-Part | Vertical 6-Part | Choker 8-Part | Choker 6-Part | Basket, vertical to 30°[1] 8-Part | Basket, vertical to 30°[1] 6-Part |
|---|---|---|---|---|---|---|---|
| 3/32 | 6 × 7 | 0.42 | 0.32 | 0.32 | 0.24 | 0.74 | 0.55 |
| 1/8 | 6 × 7 | 0.75 | 0.57 | 0.57 | 0.42 | 1.3 | 0.98 |
| 3/16 | 6 × 7 | 1.7 | 1.3 | 1.3 | 0.94 | 2.9 | 2.2 |
| 3/32 | 7 × 7 | 0.51 | 0.39 | 0.38 | 0.29 | 0.89 | 0.67 |
| 1/8 | 7 × 7 | 0.95 | 0.7 | 0.71 | 0.53 | 1.6 | 1.2 |
| 3/16 | 7 × 7 | 2.1 | 1.5 | 1.5 | 1.2 | 3.6 | 2.7 |
| 3/16 | 6 × 19 | 1.7 | 1.3 | 1.3 | 0.98 | 3.0 | 2.2 |
| 1/4 | 6 × 19 | 3.1 | 2.3 | 2.3 | 1.7 | 5.3 | 4.0 |
| 5/16 | 6 × 19 | 4.8 | 3.6 | 3.6 | 2.7 | 8.3 | 6.2 |
| 3/8 | 6 × 19 | 6.8 | 5.1 | 5.1 | 3.8 | 12.0 | 8.9 |
| 7/16 | 6 × 19 | 9.3 | 6.9 | 6.9 | 5.2 | 16.0 | 12.0 |
| 1/2 | 6 × 19 | 12.0 | 9.0 | 9.0 | 6.7 | 21.0 | 15.0 |
| 9/16 | 6 × 19 | 15.0 | 11.0 | 11.0 | 8.5 | 26.0 | 20.0 |
| 5/8 | 6 × 19 | 19.0 | 14.0 | 14.0 | 10.0 | 32.0 | 24.0 |
| 3/4 | 6 × 19 | 27.0 | 20.0 | 20.0 | 15.0 | 46.0 | 35.0 |
| 7/8 | 6 × 19 | 36.0 | 27.0 | 27.0 | 20.0 | 62.0 | 47.0 |
| 1 | 6 × 19 | 47.0 | 35.0 | 35.0 | 26.0 | 81.0 | 61.0 |

[1] These values only apply when the D/d ratio is 20 or greater where: D = Diameter of curvature around which the body of the sling is bent; d = Diameter of component rope.

## TABLE N-184-7—RATED CAPACITIES FOR 2-LEG AND 3-LEG BRIDLE SLINGS
6 × 19 and 6 × 37 Classification Improved Plow Steel Grade Rope With Fiber Core (FC)
[Horizontal angles shown in parentheses]

| Rope Diameter (inches) | Constr | 2-Leg 30°(60°) HT | 2-Leg 30°(60°) MS | 2-Leg 45° HT | 2-Leg 45° MS | 2-Leg 60°(30°) HT | 2-Leg 60°(30°) MS | 3-Leg 30°(60°) HT | 3-Leg 30°(60°) MS | 3-Leg 45° HT | 3-Leg 45° MS | 3-Leg 60°(30°) HT | 3-Leg 60°(30°) MS |
|---|---|---|---|---|---|---|---|---|---|---|---|---|---|
| 1/4 | 6 × 19 | 0.85 | 0.83 | 0.70 | 0.72 | 0.49 | 0.51 | 1.3 | 1.3 | 1.0 | 1.1 | 0.74 | 0.76 |
| 5/16 | 6 × 19 | 1.3 | 1.4 | 1.1 | 1.1 | 0.76 | 0.79 | 2.0 | 2.0 | 1.6 | 1.7 | 1.1 | 1.2 |
| 3/8 | 6 × 19 | 1.8 | 1.9 | 1.5 | 1.6 | 1.1 | 1.1 | 2.8 | 2.9 | 2.3 | 2.4 | 1.6 | 1.7 |
| 7/16 | 6 × 19 | 2.5 | 2.6 | 2.0 | 2.2 | 1.4 | 1.5 | 3.7 | 4.0 | 3.0 | 3.2 | 2.1 | 2.3 |
| 1/2 | 6 × 19 | 3.2 | 3.4 | 2.6 | 2.8 | 1.8 | 2.0 | 4.8 | 5.1 | 3.9 | 4.2 | 2.8 | 3.0 |
| 9/16 | 6 × 19 | 4.0 | 4.3 | 3.2 | 3.5 | 2.3 | 2.5 | 6.0 | 6.5 | 4.9 | 5.3 | 3.4 | 3.7 |
| 5/8 | 6 × 19 | 4.8 | 5.3 | 4.0 | 4.4 | 2.8 | 3.1 | 7.3 | 8.0 | 5.9 | 6.5 | 4.2 | 4.6 |
| 3/4 | 6 × 19 | 6.8 | 7.6 | 5.5 | 6.2 | 3.9 | 4.4 | 10.0 | 11.0 | 8.3 | 9.3 | 5.8 | 6.6 |
| 7/8 | 6 × 19 | 8.9 | 10.0 | 7.3 | 8.4 | 5.1 | 5.9 | 13.0 | 15.0 | 11.0 | 13.0 | 7.7 | 8.9 |
| 1 | 6 × 19 | 11.0 | 13.0 | 9.4 | 11.0 | 6.7 | 7.7 | 17.0 | 20.0 | 14.0 | 16.0 | 10.0 | 11.0 |
| 1 1/8 | 6 × 19 | 14.0 | 16.0 | 12.0 | 13.0 | 8.4 | 9.3 | 22.0 | 24.0 | 18.0 | 20.0 | 13.0 | 14.0 |
| 1 1/4 | 6 × 37 | 17.0 | 19.0 | 14.0 | 16.0 | 9.8 | 11.0 | 25.0 | 29.0 | 21.0 | 23.0 | 15.0 | 17.0 |
| 1 3/8 | 6 × 37 | 20.0 | 23.0 | 17.0 | 19.0 | 12.0 | 13.0 | 31.0 | 35.0 | 25.0 | 28.0 | 18.0 | 20.0 |
| 1 1/2 | 6 × 37 | 24.0 | 27.0 | 20.0 | 22.0 | 14.0 | 16.0 | 36.0 | 41.0 | 30.0 | 33.0 | 21.0 | 24.0 |
| 1 5/8 | 6 × 37 | 28.0 | 32.0 | 23.0 | 26.0 | 16.0 | 18.0 | 43.0 | 48.0 | 35.0 | 39.0 | 25.0 | 28.0 |
| 1 3/4 | 6 × 37 | 33.0 | 37.0 | 27.0 | 30.0 | 19.0 | 21.0 | 49.0 | 56.0 | 40.0 | 45.0 | 28.0 | 32.0 |
| 2 | 6 × 37 | 43.0 | 48.0 | 35.0 | 39.0 | 25.0 | 28.0 | 64.0 | 72.0 | 52.0 | 59.0 | 37.0 | 41.0 |

HT = Hand Tucked Splice.
MS = Mechanical Splice.

### TABLE N-184-8—RATED CAPACITIES FOR 2-LEG AND 3-LEG BRIDLE SLINGS

6 × 19 and 6 × 37 Classification Improved Plow Steel Grade Rope With Independent Wire Rope Core (IWRC)

[Horizontal angles shown in parentheses]

| Rope | | Rated capacities, tons (2,000 lb) | | | | | | | | | | | |
|---|---|---|---|---|---|---|---|---|---|---|---|---|---|
| | | 2-Leg bridle slings | | | | | | 3-Leg bridle slings | | | | | |
| | | 30° (60°) | | 45° angle | | 60° (30°) | | 30° (60°) | | 45° angle | | 60° (30°) | |
| Dia (inches) | Constr | HT | MS | HT | MS | HT | MS | HT | MS | HT | MS | HT | MS |
| 1/4 | 6 × 19 | 0.92 | 0.97 | 0.75 | 0.79 | 0.53 | 0.56 | 1.4 | 1.4 | 1.1 | 1.2 | 0.79 | 0.84 |
| 5/16 | 6 × 19 | 1.4 | 1.5 | 1.1 | 1.2 | 0.81 | 0.87 | 2.1 | 2.3 | 1.7 | 1.8 | 1.2 | 1.3 |
| 3/8 | 6 × 19 | 2.0 | 2.1 | 1.6 | 1.8 | 1.1 | 1.2 | 3.0 | 3.2 | 2.4 | 2.6 | 1.7 | 1.9 |
| 7/16 | 6 × 19 | 2.7 | 2.9 | 2.2 | 2.4 | 1.5 | 1.7 | 4.0 | 4.4 | 3.3 | 3.6 | 2.3 | 2.5 |
| 1/2 | 6 × 19 | 3.4 | 3.8 | 2.8 | 3.1 | 2.0 | 2.2 | 5.1 | 5.7 | 4.2 | 4.6 | 3.0 | 3.3 |
| 9/16 | 6 × 19 | 4.3 | 4.8 | 3.5 | 3.9 | 2.5 | 2.7 | 6.4 | 7.1 | 5.2 | 5.8 | 3.7 | 4.1 |
| 5/8 | 6 × 19 | 5.2 | 5.9 | 4.2 | 4.8 | 3.0 | 3.4 | 7.8 | 8.8 | 6.4 | 7.2 | 4.5 | 5.1 |
| 3/4 | 6 × 19 | 7.3 | 8.4 | 5.9 | 6.9 | 4.2 | 4.9 | 11.0 | 13.0 | 8.9 | 10.0 | 6.3 | 7.3 |
| 7/8 | 6 × 19 | 9.6 | 11.0 | 7.8 | 9.3 | 5.5 | 6.6 | 14.0 | 17.0 | 12.0 | 14.0 | 8.3 | 9.9 |
| 1 | 6 × 19 | 12.0 | 15.0 | 10.0 | 12.0 | 7.2 | 8.5 | 19.0 | 22.0 | 15.0 | 18.0 | 11.0 | 13.0 |
| 1 1/8 | 6 × 19 | 16.0 | 18.0 | 13.0 | 15.0 | 9.0 | 10.0 | 23.0 | 27.0 | 19.0 | 22.0 | 13.0 | 16.0 |
| 1 1/4 | 6 × 37 | 18.0 | 21.0 | 15.0 | 17.0 | 10.0 | 12.0 | 27.0 | 32.0 | 22.0 | 26.0 | 16.0 | 18.0 |
| 1 3/8 | 6 × 37 | 22.0 | 25.0 | 18.0 | 21.0 | 13.0 | 15.0 | 33.0 | 38.0 | 27.0 | 31.0 | 19.0 | 22.0 |
| 1 1/2 | 6 × 37 | 26.0 | 30.0 | 21.0 | 25.0 | 15.0 | 17.0 | 39.0 | 45.0 | 32.0 | 37.0 | 23.0 | 26.0 |
| 1 5/8 | 6 × 37 | 31.0 | 35.0 | 25.0 | 29.0 | 18.0 | 20.0 | 46.0 | 53.0 | 38.0 | 43.0 | 27.0 | 31.0 |
| 1 3/4 | 6 × 37 | 35.0 | 41.0 | 29.0 | 33.0 | 20.0 | 24.0 | 53.0 | 61.0 | 43.0 | 50.0 | 31.0 | 35.0 |
| 2 | 6 × 37 | 46.0 | 53.0 | 37.0 | 43.0 | 26.0 | 30.0 | 68.0 | 79.0 | 56.0 | 65.0 | 40.0 | 46.0 |

HT = Hand Tucked Splice.
MS = Mechanical Splice.

### TABLE N-184-9—RATED CAPACITIES FOR 2-LEG AND 3-LEG BRIDLE SLINGS

Cable Laid Rope—Mechanical Splice Only
7 × 7 × 7 and 7 × 7 × 19 Constructions Galvanized Aircraft Grade Rope
7 × 6 × 19 IWRC Construction Improved Plow Steel Grade Rope

[Horizontal angles shown in parentheses]

| Rope | | Rated capacities, tons (2,000 lb) | | | | | |
|---|---|---|---|---|---|---|---|
| | | 2-Leg bridle sling | | | 3-Leg bridle sling | | |
| Dia (inches) | Constr | 30° (60°) | 45° angle | 60° (30°) | 30° (60°) | 45° angle | 60° (30°) |
| 1/4 | 7 × 7 × 7 | 0.87 | 0.71 | 0.50 | 1.3 | 1.1 | 0.75 |
| 3/8 | 7 × 7 × 7 | 1.9 | 1.5 | 1.1 | 2.8 | 2.3 | 1.6 |
| 1/2 | 7 × 7 × 7 | 3.2 | 2.6 | 1.8 | 4.8 | 3.9 | 2.8 |
| 5/8 | 7 × 7 × 7 | 4.8 | 3.9 | 2.8 | 7.2 | 5.9 | 4.2 |
| 3/4 | 7 × 7 × 7 | 6.6 | 5.4 | 3.8 | 9.9 | 8.1 | 3.7 |
| 5/8 | 7 × 7 × 19 | 5.0 | 4.1 | 2.9 | 7.5 | 6.1 | 4.3 |
| 3/4 | 7 × 7 × 19 | 7.0 | 5.7 | 4.1 | 10.0 | 8.6 | 6.1 |
| 7/8 | 7 × 7 × 19 | 9.3 | 7.6 | 5.4 | 14.0 | 11.0 | 8.1 |
| 1 | 7 × 7 × 19 | 12.0 | 9.7 | 6.9 | 18.0 | 14.0 | 10.0 |
| 1 1/8 | 7 × 7 × 19 | 14.0 | 12.0 | 8.2 | 21.0 | 17.0 | 12.0 |
| 1 1/4 | 7 × 7 × 19 | 17.0 | 14.0 | 9.9 | 26.0 | 21.0 | 15.0 |
| 3/4 | 7 × 6 × 19 IWRC | 6.6 | 5.4 | 3.8 | 9.9 | 8.0 | 5.7 |
| 7/8 | 7 × 6 × 19 IWRC | 8.7 | 7.1 | 5.0 | 13.0 | 11.0 | 7.5 |
| 1 | 7 × 6 × 19 IWRC | 11.0 | 9.0 | 6.4 | 17.0 | 13.0 | 9.6 |
| 1 1/8 | 7 × 6 × 19 IWRC | 13.0 | 11.0 | 7.7 | 20.0 | 16.0 | 11.0 |
| 1 1/4 | 7 × 6 × 19 IWRC | 16.0 | 13.0 | 9.2 | 24.0 | 20.0 | 14.0 |
| 1 5/16 | 7 × 6 × 19 IWRC | 17.0 | 14.0 | 10.0 | 26.0 | 21.0 | 15.0 |
| 1 3/8 | 7 × 6 × 19 IWRC | 19.0 | 15.0 | 11.0 | 28.0 | 23.0 | 16.0 |
| 1 1/2 | 7 × 6 × 19 IWRC | 22.0 | 18.0 | 13.0 | 33.0 | 7.0 | 19.0 |

### TABLE N-184-10—RATED CAPACITIES FOR 2-LEG AND 3-LEG BRIDLE SLINGS
8-Part and 6-Part Braided Rope
6 × 7 and 6 × 19 Construction Improved Plow Steel Grade Rope
7 × 7 Construction Galvanized Aircraft Grade Rope
[Horizontal angles shown in parentheses]

| Rope | | Rated capacities, tons (2,000 lb) | | | | | | | | | | | |
|---|---|---|---|---|---|---|---|---|---|---|---|---|---|
| | | 2-Leg bridle slings | | | | | | 3-Leg bridle slings | | | | | |
| | | 30° (60°) | | 45° angle | | 60° (30°) | | 30° (60°) | | 45° angle | | 60° (30°) | |
| Dia (inches) | Constr | 8-Part | 6-Part | 8-Part | 6-Part | 8-Part | 6-Part | 8-Part | 6-Part | 8-Part | 6-Part | 8-Part | 6-Part |
| 3/32 | 6 × 7 | 0.74 | 0.55 | 0.60 | 0.45 | 0.42 | 0.32 | 1.1 | 0.83 | 0.90 | 0.68 | 0.64 | 0.48 |
| 1/8 | 6 × 7 | 1.3 | 0.98 | 1.1 | 0.80 | 0.76 | 0.57 | 2.0 | 1.5 | 1.6 | 1.2 | 1.1 | 0.85 |
| 3/16 | 6 × 7 | 2.9 | 2.2 | 2.4 | 1.8 | 1.7 | 1.3 | 4.4 | 3.3 | 3.6 | 2.7 | 2.5 | 1.9 |
| 3/32 | 7 × 7 | 0.89 | 0.67 | 0.72 | 0.55 | 0.51 | 0.39 | 1.3 | 1.0 | 1.1 | 0.82 | 0.77 | 0.58 |
| 1/8 | 7 × 7 | 1.6 | 1.2 | 1.3 | 1.0 | 0.95 | 0.71 | 2.5 | 1.8 | 2.0 | 1.5 | 1.4 | 1.1 |
| 3/16 | 7 × 7 | 3.6 | 2.7 | 2.9 | 2.2 | 2.1 | 1.5 | 5.4 | 4.0 | 4.4 | 3.3 | 3.1 | 2.3 |
| 3/16 | 6 × 19 | 3.0 | 2.2 | 2.4 | 1.8 | 1.7 | 1.3 | 4.5 | 3.4 | 3.7 | 2.8 | 2.6 | 1.9 |
| 1/4 | 6 × 19 | 5.3 | 4.0 | 4.3 | 3.2 | 3.1 | 2.3 | 8.0 | 6.0 | 6.5 | 4.9 | 4.6 | 3.4 |
| 5/16 | 6 × 19 | 8.3 | 6.2 | 6.7 | 5.0 | 4.8 | 3.6 | 12.0 | 9.3 | 10.0 | 7.6 | 7.1 | 5.4 |
| 3/8 | 6 × 19 | 12.0 | 8.9 | 9.7 | 7.2 | 6.8 | 5.1 | 18.0 | 13.0 | 14.0 | 11.0 | 10.0 | 7.7 |
| 7/16 | 6 × 19 | 16.0 | 12.0 | 13.0 | 9.8 | 9.3 | 6.9 | 24.0 | 18.0 | 20.0 | 15.0 | 14.0 | 10.0 |
| 1/2 | 6 × 19 | 21.0 | 15.0 | 17.0 | 13.0 | 12.0 | 9.0 | 31.0 | 23.0 | 25.0 | 19.0 | 18.0 | 13.0 |
| 9/16 | 6 × 19 | 26.0 | 20.0 | 21.0 | 16.0 | 15.0 | 11.0 | 39.0 | 29.0 | 32.0 | 24.0 | 23.0 | 17.0 |
| 5/8 | 6 × 19 | 32.0 | 24.0 | 26.0 | 20.0 | 10.0 | 14.0 | 48.0 | 36.0 | 40.0 | 30.0 | 28.0 | 21.0 |
| 3/4 | 6 × 19 | 46.0 | 35.0 | 38.0 | 28.0 | 27.0 | 20.0 | 69.0 | 52.0 | 56.0 | 42.0 | 40.0 | 30.0 |
| 7/8 | 6 × 19 | 62.0 | 47.0 | 51.0 | 38.0 | 36.0 | 27.0 | 94.0 | 70.0 | 76.0 | 57.0 | 54.0 | 40.0 |
| 1 | 6 × 19 | 81.0 | 61.0 | 66.0 | 50.0 | 47.0 | 35.0 | 122.0 | 91.0 | 99.0 | 74.0 | 70.0 | 53.0 |

### TABLE N-184-11—RATED CAPACITIES FOR STRAND LAID GROMMET—HAND TUCKED
Improved Plow Steel Grade Rope

| Rope body | | Rated capacities, tons (2,000 lb) | | |
|---|---|---|---|---|
| Dia (inches) | Constr | Vertical | Choker | Vertical basket[1] |
| 1/4 | 7 × 19 | 0.85 | 0.64 | 1.7 |
| 5/16 | 7 × 19 | 1.3 | 1.0 | 2.6 |
| 3/8 | 7 × 19 | 1.9 | 1.4 | 3.8 |
| 7/16 | 7 × 19 | 2.6 | 1.9 | 5.2 |
| 1/2 | 7 × 19 | 3.3 | 2.5 | 6.7 |
| 9/16 | 7 × 19 | 4.2 | 3.1 | 8.4 |
| 5/8 | 7 × 19 | 5.2 | 3.9 | 10.0 |
| 3/4 | 7 × 19 | 7.4 | 5.6 | 15.0 |
| 7/8 | 7 × 19 | 10.0 | 7.5 | 20.0 |
| 1 | 7 × 19 | 13.0 | 9.7 | 26.0 |
| 1 1/8 | 7 × 19 | 16.0 | 12.0 | 32.0 |
| 1 1/4 | 7 × 37 | 18.0 | 14.0 | 37.0 |
| 1 3/8 | 7 × 37 | 22.0 | 16.0 | 44.0 |
| 1 1/2 | 7 × 37 | 26.0 | 19.0 | 52.0 |

[1] These values only apply when the D/d ratio is 5 or greater where: D = Diameter of curvature around which rope is bent. d = Diameter of rope body.

### TABLE N-184-12—RATED CAPACITIES FOR CABLE LAID GROMMET—HAND TUCKED
7 × 6 × 7 and 7 × 6 × 19 Constructions Improved Plow Steel Grade Rope
7 × 7 × 7 Construction Galvanized Aircraft Grade Rope

| Cable body | | Rated capacities, tons (2,000 lb) | | |
|---|---|---|---|---|
| Dia (inches) | Constr | Vertical | Choker | Vertical basket[1] |
| 3/8 | 7 × 6 × 7 | 1.3 | 0.95 | 2.5 |
| 9/16 | 7 × 6 × 7 | 2.8 | 2.1 | 5.6 |
| 5/8 | 7 × 6 × 7 | 3.8 | 2.8 | 7.6 |
| 3/8 | 7 × 7 × 7 | 1.6 | 1.2 | 3.2 |
| 9/16 | 7 × 7 × 7 | 3.5 | 2.6 | 6.9 |
| 5/8 | 7 × 7 × 7 | 4.5 | 3.4 | 9.0 |
| 5/8 | 7 × 6 × 19 | 3.9 | 3.0 | 7.9 |
| 3/4 | 7 × 6 × 19 | 5.1 | 3.8 | 10.0 |
| 15/16 | 7 × 6 × 19 | 7.9 | 5.9 | 16.0 |
| 1 1/8 | 7 × 6 × 19 | 11.0 | 8.4 | 22.0 |
| 1 5/16 | 7 × 6 × 19 | 15.0 | 11.0 | 30.0 |
| 1 1/2 | 7 × 6 × 19 | 19.0 | 14.0 | 39.0 |
| 1 11/16 | 7 × 6 × 19 | 24.0 | 18.0 | 49.0 |
| 1 7/8 | 7 × 6 × 19 | 30.0 | 22.0 | 60.0 |
| 2 1/4 | 7 × 6 × 19 | 42.0 | 31.0 | 84.0 |
| 2 5/8 | 7 × 6 × 19 | 56.0 | 42.0 | 112.0 |

[1] These values only apply when the D/d ratio is 5 or greater where: D = Diameter of curvature around which cable body is bent. d = Diameter of cable body.

### TABLE N-184-13—RATED CAPACITIES FOR STRAND LAID ENDLESS SLINGS—MECHANICAL JOINT
Improved Plow Steel Grade Rope

| Rope body |   | Rated capacities, tons (2,000 lb) |   |   |
|---|---|---|---|---|
| Dia (inches) | Constr | Vertical | Choker | Vertical basket[1] |
| 1/4 | [2]6 × 19 | 0.92 | 0.69 | 1.8 |
| 3/8 | [2]6 × 19 | 2.0 | 1.5 | 4.1 |
| 1/2 | [2]6 × 19 | 3.6 | 2.7 | 7.2 |
| 5/8 | [2]6 × 19 | 5.6 | 4.2 | 11.0 |
| 3/4 | [2]6 × 19 | 8.0 | 6.0 | 16.0 |
| 7/8 | [2]6 × 19 | 11.0 | 8.1 | 21.0 |
| 1 | [2]6 × 19 | 14.0 | 10.0 | 28.0 |
| 1 1/8 | [2]6 × 19 | 18.0 | 13.0 | 35.0 |
| 1 1/4 | [2]6 × 37 | 21.0 | 15.0 | 41.0 |
| 1 3/8 | [2]6 × 37 | 25.0 | 19.0 | 50.0 |
| 1 1/2 | [2]6 × 37 | 29.0 | 22.0 | 59.0 |

[1] These values only apply when the D/d ratio is 5 or greater where: D = Diameter of curvature around which rope is bent. d = Diameter of rope body.

[2] IWRC.

### TABLE N-184-14—RATED CAPACITIES FOR CABLE LAID ENDLESS SLINGS—MECHANICAL JOINT
7 × 7 × 7 and 7 × 7 × 19 Constructions Galvanized Aircraft Grade Rope
7 × 6 × 19 IWRC Construction Improved Plow Steel Grade Rope

| Cable body |   | Rated capacities, tons (2,000 lb) |   |   |
|---|---|---|---|---|
| Dia (inches) | Constr | Vertical | Choker | Vertical basket[1] |
| 1/4 | 7 × 7 × 7 | 0.83 | 0.62 | 1.6 |
| 3/8 | 7 × 7 × 7 | 1.8 | 1.3 | 3.5 |
| 1/2 | 7 × 7 × 7 | 3.0 | 2.3 | 6.1 |
| 5/8 | 7 × 7 × 7 | 4.5 | 3.4 | 9.1 |
| 3/4 | 7 × 7 × 7 | 6.3 | 4.7 | 12.0 |
| 5/8 | 7 × 7 × 19 | 4.7 | 3.5 | 9.5 |
| 3/4 | 7 × 7 × 19 | 6.7 | 5.0 | 13.0 |
| 7/8 | 7 × 7 × 19 | 8.9 | 6.6 | 18.0 |
| 1 | 7 × 7 × 19 | 11.0 | 8.5 | 22.0 |
| 1 1/8 | 7 × 7 × 19 | 14.0 | 10.0 | 28.0 |
| 1 1/4 | 7 × 7 × 19 | 17.0 | 12.0 | 33.0 |
| 3/4 | [2]7 × 6 × 19 | 6.2 | 4.7 | 12.0 |
| 7/8 | [2]7 × 6 × 19 | 8.3 | 6.2 | 16.0 |
| 1 | [2]7 × 6 × 19 | 10.0 | 7.9 | 21.0 |
| 1 1/8 | [2]7 × 6 × 19 | 13.0 | 9.7 | 26.0 |
| 1 1/4 | [2]7 × 6 × 19 | 16.0 | 12.0 | 31.0 |
| 1 3/8 | [2]7 × 6 × 19 | 18.0 | 14.0 | 37.0 |
| 1 1/2 | [2]7 × 6 × 19 | 22.0 | 16.0 | 43.0 |

[1] These values only apply when the D/d value is 5 or greater where: D = Diameter of curvature around which cable body is bent. d = Diameter of cable body.

[2] IWRC.

**1910.184(f)(5)** *Removal from service.* Wire rope slings shall be immediately removed from service if any of the following conditions are present:

**1910.184(f)(5)(i)** Ten randomly distributed broken wires in one rope lay, or five broken wires in one strand in one rope lay.

**1910.184(f)(5)(ii)** Wear or scraping of one-third the original diameter of outside individual wires.

**1910.184(f)(5)(iii)** Kinking, crushing, bird caging or any other damage resulting in distortion of the wire rope structure.

**1910.184(f)(5)(iv)** Evidence of heat damage.

**1910.184(f)(5)(v)** End attachments that are cracked, deformed or worn.

**1910.184(f)(5)(vi)** Hooks that have been opened more than 15 percent of the normal throat opening measured at the narrowest point or twisted more than 10 degrees from the plane of the unbent hook.

**1910.184(f)(5)(vii)** Corrosion of the rope or end attachments.

**1910.184(g)** *Metal mesh slings*—**(1)** *Sling marking.* Each metal mesh sling shall have permanently affixed to it a durable marking that states the rated capacity for vertical basket hitch and choker hitch loadings.

**1910.184(g)(2)** *Handles.* Handles shall have a rated capacity at least equal to the metal fabric and exhibit no deformation after proof testing.

**1910.184(g)(3)** *Attachments of handles to fabric.* The fabric and handles shall be joined so that:

**1910.184(g)(3)(i)** The rated capacity of the sling is not reduced.

**1910.184(g)(3)(ii)** The load is evenly distributed across the width of the fabric.

**1910.184(g)(3)(iii)** Sharp edges will not damage the fabric.

**1910.184(g)(4)** *Sling coatings.* Coatings which diminish the rated capacity of a sling shall not be applied.

**1910.184(g)(5)** *Sling testing.* All new and repaired metal mesh slings, including handles, shall not be used unless proof tested by the manufacturer or equivalent entity at a minimum of 1 1/2 times their rated capacity. Elastomer impregnated slings shall be proof tested before coating.

**1910.184(g)(6)** *Proper use of metal mesh slings.* Metal mesh slings shall not be used to lift loads in excess of their rated capacities as prescribed in Table N-184-15. Slings not included in this table shall be used only in accordance with the manufacturer's recommendations.

**1910.184(g)(7)** *Safe operating temperatures.* Metal mesh slings which are not impregnated with elastomers may be used in a temperature range from minus 20 °F. to plus 550 °F. without decreasing the working load limit. Metal mesh slings impregnated with polyvinyl chloride or neoprene may be used only in a temperature range from zero degrees to plus 200 °F. For operations outside these temperature ranges or for metal mesh slings impregnated with other materials, the sling manufacturer's recommendations shall be followed.

**1910.184(g)(8)** *Repairs.* **(i)** Metal mesh slings which are repaired shall not be used unless repaired by a metal mesh sling manufacturer or an equivalent entity.

**1910.184(g)(8)(ii)** Once repaired, each sling shall be permanently marked or tagged, or a written record maintained, to indicate the date and nature of the repairs and the person or organization that performed the repairs. Records of repairs shall be made available for examination.

**1910.184(g)(9)** *Removal from service.* Metal mesh slings shall be immediately removed from service if any of the following conditions are present:

**1910.184(g)(9)(i)** A broken weld or broken brazed joint along the sling edge.

**1910.184(g)(9)(ii)** Reduction in wire diameter of 25 per cent due to abrasion or 15 per cent due to corrosion.

**1910.184(g)(9)(iii)** Lack of flexibility due to distortion of the fabric.

### TABLE N-184-15—RATED CAPACITIES
Carbon Steel and Stainless Steel Metal Mesh Slings

[Horizontal angles shown in parentheses]

| Sling width in inches | Vertical or choker | Vertical basket | 30° (60°) | 45° (45°) | 60° (30°) |
|---|---|---|---|---|---|
| \multicolumn{6}{c}{Heavy Duty-10 Ga 35 Spirals/Ft of sling width} ||||||
| 2 | 1,500 | 3,000 | 2,600 | 2,100 | 1,500 |
| 3 | 2,700 | 5,400 | 4,700 | 3,800 | 2,700 |
| 4 | 4,000 | 8,000 | 6,900 | 5,600 | 4,000 |
| 6 | 6,000 | 12,000 | 10,400 | 8,400 | 6,000 |
| 8 | 8,000 | 16,000 | 13,800 | 11,300 | 8,000 |
| 10 | 10,000 | 20,000 | 17,000 | 14,100 | 10,000 |
| 12 | 12,000 | 24,000 | 20,700 | 16,900 | 12,000 |
| 14 | 14,000 | 28,000 | 24,200 | 19,700 | 14,000 |
| 16 | 16,000 | 32,000 | 27,700 | 22,600 | 16,000 |
| 18 | 18,000 | 36,000 | 31,100 | 25,400 | 18,000 |
| 20 | 20,000 | 40,000 | 34,600 | 28,200 | 20,000 |
| \multicolumn{6}{c}{Medium Duty-12 Ga 43 Spirals/Ft of sling width} ||||||
| 2 | 1,350 | 2,700 | 2,300 | 1,900 | 1,400 |
| 3 | 2,000 | 4,000 | 3,500 | 2,800 | 2,000 |
| 4 | 2,700 | 5,400 | 4,700 | 3,800 | 2,700 |
| 6 | 4,500 | 9,000 | 7,800 | 6,400 | 4,500 |
| 8 | 6,000 | 12,000 | 10,400 | 8,500 | 6,000 |
| 10 | 7,500 | 15,000 | 13,000 | 10,600 | 7,500 |
| 12 | 9,000 | 18,000 | 15,600 | 12,700 | 9,000 |
| 14 | 10,500 | 21,000 | 18,200 | 14,800 | 10,500 |
| 16 | 12,000 | 24,000 | 20,800 | 17,000 | 12,000 |
| 18 | 13,500 | 27,000 | 23,400 | 19,100 | 13,500 |
| 20 | 15,000 | 30,000 | 26,000 | 21,200 | 15,000 |
| \multicolumn{6}{c}{Light Duty-14 Ga 59 Spirals/Ft of sling width} ||||||
| 2 | 900 | 1,800 | 1,600 | 1,300 | 900 |
| 3 | 1,400 | 2,800 | 2,400 | 2,000 | 1,400 |
| 4 | 2,000 | 4,000 | 3,500 | 2,800 | 2,000 |
| 6 | 3,000 | 6,000 | 5,200 | 4,200 | 3,000 |
| 8 | 4,000 | 8,000 | 6,900 | 5,700 | 4,000 |
| 10 | 5,000 | 10,000 | 8,600 | 7,100 | 5,000 |
| 12 | 6,000 | 12,000 | 10,400 | 8,500 | 6,000 |
| 14 | 7,000 | 14,000 | 12,100 | 9,900 | 7,000 |
| 16 | 8,000 | 16,000 | 13,900 | 11,300 | 8,000 |
| 18 | 9,000 | 18,000 | 15,600 | 12,700 | 9,000 |
| 20 | 10,000 | 20,000 | 17,300 | 14,100 | 10,000 |

**1910.184(g)(9)(iv)** Distortion of the female handle so that the depth of the slot is increased more than 10 per cent.

**1910.184(g)(9)(v)** Distortion of either handle so that the width of the eye is decreased more than 10 per cent.

**1910.184(g)(9)(vi)** A 15 percent reduction of the original cross sectional area of metal at any point around the handle eye.

**1910.184(g)(9)(vii)** Distortion of either handle out of its plane.

**1910.184(h)** *Natural and synthetic fiber rope slings*—**(1)** *Sling use.* **(i)** Fiber rope slings made from conventional three strand construction fiber rope shall not be used with loads in excess of the rated capacities prescribed in Tables N-184-16 through N-184-19.

**1910.184(h)(1)(ii)** Fiber rope slings shall have a diameter of curvature meeting at least the minimums specified in Figs. N-184-4 and N-184-5.

**1910.184(h)(1)(iii)** Slings not included in these tables shall be used only in accordance with the manufacturer's recommendations.

# OSHA Standards for General Industry

## FIGURE N-184.4—BASIC SLING CONFIGURATIONS WITH VERTICAL LEGS

**NOTES:** Angles 5° or less from the vertical may be considered vertical angles.

For slings with legs more than 5° off vertical, the actual angle as shown in Figure N-184-5 must be considered.

**EXPLANATION OF SYMBOLS: MINIMUM DIAMETER OF CURVATURE**

- ⊖ Represents a contact surface which shall have a diameter of curvature at least double the diameter of the rope from which the sling is made.
- ⊗ Represents a contact surface which shall have a diameter of curvature at least 8 times the diameter of the rope.
- ○ Represents a load in a choker hitch and illustrates the rotary force on the load and/or the slippage of the rope in contact with the load. Diameter of curvature of load surface shall be at least double the diameter of the rope.

## FIGURE N-184.5—SLING CONFIGURATIONS WITH ANGLED LEGS

**NOTES:** For vertical angles of 5° or less, refer to Figure N-184-4 "Basic Sling Configuration with Vertical Legs".

See Figure N-184-4 for explanation of symbols.

## TABLE N-184-16—MANILA ROPE SLINGS
[Angle of rope to vertical shown in parentheses]

| Rope dia. nominal in inches | Nominal wt. per 100 ft in pounds | Eye and eye sling ||||||| Endless sling |||||||
|---|---|---|---|---|---|---|---|---|---|---|---|---|---|
| | | Vertical hitch | Choker hitch | Basket hitch; Angle of rope to horizontal |||| Vertical hitch | Choker hitch | Basket hitch; Angle of rope to horizontal ||||
| | | | | 90° (0°) | 60° (30°) | 45° (45°) | 30° (60°) | | | 90° (0°) | 60° (30°) | 45° (45°) | 30° (60°) |
| 1/2 | 7.5 | 480 | 240 | 960 | 830 | 680 | 480 | 865 | 430 | 1,730 | 1,500 | 1,220 | 865 |
| 9/16 | 10.4 | 620 | 310 | 1,240 | 1,070 | 875 | 620 | 1,120 | 560 | 2,230 | 1,930 | 1,580 | 1,120 |
| 5/8 | 13.3 | 790 | 395 | 1,580 | 1,370 | 1,120 | 790 | 1,420 | 710 | 2,840 | 2,460 | 2,010 | 1,420 |
| 3/4 | 16.7 | 970 | 485 | 1,940 | 1,680 | 1,370 | 970 | 1,750 | 875 | 3,490 | 3,020 | 2,470 | 1,750 |
| 13/16 | 19.5 | 1,170 | 585 | 2,340 | 2,030 | 1,650 | 1,170 | 2,110 | 1,050 | 4,210 | 3,650 | 2,980 | 2,110 |
| 7/8 | 22.5 | 1,390 | 695 | 2,780 | 2,410 | 1,970 | 1,390 | 2,500 | 1,250 | 5,000 | 4,330 | 3,540 | 2,500 |
| 1 | 27.0 | 1,620 | 810 | 3,240 | 2,810 | 2,290 | 1,620 | 2,920 | 1,460 | 5,830 | 5,050 | 4,120 | 2,920 |
| 1/16 | 31.3 | 1,890 | 945 | 3,780 | 3,270 | 2,670 | 1,890 | 3,400 | 1,700 | 6,800 | 5,890 | 4,810 | 3,400 |
| 1 1/8 | 36.0 | 2,160 | 1,080 | 4,320 | 3,740 | 3,050 | 2,160 | 3,890 | 1,940 | 7,780 | 6,730 | 5,500 | 3,890 |
| 1 1/4 | 41.7 | 2,430 | 1,220 | 4,860 | 4,210 | 3,440 | 2,430 | 4,370 | 2,190 | 8,750 | 7,580 | 6,190 | 4,370 |
| 1 5/16 | 47.9 | 2,700 | 1,350 | 5,400 | 4,680 | 3,820 | 2,700 | 4,860 | 2,430 | 9,720 | 8,420 | 6,870 | 4,860 |
| 1 1/2 | 59.9 | 3,330 | 1,670 | 6,660 | 5,770 | 4,710 | 3,330 | 5,990 | 3,000 | 12,000 | 10,400 | 8,480 | 5,990 |
| 1 5/8 | 74.6 | 4,050 | 2,030 | 8,100 | 7,010 | 5,730 | 4,050 | 7,290 | 3,650 | 14,600 | 12,600 | 10,300 | 7,290 |
| 1 3/4 | 89.3 | 4,770 | 2,390 | 9,540 | 8,260 | 6,740 | 4,770 | 8,590 | 4,290 | 17,200 | 14,900 | 12,100 | 8,590 |
| 2 | 107.5 | 5,580 | 2,790 | 11,200 | 9,660 | 7,890 | 5,580 | 10,000 | 5,020 | 20,100 | 17,400 | 14,200 | 10,000 |
| 2 1/8 | 125.0 | 6,480 | 3,240 | 13,000 | 11,200 | 9,160 | 6,480 | 11,700 | 5,830 | 23,300 | 20,200 | 16,500 | 11,700 |
| 2 1/4 | 146.0 | 7,380 | 3,690 | 14,800 | 12,800 | 10,400 | 7,380 | 13,300 | 6,640 | 26,600 | 23,000 | 18,800 | 13,300 |
| 2 1/2 | 166.7 | 8,370 | 4,190 | 16,700 | 14,500 | 11,800 | 8,370 | 15,100 | 7,530 | 30,100 | 26,100 | 21,300 | 15,100 |
| 2 5/8 | 190.8 | 9,360 | 4,680 | 18,700 | 16,200 | 13,200 | 9,360 | 16,800 | 8,420 | 33,700 | 29,200 | 23,800 | 16,800 |

See Figs. N-184-4 and N-184-5 for sling configuration descriptions.

## TABLE N-184-17—NYLON ROPE SLINGS
[Angle of rope to vertical shown in parentheses]

| Rope dia. nominal in inches | Nominal wt. per 100 ft in pounds | Eye and eye sling ||||||| Endless sling |||||||
|---|---|---|---|---|---|---|---|---|---|---|---|---|---|
| | | Vertical hitch | Choker hitch | Basket hitch; Angle of rope to horizontal |||| Vertical hitch | Choker hitch | Basket hitch; Angle of rope to horizontal ||||
| | | | | 90° (0°) | 60° (30°) | 45° (45°) | 30° (60°) | | | 90° (0°) | 60° (30°) | 45° (45°) | 30° (60°) |
| 1/2 | 6.5 | 635 | 320 | 1,270 | 1,100 | 900 | 635 | 1,140 | 570 | 2,290 | 1,980 | 1,620 | 1,140 |
| 9/16 | 8.3 | 790 | 395 | 1,580 | 1,370 | 1,120 | 790 | 1,420 | 710 | 2,840 | 2,460 | 2,010 | 1,420 |
| 5/8 | 10.5 | 1,030 | 515 | 2,060 | 1,780 | 1,460 | 1,030 | 1,850 | 925 | 3,710 | 3,210 | 2,620 | 1,850 |
| 3/4 | 14.5 | 1,410 | 705 | 2,820 | 2,440 | 1,990 | 1,410 | 2,540 | 1,270 | 5,080 | 4,400 | 3,590 | 2,540 |
| 13/16 | 17.0 | 1,680 | 840 | 3,360 | 2,910 | 2,380 | 1,680 | 3,020 | 1,510 | 6,050 | 5,240 | 4,280 | 3,020 |
| 7/8 | 20.0 | 1,980 | 990 | 3,960 | 3,430 | 2,800 | 1,980 | 3,560 | 1,780 | 7,130 | 6,170 | 5,040 | 3,560 |
| 1 | 26.0 | 2,480 | 1,240 | 4,960 | 4,300 | 3,510 | 2,480 | 4,460 | 2,230 | 8,930 | 7,730 | 6,310 | 4,460 |
| 1 1/16 | 29.0 | 2,850 | 1,430 | 5,700 | 4,940 | 4,030 | 2,850 | 5,130 | 2,570 | 10,300 | 8,890 | 7,260 | 5,130 |
| 1 1/8 | 34.0 | 3,270 | 1,640 | 6,540 | 5,660 | 4,620 | 3,270 | 5,890 | 2,940 | 11,800 | 10,200 | 8,330 | 5,890 |
| 1 1/4 | 40.0 | 3,710 | 1,860 | 7,420 | 6,430 | 5,250 | 3,710 | 6,680 | 3,340 | 13,400 | 11,600 | 9,450 | 6,680 |
| 1 5/16 | 45.0 | 4,260 | 2,130 | 8,520 | 7,380 | 6,020 | 4,260 | 7,670 | 3,830 | 15,300 | 13,300 | 10,800 | 7,670 |
| 1 1/2 | 55.0 | 5,250 | 2,630 | 10,500 | 9,090 | 7,420 | 5,250 | 9,450 | 4,730 | 18,900 | 16,400 | 13,400 | 9,450 |
| 1 5/8 | 68.0 | 6,440 | 3,220 | 12,900 | 11,200 | 9,110 | 6,440 | 11,600 | 5,800 | 23,200 | 20,100 | 16,400 | 11,600 |
| 1 3/4 | 83.0 | 7,720 | 3,860 | 15,400 | 13,400 | 10,900 | 7,720 | 13,900 | 6,950 | 27,800 | 24,100 | 19,700 | 13,900 |
| 2 | 95.0 | 9,110 | 4,560 | 18,200 | 15,800 | 12,900 | 9,110 | 16,400 | 8,200 | 32,800 | 28,400 | 23,200 | 16,400 |
| 2 1/8 | 109.0 | 10,500 | 5,250 | 21,000 | 18,200 | 14,800 | 10,500 | 18,900 | 9,450 | 37,800 | 32,700 | 26,700 | 18,900 |
| 2 1/4 | 129.0 | 12,400 | 6,200 | 24,800 | 21,500 | 17,500 | 12,400 | 22,300 | 11,200 | 44,600 | 38,700 | 31,600 | 22,300 |
| 2 1/2 | 149.0 | 13,900 | 6,950 | 27,800 | 24,100 | 19,700 | 13,900 | 25,000 | 12,500 | 50,000 | 43,300 | 35,400 | 25,000 |
| 2 5/8 | 168.0 | 16,000 | 8,000 | 32,000 | 27,700 | 22,600 | 16,000 | 28,800 | 14,400 | 57,600 | 49,900 | 40,700 | 28,800 |

See Figs. N-184-4 and N-184-5 for sling configuration descriptions.

## TABLE N-184-18—POLYESTER ROPE SLINGS
[Angle of rope to vertical shown in parentheses]

| Rope dia. nominal in inches | Nominal wt. per 100 ft in pounds | Eye and eye sling ||||||| Endless sling |||||||
|---|---|---|---|---|---|---|---|---|---|---|---|---|---|
| | | Vertical hitch | Choker hitch | Basket hitch; Angle of rope to horizontal |||| Vertical hitch | Choker hitch | Basket hitch; Angle of rope to horizontal ||||
| | | | | 90° (0°) | 60° (30°) | 45° (45°) | 30° (60°) | | | 90° (0°) | 60° (30°) | 45° (45°) | 30° (60°) |
| 1/2 | 8.0 | 635 | 320 | 1,270 | 1,100 | 900 | 635 | 1,140 | 570 | 2,290 | 1,980 | 1,620 | 1,140 |
| 9/16 | 10.2 | 790 | 395 | 1,580 | 1,370 | 1,120 | 790 | 1,420 | 710 | 2,840 | 2,460 | 2,010 | 1,420 |
| 5/8 | 13.0 | 990 | 495 | 1,980 | 1,710 | 1,400 | 990 | 1,780 | 890 | 3,570 | 3,090 | 2,520 | 1,780 |
| 3/4 | 17.5 | 1,240 | 620 | 2,480 | 2,150 | 1,750 | 1,240 | 2,230 | 1,120 | 4,470 | 3,870 | 3,160 | 2,230 |
| 13/16 | 21.0 | 1,540 | 770 | 3,080 | 2,670 | 2,180 | 1,540 | 2,770 | 1,390 | 5,540 | 4,800 | 3,920 | 2,770 |
| 7/8 | 25.0 | 1,780 | 890 | 3,560 | 3,080 | 2,520 | 1,780 | 3,200 | 1,600 | 6,410 | 5,550 | 4,530 | 3,200 |
| 1 | 30.5 | 2,180 | 1,090 | 4,360 | 3,780 | 3,080 | 2,180 | 3,920 | 2,960 | 7,850 | 6,800 | 5,550 | 3,920 |
| 1 1/16 | 34.5 | 2,530 | 1,270 | 5,060 | 4,380 | 3,580 | 2,530 | 4,550 | 2,280 | 9,110 | 7,990 | 6,440 | 4,550 |
| 1 1/8 | 40.0 | 2,920 | 1,460 | 5,840 | 5,060 | 4,130 | 2,920 | 5,260 | 2,630 | 10,500 | 9,100 | 7,440 | 5,260 |
| 1 1/4 | 46.3 | 3,290 | 1,650 | 6,580 | 5,700 | 4,650 | 3,290 | 5,920 | 2,960 | 11,800 | 10,300 | 8,380 | 5,920 |
| 1 5/16 | 52.5 | 3,710 | 1,860 | 7,420 | 6,430 | 5,250 | 3,710 | 6,680 | 3,340 | 13,400 | 11,600 | 9,450 | 6,680 |
| 1 1/2 | 66.8 | 4,630 | 2,320 | 9,260 | 8,020 | 6,550 | 4,630 | 8,330 | 4,170 | 16,700 | 14,400 | 11,800 | 8,330 |
| 1 5/8 | 82.0 | 5,640 | 2,820 | 11,300 | 9,770 | 7,980 | 5,640 | 10,200 | 5,080 | 20,300 | 17,600 | 14,400 | 10,200 |
| 1 3/4 | 98.0 | 6,710 | 3,360 | 13,400 | 11,600 | 9,490 | 6,710 | 12,100 | 6,040 | 24,200 | 20,900 | 17,100 | 12,100 |
| 2 | 118.0 | 7,920 | 3,960 | 15,800 | 13,700 | 11,200 | 7,920 | 14,300 | 7,130 | 28,500 | 24,700 | 20,200 | 14,300 |
| 2 1/8 | 135.0 | 9,110 | 4,460 | 18,200 | 15,800 | 12,900 | 9,110 | 16,400 | 8,200 | 32,800 | 28,400 | 23,200 | 16,400 |
| 2 1/4 | 157.0 | 10,600 | 5,300 | 21,200 | 18,400 | 15,000 | 10,600 | 19,100 | 9,540 | 38,200 | 33,100 | 27,000 | 19,100 |
| 2 1/2 | 181.0 | 12,100 | 6,050 | 24,200 | 21,000 | 17,100 | 12,100 | 21,800 | 10,900 | 43,600 | 37,700 | 30,800 | 21,800 |
| 2 5/8 | 205.0 | 13,600 | 6,800 | 27,200 | 23,600 | 19,200 | 13,600 | 24,500 | 12,200 | 49,000 | 42,400 | 34,600 | 24,500 |

See Figs. N-184-4 and N-184-5 for sling configuration descriptions.

# OSHA Standards for General Industry

## TABLE N-184-19—POLYPROPYLENE ROPE SLINGS
[Angle of rope to vertical shown in parentheses]

| Rope dia. nominal in inches | Nominal wt. per 100 ft in pounds | Eye and eye sling ||||||| Endless sling |||||||
|---|---|---|---|---|---|---|---|---|---|---|---|---|---|
| | | Vertical hitch | Choker hitch | Basket hitch; Angle of rope to horizontal |||| Vertical hitch | Choker hitch | Basket hitch; Angle of rope to horizontal ||||
| | | | | 90° (0°) | 60° (30°) | 45° (45°) | 30° (60°) | | | 90° (0°) | 60° (30°) | 45° (45°) | 30° (60°) |
| 1/2 | 4.7 | 645 | 325 | 1,290 | 1,120 | 910 | 645 | 1,160 | 580 | 2,320 | 2,010 | 1,640 | 1,160 |
| 9/16 | 6.1 | 780 | 390 | 1,560 | 1,350 | 1,100 | 780 | 1,400 | 700 | 2,810 | 2,430 | 1,990 | 1,400 |
| 5/8 | 7.5 | 950 | 475 | 1,900 | 1,650 | 1,340 | 950 | 1,710 | 855 | 3,420 | 2,960 | 2,420 | 1,710 |
| 3/4 | 10.7 | 1,300 | 650 | 2,600 | 2,250 | 1,840 | 1,300 | 2,340 | 1,170 | 4,680 | 4,050 | 3,310 | 2,340 |
| 13/16 | 12.7 | 1,520 | 760 | 3,040 | 2,630 | 2,150 | 1,520 | 2,740 | 1,370 | 5,470 | 4,740 | 3,870 | 2,740 |
| 7/8 | 15.0 | 1,760 | 880 | 3,520 | 3,050 | 2,490 | 1,760 | 3,170 | 1,580 | 6,340 | 5,490 | 4,480 | 3,170 |
| 1 | 18.0 | 2,140 | 1,070 | 4,280 | 3,700 | 3,030 | 2,140 | 3,850 | 1,930 | 7,700 | 6,670 | 5,450 | 3,860 |
| 1 1/16 | 20.4 | 2,450 | 1,230 | 4,900 | 4,240 | 3,460 | 2,450 | 4,410 | 2,210 | 8,820 | 7,640 | 6,240 | 4,410 |
| 1 1/8 | 23.7 | 2,800 | 1,400 | 5,600 | 4,850 | 3,960 | 2,800 | 5,040 | 2,520 | 10,100 | 8,730 | 7,130 | 5,400 |
| 1 1/4 | 27.0 | 3,210 | 1,610 | 6,420 | 5,560 | 4,540 | 3,210 | 5,780 | 2,890 | 11,600 | 10,000 | 8,170 | 5,780 |
| 1 5/16 | 30.5 | 3,600 | 1,800 | 7,200 | 6,240 | 5,090 | 3,600 | 6,480 | 3,240 | 13,000 | 11,200 | 9,170 | 6,480 |
| 1 1/2 | 38.5 | 4,540 | 2,270 | 9,080 | 7,860 | 6,420 | 4,540 | 8,170 | 4,090 | 16,300 | 14,200 | 11,600 | 8,170 |
| 1 5/8 | 47.5 | 5,510 | 2,760 | 11,000 | 9,540 | 7,790 | 5,510 | 9,920 | 4,960 | 19,800 | 17,200 | 14,000 | 9,920 |
| 1 3/4 | 57.0 | 6,580 | 3,290 | 13,200 | 11,400 | 9,300 | 6,580 | 11,800 | 5,920 | 23,700 | 20,500 | 16,800 | 11,800 |
| 2 | 69.0 | 7,960 | 3,980 | 15,900 | 13,800 | 11,300 | 7,960 | 14,300 | 7,160 | 28,700 | 24,800 | 20,300 | 14,300 |
| 2 1/8 | 80.0 | 9,330 | 4,670 | 18,700 | 16,200 | 13,200 | 9,330 | 16,800 | 8,400 | 33,600 | 29,100 | 23,800 | 16,800 |
| 2 1/4 | 92.0 | 10,600 | 5,300 | 21,200 | 18,400 | 15,000 | 10,600 | 19,100 | 9,540 | 38,200 | 33,100 | 27,000 | 19,100 |
| 2 1/2 | 107.0 | 12,200 | 6,100 | 24,400 | 21,100 | 17,300 | 12,200 | 22,000 | 11,000 | 43,900 | 38,000 | 31,100 | 22,000 |
| 2 5/8 | 120.0 | 13,800 | 6,900 | 27,600 | 23,900 | 19,600 | 13,800 | 24,800 | 12,400 | 49,700 | 43,000 | 35,100 | 24,800 |

See Figs. N-184-4 and N-184-5 for sling configuration descriptions.

**1910.184(h)(2)** *Safe operating temperatures.* Natural and synthetic fiber rope slings, except for wet frozen slings, may be used in a temperature range from minus 20 °F. to plus 180 °F. without decreasing the working load limit. For operations outside this temperature range and for wet frozen slings, the sling manufacturer's recommendations shall be followed.

**1910.184(h)(3)** *Splicing.* Spliced fiber rope slings shall not be used unless they have been spliced in accordance with the following minimum requirements and in accordance with any additional recommendations of the manufacturer:

**1910.184(h)(3)(i)** In manila rope, eye splices shall consist of at least three full tucks, and short splices shall consist of at least six full tucks, three on each side of the splice center line.

**1910.184(h)(3)(ii)** In synthetic fiber rope, eye splices shall consist of at least four full tucks, and short splices shall consist of at least eight full tucks, four on each side of the center line.

**1910.184(h)(3)(iii)** Strand end tails shall not be trimmed flush with the surface of the rope immediately adjacent to the full tucks. This applies to all types of fiber rope and both eye and short splices. For fiber rope under one inch in diameter, the tail shall project at least six rope diameters beyond the last full tuck. For fiber rope one inch in diameter and larger, the tail shall project at least six inches beyond the last full tuck. Where a projecting tail interferes with the use of the sling, the tail shall be tapered and spliced into the body of the rope using at least two additional tucks (which will require a tail length of approximately six rope diameters beyond the last full tuck).

**1910.184(h)(3)(iv)** Fiber rope slings shall have a minimum clear length of rope between eye splices equal to 10 times the rope diameter.

**1910.184(h)(3)(v)** Knots shall not be used in lieu of splices.

**1910.184(h)(3)(vi)** Clamps not designed specifically for fiber ropes shall not be used for splicing.

**1910.184(h)(3)(vii)** For all eye splices, the eye shall be of such size to provide an included angle of not greater than 60 degrees at the splice when the eye is placed over the load or support.

**1910.184(h)(4)** *End attachments.* Fiber rope slings shall not be used if end attachments in contact with the rope have sharp edges or projections.

**1910.184(h)(5)** *Removal from service.* Natural and synthetic fiber rope slings shall be immediately removed from service if any of the following conditions are present:

**1910.184(h)(5)(i)** Abnormal wear.

**1910.184(h)(5)(ii)** Powdered fiber between strands.

**1910.184(h)(5)(iii)** Broken or cut fibers.

**1910.184(h)(5)(iv)** Variations in the size or roundness of strands.

**1910.184(h)(5)(v)** Discoloration or rotting.

**1910.184(h)(5)(vi)** Distortion of hardware in the sling.

**1910.184(h)(6)** *Repairs.* Only fiber rope slings made from new rope shall be used. Use of repaired or reconditioned fiber rope slings is prohibited.

**1910.184(i)** **Synthetic web slings**—**(1)** *Sling identification.* Each sling shall be marked or coded to show the rated capacities for each type of hitch and type of synthetic web material.

**1910.184(i)(2)** *Webbing.* Synthetic webbing shall be of uniform thickness and width and selvage edges shall not be split from the webbing's width.

**1910.184(i)(3)** *Fittings.* Fittings shall be:

**1910.184(i)(3)(i)** Of a minimum breaking strength equal to that of the sling; and

**1910.184(i)(3)(ii)** Free of all sharp edges that could in any way damage the webbing.

**1910.184(i)(4)** *Attachment of end fittings to webbing and formation of eyes.* Stitching shall be the only method used to attach end fittings to webbing and to form eyes. The thread shall be in an even pattern and contain a sufficient number of stitches to develop the full breaking strength of the sling.

**1910.184(i)(5)** *Sling use.* Synthetic web slings illustrated in Fig. N-184-6 shall not be used with loads in excess of the rated capacities specified in Tables N-184-20 through N-184-22. Slings not included in these tables shall be used only in accordance with the manufacturer's recommendations.

**1910.184(i)(6)** *Environmental conditions.* When synthetic web slings are used, the following precautions shall be taken:

**1910.184(i)(6)(i)** Nylon web slings shall not be used where fumes, vapors, sprays, mists or liquids of acids or phenolics are present.

**1910.184(i)(6)(ii)** Polyester and polypropylene web slings shall not be used where fumes, vapors, sprays, mists or liquids of caustics are present.

**1910.184(i)(6)(iii)** Web slings with aluminum fittings shall not be used where fumes, vapors, sprays, mists or liquids of caustics are present.

### FIGURE N-184-6
### BASIC SYNTHETIC WEB SLING CONSTRUCTIONS

- Triangle - Choker (Type I)
- Triangle - Triangle (Type II)
- Eye and Eye With Flat Eyes (Type III)
- Eye and Eye With Twisted Eyes (Type IV)
- Endless Type (Type V)
- Return Eye (Type VI)

### TABLE N-184-20—SYNTHETIC WEB SLINGS—1,000 POUNDS PER INCH OF WIDTH—SINGLE-PLY
[Rated capacity in pounds]

| Sling body width, inches | Triangle—Choker slings, type I; Triangle—Triangle slings, type II; Eye and eye with flat eye slings, type III; Eye and eye with twisted eye slings, type IV ||||||  Endless slings, type V ||||||  Return eye slings, type VI ||||||
|---|---|---|---|---|---|---|---|---|---|---|---|---|---|---|---|---|---|---|
| | Vert. | Choker | Vert. basket | 30° basket | 45° basket | 60° basket | Vert. | Choker | Vert. basket | 30° basket | 45° basket | 60° basket | Vert. | Choker | Vert. basket | 30° basket | 45° basket | 60° basket |
| 1 | 1,000 | 750 | 2,000 | 1,700 | 1,400 | 1,000 | 1,600 | 1,300 | 3,200 | 2,800 | 2,300 | 1,600 | 800 | 650 | 1,600 | 1,400 | 1,150 | 800 |
| 2 | 2,000 | 1,500 | 4,000 | 3,500 | 2,800 | 2,000 | 3,200 | 2,600 | 6,400 | 5,500 | 4,500 | 3,200 | 1,600 | 1,300 | 3,200 | 2,800 | 2,300 | 1,600 |
| 3 | 3,000 | 2,200 | 6,000 | 5,200 | 4,200 | 3,000 | 4,800 | 3,800 | 9,600 | 8,300 | 6,800 | 4,800 | 2,400 | 1,950 | 4,800 | 4,150 | 3,400 | 2,400 |
| 4 | 4,000 | 3,000 | 8,000 | 6,900 | 5,700 | 4,000 | 6,400 | 5,100 | 12,800 | 11,100 | 9,000 | 6,400 | 3,200 | 2,600 | 6,400 | 5,500 | 4,500 | 3,200 |
| 5 | 5,000 | 3,700 | 10,000 | 8,700 | 7,100 | 5,000 | 8,000 | 6,400 | 16,000 | 13,900 | 11,300 | 8,000 | 4,000 | 3,250 | 8,000 | 6,900 | 5,650 | 4,000 |
| 6 | 6,000 | 4,500 | 12,000 | 10,400 | 8,500 | 6,000 | 9,600 | 7,700 | 19,200 | 16,600 | 13,600 | 9,600 | 4,800 | 3,800 | 9,600 | 8,300 | 6,800 | 4,800 |

Notes: 1. All angles shown are measured from the vertical.
2. Capacities for intermediate widths not shown may be obtained by interpolation.

# OSHA Standards for General Industry

### TABLE N-184-21—SYNTHETIC WEB SLINGS—1,200 POUNDS PER INCH OF WIDTH—SINGLE-PLY
[Rated capacity in pounds]

| Sling body width, inches | Triangle—Choker slings, type I; Triangle—Triangle slings, type II; Eye and eye with flat eye slings, type III; Eye and eye with twisted eye slings, type IV ||||||| Endless slings, type V ||||||| Return eye slings, type VI |||||||
|---|---|---|---|---|---|---|---|---|---|---|---|---|---|---|---|---|---|---|
| | Vert. | Choker | Vert. basket | 30° basket | 45° basket | 60° basket | Vert. | Choker | Vert. basket | 30° basket | 45° basket | 60° basket | Vert. | Choker | Vert. basket | 30° basket | 45° basket | 60° basket |
| 1 | 1,200 | 900 | 2,400 | 2,100 | 1,700 | 1,200 | 1,900 | 1,500 | 3,800 | 3,300 | 2,700 | 1,900 | 950 | 750 | 1,900 | 1,650 | 1,350 | 950 |
| 2 | 2,400 | 1,800 | 4,800 | 4,200 | 3,400 | 2,400 | 3,800 | 3,000 | 7,600 | 6,600 | 5,400 | 3,800 | 1,900 | 1,500 | 3,800 | 3,300 | 2,700 | 1,900 |
| 3 | 3,600 | 2,700 | 7,200 | 6,200 | 5,100 | 3,600 | 5,800 | 4,600 | 11,600 | 10,000 | 8,200 | 5,800 | 2,850 | 2,250 | 5,700 | 4,950 | 4,050 | 2,850 |
| 4 | 4,800 | 3,600 | 9,600 | 8,300 | 6,800 | 4,800 | 7,700 | 6,200 | 15,400 | 13,300 | 10,900 | 7,700 | 3,800 | 3,000 | 7,600 | 6,600 | 5,400 | 3,800 |
| 5 | 6,000 | 4,500 | 12,000 | 10,400 | 8,500 | 6,000 | 9,600 | 7,700 | 19,200 | 16,600 | 13,600 | 9,600 | 4,750 | 3,750 | 9,500 | 8,250 | 6,750 | 4,750 |
| 6 | 7,200 | 5,400 | 14,400 | 12,500 | 10,200 | 7,200 | 11,500 | 9,200 | 23,000 | 19,900 | 16,300 | 11,500 | 5,800 | 4,600 | 11,600 | 10,000 | 8,200 | 5,800 |

Notes: 1. All angles shown are measured from the vertical.
2. Capacities for intermediate widths not shown may be obtained by interpolation.

### TABLE N-184-22—SYNTHETIC WEB SLINGS—1,600 POUNDS PER INCH OF WIDTH—SINGLE-PLY
[Rated capacity in pounds]

| Sling body width, inches | Triangle—Choker slings, type I; Triangle—Triangle slings, type II; Eye and eye with flat eye slings, type III; Eye and eye with twisted eye slings, type IV ||||||| Endless slings, type V ||||||| Return eye slings, type VI |||||||
|---|---|---|---|---|---|---|---|---|---|---|---|---|---|---|---|---|---|---|
| | Vert. | Choker | Vert. basket | 30° basket | 45° basket | 60° basket | Vert. | Choker | Vert. basket | 30° basket | 45° basket | 60° basket | Vert. | Choker | Vert. basket | 30° basket | 45° basket | 60° basket |
| 1 | 1,600 | 1,200 | 3,200 | 2,800 | 2,300 | 1,600 | 2,600 | 2,100 | 5,200 | 4,500 | 3,700 | 2,600 | 1,050 | 1,050 | 2,600 | 2,250 | 1,850 | 1,300 |
| 2 | 3,200 | 2,400 | 6,400 | 5,500 | 4,500 | 3,200 | 5,100 | 4,100 | 10,200 | 8,800 | 7,200 | 5,100 | 2,600 | 2,100 | 5,200 | 4,500 | 3,700 | 2,600 |
| 3 | 4,800 | 3,600 | 9,600 | 8,300 | 6,800 | 4,800 | 7,700 | 6,200 | 15,400 | 13,300 | 10,900 | 7,700 | 3,900 | 3,150 | 7,800 | 6,750 | 5,500 | 3,900 |
| 4 | 6,400 | 4,800 | 12,800 | 11,100 | 9,000 | 6,400 | 10,100 | 8,200 | 20,400 | 17,700 | 14,400 | 10,200 | 5,100 | 4,100 | 10,200 | 8,800 | 7,200 | 5,100 |
| 5 | 8,000 | 6,000 | 16,000 | 13,800 | 11,300 | 8,000 | 12,800 | 10,200 | 25,600 | 22,200 | 18,100 | 12,800 | 6,400 | 5,150 | 12,800 | 11,050 | 9,050 | 6,400 |
| 6 | 9,600 | 7,200 | 19,200 | 16,600 | 13,600 | 9,600 | 15,400 | 12,300 | 30,800 | 26,700 | 21,800 | 15,400 | 7,700 | 6,200 | 15,400 | 13,300 | 10,900 | 7,700 |

Notes: 1. All angles shown are measured from the vertical.
2. Capacities for intermediate widths not shown may be obtained by interpolation.

**1910.184(i)(7)** *Safe operating temperatures.* Synthetic web slings of polyester and nylon shall not be used at temperatures in excess of 180 °F. Polypropylene web slings shall not be used at temperatures in excess of 200 °F.

**1910.184(i)(8)** *Repairs.* **(i)** Synthetic web slings which are repaired shall not be used unless repaired by a sling manufacturer or an equivalent entity.

**1910.184(i)(8)(ii)** Each repaired sling shall be proof tested by the manufacturer or equivalent entity to twice the rated capacity prior to its return to service. The employer shall retain a certificate of the proof test and make it available for examination.

**1910.184(i)(8)(iii)** Slings, including webbing and fittings, which have been repaired in a temporary manner shall not be used.

**1910.184(i)(9)** *Removal from service.* Synthetic web slings shall be immediately removed from service if any of the following conditions are present:

**1910.184(i)(9)(i)** Acid or caustic burns;

**1910.184(i)(9)(ii)** Melting or charring of any part of the sling surface;

**1910.184(i)(9)(iii)** Snags, punctures, tears or cuts;

**1910.184(i)(9)(iv)** Broken or worn stitches; or

**1910.184(i)(9)(v)** Distortion of fittings.

[40 FR 27369, June 27, 1975, as amended at 40 FR 31598, July 28, 1975; 41 FR 13353, Mar. 30, 1976; 58 FR 35309, June 30, 1993; 61 FR 9240, Mar. 7, 1996]

### §1910.189 Sources of standards [Removed].
[§1910.189 removed at 61 FR 9240, Mar. 7, 1996, effective May 6, 1996.]

### §1910.190 Standards organizations [Removed].
[§1910.190 removed at 61 FR 9240, Mar. 7, 1996, effective May 6, 1996.]

# Subpart O—Machinery and Machine Guarding

**Authority:** Sections 4, 6, and 8 of the Occupational Safety and Health Act of 1970 (29 U.S.C. 653, 655, 657); Secretary of Labor's Order No. 12-71 (36 FR 8754), 8-76 (41 FR 25059), 9-83 (48 FR 35736), or 1-90 (55 FR 9033), as applicable; 29 CFR part 1911.

### §1910.211 Definitions.

**1910.211(a)** As used in §§1910.213 and 1910.214 unless the context clearly requires otherwise, the following woodworking machinery terms shall have the meaning prescribed in this paragraph.

**1910.211(a)(1)** *Point of operations* means that point at which cutting, shaping, boring, or forming is accomplished upon the stock.

**1910.211(a)(2)** *Push stick* means a narrow strip of wood or other soft material with a notch cut into one end and which is used to push short pieces of material through saws.

**1910.211(a)(3)** *Block* means a short block of wood, provided with a handle similar to that of a plane and a shoulder at the rear end, which is used for pushing short stock over revolving cutters.

**1910.211(b)** As used in §1910.215 unless the context clearly requires otherwise, the following abrasive wheel machinery terms shall have the meanings prescribed in this paragraph.

**1910.211(b)(1)** *Type 1 straight wheels* means wheels having diameter, thickness, and hole size dimensions, and they should be used only on the periphery. Type 1 wheels shall be mounted between flanges.

LIMITATION: Hole dimension (H) should not be greater than two-thirds of wheel diameter dimension (D) for precision, cylindrical, centerless, or surface grinding applications. Maximum hole size for all other applications should not exceed one-half wheel diameter.

**FIGURE NO. O-1**
**TYPE 1 STRAIGHT WHEELS**

TYPE 1 — STRAIGHT WHEEL.
Peripheral grinding wheel having a diameter, thickness and hole.

**1910.211(b)(2)** *Type 2 cylinder wheels* means wheels having diameter, wheel thickness, and rim thickness dimensions. Grinding is performed on the rim face only, dimension W. Cylinder wheels may be plain, plate mounted, inserted nut, or of the projecting stud type.

LIMITATION: Rim height, T dimension, is generally equal to or greater than rim thickness, W dimension.

**FIGURE NO. O-2**
**TYPE 2 CYLINDER WHEELS**

TYPE 2 — CYLINDER WHEEL
Side grinding wheel having a diameter, thickness and wall—wheel is mounted on the diameter.

**1910.211(b)(3)** *Type 6 straight cup wheels* means wheels having diameter, thickness, hole size, rim thickness, and back thickness dimensions. Grinding is always performed on rim face, W dimension.

LIMITATION: Minimum back thickness, E dimension, should not be less than one-fourth T dimension. In addition, when unthreaded hole wheels are specified, the inside flat, K dimension, must be large enough to accommodate a suitable flange.

**FIGURE NO. O-3**
**TYPE 6 STRAIGHT CUP WHEELS**

TYPE 6 — STRAIGHT-CUP WHEEL
Side grinding wheel having a diameter, thickness and hole with one side straight or flat and the opposite side recessed. This type, however, differs from Type 5 in that the grinding is performed on the wall of the abrasive created by the difference between the diameter of the recess and the outside diameter of the wheel. Therefore, the wall dimension "W" takes precedence over the diameter of the recess as an essential intermediate dimension to describe this shape type.

**1910.211(b)(4)** *Type 11 flaring cup wheels* mean wheels having double diameter dimensions D and J, and in addition have thickness, hole size, rim and back thickness dimensions. Grinding is always performed on rim face, W dimension. Type 11 wheels are subject to all limitations of use and mounting listed for type 6 straight sided cup wheels definition.

LIMITATION: Minimum back thickness, E dimension, should not be less than one-fourth T dimension. In addition when unthreaded hole wheels are specified the inside flat, K dimension, shall be large enough to accommodate a suitable flange.

**FIGURE NO. O-4**
**TYPE 11 FLARING CUP WHEELS**

TYPE 11 — FLARING-CUP WHEEL
Side grinding wheel having a wall flared or tapered outward from the back. Wall thickness at the back is normally greater than at the grinding face (W).

**1910.211(b)(5)** *Modified types 6 and 11 wheels (terrazzo)* mean some type 6 and 11 cup wheels used in the terrazzo trade having tapered K dimensions to match a special tapered flange furnished by the machine builder.

LIMITATION: These wheels shall be mounted only with a special tapered flange.

**FIGURE NO. O-5**

Type 6 Wheel (Terrazzo)    Type 11 Wheel (Terrazzo)

Typical examples of modified types 6 and 11 wheels (terrazzo) showing tapered K dimensions.

**1910.211(b)(6)** *Types 27 and 28 depressed center wheels* mean wheels having diameter, thickness, and hole size dimensions. Both types are reinforced, organic bonded wheels having offset hubs which permit side and peripheral grinding operations without interference with the mounting. Type 27 wheels are manufactured with flat grinding rims permitting notching and cutting operations. Type 28 wheels have saucer shaped grinding rims.

**1910.211(b)(6)(i)** Limitations: Special supporting, back adapter and inside flange nuts are required for the proper mounting of these types of wheels subject to limitations of §1910.215(c)(4)(i) and (ii).

**1910.211(b)(6)(ii)** Mounts which are affixed to the wheel by the manufacturer may not require an inside nut and shall not be reused.

**1910.211(b)(7)** *Type 27A depressed center, cutting-off wheels* mean wheels having diameter, thickness, and hole size dimensions. They are reinforced, organic bonded, offset hub type wheels, usually 16 inches diameter and larger, specially designed for use on cutting-off machines where mounting nut or outer flange interference cannot be tolerated.

LIMITATIONS: See §1910.215(c)(1).

**1910.211(b)(8)** *Surface feet per minute* (s.f.p.m.) means the distance in feet any one abrasive grain on the peripheral surface of a grinding wheel travels in 1 minute.

**Surface Feet Per Minute = 3.1416 × diameter in inches × r.p.m. ÷ 12 or .262 × diameter in inches × r.p.m.**

**Examples:** (a) 24-inch diameter wheel, 1,000 revolutions per minute. Surface Feet per minute .262 × 24 × 1,000 = 6,288 s.f.p.m.

(b) 12-inch diameter wheel, 1,000 revolutions per minute. Surface Feet per minute .262 × 12 × 1,000 = 3,144 s.f.p.m.

**1910.211(b)(9)** *Flanges* means collars, discs or plates between which wheels are mounted and are referred to as adaptor, sleeve, or back up type. See paragraph (c) of §1910.215 for full description.

**1910.211(b)(10)** *Snagging* means grinding which removes relatively large amounts of material without regard to close tolerances or surface finish requirements.

**1910.211(b)(11)** *Off-hand grinding* means the grinding of any material or part which is held in the operator's hand.

**1910.211(b)(12)** *Safety guard* means an enclosure designed to restrain the pieces of the grinding wheel and furnish all possible protection in the event that the wheel is broken in operation. See paragraph (b) of §1910.215.

**1910.211(b)(13)** *Cutting off wheels* means wheels having diameter thickness and hole size dimensions and are subject to all limitations of mounting and use listed for type 1 wheels, the definition in subparagraph (1) of this paragraph and paragraph (d) of §1910.215. They may be steel centered, diamond abrasive or organic bonded abrasive of the plain or reinforced type.

**1910.211(b)(13)(i)** Limitation: Cutting off wheels are recommended only for use on specially designed and fully guarded machines and are subject to the following maximum thickness and hole size limitations.

| Wheel diameter (inch) | Max. thickness |
|---|---|
| 6 inch and smaller | 3/18 |
| Larger than 6 inches to 12 inches | 1/4 |
| Larger than 12 inches to 23 inches | 3/8 |
| Larger than 23 inches | 1/2 |

**1910.211(b)(13)(ii)** Maximum hole size for cutting-off wheels should not be larger than 1/4-wheel diameter.

**1910.211(b)(14)** *Abrasive wheel* means a cutting tool consisting of abrasive grains held together by organic or inorganic bonds. Diamond and reinforced wheels are included.

**1910.211(b)(15)** *Organic wheels* means wheels which are bonded by means of an organic material such as resin, rubber, shellac, or other similar bonding agent.

**1910.211(b)(16)** *Inorganic wheels* means wheels which are bonded by means of inorganic material such as clay, glass, porcelain, sodium silicate, magnesium oxychloride, or metal. Wheels bonded with clay, glass, porcelain or related ceramic materials are characterized as *vitrified bonded wheels*.

**1910.211(c)** As used in §1910.216, unless the context clearly requires otherwise, the following mills and calenders in the rubber and plastic industries terms shall have the meanings prescribed in this paragraph.

**1910.211(c)(1)** *Bite* means the nip point between any two inrunning rolls.

**1910.211(c)(2)** *Calender* means a machine equipped with two or more metal rolls revolving in opposite directions and used for continuously sheeting or plying up rubber and plastics compounds and for frictioning or coating materials with rubber and plastics compounds.

**1910.211(c)(3)** *Mill* means a machine consisting of two adjacent metal rolls, set horizontally, which revolve in opposite directions (i.e., toward each other as viewed from above) used for the mechanical working of rubber and plastics compounds.

**1910.211(d)** As used in §1910.217, unless the context clearly requires otherwise, the following power press terms shall have the meaning prescribed in this paragraph.

**1910.211(d)(1)** *Antirepeat* means the part of the clutch/brake control system designed to limit the press to a single stroke if the tripping means is held operated. Antirepeat requires release of all tripping mechanisms before another stroke can be initiated. *Antirepeat* is also called single stroke reset or reset circuit.

**1910.211(d)(2)** *Brake* means the mechanism used on a mechanical power press to stop and/or hold the crankshaft, either directly or through a gear train, when the clutch is disengaged.

**1910.211(d)(3)** *Bolster plate* means the plate attached to the top of the bed of the press having drilled holes or T-slots for attaching the lower die or die shoe.

**1910.211(d)(4)** *Clutch* means the coupling mechanism used on a mechanical power press to couple the flywheel to the crankshaft, either directly or through a gear train.

**1910.211(d)(5)** *Full revolution clutch* means a type of clutch that, when tripped, cannot be disengaged until the crankshaft has completed a full revolution and the press slide a full stroke.

**1910.211(d)(6)** *Part revolution clutch* means a type of clutch that can be disengaged at any point before the crankshaft has completed a full revolution and the press slide a full stroke.

**1910.211(d)(7)** *Direct drive* means the type of driving arrangement wherein no clutch is used; coupling and decoupling of the driving torque is accomplished by energization and deenergization of a motor. Even though not employing a clutch, direct drives match the operational characteristics of "part revolution clutches" because the driving power may be disengaged during the stroke of the press.

**1910.211(d)(8)** *Concurrent* means acting in conjunction, and is used to describe a situation wherein two or more controls exist in an operated condition at the same time.

# OSHA Standards for General Industry

**1910.211(d)(9)** *Continuous* means uninterrupted multiple strokes of the slide without intervening stops (or other clutch control action) at the end of individual strokes.

**1910.211(d)(10)** *Counterbalance* means the mechanism that is used to balance or support the weight of the connecting rods, slide, and slide attachments.

**1910.211(d)(11)** *Device* means a press control or attachment that:

**1910.211(d)(11)(i)** Restrains the operator from inadvertently reaching into the point of operation, or

**1910.211(d)(11)(ii)** Prevents normal press operation if the operator's hands are inadvertently within the point of operation, or

**1910.211(d)(11)(iii)** Automatically withdraws the operator's hands if the operator's hands are inadvertently within the point of operation as the dies close, or

**1910.211(d)(11)(iv)** Prevents the initiation of a stroke, or stops of stroke in progress, when there is an intrusion through the sensing field by any part of the operator's body or by any other object.

**1910.211(d)(12)** *Presence sensing device* means a device designed, constructed and arranged to create a sensing field or area that signals the clutch/brake control to deactivate the clutch and activate the brake of the press when any part of the operator's body or a hand tool is within such field or area.

**1910.211(d)(13)** *Gate or movable barrier device* means a movable barrier arranged to enclose the point of operation before the press stroke can be started.

**1910.211(d)(14)** *Holdout or restraint device* means a mechanism, including attachments for operator's hands, that when anchored and adjusted prevent the operator's hands from entering the point of operation.

**1910.211(d)(15)** *Pull-out device* means a mechanism attached to the operator's hands and connected to the upper die or slide of the press, that is designed, when properly adjusted, to withdraw the operator's hands as the dies close, if the operator's hands are inadvertently within the point of operation.

**1910.211(d)(16)** *Sweep device* means a single or double arm (rod) attached to the upper die or slide of the press and designed to move the operator's hands to a safe position as the dies close, if the operator's hands are inadvertently within the point of operation.

**1910.211(d)(17)** *Two hand control device* means a two hand trip that further requires concurrent pressure from both hands of the operator during a substantial part of the die-closing portion of the stroke of the press.

**1910.211(d)(18)** *Die* means the tooling used in a press for cutting or forming material. An upper and a lower die make a complete set.

**1910.211(d)(19)** *Die builder* means any person who builds dies for power presses.

**1910.211(d)(20)** *Die set* means a tool holder held in alignment by guide posts and bushings and consisting of a lower shoe, an upper shoe or punch holder, and guide posts and bushings.

**1910.211(d)(21)** *Die setter* means an individual who places or removes dies in or from mechanical power presses, and who, as a part of his duties, makes the necessary adjustments to cause the tooling to function properly and safely.

**1910.211(d)(22)** *Die setting* means the process of placing or removing dies in or from a mechanical power press, and the process of adjusting the dies, other tooling and safeguarding means to cause them to function properly and safely.

**1910.211(d)(23)** *Die shoe* means a plate or block upon which a die holder is mounted. A die shoe functions primarily as a base for the complete die assembly, and, when used, is bolted or clamped to the bolster plate or the face of slide.

**1910.211(d)(24)** *Ejector* means a mechanism for removing work or material from between the dies.

**1910.211(d)(25)** *Face of slide* means the bottom surface of the slide to which the punch or upper die is generally attached.

**1910.211(d)(26)** *Feeding* means the process of placing or removing material within or from the point of operation.

**1910.211(d)(27)** *Automatic feeding* means feeding wherein the material or part being processed is placed within or removed from the point of operation by a method or means not requiring action by an operator on each stroke of the press.

**1910.211(d)(28)** *Semiautomatic feeding* means feeding wherein the material or part being processed is placed within or removed from the point of operation by an auxiliary means controlled by operator on each stroke of the press.

**1910.211(d)(29)** *Manual feeding* means feeding wherein the material or part being processed is handled by the operator on each stroke of the press.

**1910.211(d)(30)** *Foot control* means the foot operated control mechanism designed to be used with a clutch or clutch/brake control system.

**1910.211(d)(31)** *Foot pedal* means the foot operated lever designed to operate the mechanical linkage that trips a full revolution clutch.

**1910.211(d)(32)** *Guard* means a barrier that prevents entry of the operator's hands or fingers into the point of operation.

**1910.211(d)(33)** *Die enclosure guard* means an enclosure attached to the die shoe or stripper, or both, in a fixed position.

**1910.211(d)(34)** *Fixed barrier guard* means a die space barrier attached to the press frame.

**1910.211(d)(35)** *Interlocked press barrier guard* means a barrier attached to the press frame and interlocked so that the press stroke cannot be started normally unless the guard itself, or its hinged or movable sections, enclose the point of operation.

**1910.211(d)(36)** *Adjustable barrier guard* means a barrier requiring adjustment for each job or die setup.

**1910.211(d)(37)** *Guide post* means the pin attached to the upper or lower die shoe operating within the bushing on the opposing die shoe, to maintain the alignment of the upper and lower dies.

**1910.211(d)(38)** *Hand feeding tool* means any hand held tool designed for placing or removing material or parts to be processed within or from the point of operation.

**1910.211(d)(39)** *Inch* means an intermittent motion imparted to the slide (on machines using part revolution clutches) by momentary operation of the *Inch* operating means. Operation of the *Inch* operating means engages the driving clutch so that a small portion of one stroke or indefinite stroking can occur, depending upon the length of time the *Inch* operating means is held operated. *Inch* is a function used by the die setter for setup of dies and tooling, but is not intended for use during production operations by the operator.

**1910.211(d)(40)** *Jog* means an intermittent motion imparted to the slide by momentary operation of the drive motor, after the clutch is engaged with the flywheel at rest.

**1910.211(d)(41)** *Knockout* means a mechanism for releasing material from either die.

**1910.211(d)(42)** *Liftout* means the mechanism also known as knockout.

**1910.211(d)(43)** *Operator's station* means the complete complement of controls used by or available to an operator on a given operation for stroking the press.

**1910.211(d)(44)** *Pinch point* means any point other than the point of operation at which it is possible for a part of the body to be caught between the moving parts of a press or auxiliary equipment, or between moving and stationary parts of a press or auxiliary equipment or between the material and moving part or parts of the press or auxiliary equipment.

**1910.211(d)(45)** *Point of operation* means the area of the press where material is actually positioned and work is being performed during any process such as shearing, punching, forming, or assembling.

**1910.211(d)(46)** *Press* means a mechanically powered machine that shears, punches, forms or assembles metal or other material by means of cutting, shaping, or combination dies attached to slides. A press consists of a stationary bed or anvil, and a slide (or slides) having a controlled reciprocating motion toward and away from the bed surface, the slide being guided in a definite path by the frame of the press.

**1910.211(d)(47)** *Repeat* means an unintended or unexpected successive stroke of the press resulting from a malfunction.

**1910.211(d)(48)** *Safety block* means a prop that, when inserted between the upper and lower dies or between the bolster plate and the face of the slide, prevents the slide from falling of its own deadweight.

**1910.211(d)(49)** *Single stroke* means one complete stroke of the slide, usually initiated from a full open (or up) position, followed by closing (or down), and then a return to the full open position.

**1910.211(d)(50)** *Single stroke mechanism* means an arrangement used on a full revolution clutch to limit the travel of the slide to one complete stroke at each engagement of the clutch.

**1910.211(d)(51)** *Slide* means the main reciprocating press member. A slide is also called a ram, plunger, or platen.

**1910.211(d)(52)** *Stop control* means an operator control designed to immediately deactivate the clutch control and activate the brake to stop slide motion.

**1910.211(d)(53)** *Stripper* means a mechanism or die part for removing the parts or material from the punch.

**1910.211(d)(54)** *Stroking selector* means the part of the clutch/brake control that determines the type of stroking when the operating means is actuated. The stroking selector generally includes positions for "Off" (Clutch Control), "Inch," "Single Stroke," and "Continuous" (when Continuous is furnished).

**1910.211(d)(55)** *Trip or (tripping)* means activation of the clutch to "run" the press.

**1910.211(d)(56)** *Turnover bar* means a bar used in die setting to manually turn the crankshaft of the press.

**1910.211(d)(57)** *Two-hand trip* means a clutch actuating means requiring the concurrent use of both hands of the operator to trip the press.

**1910.211(d)(58)** *Unitized tooling* means a type of die in which the upper and lower members are incorporated into a selfcontained unit so arranged as to hold the die members in alignment.

**1910.211(d)(59)** *Control system* means sensors, manual input and mode selection elements, interlocking and decision-making circuitry, and output elements to the press operating mechanism.

**1910.211(d)(60)** *Brake monitor* means a sensor designed, constructed, and arranged to monitor the effectiveness of the press braking system.

**1910.211(d)(61)** *Presence sensing device initiation* means an operating mode of indirect manual initiation of a single stroke by a presence sensing device when it senses that work motions of the operator, related to feeding and/or removing parts, are completed and all parts of the operator's body or hand tools are safely clear of the point of operation.

**1910.211(d)(62)** *Safety system* means the integrated total system, including the pertinent elements of the press, the controls, the safeguarding and any required supplemental safeguarding, and their interfaces with the operator, and the environment, designed, constructed and arranged to operate together as a unit, such that a single failure or single operating error will not cause injury to personnel due to point of operation hazards.

**1910.211(d)(63)** *Authorized person* means one to whom the authority and responsibility to perform a specific assignment has been given by the employer.

**1910.211(d)(64)** *Certification* or *certify* means, in the case of design certification/validation, that the manufacturer has reviewed and tested the design and manufacture, and in the case of installation certification/validation and annual recertification/revalidation, that the employer has reviewed and tested the installation, and concludes in both cases that the requirements of §1910.217(a) through (h) and appendix A have been met. The certifications are made to the validation organization.

**1910.211(d)(65)** *Validation* or *validate* means for PSDI safety systems that an OSHA recognized third-party validation organization:

**1910.211(d)(65)(i)** For design certification/validation has reviewed the manufacturer's certification that the PSDI safety system meets the requirements of §1910.217(a) through (h) and appendix A and the underlying tests and analyses performed by the manufacturer, has performed additional tests and analyses which may be required by §1910.217(a) through (h) and appendix A, and concludes that the requirements of §1910.217(a) through (h) and appendix A have been met; and

**1910.211(d)(65)(ii)** For installation certification/validation and annual recertification/revalidation has reviewed the employer's certification that the PSDI safety system meets the requirements of §1910.217(a) through (h) and appendix A and the underlying tests performed by the employer, has performed additional tests and analyses which may be required by §1910.217(a) through (h) and appendix A, and concludes that the requirements of §1910.217(a) through (h) and appendix A have been met.

**1910.211(d)(66)** *Certification/validation* and *certify/validate* means the combined process of certification and validation.

**1910.211(e)** As used in §1910.218, unless the context clearly requires otherwise, the following forging and hot metal terms shall have the meaning prescribed in this paragraph.

**1910.211(e)(1)** *Forging* means the product of work on metal formed to a desired shape by impact or pressure in hammers, forging machines (upsetters), presses, rolls, and related forming equipment. Forging hammers, counterblow equipment and high-energy-rate forging machines impart impact to the workpiece, while most other types of forging equipment impart squeeze pressure in shaping the stock. Some metals can be forged at room temperature, but the majority of metals are made more plastic for forging by heating.

**1910.211(e)(2)** *Open framehammers (or blacksmith hammers)* mean hammers used primarily for the shaping of forgings by means of impact with flat dies. Open frame hammers generally are so constructed that the anvil assembly is separate from the operating mechanism and machine supports; it rests on its own independent foundation. Certain exceptions are forging hammers made with frame mounted on the anvil; e.g., the smaller, single-frame hammers are usually made with the anvil and frame in one piece.

**1910.211(e)(3)** *Steam hammers* mean a type of drop hammer where the ram is raised for each stroke by a double-action steam cylinder and the energy delivered to the workpiece is supplied by the velocity and weight of the ram and attached upper die driven downward by steam pressure. Energy delivered during each stroke may be varied.

**1910.211(e)(4)** *Gravity hammers* mean a class of forging hammer wherein energy for forging is obtained by the mass and velocity of a freely falling ram and the attached upper die. Examples: board hammers and air-lift hammers.

**1910.211(e)(5)** *Forging presses* mean a class of forging equipment wherein the shaping of metal between dies is performed by mechanical or hydraulic pressure, and usually is accomplished with a single workstroke of the press for each die station.

**1910.211(e)(6)** *Trimming presses* mean a class of auxiliary forging equipment which removes flash or excess metal from a forging. This trimming operation can also be done cold, as can coining, a product sizing operation.

**1910.211(e)(7)** *High-energy-rate forging machines* mean a class of forging equipment wherein high ram velocities resulting from the sudden release of a compressed gas against a free piston impart impact to the workpiece.

**1910.211(e)(8)** *Forging rolls* mean a class of auxiliary forging equipment wherein stock is shaped between power driven rolls bearing contoured dies. Usually used for preforming, roll forging is often employed to reduce thickness and increase length of stock.

**1910.211(e)(9)** *Ring rolls* mean a class for forging equipment used for shaping weldless rings from pierced discs or thick-walled, ring-shaped blanks between rolls which control wall thickness, ring diameter, height and contour.

**1910.211(e)(10)** *Bolt-headers* mean the same as an upsetter or forging machine except that the diameter of stock fed into the machine is much smaller, i.e., commonly three-fourths inch or less.

**1910.211(e)(11)** *Rivet making machines* mean the same as upsetters and boltheaders when producing rivets with stock diameter of 1-inch or more. Rivet making with less than 1-inch diameter is usually a cold forging operation, and therefore not included in this subpart.

**1910.211(e)(12)** *Upsetters (or forging machines, or headers)* type of forging equipment, related to the mechanical press, in which the main forming energy is applied horizontally to the workpiece which is gripped and held by prior action of the dies.

**1910.211(f)** As used in §1910.219, unless the context clearly requires otherwise, the following mechanical power-transmission guarding terms shall have the meaning prescribed in this paragraph.

**1910.211(f)(1)** *Belts* include all power transmission belts, such as flat belts, round belts, V-belts, etc., unless otherwise specified.

**1910.211(f)(2)** *Belt shifter* means a device for mechanically shifting belts from tight to loose pulleys or vice versa, or for shifting belts on cones of speed pulleys.

**1910.211(f)(3)** *Belt pole* (sometimes called a *belt shipper* or *shipper pole,*) means a device used in shifting belts on and off fixed pulleys on line or countershaft where there are no loose pulleys.

**1910.211(f)(4)** *Exposed to contact* means that the location of an object is such that a person is likely to come into contact with it and be injured.

**1910.211(f)(5)** *Flywheels* include flywheels, balance wheels, and flywheel pulleys mounted and revolving on crankshaft of engine or other shafting.

**1910.211(f)(6)** *Maintenance runway* means any permanent runway or platform used for oiling, maintenance, running adjustment, or repair work, but not for passageway.

**1910.211(f)(7)** *Nip-point belt and pulley guard* means a device which encloses the pulley and is provided with rounded or rolled edge slots through which the belt passes.

**1910.211(f)(8)** *Point of operation* means that point at which cutting, shaping, or forming is accomplished upon the stock and shall include such other points as may offer a hazard to the operator in inserting or manipulating the stock in the operation of the machine.

**1910.211(f)(9)** *Prime movers* include steam, gas, oil, and air engines, motors, steam and hydraulic turbines, and other equipment used as a source of power.

**1910.211(f)(10)** *Sheaves* mean grooved pulleys, and shall be so classified unless used as flywheels.

[39 FR 23502, June 27, 1974, as amended at 39 FR 41846, Dec. 3, 1974; 53 FR 8353, Mar. 14, 1988]

## §1910.212 General requirements for all machines.

**1910.212(a)** *Machine guarding*—**(1)** *Types of guarding.* One or more methods of machine guarding shall be provided to protect the operator and other employees in the machine area from hazards such as those created by point of operation, ingoing nip points, rotating parts, flying chips and sparks. Examples of guarding methods are—barrier guards, two-hand tripping devices, electronic safety devices, etc.

**1910.212(a)(2)** *General requirements for machine guards.* Guards shall be affixed to the machine where possible and secured elsewhere if for any reason attachment to the machine is not possible. The guard shall be such that it does not offer an accident hazard in itself.

**1910.212(a)(3)** *Point of operation guarding.* **(i)** Point of operation is the area on a machine where work is actually performed upon the material being processed.

**1910.212(a)(3)(ii)** The point of operation of machines whose operation exposes an employee to injury, shall be guarded. The guarding device shall be in conformity with any appropriate standards therefor, or, in the absence of applicable specific standards, shall be so designed and constructed as to prevent the operator from having any part of his body in the danger zone during the operating cycle.

**1910.212(a)(3)(iii)** Special handtools for placing and removing material shall be such as to permit easy handling of material without the operator placing a hand in the danger zone. Such tools shall not be in lieu of other guarding required by this section, but can only be used to supplement protection provided.

**1910.212(a)(3)(iv)** The following are some of the machines which usually require point of operation guarding:

**1910.212(a)(3)(iv)(a)** Guillotine cutters.

**1910.212(a)(3)(iv)(b)** Shears.

**1910.212(a)(3)(iv)(c)** Alligator shears.

**1910.212(a)(3)(iv)(d)** Power presses.

**1910.212(a)(3)(iv)(e)** Milling machines.

**1910.212(a)(3)(iv)(f)** Power saws.

**1910.212(a)(3)(iv)(g)** Jointers.

**1910.212(a)(3)(iv)(h)** Portable power tools.

**1910.212(a)(3)(iv)(i)** Forming rolls and calenders.

**1910.212(a)(4)** *Barrels, containers, and drums.* Revolving drums, barrels, and containers shall be guarded by an enclosure which is interlocked with the drive mechanism, so that the barrel, drum, or container cannot revolve unless the guard enclosure is in place.

**1910.212(a)(5)** *Exposure of blades.* When the periphery of the blades of a fan is less than seven (7) feet above the floor or working level, the blades shall be guarded. The guard shall have openings no larger than one-half (1/2) inch.

**1910.212(b)** *Anchoring fixed machinery.* Machines designed for a fixed location shall be securely anchored to prevent walking or moving.

## §1910.213 Woodworking machinery requirements.

**1910.213(a)** *Machine construction general.* **(1)** Each machine shall be so constructed as to be free from sensible vibration when the largest size tool is mounted and run idle at full speed.

**1910.213(a)(2)** Arbors and mandrels shall be constructed so as to have firm and secure bearing and be free from play.

**1910.213(a)(3)** [Reserved]

**1910.213(a)(4)** Any automatic cutoff saw that strokes continuously without the operator being able to control each stroke shall not be used.

**1910.213(a)(5)** Saw frames or tables shall be constructed with lugs cast on the frame or with an equivalent means to limit the size of the saw blade that can be mounted, so as to avoid overspeed caused by mounting a saw larger than intended.

**1910.213(a)(6)** Circular saw fences shall be so constructed that they can be firmly secured to the table or table assembly without changing their alignment with the saw. For saws with tilting tables or tilting arbors the fence shall be so constructed that it will remain in a line parallel with the saw, regardless of the angle of the saw with the table.

**1910.213(a)(7)** Circular saw gages shall be so constructed as to slide in grooves or tracks that are accurately machined, to insure exact alignment with the saw for all positions of the guide.

**1910.213(a)(8)** Hinged saw tables shall be so constructed that the table can be firmly secured in any position and in true alignment with the saw.

**1910.213(a)(9)** All belts, pulleys, gears, shafts, and moving parts shall be guarded in accordance with the specific requirements of §1910.219.

**1910.213(a)(10)** It is recommended that each power-driven woodworking machine be provided with a disconnect switch that can be locked in the off position.

**1910.213(a)(11)** The frames and all exposed, noncurrent-carrying metal parts of portable electric woodworking machinery operated at more than 90 volts to ground shall be grounded and other portable motors driving electric tools which are held in the hand while being operated shall be grounded if they operate at more than 90 volts to ground. The ground shall be provided through use of a separate ground wire and polarized plug and receptacle.

**1910.213(a)(12)** For all circular saws where conditions are such that there is a possibility of contact with the portion of the saw either beneath or behind the table, that portion of the saw shall be covered with an exhaust hood, or, if no exhaust system is required, with a guard that shall be so arranged as to prevent accidental contact with the saw.

**1910.213(a)(13)** Revolving double arbor saws shall be fully guarded in accordance with all the requirements for circular crosscut saws or with all the requirements for circular ripsaws, according to the kind of saws mounted on the arbors.

**1910.213(a)(14)** No saw, cutter head, or tool collar shall be placed or mounted on a machine arbor unless the tool has been accurately machined to size and shape to fit the arbor.

**1910.213(a)(15)** Combs (featherboards) or suitable jigs shall be provided at the workplace for use when a standard guard cannot be used, as in dadoing, grooving, jointing, moulding, and rabbeting.

**1910.213(b)** *Machine controls and equipment.* **(1)** A mechanical or electrical power control shall be provided on each machine to make it possible for the operator to cut off the power from each machine without leaving his position at the point of operation.

**1910.213(b)(2)** On machines driven by belts and shafting, a locking-type belt shifter or an equivalent positive device shall be used.

**1910.213(b)(3)** On applications where injury to the operator might result if motors were to restart after power failures, provision shall be made to prevent machines from automatically restarting upon restoration of power.

**1910.213(b)(4)** Power controls and operating controls should be located within easy reach of the operator while he is at his regular work location, making it unnecessary for him to reach over the cutter to make adjustments. This does not apply to constant pressure controls used only for setup purposes.

**1910.213(b)(5)** On each machine operated by electric motors, positive means shall be provided for rendering such controls or devices inoperative while repairs or adjustments are being made to the machines they control.

**1910.213(b)(6)** Each operating treadle shall be protected against unexpected or accidental tripping.

**1910.213(b)(7)** Feeder attachments shall have the feed rolls or other moving parts so covered or guarded as to protect the operator from hazardous points.

**1910.213(c)** *Hand-fed ripsaws.* **(1)** Each circular hand-fed ripsaw shall be guarded by a hood which shall completely enclose that portion of the saw above the table and that portion of the saw above the material being cut. The hood and mounting shall be arranged so that the hood will automatically adjust itself to the thickness of and remain in contact with the material being cut but it shall not offer any considerable resistance to insertion of material to saw or to passage of the material being sawed. The hood shall be made of adequate strength to resist blows and strains incidental to reasonable operation, adjusting, and handling, and shall be so designed as to protect the operator from flying splinters and broken saw teeth. It shall be made of material that is soft enough so that it will be unlikely to cause tooth breakage. The hood shall be so mounted as to insure that its operation will be positive, reliable, and in true alignment with the saw; and the mounting shall be adequate in strength to resist any reasonable side thrust or other force tending to throw it out of line.

**1910.213(c)(2)** Each hand-fed circular ripsaw shall be furnished with a spreader to prevent material from squeezing the saw or being thrown back on the operator. The spreader shall be made of hard tempered steel, or its equivalent, and shall be thinner than the saw kerf. It shall be of sufficient width to provide adequate stiffness or rigidity to resist any reasonable side thrust or blow tending to bend or throw it out of position. The spreader shall be attached so that it will remain in true alignment with the saw even when either the saw or table is tilted. The provision of a spreader in connection with grooving, dadoing, or rabbeting is not required. On the completion of such operations, the spreader shall be immediately replaced.

**1910.213(c)(3)** Each hand-fed circular ripsaw shall be provided with non-kickback fingers or dogs so located as to oppose the thrust or tendency of the saw to pick up the material or to throw it back toward the operator. They shall be designed to provide adequate holding power for all the thicknesses of materials being cut.

**1910.213(d)** *Hand-fed crosscut table saws.* **(1)** Each circular crosscut table saw shall be guarded by a hood which shall meet all the requirements of paragraph (c)(1) of this section for hoods for circular ripsaws.

**1910.213(d)(2)** [Reserved]

**1910.213(e)** *Circular resaws.* **(1)** Each circular resaw shall be guarded by a hood or shield of metal above the saw. This hood or shield shall be so designed as to guard against danger from flying splinters or broken saw teeth.

**1910.213(e)(2)** Each circular resaw (other than self-feed saws with a roller or wheel at back of the saw) shall be provided with a spreader fastened securely behind the saw. The spreader shall be slightly thinner than the saw kerf and slightly thicker than the saw disk.

**1910.213(f)** *Self-feed circular saws.* **(1)** Feed rolls and saws shall be protected by a hood or guard to prevent the hands of the operator from coming in contact with the in-running rolls at any point. The guard shall be constructed of heavy material, preferably metal, and the bottom of the guard shall come down to within three-eighths inch of the plane formed by the bottom or working surfaces of the feed rolls. This distance (three-eighths inch) may be increased to three-fourths inch, provided the lead edge of the hood is extended to be not less than 5 1/2 inches in front of the nip point between the front roll and the work.

**1910.213(f)(2)** Each self-feed circular ripsaw shall be provided with sectional non-kickback fingers for the full width of the feed rolls. They shall be located in front of the saw and so arranged as to be in continual contact with the wood being fed.

**1910.213(g)** *Swing cutoff saws.* The requirements of this paragraph are also applicable to sliding cutoff saws mounted above the table.

**1910.213(g)(1)** Each swing cutoff saw shall be provided with a hood that will completely enclose the upper half of the saw, the arbor end, and the point of operation at all positions of the saw. The hood shall be constructed in such a manner and of such material that it will protect the operator from flying splinters and broken saw teeth. Its hood shall be so designed that it will automatically cover the lower portion of the blade, so that when the saw is returned to the back of the table the hood will rise on top of the fence, and when the saw is moved forward the hood will drop on top of and remain in contact with the table or material being cut.

**1910.213(g)(2)** Each swing cutoff saw shall be provided with an effective device to return the saw automatically to the back of the table when released at any point of its travel. Such a device shall not depend for its proper functioning upon any rope, cord, or spring. If there is a counterweight, the bolts supporting the bar and counterweight shall be provided with cotter pins; and the counterweight shall be prevented from dropping by either a bolt passing through both the bar and counterweight, or a bolt put through the extreme end of the bar, or, where the counterweight does not encircle the bar, a safety chain attached to it.

**1910.213(g)(3)** Limit chains or other equally effective devices shall be provided to prevent the saw from swinging beyond the front or back edges of the table, or beyond a forward position where the gullets of the lowest saw teeth will rise above the table top.

**1910.213(g)(4)** Inverted swing cutoff saws shall be provided with a hood that will cover the part of the saw that protrudes above the top of the table or above the material being cut. It shall automatically adjust itself to the thickness of and remain in contact with the material being cut.

**1910.213(h)** *Radial saws.* **(1)** The upper hood shall completely enclose the upper portion of the blade down to a point that will include the end of the saw arbor. The upper hood shall be constructed in such a manner and of such material that it will protect the operator from flying splinters, broken saw teeth, etc., and will deflect sawdust away from the operator. The sides of the lower exposed portion of the blade shall be guarded to the full diameter of the blade by a device that will automatically adjust itself to the thickness of the stock and remain in contact with stock being cut to give maximum protection possible for the operation being performed.

**1910.213(h)(2)** Each radial saw used for ripping shall be provided with nonkickback fingers or dogs located on both sides of the saw so as to oppose the thrust or tendency of the saw to pick up the material or to throw it back toward the operator. They shall be designed to provide adequate holding power for all the thicknesses of material being cut.

**1910.213(h)(3)** An adjustable stop shall be provided to prevent the forward travel of the blade beyond the position necessary to complete the cut in repetitive operations.

**1910.213(h)(4)** Installation shall be in such a manner that the front end of the unit will be slightly higher than the rear, so as to cause the cutting head to return gently to the starting position when released by the operator.

**1910.213(h)(5)** Ripping and ploughing shall be against the direction in which the saw turns. The direction of the saw rotation shall be conspicuously marked on the hood. In addition, a permanent label not less than 1 1/2 inches by 3/4 inch shall be affixed to the rear of the guard at approximately the level of the arbor, reading as follows: "Danger: Do Not Rip or Plough From This End".

**1910.213(i)** *Bandsaws and band resaws.* **(1)** All portions of the saw blade shall be enclosed or guarded, except for the working portion of the blade between the bottom of the guide rolls and the table. Bandsaw wheels shall be fully encased. The outside periphery of the enclosure shall be solid. The front and back of the band wheels shall be either enclosed by solid material or by wire mesh or perforated metal. Such mesh or perforated metal shall be not less than 0.037 inch (U.S. Gage No. 20), and the openings shall be not greater than three-eighths inch. Solid material used for this purpose shall be of an equivalent strength and firmness. The guard for the portion of the blade between the sliding guide and the upper-saw-wheel guard shall protect the saw blade at the front and outer side. This portion of the guard shall be self-adjusting to raise and lower with the guide. The upper-wheel guard shall be made to conform to the travel of the saw on the wheel.

**1910.213(i)(2)** Each bandsaw machine shall be provided with a tension control device to indicate a proper tension for the standard saws used on the machine, in order to assist in the elimination of saw breakage due to improper tension.

**1910.213(i)(3)** Feed rolls of band resaws shall be protected with a suitable guard to prevent the hands of the operator from coming in contact with the in-running rolls at any point. The guard shall be constructed of heavy material, preferably metal, and the edge of the guard shall come to within three-eighths inch of the plane formed by the inside face of the feed roll in contact with the stock being cut.

**1910.213(j)** *Jointers.* **(1)** Each hand-fed planer and jointer with horizontal head shall be equipped with a cylindrical cutting head, the knife projection of which shall not exceed one-eighth inch beyond the cylindrical body of the head.

# OSHA Standards for General Industry

**1910.213(j)(2)** The opening in the table shall be kept as small as possible. The clearance between the edge of the rear table and the cutter head shall be not more than one-eighth inch. The table throat opening shall be not more than 2 1/2 inches when tables are set or aligned with each other for zero cut.

**1910.213(j)(3)** Each hand-fed jointer with a horizontal cutting head shall have an automatic guard which will cover all the section of the head on the working side of the fence or gage. The guard shall effectively keep the operator's hand from coming in contact with the revolving knives. The guard shall automatically adjust itself to cover the unused portion of the head and shall remain in contact with the material at all times.

**1910.213(j)(4)** Each hand-fed jointer with horizontal cutting head shall have a guard which will cover the section of the head back of the gage or fence.

**1910.213(j)(5)** Each wood jointer with vertical head shall have either an exhaust hood or other guard so arranged as to enclose completely the revolving head, except for a slot of such width as may be necessary and convenient for the application of the material to be jointed.

**1910.213(k)** *Tenoning machines.* **(1)** Feed chains and sprockets of all double end tenoning machines shall be completely enclosed, except for that portion of chain used for conveying the stock.

**1910.213(k)(2)** At the rear ends of frames over which feed conveyors run, sprockets and chains shall be guarded at the sides by plates projecting beyond the periphery of sprockets and the ends of lugs.

**1910.213(k)(3)** Each tenoning machine shall have all cutting heads, and saws if used, covered by metal guards. These guards shall cover at least the unused part of the periphery of the cutting head. If such a guard is constructed of sheet metal, the material used shall be not less than one-sixteenth inch in thickness, and if cast iron is used, it shall be not less than three-sixteenths inch in thickness.

**1910.213(k)(4)** Where an exhaust system is used, the guard shall form part or all of the exhaust hood and shall be constructed of metal of a thickness not less than that specified in subparagraph (3) of this paragraph.

**1910.213(l)** *Boring and mortising machines.* **(1)** Safety-bit chucks with no projecting set screws shall be used.

**1910.213(l)(2)** Boring bits should be provided with a guard that will enclose all portions of the bit and chuck above the material being worked.

**1910.213(l)(3)** The top of the cutting chain and driving mechanism shall be enclosed.

**1910.213(l)(4)** If there is a counterweight, one of the following or equivalent means shall be used to prevent its dropping:

**1910.213(l)(4)(i)** It shall be bolted to the bar by means of a bolt passing through both bar and counterweight;

**1910.213(l)(4)(ii)** A bolt shall be put through the extreme end of the bar;

**1910.213(l)(4)(iii)** Where the counterweight does not encircle the bar, a safety chain shall be attached to it;

**1910.213(l)(4)(iv)** Other types of counterweights shall be suspended by chain or wire rope and shall travel in a pipe or other suitable enclosure wherever they might fall and cause injury.

**1910.213(l)(5)** Universal joints on spindles of boring machines shall be completely enclosed in such a way as to prevent accidental contact by the operator.

**1910.213(l)(6)** Each operating treadle shall be covered by an inverted U-shaped metal guard, fastened to the floor, and of adequate size to prevent accidental tripping.

**1910.213(m)** *Wood shapers and similar equipment.* **(1)** The cutting heads of each wood shaper, hand-fed panel raiser, or other similar machine not automatically fed, shall be enclosed with a cage or adjustable guard so designed as to keep the operator's hand away from the cutting edge. The diameter of circular shaper guards shall be not less than the greatest diameter of the cutter. In no case shall a warning device of leather or other material attached to the spindle be acceptable.

**1910.213(m)(2)** [Reserved]

**1910.213(m)(3)** All double-spindle shapers shall be provided with a spindle starting and stopping device for each spindle.

**1910.213(n)** *Planing, molding, sticking, and matching machines.* **(1)** Each planing, molding, sticking, and matching machine shall have all cutting heads, and saws if used, covered by a metal guard. If such guard is constructed of sheet metal, the material used shall be not less than 1/16 inch in thickness, and if cast iron is used, it shall be not less than three-sixteenths inch in thickness.

**1910.213(n)(2)** Where an exhaust system is used, the guards shall form part or all of the exhaust hood and shall be constructed of metal of a thickness not less than that specified in paragraph (h)(1) of this section.

**1910.213(n)(3)** Feed rolls shall be guarded by a hood or suitable guard to prevent the hands of the operator from coming in contact with the in-running rolls at any point. The guard shall be fastened to the frame carrying the rolls so as to remain in adjustment for any thickness of stock.

**1910.213(n)(4)** Surfacers or planers used in thicknessing multiple pieces of material simultaneously shall be provided with sectional infeed rolls having sufficient yield in the construction of the sections to provide feeding contact pressure on the stock, over the permissible range of variation in stock thickness specified or for which the machine is designed. In lieu of such yielding sectional rolls, suitable section kickback finger devices shall be provided at the infeed end.

**1910.213(o)** *Profile and swing-head lathes and wood heel turning machine.* **(1)** Each profile and swing-head lathe shall have all cutting heads covered by a metal guard. If such a guard is constructed of sheet metal, the material used shall be not less than one-sixteenth inch in thickness; and if cast iron is used, it shall not be less than three-sixteenths inch in thickness.

**1910.213(o)(2)** Cutting heads on wood-turning lathes, whether rotating or not, shall be covered as completely as possible by hoods or shields.

**1910.213(o)(3)** Shoe last and spoke lathes, doweling machines, wood heel turning machines, and other automatic wood-turning lathes of the rotating knife type shall be equipped with hoods enclosing the cutter blades completely except at the contact points while the stock is being cut.

**1910.213(o)(4)** Lathes used for turning long pieces of wood stock held only between the two centers shall be equipped with long curved guards extending over the tops of the lathes in order to prevent the work pieces from being thrown out of the machines if they should become loose.

**1910.213(o)(5)** Where an exhaust system is used, the guard shall form part or all of the exhaust hood and shall be constructed of metal of a thickness not less than that specified in subparagraph (1) of this paragraph.

**1910.213(p)** *Sanding machines.* **(1)** Feed rolls of self-feed sanding machines shall be protected with a semicylindrical guard to prevent the hands of the operator from coming in contact with the in-running rolls at any point. The guard shall be constructed of heavy material, preferably metal, and firmly secured to the frame carrying the rolls so as to remain in adjustment for any thickness of stock. The bottom of the guard should come down to within three-eighths inch of a plane formed by the bottom or contact face of the feed roll where it touches the stock.

**1910.213(p)(2)** Each drum sanding machine shall have an exhaust hood, or other guard if no exhaust system is required, so arranged as to enclose the revolving drum, except for that portion of the drum above the table, if a table is used, which may be necessary and convenient for the application of the material to be finished.

**1910.213(p)(3)** Each disk sanding machine shall have the exhaust hood, or other guard if no exhaust system is required, so arranged as to enclose the revolving disk, except for that portion of the disk above the table, if a table is used, which may be necessary for the application of the material to be finished.

**1910.213(p)(4)** Belt sanding machines shall be provided with guards at each nip point where the sanding belt runs on to a pulley. These guards shall effectively prevent the hands or fingers of the operator from coming in contact with the nip points. The unused run of the sanding belt shall be guarded against accidental contact.

**1910.213(q)** *Veneer cutters and wringers.* **(1)** Veneer slicer knives shall be guarded to prevent accidental contact with knife edge, at both front and rear.

**1910.213(q)(2)** Veneer clippers shall have automatic feed or shall be provided with a guard which will make it impossible to place a finger or fingers under the knife while feeding or removing the stock.

**1910.213(q)(3)** Sprockets on chain or slat-belt conveyors shall be enclosed.

**1910.213(q)(4)** Where practicable, hand and footpower guillotine veneer cutters shall be provided with rods or plates or other satisfactory means, so arranged on the feeding side that the hands cannot reach the cutting edge of the knife while feeding or holding the stock in place.

**1910.213(q)(5)** Power-driven guillotine veneer cutters, except continuous feed trimmers, shall be equipped with:

**1910.213(q)(5)(i)** Starting devices which require the simultaneous action of both hands to start the cutting motion and of at least one hand on a control during the complete stroke of the knife; or

**1910.213(q)(5)(ii)** An automatic guard which will remove the hands of the operator from the danger zone at every descent of the blade, used in conjunction with one-hand starting devices which require two distinct movements of the device to start the cutting motion, and so designed as to return positively to the nonstarting position after each complete cycle of the knife.

**1910.213(q)(6)** Where two or more workers are employed at the same time on the same power-driven guillotine veneer cutter equipped with two-hand control, the device shall be so arranged that each worker shall be required to use both hands simultaneously on the controls to start the cutting motion, and at least one hand on a control to complete the cut.

**1910.213(q)(7)** Power-driven guillotine veneer cutters, other than continuous trimmers, shall be provided, in addition to the brake or other stopping mechanism, with an emergency device which will prevent the machine from operating in the event of failure of the brake when the starting mechanism is in the nonstarting position.

**1910.213(r)** *Miscellaneous woodworking machines.* **(1)** The feed rolls of roll type glue spreaders shall be guarded by a semicylindrical guard. The bottom of the guard shall come to within three-eighths inch of a plane formed by bottom or contact face of the feed roll where it touches the stock.

**1910.213(r)(2)** Drag saws shall be so located as to give at least a 4-foot clearance for passage when the saw is at the extreme end of the stroke; or if such clearance is not obtainable, the saw and its driving mechanism shall be provided with a standard enclosure.

**1910.213(r)(3)** For combination or universal woodworking machines each point of operation of any tool shall be guarded as required for such a tool in a separate machine.

**1910.213(r)(4)** The mention of specific machines in paragraphs (a) through (q) and this paragraph (r) of this section, inclusive, is not intended to exclude other woodworking machines from the requirement that suitable guards and exhaust hoods be provided to reduce to a minimum the hazard due to the point of operation of such machines.

**1910.213(s)** *Inspection and maintenance of woodworking machinery.* **(1)** Dull, badly set, improperly filed, or improperly tensioned saws shall be immediately removed from service, before they begin to cause the material to stick, jam, or kick back when it is fed to the saw at normal speed. Saws to which gum has adhered on the sides shall be immediately cleaned.

**1910.213(s)(2)** All knives and cutting heads of woodworking machines shall be kept sharp, properly adjusted, and firmly secured. Where two or more knives are used in one head, they shall be properly balanced.

**1910.213(s)(3)** Bearings shall be kept free from lost motion and shall be well lubricated.

**1910.213(s)(4)** Arbors of all circular saws shall be free from play.

**1910.213(s)(5)** Sharpening or tensioning of saw blades or cutters shall be done only by persons of demonstrated skill in this kind of work.

**1910.213(s)(6)** Emphasis is placed upon the importance of maintaining cleanliness around woodworking machinery, particularly as regards the effective functioning of guards and the prevention of fire hazards in switch enclosures, bearings, and motors.

**1910.213(s)(7)** All cracked saws shall be removed from service.

**1910.213(s)(8)** The practice of inserting wedges between the saw disk and the collar to form what is commonly known as a "wobble saw" shall not be permitted.

**1910.213(s)(9)** Push sticks or push blocks shall be provided at the work place in the several sizes and types suitable for the work to be done.

**1910.213(s)(10)-(11)** [Reserved]

**1910.213(s)(12)** The knife blade of jointers shall be so installed and adjusted that it does not protrude more than one-eighth inch beyond the cylindrical body of the head. Push sticks or push blocks shall be provided at the work place in the several sizes and types suitable for the work to be done.

**1910.213(s)(13)** Whenever veneer slicers or rotary veneer-cutting machines have been shutdown for the purpose of inserting logs or to make adjustments, operators shall make sure that machine is clear and other workmen are not in a hazardous position before starting the machine.

**1910.213(s)(14)** Operators shall not ride the carriage of a veneer slicer.

[39 FR 23502, June 27, 1974, as amended at 43 FR 49750, Oct. 24, 1978; 49 FR 5323, Feb. 10, 1984]

## §1910.214 Cooperage machinery [Reserved].

## §1910.215 Abrasive wheel machinery.

**1910.215(a)** *General requirements*—**(1)** *Machine guarding.* Abrasive wheels shall be used only on machines provided with safety guards as defined in the following paragraphs of this section, except:

**1910.215(a)(1)(i)** Wheels used for internal work while within the work being ground;

**1910.215(a)(1)(ii)** Mounted wheels, used in portable operations, 2 inches and smaller in diameter; and

**1910.215(a)(1)(iii)** Types 16, 17, 18, 18R, and 19 cones, plugs, and threaded hole pot balls where the work offers protection.

**1910.215(a)(2)** *Guard design.* The safety guard shall cover the spindle end, nut, and flange projections. The safety guard shall be mounted so as to maintain proper alignment with the wheel, and the strength of the fastenings shall exceed the strength of the guard, except:

**1910.215(a)(2)(i)** Safety guards on all operations where the work provides a suitable measure of protection to the operator, may be so constructed that the spindle end, nut, and outer flange are exposed; and where the nature of the work is such as to entirely cover the side of the wheel, the side covers of the guard may be omitted; and

**1910.215(a)(2)(ii)** The spindle end, nut, and outer flange may be exposed on machines designed as portable saws.

**1910.215(a)(3)** *Flanges.* Grinding machines shall be equipped with flanges in accordance with paragraph (c) of this section.

**1910.215(a)(4)** *Work rests.* On offhand grinding machines, work rests shall be used to support the work. They shall be of rigid construction and designed to be adjustable to compensate for wheel wear. Work rests shall be kept adjusted closely to the wheel with a maximum opening of one-eighth inch to prevent the work from being jammed between the wheel and the rest, which may cause wheel breakage. The work rest shall be securely clamped after each adjustment. The adjustment shall not be made with the wheel in motion.

**1910.215(a)(5)** *Excluded machinery.* Natural sandstone wheels and metal, wooden, cloth, or paper discs, having a layer of abrasive on the surface are not covered by this section.

**1910.215(b)** *Guarding of abrasive wheel machinery*—**(1)** *Cup wheels.* Cup wheels (Types 6 and 11) shall be protected by:

**1910.215(b)(1)(i)** Safety guards as specified in paragraphs (b) (1) through (10) of this section;

**1910.215(b)(1)(ii)** Band type guards as specified in paragraph (b)(11) of this section; and

**1910.215(b)(1)(iii)** Special "Revolving Cup Guards" which mount behind the wheel and turn with it. They shall be made of steel or other material with adequate strength and shall enclose the wheel sides upward from the back for one-third of the wheel thickness. The mounting features shall conform with all requirements of this section. It is necessary to maintain clearance between the wheel side and the guard. This clearance shall not exceed one-sixteenth inch.

**1910.215(b)(2)** *Guard exposure angles.* The maximum exposure angles specified in paragraphs (b) (3) through (8) of this section shall not be exceeded. Visors or other accessory equipment shall not be included as a part of the guard when measuring the guard opening, unless such equipment has strength equal to that of the guard.

**1910.215(b)(3)** *Bench and floor stands.* The angular exposure of the grinding wheel periphery and sides for safety guards used on machines known as bench and floor stands should not exceed 90° or one-fourth of the periphery. This exposure shall begin at a point not more than 65° above

# OSHA Standards for General Industry

the horizontal plane of the wheel spindle. (See Figures O-6 and O-7 and paragraph (b)(9) of this section.)

**FIGURE NO. O-6**   **FIGURE NO. O-7**

Wherever the nature of the work requires contact with the wheel below the horizontal plane of the spindle, the exposure shall not exceed 125°. (See Figures O-8 and O-9.)

**FIGURE NO. O-8**   **FIGURE NO. O-9**

**1910.215(b)(4)** *Cylindrical grinders.* The maximum angular exposure of the grinding wheel periphery and sides for safety guards used on cylindrical grinding machines shall not exceed 180°. This exposure shall begin at a point not more than 65° above the horizontal plane of the wheel spindle. (See Figures O-10 and O-11 and subparagraph (9) of this paragraph.)

**FIGURE NO. O-10**   **FIGURE NO. O-11**

**1910.215(b)(5)** *Surface grinders and cutting-off machines.* The maximum angular exposure of the grinding wheel periphery and sides for safety guards used on cutting-off machines and on surface grinding machines which employ the wheel periphery shall not exceed 150°. This exposure shall begin at a point not less than 15° below the horizontal plane of the wheel spindle. (See Figures O-12 and O-13)

**FIGURE NO. O-12**   **FIGURE NO. O-13**

**1910.215(b)(6)** *Swing frame grinders.* The maximum angular exposure of the grinding wheel periphery and sides for safety guards used on machines known as swing frame grinding machines shall not exceed 180°, and the top half of the wheel shall be enclosed at all times. (See Figures O-14 and O-15.)

**FIGURE NO. O-14**   **FIGURE NO. O-15**

**1910.215(b)(7)** *Automatic snagging machines.* The maximum angular exposure of the grinding wheel periphery and sides for safety guards used on grinders known as automatic snagging machines shall not exceed 180° and the top half of the wheel shall be enclosed at all times. (See Figures O-14 and O-15.)

**1910.215(b)(8)** *Top grinding.* Where the work is applied to the wheel above the horizontal centerline, the exposure of the grinding wheel periphery shall be as small as possible and shall not exceed 60°. (See Figures O-16 and O-17.)

**FIGURE NO. O-16**   **FIGURE NO. O-17**

**1910.215(b)(9)** *Exposure adjustment.* Safety guards of the types described in subparagraphs (3) and (4) of this paragraph, where the operator stands in front of the opening, shall be constructed so that the peripheral protecting member can be adjusted to the constantly decreasing diameter of the wheel. The maximum angular exposure above the horizontal plane of the wheel spindle as specified in paragraphs (b) (3) and (4) of this section shall never be exceeded, and the distance between the wheel periphery and the adjustable tongue or the end of the peripheral member at the top shall never exceed one-fourth inch. (See Figures O-18, O-19, O-20, O-21, O-22, and O-23.)

**1910.215(b)(10)** *Material requirements and minimum dimensions.* (i) See Figures O-36 and O-37 and Table O-9 for minimum basic thickness of peripheral and side members for various types of safety guards and classes of service.

**1910.215(b)(10)(ii)** If operating speed does not exceed 8,000 surface feet per minute cast iron safety guards, malleable iron guards or other guards as described in paragraph (b)(10)(iii) of this section shall be used.

**FIGURE O-18**  **FIGURE O-19**

**CORRECT**

*Showing adjustable tongue giving required angular protection for all sizes of wheel used.*

**FIGURE NO. O-20**  **FIGURE NO. O-21**

**CORRECT**

*Showing movable guard with opening small enough to give required protection for smallest size wheel used.*

**FIGURE NO. O-22**  **FIGURE NO. O-23**

**INCORRECT**

*Showing movable guard with size of opening correct for full size wheel but too large for smaller wheels.*

**1910.215(b)(10)(iii)** Cast steel, or structural steel, safety guards as specified in Figures O-36 and O-37 and Table O-9 shall be used where operating speeds of wheels are faster than 8,000 surface feet per minute up to a maximum of 16,000 surface feet per minute.

**1910.215(b)(10)(iv)** For cutting-off wheels 16 inches diameter and smaller and where speed does not exceed 16,000 surface feet per minute, cast iron or malleable iron safety guards as specified in Figures O- 36 and O-37, and in Table O-9 shall be used.

**1910.215(b)(10)(v)** For cutting-off wheels larger than 16 inches diameter and where speed does not exceed 14,200 surface feet per minute, safety guards as specified in Figures O-27 and O-28, and in Table O-1 shall be used.

**1910.215(b)(10)(vi)** For thread grinding wheels not exceeding 1 inch in thickness cast iron or malleable iron safety guards as specified in Figures O-36 and O-37, and in Table O-9 shall be used.

**1910.215(b)(11)** *Band type guards—general specifications.* Band type guards shall conform to the following general specifications:

**1910.215(b)(11)(i)** The bands shall be of steel plate or other material of equal or greater strength. They shall be continuous, the ends being either riveted, bolted, or welded together in such a manner as to leave the inside free from projections.

**1910.215(b)(11)(ii)** The inside diameter of the band shall not be more than 1 inch larger than the outside diameter of the wheel, and shall be mounted as nearly concentric with the wheel as practicable.

**1910.215(b)(11)(iii)** The band shall be of sufficient width and its position kept so adjusted that at no time will the wheel protrude beyond the edge of the band a distance greater than that indicated in Figure O-29 and in Table O-2 or the wall thickness (W), whichever is smaller.

**1910.215(b)(12)** *Guard design specifications.* Abrasive wheel machinery guards shall meet the design specifications of the American National Standard Safety Code for the Use, Care, and Protection of Abrasive Wheels, ANSI B7.1-1970, which is incorporated by reference as specified in §1910.6. This requirement shall not apply to natural sandstone wheels or metal, wooden, cloth, or paper discs, having a layer of abrasive on the surface.

**1910.215(c)** *Flanges*—(1) *General requirements.* All abrasive wheels shall be mounted between flanges which shall not be less than one-third the diameter of the wheel.

**1910.215(c)(1)(i)** Exceptions:

**1910.215(c)(1)(i)***(a)* Mounted wheels.

**1910.215(c)(1)(i)***(b)* Portable wheels with threaded inserts or projecting studs.

**1910.215(c)(1)(i)***(c)* Abrasive discs (inserted nut, inserted washer and projecting stud type).

**1910.215(c)(1)(i)***(d)* Plate mounted wheels.

**1910.215(c)(1)(i)***(e)* Cylinders, cup, or segmental wheels that are mounted in chucks.

**1910.215(c)(1)(i)***(f)* Types 27 and 28 wheels.

**1910.215(c)(1)(i)***(g)* Certain internal wheels.

**1910.215(c)(1)(i)***(h)* Modified types 6 and 11 wheels (terrazzo).

**1910.215(c)(1)(i)***(i)* Cutting-off wheels, Types 1 and 27A (see paragraphs (c)(1) (ii) and (iii) of this section).

**1910.215(c)(1)(ii)** Type 1 cutting-off wheels are to be mounted between properly relieved flanges which have matching bearing surfaces. Such flanges shall be at least one-fourth the wheel diameter.

**1910.215(c)(1)(iii)** Type 27A cutting-off wheels are designed to be mounted by means of flat, not relieved, flanges having matching bearing surfaces and which may be less than one-third but shall not be less than one-fourth the wheel diameter. (See Figure O-24 for one such type of mounting.)

**1910.215(c)(1)(iv)** There are three general types of flanges:

**1910.215(c)(1)(iv)***(a)* Straight relieved flanges (see Figure O-32);

**1910.215(c)(1)(iv)***(b)* Straight unrelieved flanges (see Figure O-30);

**1910.215(c)(1)(iv)***(c)* Adaptor flanges (see Figures O-33 and O-34);

**1910.215(c)(1)(v)** Regardless of flange type used, the wheel shall always be guarded. Blotters shall be used in accordance with paragraph (c)(6) of this section.

**FIGURE NO. O-24**

The Type 27 A Wheel is mounted between flat non-relieved flanges of equal bearing surfaces

# OSHA Standards for General Industry

**1910.215(c)(2)** [Reserved]

**1910.215(c)(3)** *Finish and balance.* Flanges shall be dimensionally accurate and in good balance. There shall be no rough surfaces or sharp edges.

**1910.215(c)(4)** *Uniformity of diameter.* **(i)** Both flanges, of any type, between which a wheel is mounted, shall be of the same diameter and have equal bearing surface. Exceptions are set forth in the remaining subdivisions of this subparagraph.

**1910.215(c)(4)(ii)** Type 27 and Type 28 wheels, because of their shape and usage, require specially designed adaptors. The back flange shall extend beyond the central hub or raised portion and contact the wheel to counteract the side pressure on the wheel in use. The adaptor nut which is less than the minimum one-third diameter of wheel fits in the depressed side of wheel to prevent interference in side grinding and serves to drive the wheel by its clamping force against the depressed portion of the back flange. The variance in flange diameters, the adaptor nut being less than one-third wheel diameter, and the use of side pressure in wheel operation limits the use to reinforced organic bonded wheels. Mounts which are affixed to the wheel by the manufacturer shall not be reused. Type 27 and Type 28 wheels shall be used only with a safety guard located between wheel and operator during use. (See Figure O-24-A.)

**FIGURE NO. O–24–A**

Types 27 and 28 wheels, because of their shape, require specially designed adaptors

**1910.215(c)(4)(iii)** Modified Types 6 and 11 wheels (terrazzo) with tapered K dimension.

**1910.215(c)(5)** *Recess and undercut.* **(i)** Straight relieved flanges made according to Table O-6 and Figure O-32 shall be recessed at least one-sixteenth inch on the side next to the wheel for a distance as specified in Table O-6.

**1910.215(c)(5)(ii)** Straight flanges of the adaptor or sleeve type (Table O-7 and Figures O-33 and O-34) shall be undercut so that there will be no bearing on the sides of the wheel within one-eighth inch of the arbor hole.

**1910.215(c)(6)** *Blotters.* **(i)** Blotters (compressible washers) shall always be used between flanges and abrasive wheel surfaces to insure uniform distribution of flange pressure. (See paragraph (d)(5) of this section.)

**1910.215(c)(6)(ii)** Exception:

**1910.215(c)(6)(ii)(a)** Mounted wheels.

**1910.215(c)(6)(ii)(b)** Abrasive discs (inserted nut, inserted washer, and projecting stud type).

**1910.215(c)(6)(ii)(c)** Plate mounted wheels.

**1910.215(c)(6)(ii)(d)** Cylinders, cups, or segmental wheels that are mounted in chucks.

**1910.215(c)(6)(ii)(e)** Types 27 and 28 wheels.

**1910.215(c)(6)(ii)(f)** Certain Type 1 and Type 27A cutting-off wheels.

**1910.215(c)(6)(ii)(g)** Certain internal wheels.

**1910.215(c)(6)(ii)(h)** Type 4 tapered wheels.

**1910.215(c)(6)(ii)(i)** Diamond wheels, except certain vitrified diamond wheels.

**1910.215(c)(6)(ii)(j)** Modified Types 6 and 11 wheel (terrazzo)—blotters applied flat side of wheel only.

**1910.215(c)(7)** *Driving flange.* The driving flange shall be securely fastened to the spindle and the bearing surface shall run true. When more than one wheel is mounted between a single set of flanges, wheels may be cemented together or separated by specially designed spacers. Spacers shall be equal in diameter to the mounting flanges and have equal bearing surfaces. (See paragraph (d)(6) of this section.)

**1910.215(c)(8)** *Dimensions.* **(i)** Tables O-4 and O-6 and Figures O-30 and O-32 show minimum dimensions for straight relieved and unrelieved flanges for use with wheels with small holes that fit directly on the machine spindle. Dimensions of such flanges shall never be less than indicated.

**1910.215(c)(8)(ii)** Table O-5, and Table O-7 and Figures O-31, O-33, O-34 show minimum dimensions for straight adaptor flanges for use with wheels having holes larger than the spindle. Dimensions of such adaptor flanges shall never be less than indicated.

**1910.215(c)(8)(iii)** Table O-8 and Figure O-35 show minimum dimensions for straight flanges that are an integral part of wheel sleeves which are frequently used on precision grinding machines. Dimensions of such flanges shall never be less than indicated.

**1910.215(c)(9)** *Repairs and maintenance.* All flanges shall be maintained in good condition. When the bearing surfaces become worn, warped, sprung, or damaged they should be trued or refaced. When refacing or truing, care shall be exercised to make sure that proper relief and rigidity is maintained as specified in paragraphs (c)(2) and (5) of this section and they shall be replaced when they do not conform to these subparagraphs and Table O-4, Figure O-30, Table O-5, Figure O-31, Table O-6, Figure O-32, and Table O-8, Figure O-35. Failure to observe these rules might cause excessive flange pressure around the hole of the wheel. This is especially true of wheel-sleeve or adaptor flanges.

**1910.215(d)** *Mounting*—**(1)** *Inspection.* Immediately before mounting, all wheels shall be closely inspected and sounded by the user (ring test) to make sure they have not been damaged in transit, storage, or otherwise. The spindle speed of the machine shall be checked before mounting of the wheel to be certain that it does not exceed the maximum operating speed marked on the wheel. Wheels should be tapped gently with a light non-metallic implement, such as the handle of a screwdriver for light wheels, or a wooden mallet for heavier wheels. If they sound cracked (dead), they shall not be used. This is known as the "Ring Test".

**1910.215(d)(1)(i)** Wheels must be dry and free from sawdust when applying the ring test, otherwise the sound will be deadened. It should also be noted that organic bonded wheels do not emit the same clear metallic ring as do vitrified and silicate wheels.

**FIGURE NO. O-25**

Light Wheels
Suspend from hole by small pin or finger

**FIGURE NO. O-26**

Heavy Wheels
Support on clean hard floor

**1910.215(d)(1)(ii)** "Tap" wheels about 45° each side of the vertical centerline and about 1 or 2 inches from the periphery as indicated by the spots in Figure O-25 and Figure O-26. Then rotate the wheel 45° and repeat the test. A sound and undamaged wheel will give a clear metallic tone. If cracked, there will be a dead sound and not a clear "ring."

**1910.215(d)(2)** *Arbor size.* Grinding wheels shall fit freely on the spindle and remain free under all grinding conditions. A controlled clearance between the wheel hole and the machine spindle (or wheel sleeves or adaptors) is essential to avoid excessive pressure from mounting and spindle expansion. To accomplish this, the machine spindle shall be made to nominal (standard) size plus zero minus .002 inch, and the wheel hole shall be made suitably oversize to assure safety clearance under the conditions of operating heat and pressure.

**1910.215(d)(3)** *Surface condition.* All contact surfaces of wheels, blotters and flanges shall be flat and free of foreign matter.

**1910.215(d)(4)** *Bushing.* When a bushing is used in the wheel hole it shall not exceed the width of the wheel and shall not contact the flanges.

**1910.215(d)(5)** *Blotters.* When blotters or flange facings of compressible material are required, they shall cover entire contact area of wheel flanges. Blotters need not be used with the following types of wheels:

**1910.215(d)(5)(i)** Mounted wheels.

**1910.215(d)(5)(ii)** Abrasive discs (inserted nut, inserted washer, and projecting-stud type).

**1910.215(d)(5)(iii)** Plate mounted wheels.

**1910.215(d)(5)(iv)** Cylinders, cups, or segmental wheels that are mounted in chucks.

**1910.215(d)(5)(v)** Types 27 and 28 wheels.

**1910.215(d)(5)(vi)** Certain Type 1 and Type 27A cutting-off wheels.

**1910.215(d)(5)(vii)** Certain internal wheels.

**1910.215(d)(5)(viii)** Type 4 tapered wheels.

**1910.215(d)(5)(ix)** Diamond wheels, except certain vitrified diamond wheels.

**1910.215(d)(6)** *Multiple wheel mounting.* When more than one wheel is mounted between a single set of flanges, wheels may be cemented together or separated by specially designed spacers. Spacers shall be equal in diameter to the mounting flanges and have equal bearing surfaces. When mounting wheels which have not been cemented together, or ones which do not utilize separating spacers, care must be exercised to use wheels specially manufactured for that purpose.

**FIGURE NO. O-27**

**FIGURE NO. O-28**

### TABLE O-1—MINIMUM BASIC THICKNESS FOR PERIPHERAL AND SIDE MEMBERS FOR SAFETY GUARDS USED WITH CUTTING-OFF WHEELS

| Material used in construction of guard | Maximum thickness of cutting off wheel | Speed not to exceed | 6 to 11 inches A | 6 to 11 inches B | Over 11 to 20 inches A | Over 11 to 20 inches B | Over 20 to 30 inches A | Over 20 to 30 inches B | Over 30 to 48 inches A | Over 30 to 48 inches B | Over 48 to 72 inches A | Over 48 to 72 inches B |
|---|---|---|---|---|---|---|---|---|---|---|---|---|
| Structural steel (min. tensile strength 60,000 p.s.i.). | 1/2 inch or less | 14,200 SFPM | 1/16 | 1/16 | 3/32 | 3/32 | 1/8 | 1/8 | 3/16 | 3/16 | 1/4 | 1/4 |
|  | 1/2 inch or less | 16,000 SFPM | 3/32 | 1/8 | 1/8 | 1/8 | 3/16 | 1/8 | 1/4 | 3/16 | 5/16 | 1/4 |

**FIGURE NO. O-29**

## TABLE O-3—GUIDE FOR CONSTRUCTION OF BAND TYPE GUARDS
[Maximum Wheel Speed 7,000 SFPM]

| Minimum material specifications | Diameter of wheel | Minimum thickness of band A | Minimum diameter of rivets | Maximum distance between centers of rivets |
|---|---|---|---|---|
| | | Inches | | |
| Hot rolled steel SAE 1008 | Under 8 | 1/16 | 3/16 | 3/4 |
| | 8 to 24 | 1/8 | 1/4 | 1 |
| | Over 24 to 30 | 1/4 | 3/8 | 1 1/4 |

**FIGURE NO. O-30**

Driving flange secured to spindle for use only on portable wheels with threaded inserts or projecting studs.

**FIGURE NO. O-31**

F = D - E
Blotters
Corner Undercut

## TABLE O-5—MINIMUM DIMENSIONS FOR STRAIGHT ADAPTOR FLANGE—FOR ORGANIC BONDED WHEELS OVER 1 1/4 INCHES THICK[1]
[In inches]

| Wheel diameter | Wheel hole diameter | B Minimum flange diameter | D Minimum thickness of flange at bore | E Minimum thickness of flange at edge of undercut | F[1] (D-E) minimum thickness |
|---|---|---|---|---|---|
| 12 to 14 | 4 | 6 | 7/8 | 3/8 | 1/2 |
| | 5 | 7 | 7/8 | 3/8 | 1/2 |
| | 6 | 8 | 7/8 | 3/8 | 1/2 |
| Larger than 14 to 18 | 4 | 6 | 7/8 | 3/8 | 1/2 |
| | 5 | 7 | 7/8 | 3/8 | 1/2 |
| | 6 | 8 | 7/8 | 3/8 | 1/2 |
| | 7 | 9 | 7/8 | 3/8 | 1/2 |
| | 8 | 10 | 7/8 | 3/8 | 1/2 |
| Larger than 18 to 24 | 6 | 8 | 1 | 1/2 | 1/2 |
| | 7 | 9 | 1 | 1/2 | 1/2 |
| | 8 | 10 | 1 | 1/2 | 1/2 |
| | 10 | 12 | 1 | 1/2 | 1/2 |
| | 12 | 14 | 1 | 1/2 | 1/2 |
| Larger than 24 to 30 | 12 | 15 | 1 | 1/2 | 1/2 |
| Larger than 30 to 36 | 12 | 15 | 1 3/8 | 7/8 | 1/2 |

[1] For wheels under 1 1/4 inches thick F dimension shall not exceed 40 percent of wheel thickness.

**FIGURE NO. O-32**

Driving flange secured to spindle.

### TABLE O-6—MINIMUM DIMENSIONS FOR STRAIGHT RELIEVED FLANGES[1]
[In inches]

| A | B | C | | D | E |
|---|---|---|---|---|---|
| | | Radial width of bearing surface | | | |
| Diameter of wheel | Minimum outside diameter of flanges | Minimum | Maximum | Minimum thickness of flange at bore | Minimum thickness of flange at edge of recess |
| 1 | 3/8 | 1/16 | 1/8 | 1/16 | 1/16 |
| 2 | 3/4 | 1/8 | 3/16 | 1/8 | 3/32 |
| 3 | 1 | 1/8 | 3/16 | 3/16 | 3/32 |
| 4 | 1 3/8 | 1/8 | 3/16 | 3/16 | 1/8 |
| 5 | 1 3/4 | 3/16 | 1/4 | 1/4 | 1/8 |
| 6 | 2 | 1/4 | 1/2 | 3/8 | 3/16 |
| 7 | 2 1/2 | 1/4 | 1/2 | 3/8 | 3/16 |
| 8 | 3 | 1/4 | 1/2 | 3/8 | 3/16 |
| 10 | 3 1/2 | 5/16 | 5/8 | 3/8 | 1/4 |
| 12 | 4 | 5/16 | 5/8 | 1/2 | 5/16 |
| 14 | 4 1/2 | 3/8 | 3/4 | 1/2 | 5/16 |
| 16 | 5 1/2 | 1/2 | 1 | 1/2 | 5/16 |
| 18 | 6 | 1/2 | 1 | 5/8 | 3/8 |
| 20 | 7 | 5/8 | 1 1/4 | 5/8 | 3/8 |
| 22 | 7 1/2 | 5/8 | 1 1/4 | 5/8 | 7/16 |
| 24 | 8 | 3/4 | 1 1/4 | 5/8 | 7/16 |
| 26 | 8 1/2 | 3/4 | 1 1/4 | 5/8 | 1/2 |
| 28 | 10 | 7/8 | 1 1/2 | 3/4 | 1/2 |
| 30 | 10 | 7/8 | 1 1/2 | 3/4 | 5/8 |
| 36 | 12 | 1 | 2 | 7/8 | 3/4 |
| 42 | 14 | 1 | 2 | 7/8 | 3/4 |
| 48 | 16 | 1 1/4 | 2 | 1 1/8 | 1 |
| 60 | 20 | 1 1/4 | 2 | 1 1/4 | 1 1/8 |
| 72 | 24 | 1 1/2 | 2 1/2 | 1 3/8 | 1 1/4 |

[1] Flanges for wheels under 2 inches diameter may be unrelieved and shall be maintained flat and true.

**FIGURE NO. O-33**
Central Nut Mounting
*Driving flange secured to spindle.*

**FIGURE NO. O-34**
Multiple Screw Mounting
*Driving flange secured to spindle.*

### TABLE O-7—MINIMUM DIMENSIONS FOR STRAIGHT FLANGES— FOR MECHANICAL GRINDERS 12,500 S.F.P.M. TO 16,5 S.F.P.M.[1]

| Wheel diameter | Wheel hole diameter | B—Minimum flange diameter | D—Minimum thickness of flange at bore | E—Minimum thickness of flange at edge of undercut | F[2]—(D-E) minimum thickness |
|---|---|---|---|---|---|
| 20 | 6 | 8 | 1 | 1/2 | 1/2 |
| 20 | 8 | 10 | 1 1/2 | 3/4 | 3/4 |
| 24 | 12 | 15 | 2 | 1 | 1 |
| 30 | 12 | 15 | 2 | 1 | 1 |
| 36 | 12 | 15 | 2 | 1 | |

[1] Flanges shall be of steel, quality SAE 1040 or equivalent, annealed plate, heat treated to R. 25-30.
[2] For wheels under 1 1/4 inch thick F dimension shall not exceed 40 percent of wheel thickness.

**FIGURE NO. O-35**
*Driving flange secured to spindle.*

## TABLE O-8—MINIMUM DIMENSIONS FOR STRAIGHT FLANGES USED AS WHEEL SLEEVES FOR PRECISION GRINDING ONLY

[In inches]

| Wheel diameter | Wheel hole diameter | B — Minimum outside diameter of flange | D — Minimum thickness of flange at bore | E — Minimum thickness of flange at edge of undercut |
|---|---|---|---|---|
| | Inches | Inches | Inches | Inches |
| 12 to 14 | 5 | 7 | 1/2 | 7/16 |
| Larger than 14 to 20 | 5 | 7 | 5/8 | 7/16 |
| | 6 | 8 | 5/8 | 7/16 |
| | 8 | 10 | 5/8 | 7/16 |
| | 10 | 11 1/2 | 5/8 | 7/16 |
| | 12 | 13 1/2 | 5/8 | 7/16 |
| Larger than 20 to 30 | 8 | 10 | 3/4 | 1/2 |
| | 10 | 11 1/2 | 3/4 | 1/2 |
| | 12 | 13 1/2 | 3/4 | 1/2 |
| | 16 | 17 1/2 | 3/4 | 1/2 |
| Larger than 30 to 42 | 12 | 13 1/2 | 3/4 | 1/2 |
| | 16 | 17 1/2 | 3/4 | 1/2 |
| | 18 | 19 1/2 | 3/4 | 1/2 |
| | 20 | 21 1/2 | 3/4 | 1/2 |
| Larger than 42 to 60 | 16 | 20 | 1 | 3/4 |
| | 20 | 24 | 1 | 3/4 |
| | 24 | 29 | 1 1/8 | 7/8 |

NOTE: These flanges may be clamped together by means of a central nut, or by a series of bolts or some other equivalent means of fastening. For hole sizes smaller than shown in this table, use table 12.

**FIGURE NO. O-36**

**FIGURE NO. O-37**

## TABLE O-9—MINIMUM BASIC THICKNESSES OF PERIPHERAL AND SIDE MEMBERS FOR SAFETY GUARDS
[In inches]

| Material used in construction of guard | Maximum thickness of grinding wheel | 3 to 6 inches A | 3 to 6 inches B | Over 6 to 12 inches A | Over 6 to 12 inches B | Over 12 to 16 inches A | Over 12 to 16 inches B | Over 16 to 20 inches A | Over 16 to 20 inches B | Over 20 to 24 inches A | Over 20 to 24 inches B | Over 24 to 30 inches A | Over 24 to 30 inches B | Over 24 to 30 inches A | Over 24 to 30 inches B |
|---|---|---|---|---|---|---|---|---|---|---|---|---|---|---|---|
| Material satisfactory[1] for speeds up to 8,000 SFPM. | 2 | 1/4 | 1/4 | 3/8 | 5/16 | 1/2 | 3/8 | 5/8 | 1/2 | 7/8 | 5/8 | 1 | 3/4 | 1 1/4 | 1 |
|  | 4 | 5/16 | 5/16 | 3/8 | 5/16 | 1/2 | 3/8 | 3/4 | 5/8 | 1 | 5/8 | 1 1/8 | 3/4 | 1 3/8 | 1 |
|  | 6 | 3/8 | 5/16 | 1/2 | 7/16 | 5/8 | 1/2 | 1 | 5/8 | 1 1/8 | 3/4 | 1 1/4 | 7/8 | 1 1/2 | 1 1/8 |
|  | 8 |  |  | 5/8 | 9/16 | 7/8 | 3/4 | 1 | 3/4 | 1 1/8 | 3/4 | 1 1/4 | 7/8 | 1 1/2 | 1 1/8 |
| Cast iron (min. tensile strength 20,000 p.s.i.) Class 20. | 10 |  |  | 3/4 | 11/16 | 7/8 | 3/4 | 1 | 3/4 | 1 1/8 | 3/4 | 1 1/4 | 7/8 | 1 1/2 | 1 1/8 |
|  | 16 |  |  |  |  | 1 1/8 | 1 | 1 1/4 | 1 | 1 5/16 | 1 | 1 7/16 | 1 1/16 | 1 3/4 | 1 3/8 |
|  | 20 |  |  |  |  |  |  | 1 3/8 | 1 1/8 | 1 3/8 | 1 1/8 | 1 1/2 | 1 3/8 | 2 | 1 5/8 |
| Material satisfactory[1] for speeds up to 9,000 SFPM. | 2 | 1/4 | 1/4 | 3/8 | 5/16 | 1/2 | 3/8 | 5/8 | 1/2 | 3/4 | 5/8 | 7/8 | 3/4 | 1 | 7/8 |
|  | 4 | 5/16 | 5/16 | 3/8 | 5/16 | 1/2 | 3/8 | 5/8 | 1/2 | 3/4 | 5/8 | 7/8 | 3/4 | 1 1/8 | 7/8 |
|  | 6 | 3/8 | 5/16 | 1/2 | 7/16 | 5/8 | 1/2 | 3/4 | 5/8 | 7/8 | 3/4 | 1 | 3/4 | 1 1/4 | 7/8 |
|  | 8 |  |  | 1/2 | 7/16 | 5/8 | 1/2 | 3/4 | 5/8 | 7/8 | 5/8 | 1 | 3/4 | 1 1/4 | 7/8 |
| Malleable iron (min. tensile strength 50,000 p.s.i.) Grade 32510. | 10 |  |  | 1/2 | 7/16 | 5/8 | 1/2 | 3/4 | 5/8 | 7/8 | 5/8 | 1 | 3/4 | 1 1/4 | 7/8 |
|  | 16 |  |  |  |  | 13/16 | 11/16 | 13/16 | 11/16 | 1 | 3/4 | 1 1/8 | 7/8 | 1 3/8 | 1 |
|  | 20 |  |  |  |  | 7/8 | 3/4 | 7/8 | 3/4 | 1 | 3/4 | 1 1/8 | 7/8 | 1 1/2 | 1 1/8 |
| Materials satisfactory[1] for speeds up to 16,000 SFPM. | 2 | 1/4 | 1/4 | 5/16 | 5/16 | 3/8 | 3/8 | 1/2 | 7/16 | 5/8 | 1/2 | 3/4 | 5/8 | 7/8 | 3/4 |
|  | 4 | 1/4 | 1/4 | 1/2 | 1/2 | 1/2 | 1/2 | 9/16 | 1/2 | 5/8 | 1/2 | 3/4 | 5/8 | 1 | 3/4 |
|  | 6 | 3/8 | 1/4 | 3/4 | 5/8 | 5/8 | 5/8 | 3/4 | 5/8 | 13/16 | 11/16 | 13/16 | 11/16 | 1 1/8 | 3/4 |
|  | 8 |  |  | 7/8 | 3/4 | 7/8 | 3/4 | 7/8 | 3/4 | 7/8 | 3/4 | 15/16 | 13/16 | 1 3/8 | 1 |
| Steel castings (min. tensile strength 60,000 p.s.i.) Grade V60-30. | 10 |  |  | 1 | 7/8 | 1 | 7/8 | 1 | 7/8 | 1 1/8 | 15/16 | 1 1/8 | 1 | 1 7/16 | 1 1/16 |
|  | 16 |  |  |  |  | 1 1/4 | 1 1/8 | 1 1/4 | 1 1/8 | 1 1/4 | 1 1/8 | 1 1/4 | 1 1/8 | 1 13/16 | 1 7/16 |
|  | 20 |  |  |  |  | 1 3/8 | 1 1/4 | 1 3/8 | 1 1/4 | 1 3/8 | 1 1/4 | 1 7/16 | 1 15/16 | 2 1/16 | 1 11/16 |
| Structural steel (min. tensile strength 60,000 p.s.i.) | 2 | 1/8 | 1/16 | 5/16 | 1/4 | 5/16 | 1/4 | 5/16 | 1/4 | 5/16 | 1/4 | 3/8 | 5/16 | 1/2 | 3/8 |
|  | 4 | 1/8 | 1/16 | 3/8 | 5/16 | 3/8 | 5/16 | 3/8 | 5/16 | 3/8 | 5/16 | 3/8 | 5/16 | 1/2 | 3/8 |
|  | 6 | 3/16 | 1/16 | 1/2 | 3/8 | 7/16 | 3/8 | 7/16 | 3/8 | 7/16 | 3/8 | 7/16 | 3/8 | 3/4 | 1/2 |
|  | 8 |  |  | 1/2 | 3/8 | 9/16 | 7/16 | 9 |  |  |  |  |  |  |  |
|  | 10 | 9/16 | 7/16 | 5/8 | 1/2 | 5/8 | 1/2 | 5/8 | 1/2 | 7/8 | 5/8 | 1 1/16 | 13/16 |
|  | 16 |  |  |  |  | 5/8 | 1/2 | 3/4 | 5/8 | 3/4 | 5/8 | 13/16 | 11/16 | 1 1/16 | 13/16 |
|  | 20 |  |  |  |  | 13/16 | 11/16 | 13/16 | 11/16 | 13/16 | 11/16 | 7/8 | 3/4 | 1 13/16 | 15/16 |

[1] The recommendations listed in the above table are guides for the conditions stated. Other material, designs or dimensions affording equal or superior protection are also acceptable.

### TABLE O-2—EXPOSURE VERSUS WHEEL THICKNESS
[In inches]

| Overall thickness of wheel (T) | Maximum exposure of wheel (C) |
|---|---|
| 1/2 | 1/4 |
| 1 | 1/2 |
| 2 | 3/4 |
| 3 | 1 |
| 4 | 1 1/2 |
| 5 and over | 2 |

### TABLE O-4—MINIMUM DIMENSIONS FOR STRAIGHT UNRELIEVED FLANGES FOR WHEELS WITH THREADED INSERTS OR PROJECTING STUDS

| A Diameter of wheel | B[1] Minimum outside diameter of flange | T Minimum thickness of flange |
|---|---|---|
| 1 | 5/8 | 1/8 |
| 2 | 1 | 1/8 |
| 3 | 1 | 3/16 |
| 4 | 1 3/8 | 3/16 |
| 5 | 1 3/4 | 1/4 |
| 6 | 2 | 3/8 |

[1] Note: Must be large enough to extend beyond the bushing. Where prong anchor or cupback bushing are used, this footnote does not apply.

[39 FR 23502, June 27, 1974, as amended at 43 FR 49750, Oct. 24, 1978; 49 FR 5323, Feb. 10, 1984; 61 FR 9240, Mar. 7, 1996]

## §1910.216 Mills and calenders in the rubber and plastics industries.

**1910.216(a)** *General requirements—*

**1910.216(a)(1)-(2)** [Reserved]

**1910.216(a)(3)** *Auxiliary equipment.* Mechanical and electrical equipment and auxiliaries shall be installed in accordance with this section and Subpart S of this part.

**1910.216(a)(4)** *Mill roll heights.* All new mill installations shall be installed so that the top of the operating rolls is not less than 50 inches above the level on which the operator stands, irrespective of the size of the mill. This distance shall apply to the actual working level, whether it be at the general floor level, in a pit, or on a platform.

**1910.216(b)** *Mill safety controls—***(1)** *Safety trip control.* A safety trip control shall be provided in front and in back of each mill. It shall be accessible and shall operate readily on contact. The safety trip control shall be one of the following types or a combination thereof:

**1910.216(b)(1)(i)** *Pressure-sensitive body bars.* Installed at front and back of each mill having a 46-inch roll height or over. These bars shall operate readily by pressure of the mill operator's body.

**1910.216(b)(1)(ii)** *Safety triprod.* Installed in the front and in the back of each mill and located within 2 inches of a vertical plane tangent to the front and rear rolls. The top rods shall be not more than 72 inches above the level on which the operator stands. The triprods shall be accessible and shall operate readily whether the rods are pushed or pulled.

**1910.216(b)(1)(iii)** *Safety tripwire cable or wire center cord.* Installed in the front and in the back of each mill and located within 2 inches of a vertical plane tangent to the front and rear rolls. The cables shall not be more than 72 inches above the level on which the operator stands. The tripwire cable or wire center cord shall operate readily whether cable or cord is pushed or pulled.

**1910.216(b)(2)** [Reserved]

**1910.216(b)(3)** *Auxiliary equipment.* All auxiliary equipment such as mill divider, support bars, spray pipes, feed conveyors, strip knives, etc., shall be located in such a manner as to avoid interference with access to and operation of safety devices.

**1910.216(c)** *Calender safety controls—***(1)** *Safety trip, face.* A safety triprod, cable, or wire center cord shall be provided across each pair of in-running rolls extending the length of the face of the rolls. It shall be readily accessible and operate whether pushed or pulled. The safety tripping devices shall be located within reach of the operator and the bite.

**1910.216(c)(2)** *Safety trip, side.* On both sides of the calender and near each end of the face of the roll, there shall be a cable or wire center cord connected to the safety trip. They shall operate readily when pushed or pulled.

**1910.216(d)** *Protection by location—***(1)** *Mills.* Where a mill is so installed that persons cannot normally reach through, over, under, or around to come in contact with the roll bite or be caught between a roll and an adjacent object, then, provided such elements are made a fixed part of a mill, safety control devices listed in paragraph (b) of this section shall not apply.

**1910.216(d)(2)** *Calenders.* Where a calender is so installed that persons cannot normally reach through, over, under, or around to come in contact with the roll bite or be caught between a roll and an adjacent object, then, provided such elements are made a fixed part of a calender, safety control devices listed in paragraph (c) of this section shall not apply.

**1910.216(e)** *Trip and emergency switches.* All trip and emergency switches shall not be of the automatically resetting type, but shall require manual resetting.

**1910.216(f)** *Stopping limits—***(1)** *Determination of distance of travel.* All measurements on mills and calenders shall be taken with the rolls running empty at maximum operating speed. Stopping distances shall be expressed in inches of surface travel of the roll from the instant the emergency stopping device is actuated.

**1910.216(f)(2)** *Stopping limits for mills.* All mills irrespective of the size of the rolls or their arrangement (individually or group-driven) shall be stopped within a distance, as measured in inches of surface travel, not greater than 1 1/2 percent of the peripheral no-load surface speeds of the respective rolls as determined in feet per minute.

**1910.216(f)(3)** *Stopping limits for calenders.* **(i)** All calenders, irrespective of size of the rolls or their configuration, shall be stopped within a distance, as measured in inches of surface travel, not greater than 1 3/4 percent of the peripheral no-load surface speeds of the respective calender rolls as determined in feet per minute.

**1910.216(f)(3)(ii)** Where speeds above 250 feet per minute as measured on the surface of the drive roll are used, stopping distances of more than 1 3/4 percent are permissible. Such stopping distances shall be subject to engineering determination.

[39 FR 23502, June 27, 1974, as amended at 49 FR 5323, Feb. 10, 1984; 61 FR 9240, Mar. 7, 1996]

## §1910.217 Mechanical power presses.

**1910.217(a)** *General requirements.*

**1910.217(a)(1)-(3)** [Reserved]

**1910.217(a)(4)** *Reconstruction and modification.* It shall be the responsibility of any person reconstructing, or modifying a mechanical power press to do so in accordance with paragraph (b) of this section.

**1910.217(a)(5)** *Excluded machines.* Press brakes, hydraulic and pneumatic power presses, bulldozers, hot bending and hot metal presses, forging presses and hammers, riveting machines and similar types of fastener applicators are excluded from the requirements of this section.

**1910.217(b)** *Mechanical power press guarding and construction, general—***(1)** *Hazards to personnel associated with broken or falling machine components.* Machine components shall be designed, secured, or covered to minimize hazards caused by breakage, or loosening and falling or release of mechanical energy (i.e. broken springs).

**1910.217(b)(2)** *Brakes.* Friction brakes provided for stopping or holding a slide movement shall be inherently self-engaging by requiring power or force from an external source to cause disengagement. Brake capacity shall be sufficient to stop the motion of the slide quickly and capable of holding the slide and its attachments at any point in its travel.

**1910.217(b)(3)** *Machines using full revolution positive clutches.* **(i)** Machines using full revolution clutches shall incorporate a single-stroke mechanism.

**1910.217(b)(3)(ii)** If the single-stroke mechanism is dependent upon spring action, the spring(s) shall be of the compression type, operating on a rod or guided within a hole or tube, and designed to prevent interleaving of the spring coils in event of breakage.

**1910.217(b)(4)** *Foot pedals (treadle).* **(i)** The pedal mechanism shall be protected to prevent unintended operation from falling or moving objects or by accidental stepping onto the pedal.

**1910.217(b)(4)(ii)** A pad with a nonslip contact area shall be firmly attached to the pedal.

**1910.217(b)(4)(iii)** The pedal return spring(s) shall be of the compression type, operating on a rod or guided within a hole or tube, or designed to prevent interleaving of spring coils in event of breakage.

**1910.217(b)(4)(iv)** If pedal counterweights are provided, the path of the travel of the weight shall be enclosed.

**1910.217(b)(5)** *Hand operated levers.* **(i)** Hand-lever-operated power presses shall be equipped with a spring latch on the operating lever to prevent premature or accidental tripping.

**1910.217(b)(5)(ii)** The operating levers on hand-tripped presses having more than one operating station shall be interlocked to prevent the tripping of the press except by the "concurrent" use of all levers.

**1910.217(b)(6)** *Two-hand trip.* **(i)** A two-hand trip shall have the individual operator's hand controls protected against unintentional operation and have the individual operator's hand controls arranged by design and construction and/or separation to require the use of both hands to trip the press and use a control arrangement requiring concurrent operation of the individual operator's hand controls.

**1910.217(b)(6)(ii)** Two-hand trip systems on full revolution clutch machines shall incorporate an antirepeat feature.

**1910.217(b)(6)(iii)** If two-hand trip systems are used on multiple operator presses, each operator shall have a separate set of controls.

**1910.217(b)(7)** *Machines using part revolution clutches.* **(i)** The clutch shall release and the brake shall be applied when the external clutch engaging means is removed, deactivated, or deenergized.

**1910.217(b)(7)(ii)** A red color stop control shall be provided with the clutch/brake control system. Momentary operation of the stop control shall immediately deactivate the clutch and apply the brake. The stop control shall override any other control, and reactuation of the clutch shall require use of the operating (tripping) means which has been selected.

**1910.217(b)(7)(iii)** A means of selecting Off, "Inch," Single Stroke, and Continuous (when the continuous function is furnished) shall be supplied with the clutch/brake control to select type of operation of the press. Fixing of selection shall be by means capable of supervision by the employer.

**1910.217(b)(7)(iv)** The "Inch" operating means shall be designed to prevent exposure of the workers hands within the point of operation by:

**1910.217(b)(7)(iv)***(a)* Requiring the concurrent use of both hands to actuate the clutch, or

**1910.217(b)(7)(iv)***(b)* Being a single control protected against accidental actuation and so located that the worker cannot reach into the point of operation while operating the single control.

**1910.217(b)(7)(v)** Two-hand controls for single stroke shall conform to the following requirements:

**1910.217(b)(7)(v)***(a)* Each hand control shall be protected against unintended operation and arranged by design, construction, and/or separation so that the concurrent use of both hands is required to trip the press.

**1910.217(b)(7)(v)***(b)* The control system shall be designed to permit an adjustment which will require concurrent pressure from both hands during the die closing portion of the stroke.

**1910.217(b)(7)(v)***(c)* The control system shall incorporate an antirepeat feature.

**1910.217(b)(7)(v)***(d)* The control systems shall be designed to require release of all operators' hand controls before an interrupted stroke can be resumed. This requirement pertains only to those single-stroke, two-hand controls manufactured and installed on or after August 31, 1971.

**1910.217(b)(7)(vi)** [Reserved]

**1910.217(b)(7)(vii)** Controls for more than one operating station shall be designed to be activated and deactivated in complete sets of two operator's hand controls per operating station by means capable of being supervised by the employer. The clutch/brake control system shall be designed and constructed to prevent actuation of the clutch if all operating stations are bypassed.

**1910.217(b)(7)(viii)** Those clutch/brake control systems which contain both single and continuous functions shall be designed so that completion of continuous circuits may be supervised by the employer. The initiation of continuous run shall require a prior action or decision by the operator in addition to the selection of Continuous on the stroking selector, before actuation of the operating means will result in continuous stroking.

**1910.217(b)(7)(ix)** If foot control is provided, the selection method between hand and foot control shall be separate from the stroking selector and shall be designed so that the selection may be supervised by the employer.

**1910.217(b)(7)(x)** Foot operated tripping controls, if used, shall be protected so as to prevent operation from falling or moving objects, or from unintended operation by accidental stepping onto the foot control.

**1910.217(b)(7)(xi)** The control of air-clutch machines shall be designed to prevent a significant increase in the normal stopping time due to a failure within the operating value mechanism, and to inhibit further operation if such failure does occur. This requirement shall apply only to those clutch/brake air-valve controls manufactured and installed on or after August 31, 1971, but shall not apply to machines intended only for continuous, automatic feeding applications.

**1910.217(b)(7)(xii)** The clutch/brake control shall incorporate an automatic means to prevent initiation or continued activation of the Single Stroke or Continuous functions unless the press drive motor is energized and in the forward direction.

**1910.217(b)(7)(xiii)** The clutch/brake control shall automatically deactivate in event of failure of the power or pressure supply for the clutch engaging means. Reactivation of the clutch shall require restoration of normal supply and the use of the tripping mechanism(s).

**1910.217(b)(7)(xiv)** The clutch/brake control shall automatically deactivate in event of failure of the counterbalance(s) air supply. Reactivation of the clutch shall require restoration of normal air supply and use of the tripping mechanism(s).

**1910.217(b)(7)(xv)** Selection of bar operation shall be by means capable of being supervised by the employer. A separate pushbutton shall be employed to activate the clutch, and the clutch shall be activated only if the driver motor is deenergized.

**1910.217(b)(8)** *Electrical.* **(i)** A main power disconnect switch capable of being locked only in the Off position shall be provided with every power press control system.

**1910.217(b)(8)(ii)** The motor start button shall be protected against accidental operation.

**1910.217(b)(8)(iii)** All mechanical power press controls shall incorporate a type of drive motor starter that will disconnect the drive motor from the power source in event of control voltage or power source failure, and require operation of the motor start button to restart the motor when voltage conditions are restored to normal.

**1910.217(b)(8)(iv)** All a.c. control circuits and solenoid value coils shall be powered by not more than a nominal 120-volt a.c. supply obtained from a transformer with an isolated secondary. Higher voltages that may be necessary for operation of machine or control mechanisms shall be isolated from any control mechanism handled by the operator, but motor starters with integral Start-Stop buttons may utilize line voltage control. All d.c. control circuits shall be powered by not more than a nominal 240-volt d.c. supply isolated from any higher voltages.

**1910.217(b)(8)(v)** All clutch/brake control electrical circuits shall be protected against the possibility of an accidental ground in the control circuit causing false operation of the press.

**1910.217(b)(8)(vi)** Electrical clutch/brake control circuits shall incorporate features to minimize the possibility of an unintended stroke in the event of the failure of a control component to function properly, including relays, limit switches, and static output circuits.

**1910.217(b)(9)** *Slide counterbalance systems.* **(i)** Spring counterbalance systems when used shall incorporate means to retain system parts in event of breakage.

**1910.217(b)(9)(ii)** Spring counterbalances when used shall have the capability to hold the slide and its attachments at midstroke, without brake applied.

**1910.217(b)(9)(iii)** Air counterbalance cylinders shall incorporate means to retain the piston and rod in case of breakage or loosening.

**1910.217(b)(9)(iv)** Air counterbalance cylinders shall have adequate capability to hold the slide and its attachments at any point in stroke, without brake applied.

**1910.217(b)(9)(v)** Air counterbalance cylinders shall incorporate means to prevent failure of capability (sudden loss of pressure) in event of air supply failure.

**1910.217(b)(10)** *Air controlling equipment.* Air controlling equipment shall be protected against foreign material and water entering the pneumatic system of the press. A means of air lubrication shall be provided when needed.

**1910.217(b)(11)** *Hydraulic equipment.* The maximum anticipated working pressures in any hydraulic system on a mechanical power press shall not exceed the safe working pressure rating of any component used in that system.

**1910.217(b)(12)** *Pressure vessels.* All pressure vessels used in conjunction with power presses shall conform to the American Society of Mechanical Engineers Code for Pressure Vessels, 1968 Edition, which is incorporated by reference as specified in §1910.6.

**1910.217(b)(13)** *Control reliability.* When required by paragraph (c)(5) of this section, the control system shall be constructed so that a failure within the system does not prevent the normal stopping action from being applied to the press when required, but does prevent initiation of a successive stroke until the failure is corrected. The failure shall be detectable by a simple test, or indicated by the control system. This requirement does not apply to those elements of the control system which have no effect on the protection against point of operation injuries.

**1910.217(b)(14)** *Brake system monitoring.* When required by paragraph (c)(5) of this section, the brake monitor shall meet the following requirements:

**1910.217(b)(14)(i)** Be so constructed as to automatically prevent the activation of a successive stroke if the stopping time or braking distance deteriorates to a point where the safety distance being utilized does not meet the requirements set forth in paragraph (c)(3)(iii)(*e*) or (c)(3)(vii)(*c*) of this section. The brake monitor used with the Type B gate or movable barrier device shall be installed in a manner to detect slide top-stop overrun beyond the normal limit reasonably established by the employer.

**1910.217(b)(14)(ii)** Be installed on a press such that it indicates when the performance of the braking system has deteriorated to the extent described in paragraph (b)(14)(i) of this section; and

**1910.217(b)(14)(iii)** Be constructed and installed in a manner to monitor brake system performance on each stroke.

**1910.217(c)** *Safeguarding the point of operation*—**(1)** *General requirements.* **(i)** It shall be the responsibility of the employer to provide and insure the usage of "point of operation guards" or properly applied and adjusted point of operation devices on every operation performed on a mechanical power press. See Table O-10.

**1910.217(c)(1)(ii)** The requirement of paragraph (c)(1)(i) of this section shall not apply when the point of operation opening is one-fourth inch or less. See Table O-10.

**1910.217(c)(2)** *Point of operation guards.* **(i)** Every point of operation guard shall meet the following design, construction, application, and adjustment requirements:

**1910.217(c)(2)(i)***(a)* It shall prevent entry of hands or fingers into the point of operation by reaching through, over, under or around the guard;

**1910.217(c)(2)(i)***(b)* It shall conform to the maximum permissible openings of Table O-10;

**1910.217(c)(2)(i)***(c)* It shall, in itself, create no pinch point between the guard and moving machine parts;

**1910.217(c)(2)(i)***(d)* It shall utilize fasteners not readily removable by operator, so as to minimize the possibility of misuse or removal of essential parts;

**1910.217(c)(2)(i)***(e)* It shall facilitate its inspection, and

**1910.217(c)(2)(i)***(f)* It shall offer maximum visibility of the point of operation consistent with the other requirements.

**1910.217(c)(2)(ii)** A die enclosure guard shall be attached to the die shoe or stripper in a fixed position.

**1910.217(c)(2)(iii)** A fixed barrier guard shall be attached securely to the frame of the press or to the bolster plate.

**1910.217(c)(2)(iv)** An interlocked press barrier guard shall be attached to the press frame or bolster and shall be interlocked with the press clutch control so that the clutch cannot be activated unless the guard itself, or the hinged or movable sections of the guard are in position to conform to the requirements of Table O-10.

**1910.217(c)(2)(v)** The hinged or movable sections of an interlocked press barrier guard shall not be used for manual feeding. The guard shall prevent opening of the interlocked section and reaching into the point of operation prior to die closure or prior to the cessation of slide motion. See paragraph (c)(3)(ii) of this section regarding manual feeding through interlocked press barrier devices.

**1910.217(c)(2)(vi)** The adjustable barrier guard shall be securely attached to the press bed, bolster plate, or die shoe, and shall be adjusted and operated in conformity with Table O-10 and the requirements of this subparagraph. Adjustments shall be made only by authorized personnel whose qualifications include a knowledge of the provisions of Table O-10 and this subparagraph.

**1910.217(c)(2)(vii)** A point of operation enclosure which does not meet the requirements of this subparagraph and Table O-10 shall be used only in conjunction with point of operation devices.

**1910.217(c)(3)** *Point of operation devices.* (i) Point of operation devices shall protect the operator by:

**1910.217(c)(3)(i)***(a)* Preventing and/or stopping normal stroking of the press if the operator's hands are inadvertently placed in the point of operation; or

**1910.217(c)(3)(i)***(b)* Preventing the operator from inadvertently reaching into the point of operation, or withdrawing his hands if they are inadvertently located in the point of operation, as the dies close; or

**1910.217(c)(3)(i)***(c)* Preventing the operator from inadvertently reaching into the point of operation at all times; or

**1910.217(c)(3)(i)***(d)* [Reserved]

**1910.217(c)(3)(i)***(e)* Requiring application of both of the operator's hands to machine operating controls and locating such controls at such a safety distance from the point of operation that the slide completes the downward travel or stops before the operator can reach into the point of operation with his hands; or

**1910.217(c)(3)(i)***(f)* Enclosing the point of operation before a press stroke can be initiated, and maintaining this closed condition until the motion of the slide had ceased; or

**1910.217(c)(3)(i)***(g)* Enclosing the point of operation before a press stroke can be initiated, so as to prevent an operator from reaching into the point of operation prior to die closure or prior to cessation of slide motion during the downward stroke.

**1910.217(c)(3)(ii)** A gate or movable barrier device shall protect the operator as follows:

**1910.217(c)(3)(ii)***(a)* A Type A gate or movable barrier device shall protect the operator in the manner specified in paragraph (c)(3)(i)(*f*) of this section, and

**1910.217(c)(3)(ii)***(b)* A Type B gate or movable barrier device shall protect the operator in the manner specified in paragraph (c)(3)(i)(*g*) of this section.

**1910.217(c)(3)(iii)** A presence sensing point of operation device shall protect the operator as provided in paragraph (c)(3)(i)(*a*) of this section, and shall be interlocked into the control circuit to prevent or stop slide motion if the operator's hand or other part of his body is within the sensing field of the device during the downstroke of the press slide.

**1910.217(c)(3)(iii)***(a)* The device may not be used on machines using full revolution clutches.

**1910.217(c)(3)(iii)***(b)* The device may not be used as a tripping means to initiate slide motion, except when used in total conformance with paragraph (h) of this section.

**1910.217(c)(3)(iii)***(c)* The device shall be constructed so that a failure within the system does not prevent the normal stopping action from being applied to the press when required, but does prevent the initiation of a successive stroke until the failure is corrected. The failure shall be indicated by the system.

**1910.217(c)(3)(iii)(d)** Muting (bypassing of the protective function) of such device, during the upstroke of the press slide, is permitted for the purpose of parts ejection, circuit checking, and feeding.

**1910.217(c)(3)(iii)(e)** The safety distance ($D_s$) from the sensing field to the point of operation shall be greater than the distance determined by the following formula:

$$D_s = 63 \text{ inches/second} \times T_s$$

where:

$D_s$ = minimum safety distance (inches); 63 inches/second = hand speed constant;

and

$T_s$ = stopping time of the press measured at approximately 90° position of crankshaft rotation (seconds).

**1910.217(c)(3)(iii)(f)** Guards shall be used to protect all areas of entry to the point of operation not protected by the presence sensing device.

**1910.217(c)(3)(iv)** The pull-out device shall protect the operator as specified in paragraph (c)(3)(i)(b) of this section, and shall include attachments for each of the operator's hands.

**1910.217(c)(3)(iv)(a)** Attachments shall be connected to and operated only by the press slide or upper die.

**1910.217(c)(3)(iv)(b)** Attachments shall be adjusted to prevent the operator from reaching into the point of operation or to withdraw the operator's hands from the point of operation before the dies close.

**1910.217(c)(3)(iv)(c)** A separate pull-out device shall be provided for each operator if more than one operator is used on a press.

**1910.217(c)(3)(iv)(d)** Each pull-out device in use shall be visually inspected and checked for proper adjustment at the start of each operator shift, following a new die set-up, and when operators are changed. Necessary maintenance or repair or both shall be performed and completed before the press is operated. Records of inspections and maintenance shall be kept in accordance with paragraph (e) of this section.

**1910.217(c)(3)(v)** The sweep device may not be used for point of operation safeguarding.

**1910.217(c)(3)(vi)** A holdout or a restraint device shall protect the operator as specified in paragraph (c)(3)(i)(c) of this section and shall include attachments for each of the operator's hands. Such attachments shall be securely anchored and adjusted in such a way that the operator is restrained from reaching into the point of operation. A separate set of restraints shall be provided for each operator if more than one operator is required on a press.

**1910.217(c)(3)(vii)** The two hand control device shall protect the operator as specified in paragraph (c)(3)(i)(e) of this section.

**1910.217(c)(3)(vii)(a)** When used in press operations requiring more than one operator, separate two hand controls shall be provided for each operator, and shall be designed to require concurrent application of all operators' controls to activate the slide. The removal of a hand from any control button shall cause the slide to stop.

**1910.217(c)(3)(vii)(b)** Each two hand control shall meet the construction requirements of paragraph (b)(7)(v) of this section.

**1910.217(c)(3)(vii)(c)** The safety distance ($D_s$) between each two hand control device and the point of operation shall be greater than the distance determined by the following formula:

$$D_s = 63 \text{ inches/second} \times T_s;$$

where:

$D_s$ = minimum safety distance (inches); 63 inches/second = hand speed constant;

and

$T_s$ = stopping time of the press measured at approximately 90° position of crankshaft rotation (seconds).

**1910.217(c)(3)(vii)(d)** Two hand controls shall be fixed in position so that only a supervisor or safety engineer is capable of relocating the controls.

**1910.217(c)(3)(viii)** The two hand trip device shall protect the operator as specified in paragraph (c)(3)(i)(e) of this section.

**1910.217(c)(3)(viii)(a)** When used in press operations requiring more than one operator, separate two hand trips shall be provided for each operator, and shall be designed to require concurrent application of all operators' to activate the slide.

**1910.217(c)(3)(viii)(b)** Each two hand trip shall meet the construction requirements of paragraph (b)(6) of this section.

**1910.217(c)(3)(viii)(c)** The safety distance ($D_m$) between the two hand trip and the point of operation shall be greater than the distance determined by the following formula:

$$D_m = 63 \text{ inches/second} \times T_m;$$

where:

$D_m$ = minimum safety distance (inches); 63 inches/second = hand speed constant;

and

$T_m$ = the maximum time the press takes for the die closure after it has been tripped (seconds). For full revolution clutch presses with only one engaging point $T_m$ is equal to the time necessary for one and one-half revolutions of the crankshaft. For full revolution clutch presses with more than one engaging point, $T_m$ shall be calculated as follows:

$T_m$ = [1/2 + (1 ÷ Number of engaging points per revolution)] × time necessary to complete one revolution of the crankshaft (seconds).

**1910.217(c)(3)(viii)(d)** Two hand trips shall be fixed in position so that only a supervisor or safety engineer is capable of relocating the controls.

**1910.217(c)(4)** *Hand feeding tools.* Hand feeding tools are intended for placing and removing materials in and from the press. Hand feeding tools are not a point of operation guard or protection device and shall not be used in lieu of the "guards" or devices required in this section.

**1910.217(c)(5)** *Additional requirements for safe-guarding.* Where the operator feeds or removes parts by placing one or both hands in the point of operation, and a two hand control, presence sensing device, Type B gate or movable barrier (on a part revolution clutch) is used for safeguarding:

**1910.217(c)(5)(i)** The employer shall use a control system and a brake monitor which comply with paragraphs (b) (13) and (14) of this section;

**1910.217(c)(5)(ii)** The exception in paragraph (b)(7)(v)(d) of this section for two hand controls manufactured and installed before August 31, 1971 is not applicable under this paragraph (c)(5);

**1910.217(c)(5)(iii)** The control of air clutch machines shall be designed to prevent a significant increase in the normal stopping time due to a failure within the operating valve mechanism, and to inhibit further operation if such failure does occur, where a part revolution clutch is employed. The exception in paragraph (b)(7)(xi) of this section for controls manufactured and installed before August 31, 1971, is not applicable under this paragraph (c)(5).

**1910.217(d)** *Design, construction, setting and feeding of dies*—**(1)** *General requirements.* The employer shall: (i) Use dies and operating methods designed to control or eliminate hazards to operating personnel, and (ii) furnish and enforce the use of hand tools for freeing and removing stuck work or scrap pieces from the die, so that no employee need reach into the point of operation for such purposes.

**1910.217(d)(2)** [Reserved]

**1910.217(d)(3)** *Scrap handling.* The employer shall provide means for handling scrap from roll feed or random length stock operations. Scrap cutters used in conjunction with scrap handling systems shall be safeguarded in accordance with paragraph (c) of this section and with §1910.219.

**1910.217(d)(4)** *Guide post hazard.* The hazard created by a guide post (when it is located in the immediate vicinity of the operator) when separated from its bushing by more than one-fourth inch shall be considered as a point of operation hazard and be protected in accordance with paragraph (c) of this section.

**1910.217(d)(5)** *Unitized tooling.* If unitized tooling is used, the opening between the top of the punch holder and the face of the slide, or striking pad, shall be safeguarded in accordance with the requirements of paragraph (c) of this section.

**1910.217(d)(6)** *Tonnage, stroke, and weight designation.* All dies shall be:

**1910.217(d)(6)(i)** Stamped with the tonnage and stroke requirements, or have these characteristics recorded if these records are readily available to the die setter;

**1910.217(d)(6)(ii)** Stamped to indicate upper die weight when necessary for air counterbalance pressure adjustment; and

# OSHA Standards for General Industry

**1910.217(d)(6)(iii)** Stamped to indicate complete die weight when handling equipment may become overloaded.

**1910.217(d)(7)** *Die fastening.* Provision shall be made in both the upper and lower shoes for securely mounting the die to the bolster and slide. Where clamp caps or setscrews are used in conjunction with punch stems, additional means of securing the upper shoe to the slide shall be used.

**1910.217(d)(8)** *Die handling.* Handling equipment attach points shall be provided on all dies requiring mechanical handling.

**1910.217(d)(9)** *Diesetting.* **(i)** The employer shall establish a diesetting procedure that will insure compliance with paragraph (c) of this section.

**1910.217(d)(9)(ii)** The employer shall provide spring loaded turnover bars, for presses designed to accept such turnover bars.

**1910.217(d)(9)(iii)** The employer shall provide die stops or other means to prevent losing control of the die while setting or removing dies in presses which are inclined.

**1910.217(d)(9)(iv)** The employer shall provide and enforce the use of safety blocks for use whenever dies are being adjusted or repaired in the press.

**1910.217(d)(9)(v)** The employer shall provide brushes, swabs, lubricating rolls, and automatic or manual pressure guns so that operators and diesetters shall not be required to reach into the point of operation or other hazard areas to lubricate material, punches or dies.

**1910.217(e)** *Inspection, maintenance, and modification of presses*—**(1)** *Inspection and maintenance records.* **(i)** It shall be the responsibility of the employer to establish and follow a program of periodic and regular inspections of his power presses to ensure that all their parts, auxiliary equipment, and safeguards are in a safe operating condition and adjustment. The employer shall maintain a certification record of inspections which includes the date of inspection, the signature of the person who performed the inspection and the serial number, or other identifier, of the power press that was inspected.

**1910.217(e)(1)(ii)** Each press shall be inspected and tested no less than weekly to determine the condition of the clutch/brake mechanism, anti-repeat feature and single stroke mechanism. Necessary maintenance or repair or both shall be performed and completed before the press is operated. These requirements do not apply to those presses which comply with paragraphs (b) (13) and (14) of this section. The employer shall maintain a certification record of inspections, tests and maintenance work which includes the date of the inspection, test or maintenance; the signature of the person who performed the inspection, test, or maintenance; and the serial number or other identifier of the press that was inspected, tested or maintained.

**1910.217(e)(2)** *Modification.* It shall be the responsibility of any person modifying a power press to furnish instructions with the modification to establish new or changed guidelines for use and care of the power press so modified.

**1910.217(e)(3)** *Training of maintenance personnel.* It shall be the responsibility of the employer to insure the original and continuing competence of personnel caring for, inspecting, and maintaining power presses.

**1910.217(f)** *Operation of power presses*—**(1)** [Reserved]

**1910.217(f)(2)** *Instruction to operators.* The employer shall train and instruct the operator in the safe method of work before starting work on any operation covered by this section. The employer shall insure by adequate supervision that correct operating procedures are being followed.

**1910.217(f)(3)** *Work area.* The employer shall provide clearance between machines so that movement of one operator will not interfere with the work of another. Ample room for cleaning machines, handling material, work pieces, and scrap shall also be provided. All surrounding floors shall be kept in good condition and free from obstructions, grease, oil, and water.

**1910.217(f)(4)** *Overloading.* The employer shall operate his presses within the tonnage and attachment weight ratings specified by the manufacturer.

Explanation of above diagram:

This diagram shows the accepted safe openings between the bottom edge of a guard and feed table at various distances from the danger line (point of operation).

The *clearance line* marks the distance required to prevent contact between guard and moving parts.

The *minimum guarding line* is the distance between the infeed side of the guard and the danger line which is one-half inch from the danger line.

The various openings are such that for average size hands an operator's fingers won't reach the point of operation.

After installation of point of operation guards and before a job is released for operation a check should be made to verify that the guard will prevent the operator's hands from reaching the point of operation.

### TABLE O-10
[In inches]

| Distance of opening from point of operation hazard | Maximum width of opening |
|---|---|
| 1/2 to 1 1/2 | 1/4 |
| 1 1/2 to 2 1/2 | 3/8 |
| 2 1/2 to 3 1/2 | 1/2 |
| 3 1/2 to 5 1/2 | 5/8 |
| 5 1/2 to 6 1/2 | 3/4 |
| 6 1/2 to 7 1/2 | 7/8 |
| 7 1/2 to 12 1/2 | 1 1/4 |
| 12 1/2 to 15 1/2 | 1 1/2 |
| 15 1/2 to 17 1/2 | 1 7/8 |
| 17 1/2 to 31 1/2 | 2 1/8 |

This table shows the distances that guards shall be positioned from the danger line in accordance with the required openings.

**1910.217(g)** *Reports of injuries to employees operating mechanical power presses.* **(1)** The employer shall, within 30 days of the occurrence, report to either the Director of the Directorate of Safety Standards Programs, OSHA, U.S. Department of Labor, Washington, D.C. 20210, or the State agency administering a plan approved by the Assistant Secretary of Labor for Occupational Safety and Health, all point of operation injuries to operators or other employees. The following information shall be included in the report:

**1910.217(g)(1)(i)** Employer's name, address and location of the workplace (establishment).

**1910.217(g)(1)(ii)** Employee's name, injury sustained, and the task being performed (operation, set-up, maintenance, or other).

**1910.217(g)(1)(iii)** Type of clutch used on the press (full revolution, part revolution, or direct drive).

**1910.217(g)(1)(iv)** Type of safeguard(s) being used (two hand control, two hand trip, pull-outs, sweeps, or other). If the safeguard is not described in this section, give a complete description.

**1910.217(g)(1)(v)** Cause of the accident (repeat of press, safeguard failure, removing stuck part or scrap, no safeguard provided, no safeguard in use, or other).

**1910.217(g)(1)(vi)** Type of feeding (manual with hands in dies or with hands out of dies, semiautomatic, automatic, or other).

**1910.217(g)(1)(vii)** Means used to actuate press stroke (foot trip, foot control, hand trip, hand control, or other).

**1910.217(g)(1)(viii)** Number of operators required for the operation and the number of operators provided with controls and safeguards.

**1910.217(h)** *Presence sensing device initiation (PSDI)—***(1)** *General.* **(i)** The requirements of paragraph (h) shall apply to all part revolution mechanical power presses used in the PSDI mode of operation.

**1910.217(h)(1)(ii)** The relevant requirements of paragraphs (a) through (g) of this section also shall apply to all presses used in the PSDI mode of operation, whether or not cross referenced in this paragraph (h). Such cross-referencing of specific requirements from paragraphs (a) through (g) of this section is intended only to enhance convenience and understanding in relating to the new provisions to the existing standard, and is not to be construed as limiting the applicability of other provisions in paragraphs (a) through (g) of this section.

**1910.217(h)(1)(iii)** Full revolution mechanical power presses shall not be used in the PSDI mode of operation.

**1910.217(h)(1)(iv)** Mechanical power presses with a configuration which would allow a person to enter, pass through, and become clear of the sensing field into the hazardous portion of the press shall not be used in the PSDI mode of operation.

**1910.217(h)(1)(v)** The PSDI mode of operation shall be used only for normal production operations. Die-setting and maintenance procedures shall comply with paragraphs (a) through (g) of this section, and shall not be done in the PSDI mode.

**1910.217(h)(2)** *Brake and clutch requirements.* **(i)** Presses with flexible steel band brakes or with mechanical linkage actuated brakes or clutches shall not be used in the PSDI mode.

**1910.217(h)(2)(ii)** Brake systems on presses used in the PSDI mode shall have sufficient torque so that each average value of stopping times (Ts) for stops initiated at approximately 45 degrees, 60 degrees, and 90 degrees, respectively, of crankshaft angular position, shall not be more than 125 percent of the average value of the stopping time at the top crankshaft position. Compliance with this requirement shall be determined by using the heaviest upper die to be used on the press, and operating at the fastest press speed if there is speed selection.

**1910.217(h)(2)(iii)** Where brake engagement and clutch release is effected by spring action, such spring(s) shall operate in compression on a rod or within a hole or tube, and shall be of non-interleaving design.

**1910.217(h)(3)** *Pneumatic systems.* **(i)** Air valve and air pressure supply/control.

**1910.217(h)(3)(i)(A)** The requirements of paragraphs (b)(7)(xiii), (b)(7)(xiv), (b)(10), (b)(12) and (c)(5)(iii) of this section apply to the pneumatic systems of machines used in the PSDI mode.

**1910.217(h)(3)(i)(B)** The air supply for pneumatic clutch/brake control valves shall incorporate a filter, an air regulator, and, when necessary for proper operation, a lubricator.

**1910.217(h)(3)(i)(C)** The air pressure supply for clutch/brake valves on machines used in the PSDI mode shall be regulated to pressures less than or equal to the air pressure used when making the stop time measurements required by paragraph (h)(2)(ii) of this section.

**1910.217(h)(3)(ii)** Air counterbalance systems.

**1910.217(h)(3)(ii)(A)** Where presses that have slide counterbalance systems are used in the PSDI mode, the counterbalance system shall also meet the requirements of paragraph (b)(9) of this section.

**1910.217(h)(3)(ii)(B)** Counterbalances shall be adjusted in accordance with the press manufacturer's recommendations to assure correct counterbalancing of the slide attachment (upper die) weight for all operations performed on presses used in the PSDI mode. The adjustments shall be made before performing the stopping time measurements required by paragraphs (h)(2)(ii), (h)(5)(iii), and (h)(9)(v) of this section.

**1910.217(h)(4)** *Flywheels and bearings.* Presses whose designs incorporate flywheels running on journals on the crankshaft or back shaft, or bull gears running on journals mounted on the crankshaft, shall be inspected, lubricated, and maintained as provided in paragraph (h)(10) of this section to reduce the possibility of unintended and uncontrolled press strokes caused by bearing seizure.

**1910.217(h)(5)** *Brake monitoring.* **(i)** Presses operated in the PSDI mode shall be equipped with a brake monitor that meets the requirements of paragraphs (b)(13) and (b)(14) of this section. In addition, the brake monitor shall be adjusted during installation certification to prevent successive stroking of the press if increases in stopping time cause an increase in the safety distance above that required by paragraph (h)(9)(v) of this section.

**1910.217(h)(5)(ii)** Once the PSDI safety system has been certified/validated, adjustment of the brake monitor shall not be done without prior approval of the validation organization for both the brake monitor adjustment and the corresponding adjustment of the safety distance. The validation organization shall in its installation validation, state that in what circumstances, if any, the employer has advance approval for adjustment, when prior oral approval is appropriate and when prior approval must be in writing. The adjustment shall be done under the supervision of an authorized person whose qualifications include knowledge of safety distance requirements and experience with the brake system and its adjustment. When brake wear or other factors extend press stopping time beyond the limit permitted by the brake monitor, adjustment, repair, or maintenance shall be performed on the brake or other press system element that extends the stopping time.

**1910.217(h)(5)(iii)** The brake monitor setting shall allow an increase of no more than 10 percent of the longest stopping time for the press, or 10 milliseconds, whichever is longer, measured at the top of the stroke.

**1910.217(h)(6)** *Cycle control and control systems.* **(i)** The control system on presses used in the PSDI mode shall meet the applicable requirements of paragraphs (b)(7), (b)(8), (b)(13), and (c)(5) of this section.

**1910.217(h)(6)(ii)** The control system shall incorporate a means of dynamically monitoring for decoupling of the rotary position indicating mechanism drive from the crankshaft. This monitor shall stop slide motion and prevent successive press strokes if decoupling occurs, or if the monitor itself fails.

# OSHA Standards for General Industry

**1910.217(h)(6)(iii)** The mode selection means of paragraph (b)(7)(iii) of this section shall have at least one position for selection of the PSDI mode. Where more than one interruption of the light sensing field is used in the initiation of a stroke, either the mode selection means must have one position for each function, or a separate selection means shall be provided which becomes operable when the PSDI mode is selected. Selection of PSDI mode and the number of interruptions/withdrawals of the light sensing field required to initiate a press cycle shall be by means capable of supervision by the employer.

**1910.217(h)(6)(iv)** A PSDI set-up/reset means shall be provided which requires an overt action by the operator, in addition to PSDI mode selection, before operation of the press by means of PSDI can be started.

**1910.217(h)(6)(v)** An indicator visible to the operator and readily seen by the employer shall be provided which shall clearly indicate that the system is set-up for cycling in the PSDI mode.

**1910.217(h)(6)(vi)** The control system shall incorporate a timer to deactivate PSDI when the press does not stroke within the period of time set by the timer. The timer shall be manually adjustable, to a maximum time of 30 seconds. For any timer setting greater than 15 seconds, the adjustment shall be made by the use of a special tool available only to authorized persons. Following a deactivation of PSDI by the timer, the system shall make it necessary to reset the set-up/reset means in order to reactivate the PSDI mode.

**1910.217(h)(6)(vii)** Reactivation of PSDI operation following deactivation of the PSDI mode from any other cause, such as activation of the red color stop control required by paragraph (b)(7)(ii) of this section, interruption of the presence sensing field, opening of an interlock, or reselection of the number of sensing field interruptions/withdrawals required to cycle the press, shall require resetting of the set-up/reset means.

**1910.217(h)(6)(viii)** The control system shall incorporate an automatic means to prevent initiation or continued operation in the PSDI mode unless the press drive motor is energized in the forward direction of crankshaft rotation.

**1910.217(h)(6)(ix)** The control design shall preclude any movement of the slide caused by operation of power on, power off, or selector switches, or from checks for proper operations as required by paragraph (h)(6)(xiv) of this section.

**1910.217(h)(6)(x)** All components and subsystems of the control system shall be designed to operate together to provide total control system compliance with the requirements of this section.

**1910.217(h)(6)(xi)** Where there is more than one operator of a press used for PSDI, each operator shall be protected by a separate, independently functioning, presence sensing device. The control system shall require that each sensing field be interrupted the selected number of times prior to initiating a stroke. Further, each operator shall be provided with a set-up/reset means that meets the requirements of paragraph (h)(6) of this section, and which must be actuated to initiate operation of the press in the PSDI mode.

**1910.217(h)(6)(xii)** [Reserved]

**1910.217(h)(6)(xiii)** The Control system shall incorporate interlocks for supplemental guards, if used, which will prevent stroke initiation or will stop a stroke in progress if any supplemental guard fails or is deactivated.

**1910.217(h)(6)(xiv)** The control system shall perform checks for proper operation of all cycle control logic element switches and contacts at least once each cycle. Control elements shall be checked for correct status after power "on" and before the initial PSDI stroke.

**1910.217(h)(6)(xv)** The control system shall have provisions for an "inch" operating means meeting the requirements of paragraph (b)(7)(iv) of this section. Die-setting shall not be done in the PSDI mode. Production shall not be done in the "inch" mode.

**1910.217(h)(6)(xvi)** The control system shall permit only a single stroke per initiation command.

**1910.217(h)(6)(xvii)** Controls with internally stored programs (e.g., mechanical, electro-mechanical, or electronic) shall meet the requirements of paragraph (b)(13) of this section, and shall default to a predetermined safe condition in the event of any single failure within the system. Programmable controllers which meet the requirements for controls with internally stored programs stated above shall be permitted only if all logic elements affecting the safety system and point of operation safety are internally stored and protected in such a manner that they cannot be altered or manipulated by the user to an unsafe condition.

**1910.217(h)(7)** *Environmental requirements.* Control components shall be selected, constructed, and connected together in such a way as to withstand expected operational and environmental stresses, at least including those outlined in appendix A. Such stresses shall not so affect the control system as to cause unsafe operation.

**1910.217(h)(8)** *Safety system.* **(i)** Mechanical power presses used in the PSDI mode shall be operated under the control of a safety system which, in addition to meeting the applicable requirements of paragraphs (b)(13) and (c)(5) and other applicable provisions of this section, shall function such that a single failure or single operating error shall not cause injury to personnel from point of operation hazards.

**1910.217(h)(8)(ii)** The safety system shall be designed, constructed, and arranged as an integral total system, including all elements of the press, the controls, the safeguarding and any required supplemental safeguarding, and their interfaces with the operator and that part of the environment which has effect on the protection against point of operation hazards.

**1910.217(h)(9)** *Safeguarding the point of operation.* **(i)** The point of operation of presses operated in the PSDI mode shall be safeguarded in accordance with the requirements of paragraph (c) of this section, except that the safety distance requirements of paragraph (h)(9)(v) of this section shall be used for PSDI operation.

**1910.217(h)(9)(ii)(A)** PSDI shall be implemented only by use of light curtain (photo-electric) presence sensing devices which meet the requirements of paragraph (c)(3)(iii)(c) of this section unless the requirements of the following paragraph have been met.

**1910.217(h)(9)(ii)(B)** Alternatives to photo-electric light curtains may be used for PSDI when the employer can demonstrate, through tests and analysis by the employer or the manufacturer, that the alternative is as safe as the photo-electric light curtain, that the alternative meets the conditions of this section, has the same long term reliability as light curtains and can be integrated into the entire safety system as provided for in this section. Prior to use, both the employer and manufacturer must certify that these requirements and all the other applicable requirements of this section are met and these certifications must be validated by an OSHA-recognized third-party validation organization to meet these additional requirements and all the other applicable requirements of paragraphs (a) through (h) and appendix A of this section. Three months prior to the operation of any alternative system, the employer must notify the OSHA Directorate of Safety Standards Programs of the name of the system to be installed, the manufacturer and the OSHA-recognized third-party validation organization immediately. Upon request, the employer must make available to that office all tests and analyses for OSHA review.

**1910.217(h)(9)(iii)** Individual sensing fields of presence sensing devices used to initiate strokes in the PSDI mode shall cover only one side of the press.

**1910.217(h)(9)(iv)** Light curtains used for PSDI operation shall have minimum object sensitivity not to exceed one and one-fourth inches (31.75 mm). Where light curtain object sensitivity is user-adjustable, either discretely or continuously, design features shall limit the minimum object sensitivity adjustment not to exceed one and one-fourth inches (31.75 mm). Blanking of the sensing field is not permitted.

**1910.217(h)(9)(v)** The safety distance (Ds) from the sensing field of the presence sensing device to the point of operation shall be greater than or equal to the distance determined by the formula:

$$D_s = H_s \times (T_s + T_p + T_r + 2T_m) + D_p$$

Where:

$D_s$ = Minimum safety distance.

$H_s$ = Hand speed constant of 63 inches per second (1.6 m/s).

$T_s$ = Longest press stopping time, in seconds, computed by taking averages of multiple measurements at each of three positions (45 degrees, 60 degrees, and 90 degrees) of crankshaft angular position; the longest of the three averages is the stopping time to use. ($T_s$ is defined as the sum of the kinetic energy dissipation time plus the pneumatic/magnetic/hydraulic reaction time of the clutch/brake operating mechanism(s).)

$T_p$ = Longest presence sensing device response time, in seconds.

$T_r$ = Longest response time, in seconds, of all interposing control elements between the presence sensing device and the clutch/brake operating mechanism(s).

Tm = Increase in the press stopping time at the top of the stroke, in seconds, allowed by the brake monitor for brake wear. The time increase allowed shall be limited to no more than 10 percent of the longest press stopping time measured at the top of the stroke, or 10 milliseconds, whichever is longer.

Dp = Penetration depth factor, required to provide for possible penetration through the presence sensing field by fingers or hand before detection occurs. The penetration depth factor shall be determined from Graph h-1 using the minimum object sensitivity size.

**PENETRATION DEPTH FACTOR CALCULATION**
$$Dp = 3.4 (S - 0.276)$$

PENETRATION DEPTH FACTOR Dp - (INCHES)

MINIMUM OBJECT SENSITIVITY LIMIT 1.25 INCHES

OBJECT SENSITIVITY - S (INCHES)

**1910.217(h)(9)(vi)** The presence sensing device location shall either be set at each tool change and set-up to provide at least the minimum safety distance, or fixed in location to provide a safety distance greater than or equal to the minimum safety distance for all tooling set-ups which are to be used on that press.

**1910.217(h)(9)(vii)** Where presence sensing device location is adjustable, adjustment shall require the use of a special tool available only to authorized persons.

**1910.217(h)(9)(viii)** Supplemental safeguarding shall be used to protect all areas of access to the point of operation which are unprotected by the PSDI presence sensing device. Such supplemental safeguarding shall consist of either additional light curtain (photo-electric) presence sensing devices or other types of guards which meet the requirements of paragraphs (c) and (h) of this section.

**1910.217(h)(9)(viii)(A)** Presence sensing devices used as supplemental safeguarding shall not initiate a press stroke, and shall conform to the requirements of paragraph (c)(3)(iii) and other applicable provisions of this section, except that the safety distance shall comply with paragraph (h)(9)(v) of this section.

**1910.217(h)(9)(viii)(B)** Guards used as supplemental safeguarding shall conform to the design, construction and application requirements of paragraph (c)(2) of this section, and shall be interlocked with the press control to prevent press PSDI operation if the guard fails, is removed, or is out of position.

**1910.217(h)(9)(ix)** Barriers shall be fixed to the press frame or bolster to prevent personnel from passing completely through the sensing field, where safety distance or press configuration is such that personnel could pass through the PSDI presence sensing field and assume a position where the point of operation could be accessed without detection by the PSDI presence sensing device. As an alternative, supplemental presence sensing devices used only in the safeguard mode may be provided. If used, these devices shall be located so as to detect all operator locations and positions not detected by the PSDI sensing field, and shall prevent stroking or stop a stroke in process when any supplemental sensing field(s) are interrupted.

**1910.217(h)(9)(x)** Hand tools. Where tools are used for feeding, removal of scrap, lubrication of parts, or removal of parts that stick on the die in PSDI operations:

**1910.217(h)(9)(x)(A)** The minimum diameter of the tool handle extension shall be greater than the minimum object sensitivity of the presence sensing device(s) used to initiate press strokes; or

**1910.217(h)(9)(x)(B)** The length of the hand tool shall be such as to ensure that the operator's hand will be detected for any safety distance required by the press set-ups.

**1910.217(h)(10)** *Inspection and maintenance.* **(i)** Any press equipped with presence sensing devices for use in PSDI, or for supplemental safeguarding on presses used in the PSDI mode, shall be equipped with a test rod of diameter specified by the presence sensing device manufacturer to represent the minimum object sensitivity of the sensing field. Instructions for use of the test rod shall be noted on a label affixed to the presence sensing device.

**1910.217(h)(10)(ii)** The following checks shall be made at the beginning of each shift and whenever a die change is made.

**1910.217(h)(10)(ii)(A)** A check shall be performed using the test rod according to the presence sensing device manufacturer's instructions to determine that the presence sensing device used for PSDI is operational.

**1910.217(h)(10)(ii)(B)** The safety distance shall be checked for compliance with (h)(9)(v) of this section.

**1910.217(h)(10)(ii)(C)** A check shall be made to determine that all supplemental safeguarding is in place. Where presence sensing devices are used for supplemental safeguarding, a check for proper operation shall be performed using the test rod according to the presence sensing device manufacturer's instructions.

**1910.217(h)(10)(ii)(D)** A check shall be made to assure that the barriers and/or supplemental presence sensing devices required by paragraph (h)(9)(ix) of this section are operating properly.

**1910.217(h)(10)(ii)(E)** A system or visual check shall be made to verify correct counterbalance adjustment for die weight according to the press manufacturer's instructions, when a press is equipped with a slide counterbalance system.

**1910.217(h)(10)(iii)** When presses used in the PSDI mode have flywheel or bullgear running on crankshaft mounted journals and bearings, or a flywheel mounted on back shaft journals and bearings, periodic inspections following the press manufacturer's recommendations shall be made to ascertain that bearings are in good working order, and that automatic lubrication systems for these bearings (if automatic lubrication is provided) are supplying proper lubrication. On presses with provision for manual lubrication of flywheel or bullgear bearings, lubrication shall be provided according to the press manufacturer's recommendations.

**1910.217(h)(10)(iv)** Periodic inspections of clutch and brake mechanisms shall be performed to assure they are in proper operating condition. The press manufacturer's recommendations shall be followed.

**1910.217(h)(10)(v)** When any check of the press, including those performed in accordance with the requirements of paragraphs (h)(10)(ii), (iii) or (iv) of this section, reveals a condition of noncompliance, improper adjustment, or failure, the press shall not be operated until the condition has been corrected by adjustment, replacement, or repair.

**1910.217(h)(10)(vi)** It shall be the responsibility of the employer to ensure the competence of personnel caring for, inspecting, and maintaining power presses equipped for PSDI operation, through initial and periodic training.

**1910.217(h)(11)** *Safety system certification/validation.* **(i)** Prior to the initial use of any mechanical press in the PSDI mode, two sets of certification and validation are required:

**1910.217(h)(11)(i)(A)** The design of the safety system required for the use of a press in the PSDI mode shall be certified and validated prior to installation. The manufacturer's certification shall be validated by an OSHA-recognized third-party validation organization to meet all applicable requirements of paragraphs (a) through (h) and appendix A of this section.

**1910.217(h)(11)(i)(B)** After a press has been equipped with a safety system whose design has been certified and validated in accordance with paragraph (h)(11)(i) of this section, the safety system installation shall be certified by the employer, and then shall be validated by an OSHA-recognized third-party validation organization to meet all applicable requirements of paragraphs (a) through (h) and appendix A of this section.

# OSHA Standards for General Industry

**1910.217(h)(11)(ii)** At least annually thereafter, the safety system on a mechanical power press used in the PSDI mode shall be recertified by the employer and revalidated by an OSHA-recognized third-party validation organization to meet all applicable requirements of paragraphs (a) through (h) and appendix A of this section. Any press whose safety system has not been recertified and revalidated within the preceding 12 months shall be removed from service in the PSDI mode until the safety system is recertified and revalidated.

**1910.217(h)(11)(iii)** A label shall be affixed to the press as part of each installation certification/validation and the most recent recertification/revalidation. The label shall indicate the press serial number, the minimum safety distance (Ds) required by paragraph (h)(9)(v) of this section, the fulfillment of design certification/validation, the employer's signed certification, the identification of the OSHA-recognized third-party validation organization, its signed validation, and the date the certification/validation and recertification/revalidation are issued.

**1910.217(h)(11)(iv)** Records of the installation certification and validation and the most recent recertification and revalidation shall be maintained for each safety system equipped press by the employer as long as the press is in use. The records shall include the manufacture and model number of each component and subsystem, the calculations of the safety distance as required by paragraph (h)(9)(v) of this section, and the stopping time measurements required by paragraph (h)(2)(ii) of this section. The most recent records shall be made available to OSHA upon request.

**1910.217(h)(11)(v)** The employer shall notify the OSHA-recognized third-party validation organization within five days whenever a component or a subsystem of the safety system fails or modifications are made which may affect the safety of the system. The failure of a critical component shall necessitate the removal of the safety system from service until it is recertified and revalidated, except recertification by the employer without revalidation is permitted when a non-critical component or subsystem is replaced by one of the same manufacture and design as the original, or determined by the third-party validation organization to be equivalent by similarity analysis, as set forth in appendix A.

**1910.217(h)(11)(vi)** The employer shall notify the OSHA-recognized third-party validation organization within five days of the occurrence of any point of operation injury while a press is used in the PSDI mode. This is in addition to the report of injury required by paragraph (g) of this section; however, a copy of that report may be used for this purpose.

**1910.217(h)(12)** *Die setting and work set-up.* **(i)** Die setting on presses used in the PSDI mode shall be performed in accordance with paragraphs (d) and (h) of this section.

**1910.217(h)(12)(ii)** The PSDI mode shall not be used for die setting or set-up. An alternative manual cycle initiation and control means shall be supplied for use in die setting which meets the requirements of paragraph (b)(7) of this section.

**1910.217(h)(12)(iii)** Following a die change, the safety distance, the proper application of supplemental safeguarding, and the slide counterbalance adjustment (if the press is equipped with a counterbalance) shall be checked and maintained by authorized persons whose qualifications include knowledge of the safety distance, supplemental safeguarding requirements, and the manufacturer's specifications for counterbalance adjustment. Adjustment of the location of the PSDI presence sensing device shall require use of a special tool available only to the authorized persons.

**1910.217(h)(13)** *Operator training.* **(i)** The operator training required by paragraph (f)(2) of this section shall be provided to the employee before the employee initially operates the press and as needed to maintain competence, but not less than annually thereafter. It shall include instruction relative to the following items for presses used in the PSDI mode.

**1910.217(h)(13)(i)(A)** The manufacturer's recommended test procedures for checking operation of the presence sensing device. This shall include the use of the test rod required by paragraph (h)(10)(i) of this section.

**1910.217(h)(13)(i)(B)** The safety distance required.

**1910.217(h)(13)(i)(C)** The operation, function and performance of the PSDI mode.

**1910.217(h)(13)(i)(D)** The requirements for hand tools that may be used in the PSDI mode.

**1910.217(h)(13)(i)(E)** The severe consequences that can result if he or she attempts to circumvent or by-pass any of the safeguard or operating functions of the PSDI system.

**1910.217(h)(13)(ii)** The employer shall certify that employees have been trained by preparing a certification record which includes the identity of the person trained, the signature of the employer or the person who conducted the training, and the date the training was completed. The certification record shall be prepared at the completion of training and shall be maintained on file for the duration of the employee's employment. The certification record shall be made available upon request to the Assistant Secretary for Occupational Safety and Health.

## APPENDIX A TO §1910.217—MANDATORY REQUIREMENTS FOR CERTIFICATION/VALIDATION OF SAFETY SYSTEMS FOR PRESENCE SENSING DEVICE INITIATION OF MECHANICAL POWER PRESSES

### *Purpose*

The purpose of the certification/validation of safety systems for presence sensing device initiation (PSDI) of mechanical power presses is to ensure that the safety systems are designed, installed, and maintained in accordance with all applicable requirements of 29 CFR 1910.217(a) through (h) and this appendix A.

### *General*

The certification/validation process shall utilize an independent third-party validation organization recognized by OSHA in accordance with the requirements specified in appendix C of this section.

While the employer is responsible for assuring that the certification/validation requirements in §1910.217(h)(11) are fulfilled, the design certification of PSDI safety systems may be initiated by manufacturers, employers, and/or their representatives. The term *manufacturers* refers to the manufacturer of any of the components of the safety system. An employer who assembles a PSDI safety system would be a manufacturer as well as employer for purposes of this standard and appendix.

The certification/validation process includes two stages. For design certification, in the first stage, the manufacturer (which can be an employer) certifies that the PSDI safety system meets the requirements of 29 CFR 1910.217(a) through (h) and this appendix A, based on appropriate design criteria and tests. In the second stage, the OSHA-recognized third-party validation organization validates that the PSDI safety system meets the requirements of 29 CFR 1910.217(a) through (h) and this appendix A and the manufacturer's certification by reviewing the manufacturer's design and test data and performing any additional reviews required by this standard or which it believes appropriate.

For installation certification/validation and annual recertification/revalidation, in the first stage the employer certifies or recertifies that the employer is installing or utilizing a PSDI safety system validated as meeting the design requirements of 29 CFR 1910.217(a) through (h) and this appendix A by an OSHA-recognized third-party validation organization and that the installation, operation and maintenance meet the requirements of 29 CFR 1910.217(a) through (h) and this appendix A. In the second stage, the OSHA-recognized third-party validation organization validates or revalidates that the PSDI safety system installation meets the requirements of 29 CFR 1910.217(a) through (h) and this appendix A and the employer's certification, by reviewing that the PSDI safety system has been certified; the employer's certification, designs and tests, if any; the installation, operation, maintenance and training; and by performing any additional tests and reviews which the validation organization believes is necessary.

### *Summary*

The certification/validation of safety systems for PSDI shall consider the press, controls, safeguards, operator, and environment as an integrated system which shall comply with all of the requirements in 29 CFR 1910.217(a) through (h) and this appendix A. The certification/validation process shall verify that the safety system complies with the OSHA safety requirements as follows:

### A. Design Certification/Validation

1. The major parts, components and subsystems used shall be defined by part number or serial number, as appropriate, and by manufacturer to establish the configuration of the system.

2. The identified parts, components and subsystems shall be certified by the manufacturer to be able to withstand the functional and operational environments of the PSDI safety system.

3. The total system design shall be certified by the manufacturer as complying with all requirements in 29 CFR 1910.217(a) through (h) and this appendix A.

4. The third-party validation organization shall validate the manufacturer's certification under paragraphs 2 and 3.

### B. Installation Certification/Validation

1. The employer shall certify that the PSDI safety system has been design certified and validated, that the installation meets the operational and environmental requirements specified by the manufacturer, that the installation drawings are accurate, and that the installation meets the requirements of 29 CFR 1910.217(a) through (h) and this appendix A. (The operational and installation requirements of the PSDI safety system may vary for different applications.)

2. The third-party validation organization shall validate the employer's certifications that the PSDI safety system is design certified and validated, that the installation meets the installation and environmental requirements specified by the manufacturer, and that the installation meets the requirements of 29 CFR 1910.217(a) through (h) and this appendix A.

### C. Recertification/Revalidation

1. The PSDI safety system shall remain under certification/validation for the shorter of one year or until the system hardware is changed, modified or refurbished, or operating conditions are changed (including environmental, application or facility changes), or a failure of a critical component has occurred.

2. Annually, or after a change specified in paragraph 1., the employer shall inspect and recertify the installation as meeting the requirements set forth under B., Installation Certification/Validation.

3. The third-party validation organization, annually or after a change specified in paragraph 1., shall validate the employer's certification that the requirements of paragraph B., Installation Certification/Validation have been met.

(**NOTE:** Such changes in operational conditions as die changes or press relocations not involving disassembly or revision to the safety system would not require recertification/revalidation.)

### Certification/Validation Requirements

### A. General Design Certification/Validation Requirements

1. *Certification/Validation Program Requirements.* The manufacturer shall certify and the OSHA-recognized third-party validation organization shall validate that:

(a) The design of components, subsystems, software and assemblies meets OSHA performance requirements and are ready for the intended use; and

(b) The performance of combined subsystems meets OSHA's operational requirements.

2. *Certification/Validation Program Level of Risk Evaluation Requirements.* The manufacturer shall evaluate and certify, and the OSHA-recognized third-party validation organization shall validate, the design and operation of the safety system by determining conformance with the following:

a. The safety system shall have the ability to sustain a single failure or a single operating error and not cause injury to personnel from point of operation hazards. Acceptable design features shall demonstrate, in the following order or precedence, that:

(1) No single failure points may cause injury; or

(2) Redundancy, and comparison and/or diagnostic checking, exist for the critical items that may cause injury, and the electrical, electronic, electromechanical and mechanical parts and components are selected so that they can withstand operational and external environments. The safety factor and/or derated percentage shall be specifically noted and complied with.

b. The manufacturer shall design, evaluate, test and certify, and the third-party validation organization shall evaluate and validate, that the PSDI safety system meets appropriate requirements in the following areas.

(1) Environmental Limits

(a) Temperature

(b) Relative humidity

(c) Vibration

(d) Fluid compatibility with other materials

(2) Design Limits

(a) Power requirements

(b) Power transient tolerances

(c) Compatibility of materials used

(d) Material stress tolerances and limits

(e) Stability to long term power fluctuations

(f) Sensitivity to signal acquisition

(g) Repeatability of measured parameter without inadvertent initiation of a press stroke

(h) Operational life of components in cycles, hours, or both

(i) Electromagnetic tolerance to:

(*1*) Specific operational wave lengths; and

(*2*) Externally generated wave lengths

(*3*) *New Design Certification/Validation.* Design certification/validation for a new safety system, i.e., a new design or new integration of specifically identified components and subsystems, would entail a single certification/validation which would be applicable to all identical safety systems. It would not be necessary to repeat the tests on individual safety systems of the same manufacture or design. Nor would it be necessary to repeat these tests in the case of modifications where determined by the manufacturer and validated by the third-party validation organization to be equivalent by similarity analysis. Minor modifications not affecting the safety of the system may be made by the manufacturer without revalidation.

Substantial modifications would require testing as a new safety system, as deemed necessary by the validation organization.

### B. Additional Detailed Design Certification/Validation Requirements

1. *General.* The manufacturer or the manufacturer's representative shall certify to and submit to an OSHA-recognized third-party validation organization the documentation necessary to demonstrate that the PSDI safety system design is in full compliance with the requirements of 29 CFR 1910.217(a)-(h) and this appendix A, as applicable, by means of analysis, tests, or combination of both, establishing that the following additional certification/validation requirements are fulfilled.

2. *Reaction Times.* For the purpose of demonstrating compliance with the reaction time required by §1910.217(h), the tests shall use the following definitions and requirements:

a. *Reaction time* means the time, in seconds, it takes the signal, required to activate/deactivate the system, to travel through the system, measured from the time of signal initiation to the time the function being measured is completed.

b. *Full stop* or *No movement of the slide or ram* means when the crankshaft rotation has slowed to two or less revolutions per minute, just before stopping completely.

c. *Function completion* means for, electrical, electromechanical and electronic devices, when the circuit produces a change of state in the output element of the device.

d. When the change of state is motion, the measurement shall be made at the completion of the motion.

e. The generation of the test signal introduced into the system for measuring reaction time shall be such that the initiation time can be established with an error of less than 0.5 percent of the reaction time measured.

f. The instrument used to measure reaction time shall be calibrated to be accurate to within 0.001 second.

3. *Compliance with §1910.217(h)(2)(ii).* For compliance with these requirements, the average value of the stopping time, $T_s$, shall be the arithmetic mean of at least 25 stops for each stop angle initiation measured with the brake and/or clutch unused, 50 percent worn, and 90 percent worn. The recommendations of the brake system manufacturer shall be used to simulate or estimate the brake wear. The manufacturer's recommended minimum lining depth shall be identified and documented, and an evaluation made that the minimum depth will not be exceeded before the next (annual) recertification/revalidation. A correlation of the brake and/or clutch degradation based on the above tests and/or estimates shall be made and documented. The results shall document the conditions under which the

brake and/or clutch will and will not comply with the requirement. Based upon this determination, a scale shall be developed to indicate the allowable 10 percent of the stopping time at the top of the stroke for slide or ram overtravel due to brake wear. The scale shall be marked to indicate that brake adjustment and/or replacement is required. The explanation and use of the scale shall be documented.

The test specification and procedure shall be submitted to the validation organization for review and validation prior to the test. The validation organization representative shall witness at least one set of tests.

4. *Compliance with §§1910.217(h)(5)(iii) and (h)(9)(v).* Each reaction time required to calculate the Safety Distance, including the brake monitor setting, shall be documented in separate reaction time tests. These tests shall specify the acceptable tolerance band sufficient to assure that tolerance build-up will not render the safety distance unsafe.

a. Integrated test of the press fully equipped to operate in the PSDI mode shall be conducted to establish the total system reaction time.

b. Brakes which are the adjustable type shall be adjusted properly before the test.

5. *Compliance with §1910.217(h)(2)(iii).* a. Prior to conducting the brake system test required by paragraph (h)(2)(ii), a visual check shall be made of the springs. The visual check shall include a determination that the spring housing or rod does not show damage sufficient to degrade the structural integrity of the unit, and the spring does not show any tendency to interleave.

b. Any detected broken or unserviceable springs shall be replaced before the test is conducted. The test shall be considered successful if the stopping time remains within that which is determined by paragraph (h)(9)(v) for the safety distance setting. If the increase in press stopping time exceeds the brake monitor setting limit defined in paragraph (h)(5)(iii), the test shall be considered unsuccessful, and the cause of the excessive stopping time shall be investigated. It shall be ascertained that the springs have not been broken and that they are functioning properly.

6. *Compliance with §1910.217(h)(7).* a. Tests which are conducted by the manufacturers of electrical components to establish stress, life, temperature and loading limits must be tests which are in compliance with the provisions of the National Electrical Code.

b. Electrical and/or electronic cards or boards assembled with discreet components shall be considered a subsystem and shall require separate testing that the subsystems do not degrade in any of the following conditions:

(1) Ambient temperature variation from -20 °C to +50 °C.

(2) Ambient relative humidity of 99 percent.

(3) Vibration of 45G for one millisecond per stroke when the item is to be mounted on the press frame.

(4) Electromagnetic interference at the same wavelengths used for the radiation sensing field, at the power line frequency fundamental and harmonics, and also from outogenous radiation due to system switching.

(5) Electrical power supply variations of ±15 percent.

c. The manufacturer shall specify the test requirements and procedures from existing consensus tests in compliance with the provisions of the National Electrical Code.

d. Tests designed by the manufacturer shall be made available upon request to the validation organization. The validation organization representative shall witness at least one set of each of these tests.

7. *Compliance with §1910.217(h)(9)(iv).* a. The manufacturer shall design a test to demonstrate that the prescribed minimum object sensitivity of the presence sensing device is met.

b. The test specifications and procedures shall be made available upon request to the validation organization.

8. *Compliance with §1910.217(h)(9)(x).* a. The manufacturer shall design a test(s) to establish the hand tool extension diameters allowed for variations in minimum object sensitivity response.

b. The test(s) shall document the range of object diameter sizes which will produce both single and double break conditions.

c. The test(s) specifications and procedures shall be made available upon request to the validation organization.

9. *Integrated Tests Certification/Validation.* a. The manufacturer shall design a set of integrated tests to demonstrate compliance with the following requirements:

Sections 1910.217(h)(6) (ii); (iii); (iv); (v); (vi); (vii); (viii); (ix); (xi); (xii); (xiii); (xiv); (xv); and (xvii).

b. The integrated test specifications and procedures shall be made available to the validation organization.

10. *Analysis.* a. The manufacturer shall submit to the validation organization the technical analysis such as Hazard Analysis, Failure Mode and Effect Analysis, Stress Analysis, Component and Material Selection Analysis, Fluid Compatibility, and/or other analyses which may be necessary to demonstrate, compliance with the following requirements:

Sections 1910.217(h)(8)(i) and (ii); (h)(2) (ii) and (iii); (h)(3)(i) (A) and (C), and (ii); (h)(5) (i), (ii) and (iii); (h)(6) (i), (iii), (iv), (vi), (vii), (viii), (ix), (x), (xi), (xiii), (xiv), (xv), (xvi), and (xvii); (h)(7) (i) and (ii); (h)(9) (iv), (v), (viii), (ix) and (x); (h)(10) (i) and (ii).

11. *Types of Tests Acceptable for Certification/Validation.* a. Test results obtained from development testing may be used to certify/validate the design.

b. The test results shall provide the engineering data necessary to establish confidence that the hardware and software will meet specifications, the manufacturing process has adequate quality control and the data acquired was used to establish processes, procedures, and test levels supporting subsequent hardware design, production, installation and maintenance.

12. *Validation for Design Certification/Validation.* If, after review of all documentation, tests, analyses, manufacturer's certifications, and any additional tests which the third-party validation organization believes are necessary, the third-party validation organization determines that the PSDI safety system is in full compliance with the applicable requirements of 29 CFR 1910.217(a) through (h) and this appendix A, it shall validate the manufacturer's certification that it so meets the stated requirements.

### C. Installation Certification/Validation Requirements

1. The employer shall evaluate and test the PSDI system installation, shall submit to the OSHA-recognized third-party validation organization the necessary supporting documentation, and shall certify that the requirements of §1910.217(a) through (h) and this appendix A have been met and that the installation is proper.

2. The OSHA-recognized third-party validation organization shall conduct tests, and/or review and evaluate the employer's installation tests, documentation and representations. If it so determines, it shall validate the employer's certification that the PSDI safety system is in full conformance with all requirements of 29 CFR 1910.217(a) through (h) and this appendix A.

### D. Recertification/Revalidation Requirements

1. A PSDI safety system which has received installation certification/validation shall undergo recertification/revalidation the earlier of:

a. Each time the systems hardware is significantly changed, modified, or refurbished;

b. Each time the operational conditions are significantly changed (including environmental, application or facility changes, but excluding such changes as die changes or press relocations not involving revision to the safety system);

c. When a failure of a significant component has occurred or a change has been made which may affect safety; or

d. When one year has elapsed since the installation certification/validation or the last recertification/revalidation.

2. *Conduct or recertification/revalidation.* The employer shall evaluate and test the PSDI safety system installation, shall submit to the OSHA-recognized third-party validation organization the necessary supporting documentation, and shall recertify that the requirements of §1910.217(a) through (h) and this appendix are being met. The documentation shall include, but not be limited to, the following items:

a. Demonstration of a thorough inspection of the entire press and PSDI safety system to ascertain that the installation, components and safeguarding have not been changed, modified or tampered with since the installation certification/validation or last recertification/revalidation was made.

b. Demonstrations that such adjustments as may be needed (such as to the brake monitor setting) have been accomplished with proper changes made in the records and on such notices as are located on the press and safety system.

c. Demonstration that review has been made of the reports covering the design certification/validation, the installation certification/validation, and all recertification/revalidations, in order to detect any degradation to an unsafe condition, and that necessary changes have been made to restore the safety system to previous certification/validation levels.

3. The OSHA-recognized third-party validation organization shall conduct tests, and/or review and evaluate the employer's installation, tests, documentation and representations. If it so determines, it shall revalidate the employer's recertification that the PSDI system is in full conformance with all requirements of 29 CFR 1910.217(a) through (h) and this appendix A.

### APPENDIX B TO §1910.217—NON-MANDATORY GUIDELINES FOR CERTIFICATION/VALIDATION OF SAFETY SYSTEMS FOR PRESENCE SENSING DEVICE INITIATION OF MECHANICAL POWER PRESSES

#### Objectives

This appendix provides employers, manufacturers, and their representatives, with nonmandatory guidelines for use in developing certification documents. Employers and manufacturers are encouraged to recommend other approaches if there is a potential for improving safety and reducing cost. The guidelines apply to certification/validation activity from design evaluation through the completion of the installation test and the annual recertification/revalidation tests.

#### General Guidelines

A. The certification/validation process should confirm that hazards identified by hazard analysis, (HA), failure mode effect analysis (FMEA), and other system analyses have been eliminated by design or reduced to an acceptable level through the use of appropriate design features, safety devices, warning devices, or special procedures. The certification/validation process should also confirm that residual hazards identified by operational analysis are addressed by warning, labeling safety instructions or other appropriate means.

B. The objective of the certification/validation program is to demonstrate and document that the system satisfies specification and operational requirements for safe operations.

#### Quality Control

The safety attributes of a certified/validated PSDI safety system are more likely to be maintained if the quality of the system and its parts, components and subsystem is consistently controlled. Each manufacturer supplying parts, components, subsystems, and assemblies needs to maintain the quality of the product, and each employer needs to maintain the system in a non-degraded condition.

#### Analysis Guidelines

A. Certification/validation of hardware design below the system level should be accomplished by test and/or analysis.

B. Analytical methods may be used in lieu of, in combination with, or in support of tests to satisfy specification requirements.

C. Analyses may be used for certification/validation when existing data are available or when test is not feasible.

D. Similarity analysis may be used in lieu of tests where it can be shown that the article is similar in design, manufacturing process, and quality control to another article that was previously certified/validated in accordance with equivalent or more stringent criteria. If previous design, history and application are considered to be similar, but not equal to or more exacting than earlier experiences, the additional or partial certification/validation tests should concentrate on the areas of changed or increased requirements.

#### Analysis Reports

The analysis reports should identify: (1) The basis for the analysis; (2) the hardware or software items analyzed; (3) conclusions; (4) safety factors; and (5) limit of the analysis. The assumptions made during the analysis should be clearly stated and a description of the effects of these assumptions on the conclusions and limits should be included.

Certification/validation by similarity analysis reports should identify, in addition to the above, application of the part, component or subsystem for which certification/validation is being sought as well as data from previous usage establishing adequacy of the item. Similarity analysis should not be accepted when the internal and external stresses on the item being certified/validated are not defined.

Usage experience should also include failure data supporting adequacy of the design.

### APPENDIX C TO §1910.217—MANDATORY REQUIREMENTS FOR OSHA RECOGNITION OF THIRD-PARTY VALIDATION ORGANIZATIONS FOR THE PSDI STANDARD

This appendix prescribes mandatory requirements and procedures for OSHA recognition of third-party validation organizations to validate employer and manufacturer certifications that their equipment and practices meet the requirements of the PSDI standard. The scope of the appendix includes the three categories of certification/validation required by the PSDI standard: Design Certification/Validation, Installation Certification/Validation, and Annual Recertification/Revalidation.

If further detailing of these provisions will assist the validation organization or OSHA in this activity, this detailing will be done through appropriate OSHA Program Directives.

#### I. Procedure for OSHA Recognition of Validation Organizations

#### A. Applications

1. *Eligibility.* a. Any person or organization considering itself capable of conducting a PSDI-related third-party validation function may apply for OSHA recognition.

b. However, in determining eligibility for a foreign-based third-party validation organization, OSHA shall take into consideration whether there is reciprocity of treatment by the foreign government after consultation with relevant U.S. government agencies.

2. *Content of application.* a. The application shall identify the scope of the validation activity for which the applicant wishes to be recognized, based on one of the following alternatives:

(1) Design Certification/Validation, Installation Certification/Validation, and Annual Recertification/Revalidation;

(2) Design Certification/Validation only; or

(3) Installation/Certification/Validation and Annual Recertification/Revalidation.

b. The application shall provide information demonstrating that it and any validating laboratory utilized meet the qualifications set forth in section II of this appendix.

c. The applicant shall provide information demonstrating that it and any validating laboratory utilized meet the program requirements set forth in section III of this appendix.

d. The applicant shall identify the test methods it or the validating laboratory will use to test or judge the components and operations of the PSDI safety system required to be tested by the PSDI standard and appendix A, and shall specify the reasons the test methods are appropriate.

e. The applicant may include whatever enclosures, attachments, or exhibits the applicant deems appropriate. The application need not be submitted on a Federal form.

f. The applicant shall certify that the information submitted is accurate.

3. *Filing office location.* The application shall be filed with: PSDI Certification/Validation Program, Office of Variance Determination, Occupational Safety and Health Administration, U.S. Department of Labor, Room N3653, 200 Constitution Avenue, NW., Washington, DC 20210.

4. *Amendments and withdrawals.* a. An application may be revised by an applicant at any time prior to the completion of the final staff recommendation.

b. An application may be withdrawn by an applicant, without prejudice, at any time prior to the final decision by the Assistant Secretary in paragraph I.B.8.b.(4) of this appendix.

#### B. Review and Decision Process

1. *Acceptance and field inspection.* All applications submitted will be accepted by OSHA, and their receipt acknowledged in writing. After receipt of an appli-

cation, OSHA may request additional information if it believes information relevant to the requirements for recognition have been omitted. OSHA may inspect the facilities of the third-party validation organization and any validating laboratory, and while there shall review any additional documentation underlying the application. A report shall be made of each field inspection.

2. *Requirements for recognition.* The requirements for OSHA recognition of a third-party validation organization for the PSDI standard are that the program has fulfilled the requirements of section II of this appendix for qualifications and of section III of this appendix for program requirements, and the program has identified appropriate test and analysis methods to meet the requirements of the PSDI standard and appendix A.

3. *Preliminary approval.* If, after review of the application, any additional information, and the inspection report, the applicant and any validating laboratory appear to have met the requirements for recognition, a written recommendation shall be submitted by the responsible OSHA personnel to the Assistant Secretary to approve the application with a supporting explanation.

4. *Preliminary disapproval.* If, after review of the application, additional information, and inspection report, the applicant does not appear to have met the requirements for recognition, the Director of the PSDI certification/validation program shall notify the applicant in writing, listing the specific requirements of this appendix which the applicant has not met, and the reasons.

5. *Revision of application.* After receipt of a notification of preliminary disapproval, the applicant may submit a revised application for further review by OSHA pursuant to subsection I.B. of this appendix or may request that the original application be submitted to the Assistant Secretary with a statement of reasons supplied by the applicant as to why the application should be approved.

6. *Preliminary decision by Assistant Secretary.* a. The Assistant Secretary, or a special designee for this purpose, will make a preliminary decision whether the applicant has met the requirements for recognition based on the completed application file and the written staff recommendation, as well as the statement of reasons by the applicant if there is a recommendation of disapproval.

b. This preliminary decision will be sent to the applicant and subsequently published in the **FEDERAL REGISTER**.

7. *Public review and comment period.* a. The **FEDERAL REGISTER** notice of preliminary decision will provide a period of not less than 60 calendar days for the written comments on the applicant's fulfillment of the requirements for recognition. The application, supporting documents, staff recommendation, statement of applicant's reasons, and any comments received, will be available for public inspection in the OSHA Docket Office.

b. If the preliminary decision is in favor of recognition, a member of the public, or if the preliminary decision is against recognition, the applicant may request a public hearing by the close of the comment period, if it supplies detailed reasons and evidence challenging the basis of the Assistant Secretary's preliminary decision and justifying the need for a public hearing to bring out evidence which could not be effectively supplied through written submissions.

8. *Final decision by Assistant Secretary*—a. *Without hearing.* If there are no valid requests for a hearing, based on the application, supporting documents, staff recommendation, evidence and public comment, the Assistant Secretary shall issue the final decision (including reasons) of the Department of Labor on whether the applicant has demonstrated by a preponderance of the evidence that it meets the requirements for recognition.

b. *After hearing.* If there is a valid request for a hearing pursuant to paragraph I.B.7.b. of this appendix, the following procedures will be used:

(1) The Assistant Secretary will issue a notice of hearing before an administrative law judge of the Department of Labor pursuant to the rules specified in 29 CFR part 1905, subpart C.

(2) After the hearing, pursuant to subpart C, the administrative law judge shall issue a decision (including reasons) based on the application, the supporting documentation, the staff recommendation, the public comments and the evidence submitted during the hearing (the record), stating whether it has been demonstrated, based on a preponderance of evidence, that the applicant meets the requirements for recognition. If no exceptions are filed, this is the final decision of the Department of Labor.

(3) Upon issuance of the decision, any party to the hearing may file exceptions within 20 days pursuant to subpart C. If exceptions are filed, the administrative law judge shall forward the decision, exceptions and record to the Assistant Secretary for the final decision on the application.

(4) The Assistant Secretary shall review the record, the decision by the administrative law judge, and the exceptions. Based on this, the Assistant Secretary shall issue the final decision (including reasons) of the Department of Labor stating whether the applicant has demonstrated by a preponderance of evidence that it meets the requirements for recognition.

b. *Publication.* A notification of the final decision shall be published in the **FEDERAL REGISTER**.

### C. Terms and Conditions of Recognition, Renewal and Revocation

1. The following terms and conditions shall be part of every recognition:

a. The recognition of any validation organization will be evidenced by a letter of recognition from OSHA. The letter will provide the specific details of the scope of the OSHA recognition as well as any conditions imposed by OSHA, including any Federal monitoring requirements.

b. The recognition of each validation organization will be valid for five years, unless terminated before or renewed after the expiration of the period. The dates of the period of recognition will be stated in the recognition letter.

c. The recognized validation organization shall continue to satisfy all the requirements of this appendix and the letter of recognition during the period of recognition.

2. A recognized validation organization may change a test method of the PSDI safety system certification/validation program by notifying the Assistant Secretary of the change, certifying that the revised method will be at least as effective as the prior method, and providing the supporting data upon which its conclusions are based.

3. A recognized validation organization may renew its recognition by filing a renewal request at the address in paragraph I.A.3. of this appendix, above, not less than 180 calendar days, nor more than one year, before the expiration date of its current recognition. When a recognized validation organization has filed such a renewal request, its current recognition will not expire until a final decision has been made on the request. The renewal request will be processed in accordance with subsection I.B. of this appendix, above, except that a reinspection is not required but may be performed by OSHA. A hearing will be granted to an objecting member of the public if evidence of failure to meet the requirements of this appendix is supplied to OSHA.

4. A recognized validation organization may apply to OSHA for an expansion of its current recognition to cover other categories of PSDI certification/validation in addition to those included in the current recognition. The application for expansion will be acted upon and processed by OSHA in accordance with subsection I.B. of this appendix, subject to the possible reinspection exception. If the validation organization has been recognized for more than one year, meets the requirements for expansion of recognition, and there is no evidence that the recognized validation organization has not been following the requirements of this appendix and the letter of recognition, an expansion will normally be granted. A hearing will be granted to an objecting member of the public only if evidence of failure to meet the requirements of this appendix is supplied to OSHA.

5. A recognized validation organization may voluntarily terminate its recognition, either in its entirety or with respect to any area covered in its recognition, by giving written notice to OSHA at any time. The written notice shall indicate the termination date. A validation organization may not terminate its installation certification and recertification validation functions earlier than either one year from the date of the written notice, or the date on which another recognized validation organization is able to perform the validation of installation certification and recertification.

6.a. OSHA may revoke its recognition of a validation organization if its program either has failed to continue to satisfy the requirements of this appendix or its letter of recognition, has not been performing the validation functions required by the PSDI standard and appendix A, or has misrepresented itself in its applications. Before proposing to revoke recognition, the Agency will notify the recognized validation organization of the basis of the proposed revocation and will allow rebuttal or correction of the alleged deficiencies. If the deficiencies are not corrected, OSHA may revoke recognition, effective in 60 days, unless the validation organization requests a hearing within that time.

b. If a hearing is requested, it shall be held before an administrative law judge of the Department of Labor pursuant to the rules specified in 29 CFR part 1905, subpart C.

c. The parties shall be OSHA and the recognized validation organization. The decision shall be made pursuant to the procedures specified in paragraphs I.B.8.b.(2) through (4) of this appendix except that the burden of proof shall be on OSHA to demonstrate by a preponderance of the evidence that the recognition should be revoked because the validation organization either is not meeting the requirements for recognition, has not been performing the validation functions required by the PSDI standard and appendix A, or has misrepresented itself in its applications.

### D. Provisions of OSHA Recognition

Each recognized third-party validation organization and its validating laboratories shall:

1. Allow OSHA to conduct unscheduled reviews or on-site audits of it or the validating laboratories on matters relevant to PSDI, and cooperate in the conduct of these reviews and audits;

2. Agree to terms and conditions established by OSHA in the grant of recognition on matters such as exchange of data, submission of accident reports, and assistance in studies for improving PSDI or the certification/validation process.

## II. Qualifications

The third-party validation organization, the validating laboratory, and the employees of each shall meet the requirements set forth in this section of this appendix.

### A. Experience of Validation Organization

1. The third-party validation organization shall have legal authority to perform certification/validation activities.

2. The validation organization shall demonstrate competence and experience in either power press design, manufacture or use, or testing, quality control or certification/validation of equipment comparable to power presses and associated control systems.

3. The validation organization shall demonstrate a capability for selecting, reviewing, and/or validating appropriate standards and test methods to be used for validating the certification of PSDI safety systems, as well as for reviewing judgements on the safety of PSDI safety systems and their conformance with the requirements of this section.

4. The validating organization may utilize the competence, experience, and capability of its employees to demonstrate this competence, experience and capability.

### B. Independence of Validation Organization

1. The validation organization shall demonstrate that:

a. It is financially capable to conduct the work;

b. It is free of direct influence or control by manufacturers, suppliers, vendors, representatives of employers and employees, and employer or employee organizations; and

c. Its employees are secure from discharge resulting from pressures from manufacturers, suppliers, vendors, employers or employee representatives.

2. A validation organization may be considered independent even if it has ties with manufacturers, employers or employee representatives if these ties are with at least two of these three groups; it has a board of directors (or equivalent leadership responsible for the certification/validation activities) which includes representatives of the three groups; and it has a binding commitment of funding for a period of three years or more.

### C. Validating Laboratory

The validation organization's laboratory (which organizationally may be a part of the third-party validation organization):

1. Shall have legal authority to perform the validation of certification;

2. Shall be free of operational control and influence of manufacturers, suppliers, vendors, employers, or employee representatives that would impair its integrity of performance; and

3. Shall not engage in the design, manufacture, sale, promotion, or use of the certified equipment.

### D. Facilities and Equipment

The validation organization's validating laboratory shall have available all testing facilities and necessary test and inspection equipment relevant to the validation of the certification of PSDI safety systems, installations and operations.

### E. Personnel

The validation organization and the validating laboratory shall be adequately staffed by personnel who are qualified by technical training and/or experience to conduct the validation of the certification of PSDI safety systems.

1. The validation organization shall assign overall responsibility for the validation of PSDI certification to an Administrative Director. Minimum requirements for this position are a Bachelor's degree and five years professional experience, at least one of which shall have been in responsible charge of a function in the areas of power press design or manufacture or a broad range of power press use, or in the areas of testing, quality control, or certification/validation of equipment comparable to power presses or their associated control systems.

2. The validating laboratory, if a separate organization from the validation organization, shall assign technical responsibility for the validation of PSDI certification to a Technical Director. Minimum requirements for this position are a Bachelor's degree in a technical field and five years of professional experience, at least one of which shall have been in responsible charge of a function in the area of testing, quality control or certification/validation of equipment comparable to power presses or their associated control systems.

3. If the validation organization and the validating laboratory are the same organization, the administrative and technical responsibilities may be combined in a single position, with minimum requirements as described in E.1. and 2. for the combined position.

4. The validation organization and validating laboratory shall have adequate administrative and technical staffs to conduct the validation of the certification of PSDI safety systems.

### F. Certification/Validation Mark or Logo

1. The validation organization or the validating laboratory shall own a registered certification/validation mark or logo.

2. The mark or logo shall be suitable for incorporation into the label required by paragraph (h)(11)(iii) of this section.

## III. Program Requirements

### A. Test and Certification/Validation Procedures

1. The validation organization and/or validating laboratory shall have established written procedures for test and certification/validation of PSDI safety systems. The procedures shall be based on pertinent OSHA standards and test methods, or other publicly available standards and test methods generally recognized as appropriate in the field, such as national consensus standards or published standards of professional societies or trade associations.

2. The written procedures for test and certification/validation of PSDI systems, and the standards and test methods on which they are based, shall be reproducible and be available to OSHA and to the public upon request.

### B. Test Reports

1. A test report shall be prepared for each PSDI safety system that is tested. The test report shall be signed by a technical staff representative and the Technical Director.

2. The test report shall include the following:

a. Name of manufacturer and catalog or model number of each subsystem or major component.

b. Identification and description of test methods or procedures used. (This may be through reference to published sources which describe the test methods or procedures used.)

c. Results of all tests performed.

d. All safety distance calculations.

3. A copy of the test report shall be maintained on file at the validation organization and/or validating laboratory, and shall be available to OSHA upon request.

### C. Certification/Validation Reports

1. A certification/validation report shall be prepared for each PSDI safety system for which the certification is validated. The certification/validation report shall be signed by the Administrative Director and the Technical Director.

2. The certification/validation report shall include the following:

a. Name of manufacturer and catalog or model number of each subsystem or major component.

b. Results of all tests which serve as the basis for the certification.

c. All safety distance calculations.

d. Statement that the safety system conforms with all requirements of the PSDI standard and appendix A.

3. A copy of the certification/validation report shall be maintained on file at the validation organization and/or validating laboratory, and shall be available to the public upon request.

4. A copy of the certification/validation report shall be submitted to OSHA within 30 days of its completion.

### D. Publications System

The validation organization shall make available upon request a list of PSDI safety systems which have been certified/validated by the program.

### E. Follow-up Activities

1. The validation organization or validating laboratory shall have a follow-up system for inspecting or testing manufacturer's production of design certified/validated PSDI safety system components and subassemblies where deemed appropriate by the validation organization.

2. The validation organization shall notify the appropriate product manufacturer(s) of any reports from employers of point of operation injuries which occur while a press is operated in a PSDI mode.

### F. Records

The validation organization or validating laboratory shall maintain a record of each certification/validation of a PSDI safety system, including manufacturer and/or employer certification documentation, test and working data, test report, certification/validation report, any follow-up inspections or testing, and reports of equipment failures, any reports of accidents involving the equipment, and any other pertinent information. These records shall be available for inspection by OSHA and OSHA State Plan offices.

### G. Dispute Resolution Procedures

1. The validation organization shall have a reasonable written procedure for acknowledging and processing appeals or complaints from program participants (manufacturers, producers, suppliers, vendors and employers) as well as other interested parties (employees or their representatives, safety personnel, government agencies, etc.), concerning certification or validation.

2. The validation organization may charge any complainant the reasonable charge for repeating tests needed for the resolution of disputes.

## APPENDIX D TO §1910.217— NONMANDATORY SUPPLEMENTARY INFORMATION

This appendix provides nonmandatory supplementary information and guidelines to assist in the understanding and use of 29 CFR 1910.217(h) to allow presence sensing device initiation (PSDI) of mechanical power presses. Although this appendix as such is not mandatory, it references sections and requirements which are made mandatory by other parts of the PSDI standard and appendices.

### 1. General

OSHA intends that PSDI continue to be prohibited where present state-of-the-art technology will not allow it to be done safely. Only *part revolution* type mechanical power presses are approved for PSDI. Similarly, only presses with a configuration such that a person's body cannot completely enter the bed area are approved for PSDI.

### 2. Brake and Clutch

Flexible steel band brakes do not possess a long-term reliability against structural failure as compared to other types of brakes, and therefore are not acceptable on presses used in the PSDI mode of operation.

Fast and consistent stopping times are important to safety for the PSDI mode of operation. Consistency of braking action is enhanced by high brake torque. The requirement in paragraph (h)(2)(ii) defines a high torque capability which should ensure fast and consistent stopping times.

Brake design parameters important to PSDI are high torque, low moment of inertia, low air volume (if pneumatic) mechanisms, non-interleaving engagement springs, and structural integrity which is enhanced by over-design. The requirement in paragraph (h)(2)(iii) reduces the possibility of significantly increased stopping time if a spring breaks.

As an added precaution to the requirements in paragraph (h)(2)(iii), brake adjustment locking means should be secured. Where brake springs are externally accessible, lock nuts or other means may be provided to reduce the possibility of backing off of the compression nut which holds the springs in place.

### 3. Pneumatic Systems

Elevated clutch/brake air pressure results in longer stopping time. The requirement in paragraph (h)(3)(i)(C) is intended to prevent degradation in stopping speed from higher air pressure. Higher pressures may be permitted, however, to increase clutch torque to free "jammed" dies, provided positive measures are provided to prevent the higher pressure at other times.

### 4. Flywheels and Bearings

Lubrication of bearings is considered the single greatest deterrent to their failure. The manufacturer's recommended procedures for maintenance and inspection should be closely followed.

### 5. Brake Monitoring

The approval of brake monitor adjustments, as required in paragraph (h)(5)(ii), is not considered a *recertification*, and does not necessarily involve an on-site inspection by a representative of the validation organization. It is expected that the brake monitor adjustment normally could be evaluated on the basis of the effect on the safety system certification/validation documentation retained by the validation organization.

Use of a brake monitor does not eliminate the need for periodic brake inspection and maintenance to reduce the possibility of catastrophic failures.

### 6. Cycle Control and Control Systems

The PSDI set-up/reset means required by paragraph (h)(6)(iv) may be initiated by the actuation of a special momentary pushbutton or by the actuation of a special momentary pushbutton and the initiation of a first stroke with two hand controls.

It would normally be preferable to limit the adjustment of the time required in paragraph (h)(6)(vi) to a maximum of 15 seconds. However, where an operator must do many operations outside the press, such as lubricating, trimming, deburring, etc., a longer interval up to 30 seconds is permitted.

When a press is equipped for PSDI operation, it is recommended that the presence sensing device be active as a guarding device in other production modes. This should enhance the reliability of the device and ensure that it remains operable.

An acceptable method for interlocking supplemental guards as required by paragraph (h)(6)(xiii) would be to incorporate the supplemental guard and the PSDI presence sensing device into a hinged arrangement in which the alignment of the presence sensing device serves, in effect, as the interlock. If the supplemental guards are moved, the presence sensing device would become misaligned and the press control would be deactivated. No extra microswitches or interlocking sensors would be required.

Paragraph (h)(6)(xv) of the standard requires that the control system have provisions for an "inch" operating means; that die-setting not be done in the PSDI mode; and that production not be done in the "inch" mode. It should be noted that the sensing device would be by-passed in the "inch" mode. For that reason, the prohibitions against die-setting in the PSDI mode, and against production in the "inch" mode are cited to emphasize that "inch" operation is of reduced safety and is not compatible with PSDI or other production modes.

### 7. Environmental Requirements

It is the intent of paragraph (h)(7) that control components be provided with inherent design protection against operating stresses and environmental factors affecting safety and reliability.

### 8. Safety system

The safety system provision continues the concept of paragraph (b)(13) that the probability of two independent failures in the length of time required to make one press cycle is so remote as to be a negligible risk factor in the total array of equipment and human factors. The emphasis is on an integrated total system including all elements affecting point of operation safety.

It should be noted that this does not require redundancy for press components such as structural elements, clutch/brake mechanisms, plates, etc., for which adequate reliability may be achieved by proper design, maintenance, and inspection.

### 9. Safeguarding the Point of Operation

The intent of paragraph (h)(9)(iii) is to prohibit use of mirrors to "bend" a single light curtain sensing field around corners to cover more than one side of a press. This prohibition is needed to increase the reliability of the presence sensing device in initiating a stroke only when the desired work motion has been completed.

*Object sensitivity* describes the capability of a presence sensing device to detect an object in the sensing field, expressed as the linear measurement of the smallest interruption which can be detected at any point in the field. Minimum object sensitivity describes the largest acceptable size of the interruption in the sensing field. A minimum object sensitivity of one and one-fourth inches (31.75 mm) means that a one and one-fourth inch (31.75 mm) diameter object will be continuously detected at all locations in the sensing field.

In deriving the safety distance required in paragraph (h)(9)(v), all stopping time measurements should be made with clutch/brake air pressure regulated to the press manufacturer's recommended value for full clutch torque capability. The stopping time measurements should be made with the heaviest upper die that is planned for use in the press. If the press has a slide counterbalance system, it is important that the counterbalance be adjusted correctly for upper die weight according to the manufacturer's instructions. While the brake monitor setting is based on the stopping time it actually measures, i.e., the normal stopping time at the top of the stroke, it is important that the safety distance be computed from the longest stopping time measured at any of the indicated three downstroke stopping positions listed in the explanation of Ts. The use in the formula of twice the stopping time increase, Tm, allowed by the brake monitor for brake wear allows for greater increases in the downstroke stopping time than occur in normal stopping time at the top of the stroke.

### 10. Inspection and Maintenance. [Reserved]
### 11. Safety System Certification/Validation

Mandatory requirements for certification/validation of the PSDI safety system are provided in appendix A and appendix C to this standard. Nonmandatory supplementary information and guidelines relating to certification/validation of the PSDI safety system are provided to appendix B to this standard.

[39 FR 32502, June 27, 1974, as amended at 39 FR 41846, Dec. 23, 1974; 40 FR 3982, Jan. 27, 1975; 43 FR 49750, Oct. 24, 1978; 45 FR 8594, Feb. 8, 1980; 49 FR 18295, Apr. 30, 1984; 51 FR 34561, Sept. 29, 1986; 53 FR 8353, 8358 Mar. 14, 1988; 54 FR 24333, June 7, 1989; 61 FR 9240, Mar. 7, 1996]

### §1910.218 Forging machines.

**1910.218(a)** *General requirements*—**(1)** *Use of lead.* The safety requirements of this subparagraph apply to lead casts or other use of lead in the forge shop or die shop.

**1910.218(a)(1)(i)** Thermostatic control of heating elements shall be provided to maintain proper melting temperature and prevent overheating.

**1910.218(a)(1)(ii)** Fixed or permanent lead pot installations shall be exhausted.

**1910.218(a)(1)(iii)** Portable units shall be used only in areas where good, general room ventilation is provided.

**1910.218(a)(1)(iv)** Personal protective equipment (gloves, goggles, aprons, and other items) shall be worn.

**1910.218(a)(1)(v)** A covered container shall be provided to store dross skimmings.

**1910.218(a)(1)(vi)** Equipment shall be kept clean, particularly from accumulations of yellow lead oxide.

**1910.218(a)(2)** *Inspection and maintenance.* It shall be the responsibility of the employer to maintain all forge shop equipment in a condition which will insure continued safe operation. This responsibility includes:

**1910.218(a)(2)(i)** Establishing periodic and regular maintenance safety checks and keeping certification records of these inspections which include the date of inspection, the signature of the person who performed the inspection and the serial number, or other identifier, for the forging machine which was inspected.

**1910.218(a)(2)(ii)** Scheduling and recording the inspection of guards and point of operation protection devices at frequent and regular intervals. Recording of inspections shall be in the form of a certification record which includes the date the inspection was performed, the signature of the person who performed the inspection and the serial number, or other identifier, of the equipment inspected.

**1910.218(a)(2)(iii)** Training personnel for the proper inspection and maintenance of forging machinery and equipment.

**1910.218(a)(2)(iv)** All overhead parts shall be fastened or protected in such a manner that they will not fly off or fall in event of failure.

**1910.218(a)(3)** *Hammers and presses.* **(i)** All hammers shall be positioned or installed in such a manner that they remain on or are anchored to foundations sufficient to support them according to applicable engineering standards.

**1910.218(a)(3)(ii)** All presses shall be installed in such a manner that they remain where they are positioned or they are anchored to foundations sufficient to support them according to applicable engineering standards.

### TABLE O-11—STRENGTH AND DIMENSIONS FOR WOOD RAM PROPS

| Size of timber, inches[1] | Square inches in cross section | Minimum allowable crushing strength parallel to grain, p.s.i.[2] | Maximum static load within short column range[3] | Safety factor | Maximum recommended weight of forging hammer for timber used | Maximum allowable length of timber, inches |
|---|---|---|---|---|---|---|
| 4 × 4 | 16 | 5,000 | 80,000 | 10 | 8,000 | 44 |
| 6 × 6 | 36 | 5,000 | 180,000 | 10 | 18,000 | 66 |
| 8 × 8 | 64 | 5,000 | 320,000 | 10 | 32,000 | 88 |
| 10 × 10 | 100 | 5,000 | 500,000 | 10 | 50,000 | 100 |
| 12 × 12 | 144 | 5,000 | 720,000 | 10 | 72,000 | 132 |

[1] Actual dimension.
[2] Adapted from U.S. Department of Agriculture Technical Bulletin 479. Hardwoods recommended are those whose ultimate crushing strengths in compression parallel to grain are 5,000 p.s.i. (pounds per square inch) or greater.
[3] Slenderness ratio formula for short columns is L/d = 11, where L = length of timber in inches and d = least dimension in inches; this ratio should not exceed 11.

**1910.218(a)(3)(iii)** Means shall be provided for disconnecting the power to the machine and for locking out or rendering cycling controls inoperable.

**1910.218(a)(3)(iv)** The ram shall be blocked when dies are being changed or other work is being done on the hammer. Blocks or wedges shall be made of material the strength and construction of which should meet or exceed the specifications and dimensions shown in Table O-11.

**1910.218(a)(3)(v)** Tongs shall be of sufficient length to clear the body of the worker in case of kickback, and shall not have sharp handle ends.

**1910.218(a)(3)(vi)** Oil swabs, or scale removers, or other devices to remove scale shall be provided. These devices shall be long enough to enable a man to reach the full length of the die without placing his hand or arm between the dies.

**1910.218(a)(3)(vii)** Material handling equipment shall be of adequate strength, size, and dimension to handle diesetting operations safely.

**1910.218(a)(3)(viii)** A scale guard of substantial construction shall be provided at the back of every hammer, so arranged as to stop flying scale.

**1910.218(a)(3)(ix)** A scale guard of substantial construction shall be provided at the back of every press, so arranged as to stop flying scale.

**1910.218(b)** *Hammers, general*—**(1)** *Keys.* Die keys and shims shall be made from a grade of material that will not unduly crack or splinter.

**1910.218(b)(2)** *Foot operated devices.* All foot operated devices (i.e., treadles, pedals, bars, valves, and switches) shall be substantially and effectively protected from unintended operation.

**1910.218(c)** *Presses.* All manually operated valves and switches shall be clearly identified and readily accessible.

**1910.218(d)** *Power-driven hammers*—**(1)** *Safety cylinder head.* Every steam or airhammer shall have a safety cylinder head to act as a cushion if the rod should break or pullout of the ram.

**1910.218(d)(2)** *Shutoff valve.* Steam hammers shall be provided with a quick closing emergency valve in the admission pipeline at a convenient location. This valve shall be closed and locked in the off position while the hammer is being adjusted, repaired, or serviced, or when the dies are being changed.

**1910.218(d)(3)** *Cylinder draining.* Steam hammers shall be provided with a means of cylinder draining, such as a self-draining arrangement or a quick-acting drain cock.

**1910.218(d)(4)** *Pressure pipes.* Steam or air piping shall conform to the specifications of American National Standard ANSI B31.1.0-1967, Power Piping with Addenda issued before April 28, 1971, which is incorporated by reference as specified in §1910.6.

**1910.218(e)** *Gravity hammers*—**(1)** *Air-lift hammers.* **(i)** Air-lift hammers shall have a safety cylinder head as required in paragraph (d)(1) of this section.

**1910.218(e)(1)(ii)** Air-lift hammers shall have an air shutoff valve as required in paragraph (d)(2) of this section.

**1910.218(e)(1)(iii)** Air-lift hammers shall be provided with two drain cocks: one on main head cylinder, and one on clamp cylinder.

**1910.218(e)(1)(iv)** Air piping shall conform to the specifications of the ANSI B31.1.0-1967, Power Piping with Addenda issued before April 28, 1971, which is incorporated by reference as specified in §1910.6.

**1910.218(e)(2)** *Board drophammers.* **(i)** A suitable enclosure shall be provided to prevent damaged or detached boards from falling. The board enclosure shall be securely fastened to the hammer.

**1910.218(e)(2)(ii)** All major assemblies and fittings which can loosen and fall shall be properly secured in place.

**1910.218(f)** *Forging presses*—**(1)** *Mechanical forging presses.* When dies are being changed or maintenance is being performed on the press, the following shall be accomplished:

**1910.218(f)(1)(i)** The power to the press shall be locked out.

**1910.218(f)(1)(ii)** The flywheel shall be at rest.

**1910.218(f)(1)(iii)** The ram shall be blocked with a material the strength of which shall meet or exceed the specifications or dimensions shown in Table O-11.

**1910.218(f)(2)** *Hydraulic forging presses.* When dies are being changed or maintenance is being performed on the press, the following shall be accomplished:

**1910.218(f)(2)(i)** The hydraulic pumps and power apparatus shall be locked out.

**1910.218(f)(2)(ii)** The ram shall be blocked with a material the strength of which shall meet or exceed the specifications or dimensions shown in Table O-11.

**1910.218(g)** *Trimming presses*—**(1)** *Hot trimming presses.* The requirements of paragraph (f)(1) of this section shall also apply to hot trimming presses.

**1910.218(g)(2)** *Cold trimming presses.* Cold trimming presses shall be safeguarded in accordance with §1910.217(c).

**1910.218(h)** *Upsetters*—**(1)** *General requirements.* All upsetters shall be installed so that they remain on their supporting foundations.

**1910.218(h)(2)** *Lockouts.* Upsetters shall be provided with a means for locking out the power at its entry point to the machine and rendering its cycling controls inoperable.

**1910.218(h)(3)** *Manually operated controls.* All manually operated valves and switches shall be clearly identified and readily accessible.

**1910.218(h)(4)** *Tongs.* Tongs shall be of sufficient length to clear the body of the worker in case of kickback, and shall not have sharp handle ends.

**1910.218(h)(5)** *Changing dies.* When dies are being changed, maintenance performed, or any work done on the machine, the power to the upsetter shall be locked out, and the flywheel shall be at rest.

**1910.218(i)** *Other forging equipment*—**(1)** *Boltheading.* The provisions of paragraph (h) of this section shall apply to boltheading.

**1910.218(i)(2)** *Rivet making.* The provisions of paragraph (h) of this section shall apply to rivet making.

**1910.218(j)** *Other forge facility equipment*—**(1)** *Billet shears.* A positive-type lockout device for disconnecting the power to the shear shall be provided.

**1910.218(j)(2)** *Saws.* Every saw shall be provided with a guard of not less than one-eighth inch sheet metal positioned to stop flying sparks.

**1910.218(j)(3)** *Conveyors.* Conveyor power transmission equipment shall be guarded in accordance with ANSI B20.1-1957, Safety Code for Conveyors, Cableways, and Related Equipment, which is incorporated by reference as specified in §1910.6.

**1910.218(j)(4)** *Shot blast.* The cleaning chamber shall have doors or guards to protect operators.

**1910.218(j)(5)** *Grinding.* Personal protective equipment shall be used in grinding operations, and equipment shall be used and maintained in accordance with ANSI B7.1-1970, Safety Code for the Use, Care, and Protection of Abrasive Wheels, which is incorporated by reference as specified in §1910.6, and with §1910.215.

[39 FR 23502, June 27, 1974, as amended at 49 FR 5323, Feb. 10, 1984; 51 FR 34561, Sept. 29, 1986; 61 FR 9240, Mar. 7, 1996]

## §1910.219 Mechanical power-transmission apparatus.

**1910.219(a)** *General requirements.* **(1)** This section covers all types and shapes of power-transmission belts, except the following when operating at two hundred and fifty (250) feet per minute or less: (i) Flat belts one (1) inch or less in width, (ii) flat belts two (2) inches or less in width which are free from metal lacings or fasteners, (iii) round belts one-half ( 1/2) inch or less in diameter; and (iv) single strand V-belts, the width of which is thirteen thirty-seconds (13/32) inch or less.

**1910.219(a)(2)** Vertical and inclined belts (paragraphs (e) (3) and (4) of this section) if not more than two and one-half (2 1/2) inches wide and running at a speed of less than one thousand (1,000) feet per minute, and if free from metal lacings or fastenings may be guarded with a nip-point belt and pulley guard.

**1910.219(a)(3)** For the Textile Industry, because of the presence of excessive deposits of lint, which constitute a serious fire hazard, the sides and face sections only of nip-point belt and pulley guards are required, provided the guard shall extend at least six (6) inches beyond the rim of the pulley on the in-running and off-running sides of the belt and at least two (2) inches away from the rim and face of the pulley in all other directions.

**1910.219(a)(4)** This section covers the principal features with which power transmission safeguards shall comply.

**1910.219(b)** *Prime-mover guards*—**(1)** *Flywheels.* Flywheels located so that any part is seven (7) feet or less above floor or platform shall be guarded in accordance with the requirements of this subparagraph:

**1910.219(b)(1)(i)** With an enclosure of sheet, perforated, or expanded metal, or woven wire;

**1910.219(b)(1)(ii)** With guard rails placed not less than fifteen (15) inches nor more than twenty (20) inches from rim. When flywheel extends into pit or is within 12 inches of floor, a standard toeboard shall also be provided;

**1910.219(b)(1)(iii)** When the upper rim of flywheel protrudes through a working floor, it shall be entirely enclosed or surrounded by a guardrail and toeboard.

**1910.219(b)(1)(iv)** For flywheels with smooth rims five (5) feet or less in diameter, where the preceding methods cannot be applied, the following may be used: A disk attached to the flywheel in such manner as to cover the spokes of the wheel on the exposed side and present a smooth surface and edge, at the same time providing means for periodic inspection. An open space, not exceeding four (4) inches in width, may be left between the outside edge of the disk and the rim of the wheel if desired, to facilitate turning the wheel over. Where a disk is used, the keys or other dangerous projections not covered by disk shall be cut off or covered. This subdivision does not apply to flywheels with solid web centers.

**1910.219(b)(1)(v)** Adjustable guard to be used for starting engine or for running adjustment may be provided at the flywheel of gas or oil engines. A slot opening for jack bar will be permitted.

**1910.219(b)(1)(vi)** Wherever flywheels are above working areas, guards shall be installed having sufficient strength to hold the weight of the flywheel in the event of a shaft or wheel mounting failure.

**1910.219(b)(2)** *Cranks and connecting rods.* Cranks and connecting rods, when exposed to contact, shall be guarded in accordance with paragraphs (m) and (n) of this section, or by a guardrail as described in paragraph (o)(5) of this section.

**1910.219(b)(3)** *Tail rods or extension piston rods.* Tail rods or extension piston rods shall be guarded in accordance with paragraphs (m) and (o) of this section, or by a guardrail on sides and end, with a clearance of not less than fifteen (15) nor more than twenty (20) inches when rod is fully extended.

**1910.219(c)** *Shafting*—**(1)** *Installation.* **(i)** Each continuous line of shafting shall be secured in position against excessive endwise movement.

**1910.219(c)(1)(ii)** Inclined and vertical shafts, particularly inclined idler shafts, shall be securely held in position against endwise thrust.

**1910.219(c)(2)** *Guarding horizontal shafting.* **(i)** All exposed parts of horizontal shafting seven (7) feet or less from floor or working platform, excepting runways used exclusively for oiling, or running adjustments, shall be protected by a stationary casing enclosing shafting completely or by a trough enclosing sides and top or sides and bottom of shafting as location requires.

**1910.219(c)(2)(ii)** Shafting under bench machines shall be enclosed by a stationary casing, or by a trough at sides and top or sides and bottom, as location requires. The sides of the trough shall come within at least six (6) inches of the underside of table, or if shafting is located near floor within six (6) inches of floor. In every case the sides of trough shall extend at least two (2) inches beyond the shafting or protuberance.

**1910.219(c)(3)** *Guarding vertical and inclined shafting.* Vertical and inclined shafting seven (7) feet or less from floor or working platform, excepting maintenance runways, shall be enclosed with a stationary casing in accordance with requirements of paragraphs (m) and (o) of this section.

**1910.219(c)(4)** *Projecting shaft ends.* **(i)** Projecting shaft ends shall present a smooth edge and end and shall not project more than one-half the diameter of the shaft unless guarded by nonrotating caps or safety sleeves.

**1910.219(c)(4)(ii)** Unused keyways shall be filled up or covered.

**1910.219(c)(5)** *Power-transmission apparatus located in basements.* All mechanical power transmission apparatus located in basements, towers, and rooms used exclusively for power transmission equipment shall be guarded in accordance with this section, except that the requirements for safeguarding belts, pulleys, and shafting need not be complied with when the following requirements are met:

**1910.219(c)(5)(i)** The basement, tower, or room occupied by transmission equipment is locked against unauthorized entrance.

**1910.219(c)(5)(ii)** The vertical clearance in passageways between the floor and power transmission beams, ceiling, or any other objects, is not less than five feet six inches (5 ft. 6 in.).

**1910.219(c)(5)(iii)** The intensity of illumination conforms to the requirements of ANSI A11.1-1965 (R-1970), which is incorporated by reference as specified in §1910.6.

**1910.219(c)(5)(iv)** [Reserved]

**1910.219(c)(5)(v)** The route followed by the oiler is protected in such manner as to prevent accident.

**1910.219(d)** *Pulleys*—**(1)** *Guarding.* Pulleys, any parts of which are seven (7) feet or less from the floor or working platform, shall be guarded in accordance with the standards specified in paragraphs (m) and (o) of this section. Pulleys serving as balance wheels (e.g., punch presses) on which the point of contact between belt and pulley is more than six feet six inches (6 ft. 6 in.) from the floor or platform may be guarded with a disk covering the spokes.

**1910.219(d)(2)** *Location of pulleys.* **(i)** Unless the distance to the nearest fixed pulley, clutch, or hanger exceeds the width of the belt used, a guide shall be provided to prevent the belt from leaving the pulley on the side where insufficient clearance exists.

**1910.219(d)(2)(ii)** [Reserved]

**1910.219(d)(3)** *Broken pulleys.* Pulleys with cracks, or pieces broken out of rims, shall not be used.

**1910.219(d)(4)** *Pulley speeds.* Pulleys intended to operate at rim speed in excess of manufacturers normal recommendations shall be specially designed and carefully balanced for the speed at which they are to operate.

**1910.219(e)** **Belt, rope, and chain drives**—**(1)** *Horizontal belts and ropes.* **(i)** Where both runs of horizontal belts are seven (7) feet or less from the floor level, the guard shall extend to at least fifteen (15) inches above the belt or to a standard height, except that where both runs of a horizontal belt are 42 inches or less from the floor, the belt shall be fully enclosed in accordance with paragraphs (m) and (o) of this section.

**1910.219(e)(1)(ii)** In powerplants or power-development rooms, a guardrail may be used in lieu of the guard required by subdivision (i) of this subparagraph.

**1910.219(e)(2)** *Overhead horizontal belts.* **(i)** Overhead horizontal belts, with lower parts seven (7) feet or less from the floor or platform, shall be guarded on sides and bottom in accordance with paragraph (o)(3) of this section.

**1910.219(e)(2)(ii)** Horizontal overhead belts more than seven (7) feet above floor or platform shall be guarded for their entire length under the following conditions:

**1910.219(e)(2)(ii)(a)** If located over passageways or work places and traveling 1,800 feet or more per minute.

**1910.219(e)(2)(ii)(b)** If center to center distance between pulleys is ten (10) feet or more.

**1910.219(e)(2)(ii)(c)** If belt is eight (8) inches or more in width.

**1910.219(e)(2)(iii)** Where the upper and lower runs of horizontal belts are so located that passage of persons between them would be possible, the passage shall be either:

**1910.219(e)(2)(iii)(a)** Completely barred by a guardrail or other barrier in accordance with paragraphs (m) and (o) of this section; or

**1910.219(e)(2)(iii)(b)** Where passage is regarded as necessary, there shall be a platform over the lower run guarded on either side by a railing completely filled in with wire mesh or other filler, or by a solid barrier. The upper run shall be so guarded as to prevent contact therewith either by the worker or by objects carried by him. In powerplants only the lower run of the belt need be guarded.

**1910.219(e)(2)(iv)** Overhead chain and link belt drives are governed by the same rules as overhead horizontal belts and shall be guarded in the same manner as belts.

**1910.219(e)(3)** *Vertical and inclined belts.* **(i)** Vertical and inclined belts shall be enclosed by a guard conforming to standards in paragraphs (m) and (o) of this section.

**1910.219(e)(3)(ii)** All guards for inclined belts shall be arranged in such a manner that a minimum clearance of seven (7) feet is maintained between belt and floor at any point outside of guard.

**1910.219(e)(4)** *Vertical belts.* Vertical belts running over a lower pulley more than seven (7) feet above floor or platform shall be guarded at the bottom in the same manner as horizontal overhead belts, if conditions are as stated in paragraphs (e)(2)(ii) (a) and (c) of this section.

**1910.219(e)(5)** *Cone-pulley belts.* **(i)** The cone belt and pulley shall be equipped with a belt shifter so constructed as to adequately guard the nip point of the belt and pulley. If the frame of the belt shifter does not adequately guard the nip point of the belt and pulley, the nip point shall be further protected by means of a vertical guard placed in front of the pulley and extending at least to the top of the largest step of the cone.

**1910.219(e)(5)(ii)** If the belt is of the endless type or laced with rawhide laces, and a belt shifter is not desired, the belt will be considered guarded if the nip point of the belt and pulley is protected by a nip point guard located in front of the cone extending at least to the top of the largest step of the cone, and formed to show the contour of the cone in order to give the nip point of the belt and pulley the maximum protection.

**1910.219(e)(5)(iii)** If the cone is located less than 3 feet from the floor or working platform, the cone pulley and belt shall be guarded to a height of 3 feet regardless of whether the belt is endless or laced with rawhide.

# OSHA Standards for General Industry

**1910.219(e)(6)** *Belt tighteners.* **(i)** Suspended counterbalanced tighteners and all parts thereof shall be of substantial construction and securely fastened; the bearings shall be securely capped. Means must be provided to prevent tightener from falling, in case the belt breaks.

**1910.219(e)(6)(ii)** Where suspended counterweights are used and not guarded by location, they shall be so encased as to prevent accident.

**1910.219(f)** *Gears, sprockets, and chains*—**(1)** *Gears.* Gears shall be guarded in accordance with one of the following methods:

**1910.219(f)(1)(i)** By a complete enclosure; or

**1910.219(f)(1)(ii)** By a standard guard as described in paragraph (o) of this section, at least seven (7) feet high extending six (6) inches above the mesh point of the gears; or

**1910.219(f)(1)(iii)** By a band guard covering the face of gear and having flanges extended inward beyond the root of the teeth on the exposed side or sides. Where any portion of the train of gears guarded by a band guard is less than six (6) feet from the floor a disk guard or a complete enclosure to the height of six (6) feet shall be required.

**1910.219(f)(2)** *Hand-operated gears.* Paragraph (f)(1) of this section does not apply to hand-operated gears used only to adjust machine parts and which do not continue to move after hand power is removed. However, the guarding of these gears is highly recommended.

**1910.219(f)(3)** *Sprockets and chains.* All sprocket wheels and chains shall be enclosed unless they are more than seven (7) feet above the floor or platform. Where the drive extends over other machine or working areas, protection against falling shall be provided. This subparagraph does not apply to manually operated sprockets.

**1910.219(f)(4)** *Openings for oiling.* When frequent oiling must be done, openings with hinged or sliding self-closing covers shall be provided. All points not readily accessible shall have oil feed tubes if lubricant is to be added while machinery is in motion.

**1910.219(g)** *Guarding friction drives.* The driving point of all friction drives when exposed to contact shall be guarded, all arm or spoke friction drives and all web friction drives with holes in the web shall be entirely enclosed, and all projecting belts on friction drives where exposed to contact shall be guarded.

**1910.219(h)** *Keys, setscrews, and other projections.* **(1)** All projecting keys, setscrews, and other projections in revolving parts shall be removed or made flush or guarded by metal cover. This subparagraph does not apply to keys or setscrews within gear or sprocket casings or other enclosures, nor to keys, setscrews, or oilcups in hubs of pulleys less than twenty (20) inches in diameter where they are within the plane of the rim of the pulley.

**1910.219(h)(2)** It is recommended, however, that no projecting setscrews or oilcups be used in any revolving pulley or part of machinery.

**1910.219(i)** *Collars and couplings*—**(1)** *Collars.* All revolving collars, including split collars, shall be cylindrical, and screws or bolts used in collars shall not project beyond the largest periphery of the collar.

**1910.219(i)(2)** *Couplings.* Shaft couplings shall be so constructed as to present no hazard from bolts, nuts, setscrews, or revolving surfaces. Bolts, nuts, and setscrews will, however, be permitted where they are covered with safety sleeves or where they are used parallel with the shafting and are countersunk or else do not extend beyond the flange of the coupling.

**1910.219(j)** *Bearings and facilities for oiling.* All drip cups and pans shall be securely fastened.

**1910.219(k)** *Guarding of clutches, cutoff couplings, and clutch pulleys*—**(1)** *Guards.* Clutches, cutoff couplings, or clutch pulleys having projecting parts, where such clutches are located seven (7) feet or less above the floor or working platform, shall be enclosed by a stationary guard constructed in accordance with this section. A "U" type guard is permissible.

**1910.219(k)(2)** *Engine rooms.* In engine rooms a guardrail, preferably with toeboard, may be used instead of the guard required by paragraph (k)(1) of this section, provided such a room is occupied only by engine room attendants.

**1910.219(l)** *Belt shifters, clutches, shippers, poles, perches, and fasteners*—**(1)** *Belt shifters.* **(i)** Tight and loose pulleys on all new installations made on or after August 31, 1971, shall be equipped with a permanent belt shifter provided with mechanical means to prevent belt from creeping from loose to tight pulley. It is recommended that old installations be changed to conform to this rule.

**1910.219(l)(1)(ii)** Belt shifter and clutch handles shall be rounded and be located as far as possible from danger of accidental contact, but within easy reach of the operator. Where belt shifters are not directly located over a machine or bench, the handles shall be cut off six feet six inches (6 ft. 6 in.) above floor level.

**1910.219(l)(2)** *Belt shippers and shipper poles.* The use of belt poles as substitutes for mechanical shifters is not recommended.

**1910.219(l)(3)** *Belt perches.* Where loose pulleys or idlers are not practicable, belt perches in form of brackets, rollers, etc., shall be used to keep idle belts away from the shafts.

**1910.219(l)(4)** *Belt fasteners.* Belts which of necessity must be shifted by hand and belts within seven (7) feet of the floor or working platform which are not guarded in accordance with this section shall not be fastened with metal in any case, nor with any other fastening which by construction or wear will constitute an accident hazard.

**1910.219(m)** *Standard guards—general requirements*—**(1)** *Materials.* **(i)** Standard conditions shall be secured by the use of the following materials. Expanded metal, perforated or solid sheet metal, wire mesh on a frame of angle iron, or iron pipe securely fastened to floor or to frame of machine.

**1910.219(m)(1)(ii)** All metal should be free from burrs and sharp edges.

**1910.219(m)(2)** *Methods of manufacture.* **(i)** Expanded metal, sheet or perforated metal, and wire mesh shall be securely fastened to frame.

**1910.219(m)(2)(ii)** [Reserved]

**1910.219(n)** [Reserved]

**1910.219(o)** *Approved materials*—**(1)** *Minimum requirements.* The materials and dimensions specified in this paragraph shall apply to all guards, except horizontal overhead belts, rope, cable, or chain guards more than seven (7) feet above floor, or platform.

**1910.219(o)(1)(i)** [Reserved]

**1910.219(o)(1)(i)***(a)* All guards shall be rigidly braced every three (3) feet or fractional part of their height to some fixed part of machinery or building structure. Where guard is exposed to contact with moving equipment additional strength may be necessary.

**1910.219(o)(1)(i)***(b)* [Reserved]

**1910.219(o)(1)(ii)** [Reserved]

**1910.219(o)(2)** *Wood guards.* **(i)** Wood guards may be used in the woodworking and chemical industries, in industries where the presence of fumes or where manufacturing conditions would cause the rapid deterioration of metal guards; also in construction work and in locations outdoors where extreme cold or extreme heat make metal guards and railings undesirable. In all other industries, wood guards shall not be used.

**1910.219(o)(2)(ii)** [Reserved]

**1910.219(o)(3)** *Guards for horizontal overhead belts.* **(i)** Guards for horizontal overhead belts shall run the entire length of the belt and follow the line of the pulley to the ceiling or be carried to the nearest wall, thus enclosing the belt effectively. Where belts are so located as to make it impracticable to carry the guard to wall or ceiling, construction of guard shall be such as to enclose completely the top and bottom runs of belt and the face of pulleys.

**1910.219(o)(3)(ii)** [Reserved]

**1910.219(o)(3)(iii)** Suitable reinforcement shall be provided for the ceiling rafters or overhead floor beams, where such is necessary, to sustain safely the weight and stress likely to be imposed by the guard. The interior surface of all guards, by which is meant the surface of the guard with which a belt will come in contact, shall be smooth and free from all projections of any character, except where construction demands it; protruding shallow roundhead rivets may be used. Overhead belt guards shall be at least one-quarter wider than belt which they protect, except that this clearance need not in any case exceed six (6) inches on each side. Overhead rope drive and block and roller-chain-drive guards shall be not less than six (6) inches wider than the drive on each side. In overhead silent chain-drive guards where the chain is held from lateral displacement on the sprockets, the side clearances required on drives of twenty (20) inch centers or under shall be not less than one-fourth inch from the nearest moving chain part, and on drives of over twenty (20) inch centers a minimum of one-half inch from the nearest moving chain part.

**1910.219(o)(4)** *Guards for horizontal overhead rope and chain drives.* Overhead-rope and chain-drive guard construction shall conform to the rules for overhead-belt guard.

**1910.219(o)(5)** *Guardrails and toeboards.* **(i)** Guardrail shall be forty-two (42) inches in height, with midrail between top rail and floor.

**1910.219(o)(5)(ii)** Posts shall be not more than eight (8) feet apart; they are to be permanent and substantial, smooth, and free from protruding nails, bolts, and splinters. If made of pipe, the post shall be one and one-fourth (1 1/4) inches inside diameter, or larger. If made of metal shapes or bars, their section shall be equal in strength to that of one and one-half (1 1/2) by one and one-half (1 1/2) by three-sixteenths (3/16) inch angle iron. If made of wood, the posts shall be two by four (2 × 4) inches or larger. The upper rail shall be two by four (2 × 4) inches, or two one by four (1 × 4) strips, one at the top and one at the side of posts. The midrail may be one by four (1 × 4) inches or more. Where panels are fitted with expanded metal or wire mesh, the middle rails may be omitted. Where guard is exposed to contact with moving equipment, additional strength may be necessary.

**1910.219(o)(5)(iii)** Toeboards shall be four (4) inches or more in height, of wood, metal, or of metal grill not exceeding one (1) inch mesh.

**1910.219(p)** *Care of equipment*—**(1)** *General.* All power-transmission equipment shall be inspected at intervals not exceeding 60 days and be kept in good working condition at all times.

**1910.219(p)(2)** *Shafting.* **(i)** Shafting shall be kept in alignment, free from rust and excess oil or grease.

**1910.219(p)(2)(ii)** Where explosives, explosive dusts, flammable vapors or flammable liquids exist, the hazard of static sparks from shafting shall be carefully considered.

**1910.219(p)(3)** *Bearings.* Bearings shall be kept in alignment and properly adjusted.

**1910.219(p)(4)** *Hangers.* Hangers shall be inspected to make certain that all supporting bolts and screws are tight and that supports of hanger boxes are adjusted properly.

**1910.219(p)(5)** *Pulleys.* **(i)** Pulleys shall be kept in proper alignment to prevent belts from running off.

**1910.219(p)(5)(ii)** [Reserved]

**1910.219(p)(6)** *Care of belts.*

**1910.219(p)(6)(i)** [Reserved]

**1910.219(p)(6)(ii)** Inspection shall be made of belts, lacings, and fasteners and such equipment kept in good repair.

**1910.219(p)(7)** *Lubrication.* The regular oilers shall wear tight-fitting clothing. Machinery shall be oiled when not in motion, wherever possible.

[39 FR 23502, June 27, 1974, as amended at 43 FR 49750, Oct. 24, 1978; 43 FR 51760; Nov. 7, 1978; 49 FR 5323, Feb. 10, 1984; 61 FR 9240, Mar. 7, 1996]

### §1910.220 Effective dates [Removed].

[§1910.220 removed at 61 FR 9240, Mar. 7, 1996, effective May 6, 1996.]

### §1910.221 Sources of standards [Removed].

[§1910.221 removed at 61 FR 9240, Mar. 7, 1996, effective May 6, 1996.]

### §1910.222 Standards organizations [Removed].

[§1910.222 removed at 61 FR 9240, Mar. 7, 1996, effective May 6, 1996.]

# Subpart P—Hand and Portable Powered Tools and Other Hand-Held Equipment

**Authority:** Sections 4, 6, and 8 of the Occupational Safety and Health Act of 1970 (29 U.S.C. 653, 655, 657); Secretary of Labor's Order No. 12-71 (36 FR 8754), 8-76 (41 FR 25059), 9-83 (48 FR 35736), or 1-90 (55 FR 9033), as applicable; 29 CFR part 1911.

Section 1910.243 also issued under 29 CFR part 1910.

## §1910.241 Definitions.

As used in this subpart:

**1910.241(a)** *Explosive-actuated fastening tool terms—*

**1910.241(a)(1)** *Hammer-operated piston tool—low-velocity type.* A tool which, by means of a heavy mass hammer supplemented by a load, moves a piston designed to be captive to drive a stud, pin, or fastener into a work surface, always starting the fastener at rest and in contact with the work surface. It shall be so designed that when used with any load that accurately chambers in it and that is commercially available at the time the tool is submitted for approval, it will not cause such stud, pin, or fastener to have a mean velocity in excess of 300 feet per second when measured 6.5 feet from the muzzle end of the barrel.

**1910.241(a)(2)** *High-velocity tool.* A tool or machine which, when used with a load, propels or discharges a stud, pin, or fastener, at velocities in excess of 300 feet per second when measured 6.5 feet from the muzzle end of the barrel, for the purpose of impinging it upon, affixing it to, or penetrating another object or material.

**1910.241(a)(3)** *Low-velocity piston tool.* A tool that utilizes a piston designed to be captive to drive a stud, pin, or fastener into a work surface. It shall be so designed that when used with any load that accurately chambers in it and that is commercially available at the time the tool is submitted for approval, it will not cause such stud, pin, or fastener to have a mean velocity in excess of 300 feet per second when measured 6.5 feet from the muzzle end of the barrel.

**1910.241(a)(4)** *Stud, pin, or fastener.* A fastening device specifically designed and manufactured for use in explosive-actuated fastening tools.

**1910.241(a)(5)** *To chamber.* To fit properly without the use of excess force, the case being duly supported.

**1910.241(a)(6)** *Explosive powerload, also known as load.* Any substance in any form capable of producing a propellant force.

**1910.241(a)(7)** *Tool.* An explosive-actuated fastening tool, unless otherwise indicated, and all accessories pertaining thereto.

**1910.241(a)(8)** *Protective shield or guard.* A device or guard attached to the muzzle end of the tool, which is designed to confine flying particles.

**1910.241(b)** *Abrasive wheel terms—***(1)** *Mounted wheels.* Mounted wheels, usually 2-inch diameter or smaller, and of various shapes, may be either organic or inorganic bonded abrasive wheels. They are secured to plain or threaded steel mandrels.

**1910.241(b)(2)** *Tuck pointing.* Removal, by grinding, of cement, mortar, or other nonmetallic jointing material.

**1910.241(b)(3)** *Tuck pointing wheels.* Tuck pointing wheels, usually Type 1, reinforced organic bonded wheels have diameter, thickness and hole size dimension. They are subject to the same limitations of use and mounting as Type 1 wheels defined in subparagraph (10) of this paragraph.

LIMITATION: Wheels used for tuck pointing should be reinforced, organic bonded. (See §1910.243(c)(1)(ii)(c).)

**1910.241(b)(4)** *Portable grinding.* A grinding operation where the grinding machine is designed to be hand held and may be easily moved from one location to another.

**1910.241(b)(5)** *Organic bonded wheels.* Organic wheels are wheels which are bonded by means of an organic material such as resin, rubber, shellac, or other similar bonding agent.

**1910.241(b)(6)** *Safety guard.* A safety guard is an enclosure designed to restrain the pieces of the grinding wheel and furnish all possible protection in the event that the wheel is broken in operation.

**1910.241(b)(7)** *Reinforced wheels.* The term *reinforced* as applied to grinding wheels shall define a class of organic wheels which contain strengthening fabric or filament. The term *reinforced* does not cover wheels using such mechanical additions as steel rings, steel cup backs or wire or tape winding.

**1910.241(b)(8)** *Type 11 flaring cup wheels.* Type 11 flaring cup wheels have double diameter dimensions D and J, and in addition have thickness, hole size, rim and back thickness dimensions. Grinding is always performed on rim face, W dimension. Type 11 wheels are subject to all limitations of use and mounting listed for Type 6 straight sided cup wheels definition in subparagraph (9) of this paragraph.

**TYPE 11—FLARING-CUP WHEEL**

Side grinding wheel having a wall flared or tapered outward from the back. Wall thickness at the back is normally greater than at the grinding face (W).

**FIGURE P-1**

LIMITATION: Minimum back thickness, E dimension, should not be less than one-fourth T dimension. In addition when unthreaded hole wheels are specified the inside flat, K dimension, shall be large enough to accommodate a suitable flange.

**1910.241(b)(9)** *Type 6 straight cup wheels.* Type 6 cup wheels have diameter, thickness, hole size, rim thickness, and back thickness dimensions. Grinding is always performed on rim face, W dimension.

LIMITATION: Minimum back thickness, E dimension, should not be less than one-fourth T dimension. In addition, when unthreaded hole wheels are specified, the inside flat, K dimension, must be large enough to accommodate a suitable flange.

**TYPE 6—STRAIGHT CUP WHEEL**

Side grinding wheel having a diameter, thickness and hole with one side straight or flat and the opposite side recessed. This type, however, differs from Type 5 in that the grinding is performed on the wall of the abrasive created by the difference between the diameter of the recess and the outside diameter of the wheel. Therefore, the wall dimension "W" takes precedence over the diameter of the recess as an essential intermediate dimension to describe this shape type.

**FIGURE P-2**

**1910.241(b)(10)** *Type 1 straight wheels.* Type 1 straight wheels have diameter, thickness, and hole size dimensions and should be used only on the periphery. Type 1 wheels shall be mounted between flanges.

LIMITATION: Hole dimension (H) should not be greater than two-thirds of wheel diameter dimension (D) for precision, cylindrical, centerless, or surface grinding applications. Maximum hole size for all other applications should not exceed one-half wheel diameter.

**TYPE 1—STRAIGHT WHEEL**

Peripheral grinding wheel having a diameter, thickness and hole.

**FIGURE P-3**

**1910.241(c)** [Reserved]

**1910.241(d)** *Jack terms*—**(1)** *Jack.* A jack is an appliance for lifting and lowering or moving horizontally a load by application of a pushing force.

**NOTE:** Jacks may be of the following types: Lever and ratchet, screw and hydraulic.

**1910.241(d)(2)** *Rating.* The rating of a jack is the maximum working load for which it is designed to lift safely that load throughout its specified amount of travel.

**NOTE:** To raise the rated load of a jack, the point of application of the load, the applied force, and the length of lever arm should be those designated by the manufacturer for the particular jack considered.

[39 FR 23502, June 27, 1974, as amended at 43 FR 49750, Oct. 24, 1978]

## §1910.242 Hand and portable powered tools and equipment, general.

**1910.242(a)** *General requirements.* Each employer shall be responsible for the safe condition of tools and equipment used by employees, including tools and equipment which may be furnished by employees.

**1910.242(b)** *Compressed air used for cleaning.* Compressed air shall not be used for cleaning purposes except where reduced to less than 30 p.s.i. and then only with effective chip guarding and personal protective equipment.

## §1910.243 Guarding of portable powered tools.

**1910.243(a)** *Portable powered tool*—**(1)** *Portable circular saws.* **(i)** All portable, power- driven [sic] circular saws having a blade diameter greater than 2 in. shall be equipped with guards above and below the base plate or shoe. The upper guard shall cover the saw to the depth of the teeth, except for the minimum arc required to permit the base to be tilted for bevel cuts. The lower guard shall cover the saw to the depth of the teeth, except for the minimum arc required to allow proper retraction and contact with the work. When the tool is withdrawn from the work, the lower guard shall automatically and instantly return to covering position.

**1910.243(a)(1)(ii)** Paragraph (a)(1)(i) of this section does not apply to circular saws used in the meat industry for meat cutting purposes.

**1910.243(a)(2)** *Switches and controls.* **(i)** All hand-held powered circular saws having a blade diameter greater than 2 inches, electric, hydraulic or pneumatic chain saws, and percussion tools without positive accessory holding means shall be equipped with a constant pressure switch or control that will shut off the power when the pressure is released. All hand-held gasoline powered chain saws shall be equipped with a constant pressure throttle control that will shut off the power to the saw chain when the pressure is released.

**1910.243(a)(2)(ii)** All hand-held powered drills, tappers, fastener drivers, horizontal, vertical, and angle grinders with wheels greater than 2 inches in diameter, disc sanders with discs greater than 2 inches in diameter, belt sanders, reciprocating saws, saber, scroll, and jig saws with blade shanks greater than a nominal one-fourth inch, and other similarly operating powered tools shall be equipped with a constant pressure switch or control, and may have a lock-on control provided that turnoff can be accomplished by a single motion of the same finger or fingers that turn it on.

**1910.243(a)(2)(iii)(a)** All other hand-held powered tools, such as, but not limited to, platen sanders, grinders with wheels 2 inches in diameter or less, disc sanders with discs 2 inches in diameter or less, routers, planers, laminate trimmers, nibblers, shears, saber, scroll, and jig saws with blade shanks a nominal one-fourth of an inch wide or less, may be equipped with either a positive "on-off" control, or other controls as described by paragraph (a)(2)(i) and (ii) of this section.

**1910.243(a)(2)(iii)(b)** Saber, scroll, and jig saws with nonstandard blade holders may use blades with shanks which are nonuniform in width, provided the narrowest portion of the blade shank is an integral part in mounting the blade.

**1910.243(a)(2)(iii)(c)** Blade shank width shall be measured at the narrowest portion of the blade shank when saber, scroll, and jig saws have nonstandard blade holders.

**1910.243(a)(2)(iii)(d)** *Nominal* in this subparagraph means ± 0.05 inch.

**1910.243(a)(2)(iv)** The operating control on hand-held power tools shall be so located as to minimize the possibility of its accidental operation, if such accidental operation would constitute a hazard to employees.

**1910.243(a)(2)(v)** This subparagraph does not apply to concrete vibrators, concrete breakers, powered tampers, jack hammers, rock drills, garden appliances, household and kitchen appliances, personal care appliances, medical or dental equipment, or to fixed machinery.

**1910.243(a)(3)** *Portable belt sanding machines.* Belt sanding machines shall be provided with guards at each nip point where the sanding belt runs onto a pulley. These guards shall effectively prevent the hands or fingers of the operator from coming in contact with the nip points. The unused run of the sanding belt shall be guarded against accidental contact.

**1910.243(a)(4)** *Cracked saws.* All cracked saws shall be removed from service.

**1910.243(a)(5)** *Grounding.* Portable electric powered tools shall meet the electrical requirements of subpart S of this part.

**1910.243(b)** *Pneumatic powered tools and hose*—**(1)** *Tool retainer.* A tool retainer shall be installed on each piece of utilization equipment which, without such a retainer, may eject the tool.

**1910.243(b)(2)** *Airhose.* Hose and hose connections used for conducting compressed air to utilization equipment shall be designed for the pressure and service to which they are subjected.

**1910.243(c)** *Portable abrasive wheels*—**(1)** *General requirements.* Abrasive wheels shall be used only on machine provided with safety guards as defined in paragraph (c) (1) through (4) of this section.

**1910.243(c)(1)(i)** *Exceptions.* The requirements of this paragraph (c)(1) shall not apply to the following classes of wheels and conditions.

**1910.243(c)(1)(i)(a)** Wheels used for internal work while within the work being ground;

**1910.243(c)(1)(i)(b)** Mounted wheels used in portable operations 2 inches and smaller in diameter; (see definition §1910.241(b)(1)); and

**1910.243(c)(1)(i)(c)** Types 16, 17, 18, 18R, and 19 cones, and plugs, and threaded hole pot balls where the work offers protection.

**1910.243(c)(1)(ii)(a)** A safety guard shall cover the spindle end, nut and flange projections. The safety guard shall be mounted so as to maintain proper alignment with the wheel, and the strength of the fastenings shall exceed the strength of the guard.

**1910.243(c)(1)(ii)(b)** *Exception.* Safety guards on all operations where the work provides a suitable measure of protection to the operator may be so constructed that the spindle end, nut and outer flange are exposed. Where the nature of the work is such as to entirely cover the side of the wheel, the side covers of the guard may be omitted.

**1910.243(c)(1)(ii)(c)** *Exception.* The spindle end, nut, and outer flange may be exposed on portable machines designed for, and used with, type 6, 11, 27, and 28 abrasive wheels, cutting off wheels, and tuck pointing wheels.

**1910.243(c)(2)** *Cup wheels.* Cup wheels (Types 6 and 11) shall be protected by:

**1910.243(c)(2)(i)** Safety guards as specified in paragraph (c)(1) of this section; or,

**1910.243(c)(2)(ii)** Special "revolving cup guards" which mount behind the wheel and turn with it. They shall be made of steel or other material with adequate strength and shall enclose the wheel sides upward from the back for one-third of the wheel thickness. The mounting features shall conform with all regulations. (See paragraph (c)(5) of this section.) It is necessary to maintain clearance between the wheel side and the guard. The clearance shall not exceed one-sixteenth inch; or,

**1910.243(c)(2)(iii)** Some other form of guard that will insure as good protection as that which would be provided by the guards specified in paragraph (c)(1) (i) or (ii) of this subparagraph.

**1910.243(c)(3)** *Vertical portable grinders.* Safety guards used on machines known as right angle head or vertical portable grinders shall have a maximum exposure angle of 180°, and the guard shall be so located so as to be between the operator and the wheel during use. Adjustment of guard shall be such that pieces of an accidentally broken wheel will be deflected away from the operator. (See Figure P-4.)

**FIGURE NO. P-4**

**1910.243(c)(4)** *Other portable grinders.* The maximum angular exposure of the grinding wheel periphery and sides for safety guards used on other portable grinding machines shall not exceed 180° and the top half of the wheel shall be enclosed at all times. (See Figures P-5 and P-6.)

**FIGURE NO. P-5**    **FIGURE NO. P-6**

**1910.243(c)(5)** *Mounting and inspection of abrasive wheels.* **(i)** Immediately before mounting, all wheels shall be closely inspected and sounded by the user (ring test, see subpart O, §1910.215(d)(1)) to make sure they have not been damaged in transit, storage, or otherwise. The spindle speed of the machine shall be checked before mounting of the wheel to be certain that it does not exceed the maximum operating speed marked on the wheel.

**1910.243(c)(5)(ii)** Grinding wheels shall fit freely on the spindle and remain free under all grinding conditions. A controlled clearance between the wheel hole and the machine spindle (or wheel sleeves or adaptors) is essential to avoid excessive pressure from mounting and spindle expansion. To accomplish this, the machine spindle shall be made to nominal (standard) size plus zero minus .002 inch, and the wheel hole shall be made suitably oversize to assure safety clearance under the conditions of operating heat and pressure.

**1910.243(c)(5)(iii)** All contact surfaces of wheels, blotters, and flangers shall be flat and free of foreign matter.

**1910.243(c)(5)(iv)** When a bushing is used in the wheel hole it shall not exceed the width of the wheel and shall not contact the flanges.

**1910.243(c)(5)(v)** Requirements for the use of flanges and blotters, see subpart O, §1910.215(c).

**1910.243(c)(6)** *Excluded machinery.* Natural sandstone wheels and metal, wooden, cloth, or paper discs, having a layer of abrasive on the surface are not covered by this paragraph.

**1910.243(d)** *Explosive actuated fastening tools*—**(1)** *General requirements.* **(i)** Explosive-actuated fastening tools which are actuated by explosives or any similar means and propel a stud, pin, fastener, or other object for the purpose of affixing it by penetration to any other object shall meet the design requirements in "American National Standard Safety Requirements for Explosive-Actuated Fastening Tools," ANSI A10.3-1970, which is incorporated by reference as specified in §1910.6. This requirement does not apply to devices designed for attaching objects to soft construction materials, such as wood, plaster, tar, dry wallboard, and the like, or to stud welding equipment.

**1910.243(d)(1)(ii)** Operators and assistants using tools shall be safeguarded by means of eye protection. Head and face protection shall be used, as required by working conditions, as set forth in subpart I.

**1910.243(d)(2)** *Inspection, maintenance, and tool handling*—**(i)** *High-velocity tools.* Tools of this type shall have the characteristics outlined in (*a*) through (*h*) of this section.

**1910.243(d)(2)(i)(a)** The muzzle end of the tool shall have a protective shield or guard at least 3 1/2 inches in diameter, mounted perpendicular to and concentric with the barrel, and designed to confine any flying fragments or particles that might otherwise create a hazard at the time of firing.

**1910.243(d)(2)(i)(b)** Where a standard shield or guard cannot be used, or where it does not cover all apparent avenues through which flying particles might escape, a special shield, guard, fixture, or jig designed and built by the manufacturer of the tool being used, which provides this degree of protection, shall be used as a substitute.

**1910.243(d)(2)(i)(c)** The tool shall be so designed that it cannot be fired unless it is equipped with a standard protective shield or guard, or a special shield, guard, fixture, or jig.

**1910.243(d)(2)(i)(d)(1)** The firing mechanism shall be so designed that the tool cannot fire during loading or preparation to fire, or if the tool should be dropped while loaded.

**1910.243(d)(2)(i)(d)(2)** Firing of the tool shall be dependent upon at least two separate and distinct operations of the operator, with the final firing movement being separate from the operation of bringing the tool into the firing position.

**1910.243(d)(2)(i)(e)** The tool shall be so designed as not to be operable other than against a work surface, and unless the operator is holding the tool against the work surface with a force at least 5 pounds greater than the total weight of the tool.

**1910.243(d)(2)(i)(f)** The tool shall be so designed that it will not operate when equipped with the standard guard indexed to the center position if any bearing surface of the guard is tilted more than 8° from contact with the work surface.

**1910.243(d)(2)(i)(g)** The tool shall be so designed that positive means of varying the power are available or can be made available to the operator as part of the tool, or as an auxiliary, in order to make it possible for the operator to select a power level adequate to perform the desired work without excessive force.

**1910.243(d)(2)(i)(h)** The tool shall be so designed that all breeching parts will be reasonably visible to allow a check for any foreign matter that may be present.

**1910.243(d)(2)(ii)** Tools of the low-velocity-piston type shall have the characteristics outlined in paragraphs (d)(2)(ii) (*a*) through (*e*) of this section and any additional safety features he may wish to incorporate.

**1910.243(d)(2)(ii)(a)** The muzzle end of the tool shall be designed so that suitable protective shields, guards, jigs, or fixtures, designed and built by the manufacturer of the tool being used, can be mounted perpendicular to the barrel. A standard spall shield shall be supplied with each tool.

**1910.243(d)(2)(ii)(b)(1)** The tool shall be designed so that it shall not in ordinary usage propel or discharge a stud, pin, or fastener while loading or during preparation to fire, or if the tool should be dropped while loaded.

**1910.243(d)(2)(ii)(b)(2)** Firing of the tool shall be dependent upon at least two separate and distinct operations of the operator, with the final firing movement being separate from the operation of bringing the tool into the firing position.

**1910.243(d)(2)(ii)(c)** The tool shall be so designed as not to be operable other than against a work surface, and unless the operator is holding the tool against the work surface with a force at least 5 pounds greater than the total weight of the tool.

**1910.243(d)(2)(ii)(d)** The tool shall be so designed that positive means of varying the power are available or can be made available to the operator as part of the tool, or as an auxiliary, in order to make it possible for the operator to select a power level adequate to perform the desired work without excessive force.

**1910.243(d)(2)(ii)(e)** The tool shall be so designed that all breeching parts will be reasonably visible to allow a check for any foreign matter that may be present.

**1910.243(d)(2)(iii)** Tools of the hammer-operated piston tools—low-velocity type shall have the characteristics outlined in paragraphs (d)(2)(iii) (a) through (e) of this section.

**1910.243(d)(2)(iii)(a)** The muzzle end of the tool shall be so designed that suitable protective shields, guards, jigs, or fixtures, designed and built by the manufacturer of the tool being used, can be mounted perpendicular to the barrel. A standard spall shield shall be supplied with each tool.

**1910.243(d)(2)(iii)(b)** The tool shall be so designed that it shall not in ordinary usage propel or discharge a stud, pin, or fastener while loading, or during preparation to fire, or if the tool should be dropped while loaded.

**1910.243(d)(2)(iii)(c)** Firing of the tool shall be dependent upon at least two separate and distinct operations of the operator, with the final firing movement being separate from the operation of bringing the tool into the firing position.

**1910.243(d)(2)(iii)(d)** The tool shall be so designed that positive means of varying the power are available or can be made available to the operator as part of the tool, or as an auxiliary, in order to make it possible for the operator to select a power level adequate to perform the desired work without excessive force.

**1910.243(d)(2)(iii)(e)** The tool shall be so designed that all breeching parts will be reasonably visible to allow a check for any foreign matter that may be present.

**1910.243(d)(3)** *Requirements for loads and fasteners.* (i) There shall be a standard means of identifying the power levels of loads used in tools.

**1910.243(d)(3)(ii)** [Reserved]

**1910.243(d)(3)(iii)** No load (cased or caseless) shall be used if it will accurately chamber in any existing approved commercially available low-velocity piston tool or hammer operated piston tool —low-velocity type and will cause a fastener to have a mean velocity in excess of 300 feet per second when measured 6.5 feet from the muzzle end of the barrel. No individual test firing of a series shall exceed 300 feet per second by more than 8 percent.

**1910.243(d)(3)(iv)** Fasteners used in tools shall be only those specifically manufactured for use in such tools.

**1910.243(d)(4)** *Operating requirements.* (i) Before using a tool, the operator shall inspect it to determine to his satisfaction that it is clean, that all moving parts operate freely, and that the barrel is free from obstructions.

**1910.243(d)(4)(ii)** When a tool develops a defect during use, the operator shall immediately cease to use it, until it is properly repaired.

**1910.243(d)(4)(iii)** Tools shall not be loaded until just prior to the intended firing time. Neither loaded nor empty tools are to be pointed at any workmen.

**1910.243(d)(4)(iv)** No tools shall be loaded unless being prepared for immediate use, nor shall an unattended tool be left loaded.

**1910.243(d)(4)(v)** In case of a misfire, the operator shall hold the tool in the operating position for at least 30 seconds. He shall then try to operate the tool a second time. He shall wait another 30 seconds, holding the tool in the operating position; then he shall proceed to remove the explosive load in strict accordance with the manufacturer's instructions.

**1910.243(d)(4)(vi)** A tool shall never be left unattended in a place where it would be available to unauthorized persons.

**1910.243(d)(4)(vii)** Fasteners shall not be driven into very hard or brittle materials including, but not limited to, cast iron, glazed tile, surface-hardened steel, glass block, live rock, face brick, or hollow tile.

**1910.243(d)(4)(viii)** Driving into materials easily penetrated shall be avoided unless such materials are backed by a substance that will prevent the pin or fastener from passing completely through and creating a flying-missile hazard on the other side.

**1910.243(d)(4)(ix)(a)** Fasteners shall not be driven directly into materials such as brick or concrete closer than 3 inches from the unsupported edge or corner, or into steel surfaces closer than one-half inch from the unsupported edge or corner, unless a special guard, fixture, or jig is used. (Exception: Low-velocity tools may drive no closer than 2 inches from an edge in concrete or one-fourth inch in steel.)

**1910.243(d)(4)(ix)(b)** When fastening other materials, such as a 2- by 4-inch wood section to a concrete surface, it is permissible to drive a fastener of no greater than 7/32-inch shank diameter not closer than 2 inches from the unsupported edge or corner of the work surface.

**1910.243(d)(4)(x)** Fasteners shall not be driven through existing holes unless a positive guide is used to secure accurate alignment.

**1910.243(d)(4)(xi)** No fastener shall be driven into a spalled area caused by an unsatisfactory fastening.

**1910.243(d)(4)(xii)** Tools shall not be used in an explosive or flammable atmosphere.

**1910.243(d)(4)(xiii)** All tools shall be used with the correct shield, guard, or attachment recommended by the manufacturer.

**1910.243(d)(4)(xiv)** Any tool found not in proper working order shall be immediately removed from service. The tool shall be inspected at regular intervals and shall be repaired in accordance with the manufacturer's specifications.

**1910.243(e)** *Power lawnmowers*—(1) *General requirements.* (i) Power lawnmowers of the walk-behind, riding-rotary, and reel power lawnmowers shall be guarded in accordance with the machine guarding requirements in 29 CFR 1910.212, General requirements for all machines.

**1910.243(e)(1)(ii)** All power-driven chains, belts, and gears shall be so positioned or otherwise guarded to prevent the operator's accidental contact therewith, during normal starting, mounting, and operation of the machine.

**1910.243(e)(1)(iii)** A shutoff device shall be provided to stop operation of the motor or engine. This device shall require manual and intentional reactivation to restart the motor or engine.

**1910.243(e)(1)(iv)** All positions of the operating controls shall be clearly identified.

**1910.243(e)(1)(v)** The words, "Caution. Be sure the operating control(s) is in neutral before starting the engine," or similar wording shall be clearly visible at an engine starting control point on self-propelled mowers.

**1910.243(e)(2)** *Walk-behind and riding rotary mowers.* (i) The mower blade shall be enclosed except on the bottom and the enclosure shall extend to or below the lowest cutting point of the blade in the lowest blade position.

**1910.243(e)(2)(ii)** Guards which must be removed to install a catcher assembly shall comply with the following:

**1910.243(e)(2)(ii)(a)** Warning instructions shall be affixed to the mower near the opening stating that the mower shall not be used without either the catcher assembly or the guard in place.

**1910.243(e)(2)(ii)(b)** The catcher assembly or the guard shall be shipped and sold as part of the mower.

**1910.243(e)(2)(ii)(c)** The instruction manual shall state that the mower shall not be used without either the catcher assembly or the guard in place.

**1910.243(e)(2)(ii)(d)** The catcher assembly, when properly and completely installed, shall not create a condition which violates the limits given for the guarded opening.

**1910.243(e)(2)(iii)** Openings in the blade enclosure, intended for the discharge of grass, shall be limited to a maximum vertical angle of the opening of 30°. Measurements shall be taken from the lowest blade position.

**1910.243(e)(2)(iv)** The total effective opening area of the grass discharge opening(s) shall not exceed 1,000 square degrees on units having a width of cut less than 27 1/2 inches, or 2,000 square degrees on units having a width of cut 27 1/2 inches or over.

**1910.243(e)(2)(v)** The word "Caution." or stronger wording, shall be placed on the mower at or near each discharge opening.

**1910.243(e)(2)(vi)** [Reserved]

**1910.243(e)(2)(vii)** Blade(s) shall stop rotating from the manufacturer's specified maximum speed within 15 seconds after declutching, or shutting off power.

**1910.243(e)(2)(viii)** In a multipiece blade, the means of fastening the cutting members to the body of the blade or disc shall be so designed that they will not become worn to a hazardous condition before the cutting members themselves are worn beyond use.

**1910.243(e)(2)(ix)** The maximum tip speed of any blade shall be 19,000 feet per minute.

**1910.243(e)(3)** *Walk-behind rotary mowers.* **(i)** The horizontal angle of the opening(s) in the blade enclosure, intended for the discharge of grass, shall not contact the operator area.

**1910.243(e)(3)(ii)** There shall be one of the following at all openings in the blade enclosure intended for the discharge of grass:

**1910.243(e)(3)(ii)(a)** A minimum unobstructed horizontal distance of 3 inches from the end of the discharge chute to the blade tip circle.

**1910.243(e)(3)(ii)(b)** A rigid bar fastened across the discharge opening, secured to prevent removal without the use of tools. The bottom of the bar shall be no higher than the bottom edge of the blade enclosure.

**1910.243(e)(3)(iii)** The highest point(s) of the front of the blade enclosure, except discharge openings, shall be such that any line extending a maximum of 15° downward from the horizontal toward the blade shaft axis (axes) shall not intersect the horizontal plane within the blade tip circle. The highest point(s) on the blade enclosure front, except discharge-openings, shall not exceed 1 1/4 inches above the lowest cutting point of the blade in the lowest blade position. Mowers with a swingover handle are to be considered as having no front in the blade enclosure and therefore shall comply with paragraph (e)(2)(i) of this section.

**1910.243(e)(3)(iv)** The mower handle shall be fastened to the mower so as to prevent loss of control by unintentional uncoupling while in operation.

**1910.243(e)(3)(v)** A positive upstop or latch shall be provided for the mower handle in the normal operating position(s). The upstop shall not be subject to unintentional disengagement during normal operation of the mower. The upstop or latch shall not allow the center or the handle grips to come closer than 17 inches horizontally behind the closest path of the mower blade(s) unless manually disengaged.

**1910.243(e)(3)(vi)** A swing-over handle, which complies with the above requirements, will be permitted.

**1910.243(e)(3)(vii)** Wheel drive disengaging controls, except deadman controls, shall move opposite to the direction of the vehicle motion in order to disengage the drive. Deadman controls shall automatically interrupt power to a drive when the operator's actuating force is removed, and may operate in any direction to disengage the drive.

**1910.243(e)(4)** *Riding rotary mowers.* **(i)** The highest point(s) of all openings in the blade enclosure, front shall be limited by a vertical angle of opening of 15° and a maximum distance of 1 1/4 inches above the lowest cutting point of the blade in the lowest blade position.

**1910.243(e)(4)(ii)** Opening(s) shall be placed so that grass or debris will not discharge directly toward any part of an operator seated in a normal operator position.

**1910.243(e)(4)(iii)** There shall be one of the following at all openings in the blade enclosure intended for the discharge of grass:

**1910.243(e)(4)(iii)(a)** A minimum unobstructed horizontal distance of 6 inches from the end of the discharge chute to the blade tip circle.

**1910.243(e)(4)(iii)(b)** A rigid bar fastened across the discharge opening, secured to prevent removal without the use of tools. The bottom of the bar shall be no higher than the bottom edge of the blade enclosure.

**1910.243(e)(4)(iv)** Mowers shall be provided with stops to prevent jack-knifing or locking of the steering mechanism.

**1910.243(e)(4)(v)** Vehicle stopping means shall be provided.

**1910.243(e)(4)(vi)** Hand-operated wheel drive disengaging controls shall move opposite to the direction of vehicle motion in order to disengage the drive. Foot-operated wheel drive disengaging controls shall be depressed to disengage the drive. Deadman controls, both hand and foot operated, shall automatically interrupt power to a drive when the operator's actuating force is removed, and may operate in any direction to disengage the drive.

[39 FR 23502, June 27, 1974, as amended at 43 FR 49750, Oct. 24, 1978; 49 FR 5323, Feb. 10, 1984; 50 FR 4649, Feb. 1, 1985; 61 FR 9240, Mar. 7, 1996]

## §1910.244 Other portable tools and equipment.

**1910.244(a)** *Jacks*—**(1)** *Loading and marking.*

**1910.244(a)(1)(i)** The operator shall make sure that the jack used has a rating sufficient to lift and sustain the load.

**1910.244(a)(1)(ii)** The rated load shall be legibly and permanently marked in a prominent location on the jack by casting, stamping, or other suitable means.

**1910.244(a)(2)** *Operation and maintenance.* **(i)** In the absence of a firm foundation, the base of the jack shall be blocked. If there is a possibility of slippage of the cap, a block shall be placed in between the cap and the load.

**1910.244(a)(2)(ii)** The operator shall watch the stop indicator, which shall be kept clean, in order to determine the limit of travel. The indicated limit shall not be overrun.

**1910.244(a)(2)(iii)** After the load has been raised, it shall be cribbed, blocked, or otherwise secured at once.

**1910.244(a)(2)(iv)** Hydraulic jacks exposed to freezing temperatures shall be supplied with an adequate antifreeze liquid.

**1910.244(a)(2)(v)** All jacks shall be properly lubricated at regular intervals.

**1910.244(a)(2)(vi)** Each jack shall be thoroughly inspected at times which depend upon the service conditions. Inspections shall be not less frequent than the following:

**1910.244(a)(2)(vi)(a)** For constant or intermittent use at one locality, once every 6 months,

**1910.244(a)(2)(vi)(b)** For jacks sent out of shop for special work, when sent out and when returned,

**1910.244(a)(2)(vi)(c)** For a jack subjected to abnormal load or shock, immediately before and immediately thereafter.

**1910.244(a)(2)(vii)** Repair or replacement parts shall be examined for possible defects.

**1910.244(a)(2)(viii)** Jacks which are out of order shall be tagged accordingly, and shall not be used until repairs are made.

**1910.244(b)** *Abrasive blast cleaning nozzles.* The blast cleaning nozzles shall be equipped with an operating valve which must be held open manually. A support shall be provided on which the nozzle may be mounted when it is not in use.

[39 FR 23502, June 27, 1974, as amended at 49 FR 5323, Feb. 10, 1984]

## §1910.245 Effective dates [Removed].

[§1910.245 removed at 61 FR 9240, Mar. 7, 1996, effective May 6, 1996.]

## §1910.246 Sources of standards [Removed].

[§1910.246 removed at 61 FR 9240, Mar. 7, 1996, effective May 6, 1996.]

## §1910.247 Standards organizations [Removed].

[§1910.247 removed at 61 FR 9240, Mar. 7, 1996, effective May 6, 1996.]

# Subpart Q—Welding, Cutting and Brazing

> **Authority:** Secs. 4, 6, and 8 of the Occupational Safety and Health Act of 1970 (29 U.S.C. 653, 655, 657); Secretary of Labor's Orders 12-71 (36 FR 8754), 8-76 (41 FR 25059), 9-83 (48 FR 35736), 1-90 (55 FR 9033), or 6-96 (62 FR 111), as applicable; and 29 CFR part 1911.
>
> **Source:** 55 FR 13696, Apr. 11, 1990, unless otherwise noted.

### §1910.251 Definitions.

As used in this subpart:

**1910.251(a)** *Welder* and *welding operator* mean any operator of electric or gas welding and cutting equipment.

**1910.251(b)** *Approved* means listed or approved by a nationally recognized testing laboratory. Refer to §1910.155(c)(3) for definitions of listed and approved, and §1910.7 for nationally recognized testing laboratory.

**1910.251(c)** All other welding terms are used in accordance with American Welding Society—Terms and Definitions—A3.0—1969, which is incorporated by reference as specified in §1910.6.

[55 FR 13696, Apr. 11, 1990, as amended at 61 FR 9240, Mar. 7, 1996]

### §1910.252 General requirements.

**1910.252(a)** *Fire prevention and protection*—**(1)** *Basic precautions.* For elaboration of these basic precautions and of the special precautions of paragraph (a)(2) of this section as well as a delineation of the fire protection and prevention responsibilities of welders and cutters, their supervisors (including outside contractors) and those in management on whose property cutting and welding is to be performed, see Standard for Fire Prevention in Use of Cutting and Welding Processes, NFPA Standard 51B, 1962, which is incorporated by reference as specified in §1910.6. The basic precautions for fire prevention in welding or cutting work are:

**1910.252(a)(1)(i)** *Fire hazards.* If the object to be welded or cut cannot readily be moved, all movable fire hazards in the vicinity shall be taken to a safe place.

**1910.252(a)(1)(ii)** *Guards.* If the object to be welded or cut cannot be moved and if all the fire hazards cannot be removed, then guards shall be used to confine the heat, sparks, and slag, and to protect the immovable fire hazards.

**1910.252(a)(1)(iii)** *Restrictions.* If the requirements stated in paragraphs (a)(1)(i) and (a)(1)(ii) of this section cannot be followed then welding and cutting shall not be performed.

**1910.252(a)(2)** *Special precautions.* When the nature of the work to be performed falls within the scope of paragraph (a)(1)(ii) of this section certain additional precautions may be necessary:

**1910.252(a)(2)(i)** *Combustible material.* Wherever there are floor openings or cracks in the flooring that cannot be closed, precautions shall be taken so that no readily combustible materials on the floor below will be exposed to sparks which might drop through the floor. The same precautions shall be observed with regard to cracks or holes in walls, open doorways and open or broken windows.

**1910.252(a)(2)(ii)** *Fire extinguishers.* Suitable fire extinguishing equipment shall be maintained in a state of readiness for instant use. Such equipment may consist of pails of water, buckets of sand, hose or portable extinguishers depending upon the nature and quantity of the combustible material exposed.

**1910.252(a)(2)(iii)** *Fire watch.* **(A)** Fire watchers shall be required whenever welding or cutting is performed in locations where other than a minor fire might develop, or any of the following conditions exist:

**1910.252(a)(2)(iii)(A)***(1)* Appreciable combustible material, in building construction or contents, closer than 35 feet (10.7 m) to the point of operation.

**1910.252(a)(2)(iii)(A)***(2)* Appreciable combustibles are more than 35 feet (10.7 m) away but are easily ignited by sparks.

**1910.252(a)(2)(iii)(A)***(3)* Wall or floor openings within a 35-foot (10.7 m) radius expose combustible material in adjacent areas including concealed spaces in walls or floors.

**1910.252(a)(2)(iii)(A)***(4)* Combustible materials are adjacent to the opposite side of metal partitions, walls, ceilings, or roofs and are likely to be ignited by conduction or radiation.

**1910.252(a)(2)(iii)(B)** Fire watchers shall have fire extinguishing equipment readily available and be trained in its use. They shall be familiar with facilities for sounding an alarm in the event of a fire. They shall watch for fires in all exposed areas, try to extinguish them only when obviously within the capacity of the equipment available, or otherwise sound the alarm. A fire watch shall be maintained for at least a half hour after completion of welding or cutting operations to detect and extinguish possible smoldering fires.

**1910.252(a)(2)(iv)** *Authorization.* Before cutting or welding is permitted, the area shall be inspected by the individual responsible for authorizing cutting and welding operations. He shall designate precautions to be followed in granting authorization to proceed preferably in the form of a written permit.

**1910.252(a)(2)(v)** *Floors.* Where combustible materials such as paper clippings, wood shavings, or textile fibers are on the floor, the floor shall be swept clean for a radius of 35 feet (10.7 m). Combustible floors shall be kept wet, covered with damp sand, or protected by fire-resistant shields. Where floors have been wet down, personnel operating arc welding or cutting equipment shall be protected from possible shock.

**1910.252(a)(2)(vi)** *Prohibited areas.* Cutting or welding shall not be permitted in the following situations:

**1910.252(a)(2)(vi)(A)** In areas not authorized by management.

**1910.252(a)(2)(vi)(B)** In sprinklered buildings while such protection is impaired.

**1910.252(a)(2)(vi)(C)** In the presence of explosive atmospheres (mixtures of flammable gases, vapors, liquids, or dusts with air), or explosive atmospheres that may develop inside uncleaned or improperly prepared tanks or equipment which have previously contained such materials, or that may develop in areas with an accumulation of combustible dusts.

**1910.252(a)(2)(vi)(D)** In areas near the storage of large quantities of exposed, readily ignitable materials such as bulk sulfur, baled paper, or cotton.

**1910.252(a)(2)(vii)** *Relocation of combustibles.* Where practicable, all combustibles shall be relocated at least 35 feet (10.7 m) from the work site. Where relocation is impracticable, combustibles shall be protected with flameproofed covers or otherwise shielded with metal or asbestos guards or curtains.

**1910.252(a)(2)(viii)** *Ducts.* Ducts and conveyor systems that might carry sparks to distant combustibles shall be suitably protected or shut down.

**1910.252(a)(2)(ix)** *Combustible walls.* Where cutting or welding is done near walls, partitions, ceiling or roof of combustible construction, fire-resistant shields or guards shall be provided to prevent ignition.

**1910.252(a)(2)(x)** *Noncombustible walls.* If welding is to be done on a metal wall, partition, ceiling or roof, precautions shall be taken to prevent ignition of combustibles on the other side, due to conduction or radiation, preferably by relocating combustibles. Where combustibles are not relocated, a fire watch on the opposite side from the work shall be provided.

**1910.252(a)(2)(xi)** *Combustible cover.* Welding shall not be attempted on a metal partition, wall, ceiling or roof having a combustible covering nor on walls or partitions of combustible sandwich-type panel construction.

**1910.252(a)(2)(xii)** *Pipes.* Cutting or welding on pipes or other metal in contact with combustible walls, partitions, ceilings or roofs shall not be undertaken if the work is close enough to cause ignition by conduction.

**1910.252(a)(2)(xiii)** *Management.* Management shall recognize its responsibility for the safe usage of cutting and welding equipment on its property and:

**1910.252(a)(2)(xiii)(A)** Based on fire potentials of plant facilities, establish areas for cutting and welding, and establish procedures for cutting and welding, in other areas.

**1910.252(a)(2)(xiii)(B)** Designate an individual responsible for authorizing cutting and welding operations in areas not specifically designed for such processes.

**1910.252(a)(2)(xiii)(C)** Insist that cutters or welders and their supervisors are suitably trained in the safe operation of their equipment and the safe use of the process.

**1910.252(a)(2)(xiii)(D)** Advise all contractors about flammable materials or hazardous conditions of which they may not be aware.

**1910.252(a)(2)(xiv)** *Supervisor.* The Supervisor:

**Subpart Q—Welding, Cutting and Brazing**

**1910.252(a)(2)(xiv)(A)** Shall be responsible for the safe handling of the cutting or welding equipment and the safe use of the cutting or welding process.

**1910.252(a)(2)(xiv)(B)** Shall determine the combustible materials and hazardous areas present or likely to be present in the work location.

**1910.252(a)(2)(xiv)(C)** Shall protect combustibles from ignition by the following:

**1910.252(a)(2)(xiv)(C)***(1)* Have the work moved to a location free from dangerous combustibles.

**1910.252(a)(2)(xiv)(C)***(2)* If the work cannot be moved, have the combustibles moved to a safe distance from the work or have the combustibles properly shielded against ignition.

**1910.252(a)(2)(xiv)(C)***(3)* See that cutting and welding are so scheduled that plant operations that might expose combustibles to ignition are not started during cutting or welding.

**1910.252(a)(2)(xiv)(D)** Shall secure authorization for the cutting or welding operations from the designated management representative.

**1910.252(a)(2)(xiv)(E)** Shall determine that the cutter or welder secures his approval that conditions are safe before going ahead.

**1910.252(a)(2)(xiv)(F)** Shall determine that fire protection and extinguishing equipment are properly located at the site.

**1910.252(a)(2)(xiv)(G)** Where fire watches are required, he shall see that they are available at the site.

**1910.252(a)(2)(xv)** *Fire prevention precautions.* Cutting or welding shall be permitted only in areas that are or have been made fire safe. When work cannot be moved practically, as in most construction work, the area shall be made safe by removing combustibles or protecting combustibles from ignition sources.

**1910.252(a)(3)** *Welding or cutting containers—*(i) *Used containers.* No welding, cutting, or other hot work shall be performed on used drums, barrels, tanks or other containers until they have been cleaned so thoroughly as to make absolutely certain that there are no flammable materials present or any substances such as greases, tars, acids, or other materials which when subjected to heat, might produce flammable or toxic vapors. Any pipe lines or connections to the drum or vessel shall be disconnected or blanked.

**1910.252(a)(3)(ii)** *Venting and purging.* All hollow spaces, cavities or containers shall be vented to permit the escape of air or gases before preheating, cutting or welding. Purging with inert gas is recommended.

**1910.252(a)(4)** *Confined spaces—*(i) *Accidental contact.* When arc welding is to be suspended for any substantial period of time, such as during lunch or overnight, all electrodes shall be removed from the holders and the holders carefully located so that accidental contact cannot occur and the machine be disconnected from the power source.

**1910.252(a)(4)(ii)** *Torch valve.* In order to eliminate the possibility of gas escaping through leaks or improperly closed valves, when gas welding or cutting, the torch valves shall be closed and the gas supply to the torch positively shut off at some point outside the confined area whenever the torch is not to be used for a substantial period of time, such as during lunch hour or overnight. Where practicable, the torch and hose shall also be removed from the confined space.

**1910.252(b)** *Protection of personnel—*(1) *General—*(i) *Railing.* A welder or helper working on platforms, scaffolds, or runways shall be protected against falling. This may be accomplished by the use of railings, safety belts, life lines, or some other equally effective safeguards.

**1910.252(b)(1)(ii)** *Welding cable.* Welders shall place welding cable and other equipment so that it is clear of passageways, ladders, and stairways.

**1910.252(b)(2)** *Eye protection—*(i) *Selection.* (A) Helmets or hand shields shall be used during all arc welding or arc cutting operations, excluding submerged arc welding. Helpers or attendants shall be provided with proper eye protection.

**1910.252(b)(2)(i)(B)** Goggles or other suitable eye protection shall be used during all gas welding or oxygen cutting operations. Spectacles without side shields, with suitable filter lenses are permitted for use during gas welding operations on light work, for torch brazing or for inspection.

**1910.252(b)(2)(i)(C)** All operators and attendants of resistance welding or resistance brazing equipment shall use transparent face shields or goggles, depending on the particular job, to protect their faces or eyes, as required.

**1910.252(b)(2)(i)(D)** Eye protection in the form of suitable goggles shall be provided where needed for brazing operations not covered in paragraphs (b)(2)(i)(A) through (b)(2)(i)(C) of this section.

**1910.252(b)(2)(ii)** *Specifications for protectors.* (A) Helmets and hand shields shall be made of a material which is an insulator for heat and electricity. Helmets, shields and goggles shall be not readily flammable and shall be capable of withstanding sterilization.

**1910.252(b)(2)(ii)(B)** Helmets and hand shields shall be arranged to protect the face, neck and ears from direct radiant energy from the arc.

**1910.252(b)(2)(ii)(C)** Helmets shall be provided with filter plates and cover plates designed for easy removal.

**1910.252(b)(2)(ii)(D)** All parts shall be constructed of a material which will not readily corrode or discolor the skin.

**1910.252(b)(2)(ii)(E)** Goggles shall be ventilated to prevent fogging of the lenses as much as practicable.

**1910.252(b)(2)(ii)(F)** All glass for lenses shall be tempered, substantially free from striae, air bubbles, waves and other flaws. Except when a lens is ground to provide proper optical correction for defective vision, the front and rear surfaces of lenses and windows shall be smooth and parallel.

**1910.252(b)(2)(ii)(G)** Lenses shall bear some permanent distinctive marking by which the source and shade may be readily identified.

**1910.252(b)(2)(ii)(H)** The following is a guide for the selection of the proper shade numbers. These recommendations may be varied to suit the individual's needs.

| Welding operation | Shade No. |
|---|---|
| Shielded metal-arc welding—1/16-, 3/32-, 1/8-, 5/32-inch electrodes | 10 |
| Gas-shielded arc welding (nonferrous)—1/16-, 3/32-, 1/8-, 5/32-inch electrodes | 11 |
| Gas-shielded arc welding (ferrous)—1/16-, 3/32-, 1/8-, 5/32-inch electrodes | 12 |
| Shielded metal-arc welding: 3/16-, 7/32-, 1/4-inch electrodes | 12 |
| 5/16-, 3/8-inch electrodes | 14 |
| Atomic hydrogen welding | 10-14 |
| Carbon arc welding | 14 |
| Soldering | 2 |
| Torch brazing | 3 or 4 |
| Light cutting, up to 1 inch | 3 or 4 |
| Medium cutting, 1 inch to 6 inches | 4 or 5 |
| Heavy cutting, 6 inches and over | 5 or 6 |
| Gas welding (light) up to 1/8 inch | 4 or 5 |
| Gas welding (medium) 1/8 inch to 1/2 inch | 5 or 6 |
| Gas welding (heavy) 1/2 inch and over | 6 or 8 |

Note: In gas welding or oxygen cutting where the torch produces a high yellow light, it is desirable to use a filter or lens that absorbs the yellow or sodium line in the visible light of the operation.

**1910.252(b)(2)(ii)(I)** All filter lenses and plates shall meet the test for transmission of radiant energy prescribed in ANSI Z87.1—1968—American National Standard Practice for Occupational and Educational Eye and Face Protection, which is incorporated by reference as specified in §1910.6.

**1910.252(b)(2)(iii)** *Protection from arc welding rays.* Where the work permits, the welder should be enclosed in an individual booth painted with a finish of low reflectivity such as zinc oxide (an important factor for absorbing ultraviolet radiations) and lamp black, or shall be enclosed with noncombustible screens similarly painted. Booths and screens shall permit circulation of air at floor level. Workers or other persons adjacent to the welding areas shall be protected from the rays by noncombustible or flameproof screens or shields or shall be required to wear appropriate goggles.

**1910.252(b)(3)** *Protective clothing—General requirements.* Employees exposed to the hazards created by welding, cutting, or brazing operations shall be protected by personal protective equipment in accordance with the requirements of §1910.132 of this part. Appropriate protective clothing required for any welding operation will vary with the size, nature and location of the work to be performed.

**1910.252(b)(4)** *Work in confined spaces—(i) General.* As used herein confined space is intended to mean a relatively small or restricted space such as a tank, boiler, pressure vessel, or small compartment of a ship.

**1910.252(b)(4)(ii)** *Ventilation.* Ventilation is a prerequisite to work in confined spaces. For ventilation requirements see paragraph (c) of this section.

**1910.252(b)(4)(iii)** *Securing cylinders and machinery.* When welding or cutting is being performed in any confined spaces the gas cylinders and welding machines shall be left on the outside. Before operations are started, heavy portable equipment mounted on wheels shall be securely blocked to prevent accidental movement.

**1910.252(b)(4)(iv)** *Lifelines.* Where a welder must enter a confined space through a manhole or other small opening, means shall be provided for quickly removing him in case of emergency. When safety belts and lifelines are used for this purpose they shall be so attached to the welder's body that his body cannot be jammed in a small exit opening. An attendant with a pre-planned rescue procedure shall be stationed outside to observe the welder at all times and be capable of putting rescue operations into effect.

**1910.252(b)(4)(v)** *Electrode removal.* When arc welding is to be suspended for any substantial period of time, such as during lunch or overnight, all electrodes shall be removed from the holders and the holders carefully located so that accidental contact cannot occur and the machine disconnected from the power source.

**1910.252(b)(4)(vi)** *Gas cylinder shutoff.* In order to eliminate the possibility of gas escaping through leaks of improperly closed valves, when gas welding or cutting, the torch valves shall be closed and the fuel-gas and oxygen supply to the torch positively shut off at some point outside the confined area whenever the torch is not to be used for a substantial period of time, such as during lunch hour or overnight. Where practicable the torch and hose shall also be removed from the confined space.

**1910.252(b)(4)(vii)** *Warning sign.* After welding operations are completed, the welder shall mark the hot metal or provide some other means of warning other workers.

**1910.252(c)** *Health protection and ventilation—(1) General—(i) Contamination.* The requirements in this paragraph have been established on the basis of the following three factors in arc and gas welding which govern the amount of contamination to which welders may be exposed:

**1910.252(c)(1)(i)(A)** Dimensions of space in which welding is to be done (with special regard to height of ceiling).

**1910.252(c)(1)(i)(B)** Number of welders.

**1910.252(c)(1)(i)(C)** Possible evolution of hazardous fumes, gases, or dust according to the metals involved.

**1910.252(c)(1)(ii)** *Screens.* When welding must be performed in a space entirely screened on all sides, the screens shall be so arranged that no serious restriction of ventilation exists. It is desirable to have the screens so mounted that they are about 2 feet (0.61 m) above the floor unless the work is performed at so low a level that the screen must be extended nearer to the floor to protect nearby workers from the glare of welding.

**1910.252(c)(1)(iii)** *Maximum allowable concentration.* Local exhaust or general ventilating systems shall be provided and arranged to keep the amount of toxic fumes, gases, or dusts below the maximum allowable concentration as specified in §1910.1000 of this part.

**1910.252(c)(1)(iv)** *Precautionary labels.* A number of potentially hazardous materials are employed in fluxes, coatings, coverings, and filler metals used in welding and cutting or are released to the atmosphere during welding and cutting. These include but are not limited to the materials itemized in paragraphs (c)(5) through (c)(12) of this section. The suppliers of welding materials shall determine the hazard, if any, associated with the use of their materials in welding, cutting, etc.

**1910.252(c)(1)(iv)(A)** All filler metals and fusible granular materials shall carry the following notice, as a minimum, on tags, boxes, or other containers:

**CAUTION**

Welding may produce fumes and gases hazardous to health. Avoid breathing these fumes and gases. Use adequate ventilation. See ANSI Z49.1 - 1967 Safety in Welding and Cutting published by the American Welding Society.

**1910.252(c)(1)(iv)(B)** Brazing (welding) filler metals containing cadmium in significant amounts shall carry the following notice on tags, boxes, or other containers:

**WARNING**

**CONTAINS CADMIUM—**

**POISONOUS FUMES MAY BE FORMED ON HEATING**

Do not breathe fumes. Use only with adequate ventilation such as fume collectors, exhaust ventilators, or air-supplied respirators. See ANSI Z49.1 - 1967. If chest pain, cough, or fever develops after use call physician immediately.

**1910.252(c)(1)(iv)(C)** Brazing and gas welding fluxes containing fluorine compounds shall have a cautionary wording to indicate that they contain fluorine compounds. One such cautionary wording recommended by the American Welding Society for brazing and gas welding fluxes reads as follows:

**CAUTION**

**CONTAINS FLUORIDES**

This flux when heated gives off fumes that may irritate eyes, nose and throat.

1. Avoid fumes—use only in well-ventilated spaces.
2. Avoid contact of flux with eyes or skin.
3. Do not take internally.

**1910.252(c)(2)** *Ventilation for general welding and cutting—*

**1910.252(c)(2)(i)** *General.* Mechanical ventilation shall be provided when welding or cutting is done on metals not covered in paragraphs (c)(5) through (c)(12) of this section. (For specific materials, see the ventilation requirements of paragraphs (c)(5) through (c)(12) of this section.)

**1910.252(c)(2)(i)(A)** In a space of less than 10,000 cubic feet (284 m$^3$) per welder.

**1910.252(c)(2)(i)(B)** In a room having a ceiling height of less than 16 feet (5 m).

**1910.252(c)(2)(i)(C)** In confined spaces or where the welding space contains partitions, balconies, or other structural barriers to the extent that they significantly obstruct cross ventilation.

**1910.252(c)(2)(ii)** *Minimum rate.* Such ventilation shall be at the minimum rate of 2,000 cubic feet (57 m$^3$) per minute per welder, except where local exhaust hoods and booths as per paragraph (c)(3) of this section, or airline respirators approved by the Mine Safety and Health Administration and the National Institute for Occupational Safety and Health, pursuant to the provisions of 30 CFR part 11, are provided. Natural ventilation is considered sufficient for welding or cutting operations where the restrictions in paragraph (c)(2)(i) of this section are not present.

**1910.252(c)(3)** *Local exhaust hoods and booths.* Mechanical local exhaust ventilation may be by means of either of the following:

**1910.252(c)(3)(i)** *Hoods.* Freely movable hoods intended to be placed by the welder as near as practicable to the work being welded and provided with a rate of air-flow sufficient to maintain a velocity in the direction of the hood of 100 linear feet (30 m) per minute in the zone of welding when the hood is at its most remote distance from the point of welding. The rates of ventilation required to accomplish this control velocity using a 3-inch (7.6 cm) wide flanged suction opening are shown in the following table:

| Welding zone | Minimum air flow[1] cubic feet/minute | Duct diameter, inches[2] |
|---|---|---|
| 4 to 6 inches from arc or torch | 150 | 3 |
| 6 to 8 inches from arc or torch | 275 | 3 1/2 |
| 8 to 10 inches from arc or torch | 425 | 4 1/2 |
| 10 to 12 inches from arc or torch | 600 | 5 1/2 |

[1] When brazing with cadmium bearing materials or when cutting on such materials increased rates of ventilation may be required.

[2] Nearest half-inch duct diameter based on 4,000 feet per minute velocity in pipe.

**1910.252(c)(3)(ii)** *Fixed enclosure.* A fixed enclosure with a top and not less than two sides which surround the welding or cutting operations and with a rate of airflow sufficient to maintain a velocity away from the welder of not less than 100 linear feet (30 m) per minute.

**1910.252(c)(4)** *Ventilation in confined spaces*—(i) *Air replacement.* All welding and cutting operations carried on in confined spaces shall be adequately ventilated to prevent the accumulation of toxic materials or possible oxygen deficiency. This applies not only to the welder but also to helpers and other personnel in the immediate vicinity. All air replacing that withdrawn shall be clean and respirable.

**1910.252(c)(4)(ii)** *Airline respirators.* In circumstances for which it is impossible to provide such ventilation, airline respirators or hose masks approved for this purpose by the National Institute for Occupational Safety and Health (NIOSH) under 42 CFR part 84 must be used.

**1910.252(c)(4)(iii)** *Self-contained units.* In areas immediately hazardous to life, a full-facepiece, pressure-demand, self-contained breathing apparatus or a combination full-facepiece, pressure-demand supplied-air respirator with an auxiliary, self-contained air supply approved by NIOSH under 42 CFR part 84 must be used.

**1910.252(c)(4)(iv)** *Outside helper.* Where welding operations are carried on in confined spaces and where welders and helpers are provided with hose masks, hose masks with blowers or self-contained breathing equipment approved by the Mine Safety and Health Administration and the National Institute for Occupational Safety and Health, a worker shall be stationed on the outside of such confined spaces to insure the safety of those working within.

**1910.252(c)(4)(v)** *Oxygen for ventilation.* Oxygen shall never be used for ventilation.

**1910.252(c)(5)** *Fluorine compounds*—(i) *General.* In confined spaces, welding or cutting involving fluxes, coverings, or other materials which contain fluorine compounds shall be done in accordance with paragraph (c)(4) of this section. A fluorine compound is one that contains fluorine, as an element in chemical combination, not as a free gas.

**1910.252(c)(5)(ii)** *Maximum allowable concentration.* The need for local exhaust ventilation or airline respirators for welding or cutting in other than confined spaces will depend upon the individual circumstances. However, experience has shown such protection to be desirable for fixed-location production welding and for all production welding on stainless steels. Where air samples taken at the welding location indicate that the fluorides liberated are below the maximum allowable concentration, such protection is not necessary.

**1910.252(c)(6)** *Zinc*—(i) *Confined spaces.* In confined spaces welding or cutting involving zinc-bearing base or filler metals or metals coated with zinc-bearing materials shall be done in accordance with paragraph (c)(4) of this section.

**1910.252(c)(6)(ii)** *Indoors.* Indoors, welding or cutting involving zinc-bearing base or filler metals coated with zinc-bearing materials shall be done in accordance with paragraph (c)(3) of this section.

**1910.252(c)(7)** *Lead*—(i) *Confined spaces.* In confined spaces, welding involving lead-base metals (erroneously called lead-burning) shall be done in accordance with paragraph (c)(4) of this section.

**1910.252(c)(7)(ii)** *Indoors.* Indoors, welding involving lead-base metals shall be done in accordance with paragraph (c)(3) of this section.

**1910.252(c)(7)(iii)** *Local ventilation.* In confined spaces or indoors, welding or cutting operations involving metals containing lead, other than as an impurity, or metals coated with lead-bearing materials, including paint, must be done using local exhaust ventilation or airline respirators. Such operations, when done outdoors, must be done using respirators approved for this purpose by NIOSH under 42 CFR part 84. In all cases, workers in the immediate vicinity of the cutting operation must be protected by local exhaust ventilation or airline respirators.

**1910.252(c)(8)** *Beryllium.* Welding or cutting indoors, outdoors, or in confined spaces involving beryllium-containing base or filler metals shall be done using local exhaust ventilation and airline respirators unless atmospheric tests under the most adverse conditions have established that the workers' exposure is within the acceptable concentrations defined by §1910.1000 of this part. In all cases, workers in the immediate vicinity of the welding or cutting operations shall be protected as necessary by local exhaust ventilation or airline respirators.

**1910.252(c)(9)** *Cadmium*—(i) *General.* In confined spaces or indoors, welding or cutting operations involving cadmium-bearing or cadmium-coated base metals must be done using local exhaust ventilation or airline respirators unless atmospheric tests under the most adverse conditions show that employee exposure is within the acceptable concentrations specified by 29 CFR 1910.1000. Such operations, when done outdoors, must be done using respirators, such as fume respirators, approved for this purpose by NIOSH under 42 CFR part 84.

**1910.252(c)(9)(ii)** *Confined space.* Welding (brazing) involving cadmium-bearing filler metals shall be done using ventilation as prescribed in paragraph (c)(3) or (c)(4) of this section if the work is to be done in a confined space.

**1910.252(c)(10)** *Mercury.* In confined spaces or indoors, welding or cutting operations involving metals coated with mercury-bearing materials, including paint, must be done using local exhaust ventilation or airline respirators unless atmospheric tests under the most adverse conditions show that employee exposure is within the acceptable concentrations specified by 29 CFR 1910.1000. Such operations, when done outdoors, must be done using respirators approved for this purpose by NIOSH under 42 CFR part 84.

**1910.252(c)(11)** *Cleaning compounds*—(i) *Manufacturer's instructions.* In the use of cleaning materials, because of their possible toxicity or flammability, appropriate precautions such as manufacturers instructions shall be followed.

**1910.252(c)(11)(ii)** *Degreasing.* Degreasing and other cleaning operations involving chlorinated hydrocarbons shall be so located that no vapors from these operations will reach or be drawn into the atmosphere surrounding any welding operation. In addition, trichloroethylene and perchlorethylene should be kept out of atmospheres penetrated by the ultraviolet radiation of gas-shielded welding operations.

**1910.252(c)(12)** *Cutting of stainless steels.* Oxygen cutting, using either a chemical flux or iron powder or gas-shielded arc cutting of stainless steel, shall be done using mechanical ventilation adequate to remove the fumes generated.

**1910.252(c)(13)** *First-aid equipment.* First-aid equipment shall be available at all times. All injuries shall be reported as soon as possible for medical attention. First aid shall be rendered until medical attention can be provided.

**1910.252(d)** **Industrial applications**—(1) *Transmission pipeline*—(i) *General.* The requirements of paragraphs (b) and (c) of this section and §1910.254 of this part shall be observed.

**1910.252(d)(1)(ii)** *Field shop operations.* Where field shop operations are involved for fabrication of fittings, river crossings, road crossings, and pumping and compressor stations the requirements of paragraphs (a), (b), and (c) of this section and §§1910.253 and 1910.254 of this part shall be observed.

**1910.252(d)(1)(iii)** *Electric shock.* When arc welding is performed in wet conditions, or under conditions of high humidity, special protection against electric shock shall be supplied.

**1910.252(d)(1)(iv)** *Pressure testing.* In pressure testing of pipelines, the workers and the public shall be protected against injury by the blowing out of closures or other pressure restraining devices. Also, protection shall be provided against expulsion of loose dirt that may have become trapped in the pipe.

**1910.252(d)(1)(v)** *Construction standards.* The welded construction of transmission pipelines shall be conducted in accordance with the Standard for Welding Pipe Lines and Related Facilities, API Std. 1104—1968, which is incorporated by reference as specified in §1910.6.

**1910.252(d)(1)(vi)** *Flammable substance lines.* The connection, by welding, of branches to pipelines carrying flammable substances shall be performed in accordance with Welding or Hot Tapping on Equipment Containing Flammables, API Std. PSD No. 2201—1963, which is incorporated by reference as specified in §1910.6.

**1910.252(d)(1)(vii)** *X-ray inspection.* The use of X-rays and radioactive isotopes for the inspection of welded pipeline joints shall be carried out in conformance with the American National Standard Safety Standard for Non-Medical X-ray and Sealed Gamma-Ray Sources, ANSI Z54.1—1963, which is incorporated by reference as specified in §1910.6.

**1910.252(d)(2)** *Mechanical piping systems*—(i) *General.* The requirements of paragraphs (a), (b), and (c) of this section and §§1910.253 and 1910.254 of this part shall be observed.

**1910.252(d)(2)(ii)** *X-ray inspection.* The use of X-rays and radioactive isotopes for the inspection of welded piping joints shall be in conformance with the American National Standard Safety Standard for Non-Medical X-ray and Sealed Gamma-Ray Sources, ANSI Z54.1—1963.

[55 FR 13696, Apr. 11, 1990, as amended at 61 FR 9240, Mar. 7, 1996; 63 FR 1284, Jan. 8, 1998]

### §1910.253 Oxygen-fuel gas welding and cutting.

**1910.253(a)** *General requirements*—**(1)** *Flammable mixture.* Mixtures of fuel gases and air or oxygen may be explosive and shall be guarded against. No device or attachment facilitating or permitting mixtures of air or oxygen with flammable gases prior to consumption, except at the burner or in a standard torch, shall be allowed unless approved for the purpose.

**1910.253(a)(2)** *Maximum pressure.* Under no condition shall acetylene be generated, piped (except in approved cylinder manifolds) or utilized at a pressure in excess of 15 psig (103 kPa gauge pressure) or 30 psia (206 kPa absolute. The 30 psia (206 kPa absolute) limit is intended to prevent unsafe use of acetylene in pressurized chambers such as caissons, underground excavations or tunnel construction.) This requirement is not intended to apply to storage of acetylene dissolved in a suitable solvent in cylinders manufactured and maintained according to U.S. Department of Transportation requirements, or to acetylene for chemical use. The use of liquid acetylene shall be prohibited.

**1910.253(a)(3)** *Apparatus.* Only approved apparatus such as torches, regulators or pressure-reducing valves, acetylene generators, and manifolds shall be used.

**1910.253(a)(4)** *Personnel.* Workmen in charge of the oxygen or fuel-gas supply equipment, including generators, and oxygen or fuel-gas distribution piping systems shall be instructed and judged competent by their employers for this important work before being left in charge. Rules and instructions covering the operation and maintenance of oxygen or fuel-gas supply equipment including generators, and oxygen or fuel-gas distribution piping systems shall be readily available.

**1910.253(b)** *Cylinders and containers*—**(1)** *Approval and marking.* **(i)** All portable cylinders used for the storage and shipment of compressed gases shall be constructed and maintained in accordance with the regulations of the U.S. Department of Transportation, 49 CFR parts 171-179.

**1910.253(b)(1)(ii)** Compressed gas cylinders shall be legibly marked, for the purpose of identifying the gas content, with either the chemical or the trade name of the gas. Such marking shall be by means of stenciling, stamping, or labeling, and shall not be readily removable. Whenever practical, the marking shall be located on the shoulder of the cylinder. This method conforms to the American National Standard Method for Marking Portable Compressed Gas Containers to Identify the Material Contained, ANSI Z48.1—1954, which is incorporated by reference as specified in §1910.6.

**1910.253(b)(1)(iii)** Compressed gas cylinders shall be equipped with connections complying with the American National Standard Compressed Gas Cylinder Valve Outlet and Inlet Connections, ANSI B57.1—1965, which is incorporated by reference as specified in §1910.6.

**1910.253(b)(1)(iv)** All cylinders with a water weight capacity of over 30 pounds (13.6 kg) shall be equipped with means of connecting a valve protection cap or with a collar or recess to protect the valve.

**1910.253(b)(2)** *Storage of cylinders—general.* **(i)** Cylinders shall be kept away from radiators and other sources of heat.

**1910.253(b)(2)(ii)** Inside of buildings, cylinders shall be stored in a well-protected, well-ventilated, dry location, at least 20 feet (6.1 m) from highly combustible materials such as oil or excelsior. Cylinders should be stored in definitely assigned places away from elevators, stairs, or gangways. Assigned storage spaces shall be located where cylinders will not be knocked over or damaged by passing or falling objects, or subject to tampering by unauthorized persons. Cylinders shall not be kept in unventilated enclosures such as lockers and cupboards.

**1910.253(b)(2)(iii)** Empty cylinders shall have their valves closed.

**1910.253(b)(2)(iv)** Valve protection caps, where cylinder is designed to accept a cap, shall always be in place, hand-tight, except when cylinders are in use or connected for use.

**1910.253(b)(3)** *Fuel-gas cylinder storage.* Inside a building, cylinders, except those in actual use or attached ready for use, shall be limited to a total gas capacity of 2,000 cubic feet (56 m³) or 300 pounds (135.9 kg) of liquefied petroleum gas.

**1910.253(b)(3)(i)** For storage in excess of 2,000 cubic feet (56 m³) total gas capacity of cylinders or 300 (135.9 kg) pounds of liquefied petroleum gas, a separate room or compartment conforming to the requirements specified in paragraphs (f)(6)(i)(H) and (f)(6)(i)(I) of this section shall be provided, or cylinders shall be kept outside or in a special building. Special buildings, rooms or compartments shall have no open flame for heating or lighting and shall be well ventilated. They may also be used for storage of calcium carbide in quantities not to exceed 600 (271.8 kg) pounds, when contained in metal containers complying with paragraphs (g)(1)(i) and (g)(1)(ii) of this section.

**1910.253(b)(3)(ii)** Acetylene cylinders shall be stored valve end up.

**1910.253(b)(4)** *Oxygen storage.* **(i)** Oxygen cylinders shall not be stored near highly combustible material, especially oil and grease; or near reserve stocks of carbide and acetylene or other fuel-gas cylinders, or near any other substance likely to cause or accelerate fire; or in an acetylene generator compartment.

**1910.253(b)(4)(ii)** Oxygen cylinders stored in outside generator houses shall be separated from the generator or carbide storage rooms by a noncombustible partition having a fire-resistance rating of at least 1 hour. This partition shall be without openings and shall be gastight.

**1910.253(b)(4)(iii)** Oxygen cylinders in storage shall be separated from fuel-gas cylinders or combustible materials (especially oil or grease), a minimum distance of 20 feet (6.1 m) or by a noncombustible barrier at least 5 feet (1.5 m) high having a fire-resistance rating of at least one-half hour.

**1910.253(b)(4)(iv)** Where a liquid oxygen system is to be used to supply gaseous oxygen for welding or cutting and the system has a storage capacity of more than 13,000 cubic feet (364 m³) of oxygen (measured at 14.7 psia (101 kPa) and 70 °F (21.1 °C)), connected in service or ready for service, or more than 25,000 cubic feet (700 m³) of oxygen (measured at 14.7 psia (101 kPa) and 70 °F (21.1 °C)), including unconnected reserves on hand at the site, it shall comply with the provisions of the Standard for Bulk Oxygen Systems at Consumer Sites, NFPA No. 566—1965, which is incorporated by reference as specified in §1910.6.

**1910.253(b)(5)** *Operating procedures.* **(i)** Cylinders, cylinder valves, couplings, regulators, hose, and apparatus shall be kept free from oily or greasy substances. Oxygen cylinders or apparatus shall not be handled with oily hands or gloves. A jet of oxygen must never be permitted to strike an oily surface, greasy clothes, or enter a fuel oil or other storage tank.

**1910.253(b)(5)(ii)(A)** When transporting cylinders by a crane or derrick, a cradle, boat, or suitable platform shall be used. Slings or electric magnets shall not be used for this purpose. Valve-protection caps, where cylinder is designed to accept a cap, shall always be in place.

**1910.253(b)(5)(ii)(B)** Cylinders shall not be dropped or struck or permitted to strike each other violently.

**1910.253(b)(5)(ii)(C)** Valve-protection caps shall not be used for lifting cylinders from one vertical position to another. Bars shall not be used under valves or valve-protection caps to pry cylinders loose when frozen to the ground or otherwise fixed; the use of warm (not boiling) water is recommended. Valve-protection caps are designed to protect cylinder valves from damage.

**1910.253(b)(5)(ii)(D)** Unless cylinders are secured on a special truck, regulators shall be removed and valve-protection caps, when provided for, shall be put in place before cylinders are moved.

**1910.253(b)(5)(ii)(E)** Cylinders not having fixed hand wheels shall have keys, handles, or nonadjustable wrenches on valve stems while these cylinders are in service. In multiple cylinder installations only one key or handle is required for each manifold.

**1910.253(b)(5)(ii)(F)** Cylinder valves shall be closed before moving cylinders.

**1910.253(b)(5)(ii)(G)** Cylinder valves shall be closed when work is finished.

**1910.253(b)(5)(ii)(H)** Valves of empty cylinders shall be closed.

**1910.253(b)(5)(ii)(I)** Cylinders shall be kept far enough away from the actual welding or cutting operation so that sparks, hot slag, or flame will not reach them, or fire-resistant shields shall be provided.

**1910.253(b)(5)(ii)(J)** Cylinders shall not be placed where they might become part of an electric circuit. Contacts with third rails, trolley wires, etc., shall be avoided. Cylinders shall be kept away from radiators, piping systems, layout tables, etc., that may be used for grounding electric circuits such as for arc welding machines. Any practice such as the tapping of an electrode against a cylinder to strike an arc shall be prohibited.

**1910.253(b)(5)(ii)(K)** Cylinders shall never be used as rollers or supports, whether full or empty.

**1910.253(b)(5)(ii)(L)** The numbers and markings stamped into cylinders shall not be tampered with.

**1910.253(b)(5)(ii)(M)** No person, other than the gas supplier, shall attempt to mix gases in a cylinder. No one, except the owner of the cylinder or person authorized by him, shall refill a cylinder.

**1910.253(b)(5)(ii)(N)** No one shall tamper with safety devices in cylinders or valves.

**1910.253(b)(5)(ii)(O)** Cylinders shall not be dropped or otherwise roughly handled.

**1910.253(b)(5)(ii)(P)** Unless connected to a manifold, oxygen from a cylinder shall not be used without first attaching an oxygen regulator to the cylinder valve. Before connecting the regulator to the cylinder valve, the valve shall be opened slightly for an instant and then closed. Always stand to one side of the outlet when opening the cylinder valve.

**1910.253(b)(5)(ii)(Q)** A hammer or wrench shall not be used to open cylinder valves. If valves cannot be opened by hand, the supplier shall be notified.

**1910.253(b)(5)(ii)(R)(1)** Cylinder valves shall not be tampered with nor should any attempt be made to repair them. If trouble is experienced, the supplier should be sent a report promptly indicating the character of the trouble and the cylinder's serial number. Supplier's instructions as to its disposition shall be followed.

**1910.253(b)(5)(ii)(R)(2)** Complete removal of the stem from a diaphragm-type cylinder valve shall be avoided.

**1910.253(b)(5)(iii)(A)** Fuel-gas cylinders shall be placed with valve end up whenever they are in use. Liquefied gases shall be stored and shipped with the valve end up.

**1910.253(b)(5)(iii)(B)** Cylinders shall be handled carefully. Rough handling, knocks, or falls are liable to damage the cylinder, valve or safety devices and cause leakage.

**1910.253(b)(5)(iii)(C)** Before connecting a regulator to a cylinder valve, the valve shall be opened slightly and closed immediately. The valve shall be opened while standing to one side of the outlet; never in front of it. Never crack a fuel-gas cylinder valve near other welding work or near sparks, flame, or other possible sources of ignition.

**1910.253(b)(5)(iii)(D)** Before a regulator is removed from a cylinder valve, the cylinder valve shall be closed and the gas released from the regulator.

**1910.253(b)(5)(iii)(E)** Nothing shall be placed on top of an acetylene cylinder when in use which may damage the safety device or interfere with the quick closing of the valve.

**1910.253(b)(5)(iii)(F)** If cylinders are found to have leaky valves or fittings which cannot be stopped by closing of the valve, the cylinders shall be taken outdoors away from sources of ignition and slowly emptied.

**1910.253(b)(5)(iii)(G)** A warning should be placed near cylinders having leaking fuse plugs or other leaking safety devices not to approach them with a lighted cigarette or other source of ignition. Such cylinders should be plainly tagged; the supplier should be promptly notified and his instructions followed as to their return.

**1910.253(b)(5)(iii)(H)** Safety devices shall not be tampered with.

**1910.253(b)(5)(iii)(I)** Fuel-gas shall never be used from cylinders through torches or other devices equipped with shutoff valves without reducing the pressure through a suitable regulator attached to the cylinder valve or manifold.

**1910.253(b)(5)(iii)(J)** The cylinder valve shall always be opened slowly.

**1910.253(b)(5)(iii)(K)** An acetylene cylinder valve shall not be opened more than one and one-half turns of the spindle, and preferably no more than three-fourths of a turn.

**1910.253(b)(5)(iii)(L)** Where a special wrench is required it shall be left in position on the stem of the valve while the cylinder is in use so that the fuel-gas flow can be quickly turned off in case of emergency. In the case of manifolded or coupled cylinders at least one such wrench shall always be available for immediate use.

**1910.253(c)** *Manifolding of cylinders*—(1) *Fuel-gas manifolds.* (i) Manifolds shall be approved either separately for each component part or as an assembled unit.

**1910.253(c)(1)(ii)** Except as provided in paragraph (c)(1)(iii) of this section fuel-gas cylinders connected to one manifold inside a building shall be limited to a total capacity not exceeding 300 pounds (135.9 kg) of liquefied petroleum gas or 3,000 cubic feet (84 m$^3$) of other fuel-gas. More than one such manifold with connected cylinders may be located in the same room provided the manifolds are at least 50 feet (15 m) apart or separated by a noncombustible barrier at least 5 feet (1.5 m) high having a fire-resistance rating of at least one-half hour.

**1910.253(c)(1)(iii)** Fuel-gas cylinders connected to one manifold having an aggregate capacity exceeding 300 pounds (135.9 kg) of liquefied petroleum gas or 3,000 cubic feet (84 m$^3$) of other fuel-gas shall be located outdoors, or in a separate building or room constructed in accordance with paragraphs (f)(6)(i)(H) and (f)(6)(i)(I) of this section.

**1910.253(c)(1)(iv)** Separate manifold buildings or rooms may also be used for the storage of drums of calcium carbide and cylinders containing fuel gases as provided in paragraph (b)(3) of this section. Such buildings or rooms shall have no open flames for heating or lighting and shall be well-ventilated.

**1910.253(c)(1)(v)** High-pressure fuel-gas manifolds shall be provided with approved pressure regulating devices.

**1910.253(c)(2)** *High-pressure oxygen manifolds (for use with cylinders having a Department of Transportation service pressure above 200 psig (1.36 MPa)).* (i) Manifolds shall be approved either separately for each component part or as an assembled unit.

**1910.253(c)(2)(ii)** Oxygen manifolds shall not be located in an acetylene generator room. Oxygen manifolds shall be separated from fuel-gas cylinders or combustible materials (especially oil or grease), a minimum distance of 20 feet (6.1 m) or by a noncombustible barrier at least 5 feet (1.5 m) high having a fire-resistance rating of at least one-half hour.

**1910.253(c)(2)(iii)** Except as provided in paragraph (c)(2)(iv) of this section, oxygen cylinders connected to one manifold shall be limited to a total gas capacity of 6,000 cubic feet (168 m$^3$). More than one such manifold with connected cylinders may be located in the same room provided the manifolds are at least 50 feet (15 m) apart or separated by a noncombustible barrier at least 5 feet (1.5 m) high having a fire-resistance rating of at least one-half hour.

**1910.253(c)(2)(iv)** An oxygen manifold, to which cylinders having an aggregate capacity of more than 6,000 cubic feet (168 m$^3$) of oxygen are connected, should be located outdoors or in a separate noncombustible building. Such a manifold, if located inside a building having other occupancy, shall be located in a separate room of noncombustible construction having a fire-resistance rating of at least one-half hour or in an area with no combustible material within 20 feet (6.1 m) of the manifold.

**1910.253(c)(2)(v)** An oxygen manifold or oxygen bulk supply system which has storage capacity of more than 13,000 cubic feet (364 m$^3$) of oxygen (measured at 14.7 psia (101 kPa) and 70 °F (21.1 °C)), connected in service or ready for service, or more than 25,000 cubic feet (700 m$^3$) of oxygen (measured at 14.7 psia (101 kPa) and 70 °F (21.1 °C)), including unconnected reserves on hand at the site, shall comply with the provisions of the Standard for Bulk Oxygen Systems at Consumer Sites, NFPA No. 566-1965.

**1910.253(c)(2)(vi)** High-pressure oxygen manifolds shall be provided with approved pressure-regulating devices.

**1910.253(c)(3)** *Low-pressure oxygen manifolds (for use with cylinders having a Department of Transportation service pressure not exceeding 200 psig (1.36 MPa)).* (i) Manifolds shall be of substantial construction suitable for use with oxygen at a pressure of 250 psig (1.7 MPa). They shall have a minimum bursting pressure of 1,000 psig (6.8 MPa) and shall be protected by a safety relief device which will relieve at a maximum pressure of 500 psig (3.4 MPa). DOT-4L21910.251(c)1910.252(d)1910.253(c)00 cylinders have safety devices which relieve at a maximum pressure of 250 psig (1.7 MPa) (or 235 psig (1.6 MPa) if vacuum insulation is used).

**1910.253(c)(3)(ii)** Hose and hose connections subject to cylinder pressure shall comply with paragraph (e)(5) of this section. Hose shall have a minimum bursting pressure of 1,000 psig (6.8 MPa).

**1910.253(c)(3)(iii)** The assembled manifold including leads shall be tested and proven gas-tight at a pressure of 300 psig (2.04 MPa). The fluid used for testing oxygen manifolds shall be oil-free and not combustible.

**1910.253(c)(3)(iv)** The location of manifolds shall comply with paragraphs (c)(2)(ii), (c)(2)(iii), (c)(2)(iv), and (c)(2)(v) of this section.

**1910.253(c)(3)(v)** The following sign shall be conspicuously posted at each manifold:

# OSHA Standards for General Industry

**Low-Pressure Manifold**

**Do Not Connect High-Pressure Cylinders**

**Maximum Pressure—250 psig (1.7 MPa)**

**1910.253(c)(4)** *Portable outlet headers.* **(i)** Portable outlet headers shall not be used indoors except for temporary service where the conditions preclude a direct supply from outlets located on the service piping system.

**1910.253(c)(4)(ii)** Each outlet on the service piping from which oxygen or fuel-gas is withdrawn to supply a portable outlet header shall be equipped with a readily accessible shutoff valve.

**1910.253(c)(4)(iii)** Hose and hose connections used for connecting the portable outlet header to the service piping shall comply with paragraph (e)(5) of this section.

**1910.253(c)(4)(iv)** Master shutoff valves for both oxygen and fuel-gas shall be provided at the entry end of the portable outlet header.

**1910.253(c)(4)(v)** Portable outlet headers for fuel-gas service shall be provided with an approved hydraulic back-pressure valve installed at the inlet and preceding the service outlets, unless an approved pressure-reducing regulator, an approved back-flow check valve, or an approved hydraulic back-pressure valve is installed at each outlet. Outlets provided on headers for oxygen service may be fitted for use with pressure-reducing regulators or for direct hose connection.

**1910.253(c)(4)(vi)** Each service outlet on portable outlet headers shall be provided with a valve assembly that includes a detachable outlet seal cap, chained or otherwise attached to the body of the valve.

**1910.253(c)(4)(vii)** Materials and fabrication procedures for portable outlet headers shall comply with paragraphs (d)(1), (d)(2), and (d)(5) of this section.

**1910.253(c)(4)(viii)** Portable outlet headers shall be provided with frames which will support the equipment securely in the correct operating position and protect them from damage during handling and operation.

**1910.253(c)(5)** *Manifold operation procedures.* **(i)** Cylinder manifolds shall be installed under the supervision of someone familiar with the proper practices with reference to their construction and use.

**1910.253(c)(5)(ii)** All manifolds and parts used in methods of manifolding shall be used only for the gas or gases for which they are approved.

**1910.253(c)(5)(iii)** When acetylene cylinders are coupled, approved flash arresters shall be installed between each cylinder and the coupler block. For outdoor use only, and when the number of cylinders coupled does not exceed three, one flash arrester installed between the coupler block and regulator is acceptable.

**1910.253(c)(5)(iv)** The aggregate capacity of fuel-gas cylinders connected to a portable manifold inside a building shall not exceed 3,000 cubic feet (84 m$^3$) of gas.

**1910.253(c)(5)(v)** Acetylene and liquefied fuel-gas cylinders shall be manifolded in a vertical position.

**1910.253(c)(5)(vi)** The pressure in the gas cylinders connected to and discharged simultaneously through a common manifold shall be approximately equal.

**1910.253(d)** *Service piping systems*—**(1)** *Materials and design.* **(i)(A)** Piping and fittings shall comply with section 2, Industrial Gas and Air Piping Systems, of the American National Standard Code for Pressure Piping ANSI B31.1, 1967, which is incorporated by reference as specified in §1910.6, insofar as it does not conflict with paragraphs (d)(1)(i)(A)(*1*) and (d)(1)(i)(A)(*2*) of this section:

**1910.253(d)(1)(i)(A)***(1)* Pipe shall be at least Schedule 40 and fittings shall be at least standard weight in sizes up to and including 6-inch nominal.

**1910.253(d)(1)(i)(A)***(2)* Copper tubing shall be Types K or L in accordance with the Standard Specification for Seamless Copper Water Tube, ASTM B88-66a, which is incorporated by reference as specified in §1910.6.

**1910.253(d)(1)(i)(B)** Piping shall be steel, wrought iron, brass or copper pipe, or seamless copper, brass or stainless steel tubing, except as provided in paragraphs (d)(1)(ii) and (d)(1)(iii) of this section.

**1910.253(d)(1)(ii)(A)** Oxygen piping and fittings at pressures in excess of 700 psi (4.8 MPa), shall be stainless steel or copper alloys.

**1910.253(d)(1)(ii)(B)** Hose connections and hose complying with paragraph (e)(5) of this section may be used to connect the outlet of a manifold pressure regulator to piping providing the working pressure of the piping is 250 psi (1.7 MPa) or less and the length of the hose does not exceed 5 feet (1.5 m). Hose shall have a minimum bursting pressure of 1,000 psig (6.8 MPa).

**1910.253(d)(1)(ii)(C)** When oxygen is supplied to a service piping system from a low-pressure oxygen manifold without an intervening pressure regulating device, the piping system shall have a minimum design pressure of 250 psig (1.7 MPa). A pressure regulating device shall be used at each station outlet when the connected equipment is for use at pressures less than 250 psig (1.7 MPa).

**1910.253(d)(1)(iii)(A)** Piping for acetylene or acetylenic compounds shall be steel or wrought iron.

**1910.253(d)(1)(iii)(B)** Unalloyed copper shall not be used for acetylene or acetylenic compounds except in listed equipment.

**1910.253(d)(2)** *Piping joints.* **(i)** Joints in steel or wrought iron piping shall be welded, threaded or flanged. Fittings, such as ells, tees, couplings, and unions, may be rolled, forged or cast steel, malleable iron or nodular iron. Gray or white cast iron fittings are prohibited.

**1910.253(d)(2)(ii)** Joints in brass or copper pipe shall be welded, brazed, threaded, or flanged. If of the socket type, they shall be brazed with silver-brazing alloy or similar high melting point (not less than 800 °F (427 °C)) filler metal.

**1910.253(d)(2)(iii)** Joints in seamless copper, brass, or stainless steel tubing shall be approved gas tubing fittings or the joints shall be brazed. If of the socket type, they shall be brazed with silver-brazing alloy or similar high melting point (not less than 800 °F (427 °C) filler metal.

**1910.253(d)(3)** *Installation.* **(i)** Distribution lines shall be installed and maintained in a safe operating condition.

**1910.253(d)(3)(ii)** All piping shall be run as directly as practicable, protected against physical damage, proper allowance being made for expansion and contraction, jarring and vibration. Pipe laid underground in earth shall be located below the frost line and protected against corrosion. After assembly, piping shall be thoroughly blown out with air, nitrogen, or carbon dioxide to remove foreign materials. For oxygen piping, only oil-free air, oil-free nitrogen, or oil-free carbon dioxide shall be used.

**1910.253(d)(3)(iii)** Only piping which has been welded or brazed shall be installed in tunnels, trenches or ducts. Shutoff valves shall be located outside such conduits. Oxygen piping may be placed in the same tunnel, trench or duct with fuel-gas pipelines, provided there is good natural or forced ventilation.

**1910.253(d)(3)(iv)** Low points in piping carrying moist gas shall be drained into drip pots constructed so as to permit pumping or draining out the condensate at necessary intervals. Drain valves shall be installed for this purpose having outlets normally closed with screw caps or plugs. No open end valves or petcocks shall be used, except that in drips located out of doors, underground, and not readily accessible, valves may be used at such points if they are equipped with means to secure them in the closed position. Pipes leading to the surface of the ground shall be cased or jacketed where necessary to prevent loosening or breaking.

**1910.253(d)(3)(v)** Gas cocks or valves shall be provided for all buildings at points where they will be readily accessible for shutting off the gas supply to these buildings in any emergency. There shall also be provided a shutoff valve in the discharge line from the generator, gas holder, manifold or other source of supply.

**1910.253(d)(3)(vi)** Shutoff valves shall not be installed in safety relief lines in such a manner that the safety relief device can be rendered ineffective.

**1910.253(d)(3)(vii)** Fittings and lengths of pipe shall be examined internally before assembly and, if necessary freed from scale or dirt. Oxygen piping and fittings shall be washed out with a suitable solution which will effectively remove grease and dirt but will not react with oxygen. Hot water solutions of caustic soda or trisodium phosphate are effective cleaning agents for this purpose.

**1910.253(d)(3)(viii)** Piping shall be thoroughly blown out after assembly to remove foreign materials. For oxygen piping, oil-free air, oil-free nitrogen, or oil-free carbon dioxide shall be used. For other piping, air or inert gas may be used.

**1910.253(d)(3)(ix)** When flammable gas lines or other parts of equipment are being purged of air or gas, open lights or other sources of ignition shall not be permitted near uncapped openings.

**1910.253(d)(3)(x)** No welding or cutting shall be performed on an acetylene or oxygen pipeline, including the attachment of hangers or supports, until the line has been purged. Only oil-free air, oil-free nitrogen, or oil-free carbon dioxide shall be used to purge oxygen lines.

**1910.253(d)(4)** *Painting and signs.* **(i)** Underground pipe and tubing and outdoor ferrous pipe and tubing shall be covered or painted with a suitable material for protection against corrosion.

**1910.253(d)(4)(ii)** Aboveground piping systems shall be marked in accordance with the American National Standard Scheme for the Identification of Piping Systems, ANSI A13.1-1956, which is incorporated by reference as specified in §1910.6.

**1910.253(d)(4)(iii)** Station outlets shall be marked to indicate the name of the gas.

**1910.253(d)(5)** *Testing.* **(i)** Piping systems shall be tested and proved gastight at 1 1/2 times the maximum operating pressure, and shall be thoroughly purged of air before being placed in service. The material used for testing oxygen lines shall be oil free and noncombustible. Flames shall not be used to detect leaks.

**1910.253(d)(5)(ii)** When flammable gas lines or other parts of equipment are being purged of air or gas, sources of ignition shall not be permitted near uncapped openings.

**1910.253(e)** *Protective equipment, hose, and regulators*—**(1)** *General.* Equipment shall be installed and used only in the service for which it is approved and as recommended by the manufacturer.

**1910.253(e)(2)** *Pressure relief devices.* Service piping systems shall be protected by pressure relief devices set to function at not more than the design pressure of the systems and discharging upwards to a safe location.

**1910.253(e)(3)** *Piping protective equipment.* **(i)** The fuel-gas and oxygen piping systems, including portable outlet headers shall incorporate the protective equipment shown in Figures Q-1, Q-2, and Q-3. When only a portion of a fuel-gas system is to be used with oxygen, only that portion need comply with this paragraph (e)(3)(i).

FIG. Q-1   FIG. Q-2   FIG. Q-3

LEGEND
$P_F$ — Protective equipment in fuel gas piping
$V_F$ — Fuel gas station outlet valve
$V_O$ — Oxygen station outlet valve
$S_F$ — Backflow prevention device(s) at fuel gas station outlet
$S_O$ — Backflow prevention device(s) at oxygen station outlet

**1910.253(e)(3)(ii)** Approved protective equipment (designated $P_F$ in Figures Q-1, Q-2, and Q-3) shall be installed in fuel-gas piping to prevent:

**1910.253(e)(3)(ii)(A)** Backflow of oxygen into the fuel-gas supply system;

**1910.253(e)(3)(ii)(B)** Passage of a flash back into the fuel-gas supply system; and

**1910.253(e)(3)(ii)(C)** Excessive back pressure of oxygen in the fuel-gas supply system. The three functions of the protective equipment may be combined in one device or may be provided by separate devices.

**1910.253(e)(3)(ii)(C)(1)** The protective equipment shall be located in the main supply line, as in Figure Q-1 or at the head of each branch line, as in Figure Q-2 or at each location where fuel-gas is withdrawn, as in Figure Q-3. Where branch lines are of 2-inch pipe size or larger or of substantial length, protective equipment (designated as $P_F$) shall be located as shown in either Q-2 and Q-3.

**1910.253(e)(3)(ii)(C)(2)** Backflow protection shall be provided by an approved device that will prevent oxygen from flowing into the fuel-gas system or fuel from flowing into the oxygen system (see $S_F$, Figures Q-1 and Q-2).

**1910.253(e)(3)(ii)(C)(3)** Flash-back protection shall be provided by an approved device that will prevent flame from passing into the fuel-gas system.

**1910.253(e)(3)(ii)(C)(4)** Back-pressure protection shall be provided by an approved pressure-relief device set at a pressure not greater than the pressure rating of the backflow or the flashback protection device, whichever is lower. The pressure-relief device shall be located on the downstream side of the backflow and flashback protection devices. The vent from the pressure-relief device shall be at least as large as the relief device inlet and shall be installed without low points that may collect moisture. If low points are unavoidable, drip pots with drains closed with screw plugs or caps shall be installed at the low points. The vent terminus shall not endanger personnel or property through gas discharge; shall be located away from ignition sources; and shall terminate in a hood or bend.

**1910.253(e)(3)(iii)** If pipeline protective equipment incorporates a liquid, the liquid level shall be maintained, and a suitable antifreeze may be used to prevent freezing.

**1910.253(e)(3)(iv)** Fuel gas for use with equipment not requiring oxygen shall be withdrawn upstream of the piping protective devices.

**1910.253(e)(4)** *Station outlet protective equipment.* **(i)** A check valve, pressure regulator, hydraulic seal, or combination of these devices shall be provided at each station outlet, including those on portable headers, to prevent backflow, as shown in Figures Q-1, Q-2, and Q-3 and designated as $S_F$ and $S_O$.

**1910.253(e)(4)(ii)** When approved pipeline protective equipment (designated $P_F$) is located at the station outlet as in Figure Q-3, no additional check valve, pressure regulator, or hydraulic seal is required.

**1910.253(e)(4)(iii)** A shutoff valve (designated $V_F$ and $V_O$) shall be installed at each station outlet and shall be located on the upstream side of other station outlet equipment.

**1910.253(e)(4)(iv)** If the station outlet is equipped with a detachable regulator, the outlet shall terminate in a union connection that complies with the Regulator Connection Standards, 1958, Compressed Gas Association, which is incorporated by reference as specified in §1910.6.

**1910.253(e)(4)(v)** If the station outlet is connected directly to a hose, the outlet shall terminate in a union connection complying with the Standard Hose Connection Specifications, 1957, Compressed Gas Association, which is incorporated by reference as specified in §1910.6.

**1910.253(e)(4)(vi)** Station outlets may terminate in pipe threads to which permanent connections are to be made, such as to a machine.

**1910.253(e)(4)(vii)** Station outlets shall be equipped with a detachable outlet seal cap secured in place. This cap shall be used to seal the outlet except when a hose, a regulator, or piping is attached.

**1910.253(e)(4)(viii)** Where station outlets are equipped with approved backflow and flashback protective devices, as many as four torches may be supplied from one station outlet through rigid piping, provided each outlet from such piping is equipped with a shutoff valve and provided the fuel-gas capacity of any one torch does not exceed 15 cubic feet (0.42 m$^3$) per hour. This paragraph (e)(4)(viii) does not apply to machines.

**1910.253(e)(5)** *Hose and hose connections.* **(i)** Hose for oxy-fuel gas service shall comply with the Specification for Rubber Welding Hose, 1958, Compressed Gas Association and Rubber Manufacturers Association, which is incorporated by reference as specified in §1910.6.

**1910.253(e)(5)(ii)** When parallel lengths of oxygen and acetylene hose are taped together for convenience and to prevent tangling, not more than 4 inches (10.2 cm) out of 12 inches (30.5 cm) shall be covered by tape.

**1910.253(e)(5)(iii)** Hose connections shall comply with the Standard Hose Connection Specifications, 1957, Compressed Gas Association.

**1910.253(e)(5)(iv)** Hose connections shall be clamped or otherwise securely fastened in a manner that will withstand, without leakage, twice the pressure to which they are normally subjected in service, but in no case less than a pressure of 300 psi (2.04 MPa). Oil-free air or an oil-free inert gas shall be used for the test.

**1910.253(e)(5)(v)** Hose showing leaks, burns, worn places, or other defects rendering it unfit for service shall be repaired or replaced.

**1910.253(e)(6)** *Pressure-reducing regulators.* **(i)** Pressure-reducing regulators shall be used only for the gas and pressures for which they are intended. The regulator inlet connections shall comply with Regulator Connection Standards, 1958, Compressed Gas Association.

**1910.253(e)(6)(ii)** When regulators or parts of regulators, including gages, need repair, the work shall be performed by skilled mechanics who have been properly instructed.

**1910.253(e)(6)(iii)** Gages on oxygen regulators shall be marked "USE NO OIL."

**1910.253(e)(6)(iv)** Union nuts and connections on regulators shall be inspected before use to detect faulty seats which may cause leakage of gas when the regulators are attached to the cylinder valves.

**1910.253(f)** *Acetylene generators*—**(1)** *Approval and marking.* **(i)** Generators shall be of approved construction and shall be plainly marked with the maximum rate of acetylene in cubic feet per hour for which they are designed; the weight and size of carbide necessary for a single charge; the manufacturer's name and address; and the name or number of the type of generator.

**1910.253(f)(1)(ii)** Carbide shall be of the size marked on the generator nameplate.

**1910.253(f)(2)** *Rating and pressure limitations.* **(i)** The total hourly output of a generator shall not exceed the rate for which it is approved and marked. Unless specifically approved for higher ratings, carbide-feed generators shall be rated at 1 cubic foot (0.028 m³) per hour per pound of carbide required for a single complete charge.

**1910.253(f)(2)(ii)** Relief valves shall be regularly operated to insure proper functioning. Relief valves for generating chambers shall be set to open at a pressure not in excess of 15 psig (103 kPa gauge pressure). Relief valves for hydraulic back pressure valves shall be set to open at a pressure not in excess of 20 psig (137 kPa gauge pressure).

**1910.253(f)(2)(iii)** Nonautomatic generators shall not be used for generating acetylene at pressures exceeding l psig (7 kPa gauge pressure), and all water overflows shall be visible.

**1910.253(f)(3)** *Location.* The space around the generator shall be ample for free, unobstructed operation and maintenance and shall permit ready adjustment and charging.

**1910.253(f)(4)** *Stationary acetylene generators (automatic and nonautomatic).* **(i)(A)** The foundation shall be so arranged that the generator will be level and so that no excessive strain will be placed on the generator or its connections. Acetylene generators shall be grounded.

**1910.253(f)(4)(i)(B)** Generators shall be placed where water will not freeze. The use of common salt (sodium chloride) or other corrosive chemicals for protection against freezing is not permitted. (For heating systems see paragraph (f)(6)(iii) of this section.)

**1910.253(f)(4)(i)(C)** Except when generators are prepared in accordance with paragraph (f)(7)(v) of this section, sources of ignition shall be prohibited in outside generator houses or inside generator rooms.

**1910.253(f)(4)(i)(D)** Water shall not be supplied through a continuous connection to the generator except when the generator is provided with an adequate open overflow or automatic water shutoff which will effectively prevent overfilling of the generator. Where a noncontinuous connection is used, the supply line shall terminate at a point not less than 2 inches (5 cm) above the regularly provided opening for filling so that the water can be observed as it enters the generator.

**1910.253(f)(4)(i)(E)** Unless otherwise specifically approved, generators shall not be fitted with continuous drain connections leading to sewers, but shall discharge through an open connection into a suitably vented outdoor receptacle or residue pit which may have such connections. An open connection for the sludge drawoff is desirable to enable the generator operator to observe leakage of generating water from the drain valve or sludge cock.

**1910.253(f)(4)(ii)(A)** Each generator shall be provided with a vent pipe.

**1910.253(f)(4)(ii)(B)** The escape or relief pipe shall be rigidly installed without traps and so that any condensation will drain back to the generator.

**1910.253(f)(4)(ii)(C)** The escape or relief pipe shall be carried full size to a suitable point outside the building. It shall terminate in a hood or bend located at least 12 feet (3.7 m) above the ground, preferably above the roof, and as far away as practicable from windows or other openings into buildings and as far away as practicable from sources of ignition such as flues or chimneys and tracks used by locomotives. Generating chamber relief pipes shall not be inter-connected but shall be separately led to the outside air. The hood or bend shall be so constructed that it will not be obstructed by rain, snow, ice, insects, or birds. The outlet shall be at least 3 feet (0.9 m) from combustible construction.

**1910.253(f)(4)(iii)(A)** Gas holders shall be constructed on the gasometer principle, the bell being suitably guided. The gas bell shall move freely without tendency to bind and shall have a clearance of at least 2 inches (5 cm) from the shell.

**1910.253(f)(4)(iii)(B)** The gas holder may be located in the generator room, in a separate room or out of doors. In order to prevent collapse of the gas bell or infiltration of air due to a vacuum caused by the compressor or booster pump or cooling of the gas, a compressor or booster cutoff shall be provided at a point 12 inches (0.3 m) or more above the landing point of the bell. When the gas holder is located indoors, the room shall be ventilated in accordance with paragraph (f)(6)(ii) of this section and heated and lighted in accordance with paragraphs (f)(6)(iii) and (f)(6)(iv) of this section.

**1910.253(f)(4)(iii)(C)** When the gas holder is not located within a heated building, gas holder seals shall be protected against freezing.

**1910.253(f)(4)(iii)(D)** Means shall be provided to stop the generator-feeding mechanism before the gas holder reaches the upper limit of its travel.

**1910.253(f)(4)(iii)(E)** When the gas holder is connected to only one generator, the gas capacity of the holder shall be not less than one-third of the hourly rating of the generator.

**1910.253(f)(4)(iii)(F)** If acetylene is used from the gas holder without increase in pressure at some points but with increase in pressure by a compressor or booster pump at other points, approved piping protective devices shall be installed in each supply line. The low-pressure protective device shall be located between the gas holder and the shop piping, and the medium-pressure protective device shall be located between the compressor or booster pump and the shop piping (see Figure Q-4). Approved protective equipment (designated $P_F$) is used to prevent: Backflow of oxygen into the fuel-gas supply system; passage of a flashback into the fuel-gas supply system; and excessive back pressure of oxygen in the fuel-gas supply system. The three functions of the protective equipment may be combined in one device or may be provided by separate devices.

**FIGURE Q-4**

**1910.253(f)(4)(iv)(A)** The compressor or booster system shall be of an approved type.

**1910.253(f)(4)(iv)(B)** Wiring and electric equipment in compressor or booster pump rooms or enclosures shall conform to the provisions of subpart S of this part for Class I, Division 2 locations.

**1910.253(f)(4)(iv)(C)** Compressors and booster pump equipment shall be located in well-ventilated areas away from open flames, electrical or mechanical sparks, or other ignition sources.

**1910.253(f)(4)(iv)(D)** Compressor or booster pumps shall be provided with pressure relief valves which will relieve pressure exceeding 15 psig (103 kPa gauge pressure) to a safe outdoor location as provided in paragraph (f)(4)(ii) of this section, or by returning the gas to the inlet side or to the gas supply source.

**1910.253(f)(4)(iv)(E)** Compressor or booster pump discharge outlets shall be provided with approved protective equipment. (See paragraph (e) of this section.)

**1910.253(f)(5)** *Portable acetylene generators.* **(i)(A)** All portable generators shall be of a type approved for portable use.

**1910.253(f)(5)(i)(B)** Portable generators shall not be used within 10 feet (3 m) of combustible material other than the floor.

**1910.253(f)(5)(i)(C)** Portable generators shall not be used in rooms of total volume less than 35 times the total gas-generating capacity per charge of all generators in the room. Generators shall not be used in rooms having a ceiling height of less than 10 feet (3 m). (To obtain the gas-generating capacity in cubic feet per charge, multiply the pounds of carbide per charge by 4.5.)

**1910.253(f)(5)(i)(D)** Portable generators shall be protected against freezing. The use of salt or other corrosive chemical to prevent freezing is prohibited.

**1910.253(f)(5)(ii)(A)** Portable generators shall be cleaned and recharged and the air mixture blown off outside buildings.

**1910.253(f)(5)(ii)(B)** When charged with carbide, portable generators shall not be moved by crane or derrick.

**1910.253(f)(5)(ii)(C)** When not in use, portable generators shall not be stored in rooms in which open flames are used unless the generators contain no carbide and have been thoroughly purged of acetylene. Storage rooms shall be well ventilated.

**1910.253(f)(5)(ii)(D)** When portable acetylene generators are to be transported and operated on vehicles, they shall be securely anchored to the vehicles. If transported by truck, the motor shall be turned off during charging, cleaning, and generating periods.

**1910.253(f)(5)(ii)(E)** Portable generators shall be located at a safe distance from the welding position so that they will not be exposed to sparks, slag, or misdirection of the torch flame or overheating from hot materials or processes.

**1910.253(f)(6)** *Outside generator houses and inside generator rooms for stationary acetylene generators.* **(i)(A)** No opening in any outside generator house shall be located within 5 feet (1.5 m) of any opening in another building.

**1910.253(f)(6)(i)(B)** Walls, floors, and roofs of outside generator houses shall be of noncombustible construction.

**1910.253(f)(6)(i)(C)** When a part of the generator house is to be used for the storage or manifolding of oxygen cylinders, the space to be so occupied shall be separated from the generator or carbide storage section by partition walls continuous from floor to roof or ceiling, of the type of construction stated in paragraph (f)(6)(i)(H) of this section. Such separation walls shall be without openings and shall be joined to the floor, other walls and ceiling or roof in a manner to effect a permanent gas-tight joint.

**1910.253(f)(6)(i)(D)** Exit doors shall be located so as to be readily accessible in case of emergency.

**1910.253(f)(6)(i)(E)** Explosion venting for outside generator houses and inside generator rooms shall be provided in exterior walls or roofs. The venting areas shall be equal to not less than 1 square foot (0.09 m$^2$) per 50 cubic feet (1.4 m$^3$) of room volume and may consist of any one or any combination of the following: Walls of light, noncombustible material preferably single-thickness, single-strength glass; lightly fastened hatch covers; lightly fastened swinging doors in exterior walls opening outward; lightly fastened walls or roof designed to relieve at a maximum pressure of 25 pounds per square foot (0.001 MPa).

**1910.253(f)(6)(i)(F)** The installation of acetylene generators within buildings shall be restricted to buildings not exceeding one story in height; *provided, however,* that this will not be construed as prohibiting such installations on the roof or top floor of a building exceeding such height.

**1910.253(f)(6)(i)(G)** Generators installed inside buildings shall be enclosed in a separate room.

**1910.253(f)(6)(i)(H)** The walls, partitions, floors, and ceilings of inside generator rooms shall be of noncombustible construction having a fire-resistance rating of at least 1 hour. The walls or partitions shall be continuous from floor to ceiling and shall be securely anchored. At least one wall of the room shall be an exterior wall.

**1910.253(f)(6)(i)(I)** Openings from an inside generator room to other parts of the building shall be protected by a swinging type, self-closing fire door for a Class B opening and having a rating of at least 1 hour. Windows in partitions shall be wired glass and approved metal frames with fixed sash. Installation shall be in accordance with the Standard for the Installation of Fire Doors and Windows, NFPA 80-1970, which is incorporated by reference as specified in §1910.6.

**1910.253(f)(6)(ii)** Inside generator rooms or outside generator houses shall be well ventilated with vents located at floor and ceiling levels.

**1910.253(f)(6)(iii)** Heating shall be by steam, hot water, enclosed electrically heated elements or other indirect means. Heating by flames or fires shall be prohibited in outside generator houses or inside generator rooms, or in any enclosure communicating with them.

**1910.253(f)(6)(iv)(A)** Generator houses or rooms shall have natural light during daylight hours. Where artificial lighting is necessary it shall be restricted to electric lamps installed in a fixed position. Unless specifically approved for use in atmospheres containing acetylene, such lamps shall be provided with enclosures of glass or other noncombustible material so designed and constructed as to prevent gas vapors from reaching the lamp or socket and to resist breakage. Rigid conduit with threaded connections shall be used.

**1910.253(f)(6)(iv)(B)** Lamps installed outside of wired-glass panels set in gas-tight frames in the exterior walls or roof of the generator house or room are acceptable.

**1910.253(f)(6)(v)** Electric switches, telephones, and all other electrical apparatus which may cause a spark, unless specifically approved for use inside acetylene generator rooms, shall be located outside the generator house or in a room or space separated from the generator room by a gas-tight partition, except that where the generator system is designed so that no carbide fill opening or other part of the generator is open to the generator house or room during the operation of the generator, and so that residue is carried in closed piping from the residue discharge valve to a point outside the generator house or room, electrical equipment in the generator house or room shall conform to the provisions of subpart S of this part for Class I, Division 2 locations.

**1910.253(f)(7)** *Maintenance and operation.* **(i)** Unauthorized persons shall not be permitted in outside generator houses or inside generator rooms.

**1910.253(f)(7)(i)(A)** Operating instructions shall be posted in a conspicuous place near the generator or kept in a suitable place available for ready reference.

**1910.253(f)(7)(i)(B)** When recharging generators the order of operations specified in the instructions supplied by the manufacturer shall be followed.

**1910.253(f)(7)(i)(C)** In the case of batch-type generators, when the charge of carbide is exhausted and before additional carbide is added, the generating chamber shall always be flushed out with water, renewing the water supply in accordance with the instruction card furnished by the manufacturer.

**1910.253(f)(7)(i)(D)** The water-carbide residue mixture drained from the generator shall not be discharged into sewer pipes or stored in areas near open flames. Clear water from residue settling pits may be discharged into sewer pipes.

**1910.253(f)(7)(ii)** The carbide added each time the generator is recharged shall be sufficient to refill the space provided for carbide without ramming the charge. Steel or other ferrous tools shall not be used in distributing the charge.

**1910.253(f)(7)(iii)** Generator water chambers shall be kept filled to proper level at all times except while draining during the recharging operation.

**1910.253(f)(7)(iv)** Whenever repairs are to be made or the generator is to be charged or carbide is to be removed, the water chamber shall be filled to the proper level.

**1910.253(f)(7)(v)** Previous to making repairs involving welding, soldering, or other hot work or other operations which produce a source of ignition, the carbide charge and feed mechanism shall be completely removed. All acetylene shall be expelled by completely flooding the generator shell with water and the generator shall be disconnected from the piping system. The generator shall be kept filled with water, if possible, or positioned to hold as much water as possible.

**1910.253(f)(7)(vi)** Hot repairs shall not be made in a room where there are other generators unless all the generators and piping have been purged of acetylene.

**1910.253(g)** *Calcium carbide storage*—**(1)** *Packaging.* **(i)** Calcium carbide shall be contained in metal packages of sufficient strength to prevent rupture. The packages shall be provided with a screw top or equivalent. These packages shall be constructed water- and air-tight. Solder shall not be used in such a manner that the package would fail if exposed to fire.

**1910.253(g)(1)(ii)** Packages containing calcium carbide shall be conspicuously marked "Calcium Carbide—Dangerous If Not Kept Dry" or with equivalent warning.

**1910.253(g)(1)(iii)** Caution: Metal tools, even the so-called spark resistant type may cause ignition of an acetylene and air mixture when opening carbide containers.

**1910.253(g)(1)(iv)** Sprinkler systems shall not be installed in carbide storage rooms.

**1910.253(g)(2)** *Storage indoors.* **(i)** Calcium carbide in quantities not to exceed 600 pounds (272.2 kg) may be stored indoors in dry, waterproof, and well-ventilated locations.

**1910.253(g)(2)(i)(A)** Calcium carbide not exceeding 600 pounds (272.2 kg) may be stored indoors in the same room with fuel-gas cylinders.

**1910.253(g)(2)(i)(B)** Packages of calcium carbide, except for one of each size, shall be kept sealed. The seals shall not be broken when there is carbide in excess of 1 pound (0.5 kg) in any other unsealed package of the same size of carbide in the room.

**1910.253(g)(2)(ii)** Calcium carbide exceeding 600 pounds (272.2 kg) but not exceeding 5,000 pounds (2,268 kg) shall be stored:

**1910.253(g)(2)(ii)(A)** In accordance with paragraph (g)(2)(iii) of this section;

**1910.253(g)(2)(ii)(B)** In an inside generator room or outside generator house; or

**1910.253(g)(2)(ii)(C)** In a separate room in a one-story building which may contain other occupancies, but without cellar or basement beneath the carbide storage section. Such rooms shall be constructed in accordance with paragraphs (f)(6)(i)(H) and (f)(6)(i)(I) of this section and ventilated in accordance with paragraph (f)(6)(ii) of this section. These rooms shall be used for no other purpose.

**1910.253(g)(2)(iii)** Calcium carbide in excess of 5,000 pounds (2,268 kg) shall be stored in one-story buildings without cellar or basement and used for no other purpose, or in outside generator houses. If the storage building is of noncombustible construction, it may adjoin other one-story buildings if separated therefrom by unpierced firewalls; if it is detached less than 10 feet (3 m) from such building or buildings, there shall be no opening in any of the mutually exposing sides of such buildings within 10 feet (3 m). If the storage building is of combustible construction, it shall be at least 20 feet (6.1 m) from any other one- or two-story building, and at least 30 feet (9.1 m) from any other building exceeding two stories.

**1910.253(g)(3)** *Storage outdoors.* **(i)** Calcium carbide in unopened metal containers may be stored outdoors.

**1910.253(g)(3)(ii)** Carbide containers to be stored outdoors shall be examined to make sure that they are in good condition. Periodic reexaminations shall be made for rusting or other damage to a container that might affect its water or air tightness.

**1910.253(g)(3)(iii)** The bottom tier of each row shall be placed on wooden planking or equivalent, so that the containers will not come in contact with the ground or ground water.

**1910.253(g)(3)(iv)** Containers of carbide which have been in storage the longest shall be used first.

[55 FR 13696, Apr. 11, 1990, as amended at 55 FR 32015, Aug 6, 1990; 55 FR 46053, Nov. 1, 1990; 61 FR 9241, Mar. 7, 1996]

### §1910.254 Arc welding and cutting.

**1910.254(a)** *General*—**(1)** *Equipment selection.* Welding equipment shall be chosen for safe application to the work to be done as specified in paragraph (b) of this section.

**1910.254(a)(2)** *Installation.* Welding equipment shall be installed safely as specified by paragraph (c) of this section.

**1910.254(a)(3)** *Instruction.* Workmen designated to operate arc welding equipment shall have been properly instructed and qualified to operate such equipment as specified in paragraph (d) of this section.

**1910.254(b)** *Application of arc welding equipment*—**(1)** *General.* Assurance of consideration of safety in design is obtainable by choosing apparatus complying with the Requirements for Electric Arc-Welding Apparatus, NEMA EW-1-1962, National Electrical Manufacturers Association or the Safety Standard for Transformer-Type Arc-Welding Machines, ANSI C33.2—1956, Underwriters' Laboratories, both of which are incorporated by reference as specified in §1910.6.

**1910.254(b)(2)** *Environmental conditions.* **(i)** Standard machines for arc welding service shall be designed and constructed to carry their rated load with rated temperature rises where the temperature of the cooling air does not exceed 40 °C (104 °F) and where the altitude does not exceed 3,300 feet (1,005.8 m), and shall be suitable for operation in atmospheres containing gases, dust, and light rays produced by the welding arc.

**1910.254(b)(2)(ii)** Unusual service conditions may exist, and in such circumstances machines shall be especially designed to safely meet the requirements of the service. Chief among these conditions are:

**1910.254(b)(2)(ii)(A)** Exposure to unusually corrosive fumes.

**1910.254(b)(2)(ii)(B)** Exposure to steam or excessive humidity.

**1910.254(b)(2)(ii)(C)** Exposure to excessive oil vapor.

**1910.254(b)(2)(ii)(D)** Exposure to flammable gases.

**1910.254(b)(2)(ii)(E)** Exposure to abnormal vibration or shock.

**1910.254(b)(2)(ii)(F)** Exposure to excessive dust.

**1910.254(b)(2)(ii)(G)** Exposure to weather.

**1910.254(b)(2)(ii)(H)** Exposure to unusual seacoast or shipboard conditions.

**1910.254(b)(3)** *Voltage.* The following limits shall not be exceeded:

**1910.254(b)(3)(i)** Alternating-current machines

**1910.254(b)(3)(i)(A)** Manual arc welding and cutting—80 volts.

**1910.254(b)(3)(i)(B)** Automatic (machine or mechanized) arc welding and cutting—100 volts.

**1910.254(b)(3)(ii)** Direct-current machines

**1910.254(b)(3)(ii)(A)** Manual arc welding and cutting—100 volts.

**1910.254(b)(3)(ii)(B)** Automatic (machine or mechanized) arc welding and cutting—100 volts.

**1910.254(b)(3)(iii)** When special welding and cutting processes require values of open circuit voltages higher than the above, means shall be provided to prevent the operator from making accidental contact with the high voltage by adequate insulation or other means.

**1910.254(b)(3)(iv)** For a.c. welding under wet conditions or warm surroundings where perspiration is a factor, the use of reliable automatic controls for reducing no load voltage is recommended to reduce the shock hazard.

**1910.254(b)(4)** *Design.* **(i)** A controller integrally mounted in an electric motor driven welder shall have capacity for carrying rated motor current, shall be capable of making and interrupting stalled rotor current of the motor, and may serve as the running overcurrent device if provided with the number of overcurrent units as specified by subpart S of this part.

**1910.254(b)(4)(ii)** On all types of arc welding machines, control apparatus shall be enclosed except for the operating wheels, levers, or handles.

**1910.254(b)(4)(iii)** Input power terminals, tap change devices and live metal parts connected to input circuits shall be completely enclosed and accessible only by means of tools.

**1910.254(b)(4)(iv)** Terminals for welding leads should be protected from accidental electrical contact by personnel or by metal objects, i.e., vehicles, crane hooks, etc. Protection may be obtained by use of: Dead-front receptacles for plug connections; recessed openings with nonremovable hinged covers; heavy insulating sleeving or taping or other equivalent electrical and mechanical protection. If a welding lead terminal which is intended to be used exclusively for connection to the work is connected to the grounded enclosure, it must be done by a conductor at least two AWG sizes smaller than the grounding conductor and the terminal shall be marked to indicate that it is grounded.

**1910.254(b)(4)(v)** No connections for portable control devices such as push buttons to be carried by the operator shall be connected to an a.c. circuit of higher than 120 volts. Exposed metal parts of portable control devices operating on circuits above 50 volts shall be grounded by a grounding conductor in the control cable.

**1910.254(b)(4)(vi)** Auto transformers or a.c. reactors shall not be used to draw welding current directly from any a.c. power source having a voltage exceeding 80 volts.

**1910.254(c)** *Installation of arc welding equipment*—**(1)** *General.* Installation including power supply shall be in accordance with the requirements of subpart S of this part.

**1910.254(c)(2)** *Grounding.* **(i)** The frame or case of the welding machine (except engine-driven machines) shall be grounded under the conditions and according to the methods prescribed in subpart S of this part.

**1910.254(c)(2)(ii)** Conduits containing electrical conductors shall not be used for completing a work-lead circuit. Pipelines shall not be used as a permanent part of a work-lead circuit, but may be used during construction, extension or repair providing current is not carried through threaded joints, flanged bolted joints, or caulked joints and that special precautions are used to avoid sparking at connection of the work-lead cable.

**1910.254(c)(2)(iii)** Chains, wire ropes, cranes, hoists, and elevators shall not be used to carry welding current.

**1910.254(c)(2)(iv)** Where a structure, conveyor, or fixture is regularly employed as a welding current return circuit, joints shall be bonded or provided with adequate current collecting devices.

**1910.254(c)(2)(v)** All ground connections shall be checked to determine that they are mechanically strong and electrically adequate for the required current.

**1910.254(c)(3)** *Supply connections and conductors.* **(i)** A disconnecting switch or controller shall be provided at or near each welding machine which is not equipped with such a switch or controller mounted as an integral part of the machine. The switch shall be in accordance with subpart S of this part. Overcurrent protection shall be provided as specified in subpart S of this part. A disconnect switch with overload protection or equivalent disconnect and protection means, permitted by subpart S of this part, shall be provided for each outlet intended for connection to a portable welding machine.

**1910.254(c)(3)(ii)** For individual welding machines, the rated current-carrying capacity of the supply conductors shall be not less than the rated primary current of the welding machines.

**1910.254(c)(3)(iii)** For groups of welding machines, the rated current-carrying capacity of conductors may be less than the sum of the rated primary currents of the welding machines supplied. The conductor rating shall be determined in each case according to the machine loading based on the use to be made of each welding machine and the allowance permissible in the event that all the welding machines supplied by the conductors will not be in use at the same time.

**1910.254(c)(3)(iv)** In operations involving several welders on one structure, d.c. welding process requirements may require the use of both polarities; or supply circuit limitations for a.c. welding may require distribution of machines among the phases of the supply circuit. In such cases no load voltages between electrode holders will be 2 times normal in d.c. or 1, 1.41, 1.73, or 2 times normal on a.c. machines. Similar voltage differences will exist if both a.c. and d.c. welding are done on the same structure.

**1910.254(c)(3)(iv)(A)** All d.c. machines shall be connected with the same polarity.

**1910.254(c)(3)(iv)(B)** All a.c. machines shall be connected to the same phase of the supply circuit and with the same instantaneous polarity.

**1910.254(d)** *Operation and maintenance*—**(1)** *General.* Workers assigned to operate or maintain arc welding equipment shall be acquainted with the requirements of this section and with §1910.252(a), (b) and (c) of this part.

**1910.254(d)(2)** *Machine hook up.* Before starting operations all connections to the machine shall be checked to make certain they are properly made. The work lead shall be firmly attached to the work; magnetic work clamps shall be freed from adherent metal particles of spatter on contact surfaces. Coiled welding cable shall be spread out before use to avoid serious overheating and damage to insulation.

**1910.254(d)(3)** *Grounding.* Grounding of the welding machine frame shall be checked. Special attention shall be given to safety ground connections of portable machines.

**1910.254(d)(4)** *Leaks.* There shall be no leaks of cooling water, shielding gas or engine fuel.

**1910.254(d)(5)** *Switches.* It shall be determined that proper switching equipment for shutting down the machine is provided.

**1910.254(d)(6)** *Manufacturers' instructions.* Printed rules and instructions covering operation of equipment supplied by the manufacturers shall be strictly followed.

**1910.254(d)(7)** *Electrode holders.* Electrode holders when not in use shall be so placed that they cannot make electrical contact with persons, conducting objects, fuel or compressed gas tanks.

**1910.254(d)(8)** *Electric shock.* Cables with splices within 10 feet (3 m) of the holder shall not be used. The welder should not coil or loop welding electrode cable around parts of his body.

**1910.254(d)(9)** *Maintenance.* **(i)** The operator should report any equipment defect or safety hazard to his supervisor and the use of the equipment shall be discontinued until its safety has been assured. Repairs shall be made only by qualified personnel.

**1910.254(d)(9)(ii)** Machines which have become wet shall be thoroughly dried and tested before being used.

**1910.254(d)(9)(iii)** Cables with damaged insulation or exposed bare conductors shall be replaced. Joining lengths of work and electrode cables shall be done by the use of connecting means specifically intended for the purpose. The connecting means shall have insulation adequate for the service conditions.

[55 FR 13696, Apr. 11, 1990, as amended at 61 FR 9241, Mar. 7, 1996]

### §1910.255 Resistance welding.

**1910.255(a)** *General*—**(1)** *Installation.* All equipment shall be installed by a qualified electrician in conformance with subpart S of this part. There shall be a safety-type disconnecting switch or a circuit breaker or circuit interrupter to open each power circuit to the machine, conveniently located at or near the machine, so that the power can be shut off when the machine or its controls are to be serviced.

**1910.255(a)(2)** *Thermal protection.* Ignitron tubes used in resistance welding equipment shall be equipped with a thermal protection switch.

**1910.255(a)(3)** *Personnel.* Workmen designated to operate resistance welding equipment shall have been properly instructed and judged competent to operate such equipment.

**1910.255(a)(4)** *Guarding.* Controls of all automatic or air and hydraulic clamps shall be arranged or guarded to prevent the operator from accidentally activating them.

**1910.255(b)** *Spot and seam welding machines (nonportable)*—**(1)** *Voltage.* All external weld initiating control circuits shall operate on low voltage, not over 120 volts, for the safety of the operators.

**1910.255(b)(2)** *Capacitor welding.* Stored energy or capacitor discharge type of resistance welding equipment and control panels involving high voltage (over 550 volts) shall be suitably insulated and protected by complete enclosures, all doors of which shall be provided with suitable interlocks and contacts wired into the control circuit (similar to elevator interlocks). Such interlocks or contacts shall be so designed as to effectively interrupt power and short circuit all capacitors when the door or panel is open. A manually operated switch or suitable positive device shall be installed, in addition to the mechanical interlocks or contacts, as an added safety measure assuring absolute discharge of all capacitors.

**1910.255(b)(3)** *Interlocks.* All doors and access panels of all resistance welding machines and control panels shall be kept locked and interlocked to prevent access, by unauthorized persons, to live portions of the equipment.

**1910.255(b)(4)** *Guarding.* All press welding machine operations, where there is a possibility of the operator's fingers being under the point of operation, shall be effectively guarded by the use of a device such as an electronic eye safety circuit, two hand controls or protection similar to that prescribed for punch press operation, §1910.217 of this part. All chains, gears, operating bus linkage, and belts shall be protected by adequate guards, in accordance with §1910.219 of this part.

**1910.255(b)(5)** *Shields.* The hazard of flying sparks shall be, wherever practical, eliminated by installing a shield guard of safety glass or suitable fire-resistant plastic at the point of operation. Additional shields or curtains shall be installed as necessary to protect passing persons from flying sparks. (See §1910.252(b)(2)(i)(C) of this part.)

**1910.255(b)(6)** *Foot switches.* All foot switches shall be guarded to prevent accidental operation of the machine.

**1910.255(b)(7)** *Stop buttons.* Two or more safety emergency stop buttons shall be provided on all special multispot welding machines, including 2-post and 4-post weld presses.

**1910.255(b)(8)** *Safety pins.* On large machines, four safety pins with plugs and receptacles (one in each corner) shall be provided so that when safety pins are removed and inserted in the ram or platen, the press becomes inoperative.

**1910.255(b)(9)** *Grounding.* Where technically practical, the secondary of all welding transformers used in multispot, projection and seam welding machines shall be grounded. This may be done by permanently grounding one side of the welding secondary current circuit. Where not technically practical, a center tapped grounding reactor connected across the secondary or the use of a safety disconnect switch in conjunction with the welding control are acceptable alternates. Safety disconnect shall be arranged to open both sides of the line when welding current is not present.

**1910.255(c) Portable welding machines—(1)** *Counterbalance.* All portable welding guns shall have suitable counterbalanced devices for supporting the guns, including cables, unless the design of the gun or fixture makes counterbalancing impractical or unnecessary.

**1910.255(c)(2)** *Safety chains.* All portable welding guns, transformers and related equipment that is suspended from overhead structures, eye beams, trolleys, etc., shall be equipped with safety chains or cables. Safety chains or cables shall be capable of supporting the total shock load in the event of failure of any component of the supporting system.

**1910.255(c)(3)** *Clevis.* Each clevis shall be capable of supporting the total shock load of the suspended equipment in the event of trolley failure.

**1910.255(c)(4)** *Switch guards.* All initiating switches, including retraction and dual schedule switches, located on the portable welding gun shall be equipped with suitable guards capable of preventing accidental initiation through contact with fixturing, operator's clothing, etc. Initiating switch voltage shall not exceed 24 volts.

**1910.255(c)(5)** *Moving holder.* The movable holder, where it enters the gun frame, shall have sufficient clearance to prevent the shearing of fingers carelessly placed on the operating movable holder.

**1910.255(c)(6)** *Grounding.* The secondary and case of all portable welding transformers shall be grounded. Secondary grounding may be by center tapped secondary or by a center tapped grounding reactor connected across the secondary.

**1910.255(d) Flash welding equipment—(1)** *Ventilation and flash guard.* Flash welding machines shall be equipped with a hood to control flying flash. In cases of high production, where materials may contain a film of oil and where toxic elements and metal fumes are given off, ventilation shall be provided in accordance with §1910.252(c) of this part.

**1910.255(d)(2)** *Fire curtains.* For the protection of the operators of nearby equipment, fire-resistant curtains or suitable shields shall be set up around the machine and in such a manner that the operators movements are not hampered.

**1910.255(e)** *Maintenance.* Periodic inspection shall be made by qualified maintenance personnel, and a certification record maintained. The certification record shall include the date of inspection, the signature of the person who performed the inspection and the serial number, or other identifier, for the equipment inspected. The operator shall be instructed to report any equipment defects to his supervisor and the use of the equipment shall be discontinued until safety repairs have been completed.

### §1910.256 Sources of standards [Removed].

[§1910.256 removed at 61 FR 9240, Mar. 7, 1996, effective May 6, 1996.]

### §1910.257 Standards organizations [Removed].

[§1910.257 removed at 61 FR 9240, Mar. 7, 1996, effective May 6, 1996.]

# Subpart R—Special Industries

**Authority:** Sections 4, 6, and 8 of the Occupational Safety and Health Act of 1970 (29 U.S.C. 653, 655, 657); Secretary of Labor's Order No. 12-71 (36 FR 8754), 8-76 (41 FR 25059), 9-83 (48 FR 35736), or 6-96 (62 FR 111), as applicable; and 29 CFR part 1911.

### §1910.261 Pulp, paper, and paperboard mills.

**1910.261(a)** *General requirements*—(1) *Application.* This section applies to establishments where pulp, paper, and paperboard are manufactured and converted. This section does not apply to logging and the transportation of logs to pulp, paper, and paperboard mills.

**1910.261(a)(2)** *Standards incorporated by reference.* Standards covering issues of occupational safety and health which have general application without regard to any specific industry are incorporated by reference in paragraphs (b) through (m) of this section and in subparagraphs (3) and (4) of this paragraph and made applicable under this section. Such standards shall be construed according to the rules set forth in §1910.5.

**1910.261(a)(3)** *General incorporation of standards.* Establishments subject to this section shall comply with the following standards of the American National Standards Institute, which are incorporated by reference as specified in §1910.6:

**1910.261(a)(3)(i)** Practice for Industrial Lighting, A11.1—1965 (R-1970).

**1910.261(a)(3)(ii)** Scheme for the Identification of Piping Systems, A13.1—1956.

**1910.261(a)(3)(iii)** Safety Code for Elevators, Dumbwaiters, and Moving Walks, A17.1—1965, including Supplements A17.1a—1967, A17.1b—1968, A17.1c—1969, and A17.1d—1970.

**1910.261(a)(3)(iv)** Practice for the Inspection of Elevators (Inspector's Manual), A17.2—1960, including Suppelements A17.2a—1965 and A17.2b—1967.

**1910.261(a)(3)(v)** Safety Code for Conveyors, Cableways, and Related Equipment, B20.1—1957.

**1910.261(a)(3)(vi)** Power Piping, B31.1.0—1967 and addenda B31.10a—1969. Fuel Gas Piping, B31.2—1968.

**1910.261(a)(3)(vii)** Identification of Gas-Mask Canisters, K13.1—1967.

**1910.261(a)(3)(viii)** Prevention of Sulfur Fires and Explosions, Z12.12—1968.

**1910.261(a)(3)(ix)** Installation of Blower and Exhaust Systems for Dust, Stock, and Vapor Removal or Conveying, Z33.1—1961.

**1910.261(a)(4)** *Other standards.* The following standards, which are incorporated by reference as specified in §1910.6, shall be considered standards under this section:

**1910.261(a)(4)(i)** ASME Boiler and Pressure Vessel Code, Section VIII, Unfired Pressure Vessels, including addenda 1969.

**1910.261(a)(4)(ii)** Building Exits Code for Life Safety from Fire, NFPA 101—1970.

**1910.261(a)(4)(iii)** Safety in the Handling and Use of Explosives, IME Pamphlet No. 17, July 1960, Institute of Makers of Explosives.

**1910.261(b)** *Safe practices*—(1) *Lockouts.* Devices such as padlocks shall be provided for locking out the source of power at the main disconnect switch. Before any maintenance, inspection, cleaning, adjusting, or servicing of equipment (electrical, mechanical, or other) that requires entrance into or close contact with the machinery or equipment, the main power disconnect switch or valve, or both, controlling its source of power or flow of material, shall be locked out or blocked off with padlock, blank flange, or similar device.

**1910.261(b)(2)** *Emergency lighting.* Emergency lighting shall be provided wherever it is necessary for employees to remain at their machines or stations to shut down equipment in case of power failure. Emergency lighting shall be provided at stairways and passageways or aisleways used by employees for emergency exit in case of power failure. Emergency lighting shall be provided in all plant first aid and medical facilities.

**1910.261(c)** *Handling and storage of pulpwood and pulp chips*—(1) *Handling pulpwood with forklift trucks.* Where large forklift trucks, or lift trucks with clam-jaws, are used in the yard, the operator's enclosed cab shall be provided with an escape hatch, whenever the hydraulic arm blocks escape through the side doors.

**1910.261(c)(2)** *Handling pulpwood with cranes or stackers.* **(i)** Where locomotive cranes are used for loading or unloading pulpwood, the pulpwood shall be piled so as to allow a clearance of not less than 24 inches between the pile and the end of the cab of any locomotive crane in use, when the cab is turned in any working position.

**1910.261(c)(2)(ii)** The minimum distance of the pulpwood pile from the centerline of a standard-gage track shall be maintained at not less than 8 1/2 feet.

**1910.261(c)(2)(iii)** Logs shall be piled in an orderly and stable manner, with no projection into walkways or roadways.

**1910.261(c)(2)(iv)** Railroad cars shall not be spotted on tracks adjacent to the locomotive cranes unless a 24-inch clearance is maintained, as required in paragraph (c)(2)(i) of this section.

**1910.261(c)(2)(v)** The handling and storage of other materials shall conform to paragraphs (c)(2) (i) and (ii) of this section with respect to clearance.

**1910.261(c)(2)(vi)** No person shall be permitted to walk beneath a suspended load, bucket, or hook.

**1910.261(c)(3)** *Handling pulpwood from ships.*

**1910.261(c)(3)(i)** [Reserved]

**1910.261(c)(3)(ii)** The hatch tender shall be required to signal the hoisting engineer to move the load only after the men working in the hold are in the clear.

**1910.261(c)(3)(iii)** The air in the ship's hold, tanks, or closed vessels shall be tested for oxygen deficiency and for both toxic and explosive gases and vapors.

**1910.261(c)(4)** *Handling pulpwood from flatcars and all other railway cars.* **(i)** Railroad flatcars for the conveyance of pulpwood loaded parallel to the length of the car shall be equipped with safety-stake pockets.

**1910.261(c)(4)(ii)** Where pulpwood is loaded crosswise on a flatcar sufficient stakes of sizes not smaller than 4 by 4 inches shall be used to prevent the load from shifting.

**1910.261(c)(4)(iii)** When it is necessary to cut stakes, those on the unloading side should be partially cut through first, and then the binder wires cut on the opposite side. Wire cutters equipped with long extension handles shall be used. No person shall be permitted along the dumping side of the car after the stakes have been cut.

**1910.261(c)(4)(iv)** When steel straps without stakes are used, the steel straps shall be cut from a safe area to prevent employees from being struck by the falling logs.

**1910.261(c)(4)(v)** Flatcars and all other cars shall be chocked during unloading. Where equipment is not provided with hand brakes, rail clamping chocks shall be used.

**1910.261(c)(4)(vi)** A derail shall be used to prevent movement of other rail equipment into cars where persons are working.

**1910.261(c)(5)** *Handling pulpwood from trucks.* **(i)** Cutting of stakes and binder wires shall be done in accordance with paragraph (c)4(iii) of this section.

**1910.261(c)(5)(ii)** Where binder chain and steel stakes are used, the binder chains shall be released and the stakes tripped from the opposite side of the load spillage.

**1910.261(c)(5)(iii)** Where binder chains and crane slings are used, the crane slings shall be attached and taut before the binder chains are released. The hooker shall see that the helper is clear before signaling for the movement of the load.

**1910.261(c)(6)** *Handling pulp chips from railway cars.* All cars shall be securely fastened in place and all employees in the clear before dumping is started.

**1910.261(c)(7)** *Handling pulp chips from trucks and trailers.* All trucks and trailers shall be securely fastened in place and all employees in the clear before dumping is started.

**1910.261(c)(8)** *Cranes.*

**1910.261(c)(8)(i)** [Reserved]

**1910.261(c)(8)(ii)** A safety device such as a heavy chain or cable at least equal in strength to the lifting cables shall be fastened to the boom and to the frame of the boom crane (if it is other than locomotive) at the base. Alternatively, a telescoping safety device shall be fastened to the boom and to the cab frame, so as to prevent the boom from snapping back over the cab in the event of lifting cable breakage.

**1910.261(c)(8)(iii)** A crane shall not be operated where any part thereof may come within 10 feet of overhead powerlines (or other overhead obstructions) unless the powerlines have been deenergized. The boom shall be painted bright yellow from and including the head sheave to a point 6 feet down the boom towards the cab.

**1910.261(c)(8)(iv)** Standard signals for the operation of cranes shall be established for all movements of the crane, in accordance with American National Standards B30.2—1943 (reaffirmed 1968) and B30.2.0—1967.

**1910.261(c)(8)(v)** Only one member of the crew shall be authorized to give signals to the crane operator.

**1910.261(c)(8)(vi)** All cranes shall be equipped with a suitable warning device such as a horn or whistle.

**1910.261(c)(8)(vii)** A sheave guard shall be provided beneath the head sheave of the boom.

**1910.261(c)(9)** *Traffic warning signs or signals.* **(i)** A flagman shall direct the movement of cranes or locomotives being moved across railroad tracks or roads, and at any points where the vision of the operator is restricted. The flagman must always remain in sight of the operator when the crane or locomotive is in motion. The blue flag policy shall be used to mark stationary cars day and night. This policy shall include marking the track in advance of the spotted cars (flag for daytime, light for darkness).

**1910.261(c)(9)(ii)** After cars are spotted for loading or unloading, warning flags or signs shall be placed in the center of the track at least 50 feet away from the cars and a derail set to protect workmen in the car.

**1910.261(c)(10)** *Illumination.* Artificial illumination shall be provided when loading or unloading is performed after dark, in accordance with American National Standard A11.1—1965 (R—1970).

**1910.261(c)(11)** [Reserved]

**1910.261(c)(12)** *Barking devices.* When barking drums are employed in the yard, the requirements of paragraph (e)(12) of this section shall apply.

**1910.261(c)(13)** *Hand tools.* Handles of wood hooks shall be locked to the shank to prevent them from rotating.

**1910.261(c)(14)** *Removal of pulpwood.* **(i)** The ends of a woodpile shall be properly sloped and cross-tiered into the pile. Upright poles shall not be used at the ends of woodpiles. To knock down wood from the woodpile, mechanical equipment shall be used to permit employees to keep in the clear of loosened wood.

**1910.261(c)(14)(ii)** If dynamite is used to loosen the pile, only authorized personnel shall be permitted to handle and discharge the explosive. An electric detonator is preferable for firing; if a fuse is used, it shall be an approved safety fuse with a burning rate of not less than 120 seconds per yard and a minimum length of 3 feet, in accordance with Safety in the Handling and Use of Explosives, IME Pamphlet No. 17, July 1960.

**1910.261(c)(15)** *Belt conveyors.* **(i)** The sides of the conveyor shall be constructed so that the wood will not fall off.

**1910.261(c)(15)(ii)** Where conveyors cross passageways or roadways, a horizontal platform shall be provided under the conveyor extending out from the sides of the conveyor a distance equal to 1 1/2 times the length of the wood handled. The platform shall extend the width of the road plus 2 feet on each side and shall be kept free of wood and rubbish. The edges of the platform shall be provided with toeboards or other protection to prevent wood from falling, in accordance with American National Standard A12.1—1967.

**1910.261(c)(15)(iii)** All conveyors for pulpwood shall have the inrunning nips between chain and sprockets guarded; also, turning drums shall be guarded.

**1910.261(c)(15)(iv)** Every belt conveyor shall have an emergency stop cable extending the length of the conveyor so that it may be stopped from any location along the line, or conveniently located stop buttons within 10 feet of each work station, in accordance with American National Standard B20.1—1957.

**1910.261(c)(16)** *Signs.* Where conveyors cross walkways or roadways in the yards, signs reading "Danger—Overhead Conveyor" or an equivalent warning shall be erected, in accordance with American National Standard Z35.1—1968.

**1910.261(d)** *Handling and storage of raw materials other than pulpwood or pulp chips*—**(1)** *Personal protective equipment.* Whenever possible, all dust, fumes, and gases incident to handling materials shall be controlled at the source, in accordance with American National Standard Z9.2—1960. Where control at the source is not possible, respirators with goggles or protective masks shall be provided, and employees shall wear them when handling alum, clay, soda ash, lime, bleach powder, ±sulfur, chlorine, and similar materials, and when opening rag bales.

**1910.261(d)(2)** *Clearance.* **(i)** When materials are being piled inside a building and upon platforms, an aisle clearance at least 3 feet greater than the widest truck in use shall be provided.

**1910.261(d)(2)(ii)** Baled paper and rags stored inside a building shall not be piled closer than 18 inches to walls, partitions, or sprinkler heads.

**1910.261(d)(3)** *Piling and unpiling pulp.* **(i)** Piles of wet lap pulp (unless palletized) shall be stepped back one-half the width of the sheet for each 8 feet of pile height. Sheets of pulp shall be interlapped to make the pile secure. Pulp shall not be piled over pipelines to jeopardize pipes, or so as to cause overloading of floors, or to within 18 inches below sprinkler heads.

**1910.261(d)(3)(ii)** Piles of pulp shall not be undermined when being unpiled.

**1910.261(d)(3)(iii)** Floor capacities shall be clearly marked on all floors.

**1910.261(d)(4)(i)** [Reserved]

**1910.261(d)(4)(ii)** Where rolls are pyramided two or more high, chocks shall be installed between each roll on the floor and at every row. Where pulp and paper rolls are stored on smooth floors in processing areas, rubber chocks with wooden core shall be used.

**1910.261(d)(4)(iii)** When rolls are decked two or more high, the bottom rolls shall be chocked on each side to prevent shifting in either direction.

**1910.261(e)** *Preparing pulpwood*—**(1)** *Gang and slasher saws.* A guard shall be provided in front of all gang and slasher saws to protect workers from wood thrown by saws. A guard shall be placed over tail sprockets.

**1910.261(e)(2)** *Slasher tables.* Saws shall be stopped and power switches shall be locked out and tagged whenever it is necessary for any person to be on the slasher table.

**1910.261(e)(3)** [Reserved]

**1910.261(e)(4)** *Runway to the jack ladder.* The runway from the pond or unloading dock to the table shall be protected with standard handrails and toeboards. Inclined portions shall have cleats or equivalent nonslip surfacing, in accordance with American National Standard A12.1—1967. Protective equipment shall be provided for persons working over water.

**1910.261(e)(5)** *Guards below table.* Where not protected by the frame of the machine, the underside of the slasher saws shall be enclosed with guards.

**1910.261(e)(6)** *Conveyors.* The requirements of paragraph (c)(15)(iv) of this section shall apply.

**1910.261(e)(7)** [Reserved]

**1910.261(e)(8)** *Barker feed.* Each barker shall be equipped with a feed and turnover device which will make it unnecessary for the operator to hold a bolt or log by hand during the barking operation. Eye, ear, and head protection shall be provided for the operator, in accordance with paragraph (b)(2) of this section.

**1910.261(e)(9)** [Reserved]

**1910.261(e)(10)** *Stops.* All control devices shall be locked out and tagged when knives are being changed.

**1910.261(e)(11)** *Speed governor.* Water wheels, when directly connected to barker disks or grinders, shall be provided with speed governors, if operated with gate wide open.

**1910.261(e)(12)** *Continuous barking drums.* **(i)** When platforms or floors allow access to the sides of the drums, a standard railing shall be constructed around the drums. When two or more drums are arranged side by side, proper walkways with standard handrails shall be provided between each set, in accordance with paragraph (b)(3) of this section.

**1910.261(e)(12)(ii)** Sprockets and chains, gears, and trunnions shall have standard guards, in accordance with paragraph (b)(1) of this section.

# OSHA Standards for General Industry

**1910.261(e)(12)(iii)** Whenever it becomes necessary for a workman to go within a drum, the driving mechanism shall be locked and tagged, at the main disconnect switch, in accordance with paragraph (b)(4) of this section.

**1910.261(e)(13)** *Intermittent barking drums.* In addition to motor switch, clutch, belt shifter, or other power disconnecting device, intermittent barking drums shall be equipped with a device which may be locked to prevent the drum from moving while it is being emptied or filled.

**1910.261(e)(14)** *Hydraulic barkers.* Hydraulic barkers shall be enclosed with strong baffles at the inlet and the outlet. The operator shall be protected by at least five-ply laminated glass.

**1910.261(e)(15)** *Splitter block.* The block upon or against which the wood is rested shall have a corrugated surface or other means provided that the wood will not slip. Wood to be split, and also the splitting block, shall be free of ice, snow, or chips. The operator shall be provided with eye and foot protection. A clear and unobstructed view shall be maintained between equipment and workers around the block and the workers' help area.

**1910.261(e)(16)** *Power control.* Power for the operation of the splitter shall be controlled by a clutch or equivalent device.

**1910.261(e)(17)** *Knot cleaners.* The operators of knot cleaners of the woodpecker type shall wear eye protection equipment.

**1910.261(e)(18)** *Chipper spout.* The feed system to the chipper spout shall be arranged in such a way that the operator does not stand in a direct line with the chipper spout. All chipper spouts shall be enclosed to a height of at least 42 inches from the floor or operator's platform. When other protection is not sufficient, the operator shall wear a safety belt line. The safety belt line shall be fastened in such a manner as to make it impossible for the operator to fall into the throat of the chipper. Ear protection equipment shall be worn by the operator and others in the immediate area if there is any possibility that the noise level may be harmful (see §1910.95).

**1910.261(e)(19)** *Carriers for knives.* Carriers shall be provided and used for transportation of knives.

**1910.261(f)** **Rag and old paper preparation**—**(1)** *Ripping and trimming tools.* **(i)** Hand knives and scissors shall have blunt points, shall be fastened to the table with chain or thong, and shall not be carried on the person but placed safely in racks or sheaths when not in use.

**1910.261(f)(1)(ii)** Hand knives and sharpening steels shall be provided with guards at the junction of the handle and the blade.

**1910.261(f)(2)** *Shredders, cutters, and dusters.* **(i)** Rotating heads or cylinders shall be completely enclosed except for an opening at the feed side sufficient to permit only the entry of stock. The enclosure shall extend over the top of the feed rolls. It shall be constructed either of solid material or with mesh or openings not exceeding one-half inch and substantial enough to contain flying particles and prevent accidental contact with moving parts. The enclosure shall be bolted or locked into place.

**1910.261(f)(2)(ii)** A smooth-pivoted idler roll resting on the stock or feed table shall be provided in front of feed rolls except when arrangements prevent the operator from standing closer than 36 inches to any part of the feed rolls.

**1910.261(f)(2)(iii)** Any manually fed cutter, shredder, or duster shall be provided with an idler roll as per subdivision (ii) of this subparagraph or the operator shall use special hand-feeding tools.

**1910.261(f)(2)(iv)** Hoods of cutters, shredders, and dusters shall have exhaust ventilation, in accordance with American National Standard Z9.2—1960.

**1910.261(f)(3)** *Blowers.* **(i)** Blowers used for transporting rags shall be provided with feed hoppers having outer edges located not less than 48 inches from the fan.

**1910.261(f)(3)(ii)** The arrangement of the blower discharge outlets and work areas shall be such as to prevent material from falling on workers.

**1910.261(f)(4)** *Conveyors.* Conveyors and conveyor drive belts and pulleys shall be fully enclosed or, if open and within 7 feet of the floor, shall be constructed and guarded in accordance with paragraph (c)(15) of this section and American National Standards B15.1—1953 (Reaffirmed 1958) and B20.1—1957.

**1910.261(f)(5)** *Dust.* Measures for the control of dust shall be provided, in accordance with American National Standards Z33.1—1961, Z87.1—1968, and Z88.2—1969.

**1910.261(f)(6)** *Rag cookers.* **(i)** When cleaning, inspection, or other work requires that persons enter rag cookers, all steam and water valves, or other control devices, shall be locked and tagged in the closed or "off" position. Blank flanging of pipelines is acceptable in place of closed and locked valves.

**1910.261(f)(6)(ii)** When cleaning, inspection, or other work requires that persons must enter the cooker, one person shall be stationed outside in a position to observe and assist in case of emergency, in accordance with paragraph (b)(5) of this section.

**1910.261(f)(6)(iii)** [Reserved]

**1910.261(f)(6)(iv)** Rag cookers shall be provided with safety valves in accordance with the ASME Boiler and Pressure Vessel Code, Section VIII, Unfired Pressure Vessels—1968, with Addenda.

**1910.261(g)** **Chemical processes of making pulp**—**(1)** *Sulfur burners.* **(i)** Sulfur-burner houses shall be safely and adequately ventilated, and every precaution shall be taken to guard against dust explosion hazards and fires, in accordance with American National Standards Z9.2—1960 and Z12.12—1968.

**1910.261(g)(1)(ii)** Nonsparking tools and equipment shall be used in handling dry sulfur.

**1910.261(g)(1)(iii)** Sulfur storage bins shall be kept free of sulfur dust accumulation, in accordance with American National Standard Z9.2—1960.

**1910.261(g)(1)(iv)** Sulfur-melting equipment shall not be located in the burner room.

**1910.261(g)(2)** *Protection for employees (acid plants).* **(i)** Supplied air respirators shall be strategically located for emergency and rescue use.

**1910.261(g)(2)(ii)** During inspection, repairs, or maintenance of acid towers, the workman shall be provided with eye protection, a supplied air respirator, a safety belt, and an attached lifeline. The line shall be extended to an attendant stationed outside the tower opening.

**1910.261(g)(3)** *Acid tower structure.* Outside elevators shall be inspected daily during winter months when ice materially affects safety. Elevators, runways, stairs, etc., for the acid tower shall be inspected monthly for defects that may occur because of exposure to acid or corrosive gases.

**1910.261(g)(4)** *Tanks (acid).* **(i)** Tanks shall be free of acid and shall be washed out with water, and fresh air shall be blown into them before allowing men to enter. Men entering the tanks shall be provided with supplied air respirators, lifebelts, and attached lifelines.

**1910.261(g)(4)(ii)** A man shall be stationed outside to summon assistance if necessary. All intake valves to a tank shall be blanked off or disconnected.

**1910.261(g)(5)** *Clothing.* Where lime slaking takes place, employees shall be provided with rubber boots, rubber gloves, protective aprons, and eye protection. A deluge shower and eye fountain shall be provided to flush the skin and eyes to counteract lime or acid burns.

**1910.261(g)(6)** *Lead burning.* When lead burning is being done within tanks, fresh air shall be forced into the tanks so that fresh air will reach the face of the worker first and the direction of the current will never be from the source of the fumes toward the face of the workers. Supplied air respirators (constant-flow type) shall be provided.

**1910.261(g)(7)** *Hoops for acid storage tanks.* Hoops of tanks shall be made of rods rather than flat strips and shall be safely maintained by scheduled inspections.

**1910.261(g)(8)** *Chip and sawdust bins.* Steam or compressed-air lances, or other facilities, shall be used for breaking down the arches caused by jamming in chip lofts. No worker shall be permitted to enter a bin unless provided with a safety belt, with line attached, and an attendant stationed at the bin to summon assistance.

**1910.261(g)(9)** *Exits (digester building).* At least one unobstructed exit at each end of the room shall be provided on each floor of a digester building.

**1910.261(g)(10)** *Gas masks (digester building).* Gas masks must be available, and they must furnish adequate protection against sulfurous acid and chlorine gases and be inspected and repaired in accordance with 29 CFR 1910.134.

**1910.261(g)(11)** *Elevators.* **(i)** Elevators shall be constructed in accordance with American National Standard A17.1—1965.

**1910.261(g)(11)(ii)** Elevators shall be equipped with gas masks for the maximum number of passengers.

**1910.261(g)(11)(iii)** Elevators shall be equipped with an alarm system to advise of failure.

**1910.261(g)(12)** *Blowoff valves and piping.* **(i)** The blowoff valve of a digester shall be arranged so as to be operated from another room, remote from safety valves.

**1910.261(g)(12)(ii)** Through bolts instead of cap bolts shall be used on all digester pipings.

**1910.261(g)(12)(iii)** Heavy duty pipe, valves, and fittings shall be used between the digester and blow pit. These valves, fittings, and pipes shall be inspected at least semiannually to determine the degree of deterioration.

**1910.261(g)(12)(iv)** Digester blow valves shall be pinned or locked in closed position throughout the entire cooking period.

**1910.261(g)(13)** *Blow pits and blow tanks.* **(i)** Blow-pit openings shall be preferably on the side of the pit instead of on top. When located on top, openings shall be as small as possible and shall be provided with railings, in accordance with American National Standard A12.1—1967.

**1910.261(g)(13)(ii)** A specially constructed ladder shall be used for access to blow pits, to be constructed so that the door of the blow pit cannot be closed when the ladder is in place; other means shall be provided to prevent the closing of the pit door when anyone is in the pit.

**1910.261(g)(13)(iii)** A signaling device shall be installed in the digester and blow-pit rooms and chip bins to be operated as a warning before and while digesters are being blown.

**1910.261(g)(13)(iv)** Blow-pit hoops shall be maintained in a safe condition.

**1910.261(g)(14)** *Blowing digester.* **(i)** Blowoff valves shall be opened slowly.

**1910.261(g)(14)(ii)** After the digester has started to be blown, the blowoff valve shall be left open, and the hand plate shall not be removed until the digester cook signals the blow-pit man that the blow is completed. Whenever it becomes necessary to remove the hand plate to clear stock, operators shall wear eye protection equipment and protective clothing to guard against burns from hot stock.

**1910.261(g)(14)(iii)** Means shall be provided whereby the digester cook shall signal the man in the chip bin before starting to load the digester.

**1910.261(g)(15)** *Inspecting and repairing digester.* **(i)** Valves controlling lines leading into a digester shall be locked out and tagged. The keys to the locks shall be in the possession of a person or persons doing the inspecting or making repairs.

**1910.261(g)(15)(ii)** Fresh air shall be blown into the digester constantly while workmen are inside. Supplied air respirators shall be available in the event the fresh air supply fails or is inadequate.

**1910.261(g)(15)(iii)** No inspector shall enter a digester unless a lifeline is securely fastened to his body by means of a safety belt and at least one other experienced employee is stationed outside the digester to handle the line and to summon assistance. All ladders and lifelines shall be inspected before each use.

**1910.261(g)(15)(iv)** All employees entering digesters for inspection or repair work shall be provided with protective headgear. Eye protection and dust masks shall be provided to workmen while the old brick lining is being removed, in accordance with American National Standards, Z87.1—1968, Z88.2—1969, and Z99.1—1969.

**1910.261(g)(16)** *Pressure tanks-accumulators (acid).* **(i)** Safety regulations governing inspection and repairing of pressure tanks-accumulators (acid) shall be the same as those specified in subparagraph (15) of this paragraph.

**1910.261(g)(16)(ii)** The pressure tanks-accumulators shall be inspected twice annually. (See the ASME Boiler and Pressure Vessel Code, Section VIII, Unfired Pressure Vessels—1968, with Addenda.)

**1910.261(g)(17)** *Pressure vessels (safety devices).* **(i)** A safety valve shall be installed in a separate line from each pressure vessel; no hand valve shall be installed between this safety valve and the pressure vessel. Safety valves shall be checked between each cook to be sure they have not become plugged or corroded to the point of being inoperative. (See the ASME Boiler and Pressure Vessel Code, Section VIII, Unfired Pressure Vessels—1968, with Addenda.)

**1910.261(g)(17)(ii)** All safety devices shall conform to Paragraph U-2 in the ASME Boiler and Pressure Vessel Code, Section VIII, Unfired Pressure Vessels—1968, with Addenda.

**1910.261(g)(18)** *Miscellaneous.* Insofar as the processes of the sulfate and soda operations are similar to those of the sulfite processes, the standard of paragraphs (g) (1) through (17) of this section shall apply.

**1910.261(g)(18)(i)** Quick operating showers, bubblers, etc., shall be available for emergency use in case of caustic soda burns.

**1910.261(g)(18)(ii)** Rotary tenders, smelter operators, and those cleaning smelt spouts shall be provided with eye protection equipment (fitted with lenses that filter out the harmful rays emanating from the light source) when actively engaged in their duties, in accordance with American National Standard Z87.1—1968.

**1910.261(g)(18)(iii)** Heavy-duty pipe, valves, and fittings shall be used between digester and blow pit. These shall be inspected at least semiannually to determine the degree of deterioration and repaired or replaced when necessary, in accordance with American National Standards B31.1—1955, B31.1a—1963, B31.1.0—1967, and B31.2—1968.

**1910.261(g)(18)(iv)** Smelt-dissolving tanks shall be covered and the cover kept closed, except when samples are being taken.

**1910.261(g)(18)(v)** Smelt tanks shall be provided with vent stacks and explosion doors, in accordance with American National Standard Z9.1—1951.

**1910.261(g)(19)** *Blow lines.*

**1910.261(g)(19)(i)-(ii)** [Reserved]

**1910.261(g)(19)(iii)** When blow lines from more than one digester lead into one pipe, the cock or valve of the blow line from the tank being inspected or repaired shall be locked or tagged out, or the line shall be disconnected and blocked off.

**1910.261(g)(20)** *Furnace room.* Exhaust ventilation shall be provided where niter cake is fed into a rotary furnace and shall be so designed and maintained as to keep the concentration of hydrogen sulfide gas below the parts per million listed in §1910.1000.

**1910.261(g)(21)** *Inspection and repair of tanks.* All piping leading to tanks shall be blanked off or valved and locked or tagged. Any lines to sewers shall be blanked off to protect workers from air contaminants.

**1910.261(g)(22)** *Welding.* Welding on blow tanks, accumulator tanks, or any other vessels where turpentine vapor or other combustible vapor could gather shall be done only after the vessel has been completely purged of fumes. Fresh air shall be supplied workers inside of vessels.

**1910.261(g)(23)** *Turpentine systems and storage tanks.* Nonsparking tools and ground hose shall be used when pumping out the tank. The tank shall be surrounded by a berm or moat.

**1910.261(h)** **Bleaching**—**(1)** *Bleaching engines.* Bleaching engines, except the Bellmer type, shall be completely covered on the top, with the exception of one small opening large enough to allow filling but too small to admit a man. Platforms leading from one engine to another shall have standard guardrails, in accordance with American National Standard A12.1—1967.

**1910.261(h)(2)** *Bleach mixing rooms.* **(i)** The room in which the bleach powder is mixed shall be provided with adequate exhaust ventilation, located at the floor level, in accordance with American National Standard Z9.1—1951.

**1910.261(h)(2)(ii)** Chlorine gas shall be carried away from the work place and breathing area by an exhaust system. The gas shall be rendered neutral or harmless before being discharged into the atmosphere. The requirements of American National Standard Z9.2—1960 shall apply to this subdivision.

**1910.261(h)(2)(iii)** For emergency and rescue operations, the employer must provide employees with self-contained breathing apparatuses or supplied-air respirators, and ensure that employees use these respirators, in accordance with the requirements of 29 CFR 1910.134.

**1910.261(h)(3)** *Liquid chlorine.* **(i)** Tanks of liquid chlorine shall be stored in an adequately ventilated unoccupied room, where their possible leakage cannot affect workers.

**1910.261(h)(3)(ii)** Gas masks capable of absorbing chlorine shall be supplied, conveniently placed, and regularly inspected, and workers who may be exposed to chlorine gas shall be instructed in their use.

**1910.261(h)(3)(iii)** For emergency and rescue work, independent self-contained oxygen-type masks or supplied air equipment shall be provided.

**1910.261(h)(3)(iv)** At least two exits, remote from each other, shall be provided for all rooms in which chlorine is stored.

**1910.261(h)(3)(v)** Spur tracks upon which tank cars containing chlorine and caustic are spotted and connected to pipelines shall be protected by means of a derail in front of the cars.

**1910.261(h)(3)(vi)** All chlorine, caustic, and acid lines shall be marked for positive identification, in accordance with American National Standard A13.1—1967.

**1910.261(h)(4)** *Bagged or drummed chemicals.* Bagged or drummed chemicals require efficient handling to prevent damage and spillage. Certain oxidizing chemicals used in bleaching pulp and also in some sanitizing work require added precautions for safety in storage and handling. In storage, these chemicals must be isolated from combustible materials and other chemicals with which they will react such as acids. They must also be kept dry, clean and uncontaminated.

**1910.261(i)** **Mechanical pulp process**—**(1)** *Pulp grinders.* **(i)** Water wheels directly connected to pulp grinders shall be provided with speed governors limiting the peripheral speed of the grinder to that recommended by the manufacturer.

**1910.261(i)(1)(ii)** Doors of pocket grinders shall be arranged so as to keep them from closing accidentally.

**1910.261(i)(2)** *Butting saws.* Hood guards shall be provided on butting saws, in accordance with American National Standard O1.1—1954 (reaffirmed 1961).

**1910.261(i)(3)** *Floors and platforms.* The requirements of paragraph (b)(3) of this section shall apply.

**1910.261(i)(4)** *Personal protection.* Persons exposed to falling material shall wear eye, head, foot, and shin protection equipment, in accordance with American National Standards Z87.1—1968, Z88.2—1969, Z89.1—1969, and Z41.1—1967.

**1910.261(j)** **Stock preparation**—**(1)** *Pulp shredders.* **(i)** Cutting heads shall be completely enclosed except for an opening at the feed side sufficient to permit only entry of stock. The enclosure shall be bolted or locked in place. The enclosure shall be of solid material or with mesh or other openings not exceeding one-half inch.

**1910.261(j)(1)(ii)** Either a slanting feed table with its outer edge not less than 36 inches from the cutting head or an automatic feeding device shall be provided.

**1910.261(j)(1)(iii)** Repairs for cleaning of blockage shall be done only when the shredder is shutdown and control devices locked.

**1910.261(j)(2)** *Pulp conveyors.* Pulp conveyors and conveyor drive belts and pulleys shall be fully enclosed, or if open and within 7 feet of the floor, shall be constructed and guarded in accordance with American National Standard B20.1—1957.

**1910.261(j)(3)** [Reserved]

**1910.261(j)(4)** *Beaters.* **(i)** Beater rolls shall be provided with covers.

**1910.261(j)(4)(ii)** When cleaning, inspecting, or other work requires that persons enter the beaters, all control devices shall be locked or tagged out, in accordance with paragraph (b)(4) of this section.

**1910.261(j)(4)(iii)** When beaters are fed from floor above, the chute opening, if less than 42 inches from the floor, shall be provided with a complete rail or other enclosure. Openings for manual feeding shall be sufficient only for entry of stock and shall be provided with at least two permanently secured crossrails, in accordance American National Standard A12.1—1967.

**1910.261(j)(4)(iv)** [Reserved]

**1910.261(j)(4)(v)** Floors around beaters shall be provided with sufficient drainage to remove wastes.

**1910.261(j)(5)** *Pulpers.* **(i)** All pulpers having the top or any other opening of vessel less than 42 inches from the floor or work platform shall have such openings guarded by railed or other enclosures. For manual charging, openings shall be sufficient only to permit the entry of stock and shall be provided with at least two permanently secured crossrails, in accordance with American National Standard A12.1—1967.

**1910.261(j)(5)(ii)** When cleaning, inspecting, or other work requires that persons enter the pulpers, they shall be equipped with safety belt and lifeline, and one person shall be stationed outside at a position to observe and assist in case of emergency.

**1910.261(j)(5)(iii)** When cleaning, inspecting, or other work requires that persons enter pulpers, all steam, water, or other control devices shall be locked or tagged out. Blank flanging and tagging of pipe lines is acceptable in place of closed and locked or tagged valves. Blank flanging of steam and water lines shall be acceptable in place of valve locks.

**1910.261(j)(6)** *Stock chests.* (i) All control devices shall be locked or tagged out when persons enter stock chests, in accordance with paragraph (b)(4) of this section.

**1910.261(j)(6)(ii)** When cleaning, inspecting, or other work requires that persons enter stock chests, they shall be provided with a low-voltage extension light.

**1910.261(k)** **Machine room**—**(1)** *Emergency stops.* Paper machines shall be equipped with devices that will stop the machine quickly in an emergency. The devices shall consist of push buttons for electric motive power (or electrically operated engine stops), pull cords connected directly to the prime mover, control clutches, or other devices, interlocked with adequate braking action. The devices shall be tested periodically by making use of them when stopping the machine and shall be so located that any person working on the machine can quickly disconnect the machine from the source of power in case of emergency.

**1910.261(k)(2)** *Drives.* (i) All drives shall be provided with lockout devices at the power switch which interrupts the flow of current to the unit.

**1910.261(k)(2)(ii)** All ends of rotating shafts including dryer drum shafts shall be completely guarded.

**1910.261(k)(2)(iii)** All accessible disengaged doctor blades should be covered.

**1910.261(k)(2)(iv)** All exposed shafts shall be guarded. Crossovers shall be provided.

**1910.261(k)(2)(v)** Oil cups and grease fittings shall be placed in a safe area remote from nip and heat hazards.

**1910.261(k)(3)** *Protective equipment.* Face shields, aprons, and rubber gloves shall be provided for workmen handling acids in accordance with paragraphs (b)(2) and (d)(1) of this section.

**1910.261(k)(4)-(5)** [Reserved]

**1910.261(k)(6)** *Steps.* Steps of uniform rise and tread with nonslip surfaces shall be provided at each press in accordance with American National Standard A12.1—1967.

**1910.261(k)(7)** *Plank walkways.* A removable plank shall be provided along each press, with standard guardrails installed. The planks shall have nonslip surfaces in accordance with paragraph (b)(3) of this section.

**1910.261(k)(8)** *Dryer lubrication.* If a gear bearing must be oiled while the machine is in operation, an automatic oiling device to protect the oiler shall be provided, or oil cups and grease fittings shall be placed along the walkways out of reach of hot pipes and dryer gears.

**1910.261(k)(9)** *Levers.* All levers carrying weights shall be constructed so that weights will not slip or fall off.

**1910.261(k)(10)** *First dryer.* Either a permanent guardrail or apron guard or both shall be installed in front of the first dryer in each section in accordance with paragraph (b)(1) of this section.

**1910.261(k)(11)** *Steam and hot-water pipes.* All exposed steam and hot-water pipes within 7 feet of the floor or working platform or within 15 inches measured horizontally from stairways, ramps, or fixed ladders shall be covered with an insulating material, or guarded in such manner as to prevent contact.

**1910.261(k)(12)** *Dryer gears.* Dryer gears shall be guarded excepting where the oilers' walkway is removed out of reach of the gears' nips and spokes and hot pipes in accordance with American National Standard B15.1—1953 (reaffirmed 1958).

**1910.261(k)(13)** *Broke hole.* (i) A guardrail shall be provided at broke holes in accordance with American National Standard A12.1—1967.

**1910.261(k)(13)(ii)** Where pulpers are located directly below the broke hole on a paper machine and where the broke hole opening is large enough to permit a worker to fall through, any employee pushing broke down the hole shall wear a safety belt attached to a safety belt line. The safety belt line shall be fastened in such a manner that it is impossible for the person to fall into the pulper.

**1910.261(k)(13)(iii)** An alarm bell or a flashing light shall be actuated before dropping material through the broke hole.

**1910.261(k)(14)** *Feeder belt.* A feeder belt or other effective device shall be provided for starting paper through the calender stack.

**1910.261(k)(15)** *Steps.* Steps or ladders of uniform rise and tread with nonslip surfaces shall be provided at each calender stack. Handrails and hand grips shall be provided at each calender stack in accordance with American National Standard A12.1—1967.

**1910.261(k)(16)** [Reserved]

**1910.261(k)(17)** *Sole plates.* All exposed sole plates between dryers, calenders, reels, and rewinders shall have a nonskid surface.

**1910.261(k)(18)** *Nip points.* The hazard of the nip points on all calender rolls shall be eliminated or minimized by means of an effective barrier device, or by feeding the paper into the rolls by means of a rope carrier, air jets, or hand feeding devices.

**1910.261(k)(19)** *Platforms.* [Reserved]

**1910.261(k)(20)** *Scrapers.* Alloy steel scrapers with pullthrough blades approximately 3 by 5 inches in size shall be used to remove "scabs" from calender rolls.

**1910.261(k)(21)** *Illumination.* Permanent lighting shall be installed in all areas where employees are required to make machine adjustments and sheet transfers in accordance with the American National Standard A11.1—1965 (R 1970).

**1910.261(k)(22)** *Control panels.* All control panel handles and buttons shall be protected from accidental contact.

**1910.261(k)(23)** [Reserved]

**1910.261(k)(24)** *Lifting reels.* **(i)** The reels shall stop rotating before being lifted from bearings.

**1910.261(k)(24)(ii)** All lifting equipment (clamps, cables, and slings) shall be maintained in a safe condition and inspected regularly.

**1910.261(k)(24)(iii)** Reel shafts with square block ends shall be guarded.

**1910.261(k)(25)** *Feeder belts.* Feeder belts, carrier ropes, air carriage, or other equally effective means shall be provided for starting paper into the nip or drum-type reels.

**1910.261(k)(26)** *Inrunning nip.* **(i)** Where the nipping points of all drum winders and rewinders is on the operator's side, it shall be guarded by barrier guards interlocked with the drive mechanism.

**1910.261(k)(26)(ii)** [Reserved]

**1910.261(k)(27)** *Core collars.* Set screws for securing core collars to winding and unwinding shafts shall not protrude above the face of the collar. All edges of the collar with which an operator's hand comes in contact shall be beveled to remove all sharp corners.

**1910.261(k)(28)** *Slitter knives.* Slitter knives shall be guarded so as to prevent accidental contact. Carriers shall be provided and used for transportation of slitter knives.

**1910.261(k)(29)** *Winder shaft.* The winder shall have a guide rail to align the shaft for easy entrance into the opened rewind shaft bearing housings.

**1910.261(k)(30)** *Core shaft.* When the core shaft weighs in excess of the safe standard, a mechanical device such as a dolly shall be provided for carrying all or part of the weight when it is being removed from the set of paper and placed in the dressing brackets on the winder.

**1910.261(k)(31)** *Winder area.* A nonskid surface shall be provided in the front vicinity of the winder to prevent accidental slipping.

**1910.261(k)(32)** *Radiation.* Special standards regarding the use of radiation equipment shall be posted and followed as required by §1910.96.

**1910.261(l)** *Finishing room*—**(1)** *Cleaning rolls.* Rolls shall be cleaned only on the outrunning side.

**1910.261(l)(2)** *Emergency stops.* Electrically or manually operated quick power disconnecting devices, interlocked with braking action, shall be provided on all operating sides of the machine within easy reach of all employees. These devices shall be tested by making use of them when stopping the machine.

**1910.261(l)(3)** *Core collars.* The requirements of paragraph (k)(27) of this section and the American National Standard B15.1—1953 (reaffirmed 1958) shall apply.

**1910.261(l)(4)** *Elevators.* These shall be in accordance with American National Standard A17.1—1965.

**1910.261(l)(5)** *Control panels.* The requirements of paragraph (k)(22) of this section shall apply.

**1910.261(l)(6)** *Guillotine-type cutters.* **(i)** Each guillotine-type cutter shall be equipped with a control which requires the operator and his helper, if any, to use both hands to engage the clutch.

**1910.261(l)(6)(ii)** Each guillotine-type cutter shall be equipped with a non-repeat device.

**1910.261(l)(6)(iii)** Carriers shall be provided and used for transportation of guillotine-type cutter knives.

**1910.261(l)(7)** *Rotary cutter.* **(i)** On single-knife machines a guard shall be provided at a point of contact to the knife.

**1910.261(l)(7)(ii)** On duplex cutters the protection required for single-knife machines shall be provided for the first knife, and a hood shall be provided for the second knife.

**1910.261(l)(7)(iii)** Safe access shall be provided to the knives of a rotary cutter by means of catwalks with nonslip surfaces, railings, and toeboards in accordance with paragraph (b)(3) of this section.

**1910.261(l)(7)(iv)** A guard shall be provided for the spreader or squeeze roll at the nip side on sheet cutters.

**1910.261(l)(7)(v)** Electrically or manually operated quick power disconnecting devices with adequate braking action shall be provided on all operating sides of the machine within easy reach of all operators.

**1910.261(l)(7)(vi)** The outside slitters shall be guarded.

**1910.261(l)(8)** *Platers.* **(i)** A guard shall be arranged across the face of the rolls to serve as a warning that the operator's hand is approaching the danger zone.

**1910.261(l)(8)(ii)** A quick power disconnecting device shall be installed on each machine within easy reach of the operator.

**1910.261(l)(9)** *Finishing room rewinders.* **(i)** The nipping points of all drum winders and rewinders located on the operator's side shall be guarded by either automatic or manually operated barrier guards of sufficient height to protect fully anyone working around them. The barrier guard shall be interlocked with the drive mechanism to prevent operating above jog speed without the guard in place.

A zero speed switch should be installed to prevent the guard from being raised while the roll is turning.

**1910.261(l)(9)(ii)** A nonskid surface shall be provided in front of the rewinder to prevent an employee from slipping in accordance with paragraph (b)(3) of this section.

**1910.261(l)(9)(iii)** Mechanical lifting devices shall be provided for placing and removing rolls from the machine.

**1910.261(l)(10)** *Control panels.* The requirements of paragraph (k)(22) of this section shall apply.

**1910.261(l)(11)** *Roll-type embosser.* The nipping point located on the operator's side shall be guarded by either automatic or manually operated barrier guards interlocked with the drive.

**1910.261(l)(12)** *Sorting and counting tables.* **(i)** Tables shall be smooth and free from splinters, with edges and corners rounded.

**1910.261(l)(12)(ii)** Paddles shall be smooth and free from splinters.

**1910.261(l)(13)** *Roll splitters.* The nip point and cutter knife shall be guarded by either automatic or manually operated barrier guards.

**1910.261(m)** **Materials handling**—**(1)** *Hand trucks.* No person shall be permitted to ride on a powered hand truck unless it is so designed by the manufacturer. A limit switch shall be on operating handle—30 degrees each way from a 45-degree angle up and down.

**1910.261(m)(2)** [Reserved]

**1910.261(m)(3)** *Cartons.* The carton-stitching machine shall be guarded to prevent the operator from coming in contact with the stitching head.

**1910.261(m)(4)** [Reserved]

**1910.261(m)(5)** *Unloading cars.* Flag signals, derails, or other protective devices shall be used to protect men during switching operations.

The blue flag policy shall be invoked according to paragraph (c)(9)(i) of this section.

[39 FR 23502, June 27, 1974, as amended at 40 FR 23073, May 28, 1975; 43 FR 49751, Oct. 24, 1978; 49 FR 5323, Feb. 10, 1984; 55 FR 32015, Aug. 6, 1990; 61 FR 9241, Mar. 7, 1996; 63 FR 1285, Jan. 8, 1998; 63 FR 33467, June 18, 1998]

### §1910.262 Textiles.

**1910.262(a)** *Application requirements*—**(1)** *Application.* The requirements of this subpart for textile safety apply to the design, installation, processes, operation, and maintenance of textile machinery, equipment, and other plant facilities in all plants engaged in the manufacture and processing of textiles, except those processes used exclusively in the manufacture of synthetic fibers.

**1910.262(a)(2)** *Standards incorporated by reference.* Standards covering issues of occupational safety and health which are of general application without regard to any specific industry are incorporated by reference in paragraphs of this section and made applicable to textiles. All such standards shall be construed according to the rules of construction set out in §1910.5.

**1910.262(b)** *Definitions applicable to this section*—**(1)** *Belt shifter.* A *belt shifter* is a device for mechanically shifting a belt from one pulley to another.

**1910.262(b)(2)** *Belt shifter lock.* A *belt shifter lock* is a device for positively locking the belt shifter in position while the machine is stopped and the belt is idling on the loose pulleys.

**1910.262(b)(3)** *Calender.* A *calender* in essence consists of a set of heavy rollers mounted on vertical side frames and arranged to pass cloth between them. Calenders may have two to ten rollers, or bowls, some of which can be heated.

**1910.262(b)(4)** *Embossing calender.* An *embossing calender* is a calender with two or more rolls, one of which is engraved for producing figured effects of various kinds on a fabric.

**1910.262(b)(5)** *Cans (drying).* Drying *cans* are hollow cylindrical drums mounted in a frame so they can rotate. They are heated with steam and are used to dry fabrics or yarn as it passes around the perimeter of the can.

**1910.262(b)(6)** *Carbonizing. Carbonizing* means the removing of vegetable matter such as burns, straws, etc., from wool by treatment with acid, followed by heat. The undesired matter is reduced to a carbon-like form which may be removed by dusting or shaking.

**1910.262(b)(7)** *Card.* A *card* machine consists of cylinders of various sizes—and in certain cases flats—covered with card clothing and set in relation to each other so that fibers in staple form may be separated into individual relationship. The speed of the cylinders and their direction of rotation varies. The finished product is delivered as a sliver. Cards of different types are: The revolving flat card, the roller-and-clearer card, etc.

**1910.262(b)(8)** *Card clothing. Card clothing* is the material with which many of the surfaces of a card are covered; e.g., the cylinder, doffer, etc. It consists of a thick foundation material, usually made of textile fabrics, through which are pressed many fine, closely spaced, specially bent wires.

**1910.262(b)(9)** *Comber.* A *comber* is a machine for combing fibers of cotton, wool, etc. The essential parts are a device for feeding forward a fringe of fibers at regular intervals and an arrangement of combs or pins which, at the right time, pass through the fringe. All tangled fibers, short fibers, and neps are removed and the long fibers are laid parallel.

**1910.262(b)(10)** *Combing machinery. Combing machinery* is a general classification, including combers, sliver lap machines, ribbon lap machines, and gill boxes, but excluding cards.

**1910.262(b)(11)** *Cutter (rotary staple).* A rotary staple *cutter* is a machine consisting of one or more rotary blades used for the purpose of cutting textile fibers into staple lengths.

**1910.262(b)(12)** *Exposed to contact. Exposed to contact* shall mean that the location of an object, material, nip point, or point of operation is such that a person is liable to come in contact with it in his normal course of employment.

**1910.262(b)(13)** *Garnett machine.* A *Garnett machine* means any of a number of types of machines for opening hard twisted waste of wool, cotton, silk, etc. Essentially, such machines consist of a lickerin; one or more cylinders, each having a complement worker and stripper rolls; and a fancy roll and doffer. The action of such machines is somewhat like that of a wool card, but it is much more severe in that the various rolls are covered with garnett wire instead of card clothing.

**1910.262(b)(14)** *Gill box.* A *gill box* is a machine used in the worsted system of manufacturing yarns. Its function is to arrange the fibers in parallel order. Essentially, it consists of a pair of feed rolls and a series of followers where the followers move at a faster surface speed and perform a combing action.

**1910.262(b)(15)** *Interlock.* An *interlock* is a device that operates to prevent the operation of machine while the cover or door of the machine is open or unlocked, and which will also hold the cover or door closed and locked while the machine is in motion.

**1910.262(b)(16)** *Jig (dye).* A dye *jig* is a machine for dyeing piece goods. The cloth, at full width, passes from a roller through the dye liquor in an open vat and is then wound on another roller. The operation is repeated until the desired shade is obtained.

**1910.262(b)(17)** *Kier.* A *kier* is a large metal vat, usually a pressure type, in which fabrics may be boiled out, bleached, etc.

**1910.262(b)(18)** *Lapper (ribbon).* A ribbon *lapper* is a machine used to prepare laps for feeding a cotton comb; its purpose is to provide a uniform lap in which the fibers have been straightened as much as possible.

**1910.262(b)(19)** *Lapper (sliver).* A sliver *lapper* is a machine in which a number of parallel card slivers are drafted slightly, laid side by side in a compact sheet, and wound into a cylindrical package.

**1910.262(b)(20)** *Loom.* A *loom* is a machine for effecting the interlacing of two series of yarns crossing one another at right angles. The warp yarns are wound on a warp beam and pass through heddles and reed. The filling is shot across in a shuttle and settled in place by reed and lay, and the fabric is wound on a cloth beam.

**1910.262(b)(21)** *Mangle (starch).* A starch *mangle* is a mangle that is used specifically for starching cotton goods. It commonly consists of two large rolls and a shallow open vat with several immersion rolls. The vat contains the starch solution.

**1910.262(b)(22)** *Mangle (water).* A *water mangle* is a calender having two or more rolls used for squeezing water from fabrics before drying. Water mangles also may be used in other ways during the finishing of various fabrics.

**1910.262(b)(23)** *Mule.* A *mule* is a type of spinning frame having a head stock and a carriage as its two main sections. The head stock is stationary. The carriage is movable and it carries the spindles which draft and spin the roving into the yarn. The carriage extends over the whole width of the machine and moves slowly toward and away from the head stock during the spinning operation.

**1910.262(b)(24)** *Nip. Nip* shall mean the point of contact between two in-running rolls.

**1910.262(b)(25)** *Openers and pickers. Openers and pickers* means a general classification which includes breaker pickers, intermediate pickers, finisher pickers, single process pickers, multiple process pickers, willow machines, card and picker waste cleaners, thread extractors, shredding machines, roving waste openers, shoddy pickers, bale breakers, feeders, vertical openers, lattice cleaners, horizontal cleaners, and any similar machinery equipped with either cylinders, screen section, calender section, rolls, or beaters used for the preparation of stock for further processing.

**1910.262(b)(26)** *Paddler.* A *paddler* consists of a trough for a solution and two or more squeeze rolls between which cloth passes after being passed through a mordant or dye bath.

**1910.262(b)(27)** *Point of operation. Point of operation* shall mean that part of the machine where the work of cutting, shearing, squeezing, drawing, or manipulating the stock in any other way is done.

**1910.262(b)(28)** *Printing machine (roller type).* A *roller printing machine* is a machine consisting of a large central cylinder, or pressure bowl, around the lower part of the perimeter of which is placed a series of engraved color rollers (each having a color trough), a furnisher roller, doctor blades, etc. The machine is used for printing fabrics.

**1910.262(b)(29)** *Ranges (bleaching continuous). Continuous bleaching ranges* are of several types and may be made for cloth in rope or open-width form. The goods, after wetting out, pass through a squeeze roll into a saturator containing a solution of caustic soda and then to an enclosed J-box. A V-shaped arrangement is attached to the front part of the J-box for uniform and rapid saturation of the cloth with steam before it is packed down in the J-box. The cloth, in a single strand rope form, passes over a guide roll down the first arm of the "V" and up the second. Steam is injected into the "V" at the upper end of the second arm so that the cloth is rapidly saturated with steam at this point. The J-box capacity is such that cloth will remain hot for a sufficient time to complete the scouring action. It then passes a series of washers with a squeeze roll in between. The cloth then passes through a second set of saturator, J-box, and washer, where it is treated with the peroxide solution. By slight modification of the form of the unit, the same process can be applied to open-width cloth.

**1910.262(b)(30)** *Range (mercerizing).* A *mercerizing range* consists generally of a 3-bowl mangle, a tenter frame, and a number of boxes for washing and scouring. The whole setup is in a straight line and all parts operate continuously. The combination is used to saturate the cloth with sodium hydroxide, stretch it while saturated, and washing out most of the caustic before releasing tension.

**1910.262(b)(31)** *Sanforizing machine.* A *sanforizing machine* is a machine consisting of a large steam-heated cylinder, an endless, thick, woolen felt blanket which is in close contact with the cylinder for most of its perimeter, and an electrically heated shoe which presses the cloth against the blanket while the latter is in a stretched condition as it curves around feed-in roll.

**1910.262(b)(32)** *Shearing machine.* A *shearing machine* is a machine used in shearing cloth. Cutting action is provided by a number of steel blades spirally mounted on a roller. The roller rotates in close contact with a fixed ledger blade. There may be from one to six such rollers on a machine.

**1910.262(b)(33)** *Singeing machine.* A *singeing machine* is a machine used particularly with cotton; it comprises of a heated roller, plate, or an open gas flame. The material is rapidly passed over the roller or the plate or through the open gas flame to remove, fuzz or hairiness on yarn or cloth by burning.

**1910.262(b)(34)** *Slasher.* A *slasher* is a machine used for applying a size mixture to warp yarns. Essentially, it consists of a stand for holding section beams, a size box, one or more cylindrical dryers or an enclosed hot air dryer, and a beaming end for finding the yarn on the loom beams.

**1910.262(b)(35)** *Solvent (industrial organic).* Industrial organic solvent means any organic volatile liquid or compound, or any combination of these substances which are used to dissolve or suspend a nonvolatile or slightly volatile substance for industrial utilization. It shall also apply to such substances when used as detergents or cleansing agents. It shall not apply to petroleum products when such products are used as fuel.

**1910.262(b)(36)** *Tenter frame.* A *tenter frame* is a machine for drying cloth under tension. It essentially consists of a pair of endless traveling chains fitted with clips of fine pins and carried on tracks. The cloth is firmly held at the selvages by the two chains which diverge as they move forward so that the cloth is brought to the desired width.

**1910.262(b)(37)** *Warper.* A *warper* is any machine for preparing and arranging the yarns intended for the warp of a fabric, specifically, a beam warper.

**1910.262(c)** **General safety requirements**—**(1)** *Means of stopping machines.* Every textile machine shall be provided with individual mechanical or electrical means for stopping such machines. On machines driven by belts and shafting, a locking-type shifter or an equivalent positive device shall be used. On operations where injury to the operator might result if motors were to restart after power failures, provision shall be made to prevent machines from automatically restarting upon restoration of power.

**1910.262(c)(2)** *Handles.* Stopping and starting handles shall be designed to the proper length to prevent the worker's hand or fingers from striking against any revolving part, gear guard, or any other part of the machine.

**1910.262(c)(3)-(4)** [Reserved]

**1910.262(c)(5)** *Inspection and maintenance.* All guards and other safety devices, including starting and stopping devices, shall be properly maintained.

**1910.262(c)(6)** *Lighting.* Lighting shall conform to American National Standard A11.1—1965, which is incorporated by reference as specified in §1910.6.

**1910.262(c)(7)** *Identification of piping systems.* Identification of piping systems shall conform to American National Standard A13.1—1956, which is incorporated by reference as specified in §1910.6.

**1910.262(c)(8)** *Identification of physical hazards.* Identification of physical hazards shall be in accordance with the requirements of §1910.144.

**1910.262(c)(9)** *Steam pipes.* All pipes carrying steam or hot water for process or servicing machinery, when exposed to contact and located within seven feet of the floor or working platform shall be covered with a heat-insulating material, or otherwise properly guarded.

**1910.262(d)** **Openers and pickers**—**(1)** *Beater guards.* When any opening or picker machinery is equipped with a beater, such beater shall be provided with metal covers which will prevent contact with the beater. Such covers shall be provided with an interlock which will prevent the cover from being raised while the machine is in motion and prevent the operation of the machine while the cover is open.

**1910.262(d)(2)** *Cleanout holes.* Cleanout holes within reaching distance of the fan or picker beater shall have their covers securely fastened and they shall not be opened while the machine is in motion.

**1910.262(d)(3)** *Feed rolls.* The feed rolls on all opening and picking machinery shall be covered with a guard designed to prevent the operator from reaching the nip while the machinery is in operation.

**1910.262(d)(4)** *Removal of foreign ferrous material.* All textile opener lines shall be equipped with magnetic separators, tramp iron separators, or other means for the removal of foreign ferrous material.

**1910.262(e)** **Cotton cards**—**(1)** *Enclosures.* Cylinder and lickerins shall be completely protected and the doffers should be enclosed.

**1910.262(e)(2)** *Enclosure fastenings.* The enclosures or covers shall be kept in place while the machine is in operation, except when stripping or grinding.

**1910.262(e)(3)** *Stripping rolls.* On operations calling for flat strippings which are allowed to fall on the doffer cover, where such strippings are removed by hand, the doffer cover shall be kept closed and securely fastened to prevent the opening of the cover while the machine is in operation. When it becomes necessary to clean the cards while they are in motion, a long-handled brush or dust mop shall be used.

**1910.262(f)** **Garnett machines**—**(1)** *Lickerin.* Garnett lickerins shall be enclosed.

**1910.262(f)(2)** *Fancy rolls.* Garnett fancy rolls shall be enclosed by covers. These shall be installed in a way that keeps worker rolls reasonably accessible for removal or adjustment.

**1910.262(f)(3)** *Underside of machine.* The underside of the garnett shall be guarded by a screen mesh or other form of enclosure to prevent access.

**1910.262(g)** **Spinning mules**—A substantial fender of metal or hardwood shall be installed in front of the carriage wheels, the fender to extend to within one-fourth inch of the rail.

**1910.262(h)** **Slashers**—**(1)** *Cylinder dryers*—**(i)** *Reducing valves, safety valves, and pressure gages.* Reducing valves, safety valves, and pressure gages shall conform to the ASME Pressure Vessel Code, Section VIII, Unfired Pressure Vessels, 1968, which is incorporated by reference as specified in §1910.6.

**1910.262(h)(1)(ii)** *Vacuum relief valves.* Vacuum relief valves shall conform to the ASME Code for Pressure Vessels, Section VIII, Unfired Pressure Vessels, 1968.

**1910.262(h)(1)(iii)** *Lever control.* When slashers are operated by control levers, these levers shall be connected to a horizontal bar or treadle located not more than 69 inches above the floor to control the operation from any point.

**1910.262(h)(1)(iv)** *Pushbutton control.* Slashers operated by pushbutton control shall have stop and start buttons located at each end of the machine, and additional buttons located on both sides of the machine, at the size box and the delivery end. If calender rolls are used, additional buttons shall be provided at both sides of the machine at points near the nips, except when slashers are equipped with an enclosed dryer.

**1910.262(h)(1)(v)** *Nip guards.* All nip guards shall comply with the requirements of paragraph (h)(2)(iv) of this section.

**1910.262(h)(1)(vi)** *Cylinder enclosure.* When enclosures or hoods are used over cylinder drying rolls, such enclosures or hoods shall be provided with an exhaust system which will effectively prevent wet air and steam from escaping into the workroom.

**1910.262(h)(1)(vii)** *Expansion chambers.* Slasher kettles and cookers shall be provided with expansion chambers in the covers, or drains, to prevent surging over. Steam-control valves shall be so located that they can be operated without exposing the worker to moving parts, hot surfaces, or steam.

**1910.262(h)(2)** *Enclosed hot air dryer*—**(i)** *Lever control.* When slashers are operated by control levers, these levers shall be connected to a horizontal

# OSHA Standards for General Industry

bar or treadle located not more than 69 inches above the floor to control the operation from any point.

**1910.262(h)(2)(ii)** *Push-button control.* Slashers operated by push-button control shall have one start button at each end of the machine and stop buttons shall be located on both sides of the machines at intervals spaced not more than 6 feet on centers. Inching buttons should be installed.

**1910.262(h)(2)(iii)** *Dryer enclosure.* The dryer enclosure shall be provided with an exhaust system which will effectively prevent wet air and steam from escaping into the workroom.

**1910.262(h)(2)(iv)** *Nip guards.* All nip guards shall comply with Table R-1.

### TABLE R-1—GUARD OPENINGS
[Openings in the guard or between the guard and working surface shall not be greater than the following]

| Distance of opening from nip point | Maximum width of opening |
|---|---|
| 0 to 1 1/2 | 1/4 |
| 1 1/2 to 2 1/2 | 3/8 |
| 2 1/2 to 3 1/2 | 1/2 |
| 3 1/2 to 5 1/2 | 5/8 |
| 5 1/2 to 6 1/2 | 3/4 |
| 6 1/2 to 7 1/2 | 7/8 |
| 7 1/2 to 8 1/2 | 1 1/4 |

The measurements in Table R-1 are all in inches.

**1910.262(h)(2)(v)** *Expansion chambers.* Slasher kettles and cookers shall be provided with expansion chambers in the covers, or drains, to prevent surging over. Steam control valves shall be so located that they can be operated without exposing the worker to moving parts, hot surfaces, or steam.

**1910.262(i)** *Warpers*—**(1)** *Swiveled double-bar gates.* Swiveled double-bar gates shall be installed on all warpers operating in excess of 450 yards per minute. These gates shall be so interlocked that the machine cannot be operated until the gate is in the "closed position," except for the purpose of inching or jogging.

**1910.262(i)(2)** *Closed position. Closed position* shall mean that the top bar of he gate shall be at least 42 inches from the floor or working platform; and the lower bar shall be at least 21 inches from the floor or working platform; and the gate shall be located 15 inches from the vertical tangent to the beam head.

**1910.262(j)** *Drawing frames, slubbers, roving parts, cotton combers, ring spinning frames, twisters.* Gear housing covers on all installations of drawing frames, slubbers, roving frames, cotton combers, ring spinning frames, and twisters shall be equipped with interlocks.

**1910.262(k)** *Gill boxes*—**(1)** *Pin guard.* A guard shall be placed ahead of the feed end and shall be so designed that it will prevent the worker's fingers from being caught in the pins of the intersecting fallers.

**1910.262(k)(2)** *Nip guards.* All nip guards shall comply with the requirements of paragraph (h)(2)(iv) of this section.

**1910.262(l)** *Heavy draw boxes, finishers, and speeders used in worsted drawing*—**(1)** *Band pulley covers.* Covers for band pulleys shall be closed when the machine is in motion.

**1910.262(l)(2)** *Benches or working platforms.* Branches or working platforms approximately 10 inches in height and 8 inches in width should be installed along the entire running length of the machine for the worker to stand on while creeling the machine. Such benches or platforms shall be covered with an abrasive or nonslip material.

**1910.262(m)** *Sliver and ribbon lappers (cotton).* Cover guard. An interlocking cover guard shall be installed over the large calender drums and the lap spool, designed to prevent the operator from coming in contact with the nip.

**1910.262(n)** *Looms*—**(1)** *Shuttle guard.* Each loom shall be equipped with a guard designed to minimize the danger of the shuttle flying out of the shed.

**1910.262(n)(2)** *Protection for loom fixer.* Provisions shall be made so that every loom fixer can prevent the loom from being started while he is at work on the loom. This may be accomplished by means of a lock, the key to which is retained in the possession of the loom fixer, or by some other effective means to prevent starting the loom.

**1910.262(o)** *Shearing machines.* All revolving blades on shearing machines shall be guarded so that the opening between the cloth surface and the bottom of the guard will not exceed three-eighths inch.

**1910.262(p)** *Continuous bleach range (cotton and rayon)*—**(1)** *J-box protection.* Each valve controlling the flow of steam, injurious gases, or liquids into a J-box shall be equipped with a chain, lock, and key, so that any worker who enters the J-box can lock the valve and retain the key in his possession. Any other method which will prevent steam, injurious gases, or liquids from entering the J-box while the worker is in it will be acceptable.

**1910.262(p)(2)** *Open-width bleaching.* The nip of all in-running rolls on open-width bleaching machine rolls shall be protected with a guard to prevent the worker from being caught at the nip. The guard shall extend across the entire length of the nip.

**1910.262(q)** *Kiers*—**(1)** *Reducing valves, safety valves, and pressure gages.* Reducing valves, safety valves, and pressure gages shall conform to the ASME Code for Unfired Pressure Vessels, Section VIII, Unfired/Pressure Vessels, 1968.

**1910.262(q)(2)** *Kier valve protection.* Each valve controlling the flow of steam, injurious gases, or liquids into a kier shall be equipped with a chain, lock, and key, so that any worker who enters the kier can lock the valve and retain the key in his possession. Any other method which will prevent steam, injurious gases, or liquids from entering the kier while the worker is in it will be acceptable.

**1910.262(r)** *Gray and white bins.* On new installations guard rails conforming to §1910.23 shall be provided where workers are required to plait by hand from the top of the bin so as to protect the worker from falling to a lower level.

**1910.262(s)** *Mercerizing range (piece goods)*—**(1)** *Stopping devices.* A stopping device shall be provided at each end of the machine.

**1910.262(s)(2)** *Frame ends.* A guard shall be installed at each end of the frame between the in-running chain and the clip opener, to prevent the worker's fingers from being caught.

**1910.262(s)(3)** *Mangle and washers.* The nip at the in-running rolls shall conform to §1910.264.

**1910.262(t)** *Tenter frames*—**(1)** *Stopping devices.* A stopping device shall be provided at each end of the machine.

**1910.262(t)(2)** *Frame ends.* A guard shall be installed at each end of the frame at the in-running chain and clip opener.

**1910.262(t)(3)** *Oil cups.* Oil cups shall be safely located to permit easy access.

**1910.262(u)** *Dyeing jigs*—**(1)** *Stopping devices.* Each dye jig shall be equipped with individual mechanical or electrical means for stopping the machine.

**1910.262(u)(2)** *Roll arms.* Roll arms on jigs shall be built to allow for extra large batches, and to prevent the center bar from being forced off, causing the batch to fall.

**1910.262(v)** *Padders*—*Nip guards.* All nip guards shall comply with the requirements of paragraph (h)(2)(iv) of this section.

**1910.262(w)** *Drying cans*—**(1)** *Pressure reducing valves and pressure gages.* Pressure reducing valves and pressure gages shall conform to the ASME Code for Pressure Vessels, Section VIII, 1968, Unfired Pressure Vessels.

**1910.262(w)(2)** *Vacuum collapse.* If cans are not designed to prevent vacuum collapse, each can shall be equipped with one or more vacuum relief valves with openings of sufficient size to prevent the collapse of the can if vacuum occurs.

**1910.262(x)** *Flat-work ironer*—**(1)** *Feed rolls.* The feed rolls shall be guarded to conform to §1910.264.

**1910.262(x)(2)** *Pressure rolls.* Pressure rolls shall be covered or guarded to conform to §1910.264.

**1910.262(y)** *Extractors*—**(1)** *Centrifugal extractor*—**(i)** *Cover.* Each extractor shall be equipped with a metal cover.

**1910.262(y)(1)(ii)** *Interlocking device.* Each extractor shall be equipped with an interlocking device that will prevent the cover from being opened while the basket is in motion, and also prevent the power operation of the basket while the cover is open.

**1910.262(y)(1)(iii)** *Brakes.* Each extractor shall be equipped with a mechanically or electrically operated brake to quickly stop the basket when the power driving the basket is shut off.

**1910.262(y)(1)(iv)** *Maximum allowable speed.* Each centrifugal extractor shall be effectively secured in position on the floor or foundation so as to eliminate unnecessary vibration, and should not be operated at a speed greater than the manufacturer's rating, which shall be stamped where easily visible in letters not less than one-quarter inch in height. The maximum allowable speed shall be given in revolutions per minute (rpm).

**1910.262(y)(2)** *Engine drum extractor—Over-speed governor.* Each engine individually driving an extractor shall be provided with an approved engine stop and speed limit governor.

**1910.262(y)(3)** *Squeezer or wringer extractor—Nip guards.* All nip guards shall comply with the requirements of paragraph (h)(2)(iv) of this section.

**1910.262(z)** *Nip guards.* All nip guards for water mangle, starch mangle, back-washer (worsted yarn) crabbing machines, decating machines, shall comply with the requirements of paragraph (h)(2)(iv).

**1910.262(aa)** *Sanforizing and palmer machine.* A safety trip rod, cable, or wire center cord shall be provided across the front and back of all palmer cylinders extending the length of the face of the cylinder. It shall operate readily whether pushed or pulled. This safety trip shall be not more than 72 inches above the level on which the operator stands and shall be readily accessible.

**1910.262(bb)** *Rope washers*—**(1)** *Splash guard.* Splash guards shall be installed on all rope washers unless the machine is so designed as to prevent the water or liquid from splashing the operator, the floor, or working surface.

**1910.262(bb)(2)** *Safety stop bar.* A safety trip rod, cable or wire center cord shall be provided across the front and back of all rope washers extending the length of the face of the washer. It shall operate readily whether pushed or pulled. This safety trip shall be not more than 72 inches above the level on which the operator stands and shall be readily accessible.

**1910.262(cc)** *Laundry washer tumbler or shaker*—**(1)** *Interlocking device.* Each drying tumbler, each double cylinder shaker or clothes tumbler, and each washing machine shall be equipped with an interlock device which will prevent the power operation of the inside cylinder when the outer door on the case or shell is open, and which will also prevent the outer door on the case or shell from being opened without shutting off the power.

**1910.262(cc)(2)** *Means of holding covers or doors in open position.* Each enclosed barrel shall also be equipped with adequate means for holding open the doors or covers of the inner and outer cylinders or shells while it is being loaded or unloaded.

**1910.262(dd)** *Printing machine (roller type)*—**(1)** *Nip guards.* All nip guards shall comply with the requirements of paragraph (h)(2)(iv) of this section.

**1910.262(dd)(2)** *Crown wheel and roller gear nip protection.* The engraved roller gears and the large crown wheel shall be provided with a protective disc which will enclose the nips of the in running gears. Individual discs for each nip will be acceptable.

**1910.262(ee)** *Calenders.* The nip at the in-running side of the rolls shall be provided with a guard extending across the entire length of the nip and arranged to prevent the fingers of the workers from being pulled in between the rolls or between the guard and the rolls, and constructed so that the cloth can be fed into the rolls safely.

**1910.262(ff)** *Rotary staple cutters.* A guard shall be installed completely enclosing the cutters to prevent the hands of the operator from reaching the cutting zone.

**1910.262(gg)** [Reserved]

**1910.262(hh)** *Hand bailing machine.* An angle-iron-handle stop guard shall be installed at the right angle to the frame of the machine. The stop guard shall be so designed and so located that it will prevent the handle from traveling beyond the vertical position should the handle slip from the operator's hand when the pawl has been released from the teeth of the takeup gear.

**1910.262(ii)** *Roll bench.* Cleats shall be installed on the ends of roll benches.

**1910.262(jj)** *Cuttle or swing folder (overhead type).* The bottom of the overhead folders shall be located not less than 7 feet from the floor or working surface.

**1910.262(kk)** *Color-mixing room.* Floors in color-mixing rooms shall be constructed to drain easily.

**1910.262(ll)** *Open tanks and vats for mixing and storage of hot or corrosive liquids—Shutoff valves.* Boiling tanks, caustic tanks, and hot liquid containers, so located that the operator cannot see the contents from the floor or working area, shall have emergency shutoff valves controlled from a point not subject to danger of splash. Valves shall conform to the ASME Pressure Vessel Code, section VIII, Unfired Pressure Vessels, 1968.

**1910.262(mm)** *Dye kettles and vats*—Pipes or drains of sufficient capacity to carry the contents safely away from the working area shall be installed where there are dye kettles and vats which may at any time contain hot or corrosive liquids. These shall not empty directly onto the floor.

**1910.262(nn)** *Acid carboys.* Carboys shall be provided with inclinators, or the acid shall be withdrawn from the carboys by means of pumping without pressure in the carboy, or by means of hand operated siphons.

**1910.262(oo)** *Handling caustic soda and caustic potash.* Means shall be provided for handling and emptying caustic soda and caustic potash containers to prevent workers from coming in contact with the caustic (see paragraph (qq) of this section).

**1910.262(pp)** *First aid.* Wherever acids or caustics are used, provision shall be made for a copious and flowing supply of fresh, clean water.

[39 FR 23502, June 27, 1974, as amended at 40 FR 23073, May 28, 1975; 49 FR 5324, Feb. 10, 1984; 61 FR 9241, Mar. 7, 1996; 63 FR 33467, June 18, 1998]

### §1910.263 Bakery equipment.

**1910.263(a)** *General requirements*—**(1)** *Application.* The requirements of this section shall apply to the design, installation, operation and maintenance of machinery and equipment used within a bakery.

**1910.263(a)(2)** [Reserved]

**1910.263(b)** [Reserved]

**1910.263(c)** *General machine guarding.*

**1910.263(c)(1)** [Reserved]

**1910.263(c)(2)** *Gears.* All gears shall be completely enclosed regardless of location.

**1910.263(c)(3)** *Sprockets and V-belt drives.* Sprockets and V-belt drives located within reach from platforms or pasageways or located within 8 feet 6 inches from the floor shall be completely enclosed.

**1910.263(c)(4)** [Reserved]

**1910.263(c)(5)** *Lubrication.* Where machinery must be lubricated while in motion, stationary lubrication fittings inside a machine shall be provided with extension piping to a point of safety so that the employee will not have to reach into any dangerous part of the machine when lubricating.

**1910.263(c)(6)-(7)** [Reserved]

**1910.263(c)(8)** *Hot pipes.* Exposed hot water and steam pipes shall be covered with insulating material wherever necessary to protect employee from contact.

**1910.263(d)** *Flour-handling equipment*—**(1)** *General requirements for flour handling.* (i) Wherever any of the various pieces of apparatus comprising a flour-handling system are run in electrical unity with one another the following safeguards shall apply:

**1910.263(d)(1)(i)***a* [Reserved]

**1910.263(d)(1)(i)***b* Wherever a flour-handling system is of such size that the beginning of its operation is far remote from its final delivery end, all electric motors operating each apparatus comprising this system shall be controlled at each of two points, one located at each remote end, either of which will stop all motors.

**1910.263(d)(1)(i)***c* [Reserved]

**1910.263(d)(1)(i)***d* Control circuits for magnetic controllers shall be so arranged that the opening of any one of several limit switches, which may be on an individual unit, will serve to de-energize all of the motors of that unit.

**1910.263(d)(1):***(ii)* [Reserved]

**1910.263(d)(2)** *Bag chutes and bag lifts (bag-arm elevators).* (i) Bag chutes (gravity chutes for handling flour bags) shall be so designed so as to keep to a minimum the speed of flour bags. If the chute inclines more than 30°

# OSHA Standards for General Industry

from the horizontal, there shall be an upturn at the lower end of the chute to slow down the bags.

**1910.263(d)(2)(ii)** Bag-arm elevators with manual takeoff shall be designed to operate at a capacity not exceeding seven bags per minute. The arms on the conveyor chain shall be so spaced as to obtain the full capacity of the elevator with the lowest possible chain speed. There shall be an electric limit switch at the unloading end of the bag-arm elevator so installed as to automatically stop the conveyor chain if any bag fails to clear the conveyor arms.

**1910.263(d)(2)(iii)** [Reserved]

**1910.263(d)(2)(iv)** Man lifts shall be prohibited in bakeries. Bag or barrel lifts shall not be used as man lifts.

**1910.263(d)(3)** *Dumpbin and blender.*

**1910.263(d)(3)(i)-(iv)** [Reserved]

**1910.263(d)(3)(v)** All dumpbin and blender hoods shall be of sufficient capacity to prevent circulation of flour dust outside the hoods.

**1910.263(d)(3)(vi)** All dumpbins shall be of a suitable height from floor to enable the operator to dump flour from bags, without causing undue strain or fatigue. Where the edge of any bin is more than 24 inches above the floor, a bag rest step shall be provided.

**1910.263(d)(3)(vii)** A control device for stopping the dumpbin and blender shall be provided close to the normal location of the operator.

**1910.263(d)(4)-(5)** [Reserved]

**1910.263(d)(6)** *Storage bins.*

**1910.263(d)(6)(i)** [Reserved]

**1910.263(d)(6)(ii)** Storage bins shall be provided with gaskets and locks or latches to keep the cover closed, or other equivalent devices in order to insure the dust tightness of the cover. Covers at openings where an employee may enter the bin shall also be provided with a hasp and a lock, so located that the employee may lock the cover in the open position whenever it is necessary to enter the bin.

**1910.263(d)(6)(iii)** Storage bins where the side is more than 5 feet in depth shall be provided with standard stationary safety ladders, both inside and outside, to reach from floor level to top of bin and from top of bin to inside bottom, keeping the ladder end away from the moving screw conveyor.

**1910.263(d)(6)(iv)-(v)** [Reserved]

**1910.263(d)(6)(vi)** The main entrance cover of large storage bins located at the interior exit ladder shall be provided with an electric interlock for motors operating both feed and unloading screw, so that these motors cannot operate while the cover is open.

**1910.263(d)(7)** *Screw conveyors.*

**1910.263(d)(7)(i)-(ii)** [Reserved]

**1910.263(d)(7)(iii)** The covers of all screw conveyors shall be made removable in convenient sections, held on with stationary clamps located at proper intervals keeping all covers dust-tight. Where drop or hinged bottom sections are provided this provision shall not apply.

**1910.263(d)(8)** *Sifters.* (i) Enclosures of all types of flour sifters shall be so constructed that they are dust-tight but readily accessible for interior inspection.

**1910.263(d)(8)(ii)** [Reserved]

**1910.263(d)(9)** *Flour scales.*

**1910.263(d)(9)(i)-(ii)** [Reserved]

**1910.263(d)(9)(iii)** Traveling or track-type flour scales shall be equipped with bar handles for moving same. The bar should be at least 1 inch in diameter and well away from trolley track wheels.

**1910.263(e)** *Mixers*—**(1)** *Horizontal dough mixers.* **(i)** Mixers with external power application shall have all belts, chains, gears, pulleys, sprockets, clutches, and other moving parts completely enclosed.

**1910.263(e)(1)(ii)** [Reserved]

**1910.263(e)(1)(iii)** Each mixer shall be equipped with an individual motor and control, and with a conveniently located manual switch to prevent the mixer from being started in the usual manner while the machine is being serviced and cleaned.

**1910.263(e)(1)(iv)** All electrical control stations shall be so located that the operator must be in full view of the bowl in its open position. No duplication of such controls other than a stop switch shall be permitted.

**1910.263(e)(1)(v)** All mixers with power and manual dumping arrangements shall be equipped with safety devices which shall:

**1910.263(e)(1)(v)(a)** Engage both hands of the operator, when the agitator is in motion under power, and while the bowl is opened more than one-fifth of its total opening.

**1910.263(e)(1)(v)(b)** Prevent the agitator from being started, while the bowl is more than one-fifth open, without engaging both hands of the operator;

**1910.263(e)(1)(vi)-(vii)** [Reserved]

**1910.263(e)(1)(viii)** Every mixer shall be equipped with a full enclosure over the bowl which is closed at all times while the agitator is in motion. Only minor openings in this enclosure, such as ingredient doors, flour inlets, etc., each representing less than 1 1/2 square feet in area, shall be capable of being opened while the mixer is in operation.

**1910.263(e)(1)(ix)** [Reserved]

**1910.263(e)(1)(x)** Overhead covers or doors which are subject to accidental closure shall be counterbalanced to remain in an open position or provided with means to hold them open until positively released by the operator.

**1910.263(e)(1)(xi)-(xvii)** [Reserved]

**1910.263(e)(1)(xviii)** Valves and controls to regulate the coolant in mixer jackets shall be located so as to permit access by the operator without jeopardizing his safety.

**1910.263(e)(2)** *Vertical mixers.* **(i)** Vertical mixers shall comply with paragraphs (e)(1) (i), (iii), (ix) and (x), of this section.

**1910.263(e)(2)(ii)** [Reserved]

**1910.263(e)(2)(iii)** Bowl locking devices shall be of a positive type which require the attention of the operator for unlocking.

**1910.263(e)(2)(iv)** Devices shall be made available for moving bowls weighing more than 80 pounds, with contents, into and out of the mixing position on the machine.

**1910.263(f)** *Dividers.*

**1910.263(f)(1)-(2)** [Reserved]

**1910.263(f)(3)** *Rear of divider.* The back of the divider shall have a complete cover to enclose all of the moving parts, or each individual part shall be enclosed or guarded to remove the separate hazards. The rear cover shall be provided with a limit switch in order that the machine cannot operate when this cover is open. The guard on the back shall be hinged so that it cannot be completely removed and if a catch or brace is provided for holding the cover open, it shall be designed so that it will not release due to vibrations or minor bumping whereby the cover may drop on an employee.

**1910.263(g)** *Moulders*—**(1)** *Hoppers.* Mechanical feed moulders shall be provided with hoppers so designed and connected to the proofer that an employee's hands cannot get into the hopper where they will come in contact with the in-running rolls.

**1910.263(g)(2)** *Hand-fed moulders.* Hand-fed moulders shall be provided with a belt-feed device or the hopper shall be extended high enough so that the hands of the operator cannot get into the feed rolls. The top edge of such a hopper shall be well rounded to prevent injury when it is struck or bumped by the employee's hand.

**1910.263(g)(3)** *Stopping devices.* There shall be a stopping device within easy reach of the operator who feeds the moulder and another stopping device within the reach of the employee taking the dough away from the moulder.

**1910.263(h)** *Manually fed dough brakes*—**(1)** *Top-roll protection.* The top roll shall be protected by a heavy gage metal shield extending over the roll to go within 6 inches of the hopper bottom board. The shield may be perforated to permit observation of the dough entering the rolls.

**1910.263(h)(2)** *Emergency stop bar*—An emergency stop bar shall be provided, and so located that the body of the operator will press against the bar if the operator slips and falls toward the rolls, or if the operator gets his hand caught in the rolls. The bar shall apply the body pressure to open positively a circuit that will deenergize the drive motor. In addition, a brake which is inherently self-engaging by requiring power or force from an external source to cause disengagement shall be activated at the same time causing the rolls to stop instantly. The emergency stop bar shall be checked for proper operation every 30 days.

**1910.263(i)** *Miscellaneous equipment*—**(1)** *Proof boxes.* All door locks shall be operable both from within and outside the box. Guide rails shall be installed to center the rack as it enters, passes through, and leaves the proof box.

**1910.263(i)(2)** *Fermentation room.* Fermentation room doors shall have nonshatterable wire glass or plastic panels for vision through doors.

**1910.263(i)(3)** *Troughs.* Troughs shall be mounted on antifriction bearing casters thus making it possible for the operator to move and direct the motion of the trough with a minimum of effort.

**1910.263(i)(4)** *Hand trucks.* **(i)** Casters shall be set back from corners to be out of the way of toes and heels, but not far enough back to cause the truck to be unstable.

**1910.263(i)(4)(ii)** A lock or other device shall be provided to hold the handle in vertical position when the truck is not in use.

**1910.263(i)(5)** *Lift trucks.* A lock or other device shall be provided to hold the handle in vertical position when the truck is not in use.

**1910.263(i)(6)** *Racks.*

**1910.263(i)(6)(i)** [Reserved]

**1910.263(i)(6)(ii)** Racks shall be equipped with handles so located with reference to the frame of the rack that no part of the operator's hands extends beyond the outer edge of the frame when holding onto the handles.

**1910.263(i)(6)(iii)** Antifriction bearing casters shall be used to give the operator better control of the rack.

**1910.263(i)(7)** *Conveyors.* **(i)** Wherever a conveyor passes over a main aisleway, regularly occupied work area, or passageway, the underside of the conveyor shall be completely enclosed to prevent broken chains or other material from falling in the passageway.

**1910.263(i)(7)(ii)** Stop bumpers shall be installed on all delivery ends of conveyors, wherever manual removal of the product carried is practiced.

**1910.263(i)(7)(iii)** Where hazard of getting caught exists a sufficient number of stop buttons shall be provided to enable quick stopping of the conveyor.

**1910.263(i)(8)-(10)** [Reserved]

**1910.263(i)(11)** *Ingredient premixers, emulsifiers, etc.* **(i)** All top openings shall be provided with covers attached to the machines. These covers should be so arranged and interlocked that power will be shut off whenever the cover is opened to a point where the operator's fingers might come in contact with the beaters.

**1910.263(i)(11)(ii)** [Reserved]

**1910.263(i)(12)** *Chain tackle.* **(i)** All chain tackle shall be marked prominently, permanently, and legibly with maximum load capacity.

**1910.263(i)(12)(ii)** All chain tackle shall be marked permanently and legibly with minimum support specification.

**1910.263(i)(12)(iii)** Safety hooks shall be used.

**1910.263(i)(13)** *Trough hoists, etc.* **(i)** All hoists shall be marked prominently, permanently, and legibly with maximum load capacity.

**1910.263(i)(13)(ii)** All hoists shall be marked permanently and legibly with minimum support specifications.

**1910.263(i)(13)(iii)** Safety catches shall be provided for the chain so that the chain will hold the load in any position.

**1910.263(i)(13)(iv)** Safety hooks shall be used.

**1910.263(i)(14)** *Air-conditioning units.*

**1910.263(i)(14)(i)** [Reserved]

**1910.263(i)(14)(ii)** On large units with doors to chambers large enough to be entered, all door locks shall be operable from both inside and outside.

**1910.263(i)(15)** *Pan washing tanks.*

**1910.263(i)(15)(i)** [Reserved]

**1910.263(i)(15)(ii)** The surface of the floor of the working platform shall be maintained in nonslip condition.

**1910.263(i)(15)(iii)-(iv)** [Reserved]

**1910.263(i)(15)(v)** Power ventilated exhaust hoods shall be provided over the tanks.

**1910.263(i)(16)-(19)** [Reserved]

**1910.263(i)(20)** *Bread coolers, rack type.*

**1910.263(i)(20)(i)** [Reserved]

**1910.263(i)(20)(ii)** All door locks shall be operable from both within and outside the cooler.

**1910.263(i)(21)** [Reserved]

**1910.263(i)(22)** *Doughnut machines.* Separate flues shall be provided, (i) for venting vapors from the frying section, and (ii) for venting products of combustion from the combustion chamber used to heat the fat.

**1910.263(i)(23)** *Open fat kettles.* **(i)** The floor around kettles shall be maintained in nonslip condition.

**1910.263(i)(23)(ii)-(iii)** [Reserved]

**1910.263(i)(23)(iv)** The top of the kettle shall be not less than 36 inches above floor or working level.

**1910.263(i)(24)** *Steam kettles.* **(i)** Positive locking devices shall be provided to hold kettles in the desired position.

**1910.263(i)(24)(ii)** Kettles with steam jackets shall be provided with safety valves in accordance with the ASME Pressure Vessel Code, Section VIII, Unfired Pressure Vessels, 1968, which is incorporated by reference as specified in §1910.6.

**1910.263(j)** *Slicers and wrappers*—**(1)** *Slicers.*

**1910.263(j)(1)(i)-(ii)** [Reserved]

**1910.263(j)(1)(iii)** The cover over the knife head of reciprocating-blade slicers shall be provided with an interlocking arrangement so that the machine cannot operate unless the cover is in place.

**1910.263(j)(1)(iv)** On slicers with endless band knives, each motor shall be equipped with a magnet brake which operates whenever the motor is not energized. Each door, panel, or other point of access to the cutting blades shall be arranged by means of mechanical or electric interlocks so that the motor will be deenergized if all such access doors, panels, or access points are not closed.

**1910.263(j)(1)(v)** When it is necessary to sharpen slicer blades on the machine, a barrier shall be provided leaving only sufficient opening for the sharpening stone to reach the knife blades.

**1910.263(j)(1)(vi)** [Reserved]

**1910.263(j)(1)(vii)** *Slicer wrapper conditions.*

**1910.263(j)(1)(vii)(a)-(b)** [Reserved]

**1910.263(j)(1)(vii)(c)** Mechanical control levers for starting and stopping both slicing machine conveyors and wrapping machines shall be extended or so located that an operator in one location can control both machines. Such levers should be provided wherever necessary, but these should be so arranged that there is only one station capable of starting the wrapping machine and conveyor assembly, and this starting station should be so arranged or guarded as to prevent accidental starting. The electric control station for starting and stopping the electric motor driving the wrapping machine and conveyor should be located near the clutch starting lever.

**1910.263(j)(2)** *Wrappers.*

**1910.263(j)(2)(i)-(ii)** [Reserved]

**1910.263(j)(2)(iii)** Electrical heaters on wrappers shall be protected by a cover plate properly separated or insulated from the heaters in order that accidental contact with this cover plate will not cause a burn to the operator.

**1910.263(k)** *Biscuit and cracker equipment*—**(1)** *Meal, peanut, and fig grinders.* **(i)** If the hopper is removable it shall be provided with an electric interlock so that the machine cannot be put in operation when the hopper is removed.

**1910.263(k)(1)(ii)** Where grid guards cannot be used, feed conveyors to hoppers, or baffle-type hoppers, shall be provided. Hoppers in such cases shall be enclosed and provided with hinged covers, and equipped with electric interlock to prevent operation of the machine with the cover open.

**1910.263(k)(2)** *Sugar and spice pulverizers.* **(i)** All drive belts used in connection with sugar and spice pulverizers shall be grounded by means of metal combs or other effective means of removing static electricity. All pulverizing of sugar or spice grinding shall be done in accordance with NFPA 62—1967 (Standard for Dust Hazards of Sugar and Cocoa) and NFPA 656—1959 (Standard for Dust Hazards in Spice Grinding Plants), which are incorporated by reference as specified in §1910.6.

**1910.263(k)(2)(ii)** Magnetic separators shall be provided to reduce fire and explosion hazards.

**1910.263(k)(3)** *Cheese, fruit, and food cutters.* These machines shall be protected in accordance with the requirements of paragraph (k)(1) of this section.

**1910.263(k)(4)** [Reserved]

**1910.263(k)(5)** *Reversible dough brakes.* Reversible brakes shall be provided with a guard or tripping mechanism on each side of the rolls. These guards shall be so arranged as to stop the machine or reverse the direction of the rolls so that they are outrunning if the guard is moved by contact of the operator.

**1910.263(k)(6)** *Cross-roll brakes.* Cross-roll brakes shall be provided with guards that are similar in number and equal in effectiveness to guards on hand-fed brakes.

**1910.263(k)(7)** *Box- and roll-type dough sheeters.*

**1910.263(k)(7)(i)** [Reserved]

**1910.263(k)(7)(ii)** Hoppers for sheeters shall have an automatic stop bar or automatic stopping device along the back edge of the hopper. If construction does not permit location at the back edge, the automatic stop bar or automatic stopping device shall be located where it will be most effective to accomplish the desired protection.

**1910.263(k)(8)** [Reserved]

**1910.263(k)(9)** *Rotary, die machines, pretzel rolling, and pretzel-stick extruding machines.* Dough hoppers shall have the entire opening protected with substantial grid-type guards to prevent the employee from getting his hands caught in moving parts, or the hopper shall be extended high enough so that the operator's hands cannot get into moving parts.

**1910.263(k)(10)-(11)** [Reserved]

**1910.263(k)(12)** *Pan cooling towers.* **(i)** Where pan cooling towers extend to two or more floors, a lockout switch shall be provided on each floor in order that mechanics working on the tower may positively lock the mechanism against starting. Only one start switch shall be used in the motor control circuit.

**1910.263(k)(12)(ii)** [Reserved]

**1910.263(k)(13)** *Chocolate melting, refining, and mixing kettles.* Each kettle shall be provided with a cover to enclose the top of the kettle. The bottom outlet of each kettle shall be of such size and shape that the operator cannot reach in to touch the revolving paddle or come in contact with the shear point between the paddle and the side of the kettle.

**1910.263(k)(14)-(16)** [Reserved]

**1910.263(k)(17)** *Peanut cooling trucks.* Mechanically operated peanut cooling trucks shall have a grid-type cover over the entire top.

**1910.263(l)** **Ovens**—**(1)** *General location.*

**1910.263(l)(1)(i)-(vi)** [Reserved]

**1910.263(l)(1)(vii)** Ovens shall be located so that possible fire or explosion will not expose groups of persons to possible injury. For this reason ovens shall not adjoin lockers, lunch or sales rooms, main passageways, or exits.

**1910.263(l)(2)** [Reserved]

**1910.263(l)(3)** *Safeguards of mechanical parts.* **(i)** Emergency stop buttons shall be provided on mechanical ovens near the point where operators are stationed.

**1910.263(l)(3)(ii)** All piping at ovens shall be tested to be gastight.

**1910.263(l)(3)(iii)** Main shutoff valves, operable separately from any automatic valve, shall be provided to permit turning off the fuel or steam in case of an emergency.

**1910.263(l)(3)(iii)*(a)*** Main shutoff valves shall be located so that explosions, fires, etc. will not prevent access to these valves.

**1910.263(l)(3)(iii)*(b)*** Main shutoff valves shall be locked in the closed position when men must enter the oven or when the oven is not in service.

**1910.263(l)(4)-(7)** [Reserved]

**1910.263(l)(8)** *Electrical heating equipment.*

**1910.263(l)(8)(i)-(ii)** [Reserved]

**1910.263(l)(8)(iii)** A main disconnect switch or circuit breaker shall be provided. This switch or circuit breaker shall be so located that it can be reached quickly and safely. The main switch or circuit breaker shall have provisions for locking it in the open position if any work on the electrical equipment or inside the oven must be performed.

**1910.263(l)(9)** *General requirements.* **(i)** Protecting devices shall be properly maintained and kept in working order.

**1910.263(l)(9)(ii)** All safety devices on ovens shall be inspected at intervals of not less than twice a month by an especially appointed, properly instructed bakery employee, and not less than once a year by representatives of the oven manufacturers.

**1910.263(l)(9)(iii)*(a)*** Protection of gas pilot lights shall be provided when it is impracticable to protect the main flame of the burner and where the pilot flame cannot contact the flame electrode without being in the path of the main flame of the burner. Failure of any gas pilot shall automatically shut off the fuel supply to the burner.

**1910.263(l)(9)(iii)*(b)*** Ovens with multiple burners shall be equipped with individual atmospheric pilot lights where there is sufficient secondary air in the baking chamber and where gas is available; or else each burner shall be equipped with an electric spark-type ignition device.

**1910.263(l)(9)(iv)** Burners of a capacity exceeding 150,000 B.t.u. per hour equipped with electric ignition shall be protected in addition by quick-acting combustion safeguards.

**1910.263(l)(9)(iv)*(a)*** The high-tension current for any electric spark-type ignition device shall originate in a power supply line which is interlocked with the fuel supply for the oven in such a way that in case of current failure both the source of electricity to the high-tension circuits and the fuel supply shall be turned off simultaneously.

**1910.263(l)(9)(iv)*(b)*** [Reserved]

**1910.263(l)(9)(iv)*(c)*** Combustion safeguards used in connection with electric ignition systems on ovens shall be so designed as to prevent an explosive mixture from accumulating inside the oven before ignition has taken place.

**1910.263(l)(9)(v)** When fuel is supplied and used at line pressure, safety shutoff valves shall be provided in the fuel line leading to the burner.

**1910.263(l)(9)(v)*(a)*** When fuel is supplied in excess of line pressure, safety shutoff valves shall be provided in the fuel line leading to the burners, unless the fuel supply lines are equipped with other automatic valves which will prevent the flow of fuel when the compressing equipment is stopped.

**1910.263(l)(9)(v)*(b)*** The safety shutoff valve shall be positively tight and shall be tested at least twice monthly.

**1910.263(l)(9)(v)*(c)-(d)*** [Reserved]

**1910.263(l)(9)(v)*(e)*** A safety shutoff valve shall require manual operation for reopening after it has closed, or the electric circuit shall be so arranged that it will require a manual operation for reopening the safety shutoff valve.

**1910.263(l)(9)(v)*(f)*** Manual reset-type safety shutoff valves shall be so arranged that they cannot be locked in an open position by external means.

**1910.263(l)(9)(v)*(g)*** Where blowers are used for supplying the air for combustion the safety shutoff valve shall be interlocked so that it will close in case of air failure.

**1910.263(l)(9)(v)*(h)*** Where gas or electric ignition is used, the safety shutoff valve shall close in case of ignition failure. On burners equipped with combustion safeguards, the valve shall close in case of burner flame failure.

**1910.263(l)(9)(vi)** One main, manually operated, fuel shutoff valve shall be provided on each oven, and shall be located ahead of all other valves in the system.

**1910.263(l)(9)(vii)** All individual gas or oil burners with a heating capacity over 150,000 B.t.u. per hour shall be protected by a safeguard which is actuated by the flame and which will react to flame failure in a time interval not to exceed 2 seconds. All safeguards, once having shut down a gas or oil burner, shall require manual resetting and starting of the burner or burners.

**1910.263(l)(9)(viii)** Any space in an oven (except direct fired ovens) which could be filled with an explosive mixture shall be protected by explosion vents. Explosion vents shall be made of minimum weight consistent with adequate insulation.

**1910.263(l)(9)(viii)(a)** Explosion doors which have a substantial weight shall be attached by chains or similar means to prevent flying parts from injuring the personnel in case of an explosion.

**1910.263(l)(9)(viii)(b)** Where explosion vents are so located that flying parts or gases might endanger the personnel working on or near the oven, internal or external protecting means shall be provided in the form of heavily constructed shields or deflectors made from noncombustible material.

**1910.263(l)(9)(viii)(c)** Specifically exempted from the provisions of paragraph paragraph (l)(8)(viii) of this section are heating systems on ovens in which the fuel is admitted only to enclosed spaces which shall have been tested to prove that their construction will resist repeated explosions without deformation are exempt from the requirements of paragraph (l)(8)(viii) (*a*) and (*b*) of this section.

**1910.263(l)(9)(ix)-(x)** [Reserved]

**1910.263(l)(9)(xi)** Where the gas supply pressure is substantially higher than that at which the burners of an oven are designed to operate, a gas pressure regulator shall be employed.

**1910.263(l)(9)(xi)(a)-(c)** [Reserved]

**1910.263(l)(9)(xi)(d)** A relief valve shall be placed on the outlet side of gas pressure regulators where gas is supplied at high pressure. The discharge from this valve shall be piped to the outside of the building.

**1910.263(l)(10)** *Direct-fired ovens.* **(i)** Direct-fired ovens shall be safeguarded against failure of fuel, air, or ignition.

**1910.263(l)(10)(ii)** To prevent the possible accumulation of explosive gases from being ignited after a shutdown, all direct-fired ovens with a heating capacity over 150,000 B.t.u. per hour shall be ventilated before the ignition system, combustion air blower, and the fuel can be turned on. The preventilation shall insure at least four complete changes of atmosphere in the baking chamber by discharging the oven atmosphere to the outside of the building and entraining fresh air into it. The preventilation shall be repeated whenever the heating equipment is shut down by a safety device.

**1910.263(l)(11)** *Direct recirculating ovens.* **(i)** Each circulating fan in direct recirculating ovens shall be interconnected with the burner in such a manner that the fuel is shut off by a safety valve when the fan is not running.

**1910.263(l)(11)(ii)** The flame of the burner or burners in direct recirculating ovens shall be protected by a quick-acting flame-sensitive safeguard which will automatically shut off the fuel supply in case of burner failure.

**1910.263(l)(12)-(14)** [Reserved]

**1910.263(l)(15)** *Indirect recirculating ovens.*

**1910.263(l)(15)(i)-(ii)** [Reserved]

**1910.263(l)(15)(iii)** Duct systems (in ovens) operating under pressure shall be tested for tightness in the initial starting of the oven and also at intervals not farther apart than 6 months.

[39 FR 23502, June 27, 1974, as amended at 43 FR 49765, Oct. 24, 1978; 43 FR 51760, Nov. 7, 1978; 61 FR 9241, Mar. 7, 1996]

## §1910.264 Laundry machinery and operations.

**1910.264(a)** [Reserved]

**1910.264(b)** *General requirements.* This section applies to moving parts of equipment used in laundries and to conditions peculiar to this industry, with special reference to the point of operation of laundry machines. This section does not apply to dry-cleaning operations.

**1910.264(c)** *Point-of-operation guards*—**(1)** *Washroom machines.*

**1910.264(c)(1)(i)** [Reserved]

**1910.264(c)(1)(ii)** *Washing machine.*

**1910.264(c)(1)(ii)(a)** [Reserved]

**1910.264(c)(1)(ii)(b)** Each washing machine shall be provided with means for holding open the doors or covers of inner and outer cylinders or shells while being loaded or unloaded.

**1910.264(c)(2)** *Starching and drying machines.*

**1910.264(c)(2)(i)-(ii)** [Reserved]

**1910.264(c)(2)(iii)** *Drying tumbler.*

**1910.264(c)(2)(iii)(a)** [Reserved]

**1910.264(c)(2)(iii)(b)** Each drying tumbler shall be provided with means for holding open the doors or covers of inner and outer cylinders or shells while being loaded or unloaded.

**1910.264(c)(2)(iv)** *Shaker (clothes tumbler).*

**1910.264(c)(2)(iv)(a)** [Reserved]

**1910.264(c)(2)(iv)(b)(1)** [Reserved]

**1910.264(c)(2)(iv)(b)(2)** Each shaker or clothes tumbler of the double-cylinder type shall be provided with means for holding open the doors or covers of inner and outer cylinders or shells while being loaded or unloaded.

**1910.264(c)(2)(v)** *Exception.* Provisions of paragraph (c)(2) (iii), (iv)(*a*)(*1*), and (iv)(*b*) of this section shall not apply to shakeout or conditioning tumblers where the clothes are loaded into the open end of the revolving cylinder and are automatically discharged out of the opposite end.

**1910.264(c)(3)** [Reserved]

**1910.264(c)(4)** *Miscellaneous machines and equipment.*

**1910.264(c)(4)(i)-(ii)** [Reserved]

**1910.264(c)(4)(iii)** *Steam pipes.* **(a)** All steam pipes that are within 7 feet of the floor or working platform, and with which the worker may come into contact, shall be insulated or covered with a heat-resistive material or shall be otherwise properly guarded.

**1910.264(c)(4)(iii)(b)** Where pressure-reducing valves are used, one or more relief or safety valves shall be provided on the low-pressure side of the reducing valve, in case the piping or equipment on the low-pressure side does not meet the requirements for full initial pressure. The relief or safety valve shall be located adjacent to, or as close as possible to, the reducing valve. Proper protection shall be provided to prevent injury or damage caused by fluid escaping from relief or safety valves if vented to the atmosphere. The vents shall be of ample size and as short and direct as possible. The combined discharge capacity of the relief valves shall be such that the pressure rating of the lower-pressure piping and equipment will not be exceeded if the reducing valve sticks or fails to open.

**1910.264(d)** *Operating rules*—**(1)** *General.*

**1910.264(d)(1)(i)-(ii)** [Reserved]

**1910.264(d)(1)(iii)** *Markers.* Markers and others handling soiled clothes shall be warned against touching the eyes, mouth, or any part of the body on which the skin has been broken by a scratch or abrasion; and they shall be cautioned not to touch or eat food until their hands have been thoroughly washed.

**1910.264(d)(1)(iv)** [Reserved]

**1910.264(d)(1)(v)** *Instruction of employees.* Employees shall be properly instructed as to the hazards of their work and be instructed in safe practices, by bulletins, printed rules, and verbal instructions.

**1910.264(d)(2)** *Mechanical*—**(i)** *Safety guards.* **(a)** No safeguard, safety appliance, or device attached to, or forming an integral part of any machinery shall be removed or made ineffective except for the purpose of making immediate repairs or adjustments. Any such safeguard, safety appliance, or device removed or made ineffective during the repair or adjustment of such machinery shall be replaced immediately upon the completion of such repairs or adjustments.

**1910.264(b)** [Reserved]

[39 FR 23502, June 27, 1974, as amended at 43 FR 49767, Oct. 24, 1978; 43 FR 51760, Nov. 7, 1978]

## §1910.265 Sawmills.

**1910.265(a)** *General requirements*—*Application.* This section includes safety requirements for sawmill operations including, but not limited to, log and lumber handling, sawing, trimming, and planing; waste disposal; operation of dry kilns; finishing; shipping; storage; yard and yard equipment; and for power tools and affiliated equipment used in connection with such operations, but excluding the manufacture of plywood, cooperage, and veneer.

**1910.265(b)** *Definitions applicable to this section*—**(1)** *A-frame.* The term *A-frame* means a structure made of two independent columns fastened together at the top and separated at the bottom for stability.

# OSHA Standards for General Industry

**1910.265(b)(2)** *Annealing.* The term *annealing* means heating then cooling to soften and render less brittle.

**1910.265(b)(3)** *Binder.* The term *binder* means a chain, cable, rope, or other approved material used for binding loads.

**1910.265(b)(4)** *Boom.* The term *boom* means logs or timbers fastened together end to end and used to contain floating logs. The term includes enclosed logs.

**1910.265(b)(5)** *Brow log.* The term *brow log* means a log placed parallel to a roadway at a landing or dump to protect vehicles while loading or unloading.

**1910.265(b)(6)** *Bunk.* The term *bunk* means a cross support for a load.

**1910.265(b)(7)** *Cant.* The term *cant* means a log slabbed on one or more sides.

**1910.265(b)(8)** *Carriage (log carriage).* The term *carriage* means a framework mounted on wheels which runs on tracks or in grooves in a direction parallel to the face of the saw, and which contains apparatus to hold a log securely and advance it towards the saw.

**1910.265(b)(9)** *Carrier.* The term *carrier* means an industrial truck so designed and constructed that it straddles the load to be transported with mechanisms to pick up the load and support it during transportation.

**1910.265(b)(10)** *Chipper.* The term *chipper* means a machine which cuts material into chips.

**1910.265(b)(11)** *Chock (bunk block) (cheese block).* The terms *chock*, *bunk block*, and *cheese block* mean a wedge that prevents logs or loads from moving.

**1910.265(b)(12)** *Cold deck.* The term *cold deck* means a pile of logs stored for future removal.

**1910.265(b)(13)** *Crotch lines.* The term *crotch lines* means two short lines attached to a hoisting line by a ring or shackle, the lower ends being attached to loading hooks.

**1910.265(b)(14)** *Dog (carriage dog).* The term *dog* means a steel tooth, one or more of which are attached to each carriage knee to hold log firmly in place on carriage.

**1910.265(b)(15)** *Drag saw.* The term *drag saw* means a power-driven, reciprocating crosscut saw mounted on suitable frame and used for bucking logs.

**1910.265(b)(16)** *Head block.* The term *head block* means that part of a carriage which holds the log and upon which it rests. It generally consists of base, knee, taper set, and mechanism.

**1910.265(b)(17)** *Head rig.* The term *head rig* means a combination of head saw and log carriage used for the initial breakdown of logs into timbers, cants, and boards.

**1910.265(b)(18)** *Hog.* The term *hog* means a machine for cutting or grinding slabs and other coarse residue from the mill.

**1910.265(b)(19)** *Husk.* The term *husk* means a head saw framework on a circular mill.

**1910.265(b)(20)** *Industrial truck.* The term *industrial truck* means a mobile powerdriven truck or tractor.

**1910.265(b)(21)** *Kiln tender.* The term *kiln tender* means the operator of a kiln.

**1910.265(b)(22)** *Lift truck.* The term *lift truck* means an industrial truck used for lateral transportation and equipped with a power-operated lifting device, usually in the form of forks, for piling or unpiling lumber units or packages.

**1910.265(b)(23)** *Live rolls.* The term *live rolls* means cylinders of wood or metal mounted on horizontal axes and rotated by power, which are used to convey slabs, lumber, and other wood products.

**1910.265(b)(24)** *Loading boom.* The term *loading boom* means any structure projecting from a pivot point to guide a log when lifted.

**1910.265(b)(25)** *Log deck.* The term *Log deck* means a platform in the sawmill on which the logs remain until needed for sawing.

**1910.265(b)(26)** *Lumber hauling truck.* The term *lumber hauling truck* means an industrial truck, other than a lift truck or a carrier, used for the transport of lumber.

**1910.265(b)(27)** *Log haul.* The term *log haul* means a conveyor for transferring logs to mill.

**1910.265(b)(28)** *Package.* The term *package* means a unit of lumber.

**1910.265(b)(29)** *Peavy.* The term *peavy* means a stout wooden handle fitted with a spike and hook and used for rolling logs.

**1910.265(b)(30)** *Pike pole.* The term *pike pole* means a long pole whose end is shod with a sharp pointed spike.

**1910.265(b)(31)** *Pitman rod.* The term *pitman rod* means connecting rod.

**1910.265(b)(32)** *Resaw.* The term *resaw* means band, circular, or sash gang saws used to break down slabs, cants, or flitches into lumber.

**1910.265(b)(33)** *Running line.* The term *running line* means any moving rope as distinguished from a stationary rope such as a guyline.

**1910.265(b)(34)** *Safety factor.* The term *safety factor* means a calculated reduction factor which may be applied to laboratory test values to obtain safe working stresses for wooden beams and other mechanical members; ratio of breaking load to safe load.

**1910.265(b)(35)** *Saw guide.* The term *saw guide* means a device for steadying a circular or bandsaw.

**1910.265(b)(36)** *Setwork.* The term *setwork* means a mechanism on a sawmill carriage which enables an operator to move the log into position for another cut.

**1910.265(b)(37)** *Sorting gaps.* The term *sorting gaps* means the areas on a log pond enclosed by boom sticks into which logs are sorted.

**1910.265(b)(38)** *Spreader wheel.* The term *spreader wheel* means a metal wheel that separates the board from the log in back of circular saws to prevent binding.

**1910.265(b)(39)** *Splitter.* The term *splitter* means a knife-type, nonrotating spreader.

**1910.265(b)(40)** *Sticker.* The term *sticker* means a strip of wood or other material used to separate layers of lumber.

**1910.265(b)(41)** *Stiff boom.* The term *stiff boom* means the anchored, stationary boom sticks which are tied together and on which boom men work.

**1910.265(b)(42)** *Swifter.* The term *swifter* is a means of tying boom sticks together to prevent them from spreading while being towed.

**1910.265(b)(43)** *Telltale.* The term *telltale* means a device used to serve as a warning for overhead objects.

**1910.265(b)(44)** *Top saw.* The term *top saw* means the upper of two circular saws on a head rig, both being on the same husk.

**1910.265(b)(45)** *Tramway.* The term *tramway* means a way for trams, usually consisting of parallel tracks laid on wooden beams.

**1910.265(b)(46)** *Trestle.* The term *trestle* means a braced framework of timbers, piles or steelwork for carrying a road or railroad over a depression.

**1910.265(c) Building facilities, and isolated equipment—(1)** *Safety factor.* All buildings, docks, tramways, walkways, log dumps, and other structures shall be designed, constructed and maintained so as to support the imposed load in accordance with a safety factor.

**1910.265(c)(2)** *Work areas.* Work areas under mills shall be as evenly surfaced as local conditions permit. They shall be free from unnecessary obstructions and provided with lighting facilities in accordance with American National Standard for Industrial Lighting A11.1—1965, which is incorporated by reference as specified in §1910.6.

**1910.265(c)(3)** *Floors.* Flooring in buildings and on ramps and walkways shall be constructed and installed in accordance with established principles of mechanics and sound engineering practices. They shall be of adequate strength to support the estimated or actual dead and live loads acting on them with the resultant stress not exceeding the allowable stress for the material being used.

**1910.265(c)(3)(i)** [Reserved]

**1910.265(c)(3)(ii)** *Areas beneath floor openings.* Areas under floor openings shall, where practical, be fenced off. When this is not practical, they shall be plainly marked and telltales shall be installed to hang over these areas.

**1910.265(c)(3)(iii)** *Floor maintenance.* The flooring of buildings, docks, and passageways shall be kept in good repair. When a hazardous condition develops that cannot be immediately repaired, the area shall be guarded until adequate repairs are made.

**1910.265(c)(3)(iv)** *Nonslip floors.* Floors, footwalks, and passageways in the work area around machines or other places where a person is required to stand or walk shall be provided with effective means to minimize slipping.

**1910.265(c)(4)** *Walkways, docks, and platforms—(i) Width.* Walkways, docks, and platforms shall be of sufficient width to provide adequate passage and working areas.

**1910.265(c)(4)(ii)** *Maintenance.* Walkways shall be evenly floored and kept in good repair.

**1910.265(c)(4)(iii)** *Docks.* Docks and runways used for the operation of lift trucks and other vehicles shall have a substantial guard or shear timber except where loading and unloading are being performed.

**1910.265(c)(4)(iv)** *Elevated walks.* All elevated walks, runways, or platforms, if 4 feet or more from the floor level, shall be provided with a standard railing except on loading or unloading sides of platforms. If height exceeds 6 feet, a standard toe board also shall be provided to prevent material from rolling or falling off.

**1910.265(c)(4)(v)** *Elevated platforms.* Where elevated platforms are used routinely on a daily basis they shall be equipped with stairways or fixed ladders in accordance with §1910.27.

**1910.265(c)(4)(vi)** *Hazardous locations.* Where required, walkways and stairways with standard handrails shall be provided in elevated and hazardous locations. Where such passageways are over walkways or work areas, standard toe boards shall be provided.

**1910.265(c)(5)** *Stairways—(i) Construction.* Stairways shall be constructed in accordance with §1910.24.

**1910.265(c)(5)(ii)** *Handrails.* Stairways shall be provided with a standard handrail on at least one side or on any open side. Where stairs are more than four feet wide there shall be a standard handrail at each side, and where more than eight feet wide, a third standard handrail shall be erected in the center of the stairway.

**1910.265(c)(5)(iii)** *Lighting.* All stairways shall be adequately lighted as prescribed in paragraph (c)(9) of this section.

**1910.265(c)(6)** *Emergency exits including doors and fire escapes—(i) Opening.* Doors shall not open directly on or block a flight of stairs, and shall swing in the direction of exit travel.

**1910.265(c)(6)(ii)** *Identification.* Exits shall be located and identified in a manner that affords ready exit from all work areas.

**1910.265(c)(6)(iii)** *Swinging doors.* All swinging doors shall be provided with windows; with one window for each section of double swinging doors. Such windows shall be of shatterproof or safety glass unless otherwise protected against breakage.

**1910.265(c)(6)(iv)** *Sliding doors.* Where sliding doors are used as exits, an inner door shall be cut inside each of the main doors and arranged to open outward.

**1910.265(c)(6)(v)** *Barriers and warning signs.* Where a doorway opens upon a railroad track or upon a tramway or dock over which vehicles travel, a barrier or other warning device shall be placed to prevent workmen from stepping into moving traffic.

**1910.265(c)(7)** *Air requirements.* Ventilation shall be provided to supply adequate fresh healthful air to rooms, buildings, and work areas.

**1910.265(c)(8)** *Vats and tanks.* All open vats and tanks into which workmen could fall shall be guarded.

**1910.265(c)(9)** *Lighting—(i) Adequacy.* Illumination shall be provided and designed to supply adequate general and local lighting to rooms, buildings, and work areas during the time of use.

**1910.265(c)(9)(ii)** *Effectiveness.* Factors upon which the adequacy and effectiveness of illumination will be judged, include the following:

**1910.265(c)(9)(ii)***(a)* The quantity of light in foot-candle intensity shall be sufficient for the work being done.

**1910.265(c)(9)(ii)***(b)* The quality of the light shall be such that it is free from glare, and has correct direction, diffusion, and distribution.

**1910.265(c)(9)(ii)***(c)* Shadows and extreme contrasts shall be avoided or kept to a minimum.

**1910.265(c)(10)** [Reserved]

**1910.265(c)(11)** *Hazard marking.* Physical hazard marking shall be as specified in §1910.144 of this part.

**1910.265(c)(12)** [Reserved]

**1910.265(c)(13)** *Hydraulic systems.* Means shall be provided to block, chain, or otherwise secure equipment normally supported by hydraulic pressure so as to provide for safe maintenance.

**1910.265(c)(14)** [Reserved]

**1910.265(c)(15)** *Gas piping and appliances.* All gas piping and appliances shall be installed in accordance with the American National Standard Requirements for the Installation of Gas Appliances and Gas Piping Z21.30—1964, which is incorporated by reference as specified in §1910.6.

**1910.265(c)(16)-(17)** [Reserved]

**1910.265(c)(18)** *Conveyors—(i) Standards.* Construction, operation, and maintenance of conveyors shall be in accordance with American National Standard B20.1—1957, which is incorporated by reference as specified in §1910.6.

**1910.265(c)(18)(ii)** *Guarding.* Spiked live rolls shall be guarded.

**1910.265(c)(19)** *Stationary tramways and trestles—(i) Foundations and walkways.* Tramways and trestles shall have substantial mud sills or foundations which shall be frequently inspected and kept in repair. When vehicles are operated on tramways and trestles which are used for foot passage, traffic shall be controlled or a walkway with standard handrails at the outer edge and shear timber on the inner edge shall be provided. This walkway shall be wide enough to allow adequate clearance to vehicles. When walkways cross over other thoroughfares, they shall be solidly fenced at the outer edge to a height of 42 inches over such thoroughfares.

**1910.265(c)(19)(ii)** *Clearance.* Stationary tramways and trestles shall have a vertical clearance of 22 feet over railroad rails. When constructed over carrier docks or roads, they shall have a clearance of 6 feet above the driver's foot rest on the carrier, and in no event shall this clearance be less than 12 feet from the roadway. In existing operations where it is impractical to obtain such clearance, telltales, electric signals, signs or other precautionary measures shall be installed.

**1910.265(c)(20)** *Blower, collecting, and exhaust systems—(i) Design, construction, and maintenance.* Blower collecting, and exhaust systems should be designed, constructed, and maintained in accordance with American National Standards Z33.1—1961 (For the Installation of Blower and Exhaust Systems for Dust, Stock, and Vapor Removal or Conveying) and Z12.2—1962 (R1969) (Code for the Prevention of Dust Explosion in Woodworking and Wood Flour Manufacturing Plants), which are incorporated by reference as specified in §1910.6.

**1910.265(c)(20)(ii)** *Collecting systems.* All mills containing one or more machines that create dust, shavings, chips, or slivers during a period of time equal to or greater than one-fourth of the working day, shall be equipped with a collecting system. It may be either continuous or automatic, and shall be of sufficient strength and capacity to enable it to remove such refuse from points of operation and immediate vicinities of machines and work areas.

**1910.265(c)(20)(iii)** *Exhaust or conveyor systems.* Each woodworking machine that creates dust, shavings, chips, or slivers shall be equipped with an exhaust or conveyor system located and adjusted to remove the maximum amount of refuse from the point of operation and immediate vicinity.

**1910.265(c)(20)(iv)** [Reserved]

**1910.265(c)(20)(v)** *Dust chambers.* Exhaust pipes shall not discharge into an unconfined outside pile if uncontrolled fire or explosion hazards are created. They may empty into settling or dust chambers, designed to prevent the dust or refuse from entering any work area. Such chambers shall be constructed and operated to minimize the danger of fire or dust explosion.

**1910.265(c)(20)(vi)** *Hand removal of refuse.* Provision for the daily removal of refuse shall be made in all operations not required to have an exhaust system or having refuse too heavy, bulky, or otherwise unsuitable to be handled by the exhaust system.

**1910.265(c)(21)** *Chippers—(i) Whole-log chippers.* The feed system to the chipper shall be arranged so the operator does not stand in direct line with the chipper spout (hopper). The chipper spout shall be enclosed to a height of not less than 36 inches from the floor or the operator's platform. A safety belt and lifeline shall be worn by workmen when working at or near the spout unless the spout is guarded. The lifeline shall be short enough to prevent workers from falling into the chipper.

# OSHA Standards for General Industry

**1910.265(c)(21)(ii)** *Hogs.* **(a)** Hog mills shall be so designed and arranged that from no position on the rim of the chute shall the distance to the cutter knives be less than 40 inches.

**1910.265(c)(21)(ii)(b)** Hog feed chutes shall be provided with suitable and approved baffles, which shall minimize material from being thrown from the mill.

**1910.265(c)(21)(ii)(c)** Employees feeding hog mills shall be provided with safety belts and lines unless guarded.

**1910.265(c)(22)** [Reserved]

**1910.265(c)(23)** *Bins, bunkers, hoppers, and fuel houses—(i) Guarding.* Open bins, bunkers, and hoppers whose upper edges extend less than 3 feet above working level shall be equipped with standard handrails and toe boards, or have their tops covered by a substantial grill or grating with openings small enough to prevent a man from falling through.

**1910.265(c)(23)(ii)** *Use of wheeled equipment to load bins.* Where automotive or other wheeled equipment is used to move materials into bins, bunkers, and hoppers, adequate guard rails shall be installed along each side of the runway, and a substantial bumper stop provided when necessary.

**1910.265(c)(23)(iii)** *Exits, lighting, and safety devices.* Fuel houses and bins shall have adequate exits and lighting, and all necessary safety devices shall be provided and shall be used by persons entering these structures.

**1910.265(c)(23)(iv)** *Walkways.* Where needed, fuel houses and bins shall have a standard railed platform or walkway near the top.

**1910.265(c)(24)** *Ropes, cables, slings, and chains—(i) Safe usage.* Ropes, cables, slings, and chains shall be used in accordance with safe use practices recommended by the manufacturer or within safe limits recommended by the equipment manufacturer when used in conjunction with it.

**1910.265(c)(24)(ii)** *Hooks.* No open hook shall be used in rigging to lift any load where there is hazard from relieving the tension on the hook from the load or hook catching or fouling.

**1910.265(c)(24)(iii)** *Work by qualified persons.* Installation, inspection, maintenance, repair, and testing of ropes, cables, slings, and chains shall be done only by persons qualified to do such work.

**1910.265(c)(24)(iv)** *Slings.* Proper storage shall be provided for slings while not in use.

**1910.265(c)(24)(v)** *Ropes or cables.* **(a)** Wire rope or cable shall be inspected when installed and once each week thereafter, when in use. It shall be removed from hoisting or load-carrying service when kinked or when one of the following conditions exists:

**1910.265(c)(24)(v)(a)(1)** When three broken wires are found in one lay of 6 by 6 wire rope.

**1910.265(c)(24)(v)(a)(2)** When six broken wires are found in one lay of 6 by 19 wire rope.

**1910.265(c)(24)(v)(a)(3)** When nine broken wires are found in one lay of 6 by 37 wire rope.

**1910.265(c)(24)(v)(a)(4)** When eight broken wires are found in one lay of 8 by 19 wire rope.

**1910.265(c)(24)(v)(a)(5)** When marked corrosion appears.

**1910.265(c)(24)(v)(a)(6)** Wire rope of a type not described herein shall be removed from service when 4 percent of the total number of wires composing such rope are found to be broken in one lay.

**1910.265(c)(24)(v)(b)** Wire rope removed from service due to defects shall be plainly marked or identified as being unfit for further use on cranes, hoists, and other load-carrying devices.

**1910.265(c)(24)(v)(c)** The ratio between the rope diameter and the drum, block, sheave, or pulley tread diameter shall be such that the rope will adjust itself to the bend without excessive wear, deformation, or injury. In no case shall the safe value of drums, blocks, sheaves, or pulleys be reduced when replacing such items unless compensating changes are made for rope used and for safe loading limits.

**1910.265(c)(24)(vi)** *Drums, sheaves, and pulleys.* Drums, sheaves, and pulleys shall be smooth and free from surface defects liable to injure rope. Drums, sheaves, or pulleys having eccentric bores or cracked hubs, spokes, or flanges shall be removed from service.

**1910.265(c)(24)(vii)** *Connections.* Connections, fittings, fastenings, and other parts used in connection with ropes and cables shall be of good quality and of proper size and strength, and shall be installed in accordance with the manufacturer's recommendations.

**1910.265(c)(24)(viii)** *Socketing, splicing, and seizing.* **(a)** Socketing, splicing, and seizing of cables shall be performed only by qualified persons.

**1910.265(c)(24)(viii)(b)** All eye splices shall be made in an approved manner and wire rope thimbles of proper size shall be fitted in the eye, except that in slings the use of thimbles shall be optional.

**1910.265(c)(24)(viii)(c)** Wire rope clips attached with U-bolts shall have these bolts on the dead or short end of the rope. The U-bolt nuts shall be retightened immediately after initial load carrying use and at frequent intervals thereafter.

**1910.265(c)(24)(viii)(d)** When a wedge socket-type fastening is used, the dead or short end of the cable shall be clipped with a U-bolt or otherwise made secure against loosening.

**1910.265(c)(24)(viii)(e)** *Fittings.* Hooks, shackles, rings, pad eyes, and other fittings that show excessive wear or that have been bent, twisted, or otherwise damaged shall be removed from service.

**1910.265(c)(24)(viii)(f)** *Running lines.* Running lines of hoisting equipment located within 6 feet 6 inches of the ground or working level shall be boxed off or otherwise guarded, or the operating area shall be restricted.

**1910.265(c)(24)(viii)(g)** *Number of wraps on drum.* There shall be not less than two full wraps of hoisting cable on the drum of cranes and hoists at all times of operation.

**1910.265(c)(24)(viii)(h)** *Drum flanges.* Drums shall have a flange at each end to prevent the cable from slipping off.

**1910.265(c)(24)(viii)(i)** *Sheave guards.* Bottom sheaves shall be protected by close fitting guards to prevent cable from jumping the sheave.

**1910.265(c)(24)(viii)(j)** *Preventing abrasion.* The reeving of a rope shall be so arranged as to minimize chafing or abrading while in use.

**1910.265(c)(24)(ix)** *Chains.* **(a)** Chains used in load carrying service shall be inspected before initial use and weekly thereafter.

**1910.265(c)(24)(ix)(b)** Chain shall be normalized or annealed periodically as recommended by the manufacturer.

**1910.265(c)(24)(ix)(c)** If at any time any 3-foot length of chain is found to have stretched one-third the length of a link it shall be discarded.

**1910.265(c)(24)(ix)(d)** Bolts or nails shall not be placed between two links to shorten or join chains.

**1910.265(c)(24)(ix)(e)** Broken chains shall not be spliced by inserting a bolt between two links with the head of the bolt and nut sustaining the load, or by passing one link through another and inserting a bolt or nail to hold it.

**1910.265(c)(24)(x)** *Fiber rope.* **(a)** Frozen fiber rope shall not be used in load carrying service.

**1910.265(c)(24)(x)(b)** Fiber rope that has been subjected to acid or excessive heat shall not be used for load carrying purposes.

**1910.265(c)(24)(x)(c)** Fiber rope shall be protected from abrasion by padding where it is fastened or drawn over square corners or sharp or rough surfaces.

**1910.265(c)(25)** [Reserved]

**1910.265(c)(26)** *Mechanical stackers and unstackers.*

**1910.265(c)(26)(i)** [Reserved]

**1910.265(c)(26)(ii)** *Lumber lifting devices.* Lumber lifting devices on all stackers shall be designed and arranged so as to minimize the possibility of lumber falling from such devices.

**1910.265(c)(26)(iii)** *Blocking hoisting platform.* Means shall be provided to positively block the hoisting platform when employees must go beneath the stacker or unstacker hoist.

**1910.265(c)(26)(iv)** *Identifying controls.* Every manually operated control switch shall be properly identified and so located as to be readily accessible to the operator.

**1910.265(c)(26)(v)** *Locking main control switches.* Main control switches shall be so designed that they can be locked in the open position.

**1910.265(c)(26)(vi)** *Guarding side openings.* The hoistway side openings at the top level of the stacker and unstacker shall be protected by enclosures of standard railings.

**1910.265(c)(26)(vii)** *Guarding hoistway openings.* When the hoist platform or top of the load is below the working platform, the hoistway openings shall be guarded.

**1910.265(c)(26)(viii)** *Guarding lower landing area.* The lower landing area of stackers and unstackers shall be guarded by enclosures that prevent entrance to the area or pit below the hoist platform. Entrances should be protected by electrically interlocked gates which, when open, will disconnect the power and set the hoist brakes. When the interlock is not installed, other positive means of protecting the entrance shall be provided.

**1910.265(c)(26)(ix)** *Inspection.* Every stacker and unstacker shall be inspected at frequent intervals and all defective parts shall be immediately repaired or replaced.

**1910.265(c)(26)(x)** *Cleaning pits.* Safe means of entrance and exit shall be provided to permit cleaning of pits.

**1910.265(c)(26)(xi)** *Preventing entry to hazardous area.* Where the return of trucks from unstacker to stacker is by mechanical power or gravity, adequate signs, warning devices, or barriers shall be erected to prevent entry into the hazardous area.

**1910.265(c)(27)** *Lumber piling and storage—(i) Pile foundations.* In stacking units of lumber, pile foundations shall be designed and arranged to support maximum loads without sinking, sagging, or permitting the piles to topple. In unit package piles, substantial bolsters or unit separators shall be placed between each package directly over the stickers.

**1910.265(c)(27)(ii)** *Stacking dissimilar unit packages.* Long units of lumber shall not be stacked upon shorter packages except where a stable pile can be made with the use of package separators.

**1910.265(c)(27)(iii)** *Unstable piles.* Piles of lumber which have become unstable shall be immediately made safe, or the area into which they might fall shall be fenced or barricaded and employees prohibited from entering it.

**1910.265(c)(27)(iv)** *Stickers.* Unit packages of lumber shall be provided with stickers as necessary to insure stability under ordinary operating conditions.

**1910.265(c)(27)(v)** *Sticker alignment.* Stickers shall extend the full width of the package, shall be uniformly spaced, and shall be aligned one above the other. Stickers may be lapped with a minimum overlapping of 12 inches. Stickers shall not protrude more than 2 inches beyond the sides of the package.

**1910.265(c)(27)(vi)** *Pile height.* The height of unit package piles shall be dependent on the dimensions of the packages and shall be such as to provide stability under normal operating conditions. Adjacent lumber piles may be tied together with separators to increase stability.

**1910.265(c)(28)** *Lumber loading.* Loads shall be built and secured to insure stability in transit.

**1910.265(c)(29)** *Burners—(i) Guying.* If the burner stack is not self-supporting, it shall be guyed or otherwise supported.

**1910.265(c)(29)(ii)** *Runway.* The conveyor runway to the burner shall be equipped with a standard handrail. If the runway crosses a roadway or thoroughfare, standard toe boards shall be provided in addition.

**1910.265(c)(30)** *Vehicles—(I) Scope.* Vehicles shall include all mobile equipment normally used in sawmill, planing mill, storage, shipping, and yard operations.

**1910.265(c)(30)(ii)** *Warning signals and spark arrestors.* All vehicles shall be equipped with audible warning signals and where practicable shall have spark arrestors.

**1910.265(c)(30)(iii)** *Lights.* All vehicles operated in the dark or in poorly lighted areas shall be equipped with head and tail lights.

**1910.265(c)(30)(iv)** *Overhead guard.* All vehicles operated in areas where overhead hazards exist shall be equipped with an approved overhead guard. See American National Standard Safety Code for Powered Industrial Trucks, B56.1—1969, which is incorporated by reference as specified in §1910.6.

**1910.265(c)(30)(v)** *Platform guard.* Where the operator is exposed to hazard from backing the vehicle into objects, an approved platform guard shall be provided and so arranged as to not impede exit of driver from vehicle.

**1910.265(c)(30)(vi)** [Reserved]

**1910.265(c)(30)(vii)** *Operation in buildings.* Vehicles powered by internal combustion engines shall not operate in buildings unless the buildings are adequately ventilated.

**1910.265(c)(30)(viii)** *Load limits.* No vehicle shall be operated with loads exceeding its safe load capacity.

**1910.265(c)(30)(ix)** *Brakes.* All vehicles shall be equipped with brakes capable of holding and controlling the vehicle and capacity load upon any incline or grade over which they may be operated.

**1910.265(c)(30)(x)** [Reserved]

**1910.265(c)(30)(xi)** *Carriers.* (a) Carriers shall be so designed and constructed that the operator's field of vision shall not be unnecessarily restricted.

**1910.265(c)(30)(xi)(b)** Carriers shall be provided with an access ladder or equivalent.

**1910.265(c)(30)(xii)** *Lumber hauling trucks.* (a) On trucks where movement of load on stopping would endanger the operator, a substantial bulkhead shall be installed behind the operator's seat. This shall extend to the top of the operator's compartment.

**1910.265(c)(30)(xii)(b)** Stakes, stake pockets, racks, tighteners, and binders shall provide adequate means to secure the load against any movement during transit.

**1910.265(c)(30)(xii)(c)** Where rollers are used, at least two shall be equipped with locks which shall be locked when supporting loads during transit.

**1910.265(c)(31)** *Traffic control and flow—(i) Hazardous crossings.* Railroad tracks and other hazardous crossings shall be plainly posted.

**1910.265(c)(31)(ii)** *Restricted overhead clearance.* All areas of restricted side or overhead clearance shall be plainly marked.

**1910.265(c)(31)(iii)** *Pickup and unloading points.* Pickup and unloading points and paths for lumber packages on conveyors and transfers and other areas where accurate spotting is required, shall be plainly marked and wheel stops provided where necessary.

**1910.265(c)(31)(iv)** *Aisles, passageways, and roadways.* Aisles, passageways, and roadways shall be sufficiently wide to provide safe side clearance. One-way aisles may be used for two-way traffic if suitable turnouts are provided.

**1910.265(d)** **Log handling, sorting, and storage—(1)** *Log unloading methods, equipment, and facilities—(i) Unloading methods.* (a) Stakes and chocks which trip shall be constructed in such manner that the tripping mechanism that releases the stake or chocks is activated at the opposite side of the load being tripped.

**1910.265(d)(1)(i)(b)** Binders on logs shall not be released prior to securing with unloading lines or other unloading device.

**1910.265(d)(1)(i)(c)** Binders shall be released only from the side on which the unloader operates, except when released by remote control devices or except when person making release is protected by racks or stanchions or other equivalent means.

**1910.265(d)(1)(i)(d)** Loads on which a binder is fouled by the unloading machine shall have an extra binder or metal band of equal strength placed around the load, or the load shall be otherwise secured so the fouled binder can be safely removed.

**1910.205(d)(1)(ii)** *Unloading equipment and facilities.* (a) Machines used for hoisting, unloading, or lowering logs shall be equipped with brakes capable of controlling or holding the maximum load in midair.

**1910.265(d)(1)(ii)(b)** The lifting cylinders of all hydraulically operated log handling machines shall be equipped with a positive device for preventing the uncontrolled lowering of the load or forks in case of a failure in the hydraulic system.

**1910.265(d)(1)(ii)(c)** A limit switch shall be installed on powered log handling machines to prevent the lift arms from traveling too far in the event the control switch is not released in time.

**1910.265(d)(1)(ii)(d)** When forklift-type machines are used to load trailers, a means of securing the loading attachment to the fork shall be installed and used.

**1910.265(d)(1)(ii)(e)** A-frames and similar log unloading devices shall have adequate height to provide safe clearance for swinging loads and to provide for adequate crotch lines and spreader bar devices.

**1910.265(d)(1)(ii)(f)** Log handling machines used to stack logs or lift loads above operator's head shall be equipped with adequate overhead protection.

**1910.265(d)(1)(ii)(g)** All mobile log handling machines shall be equipped with headlights and backup lights.

# OSHA Standards for General Industry

**1910.265(d)(1)(ii)(h)** Unloading devices shall be equipped with a horn or other plainly audible signaling device.

**1910.265(d)(1)(ii)(i)** Movement of unloading equipment shall be coordinated by audible or hand signals when operator's vision is impaired or operating in the vicinity of other employees.

**1910.265(d)(1)(ii)(j)** Wood pike poles shall be made of straight-grained, select material. Metal or conductive pike poles shall not be used around exposed energized electrical conductors. Defective, blunt, or dull pike poles shall not be used.

**1910.265(d)(2)** *Log unloading and storage areas—(i) General. (a)* Log dumps, booms, ponds, or storage areas used at night shall be illuminated in accordance with the requirements of American National Standard A11.1-1965 (R-1970) Standard Practice for Industrial Lighting, which is incorporated by reference as specified in §1910.6.

**1910.265(d)(2)(i)(b)** Log unloading areas shall be arranged and maintained to provide a safe working area.

**1910.265(d)(2)(i)(c)** Where skids are used, space adequate to clear a man's body shall be maintained between the top of the skids and the ground.

**1910.265(d)(2)(i)(d)** Signs prohibiting unauthorized foot or vehicle traffic in log unloading and storage areas shall be posted.

**1910.265(d)(2)(ii)** *Water log dumps. (a)* Ungrounded electrically powered hoists using handheld remote control in grounded locations, such as log dumps or mill log lifts, shall be actuated by circuits operating at less than 50 volts to ground.

**1910.265(d)(2)(ii)(b)** Roadbeds at log dumps shall be of sufficient width and evenness to insure safe operation of equipment.

**1910.265(d)(2)(ii)(c)** An adequate brow log or skid timbers or the equivalent shall be provided where necessary. Railroad-type dumps, when located where logs are dumped directly into water or where entire loads are lifted from vehicle, may be exempted providing such practice does not create a hazardous exposure of personnel or equipment.

**1910.265(d)(2)(ii)(d)** Unloading lines shall be arranged so that it is not necessary for the employees to attach them from the pond or dump side of the load except when entire loads are lifted from the log-transporting vehicle.

**1910.265(d)(2)(ii)(e)** Unloading lines, crotch lines, or equally effective means shall be arranged and used in a manner to minimize the possibility of any log from swinging or rolling back.

**1910.265(d)(2)(ii)(f)** When logs are unloaded with peavys or similar manual methods, means shall be provided and used that will minimize the danger from rolling or swinging logs.

**1910.265(d)(2)(ii)(g)** Guardrails, walkways, and standard handrails shall be installed

**1910.265(d)(2)(ii)(h)** Approved life rings (see: 46 CFR 160.099 and 46 CFR 160.050) with line attached and maintained to retain buoyancy shall be provided.

**1910.265(d)(2)(iii)** *Log booms and ponds. (a)* Walkways and floats shall be installed and securely anchored to provide adequate passageway for employees.

**1910.265(d)(2)(iii)(b)** All regular boom sticks and foot logs shall be reasonably straight, with no protruding knots and bark, and shall be capable of supporting, above the water line at either end, the weight of an employee and equipment.

**1910.265(d)(2)(iii)(c)** Permanent cable swifters shall be so arranged that it will not be necessary to roll boom sticks in order to attach or detach them.

**1910.265(d)(2)(iii)(d)** Periodic inspection of cable or dogging lines shall be made to determine when repair or removal from service is necessary.

**1910.265(d)(2)(iii)(e)** The banks of the log pond in the vicinity of the log haul shall be reinforced to prevent caving in.

**1910.265(d)(2)(iii)(f)** Artificial log ponds shall be drained, cleaned, and refilled when unhealthy stagnation or pollution occurs.

**1910.265(d)(2)(iii)(g)** Employees whose duties require them to work from boats, floating logs, boom sticks, or walkways along or on water shall be provided with and shall wear appropriate buoyant devices while performing such duties.

**1910.265(d)(2)(iii)(h)** Stiff booms shall be two float logs wide secured by boom chains or other connecting devices, and of a width adequate for the working needs. Walking surfaces shall be free of loose material and maintained in good repair.

**1910.265(d)(2)(iii)(i)** Boom sticks shall be fastened together with adequate crossties or couplings.

**1910.265(d)(2)(iii)(j)** Floating donkeys or other power-driven machinery used on booms shall be placed on a raft or float with enough buoyancy to keep the deck well above water.

**1910.265(d)(2)(iii)(k)** All sorting gaps shall have a substantial stiff boom on each side.

**1910.265(d)(2)(iv)** *Pond boats and rafts.* The applicable provisions of the Standard for Fire Protection for Motorcraft, NFPA No. 302—1968, which is incorporated by reference as specified in §1910.6, shall be complied with.

**1910.265(d)(2)(iv)(a)** Decks of pond boats shall be covered with non-slip material.

**1910.265(d)(2)(iv)(b)** Powered pond boats or rafts shall be provided with at least one approved fire extinguisher, and one lifering with line attached.

**1910.265(d)(2)(iv)(c)** Boat fuel shall be transported and stored in approved safety containers. Refer to §1910.155(c)(3) for definition of approved.

**1910.265(d)(2)(iv)(e)** Adequate ventilation shall be provided for the cabin area on enclosed cabin-type boats to prevent accumulation of harmful gases or vapors.

**1910.265(d)(2)(v)** *Dry deck storage. (a)* Dry deck storage areas shall be kept orderly and shall be maintained in a condition which is conducive to safe operation of mobile equipment.

**1910.265(d)(2)(v)(b)** Logs shall be stored in a safe and orderly manner, and roadways and traffic lanes shall be maintained at a width adequate for safe travel of log handling equipment.

**1910.265(d)(2)(v)(c)** Logs shall be arranged to minimize the chance of accidentally rolling from the deck.

**1910.265(d)(2)(vi)** *Log hauls and slips. (a)* Walkways along log hauls shall have a standard handrail on the outer edge, and cleats or other means to assure adequate footing and enable employees to walk clear of the log chute.

**1910.265(d)(2)(vi)(b)** Log haul bull chains or cable shall be designed, installed, and maintained to provide adequate safety for the work need.

**1910.265(d)(2)(vi)(c)** Log haul gear and bull chain drive mechanism shall be guarded.

**1910.265(d)(2)(vi)(d)** Substantial troughs for the return strand of log haul chains shall be provided over passageways.

**1910.265(d)(2)(vi)(e)** Log haul controls shall be located and identified to operate from a position where the operator will, at all times, be in the clear of logs, machinery, lines, and rigging. In operations where control is by lever exposed to incoming logs, the lever shall be arranged to operate the log haul only when moved toward the log slip or toward the log pond.

**1910.265(d)(2)(vi)(f)** A positive stop shall be installed on all log hauls to prevent logs from traveling too far ahead in the mill.

**1910.265(d)(2)(vi)(g)** Overhead protection shall be provided for employees working below logs being moved to the log deck.

**1910.265(d)(2)(vi)(h)** Log wells shall be provided with safeguards to minimize the possibility of logs rolling back into well from log deck.

**1910.265(d)(3)** *Log decks—(i) Access.* Safe access to the head rig shall be provided.

**1910.265(d)(3)(ii)** *Stops.* Log decks shall be provided with adequate stops, chains, or other safeguards to prevent logs from rolling down the deck onto the carriage or its runway.

**1910.265(d)(3)(iii)** *Barricade.* A barricade or other positive stop of sufficient strength to stop any log shall be erected between the sawyer's stand and the log deck.

**1910.265(d)(3)(iv)** *Loose chains.* Loose chains from overhead canting devices or other equipment shall not be allowed to hang over the log deck in such manner as to strike employees.

**1910.265(d)(3)(v)** *Swing saws.* Swing saws on log decks shall be equipped with a barricade and stops for protection of employees who may be on the opposite side of the log haul chute.

**1910.265(d)(3)(vi)** *Drag saws.* Where reciprocating log cutoff saws (drag saws) are provided, they shall not project into walkway or aisle.

**1910.265(d)(3)(vii)** *Circular cutoff saws.* Circular log bucking or cutoff saws shall be so located and guarded as to allow safe entrance to and exit from the building.

**1910.265(d)(3)(viii)** *Entrance doorway.* Where the cutoff saw partially blocks the entrance from the log haul runway, the entrance shall be guarded.

**1910.265(d)(4)** *Mechanical barkers*—**(i)** *Rotary barkers.* Rotary barking devices shall be so guarded as to protect employees from flying chips, bark, or other extraneous material.

**1910.265(d)(4)(ii)** *Elevating ramp.* If an elevating ramp or gate is used, it shall be provided with a safety chain, hook, or other means of suspension while employees are underneath.

**1910.265(d)(4)(iii)** *Area around barkers.* The hazardous area around ring barkers and their conveyors shall be fenced off or posted as a prohibited area for unauthorized persons.

**1910.265(d)(4)(iv)** *Enclosing hydraulic barkers.* Hydraulic barkers shall be enclosed with strong baffles at the inlet and outlet. The operator shall be protected by adequate safety glass or equivalent.

**1910.265(d)(4)(v)** *Holddown rolls.* Holddown rolls shall be installed at the infeed and outfeed sections of mechanical ring barkers to control the movement of logs.

**1910.265(e)** *Log breakdown and related machinery and facilities*—**(1)** *Log carriages and carriage runways*—**(i)** *Bumpers.* A substantial stop or bumper with adequate shock-absorptive qualities shall be installed at each end of the carriage runway.

**1910.265(e)(1)(ii)** *Footing.* Rider-type carriages shall be floored to provide secure footing and a firm working platform for the block setter.

**1910.265(e)(1)(iii)** *Sheave housing.* Sheaves on rope-driven carriages shall be guarded at floor line with substantial housings.

**1910.265(e)(1)(iv)** *Carriage control.* A positive means shall be provided to prevent unintended movement of the carriage. This may involve a control locking device, a carriage tie-down, or both.

**1910.265(e)(1)(v)** *Barriers and warning signs.* A barrier shall be provided to prevent employees from entering the space necessary for travel of the carriage, with headblocks fully receded, for the full length and extreme ends of carriage runways. Warning signs shall be posted at possible entry points to this area.

**1910.265(e)(1)(vi)** *Overhead clearance.* For a rider-type carriage adequate overhead clear space above the carriage deck shall be provided for the full carriage runway length.

**1910.265(e)(1)(vii)** *Sweeping devices.* Carriage track sweeping devices shall be used to keep track rails clear of debris.

**1910.265(e)(1)(viii)** *Dogs.* Dogging devices shall be adequate to secure logs, cants, or boards, during sawing operations.

**1910.265(e)(2)** *Head saws*—**(i)** *Band head saws.* **(a)** Band head saws shall not be operated at speeds in excess of those recommended by the manufacturer.

**1910.265(e)(2)(i)***(b)* Band head saws shall be thoroughly inspected for cracks, splits, broken teeth, and other defects. A bandsaw with a crack greater than one-tenth the width of the saw shall not be placed in service until width of saw is reduced to eliminate crack, until cracked section is removed, or crack development is stopped.

**1910.265(e)(2)(i)***(c)* Provisions shall be made for alerting and warning employees before starting band head saws, and measures shall be taken to insure that all persons are in the clear.

**1910.265(e)(2)(ii)** *Bandsaw wheels.* **(a)** No bandsaw wheel shall be run at a peripheral speed in excess of that recommended by the manufacturer. The manufacturer's recommended maximum speed shall be stamped in plainly legible figures on some portion of the wheel.

**1910.265(e)(2)(ii)***(b)* Band head saw wheels shall be subjected to monthly inspections. Hubs, spokes, rims, bolts, and rivets shall be thoroughly examined in the course of such inspections. A loose or damaged hub, a rim crack, or loose spokes shall make the wheel unfit for service.

**1910.265(e)(2)(ii)***(c)* Band wheels shall be completely encased or guarded, except for a portion of the upper wheel immediately around the point where the blade leaves the wheel, to permit operator to observe movement of equipment. Necessary ventilating and observation ports may be permitted. Substantial doors or gates are allowed for repair, lubrication, and saw changes; such doors or gates shall be closed securely during operation. Band head rigs shall be equipped with a saw catcher or guard of substantial construction.

**1910.265(e)(2)(iii)** *Single circular head saws.* **(a)** Circular head saws shall not be operated at speeds in excess of those specified by the manufacturer. Maximum speed shall be etched on the saw.

**1910.265(e)(2)(iii)***(b)* Circular head saws shall be equipped with safety guides which can be readily adjusted without use of hand tools.

**1910.265(e)(2)(iii)***(c)* The upper saw of a double circular mill shall be provided with a substantial hood or guard. A screen or other suitable device shall be placed so as to protect the sawyer from flying particles.

**1910.265(e)(2)(iii)***(d)* All circular sawmills where live rolls are not used behind the head saw shall be equipped with a spreader wheel or splitter.

**1910.265(e)(2)(iv)** *Twin circular head saws.* Twin circular head saw rigs such as scrag saws shall meet the specifications for single circular head saws in paragraph (e)(1)(iii) of this section where applicable.

**1910.265(e)(2)(v)** *Whole-log sash gang saws (Swedish gangs).* **(a)** Cranks, pitman rods, and other moving parts shall be adequately guarded.

**1910.265(e)(2)(v)***(b)* Feed rolls shall be enclosed by a cover over the top, front, and open ends except where guarded by location. Drive mechanism to feed rolls shall be enclosed.

**1910.265(e)(2)(v)***(c)* Carriage cradles of whole-log sash gang saws (Swedish gangs), shall be of adequate height to prevent logs from kicking out while being loaded.

**1910.265(e)(3)** *Resaws*—**(i)** *Band resaws.* Band resaws shall meet the specifications for band head saws as required by paragraph (e)(2)(i) of this section.

**1910.265(e)(3)(ii)** *Circular gang resaws.* **(a)** Banks of circular gang resaws shall be guarded by a hood.

**1910.265(e)(3)(ii)***(b)* Circular gang resaws shall be provided with safety fingers or other antikickback devices.

**1910.265(e)(3)(ii)***(c)* Circular gang resaws shall not be operated at speeds exceeding those recommended by the manufacturer.

**1910.265(e)(3)(ii)***(d)* [Reserved]

**1910.265(e)(3)(ii)***(e)* Feed rolls shall be guarded.

**1910.265(e)(3)(ii)***(f)* Each circular gang resaw, except self-feed saws with a live roll or wheel at back of saw, shall be provided with spreaders.

**1910.265(e)(3)(iii)** *Sash gang resaws.* Sash gang resaws shall meet the safety specifications of whole-log sash gang saws in accordance with the requirements of paragraph (e)(2)(v) of this section.

**1910.265(e)(4)** *Trimmer saws*—**(i)** *Maximum speed.* Trimmer saws shall not be run at peripheral speeds in excess of those recommended by the manufacturer.

**1910.265(e)(4)(ii)** *Guards.* **(a)** Trimmer saws shall be guarded in front by adequate baffles to protect against flying debris and they shall be securely bolted to a substantial frame. These guards for a series of saws shall be set as close to the top of the trimmer table as is practical.

**1910.265(e)(4)(ii)***(b)* The end saws on trimmer shall be guarded.

**1910.265(e)(4)(ii)***(c)* The rear of trimmer saws shall have a guard the full width of the saws and as much wider as practical.

**1910.265(e)(4)(iii)** *Safety stops.* Automatic trimmer saws shall be provided with safety stops or hangers to prevent saws from dropping on table.

**1910.265(e)(5)** *Edgers*—**(i)** *Location.* **(a)** Where vertical arbor edger saws are located ahead of the main saw, they shall be so guarded that an employee cannot contact any part of the edger saw from his normal position.

**1910.265(e)(5)(i)***(b)* Edgers shall not be located in the main roll case behind the head saws.

**1910.265(e)(5)(ii)** *Guards.* **(a)** The top and the openings in end and side frames of edgers shall be adequately guarded and gears and chains shall be fully housed. Guards may be hinged or otherwise arranged to permit oiling and the removal of saws.

**1910.265(e)(5)(ii)***(b)* All edgers shall be equipped with pressure feed rolls.

**1910.265(e)(5)(ii)***(c)* Pressure feed rolls on edgers shall be guarded against accidental contact.

# OSHA Standards for General Industry

**1910.265(e)(5)(iii)** *Antikickback devices.* **(a)** Edgers shall be provided with safety fingers or other approved methods of preventing kickbacks or guarding against them. A barricade in line with the edger, if properly fenced off, may be used if safety fingers are not feasible to install.

**1910.265(e)(5)(iii)(b)** A controlling device shall be installed and located so that the operator can stop the feed mechanism without releasing the tension of the pressure rolls.

**1910.265(e)(5)(iv)** *Operating speed of live rolls.* Live rolls and tailing devices in back of edger shall operate at a speed not less than the speed of the edger feed rolls.

**1910.265(e)(6)** *Planers*—**(i)** *Guards.* **(a)** All cutting heads shall be guarded.

**1910.265(e)(6)(i)(b)** Side head hoods shall be of sufficient height to safeguard the head setscrew.

**1910.265(e)(6)(i)(c)** Pressure feed rolls and "pineapples" shall be guarded.

**1910.265(e)(6)(i)(d)** Levers or controls shall be so arranged or guarded as to reduce the possibility of accidental operation.

**1910.265(f)** *Dry kilns and facilities*—**(1)** *Kiln foundations.* Dry kilns shall be constructed upon solid foundations to prevent tracks from sagging.

**1910.265(f)(2)** *Passageways.* A passageway shall be provided to give adequate clearance on at least one side or in the center of end-piled kilns and on two sides of cross-piled kilns.

**1910.265(f)(3)** *Doors*—**(i)** *Main kiln doors.* **(a)** Main kiln doors shall be provided with a method of holding them open while kiln is being loaded.

**1910.265(f)(3)(i)(b)** Counterweights on vertical lift doors shall be boxed or otherwise guarded.

**1910.265(f)(3)(i)(c)** Adequate means shall be provided to firmly secure main doors, when they are disengaged from carriers and hangers, to prevent toppling.

**1910.265(f)(3)(ii)** *Escape doors.* **(a)** If operating procedures require access to kilns, kilns shall be provided with escape doors that operate easily from the inside, swing in the direction of exit, and are located in or near the main door at the end of the passageway.

**1910.265(f)(3)(ii)(b)** Escape doors shall be of adequate height and width to accommodate an average size man.

**1910.265(f)(4)** *Pits.* Pits shall be well ventilated, drained, and lighted, and shall be large enough to safely accommodate the kiln operator together with operating devices such as valves, dampers, damper rods, and traps.

**1910.265(f)(5)** *Steam mains.* All high-pressure steam mains located in or adjacent to an operating pit shall be covered with heat-insulating material.

**1910.265(f)(6)** *Ladders.* A fixed ladder, in accordance with the requirements of §1910.27 or other adequate means shall be provided to permit access to the roof. Where controls and machinery are mounted on the roof, a permanent stairway with standard handrail shall be installed in accordance with the requirements of §1910.24.

**1910.265(f)(7)** *Chocks.* A means shall be provided for chocking or blocking cars.

**1910.265(f)(8)** *Kiln tender room.* A warm room shall be provided for kiln employees to stay in during cold weather after leaving a hot kiln.

[39 FR 23502, June 27, 1974, as amended at 40 FR 23073, May 28, 1975; 43 FR 49751, Oct. 24, 1978; 43 FR 51760, Nov. 7, 1978; 53 FR 12123, Apr. 12, 1988; 55 FR 32015, Aug. 6, 1990; 61 FR 9241, Mar. 7, 1996; 63 FR 33467, June 18, 1998]

## §1910.266 Logging operations.

**1910.266(a)** *Table of contents.*

This paragraph contains the list of paragraphs and appendices contained in this section.

a. Table of contents
b. Scope and application
c. Definitions
d. General requirements
   1. Personal protective equipment
   2. First-aid kits
   3. Seat belts
   4. Fire extinguishers
   5. Environmental conditions
   6. Work areas
   7. Signaling and signal equipment
   8. Overhead electric lines
   9. Flammable and combustible liquids
   10. Explosives and blasting agents
e. Hand and portable powered tools
   1. General requirements
   2. Chain saws
f. Machines
   1. General requirements
   2. Machine operation
   3. Protective structures
   4. Overhead guards
   5. Machine access
   6. Exhaust systems
   7. Brakes
   8. Guarding
g. Vehicles
h. Tree harvesting
   1. General requirements
   2. Manual felling
   3. Bucking and limbing
   4. Chipping
   5. Yarding
   6. Loading and unloading
   7. Transport
   8. Storage
i. Training
j. Appendices

APPENDIX A—MINIMUM FIRST-AID SUPPLIES
APPENDIX B—MINIMUM FIRST-AID TRAINING
APPENDIX C—CORRESPONDING ISO AGREEMENTS

**1910.266(b)** *Scope and application.*

**1910.266(b)(1)** This standard establishes safety practices, means, methods and operations for all types of logging, regardless of the end use of the wood. These types of logging include, but are not limited to, pulpwood and timber harvesting and the logging of sawlogs, veneer bolts, poles, pilings and other forest products. This standard does not cover the construction or use of cable yarding systems.

**1910.266(b)(2)** This standard applies to all logging operations as defined by this section.

**1910.266(b)(3)** Hazards and working conditions not specifically addressed by this section are covered by other applicable sections of part 1910.

**1910.266(c)** *Definitions applicable to this section.*

*Arch.* An open-framed trailer or built-up framework used to suspend the leading ends of trees or logs when they are skidded.

*Backcut (felling cut).* The final cut in a felling operation.

*Ballistic nylon.* A nylon fabric of high tensile properties designed to provide protection from lacerations.

*Buck.* To cut a felled tree into logs.

*Butt.* The bottom of the felled part of a tree.

*Cable yarding.* The movement of felled trees or logs from the area where they are felled to the landing on a system composed of a cable suspended from spars and/or towers. The trees or logs may be either dragged across the ground on the cable or carried while suspended from the cable.

*Chock.* A block, often wedge shaped, which is used to prevent movement; e.g., a log from rolling, a wheel from turning.

*Choker.* A sling used to encircle the end of a log for yarding. One end is passed around the load, then through a loop eye, end fitting or other device at the other end of the sling. The end that passed through the end fitting or other device is then hooked to the lifting or pulling machine.

*Danger tree.* A standing tree that presents a hazard to employees due to conditions such as, but not limited to, deterioration or physical damage to the root system, trunk, stem or limbs, and the direction and lean of the tree.

*Debark.* To remove bark from trees or logs.

*Deck.* A stack of trees or logs.

*Designated person.* An employee who has the requisite knowledge, training and experience to perform specific duties.

*Domino felling.* The partial cutting of multiple trees which are left standing and then pushed over with a pusher tree.

*Fell (fall).* To cut down trees.

*Feller (faller).* An employee who fells trees.

*Grounded.* The placement of a component of a machine on the ground or on a device where it is firmly supported.

*Guarded.* Covered, shielded, fenced, enclosed, or otherwise protected by means of suitable enclosures, covers, casings, shields, troughs, railings, screens, mats, or platforms, or by location, to prevent injury.

*Health care provider.* A health care practitioner operating with the scope of his/her license, certificate, registration or legally authorized practice.

*Landing.* Any place where logs are laid after being yarded, and before transport from the work site.

*Limbing.* To cut branches off felled trees.

*Lodged tree (hung tree).* A tree leaning against another tree or object which prevents it from falling to the ground.

*Log.* A segment sawed or split from a felled tree, such as, but not limited to, a section, bolt, or tree length.

*Logging operations.* Operations associated with felling and moving trees and logs from the stump to the point of delivery, such as, but not limited to, marking danger trees and trees/logs to be cut to length, felling, limbing, bucking, debarking, chipping, yarding, loading, unloading, storing, and transporting machines, equipment and personnel to, from and between logging sites.

*Machine.* A piece of stationary or mobile equipment having a self-contained powerplant, that is operated off-road and used for the movement of material. Machines include, but are not limited to, tractors, skidders, front-end loaders, scrapers, graders, bulldozers, swing yarders, log stackers, log loaders, and mechanical felling devices, such as tree shears and feller-bunchers. Machines do not include airplanes or aircraft (e.g., helicopters).

*Rated capacity.* The maximum load a system, vehicle, machine or piece of equipment was designed by the manufacturer to handle.

*Root wad.* The ball of a tree root and dirt that is pulled from the ground when a tree is uprooted.

*Serviceable condition.* A state or ability of a tool, machine, vehicle or other device to operate as it was intended by the manufacturer to operate.

*Skidding.* The yarding of trees or logs by pulling or towing them across the ground.

*Slope (grade).* The increase or decrease in altitude over a horizontal distance expressed as a percentage. For example, a change of altitude of 20 feet (6 m) over a horizontal distance of 100 feet (30 m) is expressed as a 20 percent slope.

*Snag.* Any standing dead tree or portion thereof.

*Spring pole.* A tree, segment of a tree, limb, or sapling which is under stress or tension due to the pressure or weight of another object.

*Tie down.* Chain, cable, steel strips or fiber webbing and binders attached to a truck, trailer or other conveyance as a means to secure loads and to prevent them from shifting or moving when they are being transported.

*Undercut.* A notch cut in a tree to guide the direction of the tree fall and to prevent splitting or kickback.

*Vehicle.* A car, bus, truck, trailer or semi-trailer owned, leased or rented by the employer that is used for transportation of employees or movement of material.

*Winching.* The winding of cable or rope onto a spool or drum.

*Yarding.* The movement of logs from the place they are felled to a landing.

**1910.266(d) General requirements—(1)** *Personal protective equipment.* **(i)** The employer shall assure that personal protective equipment, including any personal protective equipment provided by an employee, is maintained in a serviceable condition.

**1910.266(d)(1)(ii)** The employer shall assure that personal protective equipment, including any personal protective equipment provided by an employee, is inspected before initial use during each workshift. Defects or damage shall be repaired or the unserviceable personal protective equipment shall be replaced before work is commenced.

**1910.266(d)(1)(iii)** The employer shall provide, at no cost to the employee, and assure that each employee handling wire rope wears, hand protection which provides adequate protection from puncture wounds, cuts and lacerations.

**1910.266(d)(1)(iv)** The employer shall provide, at no cost to the employee, and assure that each employee who operates a chain saw wears leg protection constructed with cut-resistant material, such as ballistic nylon. The leg protection shall cover the full length of the thigh to the top of the boot on each leg to protect against contact with a moving chain saw. *Exception:* This requirement does not apply when an employee is working as a climber if the employer demonstrates that a greater hazard is posed by wearing leg protection in the particular situation, or when an employee is working from a vehicular mounted elevating and rotating work platform meeting the requirements of 29 CFR 1910.68.

**1910.266(d)(1)(v)** The employer shall assure that each employee wears foot protection, such as heavy-duty logging boots that are waterproof or water repellant, cover and provide support to the ankle. The employer shall assure that each employee who operates a chain saw wears foot protection that is constructed with cut-resistant material which will protect the employee against contact with a running chain saw. Sharp, calk-soled boots or other slip-resistant type boots may be worn where the employer demonstrates that they are necessary for the employee's job, the terrain, the timber type, and the weather conditions, provided that foot protection otherwise required by this paragraph is met.

**1910.266(d)(1)(vi)** The employer shall provide, at no cost to the employee, and assure that each employee who works in an area where there is potential for head injury from falling or flying objects wears head protection meeting the requirements of subpart I of part 1910.

**1910.266(d)(1)(vii)** The employer shall provide, at no cost to the employee, and assure that each employee wears the following:

**1910.266(d)(1)(vii)(A)** Eye protection meeting the requirements of subpart I of part 1910 where there is potential for eye injury due to falling or flying objects; and

**1910.266(d)(1)(vii)(B)** Face protection meeting the requirements of subpart I of part 1910 where there is potential for facial injury such as, but not limited to, operating a chipper. Logger-type mesh screens may be worn by employees performing chain-saw operations and yarding.

**NOTE to paragraph (d)(1)(vii):** The employee does not have to wear a separate eye protection device where face protection covering both the eyes and face is worn.

**1910.266(d)(2)** *First-aid kits.* **(i)** The employer shall provide first-aid kits at each work site where trees are being cut (e.g., felling, bucking, limbing), at each active landing, and on each employee transport vehicle. The number of first-aid kits and the content of each kit shall reflect the degree of isolation, the number of employees, and the hazards reasonably anticipated at the work site.

**1910.266(d)(2)(ii)** At a minimum, each first-aid kit shall contain the items listed in appendix A at all times.

**1910.266(d)(2)(iii)** The employer also may have the number and content of first-aid kits reviewed and approved annually by a health care provider.

**1910.266(d)(2)(iv)** The employer shall maintain the contents of each first-aid kit in a serviceable condition.

**1910.266(d)(3)** *Seat belts.* For each vehicle or machine (equipped with ROPS/FOPS or overhead guards), including any vehicle or machine provided by an employee, the employer shall assure:

# OSHA Standards for General Industry

**1910.266(d)(3)(i)** That a seat belt is provided for each vehicle or machine operator;

**1910.266(d)(3)(ii)** That each employee uses the available seat belt while the vehicle or machine is being operated;

**1910.266(d)(3)(iii)** That each employee securely and tightly fastens the seat belt to restrain the employee within the vehicle or machine cab;

**1910.266(d)(3)(iv)** That each machine seat belt meets the requirements of the Society of Automotive Engineers Standard SAE J386, June 1985, "Operator Restraint Systems for Off-Road Work Machines", which is incorporated by reference as specified in §1910.6.

**1910.266(d)(3)(v)** That seat belts are not removed from any vehicle or machine. The employer shall replace each seat belt which has been removed from any vehicle or machine that was equipped with seat belts at the time of manufacture; and

**1910.266(d)(3)(vi)** That each seat belt is maintained in a serviceable condition.

**1910.266(d)(4)** *Fire extinguishers.* The employer shall provide and maintain portable fire extinguishers on each machine and vehicle in accordance with the requirements of subpart L of part 1910.

**1910.266(d)(5)** *Environmental conditions.* All work shall terminate and each employee shall move to a place of safety when environmental conditions, such as but not limited to, electrical storms, strong winds which may affect the fall of a tree, heavy rain or snow, extreme cold, dense fog, fires, mudslides, and darkness, create a hazard for the employee in the performance of the job.

**1910.266(d)(6)** *Work areas.* **(i)** Employees shall be spaced and the duties of each employee shall be organized so the actions of one employee will not create a hazard for any other employee.

**1910.266(d)(6)(ii)** Work areas shall be assigned so that trees cannot fall into an adjacent occupied work area. The distance between adjacent occupied work areas shall be at least two tree lengths of the trees being felled. The distance between adjacent occupied work areas shall reflect the degree of slope, the density of the growth, the height of the trees, the soil structure and other hazards reasonably anticipated at that work site. A distance of greater than two tree lengths shall be maintained between adjacent occupied work areas on any slope where rolling or sliding of trees or logs is reasonably foreseeable.

**1910.266(d)(6)(iii)** Each employee performing a logging operation at a logging work site shall work in a position or location that is within visual or audible contact with another employee.

**1910.266(d)(6)(iv)** The employer shall account for each employee at the end of each workshift.

**1910.266(d)(7)** *Signaling and signal equipment.* **(i)** Hand signals or audible contact, such as but not limited to, whistles, horns, or radios, shall be utilized whenever noise, distance, restricted visibility, or other factors prevent clear understanding of normal voice communications between employees.

**1910.266(d)(7)(ii)** Engine noise, such as from a chain saw, is not an acceptable means of signaling. Other locally and regionally recognized signals may be used.

**1910.266(d)(7)(iii)** Only a designated person shall give signals, except in an emergency.

**1910.266(d)(8)** *Overhead electric lines.* **(i)** Logging operations near overhead electric lines shall be done in accordance with the requirements of 29 CFR 1910.333(c)(3).

**1910.266(d)(8)(ii)** The employer shall notify the power company immediately if a felled tree makes contact with any power line. Each employee shall remain clear of the area until the power company advises that there are no electrical hazards.

**1910.266(d)(9)** *Flammable and combustible liquids.* **(i)** Flammable and combustible liquids shall be stored, handled, transported, and used in accordance with the requirements of subpart H of part 1910.

**1910.266(d)(9)(ii)** Flammable and combustible liquids shall not be transported in the driver compartment or in any passenger-occupied area of a machine or vehicle.

**1910.266(d)(9)(iii)** Each machine, vehicle, and portable powered tool shall be shut off during fueling. Diesel-powered machines and vehicles may be fueled while they are at idle, provided that continued operation is intended and that the employer follows safe fueling and operating procedures.

**1910.266(d)(9)(iv)** Flammable and combustible liquids, including chain-saw and diesel fuel, may be used to start a fire, provided the employer assures that in the particular situation its use does not create a hazard for an employee.

**1910.266(d)(10)** *Explosives and blasting agents.* **(i)** Explosives and blasting agents shall be stored, handled, transported, and used in accordance with the requirements of subpart H of part 1910.

**1910.266(d)(10)(ii)** Only a designated person shall handle or use explosives and blasting agents.

**1910.266(d)(10)(iii)** Explosives and blasting agents shall not be transported in the driver compartment or in any passenger-occupied area of a machine or vehicle.

**1910.266(e)** **Hand and portable powered tools**—**(1)** *General requirements.* **(i)** The employer shall assure that each hand and portable powered tool, including any tool provided by an employee, is maintained in serviceable condition.

**1910.266(e)(1)(ii)** The employer shall assure that each tool, including any tool provided by an employee, is inspected before initial use during each workshift. At a minimum, the inspection shall include the following:

**1910.266(e)(1)(ii)(A)** Handles and guards, to assure that they are sound, tight-fitting, properly shaped, free of splinters and sharp edges, and in place;

**1910.266(e)(1)(ii)(B)** Controls, to assure proper function;

**1910.266(e)(1)(ii)(C)** Chain-saw chains, to assure proper adjustment;

**1910.266(e)(1)(ii)(D)** Chain-saw mufflers, to assure that they are operational and in place;

**1910.266(e)(1)(ii)(E)** Chain brakes and nose shielding devices, to assure that they are in place and function properly;

**1910.266(e)(1)(ii)(F)** Heads of shock, impact-driven and driving tools, to assure that there is no mushrooming;

**1910.266(e)(1)(ii)(G)** Cutting edges, to assure that they are sharp and properly shaped; and

**1910.266(e)(1)(ii)(H)** All other safety devices, to assure that they are in place and function properly.

**1910.266(e)(1)(iii)** The employer shall assure that each tool is used only for purposes for which it has been designed.

**1910.266(e)(1)(iv)** When the head of any shock, impact-driven or driving tool begins to chip, it shall be repaired or removed from service.

**1910.266(e)(1)(v)** The cutting edge of each tool shall be sharpened in accordance with manufacturer's specifications whenever it becomes dull during the workshift.

**1910.266(e)(1)(vi)** Each tool shall be stored in the provided location when not being used at a work site.

**1910.266(e)(1)(vii)** Racks, boxes, holsters or other means shall be provided, arranged and used for the transportation of tools so that a hazard is not created for any vehicle operator or passenger.

**1910.266(e)(2)** *Chain saws.* **(i)** Each chain saw placed into initial service after the effective date of this section shall be equipped with a chain brake and shall otherwise meet the requirements of the ANSI B175.1-1991 "Safety Requirements for Gasoline-Powered Chain Saws", which is incorporated by reference as specified in §1910.6. Each chain saw placed into service before the effective date of this section shall be equipped with a protective device that minimizes chain-saw kickback. No chain-saw kickback device shall be removed or otherwise disabled.

**1910.266(e)(2)(ii)** Each gasoline-powered chain saw shall be equipped with a continuous pressure throttle control system which will stop the chain when pressure on the throttle is released.

**1910.266(e)(2)(iii)** The chain saw shall be operated and adjusted in accordance with the manufacturer's instructions.

**1910.266(e)(2)(iv)** The chain saw shall be fueled at least 10 feet (3 m) from any open flame or other source of ignition.

**1910.266(e)(2)(v)** The chain saw shall be started at least 10 feet (3 m) from the fueling area.

**1910.266(e)(2)(vi)** The chain saw shall be started on the ground or where otherwise firmly supported. Drop starting a chain saw is prohibited.

**1910.266(e)(2)(vii)** The chain saw shall be started with the chain brake engaged.

**1910.266(e)(2)(viii)** The chain saw shall be held with the thumbs and fingers of both hands encircling the handles during operation unless the employer demonstrates that a greater hazard is posed by keeping both hands on the chain saw in that particular situation.

**1910.266(e)(2)(ix)** The chain-saw operator shall be certain of footing before starting to cut. The chain saw shall not be used in a position or at a distance that could cause the operator to become off-balance, to have insecure footing, or to relinquish a firm grip on the saw.

**1910.266(e)(2)(x)** Prior to felling any tree, the chain-saw operator shall clear away brush or other potential obstacles which might interfere with cutting the tree or using the retreat path.

**1910.266(e)(2)(xi)** The chain saw shall not be used to cut directly overhead.

**1910.266(e)(2)(xii)** The chain saw shall be carried in a manner that will prevent operator contact with the cutting chain and muffler.

**1910.266(e)(2)(xiii)** The chain saw shall be shut off or the throttle released before the feller starts his retreat.

**1910.266(e)(2)(xiv)** The chain saw shall be shut down or the chain brake shall be engaged whenever a saw is carried further than 50 feet (15.2 m). The chain saw shall be shut down or the chain brake shall be engaged when a saw is carried less than 50 feet if conditions such as, but not limited to, the terrain, underbrush and slippery surfaces, may create a hazard for an employee.

**1910.266(f)** *Machines*—**(1)** *General requirements.* **(i)** The employer shall assure that each machine, including any machine provided by an employee, is maintained in serviceable condition.

**1910.266(f)(1)(ii)** The employer shall assure that each machine, including any machine provided by an employee, is inspected before initial use during each workshift. Defects or damage shall be repaired or the unserviceable machine shall be replaced before work is commenced.

**1910.266(f)(1)(iii)** The employer shall assure that operating and maintenance instructions are available on the machine or in the area where the machine is being operated. Each machine operator and maintenance employee shall comply with the operating and maintenance instructions.

**1910.266(f)(2)** *Machine operation.* **(i)** The machine shall be started and operated only by a designated person.

**1910.266(f)(2)(ii)** Stationary logging machines and their components shall be anchored or otherwise stabilized to prevent movement during operation.

**1910.266(f)(2)(iii)** The rated capacity of any machine shall not be exceeded.

**1910.266(f)(2)(iv)** To maintain stability, the machine must be operated within the limitations imposed by the manufacturer as described in the operating and maintenance instructions for that machine.

**1910.266(f)(2)(v)** Before starting or moving any machine, the operator shall determine that no employee is in the path of the machine.

**1910.266(f)(2)(vi)** The machine shall be operated only from the operator's station or as otherwise recommended by the manufacturer.

**1910.266(f)(2)(vii)** The machine shall be operated at such a distance from employees and other machines such that operation will not create a hazard for an employee.

**1910.266(f)(2)(viii)** No employee other than the operator shall ride on any mobile machine unless seating, seat belts and other protection equivalent to that provided for the operator are provided.

**1910.266(f)(2)(ix)** No employee shall ride on any load.

**1910.266(f)(2)(x)** Before the operator leaves the operator's station of a machine, it shall be secured as follows:

**1910.266(f)(2)(x)(A)** The parking brake or brake locks shall be applied;

**1910.266(f)(2)(x)(B)** The transmission shall be placed in the manufacturer's specified park position; and

**1910.266(f)(2)(x)(C)** Each moving element such as, but not limited to, blades, buckets, saws and shears, shall be lowered to the ground or otherwise secured.

**1910.266(f)(2)(xi)** If a hydraulic or pneumatic storage device can move the moving elements such as, but not limited to, blades, buckets, saws and shears, after the machine is shut down, the pressure or stored energy from the element shall be discharged as specified by the manufacturer.

**1910.266(f)(2)(xii)** The rated capacity of any vehicle transporting a machine shall not be exceeded.

**1910.266(f)(2)(xiii)** The machine shall be loaded, secured and unloaded so that it will not create a hazard for any employee.

**1910.266(f)(3)** *Protective structures.* **(i)** Each tractor, skidder, swing yarder, log stacker, log loader and mechanical felling device, such as tree shears or feller-buncher, placed into initial service after February 9, 1995, shall be equipped with falling object protective structure (FOPS) and/or rollover protective structure (ROPS). The employer shall replace FOPS or ROPS which have been removed from any machine. *Exception:* This requirement does not apply to machines which are capable of 360 degree rotation.

**1910.266(f)(3)(ii)(A)** ROPS shall be tested, installed, and maintained in serviceable condition.

**1910.266(f)(3)(ii)(B)** Each machine manufactured after August 1, 1996, shall have ROPS tested, installed, and maintained in accordance with the Society of Automotive Engineers SAE J1040, April 1988, "Performance Criteria for Rollover Protective Structures (ROPS) for Construction, Earthmoving, Forestry, and Mining Machines", which is incorporated by reference as specified in §1910.6.

**1910.266(f)(3)(ii)(C)** This incorporation by reference was approved by the Director of the Federal Register in accordance with 5 U.S.C. 552(a) and 1 CFR part 51. Copies may be obtained from the Society of Automotive Engineers, 400 Commonwealth Drive, Warrendale, PA 15096. Copies may be inspected at the Docket Office, Occupational Safety and Health Administration, U.S. Department of Labor, 200 Constitution Avenue NW., room N2625, Washington, DC 20210, or at the Office of the Federal Register, 800 North Capitol Street NW., suite 700, Washington, DC.

**1910.266(f)(3)(iii)** FOPS shall be installed, tested and maintained in accordance with the Society of Automotive Engineers SAE J231, January 1981, "Minimum Performance Criteria for Falling Object Protective Structures (FOPS)", which is incorporated by reference as specified in §1910.6.

**1910.266(f)(3)(iv)** ROPS and FOPS shall meet the requirements of the Society of Automotive Engineers SAE J397, April 1988, "Deflection Limiting Volume-ROPS/FOPS Laboratory Evaluation", which is incorporated by reference as specified in §1910.6.

**1910.266(f)(3)(v)** Each protective structure shall be of a size that does not impede the operator's normal movements.

**1910.266(f)(3)(vi)** The overhead covering of each cab shall be of solid material and shall extend over the entire canopy.

**1910.266(f)(3)(vii)** Each machine manufactured after August 1, 1996, shall have a cab that is fully enclosed with mesh material with openings no greater than 2 inches (5.08 cm) at its least dimension. The cab may be enclosed with other material(s) where the employer demonstrates such material(s) provides equivalent protection and visibility. Exception: Equivalent visibility is not required for the lower portion of the cab where there are control panels or similar obstructions in the cab, or where visibility is not necessary for safe operation of the machine.

**1910.266(f)(3)(viii)** Each machine manufactured on or before August 1, 1996 shall have a cab which meets the requirements specified in paragraph (f)(3)(vii) or a protective canopy for the operator which meets the following requirements:

**1910.266(f)(3)(viii)(A)** The protective canopy shall be constructed to protect the operator from injury due to falling trees, limbs, saplings or branches which might enter the compartment side areas and from snapping winch lines or other objects;

**1910.266(f)(3)(viii)(B)** The lower portion of the cab shall be fully enclosed with solid material, except at entrances, to prevent the operator from being injured from obstacles entering the cab;

**1910.266(f)(3)(viii)(C)** The upper rear portion of the cab shall be fully enclosed with open mesh material with openings of such size as to reject the entrance of an object larger than 2 inches in diameter. It shall provide maximum rearward visibility; and

**1910.266(f)(3)(viii)(D)** Open mesh shall be extended forward as far as possible from the rear corners of the cab sides so as to give the maximum protection against obstacles, branches, etc., entering the cab area.

**1910.266(f)(3)(ix)** The enclosure of the upper portion of each cab shall allow maximum visibility.

# OSHA Standards for General Industry

**1910.266(f)(3)(x)** When transparent material is used to enclose the upper portion of the cab, it shall be made of safety glass or other material that the employer demonstrates provides equivalent protection and visibility.

**1910.266(f)(3)(xi)** Transparent material shall be kept clean to assure operator visibility.

**1910.266(f)(3)(xii)** Transparent material that may create a hazard for the operator, such as but not limited to, cracked, broken or scratched safety glass, shall be replaced.

**1910.266(f)(3)(xiii)** Deflectors shall be installed in front of each cab to deflect whipping saplings and branches. Deflectors shall be located so as not to impede visibility and access to the cab.

**1910.266(f)(3)(xiv)** The height of each cab entrance shall be at least 52 inches (1.3 meters) from the floor of the cab.

**1910.266(f)(3)(xv)** Each machine operated near cable yarding operations shall be equipped with sheds or roofs of sufficient strength to provide protection from breaking lines.

**1910.266(f)(4)** *Overhead guards.* Each forklift shall be equipped with an overhead guard meeting the requirements of the American Society of Mechanical Engineers, ASME B56.6-1992 (with addenda), "Safety Standard for Rough Terrain Forklift Trucks", which is incorporated by reference as specified in §1910.6.

**1910.266(f)(5)** *Machine access.* **(i)** Machine access systems, meeting the specifications of the Society of Automotive Engineers, SAE J185, June 1988, "Recommended Practice for Access Systems for Off-Road Machines", which is incorporated by reference as specified in §1910.6, shall be provided for each machine where the operator or any other employee must climb onto the machine to enter the cab or to perform maintenance.

**1910.266(f)(5)(ii)** Each machine cab shall have a second means of egress.

**1910.266(f)(5)(iii)** Walking and working surfaces of each machine and machine work station shall have a slip resistant surface to assure safe footing.

**1910.266(f)(5)(iv)** The walking and working surface of each machine shall be kept free of waste, debris and any other material which might result in fire, slipping, or falling.

**1910.266(f)(6)** *Exhaust systems.* **(i)** The exhaust pipes on each machine shall be located so exhaust gases are directed away from the operator.

**1910.266(f)(6)(ii)** The exhaust pipes on each machine shall be mounted or guarded to protect each employee from accidental contact.

**1910.266(f)(6)(iii)** The exhaust pipes shall be equipped with spark arresters. Engines equipped with turbochargers do not require spark arresters.

**1910.266(f)(6)(iv)** Each machine muffler provided by the manufacturer, or their equivalent, shall be in place at all times the machine is in operation.

**1910.266(f)(7)** *Brakes.* **(i)** Service brakes shall be sufficient to stop and hold each machine and its rated load capacity on the slopes over which it is being operated.

**1910.266(f)(7)(ii)** Each machine placed into initial service on or after September 8, 1995 shall also be equipped with: back-up or secondary brakes that are capable of stopping the machine regardless of the direction of travel or whether the engine is running; and parking brakes that are capable of continuously holding a stopped machine stationary.

**1910.266(f)(8)** *Guarding.* **(i)** Each machine shall be equipped with guarding to protect employees from exposed moving elements, such as but not limited to, shafts, pulleys, belts on conveyors, and gears, in accordance with the requirements of subpart O of part 1910.

**1910.266(f)(8)(ii)** Each machine used for debarking, limbing and chipping shall be equipped with guarding to protect employees from flying wood chunks, logs, chips, bark, limbs and other material in accordance with the requirements of subpart O of part 1910.

**1910.266(f)(8)(iii)** The guarding on each machine shall be in place at all times the machine is in operation.

**1910.266(g)** *Vehicles.* **(1)** The employer shall assure that each vehicle used to perform any logging operation is maintained in serviceable condition.

**1910.266(g)(2)** The employer shall assure that each vehicle used to perform any logging operation is inspected before initial use during each workshift. Defects or damage shall be repaired or the unserviceable vehicle shall be replaced before work is commenced.

**1910.266(g)(3)** The employer shall assure that operating and maintenance instructions are available in each vehicle. Each vehicle operator and maintenance employee shall comply with the operating and maintenance instructions.

**1910.266(g)(4)** The employer shall assure that each vehicle operator has a valid operator's license for the class of vehicle being operated.

**1910.266(g)(5)** Mounting steps and handholds shall be provided for each vehicle wherever it is necessary to prevent an employee from being injured when entering or leaving the vehicle.

**1910.266(g)(6)** The seats of each vehicle shall be securely fastened.

**1910.266(g)(7)** The requirements of paragraphs (f)(2)(iii), (f)(2)(v), (f)(2)(vii), (f)(2)(x), (f)(2)(xiii), and (f)(7) of this section shall also apply to each vehicle used to transport any employee off public roads or to perform any logging operation, including any vehicle provided by an employee.

**1910.266(h)** *Tree harvesting—***(1)** *General requirements.* **(i)** Trees shall not be felled in a manner that may create a hazard for an employee, such as but not limited to, striking a rope, cable, power line, or machine.

**1910.266(h)(1)(ii)** The immediate supervisor shall be consulted when unfamiliar or unusually hazardous conditions necessitate the supervisor's approval before cutting is commenced.

**1910.266(h)(1)(iii)** While manual felling is in progress, no yarding machine shall be operated within two tree lengths of trees being manually felled. Exception: This provision does not apply to yarding machines performing tree pulling operations.

**1910.266(h)(1)(iv)** No employee shall approach a feller closer than two tree lengths of trees being felled until the feller has acknowledged that it is safe to do so, unless the employer demonstrates that a team of employees is necessary to manually fell a particular tree.

**1910.266(h)(1)(v)** No employee shall approach a mechanical felling operation closer than two tree lengths of the trees being felled until the machine operator has acknowledged that it is safe to do so.

**1910.266(h)(1)(vi)** Each danger tree shall be felled, removed or avoided. Each danger tree, including lodged trees and snags, shall be felled or removed using mechanical or other techniques that minimize employee exposure before work is commenced in the area of the danger tree. If the danger tree is not felled or removed, it shall be marked and no work shall be conducted within two tree lengths of the danger tree unless the employer demonstrates that a shorter distance will not create a hazard for an employee.

**1910.266(h)(1)(vii)** Each danger tree shall be carefully checked for signs of loose bark, broken branches and limbs or other damage before they are felled or removed. Accessible loose bark and other damage that may create a hazard for an employee shall be removed or held in place before felling or removing the tree.

**1910.266(h)(1)(viii)** Felling on any slope where rolling or sliding of trees or logs is reasonably foreseeable shall be done uphill from, or on the same level as, previously felled trees.

**1910.266(h)(1)(ix)** Domino felling of trees is prohibited.

**NOTE to paragraph (h)(1)(ix):** The definition of domino felling does not include the felling of a single danger tree by felling another single tree into it.

**1910.266(h)(2)** *Manual felling.* **(i)** Before felling is started, the feller shall plan and clear a retreat path. The retreat path shall extend diagonally away from the expected felling line unless the employer demonstrates that such a retreat path poses a greater hazard than an alternate path. Once the backcut has been made the feller shall immediately move a safe distance away from the tree on the retreat path.

**1910.266(h)(2)(ii)** Before each tree is felled, conditions such as, but not limited to, snow and ice accumulation, the wind, the lean of tree, dead limbs, and the location of other trees, shall be evaluated by the feller and precautions taken so a hazard is not created for an employee.

**1910.266(h)(2)(iii)** Each tree shall be checked for accumulations of snow and ice. Accumulations of snow and ice that may create a hazard for an employee shall be removed before felling is commenced in the area or the area shall be avoided.

**1910.266(h)(2)(iv)** When a spring pole or other tree under stress is cut, no employee other than the feller shall be closer than two trees lengths when the stress is released.

**1910.266(h)(2)(v)** An undercut shall be made in each tree being felled unless the employer demonstrates that felling the particular tree without an undercut will not create a hazard for an employee. The undercut shall be of a size so the tree will not split and will fall in the intended direction.

**1910.266(h)(2)(vi)** A backcut shall be made in each tree being felled. The backcut shall leave sufficient hinge wood to hold the tree to the stump during most of its fall so that the hinge is able to guide the tree's fall in the intended direction.

**1910.266(h)(2)(vii)** The backcut shall be above the level of the horizontal facecut in order to provide an adequate platform to prevent kickback. Exception: The backcut may be at or below the horizontal facecut in tree pulling operations.

**NOTE to paragraph (h)(2)(vii):** This requirement does not apply to open face felling where two angled facecuts rather than a horizontal facecut are used.

**1910.266(h)(3)** *Limbing and bucking.* **(i)** Limbing and bucking on any slope where rolling or sliding of trees or logs is reasonably foreseeable shall be done on the uphill side of each tree or log.

**1910.266(h)(3)(ii)** Before bucking or limbing wind-thrown trees, precautions shall be taken to prevent the root wad, butt or logs from striking an employee. These precautions include, but are not limited to, chocking or moving the tree to a stable position.

**1910.266(h)(4)** *Chipping (in-woods locations).* **(i)** Chipper access covers or doors shall not be opened until the drum or disc is at a complete stop.

**1910.266(h)(4)(ii)** Infeed and discharge ports shall be guarded to prevent contact with the disc, knives, or blower blades.

**1910.266(h)(4)(iii)** The chipper shall be shut down and locked out in accordance with the requirements of 29 CFR 1910.147 when an employee performs any servicing or maintenance.

**1910.266(h)(4)(iv)** Detached trailer chippers shall be chocked during usage on any slope where rolling or sliding of the chipper is reasonably foreseeable.

**1910.266(h)(5)** *Yarding.* **(i)** No log shall be moved until each employee is in the clear.

**1910.266(h)(5)(ii)** Each choker shall be hooked and unhooked from the uphill side or end of the log, unless the employer demonstrates that is it not feasible in the particular situation to hook or unhook the choker from the uphill side. Where the choker is hooked or unhooked from the downhill side or end of the log, the log shall be securely chocked to prevent rolling, sliding or swinging.

**1910.266(h)(5)(iii)** Each choker shall be positioned near the end of the log or tree length.

**1910.266(h)(5)(iv)** Each machine shall be positioned during winching so the machine and winch are operated within their design limits.

**1910.266(h)(5)(v)** No yarding line shall be moved unless the yarding machine operator has clearly received and understood the signal to do so. When in doubt, the yarding machine operator shall repeat the signal and wait for a confirming signal before moving any line.

**1910.266(h)(5)(vi)** No load shall exceed the rated capacity of the pallet, trailer, or other carrier.

**1910.266(h)(5)(vii)** Towed equipment, such as but not limited to, skid pans, pallets, arches, and trailers, shall be attached to each machine or vehicle in such a manner as to allow a full 90 degree turn; to prevent overrunning of the towing machine or vehicle; and to assure that the operator is always in control of the towed equipment.

**1910.266(h)(5)(viii)** The yarding machine or vehicle, including its load, shall be operated with safe clearance from all obstructions that may create a hazard for an employee.

**1910.266(h)(5)(ix)** Each yarded tree shall be placed in a location that does not create a hazard for an employee and an orderly manner so that the trees are stable before bucking or limbing is commenced.

**1910.266(h)(6)** *Loading and unloading.* **(i)** The transport vehicle shall be positioned to provide working clearance between the vehicle and the deck.

**1910.266(h)(6)(ii)** Only the loading or unloading machine operator and other personnel the employer demonstrates are essential shall be in the loading or unloading work area during this operation.

**1910.266(h)(6)(iii)** No transport vehicle operator shall remain in the cab during loading and unloading if the logs are carried or moved over the truck cab, unless the employer demonstrates that it is necessary for the operator to do so. Where the transport vehicle operator remains in the cab, the employer shall provide operator protection, such as but not limited to, reinforcement of the cab.

**1910.266(h)(6)(iv)** Each log shall be placed on a transport vehicle in an orderly manner and tightly secured.

**1910.266(h)(6)(v)** The load shall be positioned to prevent slippage or loss during handling and transport.

**1910.266(h)(6)(vi)** Each stake and chock which is used to trip loads shall be so constructed that the tripping mechanism is activated on the side opposite the release of the load.

**1910.266(h)(6)(vii)** Each tie down shall be left in place over the peak log to secure all logs until the unloading lines or other protection the employer demonstrates is equivalent has been put in place. A stake of sufficient strength to withstand the forces of shifting or moving logs, shall be considered equivalent protection provided that the logs are not loaded higher than the stake.

**1910.266(h)(6)(viii)** Each tie down shall be released only from the side on which the unloading machine operates, except as follows:

**1910.266(h)(6)(viii)(A)** When the tie down is released by a remote control device; and

**1910.266(h)(6)(viii)(B)** When the employee making the release is protected by racks, stanchions or other protection the employer demonstrates is capable of withstanding the force of the logs.

**1910.266(h)(7)** *Transport.* The transport vehicle operator shall assure that each tie down is tight before transporting the load. While enroute, the operator shall check and tighten the tie downs whenever there is reason to believe that the tie downs have loosened or the load has shifted.

**1910.266(h)(8)** *Storage.* Each deck shall be constructed and located so it is stable and provides each employee with enough room to safely move and work in the area.

**1910.266(i)** *Training.* **(1)** The employer shall provide training for each employee, including supervisors, at no cost to the employee.

**1910.266(i)(2)** *Frequency.* Training shall be provided as follows:

**1910.266(i)(2)(i)** As soon as possible but not later than the effective date of this section for initial training for each current and new employee;

**1910.266(i)(2)(ii)** Prior to initial assignment for each new employee;

**1910.266(i)(2)(iii)** Whenever the employee is assigned new work tasks, tools, equipment, machines or vehicles; and

**1910.266(i)(2)(iv)** Whenever an employee demonstrates unsafe job performance.

**1910.266(i)(3)** *Content.* At a minimum, training shall consist of the following elements:

**1910.266(i)(3)(i)** Safe performance of assigned work tasks;

**1910.266(i)(3)(ii)** Safe use, operation and maintenance of tools, machines and vehicles the employee uses or operates, including emphasis on understanding and following the manufacturer's operating and maintenance instructions, warnings and precautions;

**1910.266(i)(3)(iii)** Recognition of safety and health hazards associated with the employee's specific work tasks, including the use of measures and work practices to prevent or control those hazards;

**1910.266(i)(3)(iv)** Recognition, prevention and control of other safety and health hazards in the logging industry;

**1910.266(i)(3)(v)** Procedures, practices and requirements of the employer's work site; and

**1910.266(i)(3)(vi)** The requirements of this standard.

**1910.266(i)(4)** Training of an employee due to unsafe job performance, or assignment of new work tasks, tools, equipment, machines, or vehicles; may be limited to those elements in paragraph (i)(3) of this section which are relevant to the circumstances giving rise to the need for training.

**1910.266(i)(5)** *Portability of training.* **(i)** Each current employee who has received training in the particular elements specified in paragraph (i)(3) of this section shall not be required to be retrained in those elements.

**1910.266(i)(5)(ii)** Each new employee who has received training in the particular elements specified in paragraph (i)(3) of this section shall not be required to be retrained in those elements prior to initial assignment.

**1910.266(i)(5)(iii)** The employer shall train each current and new employee in those elements for which the employee has not received training.

**1910.266(i)(5)(iv)** The employer is responsible for ensuring that each current and new employee can properly and safely perform the work tasks and operate the tools, equipment, machines, and vehicles used in their job.

**1910.266(i)(6)** Each new employee and each employee who is required to be trained as specified in paragraph (i)(2) of this section, shall work under the close supervision of a designated person until the employee demonstrates to the employer the ability to safely perform their new duties independently.

**1910.266(i)(7)** *First-aid training.* **(i)** The employer shall assure that each employee, including supervisors, receives or has received first-aid and CPR training meeting at least the requirements specified in appendix B.

**1910.266(i)(7)(ii)** The employer shall assure that each employee's first-aid and CPR training and/or certificate of training remain current.

**1910.266(i)(8)** All training shall be conducted by a designated person.

**1910.266(i)(9)** The employer shall assure that all training required by this section is presented in a manner that the employee is able to understand. The employer shall assure that all training materials used are appropriate in content and vocabulary to the educational level, literacy, and language skills of the employees being trained.

**1910.266(i)(10)** *Certification of training.* **(i)** The employer shall verify compliance with paragraph (i) of this section by preparing a written certification record. The written certification record shall contain the name or other identity of the employee trained, the date(s) of the training, and the signature of the person who conducted the training or the signature of the employer. If the employer relies on training conducted prior to the employee's hiring or completed prior to the effective date of this section, the certification record shall indicate the date the employer determined the prior training was adequate.

**1910.266(i)(10)(ii)** The most recent training certification shall be maintained.

**1910.266(i)(11)** *Safety and health meetings.* The employer shall hold safety and health meetings as necessary and at least each month for each employee. Safety and health meetings may be conducted individually, in crew meetings, in larger groups, or as part of other staff meetings.

**1910.266(j)** *Appendices.* Appendices A and B of this section are mandatory. The information contained in appendix C of this section is informational and is not intended to create any additional obligations not otherwise imposed or to detract from existing regulations.

**NOTE:** In the FEDERAL REGISTER of August 9, 1995, OSHA extended the stay of the following paragraphs of §1910.266 until September 8, 1995. The remaining requirements of §1910.266, which became effective on February 9, 1995, are unaffected by the extension of the partial stay:

1. (d)(1)(v)—insofar as it requires foot protection to be chain-saw resistant.
2. (d)(1)(vii)—insofar as it required face protection.
3. (d)(2)(iii).
4. (f)(2)(iv).
5. (f)(2)(xi).
6. (f)(3)(ii).
7. (f)(3)(vii).
8. (f)(3)(viii).
9. (f)(7)(ii)—insofar as it requires parking brakes to be able to stop a moving machine.
10. (g)(1) and (g)(2) insofar as they require inspection and maintenance of employee-owned vehicles.
11. (h)(2)(vii)—insofar as it precludes backcuts at the level of the horizontal cut of the undercut when the Humboldt cutting method is used.

## APPENDIX A TO §1910.266—FIRST-AID KITS (MANDATORY)

The following list sets forth the minimally acceptable number and type of first-aid supplies for first-aid kits required under paragraph (d)(2) of the logging standard. The contents of the first-aid kit listed should be adequate for small work sites, consisting of approximately two to three employees. When larger operations or multiple operations are being conducted at the same location, additional first-aid kits should be provided at the work site or additional quantities of supplies should be included in the first-aid kits:

1. Gauze pads (at least 4×4 inches).
2. Two large gauze pads (at least 8×10 inches).
3. Box adhesive bandages (band-aids).
4. One package gauze roller bandage at least 2 inches wide.
5. Two triangular bandages.
6. Wound cleaning agent such as sealed moistened towelettes.
7. Scissors.
8. At least one blanket.
9. Tweezers.
10. Adhesive tape.
11. Latex gloves.
12. Resuscitation equipment such as resuscitation bag, airway, or pocket mask.
13. Two elastic wraps.
14. Splint.
15. Directions for requesting emergency assistance.

## APPENDIX B TO §1910.266—FIRST-AID AND CPR TRAINING (MANDATORY)

The following is deemed to be the minimal acceptable first-aid and CPR training program for employees engaged in logging activities.

First-aid and CPR training shall be conducted using the conventional methods of training such as lecture, demonstration, practical exercise and examination (both written and practical). The length of training must be sufficient to assure that trainees understand the concepts of first aid and can demonstrate their ability to perform the various procedures contained in the outline below.

At a minimum, first-aid and CPR training shall consist of the following:

1. The definition of first aid.
2. Legal issues of applying first aid (Good Samaritan Laws).
3. Basic anatomy.
4. Patient assessment and first aid for the following:
   a. Respiratory arrest.
   b. Cardiac arrest.
   c. Hemorrhage.
   d. Lacerations/abrasions.
   e. Amputations.
   f. Musculoskeletal injuries.
   g. Shock.
   h. Eye injuries.
   i. Burns.
   j. Loss of consciousness.
   k. Extreme temperature exposure (hypothermia/hyperthermia).
   l. Paralysis.
   m. Poisoning.
   n. Loss of mental functioning (psychosis/hallucinations, etc.). Artificial ventilation.
   o. Drug overdose.

5. CPR.
6. Application of dressings and slings.
7. Treatment of strains, sprains, and fractures.
8. Immobilization of injured persons.
9. Handling and transporting injured persons.
10. Treatment of bites, stings, or contact with poisonous plants or animals.

### APPENDIX C TO §1910.266—COMPARABLE ISO STANDARDS (NON-MANDATORY)

The following International Labor Organization (ISO) standards are comparable to the corresponding Society of Automotive Engineers (Standards that are referenced in this standard.)

Utilization of the ISO standards in lieu of the corresponding SAE standards should result in a machine that meets the OSHA standard.

| SAE standard | ISO standard | Subject |
| --- | --- | --- |
| SAE J1040 | ISO 3471-1 | Performance Criteria for Rollover Protective Structures (ROPS) for Construction, Earthmoving, Forestry and Mining Machines. |
| SAE J397 | ISO 3164 | Deflection Limiting Volume—ROPS/FOPS Laboratory Evaluation. |
| SAE J231 | ISO 3449 | Minimum Performance Criteria for Falling Object Protective Structures (FOPS). |
| SAE J386 | ISO 6683 | Operator Restraint Systems for Off-Road Work Machines. |
| SAE J185 | ISO 2897 | Access Systems for Off-Road Machines. |

[59 FR 51741, Oct. 12, 1994, as amended at 60 FR 7449, Feb. 8, 1995; 60 FR 40458, Aug. 9, 1996; 60 FR 47035-47037, Sept. 8, 1995; 61 FR 9241, 9242, Mar. 7, 1996]

### §1910.267 [Reserved].

[§1910.267 removed and reserved at 63 FR 33467, June 18, 1998, effective August 17, 1998.]

### §1910.268 Telecommunications.

**1910.268(a)** *Application.* **(1)** This section sets forth safety and health standards that apply to the work conditions, practices, means, methods, operations, installations and processes performed at telecommunications centers and at telecommunications field installations, which are located outdoors or in building spaces used for such field installations. *Center* work includes the installation, operation, maintenance, rearrangement, and removal of communications equipment and other associated equipment in telecommunications switching centers. *Field* work includes the installation, operation, maintenance, rearrangement, and removal of conductors and other equipment used for signal or communication service, and of their supporting or containing structures, overhead or underground, on public or private rights of way, including buildings or other structures.

**1910.268(a)(2)** These standards do not apply: **(i)** To construction work, as defined in §1910.12, nor (ii) to installations under the exclusive control of electric utilities used for the purpose of communications or metering, or for generation, control, transformation, transmission, and distribution of electric energy, which are located in buildings used exclusively by the electric utilities for such purposes, or located outdoors on property owned or leased by the electric utilities or on public highways, streets, roads, etc., or outdoors by established rights on private property.

**1910.268(a)(3)** Operations or conditions not specifically covered by this section are subject to all the applicable standards contained in this part 1910. See §1910.5(c). Operations which involve construction work, as defined in §1910.12 are subject to all the applicable standards contained in part 1926 of this chapter.

**1910.268(b)** *General*—**(1)** *Buildings containing telecommunications centers*—**(i)** *Illumination.* Lighting in telecommunication centers shall be provided in an adequate amount such that continuing work operations, routine observations, and the passage of employees can be carried out in a safe and healthful manner. Certain specific tasks in centers, such as splicing cable and the maintenance and repair of equipment frame lineups, may require a higher level of illumination. In such cases, the employer shall install permanent lighting or portable supplemental lighting to attain a higher level of illumination shall be provided as needed to permit safe performance of the required task.

**1910.268(b)(1)(ii)** *Working surfaces.* Guard rails and toe boards may be omitted on distribution frame mezzanine platforms to permit access to equipment. This exemption applies only on the side or sides of the platform facing the frames and only on those portions of the platform adjacent to equipped frames.

**1910.268(b)(1)(iii)** *Working spaces. Maintenance aisles,* or *wiring aisles,* between equipment frame lineups are working spaces and are not a means of egress for purposes of §1910.34.

**1910.268(b)(1)(iv)** *Special doors.* When blastproof or power actuated doors are installed in specially designed hardsite security buildings and spaces, they shall be designed and installed so that they can be used as a means of egress in emergencies.

**1910.268(b)(1)(v)** *Equipment, machinery and machine guarding.* When power plant machinery in telecommunications centers is operated with commutators and couplings uncovered, the adjacent housing shall be clearly marked to alert personnel to the rotating machinery.

**1910.268(b)(2)** *Battery handling.* **(i)** Eye protection devices which provide side as well as frontal eye protection for employees shall be provided when measuring storage battery specific gravity or handling electrolyte, and the employer shall ensure that such devices are used by the employees. The employer shall also ensure that acid resistant gloves and aprons shall be worn for protection against spattering. Facilities for quick drenching or flushing of the eyes and body shall be provided unless the storage batteries are of the enclosed type and equipped with explosion proof vents, in which case sealed water rinse or neutralizing packs may be substituted for the quick drenching or flushing facilities. Employees assigned to work with storage batteries shall be instructed in emergency procedures such as dealing with accidental acid spills.

**1910.268(b)(2)(ii)** Electrolyte (acid or base, and distilled water) for battery cells shall be mixed in a well ventilated room. Acid or base shall be poured gradually, while stirring, into the water. Water shall never be poured into concentrated (greater than 75 percent) acid solutions. Electrolyte shall never be placed in metal containers nor stirred with metal objects.

**1910.268(b)(2)(iii)** When taking specific gravity readings, the open end of the hydrometer shall be covered with an acid resistant material while moving it from cell to cell to avoid splashing or throwing the electrolyte.

**1910.268(b)(3)** *Medical and first aid.* Employers must provide employees with readily accessible, adequate, and appropriate first aid supplies. A non-mandatory example of appropriate supplies is listed in Appendix A to 29 CFR 1910.151.

**1910.268(b)(4)** *Hazardous materials.* Highway mobile vehicles and trailers stored in garages in accordance with §1910.110 may be equipped to carry more than one LP-gas container, but the total capacity of LP-gas containers per work vehicle stored in garages shall not exceed 100 pounds of LP-gas. All container valves shall be closed when not in use.

**1910.268(b)(5)** *Compressed gas.* When using or transporting nitrogen cylinders in a horizontal position, special compartments, racks, or blocking shall be provided to prevent cylinder movement. Regulators shall be removed or guarded before a cylinder is transported.

**1910.268(b)(6)** *Support structures.* No employee, or any material or equipment, may be supported or permitted to be supported on any portion of a pole structure, platform, ladder, walkway or other elevated structure or aerial device unless the employer ensures that the support structure is first inspected by a competent person and it is determined to be adequately strong, in good working condition and properly secured in place.

**1910.268(b)(7)** *Approach distances to exposed energized overhead power lines and parts.* The employer shall ensure that no employee approaches or takes any conductive object closer to any electrically energized overhead power lines and parts than prescribed in Table R-2, unless:

**1910.268(b)(7)(i)** The employee is insulated or guarded from the energized parts (insulating gloves rated for the voltage involved shall be considered adequate insulation), or

**1910.268(b)(7)(ii)** The energized parts are insulated or guarded from the employee and any other conductive object at a different potential, or

**1910.268(b)(7)(iii)** The power conductors and equipment are deenergized and grounded.

## OSHA Standards for General Industry

### TABLE R-2—APPROACH DISTANCES TO EXPOSED ENERGIZED OVERHEAD POWER LINES AND PARTS

| Voltage range (phase to phase, RMS) | Approach distance (inches) |
|---|---|
| 300 V and less | (1) |
| Over 300V, not over 750V | 12 |
| Over 750V not over 2 kV | 18 |
| Over 2 kV, not over 15 kV | 24 |
| Over 15 kV, not over 37 kV | 36 |
| Over 37 kV, not over 87.5 kV | 42 |
| Over 87.5 kV, not over 121 kV | 48 |
| Over 121 kV, not over 140 kV | 54 |

[1] Avoid contact.

**1910.268(b)(8)** *Illumination of field work.* Whenever natural light is insufficient to adequately illuminate the worksite, artificial illumination shall be provided to enable the employee to perform the work safely.

**1910.268(c)** *Training.* Employers shall provide training in the various precautions and safe practices described in this section and shall insure that employees do not engage in the activities to which this section applies until such employees have received proper training in the various precautions and safe practices required by this section. However, where the employer can demonstrate that an employee is already trained in the precautions and safe practices required by this section prior to his employment, training need not be provided to that employee in accordance with this section. Where training is required, it shall consist of on-the-job training or classroom-type training or a combination of both. The employer shall certify that employees have been trained by preparing a certification record which includes the identity of the person trained, the signature of the employer or the person who conducted the training, and the date the training was completed. The certification record shall be prepared at the completion of training and shall be maintained on file for the duration of the employee's employment. The certification record shall be made available upon request to the Assistant Secretary for Occupational Safety and Health. Such training shall, where appropriate, include the following subjects:

**1910.268(c)(1)** Recognition and avoidance of dangers relating to encounters with harmful substances and animal, insect, or plant life;

**1910.268(c)(2)** Procedures to be followed in emergency situations; and,

**1910.268(c)(3)** First aid training, including instruction in artificial respiration.

**1910.268(d)** *Employee protection in public work areas.* **(1)** Before work is begun in the vicinity of vehicular or pedestrian traffic which may endanger employees, warning signs and/or flags or other traffic control devices shall be placed conspicuously to alert and channel approaching traffic. Where further protection is needed, barriers shall be utilized. At night, warning lights shall be prominently displayed, and excavated areas shall be enclosed with protective barricades.

**1910.268(d)(2)** If work exposes energized or moving parts that are normally protected, danger signs shall be displayed and barricades erected, as necessary, to warn other personnel in the area.

**1910.268(d)(3)** The employer shall insure that an employee finding any crossed or fallen wires which create or may create a hazardous situation at the work area: (i) Remains on guard or adopts other adequate means to warn other employees of the danger and (ii) has the proper authority notified at the earliest practical moment.

**1910.268(e)** *Tools and personal protective equipment—Generally.* Personal protective equipment, protective devices and special tools needed for the work of employees shall be provided and the employer shall ensure that they are used by employees. Before each day's use the employer shall ensure that these personal protective devices, tools, and equipment are carefully inspected by a competent person to ascertain that they are in good condition.

**1910.268(f)** *Rubber insulating equipment.* **(1)** Rubber insulating equipment designed for the voltage levels to be encountered shall be provided and the employer shall ensure that they are used by employees as required by this section. The requirements of §1910.137, Electrical Protective Equipment, shall be followed except for Table I-6.

**1910.268(f)(2)** The employer is responsible for the periodic retesting of all insulating gloves, blankets, and other rubber insulating equipment. This retesting shall be electrical, visual and mechanical. The following maximum retesting intervals shall apply:

| Gloves, blankets, and other insulating equipment | Natural rubber | Synthetic rubber |
|---|---|---|
| | Months | Months |
| New | 12 | 18 |
| Re-issued | 9 | 15 |

**1910.268(f)(3)** Gloves and blankets shall be marked to indicate compliance with the retest schedule, and shall be marked with the date the next test is due. Gloves found to be defective in the field or by the tests set forth in paragraph (f)(5) of this section shall be destroyed by cutting them open from the finger to the gauntlet.

**1910.268(g)** *Personal climbing equipment*—**(1)** *General.* Safety belts and straps shall be provided and the employer shall ensure their use when work is performed at positions more than 4 feet above ground, on poles, and on towers, except as provided in paragraphs (n)(7) and (n)(8) of this section. No safety belts, safety straps or lanyards acquired after July 1, 1975 may be used unless they meet the tests set forth in paragraph (g)(2) of this section. The employer shall ensure that all safety belts and straps are inspected by a competent person prior to each day's use to determine that they are in safe working condition.

**1910.268(g)(2)** *Telecommunication lineman's body belts, safety straps, and lanyards*—**(i)** *General requirements.* **(A)** Hardware for lineman's body belts, safety straps, and lanyards shall be drop forged or pressed steel and shall have a corrosion resistant finish tested to meet the requirements of the American Society for Testing and Materials B117-64, which is incorporated by reference as specified in §1910.6 (50-hour test). Surfaces shall be smooth and free of sharp edges. Production samples of lineman's safety straps, body belts and lanyards shall be approved by a nationally recognized testing laboratory, as having been tested in accordance with and as meeting the requirements of this paragraph.

**1910.268(g)(2)(i)(B)** All buckles shall withstand a 2,000-pound tensile test with a maximum permanent deformation no greater than one sixty-fourth inch.

**1910.268(g)(2)(i)(C)** D rings shall withstand a 5,000-pound tensile test without cracking or breaking.

**1910.268(g)(2)(i)(D)** Snaphooks shall withstand a 5,000-pound tensile test, or shall withstand a 3,000-pound tensile test and a 180° bend test. Tensile failure is indicated by distortion of the snaphook sufficient to release the keeper; bend test failure is indicated by cracking of the snaphook.

**1910.268(g)(2)(ii)** *Specific requirements.* **(A)(1)** All fabric used for safety straps shall be capable of withstanding an A.C. dielectric test of not less than 25,000 volts per foot "dry" for 3 minutes, without visible deterioration.

**1910.268(g)(2)(ii)(A)(2)** All fabric and leather used shall be tested for leakage current. Fabric or leather may not be used if the leakage current exceeds 1 milliampere when a potential of 3,000 volts is applied to the electrodes positioned 12 inches apart.

**1910.268(g)(2)(ii)(A)(3)** In lieu of alternating current tests, equivalent direct current tests may be performed.

**1910.268(g)(2)(ii)(B)** The cushion part of the body belt shall:

**1910.268(g)(2)(ii)(B)(1)** Contain no exposed rivets on the inside. This provision does not apply to belts used by craftsmen not engaged in line work.

**1910.268(g)(2)(ii)(B)(2)** Be at least three inches in width;

**1910.268(g)(2)(ii)(B)(3)** Be at least five thirty-seconds (5/32) inch thick, if made of leather; and

**1910.268(g)(2)(ii)(C)** [Reserved]

**1910.268(g)(2)(ii)(D)** Suitable copper, steel, or equivalent liners shall be used around the bars of D rings to prevent wear between these members and the leather or fabric enclosing them.

**1910.268(g)(2)(ii)(E)** All stitching shall be done with a minimum 42-pound weight nylon or equivalent thread and shall be lock stitched. Stitching parallel to an edge may not be less than three-sixteenths (3/16) inch from the edge of the narrowest member caught by the thread. The use of cross stitching on leather is prohibited.

**1910.268(g)(2)(ii)(F)** The keepers of snaphooks shall have a spring tension that will not allow the keeper to begin to open when a weight of 2 1/2 pounds or less is applied, but the keepers shall begin to open when a weight of four pounds is applied. In making this determination, the weight shall be supported on the keeper against the end of the nose.

**1910.268(g)(2)(ii)(G)** Safety straps, lanyards, and body belts shall be tested in accordance with the following procedure:

**1910.268(g)(2)(ii)(G)*1*** Attach one end of the safety strap or lanyard to a rigid support, and the other end to a 250-pound canvas bag of sand;

**1910.268(g)(2)(ii)(G)*2*** Allow the 250-pound canvas bag of sand to free fall 4 feet when testing safety straps and 6 feet when testing lanyards. In each case, the strap or lanyard shall stop the fall of the 250-pound bag;

**1910.268(g)(2)(ii)(G)*3*** Failure of the strap or lanyard shall be indicated by any breakage or slippage sufficient to permit the bag to fall free from the strap or lanyard.

**1910.268(g)(2)(ii)(G)*4*** The entire "body belt assembly" shall be tested using on D ring. A safety strap or lanyard shall be used that is capable of passing the "impact loading test" described in paragraph (g)(2)(ii)(G)*2* of this section and attached as required in paragraph (g)(2)(ii)(G)*1* of this section. The body belt shall be secured to the 250-pound bag of sand at a point which simulates the waist of a man and shall be dropped as stated in paragraph (g)(2)(ii)(G)*2* of this section. Failure of the body belt shall be indicated by any breakage or slippage sufficient to permit the bag to fall free from the body belt.

**1910.268(g)(3)** *Pole climbers.* **(i)** Pole climbers may not be used if the gaffs are less than 1 1/4 inches in length as measured on the underside of the gaff. The gaffs of pole climbers shall be covered with safety caps when not being used for their intended use.

**1910.268(g)(3)(ii)** The employer shall ensure that pole climbers are inspected by a competent person for the following conditions: Fractured or cracked gaffs or leg irons, loose or dull gaffs, broken straps or buckles. If any of these conditions exist, the defect shall be corrected before the climbers are used.

**1910.268(g)(3)(iii)** Pole climbers shall be inspected as required in this paragraph (g)(3) before each day's use and a gaff cut-out test performed at least weekly when in use.

**1910.268(g)(3)(iv)** Pole climbers may not be worn when:

**1910.268(g)(3)(iv)(A)** Working in trees (specifically designed tree climbers shall be used for tree climbing),

**1910.268(g)(3)(iv)(B)** Working on ladders,

**1910.268(g)(3)(iv)(C)** Working in an aerial lift,

**1910.268(g)(3)(iv)(D)** Driving a vehicle, nor

**1910.268(g)(3)(iv)(E)** Walking on rocky, hard, frozen, brushy or hilly terrain.

**1910.268(h)** *Ladders.* **(1)** The employer shall ensure that no employee nor any material or equipment may be supported or permitted to be supported on any portion of a ladder unless it is first determined, by inspections and checks conducted by a competent person that such ladder is adequately strong, in good condition, and properly secured in place, as required in subpart D of this part and as required in this section.

**1910.268(h)(2)** The spacing between steps or rungs permanently installed on poles and towers shall be no more than 18 inches (36 inches on any one side). This requirement also applies to fixed ladders on towers, when towers are so equipped. Spacing between steps shall be uniform above the initial unstepped section, except where working, standing, or access steps are required. Fixed ladder rungs and step rungs for poles and towers shall have a minimum diameter of 5/8". Fixed ladder rungs shall have a minimum clear width of 12 inches. Steps for poles and towers shall have a minimum clear width of 4 1/2 inches. The spacing between detachable steps may not exceed 30 inches on any one side, and these steps shall be properly secured when in use.

**1910.268(h)(3)** Portable wood ladders intended for general use may not be painted but may be coated with a translucent nonconductive coating. Portable wood ladders may not be longitudinally reinforced with metal.

**1910.268(h)(4)** Portable wood ladders that are not being carried on vehicles and are not in active use shall be stored where they will not be exposed to the elements and where there is good ventilation.

**1910.268(h)(5)** The provisions of §1910.25(c)(5) shall apply to rolling ladders used in telecommunications centers, except that such ladders shall have a minimum inside width, between the side rails, of at least eight inches.

**1910.268(h)(6)** Climbing ladders or stairways on scaffolds used for access and egress shall be affixed or built into the scaffold by proper design and engineering, and shall be so located that their use will not disturb the stability of the scaffold. The rungs of the climbing device shall be equally spaced, but may not be less than 12 inches nominal nor more than 16 inches nominal apart. Horizontal end rungs used for platform support may also be utilized as a climbing device if such rungs meet the spacing requirement of this paragraph (h)(6), and if there is sufficient clearance between the rung and the edge of the platform to afford an adequate handhold. If a portable ladder is affixed to the scaffold, it shall be securely attached and shall have rungs meeting the spacing requirements of this paragraph (h)(6). Clearance shall be provided in the back of the ladder of not less than 6 inches from center of rung to the nearest scaffold structural member.

**1910.268(h)(7)** When a ladder is supported by an aerial strand, and ladder hooks or other supports are not being used, the ladder shall be extended at least 2 feet above the strand and shall be secured to it (e.g. lashed or held by a safety strap around the strand and ladder side rail). When a ladder is supported by a pole, it shall be securely lashed to the pole unless the ladder is specifically designed to prevent movement when used in this application.

**1910.268(h)(8)** The following requirements apply to metal manhole ladders. (i) Metal manhole ladders shall be free of structural defects and free of accident hazards such as sharp edges and burrs. The metal shall be protected against corrosion unless inherently corrosion-resistant.

**1910.268(h)(8)(ii)** These ladders may be designed with parallel side rails, or with side rails varying uniformly in separation along the length (tapered), or with side rails flaring at the base to increase stability.

**1910.268(h)(8)(iii)** The spacing of rungs or steps shall be on 12-inch centers.

**1910.268(h)(8)(iv)** Connections between rungs or steps and siderails shall be constructed to insure rigidity as well as strength.

**1910.268(h)(8)(v)** Rungs and steps shall be corrugated, knurled, dimpled, coated with skid-resistant material, or otherwise treated to minimize the possibility of slipping.

**1910.268(h)(8)(vi)** Ladder hardware shall meet the strength requirements of the ladder's component parts and shall be of a material that is protected against corrosion unless inherently corrosion-resistant. Metals shall be so selected as to avoid excessive galvanic action.

**1910.268(i)** *Other tools and personal protective equipment*—**(1)** *Head protection.* Head protection meeting the requirements of ANSI Z89.2-1971, "Safety Requirements for Industrial Protective Helmets for Electrical Workers, Class B" shall be provided whenever there is exposure to possible high voltage electrical contact, and the employer shall ensure that the head protection is used by employees. ANSI Z89.2-1971 is incorporated by reference as specified in §1910.6.

**1910.268(i)(2)** *Eye protection.* Eye protection meeting the requirements of §1910.133(a)(2) thru (a)(6) shall be provided and the employer shall ensure its use by employees where foreign objects may enter the eyes due to work operations such as but not limited to:

**1910.268(i)(2)(i)** Drilling or chipping stone, brick or masonry, breaking concrete or pavement, etc. by hand tools (sledgehammer, etc.) or power tools such as pneumatic drills or hammers;

**1910.268(i)(2)(ii)** Working on or around high speed emery or other grinding wheels unprotected by guards;

**1910.268(i)(2)(iii)** Cutting or chipping terra cotta ducts, tile, etc.;

**1910.268(i)(2)(iv)** Working under motor vehicles requiring hammering;

**1910.268(i)(2)(v)** Cleaning operations using compressed air, steam, or sand blast;

**1910.268(i)(2)(vi)** Acetylene welding or similar operations where sparks are thrown off;

**1910.268(i)(2)(vii)** Using powder actuated stud drivers;

**1910.268(i)(2)(viii)** Tree pruning or cutting underbrush;

# OSHA Standards for General Industry

**1910.268(i)(2)(ix)** Handling battery cells and solutions, such as taking battery readings with a hydrometer and thermometer;

**1910.268(i)(2)(x)** Removing or rearranging strand or open wire; and

**1910.268(i)(2)(xi)** Performing lead sleeve wiping and while soldering.

**1910.268(i)(3)** *Tent heaters.* Flame-type heaters may not be used within ground tents or on platforms within aerial tents unless:

**1910.268(i)(3)(i)** The tent covers are constructed of fire resistant materials, and

**1910.268(i)(3)(ii)** Adequate ventilation is provided to maintain safe oxygen levels and avoid harmful buildup of combustion products and combustible gases.

**1910.268(i)(4)** *Torches.* Torches may be used on aerial splicing platforms or in buckets enclosed by tents provided the tent material is constructed of fire resistant material and the torch is turned off when not in actual use. Aerial tents shall be adequately ventilated while the torch is in operation.

**1910.268(i)(5)** *Portable power equipment.* Nominal 120V, or less, portable generators used for providing power at work locations do not require grounding if the output circuit is completely isolated from the frame of the unit.

**1910.268(i)(6)** *Vehicle-mounted utility generators.* Vehicle-mounted utility generators used for providing nominal 240V AC or less for powering portable tools and equipment need not be grounded to earth if all of the following conditions are met:

**1910.268(i)(6)(i)** One side of the voltage source is solidly strapped to the metallic structure of the vehicle;

**1910.268(i)(6)(ii)** Grounding-type outlets are used, with a "grounding" conductor between the outlet grounding terminal and the side of the voltage source that is strapped to the vehicle;

**1910.268(i)(6)(iii)** All metallic encased tools and equipment that are powered from this system are equipped with three-wire cords and grounding-type attachment plugs, except as designated in paragraph (i)(7) of this section.

**1910.268(i)(7)** *Portable lights, tools, and appliances.* Portable lights, tools, and appliances having noncurrent-carrying external metal housing may be used with power equipment described in paragraph (i)(5) of this section without an equipment grounding conductor. When operated from commercial power such metal parts of these devices shall be grounded, unless these tools or appliances are protected by a system of double insulation, or its equivalent. Where such a system is employed, the equipment shall be distinctively marked to indicate double insulation.

**1910.268(i)(8)** *Soldering devices.* Grounding shall be omitted when using soldering irons, guns or wire-wrap tools on telecommunications circuits.

**1910.268(i)(9)** *Lead work.* The wiping of lead joints using melted solder, gas fueled torches, soldering irons or other appropriate heating devices, and the soldering of wires or other electrical connections do not constitute the welding, cutting and brazing described in subpart Q of this part. When operated from commercial power the metal housing of electric solder pots shall be grounded. Electric solder pots may be used with the power equipment described in paragraph (i)(5) of this section without a grounding conductor. The employer shall ensure that wiping gloves or cloths and eye protection are used in lead wiping operations. A drip pan to catch hot lead drippings shall also be provided and used.

**1910.268(j)** *Vehicle-mounted material handling devices and other mechanical equipment*—**(1)** *General.* **(i)** The employer shall ensure that visual inspections are made of the equipment by a competent person each day the equipment is to be used to ascertain that it is in good condition.

**1910.268(j)(1)(ii)** The employer shall ensure that tests shall be made at the beginning of each shift by a competent person to insure the vehicle brakes and operating systems are in proper working condition.

**1910.268(j)(2)** *Scrapers, loaders, dozers, graders and tractors.* **(i)** All rubber-tired, self-propelled scrapers, rubber-tired front end loaders, rubber-tired dozers, agricultural and industrial tractors, crawler tractors, crawler-type loaders, and motor graders, with or without attachments, that are used in telecommunications work shall have rollover protective structures that meet the requirements of subpart W of part 1926 of this Title.

**1910.268(j)(2)(ii)** Eye protection shall be provided and the employer shall ensure that it is used by employees when working in areas where flying material is generated.

**1910.268(j)(3)** *Vehicle-mounted elevating and rotating work platforms.* These devices shall not be operated with any conductive part of the equipment closer to exposed energized power lines than the clearances set forth in Table R-2 of this section.

**1910.268(j)(4)** *Derrick trucks and similar equipment.* **(i)** This equipment shall not be operated with any conductive part of the equipment closer to exposed energized power lines than the clearances set forth in Table R-2 of this section.

**1910.268(j)(4)(ii)** When derricks are used to handle poles near energized power conductors, these operations shall comply with the requirements contained in paragraphs (b)(7) and (n)(11) of this section.

**1910.268(j)(4)(iii)** Moving parts of equipment and machinery carried on or mounted on telecommunications line trucks shall be guarded. This may be done with barricades as specified in paragraph (d)(2) of this section.

**1910.268(j)(4)(iv)** Derricks and the operation of derricks shall comply with the following requirements: (A) Manufacturer's specifications, load ratings and instructions for derrick operation shall be strictly observed.

**1910.268(j)(4)(iv)(B)** Rated load capacities and instructions related to derrick operation shall be conspicuously posted on a permanent weather-resistant plate or decal in a location on the derrick that is plainly visible to the derrick operator.

**1910.268(j)(4)(iv)(C)** Prior to derrick operation the parking brake must be set and the stabilizers extended if the vehicle is so equipped. When the vehicle is situated on a grade, at least two wheels must be chocked on the downgrade side.

**1910.268(j)(4)(iv)(D)** Only persons trained in the operation of the derrick shall be permitted to operate the derrick.

**1910.268(j)(4)(iv)(E)** Hand signals to derrick operators shall be those prescribed by ANSI B30.6-1969, "Safety Code for Derricks", which is incorporated by reference as specified in §1910.6.

**1910.268(j)(4)(iv)(F)** The employer shall ensure that the derrick and its associated equipment are inspected by a competent person at intervals set by the manufacturer but in no case less than once per year. Records shall be maintained including the dates of inspections, and necessary repairs made, if corrective action was required.

**1910.268(j)(4)(iv)(G)** Modifications or additions to the derrick and its associated equipment that alter its capacity or affect its safe operation shall be made only with written certification from the manufacturer, or other equivalent entity, such as a nationally recognized testing laboratory, that the modification results in the equipment being safe for its intended use. Such changes shall require the changing and posting of revised capacity and instruction decals or plates. These new ratings or limitations shall be as provided by the manufacturer or other equivalent entity.

**1910.268(j)(4)(iv)(H)** Wire rope used with derricks shall be of improved plow steel or equivalent. Wire rope safety factors shall be in accordance with American National Standards Institute B30.6-1969.

**1910.268(j)(4)(iv)(I)** Wire rope shall be taken out of service, or the defective portion removed, when any of the following conditions exist: (*1*) The rope strength has been significantly reduced due to corrosion, pitting, or excessive heat, or

**1910.268(j)(4)(iv)(I)***(2)* The thickness of the outer wires of the rope has been reduced to two-thirds or less of the original thickness, or

**1910.268(j)(4)(iv)(I)***(3)* There are more than six broken wires in any one rope lay, or

**1910.268(j)(4)(iv)(I)***(4)* There is excessive permanent distortion caused by kinking, crushing, or severe twisting of the rope.

**1910.268(k)** *Materials handling and storage*—**(1)** *Poles.* When working with poles in piles or stacks, work shall be performed from the ends of the poles as much as possible, and precautions shall be taken for the safety of employees at the other end of the pole. During pole hauling operations, all loads shall be secured to prevent displacement. Lights, reflectors and/or flags shall be displayed on the end and sides of the load as necessary. The requirements for installation, removal, or other handling of poles in pole lines are prescribed in paragraph (n) of this section which pertains to overhead lines. In the case of hoisting machinery equipped with a positive stop loadholding device, it shall be permissible for the operator to leave his position at the controls (while a load is suspended) for the sole purpose of assisting in positioning the load prior to landing it. Prior to unloading steel, poles, crossarms, and similar material, the load shall be thoroughly examined to ascertain that the load has not shifted, that binders or stakes have not broken, and that the load is not otherwise hazardous to employees.

**1910.268(k)(2)** *Cable reels.* Cable reels in storage shall be checked or otherwise restrained when there is a possibility that they might accidentally roll from position.

**1910.268(l)** *Cable fault locating and testing.* **(1)** Employees involved in using high voltages to locate trouble or test cables shall be instructed in the precautions necessary for their own safety and the safety of other employees.

**1910.268(l)(2)** Before the voltage is applied, cable conductors shall be isolated to the extent practicable. Employees shall be warned, by such techniques as briefing and tagging at all affected locations, to stay clear while the voltage is applied.

**1910.268(m)** *Grounding for employee protection—pole lines*—**(1)** *Power conductors.* Electric power conductors and equipment shall be considered as energized unless the employee can visually determine that they are bonded to one of the grounds listed in paragraph (m)(4) of this section.

**1910.268(m)(2)** *Nonworking open wire.* Nonworking open wire communications lines shall be bonded to one of the grounds listed in paragraph (m)(4) of this section.

**1910.268(m)(3)** *Vertical power conduit, power ground wires and street light fixtures.* **(i)** Metal power conduit on joint use poles, exposed vertical power ground wires, and street light fixtures which are below communications attachments or less than 20 inches above these attachments, shall be considered energized and shall be tested for voltage unless the employee can visually determine that they are bonded to the communications suspension strand or cable sheath.

**1910.268(m)(3)(ii)** If no hazardous voltage is shown by the voltage test, a temporary bond shall be placed between such street light fixture, exposed vertical power grounding conductor, or metallic power conduit and the communications cable strand. Temporary bonds used for this purpose shall have sufficient conductivity to carry at least 500 amperes for a period of one second without fusing.

**1910.268(m)(4)** *Suitable protective grounding.* Acceptable grounds for protective grounding are as follows:

**1910.268(m)(4)(i)** A vertical ground wire which has been tested, found safe, and is connected to a power system multigrounded neutral or the grounded neutral of a power secondary system where there are at least three services connected;

**1910.268(m)(4)(ii)** Communications cable sheath or shield and its supporting strand where the sheath or shield is:

**1910.268(m)(4)(ii)(A)** Bonded to an underground or buried cable which is connected to a central office ground, or

**1910.268(m)(4)(ii)(B)** Bonded to an underground metallic piping system, or

**1910.268(m)(4)(ii)(C)** Bonded to a power system multigrounded neutral or grounded neutral of a power secondary system which has at least three services connected;

**1910.268(m)(4)(iii)** Guys which are bonded to the grounds specified in paragraphs (m)(4) (i) and (ii) of this section and which have continuity uninterrupted by an insulator; and

**1910.268(m)(4)(iv)** If all of the preceding grounds are not available, arrays of driven ground rods where the resultant resistance to ground will be low enough to eliminate danger to personnel or permit prompt operation of protective devices.

**1910.268(m)(5)** *Attaching and removing temporary bonds.* When attaching grounds (bonds), the first attachment shall be made to the protective ground. When removing bonds, the connection to the line or equipment shall be removed first. Insulating gloves shall be worn during these operations.

**1910.268(m)(6)** *Temporary grounding of suspension strand.* **(i)** The suspension strand shall be grounded to the existing grounds listed in paragraph (m)(4) of this section when being placed on jointly used poles or during thunderstorm activity.

**1910.268(m)(6)(ii)** Where power crossings are encountered on nonjoint lines, the strand shall be bonded to an existing ground listed in paragraph (m)(4) of this section as close as possible to the crossing. This bonding is not required where crossings are made on a common crossing pole unless there is an upward change in grade at the pole.

**1910.268(m)(6)(iii)** Where roller-type bonds are used, they shall be restrained so as to avoid stressing the electrical connections.

**1910.268(m)(6)(iv)** Bonds between the suspension strand and the existing ground shall be at least No. 6AWG copper.

**1910.268(m)(6)(v)** Temporary bonds shall be left in place until the strand has been tensioned, dead-ended, and permanently grounded.

**1910.268(m)(6)(vi)** The requirements of paragraphs (m)(6)(i) through (m)(6)(v) of this section do not apply to the installation of insulated strand.

**1910.268(m)(7)** *Antenna work-radio transmitting stations 3-30 MHZ.* **(i)** Prior to grounding a radio transmitting station antenna, the employer shall insure that the rigger in charge:

**1910.268(m)(7)(i)(A)** Prepares a danger tag signed with his signature,

**1910.268(m)(7)(i)(B)** Requests the transmitting technician to shutdown the transmitter and to ground the antenna with its grounding switch,

**1910.268(m)(7)(i)(C)** Is notified by the transmitting technician that the transmitter has been shutdown, and

**1910.268(m)(7)(i)(D)** Tags the antenna ground switch personally in the presence of the transmitting technician after the antenna has been grounded by the transmitting technician.

**1910.268(m)(7)(ii)** Power shall not be applied to the antenna, nor shall the grounding switch be opened under any circumstances while the tag is affixed.

**1910.268(m)(7)(iii)(A)** Where no grounding switches are provided, grounding sticks shall be used, one on each side of line, and tags shall be placed on the grounding sticks, antenna switch, or plate power switch in a conspicuous place.

**1910.268(m)(7)(iii)(B)** When necessary to further reduce excessive radio frequency pickup, ground sticks or short circuits shall be placed directly on the transmission lines near the transmitter in addition to the regular grounding switches.

**1910.268(m)(7)(iii)(C)** In other cases, the antenna lines may be disconnected from ground and the transmitter to reduce pickup at the point in the field.

**1910.268(m)(7)(iv)** All radio frequency line wires shall be tested for pickup with an insulated probe before they are handled either with bare hands or with metal tools.

**1910.268(m)(7)(v)** The employer shall insure that the transmitting technician warn the riggers about adjacent lines which are, or may become energized.

**1910.268(m)(7)(vi)** The employer shall insure that when antenna work has been completed, the rigger in charge of the job returns to the transmitter, notifies the transmitting technician in charge that work has been completed, and personally removes the tag from the antenna ground switch.

**1910.268(n)** *Overhead lines*—**(1)** *Handling suspension strand.* **(i)** The employer shall insure that when handling cable suspension strand which is being installed on poles carrying exposed energized power conductors, employees shall wear insulating gloves and shall avoid body contact with the strand until after it has been tensioned, dead-ended and permanently grounded.

**1910.268(n)(1)(ii)** The strand shall be restrained against upward movement during installation:

**1910.268(n)(1)(ii)(A)** On joint-use poles, where there is an upward change in grade at the pole, and

**1910.268(n)(1)(ii)(B)** On non-joint-use poles, where the line crosses under energized power conductors.

**1910.268(n)(2)** *Need for testing wood poles.* Unless temporary guys or braces are attached, the following poles shall be tested in accordance with paragraph (n)(3) of this section and determined to be safe before employees are permitted to climb them:

**1910.268(n)(2)(i)** Dead-end poles, except properly braced or guyed "Y" or "T" cable junction poles,

**1910.268(n)(2)(ii)** Straight line poles which are not storm guyed and where adjacent span lengths exceed 165 feet,

**1910.268(n)(2)(iii)** Poles at which there is a downward change in grade and which are not guyed or braced corner poles or cable junction poles,

**1910.268(n)(2)(iv)** Poles which support only telephone drop wire, and

# OSHA Standards for General Industry

**1910.268(n)(2)(v)** Poles which carry less than ten communication line wires. On joint use poles, one power line wire shall be considered as two communication wires for purposes of this paragraph (n)(2)(v).

**1910.268(n)(3)** *Methods for testing wood poles.* One of the following methods or an equivalent method shall be used for testing wood poles:

**1910.268(n)(3)(i)** Rap the pole sharply with a hammer weighing about 3 pounds, starting near the ground line and continuing upwards circumferentially around the pole to a height of approximately 6 feet. The hammer will produce a clear sound and rebound sharply when striking sound wood. Decay pockets will be indicated by a dull sound and/or a less pronounced hammer rebound. When decay pockets are indicated, the pole shall be considered unsafe. Also, prod the pole as near the ground line as possible using a pole prod or a screwdriver with a blade at least 5 inches long. If substantial decay is encountered, the pole shall be considered unsafe.

**1910.268(n)(3)(ii)** Apply a horizontal force to the pole and attempt to rock it back and forth in a direction perpendicular to the line. Caution shall be exercised to avoid causing power wires to swing together. The force may be applied either by pushing with a pike pole or pulling with a rope. If the pole cracks during the test, it shall be considered unsafe.

**1910.268(n)(4)** *Unsafe poles or structures.* Poles or structures determined to be unsafe by test or observation may not be climbed until made safe by guying, bracing or other adequate means. Poles determined to be unsafe to climb shall, until they are made safe, be tagged in a conspicuous place to alert and warn all employees of the unsafe condition.

**1910.268(n)(5)** *Test requirements for cable suspension strand.* **(i)** Before attaching a splicing platform to a cable suspension strand, the strand shall be tested and determined to have strength sufficient to support the weight of the platform and the employee. Where the strand crosses above power wires or railroad tracks it may not be tested but shall be inspected in accordance with paragraph (n)(6) of this section.

**1910.268(n)(5)(ii)** The following method or an equivalent method shall be used for testing the strength of the strand: A rope, at least three-eighths inch in diameter, shall be thrown over the strand. On joint lines, the rope shall be passed over the strand using tree pruner handles or a wire raising tool. If two employees are present, both shall grip the double rope and slowly transfer their entire weight to the rope and attempt to raise themselves off the ground. If only one employee is present, one end of the rope which has been passed over the strand shall be tied to the bumper of the truck, or other equally secure anchorage. The employee then shall grasp the other end of the rope and attempt to raise himself off the ground.

**1910.268(n)(6)** *Inspection of strand.* Where strand passes over electric power wires or railroad tracks, it shall be inspected from an elevated working position at each pole supporting the span in question. The strand may not be used to support any splicing platform, scaffold or cable car, if any of the following conditions exist:

**1910.268(n)(6)(i)** Corrosion so that no galvanizing can be detected,

**1910.268(n)(6)(ii)** One or more wires of the strand are broken,

**1910.268(n)(6)(iii)** Worn spots, or

**1910.268(n)(6)(iv)** Burn marks such as those caused by contact with electric power wires.

**1910.268(n)(7)** *Outside work platforms.* Unless adequate railings are provided, safety straps and body belts shall be used while working on elevated work platforms such as aerial splicing platforms, pole platforms, ladder platforms and terminal balconies.

**1910.268(n)(8)** *Other elevated locations.* Safety straps and body belts shall be worn when working at elevated positions on poles, towers or similar structures, which do not have adequately guarded work areas.

**1910.268(n)(9)** *Installing and removing wire and cable.* Before installing or removing wire or cable, the pole or structure shall be guyed, braced, or otherwise supported, as necessary, to prevent failure of the pole or structure.

**1910.268(n)(10)** *Avoiding contact with energized power conductors or equipment.* When cranes, derricks, or other mechanized equipment are used for setting, moving, or removing poles, all necessary precautions shall be taken to avoid contact with energized power conductors or equipment.

**1910.268(n)(11)** *Handling poles near energized power conductors.* **(i)** Joint use poles may not be set, moved, or removed where the nominal voltage of open electrical power conductors exceeds 34.5kV phase to phase (20kV to ground).

**1910.268(n)(11)(ii)** Poles that are to be placed, moved or removed during heavy rains, sleet or wet snow in joint lines carrying more than 8.7kV phase to phase voltage (5kV to ground) shall be guarded or otherwise prevented from direct contact with overhead energized power conductors.

**1910.268(n)(11)(iii)(A)** In joint lines where the power voltage is greater than 750 volts but less than 34.5kV phase to phase (20 kV to ground), wet poles being placed, moved or removed shall be insulated with either a rubber insulating blanket, a fiberglass box guide, or equivalent protective equipment.

**1910.268(n)(11)(iii)(B)** In joint lines where the power voltage is greater than 8.7 kV phase to phase (5kV to ground) but less than 34.5kV phase to phase (20 kV to ground), dry poles being placed, moved, or removed shall be insulated with either a rubber insulating blanket, a fiberglass box guide, or equivalent protective equipment.

**1910.268(n)(11)(iii)(C)** Where wet or dry poles are being removed, insulation of the pole is not required if the pole is cut off 2 feet or more below the lowest power wire and also cut off near the ground line.

**1910.268(n)(11)(iv)** Insulating gloves shall be worn when handling the pole with either hands or tools, when there exists a possibility that the pole may contact a power conductor. Where the voltage to ground of the power conductor exceeds 15kV to ground, Class II gloves (as defined in ANSI J6.6-1971) shall be used. For voltages not exceeding 15kV to ground, insulating gloves shall have a breakdown voltage of at least 17kV.

**1910.268(n)(11)(v)** The guard or insulating material used to protect the pole shall meet the appropriate 3 minute proof test voltage requirements contained in the ANSI J6.4-1971.

**1910.268(n)(11)(vi)** When there exists a possibility of contact between the pole or the vehicle-mounted equipment used to handle the pole, and an energized power conductor, the following precautions shall be observed:

**1910.268(n)(11)(vi)(A)** When on the vehicle which carries the derrick, avoid all contact with the ground, with persons standing on the ground, and with all grounded objects such as guys, tree limbs, or metal sign posts. To the extent feasible, remain on the vehicle as long as the possibility of contact exists.

**1910.268(n)(11)(vi)(B)** When it is necessary to leave the vehicle, step onto an insulating blanket and break all contact with the vehicle before stepping off the blanket and onto the ground. As a last resort, if a blanket is not available, the employee may jump cleanly from the vehicle.

**1910.268(n)(11)(vi)(C)** When it is necessary to enter the vehicle, first step onto an insulating blanket and break all contact with the ground, grounded objects and other persons before touching the truck or derrick.

**1910.268(n)(12)** *Working position on poles.* Climbing and working are prohibited above the level of the lowest electric power conductor on the pole (exclusive of vertical runs and street light wiring), except:

**1910.268(n)(12)(i)** Where communications facilities are attached above the electric power conductors, and a rigid fixed barrier is installed between the electric power facility and the communications facility, or

**1910.268(n)(12)(ii)** Where the electric power conductors are cabled secondary service drops carrying less than 300 volts to ground and are attached 40 inches or more below the communications conductors or cables.

**1910.268(n)(13)** *Metal tapes and ropes.* **(i)** Metal measuring tapes, metal measuring ropes, or tapes containing conductive strands may not be used when working near exposed energized parts.

**1910.268(n)(13)(ii)** Where it is necessary to measure clearances from energized parts, only nonconductive devices shall be used.

**1910.268(o)** **Underground lines.** The provisions of this paragraph apply to the guarding of manholes and street openings, and to the ventilation and testing for gas in manholes and unvented vaults, where telecommunications field work is performed on or with underground lines.

**1910.268(o)(1)** *Guarding manholes and street openings.* **(i)** When covers of manholes or vaults are removed, the opening shall be promptly guarded by a railing, temporary cover, or other suitable temporary barrier which is appropriate to prevent an accidental fall through the opening and to protect employees working in the manhole from foreign objects entering the manhole.

**1910.268(o)(1)(ii)** While work is being performed in the manhole, a person with basic first aid training shall be immediately available to render assistance if there is cause for believing that a safety hazard exists, and if the requirements contained in paragraphs (d)(1) and (o)(1)(i) of this section do not adequately protect the employee(s). Examples of manhole worksite hazards which shall be considered to constitute a safety hazard include, but are not limited to:

**1910.268(o)(1)(ii)(A)** Manhole worksites where safety hazards are created by traffic patterns that cannot be corrected by provisions of paragraph (d)(1) of this section.

**1910.268(o)(1)(ii)(B)** Manhole worksites that are subject to unusual water hazards that cannot be abated by conventional means.

**1910.268(o)(1)(ii)(C)** Manhole worksites that are occupied jointly with power utilities as described in paragraph (o)(3) of this section.

**1910.268(o)(2)** *Requirements prior to entering manholes and unvented vaults.* **(i)** Before an employee enters a manhole, the following steps shall be taken:

**1910.268(o)(2)(i)(A)** The internal atmosphere shall be tested for combustible gas and, except when continuous forced ventilation is provided, the atmosphere shall also be tested for oxygen deficiency.

**1910.268(o)(2)(i)(B)** When unsafe conditions are detected by testing or other means, the work area shall be ventilated and otherwise made safe before entry.

**1910.268(o)(2)(ii)** An adequate continuous supply of air shall be provided while work is performed in manholes under any of the following conditions:

**1910.268(o)(2)(ii)(A)** Where combustible or explosive gas vapors have been initially detected and subsequently reduced to a safe level by ventilation,

**1910.268(o)(2)(ii)(B)** Where organic solvents are used in the work procedure,

**1910.268(o)(2)(ii)(C)** Where open flame torches are used in the work procedure,

**1910.268(o)(2)(ii)(D)** Where the manhole is located in that portion of a public right of way open to vehicular traffic and/or exposed to a seepage of gas or gases, or

**1910.268(o)(2)(ii)(E)** Where a toxic gas or oxygen deficiency is found.

**1910.268(o)(2)(iii)(A)** The requirements of paragraphs (o)(2) (i) and (ii) of this section do not apply to work in central office cable vaults that are adequately ventilated.

**1910.268(o)(2)(iii)(B)** The requirements of paragraphs (o)(2) (i) and (ii) of this section apply to work in unvented vaults.

**1910.268(o)(3)** *Joint power and telecommunication manholes.* While work is being performed in a manhole occupied jointly by an electric utility and a telecommunication utility, an employee with basic first aid training shall be available in the immediate vicinity to render emergency assistance as may be required. The employee whose presence is required in the immediate vicinity for the purposes of rendering emergency assistance is not to be precluded from occasionally entering a manhole to provide assistance other than in an emergency. The requirement of this paragraph (o)(3) does not preclude a qualified employee, working alone, from entering for brief periods of time, a manhole where energized cables or equipment are in service, for the purpose of inspection, housekeeping, taking readings, or similar work if such work can be performed safely.

**1910.268(o)(4)** *Ladders.* Ladders shall be used to enter and exit manholes exceeding 4 feet in depth.

**1910.268(o)(5)** *Flames.* When open flames are used in manholes, the following precautions shall be taken to protect against the accumulation of combustible gas:

**1910.268(o)(5)(i)** A test for combustible gas shall be made immediately before using the open flame device, and at least once per hour while using the device; and

**1910.268(o)(5)(ii)** a fuel tank (e.g., acetylene) may not be in the manhole unless in actual use.

**1910.268(p)** *Microwave transmission*—**(1)** *Eye protection.* Employers shall insure that employees do not look into an open waveguide which is connected to an energized source of microwave radiation.

**1910.268(p)(2)** *Hazardous area.* Accessible areas associated with microwave communication systems where the electromagnetic radiation level exceeds the radiation protection guide given in §1910.97 shall be posted as described in that section. The lower half of the warning symbol shall include the following:

Radiation in this area may exceed hazard limitations and special precautions are required. Obtain specific instruction before entering.

**1910.268(p)(3)** *Protective measures.* When an employee works in an area where the electromagnetic radiation exceeds the radiation protection guide, the employer shall institute measures that insure that the employee's exposure is not greater than that permitted by the radiation guide. Such measures shall include, but not be limited to those of an administrative or engineering nature or those involving personal protective equipment.

**1910.268(q)** *Tree trimming—electrical hazards*—**(1)** *General.* **(i)** Employees engaged in pruning, trimming, removing, or clearing trees from lines shall be required to consider all overhead and underground electrical power conductors to be energized with potentially fatal voltages, never to be touched (contacted) either directly or indirectly.

**1910.268(q)(1)(ii)** Employees engaged in line-clearing operations shall be instructed that:

**1910.268(q)(1)(ii)(A)** A direct contact is made when any part of the body touches or contacts an energized conductor, or other energized electrical fixture or apparatus.

**1910.268(q)(1)(ii)(B)** An indirect contact is made when any part of the body touches any object in contact with an energized electrical conductor, or other energized fixture or apparatus.

**1910.268(q)(1)(ii)(C)** An indirect contact can be made through conductive tools, tree branches, trucks, equipment, or other objects, or as a result of communications wires, cables, fences, or guy wires being accidentally energized.

**1910.268(q)(1)(ii)(D)** Electric shock will occur when an employee, by either direct or indirect contact with an energized conductor, energized tree limb, tool, equipment, or other object, provides a path for the flow of electricity to a grounded object or to the ground itself. Simultaneous contact with two energized conductors will also cause electric shock which may result in serious or fatal injury.

**1910.268(q)(1)(iii)** Before any work is performed in proximity to energized conductors, the system operator/owner of the energized conductors shall be contacted to ascertain if he knows of any hazards associated with the conductors which may not be readily apparent. This rule does not apply when operations are performed by or on behalf of, the system operator/owner.

**1910.268(q)(2)** *Working in proximity to electrical hazards.* **(i)** Employers shall ensure that a close inspection is made by the employee and by the foremen or supervisor in charge before climbing, entering, or working around any tree, to determine whether an electrical power conductor passes through the tree, or passes within reaching distance of an employee working in the tree. If any of these conditions exist either directly or indirectly, an electrical hazard shall be considered to exist unless the system operator/owner has caused the hazard to be removed by deenergizing the lines, or installing protective equipment.

**1910.268(q)(2)(ii)** Only qualified employees or trainees, familiar with the special techniques and hazards involved in line clearance, shall be permitted to perform the work if it is found that an electrical hazard exists.

**1910.268(q)(2)(iii)** During all tree working operations aloft where an electrical hazard of more than 750V exists, there shall be a second employee or trainee qualified in line clearance tree trimming within normal voice communication.

**1910.268(q)(2)(iv)** Where tree work is performed by employees qualified in line-clearance tree trimming and trainees qualified in line-clearance tree trimming, the clearances from energized conductors given in Table R-3 shall apply.

### TABLE R-3—MINIMUM WORKING DISTANCES FROM ENERGIZED CONDUCTORS FOR LINE-CLEARANCE TREE TRIMMERS AND LINE-CLEARANCE TREE-TRIMMER TRAINEES

| Voltage range (phase to phase) (kilovolts) | Minimum working distance |
|---|---|
| 2.1 to 15.0 | 2 ft. 0 in. |
| 15.1 to 35.0 | 2 ft. 4 in. |
| 35.1 to 46.0 | 2 ft. 6 in. |
| 46.1 to 72.5 | 3 ft. 0 in. |
| 72.6 to 121.0 | 3 ft. 4 in. |
| 138.0 to 145.0 | 3 ft. 6 in. |
| 161.0 to 169.0 | 3 ft. 8 in. |
| 230.0 to 242.0 | 5 ft. 0 in. |
| 345.0 to 362.0 | 7 ft. 0 in. |
| 500.0 to 552.0 | 11 ft. 0 in. |
| 700.0 to 765.0 | 15 ft. 0 in. |

**1910.268(q)(2)(v)** Branches hanging on an energized conductor may only be removed using appropriately insulated equipment.

**1910.268(q)(2)(vi)** Rubber footwear, including lineman's overshoes, shall not be considered as providing any measure of safety from electrical hazards.

**1910.268(q)(2)(vii)** Ladders, platforms, and aerial devices, including insulated aerial devices, may not be brought in contact with an electrical conductor. Reliance shall not be placed on their dielectric capabilities.

**1910.268(q)(2)(viii)** When an aerial lift device contacts an electrical conductor, the truck supporting the aerial lift device shall be considered as energized.

**1910.268(q)(3)** *Storm work and emergency conditions.* **(i)** Since storm work and emergency conditions create special hazards, only authorized representatives of the electric utility system operator/owner and not telecommunication workers may perform tree work in these situations where energized electrical power conductors are involved.

**1910.268(q)(3)(ii)** When an emergency condition develops due to tree operations, work shall be suspended and the system operator/owner shall be notified immediately.

**1910.268(r)** *Buried facilities*—*Communications lines and power lines in the same trench.* [Reserved]

**1910.268(s)** *Definitions*—**(1)** *Aerial lifts.* Aerial lifts include the following types of vehicle-mounted aerial devices used to elevate personnel to jobsites above ground:

**1910.268(s)(1)(i)** Extensible boom platforms,

**1910.268(s)(1)(ii)** Aerial ladders,

**1910.268(s)(1)(iii)** Articulating boom platforms,

**1910.268(s)(1)(iv)** Vertical towers,

**1910.268(s)(1)(v)** A combination of any of the above defined in ANSI A92.2-1969, which is incorporated by reference as specified in §1910.6. These devices are made of metal, wood, fiberglass reinforced plastic (FRP), or other material; are powered or manually operated; and are deemed to be aerial lifts whether or not they are capable of rotating about a substantially vertical axis.

**1910.268(s)(2)** *Aerial splicing platform.* This consists of a platform, approximately 3 ft. × 4 ft., used to perform aerial cable work. It is furnished with fiber or synthetic ropes for supporting the platform from aerial strand, detachable guy ropes for anchoring it, and a device for raising and lowering it with a handline.

**1910.268(s)(3)** *Aerial tent.* A small tent usually constructed of vinyl coated canvas which is usually supported by light metal or plastic tubing. It is designed to protect employees in inclement weather while working on ladders, aerial splicing platforms, or aerial devices.

**1910.268(s)(4)** *Alive or live (energized).* Electrically connected to a source of potential difference, or electrically charged so as to have a potential significantly different from that of the earth in the vicinity. The term *live* is sometimes used in the place of the term *current-carrying*, where the intent is clear, to avoid repetition of the longer term.

**1910.268(s)(5)** *Barricade.* A physical obstruction such as tapes, cones, or "A" frame type wood and/or metal structure intended to warn and limit access to a work area.

**1910.268(s)(6)** *Barrier.* A physical obstruction which is intended to prevent contact with energized lines or equipment, or to prevent unauthorized access to work area.

**1910.268(s)(7)** *Bond.* An electrical connection from one conductive element to another for the purpose of minimizing potential differences or providing suitable conductivity for fault current or for mitigation of leakage current and electrolytic action.

**1910.268(s)(8)** *Cable.* A conductor with insulation, or a stranded conductor with or without insulation and other coverings (single-conductor cable), or a combination of conductors insulated from one another (multiple-conductor cable).

**1910.268(s)(9)** *Cable sheath.* A protective covering applied to cables.

**NOTE:** A cable sheath may consist of multiple layers of which one or more is conductive.

**1910.268(s)(10)** *Circuit.* A conductor or system of conductors through which an electric current is intended to flow.

**1910.268(s)(11)** *Communication lines.* The conductors and their supporting or containing structures for telephone, telegraph, railroad signal, data, clock, fire, police-alarm, community television antenna and other systems which are used for public or private signal or communication service, and which operate at potentials not exceeding 400 volts to ground or 750 volts between any two points of the circuit, and the transmitted power of which does not exceed 150 watts. When communications lines operate at less than 150 volts to ground, no limit is placed on the capacity of the system. Specifically designed communications cables may include communication circuits not complying with the preceding limitations, where such circuits are also used incidentally to supply power to communication equipment.

**1910.268(s)(12)** *Conductor.* A material, usually in the form of a wire, cable, or bus bar, suitable for carrying an electric current.

**1910.268(s)(13)** *Effectively grounded.* Intentionally connected to earth through a ground connection or connections of sufficiently low impedance and having sufficient current-carrying capacity to prevent the build-up of voltages which may result in undue hazard to connected equipment or to persons.

**1910.268(s)(14)** *Equipment.* A general term which includes materials, fittings, devices, appliances, fixtures, apparatus, and similar items used as part of, or in connection with, a supply or communications installation.

**1910.268(s)(15)** *Ground (reference).* That conductive body, usually earth, to which an electric potential is referenced.

**1910.268(s)(16)** *Ground (as a noun).* A conductive connection, whether intentional or accidental, by which an electric circuit or equipment is connected to reference ground.

**1910.268(s)(17)** *Ground (as a verb).* The connecting or establishment of a connection, whether by intention or accident, of an electric circuit or equipment to reference ground.

**1910.268(s)(18)** *Ground tent.* A small tent usually constructed of vinyl coated canvas supported by a metal or plastic frame. Its purpose is to protect employees from inclement weather while working at buried cable pedestal sites or similar locations.

**1910.268(s)(19)** *Grounded conductor.* A system or circuit conductor which is intentionally grounded.

**1910.268(s)(20)** *Grounded systems.* A system of conductors in which at least one conductor or point (usually the middle wire, or the neutral point of transformer or generator windings) is intentionally grounded, either solidly or through a current-limiting device (not a current-interrupting device).

**1910.268(s)(21)** *Grounding electrode conductor. (Grounding conductor).* A conductor used to connect equipment or the grounded circuit of a wiring system to a grounding electrode.

**1910.268(s)(22)** *Insulated.* Separated from other conducting surfaces by a dielectric substance (including air space) offering a high resistance to the passage of current.

**NOTE:** When any object is said to be insulated, it is understood to be insulated in suitable manner for the conditions to which it is subjected.

Otherwise, it is, within the purpose of these rules, uninsulated. Insulating coverings of conductors in one means of making the conductor insulated.

**1910.268(s)(23)** *Insulation (as applied to cable).* That which is relied upon to insulate the conductor from other conductors or conducting parts or from ground.

**1910.268(s)(24)** *Joint use.* The sharing of a common facility, such as a manhole, trench or pole, by two or more different kinds of utilities (e.g., power and telecommunications).

**1910.268(s)(25)** *Ladder platform.* A device designed to facilitate working aloft from an extension ladder. A typical device consists of a platform (approximately 9" × 18") hinged to a welded pipe frame. The rear edge of the platform and the bottom cross-member of the frame are equipped with latches to lock the platform to ladder rungs.

**1910.268(s)(26)** *Ladder seat.* A removable seat used to facilitate work at an elevated position on rolling ladders in telecommunication centers.

**1910.268(s)(27)** *Manhole.* A subsurface enclosure which personnel may enter and which is used for the purpose of installing, operating, and maintaining submersible equipment and/or cable.

**1910.268(s)(28)** *Manhole platform.* A platform consisting of separate planks which are laid across steel platform supports. The ends of the supports are engaged in the manhole cable racks.

**1910.268(s)(29)** *Microwave transmission.* The act of communicating or signaling utilizing a frequency between 1 GHz (gigahertz) and 300 GHz inclusively.

**1910.268(s)(30)** *Nominal voltage.* The nominal voltage of a system or circuit is the value assigned to a system or circuit of a given voltage class for the purpose of convenient designation. The actual voltage may vary above or below this value.

**1910.268(s)(31)** *Pole balcony or seat.* A balcony or seat used as a support for workmen at pole-mounted equipment or terminal boxes. A typical device consists of a bolted assembly of steel details and a wooden platform. Steel braces run from the pole to the underside of the balcony. A guard rail (approximately 30" high) may be provided.

**1910.268(s)(32)** *Pole platform.* A platform intended for use by a workman in splicing and maintenance operations in an elevated position adjacent to a pole. It consists of a platform equipped at one end with a hinged chain binder for securing the platform to a pole. A brace from the pole to the underside of the platform is also provided.

**1910.268(s)(33)** *Qualified employee.* Any worker who by reason of his training and experience has demonstrated his ability to safely perform his duties.

**1910.268(s)(34)** *Qualified line-clearance tree trimmer.* A tree worker who through related training and on-the-job experience is familar with the special techniques and hazards involved in line clearance.

**1910.268(s)(35)** *Qualified line-clearance tree-trimmer trainee.* Any worker regularly assigned to a line-clearance tree-trimming crew and undergoing on-the-job training who, in the course of such training, has demonstrated his ability to perform his duties safely at his level of training.

**1910.268(s)(36)** *System operator/owner.* The person or organization that operates or controls the electrical conductors involved.

**1910.268(s)(37)** *Telecommunications center.* An installation of communication equipment under the exclusive control of an organization providing telecommunications service, that is located outdoors or in a vault, chamber, or a building space used primarily for such installations.

**NOTE:** Telecommunication centers are facilities established, equipped and arranged in accordance with engineered plans for the purpose of providing telecommunications service. They may be located on premises owned or leased by the organization providing telecommunication service, or on the premises owned or leased by others. This definition includes switch rooms (whether electromechanical, electronic, or computer controlled), terminal rooms, power rooms, repeater rooms, transmitter and receiver rooms, switchboard operating rooms, cable vaults, and miscellaneous communications equipment rooms. Simulation rooms of telecommunication centers for training or developmental purposes are also included.

**1910.268(s)(38)** *Telecommunications derricks.* Rotating or nonrotating derrick structures permanently mounted on vehicles for the purpose of lifting, lowering, or positioning hardware and materials used in telecommunications work.

**1910.268(s)(39)** *Telecommunication line truck.* A truck used to transport men, tools, and material, and to serve as a traveling workshop for telecommunication installation and maintenance work. It is sometimes equipped with a boom and auxiliary equipment for setting poles, digging holes, and elevating material or men.

**1910.268(s)(40)** *Telecommunication service.* The furnishing of a capability to signal or communicate at a distance by means such as telephone, telegraph, police and firealarm, community antenna television, or similar system, using wire, conventional cable, coaxial cable, wave guides, microwave transmission, or other similar means.

**1910.268(s)(41)** *Unvented vault.* An enclosed vault in which the only openings are access openings.

**1910.268(s)(42)** *Vault.* An enclosure above or below ground which personnel may enter, and which is used for the purpose of installing, operating, and/or maintaining equipment and/or cable which need not be of submersible design.

**1910.268(s)(43)** *Vented vault.* An enclosure as described in paragraph(s) (42) of this section, with provision for air changes using exhaust flue stack(s) and low level air intake(s), operating on differentials of pressure and temperature providing for air flow.

**1910.268(s)(44)** *Voltage of an effectively grounded circuit.* The voltage between any conductor and ground unless otherwise indicated.

**1910.268(s)(45)** *Voltage of a circuit not effectively grounded.* The voltage between any two conductors. If one circuit is directly connected to and supplied from another circuit of higher voltage (as in the case of an autotransformer), both are considered as of the higher voltage, unless the circuit of lower voltage is effectively grounded, in which case its voltage is not determined by the circuit of higher voltage. Direct connection implies electric connection as distinguished from connection merely through electromagnetic or electrostatic induction.

[40 FR 13441, Mar. 26, 1975, as amended at 43 FR 49751, Oct. 24, 1978; 47 FR 14706, Apr. 6, 1982; 52 FR 36387, Sept. 28, 1987; 54 FR 24334, June 7, 1989; 61 FR 9242, Mar. 7, 1996; 63 FR 33467, June 18, 1998]

## §1910.269 Electric power generation, transmission, and distribution.

**NOTE:** OSHA is staying the enforcement of the following paragraphs of §1910.269 until November 1, 1994: (b)(1)(ii), (d) except for (d)(2)(i) and (d)(2)(iii), (e)(2), (e)(3), (j)(2)(iii), (l)(6)(iii), (m), (n)(3), (n)(4)(ii), (n)(8), (o) except for (o)(2)(i), (r)(1)(vi), (u)(1), (u)(4), (u)(5). OSHA is also staying the enforcement of paragraphs (n)(6) and (n)(7) of §1910.269 until November 1, 1994, but only insofar as they apply to lines and equipment operated at 600 volts or less. Further, OSHA is staying the enforcement of paragraph (v)(11)(xii) of §1910.269 until Februrary 1, 1996.

**1910.269(a)** *General*—**(1)** *Application.* **(i)** This section covers the operation and maintenance of electric power generation, control, transformation, transmission, and distribution lines and equipment. These provisions apply to:

**1910.269(a)(1)(i)(A)** Power generation, transmission, and distribution installations, including related equipment for the purpose of communication or metering, which are accessible only to qualified employees;

**NOTE:** The types of installations covered by this paragraph include the generation, transmission, and distribution installations of electric utilities, as well as equivalent installations of industrial establishments. Supplementary electric generating equipment that is used to supply a workplace for emergency, standby, or similar purposes only is covered under subpart S of this part. (See paragraph (a)(1)(ii)(B) of this section.)

**1910.269(a)(1)(i)(B)** Other installations at an electric power generating station, as follows:

**1910.269(a)(1)(i)(B)***(1)* Fuel and ash handling and processing installations, such as coal conveyors,

**1910.269(a)(1)(i)(B)***(2)* Water and steam installations, such as penstocks, pipelines, and tanks, providing a source of energy for electric generators, and

**1910.269(a)(1)(i)(B)***(3)* Chlorine and hydrogen systems;

**1910.269(a)(1)(i)(C)** Test sites where electrical testing involving temporary measurements associated with electric power generation, transmission, and distribution is performed in laboratories, in the field, in substations, and on lines, as opposed to metering, relaying, and routine line work;

**1910.269(a)(1)(i)(D)** Work on or directly associated with the installations covered in paragraphs (a)(1)(i)(A) through (a)(1)(i)(C) of this section; and

# OSHA Standards for General Industry

**1910.269(a)(1)(i)(E)** Line-clearance tree-trimming operations, as follows:

**1910.269(a)(1)(i)(E)(1)** Entire §1910.269 of this part, except paragraph (r)(1) of this section, applies to line-clearance tree-trimming operations performed by qualified employees (those who are knowledgeable in the construction and operation of electric power generation, transmission, or distribution equipment involved, along with the associated hazards).

**1910.269(a)(1)(i)(E)(2)** Paragraphs (a)(2), (b), (c), (g), (k), (p), and (r) of this section apply to line-clearance tree-trimming operations performed by line-clearance tree trimmers who are not qualified employees.

**1910.269(a)(1)(ii)** Notwithstanding paragraph (a)(1)(i) of this section, §1910.269 of this part does not apply:

**1910.269(a)(1)(ii)(A)** To construction work, as defined in §1910.12 of this part; or

**1910.269(a)(1)(ii)(B)** To electrical installations, electrical safety-related work practices, or electrical maintenance considerations covered by subpart S of this part.

**NOTE 1:** Work practices conforming to §§1910.332 through 1910.335 of this part are considered as complying with the electrical safety-related work practice requirements of this section identified in Table 1 of appendix A-2 to this section, provided the work is being performed on a generation or distribution installation meeting §§1910.303 through 1910.308 of this part. This table also identifies provisions in this section that apply to work by qualified persons directly on or associated with installations of electric power generation, transmission, and distribution lines or equipment, regardless of compliance with §§1910.332 through 1910.335 of this part.

**NOTE 2:** Work practices performed by qualified persons and conforming to §1910.269 of this part are considered as complying with §§1910.333(c) and 1910.335 of this part.

**1910.269(a)(1)(iii)** This section applies in addition to all other applicable standards contained in this part 1910. Specific references in this section to other sections of part 1910 are provided for emphasis only.

**1910.269(a)(2)** *Training.* **(i)** Employees shall be trained in and familiar with the safety-related work practices, safety procedures, and other safety requirements in this section that pertain to their respective job assignments. Employees shall also be trained in and familiar with any other safety practices, including applicable emergency procedures (such as pole top and manhole rescue), that are not specifically addressed by this section but that are related to their work and are necessary for their safety.

**1910.269(a)(2)(ii)** Qualified employees shall also be trained and competent in:

**1910.269(a)(2)(ii)(A)** The skills and techniques necessary to distinguish exposed live parts from other parts of electric equipment,

**1910.269(a)(2)(ii)(B)** The skills and techniques necessary to determine the nominal voltage of exposed live parts,

**1910.269(a)(2)(ii)(C)** The minimum approach distances specified in this section corresponding to the voltages to which the qualified employee will be exposed, and

**1910.269(a)(2)(ii)(D)** The proper use of the special precautionary techniques, personal protective equipment, insulating and shielding materials, and insulated tools for working on or near exposed energized parts of electric equipment.

**NOTE:** For the purposes of this section, a person must have this training in order to be considered a qualified person.

**1910.269(a)(2)(iii)** The employer shall determine, through regular supervision and through inspections conducted on at least an annual basis, that each employee is complying with the safety-related work practices required by this section.

**1910.269(a)(2)(iv)** An employee shall receive additional training (or retraining) under any of the following conditions:

**1910.269(a)(2)(iv)(A)** If the supervision and annual inspections required by paragraph (a)(2)(iii) of this section indicate that the employee is not complying with the safety-related work practices required by this section, or

**1910.269(a)(2)(iv)(B)** If new technology, new types of equipment, or changes in procedures necessitate the use of safety-related work practices that are different from those which the employee would normally use, or

**1910.269(a)(2)(iv)(C)** If he or she must employ safety-related work practices that are not normally used during his or her regular job duties.

**NOTE:** OSHA would consider tasks that are performed less often than once per year to necessitate retraining before the performance of the work practices involved.

**1910.269(a)(2)(v)** The training required by paragraph (a)(2) of this section shall be of the classroom or on-the-job type.

**1910.269(a)(2)(vi)** The training shall establish employee proficiency in the work practices required by this section and shall introduce the procedures necessary for compliance with this section.

**1910.269(a)(2)(vii)** The employer shall certify that each employee has received the training required by paragraph (a)(2) of this section. This certification shall be made when the employee demonstrates proficiency in the work practices involved and shall be maintained for the duration of the employee's employment.

**NOTE:** Employment records that indicate that an employee has received the required training are an acceptable means of meeting this requirement.

**1910.269(a)(3)** *Existing conditions.* Existing conditions related to the safety of the work to be performed shall be determined before work on or near electric lines or equipment is started. Such conditions include, but are not limited to, the nominal voltages of lines and equipment, the maximum switching transient voltages, the presence of hazardous induced voltages, the presence and condition of protective grounds and equipment grounding conductors, the condition of poles, environmental conditions relative to safety, and the locations of circuits and equipment, including power and communication lines and fire protective signaling circuits.

**1910.269(b)** *Medical services and first aid.* The employer shall provide medical services and first aid as required in §1910.151 of this part. In addition to the requirements of §1910.151 of this part, the following requirements also apply:

**1910.269(b)(1)** *Cardiopulmonary resuscitation and first aid training.* When employees are performing work on or associated with exposed lines or equipment energized at 50 volts or more, persons trained in first aid including cardiopulmonary resuscitation (CPR) shall be available as follows:

**1910.269(b)(1)(i)** For field work involving two or more employees at a work location, at least two trained persons shall be available. However, only one trained person need be available if all new employees are trained in first aid, including CPR, within 3 months of their hiring dates.

**1910.269(b)(1)(ii)** For fixed work locations such as generating stations, the number of trained persons available shall be sufficient to ensure that each employee exposed to electric shock can be reached within 4 minutes by a trained person. However, where the existing number of employees is insufficient to meet this requirement (at a remote substation, for example), all employees at the work location shall be trained.

**1910.269(b)(2)** *First aid supplies.* First aid supplies required by §1910.151(b) of this part shall be placed in weatherproof containers if the supplies could be exposed to the weather.

**1910.269(b)(3)** *First aid kits.* Each first aid kit shall be maintained, shall be readily available for use, and shall be inspected frequently enough to ensure that expended items are replaced but at least once per year.

**1910.269(c)** *Job briefing.* The employer shall ensure that the employee in charge conducts a job briefing with the employees involved before they start each job. The briefing shall cover at least the following subjects: hazards associated with the job, work procedures involved, special precautions, energy source controls, and personal protective equipment requirements.

**1910.269(c)(1)** *Number of briefings.* If the work or operations to be performed during the work day or shift are repetitive and similar, at least one job briefing shall be conducted before the start of the first job of each day or shift. Additional job briefings shall be held if significant changes, which might affect the safety of the employees, occur during the course of the work.

**1910.269(c)(2)** *Extent of briefing.* A brief discussion is satisfactory if the work involved is routine and if the employee, by virtue of training and experience, can reasonably be expected to recognize and avoid the hazards involved in the job. A more extensive discussion shall be conducted:

**1910.269(c)(2)(i)** If the work is complicated or particularly hazardous, or

**1910.269(c)(2)(ii)** If the employee cannot be expected to recognize and avoid the hazards involved in the job.

**NOTE:** The briefing is always required to touch on all the subjects listed in the introductory text to paragraph (c) of this section.

**1910.269(c)(3)** *Working alone.* An employee working alone need not conduct a job briefing. However, the employer shall ensure that the tasks to be performed are planned as if a briefing were required.

**1910.269(d)** *Hazardous energy control (lockout/tagout) procedures—(1) Application.* The provisions of paragraph (d) of this section apply to the use of lockout/tagout procedures for the control of energy sources in installations for the purpose of electric power generation, including related equipment for communication or metering. Locking and tagging procedures for the deenergizing of electric energy sources which are used exclusively for purposes of transmission and distribution are addressed by paragraph (m) of this section.

**NOTE 1:** Installations in electric power generation facilities that are not an integral part of, or inextricably commingled with, power generation processes or equipment are covered under §1910.147 and subpart S of this part.

**NOTE 2:** Lockout and tagging procedures that comply with paragraphs (c) through (f) of §1910.147 of this part will also be deemed to comply with paragraph (d) of this section if the procedures address the hazards covered by paragraph (d) of this section.

**1910.269(d)(2)** *General.* (i) The employer shall establish a program consisting of energy control procedures, employee training, and periodic inspections to ensure that, before any employee performs any servicing or maintenance on a machine or equipment where the unexpected energizing, start up, or release of stored energy could occur and cause injury, the machine or equipment is isolated from the energy source and rendered inoperative.

**1910.269(d)(2)(ii)** The employer's energy control program under paragraph (d)(2) of this section shall meet the following requirements:

**1910.269(d)(2)(ii)(A)** If an energy isolating device is not capable of being locked out, the employer's program shall use a tagout system.

**1910.269(d)(2)(ii)(B)** If an energy isolating device is capable of being locked out, the employer's program shall use lockout, unless the employer can demonstrate that the use of a tagout system will provide full employee protection as follows:

**1910.269(d)(2)(ii)(B)***(1)* When a tagout device is used on an energy isolating device which is capable of being locked out, the tagout device shall be attached at the same location that the lockout device would have been attached, and the employer shall demonstrate that the tagout program will provide a level of safety equivalent to that obtained by the use of a lockout program.

**1910.269(d)(2)(ii)(B)***(2)* In demonstrating that a level of safety is achieved in the tagout program equivalent to the level of safety obtained by the use of a lockout program, the employer shall demonstrate full compliance with all tagout-related provisions of this standard together with such additional elements as are necessary to provide the equivalent safety available from the use of a lockout device. Additional means to be considered as part of the demonstration of full employee protection shall include the implementation of additional safety measures such as the removal of an isolating circuit element, blocking of a controlling switch, opening of an extra disconnecting device, or the removal of a valve handle to reduce the likelihood of inadvertent energizing.

**1910.269(d)(2)(ii)(C)** After November 1, 1994, whenever replacement or major repair, renovation, or modification of a machine or equipment is performed, and whenever new machines or equipment are installed, energy isolating devices for such machines or equipment shall be designed to accept a lockout device.

**1910.269(d)(2)(iii)** Procedures shall be developed, documented, and used for the control of potentially hazardous energy covered by paragraph (d) of this section.

**1910.269(d)(2)(iv)** The procedure shall clearly and specifically outline the scope, purpose, responsibility, authorization, rules, and techniques to be applied to the control of hazardous energy, and the measures to enforce compliance including, but not limited to, the following:

**1910.269(d)(2)(iv)(A)** A specific statement of the intended use of this procedure;

**1910.269(d)(2)(iv)(B)** Specific procedural steps for shutting down, isolating, blocking and securing machines or equipment to control hazardous energy;

**1910.269(d)(2)(iv)(C)** Specific procedural steps for the placement, removal, and transfer of lockout devices or tagout devices and the responsibility for them; and

**1910.269(d)(2)(iv)(D)** Specific requirements for testing a machine or equipment to determine and verify the effectiveness of lockout devices, tagout devices, and other energy control measures.

**1910.269(d)(2)(v)** The employer shall conduct a periodic inspection of the energy control procedure at least annually to ensure that the procedure and the provisions of paragraph (d) of this section are being followed.

**1910.269(d)(2)(v)(A)** The periodic inspection shall be performed by an authorized employee who is not using the energy control procedure being inspected.

**1910.269(d)(2)(v)(B)** The periodic inspection shall be designed to identify and correct any deviations or inadequacies.

**1910.269(d)(2)(v)(C)** If lockout is used for energy control, the periodic inspection shall include a review, between the inspector and each authorized employee, of that employee's responsibilities under the energy control procedure being inspected.

**1910.269(d)(2)(v)(D)** Where tagout is used for energy control, the periodic inspection shall include a review, between the inspector and each authorized and affected employee, of that employee's responsibilities under the energy control procedure being inspected, and the elements set forth in paragraph (d)(2)(vii) of this section.

**1910.269(d)(2)(v)(E)** The employer shall certify that the inspections required by paragraph (d)(2)(v) of this section have been accomplished. The certification shall identify the machine or equipment on which the energy control procedure was being used, the date of the inspection, the employees included in the inspection, and the person performing the inspection.

**NOTE:** If normal work schedule and operation records demonstrate adequate inspection activity and contain the required information, no additional certification is required.

**1910.269(d)(2)(vi)** The employer shall provide training to ensure that the purpose and function of the energy control program are understood by employees and that the knowledge and skills required for the safe application, usage, and removal of energy controls are acquired by employees. The training shall include the following:

**1910.269(d)(2)(vi)(A)** Each authorized employee shall receive training in the recognition of applicable hazardous energy sources, the type and magnitude of energy available in the workplace, and in the methods and means necessary for energy isolation and control.

**1910.269(d)(2)(vi)(B)** Each affected employee shall be instructed in the purpose and use of the energy control procedure.

**1910.269(d)(2)(vi)(C)** All other employees whose work operations are or may be in an area where energy control procedures may be used shall be instructed about the procedures and about the prohibition relating to attempts to restart or reenergize machines or equipment that are locked out or tagged out.

**1910.269(d)(2)(vii)** When tagout systems are used, employees shall also be trained in the following limitations of tags:

**1910.269(d)(2)(vii)(A)** Tags are essentially warning devices affixed to energy isolating devices and do not provide the physical restraint on those devices that is provided by a lock.

**1910.269(d)(2)(vii)(B)** When a tag is attached to an energy isolating means, it is not to be removed without authorization of the authorized person responsible for it, and it is never to be bypassed, ignored, or otherwise defeated.

**1910.269(d)(2)(vii)(C)** Tags must be legible and understandable by all authorized employees, affected employees, and all other employees whose work operations are or may be in the area, in order to be effective.

**1910.269(d)(2)(vii)(D)** Tags and their means of attachment must be made of materials which will withstand the environmental conditions encountered in the workplace.

**1910.269(d)(2)(vii)(E)** Tags may evoke a false sense of security, and their meaning needs to be understood as part of the overall energy control program.

**1910.269(d)(2)(vii)(F)** Tags must be securely attached to energy isolating devices so that they cannot be inadvertently or accidentally detached during use.

**1910.269(d)(2)(viii)** Retraining shall be provided by the employer as follows:

**1910.269(d)(2)(viii)(A)** Retraining shall be provided for all authorized and affected employees whenever there is a change in their job assignments, a change in machines, equipment, or processes that present a new hazard or whenever there is a change in the energy control procedures.

**1910.269(d)(2)(viii)(B)** Retraining shall also be conducted whenever a periodic inspection under paragraph (d)(2)(v) of this section reveals, or whenever the employer has reason to believe, that there are deviations

# OSHA Standards for General Industry

from or inadequacies in an employee's knowledge or use of the energy control procedures.

**1910.269(d)(2)(viii)(C)** The retraining shall reestablish employee proficiency and shall introduce new or revised control methods and procedures, as necessary.

**1910.269(d)(2)(ix)** The employer shall certify that employee training has been accomplished and is being kept up to date. The certification shall contain each employee's name and dates of training.

**1910.269(d)(3)** *Protective materials and hardware.* **(i)** Locks, tags, chains, wedges, key blocks, adapter pins, self-locking fasteners, or other hardware shall be provided by the employer for isolating, securing, or blocking of machines or equipment from energy sources.

**1910.269(d)(3)(ii)** Lockout devices and tagout devices shall be singularly identified; shall be the only devices used for controlling energy; may not be used for other purposes; and shall meet the following requirements:

**1910.269(d)(3)(ii)(A)** Lockout devices and tagout devices shall be capable of withstanding the environment to which they are exposed for the maximum period of time that exposure is expected.

**1910.269(d)(3)(ii)(A)***(1)* Tagout devices shall be constructed and printed so that exposure to weather conditions or wet and damp locations will not cause the tag to deteriorate or the message on the tag to become illegible.

**1910.269(d)(3)(ii)(A)***(2)* Tagout devices shall be so constructed as not to deteriorate when used in corrosive environments.

**1910.269(d)(3)(ii)(B)** Lockout devices and tagout devices shall be standardized within the facility in at least one of the following criteria: color, shape, size. Additionally, in the case of tagout devices, print and format shall be standardized.

**1910.269(d)(3)(ii)(C)** Lockout devices shall be substantial enough to prevent removal without the use of excessive force or unusual techniques, such as with the use of bolt cutters or metal cutting tools.

**1910.269(d)(3)(ii)(D)** Tagout devices, including their means of attachment, shall be substantial enough to prevent inadvertent or accidental removal. Tagout device attachment means shall be of a non-reusable type, attachable by hand, self-locking, and non-releasable with a minimum unlocking strength of no less than 50 pounds and shall have the general design and basic characteristics of being at least equivalent to a one-piece, all-environment-tolerant nylon cable tie.

**1910.269(d)(3)(ii)(E)** Each lockout device or tagout device shall include provisions for the identification of the employee applying the device.

**1910.269(d)(3)(ii)(F)** Tagout devices shall warn against hazardous conditions if the machine or equipment is energized and shall include a legend such as the following: Do Not Start, Do Not Open, Do Not Close, Do Not Energize, Do Not Operate.

**NOTE:** For specific provisions covering accident prevention tags, see §1910.145 of this part.

**1910.269(d)(4)** *Energy isolation.* Lockout and tagout device application and removal may only be performed by the authorized employees who are performing the servicing or maintenance.

**1910.269(d)(5)** *Notification.* Affected employees shall be notified by the employer or authorized employee of the application and removal of lockout or tagout devices. Notification shall be given before the controls are applied and after they are removed from the machine or equipment.

**NOTE:** See also paragraph (d)(7) of this section, which requires that the second notification take place before the machine or equipment is reenergized.

**1910.269(d)(6)** *Lockout/tagout application.* The established procedures for the application of energy control (the lockout or tagout procedures) shall include the following elements and actions, and these procedures shall be performed in the following sequence:

**1910.269(d)(6)(i)** Before an authorized or affected employee turns off a machine or equipment, the authorized employee shall have knowledge of the type and magnitude of the energy, the hazards of the energy to be controlled, and the method or means to control the energy.

**1910.269(d)(6)(ii)** The machine or equipment shall be turned off or shut down using the procedures established for the machine or equipment. An orderly shutdown shall be used to avoid any additional or increased hazards to employees as a result of the equipment stoppage.

**1910.269(d)(6)(iii)** All energy isolating devices that are needed to control the energy to the machine or equipment shall be physically located and operated in such a manner as to isolate the machine or equipment from energy sources.

**1910.269(d)(6)(iv)** Lockout or tagout devices shall be affixed to each energy isolating device by authorized employees.

**1910.269(d)(6)(iv)(A)** Lockout devices shall be attached in a manner that will hold the energy isolating devices in a "safe" or "off" position.

**1910.269(d)(6)(iv)(B)** Tagout devices shall be affixed in such a manner as will clearly indicate that the operation or movement of energy isolating devices from the "safe" or "off" position is prohibited.

**1910.269(d)(6)(iv)(B)***(1)* Where tagout devices are used with energy isolating devices designed with the capability of being locked out, the tag attachment shall be fastened at the same point at which the lock would have been attached.

**1910.269(d)(6)(iv)(B)***(2)* Where a tag cannot be affixed directly to the energy isolating device, the tag shall be located as close as safely possible to the device, in a position that will be immediately obvious to anyone attempting to operate the device.

**1910.269(d)(6)(v)** Following the application of lockout or tagout devices to energy isolating devices, all potentially hazardous stored or residual energy shall be relieved, disconnected, restrained, or otherwise rendered safe.

**1910.269(d)(6)(vi)** If there is a possibility of reaccumulation of stored energy to a hazardous level, verification of isolation shall be continued until the servicing or maintenance is completed or until the possibility of such accumulation no longer exists.

**1910.269(d)(6)(vii)** Before starting work on machines or equipment that have been locked out or tagged out, the authorized employee shall verify that isolation and deenergizing of the machine or equipment have been accomplished. If normally energized parts will be exposed to contact by an employee while the machine or equipment is deenergized, a test shall be performed to ensure that these parts are deenergized.

**1910.269(d)(7)** *Release from lockout/tagout.* Before lockout or tagout devices are removed and energy is restored to the machine or equipment, procedures shall be followed and actions taken by the authorized employees to ensure the following:

**1910.269(d)(7)(i)** The work area shall be inspected to ensure that nonessential items have been removed and that machine or equipment components are operationally intact.

**1910.269(d)(7)(ii)** The work area shall be checked to ensure that all employees have been safely positioned or removed.

**1910.269(d)(7)(iii)** After lockout or tagout devices have been removed and before a machine or equipment is started, affected employees shall be notified that the lockout or tagout devices have been removed.

**1910.269(d)(7)(iv)** Each lockout or tagout device shall be removed from each energy isolating device by the authorized employee who applied the lockout or tagout device. However, if that employee is not available to remove it, the device may be removed under the direction of the employer, provided that specific procedures and training for such removal have been developed, documented, and incorporated into the employer's energy control program. The employer shall demonstrate that the specific procedure provides a degree of safety equivalent to that provided by the removal of the device by the authorized employee who applied it. The specific procedure shall include at least the following elements:

**1910.269(d)(7)(iv)(A)** Verification by the employer that the authorized employee who applied the device is not at the facility;

**1910.269(d)(7)(iv)(B)** Making all reasonable efforts to contact the authorized employee to inform him or her that his or her lockout or tagout device has been removed; and

**1910.269(d)(7)(iv)(C)** Ensuring that the authorized employee has this knowledge before he or she resumes work at that facility.

**1910.269(d)(8)** *Additional requirements.* **(i)** If the lockout or tagout devices must be temporarily removed from energy isolating devices and the machine or equipment must be energized to test or position the machine, equipment, or component thereof, the following sequence of actions shall be followed:

**1910.269(d)(8)(i)(A)** Clear the machine or equipment of tools and materials in accordance with paragraph (d)(7)(i) of this section;

**1910.269(d)(8)(i)(B)** Remove employees from the machine or equipment area in accordance with paragraphs (d)(7)(ii) and (d)(7)(iii) of this section;

**1910.269(d)(8)(i)(C)** Remove the lockout or tagout devices as specified in paragraph (d)(7)(iv) of this section;

**1910.269(d)(8)(i)(D)** Energize and proceed with the testing or positioning; and

**1910.269(d)(8)(i)(E)** Deenergize all systems and reapply energy control measures in accordance with paragraph (d)(6) of this section to continue the servicing or maintenance.

**1910.269(d)(8)(ii)** When servicing or maintenance is performed by a crew, craft, department, or other group, they shall use a procedure which affords the employees a level of protection equivalent to that provided by the implementation of a personal lockout or tagout device. Group lockout or tagout devices shall be used in accordance with the procedures required by paragraphs (d)(2)(iii) and (d)(2)(iv) of this section including, but not limited to, the following specific requirements:

**1910.269(d)(8)(ii)(A)** Primary responsibility shall be vested in an authorized employee for a set number of employees working under the protection of a group lockout or tagout device (such as an operations lock);

**1910.269(d)(8)(ii)(B)** Provision shall be made for the authorized employee to ascertain the exposure status of all individual group members with regard to the lockout or tagout of the machine or equipment;

**1910.269(d)(8)(ii)(C)** When more than one crew, craft, department, or other group is involved, assignment of overall job-associated lockout or tagout control responsibility shall be given to an authorized employee designated to coordinate affected work forces and ensure continuity of protection; and

**1910.269(d)(8)(ii)(D)** Each authorized employee shall affix a personal lockout or tagout device to the group lockout device, group lockbox, or comparable mechanism when he or she begins work and shall remove those devices when he or she stops working on the machine or equipment being serviced or maintained.§1910.268

**1910.269(d)(8)(iii)** Procedures shall be used during shift or personnel changes to ensure the continuity of lockout or tagout protection, including provision for the orderly transfer of lockout or tagout device protection between off-going and on-coming employees, to minimize their exposure to hazards from the unexpected energizing or start-up of the machine or equipment or from the release of stored energy.

**1910.269(d)(8)(iv)** Whenever outside servicing personnel are to be engaged in activities covered by paragraph (d) of this section, the on-site employer and the outside employer shall inform each other of their respective lockout or tagout procedures, and each employer shall ensure that his or her personnel understand and comply with restrictions and prohibitions of the energy control procedures being used.

**1910.269(d)(8)(v)** If energy isolating devices are installed in a central location and are under the exclusive control of a system operator, the following requirements apply:

**1910.269(d)(8)(v)(A)** The employer shall use a procedure that affords employees a level of protection equivalent to that provided by the implementation of a personal lockout or tagout device.

**1910.269(d)(8)(v)(B)** The system operator shall place and remove lockout and tagout devices in place of the authorized employee under paragraphs (d)(4), (d)(6)(iv), and (d)(7)(iv) of this section.

**1910.269(d)(8)(v)(C)** Provisions shall be made to identify the authorized employee who is responsible for (that is, being protected by) the lockout or tagout device, to transfer responsibility for lockout and tagout devices, and to ensure that an authorized employee requesting removal or transfer of a lockout or tagout device is the one responsible for it before the device is removed or transferred.

**1910.269(e)** *Enclosed spaces.* This paragraph covers enclosed spaces that may be entered by employees. It does not apply to vented vaults if a determination is made that the ventilation system is operating to protect employees before they enter the space. This paragraph applies to routine entry into enclosed spaces in lieu of the permit-space entry requirements contained in paragraphs (d) through (k) of §1910.146 of this part. If, after the precautions given in paragraphs (e) and (t) of this section are taken, the hazards remaining in the enclosed space endanger the life of an entrant or could interfere with escape from the space, then entry into the enclosed space shall meet the permit-space entry requirements of paragraphs (d) through (k) of §1910.146 of this part.

**NOTE:** Entries into enclosed spaces conducted in accordance with the permit-space entry requirements of paragraphs (d) through (k) of §1910.146 of this part are considered as complying with paragraph (e) of this section.

**1910.269(e)(1)** *Safe work practices.* The employer shall ensure the use of safe work practices for entry into and work in enclosed spaces and for rescue of employees from such spaces.

**1910.269(e)(2)** *Training.* Employees who enter enclosed spaces or who serve as attendants shall be trained in the hazards of enclosed space entry, in enclosed space entry procedures, and in enclosed space rescue procedures.

**1910.269(e)(3)** *Rescue equipment.* Employers shall provide equipment to ensure the prompt and safe rescue of employees from the enclosed space.

**1910.269(e)(4)** *Evaluation of potential hazards.* Before any entrance cover to an enclosed space is removed, the employer shall determine whether it is safe to do so by checking for the presence of any atmospheric pressure or temperature differences and by evaluating whether there might be a hazardous atmosphere in the space. Any conditions making it unsafe to remove the cover shall be eliminated before the cover is removed.

**NOTE:** The evaluation called for in this paragraph may take the form of a check of the conditions expected to be in the enclosed space. For example, the cover could be checked to see if it is hot and, if it is fastened in place, could be loosened gradually to release any residual pressure. A determination must also be made of whether conditions at the site could cause a hazardous atmosphere, such as an oxygen deficient or flammable atmosphere, to develop within the space.

**1910.269(e)(5)** *Removal of covers.* When covers are removed from enclosed spaces, the opening shall be promptly guarded by a railing, temporary cover, or other barrier intended to prevent an accidental fall through the opening and to protect employees working in the space from objects entering the space.

**1910.269(e)(6)** *Hazardous atmosphere.* Employees may not enter any enclosed space while it contains a hazardous atmosphere, unless the entry conforms to the generic permit-required confined spaces standard in §1910.146 of this part.

**NOTE:** The term "entry" is defined in §1910.146(b) of this part.

**1910.269(e)(7)** *Attendants.* While work is being performed in the enclosed space, a person with first aid training meeting paragraph (b) of this section shall be immediately available outside the enclosed space to render emergency assistance if there is reason to believe that a hazard may exist in the space or if a hazard exists because of traffic patterns in the area of the opening used for entry. That person is not precluded from performing other duties outside the enclosed space if these duties do not distract the attendant from monitoring employees within the space.

**NOTE:** See paragraph (t)(3) of this section for additional requirements on attendants for work in manholes.

**1910.269(e)(8)** *Calibration of test instruments.* Test instruments used to monitor atmospheres in enclosed spaces shall be kept in calibration, with a minimum accuracy of ±10 percent.

**1910.269(e)(9)** *Testing for oxygen deficiency.* Before an employee enters an enclosed space, the internal atmosphere shall be tested for oxygen deficiency with a direct-reading meter or similar instrument, capable of collection and immediate analysis of data samples without the need for off-site evaluation. If continuous forced air ventilation is provided, testing is not required provided that the procedures used ensure that employees are not exposed to the hazards posed by oxygen deficiency.

**1910.269(e)(10)** *Testing for flammable gases and vapors.* Before an employee enters an enclosed space, the internal atmosphere shall be tested for flammable gases and vapors with a direct-reading meter or similar instrument capable of collection and immediate analysis of data samples without the need for off-site evaluation. This test shall be performed after the oxygen testing and ventilation required by paragraph (e)(9) of this section demonstrate that there is sufficient oxygen to ensure the accuracy of the test for flammability.

**1910.269(e)(11)** *Ventilation and monitoring.* If flammable gases or vapors are detected or if an oxygen deficiency is found, forced air ventilation shall be used to maintain oxygen at a safe level and to prevent a hazardous concentration of flammable gases and vapors from accumulating. A continuous monitoring program to ensure that no increase in flammable gas or vapor concentration occurs may be followed in lieu of ventilation, if flammable gases or vapors are detected at safe levels.

# OSHA Standards for General Industry

**NOTE:** See the definition of hazardous atmosphere for guidance in determining whether or not a given concentration of a substance is considered to be hazardous.

**1910.269(e)(12)** *Specific ventilation requirements.* If continuous forced air ventilation is used, it shall begin before entry is made and shall be maintained long enough to ensure that a safe atmosphere exists before employees are allowed to enter the work area. The forced air ventilation shall be so directed as to ventilate the immediate area where employees are present within the enclosed space and shall continue until all employees leave the enclosed space.

**1910.269(e)(13)** *Air supply.* The air supply for the continuous forced air ventilation shall be from a clean source and may not increase the hazards in the enclosed space.

**1910.269(e)(14)** *Open flames.* If open flames are used in enclosed spaces, a test for flammable gases and vapors shall be made immediately before the open flame device is used and at least once per hour while the device is used in the space. Testing shall be conducted more frequently if conditions present in the enclosed space indicate that once per hour is insufficient to detect hazardous accumulations of flammable gases or vapors.

**NOTE:** See the definition of hazardous atmosphere for guidance in determining whether or not a given concentration of a substance is considered to be hazardous.

**1910.269(f)** **Excavations.** Excavation operations shall comply with subpart P of part 1926 of this chapter.

**1910.269(g)** **Personal protective equipment—(1)** *General.* Personal protective equipment shall meet the requirements of subpart I of this part.

**1910.269(g)(2)** *Fall protection.* **(i)** Personal fall arrest equipment shall meet the requirements of subpart M of part 1926 of this chapter.

**1910.269(g)(2)(ii)** Body belts and safety straps for work positioning shall meet the requirements of §1926.959 of this chapter.

**1910.269(g)(2)(iii)** Body belts, safety straps, lanyards, lifelines, and body harnesses shall be inspected before use each day to determine that the equipment is in safe working condition. Defective equipment may not be used.

**1910.269(g)(2)(iv)** Lifelines shall be protected against being cut or abraded.

**1910.269(g)(2)(v)** Fall arrest equipment, work positioning equipment, or travel restricting equipment shall be used by employees working at elevated locations more than 4 feet (1.2 m) above the ground on poles, towers, or similar structures if other fall protection has not been provided. Fall protection equipment is not required to be used by a qualified employee climbing or changing location on poles, towers, or similar structures, unless conditions, such as, but not limited to, ice, high winds, the design of the structure (for example, no provision for holding on with hands), or the presence of contaminants on the structure, could cause the employee to lose his or her grip or footing.

**NOTE 1:** This paragraph applies to structures that support overhead electric power generation, transmission, and distribution lines and equipment. It does not apply to portions of buildings, such as loading docks, to electric equipment, such as transformers and capacitors, nor to aerial lifts. Requirements for fall protection associated with walking and working surfaces are contained in subpart D of this part; requirements for fall protection associated with aerial lifts are contained in §1910.67 of this part.

**NOTE 2:** Employees undergoing training are not considered "qualified employees" for the purposes of this provision. Unqualified employees (including trainees) are required to use fall protection any time they are more than 4 feet (1.2 m) above the ground.

**1910.269(g)(2)(vi)** The following requirements apply to personal fall arrest systems:

**1910.269(g)(2)(vi)(A)** When stopping or arresting a fall, personal fall arrest systems shall limit the maximum arresting force on an employee to 900 pounds (4 kN) if used with a body belt.

**1910.269(g)(2)(vi)(B)** When stopping or arresting a fall, personal fall arrest systems shall limit the maximum arresting force on an employee to 1800 pounds (8 kN) if used with a body harness.

**1910.269(g)(2)(vi)(C)** Personal fall arrest systems shall be rigged such that an employee can neither free fall more than 6 feet (1.8 m) nor contact any lower level.

**1910.269(g)(2)(vii)** If vertical lifelines or droplines are used, not more than one employee may be attached to any one lifeline.

**1910.269(g)(2)(viii)** Snaphooks may not be connected to loops made in webbing-type lanyards.

**1910.269(g)(2)(ix)** Snaphooks may not be connected to each other.

**1910.269(h)** **Ladders, platforms, step bolts, and manhole steps—(1)** *General.* Requirements for ladders contained in subpart D of this part apply, except as specifically noted in paragraph (h)(2) of this section.

**1910.269(h)(2)** *Special ladders and platforms.* Portable ladders and platforms used on structures or conductors in conjunction with overhead line work need not meet paragraphs (d)(2)(i) and (d)(2)(iii) of §1910.25 of this part or paragraph (c)(3)(iii) of §1910.26 of this part. However, these ladders and platforms shall meet the following requirements:

**1910.269(h)(2)(i)** Ladders and platforms shall be secured to prevent their becoming accidentally dislodged.

**1910.269(h)(2)(ii)** Ladders and platforms may not be loaded in excess of the working loads for which they are designed.

**1910.269(h)(2)(iii)** Ladders and platforms may be used only in applications for which they were designed.

**1910.269(h)(2)(iv)** In the configurations in which they are used, ladders and platforms shall be capable of supporting without failure at least 2.5 times the maximum intended load.

**1910.269(h)(3)** *Conductive ladders.* Portable metal ladders and other portable conductive ladders may not be used near exposed energized lines or equipment. However, in specialized high-voltage work, conductive ladders shall be used where the employer can demonstrate that nonconductive ladders would present a greater hazard than conductive ladders.

**1910.269(i)** **Hand and portable power tools—(1)** *General.* Paragraph (i)(2) of this section applies to electric equipment connected by cord and plug. Paragraph (i)(3) of this section applies to portable and vehicle-mounted generators used to supply cord-and plug-connected equipment. Paragraph (i)(4) of this section applies to hydraulic and pneumatic tools.

**1910.269(i)(2)** *Cord- and plug-connected equipment.* **(i)** Cord-and plug-connected equipment supplied by premises wiring is covered by subpart S of this part.

**1910.269(i)(2)(ii)** Any cord- and plug-connected equipment supplied by other than premises wiring shall comply with one of the following in lieu of §1910.243(a)(5) of this part:

**1910.269(i)(2)(ii)(A)** It shall be equipped with a cord containing an equipment grounding conductor connected to the tool frame and to a means for grounding the other end (however, this option may not be used where the introduction of the ground into the work environment increases the hazard to an employee); or

**1910.269(i)(2)(ii)(B)** It shall be of the double-insulated type conforming to subpart S of this part; or

**1910.269(i)(2)(ii)(C)** It shall be connected to the power supply through an isolating transformer with an ungrounded secondary.

**1910.269(i)(3)** *Portable and vehicle-mounted generators.* Portable and vehicle-mounted generators used to supply cord- and plug-connected equipment shall meet the following requirements:

**1910.269(i)(3)(i)** The generator may only supply equipment located on the generator or the vehicle and cord- and plug-connected equipment through receptacles mounted on the generator or the vehicle.

**1910.269(i)(3)(ii)** The non-current-carrying metal parts of equipment and the equipment grounding conductor terminals of the receptacles shall be bonded to the generator frame.

**1910.269(i)(3)(iii)** In the case of vehicle-mounted generators, the frame of the generator shall be bonded to the vehicle frame.

**1910.269(i)(3)(iv)** Any neutral conductor shall be bonded to the generator frame.

**1910.269(i)(4)** *Hydraulic and pneumatic tools.* **(i)** Safe operating pressures for hydraulic and pneumatic tools, hoses, valves, pipes, filters, and fittings may not be exceeded.

**NOTE:** If any hazardous defects are present, no operating pressure would be safe, and the hydraulic or pneumatic equipment involved may not be used. In the absence of defects, the maximum rated operating pressure is the maximum safe pressure.

**1910.269(i)(4)(ii)** A hydraulic or pneumatic tool used where it may contact exposed live parts shall be designed and maintained for such use.

**1910.269(i)(4)(iii)** The hydraulic system supplying a hydraulic tool used where it may contact exposed live parts shall provide protection against loss of insulating value for the voltage involved due to the formation of a partial vacuum in the hydraulic line.

**NOTE:** Hydraulic lines without check valves having a separation of more than 35 feet (10.7 m) between the oil reservoir and the upper end of the hydraulic system promote the formation of a partial vacuum.

**1910.269(i)(4)(iv)** A pneumatic tool used on energized electric lines or equipment or used where it may contact exposed live parts shall provide protection against the accumulation of moisture in the air supply.

**1910.269(i)(4)(v)** Pressure shall be released before connections are broken, unless quick acting, self-closing connectors are used. Hoses may not be kinked.

**1910.269(i)(4)(vi)** Employees may not use any part of their bodies to locate or attempt to stop a hydraulic leak.

**1910.269(j)** *Live-line tools*—(1) *Design of tools.* Live-line tool rods, tubes, and poles shall be designed and constructed to withstand the following minimum tests:

**1910.269(j)(1)(i)** 100,000 volts per foot (3281 volts per centimeter) of length for 5 minutes if the tool is made of fiberglass-reinforced plastic (FRP), or

**1910.269(j)(1)(ii)** 75,000 volts per foot (2461 volts per centimeter) of length for 3 minutes if the tool is made of wood, or

**1910.269(j)(1)(iii)** Other tests that the employer can demonstrate are equivalent.

**NOTE:** Live-line tools using rod and tube that meet ASTM F711-89, Standard Specification for Fiberglass-Reinforced Plastic (FRP) Rod and Tube Used in Live-Line Tools, conform to paragraph (j)(1)(i) of this section.

**1910.269(j)(2)** *Condition of tools.* (i) Each live-line tool shall be wiped clean and visually inspected for defects before use each day.

**1910.269(j)(2)(ii)** If any defect or contamination that could adversely affect the insulating qualities or mechanical integrity of the live-line tool is present after wiping, the tool shall be removed from service and examined and tested according to paragraph (j)(2)(iii) of this section before being returned to service.

**1910.269(j)(2)(iii)** Live-line tools used for primary employee protection shall be removed from service every 2 years and whenever required under paragraph (j)(2)(ii) of this section for examination, cleaning, repair, and testing as follows:

**1910.269(j)(2)(iii)(A)** Each tool shall be thoroughly examined for defects.

**1910.269(j)(2)(iii)(B)** If a defect or contamination that could adversely affect the insulating qualities or mechanical integrity of the live-line tool is found, the tool shall be repaired and refinished or shall be permanently removed from service. If no such defect or contamination is found, the tool shall be cleaned and waxed.

**1910.269(j)(2)(iii)(C)** The tool shall be tested in accordance with paragraphs (j)(2)(iii)(D) and (j)(2)(iii)(E) of this section under the following conditions:

**1910.269(j)(2)(iii)(C)***(1)* After the tool has been repaired or refinished; and

**1910.269(j)(2)(iii)(C)***(2)* After the examination if repair or refinishing is not performed, unless the tool is made of FRP rod or foam-filled FRP tube and the employer can demonstrate that the tool has no defects that could cause it to fail in use.

**1910.269(j)(2)(iii)(D)** The test method used shall be designed to verify the tool's integrity along its entire working length and, if the tool is made of fiberglass-reinforced plastic, its integrity under wet conditions.

**1910.269(j)(2)(iii)(E)** The voltage applied during the tests shall be as follows:

**1910.269(j)(2)(iii)(E)***(1)* 75,000 volts per foot (2461 volts per centimeter) of length for 1 minute if the tool is made of fiberglass, or

**1910.269(j)(2)(iii)(E)***(2)* 50,000 volts per foot (1640 volts per centimeter) of length for 1 minute if the tool is made of wood, or

**1910.269(j)(2)(iii)(E)***(3)* Other tests that the employer can demonstrate are equivalent.

**NOTE:** Guidelines for the examination, cleaning, repairing, and in-service testing of live-line tools are contained in the Institute of Electrical and Electronics Engineers Guide for In-Service Maintenance and Electrical Testing of Live-Line Tools, IEEE Std. 978-1984.

**1910.269(k)** *Materials handling and storage*—(1) *General.* Material handling and storage shall conform to the requirements of subpart N of this part.

**1910.269(k)(2)** *Materials storage near energized lines or equipment.* (i) In areas not restricted to qualified persons only, materials or equipment may not be stored closer to energized lines or exposed energized parts of equipment than the following distances plus an amount providing for the maximum sag and side swing of all conductors and providing for the height and movement of material handling equipment:

**1910.269(k)(2)(i)(A)** For lines and equipment energized at 50 kV or less, the distance is 10 feet (305 cm).

**1910.269(k)(2)(i)(B)** For lines and equipment energized at more than 50 kV, the distance is 10 feet (305 cm) plus 4 inches (10 cm) for every 10 kV over 50 kV.

**1910.269(k)(2)(ii)** In areas restricted to qualified employees, material may not be stored within the working space about energized lines or equipment.

**NOTE:** Requirements for the size of the working space are contained in paragraphs (u)(1) and (v)(3) of this section.

**1910.269(l)** *Working on or near exposed energized parts.* This paragraph applies to work on exposed live parts, or near enough to them, to expose the employee to any hazard they present.

**1910.269(l)(1)** *General.* Only qualified employees may work on or with exposed energized lines or parts of equipment. Only qualified employees may work in areas containing unguarded, uninsulated energized lines or parts of equipment operating at 50 volts or more. Electric lines and equipment shall be considered and treated as energized unless the provisions of paragraph (d) or paragraph (m) of this section have been followed.

**1910.269(l)(1)(i)** Except as provided in paragraph (l)(1)(ii) of this section, at least two employees shall be present while the following types of work are being performed:

**1910.269(l)(1)(i)(A)** Installation, removal, or repair of lines that are energized at more than 600 volts,

**1910.269(l)(1)(i)(B)** Installation, removal, or repair of deenergized lines if an employee is exposed to contact with other parts energized at more than 600 volts,

**1910.269(l)(1)(i)(C)** Installation, removal, or repair of equipment, such as transformers, capacitors, and regulators, if an employee is exposed to contact with parts energized at more than 600 volts,

**1910.269(l)(1)(i)(D)** Work involving the use of mechanical equipment, other than insulated aerial lifts, near parts energized at more than 600 volts, and

**1910.269(l)(1)(i)(E)** Other work that exposes an employee to electrical hazards greater than or equal to those posed by operations that are specifically listed in paragraphs (l)(1)(i)(A) through (l)(1)(i)(D) of this section.

**1910.269(l)(1)(ii)** Paragraph (l)(1)(i) of this section does not apply to the following operations:

**1910.269(l)(1)(ii)(A)** Routine switching of circuits, if the employer can demonstrate that conditions at the site allow this work to be performed safely,

**1910.269(l)(1)(ii)(B)** Work performed with live-line tools if the employee is positioned so that he or she is neither within reach of nor otherwise exposed to contact with energized parts, and

**1910.269(l)(1)(ii)(C)** Emergency repairs to the extent necessary to safeguard the general public.

**1910.269(l)(2)** *Minimum approach distances.* The employer shall ensure that no employee approaches or takes any conductive object closer to exposed energized parts than set forth in Table R-6 through Table R-10, unless:

**1910.269(l)(2)(i)** The employee is insulated from the energized part (insulating gloves or insulating gloves and sleeves worn in accordance with paragraph (l)(3) of this section are considered insulation of the employee only with regard to the energized part upon which work is being performed), or

**1910.269(l)(2)(ii)** The energized part is insulated from the employee and from any other conductive object at a different potential, or

**1910.269(l)(2)(iii)** The employee is insulated from any other exposed conductive object, as during live-line bare-hand work.

# OSHA Standards for General Industry

**NOTE:** Paragraphs (u)(5)(i) and (v)(5)(i) of this section contain requirements for the guarding and isolation of live parts. Parts of electric circuits that meet these two provisions are not considered as "exposed" unless a guard is removed or an employee enters the space intended to provide isolation from the live parts.

**1910.269(l)(3)** *Type of insulation.* If the employee is to be insulated from energized parts by the use of insulating gloves (under paragraph (l)(2)(i) of this section), insulating sleeves shall also be used. However, insulating sleeves need not be used under the following conditions:

**1910.269(l)(3)(i)** If exposed energized parts on which work is not being performed are insulated from the employee and

**1910.269(l)(3)(ii)** If such insulation is placed from a position not exposing the employee's upper arm to contact with other energized parts.

**1910.269(l)(4)** *Working position.* The employer shall ensure that each employee, to the extent that other safety-related conditions at the worksite permit, works in a position from which a slip or shock will not bring the employee's body into contact with exposed, uninsulated parts energized at a potential different from the employee.

**1910.269(l)(5)** *Making connections.* The employer shall ensure that connections are made as follows:

**1910.269(l)(5)(i)** In connecting deenergized equipment or lines to an energized circuit by means of a conducting wire or device, an employee shall first attach the wire to the deenergized part;

**1910.269(l)(5)(ii)** When disconnecting equipment or lines from an energized circuit by means of a conducting wire or device, an employee shall remove the source end first; and

**1910.269(l)(5)(iii)** When lines or equipment are connected to or disconnected from energized circuits, loose conductors shall be kept away from exposed energized parts.

**1910.269(l)(6)** *Apparel.* (i) When work is performed within reaching distance of exposed energized parts of equipment, the employer shall ensure that each employee removes or renders nonconductive all exposed conductive articles, such as key or watch chains, rings, or wrist watches or bands, unless such articles do not increase the hazards associated with contact with the energized parts.

**1910.269(l)(6)(ii)** The employer shall train each employee who is exposed to the hazards of flames or electric arcs in the hazards involved.

**1910.269(l)(6)(iii)** The employer shall ensure that each employee who is exposed to the hazards of flames or electric arcs does not wear clothing that, when exposed to flames or electric arcs, could increase the extent of injury that would be sustained by the employee.

**NOTE:** Clothing made from the following types of fabrics, either alone or in blends, is prohibited by this paragraph, unless the employer can demonstrate that the fabric has been treated to withstand the conditions that may be encountered or that the clothing is worn in such a manner as to eliminate the hazard involved: acetate, nylon, polyester, rayon.

**1910.269(l)(7)** *Fuse handling.* When fuses must be installed or removed with one or both terminals energized at more than 300 volts or with exposed parts energized at more than 50 volts, the employer shall ensure that tools or gloves rated for the voltage are used. When expulsion-type fuses are installed with one or both terminals energized at more than 300 volts, the employer shall ensure that each employee wears eye protection meeting the requirements of subpart I of this part, uses a tool rated for the voltage, and is clear of the exhaust path of the fuse barrel.

**1910.269(l)(8)** *Covered (noninsulated) conductors.* The requirements of this section which pertain to the hazards of exposed live parts also apply when work is performed in the proximity of covered (noninsulated) wires.

**1910.269(l)(9)** *Noncurrent-carrying metal parts.* Noncurrent-carrying metal parts of equipment or devices, such as transformer cases and circuit breaker housings, shall be treated as energized at the highest voltage to which they are exposed, unless the employer inspects the installation and determines that these parts are grounded before work is performed.

**1910.269(l)(10)** *Opening circuits under load.* Devices used to open circuits under load conditions shall be designed to interrupt the current involved.

### TABLE R-6—AC LIVE-LINE WORK MINIMUM APPROACH DISTANCE

| Nominal voltage in kilovolts phase to phase | Phase to ground exposure (ft-in) | Phase to ground exposure (m) | Phase to phase exposure (ft-in) | Phase to phase exposure (m) |
|---|---|---|---|---|
| 0.05 to 1.0 | (4) | (4) | (4) | (4) |
| 1.1 to 15.0 | 2-1 | 0.64 | 2-2 | 0.66 |
| 15.1 to 36.0 | 2-4 | 0.72 | 2-7 | 0.77 |
| 36.1 to 46.0 | 2-7 | 0.77 | 2-10 | 0.85 |
| 46.1 to 72.5 | 3-0 | 0.90 | 3-6 | 1.05 |
| 72.6 to 121 | 3-2 | 0.95 | 4-3 | 1.29 |
| 138 to 145 | 3-7 | 1.09 | 4-11 | 1.50 |
| 161 to 169 | 4-0 | 1.22 | 5-8 | 1.71 |
| 230 to 242 | 5-3 | 1.59 | 7-6 | 2.27 |
| 345 to 362 | 8-6 | 2.59 | 12-6 | 3.80 |
| 500 to 550 | 11-3 | 3.42 | 18-1 | 5.50 |
| 765 to 800 | 14-11 | 4.53 | 26-0 | 7.91 |

NOTE 1: These distances take into consideration the highest switching surge an employee will be exposed to on any system with air as the insulating medium and the maximum voltages shown.

NOTE 2: The clear live-line tool distance shall equal or exceed the values for the indicated voltage ranges.

NOTE 3: See appendix B to this section for information on how the minimum approach distances listed in the tables were derived.

4 Avoid contact.

### TABLE R-7—AC LIVE-LINE WORK MINIMUM APPROACH DISTANCE WITH OVERVOLTAGE FACTOR PHASE-TO-GROUND EXPOSURE

| | Distance in feet-inches | | | | | | |
|---|---|---|---|---|---|---|---|
| | Maximum phase-to-phase voltage in kilovolts | | | | | | |
| Maximum anticipated per-unit transient overvoltage | 121 | 145 | 169 | 242 | 362 | 552 | 800 |
| 1.5 | ......... | ......... | ......... | ......... | ......... | 6-0 | 9-8 |
| 1.6 | ......... | ......... | ......... | ......... | ......... | 6-6 | 10-8 |
| 1.7 | ......... | ......... | ......... | ......... | ......... | 7-0 | 11-8 |
| 1.8 | ......... | ......... | ......... | ......... | ......... | 7-7 | 12-8 |
| 1.9 | ......... | ......... | ......... | ......... | ......... | 8-1 | 13-9 |
| 2.0 | 2-5 | 2-9 | 3-0 | 3-10 | 5-3 | 8-9 | 14-11 |
| 2.1 | 2-6 | 2-10 | 3-2 | 4-0 | 5-5 | 9-4 | ......... |
| 2.2 | 2-7 | 2-11 | 3-3 | 4-1 | 5-9 | 9-11 | ......... |
| 2.3 | 2-8 | 3-0 | 3-4 | 4-3 | 6-1 | 10-6 | ......... |
| 2.4 | 2-9 | 3-1 | 3-5 | 4-5 | 6-4 | 11-3 | ......... |
| 2.5 | 2-9 | 3-2 | 3-6 | 4-6 | 6-8 | ......... | ......... |
| 2.6 | 2-10 | 3-3 | 3-8 | 4-8 | 7-1 | ......... | ......... |
| 2.7 | 2-11 | 3-4 | 3-9 | 4-10 | 7-5 | ......... | ......... |
| 2.8 | 3-0 | 3-5 | 3-10 | 4-11 | 7-9 | ......... | ......... |
| 2.9 | 3-1 | 3-6 | 3-11 | 5-1 | 8-2 | ......... | ......... |
| 3.0 | 3-2 | 3-7 | 4-0 | 5-3 | 8-6 | ......... | ......... |

NOTE 1: The distance specified in this table may be applied only where the maximum anticipated per-unit transient overvoltage has been determined by engineering analysis and has been supplied by the employer. Table R-6 applies otherwise.

NOTE 2: The distances specified in this table are the air, bare-hand, and live-line tool distances.

NOTE 3: See appendix B to this section for information on how the minimum approach distances listed in the tables were derived and on how to calculate revised minimum approach distances based on the control of transient overvoltages.

### TABLE R-8—AC LIVE-LINE WORK MINIMUM APPROACH DISTANCE WITH OVERVOLTAGE FACTOR PHASE-TO-PHASE EXPOSURE

| | Distance in feet-inches | | | | | | |
|---|---|---|---|---|---|---|---|
| | Maximum phase-to-phase voltage in kilovolts | | | | | | |
| Maximum anticipated per-unit transient overvoltage | 121 | 145 | 169 | 242 | 362 | 552 | 800 |
| 1.5 | ......... | ......... | ......... | ......... | ......... | 7-4 | 12-1 |
| 1.6 | ......... | ......... | ......... | ......... | ......... | 8-9 | 14-6 |
| 1.7 | ......... | ......... | ......... | ......... | ......... | 10-2 | 17-2 |
| 1.8 | ......... | ......... | ......... | ......... | ......... | 11-7 | 19-11 |
| 1.9 | ......... | ......... | ......... | ......... | ......... | 13-2 | 22-11 |
| 2.0 | 3-7 | 4-1 | 4-8 | 6-1 | 8-7 | 14-10 | 26-0 |
| 2.1 | 3-7 | 4-2 | 4-9 | 6-3 | 8-10 | 15-7 | ......... |
| 2.2 | 3-8 | 4-3 | 4-10 | 6-4 | 9-2 | 16-4 | ......... |
| 2.3 | 3-9 | 4-4 | 4-11 | 6-6 | 9-6 | 17-2 | ......... |
| 2.4 | 3-10 | 4-5 | 5-0 | 6-7 | 9-11 | 18-1 | ......... |
| 2.5 | 3-11 | 4-6 | 5-2 | 6-9 | 10-4 | ......... | ......... |
| 2.6 | 4-0 | 4-7 | 5-3 | 6-11 | 10-9 | ......... | ......... |
| 2.7 | 4-1 | 4-8 | 5-4 | 7-0 | 11-2 | ......... | ......... |
| 2.8 | 4-1 | 4-9 | 5-5 | 7-2 | 11-7 | ......... | ......... |
| 2.9 | 4-2 | 4-10 | 5-6 | 7-4 | 12-1 | ......... | ......... |
| 3.0 | 4-3 | 4-11 | 5-8 | 7-6 | 12-6 | ......... | ......... |

NOTE 1: The distance specified in this table may be applied only where the maximum anticipated per-unit transient overvoltage has been determined by engineering analysis and has been supplied by the employer. Table R-6 applies otherwise.

NOTE 2: The distances specified in this table are the air, bare-hand, and live-line tool distances.

NOTE 3: See appendix B to this section for information on how the minimum approach distances listed in the tables were derived and on how to calculate revised minimum approach distances based on the control of transient overvoltages.

### TABLE R-9—DC LIVE-LINE WORK MINIMUM APPROACH DISTANCE WITH OVERVOLTAGE FACTOR

| | Distance in feet-inches | | | | |
|---|---|---|---|---|---|
| | Maximum line-to-ground voltage in kilovolts | | | | |
| Maximum anticipated per-unit transient overvoltage | 250 | 400 | 500 | 600 | 750 |
| 1.5 or lower | 3-8 | 5-3 | 6-9 | 8-7 | 11-10 |
| 1.6 | 3-10 | 5-7 | 7-4 | 9-5 | 13-1 |
| 1.7 | 4-1 | 6-0 | 7-11 | 10-3 | 14-4 |
| 1.8 | 4-3 | 6-5 | 8-7 | 11-2 | 15-9 |

NOTE 1: The distances specified in this table may be applied only where the maximum anticipated per-unit transient overvoltage has been determined by engineering analysis and has been supplied by the employer. However, if the transient overvoltage factor is not known, a factor of 1.8 shall be assumed.

NOTE 2: The distances specified in this table are the air, bare-hand, and live-line tool distances.

### TABLE R-10—ALTITUDE CORRECTION FACTOR

| Altitude | | | | Correction factor | |
|---|---|---|---|---|---|
| ft | ft | m | m | | |
| 3000 | 10000 | 900 | 3000 | 1.00 | 1.20 |
| 4000 | 12000 | 1200 | 3600 | 1.02 | 1.25 |
| 5000 | 14000 | 1500 | 4200 | 1.05 | 1.30 |
| 6000 | 16000 | 1800 | 4800 | 1.08 | 1.35 |
| 7000 | 18000 | 2100 | 5400 | 1.11 | 1.39 |
| 8000 | 20000 | 2400 | 6000 | 1.14 | 1.44 |
| 9000 | | 2700 | | 1.17 | |

NOTE: If the work is performed at elevations greater than 3000 ft (900 m) above mean sea level, the minimum approach distance shall be determined by multiplying the distances in Table R-6 through Table R-9 by the correction factor corresponding to the altitude at which work is performed.

**1910.269(m)** *Deenergizing lines and equipment for employee protection*—**(1)** *Application.* Paragraph (m) of this section applies to the deenergizing of transmission and distribution lines and equipment for the purpose of protecting employees. Control of hazardous energy sources used in the generation of electric energy is covered in paragraph (d) of this section. Conductors and parts of electric equipment that have been deenergized under procedures other than those required by paragraph (d) or (m) of this section, as applicable, shall be treated as energized.

**1910.269(m)(2)** *General.* **(i)** If a system operator is in charge of the lines or equipment and their means of disconnection, all of the requirements of paragraph (m)(3) of this section shall be observed, in the order given.

**1910.269(m)(2)(ii)** If no system operator is in charge of the lines or equipment and their means of disconnection, one employee in the crew shall be designated as being in charge of the clearance. All of the requirements of paragraph (m)(3) of this section apply, in the order given, except as provided in paragraph (m)(2)(iii) of this section. The employee in charge of the clearance shall take the place of the system operator, as necessary.

**1910.269(m)(2)(iii)** If only one crew will be working on the lines or equipment and if the means of disconnection is accessible and visible to and under the sole control of the employee in charge of the clearance, paragraphs (m)(3)(i), (m)(3)(iii), (m)(3)(iv), (m)(3)(viii), and (m)(3)(xii) of this section do not apply. Additionally, tags required by the remaining provisions of paragraph (m)(3) of this section need not be used.

**1910.269(m)(2)(iv)** Any disconnecting means that are accessible to persons outside the employer's control (for example, the general public) shall be rendered inoperable while they are open for the purpose of protecting employees.

**1910.269(m)(3)** *Deenergizing lines and equipment.* **(i)** A designated employee shall make a request of the system operator to have the particular section of line or equipment deenergized. The designated employee becomes the employee in charge (as this term is used in paragraph (m)(3) of this section) and is responsible for the clearance.

**1910.269(m)(3)(ii)** All switches, disconnectors, jumpers, taps, and other means through which known sources of electric energy may be supplied to the particular lines and equipment to be deenergized shall be opened. Such means shall be rendered inoperable, unless its design does not so permit, and tagged to indicate that employees are at work.

**1910.269(m)(3)(iii)** Automatically and remotely controlled switches that could cause the opened disconnecting means to close shall also be tagged at the point of control. The automatic or remote control feature shall be rendered inoperable, unless its design does not so permit.

**1910.269(m)(3)(iv)** Tags shall prohibit operation of the disconnecting means and shall indicate that employees are at work.

**1910.269(m)(3)(v)** After the applicable requirements in paragraphs (m)(3)(i) through (m)(3)(iv) of this section have been followed and the employee in charge of the work has been given a clearance by the system operator, the lines and equipment to be worked shall be tested to ensure that they are deenergized.

**1910.269(m)(3)(vi)** Protective grounds shall be installed as required by paragraph (n) of this section.

**1910.269(m)(3)(vii)** After the applicable requirements of paragraphs (m)(3)(i) through (m)(3)(vi) of this section have been followed, the lines and equipment involved may be worked as deenergized.

**1910.269(m)(3)(viii)** If two or more independent crews will be working on the same lines or equipment, each crew shall independently comply with the requirements in paragraph (m)(3) of this section.

**1910.269(m)(3)(ix)** To transfer the clearance, the employee in charge (or, if the employee in charge is forced to leave the worksite due to illness or other emergency, the employee's supervisor) shall inform the system operator; employees in the crew shall be informed of the transfer; and the new employee in charge shall be responsible for the clearance.

**1910.269(m)(3)(x)** To release a clearance, the employee in charge shall:

**1910.269(m)(3)(x)(A)** Notify employees under his or her direction that the clearance is to be released;

**1910.269(m)(3)(x)(B)** Determine that all employees in the crew are clear of the lines and equipment;

**1910.269(m)(3)(x)(C)** Determine that all protective grounds installed by the crew have been removed; and

**1910.269(m)(3)(x)(D)** Report this information to the system operator and release the clearance.

**1910.269(m)(3)(xi)** The person releasing a clearance shall be the same person that requested the clearance, unless responsibility has been transferred under paragraph (m)(3)(ix) of this section.

**1910.269(m)(3)(xii)** Tags may not be removed unless the associated clearance has been released under paragraph (m)(3)(x) of this section.

**1910.269(m)(3)(xiii)** Only after all protective grounds have been removed, after all crews working on the lines or equipment have released their clearances, after all employees are clear of the lines and equipment, and after all protective tags have been removed from a given point of disconnection, may action be initiated to reenergize the lines or equipment at that point of disconnection.

**1910.269(n)** *Grounding for the protection of employees*—**(1)** *Application.* Paragraph (n) of this section applies to the grounding of transmission and distribution lines and equipment for the purpose of protecting employees. Paragraph (n)(4) of this section also applies to the protective grounding of other equipment as required elsewhere in this section.

**1910.269(n)(2)** *General.* For the employee to work lines or equipment as deenergized, the lines or equipment shall be deenergized under the provisions of paragraph (m) of this section and shall be grounded as specified in paragraphs (n)(3) through (n)(9) of this section. However, if the employer can demonstrate that installation of a ground is impracticable or that the conditions resulting from the installation of a ground would present greater hazards than working without grounds, the lines and equipment may be treated as deenergized provided all of the following conditions are met:

**1910.269(n)(2)(i)** The lines and equipment have been deenergized under the provisions of paragraph (m) of this section.

**1910.269(n)(2)(ii)** There is no possibility of contact with another energized source.

**1910.269(n)(2)(iii)** The hazard of induced voltage is not present.

**1910.269(n)(3)** *Equipotential zone.* Temporary protective grounds shall be placed at such locations and arranged in such a manner as to prevent each employee from being exposed to hazardous differences in electrical potential.

**1910.269(n)(4)** *Protective grounding equipment.* **(i)** Protective grounding equipment shall be capable of conducting the maximum fault current that could flow at the point of grounding for the time necessary to clear the fault. This equipment shall have an ampacity greater than or equal to that of No. 2 AWG copper.

**NOTE:** Guidelines for protective grounding equipment are contained in American Society for Testing and Materials Standard Specifications for Temporary Grounding Systems to be Used on De-Energized Electric Power Lines and Equipment, ASTM F855-1990.

**1910.269(n)(4)(ii)** Protective grounds shall have an impedance low enough to cause immediate operation of protective devices in case of accidental energizing of the lines or equipment.

**1910.269(n)(5)** *Testing.* Before any ground is installed, lines and equipment shall be tested and found absent of nominal voltage, unless a previously installed ground is present.

**1910.269(n)(6)** *Order of connection.* When a ground is to be attached to a line or to equipment, the ground-end connection shall be attached first, and then the other end shall be attached by means of a live-line tool.

**1910.269(n)(7)** *Order of removal.* When a ground is to be removed, the grounding device shall be removed from the line or equipment using a live-line tool before the ground-end connection is removed.

**1910.269(n)(8)** *Additional precautions.* When work is performed on a cable at a location remote from the cable terminal, the cable may not be grounded at the cable terminal if there is a possibility of hazardous transfer of potential should a fault occur.

**1910.269(n)(9)** *Removal of grounds for test.* Grounds may be removed temporarily during tests. During the test procedure, the employer shall ensure that each employee uses insulating equipment and is isolated from any hazards involved, and the employer shall institute any additional measures as may be necessary to protect each exposed employee in case the previously grounded lines and equipment become energized.

**1910.269(o)** *Testing and test facilities*—**(1)** *Application.* Paragraph (o) of this section provides for safe work practices for high-voltage and high-power testing performed in laboratories, shops, and substations, and in the field and on electric transmission and distribution lines and equipment. It applies only to testing involving interim measurements utilizing high voltage, high power, or combinations of both, and not to testing involving continuous measurements as in routine metering, relaying, and normal line work.

**NOTE:** Routine inspection and maintenance measurements made by qualified employees are considered to be routine line work and are not included in the scope of paragraph (o) of this section, as long as the hazards related to the use of intrinsic high-voltage or high-power sources require only the normal precautions associated with routine operation and maintenance work required in the other paragraphs of this section. Two typical examples of such excluded test work procedures are "phasing-out" testing and testing for a "no-voltage" condition.

**1910.269(o)(2)** *General requirements.* **(i)** The employer shall establish and enforce work practices for the protection of each worker from the hazards of high-voltage or high-power testing at all test areas, temporary and permanent. Such work practices shall include, as a minimum, test area guarding, grounding, and the safe use of measuring and control circuits. A means providing for periodic safety checks of field test areas shall also be included. (See paragraph (o)(6) of this section.)

**1910.269(o)(2)(ii)** Employees shall be trained in safe work practices upon their initial assignment to the test area, with periodic reviews and updates provided as required by paragraph (a)(2) of this section.

**1910.269(o)(3)** *Guarding of test areas.* **(i)** Permanent test areas shall be guarded by walls, fences, or barriers designed to keep employees out of the test areas.

**1910.269(o)(3)(ii)** In field testing, or at a temporary test site where permanent fences and gates are not provided, one of the following means shall be used to prevent unauthorized employees from entering:

**1910.269(o)(3)(ii)(A)** The test area shall be guarded by the use of distinctively colored safety tape that is supported approximately waist high and to which safety signs are attached,

**1910.269(o)(3)(ii)(B)** The test area shall be guarded by a barrier or barricade that limits access to the test area to a degree equivalent, physically and visually, to the barricade specified in paragraph (o)(3)(ii)(A) of this section, or

**1910.269(o)(3)(ii)(C)** The test area shall be guarded by one or more test observers stationed so that the entire area can be monitored.

**1910.269(o)(3)(iii)** The barriers required by paragraph (o)(3)(ii) of this section shall be removed when the protection they provide is no longer needed.

**1910.269(o)(3)(iv)** Guarding shall be provided within test areas to control access to test equipment or to apparatus under test that may become energized as part of the testing by either direct or inductive coupling, in order to prevent accidental employee contact with energized parts.

**1910.269(o)(4)** *Grounding practices.* **(i)** The employer shall establish and implement safe grounding practices for the test facility.

**1910.269(o)(4)(i)(A)** All conductive parts accessible to the test operator during the time the equipment is operating at high voltage shall be maintained at ground potential except for portions of the equipment that are isolated from the test operator by guarding.

**1910.269(o)(4)(i)(B)** Wherever ungrounded terminals of test equipment or apparatus under test may be present, they shall be treated as energized until determined by tests to be deenergized.

**1910.269(o)(4)(ii)** Visible grounds shall be applied, either automatically or manually with properly insulated tools, to the high-voltage circuits after they are deenergized and before work is performed on the circuit or item or apparatus under test. Common ground connections shall be solidly connected to the test equipment and the apparatus under test.

**1910.269(o)(4)(iii)** In high-power testing, an isolated ground-return conductor system shall be provided so that no intentional passage of current, with its attendant voltage rise, can occur in the ground grid or in the earth. However, an isolated ground-return conductor need not be provided if the employer can demonstrate that both the following conditions are met:

**1910.269(o)(4)(iii)(A)** An isolated ground-return conductor cannot be provided due to the distance of the test site from the electric energy source, and

**1910.269(o)(4)(iii)(B)** Employees are protected from any hazardous step and touch potentials that may develop during the test.

**NOTE:** See appendix C to this section for information on measures that can be taken to protect employees from hazardous step and touch potentials.

**1910.269(o)(4)(iv)** In tests in which grounding of test equipment by means of the equipment grounding conductor located in the equipment power cord cannot be used due to increased hazards to test personnel or the prevention of satisfactory measurements, a ground that the employer can demonstrate affords equivalent safety shall be provided, and the safety ground shall be clearly indicated in the test set-up.

**1910.269(o)(4)(v)** When the test area is entered after equipment is deenergized, a ground shall be placed on the high-voltage terminal and any other exposed terminals.

**1910.269(o)(4)(v)(A)** High capacitance equipment or apparatus shall be discharged through a resistor rated for the available energy.

**1910.269(o)(4)(v)(B)** A direct ground shall be applied to the exposed terminals when the stored energy drops to a level at which it is safe to do so.

**1910.269(o)(4)(vi)** If a test trailer or test vehicle is used in field testing, its chassis shall be grounded. Protection against hazardous touch potentials with respect to the vehicle, instrument panels, and other conductive parts accessible to employees shall be provided by bonding, insulation, or isolation.

# OSHA Standards for General Industry

**1910.269(o)(5)** *Control and measuring circuits.* **(i)** Control wiring, meter connections, test leads and cables may not be run from a test area unless they are contained in a grounded metallic sheath and terminated in a grounded metallic enclosure or unless other precautions are taken that the employer can demonstrate as ensuring equivalent safety.

**1910.269(o)(5)(ii)** Meters and other instruments with accessible terminals or parts shall be isolated from test personnel to protect against hazards arising from such terminals and parts becoming energized during testing. If this isolation is provided by locating test equipment in metal compartments with viewing windows, interlocks shall be provided to interrupt the power supply if the compartment cover is opened.

**1910.269(o)(5)(iii)** The routing and connections of temporary wiring shall be made secure against damage, accidental interruptions and other hazards. To the maximum extent possible, signal, control, ground, and power cables shall be kept separate.

**1910.269(o)(5)(iv)** If employees will be present in the test area during testing, a test observer shall be present. The test observer shall be capable of implementing the immediate deenergizing of test circuits for safety purposes.

**1910.269(o)(6)** *Safety check.* **(i)** Safety practices governing employee work at temporary or field test areas shall provide for a routine check of such test areas for safety at the beginning of each series of tests.

**1910.269(o)(6)(ii)** The test operator in charge shall conduct these routine safety checks before each series of tests and shall verify at least the following conditions:

**1910.269(o)(6)(ii)(A)** That barriers and guards are in workable condition and are properly placed to isolate hazardous areas;

**1910.269(o)(6)(ii)(B)** That system test status signals, if used, are in operable condition;

**1910.269(o)(6)(ii)(C)** That test power disconnects are clearly marked and readily available in an emergency;

**1910.269(o)(6)(ii)(D)** That ground connections are clearly identifiable;

**1910.269(o)(6)(ii)(E)** That personal protective equipment is provided and used as required by subpart I of this part and by this section; and

**1910.269(o)(6)(ii)(F)** That signal, ground, and power cables are properly separated.

**1910.269(p)** *Mechanical equipment*—**(1)** *General requirements.* **(i)** The critical safety components of mechanical elevating and rotating equipment shall receive a thorough visual inspection before use on each shift.

**NOTE:** Critical safety components of mechanical elevating and rotating equipment are components whose failure would result in a free fall or free rotation of the boom.

**1910.269(p)(1)(ii)** No vehicular equipment having an obstructed view to the rear may be operated on off-highway jobsites where any employee is exposed to the hazards created by the moving vehicle, unless:

**1910.269(p)(1)(ii)(A)** The vehicle has a reverse signal alarm audible above the surrounding noise level, or

**1910.269(p)(1)(ii)(B)** The vehicle is backed up only when a designated employee signals that it is safe to do so.

**1910.269(p)(1)(iii)** The operator of an electric line truck may not leave his or her position at the controls while a load is suspended, unless the employer can demonstrate that no employee (including the operator) might be endangered.

**1910.269(p)(1)(iv)** Rubber-tired, self-propelled scrapers, rubber-tired front-end loaders, rubber-tired dozers, wheel-type agricultural and industrial tractors, crawler-type tractors, crawler-type loaders, and motor graders, with or without attachments, shall have roll-over protective structures that meet the requirements of subpart W of part 1926 of this chapter.

**1910.269(p)(2)** *Outriggers.* **(i)** Vehicular equipment, if provided with outriggers, shall be operated with the outriggers extended and firmly set as necessary for the stability of the specific configuration of the equipment. Outriggers may not be extended or retracted outside of clear view of the operator unless all employees are outside the range of possible equipment motion.

**1910.269(p)(2)(ii)** If the work area or the terrain precludes the use of outriggers, the equipment may be operated only within its maximum load ratings for the particular configuration of the equipment without outriggers.

**1910.269(p)(3)** *Applied loads.* Mechanical equipment used to lift or move lines or other material shall be used within its maximum load rating and other design limitations for the conditions under which the work is being performed.

**1910.269(p)(4)** *Operations near energized lines or equipment.* **(i)** Mechanical equipment shall be operated so that the minimum approach distances of Table R-6 through Table R-10 are maintained from exposed energized lines and equipment. However, the insulated portion of an aerial lift operated by a qualified employee in the lift is exempt from this requirement.

**1910.269(p)(4)(ii)** A designated employee other than the equipment operator shall observe the approach distance to exposed lines and equipment and give timely warnings before the minimum approach distance required by paragraph (p)(4)(i) is reached, unless the employer can demonstrate that the operator can accurately determine that the minimum approach distance is being maintained.

**1910.269(p)(4)(iii)** If, during operation of the mechanical equipment, the equipment could become energized, the operation shall also comply with at least one of paragraphs (p)(4)(iii)(A) through (p)(4)(iii)(C) of this section.

**1910.269(p)(4)(iii)(A)** The energized lines exposed to contact shall be covered with insulating protective material that will withstand the type of contact that might be made during the operation.

**1910.269(p)(4)(iii)(B)** The equipment shall be insulated for the voltage involved. The equipment shall be positioned so that its uninsulated portions cannot approach the lines or equipment any closer than the minimum approach distances specified in Table R-6 through Table R-10.

**1910.269(p)(4)(iii)(C)** Each employee shall be protected from hazards that might arise from equipment contact with the energized lines. The measures used shall ensure that employees will not be exposed to hazardous differences in potential. Unless the employer can demonstrate that the methods in use protect each employee from the hazards that might arise if the equipment contacts the energized line, the measures used shall include all of the following techniques:

**1910.269(p)(4)(iii)(C)(1)** Using the best available ground to minimize the time the lines remain energized,

**1910.269(p)(4)(iii)(C)(2)** Bonding equipment together to minimize potential differences,

**1910.269(p)(4)(iii)(C)(3)** Providing ground mats to extend areas of equipotential, and

**1910.269(p)(4)(iii)(C)(4)** Employing insulating protective equipment or barricades to guard against any remaining hazardous potential differences.

**NOTE:** Appendix C to this section contains information on hazardous step and touch potentials and on methods of protecting employees from hazards resulting from such potentials.

**1910.269(q)** *Overhead lines.* This paragraph provides additional requirements for work performed on or near overhead lines and equipment.

**1910.269(q)(1)** *General.* **(i)** Before elevated structures, such as poles or towers, are subjected to such stresses as climbing or the installation or removal of equipment may impose, the employer shall ascertain that the structures are capable of sustaining the additional or unbalanced stresses. If the pole or other structure cannot withstand the loads which will be imposed, it shall be braced or otherwise supported so as to prevent failure.

**NOTE:** Appendix D to this section contains test methods that can be used in ascertaining whether a wood pole is capable of sustaining the forces that would be imposed by an employee climbing the pole. This paragraph also requires the employer to ascertain that the pole can sustain all other forces that will be imposed by the work to be performed.

**1910.269(q)(1)(ii)** When the pole is set, moved, or removed near exposed energized overhead conductors, the pole may not contact the conductors.

**1910.269(q)(1)(iii)** When a pole is set, moved, or removed near an exposed energized overhead conductor, the employer shall ensure that each employee wears electrical protective equipment or uses insulated devices when handling the pole and that no employee contacts the pole with uninsulated parts of his or her body.

**1910.269(q)(1)(iv)** To protect employees from falling into holes into which poles are to be placed, the holes shall be attended by employees or physically guarded whenever anyone is working nearby.

**1910.269(q)(2)** *Installing and removing overhead lines.* The following provisions apply to the installation and removal of overhead conductors or cable.

**1910.269(q)(2)(i)** The employer shall use the tension stringing method, barriers, or other equivalent measures to minimize the possibility that conductors and cables being installed or removed will contact energized power lines or equipment.

**1910.269(q)(2)(ii)** The protective measures required by paragraph (p)(4)(iii) of this section for mechanical equipment shall also be provided for conductors, cables, and pulling and tensioning equipment when the conductor or cable is being installed or removed close enough to energized conductors that any of the following failures could energize the pulling or tensioning equipment or the wire or cable being installed or removed:

**1910.269(q)(2)(ii)(A)** Failure of the pulling or tensioning equipment,

**1910.269(q)(2)(ii)(B)** Failure of the wire or cable being pulled, or

**1910.269(q)(2)(ii)(C)** Failure of the previously installed lines or equipment.

**1910.269(q)(2)(iii)** If the conductors being installed or removed cross over energized conductors in excess of 600 volts and if the design of the circuit-interrupting devices protecting the lines so permits, the automatic-reclosing feature of these devices shall be made inoperative.

**1910.269(q)(2)(iv)** Before lines are installed parallel to existing energized lines, the employer shall make a determination of the approximate voltage to be induced in the new lines, or work shall proceed on the assumption that the induced voltage is hazardous. Unless the employer can demonstrate that the lines being installed are not subject to the induction of a hazardous voltage or unless the lines are treated as energized, the following requirements also apply:

**1910.269(q)(2)(iv)(A)** Each bare conductor shall be grounded in increments so that no point along the conductor is more than 2 miles (3.22 km) from a ground.

**1910.269(q)(2)(iv)(B)** The grounds required in paragraph (q)(2)(iv)(A) of this section shall be left in place until the conductor installation is completed between dead ends.

**1910.269(q)(2)(iv)(C)** The grounds required in paragraph (q)(2)(iv)(A) of this section shall be removed as the last phase of aerial cleanup.

**1910.269(q)(2)(iv)(D)** If employees are working on bare conductors, grounds shall also be installed at each location where these employees are working, and grounds shall be installed at all open dead-end or catch-off points or the next adjacent structure.

**1910.269(q)(2)(iv)(E)** If two bare conductors are to be spliced, the conductors shall be bonded and grounded before being spliced.

**1910.269(q)(2)(v)** Reel handling equipment, including pulling and tensioning devices, shall be in safe operating condition and shall be leveled and aligned.

**1910.269(q)(2)(vi)** Load ratings of stringing lines, pulling lines, conductor grips, load-bearing hardware and accessories, rigging, and hoists may not be exceeded.

**1910.269(q)(2)(vii)** Pulling lines and accessories shall be repaired or replaced when defective.

**1910.269(q)(2)(viii)** Conductor grips may not be used on wire rope, unless the grip is specifically designed for this application.

**1910.269(q)(2)(ix)** Reliable communications, through two-way radios or other equivalent means, shall be maintained between the reel tender and the pulling rig operator.

**1910.269(q)(2)(x)** The pulling rig may only be operated when it is safe to do so.

**NOTE:** Examples of unsafe conditions include employees in locations prohibited by paragraph (q)(2)(xi) of this section, conductor and pulling line hang-ups, and slipping of the conductor grip.

**1910.269(q)(2)(xi)** While the conductor or pulling line is being pulled (in motion) with a power-driven device, employees are not permitted directly under overhead operations or on the cross arm, except as necessary to guide the stringing sock or board over or through the stringing sheave.

**1910.269(q)(3)** *Live-line bare-hand work.* In addition to other applicable provisions contained in this section, the following requirements apply to live-line bare-hand work:

**1910.269(q)(3)(i)** Before using or supervising the use of the live-line bare-hand technique on energized circuits, employees shall be trained in the technique and in the safety requirements of paragraph (q)(3) of this section. Employees shall receive refresher training as required by paragraph (a)(2) of this section.

**1910.269(q)(3)(ii)** Before any employee uses the live-line bare-hand technique on energized high-voltage conductors or parts, the following information shall be ascertained:

**1910.269(q)(3)(ii)(A)** The nominal voltage rating of the circuit on which the work is to be performed,

**1910.269(q)(3)(ii)(B)** The minimum approach distances to ground of lines and other energized parts on which work is to be performed, and

**1910.269(q)(3)(ii)(C)** The voltage limitations of equipment to be used.

**1910.269(q)(3)(iii)** The insulated equipment, insulated tools, and aerial devices and platforms used shall be designed, tested, and intended for live-line bare-hand work. Tools and equipment shall be kept clean and dry while they are in use.

**1910.269(q)(3)(iv)** The automatic-reclosing feature of circuit-interrupting devices protecting the lines shall be made inoperative, if the design of the devices permits.

**1910.269(q)(3)(v)** Work may not be performed when adverse weather conditions would make the work hazardous even after the work practices required by this section are employed. Additionally, work may not be performed when winds reduce the phase-to-phase or phase-to-ground minimum approach distances at the work location below that specified in paragraph (q)(3)(xiii) of this section, unless the grounded objects and other lines and equipment are covered by insulating guards.

**NOTE:** Thunderstorms in the immediate vicinity, high winds, snow storms, and ice storms are examples of adverse weather conditions that are presumed to make live-line bare-hand work too hazardous to perform safely.

**1910.269(q)(3)(vi)** A conductive bucket liner or other conductive device shall be provided for bonding the insulated aerial device to the energized line or equipment.

**1910.269(q)(3)(vi)(A)** The employee shall be connected to the bucket liner or other conductive device by the use of conductive shoes, leg clips, or other means.

**1910.269(q)(3)(vi)(B)** Where differences in potentials at the worksite pose a hazard to employees, electrostatic shielding designed for the voltage being worked shall be provided.

**1910.269(q)(3)(vii)** Before the employee contacts the energized part, the conductive bucket liner or other conductive device shall be bonded to the energized conductor by means of a positive connection. This connection shall remain attached to the energized conductor until the work on the energized circuit is completed.

**1910.269(q)(3)(viii)** Aerial lifts to be used for live-line bare-hand work shall have dual controls (lower and upper) as follows:

**1910.269(q)(3)(viii)(A)** The upper controls shall be within easy reach of the employee in the bucket. On a two-bucket-type lift, access to the controls shall be within easy reach from either bucket.

**1910.269(q)(3)(viii)(B)** The lower set of controls shall be located near the base of the boom, and they shall be so designed that they can override operation of the equipment at any time.

**1910.269(q)(3)(ix)** Lower (ground-level) lift controls may not be operated with an employee in the lift, except in case of emergency.

**1910.269(q)(3)(x)** Before employees are elevated into the work position, all controls (ground level and bucket) shall be checked to determine that they are in proper working condition.

**1910.269(q)(3)(xi)** Before the boom of an aerial lift is elevated, the body of the truck shall be grounded, or the body of the truck shall be barricaded and treated as energized.

**1910.269(q)(3)(xii)** A boom-current test shall be made before work is started each day, each time during the day when higher voltage is encountered, and when changed conditions indicate a need for an additional test. This test shall consist of placing the bucket in contact with an energized source equal to the voltage to be encountered for a minimum of 3 minutes. The leakage current may not exceed 1 microampere per kilovolt of nominal phase-to-ground voltage. Work from the aerial lift shall be immediately suspended upon indication of a malfunction in the equipment.

**1910.269(q)(3)(xiii)** The minimum approach distances specified in Table R-6 through Table R-10 shall be maintained from all grounded objects and from lines and equipment at a potential different from that to which the live-line bare-hand equipment is bonded, unless such grounded objects and other lines and equipment are covered by insulating guards.

**1910.269(q)(3)(xiv)** While an employee is approaching, leaving, or bonding to an energized circuit, the minimum approach distances in Table R-6 through Table R-10 shall be maintained between the employee and any grounded parts, including the lower boom and portions of the truck.

**1910.269(q)(3)(xv)** While the bucket is positioned alongside an energized bushing or insulator string, the phase-to-ground minimum approach distances of Table R-6 through Table R-10 shall be maintained between all parts of the bucket and the grounded end of the bushing or insulator string or any other grounded surface.

**1910.269(q)(3)(xvi)** Hand lines may not be used between the bucket and the boom or between the bucket and the ground. However, non-conductive-type hand lines may be used from conductor to ground if not supported from the bucket. Ropes used for live-line bare-hand work may not be used for other purposes.

**1910.269(q)(3)(xvii)** Uninsulated equipment or material may not be passed between a pole or structure and an aerial lift while an employee working from the bucket is bonded to an energized part.

**1910.269(q)(3)(xviii)** A minimum approach distance table reflecting the minimum approach distances listed in Table R-6 through Table R-10 shall be printed on a plate of durable non-conductive material. This table shall be mounted so as to be visible to the operator of the boom.

**1910.269(q)(3)(xix)** A non-conductive measuring device shall be readily accessible to assist employees in maintaining the required minimum approach distance.

**1910.269(q)(4)** *Towers and structures.* The following requirements apply to work performed on towers or other structures which support overhead lines.

**1910.269(q)(4)(i)** The employer shall ensure that no employee is under a tower or structure while work is in progress, except where the employer can demonstrate that such a working position is necessary to assist employees working above.

**1910.269(q)(4)(ii)** Tag lines or other similar devices shall be used to maintain control of tower sections being raised or positioned, unless the employer can demonstrate that the use of such devices would create a greater hazard.

**1910.269(q)(4)(iii)** The loadline may not be detached from a member or section until the load is safely secured.

**1910.269(q)(4)(iv)** Except during emergency restoration procedures, work shall be discontinued when adverse weather conditions would make the work hazardous in spite of the work practices required by this section.

**NOTE:** Thunderstorms in the immediate vicinity, high winds, snow storms, and ice storms are examples of adverse weather conditions that are presumed to make this work too hazardous to perform, except under emergency conditions.

**1910.269(r)** *Line-clearance tree trimming operations.* This paragraph provides additional requirements for line-clearance tree-trimming operations and for equipment used in these operations.

**1910.269(r)(1)** *Electrical hazards.* This paragraph does not apply to qualified employees.

**1910.269(r)(1)(i)** Before an employee climbs, enters, or works around any tree, a determination shall be made of the nominal voltage of electric power lines posing a hazard to employees. However, a determination of the maximum nominal voltage to which an employee will be exposed may be made instead, if all lines are considered as energized at this maximum voltage.

**1910.269(r)(1)(ii)** There shall be a second line-clearance tree trimmer within normal (that is, unassisted) voice communication under any of the following conditions:

**1910.269(r)(1)(ii)(A)** If a line-clearance tree trimmer is to approach more closely than 10 feet (305 cm) any conductor or electric apparatus energized at more than 750 volts or

**1910.269(r)(1)(ii)(B)** If branches or limbs being removed are closer to lines energized at more than 750 volts than the distances listed in Table R-6, Table R-9, and Table R-10 or

**1910.269(r)(1)(ii)(C)** If roping is necessary to remove branches or limbs from such conductors or apparatus.

**1910.269(r)(1)(iii)** Line-clearance tree trimmers shall maintain the minimum approach distances from energized conductors given in Table R-6, Table R-9, and Table R-10.

**1910.269(r)(1)(iv)** Branches that are contacting exposed energized conductors or equipment or that are within the distances specified in Table R-6, Table R-9, and Table R-10 may be removed only through the use of insulating equipment.

**NOTE:** A tool constructed of a material that the employer can demonstrate has insulating qualities meeting paragraph (j)(1) of this section is considered as insulated under this paragraph if the tool is clean and dry.

**1910.269(r)(1)(v)** Ladders, platforms, and aerial devices may not be brought closer to an energized part than the distances listed in Table R-6, Table R-9, and Table R-10.

**1910.269(r)(1)(vi)** Line-clearance tree-trimming work may not be performed when adverse weather conditions make the work hazardous in spite of the work practices required by this section. Each employee performing line-clearance tree trimming work in the aftermath of a storm or under similar emergency conditions shall be trained in the special hazards related to this type of work.

**NOTE:** Thunderstorms in the immediate vicinity, high winds, snow storms, and ice storms are examples of adverse weather conditions that are presumed to make line-clearance tree trimming work too hazardous to perform safely.

**1910.269(r)(2)** *Brush chippers.* **(i)** Brush chippers shall be equipped with a locking device in the ignition system.

**1910.269(r)(2)(ii)** Access panels for maintenance and adjustment of the chipper blades and associated drive train shall be in place and secure during operation of the equipment.

**1910.269(r)(2)(iii)** Brush chippers not equipped with a mechanical infeed system shall be equipped with an infeed hopper of length sufficient to prevent employees from contacting the blades or knives of the machine during operation.

**1910.269(r)(2)(iv)** Trailer chippers detached from trucks shall be chocked or otherwise secured.

**1910.269(r)(2)(v)** Each employee in the immediate area of an operating chipper feed table shall wear personal protective equipment as required by subpart I of this part.

**1910.269(r)(3)** *Sprayers and related equipment.* **(i)** Walking and working surfaces of sprayers and related equipment shall be covered with slip-resistant material. If slipping hazards cannot be eliminated, slip-resistant footwear or handrails and stair rails meeting the requirements of subpart D may be used instead of slip-resistant material.

**1910.269(r)(3)(ii)** Equipment on which employees stand to spray while the vehicle is in motion shall be equipped with guardrails around the working area. The guardrail shall be constructed in accordance with subpart D of this part.

**1910.269(r)(4)** *Stump cutters.* **(i)** Stump cutters shall be equipped with enclosures or guards to protect employees.

**1910.269(r)(4)(ii)** Each employee in the immediate area of stump grinding operations (including the stump cutter operator) shall wear personal protective equipment as required by subpart I of this part.

**1910.269(r)(5)** *Gasoline-engine power saws.* Gasoline-engine power saw operations shall meet the requirements of §1910.266(e) and the following:

**1910.269(r)(5)(i)** Each power saw weighing more than 15 pounds (6.8 kilograms, service weight) that is used in trees shall be supported by a separate line, except when work is performed from an aerial lift and except during topping or removing operations where no supporting limb will be available.

**1910.269(r)(5)(ii)** Each power saw shall be equipped with a control that will return the saw to idling speed when released.

**1910.269(r)(5)(iii)** Each power saw shall be equipped with a clutch and shall be so adjusted that the clutch will not engage the chain drive at idling speed.

**1910.269(r)(5)(iv)** A power saw shall be started on the ground or where it is otherwise firmly supported. Drop starting of saws over 15 pounds (6.8

kg) is permitted outside of the bucket of an aerial lift only if the area below the lift is clear of personnel.

**1910.269(r)(5)(v)** A power saw engine may be started and operated only when all employees other than the operator are clear of the saw.

**1910.269(r)(5)(vi)** A power saw may not be running when the saw is being carried up into a tree by an employee.

**1910.269(r)(5)(vii)** Power saw engines shall be stopped for all cleaning, refueling, adjustments, and repairs to the saw or motor, except as the manufacturer's servicing procedures require otherwise.

**1910.269(r)(6)** *Backpack power units for use in pruning and clearing.* **(i)** While a backpack power unit is running, no one other than the operator may be within 10 feet (305 cm) of the cutting head of a brush saw.

**1910.269(r)(6)(ii)** A backpack power unit shall be equipped with a quick shutoff switch readily accessible to the operator.

**1910.269(r)(6)(iii)** Backpack power unit engines shall be stopped for all cleaning, refueling, adjustments, and repairs to the saw or motor, except as the manufacturer's servicing procedures require otherwise.

**1910.269(r)(7)** *Rope.* (i) Climbing ropes shall be used by employees working aloft in trees. These ropes shall have a minimum diameter of 0.5 inch (1.2 cm) with a minimum breaking strength of 2300 pounds (10.2 kN). Synthetic rope shall have elasticity of not more than 7 percent.

**1910.269(r)(7)(ii)** Rope shall be inspected before each use and, if unsafe (for example, because of damage or defect), may not be used.

**1910.269(r)(7)(iii)** Rope shall be stored away from cutting edges and sharp tools. Rope contact with corrosive chemicals, gas, and oil shall be avoided.

**1910.269(r)(7)(iv)** When stored, rope shall be coiled and piled, or shall be suspended, so that air can circulate through the coils.

**1910.269(r)(7)(v)** Rope ends shall be secured to prevent their unraveling.

**1910.269(r)(7)(vi)** Climbing rope may not be spliced to effect repair.

**1910.269(r)(7)(vii)** A rope that is wet, that is contaminated to the extent that its insulating capacity is impaired, or that is otherwise not considered to be insulated for the voltage involved may not be used near exposed energized lines.

**1910.269(r)(8)** *Fall protection.* Each employee shall be tied in with a climbing rope and safety saddle when the employee is working above the ground in a tree, unless he or she is ascending into the tree.

**1910.269(s)** *Communication facilities*—**(1)** *Microwave transmission.* **(i)** The employer shall ensure that no employee looks into an open waveguide or antenna that is connected to an energized microwave source.

**1910.269(s)(1)(ii)** If the electromagnetic radiation level within an accessible area associated with microwave communications systems exceeds the radiation protection guide given in §1910.97(a)(2) of this part, the area shall be posted with the warning symbol described in §1910.97(a)(3) of this part. The lower half of the warning symbol shall include the following statements or ones that the employer can demonstrate are equivalent:

Radiation in this area may exceed hazard limitations and special precautions are required. Obtain specific instruction before entering.

**1910.269(s)(1)(iii)** When an employee works in an area where the electromagnetic radiation could exceed the radiation protection guide, the employer shall institute measures that ensure that the employee's exposure is not greater than that permitted by that guide. Such measures may include administrative and engineering controls and personal protective equipment.

**1910.269(s)(2)** *Power line carrier.* Power line carrier work, including work on equipment used for coupling carrier current to power line conductors, shall be performed in accordance with the requirements of this section pertaining to work on energized lines.

**1910.269(t)** *Underground electrical installations.* This paragraph provides additional requirements for work on underground electrical installations.

**1910.269(t)(1)** *Access.* A ladder or other climbing device shall be used to enter and exit a manhole or subsurface vault exceeding 4 feet (122 cm) in depth. No employee may climb into or out of a manhole or vault by stepping on cables or hangers.

**1910.269(t)(2)** *Lowering equipment into manholes.* Equipment used to lower materials and tools into manholes or vaults shall be capable of supporting the weight to be lowered and shall be checked for defects before use. Before tools or material are lowered into the opening for a manhole or vault, each employee working in the manhole or vault shall be clear of the area directly under the opening.

**1910.269(t)(3)** *Attendants for manholes.* **(i)** While work is being performed in a manhole containing energized electric equipment, an employee with first aid and CPR training meeting paragraph (b)(1) of this section shall be available on the surface in the immediate vicinity to render emergency assistance.

**1910.269(t)(3)(ii)** Occasionally, the employee on the surface may briefly enter a manhole to provide assistance, other than emergency.

**NOTE 1:** An attendant may also be required under paragraph (e)(7) of this section. One person may serve to fulfill both requirements. However, attendants required under paragraph (e)(7) are not permitted to enter the manhole.

**NOTE 2:** Employees entering manholes containing unguarded, uninsulated energized lines or parts of electric equipment operating at 50 volts or more are required to be qualified under paragraph (l)(1) of this section.

**1910.269(t)(3)(iii)** For the purpose of inspection, housekeeping, taking readings, or similar work, an employee working alone may enter, for brief periods of time, a manhole where energized cables or equipment are in service, if the employer can demonstrate that the employee will be protected from all electrical hazards.

**1910.269(t)(3)(iv)** Reliable communications, through two-way radios or other equivalent means, shall be maintained among all employees involved in the job.

**1910.269(t)(4)** *Duct rods.* If duct rods are used, they shall be installed in the direction presenting the least hazard to employees. An employee shall be stationed at the far end of the duct line being rodded to ensure that the required minimum approach distances are maintained.

**1910.269(t)(5)** *Multiple cables.* When multiple cables are present in a work area, the cable to be worked shall be identified by electrical means, unless its identity is obvious by reason of distinctive appearance or location or by other readily apparent means of identification. Cables other than the one being worked shall be protected from damage.

**1910.269(t)(6)** *Moving cables.* Energized cables that are to be moved shall be inspected for defects.

**1910.269(t)(7)** *Defective cables.* Where a cable in a manhole has one or more abnormalities that could lead to or be an indication of an impending fault, the defective cable shall be deenergized before any employee may work in the manhole, except when service load conditions and a lack of feasible alternatives require that the cable remain energized. In that case, employees may enter the manhole provided they are protected from the possible effects of a failure by shields or other devices that are capable of containing the adverse effects of a fault in the joint.

**NOTE:** Abnormalities such as oil or compound leaking from cable or joints, broken cable sheaths or joint sleeves, hot localized surface temperatures of cables or joints, or joints that are swollen beyond normal tolerance are presumed to lead to or be an indication of an impending fault.

**1910.269(t)(8)** *Sheath continuity.* When work is performed on buried cable or on cable in manholes, metallic sheath continuity shall be maintained or the cable sheath shall be treated as energized.

**1910.269(u)** *Substations.* This paragraph provides additional requirements for substations and for work performed in them.

**1910.269(u)(1)** *Access and working space.* Sufficient access and working space shall be provided and maintained about electric equipment to permit ready and safe operation and maintenance of such equipment.

**NOTE:** Guidelines for the dimensions of access and working space about electric equipment in substations are contained in American National Standard—National Electrical Safety Code, ANSI C2-1987. Installations meeting the ANSI provisions comply with paragraph (u)(1) of this section. An installation that does not conform to this ANSI standard will, nonetheless, be considered as complying with paragraph (u)(1) of this section if the employer can demonstrate that the installation provides ready and safe access based on the following evidence:

(1) That the installation conforms to the edition of ANSI C2 that was in effect at the time the installation was made,

(2) That the configuration of the installation enables employees to maintain the minimum approach distances required by paragraph (l)(2) of this section while they are working on exposed, energized parts, and

(3) That the precautions taken when work is performed on the installation provide protection equivalent to the protection that would be provided by access and working space meeting ANSI C2-1987.

**1910.269(u)(2)** *Draw-out-type circuit breakers.* When draw-out-type circuit breakers are removed or inserted, the breaker shall be in the open position. The control circuit shall also be rendered inoperative, if the design of the equipment permits.

**1910.269(u)(3)** *Substation fences.* Conductive fences around substations shall be grounded. When a substation fence is expanded or a section is removed, fence grounding continuity shall be maintained, and bonding shall be used to prevent electrical discontinuity.

**1910.269(u)(4)** *Guarding of rooms containing electric supply equipment.* **(i)** Rooms and spaces in which electric supply lines or equipment are installed shall meet the requirements of paragraphs (u)(4)(ii) through (u)(4)(v) of this section under the following conditions:

**1910.269(u)(4)(i)(A)** If exposed live parts operating at 50 to 150 volts to ground are located within 8 feet of the ground or other working surface inside the room or space,

**1910.269(u)(4)(i)(B)** If live parts operating at 151 to 600 volts and located within 8 feet of the ground or other working surface inside the room or space are guarded only by location, as permitted under paragraph (u)(5)(i) of this section, or

**1910.269(u)(4)(i)(C)** If live parts operating at more than 600 volts are located within the room or space, unless:

**1910.269(u)(4)(i)(C)***(1)* The live parts are enclosed within grounded, metal-enclosed equipment whose only openings are designed so that foreign objects inserted in these openings will be deflected from energized parts, or

**1910.269(u)(4)(i)(C)***(2)* The live parts are installed at a height above ground and any other working surface that provides protection at the voltage to which they are energized corresponding to the protection provided by an 8-foot height at 50 volts.

**1910.269(u)(4)(i)(C)***(2)(ii)* The rooms and spaces shall be so enclosed within fences, screens, partitions, or walls as to minimize the possibility that unqualified persons will enter.

**1910.269(u)(4)(i)(C)***(2)(iii)* Signs warning unqualified persons to keep out shall be displayed at entrances to the rooms and spaces.

**1910.269(u)(4)(i)(C)***(2)(iv)* Entrances to rooms and spaces that are not under the observation of an attendant shall be kept locked.

**1910.269(u)(4)(i)(C)***(2)(v)* Unqualified persons may not enter the rooms or spaces while the electric supply lines or equipment are energized.

**1910.269(u)(5)** *Guarding of energized parts.* **(i)** Guards shall be provided around all live parts operating at more than 150 volts to ground without an insulating covering, unless the location of the live parts gives sufficient horizontal or vertical or a combination of these clearances to minimize the possibility of accidental employee contact.

**NOTE:** Guidelines for the dimensions of clearance distances about electric equipment in substations are contained in American National Standard—National Electrical Safety Code, ANSI C2-1987. Installations meeting the ANSI provisions comply with paragraph (u)(5)(i) of this section. An installation that does not conform to this ANSI standard will, nonetheless, be considered as complying with paragraph (u)(5)(i) of this section if the employer can demonstrate that the installation provides sufficient clearance based on the following evidence:

(1) That the installation conforms to the edition of ANSI C2 that was in effect at the time the installation was made,

(2) That each employee is isolated from energized parts at the point of closest approach, and

(3) That the precautions taken when work is performed on the installation provide protection equivalent to the protection that would be provided by horizontal and vertical clearances meeting ANSI C2-1987.

**1910.269(u)(5)(ii)** Except for fuse replacement and other necessary access by qualified persons, the guarding of energized parts within a compartment shall be maintained during operation and maintenance functions to prevent accidental contact with energized parts and to prevent tools or other equipment from being dropped on energized parts.

**1910.269(u)(5)(iii)** When guards are removed from energized equipment, barriers shall be installed around the work area to prevent employees who are not working on the equipment, but who are in the area, from contacting the exposed live parts.

**1910.269(u)(6)** *Substation entry.* **(i)** Upon entering an attended substation, each employee other than those regularly working in the station shall report his or her presence to the employee in charge in order to receive information on special system conditions affecting employee safety.

**1910.269(u)(6)(ii)** The job briefing required by paragraph (c) of this section shall cover such additional subjects as the location of energized equipment in or adjacent to the work area and the limits of any deenergized work area.

**1910.269(v)** **Power generation.** This paragraph provides additional requirements and related work practices for power generating plants.

**1910.269(v)(1)** *Interlocks and other safety devices.* **(i)** Interlocks and other safety devices shall be maintained in a safe, operable condition.

**1910.269(v)(1)(ii)** No interlock or other safety device may be modified to defeat its function, except for test, repair, or adjustment of the device.

**1910.269(v)(2)** *Changing brushes.* Before exciter or generator brushes are changed while the generator is in service, the exciter or generator field shall be checked to determine whether a ground condition exists. The brushes may not be changed while the generator is energized if a ground condition exists.

**1910.269(v)(3)** *Access and working space.* Sufficient access and working space shall be provided and maintained about electric equipment to permit ready and safe operation and maintenance of such equipment.

**NOTE:** Guidelines for the dimensions of access and working space about electric equipment in generating stations are contained in American National Standard—National Electrical Safety Code, ANSI C2-1987. Installations meeting the ANSI provisions comply with paragraph (v)(3) of this section. An installation that does not conform to this ANSI standard will, nonetheless, be considered as complying with paragraph (v)(3) of this section if the employer can demonstrate that the installation provides ready and safe access based on the following evidence:

(1) That the installation conforms to the edition of ANSI C2 that was in effect at the time the installation was made,

(2) That the configuration of the installation enables employees to maintain the minimum approach distances required by paragraph (l)(2) of this section while they are working on exposed, energized parts, and

(3) That the precautions taken when are working is performed on the installation provide protection equivalent to the protection that would be provided by access and working space meeting ANSI C2-1987.

**1910.269(v)(4)** *Guarding of rooms containing electric supply equipment.* **(i)** Rooms and spaces in which electric supply lines or equipment are installed shall meet the requirements of paragraphs (v)(4)(ii) through (v)(4)(v) of this section under the following conditions:

**1910.269(v)(4)(i)(A)** If exposed live parts operating at 50 to 150 volts to ground are located within 8 feet of the ground or other working surface inside the room or space,

**1910.269(v)(4)(i)(B)** If live parts operating at 151 to 600 volts and located within 8 feet of the ground or other working surface inside the room or space are guarded only by location, as permitted under paragraph (v)(5)(i) of this section, or

**1910.269(v)(4)(i)(C)** If live parts operating at more than 600 volts are located within the room or space, unless:

**1910.269(v)(4)(i)(C)***(1)* The live parts are enclosed within grounded, metal-enclosed equipment whose only openings are designed so that foreign objects inserted in these openings will be deflected from energized parts, or

**1910.269(v)(4)(i)(C)***(2)* The live parts are installed at a height above ground and any other working surface that provides protection at the voltage to which they are energized corresponding to the protection provided by an 8-foot height at 50 volts.

**1910.269(v)(4)(ii)** The rooms and spaces shall be so enclosed within fences, screens, partitions, or walls as to minimize the possibility that unqualified persons will enter.

**1910.269(v)(4)(iii)** Signs warning unqualified persons to keep out shall be displayed at entrances to the rooms and spaces.

**1910.269(v)(4)(iv)** Entrances to rooms and spaces that are not under the observation of an attendant shall be kept locked.

**1910.269(v)(4)(v)** Unqualified persons may not enter the rooms or spaces while the electric supply lines or equipment are energized.

**1910.269(v)(5)** *Guarding of energized parts.* **(i)** Guards shall be provided around all live parts operating at more than 150 volts to ground without an insulating covering, unless the location of the live parts gives sufficient horizontal or vertical or a combination of these clearances to minimize the possibility of accidental employee contact.

NOTE: Guidelines for the dimensions of clearance distances about electric equipment in generating stations are contained in American National Standard—National Electrical Safety Code, ANSI C2-1987. Installations meeting the ANSI provisions comply with paragraph (v)(5)(i) of this section. An installation that does not conform to this ANSI standard will, nonetheless, be considered as complying with paragraph (v)(5)(i) of this section if the employer can demonstrate that the installation provides sufficient clearance based on the following evidence:

(1) That the installation conforms to the edition of ANSI C2 that was in effect at the time the installation was made,

(2) That each employee is isolated from energized parts at the point of closest approach, and

(3) That the precautions taken when work is performed on the installation provide protection equivalent to the protection that would be provided by horizontal and vertical clearances meeting ANSI C2-1987.

**1910.269(v)(5)(ii)** Except for fuse replacement or other necessary access by qualified persons, the guarding of energized parts within a compartment shall be maintained during operation and maintenance functions to prevent accidental contact with energized parts and to prevent tools or other equipment from being dropped on energized parts.

**1910.269(v)(5)(iii)** When guards are removed from energized equipment, barriers shall be installed around the work area to prevent employees who are not working on the equipment, but who are in the area, from contacting the exposed live parts.

**1910.269(v)(6)** *Water or steam spaces.* The following requirements apply to work in water and steam spaces associated with boilers:

**1910.269(v)(6)(i)** A designated employee shall inspect conditions before work is permitted and after its completion. Eye protection, or full face protection if necessary, shall be worn at all times when condenser, heater, or boiler tubes are being cleaned.

**1910.269(v)(6)(ii)** Where it is necessary for employees to work near tube ends during cleaning, shielding shall be installed at the tube ends.

**1910.269(v)(7)** *Chemical cleaning of boilers and pressure vessels.* The following requirements apply to chemical cleaning of boilers and pressure vessels:

**1910.269(v)(7)(i)** Areas where chemical cleaning is in progress shall be cordoned off to restrict access during cleaning. If flammable liquids, gases, or vapors or combustible materials will be used or might be produced during the cleaning process, the following requirements also apply:

**1910.269(v)(7)(i)(A)** The area shall be posted with signs restricting entry and warning of the hazards of fire and explosion; and

**1910.269(v)(7)(i)(B)** Smoking, welding, and other possible ignition sources are prohibited in these restricted areas.

**1910.269(v)(7)(ii)** The number of personnel in the restricted area shall be limited to those necessary to accomplish the task safely.

**1910.269(v)(7)(iii)** There shall be ready access to water or showers for emergency use.

NOTE: See §1910.141 of this part for requirements that apply to the water supply and to washing facilities.

**1910.269(v)(7)(iv)** Employees in restricted areas shall wear protective equipment meeting the requirements of subpart I of this part and including, but not limited to, protective clothing, boots, goggles, and gloves.

**1910.269(v)(8)** *Chlorine systems.* **(i)** Chlorine system enclosures shall be posted with signs restricting entry and warning of the hazard to health and the hazards of fire and explosion.

NOTE: See subpart Z of this part for requirements necessary to protect the health of employees from the effects of chlorine.

**1910.269(v)(8)(ii)** Only designated employees may enter the restricted area. Additionally, the number of personnel shall be limited to those necessary to accomplish the task safely.

**1910.269(v)(8)(iii)** Emergency repair kits shall be available near the shelter or enclosure to allow for the prompt repair of leaks in chlorine lines, equipment, or containers.

**1910.269(v)(8)(iv)** Before repair procedures are started, chlorine tanks, pipes, and equipment shall be purged with dry air and isolated from other sources of chlorine.

**1910.269(v)(8)(v)** The employer shall ensure that chlorine is not mixed with materials that would react with the chlorine in a dangerously exothermic or other hazardous manner.

**1910.269(v)(9)** *Boilers.* **(i)** Before internal furnace or ash hopper repair work is started, overhead areas shall be inspected for possible falling objects. If the hazard of falling objects exists, overhead protection such as planking or nets shall be provided.

**1910.269(v)(9)(ii)** When opening an operating boiler door, employees shall stand clear of the opening of the door to avoid the heat blast and gases which may escape from the boiler.

**1910.269(v)(10)** *Turbine generators.* **(i)** Smoking and other ignition sources are prohibited near hydrogen or hydrogen sealing systems, and signs warning of the danger of explosion and fire shall be posted.

**1910.269(v)(10)(ii)** Excessive hydrogen makeup or abnormal loss of pressure shall be considered as an emergency and shall be corrected immediately.

**1910.269(v)(10)(iii)** A sufficient quantity of inert gas shall be available to purge the hydrogen from the largest generator.

**1910.269(v)(11)** *Coal and ash handling.* **(i)** Only designated persons may operate railroad equipment.

**1910.269(v)(11)(ii)** Before a locomotive or locomotive crane is moved, a warning shall be given to employees in the area.

**1910.269(v)(11)(iii)** Employees engaged in switching or dumping cars may not use their feet to line up drawheads.

**1910.269(v)(11)(iv)** Drawheads and knuckles may not be shifted while locomotives or cars are in motion.

**1910.269(v)(11)(v)** When a railroad car is stopped for unloading, the car shall be secured from displacement that could endanger employees.

**1910.269(v)(11)(vi)** An emergency means of stopping dump operations shall be provided at railcar dumps.

**1910.269(v)(11)(vii)** The employer shall ensure that employees who work in coal- or ash-handling conveyor areas are trained and knowledgeable in conveyor operation and in the requirements of paragraphs (v)(11)(viii) through (v)(11)(xii) of this section.

**1910.269(v)(11)(viii)** Employees may not ride a coal- or ash-handling conveyor belt at any time. Employees may not cross over the conveyor belt, except at walkways, unless the conveyor's energy source has been deenergized and has been locked out or tagged in accordance with paragraph (d) of this section.

**1910.269(v)(11)(ix)** What could cause injury when started may not be started until personnel in the area are alerted by a signal or by a designated person that the conveyor is about to start.

**1910.269(v)(11)(x)** If a conveyor that could cause injury when started is automatically controlled or is controlled from a remote location, an audible device shall be provided that sounds an alarm that will be recognized by each employee as a warning that the conveyor will start and that can be clearly heard at all points along the conveyor where personnel may be present. The warning device shall be actuated by the device starting the conveyor and shall continue for a period of time before the conveyor starts that is long enough to allow employees to move clear of the conveyor system. A visual warning may be used in place of the audible device if the employer can demonstrate that it will provide an equally effective warning in the particular circumstances involved.

*Exception:* If the employer can demonstrate that the system's function would be seriously hindered by the required time delay, warning signs may be provided in place of the audible warning device. If the system was installed before January 31, 1995, warning signs may be provided in place of the audible warning device until such time as the conveyor or its control system is rebuilt or rewired. These warning signs shall be clear, concise, and legible and shall indicate that conveyors and allied equipment may be started at any time, that danger exists, and that personnel must keep clear. These warning signs shall be provided along the conveyor at areas not guarded by position or location.

# OSHA Standards for General Industry

**1910.269(v)(11)(xi)** Remotely and automatically controlled conveyors, and conveyors that have operating stations which are not manned or which are beyond voice and visual contact from drive areas, loading areas, transfer points, and other locations on the conveyor path not guarded by location, position, or guards shall be furnished with emergency stop buttons, pull cords, limit switches, or similar emergency stop devices. However, if the employer can demonstrate that the design, function, and operation of the conveyor do not expose an employee to hazards, an emergency stop device is not required.

**1910.269(v)(11)(xi)(A)** Emergency stop devices shall be easily identifiable in the immediate vicinity of such locations.

**1910.269(v)(11)(xi)(B)** An emergency stop device shall act directly on the control of the conveyor involved and may not depend on the stopping of any other equipment.

**1910.269(v)(11)(xi)(C)** Emergency stop devices shall be installed so that they cannot be overridden from other locations.

**1910.269(v)(11)(xii)** Where coal-handling operations may produce a combustible atmosphere from fuel sources or from flammable gases or dust, sources of ignition shall be eliminated or safely controlled to prevent ignition of the combustible atmosphere.

**NOTE:** Locations that are hazardous because of the presence of combustible dust are classified as Class II hazardous locations. See §1910.307 of this part.

**1910.269(v)(11)(xiii)** An employee may not work on or beneath overhanging coal in coal bunkers, coal silos, or coal storage areas, unless the employee is protected from all hazards posed by shifting coal.

**1910.269(v)(11)(xiv)** An employee entering a bunker or silo to dislodge the contents shall wear a body harness with lifeline attached. The lifeline shall be secured to a fixed support outside the bunker and shall be attended at all times by an employee located outside the bunker or facility.

**1910.269(v)(12)** *Hydroplants and equipment.* Employees working on or close to water gates, valves, intakes, forebays, flumes, or other locations where increased or decreased water flow or levels may pose a significant hazard shall be warned and shall vacate such dangerous areas before water flow changes are made.

**1910.269(w)** *Special conditions*—**(1)** *Capacitors.* The following additional requirements apply to work on capacitors and on lines connected to capacitors.

**NOTE:** See paragraphs (m) and (n) of this section for requirements pertaining to the deenergizing and grounding of capacitor installations.

**1910.269(w)(1)(i)** Before employees work on capacitors, the capacitors shall be disconnected from energized sources and, after a wait of at least 5 minutes from the time of disconnection, short-circuited.

**1910.269(w)(1)(ii)** Before the units are handled, each unit in series-parallel capacitor banks shall be short-circuited between all terminals and the capacitor case or its rack. If the cases of capacitors are on ungrounded substation racks, the racks shall be bonded to ground.

**1910.269(w)(1)(iii)** Any line to which capacitors are connected shall be short-circuited before it is considered deenergized.

**1910.269(w)(2)** *Current transformer secondaries.* The secondary of a current transformer may not be opened while the transformer is energized. If the primary of the current transformer cannot be deenergized before work is performed on an instrument, a relay, or other section of a current transformer secondary circuit, the circuit shall be bridged so that the current transformer secondary will not be opened.

**1910.269(w)(3)** *Series streetlighting.* **(i)** If the open-circuit voltage exceeds 600 volts, the series streetlighting circuit shall be worked in accordance with paragraph (q) or (t) of this section, as appropriate.

**1910.269(w)(3)(ii)** A series loop may only be opened after the streetlighting transformer has been deenergized and isolated from the source of supply or after the loop is bridged to avoid an open-circuit condition.

**1910.269(w)(4)** *Illumination.* Sufficient illumination shall be provided to enable the employee to perform the work safely.

**1910.269(w)(5)** *Protection against drowning.* **(i)** Whenever an employee may be pulled or pushed or may fall into water where the danger of drowning exists, the employee shall be provided with and shall use U.S. Coast Guard approved personal flotation devices.

**1910.269(w)(5)(ii)** Each personal flotation device shall be maintained in safe condition and shall be inspected frequently enough to ensure that it does not have rot, mildew, water saturation, or any other condition that could render the device unsuitable for use.

**1910.269(w)(5)(iii)** An employee may cross streams or other bodies of water only if a safe means of passage, such as a bridge, is provided.

**1910.269(w)(6)** *Employee protection in public work areas.* **(i)** Traffic control signs and traffic control devices used for the protection of employees shall meet the requirements of §1926.200(g)(2) of this chapter.

**1910.269(w)(6)(ii)** Before work is begun in the vicinity of vehicular or pedestrian traffic that may endanger employees, warning signs or flags and other traffic control devices shall be placed in conspicuous locations to alert and channel approaching traffic.

**1910.269(w)(6)(iii)** Where additional employee protection is necessary, barricades shall be used.

**1910.269(w)(6)(iv)** Excavated areas shall be protected with barricades.

**1910.269(w)(6)(v)** At night, warning lights shall be prominently displayed.

**1910.269(w)(7)** *Backfeed.* If there is a possibility of voltage backfeed from sources of cogeneration or from the secondary system (for example, backfeed from more than one energized phase feeding a common load), the requirements of paragraph (l) of this section apply if the lines or equipment are to be worked as energized, and the requirements of paragraphs (m) and (n) of this section apply if the lines or equipment are to be worked as deenergized.

**1910.269(w)(8)** *Lasers.* Laser equipment shall be installed, adjusted, and operated in accordance with §1926.54 of this chapter.

**1910.269(w)(9)** *Hydraulic fluids.* Hydraulic fluids used for the insulated sections of equipment shall provide insulation for the voltage involved.

**1910.269(x)** *Definitions.*

*Affected employee.* An employee whose job requires him or her to operate or use a machine or equipment on which servicing or maintenance is being performed under lockout or tagout, or whose job requires him or her to work in an area in which such servicing or maintenance is being performed.

*Attendant.* An employee assigned to remain immediately outside the entrance to an enclosed or other space to render assistance as needed to employees inside the space.

*Authorized employee.* An employee who locks out or tags out machines or equipment in order to perform servicing or maintenance on that machine or equipment. An affected employee becomes an authorized employee when that employee's duties include performing servicing or maintenance covered under this section.

*Automatic circuit recloser.* A self-controlled device for interrupting and reclosing an alternating current circuit with a predetermined sequence of opening and reclosing followed by resetting, hold-closed, or lockout operation.

*Barricade.* A physical obstruction such as tapes, cones, or A-frame type wood or metal structures intended to provide a warning about and to limit access to a hazardous area.

*Barrier.* A physical obstruction which is intended to prevent contact with energized lines or equipment or to prevent unauthorized access to a work area.

*Bond.* The electrical interconnection of conductive parts designed to maintain a common electrical potential.

*Bus.* A conductor or a group of conductors that serve as a common connection for two or more circuits.

*Bushing.* An insulating structure, including a through conductor or providing a passageway for such a conductor, with provision for mounting on a barrier, conducting or otherwise, for the purposes of insulating the conductor from the barrier and conducting current from one side of the barrier to the other.

*Cable.* A conductor with insulation, or a stranded conductor with or without insulation and other coverings (single-conductor cable), or a combination of conductors insulated from one another (multiple-conductor cable).

*Cable sheath.* A conductive protective covering applied to cables.

**NOTE:** A cable sheath may consist of multiple layers of which one or more is conductive.

*Circuit.* A conductor or system of conductors through which an electric current is intended to flow.

*Clearance (between objects).* The clear distance between two objects measured surface to surface.

*Clearance (for work).* Authorization to perform specified work or permission to enter a restricted area.

*Communication lines.* (See *Lines, communication.*)

*Conductor.* A material, usually in the form of a wire, cable, or bus bar, used for carrying an electric current.

*Covered conductor.* A conductor covered with a dielectric having no rated insulating strength or having a rated insulating strength less than the voltage of the circuit in which the conductor is used.

*Current-carrying part.* A conducting part intended to be connected in an electric circuit to a source of voltage. Non-current-carrying parts are those not intended to be so connected.

*Deenergized.* Free from any electrical connection to a source of potential difference and from electric charge; not having a potential different from that of the earth.

**NOTE:** The term is used only with reference to current-carrying parts, which are sometimes energized (alive).

*Designated employee (designated person).* An employee (or person) who is designated by the employer to perform specific duties under the terms of this section and who is knowledgeable in the construction and operation of the equipment and the hazards involved.

*Electric line truck.* A truck used to transport personnel, tools, and material for electric supply line work.

*Electric supply equipment.* Equipment that produces, modifies, regulates, controls, or safeguards a supply of electric energy.

*Electric supply lines.* (See *Lines, electric supply.*)

*Electric utility.* An organization responsible for the installation, operation, or maintenance of an electric supply system.

*Enclosed space.* A working space, such as a manhole, vault, tunnel, or shaft, that has a limited means of egress or entry, that is designed for periodic employee entry under normal operating conditions, and that under normal conditions does not contain a hazardous atmosphere, but that may contain a hazardous atmosphere under abnormal conditions.

**NOTE:** Spaces that are enclosed but not designed for employee entry under normal operating conditions are not considered to be enclosed spaces for the purposes of this section. Similarly, spaces that are enclosed and that are expected to contain a hazardous atmosphere are not considered to be enclosed spaces for the purposes of this section. Such spaces meet the definition of permit spaces in §1910.146 of this part, and entry into them must be performed in accordance with that standard.

*Energized (alive, live).* Electrically connected to a source of potential difference, or electrically charged so as to have a potential significantly different from that of earth in the vicinity.

*Energy isolating device.* A physical device that prevents the transmission or release of energy, including, but not limited to, the following: a manually operated electric circuit breaker, a disconnect switch, a manually operated switch, a slide gate, a slip blind, a line valve, blocks, and any similar device with a visible indication of the position of the device. (Push buttons, selector switches, and other control-circuit-type devices are not energy isolating devices.)

*Energy source.* Any electrical, mechanical, hydraulic, pneumatic, chemical, nuclear, thermal, or other energy source that could cause injury to personnel.

*Equipment (electric).* A general term including material, fittings, devices, appliances, fixtures, apparatus, and the like used as part of or in connection with an electrical installation.

*Exposed.* Not isolated or guarded.

*Ground.* A conducting connection, whether intentional or accidental, between an electric circuit or equipment and the earth, or to some conducting body that serves in place of the earth.

*Grounded.* Connected to earth or to some conducting body that serves in place of the earth.

*Guarded.* Covered, fenced, enclosed, or otherwise protected, by means of suitable covers or casings, barrier rails or screens, mats, or platforms, designed to minimize the possibility, under normal conditions, of dangerous approach or accidental contact by persons or objects.

**NOTE:** Wires which are insulated, but not otherwise protected, are not considered as guarded.

*Hazardous atmosphere* means an atmosphere that may expose employees to the risk of death, incapacitation, impairment of ability to self-rescue (that is, escape unaided from an enclosed space), injury, or acute illness from one or more of the following causes:

**1910.269(x)(1)** Flammable gas, vapor, or mist in excess of 10 percent of its lower flammable limit (LFL);

**1910.269(x)(2)** Airborne combustible dust at a concentration that meets or exceeds its LFL;

**NOTE:** This concentration may be approximated as a condition in which the dust obscures vision at a distance of 5 feet (1.52 m) or less.

**1910.269(x)(3)** Atmospheric oxygen concentration below 19.5 percent or above 23.5 percent;

**1910.269(x)(4)** Atmospheric concentration of any substance for which a dose or a permissible exposure limit is published in Subpart G, *Occupational Health and Environmental Control*, or in Subpart Z, *Toxic and Hazardous Substances*, of this part and which could result in employee exposure in excess of its dose or permissible exposure limit;

**NOTE:** An atmospheric concentration of any substance that is not capable of causing death, incapacitation, impairment of ability to self-rescue, injury, or acute illness due to its health effects is not covered by this provision.

**1910.269(x)(5)** Any other atmospheric condition that is immediately dangerous to life or health.

**NOTE:** For air contaminants for which OSHA has not determined a dose or permissible exposure limit, other sources of information, such as Material Safety Data Sheets that comply with the Hazard Communication Standard, §1910.1200 of this part, published information, and internal documents can provide guidance in establishing acceptable atmospheric conditions.

*High-power tests.* Tests in which fault currents, load currents, magnetizing currents, and line-dropping currents are used to test equipment, either at the equipment's rated voltage or at lower voltages.

*High-voltage tests.* Tests in which voltages of approximately 1000 volts are used as a practical minimum and in which the voltage source has sufficient energy to cause injury.

*High wind.* A wind of such velocity that the following hazards would be present:

(1) An employee would be exposed to being blown from elevated locations, or

(2) An employee or material handling equipment could lose control of material being handled, or

(3) An employee would be exposed to other hazards not controlled by the standard involved.

**NOTE:** Winds exceeding 40 miles per hour (64.4 kilometers per hour), or 30 miles per hour (48.3 kilometers per hour) if material handling is involved, are normally considered as meeting this criteria unless precautions are taken to protect employees from the hazardous effects of the wind.

*Immediately dangerous to life or health* (IDLH) means any condition that poses an immediate or delayed threat to life or that would cause irreversible adverse health effects or that would interfere with an individual's ability to escape unaided from a permit space.

**NOTE:** Some materials—hydrogen fluoride gas and cadmium vapor, for example—may produce immediate transient effects that, even if severe, may pass without medical attention, but are followed by sudden, possibly fatal collapse 12-72 hours after exposure. The victim "feels normal" from recovery from transient effects until collapse. Such materials in hazardous quantities are considered to be "immediately" dangerous to life or health.

*Insulated.* Separated from other conducting surfaces by a dielectric (including air space) offering a high resistance to the passage of current.

# OSHA Standards for General Industry

**NOTE:** When any object is said to be insulated, it is understood to be insulated for the conditions to which it is normally subjected. Otherwise, it is, within the purpose of this section, uninsulated.

*Insulation (cable).* That which is relied upon to insulate the conductor from other conductors or conducting parts or from ground.

*Line-clearance tree trimmer.* An employee who, through related training or on-the-job experience or both, is familiar with the special techniques and hazards involved in line-clearance tree trimming.

**NOTE 1:** An employee who is regularly assigned to a line-clearance tree-trimming crew and who is undergoing on-the-job training and who, in the course of such training, has demonstrated an ability to perform duties safely at his or her level of training and who is under the direct supervision of a line-clearance tree trimmer is considered to be a line-clearance tree trimmer for the performance of those duties.

**NOTE 2:** A line-clearance tree trimmer is not considered to be a "qualified employee" under this section unless he or she has the training required for a qualified employee under paragraph (a)(2)(ii) of this section. However, under the electrical safety-related work practices standard in subpart S of this part, a line-clearance tree trimmer *is* considered to be a "qualified employee". Tree trimming performed by such "qualified employees" is not subject to the electrical safety-related work practice requirements contained in §§1910.331 through 1910.335 of this part. (See also the note following §1910.332(b)(3) of this part for information regarding the training an employee must have to be considered a qualified employee under §§1910.331 through 1910.335 of this part.)

*Line-clearance tree trimming.* The pruning, trimming, repairing, maintaining, removing, or clearing of trees or the cutting of brush that is within 10 feet (305 cm) of electric supply lines and equipment.

*Lines—(1) Communication lines.* The conductors and their supporting or containing structures which are used for public or private signal or communication service, and which operate at potentials not exceeding 400 volts to ground or 750 volts between any two points of the circuit, and the transmitted power of which does not exceed 150 watts. If the lines are operating at less than 150 volts, no limit is placed on the transmitted power of the system. Under certain conditions, communication cables may include communication circuits exceeding these limitations where such circuits are also used to supply power solely to communication equipment.

**NOTE:** Telephone, telegraph, railroad signal, data, clock, fire, police alarm, cable television, and other systems conforming to this definition are included. Lines used for signaling purposes, but not included under this definition, are considered as electric supply lines of the same voltage.

*(2) Electric supply lines.* Conductors used to transmit electric energy and their necessary supporting or containing structures. Signal lines of more than 400 volts are always supply lines within this section, and those of less than 400 volts are considered as supply lines, if so run and operated throughout.

*Manhole.* A subsurface enclosure which personnel may enter and which is used for the purpose of installing, operating, and maintaining submersible equipment or cable.

*Manhole steps.* A series of steps individually attached to or set into the walls of a manhole structure.

*Minimum approach distance.* The closest distance an employee is permitted to approach an energized or a grounded object.

*Qualified employee (qualified person).* One knowledgeable in the construction and operation of the electric power generation, transmission, and distribution equipment involved, along with the associated hazards.

**NOTE 1:** An employee must have the training required by paragraph (a)(2)(ii) of this section in order to be considered a qualified employee.

**NOTE 2:** Except under paragraph (g)(2)(v) of this section, an employee who is undergoing on-the-job training and who, in the course of such training, has demonstrated an ability to perform duties safely at his or her level of training and who is under the direct supervision of a qualified person is considered to be a qualified person for the performance of those duties.

*Step bolt.* A bolt or rung attached at intervals along a structural member and used for foot placement during climbing or standing.

*Switch.* A device for opening and closing or for changing the connection of a circuit. In this section, a switch is understood to be manually operable, unless otherwise stated.

*System operator.* A qualified person designated to operate the system or its parts.

*Vault.* An enclosure, above or below ground, which personnel may enter and which is used for the purpose of installing, operating, or maintaining equipment or cable.

*Vented vault.* A vault that has provision for air changes using exhaust flue stacks and low level air intakes operating on differentials of pressure and temperature providing for airflow which precludes a hazardous atmosphere from developing.

*Voltage.* The effective (rms) potential difference between any two conductors or between a conductor and ground. Voltages are expressed in nominal values unless otherwise indicated. The nominal voltage of a system or circuit is the value assigned to a system or circuit of a given voltage class for the purpose of convenient designation. The operating voltage of the system may vary above or below this value.

### APPENDIX A TO §1910.269—FLOW CHARTS

This appendix presents information, in the form of flow charts, that illustrates the scope and application of §1910.269. This appendix addresses the interface between §1910.269 and subpart S of this part (*Electrical*), between §1910.269 and §1910.146 of this part (*Permit-required confined spaces*), and between §1910.269 and §1910.147 of this part (*The control of hazardous energy (lockout/tagout)*). These flow charts provide guidance for employers trying to implement the requirements of §1910.269 in combination with other General Industry Standards contained in part 1910.

### APPENDIX A-1 TO §1910.269—APPLICATION OF §1910.269 AND SUBPART S OF THIS PART TO ELECTRICAL INSTALLATIONS

Is this an electric power generation, transmission, or distribution installation?[1]
- YES → Is it a generation installation?
  - YES → §1910.269(v)[2]
  - NO → §1910.269(u)[2]
- NO → §§1910.302 through 1910.308

[1] Electrical installation design requirements only. See Appendix 1B for electrical safety-related work practices. Supplementary electric generating equipment that is used to supply a workplace for emergency, standby, or similar purposes only is not considered to be an electric power generation installation.

[2] See Table 1 of Appendix A-2 for requirements that can be met through compliance with Subpart S.

## APPENDIX A-2 TO §1910.269—APPLICATION OF §1910.269 AND SUBPART S OF THIS PART TO ELECTRICAL SAFETY-RELATED WORK PRACTICES

[Flowchart:]
- Are the employees "qualified" as defined in §1910.269(x)?
  - NO → §§1910.332 through 1910.335
  - YES → Is this an electric power generation, transmission, or distribution installation?
    - NO → Is it a commingled[1] installation?
      - NO → §§1910.332 through 1910.335
      - YES → §1910.269 plus §1910.332, §1910.333(a) & (b), and §1910.334
    - YES → Does the installation conform to §§1910.302 through 1910.308?
      - NO → §1910.269
      - YES → §1910.269 OR §§1910.332 through 1910.335 plus the supplementary requirements of §1910.269 identified in Appendix A-2, Table 1

[1] Commingled to the extent that the electric power generation, transmission, or distribution installation poses the greater hazard.

### TABLE 1—ELECTRICAL SAFETY-RELATED WORK PRACTICES IN §1910.269

| Compliance with subpart S is considered as compliance with Sec. 1910.269[1] | Paragraphs that apply regardless of compliance with subpart S |
|---|---|
| (d), electric shock hazards only | (a)(2)[2] and (a)(3)[2]. |
| (h)(3) | (b)[2]. |
| (i)(2) | (c)[2]. |
| (k) | (d), other than electric shock hazards. |
| (l)(1) through (l)(4), (l)(6)(i), and (l)(8) through (l)(10) | (e). |
| (m) | (f). |
| (p)(4) | (g). |
| (s)(2) | (h)(1) and (h)(2). |
| (u)(1) and (u)(3) through (u)(5) | (i)(3)[2] and (i)(4)[2]. |
| (v)(3) through (v)(5) | (j)[2]. |
| (w)(1) and (w)(7) | (l)(5)[2], (l)(6)(ii)[2], (l)(6)(iii)[2], and (l)(7)[2]. |
|  | (n)[2]. |
|  | (o)[2]. |
|  | (p)(1) through (p)(3). |
|  | (q)[2]. |
|  | (r)[2]. |
|  | (s)(1). |
|  | (t)[2]. |
|  | (u)(2)[2] and (u)(6)[2]. |
|  | (v)(1), (v)(2)[2], and (v)(6) through (v)(12). |
|  | (w)(2) through (w)(6)[2], (w)(8), and (w)(9)[2]. |

[1] If the electrical installation meets the requirements of §§1910.303 through 1910.308 of this part, then the electrical installation and any associated electrical safety-related work practices conforming to §§1910.332 through 1910.335 of this part are considered to comply with these provisions of §1910.269 of this part.

[2] These provisions include electrical safety requirements that must be met regardless of compliance with subpart S of this part.

## APPENDIX A-3 TO §1910.269—APPLICATION OF §1910.269 AND SUBPART S OF THIS PART TO TREE-TRIMMING OPERATIONS

[Flowchart:]
- Is the tree within 10 feet[1] of an overhead line?
  - NO → Neither §1910.269 nor Subpart S applies.
  - YES → Is the employee a line-clearance tree trimmer?
    - NO → Subpart S applies. (Employee may not trim branch within 10 feet[1] of line.)
    - YES → §1910.269 applies. (Clearances are specified in §1910.269(r)(1)(iii).)

[1] 10 feet plus 4 inches for every 10 kilovolts over 50 kilovolts

## APPENDIX A-4 TO §1910.269—APPLICATION OF §§1910.147, 1910.269 AND 1910.333 TO HAZARDOUS ENERGY CONTROL PROCEDURES (LOCKOUT/TAGOUT)

[Flowchart:]
- Is this an electric power generation, transmission, or distribution installation?[1]
  - YES → Is it a generation installation?
    - YES → §1910.269(d) or §1910.147
    - NO → §1910.269(m)
  - NO → Is it a commingled[2] installation?
    - YES → Is it a generation installation? (loop)
    - NO → Is there a hazard of electric shock?
      - YES → §1910.333(b) or §1910.147[3]
      - NO → §1910.147

[1] If the installation conforms to §§1910.303 through 1910.308, the lockout and tagging procedures of 1910.333(b) may be followed for electric shock hazards

[2] Commingled to the extent that the electric power generation, transmission, or distribution installation poses the greater hazard.

[3] §1910.333(b)(2)(iii)(D) and (b)(2)(iv)(B) still apply.

## APPENDIX A-5 TO §1910.269—APPLICATION OF §§1910.146 AND 1910.269 TO PERMIT-REQUIRED CONFINED SPACES

[Flowchart:
- Is this a confined space as defined in §1910.146(b)?[1] — NO → Neither §1910.146 nor §1910.269(e) apply to entry.
- YES → Is it a permit space as defined in §1910.146(b)? — NO → (Neither applies)
- YES → Does the work performed fall within the scope of §1910.269? — NO → §1910.146
- YES → Is this space an enclosed space as defined in §1910.269(x)? — NO → §1910.146
- YES → Are hazards controlled through measures required by §1910.269? — YES → §1910.269(e) or §1910.146 — NO → §1910.146]

[1] See §1910.146(c) for general non-entry requirements that apply to all confined spaces.

## APPENDIX B TO §1910.269—WORKING ON EXPOSED ENERGIZED PARTS

### I. Introduction

Electric transmission and distribution line installations have been designed to meet National Electrical Safety Code (NESC), ANSI C2, requirements and to provide the level of line outage performance required by system reliability criteria. Transmission and distribution lines are also designed to withstand the maximum overvoltages expected to be impressed on the system. Such overvoltages can be caused by such conditions as switching surges, faults, or lightning. Insulator design and lengths and the clearances to structural parts (which, for low voltage through extra-high voltage, or EHV, facilities, are generally based on the performance of the line as a result of contamination of the insulation or during storms) have, over the years, come closer to the minimum approach distances used by workers (which are generally based on non-storm conditions). Thus, as minimum approach (working) distances and structural distances (clearances) converge, it is increasingly important that basic considerations for establishing safe approach distances for performing work be understood by the designers and the operating and maintenance personnel involved.

The information in this appendix will assist employers in complying with the minimum approach distance requirements contained in paragraphs (l)(2) and (q)(3) of this section. The technical criteria and methodology presented herein is mandatory for employers using reduced minimum approach distances as permitted in Table R-7 and Table R-8. This appendix is intended to provide essential background information and technical criteria for the development or modification, if possible, of the safe minimum approach distances for electric transmission and distribution live-line work. The development of these safe distances must be undertaken by persons knowledgeable in the techniques discussed in this appendix and competent in the field of electric transmission and distribution system design.

### II. General

#### A. Definitions

The following definitions from §1910.269(x) relate to work on or near transmission and distribution lines and equipment and the electrical hazards they present.

*Exposed.* Not isolated or guarded.

*Guarded.* Covered, fenced, enclosed, or otherwise protected, by means of suitable covers or casings, barrier rails or screens, mats, or platforms, designed to minimize the possibility, under normal conditions, of dangerous approach or accidental contact by persons or objects.

**NOTE:** Wires which are insulated, but not otherwise protected, are not considered as guarded.

*Insulated.* Separated from other conducting surfaces by a dielectric (including air space) offering a high resistance to the passage of current.

**NOTE:** When any object is said to be insulated, it is understood to be insulated for the conditions to which it is normally subjected. Otherwise, it is, within the purpose of this section, uninsulated.

#### B. Installations Energized at 50 to 300 Volts

The hazards posed by installations energized at 50 to 300 volts are the same as those found in many other workplaces. That is not to say that there is no hazard, but the complexity of electrical protection required does not compare to that required for high voltage systems. The employee must avoid contact with the exposed parts, and the protective equipment used (such as rubber insulating gloves) must provide insulation for the voltages involved.

#### C. Exposed Energized Parts Over 300 Volts AC

Table R-6, Table R-7, and Table R-8 of §1910.269 provide safe approach and working distances in the vicinity of energized electric apparatus so that work can be done safely without risk of electrical flashover.

The working distances must withstand the maximum transient overvoltage that can reach the work site under the working conditions and practices in use. Normal system design may provide or include a means to control transient overvoltages, or temporary devices may be employed to achieve the same result. The use of technically correct practices or procedures to control overvoltages (for example, portable gaps or preventing the automatic control from initiating breaker reclosing) enables line design and operation to be based on reduced transient overvoltage values. Technical information for U.S. electrical systems indicates that current design provides for the following maximum transient overvoltage values (usually produced by switching surges): 362 kV and less—3.0 per unit; 552 kV—2.4 per unit; 800 kV—2.0 per unit.

Additional discussion of maximum transient overvoltages can be found in paragraph IV.A.2, later in this appendix.

### III. Determination of the Electrical Component of Minimum Approach Distances

#### A. Voltages of 1.1 kV to 72.5 kV

For voltages of 1.1 kV to 72.5 kV, the electrical component of minimum approach distances is based on American National Standards Institute (ANSI)/American Institute of Electrical Engineers (AIEE) Standard No.4, March 1943, Tables III and IV. (AIEE is the predecessor technical society to the Institute of Electrical and Electronic Engineers (IEEE).) These distances are calculated by the following formula:

Equation (1)—For voltages of 1.1 kV to 72.5 kV

$$D = \left(\frac{V_{max} \times pu}{124}\right)^{1.63}$$

Where:

D = Electrical component of the minimum approach distance in air in feet

$V_{max}$ = Maximum rated line-to-ground rms voltage in kV

pu = Maximum transient overvoltage factor in per unit

SOURCE: AIEE Standard No. 4, 1943.

This formula has been used to generate Table 1.

### TABLE 1—AC ENERGIZED LINE-WORK PHASE-TO-GROUND ELECTRICAL COMPONENT OF THE MINIMUM APPROACH DISTANCE—1.1 TO 72.5 KV

| Maximum anticipated per-unit transient overvoltage | Phase to phase voltage ||||
|---|---|---|---|---|
| | 15,000 | 36,000 | 46,000 | 72,500 |
| 3.0 | 0.08 | 0.33 | 0.49 | 1.03 |

NOTE: The distances given (in feet) are for air as the insulating medium and provide no additional clearance for inadvertent movement.

### B. Voltages of 72.6 kV to 800 kV

For voltages of 72.6 kV to 800 kV, the electrical component of minimum approach distances is based on ANSI/IEEE Standard 516-1987, "IEEE Guide for Maintenance Methods on Energized Power Lines." This standard gives the electrical component of the minimum approach distance based on power frequency rod-gap data, supplemented with transient overvoltage information and a saturation factor for high voltages. The distances listed in ANSI/IEEE Standard 516 have been calculated according to the following formula:

*Equation (2)—For voltages of 72.6 kV to 800 kV*

$$D = (C+a)puV_{max}$$

Where:

D = Electrical component of the minimum approach distance in air in feet

C = 0.01 to take care of correction factors associated with the variation of gap sparkover with voltage

a = A factor relating to the saturation of air at voltages of 345 kV or higher

pu = Maximum anticipated transient overvoltage, in per unit (p.u.)

$V_{max}$ = Maximum rms system line-to-ground voltage in kilovolts—it should be the "actual" maximum, or the normal highest voltage for the range (for example, 10 percent above the nominal voltage)

SOURCE: Formula developed from ANSI/IEEE Standard No. 516, 1987.

This formula is used to calculate the electrical component of the minimum approach distances in air and is used in the development of Table 2 and Table 3.

### TABLE 2—AC ENERGIZED LINE-WORK PHASE-TO-GROUND ELECTRICAL COMPONENT OF THE MINIMUM APPROACH DISTANCE—121 TO 242 KV

| Maximum anticipated per-unit transient overvoltage | Phase to phase voltage ||||
|---|---|---|---|---|
| | 121,000 | 145,000 | 169,000 | 242,000 |
| 2.0 | 1.40 | 1.70 | 2.00 | 2.80 |
| 2.1 | 1.47 | 1.79 | 2.10 | 2.94 |
| 2.2 | 1.54 | 1.87 | 2.20 | 3.08 |
| 2.3 | 1.61 | 1.96 | 2.30 | 3.22 |
| 2.4 | 1.68 | 2.04 | 2.40 | 3.35 |
| 2.5 | 1.75 | 2.13 | 2.50 | 3.50 |
| 2.6 | 1.82 | 2.21 | 2.60 | 3.64 |
| 2.7 | 1.89 | 2.30 | 2.70 | 3.76 |
| 2.8 | 1.96 | 2.38 | 2.80 | 3.92 |
| 2.9 | 2.03 | 2.47 | 2.90 | 4.05 |
| 3.0 | 2.10 | 2.55 | 3.00 | 4.29 |

NOTE: The distances given (in feet) are for air as the insulating medium and provide no additional clearance for inadvertent movement.

### TABLE 3—AC ENERGIZED LINE-WORK PHASE-TO-GROUND ELECTRICAL COMPONENT OF THE MINIMUM APPROACH DISTANCE—362 TO 800 KV

| Maximum anticipated per-unit transient overvoltage | Phase to phase voltage |||
|---|---|---|---|
| | 362,000 | 552,000 | 800,000 |
| 1.5 | ......... | 4.97 | 8.66 |
| 1.6 | ......... | 5.46 | 9.60 |
| 1.7 | ......... | 5.98 | 10.60 |
| 1.8 | ......... | 6.51 | 11.64 |
| 1.9 | ......... | 7.08 | 12.73 |
| 2.0 | 4.20 | 7.68 | 13.86 |
| 2.1 | 4.41 | 8.27 | ......... |
| 2.2 | 4.70 | 8.87 | ......... |
| 2.3 | 5.01 | 9.49 | ......... |
| 2.4 | 5.34 | 10.21 | ......... |
| 2.5 | 5.67 | ......... | ......... |
| 2.6 | 6.01 | ......... | ......... |
| 2.7 | 6.36 | ......... | ......... |
| 2.8 | 6.73 | ......... | ......... |
| 2.9 | 7.10 | ......... | ......... |
| 3.0 | 7.48 | ......... | ......... |

NOTE: The distances given (in feet) are for air as the insulating medium and provide no additional clearance for inadvertent movement.

### C. Provisions for Inadvertent Movement

The minimum approach distances (working distances) must include an "adder" to compensate for the inadvertent movement of the worker relative to an energized part or the movement of the part relative to the worker. A certain allowance must be made to account for this possible inadvertent movement and to provide the worker with a comfortable and safe zone in which to work. A distance for inadvertent movement (called the "ergonomic component of the minimum approach distance") must be added to the electrical component to determine the total safe minimum approach distances used in live-line work.

One approach that can be used to estimate the ergonomic component of the minimum approach distance is response time-distance analysis. When this technique is used, the total response time to a hazardous incident is estimated and converted to distance travelled. For example, the driver of a car takes a given amount of time to respond to a "stimulus" and stop the vehicle. The elapsed time involved results in a distance being travelled before the car comes to a complete stop. This distance is dependent on the speed of the car at the time the stimulus appears.

In the case of live-line work, the employee must first perceive that he or she is approaching the danger zone. Then, the worker responds to the danger and must decelerate and stop all motion toward the energized part. During the time it takes to stop, a distance will have been traversed. It is this distance that must be added to the electrical component of the minimum approach distance to obtain the total safe minimum approach distance.

At voltages below 72.5 kV, the electrical component of the minimum approach distance is smaller than the ergonomic component. At 72.5 kV the electrical component is only a little more than 1 foot. An ergonomic component of the minimum approach distance is needed that will provide for all the worker's unexpected movements. The usual live-line work method for these voltages is the use of rubber insulating equipment, frequently rubber gloves. The energized object needs to be far enough away to provide the worker's face with a safe approach distance, as his or her hands and arms are insulated. In this case, 2 feet has been accepted as a sufficient and practical value.

For voltages between 72.6 and 800 kV, there is a change in the work practices employed during energized line work. Generally, live-line tools (hot sticks) are employed to perform work while equipment is energized. These tools, by design, keep the energized part at a constant distance from the employee and thus maintain the appropriate minimum approach distance automatically.

The length of the ergonomic component of the minimum approach distance is also influenced by the location of the worker and by the nature of the work. In these higher voltage ranges, the employees use work methods that more tightly control their movements than when the workers perform rubber glove work. The worker is farther from energized line or equipment and needs to be more precise in his or her movements just to perform the work.

For these reasons, a smaller ergonomic component of the minimum approach distance is needed, and a distance of 1 foot has been selected for voltages between 72.6 and 800 kV.

Table 4 summarizes the ergonomic component of the minimum approach distance for the two voltage ranges.

### TABLE 4—ERGONOMIC COMPONENT OF MINIMUM APPROACH DISTANCE

| Voltage range (kV) | Distance (feet) |
|---|---|
| 1.1 to 72.5 | 2.0 |
| 72.6 to 800 | 1.0 |

NOTE: This distance must be added to the electrical component of the minimum approach distance to obtain the full minimum approach distance.

### D. Bare-Hand Live-Line Minimum Approach Distances

Calculating the strength of phase-to-phase transient overvoltages is complicated by the varying time displacement between overvoltages on parallel conductors (electrodes) and by the varying ratio between the positive and negative voltages on the two electrodes. The time displacement causes the maximum voltage between phases to be less than the sum of the phase-to-ground voltages. The International Electrotechnical Commission (IEC) Technical Committee 28, Working Group 2, has developed the following formula for determining the phase-to-phase maximum transient overvoltage, based on the per unit (p.u.) of the system nominal voltage phase-to-ground crest:

$$pu_p = pu_g + 1.6.$$

Where:

$pu_g$ = p.u. phase-to-ground maximum transient overvoltage

$pu_p$ = p.u. phase-to-phase maximum transient overvoltage

This value of maximum anticipated transient overvoltage must be used in Equation (2) to calculate the phase-to-phase minimum approach distances for live-line bare-hand work.

### E. Compiling the Minimum Approach Distance Tables

For each voltage involved, the distance in Table 4 in this appendix has been added to the distance in Table 1, Table 2 or Table 3 in this appendix to determine the resulting minimum approach distances in Table R-6, Table R-7, and Table R-8 in §1910.269.

### F. Miscellaneous Correction Factors

The strength of an air gap is influenced by the changes in the air medium that forms the insulation. A brief discussion of each factor follows, with a summary at the end.

1. *Dielectric strength of air.* The dielectric strength of air in a uniform electric field at standard atmospheric conditions is approximately 31 kV (crest) per cm at 60 Hz. The disruptive gradient is affected by the air pressure, temperature, and humidity, by the shape, dimensions, and separation of the electrodes, and by the characteristics of the applied voltage (wave shape).

2. *Atmospheric effect.* Flashover for a given air gap is inhibited by an increase in the density (humidity) of the air. The empirically determined electrical strength of a given gap is normally applicable at standard atmospheric conditions (20 °C, 101.3 kPa, 11 g/cm$^3$ humidity).

The combination of temperature and air pressure that gives the lowest gap flashover voltage is high temperature and low pressure. These are conditions not likely to occur simultaneously. Low air pressure is generally associated with high humidity, and this causes increased electrical strength. An average air pressure is more likely to be associated with low humidity. Hot and dry working conditions are thus normally associated with reduced electrical strength.

The electrical component of the minimum approach distances in Table 1, Table 2, and Table 3 has been calculated using the maximum transient overvoltages to determine withstand voltages at standard atmospheric conditions.

3. *Altitude.* The electrical strength of an air gap is reduced at high altitude, due principally to the reduced air pressure. An increase of about 3 percent per 300 meters in the minimum approach distance for altitudes above 900 meters is required. Table R-10 of §1910.269 presents this information in tabular form.

*Summary.* After taking all these correction factors into account and after considering their interrelationships relative to the air gap insulation strength and the conditions under which live work is performed, one finds that only a correction for altitude need be made. An elevation of 900 meters is established as the base elevation, and the values of the electrical component of the minimum approach distances has been derived with this correction factor in mind. Thus, the values used for elevations below 900 meters are conservative without any change; corrections have to be made only above this base elevation.

### IV. Determination of Reduced Minimum Approach Distances

#### A. Factors Affecting Voltage Stress at the Work Site

1. *System voltage (nominal).* The nominal system voltage range sets the absolute lower limit for the minimum approach distance. The highest value within the range, as given in the relevant table, is selected and used as a reference for per unit calculations.

2. *Transient overvoltages.* Transient overvoltages may be generated on an electrical system by the operation of switches or breakers, by the occurrence of a fault on the line or circuit being worked or on an adjacent circuit, and by similar activities. Most of the overvoltages are caused by switching, and the term "switching surge" is often used to refer generically to all types of overvoltages. However, each overvoltage has an associated transient voltage wave shape. The wave shape arriving at the site and its magnitude vary considerably.

The information used in the development of the minimum approach distances takes into consideration the most common wave shapes; thus, the required minimum approach distances are appropriate for any transient overvoltage level usually found on electric power generation, transmission, and distribution systems. The values of the per unit (p.u.) voltage relative to the nominal maximum voltage are used in the calculation of these distances.

3. *Typical magnitude of overvoltages.* The magnitude of typical transient overvoltages is given in Table 5.

4. *Standard deviation—air-gap withstand.* For each air gap length, and under the same atmospheric conditions, there is a statistical variation in the breakdown voltage. The probability of the breakdown voltage is assumed to have a normal (Gaussian) distribution. The standard deviation of this distribution varies with the wave shape, gap geometry, and atmospheric conditions. The withstand voltage of the air gap used in calculating the electrical component of the minimum approach distance has been set at three standard deviations ($3\sigma^1$) below the critical flashover voltage. (The critical flashover voltage is the crest value of the impulse wave that, under specified conditions, causes flashover on 50 percent of the applications. An impulse wave of three standard deviations below this value, that is, the withstand voltage, has a probability of flashover of approximately 1 in 1000.)

### TABLE 5—MAGNITUDE OF TYPICAL TRANSIENT OVERVOLTAGES

| Cause | Magnitude (per unit) |
|---|---|
| Energized 200 mile line without closing resistors | 3.5 |
| Energized 200 mile line with one step closing resistor | 2.1 |
| Energized 200 mile line with multi-step resistor | 2.5 |
| Reclosed with trapped charge one step resistor | 2.2 |
| Opening surge with single restrike | 3.0 |
| Fault initiation unfaulted phase | 2.1 |
| Fault initiation adjacent circuit | 2.5 |
| Fault clearing | 1.7-1.9 |

Source: ANSI/IEEE Standard No. 516, 1987.

5. *Broken Insulators.* Tests have shown that the insulation strength of an insulator string with broken skirts is reduced. Broken units may have lost up to 70% of their withstand capacity. Because the insulating capability of a broken unit cannot be determined without testing it, damaged units in an insulator are usually considered to have no insulating value. Additionally, the overall insulating strength of a string with broken units may be further reduced in the presence of a live-line tool alongside it. The number of good units that must be present in a string is based on the maximum overvoltage possible at the worksite.

### B. Minimum Approach Distances Based on Known Maximum Anticipated Per-Unit Transient Overvoltages

1. *Reduction of the minimum approach distance for AC systems.* When the transient overvoltage values are known and supplied by the employer, Table R-7 and Table R-8 of §1910.269 allow the minimum approach distances from energized parts to be reduced. In order to determine what this maximum overvoltage is, the employer must undertake an engineering analysis of the system. As a result of this engineering study, the employer must provide new live work procedures, reflecting the new minimum approach distances, the conditions and limitations of application of the new minimum approach distances, and the specific practices to be used when these procedures are implemented.

2. *Calculation of reduced approach distance values.* The following method of calculating reduced minimum approach distances is based on ANSI/IEEE Standard 516:

*Step 1.* Determine the maximum voltage (with respect to a given nominal voltage range) for the energized part.

*Step 2.* Determine the maximum transient overvoltage (normally a switching surge) that can be present at the work site during work operation.

*Step 3.* Determine the technique to be used to control the maximum transient overvoltage. (See paragraphs IV.C and IV.D of this appendix.) Determine the maximum voltage that can exist at the work site with that form of control in place and with a confidence level of $3\sigma$. This voltage is considered to be the withstand voltage for the purpose of calculating the appropriate minimum approach distance.

*Step 4.* Specify in detail the control technique to be used, and direct its implementation during the course of the work.

*Step 5.* Using the new value of transient overvoltage in per unit (p.u.), determine the required phase-to-ground minimum approach distance from Table R-7 or Table R-8 of §1910.269.

### C. Methods of Controlling Possible Transient Overvoltage Stress Found on a System

1. *Introduction.* There are several means of controlling overvoltages that occur on transmission systems. First, the operation of circuit breakers or other switching devices may be modified to reduce switching transient overvoltages. Second, the overvoltage itself may be forcibly held to an acceptable level by means of installation of surge arresters at the specific location to be protected. Third, the transmission system may be changed to minimize the effect of switching operations.

2. *Operation of circuit breakers.*[2] The maximum transient overvoltage that can reach the work site is often due to switching on the line on which work is being performed. If the automatic-reclosing is removed during energized line work so that the line will not be re-energized after being opened for any reason, the maximum switching surge overvoltage is then limited to the larger of the opening surge or the greatest possible fault-generated surge, provided that the devices (for example, insertion resistors) are operable and will function to limit the transient overvoltage. It is essential that the operating ability of such devices be assured when they are employed to limit the overvoltage level. If it is prudent not to remove the reclosing feature (because of system operating conditions), other methods of controlling the switching surge level may be necessary.

Transient surges on an adjacent line, particularly for double circuit construction, may cause a significant overvoltage on the line on which work is being performed. The coupling to adjacent lines must be accounted for when minimum approach distances are calculated based on the maximum transient overvoltage.

3. *Surge arresters.* The use of modern surge arresters has permitted a reduction in the basic impulse-insulation levels of much transmission system equipment. The primary function of early arresters was to protect the system insulation from the effects of lightning. Modern arresters not only dissipate lightning-caused transients, but may also control many other system transients that may be caused by switching or faults.

It is possible to use properly designed arresters to control transient overvoltages along a transmission line and thereby reduce the requisite length of the insulator string. On the other hand, if the installation of arresters has not been used to reduce the length of the insulator string, it may be used to reduce the minimum approach distance instead.[3]

4. *Switching Restrictions.* Another form of overvoltage control is the establishment of switching restrictions, under which breakers are not permitted to be operated until certain system conditions are satisfied. Restriction of switching is achieved by the use of a tagging system, similar to that used for a "permit", except that the common term used for this activity is a "hold-off" or "restriction". These terms are used to indicate that operation is not prevented, but only modified during the live-work activity.

### D. Minimum Approach Distance Based on Control of Voltage Stress (Overvoltages) at the Work Site.

Reduced minimum approach distances can be calculated as follows:

1. *First Method—Determining the reduced minimum approach distance from a given withstand voltage.*[4]

*Step 1.* Select the appropriate withstand voltage for the protective gap based on system requirements and an acceptable probability of actual gap flashover.

*Step 2.* Determine a gap distance that provides a withstand voltage[5] greater than or equal to the one selected in the first step.[6]

*Step 3.* Using 110 percent of the gap's critical flashover voltage, determine the electrical component of the minimum approach distance from Equation (2) or Table 6, which is a tabulation of distance vs. withstand voltage based on Equation (2).

*Step 4.* Add the 1-foot ergonomic component to obtain the total minimum approach distance to be maintained by the employee.

2. *Second Method—Determining the necessary protective gap length from a desired (reduced) minimum approach distance.*

*Step 1.* Determine the desired minimum approach distance for the employee. Subtract the 1-foot ergonomic component of the minimum approach distance.

*Step 2.* Using this distance, calculate the air gap withstand voltage from Equation (2). Alternatively, find the voltage corresponding to the distance in Table 6.[7]

---

[1] Sigma, $\sigma$, is the symbol for standard deviation.

[2] The detailed design of a circuit interrupter, such as the design of the contacts, of resistor insertion, and of breaker timing control, are beyond the scope of this appendix. These features are routinely provided as part of the design for the system. Only features that can limit the maximum switching transient overvoltage on a system are discussed in this appendix.

[3] Surge arrestor application is beyond the scope of this appendix. However, if the arrester is installed near the work site, the application would be similar to protective gaps as discussed in paragraph IV.D of this appendix.

[4] Since a given rod gap of a given configuration corresponds to a certain withstand voltage, this method can also be used to determine the minimum approach distance for a known gap.

*Step 3.* Select a protective gap distance corresponding to a critical flashover voltage that, when multiplied by 110 percent, is less than or equal to the withstand voltage from Step 2.

*Step 4.* Calculate the withstand voltage of the protective gap (85 percent of the critical flashover voltage) to ensure that it provides an acceptable risk of flashover during the time the gap is installed.

### TABLE 6—WITHSTAND DISTANCES FOR TRANSIENT OVERVOLTAGES

| Crest voltage (kV) | Withstand distance (in feet) air gap |
|---|---|
| 100 | 0.71 |
| 150 | 1.06 |
| 200 | 1.41 |
| 250 | 1.77 |
| 300 | 2.12 |
| 350 | 2.47 |
| 400 | 2.83 |
| 450 | 3.18 |
| 500 | 3.54 |
| 550 | 3.89 |
| 600 | 4.24 |
| 650 | 4.60 |
| 700 | 5.17 |
| 750 | 5.73 |
| 800 | 6.31 |
| 850 | 6.91 |
| 900 | 7.57 |
| 950 | 8.23 |
| 1000 | 8.94 |
| 1050 | 9.65 |
| 1100 | 10.42 |
| 1150 | 11.18 |
| 1200 | 12.05 |
| 1250 | 12.90 |
| 1300 | 13.79 |
| 1350 | 14.70 |
| 1400 | 15.64 |
| 1450 | 16.61 |
| 1500 | 17.61 |
| 1550 | 18.63 |

Source: Calculations are based on Equation (2).
NOTE: The air gap is based on the 60-Hz rod-gap withstand distance.

3. *Sample protective gap calculations.*

*Problem 1:* Work is to be performed on a 500-kV transmission line that is subject to transient overvoltages of 2.4 p.u. The maximum operating voltage of the line is 552 kV. Determine the length of the protective gap that will provide the minimum practical safe approach distance. Also, determine what that minimum approach distance is.

*Step 1.* Calculate the smallest practical maximum transient overvoltage (1.25 times the crest line-to-ground voltage):[8]

$$552 \text{ kV} \times \frac{\sqrt{2}}{\sqrt{3}} \times 1.25 = 563 \text{ kV.}$$

This will be the withstand voltage of the protective gap.

*Step 2.* Using test data for a particular protective gap, select a gap that has a critical flashover voltage greater than or equal to:

$$563 \text{ kV} \div 0.85 = 662 \text{ kV.}$$

For example, if a protective gap with a 4.0-foot spacing tested to a critical flashover voltage of 665 kV, crest, select this gap spacing.

*Step 3.* This protective gap corresponds to a 110 percent of critical flashover voltage value of:

$$665 \text{ kV} \times 1.10 = 732 \text{ kV.}$$

This corresponds to the withstand voltage of the electrical component of the minimum approach distance.

*Step 4.* Using this voltage in Equation (2) results in an electrical component of the minimum approach distance of:

$$D = (0.01 + 0.0006) \times \frac{552 \text{ kV}}{\sqrt{3}} = 5.5 \text{ ft}$$

*Step 5.* Add 1 foot to the distance calculated in step 4, resulting in a total minimum approach distance of 6.5 feet.

*Problem 2:* For a line operating at a maximum voltage of 552 kV subject to a maximum transient overvoltage of 2.4 p.u., find a protective gap distance that will permit the use of a 9.0-foot minimum approach distance. (A minimum approach distance of 11 feet, 3 inches is normally required.)

*Step 1.* The electrical component of the minimum approach distance is 8.0 feet (9.0-1.0).

*Step 2.* From Table 6, select the withstand voltage corresponding to a distance of 8.0 feet. By interpolation:

$$900 \text{ kV} + \left(50 \times \frac{(8.00 - 7.57)}{(8.23 - 7.57)}\right) = 933 \text{ kV}$$

*Step 3.* The voltage calculated in Step 2 corresponds to 110 percent of the critical flashover voltage of the gap that should be employed. Using test data for a particular protective gap, select a gap that has a critical flashover voltage less than or equal to:

$$D = (0.01 + 0.0006) \times \frac{732 \text{ kV}}{\sqrt{2}}$$

For example, if a protective gap with a 5.8-foot spacing tested to a critical flashover voltage of 820 kV, crest, select this gap spacing.

*Step 4.* The withstand voltage of this protective gap would be:

$$820 \text{ kV} \times 0.85 = 697 \text{ kV.}$$

The maximum operating crest voltage would be:

$$552 \text{ kV} \times \frac{\sqrt{2}}{\sqrt{3}} = 449 \text{ kV.}$$

The crest withstand voltage of the protective gap in per unit is thus:

$$697 \text{ kV} + 449 \text{ kV} = 1.55 \text{ p.u.}$$

If this is acceptable, the protective gap could be installed with a 5.8-foot spacing, and the minimum approach distance could then be reduced to 9.0 feet.

4. *Comments and variations.* The 1-foot ergonomic component of the minimum approach distance must be added to the electrical component of the minimum approach distance calculated under paragraph IV.D of this appendix. The calculations may be varied by starting with the protective gap distance or by starting with the minimum approach distance.

### E. Location of Protective Gaps

1. Installation of the protective gap on a structure adjacent to the work site is an acceptable practice, as this does not significantly reduce the protection afforded by the gap.

---

[5] The withstand voltage for the gap is equal to 85 percent of its critical flashover voltage.

[6] Switch steps 1 and 2 if the length of the protective gap is known. The withstand voltage must then be checked to ensure that it provides an acceptable probability of gap flashover. In general, it should be at least 1.25 times the maximum crest operating voltage.

[7] Since the value of the saturation factor, a, in Equation (2) is dependent on the maximum voltage, several iterative computations may be necessary to determine the correct withstand voltage using the equation. A graph of withstand voltage vs. distance is given in ANSI/IEEE Std. 516, 1987. This graph could also be used to determine the appropriate withstand voltage for the minimum approach distance involved.

[8] To eliminate unwanted flashovers due to minor system disturbances, it is desirable to have the crest withstand voltage no lower than 1.25 p.u.

2. Gaps installed at terminal stations of lines or circuits provide a given level of protection. The level may not, however, extend throughout the length of the line to the worksite. The use of gaps at terminal stations must be studied in depth. The use of substation terminal gaps raises the possibility that separate surges could enter the line at opposite ends, each with low enough magnitude to pass the terminal gaps without flashover. When voltage surges are initiated simultaneously at each end of a line and travel toward each other, the total voltage on the line at the point where they meet is the arithmetic sum of the two surges. A gap that is installed within 0.5 mile of the work site will protect against such intersecting waves. Engineering studies of a particular line or system may indicate that adequate protection can be provided by even more distant gaps.

3. If protective gaps are used at the work site, the work site impulse insulation strength is established by the gap setting. Lightning strikes as much as 6 miles away from the worksite may cause a voltage surge greater than the insulation withstand voltage, and a gap flashover may occur. The flashover will not occur between the employee and the line, but across the protective gap instead.

4. There are two reasons to disable the automatic-reclosing feature of circuit-interrupting devices while employees are performing live-line maintenance:

- To prevent the reenergizing of a circuit faulted by actions of a worker, which could possibly create a hazard or compound injuries or damage produced by the original fault;
- To prevent any transient overvoltage caused by the switching surge that would occur if the circuit were reenergized.

However, due to system stability considerations, it may not always be feasible to disable the automatic-reclosing feature.

### APPENDIX C TO §1910.269—
### PROTECTION FROM STEP AND TOUCH POTENTIALS

#### I. Introduction

When a ground fault occurs on a power line, voltage is impressed on the "grounded" object faulting the line. The voltage to which this object rises depends largely on the voltage on the line, on the impedance of the faulted conductor, and on the impedance to "true," or "absolute," ground represented by the object. If the object causing the fault represents a relatively large impedance, the voltage impressed on it is essentially the phase-to-ground system voltage. However, even faults to well grounded transmission towers or substation structures can result in hazardous voltages.[1] The degree of the hazard depends upon the magnitude of the fault current and the time of exposure.

#### II. Voltage-Gradient Distribution

##### A. Voltage-Gradient Distribution Curve

The dissipation of voltage from a grounding electrode (or from the grounded end of an energized grounded object) is called the ground potential gradient. Voltage drops associated with this dissipation of voltage are called ground potentials. Figure 1 is a typical voltage-gradient distribution curve (assuming a uniform soil texture). This graph shows that voltage decreases rapidly with increasing distance from the grounding electrode.

**FIGURE 1—TYPICAL VOLTAGE-GRADIENT DISTRIBUTION CURVE**

##### B. Step and Touch Potentials

"Step potential" is the voltage between the feet of a person standing near an energized grounded object. It is equal to the difference in voltage, given by the voltage distribution curve, between two points at different distances from the "electrode". A person could be at risk of injury during a fault simply by standing near the grounding point.

"Touch potential" is the voltage between the energized object and the feet of a person in contact with the object. It is equal to the difference in voltage between the object (which is at a distance of 0 feet) and a point some distance away. It should be noted that the touch potential could be nearly the full voltage across the grounded object if that object is grounded at a point remote from the place where the person is in contact with it. For example, a crane that was grounded to the system neutral and that contacted an energized line would expose any person in contact with the crane or its uninsulated load line to a touch potential nearly equal to the full fault voltage.

---

[1] This appendix provides information primarily with respect to employee protection from contact between equipment being used and an energized power line. The information presented is also relevant to ground faults to transmission towers and substation structures; however, grounding systems for these structures should be designed to minimize the step and touch potentials involved.

**FIGURE 2—STEP AND TOUCH POTENTIALS**

**FIGURE 3—PROTECTION FROM GROUND-POTENTIAL GRADIENTS**

### C. Protection From the Hazards of Ground-Potential Gradients.

An engineering analysis of the power system under fault conditions can be used to determine whether or not hazardous step and touch voltages will develop. The result of this analysis can ascertain the need for protective measures and can guide the selection of appropriate precautions.

Several methods may be used to protect employees from hazardous ground-potential gradients, including equipotential zones, insulating equipment, and restricted work areas.

1. The creation of an equipotential zone will protect a worker standing within it from hazardous step and touch potentials. (See Figure 3.) Such a zone can be produced through the use of a metal mat connected to the grounded object. In some cases, a grounding grid can be used to equalize the voltage within the grid. Equipotential zones will not, however, protect employees who are either wholly or partially outside the protected area. Bonding conductive objects in the immediate work area can also be used to minimize the potential between the objects and between each object and ground. (Bonding an object outside the work area can increase the touch potential to that object in some cases, however.)

2. The use of insulating equipment, such as rubber gloves, can protect employees handling grounded equipment and conductors from hazardous touch potentials. The insulating equipment must be rated for the highest voltage that can be impressed on the grounded objects under fault conditions (rather than for the full system voltage).

3. Restricting employees from areas where hazardous step or touch potentials could arise can protect employees not directly involved in the operation being performed. Employees on the ground in the vicinity of transmission structures should be kept at a distance where step voltages would be insufficient to cause injury. Employees should not handle grounded conductors or equipment likely to become energized to hazardous voltages unless the employees are within an equipotential zone or are protected by insulating equipment.

## APPENDIX D TO §1910.269—METHODS OF INSPECTING AND TESTING WOOD POLES

### I. Introduction

When work is to be performed on a wood pole, it is important to determine the condition of the pole before it is climbed. The weight of the employee, the weight of equipment being installed, and other working stresses (such as the removal or retensioning of conductors) can lead to the failure of a defective pole or one that is not designed to handle the additional stresses.[1] For these reasons, it is essential that an inspection and test of the condition of a wood pole be performed before it is climbed.

If the pole is found to be unsafe to climb or to work from, it must be secured so that it does not fail while an employee is on it. The pole can be secured by a line truck boom, by ropes or guys, or by lashing a new pole alongside it. If a new one is lashed alongside the defective pole, work should be performed from the new one.

### II. Inspection of Wood Poles

Wood poles should be inspected by a qualified employee for the following conditions:[2]

### A. General Condition

The pole should be inspected for buckling at the ground line and for an unusual angle with respect to the ground. Buckling and odd angles may indicate that the pole has rotted or is broken.

### B. Cracks

The pole should be inspected for cracks. Horizontal cracks perpendicular to the grain of the wood may weaken the pole. Vertical ones, although not considered to be a sign of a defective pole, can pose a hazard to the climber, and the employee should keep his or her gaffs away from them while climbing.

---

[1] A properly guyed pole in good condition should, at a minimum, be able to handle the weight of an employee climbing it.

[2] The presence of any of these conditions is an indication that the pole may not be safe to climb or to work from. The employee performing the inspection must be qualified to make a determination as to whether or not it is safe to perform the work without taking additional precautions.

### C. Holes
Hollow spots and woodpecker holes can reduce the strength of a wood pole.

### D. Shell Rot and Decay
Rotting and decay are cutout hazards and are possible indications of the age and internal condition of the pole.

### E. Knots
One large knot or several smaller ones at the same height on the pole may be evidence of a weak point on the pole.

### F. Depth of Setting
Evidence of the existence of a former ground line substantially above the existing ground level may be an indication that the pole is no longer buried to a sufficient extent.

### G. Soil Conditions
Soft, wet, or loose soil may not support any changes of stress on the pole.

### H. Burn Marks
Burning from transformer failures or conductor faults could damage the pole so that it cannot withstand mechanical stress changes.

## III. Testing of Wood Poles
The following tests, which have been taken from §1910.268(n)(3), are recognized as acceptable methods of testing wood poles:

### A. Hammer Test
Rap the pole sharply with a hammer weighing about 3 pounds, starting near the ground line and continuing upwards circumferentially around the pole to a height of approximately 6 feet. The hammer will produce a clear sound and rebound sharply when striking sound wood. Decay pockets will be indicated by a dull sound or a less pronounced hammer rebound. Also, prod the pole as near the ground line as possible using a pole prod or a screwdriver with a blade at least 5 inches long. If substantial decay is encountered, the pole is considered unsafe.

### B. Rocking Test
Apply a horizontal force to the pole and attempt to rock it back and forth in a direction perpendicular to the line. Caution must be exercised to avoid causing power lines to swing together. The force may be applied either by pushing with a pike pole or pulling with a rope. If the pole cracks during the test, it shall be considered unsafe.

## APPENDIX E TO §1910.269—REFERENCE DOCUMENTS

The references contained in this appendix provide information that can be helpful in understanding and complying with the requirements contained in §1910.269. The national consensus standards referenced in this appendix contain detailed specifications that employers may follow in complying with the more performance-oriented requirements of OSHA's final rule. Except as specifically noted in §1910.269, however, compliance with the national consensus standards is not a substitute for compliance with the provisions of the OSHA standard.

ANSI/SIA A92.2-1990, American National Standard for Vehicle-Mounted Elevating and Rotating Aerial Devices.

ANSI C2-1993, National Electrical Safety Code.

ANSI Z133.1-1988, American National Standard Safety Requirements for Pruning, Trimming, Repairing, Maintaining, and Removing Trees, and for Cutting Brush.

ANSI/ASME B20.1-1990, Safety Standard for Conveyors and Related Equipment.

ANSI/IEEE Std. 4-1978 (Fifth Printing), IEEE Standard Techniques for High-Voltage Testing.

ANSI/IEEE Std. 100-1988, IEEE Standard Dictionary of Electrical and Electronic Terms.

ANSI/IEEE Std. 516-1987, IEEE Guide for Maintenance Methods on Energized Power-Lines.

ANSI/IEEE Std. 935-1989, IEEE Guide on Terminology for Tools and Equipment to Be Used in Live Line Working.

ANSI/IEEE Std. 957-1987, IEEE Guide for Cleaning Insulators.

ANSI/IEEE Std. 978-1984 (R1991), IEEE Guide for In-Service Maintenance and Electrical Testing of Live-Line Tools.

ASTM D 120-87, Specification for Rubber Insulating Gloves.

ASTM D 149-92, Test Method for Dielectric Breakdown Voltage and Dielectric Strength of Solid Electrical Insulating Materials at Commercial Power Frequencies.

ASTM D 178-93, Specification for Rubber Insulating Matting.

ASTM D 1048-93, Specification for Rubber Insulating Blankets.

ASTM D 1049-93, Specification for Rubber Insulating Covers.

ASTM D 1050-90, Specification for Rubber Insulating Line Hose.

ASTM D 1051-87, Specification for Rubber Insulating Sleeves.

ASTM F 478-92, Specification for In-Service Care of Insulating Line Hose and Covers.

ASTM F 479-93, Specification for In-Service Care of Insulating Blankets.

ASTM F 496-93b, Specification for In-Service Care of Insulating Gloves and Sleeves.

ASTM F 711-89, Specification for Fiberglass-Reinforced Plastic (FRP) Rod and Tube Used in Live Line Tools.

ASTM F 712-88, Test Methods for Electrically Insulating Plastic Guard Equipment for Protection of Workers.

ASTM F 819-83a (1988), Definitions of Terms Relating to Electrical Protective Equipment for Workers.

ASTM F 855-90, Specifications for Temporary Grounding Systems to Be Used on De-Energized Electric Power Lines and Equipment.

ASTM F 887-91a, Specifications for Personal Climbing Equipment.

ASTM F 914-91, Test Method for Acoustic Emission for Insulated Aerial Personnel Devices.

ASTM F 968-93, Specification for Electrically Insulating Plastic Guard Equipment for Protection of Workers.

ASTM F 1116-88, Test Method for Determining Dielectric Strength of Overshoe Footwear.

ASTM F 1117-87, Specification for Dielectric Overshoe Footwear.

ASTM F 1236-89, Guide for Visual Inspection of Electrical Protective Rubber Products.

ASTM F 1505-94, Standard Specification for Insulated and Insulating Hand Tools.

ASTM F 1506-94, Standard Performance Specification for Textile Materials for Wearing Apparel for Use by Electrical Workers Exposed to Momentary Electric Arc and Related Thermal Hazards.

IEEE Std. 62-1978, IEEE Guide for Field Testing Power Apparatus Insulation.

IEEE Std. 524-1992, IEEE Guide to the Installation of Overhead Transmission Line Conductors.

IEEE Std. 1048-1990, IEEE Guide for Protective Grounding of Power Lines.

IEEE Std. 1067-1990, IEEE Guide for the In-Service Use, Care, Maintenance, and Testing of Conductive Clothing for Use on Voltages up to 765 kV AC.

[59 FR 4437, Jan. 31, 1994; 59 FR 33658, June 30, 1994, as amended at 59 FR 4458, Jan. 31, 1994; 59 FR 40729, Aug. 9, 1994; 59 FR 51748, Oct. 12, 1994]

## §1910.272 Grain handling facilities.

**1910.272(a)** *Scope.* This section contains requirements for the control of grain dust fires and explosions, and certain other safety hazards associated with grain handling facilities. It applies in addition to all other relevant provisions of part 1910 (or part 1917 at marine terminals).

**1910.272(b)** *Application.* **(1)** Paragraphs (a) through (n) of this section apply to grain elevators, feed mills, flour mills, rice mills, dust pelletizing plants, dry corn mills, soybean flaking operations, and the dry grinding operations of soycake.

**1910.272(b)(2)** Paragraphs (o), (p), and (q) of this section apply only to grain elevators.

## 1910.272(c) Definitions.

*Choked leg* means a condition of material buildup in the bucket elevator that results in the stoppage of material flow and bucket movement. A bucket elevator is not considered choked that has the up-leg partially or fully loaded and has the boot and discharge cleared allowing bucket movement.

*Flat storage structure* means a grain storage building or structure that will not empty completely by gravity, has an unrestricted ground level opening for entry, and must be entered to reclaim the residual grain using powered equipment or manual means.

*Fugitive grain dust* means combustible dust particles, emitted from the stock handling system, of such size as will pass through a U.S. Standard 40 mesh sieve (425 microns or less).

*Grain elevator* means a facility engaged in the receipt, handling, storage, and shipment of bulk raw agricultural commodities such as corn, wheat, oats, barley, sunflower seeds, and soybeans.

*Hot work* means work involving electric or gas welding, cutting, brazing, or similar flame producing operations.

*Inside bucket elevator* means a bucket elevator that has the boot and more than 20 percent of the total leg height (above grade or ground level) inside the grain elevator structure. Bucket elevators with leg casings that are inside (and pass through the roofs) of rail or truck dump sheds with the remainder of the leg outside of the grain elevator structure, are not considered inside bucket elevators.

*Jogging* means repeated starting and stopping of drive motors in an attempt to clear choked legs.

*Lagging* means a covering on drive pulleys used to increase the coefficient of friction between the pulley and the belt.

*Permit* means the written certification by the employer authorizing employees to perform identified work operations subject to specified precautions.

**1910.272(d) Emergency action plan.** The employer shall develop and implement an emergency action plan meeting the requirements contained in §1910.38.

**1910.272(e) Training. (1)** The employer shall provide training to employees at least annually and when changes in job assignment will expose them to new hazards. Current employees, and new employees prior to starting work, shall be trained in at least the following:

**1910.272(e)(1)(i)** General safety precautions associated with the facility, including recognition and preventive measures for the hazards related to dust accumulations and common ignition sources such as smoking; and,

**1910.272(e)(1)(ii)** Specific procedures and safety practices applicable to their job tasks including but not limited to, cleaning procedures for grinding equipment, clearing procedures for choked legs, housekeeping procedures, hot work procedures, preventive maintenance procedures and lock-out/tag-out procedures.

**1910.272(e)(2)** Employees assigned special tasks, such as bin entry and handling of flammable or toxic substances, shall be provided training to perform these tasks safely.

**NOTE to paragraph (e)(2):** Training for an employee who enters grain storage structures includes training about engulfment and mechanical hazards and how to avoid them.

**1910.272(f) Hot work permit. (1)** The employer shall issue a permit for all hot work, with the following exceptions:

**1910.272(f)(1)(i)** Where the employer or the employer's representative (who would otherwise authorize the permit) is present while the hot work is being performed;

**1910.272(f)(1)(ii)** In welding shops authorized by the employer;

**1910.272(f)(1)(iii)** In hot work areas authorized by the employer which are located outside of the grain handling structure.

**1910.272(f)(2)** The permit shall certify that the requirements contained in §1910.252(a) have been implemented prior to beginning the hot work operations. The permit shall be kept on file until completion of the hot work operations.

**1910.272(g) Entry into grain storage structures.** This paragraph applies to employee entry into bins, silos, tanks, and other grain storage structures. *Exception:* Entry through unrestricted ground level openings into flat storage structures in which there are no toxicity, flammability, oxygen-deficiency, or other atmospheric hazards is covered by paragraph (h) of this section. For the purposes of this paragraph (g), the term "grain" includes raw and processed grain and grain products in facilities within the scope of paragraph (b)(1) of this section.

**1910.272(g)(1)** The following actions shall be taken before employees enter bins, silos, or tanks:

**1910.272(g)(1)(i)** The employer shall issue a permit for entering bins, silos, or tanks unless the employer or the employer's representative (who would otherwise authorize the permit) is present during the entire operation. The permit shall certify that the precautions contained in this paragraph (§1910.272(g)) have been implemented prior to employees entering bins, silos or tanks. The permit shall be kept on file until completion of the entry operations.

**1910.272(g)(1)(ii)** All mechanical, electrical, hydraulic, and pneumatic equipment which presents a danger to employees inside grain storage structures shall be deenergized and shall be disconnected, locked-out and tagged, blocked-off, or otherwise prevented from operating by other equally effective means or methods.

**1910.272(g)(1)(iii)** The atmosphere within a bin, silo, or tank shall be tested for the presence of combustible gases, vapors, and toxic agents when the employer has reason to believe they may be present. Additionally, the atmosphere within a bin, silo, or tank shall be tested for oxygen content unless there is continuous natural air movement or continuous forced-air ventilation before and during the period employees are inside. If the oxygen level is less than 19.5%, or if combustible gas or vapor is detected in excess of 10% of the lower flammable limit, or if toxic agents are present in excess of the ceiling values listed in subpart Z of 29 CFR part 1910, or if toxic agents are present in concentrations that will cause health effects which prevent employees from effecting self-rescue or communication to obtain assistance, the following provisions apply.

**1910.272(g)(1)(iii)(A)** Ventilation shall be provided until the unsafe condition or conditions are eliminated, and the ventilation shall be continued as long as there is a possibility of recurrence of the unsafe condition while the bin, silo, or tank is occupied by employees.

**1910.272(g)(1)(iii)(B)** If toxicity or oxygen deficiency cannot be eliminated by ventilation, employees entering the bin, silo, or tank shall wear an appropriate respirator. Respirator use shall be in accordance with the requirements of §1910.134.

**1910.272(g)(1)(iv)** "Walking down grain" and similar practices where an employee walks on grain to make it flow within or out from a grain storage structure, or where an employee is on moving grain, are prohibited.

**1910.272(g)(2)** Whenever an employee enters a grain storage structure from a level at or above the level of the stored grain or grain products, or whenever an employee walks or stands on or in stored grain of a depth which poses an engulfment hazard, the employer shall equip the employee with a body harness with lifeline, or a boatswain's chair that meets the requirements of subpart D of this part. The lifeline shall be so positioned, and of sufficient length, to prevent the employee from sinking further than waist-deep in the grain. *Exception:* Where the employer can demonstrate that the protection required by this paragraph is not feasible or creates a greater hazard, the employer shall provide an alternative means of protection which is demonstrated to prevent the employee from sinking further than waist-deep in the grain.

**NOTE to paragraph (g)(2):** When the employee is standing or walking on a surface which the employer demonstrates is free from engulfment hazards, the lifeline or alternative means may be disconnected or removed.

**1910.272(g)(3)** An observer, equipped to provide assistance, shall be stationed outside the bin, silo, or tank being entered by an employee. Communications (visual, voice, or signal line) shall be maintained between the observer and employee entering the bin, silo, or tank.

**1910.272(g)(4)** The employer shall provide equipment for rescue operations which is specifically suited for the bin, silo, or tank being entered.

**1910.272(g)(5)** The employee acting as observer shall be trained in rescue procedures, including notification methods for obtaining additional assistance.

**1910.272(g)(6)** Employees shall not enter bins, silos, or tanks underneath a bridging condition, or where a buildup of grain products on the sides could fall and bury them.

**1910.272(h)** *Entry into flat storage structures.* For the purposes of this paragraph (h), the term "grain" means raw and processed grain and grain products in facilities within the scope of paragraph (b)(1) of this section.

**1910.272(h)(1)** Each employee who walks or stands on or in stored grain, where the depth of the grain poses an engulfment hazard, shall be equipped with a lifeline or alternative means which the employer demonstrates will prevent the employee from sinking further than waist-deep into the grain.

**NOTE to paragraph (h)(1):** When the employee is standing or walking on a surface which the employer demonstrates is free from engulfment hazards, the lifeline or alternative means may be disconnected or removed.

**1910.272(h)(2)(i)** Whenever an employee walks or stands on or in stored grain or grain products of a depth which poses an engulfment hazard, all equipment which presents a danger to that employee (such as an auger or other grain transport equipment) shall be deenergized, and shall be disconnected, locked-out and tagged, blocked-off, or otherwise prevented from operating by other equally effective means or methods.

**1910.272(h)(2)(ii)** "Walking down grain" and similar practices where an employee walks on grain to make it flow within or out from a grain storage structure, or where an employee is on moving grain, are prohibited.

**1910.272(h)(3)** No employee shall be permitted to be either underneath a bridging condition, or in any other location where an accumulation of grain on the sides or elsewhere could fall and engulf that employee.

**1910.272(i)** *Contractors.* **(1)** The employer shall inform contractors performing work at the grain handling facility of known potential fire and explosion hazards related to the contractor's work and work area. The employer shall also inform contractors of the applicable safety rules of the facility.

**1910.272(i)(2)** The employer shall explain the applicable provisions of the emergency action plan to contractors.

**1910.272(j)** *Housekeeping.* **(1)** The employer shall develop and implement a written housekeeping program that establishes the frequency and method(s) determined best to reduce accumulations of fugitive grain dust on ledges, floors, equipment, and other exposed surfaces.

**1910.272(j)(2)** In addition, the housekeeping program for *grain elevators* shall address fugitive grain dust accumulations at priority housekeeping areas.

**1910.272(j)(2)(i)** Priority housekeeping areas shall include *at least* the following:

**1910.272(j)(2)(i)(A)** Floor areas within 35 feet (10.7 m) of inside bucket elevators;

**1910.272(j)(2)(i)(B)** Floors of enclosed areas containing grinding equipment;

**1910.272(j)(2)(i)(C)** Floors of enclosed areas containing grain dryers located inside the facility.

**1910.272(j)(2)(ii)** The employer shall immediately remove any fugitive grain dust accumulations whenever they exceed 1/8 inch (.32 cm) at priority housekeeping areas, pursuant to the housekeeping program, or shall demonstrate and assure, through the development and implementation of the housekeeping program, that equivalent protection is provided.

**1910.272(j)(3)** The use of compressed air to blow dust from ledges, walls, and other areas shall only be permitted when all machinery that presents an ignition source in the area is shut-down, and all other known potential ignition sources in the area are removed or controlled.

**1910.272(j)(4)** Grain and product spills shall not be considered fugitive grain dust accumulations. However, the housekeeping program shall address the procedures for removing such spills from the work area.

**1910.272(k)** *Grate openigs [sic openings].* Receiving-pit feed openings, such as truck or railcar receiving-pits, shall be covered by grates. The width of openings in the grates shall be a maximum of 2 1/2 inches (6.35 cm).

**1910.272(l)** *Filter collectors.* **(1)** All fabric dust filter collectors which are a part of a pneumatic dust collection system shall be equipped with a monitoring device that will indicate a pressure drop across the surface of the filter.

**1910.272(l)(2)** Filter collectors installed after March 30, 1988 shall be:

**1910.272(l)(2)(i)** Located outside the facility; or

**1910.272(l)(2)(ii)** Located in an area inside the facility protected by an explosion suppression system; or

**1910.272(l)(2)(iii)** Located in an area inside the facility that is separated from other areas of the facility by construction having at least a one hour fire-resistance rating, and which is adjacent to an exterior wall and vented to the outside. The vent and ductwork shall be designed to resist rupture due to deflagration.

**1910.272(m)** *Preventive maintenance.* **(1)** The employer shall implement preventive maintenance procedures consisting of:

**1910.272(m)(1)(i)** Regularly scheduled inspections of at least the mechanical and safety control equipment associated with dryers, grain stream processing equipment, dust collection equipment including filter collectors, and bucket elevators;

**1910.272(m)(1)(ii)** Lubrication and other appropriate maintenance in accordance with manufacturers' recommendations, or as determined necessary by prior operating records.

**1910.272(m)(2)** The employer shall promptly correct dust collection systems which are malfunctioning or which are operating below designed efficiency. Additionally, the employer shall promptly correct, or remove from service, overheated bearings and slipping or misaligned belts associated with inside bucket elevators.

**1910.272(m)(3)** A certification record shall be maintained of each inspection, performed in accordance with this paragraph (m), containing the date of the inspection, the name of the person who performed the inspection and the serial number, or other identifier, of the equipment specified in paragraph (m)(1)(i) of this section that was inspected.

**1910.272(m)(4)** The employer shall implement procedures for the use of tags and locks which will prevent the inadvertent application of energy or motion to equipment being repaired, serviced, or adjusted, which could result in employee injury. Such locks and tags shall be removed in accordance with established procedures only by the employee installing them or, if unavailable, by his or her supervisor.

**1910.272(n)** *Grain stream processing equipment.* The employer shall equip grain stream processing equipment (such as hammer mills, grinders, and pulverizers) with an effective means of removing ferrous material from the incoming grain stream.

**1910.272(o)** *Emergency escape.* **(1)** The employer shall provide at least two means of emergency escape from galleries (bin decks).

**1910.272(o)(2)** The employer shall provide at least one means of emergency escape in tunnels of existing grain elevators. Tunnels in grain elevators constructed after the effective date of this standard shall be provided with at least two means of emergency escape.

**1910.272(p)** *Continuous-flow bulk raw grain dryers.* **(1)** All direct-heat grain dryers shall be equipped with automatic controls that:

**1910.272(p)(1)(i)** Will shut-off the fuel supply in case of power or flame failure or interruption of air movement through the exhaust fan; and,

**1910.272(p)(1)(ii)** Will stop the grain from being fed into the dryer if excessive temperature occurs in the exhaust of the drying section.

**1910.272(p)(2)** Direct-heat grain dryers installed after March 30, 1988 shall be:

**1910.272(p)(2)(i)** Located outside the grain elevator; or

**1910.272(p)(2)(ii)** Located in an area inside the grain elevator protected by a fire or explosion suppression system; or

**1910.272(p)(2)(iii)** Located in an area inside the grain elevator which is separated from other areas of the facility by construction having at least a one hour fire-resistance rating.

**1910.272(q)** *Inside bucket elevators.* **(1)** Bucket elevators shall not be jogged to free a choked leg.

**1910.272(q)(2)** All belts and lagging purchased after March 30, 1988 shall be conductive. Such belts shall have a surface electrical resistance not to exceed 300 megohms.

**1910.272(q)(3)** All bucket elevators shall be equipped with a means of access to the head pulley section to allow inspection of the head pulley, lagging, belt, and discharge throat of the elevator head. The boot section shall also be provided with a means of access for clean-out of the boot and for inspection of the boot, pulley, and belt.

**1910.272(q)(4)** All the employer shall:

**1910.272(q)(4)(i)** Mount bearings externally to the leg casing; or,

**1910.272(q)(4)(ii)** Provide vibration monitoring, temperature monitoring, or other means to monitor the condition of those bearings mounted inside or partially-inside the leg casing.

**1910.272(q)(5)** All the employer shall equip bucket elevators with a motion detection device which will shut-down the bucket elevator when the belt speed is reduced by no more than 20% of the normal operating speed.

**1910.272(q)(6)** All the employer shall:

**1910.272(q)(6)(i)** Equip bucket elevators with a belt alignment monitoring device which will initiate an alarm to employees when the belt is not tracking properly; or,

**1910.272(q)(6)(ii)** Provide a means to keep the belt tracking properly, such as a system that provides constant alignment adjustment of belts.

**1910.272(q)(7)** Paragraphs (q)(5) and (q)(6) of this section do not apply to grain elevators having a permanent storage capacity of less than one million bushels, provided that daily visual inspection is made of bucket movement and tracking of the belt.

**1910.272(q)(8)** Paragraphs (q)(4), (q)(5), and (q)(6) of this section do not apply to the following:

**1910.272(q)(8)(i)** Bucket elevators which are equipped with an operational fire and explosion suppression system capable of protecting at least the head and boot section of the bucket elevator; or,

**1910.272(q)(8)(ii)** Bucket elevators which are equipped with pneumatic or other dust control systems or methods that keep the dust concentration inside the bucket elevator at least 25% below the lower explosive limit at all times during operations.

**NOTE:** The following appendices to §1910.272 serve as nonmandatory guidelines to assist employers and employees in complying with the requirements of this section, as well as to provide other helpful information.

No additional burdens are imposed through these appendices.

## APPENDIX A TO §1910.272— GRAIN HANDLING FACILITIES

Examples presented in this appendix may not be the only means of achieving the performance goals in the standard.

### 1. Scope and Application

The provisions of this standard apply in addition to any other applicable requirements of this part 1910 (or part 1917 at marine terminals). The standard contains requirements for new and existing grain handling facilities. The standard does not apply to seed plants which handle and prepare seeds for planting of future crops, nor to on-farm storage or feed lots.

### 2. Emergency Action Plan

The standard requires the employer to develop and implement an emergency action plan. The emergency action plan (§1910.38) covers those designated actions employers and employees are to take to ensure employee safety from fire and other emergencies. The plan specifies certain minimum elements which are to be addressed. These elements include the establishment of an employee alarm system, the development of evacuation procedures, and training employees in those actions they are to take during an emergency.

The standard does not specify a particular method for notifying employees of an emergency. Public announcement systems, air horns, steam whistles, a standard fire alarm system, or other types of employee alarm may be used. However, employers should be aware that employees in a grain facility may have difficulty hearing an emergency alarm, or distinguishing an emergency alarm from other audible signals at the facility, or both. Therefore, it is important that the type of employee alarm used be distinguishable and distinct.

The use of floor plans or workplace maps which clearly show the emergency escape routes should be included in the emergency action plan; color coding will aid employees in determining their route assignments. The employer should designate a safe area, outside the facility, where employees can congregate after evacuation, and implement procedures to account for all employees after emergency evacuation has been completed.

It is also recommended that employers seek the assistance of the local fire department for the purpose of preplanning for emergencies. Preplanning is encouraged to facilitate coordination and cooperation between facility personnel and those who may be called upon for assistance during an emergency. It is important for emergency service units to be aware of the usual work locations of employees at the facility.

### 3. Training

It is important that employees be trained in the recognition and prevention of hazards associated with grain facilities, especially those hazards associated with their own work tasks. Employees should understand the factors which are necessary to produce a fire or explosion, i.e., fuel (such as grain dust), oxygen, ignition source, and (in the case of explosions) confinement. Employees should be made aware that any efforts they make to keep these factors from occurring simultaneously will be an important step in reducing the potential for fires and explosions.

The standard provides flexibility for the employer to design a training program which fulfills the needs of a facility. The type, amount, and frequency of training will need to reflect the tasks that employees are expected to perform. Although training is to be provided to employees at least annually, it is recommended that safety meetings or discussions and drills be conducted at more frequent intervals.

The training program should include those topics applicable to the particular facility, as well as topics such as: Hot work procedures; lockout/tag-out procedures; bin entry procedures; bin cleaning procedures; grain dust explosions; fire prevention; procedures for handling "hot grain"; housekeeping procedures, including methods and frequency of dust removal; pesticide and fumigant usage; proper use and maintenance of personal protective equipment; and, preventive maintenance. The types of work clothing should also be considered in the program at least to caution against using polyester clothing that easily melts and increases the severity of burns, as compared to wool or fire retardant cotton.

In implementing the training program, it is recommended that the employer utilize films, slide-tape presentations, pamphlets, and other information which can be obtained from such sources as the Grain Elevator and Processing Society, the Cooperative Extension Service of the U.S. Department of Agriculture, Kansas State University's Extension Grain Science and Industry, and other state agriculture schools, industry associations, union organizations, and insurance groups.

### 4. Hot Work Permit

The implementation of a permit system for hot work is intended to assure that employers maintain control over operations involving hot work and to assure that employees are aware of and utilize appropriate safeguards when conducting these activities.

Precautions for hot work operations are specified in 29 CFR 1910.252(a), and include such safeguards as relocating the hot work operation to a safe location if possible, relocating or covering combustible material in the vicinity, providing fire extinguishers, and provisions for establishing a fire watch. Permits are not required for hot work operations conducted in the presence of the employer or the employer's authorized representative who would otherwise issue the permit, or in an employer authorized welding shop or when work is conducted outside and away from the facility.

It should be noted that the permit is not a record, but is an authorization of the employer certifying that certain safety precautions have been implemented prior to the beginning of work operations.

### 5. Entry Into Bins, Silos, And Tanks

In order to assure that employers maintain control over employee entry into bins, silos, and tanks, OSHA is requiring that the employer issue a permit for entry into bins, silos, and tanks unless the employer (or the employer's representative who would otherwise authorize the permit) is present at the entry and during the entire operation.

Employees should have a thorough understanding of the hazards associated with entry into bins, silos, and tanks. Employees are not to be permitted to enter these spaces from the bottom when grain or other agricultural products are hung up or sticking to the sides which might fall and injure or kill an employee. Employees should be made aware that the atmosphere in bins, silos, and tanks can be oxygen deficient or toxic. Employees should be trained in the proper methods of testing the atmosphere, as well as in the appropriate procedures to be taken if the atmosphere is found to be oxygen deficient or toxic. When a fumigant has been recently applied in these areas and entry must be made, aeration fans should be running continuously to assure a safe atmosphere for those inside. Periodic monitoring of toxic levels shuld be done by direct reading instruments to measure the levels, and, if there is an increase in these readings, appropriate actions should be promptly taken.

Employees have been buried and suffocated in grain or other agricultural products because they sank into the material. Therefore, it is suggested that employees not be permitted to walk or stand on the grain or other grain product where the depth is greater than waist high. In this regard, employees must use a full body harness or boatswain's chair with a lifeline when entering from the top. A winch system with mechanical advantage (either powered or manual) would allow better control of the employee than just using a hand held hoist line, and such a system would allow the observer to remove the employee easily without having to enter the space.

It is important that employees be trained in the proper selection and use of any personal protective equipment which is to be worn. Equally important is the training of employees in the planned emergency rescue procedures. Employers should carefully read §1910.134(e)(3) and assure that their procedures follow these requirements. The employee acting as observer is to be equipped to provide assistance and is to know procedures for obtaining additional assistance. The observer should not enter a space until adequate assistance is available. It is recommended that an employee trained in CPR be readily available to provide assistance to those employees entering bins, silos, or tanks.

### 6. Contractors

These provisions of the standard are intended to ensure that outside contractors are cognizant of the hazards associated with grain handling facilities, particularly in relation to the work they are to perform for the employer. Also, in the event of an emergency, contractors should be able to take appropriate action as a part of the overall facility emergency action plan. Contractors should also be aware of the employer's permit systems. Contractors should develop specified procedures for performing hot work and for entry into bins, silos, and tanks and these activities should be coordinated with the employer. Contractors are responsible for informing their own employees.

This coordination will help to ensure that employers know what work is being performed at the facility by contractors; where it is being performed; and, that it is being performed in a manner that will not endanger employees.

### 7. Housekeeping

The housekeeping program is to be designed to keep dust accumulations and emissions under control inside grain facilities. The housekeeping program, which is to be written, is to specify the frequency and method(s) used to best reduce dust accumulations.

Ship, barge, and rail loadout and receiving areas which are located outside the facility need not be addressed in the housekeeping program. Additionally, truck dumps which are open on two or more sides need not be addressed by the housekeeping program. Other truck dumps should be addressed in the housekeeping program to provide for regular cleaning during periods of receiving grain or agricultural products. The housekeeping program should provide coverage for all workspaces in the facility and include walls, beams, etc., especially in relation to the extent that dust could accumulate.

#### Dust Accumulations

Almost all facilities will require some level of manual housekeeping. Manual housekeeping methods, such as vacuuming or sweeping with soft bristle brooms, should be used which will minimize the possibility of layered dust being suspended in the air when it is being removed.

The housekeeping program should include a contingency plan to respond to situations where dust accumulates rapidly due to a failure of a dust enclosure hood, an unexpected breakdown of the dust control system, a dust-tight connection inadvertently knocked open, etc.

The housekeeping program should also specify the manner of handling spills. Grain spills are not considered to be dust accumulations.

A fully enclosed horizontal belt conveying system where the return belt is inside the enclosure should have inspection access such as sliding panels or doors to permit checking of equipment, checking for dust accumulations and facilitate cleaning if needed.

#### Dust Emissions

Employers should analyze the entire stock handling system to determine the location of dust emissions and effective methods to control or to eliminate them. The employer should make sure that holes in spouting, casings of bucket elevators, pneumatic conveying pipes, screw augers, or drag conveyor casings, are patched or otherwise properly repaired to prevent leakage. Minimizing free falls of grain or grain products by using choke feeding techniques, and utilization of dust-tight enclosures at transfer points, can be effective in reducing dust emissions.

Each housekeeping program should specify the schedules and control measures which will be used to control dust emitted from the stock handling system. The housekeeping program should address the schedules to be used for cleaning dust accumulations from motors, critical bearings and other potential ignition sources in the working areas. Also, the areas around bucket elevator legs, milling machinery and similar equipment should be given priority in the cleaning schedule. The method of disposal of the dust which is swept or vacuumed should also be planned.

Dust may accumulate in somewhat inaccessible areas, such as those areas where ladders or scaffolds might be necessary to reach them. The employer may want to consider the use of compressed air and long lances to blow down these areas frequently. The employer may also want to consider the periodic use of water and hoselines to wash down these areas. If these methods are used, they are to be specified in the housekeeping program along with the appropriate safety precautions, including the use of personal protective equipment such as eyewear and dust respirators.

Several methods have been effective in controlling dust emissions. A frequently used method of controlling dust emissions is a pneumatic dust collection system. However, the installation of a poorly designed pneumatic dust collection system has fostered a false sense of security and has often led to an inappropriate reduction in manual housekeeping. Therefore, it is imperative that the system be designed properly and installed by a competent contractor. Those employers who have a pneumatic dust control system that is not working according to expectations should request the engineering design firm, or the manufacturer of the filter and related equipment, to conduct an evaluation of the system to determine the corrections necessary for proper operation of the system. If the design firm or manufacturer of the equipment is not known, employers should contact their trade association for recommendations of competent designers of pneumatic dust control systems who could provide assistance.

When installing a new or upgraded pneumatic control system, the employer should insist on an acceptance test period of 30 to 45 days of operation to ensure that the system is operating as intended and designed. The employer should also obtain maintenance, testing, and inspection information from the manufacturer to ensure that the system will continue to operate as designed.

Aspiration of the leg, as part of a pneumatic dust collection system, is another effective method of controlling dust emissions. Aspiration of the leg consists of a flow of air across the entire boot, which entrains the liberated dust and carries it up the up-leg to take-off points. With proper aspiration, dust concentrations in the leg can be lowered below the lower explosive limit. Where a prototype leg installation has been instrumented and shown to be effective in keeping the dust level 25% below the lower explosive limit during normal operations for the various products handled, then other legs of similar size, capacity and products being handled which have the same design criteria for the air aspiration would be acceptable to OSHA, provided the prototype test report is available on site.

Another method of controlling dust emissions is enclosing the conveying system, pressurizing the general work area, and providing a lower pressure inside the enclosed conveying system. Although this method is effective in controlling dust emissions from the conveying system, adequate access to the inside of the enclosure is necessary to facilitate frequent removal of dust accumulations. This is also necessary for those systems called "self-cleaning."

The use of edible oil sprayed on or into a moving stream of grain is another method which has been used to control dust emissions. Tests performed using this method have shown that the oil treatment can reduce dust emissions. Repeated handling of the grain may necessitate additional oil treatment to prevent liberation of dust. However, before using this method, operators of grain handling facilities should be aware that the Food and Drug Administration must approve the specific oil treatment used on products for food or feed.

As a part of the housekeeping program, grain elevators are required to address accumulations of dust at priority areas using the action level. The standard specifies a maximum accumulation of 1/8 inch dust, measurable by a ruler or other measuring device, anywhere within a priority area as the upper limit at which time employers must initiate action to remove the accumulations using designated means or methods. Any accumulation in excess of this amount and where no action has been initiated to implement cleaning would constitute a violation of the standard, unless the employer can demonstrate equivalent protection. Employers should make every effort to minimize dust accumulations on exposed surfaces since dust is the fuel for a fire or explosion, and it is recognized that a 1/8 inch dust accumulation is more than enough to fuel such occurrences.

### 8. Filter Collectors

Proper sizing of filter collectors for the pneumatic dust control system they serve is very important for the overall effectiveness of the system. The air to cloth ratio of the system should be in accordance with the manufacturer's recommendations. If higher ratios are used, they can result in more maintenance on the filter, shorter bag or sock life, increased differential pressure resulting in higher energy costs, and an increase in operational problems.

A photohelic gauge, magnehelic gauge, or manometer, may be used to indicate the pressure rise across the inlet and outlet of the filter. When the pressure exceeds the design value for the filter, the air volume will start to drop, and maintenance will be required. Any of these three monitoring devices is acceptable as meeting paragraph (l)(1) of the standard.

The employer should establish a level or target reading on the instrument which is consistent with the manufacturer's recommendations that will indicate when the filter should be serviced. This target reading on the instrument and the accompanying procedures should be in the preventive maintenance program. These efforts would minimize the blinding of the filter and the subsequent failure of the pneumatic dust control system.

There are other instruments that the employer may want to consider using to monitor the operation of the filter. One instrument is a zero motion switch for detecting a failure of motion by the rotary discharge valve on the hopper. If the rotary discharge valve stops turning, the dust released by the bag or sock will accumulate in the filter hopper until the filter becomes clogged. Another instrument is a level indicator which is installed in the hopper of the filter to detect the buildup of dust that would otherwise cause the filter hopper to be plugged. The installation of these instruments should be in accordance with manufacturer's recommendations.

All of these monitoring devices and instruments are to be capable of being read at an accessible location and checked as frequently as specified in the preventive maintenance program.

Filter collectors on portable vacuum cleaners, and those used where fans are not part of the system, are not covered by requirements of paragraph (l) of the standard.

### 9. Preventive Maintenance

The control of dust and the control of ignition sources are the most effective means for reducing explosion hazards. Preventive maintenance is related to ignition sources in the same manner as housekeeping is related to dust control and should be treated as a major function in a facility. Equipment such as critical bearings, belts, buckets, pulleys, and milling machinery are potential ignition sources, and periodic inspection and lubrication of such equipment through a scheduled preventive maintenance program is an effective method for keeping equipment functioning properly and safely. The use of vibration detection methods, heat sensitive tape or other heat detection methods that can be seen by the inspector or maintenance person will allow for a quick, accurate, and consistent evaluation of bearings and will help in the implementation of the program.

The standard does not require a specific frequency for preventive maintenance. The employer is permitted flexibility in determining the appropriate interval for maintenance provided that the effectiveness of the maintenance program can be demonstrated. Scheduling of preventive maintenance should be based on manufacturer's recommendations for effective operation, as well as from the employer's previous experience with the equipment. However, the employer's schedule for preventive maintenance should be frequent enough to allow for both prompt identification and correction of any problems concerning the failure or malfunction of the mechanical and safety control equipment associated with bucket elevators, dryers, filter collectors and magnets. The pressure-drop monitoring device for a filter collector, and the condition of the lagging on the head pulley, are examples of items that require regularly scheduled inspections. A system of identifying the date, the equipment inspected and the maintenance performed, if any, will assist employers in continually refining their preventive maintenance schedules and identifying equipment problem areas. Open work orders where repair work or replacement is to be done at a designated future date as scheduled, would be an indication of an effective preventive maintenance program.

It is imperative that the prearranged schedule of maintenance be adhered to regardless of other facility constraints. The employer should give priority to the maintenance or repair work associated with safety control equipment, such as that on dryers, magnets, alarm and shut-down systems on bucket elevators, bearings on bucket elevators, and the filter collectors in the dust control system. Benefits of a strict preventive maintenance program can be a reduction of unplanned downtime, improved equipment performance, planned use of resources, more efficient operations, and, most importantly, safer operations.

The standard also requires the employer to develop and implement procedures consisting of locking out and tagging equipment to prevent the inadvertent application of energy or motion to equipment being repaired, serviced, or adjusted, which could result in employee injury. All employees who have responsibility for repairing or servicing equipment, as well as those who operate the equipment, are to be familiar with the employer's lock and tag procedures. A lock is to be used as the positive means to prevent operation of the disconnected equipment. Tags are to be used to inform employees why equipment is locked out. Tags are to meet requirements in §1910.145(f). Locks and tags may only be removed by employees that placed them, or by their supervisor, to ensure the safety of the operation.

### 10. Grain Stream Processing Equipment

The standard requires an effective means of removing ferrous material from grain streams so that such material does not enter equipment such as hammer mills, grinders and pulverizers. Large foreign objects, such as stones, should have been removed at the receiving pit. Introduction of foreign objects and ferrous material into such equipment can produce sparks which can create an explosion hazard. Acceptable means for removal of ferrous materials include the use of permanent or electromagnets. Means used to separate foreign objects and ferrous material should be cleaned regularly and kept in good repair as part of the preventive maintenance program in order to maximize their effectiveness.

### 11. Emergency Escape

The standard specifies that at least two means of escape must be provided from galleries (bin decks). Means of emergency escape may include any available means of egress (consisting of three components, exit access, exit, and exit discharge as defined in §1910.35), the use of controlled descent devices with landing velocities not to exceed 15 ft/sec., or emergency escape ladders from galleries. Importantly, the means of emergency escape are to be addressed in the facility emergency action plan. Employees are to know the location of the nearest means of emergency escape and the action they must take during an emergency.

### 12. Dryers

Liquefied petroleum gas fired dryers should have the vaporizers installed at least ten feet from the dryer. The gas piping system should be protected from mechanical damage. The employer should establish procedures for locating and repairing leaks when there is a strong odor of gas or other signs of a leak.

### 13. Inside Bucket Elevators

Hazards associated with inside bucket elevator legs are the source of many grain elevator fires and explosions. Therefore, to mitigate these hazards, the standard requires the implementation of special safety precautions and procedures, as well as the installation of safety control devices. The standard provides for a phase-in period for many of the requirements to provide the employer time for planning the implementation of the requirements. Additionally, for elevators with a permanent storage capacity of less than one million bushels, daily visual inspection of belt alignment and bucket movement can be substituted for alignment monitoring devices and motion detection devices.

The standard requires that belts (purchased after the effective date of the standard) have surface electrical resistance not to exceed 300 megohms. Test methods available regarding electrical resistance of belts are: The American Society for Testing and Materials D257-76, "Standard Test Methods for D-C Resistance or Conductance of Insulating Materials"; and, the International Standards Organization's #284, "Conveyor Belts-Electrical Conductivity-Specification and Method of Test." When an employer has a written certification from the manufacturer that a belt has been tested using one of the above test methods, and meets the 300 megohm criteria, the belt is acceptable as meeting this standard. When using conductive belts, the employer should make certain that the head pulley and shaft are grounded through the drive motor ground or by some other equally effective means. When V-type belts are used to transmit power to the head pulley assembly from the motor drive shaft, it will be necessary to provide electrical continuity from the head pulley assembly to ground, e.g., motor grounds.

Employers should also consider purchasing new belts that are flame retardant or fire resistive. A flame resistance test for belts is contained in 30 CFR 18.65.

## APPENDIX B TO §1910.272—GRAIN HANDLING FACILITIES

### National Consensus Standards

The following table contains a cross-reference listing of current national consensus standards which provide information that may be of assistance to grain handling operations. Employers who comply with provisions in these national consensus standards that provide equal or greater protection than those in §1910.272 will be considered in compliance with the corresponding requirements in §1910.272.

| Subject | National consensus standards |
|---|---|
| Grain elevators and facilities handling bulk raw agricultural commodities............................. | ANSI/NFPA 61B |
| Feed mills............................................................ | ANSI/NFPA 61C |
| Facilities handling agricultural commodities for human consumption. ....................... | ANSI/NFPA 61D |
| Pneumatic conveying systems for agricultural commodities ..................... | ANSI/NFPA 66 |
| Guide for explosion venting........................... | ANSI/NFPA 68 |
| Explosion prevention systems ....................... | ANSI/NFPA 69 |
| Dust removal and exhaust systems.............. | ANSI/NFPA 91 |

## APPENDIX C TO §1910.272—GRAIN HANDLING FACILITIES

### References for Further Information

The following references provide information which can be helpful in understanding the requirements contained in various provisions of the standard, as well as provide other helpful information.

1. *Accident Prevention Manual for Industrial Operations;* National Safety Council, 425 North Michigan Avenue, Chicago, Illinois 60611.

2. *Practical Guide to Elevator Design;* National Grain and Feed Association, P.O. Box 28328, Washington, DC 20005.

3. *Dust Control for Grain Elevators;* National Grain and Feed Association, P.O. Box 28328, Washington, DC 20005.

4. *Prevention of Grain Elevator and Mill Explosions;* National Academy of Sciences, Washington, DC. (Available from National Technical Information Service, Springfield, Virginia 22151.)

5. *Standard for the Prevention of Fires and Explosions in Grain Elevators and Facilities Handling Bulk Raw Agricultural Commodities,* NFPA 61B; National Fire Protection Association, Batterymarch Park, Quincy, Massachusetts 02269.

6. *Standard for the Prevention of Fire and Dust Explosions in Feed Mills,* NFPA 61C; National Fire Protection Association, Batterymarch Park, Quincy, Massachusetts 02269.

7. *Standard for the Prevention of Fire and Dust Explosions in the Milling of Agricultural Commodities for Human Consumption,* NFPA 61D; National Fire Protection Association, Batterymarch Park, Quincy, Massachusetts 02269.

8. *Standard for Pneumatic Conveying Systems for Handling Feed, Flour, Grain and Other Agricultural Dusts,* NFPA 66; National Fire Protection Association, Batterymarch Park, Quincy, Massachusetts 02269.

9. *Guide for Explosion Venting,* NFPA 68; National Fire Protection Association, Batterymarch Park, Quincy, Massachusetts 02269.

10. *Standard on Explosion Prevention Systems,* NFPA 69; National Fire Protection Association, Batterymarch Park, Quincy, Massachusetts 02269.

11. *Safety-Operations Plans;* U.S. Department of Agriculture, Washington, DC 20250.

12. *Inplant Fire Prevention Control Programs;* Mill Mutual Fire Prevention Mutual Fire Prevention Bureau, 1 Pierce Place, Suite 1260 West, Itasca, Illinois 60143-1269.

13. *Guidelines for Terminal Elevators;* Mill Mutual Fire Prevention Bureau, 1 Pierce Place, Suite 1260 West, Itasca, Illinois 60143-1269.

14. *Standards for Preventing the Horizontal and Vertical Spread of Fires in Grain Handling Properties;* Mill Mutual Fire Mutual Fire Prevention Bureau, 1 Pierce Place, Suite 1260 West, Itasca, Illinois 60143-1269.

15. *Belt Conveyors for Bulk Materials,* Part I and Part II, Data Sheet 570, Revision A; National Safety Council, 425 North Michigan Avenue, Chicago, Illinois 60611.

16. *Suggestions for Precautions and Safety Practices in Welding and Cutting;* Mill Mutual Fire Prevention Bureau, 1 Pierce Place, Suite 1260 West, Itasca, Illinois 60143-1269.

17. *Food Bins and Tanks,* Data Sheet 524; National Safety Council, 425 North Michigan Avenue, Chicago, Illinois 60611.

18. *Pneumatic Dust Control in Grain Elevators;* National Academy of Sciences, Washington, DC. (Available from National Technical Information Service, Springfield, Virginia 22151.)

19. *Dust Control Analysis and Layout Procedures for Grain Storage and Processing Plants;* Mill Mutual Fire Prevention Bureau, 1 Pierce Place, Suite 1260 West, Itasca, Illinois 60143-1269.

20. *Standard for the Installation of Blower and Exhaust Systems for Dust, Stock and Vapor Removal,* NFPA 91; National Fire Protection Association, Batterymarch Park, Quincy, Massachusetts 02269.

21. *Standards for the Installation of Direct Heat Grain Driers in Grain and Milling Properties;* Mill Mutual Fire Prevention Bureau, 1 Pierce Place, Suite 1260 West, Itasca, Illinois 60143-1269.

22. *Guidelines for Lubrication and Bearing Maintenance;* Mill Mutual Fire Prevention Bureau, 1 Pierce Place, Suite 1260 West, Itasca, Illinois 60143-1269.

23. *Organized Maintenance in Grain and Milling Properties;* Mill Mutual Fire Prevention Bureau, 1 Pierce Place, Suite 1260 West, Itasca, Illinois 60143-1269.

24. *Safe and Efficient Elevator Legs for Grain and Milling Properties;* Mill Mutual Fire Prevention Bureau, 1 Pierce Place, Suite 1260 West, Itasca, Illinois 60143-1269.

25. *Explosion Venting and Supression of Bucket Elevators;* National Grain and Feed Association, P.O. Box 28328, Washington, DC 20005.

26. *Lightning Protection Code,* NFPA 78; National Fire Protection Association, Batterymarch Park, Quincy, Massachusetts 02269.

27. *Occupational Safety in Grain Elevators,* DHHS (NIOSH) Publication No. 83-126); National Institute for Occupational Safety and Health, Morgantown, West Virginia 26505.

28. *Retrofitting and Constructing Grain Elevators;* National Grain and Feed Association, P.O. Box 28328, Washington, DC 20005.

29. *Grain Industry Safety and Health Center—Training Series* (Preventing grain dust explosions, operations maintenance safety, transportation safety, occupational safety and health); Grain Elevator and Processing Society, P.O. Box 15026, Commerce Station, Minneapolis, Minnesota 55415-0026.

30. *Suggestions for Organized Maintenance;* The Mill Mutuals Loss Control Department, 1 Pierce Place, Suite 1260 West, Itasca, Illinois 60143-1269.

31. *Safety—The First Step to Success;* The Mill Mutual Loss Control Department, 1 Pierce Place, Suite 1260 West, Itasca, Illinois 60143-1269.

32. *Emergency Plan Notebook;* Schoeff, Robert W. and James L. Balding, Kansas State University, Cooperative Extension Service, Extension Grain Science and Industry, Shellenberger Hall, Manhattan, Kansas 66506.

[52 FR 49625, Dec. 31, 1987, as amended at 53 FR 17696, May 18, 1988; 54 FR 24334, June 7, 1989; 55 FR 25094, June 20, 1990; 61 FR 9242, Mar. 7, 1996; 61 FR 9584, Mar. 8, 1996]

### §1910.274 Sources of standards [Removed].

[§1910.274 removed at 61 FR 9242, Mar. 7, 1996, effective May 6, 1996.]

### §1910.275 Standards organizations [Removed].

[§1910.275 removed at 61 FR 9242, Mar. 7, 1996, effective May 6, 1996.]

# Subpart S—Electrical

**Authority:** Secs. 4, 6, 8, Occupational Safety and Health Act of 1970 (29 U.S.C. 653, 655, 657; Secretary of Labor's Order No. 8-76 (41 FR 25059) or 1-90 (55 FR 9033), as applicable; 29 CFR part 1911.

**Source:** 46 FR 4056, Jan. 16, 1981, unless otherwise noted.

## General

### §1910.301 Introduction.

This subpart addresses electrical safety requirements that are necessary for the practical safeguarding of employees in their workplaces and is divided into four major divisions as follows:

**1910.301(a)** *Design safety standards for electrical systems.* These regulations are contained in §§1910.302 through 1910.330. Sections 1910.302 through 1910.308 contain design safety standards for electric utilization systems. Included in this category are all electric equipment and installations used to provide electric power and light for employee workplaces. Sections 1910.309 through 1910.330 are reserved for possible future design safety standards for other electrical systems.

**1910.301(b)** *Safety-related work practices.* These regulations will be contained in §§1910.331 through 1910.360.

**1910.301(c)** *Safety-related maintenance requirements.* These regulations will be contained in §§1910.361 through 1910.380.

**1910.301(d)** *Safety requirements for special equipment.* These regulations will be contained in §§1910.381 through 1910.398.

**1910.301(e)** *Definitions.* Definitions applicable to each division are contained in §1910.399.

[46 FR 4056, Jan. 16, 1982; 46 FR 40185, Aug. 7, 1981]

## Design Safety Standards for Electrical Systems

### §1910.302 Electric utilization systems.

Sections 1910.302 through 1910.308 contain design safety standards for electric utilization systems.

**1910.302(a)** *Scope*—(1) *Covered.* The provisions of §§1910.302 through 1910.308 of this subpart cover electrical installations and utilization equipment installed or used within or on buildings, structures, and other premises including:

**1910.302(a)(1)(i)** Yards,

**1910.302(a)(1)(ii)** Carnivals,

**1910.302(a)(1)(iii)** Parking and other lots,

**1910.302(a)(1)(iv)** Mobile homes,

**1910.302(a)(1)(v)** Recreational vehicles,

**1910.302(a)(1)(vi)** Industrial substations,

**1910.302(a)(1)(vii)** Conductors that connect the installations to a supply of electricity, and

**1910.302(a)(1)(viii)** Other outside conductors on the premises.

**1910.302(a)(2)** *Not covered.* The provisions of §§1910.302 through 1910.308 of this subpart do not cover:

**1910.302(a)(2)(i)** Installations in ships, watercraft, railway rolling stock, aircraft, or automotive vehicles other than mobile homes and recreational vehicles.

**1910.302(a)(2)(ii)** Installations underground in mines.

**1910.302(a)(2)(iii)** Installations of railways for generation, transformation, transmission, or distribution of power used exclusively for operation of rolling stock or installations used exclusively for signaling and communication purposes.

**1910.302(a)(2)(iv)** Installations of communication equipment under the exclusive control of communication utilities, located outdoors or in building spaces used exclusively for such installations.

**1910.302(a)(2)(v)** Installations under the exclusive control of electric utilities for the purpose of communication or metering; or for the generation, control, transformation, transmission, and distribution of electric energy located in buildings used exclusively by utilities for such purposes or located outdoors on property owned or leased by the utility or on public highways, streets, roads, etc., or outdoors by established rights on private property.

**1910.302(b)** *Extent of application.* **(1)** The requirements contained in the sections listed below shall apply to all electrical installations and utilization equipment, regardless of when they were designed or installed.

Sections:

| | |
|---|---|
| 1910.303(b) | Examination, installation, and use of equipment. |
| 1910.303(c) | Splices. |
| 1910.303(d) | Arcing parts. |
| 1910.303(e) | Marking. |
| 1910.303(f) | Identification of disconnecting means. |
| 1910.303(g)(2) | Guarding of live parts. |
| 1910.304(e)(l)(i) | Protection of conductors and equipment. |
| 1910.304(e)(l)(iv) | Location in or on premises. |
| 1910.304(e)(l)(v) | Arcing or suddenly moving parts. |
| 1910.304(f)(l)(ii) | 2-Wire DC systems to be grounded: |
| 1910.304(f)(l)(iii) and 1910.304(f)(l)(iv) | AC Systems to be grounded. |
| 1910.304(f)(l)(v) | AC Systems 50 to 1000 volts not required to be grounded. |
| 1910.304(f)(3) | Grounding connections. |
| 1910.304(f)(4) | Grounding path. |
| 1910.304(f)(5)(iv)(a) through 1910.304(f)(5)(iv)(d) | Fixed equipment required to be grounded. |
| 1910.304(f)(5)(v) | Grounding of equipment connected by cord and plug. |
| 1910.304(f)(5)(vi) | Grounding of nonelectrical equipment. |
| 1910.304(f)(6)(i) | Methods of grounding fixed equipment. |
| 1910.305(g)(l)(i) and 1910.305(g)(1)(ii) | Flexible cords and cables, uses. |
| 1910.305(g)(l)(iii) | Flexible cords and cables prohibited. |
| 1910.305(g)(2)(ii) | Flexible cords and cables, splices. |
| 1910.305(g)(2)(iii) | Pull at joints and terminals of flexible cords and cables. |
| 1910.307 | Hazardous (classified) locations. |

**1910.302(b)(2)** Every electric utilization system and all utilization equipment installed after March 15, 1972, and every major replacement, modification, repair, or rehabilitation, after March 15, 1972, of any part of any electric utilization system or utilization equipment installed before March 15, 1972, shall comply with the provisions of §§1910.302 through 1910.308.

**NOTE:** "Major replacements, modifications, repairs, or rehabilitations" include work similar to that involved when a new building or facility is built, a new wing is added, or an entire floor is renovated.

**1910.302(b)(3)** The following provisions apply to electric utilization systems and utilization equipment installed after April 16, 1981:

| | |
|---|---|
| §1910.303(h)(4)(i) and (ii) | Entrance and access to workspace (over 600 volts). |
| §1910.304(e)(1)(vi)(b) | Circuit breakers operated vertically. |
| §1910.304(e)(1)(vi)(c) | Circuit breakers used as switches. |
| §1910.304(f)(7)(ii) | Grounding of systems of 1000 volts or more supplying portable or mobile equipment. |
| §1910.305(j)(6)(ii)(b) | Switching series capacitors over 600 volts. |
| §1910.306(c)(2) | Warning signs for elevators and escalators. |
| §1910.306(i) | Electrically controlled irrigation machines. |
| §1910.306(j)(5) | Ground-fault circuit interrupters for fountains. |
| §1910.308(a)(1)(ii) | Physical protection of conductors over 600 volts. |
| §1910.308(c)(2) | Marking of Class 2 and Class 3 power supplies. |
| §1910.308(d) | Fire protective signaling circuits. |

[46 FR 4056, Jan. 16, 1981; 46 FR 40185, Aug. 7, 1981]

## §1910.303 General requirements.

**1910.303(a)** *Approval.* The conductors and equipment required or permitted by this subpart shall be acceptable only if approved.

**1910.303(b)** *Examination, installation, and use of equipment*—**(1)** *Examination.* Electrical equipment shall be free from recognized hazards that are likely to cause death or serious physical harm to employees. Safety of equipment shall be determined using the following considerations:

**1910.303(b)(i)** Suitability for installation and use in conformity with the provisions of this subpart. Suitability of equipment for an identified purpose may be evidenced by listing or labeling for that identified purpose.

**1910.303(b)(ii)** Mechanical strength and durability, including, for parts designed to enclose and protect other equipment, the adequacy of the protection thus provided.

**1910.303(b)(iii)** Electrical insulation.

**1910.303(b)(iv)** Heating effects under conditions of use.

**1910.303(b)(v)** Arcing effects.

**1910.303(b)(vi)** Classification by type, size, voltage, current capacity, specific use.

**1910.303(b)(vii)** Other factors which contribute to the practical safeguarding of employees using or likely to come in contact with the equipment.

**1910.303(b)(2)** *Installation and use.* Listed or labeled equipment shall be used or installed in accordance with any instructions included in the listing or labeling.

**1910.303(c)** *Splices.* Conductors shall be spliced or joined with splicing devices suitable for the use or by brazing, welding, or soldering with a fusible metal or alloy. Soldered splices shall first be so spliced or joined as to be mechanically and electrically secure without solder and then soldered. All splices and joints and the free ends of conductors shall be covered with an insulation equivalent to that of the conductors or with an insulating device suitable for the purpose.

**1910.303(d)** *Arcing parts.* Parts of electric equipment which in ordinary operation produce arcs, sparks, flames, or molten metal shall be enclosed or separated and isolated from all combustible material.

**1910.303(e)** *Marking.* Electrical equipment may not be used unless the manufacturer's name, trademark, or other descriptive marking by which the organization responsible for the product may be identified is placed on the equipment. Other markings shall be provided giving voltage, current, wattage, or other ratings as necessary. The marking shall be of sufficient durability to withstand the environment involved.

**1910.303(f)** *Identification of disconnecting means and circuits.* Each disconnecting means required by this subpart for motors and appliances shall be legibly marked to indicate its purpose, unless located and arranged so the purpose is evident. Each service, feeder, and branch circuit, at its disconnecting means or overcurrent device, shall be legibly marked to indicate its purpose, unless located and arranged so the purpose is evident. These markings shall be of sufficient durability to withstand the environment involved.

**1910.303(g)** *600 Volts, nominal, or less*—**(1)** *Working space about electric equipment.* Sufficient access and working space shall be provided and maintained about all electric equipment to permit ready and safe operation and maintenance of such equipment.

**1910.303(g)(1)(i)** *Working clearances.* Except as required or permitted elsewhere in this subpart, the dimension of the working space in the direction of access to live parts operating at 600 volts or less and likely to require examination, adjustment, servicing, or maintenance while alive may not be less than indicated in Table S-1. In addition to the dimensions shown in Table S-1, workspace may not be less than 30 inches wide in front of the electric equipment. Distances shall be measured from the live parts if they are exposed, or from the enclosure front or opening if the live parts are enclosed. Concrete, brick, or tile walls are considered to be grounded. Working space is not required in back of assemblies such as dead-front switchboards or motor control centers where there are no renewable or adjustable parts such as fuses or switches on the back and where all connections are accessible from locations other than the back.

### TABLE S-1—WORKING CLEARANCES

| Nominal voltage to ground | Minimum clear distance for condition[2] (ft) | | |
|---|---|---|---|
| | (a) | (b) | (c) |
| 0-150 | [1]3 | [1]3 | 3 |
| 151-600 | [1]3 | 3 1/2 | 4 |

[1] Minimum clear distances may be 2 feet 6 inches for installations built prior to April 16, 1981.

[2] Conditions (a), (b), and (c), are as follows: (a) Exposed live parts on one side and no live or grounded parts on the other side of the working space, or exposed live parts on both sides effectively guarded by suitable wood or other insulating material. Insulated wire or insulated busbars operating at not over 300 volts are not considered live parts. (b) Exposed live parts on one side and grounded parts on the other side. (c) Exposed live parts on both sides of the workspace [not guarded as provided in Condition (a)] with the operator between.

**1910.303(g)(1)(ii)** *Clear spaces.* Working space required by this subpart may not be used for storage. When normally enclosed live parts are exposed for inspection or servicing, the working space, if in a passageway or general open space, shall be suitably guarded.

**1910.303(g)(1)(iii)** *Access and entrance to working space.* At least one entrance of sufficient area shall be provided to give access to the working space about electric equipment.

**1910.303(g)(1)(iv)** *Front working space.* Where there are live parts normally exposed on the front of switchboards or motor control centers, the working space in front of such equipment may not be less than 3 feet.

**1910.303(g)(1)(v)** *Illumination.* Illumination shall be provided for all working spaces about service equipment, switchboards, panelboards, and motor control centers installed indoors.

**1910.303(g)(1)(vi)** *Headroom.* The minimum headroom of working spaces about service equipment, switchboards, panel-boards, or motor control centers shall be 6 feet 3 inches.

**NOTE:** As used in this section a motor control center is an assembly of one or more enclosed sections having a common power bus and principally containing motor control units.

**1910.303(g)(2)** *Guarding of live parts.* **(i)** Except as required or permitted elsewhere in this subpart, live parts of electric equipment operating at 50 volts or more shall be guarded against accidental contact by approved cabinets or other forms of approved enclosures, or by any of the following means:

**1910.303(g)(2)(i)(A)** By location in a room, vault, or similar enclosure that is accessible only to qualified persons.

**1910.303(g)(2)(i)(B)** By suitable permanent, substantial partitions or screens so arranged that only qualified persons will have access to the space within reach of the live parts. Any openings in such partitions or screens shall be so sized and located that persons are not likely to come into accidental contact with the live parts or to bring conducting objects into contact with them.

**1910.303(g)(2)(i)(C)** By location on a suitable balcony, gallery, or platform so elevated and arranged as to exclude unqualified persons.

**1910.303(g)(2)(i)(D)** By elevation of 8 feet or more above the floor or other working surface.

**1910.303(g)(2)(ii)** In locations where electric equipment would be exposed to physical damage, enclosures or guards shall be so arranged and of such strength as to prevent such damage.

**1910.303(g)(2)(iii)** Entrances to rooms and other guarded locations containing exposed live parts shall be marked with conspicuous warning signs forbidding unqualified persons to enter.

**1910.303(h)** *Over 600 volts, nominal*—**(1)** *General.* Conductors and equipment used on circuits exceeding 600 volts, nominal, shall comply with all applicable provisions of paragraphs (a) through (g) of this section and with the following provisions which supplement or modify those requirements. The provisions of paragraphs (h)(2), (h)(3), and (h)(4) of this section do not apply to equipment on the supply side of the service conductors.

**1910.303(h)(2)** *Enclosure for electrical installations.* Electrical installations in a vault, room, closet or in an area surrounded by a wall, screen, or fence, access

# OSHA Standards for General Industry

to which is controlled by lock and key or other approved means, are considered to be accessible to qualified persons only. A wall, screen, or fence less than 8 feet in height is not considered to prevent access unless it has other features that provide a degree of isolation equivalent to an 8 foot fence. The entrances to all buildings, rooms, or enclosures containing exposed live parts or exposed conductors operating at over 600 volts, nominal, shall be kept locked or shall be under the observation of a qualified person at all times.

**1910.303(h)(2)(i)** *Installations accessible to qualified persons only.* Electrical installations having exposed live parts shall be accessible to qualified persons only and shall comply with the applicable provisions of paragraph (h)(3) of this section.

**1910.303(h)(2)(ii)** *Installations accessible to unqualified persons.* Electrical installations that are open to unqualified persons shall be made with metal-enclosed equipment or shall be enclosed in a vault or in an area, access to which is controlled by a lock. If metal-enclosed equipment is installed so that the bottom of the enclosure is less than 8 feet above the floor, the door or cover shall be kept locked. Metal-enclosed switchgear, unit substations, transformers, pull boxes, connection boxes, and other similar associated equipment shall be marked with appropriate caution signs. If equipment is exposed to physical damage from vehicular traffic, suitable guards shall be provided to prevent such damage. Ventilating or similar openings in metal-enclosed equipment shall be designed so that foreign objects inserted through these openings will be deflected from energized parts.

**1910.303(h)(3)** *Workspace about equipment.* Sufficient space shall be provided and maintained about electric equipment to permit ready and safe operation and maintenance of such equipment. Where energized parts are exposed, the minimum clear workspace may not be less than 6 feet 6 inches high (measured vertically from the floor or platform), or less than 3 feet wide (measured parallel to the equipment). The depth shall be as required in Table S-2. The workspace shall be adequate to permit at least a 90-degree opening of doors or hinged panels.

**1910.303(h)(3)(i)** *Working space.* The minimum clear working space in front of electric equipment such as switchboards, control panels, switches, circuit breakers, motor controllers, relays, and similar equipment may not be less than specified in Table S-2 unless otherwise specified in this subpart. Distances shall be measured from the live parts if they are exposed, or from the enclosure front or opening if the live parts are enclosed. However, working space is not required in back of equipment such as deadfront switchboards or control assemblies where there are no renewable or adjustable parts (such as fuses or switches) on the back and where all connections are accessible from locations other than the back. Where rear access is required to work on de-energized parts on the back of enclosed equipment, a minimum working space of 30 inches horizontally shall be provided.

### TABLE S-2—MINIMUM DEPTH OF CLEAR WORKING SPACE IN FRONT OF ELECTRIC EQUIPMENT

| Nominal voltage to ground | Conditions[2] (ft) (a) | (b) | (c) |
|---|---|---|---|
| 601 to 2,500 | 3 | 4 | 5 |
| 2,501 to 9,000 | 4 | 5 | 6 |
| 9,001 to 25,000 | 5 | 6 | 9 |
| 25,001 to 75kV[1] | 6 | 8 | 10 |
| Above 75kV[1] | 8 | 10 | 12 |

[1] Minimum depth of clear working space in front of electric equipment with a nominal voltage to ground above 25,000 volts may be the same as for 25,000 volts under Conditions (a), (b), and (c) for installations built prior to April 16, 1981.

[2] Conditions (a), (b), and (c) are as follows: (a) Exposed live parts on one side and no live or grounded parts on the other side of the working space, or exposed live parts on both sides effectively guarded by suitable wood or other insulating materials. Insulated wire or insulated busbars operating at not over 300 volts are not considered live parts. (b) Exposed live parts on one side and grounded parts on the other side. Concrete, brick, or tile walls will be considered as grounded surfaces. (c) Exposed live parts on both sides of the workspace not guarded as provided in Condition (a) with the operator between.

**1910.303(h)(3)(ii)** *Illumination.* Adequate illumination shall be provided for all working spaces about electric equipment. The lighting outlets shall be so arranged that persons changing lamps or making repairs on the lighting system will not be endangered by live parts or other equipment. The points of control shall be so located that persons are not likely to come in contact with any live part or moving part of the equipment while turning on the lights.

**1910.303(h)(3)(iii)** *Elevation of unguarded live parts.* Unguarded live parts above working space shall be maintained at elevations not less than specified in Table S-3.

### TABLE S-3—ELEVATION OF UNGUARDED ENERGIZED PARTS ABOVE WORKING SPACE

| Nominal voltage between phases | Minimum elevation |
|---|---|
| 601 to 7,500 | *8 feet 6 inches. |
| 7,501 to 35,000 | 9 feet. |
| Over 35kV | 9 feet + 0.37 inches per kV above 35kV. |

* Note: Minimum elevation may be 8 feet 0 inches for installations built prior to April 16, 1981 if the nominal voltage between phases is in the range of 601-6600 volts.

**1910.303(h)(4)** *Entrance and access to workspace.* (See §1910.302(b)(3).)

**1910.303(h)(4)(i)** At least one entrance not less than 24 inches wide and 6 feet 6 inches high shall be provided to give access to the working space about electric equipment. On switchboard and control panels exceeding 48 inches in width, there shall be one entrance at each end of such board where practicable. Where bare energized parts at any voltage or insulated energized parts above 600 volts are located adjacent to such entrance, they shall be suitably guarded.

**1910.303(h)(4)(ii)** Permanent ladders or stairways shall be provided to give safe access to the working space around electric equipment installed on platforms, balconies, mezzanine floors, or in attic or roof rooms or spaces.

[46 FR 4056, Jan. 16, 1981; 46 FR 40185, Aug. 7, 1981]

## §1910.304 Wiring design and protection.

**1910.304(a)** *Use and identification of grounded and grounding conductors.* **(1)** *Identification of conductors.* A conductor used as a grounded conductor shall be identifiable and distinguishable from all other conductors. A conductor used as an equipment grounding conductor shall be identifiable and distinguishable from all other conductors.

**1910.304(a)(2)** *Polarity of connections.* No grounded conductor may be attached to any terminal or lead so as to reverse designated polarity.

**1910.304(a)(3)** *Use of grounding terminals and devices.* A grounding terminal or grounding-type device on a receptacle, cord connector, or attachment plug may not be used for purposes other than grounding.

**1910.304(b)** *Branch circuits*—**(1)** [Reserved]

**1910.304(b)(2)** *Outlet devices.* Outlet devices shall have an ampere rating not less than the load to be served.

**1910.304(c)** *Outside conductors, 600 volts, nominal, or less.* Paragraphs (c)(1), (c)(2), (c)(3), and (c)(4) of this section apply to branch circuit, feeder, and service conductors rated 600 volts, nominal, or less and run outdoors as open conductors. Paragraph (c)(5) applies to lamps installed under such conductors.

**1910.304(c)(1)** *Conductors on poles.* Conductors supported on poles shall provide a horizontal climbing space not less than the following:

**1910.304(c)(1)(i)** Power conductors below communication conductors—30 inches.

**1910.304(c)(1)(ii)** Power conductors alone or above communication conductors: 300 volts or less—24 inches; more than 300 volts—30 inches.

**1910.304(c)(1)(iii)** Communication conductors below power conductors with power conductors 300 volts or less—24 inches; more than 300 volts—30 inches.

**1910.304(c)(2)** *Clearance from ground.* Open conductors shall conform to the following minimum clearances:

**1910.304(c)(2)(i)** 10 feet—above finished grade, sidewalks, or from any platform or projection from which they might be reached.

**1910.304(c)(2)(ii)** 12 feet—over areas subject to vehicular traffic other than truck traffic.

**1910.304(c)(2)(iii)** 15 feet—over areas other than those specified in paragraph (c)(2)(iv) of this section that are subject to truck traffic.

**1910.304(c)(2)(iv)** 18 feet—over public streets, alleys, roads, and driveways.

**1910.304(c)(3)** *Clearance from building openings.* Conductors shall have a clearance of at least 3 feet from windows, doors, porches, fire escapes, or similar locations. Conductors run above the top level of a window are considered to be out of reach from that window and, therefore, do not have to be 3 feet away.

**1910.304(c)(4)** *Clearance over roofs.* Conductors shall have a clearance of not less than 8 feet from the highest point of roofs over which they pass, except that:

**1910.304(c)(4)(i)** Where the voltage between conductors is 300 volts or less and the roof has a slope of not less than 4 inches in 12, the clearance from roofs shall be at least 3 feet, or

**1910.304(c)(4)(ii)** Where the voltage between conductors is 300 volts or less and the conductors do not pass over more than 4 feet of the overhang portion of the roof and they are terminated at a through-the-roof raceway or approved support, the clearance from roofs shall be at least 18 inches.

**1910.304(c)(5)** *Location of outdoor lamps.* Lamps for outdoor lighting shall be located below all live conductors, transformers, or other electric equipment, unless such equipment is controlled by a disconnecting means that can be locked in the open position or unless adequate clearances or other safeguards are provided for relamping operations.

**1910.304(d)** *Services—***(1)** *Disconnecting means—***(i)** *General.* Means shall be provided to disconnect all conductors in a building or other structure from the service-entrance conductors. The disconnecting means shall plainly indicate whether it is in the open or closed position and shall be installed at a readily accessible location nearest the point of entrance of the service-entrance conductors.

**1910.304(d)(1)(ii)** *Simultaneous opening of poles.* Each service disconnecting means shall simultaneously disconnect all ungrounded conductors.

**1910.304(d)(2)** *Services over 600 volts, nominal.* The following additional requirements apply to services over 600 volts, nominal.

**1910.304(d)(2)(i)** *Guarding.* Service-entrance conductors installed as open wires shall be guarded to make them accessible only to qualified persons.

**1910.304(d)(2)(ii)** *Warning signs.* Signs warning of high voltage shall be posted where other than qualified employees might come in contact with live parts.

**1910.304(e)** *Overcurrent protection.* **(1)** *600 volts, nominal, or less.* The following requirements apply to overcurrent protection of circuits rated 600 volts, nominal, or less.

**1910.304(e)(1)(i)** *Protection of conductors and equipment.* Conductors and equipment shall be protected from overcurrent in accordance with their ability to safely conduct current.

**1910.304(e)(1)(ii)** *Grounded conductors.* Except for motor running overload protection, overcurrent devices may not interrupt the continuity of the grounded conductor unless all conductors of the circuit are opened simultaneously.

**1910.304(e)(1)(iii)** *Disconnection of fuses and thermal cutouts.* Except for service fuses, all cartridge fuses which are accessible to other than qualified persons and all fuses and thermal cutouts on circuits over 150 volts to ground shall be provided with disconnecting means. This disconnecting means shall be installed so that the fuse or thermal cutout can be disconnected from its supply without disrupting service to equipment and circuits unrelated to those protected by the overcurrent device.

**1910.304(e)(1)(iv)** *Location in or on premises.* Overcurrent devices shall be readily accessible to each employee or authorized building management personnel. These overcurrent devices may not be located where they will be exposed to physical damage nor in the vicinity of easily ignitible material.

**1910.304(e)(1)(v)** *Arcing or suddenly moving parts.* Fuses and circuit breakers shall be so located or shielded that employees will not be burned or otherwise injured by their operation.

**1910.304(e)(1)(vi)** *Circuit breakers.* **(A)** Circuit breakers shall clearly indicate whether they are in the open (off) or closed (on) position.

**1910.304(e)(1)(vi)(B)** Where circuit breaker handles on switchboards are operated vertically rather than horizontally or rotationally, the up position of the handle shall be the closed (on) position. (See §1910.302(b)(3).)

**1910.304(e)(1)(vi)(C)** If used as switches in 120-volt, fluorescent lighting circuits, circuit breakers shall be approved for the purpose and marked "SWD." (See §1910.302(b)(3).)

**1910.304(e)(2)** *Over 600 volts, nominal.* Feeders and branch circuits over 600 volts, nominal, shall have short-circuit protection.

**1910.304(f)** **Grounding.** Paragraphs (f)(1) through (f)(7) of this section contain grounding requirements for systems, circuits, and equipment.

**1910.304(f)(1)** *Systems to be grounded.* The following systems which supply premises wiring shall be grounded:

**1910.304(f)(1)(i)** All 3-wire DC systems shall have their neutral conductor grounded.

**1910.304(f)(1)(ii)** Two-wire DC systems operating at over 50 volts through 300 volts between conductors shall be grounded unless:

**1910.304(f)(1)(ii)(A)** They supply only industrial equipment in limited areas and are equipped with a ground detector; or

**1910.304(f)(1)(ii)(B)** They are rectifier-derived from an AC system complying with paragraphs (f)(1)(iii), (f)(1)(iv), and (f)(1)(v) of this section; or

**1910.304(f)(1)(ii)(C)** They are fire-protective signaling circuits having a maximum current of 0.030 amperes.

**1910.304(f)(1)(iii)** AC circuits of less than 50 volts shall be grounded if they are installed as overhead conductors outside of buildings or if they are supplied by transformers and the transformer primary supply system is ungrounded or exceeds 150 volts to ground.

**1910.304(f)(1)(iv)** AC systems of 50 volts to 1000 volts shall be grounded under any of the following conditions, unless exempted by paragraph (f)(1)(v) of this section:

**1910.304(f)(1)(iv)(A)** If the system can be so grounded that the maximum voltage to ground on the ungrounded conductors does not exceed 150 volts;

**1910.304(f)(1)(iv)(B)** If the system is nominally rated 480Y/277 volt, 3-phase, 4-wire in which the neutral is used as a circuit conductor;

**1910.304(f)(1)(iv)(C)** If the system is nominally rated 240/120 volt, 3-phase, 4-wire in which the midpoint of one phase is used as a circuit conductor; or

**1910.304(f)(1)(iv)(D)** If a service conductor is uninsulated.

**1910.304(f)(1)(v)** AC systems of 50 volts to 1000 volts are not required to be grounded under any of the following conditions:

**1910.304(f)(1)(v)(A)** If the system is used exclusively to supply industrial electric furnaces for melting, refining, tempering, and the like.

**1910.304(f)(1)(v)(B)** If the system is separately derived and is used exclusively for rectifiers supplying only adjustable speed industrial drives.

**1910.304(f)(1)(v)(C)** If the system is separately derived and is supplied by a transformer that has a primary voltage rating less than 1000 volts, provided all of the following conditions are met:

**1910.304(f)(1)(v)(C)(1)** The system is used exclusively for control circuits,

**1910.304(f)(1)(v)(C)(2)** The conditions of maintenance and supervision assure that only qualified persons will service the installation,

**1910.304(f)(1)(v)(C)(3)** Continuity of control power is required, and

**1910.304(f)(1)(v)(C)(4)** Ground detectors are installed on the control system.

**1910.304(f)(1)(v)(D)** If the system is an isolated power system that supplies circuits in health care facilities.

**1910.304(f)(2)** *Conductors to be grounded.* For AC premises wiring systems the identified conductor shall be grounded.

**1910.304(f)(3)** *Grounding connections.* **(i)** For a grounded system, a grounding electrode conductor shall be used to connect both the equipment grounding conductor and the grounded circuit conductor to the grounding electrode. Both the equipment grounding conductor and the grounding electrode conductor shall be connected to the grounded circuit conductor on the supply side of the service disconnecting means, or on the supply side of the system disconnecting means or overcurrent devices if the system is separately derived.

**1910.304(f)(3)(ii)** For an ungrounded service-supplied system, the equipment grounding conductor shall be connected to the grounding electrode conductor at the service equipment. For an ungrounded separately derived system, the equipment grounding conductor shall be connected to the grounding electrode conductor at, or ahead of, the system disconnecting means or overcurrent devices.

**1910.304(f)(3)(iii)** On extensions of existing branch circuits which do not have an equipment grounding conductor, grounding-type receptacles may be grounded to a grounded cold water pipe near the equipment.

# OSHA Standards for General Industry

**1910.304(f)(4)** *Grounding path.* The path to ground from circuits, equipment, and enclosures shall be permanent and continuous.

**1910.304(f)(5)** *Supports, enclosures, and equipment to be grounded—(i) Supports and enclosures for conductors.* Metal cable trays, metal raceways, and metal enclosures for conductors shall be grounded, except that:

**1910.304(f)(5)(i)(A)** Metal enclosures such as sleeves that are used to protect cable assemblies from physical damage need not be grounded; or

**1910.304(f)(5)(i)(B)** Metal enclosures for conductors added to existing installations of open wire, knob-and-tube wiring, and nonmetallic-sheathed cable need not be grounded if all of the following conditions are met: (*1*) Runs are less than 25 feet; (*2*) enclosures are free from probable contact with ground, grounded metal, metal laths, or other conductive materials; and (*3*) enclosures are guarded against employee contact.

**1910.304(f)(5)(ii)** *Service equipment enclosures.* Metal enclosures for service equipment shall be grounded.

**1910.304(f)(5)(iii)** *Frames of ranges and clothes dryers.* Frames of electric ranges, wall-mounted ovens, counter-mounted cooking units, clothes dryers, and metal outlet or junction boxes which are part of the circuit for these appliances shall be grounded.

**1910.304(f)(5)(iv)** *Fixed equipment.* Exposed non-current-carrying metal parts of fixed equipment which may become energized shall be grounded under any of the following conditions:

**1910.304(f)(5)(iv)(A)** If within 8 feet vertically or 5 feet horizontally of ground or grounded metal objects and subject to employee contact.

**1910.304(f)(5)(iv)(B)** If located in a wet or damp location and not isolated.

**1910.304(f)(5)(iv)(C)** If in electrical contact with metal.

**1910.304(f)(5)(iv)(D)** If in a hazardous (classified) location.

**1910.304(f)(5)(iv)(E)** If supplied by a metal-clad, metal-sheathed, or grounded metal raceway wiring method.

**1910.304(f)(5)(iv)(F)** If equipment operates with any terminal at over 150 volts to ground; however, the following need not be grounded:

**1910.304(f)(5)(iv)(F)***(1)* Enclosures for switches or circuit breakers used for other than service equipment and accessible to qualified persons only;

**1910.304(f)(5)(iv)(F)***(2)* Metal frames of electrically heated appliances which are permanently and effectively insulated from ground; and

**1910.304(f)(5)(iv)(F)***(3)* The cases of distribution apparatus such as transformers and capacitors mounted on wooden poles at a height exceeding 8 feet above ground or grade level.

**1910.304(f)(5)(v)** *Equipment connected by cord and plug.* Under any of the conditions described in paragraphs (f)(5)(v)(A) through (f)(5)(v)(C) of this section, exposed non-current-carrying metal parts of cord- and plug-connected equipment which may become energized shall be grounded.

**1910.304(f)(5)(v)(A)** If in hazardous (classified) locations (see §1910.307).

**1910.304(f)(5)(v)(B)** If operated at over 150 volts to ground, except for guarded motors and metal frames of electrically heated appliances if the appliance frames are permanently and effectively insulated from ground.

**1910.304(f)(5)(v)(C)** If the equipment is of the following types:

**1910.304(f)(5)(v)(C)***(1)* Refrigerators, freezers, and air conditioners;

**1910.304(f)(5)(v)(C)***(2)* Clothes-washing, clothes-drying and dishwashing machines, sump pumps, and electrical aquarium equipment;

**1910.304(f)(5)(v)(C)***(3)* Hand-held motor-operated tools;

**1910.304(f)(5)(v)(C)***(4)* Motor-operated appliances of the following types: hedge clippers, lawn mowers, snow blowers, and wet scrubbers;

**1910.304(f)(5)(v)(C)***(5)* Cord- and plug-connected appliances used in damp or wet locations or by employees standing on the ground or on metal floors or working inside of metal tanks or boilers;

**1910.304(f)(5)(v)(C)***(6)* Portable and mobile X-ray and associated equipment;

**1910.304(f)(5)(v)(C)***(7)* Tools likely to be used in wet and conductive locations; and

**1910.304(f)(5)(v)(C)***(8)* Portable hand lamps.

Tools likely to be used in wet and conductive locations need not be grounded if supplied through an isolating transformer with an ungrounded secondary of not over 50 volts. Listed or labeled portable tools and appliances protected by an approved system of double insulation, or its equivalent, need not be grounded. If such a system is employed, the equipment shall be distinctively marked to indicate that the tool or appliance utilizes an approved system of double insulation.

**1910.304(f)(5)(vi)** *Nonelectrical equipment.* The metal parts of the following nonelectrical equipment shall be grounded: frames and tracks of electrically operated cranes; frames of nonelectrically driven elevator cars to which electric conductors are attached; hand operated metal shifting ropes or cables of electric elevators, and metal partitions, grill work, and similar metal enclosures around equipment of over 750 volts between conductors.

**1910.304(f)(6)** *Methods of grounding fixed equipment.* **(i)** Non-current-carrying metal parts of fixed equipment, if required to be grounded by this subpart, shall be grounded by an equipment grounding conductor which is contained within the same raceway, cable, or cord, or runs with or encloses the circuit conductors. For DC circuits only, the equipment grounding conductor may be run separately from the circuit conductors.

**1910.304(f)(6)(ii)** Electric equipment is considered to be effectively grounded if it is secured to, and in electrical contact with, a metal rack or structure that is provided for its support and the metal rack or structure is grounded by the method specified for the non-current-carrying metal parts of fixed equipment in paragraph (f)(6)(i) of this section. For installations made before April 16, 1981, only, electric equipment is also considered to be effectively grounded if it is secured to, and in metallic contact with, the grounded structural metal frame of a building. Metal car frames supported by metal hoisting cables attached to or running over metal sheaves or drums of grounded elevator machines are also considered to be effectively grounded.

**1910.304(f)(7)** *Grounding of systems and circuits of 1000 volts and over (high voltage)—(i) General.* If high voltage systems are grounded, they shall comply with all applicable provisions of paragraphs (f)(1) through (f)(6) of this section as supplemented and modified by this paragraph (f)(7).

**1910.304(f)(7)(ii)** *Grounding of systems supplying portable or mobile equipment.* (See §1910.302(b)(3).) Systems supplying portable or mobile high voltage equipment, other than substations installed on a temporary basis, shall comply with the following:

**1910.304(f)(7)(ii)(A)** Portable and mobile high voltage equipment shall be supplied from a system having its neutral grounded through an impedance. If a delta-connected high voltage system is used to supply the equipment, a system neutral shall be derived.

**1910.304(f)(7)(ii)(B)** Exposed non-current-carrying metal parts of portable and mobile equipment shall be connected by an equipment grounding conductor to the point at which the system neutral impedance is grounded.

**1910.304(f)(7)(ii)(C)** Ground-fault detection and relaying shall be provided to automatically de-energize any high voltage system component which has developed a ground fault. The continuity of the equipment grounding conductor shall be continuously monitored so as to de-energize automatically the high voltage feeder to the portable equipment upon loss of continuity of the equipment grounding conductor.

**1910.304(f)(7)(ii)(D)** The grounding electrode to which the portable or mobile equipment system neutral impedance is connected shall be isolated from and separated in the ground by at least 20 feet from any other system or equipment grounding electrode, and there shall be no direct connection between the grounding electrodes, such as buried pipe, fence, etc.

**1910.304(f)(7)(iii)** *Grounding of equipment.* All non-current-carrying metal parts of portable equipment and fixed equipment including their associated fences, housings, enclosures, and supporting structures shall be grounded. However, equipment which is guarded by location and isolated from ground need not be grounded. Additionally, pole-mounted distribution apparatus at a height exceeding 8 feet above ground or grade level need not be grounded.

[46 FR 4056, Jan. 16, 1981; 46 FR 40185, Aug. 7, 1981, as amended at 55 FR 32015, Aug. 6, 1990]

## §1910.305 Wiring methods, components, and equipment for general use.

**1910.305(a)** *Wiring methods.* The provisions of this section do not apply to the conductors that are an integral part of factory-assembled equipment.

**1910.305(a)(1)** *General requirements—(i) Electrical continuity of metal raceways and enclosures.* Metal raceways, cable armor, and other metal enclosures for conductors shall be metallically joined together into a continuous electric conductor and shall be so connected to all boxes, fittings, and cabinets as to provide effective electrical continuity.

**1910.305(a)(1)(ii)** *Wiring in ducts.* No wiring systems of any type shall be installed in ducts used to transport dust, loose stock or flammable vapors. No wiring system of any type may be installed in any duct used for vapor removal or for ventilation of commercial-type cooking equipment, or in any shaft containing only such ducts.

**1910.305(a)(2)** *Temporary wiring.* Temporary electrical power and lighting wiring methods may be of a class less than would be required for a permanent installation. Except as specifically modified in this paragraph, all other requirements of this subpart for permanent wiring shall apply to temporary wiring installations.

**1910.305(a)(2)(i)** *Uses permitted, 600 volts, nominal, or less.* Temporary electrical power and lighting installations 600 volts, nominal, or less may be used only:

**1910.305(a)(2)(i)(A)** During and for remodeling, maintenance, repair, or demolition of buildings, structures, or equipment, and similar activities;

**1910.305(a)(2)(i)(B)** For experimental or development work, and

**1910.305(a)(2)(i)(C)** For a period not to exceed 90 days for Christmas decorative lighting, carnivals, and similar purposes.

**1910.305(a)(2)(ii)** *Uses permitted, over 600 volts, nominal.* Temporary wiring over 600 volts, nominal, may be used only during periods of tests, experiments, or emergencies.

**1910.305(a)(2)(iii)** *General requirements for temporary wiring.* **(A)** Feeders shall originate in an approved distribution center. The conductors shall be run as multiconductor cord or cable assemblies, or, where not subject to physical damage, they may be run as open conductors on insulators not more than 10 feet apart.

**1910.305(a)(2)(iii)(B)** Branch circuits shall originate in an approved power outlet or panelboard. Conductors shall be multiconductor cord or cable assemblies or open conductors. If run as open conductors they shall be fastened at ceiling height every 10 feet. No branch-circuit conductor may be laid on the floor. Each branch circuit that supplies receptacles or fixed equipment shall contain a separate equipment grounding conductor if run as open conductors.

**1910.305(a)(2)(iii)(C)** Receptacles shall be of the grounding type. Unless installed in a complete metallic raceway, each branch circuit shall contain a separate equipment grounding conductor and all receptacles shall be electrically connected to the grounding conductor.

**1910.305(a)(2)(iii)(D)** No bare conductors nor earth returns may be used for the wiring of any temporary circuit.

**1910.305(a)(2)(iii)(E)** Suitable disconnecting switches or plug connectors shall be installed to permit the disconnection of all ungrounded conductors of each temporary circuit.

**1910.305(a)(2)(iii)(F)** Lamps for general illumination shall be protected from accidental contact or breakage. Protection shall be provided by elevation of at least 7 feet from normal working surface or by a suitable fixture or lampholder with a guard.

**1910.305(a)(2)(iii)(G)** Flexible cords and cables shall be protected from accidental damage. Sharp corners and projections shall be avoided. Where passing through doorways or other pinch points, flexible cords and cables shall be provided with protection to avoid damage.

**1910.305(a)(3)** *Cable trays—(i) Uses permitted.* **(a)** Only the following may be installed in cable tray systems:

**1910.305(a)(3)(i)(a)(1)** Mineral-insulated metal-sheathed cable (Type MI);

**1910.305(a)(3)(i)(a)(2)** Armored cable (Type AC);

**1910.305(a)(3)(i)(a)(3)** Metal-clad cable (Type MC);

**1910.305(a)(3)(i)(a)(4)** Power-limited tray cable (Type PLTC);

**1910.305(a)(3)(i)(a)(5)** Nonmetallic-sheathed cable (Type NM or NMC);

**1910.305(a)(3)(i)(a)(6)** Shielded nonmetallic-sheathed cable (Type SNM);

**1910.305(a)(3)(i)(a)(7)** Multiconductor service-entrance cable (Type SE or USE);

**1910.305(a)(3)(i)(a)(8)** Multiconductor underground feeder and branch-circuit cable (Type UF);

**1910.305(a)(3)(i)(a)(9)** Power and control tray cable (Type TC);

**1910.305(a)(3)(i)(a)(10)** Other factory-assembled, multiconductor control, signal, or power cables which are specifically approved for installation in cable trays; or

**1910.305(a)(3)(i)(a)(11)** Any approved conduit or raceway with its contained conductors.

**1910.305(a)(3)(i)(b)** In industrial establishments only, where conditions of maintenance and supervision assure that only qualified persons will service the installed cable tray system, the following cables may also be installed in ladder, ventilated trough, or 4 inch ventilated channel-type cable trays:

**1910.305(a)(3)(i)(b)(1)** Single conductor cables which are 250 MCM or larger and are Types RHH, RHW, MV, USE, or THW, and other 250 MCM or larger single conductor cables if specifically approved for installation in cable trays. Where exposed to direct rays of the sun, cables shall be sunlight-resistant.

**1910.305(a)(3)(i)(b)(2)** Type MV cables, where exposed to direct rays of the sun, shall be sunlight-resistant.

**1910.305(a)(3)(i)(c)** Cable trays in hazardous (classified) locations shall contain only the cable types permitted in such locations.

**1910.305(a)(3)(ii)** *Uses not permitted.* Cable tray systems may not be used in hoistways or where subjected to severe physical damage.

**1910.305(a)(4)** *Open wiring on insulators—(i) Uses permitted.* Open wiring on insulators is only permitted on systems of 600 volts, nominal, or less for industrial or agricultural establishments and for services.

**1910.305(a)(4)(ii)** *Conductor supports.* Conductors shall be rigidly supported on noncombustible, nonabsorbent insulating materials and may not contact any other objects.

**1910.305(a)(4)(iii)** *Flexible nonmetallic tubing.* In dry locations where not exposed to severe physical damage, conductors may be separately enclosed in flexible nonmetallic tubing. The tubing shall be in continuous lengths not exceeding 15 feet and secured to the surface by straps at intervals not exceeding 4 feet 6 inches.

**1910.305(a)(4)(iv)** *Through walls, floors, wood cross members, etc.* Open conductors shall be separated from contact with walls, floors, wood cross members, or partitions through which they pass by tubes or bushings of noncombustible, nonabsorbent insulating material. If the bushing is shorter than the hole, a waterproof sleeve of nonconductive material shall be inserted in the hole and an insulating bushing slipped into the sleeve at each end in such a manner as to keep the conductors absolutely out of contact with the sleeve. Each conductor shall be carried through a separate tube or sleeve.

**1910.305(a)(4)(v)** *Protection from physical damage.* Conductors within 7 feet from the floor are considered exposed to physical damage. Where open conductors cross ceiling joints and wall studs and are exposed to physical damage, they shall be protected.

**1910.305(b)** *Cabinets, boxes, and fittings—(1) Conductors entering boxes, cabinets, or fittings.* Conductors entering boxes, cabinets, or fittings shall also be protected from abrasion, and openings through which conductors enter shall be effectively closed. Unused openings in cabinets, boxes, and fittings shall be effectively closed.

**1910.305(b)(2)** *Covers and canopies.* All pull boxes, junction boxes, and fittings shall be provided with covers approved for the purpose. If metal covers are used they shall be grounded. In completed installations each outlet box shall have a cover, faceplate, or fixture canopy. Covers of outlet boxes having holes through which flexible cord pendants pass shall be provided with bushings designed for the purpose or shall have smooth, well-rounded surfaces on which the cords may bear.

**1910.305(b)(3)** *Pull and junction boxes for systems over 600 volts, nominal.* In addition to other requirements in this section for pull and junction boxes, the following shall apply to these boxes for systems over 600 volts, nominal:

**1910.305(b)(3)(i)** Boxes shall provide a complete enclosure for the contained conductors or cables.

# OSHA Standards for General Industry

**1910.305(b)(3)(ii)** Boxes shall be closed by suitable covers securely fastened in place. Underground box covers that weigh over 100 pounds meet this requirement. Covers for boxes shall be permanently marked "HIGH VOLTAGE." The marking shall be on the outside of the box cover and shall be readily visible and legible.

**1910.305(c)** *Switches*—**(1)** *Knife switches.* Single-throw knife switches shall be so connected that the blades are dead when the switch is in the open position. Single-throw knife switches shall be so placed that gravity will not tend to close them. Single-throw knife switches approved for use in the inverted position shall be provided with a locking device that will ensure that the blades remain in the open position when so set. Double-throw knife switches may be mounted so that the throw will be either vertical or horizontal. However, if the throw is vertical a locking device shall be provided to ensure that the blades remain in the open position when so set.

**1910.305(c)(2)** *Faceplates for flush-mounted snap switches.* Flush snap switches that are mounted in ungrounded metal boxes and located within reach of conducting floors or other conducting surfaces shall be provided with faceplates of nonconducting, noncombustible material.

**1910.305(d)** *Switchboards and panelboards.* Switchboards that have any exposed live parts shall be located in permanently dry locations and accessible only to qualified persons. Panelboards shall be mounted in cabinets, cutout boxes, or enclosures approved for the purpose and shall be dead front. However, panelboards other than the dead front externally-operable type are permitted where accessible only to qualified persons. Exposed blades of knife switches shall be dead when open.

**1910.305(e)** *Enclosures for damp or wet locations.* **(1)** Cabinets, cutout boxes, fittings, boxes, and panelboard enclosures in damp or wet locations shall be installed so as to prevent moisture or water from entering and accumulating within the enclosures. In wet locations the enclosures shall be weatherproof.

**1910.305(e)(2)** Switches, circuit breakers, and switchboards installed in wet locations shall be enclosed in weatherproof enclosures.

**1910.305(f)** *Conductors for general wiring.* All conductors used for general wiring shall be insulated unless otherwise permitted in this Subpart. The conductor insulation shall be of a type that is approved for the voltage, operating temperature, and location of use. Insulated conductors shall be distinguishable by appropriate color or other suitable means as being grounded conductors, ungrounded conductors, or equipment grounding conductors.

**1910.305(g)** *Flexible cords and cables*—**(1)** *Use of flexible cords and cables.* **(i)** Flexible cords and cables shall be approved and suitable for conditions of use and location. Flexible cords and cables shall be used only for:

**1910.305(g)(1)(i)(A)** Pendants;

**1910.305(g)(1)(i)(B)** Wiring of fixtures;

**1910.305(g)(1)(i)(C)** Connection of portable lamps or appliances;

**1910.305(g)(1)(i)(D)** Elevator cables;

**1910.305(g)(1)(i)(E)** Wiring of cranes and hoists;

**1910.305(g)(1)(i)(F)** Connection of stationary equipment to facilitate their frequent interchange;

**1910.305(g)(1)(i)(G)** Prevention of the transmission of noise or vibration;

**1910.305(g)(1)(i)(H)** Appliances where the fastening means and mechanical connections are designed to permit removal for maintenance and repair; or

**1910.305(g)(1)(i)(I)** Data processing cables approved as a part of the data processing system.

**1910.305(g)(1)(ii)** If used as permitted in paragraphs (g)(1)(i)(c), (g)(1)(i)(f), or (g)(1)(i)(h) of this section, the flexible cord shall be equipped with an attachment plug and shall be energized from an approved receptacle outlet.

**1910.305(g)(1)(iii)** Unless specifically permitted in paragraph (g)(1)(i) of this section, flexible cords and cables may not be used:

**1910.305(g)(1)(iii)(A)** As a substitute for the fixed wiring of a structure;

**1910.305(g)(1)(iii)(B)** Where run through holes in walls, ceilings, or floors;

**1910.305(g)(1)(iii)(C)** Where run through doorways, windows, or similar openings;

**1910.305(g)(1)(iii)(D)** Where attached to building surfaces; or

**1910.305(g)(1)(iii)(E)** Where concealed behind building walls, ceilings, or floors.

**1910.305(g)(1)(iv)** Flexible cords used in show windows and showcases shall be Type S, SO, SJ, SJO, ST, STO, SJT, SJTO, or AFS except for the wiring of chain-supported lighting fixtures and supply cords for portable lamps and other merchandise being displayed or exhibited.

**1910.305(g)(2)** *Identification, splices, and terminations.* **(i)** A conductor of a flexible cord or cable that is used as a grounded conductor or an equipment grounding conductor shall be distinguishable from other conductors. Types SJ, SJO, SJT, SJTO, S, SO, ST, and STO shall be durably marked on the surface with the type designation, size, and number of conductors.

**1910.305(g)(2)(ii)** Flexible cords shall be used only in continuous lengths without splice or tap. Hard service flexible cords No. 12 or larger may be repaired if spliced so that the splice retains the insulation, outer sheath properties, and usage characteristics of the cord being spliced.

**1910.305(g)(2)(iii)** Flexible cords shall be connected to devices and fittings so that strain relief is provided which will prevent pull from being directly transmitted to joints or terminal screws.

**1910.305(h)** *Portable cables over 600 volts, nominal.* Multiconductor portable cable for use in supplying power to portable or mobile equipment at over 600 volts, nominal, shall consist of No. 8 or larger conductors employing flexible stranding. Cables operated at over 2,000 volts shall be shielded for the purpose of confining the voltage stresses to the insulation. Grounding conductors shall be provided. Connectors for these cables shall be of a locking type with provisions to prevent their opening or closing while energized. Strain relief shall be provided at connections and terminations. Portable cables may not be operated with splices unless the splices are of the permanent molded, vulcanized, or other approved type. Termination enclosures shall be suitably marked with a high voltage hazard warning, and terminations shall be accessible only to authorized and qualified personnel.

**1910.305(i)** *Fixture wires*—**(1)** *General.* Fixture wires shall be approved for the voltage, temperature, and location of use. A fixture wire which is used as a grounded conductor shall be identified.

**1910.305(i)(2)** *Uses permitted.* Fixture wires may be used:

**1910.305(i)(2)(i)** For installation in lighting fixtures and in similar equipment where enclosed or protected and not subject to bending or twisting in use; or

**1910.305(i)(2)(ii)** For connecting lighting fixtures to the branch-circuit conductors supplying the fixtures.

**1910.305(i)(3)** *Uses not permitted.* Fixture wires may not be used as branch-circuit conductors except as permitted for Class 1 power limited circuits.

**1910.305(j)** *Equipment for general use*—**(1)** *Lighting fixtures, lampholders, lamps, and receptacles.* **(i)** Fixtures, lampholders, lamps, rosettes, and receptacles may have no live parts normally exposed to employee contact. However, rosettes and cleat-type lampholders and receptacles located at least 8 feet above the floor may have exposed parts.

**1910.305(j)(1)(ii)** Handlamps of the portable type supplied through flexible cords shall be equipped with a handle of molded composition or other material approved for the purpose, and a substantial guard shall be attached to the lampholder or the handle.

**1910.305(j)(1)(iii)** Lampholders of the screw-shell type shall be installed for use as lampholders only. Lampholders installed in wet or damp locations shall be of the weatherproof type.

**1910.305(j)(1)(iv)** Fixtures installed in wet or damp locations shall be approved for the purpose and shall be so constructed or installed that water cannot enter or accumulate in wireways, lampholders, or other electrical parts.

**1910.305(j)(2)** *Receptacles, cord connectors, and attachment plugs (caps).* **(i)** Receptacles, cord connectors, and attachment plugs shall be constructed so that no receptacle or cord connector will accept an attachment plug with a different voltage or current rating than that for which the device is intended. However, a 20-ampere T-slot receptacle or cord connector may accept a 15-ampere attachment plug of the same voltage rating.

**1910.305(j)(2)(ii)** A receptacle installed in a wet or damp location shall be suitable for the location.

**1910.305(j)(3)** *Appliances.* **(i)** Appliances, other than those in which the current-carrying parts at high temperatures are necessarily exposed, may have no live parts normally exposed to employee contact.

**1910.305(j)(3)(ii)** A means shall be provided to disconnect each appliance.

**1910.305(j)(3)(iii)** Each appliance shall be marked with its rating in volts and amperes or volts and watts.

**1910.305(j)(4)** *Motors.* This paragraph applies to motors, motor circuits, and controllers.

**1910.305(j)(4)(i)** *In sight from.* If specified that one piece of equipment shall be "in sight from" another piece of equipment, one shall be visible and not more than 50 feet from the other.

**1910.305(j)(4)(ii)** *Disconnecting means.* **(A)** A disconnecting means shall be located in sight from the controller location. However, a single disconnecting means may be located adjacent to a group of coordinated controllers mounted adjacent to each other on a multi-motor continuous process machine. The controller disconnecting means for motor branch circuits over 600 volts, nominal, may be out of sight of the controller, if the controller is marked with a warning label giving the location and identification of the disconnecting means which is to be locked in the open position.

**1910.305(j)(4)(ii)(B)** The disconnecting means shall disconnect the motor and the controller from all ungrounded supply conductors and shall be so designed that no pole can be operated independently.

**1910.305(j)(4)(ii)(C)** If a motor and the driven machinery are not in sight from the controller location, the installation shall comply with one of the following conditions:

**1910.305(j)(4)(ii)(C)***(1)* The controller disconnecting means shall be capable of being locked in the open position.

**1910.305(j)(4)(ii)(C)***(2)* A manually operable switch that will disconnect the motor from its source of supply shall be placed in sight from the motor location.

**1910.305(j)(4)(ii)(D)** The disconnecting means shall plainly indicate whether it is in the open (off) or closed (on) position.

**1910.305(j)(4)(ii)(E)** The disconnecting means shall be readily accessible. If more than one disconnect is provided for the same equipment, only one need be readily accessible.

**1910.305(j)(4)(ii)(F)** An individual disconnecting means shall be provided for each motor, but a single disconnecting means may be used for a group of motors under any one of the following conditions:

**1910.305(j)(4)(ii)(F)***(1)* If a number of motors drive special parts of a single machine or piece of apparatus, such as a metal or woodworking machine, crane, or hoist;

**1910.305(j)(4)(ii)(F)***(2)* If a group of motors is under the protection of one set of branch-circuit protective devices; or

**1910.305(j)(4)(ii)(F)***(3)* If a group of motors is in a single room in sight from the location of the disconnecting means.

**1910.305(j)(4)(iii)** *Motor overload, short-circuit, and ground-fault protection.* Motors, motor-control apparatus, and motor branch-circuit conductors shall be protected against overheating due to motor overloads or failure to start, and against short-circuits or ground faults. These provisions shall not require overload protection that will stop a motor where a shutdown is likely to introduce additional or increased hazards, as in the case of fire pumps, or where continued operation of a motor is necessary for a safe shutdown of equipment or process and motor overload sensing devices are connected to a supervised alarm.

**1910.305(j)(4)(iv)** *Protection of live parts—all voltages.* **(A)** Stationary motors having commutators, collectors, and brush rigging located inside of motor end brackets and not conductively connected to supply circuits operating at more than 150 volts to ground need not have such parts guarded. Exposed live parts of motors and controllers operating at 50 volts or more between terminals shall be guarded against accidental contact by any of the following:

**1910.305(j)(4)(iv)(A)***(1)* By installation in a room or enclosure that is accessible only to qualified persons;

**1910.305(j)(4)(iv)(A)***(2)* By installation on a suitable balcony, gallery, or platform, so elevated and arranged as to exclude unqualified persons; or

**1910.305(j)(4)(iv)(A)***(3)* By elevation 8 feet or more above the floor.

**1910.305(j)(4)(iv)(B)** Where live parts of motors or controllers operating at over 150 volts to ground are guarded against accidental contact only by location, and where adjustment or other attendance may be necessary during the operation of the apparatus, suitable insulating mats or platforms shall be provided so that the attendant cannot readily touch live parts unless standing on the mats or platforms.

**1910.305(j)(5)** *Transformers.* **(i)** The following paragraphs cover the installation of all transformers except the following:

**1910.305(j)(5)(i)(A)** Current transformers;

**1910.305(j)(5)(i)(B)** Dry-type transformers installed as a component part of other apparatus;

**1910.305(j)(5)(i)(C)** Transformers which are an integral part of an X-ray, high frequency, or electrostatic-coating apparatus;

**1910.305(j)(5)(i)(D)** Transformers used with Class 2 and Class 3 circuits, sign and outline lighting, electric discharge lighting, and power-limited fire-protective signalling circuits; and

**1910.305(j)(5)(i)(E)** Liquid-filled or dry-type transformers used for research, development, or testing, where effective safeguard arrangements are provided.

**1910.305(j)(5)(ii)** The operating voltage of exposed live parts of transformer installations shall be indicated by warning signs or visible markings on the equipment or structure.

**1910.305(j)(5)(iii)** Dry-type, high fire point liquid-insulated, and askarel-insulated transformers installed indoors and rated over 35kV shall be in a vault.

**1910.305(j)(5)(iv)** If they present a fire hazard to employees, oil-insulated transformers installed indoors shall be in a vault.

**1910.305(j)(5)(v)** Combustible material, combustible buildings and parts of buildings, fire escapes, and door and window openings shall be safeguarded from fires which may originate in oil-insulated transformers attached to or adjacent to a building or combustible material.

**1910.305(j)(5)(vi)** Transformer vaults shall be constructed so as to contain fire and combustible liquids within the vault and to prevent unauthorized access. Locks and latches shall be so arranged that a vault door can be readily opened from the inside.

**1910.305(j)(5)(vii)** Any pipe or duct system foreign to the vault installation may not enter or pass through a transformer vault.

**1910.305(j)(5)(viii)** Materials may not be stored in transformer vaults.

**1910.305(j)(6)** *Capacitors.* **(i)** All capacitors, except surge capacitors or capacitors included as a component part of other apparatus, shall be provided with an automatic means of draining the stored charge after the capacitor is disconnected from its source of supply.

**1910.305(j)(6)(ii)** Capacitors rated over 600 volts, nominal, shall comply with the following additional requirements:

**1910.305(j)(6)(ii)(A)** Isolating or disconnecting switches (with no interrupting rating) shall be interlocked with the load interrupting device or shall be provided with prominently displayed caution signs to prevent switching load current.

**1910.305(j)(6)(ii)(B)** For series capacitors (see §1910.302(b)(3)), the proper switching shall be assured by use of at least one of the following:

**1910.305(j)(6)(ii)(B)***(1)* Mechanically sequenced isolating and bypass switches,

**1910.305(j)(6)(ii)(B)***(2)* Interlocks, or

**1910.305(j)(6)(ii)(B)***(3)* Switching procedure prominently displayed at the switching location.

**1910.305(j)(7)** *Storage batteries.* Provisions shall be made for sufficient diffusion and ventilation of gases from storage batteries to prevent the accumulation of explosive mixtures.

[46 FR 4056, Jan. 16, 1981; 46 FR 40185, Aug. 7, 1981]

### §1910.306 Specific purpose equipment and installations.

**1910.306(a)** *Electric signs and outline lighting—***(1)** *Disconnecting means.* Signs operated by electronic or electromechanical controllers located outside the sign shall have a disconnecting means located inside the controller enclosure or within sight of the controller location, and it shall be capable of being locked in the open position. Such disconnecting means shall have no pole that can be operated independently, and it shall open all ungrounded conductors that supply the controller and sign. All other signs, except the portable type, and all outline lighting installations shall have an externally operable disconnecting means which can open all ungrounded conductors and is within the sight of the sign or outline lighting it controls.

**1910.306(a)(2)** Doors or covers giving access to uninsulated parts of indoor signs or outline lighting exceeding 600 volts and accessible to other than qualified persons shall either be provided with interlock switches to disconnect the primary circuit or shall be so fastened that the use of other than ordinary tools will be necessary to open them.

**1910.306(b)** *Cranes and hoists.* This paragraph applies to the installation of electric equipment and wiring used in connection with cranes, monorail hoists, hoists, and all runways.

**1910.306(b)(1)** *Disconnecting means.* A readily accessible disconnecting means (i) shall be provided between the runway contact conductors and the power supply.

**1910.306(b)(1)(ii)** Another disconnecting means, capable of being locked in the open position, shall be provided in the leads from the runway contact conductors or other power supply on any crane or monorail hoist.

**1910.306(b)(1)(ii)(A)** If this additional disconnecting means is not readily accessible from the crane or monorail hoist operating station, means shall be provided at the operating station to open the power circuit to all motors of the crane or monorail hoist.

**1910.306(b)(1)(ii)(B)** The additional disconnect may be omitted if a monorail hoist or hand-propelled crane bridge installation meets all of the following:

**1910.306(b)(1)(ii)(B)***(1)* The unit is floor controlled;

**1910.306(b)(1)(ii)(B)***(2)* The unit is within view of the power supply disconnecting means; and

**1910.306(b)(1)(ii)(B)***(3)* No fixed work platform has been provided for servicing the unit.

**1910.306(b)(2)** *Control.* A limit switch or other device shall be provided to prevent the load block from passing the safe upper limit of travel of any hoisting mechanism.

**1910.306(b)(3)** *Clearance.* The dimension of the working space in the direction of access to live parts which may require examination, adjustment, servicing, or maintenance while alive shall be a minimum of 2 feet 6 inches. Where controls are enclosed in cabinets, the door(s) shall either open at least 90 degrees or be removable.

**1910.306(c)** *Elevators, dumbwaiters, escalators, and moving walks—* **(1)** *Disconnecting means.* Elevators, dumbwaiters, escalators, and moving walks shall have a single means for disconnecting all ungrounded main power supply conductors for each unit.

**1910.306(c)(2)** *Warning signs.* If interconnections between control panels are necessary for operation of the system on a multicar installation that remains energized from a source other than the disconnecting means, a warning sign shall be mounted on or adjacent to the disconnecting means. The sign shall be clearly legible and shall read "Warning—Parts of the control panel are not de-energized by this switch." (See §1910.302(b)(3).)

**1910.306(c)(3)** *Control panels.* If control panels are not located in the same space as the drive machine, they shall be located in cabinets with doors or panels capable of being locked closed.

**1910.306(d)** *Electric welders—disconnecting means.* **(1)** A disconnecting means shall be provided in the supply circuit for each motor-generator arc welder, and for each AC transformer and DC rectifier arc welder which is not equipped with a disconnect mounted as an intergral part of the welder.

**1910.306(d)(2)** A switch or circuit breaker shall be provided by which each resistance welder and its control equipment can be isolated from the supply circuit. The ampere rating of this disconnecting means may not be less than the supply conductor ampacity.

**1910.306(e)** *Data processing systems—disconnecting means.* A disconnecting means shall be provided to disconnect the power to all electronic equipment in data processing or computer rooms. This disconnecting means shall be controlled from locations readily accessible to the operator at the principal exit doors. There shall also be a similar disconnecting means to disconnect the air conditioning system serving this area.

**1910.306(f)** *X-Ray equipment.* This paragraph applies to X-Ray equipment for other than medical or dental use.

**1910.306(f)(1)** *Disconnecting means.* **(i)** A disconnecting means shall be provided in the supply circuit. The disconnecting means shall be operable from a location readily accessible from the X-ray control. For equipment connected to a 120-volt branch circuit of 30 amperes or less, a grounding-type attachment plug cap and receptacle of proper rating may serve as a disconnecting means.

**1910.306(f)(1)(ii)** If more than one piece of equipment is operated from the same high-voltage circuit, each piece or each group of equipment as a unit shall be provided with a high-voltage switch or equivalent disconnecting means. This disconnecting means shall be constructed, enclosed, or located so as to avoid contact by employees with its live parts.

**1910.306(f)(2)** *Control—***(i)** *Radiographic and fluoroscopic types.* Radiographic and fluoroscopic-type equipment shall be effectively enclosed or shall have interlocks that de-energize the equipment automatically to prevent ready access to live current-carrying parts.

**1910.306(f)(2)(ii)** *Diffraction and irradiation types.* Diffraction-and irradiation-type equipment shall be provided with a means to indicate when it is energized unless the equipment or installation is effectively enclosed or is provided with interlocks to prevent access to live current-carrying parts during operation.

**1910.306(g)** *Induction and dielectric heating equipment—***(1)** *Scope.* Paragraphs (g)(2) and (g)(3) of this section cover induction and dielectric heating equipment and accessories for industrial and scientific applications, but not for medical or dental applications or for appliances.

**1910.306(g)(2)** *Guarding and grounding.* **(i)** *Enclosures.* The converting apparatus (including the DC line) and high-frequency electric circuits (excluding the output circuits and remote-control circuits) shall be completely contained within enclosures of noncombustible material.

**1910.306(g)(2)(ii)** *Panel controls.* All panel controls shall be of dead-front construction.

**1910.306(g)(2)(iii)** *Access to internal equipment.* Where doors are used for access to voltages from 500 to 1000 volts AC or DC, either door locks or interlocks shall be provided. Where doors are used for access to voltages of over 1000 volts AC or DC, either mechanical lockouts with a disconnecting means to prevent access until voltage is removed from the cubicle, or both door interlocking and mechanical door locks, shall be provided.

**1910.306(g)(2)(iv)** *Warning labels.* "Danger" labels shall be attached on the equipment and shall be plainly visible even when doors are open or panels are removed from compartments containing voltages of over 250 volts AC or DC.

**1910.306(g)(2)(v)** *Work applicator shielding.* Protective cages or adequate shielding shall be used to guard work applicators other than induction heating coils. Induction heating coils shall be protected by insulation and/or refractory materials. Interlock switches shall be used on all hinged access doors, sliding panels, or other such means of access to the applicator. Interlock switches shall be connected in such a manner as to remove all power from the applicator when any one of the access doors or panels is open. Interlocks on access doors or panels are not required if the applicator is an induction heating coil at DC ground potential or operating at less than 150 volts AC.

**1910.306(g)(2)(vi)** *Disconnecting means.* A readily accessible disconnecting means shall be provided by which each unit of heating equipment can be isolated from its supply circuit.

**1910.306(g)(3)** *Remote control.* If remote controls are used for applying power, a selector switch shall be provided and interlocked to provide power from only one control point at a time. Switches operated by foot pressure shall be provided with a shield over the contact button to avoid accidental closing of the switch.

**1910.306(h)** *Electrolytic cells.* **(1)** *Scope.* These provisions for electrolytic cells apply to the installation of the electrical components and accessory equipment of electrolytic cells, electrolytic cell lines, and process power supply for the production of aluminum, cadmium, chlorine, copper, fluorine, hydrogen peroxide, magnesium, sodium, sodium chlorate, and zinc. Cells used as a source of electric energy and for electroplating processes and cells used for production of hydrogen are not covered by these provisions.

**1910.306(h)(2)** *Definitions applicable to this paragraph.*

*Cell line:* An assembly of electrically interconnected electrolytic cells supplied by a source of direct-current power.

*Cell line attachments and auxiliary equipment:* Cell line attachments and auxiliary equipment include, but are not limited to: auxiliary tanks; process piping; duct work; structural supports; exposed cell line conductors; conduits and other raceways; pumps; positioning equipment and cell cutout or by-pass electrical devices. Auxiliary equipment also includes tools, welding machines, crucibles, and other portable equipment used for operation and maintenance within the electrolytic cell line working zone. In the cell line working zone, auxiliary equipment includes the exposed conductive surfaces of ungrounded cranes and crane-mounted cell-servicing equipment.

*Cell line working zone:* The cell line working zone is the space envelope wherein operation or maintenance is normally performed on or in the vicinity of exposed energized surfaces of cell lines or their attachments.

*Electrolytic Cells:* A receptacle or vessel in which electrochemical reactions are caused by applying energy for the purpose of refining or producing usable materials.

**1910.306(h)(3)** *Application.* Installations covered by paragraph (h) of this section shall comply with all applicable provisions of this subpart, except as follows:

**1910.306(h)(3)(i)** Overcurrent protection of electrolytic cell DC process power circuits need not comply with the requirements of §1910.304(e).

**1910.306(h)(3)(ii)** Equipment located or used within the cell line working zone or associated with the cell line DC power circuits need not comply with the provisions of §1910.304(f).

**1910.306(h)(3)(iii)** Electrolytic cells, cell line conductors, cell line attachments, and the wiring of auxiliary equipment and devices within the cell line working zone need not comply with the provisions of §§1910.303, and 1910.304(b) and (c).

**1910.306(h)(4)** *Disconnecting means.* **(i)** If more than one DC cell line process power supply serves the same cell line, a disconnecting means shall be provided on the cell line circuit side of each power supply to disconnect it from the cell line circuit.

**1910.306(h)(4)(ii)** Removable links or removable conductors may be used as the disconnecting means.

**1910.306(h)(5)** *Portable electric equipment.* **(i)** The frames and enclosures of portable electric equipment used within the cell line working zone may not be grounded. However, these frames and enclosures may be grounded if the cell line circuit voltage does not exceed 200 volts DC or if the frames are guarded.

**1910.306(h)(5)(ii)** Ungrounded portable electric equipment shall be distinctively marked and may not be interchangeable with grounded portable electric equipment.

**1910.306(h)(6)** *Power supply circuits and receptacles for portable electric equipment.* **(i)** Circuits supplying power to ungrounded receptacles for hand-held, cord- and plug-connected equipment shall be electrically isolated from any distribution system supplying areas other than the cell line working zone and shall be ungrounded. Power for these circuits shall be supplied through isolating transformers.

**1910.306(h)(6)(ii)** Receptacles and their mating plugs for ungrounded equipment may not have provision for a grounding conductor and shall be of a configuration which prevents their use for equipment required to be grounded.

**1910.306(h)(6)(iii)** Receptacles on circuits supplied by an isolating transformer with an ungrounded secondary shall have a distinctive configuration, shall be distinctively marked, and may not be used in any other location in the plant.

**1910.306(h)(7)** *Fixed and portable electric equipment.* **(i)** AC systems supplying fixed and portable electric equipment within the cell line working zone need not be grounded.

**1910.306(h)(7)(ii)** Exposed conductive surfaces, such as electric equipment housings, cabinets, boxes, motors, raceways and the like that are within the cell line working zone need not be grounded.

**1910.306(h)(7)(iii)** Auxiliary electrical devices, such as motors, transducers, sensors, control devices, and alarms, mounted on an electrolytic cell or other energized surface, shall be connected by any of the following means:

**1910.306(h)(7)(iii)(A)** Multiconductor hard usage or extra hard usage flexible cord;

**1910.306(h)(7)(iii)(B)** Wire or cable in suitable raceways; or

**1910.306(h)(7)(iii)(C)** Exposed metal conduit, cable tray, armored cable, or similar metallic systems installed with insulating breaks such that they will not cause a potentially hazardous electrical condition.

**1910.306(h)(7)(iv)** Fixed electric equipment may be bonded to the energized conductive surfaces of the cell line, its attachments, or auxiliaries. If fixed electric equipment is mounted on an energized conductive surface, it shall be bonded to that surface.

**1910.306(h)(8)** *Auxiliary nonelectric connections.* Auxiliary nonelectric connections, such as air hoses, water hoses, and the like, to an electrolytic cell, its attachments, or auxiliary equipment may not have continuous conductive reinforcing wire, armor, braids, and the like. Hoses shall be of a nonconductive material.

**1910.306(h)(9)** *Cranes and hoists.* **(i)** The conductive surfaces of cranes and hoists that enter the cell line working zone need not be grounded. The portion of an overhead crane or hoist which contacts an energized electrolytic cell or energized attachments shall be insulated from ground.

**1910.306(h)(9)(ii)** Remote crane or hoist controls which may introduce hazardous electrical conditions into the cell line working zone shall employ one or more of the following systems:

**1910.306(h)(9)(ii)(A)** Insulated and ungrounded control circuit;

**1910.306(h)(9)(ii)(B)** Nonconductive rope operator;

**1910.306(h)(9)(ii)(C)** Pendant pushbutton with nonconductive supporting means and having nonconductive surfaces or ungrounded exposed conductive surfaces; or

**1910.306(h)(9)(ii)(D)** Radio.

**1910.306(i)** *Electrically driven or controlled irrigation machines.* (See §1910.302(b)(3).)

**1910.306(i)(1)** *Lightning protection.* If an electrically driven or controlled irrigation machine has a stationary point, a driven ground rod shall be connected to the machine at the stationary point for lightning protection.

**1910.306(i)(2)** *Disconnecting means.* The main disconnecting means for a center pivot irrigation machine shall be located at the point of connection of electrical power to the machine and shall be readily accessible and capable of being locked in the open position. A disconnecting means shall be provided for each motor and controller.

**1910.306(j)** *Swimming pools, fountains, and similar installations*—**(1)** *Scope.* Paragraphs (j)(2) through (j)(5) of this section apply to electric wiring for and equipment in or adjacent to all swimming, wading, therapeutic, and decorative pools and fountains, whether permanently installed or storable, and to metallic auxiliary equipment, such as pumps, filters, and similar equipment. Therapeutic pools in health care facilities are exempt from these provisions.

**1910.306(j)(2)** *Lighting and receptacles*—**(i)** *Receptacles.* A single receptacle of the locking and grounding type that provides power for a permanently installed swimming pool recirculating pump motor may be located not less than 5 feet from the inside walls of a pool. All other receptacles on the property shall be located at least 10 feet from the inside walls of a pool. Receptacles which are located within 15 feet of the inside walls of the pool shall be protected by ground-fault circuit interrupters.

**NOTE:** In determining these dimensions, the distance to be measured is the shortest path the supply cord of an appliance connected to the receptacle would follow without piercing a floor, wall, or ceiling of a building or other effective permanent barrier.

**1910.306(j)(2)(ii)** *Lighting fixtures and lighting outlets.* **(A)** Unless they are 12 feet above the maximum water level, lighting fixtures and lighting outlets may not be installed over a pool or over the area extending 5 feet horizontally from the inside walls of a pool. However, a lighting fixture or lighting outlet which has been installed before April 16, 1981, may be located less than 5 feet measured horizontally from the inside walls of a pool if it is at least 5 feet above the surface of the maximum water level and shall be rigidly attached to the existing structure. It shall also be protected by a ground-fault circuit interrupter installed in the branch circuit supplying the fixture.

**1910.306(j)(2)(ii)(B)** Unless installed 5 feet above the maximum water level and rigidly attached to the structure adjacent to or enclosing the pool, lighting fixtures and lighting outlets installed in the area extending between 5 feet and 10 feet horizontally from the inside walls of a pool shall be protected by a ground-fault circuit interrupter.

# OSHA Standards for General Industry

**1910.306(j)(3)** *Cord- and plug-connected equipment.* Flexible cords used with the following equipment may not exceed 3 feet in length and shall have a copper equipment grounding conductor with a grounding-type attachment plug.

**1910.306(j)(3)(i)** Cord- and plug-connected lighting fixtures installed within 16 feet of the water surface of permanently installed pools.

**1910.306(j)(3)(ii)** Other cord- and plug-connected, fixed or stationary equipment used with permanently installed pools.

**1910.306(j)(4)** *Underwater equipment.* **(i)** A ground-fault circuit interrupter shall be installed in the branch circuit supplying underwater fixtures operating at more than 15 volts. Equipment installed underwater shall be approved for the purpose.

**1910.306(j)(4)(ii)** No underwater lighting fixtures may be installed for operation at over 150 volts between conductors.

**1910.306(j)(5)** *Fountains.* All electric equipment operating at more than 15 volts, including power supply cords, used with fountains shall be protected by ground-fault circuit interrupters. (See §1910.302(b)(3).)

[46 FR 4056, Jan. 16, 1981; 46 FR 40185, Aug. 7, 1981]

## §1910.307 Hazardous (classified) locations.

**1910.307(a)** *Scope.* This section covers the requirements for electric equipment and wiring in locations which are classified depending on the properties of the flammable vapors, liquids or gases, or combustible dusts or fibers which may be present therein and the likelihood that a flammable or combustible concentration or quantity is present. Hazardous (classified) locations may be found in occupancies such as, but not limited to, the following: aircraft hangars, gasoline dispensing and service stations, bulk storage plants for gasoline or other volatile flammable liquids, paint-finishing process plants, health care facilities, agricultural or other facilities where excessive combustible dusts may be present, marinas, boat yards, and petroleum and chemical processing plants. Each room, section or area shall be considered individually in determining its classification. These hazardous (classified) locations are assigned six designations as follows:

Class I, Division 1

Class I, Division 2

Class II, Division 1

Class II, Division 2

Class III, Division 1

Class III, Division 2

For definitions of these locations see §1910.399(a). All applicable requirements in this subpart shall apply to hazardous (classified) locations, unless modified by provisions of this section.

**1910.307(b)** *Electrical installations.* Equipment, wiring methods, and installations of equipment in hazardous (classified) locations shall be intrinsically safe, approved for the hazardous (classified) location, or safe or for the hazardous (classified) location. Requirements for each of these options are as follows:

**1910.307(b)(1)** *Intrinsically safe.* Equipment and associated wiring approved as intrinsically safe shall be permitted in any hazardous (classified) location for which it is approved.

**1910.307(b)(2)** *Approved for the hazardous (classified) location.* **(i)** Equipment shall be approved not only for the class of location but also for the ignitible or combustible properties of the specific gas, vapor, dust, or fiber that will be present.

**NOTE:** NFPA 70, the National Electrical Code, lists or defines hazardous gases, vapors, and dusts by "Groups" characterized by their ignitible or combustible properties.

**1910.307(b)(2)(ii)** Equipment shall be marked to show the class, group, and operating temperature or temperature range, based on operation in a 40 degrees C ambient, for which it is approved. The temperature marking may not exceed the ignition temperature of the specific gas or vapor to be encountered. However, the following provisions modify this marking requirement for specific equipment:

**1910.307(b)(2)(ii)(A)** Equipment of the non-heat-producing type, such as junction boxes, conduit, and fittings, and equipment of the heat-producing type having a maximum temperature not more than 100 °C (212 °F) need not have a marked operating temperature or temperature range.

**1910.307(b)(2)(ii)(B)** Fixed lighting fixtures marked for use in Class I, Division 2 locations only, need not be marked to indicate the group.

**1910.307(b)(2)(ii)(C)** Fixed general-purpose equipment in Class I locations, other than lighting fixtures, which is acceptable for use in Class I, Division 2 locations need not be marked with the class, group, division, or operating temperature.

**1910.307(b)(2)(ii)(D)** Fixed dust-tight equipment, other than lighting fixtures, which is acceptable for use in Class II, Division 2 and Class III locations need not be marked with the class, group, division, or operating temperature.

**1910.307(b)(3)** *Safe for the hazardous (classified) location.* Equipment which is safe for the location shall be of a type and design which the employer demonstrates will provide protection from the hazards arising from the combustibility and flammability of vapors, liquids, gases, dusts, or fibers.

**NOTE:** The National Electrical Code, NFPA 70, contains guidelines for determining the type and design of equipment and installations which will meet this requirement. The guidelines of this document address electric wiring, equipment, and systems installed in hazardous (classified) locations and contain specific provisions for the following: wiring methods, wiring connections; conductor insulation, flexible cords, sealing and drainage, transformers, capacitors, switches, circuit breakers, fuses, motor controllers, receptacles, attachment plugs, meters, relays, instruments, resistors, generators, motors, lighting fixtures, storage battery charging equipment, electric cranes, electric hoists and similar equipment, utilization equipment, signaling systems, alarm systems, remote control systems, local loud speaker and communication systems, ventilation piping, live parts, lightning surge protection, and grounding. Compliance with these guidelines will constitute one means, but not the only means, of compliance with this paragraph.

**1910.307(c)** *Conduits.* All conduits shall be threaded and shall be made wrench-tight. Where it is impractical to make a threaded joint tight, a bonding jumper shall be utilized.

**1910.307(d)** *Equipment in Division 2 locations.* Equipment that has been approved for a Division 1 location may be installed in a Division 2 location of the same class and group. General-purpose equipment or equipment in general-purpose enclosures may be installed in Division 2 locations if the equipment does not constitute a source of ignition under normal operating conditions.

[46 FR 4056, Jan. 16, 1981; 46 FR 40185, Aug. 7, 1981]

## §1910.308 Special systems.

**1910.308(a)** *Systems over 600 volts, nominal.* Paragraphs (a) (1) through (4) of this section cover the general requirements for all circuits and equipment operated at over 600 volts.

**1910.308(a)(1)** *Wiring methods for fixed installations.* **(i)** Above-ground conductors shall be installed in rigid metal conduit, in intermediate metal conduit, in cable trays, in cablebus, in other suitable raceways, or as open runs of metal-clad cable suitable for the use and purpose. However, open runs of non-metallic-sheathed cable or of bare conductors or busbars may be installed in locations accessible only to qualified persons. Metallic shielding components, such as tapes, wires, or braids for conductors, shall be grounded. Open runs of insulated wires and cables having a bare lead sheath or a braided outer covering shall be supported in a manner designed to prevent physical damage to the braid or sheath.

**1910.308(a)(1)(ii)** Conductors emerging from the ground shall be enclosed in approved raceways. (See §1910.302(b)(3).)

**1910.308(a)(2)** *Interrupting and isolating devices.* **(i)** Circuit breaker installations located indoors shall consist of metal-enclosed units or fire-resistant cell-mounted units. In locations accessible only to qualified personnel, open mounting of circuit breakers is permitted. A means of indicating the open and closed position of circuit breakers shall be provided.

**1910.308(a)(2)(ii)** Fused cutouts installed in buildings or tranformer vaults shall be of a type approved for the purpose. They shall be readily accessible for fuse replacement.

**1910.308(a)(2)(iii)** A means shall be provided to completely isolate equipment for inspection and repairs. Isolating means which are not designed to interrupt the load current of the circuit shall be either interlocked with an approved circuit interrupter or provided with a sign warning against opening them under load.

**1910.308(a)(3)** *Mobile and portable equipment.* **(i)** *Power cable connections to mobile machines.* A metallic enclosure shall be provided on the mobile machine for enclosing the terminals of the power cable. The enclosure shall include provisions for a solid connection for the ground wire(s) terminal to effectively ground the machine frame. The method of cable termination used shall prevent any strain or pull on the cable from stressing the electrical connections. The enclosure shall have provision for locking so only authorized qualified persons may open it and shall be marked with a sign warning of the presence of energized parts.

**1910.308(a)(3)(ii)** *Guarding live parts.* All energized switching and control parts shall be enclosed in effectively grounded metal cabinets or enclosures. Circuit breakers and protective equipment shall have the operating means projecting through the metal cabinet or enclosure so these units can be reset without locked doors being opened. Enclosures and metal cabinets shall be locked so that only authorized qualified persons have access and shall be marked with a sign warning of the presence of energized parts. Collector ring assemblies on revolving-type machines (shovels, draglines, etc.) shall be guarded.

**1910.308(a)(4)** *Tunnel installation—***(i)** *Application.* The provisions of this paragraph apply to installation and use of high-voltage power distribution and utilization equipment which is portable and/or mobile, such as substations, trailers, cars, mobile shovels, draglines, hoists, drills, dredges, compressors, pumps, conveyors, and underground excavators.

**1910.308(a)(4)(ii)** *Conductors.* Conductors in tunnels shall be installed in one or more of the following:

**1910.308(a)(4)(ii)(A)** Metal conduit or other metal raceway,

**1910.308(a)(4)(ii)(B)** Type MC cable, or

**1910.308(a)(4)(ii)(C)** Other approved multiconductor cable.

Conductors shall also be so located or guarded as to protect them from physical damage. Multiconductor portable cable may supply mobile equipment. An equipment grounding conductor shall be run with circuit conductors inside the metal raceway or inside the multiconductor cable jacket. The equipment grounding conductor may be insulated or bare.

**1910.308(a)(4)(iii)** *Guarding live parts.* Bare terminals of transformers, switches, motor controllers, and other equipment shall be enclosed to prevent accidental contact with energized parts. Enclosures for use in tunnels shall be drip-proof, weatherproof, or submersible as required by the environmental conditions.

**1910.308(a)(4)(iv)** *Disconnecting means.* A disconnecting means that simultaneously opens all ungrounded conductors shall be installed at each transformer or motor location.

**1910.308(a)(4)(v)** *Grounding and bonding.* All nonenergized metal parts of electric equipment and metal raceways and cable sheaths shall be effectively grounded and bonded to all metal pipes and rails at the portal and at intervals not exceeding 1000 feet throughout the tunnel.

**1910.308(b)** *Emergency power systems—***(1)** *Scope.* The provisions for emergency systems apply to circuits, systems, and equipment intended to supply power for illumination and special loads, in the event of failure of the normal supply.

**1910.308(b)(2)** *Wiring methods.* Emergency circuit wiring shall be kept entirely independent of all other wiring and equipment and may not enter the same raceway, cable, box, or cabinet or other wiring except either where common circuit elements suitable for the purpose are required, or for transferring power from the normal to the emergency source.

**1910.308(b)(3)** *Emergency illumination.* Where emergency lighting is necessary, the system shall be so arranged that the failure of any individual lighting element, such as the burning out of a light bulb, cannot leave any space in total darkness.

**1910.308(c)** *Class 1, Class 2, and Class 3 remote control, signaling, and power-limited circuits—***(1)** *Classification.* Class 1, Class 2, or Class 3 remote control, signaling, or power-limited circuits are characterized by their usage and electrical power limitation which differentiates them from light and power circuits. These circuits are classified in accordance with their respective voltage and power limitations as summarized in paragraphs (c)(1)(i) through (c)(1)(iii) of this section.

**1910.308(c)(1)(i)** *Class 1 circuits.* **(A)** A Class 1 power-limited circuit is supplied from a source having a rated output of not more than 30 volts and 1000 volt-amperes.

**1910.308(c)(1)(i)(B)** A Class 1 remote control circuit or a Class 1 signaling circuit has a voltage which does not exceed 600 volts; however, the power output of the source need not be limited.

**1910.308(c)(1)(ii)** *Class 2 and Class 3 circuits.* **(A)** Power for Class 2 and Class 3 circuits is limited either inherently (in which no overcurrent protection is required) or by a combination of a power source and overcurrent protection.

**1910.308(c)(1)(ii)(B)** The maximum circuit voltage is 150 volts AC or DC for a Class 2 inherently limited power source, and 100 volts AC or DC for a Class 3 inherently limited power source.

**1910.308(c)(1)(ii)(C)** The maximum circuit voltage is 30 volts AC and 60 volts DC for a Class 2 power source limited by overcurrent protection, and 150 volts AC or DC for a Class 3 power source limited by overcurrent protection.

**1910.308(c)(1)(iii)** The maximum circuit voltages in paragraphs (c)(1)(i) and (c)(1)(ii) of this section apply to sinusoidal AC or continuous DC power sources, and where wet contact occurence is not likely.

**1910.308(c)(2)** *Marking.* A Class 2 or Class 3 power supply unit shall be durably marked where plainly visible to indicate the class of supply and its electrical rating. (See §1910.302(b)(3).)

**1910.308(d)** *Fire protective signaling systems.* (See §1910.302(b)(3).)

**1910.308(d)(1)** *Classifications.* Fire protective signaling circuits shall be classified either as non-power limited or power limited.

**1910.308(d)(2)** *Power sources.* The power sources for use with fire protective signaling circuits shall be either power limited or nonlimited as follows:

**1910.308(d)(2)(i)** The power supply of non-power-limited fire protective signaling circuits shall have an output voltage not in excess of 600 volts.

**1910.308(d)(2)(ii)** The power for power-limited fire protective signaling circuits shall be either inherently limited, in which no overcurrent protection is required, or limited by a combination of a power source and overcurrent protection.

**1910.308(d)(3)** *Non-power-limited conductor location.* Non-power-limited fire protective signaling circuits and Class 1 circuits may occupy the same enclosure, cable, or raceway provided all conductors are insulated for maximum voltage of any conductor within the enclosure, cable, or raceway. Power supply and fire protective signaling circuit conductors are permitted in the same enclosure, cable, or raceway only if connected to the same equipment.

**1910.308(d)(4)** *Power-limited conductor location.* Where open conductors are installed, power-limited fire protective signaling circuits shall be separated at least 2 inches from conductors of any light, power, Class 1, and non-power-limited fire protective signaling circuits unless a special and equally protective method of conductor separation is employed. Cables and conductors of two or more power-limited fire protective signaling circuits or Class 3 circuits are permitted in the same cable, enclosure, or raceway. Conductors of one or more Class 2 circuits are permitted within the same cable, enclosure, or raceway with conductors of power-limited fire protective signaling circuits provided that the insulation of Class 2 circuit conductors in the cable, enclosure, or raceway is at least that needed for the power-limited fire protective signaling circuits.

**1910.308(d)(5)** *Identification.* Fire protective signaling circuits shall be identified at terminal and junction locations in a manner which will prevent unintentional interference with the signaling circuit during testing and servicing. Power-limited fire protective signaling circuits shall be durably marked as such where plainly visible at terminations.

**1910.308(e)** *Communications systems—***(1)** *Scope.* These provisions for communication systems apply to such systems as central-station-connected and non-central-station-connected telephone circuits, radio and television receiving and transmitting equipment, including community antenna television and radio distribution systems, telegraph, district messenger, and outside wiring for fire and burglar alarm, and similar central station systems. These installations need not comply with the provisions of §§1910.303 through 1910.308(d), except §1910.304(c)(1) and §1910.307(b).

**1910.308(e)(2)** *Protective devices.* **(i)** Communication circuits so located as to be exposed to accidental contact with light or power conductors operating at over 300 volts shall have each circuit so exposed provided with a protector approved for the purpose.

**1910.308(e)(2)(ii)** Each conductor of a lead-in from an outdoor antenna shall be provided with an antenna discharge unit or other suitable means that will drain static charges from the antenna system.

# OSHA Standards for General Industry

**1910.308(e)(3)** *Conductor location*—**(i)** *Outside of buildings.* **(a)** Receiving distribution lead-in or aerial-drop cables attached to buildings and lead-in conductors to radio transmitters shall be so installed as to avoid the possibility of accidental contact with electric light or power conductors.

**1910.308(e)(3)(i)(b)** The clearance between lead-in conductors and any lightning protection conductors may not be less than 6 feet.

**1910.308(e)(3)(ii)** *On poles.* Where practicable, communication conductors on poles shall be located below the light or power conductors. Communications conductors may not be attached to a crossarm that carries light or power conductors.

**1910.308(e)(3)(iii)** *Inside of buildings.* Indoor antennas, lead-ins, and other communication conductors attached as open conductors to the inside of buildings shall be located at least 2 inches from conductors of any light or power or Class 1 circuits unless a special and equally protective method of conductor separation, approved for the purpose, is employed.

**1910.308(e)(4)** *Equipment location.* Outdoor metal structures supporting antennas, as well as self-supporting antennas such as vertical rods or dipole structures, shall be located as far away from overhead conductors of electric light and power circuits of over 150 volts to ground as necessary to avoid the possibility of the antenna or structure falling into or making accidental contact with such circuits.

**1910.308(e)(5)** *Grounding*—**(i)** *Lead-in conductors.* If exposed to contact with electric light and power conductors, the metal sheath of aerial cables entering buildings shall be grounded or shall be interrupted close to the entrance to the building by an insulating joint or equivalent device. Where protective devices are used, they shall be grounded in an approved manner.

**1910.308(e)(5)(ii)** *Antenna structures.* Masts and metal structures supporting antennas shall be permanently and effectively grounded without splice or connection in the grounding conductor.

**1910.308(e)(5)(iii)** *Equipment enclosures.* Transmitters shall be enclosed in a metal frame or grill or separated from the operating space by a barrier, all metallic parts of which are effectively connected to ground. All external metal handles and controls accessible to the operating personnel shall be effectively grounded. Unpowered equipment and enclosures shall be considered grounded where connected to an attached coaxial cable with an effectively grounded metallic shield.

[46 FR 4056, Jan. 16, 1981; 46 FR 40185, Aug. 7, 1981]

## §§1910.309-1910.330 [Reserved]

### Safety-Related Work Practices

### §1910.331 Scope.

**1910.331(a)** ***Covered work by both qualified and unqualified persons.*** The provisions of §§1910.331 through 1910.335 cover electrical safety-related work practices for both qualified persons (those who have training in avoiding the electrical hazards of working on or near exposed energized parts) and unqualified persons (those with little or no such training) working on, near, or with the following installations:

**1910.331(a)(1)** *Premises wiring.* Installations of electric conductors and equipment within or on buildings or other structures, and on other premises such as yards, carnival, parking, and other lots, and industrial substations;

**1910.331(a)(2)** *Wiring for connection to supply.* Installations of conductors that connect to the supply of electricity; and

**1910.331(a)(3)** *Other wiring.* Installations of other outside conductors on the premises.

**1910.331(a)(4)** *Optical fiber cable.* Installations of optical fiber cable where such installations are made along with electric conductors.

**NOTE:** See §1910.399 for the definition of "qualified person." See §1910.332 for training requirements that apply to qualified and unqualified persons.

**1910.331(b)** ***Other covered work by unqualified persons.*** The provisions of §§1910.331 through 1910.335 also cover work performed by unqualified persons on, near, or with the installations listed in paragraphs (c)(1) through (c)(4) of this section.

**1910.331(c)** ***Excluded work by qualified persons.*** The provisions of §§1910.331 through 1910.335 do not apply to work performed by qualified persons on or directly associated with the following installations:

**1910.331(c)(1)** *Generation, transmission, and distribution installations.* Installations for the generation, control, transformation, transmission, and distribution of electric energy (including communication and metering) located in buildings used for such purposes or located outdoors.

**NOTE 1:** Work on or directly associated with installations of utilization equipment used for purposes other than generating, transmitting, or distributing electric energy (such as installations which are in office buildings, warehouses, garages, machine shops, or recreational buildings, or other utilization installations which are not an integral part of a generating installation, substation, or control center) is covered under paragraph (a)(1) of this section.

**NOTE 2:** For work on or directly associated with utilization installations, an employer who complies with the work practices of §1910.269 (electric power generation, transmission, and distribution) will be deemed to be in compliance with §1910.333(c) and §1910.335. However, the requirements of §1910.332, §1910.333(a), §1910.333(b), and §1910.334 apply to *all* work on or directly associated with utilization installations, regardless of whether the work is performed by qualified or unqualified persons.

**NOTE 3:** Work on or directly associated with generation, transmission, or distribution installations includes:

**1910.331(c)(1)(1)** Work performed directly on such installations, such as repairing overhead or underground distribution lines or repairing a feed-water pump for the boiler in a generating plant.

**1910.331(c)(1)(2)** Work directly associated with such installations, such as line-clearance tree trimming and replacing utility poles.

**1910.331(c)(1)(3)** Work on electric utilization circuits in a generating plant provided that:

**1910.331(c)(1)(3)(A)** Such circuits are commingled with installations of power generation equipment or circuits, and

**1910.331(c)(1)(3)(B)** The generation equipment or circuits present greater electrical hazards than those posed by the utilization equipment or circuits (such as exposure to higher voltages or lack of overcurrent protection).

This work is covered by §1910.269 of this Part.

**1910.331(c)(2)** *Communications installations.* Installations of communication equipment to the extent that the work is covered under §1910.268.

**1910.331(c)(3)** *Installations in vehicles.* Installations in ships, watercraft, railway rolling stock, aircraft, or automotive vehicles other than mobile homes and recreational vehicles.

**1910.331(c)(4)** *Railway installations.* Installations of railways for generation, transformation, transmission, or distribution of power used exclusively for operation of rolling stock or installations of railways used exclusively for signaling and communication purposes.

[55 FR 32016, Aug. 6, 1990, as amended at 59 FR 4476, Jan. 31, 1994]

### §1910.332 Training.

**1910.332(a)** *Scope.* The training requirements contained in this section apply to employees who face a risk of electric shock that is not reduced to a safe level by the electrical installation requirements of §§1910.303 through 1910.308.

**NOTE:** Employees in occupations listed in Table S-4 face such a risk and are required to be trained. Other employees who also may reasonably be expected to face a comparable risk of injury due to electric shock or other electrical hazards must also be trained.

**1910.332(b)** ***Content of training.*** **(1)** *Practices addressed in this standard.* Employees shall be trained in and familiar with the safety-related work practices required by §§1910.331 through 1910.335 that pertain to their respective job assignments.

**1910.332(b)(2)** *Additional requirements for unqualified persons.* Employees who are covered by paragraph (a) of this section but who are not qualified persons shall also be trained in and familiar with any electrically related safety practices not specifically addressed by §§1910.331 through 1910.335 but which are necessary for their safety.

**1910.332(b)(3)** *Additional requirements for qualified persons.* Qualified persons (i.e., those permitted to work on or near exposed energized parts) shall, at a minimum, be trained in and familiar with the following:

**1910.332(b)(3)(i)** The skills and techniques necessary to distinguish exposed live parts from other parts of electric equipment,

**1910.332(b)(3)(ii)** The skills and techniques necessary to determine the nominal voltage of exposed live parts, and

**1910.332(b)(3)(iii)** The clearance distances specified in §1910.333(c) and the corresponding voltages to which the qualified person will be exposed.

**NOTE 1:** For the purposes of §§1910.331 through 1910.335, a person must have the training required by paragraph (b)(3) of this section in order to be considered a qualified person.

**NOTE 2:** Qualified persons whose work on energized equipment involves either direct contact or contact by means of tools or materials must also have the training needed to meet §1910.333(c)(2).

**1910.332(c)** *Type of training.* The training required by this section shall be of the classroom or on-the-job type. The degree of training provided shall be determined by the risk to the employee.

| Occupation |
|---|
| Blue collar supervisors.[1] |
| Electrical and electronic engineers.[1] |
| Electrical and electronic equipment assemblers.[1] |
| Electrical and electronic technicians.[1] |
| Electricians. |
| Industrial machine operators.[1] |
| Material handling equipment operators.[1] |
| Mechanics and repairers.[1] |
| Painters.[1] |
| Riggers and roustabouts.[1] |
| Stationary engineers.[1] |
| Welders. |

[1] Workers in these groups do not need to be trained if their work or the work of those they supervise does not bring them or the employees they supervise close enough to exposed parts of electric circuits operating at 50 volts or more to ground for a hazard to exist.

[55 FR 32016, Aug. 6, 1990]

### §1910.333 Selection and use of work practices.

**1910.333(a)** *General.* Safety-related work practices shall be employed to prevent electric shock or other injuries resulting from either direct or indirect electrical contacts, when work is performed near or on equipment or circuits which are or may be energized. The specific safety-related work practices shall be consistent with the nature and extent of the associated electrical hazards.

**1910.333(a)(1)** *Deenergized parts.* Live parts to which an employee may be exposed shall be deenergized before the employee works on or near them, unless the employer can demonstrate that deenergizing introduces additional or increased hazards or is infeasible due to equipment design or operational limitations. Live parts that operate at less than 50 volts to ground need not be deenergized if there will be no increased exposure to electrical burns or to explosion due to electric arcs.

**NOTE 1:** Examples of increased or additional hazards include interruption of life support equipment, deactivation of emergency alarm systems, shutdown of hazardous location ventilation equipment, or removal of illumination for an area.

**NOTE 2:** Examples of work that may be performed on or near energized circuit parts because of infeasibility due to equipment design or operational limitations include testing of electric circuits that can only be performed with the circuit energized and work on circuits that form an integral part of a continuous industrial process in a chemical plant that would otherwise need to be completely shut down in order to permit work on one circuit or piece of equipment.

**NOTE 3:** Work on or near deenergized parts is covered by paragraph (b) of this section.

**1910.333(a)(2)** *Energized parts.* If the exposed live parts are not deenergized (i.e., for reasons of increased or additional hazards or infeasibility), other safety-related work practices shall be used to protect employees who may be exposed to the electrical hazards involved. Such work practices shall protect employees against contact with energized circuit parts directly with any part of their body or indirectly through some other conductive object. The work practices that are used shall be suitable for the conditions under which the work is to be performed and for the voltage level of the exposed electric conductors or circuit parts. Specific work practice requirements are detailed in paragraph (c) of this section.

**1910.333(b)** *Working on or near exposed deenergized parts* —**(1)** *Application.* This paragraph applies to work on exposed deenergized parts or near enough to them to expose the employee to any electrical hazard they present. Conductors and parts of electric equipment that have been deenergized but have not been locked out or tagged in accordance with paragraph (b) of this section shall be treated as energized parts, and paragraph (c) of this section applies to work on or near them.

**1910.333(b)(2)** *Lockout and tagging.* While any employee is exposed to contact with parts of fixed electric equipment or circuits which have been deenergized, the circuits energizing the parts shall be locked out or tagged or both in accordance with the requirements of this paragraph. The requirements shall be followed in the order in which they are presented (i.e., paragraph (b)(2)(i) first, then paragraph (b)(2)(ii), etc.).

**NOTE 1:** As used in this section, fixed equipment refers to equipment fastened in place or connected by permanent wiring methods.

**NOTE 2:** Lockout and tagging procedures that comply with paragraphs (c) through (f) of §1910.147 will also be deemed to comply with paragraph (b)(2) of this section provided that:

**1910.333(b)(2)(1)** The procedures address the electrical safety hazards covered by this Subpart; and

**1910.333(b)(2)(2)** The procedures also incorporate the requirements of paragraphs (b)(2)(iii)(D) and (b)(2)(iv)(B) of this section.

**1910.333(b)(2)(i)** *Procedures.* The employer shall maintain a written copy of the procedures outlined in paragraph (b)(2) and shall make it available for inspection by employees and by the Assistant Secretary of Labor and his or her authorized representatives.

**NOTE:** The written procedures may be in the form of a copy of paragraph (b) of this section.

**1910.333(b)(2)(ii)** *Deenergizing equipment.* **(A)** Safe procedures for deenergizing circuits and equipment shall be determined before circuits or equipment are deenergized.

**1910.333(b)(2)(ii)(B)** The circuits and equipment to be worked on shall be disconnected from all electric energy sources. Control circuit devices, such as push buttons, selector switches, and interlocks, may not be used as the sole means for deenergizing circuits or equipment. Interlocks for electric equipment may not be used as a substitute for lockout and tagging procedures.

**1910.333(b)(2)(ii)(C)** Stored electric energy which might endanger personnel shall be released. Capacitors shall be discharged and high capacitance elements shall be short-circuited and grounded, if the stored electric energy might endanger personnel.

**NOTE:** If the capacitors or associated equipment are handled in meeting this requirement, they shall be treated as energized.

**1910.333(b)(2)(ii)(D)** Stored non-electrical energy in devices that could re-energize electric circuit parts shall be blocked or relieved to the extent that the circuit parts could not be accidentally energized by the device.

**1910.333(b)(2)(iii)** *Application of locks and tags.* **(A)** A lock and a tag shall be placed on each disconnecting means used to deenergize circuits and equipment on which work is to be performed, except as provided in paragraphs (b)(2)(iii)(C) and (b)(2)(iii)(E) of this section. The lock shall be attached so as to prevent persons from operating the disconnecting means unless they resort to undue force or the use of tools.

**1910.333(b)(2)(iii)(B)** Each tag shall contain a statement prohibiting unauthorized operation of the disconnecting means and removal of the tag.

**1910.333(b)(2)(iii)(C)** If a lock cannot be applied, or if the employer can demonstrate that tagging procedures will provide a level of safety equivalent to that obtained by the use of a lock, a tag may be used without a lock.

**1910.333(b)(2)(iii)(D)** A tag used without a lock, as permitted by paragraph (b)(2)(iii)(C) of this section, shall be supplemented by at least one additional safety measure that provides a level of safety equivalent to that obtained by the use of a lock. Examples of additional safety measures include the removal of an isolating circuit element, blocking of a controlling switch, or opening of an extra disconnecting device.

**1910.333(b)(2)(iii)(E)** A lock may be placed without a tag only under the following conditions:

# OSHA Standards for General Industry

**1910.333(b)(2)(iii)(E)***(1)* Only one circuit or piece of equipment is deenergized, and

**1910.333(b)(2)(iii)(E)***(2)* The lockout period does not extend beyond the work shift, and

**1910.333(b)(2)(iii)(E)***(3)* Employees exposed to the hazards associated with reenergizing the circuit or equipment are familiar with this procedure.

**1910.333(b)(2)(iv)** Verification of deenergized condition. The requirements of this paragraph shall be met before any circuits or equipment can be considered and worked as deenergized.

**1910.333(b)(2)(iv)(A)** A qualified person shall operate the equipment operating controls or otherwise verify that the equipment cannot be restarted.

**1910.333(b)(2)(iv)(B)** A qualified person shall use test equipment to test the circuit elements and electrical parts of equipment to which employees will be exposed and shall verify that the circuit elements and equipment parts are deenergized. The test shall also determine if any energized condition exists as a result of inadvertently induced voltage or unrelated voltage backfeed even though specific parts of the circuit have been deenergized and presumed to be safe. If the circuit to be tested is over 600 volts, nominal, the test equipment shall be checked for proper operation immediately before and immediately after this test.

**1910.333(b)(2)(v)** *Reenergizing equipment.* These requirements shall be met, in the order given, before circuits or equipment are reenergized, even temporarily.

**1910.333(b)(2)(v)(A)** A qualified person shall conduct tests and visual inspections, as necessary, to verify that all tools, electrical jumpers, shorts, grounds, and other such devices have been removed, so that the circuits and equipment can be safely energized.

**1910.333(b)(2)(v)(B)** Employees exposed to the hazards associated with reenergizing the circuit or equipment shall be warned to stay clear of circuits and equipment.

**1910.333(b)(2)(v)(C)** Each lock and tag shall be removed by the employee who applied it or under his or her direct supervision. However, if this employee is absent from the workplace, then the lock or tag may be removed by a qualified person designated to perform this task provided that:

**1910.333(b)(2)(v)(C)***(1)* The employer ensures that the employee who applied the lock or tag is not available at the workplace, and

**1910.333(b)(2)(v)(C)***(2)* The employer ensures that the employee is aware that the lock or tag has been removed before he or she resumes work at that workplace.

**1910.333(b)(2)(v)(D)** There shall be a visual determination that all employees are clear of the circuits and equipment.

**1910.333(c)** *Working on or near exposed energized parts*—**(1)** Application. This paragraph applies to work performed on exposed live parts (involving either direct contact or contact by means of tools or materials) or near enough to them for employees to be exposed to any hazard they present.

**1910.333(c)(2)** *Work on energized equipment.* Only qualified persons may work on electric circuit parts or equipment that have not been deenergized under the procedures of paragraph (b) of this section. Such persons shall be capable of working safely on energized circuits and shall be familiar with the proper use of special precautionary techniques, personal protective equipment, insulating and shielding materials, and insulated tools.

**1910.333(c)(3)** *Overhead lines.* If work is to be performed near overhead lines, the lines shall be deenergized and grounded, or other protective measures shall be provided before work is started. If the lines are to be deenergized, arrangements shall be made with the person or organization that operates or controls the electric circuits involved to deenergize and ground them. If protective measures, such as guarding, isolating, or insulating are provided, these precautions shall prevent employees from contacting such lines directly with any part of their body or indirectly through conductive materials, tools, or equipment.

**NOTE:** The work practices used by qualified persons installing insulating devices on overhead power transmission or distribution lines are covered by §1910.269 of this Part, not by §§1910.332 through 1910.335 of this Part. Under paragraph (c)(2) of this section, unqualified persons are prohibited from performing this type of work.

**1910.333(c)(3)(i)** *Unqualified persons.* **(A)** When an unqualified person is working in an elevated position near overhead lines, the location shall be such that the person and the longest conductive object he or she may contact cannot come closer to any unguarded, energized overhead line than the following distances:

**1910.333(c)(3)(i)(A)***(1)* For voltages to ground 50kV or below—10 ft. (305 cm);

**1910.333(c)(3)(i)(A)***(2)* For voltages to ground over 50kV—10 ft. (305 cm) plus 4 in. (10 cm) for every 10kV over 50kV.

**1910.333(c)(3)(i)(B)** When an unqualified person is working on the ground in the vicinity of overhead lines, the person may not bring any conductive object closer to unguarded, energized overhead lines than the distances given in paragraph (c)(3)(i)(A) of this section.

**NOTE:** For voltages normally encountered with overhead power lines, objects which do not have an insulating rating for the voltage involved are considered to be conductive.

**1910.333(c)(3)(ii)** *Qualified persons.* When a qualified person is working in the vicinity of overhead lines, whether in an elevated position or on the ground, the person may not approach or take any conductive object without an approved insulating handle closer to exposed energized parts than shown in Table S-5 unless:

**1910.333(c)(3)(ii)(A)** The person is insulated from the energized part (gloves, with sleeves if necessary, rated for the voltage involved are considered to be insulation of the person from the energized part on which work is performed), or

**1910.333(c)(3)(ii)(B)** The energized part is insulated both from all other conductive objects at a different potential and from the person, or

**1910.333(c)(3)(ii)(C)** The person is insulated from all conductive objects at a potential different from that of the energized part.

### TABLE S-5—APPROACH DISTANCES FOR QUALIFIED EMPLOYEES—ALTERNATING CURRENT

| Voltage range (phase to phase) | Minimum approach distance |
|---|---|
| 300V and less | Avoid contact. |
| Over 300V, not over 750V | 1 ft. 0 in. (30.5 cm). |
| Over 750V, not over 2kV | 1 ft. 6 in. (46 cm). |
| Over 2kV, not over 15kV | 2 ft. 0 in. (61 cm). |
| Over 15kV, not over 37kV | 3 ft. 0 in. (91 cm). |
| Over 37kV, not over 87.5kV | 3 ft. 6 in. (107 cm). |
| Over 87.5kV, not over 121kV | 4 ft. 0 in. (122 cm). |
| Over 121kV, not over 140kV | 4 ft. 6 in. (137 cm). |

**1910.333(c)(3)(iii)** *Vehicular and mechanical equipment.* **(A)** Any vehicle or mechanical equipment capable of having parts of its structure elevated near energized overhead lines shall be operated so that a clearance of 10 ft. (305 cm) is maintained. If the voltage is higher than 50kV, the clearance shall be increased 4 in. (10 cm) for every 10kV over that voltage. However, under any of the following conditions, the clearance may be reduced:

**1910.333(c)(3)(iii)(A)***(1)* If the vehicle is in transit with its structure lowered, the clearance may be reduced to 4 ft. (122 cm). If the voltage is higher than 50kV, the clearance shall be increased 4 in. (10 cm) for every 10kV over that voltage.

**1910.333(c)(3)(iii)(A)***(2)* If insulating barriers are installed to prevent contact with the lines, and if the barriers are rated for the voltage of the line being guarded and are not a part of or an attachment to the vehicle or its raised structure, the clearance may be reduced to a distance within the designed working dimensions of the insulating barrier.

**1910.333(c)(3)(iii)(A)***(3)* If the equipment is an aerial lift insulated for the voltage involved, and if the work is performed by a qualified person, the clearance (between the uninsulated portion of the aerial lift and the power line) may be reduced to the distance given in Table S-5.

**1910.333(c)(3)(iii)(B)** Employees standing on the ground may not contact the vehicle or mechanical equipment or any of its attachments, unless:

**1910.333(c)(3)(iii)(B)***(1)* The employee is using protective equipment rated for the voltage; or

**1910.333(c)(3)(iii)(B)***(2)* The equipment is located so that no uninsulated part of its structure (that portion of the structure that provides a conductive path to employees on the ground) can come closer to the line than permitted in paragraph (c)(3)(iii) of this section.

**1910.333(c)(3)(iii)(C)** If any vehicle or mechanical equipment capable of having parts of its structure elevated near energized overhead lines is intentionally grounded, employees working on the ground near the point of grounding may not stand at the grounding location whenever there is a possibility of overhead line contact. Additional precautions, such as the use of barricades or insulation, shall be taken to protect employees from hazardous ground potentials, depending on earth resistivity and fault currents, which can develop within the first few feet or more outward from the grounding point.

**1910.333(c)(4)** *Illumination.* **(i)** Employees may not enter spaces containing exposed energized parts, unless illumination is provided that enables the employees to perform the work safely.

**1910.333(c)(4)(ii)** Where lack of illumination or an obstruction precludes observation of the work to be performed, employees may not perform tasks near exposed energized parts. Employees may not reach blindly into areas which may contain energized parts.

**1910.333(c)(5)** *Confined or enclosed work spaces.* When an employee works in a confined or enclosed space (such as a manhole or vault) that contains exposed energized parts, the employer shall provide, and the employee shall use, protective shields, protective barriers, or insulating materials as necessary to avoid inadvertent contact with these parts. Doors, hinged panels, and the like shall be secured to prevent their swinging into an employee and causing the employee to contact exposed energized parts.

**1910.333(c)(6)** *Conductive materials and equipment.* Conductive materials and equipment that are in contact with any part of an employee's body shall be handled in a manner that will prevent them from contacting exposed energized conductors or circuit parts. If an employee must handle long dimensional conductive objects (such as ducts and pipes) in areas with exposed live parts, the employer shall institute work practices (such as the use of insulation, guarding, and material handling techniques) which will minimize the hazard.

**1910.333(c)(7)** *Portable ladders.* Portable ladders shall have nonconductive siderails if they are used where the employee or the ladder could contact exposed energized parts.

**1910.333(c)(8)** *Conductive apparel.* Conductive articles of jewelry and clothing (such as watch bands, bracelets, rings, key chains, necklaces, metalized aprons, cloth with conductive thread, or metal headgear) may not be worn if they might contact exposed energized parts. However, such articles may be worn if they are rendered nonconductive by covering, wrapping, or other insulating means.

**1910.333(c)(9)** *Housekeeping duties.* Where live parts present an electrical contact hazard, employees may not perform housekeeping duties at such close distances to the parts that there is a possibility of contact, unless adequate safeguards (such as insulating equipment or barriers) are provided. Electrically conductive cleaning materials (including conductive solids such as steel wool, metalized cloth, and silicon carbide, as well as conductive liquid solutions) may not be used in proximity to energized parts unless procedures are followed which will prevent electrical contact.

**1910.333(c)(10)** *Interlocks.* Only a qualified person following the requirements of paragraph (c) of this section may defeat an electrical safety interlock, and then only temporarily while he or she is working on the equipment. The interlock system shall be returned to its operable condition when this work is completed.

[55 FR 32016, Aug. 6, 1990; 55 FR 42053, Nov. 1, 1990, as amended at 59 FR 4476, Jan. 31, 1994]

## §1910.334 Use of equipment.

**1910.334(a)** *Portable electric equipment.* This paragraph applies to the use of cord- and plug-connected equipment, including flexible cord sets (extension cords).

**1910.334(a)(1)** *Handling.* Portable equipment shall be handled in a manner which will not cause damage. Flexible electric cords connected to equipment may not be used for raising or lowering the equipment. Flexible cords may not be fastened with staples or otherwise hung in such a fashion as could damage the outer jacket or insulation.

**1910.334(a)(2)** *Visual inspection.* **(i)** Portable cord- and plug-connected equipment and flexible cord sets (extension cords) shall be visually inspected before use on any shift for external defects (such as loose parts, deformed and missing pins, or damage to outer jacket or insulation) and for evidence of possible internal damage (such as pinched or crushed outer jacket). Cord- and plug-connected equipment and flexible cord sets (extension cords) which remain connected once they are put in place and are not exposed to damage need not be visually inspected until they are relocated.

**1910.334(a)(2)(ii)** If there is a defect or evidence of damage that might expose an employee to injury, the defective or damaged item shall be removed from service, and no employee may use it until repairs and tests necessary to render the equipment safe have been made.

**1910.334(a)(2)(iii)** When an attachment plug is to be connected to a receptacle (including any on a cord set), the relationship of the plug and receptacle contacts shall first be checked to ensure that they are of proper mating configurations.

**1910.334(a)(3)** *Grounding-type equipment.* **(i)** A flexible cord used with grounding-type equipment shall contain an equipment grounding conductor.

**1910.334(a)(3)(ii)** Attachment plugs and receptacles may not be connected or altered in a manner which would prevent proper continuity of the equipment grounding conductor at the point where plugs are attached to receptacles. Additionally, these devices may not be altered to allow the grounding pole of a plug to be inserted into slots intended for connection to the current-carrying conductors.

**1910.334(a)(3)(iii)** Adapters which interrupt the continuity of the equipment grounding connection may not be used.

**1910.334(a)(4)** *Conductive work locations.* Portable electric equipment and flexible cords used in highly conductive work locations (such as those inundated with water or other conductive liquids), or in job locations where employees are likely to contact water or conductive liquids, shall be approved for those locations.

**1910.334(a)(5)** *Connecting attachment plugs.* **(i)** Employees' hands may not be wet when plugging and unplugging flexible cords and cord- and plug-connected equipment, if energized equipment is involved.

**1910.334(a)(5)(ii)** Energized plug and receptacle connections may be handled only with insulating protective equipment if the condition of the connection could provide a conducting path to the employee's hand (if, for example, a cord connector is wet from being immersed in water).

**1910.334(a)(5)(iii)** Locking-type connectors shall be properly secured after connection.

**1910.334(b)** *Electric power and lighting circuits.* **(1)** *Routine opening and closing of circuits.* Load rated switches, circuit breakers, or other devices specifically designed as disconnecting means shall be used for the opening, reversing, or closing of circuits under load conditions. Cable connectors not of the load-break type, fuses, terminal lugs, and cable splice connections may not be used for such purposes, except in an emergency.

**1910.334(b)(2)** *Reclosing circuits after protective device operation.* After a circuit is deenergized by a circuit protective device, the circuit may not be manually reenergized until it has been determined that the equipment and circuit can be safely energized. The repetitive manual reclosing of circuit breakers or reenergizing circuits through replaced fuses is prohibited.

NOTE: When it can be determined from the design of the circuit and the overcurrent devices involved that the automatic operation of a device was caused by an overload rather than a fault condition, no examination of the circuit or connected equipment is needed before the circuit is reenergized.

**1910.334(b)(3)** *Overcurrent protection modification.* Overcurrent protection of circuits and conductors may not be modified, even on a temporary basis, beyond that allowed by §1910.304(e), the installation safety requirements for overcurrent protection.

**1910.334(c)** *Test instruments and equipment.* **(1)** *Use.* Only qualified persons may perform testing work on electric circuits or equipment.

**1910.334(c)(2)** *Visual inspection.* Test instruments and equipment and all associated test leads, cables, power cords, probes, and connectors shall be visually inspected for external defects and damage before the equipment is used. If there is a defect or evidence of damage that might expose an employee to injury, the defective or damaged item shall be removed from service, and no employee may use it until repairs and tests necessary to render the equipment safe have been made.

**1910.334(c)(3)** *Rating of equipment.* Test instruments and equipment and their accessories shall be rated for the circuits and equipment to which they will be connected and shall be designed for the environment in which they will be used.

**1910.334(d)** *Occasional use of flammable or ignitible materials.* Where flammable materials are present only occasionally, electric equipment capable of igniting them shall not be used, unless measures are taken to prevent hazardous conditions from developing. Such materials include, but are not limited to: flammable gases, vapors, or liquids; combustible dust; and ignitible fibers or flyings.

**NOTE:** Electrical installation requirements for locations where flammable materials are present on a regular basis are contained in §1910.307.

[55 FR 32019, Aug. 6, 1990]

## §1910.335 Safeguards for personnel protection.

**1910.335(a)** *Use of protective equipment—***(1)** *Personal protective equipment.* (i) Employees working in areas where there are potential electrical hazards shall be provided with, and shall use, electrical protective equipment that is appropriate for the specific parts of the body to be protected and for the work to be performed.

**NOTE:** Personal protective equipment requirements are contained in subpart I of this part.

**1910.335(a)(1)(ii)** Protective equipment shall be maintained in a safe, reliable condition and shall be periodically inspected or tested, as required by §1910.137.

**1910.335(a)(1)(iii)** If the insulating capability of protective equipment may be subject to damage during use, the insulating material shall be protected. (For example, an outer covering of leather is sometimes used for the protection of rubber insulating material.)

**1910.335(a)(1)(iv)** Employees shall wear nonconductive head protection wherever there is a danger of head injury from electric shock or burns due to contact with exposed energized parts.

**1910.335(a)(1)(v)** Employees shall wear protective equipment for the eyes or face wherever there is danger of injury to the eyes or face from electric arcs or flashes or from flying objects resulting from electrical explosion.

**1910.335(a)(2)** *General protective equipment and tools.* (i) When working near exposed energized conductors or circuit parts, each employee shall use insulated tools or handling equipment if the tools or handling equipment might make contact with such conductors or parts. If the insulating capability of insulated tools or handling equipment is subject to damage, the insulating material shall be protected.

**1910.335(a)(2)(i)(A)** Fuse handling equipment, insulated for the circuit voltage, shall be used to remove or install fuses when the fuse terminals are energized.

**1910.335(a)(2)(i)(B)** Ropes and handlines used near exposed energized parts shall be nonconductive.

**1910.335(a)(2)(ii)** Protective shields, protective barriers, or insulating materials shall be used to protect each employee from shock, burns, or other electrically related injuries while that employee is working near exposed energized parts which might be accidentally contacted or where dangerous electric heating or arcing might occur. When normally enclosed live parts are exposed for maintenance or repair, they shall be guarded to protect unqualified persons from contact with the live parts.

**1910.335(b)** *Alerting techniques.* The following alerting techniques shall be used to warn and protect employees from hazards which could cause injury due to electric shock, burns, or failure of electric equipment parts:

**1910.335(b)(1)** *Safety signs and tags.* Safety signs, safety symbols, or accident prevention tags shall be used where necessary to warn employees about electrical hazards which may endanger them, as required by §1910.145.

**1910.335(b)(2)** *Barricades.* Barricades shall be used in conjunction with safety signs where it is necessary to prevent or limit employee access to work areas exposing employees to uninsulated energized conductors or circuit parts. Conductive barricades may not be used where they might cause an electrical contact hazard.

**1910.335(b)(3)** *Attendants.* If signs and barricades do not provide sufficient warning and protection from electrical hazards, an attendant shall be stationed to warn and protect employees.

[55 FR 32020, Aug. 6, 1990]

## §§1910.336—1910.360 [Reserved]

### Safety-Related Maintenance Requirements

## §§1910.361—1910.380 [Reserved]

### Safety Requirements for Special Equipment

## §§1910.381—1910.398 [Reserved]

### Definitions

## §1910.399 Definitions applicable to this subpart.

*Acceptable.* An installation or equipment is acceptable to the Assistant Secretary of Labor, and approved within the meaning of this Subpart S:

(i) If it is accepted, or certified, or listed, or labeled, or otherwise determined to be safe by a nationally recognized testing laboratory; or

(ii) With respect to an installation or equipment of a kind which no nationally recognized testing laboratory accepts, certifies, lists, labels, or determines to be safe, if it is inspected or tested by another Federal agency, or by a State, municipal, or other local authority responsible for enforcing occupational safety provisions of the National Electrical Code and found in compliance with the provisions of the National Electrical Code as applied in this subpart; or

(iii) With respect to custom-made equipment or related installations which are designed, fabricated for, and intended for use by a particular customer, if it is determined to be safe for its intended use by its manufacturer on the basis of test data which the employer keeps and makes available for inspection to the Assistant Secretary and his authorized representatives. Refer to §1910.7 for definition of nationally recognized testing laboratory.

*Accepted.* An installation is "accepted" if it has been inspected and found by a nationally recognized testing laboratory to conform to specified plans or to procedures of applicable codes.

*Accessible.* (As applied to wiring methods.) Capable of being removed or exposed without damaging the building structure or finish, or not permanently closed in by the structure or finish of the building. (See "concealed" and "exposed.")

*Accessible.* (As applied to equipment.) Admitting close approach; not guarded by locked doors, elevation, or other effective means. (See "Readily accessible.")

*Ampacity.* Current-carrying capacity of electric conductors expressed in amperes.

*Appliances.* Utilization equipment, generally other than industrial, normally built in standardized sizes or types, which is installed or connected as a unit to perform one or more functions such as clothes washing, air conditioning, food mixing, deep frying, etc.

*Approved.* Acceptable to the authority enforcing this subpart. The authority enforcing this subpart is the Assistant Secretary of Labor for Occupational Safety and Health. The definition of "acceptable" indicates what is acceptable to the Assistant Secretary of Labor, and therefore approved within the meaning of this Subpart.

*Approved for the purpose.* Approved for a specific purpose, environment, or application described in a particular standard requirement.

Suitability of equipment or materials for a specific purpose, environment or application may be determined by a nationally recognized testing laboratory, inspection agency or other organization concerned with product evaluation as part of its listing and labeling program. (See "Labeled" or "Listed.")

*Armored cable.* Type AC armored cable is a fabricated assembly of insulated conductors in a flexible metallic enclosure.

*Askarel.* A generic term for a group of nonflammable synthetic chlorinated hydrocarbons used as electrical insulating media. Askarels of various compositional types are used. Under arcing conditions the gases produced, while consisting predominantly of noncombustible hydrogen chloride, can include varying amounts of combustible gases depending upon the askarel type.

*Attachment plug (Plug cap) (Cap).* A device which, by insertion in a receptacle, establishes connection between the conductors of the attached flexible cord and the conductors connected permanently to the receptacle.

*Automatic.* Self-acting, operating by its own mechanism when actuated by some impersonal influence, as, for example, a change in current strength, pressure, temperature, or mechanical configuration.

*Bare conductor.* See *"Conductor."*

*Bonding.* The permanent joining of metallic parts to form an electrically conductive path which will assure electrical continuity and the capacity to conduct safely any current likely to be imposed.

*Bonding jumper.* A reliable conductor to assure the required electrical conductivity between metal parts required to be electrically connected.

*Branch circuit.* The circuit conductors between the final overcurrent device protecting the circuit and the outlet(s).

*Building.* A structure which stands alone or which is cut off from adjoining structures by fire walls with all openings therein protected by approved fire doors.

*Cabinet.* An enclosure designed either for surface or flush mounting, and provided with a frame, mat, or trim in which a swinging door or doors are or may be hung.

*Cable tray system.* A cable tray system is a unit or assembly of units or sections, and associated fittings, made of metal or other noncombustible materials forming a rigid structural system used to support cables. Cable tray systems include ladders, troughs, channels, solid bottom trays, and other similar structures.

*Cablebus.* Cablebus is an approved assembly of insulated conductors with fittings and conductor terminations in a completely enclosed, ventilated, protective metal housing.

*Center pivot irrigation machine.* A center pivot irrigation machine is a multimotored irrigation machine which revolves around a central pivot and employs alignment switches or similar devices to control individual motors.

*Certified.* Equipment is "certified" if it (a) has been tested and found by a nationally recognized testing laboratory to meet nationally recognized standards or to be safe for use in a specified manner, or (b) is of a kind whose production is periodically inspected by a nationally recognized testing laboratory, and (c) it bears a label, tag, or other record of certification.

*Circuit breaker.* (i) (600 volts nominal, or less). A device designed to open and close a circuit by nonautomatic means and to open the circuit automatically on a predetermined overcurrent without injury to itself when properly applied within its rating.

(ii) (Over 600 volts, nominal). A switching device capable of making, carrying, and breaking currents under normal circuit conditions, and also making, carrying for a specified time, and breaking currents under specified abnormal circuit conditions, such as those of short circuit.

*Class I locations.* Class I locations are those in which flammable gases or vapors are or may be present in the air in quantities sufficient to produce explosive or ignitible mixtures. Class I locations include the following:

(i) *Class I, Division 1.* A Class I, Division 1 location is a location: (a) in which hazardous concentrations of flammable gases or vapors may exist under normal operating conditions; or (b) in which hazardous concentrations of such gases or vapors may exist frequently because of repair or maintenance operations or because of leakage; or (c) in which breakdown or faulty operation of equipment or processes might release hazardous concentrations of flammable gases or vapors, and might also cause simultaneous failure of electric equipment.

**NOTE:** This classification usually includes locations where volatile flammable liquids or liquefied flammable gases are transferred from one container to another; interiors of spray booths and areas in the vicinity of spraying and painting operations where volatile flammable solvents are used; locations containing open tanks or vats of volatile flammable liquids; drying rooms or compartments for the evaporation of flammable solvents; locations containing fat and oil extraction equipment using volatile flammable solvents; portions of cleaning and dyeing plants where flammable liquids are used; gas generator rooms and other portions of gas manufacturing plants where flammable gas may escape; inadequately ventilated pump rooms for flammable gas or for volatile flammable liquids; the interiors of refrigerators and freezers in which volatile flammable materials are stored in open, lightly stoppered, or easily ruptured containers; and all other locations where ignitible concentrations of flammable vapors or gases are likely to occur in the course of normal operations.

(ii) *Class I, Division 2.* A Class I, Division 2 location is a location: (a) in which volatile flammable liquids or flammable gases are handled, processed, or used, but in which the hazardous liquids, vapors, or gases will normally be confined within closed containers or closed systems from which they can escape only in case of accidental rupture or breakdown of such containers or systems, or in case of abnormal operation of equipment; or (b) in which hazardous concentrations of gases or vapors are normally prevented by positive mechanical ventilation, and which might become hazardous through failure or abnormal operations of the ventilating equipment; or (c) that is adjacent to a Class I, Division 1 location, and to which hazardous concentrations of gases or vapors might occasionally be communicated unless such communication is prevented by adequate positive-pressure ventilation from a source of clean air, and effective safeguards against ventilation failure are provided.

**NOTE:** This classification usually includes locations where volatile flammable liquids or flammable gases or vapors are used, but which would become hazardous only in case of an accident or of some unusual operating condition. The quantity of flammable material that might escape in case of accident, the adequacy of ventilating equipment, the total area involved, and the record of the industry or business with respect to explosions or fires are all factors that merit consideration in determining the classification and extent of each location.

Piping without valves, checks, meters, and similar devices would not ordinarily introduce a hazardous condition even though used for flammable liquids or gases. Locations used for the storage of flammable liquids or a liquefied or compressed gases in sealed containers would not normally be considered hazardous unless also subject to other hazardous conditions.

Electrical conduits and their associated enclosures separated from process fluids by a single seal or barrier are classed as a Division 2 location if the outside of the conduit and enclosures is a nonhazardous location.

*Class II locations.* Class II locations are those that are hazardous because of the presence of combustible dust. Class II locations include the following:

(i) *Class II, Division 1.* A Class II, Division 1 location is a location: (a) In which combustible dust is or may be in suspension in the air under normal operating conditions, in quantities sufficient to produce explosive or ignitible mixtures; or (b) where mechanical failure or abnormal operation of machinery or equipment might cause such explosive or ignitible mixtures to be produced, and might also provide a source of ignition through simultaneous failure of electric equipment, operation of protection devices, or from other causes, or (c) in which combustible dusts of an electrically conductive nature may be present.

**NOTE:** This classification may include areas of grain handling and processing plants, starch plants, sugar-pulverizing plants, malting plants, hay-grinding plants, coal pulverizing plants, areas where metal dusts and powders are produced or processed, and other similar locations which contain dust producing machinery and equipment (except where the equipment is dust-tight or vented to the outside). These areas would have combustible dust in the air, under normal operating conditions, in quantities sufficient to produce explosive or ignitible mixtures. Combustible dusts which are electrically nonconductive include dusts produced in the handling and processing of grain and grain products, pulverized sugar and cocoa, dried egg and milk powders, pulverized spices, starch and pastes, potato and woodflour, oil meal from beans and seed, dried hay, and other organic materials which may produce combustible dusts when processed or handled. Dusts containing magnesium or aluminum are particularly hazardous and the use of extreme caution is necessary to avoid ignition and explosion.

(ii) *Class II, Division 2.* A Class II, Division 2 location is a location in which: (a) combustible dust will not normally be in suspension in the air in quantities sufficient to produce explosive or ignitible mixtures, and dust accumulations are normally insufficient to interfere with the normal operation of electrical equipment or other apparatus; or (b) dust may be in suspension in the air as a result of infrequent malfunctioning of handling or processing equipment, and dust accumulations resulting therefrom may be ignitible by abnormal operation or failure of electrical equipment or other apparatus.

**NOTE:** This classification includes locations where dangerous concentrations of suspended dust would not be likely but where dust accumulations might form on or in the vicinity of electric equipment. These areas may contain equipment from which appreciable quantities of dust would escape under abnormal operating conditions or be adjacent to a Class II Division 1 location, as described above, into which an explosive or ignitible concentration of dust may be put into suspension under abnormal operating conditions.

*Class III locations.* Class III locations are those that are hazardous because of the presence of easily ignitible fibers or flyings but in which such fibers or flyings are not likely to be in suspension in the air in quantities sufficient to produce ignitible mixtures. Class III locations include the following:

(i) *Class III, Division 1.* A Class III, Division 1 location is a location in which easily ignitible fibers or materials producing combustible flyings are handled, manufactured, or used.

**NOTE:** Such locations usually include some parts of rayon, cotton, and other textile mills; combustible fiber manufacturing and processing plants; cotton

# OSHA Standards for General Industry

gins and cotton-seed mills; flax-processing plants; clothing manufacturing plants; woodworking plants, and establishments; and industries involving similar hazardous processes or conditions.

Easily ignitible fibers and flyings include rayon, cotton (including cotton linters and cotton waste), sisal or henequen, istle, jute, hemp, tow, cocoa fiber, oakum, baled waste kapok, Spanish moss, excelsior, and other materials of similar nature.

(ii) *Class III, Division 2.* A Class III, Division 2 location is a location in which easily ignitible fibers are stored or handled, except in process of manufacture.

*Collector ring.* A collector ring is an assembly of slip rings for transferring electrical energy from a stationary to a rotating member.

*Concealed.* Rendered inaccessible by the structure or finish of the building. Wires in concealed raceways are considered concealed, even though they may become accessible by withdrawing them. [See *Accessible.* (As applied to wiring methods.)]

*Conductor.* (i) *Bare.* A conductor having no covering or electrical insulation whatsoever.

(ii) *Covered.* A conductor encased within material of composition or thickness that is not recognized as electrical insulation.

(iii) *Insulated.* A conductor encased within material of composition and thickness that is recognized as electrical insulation.

*Conduit body.* A separate portion of a conduit or tubing system that provides access through a removable cover(s) to the interior of the system at a junction of two or more sections of the system or at a terminal point of the system. Boxes such as FS and FD or larger cast or sheet metal boxes are not classified as conduit bodies.

*Controller.* A device or group of devices that serves to govern, in some predetermined manner, the electric power delivered to the apparatus to which it is connected.

*Cooking unit, counter-mounted.* A cooking appliance designed for mounting in or on a counter and consisting of one or more heating elements, internal wiring, and built-in or separately mountable controls. (See *Oven, wall-mounted.*)

*Covered conductor.* See *Conductor.*

*Cutout.* (Over 600 volts, nominal.) An assembly of a fuse support with either a fuseholder, fuse carrier, or disconnecting blade. The fuseholder or fuse carrier may include a conducting element (fuse link), or may act as the disconnecting blade by the inclusion of a nonfusible member.

*Cutout box.* An enclosure designed for surface mounting and having swinging doors or covers secured directly to and telescoping with the walls of the box proper. (See *Cabinet.*)

*Damp location.* See *Location.*

*Dead front.* Without live parts exposed to a person on the operating side of the equipment.

*Device.* A unit of an electrical system which is intended to carry but not utilize electric energy.

*Dielectric heating.* Dielectric heating is the heating of a nominally insulating material due to its own dielectric losses when the material is placed in a varying electric field.

*Disconnecting means.* A device, or group of devices, or other means by which the conductors of a circuit can be disconnected from their source of supply.

*Disconnecting (or Isolating) switch.* (Over 600 volts, nominal.) A mechanical switching device used for isolating a circuit or equipment from a source of power.

*Dry location.* See *Location.*

*Electric sign.* A fixed, stationary, or portable self-contained, electrically illuminated utilization equipment with words or symbols designed to convey information or attract attention.

*Enclosed.* Surrounded by a case, housing, fence or walls which will prevent persons from accidentally contacting energized parts.

*Enclosure.* The case or housing of apparatus, or the fence or walls surrounding an installation to prevent personnel from accidentally contacting energized parts, or to protect the equipment from physical damage.

*Equipment.* A general term including material, fittings, devices, appliances, fixtures, apparatus, and the like, used as a part of, or in connection with, an electrical installation.

*Equipment grounding conductor.* See *Grounding conductor, equipment.*

*Explosion-proof apparatus.* Apparatus enclosed in a case that is capable of withstanding an explosion of a specified gas or vapor which may occur within it and of preventing the ignition of a specified gas or vapor surrounding the enclosure by sparks, flashes, or explosion of the gas or vapor within, and which operates at such an external temperature that it will not ignite a surrounding flammable atmosphere.

*Exposed.* (As applied to live parts.) Capable of being inadvertently touched or approached nearer than a safe distance by a person. It is applied to parts not suitably guarded, isolated, or insulated. (See *Accessible.* and *Concealed.*)

*Exposed.* (As applied to wiring methods.) On or attached to the surface or behind panels designed to allow access. [See *Accessible.* (As applied to wiring methods.)]

*Exposed.* (For the purposes of §1910.308(e), *Communications systems.*) Where the circuit is in such a position that in case of failure of supports or insulation, contact with another circuit may result.

*Externally operable.* Capable of being operated without exposing the operator to contact with live parts.

*Feeder.* All circuit conductors between the service equipment, or the generator switchboard of an isolated plant, and the final branch-circuit overcurrent device.

*Fitting.* An accessory such as a locknut, bushing, or other part of a wiring system that is intended primarily to perform a mechanical rather than an electrical function.

*Fuse.* (Over 600 volts, nominal.) An overcurrent protective device with a circuit opening fusible part that is heated and severed by the passage of overcurrent through it. A fuse comprises all the parts that form a unit capable of performing the prescribed functions. It may or may not be the complete device necessary to connect it into an electrical circuit.

*Ground.* A conducting connection, whether intentional or accidental, between an electrical circuit or equipment and the earth, or to some conducting body that serves in place of the earth.

*Grounded.* Connected to earth or to some conducting body that serves in place of the earth.

*Grounded, effectively.* (Over 600 volts, nominal.) Permanently connected to earth through a ground connection of sufficiently low impedance and having sufficient ampacity that ground fault current which may occur cannot build up to voltages dangerous to personnel.

*Grounded conductor.* A system or circuit conductor that is intentionally grounded.

*Grounding conductor.* A conductor used to connect equipment or the grounded circuit of a wiring system to a grounding electrode or electrodes.

*Grounding conductor, equipment.* The conductor used to connect the non-current-carrying metal parts of equipment, raceways, and other enclosures to the system grounded conductor and/or the grounding electrode conductor at the service equipment or at the source of a separately derived system.

*Grounding electrode conductor.* The conductor used to connect the grounding electrode to the equipment grounding conductor and/or to the grounded conductor of the circuit at the service equipment or at the source of a separately derived system.

*Ground-fault circuit-interrupter.* A device whose function is to interrupt the electric circuit to the load when a fault current to ground exceeds some predetermined value that is less than that required to operate the overcurrent protective device of the supply circuit.

*Guarded.* Covered, shielded, fenced, enclosed, or otherwise protected by means of suitable covers, casings, barriers, rails, screens, mats, or platforms to remove the likelihood of approach to a point of danger or contact by persons or objects.

*Health care facilities.* Buildings or portions of buildings and mobile homes that contain, but are not limited to, hospitals, nursing homes, extended care facilities, clinics, and medical and dental offices, whether fixed or mobile.

*Heating equipment.* For the purposes of §1910.306(g), the term *heating equipment* includes any equipment used for heating purposes if heat is generated by induction or dielectric methods.

*Hoistway.* Any shaftway, hatchway, well hole, or other vertical opening or space in which an elevator or dumbwaiter is designed to operate.

*Identified.* Identified, as used in reference to a conductor or its terminal, means that such conductor or terminal can be readily recognized as grounded.

*Induction heating.* Induction heating is the heating of a nominally conductive material due to its own $I^2R$ losses when the material is placed in a varying electromagnetic field.

*Insulated conductor.* See *Conductor*.

*Interrupter switch.* (Over 600 volts, nominal.) A switch capable of making, carrying, and interrupting specified currents.

*Irrigation machine.* An irrigation machine is an electrically driven or controlled machine, with one or more motors, not hand portable, and used primarily to transport and distribute water for agricultural purposes.

*Isolated.* Not readily accessible to persons unless special means for access are used.

*Isolated power system.* A system comprising an isolating transformer or its equivalent, a line isolation monitor, and its ungrounded circuit conductors.

*Labeled.* Equipment is *labeled* if there is attached to it a label, symbol, or other identifying mark of a nationally recognized testing laboratory which, (a) makes periodic inspections of the production of such equipment, and (b) whose labeling indicates compliance with nationally recognized standards or tests to determine safe use in a specified manner.

*Lighting outlet.* An outlet intended for the direct connection of a lampholder, a lighting fixture, or a pendant cord terminating in a lampholder.

*Line-clearance tree trimming.* The pruning, trimming, repairing, maintaining, removing, or clearing of trees or cutting of brush that is within 10 feet (305 cm) of electric supply lines and equipment.

*Listed.* Equipment is *listed* if it is of a kind mentioned in a list which, (a) is published by a nationally recognized laboratory which makes periodic inspection of the production of such equipment, and (b) states such equipment meets nationally recognized standards or has been tested and found safe for use in a specified manner.

*Location*—(i) *Damp location.* Partially protected locations under canopies, marquees, roofed open porches, and like locations, and interior locations subject to moderate degrees of moisture, such as some basements, some barns, and some cold-storage warehouses.

(ii) *Dry location.* A location not normally subject to dampness or wetness. A location classified as dry may be temporarily subject to dampness or wetness, as in the case of a building under construction.

(iii) *Wet location.* Installations underground or in concrete slabs or masonry in direct contact with the earth, and locations subject to saturation with water or other liquids, such as vehicle-washing areas, and locations exposed to weather and unprotected.

*May.* If a discretionary right, privilege, or power is conferred, the word "may" is used. If a right, privilege, or power is abridged or if an obligation to abstain from acting is imposed, the word "may" is used with a restrictive "no," "not," or "only." (E.g., no employer may … ; an employer may not … ; only qualified persons may….)

*Medium voltage cable.* Type MV medium voltage cable is a single or multiconductor solid dielectric insulated cable rated 2000 volts or higher.

*Metal-clad cable.* Type MC cable is a factory assembly of one or more conductors, each individually insulated and enclosed in a metallic sheath of interlocking tape, or a smooth or corrugated tube.

*Mineral-insulated metal-sheathed cable.* Type MI mineral-insulated metal-sheathed cable is a factory assembly of one or more conductors insulated with a highly compressed refractory mineral insulation and enclosed in a liquidtight and gastight continuous copper sheath.

*Mobile X-ray.* X-ray equipment mounted on a permanent base with wheels and/or casters for moving while completely assembled.

*Nonmetallic-sheathed cable.* Nonmetallic-sheathed cable is a factory assembly of two or more insulated conductors having an outer sheath of moisture resistant, flame-retardant, nonmetallic material. Nonmetallic sheathed cable is manufactured in the following types:

(i) *Type NM.* The overall covering has a flame-retardant and moisture-resistant finish.

(ii) *Type NMC.* The overall covering is flame-retardant, moisture-resistant, fungus-resistant, and corrosion-resistant.

*Oil (filled) cutout.* (Over 600 volts, nominal.) A cutout in which all or part of the fuse support and its fuse link or disconnecting blade are mounted in oil with complete immersion of the contacts and the fusible portion of the conducting element (fuse link), so that arc interruption by severing of the fuse link or by opening of the contacts will occur under oil.

*Open wiring on insulators.* Open wiring on insulators is an exposed wiring method using cleats, knobs, tubes, and flexible tubing for the protection and support of single insulated conductors run in or on buildings, and not concealed by the building structure.

*Outlet.* A point on the wiring system at which current is taken to supply utilization equipment.

*Outline lighting.* An arrangement of incandescent lamps or electric discharge tubing to outline or call attention to certain features such as the shape of a building or the decoration of a window.

*Oven, wall-mounted.* An oven for cooking purposes designed for mounting in or on a wall or other surface and consisting of one of more heating elements, internal wiring, and built-in or separately mountable controls. (See *Cooking unit, counter-mounted*.)

*Overcurrent.* Any current in excess of the rated current of equipment or the ampacity of a conductor. It may result from overload (see definition), short circuit, or ground fault. A current in excess of rating may be accommodated by certain equipment and conductors for a given set of conditions. Hence the rules for overcurrent protection are specific for particular situations.

*Overload.* Operation of equipment in excess of normal, full load rating, or of a conductor in excess of rated ampacity which, when it persists for a sufficient length of time, would cause damage or dangerous overheating. A fault, such as a short circuit or ground fault, is not an overload. (See *Overcurrent*.)

*Panelboard.* A single panel or group of panel units designed for assembly in the form of a single panel; including buses, automatic overcurrent devices, and with or without switches for the control of light, heat, or power circuits; designed to be placed in a cabinet or cutout box placed in or against a wall or partition and accessible only from the front. (See *Switchboard*.)

*Permanently installed decorative fountains and reflection pools.* Those that are constructed in the ground, on the ground, or in a building in such a manner that the pool cannot be readily disassembled for storage and are served by electrical circuits of any nature. These units are primarily constructed for their aesthetic value and not intended for swimming or wading.

*Permanently installed swimming pools, wading and therapeutic pools.* Those that are constructed in the ground, on the ground, or in a building in such a manner that the pool cannot be readily disassembled for storage whether or not served by electrical circuits of any nature.

*Portable X-ray.* X-ray equipment designed to be hand-carried.

*Power and control tray cable.* Type TC power and control tray cable is a factory assembly of two or more insulated conductors, with or without associated bare or covered grounding conductors under a nonmetallic sheath, approved for installation in cable trays, in raceways, or where supported by a messenger wire.

*Power fuse.* (Over 600 volts, nominal.) See *Fuse*.

*Power-limited tray cable.* Type PLTC nonmetallic-sheathed power limited tray cable is a factory assembly of two or more insulated conductors under a nonmetallic jacket.

*Power outlet.* An enclosed assembly which may include receptacles, circuit breakers, fuseholders, fused switches, buses and watt-hour meter mounting means; intended to supply and control power to mobile homes, recreational vehicles or boats, or to serve as a means for distributing power required to operate mobile or temporarily installed equipment.

*Premises wiring system.* That interior and exterior wiring, including power, lighting, control, and signal circuit wiring together with all of its associated hardware, fittings, and wiring devices, both permanently and temporarily installed, which extends from the load end of the service drop, or load end of the service lateral conductors to the outlet(s). Such wiring does not include wiring internal to appliances, fixtures, motors, controllers, motor control centers, and similar equipment.

## OSHA Standards for General Industry

*Qualified person.* One familiar with the construction and operation of the equipment and the hazards involved.

**NOTE 1:** Whether an employee is considered to be a "qualified person" will depend upon various circumstances in the workplace. It is possible and, in fact, likely for an individual to be considered "qualified" with regard to certain equipment in the workplace, but "unqualified" as to other equipment. (See §1910.332(b)(3) for training requirements that specifically apply to qualified persons.)

**NOTE 2:** An employee who is undergoing on-the-job training and who, in the course of such training, has demonstrated an ability to perform duties safely at his or her level of training and who is under the direct supervision of a qualified person is considered to be a qualified person for the performance of those duties.

*Raceway.* A channel designed expressly for holding wires, cables, or busbars, with additional functions as permitted in this subpart. Raceways may be of metal or insulating material, and the term includes rigid metal conduit, rigid nonmetallic conduit, intermediate metal conduit, liquidtight flexible metal conduit, flexible metallic tubing, flexible metal conduit, electrical metallic tubing, underfloor raceways, cellular concrete floor raceways, cellular metal floor raceways, surface raceways, wireways, and busways.

*Readily accessible.* Capable of being reached quickly for operation, renewal, or inspections, without requiring those to whom ready access is requisite to climb over or remove obstacles or to resort to portable ladders, chairs, etc. (See *Accessible*.)

*Receptacle.* A receptacle is a contact device installed at the outlet for the connection of a single attachment plug. A single receptacle is a single contact device with no other contact device on the same yoke. A multiple receptacle is a single device containing two or more receptacles.

*Receptacle outlet.* An outlet where one or more receptacles are installed.

*Remote-control circuit.* Any electric circuit that controls any other circuit through a relay or an equivalent device.

*Sealable equipment.* Equipment enclosed in a case or cabinet that is provided with a means of sealing or locking so that live parts cannot be made accessible without opening the enclosure. The equipment may or may not be operable without opening the enclosure.

*Separately derived system.* A premises wiring system whose power is derived from generator, transformer, or converter winding and has no direct electrical connection, including a solidly connected grounded circuit conductor, to supply conductors originating in another system.

*Service.* The conductors and equipment for delivering energy from the electricity supply system to the wiring system of the premises served.

*Service cable.* Service conductors made up in the form of a cable.

*Service conductors.* The supply conductors that extend from the street main or from transformers to the service equipment of the premises supplied.

*Service drop.* The overhead service conductors from the last pole or other aerial support to and including the splices, if any, connecting to the service-entrance conductors at the building or other structure.

*Service-entrance cable.* Service-entrance cable is a single conductor or multiconductor assembly provided with or without an overall covering, primarily used for services and of the following types:

(i) *Type SE*, having a flame-retardant, moisture-resistant covering, but not required to have inherent protection against mechanical abuse.

(ii) *Type USE*, recognized for underground use, having a moisture-resistant covering, but not required to have a flame-retardant covering or inherent protection against mechanical abuse. Single-conductor cables having an insulation specifically approved for the purpose do not require an outer covering.

*Service-entrance conductors, overhead system.* The service conductors between the terminals of the service equipment and a point usually outside the building, clear of building walls, where joined by tap or splice to the service drop.

*Service entrance conductors, underground system.* The service conductors between the terminals of the service equipment and the point of connection to the service lateral. Where service equipment is located outside the building walls, there may be no service-entrance conductors, or they may be entirely outside the building.

*Service equipment.* The necessary equipment, usually consisting of a circuit breaker or switch and fuses, and their accessories, located near the point of entrance of supply conductors to a building or other structure, or an otherwise defined area, and intended to consititute the main control and means of cutoff of the supply.

*Service raceway.* The raceway that encloses the service-entrance conductors.

*Shielded nonmetallic-sheathed cable.* Type SNM, shielded nonmetallic-sheathed cable is a factory assembly of two or more insulated conductors in an extruded core of moisture-resistant, flame-resistant nonmetallic material, covered with an overlapping spiral metal tape and wire shield and jacketed with an extruded moisture-, flame-, oil-, corrosion-, fungus-, and sunlight-resistant nonmetallic material.

*Show window.* Any window used or designed to be used for the display of goods or advertising material, whether it is fully or partly enclosed or entirely open at the rear and whether or not it has a platform raised higher than the street floor level.

*Sign.* See *Electric sign*.

*Signaling circuit.* Any electric circuit that energizes signaling equipment.

*Special permission.* The written consent of the authority having jurisdiction.

*Storable swimming or wading pool.* A pool with a maximum dimension of 15 feet and a maximum wall height of 3 feet and is so constructed that it may be readily disassembled for storage and reassembled to its original integrity.

*Switchboard.* A large single panel, frame, or assembly of panels which have switches, buses, instruments, overcurrent and other protective devices mounted on the face or back or both. Switchboards are generally accessible from the rear as well as from the front and are not intended to be installed in cabinets. (See *Panelboard*.)

*Switches.*

(i) *General-use switch.* A switch intended for use in general distribution and branch circuits. It is rated in amperes, and it is capable of interrupting its rated current at its rated voltage.

(ii) *General-use snap switch.* A form of general-use switch so constructed that it can be installed in flush device boxes or on outlet box covers, or otherwise used in conjunction with wiring systems recognized by this subpart.

(iii) *Isolating switch.* A switch intended for isolating an electric circuit from the source of power. It has no interrupting rating, and it is intended to be operated only after the circuit has been opened by some other means.

(iv) *Motor-circuit switch.* A switch, rated in horsepower, capable of interrupting the maximum operating overload current of a motor of the same horsepower rating as the switch at the rated voltage.

*Switching devices.* (Over 600 volts, nominal.) Devices designed to close and/or open one or more electric circuits. Included in this category are circuit breakers, cutouts, disconnecting (or isolating) switches, disconnecting means, interrupter switches, and oil (filled) cutouts.

*Transportable X-ray.* X-ray equipment installed in a vehicle or that may readily be disassembled for transport in a vehicle.

*Utilization equipment.* Utilization equipment means equipment which utilizes electric energy for mechanical, chemical, heating, lighting, or similar useful purpose.

*Utilization system.* A utilization system is a system which provides electric power and light for employee workplaces, and includes the premises wiring system and utilization equipment.

*Ventilated.* Provided with a means to permit circulation of air sufficient to remove an excess of heat, fumes, or vapors.

*Volatile flammable liquid.* A flammable liquid having a flash point below 38 degrees C (100 degrees F) or whose temperature is above its flash point.

*Voltage (of a circuit).* The greatest root-mean-square (effective) difference of potential between any two conductors of the circuit concerned.

*Voltage, nominal.* A nominal value assigned to a circuit or system for the purpose of conveniently designating its voltage class (as 120/240, 480Y/277, 600, etc.). The actual voltage at which a circuit operates can vary from the nominal within a range that permits satisfactory operation of equipment.

*Voltage to ground.* For grounded circuits, the voltage between the given conductor and that point or conductor of the circuit that is grounded;

for ungrounded circuits, the greatest voltage between the given conductor and any other conductor of the circuit.

*Watertight.* So constructed that moisture will not enter the enclosure.

*Weatherproof.* So constructed or protected that exposure to the weather will not interfere with successful operation. Rainproof, raintight, or watertight equipment can fulfill the requirements for weatherproof where varying weather conditions other than wetness, such as snow, ice, dust, or temperature extremes, are not a factor.

*Wet location.* See *Location.*

*Wireways.* Wireways are sheet-metal troughs with hinged or removable covers for housing and protecting electric wires and cable and in which conductors are laid in place after the wireway has been installed as a complete system.

[46 FR 4056, Jan. 16, 1981; 46 FR 40185, Aug. 7, 1981, as amended at 53 FR 12123, Apr. 12, 1988; 55 FR 32020, Aug. 6, 1990; 55 FR 46054, Nov. 1, 1990]

### Appendix A to Subpart S to Part 1910—Reference Documents

The following references provide information which can be helpful in understanding and complying with the requirements contained in Subpart S:

ANSI A17.1-71 Safety Code for Elevators, Dumbwaiters, Escalators and Moving Walks.

ANSI B9.1-71 Safety Code for Mechanical Refrigeration.

ANSI B30.2-76 Safety Code for Overhead and Gantry Cranes.

ANSI B30.3-75 Hammerhead Tower Cranes.

ANSI B30.4-73 Safety Code for Portal, Tower, and Pillar Cranes.

ANSI B30.5-68 Safety Code for Crawler, Locomotive, and Truck Cranes.

ANSI B30.6-77 Derricks.

ANSI B30.7-77 Base Mounted Drum Hoists.

ANSI B30.8-71 Safety Code for Floating Cranes and Floating Derricks.

ANSI B30.11-73 Monorail Systems and Underhung Cranes.

ANSI B30.12-75 Handling Loads Suspended from Rotorcraft.

ANSI B30.13-77 Controlled Mechanical Storage Cranes.

ANSI B30.15-73 Safety Code for Mobile Hydraulic Cranes.

ANSI B30.16-73 Overhead Hoists.

ANSI C2-81 National Electrical Safety Code.

ANSI C33.27-74 Safety Standard for Outlet Boxes and Fittings for Use in Hazardous Locations, Class I, Groups A, B, C, and D, and Class II, Groups E, F, and G.

ANSI K61.1-72 Safety Requirements for the Storage and Handling of Anhydrous Ammonia.

ASTM D2155-66 Test Method for Autoignition Temperature of Liquid Petroleum Products.

ASTM D3176-74 Method for Ultimate Analysis of Coal and Coke.

ASTM D3180-74 Method for Calculating Coal and Coke Analyses from As Determined to Different Bases.

IEEE 463-77 Standard for Electrical Safety Practices in Electrolytic Cell Line Working Zones.

NFPA 20-76 Standard for the Installation of Centrifugal Fire Pumps.

NFPA 30-78 Flammable and Combustible Liquids Code.

NFPA 32-74 Standard for Drycleaning Plants.

NFPA 33-73 Standard for Spray Application Using Flammable and Combustible Materials.

NFPA 34-74 Standard for Dip Tanks Containing Flammable or Combustible Liquids.

NFPA 35-76 Standard for the Manufacture of Organic Coatings.

NFPA 36-74 Standard for Solvent Extraction Plants.

NFPA 40-74 Standard for the Storage and Handling of Cellulose Nitrate Motion Picture Film.

NFPA 56A-73 Standard for the Use of Inhalation Anesthetics (Flammable and Nonflammable).

NFPA 56F-74 Standard for Nonflammable Medical Gas Systems.

NFPA 58-76 Standard for the Storage and Handling of Liquefied Petroleum Gases.

NFPA 59-76 Standard for the Storage and Handling of Liquefied Petroleum Gases at Utility Gas Plants.

NFPA 70-78 National Electrical Code.

NFPA 70C-74 Hazardous Locations Classification.

NFPA 70E Standard for the Electrical Safety Requirements for Employee Workplaces.

NFPA 71-77 Standard for the Installation, Maintenance, and Use of Central Station Signaling Systems.

NFPA 72A-75 Standard for the Installation, Maintenance, and Use of Local Protective Signaling Systems for Watchman, Fire Alarm, and Supervisory Service.

NFPA 72B-75 Standard for the Installation, Maintenance, and Use of Auxiliary Protective Signaling Systems for Fire Alarm Service.

NFPA 72C-75 Standard for the Installation, Maintenance, and Use of Remote Station Protective Signaling Systems.

NFPA 72D-75 Standard for the Installation, Maintenance, and Use of Proprietary Protective Signaling Systems for Watchman, Fire Alarm, and Supervisory Service.

NFPA 72E-74 Standard for Automatic Fire Detectors.

NFPA 74-75 Standard for Installation, Maintenance, and Use of Household Fire Warning Equipment.

NFPA 76A-73 Standard for Essential Electrical Systems for Health Care Facilities.

NFPA 77-72 Recommended Practice on Static Electricity.

NFPA 80-77 Standard for Fire Doors and Windows.

NFPA 86A-73 Standard for Ovens and Furnaces; Design, Location and Equipment.

NFPA 88A-73 Standard for Parking Structures.

NFPA 88B-73 Standard for Repair Garages.

NFPA 91-73 Standard for the Installation of Blower and Exhaust Systems for Dust, Stock, and Vapor Removal, or Conveying.

NFPA 101-78 Code for Safety to Life from Fire in Buildings and Structures. (Life Safety Code.)

NFPA 325M-69 Fire-Hazard Properties of Flammable Liquids, Gases, and Volatile Solids.

NFPA 493-75 Standard for Intrinsically Safe Apparatus for Use in Class I Hazardous Locations and Its Associated Apparatus.

NFPA 496-74 Standard for Purged and Pressurized Enclosures for Electrical Equipment in Hazardous Locations.

NFPA 497-75 Recommended Practice for Classification of Class I Hazardous Locations for Electrical Installations in Chemical Plants.

NFPA 505-75 Fire Safety Standard for Powered Industrial Trucks Including Type Designations and Areas of Use.

NMAB 353-1-79 Matrix of Combustion-Relevant Properties and Classification of Gases, Vapors, and Selected Solids.

NMAB 353-2-79 Test Equipment for Use in Determining Classifications of Combustible Dusts.

NMAB 353-3-80 Classification of Combustible Dusts in Accordance with the National Electrical Code.

[46 FR 4056, Jan. 16, 1981; 46 FR 40185, Aug. 7, 1981]

### Appendix B to Subpart S to Part 1910—Explanatory Data [Reserved]

### Appendix C to Subpart S to Part 1910—Tables, Notes, and Charts [Reserved]

# Subpart T—Commercial Diving Operations

> **Authority:** Sections 4, 6, and 8 of the Occupational Safety and Health Act of 1970 (29 U.S.C. 653, 655, and 657); Section 107, Contract Work Hours and Safety Standards Act (the Construction Safety Act) (40 U.S.C. 333); Section 41, Longshore and Harbor Workers' Compensation Act (33 U.S.C. 941); Secretary of Labor's Order No. 8-76 (41 FR 25059), 9-83 (48 FR 35736), 1-90 (55 FR 9033), 6-96 (62 FR 111), 3-2000 (65 FR 50017), or 5-2002 (67 FR 65008), as applicable; 29 CFR part 1911.
>
> **Source:** 42 FR 37668, July 22, 1977, unless otherwise noted. [58 FR 35310, June 30, 1993; 61 FR 9227, March 7, 1996; 69 FR 7363, Feb. 17, 2004]

## General

### §1910.401 Scope and application.

**1910.401(a)** *Scope.* **(1)** This subpart (standard) applies to every place of employment within the waters of the United States, or within any State, the District of Columbia, the Commonwealth of Puerto Rico, the Virgin Islands, American Samoa, Guam, the Trust Territory of the Pacific Islands, Wake Island, Johnston Island, the Canal Zone, or within the Outer Continental Shelf lands as defined in the Outer Continental Shelf Lands Act (67 Stat. 462, 43 U.S.C. 1331), where diving and related support operations are performed.

**1910.401(a)(2)** This standard applies to diving and related support operations conducted in connection with all types of work and employments, including general industry, construction, ship repairing, shipbuilding, shipbreaking and longshoring. However, this standard does not apply to any diving operation:

**1910.401(a)(2)(i)** Performed solely for instructional purposes, using open-circuit, compressed-air SCUBA and conducted within the no-decompression limits;

**1910.401(a)(2)(ii)** Performed solely for search, rescue, or related public safety purposes by or under the control of a governmental agency; or

**1910.401(a)(2)(iii)** Governed by 45 CFR part 46 (Protection of Human Subjects, U.S. Department of Health and Human Services) or equivalent rules or regulations established by another federal agency, which regulate research, development, or related purposes involving human subjects.

**1910.401(a)(2)(iv)** Defined as scientific diving and which is under the direction and control of a diving program containing at least the following elements:

**1910.401(a)(2)(iv)(A)** Diving safety manual which includes at a minimum: Procedures covering all diving operations specific to the program; procedures for emergency care, including recompression and evacuation; and criteria for diver training and certification.

**1910.401(a)(2)(iv)(B)** Diving control (safety) board, with the majority of its members being active divers, which shall at a minimum have the authority to: Approve and monitor diving projects; review and revise the diving safety manual; assure compliance with the manual; certify the depths to which a diver has been trained; take disciplinary action for unsafe practices; and, assure adherence to the buddy system (a diver is accompanied by and is in continuous contact with another diver in the water) for SCUBA diving.

**1910.401(a)(3)** *Alternative requirements for recreational diving instructors and diving guides.* Employers of recreational diving instructors and diving guides are not required to comply with the decompression-chamber requirements specified by paragraphs (b)(2) and (c)(3)(iii) of § 1910.423 and paragraph (b)(1) of § 1910.426 when they meet all of the following conditions:

**1910.401(a)(3)(i)** The instructor or guide is engaging solely in recreational diving instruction or dive-guiding operations;

**1910.401(a)(3 )(ii)** The instructor or guide is diving within the no-decompression limits in these operations;

**1910.401(a)(3 )(iii)** The instructor or guide is using a nitrox breathing-gas mixture consisting of a high percentage of oxygen (more than 22% by volume) mixed with nitrogen;

**1910.401(a)(3 )(iv)** The instructor or guide is using an open-circuit, semi-closed-circuit, or closed-circuit self-contained underwater breathing apparatus (SCUBA); and

**1910.401(a)(3) (v)** The employer of the instructor or guide is complying with all requirements of Appendix C of this subpart.

**1910.401(b)** *Application in emergencies.* An employer may deviate from the requirements of this standard to the extent necessary to prevent or minimize a situation which is likely to cause death, serious physical harm, or major environmental damage, provided that the employer:

**1910.401(b)(1)** Notifies the Area Director, Occupational Safety and Health Administration within 48 hours of the onset of the emergency situation indicating the nature of the emergency and extent of the deviation from the prescribed regulations; and

**1910.401(b)(2)** Upon request from the Area Director, submits such information in writing.

**1910.401(c)** *Employer obligation.* The employer shall be responsible for compliance with:

**1910.401(c)(1)** All provisions of this standard of general applicability; and

**1910.401(c)(2)** All requirements pertaining to specific diving modes to the extent diving operations in such modes are conducted.

[42 FR 37668, July 22, 1977, as amended at 47 FR 53365, Nov. 26, 1982; 58 FR 35310, June 30, 1993]

### §1910.402 Definitions.

As used in this standard, the listed terms are defined as follows:

*Acfm:* Actual cubic feet per minute.

*ASME Code or equivalent:* ASME (American Society of Mechanical Engineers) Boiler and Pressure Vessel Code, Section VIII, or an equivalent code which the employer can demonstrate to be equally effective.

*ATA:* Atmosphere absolute.

*Bell:* An enclosed compartment, pressurized (closed bell) or unpressurized (open bell), which allows the diver to be transported to and from the underwater work area and which may be used as a temporary refuge during diving operations.

*Bottom time:* The total elasped time measured in minutes from the time when the diver leaves the surface in descent to the time that the diver begins ascent.

*Bursting pressure:* The pressure at which a pressure containment device would fail structurally.

*Cylinder:* A pressure vessel for the storage of gases.

*Decompression chamber:* A pressure vessel for human occupancy such as a surface decompression chamber, closed bell, or deep diving system used to decompress divers and to treat decompression sickness.

*Decompression sickness:* A condition with a variety of symptoms which may result from gas or bubbles in the tissues of divers after pressure reduction.

*Decompression table:* A profile or set of profiles of depth-time relationships for ascent rates and breathing mixtures to be followed after a specific depth-time exposure or exposures.

*Dive-guiding operations* means leading groups of sports divers, who use an open-circuit, semi-closed-circuit, or closed-circuit self-contained underwater breathing apparatus, to local undersea diving locations for recreational purposes.

*Dive location:* A surface or vessel from which a diving operation is conducted.

*Dive-location reserve breathing gas:* A supply system of air or mixed-gas (as appropriate) at the dive location which is independent of the primary supply system and sufficient to support divers during the planned decompression.

*Dive team:* Divers and support employees involved in a diving operation, including the designated person-in-charge.

*Diver:* An employee working in water using underwater apparatus which supplies compressed breathing gas at the ambient pressure.

*Diver-carried reserve breathing gas:* A diver-carried supply of air or mixed gas (as appropriate) sufficient under standard operating conditions to allow the diver to reach the surface, or another source of breathing gas, or to be reached by a standby diver.

*Diving mode:* A type of diving requiring specific equipment, procedures and techniques (SCUBA, surface-supplied air, or mixed gas).

*Fsw:* Feet of seawater (or equivalent static pressure head).

*Heavy gear:* Diver-worn deep-sea dress including helmet, breastplate, dry suit, and weighted shoes.

*Hyperbaric conditions:* Pressure conditions in excess of surface pressure.

*Inwater stage:* A suspended underwater platform which supports a diver in the water.

*Liveboating:* The practice of supporting a surfaced-supplied air or mixed gas diver from a vessel which is underway.

*Mixed-gas diving:* A diving mode in which the diver is supplied in the water with a breathing gas other than air.

*No-decompression limits:* The depth-time limits of the "no-decompression limits and repetitive dive group designation table for no-decompression air dives", U.S. Navy Diving Manual or equivalent limits which the employer can demonstrate to be equally effective.

*Psi(g):* Pounds per square inch (gauge).

*Recreational diving* instruction means training diving students in the use of recreational diving procedures and the safe operation of diving equipment, including an open-circuit, semi-closed-circuit, or closed-circuit self-contained underwater breathing apparatus, during dives.

*Scientific diving* means diving performed solely as a necessary part of a scientific, research, or educational activity by employees whose sole purpose for diving is to perform scientific research tasks. Scientific diving does not include performing any tasks usually associated with commercial diving such as: Placing or removing heavy objects underwater; inspection of pipelines and similar objects; construction; demolition; cutting or welding; or the use of explosives.

*SCUBA diving:* A diving mode independent of surface supply in which the diver uses open circuit self-contained underwater breathing apparatus.

*Standby diver:* A diver at the dive location available to assist a diver in the water.

*Surface-supplied air diving:* A diving mode in which the diver in the water is supplied from the dive location with compressed air for breathing.

*Treatment table:* A depth-time and breathing gas profile designed to treat decompression sickness.

*Umbilical:* The composite hose bundle between a dive location and a diver or bell, or between a diver and a bell, which supplies the diver or bell with breathing gas, communications, power, or heat as appropriate to the diving mode or conditions, and includes a safety line between the diver and the dive location.

*Volume tank:* A pressure vessel connected to the outlet of a compressor and used as an air reservoir.

*Working pressure:* The maximum pressure to which a pressure containment device may be exposed under standard operating conditions.

[42 FR 37668, July 22, 1977, as amended at 47 FR 53365, Nov. 26, 1982; 69 FR 7363, Feb. 17 2004]

## Personnel Requirements
### §1910.410 Qualifications of dive team.

**1910.410(a)** *General.* **(1)** Each dive team member shall have the experience or training necessary to perform assigned tasks in a safe and healthful manner.

**1910.410(a)(2)** Each dive team member shall have experience or training in the following:

**1910.410(a)(2)(i)** The use of tools, equipment and systems relevant to assigned tasks;

**1910.410(a)(2)(ii)** Techniques of the assigned diving mode: and

**1910.410(a)(2)(iii)** Diving operations and emergency procedures.

**1910.410(a)(3)** All dive team members shall be trained in cardiopulmonary resuscitation and first aid (American Red Cross standard course or equivalent).

**1910.410(a)(4)** Dive team members who are exposed to or control the exposure of others to hyperbaric conditions shall be trained in diving-related physics and physiology.

**1910.410(b)** *Assignments.* **(1)** Each dive team member shall be assigned tasks in accordance with the employee's experience or training, except that limited additional tasks may be assigned to an employee undergoing training provided that these tasks are performed under the direct supervision of an experienced dive team member.

**1910.410(b)(2)** The employer shall not require a dive team member to be exposed to hyperbaric conditions against the employee's will, except when necessary to complete decompression or treatment procedures.

**1910.410(b)(3)** The employer shall not permit a dive team member to dive or be otherwise exposed to hyperbaric conditions for the duration of any temporary physical impairment or condition which is known to the employer and is likely to affect adversely the safety or health of a dive team member.

**1910.410(c)** *Designated person-in-charge.* **(1)** The employer or an employee designated by the employer shall be at the dive location in charge of all aspects of the diving operation affecting the safety and health of dive team members.

**1910.410(c)(2)** The designated person-in-charge shall have experience and training in the conduct of the assigned diving operation.

## General Operations Procedures
### §1910.420 Safe practices manual.

**1910.420(a)** *General.* The employer shall develop and maintain a safe practices manual which shall be made available at the dive location to each dive team member.

**1910.420(b)** *Contents.* **(1)** The safe practices manual shall contain a copy of this standard and the employer's policies for implementing the requirements of this standard.

**1910.420(b)(2)** For each diving mode engaged in, the safe practices manual shall include:

**1910.420(b)(2)(i)** Safety procedures and checklists for diving operations;

**1910.420(b)(2)(ii)** Assignments and responsibilities of the dive team members;

**1910.420(b)(2)(iii)** Equipment procedures and checklists; and

**1910.420(b)(2)(iv)** Emergency procedures for fire, equipment failure, adverse environmental conditions, and medical illness and injury.

[42 FR 37668, July 22, 1977, as amended at 49 FR 18295, Apr. 30, 1984; 61 FR 5507, Feb. 13, 1996]

### §1910.421 Pre-dive procedures.

**1910.421(a)** *General.* The employer shall comply with the following requirements prior to each diving operation, unless otherwise specified.

**1910.421(b)** *Emergency aid.* A list shall be kept at the dive location of the telephone or call numbers of the following:

**1910.421(b)(1)** An operational decompression chamber (if not at the dive location);

**1910.421(b)(2)** Accessible hospitals;

**1910.421(b)(3)** Available physicians;

**1910.421(b)(4)** Available means of transportation; and

**1910.421(b)(5)** The nearest U.S. Coast Guard Rescue Coordination Center.

**1910.421(c)** *First aid supplies.* **(1)** A first aid kit appropriate for the diving operation and approved by a physician shall be available at the dive location.

**1910.421(c)(2)** When used in a decompression chamber or bell, the first aid kit shall be suitable for use under hyperbaric conditions.

**1910.421(c)(3)** In addition to any other first aid supplies, an American Red Cross standard first aid handbook or equivalent, and a bag-type manual resuscitator with transparent mask and tubing shall be available at the dive location.

**1910.421(d)** *Planning and assessment.* Planning of a diving operation shall include an assessment of the safety and health aspects of the following:

**1910.421(d)(1)** Diving mode;

**1910.421(d)(2)** Surface and underwater conditions and hazards;

**1910.421(d)(3)** Breathing gas supply (including reserves);

**1910.421(d)(4)** Thermal protection;

**1910.421(d)(5)** Diving equipment and systems;

**1910.421(d)(6)** Dive team assignments and physical fitness of dive team members (including any impairment known to the employer);

**1910.421(d)(7)** Repetitive dive designation or residual inert gas status of dive team members;

**1910.421(d)(8)** Decompression and treatment procedures (including altitude corrections); and

# OSHA Standards for General Industry

**1910.421(d)(9)** Emergency procedures.

**1910.421(e)** *Hazardous activities.* To minimize hazards to the dive team, diving operations shall be coordinated with other activities in the vicinity which are likely to interfere with the diving operation.

**1910.421(f)** *Employee briefing.* **(1)** Dive team members shall be briefed on:

**1910.421(f)(1)(i)** The tasks to be undertaken;

**1910.421(f)(1)(ii)** Safety procedures for the diving mode;

**1910.421(f)(1)(iii)** Any unusual hazards or environmental conditions likely to affect the safety of the diving operation; and

**1910.421(f)(1)(iv)** Any modifications to operating procedures necessitated by the specific diving operation.

**1910.421(f)(2)** Prior to making individual dive team member assignments, the employer shall inquire into the dive team member's current state of physical fitness, and indicate to the dive team member the procedure for reporting physical problems or adverse physiological effects during and after the dive.

**1910.421(g)** *Equipment inspection.* The breathing gas supply system including reserve breathing gas supplies, masks, helmets, thermal protection, and bell handling mechanism (when appropriate) shall be inspected prior to each dive.

**1910.421(h)** *Warning signal.* When diving from surfaces other than vessels in areas capable of supporting marine traffic, a rigid replica of the international code flag "A" at least one meter in height shall be displayed at the dive location in a manner which allows all-round visibility, and shall be illuminated during night diving operations.

[42 FR 37668, July 22, 1977, as amended at 47 FR 14706, Apr. 6, 1982; 54 FR 24334, June 7, 1989; 61 FR 5507, Feb. 13, 1996]

## §1910.422 Procedures during dive.

**1910.422(a)** *General.* The employer shall comply with the following requirements which are applicable to each diving operation unless otherwise specified.

**1910.422(b)** *Water entry and exit.* **(1)** A means capable of supporting the diver shall be provided for entering and exiting the water.

**1910.422(b)(2)** The means provided for exiting the water shall extend below the water surface.

**1910.422(b)(3)** A means shall be provided to assist an injured diver from the water or into a bell.

**1910.422(c)** *Communications.* **(1)** An operational two-way voice communication system shall be used between:

**1910.422(c)(1)(i)** Each surface-supplied air or mixed-gas diver and a dive team member at the dive location or bell (when provided or required); and

**1910.422(c)(1)(ii)** The bell and the dive location.

**1910.422(c)(2)** An operational, two-way communication system shall be available at the dive location to obtain emergency assistance.

**1910.422(d)** *Decompression tables.* Decompression, repetitive, and no-decompression tables (as appropriate) shall be at the dive location.

**1910.422(e)** *Dive profiles.* A depth-time profile, including when appropriate any breathing gas changes, shall be maintained for each diver during the dive including decompression.

**1910.422(f)** *Hand-held power tools and equipment.* **(1)** Hand-held electrical tools and equipment shall be de-energized before being placed into or retrieved from the water.

**1910.422(f)(2)** Hand-held power tools shall not be supplied with power from the dive location until requested by the diver.

**1910.422(g)** *Welding and burning.* **(1)** A current supply switch to interrupt the current flow to the welding or burning electrode shall be:

**1910.422(g)(1)(i)** Tended by a dive team member in voice communication with the diver performing the welding or burning; and

**1910.422(g)(1)(ii)** Kept in the open position except when the diver is welding or burning.

**1910.422(g)(2)** The welding machine frame shall be grounded.

**1910.422(g)(3)** Welding and burning cables, electrode holders, and connections shall be capable of carrying the maximum current required by the work, and shall be properly insulated.

**1910.422(g)(4)** Insulated gloves shall be provided to divers performing welding and burning operations.

**1910.422(g)(5)** Prior to welding or burning on closed compartments, structures or pipes, which contain a flammable vapor or in which a flammable vapor may be generated by the work, they shall be vented, flooded, or purged with a mixture of gases which will not support combustion.

**1910.422(h)** *Explosives.* **(1)** Employers shall transport, store, and use explosives in accordance with this section and the applicable provisions of §1910.109 and §1926.912 of Title 29 of the Code of Federal Regulations.

**1910.422(h)(2)** Electrical continuity of explosive circuits shall not be tested until the diver is out of the water.

**1910.422(h)(3)** Explosives shall not be detonated while the diver is in the water.

**1910.422(i)** *Termination of dive.* The working interval of a dive shall be terminated when:

**1910.422(i)(1)** A diver requests termination;

**1910.422(i)(2)** A diver fails to respond correctly to communications or signals from a dive team member;

**1910.422(i)(3)** Communications are lost and can not be quickly re-established between the diver and a dive team member at the dive location, and between the designated person-in-charge and the person controlling the vessel in liveboating operations; or

**1910.422(i)(4)** A diver begins to use diver-carried reserve breathing gas or the dive-location reserve breathing gas.

## §1910.423 Post-dive procedures.

**1910.423(a)** *General.* The employer shall comply with the following requirements which are applicable after each diving operation, unless otherwise specified.

**1910.423(b)** *Precautions.* **(1)** After the completion of any dive, the employer shall:

**1910.423(b)(1)(i)** Check the physical condition of the diver;

**1910.423(b)(1)(ii)** Instruct the diver to report any physical problems or adverse physiological effects including symptoms of decompression sickness;

**1910.423(b)(1)(iii)** Advise the diver of the location of a decompression chamber which is ready for use; and

**1910.423(b)(1)(iv)** Alert the diver to the potential hazards of flying after diving.

**1910.423(b)(2)** For any dive outside the no-decompression limits, deeper than 100 fsw or using mixed gas as a breathing mixture, the employer shall instruct the diver to remain awake and in the vicinity of the decompression chamber which is at the dive location for at least one hour after the dive (including decompression or treatment as appropriate).

**1910.423(c)** *Recompression capability.* **(1)** A decompression chamber capable of recompressing the diver at the surface to a minimum of 165 fsw (6 ATA) shall be available at the dive location for:

**1910.423(c)(1)(i)** Surface-supplied air diving to depths deeper than 100 fsw and shallower than 220 fsw;

**1910.423(c)(1)(ii)** Mixed gas diving shallower than 300 fsw; or

**1910.423(c)(1)(iii)** Diving outside the no-decompression limits shallower than 300 fsw.

**1910.423(c)(2)** A decompression chamber capable of recompressing the diver at the surface to the maximum depth of the dive shall be available at the dive location for dives deeper than 300 fsw.

**1910.423(c)(3)** The decompression chamber shall be:

**1910.423(c)(3)(i)** Dual-lock;

**1910.423(c)(3)(ii)** Multiplace; and

**1910.423(c)(3)(iii)** Located within 5 minutes of the dive location.

**1910.423(c)(4)** The decompression chamber shall be equipped with:

**1910.423(c)(4)(i)** A pressure gauge for each pressurized compartment designed for human occupancy;

**1910.423(c)(4)(ii)** A built-in-breathing-system with a minimum of one mask per occupant;

**1910.423(c)(4)(iii)** A two-way voice communication system between occupants and a dive team member at the dive location;

**1910.423(c)(4)(iv)** A viewport; and

**1910.423(c)(4)(v)** Illumination capability to light the interior.

**1910.423(c)(5)** Treatment tables, treatment gas appropriate to the diving mode, and sufficient gas to conduct treatment shall be available at the dive location.

**1910.423(c)(6)** A dive team member shall be available at the dive location during and for at least one hour after the dive to operate the decompression chamber (when required or provided).

**1910.423(d)** *Record of dive.* **(1)** The following information shall be recorded and maintained for each diving operation:

**1910.423(d)(1)(i)** Names of dive team members including designated person-in-charge;

**1910.423(d)(1)(ii)** Date, time, and location;

**1910.423(d)(1)(iii)** Diving modes used;

**1910.423(d)(1)(iv)** General nature of work performed;

**1910.423(d)(1)(v)** Approximate underwater and surface conditions (visibility, water temperature and current); and

**1910.423(d)(1)(vi)** Maximum depth and bottom time for each diver.

**1910.423(d)(2)** For each dive outside the no-decompression limits, deeper than 100 fsw or using mixed gas, the following additional information shall be recorded and maintained:

**1910.423(d)(2)(i)** Depth-time and breathing gas profiles;

**1910.423(d)(2)(ii)** Decompression table designation (including modification); and

**1910.423(d)(2)(iii)** Elapsed time since last pressure exposure if less than 24 hours or repetitive dive designation for each diver.

**1910.423(d)(3)** For each dive in which decompression sickness is suspected or symptoms are evident, the following additional information shall be recorded and maintained:

**1910.423(d)(3)(i)** Description of decompression sickness symptoms (including depth and time of onset); and

**1910.423(d)(3)(ii)** Description and results of treatment.

**1910.423(e)** *Decompression procedure assessment.* The employer shall:

**1910.423(e)(1)** Investigate and evaluate each incident of decompression sickness based on the recorded information, consideration of the past performance of decompression table used, and individual susceptibility;

**1910.423(e)(2)** Take appropriate corrective action to reduce the probability of recurrence of decompression sickness; and

**1910.423(e)(3)** Prepare a written evaluation of the decompression procedure assessment, including any corrective action taken, within 45 days of the incident of decompression sickness.

[42 FR 37668, July 22, 1977, as amended at 49 FR 18295, Apr. 30, 1984; 61 FR 5507, Feb. 13, 1996]

### Specific Operations Procedures

### §1910.424 SCUBA diving.

**1910.424(a)** *General.* Employers engaged in SCUBA diving shall comply with the following requirements, unless otherwise specified.

**1910.424(b)** *Limits.* SCUBA diving shall not be conducted:

**1910.424(b)(1)** At depths deeper than 130 fsw;

**1910.424(b)(2)** At depths deeper than 100 fsw or outside the no-decompression limits unless a decompression chamber is ready for use;

**1910.424(b)(3)** Against currents exceeding one (1) knot unless line-tended; or

**1910.424(b)(4)** In enclosed or physically confining spaces unless line-tended.

**1910.424(c)** *Procedures.* **(1)** A standby diver shall be available while a diver is in the water.

**1910.424(c)(2)** A diver shall be line-tended from the surface, or accompanied by another diver in the water in continuous visual contact during the diving operations.

**1910.424(c)(3)** A diver shall be stationed at the underwater point of entry when diving is conducted in enclosed or physically confining spaces.

**1910.424(c)(4)** A diver-carried reserve breathing gas supply shall be provided for each diver consisting of:

**1910.424(c)(4)(i)** A manual reserve (J valve); or

**1910.424(c)(4)(ii)** An independent reserve cylinder with a separate regulator or connected to the underwater breathing apparatus.

**1910.424(c)(5)** The valve of the reserve breathing gas supply shall be in the closed position prior to the dive.

### §1910.425 Surface-supplied air diving.

**1910.425(a)** *General.* Employers engaged in surface-supplied air diving shall comply with the following requirements, unless otherwise specified.

**1910.425(b)** *Limits.* **(1)** Surface-supplied air diving shall not be conducted at depths deeper than 190 fsw, except that dives with bottom times of 30 minutes or less may be conducted to depths of 220 fsw.

**1910.425(b)(2)** A decompression chamber shall be ready for use at the dive location for any dive outside the no-decompression limits or deeper than 100 fsw.

**1910.425(b)(3)** A bell shall be used for dives with an inwater decompression time greater than 120 minutes, except when heavy gear is worn or diving is conducted in physically confining spaces.

**1910.425(c)** *Procedures.* **(1)** Each diver shall be continuously tended while in the water.

**1910.425(c)(2)** A diver shall be stationed at the underwater point of entry when diving is conducted in enclosed or physically confining spaces.

**1910.425(c)(3)** Each diving operation shall have a primary breathing gas supply sufficient to support divers for the duration of the planned dive including decompression.

**1910.425(c)(4)** For dives deeper than 100 fsw or outside the no-decompression limits:

**1910.425(c)(4)(i)** A separate dive team member shall tend each diver in the water;

**1910.425(c)(4)(ii)** A standby diver shall be available while a diver is in the water;

**1910.425(c)(4)(iii)** A diver-carried reserve breathing gas supply shall be provided for each diver except when heavy gear is worn; and

**1910.425(c)(4)(iv)** A dive-location reserve breathing gas supply shall be provided.

**1910.425(c)(5)** For heavy-gear diving deeper than 100 fsw or outside the no-decompression limits:

**1910.425(c)(5)(i)** An extra breathing gas hose capable of supplying breathing gas to the diver in the water shall be available to the standby diver.

**1910.425(c)(5)(ii)** An inwater stage shall be provided to divers in the water.

**1910.425(c)(6)** Except when heavy gear is worn or where physical space does not permit, a diver-carried reserve breathing gas supply shall be provided whenever the diver is prevented by the configuration of the dive area from ascending directly to the surface.

### §1910.426 Mixed-gas diving.

**1910.426(a)** *General.* Employers engaged in mixed-gas diving shall comply with the following requirements, unless otherwise specified.

**1910.426(b)** *Limits.* Mixed-gas diving shall be conducted only when:

**1910.426(b)(1)** A decompression chamber is ready for use at the dive location; and

**1910.426(b)(1)(i)** A bell is used at depths greater than 220 fsw or when the dive involves inwater decompression time of greater than 120 minutes, except when heavy gear is worn or when diving in physically confining spaces; or

**1910.426(b)(1)(ii)** A closed bell is used at depths greater than 300 fsw, except when diving is conducted in physically confining spaces.

**1910.426(c)** *Procedures.* **(1)** A separate dive team member shall tend each diver in the water.

**1910.426(c)(2)** A standby diver shall be available while a diver is in the water.

**1910.426(c)(3)** A diver shall be stationed at the underwater point of entry when diving is conducted in enclosed or physically confining spaces.

**1910.426(c)(4)** Each diving operation shall have a primary breathing gas supply sufficient to support divers for the duration of the planned dive including decompression.

**1910.426(c)(5)** Each diving operation shall have a dive-location reserve breathing gas supply.

**1910.426(c)(6)** When heavy gear is worn:

**1910.426(c)(6)(i)** An extra breathing gas hose capable of supplying breathing gas to the diver in the water shall be available to the standby diver; and

**1910.426(c)(6)(ii)** An inwater stage shall be provided to divers in the water.

**1910.426(c)(7)** An inwater stage shall be provided for divers without access to a bell for dives deeper than 100 fsw or outside the no-decompression limits.

**1910.426(c)(8)** When a closed bell is used, one dive team member in the bell shall be available and tend the diver in the water.

**1910.426(c)(9)** Except when heavy gear is worn or where physical space does not permit, a diver-carried reserve breathing gas supply shall be provided for each diver:

**1910.426(c)(9)(i)** Diving deeper than 100 fsw or outside the no-decompression limits; or

**1910.426(c)(9)(ii)** Prevented by the configuration of the dive area from directly ascending to the surface.

## §1910.427 Liveboating.

**1910.427(a)** *General.* Employers engaged in diving operations involving liveboating shall comply with the following requirements.

**1910.427(b)** *Limits.* Diving operations involving liveboating shall not be conducted:

**1910.427(b)(1)** With an inwater decompression time of greater than 120 minutes;

**1910.427(b)(2)** Using surface-supplied air at depths deeper than 190 fsw, except that dives with bottom times of 30 minutes or less may be conducted to depths of 220 fsw;

**1910.427(b)(3)** Using mixed gas at depths greater than 220 fsw;

**1910.427(b)(4)** In rough seas which significantly inpede diver mobility or work function; or

**1910.427(b)(5)** In other than daylight hours.

**1910.427(c)** *Procedures.* **(1)** The propeller of the vessel shall be stopped before the diver enters or exits the water.

**1910.427(c)(2)** A device shall be used which minimizes the possibility of entanglement of the diver's hose in the propeller of the vessel.

**1910.427(c)(3)** Two-way voice communication between the designated person-in-charge and the person controlling the vessel shall be available while the diver is in the water.

**1910.427(c)(4)** A standby diver shall be available while a diver is in the water.

**1910.427(c)(5)** A diver-carried reserve breathing gas supply shall be carried by each diver engaged in liveboating operations.

### Equipment Procedures and Requirements

## §1910.430 Equipment.

**1910.430(a)** *General.* **(1)** All employers shall comply with the following requirements, unless otherwise specified.

**1910.430(a)(2)** Each equipment modification, repair, test, calibration or maintenance service shall be recorded by means of a tagging or logging system, and include the date and nature of work performed, and the name or initials of the person performing the work.

**1910.430(b)** *Air compressor system.* **(1)** Compressors used to supply air to the diver shall be equipped with a volume tank with a check valve on the inlet side, a pressure gauge, a relief valve, and a drain valve.

**1910.430(b)(2)** Air compressor intakes shall be located away from areas containing exhaust or other contaminants.

**1910.430(b)(3)** Respirable air supplied to a diver shall not contain:

**1910.430(b)(3)(i)** A level of carbon monoxide (CO) greater than 20 p/m;

**1910.430(b)(3)(ii)** A level of carbon dioxide ($CO_2$) greater than 1,000 p/m;

**1910.430(b)(3)(iii)** A level of oil mist greater than 5 milligrams per cubic meter; or

**1910.430(b)(3)(iv)** A noxious or pronounced odor.

**1910.430(b)(4)** The output of air compressor systems shall be tested for air purity every 6 months by means of samples taken at the connection to the distribution system, except that non-oil lubricated compressors need not be tested for oil mist.

**1910.430(c)** *Breathing gas supply hoses.* **(1)** Breathing gas supply hoses shall:

**1910.430(c)(1)(i)** Have a working pressure at least equal to the working pressure of the total breathing gas system;

**1910.430(c)(1)(ii)** Have a rated bursting pressure at least equal to 4 times the working pressure;

**1910.430(c)(1)(iii)** Be tested at least annually to 1.5 times their working pressure; and

**1910.430(c)(1)(iv)** Have their open ends taped, capped or plugged when not in use.

**1910.430(c)(2)** Breathing gas supply hose connectors shall:

**1910.430(c)(2)(i)** Be made of corrosion-resistant materials;

**1910.430(c)(2)(ii)** Have a working pressure at least equal to the working pressure of the hose to which they are attached; and

**1910.430(c)(2)(iii)** Be resistant to accidental disengagement.

**1910.430(c)(3)** Umbilicals shall:

**1910.430(c)(3)(i)** Be marked in 10-ft. increments to 100 feet beginning at the diver's end, and in 50 ft. increments thereafter;

**1910.430(c)(3)(ii)** Be made of kink-resistant materials; and

**1910.430(c)(3)(iii)** Have a working pressure greater than the pressure equivalent to the maximum depth of the dive (relative to the supply source) plus 100 psi.

**1910.430(d)** *Buoyancy control.* **(1)** Helmets or masks connected directly to the dry suit or other buoyancy-changing equipment shall be equipped with an exhaust valve.

**1910.430(d)(2)** A dry suit or other buoyancy-changing equipment not directly connected to the helmet or mask shall be equipped with an exhaust valve.

**1910.430(d)(3)** When used for SCUBA diving, a buoyancy compensator shall have an inflation source separate from the breathing gas supply.

**1910.430(d)(4)** An inflatable flotation device capable of maintaining the diver at the surface in a face-up position, having a manually activated inflation source independent of the breathing supply, an oral inflation device, and an exhaust valve shall be used for SCUBA diving.

**1910.430(e)** *Compressed gas cylinders.* Compressed gas cylinders shall:

**1910.430(e)(1)** Be designed, constructed and maintained in accordance with the applicable provisions of 29 CFR 1910.101 and 1910.169 through 1910.171.

**1910.430(e)(2)** Be stored in a ventilated area and protected from excessive heat;

**1910.430(e)(3)** Be secured from falling; and

**1910.430(e)(4)** Have shut-off valves recessed into the cylinder or protected by a cap, except when in use or manifolded, or when used for SCUBA diving.

**1910.430(f)** *Decompression chambers.* **(1)** Each decompression chamber manufactured after the effective date of this standard, shall be built and maintained in accordance with the ASME Code or equivalent.

**1910.430(f)(2)** Each decompression chamber manufactured prior to the effective date of this standard shall be maintained in conformity with the code requirements to which it was built, or equivalent.

**1910.430(f)(3)** Each decompression chamber shall be equipped with:

**1910.430(f)(3)(i)** Means to maintain the atmosphere below a level of 25 percent oxygen by volume;

**1910.430(f)(3)(ii)** Mufflers on intake and exhaust lines, which shall be regularly inspected and maintained;

**1910.430(f)(3)(iii)** Suction guards on exhaust line openings; and

**1910.430(f)(3)(iv)** A means for extinguishing fire, and shall be maintained to minimize sources of ignition and combustible material.

**1910.430(g)** *Gauges and timekeeping devices.* **(1)** Gauges indicating diver depth which can be read at the dive location shall be used for all dives except SCUBA.

**1910.430(g)(2)** Each depth gauge shall be deadweight tested or calibrated against a master reference gauge every 6 months, and when there is a discrepancy greater than two percent (2 percent) of full scale between any two equivalent gauges.

**1910.430(g)(3)** A cylinder pressure gauge capable of being monitored by the diver during the dive shall be worn by each SCUBA diver.

**1910.430(g)(4)** A timekeeping device shall be available at each dive location.

**1910.430(h)** *Masks and helmets.* **(1)** Surface-supplied air and mixed-gas masks and helmets shall have:

**1910.430(h)(1)(i)** A non-return valve at the attachment point between helmet or mask and hose which shall close readily and positively; and

**1910.430(h)(1)(ii)** An exhaust valve.

**1910.430(h)(2)** Surface-supplied air masks and helmets shall have a minimum ventilation rate capability of 4.5 acfm at any depth at which they are operated or the capability of maintaining the diver's inspired carbon dioxide partial pressure below 0.02 ATA when the diver is producing carbon dioxide at the rate of 1.6 standard liters per minute.

**1910.430(i)** *Oxygen safety.* **(1)** Equipment used with oxygen or mixtures containing over forty percent (40%) by volume oxygen shall be designed for oxygen service.

**1910.430(i)(2)** Components (except umbilicals) exposed to oxygen or mixtures containing over forty percent (40%) by volume oxygen shall be cleaned of flammable materials before use.

**1910.430(i)(3)** Oxygen systems over 125 psig and compressed air systems over 500 psig shall have slow-opening shut-off valves.

**1910.430(j)** *Weights and harnesses.* **(1)** Except when heavy gear is worn, divers shall be equipped with a weight belt or assembly capable of quick release.

**1910.430(j)(2)** Except when heavy gear is worn or in SCUBA diving, each diver shall wear a safety harness with:

**1910.430(j)(2)(i)** A positive buckling device;

**1910.430(j)(2)(ii)** An attachment point for the umbilical to prevent strain on the mask or helmet; and

**1910.430(j)(2)(iii)** A lifting point to distribute the pull force of the line over the diver's body.

[39 FR 23502, June 27, 1974, as amended at 49 FR 18295, Apr. 30, 1984; 51 FR 33033, Sept. 18, 1986; 61 FR 5507, Feb. 13, 1996]

## Recordkeeping

### §1910.440 Recordkeeping requirements.

**1910.440(a)(1)** [Reserved]

**1910.440(a)(2)** The employer shall record the occurrence of any diving-related injury or illness which requires any dive team member to be hospitalized for 24 hours or more, specifying the circumstances of the incident and the extent of any injuries or illnesses.

**1910.440(b)** *Availability of records.* **(1)** Upon the request of the Assistant Secretary of Labor for Occupational Safety and Health, or the Director, National Institute for Occupational Safety and Health, Department of Health and Human Services of their designees, the employer shall make available for inspection and copying any record or document required by this standard.

**1910.440(b)(2)** Records and documents required by this standard shall be provided upon request to employees, designated representatives, and the Assistant Secretary in accordance with 29 CFR 1910.1020(a)-(e) and (g)-(i). Safe practices manuals (§1910.420), depth-time profiles (§1910.422), recordings of dives (§1910.423), decompression procedure assessment evaluations (§1910.423), and records of hospitalizations (§1910.440) shall be provided in the same manner as employee exposure records or analyses using exposure or medical records. Equipment inspections and testing records which pertain to employees (§1910.430) shall also be provided upon request to employees and their designated representatives.

**1910.440(b)(3)** Records and documents required by this standard shall be retained by the employer for the following period:

**1910.440(b)(3)(i)** Dive team member medical records (physician's reports) (§1910.411)—5 years;

**1910.440(b)(3)(ii)** Safe practices manual (§1910.420)—current document only;

**1910.440(b)(3)(iii)** Depth-time profile (§1910.422)—until completion of the recording of dive, or until completion of decompression procedure assessment where there has been an incident of decompression sickness;

**1910.440(b)(3)(iv)** Recording of dive (§1910.423)—1 year, except 5 years where there has been an incident of decompression sickness;

**1910.440(b)(3)(v)** Decompression procedure assessment evaluations (§1910.423)—5 years;

**1910.440(b)(3)(vi)** Equipment inspections and testing records (§1910.430)—current entry or tag, or until equipment is withdrawn from service;

**1910.440(b)(3)(vii)** Records of hospitalizations (§1910.440)—5 years.

**1910.440(b)(4)** After the expiration of the retention period of any record required to be kept for five (5) years, the employer shall forward such records to the National Institute for Occupational Safety and Health, Department of Health and Human Services. The employer shall also comply with any additional requirements set forth at 29 CFR 1910.20(h).

**1910.440(b)(5)** In the event the employer ceases to do business:

**1910.440(b)(5)(i)** The successor employer shall receive and retain all dive and employee medical records required by this standard; or

**1910.440(b)(5)(ii)** If there is no successor employer, dive and employee medical records shall be forwarded to the National Institute for Occupational Safety and Health, Department of Health and Human Services.

[42 FR 37668, July 22, 1977, as amended at 45 FR 35281, May 23, 1980; 47 FR 14706, Apr. 6, 1982; 51 FR 34562, Sept. 29, 1986; 61 FR 5507, Feb. 13, 1996; 61 FR 9227, Mar. 7, 1996]

### Appendix A to Subpart T to Part 1910—Examples of Conditions Which May Restrict or Limit Exposure to Hyperbaric Conditions

The following disorders may restrict or limit occupational exposure to hyperbaric conditions depending on severity, presence of residual effects, response to therapy, number of occurrences, diving mode, or degree and duration of isolation.

History of seizure disorder other than early febrile convulsions.

Malignancies (active) unless treated and without recurrence for 5 yrs.

Chronic inability to equalize sinus and/or middle ear pressure.

Cystic or cavitary disease of the lungs.

Impaired organ function caused by alcohol or drug use.

Conditions requiring continuous medication for control (e.g., antihistamines, steroids, barbiturates, moodaltering drugs, or insulin).

Meniere's disease.

Hemoglobinopathies.

Obstructive or restrictive lung disease.

Vestibular end organ destruction.

Pneumothorax.

Cardiac abnormalities (e.g., pathological heart block, valvular disease, intraventricular conduction defects other than isolated right bundle branch block, angina pectoris, arrhythmia, coronary artery disease).

Juxta-articular osteonecrosis.

# OSHA Standards for General Industry

### Appendix B to Subpart T to Part 1910— Guidelines for Scientific Diving

This appendix contains guidelines that will be used in conjunction with §1910.401(a)(2)(iv) to determine those scientific diving programs which are exempt from the requirements for commercial diving. The guidelines are as follows:

1. The Diving Control Board consists of a majority of active scientific divers and has autonomous and absolute authority over the scientific diving program's operations.

2. The purpose of the project using scientific diving is the advancement of science; therefore, information and data resulting from the project are non-proprietary.

3. The tasks of a scientific diver are those of an observer and data gatherer. Construction and trouble-shooting tasks traditionally associated with commercial diving are not included within scientific diving.

4. Scientific divers, based on the nature of their activities, must use scientific expertise in studying the underwater environment and, therefore, are scientists or scientists in training.

[50 FR 1050, Jan. 9, 1985]

### Appendix C to Subpart T of Part 1910—Alternative Conditions Under § 1910.401(a)(3) for Recreational Diving Instructors and Diving Guides (Mandatory)

Paragraph (a)(3) of § 1910.401 specifies that an employer of recreational diving instructors and diving guides (hereafter, "divers" or "employees") who complies with all of the conditions of this appendix need not provide a decompression chamber for these divers as required under §§ 1910.423(b)(2) or (c)(3) or 1910.426(b)(1).

#### 1. Equipment Requirements for Rebreathers

(a) The employer must ensure that each employee operates the rebreather (i.e., semi-closed-circuit and closed-circuit self-contained underwater breathing apparatuses (hereafter, "SCUBAs")) according to the rebreather manufacturer's instructions.

(b) The employer must ensure that each rebreather has a counterlung that supplies a sufficient volume of breathing gas to their divers to sustain the divers' respiration rates, and contains a baffle system and/or other moisture separating system that keeps moisture from entering the scrubber.

(c) The employer must place a moisture trap in the breathing loop of the rebreather, and ensure that:

(i) The rebreather manufacturer approves both the moisture trap and its location in the breathing loop; and

(ii) Each employee uses the moisture trap according to the rebreather manufacturer's instructions.

(d) The employer must ensure that each rebreather has a continuously functioning moisture sensor, and that:

(i) The moisture sensor connects to a visual (e.g., digital, graphic, analog) or auditory (e.g., voice, pure tone) alarm that is readily detectable by the diver under the diving conditions in which the diver operates, and warns the diver of moisture in the breathing loop in sufficient time to terminate the dive and return safely to the surface; and

(ii) Each diver uses the moisture sensor according to the rebreather manufacturer's instructions.

(e) The employer must ensure that each rebreather contains a continuously functioning $CO_2$ sensor in the breathing loop, and that:

(i) The rebreather manufacturer approves the location of the $CO_2$ sensor in the breathing loop;

(ii) The $CO_2$ sensor is integrated with an alarm that operates in a visual (e.g., digital, graphic, analog) or auditory (e.g., voice, pure tone) mode that is readily detectable by each diver under the diving conditions in which the diver operates; and

(iii) The $CO_2$ alarm remains continuously activated when the inhaled $CO_2$ level reaches and exceeds 0.005 atmospheres absolute (ATA).

(f) Before each day's diving operations, and more often when necessary, the employer must calibrate the $CO_2$ sensor according to the sensor manufacturer's instructions, and ensure that:

(i) The equipment and procedures used to perform this calibration are accurate to within 10% of a $CO_2$ concentration of 0.005 ATA or less;

(ii) The equipment and procedures maintain this accuracy as required by the sensor manufacturer's instructions; and

(iii) The calibration of the $CO_2$ sensor is accurate to within 10% of a $CO_2$ concentration of 0.005 ATA or less.

(g) The employer must replace the $CO_2$ sensor when it fails to meet the accuracy requirements specified in paragraph 1(f)(iii) of this appendix, and ensure that the replacement $CO_2$ sensor meets the accuracy requirements specified in paragraph 1(f)(iii) of this appendix before placing the rebreather in operation.

(h) As an alternative to using a continuously functioning $CO_2$ sensor, the employer may use a schedule for replacing $CO_2$-sorbent material provided by the rebreather manufacturer. The employer may use such a schedule only when the rebreather manufacturer has developed it according to the canister- testing protocol specified below in Condition 11, and must use the canister within the temperature range for which the manufacturer conducted its scrubber canister tests following that protocol. Variations above or below the range are acceptable only after the manufacturer adds that lower or higher temperature to the protocol.

(i) When using $CO_2$-sorbent replacement schedules, the employer must ensure that each rebreather uses a manufactured (i.e., commercially pre-packed), disposable scrubber cartridge containing a $CO_2$-sorbent material that:

(i) Is approved by the rebreather manufacturer;

(ii) Removes $CO_2$ from the diver's exhaled gas; and

(iii) Maintains the $CO_2$ level in the breathable gas (i.e., the gas that a diver inhales directly from the regulator) below a partial pressure of 0.01 ATA.

(j) As an alternative to manufactured, disposable scrubber cartridges, the employer may fill $CO_2$ scrubber cartridges manually with $CO_2$-sorbent material when:

(i) The rebreather manufacturer permits manual filling of scrubber cartridges;

(ii) The employer fills the scrubber cartridges according to the rebreather manufacturer's instructions;

(iii) The employer replaces the $CO_2$-sorbent material using a replacement schedule developed under paragraph 1(h) of this appendix; and

(iv) The employer demonstrates that manual filling meets the requirements specified in paragraph 1(i) of this appendix.

(k) The employer must ensure that each rebreather has an information module that provides:

(i) A visual (e.g., digital, graphic, analog) or auditory (e.g., voice, pure tone) display that effectively warns the diver of solenoid failure (when the rebreather uses solenoids) and other electrical weaknesses or failures (e.g., low battery voltage);

(ii) For a semi-closed circuit rebreather, a visual display for the partial pressure of $CO_2$, or deviations above and below a preset $CO_2$ partial pressure of 0.005 ATA; and

(iii) For a closed-circuit rebreather, a visual display for: partial pressures of $O_2$ and $CO_2$, or deviations above and below a preset $CO_2$ partial pressure of 0.005 ATA and a preset $O_2$ partial pressure of 1.40 ATA or lower; gas temperature in the breathing loop; and water temperature.

(l) Before each day's diving operations, and more often when necessary, the employer must ensure that the electrical power supply and electrical and electronic circuits in each rebreather are operating as required by the rebreather manufacturer's instructions.

#### 2. Special Requirements for Closed-Circuit Rebreathers

(a) The employer must ensure that each closed-circuit rebreather uses supply-pressure sensors for the $O_2$ and diluent (i.e., air or nitrogen) gases and continuously functioning sensors for detecting temperature in the inhalation side of the gas-loop and the ambient water.

(b) The employer must ensure that:

(i) At least two $O_2$ sensors are located in the inhalation side of the breathing loop; and

(ii) The O2 sensors are: functioning continuously; temperature compensated; and approved by the rebreather manufacturer.

(c) Before each day's diving operations, and more often when necessary, the employer must calibrate O2 sensors as required by the sensor manufacturer's instructions. In doing so, the employer must:

(i) Ensure that the equipment and procedures used to perform the calibration are accurate to within 1% of the O2 fraction by volume;

(ii) Maintain this accuracy as required by the manufacturer of the calibration equipment;

(iii) Ensure that the sensors are accurate to within 1% of the O2 fraction by volume;

(iv) Replace O2 sensors when they fail to meet the accuracy requirements specified in paragraph 2(c)(iii) of this appendix; and

(v) Ensure that the replacement O2 sensors meet the accuracy requirements specified in paragraph 2(c)(iii) of this appendix before placing a rebreather in operation.

(d) The employer must ensure that each closed-circuit rebreather has:

(i) A gas-controller package with electrically operated solenoid O2-supply valves;

(ii) A pressure-activated regulator with a second-stage diluent-gas addition valve;

(iii) A manually operated gas-supply bypass valve to add O2 or diluent gas to the breathing loop; and

(iv) Separate O2 and diluent-gas cylinders to supply the breathing-gas mixture.

### 3. O2 Concentration in the Breathing Gas

The employer must ensure that the fraction of O2 in the nitrox breathing-gas mixture:

(a) Is greater than the fraction of O2 in compressed air (*i.e.*, exceeds 22% by volume);

(b) For open-circuit SCUBA, never exceeds a maximum fraction of breathable O2 of 40% by volume or a maximum O2 partial pressure of 1.40 ATA, whichever exposes divers to less O2; and

(c) For a rebreather, never exceeds a maximum O2 partial pressure of 1.40 ATA.

### 4. Regulating O2 Exposures and Diving Depth

(a) Regarding O2 exposure, the employer must:

(i) Ensure that the exposure of each diver to partial pressures of O2 between 0.60 and 1.40 ATA does not exceed the 24-hour single-exposure time limits specified either by the 2001 National Oceanic and Atmospheric Administration Diving Manual (the "2001 NOAA Diving Manual"), or by the report entitled "Enriched Air Operations and Resource Guide" published in 1995 by the Professional Association of Diving Instructors (known commonly as the "1995 DSAT Oxygen Exposure Table"); and

(ii) Determine a diver's O2-exposure duration using the diver's maximum O2 exposure (partial pressure of O2) during the dive and the total dive time (*i.e.*, from the time the diver leaves the surface until the diver returns to the surface).

(b) Regardless of the diving equipment used, the employer must ensure that no diver exceeds a depth of 130 feet of sea water ("fsw") or a maximum O2 partial pressure of 1.40 ATA, whichever exposes the diver to less O2.

### 5. Use of No-Decompression Limits

(a) For diving conducted while using nitrox breathing-gas mixtures, the employer must ensure that each diver remains within the no-decompression limits specified for single and repetitive air diving and published in the 2001 NOAA Diving Manual or the report entitled "Development and Validation of No-Stop Decompression Procedures for Recreational Diving: The DSAT Recreational Dive Planner," published in 1994 by Hamilton Research Ltd. (known commonly as the "1994 DSAT No-Decompression Tables").

(b) An employer may permit a diver to use a dive-decompression computer designed to regulate decompression when the dive-decompression computer uses the no-decompression limits specified in paragraph 5(a) of this appendix, and provides output that reliably represents those limits.

### 6. Mixing and Analyzing the Breathing Gas

(a) The employer must ensure that:

(i) Properly trained personnel mix nitrox-breathing gases, and that nitrogen is the only inert gas used in the breathing-gas mixture; and

(ii) When mixing nitrox-breathing gases, they mix the appropriate breathing gas before delivering the mixture to the breathing-gas cylinders, using the continuous-flow or partial-pressure mixing techniques specified in the 2001 NOAA Diving Manual, or using a filter-membrane system.

(b) Before the start of each day's diving operations, the employer must determine the O2 fraction of the breathing-gas mixture using an O2 analyzer. In doing so, the employer must:

(i) Ensure that the O2 analyzer is accurate to within 1% of the O2 fraction by volume.

(ii) Maintain this accuracy as required by the manufacturer of the analyzer.

(c) When the breathing gas is a commercially supplied nitrox breathing-gas mixture, the employer must ensure that the O2 meets the medical USP specifications (Type I, Quality Verification Level A) or aviator's breathing-oxygen specifications (Type I, Quality Verification Level E) of CGA G-4.3-2000 ("Commodity Specification for Oxygen"). In addition, the commercial supplier must:

(i) Determine the O2 fraction in the breathing-gas mixture using an analytic method that is accurate to within 1% of the O2 fraction by volume;

(ii) Make this determination when the mixture is in the charged tank and after disconnecting the charged tank from the charging apparatus;

(iii) Include documentation of the O2-analysis procedures and the O2 fraction when delivering the charged tanks to the employer.

(d) Before producing nitrox breathing-gas mixtures using a compressor in which the gas pressure in any system component exceeds 125 pounds per square inch (psi), the:

(i) Compressor manufacturer must provide the employer with documentation that the compressor is suitable for mixing high-pressure air with the highest O2 fraction used in the nitrox breathing-gas mixture when operated according to the manufacturer's operating and maintenance specifications;

(ii) Employer must comply with paragraph 6(e) of this appendix, unless the compressor is rated for O2 service and is oil-less or oil-free; and

(iii) Employer must ensure that the compressor meets the requirements specified in paragraphs (i)(1) and (i)(2) of § 1910.430 whenever the highest O2 fraction used in the mixing process exceeds 40%.

(e) Before producing nitrox breathing-gas mixtures using an oil-lubricated compressor to mix high-pressure air with O2, and regardless of the gas pressure in any system component, the:

(i) Employer must use only uncontaminated air (*i.e.*, air containing no hydrocarbon particulates) for the nitrox breathing-gas mixture;

(ii) Compressor manufacturer must provide the employer with documentation that the compressor is suitable for mixing the high-pressure air with the highest O2 fraction used in the nitrox breathing-gas mixture when operated according to the manufacturer's operating and maintenance specifications;

(iii) Employer must filter the high-pressure air to produce O2-compatible air;

(iv) The filter-system manufacturer must provide the employer with documentation that the filter system used for this purpose is suitable for producing O2-compatible air when operated according to the manufacturer's operating and maintenance specifications; and

(v) Employer must continuously monitor the air downstream from the filter for hydrocarbon contamination.

(f) The employer must ensure that diving equipment using nitrox breathing-gas mixtures or pure O2 under high pressure (*i.e.*, exceeding 125 psi) conforms to the O2-service requirements specified in paragraphs (i)(1) and (i)(2) of § 1910.430.

### 7. Emergency Egress

(a) Regardless of the type of diving equipment used by a diver (*i.e.*, open-circuit SCUBA or rebreathers), the employer must ensure that the equipment contains (or incorporates) an open-circuit emergency-egress system (a "bail-out" system) in which the second stage of the regulator connects to a separate supply of emergency breathing gas, and the emergency breathing gas consists of air or the same nitrox breathing-gas mixture used during the dive.

(b) As an alternative to the "bail-out" system specified in paragraph 7(a) of this appendix, the employer may use:

(i) For open-circuit SCUBA, an emergency-egress system as specified in § 1910.424(c)(4); or

(ii) For a semi-closed-circuit and closed-circuit rebreather, a system configured so that the second stage of the regulator connects to a reserve supply of emergency breathing gas.

(c) The employer must obtain from the rebreather manufacturer sufficient information to ensure that the bail-out system performs reliably and has sufficient capacity to enable the diver to terminate the dive and return safely to the surface.

### 8. Treating Diving-Related Medical Emergencies

(a) Before each day's diving operations, the employer must:

(i) Verify that a hospital, qualified health-care professionals, and the nearest Coast Guard Coordination Center (or an equivalent rescue service operated by a state, county, or municipal agency) are available to treat diving-related medical emergencies;

(ii) Ensure that each dive site has a means to alert these treatment resources in a timely manner when a diving-related medical emergency occurs; and

(iii) Ensure that transportation to a suitable decompression chamber is readily available when no decompression chamber is at the dive site, and that this transportation can deliver the injured diver to the decompression chamber within four (4) hours travel time from the dive site.

(b) The employer must ensure that portable $O_2$ equipment is available at the dive site to treat injured divers. In doing so, the employer must ensure that:

(i) The equipment delivers medical-grade $O_2$ that meets the requirements for medical USP oxygen (Type I, Quality Verification Level A) of CGA G-4.3-2000 ("Commodity Specification for Oxygen");

(ii) The equipment delivers this $O_2$ to a transparent mask that covers the injured diver's nose and mouth; and

(iii) Sufficient $O_2$ is available for administration to the injured diver from the time the employer recognizes the symptoms of a diving-related medical emergency until the injured diver reaches a decompression chamber for treatment.

(c) Before each day's diving operations, the employer must:

(i) Ensure that at least two attendants, either employees or non-employees, qualified in first-aid and administering $O_2$ treatment, are available at the dive site to treat diving-related medical emergencies; and

(ii) Verify their qualifications for this task.

### 9. Diving Logs and No-Decompression Tables

(a) Before starting each day's diving operations, the employer must:

(i) Designate an employee or a non-employee to make entries in a diving log; and

(ii) Verify that this designee understands the diving and medical terminology, and proper procedures, for making correct entries in the diving log.

(b) The employer must:

(i) Ensure that the diving log conforms to the requirements specified by paragraph (d) ("Record of dive") of § 1910.423; and

(ii) Maintain a record of the dive according to § 1910.440 ("Recordkeeping requirements").

(c) The employer must ensure that a hard-copy of the no-decompression tables used for the dives (as specified in paragraph 6(a) of this appendix) is readily available at the dive site, whether or not the divers use dive-decompression computers.

### 10. Diver Training

The employer must ensure that each diver receives training that enables the diver to perform work safely and effectively while using open-circuit SCUBAs or rebreathers supplied with nitrox breathing-gas mixtures. Accordingly, each diver must be able to demonstrate the ability to perform critical tasks safely and effectively, including, but not limited to: recognizing the effects of breathing excessive $CO_2$ and $O_2$; taking appropriate action after detecting excessive levels of $CO_2$ and $O_2$; and properly evaluating, operating, and maintaining their diving equipment under the diving conditions they encounter.

### 11. Testing Protocol for Determining the CO2 Limits of Rebreather Canisters

(a) The employer must ensure that the rebreather manufacturer has used the following procedures for determining that the $CO_2$-sorbent material meets the specifications of the sorbent material's manufacturer:

(i) The North Atlantic Treating Organization $CO_2$ absorbent-activity test;

(ii) The RoTap shaker and nested-sieves test;

(iii) The Navy Experimental Diving Unit ("NEDU")-derived Schlegel test; and

(iv) The NEDU MeshFit software.

(b) The employer must ensure that the rebreather manufacturer has applied the following canister-testing materials, methods, procedures, and statistical analyses:

(i) Use of a nitrox breathing-gas mixture that has an $O_2$ fraction maintained at 0.28 (equivalent to 1.4 ATA of $O_2$ at 130 fsw, the maximum $O_2$ concentration permitted at this depth);

(ii) While operating the rebreather at a maximum depth of 130 fsw, use of a breathing machine to continuously ventilate the rebreather with breathing gas that is at 100% humidity and warmed to a temperature of 98.6 degrees F (37 degrees C) in the heating-humidification chamber;

(iii) Measurement of the $O_2$ concentration of the inhalation breathing gas delivered to the mouthpiece;

(iv) Testing of the canisters using the three ventilation rates listed in Table I below (with the required breathing-machine tidal volumes and frequencies, and $CO_2$-injection rates, provided for each ventilation rate):

#### Table I. -- Canister Testing Parameters

| Ventilation rates (Lpm, ATPS[1]) | Breathing machine tidal volumes (L) | Breathing machine frequencies (breaths per min.) | $CO_2$ injection rates (Lpm, STPD \2\) |
|---|---|---|---|
| 22.5 | 1.5 | 15 | 0.90 |
| 40.0 | 2.0 | 20 | 1.35 |
| 62.5 | 2.5 | 25 | 2.25 |

1 ATPS means ambient temperature and pressure, saturated with water.

2 STPD means standard temperature and pressure, dry; the standard temperature is 32 degrees F (0 degrees C).

(v) When using a work rate (*i.e.*, breathing-machine tidal volume and frequency) other than the work rates listed in the table above, addition of the appropriate combinations of ventilation rates and $CO_2$-injection rates;

(vi) Performance of the $CO_2$ injection at a constant (steady) and continuous rate during each testing trial;

(vii) Determination of canister duration using a minimum of four (4) water temperatures, including 40, 50, 70, and 90 degrees F (4.4, 10.0, 21.1, and 32.2 degrees C, respectively);

(viii) Monitoring of the breathing-gas temperature at the rebreather mouthpiece (at the "chrome T" connector), and ensuring that this temperature conforms to the temperature of a diver's exhaled breath at the water temperature and ventilation rate used during the testing trial; [1]

(ix) Implementation of at least eight (8) testing trials for each combination of temperature and ventilation-$CO_2$-injection rates (for example, eight testing trials at 40 degrees F using a ventilation rate of 22.5 Lpm at a $CO_2$-injection rate of 0.90 Lpm);

(x) Allowing the water temperature to vary no more than 2.0 degrees F ( 1.0 degree C) *between* each of the eight testing trials, and no more than 1.0 degree F ( 0.5 degree C) within each testing trial;

(xi) Use of the average temperature for each set of eight testing trials in the statistical analysis of the testing-trial results, with the testing-trial results being the time taken for the inhaled breathing gas to reach 0.005 ATA of $CO_2$ (*i.e.*, the canister-duration results);

(xii) Analysis of the canister-duration results using the repeated-measures statistics described in NEDU Report 2-99;

(xiii) Specification of the replacement schedule for the $CO_2$-sorbent materials in terms of the lower prediction line (or limit) of the 95% confidence interval; and

(xiv) Derivation of replacement schedules only by interpolating among, but not by extrapolating beyond, the depth, water temperatures, and exercise levels used during canister testing.

[FR Doc. 04-3289 Filed 2-13-04; 8:45 am]

[1] ATA is the partial pressure of a constituent gas in the total pressure of a breathing gas. When the percentage of the constituent gas in the breathing gas remains constant throughout a dive, its partial pressure, or ATA, increases in direct proportion to increases in diving depth.

[2] "Channeling" describes open spaces (or channels) that form in the sorbent material, and that permit exhaled breathing gas to pass through that part of the material to the inhalation side of the breathing apparatus with little or no absorption of the $CO_2$ contained in the exhaled breathing gas. Channeling can be prevented by compressing the sorbent material uniformly in the canister (*e.g.*, by shaking the canister vigorously).

[3] Although low $O_2$ levels are rare under nitrox breathing conditions, the sensors also would detect levels of $O_2$ less than 22% by volume (see Condition 3 of Appendix C below).

[4] By definition, a nitrox breathing-gas mixture must contain a higher percentage of oxygen than is found in normal air (*i.e.*, 21%), usually 32% and 36% oxygen (Ex. 3-12).

[5] Note that the final rule reverses the designations of proposed Conditions 5 and 6 to Conditions 6 and 5, respectively. Redesignating proposed Condition 6 as Condition 5 groups it with the other conditions (*i.e.*, 3 and 4) that address procedures for protecting divers from $O_2$ toxicity and DCS.

[6] The aviator's oxygen specification is Type I, Quality Verification Level E (Aviator's Breathing Oxygen), while the specification for medical-grade oxygen is Type I, Quality Verification Level A (Medical USP); see Ex. 3-15, p. 2.

[7] In addition, employers already are required to comply with other OSHA standards that provide for accurate mixing and decontamination (especially hydrocarbon removal) of breathing gases, and they must assure that employees are properly protected during these activities. These standards include the appropriate provisions of §§ 1910.101 ("Compressed gases (general requirements)"), 1910.169 ("Air receivers"), and 1910.134 ("Respiratory protection").

[8] Paragraph (c)(4) of § 1910.424 is an emergency-escape provision in OSHA's existing CDO standards that applies to divers using air-supplied open-circuit SCUBA.

[9] Dr. Raymond made these comments in reference to an OSHA news release (dated January 10, 2003) that stated erroneously that nitrox breathing-gas mixtures prevented AGE. Dr. Raymond did indicate correctly that nitrox breathing-gas mixtures do not, in fact, prevent AGE. OSHA subsequently corrected this news release (with the same issue date of January 10, 2003).

[1] NEDU can provide the manufacturer with information on the temperature of a diver's exhaled breath at various water temperatures and ventilation rates, as well as techniques and procedures used to maintain these temperatures during the testing trials.

# Subpart U—V [Reserved]

§1910.442—1910.899 [Reserved]

# Subpart W—Program Standards

**§1910.900 Ergonomics program standard. [Repealed and Reserved.]**

[Subpart W removed and reserved, at 66 FR 20403, April 23, 2001, effective April 23, 2001.]

# Subparts X—Y [Reserved]

§§1910.901—1910.999 [Reserved]

# Subpart Z—Toxic and Hazardous Substances

**Authority:** Sections 4, 6, and 8 of the Occupational Safety and Health Act of 1970 (29 U.S.C. 653, 655, 657); Secretary of Labor's Order No. 12-71 (36 FR 8754), 8-76 (41 FR 25059), 9-83 (48 FR 35736), 1-90 (55 FR 9033), 6-96 (62 FR 111) or 3-2000 (65 FR 50017), as applicable; and 29 CFR part 1911.

All of subpart Z issued under Sec. 6(b) of the Occupational Safety and Health Act, except those substances that have exposure limits listed in Tables Z-1, Z-2, and Z-3 of 29 CFR 1910.1000. The latter were issued under Sec. 6(a) (29 U.S.C. 655(a)).

Section 1910.1000, Tables Z-1, Z-2, and Z-3 also issued under 5 U.S.C. 533, Section 1910.1000 Tables Z-1, Z-2, and Z-3 not issued under 29 CFR part 1911 except for the arsenic (organic compounds), benzene, and cotton dust listings.

Section 1910.1001 also issued under section 107 of the Contract Work Hours and Safety Standards Act (40 U.S.C. 333) and 5 U.S.C. 553.

Section 1910.1002 not issued under 29 U.S.C. 655 or 29 CFR part 1911; also issued under 5 U.S.C. 553.

Sections 1910.1018, 1910.1029 and 1910.1200 are also issued under 29 U.S.C. 653.

Section 1910.1030 is also issued under Pub. L. 106-430, 114 Stat. 1901.

**Source:** 39 FR 23502, June 27, 1974, unless otherwise noted. Redesignated at 40 FR 23072, May 28, 1975. Amended at 66 FR 5317, Jan. 18, 2001.

## §1910.1000 Air contaminants.

An employee's exposure to any substance listed in Tables Z-1, Z-2, or Z-3 of this section shall be limited in accordance with the requirements of the following paragraphs of this section.

**1910.1000(a)** *Table Z-1*—(1) *Substances with limits preceded by "C"—Ceiling Values.* An employee's exposure to any substance in Table Z-1, the exposure limit of which is preceded by a "C", shall at no time exceed the exposure limit given for that substance. If instantaneous monitoring is not feasible, then the ceiling shall be assessed as a 15-minute time weighted average exposure which shall not be exceeded at any time during the working day.

**1910.1000(a)(2)** *Other substances—8-hour Time Weighted Averages.* An employee's exposure to any substance in Table Z-1, the exposure limit of which is not preceded by a "C", shall not exceed the 8-hour Time Weighted Average given for that substance in any 8-hour work shift of a 40-hour work week.

**1910.1000(b)** *Table Z-2.* An employee's exposure to any substance listed in Table Z-2 shall not exceed the exposure limits specified as follows:

**1910.1000(b)(1)** *8-hour time weighted averages.* An employee's exposure to any substance listed in Table Z-2, in any 8-hour work shift of a 40-hour work week, shall not exceed the 8-hour time weighted average limit given for that substance in Table Z-2.

**1910.1000(b)(2)** *Acceptable ceiling concentrations.* An employee's exposure to a substance listed in Table Z-2 shall not exceed at any time during an 8-hour shift the acceptable ceiling concentration limit given for the substance in the table, except for a time period, and up to a concentration not exceeding the maximum duration and concentration allowed in the column under "acceptable maximum peak above the acceptable ceiling concentration for an 8-hour shift."

**1910.1000(b)(3)** *Example.* During an 8-hour work shift, an employee may be exposed to a concentration of Substance A (with a 10 ppm TWA, 25 ppm ceiling and 50 ppm peak) above 25 ppm (but never above 50 ppm) only for a maximum period of 10 minutes. Such exposure must be compensated by exposures to concentrations less than 10 ppm so that the cumulative exposure for the entire 8-hour work shift does not exceed a weighted average of 10 ppm.

**1910.1000(c)** *Table Z-3.* An employee's exposure to any substance listed in Table Z-3, in any 8-hour work shift of a 40-hour work week, shall not exceed the 8-hour time weighted average limit given for that substance in the table.

**1910.1000(d)** *Computation formulae.* The computation formula which shall apply to employee exposure to more than one substance for which 8-hour time weighted averages are listed in subpart Z of 29 CFR part 1910 in order to determine whether an employee is exposed over the regulatory limit is as follows:

**1910.1000(d)(1)(i)** The cumulative exposure for an 8-hour work shift shall be computed as follows:

$$E = (C_a T_a + C_b T_b + \ldots C_n T_n) \div 8$$

Where:

E is the equivalent exposure for the working shift.

C is the concentration during any period of time T where the concentration remains constant.

T is the duration in hours of the exposure at the concentration C.

The value of E shall not exceed the 8-hour time weighted average specified in subpart Z of 29 CFR part 1910 for the substance involved.

**1910.1000(d)(1)(ii)** To illustrate the formula prescribed in paragraph (d)(1)(i) of this section, assume that Substance A has an 8-hour time weighted average limit of 100 ppm noted in Table Z-1. Assume that an employee is subject to the following exposure:

Two hours exposure at 150 ppm

Two hours exposure at 75 ppm

Four hours exposure at 50 ppm

Substituting this information in the formula, we have

$(2 \times 150 + 2 \times 75 + 4 \times 50) \div 8 = 81.25$ ppm

Since 81.25 ppm is less than 100 ppm, the 8-hour time weighted average limit, the exposure is acceptable.

**1910.1000(d)(2)(i)** In case of a mixture of air contaminants an employer shall compute the equivalent exposure as follows:

$$E_m = (C_1 \div L_1 + C_2 \div L_2) + \ldots (C_n \div L_n)$$

Where:

$E_m$ is the equivalent exposure for the mixture.

C is the concentration of a particular contaminant.

L is the exposure limit for that substance specified in subpart Z of 29 CFR part 1910.

The value of $E_m$ shall not exceed unity (1).

**1910.1000(d)(2)(ii)** To illustrate the formula prescribed in paragraph (d)(2)(i) of this section, consider the following exposures:

| Substance | Actual concentration of 8-hour exposure (ppm) | 8-hour TWA PEL (ppm) |
|---|---|---|
| B | 500 | 1,000 |
| C | 45 | 200 |
| D | 40 | 200 |

Substituting in the formula, we have:

$E_m = 500 \div 1,000 + 45 \div 200 + 40 \div 200$

$E_m = 0.500 + 0.225 + 0.200$

$E_m = 0.925$

Since $E_m$ is less than unity (1), the exposure combination is within acceptable limits.

**1910.1000(e)** To achieve compliance with paragraphs (a) through (d) of this section, administrative or engineering controls must first be determined and implemented whenever feasible. When such controls are not feasible to achieve full compliance, protective equipment or any other protective measures shall be used to keep the exposure of employees to air contaminants within the limits prescribed in this section. Any equipment and/or technical measures used for this purpose must be approved for each particular use by a competent industrial hygienist or other technically qualified person. Whenever respirators are used, their use shall comply with 1910.134.

## TABLE Z-1—LIMITS FOR AIR CONTAMINANTS

| Substance | CAS No. (c) | ppm (a)[1] | mg/m³ (b)[1] | Skin designation |
|---|---|---|---|---|
| Acetaldehyde | 75-07-0 | 200 | 360 | |
| Acetic acid | 64-19-7 | 10 | 25 | |
| Acetic anhydride | 108-24-7 | 5 | 20 | |
| Acetone | 67-64-1 | 1000 | 2400 | |
| Acetonitrile | 75-05-8 | 40 | 70 | |
| 2-Acetylaminofluorine; see 1910.1014. | 53-96-3 | | | |
| Acetylene dichloride; see 1,2-Dichloroethylene. | | | | |
| Acetylene tetrabromide | 79-27-6 | 1 | 14 | |
| Acrolein | 107-02-8 | 0.1 | 0.25 | |
| Acrylamide | 79-06-1 | ........ | 0.3 | X |
| Acrylonitrile; see 1910.1045. | 107-13-1 | | | |
| Aldrin | 309-00-2 | ........ | 0.25 | X |
| Allyl alcohol | 107-18-6 | 2 | 5 | X |
| Allyl chloride | 107-05-1 | 1 | 3 | |
| Allyl glycidyl ether (AGE) | 106-92-3 | (C)10 | (C)45 | |
| Allyl propyl disulfide | 2179-59-1 | 2 | 12 | |
| Alpha-Alumina | 1344-28-1 | | | |
| Total dust | | ........ | 15 | |
| Respirable fraction | | ........ | 5 | |
| Aluminum, metal (as Al) | 7429-90-5 | | | |
| Total dust | | ........ | 15 | |
| Respirable fraction | | ........ | 5 | |
| 4-Aminodiphenyl; see 1910.1011. | 92-67-1 | | | |
| 2-Aminoethanol; see Ethanolamine. | | | | |
| 2-Aminopyridine | 504-29-0 | 0.5 | 2 | |
| Ammonia | 7664-41-7 | 50 | 35 | |
| Ammonium sulfamate | 7773-06-0 | | | |
| Total dust | | ........ | 15 | |
| Respirable fraction | | ........ | 5 | |
| n-Amyl acetate | 628-63-7 | 100 | 525 | |
| sec-Amyl acetate | 626-38-0 | 125 | 650 | |
| Aniline and homologs | 62-53-3 | 5 | 19 | X |
| Anisidine (o-, p-isomers) | 29191-52-4 | ........ | 0.5 | X |
| Antimony and compounds (as Sb) | 7440-36-0 | ........ | 0.5 | |
| ANTU (alpha Naphthylthiourea) | 86-88-4 | ........ | 0.3 | |
| Arsenic, inorganic compounds (as As); see 1910.1018. | 7440-38-2 | | | |
| Arsenic, organic compounds (as As) | 7440-38-2 | ........ | 0.5 | |
| Arsine | 7784-42-1 | 0.05 | 0.2 | |
| Asbestos; see 1910.1001. | (⁴) | | | |
| Azinphos-methyl | 86-50-0 | ........ | 0.2 | X |
| Barium, soluble compounds (as Ba) | 7440-39-3 | ........ | 0.5 | |
| Barium sulfate | 7727-43-7 | | | |
| Total dust | | ........ | 15 | |
| Respirable fraction | | ........ | 5 | |
| Benomyl | 17804-35-2 | | | |
| Total dust | | ........ | 15 | |
| Respirable fraction | | ........ | 5 | |
| Benzene; see 1910.1028 | 71-43-2 | | | |
| See Table Z-2 for the limits applicable in the operations or sectors excluded in 1910.1028[d] | | | | |
| Benzidine; see 1910.1010 | 92-87-5 | | | |
| p-Benzoquinone; see Quinone. | | | | |
| Benzo(a)pyrene; see Coal tar pitch volatiles. | | | | |
| Benzoyl peroxide | 94-36-0 | ........ | 5 | |
| Benzyl chloride | 100-44-7 | 1 | 5 | |
| Beryllium and beryllium compounds (as Be) | 7440-41-7 | | (²) | |
| Biphenyl; see Diphenyl. | | | | |
| Bismuth telluride, Undoped | 1304-82-1 | | | |
| Total dust | | ........ | 15 | |
| Respirable fraction | | ........ | 5 | |
| Boron oxide | 1303-86-2 | | | |
| Total dust | | ........ | 15 | |
| Boron trifluoride | 7637-07-2 | (C)1 | (C)3 | |
| Bromine | 7726-95-6 | 0.1 | 0.7 | |
| Bromoform | 75-25-2 | 0.5 | 5 | X |
| Butadiene (1,3-Butadiene);See 29 CFR 1910.1051; 29 CFR 1910.19(l). | 106-99-0 | 1 ppm/5 ppm STEL | | |
| Butanethiol; see Butyl mercaptan. | | | | |
| 2-Butanone (Methyl ethyl ketone) | 78-93-3 | 200 | 590 | |
| 2-Butoxyethanol | 111-76-2 | 50 | 240 | X |
| n-Butyl-acetate | 123-86-4 | 150 | 710 | |
| sec-Butyl acetate | 105-46-4 | 200 | 950 | |
| tert-Butyl acetate | 540-88-5 | 200 | 950 | |
| n-Butyl alcohol | 71-36-3 | 100 | 300 | |
| sec-Butyl alcohol | 78-92-2 | 150 | 450 | |
| tert-Butyl alcohol | 75-65-0 | 100 | 300 | |
| Butylamine | 109-73-9 | (C)5 | (C)15 | X |
| tert-Butyl chromate (as CrO₃); see 1910.1026 | 1189-85-1 | ........ | | |

# OSHA Standards for General Industry

| Substance | CAS No. (c) | ppm (a)[1] | mg/m³ (b)[1] | Skin designation |
|---|---|---|---|---|
| n-Butyl glycidyl ether (BGE) | 2426-08-6 | 50 | 270 | |
| Butyl mercaptan | 109-79-5 | 10 | 35 | |
| p-tert-Butyltoluene | 98-51-1 | 10 | 60 | |
| Cadmium (as Cd); see 1910.1027 | 7440-43-9 | | | |
| Calcium carbonate | 1317-65-3 | | | |
| Total dust | | ........ | 15 | |
| Respirable fraction | | ........ | 5 | |
| Calcium hydroxide | 1305-62-0 | | | |
| Total dust | | ........ | 15 | |
| Respirable fraction | | ........ | 5 | |
| Calcium oxide | 1305-78-8 | ........ | 5 | |
| Calcium silicate | 1344-95-2 | | | |
| Total dust | | ........ | 15 | |
| Respirable fraction | | ........ | 5 | |
| Calcium sulfate | 7778-18-9 | | | |
| Total dust | | ........ | 15 | |
| Respirable fraction | | ........ | 5 | |
| Camphor, synthetic | 76-22-2 | ........ | 2 | |
| Carbaryl (Sevin) | 63-25-2 | ........ | 5 | |
| Carbon black | 1333-86-4 | ........ | 3.5 | |
| Carbon dioxide | 124-38-9 | 5000 | 9000 | |
| Carbon disulfide | 75-15-0 | | ([2]) | |
| Carbon monoxide | 630-08-0 | 50 | 55 | |
| Carbon tetrachloride | 56-23-5 | | ([2]) | |
| Cellulose | 9004-34-6 | | | |
| Total dust | | ........ | 15 | |
| Respirable fraction | | ........ | 5 | |
| Chlordane | 57-74-9 | ........ | 0.5 | X |
| Chlorinated camphene | 8001-35-2 | ........ | 0.5 | X |
| Chlorinated diphenyl oxide | 55720-99-5 | ........ | 0.5 | |
| Chlorine | 7782-50-5 | (C)1 | (C)3 | |
| Chlorine dioxide | 10049-04-4 | 0.1 | 0.3 | |
| Chlorine trifluoride | 7790-91-2 | (C)0.1 | (C)0.4 | |
| Chloroacetaldehyde | 107-20-0 | (C)1 | (C)3 | |
| a-Chloroacetophenone (Phenacyl chloride) | 532-27-4 | 0.05 | 0.3 | |
| Chlorobenzene | 108-90-7 | 75 | 350 | |
| o-Chlorobenzylidene malononitrile | 2698-41-1 | 0.05 | 0.4 | |
| Chlorobromomethane | 74-97-5 | 200 | 1050 | |
| 2-Chloro-1,3-butadiene; see beta-Chloroprane. | | | | |
| Chlorodiphenyl (42% Chlorine) (PCB) | 53469-21-9 | ........ | 1 | X |
| Chlorodiphenyl (54% Chlorine) (PCB) | 11097-69-1 | ........ | 0.5 | X |
| 1-Chloro-2,3-epoxypropane; see Epichlorohydrin. | | | | |
| 2-Chloroethanol; see Ethylene chlorohydrin. | | | | |
| Chloroethylene; see Vinyl chloride. | | | | |
| Chloroform (Trichloromethane) | 67-66-3 | (C)50 | (C)240 | |
| bis(Chloromethyl) ether; see 1910.1008 | 542-88-1 | | | |
| Chloromethyl methyl ether; see 1910.1006 | 107-30-2 | | | |
| 1-Chloro-1-nitropropane | 600-25-9 | 20 | 100 | |
| Chloropicrin | 76-06-2 | 0.1 | 0.7 | |
| beta-Chloroprene | 126-99-8 | 25 | 90 | X |
| 2-Chloro-6-(trichloromethyl) pyridine | 1929-82-4 | | | |
| Total dust | | ........ | 15 | |
| Respirable fraction | | ........ | 5 | |
| Chromium (II) compounds. (as Cr) | 7440-47-3 | ........ | 0.5 | |
| Chromium (III) compounds. (as Cr) | 7440-47-3 | ........ | 0.5 | |
| Chromium (VI) compounds; see 1910.1026[5] | 7440-47-3 | | 0.5 | |
| Chromium metal and insol. salts (as Cr) | 7440-47-3 | ........ | 1 | |
| Chrysene; see Coal tar pitch volatiles. | | | | |
| Clopidol | 2971-90-6 | | | |
| Total dust | | ........ | 15 | |
| Respirable fraction | | ........ | 5 | |
| Coal dust (less than 5% SiO₂), respirable fraction | | | ([3]) | |
| Coal dust (greater than or equal to 5% SiO₂), respirable fraction | | | ([3]) | |
| Coal tar pitch volatiles (benzene soluble fraction), anthracene, BaP, phenanthrene, acridine, chrysene, pyrene | 65966-93-2 | ........ | 0.2 | |
| Cobalt metal, dust, and fume (as Co) | 7440-48-4 | ........ | 0.1 | |
| Coke oven emissions; see 1910.1029. | | | | |
| Copper | 7440-50-8 | | | |
| Fume (as Cu) | | ........ | 0.1 | |
| Dusts and mists (as Cu) | | ........ | 1 | |
| Cotton dustᵉ; see 1910.1043 | | ........ | 1 | |
| Crag herbicide (Sesone) | 136-78-7 | | | |
| Total dust | | ........ | 15 | |
| Respirable fraction | | ........ | 5 | |
| Cresol, all isomers | 1319-77-3 | 5 | 22 | X |
| Crotonaldehyde | 123-73-9; 4170-30-3 | 2 | 6 | |
| Cumene | 98-82-8 | 50 | 245 | X |
| Cyanides (as CN) | ([4]) | ........ | 5 | X |
| Cyclohexane | 110-82-7 | 300 | 1050 | |

§1910.1000

Toxic and Hazardous Substances

| Substance | CAS No. (c) | ppm (a)[1] | mg/m³ (b)[1] | Skin designation |
|---|---|---|---|---|
| Cyclohexanol | 108-93-0 | 50 | 200 | |
| Cyclohexanone | 108-94-1 | 50 | 200 | |
| Cyclohexene | 110-83-8 | 300 | 1015 | |
| Cyclopentadiene | 542-92-7 | 75 | 200 | |
| 2,4-D (Dichlorophenoxyacetic acid) | 94-75-7 | ........ | 10 | |
| Decaborane | 17702-41-9 | 0.05 | 0.3 | X |
| Demeton (Systox) | 8065-48-3 | ........ | 0.1 | X |
| Diacetone alcohol (4-Hydroxy-4 methyl-2-pentanone) | 123-42-2 | 50 | 240 | |
| 1,2-Diaminoethane; see Ethylenediamine. | | | | |
| Diazomethane | 334-88-3 | 0.2 | 0.4 | |
| Diborane | 19287-45-7 | 0.1 | 0.1 | |
| 1,2-Dibromo-3-chloropropane (DBCP); see 1910.1044. | 96-12-8 | | | |
| 1,2-Dibromoethane; see Ethylene dibromide. | | | | |
| Dibutyl phosphate | 107-66-4 | 1 | 5 | |
| Dibutyl phthalate | 84-74-2 | ........ | 5 | |
| o-Dichlorobenzene | 95-50-1 | (C)50 | (C)300 | |
| p-Dichlorobenzene | 106-46-7 | 75 | 450 | |
| 3,-Dichlorobenzidine; see 1910.1007. | 91-94-1 | | | |
| Dichlorodifluoromethane | 75-71-8 | 1000 | 4950 | |
| 1,3-Dichloro-5,5-dimethyl hydantoin | 118-52-5 | ........ | 0.2 | |
| Dichlorodiphenyltrichloroethane (DDT) | 50-29-3 | ........ | 1 | X |
| 1,1 Dichloroethane | 75-34-3 | 100 | 400 | |
| 1,2-Dichloroethane; see Ethylene dichloride. | | | | |
| 1,2-Dichloroethylene | 540-59-0 | 200 | 790 | |
| Dichloroethyl ether | 111-44-4 | (C)15 | (C)90 | X |
| Dichloromethane; see Methylene chloride. | | | | |
| Dichloromonofluoromethane | 75-43-4 | 1000 | 4200 | |
| 1,1-Dichloro-1-nitroethane | 594-72-9 | (C)10 | (C)60 | |
| 1,2-Dichloropropane; see Propylene dichloride. | | | | |
| Dichlorotetrafluoroethane | 76-14-2 | 1000 | 7000 | |
| Dichlorvos (DDVP) | 62-73-7 | ........ | 1 | X |
| Dicyclopentadienyl iron | 102-54-5 | | | |
| Total dust | | ........ | 15 | |
| Respirable fraction | | ........ | 5 | |
| Dieldrin | 60-57-1 | ........ | 0.25 | X |
| Diethylamine | 109-89-7 | 25 | 75 | |
| 2-Diethylaminoethanol | 100-37-8 | 10 | 50 | X |
| Diethyl ether; see Ethyl ether. | | | | |
| Difluorodibromomethane | 75-61-6 | 100 | 860 | |
| Diglycidyl ether (DGE) | 2238-07-5 | (C)0.5 | (C)2.8 | |
| Dihydroxybenzene; see Hydroquinone. | | | | |
| Diisobutyl ketone | 108-83-8 | 50 | 290 | |
| Diisopropylamine | 108-18-9 | 5 | 20 | X |
| 4-Dimethylaminoazobenzene; see 1910.1015 | 60-11-7 | | | |
| Dimethoxymethane; see Methylal. | | | | |
| Dimethyl acetamide | 127-19-5 | 10 | 35 | X |
| Dimethylamine | 124-40-3 | 10 | 18 | |
| Dimethylaminobenzene; see Xylidine. | | | | |
| Dimethylaniline (N,N-Dimethylaniline) | 121-69-7 | 5 | 25 | X |
| Dimethylbenzene; see Xylene. | | | | |
| Dimethyl-1,2-dibromo-2,2-dichloroethyl phosphate | 300-76-5 | ........ | 3 | |
| Dimethylformamide | 68-12-2 | 10 | 30 | X |
| 2,6-Dimethyl-4-heptanone; see Diisobutyl ketone. | | | | |
| 1,1-Dimethylhydrazine | 57-14-7 | 0.5 | 1 | X |
| Dimethylphthalate | 131-11-3 | ........ | 5 | |
| Dimethyl sulfate | 77-78-1 | 1 | 5 | X |
| Dinitrobenzene (all isomers) | | | 1 | X |
| (ortho) | 528-29-0 | | | |
| (meta) | 99-65-0 | | | |
| (para) | 100-25-4 | | | |
| Dinitro-o-cresol | 534-52-1 | ........ | 0.2 | X |
| Dinitrotoluene | 25321-14-6 | | 1.5 | X |
| Dioxane (Diethylene dioxide) | 123-91-1 | 100 | 360 | X |
| Diphenyl (Biphenyl) | 92-52-4 | 0.2 | 1 | |
| Diphenylmethane diisocyanate; see Methylene bisphenyl isocyanate. | | | | |
| Dipropylene glycol methyl ether | 34590-94-8 | 100 | 600 | X |
| Di-sec octyl phthalate (Di-(2-ethylhexyl) phthalate) | 117-81-7 | ........ | 5 | |
| Emery | 12415-34-8 | | | |
| Total dust | | ........ | 15 | |
| Respirable fraction | | ........ | 5 | |
| Endrin | 72-20-8 | ........ | 0.1 | X |
| Epichlorohydrin | 106-89-8 | 5 | 19 | X |
| EPN | 2104-64-5 | | 0.5 | X |
| 1,2-Epoxypropane; see Propylene oxide. | | | | |
| 2,3-Epoxy-1-propanol; see Glycidol. | | | | |
| Ethanethiol; see Ethyl mercaptan. | | | | |
| Ethanolamine | 141-43-5 | 3 | 6 | |
| 2-Ethoxyethanol (Cellosolve) | 110-80-5 | 200 | 740 | X |
| 2-Ethoxyethyl acetate (Cellosolve acetate) | 111-15-9 | 100 | 540 | X |
| Ethyl acetate | 141-78-6 | 400 | 1400 | |

# OSHA Standards for General Industry

| Substance | CAS No. (c) | ppm (a)[1] | mg/m³ (b)[1] | Skin designation |
|---|---|---|---|---|
| Ethyl acrylate | 140-88-5 | 25 | 100 | X |
| Ethyl alcohol (Ethanol) | 64-17-5 | 1000 | 1900 | |
| Ethylamine | 75-04-7 | 10 | 18 | |
| Ethyl amyl ketone (5-Methyl-3-heptanone) | 541-85-5 | 25 | 130 | |
| Ethyl benzene | 100-41-4 | 100 | 435 | |
| Ethyl bromide | 74-96-4 | 200 | 890 | |
| Ethyl butyl ketone (3-Heptanone) | 106-35-4 | 50 | 230 | |
| Ethyl chloride | 75-00-3 | 1000 | 2600 | |
| Ethyl ether | 60-29-7 | 400 | 1200 | |
| Ethyl formate | 109-94-4 | 100 | 300 | |
| Ethyl mercaptan | 75-08-1 | (C)10 | (C)25 | |
| Ethyl silicate | 78-10-4 | 100 | 850 | |
| Ethylene chlorohydrin | 107-07-3 | 5 | 16 | X |
| Ethylenediamine | 107-15-3 | 10 | 25 | |
| Ethylene dibromide | 106-93-4 | | ([2]) | |
| Ethylene dichloride (1,2-Dichloroethane) | 107-06-2 | | ([2]) | |
| Ethylene glycol dinitrate | 628-96-6 | (C)0.2 | (C)1 | X |
| Ethylene glycol methyl acetate; see Methyl cellosolve acetate. | | | | |
| Ethyleneimine; see 1910.1012 | 151-56-4 | | | |
| Ethylene oxide; see 1910.1047 | 75-21-8 | | | |
| Ethylidene chloride; see 1,1-Dichloroethane. | | | | |
| N-Ethylmorpholine | 100-74-3 | 20 | 94 | X |
| Ferbam | 14484-64-1 | | | |
|    Total dust | | ........ | 15 | |
| Ferrovanadium dust | 12604-58-9 | ........ | 1 | |
| Fluorides (as F) | ([4]) | ........ | 2.5 | |
| Fluorine | 7782-41-4 | 0.1 | 0.2 | |
| Fluorotrichloromethane (Trichlorofluoromethane) | 75-69-4 | 1000 | 5600 | |
| Formaldehyde; see 1910.1048 | 50-00-0 | | | |
| Formic acid | 64-18-6 | 5 | 9 | |
| Furfural | 98-01-1 | 5 | 20 | X |
| Furfuryl alcohol | 98-00-0 | 50 | 200 | |
| Grain dust (oat, wheat, barley) | ........ | ........ | 10 | |
| Glycerin (mist) | 56-81-5 | | | |
|    Total dust | | ........ | 15 | |
|    Respirable fraction | | ........ | 5 | |
| Glycidol | 556-52-5 | 50 | 150 | |
| Glycol monoethyl ether; see 2-Ethoxyethanol. | | | | |
| Graphite, natural, respirable dust | 7782-42-5 | | ([3]) | |
| Graphite, synthetic | | | | |
|    Total dust | | ........ | 15 | |
|    Respirable fraction | | ........ | 5 | |
| Guthion; see Azinphos methyl. | | | | |
| Gypsum | 13397-24-5 | | | |
|    Total dust | | ........ | 15 | |
|    Respirable fraction | | ........ | 5 | |
| Hafnium | 7440-58-6 | | 0.5 | |
| Heptachlor | 76-44-8 | ........ | 0.5 | X |
| Heptane (n-Heptane) | 142-82-5 | 500 | 2000 | |
| Hexachloroethane | 67-72-1 | 1 | 10 | X |
| Hexachloronaphthalene | 1335-87-1 | ........ | 0.2 | X |
| n-Hexane | 110-54-3 | 500 | 1800 | |
| 2-Hexanone (Methyl n-butyl ketone) | 591-78-6 | 100 | 410 | |
| Hexone (Methyl isobutyl ketone) | 108-10-1 | 100 | 410 | |
| sec-Hexyl acetate | 108-84-9 | 50 | 300 | |
| Hydrazine | 302-01-2 | 1 | 1.3 | X |
| Hydrogen bromide | 10035-10-6 | 3 | 10 | |
| Hydrogen chloride | 7647-01-0 | (C)5 | (C)7 | |
| Hydrogen cyanide | 74-90-8 | 10 | 11 | X |
| Hydrogen fluoride (as F) | 7664-39-3 | | ([2]) | |
| Hydrogen peroxide | 7722-84-1 | 1 | 1.4 | |
| Hydrogen selenide (as Se) | 7783-07-5 | 0.05 | 0.2 | |
| Hydrogen sulfide | 7783-06-4 | | ([2]) | |
| Hydroquinone | 123-31-9 | ........ | 2 | |
| Iodine | 7553-56-2 | (C)0.1 | (C)1 | |
| Iron oxide fume | 1309-37-1 | ........ | 10 | |
| Isoamyl acetate | 123-92-2 | 100 | 525 | |
| Isoamyl alcohol (primary and secondary) | 123-51-3 | 100 | 360 | |
| Isobutyl acetate | 110-19-0 | 150 | 700 | |
| Isobutyl alcohol | 78-83-1 | 100 | 300 | |
| Isophorone | 78-59-1 | 25 | 140 | |
| Isopropyl acetate | 108-21-4 | 250 | 950 | |
| Isopropyl alcohol | 67-63-0 | 400 | 980 | |
| Isopropylamine | 75-31-0 | 5 | 12 | |
| Isopropyl ether | 108-20-3 | 500 | 2100 | |
| Isopropyl glycidyl ether (IGE) | 4016-14-2 | 50 | 240 | |
| Kaolin | 1332-58-7 | | | |
|    Total dust | | ........ | 15 | |
|    Respirable fraction | | ........ | 5 | |
| Ketene | 463-51-4 | 0.5 | 0.9 | |

§1910.1000   Toxic and Hazardous Substances

## Subpart Z—Toxic and Hazardous Substances

| Substance | CAS No. (c) | ppm (a)[1] | mg/m³ (b)[1] | Skin designation |
|---|---|---|---|---|
| Lead, inorganic (as Pb); see 1910.1025. | 7439-92-1 | | | |
| Limestone | 1317-65-3 | | | |
|   Total dust | | ........ | 15 | |
|   Respirable fraction | | ........ | 5 | |
| Lindane | 58-89-9 | ........ | 0.5 | X |
| Lithium hydride | 7580-67-8 | ........ | 0.025 | |
| L.P.G. (Liquefied petroleum gas) | 68476-85-7 | 1000 | 1800 | |
| Magnesite | 546-93-0 | | | |
|   Total dust | | ........ | 15 | |
|   Respirable fraction | | ........ | 5 | |
| Magnesium oxide fume | 1309-48-4 | | | |
|   Total particulate | | ........ | 15 | |
| Malathion | 121-75-5 | | | |
|   Total dust | | ........ | 15 | X |
| Maleic anhydride | 108-31-6 | 0.25 | 1 | |
| Manganese compounds (as Mn) | 7439-96-5 | ........ | (C)5 | |
| Manganese fume (as Mn) | 7439-96-5 | ........ | (C)5 | |
| Marble | 1317-65-3 | | | |
|   Total dust | | ........ | 15 | |
|   Respirable fraction | | ........ | 5 | |
| Mercury (aryl and inorganic) (as Hg) | 7439-97-6 | | (²) | |
| Mercury (organo) alkyl compounds (as Hg) | 7439-97-6 | | (²) | |
| Mercury (vapor) (as Hg) | 7439-97-6 | | (²) | |
| Mesityl oxide | 141-79-7 | 25 | 100 | |
| Methanethiol; see Methyl mercaptan. | | | | |
| Methoxychlor | 72-43-5 | | | |
|   Total dust | | ........ | 15 | |
| 2-Methoxyethanol (Methyl cellosolve) | 109-86-4 | 25 | 80 | X |
| 2-Methoxyethyl acetate (Methyl cellosolve acetate) | 110-49-6 | 25 | 120 | X |
| Methyl acetate | 79-20-9 | 200 | 610 | |
| Methyl acetylene (Propyne) | 74-99-7 | 1000 | 1650 | |
| Methyl acetylene-propadiene mixture (MAPP) | | 1000 | 1800 | |
| Methyl acrylate | 96-33-3 | 10 | 35 | X |
| Methylal (Dimethoxy-methane) | 109-87-5 | 1000 | 3100 | |
| Methyl alcohol | 67-56-1 | 200 | 260 | |
| Methylamine | 74-89-5 | 10 | 12 | |
| Methyl amyl alcohol; see Methyl isobutyl carbinol. | | | | |
| Methyl n-amyl ketone | 110-43-0 | 100 | 465 | |
| Methyl bromide | 74-83-9 | (C)20 | (C)80 | X |
| Methyl butyl ketone; see 2-Hexanone. | | | | |
| Methyl cellosolve; see 2-Methoxyethanol. | | | | |
| Methyl cellosolve acetate; see 2-Methoxyethyl acetate. | | | | |
| Methyl chloride | 74-87-3 | | (²) | |
| Methyl chloroform (1,1,1-Trichloroethane) | 71-55-6 | 350 | 1900 | |
| Methylcyclohexane | 108-87-2 | 500 | 2000 | |
| Methylcyclohexanol | 25639-42-3 | 100 | 470 | |
| o-Methylcyclohexanone | 583-60-8 | 100 | 460 | X |
| Methylene chloride | 75-09-2 | | (²) | |
| Methyl ethyl ketone (MEK); see 2-Butanone. | | | | |
| Methyl formate | 107-31-3 | 100 | 250 | |
| Methyl hydrazine (Monomethyl hydrazine) | 60-34-4 | (C)0.2 | (C)0.35 | X |
| Methyl iodide | 74-88-4 | 5 | 28 | X |
| Methyl isoamyl ketone | 110-12-3 | 100 | 475 | |
| Methyl isobutyl carbinol | 108-11-2 | 25 | 100 | X |
| Methyl isobutyl ketone; see Hexone. | | | | |
| Methyl isocyanate | 624-83-9 | 0.02 | 0.05 | X |
| Methyl mercaptan | 74-93-1 | (C)10 | (C)20 | |
| Methyl methacrylate | 80-62-6 | 100 | 410 | |
| Methyl propyl ketone; see 2-Pentanone. | | | | |
| alpha-Methyl styrene | 98-83-9 | (C)100 | (C)480 | |
| Methylene bisphenyl isocyanate (MDI) | 101-68-8 | (C)0.02 | (C)0.2 | |
| Mica; see Silicates. | | | | |
| Molybdenum (as Mo) | 7439-98-7 | | | |
|   Soluble compounds | | ........ | 5 | |
|   Insoluble compounds. | | | | |
|     Total dust | | ........ | 15 | |
| Monomethyl aniline | 100-61-8 | 2 | 9 | X |
| Monomethyl hydrazine; see Methyl hydrazine. | | | | |
| Morpholine | 110-91-8 | 20 | 70 | X |
| Naphtha (Coal tar) | 8030-30-6 | 100 | 400 | |
| Naphthalene | 91-20-3 | 10 | 50 | |
| alpha-Naphthylamine; see 1910.1004 | 134-32-7 | | | |
| beta-Naphthylamine; see 1910.1009 | 91-59-8 | | | |
| Nickel carbonyl (as Ni) | 13463-39-3 | 0.001 | 0.007 | |
| Nickel, metal and insoluble compounds (as Ni) | 7440-02-0 | ........ | 1 | |
| Nickel, soluble compounds (as Ni) | 7440-02-0 | ........ | 1 | |
| Nicotine | 54-11-5 | ........ | 0.5 | X |
| Nitric acid | 7697-37-2 | 2 | 5 | |
| Nitric oxide | 10102-43-9 | 25 | 30 | |
| p-Nitroaniline | 100-01-6 | 1 | 6 | X |

# OSHA Standards for General Industry

| Substance | CAS No. (c) | ppm (a)[1] | mg/m³ (b)[1] | Skin designation |
|---|---|---|---|---|
| Nitrobenzene | 98-95-3 | 1 | 5 | X |
| p-Nitrochlorobenzene | 100-00-5 | ........ | 1 | X |
| 4-Nitrodiphenyl; see 1910.1003 | 92-93-3 | | | |
| Nitroethane | 79-24-3 | 100 | 310 | |
| Nitrogen dioxide | 10102-44-0 | (C)5 | (C)9 | |
| Nitrogen trifluoride | 7783-54-2 | 10 | 29 | |
| Nitroglycerin | 55-63-0 | (C)0.2 | (C)2 | X |
| Nitromethane | 75-52-5 | 100 | 250 | |
| 1-Nitropropane | 108-03-2 | 25 | 90 | |
| 2-Nitropropane | 79-46-9 | 25 | 90 | |
| N-Nitrosodimethylamine; see 1910.1016. | | | | |
| Nitrotoluene (all isomers) | | 5 | 30 | X |
|     o-isomer | 88-72-2 | | | |
|     m-isomer | 99-08-1 | | | |
|     p-isomer | 99-99-0 | | | |
| Nitrotrichloromethane; see Chloropicrin. | | | | |
| Oclachloronaphthalene | 2234-13-1 | ........ | 0.1 | X |
| Octane | 111-65-9 | 500 | 2350 | |
| Oil mist, mineral | 8012-95-1 | ........ | 5 | |
| Osmium tetroxide (as Os) | 20816-12-0 | ........ | 0.002 | |
| Oxalic acid | 144-62-7 | ........ | 1 | |
| Oxygen difluoride | 7783-41-7 | 0.05 | 0.1 | |
| Ozone | 10028-15-6 | 0.1 | 0.2 | |
| Paraquat, respirable dust | 4685-14-7; 1910-42-5; 2074-50-2 | ........ | 0.5 | X |
| Parathion | 56-38-2 | ........ | 0.1 | X |
| Particulates not otherwise regulated (PNOR)[f]. | | | | |
|     Total dust | | ........ | 15 | |
|     Respirable fraction | | ........ | 5 | |
| PCB; see Chlorodiphenyl (42% and 54% chlorine). | | | | |
| Pentaborane | 19624-22-7 | 0.005 | 0.01 | |
| Pentachloronaphthalene | 1321-64-8 | | 0.5 | X |
| Pentachlorophenol | 87-86-5 | | 0.5 | X |
| Pentaerythritol | 115-77-5 | | | |
|     Total dust | | ........ | 15 | |
|     Respirable fraction | | ........ | 5 | |
| Pentane | 109-66-0 | 1000 | 2950 | |
| 2-Pentanone (Methyl propyl ketone) | 107-87-9 | 200 | 700 | |
| Perchloroethylene (Tetrachloroethylene) | 127-18-4 | | (²) | |
| Perchloromethyl mercaptan | 594-42-3 | 0.1 | 0.8 | |
| Perchloryl fluoride | 7616-94-6 | 3 | 13.5 | |
| Petroleum distillates (Naphtha) (Rubber Solvent) | | 500 | 2000 | |
| Phenol | 108-95-2 | 5 | 19 | X |
| p-Phenylene diamine | 106-50-3 | ........ | 0.1 | X |
| Phenyl ether, vapor | 101-84-8 | 1 | 7 | |
| Phenyl ether-biphenyl mixture, vapor | | 1 | 7 | |
| Phenylethylene; see Styrene. | | | | |
| Phenyl glycidyl ether (PGE) | 122-60-1 | 10 | 60 | |
| Phenythydrazine | 100-63-0 | 5 | 22 | X |
| Phosdrin (Mevinphos) | 7786-34-7 | ........ | 0.1 | X |
| Phosgene (Carbonyl chloride) | 75-44-5 | 0.1 | 0.4 | |
| Phosphine | 7803-51-2 | 0.3 | 0.4 | |
| Phosphoric acid | 7664-38-2 | ........ | 1 | |
| Phosphorus (yellow) | 7723-14-0 | ........ | 0.1 | |
| Phosphorus pentachloride | 10026-13-8 | ........ | 1 | |
| Phosphorus pentasulfide | 1314-80-3 | ........ | 1 | |
| Phosphorus trichloride | 7719-12-2 | 0.5 | 3 | |
| Phthalic anhydride | 85-44-9 | 2 | 12 | |
| Picloram | 1918-02-1 | | | |
|     Total dust | | ........ | 15 | |
|     Respirable fraction | | ........ | 5 | |
| Picric acid | 88-89-1 | ........ | 0.1 | X |
| Pindone (2-Pivalyl-1,3-indandione) | 83-26-1 | ........ | 0.1 | |
| Plaster of Paris | 26499-65-0 | | | |
|     Total dust | | ........ | 15 | |
|     Respirable fraction | | ........ | 5 | |
| Platinum (as Pt) | 7440-06-4 | | | |
|     Metal | | ........ | ........ | |
|     Soluble salts | | ........ | 0.002 | |
| Portland cement | 65997-15-1 | | | |
|     Total dust | | | 15 | |
|     Respirable fraction | | ........ | 5 | |
| Propane | 74-98-6 | 1000 | 1800 | |
| beta-Propriolactone; see 1910.1013. | 57-57-8 | | | |
| n-Propyl acetate | 109-60-4 | 200 | 840 | |
| n-Propyl alcohol | 71-23-8 | 200 | 500 | |
| n-Propyl nitrate | 627-13-4 | 25 | 110 | |
| Propylene dichloride | 78-87-5 | 75 | 350 | |
| Propylene imine | 75-55-8 | 2 | 5 | X |

## Subpart Z—Toxic and Hazardous Substances

| Substance | CAS No. (c) | ppm (a)[1] | mg/m³ (b)[1] | Skin designation |
|---|---|---|---|---|
| Propylene oxide | 75-56-9 | 100 | 240 | |
| Propyne; see Methyl acetylene. | | | | |
| Pyrethrum | 8003-34-7 | ........ | 5 | |
| Pyridine | 110-86-1 | 5 | 15 | |
| Quinone | 106-51-4 | 0.1 | 0.4 | |
| RDX; see Cyclonite. | | | | |
| Rhodium (as Rh), metal fume and insoluble compounds | 7440-16-6 | ........ | 0.1 | |
| Rhodium (as Rh), soluble compounds | 7440-16-6 | ........ | 0.001 | |
| Ronnel | 299-84-3 | ........ | 15 | |
| Rotenone | 83-79-4 | ........ | 5 | |
| Rouge | | | | |
|   Total dust | | ........ | 15 | |
|   Respirable fraction | | ........ | 5 | |
| Selenium compounds (as Se) | 7782-49-2 | ........ | 0.2 | |
| Selenium hexafluoride (as Se) | 7783-79-1 | 0.05 | 0.4 | |
| Silica, amorphous, precipitated and gel | 112926-00-8 | | (³) | |
| Silica, amorphous, diatomaceous earth, containing less than 1% crystalline silica | 61790-53-2 | | (³) | |
| Silica, crystalline cristobalite, respirable dust | 14464-46-1 | | (³) | |
| Silica, crystalline quartz, respirable dust | 14808-60-7 | | (³) | |
| Silica, crystalline tripoli (as quartz), respirable dust | 1317-95-9 | | (³) | |
| Silica, crystalline tridymite, respirable dust | 15468-32-3 | | (³) | |
| Silica, fused, respirable dust | 60676-86-0 | | (³) | |
| Silicates (less than 1% crystalline silica) | | | | |
|   Mica (respirable dust) | 12001-26-2 | | (³) | |
|   Soapstone, total dust | ........ | | (³) | |
|   Soapstone, respirable dust | ........ | | (³) | |
|   Talc (containing asbestos); use asbestos limit; see 29 CFR 1910.1001 | ........ | | (³) | |
|   Talc (containing no asbestos), respirable dust | 14807-96-6 | | (³) | |
|   Tremolite, asbestiform; see 1910.1001. | | | | |
| Silicon | 7440-21-3 | | | |
|   Total dust | | ........ | 15 | |
|   Respirable fraction | | ........ | 5 | |
| Silicon carbide | 409-21-2 | | | |
|   Total dust | | ........ | 15 | |
|   Respirable fraction | | ........ | 5 | |
| Silver, metal and soluble compounds (as Ag) | 7440-22-4 | ........ | 0.01 | |
| Soapstone; see Silicates. | | | | |
| Sodium fluoroacetate | 62-74-8 | ........ | 0.05 | X |
| Sodium hydroxide | 1310-73-2 | ........ | 2 | |
| Starch | 9005-25-8 | | | |
|   Total dust | | ........ | 15 | |
|   Respirable fraction | | ........ | 5 | |
| Stibine | 7803-52-3 | 0.1 | 0.5 | |
| Stoddard solvent | 8052-41-3 | 500 | 2900 | |
| Strychnine | 57-24-9 | ........ | 0.15 | |
| Styrene | 100-42-5 | | (²) | |
| Sucrose | 57-50-1 | | | |
|   Total dust | | ........ | 15 | |
|   Respirable fraction | | ........ | 5 | |
| Sulfur dioxide | 7446-09-5 | 5 | 13 | |
| Sulfur hexafluoride | 2551-62-4 | 1000 | 6000 | |
| Sulfuric acid | 7664-93-9 | ........ | 1 | |
| Sulfur monochloride | 10025-67-9 | 1 | 6 | |
| Sulfur pentafluoride | 5714-22-7 | 0.025 | 0.25 | |
| Sulfuryl fluoride | 2699-79-8 | 5 | 20 | |
| Systox; see Demeton. | | | | |
| 2,4,5-T (2,4,5-trichlorophenoxyacetic acid) | 93-76-5 | ........ | 10 | |
| Tato; see Silicates. | | | | |
| Tantalum, metal and oxide dust | 7440-25-7 | ........ | 5 | |
| TEDP (Sulfotep) | 3689-24-5 | ........ | 0.2 | X |
| Tellurium and compounds (as Te) | 13494-80-9 | ........ | 0.1 | |
| Tellurium hexafluoride (as Te) | 7783-80-4 | 0.02 | 0.2 | |
| Temephos | 3383-96-8 | | | |
|   Total dust | | ........ | 15 | |
|   Respirable fraction | | ........ | 5 | |
| TEPP (Tetraethyl pyrophosphate) | 107-49-3 | ........ | 0.05 | X |
| Terphenyls | 26140-60-3 | (C)1 | (C)9 | |
| 1,1,1,2-Tetrachloro-2,2-difluoroethane | 76-11-9 | 500 | 4170 | |
| 1,1,2,2-Tetrachloro-1,2-difluoroethane | 76-12-0 | 500 | 4170 | |
| 1,1,2,2-Tetrachloroethane | 79-34-5 | 5 | 35 | X |
| Tetrachloroethylene; see Perchloroethylene. | | | | |
| Tetrachloromethane; see Carbon tetrachloride. | | | | |
| Tetrachloronaphthalene | 1335-88-2 | ........ | 2 | X |
| Tetraethyl lead (as Pb) | 78-00-2 | ........ | 0.075 | X |
| Tetrahydrofuran | 109-99-9 | 200 | 590 | |
| Tetramethyl lead (as Pb) | 75-74-1 | ........ | 0.075 | X |
| Tetramethyl succinonitrile | 3333-52-6 | 0.5 | 3 | X |
| Tetranitromethane | 509-14-8 | 1 | 8 | |
| Tetryl (2,4,6-Trinitrophenylmethylnitramine) | 479-45-8 | ........ | 1.5 | X |
| Thallium, soluble compounds (as Tl) | 7440-28-0 | ........ | 0.1 | X |

# OSHA Standards for General Industry

461

§1910.1000 — Toxic and Hazardous Substances

| Substance | CAS No. (c) | ppm (a)[1] | mg/m³ (b)[1] | Skin designation |
|---|---|---|---|---|
| 4,4'-Thiobis (6-tert, Butyl-m-cresol) | 96-69-5 | | | |
| Total dust | | ........ | 15 | |
| Respirable fraction | | ........ | 5 | |
| Thiram | 137-26-8 | ........ | 5 | |
| Tin, inorganic compounds (except oxides) (as Sn) | 7440-31-5 | ........ | 2 | |
| Tin, organic compounds (as Sn) | 7440-31-5 | ........ | 0.1 | |
| Titanium dioxide | 13463-67-7 | | | |
| Total dust | | ........ | 15 | |
| Toluene | 108-88-3 | | (²) | |
| Toluene-2,4-diisocyanate (TDI) | 584-84-9 | (C)0.02 | (C)0.14 | |
| o-Toluidine | 95-53-4 | 5 | 22 | X |
| Toxaphene; see Chlorinated camphene. | | | | |
| Tremolite; see Silicates. | | | | |
| Tributyl phosphate | 126-73-8 | ........ | 5 | |
| 1,1,1-Trichloroethane; see Methyl chloroform. | | | | |
| 1,1,2-Trichloroethane | 79-00-5 | 10 | 45 | X |
| Trichloroethylene | 79-01-6 | | (²) | |
| Trichloromethane; see Chloroform. | | | | |
| Trichloronaphthalene | 1321-65-9 | ........ | 5 | X |
| 1,2,3-Trichloropropane | 96-18-4 | 50 | 300 | |
| 1,1,2-Trichloro-1,2,2-trifluoroethane | 76-13-1 | 1000 | 7600 | |
| Triethylamine | 121-44-8 | 25 | 100 | |
| Trifluorobromomethane | 75-63-8 | 1000 | 6100 | |
| 2,4,6-Trinitrophenol; see Picric acid. | | | | |
| 2,4,6-Trinitrophenylmethylnitramine; see Tetryl. | | | | |
| 2,4,6-Trinitrotoluene (TNT) | 118-96-7 | ........ | 1.5 | X |
| Triorthocresyl phosphate | 78-30-8 | ........ | 0.1 | |
| Triphenyl phosphate | 115-86-6 | ........ | 3 | |
| Turpentine | 8006-64-2 | 100 | 560 | |
| Uranium (as U) | 7440-61-1 | | | |
| Soluble compounds | | ........ | 0.05 | |
| Insoluble compounds | | ........ | 0.25 | |
| Vanadium | 1314-62-1 | | | |
| Respirable dust (as $V_2O_5$) | | ........ | (C)0.5 | |
| Fume (as $V_2O_5$) | | ........ | (C)0.1 | |
| Vegetable oil mist | | | | |
| Total dust | | ........ | 15 | |
| Respirable fraction | | ........ | 5 | |
| Vinyl benzene; see Styrene. | | | | |
| Vinyl chloride; see 1910.1017 | 75-01-4 | | | |
| Vinyl cyanide; see Acrylonitrile. | | | | |
| Vinyl toluene | 25013-15-4 | 100 | 480 | |
| Warfarin | 81-81-2 | ........ | 0.1 | |
| Xylenes (o-, m-, p-isomers) | 1330-20-7 | 100 | 435 | |
| Xylidine | 1300-73-8 | 5 | 25 | X |
| Yttrium | 7440-65-5 | ........ | 1 | |
| Zinc chloride fume | 7646-85-7 | ........ | 1 | |
| Zinc oxide fume | 1314-13-2 | ........ | 5 | |
| Zinc oxide | 1314-13-2 | | | |
| Total dust | | ........ | 15 | |
| Respirable fraction | | ........ | 5 | |
| Zinc stearate | 557-05-1 | | | |
| Total dust | | ........ | 15 | |
| Respirable fraction | | ........ | 5 | |
| Zirconium compounds (as Zr) | 7440-67-7 | ........ | 5 | |

[1] The PELs are 8-hour TWAs unless otherwise noted; a (C) designation denotes a ceiling limit. They are to be determined from breathing-zone air samples.

(a) Parts of vapor or gas per million parts of contaminated air by volume at 25 °C and 760 torr.

(b) Milligrams of substance per cubic meter of air. When entry is in this column only, the value is exact; when listed with a ppm entry, it is approximate.

(c) The CAS number is for information only. Enforcement is based on the substance name. For an entry covering more than one metal compound, measured as the metal, the CAS number for the metal is given—not CAS numbers for the individual compounds.

(d) The final benzene standard in 1910.1028 applies to all occupational exposures to benzene except in some circumstances the distribution and sale of fuels, sealed containers and pipelines, coke production, oil and gas drilling and production, natural gas processing, and the percentage exclusion for liquid mixtures; for the excepted subsegments, the benzene limits in Table Z-2 apply. See 1910.1028 for specific circumstances.

(e) This 8-hour TWA applies to respirable dust as measured by a vertical elutriator cotton dust sampler or equivalent instrument. The time-weighted average applies to the cottom waste processing operations of waste recycling (sorting, blending, cleaning and willowing) and garnetting. See also 1910.1043 for cotton dust limits applicable to other sectors.

(f) All inert or nuisance dusts, whether mineral, inorganic, or organic, not listed specifically by substance name are covered by the Particulates Not Otherwise Regulated (PNOR) limit which is the same as the inert or nuisance dust limit of Table Z-3.

[2] See Table Z-2.

[3] See Table Z-3.

[4] Varies with compound.

[5] See Table Z-2 for the exposure limits for any operations or sections where the exposure limits in §1910.1026 are stayed or are otherwise not in effect.

## TABLE Z-2

| Substance | 8-hour time weighted average | Acceptable ceiling concentration | Acceptable maximum peak above the acceptable ceiling concentration for an 8-hr shift ||
|---|---|---|---|---|
| | | | Concentration | Maximum duration |
| Benzene[a] (Z37.40-1969) | 10 ppm | 25 ppm | 50 ppm | 10 minutes. |
| Beryllium and beryllium compounds (Z37.29-1970) | 2 µg/m³ | 5 µg/m³ | 25 µg/m³ | 30 minutes. |
| Cadmium fume[b] (Z37.5-1970) | 0.1 mg/m³ | 0.3 mg/m³ | — | |
| Cadmium dust[b] (Z37.5-1970) | 0.2 mg/m³ | 0.6 mg/m³ | — | |
| Carbon disulfide (Z37.3-1968) | 20 ppm | 30 ppm | 100 ppm | 30 minutes. |
| Carbon tetrachloride (Z37.17-1967) | 10 ppm | 25 ppm | 200 ppm | 5 min. in any 4 hrs. |
| Chromic acid and chromates (Z37.7-1971) (as $CrO_3$)[c] | — | 1 mg/10m³ | | |
| Ethylene dibromide (Z37.31-1970) | 20 ppm | 30 ppm | 50 ppm | 5 minutes. |
| Ethylene dichloride (Z37.21-1969) | 50 ppm | 100 ppm | 200 ppm | 5 min. in any 3 hrs. |
| Fluoride as dust (Z37.28-1969) | 2.5 mg/m³ | — | — | |
| Formaldehyde; see 1910.1048 | — | — | — | |
| Hydrogen fluoride (Z37.28-1969) | 3 ppm | — | — | |
| Hydrogen sulfide (Z37.2-1966) | | 20 ppm | 50 ppm | 10 mins. once, only if no other meas. exp. occurs. |
| Mercury (Z37.8-1971) | — | 1 mg/10m³ | — | |
| Methyl chloride (Z37.18-1969) | 100 ppm | 200 ppm | 300 ppm | 5 mins. in any 3 hrs. |
| Methylene Chloride: See §1919.52. | | | | |
| Organo (alkyl) mercury (Z37.30-1969) | 0.01 mg/m³ | 0.04 mg/m³ | — | |
| Styrene (Z37.15-1969) | 100 ppm | 200 ppm | 600 ppm | 5 mins. in any 3 hrs. |
| Tetrachloroethylene (Z37.22-1967) | 100 ppm | 200 ppm | 300 ppm | 5 mins. in any 3 hrs. |
| Toluene (Z37.12-1967) | 200 ppm | 300 ppm | 500 ppm | 10 minutes. |
| Trichloroethylene (Z37.19-1967) | 100 ppm | 200 ppm | 300 ppm | 5 mins. in any 2 hrs. |

[a] This standard applies to the industry segments exempt from the 1 ppm 8-hour TWA and 5 ppm STEL of the benzene standard at 1910.1028.

[b] This standard applies to any operations or sectors for which the Cadmium standard, 1910.1027, is stayed or otherwise not in effect.

[c] The standard applies to any operations or sectors for which the Hexavalent Chromium standard, 1910.1026 is stayed or otherwise is not in effect.

### TABLE Z-3—MINERAL DUSTS

| Substance | mppcf[a] | mg/m³ |
|---|---|---|
| Silica: | | |
| Crystalline | | |
| Quartz (Respirable) | $\frac{250^{b}}{\%SiO_2 + 5}$ | $\frac{10 \text{ mg/m}^{3e}}{\% SiO_2 + 2}$ |
| Quartz (Total Dust) | ........ | $\frac{30 \text{ mg/m}^3}{\% SiO_2 + 2}$ |
| Cristobalite: Use 1/2 the value calculated from the count or mass formulae for quartz. | | |
| Tridymite: Use 1/2 the value calculated from the formulae for quartz. | | |
| Amorphous, including natural diatomaceous earth | 20 | $\frac{80 \text{ mg/m}^3}{\%SiO_2}$ |
| Silicates (less than 1% crystalline silica): | | |
| Mica | 20 | |
| Soapstone | 20 | |
| Talc (not containing asbestos) | 20[c] | |
| Talc (containing asbestos) Use asbestos limit. | | |
| Tremolite, asbestiform (see 29 CFR 1910.1001). | | |
| Portland cement | 50 | |
| Graphite (Natural) | 15 | |
| Coal Dust: | | |
| Respirable fraction less than 5% $SiO_2$ | ........ | 2.4 mg/m³[e] |
| Respirable fraction greater than 5% $SiO_2$ | ........ | $\frac{10 \text{ mg/m}^{3e}}{\%SiO_2 + 2}$ |
| Inert or Nuisance Dust:[d] | | |
| Respirable fraction | 15 | 5 mg/m³ |
| Total dust | 50 | 15 mg/m³ |

**NOTE:** Conversion factors - mppcf x 35.3 = million particles per cubic meter = particles per c.c.

[a] Millions of particles per cubic foot of air, based on impinger samples counted by light-field techniques.

[b] The percentage of crystalline silica in the formula is the amount determined from airborne samples, except in those instances in which other methods have been shown to be applicable.

[c] Containing less than 1% quartz; if 1% quartz or more, use quartz limit.

[d] All inert or nuisance dusts, whether mineral, inorganic, or organic, not listed specifically by substance name are covered by this limit, which is the same as the Particulates Not Otherwise Regulated (PNOR) limit in Table Z-1.

[e] Both concentration and percent quartz for the application of this limit are to be determined from the fraction passing a size-selector with the following characteristics:

| Aerodynamic diameter (unit density sphere) | Percent passing selector |
|---|---|
| 2 | 90 |
| 2.5 | 75 |
| 3.5 | 50 |
| 5.0 | 25 |
| 10 | 0 |

The measurements under this note refer to the use of an AEC (now NRC) instrument. The respirable fraction of coal dust is determined with an MRE; the figure corresponding to that of 2.4 mg/m³ in the table for coal dust is 4.5 mg/m³K.

[58 FR 35340, June 30. 1993; 58 FR 40191, July 27, 1993, as amended at 61 FR 56831, Nov. 4, 1996; 62 FR 1600, Jan. 10, 1997; 62 FR 42018, Aug. 4, 1997]

### §1910.1001 Asbestos.

**1910.1001(a)** *Scope and application.* **(1)** This section applies to all occupational exposures to asbestos in all industries covered by the Occupational Safety and Health Act, except as provided in paragraph (a)(2) and (3) of this section.

**1910.1001(a)(2)** This section does not apply to construction work as defined in 29 CFR 1910.12(b). (Exposure to asbestos in construction work is covered by 29 CFR 1926.1101).

# OSHA Standards for General Industry

**1910.1001(a)(3)** This section does not apply to ship repairing, shipbuilding and shipbreaking employments and related employments as defined in 29 CFR 1915.4. (Exposure to asbestos in these employments is covered by 29 CFR 1915.1001).

**1910.1001(b) Definitions.**

*Asbestos* includes chrysotile, amosite, crocidolite, tremolite asbestos, anthophyllite asbestos, actinolite asbestos, and any of these minerals that have been chemically treated and/or altered.

*Asbestos-containing material* (ACM) means any material containing more than 1% asbestos.

*Assistant Secretary* means the Assistant Secretary of Labor for Occupational Safety and Health, U.S. Department of Labor, or designee.

*Authorized person* means any person authorized by the employer and required by work duties to be present in regulated areas.

*Building/facility owner* is the legal entity, including a lessee, which exercises control over management and record keeping functions relating to a building and/or facility in which activities covered by this standard take place.

*Certified industrial hygienist (CIH)* means one certified in the practice of industrial hygiene by the American Board of Industrial Hygiene.

*Director* means the Director of the National Institute for Occupational Safety and Health, U.S. Department of Health and Human Services, or designee.

*Employee exposure* means that exposure to airborne asbestos that would occur if the employee were not using respiratory protective equipment.

*Fiber* means a particulate form of asbestos 5 micrometers or longer, with a length-to-diameter ratio of at least 3 to 1.

*High-efficiency particulate air (HEPA) filter* means a filter capable of trapping and retaining at least 99.97 percent of 0.3 micrometer diameter mono-disperse particles.

*Homogeneous area* means an area of surfacing material or thermal system insulation that is uniform in color and texture.

*Industrial hygienist* means a professional qualified by education, training, and experience to anticipate, recognize, evaluate and develop controls for occupational health hazards.

*PACM* means "presumed asbestos containing material."

*Presumed asbestos containing material* means thermal system insulation and surfacing material found in buildings constructed no later than 1980. The designation of a material as "PACM" may be rebutted pursuant to paragraph (j)(8) of this section.

*Regulated area* means an area established by the employer to demarcate areas where airborne concentrations of asbestos exceed, or there is a reasonable possibility they may exceed, the permissible exposure limits.

*Surfacing ACM* means surfacing material which contains more than 1% asbestos.

*Surfacing material* means material that is sprayed, troweled-on or otherwise applied to surfaces (such as acoustical plaster on ceilings and fireproofing materials on structural members, or other materials on surfaces for acoustical, fireproofing, and other purposes).

*Thermal System Insulation (TSI)* means ACM applied to pipes, fittings, boilers, breeching, tanks, ducts or other structural components to prevent heat loss or gain.

*Thermal System Insulation ACM* means thermal system insulation which contains more than 1% asbestos.

**1910.1001(c)** *Permissible exposure limit (PELS)*—**(1)** *Time-weighted average limit (TWA).* The employer shall ensure that no employee is exposed to an airborne concentration of asbestos in excess of 0.1 fiber per cubic centimeter of air as an eight (8)-hour time-weighted average (TWA) as determined by the method prescribed in Appendix A to this section, or by an equivalent method.

**1910.1001(c)(2)** *Excursion limit.* The employer shall ensure that no employee is exposed to an airborne concentration of asbestos in excess of 1.0 fiber per cubic centimeter of air (1 f/cc) as averaged over a sampling period of thirty (30) minutes as determined by the method prescribed in Appendix A to this section, or by an equivalent method.

**1910.1001(d)** *Exposure monitoring*—**(1)** *General.* **(i)** Determinations of employee exposure shall be made from breathing zone air samples that are representative of the 8-hour TWA and 30-minute short-term exposures of each employee.

**1910.1001(d)(1)(ii)** Representative 8-hour TWA employee exposures shall be determined on the basis of one or more samples representing full-shift exposures for each shift for each employee in each job classification in each work area. Representative 30-minute short-term employee exposures shall be determined on the basis of one or more samples representing 30 minute exposures associated with operations that are most likely to produce exposures above the excursion limit for each shift for each job classification in each work area.

**1910.1001(d)(2)** *Initial monitoring.* **(i)** Each employer who has a workplace or work operation covered by this standard, except as provided for in paragraphs (d)(2)(ii) and (d)(2)(iii) of this section, shall perform initial monitoring of employees who are, or may reasonably be expected to be exposed to airborne concentrations at or above the TWA permissible exposure limit and/or excursion limit.

**1910.1001(d)(2)(ii)** Where the employer has monitored after March 31, 1992, for the TWA permissible exposure limit and/or the excursion limit, and the monitoring satisfies all other requirements of this section, the employer may rely on such earlier monitoring results to satisfy the requirements of paragraph (d)(2)(i) of this section.

**1910.1001(d)(2)(iii)** Where the employer has relied upon objective data that demonstrate that asbestos is not capable of being released in airborne concentrations at or above the TWA permissible exposure limit and/or excursion limit under the expected conditions of processing, use, or handling, then no initial monitoring is required.

**1910.1001(d)(3)** *Monitoring frequency (periodic monitoring) and patterns.* After the initial determinations required by paragraph (d)(2)(i) of this section, samples shall be of such frequency and pattern as to represent with reasonable accuracy the levels of exposure of the employees. In no case shall sampling be at intervals greater than six months for employees whose exposures may reasonably be foreseen to exceed the TWA permissible exposure limit and/or excursion limit.

**1910.1001(d)(4)** *Changes in monitoring frequency.* If either the initial or the periodic monitoring required by paragraphs (d)(2) and (d)(3) of this section statistically indicates that employee exposures are below the TWA permissible exposure limit and/or excursion limit, the employer may discontinue the monitoring for those employees whose exposures are represented by such monitoring.

**1910.1001(d)(5)** *Additional monitoring.* Notwithstanding the provisions of paragraphs (d)(2)(ii) and (d)(4) of this section, the employer shall institute the exposure monitoring required under paragraphs (d)(2)(i) and (d)(3) of this section whenever there has been a change in the production, process, control equipment, personnel or work practices that may result in new or additional exposures above the TWA permissible exposure limit and/or excursion limit or when the employer has any reason to suspect that a change may result in new or additional exposures above the PEL and/or excursion limit.

**1910.1001(d)(6)** *Method of monitoring.* **(i)** All samples taken to satisfy the monitoring requirements of paragraph (d) of this section shall be personal samples collected following the procedures specified in Appendix A.

**1910.1001(d)(6)(ii)** All samples taken to satisfy the monitoring requirements of paragraph (d) of this section shall be evaluated using the OSHA Reference Method (ORM) specified in Appendix A of this section, or an equivalent counting method.

**1910.1001(d)(6)(iii)** If an equivalent method to the ORM is used, the employer shall ensure that the method meets the following criteria:

**1910.1001(d)(6)(iii)(A)** Replicate exposure data used to establish equivalency are collected in side-by-side field and laboratory comparisons; and

**1910.1001(d)(6)(iii)(B)** The comparison indicates that 90% of the samples collected in the range 0.5 to 2.0 times the permissible limit have an accuracy range of plus or minus 25 percent of the ORM results at a 95% confidence level as demonstrated by a statistically valid protocol; and

**1910.1001(d)(6)(iii)(C)** The equivalent method is documented and the results of the comparison testing are maintained.

**1910.1001(d)(6)(iv)** To satisfy the monitoring requirements of paragraph (d) of this section, employers must use the results of monitoring analysis performed by laboratories which have instituted quality assurance programs that include the elements as prescribed in Appendix A of this section.

**1910.1001(d)(7)** *Employee notification of monitoring results.* **(i)** The employer must, within 15 working days after the receipt of the results of any monitoring performed under this sections, notify each affected employee of these results either individually in writing or by posting the results in an appropriate location that is accessible to affected employees.

**1910.1001(d)(7)(ii)** The written notification required by paragraph (d)(7)(i) of this section shall contain the corrective action being taken by the employer to reduce employee exposure to or below the TWA and/or excursion limit, wherever monitoring results indicated that the TWA and/or excursion limit had been exceeded.

**1910.1001(e)** *Regulated Areas*—**(1)** *Establishment.* The employer shall establish regulated areas wherever airborne concentrations of asbestos and/or PACM are in excess of the TWA and/or excursion limit prescribed in paragraph (c) of this section.

**1910.1001(e)(2)** *Demarcation.* Regulated areas shall be demarcated from the rest of the workplace in any manner that minimizes the number of persons who will be exposed to asbestos.

**1910.1001(e)(3)** *Access.* Access to regulated areas shall be limited to authorized persons or to persons authorized by the Act or regulations issued pursuant thereto.

**1910.1001(e)(4)** *Provision of respirators.* Each person entering a regulated area shall be supplied with and required to use a respirator, selected in accordance with paragraph (g)(2) of this section.

**1910.1001(e)(5)** *Prohibited activities.* The employer shall ensure that employees do not eat, drink, smoke, chew tobacco or gum, or apply cosmetics in the regulated areas.

**1910.1001(f)** *Methods of compliance*—**(1)** *Engineering controls and work practices.* **(i)** The employer shall institute engineering controls and work practices to reduce and maintain employee exposure to or below the TWA and/or excursion limit prescribed in paragraph (c) of this section, except to the extent that such controls are not feasible.

**1910.1001(f)(1)(ii)** Wherever the feasible engineering controls and work practices that can be instituted are not sufficient to reduce employee exposure to or below the TWA and/or excursion limit prescribed in paragraph (c) of this section, the employer shall use them to reduce employee exposure to the lowest levels achievable by these controls and shall supplement them by the use of respiratory protection that complies with the requirements of paragraph (g) of this section.

**1910.1001(f)(1)(iii)** For the following operations, wherever feasible engineering controls and work practices that can be instituted are not sufficient to reduce the employee exposure to or below the TWA and/or excursion limit prescribed in paragraph (c) of this section, the employer shall use them to reduce employee exposure to or below 0.5 fiber per cubic centimeter of air (as an eight-hour time-weighted average) or 2.5 fibers/cc for 30 minutes (short-term exposure) and shall supplement them by the use of any combination of respiratory protection that complies with the requirements of paragraph (g) of this section, work practices and feasible engineering controls that will reduce employee exposure to or below the TWA and to or below the excursion limit permissible prescribed in paragraph (c) of this section: Coupling cutoff in primary asbestos cement pipe manufacturing; sanding in primary and secondary asbestos cement sheet manufacturing; grinding in primary and secondary friction product manufacturing; carding and spinning in dry textile processes; and grinding and sanding in primary plastics manufacturing.

**1910.1001(f)(1)(iv)** *Local exhaust ventilation.* Local exhaust ventilation and dust collection systems shall be designed, constructed, installed, and maintained in accordance with good practices such as those found in the American National Standard Fundamentals Governing the Design and Operation of Local Exhaust Systems, ANSI Z9.2-1979.

**1910.1001(f)(1)(v)** *Particular tools.* All hand-operated and power-operated tools which would produce or release fibers of asbestos, such as, but not limited to, saws, scorers, abrasive wheels, and drills, shall be provided with local exhaust ventilation systems which comply with paragraph (f)(1)(iv) of this section.

**1910.1001(f)(1)(vi)** *Wet methods.* Insofar as practicable, asbestos shall be handled, mixed, applied, removed, cut, scored, or otherwise worked in a wet state sufficient to prevent the emission of airborne fibers so as to expose employees to levels in excess of the TWA and/or excursion limit, prescribed in paragraph (c) of this section, unless the usefulness of the product would be diminished thereby.

**1910.1001(f)(1)(vii)** [Reserved]

**1910.1001(f)(1)(viii)** *Particular products and operations.* No asbestos cement, mortar, coating, grout, plaster, or similar material containing asbestos, shall be removed from bags, cartons, or other containers in which they are shipped, without being either wetted, or enclosed, or ventilated so as to prevent effectively the release of airborne fibers.

**1910.1001(f)(1)(ix)** *Compressed air.* Compressed air shall not be used to remove asbestos or materials containing asbestos unless the compressed air is used in conjunction with a ventilation system which effectively captures the dust cloud created by the compressed air.

**1910.1001(f)(1)(x)** *Flooring.* Sanding of asbestos-containing flooring material is prohibited.

**1910.1001(f)(2)** *Compliance program.* **(i)** Where the TWA and/or excursion limit is exceeded, the employer shall establish and implement a written program to reduce employee exposure to or below the TWA and to or below the excursion limit by means of engineering and work practice controls as required by paragraph (f)(1) of this section, and by the use of respiratory protection where required or permitted under this section.

**1910.1001(f)(2)(ii)** Such programs shall be reviewed and updated as necessary to reflect significant changes in the status of the employer's compliance program.

**1910.1001(f)(2)(iii)** Written programs shall be submitted upon request for examination and copying to the Assistant Secretary, the Director, affected employees and designated employee representatives.

**1910.1001(f)(2)(iv)** The employer shall not use employee rotation as a means of compliance with the TWA and/or excursion limit.

**1910.1001(f)(3)** Specific compliance methods for brake and clutch repair:

**1910.1001(f)(3)(i)** Engineering controls and work practices for brake and clutch repair and service. During automotive brake and clutch inspection, disassembly, repair and assembly operations, the employer shall institute engineering controls and work practices to reduce employee exposure to materials containing asbestos using a negative pressure enclosure/HEPA vacuum system method or low pressure/wet cleaning method, which meets the detailed requirements set out in Appendix F to this section. The employer may also comply using an equivalent method which follows written procedures which the employer demonstrates can achieve results equivalent to Method A in Appendix F to this section. For facilities in which no more than 5 pair of brakes or 5 clutches are inspected, disassembled, repaired, or assembled per week, the method set forth in paragraph [D] of Appendix F to this section may be used.

**1910.1001(f)(3)(ii)** The employer may also comply by using an equivalent method which follows written procedures, which the employer demonstrates can achieve equivalent exposure reductions as do the two "preferred methods." Such demonstration must include monitoring data conducted under workplace conditions closely resembling the process, type of asbestos containing materials, control method, work practices and environmental conditions which the equivalent method will be used, or objective data, which document that under all reasonably foreseeable conditions of brake and clutch repair applications, the method results in exposures which are equivalent to the methods set out in Appendix F to this section.

**1910.1001(g)** *Respiratory protection*—**(1)** *General.* For employees who use respirators required by this section, the employer must provide respirators that comply with the requirements of this paragraph. Respirators must be used during:

**1910.1001(g)(1)(i)** Periods necessary to install or implement feasible engineering and work-practice controls.

**1910.1001(g)(1)(ii)** Work operations, such as maintenance and repair activities, for which engineering and work-practice controls are not feasible.

**1910.1001(g)(1)(iii)** Work operations for which feasible engineering and work-practice controls are not yet sufficient to reduce employee exposure to or below the TWA and/or excursion limit.

**1910.1001(g)(1)(iv)** Emergencies.

**1910.1001(g)(2)** *Respirator program.* **(i)** The employer must implement a respiratory protection program in accordance with 29 CFR 1910.134(b) through (d) (except (d)(1)(iii)), and (f) through (m).

**1910.1001(g)(2)(ii)** The employer must provide a tight-fitting, powered, air-purifying respirator instead of any negative-pressure respirator specified in Table 1 of this section when an employee chooses to use this type of respirator and the respirator provides adequate protection to the employee.

**1910.1001(g)(2)(iii)** No employee must be assigned to tasks requiring the use of respirators if, based on their most recent medical examination, the examining physician determines that the employee will be unable to function normally using a respirator, or that the safety or health of the employee or other employees will be impaired by the use of a respirator. Such employees must be assigned to another job or given the opportunity to transfer to a different position, the duties of which they can perform. If such a transfer position is available, the position must be with the same employer, in the same geographical area, and with the same seniority, status, and rate of pay the employee had just prior to such transfer.

**1910.1001(g)(3)** *Respirator selection.* The employer must select and provide the appropriate respirator from Table 1 of this section.

### TABLE 1—RESPIRATORY PROTECTION FOR ASBESTOS FIBERS

| Airborne concentration of asbestos or conditions of use | Required respirator |
| --- | --- |
| Not in excess of 1 f/cc (10 X PEL). | Half-mask air purifying respirator other than a disposable respirator, equipped with high efficiency filters. |
| Not in excess of 5 f/cc (50 X PEL). | Full facepiece air-purifying respirator equipped with high efficiency filters. |
| Not in excess of 10 f/cc (100 X PEL) | Any powered air-purifying respirator equipped with high efficiency filters or any supplied air respirator operated in continuous flow mode. |
| Not in excess of 100 f/cc (1,000 X PEL). | Full facepiece supplied air respirator operated in pressure demand mode. |
| Greater than 100 f/cc (1,000 X PEL) or unknown concentration. | Full facepiece supplied air respirator operated in pressure demand mode, equipped with an auxiliary positive pressure self-contained breathing apparatus. |

NOTE:

a. Respirators assigned for high environmental concentrations may be used at lower concentrations, or when required respirator use is independent of concentration.

b. A high efficiency filter means a filter that is at least 99.97 percent efficient against mono-dispersed particles of 0.3 micrometers in diameter or larger.

**1910.1001(h)** *Protective work clothing and equipment*—**(1)** *Provision and use.* If an employee is exposed to asbestos above the TWA and/or excursion limit, or where the possibility of eye irritation exists, the employer shall provide at no cost to the employee and ensure that the employee uses appropriate protective work clothing and equipment such as, but not limited to:

**1910.1001(h)(1)(i)** Coveralls or similar full-body work clothing;

**1910.1001(h)(1)(ii)** Gloves, head coverings, and foot coverings; and

**1910.1001(h)(1)(iii)** Face shields, vented goggles, or other appropriate protective equipment which complies with 1910.133 of this Part.

**1910.1001(h)(2)** *Removal and storage.* **(i)** The employer shall ensure that employees remove work clothing contaminated with asbestos only in change rooms provided in accordance with paragraph (i)(1) of this section.

**1910.1001(h)(2)(ii)** The employer shall ensure that no employee takes contaminated work clothing out of the change room, except those employees authorized to do so for the purpose of laundering, maintenance, or disposal.

**1910.1001(h)(2)(iii)** Contaminated work clothing shall be placed and stored in closed containers which prevent dispersion of the asbestos outside the container.

**1910.1001(h)(2)(iv)** Containers of contaminated protective devices or work clothing which are to be taken out of change rooms or the workplace for cleaning, maintenance or disposal, shall bear labels in accordance with paragraph (j)(4) of this section.

**1910.1001(h)(3)** *Cleaning and replacement.* **(i)** The employer shall clean, launder, repair, or replace protective clothing and equipment required by this paragraph to maintain their effectiveness. The employer shall provide clean protective clothing and equipment at least weekly to each affected employee.

**1910.1001(h)(3)(ii)** The employer shall prohibit the removal of asbestos from protective clothing and equipment by blowing or shaking.

**1910.1001(h)(3)(iii)** Laundering of contaminated clothing shall be done so as to prevent the release of airborne fibers of asbestos in excess of the permissible exposure limits prescribed in paragraph (c) of this section.

**1910.1001(h)(3)(iv)** Any employer who gives contaminated clothing to another person for laundering shall inform such person of the requirement in paragraph (h)(3)(iii) of this section to effectively prevent the release of airborne fibers of asbestos in excess of the permissible exposure limits.

**1910.1001(h)(3)(v)** The employer shall inform any person who launders or cleans protective clothing or equipment contaminated with asbestos of the potentially harmful effects of exposure to asbestos.

**1910.1001(h)(3)(vi)** Contaminated clothing shall be transported in sealed impermeable bags, or other closed, impermeable containers, and labeled in accordance with paragraph (j) of this section.

**1910.1001(i)** *Hygiene facilities and practices*—**(1)** *Change rooms.* **(i)** The employer shall provide clean change rooms for employees who work in areas where their airborne exposure to asbestos is above the TWA and/or excursion limit.

**1910.1001(i)(1)(ii)** The employer shall ensure that change rooms are in accordance with 1910.141(e) of this part, and are equipped with two separate lockers or storage facilities, so separated as to prevent contamination of the employee's street clothes from his protective work clothing and equipment.

**1910.1001(i)(2)** *Showers.* **(i)** The employer shall ensure that employees who work in areas where their airborne exposure is above the TWA and/or excursion limit, shower at the end of the work shift.

**1910.1001(i)(2)(ii)** The employer shall provide shower facilities which comply with 1910.141(d)(3) of this part.

**1910.1001(i)(2)(iii)** The employer shall ensure that employees who are required to shower pursuant to paragraph (i)(2)(i) of this section do not leave the workplace wearing any clothing or equipment worn during the work shift.

**1910.1001(i)(3)** *Lunchrooms.* **(i)** The employer shall provide lunchroom facilities for employees who work in areas where their airborne exposure is above the TWA and/or excursion limit.

**1910.1001(i)(3)(ii)** The employer shall ensure that lunchroom facilities have a positive pressure, filtered air supply, and are readily accessible to employees.

**1910.1001(i)(3)(iii)** The employer shall ensure that employees who work in areas where their airborne exposure is above the PEL and/or excursion limit wash their hands and faces prior to eating, drinking or smoking.

**1910.1001(i)(3)(iv)** The employer shall ensure that employees do not enter lunchroom facilities with protective work clothing or equipment unless surface asbestos fibers have been removed from the clothing or equipment by vacuuming or other method that removes dust without causing the asbestos to become airborne.

**1910.1001(i)(4)** *Smoking in work areas.* The employer shall ensure that employees do not smoke in work areas where they are occupationally exposed to asbestos because of activities in that work area.

**1910.1001(j)** *Communication of hazards to employees*—*Introduction.* This section applies to the communication of information concerning asbestos hazards in general industry to facilitate compliance with this standard. Asbestos exposure in general industry occurs in a wide variety of industrial and commercial settings. Employees who manufacture asbestos-containing products may be exposed to asbestos fibers. Employees who repair and replace automotive brakes and clutches may be exposed to asbestos fibers. In addition, employees engaged in housekeeping activities in industrial facilities with asbestos product manufacturing operations, and in public and commercial buildings with installed asbestos containing materials may be exposed to asbestos fibers. Most of these workers are covered by this general industry standard, with the exception of state or local governmental employees in non-state plan states. It should be noted that employees who perform housekeeping activities during and after construction activities are covered by the asbestos construction standard, 29 CFR 1926.1101, formerly 1926.58. However, housekeeping employees, regardless of industry designation, should know whether building components they maintain may expose them to asbestos. The same hazard communication provisions will protect employees who perform housekeeping operations in all three asbestos standards; general industry, construction, and shipyard employment. As noted in the construction standard, building owners are often the only and/or best source of information concerning the presence of previously installed asbestos containing building materials. Therefore they, along with employers of potentially exposed employees, are assigned specific information conveying and retention duties under this section.

**1910.1001(j)(1)** *Installed Asbestos Containing Material.* Employers and building owners are required to treat installed TSI and sprayed on and troweled-on surfacing materials as ACM in buildings constructed no later than 1980 for purposes of this standard. These materials are designated "presumed ACM or PACM", and are defined in paragraph (b) of this section. Asphalt and vinyl flooring material installed no later than 1980 also must be treated as asbestos-containing. The employer or building owner may demonstrate that PACM and flooring material do not contain asbestos by complying with paragraph (j)(8)(iii) of this section.

**1910.1001(j)(2)** *Duties of employers and building and facility owners.* **(i)** Building and facility owners shall determine the presence, location, and quantity of ACM and/or PACM at the work site. Employers and building and facility owners shall exercise due diligence in complying with these requirements to inform employers and employees about the presence and location of ACM and PACM.

**1910.1001(j)(2)(ii)** Building and facility owners shall maintain records of all information required to be provided pursuant to this section and/or otherwise known to the building owner concerning the presence, location and quantity of ACM and PACM in the building/facility. Such records shall be kept for the duration of ownership and shall be transferred to successive owners.

**1910.1001(j)(2)(iii)** Building and facility owners shall inform employers of employees, and employers shall inform employees who will perform housekeeping activities in areas which contain ACM and/or PACM of the presence and location of ACM and/or PACM in such areas which may be contacted during such activities.

**1910.1001(j)(3)** *Warning signs—(i) Posting.* Warning signs shall be provided and displayed at each regulated area. In addition, warning signs shall be posted at all approaches to regulated areas so that an employee may read the signs and take necessary protective steps before entering the area.

**1910.1001(j)(3)(ii)** *Sign specifications.* **(A)** The warning signs required by paragraph (j)(3) of this section shall bear the following information:

**DANGER**

**ASBESTOS**

**CANCER AND LUNG DISEASE HAZARD**

**AUTHORIZED PERSONNEL ONLY**

**1910.1001(j)(3)(ii)(B)** In addition, where the use of respirators and protective clothing is required in the regulated area under this section, the warning signs shall include the following:

**RESPIRATORS AND PROTECTIVE CLOTHING ARE REQUIRED IN THIS AREA**

**1910.1001(j)(3)(iii)** [Reserved]

**1910.1001(j)(3)(iv)** The employer shall ensure that employees working in and contiguous to regulated areas comprehend the warning signs required to be posted by paragraph (j)(3)(i) of this section. Means to ensure employee comprehension may include the use of foreign languages, pictographs and graphics.

**1910.1001(j)(3)(v)** At the entrance to mechanical rooms/areas in which employees reasonably can be expected to enter and which contain ACM and/or PACM, the building owner shall post signs which identify the material which is present, its location, and appropriate work practices which, if followed, will ensure that ACM and/or PACM will not be disturbed. The employer shall ensure, to the extent feasible, that employees who come in contact with these signs can comprehend them. Means to ensure employee comprehension may include the use of foreign languages, pictographs, graphics, and awareness training.

**1910.1001(j)(4)** *Warning labels—(i) Labeling.* Warning labels shall be affixed to all raw materials, mixtures, scrap, waste, debris, and other products containing asbestos fibers, or to their containers. When a building owner or employer identifies previously installed ACM and/or PACM, labels or signs shall be affixed or posted so that employees will be notified of what materials contain ACM and/or PACM. The employer shall attach such labels in areas where they will clearly be noticed by employees who are likely to be exposed, such as at the entrance to mechanical room/areas. Signs required by paragraph (j)(3) of this section may be posted in lieu of labels so long as they contain information required for labelling.

**1910.1001(j)(4)(ii)** *Label specifications.* The labels shall comply with the requirements of 29 CFR 1910.1200(f) of OSHA's Hazard Communication standard, and shall include the following information:

**DANGER**

**CONTAINS ASBESTOS FIBERS**

**AVOID CREATING DUST**

**CANCER AND LUNG DISEASE HAZARD**

**1910.1001(j)(5)** *Material safety data sheets.* Employers who are manufacturers or importers of asbestos or asbestos products shall comply with the requirements regarding development of material safety data sheets as specified in 29 CFR 1910.1200(g) of OSHA's Hazard Communication standard, except as provided by paragraph (j)(6) of this section.

**1910.1001(j)(6)** The provisions for labels required by paragraph (j)(4) of this section or for material safety data sheets required by paragraph (j)(5) of this section do not apply where:

**1910.1001(j)(6)(i)** Asbestos fibers have been modified by a bonding agent, coating, binder, or other material provided that the manufacturer can demonstrate that during any reasonably foreseeable use, handling, storage, disposal, processing, or transportation, no airborne concentrations of fibers of asbestos in excess of the TWA permissible exposure level and/or excursion limit will be released or

**1910.1001(j)(6)(ii)** Asbestos is present in a product in concentrations less than 1.0%.

**1910.1001(j)(7)** *Employee information and training.* **(i)** The employer shall institute a training program for all employees who are exposed to airborne concentrations of asbestos at or above the PEL and/or excursion limit and ensure their participation in the program.

**1910.1001(j)(7)(ii)** Training shall be provided prior to or at the time of initial assignment and at least annually thereafter.

**1910.1001(j)(7)(iii)** The training program shall be conducted in a manner which the employee is able to understand. The employer shall ensure that each employee is informed of the following:

**1910.1001(j)(7)(iii)(A)** The health effects associated with asbestos exposure;

**1910.1001(j)(7)(iii)(B)** The relationship between smoking and exposure to asbestos producing lung cancer:

**1910.1001(j)(7)(iii)(C)** The quantity, location, manner of use, release, and storage of asbestos, and the specific nature of operations which could result in exposure to asbestos;

**1910.1001(j)(7)(iii)(D)** The engineering controls and work practices associated with the employee's job assignment;

**1910.1001(j)(7)(iii)(E)** The specific procedures implemented to protect employees from exposure to asbestos, such as appropriate work practices, emergency and clean-up procedures, and personal protective equipment to be used;

**1910.1001(j)(7)(iii)(F)** The purpose, proper use, and limitations of respirators and protective clothing, if appropriate;

**1910.1001(j)(7)(iii)(G)** The purpose and a description of the medical surveillance program required by paragraph (l) of this section;

**1910.1001(j)(7)(iii)(H)** The content of this standard, including appendices.

**1910.1001(j)(7)(iii)(I)** The names, addresses and phone numbers of public health organizations which provide information, materials, and/or conduct programs concerning smoking cessation. The employer may distribute the list of such organizations contained in Appendix I to this section, to comply with this requirement.

**1910.1001(j)(7)(iii)(J)** The requirements for posting signs and affixing labels and the meaning of the required legends for such signs and labels.

**1910.1001(j)(7)(iv)** The employer shall also provide, at no cost to employees who perform housekeeping operations in an area which contains ACM or PACM, an asbestos awareness training course, which shall at a minimum contain the following elements: health effects of asbestos, locations of ACM and PACM in the building/facility, recognition of ACM and PACM damage and deterioration, requirements in this standard relating to housekeeping, and proper response to fiber release episodes, to all employees who perform housekeeping work in areas where ACM and/or PACM is present. Each such employee shall be so trained at least once a year.

**1910.1001(j)(7)(v)** Access to information and training materials.

**1910.1001(j)(7)(v)(A)** The employer shall make a copy of this standard and its appendices readily available without cost to all affected employees.

**1910.1001(j)(7)(v)(B)** The employer shall provide, upon request, all materials relating to the employee information and training program to the Assistant Secretary and the training program to the Assistant Secretary and the Director.

**1910.1001(j)(7)(v)(C)** The employer shall inform all employees concerning the availability of self-help smoking cessation program material. Upon employee request, the employer shall distribute such material, consisting of NIH Publication No. 89-1647, or equivalent self-help material, which is approved or published by a public health organization listed in Appendix I to this section.

**1910.1001(j)(8)** *Criteria to rebut the designation of installed material as PACM.* **(i)** At any time, an employer and/or building owner may demonstrate, for purposes of this standard, that PACM does not contain asbestos. Building owners and/or employers are not required to communicate information about the presence of building material for which such a demonstration pursuant to the requirements of paragraph (j)(8)(ii) of this section has been made. However, in all such cases, the information, data and analysis supporting the determination that PACM does not contain asbestos, shall be retained pursuant to paragraph (m) of this section.

**1910.1001(j)(8)(ii)** An employer or owner may demonstrate that PACM does not contain asbestos by the following:

**1910.1001(j)(8)(ii)(A)** Having a completed inspection conducted pursuant to the requirements of AHERA (40 CFR 763, Subpart E) which demonstrates that no ACM is present in the material; or

**1910.1001(j)(8)(ii)(B)** Performing tests of the material containing PACM which demonstrate that no ACM is present in the material. Such tests shall include analysis of bulk samples collected in the manner described in 40 CFR 763.86. The tests, evaluation and sample collection shall be conducted by an accredited inspector or by a CIH. Analysis of samples shall be performed by persons or laboratories with proficiency demonstrated by current successful participation in a nationally recognized testing program such as the National Voluntary Laboratory Accreditation Program (NVLAP) or the National Institute for Standards and Technology (NIST) or the Round Robin for bulk samples administered by the American Industrial Hygiene Association (AIHA) or an equivalent nationally-recognized round robin testing program.

**1910.1001(j)(8)(iii)** The employer and/or building owner may demonstrate that flooring material including associated mastic and backing does not contain asbestos, by a determination of an industrial hygienist based upon recognized analytical techniques showing that the material is not ACM.

**1910.1001(k)** *Housekeeping.* **(1)** All surfaces shall be maintained as free as practicable of ACM waste and debris and accompanying dust.

**1910.1001(k)(2)** All spills and sudden releases of material containing asbestos shall be cleaned up as soon as possible.

**1910.1001(k)(3)** Surfaces contaminated with asbestos may not be cleaned by the use of compressed air.

**1910.1001(k)(4)** *Vacuuming.* HEPA-filtered vacuuming equipment shall be used for vacuuming asbestos containing waste and debris. The equipment shall be used and emptied in a manner which minimizes the reentry of asbestos into the workplace.

**1910.1001(k)(5)** Shoveling, dry sweeping and dry clean-up of asbestos may be used only where vacuuming and/or wet cleaning are not feasible.

**1910.1001(k)(6)** *Waste disposal.* Waste, scrap, debris, bags, containers, equipment, and clothing contaminated with asbestos consigned for disposal, shall be collected, recycled and disposed of in sealed impermeable bags, or other closed, impermeable containers.

**1910.1001(k)(7)** Care of asbestos-containing flooring material.

**1910.1001(k)(7)(i)** Sanding of asbestos-containing floor material is prohibited.

**1910.1001(k)(7)(ii)** Stripping of finishes shall be conducted using low abrasion pads at speeds lower than 300 rpm and wet methods.

**1910.1001(k)(7)(iii)** Burnishing or dry buffing may be performed only on asbestos-containing flooring which has sufficient finish so that the pad cannot contact the asbestos-containing material.

**1910.1001(k)(8)** Waste and debris and accompanying dust in an area containing accessible ACM and/or PACM or visibly deteriorated ACM, shall not be dusted or swept dry, or vacuumed without using a HEPA filter.

**1910.1001(l)** *Medical surveillance*—**(1)** *General*—**(i)** *Employees covered.* The employer shall institute a medical surveillance program for all employees who are or will be exposed to airborne concentrations of fibers of asbestos at or above the TWA and/or excursion limit.

**1910.1001(l)(1)(ii)** *Examination by a physician.* **(A)** The employer shall ensure that all medical examinations and procedures are performed by or under the supervision of a licensed physician, and shall be provided without cost to the employee and at a reasonable time and place.

**1910.1001(l)(1)(ii)(B)** Persons other than licensed physicians, who administer the pulmonary function testing required by this section, shall complete a training course in spirometry sponsored by an appropriate academic or professional institution.

**1910.1001(l)(2)** *Pre-placement examinations.* **(i)** Before an employee is assigned to an occupation exposed to airborne concentrations of asbestos fibers at or above the TWA and/or excursion limit, a pre-placement medical examination shall be provided or made available by the employer.

**1910.1001(l)(2)(ii)** Such examination shall include, as a minimum, a medical and work history; a complete physical examination of all systems with emphasis on the respiratory system, the cardiovascular system and digestive tract; completion of the respiratory disease standardized questionnaire in Appendix D to this section, Part 1; a chest roentgenogram (posterior-anterior 14×17 inches); pulmonary function tests to include forced vital capacity (FVC) and forced expiratory volume at 1 second (FEV (1.0)); and any additional tests deemed appropriate by the examining physician. Interpretation and classification of chest roentgenogram shall be conducted in accordance with Appendix E to this section.

**1910.1001(l)(3)** *Periodic examinations.* **(i)** Periodic medical examinations shall be made available annually.

**1910.1001(l)(3)(ii)** The scope of the medical examination shall be in conformance with the protocol established in paragraph (l)(2)(ii) of this section, except that the frequency of chest roentgenogram shall be conducted in accordance with Table 2, and the abbreviated standardized questionnaire contained in, Part 2 of Appendix D to this section shall be administered to the employee.

### TABLE 2—FREQUENCY OF CHEST ROENTGENOGRAM

| Years since first exposure | Age of employee |||
|---|---|---|---|
| | 15 to 35 | 35 + to 45 | 45+ |
| 0 to 10 | Every 5 years | Every 5 years | Every 5 years |
| 10+ | Every 5 years | Every 2 years | Every 1 year |

**1910.1001(l)(4)** *Termination of employment examinations.* **(i)** The employer shall provide, or make available, a termination of employment medical examination for any employee who has been exposed to airborne concentrations of fibers of asbestos at or above the TWA and/or excursion limit.

**1910.1001(l)(4)(ii)** The medical examination shall be in accordance with the requirements of the periodic examinations stipulated in paragraph (l)(3) of this section, and shall be given within 30 calendar days before or after the date of termination of employment.

**1910.1001(l)(5)** *Recent examinations.* No medical examination is required of any employee, if adequate records show that the employee has been examined in accordance with any of paragraphs ((l)(2) through (l)(4) of this section within the past 1 year period. A pre- employment medical examination which was required as a condition of employment by the employer, may not be used by that employer to meet the requirements of this paragraph, unless the cost of such examination is borne by the employer.

**1910.1001(l)(6)** *Information provided to the physician.* The employer shall provide the following information to the examining physician:

**1910.1001(l)(6)(i)** A copy of this standard and Appendices D and E.

**1910.1001(l)(6)(ii)** A description of the affected employee's duties as they relate to the employee's exposure.

**1910.1001(l)(6)(iii)** The employee's representative exposure level or anticipated exposure level.

**1910.1001(l)(6)(iv)** A description of any personal protective and respiratory equipment used or to be used.

**1910.1001(l)(6)(v)** Information from previous medical examinations of the affected employee that is not otherwise available to the examining physician.

**1910.1001(l)(7)** *Physician's written opinion.* **(i)** The employer shall obtain a written opinion from the examining physician. This written opinion shall contain the results of the medical examination and shall include:

**1910.1001(l)(7)(i)(A)** The physician's opinion as to whether the employee has any detected medical conditions that would place the employee at an increased risk of material health impairment from exposure to asbestos;

**1910.1001(l)(7)(i)(B)** Any recommended limitations on the employee or upon the use of personal protective equipment such as clothing or respirators;

**1910.1001(l)(7)(i)(C)** A statement that the employee has been informed by the physician of the results of the medical examination and of any medical conditions resulting from asbestos exposure that require further explanation or treatment; and

**1910.1001(l)(7)(i)(D)** A statement that the employee has been informed by the physician of the increased risk of lung cancer attributable to the combined effect of smoking and asbestos exposure.

**1910.1001(l)(7)(ii)** The employer shall instruct the physician not to reveal in the written opinion given to the employer specific findings or diagnoses unrelated to occupational exposure to asbestos.

**1910.1001(l)(7)(iii)** The employer shall provide a copy of the physician's written opinion to the affected employee within 30 days from its receipt.

**1910.1001(m)** *Recordkeeping*—**(1)** *Exposure measurements.*

**NOTE:** The employer may utilize the services of competent organizations such as industry trade associations and employee associations to maintain the records required by this section.

**1910.1001(m)(1)(i)** The employer shall keep an accurate record of all measurements taken to monitor employee exposure to asbestos as prescribed in paragraph (d) of this section.

**1910.1001(m)(1)(ii)** This record shall include at least the following information:

**1910.1001(m)(1)(ii)(A)** The date of measurement;

**1910.1001(m)(1)(ii)(B)** The operation involving exposure to asbestos which is being monitored;

**1910.1001(m)(1)(ii)(C)** Sampling and analytical methods used and evidence of their accuracy;

**1910.1001(m)(1)(ii)(D)** Number, duration, and results of samples taken;

**1910.1001(m)(1)(ii)(E)** Type of respiratory protective devices worn, if any; and

**1910.1001(m)(1)(ii)(F)** Name, social security number and exposure of the employees whose exposure are represented.

**1910.1001(m)(1)(iii)** The employer shall maintain this record for at least thirty (30) years, in accordance with 29 CFR 1910.20.

**1910.1001(m)(2)** *Objective data for exempted operations.* **(i)** Where the processing, use, or handling of products made from or containing asbestos is exempted from other requirements of this section under paragraph (d)(2)(iii) of this section, the employer shall establish and maintain an accurate record of objective data reasonably relied upon in support of the exemption.

**1910.1001(m)(2)(ii)** The record shall include at least the following:

**1910.1001(m)(2)(ii)(A)** The product qualifying for exemption;

**1910.1001(m)(2)(ii)(B)** The source of the objective data;

**1910.1001(m)(2)(ii)(C)** The testing protocol, results of testing, and/or analysis of the material for the release of asbestos;

**1910.1001(m)(2)(ii)(D)** A description of the operation exempted and how the data support the exemption; and

**1910.1001(m)(2)(ii)(E)** Other data relevant to the operations, materials, processing, or employee exposures covered by the exemption.

**1910.1001(m)(2)(iii)** The employer shall maintain this record for the duration of the employer's reliance upon such objective data.

**1910.1001(m)(3)** *Medical surveillance.* **(i)** The employer shall establish and maintain an accurate record for each employee subject to medical surveillance by paragraph (l)(1)(i) of this section, in accordance with 29 CFR 1910.1020.

**1910.1001(m)(3)(ii)** The record shall include at least the following information:

**1910.1001(m)(3)(ii)(A)** The name and social security number of the employee;

**1910.1001(m)(3)(ii)(B)** Physician's written opinions;

**1910.1001(m)(3)(ii)(C)** Any employee medical complaints related to exposure to asbestos; and

**1910.1001(m)(3)(ii)(D)** A copy of the information provided to the physician as required by paragraph (l)(6) of this section.

**1910.1001(m)(3)(iii)** The employer shall ensure that this record is maintained for the duration of employment plus thirty (30) years, in accordance with 29 CFR 1910.1020.

**1910.1001(m)(4)** *Training.* The employer shall maintain all employee training records for one (1) year beyond the last date of employment of that employee.

**1910.1001(m)(5)** *Availability.* **(i)** The employer, upon written request, shall make all records required to be maintained by this section available to the Assistant Secretary and the Director for examination and copying.

**1910.1001(m)(5)(ii)** The employer, upon request shall make any exposure records required by paragraph (m)(1) of this section available for examination and copying to affected employees, former employees, designated representatives and the Assistant Secretary, in accordance with 29 CFR 1910.1020(a) through (e) and (g) through (i).

**1910.1001(m)(5)(iii)** The employer, upon request, shall make employee medical records required by paragraph (m)(3) of this section available for examination and copying to the subject employee, to anyone having the specific written consent of the subject employee, and the Assistant Secretary, in accordance with 29 CFR 1910.1020.

**1910.1001(m)(6)** *Transfer of records.* **(i)** The employer shall comply with the requirements concerning transfer of records set forth in 29 CFR 1910.1020(h).

**1910.1001(m)(6)(ii)** Whenever the employer ceases to do business and there is no successor employer to receive and retain the records for the prescribed period, the employer shall notify the Director at least 90 days prior to disposal of records and, upon request, transmit them to the Director.

**1910.1001(n)** *Observation of monitoring*—**(1)** *Employee observation.* The employer shall provide affected employees or their designated representatives an opportunity to observe any monitoring of employee exposure to asbestos conducted in accordance with paragraph (d) of this section.

**1910.1001(n)(2)** *Observation procedures.* When observation of the monitoring of employee exposure to asbestos requires entry into an area where the use of protective clothing or equipment is required, the observer shall be provided with and be required to use such clothing and equipment and shall comply with all other applicable safety and health procedures.

**1910.1001(o)** *Appendices.* **(1)** Appendices A, C, D, E, and F to this section are incorporated as part of this section and the contents of these Appendices are mandatory.

**1910.1001(o)(2)** Appendices B, G, H, I, and J to this section are informational and are not intended to create any additional obligations not otherwise imposed or to detract from any existing obligations.

## APPENDIX A TO §1910.1001—OSHA REFERENCE METHOD—MANDATORY

This mandatory appendix specifies the procedure for analyzing air samples for asbestos and specifies quality control procedures that must be implemented by laboratories performing the analysis. The sampling and analytical methods described below represent the elements of the available monitoring methods (such as Appendix B of their regulation, the most current version of the OSHA method ID-160, or the most current version of the NIOSH Method 7400). All employers who are required to conduct air monitoring under paragraph (d) of the standard are required to utilize analytical laboratories that use this procedure, or an equivalent method, for collecting and analyzing samples.

### Sampling and Analytical Procedure

1. The sampling medium for air samples shall be mixed cellulose ester filter membranes. These shall be designated by the manufacturer as suitable for asbestos counting. See below for rejection of blanks.

2. The preferred collection device shall be the 25-mm diameter cassette with an open-faced 50-mm electrically conductive extension cowl. The 37-mm cassette may be used if necessary but only if written justification for the need to use the 37-mm filter cassette accompanies the sample results in the employee's exposure monitoring record. Do not reuse or reload cassettes for asbestos sample collection.

3. An air flow rate between 0.5 liter/min and 2.5 liters/min shall be selected for the 25-mm cassette. If the 37-mm cassette is used, an air flow rate between 1 liter/min and 2.5 liters/min shall be selected.

4. Where possible, a sufficient air volume for each air sample shall be collected to yield between 100 and 1,300 fibers per square millimeter on the membrane filter. If a filter darkens in appearance or if loose dust is seen on the filter, a second sample shall be started.

5. Ship the samples in a rigid container with sufficient packing material to prevent dislodging the collected fibers. Packing material that has a high electrostatic charge on its surface (e.g., expanded polystyrene) cannot be used because such material can cause loss of fibers to the sides of the cassette.

6. Calibrate each personal sampling pump before and after use with a representative filter cassette installed between the pump and the calibration devices.

7. Personal samples shall be taken in the "breathing zone" of the employee (i.e., attached to or near the collar or lapel near the worker's face).

8. Fiber counts shall be made by positive phase contrast using a microscope with an 8 to 10 X eyepiece and a 40 to 45 X objective for a total magnification of approximately 400 X and a numerical aperture of 0.65 to 0.75. The microscope shall also be fitted with a green or blue filter.

9. The microscope shall be fitted with a Walton-Beckett eyepiece graticule calibrated for a field diameter of 100 micrometers (±2 micrometers).

10. The phase-shift detection limit of the microscope shall be about 3 degrees measured using the HSE phase shift test slide as outlined below.

a. Place the test slide on the microscope stage and center it under the phase objective.

b. Bring the blocks of grooved lines into focus.

**NOTE:** The slide consists of seven sets of grooved lines (ca. 20 grooves to each block) in descending order of visibility from sets 1 to 7, seven being the least visible. The requirements for asbestos counting are that the microscope optics must resolve the grooved lines in set 3 completely, although they may appear somewhat faint, and that the grooved lines in sets 6 and 7 must be invisible. Sets 4 and 5 must be at least partially visible but may vary slightly in visibility between microscopes. A microscope that fails to meet these requirements has either too low or too high a resolution to be used for asbestos counting.

c. If the image deteriorates, clean and adjust the microscope optics. If the problem persists, consult the microscope manufacturer.

11. Each set of samples taken will include 10% field blanks or a minimum of 2 field blanks. These blanks must come from the same lot as the filters used for sample collection. The field blank results shall be averaged and subtracted from the analytical results before reporting. A set consists of any sample or group of samples for which an evaluation for this standard must be made. Any samples represented by a field blank having a fiber count in excess of the detection limit of the method being used shall be rejected.

12. The samples shall be mounted by the acetone/triacetin method or a method with an equivalent index of refraction and similar clarity.

13. Observe the following counting rules.

a. Count only fibers equal to or longer than 5 micrometers. Measure the length of curved fibers along the curve.

b. In the absence of other information, count all particles as asbestos that have a length-to-width ratio (aspect ratio) of 3:1 or greater.

c. Fibers lying entirely within the boundary of the Walton-Beckett graticule field shall receive a count of 1. Fibers crossing the boundary once, having one end within the circle, shall receive the count of one half (1/2). Do not count any fiber that crosses the graticule boundary more than once. Reject and do not count any other fibers even though they may be visible outside the graticule area.

d. Count bundles of fibers as one fiber unless individual fibers can be identified by observing both ends of an individual fiber.

e. Count enough graticule fields to yield 100 fibers. Count a minimum of 20 fields; stop counting at 100 fields regardless of fiber count.

14. Blind recounts shall be conducted at the rate of 10 percent.

### Quality Control Procedures

1. Intralaboratory program. Each laboratory and/or each company with more than one microscopist counting slides shall establish a statistically designed quality assurance program involving blind recounts and comparisons between microscopists to monitor the variability of counting by each microscopist and between microscopists. In a company with more than one laboratory, the program shall include all laboratories and shall also evaluate the laboratory-to-laboratory variability.

2.a. Interlaboratory program. Each laboratory analyzing asbestos samples for compliance determination shall implement an interlaboratory quality assurance program that as a minimum includes participation of at least two other independent laboratories. Each laboratory shall participate in round robin testing at least once every 6 months with at least all the other laboratories in its interlaboratory quality assurance group. Each laboratory shall submit slides typical of its own work load for use in this program. The round robin shall be designed and results analyzed using appropriate statistical methodology.

2.b. All laboratories should also participate in a national sample testing scheme such as the Proficiency Analytical Testing Program (PAT), or the Asbestos Registry sponsored by the American Industrial Hygiene Association (AIHA).

3. All individuals performing asbestos analysis must have taken the NIOSH course for sampling and evaluating airborne asbestos dust or an equivalent course.

4. When the use of different microscopes contributes to differences between counters and laboratories, the effect of the different microscope shall be evaluated and the microscope shall be replaced, as necessary.

5. Current results of these quality assurance programs shall be posted in each laboratory to keep the microscopists informed.

### APPENDIX B TO § 1910.1001—DETAILED PROCEDURES FOR ASBESTOS SAMPLING AND ANALYSIS—NON-MANDATORY

Matrix Air:
OSHA Permissible Exposure Limits:
  Time Weighted Average ............... 0.1 fiber/cc
  Excursion Level (30 minutes) ...... 1.0 fiber/cc
Collection Procedure:

A known volume of air is drawn through a 25-mm diameter cassette containing a mixed-cellulose ester filter. The cassette must be equipped with an electrically conductive 50-mm extension cowl. The sampling time and rate are chosen to give a fiber density of between 100 to 1,300 fibers/mm$^2$ on the filter.

Recommended Sampling Rate ............... 0.5 to 5.0 liters/minute (L/min)
Recommended Air Volumes:
  Minimum .......................... 25 L
  Maximum .......................... 2,400 L

Analytical Procedure: A portion of the sample filter is cleared and prepared for asbestos fiber counting by Phase Contrast Microscopy (PCM) at 400X.

Commercial manufacturers and products mentioned in this method are for descriptive use only and do not constitute endorsements by USDOL-OSHA. Similar products from other sources can be substituted.

#### 1. Introduction

This method describes the collection of airborne asbestos fibers using calibrated sampling pumps with mixed-cellulose ester (MCE) filters and analysis by phase contrast microscopy (PCM). Some terms used are unique to this method and are defined below:

*Asbestos:* A term for naturally occurring fibrous minerals. Asbestos includes chrysotile, crocidolite, amosite (cummingtonite-grunerite asbestos), tremolite asbestos, actinolite asbestos, anthophyllite asbestos, and any of these minerals that have been chemically treated and/or altered. The precise chemical formulation of each species will vary with the location from which it was mined. Nominal compositions are listed:

| | |
|---|---|
| Chrysotile | $Mg_3 Si_2 O_5 (OH)_4$ |
| Crocidolite | $Na_2 Fe_3 2 + Fe_2 3 + Si_8 O_{22} (OH)^2$ |
| Amosite | $(Mg,Fe)_7 Si_8 O_{22} (OH)_2$ |
| Tremolite-actinolite | $Ca_2(Mg,Fe)_5 Si_8 O_{22} (OH)_2$ |
| Anthophyllite | $(Mg,Fe)_7 Si_8 O_{22} (OH)_2$ |

*Asbestos Fiber:* A fiber of asbestos which meets the criteria specified below for a fiber.

*Aspect Ratio:* The ratio of the length of a fiber to it's diameter (e.g., 3:1, 5:1 aspect ratios).

*Cleavage Fragments:* Mineral particles formed by comminution of minerals, especially those characterized by parallel sides and a moderate aspect ratio (usually less than 20:1).

*Detection Limit:* The number of fibers necessary to be 95% certain that the result is greater than zero.

*Differential Counting:* The term applied to the practice of excluding certain kinds of fibers from the fiber count because they do not appear to be asbestos.

*Fiber:* A particle that is 5 µm or longer, with a length-to-width ratio of 3 to 1 or longer.

*Field:* The area within the graticule circle that is superimposed on the microscope image.

*Set:* The samples which are taken, submitted to the laboratory, analyzed, and for which, interim or final result reports are generated.

*Tremolite, Anthophyllite, and Actinolite:* The non-asbestos form of these minerals which meet the definition of a fiber. It includes any of these minerals that have been chemically treated and/or altered.

*Walton-Beckett Graticule:* An eyepiece graticule specifically designed for asbestos fiber counting. It consists of a circle with a projected diameter of 100 2 µm (area of about 0.00785 mm²) with a crosshair having tic-marks at 3-µm intervals in one direction and 5-µm in the orthogonal direction. There are marks around the periphery of the circle to demonstrate the proper sizes and shapes of fibers. This design is reproduced in Figure 1. The disk is placed in one of the microscope eyepieces so that the design is superimposed on the field of view.

### 1.1. History

Early surveys to determine asbestos exposures were conducted using impinger counts of total dust with the counts expressed as million particles per cubic foot. The British Asbestos Research Council recommended filter membrane counting in 1969. In July 1969, the Bureau of Occupational Safety and Health published a filter membrane method for counting asbestos fibers in the United States. This method was refined by NIOSH and published as P CAM 239. On May 29, 1971, OSHA specified filter membrane sampling with phase contrast counting for evaluation of asbestos exposures at work sites in the United States. The use of this technique was again required by OSHA in 1986. Phase contrast microscopy has continued to be the method of choice for the measurement of occupational exposure to asbestos.

### 1.2. Principle

Air is drawn through a MCE filter to capture airborne asbestos fibers. A wedge shaped portion of the filter is removed, placed on a glass microscope slide and made transparent. A measured area (field) is viewed by PCM. All the fibers meeting defined criteria for asbestos are counted and considered a measure of the airborne asbestos concentration.

### 1.3. Advantages and Disadvantages

There are four main advantages of PCM over other methods:

(1) The technique is specific for fibers. Phase contrast is a fiber counting technique which excludes non-fibrous particles from the analysis.

(2) The technique is inexpensive and does not require specialized knowledge to carry out the analysis for total fiber counts.

(3) The analysis is quick and can be performed on-site for rapid determination of air concentrations of asbestos fibers.

(4) The technique has continuity with historical epidemiological studies so that estimates of expected disease can be inferred from long-term determinations of asbestos exposures.

The main disadvantage of PCM is that it does not positively identify asbestos fibers. Other fibers which are not asbestos may be included in the count unless differential counting is performed. This requires a great deal of experience to adequately differentiate asbestos from non-asbestos fibers. Positive identification of asbestos must be performed by polarized light or electron microscopy techniques. A further disadvantage of PCM is that the smallest visible fibers are about 0.2 µm in diameter while the finest asbestos fibers may be as small as 0.02 µm in diameter. For some exposures, substantially more fibers may be present than are actually counted.

### 1.4. Workplace Exposure

Asbestos is used by the construction industry in such products as shingles, floor tiles, asbestos cement, roofing felts, insulation and acoustical products. Non-construction uses include brakes, clutch facings, paper, paints, plastics, and fabrics. One of the most significant exposures in the workplace is the removal and encapsulation of asbestos in schools, public buildings, and homes. Many workers have the potential to be exposed to asbestos during these operations.

About 95% of the asbestos in commercial use in the United States is chrysotile. Crocidolite and amosite make up most of the remainder. Anthophyllite and tremolite or actinolite are likely to be encountered as contaminants in various industrial products.

### 1.5. Physical Properties

Asbestos fiber possesses a high tensile strength along its axis, is chemically inert, non-combustible, and heat resistant. It has a high electrical resistance and good sound absorbing properties. It can be weaved into cables, fabrics or other textiles, and also matted into asbestos papers, felts, or mats.

## 2. Range and Detection Limit

2.1. The ideal counting range on the filter is 100 to 1,300 fibers/mm². With a Walton-Beckett graticule this range is equivalent to 0.8 to 10 fibers/field. Using NIOSH counting statistics, a count of 0.8 fibers/field would give an approximate coefficient of variation (CV) of 0.13.

2.2. The detection limit for this method is 4.0 fibers per 100 fields or 5.5 fibers/mm². This was determined using an equation to estimate the maximum CV possible at a specific concentration (95% confidence) and a Lower Control Limit of zero. The CV value was then used to determine a corresponding concentration from historical CV vs fiber relationships. As an example:

**Lower Control Limit (95% Confidence) = AC - 1.645(CV)(AC)**

Where:

AC = Estimate of the airborne fiber concentration (fibers/cc) Setting the Lower Control Limit = 0 and solving for CV:

0 = AC - 1.645(CV)(AC)

CV = 0.61

This value was compared with CV vs. count curves. The count at which CV = 0.61 for Leidel-Busch counting statistics or for an OSHA Salt Lake Technical Center (OSHA-SLTC) CV curve (see Appendix A for further information) was 4.4 fibers or 3.9 fibers per 100 fields, respectively. Although a lower detection limit of 4 fibers per 100 fields is supported by the OSHA-SLTC data, both data sets support the 4.5 fibers per 100 fields value.

## 3. Method Performance—Precision and Accuracy

Precision is dependent upon the total number of fibers counted and the uniformity of the fiber distribution on the filter. A general rule is to count at least 20 and not more than 100 fields. The count is discontinued when 100 fibers are counted, provided that 20 fields have already been counted. Counting more than 100 fibers results in only a small gain in precision. As the total count drops below 10 fibers, an accelerated loss of precision is noted.

At this time, there is no known method to determine the absolute accuracy of the asbestos analysis. Results of samples prepared through the Proficiency Analytical Testing (PAT) Program and analyzed by the OSHA-SLTC showed no significant bias when compared to PAT reference values. The PAT samples were analyzed from 1987 to 1989 (N=36) and the concentration range was from 120 to 1,300 fibers/mm².

## 4. Interferences

Fibrous substances, if present, may interfere with asbestos analysis. Some common fibers are:

fiberglass

anhydrite

plant fibers

perlite veins

gypsum

some synthetic fibers

membrane structures

sponge spicules

# OSHA Standards for General Industry

diatoms

microorganisms

wollastonite

The use of electron microscopy or optical tests such as polarized light, and dispersion staining may be used to differentiate these materials from asbestos when necessary.

## 5. Sampling
### 5.1. Equipment

*5.1.1.* Sample assembly (The assembly is shown in Figure 3). Conductive filter holder consisting of a 25-mm diameter, 3-piece cassette having a 50-mm long electrically conductive extension cowl. Backup pad, 25-mm, cellulose. Membrane filter, mixed-cellulose ester (MCE), 25-mm, plain, white, 0.4 to 1.2-μm pore size.

**NOTES:** (a) Do not re-use cassettes.

(b) Fully conductive cassettes are required to reduce fiber loss to the sides of the cassette due to electrostatic attraction.

(c) Purchase filters which have been selected by the manufacturer for asbestos counting or analyze representative filters for fiber background before use. Discard the filter lot if more than 4 fibers/100 fields are found.

(d) To decrease the possibility of contamination, the sampling system (filter-backup pad-cassette) for asbestos is usually preassembled by the manufacturer.

(e) Other cassettes, such as the Bell-mouth, may be used within the limits of their validation.

*5.1.2.* Gel bands for sealing cassettes.

*5.1.3.* Sampling pump.

Each pump must be a battery operated, self-contained unit small enough to be placed on the monitored employee and not interfere with the work being performed. The pump must be capable of sampling at the collection rate for the required sampling time.

*5.1.4.* Flexible tubing, 6-mm bore.

*5.1.5.* Pump calibration.

Stopwatch and bubble tube/burette or electronic meter.

### 5.2. Sampling Procedure

*5.2.1.* Seal the point where the base and cowl of each cassette meet with a gel band or tape.

*5.2.2.* Charge the pumps completely before beginning.

*5.2.3.* Connect each pump to a calibration cassette with an appropriate length of 6-mm bore plastic tubing. Do not use luer connectors—the type of cassette specified above has built-in adapters.

*5.2.4.* Select an appropriate flow rate for the situation being monitored. The sampling flow rate must be between 0.5 and 5.0 l/min for personal sampling and is commonly set between 1 and 2 l/min. Always choose a flow rate that will not produce overloaded filters.

*5.2.5.* Calibrate each sampling pump before and after sampling with a calibration cassette in-line (NOTE: This calibration cassette should be from the same lot of cassettes used for sampling). Use a primary standard (e.g., bubble burette) to calibrate each pump. If possible, calibrate at the sampling site.

**NOTE:** If sampling site calibration is not possible, environmental influences may affect the flow rate. The extent is dependent on the type of pump used. Consult with the pump manufacturer to determine dependence on environmental influences. If the pump is affected by temperature and pressure changes, correct the flow rate using the formula shown in the section "Sampling Pump Flow Rate Corrections" at the end of this appendix.

*5.2.6.* Connect each pump to the base of each sampling cassette with flexible tubing. Remove the end cap of each cassette and take each air sample open face. Assure that each sample cassette is held open side down in the employee's breathing zone during sampling. The distance from the nose/mouth of the employee to the cassette should be about 10 cm. Secure the cassette on the collar or lapel of the employee using spring clips or other similar devices.

*5.2.7.* A suggested minimum air volume when sampling to determine TWA compliance is 25 L. For Excursion Limit (30 min sampling time) evaluations, a minimum air volume of 48 L is recommended.

*5.2.8.* The most significant problem when sampling for asbestos is overloading the filter with non-asbestos dust. Suggested maximum air sample volumes for specific environments are:

| Environment | Air vol. (L) |
|---|---|
| Asbestos removal operations (visible dust) | 100 |
| Asbestos removal operations (little dust) | 240 |
| Office environments | 400 to 2,400 |

**CAUTION:** Do not overload the filter with dust. High levels of non-fibrous dust particles may obscure fibers on the filter and lower the count or make counting impossible. If more than about 25 to 30% of the field area is obscured with dust, the result may be biased low. Smaller air volumes may be necessary when there is excessive non-asbestos dust in the air.

While sampling, observe the filter with a small flashlight. If there is a visible layer of dust on the filter, stop sampling, remove and seal the cassette, and replace with a new sampling assembly. The total dust loading should not exceed 1 mg.

*5.2.9.* Blank samples are used to determine if any contamination has occurred during sample handling. Prepare two blanks for the first 1 to 20 samples. For sets containing greater than 20 samples, prepare blanks as 10% of the samples. Handle blank samples in the same manner as air samples with one exception: Do not draw any air through the blank samples. Open the blank cassette in the place where the sample cassettes are mounted on the employee. Hold it open for about 30 seconds. Close and seal the cassette appropriately. Store blanks for shipment with the sample cassettes.

*5.2.10.* Immediately after sampling, close and seal each cassette with the base and plastic plugs. Do not touch or puncture the filter membrane as this will invalidate the analysis.

*5.2.11* Attach and secure a sample seal around each sample cassette in such a way as to assure that the end cap and base plugs cannot be removed without destroying the seal. Tape the ends of the seal together since the seal is not long enough to be wrapped end-to-end. Also wrap tape around the cassette at each joint to keep the seal secure.

### 5.3. Sample Shipment

*5.3.1.* Send the samples to the laboratory with paperwork requesting asbestos analysis. List any known fibrous interferences present during sampling on the paperwork. Also, note the workplace operation(s) sampled.

*5.3.2.* Secure and handle the samples in such that they will not rattle during shipment nor be exposed to static electricity. Do not ship samples in expanded polystyrene peanuts, vermiculite, paper shreds, or excelsior. Tape sample cassettes to sheet bubbles and place in a container that will cushion the samples in such a manner that they will not rattle.

*5.3.3.* To avoid the possibility of sample contamination, always ship bulk samples in separate mailing containers.

## 6. Analysis
### 6.1. Safety Precautions

*6.1.1.* Acetone is extremely flammable and precautions must be taken not to ignite it. Avoid using large containers or quantities of acetone. Transfer the solvent in a ventilated laboratory hood. Do not use acetone near any open flame. For generation of acetone vapor, use a spark free heat source.

*6.1.2.* Any asbestos spills should be cleaned up immediately to prevent dispersal of fibers. Prudence should be exercised to avoid contamination of laboratory facilities or exposure of personnel to asbestos. Asbestos spills should be cleaned up with wet methods and/or a High Efficiency Particulate-Air (HEPA) filtered vacuum.

*Caution:* Do not use a vacuum without a HEPA filter—It will disperse fine asbestos fibers in the air.

### 6.2. Equipment

*6.2.1.* Phase contrast microscope with binocular or trinocular head.

*6.2.2.* Widefield or Huygenian 10X eyepieces (NOTE: The eyepiece containing the graticule must be a focusing eyepiece. Use a 40X phase objective with a numerical aperture of 0.65 to 0.75).

6.2.3. Kohler illumination (if possible) with green or blue filter.

6.2.4. Walton-Beckett Graticule, type G-22 with 100 ±2 µm projected diameter.

6.2.5. Mechanical stage.

A rotating mechanical stage is convenient for use with polarized light.

6.2.6. Phase telescope.

6.2.7. Stage micrometer with 0.01-mm subdivisions.

6.2.8. Phase-shift test slide, mark II (Available from PTR optics Ltd., and also McCrone).

6.2.9. Precleaned glass slides, 25 mm X 75 mm. One end can be frosted for convenience in writing sample numbers, etc., or paste-on labels can be used.

6.2.10. Cover glass #1 1/2.

6.2.11. Scalpel (#10, curved blade).

6.2.12. Fine tipped forceps.

6.2.13. Aluminum block for clearing filter (see Appendix D and Figure 4).

6.2.14. Automatic adjustable pipette, 100- to 500-µL.

6.2.15. Micropipette, 5 µL.

### 6.3. Reagents

6.3.1. Acetone (HPLC grade).

6.3.2. Triacetin (glycerol triacetate).

6.3.3. Lacquer or nail polish.

### 6.4. Standard Preparation

A way to prepare standard asbestos samples of known concentration has not been developed. It is possible to prepare replicate samples of nearly equal concentration. This has been performed through the PAT program. These asbestos samples are distributed by the AIHA to participating laboratories.

Since only about one-fourth of a 25-mm sample membrane is required for an asbestos count, any PAT sample can serve as a "standard" for replicate counting.

### 6.5. Sample Mounting

**NOTE:** See Safety Precautions in Section 6.1. before proceeding. The objective is to produce samples with a smooth (non-grainy) background in a medium with a refractive index of approximately 1.46. The technique below collapses the filter for easier focusing and produces permanent mounts which are useful for quality control and interlaboratory comparison.

An aluminum block or similar device is required for sample preparation.

6.5.1. Heat the aluminum block to about 70 °C. The hot block should not be used on any surface that can be damaged by either the heat or from exposure to acetone.

6.5.2. Ensure that the glass slides and cover glasses are free of dust and fibers.

6.5.3. Remove the top plug to prevent a vacuum when the cassette is opened. Clean the outside of the cassette if necessary. Cut the seal and/or tape on the cassette with a razor blade. Very carefully separate the base from the extension cowl, leaving the filter and backup pad in the base.

6.5.4. With a rocking motion cut a triangular wedge from the filter using the scalpel. This wedge should be one-sixth to one-fourth of the filter. Grasp the filter wedge with the forceps on the perimeter of the filter which was clamped between the cassette pieces. DO NOT TOUCH the filter with your finger. Place the filter on the glass slide sample side up. Static electricity will usually keep the filter on the slide until it is cleared.

6.5.5. Place the tip of the micropipette containing about 200 µL acetone into the aluminum block. Insert the glass slide into the receiving slot in the aluminum block. Inject the acetone into the block with slow, steady pressure on the plunger while holding the pipette firmly in place. Wait 3 to 5 seconds for the filter to clear, then remove the pipette and slide from the aluminum block.

6.5.6. Immediately (less than 30 seconds) place 2.5 to 3.5 µL of triacetin on the filter (NOTE: Waiting longer than 30 seconds will result in increased index of refraction and decreased contrast between the fibers and the preparation. This may also lead to separation of the cover slip from the slide).

6.5.7. Lower a cover slip gently onto the filter at a slight angle to reduce the possibility of forming air bubbles. If more than 30 seconds have elapsed between acetone exposure and triacetin application, glue the edges of the cover slip to the slide with lacquer or nail polish.

6.5.8. If clearing is slow, warm the slide for 15 min on a hot plate having a surface temperature of about 50 °C to hasten clearing. The top of the hot block can be used if the slide is not heated too long.

6.5.9. Counting may proceed immediately after clearing and mounting are completed.

### 6.6. Sample Analysis

Completely align the microscope according to the manufacturer's instructions. Then, align the microscope using the following general alignment routine at the beginning of every counting session and more often if necessary.

#### 6.6.1. Alignment

(1) Clean all optical surfaces. Even a small amount of dirt can significantly degrade the image.

(2) Rough focus the objective on a sample.

(3) Close down the field iris so that it is visible in the field of view. Focus the image of the iris with the condenser focus. Center the image of the iris in the field of view.

(4) Install the phase telescope and focus on the phase rings. Critically center the rings. Misalignment of the rings results in astigmatism which will degrade the image.

(5) Place the phase-shift test slide on the microscope stage and focus on the lines. The analyst must see line set 3 and should see at least parts of 4 and 5 but, not see line set 6 or 6. A microscope/microscopist combination which does not pass this test may not be used.

#### 6.6.2. Counting Fibers

(1) Place the prepared sample slide on the mechanical stage of the microscope. Position the center of the wedge under the objective lens and focus upon the sample.

(2) Start counting from one end of the wedge and progress along a radial line to the other end (count in either direction from perimeter to wedge tip). Select fields randomly, without looking into the eyepieces, by slightly advancing the slide in one direction with the mechanical stage control.

(3) Continually scan over a range of focal planes (generally the upper 10 to 15 mm of the filter surface) with the fine focus control during each field count. Spend at least 5 to 15 seconds per field.

(4) Most samples will contain asbestos fibers with fiber diameters less than 1 mm. Look carefully for faint fiber images. The small diameter fibers will be very hard to see. However, they are an important contribution to the total count.

(5) Count only fibers equal to or longer than 5 mm. Measure the length of curved fibers along the curve.

(6) Count fibers which have a length to width ratio of 3:1 or greater.

(7) Count all the fibers in at least 20 fields. Continue counting until either 100 fibers are counted or 100 fields have been viewed; whichever occurs first. Count all the fibers in the final field.

(8) Fibers lying entirely within the boundary of the Walton-Beckett graticule field shall receive a count of 1. Fibers crossing the boundary once, having one end within the circle shall receive a count of 1/2. Do not count any fiber that crosses the graticule boundary more than once. Reject and do not count any other fibers even though they may be visible outside the graticule area. If a fiber touches the circle, it is considered to cross the line.

(9) Count bundles of fibers as one fiber unless individual fibers can be clearly identified and each individual fiber is clearly not connected to another counted fiber. See Figure 1 for counting conventions.

(10) Record the number of fibers in each field in a consistent way such that filter non-uniformity can be assessed.

(11) Regularly check phase ring alignment.

(12) When an agglomerate (mass of material) covers more than 25% of the field of view, reject the field tand select another. Do not include it in the number of fields counted.

(13) Perform a "blind recount" of 1 in every 10 filter wedges (slides). Re-label the slides using a person other than the original counter.

# OSHA Standards for General Industry

## 6.7. Fiber Identification

As previously mentioned in Section 1.3, PCM does not provide positive confirmation of asbestos fibers. Alternate differential counting techniques should be used if discrimination is desirable. Differential counting may include primary discrimination based on morphology, polarized light analysis of fibers, or modification of PCM data by Scanning Electron or Transmission Electron Microscopy.

A great deal of experience is required to routinely and correctly perform differential counting. It is discouraged unless it is legally necessary. Then, only if a fiber is obviously not asbestos should it be excluded from the count. Further discussion of this technique can be found in reference 8.10.

If there is a question whether a fiber is asbestos or not, follow the rule:

"WHEN IN DOUBT, COUNT."

## 6.8. Analytical Recommendations— Quality Control System

*6.8.1.* All individuals performing asbestos analysis must have taken the NIOSH course for sampling and evaluating airborne asbestos or an equivalent course.

*6.8.2.* Each laboratory engaged in asbestos counting shall set up a slide trading arrangement with at least two other laboratories in order to compare performance and eliminate inbreeding of error. The slide exchange occurs at least semiannually. The round robin results shall be posted where all analysts can view individual analyst's results.

*6.8.3.* Each laboratory engaged in asbestos counting shall participate in the Proficiency Analytical Testing Program, the Asbestos Analyst Registry or equivalent.

*6.8.4.* Each analyst shall select and count prepared slides from a "slide bank". These are quality assurance counts. The slide bank shall be prepared using uniformly distributed samples taken from the workload. Fiber densities should cover the entire range routinely analyzed by the laboratory. These slides are counted blind by all counters to establish an original standard deviation. This historical distribution is compared with the quality assurance counts. A counter must have 95% of all quality control samples counted within three standard deviations of the historical mean. This count is then integrated into a new historical mean and standard deviation for the slide.

The analyses done by the counters to establish the slide bank may be used for an interim quality control program if the data are treated in a proper statistical fashion.

## 7. Calculations

7.1. Calculate the estimated airborne asbestos fiber concentration on the filter sample using the following formula:

where:

AC = Airborne fiber concentration

$$AC = \frac{\left[\left(\frac{FB}{FL}\right) - \left(\frac{BFB}{BFL}\right)\right] \times ECA}{1000 \times FR \times T \times MFA}$$

FB = Total number of fibers greater than 5 μm counted

FL = Total number of fields counted on the filter

BFB = Total number of fibers greater than 5 μm counted in the blank

BFL = Total number of fields counted on the blank

ECA = Effective collecting area of filter (385 mm² nominal for a 25-mm filter.)

FR = Pump flow rate (l/min)

MFA = Microscope count field area (mm²). This is 0.00785 mm² for a Walton-Beckett Graticule.

T = Sample collection time (min)

1,000 = Conversion of L to cc

**NOTE:** The collection area of a filter is seldom equal to 385 mm². It is appropriate for laboratories to routinely monitor the exact diameter using an inside micrometer. The collection area is calculated according to the formula:

$$\text{Area} = \pi(d/2)^2$$

## 7.2. Short-cut Calculation

Since a given analyst always has the same interpupillary distance, the number of fields per filter for a particular analyst will remain constant for a given size filter. The field size for that analyst is constant (i.e., the analyst is using an assigned microscope and is not changing the reticle).

For example, if the exposed area of the filter is always 385 mm² and the size of the field is always 0.00785 mm², the number of fields per filter will always be 49,000. In addition it is necessary to convert liters of air to cc. These three constants can then be combined such that ECA/(1,000 × MFA) = 49. The previous equation simplifies to:

$$AC = \frac{\left(\frac{FB}{FL}\right) - \left(\frac{BFB}{BFL}\right) \times 49}{FR \times T}$$

## 7.3. Recount Calculations

As mentioned in step 13 of Section 6.6.2., a "blind recount" of 10% of the slides is performed. In all cases, differences will be observed between the first and second counts of the same filter wedge. Most of these differences will be due to chance alone, that is, due to the random variability (precision) of the count method. Statistical recount criteria enables one to decide whether observed differences can be explained due to chance alone or are probably due to systematic differences between analysts, microscopes, or other biasing factors.

The following recount criterion is for a pair of counts that estimate AC in fibers/cc. The criterion is given at the type-I error level. That is, there is 5% maximum risk that we will reject a pair of counts for the reason that one might be biased, when the large observed difference is really due to chance.

Reject a pair of counts if:

$$|\sqrt{AC_2} - \sqrt{AC_1}| > 2.78 \times \left(\sqrt{AC_{AVG}}\right) \times CV_{FB}$$

Where:

$AC_1$ = lower estimated airborne fiber concentration

$AC_2$ = higher estimated airborne fiber concentration

$AC_{AVG}$ = average of the two concentration estimates

$CV_{FB}$ = CV for the average of the two concentration estimates

If a pair of counts are rejected by this criterion then, recount the rest of the filters in the submitted set. Apply the test and reject any other pairs failing the test. Rejection shall include a memo to the industrial hygienist stating that the sample failed a statistical test for homogeneity and the true air concentration may be significantly different than the reported value.

## 7.4. Reporting Results

Report results to the industrial hygienist as fibers/cc. Use two significant figures. If multiple analyses are performed on a sample, an average of the results is to be reported unless any of the results can be rejected for cause.

## 8. References

8.1. Dreesen, W.C., et al, *U.S. Public Health Service: A Study of Asbestosis in the Asbestos Textile Industry*, (Public Health Bulletin No. 241), US Treasury Dept., Washington, DC, 1938.

8.2. *Asbestos Research Council: The Measurement of Airborne Asbestos Dust by the Membrane Filter Method* (Technical Note), Asbestos Research Council, Rockdale, Lancashire, Great Britain, 1969.

8.3. Bayer, S.G., Zumwalde, R.D., Brown, T.A., *Equipment and Procedure for Mounting Millipore Filters and Counting Asbestos Fibers by Phase Contrast Microscopy*, Bureau of Occupational Health, U.S. Dept. of Health, Education and Welfare, Cincinnati, OH, 1969.

8.4. *NIOSH Manual of Analytical Methods*, 2nd ed., Vol. 1 (DHEW/NIOSH Pub. No. 77-157-A). National Institute for Occupational Safety and Health, Cincinnati, OH, 1977. pp. 239-1–239-21.

8.5. *Asbestos*, Code of Federal Regulations 29 CFR 1910.1001. 1971.

8.6. *Occupational Exposure to Asbestos, Tremolite, Anthophyllite, and Actinolite. Final Rule*, FEDERAL REGISTER 51:119 (20 June 1986). pp.22612-22790.

8.7. *Asbestos, Tremolite, Anthophyllite, and Actinolite,* Code of Federal Regulations 1910.1001. 1988. pp 711-752.

8.8. *Criteria for a Recommended Standard—Occupational Exposure to Asbestos* (DHEW/NIOSH Pub. No. HSM 72-10267), National Institute for Occupational Safety and Health NIOSH, Cincinnati, OH, 1972. pp. III-1-III-24.

8.9. Leidel, N.A., Bayer, S.G., Zumwalde, R.D., Busch, K.A., *USPHS/NIOSH Membrane Filter Method for Evaluating Airborne Asbestos Fibers* (DHEW/NIOSH Pub. No. 79-127). National Institute for Occupational Safety and Health, Cincinnati, OH, 1979.

8.10. Dixon, W.C., *Applications of Optical Microscopy in Analysis of Asbestos and Quartz,* Analytical Techniques in Occupational Health Chemistry, edited by D.D. Dollberg and A.W. Verstuyft. Wash. DC: American Chemical Society, (ACS Symposium Series 120) 1980. pp. 13-41.

### Quality Control

The OSHA asbestos regulations require each laboratory to establish a quality control program. The following is presented as an example of how the OSHA-SLTC constructed its internal CV curve as part of meeting this requirement. Data is from 395 samples collected during OSHA compliance inspections and analyzed from October 1980 through April 1986.

Each sample was counted by 2 to 5 different counters independently of one another. The standard deviation and the CV statistic was calculated for each sample. This data was then plotted on a graph of CV vs. fibers/mm². A least squares regression was performed using the following equation:

$$CV = \text{antilog}_{10}[A(\log_{10}(x))^2 + B(\log_{10}(x)) + C]$$

where:

$x$ = the number of fibers/mm²

Application of least squares gave:

$A = 0.182205$

$B = -0.973343$

$C = 0.327499$

Using these values, the equation becomes:

$CV = \text{antilog}_{10}[0.182205(\log_{10}(x))^2 - 0.973343(\log_{10}(x)) + 0.327499]$

*Sampling Pump Flow Rate Corrections*

This correction is used if a difference greater than 5% in ambient temperature and/or pressure is noted between calibration and sampling sites and the pump does not compensate for the differences.

$$Q_{act} = Q_{cal} \times \sqrt{\left(\frac{P_{cal}}{P_{act}}\right) \times \left(\frac{T_{act}}{T_{cal}}\right)}$$

Where:

$Q_{act}$ = actual flow rate

$Q_{cal}$ = calibrated flow rate (if a rotameter was used, the rotameter value)

$P_{cal}$ = uncorrected air pressure at calibration

$P_{act}$ = uncorrected air pressure at sampling site

$T_{act}$ = temperature at sampling site (K)

$T_{cal}$ = temperature at calibration (K)

### Walton-Beckett Graticule

When ordering the Graticule for asbestos counting, specify the exact disc diameter needed to fit the ocular of the microscope and the diameter (mm) of the circular counting area. Instructions for measuring the dimensions necessary are listed:

(1) Insert any available graticule into the focusing eyepiece and focus so that the graticule lines are sharp and clear.

(2) Align the microscope.

(3) Place a stage micrometer on the microscope object stage and focus the microscope on the graduated lines.

(4) Measure the magnified grid length, PL (µm), using the stage micrometer.

(5) Remove the graticule from the microscope and measure its actual grid length, AL (mm). This can be accomplished by using a mechanical stage fitted with verniers, or a jeweler's loupe with a direct reading scale.

(6) Let D = 100 µm. Calculate the circle diameter, $d_c$ (mm), for the Walton-Beckett graticule and specify the diameter when making a purchase:

$$d_c = \frac{AL \times D}{PL}$$

Example: If PL = 108 µm, AL = 2.93 mm and D = 100 µm, then,

$$d_c = \frac{2.93 \times 100}{108} = 2.71 \text{ mm}$$

(7) Each eyepiece-objective-reticle combination on the microscope must be calibrated. Should any of the three be changed (by zoom adjustment, disassembly, replacement, etc.), the combination must be recalibrated. Calibration may change if interpupillary distance is changed. Measure the field diameter, D (acceptable range: 100 ±2 µm) with a stage micrometer upon receipt of the graticule from the manufacturer. Determine the field area (mm²).

Field Area = $\pi(D/2)^2$

If D = 100 µm = 0.1 mm, then

Field Area = $\pi(0.1 \text{ mm}/2)^2 = 0.00785$ mm²

The Graticule is available from: Graticules Ltd., Morley Road, Tonbridge TN9 1RN, Kent, England (Telephone 011-44-732-359061). Also available from PTR Optics Ltd., 145 Newton Street, Waltham, MA 02154 [telephone (617) 891-6000] or McCrone Accessories and Components, 2506 S. Michigan Ave., Chicago, IL 60616 [phone (312)-842-7100]. The graticule is custom made for each microscope.

### COUNTS FOR THE FIBERS IN THE FIGURE

| Structure No. | Count | Explanation |
| --- | --- | --- |
| 1 to 6 | 1 | Single fibers all contained within the circle. |
| 7 | 1/2 | Fiber crosses circle once. |
| 8 | 0 | Fiber too short. |
| 9 | 2 | Two crossing fibers. |
| 10 | 0 | Fiber outside graticule. |
| 11 | 0 | Fiber crosses graticule twice. |
| 12 | 1/2 | Although split, fiber only crosses once. |

# OSHA Standards for General Industry

**FIGURE 1—WALTON-BECKETT GRATICULE WITH SOME EXPLANATORY FIBERS**

## APPENDIX C TO §1910.1001 [Reserved].

## APPENDIX D TO §1910.1001— MEDICAL QUESTIONNAIRES; MANDATORY

This mandatory appendix contains the medical questionnaires that must be administered to all employees who are exposed to asbestos above the permissible exposure limit, and who will therefore be included in their employer's medical surveillance program. Part 1 of the appendix contains the Initial Medical Questionnaire, which must be obtained for all new hires who will be covered by the medical surveillance requirements. Part 2 includes the abbreviated Periodical Medical Questionnaire, which must be administered to all employees who are provided periodic medical examinations under the medical surveillance provisions of the standard.

## Part 1
### INITIAL MEDICAL QUESTIONNAIRE

1. NAME _____
2. SOCIAL SECURITY # _____
3. CLOCK NUMBER _____
4. PRESENT OCCUPATION _____
5. PLANT _____
6. ADDRESS _____
7. _____ (Zip Code)
8. TELEPHONE NUMBER _____
9. INTERVIEWER _____
10. Date _____
11. Date of Birth _____ Month _____ Day _____ Year
12. Place of Birth _____
13. Sex   1. Male _____   2. Female _____
14. What is your marital status?   1. Single _____   2. Married _____   3. Widowed _____   4. Separated/Divorced _____
15. Race   1. White _____   2. Black _____   3. Asian _____   4. Hispanic _____   5. Indian _____   6. Other _____
16. What is the highest grade completed in school? _____ (For example 12 years is completion of high school)

**OCCUPATIONAL HISTORY**

17. A. Have you ever worked full time (30 hours per week or more) for 6 months or more?   1. Yes _____   2. No _____
    IF YES TO 17A:
    B. Have you ever worked for a year or more in any dusty job?   1. Yes _____   2. No _____   3. Does Not Apply _____
       Specify job/Industry _____   Total Years Worked _____
       Was dust exposure:   1. Mild _____   2. Moderate _____   3. Severe _____
    C. Have you ever been exposed to gas or chemical fumes in your work?   1. Yes _____   2. No _____
       Specify job/Industry _____   Total Years Worked _____
       Was dust exposure:   1. Mild _____   2. Moderate _____   3. Severe _____
    D. What has been your usual occupation or job—the one you have worked at the longest?
       1. Job occupation _____
       2. Number of years employed in this occupation _____
       3. Position/job title _____
       4. Business, field or industry _____
    (Record on lines the years in which you have worked in any of these industries, e.g. 1960-1969)

    Have you ever worked:
    
    |  | YES | NO |
    |---|---|---|
    | E. In a mine? | ☐ | ☐ |
    | F. In a quarry? | ☐ | ☐ |
    | G. In a foundry? | ☐ | ☐ |
    | H. In a pottery? | ☐ | ☐ |
    | I. In a cotton, flax or hemp mill? | ☐ | ☐ |
    | J. With asbestos? | ☐ | ☐ |

18. **PAST MEDICAL HISTORY**

    |  | YES | NO |
    |---|---|---|
    | A. Do you consider yourself to be in good health? | ☐ | ☐ |
    
    If "NO" state reason _____
    
    |  | YES | NO |
    |---|---|---|
    | B. Have you any defect of vision? | ☐ | ☐ |
    
    If "YES" state nature of defect _____
    
    |  | YES | NO |
    |---|---|---|
    | C. Have you any hearing defect? | ☐ | ☐ |
    
    If "YES" state nature of defect _____
    
    D. Are you suffering from or have you ever suffered from:
    
    |  | YES | NO |
    |---|---|---|
    | a. Epilepsy (or fits, seizures, convulsions)? | ☐ | ☐ |
    | b. Rheumatic fever? | ☐ | ☐ |
    | c. Kidney disease? | ☐ | ☐ |
    | d. Bladder disease? | ☐ | ☐ |
    | e. Diabetes? | ☐ | ☐ |
    | f. Jaundice? | ☐ | ☐ |

## Subpart Z—Toxic and Hazardous Substances

### CHEST COLDS AND CHEST ILLNESSES

19. A. If you get a cold, does it usually go to your chest? (Usually means more than 1/2 the time)
    1. Yes ___ 2. No ___ 3. Don't get colds ___

20. A. During the past 3 years, have you had any chest illnesses that have kept you off work, indoors at home, or in bed?
    1. Yes ___ 2. No ___
    IF YES TO 20A:
    B. Did you produce phlegm with any of these chest illnesses?
    1. Yes ___ 2. No ___ 3. Does Not Apply ___
    C. In the last 3 years, how many such illnesses with (increased) phlegm did you have which lasted a week or more?
    Number of illnesses ___ No such illnesses ___

21. Did you have any lung trouble before the age of 16? 1. Yes ___ 2. No ___

22. Have you ever had any of the following?
    1 A. Attacks of bronchitis? 1. Yes ___ 2. No ___
    IF YES TO 1A:
    B. Was it confirmed by a doctor? 1. Yes ___ 2. No ___ 3. Does Not Apply ___
    C. At what age was your first attack? Age in Years ___ Does Not Apply ___

    2 A. Pneumonia (include bronchopneumonia)? 1. Yes ___ 2. No ___
    IF YES TO 2A:
    B. Was it confirmed by a doctor? 1. Yes ___ 2. No ___ 3. Does Not Apply ___
    C. At what age did you first have it? Age in Years ___ Does Not Apply ___

    3 A. Hay Fever? 1. Yes ___ 2. No ___
    IF YES TO 3A:
    B. Was it confirmed by a doctor? 1. Yes ___ 2. No ___ 3. Does Not Apply ___
    C. At what age did it start? Age in Years ___ Does Not Apply ___

23. A. Have you ever had chronic bronchitis? 1. Yes ___ 2. No ___
    IF YES TO 23A:
    B. Do you still have it? 1. Yes ___ 2. No ___ 3. Does Not Apply ___
    C. Was it confirmed by a doctor? 1. Yes ___ 2. No ___ 3. Does Not Apply ___
    D. At what age did it start? Age in Years ___ Does Not Apply ___

24. A. Have you ever had emphysema? 1. Yes ___ 2. No ___
    IF YES TO 24A:
    B. Do you still have it? 1. Yes ___ 2. No ___ 3. Does Not Apply ___
    C. Was it confirmed by a doctor? 1. Yes ___ 2. No ___ 3. Does Not Apply ___
    D. At what age did it start? Age in Years ___ Does Not Apply ___

25. A. Have you ever had asthma? 1. Yes ___ 2. No ___
    IF YES TO 25A:
    B. Do you still have it? 1. Yes ___ 2. No ___ 3. Does Not Apply ___
    C. Was it confirmed by a doctor? 1. Yes ___ 2. No ___ 3. Does Not Apply ___
    D. At what age did it start? Age in Years ___ Does Not Apply ___
    E. If you no longer have it, at what age did it stop? Age Stopped ___ Does Not Apply ___

26. Have you ever had:
    A. Any other chest illness? 1. Yes ___ 2. No ___
    If yes, please specify ___
    B. Any chest operations? 1. Yes ___ 2. No ___
    If yes, please specify ___
    C. Any chest injuries? 1. Yes ___ 2. No ___
    If yes, please specify ___

27. A. Has a doctor ever told you that you had heart trouble? 1. Yes ___ 2. No ___
    IF YES TO 27A:
    B. Have you ever had treatment for heart trouble in the past 10 years? 1. Yes ___ 2. No ___ 3. Does Not Apply ___

28. A. Has a doctor ever told you that you had high blood pressure? 1. Yes ___ 2. No ___
    IF YES TO 28A:
    B. Have you had any treatment for high blood pressure (hypertension) in the past 10 years? 1. Yes ___ 2. No ___ 3. Does Not Apply ___

29. When did you last have your chest X-rayed? (Year) ___

30. Where did you last have your chest X-rayed (if known)? ___
    What was the outcome? ___

### FAMILY HISTORY

31. Were either of your natural parents ever told by a doctor that they had a chronic lung condition such as:

    |  | FATHER | MOTHER |
    |---|---|---|
    | A. Chronic Bronchitis? | 1. Yes ___ 2. No ___ 3. Don't Know ___ | 1. Yes ___ 2. No ___ 3. Don't Know ___ |
    | B. Emphysema? | | |
    | C. Asthma? | | |
    | D. Lung cancer? | | |
    | E. Other chest conditions? | | |
    | F. Is parent currently alive? | Age if Living ___ Age at Death ___ Don't Know ___ | Age if Living ___ Age at Death ___ Don't Know ___ |
    | G. Please Specify | | |
    | H. Please specify cause of death | | |

### COUGH

32. A. Do you usually have a cough? (Count a cough with first smoke or on first going out of doors. Exclude clearing of throat.) [If no, skip to question 32C.] 1. Yes ___ 2. No ___
    B. Do you usually cough as much as 4 to 6 times a day 4 or more days out of the week? 1. Yes ___ 2. No ___
    C. Do you usually cough at all on getting up or first thing in the morning? 1. Yes ___ 2. No ___
    D. Do you usually cough at all during the rest of the day or at night? 1. Yes ___ 2. No ___
    IF YES TO ANY OF ABOVE (32A, B, C, or D), ANSWER THE FOLLOWING. IF NO TO ALL, CHECK DOES NOT APPLY AND SKIP TO NEXT PAGE.
    E. Do you usually cough like this on most days for 3 consecutive months or more during the year?
    1. Yes ___ 2. No ___ 3. Does Not Apply ___
    F. For how many years have you had the cough? Number of Years ___ Does Not Apply ___

33. A. Do you usually bring up phlegm from your chest? 1. Yes ___ 2. No ___
    (Count phlegm with the first smoke or on first going out of doors. Exclude phlegm from the nose. Count swallowed phlegm.) [If no, skip to 33C]
    B. Do you usually bring up phlegm like this as much as twice a day 4 or more days out of the week?
    1. Yes ___ 2. No ___
    C. Do you usually bring up phlegm at all on getting up or first thing in the morning?
    1. Yes ___ 2. No ___
    D. Do you usually bring up phlegm at all during the rest of the day or at night?
    1. Yes ___ 2. No ___
    IF YES TO ANY OF ABOVE (33A, B, C, or D), ANSWER THE FOLLOWING. IF NO TO ALL, CHECK DOES NOT APPLY AND SKIP TO 34A.
    E. Do you bring up phlegm like this on most days for 3 consecutive months or more during the year?
    1. Yes ___ 2. No ___ 3. Does Not Apply ___
    F. For how many years have you had trouble with phlegm? Number of years ___ Does not apply ___

### EPISODES OF COUGH AND PHLEGM

34. A. Have you had periods or episodes of (increased *) cough and phlegm lasting for 3 weeks or more each year?
    * (For persons who usually have cough and/or phlegm) 1. Yes ___ 2. No ___ Does not apply ___
    IF YES TO 34A:
    B. For how long have you had at least 1 such episode per year? Number of years ___ Does not apply ___

### WHEEZING

35. Does your chest ever sound wheezy or whistling
    1. When you have a cold? 1. Yes ___ 2. No ___
    2. Occasionally apart from colds? 1. Yes ___ 2. No ___
    3. Most days or nights? 1. Yes ___ 2. No ___
    IF YES TO 1, 2, or 3 in 35A:
    B. For how many years has this been present? Number of years ___ Does not apply ___

36. A. Have you ever had an attack of wheezing that has made you feel short of breath? 1. Yes ___ 2. No ___
    IF YES TO 36A
    B. How old were you when you had your first such attack? Age in years ___ Does not apply ___
    C. Have you had 2 or more such episodes? 1. Yes ___ 2. No ___ 3. Does not apply ___
    D. Have you ever required medicine or treatment for the(se) attack(s)? 1. Yes ___ 2. No ___ 3. Does not apply ___

# OSHA Standards for General Industry

## Part 2
## PERIODIC MEDICAL QUESTIONNAIRE

1. NAME _____
2. SOCIAL SECURITY # _____   3. CLOCK NUMBER _____
4. PRESENT OCCUPATION _____   5. PLANT _____
6. ADDRESS _____
   _____ (Zip Code)
   7. TELEPHONE NUMBER _____
8. Date _____   9. INTERVIEWER _____
10. Date _____
11. What is your marital status?   1. Single ___   2. Married ___   3. Widowed ___   4. Separated/Divorced ___

### 12 OCCUPATIONAL HISTORY
12 A. In the past year, did you work full time (30 hours per week or more) for 6 months or more?
   1. Yes ___   2. No ___
IF YES TO 12A:
12 B. In the past year, did you work in a dusty job?   1. Yes ___   2. No ___   3. Does Not Apply ___
12 C. Was dust exposure:   1. Mild ___   2. Moderate ___   3. Severe ___
12 D. In the past year, were you exposed to gas or chemical fumes in your work?   1. Yes ___   2. No ___
12 E. Was exposure:   1. Mild ___   2. Moderate ___   3. Severe ___
12 F. In the past year, what was your:   1. Job/occupation? ___   2. Position/job title? ___

### 13. RECENT MEDICAL HISTORY
13 A. Do you consider yourself to be in good health?   Yes ___   No ___
   If NO, state reason _____
13 B. In the past year, have you developed:

|  | Yes | No |
|---|---|---|
| Epilepsy? | | |
| Rheumatic fever? | | |
| Kidney disease? | | |
| Bladder disease? | | |
| Diabetes? | | |
| Jaundice? | | |
| Cancer? | | |

### 14. CHEST COLDS AND CHEST ILLNESSES
14 A. If you get a cold, does it usually go to your chest? (Usually means more than 1/2 the time)
   1. Yes ___   2. No ___   3. Don't get colds ___
15 A. During the past year, have you had any chest illnesses that have kept you off work, indoors at home, or in bed?
   1. Yes ___   2. No ___   3. Does not apply ___
IF YES TO 15A:
15 B. Did you produce phlegm with any of these chest illnesses?
   1. Yes ___   2. No ___   Does not apply ___
15 C. In the past year, how many such illnesses with (increased) phlegm did you have which lasted a week or more?
   Number of illnesses ___   No such illnesses ___

### 16. RESPIRATORY SYSTEM
In the past year have you had:

| | Yes or No | Further Comment on Positive Answers |
|---|---|---|
| Asthma | | |
| Bronchitis | | |
| Hay Fever | | |
| Other Allergies | | |
| Pneumonia | | |
| Tuberculosis | | |
| Chest Surgery | | |
| Other Lung Problems | | |
| Heart Disease | | |

Do you have: | Yes or No | Further Comment on Positive Answers
Frequent colds
Chronic cough
Shortness of breath when walking or climbing one flight of stairs

Do you:
Wheeze
Cough up phlegm
Smoke cigarettes ___   Packs per day ___   How many years ___

Date _____   Signature _____

---

### BREATHLESSNESS
37. If disabled from walking by any condition other than heart or lung disease, please describe and proceed to question 39A.
   Nature of condition(s) _____
38 A. Are you troubled by shortness of breath when hurrying on the level or walking up a slight hill?
   1. Yes ___   2. No ___
IF YES TO 38A:
B. Do you have to walk slower than people of your age on the level because of breathlessness?
   1. Yes ___   2. No ___   3. Does not apply ___
C. Do you ever have to stop for breath when walking at your own pace on the level?
   1. Yes ___   2. No ___   3. Does not apply ___
D. Do you ever have to stop for breath after walking about 100 yards (or after a few minutes) on the level?
   1. Yes ___   2. No ___   3. Does not apply ___
E. Are you too breathless to leave the house or breathless on dressing or climbing one flight of stairs?
   1. Yes ___   2. No ___   3. Does not apply ___

### TOBACCO SMOKING
39 A. Have you ever smoked cigarettes? (No means less than 20 packs of cigarettes or 12 oz. of tobacco in a lifetime or less than 1 cigarette a day for 1 year.)   1. Yes ___   2. No ___
IF YES TO 39A:
B. Do you now smoke cigarettes (as of one month ago)?   1. Yes ___   2. No ___   3. Does not apply ___
C. How old were you when you first started regular cigarette smoking? ___ Age in Years ___   Does not apply ___
D. If you have stopped smoking cigarettes completely, how old were you when you stopped?
   Age stopped ___   Check if still smoking ___   Does not apply ___
E. How many cigarettes do you smoke per day now? ___   Cigarettes per day ___   Does not apply ___
F. On the average of the entire time you smoked, how many cigarettes did you smoke per day?
   Cigarettes per day ___   Does not apply ___
G. Do you or did you inhale the cigarette smoke?
   1. Does not apply ___   2. Not at all ___   3. Slightly ___   4. Moderately ___   5. Deeply ___
40 A. Have you ever smoked a pipe regularly?   1. Yes ___   2. No ___
   (Yes means more than 12 oz. of tobacco in a lifetime.)
IF YES TO 40A:

### FOR PERSONS WHO HAVE EVER SMOKED A PIPE
B. 1. How old were you when you started to smoke a pipe regularly?   Age ___
   2. If you have stopped smoking a pipe completely, how old were you when you stopped?
      Age stopped ___   Check if still smoking pipe ___   Does not apply ___
C. On the average over the entire time you smoked a pipe, how much pipe tobacco did you smoke per week?
   ___ oz. per week (a standard pouch of tobacco contains 1 1/2 oz.)   Does not apply ___
D. How much pipe tobacco are you smoking now? ___ oz. per week ___   Not currently smoking a pipe ___
E. Do you or did you inhale the pipe smoke?
   1. Never smoked ___   2. Not at all ___   3. Slightly ___   4. Moderately ___   5. Deeply ___
41 A. Have you ever smoked cigars regularly?   1. Yes ___   2. No ___
   (Yes means more than 1 cigar a week for a year)
IF YES TO 41A:

### FOR PERSONS WHO HAVE EVER SMOKED CIGARS
B. 1. How old were you when you started smoking cigars regularly?   Age ___
   2. If you have stopped smoking cigars completely, how old were you when you stopped?
      Age stopped ___   Check if still smoking cigars ___   Does not apply ___
C. On the average over the entire time you smoked cigars, how many cigars did you smoke per week?
   Cigars per week ___   Does not apply ___
D. How many cigars are you smoking per week now?
   Cigars per week ___   Check if not smoking cigars currently ___
E. Do or did you inhale the cigar smoke?
   1. Never smoked ___   2. Not at all ___   3. Slightly ___   4. Moderately ___   5. Deeply ___

Signature _____   Date _____

# APPENDIX E TO §1910.1001—
# INTERPRETATION AND CLASSIFICATION OF CHEST ROENTGENOGRAMS—MANDATORY

(a) Chest roentgenograms shall be interpreted and classified in accordance with a professionally accepted Classification system and recorded on an interpretation form following the format of the CDC/NIOSH (M) 2.8 form. As a minimum, the content within the bold lines of this form (items 1 though 4) shall be included. This form is not to be submitted to NIOSH.

(b) Roentgenograms shall be interpreted and classified only by a B-reader, a board eligible/certified radiologist, or an experienced physician with known expertise in pneumoconioses.

(c) All interpreters, whenever interpreting chest roentgenograms made under this section, shall have immediately available for reference a complete set of the ILO-U/C International Classification of Radiographs for Pneumoconioses, 1980.

# APPENDIX F TO §1910.1001—
# WORK PRACTICES AND ENGINEERING CONTROLS FOR AUTOMOTIVE BRAKE AND CLUTCH INSPECTION, DISASSEMBLY, REPAIR AND ASSEMBLY—MANDATORY

This mandatory appendix specifies engineering controls and work practices that must be implemented by the employer during automotive brake and clutch inspection, disassembly, repair, and assembly operations. Proper use of these engineering controls and work practices by trained employees will reduce employees' asbestos exposure below the permissible exposure level during clutch and brake inspection, disassembly, repair, and assembly operations. The employer shall institute engineering controls and work practices using either the method set forth in paragraph [A] or paragraph [B] of this appendix, or any other method which the employer can demonstrate to be equivalent in terms of reducing employee exposure to asbestos as defined and which meets the requirements described in paragraph [C] of this appendix, for those facilities in which no more than 5 pairs of brakes or 5 clutches are inspected, disassembled, reassembled and/or repaired per week, the method set forth in paragraph [D] of this appendix may be used:

[A] *Negative Pressure Enclosure/HEPA Vacuum System Method*

(1) The brake and clutch inspection, disassembly, repair, and assembly operations shall be enclosed to cover and contain the clutch or brake assembly and to prevent the release of asbestos fibers into the worker's breathing zone.

(2) The enclosure shall be sealed tightly and thoroughly inspected for leaks before work begins on brake and clutch inspection, disassembly, repair, and assembly.

(3) The enclosure shall be such that the worker can clearly see the operation and shall provide impermeable sleeves through which the worker can handle the brake and clutch inspection, disassembly, repair and assembly. The integrity of the sleeves and ports shall be examined before work begins.

(4) A HEPA-filtered vacuum shall be employed to maintain the enclosure under negative pressure throughout the operation. Compressed-air may be used to remove asbestos fibers or particles from the enclosure.

(5) The HEPA vacuum shall be used first to loosen the asbestos containing residue from the brake and clutch parts and then to evacuate the loosened asbestos containing material from the enclosure and capture the material in the vacuum filter.

(6) The vacuum's filter, when full, shall be first wetted with a fine mist of water, then removed and placed immediately in an impermeable container, labeled according to paragraph (j)(4) of this section and disposed of according to paragraph (k) of this section.

(7) Any spills or releases of asbestos containing waste material from inside of the enclosure or vacuum hose or vacuum filter shall be immediately cleaned up and disposed of according to paragraph (k) of this section.

[B] *Low Pressure/Wet Cleaning Method*

(1) A catch basin shall be placed under the brake assembly, positioned to avoid splashes and spills.

(2) The reservoir shall contain water containing an organic solvent or wetting agent. The flow of liquid shall be controlled such that the brake assembly is gently flooded to prevent the asbestos-containing brake dust from becoming airborne.

(3) The aqueous solution shall be allowed to flow between the brake drum and brake support before the drum is removed.

(4) After removing the brake drum, the wheel hub and back of the brake assembly shall be thoroughly wetted to suppress dust.

(5) The brake support plate, brake shoes and brake components used to attach the brake shoes shall be thoroughly washed before removing the old shoes.

(6) In systems using filters, the filters, when full, shall be first wetted with a fine mist of water, then removed and placed immediately in an impermeable container, labeled according to paragraph (j)(4) of this section and disposed of according to paragraph (k) of this section.

(7) Any spills of asbestos-containing aqueous solution or any asbestos-containing waste material shall be cleaned up immediately and disposed of according to paragraph (k) of this section.

(8) The use of dry brushing during low pressure/wet cleaning operations is prohibited.

[C] *Equivalent Methods*

An equivalent method is one which has sufficient written detail so that it can be reproduced and has been demonstrated that the exposures resulting from the equivalent method are equal to or less than the exposures which would result from the use of the method described in paragraph [A] of this appendix. For purposes of making this comparison, the employer shall assume that exposures resulting from the use of the method described in paragraph [A] of this appendix shall not exceed 0.016 f/cc, as measured by the OSHA reference method and as averaged over at least 18 personal samples.

[D] *Wet Method.*

(1) A spray bottle, hose nozzle, or other implement capable of delivering a fine mist of water or amended water or other delivery system capable of delivering water at low pressure, shall be used to first thoroughly wet the brake and clutch parts. Brake and clutch components shall then be wiped clean with a cloth.

(2) The cloth shall be placed in an impermeable container, labelled according to paragraph (j)(4) of this section and then disposed of according to paragraph (k) of this section, or the cloth shall be laundered in a way to prevent the release of asbestos fibers in excess of 0.1 fiber per cubic centimeter of air.

(3) Any spills of solvent or any asbestos containing waste material shall be cleaned up immediately according to paragraph (k) of this section.

(4) The use of dry brushing during the wet method operations is prohibited.

# APPENDIX G TO §1910.1001—
# SUBSTANCE TECHNICAL INFORMATION FOR ASBESTOS—NON-MANDATORY

I. Substance Identification

A. Substance: "Asbestos" is the name of a class of magnesium-silicate minerals that occur in fibrous form. Minerals that are included in this group are chrysotile, crocidolite, amosite, tremolite asbestos, anthophyllite asbestos, and actinolite asbestos.

B. Asbestos is used in the manufacture of heat-resistant clothing, automotive brake and clutch linings, and a variety of building materials including floor tiles, roofing felts, ceiling tiles, asbestos-cement pipe and sheet, and fire-resistant drywall. Asbestos is also present in pipe and boiler insulation materials, and in sprayed-on materials located on beams, in crawlspaces, and between walls.

C. The potential for a product containing asbestos to release breathable fibers depends on its degree of friability. Friable means that the material can be crumbled with hand pressure and is therefore likely to emit fibers. The fibrous or fluffy sprayed-on materials used for fireproofing, insulation, or sound proofing are considered to be friable, and they readily release airborne fibers if disturbed. Materials such as vinyl-asbestos floor tile or roofing felts are considered nonfriable and generally do not emit airborne fibers unless subjected to sanding or sawing operations. Asbestos-cement pipe or sheet can emit airborne fibers if the materials are cut or sawed, or if they are broken during demolition operations.

D. Permissible exposure: Exposure to airborne asbestos fibers may not exceed 0.2 fibers per cubic centimeter of air (0.1 f/cc) averaged over the 8-hour workday.

II. Health Hazard Data

A. Asbestos can cause disabling respiratory disease and various types of cancers if the fibers are inhaled. Inhaling or ingesting fibers from contaminated clothing or skin can also result in these diseases. The symptoms of these diseases generally do not appear for 20 or more years after initial exposure.

B. Exposure to asbestos has been shown to cause lung cancer, mesothelioma, and cancer of the stomach and colon. Mesothelioma is a rare cancer of the thin membrane lining of the chest and abdomen. Symptoms of mesothelioma include shortness of breath, pain in the walls of the chest, and/or abdominal pain.

III. Respirators and Protective Clothing

A. Respirators: You are required to wear a respirator when performing tasks that result in asbestos exposure that exceeds the permissible exposure limit (PEL) of 0.1 f/cc. These conditions can occur while your employer is in the process of installing engineering controls to reduce asbestos exposure, or where engineering controls are not feasible to reduce asbestos exposure. Air-purifying respirators equipped with a high-efficiency particulate air (HEPA) filter can be used where airborne asbestos fiber concentrations do not exceed 2 f/cc; otherwise, air-supplied, positive-pressure, full facepiece respirators must be used. Disposable respirators or dust masks are not permitted to be used for asbestos work. For effective protection, respirators must fit your face and head snugly. Your employer is required to conduct fit tests when you are first assigned a respirator and every 6 months thereafter. Respirators should not be loosened or removed in work situations where their use is required.

B. Protective clothing: You are required to wear protective clothing in work areas where asbestos fiber concentrations exceed the permissible exposure limit.

IV. Disposal Procedures and Cleanup

A. Wastes that are generated by processes where asbestos is present include:

1. Empty asbestos shipping containers.

2. Process wastes such as cuttings, trimmings, or reject material.

3. Housekeeping waste from sweeping or vacuuming.

4. Asbestos fireproofing or insulating material that is removed from buildings.

5. Building products that contain asbestos removed during building renovation or demolition.

6. Contaminated disposable protective clothing.

B. Empty shipping bags can be flattened under exhaust hoods and packed into airtight containers for disposal. Empty shipping drums are difficult to clean and should be sealed.

C. Vacuum bags or disposable paper filters should not be cleaned, but should be sprayed with a fine water mist and placed into a labeled waste container.

D. Process waste and housekeeping waste should be wetted with water or a mixture of water and surfactant prior to packaging in disposable containers.

E. Material containing asbestos that is removed from buildings must be disposed of in leak-tight 6-mil thick plastic bags, plastic-lined cardboard containers, or plastic-lined metal containers. These wastes, which are removed while wet, should be sealed in containers before they dry out to minimize the release of asbestos fibers during handling.

V. Access to Information

A. Each year, your employer is required to inform you of the information contained in this standard and appendices for asbestos. In addition, your employer must instruct you in the proper work practices for handling materials containing asbestos, and the correct use of protective equipment.

B. Your employer is required to determine whether you are being exposed to asbestos. You or your representative has the right to observe employee measurements and to record the results obtained. Your employer is required to inform you of your exposure, and, if you are exposed above the permissible limit, he or she is required to inform you of the actions that are being taken to reduce your exposure to within the permissible limit.

C. Your employer is required to keep records of your exposures and medical examinations. These exposure records must be kept for at least thirty (30) years. Medical records must be kept for the period of your employment plus thirty (30) years.

D. Your employer is required to release your exposure and medical records to your physician or designated representative upon your written request.

## APPENDIX H TO §1910.1001— MEDICAL SURVEILLANCE GUIDELINES FOR ASBESTOS NON-MANDATORY

I. Route of Entry Inhalation, Ingestion

II. Toxicology

Clinical evidence of the adverse effects associated with exposure to asbestos is present in the form of several well-conducted epidemiological studies of occupationally exposed workers, family contacts of workers, and persons living near asbestos mines. These studies have shown a definite association between exposure to asbestos and an increased incidence of lung cancer, pleural and peritoneal mesothelioma, gastrointestinal cancer, and asbestosis. The latter is a disabling fibrotic lung disease that is caused only by exposure to asbestos. Exposure to asbestos has also been associated with an increased incidence of esophageal, kidney, laryngeal, pharyngeal, and buccal cavity cancers. As with other known chronic occupational diseases, disease associated with asbestos generally appears about 20 years following the first occurrence of exposure: There are no known acute effects associated with exposure to asbestos.

Epidemiological studies indicate that the risk of lung cancer among exposed workers who smoke cigarettes is greatly increased over the risk of lung cancer among non-exposed smokers or exposed nonsmokers. These studies suggest that cessation of smoking will reduce the risk of lung cancer for a person exposed to asbestos but will not reduce it to the same level of risk as that existing for an exposed worker who has never smoked.

III. Signs and Symptoms of Exposure-Related Disease

The signs and symptoms of lung cancer or gastrointestinal cancer induced by exposure to asbestos are not unique, except that a chest X-ray of an exposed patient with lung cancer may show pleural plaques, pleural calcification, or pleural fibrosis. Symptoms characteristic of mesothelioma include shortness of breath, pain in the walls of the chest, or abdominal pain. Mesothelioma has a much longer latency period compared with lung cancer (40 years versus 15-20 years), and mesothelioma is therefore more likely to be found among workers who were first exposed to asbestos at an early age. Mesothelioma is always fatal.

Asbestosis is pulmonary fibrosis caused by the accumulation of asbestos fibers in the lungs. Symptoms include shortness of breath, coughing, fatigue, and vague feelings of sickness. When the fibrosis worsens, shortness of breath occurs even at rest. The diagnosis of asbestosis is based on a history of exposure to asbestos, the presence of characteristic radiologic changes, end-inspiratory crackles (rales), and other clinical features of fibrosing lung disease. Pleural plaques and thickening are observed on X-rays taken during the early stages of the disease. Asbestosis is often a progressive disease even in the absence of continued exposure, although this appears to be a highly individualized characteristic. In severe cases, death may be caused by respiratory or cardiac failure.

IV. Surveillance and Preventive Considerations

As noted above, exposure to asbestos has been linked to an increased risk of lung cancer, mesothelioma, gastrointestinal cancer, and asbestosis among occupationally exposed workers. Adequate screening tests to determine an employee's potential for developing serious chronic diseases, such as cancer, from exposure to asbestos do not presently exist. However, some tests, particularly chest X-rays and pulmonary function tests, may indicate that an employee has been overexposed to asbestos increasing his or her risk of developing exposure-related chronic diseases. It is important for the physician to become familiar with the operating conditions in which occupational exposure to asbestos is likely to occur. This is particularly important in evaluating medical and work histories and in conducting physical examinations. When an active employee has been identified as having been overexposed to asbestos, measures taken by the employer to eliminate or mitigate further exposure should also lower the risk of serious long-term consequences.

The employer is required to institute a medical surveillance program for all employees who are or will be exposed to asbestos at or above the permissible exposure limit (0.1 fiber per cubic centimeter of air). All examinations and procedures must be performed by or under the supervision of a licensed physician, at a reasonable time and place, and at no cost to the employee.

Although broad latitude is given to the physician in prescribing specific tests to be included in the medical surveillance program, OSHA requires inclusion of the following elements in the routine examination:

(i) Medical and work histories with special emphasis directed to symptoms of the respiratory system, cardiovascular system, and digestive tract.

(ii) Completion of the respiratory disease questionnaire contained in Appendix D.

(iii) A physical examination including a chest roentgenogram and pulmonary function test that includes measurement of the employee's forced vital capacity (FVC) and forced expiratory volume at one second ($FEV_1$).

(iv) Any laboratory or other test that the examining physician deems by sound medical practice to be necessary.

The employer is required to make the prescribed tests available at least annually to those employees covered; more often than specified if recommended by the examining physician; and upon termination of employment.

The employer is required to provide the physician with the following information: A copy of this standard and appendices; a description of the employee's duties as they relate to asbestos exposure; the employee's representative level of exposure to asbestos; a description of any personal protective and respiratory equipment used; and information from previous medical examinations of the affected employee that is not otherwise available to the physician. Making this information available to the physician will aid in the evaluation of the employee's health in relation to assigned duties and fitness to wear personal protective equipment, if required.

The employer is required to obtain a written opinion from the examining physician containing the results of the medical examination; the physician's opinion as to whether the employee has any detected medical conditions that would place the employee at an increased risk of exposure-related disease; any recommended limitations on the employee or on the use of personal protective equipment; and a statement that the employee has been informed by the physician of the results of the medical examination and of any medical conditions related to asbestos exposure that require further explanation or treatment. This written opinion must not reveal specific findings or diagnoses unrelated to exposure to asbestos, and a copy of the opinion must be provided to the affected employee.

### APPENDIX I TO §1910.1001—SMOKING CESSATION PROGRAM INFORMATION FOR ASBESTOS—NON-MANDATORY

The following organizations provide smoking cessation information and program material.

1. The National Cancer Institute operates a toll-free Cancer Information Service (CIS) with trained personnel to help you. Call 1-800-4-CANCER[*] to reach the CIS office serving your area, or write: Office of Cancer Communications, National Cancer Institute, National Institutes of Health, Building 31, Room 10A24, Bethesda, Maryland 20892.

2. American Cancer Society, 3340 Peachtree Road, NE., Atlanta, Georgia 30062, (404) 320-3333.

The American Cancer Society (ACS) is a voluntary organization composed of 58 divisions and 3,100 local units. Through "The Great American Smokeout" in November, the annual Cancer Crusade in April, and numerous educational materials, ACS helps people learn about the health hazards of smoking and become successful ex-smokers.

3. American Heart Association, 7320 Greenville Avenue, Dallas, Texas 75231, (214) 750-5300.

The American Heart Association (AHA) is a voluntary organization with 130,000 members (physicians, scientists, and laypersons) in 55 state and regional groups. AHA produces a variety of publications and audiovisual materials about the effects of smoking on the heart. AHA also has developed a guidebook for incorporating a weight-control component into smoking cessation programs.

4. American Lung Association, 1740 Broadway, New York, New York 10019, (212) 245-8000.

A voluntary organization of 7,500 members (physicians, nurses, and laypersons), the American Lung Association (ALA) conducts numerous public information programs about the health effect of smoking. ALA has 59 state and 85 local units. The organization actively supports legislation and information campaigns for non-smokers' rights and provides help for smokers who want to quit, for example, through "Freedom From Smoking," a self-help smoking cessation program.

5. Office on Smoking and Health, U.S. Department of Health and Human Services, 5600 Fishers Lane, Park Building, Room 110, Rockville, Maryland 20857.

The Office on Smoking and Health (OSH) is the Department of Health and Human Services' lead agency in smoking control. OSH has sponsored distribution of publications on smoking-related topics, such as free flyers on relapse after initial quitting, helping a friend or family member quit smoking, the health hazards of smoking, and the effects of parental smoking on teenagers.

[*] In Hawaii, on Oahu call 524-1234 (call collect from neighboring islands).

Spanish-speaking staff members are available during daytime hours to callers from the following areas: California, Florida, Georgia, Illinois, New Jersey (area code 210), New York, and Texas. Consult your local telephone directory for listings of local chapters.

### APPENDIX J TO §1910.1001—POLARIZED LIGHT MICROSCOPY OF ASBESTOS—NON-MANDATORY

Method number: ID-191

Matrix: Bulk

Collection Procedure

Collect approximately 1 to 2 grams of each type of material and place into separate 20 mL scintillation vials.

Analytical Procedure

A portion of each separate phase is analyzed by gross examination, phase-polar examination, and central stop dispersion microscopy.

Commercial manufacturers and products mentioned in this method are for descriptive use only and do not constitute endorsements by USDOL-OSHA. Similar products from other sources may be substituted.

*1. Introduction*

This method describes the collection and analysis of asbestos bulk materials by light microscopy techniques including phase- polar illumination and central-stop dispersion microscopy. Some terms unique to asbestos analysis are defined below:

*Amphibole:* A family of minerals whose crystals are formed by long, thin units which have two thin ribbons of double chain silicate with a brucite ribbon in between. The shape of each unit is similar to an "I beam". Minerals important in asbestos analysis include cummingtonite-grunerite, crocidolite, tremolite-actinolite and anthophyllite.

*Asbestos:* A term for naturally occurring fibrous minerals. Asbestos includes chrysotile, cummingtonite-grunerite asbestos (amosite), anthophyllite asbestos, tremolite asbestos, crocidolite, actinolite asbestos and any of these minerals which have been chemically treated or altered. The precise chemical formulation of each species varies with the location from which it was mined. Nominal compositions are listed:

| | |
|---|---|
| Chrysotile | $Mg_3 Si_2 O_5 (OH)_4$ |
| Crocidolite (Riebeckite asbestos) | $Na_2 Fe_2^{2+} + Fe_2^{3+} + Si_8 O_{22} (OH)_2$ |
| Cummingtonite-Grunerite asbestos (Amosite) | $(Mg,Fe)_7 Si_8 O_{22} (OH)_2$ |
| Tremolite-Actinolite asbestos | $Ca_2 (Mg,Fe)_5 Si_8 O_{22} (OH)_2$ |
| Anthophyllite asbestos | $(Mg,Fe)_7 Si_8 O_{22} (OH)_2$ |

*Asbestos Fiber:* A fiber of asbestos meeting the criteria for a fiber. (See section 3.5.)

*Aspect Ratio:* The ratio of the length of a fiber to its diameter usually defined as "length : width", e.g., 3:1.

*Brucite:* A sheet mineral with the composition $Mg(OH)_2$.

*Central Stop Dispersion Staining (microscope):* This is a dark field microscope technique that images particles using only light refracted by the particle, excluding light that travels through the particle unrefracted. This is usually accomplished with a McCrone objective or other arrangement which places a circular stop with apparent aperture equal to the objective aperture in the back focal plane of the microscope.

*Cleavage Fragments:* Mineral particles formed by the comminution of minerals, especially those characterized by relatively parallel sides and moderate aspect ratio.

*Differential Counting:* The term applied to the practice of excluding certain kinds of fibers from a phase contrast asbestos count because they are not asbestos.

*Fiber:* A particle longer than or equal to 5 µm with a length to width ratio greater than or equal to 3:1. This may include cleavage fragments. (see section 3.5 of this appendix).

*Phase Contrast:* Contrast obtained in the microscope by causing light scattered by small particles to destructively interfere with unscattered light, thereby enhancing the visibility of very small particles and particles with very low intrinsic contrast.

*Phase Contrast Microscope:* A microscope configured with a phase mask pair to create phase contrast. The technique which uses this is called Phase Contrast Microscopy (PCM).

*Phase-Polar Analysis:* This is the use of polarized light in a phase contrast microscope. It is used to see the same size fibers that are visible in air filter analysis. Although fibers finer than 1 µm are visible, analysis of these is inferred from analysis of larger bundles that are usually present.

*Phase-Polar Microscope:* The phase-polar microscope is a phase contrast microscope which has an analyzer, a polarizer, a first order red plate and a rotating phase condenser all in place so that the polarized light image is enhanced by phase contrast.

*Sealing Encapsulant:* This is a product which can be applied, preferably by spraying, onto an asbestos surface which will seal the surface so that fibers cannot be released.

*Serpentine:* A mineral family consisting of minerals with the general composition $Mg_3(Si_2O_5)(OH)_4$ having the magnesium in brucite layer over a silicate layer. Minerals important in asbestos analysis included in this family are chrysotile, lizardite, antigorite.

1.1. History

Light microscopy has been used for well over 100 years for the determination of mineral species. This analysis is carried out using specialized polarizing microscopes as well as bright field microscopes. The identification of minerals is an on-going process with many new minerals described each year. The first recorded use of asbestos was in Finland about 2500 B.C. where the material was used in the mud wattle for the wooden huts the people lived in as well as strengthening for pottery. Adverse health aspects of the mineral were noted nearly 2000 years ago when Pliny the Younger wrote about the poor health of slaves in the asbestos mines. Although known to be injurious for centuries, the first modern references to its toxicity were by the British Labor Inspectorate when it banned asbestos dust from the workplace in 1898. Asbestosis cases were described in the literature after the turn of the century. Cancer was first suspected in the mid 1930's and a causal link to mesothelioma was made in 1965. Because of the public concern for worker and public safety with the use of this material, several different types of analysis were applied to the determination of asbestos content. Light microscopy requires a great deal of experience and craft. Attempts were made to apply less subjective methods to the analysis. X-ray diffraction was partially successful in determining the mineral types but was unable to separate out the fibrous portions from the non-fibrous portions. Also, the minimum detection limit for asbestos analysis by X-ray diffraction (XRD) is about 1%. Differential Thermal Analysis (DTA) was no more successful. These provide useful corroborating information when the presence of asbestos has been shown by microscopy; however, neither can determine the difference between fibrous and non-fibrous minerals when both habits are present. The same is true of Infrared Absorption (IR).

When electron microscopy was applied to asbestos analysis, hundreds of fibers were discovered present too small to be visible in any light microscope. There are two different types of electron microscope used for asbestos analysis: Scanning Electron Microscope (SEM) and Transmission Electron Microscope (TEM). Scanning Electron Microscopy is useful in identifying minerals. The SEM can provide two of the three pieces of information required to identify fibers by electron microscopy: morphology and chemistry. The third is structure as determined by Selected Area Electron Diffraction—SAED which is performed in the TEM. Although the resolution of the SEM is sufficient for very fine fibers to be seen, accuracy of chemical analysis that can be performed on the fibers varies with fiber diameter in fibers of less than 0.2 µm diameter. The TEM is a powerful tool to identify fibers too small to be resolved by light microscopy and should be used in conjunction with this method when necessary. The TEM can provide all three pieces of information required for fiber identification. Most fibers thicker than 1 µm can adequately be defined in the light microscope. The light microscope remains as the best instrument for the determination of mineral type. This is because the minerals under investigation were first described analytically with the light microscope. It is inexpensive and gives positive identification for most samples analyzed. Further, when optical techniques are inadequate, there is ample indication that alternative techniques should be used for complete identification of the sample.

1.2. Principle

Minerals consist of atoms that may be arranged in random order or in a regular arrangement. Amorphous materials have atoms in random order while crystalline materials have long range order. Many materials are transparent to light, at least for small particles or for thin sections. The properties of these materials can be investigated by the effect that the material has on light passing through it. The six asbestos minerals are all crystalline with particular properties that have been identified and cataloged. These six minerals are anisotropic. They have a regular array of atoms, but the arrangement is not the same in all directions. Each major direction of the crystal presents a different regularity. Light photons travelling in each of these main directions will encounter different electrical neighborhoods, affecting the path and time of travel. The techniques outlined in this method use the fact that light traveling through fibers or crystals in different directions will behave differently, but predictably. The behavior of the light as it travels through a crystal can be measured and compared with known or determined values to identify the mineral species. Usually, Polarized Light Microscopy (PLM) is performed with strain-free objectives on a bright-field microscope platform. This would limit the resolution of the microscope to about 0.4 µm. Because OSHA requires the counting and identification of fibers visible in phase contrast, the phase contrast platform is used to visualize the fibers with the polarizing elements added into the light path. Polarized light methods cannot identify fibers finer than about 1 µm in diameter even though they are visible. The finest fibers are usually identified by inference from the presence of larger, identifiable fiber bundles. When fibers are present, but not identifiable by light microscopy, use either SEM or TEM to determine the fiber identity.

1.3. Advantages and Disadvantages

The advantages of light microcopy are:

(a) Basic identification of the materials was first performed by light microscopy and gross analysis. This provides a large base of published information against which to check analysis and analytical technique.

(b) The analysis is specific to fibers. The minerals present can exist in asbestiform, fibrous, prismatic, or massive varieties all at the same time. Therefore, bulk methods of analysis such as X-ray diffraction, IR analysis, DTA, etc. are inappropriate where the material is not known to be fibrous.

(c) The analysis is quick, requires little preparation time, and can be performed on-site if a suitably equipped microscope is available.

The disadvantages are:

(a) Even using phase-polar illumination, not all the fibers present may be seen. This is a problem for very low asbestos concentrations where agglomerations or large bundles of fibers may not be present to allow identification by inference.

(b) The method requires a great degree of sophistication on the part of the microscopist. An analyst is only as useful as his mental catalog of images. Therefore, a microscopist's accuracy is enhanced by experience. The mineralogical training of the analyst is very important. It is the basis on which subjective decisions are made.

(c) The method uses only a tiny amount of material for analysis. This may lead to sampling bias and false results (high or low). This is especially true if the sample is severely inhomogeneous.

(d) Fibers may be bound in a matrix and not distinguishable as fibers so identification cannot be made.

1.4. Method Performance

*1.4.1.* This method can be used for determination of asbestos content from 0 to 100% asbestos. The detection limit has not been adequately determined, although for selected samples, the limit is very low, depending on the number of particles examined. For mostly homogeneous, finely divided samples, with no difficult fibrous interferences, the detection limit is below 1%. For inhomogeneous samples (most samples), the detection limit remains undefined. NIST has conducted proficiency testing of laboratories on a national scale. Although each round is reported statistically with an average, control limits, etc., the results indicate a difficulty in establishing precision especially in the low concentration range. It is suspected that there is significant bias in the low range especially near 1%. EPA tried to remedy this by requiring a mandatory point counting scheme for samples less than 10%. The point counting procedure is tedious, and may introduce significant biases of its own. It has not been incorporated into this method.

*1.4.2.* The precision and accuracy of the quantitation tests performed in this method are unknown. Concentrations are easier to determine in commercial products where asbestos was deliberately added because the amount is usually more than a few percent. An analyst's results can be "calibrated" against the known amounts added by the manufacturer. For geological samples, the degree of homogeneity affects the precision.

*1.4.3.* The performance of the method is analyst dependent. The analyst must choose carefully and not necessarily randomly the portions for analysis to assure that detection of asbestos occurs when it is present. For this reason, the analyst must have adequate training in sample preparation, and experience in the location and identification of asbestos in samples. This is usually accomplished through substantial on-the-job training as well as formal education in mineralogy and microscopy.

*1.5. Interferences*

Any material which is long, thin, and small enough to be viewed under the microscope can be considered an interference for asbestos. There are literally hundreds of interferences in workplaces. The techniques described in this method are normally sufficient to eliminate the interferences. An analyst's success in eliminating the interferences depends on proper training.

Asbestos minerals belong to two mineral families: the serpentines and the amphiboles. In the serpentine family, the only common fibrous mineral is chrysotile. Occasionally, the mineral antigorite occurs in a fibril habit with morphology similar to the amphiboles. The amphibole minerals consist of a score of different minerals of which only five are regulated by federal standard: amosite, crocidolite, anthophyllite asbestos, tremolite asbestos and actinolite asbestos. These are the only amphibole minerals that have been commercially exploited for their fibrous properties; however, the rest can and do occur occasionally in asbestiform habit.

In addition to the related mineral interferences, other minerals common in building material may present a problem for some microscopists: gypsum, anhydrite, brucite, quartz fibers, talc fibers or ribbons, wollastonite, perlite, attapulgite, etc. Other fibrous materials commonly present in workplaces are: fiberglass, mineral wool, ceramic wool, refractory ceramic fibers, kevlar, nomex, synthetic fibers, graphite or carbon fibers, cellulose (paper or wood) fibers, metal fibers, etc.

Matrix embedding material can sometimes be a negative interference. The analyst may not be able to easily extract the fibers from the matrix in order to use the method. Where possible, remove the matrix before the analysis, taking careful note of the loss of weight. Some common matrix materials are: vinyl, rubber, tar, paint, plant fiber, cement, and epoxy. A further negative interference is that the asbestos fibers themselves may be either too small to be seen in Phase contrast Microscopy (PCM) or of a very low fibrous quality, having the appearance of plant fibers. The analyst's ability to deal with these materials increases with experience.

*1.6. Uses and Occupational Exposure*

Asbestos is ubiquitous in the environment. More than 40% of the land area of the United States is composed of minerals which may contain asbestos. Fortunately, the actual formation of great amounts of asbestos is relatively rare. Nonetheless, there are locations in which environmental exposure can be severe such as in the Serpentine Hills of California.

There are thousands of uses for asbestos in industry and the home. Asbestos abatement workers are the most current segment of the population to have occupational exposure to great amounts of asbestos. If the material is undisturbed, there is no exposure. Exposure occurs when the asbestos-containing material is abraded or otherwise disturbed during maintenance operations or some other activity. Approximately 95% of the asbestos in place in the United States is chrysotile.

Amosite and crocidolite make up nearly all the difference. Tremolite and anthophyllite make up a very small percentage. Tremolite is found in extremely small amounts in certain chrysotile deposits. Actinolite exposure is probably greatest from environmental sources, but has been identified in vermiculite containing, sprayed-on insulating materials which may have been certified as asbestos-free.

*1.7. Physical and Chemical Properties*

The nominal chemical compositions for the asbestos minerals were given in Section 1. Compared to cleavage fragments of the same minerals, asbestiform fibers possess a high tensile strength along the fiber axis. They are chemically inert, non-combustible, and heat resistant. Except for chrysotile, they are insoluble in Hydrochloric acid (HCl). Chrysotile is slightly soluble in HCl. Asbestos has high electrical resistance and good sound absorbing characteristics. It can be woven into cables, fabrics or other textiles, or matted into papers, felts, and mats.

*1.8. Toxicology (This Section is for Information Only and Should Not Be Taken as OSHA Policy)*

Possible physiologic results of respiratory exposure to asbestos are mesothelioma of the pleura or peritoneum, interstitial fibrosis, asbestosis, pneumoconiosis, or respiratory cancer. The possible consequences of asbestos exposure are detailed in the NIOSH Criteria Document or in the OSHA Asbestos Standards 29 CFR 1910.1001 and 29 CFR 1926.1101 and 29 CFR 1915.1001.

*2. Sampling Procedure*

*2.1. Equipment for Sampling*

(a) Tube or cork borer sampling device

(b) Knife

(c) 20 mL scintillation vial or similar vial

(d) Sealing encapsulant

*2.2. Safety Precautions*

Asbestos is a known carcinogen. Take care when sampling. While in an asbestos-containing atmosphere, a properly selected and fit-tested respirator should be worn. Take samples in a manner to cause the least amount of dust. Follow these general guidelines:

(a) Do not make unnecessary dust.

(b) Take only a small amount (1 to 2 g).

(c) Tightly close the sample container.

(d) Use encapsulant to seal the spot where the sample was taken, if necessary.

*2.3. Sampling Procedure*

Samples of any suspect material should be taken from an inconspicuous place. Where the material is to remain, seal the sampling wound with an encapsulant to eliminate the potential for exposure from the sample site. Microscopy requires only a few milligrams of material. The amount that will fill a 20 mL scintillation vial is more than adequate. Be sure to collect samples from all layers and phases of material. If possible, make separate samples of each different phase of the material. This will aid in determining the actual hazard. *DO NOT USE ENVELOPES, PLASTIC OR PAPER BAGS OF ANY KIND TO COLLECT SAMPLES.* The use of plastic bags presents a contamination hazard to laboratory personnel and to other samples. When these containers are opened, a bellows effect blows fibers out of the container onto everything, including the person opening the container.

If a cork-borer type sampler is available, push the tube through the material all the way, so that all layers of material are sampled. Some samplers are intended to be disposable. These should be capped and sent to the laboratory. If a non-disposable cork borer is used, empty the contents into a scintillation vial and send to the laboratory. Vigorously and completely clean the cork borer between samples.

*2.4 Shipment*

Samples packed in glass vials must not touch or they might break in shipment.

(a) Seal the samples with a sample seal over the end to guard against tampering and to identify the sample.

(b) Package the bulk samples in separate packages from the air samples. They may cross-contaminate each other and will invalidate the results of the air samples.

(c) Include identifying paperwork *with* the samples, but not in contact with the suspected asbestos.

(d) To maintain sample accountability, ship the samples by certified mail, overnight express, or hand carry them to the laboratory.

*3. Analysis*

The analysis of asbestos samples can be divided into two major parts: sample preparation and microscopy. Because of the different asbestos uses that may be encountered by the analyst, each sample may need different preparation steps. The choices are outlined below. There are several different tests that are performed to identify the asbestos species and determine the percentage. They will be explained below.

*3.1. Safety*

(a) Do not create unnecessary dust. Handle the samples in HEPA-filter equipped hoods. If samples are received in bags, envelopes or other inappropriate container, open them only in a hood having a face velocity at or greater than 100 fpm. Transfer a small amount to a scintillation vial and only handle the smaller amount.

(b) Open samples in a hood, never in the open lab area.

(c) Index of refraction oils can be toxic. Take care not to get this material on the skin. Wash immediately with soap and water if this happens.

(d) Samples that have been heated in the muffle furnace or the drying oven may be hot. Handle them with tongs until they are cool enough to handle.

(e) Some of the solvents used, such as THF (tetrahydrofuran), are toxic and should only be handled in an appropriate fume hood and according to instructions given in the Material Safety Data Sheet (MSDS).

*3.2. Equipment*

(a) Phase contrast microscope with 10X, 16X and 40X objectives, 10X wide-field eyepieces, G-22 Walton-Beckett graticule, Whipple disk, polarizer, analyzer and first order red or gypsum plate, 100 Watt illuminator, rotating position condenser with oversize phase rings, central stop dispersion objective, Kohler illumination and a rotating mechanical stage.

(b) Stereo microscope with reflected light illumination, transmitted light illumination, polarizer, analyzer and first order red or gypsum plate, and rotating stage.

(c) Negative pressure hood for the stereo microscope

(d) Muffle furnace capable of 600 °C

(e) Drying oven capable of 50—150 °C

(f) Aluminum specimen pans

(g) Tongs for handling samples in the furnace

(h) High dispersion index of refraction oils (Special for dispersion staining.)

$n = 1.550$
$n = 1.585$
$n = 1.590$
$n = 1.605$
$n = 1.620$
$n = 1.670$
$n = 1.680$
$n = 1.690$

(i) A set of index of refraction oils from about n=1.350 to n=2.000 in n=0.005 increments. (Standard for Becke line analysis.)

(j) Glass slides with painted or frosted ends 1×3 inches 1 mm thick, precleaned.

(k) Cover Slips 22×22 mm, #1 1/2

(l) Paper clips or dissection needles

(m) Hand grinder

(n) Scalpel with both #10 and #11 blades

(o) 0.1 molar HCl

(p) Decalcifying solution (Baxter Scientific Products) Ethylenediaminetetraacetic Acid,

| | |
|---|---|
| Tetrasodium | 0.7 g/liter |
| Sodium Potassium Tartrate | 8.0 mg/liter |
| Hydrochloric Acid | 99.2 g/liter |
| Sodium Tartrate | 0.14 g/liter |

(q) Tetrahydrofuran (THF)

(r) Hotplate capable of 60 °C

(s) Balance

(t) Hacksaw blade

(u) Ruby mortar and pestle

*3.3. Sample Pre-Preparation*

Sample preparation begins with pre-preparation which may include chemical reduction of the matrix, heating the sample to dryness or heating in the muffle furnace. The end result is a sample which has been reduced to a powder that is sufficiently fine to fit under the cover slip. Analyze different phases of samples separately, e.g., tile and the tile mastic should be analyzed separately as the mastic may contain asbestos while the tile may not.

(a) *Wet samples*

Samples with a high water content will not give the proper dispersion colors and must be dried prior to sample mounting. Remove the lid of the scintillation vial, place the bottle in the drying oven and heat at 100°C to dryness (usually about 2 h). Samples which are not submitted to the lab in glass must be removed and placed in glass vials or aluminum weighing pans before placing them in the drying oven.

(b) *Samples With Organic Interference—Muffle Furnace*

These may include samples with tar as a matrix, vinyl asbestos tile, or any other organic that can be reduced by heating. Remove the sample from the vial and weigh in a balance to determine the weight of the submitted portion. Place the sample in a muffle furnace at 500 °C for 1 to 2 h or until all obvious organic material has been removed. Retrieve, cool and weigh again to determine the weight loss on ignition. This is necessary to determine the asbestos content of the submitted sample, because the analyst will be looking at a reduced sample.

**NOTE:** Heating above 600 °C will cause the sample to undergo a structural change which, given sufficient time, will convert the chrysotile to forsterite. Heating even at lower temperatures for 1 to 2 h may have a measurable effect on the optical properties of the minerals. If the analyst is unsure of what to expect, a sample of standard asbestos should be heated to the same temperature for the same length of time so that it can be examined for the proper interpretation.

(c) *Samples With Organic Interference—THF*

Vinyl asbestos tile is the most common material treated with this solvent, although, substances containing tar will sometimes yield to this treatment. Select a portion of the material and then grind it up if possible. Weigh the sample and place it in a test tube. Add sufficient THF to dissolve the organic matrix. This is usually about 4 to 5 mL. *Remember, THF is highly flammable.* Filter the remaining material through a tared silver membrane, dry and weigh to determine how much is left after the solvent extraction. Further process the sample to remove carbonate or mount directly.

(d) *Samples With Carbonate Interference*

Carbonate material is often found on fibers and sometimes must be removed in order to perform dispersion microscopy. Weigh out a portion of the material and place it in a test tube. Add a sufficient amount of 0.1 M HCl or decalcifying solution in the tube to react all the carbonate as evidenced by gas formation; i.e., when the gas bubbles stop, add a little more solution. If no more gas forms, the reaction is complete. Filter the material out through a tared silver membrane, dry and weigh to determine the weight lost.

*3.4. Sample Preparation*

Samples must be prepared so that accurate determination can be made of the asbestos type and amount present. The following steps are carried out in the low-flow hood (a low-flow hood has less than 50 fpm flow):

(1) If the sample has large lumps, is hard, or cannot be made to lie under a cover slip, the grain size must be reduced. Place a small amount between two slides and grind the material between them or grind a small amount in a clean mortar and pestle. The choice of whether to

use an alumina, ruby, or diamond mortar depends on the hardness of the material. Impact damage can alter the asbestos mineral if too much mechanical shock occurs. (Freezer mills can completely destroy the observable crystallinity of asbestos and should not be used). For some samples, a portion of material can be shaved off with a scalpel, ground off with a hand grinder or hack saw blade.

The preparation tools should either be disposable or cleaned thoroughly. Use vigorous scrubbing to loosen the fibers during the washing. Rinse the implements with copious amounts of water and air-dry in a dust-free environment.

(2) If the sample is powder or has been reduced as in (1) above, it is ready to mount. Place a glass slide on a piece of optical tissue and write the identification on the painted or frosted end. Place two drops of index of refraction medium n=1.550 on the slide. (The medium n=1.550 is chosen because it is the matching index for chrysotile. Dip the end of a clean paper-clip or dissecting needle into the droplet of refraction medium on *the slide* to moisten it. Then dip the probe into the powder sample. Transfer what sticks on the probe to the slide. The material on the end of the probe should have a diameter of about 3 mm for a good mount. If the material is very fine, less sample may be appropriate. For non-powder samples such as fiber mats, forceps should be used to transfer a small amount of material to the slide. Stir the material in the medium on the slide, spreading it out and making the preparation as uniform as possible. Place a cover-slip on the preparation by gently lowering onto the slide and allowing it to fall "trapdoor" fashion on the preparation to push out any bubbles. Press gently on the cover slip to even out the distribution of particulate on the slide. If there is insufficient mounting oil on the slide, one or two drops may be placed near the edge of the coverslip on the slide. Capillary action will draw the necessary amount of liquid into the preparation. Remove excess oil with the point of a laboratory wiper.

Treat at least two different areas of each phase in this fashion. Choose representative areas of the sample. It may be useful to select particular areas or fibers for analysis. This is useful to identify asbestos in severely inhomogeneous samples.

When it is determined that amphiboles may be present, repeat the above process using the appropriate high-dispersion oils until an identification is made or all six asbestos minerals have been ruled out. Note that percent determination must be done in the index medium 1.550 because amphiboles tend to disappear in their matching mediums.

3.5. Analytical Procedure

**NOTE:** This method presumes some knowledge of mineralogy and optical petrography.

The analysis consists of three parts: The determination of whether there is asbestos present, what type is present and the determination of how much is present. The general flow of the analysis is:

(1) Gross examination.

(2) Examination under polarized light on the stereo microscope.

(3) Examination by phase-polar illumination on the compound phase microscope.

(4) Determination of species by dispersion stain. Examination by Becke line analysis may also be used; however, this is usually more cumbersome for asbestos determination.

(5) Difficult samples may need to be analyzed by SEM or TEM, or the results from those techniques combined with light microscopy for a definitive identification. Identification of a particle as asbestos requires that it be asbestiform. Description of particles should follow the suggestion of Campbell. (Figure 1)

**FIGURE 1—PARTICLE DEFINITIONS SHOWING MINERAL GROWTH HABITS FROM THE U.S. BUREAU OF MINES**

Single-crystal shapes: Equant, Prismatic, Acicular, Fiber, Fibril, Filiform, Bladed, Platy, Lamellar

Crystal-aggregate patterns or arrangements: Asbestiform, Columnar, Fibrous (See "Asbestiform" above), Lamellar, Massive, Radiating, Reticulated

For the purpose of regulation, the mineral must be one of the six minerals covered and must be in the asbestos growth habit. Large specimen samples of asbestos generally have the gross appearance of wood. Fibers are easily parted from it. Asbestos fibers are very long compared with their widths. The fibers have a very high tensile strength as demonstrated by bending without breaking. Asbestos fibers exist in bundles that are easily parted, show longitudinal fine structure and may be tufted at the ends showing "bundle of sticks" morphology. In the microscope some of these properties may not be observable. Amphiboles do not always show striations along their length even when they are asbestos. Neither will they always show tufting. They generally do not show a curved nature except for very long fibers. Asbestos and asbestiform minerals are usually characterized in groups by extremely high aspect ratios (greater than 100:1). While aspect ratio analysis is useful for characterizing populations of fibers, it cannot be used to identify individual fibers of intermediate to short aspect ratio. Observation of many fibers is often necessary to determine whether a sample consists of "cleavage fragments" or of asbestos fibers.

Most cleavage fragments of the asbestos minerals are easily distinguishable from true asbestos fibers. This is because true cleavage fragments usually have larger diameters than 1 µm. Internal structure of particles larger than this usually shows them to have no internal fibrillar structure. In addition, cleavage fragments of the monoclinic amphiboles show inclined extinction under crossed polars with no compensator. Asbestos fibers usually show extinction at zero degrees or ambiguous extinction if any at all. Morphologically, the larger cleavage fragments are obvious by their blunt or stepped ends showing prismatic habit. Also, they tend to be acicular rather than filiform.

Where the particles are less than 1 µm in diameter and have an aspect ratio greater than or equal to 3:1, it is recommended that the sample be analyzed by SEM or TEM if there is any question whether the fibers are cleavage fragments or asbestiform particles.

Care must be taken when analyzing by electron microscopy because the interferences are different from those in light microscopy and may structurally be very similar to asbestos. The classic interference is between anthophyllite and biopyribole or intermediate fiber. Use the same morphological clues for electron microscopy as are used for light microscopy, e.g., fibril splitting, internal longitudinal striation, fraying, curvature, etc.

(1) Gross examination:

Examine the sample, preferably in the glass vial. Determine the presence of any obvious fibrous component. Estimate a percentage based on previous experience and current observation. Determine whether any pre-preparation is necessary. Determine the number of phases present. This step may be carried out or augmented by observation at 6 to 40× under a stereo microscope.

(2) After performing any necessary pre-preparation, prepare slides of each phase as described above. Two preparations of the same phase in the same index medium can be made side-by-side on the same glass for convenience. Examine with the polarizing stereo microscope. Estimate the percentage of asbestos based on the amount of birefringent fiber present.

(3) Examine the slides on the phase-polar microscopes at magnifications of 160 and 400×. Note the morphology of the fibers. Long, thin, very straight fibers with little curvature are indicative of fibers from the amphibole family. Curved, wavy fibers are usually indicative of chrysotile. Estimate the percentage of asbestos on the phase-polar microscope under conditions of crossed polars and a gypsum plate. Fibers smaller than 1.0 µm in thickness must be identified by inference to the presence of larger, identifiable fibers and morphology. If no larger fibers are visible, electron microscopy should be performed. At this point, only a tentative identification can be made. Full identification must be made with dispersion microscopy. Details of the tests are included in the appendices.

(4) Once fibers have been determined to be present, they must be identified. Adjust the microscope for dispersion mode and observe the fibers. The microscope has a rotating stage, one polarizing element, and a system for generating dark-field dispersion microscopy (see Section 4.6. of this appendix). Align a fiber with its length parallel to the polarizer and note the color of the Becke lines. Rotate the stage to bring the fiber length perpendicular to the polarizer and note the color. Repeat this process for every fiber or fiber bundle examined. The colors must be consistent with the colors generated by standard asbestos reference materials for a positive identification. In n=1.550, amphiboles will generally show a yellow to straw-yellow color indicating that the fiber indices of refraction are higher than the liquid. If long, thin fibers are noted and the colors are yellow, prepare further slides as above in the suggested matching liquids listed below:

| Type of asbestos | Index of refraction |
|---|---|
| Chrysotile | n=1.550. |
| Amosite | n=1.670 or 1.680. |
| Crocidolite | n=1.690. |
| Anthophyllite | n=1.605 and 1.620. |
| Tremolite | n=1.605 and 1.620. |
| Actinolite | n=1.620. |

Where more than one liquid is suggested, the first is preferred; however, in some cases this liquid will not give good dispersion color. Take care to avoid interferences in the other liquid; e.g., wollastonite in n=1.620 will give the same colors as tremolite. In n=1.605 wollastonite will appear yellow in all directions. Wollastonite may be determined under crossed polars as it will change from blue to yellow as it is rotated along its fiber axis by tapping on the cover slip. Asbestos minerals will not change in this way.

Determination of the angle of extinction may, when present, aid in the determination of anthophyllite from tremolite. True asbestos fibers usually have 0° extinction or ambiguous extinction, while cleavage fragments have more definite extinction.

Continue analysis until both preparations have been examined and all present species of asbestos are identified. If there are no fibers present, or there is less than 0.1% present, end the analysis with the minimum number of slides (2).

(5) Some fibers have a coating on them which makes dispersion microscopy very difficult or impossible. Becke line analysis or electron microscopy may be performed in those cases. Determine the percentage by light microscopy. TEM analysis tends to overestimate the actual percentage present.

(6) Percentage determination is an estimate of occluded area, tempered by gross observation. Gross observation information is used to make sure that the high magnification microscopy does not greatly over- or underestimate the amount of fiber present. This part of the analysis requires a great deal of experience. Satisfactory models for asbestos content analysis have not yet been developed, although some models based on metallurgical grain-size determination have found some utility. Estimation is more easily handled in situations where the grain sizes visible at about 160× are about the same and the sample is relatively homogeneous.

View all of the area under the cover slip to make the percentage determination. View the fields while moving the stage, paying attention to the clumps of material. These are not usually the best areas to perform dispersion microscopy because of the interference from other materials. But, they are the areas most likely to represent the accurate percentage in the sample. Small amounts of asbestos require slower scanning and more frequent analysis of individual fields.

Report the area occluded by asbestos as the concentration. This estimate does not generally take into consideration the difference in density of the different species present in the sample. For most samples this is adequate. Simulation studies with similar materials must be carried out to apply microvisual estimation for that purpose and is beyond the scope of this procedure.

(7) Where successive concentrations have been made by chemical or physical means, the amount reported is the percentage of the material in the "as submitted" or original state. The percentage determined by microscopy is multiplied by the fractions remaining after pre-preparation steps to give the percentage in the original sample. For example:

Step 1. 60% remains after heating at 550 °C for 1 h. Step 2. 30% of the residue of step 1 remains after dissolution of carbonate in 0.1 m HCl.

Step 3. Microvisual estimation determines that 5% of the sample is chrysotile asbestos.

The reported result is:

R = (Microvisual result in percent) × (Fraction remaining after step 2) × (Fraction remaining of original sample after step 1)

R = (5) × (.30) × (.60) = 0.9%

(8) Report the percent and type of asbestos present. For samples where asbestos was identified, but is less than 1.0%, report "Asbestos present, less than 1.0%." There must have been at least two observed fibers or fiber bundles in the two preparations to be reported as present. For samples where asbestos was not seen, report as "None Detected."

### 4. Auxiliary Information

Because of the subjective nature of asbestos analysis, certain concepts and procedures need to be discussed in more depth. This information will help the analyst understand why some of the procedures are carried out the way they are.

#### 4.1. Light

Light is electromagnetic energy. It travels from its source in packets called quanta. It is instructive to consider light as a plane wave. The light has a direction of travel. Perpendicular to this and mutually perpendicular to each other, are two vector components. One is the magnetic vector and the other is the electric vector. We shall only be concerned with the electric vector. In this description, the interaction of the vector and the mineral will describe all the observable phenomena. From a light source such a microscope illuminator, light travels in all different direction from the filament.

In any given direction away from the filament, the electric vector is perpendicular to the direction of travel of a light ray. While perpendicular, its orientation is random about the travel axis. If the electric vectors from all the light rays were lined up by passing the light through a filter that would only let light rays with electric vectors oriented in one direction pass, the light would then be *POLARIZED*.

Polarized light interacts with matter in the direction of the electric vector. This is the polarization direction. Using this property it is possible to use polarized light to probe different materials and identify them by how they interact with light.

The speed of light in a vacuum is a constant at about $2.99 \times 10^8$ m/s. When light travels in different materials such as air, water, minerals or oil, it does not travel at this speed. It travels slower. This slowing is a function of both the material through which the light is traveling and the wavelength or frequency of the light. In general, the more dense the material, the slower the light travels. Also, generally, the higher the frequency, the slower the light will travel. The ratio of the speed of light in a vacuum to that in a material is called the index of refraction (n). It is usually measured at 589 nm (the sodium D line). If white light (light containing all the visible wavelengths) travels through a material, rays of longer wavelengths will travel faster than those of shorter wavelengths, this separation is called dispersion. Dispersion is used as an identifier of materials as described in Section 4.6.

### 4.2. Material Properties

Materials are either amorphous or crystalline. The difference between these two descriptions depends on the positions of the atoms in them. The atoms in amorphous materials are randomly arranged with no long range order. An example of an amorphous material is glass. The atoms in crystalline materials, on the other hand, are in regular arrays and have long range order. Most of the atoms can be found in highly predictable locations. Examples of crystalline material are salt, gold, and the asbestos minerals.

It is beyond the scope of this method to describe the different types of crystalline materials that can be found, or the full description of the classes into which they can fall. However, some general crystallography is provided below to give a foundation to the procedures described.

With the exception of anthophyllite, all the asbestos minerals belong to the monoclinic crystal type. The unit cell is the basic repeating unit of the crystal and for monoclinic crystals can be described as having three unequal sides, two 90° angles and one angle not equal to 90°. The orthorhombic group, of which anthophyllite is a member has three unequal sides and three 90° angles. The unequal sides are a consequence of the complexity of fitting the different atoms into the unit cell. Although the atoms are in a regular array, that array is not symmetrical in all directions. There is long range order in the three major directions of the crystal. However, the order is different in each of the three directions. This has the effect that the index of refraction is different in each of the three directions. Using polarized light, we can investigate the index of refraction in each of the directions and identify the mineral or material under investigation. The indices $\alpha$, $\beta$, and $\gamma$ are used to identify the lowest, middle, and highest index of refraction respectively. The x direction, associated with $\alpha$ is called the fast axis. Conversely, the z direction is associated with $\gamma$ and is the slow direction. Crocidolite has $\alpha$ along the fiber length making it "length-fast". The remainder of the asbestos minerals have the $\gamma$ axis along the fiber length. They are called "length-slow". This orientation to fiber length is used to aid in the identification of asbestos.

### 4.3. Polarized Light Technique

Polarized light microscopy as described in this section uses the phase-polar microscope described in Section 3.2. A phase contrast microscope is fitted with two polarizing elements, one below and one above the sample. The polarizers have their polarization directions at right angles to each other. Depending on the tests performed, there may be a compensator between these two polarizing elements. Light emerging from a polarizing element has its electric vector pointing in the polarization direction of the element. The light will not be subsequently transmitted through a second element set at a right angle to the first element. Unless the light is altered as it passes from one element to the other, there is no transmission of light.

### 4.4. Angle of Extinction

Crystals which have different crystal regularity in two or three main directions are said to be anisotropic. They have a different index of refraction in each of the main directions. When such a crystal is inserted between the crossed polars, the field of view is no longer dark but shows the crystal in color. The color depends on the properties of the crystal. The light acts as if it travels through the crystal along the optical axes. If a crystal optical axis were lined up along one of the polarizing directions (either the polarizer or the analyzer) the light would appear to travel only in that direction, and it would blink out or go dark. The difference in degrees between the fiber direction and the angle at which it blinks out is called the angle of extinction. When this angle can be measured, it is useful in identifying the mineral. The procedure for measuring the angle of extinction is to first identify the polarization direction in the microscope. A commercial alignment slide can be used to establish the polarization directions or use anthophyllite or another suitable mineral. This mineral has a zero degree angle of extinction and will go dark to extinction as it aligns with the polarization directions. When a fiber of anthophyllite has gone to extinction, align the eyepiece reticle or graticule with the fiber so that there is a visual cue as to the direction of polarization in the field of view. Tape or otherwise secure the eyepiece in this position so it will not shift.

After the polarization direction has been identified in the field of view, move the particle of interest to the center of the field of view and align it with the polarization direction. For fibers, align the fiber along this direction. Note the angular reading of the rotating stage. Looking at the particle, rotate the stage until the fiber goes dark or "blinks out". Again note the reading of the stage. The difference in the first reading and the second is an angle of extinction.

The angle measured may vary as the orientation of the fiber changes about its long axis. Tables of mineralogical data usually report the maximum angle of extinction. Asbestos forming minerals, when they exhibit an angle of extinction, usually do show an angle of extinction close to the reported maximum, or as appropriate depending on the substitution chemistry.

### 4.5. Crossed Polars with Compensator

When the optical axes of a crystal are not lined up along one of the polarizing directions (either the polarizer or the analyzer) part of the light travels along one axis and part travels along the other visible axis. This is characteristic of birefringent materials.

The color depends on the difference of the two visible indices of refraction and the thickness of the crystal. The maximum difference available is the difference between the $\alpha$ and the $\gamma$ axes. This maximum difference is usually tabulated as the birefringence of the crystal.

For this test, align the fiber at 45° to the polarization directions in order to maximize the contribution to each of the optical axes. The colors seen are called retardation colors. They arise from the recombination of light which has traveled through the two separate directions of the crystal. One of the rays is retarded behind the other since the light in that direction travels slower. On recombination, some of the colors which make up white light are enhanced by constructive interference and some are suppressed by destructive interference. The result is a color dependent on the difference between the indices and the thickness of the crystal. The proper colors, thicknesses, and retardations are shown on a Michel-Levy chart. The three items, retardation, thickness and birefringence are related by the following relationship:

$$R = t (n_\gamma - n_\alpha)$$

**R = retardation, t = crystal thickness in μm, and**

$n_{\alpha,\gamma}$ **= indices of refraction.**

Examination of the equation for asbestos minerals reveals that the visible colors for almost all common asbestos minerals and fiber sizes are shades of gray and black. The eye is relatively poor at discriminating different shades of gray. It is very good at discriminating different colors. In order to compensate for the low retardation, a compensator is added to the light train between the polarization elements. The compensator used for this test is a gypsum plate of known thickness and birefringence. Such a compensator when oriented at 45° to the polarizer direction, provides a retardation of 530 nm of the 530 nm wavelength color. This enhances the red color and gives the background a characteristic red to red-magenta color. If this "full-wave" compensator is in place when the asbestos preparation is inserted into the light train, the colors seen on the fibers are quite different. Gypsum, like asbestos has a fast axis and a slow axis. When a fiber is aligned with its fast axis in the same direction as the fast axis of the gypsum plate, the ray vibrating in the slow direction is retarded by both the asbestos and the gypsum. This results in a higher retardation than would be present for either of the two minerals. The color seen is a second order blue. When the fiber is rotated 90° using the rotating stage, the slow direction of the fiber is now aligned with the fast direction of the gypsum and the fast direction of the fiber is aligned with the slow direction of the gypsum. Thus, one ray vibrates faster in the fast direction of the gypsum, and slower in the slow direction of the fiber; the other ray will vibrate slower in the slow direction of the gypsum and faster in the fast direction of the fiber. In this case, the effect is subtractive and the color seen is a first order yellow. As long as the fiber thickness does not add appreciably to the color, the same basic colors will be seen for all asbestos types except crocidolite. In crocidolite the colors will be weaker, may be in the opposite directions, and will be altered by the blue absorption color natural to crocidolite. Hundreds of other materials will give the same colors as asbestos, and therefore, this test is not definitive for asbestos. The test is useful in discriminating against fiberglass or other amorphous fibers such as some synthetic fibers. Certain synthetic fibers will show retardation colors different than asbestos; however, there are some forms of polyethylene and aramid which will show morphology and retardation colors similar to asbestos minerals. This test must be supplemented with a positive identification test when birefringent fibers are present which can not be excluded by morphology. This test is relatively ineffective for use on fibers less than 1 μm in diameter. For positive confirmation TEM or SEM should be used if no larger bundles or fibers are visible.

4.6. Dispersion Staining

Dispersion microscopy or dispersion staining is the method of choice for the identification of asbestos in bulk materials. Becke line analysis is used by some laboratories and yields the same results as does dispersion staining for asbestos and can be used in lieu of dispersion staining. Dispersion staining is performed on the same platform as the phase-polar analysis with the analyzer and compensator removed. One polarizing element remains to define the direction of the light so that the different indices of refraction of the fibers may be separately determined. Dispersion microscopy is a dark-field technique when used for asbestos. Particles are imaged with scattered light. Light which is unscattered is blocked from reaching the eye either by the back field image mask in a McCrone objective or a back field image mask in the phase condenser. The most convenient method is to use the rotating phase condenser to move an oversized phase ring into place. The ideal size for this ring is for the central disk to be just larger than the objective entry aperture as viewed in the back focal plane. The larger the disk, the less scattered light reaches the eye. This will have the effect of diminishing the intensity of dispersion color and will shift the actual color seen. The colors seen vary even on microscopes from the same manufacturer. This is due to the different bands of wavelength exclusion by different mask sizes. The mask may either reside in the condenser or in the objective back focal plane. It is imperative that the analyst determine by experimentation with asbestos standards what the appropriate colors should be for each asbestos type. The colors depend also on the temperature of the preparation and the exact chemistry of the asbestos. Therefore, some slight differences from the standards should be allowed. This is not a serious problem for commercial asbestos uses. This technique is used for identification of the indices of refraction for fibers by recognition of color. There is no direct numerical readout of the index of refraction. Correlation of color to actual index of refraction is possible by referral to published conversion tables. This is not necessary for the analysis of asbestos. Recognition of appropriate colors along with the proper morphology are deemed sufficient to identify the commercial asbestos minerals. Other techniques including SEM, TEM, and XRD may be required to provide additional information in order to identify other types of asbestos.

Make a preparation in the suspected matching high dispersion oil, e.g., n=1.550 for chrysotile. Perform the preliminary tests to determine whether the fibers are birefringent or not. Take note of the morphological character. Wavy fibers are indicative of chrysotile while long, straight, thin, frayed fibers are indicative of amphibole asbestos. This can aid in the selection of the appropriate matching oil. The microscope is set up and the polarization direction is noted as in Section 4.4. Align a fiber with the polarization direction. Note the color. This is the color parallel to the polarizer. Then rotate the fiber rotating the stage 90° so that the polarization direction is across the fiber. This is the perpendicular position. Again note the color. Both colors must be consistent with standard asbestos minerals in the correct direction for a positive identification of asbestos. If only one of the colors is correct while the other is not, the identification is not positive. If the colors in both directions are bluish-white, the analyst has chosen a matching index oil which is higher than the correct matching oil, e.g., the analyst has used n=1.620 where chrysotile is present. The next lower oil (Section 3.5.) should be used to prepare another specimen. If the color in both directions is yellow-white to straw-yellow-white, this indicates that the index of the oil is lower than the index of the fiber, e.g., the preparation is in n=1.550 while anthophyllite is present. Select the next higher oil (Section 3.5.) and prepare another slide. Continue in this fashion until a positive identification of all asbestos species present has been made or all possible asbestos species have been ruled out by negative results in this test. Certain plant fibers can have similar dispersion colors as asbestos. Take care to note and evaluate the morphology of the fibers or remove the plant fibers in pre- preparation. Coating material on the fibers such as carbonate or vinyl may destroy the dispersion color. Usually, there will be some outcropping of fiber which will show the colors sufficient for identification. When this is not the case, treat the sample as described in Section 3.3. and then perform dispersion staining. Some samples will yield to Becke line analysis if they are coated or electron microscopy can be used for identification.

5. References

5.1. Crane, D.T., *Asbestos in Air*, OSHA method ID160, Revised November 1992.

5.2. Ford, W.E., *Dana's Textbook of Mineralogy*; Fourth Ed.; John Wiley and Son, New York, 1950, p. vii.

5.3. Selikoff,.I.J., Lee, D.H.K., *Asbestos and Disease*, Academic Press, New York, 1978, pp. 3,20.

5.4. *Women Inspectors of Factories.* Annual Report for 1898, H.M. Statistical Office, London, p. 170 (1898).

5.5. Selikoff, I.J., Lee, D.H.K., *Asbestos and Disease*, Academic Press, New York, 1978, pp. 26,30.

5.6. Campbell, W.J., et al, *Selected Silicate Minerals and Their Asbestiform Varieties*, United States Department of the Interior, Bureau of Mines, Information Circular 8751, 1977.

5.7. *Asbestos*, Code of Federal Regulations, 29 CFR 1910.1001 and 29 CFR 1926.58.

5.8. *National Emission Standards for Hazardous Air Pollutants; Asbestos NESHAP Revision*, FEDERAL REGISTER, Vol. 55, No. 224, 20 November 1990, p. 48410.

5.9. Ross, M. *The Asbestos Minerals: Definitions, Description, Modes of Formation, Physical and Chemical Properties and Health Risk to the Mining Community*, Nation Bureau of Standards Special Publication, Washington, DC, 1977.

5.10. Lilis, R., Fibrous Zeolites and Endemic Mesothelioma in Cappadocia, Turkey, *J. Occ Medicine*, 1981, 23,(8),548-550.

5.11. *Occupational Exposure to Asbestos—1972*, U.S. Department of Health, Education and Welfare, Public Health Service, Center for Disease Control, National Institute for Occupational Safety and Health, HSM-72-10267.

5.12. Campbell, W.J., et al, *Relationship of Mineral Habit to Size Characteristics for Tremolite Fragments and Fibers*, United States Department of the Interior, Bureau of Mines, Information Circular 8367, 1979.

5.13. Mefford, D., DCM Laboratory, Denver, private communication, July 1987.

5.14. Deer, W.A., Howie, R.A., Zussman, J., *Rock Forming Minerals*, Longman, Thetford, UK, 1974.

5.15. Kerr, P.F., *Optical Mineralogy*; Third Ed. McGraw-Hill, New York, 1959.

5.16. Veblen, D.R. (Ed.), *Amphiboles and Other Hydrous Pyriboles—Mineralogy, Reviews in Mineralogy*, Vol 9A, Michigan, 1982, pp 1-102.

5.17. Dixon, W.C., *Applications of Optical Microscopy in the Analysis of Asbestos* and Quartz, ACS Symposium Series, No. 120, Analytical Techniques in Occupational Health Chemistry, 1979.

5.18. Polarized Light Microscopy, McCrone Research Institute, Chicago, 1976.

5.19. Asbestos Identification, McCrone Research Institute, G & G printers, Chicago, 1987.

5.20. McCrone, W.C., Calculation of Refractive Indices from Dispersion Staining Data, The Microscope, No 37, Chicago, 1989.

5.21. Levadie, B. (Ed.), *Asbestos and Other Health Related Silicates*, ASTM Technical Publication 834, ASTM, Philadelphia 1982.

5.22. Steel, E. and Wylie, A., Riordan, P.H. (Ed.), Mineralogical Characteristics of Asbestos, *Geology of Asbestos Deposits*, pp. 93-101, SME-AIME, 1981.

5.23. Zussman, J., The Mineralogy of Asbestos, *Asbestos: Properties, Applications and Hazards*, pp. 45-67 Wiley, 1979.

[51 FR 22733, June 20, 1986, as amended at 51 FR 37004, Oct. 17, 1986; 52 FR 17754, 17755, May 12, 1987; 53 FR 35625, September 14, 1988; 54 FR 24334, June 7, 1989; 54 FR 29546, July 13, 1989; 54 FR 52027, Dec. 20, 1989, 55 FR 3731, Feb. 5, 1990; 55 FR 34710, Aug. 24, 1990; 57 FR 24330, June 8, 1992; 59 FR 41057, Aug. 10, 1994; 60 FR 9625, Feb. 21, 1995; 60

FR 33344, June 28, 1995; 60 FR 33984-33987, June 29, 1995; 61 FR 5508, Feb. 13, 1996; 61 FR 43457, Aug. 23, 1996; 63 FR 1285, Jan. 8, 1998]

### §1910.1002 Coal tar pitch volatiles; interpretation of term.

As used in §1910.1000 (Table Z-1), coal tar pitch volatiles include the fused polycyclic hydrocarbons which volatilize from the distillation residues of coal, petroleum (excluding asphalt), wood, and other organic matter. Asphalt (CAS 8052-42-4, and CAS 64742-93-4) is not covered under the "coal tar pitch volatiles" standard.

[48 FR 2768, Jan. 21, 1983]

### §1910.1003 13 Carcinogens (4-Nitrobiphenyl, etc.).

**1910.1003(a)** *Scope and application.* **(1)** This section applies to any area in which the 13 carcinogens addressed by this section are manufactured, processed, repackaged, released, handled, or stored, but shall not apply to transshipment in sealed containers, except for the labeling requirements under paragraphs (e)(2), (3) and (4) of this section. The 13 carcinogens are the following:

4-Nitrobiphenyl, Chemical Abstracts Service Register Number (CAS No.) 92933;

alpha-Naphthylamine, CAS No. 134327;

methyl chloromethyl ether, CAS No. 107302;

3,'-Dichlorobenzidine (and its salts) CAS No. 91941;

bis-Chloromethyl ether, CAS No. 542881;

beta-Naphthylamine, CAS No. 91598;

Benzidine, CAS No. 92875;

4-Aminodiphenyl, CAS No. 92671;

Ethyleneimine, CAS No. 151564;

beta-Propiolactone, CAS No. 57578;

2-Acetylaminofluorene, CAS No. 53963;

4-Dimethylaminoazo-benzene, CAS No. 60117; and

N-Nitrosodimethylamine, CAS No. 62759.

**1910.1003(a)(2)** This section shall not apply to the following:

**1910.1003(a)(2)(i)** Solid or liquid mixtures containing less than 0.1 percent by weight or volume of 4-Nitrobiphenyl; methyl chloromethyl ether; bis-chloromethyl ether; beta-Naphthylamine; benzidine or 4-Aminodiphenyl; and

**1910.1003(a)(2)(ii)** Solid or liquid mixtures containing less than 1.0 percent by weight or volume of alpha-Naphthylamine; 3,'- Dichlorobenzidine (and its salts); Ethyleneimine; beta-Propiolactone; 2-Acetylaminofluorene; 4-Dimethylaminoazobenzene, or N-Nitrosodimethylamine.

**1910.1003(b)** *Definitions.* For the purposes of this section:

*Absolute filter* is one capable of retaining 99.97 percent of a mono disperse aerosol of 0.3 μm particles.

*Authorized employee* means an employee whose duties require him to be in the regulated area and who has been specifically assigned by the employer.

*Clean change room* means a room where employees put on clean clothing and/or protective equipment in an environment free of the 13 carcinogens addressed by this section. The clean change room shall be contiguous to and have an entry from a shower room, when the shower room facilities are otherwise required in this section.

*Closed system* means an operation involving a carcinogen addressed by this section where containment prevents the release of the material into regulated areas, non-regulated areas, or the external environment.

*Decontamination* means the inactivation of a carcinogen addressed by this section or its safe disposal.

*Director* means the Director, National Institute for Occupational Safety and Health, or any person directed by him or the Secretary of Health and Human Services to act for the Director.

*Disposal* means the safe removal of the carcinogens addressed by this section from the work environment.

*Emergency* means an unforeseen circumstance or set of circumstances resulting in the release of a carcinogen addressed by this section that may result in exposure to or contact with the material.

*External environment* means any environment external to regulated and nonregulated areas.

*Isolated system* means a fully enclosed structure other than the vessel of containment of a carcinogen addressed by this section that is impervious to the passage of the material and would prevent the entry of the carcinogen addressed by this section into regulated areas, nonregulated areas, or the external environment, should leakage or spillage from the vessel of containment occur.

*Laboratory-type hood* is a device enclosed on the three sides and the top and bottom, designed and maintained so as to draw air inward at an average linear face velocity of 150 feet per minute with a minimum of 125 feet per minute; designed, constructed, and maintained in such a way that an operation involving a carcinogen addressed by this section within the hood does not require the insertion of any portion of any employee's body other than his hands and arms.

*Nonregulated area* means any area under the control of the employer where entry and exit is neither restricted nor controlled.

*Open-vessel system* means an operation involving a carcinogen addressed by this section in an open vessel that is not in an isolated system, a laboratory-type hood, nor in any other system affording equivalent protection against the entry of the material into regulated areas, non-regulated areas, or the external environment.

*Protective clothing* means clothing designed to protect an employee against contact with or exposure to a carcinogen addressed by this section.

*Regulated area* means an area where entry and exit is restricted and controlled.

**1910.1003(c)** *Requirements for areas containing a carcinogen addressed by this section.* A regulated area shall be established by an employer where a carcinogen addressed by this section is manufactured, processed, used, repackaged, released, handled or stored. All such areas shall be controlled in accordance with the requirements for the following category or categories describing the operation involved:

**1910.1003(c)(1)** *Isolated systems.* Employees working with a carcinogen addressed by this section within an isolated system such as a "glove box" shall wash their hands and arms upon completion of the assigned task and before engaging in other activities not associated with the isolated system.

**1910.1003(c)(2)** *Closed system operation.* **(i)** Within regulated areas where the carcinogens addressed by this section are stored in sealed containers, or contained in a closed system, including piping systems, with any sample ports or openings closed while the carcinogens addressed by this section are contained within, access shall be restricted to authorized employees only.

**1910.1003(c)(2)(ii)** Employees exposed to 4-Nitrobiphenyl; alpha-Naphthylamine; 3,'-Dichlorobenzidine (and its salts); beta-Naphthylamine; benzidine; 4-Aminodiphenyl; 2-Acetylaminofluorene; 4-Dimethylaminoazo-benzene; and N-Nitrosodimethylamine shall be required to wash hands, forearms, face, and neck upon each exit from the regulated areas, close to the point of exit, and before engaging in other activities.

**1910.1003(c)(3)** *Open-vessel system operations.* Open-vessel system operations as defined in paragraph (b)(13) of this section are prohibited.

**1910.1003(c)(4)** *Transfer from a closed system, charging or discharging point operations, or otherwise opening a closed system.* In operations involving "laboratory-type hoods," or in locations where the carcinogens addressed by this section are contained in an otherwise "closed system," but is transferred, charged, or discharged into other normally closed containers, the provisions of this paragraph shall apply.

**1910.1003(c)(4)(i)** Access shall be restricted to authorized employees only.

**1910.1003(c)(4)(ii)** Each operation shall be provided with continuous local exhaust ventilation so that air movement is always from ordinary work areas to the operation. Exhaust air shall not be discharged to regulated areas, nonregulated areas or the external environment unless decontaminated. Clean makeup air shall be introduced in sufficient volume to maintain the correct operation of the local exhaust system.

**1910.1003(c)(4)(iii)** Employees shall be provided with, and required to wear, clean, full body protective clothing (smocks, coveralls, or long-sleeved shirt and pants), shoe covers and gloves prior to entering the regulated area.

**1910.1003(c)(4)(iv)** Employees engaged in handling operations involving the carcinogens addressed by this section must be provided with, and required to wear and use a half-face filter-type respirator with filters for dusts, mists, and fumes, or air-purifying canisters or cartridges. A respirator affording higher levels of protection than this respirator may be substituted.

**1910.1003(c)(4)(v)** Prior to each exit from a regulated area, employees shall be required to remove and leave protective clothing and equipment at the point of exit and at the last exit of the day, to place used clothing and equipment in impervious containers at the point of exit for purposes of decontamination or disposal. The contents of such impervious containers shall be identified, as required under paragraphs (e), (2), (3), and (4) of this section.

**1910.1003(c)(4)(vi)** Drinking fountains are prohibited in the regulated area.

**1910.1003(c)(4)(vii)** Employees shall be required to wash hands, forearms, face, and neck on each exit from the regulated area, close to the point of exit, and before engaging in other activities and employees exposed to 4-Nitrobiphenyl; alpha-Naphthylamine; 3,'-Dichlorobenzidine (and its salts); beta-Naphthylamine; Benzidine; 4-Aminodiphenyl; 2-Acetylaminofluorene; 4-Dimethylaminoazo-benzene; and N-Nitrosodimethylamine shall be required to shower after the last exit of the day.

**1910.1003(c)(5)** *Maintenance and decontamination activities.* In cleanup of leaks of spills, maintenance, or repair operations on contaminated systems or equipment, or any operations involving work in an area where direct contact with a carcinogen addressed by this section could result, each authorized employee entering that area shall:

**1910.1003(c)(5)(i)** Be provided with and required to wear clean, impervious garments, including gloves, boots, and continuous-air supplied hood in accordance with §1910.134;

**1910.1003(c)(5)(ii)** Be decontaminated before removing the protective garments and hood;

**1910.1003(c)(5)(iii)** Be required to shower upon removing the protective garments and hood.

**1910.1003(d)** *General regulated area requirements*—**(1)** *Respirator program.* The employer must implement a respiratory protection program in accordance with 29 CFR 1910.134 (b), (c), (d) (except (d)(1)(iii) and (iv), and (d)(3)), and (e) through (m).

**1910.1003(d)(2)** *Emergencies.* In an emergency, immediate measures including, but not limited to, the requirements of paragraphs (d)(2) (i) through (v) of this section shall be implemented.

**1910.1003(d)(2)(i)** The potentially affected area shall be evacuated as soon as the emergency has been determined.

**1910.1003(d)(2)(ii)** Hazardous conditions created by the emergency shall be eliminated and the potentially affected area shall be decontaminated prior to the resumption of normal operations.

**1910.1003(d)(2)(iii)** Special medical surveillance by a physician shall be instituted within 24 hours for employees present in the potentially affected area at the time of the emergency. A report of the medical surveillance and any treatment shall be included in the incident report, in accordance with paragraph (f)(2) of this section.

**1910.1003(d)(2)(iv)** Where an employee has a known contact with a carcinogen addressed by this section, such employee shall be required to shower as soon as possible, unless contraindicated by physical injuries.

**1910.1003(d)(2)(v)** An incident report on the emergency shall be reported as provided in paragraph (f)(2) of this section.

**1910.1003(d)(2)(vi)** Emergency deluge showers and eyewash fountains supplied with running potable water shall be located near, within sight of, and on the same level with locations where a direct exposure to Ethyleneimine or beta-Propiolactone only would be most likely as a result of equipment failure or improper work practice.

**1910.1003(d)(3)** *Hygiene facilities and practices.* **(i)** Storage or consumption of food, storage or use of containers of beverages, storage or application of cosmetics, smoking, storage of smoking materials, tobacco products or other products for chewing, or the chewing of such products are prohibited in regulated areas.

**1910.1003(d)(3)(ii)** Where employees are required by this section to wash, washing facilities shall be provided in accordance with §1910.141(d)(1) and (2) (ii) through (vii).

**1910.1003(d)(3)(iii)** Where employees are required by this section to shower, shower facilities shall be provided in accordance with §1910.141(d)(3).

**1910.1003(d)(3)(iv)** Where employees wear protective clothing and equipment, clean change rooms shall be provided for the number of such employees required to change clothes, in accordance with §1910.141(e).

**1910.1003(d)(3)(v)** Where toilets are in regulated areas, such toilets shall be in a separate room.

**1910.1003(d)(4)** *Contamination control.* **(i)** Except for outdoor systems, regulated areas shall be maintained under pressure negative with respect to nonregulated areas. Local exhaust ventilation may be used to satisfy this requirement. Clean makeup air in equal volume shall replace air removed.

**1910.1003(d)(4)(ii)** Any equipment, material, or other item taken into or removed from a regulated area shall be done so in a manner that does not cause contamination in nonregulated areas or the external environment.

**1910.1003(d)(4)(iii)** Decontamination procedures shall be established and implemented to remove carcinogens addressed by this section from the surfaces of materials, equipment, and the decontamination facility.

**1910.1003(d)(4)(iv)** Dry sweeping and dry mopping are prohibited for 4-Nitrobiphenyl; alpha-Naphthylamine; 3,'-Dichlorobenzidine (and its salts); beta-Naphthylamine; Benzidine; 4-Aminodiphenyl; 2-Acetylaminofluorene; 4-Dimethylaminoazo-benzene and N-Nitrosodimethylamine.

**1910.1003(e)** *Signs, information and training*—**(1)** *Signs*—**(i)** Entrances to regulated areas shall be posted with signs bearing the legend:

**CANCER-SUSPECT AGENT**

**AUTHORIZED PERSONNEL ONLY**

**1910.1003(e)(1)(ii)** Entrances to regulated areas containing operations covered in paragraph (c)(5) of this section shall be posted with signs bearing the legend:

**CANCER-SUSPECT AGENT EXPOSED IN THIS AREA**

**IMPERVIOUS SUIT INCLUDING GLOVES, BOOTS, AND AIR-SUPPLIED HOOD REQUIRED AT ALL TIMES**

**AUTHORIZED PERSONNEL ONLY**

**1910.1003(e)(1)(iii)** Appropriate signs and instructions shall be posted at the entrance to, and exit from, regulated areas, informing employees of the procedures that must be followed in entering and leaving a regulated area.

**1910.1003(e)(2)** *Container contents identification.* **(i)** Containers of a carcinogen addressed by this section and containers required under paragraphs (c)(4)(v) and (c)(6) (vii)(B) and (viii)(B) of this section that are accessible only to and handled only by authorized employees, or by other employees trained in accordance with paragraph (e)(5) of this section, may have contents identification limited to a generic or proprietary name or other proprietary identification of the carcinogen and percent.

**1910.1003(e)(2)(ii)** Containers of a carcinogen addressed by this section and containers required under paragraphs (c)(4)(v) and (c)(6) (vii)(B) and (viii)(B) of this section that are accessible to or handled by employees other than authorized employees or employees trained in accordance with paragraph (e)(5) of this section shall have contents identification that includes the full chemical name and Chemical Abstracts Service Registry number as listed in paragraph (a)(1) of this section.

**1910.1003(e)(2)(iii)** Containers shall have the warning words "CANCER-SUSPECT AGENT" displayed immediately under or adjacent to the contents identification.

**1910.1003(e)(2)(iv)** Containers whose contents are carcinogens addressed by this section with corrosive or irritating properties shall have label statements warning of such hazards noting, if appropriate, particularly sensitive or affected portions of the body.

**1910.1003(e)(3)** *Lettering.* Lettering on signs and instructions required by paragraph (e)(1) shall be a minimum letter height of 2 inches (5 cm). Labels on containers required under this section shall not be less than one-half the size of the largest lettering on the package, and not less than 8-point type in any instance. *Provided,* That no such required lettering need be more than 1 inch (2.5 cm) in height.

**1910.1003(e)(4)** *Prohibited statements.* No statement shall appear on or near any required sign, label, or instruction that contradicts or detracts from the effect of any required warning, information, or instruction.

**1910.1003(e)(5)** *Training and indoctrination.* **(i)** Each employee prior to being authorized to enter a regulated area, shall receive a training and indoctrination program including, but not necessarily limited to:

**1910.1003(e)(5)(i)(A)** The nature of the carcinogenic hazards of a carcinogen addressed by this section, including local and systemic toxicity;

**1910.1003(e)(5)(i)(B)** The specific nature of the operation involving a carcinogen addressed by this section that could result in exposure;

**1910.1003(e)(5)(i)(C)** The purpose for and application of the medical surveillance program, including, as appropriate, methods of self-examination;

**1910.1003(e)(5)(i)(D)** The purpose for and application of decontamination practices and purposes;

**1910.1003(e)(5)(i)(E)** The purpose for and significance of emergency practices and procedures;

**1910.1003(e)(5)(i)(F)** The employee's specific role in emergency procedures;

**1910.1003(e)(5)(i)(G)** Specific information to aid the employee in recognition and evaluation of conditions and situations which may result in the release of a carcinogen addressed by this section;

**1910.1003(e)(5)(i)(H)** The purpose for and application of specific first aid procedures and practices;

**1910.1003(e)(5)(i)(I)** A review of this section at the employee's first training and indoctrination program and annually thereafter.

**1910.1003(e)(5)(ii)** Specific emergency procedures shall be prescribed, and posted, and employees shall be familiarized with their terms, and rehearsed in their application.

**1910.1003(e)(5)(iii)** All materials relating to the program shall be provided upon request to authorized representatives of the Assistant Secretary and the Director.

**1910.1003(f)** Reserved.

**1910.1003(g)** *Medical surveillance.* At no cost to the employee, a program of medical surveillance shall be established and implemented for employees considered for assignment to enter regulated areas, and for authorized employees.

**1910.1003(g)(1)** *Examinations.* **(i)** Before an employee is assigned to enter a regulated area, a preassignment physical examination by a physician shall be provided. The examination shall include the personal history of the employee, family and occupational background, including genetic and environmental factors.

**1910.1003(g)(1)(ii)** Authorized employees shall be provided periodic physical examinations, not less often than annually, following the preassignment examination.

**1910.1003(g)(1)(iii)** In all physical examinations, the examining physician shall consider whether there exist conditions of increased risk, including reduced immunological competence, those undergoing treatment with steroids or cytotoxic agents, pregnancy, and cigarette smoking.

**1910.1003(g)(2)** *Records.* **(i)** Employers of employees examined pursuant to this paragraph shall cause to be maintained complete and accurate records of all such medical examinations. Records shall be maintained for the duration of the employee's employment. Upon termination of the employee's employment, including retirement or death, or in the event that the employer ceases business without a successor, records, or notarized true copies thereof, shall be forwarded by registered mail to the Director.

**1910.1003(g)(2)(ii)** Records required by this paragraph shall be provided upon request to employees, designated representatives, and the Assistant Secretary in accordance with 29 CFR 1910.1020(a) through (e) and (g) through (i). These records shall also be provided upon request to the Director.

**1910.1003(g)(2)(iii)** Any physician who conducts a medical examination required by this paragraph shall furnish to the employer a statement of the employee's suitability for employment in the specific exposure.

[61 FR 9242, Mar. 7, 1996, as amended at 63 FR 1286, Jan. 8, 1998; 63 FR 20099, Apr. 23, 1998]

### §1910.1004 alpha-Naphthylamine.

See §1910.1003, *13 carcinogens.*

[61 FR 9245, Mar. 7, 1996]

### §1910.1005 [Reserved].

### §1910.1006 Methyl chloromethyl ether.

See §1910.1003, *13 carcinogens.*

[61 FR 9245, Mar. 7, 1996]

### §1910.1007 3,3´-Dichlorobenzidine (and its salts).

See §1910.1003, *13 carcinogens.*

[61 FR 9245, Mar. 7, 1996]

### §1910.1008 bis-Chloromethyl ether.

See §1910.1003, *13 carcinogens.*

[61 FR 9245, Mar. 7, 1996]

### §1910.1009 beta-Naphthylamine.

See §1910.1003, *13 carcinogens.*

[61 FR 9245, Mar. 7, 1996]

### §1910.1010 Benzidine.

See §1910.1003, *13 carcinogens.*

[61 FR 9245, Mar. 7, 1996]

### §1910.1011 4-Aminodiphenyl.

See §1910.1003, *13 carcinogens.*

[61 FR 9245, Mar. 7, 1996]

### §1910.1012 Ethyleneimine.

See §1910.1003, *13 carcinogens.*

[61 FR 9245, Mar. 7, 1996]

### §1910.1013 beta-Propiolactone.

See §1910.1003, *13 carcinogens.*

[61 FR 9245, Mar. 7, 1996]

### §1910.1014 2-Acetylaminofluorene.

See §1910.1003, *13 carcinogens.*

[61 FR 9245, Mar. 7, 1996]

### §1910.1015 4-Dimethylaminoazo-benzene.

See §1910.1003, *13 carcinogens.*

[61 FR 9245, Mar. 7, 1996]

### §1910.1016 N-Nitrosodimethylamine.

See §1910.1003, *13 carcinogens.*

[61 FR 9245, Mar. 7, 1996]

### §1910.1017 Vinyl chloride.

**1910.1017(a)** *Scope and application.* **(1)** This section includes requirements for the control of employee exposure to vinyl chloride (chloroethene), Chemical Abstracts Service Registry No. 75014.

**1910.1017(a)(2)** This section applies to the manufacture, reaction, packaging, repackaging, storage, handling or use of vinyl chloride or polyvinyl chloride, but does not apply to the handling or use of fabricated products made of polyvinyl chloride.

**1910.1017(a)(3)** This section applies to the transportation of vinyl chloride or polyvinyl chloride except to the extent that the Department of Transportation may regulate the hazards covered by this section.

**1910.1017(b)** *Definitions.* **(1)** *Action level* means a concentration of vinyl chloride of 0.5 ppm averaged over an 8-hour work day.

# OSHA Standards for General Industry

**1910.1017(b)(2)** *Assistant Secretary* means the Assistant Secretary of Labor for Occupational Safety and Health, U.S. Department of Labor, or his designee.

**1910.1017(b)(3)** *Authorized person* means any person specifically authorized by the employer whose duties require him to enter a regulated area or any person entering such an area as a designated representative of employees for the purpose of exercising an opportunity to observe monitoring and measuring procedures.

**1910.1017(b)(4)** *Director* means the Director, National Institute for Occupational Safety and Health, U.S. Department of Health and Human Services, or his designee.

**1910.1017(b)(5)** *Emergency* means any occurrence such as, but not limited to, equipment failure, or operation of a relief device which is likely to, or does, result in massive release of vinyl chloride.

**1910.1017(b)(6)** *Fabricated product* means a product made wholly or partly from polyvinyl chloride, and which does not require further processing at temperatures, and for times, sufficient to cause mass melting of the polyvinyl chloride resulting in the release of vinyl chloride.

**1910.1017(b)(7)** *Hazardous operation* means any operation, procedure, or activity where a release of either vinyl chloride liquid or gas might be expected as a consequence of the operation or because of an accident in the operation, which would result in an employee exposure in excess of the permissible exposure limit.

**1910.1017(b)(8)** *OSHA Area Director* means the Director for the Occupational Safety and Health Administration Area Office having jurisdiction over the geographic area in which the employer's establishment is located.

**1910.1017(b)(9)** *Polyvinyl chloride* means polyvinyl chloride homopolymer or copolymer before such is converted to a fabricated product.

**1910.1017(b)(10)** *Vinyl chloride* means vinyl chloride monomer.

**1910.1017(c)** **Permissible exposure limit. (1)** No employee may be exposed to vinyl chloride at concentrations greater than 1 ppm averaged over any 8-hour period, and

**1910.1017(c)(2)** No employee may be exposed to vinyl chloride at concentrations greater than 5 ppm averaged over any period not exceeding 15 minutes.

**1910.1017(c)(3)** No employee may be exposed to vinyl chloride by direct contact with liquid vinyl chloride.

**1910.1017(d)** **Monitoring. (1)** A program of initial monitoring and measurement shall be undertaken in each establishment to determine if there is any employee exposed, without regard to the use of respirators, in excess of the action level.

**1910.1017(d)(2)** Where a determination conducted under paragraph (d)(1) of this section shows any employee exposures, without regard to the use of respirators, in excess of the action level, a program for determining exposures for each such employee shall be established. Such a program:

**1910.1017(d)(2)(i)** Must be repeated at least quarterly for any employee exposed, without regard to the use of respirators, in excess of the permissible exposure limit.

**1910.1017(d)(2)(ii)** Must be repeated not less than every 6 months for any employee exposed without regard to the use of respirators, at or above the action level.

**1910.1017(d)(2)(iii)** May be discontinued for any employee only when at least two consecutive monitoring determinations, made not less than 5 working days apart, show exposures for that employee at or below the action level.

**1910.1017(d)(3)** Whenever there has been a production, process or control change which may result in an increase in the release of vinyl chloride, or the employer has any other reason to suspect that any employee may be exposed in excess of the action level, a determination of employee exposure under paragraph (d)(1) of this section shall be performed.

**1910.1017(d)(4)** The method of monitoring and measurement shall have an accuracy (with a confidence level of 95 percent) of not less than plus or minus 50 percent from 0.25 through 0.5 ppm, plus or minus 35 percent from over 0.5 ppm through 1.0 ppm, and plus or minus 25 percent over 1.0 ppm. (Methods meeting these accuracy requirements are available in the "NIOSH Manual of Analytical Methods").

**1910.1017(d)(5)** Employees or their designated representatives shall be afforded reasonable opportunity to observe the monitoring and measuring required by this paragraph.

**1910.1017(e)** **Regulated area. (1)** A regulated area shall be established where:

**1910.1017(e)(1)(i)** Vinyl chloride or polyvinyl chloride is manufactured, reacted, repackaged, stored, handled or used; and

**1910.1017(e)(1)(ii)** Vinyl chloride concentrations are in excess of the permissible exposure limit.

**1910.1017(e)(2)** Access to regulated areas shall be limited to authorized persons.

**1910.1017(f)** **Methods of compliance.** Employee exposures to vinyl chloride shall be controlled to at or below the permissible exposure limit provided in paragraph (c) of this section by engineering, work practice, and personal protective controls as follows:

**1910.1017(f)(1)** Feasible engineering and work practice controls shall immediately be used to reduce exposures to at or below the permissible exposure limit.

**1910.1017(f)(2)** Wherever feasible engineering and work practice controls which can be instituted immediately are not sufficient to reduce exposures to at or below the permissible exposure limit, they shall nonetheless be used to reduce exposures to the lowest practicable level, and shall be supplemented by respiratory protection in accordance with paragraph (g) of this section. A program shall be established and implemented to reduce exposures to at or below the permissible exposure limit, or to the greatest extent feasible, solely by means of engineering and work practice controls, as soon as feasible.

**1910.1017(f)(3)** Written plans for such a program shall be developed and furnished upon request for examination and copying to authorized representatives of the Assistant Secretary and the Director. Such plans must be updated at least annually.

**1910.1017(g)** **Respiratory protection—(1)** *General.* For employees who use respirators required by this section, the employer must provide respirators that comply with the requirements of this paragraph.

**1910.1017(g)(2)** *Respirator program.* The employer must implement a respiratory protection program in accordance with 29 CFR 1910.134 (b) through (d) (except (d)(1)(iii), and (d)(3)(iii)(B)(*1*) and (*2*)), and (f) through (m).

**1910.1017(g)(3)** *Respirator selection.* **(i)** Respirators must be selected from the following table:

| Atmospheric concentration of vinyl chloride | Required apparatus |
|---|---|
| (i) Unknown, or above 3,600 p/m. | Open-circuit, self-contained breathing apparatus, pressure demand type, with full facepiece. |
| (ii) Not over 3,600 p/m | (A) Combination type C supplied air respirator, pressure demand type, with full or half facepiece, and auxiliary self-contained air supply; or |
| (iii) Not over 1,000 p/m | (B) Combination type, supplied air respirator continuous flow type, with full or half facepiece, and auxiliary self-contained air supply. Type C, supplied air respirator, continuous flow type, with full or half facepiece, helmet or hood. |
| (iv) Not over 100 p/m | (A) Combination type C supplied air respirator demand type, with full facepiece, and auxiliary self-contained air supply; or |
|  | (B) Open-circuit self-contained breathing apparatus with full facepiece, in demand mode; or Type (C) supplied air respirator, demand type, with full facepiece. |
| (v) Not over 25 p/m | (A) A powered air-purifying respirator with hood, helmet, full or half facepiece, and a canister which provides a service life of at least 4 hours for concentrations of vinyl chloride up to 25 p/m, or |
|  | (B) Gas mask, front- or back-mounted canister which provides a service life of at least 4 hours for concentrations of vinyl chloride up to 25 p/m. |
| (vi) Not over 10 p/m | (A) Combination type C supplied-air respirator, demand type, with half facepiece, and auxiliary self-contained air supply; or |
|  | (B) Type C supplied-air respirator, demand type, with half facepiece; or |
|  | (C) Any chemical cartridge respirator with an organic vapor cartridge which provides a service life of at least 1 hour for concentrations of vinyl chloride up to 10 p/m. |

**1910.1017(g)(3)(ii)** When air-purifying respirators are used:

**1910.1017(g)(3)(ii)(A)** Air-purifying canisters or cartridges must be replaced prior to the expiration of their service life or the end of the shift in which they are first used, whichever occurs first.

**1910.1017(g)(3)(ii)(B)** A continuous-monitoring and alarm system must be provided when concentrations of vinyl chloride could reasonably exceed the allowable concentrations for the devices in use. Such a system must be used to alert employees when vinyl chloride concentrations exceed the allowable concentrations for the devices in use.

**1910.1017(g)(3)(iii)** Respirators specified for higher concentrations may be used for lower concentrations.

**1910.1017(h)** *Hazardous operations.* **(1)** Employees engaged in hazardous operations, including entry of vessels to clean polyvinyl chloride residue from vessel walls, shall be provided and required to wear and use;

**1910.1017(h)(1)(i)** Respiratory protection in accordance with paragraphs (c) and (g) of this section; and

**1910.1017(h)(1)(ii)** Protective garments to prevent skin contact with liquid vinyl chloride or with polyvinyl chloride residue from vessel walls. The protective garments shall be selected for the operation and its possible exposure conditions.

**1910.1017(h)(2)** Protective garments shall be provided clean and dry for each use.

**1910.1017(i)** *Emergency situations.* A written operational plan for emergency situations shall be developed for each facility storing, handling, or otherwise using vinyl chloride as a liquid or compressed gas. Appropriate portions of the plan shall be implemented in the event of an emergency. The plan shall specifically provide that:

**1910.1017(i)(1)** Employees engaged in hazardous operations or correcting situations of existing hazardous releases shall be equipped as required in paragraph (h) of this section;

**1910.1017(i)(2)** Other employees not so equipped shall evacuate the area and not return until conditions are controlled by the methods required in paragraph (f) of this section and the emergency is abated.

**1910.1017(j)** *Training.* Each employee engaged in vinyl chloride or polyvinyl chloride operations shall be provided training in a program relating to the hazards of vinyl chloride and precautions for its safe use.

**1910.1017(j)(1)** The program shall include:

**1910.1017(j)(1)(i)** The nature of the health hazard from chronic exposure to vinyl chloride including specifically the carcinogenic hazard;

**1910.1017(j)(1)(ii)** The specific nature of operations which could result in exposure to vinyl chloride in excess of the permissible limit and necessary protective steps;

**1910.1017(j)(1)(iii)** The purpose for, proper use, and limitations of respiratory protective devices;

**1910.1017(j)(1)(iv)** The fire hazard and acute toxicity of vinyl chloride, and the necessary protective steps;

**1910.1017(j)(1)(v)** The purpose for and a description of the monitoring program;

**1910.1017(j)(1)(vi)** The purpose for, and a description of, the medical surveillance program;

**1910.1017(j)(1)(vii)** Emergency procedures;

**1910.1017(j)(1)(viii)** Specific information to aid the employee in recognition of conditions which may result in the release of vinyl chloride; and

**1910.1017(j)(1)(ix)** A review of this standard at the employee's first training and indoctrination program, and annually thereafter.

**1910.1017(j)(2)** All materials relating to the program shall be provided upon request to the Assistant Secretary and the Director.

**1910.1017(k)** *Medical surveillance.* A program of medical surveillance shall be instituted for each employee exposed, without regard to the use of respirators, to vinyl chloride in excess of the action level. The program shall provide each such employee with an opportunity for examinations and tests in accordance with this paragraph. All medical examinations and procedures shall be performed by or under the supervision of a licensed physician, and shall be provided without cost to the employee.

**1910.1017(k)(1)** At the time of initial assignment, or upon institution of medical surveillance;

**1910.1017(k)(1)(i)** A general physical examination shall be performed, with specific attention to detecting enlargement of liver, spleen or kidneys, or dysfunction in these organs, and for abnormalities in skin, connective tissues and the pulmonary system (See Appendix A).

**1910.1017(k)(1)(ii)** A medical history shall be taken, including the following topics:

**1910.1017(k)(1)(ii)(A)** Alcohol intake;

**1910.1017(k)(1)(ii)(B)** Past history of hepatitis;

**1910.1017(k)(1)(ii)(C)** Work history and past exposure to potential hepatotoxic agents, including drugs and chemicals;

**1910.1017(k)(1)(ii)(D)** Past history of blood transfusions; and

**1910.1017(k)(1)(ii)(E)** Past history of hospitalizations.

**1910.1017(k)(1)(iii)** A serum specimen shall be obtained and determinations made of:

**1910.1017(k)(1)(iii)(A)** Total bilirubin;

**1910.1017(k)(1)(iii)(B)** Alkaline phosphatase;

**1910.1017(k)(1)(iii)(C)** Serum glutamic oxalacetic transaminase (SGOT);

**1910.1017(k)(1)(iii)(D)** Serum glutamic pyruvic transaminase (SGPT); and

**1910.1017(k)(1)(iii)(E)** Gamma glustamyl transpeptidase.

**1910.1017(k)(2)** Examinations must be provided in accordance with this paragraph at least annually.

**1910.1017(k)(2)(i)** Every 6 months for each employee who has been employed in vinyl chloride or polyvinyl chloride manufacturing for 10 years or longer; and

**1910.1017(k)(2)(ii)** Annually for all other employees.

**1910.1017(k)(3)** Each employee exposed to an emergency shall be afforded appropriate medical surveillance.

**1910.1017(k)(4)** A statement of each employee's suitability for continued exposure to vinyl chloride including use of protective equipment and respirators, shall be obtained from the examining physician promptly after any examination. A copy of the physician's statement shall be provided each employee.

**1910.1017(k)(5)** If any employee's health would be materially impaired by continued exposure, such employee shall be withdrawn from possible contact with vinyl chloride.

**1910.1017(k)(6)** Laboratory analyses for all biological specimens included in medical examination shall be performed by accredited laboratories.

**1910.1017(k)(7)** If the examining physician determines that alternative medical examinations to those required by paragraph (k)(1) of this section will provide at least equal assurance of detecting medical conditions pertinent to the exposure to vinyl chloride, the employer may accept such alternative examinations as meeting the requirements of paragraph (k)(1) of this section, if the employer obtains a statement from the examining physician setting forth the alternative examinations and the rationale for substitution. This statement shall be available upon request for examination and copying to authorized representatives of the Assistant Secretary and the Director.

**1910.1017(l)** *Signs and labels.* **(1)** Entrances to regulated areas shall be posted with legible signs bearing the legend:

**CANCER-SUSPECT AGENT AREA**
**AUTHORIZED PERSONNEL ONLY**

**1910.1017(l)(2)** Areas containing hazardous operations or where an emergency currently exists shall be posted with legible signs bearing the legend:

**CANCER-SUSPECT AGENT IN THIS AREA PROTECTIVE EQUIPMENT REQUIRED AUTHORIZED PERSONNEL ONLY**

**1910.1017(l)(3)** Containers of polyvinyl chloride resin waste from reactors or other waste contaminated with vinyl chloride shall be legibly labeled:

**CONTAMINATED WITH VINYL CHLORIDE**

**CANCER-SUSPECT AGENT**

**1910.1017(l)(4)** Containers of polyvinyl chloride shall be legibly labeled:

**POLYVINYL CHLORIDE (OR TRADE NAME)**

**CONTAINS**

**VINYL CHLORIDE**

**VINYL CHLORIDE IS A CANCER-SUSPECT AGENT**

**1910.1017(l)(5)** Containers of vinyl chloride shall be legibly labeled either:

**1910.1017(l)(5)(i)**

**VINYL CHLORIDE**

**EXTREMELY FLAMMABLE GAS UNDER PRESSURE**

**CANCER SUSPECT AGENT**

or (ii) In accordance with 49 CFR Parts 170 through 189, with the additional legend:

**CANCER-SUSPECT AGENT**

applied near the label or placard.

**1910.1017(l)(6)** No statement shall appear on or near any required sign, label or instruction which contradicts or detracts from the effect of, any required warning, information or instruction.

**1910.1017(m)** *Records.* **(1)** All records maintained in accordance with this section shall include the name and social security number of each employee where relevant.

**1910.1017(m)(2)** Records of required monitoring and measuring and medical records shall be provided upon request to employees, designated representatives, and the Assistant Secretary in accordance with 29 CFR 1910.1020(a) through (e) and (g) through (i). These records shall be provided upon request to the Director. Authorized personnel rosters shall also be provided upon request to the Assistant Secretary and the Director.

**1910.1017(m)(2)(i)** Monitoring and measuring records shall:

**1910.1017(m)(2)(i)(A)** State the date of such monitoring and measuring and the concentrations determined and identify the instruments and methods used;

**1910.1017(m)(2)(i)(B)** Include any additional information necessary to determine individual employee exposures where such exposures are determined by means other than individual monitoring of employees; and

**1910.1017(m)(2)(i)(C)** Be maintained for not less than 30 years.

**1910.1017(m)(2)(ii)** [Reserved]

**1910.1017(m)(2)(iii)** Medical records shall be maintained for the duration of the employment of each employee plus 20 years, or 30 years, whichever is longer.

**1910.1017(m)(3)** In the event that the employer ceases to do business and there is no successor to receive and retain his records for the prescribed period, these records shall be transmitted by registered mail to the Director, and each employee individually notified in writing of this transfer. The employer shall also comply with any additional requirements set forth in 29 CFR 1910.1020(h).

**1910.1017(n)** *Employee notification of monitoring results.* The employer must, within 15 working days after the receipt of the results of any monitoring performed under this section, notify each affected employee of these results and the steps being taken to reduce exposures within the permissible exposure limit either individually in writing or by posting the results in an appropriate location that is accessible to affected employees.

## APPENDIX A TO §1910.1017— SUPPLEMENTARY MEDICAL INFORMATION

When required tests under paragraph (k)(1) of this section show abnormalities, the tests should be repeated as soon as practicable, preferably within 3 to 4 weeks. If tests remain abnormal, consideration should be given to withdrawal of the employee from contact with vinyl chloride, while a more comprehensive examination is made.

Additional tests which may be useful:

A. For kidney dysfunction: urine examination for albumin, red blood cells, and exfoliative abnormal cells.

B. Pulmonary system: Forced vital capacity, Forced expiratory volume at 1 second, and chest roentgenogram (posterior-anterior, 14 × 17 inches).

C. Additional serum tests: Lactic acid dehydrogenase, lactic acid dehydrogenase isoenzyme, protein determination, and protein electrophoresis.

D. For a more comprehensive examination on repeated abnormal serum tests: Hepatitis B antigen, and liver scanning.

[39 FR 35896, Oct. 4, 1974; 39 FR 41848, Dec. 3, 1974, as amended at 40 FR 13211, Mar. 25, 1975. Redesignated at 40 FR 23072, May 28, 1975 and amended at 43 FR 49751, Oct. 24, 1978; 45 FR 35282, May 23, 1980; 54 FR 24334, June 7, 1989; 58 FR 35310, June 30, 1993; 61 FR 5508, Feb. 13, 1996; 63 FR 1286, Jan. 8, 1998]

## §1910.1018 Inorganic arsenic.

**1910.1018(a)** *Scope and application.* This section applies to all occupational exposures to inorganic arsenic except that this section does not apply to employee exposures in agriculture or resulting from pesticide application, the treatment of wood with preservatives or the utilization of arsenically preserved wood.

**1910.1018(b)** *Definitions. Action level* means a concentration of inorganic arsenic of 5 micrograms per cubic meter of air (5 µg/m³) averaged over any eight (8) hour period.

*Assistant Secretary* means the Assistant Secretary of Labor for Occupational Safety and Health, U.S. Department of Labor, or designee.

*Authorized person* means any person specifically authorized by the employer whose duties require the person to enter a regulated area, or any person entering such an area as a designated representative of employees for the purpose of exercising the right to observe monitoring and measuring procedures under paragraph (e) of this section.

*Director* means the Director, National Institute for Occupational Safety and Health, U.S. Department of Health and Human Services, or designee.*Inorganic arsenic* means copper aceto- arsenite and all inorganic compounds containing arsenic except arsine, measured as arsenic (As).

**1910.1018(c)** *Permissible exposure limit.* The employer shall assure that no employee is exposed to inorganic arsenic at concentrations greater than 10 micrograms per cubic meter of air (10 µg/m³), averaged over any 8-hour period.

**1910.1018(d)** Reserved.

**1910.1018(e)** *Exposure monitoring*—**(1)** *General.* **(i)** Determinations of airborne exposure levels shall be made from air samples that are representative of each employee's exposure to inorganic arsenic over an eight (8) hour period.

**1910.1018(e)(1)(ii)** For the purposes of this section, employee exposure is that exposure which would occur if the employee were not using a respirator.

**1910.1018(e)(1)(iii)** The employer shall collect full shift (for at least 7 continuous hours) personal samples including at least one sample for each shift for each job classification in each work area.

**1910.1018(e)(2)** *Initial monitoring.* Each employer who has a workplace or work operation covered by this standard shall monitor each such workplace and work operation to accurately determine the airborne concentration of inorganic arsenic to which employees may be exposed.

**1910.1018(e)(3)** *Frequency.* **(i)** If the initial monitoring reveals employee exposure to be below the action level the measurements need not be repeated except as otherwise provided in paragraph (e)(4) of this section.

**1910.1018(e)(3)(ii)** If the initial monitoring, required by this section, or subsequent monitoring reveals employee exposure to be above the permissible exposure limit, the employer shall repeat monitoring at least quarterly.

**1910.1018(e)(3)(iii)** If the initial monitoring, required by this section, or subsequent monitoring reveals employee exposure to be above the action level and below the permissible exposure limit the employer shall repeat monitoring at least every six months.

**1910.1018(e)(3)(iv)** The employer shall continue monitoring at the required frequency until at least two consecutive measurements, taken at least seven (7) days apart, are below the action level at which time the employer may discontinue monitoring for that employee until such time as any of the events in paragraph (e)(4) of this section occur.

**1910.1018(e)(4)** *Additional monitoring.* Whenever there has been a production, process, control or personal change which may result in new or additional exposure to inorganic arsenic, or whenever the employer has any other reason to suspect a change which may result in new or additional exposures to inorganic arsenic, additional monitoring which complies with paragraph (e) of this section shall be conducted.

**1910.1018(e)(5)** *Employee notification.* **(i)** The employer must, within 15 working days after the receipt of the results of any monitoring performed under this section, notify each affected employee of these results either individually in writing or by posting the results in an appropriate location that is accessible to affected employees.

**1910.1018(e)(5)(ii)** Whenever the results indicate that the representative employee exposure exceeds the permissible exposure limit, the employer shall include in the written notice a statement that the permissible exposure limit was exceeded and a description of the corrective action taken to reduce exposure to or below the permissible exposure limit.

**1910.1018(e)(6)** *Accuracy of measurement.* **(i)** The employer shall use a method of monitoring and measurement which has an accuracy (with a confidence level of 95 percent) of not less than plus or minus 25 percent for concentrations of inorganic arsenic greater than or equal to 10 µg/m³.

**1910.1018(e)(6)(ii)** The employer shall use a method of monitoring and measurement which has an accuracy (with confidence level of 95 percent) of not less than plus or minus 35 percent for concentrations of inorganic arsenic greater than 5 µg/m³ but less than 10 µg/m³.

**1910.1018(f)** *Regulated area*—**(1)** *Establishment.* The employer shall establish regulated areas where worker exposures to inorganic arsenic, without regard to the use of respirators, are in excess of the permissible limit.

**1910.1018(f)(2)** *Demarcation.* Regulated areas shall be demarcated and segregated from the rest of the workplace in any manner that minimizes the number of persons who will be exposed to inorganic arsenic.

**1910.1018(f)(3)** *Access.* Access to regulated areas shall be limited to authorized persons or to persons otherwise authorized by the Act or regulations issued pursuant thereto to enter such areas.

**1910.1018(f)(4)** *Provision of respirators.* All persons entering a regulated area shall be supplied with a respirator, selected in accordance with paragraph (h)(2) of this section.

**1910.1018(f)(5)** *Prohibited activities.* The employer shall assure that in regulated areas, food or beverages are not consumed, smoking products, chewing tobacco and gum are not used and cosmetics are not applied, except that these activities may be conducted in the lunchrooms, change rooms and showers required under paragraph (m) of this section. Drinking water may be consumed in the regulated area.

**1910.1018(g)** *Methods of compliance*—**(1)** *Controls.* **(i)** The employer shall institute at the earliest possible time but not later than December 31, 1979, engineering and work practice controls to reduce exposures to or below the permissible exposure limit, except to the extent that the employer can establish that such controls are not feasible.

**1910.1018(g)(1)(ii)** Where engineering and work practice controls are not sufficient to reduce exposures to or below the permissible exposure limit, they shall nonetheless be used to reduce exposures to the lowest levels achievable by these controls and shall be supplemented by the use of respirators in accordance with paragraph (h) of this section and other necessary personal protective equipment. Employee rotation is not required as a control strategy before respiratory protection is instituted.

**1910.1018(g)(2)** *Compliance Program.* **(i)** The employer shall establish and implement a written program to reduce exposures to or below the permissible exposure limit by means of engineering and work practice controls.

**1910.1018(g)(2)(ii)** Written plans for these compliance programs shall include at least the following:

**1910.1018(g)(2)(ii)(A)** A description of each operation in which inorganic arsenic is emitted; e.g., machinery used, material processed, controls in place, crew size, operating procedures and maintenance practices;

**1910.1018(g)(2)(ii)(B)** Engineering plans and studies used to determine methods selected for controlling exposure to inorganic arsenic;

**1910.1018(g)(2)(ii)(C)** A report of the technology considered in meeting the permissible exposure limit;

**1910.1018(g)(2)(ii)(D)** Monitoring data;

**1910.1018(g)(2)(ii)(E)** A detailed schedule for implementation of the engineering controls and work practices that cannot be implemented immediately and for the adaption and implementation of any additional engineering and work practices necessary to meet the permissible exposure limit;

**1910.1018(g)(2)(ii)(F)** Whenever the employer will not achieve the permissible exposure limit with engineering controls and work practices by December 31, 1979, the employer shall include in the compliance plan an analysis of the effectiveness of the various controls, shall install engineering controls and institute work practices on the quickest schedule feasible, and shall include in the compliance plan and implement a program to minimize the discomfort and maximize the effectiveness of respirator use; and

**1910.1018(g)(2)(ii)(G)** Other relevant information.

**1910.1018(g)(2)(iii)** Written plans for such a program shall be submitted upon request to the Assistant Secretary and the Director, and shall be available at the worksite for examination and copying by the Assistant Secretary, Director, any affected employee or authorized employee representatives.

# OSHA Standards for General Industry

**1910.1018(g)(2)(iv)** The plans required by this paragraph must be revised and updated at least annually to reflect the current status of the program.

**1910.1018(h)** *Respiratory protection*—**(1)** *General.* For employees who use respirators required by this section, the employer must provide respirators that comply with the requirements of this paragraph. Respirators must be used during:

**1910.1018(h)(1)(i)** Periods necessary to install or implement feasible engineering or work-practice controls.

**1910.1018(h)(1)(ii)** Work operations, such as maintenance and repair activities, for which the employer establishes that engineering and work-practice controls are not feasible.

**1910.1018(h)(1)(iii)** Work operations for which engineering and work-practice controls are not yet sufficient to reduce employee exposures to or below the permissible exposure limit.

**1910.1018(h)(1)(iv)** Emergencies.

**1910.1018(h)(2)** *Respirator program.* **(i)** The employer must implement a respiratory protection program in accordance with 29 CFR 1910.134(b) through (d) (except (d)(1)(iii)), and (f) through (m).

**1910.1018(h)(2)(ii)** If an employee exhibits breathing difficulty during fit testing or respirator use, they must be examined by a physician trained in pulmonary medicine to determine whether they can use a respirator while performing the required duty.

**1910.1018(h)(3)** *Respirator selection.* **(i)** The employer must use Table I of this section to select the appropriate respirator or combination of respirators for inorganic arsenic compounds without significant vapor pressure, and Table II of this section to select the appropriate respirator or combination of respirators for inorganic arsenic compounds that have significant vapor pressure.

**1910.1018(h)(3)(ii)** When employee exposures exceed the permissible exposure limit for inorganic arsenic and also exceed the relevant limit for other gases (for example, sulfur dioxide), an air-purifying respirator provided to the employee as specified by this section must have a combination high-efficiency filter with an appropriate gas sorbent. (See footnote in Table 1 of this section.)

**1910.1018(h)(3)(iii)** Employees required to use respirators may choose, and the employer must provide, a powered air-purifying respirator if it will provide proper protection. In addition, the employer must provide a combination dust and acid-gas respirator to employees who are exposed to gases over the relevant exposure limits.

### TABLE I—RESPIRATORY PROTECTION FOR INORGANIC ARSENIC PARTICULATE EXCEPT FOR THOSE WITH SIGNIFICANT VAPOR PRESSURE

| Concentration of inorganic arsenic (as As) or condition of use | Required respirator |
|---|---|
| (i) Unknown or greater or lesser than 20,000 µg/m³ (20 mg/m(3)) or firefighting. | (A) Any full facepiece self-contained breathing apparatus operated in positive pressure mode. |
| (ii) Not greater than 20,000 µg/m(3) (20 mg/m(3)). | (A) Supplied air respirator with full facepiece, hood, or helmet or suit and operated in positive pressure mode. |
| (iii) Not greater than 10,000 µg/m(3) (10 mg/m(3)). | (A) Powered air-purifying respirators in all inlet face coverings with high efficiency filters.[1] (B) Half-mask supplied air respirators operated in positive pressure mode. |
| (iv) Not greater than 500 µg/m(3). | (A) Full facepiece air-purifying respirator equipped with high-efficiency filter.[1] (B) Any full facepiece supplied air respirator. (C) Any full facepiece self-contained breathing apparatus. |
| (v) Not greater than 100 µg/m(3). | (A) Half-mask air-purifying respirator equipped with high-efficiency filter.[1] (B) Any half-mask supplied air respirator. |

[1] High-efficiency filter-99.97 pct efficiency against 0.3 micrometer monodisperse diethyl-hexyl phthalate (DOP) particles.

### TABLE II—RESPIRATORY PROTECTION FOR INORGANIC ARSENICALS (SUCH AS ARSENIC TRICHLORIDE[2] AND ARSENIC PHOSPHIDE) WITH SIGNIFICANT VAPOR PRESSURE

| Concentration of inorganic arsenic (as As) or condition of use | Required respirator |
|---|---|
| (i) Unknown or greater or lesser than 20,000 µg/m(3) (20 mg/m(3)) or firefighting. | (A) Any full facepiece self-contained breathing apparatus operated in positive pressure mode. |
| (ii) Not greater than 20,000 µg/m(3) (20 mg/m(3)). | (A) Supplied air respirator with full facepiece, hood, or helmet or suit and operated in positive pressure mode. |
| (iii) Not greater than 10,000 µg/m(3) (10 mg/m(3)). | (A) Half-mask[2] supplied air respirator operated in positive pressure mode. |
| (iv) Not greater than 500 µg/m(3). | (A) Front or back mounted gas mask equipped with high-efficiency filter[1] and acid gas canister. (B) Any full facepiece supplied air respirator. (C) Any full facepiece self-contained breathing apparatus. |
| (v) Not greater than 100 µg/m(3). | (A) Half-mask air-purifying respirator equipped with high efficiency filter[1] and acid gas cartridge. (B) Any half-mask supplied air respirator. |

[1] High-efficiency filter-99.97 pct efficiency against 0.3 micrometer monodisperse diethyl-hexyl phthalate (DOP) particles.
[2] Half-mask respirators shall not be used for protection against arsenic trichloride, as it is rapidly absorbed through the skin.

**1910.1018(i)** [Reserved]

**1910.1018(j)** *Protective work clothing and equipment*—**(1)** *Provision and use.* Where the possibility of skin or eye irritation from inorganic arsenic exists, and for all workers working in regulated areas, the employer shall provide at no cost to the employee and assure that employees use appropriate and clean protective work clothing and equipment such as, but not limited to:

**1910.1018(j)(1)(i)** Coveralls or similar full-body work clothing;

**1910.1018(j)(1)(ii)** Gloves, and shoes or coverlets;

**1910.1018(j)(1)(iii)** Face shields or vented goggles when necessary to prevent eye irritation, which comply with the requirements of §1910.133(a)(2)-(6); and

**1910.1018(j)(1)(iv)** Impervious clothing for employees subject to exposure to arsenic trichloride.

**1910.1018(j)(2)** *Cleaning and replacement.* **(i)** The employer shall provide the protective clothing required in paragraph (j) (1) of this section in a freshly laundered and dry condition at least weekly, and daily if the employee works in areas where exposures are over 100 µg/m³ of inorganic arsenic or in areas where more frequent washing is needed to prevent skin irritation.

**1910.1018(j)(2)(ii)** The employer shall clean, launder, or dispose of protective clothing required by paragraph (j) (1) of this section.

**1910.1018(j)(2)(iii)** The employer shall repair or replace the protective clothing and equipment as needed to maintain their effectiveness.

**1910.1018(j)(2)(iv)** The employer shall assure that all protective clothing is removed at the completion of a work shift only in change rooms prescribed in paragraph (m) (1) of this section.

**1910.1018(j)(2)(v)** The employer shall assure that contaminated protective clothing which is to be cleaned, laundered, or disposed of, is placed in a closed container in the change-room which prevents dispersion of inorganic arsenic outside the container.

**1910.1018(j)(2)(vi)** The employer shall inform in writing any person who cleans or launders clothing required by this section, of the potentially harmful effects including the carcinogenic effects of exposure to inorganic arsenic.

**1910.1018(j)(2)(vii)** The employer shall assure that the containers of contaminated protective clothing and equipment in the workplace or which are to be removed from the workplace are labelled as follows:

**CAUTION: Clothing contaminated with inorganic arsenic; do not remove dust by blowing or shaking. Dispose of inorganic arsenic contaminated wash water in accordance with applicable local, State or Federal regulations.**

**1910.1018(j)(2)(viii)** The employer shall prohibit the removal of inorganic arsenic from protective clothing or equipment by blowing or shaking.

**1910.1018(k)** *Housekeeping*—**(1)** *Surfaces.* All surfaces shall be maintained as free as practicable of accumulations of inorganic arsenic.

**1910.1018(k)(2)** *Cleaning floors.* Floors and other accessible surfaces contaminated with inorganic arsenic may not be cleaned by the use of compressed air, and shoveling and brushing may be used only where vacuuming or other relevant methods have been tried and found not to be effective.

**1910.1018(k)(3)** *Vacuuming.* Where vacuuming methods are selected, the vacuums shall be used and emptied in a manner to minimize the reentry of inorganic arsenic into the workplace.

**1910.1018(k)(4)** *Housekeeping plan.* A written housekeeping and maintenance plan shall be kept which shall list appropriate frequencies for carrying out housekeeping operations, and for cleaning and maintaining dust collection equipment. The plan shall be available for inspection by the Assistant Secretary.

**1910.1018(k)(5)** *Maintenance of equipment.* Periodic cleaning of dust collection and ventilation equipment and checks of their effectiveness shall be carried out to maintain the effectiveness of the system and a notation kept of the last check of effectiveness and cleaning or maintenance.

**1910.1018(l)** [Reserved]

**1910.1018(m)** *Hygiene facilities and practices*—**(1)** *Change rooms.* The employer shall provide for employees working in regulated areas or subject to the possibility of skin or eye irritation from inorganic arsenic, clean change rooms equipped with storage facilities for street clothes and separate storage facilities for protective clothing and equipment in accordance with 29 CFR 1910.141(e).

**1910.1018(m)(2)** *Showers.* **(i)** The employer shall assure that employees working in regulated areas or subject to the possibility of skin or eye irritation from inorganic arsenic shower at the end of the work shift.

**1910.1018(m)(2)(ii)** The employer shall provide shower facilities in accordance with §1910.141(d)(3).

**1910.1018(m)(3)** *Lunchrooms.* **(i)** The employer shall provide for employees working in regulated areas, lunchroom facilities which have a temperature controlled, positive pressure, filtered air supply, and which are readily accessible to employees working in regulated areas.

**1910.1018(m)(3)(ii)** The employer shall assure that employees working in the regulated area or subject to the possibility of skin or eye irritation from exposure to inorganic arsenic wash their hands and face prior to eating.

**1910.1018(m)(4)** *Lavatories.* The employer shall provide lavatory facilities which comply with §1910.141(d)(1) and (2).

**1910.1018(m)(5)** *Vacuuming clothes.* The employer shall provide facilities for employees working in areas where exposure, without regard to the use of respirators, exceeds 100 µg/m³ to vacuum their protective clothing and clean or change shoes worn in such areas before entering change rooms, lunchrooms or shower rooms required by paragraph (j) of this section and shall assure that such employees use such facilities.

**1910.1018(m)(6)** *Avoidance of skin irritation.* The employer shall assure that no employee is exposed to skin or eye contact with arsenic trichloride, or to skin or eye contact with liquid or particulate inorganic arsenic which is likely to cause skin or eye irritation.

**1910.1018(n)** *Medical surveillance*—**(1)** *General*—**(i)** *Employees covered.* The employer shall institute a medical surveillance program for the following employees:

**1910.1018(n)(1)(i)(A)** All employees who are or will be exposed above the action level, without regard to the use of respirators, at least 30 days per year; and

**1910.1018(n)(1)(i)(B)** All employees who have been exposed above the action level, without regard to respirator use, for 30 days or more per year for a total of 10 years or more of combined employment with the employer or predecessor employers prior to or after the effective date of this standard. The determination of exposures prior to the effective date of this standard shall be based upon prior exposure records, comparison with the first measurements taken after the effective date of this standard, or comparison with records of exposures in areas with similar processes, extent of engineering controls utilized and materials used by that employer.

**1910.1018(n)(1)(ii)** *Examination by physician.* The employer shall assure that all medical examinations and procedures are performed by or under the supervision of a licensed physician, and shall be provided without cost to the employee, without loss of pay and at a reasonable time and place.

**1910.1018(n)(2)** *Initial examinations.* By December 1, 1978, for employees initially covered by the medical provisions of this section, or thereafter at the time of initial assignment to an area where the employee is likely to be exposed over the action level at least 30 days per year, the employer shall provide each affected employee an opportunity for a medical examination, including at least the following elements:

**1910.1018(n)(2)(i)** A work history and a medical history which shall include a smoking history and the presence and degree of respiratory symptoms such as breathlessness, cough, sputum production and wheezing.

**1910.1018(n)(2)(ii)** A standard posterior-anterior chest x-ray;

**1910.1018(n)(2)(ii)(A)** A 14″ by 17″ posterior-anterior chest X-ray and International Labor Office UICC/Cincinnati (ILO U/C) rating;

**1910.1018(n)(2)(ii)(B)** A nasal and skin examination; and

**1910.1018(n)(2)(ii)(C)** Other examinations which the physician believes appropriate because of the employees exposure to inorganic arsenic or because of required respirator use.

**1910.1018(n)(3)** *Periodic examinations.* **(i)** Examinations must be provided in accordance with this paragraph at least annually.

**1910.1018(n)(3)(ii)** Whenever a covered employee has not taken the examinations specified in paragraphs (n)(2)(i) and (n)(2)(ii) of this section within six (6) months preceding the termination of employment, the employer shall provide such examinations to the employee upon termination of employment.

**1910.1018(n)(4)** *Additional examinations.* If the employee for any reason develops signs or symptoms commonly associated with exposure to inorganic arsenic the employer shall provide an appropriate examination and emergency medical treatment.

**1910.1018(n)(5)** *Information provided to the physician.* The employer shall provide the following information to the examining physician:

**1910.1018(n)(5)(i)** A copy of this standard and its appendices;

**1910.1018(n)(5)(ii)** A description of the affected employee's duties as they relate to the employee's exposure;

**1910.1018(n)(5)(iii)** The employee's representative exposure level or anticipated exposure level;

**1910.1018(n)(5)(iv)** A description of any personal protective equipment used or to be used; and

**1910.1018(n)(5)(v)** Information from previous medical examinations of the affected employee which is not readily available to the examining physician.

**1910.1018(n)(6)** *Physician's written opinion.* **(i)** The employer shall obtain a written opinion from the examining physician which shall include:

**1910.1018(n)(6)(i)(A)** The results of the medical examination and tests performed;

**1910.1018(n)(6)(i)(B)** The physician's opinion as to whether the employee has any detected medical conditions which would place the employee at increased risk of material impairment of the employee's health from exposure to inorganic arsenic;

**1910.1018(n)(6)(i)(C)** Any recommended limitations upon the employee's exposure to inorganic arsenic or upon the use of protective clothing or equipment such as respirators; and

**1910.1018(n)(6)(i)(D)** A statement that the employee has been informed by the physician of the results of the medical examination and any medical conditions which require further explanation or treatment.

**1910.1018(n)(6)(ii)** The employer shall instruct the physician not to reveal in the written opinion specific findings or diagnoses unrelated to occupational exposure.

**1910.1018(n)(6)(iii)** The employer shall provide a copy of the written opinion to the affected employee.

**1910.1018(o)** *Employee information and training*—**(1)** *Training program.* **(i)** The employer shall institute a training program for all employees who are subject to exposure to inorganic arsenic above the action level without regard to respirator use, or for whom there is the possibility of skin or eye irritation from inorganic arsenic. The employer shall assure that those employees participate in the training program.

**1910.1018(o)(1)(ii)** The training program shall be provided by October 1, 1978, for employees covered by this provision, at the time of initial assignment for those subsequently covered by this provision, and at least annually for other covered employees thereafter; and the employer shall assure that each employee is informed of the following:

**1910.1018(o)(1)(ii)(A)** The information contained in Appendix A;

**1910.1018(o)(1)(ii)(B)** The quantity, location, manner of use, storage, sources of exposure, and the specific nature of operations which could result in exposure to inorganic arsenic as well as any necessary protective steps;

**1910.1018(o)(1)(ii)(C)** The purpose, proper use, and limitation of respirators;

**1910.1018(o)(1)(ii)(D)** The purpose and a description of the medical surveillance program as required by paragraph (n) of this section;

**1910.1018(o)(1)(ii)(E)** The engineering controls and work practices associated with the employee's job assignment; and

**1910.1018(o)(1)(ii)(F)** A review of this standard.

**1910.1018(o)(2)** *Access to training materials.* **(i)** The employer shall make readily available to all affected employees a copy of this standard and its appendices.

**1910.1018(o)(2)(ii)** The employer shall provide; upon request, all materials relating to the employee information and training program to the Assistant Secretary and the Director.

**1910.1018(p)** *Signs and labels*—**(1)** *General.* **(i)** The employer may use labels or signs required by other statutes, regulations, or ordinances in addition to, or in combination with, signs and labels required by this paragraph.

**1910.1018(p)(1)(ii)** The employer shall assure that no statement appears on or near any sign or label required by this paragraph which contradicts or detracts from the meaning of the required sign or label.

**1910.1018(p)(2)** *Signs.* **(i)** The employer shall post signs demarcating regulated areas bearing the legend;

**DANGER**

**INORGANIC ARSENIC**

**CANCER HAZARD**

**AUTHORIZED PERSONNEL ONLY**

**NO SMOKING OR EATING**

**RESPIRATOR REQUIRED**

**1910.1018(p)(2)(ii)** The employer shall assure that signs required by this paragraph are illuminated and cleaned as necessary so that the legend is readily visible.

**1910.1018(p)(3)** *Labels.* The employer shall apply precautionary labels to all shipping and storage containers of inorganic arsenic, and to all products containing inorganic arsenic except when the inorganic arsenic in the product is bound in such a manner so as to make unlikely the possibility of airborne exposure to inorganic arsenic. (Possible examples of products not requiring labels are semiconductors, light emitting diodes and glass). The label shall bear the following legend:

**DANGER**

**CONTAINS INORGANIC ARSENIC**

**CANCER HAZARD**

**HARMFUL IF INHALED OR SWALLOWED**

**USE ONLY WITH ADEQUATE VENTILATION**

**OR RESPIRATORY PROTECTION**

**1910.1018(q)** *Recordkeeping*—**(1)** *Exposure monitoring.* **(i)** The employer shall establish and maintain an accurate record of all monitoring required by paragraph (e) of this section.

**1910.1018(q)(1)(ii)** This record shall include:

**1910.1018(q)(1)(ii)(A)** The date(s), number, duration location, and results of each of the samples taken, including a description of the sampling procedure used to determine representative employee exposure where applicable;

**1910.1018(q)(1)(ii)(B)** A description of the sampling and analytical methods used and evidence of their accuracy;

**1910.1018(q)(1)(ii)(C)** The type of respiratory protective devices worn, if any;

**1910.1018(q)(1)(ii)(D)** Name, social security number, and job classification of the employees monitored and of all other employees whose exposure the measurement is intended to represent; and

**1910.1018(q)(1)(ii)(E)** The environmental variables that could affect the measurement of the employee's exposure.

**1910.1018(q)(1)(iii)** The employer shall maintain these monitoring records for at least 40 years or for the duration of employment plus 20 years, whichever, is longer.

**1910.1018(q)(2)** *Medical surveillance.* **(i)** The employer shall establish and maintain an accurate record for each employee subject to medical surveillance as required by paragraph (n) of this section.

**1910.1018(q)(2)(ii)** This record shall include:

**1910.1018(q)(2)(ii)(A)** The name, social security number, and description of duties of the employee;

**1910.1018(q)(2)(ii)(B)** A copy of the physician's written opinions;

**1910.1018(q)(2)(ii)(C)** Results of any exposure monitoring done for that employee and the representative exposure levels supplied to the physician; and

**1910.1018(q)(2)(ii)(D)** Any employee medical complaints related to exposure to inorganic arsenic.

**1910.1018(q)(2)(iii)** The employer shall in addition keep, or assure that the examining physician keeps, the following medical records;

**1910.1018(q)(2)(iii)(A)** A copy of the medical examination results including medical and work history required under paragraph (n) of this section;

**1910.1018(q)(2)(iii)(B)** A description of the laboratory procedures and a copy of any standards or guidelines used to interpret the test results or references to that information;

**1910.1018(q)(2)(iii)(C)** The initial X-ray;

**1910.1018(q)(2)(iii)(D)** The X-rays for the most recent 5 years; and

**1910.1018(q)(2)(iii)(E)** Any X-rays with a demonstrated abnormality and all subsequent X-rays;

**1910.1018(q)(2)(iv)** The employer shall maintain or assure that the physician maintains those medical records for at least 40 years, or for the duration of employment plus 20 years whichever is longer.

**1910.1018(q)(3)** *Availability.* (i) The employer shall make available upon request all records required to be maintained by paragraph (q) of this section to the Assistant Secretary and the Director for examination and copying.

**1910.1018(q)(3)(ii)** Records required by this paragraph shall be provided upon request to employees, designated representatives, and the Assistant Secretary in accordance with 29 CFR 1910.1020(a) through (e) and (g) through (i).

**1910.1018(q)(4)** *Transfer of records.* (i) Whenever the employer ceases to do business, the successor employer shall receive and retain all records required to be maintained by this section.

**1910.1018(q)(4)(ii)** Whenever the employer ceases to do business and there is no successor employer to receive and retain the records required to be maintained by this section for the prescribed period, these records shall be transmitted to the Director.

**1910.1018(q)(4)(iii)** At the expiration of the retention period for the records required to be maintained by this section, the employer shall notify the Director at least 3 months prior to the disposal of such records and shall transmit those records to the Director if he requests them within that period.

**1910.1018(q)(4)(iv)** The employer shall also comply with any additional requirements involving the transfer of records set in 29 CFR 1910.1020(h).

**1910.1018(r)** *Observation of monitoring*—**(1)** *Employee observation.* The employer shall provide affected employees or their designated representatives an opportunity to observe any monitoring of employee exposure to inorganic arsenic conducted pursuant to paragraph (e) of this section.

**1910.1018(r)(2)** *Observation procedures.* (i) Whenever observation of the monitoring of employee exposure to inorganic arsenic requires entry into an area where the use of respirators, protective clothing, or equipment is required, the employer shall provide the observer with and assure the use of such respirators, clothing, and such equipment, and shall require the observer to comply with all other applicable safety and health procedures.

**1910.1018(r)(2)(ii)** Without interfering with the monitoring, observers shall be entitled to;

**1910.1018(r)(2)(ii)(A)** Receive an explanation of the measurement procedures;

**1910.1018(r)(2)(ii)(B)** Observe all steps related to the monitoring of inorganic arsenic performed at the place of exposure; and

**1910.1018(r)(2)(ii)(C)** Record the results obtained or receive copies of the results when returned by the laboratory.

**1910.1018(s)** *Appendices.* The information contained in the appendices to this section is not intended by itself, to create any additional obligations not otherwise imposed by this standard nor detract from any existing obligation.

## APPENDIX A TO §1910.1018—INORGANIC ARSENIC SUBSTANCE INFORMATION SHEET

### I. SUBSTANCE IDENTIFICATION

A. *Substance.* Inorganic Arsenic.

B. *Definition.* Copper acetoarsenite, arsenic and all inorganic compounds containing arsenic except arsine, measured as arsenic (As).

C. *Permissible Exposure Limit.* 10 micrograms per cubic meter of air as determined as an average over an 8-hour period. No employee may be exposed to any skin or eye contact with arsenic trichloride or to skin or eye contact likely to cause skin or eye irritation.

D. *Regulated Areas.* Only employees authorized by your employer should enter a regulated area.

### II. HEALTH HAZARD DATA

A. *Comments.* The health hazard of inorganic arsenic is high.

B. *Ways in which the chemical affects your body.* Exposure to airborne concentrations of inorganic arsenic may cause lung cancer, and can be a skin irritant. Inorganic arsenic may also affect your body if swallowed. One compound in particular, arsenic trichloride, is especially dangerous because it can be absorbed readily through the skin. Because inorganic arsenic is a poison, you should wash your hands thoroughly prior to eating or smoking.

### III. PROTECTIVE CLOTHING AND EQUIPMENT

A. *Respirators.* Respirators will be provided by your employer at no cost to you for routine use if your employer is in the process of implementing engineering and work practice controls or where engineering and work practice controls are not feasible or insufficient. You must wear respirators for non-routine activities or in emergency situations where you are likely to be exposed to levels of inorganic arsenic in excess of the permissible exposure limit. Since how well your respirator fits your face is very important, your employer is required to conduct fit tests to make sure the respirator seals properly when you wear it. These tests are simple and rapid and will be explained to you during training sessions.

B. *Protective clothing.* If you work in a regulated area, your employer is required to provide at no cost to you, and you must wear, appropriate, clean, protective clothing and equipment. The purpose of this equipment is to prevent you from bringing to your home arsenic-contaminated dust and to protect your body from repeated skin contact with inorganic arsenic likely to cause skin irritation. This clothing should include such items as coveralls or similar full-body clothing, gloves, shoes or coverlets, and aprons. Protective equipment should include face shields or vented goggles, where eye irritation may occur.

### IV. HYGIENE FACILITIES AND PRACTICES

You must not eat, drink, smoke, chew gum or tobacco, or apply cosmetics in the regulated area, except that drinking water is permitted. If you work in a regulated area your employer is required to provide lunchrooms and other areas for these purposes.

If you work in a regulated area, your employer is required to provide showers, washing facilities, and change rooms. You must wash your face, and hands before eating and must shower at the end of the work shift. Do not take used protective clothing out of change rooms without your employer's permission. Your employer is required to provide for laundering or cleaning of your protective clothing.

### V. SIGNS AND LABELS

Your employer is required to post warning signs and labels for your protection. Signs must be posted in regulated areas. The signs must warn that a cancer hazard is present, that only authorized employees may enter the area, and that no smoking or eating is allowed, and that respirators must be worn.

### VI. MEDICAL EXAMINATIONS

If your exposure to arsenic is over the Action Level (5 mg/m$^3$)—(including all persons working in regulated areas) at least 30 days per year, or you have been exposed to arsenic for more than 10 years over the Action Level, your employer is required to provide you with a medical examination. The examination shall be every 6 months for employees over 45 years old or with more than 10 years exposure over the Action Level and annually for other covered employees. The medical examination must include a medical history; a chest x-ray; a skin examination and a nasal examination. The examining physician will provide a written opinion to your employer containing the results of your medical exams. You should also receive a copy of this opinion. The physician must not tell your employer any conditions he detects unrelated to occupational exposure to arsenic but must tell you those conditions.

# OSHA Standards for General Industry

## VII. OBSERVATION OF MONITORING

Your employer is required to monitor your exposure to arsenic and you or your representatives are entitled to observe the monitoring procedure. You are entitled to receive an explanation of the measurement procedure, and to record the results obtained. When the monitoring procedure is taking place in an area where respirators or personal protective clothing and equipment are required to be worn, you must also be provided with and must wear the protective clothing and equipment.

## VIII. ACCESS TO RECORDS

You or your representative are entitled to records of your exposure to inorganic arsenic and your medical examination records if you request your employer to provide them.

## IX. TRAINING AND NOTIFICATION

Additional information on all of these items plus training as to hazards of exposure to inorganic arsenic and the engineering and work practice controls associated with your job will also be provided by your employer. If you are exposed over the permissible exposure limit, your employer must inform you of that fact and the actions he is taking to reduce your exposures.

## APPENDIX B TO §1910.1018— SUBSTANCE TECHNICAL GUIDELINES

### ARSENIC, ARSENIC TRIOXIDE, ARSENIC TRICHLORIDE (THREE EXAMPLES)

#### I. Physical and chemical properties

A. Arsenic (metal).

1. Formula: As.
2. Appearance: Gray metal.
3. Melting point: Sublimes without melting at 613 °C.
4. Specific Gravity: ($H_2O=1$):5.73.
5. Solubility in water: Insoluble.

B. Arsenic Trioxide.

1. Formula: $As_2O_3$, ($As_4O_6$).
2. Appearance: White powder.
3. Melting point: 315 °C.
4. Specific Gravity ($H_2O=1$):3.74.
5. Solubility in water: 3.7 grams in 100cc of water at 20 °C.

C. Arsenic Trichloride (liquid).

1. Formula: $AsCl_3$.
2. Appearance: Colorless or pale yellow liquid.
3. Melting point: -8.5 °C.
4. Boiling point: 130.2 °C.
5. Specific Gravity ($H_2O=1$):2.16 at 20 °C.
6. Vapor Pressure: 10mm Hg at 23.5 °C.
7. Solubility in Water: Decomposes in water.

#### II. Fire, explosion and reactivity data.

A. Fire: Arsenic, arsenic Trioxide and Arsenic Trichloride are nonflammable.

B. Reactivity:

1. Conditions Contributing to instability: Heat.
2. Incompatibility: Hydrogen gas can react with inorganic arsenic to form the highly toxic gas arsine.

#### III. Monitoring and Measurement Procedures

Samples collected should be full shift (at least 7-hour) samples. Sampling should be done using a personal sampling pump at a flow rate of 2 liters per minute. Samples should be collected on 0.8 micrometer pore size membrane filter (37mm diameter). Volatile arsenicals such as arsenic trichloride can be most easily collected in a midget bubbler filled with 15 ml. of 0.1 N NaOH.

The method of sampling and analysis should have an accuracy of not less ±25 percent (with a confidence limit of 95 percent) for 10 micrograms per cubic meter of air (10 µg/m³) and ±35 percent (with a confidence limit of 95 percent) for concentrations of inorganic arsenic between 5 and 10 µg/m³.

## APPENDIX C TO §1910.1018— MEDICAL SURVEILLANCE GUIDELINES

### I. GENERAL

Medical examinations are to be provided for all employees exposed to levels of inorganic arsenic above the action level (5 µg/m³) for at least 30 days per year (which would include among others, all employees, who work in regulated areas). Examinations are also to be provided to all employees who have had 10 years or more exposure above the action level for more than 30 days per year while working for the present or predecessor employer though they may no longer be exposed above the level.

An initial medical examination is to be provided to all such employees by December 1, 1978. In addition, an initial medical examination is to be provided to all employees who are first assigned to areas in which worker exposure will probably exceed 5 µg/m³ (after the effective date of this standard) at the time of initial assignment. In addition to its immediate diagnostic usefulness, the initial examination will provide a baseline for comparing future test results. The initial examination must include as a minimum the following elements:

(1) A work and medical history, including a smoking history, and presence and degree of respiratory symptoms such as breathlessness, cough, sputum production, and wheezing;

(2) A 14″ by 17″ posterior-anterior chest X-ray;

(3) A nasal and skin examination; and

(4) Other examinations which the physician believes appropriate because of the employee's exposure to inorganic arsenic or because of required respirator use.

Periodic examinations are also to be provided to the employees listed above. The periodic examinations shall be given annually for those covered employees 45 years of age or less with fewer than 10 years employment in areas where employee exposure exceeds the action level (5 µg/m³). Periodic examinations need not include sputum cytology and only an updated medical history is required.

Periodic examinations for other covered employees, shall be provided every six (6) months. These examinations shall include all tests required in the initial examination, except that the medical history need only be updated.

The examination contents are minimum requirements. Additional tests such as lateral and oblique X-rays or pulmonary function tests may be useful. For workers exposed to three arsenicals which are associated with lymphatic cancer, copper acetoarsenite, potassium arsenite, or sodium arsenite the examination should also include palpation of superficial lymph nodes and complete blood count.

### II. NONCARCINOGENIC EFFECTS

The OSHA standard is based on minimizing risk of exposed workers dying of lung cancer from exposure to inorganic arsenic. It will also minimize skin cancer from such exposures.

The following three sections quoted from "Occupational Diseases: A Guide to Their Recognition", Revised Edition, June 1977, National Institute for Occupational Safety and Health is included to provide information on the nonneoplastic effects of exposure to inorganic arsenic. Such effects should not occur if the OSHA standards are followed.

A. *Local*—Trivalent arsenic compounds are corrosive to the skin. Brief contact has no effect but prolonged contact results in a local hyperemia and later vesicular or pustular eruption. The moist mucous membranes are most sensitive to the irritant action. Conjunctiva, moist and macerated areas of skin, the eyelids, the angles of the ears, nose, mouth, and respiratory mucosa are also vulnerable to the irritant effects. The wrists are common sites of dermatitis, as are the genitalia if personal hygiene is poor. Perforations of the nasal septum may occur. Arsenic trioxide and pentoxide are capable of producing skin sensitization and contact dermatitis. Arsenic is also capable of producing keratoses, especially of the palms and soles.

B. *Systemic*—The acute toxic effects of arsenic are generally seen following ingestion of inorganic arsenical compounds. This rarely occurs in an industrial setting. Symptoms develop within 1/2 to 4 hours following ingestion and are usually characterized by constriction of the throat followed by dysphagia, epigastric pain, vomiting, and watery diarrhea. Blood may appear in vomitus and stools. If the amount ingested is sufficiently high, shock may develop due to severe fluid loss, and death may ensue in 24 hours. If the acute effects are survived, exfoliative dermatitis and peripheral neuritis may develop.

Cases of acute arsenical poisoning due to inhalation are exceedingly rare in industry. When it does occur, respiratory tract symptoms—cough, chest pain, dyspnea—giddiness, headache, and extreme general weakness precede gastrointestinal symptoms. The acute toxic symptoms of trivalent arsenical poisoning are due to severe inflammation of the mucous membranes and greatly increased permeability of the blood capillaries.

Chronic arsenical poisoning due to ingestion is rare and generally confined to patients taking prescribed medications. However, it can be a concomitant of inhaled inorganic arsenic from swallowed sputum and improper eating habits. Symptoms are weight loss, nausea and diarrhea alternating with constipation, pigmentation and eruption of the skin, loss of hair, and peripheral neuritis. Chronic hepatitis and cirrhosis have been described. Polyneuritis may be the salient feature, but more frequently there are numbness and parasthenias of "glove and stocking" distribution. The skin lesions are usually melanotic and keratotic and may occasionally take the form of an intradermal cancer of the squamous cell type, but without infiltrative properties. Horizontal white lines (striations) on the fingernails and toenails are commonly seen in chronic arsenical poisoning and are considered to be a diagnostic accompaniment of arsenical polyneuritis.

Inhalation of inorganic arsenic compounds is the most common cause of chronic poisoning in the industrial situation. This condition is divided into three phases based on signs and symptoms.

First Phase: The worker complains of weakness, loss of appetite, some nausea, occasional vomiting, a sense of heaviness in the stomach, and some diarrhea.

Second Phase: The worker complains of conjunctivitis, a catarrhal state of the mucous membranes of the nose, larynx, and respiratory passage. Coryza, hoarseness, and mild tracheobronchitis may occur. Perforation of the nasal septum is common, and is probably the most typical lesion of the upper respiratory tract in occupational exposure to arsenical dust. Skin lesions, eczematoid and allergic in type, are common.

Third Phase: The worker complains of symptoms of peripheral neuritis, initially of hands and feet, which is essentially sensory. In more severe cases, motor paralyses occur; the first muscles affected are usually the toe extensors and the peronei. In only the most severe cases will paralysis of flexor muscles of the feet or of the extensor muscles of hands occur.

Liver damage from chronic arsenical poisoning is still debated, and as yet the question is unanswered. In cases of chronic and acute arsenical poisoning, toxic effects to the myocardium have been reported based on EKG changes. These findings, however, are now largely discounted and the EKG changes are ascribed to electrolyte disturbances concomitant with arsenicalism. Inhalation of arsenic trioxide and other inorganic arsenical dusts does not give rise to radiological evidence or pneumoconiosis. Arsenic does have a depressant effect upon the bone marrow, with disturbances of both erythropoiesis and myelopoiesis.

### BIBLIOGRAPHY

Dinman, B. D. 1960. Arsenic; chronic human intoxication. J. Occup. Med. 2:137.

Elkins, H. B. 1959. The Chemistry of Industrial Toxicology, 2nd ed. John Wiley and Sons, New York.

Holmquist, L. 1951. Occupational arsenical dermatitis; a study among employees at a copper-ore smelting works including investigations of skin reactions to contact with arsenic compounds. Acta. Derm. Venereol. (Supp. 26) 31:1.

Pinto, S. S., and C. M. McGill. 1953. Arsenic trioxide exposure in industry. Ind. Med. Surg. 22:281.

Pinto, S. S., and K. W. Nelson. 1976. Arsenic toxicology and industrial exposure. Annu. Rev. Pharmacol. Toxicol. 16:95.

Vallee, B. L., D. D. Ulmer, and W. E. C. Wacker. 1960. Arsenic toxicology and biochemistry. AMA Arch. Indust. Health 21:132.

[39 FR 23502, June 27, 1974, as amended at 43 FR 19624, May 5, 1978; 43 FR 28472, June 30, 1978; 45 FR 35282, May 23, 1980; 54 FR 24334, June 7, 1989; 58 FR 35310, June 30, 1993; 61 FR 5508, Feb. 13, 1996; 61 FR 9245, Mar. 7, 1996; 63 FR 1286, Jan. 8, 1998; 63 FR 33468, June 18, 1998]

### §1910.1020 Access to employee exposure and medical records.

**1910.1020(a)** *Purpose.* The purpose of this section is to provide employees and their designated representatives a right of access to relevant exposure and medical records; and to provide representatives of the Assistant Secretary a right of access to these records in order to fulfill responsibilities under the Occupational Safety and Health Act. Access by employees, their representatives, and the Assistant Secretary is necessary to yield both direct and indirect improvements in the detection, treatment, and prevention of occupational disease. Each employer is responsible for assuring compliance with this section, but the activities involved in complying with the access to medical records provisions can be carried out, on behalf of the employer, by the physician or other health care personnel in charge of employee medical records. Except as expressly provided, nothing in this section is intended to affect existing legal and ethical obligations concerning the maintenance and confidentiality of employee medical information, the duty to disclose information to a patient/employee or any other aspect of the medical-care relationship, or affect existing legal obligations concerning the protection of trade secret information.

**1910.1020(b)** *Scope and application.* **(1)** This section applies to each general industry, maritime, and construction employer who makes, maintains, contracts for, or has access to employee exposure or medical records, or analyses thereof, pertaining to employees exposed to toxic substances or harmful physical agents.

**1910.1020(b)(2)** This section applies to all employee exposure and medical records, and analyses thereof, of such employees, whether or not the records are mandated by specific occupational safety and health standards.

**1910.1020(b)(3)** This section applies to all employee exposure and medical records, and analyses thereof, made or maintained in any manner, including on an in-house of contractual (e.g., fee-for-service) basis. Each employer shall assure that the preservation and access requirements of this section are complied with regardless of the manner in which the records are made or maintained.

**1910.1020(c)** *Definitions.* **(1)** *Access* means the right and opportunity to examine and copy.

**1910.1020(c)(2)** *Analysis using exposure or medical records* means any compilation of data or any statistical study based at least in part on information collected from individual employee exposure or medical records or information collected from health insurance claims records, provided that either the analysis has been reported to the employer or no further work is currently being done by the person responsible for preparing the analysis.

**1910.1020(c)(3)** *Designated representative* means any individual or organization to whom an employee gives written authorization to exercise a right of access. For the purposes of access to employee exposure records and analyses using exposure or medical records, a recognized or certified collective bargaining agent shall be treated automatically as a designated representative without regard to written employee authorization.

**1910.1020(c)(4)** *Employee* means a current employee, a former employee, or an employee being assigned or transferred to work where there will be exposure to toxic substances or harmful physical agents. In the case of a deceased or legally incapacitated employee, the employee's legal representative may directly exercise all the employee's rights under this section.

**1910.1020(c)(5)** *Employee exposure record* means a record containing any of the following kinds of information:

**1910.1020(c)(5)(i)** Environmental (workplace) monitoring or measuring of a toxic substance or harmful physical agent, including personal, area, grab, wipe, or other form of sampling, as well as related collection and analytical methodologies, calculations, and other background data relevant to interpretation of the results obtained;

**1910.1020(c)(5)(ii)** Biological monitoring results which directly assess the absorption of a toxic substance or harmful physical agent by body systems (e.g., the level of a chemical in the blood, urine, breath, hair, fingernails, etc) but not including results which assess the biological effect of a substance or agent or which assess an employee's use of alcohol or drugs;

**1910.1020(c)(5)(iii)** Material safety data sheets indicating that the material may pose a hazard to human health; or

# OSHA Standards for General Industry

**1910.1020(c)(5)(iv)** In the absence of the above, a chemcial inventory or any other record which reveals where and when used and the identity (e.g., chemical, common, or trade name) of a toxic substance or harmful physical agent.

**1910.1020(c)(6)(i)** *Employee medical record* means a record concerning the health status of an employee which is made or maintained by a physician, nurse, or other health care personnel or technician, including:

**1910.1020(c)(6)(i)(A)** Medical and employment questionnaires or histories (including job description and occupational exposures),

**1910.1020(c)(6)(i)(B)** The results of medical examinations (pre-employment, pre-assignment, periodic, or episodic) and laboratory tests (including chest and other X-ray examinations taken for the purposes of establishing a base-line or detecting occupational illness, and all biological monitoring not defined as an "employee exposure record"),

**1910.1020(c)(6)(i)(C)** Medical opinions, diagnoses, progress notes, and recommendations,

**1910.1020(c)(6)(i)(D)** First aid records,

**1910.1020(c)(6)(i)(E)** Descriptions of treatments and prescriptions, and

**1910.1020(c)(6)(i)(F)** Employee medical complaints.

**1910.1020(c)(6)(ii)** "Employee medical record" does not include medical information in the form of:

**1910.1020(c)(6)(ii)(A)** Physical specimens (e.g., blood or urine samples) which are routinely discarded as a part of normal medical practice; or

**1910.1020(c)(6)(ii)(B)** Records concerning health insurance claims if maintained separately from the employer's medical program and its records, and not accessible to the employer by employee name or other direct personal identifier (e.g., social security number, payroll number, etc.); or

**1910.1020(c)(6)(ii)(C)** Records created solely in preparation for litigation which are privileged from discovery under the applicable rules of procedure or evidence; or

**1910.1020(c)(6)(ii)(D)** Records concerning voluntary employee assistance programs (alcohol, drug abuse, or personal counseling programs) if maintained separately from the employer's medical program and its records.

**1910.1020(c)(7)** *Employer* means a current employer, a former employer, or a successor employer.

**1910.1020(c)(8)** *Exposure* or *exposed* means that an employee is subjected to a toxic substance or harmful physical agent in the course of employment through any route of entry (inhalation, ingestion, skin contact or absorption, etc.), and includes past exposure and potential (e.g., accidental or possible) exposure, but does not include situations where the employer can demonstrate that the toxic substance or harmful physical agent is not used, handled, stored, generated, or present in the workplace in any manner different from typical non-occupational situations.

**1910.1020(c)(9)** *Health Professional* means a physician, occupational health nurse, industrial hygienist, toxicologist, or epidemiologist, providing medical or other occupational health services to exposed employees.

**1910.1020(c)(10)** *Record* means any item, collection, or grouping of information regardless of the form or process by which it is maintained (e.g., paper document, microfiche, microfilm, X-ray film, or automated data processing).

**1910.1020(c)(11)** *Specific chemical identity* means the chemical name, Chemical Abstracts Service (CAS) Registry Number, or any other information that reveals the precise chemical designation of the substance.

**1910.1020(c)(12)(i)** *Specific written consent* means a written authorization containing the following:

**1910.1020(c)(12)(i)(A)** The name and signature of the employee authorizing the release of medical information,

**1910.1020(c)(12)(i)(B)** The date of the written authorization,

**1910.1020(c)(12)(i)(C)** The name of the individual or organization that is authorized to release the medical information,

**1910.1020(c)(12)(i)(D)** The name of the designated representative (individual or organization) that is authorized to receive the released information,

**1910.1020(c)(12)(i)(E)** A general description of the medical information that is authorized to be released,

**1910.1020(c)(12)(i)(F)** A general description of the purpose for the release of the medical information, and

**1910.1020(c)(12)(i)(G)** A date or condition upon which the written authorization will expire (if less than one year).

**1910.1020(c)(12)(ii)** A written authorization does not operate to authorize the release of medical information not in existence on the date of written authorization, unless the release of future information is expressly authorized, and does not operate for more than one year from the date of written authorization.

**1910.1020(c)(12)(iii)** A written authorization may be revoked in writing prospectively at any time.

**1910.1020(c)(13)** *Toxic substance or harmful physical agent* means any chemical substance, biological agent (bacteria, virus, fungus, etc.), or physical stress (noise, heat, cold, vibration, repetitive motion, ionizing and non-ionizing radiation, hypo-or hyperbaric pressure, etc.) which:

**1910.1020(c)(13)(i)** Is listed in the latest printed edition of the National Institute for Occupational Safety and Health (NIOSH) Registry of Toxic Effects of Chemical Substances (RTECS), which is incorporated by reference as specified in §1910.6; or

**1910.1020(c)(13)(ii)** Has yielded positive evidence of an acute or chronic health hazard in testing conducted by, or known to, the employer; or

**1910.1020(c)(13)(iii)** Is the subject of a material safety data sheet kept by or known to the employer indicating that the material may pose a hazard to human health.

**1910.1020(c)(14)** *Trade secret* means any confidential formula, pattern, process, device, or information or compilation of information that is used in an employer's business and that gives the employer an opportunity to obtain an advantage over competitors who do not know or use it.

**1910.1020(d)** *Preservation of records.* **(1)** Unless a specific occupational safety and health standard provides a different period of time, each employer shall assure the preservation and retention of records as follows:

**1910.1020(d)(1)(i)** *Employee medical records.* The medical record for each employee shall be preserved and maintained for at least the duration of employment plus thirty (30) years, except that the following types of records need not be retained for any specified period:

**1910.1020(d)(1)(i)(A)** Health insurance claims records maintained separately from the employer's medical program and its records,

**1910.1020(d)(1)(i)(B)** First aid records (not including medical histories) of one-time treatment and subsequent observation of minor scratches, cuts, burns, splinters, and the like which do not involve medical treatment, loss of consciousness, restriction of work or motion, or transfer to another job, if made on-site by a non-physician and if maintained separately from the employer's medical program and its records, and

**1910.1020(d)(1)(i)(C)** The medical records of employees who have worked for less than (1) year for the employer need not be retained beyond the term of employment if they are provided to the employee upon the termination of employment.

**1910.1020(d)(1)(ii)** *Employee exposure records.* Each employee exposure record shall be preserved and maintained for at least thirty (30) years, except that:

**1910.1020(d)(1)(ii)(A)** Background data to environmental (workplace) monitoring or measuring, such as laboratory reports and worksheets, need only be retained for one (1) year as long as the sampling results, the collection methodology (sampling plan), a description of the analytical and mathematical methods used, and a summary of other background data relevant to interpretation of the results obtained, are retained for at least thirty (30) years; and

**1910.1020(d)(1)(ii)(B)** Material safety data sheets and paragraph (c)(5)(iv) records concerning the identity of a substance or agent need not be retained for any specified period as long as some record of the identity (chemical name if known) of the substance or agent, where it was used, and when it was used is retained for at least thirty (30) years;[1] and

---

[1] Material safety data sheets must be kept for those chemicals currently in use that are effected by the Hazard Communication Standard in accordance with 29 CFR 1910.1200(g).

**1910.1020(d)(1)(ii)(C)** Biological monitoring results designated as exposure records by specific occupational safety and health standards shall be preserved and maintained as required by the specific standard.

**1910.1020(d)(1)(iii)** *Analyses using exposure or medical records.* Each analysis using exposure or medial records shall be preserved and maintained for at least thirty (30) years.

**1910.1020(d)(2)** Nothing in this section is intended to mandate the form, manner, or process by which an employer preserves a record as long as the information contained in the record is preserved and retrievable, except that chest X-ray films shall be preserved in their original state.

**1910.1020(e)** *Access to records*—**(1)** *General.* **(i)** Whenever an employee or designated representative requests access to a record, the employer shall assure that access is provided in a reasonable time, place, and manner. If the employer cannot reasonably provide access to the record within fifteen (15) working days, the employer shall within the fifteen (15) working days apprise the employee or designated representative requesting the record of the reason for the delay and the earliest date when the record can be made available.

**1910.1020(e)(1)(ii)** The employer may require of the requester only such information as should be readily known to the requester and which may be necessary to locate or identify the records being requested (e.g., dates and locations where the employee worked during the time period in question).

**1910.1020(e)(1)(iii)** Whenever an employee or designated representative requests a copy of a record, the employer shall assure that either:

**1910.1020(e)(1)(iii)(A)** A copy of the record is provided without cost to the employee or representative,

**1910.1020(e)(1)(iii)(B)** The necessary mechanical copying facilities (e.g., photocopying) are made available without cost to the employee or representative for copying the record, or

**1910.1020(e)(1)(iii)(C)** The record is loaned to the employee or representative for a reasonable time to enable a copy to be made.

**1910.1020(e)(1)(iv)** In the case of an original X-ray, the employer may restrict access to on-site examination or make other suitable arrangements for the temporary loan of the X-ray.

**1910.1020(e)(1)(v)** Whenever a record has been previously provided without cost to an employee or designated representative, the employer may charge reasonable, non-discriminatory administrative costs (i.e., search and copying expenses but not including overhead expenses) for a request by the employee or designated representative for additional copies of the record, except that

**1910.1020(e)(1)(v)(A)** An employer shall not charge for an initial request for a copy of new information that has been added to a record which was previously provided; and

**1910.1020(e)(1)(v)(B)** An employer shall not charge for an initial request by a recognized or certified collective bargaining agent for a copy of an employee exposure record or an analysis using exposure or medical records.

**1910.1020(e)(1)(vi)** Nothing in this section is intended to preclude employees and collective bargaining agents from collectively bargaining to obtain access to information in addition to that available under this section.

**1910.1020(e)(2)** *Employee and designated representative access*—**(i)** *Employee exposure records.* (A) Except as limited by paragraph (f) of this section, each employer shall, upon request, assure the access to each employee and designated representative to employee exposure records relevant to the employee. For the purpose of this section, an exposure record relevant to the employee consists of:

**1910.1020(e)(2)(i)(A)***(1)* A record which measures or monitors the amount of a toxic substance or harmful physical agent to which the employee is or has been exposed;

**1910.1020(e)(2)(i)(A)***(2)* In the absence of such directly relevant records, such records of other employees with past or present job duties or working conditions related to or similar to those of the employee to the extent necessary to reasonably indicate the amount and nature of the toxic substances or harmful physical agents to which the employee is or has been subjected, and

**1910.1020(e)(2)(i)(A)***(3)* Exposure records to the extent necessary to reasonably indicate the amount and nature of the toxic substances or harmful physical agents at workplaces or under working conditions to which the employee is being assigned or transferred.

**1910.1020(e)(2)(i)(B)** Requests by designated representatives for unconsented access to employee exposure records shall be in writing and shall specify with reasonable particularity:

**1910.1020(e)(2)(i)(B)***(1)* The records requested to be disclosed; and

**1910.1020(e)(2)(i)(B)***(2)* The occupational health need for gaining access to these records.

**1910.1020(e)(2)(ii)** *Employee medical records.* **(A)** Each employer shall, upon request, assure the access of each employee to employee medical records of which the employee is the subject, except as provided in paragraph (e)(2)(ii)(D) of this section.

**1910.1020(e)(2)(ii)(B)** Each employer shall, upon request, assure the access of each designated representative to the employee medical records of any employee who has given the designated representative specific written consent. Appendix A to this section contains a sample form which may be used to establish specific written consent for access to employee medical records.

**1910.1020(e)(2)(ii)(C)** Whenever access to employee medical records is requested, a physician representing the employer may recommend that the employee or designated representative:

**1910.1020(e)(2)(ii)(C)***(1)* Consult with the physician for the purposes of reviewing and discussing the records requested,

**1910.1020(e)(2)(ii)(C)***(2)* Accept a summary of material facts and opinions in lieu of the records requested, or

**1910.1020(e)(2)(ii)(C)***(3)* Accept release of the requested records only to a physician or other designated representative.

**1910.1020(e)(2)(ii)(D)** Whenever an employee requests access to his or her employee medical records, and a physician representing the employer believes that direct employee access to information contained in the records regarding a specific diagnosis of a terminal illness or a psychiatric condition could be detrimental to the employee's health, the employer may inform the employee that access will only be provided to a designated representative of the employee having specific written consent, and deny the employee's request for direct access to this information only. Where a designated representative with specific written consent requests access to information so withheld, the employer shall assure the access of the designated representative to this information, even when it is known that the designated representative will give the information to the employee.

**1910.1020(e)(2)(ii)(E)** A physician, nurse, or other responsible health care personnel maintaining medical records may delete from requested medical records the identity of a family member, personal friend, or fellow employee who has provided confidential information concerning an employee's health status.

**1910.1020(e)(2)(iii)** *Analyses using exposure or medical records.* **(A)** Each employee shall, upon request, assure the access of each employee and designated representative to each analysis using exposure or medical records concerning the employee's working conditions or workplace.

**1910.1020(e)(2)(iii)(B)** Whenever access is requested to an analysis which reports the contents of employee medical records by either direct identifier (name, address, social security number, payroll number, etc.) or by information which could reasonably be used under the circumstances indirectly to identify specific employees (exact age, height, weight, race, sex, date of initial employment, job title, etc.), the employer shall assure that personal identifiers are removed before access is provided. If the employer can demonstrate that removal of personal identifiers from an analysis is not feasible, access to the personally identifiable portions of the analysis need not be provided.

**1910.1020(e)(3)** *OSHA access.* (i) Each employer shall, upon request, and without derogation of any rights under the Constitution or the Occupational Safety and Health Act of 1970, 29 U.S.C. 651 *et seq.*, that the employer chooses to exercise, assure the prompt access of representatives of the Assistant Secretary of Labor for Occupational Safety and Health to employee exposure and medical records and to analyses using exposure or medical records. Rules of agency practice and procedure governing OSHA access to employee medical records are contained in 29 CFR 1913.10.

**1910.1020(e)(3)(ii)** Whenever OSHA seeks access to personally identifiable employee medical information by presenting to the employer a written access order pursuant to 29 CFR 1913.10(d), the employer shall prominently post a copy of the written access order and its accompanying cover letter for at least fifteen (15) working days.

**1910.1020(f)** *Trade secrets.* **(1)** Except as provided in paragraph (f)(2) of this section, nothing in this section precludes an employer from deleting from records requested by a health professional, employee, or designated representative any trade secret data which discloses manufacturing processes, or discloses the percentage of a chemical substance in mixture, as long as the health professional, employee, or designated representative is notified that information has been deleted. Whenever deletion of trade secret information substantially impairs evaluation of the place where or the time when exposure to a toxic substance or harmful physical agent occurred, the employer shall provide alternative information which is sufficient to permit the requesting party to identify where and when exposure occurred.

**1910.1020(f)(2)** The employer may withhold the specific chemical identity, including the chemical name and other specific identification of a toxic substance from a disclosable record provided that:

**1910.1020(f)(2)(i)** The claim that the information withheld is a trade secret can be supported;

**1910.1020(f)(2)(ii)** All other available information on the properties and effects of the toxic substance is disclosed;

**1910.1020(f)(2)(iii)** The employer informs the requesting party that the specific chemical identity is being withheld as a trade secret; and

**1910.1020(f)(2)(iv)** The specific chemical identity is made available to health professionals, employees and designated representatives in accordance with the specific applicable provisions of this paragraph.

**1910.1020(f)(3)** Where a treating physician or nurse determines that a medical emergency exists and the specific chemical identity of a toxic substance is necessary for emergency or first-aid treatment, the employer shall immediately disclose the specific chemical identity of a trade secret chemical to the treating physician or nurse, regardless of the existence of a written statement of need or a confidentiality agreement. The employer may require a written statement of need and confidentiality agreement, in accordance with the provisions of paragraphs (f)(4) and (f)(5), as soon as circumstances permit.

**1910.1020(f)(4)** In non-emergency situations, an employer shall, upon request, disclose a specific chemical identity, otherwise permitted to be withheld under paragraph (f)(2) of this section, to a health professional, employee, or designated representative if:

**1910.1020(f)(4)(i)** The request is in writing;

**1910.1020(f)(4)(ii)** The request describes with reasonable detail one or more of the following occupational health needs for the information:

**1910.1020(f)(4)(ii)(A)** To assess the hazards of the chemicals to which employees will be exposed;

**1910.1020(f)(4)(ii)(B)** To conduct or assess sampling of the workplace atmosphere to determine employee exposure levels;

**1910.1020(f)(4)(ii)(C)** To conduct pre-assignment or periodic medical surveillance of exposed employees;

**1910.1020(f)(4)(ii)(D)** To provide medical treatment to exposed employees;

**1910.1020(f)(4)(ii)(E)** To select or assess appropriate personal protective equipment for exposed employees;

**1910.1020(f)(4)(ii)(F)** To design or assess engineering controls or other protective measures for exposed employees; and

**1910.1020(f)(4)(ii)(G)** To conduct studies to determine the health effects of exposure.

**1910.1020(f)(4)(iii)** The request explains in detail why the disclosure of the specific chemical identity is essential and that, in lieu thereof, the disclosure of the following information would not enable the health professional, employee or designated representative to provide the occupational health services described in paragraph (f)(4)(ii) of this section:

**1910.1020(f)(4)(iii)(A)** The properties and effects of the chemical;

**1910.1020(f)(4)(iii)(B)** Measures for controlling workers' exposure to the chemical;

**1910.1020(f)(4)(iii)(C)** Methods of monitoring and analyzing worker exposure to the chemical; and,

**1910.1020(f)(4)(iii)(D)** Methods of diagnosing and treating harmful exposures to the chemical;

**1910.1020(f)(4)(iv)** The request includes a description of the procedures to be used to maintain the confidentiality of the disclosed information; and,

**1910.1020(f)(4)(v)** The health professional, employee, or designated representative and the employer or contractor of the services of the health professional or designated representative agree in a written confidentiality agreement that the health professional, employee or designated representative will not use the trade secret information for any purpose other than the health need(s) asserted and agree not to release the information under any circumstances other than to OSHA, as provided in paragraph (f)(7) of this section, except as authorized by the terms of the agreement or by the employer.

**1910.1020(f)(5)** The confidentiality agreement authorized by paragraph (f)(4)(iv) of this section:

**1910.1020(f)(5)(i)** May restrict the use of the information to the health purposes indicated in the written statement of need;

**1910.1020(f)(5)(ii)** May provide for appropriate legal remedies in the event of a breach of the agreement, including stipulation of a reasonable pre-estimate of likely damages; and,

**1910.1020(f)(5)(iii)** May not include requirements for the posting of a penalty bond.

**1910.1020(f)(6)** Nothing in this section is meant to preclude the parties from pursuing non-contractual remedies to the extent permitted by law.

**1910.1020(f)(7)** If the health professional, employee or designated representative receiving the trade secret information decides that there is a need to disclose it to OSHA, the employer who provided the information shall be informed by the health professional prior to, or at the same time as, such disclosure.

**1910.1020(f)(8)** If the employer denies a written request for disclosure of a specific chemical identity, the denial must:

**1910.1020(f)(8)(i)** Be provided to the health professional, employee or designated representative within thirty days of the request;

**1910.1020(f)(8)(ii)** Be in writing;

**1910.1020(f)(8)(iii)** Include evidence to support the claim that the specific chemical identity is a trade secret;

**1910.1020(f)(8)(iv)** State the specific reasons why the request is being denied; and,

**1910.1020(f)(8)(v)** Explain in detail how alternative information may satisfy the specific medical or occupational health need without revealing the specific chemical identity.

**1910.1020(f)(9)** The health professional, employee, or designated representative whose request for information is denied under paragraph (f)(4) of this section may refer the request and the written denial of the request to OSHA for consideration.

**1910.1020(f)(10)** When a heath professional employee, or designated representative refers a denial to OSHA under paragraph (f)(9) of this section, OSHA shall consider the evidence to determine if:

**1910.1020(f)(10)(i)** The employer has supported the claim that the specific chemical identity is a trade secret;

**1910.1020(f)(10)(ii)** The health professional employee, or designated representative has supported the claim that there is a medical or occupational health need for the information; and

**1910.1020(f)(10)(iii)** The health professional, employee or designated representative has demonstrated adequate means to protect the confidentiality.

**1910.1020(f)(11)(i)** If OSHA determines that the specific chemical identity requested under paragraph (f)(4) of this section is not a *bona fide* trade secret, or that it is a trade secret but the requesting health professional, employee or designated representatives has a legitimate medical or occupational health need for the information, has executed a written confidentiality agreement, and has shown adequate means for complying with the terms of such agreement, the employer will be subject to citation by OSHA.

**1910.1020(f)(11)(ii)** If an employer demonstrates to OSHA that the execution of a confidentiality agreement would not provide sufficient protection against the potential harm from the unauthorized disclosure of a trade secret specific chemical identity, the Assistant Secretary may issue such orders or impose such additional limitations or conditions upon the disclosure of the requested chemical information as may be appropriate to assure that the occupational health needs are met without an undue risk of harm to the employer.

**1910.1020(f)(12)** Notwithstanding the existence of a trade secret claim, an employer shall, upon request, disclose to the Assistant Secretary any information which this section requires the employer to make available. Where there is a trade secret claim, such claim shall be made no later than at the time the information is provided to the Assistant Secretary so that suitable determinations of trade secret status can be made and the necessary protections can be implemented.

**1910.1020(f)(13)** Nothing in this paragraph shall be construed as requiring the disclosure under any circumstances of process or percentage of mixture information which is trade secret.

**1910.1020(g)** *Employee information.* **(1)** Upon an employee's first entering into employment, and at least annually thereafter, each employer shall inform current employees covered by this section of the following:

**1910.1020(g)(1)(i)** The existence, location, and availability of any records covered by this section;

**1910.1020(g)(1)(ii)** The person responsible for maintaining and providing access to records; and

**1910.1020(g)(1)(iii)** Each employee's rights of access to these records.

**1910.1020(g)(2)** Each employer shall keep a copy of this section and its appendices, and make copies readily available, upon request, to employees. The employer shall also distribute to current employees any informational materials concerning this section which are made available to the employer by the Assistant Secretary of Labor for Occupational Safety and Health.

**1910.1020(h)** *Transfer of records.* **(1)** Whenever an employer is ceasing to do business, the employer shall transfer all records subject to this section to the successor employer. The successor employer shall receive and maintain these records.

**1910.1020(h)(2)** Whenever an employer is ceasing to do business and there is no successor employer to receive and maintain the records subject to this standard, the employer shall notify affected current employees of their rights of access to records at least three (3) months prior to the cessation of the employer's business.

**1910.1020(h)(3)** Whenever an employer either is ceasing to do business and there is no successor employer to receive and maintain the records, or intends to dispose of any records required to be preserved for at least thirty (30) years, the employer shall:

**1910.1020(h)(3)(i)** Transfer the records to the Director of the National Institute for Occupational Safety and Health (NIOSH) if so required by a specific occupational safety and health standard; or

**1910.1020(h)(3)(ii)** Notify the Director of NIOSH in writing of the impending disposal of records at least three (3) months prior to the disposal of the records.

**1910.1020(h)(4)** Where an employer regularly disposes of records required to be preserved for at least thirty (30) years, the employer may, with at least (3) months notice, notify the Director of NIOSH on an annual basis of the records intended to be disposed of in the coming year.

**1910.1020(i)** *Appendices.* The information contained in appendices A and B to this section is not intended, by itself, to create any additional obligations not otherwise imposed by this section nor detract from any existing obligation.

## APPENDIX A TO §1910.20 [sic §1910.1020]—SAMPLE AUTHORIZATION LETTER FOR THE RELEASE OF EMPLOYEE MEDICAL RECORD INFORMATION TO A DESIGNATED REPRESENTATIVE (NON-MANDATORY)

I, _____ (full name of worker/patient), hereby authorize _____ (individual or organization holding the medical records) to release to _____ (individual or organization authorized to receive the medical information), the following medical information from my personal medical records:

(Describe generally the information desired to be released)
_____
_____
_____

I give my permission for this medical information to be used for the following purpose:
_____
_____

but I do not give permission for any other use or re-disclosure of this information.

**NOTE:** Several extra lines are provided below so that you can place additional restrictions on this authorization letter if you want to. You may, however, leave these lines blank. On the other hand, you may want to (1) specify a particular expiration date for this letter (if less than one year); (2) describe medical information to be created in the future that you intend to be covered by this authorization letter; or (3) describe portions of the medical information in your records which you do not intend to be released as a result of this letter.)
_____
_____
_____
_____

Full name of Employee or Legal Representative
_____

Signature of Employee or Legal Representative
_____

Date of Signature _____

## APPENDIX B TO §1910.20 [sic §1910.1020]—AVAILABILITY OF NIOSH REGISTRY OF TOXIC EFFECTS OF CHEMICAL SUBSTANCES (RTECS) (NON-MANDATORY)

The final regulation, 29 CFR 1910.20, applies to all employee exposure and medical records, and analyses thereof, of employees exposed to toxic substances or harmful physical agents (paragraph (b)(2)). The term *toxic substance or harmful physical agent* is defined by paragraph (c)(13) to encompass chemical substances, biological agents, and physical stresses for which there is evidence of harmful health effects. The regulation uses the latest printed edition of the National Institute for Occupational Safety and Health (NIOSH) Registry of Toxic Effects of Chemical Substances (RTECS) as one of the chief sources of information as to whether evidence of harmful health effects exists. If a substance is listed in the latest printed RTECS, the regulation applies to exposure and medical records (and analyses of these records) relevant to employees exposed to the substance.

It is appropriate to note that the final regulation does not require that employers purchase a copy of RTECS, and many employers need not consult RTECS to ascertain whether their employee exposure or medical records are subject to the rule. Employers who do not currently have the latest printed edition of the NIOSH RTECS, however, may desire to obtain a copy. The RTECS is issued in an annual printed edition as mandated by section 20(a)(6) of the Occupational Safety and Health Act (29 U.S.C. 669(a)(6)).

# OSHA Standards for General Industry

The Introduction to the 1980 printed edition describes the RTECS as follows:

"The 1980 edition of the Registry of Toxic Effects of Chemical Substances, formerly known as the Toxic Substances list, is the ninth revision prepared in compliance with the requirements of Section 20(a)(6) of the Occupational Safety and Health Act of 1970 (Public Law 91-596). The original list was completed on June 28, 1971, and has been updated annually in book format. Beginning in October 1977, quarterly revisions have been provided in microfiche. This edition of the Registry contains 168,096 listings of chemical substances: 45,156 are names of different chemicals with their associated toxicity data and 122,940 are synonyms. This edition includes approximately 5,900 new chemical compounds that did not appear in the 1979 Registry. (p. xi)

"The Registry's purposes are many, and it serves a variety of users. It is a single source document for basic toxicity information and for other data, such as chemical identifiers ad information necessary for the preparation of safety directives and hazard evaluations for chemical substances. The various types of toxic effects linked to literature citations provide researchers and occupational health scientists with an introduction to the toxicological literature, making their own review of the toxic hazards of a given substance easier. By presenting data on the lowest reported doses that produce effects by several routes of entry in various species, the Registry furnishes valuable information to those responsible for preparing safety data sheets for chemical substances in the workplace. Chemical and production engineers can use the Registry to identify the hazards which may be associated with chemical intermediates in the development of final products, and thus can more readily select substitutes or alternative processes which may be less hazardous. Some organizations, including health agencies and chemical companies, have included the NIOSH Registry accession numbers with the listing of chemicals in their files to reference toxicity information associated with those chemicals. By including foreign language chemical names, a start has been made toward providing rapid identification of substances produced in other countries. (p. xi)

"In this edition of the Registry, the editors intend to identify "all known toxic substances" which may exist in the environment and to provide pertinent data on the toxic effects from known doses entering an organism by any route described. (p xi)

"It must be reemphasized that the entry of a substance in the Registry does not automatically mean that it must be avoided. A listing does mean, however, that the substance has the documented potential of being harmful if misused, and care must be exercised to prevent tragic consequences. Thus, the Registry lists many substances that are common in everyday life and are in nearly every household in the United States. One can name a variety of such dangerous substances: prescription and non-prescription drugs; food additives; pesticide concentrates, sprays, and dusts; fungicides; herbicides; paints; glazes, dyes; bleaches and other household cleaning agents; alkalies; and various solvents and diluents. The list is extensive because chemicals have become an integral part of our existence."

The RTECS printed edition may be purchased from the Superintendent of Documents, U.S. Government Printing Office (GPO), Washington, DC 20402 (202-783-3238).

Some employers may desire to subscribe to the quarterly update to the RTECS which is published in a microfiche edition. An annual subscription to the quarterly microfiche may be purchased from the GPO (Order the "Microfiche Edition, Registry of Toxic Effects of Chemical Substances"). Both the printed edition and the microfiche edition of RTECS are available for review at many university and public libraries throughout the country. The latest RTECS editions may also be examined at the OSHA Technical Data Center, Room N2439—Rear, United States Department of Labor, 200 Constitution Avenue, NW., Washington, DC 20210 (202-523-9700), or at any OSHA Regional or Area Office (See, major city telephone directories under United States Government-Labor Department).

[53 FR 38163, Sept. 29, 1988; 53 FR 49981, Dec. 13, 1988, as amended at 54 FR 24333, June 7, 1989; 55 FR 26431, June 28, 1990; 61 FR 9735, Mar. 7, 1996. Redesignated at 61 FR 31430, June 20, 1996]

## §1910.1025 Lead.

**1910.1025(a)** *Scope and application.* **(1)** This section applies to all occupational exposure to lead, except as provided in paragraph (a)(2).

**1910.1025(a)(2)** This section does not apply to the construction industry or to agricultural operations covered by 29 CFR Part 1928.

**1910.1025(b)** *Definitions.* Action level means employee exposure, without regard to the use of respirators, to an airborne concentration of lead of 30 micrograms per cubic meter of air (30 µg/m$^3$) averaged over an 8-hour period.

*Assistant Secretary* means the Assistant Secretary of Labor for Occupational Safety and Health, U.S. Department of Labor, or designee.

*Director* means the Director, National Institute for Occupational Safety and Health (NIOSH), U.S. Department of Health, Education, and Welfare, or designee.

*Lead* means metallic lead, all inorganic lead compounds, and organic lead soaps. Excluded from this definition are all other organic lead compounds.

**1910.1025(c)** *Permissible exposure limit (PEL).* **(1)** The employer shall assure that no employee is exposed to lead at concentrations greater than fifty micrograms per cubic meter of air (50 µg/m$^3$) averaged over an 8-hour period.

**1910.1025(c)(2)** If an employee is exposed to lead for more than 8 hours in any work day, the permissible exposure limit, as a time weighted average (TWA) for that day, shall be reduced according to the following formula:

Maximum permissible limit (in µg/m$^3$) = 400÷hours worked in the day.

**1910.1025(c)(3)** When respirators are used to supplement engineering and work practice controls to comply with the PEL and all the requirements of paragraph (f) have been met, employee exposure, for the purpose of determining whether the employer has complied with the PEL, may be considered to be at the level provided by the protection factor of the respirator for those periods the respirator is worn. Those periods may be averaged with exposure levels during periods when respirators are not worn to determine the employee's daily TWA exposure.

**1910.1025(d)** *Exposure monitoring*—**(1)** *General.* **(i)** For the purposes of paragraph (d), employee exposure is that exposure which would occur if the employee were not using a respirator.

**1910.1025(d)(1)(ii)** With the exception of monitoring under paragraph (d)(3), the employer shall collect full shift (for at least 7 continuous hours) personal samples including at least one sample for each shift for each job classification in each work area.

**1910.1025(d)(1)(iii)** Full shift personal samples shall be representative of the monitored employee's regular, daily exposure to lead.

**1910.1025(d)(2)** *Initial determination.* Each employer who has a workplace or work operation covered by this standard shall determine if any employee may be exposed to lead at or above the action level.

**1910.1025(d)(3)** *Basis of initial determination.* **(i)** The employer shall monitor employee exposures and shall base initial determinations on the employee exposure monitoring results and any of the following, relevant considerations:

**1910.1025(d)(3)(i)(A)** Any information, observations, or calculations which would indicate employee exposure to lead;

**1910.1025(d)(3)(i)(B)** Any previous measurements of airborne lead; and

**1910.1025(d)(3)(i)(C)** Any employee complaints of symptoms which may be attributable to exposure to lead.

**1910.1025(d)(3)(ii)** Monitoring for the initial determination may be limited to a representative sample of the exposed employees who the employer reasonably believes are exposed to the greatest airborne concentrations of lead in the workplace.

**1910.1025(d)(3)(iii)** Measurements of airborne lead made in the preceding 12 months may be used to satisfy the requirement to monitor under paragraph (d)(3)(i) if the sampling and analytical methods used meet the accuracy and confidence levels of paragraph (d)(9) of this section.

**1910.1025(d)(4)** *Positive initial determination and initial monitoring.* **(i)** Where a determination conducted under paragraphs (d) (2) and (3) of this section shows the possibility of any employee exposure at or above the action level, the employer shall conduct monitoring which is representative of the exposure for each employee in the workplace who is exposed to lead.

**1910.1025(d)(4)(ii)** Measurements of airborne lead made in the preceding 12 months may be used to satisfy this requirement if the sampling and analytical methods used meet the accuracy and confidence levels of paragraph (d)(9) of this section.

**1910.1025(d)(5)** *Negative initial determination.* Where a determination, conducted under paragraphs (d) (2) and (3) of this section is made that no employee is exposed to airborne concentrations of lead at or above the action level, the employer shall make a written record of such determination. The record shall include at least the information specified in paragraph (d)(3) of this section and shall also include the date of determination, location within the worksite, and the name and social security number of each employee monitored.

**1910.1025(d)(6)** *Frequency.* **(i)** If the initial monitoring reveals employee exposure to be below the action level the measurements need not be repeated except as otherwise provided in paragraph (d)(7) of this section.

**1910.1025(d)(6)(ii)** If the initial determination or subsequent monitoring reveals employee exposure to be at or above the action level but below the permissible exposure limit the employer shall repeat monitoring in accordance with this paragraph at least every 6 months. The employer shall continue monitoring at the required frequency until at least two consecutive measurements, taken at least 7 days apart, are below the action level at which time the employer may discontinue monitoring for that employee except as otherwise provided in paragraph (d)(7) of this section.

**1910.1025(d)(6)(iii)** If the initial monitoring reveals that employee exposure is above the permissible exposure limit the employer shall repeat monitoring quarterly. The employer shall continue monitoring at the required frequency until at least two consecutive measurements, taken at least 7 days apart, are below the PEL but at or above the action level at which time the employer shall repeat monitoring for that employee at the frequency specified in paragraph (d)(6)(ii), except as otherwise provided in paragraph (d)(7) of this section.

**1910.1025(d)(7)** *Additional monitoring.* Whenever there has been a production, process, control or personnel change which may result in new or additional exposure to lead, or whenever the employer has any other reason to suspect a change which may result in new or additional exposures to lead, additional monitoring in accordance with this paragraph shall be conducted.

**1910.1025(d)(8)** *Employee notification.* **(i)** The employer must, within 15 working days after the receipt of the results of any monitoring performed under this section, notify each affected employee of these results either individually in writing or by posting the results in an appropriate location that is accessible to affected employees.

**1910.1025(d)(8)(ii)** Whenever the results indicate that the representative employee exposure, without regard to respirators, exceeds the permissible exposure limit, the employer shall incude in the written notice a statement that the permissible exposure limit was exceeded and a description of the corrective action taken or to be taken to reduce exposure to or below the permissible exposure limit.

**1910.1025(d)(9)** *Accuracy of measurement.* The employer shall use a method of monitoring and analysis which has an accuracy (to a confidence level of 95%) of not less than plus or minus 20 percent for airborne concentrations of lead equal to or greater than 30 µg/m$^3$.

**1910.1025(e)** **Methods of compliance**—**(1)** *Engineering and work practice controls.* **(i)** Where any employee is exposed to lead above the permissible exposure limit for more than 30 days per year, the employer shall implement engineering and work practice controls (including administrative controls) to reduce and maintain employee exposure to lead in accordance with the implementation schedule in Table I below, except to the extent that the employer can demonstrate that such controls are not feasible. Wherever the engineering and work practice controls which can be instituted are not sufficient to reduce employee exposure to or below the permissible exposure limit, the employer shall nonetheless use them to reduce exposures to the lowest feasible level and shall supplement them by the use of respiratory protection which complies with the requirements of paragraph (f) of this section.

**1910.1025(e)(1)(ii)** Where any employee is exposed to lead above the permissible exposure limit, but for 30 days or less per year, the employer shall implement engineering controls to reduce exposures to 200 µg/m$^3$, but thereafter may implement any combination of engineering, work practice (including administrative controls), and respiratory controls to reduce and maintain employee exposure to lead to or below 50 µg/m$^3$.

### TABLE I

| Industry | Compliance dates:[1] (50 µg/m$^3$) |
|---|---|
| Lead chemicals, secondary copper smelting | July 19, 1996. |
| Nonferrous foundries | July 19, 1996.[2] |
| Brass and bronze ingot manufacture | 6 years.[3] |

[1] Calculated by counting from the date the stay on implementation of paragraph (e)(1) was lifted by the U.S. Court of Appeals for the District of Columbia, the number of years specified in the 1978 lead standard and subsequent amendments for compliance with the PEL of 50 µg/m$^3$ for exposure to airborne concentrations of lead levels for the particular industry.

[2] Large nonferrous foundries (20 or more employees) are required to achieve the PEL of 50 µg/m$^3$ by means of engineering and work practice controls. Small nonferrous foundries (fewer than 20 employees) are required to achieve an 8-hour TWA of 75 µg/m$^3$ by such controls.

[3] Expressed as the number of years from the date on which the Court lifts the stay on the implementation of paragraph (e)(1) for this industry for employers to achieve a lead in air concentration of 75 µg/m$^3$. Compliance with paragraph (e) in this industry is determined by a compliance directive that incorporates elements from the settlement agreement between OSHA and representatives of the industry.

**1910.1025(e)(2)** *Respiratory protection.* Where engineering and work practice controls do not reduce employee exposure to or below the 50 µg/m$^3$ permissible exposure limit, the employer shall supplement these controls with respirators in accordance with paragraph (f).

**1910.1025(e)(3)** *Compliance program.* **(i)** Each employer shall establish and implement a written compliance program to reduce exposures to or below the permissible exposure limit, and interim levels if applicable, solely by means of engineering and work practice controls in accordance with the implementation schedule in paragraph (e)(1).

**1910.1025(e)(3)(ii)** Written plans for these compliance programs shall include at least the following:

**1910.1025(e)(3)(ii)(A)** A description of each operation in which lead is emitted; e.g., machinery used, material processed, controls in place, crew size, employee job responsibilities, operating procedures and maintenance practices;

**1910.1025(e)(3)(ii)(B)** A description of the specific means that will be employed to achieve compliance, including engineering plans and studies used to determine methods selected for controlling exposure to lead;

**1910.1025(e)(3)(ii)(C)** A report of the technology considered in meeting the permissible exposure limit;

**1910.1025(e)(3)(ii)(D)** Air monitoring data which documents the source of lead emissions;

**1910.1025(e)(3)(ii)(E)** A detailed schedule for implementation of the program, including documentation such as copies of purchase orders for equipment, construction contracts, etc.;

**1910.1025(e)(3)(ii)(F)** A work practice program which includes items required under paragraphs (g), (h) and (i) of this regulation;

**1910.1025(e)(3)(ii)(G)** An administrative control schedule required by paragraph (e)(6), if applicable;

**1910.1025(e)(3)(ii)(H)** Other relevant information.

**1910.1025(e)(3)(iii)** Written programs shall be submitted upon request to the Assistant Secretary and the Director, and shall be available at the worksite for examination and copying by the Assistant Secretary, Director, any affected employee or authorized employee representatives.

**1910.1025(e)(3)(iv)** Written programs must be revised and updated at least annually to reflect the current status of the program.

**1910.1025(e)(4)** *Mechanical ventilation.* **(i)** When ventilation is used to control exposure, measurements which demonstrate the effectiveness of the system in controlling exposure, such as capture velocity, duct velocity, or static pressure shall be made at least every 3 months. Measurements of the system's effectiveness in controlling exposure shall be made within 5 days of any change in production, process, or control which might result in a change in employee exposure to lead.

**1910.1025(e)(4)(ii)** *Recirculation of air.* If air from exhaust ventilation is recirculated into the workplace, the employer shall assure that (A) the system has a high efficiency filter with reliable back-up filter; and (B) controls

# OSHA Standards for General Industry

to monitor the concentration of lead in the return air and to bypass the recirculation system automatically if it fails are installed, operating, and maintained.

**1910.1025(e)(5)** *Administrative controls.* If administrative controls are used as a means of reducing employees TWA exposure to lead, the employer shall establish and implement a job rotation schedule which includes:

**1910.1025(e)(5)(i)** Name or identification number of each affected employee;

**1910.1025(e)(5)(ii)** Duration and exposure levels at each job or work station where each affected employee is located; and

**1910.1025(e)(5)(iii)** Any other information which may be useful in assessing the reliability of administrative controls to reduce exposure to lead.

**1910.1025(f)** *Respiratory protection*—**(1)** *General.* For employees who use respirators required by this section, the employer must provide respirators that comply with the requirements of this paragraph. Respirators must be used during:

**1910.1025(f)(1)(i)** Periods necessary to install or implement engineering or work-practice controls.

**1910.1025(f)(1)(ii)** Work operations for which engineering and work-practice controls are not sufficient to reduce employee exposures to or below the permissible exposure limit.

**1910.1025(f)(1)(iii)** Periods when an employee requests a respirator.

**1910.1025(f)(2)** *Respirator program.* **(i)** The employer must implement a respiratory protection program in accordance with 29 CFR 1910.134(b) through (d) (except (d)(1)(iii)), and (f) through (m).

**1910.1025(f)(2)(ii)** If an employee has breathing difficulty during fit testing or respirator use, the employer must provide the employee with a medical examination in accordance with paragraph (j)(3)(i)(C) of this section to determine whether or not the employee can use a respirator while performing the required duty.

### TABLE II—RESPIRATORY PROTECTION FOR LEAD AEROSOLS

| Airborne concentration of lead or condition of use | Required respirator |
| --- | --- |
| Not in excess of 0.5 mg/m$^3$ (10X PEL) | Half-mask, air-purifying respirator equipped with high efficiency filters.[2,3] |
| Not in excess of 2.5 mg/m$^3$ (50X PEL) | Full facepiece, air-purifying respirator with high efficiency filters.[3] |
| Not in excess of 50 mg/m$^3$ (1000X PEL) | (1) Any powered, air-purifying respirator with high efficiency filters[3]; or (2) Half-mask supplied-air respirator operated in positive-pressure mode.[2] |
| Not in excess of 100 mg/m$^3$ (2000X PEL) | Supplied-air respirators with full facepiece, hood, helmet, or suit, operated in positive pressure mode. |
| Greater than 100 mg/m$^3$, unknown concentration or fire fighting | Full facepiece, self-contained breathing apparatus operated in positive-pressure mode. |

[1] Respirators specified for high concentrations can be used at lower concentrations of lead.
[2] Full facepiece is required if the lead aerosols cause eye or skin irritation at the use concentrations.
[3] A high efficiency particulate filter means 99.97 percent efficient against 0.3 micron size particles.

**1910.1025(f)(3)** *Respirator selection.* **(i)** The employer must select the appropriate respirator or combination of respirators from Table II of this section.

**1910.1025(f)(3)(ii)** The employer must provide a powered air-purifying respirator instead of the respirator specified in Table II of this section when an employee chooses to use this type of respirator and such a respirator provides adequate protection to the employee.

**1910.1025(g)** *Protective work clothing and equipment*—**(1)** *Provision and use.* If an employee is exposed to lead above the PEL, without regard to the use of respirators or where the possibility of skin or eye irritation exists, the employer shall provide at no cost to the employee and assure that the employee uses appropriate protective work clothing and equipment such as, but not limited to:

**1910.1025(g)(1)(i)** Coveralls or similar full-body work clothing;

**1910.1025(g)(1)(ii)** Gloves, hats, and shoes or disposable shoe coverlets; and

**1910.1025(g)(1)(iii)** Face shields, vented goggles, or other appropriate protective equipment which complies with §1910.133 of this Part.

**1910.1025(g)(2)** *Cleaning and replacement.* **(i)** The employer shall provide the protective clothing required in paragraph (g)(1) of this section in a clean and dry condition at least weekly, and daily to employees whose exposure levels without regard to a respirator are over 200 μg/m$^3$ of lead as an 8-hour TWA.

**1910.1025(g)(2)(ii)** The employer shall provide for the cleaning, laundering, or disposal of protective clothing and equipment required by paragraph (g)(1) of this section.

**1910.1025(g)(2)(iii)** The employer shall repair or replace required protective clothing and equipment as needed to maintain their effectiveness.

**1910.1025(g)(2)(iv)** The employer shall assure that all protective clothing is removed at the completion of a work shift only in change rooms provided for that purpose as prescribed in paragraph (i)(2) of this section.

**1910.1025(g)(2)(v)** The employer shall assure that contaminated protective clothing which is to be cleaned, laundered, or disposed of, is placed in a closed container in the change-room which prevents dispersion of lead outside the container.

**1910.1025(g)(2)(vi)** The employer shall inform in writing any person who cleans or launders protective clothing or equipment of the potentially harmful effects of exposure to lead.

**1910.1025(g)(2)(vii)** The employer shall assure that the containers of contaminated protective clothing and equipment required by paragraph (g)(2)(v) are labelled as follows:

> **CAUTION: CLOTHING CONTAMINATED WITH LEAD. DO NOT REMOVE DUST BY BLOWING OR SHAKING. DISPOSE OF LEAD CONTAMINATED WASH WATER IN ACCORDANCE WITH APPLICABLE LOCAL, STATE, OR FEDERAL REGULATIONS.**

**1910.1025(g)(2)(viii)** The employer shall prohibit the removal of lead from protective clothing or equipment by blowing, shaking, or any other means which disperses lead into the air.

**1910.1025(h)** *Housekeeping*—**(1)** *Surfaces.* All surfaces shall be maintained as free as practicable of accumulations of lead.

**1910.1025(h)(2)** *Cleaning floors.* **(i)** Floors and other surfaces where lead accumulates may not be cleaned by the use of compressed air.

**1910.1025(h)(2)(ii)** Shoveling, dry or wet sweeping, and brushing may be used only where vacuuming or other equally effective methods have been tried and found not to be effective.

**1910.1025(h)(3)** *Vacuuming.* Where vacuuming methods are selected, the vacuums shall be used and emptied in a manner which minimizes the reentry of lead into the workplace.

**1910.1025(i)** *Hygiene facilities and practices.* **(1)** The employer shall assure that in areas where employees are exposed to lead above the PEL, without regard to the use of respirators, food or beverage is not present or consumed, tobacco products are not present or used, and cosmetics are not applied, except in change rooms, lunchrooms, and showers required under paragraphs (i)(2) through (i)(4) of this section.

**1910.1025(i)(2)** *Change rooms.* **(i)** The employer shall provide clean change rooms for employees who work in areas where their airborne exposure to lead is above the PEL, without regard to the use of respirators.

**1910.1025(i)(2)(ii)** The employer shall assure that change rooms are equipped with separate storage facilities for protective work clothing and equipment and for street clothes which prevent cross-contamination.

**1910.1025(i)(3)** *Showers.* **(i)** The employer shall assure that employees who work in areas where their airborne exposure to lead is above the PEL, without regard to the use of respirators, shower at the end of the work shift.

**1910.1025(i)(3)(ii)** The employer shall provide shower facilities in accordance with §1910.141(d)(3) of this part.

**1910.1025(i)(3)(iii)** The employer shall assure that employees who are required to shower pursuant to paragraph (i)(3)(i) do not leave the workplace wearing any clothing or equipment worn during the work shift.

**1910.1025(i)(4)** *Lunchrooms.* **(i)** The employer shall provide lunchroom facilities for employees who work in areas where their airborne exposure to lead is above the PEL, without regard to the use of respirators.

**1910.1025(i)(4)(ii)** The employer shall assure that lunchroom facilities have a temperature controlled, positive pressure, filtered air supply, and are readily accessible to employees.

**1910.1025(i)(4)(iii)** The employer shall assure that employees who work in areas where their airborne exposure to lead is above the PEL without regard to the use of a respirator wash their hands and face prior to eating, drinking, smoking or applying cosmetics.

**1910.1025(i)(4)(iv)** The employer shall assure that employees do not enter lunchroom facilities with protective work clothing or equipment unless surface lead dust has been removed by vacuuming, downdraft booth, or other cleaning method.

**1910.1025(i)(5)** *Lavatories.* The employer shall provide an adequate number of lavatory facilities which comply with §1910.141(d)(1) and (2) of this part.

**1910.1025(j)** **Medical surveillance**—**(1)** *General.* **(i)** The employer shall institute a medical surveillance program for all employees who are or may be exposed above the action level for more than 30 days per year.

**1910.1025(j)(1)(ii)** The employer shall assure that all medical examinations and procedures are performed by or under the supervision of a licensed physician.

**1910.1025(j)(1)(iii)** The employer shall provide the required medical surveillance including multiple physician review under paragraph (j)(3)(iii) without cost to employees and at a reasonable time and place.

**1910.1025(j)(2)** *Biological monitoring*—**(i)** *Blood lead and ZPP level sampling and analysis.* The employer shall make available biological monitoring in the form of blood sampling and analysis for lead and zinc protoporphyrin levels to each employee covered under paragraph (j)(1)(i) of this section on the following schedule:

**1910.1025(j)(2)(i)(A)** At least every 6 months to each employee covered under paragraph (j)(1)(i) of this section;

**1910.1025(j)(2)(i)(B)** At least every two months for each employee whose last blood sampling and analysis indicated a blood lead level at or above 40 µg/100 g of whole blood. This frequency shall continue until two consecutive blood samples and analyses indicate a blood lead level below 40 µg/100 g of whole blood; and

**1910.1025(j)(2)(i)(C)** At least monthly during the removal period of each employee removed from exposure to lead due to an elevated blood lead level.

**1910.1025(j)(2)(ii)** *Follow-up blood sampling tests.* Whenever the results of a blood lead level test indicate that an employee's blood lead level exceeds the numerical criterion for medical removal under paragraph (k)(1)(i)(A) of this section, the employer shall provide a second (follow-up) blood sampling test within two weeks after the employer receives the results of the first blood sampling test.

**1910.1025(j)(2)(iii)** *Accuracy of blood lead level sampling and analysis.* Blood lead level sampling and analysis provided pursuant to this section shall have an accuracy (to a confidence level of 95 percent) within plus or minus 15 percent or 6 µg/100ml, whichever is greater, and shall be conducted by a laboratory licensed by the Center for Disease Control, United States Department of Health, Education and Welfare (CDC) or which has received a satisfactory grade in blood lead proficiency testing from CDC in the prior twelve months.

**1910.1025(j)(2)(iv)** *Employee notification.* Within five working days after the receipt of biological monitoring results, the employer shall notify in writing each employee whose blood lead level exceeds 40 µg/100 g: (A) of that employee's blood lead level and (B) that the standard requires temporary medical removal with Medical Removal Protection benefits when an employee's blood lead level exceeds the numerical criterion for medical removal under paragraph (k)(1)(i) of this section.

**1910.1025(j)(3)** *Medical examinations and consultations*—**(i)** *Frequency.* The employer shall make available medical examinations and consultations to each employee covered under paragraph (j)(1)(i) of this section on the following schedule:

**1910.1025(j)(3)(i)(A)** At least annually for each employee for whom a blood sampling test conducted at any time during the preceding 12 months indicated a blood lead level at or above 40 µg/100 g;

**1910.1025(j)(3)(i)(B)** Prior to assignment for each employee being assigned for the first time to an area in which airborne concentrations of lead are at or above the action level;

**1910.1025(j)(3)(i)(C)** As soon as possible, upon notification by an employee either that the employee has developed signs or symptoms commonly associated with lead intoxication, that the employee desires medical advice concerning the effects of current or past exposure to lead on the employee's ability to procreate a healthy child, or that the employee has demonstrated difficulty in breathing during a respirator fitting test or during use; and

**1910.1025(j)(3)(i)(D)** As medically appropriate for each employee either removed from exposure to lead due to a risk of sustaining material impairment to health, or otherwise limited pursuant to a final medical determination.

**1910.1025(j)(3)(ii)** *Content.* Medical examinations made available pursuant to paragraph (j)(3)(i) (A) through (B) of this section shall include the following elements:

**1910.1025(j)(3)(ii)(A)** A detailed work history and a medical history, with particular attention to past lead exposure (occupational and non-occupational), personal habits (smoking, hygiene), and past gastrointestinal, hematologic, renal, cardiovascular, reproductive and neurological problems;

**1910.1025(j)(3)(ii)(B)** A thorough physical examination, with particular attention to teeth, gums, hematologic, gastrointestinal, renal, cardiovascular, and neurological systems. Pulmonary status should be evaluated if respiratory protection will be used;

**1910.1025(j)(3)(ii)(C)** A blood pressure measurement;

**1910.1025(j)(3)(ii)(D)** A blood sample and analysis which determines:

**1910.1025(j)(3)(ii)(D)***(1)* Blood lead level;

**1910.1025(j)(3)(ii)(D)***(2)* Hemoglobin and hematocrit determinations, red cell indices, and examination of peripheral smear morphology;

**1910.1025(j)(3)(ii)(D)***(3)* Zinc protoporphyrin;

**1910.1025(j)(3)(ii)(D)***(4)* Blood urea nitrogen; and,

**1910.1025(j)(3)(ii)(D)***(5)* Serum creatinine;

**1910.1025(j)(3)(ii)(E)** A routine urinalysis with microscopic examination; and

**1910.1025(j)(3)(ii)(F)** Any laboratory or other test which the examining physician deems necessary by sound medical practice.

The content of medical examinations made available pursuant to paragraph (j)(3)(i) (C) through (D) of this section shall be determined by an examining physician and, if requested by an employee, shall include pregnancy testing or laboratory evaluation of male fertility.

**1910.1025(j)(3)(iii)** *Multiple physician review mechanism.* **(A)** If the employer selects the initial physician who conducts any medical examination or consultation provided to an employee under this section, the employee may designate a second physician:

**1910.1025(j)(3)(iii)(A)***(1)* To review any findings, determinations or recommendations of the initial physician; and

**1910.1025(j)(3)(iii)(A)***(2)* To conduct such examinations, consultations, and laboratory tests as the second physician deems necessary to facilitate this review.

**1910.1025(j)(3)(iii)(B)** The employer shall promptly notify an employee of the right to seek a second medical opinion after each occasion that an initial physician conducts a medical examination or consultation pursuant to this section. The employer may condition its participation in, and payment for, the multiple physician review mechanism upon the employee doing the following within fifteen (15) days after receipt of the foregoing notification, or receipt of the initial physician's written opinion, whichever is later:

**1910.1025(j)(3)(iii)(B)(1)** The employee informing the employer that he or she intends to seek a second medical opinion, and

**1910.1025(j)(3)(iii)(B)(2)** The employee initiating steps to make an appointment with a second physician.

**1910.1025(j)(3)(iii)(C)** If the findings, determinations or recommendations of the second physician differ from those of the initial physician, then the employer and the employee shall assure that efforts are made for the two physicians to resolve any disagreement.

**1910.1025(j)(3)(iii)(D)** If the two physicians have been unable to quickly resolve their disagreement, then the employer and the employee through their respective physicians shall designate a third physician:

**1910.1025(j)(3)(iii)(D)(1)** To review any findings, determinations or recommendations of the prior physicians; and

**1910.1025(j)(3)(iii)(D)(2)** To conduct such examinations, consultations, laboratory tests and discussions with the prior physicians as the third physician deems necessary to resolve the disagreement of the prior physicians.

**1910.1025(j)(3)(iii)(E)** The employer shall act consistent with the findings, determinations and recommendations of the third physician, unless the employer and the employee reach an agreement which is otherwise consistent with the recommendations of at least one of the three physicians.

**1910.1025(j)(3)(iv)** *Information provided to examining and consulting physicians.* **(A)** The employer shall provide an initial physician conducting a medical examination or consultation under this section with the following information:

**1910.1025(j)(3)(iv)(A)(1)** A copy of this regulation for lead including all Appendices;

**1910.1025(j)(3)(iv)(A)(2)** A description of the affected employee's duties as they relate to the employee's exposure;

**1910.1025(j)(3)(iv)(A)(3)** The employee's exposure level or anticipated exposure level to lead and to any other toxic substance (if applicable);

**1910.1025(j)(3)(iv)(A)(4)** A description of any personal protective equipment used or to be used;

**1910.1025(j)(3)(iv)(A)(5)** Prior blood lead determinations; and

**1910.1025(j)(3)(iv)(A)(6)** All prior written medical opinions concerning the employee in the employer's possession or control.

**1910.1025(j)(3)(iv)(B)** The employer shall provide the foregoing information to a second or third physician conducting a medical examination or consultation under this section upon request either by the second or third physician, or by the employee.

**1910.1025(j)(3)(v)** *Written medical opinions.* **(A)** The employer shall obtain and furnish the employee with a copy of a written medical opinion from each examining or consulting physician which contains the following information:

**1910.1025(j)(3)(v)(A)(1)** The physician's opinion as to whether the employee has any detected medical condition which would place the employee at increased risk of material impairment of the employee's health from exposure to lead;

**1910.1025(j)(3)(v)(A)(2)** Any recommended special protective measures to be provided to the employee, or limitations to be placed upon the employee's exposure to lead;

**1910.1025(j)(3)(v)(A)(3)** Any recommended limitation upon the employee's use of respirators, including a determination of whether the employee can wear a powered air purifying respirator if a physician determines that the employee cannot wear a negative pressure respirator; and

**1910.1025(j)(3)(v)(A)(4)** The results of the blood lead determinations.

**1910.1025(j)(3)(v)(B)** The employer shall instruct each examining and consulting physician to:

**1910.1025(j)(3)(v)(B)(1)** Not reveal either in the written opinion, or in any other means of communication with the employer, findings, including laboratory results, or diagnoses unrelated to an employee's occupational exposure to lead; and

**1910.1025(j)(3)(v)(B)(2)** Advise the employee of any medical condition, occupational or nonoccupational, which dictates further medical examination or treatment.

**1910.1025(j)(3)(vi)** *Alternate Physician Determination Mechanisms.* The employer and an employee or authorized employee representative may agree upon the use of any expeditious alternate physician determination mechanism in lieu of the multiple physician review mechanism provided by this paragraph so long as the alternate mechanism otherwise satisfies the requirements contained in this paragraph.

**1910.1025(j)(4)** *Chelation.* **(i)** The employer shall assure that any person whom he retains, employs, supervises or controls does not engage in prophylactic chelation of any employee at any time.

**1910.1025(j)(4)(ii)** If therapeutic or diagnostic chelation is to be performed by any person in paragraph (j)(4)(i), the employer shall assure that it be done under the supervision of a licensed physician in a clinical setting with thorough and appropriate medical monitoring and that the employee is notified in writing prior to its occurrence.

**1910.1025(k)** **Medical Removal Protection**—**(1)** *Temporary medical removal and return of an employee*—**(i)** *Temporary removal due to elevated blood lead levels.* **(A)** The employer shall remove an employee from work having an exposure to lead at or above the action level on each occasion that a periodic and a follow-up blood sampling test conducted pursuant to this section indicate that the employee's blood lead level is at or above 60 µg/100 g of whole blood; and

**1910.1025(k)(1)(i)(B)** The employer shall remove an employee from work having an exposure to lead at or above the action level on each occasion that the average of the last three blood sampling tests conducted pursuant to this section (or the average of all blood sampling tests conducted over the previous six (6) months, whichever is longer) indicates that the employee's blood lead level is at or above 50 µg/100 g of whole blood; provided, however, that an employee need not be removed if the last blood sampling test indicates a blood lead level at or below 40 µg/100 g of whole blood.

**1910.1025(k)(1)(ii)** *Temporary removal due to a final medical determination.* **(A)** The employer shall remove an employee from work having an exposure to lead at or above the action level on each occasion that a final medical determination results in a medical finding, determination, or opinion that the employee has a detected medical condition which places the employee at increased risk of material impairment to health from exposure to lead.

**1910.1025(k)(1)(ii)(B)** For the purposes of this section, the phrase "final medical determination" shall mean the outcome of the multiple physician review mechanism or alternate medical determination mechanism used pursuant to the medical surveillance provisions of this section.

**1910.1025(k)(1)(ii)(C)** Where a final medical determination results in any recommended special protective measures for an employee, or limitations on an employee's exposure to lead, the employer shall implement and act consistent with the recommendation.

**1910.1025(k)(1)(iii)** *Return of the employee to former job status.* **(A)** The employer shall return an employee to his or her former job status:

**1910.1025(k)(1)(iii)(A)(1)** For an employee removed due to a blood lead level at or above 60 µg/100 g, or due to an average blood lead level at or above 50 µg/100 g, when two consecutive blood sampling tests indicate that the employee's blood lead level is at or below 40 µg/100 g of whole blood;

**1910.1025(k)(1)(iii)(A)(2)** For an employee removed due to a final medical determination, when a subsequent final medical determination results in a medical finding, determination, or opinion that the employee no longer has a detected medical condition which places the employee at increased risk of material impairment to health from exposure to lead.

**1910.1025(k)(1)(iii)(B)** For the purposes of this section, the requirement that an employer return an employee to his or her former job status is not intended to expand upon or restrict any rights an employee has or would have had, absent temporary medical removal, to a specific job classification or position under the terms of a collective bargaining agreement.

**1910.1025(k)(1)(iv)** *Removal of other employee special protective measure or limitations.* The employer shall remove any limitations placed on an employee or end any special protective measures provided to an employee pursuant to a final medical determination when a subsequent final medical determination indicates that the limitations or special protective measures are no longer necessary.

**1910.1025(k)(1)(v)** *Employer options pending a final medical determination.* Where the multiple physician review mechanism, or alternate medical determination mechanism used pursuant to the medical surveillance provisions of this section, has not yet resulted in a final medical determination with respect to an employee, the employer shall act as follows:

**1910.1025(k)(1)(v)(A)** *Removal.* The employer may remove the employee from exposure to lead, provide special protective measures to the employee, or place limitations upon the employee, consistent with the medical findings, determinations, or recommendations of any of the physicians who have reviewed the employee's health status.

**1910.1025(k)(1)(v)(B)** *Return.* The employer may return the employee to his or her former job status, end any special protective measures provided to the employee, and remove any limitations placed upon the employee, consistent with the medical findings, determinations, or recommendations of any of the physicians who have reviewed the employee's health status, with two exceptions. If

**1910.1025(k)(1)(v)(B)(1)** the initial removal, special protection, or limitation of the employee resulted from a final medical determination which differed from the findings, determinations, or recommendations of the initial physician or

**1910.1025(k)(1)(v)(B)(2)** The employee has been on removal status for the preceding eighteen months due to an elevated blood lead level, then the employer shall await a final medical determination.

**1910.1025(k)(2)** *Medical removal protection benefits—(i) Provision of medical removal protection benefits.* The employer shall provide to an employee up to eighteen (18) months of medical removal protection benefits on each occasion that an employee is removed from exposure to lead or otherwise limited pursuant to this section.

**1910.1025(k)(2)(ii)** *Definition of medical removal protection benefits.* For the purposes of this section, the requirement that an employer provide medical removal protection benefits means that the employer shall maintain the earnings, seniority and other employment rights and benefits of an employee as though the employee had not been removed from normal exposure to lead or otherwise limited.

**1910.1025(k)(2)(iii)** *Follow-up medical surveillance during the period of employee removal or limitation.* During the period of time that an employee is removed from normal exposure to lead or otherwise limited, the employer may condition the provision of medical removal protection benefits upon the employee's participation in follow-up medical surveillance made available pursuant to this section.

**1910.1025(k)(2)(iv)** *Workers' compensation claims.* If a removed employee files a claim for workers' compensation payments for a lead-related disability, then the employer shall continue to provide medical removal protection benefits pending disposition of the claim. To the extent that an award is made to the employee for earnings lost during the period of removal, the employer's medical removal protection obligation shall be reduced by such amount. The employer shall receive no credit for workers' compensation payments received by the employee for treatment related expenses.

**1910.1025(k)(2)(v)** *Other credits.* The employer's obligation to provide medical removal protection benefits to a removed employee shall be reduced to the extent that the employee receives compensation for earnings lost during the period of removal either from a publicly or employer-funded compensation program, or receives income from employment with another employer made possible by virtue of the employee's removal.

**1910.1025(k)(2)(vi)** *Employees whose blood lead levels do not adequately decline within 18 months of removal.* The employer shall take the following measures with respect to any employee removed from exposure to lead due to an elevated blood lead level whose blood lead level has not declined within the past eighteen (18) months of removal so that the employee has been returned to his or her former job status:

**1910.1025(k)(2)(vi)(A)** The employer shall make available to the employee a medical examination pursuant to this section to obtain a final medical determination with respect to the employee;

**1910.1025(k)(2)(vi)(B)** The employer shall assure that the final medical determination obtained indicates whether or not the employee may be returned to his or her former job status, and if not, what steps should be taken to protect the employee's health;

**1910.1025(k)(2)(vi)(C)** Where the final medical determination has not yet been obtained, or once obtained indicates that the employee may not yet be returned to his or her former job status, the employer shall continue to provide medical removal protection benefits to the employee until either the employee is returned to former job status, or a final medical determination is made that the employee is incapable of ever safely returning to his or her former job status.

**1910.1025(k)(2)(vi)(D)** Where the employer acts pursuant to a final medical determination which permits the return of the employee to his or her former job status despite what would otherwise be an unacceptable blood lead level, later questions concerning removing the employee again shall be decided by a final medical determination. The employer need not automatically remove such an employee pursuant to the blood lead level removal criteria provided by this section.

**1910.1025(k)(2)(vii)** *Voluntary Removal or Restriction of An Employee.* Where an employer, although not required by this section to do so, removes an employee from exposure to lead or otherwise places limitations on an employee due to the effects of lead exposure on the employee's medical condition, the employer shall provide medical removal protection benefits to the employee equal to that required by paragraph (k)(2)(i) of this section.

**1910.1025(l)** *Employee information and training—(1) Training program.* (i) Each employer who has a workplace in which there is a potential exposure to airborne lead at any level shall inform employees of the content of Appendices A and B of this regulation.

**1910.1025(l)(1)(ii)** The employer shall institute a training program for and assure the participation of all employees who are subject to exposure to lead at or above the action level or for whom the possibility of skin or eye irritation exists.

**1910.1025(l)(1)(iii)** The employer shall provide initial training by 180 days from the effective date for those employees covered by paragraph (l)(1) (ii) on the standard's effective date and prior to the time of initial job assignment for those employees subsequently covered by this paragraph.

**1910.1025(l)(1)(iv)** The training program shall be repeated at least annually for each employee.

**1910.1025(l)(1)(v)** The employer shall assure that each employee is informed of the following:

**1910.1025(l)(1)(v)(A)** The content of this standard and its appendices;

**1910.1025(l)(1)(v)(B)** The specific nature of the operations which could result in exposure to lead above the action level;

**1910.1025(l)(1)(v)(C)** The purpose, proper selection, fitting, use, and limitations of respirators;

**1910.1025(l)(1)(v)(D)** The purpose and a description of the medical surveillance program, and the medical removal protection program including information concerning the adverse health effects associated with excessive exposure to lead (with particular attention to the adverse reproductive effects on both males and females);

**1910.1025(l)(1)(v)(E)** The engineering controls and work practices associated with the employee's job assignment;

**1910.1025(l)(1)(v)(F)** The contents of any compliance plan in effect; and

**1910.1025(l)(1)(v)(G)** Instructions to employees that chelating agents should not routinely be used to remove lead from their bodies and should not be used at all except under the direction of a licensed physician;

**1910.1025(l)(2)** *Access to information and training materials.* (i) The employer shall make readily available to all affected employees a copy of this standard and its appendices.

**1910.1025(l)(2)(ii)** The employer shall provide, upon request, all materials relating to the employee information and training program to the Assistant Secretary and the Director.

**1910.1025(l)(2)(iii)** In addition to the information required by paragraph (l)(1)(v), the employer shall include as part of the training program, and shall distribute to employees, any materials pertaining to the Occupational Safety and Health Act, the regulations issued pursuant to that Act, and this lead standard, which are made available to the employer by the Assistant Secretary.

**1910.1025(m)** *Signs—(1) General.* (i) The employer may use signs required by other statutes, regulations or ordinances in addition to, or in combination with, signs required by this paragraph.

**1910.1025(m)(1)(ii)** The employer shall assure that no statement appears on or near any sign required by this paragraph which contradicts or detracts from the meaning of the required sign.

**1910.1025(m)(2)** *Signs.* (i) The employer shall post the following warning signs in each work area where the PEL is exceeded:

# OSHA Standards for General Industry

**WARNING**

**LEAD WORK AREA**

**POISON**

**NO SMOKING OR EATING**

**1910.1025(m)(2)(ii)** The employer shall assure that signs required by this paragraph are illuminated and cleaned as necessary so that the legend is readily visible.

**1910.1025(n)** *Recordkeeping*—**(1)** *Exposure monitoring.* **(i)** The employer shall establish and maintain an accurate record of all monitoring required in paragraph (d) of this section.

**1910.1025(n)(1)(ii)** This record shall include:

**1910.1025(n)(1)(ii)(A)** The date(s), number, duration, location and results of each of the samples taken, including a description of the sampling procedure used to determine representative employee exposure where applicable;

**1910.1025(n)(1)(ii)(B)** A description of the sampling and analytical methods used and evidence of their accuracy;

**1910.1025(n)(1)(ii)(C)** The type of respiratory protective devices worn, if any;

**1910.1025(n)(1)(ii)(D)** Name, social security number, and job classification of the employee monitored and of all other employees whose exposure the measurement is intended to represent; and

**1910.1025(n)(1)(ii)(E)** The environmental variables that could affect the measurement of employee exposure.

**1910.1025(n)(1)(iii)** The employer shall maintain these monitoring records for at least 40 years or for the duration of employment plus 20 years, whichever is longer.

**1910.1025(n)(2)** *Medical surveillance.* **(i)** The employer shall establish and maintain an accurate record for each employee subject to medical surveillance as required by paragraph (j) of this section.

**1910.1025(n)(2)(ii)** This record shall include:

**1910.1025(n)(2)(ii)(A)** The name, social security number, and description of the duties of the employee;

**1910.1025(n)(2)(ii)(B)** A copy of the physician's written opinions;

**1910.1025(n)(2)(ii)(C)** Results of any airborne exposure monitoring done for that employee and the representative exposure levels supplied to the physician; and

**1910.1025(n)(2)(ii)(D)** Any employee medical complaints related to exposure to lead.

**1910.1025(n)(2)(iii)** The employer shall keep, or assure that the examining physician keeps, the following medical records:

**1910.1025(n)(2)(iii)(A)** A copy of the medical examination results including medical and work history required under paragraph (j) of this section;

**1910.1025(n)(2)(iii)(B)** A description of the laboratory procedures and a copy of any standards or guidelines used to interpret the test results or references to that information;

**1910.1025(n)(2)(iii)(C)** A copy of the results of biological monitoring.

**1910.1025(n)(2)(iv)** The employer shall maintain or assure that the physician maintains those medical records for at least 40 years, or for the duration of employment plus 20 years, whichever is longer.

**1910.1025(n)(3)** *Medical removals.* **(i)** The employer shall establish and maintain an accurate record for each employee removed from current exposure to lead pursuant to paragraph (k) of this section.

**1910.1025(n)(3)(ii)** Each record shall include:

**1910.1025(n)(3)(ii)(A)** The name and social security number of the employee;

**1910.1025(n)(3)(ii)(B)** The date on each occasion that the employee was removed from current exposure to lead as well as the corresponding date on which the employee was returned to his or her former job status;

**1910.1025(n)(3)(ii)(C)** A brief explanation of how each removal was or is being accomplished; and

**1910.1025(n)(3)(ii)(D)** A statement with respect to each removal indicating whether or not the reason for the removal was an elevated blood lead level.

**1910.1025(n)(3)(iii)** The employer shall maintain each medical removal record for at least the duration of an employee's employment.

**1910.1025(n)(4)** *Availability.* **(i)** The employer shall make available upon request all records required to be maintained by paragraph (n) of this section to the Assistant Secretary and the Director for examination and copying.

**1910.1025(n)(4)(ii)** Environmental monitoring, medical removal, and medical records required by this paragraph shall be provided upon request to employees, designated representatives, and the Assistant Secretary in accordance with 29 CFR 1910.1020(a)-(e) and (2)-(i). Medical removal records shall be provided in the same manner as environmental monitoring records.

**1910.1025(n)(5)** *Transfer of records.* **(i)** Whenever the employer ceases to do business, the successor employer shall receive and retain all records required to be maintained by paragraph (n) of this section.

**1910.1025(n)(5)(ii)** Whenever the employer ceases to do business and there is no successor employer to receive and retain the records required to be maintained by this section for the prescribed period, these records shall be transmitted to the Director.

**1910.1025(n)(5)(iii)** At the expiration of the retention period for the records required to be maintained by this section, the employer shall notify the Director at least 3 months prior to the disposal of such records and shall transmit those records to the Director if requested within the period.

**1910.1025(n)(5)(iv)** The employer shall also comply with any additional requirements involving transfer of records set forth in 29 CFR 1910.1020(h).

**1910.1025(o)** *Observation of monitoring.* **(1)** *Employee observation.* The employer shall provide affected employees or their designated representatives an opportunity to observe any monitoring of employee exposure to lead conducted pursuant to paragraph (d) of this section.

**1910.1025(o)(2)** *Observation procedures.* **(i)** Whenever observation of the monitoring of employee exposure to lead requires entry into an area where the use of respirators, protective clothing or equipment is required, the employer shall provide the observer with and assure the use of such respirators, clothing and such equipment, and shall require the observer to comply with all other applicable safety and health procedures.

**1910.1025(o)(2)(ii)** Without interfering with the monitoring, observers shall be entitled to:

**1910.1025(o)(2)(ii)(A)** Receive an explanation of the measurement procedures;

**1910.1025(o)(2)(ii)(B)** Observe all steps related to the monitoring of lead performed at the place of exposure; and

**1910.1025(o)(2)(ii)(C)** Record the results obtained or receive copies of the results when returned by the laboratory.

**1910.1025(p)** *Appendices.* The information contained in the appendices to this section is not intended by itself, to create any additional obligations not otherwise imposed by this standard nor detract from any existing obligation.

## APPENDIX A TO §1910.1025— SUBSTANCE DATA SHEET FOR OCCUPATIONAL EXPOSURE TO LEAD

### I. SUBSTANCE IDENTIFICATION

A. *Substance:* Pure lead (Pb) is a heavy metal at room temperature and pressure and is a basic chemical element. It can combine with various other substances to form numerous lead compounds.

B. *Compounds Covered by the Standard:* The word "lead" when used in this standard means elemental lead, all inorganic lead compounds and a class of organic lead compounds called lead soaps. This standard does not apply to other organic lead compounds.

C. *Uses:* Exposure to lead occurs in at least 120 different occupations, including primary and secondary lead smelting, lead storage battery manufacturing, lead pigment manufacturing and use, solder manufacturing and use, shipbuilding and ship repairing, auto manufacturing, and printing.

D. *Permissible Exposure:* The Permissible Exposure Limit (PEL) set by the standard is 50 micrograms of lead per cubic meter of air (50 µg/m$^3$), averaged over an 8-hour workday.

E. *Action Level:* The standard establishes an action level of 30 micrograms per cubic meter of air (30 µg/m³), time weighted average, based on an 8-hour work-day. The action level initiates several requirements of the standard, such as exposure monitoring, medical surveillance, and training and education.

## II. HEALTH HAZARD DATA

A. *Ways in which lead enters your body.* When absorbed into your body in certain doses lead is a toxic substance. The object of the lead standard is to prevent absorption of harmful quantities of lead. The standard is intended to protect you not only from the immediate toxic effects of lead, but also from the serious toxic effects that may not become apparent until years of exposure have passed.

Lead can be absorbed into your body by inhalation (breathing) and ingestion (eating). Lead (except for certain organic lead compounds not covered by the standard, such as tetraethyl lead) is not absorbed through your skin. When lead is scattered in the air as a dust, fume or mist it can be inhaled and absorbed through you lungs and upper respiratory tract. Inhalation of airborne lead is generally the most important source of occupational lead absorption. You can also absorb lead through your digestive system if lead gets into your mouth and is swallowed. If you handle food, cigarettes, chewing tobacco, or make-up which have lead on them or handle them with hands contaminated with lead, this will contribute to ingestion.

A significant portion of the lead that you inhale or ingest gets into your blood stream. Once in your blood stream, lead is circulated throughout your body and stored in various organs and body tissues. Some of this lead is quickly filtered out of your body and excreted, but some remains in the blood and other tissues. As exposure to lead continues, the amount stored in your body will increase if you are absorbing more lead than your body is excreting. Even though you may not be aware of any immediate symptoms of disease, this lead stored in your tissues can be slowly causing irreversible damage, first to individual cells, then to your organs and whole body systems.

B. *Effects of overexposure to lead*—(1) *Short term (acute) overexposure.* Lead is a potent, systemic poison that serves no known useful function once absorbed by your body. Taken in large enough doses, lead can kill you in a matter of days. A condition affecting the brain called acute encephalopathy may arise which develops quickly to seizures, coma, and death from cardiorespiratory arrest. A short term dose of lead can lead to acute encephalopathy. Short term occupational exposures of this magnitude are highly unusual, but not impossible. Similar forms of encephalopathy may, however, arise from extended, chronic exposure to lower doses of lead. There is no sharp dividing line between rapidly developing acute effects of lead, and chronic effects which take longer to acquire. Lead adversely affects numerous body systems, and causes forms of health impairment and disease which arise after periods of exposure as short as days or as long as several years.

(2) *Long-term (chronic) overexposure.* Chronic overexposure to lead may result in severe damage to your blood-forming, nervous, urinary and reproductive systems. Some common symptoms of chronic overexposure include loss of appetite, metallic taste in the mouth, anxiety, constipation, nausea, pallor, excessive tiredness, weakness, insomnia, headache, nervous irritability, muscle and joint pain or soreness, fine tremors, numbness, dizziness, hyperactivity and colic. In lead colic there may be severe abdominal pain.

Damage to the central nervous system in general and the brain (encephalopathy) in particular is one of the most severe forms of lead poisoning. The most severe, often fatal, form of encephalopathy may be preceded by vomiting, a feeling of dullness progressing to drowsiness and stupor, poor memory, restlessness, irritability, tremor, and convulsions. It may arise suddenly with the onset of seizures, followed by coma, and death. There is a tendency for muscular weakness to develop at the same time. This weakness may progress to paralysis often observed as a characteristic "wrist drop" or "foot drop" and is a manifestation of a disease to the nervous system called peripheral neuropathy.

Chronic overexposure to lead also results in kidney disease with few, if any, symptoms appearing until extensive and most likely permanent kidney damage has occurred. Routine laboratory tests reveal the presence of this kidney disease only after about two-thirds of kidney function is lost. When overt symptoms of urinary dysfunction arise, it is often too late to correct or prevent worsening conditions, and progression to kidney dialysis or death is possible.

Chronic overexposure to lead impairs the reproductive systems of both men and women. Overexposure to lead may result in decreased sex drive, impotence and sterility in men. Lead can alter the structure of sperm cells raising the risk of birth defects. There is evidence of miscarriage and stillbirth in women whose husbands were exposed to lead or who were exposed to lead themselves. Lead exposure also may result in decreased fertility, and abnormal menstrual cycles in women. The course of pregnancy may be adversely affected by exposure to lead since lead crosses the placental barrier and poses risks to developing fetuses. Children born of parents either one of whom were exposed to excess lead levels are more likely to have birth defects, mental retardation, behavioral disorders or die during the first year of childhood.

Overexposure to lead also disrupts the blood-forming system resulting in decreased hemoglobin (the substance in the blood that carries oxygen to the cells) and ultimately anemia. Anemia is characterized by weakness, pallor and fatigability as a result of decreased oxygen carrying capacity in the blood.

(3) *Health protection goals of the standard.* Prevention of adverse health effects for most workers from exposure to lead throughout a working lifetime requires that worker blood lead (PbB) levels be maintained at or below forty micrograms per one hundred grams of whole blood (40 µg/100 g). The blood lead levels of workers (both male and female workers) who intend to have children should be maintained below 30 µg/100 g to minimize adverse reproductive health effects to the parents and to the developing fetus.

The measurement of your blood lead level is the most useful indicator of the amount of lead being absorbed by your body. Blood lead levels (PbB) are most often reported in units of milligrams (mg) or micrograms (µg) of lead (1 mg=1000 µg) per 100 grams (100 g), 100 milliters (100 ml) or deciliter (dl) of blood. These three units are essentially the same. Sometime PbB's are expressed in the form of mg% or µg%. This is a shorthand notation for 100g, 100 ml, or dl.

PbB measurements show the amount of lead circulating in your blood stream, but do not give any information about the amount of lead stored in your various tissues. PbB measurements merely show current absorption of lead, not the effect that lead is having on your body or the effects that past lead exposure may have already caused. Past research into lead-related diseases, however, has focused heavily on associations between PbBs and various diseases. As a result, your PbB is an important indicator of the likelihood that you will gradually acquire a lead-related health impairment or disease.

Once your blood lead level climbs above 40 µg/100 g, your risk of disease increases. There is a wide variability of individual response to lead, thus it is difficult to say that a particular PbB in a given person will cause a particular effect. Studies have associated fatal encephalopathy with PbBs as low as 150 µg/100 g. Other studies have shown other forms of diseases in some workers with PbBs well below 80 µg/100 g. Your PbB is a crucial indicator of the risks to your health, but one other factor is also extremely important. This factor is the length of time you have had elevated PbBs. The longer you have an elevated PbB, the greater the risk that large quantities of lead are being gradually stored in your organs and tissues (body burden). The greater your overall body burden, the greater the chances of substantial permanent damage.

The best way to prevent all forms of lead-related impairments and diseases—both short term and long term- is to maintain your PbB below 40 µg/100 g. The provisions of the standard are designed with this end in mind. Your employer has prime responsibility to assure that the provisions of the standard are complied with both by the company and by individual workers. You as a worker, however, also have a responsibility to assist your employer in complying with the standard. You can play a key role in protecting your own health by learning about the lead hazards and their control, learning what the standard requires, following the standard where it governs your own actions, and seeing that your employer complies with provisions governing his actions.

(4) *Reporting signs and symptoms of health problems.* You should immediately notify your employer if you develop signs or symptoms associated with lead poisoning or if you desire medical advice concerning the effects of current or past exposure to lead on your ability to have a healthy child. You should also notify your employer if you have difficulty breathing during a respirator fit test or while wearing a respirator. In each of these cases your employer must make available to you appropriate medical examinations or consultations. These must be provided at no cost to you and at a reasonable time and place.

The standard contains a procedure whereby you can obtain a second opinion by a physician of your choice if the employer selected the initial physician.

# OSHA Standards for General Industry

## APPENDIX B TO §1910.1025— EMPLOYEE STANDARD SUMMARY

This appendix summarizes key provisions of the standard that you as a worker should become familiar with.

### I. PERMISSIBLE EXPOSURE LIMIT (PEL)— PARAGRAPH (c)

The standards sets a permissible exposure limit (PEL) of fifty micrograms of lead per cubic meter of air (50 µg/m³), averaged over an 8-hour workday. This is the highest level of lead in air to which you may be permissibly exposed over an 8-hour workday. Since it is an 8-hour average it permits short exposures above the PEL so long as for each 8-hour work day your average exposure does not exceed the PEL.

This standard recognizes that your daily exposure to lead can extend beyond a typical 8-hour workday as the result of overtime or other alterations in your work schedule. To deal with this, the standard contains a formula which reduces your permissible exposure when you are exposed more than 8 hours. For example, if you are exposed to lead for 10 hours a day, the maximum permitted average exposure would be 40 µg/m³.

### II. EXPOSURE MONITORING—PARAGRAPH (d)

If lead is present in the workplace where you work in any quantity, your employer is required to make an initial determination of whether the action level is exceeded for any employee. This initial determination must include instrument monitoring of the air for the presence of lead and must cover the exposure of a representative number of employees who are reasonably believed to have the highest exposure levels. If your employer has conducted appropriate air sampling for lead in the past year he may use these results. If there have been any employee complaints of symptoms which may be attributable to exposure to lead or if there is any other information or observations which would indicate employee exposure to lead, this must also be considered as part of the initial determination. This initial determination must have been completed by March 31, 1979. If this initial determination shows that a reasonable possibility exists that *any* employee may be exposed, without regard to respirators, over the action level (30 µg/m³) your employer must set up an air monitoring program to determine the exposure level of every employee exposed to lead at your workplace.

In carrying out this air monitoring program, your employer is not required to monitor the exposure of every employee, but he must monitor a representative number of employees and job types. Enough sampling must be done to enable each employee's exposure level to be reasonably least one full shift (at least 7 hours) air sample. In addition, these air samples must be taken under conditions which represent each employee's *regular*, daily exposure to lead. All initial exposure monitoring must have been completed by May 30, 1979.

If you are exposed to lead and air sampling is performed, your employer is required to quickly notify you in writing of air monitoring results which represent your exposure. If the results indicate your exposure exceeds the PEL (without regard to your use of respirators), then your employer must also notify you of this in writing, and provide you with a description of the corrective action that will be taken to reduce your exposure.

Your exposure must be rechecked by monitoring every six months if your exposure is over the action level but below the PEL. Air monitoring must be repeated every 3 months if you are exposed over the PEL. Your employer may discontinue monitoring for you if 2 consecutive measurements, taken at least two weeks apart, are below the action level. However, whenever there is a production, process, control, or personnel change at your workplace which may result in new or additional exposure to lead, or whenever there is any other reason to suspect a change which may result in new or additional exposure to lead, your employer must perform additional monitoring.

### III. METHODS OF COMPLIANCE—PARAGRAPH (e)

Your employer is required to assure that no employee is exposed to lead in excess of the PEL. The standard establishes a priority of methods to be used to meet the PEL.

### IV. RESPIRATORY PROTECTION—PARAGRAPH (f)

Your employer is required to provide and assure your use of respirators when your exposure to lead is not controlled below the PEL by other means. The employer must pay the cost of the respirator. Whenever you request one, your employer is also required to provide you a respirator even if your air exposure level does not exceed the PEL. You might desire a respirator when, for example, you have received medical advice that your lead absorption should be decreased. Or, you may intend to have children in the near future, and want to reduce the level of lead in your body to minimize adverse reproductive effects. While respirators are the least satisfactory means of controlling your exposure, they are capable of providing significant protection if properly chosen, fitted, worn, cleaned, maintained, and replaced when they stop providing adequate protection.

Your employer is required to select respirators from the seven types listed in Table II of the Respiratory Protection section of the standard (§1910.1025(f)). Any respirator chosen must be approved by the National Institute for Occupational Safety and Health (NIOSH) under the provisions of 42 CFR part 84. This respirator selection table will enable your employer to choose a type of respirator that will give you a proper amount of protection based on your airborne lead exposure. Your employer may select a type of respirator that provides greater protection than that required by the standard; that is, one recommended for a higher concentration of lead than is present in your workplace. For example, a powered air-purifying respirator (PAPR) is much more protective than a typical negative pressure respirator, and may also be more comfortable to wear. A PAPR has a filter, cartridge, or canister to clean the air, and a power source that continuously blows filtered air into your breathing zone. Your employer might make a PAPR available to you to ease the burden of having to wear a respirator for long periods of time. The standard provides that you can obtain a PAPR upon request.

Your employer must also start a Respiratory Protection Program. This program must include written procedures for the proper selection, use, cleaning, storage, and maintenance of respirators.

Your employer must ensure that your respirator facepiece fits properly. Proper fit of a respirator facepiece is critical to your protection from airborne lead. Obtaining a proper fit on each employee may require your employer to make available several different types of respirator masks. To ensure that your respirator fits properly and that facepiece leakage is minimal, your employer must give you either a qualitative or quantitative fit test as specified in Appendix A of the Respiratory Protection standard located at 29 CFR 1910.134.

You must also receive from your employer proper training in the use of respirators. Your employer is required to teach you how to wear a respirator, to know why it is needed, and to understand its limitations.

The standard provides that if your respirator uses filter elements, you must be given an opportunity to change the filter elements whenever an increase in breathing resistance is detected. You also must be permitted to periodically leave your work area to wash your face and respirator facepiece whenever necessary to prevent skin irritation. If you ever have difficulty in breathing during a fit test or while using a respirator, your employer must make a medical examination available to you to determine whether you can safely wear a respirator. The result of this examination may be to give you a positive pressure respirator (which reduces breathing resistance) or to provide alternative means of protection.

### V. PROTECTIVE WORK CLOTHING AND EQUIPMENT— PARAGRAPH (g)

If you are exposed to lead above the PEL, or if you are exposed to lead compounds such as lead arsenate or lead azide which can cause skin and eye irritation, your employer must provide you with protective work clothing and equipment appropriate for the hazard. If work clothing is provided, it must be provided in a clean and dry condition at least weekly, and daily if your airborne exposure to lead is greater than 200 µg/m³. Appropriate protective work clothing and equipment can include coveralls or similar full-body work clothing, gloves, hats, shoes or disposable shoe coverlets, and face shields or vented goggles. Your employer is required to provide all such equipment at no cost to you. He is responsible for providing repairs and replacement as necessary, and also is responsible for the cleaning, laundering or disposal of protective clothing and equipment. Contaminated work clothing or equipment must be removed in change rooms and not worn home or you will extend your exposure and expose your family since lead from your clothing can accumulate in your house, car, etc. Contaminated clothing which is to be cleaned, laundered or disposed of must be placed in closed containers in the change room. At no time may lead be removed from protective clothing or equipment by any means which disperses lead into the workroom air.

### VI. HOUSEKEEPING—PARAGRAPH (h)

Your employer must establish a housekeeping program sufficient to maintain all surfaces as free as practicable of accumulations of lead dust. Vacuuming is the preferred method of meeting this requirement, and the use of compressed air to clean floors and other surfaces is absolutely prohibited. Dry or wet sweeping, shoveling, or brushing may not be used except where vaccuming or other equally effective methods have been tried and do not work. Vacuums must be used and emptied in a manner which minimizes the reentry of lead into the workplace.

## VII. HYGIENE FACILITIES AND PRACTICES—PARAGRAPH (i)

The standard requires that change rooms, showers, and filtered air lunchrooms be constructed and made available to workers exposed to lead above the PEL. When the PEL is exceeded the employer must assure that food and beverage is not present or consumed, tobacco products are not present or used, and cosmetics are not applied, except in these facilities. Change rooms, showers, and lunchrooms, must be used by workers exposed in excess of the PEL. After showering, *no* clothing or equipment worn during the shift may be worn home, and this includes shoes and underwear. Your own clothing worn during the shift should be carried home and cleaned carefully so that it does not contaminate your home. Lunchrooms may not be entered with protective clothing or equipment unless surface dust has been removed by vacuuming, downdraft booth, or other cleaning method. Finally, workers exposed above the PEL must wash both their hands and faces prior to eating, drinking, smoking or applying cosmetics.

All of the facilities and hygiene practices just discussed are essential to minimize additional sources of lead absorption from inhalation or ingestion of lead that may accumulate on you, your clothes, or your possessions. Strict compliance with these provisions can virtually eliminate several sources of lead exposure which significantly contribute to excessive lead absorption.

## VIII. MEDICAL SURVEILLANCE—PARAGRAPH (j)

The medical surveillance program is part of the standard's comprehensive approach to the prevention of lead-related disease. Its purpose is to supplement the main thrust of the standard which is aimed at minimizing airborne concentrations of lead and sources of ingestion. Only medical surveillance can determine if the other provisions of the standard have affectively protected you as an individual. Compliance with the standard's provision will protect most workers from the adverse effects of lead exposure, but may not be satisfactory to protect individual workers (1) who have high body burdens of lead acquired over past years, (2) who have additional uncontrolled sources of non-occupational lead exposure, (3) who exhibit unusual variations in lead absorption rates, or (4) who have specific non-work related medical conditions which could be aggravated by lead exposure (e.g., renal disease, anemia). In addition, control systems may fail, or hygiene and respirator programs may be inadequate. Periodic medical surveillance of individual workers will help detect those failures. Medical surveillance will also be important to protect your reproductive ability—regardless of whether you are a man or woman.

All medical surveillance required by the standard must be performed by or under the supervision of a licensed physician. The employer must provide required medical surveillance without cost to employees and at a reasonable time and place. The standard's medical surveillance program has two parts-periodic biological monitoring and medical examinations.

Your employer's obligation to offer you medical surveillance is triggered by the results of the air monitoring program. Medical surveillance must be made available to all employees who are exposed in excess of the action level for more than 30 days a year. The initial phase of the medical surveillance program, which includes blood lead level tests and medical examinations, must be completed for all covered employees no later than August 28, 1979. Priority within this first round of medical surveillance must be given to employees whom the employer believes to be at greatest risk from continued exposure (for example, those with the longest prior exposure to lead, or those with the highest current exposure). Thereafter, the employer must periodically make medical surveillance—both biological monitoring and medical examinations—available to all covered employees.

Biological monitoring under the standard consists of blood lead level (PbB) and zinc protoporphyrin tests at least every 6 months after the initial PbB test. A zinc protoporphyrin (ZPP) test is a very useful blood test which measures an effect of lead on your body. Thus biological monitoring under the standard is currently limited to PbB testing. If a worker's PbB exceeds 40 µg/100 g the monitoring frequency must be increased from every 6 months to at least every 2 months and not reduced until two consecutive PbBs indicate a blood lead level below 40 µg/100 g. Each time your PbB is determined to be over 40 µg/100 g, your employer must notify you of this in writing within five working days of his receipt of the test results. The employer must also inform you that the standard requires temporary medical removal with economic protection when your PbB exceeds certain criteria. (See Discussion of Medical Removal Protection—Paragraph (k).) During the first year of the standard, this removal criterion is 80 µg/100 g. Anytime your PbB exceeds 80 µg/100 g your employer must make available to you a prompt follow-up PbB test to ascertain your PbB. If the two tests both exceed 80 µg/100 g and you are temporarily removed, then your employer must make successive PbB tests available to you on a monthly basis during the period of your removal.

Medical examinations beyond the initial one must be made available on an annual basis if your blood lead level exceeds 40 µg/100 g at any time during the preceding year. The initial examination will provide information to establish a baseline to which subsequent data can be compared. An initial medical examination must also be made available (prior to assignment) for each employee being assigned for the first time to an area where the airborne concentration of lead equals or exceeds the action level. In addition, a medical examination or consultation must be made available as soon as possible if you notify your employer that you are experiencing signs or symptoms commonly associated with lead poisoning or that you have difficulty breathing while wearing a respirator or during a respirator fit test. You must also be provided a medical examination or consultation if you notify your employer that you desire medical advice concerning the effects of current or past exposure to lead on your ability to procreate a healthy child.

Finally, appropriate follow-up medical examinations or consultations may also be provided for employees who have been temporarily removed from exposure under the medical removal protection provisions of the standard. (See Part IX, below.)

The standard specifies the minimum content of pre-assignment and annual medical examinations. The content of other types of medical examinations and consultations is left up to the sound discretion of the examining physician. Pre-assignment and annual medical examinations must include (1) a detailed work history and medical history, (2) a thorough physical examination, and (3) a series of laboratory tests designed to check your blood chemistry and your kidney function. In addition, at any time upon your request, a laboratory evaluation of male fertility will be made (microscopic examination of a sperm sample), or a pregnancy test will be given.

The standard does not require that you participate in any of the medical procedures, tests, etc. which your employer is required to make available to you. Medical surveillance can, however, play a very important role in protecting your health. You are strongly encouraged, therefore, to participate in a meaningful fashion. The standard contains a multiple physician review mechanism which would give you a chance to have a physician of your choice directly participate in the medical surveillance program. If you were dissatisfied with an examination by a physician chosen by your employer, you could select a second physician to conduct an independent analysis. The two doctors would attempt to resolve any differences of opinion, and select a third physician to resolve any firm dispute. Generally your employer will choose the physician who conducts medical surveillance under the lead standard—unless you and your employer can agree on the choice of a physician or physicians. Some companies and unions have agreed in advance, for example, to use certain independent medical laboratories or panels of physicians. Any of these arrangements are acceptable so long as required medical surveillance is made available to workers.

The standard requires your employer to provide certain information to a physician to aid in his or her examination of you. This information includes (1) the standard and its appendices, (2) a description of your duties as they relate to lead exposure, (3) your exposure level, (4) a description of personal protective equipment you wear, (5) prior blood lead level results, and (6) prior written medical opinions concerning you that the employer has. After a medical examination or consultation the physician must prepare a written report which must contain (1) the physician's opinion as to whether you have any medical condition which places you at increased risk of material impairment to health from exposure to lead, (2) any recommended special protective measures to be provided to you, (3) any blood lead level determinations, and (4) any recommended limitation on your use of respirators. This last element must include a determination of whether you can wear a powered air purifying respirator (PAPR) if you are found unable to wear a negative pressure respirator.

The medical surveillance program of the lead standard may at some point in time serve to notify certain workers that they have acquired a disease or other adverse medical condition as a result of occupational lead exposure. If this is true, these workers might have legal rights to compensation from public agencies, their employers, firms that supply hazardous products to their employers, or other persons. Some states have laws, including worker compensation laws, that disallow a worker who learns of a job-related health impairment to sue, unless the worker sues within a short period of time after learning of the impairment. (This period of time may be a matter of months or years.) An attorney can be consulted about these possibilities. It should be stressed that OSHA is in no way trying to either encourage or discourage claims or lawsuits. However, since results of the standard's medical surveillance program can significantly affect the legal remedies of a worker who has acquired a job-related disease or impairment, it is proper for OSHA to make you aware of this.

The medical surveillance section of the standard also contains provisions dealing with chelation. Chelation is the use of certain drugs (administered in pill form or injected into the body) to reduce the amount of lead absorbed in body tissues. Experience accumulated by the medical and scientific communities has largely confirmed the effectiveness of this type of therapy for the treatment of very severe lead poisoning. On the other hand, it has also been established that there can be a long list of extremely harmful side effects associated with the use of chelating agents. The medical community has balanced the advantages and disadvantages resulting from the use of chelating agents in various circumstances and has established when the use of these agents is acceptable. The standard includes these accepted limitations due to a history of abuse of chelation therapy by some lead companies. The most widely used chelating agents are calcium disodium EDTA, (Ca Na$_2$EDTA), Calcium Disodium Versenate (Versenate), and d-penicillamine (pencillamine or Cupramine).

The standard prohibits "prophylactic chelation" of any employee by any person the employer retains, supervises or controls. "Prophylactic chelation" is the routine use of chelating or similarly acting drugs to *prevent* elevated blood levels in workers who are occupationally exposed to lead, or the use of these drugs to *routinely* lower blood lead levels to predesignated concentrations believed to be 'safe'. It should be emphasized that where an employer takes a worker who has no symptoms of lead poisoning and has chelation carried out by a physician (either inside or outside of a hospital) solely to reduce the worker's blood lead level, that will generally be considered prophylactic chelation. The use of a hospital and a physician does not mean that prophylactic chelation is not being performed. Routine chelation to prevent increased or reduce current blood lead levels is unacceptable whatever the setting.

The standard allows the use of "therapeutic" or "diagnostic" chelation if administered under the supervision of a licensed physician in a clinical setting with thorough and appropriate medical monitoring. Therapeutic chelation responds to severe lead poisoning where there are marked symptoms. Diagnostic chelation involved giving a patient a dose of the drug then collecting all urine excreted for some period of time as an aid to the diagnosis of lead poisoning.

In cases where the examining physician determines that chelation is appropriate, you must be notified in writing of this fact before such treatment. This will inform you of a potentially harmful treatment, and allow you to obtain a second opinion.

### IX. MEDICAL REMOVAL PROTECTION—PARAGRAPH (k)

Excessive lead absorption subjects you to increased risk of disease. Medical removal protection (MRP) is a means of protecting you when, for whatever reasons, other methods, such as engineering controls, work practices, and respirators, have failed to provide the protection you need. MRP involves the temproary removal of a worker from his or her regular job to a place of significantly lower exposure without any loss of earnings, seniority, or other employment rights or benefits. The purpose of this program is to cease further lead absorption and allow your body to naturally excrete lead which has previously been absorbed. Temporary medical removal can result from an elevated blood lead level, or a medical opinion. Up to 18 months of protection is provided as a result of either form of removal. The vast majority of removed workers, however, will return to their former jobs long before this eighteen month period expires. The standard contains special provisions to deal with the extraordinary but possible case where a longterm worker's blood lead level does not adequately decline during eighteen months of removal.

During the first year of the standard, if your blood lead level is 80 μg/100 g or above you must be removed from any exposure where your air lead level without a respirator would be 100 μg/m$^3$ or above. If you are removed from your normal job you may not be returned until your blood lead level declines to at least 60 μg/100 g. These criteria for removal and return will change according to the following schedule:

| | Removal blood lead (μg/100 g) | Air lead (μg/m$^3$) | Return blood lead (μg/100 g) |
|---|---|---|---|
| After March 1, 1980 | 70 and above | 50 and above | At or below 50. |
| After March 1, 1981 | 60 and above | 30 and above | At or below 40. |
| After March 1, 1983 | 50 and above averaged over six months | 30 and above | Do. |

You may also be removed from exposure even if your blood lead levels are below these criteria if a final medical determination indicates that you temporarily need reduced lead exposure for medical reasons. If the physician who is implementing your employers medical program makes a final written opinion recommending your removal or other special protective measures, your employer must implement the physician's recommendation. If you are removed in this manner, you may only be returned when the doctor indicates that it is safe for you to do so.

The standard does not give specific instructions dealing with what an employer must do with a removed worker. Your job assignment upon removal is a matter for you, your employer and your union (if any) to work out consistent with existing procedures for job assignments. Each removal must be accomplished in a manner consistent with existing collective bargaining relationships. Your employer is given broad discretion to implement temporary removals so long as no attempt is made to override existing agreements. Similarly, a removed worker is provided no right to veto an employer's choice which satisfies the standard.

In most cases, employers will likely transfer removed employees to other jobs with sufficiently low lead exposure. Alternatively, a worker's hours may be reduced so that the time weighted average exposure is reduced, or he or she may be temporarily laid off if no other alternative is feasible.

In all of these situation, MRP benefits must be provided during the period of removal—i.e., you continue to receive the same earnings, seniority, and other rights and benefits you would have had if you had not been removed. Earnings includes more than just your base wage; it includes overtime, shift differentials, incentives, and other compensation you would have earned if you had not been removed. During the period of removal you must also be provided with appropriate follow-up medical surveillance. If you were removed because your blood lead level was too high, you must be provided with a monthly blood test. If a medical opinion caused your removal, you must be provided medical tests or examinations that the doctor believes to be appropriate. If you do not participate in this follow up medical surveillance, you may lose your eligibility for MRP benefits.

When you are medically eligible to return to your former job, your employer must return you to your "former job status." This means that you are entitled to the position, wages, benefits, etc., you would have had if you had not been removed. If you would still be in your old job if no removal had occurred that is where you go back. If not, you are returned consistent with whatever job assignment discretion your employer would have had if no removal had occurred. MRP only seeks to maintain your rights, not expand them or diminish them.

If you are removed under MRP and you are also eligible for worker compensation or other compensation for lost wages, your employer's MRP benefits obligation is reduced by the amount that you *actually* receive from these other sources. This is also true if you obtain other employment during the time you are laid off with MRP benefits.

The standard also covers situations where an employer *voluntarily* removes a worker from exposure to lead due to the effects of lead on the employee's medical condition, even though the standard does not require removal. In these situations MRP benefits must still be provided as though the standard required removal. Finally, it is important to note that in all cases where removal is required, respirators cannot be used as a substitute. Respirators may be used before removal becomes necessary, but not as an alternative to a transfer to a low exposure job, or to a lay-off with MRP benefits.

## X. EMPLOYEE INFORMATION AND TRAINING—PARAGRAPH (l)

Your employer is required to provide an information and training program for all employees exposed to lead above the action level or who may suffer skin or eye irritation from lead. This program must inform these employees of the specific hazards associated with their work environment, protective measures which can be taken, the danger of lead to their bodies (including their reproductive systems), and their rights under the standard. In addition your employer must make readily available to all employees, including those exposed below the action level, a copy of the standard and its appendices and must distribute to all employees any materials provided to the employer by the Occupational Safety and Health Administration (OSHA).

Your employer is required to complete this training program for all employees by August 28, 1979. After this date, all new employees must be trained prior to initial assignment to areas where there is a possibility of exposure over the action level.

This training program must also be provided at least annually thereafter.

## XI. SIGNS—PARAGRAPH (m)

The standard requires that the following warning sign be posted in work areas where the exposure to lead exceeds the PEL:

**WARNING**

**LEAD WORK AREA**

**NO SMOKING OR EATING**

## XII. RECORDKEEPING—PARAGRAPH (n)

Your employer is required to keep all records of exposure monitoring for airborne lead. These records must include the name and job classification of employees measured, details of the sampling and analytic techniques, the results of this sampling, and the type of respiratory protection being worn by the person sampled. Your employer is also required to keep all records of biological monitoring and medical examination results. These must include the names of the employees, the physician's written opinion, and a copy of the results of the examination. All of the above kinds of records must be kept for 40 years, or for at least 20 years after your termination of employment, whichever is longer.

Recordkeeping is also required if you are temporarily removed from your job under the medical removal protection program. This record must include your name and social security number, the date of your removal and return, how the removal was or is being accomplished, and whether or not the reason for the removal was an elevated blood lead level. Your employer is required to keep each medical removal record only for as long as the duration of an employee's employment.

The standard requires that if you request to see or copy environmental monitoring, blood lead level monitoring, or medical removal records, they must be made available to you or to a representative that you authorize. Your union also has access to these records. Medical records other than PbB's must also be provided upon request to you, to your physician or to any other person whom you may specifically designate. Your union does not have access to your personal medical records unless you authorize their access.

## XIII. OBSERVATIONS OF MONITORING—PARAGRAPH (o)

When air monitoring for lead is performed at your workplace as required by this standard, your employer must allow you or someone you designate to act as an observer of the monitoring. Observers are entitled to an explanation of the measurement procedure, and to record the results obtained. Since results will not normally be available at the time of the monitoring, observers are entitled to record or receive the results of the monitoring when returned by the laboratory. Your employer is required to provide the observer with any personal protective devices required to be worn by employees working in the area that is being monitored. The employer must require the observer to wear all such equipment and to comply with all other applicable safety and health procedures.

## XIV. FOR ADDITIONAL INFORMATION

A. Copies of the Standard and explanatory material may be obtained by writing or calling the OSHA Docket Office, U.S. Department of Labor, room N2634, 200 Constitution Avenue, N.W., Washington, DC 20210. Telephone: (202) 219-7894.

1. The standard and summary of the statement of reasons (preamble), FEDERAL REGISTER, Volume 43, pp. 52952-53014, November 14, 1978.

2. The full statement of reasons (preamble) FEDERAL REGISTER, vol. 43, pp. 54354-54509, November 21, 1978.

3. Partial Administrative Stay and Corrections to the standard, (44 FR 5446-5448) January 26, 1979.

4. Notice of the Partial Judicial Stay (44 FR 14554-14555) March 13, 1979.

5. Corrections to the preamble, FEDERAL REGISTER, vol. 44, pp. 20680-20681, April 6, 1979.

6. Additional correction to the preamble concerning the construction industry FEDERAL REGISTER, vol. 44, p. 50338, August 28, 1979.

7. Appendices to the standard (Appendices A, B, C), FEDERAL REGISTER, Vol. 44, pp. 60980-60995, October 23, 1979.

8. Corrections to appendices, FEDERAL REGISTER, Vol. 44, 68828, November 30, 1979.

9. Revision to the standard and an additional appendix (Appendix D), FEDERAL REGISTER, Vol. 47, pp. 51117-51119, November 12, 1982.

10. Notice of reopening of lead rulemaking for nine remand industry sectors FEDERAL REGISTER, vol. 53, pp. 11511-11513, April 7, 1988.

11. Statement of reasons, FEDERAL REGISTER, vol. 54, pp. 29142-29275, July 11, 1989.

12. Statement of reasons, FEDERAL REGISTER, vol. 55, pp. 3146-3167, January 30, 1990.

13. Correction to appendix B, FEDERAL REGISTER, vol. 55, pp. 4998-4999, February 13, 1991.

14. Correction to appendices, FEDERAL REGISTER, vol. 56, p. 24686, May 31, 1991.

B. Additional information about the standard, its enforcement, and your employer's compliance can be obtained from the nearest OSHA Area Office listed in your telephone directory under United States Government/Department of Labor.

## APPENDIX C TO §1910.1025—MEDICAL SURVEILLANCE GUIDELINES

### INTRODUCTION

The primary purpose of the Occupational Safety and Health Act of 1970 is to assure, so far as possible, safe and healthful working conditions for every working man and woman. The occupational health standard for inorganic lead[1] was promulgated to protect workers exposed to inorganic lead including metallic lead, all inorganic lead compounds and organic lead soaps.

Under this final standard in effect as of March 1, 1979, occupational exposure to inorganic lead is to be limited to 50 µg/m$^3$ (micrograms per cubic meter) based on an 8 hour time-weighted average (TWA). This level of exposure eventually must be achieved through a combination of engineering, work practice and other administrative controls. Periods of time ranging from 1 to 10 years are provided for different industries to implement these controls. The schedule which is based on individual industry considerations is given in Table 1. Until these controls are in place, respirators must be used to meet the 50 µg/m$^3$ exposure limit.

The standard also provides for a program of biological monitoring and medical surveillance for all employees exposed to levels of inorganic lead above the action level of 30 µg/m$^3$ (TWA) for more than 30 days per year.

The purpose of this document is to outline the medical surveillance provisions of the standard for inorganic lead, and to provide further information to the physician regarding the examination and evaluation of workers exposed to inorganic lead.

---

[1] The term inorganic lead used throughout the medical surveillance appendices is meant to be synonymous with the definition of lead set forth in the standard.

# OSHA Standards for General Industry

Section 1 provides a detailed description of the monitoring procedure including the required frequency of blood testing for exposed workers, provisions for medical removal protection (MRP), the recommended right of the employee to a second medical opinion, and notification and recordkeeping requirements of the employer. A discussion of the requirements for respirator use and respirator monitoring and OSHA's position on prophylactic chelation therapy are also included in this section.

Section 2 discusses the toxic effects and clinical manifestations of lead poisoning and effects of lead intoxication on enzymatic pathways in heme synthesis. The adverse effects on both male and female reproductive capacity and on the fetus are also discussed.

Section 3 outlines the recommended medical evaluation of the worker exposed to inorganic lead including details of the medical history, physical examination, and recommended laboratory tests, which are based on the toxic effects of lead as discussed in Section 2.

Section 4 provides detailed information concerning the laboratory tests available for the monitoring of exposed workers. Included also is a discussion of the relative value of each test and the limitations and precautions which are necessary in the interpretation of the laboratory results.

## TABLE 1

| Permissible airborne lead levels by industry (µg/m³)[1] | Mar. 1, 1979 | Mar. 1, 1980 | Mar. 1, 1981 | Mar. 1, 1982 | Mar. 1, 1984 | Mar. 1, 1989 (final) |
|---|---|---|---|---|---|---|
| 1. Primary lead production | 200 | 200 | 200 | 100 | 100 | 50 |
| 2. Secondary lead production | 200 | 200 | 200 | 100 | 50 | 50 |
| 3. Lead-acid battery manufacturing | 200 | 200 | 100 | 100 | 50 | 50 |
| 4. Nonferrous foundries | 200 | 100 | 100 | 100 | 50 | 50 |
| 5. Lead pigment manufacturing | 200 | 200 | 200 | 100 | 50 | 50 |
| 6. All other industries | 200 | 50 | 50 | 50 | 50 | 50 |

[1] Airborne levels to be achieved without reliance or respirator protection through a combination of engineering, work practice and other administrative controls. While these controls are being implemented respirators must be used to meet the 50 µg/m³ exposure limit.

## I. MEDICAL SURVEILLANCE AND MONITORING REQUIREMENTS FOR WORKERS EXPOSED TO INORGANIC LEAD

Under the occupational health standard for inorganic lead, a program of biological monitoring and medical surveillance is to be made available to all employees exposed to lead above the action level of 30 µg/m³ TWA for more than 30 days each year. This program consists of periodic blood sampling and medical evaluation to be performed on a schedule which is defined by previous laboratory results, worker complaints or concerns, and the clinical assessment of the examining physician.

Under this program, the blood lead level of all employees who are exposed to lead above the action level of 30 µg/m³ is to be determined at least every six months. The frequency is increased to every two months for employees whose last blood lead level was between 40 µg/100 g whole blood and the level requiring employee medical removal to be discussed below. For employees who are removed from exposure to lead due to an elevated blood lead, a new blood lead level must be measured monthly. A zinc protoporphyrin (ZPP) is required on each occasion that a blood lead level measurement is made.

An annual medical examination and consultation performed under the guidelines discussed in Section 3 is to be made available to each employee for whom a blood test conducted at any time during the preceding 12 months indicated a blood lead level at or above 40 µg/100 g. Also, an examination is to be given to all employees prior to their assignment to an area in which airborne lead concentrations reach or exceed the action level. In addition, a medical examination must be provided as soon as possible after notification by an employee that the employee has developed signs or symptoms commonly associated with lead intoxication, that the employee desires medical advice regarding lead exposure and the ability to procreate a healthy child, or that the employee has demonstrated difficulty in breathing during a respirator fitting test or during respirator use. An examination is also to be made available to each employee removed from exposure to lead due to a risk of sustaining material impairment to health, or otherwise limited or specially protected pursuant to medical recommendations.

Results of biological monitoring or the recommendations of an examining physician may necessitate removal of an employee from further lead exposure pursuant to the standard's medical removal protection (MRP) program. The object of the MRP program is to provide temporary medical removal to workers either with substantially elevated blood lead levels or otherwise at risk of sustaining material health impairment from continued substantial exposure to lead. The following guidelines which are summarized in Table 2 were created under the standard for the temporary removal of an exposed employee and his or her subsequent return to work ing an exposure area.

## TABLE 2

| | | Effective date | | | | |
|---|---|---|---|---|---|---|
| | | Mar. 1, 1979 | Mar. 1, 1980 | Mar. 1, 1981 | Mar. 1, 1982 | Mar. 1, 1983 (final) |
| A | Blood lead level requiring employee medical removal. (Level must be confirmed with second follow-up blood lead level within two weeks of first report.) | ≥80 µg/100 g | ≥70 µg/100 g | ≥60 µg/100 g | ≥60 µg/100 g | ≥ 60 µg/100 g or average of last three blood samples or all blood samples over previous 6 months (whichever is over a longer time period) is 50 µg/100 g or greater unless last blood samples is 40 µg/100 g or less. |
| B | Frequency which employees exposed to action level of lead (30 µg/m³ TWA) must have blood ead level checked (ZPP is also required in each occasion that a blood lead is obtained.) | | | | | |
| | 1. Last blood lead level less than 40 µg/100g | Every 6 months | Every 6 months | Every 6 months | Every 6 months | Every 6 months |
| | 2. Last blood lead level between 40 µg/100 g and level requiring medical removal (see A above). | Every 2 months | Every 2 months | Every 2 months | Every 2 months | Every 2 months |
| | 3. Employees removed from exposure to lead because of an elevated blood lead level. | Every 1 month | Every 1 month | Every 1 month | Every 1 month | Every 1 month |
| C | Permissible airborne exposure limit for workers removed from work due to an elevated blood lead level (without regard to respirator protection). | 100 µg/m³ 8 hr TWA | 50 µg/m³ 8 hr TWA | 30 µg/m³ 8 hr TWA | 30 µg/m³ 8 hr TWA | 30 µg/m³ 8 hr TWA |
| D | Blood lead level confirmed with a second blood analysis, at which employee may return to work. Permissible exposure without regard to respirator protection is listed by industry in Table 1. | .60 µg/m100 g | .50 µg/m100 g | .40 µg/100 g | .40 µg/100 g | .40 µg/100 g. |

NOTE: When medical opinion indicates that an employee is at risk of material impairment from exposure to lead, the physician can remove an employee from exposures exceeding the action level (or less) or recommend special protective measures as deemed appropriate and necessary. Medical monitoring during the medical removal period can be more stringent than noted in the table above if the physician so specifies. Return to work or removal of limitations and special protections is permitted when the physician indicates that the worker is no longer at risk of material impairment.

Under the standard's ultimate worker removal criteria, a worker is to be removed from any work having any eight hour TWA exposure to lead of 30 µg/m³ or more whenever either of the following circumstances apply: (1) a blood lead level of 60 µg/100 g or greater is obtained and confirmed by a second follow-up blood lead level performed within two weeks after the employer receives the results of the first blood sampling test, or (2) the average of the previous three blood lead determinations or the average of all blood lead determinations conducted during the previous six months, whichever encompasses the longest time period, equals or exceeds 50 µg/100 g, unless the last blood sample indicates a blood lead level at or below 40 µg/100 g in which case the employee need not be removed. Medical removal is to continue until two consecutive blood lead levels are 40 µg/100 g or less.

During the first two years that the ultimate removal criteria are being phased in, the return criteria have been set to assure that a worker's blood lead level has substantially declined during the period of removal. From March 1, 1979 to March 1, 1980, the blood lead level requiring employee medical removal is 80 µg/100 g. Workers found to have a confirmed blood lead at this level or greater need only be removed from work having a daily 8 hour TWA exposure to lead at or above 100 µg/m³. Workers so removed are to be returned to work when their blood lead levels are at or below 60 µg/100 g of whole blood. From March 1, 1980 to March 1, 1981, the blood lead level requiring medical removal is 70 µg/100 g. During this period workers need only be removed from jobs having a daily 8 hour TWA exposure to lead at or above 50 µg/m³ and are to be returned to work when a level of 50 µg/100 g is achieved. Beginning March 1, 1981, return depends on a worker's blood lead level declining to 40 µg/100 g of whole blood.

As part of the standard, the employer is required to notify in writing each employee whose blood lead level exceeds 40 µg/100 g. In addition each such employee is to be informed that the standard requires medical removal with MRP benefits, discussed below, when an employee's blood lead level exceeds the above defined limits.

In addition to the above blood lead level criteria, temporary worker removal may also take place as a result of medical determinations and recommendations. Written medical opinions must be prepared after each examination pursuant to the standard. If the examining physician includes a medical finding, determination or opinion that the employee has a medical condition which places the employee at increased risk of material health impairment from exposure to lead, then the employee must be removed from exposure to lead at or above the action level. Alternatively, if the examining physician recommends special protective measures for an employee (e.g., use of a powered air purifying respirator) or recommends limitations on an employee's exposure to lead, then the employer must implement these recommendations. Recommendations may be more stringent than the specific provisions of the standard. The examining physician, therefore, is given broad flexibility to tailor special protective procedures to the needs of individual employees. This flexibility extends to the evaluation and management of pregnant workers and male and female workers who are planning to raise children. Based on the history, physical examination, and laboratory studies, the physician might recommend special protective measures or medical removal for an employee who is pregnant or who is planning to conceive a child when, in the physician's judgment, continued exposure to lead at the current job would pose a significant risk. The return of the employee to his or her former job status, or the removal of special protections or limitations, depends upon the examining physician determining that the employee is no longer at increased risk of material impairment or that special measures are no longer needed.

During the period of any form of special protection or removal, the employer must maintain the worker's earnings, seniority, and other employment rights and benefits (as though the worker had not been removed) for a period of up to 18 months. This economic protection will maximize meaningful worker participation in the medical surveillance program, and is appropriate as part of the employer's overall obligation to provide a safe and healthful workplace. The provisions of MRP benefits during the employee's removal period may, however, be conditioned upon participation in medical surveillance.

On rare occasions, an employee's blood lead level may not acceptably decline within 18 months of removal. This situation will arise only in unusual circumstances, thus the standard relies on an individual medical examination to determine how to protect such an employee. This medical determination is to be based on both laboratory values, including lead levels, zinc protoporphyrin levels, blood counts, and other tests felt to be warranted, as well as the physician's judgment that any symptoms or findings on physical examination are a result of lead toxicity. The medical determination may be that the employee is incapable of ever safely returning to his or her former job status. The medical determination may provide additional removal time past 18 months for some employees or specify special protective measures to be implemented.

The lead standard provides for a multiple physician review in cases where the employee wishes a second opinion concerning potential lead poisoning or toxicity. If an employee wishes a second opinion, he or she can make an appointment with a physician of his or her choice. This second physician will review the findings, recommendations or determinations of the first physician and conduct any examinations, consultations or tests deemed necessary in an attempt to make a final medical determination. If the first and second physicians do not agree in their assessment they must try to resolve their differences. If they cannot reach an agreement then they must designate a third physician to resolve the dispute.

The employer must provide examining and consulting physicians with the following specific information: a copy of the lead regulations and all appendices, a description of the employee's duties as related to exposure, the exposure level to lead and any other toxic substances (if applicable), a description of personal protective equipment used, blood lead levels, and all prior written medical opinions regarding the employee in the employer's possession or control. The employer must also obtain from the physician and provide the employee with a written medical opinion containing blood lead levels, the physicians's opinion as to whether the employee is at risk of material impairment to health, any recommended protective measures for the employee if further exposure is permitted, as well as any recommended limitations upon an employee's use of respirators.

Employers must instruct each physician not to reveal to the employer in writing or in any other way his or her findings, laboratory results, or diagnoses which are felt to be unrelated to occupational lead exposure. They must also instruct each physician to advise the employee of any occupationally or non-occupationally related medical condition requiring further treatment or evaluation.

The standard provides for the use of respirators where engineering and other primary controls have not been fully implemented. However, the use of respirator protection shall not be used in lieu of temporary medical removal due to elevated blood lead levels or findings that an employee is at risk of material health impairment. This is based on the numerous inadequacies of respirators including skin rash where the facepiece makes contact with the skin, unacceptable stress to breathing in some workers with underlying cardiopulmonary impairment, difficulty in providing adequate fit, the tendency for respirators to create additional hazards by interfering with vision, hearing, and mobility, and the difficulties of assuring the maximum effectiveness of a complicated work practice program involving respirators. Respirators do, however, serve a useful function where engineering and work practice controls are inadequate by providing supplementary, interim, or short-term protection, provided they are properly selected for the environment in which the employee will be working, properly fitted to the employee, maintained and cleaned periodically, and worn by the employee when required.

In its final standard on occupational exposure to inorganic lead, OSHA has prohibited prophylactic chelation. Diagnostic and therapeutic chelation are permitted only under the supervision of a licensed physician with appropriate medical monitoring in an acceptable clinical setting. The decision to initiate chelation therapy must be made on an individual basis and take into account the severity of symptoms felt to be a result of lead toxicity along with blood lead levels, ZPP levels, and other laboratory tests as appropriate. EDTA and penicillamine which are the primary chelating agents used in the therapy of occupational lead poisoning have significant potential side effects and their use must be justified on the basis of expected benefits to the worker. Unless frank and severe symptoms are present, therapeutic chelation is not recommended given the opportunity to remove a worker from exposure and allow the body to naturally excrete accumulated lead. As a diagnostic aid, the chelation mobilization test using CA-EDTA has limited applicability. According to some investigators, the test can differentiate between lead-induced and other nephropathies. The test may also provide an estimation of the mobile fraction of the total body lead burden.

Employers are required to assure that accurate records are maintained on exposure monitoring, medical surveillance, and medical removal for each employee. Exposure monitoring and medical surveillance records must be kept for 40 years or the duration of employment plus 20 years, whichever is longer, while medical removal records must be maintained for the duration of employment. All records required under the standard must be made available upon request to the Assistant Secretary of Labor for Occupational Safety and Health and the Director of the National Institute for Occupational Safety and Health. Employers must also make environmental and biological monitoring and medical removal records available to affected employees and to former employees or their authorized employee representatives. Employees or their specifically designated representatives have access to their entire medical surveillance records.

In addition, the standard requires that the employer inform all workers exposed to lead at or above the action level of the provisions of the standard and all its appendices, the purpose and description of medical surveillance and provisions for medical removal protection if temporary removal is required. An understanding of the potential health effects of lead exposure by all exposed employees along with full understanding of their rights under the lead standard is essential for an effective monitoring program.

### II. ADVERSE HEALTH EFFECTS OF INORGANIC LEAD

Although the toxicity of lead has been known for 2,000 years, the knowledge of the complex relationship between lead exposure and human response is still being refined. Significant research into the toxic properties of lead continues throughout the world, and it should be anticipated that our understanding of thresholds of effects and margins of safety will be improved in future years. The provisions of the lead standard are founded on two prime medical judgments: first, the prevention of adverse health effects from exposure to lead throughout a working lifetime requires that worker blood lead levels be maintained at or below 40 µg/100 g and second, the blood lead levels of workers, male or female, who intend to parent in the near future should be maintained below 30 µg/100 g to minimize adverse reproductive health effects to the parents and developing fetus. The adverse effects of lead on reproduction are being actively researched and OSHA encourages the physician to remain abreast of recent developments in the area to best advise pregnant workers or workers planning to conceive children.

The spectrum of health effects caused by lead exposure can be subdivided into five developmental stages: normal, physiological changes of uncertain significance, pathophysiological changes, overt symptoms (morbidity), and mortality. Within this process there are no sharp distinctions, but rather a continuum of effects. Boundaries between categories overlap due to the wide variation of individual responses and exposures in the working population. OSHA's development of the lead standard focused on pathophysiological changes as well as later stages of disease.

1. *Heme Synthesis Inhibition.* The earliest demonstrated effect of lead involves its ability to inhibit at least two enzymes of the heme synthesis pathway at very low blood levels. Inhibition of delta aminolevulinic acid dehydrase (ALA-D) which catalyzes the conversion of delta-aminolevulinic acid (ALA) to protoporphyrin is observed at a blood lead level below 20 µg/100 g whole blood. At a blood lead level of 40 µg/100 g, more than 20% of the population would have 70% inhibition of ALA-D. There is an exponential increase in ALA excretion at blood lead levels greater than 40 µg/100 g.

Another enzyme, ferrochelatase, is also inhibited at low blood lead levels. Inhibition of ferrochelatase leads to increased free erythrocyte protoporphyrin (FEP) in the blood which can then bind to zinc to yield zinc protoporphyrin. At a blood lead level of 50 µg/100 g or greater, nearly 100% of the population will have an increase in FEP. There is also an exponential relationship between blood lead levels greater than 40 µg/100 g and the associated ZPP level, which has led to the development of the ZPP screening test for lead exposure.

While the significance of these effects is subject to debate, it is OSHA's position that these enzyme disturbances are early stages of a disease process which may eventually result in the clinical symptoms of lead poisoning. Whether or not the effects do progress to the later stages of clinical disease, disruption of these enzyme processes over a working lifetime is considered to be a material impairment of health.

One of the eventual results of lead-induced inhibition of enzymes in the heme synthesis pathway is anemia which can be asymptomatic if mild but associated with a wide array of symptoms including dizziness, fatigue, and tachycardia when more severe. Studies have indicated that lead levels as low as 50 µg/100 g can be associated with a definite decreased hemoglobin, although most cases of lead-induced anemia, as well as shortened red-cell survival times, occur at lead levels exceeding 80 µg/100 g. Inhibited hemoglobin synthesis is more common in chronic cases whereas shortened erythrocyte life span is more common in acute cases.

In lead-induced anemias, there is usually a reticulocytosis along with the presence of basophilic stippling, and ringed sideroblasts, although none of the above are pathognomonic for lead-induced anemia.

2. *Neurological Effects.* Inorganic lead has been found to have toxic effects on both the central and peripheral nervous systems. The earliest stages of lead-induced central nervous system effects first manifest themselves in

the form of behavioral disturbances and central nervous system symptoms including irritability, restlessness, insomnia and other sleep disturbances, fatigue, vertigo, headache, poor memory, tremor, depression, and apathy. With more severe exposure, symptoms can progress to drowsiness, stupor, hallucinations, delerium, convulsions and coma.

The most severe and acute form of lead poisoning which usually follows ingestion or inhalation of large amounts of lead is acute encephalopathy which may arise precipitously with the onset of intractable seizures, coma, cardiorespiratory arrest, and death within 48 hours.

While there is disagreement about what exposure levels are needed to produce the earliest symptoms, most experts agree that symptoms definitely can occur at blood lead levels of 60 µg/100 g whole blood and therefore recommend a 40 µg/100 g maximum. The central nervous system effects frequently are not reversible following discontinued exposure or chelation therapy and when improvement does occur, it is almost always only partial.

The peripheral neuropathy resulting from lead exposure characteristically involves only motor function with minimal sensory damage and has a marked predilection for the extensor muscles of the most active extremity. The peripheral neuropathy can occur with varying degrees of severity. The earliest and mildest form which can be detected in workers with blood lead levels as low as 50 µg/100 g is manifested by slowing of motor nerve conduction velocity often without clinical symptoms. With progression of the neuropathy there is development of painless extensor muscle weakness usually involving the extensor muscles of the fingers and hand in the most active upper extremity, followed in severe cases by wrist drop or, much less commonly, foot drop.

In addition to slowing of nerve conduction, electromyographical studies in patients with blood lead levels greater than 50 µg/100 g have demonstrated a decrease in the number of acting motor unit potentials, an increase in the duration of motor unit potentials, and spontaneous pathological activity including fibrillations and fasciculations. Whether these effects occur at levels of 40 µg/100 g is undetermined.

While the peripheral neuropathies can occasionally be reversed with therapy, again such recovery is not assured particularly in the more severe neuropathies and often improvement is only partial. The lack of reversibility is felt to be due in part to segmental demyelination.

3. *Gastrointestinal.* Lead may also affect the gastrointestinal system producing abdominal colic or diffuse abdominal pain, constipation, obstipation, diarrhea, anorexia, nausea and vomiting. Lead colic rarely develops at blood lead levels below 80 µg/100 g.

4. *Renal.* Renal toxicity represents one of the most serious health effects of lead poisoning. In the early stages of disease nuclear inclusion bodies can frequently be identified in proximal renal tubular cells. Renal function remains normal and the changes in this stage are probably reversible. With more advanced disease there is progressive interstitial fibrosis and impaired renal function. Eventually extensive interstitial fibrosis ensues with sclerotic glomeruli and dilated and atrophied proximal tubules; all represent end stage kidney disease. Azotemia can be progressive, eventually resulting in frank uremia necessitating dialysis. There is occasionally associated hypertension and hyperuricemia with or without gout.

Early kidney disease is difficult to detect. The urinalysis is normal in early lead nephropathy and the blood urea nitrogen and serum creatinine increase only when two- thirds of kidney function is lost. Measurement of creatinine clearance can often detect earlier disease as can other methods of measurement of glomerular filtration rate. An abnormal Ca-EDTA mobilization test has been used to differentiate between lead-induced and other nephropathies, but this procedure is not widely accepted. A form of Fanconi syndrome with aminoaciduria, glycosuria, and hyperphosphaturia indicating severe injury to the proximal renal tubules is occasionally seen in children.

5. *Reproductive effects.* Exposure to lead can have serious effects on reproductive function in both males and females. In male workers exposed to lead there can be a decrease in sexual drive, impotence, decreased ability to produce healthy sperm, and sterility. Malformed sperm (teratospermia), decreased number of sperm (hypospermia), and sperm with decreased motility (asthenospermia) can all occur. Teratospermia has been noted at mean blood lead levels of 53 µg/100 g and hypospermia and asthenospermia at 41 µg/100 g. Furthermore, there appears to be a dose-response relationship for teratospermia in lead exposed workers.

Women exposed to lead may experience menstrual disturbances including dysmenorrhea, menorrhagia and amenorrhea. Following exposure to lead, women have a higher frequency of sterility, premature births, spontaneous miscarriages, and stillbirths.

Germ cells can be affected by lead and cause genetic damage in the egg or sperm cells before conception and result in failure to implant, miscarriage, stillbirth, or birth defects.

Infants of mothers with lead poisoning have a higher mortality during the first year and suffer from lowered birth weights, slower growth, and nervous system disorders.

Lead can pass through the placental barrier and lead levels in the mother's blood are comparable to concentrations of lead in the umbilical cord at birth. Transplacental passage becomes detectable at 12-14 weeks of gestation and increases until birth.

There is little direct data on damage to the fetus from exposure to lead but it is generally assumed that the fetus and newborn would be at least as susceptible to neurological damage as young children. Blood lead levels of 50-60 µg/100 g in children can cause significant neurobehavioral impairments and there is evidence of hyperactivity at blood levels as low as 25 µg/100 g. Given the overall body of literature concerning the adverse health effects of lead in children, OSHA feels that the blood lead level in children should be maintained below 30 µg/100 g with a population mean of 15 µg/100 g. Blood lead levels in the fetus and newborn likewise should not exceed 30 µg/100 g.

Because of lead's ability to pass through the placental barrier and also because of the demonstrated adverse effects of lead on reproductive function in both the male and female as well as the risk of genetic damage of lead on both the ovum and sperm, OSHA recommends a 30 µg/100 g maximum permissible blood lead level in both males and females who wish to bear children.

6. *Other toxic effects.* Debate and research continue on the effects of lead on the human body. Hypertension has frequently been noted in occupationally exposed individuals although it is difficult to assess whether this is due to lead's adverse effects on the kidney or if some other mechanism is involved. Vascular and electrocardiogarphic changes have been detected but have not been well characterized. Lead is thought to impair thyroid function and interfere with the pituitary-adrenal axis, but again these effects have not been well defined.

### III. MEDICAL EVALUATION

The most important principle in evaluating a worker for any occupational disease including lead poisoning is a high index of suspicion on the part of the examining physician. As discussed in Section 2, lead can affect numerous organ systems and produce a wide array of signs and symptoms, most of which are non-specific and subtle in nature at least in the early stages of disease. Unless serious concern for lead toxicity is present, many of the early clues to diagnosis may easily be overlooked.

The crucial initial step in the medical evaluation is recognizing that a worker's employment can result in exposure to lead. The worker will frequently be able to define exposures to lead and lead containing materials but often will not volunteer this information unless specifically asked. In other situations the worker may not know of any exposures to lead but the suspicion might be raised on the part of the physician because of the industry or occupation of the worker. Potential occupational exposure to lead and its compounds occur in at least 120 occupations, including lead smelting, the manufacture of lead storage batteries, the manufacture of lead pigments and products containing pigments, solder manufacture, shipbuilding and ship repair, auto manufacturing, construction, and painting.

Once the possibility for lead exposure is raised, the focus can then be directed toward eliciting information from the medical history, physical exam, and finally from laboratory data to evaluate the worker for potential lead toxicity.

A complete and detailed work history is important in the initial evaluation. A listing of all previous employment with information on work processes, exposure to fumes or dust, known exposures to lead or other toxic substances, respiratory protection used, and previous medical surveillance should all be included in the worker's record. Where exposure to lead is suspected, information concerning on-the-job personal hygiene, smoking or eating habits in work areas, laundry procedures, and use of any protective clothing or respiratory protection equipment should be noted. A complete work history is essential in the medical evaluation of a worker with suspected lead toxicity, especially when long term effects such as neurotoxicity and nephrotoxicity are considered.

The medical history is also of fundamental importance and should include a listing of all past and current medical conditions, current medications including proprietary drug intake, previous surgeries and hospitalizations, allergies, smoking history, alcohol consumption, and also non-occupational lead exposures such as hobbies (hunting, riflery). Also known childhood exposures should be elicited. Any previous history of hematological, neurological, gastrointestinal, renal, psychological, gynecological, genetic, or reproductive problems should be specifically noted.

A careful and complete review must be performed to assess both recognized complaints and subtle or slowly acquired symptoms which the worker might not appreciate as being significant. The review of symptoms should include the following:

General—weight loss, fatigue, decreased appetite.

Head, Eyes, Ears, Nose, Throat (HEENT)—headaches, visual disturbances or decreased visual acuity, hearing deficits or tinnitus, pigmentation of the oral mucosa, or metallic taste in mouth.

Cardio-pulmonary—shortness of breath, cough, chest pains, palpitations, or orthopnea.

Gastrointestinal—nausea, vomiting, heartburn, abdominal pain, constipation or diarrhea.

Neurologic—irritability, insomnia, weakness (fatigue), dizziness, loss of memory, confusion, hallucinations, incoordination, ataxia, decreased strength in hands or feet, disturbances in gait, difficulty in climbing stairs, or seizures.

Hematologic—pallor, easy fatigability, abnormal blood loss, melena.

Reproductive (male and female and spouse where relevant)—history of infertility, impotence, loss of libido, abnormal menstrual periods, history of miscarriages, stillbirths, or children with birth defects.

Musculo-skeletal—muscle and joint pains.

The physical examination should emphasize the neurological, gastrointestinal, and cardiovascular systems. The worker's weight and blood pressure should be recorded and the oral mucosa checked for pigmentation characteristic of a possible Burtonian or lead line on the gingiva. It should be noted, however, that the lead line may not be present even in severe lead poisoning if good oral hygiene is practiced.

The presence of pallor on skin examination may indicate an anemia, which if severe might also be associated with a tachycardia. If an anemia is suspected, an active search for blood loss should be undertaken including potential blood loss through the gastrointestinal tract.

A complete neurological examination should include an adequate mental status evaluation including a search for behavioral and psychological disturbances, memory testing, evaluation for irritability, insomnia, hallucinations, and mental clouding. Gait and coordination should be examined along with close observation for tremor. A detailed evaluation of peripheral nerve function including careful sensory and motor function testing is warranted. Strength testing particularly of extensor muscle groups of all extremities is of fundamental importance.

Cranial nerve evaluation should also be included in the routine examination.

The abdominal examination should include auscultation for bowel sounds and abdominal bruits and palpation for organomegaly, masses, and diffuse abdominal tenderness.

Cardiovascular examination should evaluate possible early signs of congestive heart failure. Pulmonary status should be addressed particularly if respiratory protection is contemplated.

As part of the medical evaluation, the lead standard requires the following laboratory studies:

1. Blood lead level

2. Hemoglobin and hematocrit determinations, red cell indices, and examination of the peripheral blood smear to evaluate red blood cell morphology

3. Blood urea nitrogen

4. Serum creatinine

5. Routine urinalysis with microscopic examination

6. A zinc protoporphyrin level

In addition to the above, the physician is authorized to order any further laboratory or other tests which he or she deems necessary in accordance with sound medical practice. The evaluation must also include pregnancy testing or laboratory evaluation of male fertility if requested by the employee.

Additional tests which are probably not warranted on a routine basis but may be appropriate when blood lead and ZPP levels are equivocal include delta aminolevulinic acid and coproporphyrin concentrations in the urine, and dark-field illumination for detection of basophilic stippling in red blood cells.

If an anemia is detected further studies including a careful examination of the peripheral smear, reticulocyte count, stool for occult blood, serum iron, total iron binding capacity, bilirubin, and, if appropriate, vitamin B12 and folate may be of value in attempting to identify the cause of the anemia.

If a peripheral neuropathy is suspected, nerve conduction studies are warranted both for diagnosis and as a basis to monitor any therapy.

If renal disease is questioned, a 24 hour urine collection for creatinine clearance, protein, and electrolytes may be indicated. Elevated uric acid levels may result from lead-induced renal disease and a serum uric acid level might be performed.

An electrocardiogram and chest x-ray may be obtained as deemed appropriate.

Sophisticated and highly specialized testing should not be done routinely and where indicated should be under the direction of a specialist.

### IV. LABORATORY EVALUATION

The blood lead level at present remains the single most important test to monitor lead exposure and is the test used in the medical surveillance program under the lead standard to guide employee medical removal. The ZPP has several advantages over the blood lead level. Because of its relatively recent development and the lack of extensive data concerning its interpretation, the ZPP currently remains an ancillary test.

This section will discuss the blood lead level and ZPP in detail and will outline their relative advantages and disadvantages. Other blood tests currently available to evaluate lead exposure will also be reviewed.

The blood lead level is a good index of current or recent lead absorption when there is no anemia present and when the worker has not taken any chelating agents. However, blood lead levels along with urinary lead levels do not necessarily indicate the total body burden of lead and are not adequate measures of past exposure. One reason for this is that lead has a high affinity for bone and up to 90% of the body's total lead is deposited there. A very important component of the total lead body burden is lead in soft tissue (liver, kidney, and brain). This fraction of the lead body burden, the biologically active lead, is not entirely reflected by blood lead levels since it is a function of the dynamics of lead absorption, distribution, deposition in bone and excretion. Following discontinuation of exposure to lead, the excess body burden is only slowly mobilized from bone and other relatively stable body stores and excreted. Consequently, a high blood lead level may only represent recent heavy exposure to lead without a significant total body excess and likewise a low blood lead level does not exclude an elevated total body burden of lead.

Also due to its correlation with recent exposures, the blood lead level may vary considerably over short time intervals.

To minimize laboratory error and erroneous results due to contamination, blood specimens must be carefully collected after thorough cleaning of the skin with appropriate methods using lead-free blood containers and analyzed by a reliable laboratory. Under the standard, samples must be analyzed in laboratories which are approved by the Center for Disease Control (CDC) or which have received satisfactory grades in proficiency testing by the CDC in the previous year. Analysis is to be made using atomic absorption spectrophotometry, anodic stripping voltammetry or any method which meets the accuracy requirements set forth by the standard.

The determination of lead in urine is generally considered a less reliable monitoring technique than analysis of whole blood primarily due to individual variability in urinary excretion capacity as well as the technical difficulty of obtaining accurate 24 hour urine collections. In addition, workers with renal insufficiency, whether due to lead or some other cause, may have decreased lead clearance and consequently urine lead levels may underestimate the true lead burden. Therefore, urine lead levels should not be used as a routine test.

The zinc protoporphyrin test, unlike the blood lead determination, measures an adverse metabolic effect of lead and as such is a better indicator of lead toxicity than the level of blood lead itself. The level of ZPP reflects lead absorption over the preceding 3 to 4 months, and therefore is a better indicator of lead body burden. The ZPP requires more time than the blood lead to read significantly elevated levels; the return to normal after discontinuing lead exposure is also slower. Furthermore, the ZPP test is simpler, faster, and less expensive to perform and no contamination is possible. Many investigators believe it is the most reliable means of monitoring chronic lead absorption.

Zinc protoporphyrin results from the inhibition of the enzyme ferrochelatase which catalyzes the insertion of an iron molecule into the protoporphyrin molecule, which then becomes heme. If iron is not inserted into the molecule then zinc, having a greater affinity for protoporphyrin, takes the place of the iron, forming ZPP.

An elevation in the level of circulating ZPP may occur at blood lead levels as low as 20-30 µg/100 g in some workers. Once the blood lead level has reached 40 µg/100 g there is more marked rise in the ZPP value from its normal range of less than 100 µg/100 ml. Increases in blood lead levels beyond 40 µg/100 g are associated with exponential increases in ZPP.

Whereas blood lead levels fluctuate over short time spans, ZPP levels remain relatively stable. ZPP is measured directly in red blood cells and is present for the cell's entire 120 day life-span. Therefore, the ZPP level in blood reflects the average ZPP production over the previous 3-4 months and consequently the average lead exposure during that time interval.

It is recommended that a hematocrit be determined whenever a confirmed ZPP of 50 µg/100 ml whole blood is obtained to rule out a significant underlying anemia. If the ZPP is in excess of 100 µg/100 ml and not associated with abnormal elevations in blood lead levels, the laboratory should be checked to be sure that blood leads were determined using atomic absorption spectrophotometry anodic stripping voltammetry, or any method which meets the accuracy requirements set forth by the standard by a CDC approved laboratory which is experienced in lead level determinations. Repeat periodic blood lead studies should be obtained in all individuals with elevated ZPP levels to be certain that an associated elevated blood lead level has not been missed due to transient fluctuations in blood leads.

ZPP has a characteristic fluorescence spectrum with a peak at 594 nm which is detectable with a hematofluorimeter. The hematofluorimeter is accurate and portable and can provide on-site, instantaneous results for workers who can be frequently tested via a finger prick.

However, careful attention must be given to calibration and quality control procedures. Limited data on blood lead—ZPP correlations and the ZPP levels which are associated with the adverse health effects discussed in Section 2 are the major limitations of the test. Also it is difficult to correlate ZPP levels with environmental exposure and there is some variation of response with age and sex. Nevertheless, the ZPP promises to be an important diagnostic test for the early detection of lead toxicity and its value will increase as more data is collected regarding its relationship to other manifestations of lead poisoning.

Levels of delta-aminolevulinic acid (ALA) in the urine are also used as a measure of lead exposure. Increasing concentrations of ALA are believed to result from the inhibition of the enzyme delta-aminolevulinic acid dehydrase (ALA-D). Although the test is relatively easy to perform, inexpensive, and rapid, the disadvantages include variability in results, the necessity to collect a complete 24 hour urine sample which has a specific gravity greater than 1.010, and also the fact that ALA decomposes in the presence of light.

The pattern of porphyrin excretion in the urine can also be helpful in identifying lead intoxication. With lead poisoning, the urine concentrations of coproporphyrins I and II, porphobilinogen and uroporphyrin I rise. The most important increase, however, is that of coproporphyrin III; levels may exceed 5,000 µg/l in the urine in lead poisoned individuals, but its correlation with blood lead levels and ZPP are not as good as those of ALA. Increases in urinary porphyrins are not diagnostic of lead toxicity and may be seen in porphyria, some liver diseases, and in patients with high reticulocyte counts.

*Summary.* The Occupational Safety and Health Administration's standard for inorganic lead places significant emphasis on the medical surveillance of all workers exposed to levels of inorganic lead above the action level of 30 µg/m³ TWA. The physician has a fundamental role in this surveillance program, and in the operation of the medical removal protection program.

Even with adequate worker education on the adverse health effects of lead and appropriate training in work practices, personal hygiene and other control measures, the physician has a primary responsibility for evaluating potential lead toxicity in the worker. It is only through a careful and detailed medical and work history, a complete physical examination and appropriate laboratory testing that an accurate assessment can be made. Many of the adverse health effects of lead toxicity are either irreversible or only partially reversible and therefore early detection of disease is very important.

This document outlines the medical monitoring program as defined by the occupational safety and health standard for inorganic lead. It reviews the adverse health effects of lead poisoning and describes the important elements of the history and physical examinations as they relate to these adverse effects. Finally, the appropriate laboratory testing for evaluating lead exposure and toxicity is presented.

It is hoped that this review and discussion will give the physician a better understanding of the OSHA standard with the ultimate goal of protecting the health and well-being of the worker exposed to lead under his or her care.

### APPENDIX D TO §1910.1025—QUALITATIVE FIT TEST PROTOCOLS [Removed].

[§1910.1025, App. D, removed at 63 FR 1287, Jan. 8, 1988, eff. Apr. 8, 1998.]

[43 FR 53007, Nov. 14, 1978]

### § 1910.1026 Chromium (VI).

**1910.1026(a)** *Scope.* **(1)** This standard applies to occupational exposures to chromium (VI) in all forms and compounds in general industry, except:

**1910.1026(a)(2)** Exposures that occur in the application of pesticides regulated by the Environmental Protection Agency or another Federal government agency (*e.g.*, the treatment of wood with preservatives);

**1910.1026(a)(3)** Exposures to portland cement; or

**1910.1026(a)(4)** Where the employer has objective data demonstrating that a material containing chromium or a specific process, operation, or activity involving chromium cannot release dusts, fumes, or mists of chromium (VI) in concentrations at or above 0.5 µg/m3 as an 8-hour time-weighted average (TWA) under any expected conditions of use.

**1910.1026(b)** *Definitions.* For the purposes of this section the following definitions apply:

*Action level* means a concentration of airborne chromium (VI) of 2.5 micrograms per cubic meter of air (2.5 µg/m³) calculated as an 8-hour time-weighted average (TWA).

*Assistant Secretary* means the Assistant Secretary of Labor for Occupational Safety and Health, U.S. Department of Labor, or designee.

*Chromium (VI) [hexavalent chromium or Cr(VI)]* means chromium with a valence of positive six, in any form and in any compound.

*Director* means the Director of the National Institute for Occupational Safety and Health (NIOSH), U.S. Department of Health and Human Services, or designee.

*Emergency* means any occurrence that results, or is likely to result, in an uncontrolled release of chromium (VI). If an incidental release of chromium (VI) can be controlled at the time of release by employees in the immediate release area, or by maintenance personnel, it is not an emergency.

*Employee exposure* means the exposure to airborne chromium (VI) that would occur if the employee were not using a respirator.

*High-efficiency particulate air [HEPA] filter* means a filter that is at least 99.97 percent efficient in removing mono-dispersed particles of 0.3 micrometers in diameter or larger.

*Historical monitoring data* means data from chromium (VI) monitoring conducted prior to May 30, 2006, obtained during work operations conducted under workplace conditions closely resembling the processes, types of material, control methods, work practices, and environmental conditions in the employer's current operations.

*Objective data* means information such as air monitoring data from industry-wide surveys or calculations based on the composition or chemical and physical properties of a substance demonstrating the employee exposure to chromium (VI) associated with a particular product or material or a specific process, operation, or activity. The data must reflect workplace conditions closely resembling the processes, types of material, control methods, work practices, and environmental conditions in the employer's current operations.

*Physician or other licensed health care professional [PLHCP]* is an individual whose legally permitted scope of practice (*i.e.*, license, registration, or certification) allows him or her to independently provide or be delegated the responsibility to provide some or all of the particular health care services required by paragraph (k) of this section.

*Regulated area* means an area, demarcated by the employer, where an employee's exposure to airborne concentrations of chromium (VI) exceeds, or can reasonably be expected to exceed, the PEL.

*This section* means this § 1910.1026 chromium (VI) standard.

**1910.1026(c)** *Permissible exposure limit (PEL).* The employer shall ensure that no employee is exposed to an airborne concentration of chromium (VI) in excess of 5 micrograms per cubic meter of air (5 µg/m³), calculated as an 8-hour time-weighted average (TWA).

**1910.1026(d)** *Exposure determination.* **(1)** *General.* Each employer who has a workplace or work operation covered by this section shall determine the 8-hour TWA exposure for each employee exposed to chromium (VI). This determination shall be made in accordance with either paragraph (d)(2) or paragraph (d)(3) of this section.

**1910.1026(d)(2)** *Scheduled monitoring option.* **(i)** The employer shall perform initial monitoring to determine the 8-hour TWA exposure for each employee on the basis of a sufficient number of personal breathing zone air samples to accurately characterize full shift exposure on each shift, for each job classification, in each work area. Where an employer does representative sampling instead of sampling all employees in order to meet this requirement, the employer shall sample the employee(s) expected to have the highest chromium (VI) exposures.

**1910.1026(d)(2)(ii)** If initial monitoring indicates that employee exposures are below the action level, the employer may discontinue monitoring for those employees whose exposures are represented by such monitoring.

**1910.1026(d)(2)(iii)** If monitoring reveals employee exposures to be at or above the action level, the employer shall perform periodic monitoring at least every six months.

**1910.1026(d)(2)(iv)** If monitoring reveals employee exposures to be above the PEL, the employer shall perform periodic monitoring at least every three months.

**1910.1026(d)(2)(v)** If periodic monitoring indicates that employee exposures are below the action level, and the result is confirmed by the result of another monitoring taken at least seven days later, the employer may discontinue the monitoring for those employees whose exposures are represented by such monitoring.

**1910.1026(d)(2)(vi)** The employer shall perform additional monitoring when there has been any change in the production process, raw materials, equipment, personnel, work practices, or control methods that may result in new or additional exposures to chromium (VI), or when the employer has any reason to believe that new or additional exposures have occurred.

**1910.1026(d)(3)** *Performance-oriented option.* The employer shall determine the 8-hour TWA exposure for each employee on the basis of any combination of air monitoring data, historical monitoring data, or objective data sufficient to accurately characterize employee exposure to chromium (VI).

**1910.1026(d)(4)** *Employee notification of determination results.* **(i)** Where the exposure determination indicates that employee exposure exceeds the PEL, within 15 working days the employer shall either post the results in an appropriate location that is accessible to all affected employees or shall notify each affected employee individually in writing of the results.

**1910.1026(d)(4)(ii)** Whenever the exposure determination indicates that employee exposure is above the PEL, the employer shall describe in the written notification the corrective action being taken to reduce employee exposure to or below the PEL.

**1910.1026(d)(5)** *Accuracy of measurement.* Where air monitoring is performed to comply with the requirements of this section, the employer shall use a method of monitoring and analysis that can measure chromium (VI) to within an accuracy of plus or minus 25 percent (+/- 25%) and can produce accurate measurements to within a statistical confidence level of 95 percent for airborne concentrations at or above the action level.

**1910.1026(d)(6)** *Observation of monitoring.* **(i)** Where air monitoring is performed to comply with the requirements of this section, the employer shall provide affected employees or their designated representatives an opportunity to observe any monitoring of employee exposure to chromium (VI).

**1910.1026(d)(6)(ii)** When observation of monitoring requires entry into an area where the use of protective clothing or equipment is required, the employer shall provide the observer with clothing and equipment and shall assure that the observer uses such clothing and equipment and complies with all other applicable safety and health procedures.

**1910.1026(e)** *Regulated areas.* **(1)** *Establishment.* The employer shall establish a regulated area wherever an employee's exposure to airborne concentrations of chromium (VI) is, or can reasonably be expected to be, in excess of the PEL.

**1910.1026(e)(2)** *Demarcation.* The employer shall ensure that regulated areas are demarcated from the rest of the workplace in a manner that adequately establishes and alerts employees of the boundaries of the regulated area.

**1910.1026(e)(3)** *Access.* The employer shall limit access to regulated areas to:

**1910.1026(e)(3)(i)** Persons authorized by the employer and required by work duties to be present in the regulated area;

**1910.1026(e)(3)(ii)** Any person entering such an area as a designated representative of employees for the purpose of exercising the right to observe monitoring procedures under paragraph (d) of this section; or

**1910.1026(e)(3)(iii)** Any person authorized by the Occupational Safety and Health Act or regulations issued under it to be in a regulated area.

**1910.1026(f)** *Methods of compliance.* **(1)** *Engineering and work practice controls.* **(i)** Except as permitted in paragraph (f)(1)(ii) and paragraph (f)(1)(iii) of this section, the employer shall use engineering and work practice controls to reduce and maintain employee exposure to chromium (VI) to or below the PEL unless the employer can demonstrate that such controls are not feasible. Wherever feasible engineering and work practice controls are not sufficient to reduce employee exposure to or below the PEL, the employer shall use them to reduce employee exposure to the lowest levels achievable, and shall supplement them by the use of respiratory protection that complies with the requirements of paragraph (g) of this section.

**1910.1026(f)(1)(ii)** Where painting of aircraft or large aircraft parts is performed in the aerospace industry, the employer shall use engineering and work practice controls to reduce and maintain employee exposure to chromium (VI) to or below 25 µg/m³ unless the employer can demonstrate that such controls are not feasible. The employer shall supplement such engineering and work practice controls with the use of respiratory protection that complies with the requirements of paragraph (g) of this section to achieve the PEL.

**1910.1026(f)(1)(iii)** Where the employer can demonstrate that a process or task does not result in any employee exposure to chromium (VI) above the PEL for 30 or more days per year (12 consecutive months), the requirement to implement engineering and work practice controls to achieve the PEL does not apply to that process or task.

**1910.1026(f)(2)** *Prohibition of rotation.* The employer shall not rotate employees to different jobs to achieve compliance with the PEL.

**1910.1026(g)** *Respiratory protection.* **(1)** *General.* The employer shall provide respiratory protection for employees during:

**1910.1026(g)(1)(i)** Periods necessary to install or implement feasible engineering and work practice controls;

**1910.1026(g)(1)(ii)** Work operations, such as maintenance and repair activities, for which engineering and work practice controls are not feasible;

**1910.1026(g)(1)(iii)** Work operations for which an employer has implemented all feasible engineering and work practice controls and such controls are not sufficient to reduce exposures to or below the PEL;

**1910.1026(g)(1)(iv)** Work operations where employees are exposed above the PEL for fewer than 30 days per year, and the employer has elected not to implement engineering and work practice controls to achieve the PEL; or

**1910.1026(g)(1)(v)** Emergencies.

**1910.1026(g)(2)** *Respiratory protection program.* Where respirator use is required by this section, the employer shall institute a respiratory protection program in accordance with 29 CFR 1910.134.

**1910.1026(h)** *Protective work clothing and equipment.* **(1)** *Provision and use.* Where a hazard is present or is likely to be present from skin or eye contact with chromium (VI), the employer shall provide appropriate personal protective clothing and equipment at no cost to employees, and shall ensure that employees use such clothing and equipment.

**1910.1026(h)(2)** *Removal and storage.* **(i)** The employer shall ensure that employees remove all protective clothing and equipment contaminated with chromium (VI) at the end of the work shift or at the completion of their tasks involving chromium (VI) exposure, whichever comes first.

**1910.1026(h)(2)(ii)** The employer shall ensure that no employee removes chromium (VI)-contaminated protective clothing or equipment from the workplace, except for those employees whose job it is to launder, clean, maintain, or dispose of such clothing or equipment.

**1910.1026(h)(2)(iii)** When contaminated protective clothing or equipment is removed for laundering, cleaning, maintenance, or disposal, the employer shall ensure that it is stored and transported in sealed, impermeable bags or other closed, impermeable containers.

**1910.1026(h)(2)(iv)** Bags or containers of contaminated protective clothing or equipment that are removed from change rooms for laundering, cleaning, maintenance, or disposal shall be labeled in accordance with the requirements of the Hazard Communication Standard, 29 CFR 1910.1200.

**1910.1026(h)(3)** *Cleaning and replacement.* **(i)** The employer shall clean, launder, repair and replace all protective clothing and equipment required by this section as needed to maintain its effectiveness.

**1910.1026(h)(3)(ii)** The employer shall prohibit the removal of chromium (VI) from protective clothing and equipment by blowing, shaking, or any other means that disperses chromium (VI) into the air or onto an employee's body.

**1910.1026(h)(3)(iii)** The employer shall inform any person who launders or cleans protective clothing or equipment contaminated with chromium (VI) of the potentially harmful effects of exposure to chromium (VI) and that the clothing and equipment should be laundered or cleaned in a manner that minimizes skin or eye contact with chromium (VI) and effectively prevents the release of airborne chromium (VI) in excess of the PEL.

**1910.1026(i)** *Hygiene areas and practices.* **(1)** *General.* Where protective clothing and equipment is required, the employer shall provide change rooms in conformance with 29 CFR 1910.141. Where skin contact with chromium (VI) occurs, the employer shall provide washing facilities in conformance with 29 CFR 1910.141. Eating and drinking areas provided by the employer shall also be in conformance with § 1910.141.

**1910.1026(i)(2)** *Change rooms.* The employer shall assure that change rooms are equipped with separate storage facilities for protective clothing and equipment and for street clothes, and that these facilities prevent cross-contamination.

**1910.1026(i)(3)** *Washing facilities.* **(i)** The employer shall provide readily accessible washing facilities capable of removing chromium (VI) from the skin, and shall ensure that affected employees use these facilities when necessary.

**1910.1026(i)(3)(ii)** The employer shall ensure that employees who have skin contact with chromium (VI) wash their hands and faces at the end of the work shift and prior to eating, drinking, smoking, chewing tobacco or gum, applying cosmetics, or using the toilet.

**1910.1026(i)(4)** *Eating and drinking areas.* **(i)** Whenever the employer allows employees to consume food or beverages at a worksite where chromium (VI) is present, the employer shall ensure that eating and drinking areas and surfaces are maintained as free as practicable of chromium (VI).

**1910.1026(i)(4)(ii)** The employer shall ensure that employees do not enter eating and drinking areas with protective work clothing or equipment unless surface chromium (VI) has been removed from the clothing and equipment by methods that do not disperse chromium (VI) into the air or onto an employee's body.

**1910.1026(i)(5)** *Prohibited activities.* The employer shall ensure that employees do not eat, drink, smoke, chew tobacco or gum, or apply cosmetics in regulated areas, or in areas where skin or eye contact with chromium (VI) occurs; or carry the products associated with these activities, or store such products in these areas.

**1910.1026(j)** *Housekeeping.* **(1)** *General.* The employer shall ensure that:

**1910.1026(j)(1)(i)** All surfaces are maintained as free as practicable of accumulations of chromium (VI).

**1910.1026(j)(1)(ii)** All spills and releases of chromium (VI) containing material are cleaned up promptly.

**1910.1026(j)(2)** *Cleaning methods.* **(i)** The employer shall ensure that surfaces contaminated with chromium (VI) are cleaned by HEPA-filter vacuuming or other methods that minimize the likelihood of exposure to chromium (VI).

**1910.1026(j)(2)(ii)** Dry shoveling, dry sweeping, and dry brushing may be used only where HEPA-filtered vacuuming or other methods that minimize the likelihood of exposure to chromium (VI) have been tried and found not to be effective.

**1910.1026(j)(2)(iii)** The employer shall not allow compressed air to be used to remove chromium (VI) from any surface unless:

**1910.1026(j)(2)(iii)(A)** The compressed air is used in conjunction with a ventilation system designed to capture the dust cloud created by the compressed air; or

**1910.1026(j)(2)(iii)(B)** No alternative method is feasible.

**1910.1026(j)(2)(iv)** The employer shall ensure that cleaning equipment is handled in a manner that minimizes the reentry of chromium (VI) into the workplace.

**1910.1026(j)(3)** *Disposal.* The employer shall ensure that:

**1910.1026(j)(3)(i)** Waste, scrap, debris, and any other materials contaminated with chromium (VI) and consigned for disposal are collected and disposed of in sealed, impermeable bags or other closed, impermeable containers.

**1910.1026(j)(3)(ii)** Bags or containers of waste, scrap, debris, and any other materials contaminated with chromium (VI) that are consigned for disposal are labeled in accordance with the requirements of the Hazard Communication Standard, 29 CFR 1910.1200.

**1910.1026(k)** *Medical surveillance.* **(1)** *General.* **(i)** The employer shall make medical surveillance available at no cost to the employee, and at a reasonable time and place, for all employees:

**1910.1026(k)(1)(i)(A)** Who are or may be occupationally exposed to chromium (VI) at or above the action level for 30 or more days a year;

**1910.1026(k)(1)(i)(B)** Experiencing signs or symptoms of the adverse health effects associated with chromium (VI) exposure; or

**1910.1026(k)(1)(i)(C)** Exposed in an emergency.

**1910.1026(k)(1)(ii)** The employer shall assure that all medical examinations and procedures required by this section are performed by or under the supervision of a PLHCP.

**1910.1026(k)(2)** *Frequency.* The employer shall provide a medical examination:

**1910.1026(k)(2)(i)** Within 30 days after initial assignment, unless the employee has received a chromium (VI) related medical examination that meets the requirements of this paragraph within the last twelve months;

**1910.1026(k)(2)(ii)** Annually;

**1910.1026(k)(2)(iii)** Within 30 days after a PLHCP's written medical opinion recommends an additional examination;

**1910.1026(k)(2)(iv)** Whenever an employee shows signs or symptoms of the adverse health effects associated with chromium (VI) exposure;

**1910.1026(k)(2)(v)** Within 30 days after exposure during an emergency which results in an uncontrolled release of chromium (VI); or

**1910.1026(k)(2)(vi)** At the termination of employment, unless the last examination that satisfied the requirements of paragraph (k) of this section was less than six months prior to the date of termination.

**1910.1026(k)(3)** *Contents of examination.* A medical examination consists of:

**1910.1026(k)(3)(i)** A medical and work history, with emphasis on: Past, present, and anticipated future exposure to chromium (VI); any history of respiratory system dysfunction; any history of asthma, dermatitis, skin ulceration, or nasal septum perforation; and smoking status and history;

**1910.1026(k)(3)(ii)** A physical examination of the skin and respiratory tract; and

**1910.1026(k)(3)(iii)** Any additional tests deemed appropriate by the examining PLHCP.

**1910.1026(k)(4)** *Information provided to the PLHCP.* The employer shall ensure that the examining PLHCP has a copy of this standard, and shall provide the following information:

# OSHA Standards for General Industry

**1910.1026(k)(4)(i)** A description of the affected employee's former, current, and anticipated duties as they relate to the employee's occupational exposure to chromium (VI);

**1910.1026(k)(4)(ii)** The employee's former, current, and anticipated levels of occupational exposure to chromium (VI);

**1910.1026(k)(4)(iii)** A description of any personal protective equipment used or to be used by the employee, including when and for how long the employee has used that equipment; and

**1910.1026(k)(4)(iv)** Information from records of employment-related medical examinations previously provided to the affected employee, currently within the control of the employer.

**1910.1026(k)(5)** *PLHCP's written medical opinion.* **(i)** The employer shall obtain a written medical opinion from the PLHCP, within 30 days for each medical examination performed on each employee, which contains:

**1910.1026(k)(5)(i)(A)** The PLHCP's opinion as to whether the employee has any detected medical condition(s) that would place the employee at increased risk of material impairment to health from further exposure to chromium (VI);

**1910.1026(k)(5)(i)(B)** Any recommended limitations upon the employee's exposure to chromium (VI) or upon the use of personal protective equipment such as respirators;

**1910.1026(k)(5)(i)(C)** A statement that the PLHCP has explained to the employee the results of the medical examination, including any medical conditions related to chromium (VI) exposure that require further evaluation or treatment, and any special provisions for use of protective clothing or equipment.

**1910.1026(k)(5)(ii)** The PLHCP shall not reveal to the employer specific findings or diagnoses unrelated to occupational exposure to chromium (VI).

**1910.1026(k)(5)(iii)** The employer shall provide a copy of the PLHCP's written medical opinion to the examined employee within two weeks after receiving it.

**1910.1026(l)** *Communication of chromium (VI) hazards to employees.*

**1910.1026(l)(1)** *General.* In addition to the requirements of the Hazard Communication Standard, 29 CFR 1910.1200, employers shall comply with the following requirements.

**1910.1026(l)(2)** *Employee information and training.* **(i)** The employer shall ensure that each employee can demonstrate knowledge of at least the following:

**1910.1026(l)(2)(i)(A)** The contents of this section; and

**1910.1026(l)(2)(i)(B)** The purpose and a description of the medical surveillance program required by paragraph (k) of this section.

**1910.1026(l)(2)(ii)** The employer shall make a copy of this section readily available without cost to all affected employees.

**1910.1026(m)** *Recordkeeping.* **(1)** *Air monitoring data.* **(i)** The employer shall maintain an accurate record of all air monitoring conducted to comply with the requirements of this section.

**1910.1026(m)(1)(ii)** This record shall include at least the following information:

**1910.1026(m)(1)(ii)(A)** The date of measurement for each sample taken;

**1910.1026(m)(1)(ii)(B)** The operation involving exposure to chromium (VI) that is being monitored;

**1910.1026(m)(1)(ii)(C)** Sampling and analytical methods used and evidence of their accuracy;

**1910.1026(m)(1)(ii)(D)** Number, duration, and the results of samples taken;

**1910.1026(m)(1)(ii)(E)** Type of personal protective equipment, such as respirators worn; and

**1910.1026(m)(1)(ii)(F)** Name, social security number, and job classification of all employees represented by the monitoring, indicating which employees were actually monitored.

**1910.1026(m)(1)(iii)** The employer shall ensure that exposure records are maintained and made available in accordance with 29 CFR 1910.1020.

**1910.1026(m)(2)** *Historical monitoring data.* **(i)** Where the employer has relied on historical monitoring data to determine exposure to chromium (VI), the employer shall establish and maintain an accurate record of the historical monitoring data relied upon.

**1910.1026(m)(2)(ii)** The record shall include information that reflects the following conditions:

**1910.1026(m)(2)(ii)(A)** The data were collected using methods that meet the accuracy requirements of paragraph (d)(5) of this section;

**1910.1026(m)(2)(ii)(B)** The processes and work practices that were in use when the historical monitoring data were obtained are essentially the same as those to be used during the job for which exposure is being determined;

**1910.1026(m)(2)(ii)(C)** The characteristics of the chromium (VI) containing material being handled when the historical monitoring data were obtained are the same as those on the job for which exposure is being determined;

**1910.1026(m)(2)(ii)(D)** Environmental conditions prevailing when the historical monitoring data were obtained are the same as those on the job for which exposure is being determined; and

**1910.1026(m)(2)(ii)(E)** Other data relevant to the operations, materials, processing, or employee exposures covered by the exception.

**1910.1026(m)(2)(iii)** The employer shall ensure that historical exposure records are maintained and made available in accordance with 29 CFR 1910.1020.

**1910.1026(m)(3)** *Objective data.* **(i)** The employer shall maintain an accurate record of all objective data relied upon to comply with the requirements of this section.

**1910.1026(m)(3)(ii)** This record shall include at least the following information:

**1910.1026(m)(3)(ii)(A)** The chromium containing material in question;

**1910.1026(m)(3)(ii)(B)** The source of the objective data;

**1910.1026(m)(3)(ii)(C)** The testing protocol and results of testing, or analysis of the material for the release of chromium (VI);

**1910.1026(m)(3)(ii)(D)** A description of the process, operation, or activity and how the data support the determination; and

**1910.1026(m)(3)(ii)(E)** Other data relevant to the process, operation, activity, material, or employee exposures.

**1910.1026(m)(3)(iii)** The employer shall ensure that objective data are maintained and made available in accordance with 29 CFR 1910.1020.

**1910.1026(m)(4)** *Medical surveillance.* **(i)** The employer shall establish and maintain an accurate record for each employee covered by medical surveillance under paragraph (k) of this section.

**1910.1026(m)(4)(ii)** The record shall include the following information about the employee:

**1910.1026(m)(4)(ii)(A)** Name and social security number;

**1910.1026(m)(4)(ii)(B)** A copy of the PLHCP's written opinions;

**1910.1026(m)(4)(ii)(C)** A copy of the information provided to the PLHCP as required by paragraph (k)(4) of this section.

**1910.1026(m)(4)(iii)** The employer shall ensure that medical records are maintained and made available in accordance with 29 CFR 1910.1020.

**1910.1026(n)** *Dates.* **(1)** For employers with 20 or more employees, all obligations of this section, except engineering controls required by paragraph (f) of this section, commence November 27, 2006.

**1910.1026(n)(2)** For employers with 19 or fewer employees, all obligations of this section, except engineering controls required by paragraph (f) of this section, commence May 30, 2007.

**1910.1026(n)(3)** For all employers, engineering controls required by paragraph (f) of this section shall be implemented no later than May 31, 2010.

## §1910.1027 Cadmium.

**1910.1027(a)** *Scope.* This standard applies to all occupational exposures to cadmium and cadmium compounds, in all forms, and in all industries covered by the Occupational Safety and Health Act, except the construction-related industries, which are covered under 29 CFR 1926.63.

**1910.1027(b)** *Definitions.*

*Action level* (AL) is defined as an airborne concentration of cadmium of 2.5 micrograms per cubic meter of air (2.5 μg/m$^3$), calculated as an 8-hour time-weighted average (TWA).

*Assistant Secretary* means the Assistant Secretary of Labor for Occupational Safety and Health, U.S. Department of Labor, or designee.

*Authorized person* means any person authorized by the employer and required by work duties to be present in regulated areas or any person authorized by the OSH Act or regulations issued under it to be in regulated areas.

*Director* means the Director of the National Institute for Occupational Safety and Health (NIOSH), U.S. Department of Health and Human Services, or designee.

*Employee exposure* and similar language referring to the air cadmium level to which an employee is exposed means the exposure to airborne cadmium that would occur if the employee were not using respiratory protective equipment.

*Final medical determination* is the written medical opinion of the employee's health status by the examining physician under paragraphs (l)(3)-(12) of this section or, if multiple physician review under paragraph (l)(13) of this section or the alternative physician determination under paragraph (l)(14) of this section is invoked, it is the final, written medical finding, recommendation or determination that emerges from that process.

*High-efficiency particulate air (HEPA) filter* means a filter capable of trapping and retaining at least 99.97 percent of mono-dispersed particles of 0.3 micrometers in diameter.

*Regulated area* means an area demarcated by the employer where an employee's exposure to airborne concentrations of cadmium exceeds, or can reasonably be expected to exceed the permissible exposure limit (PEL).

*This section* means this cadmium standard.

**1910.1027(c)** *Permissible Exposure Limit (PEL).* The employer shall assure that no employee is exposed to an airborne concentration of cadmium in excess of five micrograms per cubic meter of air (5 μg/m$^3$), calculated as an eight-hour time-weighted average exposure (TWA).

**1910.1027(d)** *Exposure monitoring*—**(1)** *General.* **(i)** Each employer who has a workplace or work operation covered by this section shall determine if any employee may be exposed to cadmium at or above the action level.

**1910.1027(d)(1)(ii)** Determinations of employee exposure shall be made from breathing zone air samples that reflect the monitored employee's regular, daily 8-hour TWA exposure to cadmium.

**1910.1027(d)(1)(iii)** Eight-hour TWA exposures shall be determined for each employee on the basis of one or more personal breathing zone air samples reflecting full shift exposure on each shift, for each job classification, in each work area. Where several employees perform the same job tasks, in the same job classification, on the same shift, in the same work area, and the length, duration, and level of cadmium exposures are similar, an employer may sample a representative fraction of the employees instead of all employees in order to meet this requirement. In representative sampling, the employer shall sample the employee(s) expected to have the highest cadmium exposures.

**1910.1027(d)(2)** *Specific.* **(i)** Initial monitoring. Except as provided for in paragraphs (d)(2)(ii) and (d)(2)(iii) of this section, the employer shall monitor employee exposures and shall base initial determinations on the monitoring results.

**1910.1027(d)(2)(ii)** Where the employer has monitored after September 14, 1991, under conditions that in all important aspects closely resemble those currently prevailing and where that monitoring satisfies all other requirements of this section, including the accuracy and confidence levels of paragraph (d)(6) of this section, the employer may rely on such earlier monitoring results to satisfy the requirements of paragraph (d)(2)(i) of this section.

**1910.1027(d)(2)(iii)** Where the employer has objective data, as defined in paragraph (n)(2) of this section, demonstrating that employee exposure to cadmium will not exceed the action level under the expected conditions of processing, use, or handling, the employer may rely upon such data instead of implementing initial monitoring.

**1910.1027(d)(3)** *Monitoring Frequency (periodic monitoring).* **(i)** If the initial monitoring or periodic monitoring reveals employee exposures to be at or above the action level, the employer shall monitor at a frequency and pattern needed to represent the levels of exposure of employees and where exposures are above the PEL to assure the adequacy of respiratory selection and the effectiveness of engineering and work practice controls. However, such exposure monitoring shall be performed at least every six months. The employer, at a minimum, shall continue these semi-annual measurements unless and until the conditions set out in paragraph (d)(3)(ii) of this section are met.

**1910.1027(d)(3)(ii)** If the initial monitoring or the periodic monitoring indicates that employee exposures are below the action level and that result is confirmed by the results of another monitoring taken at least seven days later, the employer may discontinue the monitoring for those employees whose exposures are represented by such monitoring.

**1910.1027(d)(4)** *Additional Monitoring.* The employer also shall institute the exposure monitoring required under paragraphs (d)(2)(i) and (d)(3) of this section whenever there has been a change in the raw materials, equipment, personnel, work practices, or finished products that may result in additional employees being exposed to cadmium at or above the action level or in employees already exposed to cadmium at or above the action level being exposed above the PEL, or whenever the employer has any reason to suspect that any other change might result in such further exposure.

**1910.1027(d)(5)** *Employee Notification of Monitoring Results.* **(i)** The employer must, within 15 working days after the receipt of the results of any monitoring performed under this section, notify each affected employee of these results either individually in writing or by posting the results in an appropriate location that is accessible to employees.528

**1910.1027(d)(5)(ii)** Wherever monitoring results indicate that employee exposure exceeds the PEL, the employer shall include in the written notice a statement that the PEL has been exceeded and a description of the corrective action being taken by the employer to reduce employee exposure to or below the PEL.

**1910.1027(d)(6)** *Accuracy of measurement.* The employer shall use a method of monitoring and analysis that has an accuracy of not less than plus or minus 25 percent (±25%), with a confidence level of 95 percent, for airborne concentrations of cadmium at or above the action level, the permissible exposure limit (PEL), and the separate engineering control air limit (SECAL).

**1910.1027(e)** *Regulated areas*—**(1)** *Establishment.* The employer shall establish a regulated area wherever an employee's exposure to airborne concentrations of cadmium is, or can reasonably be expected to be in excess of the permissible exposure limit (PEL).

**1910.1027(e)(2)** *Demarcation.* Regulated areas shall be demarcated from the rest of the workplace in any manner that adequately establishes and alerts employees of the boundaries of the regulated area.

**1910.1027(e)(3)** *Access.* Access to regulated areas shall be limited to authorized persons.

**1910.1027(e)(4)** *Provision of respirators.* Each person entering a regulated area shall be supplied with and required to use a respirator, selected in accordance with paragraph (g)(2) of this section.

**1910.1027(e)(5)** *Prohibited activities.* The employer shall assure that employees do not eat, drink, smoke, chew tobacco or gum, or apply cosmetics in regulated areas, carry the products associated with these activities into regulated areas, or store such products in those areas.

**1910.1027(f)** *Methods of compliance*—**(1)** *Compliance hierarchy.* **(i)** Except as specified in paragraphs (f)(1) (ii), (iii) and (iv) of this section the employer shall implement engineering and work practice controls to reduce and maintain employee exposure to cadmium at or below the PEL, except to the extent that the employer can demonstrate that such controls are not feasible.

**1910.1027(f)(1)(ii)** Except as specified in paragraphs (f)(1) (iii) and (iv) of this section, in industries where a separate engineering control air limit (SECAL) has been specified for particular processes (See Table 1 in this paragraph (f)(1)(ii)), the employer shall implement engineering and work practice controls to reduce and maintain employee exposure at or below the SECAL, except to the extent that the employer can demonstrate that such controls are not feasible.

# OSHA Standards for General Industry

### TABLE 1—SEPARATE ENGINEERING CONTROL AIRBORNE LIMITS (SECALS) FOR PROCESSES IN SELECTED INDUSTRIES

| Industry | Process | SECAL (µg/m³) |
|---|---|---|
| Nickel cadmium battery | Plate making, plate preparation | 50 |
| | All other processes | 15 |
| Zinc/Cadmium refining* | Cadmium refining, casting, melting, oxide production, sinter plant | 50 |
| Pigment manufacture | Calcine, crushing, milling, blending | 50 |
| | All other processes | 15 |
| Stabilizers* | Cadmium oxide charging, crushing, drying, blending | 50 |
| Lead smelting* | Sinter plant, blast furnace, baghouse, yard area | 50 |
| Plating* | Mechanical plating | 15 |

\* Processes in these industries that are not specified in this table must achieve the PEL using engineering controls and work practices as required in f(1)(i).

**1910.1027(f)(1)(iii)** The requirement to implement engineering and work practice controls to achieve the PEL or, where applicable, the SECAL does not apply where the employer demonstrates the following:

**1910.1027(f)(1)(iii)(A)** The employee is only intermittently exposed; and

**1910.1027(f)(1)(iii)(B)** The employee is not exposed above the PEL on 30 or more days per year (12 consecutive months).

**1910.1027(f)(1)(iv)** Wherever engineering and work practice controls are required and are not sufficient to reduce employee exposure to or below the PEL or, where applicable, the SECAL, the employer nonetheless shall implement such controls to reduce exposures to the lowest levels achievable. The employer shall supplement such controls with respiratory protection that complies with the requirements of paragraph (g) of this section and the PEL.

**1910.1027(f)(1)(v)** The employer shall not use employee rotation as a method of compliance.

**1910.1027(f)(2)** *Compliance program.* **(i)** Where the PEL is exceeded, the employer shall establish and implement a written compliance program to reduce employee exposure to or below the PEL by means of engineering and work practice controls, as required by paragraph (f)(1) of this section. To the extent that engineering and work practice controls cannot reduce exposures to or below the PEL, the employer shall include in the written compliance program the use of appropriate respiratory protection to achieve compliance with the PEL.

**1910.1027(f)(2)(ii)** Written compliance programs shall include at least the following:

**1910.1027(f)(2)(ii)(A)** A description of each operation in which cadmium is emitted; e.g., machinery used, material processed, controls in place, crew size, employee job responsibilities, operating procedures, and maintenance practices;

**1910.1027(f)(2)(ii)(B)** A description of the specific means that will be employed to achieve compliance, including engineering plans and studies used to determine methods selected for controlling exposure to cadmium, as well as, where necessary, the use of appropriate respiratory protection to achieve the PEL;

**1910.1027(f)(2)(ii)(C)** A report of the technology considered in meeting the PEL;

**1910.1027(f)(2)(ii)(D)** Air monitoring data that document the sources of cadmium emissions;

**1910.1027(f)(2)(ii)(E)** A detailed schedule for implementation of the program, including documentation such as copies of purchase orders for equipment, construction contracts, etc.;

**1910.1027(f)(2)(ii)(F)** A work practice program that includes items required under paragraphs (h), (i), and (j) of this section;

**1910.1027(f)(2)(ii)(G)** A written plan for emergency situations, as specified in paragraph (h) of this section; and

**1910.1027(f)(2)(ii)(H)** Other relevant information.

**1910.1027(f)(2)(iii)** The written compliance programs shall be reviewed and updated at least annually, or more often if necessary, to reflect significant changes in the employer's compliance status.

**1910.1027(f)(2)(iv)** Written compliance programs shall be provided upon request for examination and copying to affected employees, designated employee representatives as well as to the Assistant Secretary, and the Director.

**1910.1027(f)(3)** *Mechanical ventilation.* **(i)** When ventilation is used to control exposure, measurements that demonstrate the effectiveness of the system in controlling exposure, such as capture velocity, duct velocity, or static pressure shall be made as necessary to maintain its effectiveness.

**1910.1027(f)(3)(ii)** Measurements of the system's effectiveness in controlling exposure shall be made as necessary within five working days of any change in production, process, or control that might result in a significant increase in employee exposure to cadmium.

**1910.1027(f)(3)(iii)** Recirculation of air. If air from exhaust ventilation is recirculated into the workplace, the system shall have a high efficiency filter and be monitored to assure effectiveness.

**1910.1027(f)(3)(iv)** Procedures shall be developed and implemented to minimize employee exposure to cadmium when maintenance of ventilation systems and changing of filters is being conducted.

**1910.1027(g)** *Respiratory protection*—**(1)** *General.* For employees who use respirators required by this section, the employer must provide respirators that comply with the requirements of this paragraph. Respirators must be used during:

**1910.1027(g)(1)(i)** Periods necessary to install or implement feasible engineering and work-practice controls when employee exposure levels exceed the PEL.

**1910.1027(g)(1)(ii)** Maintenance and repair activities, and brief or intermittent operations, for which employee exposures exceed the PEL and engineering and work-practice controls are not feasible or are not required.

**1910.1027(g)(1)(iii)** Activities in regulated areas specified in paragraph (e) of this section.

**1910.1027(g)(1)(iv)** Work operations for which the employer has implemented all feasible engineering and work-practice controls and such controls are not sufficient to reduce employee exposures to or below the PEL.

**1910.1027(g)(1)(v)** Work operations for which an employee is exposed to cadmium at or above the action level, and the employee requests a respirator.

**1910.1027(g)(1)(vi)** Work operations for which an employee is exposed to cadmium above the PEL and engineering controls are not required by paragraph (f)(1)(ii) of this section.

**1910.1027(g)(1)(vii)** Emergencies.

**1910.1027(g)(2)** *Respirator program.* **(i)** The employer must implement a respiratory protection program in accordance with 29 CFR 1910.134(b) through (d) (except (d)(1)(iii)), and (f) through (m).

**1910.1027(g)(2)(ii)** No employees must use a respirator if, based on their most recent medical examination, the examining physician determines that they will be unable to continue to function normally while using a respirator. If the physician determines that the employee must be limited in, or removed from, their current job because of their inability to use a respirator, the limitation or removal must be in accordance with paragraphs (l)(11) and (12) of this section.

**1910.1027(g)(2)(iii)** If an employee has breathing difficulty during fit testing or respirator use, the employer must provide the employee with a medical examination in accordance with paragraph (l)(6)(ii) of this section to determine if the employee can use a respirator while performing the required duties.

**1910.1027(g)(3)** *Respirator selection.* **(i)** The employer must select the appropriate respirator from Table 2 of this section.

### TABLE 2—RESPIRATORY PROTECTION FOR CADMIUM

| Airborne concentration or condition of use[a] | Required respirator type[b] |
|---|---|
| 10 × or less | A half mask, air-purifying equipped with a HEPA[c] filter.[d] |
| 25 × or less | A powered air-purifying respirator ("PAPR") with a loose-fitting hood or helmet equipped with a HEPA filter, or a supplied-air respirator with a loose-fitting hood or helmet facepiece operated in the continuous flow mode. |
| 50 × or less | A full facepiece air-purifying respirator equipped with a HEPA filter, or a powered air-purifying respirator with a tight-fitting half mask equipped with a HEPA filter, or a supplied-air respirator with a tight-fitting half mask operated in the continuous flow mode. |
| 250 × or less | A powered air-purifying respirator with a tight fitting full facepiece equipped with a HEPA filter, or a supplied-air respirator with a tight-fitting full facepiece operated in the continuous flow mode. |
| 1000 × or less | A supplied air respirator with half mask or full facepiece operated in the pressure demand or other positive pressure mode. |
| >1000 × or unknown concentrations | A self-contained breathing apparatus with a full facepiece operated in the pressure demand or other positive pressure mode, or a supplied-air respirator with a full facepiece operated in the pressure demand or other positive pressure mode and equipped with an auxiliary escape type self-contained breathing apparatus operated in the pressure demand mode. |
| Fire fighting | A self-contained breathing apparatus with full facepiece operated in the pressure demand or other positive pressure mode. |

[a] Concentrations expressed as multiple of the PEL.
[b] Respirators assigned for higher environmental concentrations may be used at lower exposure levels. Quantitative fit testing is required for all tight-fitting air purifying respirators where airborne concentration of cadmium exceeds 10 times the TWA PEL (10 × 5 ug/m$^3$ = 50 ug/m$^3$). A full facepiece respirator is required when eye irritation is experienced.
[c] HEPA means High-efficiency Particulate Air.
[d] Fit testing, qualitative or quantitative, is required.
SOURCE: Respiratory Decision Logic, NIOSH, 1987.

**1910.1027(g)(3)(ii)** The employer must provide an employee with a powered air-purifying respirator instead of a negative-pressure respirator when an employee who is entitled to a respirator chooses to use this type of respirator and such a respirator provides adequate protection to the employee.

**1910.1027(h)** *Emergency situations.* The employer shall develop and implement a written plan for dealing with emergency situations involving substantial releases of airborne cadmium. The plan shall include provisions for the use of appropriate respirators and personal protective equipment. In addition, employees not essential to correcting the emergency situation shall be restricted from the area and normal operations halted in that area until the emergency is abated.

**1910.1027(i)** *Protective work clothing and equipment*—**(1)** *Provision and use.* If an employee is exposed to airborne cadmium above the PEL or where skin or eye irritation is associated with cadmium exposure at any level, the employer shall provide at no cost to the employee, and assure that the employee uses, appropriate protective work clothing and equipment that prevents contamination of the employee and the employee's garments. Protective work clothing and equipment includes, but is not limited to:

**1910.1027(i)(1)(i)** Coveralls or similar full-body work clothing;

**1910.1027(i)(1)(ii)** Gloves, head coverings, and boots or foot coverings; and

**1910.1027(i)(1)(iii)** Face shields, vented goggles, or other appropriate protective equipment that complies with 29 CFR 1910.133.

**1910.1027(i)(2)** *Removal and storage.* **(i)** The employer shall assure that employees remove all protective clothing and equipment contaminated with cadmium at the completion of the work shift and do so only in change rooms provided in accordance with paragraph (j)(1) of this section.

**1910.1027(i)(2)(ii)** The employer shall assure that no employee takes cadmium-contaminated protective clothing or equipment from the workplace, except for employees authorized to do so for purposes of laundering, cleaning, maintaining, or disposing of cadmium contaminated protective clothing and equipment at an appropriate location or facility away from the workplace.

**1910.1027(i)(2)(iii)** The employer shall assure that contaminated protective clothing and equipment, when removed for laundering, cleaning, maintenance, or disposal, is placed and stored in sealed, impermeable bags or other closed, impermeable containers that are designed to prevent dispersion of cadmium dust.

**1910.1027(i)(2)(iv)** The employer shall assure that bags or containers of contaminated protective clothing and equipment that are to be taken out of the change rooms or the workplace for laundering, cleaning, maintenance or disposal shall bear labels in accordance with paragraph (m)(3) of this section.

**1910.1027(i)(3)** *Cleaning, replacement, and disposal.* **(i)** The employer shall provide the protective clothing and equipment required by paragraph (i)(1) of this section in a clean and dry condition as often as necessary to maintain its effectiveness, but in any event at least weekly. The employer is responsible for cleaning and laundering the protective clothing and equipment required by this paragraph to maintain its effectiveness and is also responsible for disposing of such clothing and equipment.

**1910.1027(i)(3)(ii)** The employer also is responsible for repairing or replacing required protective clothing and equipment as needed to maintain its effectiveness. When rips or tears are detected while an employee is working they shall be immediately mended, or the worksuit shall be immediately replaced.

**1910.1027(i)(3)(iii)** The employer shall prohibit the removal of cadmium from protective clothing and equipment by blowing, shaking, or any other means that disperses cadmium into the air.

**1910.1027(i)(3)(iv)** The employer shall assure that any laundering of contaminated clothing or cleaning of contaminated equipment in the workplace is done in a manner that prevents the release of airborne cadmium in excess of the permissible exposure limit prescribed in paragraph (c) of this section.

**1910.1027(i)(3)(v)** The employer shall inform any person who launders or cleans protective clothing or equipment contaminated with cadmium of the potentially harmful effects of exposure to cadmium and that the clothing and equipment should be laundered or cleaned in a manner to effectively prevent the release of airborne cadmium in excess of the PEL.

**1910.1027(j)** *Hygiene areas and practices*—**(1)** *General.* For employees whose airborne exposure to cadmium is above the PEL, the employer shall provide clean change rooms, handwashing facilities, showers, and lunchroom facilities that comply with 29 CFR 1910.141.

**1910.1027(j)(2)** *Change rooms.* The employer shall assure that change rooms are equipped with separate storage facilities for street clothes and for protective clothing and equipment, which are designed to prevent dispersion of cadmium and contamination of the employee's street clothes.

# OSHA Standards for General Industry

**1910.1027(j)(3)** *Showers and handwashing facilities.* **(i)** The employer shall assure that employees who are exposed to cadmium above the PEL shower during the end of the work shift.

**1910.1027(j)(3)(ii)** The employer shall assure that employees whose airborne exposure to cadmium is above the PEL wash their hands and faces prior to eating, drinking, smoking, chewing tobacco or gum, or applying cosmetics.

**1910.1027(j)(4)** *Lunchroom facilities.* **(i)** The employer shall assure that the lunchroom facilities are readily accessible to employees, that tables for eating are maintained free of cadmium, and that no employee in a lunchroom facility is exposed at any time to cadmium at or above a concentration of 2.5 µg/m³.

**1910.1027(j)(4)(ii)** The employer shall assure that employees do not enter lunchroom facilities with protective work clothing or equipment unless surface cadmium has been removed from the clothing and equipment by HEPA vacuuming or some other method that removes cadmium dust without dispersing it.

**1910.1027(k)** *Housekeeping.* **(1)** All surfaces shall be maintained as free as practicable of accumulations of cadmium.

**1910.1027(k)(2)** All spills and sudden releases of material containing cadmium shall be cleaned up as soon as possible.

**1910.1027(k)(3)** Surfaces contaminated with cadmium shall, wherever possible, be cleaned by vacuuming or other methods that minimize the likelihood of cadmium becoming airborne.

**1910.1027(k)(4)** HEPA-filtered vacuuming equipment or equally effective filtration methods shall be used for vacuuming. The equipment shall be used and emptied in a manner that minimizes the reentry of cadmium into the workplace.

**1910.1027(k)(5)** Shoveling, dry or wet sweeping, and brushing may be used only where vacuuming or other methods that minimize the likelihood of cadmium becoming airborne have been tried and found not to be effective.

**1910.1027(k)(6)** Compressed air shall not be used to remove cadmium from any surface unless the compressed air is used in conjunction with a ventilation system designed to capture the dust cloud created by the compressed air.

**1910.1027(k)(7)** Waste, scrap, debris, bags, containers, personal protective equipment, and clothing contaminated with cadmium and consigned for disposal shall be collected and disposed of in sealed impermeable bags or other closed, impermeable containers. These bags and containers shall be labeled in accordance with paragraph (m)(2) of this section.

**1910.1027(l)** *Medical surveillance*—**(1)** *General*—**(i)** *Scope.* **(A)** Currently exposed—The employer shall institute a medical surveillance program for all employees who are or may be exposed to cadmium at or above the action level unless the employer demonstrates that the employee is not, and will not be, exposed at or above the action level on 30 or more days per year (twelve consecutive months); and,

**1910.1027(l)(1)(i)(B)** Previously exposed—The employer shall also institute a medical surveillance program for all employees who prior to the effective date of this section might previously have been exposed to cadmium at or above the action level by the employer, unless the employer demonstrates that the employee did not prior to the effective date of this section work for the employer in jobs with exposure to cadmium for an aggregated total of more than 60 months.

**1910.1027(l)(1)(ii)** To determine an employee's fitness for using a respirator, the employer shall provide the limited medical examination specified in paragraph (l)(6) of this section.

**1910.1027(l)(1)(iii)** The employer shall assure that all medical examinations and procedures required by this standard are performed by or under the supervision of a licensed physician, who has read and is familiar with the health effects section of appendix A to this section, the regulatory text of this section, the protocol for sample handling and laboratory selection in appendix F to this section, and the questionnaire of appendix D to this section. These examinations and procedures shall be provided without cost to the employee and at a time and place that is reasonable and convenient to employees.

**1910.1027(l)(1)(iv)** The employer shall assure that the collecting and handling of biological samples of cadmium in urine (CdU), cadmium in blood (CdB), and beta-2 microglobulin in urine ($\beta_2$-M) taken from employees under this section is done in a manner that assures their reliability and that analysis of biological samples of cadmium in urine (CdU), cadmium in blood (CdB), and beta-2 microglobulin in urine ($\beta_2$-M) taken from employees under this section is performed in laboratories with demonstrated proficiency for that particular analyte. (See appendix F to this section.)

**1910.1027(l)(2)** *Initial examination.* **(i)** The employer shall provide an initial (preplacement) examination to all employees covered by the medical surveillance program required in paragraph (l)(1)(i) of this section. The examination shall be provided to those employees within 30 days after initial assignment to a job with exposure to cadmium or no later than 90 days after the effective date of this section, whichever date is later.

**1910.1027(l)(2)(ii)** The initial (preplacement) medical examination shall include:

**1910.1027(l)(2)(ii)(A)** A detailed medical and work history, with emphasis on: Past, present, and anticipated future exposure to cadmium; any history of renal, cardiovascular, respiratory, hematopoietic, reproductive, and/or musculo-skeletal system dysfunction; current usage of medication with potential nephrotoxic side-effects; and smoking history and current status; and

**1910.1027(l)(2)(ii)(B)** Biological monitoring that includes the following tests:

**1910.1027(l)(2)(ii)(B)(1)** Cadmium in urine (CdU), standardized to grams of creatinine (g/Cr);

**1910.1027(l)(2)(ii)(B)(2)** Beta-2 microglobulin in urine ($\beta_2$-M), standardized to grams of creatinine (g/Cr), with pH specified, as described in appendix F to this section; and

**1910.1027(l)(2)(ii)(B)(3)** Cadmium in blood (CdB), standardized to liters of whole blood (lwb).

**1910.1027(l)(2)(iii)** Recent Examination: An initial examination is not required to be provided if adequate records show that the employee has been examined in accordance with the requirements of paragraph (l)(2)(ii) of this section within the past 12 months. In that case, such records shall be maintained as part of the employee's medical record and the prior exam shall be treated as if it were an initial examination for the purposes of paragraphs (l)(3) and (4) of this section.

**1910.1027(l)(3)** *Actions triggered by initial biological monitoring:* **(i)** If the results of the initial biological monitoring tests show the employee's CdU level to be at or below 3 µg/g Cr, $\beta_2$-M level to be at or below 300 µg/g Cr and CdB level to be at or below 5 µg/lwb, then:

**1910.1027(l)(3)(i)(A)** For currently exposed employees, who are subject to medical surveillance under paragraph (l)(1)(i)(A) of this section, the employer shall provide the minimum level of periodic medical surveillance in accordance with the requirements in paragraph (l)(4)(i) of this section; and

**1910.1027(l)(3)(i)(B)** For previously exposed employees, who are subject to medical surveillance under paragraph (l)(1)(i)(B) of this section, the employer shall provide biological monitoring for CdU, $\beta_2$-M, and CdB one year after the initial biological monitoring and then the employer shall comply with the requirements of paragraph (l)(4)(v) of this section.

**1910.1027(l)(3)(ii)** For all employees who are subject to medical surveillance under paragraph (l)(1)(i) of this section, if the results of the initial biological monitoring tests show the level of CdU to exceed 3 µg/g Cr, the level of $\beta_2$-M to exceed 300 µg/g Cr, or the level of CdB to exceed 5 µg/lwb, the employer shall:

**1910.1027(l)(3)(ii)(A)** Within two weeks after receipt of biological monitoring results, reassess the employee's occupational exposure to cadmium as follows:

**1910.1027(l)(3)(ii)(A)(1)** Reassess the employee's work practices and personal hygiene;

**1910.1027(l)(3)(ii)(A)(2)** Reevaluate the employee's respirator use, if any, and the respirator program;

**1910.1027(l)(3)(ii)(A)(3)** Review the hygiene facilities;

**1910.1027(l)(3)(ii)(A)(4)** Reevaluate the maintenance and effectiveness of the relevant engineering controls;

**1910.1027(l)(3)(ii)(A)(5)** Assess the employee's smoking history and status;

**1910.1027(l)(3)(ii)(B)** Within 30 days after the exposure reassessment, specified in paragraph (l)(3)(ii)(A) of this section, take reasonable steps to correct any deficiencies found in the reassessment that may be responsible for the employee's excess exposure to cadmium; and,

**1910.1027(l)(3)(ii)(C)** Within 90 days after receipt of biological monitoring results, provide a full medical examination to the employee in accordance with the requirements of paragraph (l)(4)(ii) of this section. After completing the medical examination, the examining physician shall determine in a written medical opinion whether to medically remove the employee. If the physician determines that medical removal is not necessary, then until the employee's CdU level falls to or below 3 µg/g Cr, $\beta_2$-M level falls to or below 300 µg/g Cr and CdB level falls to or below 5 µg/lwb, the employer shall:

**1910.1027(l)(3)(ii)(C)(1)** Provide biological monitoring in accordance with paragraph (l)(2)(ii)(B) of this section on a semiannual basis; and

**1910.1027(l)(3)(ii)(C)(2)** Provide annual medical examinations in accordance with paragraph (l)(4)(ii) of this section.

**1910.1027(l)(3)(iii)** For all employees who are subject to medical surveillance under paragraph (l)(1)(i) of this section, if the results of the initial biological monitoring tests show the level of CdU to be in excess of 15 µg/g Cr, or the level of CdB to be in excess of 15 µg/lwb, or the level of $\beta_2$-M to be in excess of 1,500 µg/g Cr, the employer shall comply with the requirements of paragraphs (l)(3)(ii)(A)-(B) of this section. Within 90 days after receipt of biological monitoring results, the employer shall provide a full medical examination to the employee in accordance with the requirements of paragraph (l)(4)(ii) of this section. After completing the medical examination, the examining physician shall determine in a written medical opinion whether to medically remove the employee. However, if the initial biological monitoring results and the biological monitoring results obtained during the medical examination both show that: CdU exceeds 15 µg/g Cr; or CdB exceeds 15 µg/lwb; or $\beta_2$-M exceeds 1500 µg/g Cr, and in addition CdU exceeds 3 µg/g Cr or CdB exceeds 5 µg/liter of whole blood, then the physician shall medically remove the employee from exposure to cadmium at or above the action level. If the second set of biological monitoring results obtained during the medical examination does not show that a mandatory removal trigger level has been exceeded, then the employee is not required to be removed by the mandatory provisions of this paragraph. If the employee is not required to be removed by the mandatory provisions of this paragraph or by the physician's determination, then until the employee's CdU level falls to or below 3 µg/g Cr, $\beta_2$-M level falls to or below 300 µg/g Cr and CdB level falls to or below 5 µg/lwb, the employer shall:

**1910.1027(l)(3)(iii)(A)** Periodically reassess the employee's occupational exposure to cadmium;

**1910.1027(l)(3)(iii)(B)** Provide biological monitoring in accordance with paragraph (l)(2)(ii)(B) of this section on a quarterly basis; and

**1910.1027(l)(3)(iii)(C)** Provide semiannual medical examinations in accordance with paragraph (l)(4)(ii) of this section.

**1910.1027(l)(3)(iv)** For all employees to whom medical surveillance is provided, beginning on January 1, 1999, and in lieu of paragraphs (l)(3)(i)-(iii) of this section:

**1910.1027(l)(3)(iv)(A)** If the results of the initial biological monitoring tests show the employee's CdU level to be at or below 3 µg/g Cr, $\beta_2$-M level to be at or below 300 µg/g Cr and CdB level to be at or below 5 µg/lwb, then for currently exposed employees, the employer shall comply with the requirements of paragraph (l)(3)(i)(A) of this section, and for previously exposed employees, the employer shall comply with the requirements of paragraph (l)(3)(i)(B) of this section;

**1910.1027(l)(3)(iv)(B)** If the results of the initial biological monitoring tests show the level of CdU to exceed 3 µg/g Cr, the level of $\beta_2$-M to exceed 300 µg/g Cr, or the level of CdB to exceed 5 µg/lwb, the employer shall comply with the requirements of paragraphs (l)(3)(ii)(A)-(C) of this section; and,

**1910.1027(l)(3)(iv)(C)** If the results of the initial biological monitoring tests show the level of CdU to be in excess of 7 µg/g Cr, or the level of CdB to be in excess of 10 µg/lwb, or the level of $\beta_2$-M to be in excess of 750 µg/g Cr, the employer shall: Comply with the requirements of paragraphs (l)(3)(ii)(A)-(B) of this section; and, within 90 days after receipt of biological monitoring results, provide a full medical examination to the employee in accordance with the requirements of paragraph (l)(4)(ii) of this section. After completing the medical examination, the examining physician shall determine in a written medical opinion whether to medically remove the employee. However, if the initial biological monitoring results and the biological monitoring results obtained during the medical examination both show that: CdU exceeds 7 µg/g Cr; or CdB exceeds 10 µg/lwb; or $\beta_2$-M exceeds 750 µg/g Cr, and in addition CdU exceeds 3 µg/g Cr or CdB exceeds 5 µg/liter of whole blood, then the physician shall medically remove the employee from exposure to cadmium at or above the action level. If the second set of biological monitoring results obtained during the medical examination does not show that a mandatory removal trigger level has been exceeded, then the employee is not required to be removed by the mandatory provisions of this paragraph. If the employee is not required to be removed by the mandatory provisions of this paragraph or by the physician's determination, then until the employee's CdU level falls to or below 3 µg/g Cr, $\beta_2$-M level falls to or below 300 µg/g Cr and CdB level falls to or below 5 µg/lwb, the employer shall: periodically reassess the employee's occupational exposure to cadmium; provide biological monitoring in accordance with paragraph (l)(2)(ii)(B) of this section on a quarterly basis; and provide semiannual medical examinations in accordance with paragraph (l)(4)(ii) of this section.

**1910.1027(l)(4)** *Periodic medical surveillance.* **(i)** For each employee who is covered under paragraph (l)(1)(i)(A) of this section, the employer shall provide at least the minimum level of periodic medical surveillance, which consists of periodic medical examinations and periodic biological monitoring. A periodic medical examination shall be provided within one year after the initial examination required by paragraph (l)(2) of this section and thereafter at least biennially. Biological sampling shall be provided at least annually, either as part of a periodic medical examination or separately as periodic biological monitoring.

**1910.1027(l)(4)(ii)** The periodic medical examination shall include:

**1910.1027(l)(4)(ii)(A)** A detailed medical and work history, or update thereof, with emphasis on: Past, present and anticipated future exposure to cadmium; smoking history and current status; reproductive history; current use of medications with potential nephrotoxic side-effects; any history of renal, cardiovascular, respiratory, hematopoietic, and/or musculo-skeletal system dysfunction; and as part of the medical and work history, for employees who wear respirators, questions 3-11 and 25-32 in Appendix D to this section;

**1910.1027(l)(4)(ii)(B)** A complete physical examination with emphasis on: Blood pressure, the respiratory system, and the urinary system;

**1910.1027(l)(4)(ii)(C)** A 14 inch by 17 inch, or a reasonably standard sized posterior-anterior chest X-ray (after the initial X-ray, the frequency of chest X-rays is to be determined by the examining physician);

**1910.1027(l)(4)(ii)(D)** Pulmonary function tests, including forced vital capacity (FVC) and forced expiratory volume at 1 second (FEV1);

**1910.1027(l)(4)(ii)(E)** Biological monitoring, as required in paragraph (l)(2)(ii)(B) of this section;

**1910.1027(l)(4)(ii)(F)** Blood analysis, in addition to the analysis required under paragraph (l)(2)(ii)(B) of this section, including blood urea nitrogen, complete blood count, and serum creatinine;

**1910.1027(l)(4)(ii)(G)** Urinalysis, in addition to the analysis required under paragraph (l)(2)(ii)(B) of this section, including the determination of albumin, glucose, and total and low molecular weight proteins;

**1910.1027(l)(4)(ii)(H)** For males over 40 years old, prostate palpation, or other at least as effective diagnostic test(s); and

**1910.1027(l)(4)(ii)(I)** Any additional tests deemed appropriate by the examining physician.

**1910.1027(l)(4)(iii)** Periodic biological monitoring shall be provided in accordance with paragraph (l)(2)(ii)(B) of this section.

**1910.1027(l)(4)(iv)** If the results of periodic biological monitoring or the results of biological monitoring performed as part of the periodic medical examination show the level of the employee's CdU, $\beta_2$-M, or CdB to be in excess of the levels specified in paragraphs (l)(3)(ii) or (iii); or, beginning on January 1, 1999, in excess of the levels specified in paragraphs (l)(3)(ii) or (iv) of this section, the employer shall take the appropriate actions specified in paragraphs (l)(3)(ii)-(iv) of this section.

**1910.1027(l)(4)(v)** For previously exposed employees under paragraph (l)(1)(i)(B) of this section:

**1910.1027(l)(4)(v)(A)** If the employee's levels of CdU did not exceed 3 µg/g Cr, CdB did not exceed 5 µg/lwb, and $\beta_2$-M did not exceed 300 µg/g Cr in the initial biological monitoring tests, and if the results of the followup biological monitoring required by paragraph (l)(3)(i)(B) of this section one year after the initial examination confirm the previous results, the employer may discontinue all periodic medical surveillance for that employee.

**1910.1027(l)(4)(v)(B)** If the initial biological monitoring results for CdU, CdB, or $\beta_2$-M were in excess of the levels specified in paragraph (l)(3)(i) of this section, but subsequent biological monitoring results required by paragraph (l)(3)(ii)-(iv) of this section show that the employee's CdU levels no longer exceed 3 µg/g Cr, CdB levels no longer exceed 5 µg/lwb, and $\beta_2$-M levels no longer exceed 300 µg/g Cr, the employer shall provide biological monitoring for CdU, CdB, and $\beta_2$-M one year after these most recent biological monitoring results. If the results of the followup biological monitoring, specified in this paragraph, confirm the previous results, the employer may discontinue all periodic medical surveillance for that employee.

**1910.1027(l)(4)(v)(C)** However, if the results of the follow-up tests specified in paragraph (l)(4)(v)(A) or (B) of this section indicate that the level of the employee's CdU, $\beta_2$-M, or CdB exceeds these same levels, the employer is required to provide annual medical examinations in accordance with the provisions of paragraph (l)(4)(ii) of this section until the results of biological monitoring are consistently below these levels or the examining physician determines in a written medical opinion that further medical surveillance is not required to protect the employee's health.

**1910.1027(l)(4)(vi)** A routine, biennial medical examination is not required to be provided in accordance with paragraphs (l)(3)(i) and (l)(4) of this section if adequate medical records show that the employee has been examined in accordance with the requirements of paragraph (l)(4)(ii) of this section within the past 12 months. In that case, such records shall be maintained by the employer as part of the employee's medical record, and the next routine, periodic medical examination shall be made available to the employee within two years of the previous examination.

**1910.1027(l)(5)** *Actions triggered by medical examinations.* **(i)** If the results of a medical examination carried out in accordance with this section indicate any laboratory or clinical finding consistent with cadmium toxicity that does not require employer action under paragraph (l)(2), (3) or (4) of this section, the employer, within 30 days, shall reassess the employee's occupational exposure to cadmium and take the following corrective action until the physician determines they are no longer necessary:

**1910.1027(l)(5)(i)(A)** Periodically reassess: The employee's work practices and personal hygiene; the employee's respirator use, if any; the employee's smoking history and status; the respiratory protection program; the hygiene facilities; and the maintenance and effectiveness of the relevant engineering controls;

**1910.1027(l)(5)(i)(B)** Within 30 days after the reassessment, take all reasonable steps to correct the deficiencies found in the reassessment that may be responsible for the employee's excess exposure to cadmium;

**1910.1027(l)(5)(i)(C)** Provide semiannual medical reexaminations to evaluate the abnormal clinical sign(s) of cadmium toxicity until the results are normal or the employee is medically removed; and

**1910.1027(l)(5)(i)(D)** Where the results of tests for total proteins in urine are abnormal, provide a more detailed medical evaluation of the toxic effects of cadmium on the employee's renal system.

**1910.1027(l)(6)** *Examination for respirator use.* **(i)** To determine an employee's fitness for respirator use, the employer shall provide a medical examination that includes the elements specified in paragraph (l)(6)(i)(A)-(D) of this section. This examination shall be provided prior to the employee's being assigned to a job that requires the use of a respirator or no later than 90 days after this section goes into effect, whichever date is later, to any employee without a medical examination within the preceding 12 months that satisfies the requirements of this paragraph.

**1910.1027(l)(6)(i)(A)** A detailed medical and work history, or update thereof, with emphasis on: Past exposure to cadmium; smoking history and current status; any history of renal, cardiovascular, respiratory, hematopoietic, and/or musculoskeletal system dysfunction; a description of the job for which the respirator is required; and questions 3-11 and 25-32 in appendix D to this section;

**1910.1027(l)(6)(i)(B)** A blood pressure test;

**1910.1027(l)(6)(i)(C)** Biological monitoring of the employee's levels of CdU, CdB and $\beta_2$-M in accordance with the requirements of paragraph (l)(2)(ii)(B) of this section, unless such results already have been obtained within the previous 12 months; and

**1910.1027(l)(6)(i)(D)** Any other test or procedure that the examining physician deems appropriate.

**1910.1027(l)(6)(ii)** After reviewing all the information obtained from the medical examination required in paragraph (l)(6)(i) of this section, the physician shall determine whether the employee is fit to wear a respirator.

**1910.1027(l)(6)(iii)** Whenever an employee has exhibited difficulty in breathing during a respirator fit test or during use of a respirator, the employer, as soon as possible, shall provide the employee with a periodic medical examination in accordance with paragraph (l)(4)(ii) of this section to determine the employee's fitness to wear a respirator.

**1910.1027(l)(6)(iv)** Where the results of the examination required under paragraph (l)(6)(i), (ii), or (iii) of this section are abnormal, medical limitation or prohibition of respirator use shall be considered. If the employee is allowed to wear a respirator, the employee's ability to continue to do so shall be periodically evaluated by a physician.

**1910.1027(l)(7)** *Emergency examinations.* **(i)** In addition to the medical surveillance required in paragraphs (l)(2)-(6) of this section, the employer shall provide a medical examination as soon as possible to any employee who may have been acutely exposed to cadmium because of an emergency.

**1910.1027(l)(7)(ii)** The examination shall include the requirements of paragraph (l)(4)(ii) of this section, with emphasis on the respiratory system, other organ systems considered appropriate by the examining physician, and symptoms of acute overexposure, as identified in paragraphs II (B)(1)-(2) and IV of appendix A to this section.

**1910.1027(l)(8)** *Termination of employment examination.* **(i)** At termination of employment, the employer shall provide a medical examination in accordance with paragraph (l)(4)(ii) of this section, including a chest X-ray, to any employee to whom at any prior time the employer was required to provide medical surveillance under paragraphs (l)(1)(i) or (l)(7) of this section. However, if the last examination satisfied the requirements of paragraph (l)(4)(ii) of this section and was less than six months prior to the date of termination, no further examination is required unless otherwise specified in paragraphs (l)(3) or (l)(5) of this section;

**1910.1027(l)(8)(ii)** However, for employees covered by paragraph (l)(1)(i)(B) of this section, if the employer has discontinued all periodic medical surveillance under paragraph (l)(4)(v) of this section, no termination of employment medical examination is required.

**1910.1027(l)(9)** *Information provided to the physician.* The employer shall provide the following information to the examining physician:

**1910.1027(l)(9)(i)** A copy of this standard and appendices;

**1910.1027(l)(9)(ii)** A description of the affected employee's former, current, and anticipated duties as they relate to the employee's occupational exposure to cadmium;

**1910.1027(l)(9)(iii)** The employee's former, current, and anticipated future levels of occupational exposure to cadmium;

**1910.1027(l)(9)(iv)** A description of any personal protective equipment, including respirators, used or to be used by the employee, including when and for how long the employee has used that equipment; and

**1910.1027(l)(9)(v)** relevant results of previous biological monitoring and medical examinations.

**1910.1027(l)(10)** *Physician's written medical opinion.* **(i)** The employer shall promptly obtain a written medical opinion from the examining physician for each medical examination performed on each employee. This written opinion shall contain:

**1910.1027(l)(10)(i)(A)** The physician's diagnosis for the employee;

**1910.1027(l)(10)(i)(B)** The physician's opinion as to whether the employee has any detected medical condition(s) that would place the employee at increased risk of material impairment to health from further exposure to cadmium, including any indications of potential cadmium toxicity;

**1910.1027(l)(10)(i)(C)** The results of any biological or other testing or related evaluations that directly assess the employee's absorption of cadmium;

**1910.1027(l)(10)(i)(D)** Any recommended removal from, or limitation on the activities or duties of the employee or on the employee's use of personal protective equipment, such as respirators;

**1910.1027(l)(10)(i)(E)** A statement that the physician has clearly and carefully explained to the employee the results of the medical examination, including all biological monitoring results and any medical conditions related to cadmium exposure that require further evaluation or treatment, and any limitation on the employee's diet or use of medications.

**1910.1027(l)(10)(ii)** The employer promptly shall obtain a copy of the results of any biological monitoring provided by an employer to an employee independently of a medical examination under paragraphs (l)(2) and (l)(4) of this section, and, in lieu of a written medical opinion, an explanation sheet explaining those results.

**1910.1027(l)(10)(iii)** The employer shall instruct the physician not to reveal orally or in the written medical opinion given to the employer specific findings or diagnoses unrelated to occupational exposure to cadmium.

**1910.1027(l)(11)** *Medical Removal Protection (MRP)*—**(i)** *General.* **(A)** The employer shall temporarily remove an employee from work where there is excess exposure to cadmium on each occasion that medical removal is required under paragraph (l)(3), (l)(4), or (l)(6) of this section and on each occasion that a physician determines in a written medical opinion that the employee should be removed from such exposure. The physician's determination may be based on biological monitoring results, inability to wear a respirator, evidence of illness, other signs or symptoms of cadmium-related dysfunction or disease, or any other reason deemed medically sufficient by the physician.

**1910.1027(l)(11)(i)(B)** The employer shall medically remove an employee in accordance with paragraph (l)(11) of this section regardless of whether at the time of removal a job is available into which the removed employee may be transferred.

**1910.1027(l)(11)(i)(C)** Whenever an employee is medically removed under paragraph (l)(11) of this section, the employer shall transfer the removed employee to a job where the exposure to cadmium is within the permissible levels specified in that paragraph as soon as one becomes available.

**1910.1027(l)(11)(i)(D)** For any employee who is medically removed under the provisions of paragraph (l)(11)(i) of this section, the employer shall provide follow-up biological monitoring in accordance with (l)(2)(ii)(B) of this section at least every three months and follow-up medical examinations semi-annually at least every six months until in a written medical opinion the examining physician determines that either the employee may be returned to his/her former job status as specified under paragraph (l)(11)(iv)-(v) of this section or the employee must be permanently removed from excess cadmium exposure.

**1910.1027(l)(11)(i)(E)** The employer may not return an employee who has been medically removed for any reason to his/her former job status until a physician determines in a written medical opinion that continued medical removal is no longer necessary to protect the employee's health.

**1910.1027(l)(11)(ii)** Where an employee is found unfit to wear a respirator under paragraph (l)(6)(ii) of this section, the employer shall remove the employee from work where exposure to cadmium is above the PEL.

**1910.1027(l)(11)(iii)** Where removal is based on any reason other than the employee's inability to wear a respirator, the employer shall remove the employee from work where exposure to cadmium is at or above the action level.

**1910.1027(l)(11)(iv)** Except as specified in paragraph (l)(11)(v) of this section, no employee who was removed because his/her level of CdU, CdB and/or $\beta_2$-M exceeded the medical removal trigger levels in paragraph (l)(3) or (l)(4) of this section may be returned to work with exposure to cadmium at or above the action level until the employee's levels of CdU fall to or below 3 µg/g Cr, CdB falls to or below 5 µg/lwb, and $\beta_2$-M falls to or below 300 µg/g Cr.

**1910.1027(l)(11)(v)** However, when in the examining physician's opinion continued exposure to cadmium will not pose an increased risk to the employee's health and there are special circumstances that make continued medical removal an inappropriate remedy, the physician shall fully discuss these matters with the employee, and then in a written determination may return a worker to his/her former job status despite what would otherwise be unacceptably high biological monitoring results. Thereafter, the returned employee shall continue to be provided with medical surveillance as if he/she were still on medical removal until the employee's levels of CdU fall to or below 3 µg/g Cr, CdB falls to or below 5 µg/lwb, and $\beta_2$-M falls to or below 300 µg/g Cr.

**1910.1027(l)(11)(vi)** Where an employer, although not required by paragraph (l)(11)(i)-(iii) of this section to do so, removes an employee from exposure to cadmium or otherwise places limitations on an employee due to the effects of cadmium exposure on the employee's medical condition, the employer shall provide the same medical removal protection benefits to that employee under paragraph (l)(12) of this section as would have been provided had the removal been required under paragraph (l)(11)(i)-(iii) of this section.

**1910.1027(l)(12)** *Medical Removal Protection Benefits (MRPB).* **(i)** The employer shall provide MRPB for up to a maximum of 18 months to an employee each time and while the employee is temporarily medically removed under paragraph (l)(11) of this section.

**1910.1027(l)(12)(ii)** For purposes of this section, the requirement that the employer provide MRPB means that the employer shall maintain the total normal earnings, seniority, and all other employee rights and benefits of the removed employee, including the employee's right to his/her former job status, as if the employee had not been removed from the employee's job or otherwise medically limited.

**1910.1027(l)(12)(iii)** Where, after 18 months on medical removal because of elevated biological monitoring results, the employee's monitoring results have not declined to a low enough level to permit the employee to be returned to his/her former job status:

**1910.1027(l)(12)(iii)(A)** The employer shall make available to the employee a medical examination pursuant to this section in order to obtain a final medical determination as to whether the employee may be returned to his/her former job status or must be permanently removed from excess cadmium exposure; and

**1910.1027(l)(12)(iii)(B)** The employer shall assure that the final medical determination indicates whether the employee may be returned to his/her former job status and what steps, if any, should be taken to protect the employee's health.

**1910.1027(l)(12)(iv)** The employer may condition the provision of MRPB upon the employee's participation in medical surveillance provided in accordance with this section.

**1910.1027(l)(13)** *Multiple physician review.* **(i)** If the employer selects the initial physician to conduct any medical examination or consultation provided to an employee under this section, the employee may designate a second physician to:

**1910.1027(l)(13)(i)(A)** Review any findings, determinations, or recommendations of the initial physician; and

**1910.1027(l)(13)(i)(B)** Conduct such examinations, consultations, and laboratory tests as the second physician deems necessary to facilitate this review.

**1910.1027(l)(13)(ii)** The employer shall promptly notify an employee of the right to seek a second medical opinion after each occasion that an initial physician provided by the employer conducts a medical examination or consultation pursuant to this section. The employer may condition its participation in, and payment for, multiple physician review upon the employee doing the following within fifteen (15) days after receipt of this notice, or receipt of the initial physician's written opinion, whichever is later:

**1910.1027(l)(13)(ii)(A)** Informing the employer that he or she intends to seek a medical opinion; and

**1910.1027(l)(13)(ii)(B)** Initiating steps to make an appointment with a second physician.

**1910.1027(l)(13)(iii)** If the findings, determinations, or recommendations of the second physician differ from those of the initial physician, then the employer and the employee shall assure that efforts are made for the two physicians to resolve any disagreement.

**1910.1027(l)(13)(iv)** If the two physicians have been unable to quickly resolve their disagreement, then the employer and the employee, through their respective physicians, shall designate a third physician to:

**1910.1027(l)(13)(iv)(A)** Review any findings, determinations, or recommendations of the other two physicians; and

**1910.1027(l)(13)(iv)(B)** Conduct such examinations, consultations, laboratory tests, and discussions with the other two physicians as the third physician deems necessary to resolve the disagreement among them.

**1910.1027(l)(13)(v)** The employer shall act consistently with the findings, determinations, and recommendations of the third physician, unless the employer and the employee reach an agreement that is consistent with the recommendations of at least one of the other two physicians.

**1910.1027(l)(14)** *Alternate physician determination.* The employer and an employee or designated employee representative may agree upon the use of any alternate form of physician determination in lieu of the multiple physician review provided by paragraph (l)(13) of this section, so long as the alternative is expeditious and at least as protective of the employee.

**1910.1027(l)(15)** *Information the employer must provide the employee.* **(i)** The employer shall provide a copy of the physician's written medical opinion to the examined employee within two weeks after receipt thereof.

**1910.1027(l)(15)(ii)** The employer shall provide the employee with a copy of the employee's biological monitoring results and an explanation sheet explaining the results within two weeks after receipt thereof.

**1910.1027(l)(15)(iii)** Within 30 days after a request by an employee, the employer shall provide the employee with the information the employer is required to provide the examining physician under paragraph (l)(9) of this section.

**1910.1027(l)(16)** *Reporting.* In addition to other medical events that are required to be reported on the OSHA Form No. 200, the employer shall report any abnormal condition or disorder caused by occupational exposure to cadmium associated with employment as specified in Chapter (V)(E) of the Reporting Guidelines for Occupational Injuries and Illnesses.

**1910.1027(m)** *Communication of cadmium hazards to employees*—**(1)** *General.* In communications concerning cadmium hazards, employers shall comply with the requirements of OSHA's Hazard Communication Standard, 29 CFR 1910.1200, including but not limited to the requirements concerning warning signs and labels, material safety data sheets (MSDS), and employee information and training. In addition, employers shall comply with the following requirements:

**1910.1027(m)(2)** *Warning signs.* **(i)** Warning signs shall be provided and displayed in regulated areas. In addition, warning signs shall be posted at all approaches to regulated areas so that an employee may read the signs and take necessary protective steps before entering the area.

**1910.1027(m)(2)(ii)** Warning signs required by paragraph (m)(2)(i) of this section shall bear the following information:

**DANGER**

**CADMIUM**

**CANCER HAZARD**

**CAN CAUSE LUNG AND KIDNEY DISEASE**

**AUTHORIZED PERSONNEL ONLY**

**RESPIRATORS REQUIRED IN THIS AREA**

**1910.1027(m)(2)(iii)** The employer shall assure that signs required by this paragraph are illuminated, cleaned, and maintained as necessary so that the legend is readily visible.

**1910.1027(m)(3)** *Warning labels.* **(i)** Shipping and storage containers containing cadmium, cadmium compounds, or cadmium contaminated clothing, equipment, waste, scrap, or debris shall bear appropriate warning labels, as specified in paragraph (m)(3)(ii) of this section.

**1910.1027(m)(3)(ii)** The warning labels shall include at least the following information:

**DANGER**

**CONTAINS CADMIUM**

**CANCER HAZARD**

**AVOID CREATING DUST**

**CAN CAUSE LUNG AND KIDNEY DISEASE**

**1910.1027(m)(3)(iii)** Where feasible, installed cadmium products shall have a visible label or other indication that cadmium is present.

**1910.1027(m)(4)** *Employee information and training.* **(i)** The employer shall institute a training program for all employees who are potentially exposed to cadmium, assure employee participation in the program, and maintain a record of the contents of such program.

**1910.1027(m)(4)(ii)** Training shall be provided prior to or at the time of initial assignment to a job involving potential exposure to cadmium and at least annually thereafter.

**1910.1027(m)(4)(iii)** The employer shall make the training program understandable to the employee and shall assure that each employee is informed of the following:

**1910.1027(m)(4)(iii)(A)** The health hazards associated with cadmium exposure, with special attention to the information incorporated in appendix A to this section;

**1910.1027(m)(4)(iii)(B)** The quantity, location, manner of use, release, and storage of cadmium in the workplace and the specific nature of operations that could result in exposure to cadmium, especially exposures above the PEL;

**1910.1027(m)(4)(iii)(C)** The engineering controls and work practices associated with the employee's job assignment;

**1910.1027(m)(4)(iii)(D)** The measures employees can take to protect themselves from exposure to cadmium, including modification of such habits as smoking and personal hygiene, and specific procedures the employer has implemented to protect employees from exposure to cadmium such as appropriate work practices, emergency procedures, and the provision of personal protective equipment;

**1910.1027(m)(4)(iii)(E)** The purpose, proper selection, fitting, proper use, and limitations of respirators and protective clothing;

**1910.1027(m)(4)(iii)(F)** The purpose and a description of the medical surveillance program required by paragraph (l) of this section;

**1910.1027(m)(4)(iii)(G)** The contents of this section and its appendices; and

**1910.1027(m)(4)(iii)(H)** The employee's rights of access to records under §1910.1020(e) and (g).

**1910.1027(m)(4)(iv)** Additional access to information and training program and materials.

**1910.1027(m)(4)(iv)(A)** The employer shall make a copy of this section and its appendices readily available without cost to all affected employees and shall provide a copy if requested.

**1910.1027(m)(4)(iv)(B)** The employer shall provide to the Assistant Secretary or the Director, upon request, all materials relating to the employee information and the training program.

**1910.1027(n)** *Recordkeeping*—**(1)** *Exposure monitoring.* **(i)** The employer shall establish and keep an accurate record of all air monitoring for cadmium in the workplace.

**1910.1027(n)(1)(ii)** This record shall include at least the following information:

**1910.1027(n)(1)(ii)(A)** The monitoring date, duration, and results in terms of an 8-hour TWA of each sample taken;

**1910.1027(n)(1)(ii)(B)** The name, social security number, and job classification of the employees monitored and of all other employees whose exposures the monitoring is intended to represent;

**1910.1027(n)(1)(ii)(C)** A description of the sampling and analytical methods used and evidence of their accuracy;

**1910.1027(n)(1)(ii)(D)** The type of respiratory protective device, if any, worn by the monitored employee;

**1910.1027(n)(1)(ii)(E)** A notation of any other conditions that might have affected the monitoring results.

**1910.1027(n)(1)(iii)** The employer shall maintain this record for at least thirty (30) years, in accordance with 29 CFR 1910.1020.

**1910.1027(n)(2)** *Objective data for exemption from requirement for initial monitoring.* **(i)** For purposes of this section, objective data are information demonstrating that a particular product or material containing cadmium or a specific process, operation, or activity involving cadmium cannot release dust or fumes in concentrations at or above the action level even under the worst-case release conditions. Objective data can be obtained from an industry-wide study or from laboratory product test results from manufacturers of cadmium-containing products or materials. The data the employer uses from an industry-wide survey must be obtained under workplace conditions closely resembling the processes, types of material, control methods, work practices and environmental conditions in the employer's current operations.

**1910.1027(n)(2)(ii)** The employer shall establish and maintain a record of the objective data for at least 30 years.

**1910.1027(n)(3)** *Medical surveillance.* **(i)** The employer shall establish and maintain an accurate record for each employee covered by medical surveillance under paragraph (l)(1)(i) of this section.

**1910.1027(n)(3)(ii)** The record shall include at least the following information about the employee:

**1910.1027(n)(3)(ii)(A)** Name, social security number, and description of the duties;

**1910.1027(n)(3)(ii)(B)** A copy of the physician's written opinions and an explanation sheet for biological monitoring results;

**1910.1027(n)(3)(ii)(C)** A copy of the medical history, and the results of any physical examination and all test results that are required to be provided by this section, including biological tests, X-rays, pulmonary function tests, etc., or that have been obtained to further evaluate any condition that might be related to cadmium exposure;

**1910.1027(n)(3)(ii)(D)** The employee's medical symptoms that might be related to exposure to cadmium; and

**1910.1027(n)(3)(ii)(E)** A copy of the information provided to the physician as required by paragraph (l)(9)(ii)-(v) of this section.

**1910.1027(n)(3)(iii)** The employer shall assure that this record is maintained for the duration of employment plus thirty (30) years, in accordance with 29 CFR 1910.1020.

**1910.1027(n)(4)** *Training.* The employer shall certify that employees have been trained by preparing a certification record which includes the identity of the person trained, the signature of the employer or the person who conducted the training, and the date the training was completed. The certification records shall be prepared at the completion of training and shall be maintained on file for one (1) year beyond the date of training of that employee.

**1910.1027(n)(5)** *Availability.* (i) Except as otherwise provided for in this section, access to all records required to be maintained by paragraphs (n)(1)-(4) of this section shall be in accordance with the provisions of 29 CFR 1910.1020.

**1910.1027(n)(5)(ii)** Within 15 days after a request, the employer shall make an employee's medical records required to be kept by paragraph (n)(3) of this section available for examination and copying to the subject employee, to designated representatives, to anyone having the specific written consent of the subject employee, and after the employee's death or incapacitation, to the employee's family members.

**1910.1027(n)(6)** *Transfer of records.* Whenever an employer ceases to do business and there is no successor employer to receive and retain records for the prescribed period or the employer intends to dispose of any records required to be preserved for at least 30 years, the employer shall comply with the requirements concerning transfer of records set forth in 29 CFR 1910.1020(h).

**1910.1027(o)** *Observation of monitoring*—(1) *Employee observation.* The employer shall provide affected employees or their designated representatives an opportunity to observe any monitoring of employee exposure to cadmium.

**1910.1027(o)(2)** *Observation procedures.* When observation of monitoring requires entry into an area where the use of protective clothing or equipment is required, the employer shall provide the observer with that clothing and equipment and shall assure that the observer uses such clothing and equipment and complies with all other applicable safety and health procedures.

**1910.1027(p)** *Dates*—(1) *Effective date.* This section shall become effective December 14, 1992.

**1910.1027(p)(2)** *Start-up dates.* All obligations of this section commence on the effective date except as follows:

**1910.1027(p)(2)(i)** *Exposure monitoring.* Except for small businesses (nineteen (19) or fewer employees), initial monitoring required by paragraph (d)(2) of this section shall be completed as soon as possible and in any event no later than 60 days after the effective date of this standard. For small businesses, initial monitoring required by paragraph (d)(2) of this section shall be completed as soon as possible and in any event no later than 120 days after the effective date of this standard.

**1910.1027(p)(2)(ii)** *Regulated areas.* Except for small business, defined under paragraph (p)(2)(i) of this section, regulated areas required to be established by paragraph (e) of this section shall be set up as soon as possible after the results of exposure monitoring are known and in any event no later than 90 days after the effective date of this section. For small businesses, regulated areas required to be established by paragraph (e) of this section shall be set up as soon as possible after the results of exposure monitoring are known and in any event no later than 150 days after the effective date of this section.

**1910.1027(p)(2)(iii)** *Respiratory protection.* Except for small businesses, defined under paragraph (p)(2)(i) of this section, respiratory protection required by paragraph (g) of this section shall be provided as soon as possible and in any event no later than 90 days after the effective date of this section. For small businesses, respiratory protection required by paragraph (g) of this section shall be provided as soon as possible and in any event no later than 150 days after the effective date of this section.

**1910.1027(p)(2)(iv)** *Compliance program.* Written compliance programs required by paragraph (f)(2) of this section shall be completed and available for inspection and copying as soon as possible and in any event no later than 1 year after the effective date of this section.

**1910.1027(p)(2)(v)** *Methods of compliance.* The engineering controls required by paragraph (f)(1) of this section shall be implemented as soon as possible and in any event no later than two (2) years after the effective date of this section. Work practice controls shall be implemented as soon as possible. Work practice controls that are directly related to engineering controls to be implemented in accordance with the compliance plan shall be implemented as soon as possible after such engineering controls are implemented.

**1910.1027(p)(2)(vi)** *Hygiene and lunchroom facilities.* (A) Handwashing facilities, permanent or temporary, shall be provided in accordance with 29 CFR 1910.141(d)(1) and (2) as soon as possible and in any event no later than 60 days after the effective date of this section.

**1910.1027(p)(2)(vi)(B)** Change rooms, showers, and lunchroom facilities shall be completed as soon as possible and in any event no later than 1 year after the effective date of this section.

**1910.1027(p)(2)(vii)** *Employee information and training.* Except for small businesses, defined under paragraph (p)(2)(i) of this section, employee information and training required by paragraph (m)(4) of this section shall be provided as soon as possible and in any event no later than 90 days after the effective date of this standard. For small businesses, employee information and training required by paragraph (m)(4) of this standard shall be provided as soon as possible and in any event no later than 180 days after the effective date of this standard.

**1910.1027(p)(2)(viii)** *Medical surveillance.* Except for small businesses, defined under paragraph (p)(2)(i) of this section, initial medical examinations required by paragraph (l) of this section shall be provided as soon as possible and in any event no later than 90 days after the effective date of this standard. For small businesses, initial medical examinations required by paragraph (l) of this section shall be provided as soon as possible and in any event no later than 180 days after the effective date of this standard.

**1910.1027(q)** *Appendices.* Except where portions of appendices A, B, D, E, and F to this section are expressly incorporated in requirements of this section, these appendices are purely informational and are not intended to create any additional obligations not otherwise imposed or to detract from any existing obligations.

## APPENDIX A TO §1910.1027— SUBSTANCE SAFETY DATA SHEET

### CADMIUM

#### I. Substance Identification

A. Substance: Cadmium.

B. 8-Hour, Time-weighted-average, Permissible Exposure Limit (TWA PEL): 1. TWA PEL: Five micrograms of cadmium per cubic meter of air 5 µg/m³, time-weighted average (TWA) for an 8-hour workday.

C. Appearance: Cadmium metal—soft, blue-white, malleable, lustrous metal or grayish-white powder. Some cadmium compounds may also appear as a brown, yellow, or red powdery substance.

#### II. Health Hazard Data

A. Routes of Exposure. Cadmium can cause local skin or eye irritation. Cadmium can affect your health if you inhale it or if you swallow it.

B. Effects of Overexposure.

1. Short-term (acute) exposure: Cadmium is much more dangerous by inhalation than by ingestion. High exposures to cadmium that may be immediately dangerous to life or health occur in jobs where workers handle large quantities of cadmium dust or fume; heat cadmium-containing compounds or cadmium-coated surfaces; weld with cadmium solders or cut cadmium-containing materials such as bolts.

2. Severe exposure may occur before symptoms appear. Early symptoms may include mild irritation of the upper respiratory tract, a sensation of constriction of the throat, a metallic taste and/or a cough. A period of 1-10 hours may precede the onset of rapidly progressing shortness of breath, chest pain, and flu-like symptoms with weakness, fever, headache, chills,

sweating and muscular pain. Acute pulmonary edema usually develops within 24 hours and reaches a maximum by three days. If death from asphyxia does not occur, symptoms may resolve within a week.

3. Long-term (chronic) exposure. Repeated or long-term exposure to cadmium, even at relatively low concentrations, may result in kidney damage and an increased risk of cancer of the lung and of the prostate.

C. Emergency First Aid Procedures.

1. Eye exposure: Direct contact may cause redness or pain. Wash eyes immediately with large amounts of water, lifting the upper and lower eyelids. Get medical attention immediately.

2. Skin exposure: Direct contact may result in irritation. Remove contaminated clothing and shoes immediately. Wash affected area with soap or mild detergent and large amounts of water. Get medical attention immediately.

3. Ingestion: Ingestion may result in vomiting, abdominal pain, nausea, diarrhea, headache and sore throat. Treatment for symptoms must be administered by medical personnel. Under no circumstances should the employer allow any person whom he retains, employs, supervises or controls to engage in therapeutic chelation. Such treatment is likely to translocate cadmium from pulmonary or other tissue to renal tissue. Get medical attention immediately.

4. Inhalation: If large amounts of cadmium are inhaled, the exposed person must be moved to fresh air at once. If breathing has stopped, perform cardiopulmonary resuscitation. Administer oxygen if available. Keep the affected person warm and at rest. Get medical attention immediately.

5. Rescue: Move the affected person from the hazardous exposure. If the exposed person has been overcome, attempt rescue only after notifying at least one other person of the emergency and putting into effect established emergency procedures. Do not become a casualty yourself. Understand your emergency rescue procedures and know the location of the emergency equipment before the need arises.

### III. Employee Information

A. Protective Clothing and Equipment.

1. Respirators: You may be required to wear a respirator for non-routine activities; in emergencies; while your employer is in the process of reducing cadmium exposures through engineering controls; and where engineering controls are not feasible. If respirators are worn in the future, they must have a joint Mine Safety and Health Administration (MSHA) and National Institute for Occupational Safety and Health (NIOSH) label of approval. Cadmium does not have a detectable odor except at levels well above the permissible exposure limits. If you can smell cadmium while wearing a respirator, proceed immediately to fresh air. If you experience difficulty breathing while wearing a respirator, tell your employer.

2. Protective Clothing: You may be required to wear impermeable clothing, gloves, foot gear, a face shield, or other appropriate protective clothing to prevent skin contact with cadmium. Where protective clothing is required, your employer must provide clean garments to you as necessary to assure that the clothing protects you adequately. The employer must replace or repair protective clothing that has become torn or otherwise damaged.

3. Eye Protection: You may be required to wear splash-proof or dust resistant goggles to prevent eye contact with cadmium.

B. Employer Requirements.

1. Medical: If you are exposed to cadmium at or above the action level, your employer is required to provide a medical examination, laboratory tests and a medical history according to the medical surveillance provisions under paragraph (1) of this standard. (See summary chart and tables in this appendix A.) These tests shall be provided without cost to you. In addition, if you are accidentally exposed to cadmium under conditions known or suspected to constitute toxic exposure to cadmium, your employer is required to make special tests available to you.

2. Access to Records: All medical records are kept strictly confidential. You or your representative are entitled to see the records of measurements of your exposure to cadmium. Your medical examination records can be furnished to your personal physician or designated representative upon request by you to your employer.

3. Observation of Monitoring: Your employer is required to perform measurements that are representative of your exposure to cadmium and you or your designated representative are entitled to observe the monitoring procedure. You are entitled to observe the steps taken in the measurement procedure, and to record the results obtained. When the monitoring procedure is taking place in an area where respirators or personal protective clothing and equipment are required to be worn, you or your representative must also be provided with, and must wear the protective clothing and equipment.

C. Employee Requirements—You will not be able to smoke, eat, drink, chew gum or tobacco, or apply cosmetics while working with cadmium in regulated areas. You will also not be able to carry or store tobacco products, gum, food, drinks or cosmetics in regulated areas because these products easily become contaminated with cadmium from the workplace and can therefore create another source of unnecessary cadmium exposure.

Some workers will have to change out of work clothes and shower at the end of the day, as part of their workday, in order to wash cadmium from skin and hair. Handwashing and cadmium-free eating facilities shall be provided by the employer and proper hygiene should always be performed before eating. It is also recommended that you do not smoke or use tobacco products, because among other things, they naturally contain cadmium. For further information, read the labeling on such products.

### IV. Physician Information

A. Introduction.—The medical surveillance provisions of paragraph (1) generally are aimed at accomplishing three main interrelated purposes: First, identifying employees at higher risk of adverse health effects from excess, chronic exposure to cadmium; second, preventing cadmium-induced disease; and third, detecting and minimizing existing cadmium-induced disease. The core of medical surveillance in this standard is the early and periodic monitoring of the employee's biological indicators of: (a) Recent exposure to cadmium; (b) cadmium body burden; and (c) potential and actual kidney damage associated with exposure to cadmium.

The main adverse health effects associated with cadmium overexposure are lung cancer and kidney dysfunction. It is not yet known how to adequately biologically monitor human beings to specifically prevent cadmium-induced lung cancer. By contrast, the kidney can be monitored to provide prevention and early detection of cadmium-induced kidney damage. Since, for non-carcinogenic effects, the kidney is considered the primary target organ of chronic exposure to cadmium, the medical surveillance provisions of this standard effectively focus on cadmium-induced kidney disease. Within that focus, the aim, where possible, is to prevent the onset of such disease and, where necessary, to minimize such disease as may already exist. The by-products of successful prevention of kidney disease are anticipated to be the reduction and prevention of other cadmium-induced diseases.

B. Health Effects.—The major health effects associated with cadmium overexposure are described below.

1. Kidney: The most prevalent non-malignant disease observed among workers chronically exposed to cadmium is kidney dysfunction. Initially, such dysfunction is manifested as proteinuria. The proteinuria associated with cadmium exposure is most commonly characterized by excretion of low-molecular weight proteins (15,000 to 40,000 MW) accompanied by loss of electrolytes, uric acid, calcium, amino acids, and phosphate. The compounds commonly excreted include: beta-2-microglobulin ($\beta_2$-M), retinol binding protein (RBP), immunoglobulin light chains, and lysozyme. Excretion of low molecular weight proteins are characteristic of damage to the proximal tubules of the kidney (Iwao et al., 1980).

It has also been observed that exposure to cadmium may lead to urinary excretion of high-molecular weight proteins such as albumin, immunoglobulin G, and glycoproteins (Ex. 29). Excretion of high-molecular weight proteins is typically indicative of damage to the glomeruli of the kidney. Bernard et al., (1979) suggest that damage to the glomeruli and damage to the proximal tubules of the kidney may both be linked to cadmium exposure but they may occur independently of each other.

Several studies indicate that the onset of low-molecular weight proteinuria is a sign of irreversible kidney damage (Friberg et al., 1974; Roels et al., 1982; Piscator 1984; Elinder et al., 1985; Smith et al., 1986). Above specific levels of $\beta_2$-M associated with cadmium exposure it is unlikely that $\beta_2$-M levels return to normal even when cadmium exposure is eliminated by removal of the individual from the cadmium work environment (Friberg, Ex. 29, 1990).

Some studies indicate that such proteinuria may be progressive; levels of $\beta_2$-M observed in the urine increase with time even after cadmium exposure has ceased. See, for example, Elinder et al., 1985. Such observations, however, are not universal, and it has been suggested that studies in which proteinuria has not been observed to progress may not have tracked patients for a sufficiently long time interval (Jarup, Ex. 8-661).

When cadmium exposure continues after the onset of proteinuria, chronic nephrotoxicity may occur (Friberg, Ex. 29). Uremia results from the inability of the glomerulus to adequately filter blood. This leads to severe disturbance of electrolyte concentrations and may lead to various clinical complications including kidney stones (L-140-50).

After prolonged exposure to cadmium, glomerular proteinuria, glucosuria, aminoaciduria, phosphaturia, and hypercalciuria may develop (Exs. 8-86, 4-28, 14-18). Phosphate, calcium, glucose, and amino acids are essential to life, and under normal conditions, their excretion should be regulated by the kidney. Once low molecular weight proteinuria has developed, these elements dissipate from the human body. Loss of glomerular function may also occur, manifested by decreased glomerular filtration rate and increased serum creatinine. Severe cadmium-induced renal damage may eventually develop into chronic renal failure and uremia (Ex. 55).

Studies in which animals are chronically exposed to cadmium confirm the renal effects observed in humans (Friberg et al., 1986). Animal studies also confirm problems with calcium metabolism and related skeletal effects which have been observed among humans exposed to cadmium in addition to the renal effects. Other effects commonly reported in chronic animal studies include anemia, changes in liver morphology, immunosuppression and hypertension. Some of these effects may be associated with co-factors. Hypertension, for example, appears to be associated with diet as well as cadmium exposure. Animals injected with cadmium have also shown testicular necrosis (Ex. 8-86B).

### 2. Biological Markers

It is universally recognized that the best measures of cadmium exposures and its effects are measurements of cadmium in biological fluids, especially urine and blood. Of the two, CdU is conventionally used to determine body burden of cadmium in workers without kidney disease. CdB is conventionally used to monitor for recent exposure to cadmium. In addition, levels of CdU and CdB historically have been used to predict the percent of the population likely to develop kidney disease (Thun et al., Ex. L-140-50; WHO, Ex. 8-674; ACGIH, Exs. 8-667, 140-50).

The third biological parameter upon which OSHA relies for medical surveillance is Beta-2-microglobulin in urine ($\beta_2$-M), a low molecular weight protein. Excess $\beta_2$-M has been widely accepted by physicians and scientists as a reliable indicator of functional damage to the proximal tubule of the kidney (Exs. 8-447, 144-3-C, 4-47, L-140-45, 19-43-A).

Excess $\beta_2$-M is found when the proximal tubules can no longer reabsorb this protein in a normal manner. This failure of the proximal tubules is an early stage of a kind of kidney disease that commonly occurs among workers with excessive cadmium exposure. Used in conjunction with biological test results indicating abnormal levels of CdU and CdB, the finding of excess $\beta_2$-M can establish for an examining physician that any existing kidney disease is probably cadmium-related (Trs. 6/6/90, pp. 82-86, 122, 134). The upper limits of normal levels for cadmium in urine and cadmium in blood are 3 µg Cd/gram creatinine in urine and 5 µg Cd/liter whole blood, respectively. These levels were derived from broad-based population studies.

Three issues confront the physicians in the use of $\beta_2$-M as a marker of kidney dysfunction and material impairment. First, there are a few other causes of elevated levels of $\beta_2$-M not related to cadmium exposures, some of which may be rather common diseases and some of which are serious diseases (e.g., myeloma or transient flu, Exs. 29 and 8-086). These can be medically evaluated as alternative causes (Friberg, Ex. 29). Also, there are other factors that can cause $\beta_2$-M to degrade so that low levels would result in workers with tubular dysfunction. For example, regarding the degradation of $\beta_2$-M, workers with acidic urine (pH<6) might have $\beta_2$-M levels that are within the "normal" range when in fact kidney dysfunction has occurred (Ex. L-140-1) and the low molecular weight proteins are degraded in acid urine. Thus, it is very important that the pH of urine be measured, that urine samples be buffered as necessary (See appendix F.), and that urine samples be handled correctly, i.e., measure the pH of freshly voided urine samples, then if necessary, buffer to pH>6 (or above for shipping purposes), measure pH again and then, perhaps, freeze the sample for storage and shipping. (See also appendix F.) Second, there is debate over the pathological significance of proteinuria, however, most world experts believe that $\beta_2$-M levels greater than 300 µg/g Cr are abnormal (Elinder, Ex. 55, Friberg, Ex. 29). Such levels signify kidney dysfunction that constitutes material impairment of health. Finally, detection of $\beta_2$-M at low levels has often been considered difficult, however, many laboratories have the capability of detecting excess $\beta_2$-M using simple kits, such as the Phadebas Delphia test, that are accurate to levels of 100 µg $\beta_2$-M/g Cr U (Ex. L-140-1).

Specific recommendations for ways to measure $\beta_2$-M and proper handling of urine samples to prevent degradation of $\beta_2$-M have been addressed by OSHA in appendix F, in the section on laboratory standardization. All biological samples must be analyzed in a laboratory that is proficient in the analysis of that particular analyte, under paragraph (l)(1)(iv). (See appendix F). Specifically, under paragraph (l)(1)(iv), the employer is to assure that the collecting and handling of biological samples of cadmium in urine (CdU), cadmium in blood (CdB), and beta-2 microglobulin in urine ($\beta_2$-M) taken from employees is collected in a manner that assures reliability. The employer must also assure that analysis of biological samples of cadmium in urine (CdU), cadmium in blood (CdB), and beta-2 microglobulin in urine ($\beta_2$-M) taken from employees is performed in laboratories with demonstrated proficiency for that particular analyte. (See appendix F.)

### 3. Lung and Prostate Cancer

The primary sites for cadmium-associated cancer appear to be the lung and the prostate (L-140-50). Evidence for an association between cancer and cadmium exposure derives from both epidemiological studies and animal experiments. Mortality from prostate cancer associated with cadmium is slightly elevated in several industrial cohorts, but the number of cases is small and there is not clear dose-response relationship. More substantive evidence exists for lung cancer.

The major epidemiological study of lung cancer was conducted by Thun et al., (Ex. 4-68). Adequate data on cadmium exposures were available to allow evaluation of dose-response relationships between cadmium exposure and lung cancer. A statistically significant excess of lung cancer attributed to cadmium exposure was observed in this study even when confounding variables such as co-exposure to arsenic and smoking habits were taken into consideration (Ex. L-140-50).

The primary evidence for quantifying a link between lung cancer and cadmium exposure from animal studies derives from two rat bioassay studies; one by Takenaka et al., (1983), which is a study of cadmium chloride and a second study by Oldiges and Glaser (1990) of four cadmium compounds.

Based on the above cited studies, the U.S. Environmental Protection Agency (EPA) classified cadmium as "B1", a probable human carcinogen, in 1985 (Ex. 4-4). The International Agency for Research on Cancer (IARC) in 1987 also recommended that cadmium be listed as "2A", a probable human carcinogen (Ex. 4-15). The American Conference of Governmental Industrial Hygienists (ACGIH) has recently recommended that cadmium be labeled as a carcinogen. Since 1984, NIOSH has concluded that cadmium is possibly a human carcinogen and has recommended that exposures be controlled to the lowest level feasible.

### 4. Non-carcinogenic Effects

Acute pneumonitis occurs 10 to 24 hours after initial acute inhalation of high levels of cadmium fumes with symptoms such as fever and chest pain (Exs. 30, 8-86B). In extreme exposure cases pulmonary edema may develop and cause death several days after exposure. Little actual exposure measurement data is available on the level of airborne cadmium exposure that causes such immediate adverse lung effects, nonetheless, it is reasonable to believe a cadmium concentration of approximately 1 mg/m$^3$ over an eight hour period is "immediately dangerous" (55 FR 4052, ANSI; Ex. 8-86B).

In addition to acute lung effects and chronic renal effects, long term exposure to cadmium may cause other severe effects on the respiratory system. Reduced pulmonary function and chronic lung disease indicative of emphysema have been observed in workers who have had prolonged exposure to cadmium dust or fumes (Exs. 4-29, 4-22, 4-42, 4-50, 4-63). In a study of workers conducted by Kazantzis et al., a statistically significant excess of worker deaths due to chronic bronchitis was found, which in his opinion was directly related to high cadmium exposures of 1 mg/m$^3$ or more (Tr. 6/8/90, pp. 156-157).

Cadmium need not be respirable to constitute a hazard. Inspirable cadmium particles that are too large to be respirable but small enough to enter the tracheobronchial region of the lung can lead to bronchoconstriction, chronic pulmonary disease, and cancer of that portion of the lung. All of these diseases have been associated with occupational exposure to cadmium (Ex. 8-86B). Particles that are constrained by their size to the extra-thoracic regions of the respiratory system such as the nose and maxillary sinuses can be swallowed through mucocillary clearance and be absorbed into the body (ACGIH, Ex. 8-692). The impaction of these particles in the upper airways can lead to anosmia, or loss of sense of smell, which is an early indication of overexposure among workers exposed to heavy metals. This condition is commonly reported among cadmium-exposed workers (Ex. 8-86-B).

### C. Medical Surveillance

In general, the main provisions of the medical surveillance section of the standard, under paragraphs (l)(1)-(17) of the regulatory text, are as follows:

1. Workers exposed above the action level are covered;

2. Workers with intermittent exposures are not covered;

3. Past workers who are covered receive biological monitoring for at least one year;

4. Initial examinations include a medical questionnaire and biological monitoring of cadmium in blood (CdB), cadmium in urine (CdU), and Beta-2-microglobulin in urine ($\beta_2$-M);

5. Biological monitoring of these three analytes is performed at least annually; full medical examinations are performed biennially;

6. Until five years from the effective date of the standard, medical removal is required when CdU is greater than 15 µg/gram creatinine (g Cr), or CdB is greater than 15 µg/liter whole blood (lwb), or $\beta_2$-M is greater than 1500 µg/g Cr, and CdB is greater than 5 µg/lwb or CdU is greater than 3 µg/g Cr;

7. Beginning five years after the standard is in effect, medical removal triggers will be reduced;

8. Medical removal protection benefits are to be provided for up to 18 months;

9. Limited initial medical examinations are required for respirator usage;

10. Major provisions are fully described under section (l) of the regulatory text; they are outlined here as follows:

A. Eligibility

B. Biological monitoring

C. Actions triggered by levels of CdU, CdB, and $\beta_2$-M (See Summary Charts and Tables in Attachment-1.)

D. Periodic medical surveillance

E. Actions triggered by periodic medical surveillance (See appendix A Summary Chart and Tables in Attachment-1.)

F. Respirator usage

G. Emergency medical examinations

H. Termination examination

I. Information to physician

J. Physician's medical opinion

K. Medical removal protection

L. Medical removal protection benefits

M. Multiple physician review

N. Alternate physician review

O. Information employer gives to employee

P. Recordkeeping

Q. Reporting on OSHA form 200

11. The above mentioned summary of the medical surveillance provisions, the summary chart, and tables for the actions triggered at different levels of CdU, CdB and $\beta_2$-M (in appendix A Attachment-1) are included only for the purpose of facilitating understanding of the provisions of paragraphs (l)(3) of the final cadmium standard. The summary of the provisions, the summary chart, and the tables do not add to or reduce the requirements in paragraph (l)(3).

### D. Recommendations to Physicians

1. It is strongly recommended that patients with tubular proteinuria are counseled on: The hazards of smoking; avoidance of nephrotoxins and certain prescriptions and over-the-counter medications that may exacerbate kidney symptoms; how to control diabetes and/or blood pressure; proper hydration, diet, and exercise (Ex. 19-2). A list of prominent or common nephrotoxins is attached. (See appendix A Attachment-2.)

2. DO NOT CHELATE; KNOW WHICH DRUGS ARE NEPHROTOXINS OR ARE ASSOCIATED WITH NEPHRITIS.

3. The gravity of cadmium-induced renal damage is compounded by the fact there is no medical treatment to prevent or reduce the accumulation of cadmium in the kidney (Ex. 8-619). Dr. Friberg, a leading world expert on cadmium toxicity, indicated in 1992, that there is no form of chelating agent that could be used without substantial risk. He stated that tubular proteinuria has to be treated in the same way as other kidney disorders (Ex. 29).

4. After the results of a workers' biological monitoring or medical examination are received the employer is required to provide an information sheet to the patient, briefly explaining the significance of the results. (See Attachment 3 of this appendix A.)

5. For additional information the physician is referred to the following additional resources:

a. The physician can always obtain a copy of the preamble, with its full discussion of the health effects, from OSHA's Computerized Information System (OCIS).

b. The Docket Officer maintains a record of the rulemaking. The Cadmium Docket (H-057A), is located at 200 Constitution Ave. NW., room N-2625, Washington, DC 20210; telephone: 202-219-7894.

c. The following articles and exhibits in particular from that docket (H-057A):

| Exhibit number | Author and paper title |
|---|---|
| 8-447 | Lauwerys et. al., Guide for physicians, "Health Maintenance of Workers Exposed to Cadmium," published by the Cadmium Council. |
| 4-67 | Takenaka, S., H. Oldiges, H. Konig, D. Hochrainer, G. Oberdorster. "Carcinogenicity of Cadmium Chloride Aerosols in Wistar Rats". JNCI 70:367-373, 1983. (32) |
| 4-68 | Thun, M.J., T.M. Schnoor, A.B. Smith, W.E. Halperin, R.A. Lemen. "Mortality Among a Cohort of U.S. Cadmium Production Workers—An Update." JNCI 74(2):325-33, 1985. (8) |
| 4-25 | Elinder, C.G., Kjellstrom, T., Hogstedt, C., et al., "Cancer Mortality of Cadmium Workers." Brit. J. Ind. Med. 42:651-655, 1985. (14) |
| 4-26 | Ellis, K.J. et al., "Critical Concentrations of Cadmium in Human Renal Cortex: Dose Effect Studies to Cadmium Smelter Workers." J. Toxicol. Environ. Health 7:691-703, 1981. (76) |
| 4-27 | Ellis, K.J., S.H. Cohn and T.J. Smith. "Cadmium Inhalation Exposure Estimates: Their Significance with Respect to Kidney and Liver Cadmium Burden." J. Toxicol. Environ. Health 15:173-187, 1985. |
| 4-28 | Falck, F.Y., Jr., Fine, L.J., Smith, R.G., McClatchey, K.D., Annesley, T., England, B., and Schork, A.M. "Occupational Cadmium Exposure and Renal Status." Am. J. Ind. Med. 4:541, 1983. (64) |
| 8-86A | Friberg, L., C.G. Elinder, et al., "Cadmium and Health a Toxicological and Epidemiological Appraisal, Volume I, Exposure, Dose, and Metabolism." CRC Press, Inc., Boca Raton, FL, 1986. (Available from the OSHA Technical Data Center) |
| 8-86B | Friberg, L., C.G. Elinder, et al., "Cadmium and Health: A Toxicological and Epidemiological Appraisal, Volume II, Effects and Response." CRC Press, Inc., Boca Raton, FL, 1986. (Available from the OSHA Technical Data Center) |
| L-140-45 | Elinder, C.G., "Cancer Mortality of Cadmium Workers", Brit. J. Ind. Med., 42, 651-655, 1985. |
| L-140-50 | Thun, M., Elinder, C.G., Friberg, L, "Scientific Basis for an Occupational Standard for Cadmium, Am. J. Ind. Med., 20; 629-642, 1991. |

### V. Information Sheet

The information sheet (appendix A Attachment-3.) or an equally explanatory one should be provided to you after any biological monitoring results are reviewed by the physician, or where applicable, after any medical examination.

### ATTACHMENT 1—APPENDIX A SUMMARY CHART AND TABLES A AND B OF ACTIONS TRIGGERED BY BIOLOGICAL MONITORING

### APPENDIX A SUMMARY CHART— SECTION (l)(3) MEDICAL SURVEILLANCE

Categorizing Biological Monitoring Results

(A) Biological monitoring results categories are set forth in Appendix A Table A for the periods ending December 31, 1998 and for the period beginning January 1, 1999.

(B) The results of the biological monitoring for the initial medical exam and the subsequent exams shall determine an employee's biological monitoring result category.

Actions Triggered by Biological Monitoring

(A)

(i) The actions triggered by biological monitoring for an employee are set forth in Appendix A Table B.

(ii) The biological monitoring results for each employee under section (l)(3) shall determine the actions required for that employee. That is, for any employee in biological monitoring category C, the employer will perform all of the actions for which there is an X in column C of Appendix A Table B.

(iii) An employee is assigned the alphabetical category ("A" being the lowest) depending upon the test results of the three biological markers.

(iv) An employee is assigned category A if monitoring results for all three biological markers fall at or below the levels indicated in the table listed for category A.

(v) An employee is assigned category B if any monitoring result for any of the three biological markers fall within the range of levels indicated in the table listed for category B, providing no result exceeds the levels listed for category B.

(vi) An employee is assigned category C if any monitoring result for any of the three biological markers are above the levels listed for category C.

(B) The user of Appendix A Tables A and B should know that these tables are provided only to facilitate understanding of the relevant provisions of paragraph (l)(3) of this section. Appendix A Tables A and B are not meant to add to or subtract from the requirements of those provisions.

### APPENDIX A TABLE A—CATEGORIZATION OF BIOLOGICAL MONITORING RESULTS

#### APPLICABLE THROUGH 1998 ONLY

| Biological marker | Monitoring result categories |||
|---|---|---|---|
| | A | B | C |
| Cadmium in urine (CdU) (µg/g creatinine) | ≤3 | >3 and ≤15 | >15 |
| $\beta_2$microglobulin ($\beta_2$-M) (µg/g creatinine) | ≤300 | >300 and ≤1500 | >1500* |
| Cadmium in blood (CdB) (µg/liter whole blood) | ≤5 | >5 and ≤15 | >15 |

* If an employee's $\beta_2$-M levels are above 1,500 µg/g creatinine, in order for mandatory medical removal to be required (See Appendix A Table B.), either the employee's CdU level must also be >3 µg/g creatinine or CdB level must also be >5 µg/liter whole blood.

#### APPLICABLE BEGINNING JANUARY 1, 1999

| Biological marker | Monitoring result categories |||
|---|---|---|---|
| | A | B | C |
| Cadmium in urine (CdU) (µg/g creatinine) | ≤3 | >3 and ≥7 | >7 |
| $\beta_2$microglobulin ($\beta_2$-M) (µg/g creatinine) | ≤300 | >300 and ≤750 | >750* |
| Cadmium in blood (CdB) (µg/liter whole blood) | ≤5 | >5 and ≤10 | >10 |

* If an employee's $\beta_2$-M levels are above 750 µg/g creatinine, in order for mandatory medical removal to be required (See Appendix A Table B.), either the employee's CdU level must also be >3 µg/g creatinine or CdB level must also be >5 µg/liter whole blood.

# OSHA Standards for General Industry

## APPENDIX A TABLE B—ACTIONS DETERMINED BY BIOLOGICAL MONITORING

This table presents the actions required based on the monitoring result in Appendix A Table A. Each item is a separate requirement in citing non-compliance. For example, a medical examination within 90 days for an employee in category B is separate from the requirement to administer a periodic medical examination for category B employees on an annual basis.

| Required actions | A[1] | B[1] | C[1] |
|---|---|---|---|
| **(1) Biological monitoring:** | | | |
| (a) Annual | X | | |
| (b) Semiannual | | X | |
| (c) Quarterly | | | X |
| **(2) Medical examination:** | | | |
| (a) Biennial | X | | |
| (b) Annual | | X | |
| (c) Semiannual | | | X |
| (d) Within 90 days | | X | X |
| **(3) Assess within two weeks:** | | | |
| (a) Excess cadmium exposure | | X | X |
| (b) Work practices | | X | X |
| (c) Personal hygiene | | X | X |
| (d) Respirator usage | | X | X |
| (e) Smoking history | | X | X |
| (f) Hygiene facilities | | X | X |
| (g) Engineering controls | | X | X |
| (h) Correct within 30 days | | X | X |
| (i) Periodically assess exposures | | | X |
| **(4) Discretionary medical removal** | | X | X |
| **(5) Mandatory medical removal** | | | X[2] |

[1] For all employees covered by medical surveillance exclusively because of exposures prior to the effective date of this standard, if they are in Category A, the employer shall follow the requirements of paragraphs (1)(3)(i)(B) and (1)(4)(v)(A). If they are in Category B or C, the employer shall follow the requirements of paragraphs (1)(4)(v)(B) - (C).

[2] See footnote Appendix A Table A.

## APPENDIX A—ATTACHMENT 2—LIST OF MEDICATIONS

A list of the more common medications that a physician, and the employee, may wish to review is likely to include some of the following: (1) Anticonvulsants: Paramethadione, phenytoin, trimethadone; (2) antihypertensive drugs: Captopril, methyldopa; (3) antimicrobials: Aminoglycosides, amphotericin B, cephalosporins, ethambutol; (4) antineoplastic agents: Cisplatin, methotrexate, mitomycin-C, nitrosoureas, radiation; (4) sulfonamide diuretics: Acetazolamide, chlortbalidone, furosemide, thiazides; (5) halogenated alkanes, hydrocarbons, and solvents that may occur in some settings: Carbon tetrachloride, ethylene glycol, toluene; iodinated radiographic contrast media; nonsteroidal anti-inflammatory drugs; and, (7) other miscellaneous compounds: Acetaminophen, allopurinol, amphetamines, azathioprine, cimetidine, cyclosporine, lithium, methoxyflurane, methysergide, D-penicillamine, phenacetin, phenendione. A list of drugs associated with acute interstitial nephritis includes: (1) Antimicrobial drugs: Cephalosporins, chloramphenicol, colistin, erythromycin, ethambutol, isoniazid, para-aminosalicylic acid, penicillins, polymyxin B, rifampin, sulfonamides, tetracyclines, and vancomycin; (2) other miscellaneous drugs: Allopurinol, antipyrene, azathioprine, captopril, cimetidine, clofibrate, methyldopa, phenindione, phenylpropanolamine, phenytoin, probenecid, sulfinpyrazone, sulfonamid diuretics, triamterene; and, (3) metals: Bismuth, gold.

This list have been derived from commonly available medical textbooks (e.g., Ex. 14-18). The list has been included merely to facilitate the physician's, employer's, and employee's understanding. The list does not represent an official OSHA opinion or policy regarding the use of these medications for particular employees. The use of such medications should be under physician discretion.

## ATTACHMENT 3—BIOLOGICAL MONITORING AND MEDICAL EXAMINATION RESULTS

Employee _____ Testing Date _____

Cadmium in Urine _____ µg/g Cr—Normal Levels: ≤3 µg/g Cr.

Cadmium in Blood _____ µg/lwb—Normal Levels: ≤5 µg/lwb.

Beta-2-microglobulin in Urine _____ µg/g Cr—Normal Levels: ≤300 µg/g Cr.

Physical Examination Results: N/A _____ Satisfactory _____ Unsatisfactory _____ (see physician again).

Physician's Review of Pulmonary Function Test: N/A _____

Normal _____ Abnormal _____ .

Next biological monitoring or medical examination scheduled for _____ .

The biological monitoring program has been designed for three main purposes: 1) to identify employees at risk of adverse health effects from excess, chronic exposure to cadmium; 2) to prevent cadmium-induced disease(s); and 3) to detect and minimize existing cadmium-induced disease(s).

The levels of cadmium in the urine and blood provide an estimate of the total amount of cadmium in the body. The amount of a specific protein in the urine (beta-2-microglobulin) indicates changes in kidney function. All three tests must be evaluated together. A single mildly elevated result may not be important if testing at a later time indicates that the results are normal and the workplace has been evaluated to decrease possible sources of cadmium exposure. The levels of cadmium or beta-2-microglobulin may change over a period of days to months and the time needed for those changes to occur is different for each worker.

If the results for biological monitoring are above specific "high levels" [cadmium urine greater than 10 micrograms per gram of creatinine (µg/g Cr), cadmium blood greater than 10 micrograms per liter of whole blood (µg/lwb), or beta-2-microglobulin greater than 1000 micrograms per gram of creatinine (µg/g Cr)], the worker has a much greater chance of developing other kidney diseases.

One way to measure for kidney function is by measuring beta-2-microglobulin in the urine. Beta-2-microglobulin is a protein which is normally found in the blood as it is being filtered in the kidney, and the kidney reabsorbs or returns almost all of the beta-2-microglobulin to the blood. A very small amount (less than 300 µg/g Cr in the urine) of beta-2-microglobulin is not reabsorbed into the blood, but is released in the urine. If cadmium damages the kidney, the amount of beta-2-microglobulin in the urine increases because the kidney cells are unable to reabsorb the beta-2-microglobulin normally. An increase in the amount of beta-2-microglobulin in the urine is a very early sign of kidney dysfunction. A small increase in beta-2-microglobulin in the urine will serve as an early warning sign that the worker may be absorbing cadmium from the air, cigarettes contaminated in the workplace, or eating in areas that are cadmium contaminated.

Even if cadmium causes permanent changes in the kidney's ability to reabsorb beta-2-microglobulin, and the beta-2-microglobulin is above the "high levels", the loss of kidney function may not lead to any serious health problems. Also, renal function naturally declines as people age. The risk for changes in kidney function for workers who have biological monitoring results between the "normal values" and the "high levels" is not well known. Some people are more cadmium-tolerant, while others are more cadmium-susceptible.

For anyone with even a slight increase of beta-2-microglobulin, cadmium in the urine, or cadmium in the blood, it is very important to protect the kidney from further damage. Kidney damage can come from other sources than excess cadmium-exposure so it is also recommended that if a worker's levels are "high" he/she should receive counseling about drinking more water; avoiding cadmium-tainted tobacco and certain medications (nephrotoxins, acetaminophen); controlling diet, vitamin intake, blood pressure and diabetes; etc.

## APPENDIX B TO §1910.1027—SUBSTANCE TECHNICAL GUIDELINES FOR CADMIUM

I. Cadmium Metal

A. *Physical and Chemical Data.*

1. *Substance Identification.*

*Chemical name:* Cadmium.

*Formula:* Cd.

*Molecular Weight:* 112.4.

*Chemical Abstracts Service (CAS) Registry No.:* 7740-43-9.

*Other Identifiers:* RETCS EU9800000; EPA D006; DOT 2570 53.

*Synonyms:* Colloidal Cadmium: Kadmium (German): CI 77180.

2. *Physical data.*

*Boiling point:* (760 mm Hg): 765 °C.

*Melting point:* 321 °C.

*Specific Gravity:* (H$_2$O = 1 @ 20 °C): 8.64.

*Solubility:* Insoluble in water; soluble in dilute nitric acid and in sulfuric acid.

*Appearance:* Soft, blue-white, malleable, lustrous metal or grayish-white powder.

B. *Fire, Explosion and Reactivity Data.*

1. *Fire.*

*Fire and Explosion Hazards:* The finely divided metal is pyrophoric, that is the dust is a severe fire hazard and moderate explosion hazard when exposed to heat or flame. Burning material reacts violently with extinguishing agents such as water, foam, carbon dioxide, and halons.

*Flash point:* Flammable (dust).

*Extinguishing media:* Dry sand, dry dolomite, dry graphite, or sodimum chloride.

2. *Reactivity.*

*Conditions contributing to instability:* Stable when kept in sealed containers under normal temperatures and pressure, but dust may ignite upon contact with air. Metal tarnishes in moist air.

*Incompatibilities:* Ammonium nitrate, fused: Reacts violently or explosively with cadmium dust below 20 °C. Hydrozoic acid: Violent explosion occurs after 30 minutes. Acids: Reacts violently, forms hydrogen gas. Oxidizing agents or metals: Strong reaction with cadmium dust. Nitryl fluoride at slightly elevated temperature: Glowing or white incandescence occurs. Selenium: Reacts exothermically. Ammonia: Corrosive reaction. Sulfur dioxide: Corrosive reaction. Fire extinguishing agents (water, foam, carbon dioxide, and halons): Reacts violently. Tellurium: Incandescent reaction in hydrogen atmosphere.

*Hazardous decomposition products:* The heated metal rapidly forms highly toxic, brownish fumes of oxides of cadmium.

C. *Spill, Leak and Disposal Procedures.*

1. *Steps to be taken if the materials is released or spilled.* Do not touch spilled material. Stop leak if you can do it without risk. Do not get water inside container. For large spills, dike spill for later disposal. Keep unnecessary people away. Isolate hazard area and deny entry. The Superfund Amendments and Reauthorization Act of 1986 Section 304 requires that a release equal to or greater than the reportable quantity for this substance (1 pound) must be immediately reported to the local emergency planning committee, the state emergency response commission, and the National Response Center (800) 424-8802; in Washington, DC metropolitan area (202) 426-2675.

II. Cadmium Oxide

A. *Physical and Chemical Date.*

1. *Substance identification.*

*Chemical name:* Cadmium Oxide.

*Formula:* CdO.

*Molecular Weight:* 128.4.

*CAS No.:* 1306-19-0.

*Other Identifiers:* RTECS EV1929500.

*Synonyms:* Kadmu tlenek (Polish).

2. *Physical data.*

*Boiling point (760 mm Hg):* 950 °C decomposes.

*Melting point:* 1500 °C.

*Specific Gravity:* (H$_2$O = 1 @ 20 °C): 7.0.

*Solubility:* Insoluble in water; soluble in acids and alkalines.

*Appearance:* Red or brown crystals.

B. *Fire, Explosion and Reactivity Data.*

1. *Fire.*

*Fire and Explosion Hazards:* Negligible fire hazard when exposed to heat or flame.

*Flash point:* Nonflammable.

*Extinguishing media:* Dry chemical, carbon dioxide, water spray or foam.

2. *Reactivity.*

*Conditions contributing to instability:* Stable under normal temperatures and pressures.

*Incompatibilities:* Magnesium may reduce CdO$_2$ explosively on heating.

*Hazardous decomposition products:* Toxic fumes of cadmium.

C. *Spill Leak and Disposal Procedures.*

1. *Steps to be taken if the material is released or spilled.* Do not touch spilled material. Stop leak if you can do it without risk. For small spills, take up with sand or other absorbent material and place into containers for later disposal. For small dry spills, use a clean shovel to place material into clean, dry container and then cover. Move containers from spill area. For larger spills, dike far ahead of spill for later disposal. Keep unnecessary people away. Isolate hazard area and deny entry. The Superfund Amendments and Reauthorization Act of 1986 Section 304 requires that a release equal to or greater than the reportable quantity for this substance (1 pound) must be immediately reported to the local emergency planning committee, the state emergency response commission, and the National Response Center (800) 424-8802; in Washington, DC metropolitan area (202) 426-2675.

III. Cadmium Sulfide.

A. *Physical and Chemical Data.*

1. *Substance Identification.*

*Chemical name:* Cadmium sulfide.

*Formula:* CdS.

*Molecular weight:* 144.5.

*CAS No.* 1306-23-6.

*Other Identifiers:* RTECS EV3150000.

*Synonyms:* Aurora yellow; Cadmium Golden 366; Cadmium Lemon Yellow 527; Cadmium Orange; Cadmium Primrose 819; Cadmium Sulphide; Cadmium Yellow; Cadmium Yellow 000; Cadmium Yellow Conc. Deep; Cadmium Yellow Conc. Golden; Cadmium Yellow Conc. Lemon; Cadmium Yellow Conc. Primrose; Cadmium Yellow Oz. Dark; Cadmium Yellow Primrose 47-1400; Cadmium Yellow 10G Conc.; Cadmium Yellow 892; Cadmopur Golden Yellow N; Cadmopur Yellow: Capsebon; C.I. 77199; C.I. Pigment Orange 20; CI Pigment Yellow 37; Ferro Lemon Yellow; Ferro Orange Yellow; Ferro Yellow; Greenockite; NCI-C02711.

2. *Physical data.*

*Boiling point (760 mm. Hg):* sublines in N$_2$ at 980 °C.

*Melting point:* 1750 °C (100 atm).

*Specific Gravity:* (H$_2$O = 1 @ 20 °C): 4.82.

*Solubility:* Slightly soluble in water; soluble in acid.

*Appearance:* Light yellow or yellow-orange crystals.

B. *Fire, Explosion and Reactivity Data.*

1. *Fire.*

*Fire and Explosion Hazards:* Neglible fire hazard when exposed to heat or flame.

*Flash point:* Nonflammable.

*Extinguishing media:* Dry chemical, carbon dioxide, water spray or foam.

2. *Reactivity.*

*Conditions contributing to instability:* Generally non-reactive under normal conditions. Reacts with acids to form toxic hydrogen sulfide gas.

# OSHA Standards for General Industry

*Incompatibilities:* Reacts vigorously with iodinemonochloride.

*Hazardous decomposition products:* Toxic fumes of cadmium and sulfur oxides.

C. *Spill Leak and Disposal Procedures.*

1. *Steps to be taken if the material is released or spilled.* Do not touch spilled material. Stop leak if you can do it without risk. For small, dry spills, with a clean shovel place material into clean, dry container and cover. Move containers from spill area. For larger spills, dike far ahead of spill for later disposal. Keep unnecessary people away. Isolate hazard and deny entry.

IV. Cadmium Chloride.

A. *Physical and Chemical Data.*

1. *Substance Identification.*

Chemical name: Cadmium chloride.

Formula: $CdCl_2$.

Molecular weight: 183.3.

CAS No. 10108-64-2.

Other Identifiers: RTECS EY0175000.

Synonyms: Caddy; Cadmium dichloride; NA 2570 (DOT); UI-CAD; dichlorocadmium.

2. *Physical data.*

Boiling point (760 mm Hg): 960 °C.

Melting point: 568 °C.

Specific Gravity: ($H_2O$ = 1 @ 20 °C): 4.05.

Solubility: Soluble in water (140 g/100 cc); soluble in acetone.

Appearance: Small, white crystals.

B. *Fire, Explosion and Reactivity Data.*

1. *Fire.*

*Fire and Explosion Hazards:* Negligible fire and negligible explosion hazard in dust form when exposed to heat or flame.

*Flash point:* Nonflammable.

*Extinguishing media:* Dry chemical, carbon dioxide, water spray or foam.

2. *Reactivity.*

*Conditions contributing to instability:* Generally stable under normal temperatures and pressures.

*Incompatibilities:* Bromine triflouride rapidly attacks cadmium chloride. A mixture of potassium and cadmium chloride may produce a strong explosion on impact.

*Hazardous decomposition products:* Thermal ecompostion may release toxic fumes of hydrogen chloride, chloride, chlorine or oxides of cadmium.

C. *Spill Leak and Disposal Procedures.*

1. *Steps to be taken if the materials is released or spilled.* Do not touch spilled material. Stop leak if you can do it without risk. For small, dry spills, with a clean shovel place material into clean, dry container and cover. Move containers from spill area. For larger spills, dike far ahead of spill for later disposal. Keep unnecessary people away. Isolate hazard and deny entry. The Superfund Amendments and Reauthorization Act of 1986 Section 304 requires that a release equal to or greater than the reportable quantity for this substance (100 pounds) must be immediately reported to the local emergency planning committee, the state emergency response commission, and the National Response Center (800) 424-8802; in Washington, DC Metropolitan area (202) 426-2675.

## APPENDIX C TO §1910.1027—QUALITATIVE AND QUANTITATIVE FIT TESTING PROCEDURES [Reserved].

[§1910.1027, App. C, removed and reserved at 63 FR 1288, Jan. 8, 1998, eff. Apr. 8, 1998.]

## APPENDIX D TO §1910.1027—OCCUPATIONAL HEALTH HISTORY INTERVIEW WITH REFERENCE TO CADMIUM EXPOSURE

10. What is or was the greatest number of packs per day that you have smoked? Number of packs ____
If you quit smoking cigarettes, how many years ago did you quit? ❏ Less than 1 year  Number of years ____
How many packs a day do you now smoke? Number of packs per day ____
Have you ever been told by a doctor that you had a kidney or urinary tract disease or disorder? ❏ Yes ❏ No

11. Have you ever had any of these disorders?
Kidney stones ❏ Yes ❏ No   Protein in urine ❏ Yes ❏ No   Blood in urine ❏ Yes ❏ No
Difficulty urinating ❏ Yes ❏ No   Other kidney/Urinary disorders ❏ Yes ❏ No
Please describe problems, age, treatment, and follow up for any kidney or urinary problems you have had:
____

12. Have you ever been told by a doctor or other health care provider who took your blood pressure that your blood pressure was high? ❏ Yes ❏ No

13. Have you ever been advised to take any blood pressure medication? ❏ Yes ❏ No

14. Are you presently taking any blood pressure medication? ❏ Yes ❏ No

15. Are you presently taking any other medication? ❏ Yes ❏ No

16. Please list any blood pressure or other medications and describe how long you have been taking each one:
Medicine: ____
How Long Taken ____

17. Have you ever been told by a doctor that you have diabetes? (sugar in your blood or urine) ❏ Yes ❏ No
If yes, do you presently see a doctor about your diabetes? ❏ Yes ❏ No
If yes, how do you control your blood sugar? ❏ Diet alone ❏ Diet plus oral medicine ❏ Diet plus insulin (injection)

18. Have you ever been told by a doctor that you had:
Anemia ❏ Yes ❏ No   A low blood count? ❏ Yes ❏ No

19. Do you presently feel that you tire or run out of energy sooner than normal or sooner than other people your age? ❏ Yes ❏ No
If yes, for how long have you felt that you tire easily? ❏ Less than 1 year ❏ Number of years ____

20. Have you given blood within the last year? ❏ Yes ❏ No   If yes, how many times? ❏ Number of times ____
How long ago was the last time you gave blood? ❏ Less than 1 month ❏ Number of months ____

21. Within the last year have you had any injuries with heavy bleeding? ❏ Yes ❏ No
If yes, how long ago? ❏ Less than 1 month ❏ Number of months ____
Describe: ____

22. Have you recently had any surgery? ❏ Yes ❏ No
If yes, please describe: ____

23. Have you seen any blood lately in your stool or after a bowel movement? ❏ Yes ❏ No

24. If yes, did you had a test for blood in your stool? ❏ Yes ❏ No
If yes, did the test show any blood in the stool? ❏ Yes ❏ No
What further evaluation and treatment were done? ____

## OCCUPATIONAL HEALTH HISTORY INTERVIEW WITH REFERENCE TO CADMIUM EXPOSURE

*Directions*

(To be read by employee and signed prior to the interview)

Please answer the questions you will be asked as completely and carefully as you can. These questions are asked of everyone who works with cadmium. You will also be asked to give blood and urine samples. The doctor will give your employer a written opinion on whether you are physically capable of working with cadmium. Legally, the doctor cannot share personal information you may tell him/her with your employer. The following information is considered strictly confidential. The results of the tests will go to you, your doctor and your employer. You will also receive an information sheet explaining the results of any biological monitoring or physical examinations performed.

If you are just being hired, the results of this interview and examination will be used to:

(1) Establish your health status and see if working with cadmium might be expected to cause unusual problems.
(2) Determine your health status today and see if there are changes over time.
(3) See if you can wear a respirator safely.

If you are not a new hire:

OSHA says that everyone who works with cadmium can have periodic medical examinations performed by a doctor. The reasons for this are:

(a) If there are changes in your health, either because of cadmium or some other reason, to find them early.
(b) to prevent kidney damage.

Please sign below.

I have read these directions and understand them:

Employee signature _____

Date _____

Thank you for answering these questions.

(Suggested Format)

Name _____ Age _____

Social Security # _____ Company _____

Job _____

Type of Preplacement Exam: ☐ Periodic ☐ Termination ☐ Initial ☐ Other

Blood Pressure _____ Pulse Rate _____

1. How long have you worked at the job listed above? ☐ Not yet hired   Number of months ____ Number of years ____
2. Job Duties etc.

3. Have you ever been told by a doctor that you had bronchitis? ☐ Yes ☐ No
   If yes, how long ago? ☐ Number of months ____ Number of years ____
4. Have you ever been told by a doctor that you had emphysema? ☐ Yes ☐ No
   If yes, how long ago? ☐ Number of months ____ Number of years ____
5. Have you ever been told by a doctor that you had other lung problems? ☐ Yes ☐ No
   If yes, please describe type of lung problems and when you had these problems

6. In the past year, have you had a cough? ☐ Yes ☐ No
   If yes, did you cough up sputum? ☐ Yes ☐ No
   If yes, how long did the cough with sputum production last? ☐ Less than 3 months ☐ 3 months or longer
   If yes, for how many years have you had episodes of cough with sputum production lasting this long?
   ☐ Less than one ☐ 1 ☐ 2 ☐ Longer than 2
7. Have you ever smoked cigarettes? ☐ Yes ☐ No
8. Do you now smoke cigarettes? ☐ Yes ☐ No
9. If you smoke or have smoked cigarettes, for how many years have you smoked, or did you smoke?
   ☐ Less than 1 year  Number of years ____

---

The following questions pertain to the ability to wear a respirator. Additional information for the physician can be found in The Respiratory Protective Devices Manual.

25. Have you ever been told by a doctor that you have asthma? ☐ Yes ☐ No
    If yes, are you presently taking any medication for asthma? Mark all that apply. ☐ Shots ☐ Pills ☐ Inhaler
26. Have you ever had a heart attack? ☐ Yes ☐ No
    If yes, how long ago? Number of years ____ Number of months ____
27. Have you ever had pains in your chest? ☐ Yes ☐ No
    If yes, when did it usually happen? ☐ While resting ☐ While working ☐ While exercising ☐ Activity didn't matter
28. Have you ever had a thyroid problem? ☐ Yes ☐ No
29. Have you ever had a seizure or fits? ☐ Yes ☐ No
30. Have you ever had a stroke (cerebrovascular accident)? ☐ Yes ☐ No
31. Have you ever had a ruptured eardrum or a serious hearing problem? ☐ Yes ☐ No
32. Do you now have a claustrophobia, meaning fear of crowded or closed in spaces or any psychological problems that would make it hard for you to wear a respirator? ☐ Yes ☐ No

The following questions pertain to reproductive history.

33. Have you or your partner had a problem conceiving a child? ☐ Yes ☐ No
    If yes, specify: ☐ Self ☐ Present mate ☐ Previous mate
34. Have you or your partner consulted a physician for a fertility or other reproductive problem? ☐ Yes ☐ No
    If yes, specify who consulted the physician: ☐ Self ☐ Spouse/partner ☐ Self and partner
    If yes, specify diagnosis made:

35. Have you or your partner ever conceived a child resulting in a miscarriage, still birth or deformed offspring?
    ☐ Yes ☐ No
    If yes, specify: ☐ Miscarriage ☐ Still birth ☐ Deformed offspring
    If outcome was a deformed offspring, please specify type:
36. Was this outcome a result of a pregnancy of: ☐ Yours with present partner ☐ Yours with a previous partner
37. Did the timing of any abnormal pregnancy outcome coincide with present employment? ☐ Yes ☐ No
    List dates of occurrences:
38. What is the occupation of your spouse or partner?

**For Women Only**

39. Do you have menstrual periods? ☐ Yes ☐ No   Have you had menstrual irregularities? ☐ Yes ☐ No
    If yes, specify type:
    If yes, what was the approximated date this problem began?
    Approximate date problem stopped?

**For Men Only**

40. Have you ever been diagnosed by a physician as having prostate gland problem(s)? ☐ Yes ☐ No
    If yes, please describe type of problem(s) and what was done to evaluate and treat the problem(s):

# APPENDIX E TO §1910.1027— CADMIUM IN WORKPLACE ATMOSPHERES

Method Number: ID-189

Matrix: Air

OSHA Permissible Exposure Limits: 5 µg/m$^3$ (TWA), 2.5 µg/m$^3$ (Action Level TWA)

Collection Procedure: A known volume of air is drawn through a 37-mm diameter filter cassette containing a 0.8-m mixed cellulose ester membrane filter (MCEF).

Recommended Air Volume: 960 L

Recommended Sampling Rate: 2.0 L/min

Analytical Procedure: Air filter samples are digested with nitric acid. After digestion, a small amount of hydrochloric acid is added. The samples are then diluted to volume with deionized water and analyzed by either flame atomic absorption spectroscopy (AAS) or flameless atomic absorption spectroscopy using a heated graphite furnace atomizer (AAS-HGA).

Detection Limits:

Qualitative: 0.2 µg/m$^3$ for a 200 L sample by Flame AAS, 0.007 µg/m$^3$ for a 60 L sample by AAS-HGA

Quantitative: 0.70 µg/m$^3$ for a 200 L sample by Flame AAS, 0.025 µg/m$^3$ for a 60 L sample by AAS-HGA

Precision and Accuracy: (Flame AAS Analysis and AAS-HGA Analysis):

Validation Level: 2.5 to 10 µg/m$^3$ for a 400 L air vol, 1.25 to 5.0 µg/m$^3$ for a 60 L air vol

$CV_1$ (pooled): 0.010, 0.043

Analytical Bias: +4.0%, -5.8%

Overall Analytical Error: ±6.0%, ±14.2%

Method Classification: Validated

Date: June, 1992

Inorganic Service Branch II, OSHA Salt Lake Technical Center, Salt Lake City, Utah

Commercial manufacturers and products mentioned in this method are for descriptive use only and do not constitute endorsements by USDOL-OSHA. Similar products from other sources can be substituted.

## 1. INTRODUCTION

### 1.1. Scope

This method describes the collection of airborne elemental cadmium and cadmium compounds on 0.8-µm mixed cellulose ester membrane filters and their subsequent analysis by either flame atomic absorption spectroscopy (AAS) or flameless atomic absorption spectroscopy using a heated graphite furnace atomizer (AAS-HGA). It is applicable for both TWA and Action Level TWA Permissible Exposure Level (PEL) measurements. The two atomic absorption analytical techniques included in the method do not differentiate between cadmium fume and cadmium dust samples. They also do not differentiate between elemental cadmium and its compounds.

### 1.2. Principle

Airborne elemental cadmium and cadmium compounds are collected on a 0.8-µm mixed cellulose ester membrane filter (MCEF). The air filter samples are digested with concentrated nitric acid to destroy the organic matrix and dissolve the cadmium analytes. After digestion, a small amount of concentrated hydrochloric acid is added to help dissolve other metals which may be present. The samples are diluted to volume with deionized water and then aspirated into the oxidizing air/acetylene flame of an atomic absorption spectrophotometer for analysis of elemental cadmium.

If the concentration of cadmium in a sample solution is too low for quantitation by this flame AAS analytical technique, and the sample is to be averaged with other samples for TWA calculations, aliquots of the sample and a matrix modifier are later injected onto a L'vov platform in a pyrolytically-coated graphite tube of a Zeeman atomic absorption spectrophotometer/graphite furnace assembly for analysis of elemental cadmium. The matrix modifier is added to stabilize the cadmium metal and minimize sodium chloride as an interference during the high temperature charring step of the analysis (5.1., 5.2.).

### 1.3. History

Previously, two OSHA sampling and analytical methods for cadmium were used concurrently (5.3., 5.4.). Both of these methods also required 0.8-µm mixed cellulose ester membrane filters for the collection of air samples. These cadmium air filter samples were analyzed by either flame atomic absorption spectroscopy (5.3.) or inductively coupled plasma/atomic emission spectroscopy (ICP-AES) (5.4.). Neither of these two analytical methods have adequate sensitivity for measuring workplace exposure to airborne cadmium at the new lower TWA and Action Level TWA PEL levels when consecutive samples are taken on one employee and the sample results need to be averaged with other samples to determine a single TWA.

The inclusion of two atomic absorption analytical techniques in the new sampling and analysis method for airborne cadmium permits quantitation of sample results over a broad range of exposure levels and sampling periods. The flame AAS analytical technique included in this method is similar to the previous procedure given in the General Metals Method ID-121 (5.3.) with some modifications. The sensitivity of the AAS-HGA analytical technique included in this method is adequate to measure exposure levels at 1/10 the Action Level TWA, or lower, when less than full-shift samples need to be averaged together.

### 1.4. Properties (5.5.)

Elemental cadmium is a silver-white, blue-tinged, lustrous metal which is easily cut with a knife. It is slowly oxidized by moist air to form cadmium oxide. It is insoluble in water, but reacts readily with dilute nitric acid. Some of the physical properties and other descriptive information of elemental cadmium are given below:

| | |
|---|---|
| CAS No. | 7440-43-9 |
| Atomic Number | 48 |
| Atomic Symbol | Cd |
| Atomic Weight | 112.41 |
| Melting Point | 321 °C |
| Boiling Point | 765 °C |
| Density | 8.65 g/mL (25 °C) |

The properties of specific cadmium compounds are described in reference 5.5.

### 1.5. Method Performance

A synopsis of method performance is presented below. Further information can be found in Section 4.

1.5.1. The qualitative and quantitative detection limits for the flame AAS analytical technique are 0.04 µg (0.004 µg/mL) and 0.14 µg (0.014 µg/mL) cadmium, respectively, for a 10 mL solution volume. These correspond, respectively, to 0.2 µg/m$^3$ and 0.70 µg/m$^3$ for a 200 L air volume.

1.5.2. The qualitative and quantitative detection limits for the AAS-HGA analytical technique are 0.44 ng (0.044 ng/mL) and 1.5 ng (0.15 ng/mL) cadmium, respectively, for a 10 mL solution volume. These correspond, respectively, to 0.007 µg/m$^3$ and 0.025 µg/m$^3$ for a 60 L air volume.

1.5.3. The average recovery by the flame AAS analytical technique of 17 spiked MCEF samples containing cadmium in the range of 0.5 to 2.0 times the TWA target concentration of 5 µg/m$^3$ (assuming a 400 L air volume) was 104.0% with a pooled coefficient of variation (CV1) of 0.010. The flame analytical technique exhibited a positive bias of +4.0% for the validated concentration range. The overall analytical error (OAE) for the flame AAS analytical technique was ±6.0%.

1.5.4. The average recovery by the AAS-HGA analytical technique of 18 spiked MCEF samples containing cadmium in the range of 0.5 to 2.0 times the Action Level TWA target concentration of 2.5 µg/m$^3$ (assuming a 60 L air volume) was 94.2% with a pooled coefficient of variation (CV1) of 0.043. The AAS-HGA analytical technique exhibited a negative bias of -5.8% for the validated concentration range. The overall analytical error (OAE) for the AAS-HGA analytical technique was ±14.2%.

1.5.5. Sensitivity in flame atomic absorption is defined as the characteristic concentration of an element required to produce a signal of 1% absorbance (0.0044 absorbance units). Sensitivity values are listed for each element by the atomic absorption spectrophotometer manufacturer and have proved to be a very valuable diagnostic tool to determine if instrumental parameters are optimized and if the instrument is performing up to specification. The sensitivity of the spectrophotometer used in the validation of the flame AAS analytical technique agreed with the manufacturer specifications

(5.6.); the 2 µg/mL cadmium standard gave an absorbance reading of 0.350 abs. units.

1.5.6. Sensitivity in graphite furnace atomic absorption is defined in terms of the characteristic mass, the number of picograms required to give an integrated absorbance value of 0.0044 absorbance-second (5.7.). Data suggests that under Stabilized Temperature Platform Furnace (STPF) conditions (see Section 1.6.2.), characteristic mass values are transferable between properly functioning instruments to an accuracy of about 20% (5.2.). The characteristic mass for STPF analysis of cadmium with Zeeman background correction listed by the manufacturer of the instrument used in the validation of the AAS-HGA analytical technique was 0.35 pg. The experimental characteristic mass value observed during the determination of the working range and detection limits of the AAS-HGA analytical technique was 0.41 pg.

### 1.6. Interferences

1.6.1. High concentrations of silicate interfere in determining cadmium by flame AAS (5.6.). However, silicates are not significantly soluble in the acid matrix used to prepare the samples.

1.6.2. Interferences, such as background absorption, are reduced to a minimum in the AAS-HGA analytical technique by taking full advantage of the Stabilized Temperature Platform Furnace (STPF) concept. STPF includes all of the following parameters (5.2.):

a. Integrated Absorbance,

b. Fast Instrument Electronics and Sampling Frequency,

c. Background Correction,

d. Maximum Power Heating,

e. Atomization off the L'vov platform in a pyrolytically coated graphite tube,

f. Gas Stop during Atomization,

g. Use of Matrix Modifiers.

### 1.7. Toxicology (5.14.)

Information listed within this section is synopsis of current knowledge of the physiological effects of cadmium and is not intended to be used as the basis for OSHA policy. IARC classifies cadmium and certain of its compounds as Group 2A carcinogens (probably carcinogenic to humans). Cadmium fume is intensely irritating to the respiratory tract. Workplace exposure to cadmium can cause both chronic and acute effects. Acute effects include tracheobronchitis, pneumonitis, and pulmonary edema. Chronic effects include anemia, rhinitis/anosmia, pulmonary emphysema, proteinuria and lung cancer. The primary target organs for chronic disease are the kidneys (non-carcinogenic) and the lungs (carcinogenic).

## 2. SAMPLING

### 2.1. Apparatus

2.1.1. Filter cassette unit for air sampling: A 37-mm diameter mixed cellulose ester membrane filter with a pore size of 0.8-µm contained in a 37-mm polystyrene two- or three-piece cassette filter holder (part no. MAWP 037 A0, Millipore Corp., Bedford, MA). The filter is supported with a cellulose backup pad. The cassette is sealed prior to use with a shrinkable gel band.

2.1.2. A calibrated personal sampling pump whose flow is determined to an accuracy of ±5% at the recommended flow rate with the filter cassette unit in line.

### 2.2. Procedure

2.2.1. Attach the prepared cassette to the calibrated sampling pump (the backup pad should face the pump) using flexible tubing. Place the sampling device on the employee such that air is sampled from the breathing zone.

2.2.2. Collect air samples at a flow rate of 2.0 L/min. If the filter does not become overloaded, a full-shift (at least seven hours) sample is strongly recommended for TWA and Action Level TWA measurements with a maximum air volume of 960 L. If overloading occurs, collect consecutive air samples for shorter sampling periods to cover the full workshift.

2.2.3. Replace the end plugs into the filter cassettes immediately after sampling. Record the sampling conditions.

2.2.4. Securely wrap each sample filter cassette end-to-end with an OSHA Form 21 sample seal.

2.2.5. Submit at least one blank sample with each set of air samples. The blank sample should be handled the same as the other samples except that no air is drawn through it.

2.2.6. Ship the samples to the laboratory for analysis as soon as possible in a suitable container designed to prevent damage in transit.

## 3. ANALYSIS

### 3.1. Safety Precautions

3.1.1. Wear safety glasses, protective clothing and gloves at all times.

3.1.2. Handle acid solutions with care. Handle all cadmium samples and solutions with extra care (see Sect. 1.7.). Avoid their direct contact with work area surfaces, eyes, skin and clothes. Flush acid solutions which contact the skin or eyes with copious amounts of water.

3.1.3. Perform all acid digestions and acid dilutions in an exhaust hood while wearing a face shield. To avoid exposure to acid vapors, do not remove beakers containing concentrated acid solutions from the exhaust hood until they have returned to room temperature and have been diluted or emptied.

3.1.4. Exercise care when using laboratory glassware. Do not use chipped pipets, volumetric flasks, beakers or any glassware with sharp edges exposed in order to avoid the possibility of cuts or abrasions.

3.1.5. Never pipet by mouth.

3.1.6. Refer to the instrument instruction manuals and SOPs (5.8., 5.9.) for proper and safe operation of the atomic absorption spectrophotometer, graphite furnace atomizer and associated equipment.

3.1.7. Because metallic elements and other toxic substances are vaporized during AAS flame or graphite furnace atomizer operation, it is imperative that an exhaust vent be used. Always ensure that the exhaust system is operating properly during instrument use.

### 3.2. Apparatus for Sample and Standard Preparation

3.2.1. Hot plate, capable of reaching 150 °C, installed in an exhaust hood.

3.2.2. Phillips beakers, 125 mL.

3.2.3. Bottles, narrow-mouth, polyethylene or glass with leakproof caps: used for storage of standards and matrix modifier.

3.2.4. Volumetric flasks, volumetric pipets, beakers and other associated general laboratory glassware.

3.2.5. Forceps and other associated general laboratory equipment.

### 3.3. Apparatus for Flame AAS Analysis

3.3.1. Atomic absorption spectrophotometer consisting of a(an):

Nebulizer and burner head

Pressure regulating devices capable of maintaining constant oxidant and fuel pressures

Optical system capable of isolating the desired wavelength of radiation (228.8 nm)

Adjustable slit

Light measuring and amplifying device

Display, strip chart, or computer interface for indicating the amount of absorbed radiation

Cadmium hollow cathode lamp or electrodeless discharge lamp (EDL) and power supply

3.3.2. Oxidant: compressed air, filtered to remove water, oil and other foreign substances.

3.3.3. Fuel: standard commercially available tanks of acetylene dissolved in acetone; tanks should be equipped with flash arresters.

CAUTION: Do not use grades of acetylene containing solvents other than acetone because they may damage the PVC tubing used in some instruments.

3.3.4. Pressure-reducing valves: two gauge, two-stage pressure regulators to maintain fuel and oxidant pressures somewhat higher than the controlled operating pressures of the instrument.

3.3.5. Exhaust vent installed directly above the spectrophotometer burner head.

### 3.4. Apparatus for AAS-HGA Analysis

3.4.1. Atomic absorption spectrophotometer consisting of a(an):

Heated graphite furnace atomizer (HGA) with argon purge system

# OSHA Standards for General Industry

Pressure-regulating devices capable of maintaining constant argon purge pressure

Optical system capable of isolating the desired wavelength of radiation (228.8 nm)

Adjustable slit

Light measuring and amplifying device

Display, strip chart, or computer interface for indicating the amount of absorbed radiation (as integrated absorbance, peak area)

Background corrector: Zeeman or deuterium arc. The Zeeman background corrector is recommended

Cadmium hollow cathode lamp or electrodeless discharge lamp (EDL) and power supply

Autosampler capable of accurately injecting 5 to 20 µL sample aliquots onto the L'vov Platform in a graphite tube

3.4.2. Pyrolytically coated graphite tubes containing solid, pyrolytic L'vov platforms.

3.4.3. Polyethylene sample cups, 2.0 to 2.5 mL, for use with the autosampler.

3.4.4. Inert purge gas for graphite furnace atomizer: compressed gas cylinder of purified argon.

3.4.5. Two gauge, two-stage pressure regulator for the argon gas cylinder.

3.4.6. Cooling water supply for graphite furnace atomizer.

3.4.7. Exhaust vent installed directly above the graphite furnace atomizer.

### 3.5. Reagents

All reagents should be ACS analytical reagent grade or better.

3.5.1. Deionized water with a specific conductance of less than 10 µS.

3.5.2. Concentrated nitric acid, $HNO_3$.

3.5.3. Concentrated hydrochloric acid, HCl.

3.5.4. Ammonium phosphate, monobasic, $NH_4H_2PO_4$.

3.5.5. Magnesium nitrate, $Mg(NO_3)(NO_3)_2 \cdot 6H_2O$.

3.5.6. Diluting solution (4% $HNO_3$, 0.4% HCl): Add 40 mL $HNO_3$ and 4 mL HCl carefully to approximately 500 mL deionized water and dilute to 1 L with deionized water.

3.5.7. Cadmium standard stock solution, 1,000 µg/mL: Use a commercially available certified 1,000 µg/mL cadmium standard or, alternatively, dissolve 1.0000 g of cadmium metal in a minimum volume of 1:1 HCl and dilute to 1 L with 4% $HNO_3$. Observe expiration dates of commercial standards. Properly dispose of commercial standards with no expiration dates or prepared standards one year after their receipt or preparation date.

3.5.8. Matrix modifier for AAS-HGA analysis: Dissolve 1.0 g $NH_4H_2PO_4$ and 0.15 g $Mg(NO_3)_2 \cdot 6H_2O$ in approximately 200 mL deionized water. Add 1 mL $HNO_3$ and dilute to 500 mL with deionized water.

3.5.9. Nitric Acid, 1:1 $HNO_3$/DI $H_2O$ mixture: Carefully add a measured volume of concentrated $HNO_3$ to an equal volume of DI $H_2O$.

3.5.10. Nitric acid, 10% v/v: Carefully add 100 mL of concentrated $HNO_3$ to 500 mL of DI $H_2O$ and dilute to 1 L.

### 3.6. Glassware Preparation

3.6.1. Clean Phillips beakers by refluxing with 1:1 nitric acid on a hot plate in a fume hood. Thoroughly rinse with deionized water and invert the beakers to allow them to drain dry.

3.6.2. Rinse volumetric flasks and all other glassware with 10% nitric acid and deionized water prior to use.

### 3.7. Standard Preparation for Flame AAS Analysis

3.7.1. Dilute stock solutions: Prepare 1, 5, 10 and 100 µg/mL cadmium standard stock solutions by making appropriate serial dilutions of 1,000 µg/mL cadmium standard stock solution with the diluting solution described in Section 3.5.6.

3.7.2. Working standards: Prepare cadmium working standards in the range of 0.02 to 2.0 µg/mL by making appropriate serial dilutions of the dilute stock solutions with the same diluting solution. A suggested method of preparation of the working standards is given below.

| Working standard (µg/mL) | Std solution (µg/mL) | Aliquot (mL) | Final vol. (mL) |
|---|---|---|---|
| 0.02 | 1 | 10 | 500 |
| 0.05 | 5 | 5 | 500 |
| 0.1 | 10 | 5 | 500 |
| 0.2 | 10 | 10 | 500 |
| 0.5 | 10 | 25 | 500 |
| 1 | 100 | 5 | 500 |
| 2 | 100 | 10 | 500 |

Store the working standards in 500-ml, narrow-mouth polyethylene or glass bottles with leak proof caps. Prepare every twelve months.

### 3.8. Standard Preparation for AAS-HGA Analysis

3.8.1. Dilute stock solutions: Prepare 10, 100 and 1,000 ng/mL cadmium standard stock solutions by making appropriate ten-fold serial dilutions of the 1,000 µg/mL cadmium standard stock solution with the diluting solution described in Section 3.5.6.

3.8.2. Working standards: Prepare cadmium working standards in the range of 0.2 to 20 ng/mL by making appropriate serial dilutions of the dilute stock solutions with the same diluting solution. A suggested method of preparation of the working standards is given below.

| Working standard (ng/mL) | Std solution (ng/mL) | Aliquot (mL) | Final vol. (mL) |
|---|---|---|---|
| 0.2 | 10 | 2 | 100 |
| 0.5 | 10 | 5 | 100 |
| 1 | 10 | 10 | 100 |
| 2 | 100 | 2 | 100 |
| 5 | 100 | 5 | 100 |
| 10 | 100 | 10 | 100 |
| 20 | 1,000 | 2 | 100 |

Store the working standards in narrow-mouth polyethylene or glass bottles with leakproof caps. Prepare monthly.

### 3.9. Sample Preparation

3.9.1. Carefully transfer each sample filter with forceps from its filter cassette unit to a clean, separate 125-mL Phillips beaker along with any loose dust found in the cassette. Label each Phillips beaker with the appropriate sample number.

3.9.2. Digest the sample by adding 5 mL of concentrated nitric acid ($HNO_3$) to each Phillips beaker containing an air filter sample. Place the Phillips beakers on a hot plate in an exhaust hood and heat the samples until approximately 0.5 mL remains. The sample solution in each Phillips beaker should become clear. If it is not clear, digest the sample with another portion of concentrated nitric acid.

3.9.3. After completing the $HNO_3$ digestion and cooling the samples, add 40 µL (2 drops) of concentrated HCl to each air sample solution and then swirl the contents. Carefully add about 5 mL of deionized water by pouring it down the inside of each beaker.

3.9.4. Quantitatively transfer each cooled air sample solution from each Phillips beaker to a clean 10-mL volumetric flask. Dilute each flask to volume with deionized water and mix well.

### 3.10. Flame AAS Analysis

Analyze all of the air samples for their cadmium content by flame atomic absorption spectroscopy (AAS) according to the instructions given below.

3.10.1. Set up the atomic absorption spectrophotometer for the air/acetylene flame analysis of cadmium according to the SOP (5.8.) or the manufacturer's operational instructions. For the source lamp, use the cadmium hollow cathode or electrodeless discharge lamp operated at the manufacturer's recommended rating for continuous operation. Allow the lamp to warm up 10 to 20 min or until the energy output stabilizes. Optimize conditions such as lamp position, burner head alignment, fuel and oxidant flow rates, etc. See the SOP or specific instrument manuals for details. Instrumental parameters for the Perkin-Elmer Model 603 used in the validation of this method are given in Attachment 1.

3.10.2. Aspirate and measure the absorbance of a standard solution of cadmium. The standard concentration should be within the linear range. For the instrumentation used in the validation of this method a 2 µg/mL cadmium standard gives a net absorbance reading of about 0.350 abs. units (see Section 1.5.5.) when the instrument and the source lamp are performing to manufacturer specifications.

3.10.3. To increase instrument response, scale expand the absorbance reading of the aspirated 2 µg/mL working standard approximately four times. Increase the integration time to at least 3 seconds to reduce signal noise.

3.10.4. Autozero the instrument while aspirating a deionized water blank. Monitor the variation in the baseline absorbance reading (baseline noise) for a few minutes to insure that the instrument, source lamp and associated equipment are in good operating condition.

3.10.5. Aspirate the working standards and samples directly into the flame and record their absorbance readings. Aspirate the deionized water blank immediately after every standard or sample to correct for and monitor any baseline drift and noise. Record the baseline absorbance reading of each deionized water blank. Label each standard and sample reading and its accompanying baseline reading.

3.10.6. It is recommended that the entire series of working standards be analyzed at the beginning and end of the analysis of a set of samples to establish a concentration-response curve, ensure that the standard readings agree with each other and are reproducible. Also, analyze a working standard after every five or six samples to monitor the performance of the spectrophotometer. Standard readings should agree within ±10 to 15% of the readings obtained at the beginning of the analysis.

3.10.7. Bracket the sample readings with standards during the analysis. If the absorbance reading of a sample is above the absorbance reading of the highest working standard, dilute the sample with diluting solution and reanalyze. Use the appropriate dilution factor in the calculations.

3.10.8. Repeat the analysis of approximately 10% of the samples for a check of precision.

3.10.9. If possible, analyze quality control samples from an independent source as a check on analytical recovery and precision.

3.10.10. Record the final instrument settings at the end of the analysis. Date and label the output.

### 3.11. AAS-HGA Analysis

Initially analyze all of the air samples for their cadmium content by flame atomic absorption spectroscopy (AAS) according to the instructions given in Section 3.10. If the concentration of cadmium in a sample solution is less than three times the quantitative detection limit [0.04 µg/mL (40 ng/mL) for the instrumentation used in the validation] and the sample results are to be averaged with other samples for TWA calculations, proceed with the AAS-HGA analysis of the sample as described below.

3.11.1. Set up the atomic absorption spectrophotometer and HGA for flameless atomic absorption analysis of cadmium according to the SOP (5.9.) or the manufacturer's operational instructions and allow the instrument to stabilize. The graphite furnace atomizer is equipped with a pyrolytically coated graphite tube containing a pyrolytic platform. For the source lamp, use a cadmium hollow cathode or electrodeless discharge lamp operated at the manufacturer's recommended setting for graphite furnace operation. The Zeeman background corrector and EDL are recommended for use with the L'vov platform. Instrumental parameters for the Perkin-Elmer Model 5100 spectrophotometer and Zeeman HGA-600 graphite furnace used in the validation of this method are given in Attachment 2.

3.11.2. Optimize the energy reading of the spectrophotometer at 228.8 nm by adjusting the lamp position and the wavelength according to the manufacturer's instructions.

3.11.3. Set up the autosampler to inject a 5-µL aliquot of the working standard, sample or reagent blank solution onto the L'vov platform along with a 10-µL overlay of the matrix modifier.

3.11.4. Analyze the reagent blank (diluting solution, Section 3.5.6.) and then autozero the instrument before starting the analysis of a set of samples. It is recommended that the reagent blank be analyzed several times during the analysis to assure the integrated absorbance (peak area) reading remains at or near zero.

3.11.5. Analyze a working standard approximately midway in the linear portion of the working standard range two or three times to check for reproducibility and sensitivity (see sections 1.5.5. and 1.5.6.) before starting the analysis of samples. Calculate the experimental characteristic mass value from the average integrated absorbance reading and injection volume of the analyzed working standard. Compare this value to the manufacturer's suggested value as a check of proper instrument operation.

3.11.6. Analyze the reagent blank, working standard, and sample solutions. Record and label the peak area (abs-sec) readings and the peak and background peak profiles on the printer/plotter.

3.11.7. It is recommended the entire series of working standards be analyzed at the beginning and end of the analysis of a set of samples. Establish a concentration-response curve and ensure standard readings agree with each other and are reproducible. Also, analyze a working standard after every five or six samples to monitor the performance of the system. Standard readings should agree within ±15% of the readings obtained at the beginning of the analysis.

3.11.8. Bracket the sample readings with standards during the analysis. If the peak area reading of a sample is above the peak area reading of the highest working standard, dilute the sample with the diluting solution and reanalyze. Use the appropriate dilution factor in the calculations.

3.11.9. Repeat the analysis of approximately 10% of the samples for a check of precision.

3.11.10. If possible, analyze quality control samples from an independent source as a check of analytical recovery and precision.

3.11.11. Record the final instrument settings at the end of the analysis. Date and label the output.

### 3.12. Calculations

**NOTE:** Standards used for HGA analysis are in ng/mL. Total amounts of cadmium from calculations will be in ng (not g) unless a prior conversion is made.

3.12.1. Correct for baseline drift and noise in flame AAS analysis by subtracting each baseline absorbance reading from its corresponding working standard or sample absorbance reading to obtain the net absorbance reading for each standard and sample.

3.12.2. Use a least squares regression program to plot a concentration-response curve of net absorbance reading (or peak area for HGA analysis) versus concentration (µg/mL or ng/mL) of cadmium in each working standard.

3.12.3. Determine the concentration (µg/mL or ng/mL) of cadmium in each sample from the resulting concentration-response curve. If the concentration of cadmium in a sample solution is less than three times the quantitative detection limit [0.04 µg/mL (40 ng/mL) for the instrumentation used in the validation of the method] and if consecutive samples were taken on one employee and the sample results are to be averaged with other samples to determine a single TWA, reanalyze the sample by AAS-HGA as described in Section 3.11. and report the AAS-HGA analytical results.

3.12.4. Calculate the total amount (µg or ng) of cadmium in each sample from the sample solution volume (mL):

$$W = (C)(\text{sample vol, mL})(DF)$$

Where:

$W$ = Total cadmium in sample

$C$ = Calculated concentration of cadmium

$DF$ = Dilution Factor (if applicable)

3.12.5. Make a blank correction for each air sample by subtracting the total amount of cadmium in the corresponding blank sample from the total amount of cadmium in the sample.

3.12.6. Calculate the concentration of cadmium in an air sample ($mg/m^3$ or $µg/m^3$) by using one of the following equations:

$$mg/m^3 = W_{bc}/(\text{Air vol sampled, L})$$

or

$$µg/m^3 = (W_{bc})(1{,}000 \text{ ng/µg})/(\text{Air vol sampled, L})$$

Where:

$W_{bc}$ = blank corrected total µg cadmium in the sample.
  (1 µg = 1,000 ng)

## 4. BACKUP DATA

### 4.1. Introduction

4.1.1. The purpose of this evaluation is to determine the analytical method recovery, working standard range, and qualitative and quantitative detection limits of the two atomic absorption analytical techniques included in this method. The evaluation consisted of the following experiments:

1. An analysis of 24 samples (six samples each at 0.1, 0.5, 1 and 2 times the TWA-PEL) for the analytical method recovery study of the flame AAS analytical technique.

2. An analysis of 18 samples (six samples each at 0.5, 1 and 2 times the Action Level TWA-PEL) for the analytical method recovery study of the AAS-HGA analytical technique.

3. Multiple analyses of the reagent blank and a series of standard solutions to determine the working standard range and the qualitative and quantitative detection limits for both atomic absorption analytical techniques.

4.1.2. The analytical method recovery results at all test levels were calculated from concentration-response curves and statistically examined for outliers at the 99% confidence level. Possible outliers were determined using the Treatment of Outliers test (5.10.). In addition, the sample results of the two analytical techniques, at 0.5, 1.0 and 2.0 times their target concentrations, were tested for homogeneity of variances also at the 99% confidence level. Homogeneity of the coefficients of variation was determined using the Bartlett's test (5.11.). The overall analytical error (OAE) at the 95% confidence level was calculated using the equation (5.12.):

$$OAE = \pm[|Bias| + (1.96)(CV_1(pooled))(100\%)]$$

4.1.3. A derivation of the International Union of Pure and Applied Chemistry (IUPAC) detection limit equation (5.13.) was used to determine the qualitative and quantitative detection limits for both atomic absorption analytical techniques:

$$C_{ld} = k(sd)/m \quad \text{(Equation 1)}$$

Where:

$C_{ld}$ = the smallest reliable detectable concentration an analytical instrument can determine at a given confidence level.

k = 3 for the Qualitative Detection Limit at the 99.86% Confidence Level.

= 10 for the Quantitative Detection Limit at the 99.99% Confidence Level.

sd = standard deviation of the reagent blank (Rbl) readings.

m = analytical sensitivity or slope as calculated by linear regression.

4.1.4. Collection efficiencies of metallic fume and dust atmospheres on 0.8-µm mixed cellulose ester membrane filters are well documented and have been shown to be excellent (5.11.). Since elemental cadmium and the cadmium component of cadmium compounds are nonvolatile, stability studies of cadmium spiked MCEF samples were not performed.

### 4.2. Equipment

4.2.1. A Perkin-Elmer (PE) Model 603 spectrophotometer equipped with a manual gas control system, a stainless steel nebulizer, a burner mixing chamber, a flow spoiler and a 10 cm. (one-slot) burner head was used in the experimental validation of the flame AAS analytical technique. A PE cadmium hollow cathode lamp, operated at the manufacturer's recommended current setting for continuous operation (4 mA), was used as the source lamp. Instrument parameters are listed in Attachment 1.

4.2.2. A PE Model 5100 spectrophotometer, Zeeman HGA-600 graphite furnace atomizer and AS-60 HGA autosampler were used in the experimental validation of the AAS-HGA analytical technique. The spectrophotometer was equipped with a PE Series 7700 professional computer and Model PR-310 printer. A PE System 2 cadmium electrodeless discharge lamp, operated at the manufacturer's recommended current setting for modulated operation (170 mA), was used as the source lamp. Instrument parameters are listed in Attachment 2.

### 4.3. Reagents

4.3.1. J.T. Baker Chem. Co. (Analyzed grade) concentrated nitric acid, 69.0-71.0%, and concentrated hydrochloric acid, 36.5-38.0%, were used to prepare the samples and standards.

4.3.2. Ammonium phosphate, monobasic, $NH_4H_2PO_4$ and magnesium nitrate, $Mg(NO_3)_2 6H_2O$, both manufactured by the Mallinckrodt Chem. Co., were used to prepare the matrix modifier for AAS-HGA analysis.

### 4.4. Standard Preparation for Flame AAS Analysis

4.4.1. Dilute stock solutions: Prepared 0.01, 0.1, 1, 10 and 100 µg/mL cadmium standard stock solutions by making appropriate serial dilutions of a commercially available 1,000 µg/mL cadmium standard stock solution (RICCA Chemical Co., Lot# A102) with the diluting solution (4% $HNO_3$, 0.4% HCl).

4.4.2. Analyzed Standards: Prepared cadmium standards in the range of 0.001 to 2.0 µg/mL by pipetting 2 to 10 mL of the appropriate dilute cadmium stock solution into a 100-mL volumetric flask and diluting to volume with the diluting solution. (See Section 3.7.2.)

### 4.5. Standard Preparation for AAS-HGA Analysis

4.5.1. Dilute stock solutions: Prepared 1, 10, 100 and 1,000 ng/mL cadmium standard stock solutions by making appropriate serial dilutions of a commercially available 1,000 µg/mL cadmium standard stock solution (J.T. Baker Chemical Co., Instra-analyzed, Lot# D22642) with the diluting solution (4% $HNO_3$, 0.4% HCl).

4.5.2. Analyzed Standards: Prepared cadmium standards in the range of 0.1 to 40 ng/mL by pipetting 2 to 10 mL of the appropriate dilute cadmium stock solution into a 100-mL volumetric flask and diluting to volume with the diluting solution. (See Section 3.8.2.)

### 4.6. Detection Limits and Standard Working Range for Flame AAS Analysis

4.6.1. Analyzed the reagent blank solution and the entire series of cadmium standards in the range of 0.001 to 2.0 µg/mL three to six times according to the instructions given in Section 3.10. The diluting solution (4% $HNO_3$, 0.4% HCl) was used as the reagent blank. The integration time on the PE 603 spectrophotometer was set to 3.0 seconds and a four-fold expansion of the absorbance reading of the 2.0 µg/mL cadmium standard was made prior to analysis. The 2.0 µg/mL standard gave a net absorbance reading of 0.350 abs. units prior to expansion in agreement with the manufacturer's specifications (5.6.).

4.6.2. The net absorbance readings of the reagent blank and the low concentration Cd standards from 0.001 to 0.1 µg/mL and the statistical analysis of the results are shown in Table I. The standard deviation, sd, of the six net absorbance readings of the reagent blank is 1.05 abs. units. The slope, m, as calculated by a linear regression plot of the net absorbance readings (shown in Table II) of the 0.02 to 1.0 µg/mL cadmium standards versus their concentration is 772.7 abs. units/(µg/mL).

4.6.3. If these values for sd and the slope, m, are used in Eqn. 1 (Sect. 4.1.3.), the qualitative and quantitative detection limits as determined by the IUPAC Method are:

$$C_{ld} = (3)(1.05 \text{ abs. units})/(772.7 \text{ abs. units}/(\mu g/mL))$$

= 0.0041 µg/mL for the qualitative detection limit.

$$C_{ld} = (10)(1.05 \text{ abs. units})/(772.7 \text{ abs. units}/\mu g/mL))$$

= 0.014 µg/mL for the quantitative detection limit.

The qualitative and quantitative detection limits for the flame AAS analytical technique are 0.041 µg and 0.14 µg cadmium, respectively, for a 10 mL solution volume. These correspond, respectively, to 0.2 µg/m³ and 0.70 µg/m³ for a 200 L air volume.

4.6.4. The recommended Cd standard working range for flame AAS analysis is 0.02 to 2.0 µg/mL. The net absorbance readings of the reagent blank and the recommended working range standards and the statistical analysis of the results are shown in Table II. The standard of lowest concentration in the working range, 0.02 µg/mL, is slightly greater than the calculated quantitative detection limit, 0.014 µg/mL. The standard of highest concentration in the working range, 2.0 µg/mL, is at the upper end of the linear working range suggested by the manufacturer (5.6.). Although the standard net absorbance readings are not strictly linear at concentrations above 0.5 µg/mL, the deviation from linearity is only about 10% at the upper end of the recommended standard working range. The deviation from linearity is probably caused by the four-fold expansion of the signal suggested in the method. As shown in Table II, the precision of the standard net absorbance readings are excellent throughout the recommended working range; the relative standard deviations of the readings range from 0.009 to 0.064.

### 4.7. Detection Limits and Standard Working Range for AAS-HGA Analysis

4.7.1. Analyzed the reagent blank solution and the entire series of cadmium standards in the range of 0.1 to 40 ng/mL according to the instructions given in Section 3.11. The diluting solution (4% $HNO_3$, 0.4% HCl) was used as the reagent blank. A fresh aliquot of the reagent blank and of each standard was used for every analysis. The experimental characteristic mass value was 0.41 pg, calculated from the average peak area (abs-sec) reading of the 5 ng/mL standard which is approximately midway in the linear portion of the working standard range. This agreed within 20% with the characteristic mass value, 0.35 pg, listed by the manufacturer of the instrument (5.2.).

4.7.2. The peak area (abs-sec) readings of the reagent blank and the low concentration Cd standards from 0.1 to 2.0 ng/mL and statistical analysis of the results are shown in Table III. Five of the reagent blank peak area readings were zero and the sixth reading was 1 and was an outlier. The near lack of a blank signal does not satisfy a strict interpretation of the IUPAC method for determining the detection limits. Therefore, the standard deviation of the six peak area readings of the 0.2 ng/mL cadmium standard, 0.75 abs-sec, was used to calculate the detection limits by the IUPAC method. The slope, m, as calculated by a linear regression plot of the peak area (abs-sec) readings (shown in Table IV) of the 0.2 to 10 ng/mL cadmium standards versus their concentration is 51.5 abs-sec/(ng/mL).

4.7.3. If 0.75 abs-sec (sd) and 51.5 abs-sec/(ng/mL) (m) are used in Eqn. 1 (Sect. 4.1.3.), the qualitative and quantitative detection limits as determined by the IUPAC method are:

$$C_{ld} = (3)(0.75 \text{ abs-sec})/(51.5 \text{ abs-sec}/(\text{ng/mL}))$$

$$= 0.044 \text{ ng/mL for the qualitative detection limit.}$$

$$C_{ld} = (10)(0.75 \text{ abs-sec})/(51.5 \text{ abs-sec}/(\text{ng/mL})) = 0.15 \text{ ng/mL for the quantitative detection limit.}$$

The qualitative and quantitative detection limits for the AAS-HGA analytical technique are 0.44 ng and 1.5 ng cadmium, respectively, for a 10 mL solution volume. These correspond, respectively, to 0.007 $\mu g/m^3$ and 0.025 $\mu g/m^3$ for a 60 L air volume.

4.7.4. The peak area (abs-sec) readings of the Cd standards from 0.2 to 40 ng/mL and the statistical analysis of the results are given in Table IV. The recommended standard working range for AAS-HGA analysis is 0.2 to 20 ng/mL. The standard of lowest concentration in the recommended working range is slightly greater than the calculated quantitative detection limit, 0.15 ng/mL. The deviation from linearity of the peak area readings of the 20 ng/mL standard, the highest concentration standard in the recommended working range, is approximately 10%. The deviations from linearity of the peak area readings of the 30 and 40 ng/mL standards are significantly greater than 10%. As shown in Table IV, the precision of the peak area readings are satisfactory throughout the recommended working range; the relative standard deviations of the readings range from 0.025 to 0.083.

### 4.8. Analytical Method Recovery for Flame AAS Analysis

4.8.1. Four sets of spiked MCEF samples were prepared by injecting 20 μL of 10, 50, 100 and 200 μg/mL dilute cadmium stock solutions on 37 mm diameter filters (part no. AAWP 037 00, Millipore Corp., Bedford, MA) with a calibrated micropipet. The dilute stock solutions were prepared by making appropriate serial dilutions of a commercially available 1,000 μg/mL cadmium standard stock solution (RICCA Chemical Co., Lot# A102) with the diluting solution (4% $HNO_3$, 0.4% HCl). Each set contained six samples and a sample blank. The amount of cadmium in the prepared sets were equivalent to 0.1, 0.5, 1.0 and 2.0 times the TWA PEL target concentration of 5 $\mu g/m^3$ for a 400 L air volume.

4.8.2. The air-dried spiked filters were digested and analyzed for their cadmium content by flame atomic absorption spectroscopy (AAS) following the procedure described in Section 3. The 0.02 to 2.0 μg/mL cadmium standards (the suggested working range) were used in the analysis of the spiked filters.

4.8.3. The results of the analysis are given in Table V. One result at 0.5 times the TWA PEL target concentration was an outlier and was excluded from statistical analysis. Experimental justification for rejecting it is that the outlier value was probably due to a spiking error. The coefficients of variation for the three test levels at 0.5 to 2.0 times the TWA PEL target concentration passed the Bartlett's test and were pooled.

4.8.4. The average recovery of the six spiked filter samples at 0.1 times the TWA PEL target concentration was 118.2% with a coefficient of variation ($CV_1$) of 0.128. The average recovery of the spiked filter samples in the range of 0.5 to 2.0 times the TWA target concentration was 104.0% with a pooled coefficient of variation ($CV_1$) of 0.010. Consequently, the analytical bias found in these spiked sample results over the tested concentration range was +4.0% and the OAE was ±6.0%.

### 4.9. Analytical Method Recovery for AAS-HGA Analysis

4.9.1. Three sets of spiked MCEF samples were prepared by injecting 15 μL of 5, 10 and 20 μg/mL dilute cadmium stock solutions on 37 mm diameter filters (part no. AAWP 037 00, Millipore Corp., Bedford, MA) with a calibrated micropipet. The dilute stock solutions were prepared by making appropriate serial dilutions of a commercially available certified 1,000 μg/mL cadmium standard stock solution (Fisher Chemical Co., Lot# 913438-24) with the diluting solution (4% $HNO_3$, 0.4% HCl). Each set contained six samples and a sample blank. The amount of cadmium in the prepared sets were equivalent to 0.5, 1 and 2 times the Action Level TWA target concentration of 2.5 $\mu g/m^3$ for a 60 L air volume.

4.9.2. The air-dried spiked filters were digested and analyzed for their cadmium content by flameless atomic absorption spectroscopy using a heated graphite furnace atomizer following the procedure described in Section 3. A five-fold dilution of the spiked filter samples at 2 times the Action Level TWA was made prior to their analysis. The 0.05 to 20 ng/mL cadmium standards were used in the analysis of the spiked filters.

4.9.3. The results of the analysis are given in Table VI. There were no outliers. The coefficients of variation for the three test levels at 0.5 to 2.0 times the Action Level TWA PEL passed the Bartlett's test and were pooled. The average recovery of the spiked filter samples was 94.2% with a pooled coefficient of variation ($CV_1$) of 0.043. Consequently, the analytical bias was -5.8% and the OAE was ±14.2%.

### 4.10. Conclusions

The experiments performed in this evaluation show the two atomic absorption analytical techniques included in this method to be precise and accurate and have sufficient sensitivity to measure airborne cadmium over a broad range of exposure levels and sampling periods.

## 5. REFERENCES

5.1. Slavin, W. Graphite Furnace AAS—A Source Book; Perkin-Elmer Corp., Spectroscopy Div.: Ridgefield, CT, 1984; p. 18 and pp. 83-90.

5.2. Grosser, Z., Ed.; Techniques in Graphite Furnace Atomic Absorption Spectrophotometry; Perkin-Elmer Corp., Spectroscopy Div.: Ridgefield, CT, 1985.

5.3. Occupational Safety and Health Administration Salt Lake Technical Center: Metal and Metalloid Particulate in Workplace Atmospheres (Atomic Absorption) (USDOL/OSHA Method No. ID-121). In OSHA Analytical Methods Manual 2nd ed. Cincinnati, OH: American Conference of Governmental Industrial Hygienists, 1991.

5.4. Occupational Safety and Health Administration Salt Lake Technical Center: Metal and Metalloid Particulate in Workplace Atmospheres (ICP) (USDOL/OSHA Method No. ID-125G). In OSHA Analytical Methods Manual 2nd ed. Cincinnati, OH: American Conference of Governmental Industrial Hygienists, 1991.

5.5. Windholz, M., Ed.; The Merck Index, 10th ed.; Merck & Co.: Rahway, NJ, 1983.

5.6. Analytical Methods for Atomic Absorption Spectrophotometry, The Perkin-Elmer Corporation: Norwalk, CT, 1982.

5.7. Slavin, W., D.C. Manning, G. Carnrick, and E. Pruszkowska: Properties of the Cadmium Determination with the Platform Furnace and Zeeman Background Correction. Spectrochim. Acta 38B:1157-1170 (1983).

5.8. Occupational Safety and Health Administration Salt Lake Technical Center: Standard Operating Procedure for Atomic Absorption. Salt Lake City, UT: USDOL/OSHA-SLTC, in progress.

5.9. Occupational Safety and Health Administration Salt Lake Technical Center: AAS-HGA Standard Operating Procedure. Salt Lake City, UT: USDOL/OSHA-SLTC, in progress.

5.10. Mandel, J.: Accuracy and Precision, Evaluation and Interpretation of Analytical Results, The Treatment of Outliers. In Treatise On Analytical Chemistry, 2nd ed., Vol.1, edited by I. M. Kolthoff and P. J. Elving. New York: John Wiley and Sons, 1978. pp. 282-285.

5.11. National Institute for Occupational Safety and Health: Documentation of the NIOSH Validation Tests by D. Taylor, R. Kupel, and J. Bryant (DHEW/NIOSH Pub. No. 77-185). Cincinnati, OH: National Institute for Occupational

Safety and Health, 1977.

5.12. Occupational Safety and Health Administration Analytical Laboratory: Precision and Accuracy Data Protocol for Laboratory Validations. In OSHA Analytical Methods Manual 1st ed. Cincinnati, OH: American Conference of Governmental Industrial Hygienists (Pub. No. ISBN: 0-936712-66-X), 1985.

5.13. Long, G.L. and J.D. Winefordner: Limit of Detection—A Closer Look at the IUPAC Definition. Anal.Chem. 55:712A-724A (1983).

5.14. American Conference of Governmental Industrial Hygienists: Documentation of Threshold Limit Values and Biological Exposure Indices. 5th ed. Cincinnati, OH: American Conference of Governmental Industrial Hygienists, 1986.

### TABLE I—CD DETECTION LIMIT STUDY
[Flame AAS Analysis]

| STD (µg/mL) | Absorbance reading at 228.8 nm | Statistical Analysis |
|---|---|---|
| Reagent blank | 5　2 | n = 6. |
|  | 4　3 | mean = 3.50. |
|  | 4　3 | std dev = 1.05. |
|  |  | CV = 0.30. |
| 0.001 | 6　6 | n = 6. |
|  | 2　4 | mean = 5.00. |
|  | 6　6 | std dev = 1.67. |
|  |  | CV = 0.335. |
| 0.002 | 5　7 | n = 6. |
|  | 7　3 | mean = 5.50. |
|  | 7　4 | std dev = 1.76. |
|  |  | CV = 0.320. |
| 0.005 | 7　7 | n = 6. |
|  | 8　8 | mean = 7.33. |
|  | 8　6 | std dev = 0.817. |
|  |  | CV = 0.111. |
| 0.010 | 10　9 | n = 6. |
|  | 10　13 | mean = 10.3. |
|  | 10　10 | std dev = 1.37. |
|  |  | CV = 0.133. |
| 0.020 | 20　23 | n = 6. |
|  | 20　22 | mean = 20.8. |
|  | 20　20 | std dev = 1.33. |
|  |  | CV = 0.064. |
| 0.050 | 42　42 | n = 6. |
|  | 42　42 | mean = 42.5. |
|  | 42　45 | std dev = 1.22. |
|  |  | CV = 0.029. |
| 0.10 | 84 | n = 3. |
|  | 80 | mean = 82.3. |
|  | 83 | std dev = 2.08. |
|  |  | CV = 0.025. |

### TABLE II—CD STANDARD WORKING RANGE STUDY
[Flame AAS Analysis]

| STD (µg/mL) | Absorbance reading at 228.8 nm | Statistical analysis |
|---|---|---|
| Reagent blank | 5　2 | n = 6. |
|  | 4　3 | mean = 3.50. |
|  | 4　3 | std dev = 1.05. |
|  |  | CV = 0.30. |
| 0.020 | 20　23 | n = 6. |
|  | 20　22 | mean = 20.8. |
|  | 20　20 | std dev = 1.33. |
|  |  | CV = 0.064. |
| 0.050 | 42　42 | n = 6. |
|  | 42　42 | mean = 42.5. |
|  | 42　45 | std dev = 1.22. |
|  |  | CV = 0.029. |
| 0.10 | 84 | n = 3. |
|  | 80 | mean = 82.3. |
|  | 83 | std dev = 2.08. |
|  |  | CV = 0.025. |
| 0.20 | 161 | n = 3. |
|  | 161 | mean = 160.0. |
|  | 158 | std dev = 1.73. |
|  |  | CV = 0.011. |
| 0.50 | 391 | n = 3. |
|  | 389 | mean = 391.0. |
|  | 393 | std dev = 2.00. |
|  |  | CV = 0.005. |
| 1.00 | 760 | n = 3. |
|  | 748 | mean = 753.3. |
|  | 752 | std dev = 6.11. |
|  |  | CV = 0.008. |
| 2.00 | 1416 | n = 3. |
|  | 1426 | mean = 1414.3. |
|  | 1401 | std dev = 12.6. |
|  |  | CV = 0.009. |

## TABLE III—CD DETECTION LIMIT STUDY
[AAS-HGA Analysis]

| STD (ng/mL) | Peak area readings ×10³ at 228.8 nm | | Statistical analysis |
|---|---|---|---|
| Reagent blank | 0 | 0 | n = 6. |
|  | 0 | 1 | mean = 0.167. |
|  | 0 | 0 | std dev = 0.41. |
|  |   |   | CV = 2.45. |
| 0.1 | 8 | 6 | n = 6. |
|  | 5 | 7 | mean = 7.7. |
|  | 13 | 7 | std dev = 2.8. |
|  |   |   | CV = 0.366. |
| 0.2 | 11 | 13 | n = 6. |
|  | 11 | 12 | mean = 11.8. |
|  | 12 | 12 | std dev = 0.75. |
|  |   |   | CV = 0.064. |
| 0.5 | 28 | 33 | n = 6. |
|  | 26 | 28 | mean = 28.8. |
|  | 28 | 30 | std dev = 2.4. |
|  |   |   | CV = 0.083. |
| 1.0 | 52 | 55 | n = 6. |
|  | 56 | 58 | mean = 54.8. |
|  | 54 | 54 | std dev = 2.0. |
|  |   |   | CV = 0.037. |
| 2.0 | 101 | 112 | n = 6. |
|  | 110 | 110 | mean = 108.8. |
|  | 110 | 110 | std dev = 3.9. |
|  |   |   | CV = 0.036. |

## TABLE IV—CD STANDARD WORKING RANGE STUDY
[AAS-HGA Analysis]

| STD (ng/mL) | Peak area readings ×10³ at 228.8 nm | | Statistical analysis |
|---|---|---|---|
| 0.2 | 11 | 13 | n = 6. |
|  | 11 | 12 | mean = 11.8. |
|  | 12 | 12 | std dev = 0.75. |
|  |   |   | CV = 0.064. |
| 0.5 | 28 | 33 | n = 6. |
|  | 26 | 28 | mean = 28.8. |
|  | 28 | 30 | std dev = 2.4. |
|  |   |   | CV = 0.083. |
| 1.0 | 52 | 55 | n = 6. |
|  | 56 | 58 | mean = 54.8. |
|  | 54 | 54 | std dev = 2.0. |
|  |   |   | CV = 0.037. |
| 2.0 | 101 | 112 | n = 6. |
|  | 110 | 110 | mean = 108.8. |
|  | 110 | 110 | std dev = 3.9. |
|  |   |   | CV = 0.036. |
| 5.0 | 247 | 265 | n = 6. |
|  | 268 | 275 | mean = 265.5. |
|  | 259 | 279 | std dev = 11.5. |
|  |   |   | CV = 0.044. |
| 10.0 | 495 | 520 | n = 6. |
|  | 523 | 513 | mean = 516.7. |
|  | 516 | 533 | std dev = 12.7. |
|  |   |   | CV = 0.025. |
| 20.0 | 950 | 953 | n = 6. |
|  | 951 | 958 | mean = 941.8. |
|  | 949 | 890 | std dev = 25.6. |
|  |   |   | CV = 0.027. |
| 30.0 | 1269 | 1291 | n = 6. |
|  | 1303 | 1307 | mean = 1293. |
|  | 1295 | 1290 | std dev = 13.3. |
|  |   |   | CV = 0.010. |
| 40.0 | 1505 | 1567 | n = 6. |
|  | 1535 | 1567 | mean = 1552. |
|  | 1566 | 1572 | std dev = 26.6. |
|  |   |   | CV = 0.017. |

## TABLE V—ANALYTICAL METHOD RECOVERY
[Flame AAS Analysis]

| Test level | 0.5 × | | | 1.0 × | | | 2.0 × | |
|---|---|---|---|---|---|---|---|---|
| µg taken | µg found | Percent rec. | µg taken | µg found | Percent rec. | µg taken | µg found | Percent rec. |
| 1.00 | 1.0715 | 107.2 | 2.00 | 2.0688 | 103.4 | 4.00 | 4.1504 | 103.8 |
| 1.00 | 1.0842 | 108.4 | 2.00 | 2.0174 | 100.9 | 4.00 | 4.1108 | 102.8 |
| 1.00 | 1.0842 | 108.4 | 2.00 | 2.0431 | 102.2 | 4.00 | 4.0581 | 101.5 |
| 1.00 | *1.0081 | *100.8 | 2.00 | 2.0431 | 102.2 | 4.00 | 4.0844 | 102.1 |
| 1.00 | 1.0715 | 107.2 | 2.00 | 2.0174 | 100.9 | 4.00 | 4.1504 | 103.8 |
| 1.00 | 1.0842 | 108.4 | 2.00 | 2.0045 | 100.2 | 4.00 | 4.1899 | 104.7 |

n = 5, 6, 6
mean = 107.9, 101.6, 103.1
std dev = 0.657, 1.174, 1.199
$CV_1$ = 0.006, 0.011, 0.012
$CV_1$ (pooled) = 0.010

\* Rejected as an outlier·this value did not pass the outlier T-test at the 99% confidence level.

| Test level | 0.1 × | |
|---|---|---|
| µg taken | µg found | Percent rec. |
| 0.200 | 0.2509 | 125.5 |
| 0.200 | 0.2509 | 125.5 |
| 0.200 | 0.2761 | 138.1 |
| 0.200 | 0.2258 | 112.9 |
| 0.200 | 0.2258 | 112.9 |
| 0.200 | 0.1881 | 94.1 |

n = 6
mean = 118.2
std dev = 15.1
$CV_1$ = 0.128

## TABLE VI—ANALYTICAL METHOD RECOVERY
[AAS-HGA Analysis]

| Test level | 0.5 × | | | 1.0 × | | | 2.0 × | |
|---|---|---|---|---|---|---|---|---|
| ng taken | ng found | Percent rec. | ng taken | ng found | Percent rec. | ng taken | ng found | Percent rec. |
| 75 | 71.23 | 95.0 | 150 | 138.00 | 92.0 | 300 | 258.43 | 86.1 |
| 75 | 71.47 | 95.3 | 150 | 138.29 | 92.2 | 300 | 258.46 | 86.2 |
| 75 | 70.02 | 93.4 | 150 | 136.30 | 90.9 | 300 | 280.55 | 93.5 |
| 75 | 77.34 | 103.1 | 150 | 146.62 | 97.7 | 300 | 288.34 | 96.1 |
| 75 | 78.32 | 104.4 | 150 | 145.17 | 96.8 | 300 | 261.74 | 87.2 |
| 75 | 71.96 | 95.9 | 150 | 144.88 | 96.6 | 300 | 277.22 | 92.4 |

n = 6, 6, 6
mean = 97.9, 94.4, 90.3
std dev = 4.66, 2.98, 4.30
$CV_1$ = 0.048, 0.032, 0.048
$CV_1$ (pooled) = 0.043

### ATTACHMENT 1
*Instrumental Parameters for Flame AAS Analysis*

Atomic Absorption Spectrophotometer (Perkin-Elmer Model 603)

Flame: Air/Acetylene—lean, blue

Oxidant Flow: 55

Fuel Flow: 32

Wavelength: 228.8 nm

Slit: 4 (0.7 nm)

Range: UV

Signal: Concentration (4 exp)

Integration Time: 3 sec

### ATTACHMENT 2
*Instrumental Parameters for HGA Analysis*

Atomic Absorption Spectrophotometer (Perkin-Elmer Model 5100)

Signal Type: Zeeman AA

Slitwidth: 0.7 nm

Wavelength: 228.8 nm

Measurement: Peak Area

Integration Time: 6.0 sec

BOC Time: 5 sec

BOC = Background Offset Correction.

**ZEEMAN GRAPHITE FURNACE
(PERKIN–ELMER MODEL HGA-600)**

| Step | Ramp time (sec) | Hold time (sec) | Temp. (°C) | Argon flow (mL/min) | Read (sec) |
|---|---|---|---|---|---|
| 1) Predry.............. | 5 | 10 | 90 | 300 | ..... |
| 2) Dry ................. | 30 | 10 | 140 | 300 | ..... |
| 3) Char ............... | 10 | 20 | 900 | 300 | ..... |
| 4) Cool Down ........ | 1 | 8 | 30 | 300 | ..... |
| 5) Atomize............ | 0 | 5 | 1600 | 0 | −1 |
| 6) Burnout........... | 1 | 8 | 2500 | 300 | ..... |

## APPENDIX F TO §1910.1027—NON–MANDATORY PROTOCOL FOR BIOLOGICAL MONITORING

### 1.00 Introduction

Under the final OSHA cadmium rule (29 CFR part 1910), monitoring of biological specimens and several periodic medical examinations are required for eligible employees. These medical examinations are to be conducted regularly, and medical monitoring is to include the periodic analysis of cadmium in blood (CDB), cadmium in urine (CDU) and beta-2-microglobulin in urine (B2MU). As CDU and B2MU are to be normalized to the concentration of creatinine in urine (CRTU), then CRTU must be analyzed in conjunction with CDU and B2MU analyses.

The purpose of this protocol is to provide procedures for establishing and maintaining the quality of the results obtained from the analyses of CDB, CDU and B2MU by commercial laboratories. Laboratories conforming to the provisions of this nonmandatory protocol shall be known as "participating laboratories." The biological monitoring data from these laboratories will be evaluated by physicians responsible for biological monitoring to determine the conditions under which employees may continue to work in locations exhibiting airborne-cadmium concentrations at or above defined actions levels (see paragraphs (l)(3) and (l)(4) of the final rule). These results also may be used to support a decision to remove workers from such locations.

Under the medical monitoring program for cadmium, blood and urine samples must be collected at defined intervals from workers by physicians responsible for medical monitoring; these samples are sent to commercial laboratories that perform the required analyses and report results of these analyses to the responsible physicians. To ensure the accuracy and reliability of these laboratory analyses, the laboratories to which samples are submitted should participate in an ongoing and efficacious proficiency testing program. Availability of proficiency testing programs may vary with the analyses performed.

To test proficiency in the analysis of CDB, CDU and B2MU, a laboratory should participate either in the interlaboratory comparison program operated by the Centre de Toxicologie du Quebec (CTQ) or an equivalent program. (Currently, no laboratory in the U.S. performs proficiency testing on CDB, CDU or B2MU.) Under this program, CTQ sends participating laboratories 18 samples of each analyte (CDB, CDU and/or B2MU) annually for analysis. Participating laboratories must return the results of these analyses to CTQ within four to five weeks after receiving the samples.

The CTQ program pools analytical results from many participating laboratories to derive consensus mean values for each of the samples distributed. Results reported by each laboratory then are compared against these consensus means for the analyzed samples to determine the relative performance of each laboratory. The proficiency of a participating laboratory is a function of the extent of agreement between results submitted by the participating laboratory and the consensus values for the set of samples analyzed.

Proficiency testing for CRTU analysis (which should be performed with CDU and B2MU analyses to evaluate the results properly) also is recommended. In the U.S., only the College of American Pathologists (CAP) currently conducts CRTU proficiency testing; participating laboratories should be accredited for CRTU analysis by the CAP.

Results of the proficiency evaluations will be forwarded to the participating laboratory by the proficiency-testing laboratory, as well as to physicians designated by the participating laboratory to receive this information. In addition, the participating laboratory should, on request, submit the results of their internal Quality Assurance/Quality Control (QA/QC) program for each analytic procedure (i.e., CDB, CDU and/or B2MU) to physicians designated to receive the proficiency results. For participating laboratories offering CDU and/or B2MU analyses, QA/QC documentation also should be provided for CRTU analysis. (Laboratories should provide QA/QC information regarding CRTU analysis directly to the requesting physician if they perform the analysis in-house; if CRTU analysis is performed by another laboratory under contract, this information should be provided to the physician by the contract laboratory.)

QA/QC information, along with the actual biological specimen measurements, should be provided to the responsible physician using standard formats. These physicians then may collate the QA/QC information with proficiency test results to compare the relative performance of laboratories, as well as to facilitate evaluation of the worker monitoring data. This information supports decisions made by the physician with regard to the biological monitoring program, and for mandating medical removal.

This protocol describes procedures that may be used by the responsible physicians to identify laboratories most likely to be proficient in the analysis of samples used in the biological monitoring of cadmium; also provided are procedures for record keeping and reporting by laboratories participating in proficiency testing programs, and recommendations to assist these physicians in interpreting analytical results determined by participating laboratories. As the collection and handling of samples affects the quality of the data, recommendations are made for these tasks. Specifications for analytical methods to be used in the medical monitoring program are included in this protocol as well.

In conclusion, this document is intended as a supplement to characterize and maintain the quality of medical monitoring data collected under the final cadmium rule promulgated by OSHA (29 CFR part 1910). OSHA has been granted authority under the Occupational Safety and Health Act of 1970 to protect workers from the effects of exposure to hazardous substances in the work place and to mandate adequate monitoring of workers to determine when adverse health effects may be occurring. This nonmandatory protocol is intended to provide guidelines and recommendations to improve the accuracy and reliability of the procedures used to analyze the biological samples collected as part of the medical monitoring program for cadmium.

### 2.0 Definitions

When the terms below appear in this protocol, use the following definitions.

*Accuracy:* A measure of the bias of a data set. Bias is a systematic error that is either inherent in a method or caused by some artifact or idiosyncracy of the measurement system. Bias is characterized by a consistent deviation (positive or negative) in the results from an accepted reference value.

*Arithmetic Mean:* The sum of measurements in a set divided by the number of measurements in a set.

*Blind Samples:* A quality control procedure in which the concentration of analyte in the samples should be unknown to the analyst at the time that the analysis is performed.

*Coefficient of Variation:* The ratio of the standard deviation of a set of measurements to the mean (arithmetic or geometric) of the measurements.

*Compliance Samples:* Samples from exposed workers sent to a participating laboratory for analysis.

*Control Charts:* Graphic representations of the results for quality control samples being analyzed by a participating laboratory.

*Control Limits:* Statistical limits which define when an analytic procedure exceeds acceptable parameters; control limits provide a method of assessing the accuracy of analysts, laboratories, and discrete analytic runs.

*Control Samples:* Quality control samples.

*F/T:* The measured amount of an analyte divided by the theoretical value (defined below) for that analyte in the sample analyzed; this ratio is a measure of the recovery for a quality control sample.

*Geometric Mean:* The natural antilog of the mean of a set of natural log-transformed data.

*Geometric Standard Deviation:* The antilog of the standard deviation of a set of natural log-transformed data.

*Limit of Detection:* Using a predefined level of confidence, this is the lowest measured value at which some of the measured material is likely to have come from the sample.

*Mean:* A central tendency of a set of data; in this protocol, this mean is defined as the *arithmetic mean* (see definition of *arithmetic mean* above) unless stated otherwise.

*Performance:* A measure of the overall quality of data reported by a laboratory.

*Pools:* Groups of quality-control samples to be established for each target value (defined below) of an analyte. For the protocol provided in attachment 3, for example, the theoretical value of the quality control samples of the pool must be within a range defined as plus or minus (±) 50% of the target value. Within each analyte pool, there must be quality control samples of at least 4 theoretical values.

*Precision:* The extent of agreement between repeated, independent measurements of the same quantity of an analyte.

*Proficiency:* The ability to satisfy a specified level of analyte performance.

*Proficiency Samples:* Specimens, the values of which are unknown to anyone at a participating laboratory, and which are submitted by a participating laboratory for proficiency testing.

*Quality or Data Quality:* A measure of the confidence in the measurement value.

*Quality Control (QC) Samples:* Specimens, the value of which is unknown to the analyst, but is known to the appropriate QA/QC personnel of a participating laboratory; when used as part of a laboratory QA/QC program, the theoretical values of these samples should not be known to the analyst until the analyses are complete. QC samples are to be run in sets consisting of one QC sample from each pool (see definition of "pools" above).

*Sensitivity:* For the purposes of this protocol, the limit of detection.

*Standard Deviation:* A measure of the distribution or spread of a data set about the mean; the standard deviation is equal to the positive square root of the variance, and is expressed in the same units as the original measurements in the data set.

*Standards:* Samples with values known by the analyst and used to calibrate equipment and to check calibration throughout an analytic run. In a laboratory QA/QC program, the values of the standards must exceed the values obtained for compliance samples such that the lowest standard value is near the limit of detection and the highest standard is higher than the highest compliance sample or QC sample. Standards of at least three different values are to be used for calibration, and should be constructed from at least 2 different sources.

*Target Value:* Those values of CDB, CDU or B2MU which trigger some action as prescribed in the medical surveillance section of the regulatory text of the final cadmium rule. For CDB, the target values are 5, 10 and 15 µg/l. For CDU, the target values are 3, 7, and 15 µg/g CRTU. For $B_2$ MU, the target values are 300, 750 and 1500 µg/g CRTU. (Note that target values may vary as a function of time.)

*Theoretical Value (or Theoretical Amount):* The reported concentration of a quality-control sample (or calibration standard) derived from prior characterizations of the sample.

*Value or Measurement Value:* The numerical result of a measurement.

*Variance:* A measure of the distribution or spread of a data set about the mean; the variance is the sum of the squares of the differences between the mean and each discrete measurement divided by one less than the number of measurements in the data set.

### 3.0 Protocol
This protocol provides procedures for characterizing and maintaining the quality of analytic results derived for the medical monitoring program mandated for workers under the final cadmium rule.

### 3.1 Overview
The goal of this protocol is to assure that medical monitoring data are of sufficient quality to facilitate proper interpretation. The data quality objectives (DQOs) defined for the medical monitoring program are summarized in Table 1. Based on available information, the DQOs presented in Table 1 should be achievable by the majority of laboratories offering the required analyses commercially; OSHA recommends that only laboratories meeting these DQOs be used for the analysis of biological samples collected for monitoring cadmium exposure.

**TABLE 1—RECOMMENDED DATA QUALITY OBJECTIVES (DQOs) FOR THE CADMIUM MEDICAL MONITORING PROGRAM**

| Analyte/concentration pool | Limit of detection | Precision (CV) (%) | Accuracy |
|---|---|---|---|
| Cadmium in blood | 0.5 µg/g | .......... | ±1 µg/l or 15% of the mean. |
| ≤ 2 µg/l | .......... | 40 | |
| >2 µg/l | .......... | 20 | |
| Cadmium in urine | 0.5 µg/g creatinine | .......... | ±1 µg/l or 15% of the mean. |
| ≤ 2 µg/l creatinine | .......... | 40 | |
| >2 µg/l creatinine | .......... | 20 | |
| β-2-microglobulin in urine: 100 µg/l creatine. | 100 µg/g creatinine | 5 | ±15% of the mean. |

To satisfy the DQOs presented in Table 1, OSHA provides the following guidelines:

1. Procedures for the collection and handling of blood and urine are specified (Section 3.4.1 of this protocol);

2. Preferred analytic methods for the analysis of CDB, CDU and B2MU are defined (and a method for the determination of CRTU also is specified since CDU and B2MU results are to be normalized to the level of CRTU).

3. Procedures are described for identifying laboratories likely to provide the required analyses in an accurate and reliable manner;

4. These guidelines (Sections 3.2.1 to 3.2.3, and Section 3.3) include recommendations regarding internal QA/QC programs for participating laboratories, as well as levels of proficiency through participation in an interlaboratory proficiency program;

5. Procedures for QA/QC record keeping (Section 3.3.2), and for reporting QC/QA results are described (Section 3.3.3); and,

6. Procedures for interpreting medical monitoring results are specified (Section 3.4.3).

Methods recommended for the biological monitoring of eligible workers are:

1. The method of Stoeppler and Brandt (1980) for CDB determinations (limit of detection: 0.5 µg/l);

2. The method of Pruszkowska et al. (1983) for CDU determinations (limit of detection: 0.5 µg/l of urine); and,

3. The Pharmacia Delphia test kit (Pharmacia 1990) for the determination of B2MU (limit of detection: 100 µg/l urine).

Because both CDU and B2MU should be reported in µg/g CRTU, an independent determination of CRTU is recommended. Thus, both the OSHA Salt Lake City Technical Center (OSLTC) method (OSHA, no date) and the Jaffe method (Du Pont, no date) for the determination of CRTU are specified under this protocol (i.e., either of these 2 methods may be used). Note that although detection limits are not reported for either of these CRTU methods, the range of measurements expected for CRTU (0.9-1.7 g/l) are well above the likely limit of detection for either of these methods (Harrison, 1987).

Laboratories using alternate methods should submit sufficient data to the responsible physicians demonstrating that the alternate method is capable of satisfying the defined data quality objectives of the program. Such laboratories also should submit a QA/QC plan that documents the performance of the alternate method in a manner entirely equivalent to the QA/QC plans proposed in Section 3.3.1.

### 3.2 Duties of the Responsible Physician
The responsible physician will evaluate biological monitoring results provided by participating laboratories to determine whether such laboratories are proficient and have satisfied the QA/QC recommendations. In determining which laboratories to employ for this purpose, these physicians should review proficiency and QA/QC data submitted to them by the participating laboratories.

Participating laboratories should demonstrate proficiency for each analyte (CDU, CDB and B2MU) sampled under the biological monitoring program.

Participating laboratories involved in analyzing CDU and B2MU also should demonstrate proficiency for CRTU analysis, or provide evidence of a contract with a laboratory proficient in CRTU analysis.

### 3.2.1 Recommendations for Selecting Among Existing Laboratories

OSHA recommends that existing laboratories providing commercial analyses for CDB, CDU and/or B2MU for the medical monitoring program satisfy the following criteria:

1. Should have performed commercial analyses for the appropriate analyte (CDB, CDU and/or B2MU) on a regular basis over the last 2 years;

2. Should provide the responsible physician with an internal QA/QC plan;

3. If performing CDU or B2MU analyses, the participating laboratory should be accredited by the CAP for CRTU analysis, and should be enrolled in the corresponding CAP survey (note that alternate credentials may be acceptable, but acceptability is to be determined by the responsible physician); and,

4. Should have enrolled in the CTQ interlaboratory comparison program for the appropriate analyte (CDB, CDU and/or B2MU).

Participating laboratories should submit appropriate documentation demonstrating compliance with the above criteria to the responsible physician. To demonstrate compliance with the first of the above criteria, participating laboratories should submit the following documentation for each analyte they plan to analyze (note that each document should cover a period of at least 8 consecutive quarters, and that the period designated by the term "regular analyses" is at least once a quarter):

1. Copies of laboratory reports providing results from regular analyses of the appropriate analyte (CDB, CDU and/or B2MU);

2. Copies of 1 or more signed and executed contracts for the provision of regular analyses of the appropriate analyte (CDB, CDU and/or B2MU); or,

3. Copies of invoices sent to 1 or more clients requesting payment for the provision of regular analyses of the appropriate analyte (CDB, CDU and/or B2MU). Whatever the form of documentation submitted, the specific analytic procedures conducted should be identified directly. The forms that are copied for submission to the responsible physician also should identify the laboratory which provided these analyses.

To demonstrate compliance with the second of the above criteria, a laboratory should submit to the responsible physician an internal QA/QC plan detailing the standard operating procedures to be adopted for satisfying the recommended QA/QC procedures for the analysis of each specific analyte (CDB, CDU and/or B2MU). Procedures for internal QA/QC programs are detailed in Section 3.3.1 below.

To satisfy the third of the above criteria, laboratories analyzing for CDU or B2MU also should submit a QA/QC plan for creatinine analysis (CRTU); the QA/QC plan and characterization analyses for CRTU must come from the laboratory performing the CRTU analysis, even if the CRTU analysis is being performed by a contract laboratory.

Laboratories enrolling in the CTQ program (to satisfy the last of the above criteria) must remit, with the enrollment application, an initial fee of approximately $100 per analyte. (Note that this fee is only an estimate, and is subject to revision without notice.) Laboratories should indicate on the application that they agree to have proficiency test results sent by the CTQ directly to the physicians designated by participating laboratories.

Once a laboratory's application is processed by the CTQ, the laboratory will be assigned a code number which will be provided to the laboratory on the initial confirmation form, along with identification of the specific analytes for which the laboratory is participating. Confirmation of participation will be sent by the CTQ to physicians designated by the applicant laboratory.

### 3.2.2 Recommended Review of Laboratories Selected to Perform Analyses

Six months after being selected initially to perform analyte determinations, the status of participating laboratories should be reviewed by the responsible physicians. Such reviews should then be repeated every 6 months or whenever additional proficiency or QA/QC documentation is received (whichever occurs first).

As soon as the responsible physician has received the CTQ results from the first 3 rounds of proficiency testing (i.e., 3 sets of 3 samples each for CDB, CDU and/or B2MU) for a participating laboratory, the status of the laboratory's continued participation should be reviewed. Over the same initial 6-month period, participating laboratories also should provide responsible physicians the results of their internal QA/QC monitoring program used to assess performance for each analyte (CDB, CDU and/or B2MU) for which the laboratory performs determinations. This information should be submitted using appropriate forms and documentation.

The status of each participating laboratory should be determined for each analyte (i.e., whether the laboratory satisfies minimum proficiency guidelines based on the proficiency samples sent by the CTQ and the results of the laboratory's internal QA/QC program). To maintain competency for analysis of CDB, CDU and/or B2MU during the first review, the laboratory should satisfy performance requirements for at least 2 of the 3 proficiency samples provided in each of the 3 rounds completed over the 6-month period. Proficiency should be maintained for the analyte(s) for which the laboratory conducts determinations.

To continue participation for CDU and/or B2MU analyse, laboratories also should either maintain accreditation for CRTU analysis in the CAP program and participate in the CAP surveys, or they should contract the CDU and B2MU analyses to a laboratory which satisfies these requirements (or which can provide documentation of accreditation/participation in an equivalent program).

The performance requirement for CDB analysis is defined as an analytical result within ±1 µg/l blood or 15% of the consensus mean (whichever is greater). For samples exhibiting a consensus mean less than 1 µg/l, the performance requirement is defined as a concentration between the detection limit of the analysis and a maximum of 2 µg/l. The purpose for redefining the acceptable interval for low CDB values is to encourage proper reporting of the actual values obtained during measurement; laboratories, therefore, will not be penalized (in terms of a narrow range of acceptability) for reporting measured concentrations smaller than 1 µg/l.

The performance requirement for CDU analysis is defined as an analytical result within ±1 µg/l urine or 15% of the consensus mean (whichever is greater). For samples exhibiting a consensus mean less than 1 µg/l urine, the performance requirement is defined as a concentration between the detection limit of the analysis and a maximum of 2 µg/l urine. Laboratories also should demonstrate proficiency in creatinine analysis as defined by the CAP. Note that reporting CDU results, other than for the CTQ proficiency samples (i.e., compliance samples), should be accompanied with results of analyses for CRTU, and these 2 sets of results should be combined to provide a measure of CDU in units of µg/g CRTU.

The performance requirement for B2MU is defined as analytical results within ±15% of the consensus mean. Note that reporting B2MU results, other than for CTQ proficiency samples (i.e., compliance samples), should be accompanied with results of analyses for CRTU, and these 2 sets of results should be combined to provide a measure of B2MU in units of µg/g CRTU.

There are no recommended performance checks for CRTU analyses. As stated previously, laboratories performing CRTU analysis in support of CDU or B2MU analyses should be accredited by the CAP, and participating in the CAP's survey for CRTU.

Following the first review, the status of each participating laboratory should be reevaluated at regular intervals (i.e., corresponding to receipt of results from each succeeding round of proficiency testing and submission of reports from a participating laboratory's internal QA/QC program).

After a year of collecting proficiency test results, the following proficiency criterion should be added to the set of criteria used to determine the participating laboratory's status (for analyzing CDB, CDU and/or B2MU): A participating laboratory should not fail performance requirements for more than 4 samples from the 6 most recent consecutive rounds used to assess proficiency for CDB, CDU and/or B2MU separately (i.e., a total of 18 discrete proficiency samples for each analyte). Note that this requirement does not replace, but supplements, the recommendation that a laboratory should satisfy the performance criteria for at least 2 of the 3 samples tested for each round of the program.

### 3.2.3 Recommendations for Selecting Among Newly-Formed Laboratories (or Laboratories that Previously Failed to Meet the Protocol Guidelines)

OSHA recommends that laboratories that have not previously provided commercial analyses of CDB, CDU and/or B2MU (or have done so for a period less than 2 years), or which have provided these analyses for 2 or more years but have not conformed previously with these protocol guidelines, should satisfy the following provisions for each analyte for which determinations are to be made prior to being selected to analyze biological samples under the medical monitoring program:

1. Submit to the responsible physician an internal QA/QC plan detailing the standard operating procedures to be adopted for satisfying the

# OSHA Standards for General Industry

QA/QC guidelines (guidelines for internal QA/QC programs are detailed in Section 3.3.1);

2. Submit to the responsible physician the results of the initial characterization analyses for each analyte for which determinations are to be made;

3. Submit to the responsible physician the results, for the initial 6-month period, of the internal QA/QC program for each analyte for which determinations are to be made (if no commercial analyses have been conducted previously, a minimum of 2 mock standardization trials for each analyte should be completed per month for a 6-month period);

4. Enroll in the CTQ program for the appropriate analyte for which determinations are to be made, and arrange to have the CTQ program submit the initial confirmation of participation and proficiency test results directly to the designated physicians. Note that the designated physician should receive results from 3 completed rounds from the CTQ program before approving a laboratory for participation in the biological monitoring program;

5. Laboratories seeking participation for CDU and/or B2MU analyses should submit to the responsible physician documentation of accreditation by the CAP for CRTU analyses performed in conjunction with CDU and/or B2MU determinations (if CRTU analyses are conducted by a contract laboratory, this laboratory should submit proof of CAP accreditation to the responsible physician); and,

6. Documentation should be submitted on an appropriate form.

To participate in CDB, CDU and/or B2MU analyses, the laboratory should satisfy the above criteria for a minimum of 2 of the 3 proficiency samples provided in each of the 3 rounds of the CTQ program over a 6-month period; this procedure should be completed for each appropriate analyte. Proficiency should be maintained for each analyte to continue participation. Note that laboratories seeking participation for CDU or B2MU also should address the performance requirements for CRTU, which involves providing evidence of accreditation by the CAP and participation in the CAP surveys (or an equivalent program).

The performance requirement for CDB analysis is defined as an analytical result within ±1 µg/l or 15% of the consensus mean (whichever is greater). For samples exhibiting a consensus mean less than 1 µg/l, the performance requirement is defined as a concentration between the detection limit of the analysis and a maximum of 2 µg/l. The purpose of redefining the acceptable interval for low CDB values is to encourage proper reporting of the actual values obtained during measurement; laboratories, therefore, will not be penalized (in terms of a narrow range of acceptability) for reporting measured concentrations less than 1 µg/l.

The performance requirement for CDU analysis is defined as an analytical result within ±1 µg/l urine or 15% of the consensus mean (whichever is greater). For samples exhibiting a consensus mean less than 1 µg/l urine, the performance requirement is defined as a concentration that falls between the detection limit of the analysis and a maximum of 2 µg/l urine. Performance requirements for the companion CRTU analysis (defined by the CAP) also should be met. Note that reporting CDU results, other than for CTQ proficiency testing should be accompanied with results of CRTU analyses, and these 2 sets of results should be combined to provide a measure of CDU in units of µg/g CRTU.

The performance requirement for B2MU is defined as an analytical result within ±15% of the consensus mean. Note that reporting B2MU results, other than for CTQ proficiency testing should be accompanied with results of CRTU analysis, these 2 sets of results should be combined to provide a measure of B2MU in units of µg/g CRTU.

Once a new laboratory has been approved by the responsible physician for conducting analyte determinations, the status of this approval should be reviewed periodically by the responsible physician as per the criteria presented under Section 3.2.2.

Laboratories which have failed previously to gain approval of the responsible physician for conducting determinations of 1 or more analytes due to lack of compliance with the criteria defined above for existing laboratories (Section 3.2.1), may obtain approval by satisfying the criteria for newly-formed laboratories defined under this section; for these laboratories, the second of the above criteria may be satisfied by submitting a new set of characterization analyses for each analyte for which determinations are to be made.

Reevaluation of these laboratories is discretionary on the part of the responsible physician. Reevaluation, which normally takes about 6 months, may be expedited if the laboratory can achieve 100% compliance with the proficiency test criteria using the 6 samples of each analyte submitted to the CTQ program during the first 2 rounds of proficiency testing.

For laboratories seeking reevaluation for CDU or B2MU analysis, the guidelines for CRTU analyses also should be satisfied, including accreditation for CRTU analysis by the CAP, and participation in the CAP survey program (or accreditation/participation in an equivalent program).

### 3.2.4 Future Modifications to the Protocol Guidelines

As participating laboratories gain experience with analyses for CDB, CDU and B2MU, it is anticipated that the performance achievable by the majority of laboratories should improve until it approaches that reported by the research groups which developed each method. OSHA, therefore, may choose to recommend stricter performance guidelines in the future as the overall performance of participating laboratories improves.

### 3.3 Guidelines for Record Keeping and Reporting

To comply with these guidelines, participating laboratories should satisfy the above-stated performance and proficiency recommendations, as well as the following internal QA/QC, record keeping, and reporting provisions.

If a participating laboratory fails to meet the provisions of these guidelines, it is recommended that the responsible physician disapprove further analyses of biological samples by that laboratory until it demonstrates compliance with these guidelines. On disapproval, biological samples should be sent to a laboratory that can demonstrate compliance with these guidelines, at least until the former laboratory is reevaluated by the responsible physician and found to be in compliance.

The following record keeping and reporting procedures should be practiced by participating laboratories.

### 3.3.1 Internal Quality Assurance/Quality Control Procedures

Laboratories participating in the cadmium monitoring program should develop and maintain an internal quality assurance/quality control (QA/QC) program that incorporates procedures for establishing and maintaining control for each of the analytic procedures (determinations of CDB, CDU and/or B2MU) for which the laboratory is seeking participation. For laboratories analyzing CDU and/or B2MU, a QA/QC program for CRTU also should be established.

Written documentation of QA/QC procedures should be described in a formal QA/QC plan; this plan should contain the following information: Sample acceptance and handling procedures (i.e., chain-of-custody); sample preparation procedures; instrument parameters; calibration procedures; and, calculations. Documentation of QA/QC procedures should be sufficient to identify analytical problems, define criteria under which analysis of compliance samples will be suspended, and describe procedures for corrective actions.

#### 3.3.1.1 *QA/QC procedures for establishing control of CDB and CDU analyses*

The QA/QC program for CDB and CDU should address, at a minimum, procedures involved in calibration, establishment of control limits, internal QC analyses and maintaining control, and corrective-action protocols. Participating laboratory should develop and maintain procedures to assure that analyses of compliance samples are within control limits, and that these procedures are documented thoroughly in a QA/QC plan.

A nonmandatory QA/QC protocol is presented in Attachment 1. This attachment is illustrative of the procedures that should be addressed in a proper QA/QC program.

*Calibration.* Before any analytic runs are conducted, the analytic instrument should be calibrated. Calibration should be performed at the beginning of each day on which QC and/or compliance samples are run. Once calibration is established, QC or compliance samples may be run. Regardless of the type of samples run, about every fifth sample should serve as a standard to assure that calibration is being maintained.

Calibration is being maintained if the standard is within ±15% of its theoretical value. If a standard is more than ±15% of its theoretical value, the run has exceeded control limits due to calibration error; the entire set of samples then should be reanalyzed after recalibrating or the results should be recalculated based on a statistical curve derived from that set of standards.

It is essential that the value of the highest standard analyzed be higher than the highest sample analyzed; it may be necessary, therefore, to run a high standard at the end of the run, which has been selected based on results obtained over the course of the run (i.e., higher than any standard analyzed to that point).

Standards should be kept fresh; as samples age, they should be compared with new standards and replaced if necessary.

*Internal Quality Control Analyses.* Internal QC samples should be determined interspersed with analyses of compliance samples. At a minimum, these samples should be run at a rate of 5% of the compliance samples or at least one set of QC samples per analysis of compliance samples, whichever is greater. If only 2 samples are run, they should contain different levels of cadmium.

Internal QC samples may be obtained as commercially-available reference materials and/or they may be internally prepared. Internally-prepared samples should be well characterized and traced, or compared to a reference material for which a consensus value is available.

Levels of cadmium contained in QC samples should not be known to the analyst prior to reporting the results of the analysis.

Internal QC results should be plotted or charted in a manner which describes sample recovery and laboratory control limits.

*Internal Control Limits.* The laboratory protocol for evaluating internal QC analyses per control limits should be clearly defined. Limits may be based on statistical methods (e.g., as $2\bar{\sigma}$ from the laboratory mean recovery), or on proficiency testing limits (e.g., ±1 µg or 15% of the mean, whichever is greater). Statistical limits that exceed ±40% should be reevaluated to determine the source error in the analysis.

When laboratory limits are exceeded, analytic work should terminate until the source of error is determined and corrected; compliance samples affected by the error should be reanalyzed. In addition, the laboratory protocol should address any unusual trends that develop which may be biasing the results. Numerous, consecutive results above or below laboratory mean recoveries, or outside laboratory statistical limits, indicate that problems may have developed.

*Corrective Actions.* The QA/QC plan should document in detail specific actions taken if control limits are exceeded or unusual trends develop. Corrective actions should be noted on an appropriate form, accompanied by supporting documentation.

In addition to these actions, laboratories should include whatever additional actions are necessary to assure that accurate data are reported to the responsible physicians.

*Reference Materials.* The following reference materials may be available:

### Cadmium in Blood (CDB)

1. Centre de Toxicologie du Quebec, Le Centre Hospitalier de l'Universite Laval, 2705 boul. Laurier, Quebec, Que., Canada G1V 4G2. (Prepared 6 times per year at 1-15 µg Cd/l.)

2. H. Marchandise, Community Bureau of Reference-BCR, Directorate General XII, Commission of the European Communities, 200, rue de la Loi, B-1049, Brussels, Belgium. (Prepared as Bl CBM-1 at 5.37 µg Cd/l, and Bl CBM-2 at 12.38 µg Cd/l.)

3. Kaulson Laboratories Inc., 691 Bloomfield Ave., Caldwell, NJ 07006; tel: (201) 226-9494, FAX (201) 226-3244. (Prepared as #0141 [As, Cd, Hg, Pb] at 2 levels.)

### Cadmium in Urine (CDU)

1. Centre de Toxicologie du Quebec, Le Centre Hospitalier de l'Universite Laval, 2705 boul. Laurier, Quebec, Que., Canada G1V 4G2. (Prepared 6 times per year.)

2. National Institute of Standards and Technology (NIST), Dept. of Commerce, Gaithersburg, MD; tel: (301) 975-6776. (Prepared as SRM 2670 freeze-dried urine [metals]; set includes normal and elevated levels of metals; cadmium is certified for elevated level of 88.0 µg/l in reconstituted urine.)

3. Kaulson Laboratories Inc., 691 Bloomfield Ave., Caldwell, NJ 07006; tel: (201) 226-9494, FAX (201) 226-3244. (Prepared as #0140 [As, Cd, Hg, Pb] at 2 levels.)

#### 3.3.1.2 QA/QC procedures for establishing control of B2MU

A written, detailed QA/QC plan for B2MU analysis should be developed. The QA/QC plan should contain a protocol similar to those protocols developed for the CDB/CDU analyses. Differences in analyses may warrant some differences in the QA/QC protocol, but procedures to ensure analytical integrity should be developed and followed.

Examples of performance summaries that can be provided include measurements of accuracy (i.e., the means of measured values versus target values for the control samples) and precision (i.e., based on duplicate analyses). It is recommended that the accuracy and precision measurements be compared to those reported as achievable by the Pharmacia Delphia kit (Pharmacia 1990) to determine if and when unsatisfactory analyses have arisen. If the measurement error of 1 or more of the control samples is more than 15%, the run exceeds control limits. Similarly, this decision is warranted when the average CV for duplicate samples is greater than 5%.

### 3.3.2 Procedures for Record Keeping

To satisfy reporting requirements for commercial analyses of CDB, CDU and/or B2MU performed for the medical monitoring program mandated under the cadmium rule, participating laboratories should maintain the following documentation for each analyte:

1. For each analytic instrument on which analyte determinations are made, records relating to the most recent calibration and QC sample analyses;

2. For these instruments, a tabulated record for each analyte of those determinations found to be within and outside of control limits over the past 2 years;

3. Results for the previous 2 years of the QC sample analyses conducted under the internal QA/QC program (this information should be: Provided for each analyte for which determinations are made and for each analytic instrument used for this purpose, sufficient to demonstrate that internal QA/QC programs are being executed properly, and consistent with data sent to responsible physicians.

4. Duplicate copies of monitoring results for each analyte sent to clients during the previous 5 years, as well as associated information; supporting material such as chain-of-custody forms also should be retained; and,

5. Proficiency test results and related materials received while participating in the CTQ interlaboratory program over the past 2 years; results also should be tabulated to provide a serial record of relative error (derived per Section 3.3.3 below).

### 3.3.3 Reporting Procedures

Participating laboratories should maintain these documents: QA/QC program plans; QA/QC status reports; CTQ proficiency program reports; and, analytical data reports. The information that should be included in these reports is summarized in Table 2; a copy of each report should be sent to the responsible physician.

**TABLE 2—REPORTING PROCEDURES FOR LABORATORIES PARTICIPATING IN THE CADMIUM MEDICAL MONITORING PROGRAM**

| | Report | Frequency (time frame) | Contents |
|---|---|---|---|
| 1 | QA/QC Program Plan | Once (initially). | A detailed description of the QA/QC protocol to be established by the laboratory to maintain control of analyte determinations. |
| 2 | QA/QC Status Report | Every 2 months | Results of the QC samples incorporated into regular runs for each (over the period since the last report). |
| 3 | Proficiency Report | Attached to every data report. | Results from the last full year of proficiency samples submitted to the CTQ program and Results of the 100 most recent QC samples incorporated into regular runs for each instrument. |
| 4 | Analytical Data Report | For all reports of data results. | Date the sample was received; Date the sample was analyzed; Appropriate chain-of-custody information; Types of analyses performed; Results of the requested analyses and Copy of the most current proficiency report. |

As noted in Section 3.3.1, a QA/QC program plan should be developed that documents internal QA/QC procedures (defined under Section 3.3.1) to be implemented by the participating laboratory for each analyte; this plan should provide a list identifying each instrument used in making analyte determinations.

A QA/QC status report should be written bimonthly for each analyte. In this report, the results of the QC program during the reporting period should be reported for each analyte in the following manner: The number (N) of QC samples analyzed during the period; a table of the target levels defined for each sample and the corresponding measured values; the mean of F/T value (as defined below) for the set of QC samples run during the period; and, use of $\tilde{X} \pm \tilde{\sigma}$ as defined below) for the set of QC samples run during the period as a measure of precision.

As noted in Section 2, an F/T value for a QC sample is the ratio of the measured concentration of analyte to the established (i.e., reference) concentration of analyte for that QC sample. The equation below describes the derivation of the mean for F/T values, X, (with N being the total number of samples analyzed):

$$\overline{X} = \frac{\Sigma (F/T)}{N}$$

The standard deviation, $\hat{\sigma}$, for these measurements is derived using the following equation (note that $\tilde{\sigma}$ is twice this value):

$$\hat{\sigma} = \left[ \frac{\Sigma \left( \frac{F}{T} - \overline{X} \right)^2}{N - 1} \right]^{1/2}$$

The nonmandatory QA/QC protocol (see Attachment 1) indicates that QC samples should be divided into several discrete pools, and a separate estimate of precision for each pools then should be derived. Several precision estimates should be provided for concentrations which differ in average value. These precision measures may be used to document improvements in performance with regard to the combined pool.

Participating laboratories should use the CTQ proficiency program for each analyte. Results of the this program will be sent by CTQ directly to physicians designated by the participating laboratories. Proficiency results from the CTQ program are used to establish the accuracy of results from each participating laboratory, and should be provided to responsible physicians for use in trend analysis. A proficiency report consisting of these proficiency results should accompany data reports as an attachment.

For each analyte, the proficiency report should include the results from the 6 previous proficiency rounds in the following format:

1. Number (N) of samples analyzed;

2. Mean of the target levels, $(1/N)\Sigma_i$, with $T_i$ being a consensus mean for the sample;

3. Mean of the measurements, $(1/N)\Sigma_i$, with $M_i$ being a sample measurement;

4. A measure of error defined by:

$$(1/N)\Sigma(T_i - M_i)^2$$

Analytical data reports should be submitted to responsible physicians directly. For each sample, report the following information: The date the sample was received; the date the sample was analyzed; appropriate chain-of-custody information; the type(s) of analyses performed; and, the results of the analyses. This information should be reported on a form similar to the form provided an appropriate form. The most recent proficiency program report should accompany the analytical data reports (as an attachment).

Confidence intervals for the analytical results should be reported as $X \pm \tilde{\sigma}$, with X being the measured value and $\sigma$ the standard deviation calculated as described above.

For CDU or B2MU results, which are combined with CRTU measurements for proper reporting, the 95% confidence limits are derived from the limits for CDU or B2MU, (p), and the limits for CRTU, (q), as follows:

$$\frac{X}{Y} \pm \left(\frac{1}{Y^2}\right) \left( Y^2 \times p^2 \times X^2 \times q^2 \right)^{1/2}$$

For these calculations, $X \pm p$ is the measurement and confidence limits for CDU or B2MU, and $Y \pm q$ is the measurement and confidence limit for CRTU.

Participating laboratories should notify responsible physicians as soon as they receive information indicating a change in their accreditation status with the CTQ or the CAP. These physicians should not be expected to wait until formal notice of a status change has been received from the CTQ or the CAP.

### 3.4 Instructions to Physicians

Physicians responsible for the medical monitoring of cadmium-exposed workers must collect the biological samples from workers; they then should select laboratories to perform the required analyses, and should interpret the analytic results.

### 3.4.1 Sample Collection and Holding Procedures

*Blood Samples.* The following procedures are recommended for the collection, shipment and storage of blood samples for CDB analysis to reduce analytical variablility; these recommendations were obtained primarily through personal communications with J.P. Weber of the CTQ (1991), and from reports by the Centers for Disease Control (CDC, 1986) and Stoeppler and Brandt (1980).

To the extent possible, blood samples should be collected from workers at the same time of day. Workers should shower or thoroughly wash their hands and arms before blood samples are drawn. The following materials are needed for blood sample collection: Alcohol wipes; sterile gauze sponges; band-aids; 20-gauge, 1.5-in. stainless steel needles (sterile); preprinted labels; tourniquets; vacutainer holders; 3-ml "metal free" vacutainer tubes (i.e., dark-blue caps), with EDTA as an anti-coagulant; and, styrofoam vacutainer shipping containers.

Whole blood samples are taken by venipuncture. Each blue-capped tube should be labeled or coded for the worker and company before the sample is drawn. (Blue-capped tubes are recommended instead of red-capped tubes because the latter may consist of red coloring pigment containing cadmium, which could contaminate the samples.) Immediately after sampling, the vacutainer tubes must be thoroughly mixed by inverting the tubes at least 10 times manually or mechanically using a Vortex device (for 15 sec). Samples should be refrigerated immediately or stored on ice until they can be packed for shipment to the participating laboratory for analysis.

The CDC recommends that blood samples be shipped with a "cool pak" to keep the samples cold during shipment. However, the CTQ routinely ships and receives blood samples for cadmium analysis that have not been kept cool during shipment. The CTQ has found no deterioration of cadmium in biological fluids that were shipped via parcel post without a cooling agent, even though these deliveries often take 2 weeks to reach their destination.

*Urine Samples.* The following are recommended procedures for the collection, shipment and storage of urine for CDU and B2MU analyses, and were obtained primarily through personal communications with J.P. Weber of the CTQ (1991), and from reports by the CDC (1986) and Stoeppler and Brandt (1980).

Single "spot" samples are recommended. As B2M can degrade in the bladder, workers should first empty their bladder and then drink a large glass of water at the start of the visit. Urine samples then should be collected within 1 hour. Separate samples should be collected for CDU and B2MU using the following materials: Sterile urine collection cups (250 ml); small sealable plastic bags; preprinted labels; 15-ml polypropylene or polyethylene screw-cap tubes; lab gloves ("metal free"); and, preservatives (as indicated).

The sealed collection cup should be kept in the plastic bag until collection time. The workers should wash their hands with soap and water before receiving the collection cup. The collection cup should not be opened until just before voiding and the cup should be sealed immediately after filling. It is important that the inside of the container and cap are not touched by, or come into contact with, the body, clothing or other surfaces.

For CDU analyzes, the cup is swirled gently to resuspend any solids, and the 15-ml tube is filled with 10-12 ml urine. The CDC recommends the addition of 100 μl concentrated $HNO_3$ as a preservative before sealing the tube and then freezing the sample. The CTQ recommends minimal handling and does not acidify their interlaboratory urine reference materials prior to shipment, nor do they freeze the sample for shipment. At the CTQ, if the urine sample has much sediment, the sample is acidified in the lab to free any cadmium in the precipitate.

For B2M, the urine sample should be collected directly into a polyethylene bottle previously washed with dilute nitric acid. The pH of the urine should be measured and adjusted to 8.0 with 0.1 N NaOH immediately following collection. Samples should be frozen and stored at -20 °C until testing is performed. The B2M in the samples should be stable for 2 days when stored at 2-8 °C, and for at least 2 months at -20 °C. Repeated freezing and thawing should be avoided to prevent denaturing the B2M (Pharmacia 1990).

### 3.4.2 Recommendations for Evaluating Laboratories
Using standard error data and the results of proficiency testing obtained from CTQ, responsible physicians can make an informed choice of which laboratory to select to analyze biological samples. In general, laboratories with small standard errors and little disparity between target and measured values tend to make precise and accurate sample determinations. Estimates of precision provided to the physicians with each set of monitoring results can be compared to previously-reported proficiency and precision estimates. The latest precision estimates should be at least as small as the standard error reported previously by the laboratory. Moreover, there should be no indication that precision is deteriorating (i.e., increasing values for the precision estimates). If precision is deteriorating, physicians may decide to use another laboratory for these analyses. QA/QC information provided by the participating laboratories to physicians can, therefore, assist physicians in evaluating laboratory performance.

### 3.4.3 Use and Interpretation of Results
When the responsible physician has received the CDB, CDU and/or B2MU results, these results must be compared to the action levels discussed in the final rule for cadmium. The comparison of the sample results to action levels is straightforward. The measured value reported from the laboratory can be compared directly to the action levels; if the reported value exceeds an action level, the required actions must be initiated.

### 4.0 Background
Cadmium is a naturally-occurring environmental contaminant to which humans are continually exposed in food, water, and air. The average daily intake of cadmium by the U.S. population is estimated to be 10-20 µg/day. Most of this intake is via ingestion, for which absorption is estimated at 4-7% (Kowal et al. 1979). An additional nonoccupational source of cadmium is smoking tobacco; smoking a pack of cigarettes a day adds an additional 2-4 µg cadmium to the daily intake, assuming absorption via inhalation of 25-35% (Nordberg and Nordberg 1988; Friberg and Elinder 1988; Travis and Haddock 1980).

Exposure to cadmium fumes and dusts in an occupational setting where air concentrations are 20-50 µg/m$^3$ results in an additional daily intake of several hundred micrograms (Friberg and Elinder 1988, p. 563). In such a setting, occupational exposure to cadmium occurs primarily via inhalation, although additional exposure may occur through the ingestion of material via contaminated hands if workers eat or smoke without first washing. Some of the particles that are inhaled initially may be ingested when the material is deposited in the upper respiratory tract, where it may be cleared by mucociliary transport and subsequently swallowed.

Cadmium introduced into the body through inhalation or ingestion is transported by the albumin fraction of the blood plasma to the liver, where it accumulates and is stored principally as a bound form complexed with the protein metallothionein. Metallothionein-bound cadmium is the main form of cadmium subsequently transported to the kidney; it is these 2 organs, the liver and kidney, in which the majority of the cadmium body burden accumulates. As much as one half of the total body burden of cadmium may be found in the kidneys (Nordberg and Nordberg 1988).

Once cadmium has entered the body, elimination is slow; about 0.02% of the body burden is excreted per day via urinary/fecal elimination. The whole-body half-life of cadmium is 10-35 years, decreasing slightly with increasing age (Travis and Haddock 1980).

The continual accumulation of cadmium is the basis for its chronic noncarcinogenic toxicity. This accumulation makes the kidney the target organ in which cadmium toxicity usually is first observed (Piscator 1964). Renal damage may occur when cadmium levels in the kidney cortex approach 200 µg/g wet tissue-weight (Travis and Haddock 1980).

The kinetics and internal distribution of cadmium in the body are complex, and depend on whether occupational exposure to cadmium is ongoing or has terminated. In general, cadmium in blood is related principally to recent cadmium exposure, while cadmium in urine reflects cumulative exposure (i.e., total body burden) (Lauwerys et al. 1976; Friberg and Elinder 1988).

### 4.1 Health Effects
Studies of workers in a variety of industries indicate that chronic exposure to cadmium may be linked to several adverse health effects including kidney dysfunction, reduced pulmonary function, chronic lung disease and cancer (Federal Register 1990). The primary sites for cadmium-associated cancer appear to be the lung and the prostate.

*Cancer.* Evidence for an association between cancer and cadmium exposure comes from both epidemiological studies and animal experiments. Pott (1965) found a statistically significant elevation in the incidence of prostate cancer among a cohort of cadmium workers. Other epidemiology studies also report an elevated incidence of prostate cancer; however, the increases observed in these other studies were not statistically significant (Meridian Research, Inc. 1989).

One study (Thun et al. 1985) contains sufficiently quantitative estimates of cadmium exposure to allow evaluation of dose-response relationships between cadmium exposure and lung cancer. A statistically significant excess of lung cancer attributed to cadmium exposure was found in this study, even after accounting for confounding variables such as coexposure to arsenic and smoking habits (Meridian Research, Inc. 1989).

Evidence for quantifying a link between lung cancer and cadmium exposure comes from a single study (Takenaka et al. 1983). In this study, dose-response relationships developed from animal data were extrapolated to humans using a variety of models. OSHA chose the multistage risk model for estimating the risk of cancer for humans using these animal data. Animal injection studies also suggest an association between cadmium exposure and cancer, particularly observations of an increased incidence of tumors at sites remote from the point of injection. The International Agency for Research on Cancer (IARC) (Supplement 7, 1987) indicates that this, and related, evidence is sufficient to classify cadmium as an animal carcinogen. However, the results of these injection studies cannot be used to quantify risks attendant to human occupational exposures due to differences in routes of exposure (Meridian Research, Inc. 1989).

Based on the above-cited studies, the U.S. Environmental Protection Agency (EPA) classifies cadmium as "B1," a probable human carcinogen (USEPA 1985). IARC in 1987 recommended that cadmium be listed as a probable human carcinogen.

*Kidney Dysfunction.* The most prevalent nonmalignant effect observed among workers chronically exposed to cadmium is kidney dysfunction. Initially, such dysfunction is manifested by proteinuria (Meridian Research, Inc. 1989; Roth Associates, Inc. 1989). Proteinuria associated with cadmium exposure is most commonly characterized by excretion of low-molecular weight proteins (15,000-40,000 MW), accompanied by loss of electrolytes, uric acid, calcium, amino acids, and phosphate. Proteins commonly excreted include β-2-microglobulin (B2M), retinol-binding protein (RBP), immunoglobulin light chains, and lysozyme. Excretion of low molecular weight proteins is characteristic of damage to the proximal tubules of the kidney (Iwao et al. 1980).

Exposure to cadmium also may lead to urinary excretion of high-molecular weight proteins such as albumin, immunoglobulin G, and glycoproteins (Meridian Research, Inc. 1989; Roth Associates, Inc. 1989). Excretion of high-molecular weight proteins is indicative of damage to the glomeruli of the kidney. Bernard et al. (1979) suggest that cadmium-associated damage to the glomeruli and damage to the proximal tubules of the kidney develop independently of each other, but may occur in the same individual.

Several studies indicate that the onset of low-molecular weight proteinuria is a sign of irreversible kidney damage (Friberg et al. 1974; Roels et al. 1982; Piscator 1984; Elinder et al. 1985; Smith et al. 1986). For many workers, once sufficiently elevated levels of B2M are observed in association with cadmium exposure, such levels do not appear to return to normal even when cadmium exposure is eliminated by removal of the worker from the cadmium-contaminated work environment (Friberg, exhibit 29, 1990).

Some studies indicate that cadmium-induced proteinuria may be progressive; levels of B2MU increase even after cadmium exposure has ceased (Elinder et al. 1985). Other researchers have reached similar conclusions (Frieburg testimony, OSHA docket exhibit 29, Elinder testimony, OSHA docket exhibit 55, and OSHA docket exhibits 8-86B). Such observations are not universal, however (Smith et al. 1986; Tsuchiya 1976). Studies in which proteinuria has not been observed, however, may have initiated the reassessment too early (Meridian Research, Inc. 1989; Roth Associates, Inc. 1989; Roels 1989).

A quantitative assessment of the risks of developing kidney dysfunction as a result of cadmium exposure was performed using the data from Ellis et al. (1984) and Falck et al. (1983). Meridian Research, Inc. (1989) and Roth Associates, Inc. (1989) employed several mathematical models to evaluate the data from the 2 studies, and the results indicate that cumulative cadmium exposure levels between 5 and 100 µg-years/m$^3$ correspond with a one-in-a-thousand probability of developing kidney dysfunction.

When cadmium exposure continues past the onset of early kidney damage (manifested as proteinuria), chronic nephrotoxicity may occur (Meridian Research, Inc. 1989; Roth Associates, Inc. 1989). Uremia, which is the loss of the glomerulus' ability to adequately filter blood, may result. This condition leads to severe disturbance of electrolyte concentrations, which may result in various clinical complications including atherosclerosis, hypertension, pericarditis, anemia, hemorrhagic tendencies, deficient cellular immunity, bone changes, and other problems. Progression of the disease may require dialysis or a kidney transplant.

Studies in which animals are chronically exposed to cadmium confirm the renal effects observed in humans (Friberg et al. 1986). Animal studies also confirm cadmium-related problems with calcium metabolism and associated skeletal effects, which also have been observed among humans. Other effects commonly reported in chronic animal studies include anemia, changes in liver morphology, immunosuppression and hypertension. Some of these effects may be associated with cofactors; hypertension, for example, appears to be associated with diet, as well as with cadmium exposure. Animals injected with cadmium also have shown testicular necrosis.

### 4.2 Objectives for Medical Monitoring

In keeping with the observation that renal disease tends to be the earliest clinical manifestation of cadmium toxicity, the final cadmium standard mandates that eligible workers must be medically monitored to prevent this condition (as well as cadmimum-induced cancer). The objectives of medical-monitoring, therefore, are to: Identify workers at significant risk of adverse health effects from excess, chronic exposure to cadmium; prevent future cases of cadmium-induced disease; detect and minimize existing cadmium-induced disease; and, identify workers most in need of medical intervention.

The overall goal of the medical monitoring program is to protect workers who may be exposed continuously to cadmium over a 45-year occupational lifespan. Consistent with this goal, the medical monitoring program should assure that:

1. Current exposure levels remain sufficiently low to prevent the accumulation of cadmium body burdens sufficient to cause disease in the future by monitoring CDB as an indicator of recent cadmium exposure;

2. Cumulative body burdens, especially among workers with undefined historical exposures, remain below levels potentially capable of leading to damage and disease by assessing CDU as an indicator of cumulative exposure to cadmium; and,

3. Health effects are not occurring among exposed workers by determining B2MU as an early indicator of the onset of cadmium-induced kidney disease.

### 4.3 Indicators of Cadmium Exposure and Disease

Cadmium is present in whole blood bound to albumin, in erythrocytes, and as a metallothionein-cadmium complex. The metallothionein-cadmium complex that represents the primary transport mechanism for cadmium delivery to the kidney. CDB concentrations in the general, nonexposed population average 1 µg Cd/l whole blood, with smokers exhibiting higher levels (see Section 5.1.6). Data presented in Section 5.1.6 shows that 95% of the general population not occupationally exposed to cadmium have CDB levels less than 5 µg Cd/l.

If total body burdens of cadmium remain low, CDB concentrations indicate recent exposure (i.e., daily intake). This conclusion is based on data showing that cigarette smokers exhibit CDB concentrations of 2-7 µg/l depending on the number of cigarettes smoked per day (Nordberg and Nordberg 1988), while CDB levels for those who quit smoking return to general population values (approximately 1 µg/l) within several weeks (Lauwerys et al. 1976). Based on these observations, Lauwerys et al. (1976) concluded that CDB has a biological half-life of a few weeks to less than 3 months. As indicated in Section 3.1.6, the upper 95th percentile for CDB levels observed among those who are not occupationally exposed to cadmium is 5 µg/l, which suggests that the absolute upper limit to the range reported for smokers by Nordberg and Nordberg may have been affected by an extreme value (i.e., beyond $2\sigma$ above the mean).

Among occupationally-exposed workers, the occupational history of exposure to cadmium must be evaluated to interpret CDB levels. New workers, or workers with low exposures to cadmium, exhibit CDB levels that are representative of recent exposures, similar to the general population. However, for workers with a history of chronic exposure to cadmium, who have accumulated significant stores of cadmium in the kidneys/liver, part of the CDB concentrations appear to indicate body burden. If such workers are removed from cadmium exposure, their CDB levels remain elevated, possibly for years, reflecting prior long-term accumulation of cadmium in body tissues. This condition tends to occur, however, only beyond some threshold exposure value, and possibly indicates the capacity of body tissues to accumulate cadmium which cannot be excreted readily (Friberg and Elinder 1988; Nordberg and Nordberg 1988).

CDU is widely used as an indicator of cadmium body burdens (Nordberg and Nordberg 1988). CDU is the major route of elimination and, when CDU is measured, it is commonly expressed either as µg Cd/l urine (unadjusted), µg Cd/l urine (adjusted for specific gravity), or µg Cd/g CRTU (see Section 5.2.1). The metabolic model for CDU is less complicated than CDB, since CDU is dependent in large part on the body (i.e., kidney) burden of cadmium. However, a small proportion of CDU still be attributed to recent cadmium exposure, particularly if exposure to high airborne concentrations of cadmium occurred. Note that CDU is subject to larger interindividual and day-to-day variations than CDB, so repeated measurements are recommended for CDU evaluations.

CDU is bound principally to metallothionein, regardless of whether the cadmium originates from metallothionein in plasma or from the cadmium pool accumulated in the renal tubules. Therefore, measurement of metallothionein in urine may provide information similar to CDU, while avoiding the contamination problems that may occur during collection and handling urine for cadmium analysis (Nordberg and Nordberg 1988). However, a commercial method for the determination of metallothionein at the sensitivity levels required under the final cadmium rule is not currently available; therefore, analysis of CDU is recommended.

Among the general population not occupationally exposed to cadmium, CDU levels average less than 1 µg/l (see Section 5.2.7). Normalized for creatinine (CRTU), the average CDU concentration of the general population is less than 1 µg/g CRTU. As cadmium accumulates over the lifespan, CDU increases with age. Also, cigarette smokers may eventually accumulate twice the cadmium body burden of nonsmokers, CDU is slightly higher in smokers than in nonsmokers, even several years after smoking cessation (Nordberg and Nordberg 1988). Despite variations due to age and smoking habits, 95% of those not occupationally exposed to cadmium exhibit levels of CDU less than 3 µg/g CRTU (based on the data presented in Section 5.2.7).

About 0.02% of the cadmium body burden is excreted daily in urine. When the critical cadmium concentration (about 200 ppm) in the kidney is reached, or if there is sufficient cadmium-induced kidney dysfunction, dramatic increases in CDU are observed (Nordberg and Nordberg 1988). Above 200 ppm, therefore, CDU concentrations cease to be an indicator of cadmium body burden, and are instead an index of kidney failure.

Proteinuria is an index of kidney dysfunction, and is defined by OSHA to be a material impairment. Several small proteins may be monitored as markers for proteinuria. Below levels indicative of proteinuria, these small proteins may be early indicators of increased risk of cadmium-induced renal tubular disease. Analytes useful for monitoring cadmium-induced renal tubular damage include:

1. β-2-Microglobulin (B2M), currently the most widely used assay for detecting kidney dysfunction, is the best characterized analyte available (Iwao et al. 1980; Chia et al. 1989);

2. Retinol Binding Protein (RBP) is more stable than B2M in acidic urine (i.e., B2M breakdown occurs if urinary pH is less than 5.5; such breakdown may result in false [i.e., low] B2M values [Bernard and Lauwerys, 1990]);

3. N-Acetyl-B-Glucosaminidase (NAG) is the analyte of an assay that is simple, inexpensive, reliable, and correlates with cadmium levels under 10 µg/g CRTU, but the assay is less sensitive than RBP or B2M (Kawada et al. 1989);

4. Metallothionein (MT) correlates with cadmium and B2M levels, and may be a better predictor of cadmium exposure than CDU and B2M (Kawada et al. 1989);

5. Tamm-Horsfall Glycoprotein (THG) increases slightly with elevated cadmium levels, but this elevation is small compared to increases in urinary albumin, RBP, or B2M (Bernard and Lauwerys 1990);

6. Albumin (ALB), determined by the biuret method, is not sufficiently sensitive to serve as an early indicator of the onset of renal disease (Piscator 1962);

7. Albumin (ALB), determined by the Amido Black method, is sensitive and reproducible, but involves a time-consuming procedure (Piscator 1962);

8. Glycosaminoglycan (GAG) increases among cadmium workers, but the significance of this effect is unknown because no relationship has been found between elevated GAG and other indices of tubular damage (Bernard and Lauwerys 1990);

9. Trehalase seems to increase earlier than B2M during cadmium exposure, but the procedure for analysis is complicated and unreliable (Iwata et al. 1988); and,

10. Kallikrein is observed at lower concentrations among cadmium-exposed workers than among normal controls (Roels et al. 1990).

Of the above analytes, B2M appears to be the most widely used and best characterized analyte to evaluate the presence/absence, as well as the extent of, cadmium-induced renal tubular damage (Kawada, Koyama, and Suzuki 1989; Shaikh and Smith 1984; Nogawa 1984). However, it is important that samples be collected and handled so as to minimize B2M degradation under acidic urine conditions.

The threshold value of B2MU commonly used to indicate the presence of kidney damage 300 µg/g CRTU (Kjellstrom et al. 1977a; Buchet et al. 1980; and Kowal and Zirkes 1983). This value represents the upper 95th or 97.5th percentile level of urinary excretion observed among those without tubular dysfunction (Elinder, exbt L-140-45, OSHA docket H057A). In agreement with these conclusions, the data presented in Section 5.3.7 of this protocol generally indicate that the level of 300 µg/g CRTU appears to define the boundary for kidney dysfunction. It is not clear, however, that this level represents the upper 95th percentile of values observed among those who fail to demonstrate proteinuria effects.

Although elevated B2MU levels appear to be a fairly specific indicator of disease associated with cadmium exposure, other conditions that may lead to elevated B2MU levels include high fevers from influenza, extensive physical exercise, renal disease unrelated to cadmium exposure, lymphomas, and AIDS (Iwao et al. 1980; Schardun and van Epps 1987). Elevated B2M levels observed in association with high fevers from influenza or from extensive physical exercise are transient, and will return to normal levels once the fever has abated or metabolic rates return to baseline values following exercise. The other conditions linked to elevated B2M levels can be diagnosed as part of a properly-designed medical examination. Consequently, monitoring B2M, when accompanied by regular medical examinations and CDB and CDU determinations (as indicators of present and past cadmium exposure), may serve as a specific, early indicator of cadmium-induced kidney damage.

### 4.4 Criteria for Medical Monitoring of Cadmium Workers

Medical monitoring mandated by the final cadmium rule includes a combination of regular medical examinations and periodic monitoring of 3 analytes: CDB, CDU and B2MU. As indicated above, CDB is monitored as an indicator of current cadmium exposure, while CDU serves as an indicator of the cadmium body burden; B2MU is assessed as an early marker of irreversible kidney damage and disease.

The final cadmium rule defines a series of action levels that have been developed for each of the 3 analytes to be monitored. These action levels serve to guide the responsible physician through a decision-making process. For each action level that is exceeded, a specific response is mandated. The sequence of action levels, and the attendant actions, are described in detail in the final cadmium rule.

Other criteria used in the medical decision-making process relate to tests performed during the medical examination (including a determination of the ability of a worker to wear a respirator). These criteria, however, are not affected by the results of the analyte determinations addressed in the above paragraphs and, consequently, will not be considered further in these guidelines.

### 4.5 Defining to Quality and Proficiency of the Analyte Determinations

As noted above in Sections 2 and 3, the quality of a measurement should be defined along with its value to properly interpret the results. Generally, it is necessary to know the accuracy and the precision of a measurement before it can be properly evaluated. The precision of the data from a specific laboratory indicates the extent to which the repeated measurements of the same sample vary within that laboratory. The accuracy of the data provides an indication of the extent to which these results deviate from average results determined from many laboratories performing the same measurement (i.e., in the absence of an independent determination of the true value of a measurement). Note that terms are defined operationally relative to the manner in which they will be used in this protocol. Formal definitions for the terms in italics used in this section can be found in the list of definitions (Section 2).

Another data quality criterion required to properly evaluate measurement results is the limit of detection of that measurement. For measurements to be useful, the range of the measurement which is of interest for biological monitoring purposes must lie entirely above the limit of detection defined for that measurement.

The overall quality of a laboratory's results is termed the performance of that laboratory. The degree to which a laboratory satisfies a minimum performance level is referred to as the proficiency of the laboratory. A successful medical monitoring program, therefore, should include procedures developed for monitoring and recording laboratory performance; these procedures can be used to identify the most proficient laboratories.

### 5.0 Overview of Medical Monitoring Tests for CDB, CDU, B2MU and CRTU

To evaluate whether available methods for assessing CDB, CDU, B2MU and CRTU are adequate for determining the parameters defined by the proposed action levels, it is necessary to review procedures available for sample collection, preparation and analysis. A variety of techniques for these purposes have been used historically for the determination of cadmium in biological matrices (including CDB and CDU), and for the determination of specific proteins in biological matrices (including B2MU). However, only the most recent techniques are capable of satisfying the required accuracy, precision and sensitivity (i.e., limit of detection) for monitoring at the levels mandated in the final cadmium rule, while still facilitating automated analysis and rapid processing.

### 5.1 Measuring Cadmium in Blood (CDB)

Analysis of biological samples for cadmium requires strict analytical discipline regarding collection and handling of samples. In addition to occupational settings, where cadmium contamination would be apparent, cadmium is a ubiquitous environmental contaminant, and much care should be exercised to ensure that samples are not contaminated during collection, preparation or analysis. Many common chemical reagents are contaminated with cadmium at concentrations that will interfere with cadmium analysis; because of the widespread use of cadmium compounds as colored pigments in plastics and coatings, the analyst should continually monitor each manufacturer's chemical reagents and collection containers to prevent contamination of samples.

Guarding against cadmium contamination of biological samples is particularly important when analyzing blood samples because cadmium concentrations in blood samples from nonexposed populations are generally less than 2 µg/l (2 ng/ml), while occupationally-exposed workers can be at medical risk to cadmium toxicity if blood concentrations exceed 5 µg/l (ACGIH 1991 and 1992). This narrow margin between exposed and unexposed samples requires that exceptional care be used in performing analytic determinations for biological monitoring for occupational cadmium exposure.

Methods for quantifying cadmium in blood have improved over the last 40 years primarily because of improvements in analytical instrumentation. Also, due to improvements in analytical techniques, there is less need to perform extensive multi-step sample preparations prior to analysis. Complex sample preparation was previously required to enhance method sensitivity (for cadmium), and to reduce interference by other metals or components of the sample.

### 5.1.1 Analytical Techniques Used to Monitor Cadmium in Biological Matrices

### TABLE 3—COMPARISON OF ANALYTICAL PROCEDURES/INSTRUMENTATION FOR DETERMINATION OF CADMIUM IN BIOLOGICAL SAMPLES

| Analytical procedure | Limit of detection [ng/(g or ml)] | Specified biological matrix | Reference | Comments |
|---|---|---|---|---|
| Flame Atomic Absorption Spectroscopy (FAAS). | ≥ 1.0 | Any matrix | Perkin-Elmer (1982). | Not sensitive enough for biomonitoring without extensive sample digestion, metal chelation and organic solvent extraction. |
| Graphite Furnace Atomic Absorption Spectroscopy (GFAAS). | 0.04<br>≥ 0.20 | Urine<br>Blood | Pruszkowska et al. (1983).<br>Stoeppler and Brandt (1980). | Methods of choice for routine cadmium analysis. |
| Inductively-Coupled Argon-Plasma Atomic Emission Spectroscopy (ICAP AES). | 2.0 | Any matrix | NIOSH (1984A) | Requires extensive sample preparation and concentration of metal with chelating resin. Advantage is simultaneous analyses for as many as 10 metals from 1 sample. |
| Neutron Activation Gamma Spectroscopy (NA). | 1.5 | In vivo (liver) | Ellis et al. (1983) | Only available *in vivo* method for direct determination of cadmium body tissue burdens; expensive; absolute determination of cadmium in reference materials. |
| Isotope Dilution Mass Spectroscopy (IDMS). | < 1.0 | Any matrix | Michiels and DeBievre (1986). | Suitable for absolute determination of cadmium in reference materials; expensive. |
| Diffential Pulse Anodic Stripping Voltammetry (DPASV). | < 1.0 | Any matrix | Stoeppler and Brandt (1980). | Suitable for absolute determination of cadmium in reference materials; efficient method to check accuracy of analytical method. |

A number of analytical techniques have been used for determining cadmium concentrations in biological materials. A summary of the characteristics of the most widely employed techniques is presented in Table 3. The technique most suitable for medical monitoring for cadmium is atomic absorption spectroscopy (AAS).

To obtain a measurement using AAS, a light source (i.e., hollow cathode or lectrode-free discharge lamp) containing the element of interest as the cathode, is energized and the lamp emits a spectrum that is unique for that element. This light source is focused through a sample cell, and a selected wavelength is monitored by a monochrometer and photodetector cell. Any ground state atoms in the sample that match those of the lamp element and are in the path of the emitted light may absorb some of the light and decrease the amount of light that reaches the photodetector cell. The amount of light absorbed at each characteristic wavelength is proportional to the number of ground state atoms of the corresponding element that are in the pathway of the light between the source and detector.

To determine the amount of a specific metallic element in a sample using AAS, the sample is dissolved in a solvent and aspirated into a high-temperature flame as an aerosol. At high temperatures, the solvent is rapidly evaporated or decomposed and the solute is initially solidified; the majority of the sample elements then are transformed into an atomic vapor. Next, a light beam is focused above the flame and the amount of metal in the sample can be determined by measuring the degree of absorbance of the atoms of the target element released by the flame at a characteristic wavelength.

A more refined atomic absorption technique, flameless AAS, substitutes an electrothermal, graphite furnace for the flame. An aliquot (10-100 µl) of the sample is pipetted into the cold furnace, which is then heated rapidly to generate an atomic vapor of the element.

AAS is a sensitive and specific method for the elemental analysis of metals; its main drawback is nonspecific background absorbtion and scattering of the light beam by particles of the sample as it decomposes at high temperatures; nonspecific absorbance reduces the sensitivity of the analytical method. The problem of nonspecific absorbance and scattering can be reduced by extensive sample pretreatment, such as ashing and/or acid digestion of the sample to reduce its organic content.

Current AAS instruments employ background correction devices to adjust electronically for background absorbtion and scattering. A common method to correct for background effects is to use a deuterium arc lamp as a second light source. A continuum light source, such as the deuterium lamp, emits a broad spectrum of wavelengths instead of specific wavelengths characteristic of a particular element, as with the hollow cathode tube. With this system, light from the primary source and the continuum source are passed alternately through the sample cell. The target element effectively absorbs light only from the primary source (which is much brighter than the continuum source at the characteristic wavelengths), while the background matrix absorbs and scatters light from both sources equally. Therefore, when the ratio of the two beams is measured electronically, the effect of nonspecific background absorption and scattering is eliminated. A less common, but more sophisticated, backgrond correction system is based on the Zeeman effect, which uses a magnetically-activated light polarizer to compensate electronically for nonspecific absorbtion and scattering.

Atomic emission spectroscopy with inductively-coupled argon plasma (AES-ICAP) is widely used to analyze for metals. With this instrument, the sample is aspirated into an extremely hot argon plasma flame, which excites the metal atoms; emission spectra specific for the sample element then are generated. The quanta of emitted light passing through a monochrometer are amplified by photomultiplier tubes and measured by a photodetector to determine the amount of metal in the sample. An advantage of AES-ICAP over AAS is that multi-elemental analyses of a sample can be performed by simultaneously measuring specific elemental emission energies. However, AES-ICAP lacks the sensitivity of AAS, exhibiting a limit of detection which is higher than the limit of detection for graphite-furnace AAS (Table 3).

Neutron activation (NA) analysis and isotope dilution mass spectrometry (IDMS) are 2 additional, but highly specialized, methods that have been used for cadmium determinations. These methods are expensive because they require elaborate and sophisticated instrumentation.

NA analysis has the distinct advantage over other analytical methods of being able to determine cadmium body burdens in specific organs (e.g., liver, kidney) in vivo (Ellis et al. 1983). Neutron bombardment of the target transforms cadmium-113 to cadmium-114, which promptly decays ($<10^{-14}$ sec) to its ground state, emitting gamma rays that are measured using large gamma detectors; appropriate shielding and instrumentation are required when using this method.

IDMS analysis, a definitive but laborious method, is based on the change in the ratio of 2 isotopes of cadmium (cadmium 111 and 112) that occurs when a known amount of the element (with an artificially altered ratio of the same isotopes [i.e., a cadmium 111 "spike"] is added to a weighed aliquot of the sample (Michiels and De Bievre 1986).

### 5.1.2 Methods Developed for CDB Determinations

A variety of methods have been used for preparing and analyzing CDB samples; most of these methods rely on one of the analytical techniques described above. Among the earliest reports, Princi (1947) and Smith et al. (1955) employed a colorimetric procedure to analyze for CDB and CDU. Samples were dried and digested through several cycles with concentrated mineral acids ($HNO_3$ and $H_2SO_4$) and hydrogen peroxide ($H_2O_2$). The digest was neutralized, and the cadmium was complexed with diphenylthiocarbazone and extracted with chloroform. The dithizone-cadmium complex then was quantified using a spectrometer.

Colorimetric procedures for cadmium analyses were replaced by methods based on atomic absorption spectroscopy (AAS) in the early 1960s, but many of the complex sample preparation procedures were retained. Kjellstrom (1979) reports that in Japanese, American and Swedish laboratories during the early 1970s, blood samples were wet ashed with mineral acids or ashed at high temperature and wetted with nitric acid. The cadmium in the digest was complexed with metal chelators including diethyl dithiocarbamate (DDTC), ammonium pyrrolidine dithiocarbamate (APDC) or diphenylthiocarbazone (dithizone) in ammonia-citrate buffer and extracted with methyl isobutyl ketone (MIBK). The resulting solution then was analyzed by flame AAS or graphite-furnace AAS for cadmium determinations using deuterium-lamp background correction.

In the late 1970s, researchers began developing simpler preparation procedures. Roels et al. (1978) and Roberts and Clark (1986) developed simplified digestion procedures. Using the Roberts and Clark method, a 0.5 ml aliquot of blood is collected and transferred to a digestion tube containing 1 ml concentrated $HNO_3$. The blood is then digested at 110 °C for 4 hours. The sample is reduced in volume by continued heating, and 0.5 ml 30% $H_2O_2$ is added as the sample dries. The residue is dissolved in 5 ml dilute (1%) $HNO_3$, and 20 µl of sample is then analyzed by graphite-furnace AAS with deuterium-background correction.

The current trend in the preparation of blood samples is to dilute the sample and add matrix modifiers to reduce background interference, rather than digesting the sample to reduce organic content. The method of Stoeppler and Brandt (1980), and the abbreviated procedure published in the American Public Health Association's (APHA) *Methods for Biological Monitoring* (1988), are straightforward and are nearly identical. For the APHA method, a small aliquot (50-300 µl) of whole blood that has been stabilized with ethylenediaminetetraacetate (EDTA) is added to 1.0 ml 1 M $HNO_3$, vigorously shaken and centrifuged. Aliquots (10-25 µl) of the supernatant then are then analyzed by graphite-furnace AAS with appropriate background correction.

Using the method of Stoeppler and Brandt (1980), aliquots (50-200 µl) of whole blood that have been stabilized with EDTA are pipetted into clean polystyrene tubes and mixed with 150-600 µl of 1 M $HNO_3$. After vigorous shaking, the solution is centrifuged and a 10-25 µl aliquot of the supernatant then is analyzed by graphite-furnace AAS with appropriate background correction.

Claeys-Thoreau (1982) and DeBenzo et al. (1990) diluted blood samples at a ratio of 1:10 with a matrix modifier (0.2% Triton X-100, a wetting agent) for direct determinations of CDB. DeBenzo et al. also demonstrated that aqueous standards of cadmium, instead of spiked, whole-blood samples, could be used to establish calibration curves if standards and samples are treated with additional small volumes of matrix modifiers (i.e., 1% $HNO_3$, 0.2% ammonium hydrogenphosphate and 1 mg/ml magnesium salts).

These direct dilution procedures for CDB analysis are simple and rapid. Laboratories can process more than 100 samples a day using a dedicated graphite-furnace AAS, an auto-sampler, and either a Zeeman- or a deuterium-background correction system. Several authors emphasize using optimum settings for graphite-furnace temperatures during the drying, charring, and atomization processes associated with the flameless AAS method, and the need to run frequent QC samples when performing automated analysis.

### 5.1.3 Sample Collection and Handling

Sample collection procedures are addressed primarily to identify ways to minimize the degree of variability that may be introduced by sample collection during medical monitoring. It is unclear at this point the extent to which collection procedures contribute to variability among CDB samples. Sources of variation that may result from sampling procedures include time-of-day effects and introduction of external contamination during the collection process. To minimize these sources, strict adherence to a sample collection protocol is recommended. Such a protocol must include provisions for thorough cleaning of the site from which blood will be extracted; also, every effort should be made to collect samples near the same time of day. It is also important to recognize that under the recent OSHA blood-borne pathogens standard (29 CFR 1910.1030), blood samples and certain body fluids must be handled and treated as if they are infectious.

### 5.1.4 Best Achievable Performance

The best achievable performance using a particular method for CDB determinations is assumed to be equivalent to the performance reported by research laboratories in which the method was developed.

For their method, Roberts and Clark (1986) demonstrated a limit of detection of 0.4 µg Cd/l in whole blood, with a linear response curve from 0.4 to 16.0 µg Cd/l. They report a coefficient of variation (CV) of 6.7% at 8.0 µg/l.

The APHA (1988) reports a range of 1.0-25 µg/l, with a CV of 7.3% (concentration not stated). Insufficient documentation was available to critique this method.

Stoeppler and Brandt (1980) achieved a detection limit of 0.2 µg Cd/l whole blood, with a linear range of 0.4-12.0 µg Cd/l, and a CV of 15-30%, for samples at <1.0 µg/l. Improved precision (CV of 3.8%) was reported for CDB concentrations at 9.3 µg/l.

### 5.1.5 General Method Performance

For any particular method, the performance expected from commercial laboratories may be somewhat lower than that reported by the research laboratory in which the method was developed. With participation in appropriate proficiency programs and use of a proper in-house QA/QC program incorporating provisions for regular corrective actions, the performance of commercial laboratories is expected to approach that reported by research laboratories. Also, the results reported for existing proficiency programs serve as a gauge of the likely level of performance that currently can be expected from commercial laboratories offering these analyses.

Weber (1988) reports on the results of the proficiency program run by the Centre de Toxicologie du Quebec (CTQ). As indicated previously, participants in that program receive 18 blood samples per year having cadmium concentrations ranging from 0.2-20 µg/l. Currently, 76 laboratories are participating in this program. The program is established for several analytes in addition to cadmium, and not all of these laboratories participate in the cadmium proficiency-testing program.

Under the CTQ program, cadmium results from individual laboratories are compared against the consensus mean derived for each sample. Results indicate that after receiving 60 samples (i.e., after participation for approximately three years), 60% of the laboratories in the program are able to report results that fall within ±1 µg/l or 15% of the mean, whichever is greater. (For this procedure, the 15% criterion was applied to concentrations exceeding 7 µg/l.) On any single sample of the last 20 samples, the percentage of laboratories falling within the specified range is between 55 and 80%.

The CTQ also evaluates the performance of participating laboratories against a less severe standard: ±2 µg/l or 15% of the mean, whichever is greater (Weber 1988); 90% of participating laboratories are able to satisfy this standard after approximately 3 years in the program. (The 15% criterion is used for concentrations in excess of 13 µg/l.) On any single sample of the last 15 samples, the percentage of laboratories falling within the specified range is between 80 and 95% (except for a single test for which only 60% of the laboratories achieved the desired performance).

Based on the data presented in Weber (1988), the CV for analysis of CDB is nearly constant at 20% for cadmium concentrations exceeding 5 µg/l, and increases for cadmium concentrations below 5 µg/l. At 2 µg/l, the reported CV rises to approximately 40%. At 1 µg/l, the reported CV is approximately 60%.

Participating laboratories also tend to overestimate concentrations for samples exhibiting concentrations less than 2 µg/l (see Figure 11 of Weber 1988). This problem is due in part to the proficiency evaluation criterion that allows reporting a minimum ±2.0 µg/l for evaluated CDB samples. There is currently little economic or regulatory incentive for laboratories participating in the CTQ program to achieve greater accuracy for CDB samples containing cadmium at concentrations less than 2.0 µg/l, even if the laboratory has the experience and competency to distinguish among lower concentrations in the samples obtained from the CTQ.

The collective experience of international agencies and investigators demonstrate the need for a vigorous QC program to ensure that CDB values reported by participating laboratories are indeed reasonably accurate. As Friberg (1988) stated:

"Information about the quality of published data has often been lacking. This is of concern as assessment of metals in trace concentrations in biological media are fraught with difficulties from the collection, handling, and storage of samples to the chemical analyses. This has been proven over and over again from the results of interlaboratory testing and quality control exercises. Large variations in results were reported even from 'experienced' laboratories."

The UNEP/WHO global study of cadmium biological monitoring set a limit for CDB accuracy using the maximum allowable deviation method at $Y = X \pm (0.1 X + 1)$ for a targeted concentration of 10 µg Cd/l (Friberg and Vahter 1983). The performance of participating laboratories over a concentration range of 1.5-12 µg/l was reported by Lind et al. (1987). Of the 3 QC runs conducted during 1982 and 1983, 1 or 2 of the 6 laboratories failed each run. For the years 1983 and 1985, between zero and 2 laboratories failed each of the consecutive QC runs.

In another study (Vahter and Friberg 1988), QC samples consisting of both external (unknown) and internal (stated) concentrations were distributed to laboratories participating in the epidemiology research. In this study, the maximum acceptable deviation between the regression analysis of reported results and reference values was set at $Y = X \pm (0.05X + 0.2)$ for a concentration range of 0.3-5.0 µg Cd/l. It is reported that only 2 of 5 laboratories had acceptable data after the first QC set, and only 1 of 5 laboratories had acceptable data after the second QC set. By the fourth QC set, however, all 5 laboratories were judged proficient.

The need for high quality CDB monitoring is apparent when the toxicological and biological characteristics of this metal are considered; an increase in CDB from 2 to 4 µg/l could cause a doubling of the cadmium accumulation in the kidney, a critical target tissue for selective cadmium accumulation (Nordberg and Nordberg 1988).

Historically, the CDC's internal QC program for CDB cadmium monitoring program has found achievable accuracy to be ±10% of the true value at CDB concentrations ≥5.0 µg/l (Paschal 1990). Data on the performance of laboratories participating in this program currently are not available.

### 5.1.6 Observed CDB Concentrations

As stated in Section 4.3, CDB concentrations are representative of ongoing levels of exposure to cadmium. Among those who have been exposed chronically to cadmium for extended periods, however, CDB may contain a component attributable to the general cadmium body burden.

#### 5.1.6.1 CDB Concentrations Among Unexposed Samples

Numerous studies have been conducted examining CDB concentrations in the general population, and in control groups used for comparison with cadmium-exposed workers. A number of reports have been published that present erroneously high values of CDB (Nordberg and Nordberg 1988). This problem was due to contamination of samples during sampling and analysis, and to errors in analysis. Early AAS methods were not sufficiently sensitive to accurately estimate CDB concentrations.

Table 4 presents results of recent studies reporting CDB levels for the general U.S. population not exposed occupationally to cadmium. Other surveys of tissue cadmium using U.S. samples and conducted as part of a cooperative effort among Japan, Sweden and the U.S., did not collect CDB data because standard analytical methodologies were unavailable, and because of analytic problems (Kjellstrom 1979; SWRI 1978).

### TABLE 4—BLOOD CADMIUM CONCENTRATIONS OF U.S. POPULATION NOT OCCUPATIONALLY EXPOSED TO CADMIUM[a]

| Study No. | No. in study (n) | Sex | Age | Smoking habits[b] | Arithmetic mean (±S.D.)[c] | Absolute range or (95% CI)[d] | Geometric mean (±GSD)[e] | Lower 95th percentile of distribution[f] | Upper 95th percentile of distribution[f] | Reference |
|---|---|---|---|---|---|---|---|---|---|---|
| 1 | 80 | M | 4 to 69 | NS, S | 1.13 | 0.35-3.3 | 0.98±1.71 | 0.4 | 2.4 | Kowal et al. (1979). |
|  | 88 | F | 4 to 69 | NS, S | 1.03 | 0.21-3.3 | 0.91±1.63 | 0.4 | 2.0 |  |
|  | 115 | M/F | 4 to 69 | NS | 0.95 | 0.21-3.3 | 0.85±1.59 | 0.4 | 1.8 |  |
|  | 31 | M/F | 4 to 69 | S | 1.54 | 0.4-3.3 | 1.37±1.65 | 0.6 | 3.2 |  |
| 2 | 10 | M | Adults | (?) | 2.0±2.1 | (0.5-5.0) |  | [g](0) | [g](5.8) | Ellis et al. (1983). |
| 3 | 24 | M | Adults | NS |  |  | 0.6±1/87 | 0.2 | 1.8 | Frieberg and Vahter (1983). |
|  | 20 | M | Adults | S |  |  | 1.2±2.13 | 0.3 | 4.4 |  |
|  | 64 | F | Adults | NS |  |  | 0.5±1.85 | 0.2 | 1.4 |  |
|  | 39 | F | Adults | S |  |  | 0.8±2.22 | 0.2 | 3.1 |  |
| 4 | 32 | M | Adults | S, NS |  |  | 1.2±2.0 | 0.4 | 3.9 | Thun et al. (1989). |
| 5 | 35 | M | Adults | (?) | 2.1±2.1 | (0.5-7.3) |  | [g](0) | [g](5.6) | Mueller et al. (1989). |

[a] Concentrations reported in µg Cd/l blood unless otherwise stated.
[b] NS–never smoked; S–current cigarette smoker.
[c] S.D.–Arithmetic Standard Deviation.
[d] C.I.–Confidence interval.
[e] GSD–Geometric Standard Deviation.
[f] Based on an assumed lognormal distribution.
[g] Based on an assumed normal distribution.

Arithmetic and/or geometric means and standard deviations are provided in Table 4 for measurements among the populations defined in each study listed. The range of reported measurements and/or the 95% upper and lower confidence intervals for the means are presented when this information was reported in a study. For studies reporting either an arithmetic or geometric standard deviation along with a mean, the lower and upper 95th percentile for the distribution also were derived and reported in the table.

The data provided in table 4 from Kowal et al. (1979) are from studies conducted between 1974 and 1976 evaluating CDB levels for the general population in Chicago, and are considered to be representative of the U.S. population. These studies indicate that the average CDB concentration among those not occupationally exposed to cadmium is approximately 1 µg/l.

In several other studies presented in Table 4, measurements are reported separately for males and females, and for smokers and nonsmokers. The data in this table indicate that similar CDB levels are observed among males and females in the general population, but that smokers tend to exhibit higher CDB levels than nonsmokers. Based on the Kowal et al. (1979) study, smokers not occupationally exposed to cadmium exhibit an average CDB level of 1.4 µg/l.

In general, nonsmokers tend to exhibit levels ranging to 2 µg/l, while levels observed among smokers range to 5 µg/l. Based on the data presented in Table 4, 95% of those not occupationally exposed to cadmium exhibit CDB levels less than 5 µg/l.

#### 5.1.6.2 CDB Concentrations Among Exposed Workers

Table 5 is a summary of results from studies reporting CDB levels among workers exposed to cadmium in the work place. As in Table 4, arithmetic and/or geometric means and standard deviations are provided if reported in the listed studies. The absolute range, or the 95% confidence interval around the mean, of the data in each study are provided when reported. In addition, the lower and upper 95th percentile of the distribution are presented for each study in which a mean and corresponding standard deviation were reported. Table 5 also provides estimates of the duration, and level, of exposure to cadmium in the work place if these data were reported in the listed studies. The data presented in table 5 suggest that CDB levels are dose related. Sukuri et al. (1983) show that higher CDB levels are observed among workers experiencing higher work place exposure. This trend appears to be true of the studies listed in the table.

CDB levels reported in table 5 are higher among those showing signs of cadmium-related kidney damage than those showing no such damage. Lauwerys et al. (1976) report CDB levels among workers with kidney lesions that generally are above the levels reported for workers without kidney lesions. Ellis et al. (1983) report a similar observation comparing workers with and without renal dysfunction, although they found more overlap between the 2 groups than Lauwerys et al.

## TABLE 5—BLOOD CADMIUM IN WORKERS EXPOSED TO CADMIUM IN THE WORKPLACE

| Study No. | Work environment (worker population monitored) | No. in study | Employment in years (mean) | Mean concentration of cadmium in air (µg/m³) | Concentrations of Cadmium in blood[a] — Arithmetic mean (±S.D.)[b] | Absolute range or (95% C.I.)[c] | Geometric mean (GSD)[d] | Lower 95th percentile of range[e] ( )[f] | Upper 95th percentile of range[e] ( )[f] | Reference |
|---|---|---|---|---|---|---|---|---|---|---|
| 1 | Ni-Cd battery plant and Cd production plant: | | 3-40 | ≤90 | | | | | | Lauwerys et al. 1976. |
| | (Workers without kidney lesions)... | 96 | ......... | ......... | 21.4±1.9 | ......... | ......... | (18) | (25) | |
| | (Workers with kidney lesions)........ | 25 | ......... | ......... | 38.8±3.8 | ......... | ......... | (32) | (45) | |
| 2 | Ni-Cd battery plant: | | | | | | | | | Adamsson et al. (1979). |
| | (Smokers) ..................................... | 7 | (5) | 10.1 | 22.7 | 7.3-67.2 | | | | |
| | (Nonsmokers) ................................ | 8 | (9) | 7.0 | 7.0 | 4.9-10.5 | | | | |
| 3 | Cadmium alloy plant: | | | [1,000-5 yrs; 40-5 yrs] | | | | | | Sukuri et al. 1982. |
| | (High exposure group)................... | 7 | (10.6) | | 20.8±7.1 | ......... | ......... | (7.3) | (34) | |
| | (Low exposure group)..................... | 9 | (7.3) | | 7.1±1.1 | ......... | ......... | (5.1) | (9.1) | |
| 4 | Retrospective study of workers with renal problems: | 19 | 15-41 | | | | | | ......... | Roels et al. 1982. |
| | (Before removal).............................. | | (27.2) | ......... | 39.9±3.7 | 11-179 | ......... | (34) | (46) | |
| | (After removal) ................................ | | [g](4.2) | ......... | 14.1±5.6 | 5.7-27.4 | ......... | (4.4) | (24) | |
| 5 | Cadmium production plant: | | | | | | | | | Ellis et al. 1983. |
| | (Workers without renal dysfunction). | 33 | 1-34 | ......... | 15±5.7 | 7-31 | ......... | (5.4) | (25) | |
| | (Workers with renal dysfunction) .. | 18 | 10-34 | ......... | 24±8.5 | 10-34 | ......... | (9.3) | (39) | |
| 6 | Cd-Cu alloy plant ............................ | 75 | Up to 39 | ......... | ......... | ......... | 8.8±1.1 | 7.5 | 10 | Mason et al. 1988. |
| 7 | Cadmium recovery operation—Current (19) and former (26) workers. | 45 | (19.0) | ......... | ......... | ......... | 7.9±2.0 | 2.5 | 25 | Thun et al. 1989. |
| 8 | Cadmium recovery operation | 40 | ......... | ......... | 10.2±5.3 | 2.2-18.8 | ......... | (1.3) | (19) | Mueller et al. 1989. |

[a] Concentrations reported in µg Cd/l blood unless otherwise stated.
[b] S.D.—Standard Deviation.
[c] C.I.—Confidence Interval.
[d] GSD—Geometric Standard Deviation.
[e] Based on an assumed lognormal distribution.
[f] Based on an assumed normal distribution.
[g] Years following removal.

The data in table 5 also indicate that CDB levels are higher among those experiencing current occupational exposure than those who have been removed from such exposure. Roels et al. (1982) indicate that CDB levels observed among workers experiencing ongoing exposure in the work place are almost entirely above levels observed among workers removed from such exposure. This finding suggests that CDB levels decrease once cadmium exposure has ceased.

A comparison of the data presented in tables 4 and 5 indicates that CDB levels observed among cadmium-exposed workers is significantly higher than levels observed among the unexposed groups. With the exception of 2 studies presented in table 5 (1 of which includes former workers in the sample group tested), the lower 95th percentile for CDB levels among exposed workers are greater than 5 µg/l, which is the value of the upper 95th percentile for CDB levels observed among those who are not occupationally exposed. Therefore, a CDB level of 5 µg/l represents a threshold above which significant work place exposure to cadmium may be occurring.

### 5.1.7 Conclusions and Recommendations for CDB

Based on the above evaluation, the following recommendations are made for a CDB proficiency program.

#### 5.1.7.1 Recommended method

The method of Stoeppler and Brandt (1980) should be adopted for analyzing CDB. This method was selected over other methods for its straightforward sample-preparation procedures, and because limitations of the method were described adequately. It also is the method used by a plurality of laboratories currently participating in the CTQ proficiency program. In a recent CTQ interlaboratory comparison report (CTQ 1991), analysis of the methods used by laboratories to measure CDB indicates that 46% (11 of 24) of the participating laboratories used the Stoeppler and Brandt methodology ($HNO_3$ deproteinization of blood followed by analysis of the supernatant by GF-AAS). Other CDB methods employed by participating laboratories identified in the CTQ report include dilution of blood (29%), acid digestion (12%) and miscellaneous methods (12%).

Laboratories may adopt alternate methods, but it is the responsibility of the laboratory to demonstrate that the alternate methods meet the data quality objectives defined for the Stoeppler and Brandt method (see Section 5.1.7.2 below).

#### 5.1.7.2 Data quality objectives

Based on the above evaluation, the following data quality objectives (DQOs) should facilitate interpretation of analytical results.

*Limit of Detection.* 0.5 µg/l should be achievable using the Stoeppler and Brandt method. Stoeppler and Brandt (1980) report a limit of detection equivalent to ≤0.2 µg/l in whole blood using 25 µl aliquots of deproteinized, diluted blood samples.

*Accuracy.* Initially, some of the laboratories performing CDB measurements may be expected to satisfy criteria similar to the less severe criteria specified by the CTQ program, i.e., measurements within 2 g/l or 15% (whichever is greater) of the target value. About 60% of the laboratories enrolled in the CTQ program could meet this criterion on the first proficiency test (Weber 1988).

Currently, approximately 12 laboratories in the CTQ program are achieving an accuracy for CDB analysis within the more severe constraints of ±1 µg/l or 15% (whichever is greater). Later, as laboratories gain experience, they should achieve the level of accuracy exhibited by these 12 laboratories. The experience in the CTQ program has shown that, even without incentives, laboratories benefit from the feedback of the program; after they have analyzed 40-50 control samples from the program, performance improves to the point where about 60% of the laboratories can meet the stricter criterion of ±1 µg/l or 15% (Weber 1988). Thus, this stricter target accuracy is a reasonable DQO.

*Precision.* Although Stoeppler and Brandt (1980) suggest that a coefficient of variation (CV) near 1.3% (for a 10 µg/l concentration) is achievable for within-run reproducibility, it is recognized that other factors affecting within- and between-run comparability will increase the achievable CV. Stoeppler and Brandt (1980) observed CVs that were as high as 30% for low concentrations (0.4 µg/l), and CVs of less than 5% for higher concentrations.

For internal QC samples (see Section 3.3.1), laboratories should attain an overall precision near 25%. For CDB samples with concentrations less than 2 µg/l, a target precision of 40% is reasonable, while precisions of 20% should be achievable for concentrations greater than 2 µg/l. Although these values are more strict than values observed in the CTQ interlaboratory program reported by Webber (1988), they are within the achievable limits reported by Stoeppler and Brandt (1980).

### 5.1.7.3 Quality assurance/quality control

Commercial laboratories providing measurement of CDB should adopt an internal QA/QC program that incorporates the following components: Strict adherence to the selected method, including all calibration requirements; regular incorporation of QC samples during actual runs; a protocol for corrective actions, and documentation of these actions; and, participation in an interlaboratory proficiency program. Note that the nonmandatory QA/QC program presented in Attachment 1 is based on the Stoeppler and Brandt method for CDB analysis. Should an alternate method be adopted, the laboratory should develop a QA/QC program satisfying the provisions of Section 3.3.1.

### 5.2 Measuring Cadmium in Urine (CDU)

As in the case of CDB measurement, proper determination of CDU requires strict analytical discipline regarding collection and handling of samples. Because cadmium is both ubiquitous in the environment and employed widely in coloring agents for industrial products that may be used during sample collection, preparation and analysis, care should be exercised to ensure that samples are not contaminated during the sampling procedure.

Methods for CDU determination share many of the same features as those employed for the determination of CDB. Thus, changes and improvements to methods for measuring CDU over the past 40 years parallel those used to monitor CDB. The direction of development has largely been toward the simplification of sample preparation techniques made possible because of improvements in analytic techniques.

### 5.2.1 Units of CDU Measurement

Procedures adopted for reporting CDU concentrations are not uniform. In fact, the situation for reporting CDU is more complicated than for CDB, where concentrations are normalized against a unit volume of whole blood.

Concentrations of solutes in urine vary with several biological factors (including the time since last voiding and the volume of liquid consumed over the last few hours); as a result, solute concentrations should be normalized against another characteristic of urine that represents changes in solute concentrations. The 2 most common techniques are either to standardize solute concentrations against the concentration of creatinine, or to standardize solute concentrations against the specific gravity of the urine. Thus, CDU concentrations have been reported in the literature as "uncorrected" concentrations of cadmium per volume of urine (i.e., µg Cd/l urine), "corrected" concentrations of cadmium per volume of urine at a standard specific gravity (i.e., µg Cd/l urine at a specific gravity of 1.020), or "corrected" mass concentration per unit mass of creatinine (i.e., µg Cd/g creatinine). (CDU concentrations [whether uncorrected or corrected for specific gravity, or normalized to creatinine] occasionally are reported in nanomoles [i.e., nmoles] of cadmium per unit mass or volume. In this protocol, these values are converted to µg of cadmium per unit mass or volume using 89 nmoles of cadmium=10 µg.)

While it is agreed generally that urine values of analytes should be normalized for reporting purposes, some debate exists over what correction method should be used. The medical community has long favored normalization based on creatinine concentration, a common urinary constituent. Creatinine is a normal product of tissue catabolism, is excreted at a uniform rate, and the total amount excreted per day is constant on a day-to-day basis (NIOSH 1984b). While this correction method is accepted widely in Europe, and within some occupational health circles, Kowals (1983) argues that the use of specific gravity (i.e., total solids per unit volume) is more straightforward and practical (than creatinine) in adjusting CDU values for populations that vary by age or gender.

Kowals (1983) found that urinary creatinine (CRTU) is lower in females than males, and also varies with age. Creatinine excretion is highest in younger males (20-30 years old), decreases at middle age (50-60 years), and may rise slightly in later years. Thus, cadmium concentrations may be underestimated for some workers with high CRTU levels.

Within a single void urine collection, urine concentration of any analyte will be affected by recent consumption of large volumes of liquids, and by heavy physical labor in hot environments. The absolute amount of analyte excreted may be identical, but concentrations will vary widely so that urine must be corrected for specific gravity (i.e., to normalize concentrations to the quantity of total solute) using a fixed value (e.g., 1.020 or 1.024). However,

since heavy-metal exposure may increase urinary protein excretion, there is a tendency to underestimate cadmium concentrations in samples with high specific gravities when specific-gravity corrections are applied.

Despite some shortcomings, reporting solute concentrations as a function of creatinine concentration is accepted generally; OSHA therefore recommends that CDU levels be reported as the mass of cadmium per unit mass of creatinine (µg/g CTRU).

Reporting CDU as µg/g CRTU requires an additional analytical process beyond the analysis of cadmium: Samples must be analyzed independently for creatinine so that results may be reported as the ratio of cadmium to creatinine concentrations found in the urine sample. Consequently, the overall quality of the analysis depends on the combined performance by a laboratory on these 2 determinations. The analysis used for CDU determinations is addressed below in terms of µg Cd/l, with analysis of creatinine addressed separately. Techniques for assessing creatinine are discussed in Section 5.4.

Techniques for deriving cadmium as a ratio of CRTU, and the confidence limits for independent measurements of cadmium and CRTU, are provided in Section 3.3.3.

### 5.2.2 Analytical Techniques Used to Monitor CDU

Analytical techniques used for CDU determinations are similar to those employed for CDB determinations; these techniques are summarized in Table 3. As with CDB monitoring, the technique most suitable for CDU determinations is atomic absorption spectroscopy (AAS). AAS methods used for CDU determinations typically employ a graphite furnace, with background correction made using either the deuterium-lamp or Zeeman techniques; Section 5.1.1 provides a detailed description of AAS methods.

### 5.2.3 Methods Developed for CDU Determinations

Princi (1947), Smith et al. (1955), Smith and Kench (1957), and Tsuchiya (1967) used colorimetric procedures similar to those described in the CDB section above to estimate CDU concentrations. In these methods, urine (50 ml) is reduced to dryness by heating in a sand bath and digested (wet ashed) with mineral acids. Cadmium then is complexed with dithiazone, extracted with chloroform and quantified by spectrophotometry. These early studies typically report reagent blank values equivalent to 0.3 µg Cd/l, and CDU concentrations among nonexposed control groups at maximum levels of 10 µg Cd/l—erroneously high values when compared to more recent surveys of cadmium concentrations in the general population.

By the mid-1970s, most analytical procedures for CDU analysis used either wet ashing (mineral acid) or high temperatures (>400 °C) to digest the organic matrix of urine, followed by cadmium chelation with APDC or DDTC solutions and extraction with MIBK. The resulting aliquots were analyzed by flame or graphite-furnace AAS (Kjellstrom 1979).

Improvements in control over temperature parameters with electrothermal heating devices used in conjunction with flameless AAS techniques, and optimization of temperature programs for controlling the drying, charring, and atomization processes in sample analyses, led to improved analytical detection of diluted urine samples without the need for sample digestion or ashing. Roels et al. (1978) successfully used a simple sample preparation, dilution of 1.0 ml aliquots of urine with 0.1 N $HNO_3$, to achieve accurate low-level determinations of CDU.

In the method described by Pruszkowska et al. (1983), which has become the preferred method for CDU analysis, urine samples were diluted at a ratio of 1:5 with water; diammonium hydrogenphosphate in dilute $HNO_3$ was used as a matrix modifier. The matrix modifier allows for a higher charring temperature without loss of cadmium through volatilization during preatomization. This procedure also employs a stabilized temperature platform in a graphite furnace, while nonspecific background absorbtion is corrected using the Zeeman technique. This method allows for an absolute detection limit of approximately 0.04 µg Cd/l urine.

### 5.2.4 Sample Collection and Handling

Sample collection procedures for CDU may contribute to variability observed among CDU measurements. Sources of variation attendant to sampling include time-of-day, the interval since ingestion of liquids, and the introduction of external contamination during the collection process. Therefore, to minimize contributions from these variables, strict adherence to a sample-collection protocol is recommended. This protocol should include provisions for normalizing the conditions under which urine is collected. Every effort also should be made to collect samples during the same time of day.

Collection of urine samples from an industrial work force for biological monitoring purposes usually is performed using "spot" (i.e., single-void) urine with the pH of the sample determined immediately. Logistic and sample-integrity problems arise when efforts are made to collect urine over long periods (e.g., 24 hrs). Unless single-void urines are used, there are numerous opportunities for measurement error because of poor control over sample collection, storage and environmental contamination.

To minimize the interval during which sample urine resides in the bladder, the following adaption to the "spot" collection procedure is recommended: The bladder should first be emptied, and then a large glass of water should be consumed; the sample may be collected within an hour after the water is consumed.

### 5.2.5 Best Achievable Performance

Performance using a particular method for CDU determinations is assumed to be equivalent to the performance reported by the research laboratories in which the method was developed. Pruszkowska et al. (1983) report a detection limit of 0.04 µg/l CDU, with a CV of <4% between 0-5 µg/l. The CDC reports a minimum CDU detection limit of 0.07 µg/l using a modified method based on Pruszkowska et al. (1983). No CV is stated in this protocol; the protocol contains only rejection criteria for internal QC parameters used during accuracy determinations with known standards (Attachment 8 of exhibit 106 of OSHA docket H057A). Stoeppler and Brandt (1980) report a CDU detection limit of 0.2 µ/l for their methodology.

### 5.2.6 General Method Performance

For any particular method, the expected initial performance from commercial laboratories may be somewhat lower than that reported by the research laboratory in which the method was developed. With participation in appropriate proficiency programs, and use of a proper in-house QA/QC program incorporating provisions for regular corrective actions, the performance of commercial laboratories may be expected to improve and approach that reported by a research laboratories. The results reported for existing proficiency programs serve to specify the initial level of performance that likely can be expected from commercial laboratories offering analysis using a particular method.

Weber (1988) reports on the results of the CTQ proficiency program, which includes CDU results for laboratories participating in the program. Results indicate that after receiving 60 samples (i.e., after participating in the program for approximately 3 years), approximately 80% of the participating laboratories report CDU results ranging between ±2 µg/l or 15% of the consensus mean, whichever is greater. On any single sample of the last 15 samples, the proportion of laboratories falling within the specified range is between 75 and 95%, except for a single test for which only 60% of the laboratories reported acceptable results. For each of the last 15 samples, approximately 60% of the laboratories reported results within ±1 µg or 15% of the mean, whichever is greater. The range of concentrations included in this set of samples was not reported.

Another report from the CTQ (1991) summarizes preliminary CDU results from their 1991 interlaboratory program. According to the report, for 3 CDU samples with values of 9.0, 16.8, 31.5 µg/l, acceptable results (target of ±2 µg/l or 15 % of the consensus mean, whichever is greater) were achieved by only 44-52% of the 34 laboratories participating in the CDU program. The overall CVs for these 3 CDU samples among the 34 participating laboratories were 31%, 25%, and 49%, respectively. The reason for this poor performance has not been determined.

A more recent report from the CTQ (Weber, private communication) indicates that 36% of the laboratories in the program have been able to achieve the target of ±1 µg/l or 15% for more than 75% of the samples analyzed over the last 5 years, while 45% of participating laboratories achieved a target of ±2 µg/l or 15% for more than 75% of the samples analyzed over the same period.

Note that results reported in the interlaboratory programs are in terms of µg Cd/l of urine, unadjusted for creatinine. The performance indicated, therefore, is a measure of the performance of the cadmium portion of the analyses, and does not include variation that may be introduced during the analysis of CRTU.

### 5.2.7 Observed CDU Concentrations

Prior to the onset of renal dysfunction, CDU concentrations provide a general indication of the exposure history (i.e., body burden) (see Section 4.3). Once renal dysfunction occurs, CDU levels appear to increase and are no longer indicative solely of cadmium body burden (Friberg and Elinder 1988).

#### 5.2.7.1 *Range of CDU concentrations observed among unexposed samples*

Surveys of CDU concentrations in the general population were first reported from cooperative studies among industrial countries (i.e., Japan, U.S. and Sweden) conducted in the mid-1970s. In summarizing these data, Kjellstrom (1979) reported that CDU concentrations among Dallas, Texas men (age range: <9-59 years; smokers and nonsmokers) varied from 0.11-1.12 µg/l (uncorrected for creatinine or specific gravity). These CDU concentrations are intermediate between population values found in Sweden (range: 0.11-0.80 µg/l) and Japan (range: 0.14-2.32 µg/l).

Kowal and Zirkes (1983) reported CDU concentrations for almost 1,000 samples collected during 1978-79 from the general U.S. adult population (i.e., nine states; both genders; ages 20-74 years). They report that CDU concentrations are lognormally distributed; low levels predominated, but a small proportion of the population exhibited high levels. These investigators transformed the CDU concentrations values, and reported the same data 3 different ways: µg/l urine (unadjusted), µg/l (specific gravity adjusted to 1.020), and µg/g CRTU. These data are summarized in Tables 6 and 7.

Based on further statistical examination of these data, including the lifestyle characteristics of this group, Kowal (1988) suggested increased cadmium absorption (i.e., body burden) was correlated with low dietary intakes of calcium and iron, as well as cigarette smoking.

CDU levels presented in Table 6 are adjusted for age and gender. Results suggest that CDU levels may be slightly different among men and women (i.e., higher among men when values are unadjusted, but lower among men when the values are adjusted, for specific gravity or CRTU). Mean differences among men and women are small compared to the standard deviations, and therefore may not be significant. Levels of CDU also appear to increase with age. The data in Table 6 suggest as well that reporting CDU levels adjusted for specific gravity or as a function of CRTU results in reduced variability.

**TABLE 6—URINE CADMIUM CONCENTRATIONS IN THE U.S. ADULT POPULATION: NORMAL AND CONCENTRATION-ADJUSTED VALUES BY AGE AND SEX[1]**

| | Geometric means (and geometric standard deviations) | | |
|---|---|---|---|
| | Unadjusted (µg/l) | SG-adjusted[2] µg/l at 1.020 | Creatine-adjusted (µg/g) |
| Sex: | | | |
| Male (n = 484) | 0.55 (2.9) | 0.73 (2.6) | 0.55 (2.7) |
| Female (n = 498) | 0.49 (3.0) | 0.86 (2.7) | 0.78 (2.7) |
| Age: | | | |
| 20-29 (n = 222) | 0.32 (3.0) | 0.43 (2.7) | 0.32 (2.7) |
| 30-39 (n = 141) | 0.46 (3.2) | 0.70 (2.8) | 0.54 (2.7) |
| 40-49 (n = 142) | 0.50 (3.0) | 0.81 (2.6) | 0.70 (2.7) |
| 50-59 (n = 117) | 0.61 (2.9) | 0.99 (2.4) | 0.90 (2.3) |
| 60-69 (n = 272) | 0.76 (2.6) | 1.16 (2.3) | 1.03 (2.3) |

[1] From Kowal and Zirkes 1983.

[2] SC-adjusted is adjusted for specific gravity.

**TABLE 7—URINE CADMIUM CONCENTRATIONS IN THE U.S. ADULT POPULATION: CUMULATIVE FREQUENCY DISTRIBUTION OF URINARY CADMIUM (N = 982)[1]**

| Range of concentrations | Unadjusted (µg/l) percent | SG-adjusted (µg/l at 1.020) percent | Creatine-adjusted (µg/g) percent |
|---|---|---|---|
| <0.5 | 43.9 | 28.0 | 35.8 |
| 0.6-1.0 | 71.7 | 56.4 | 65.6 |
| 1.1-1.5 | 84.4 | 74.9 | 81.4 |
| 1.6-2.0 | 91.3 | 84.7 | 88.9 |
| 2.1-3.0 | 97.3 | 94.4 | 95.8 |
| 3.1-4.0 | 98.8 | 97.4 | 97.2 |
| 4.1-5.0 | 99.4 | 98.2 | 97.9 |
| 5.1-10.0 | 99.6 | 99.4 | 99.3 |
| 10.0-20.0 | 99.8 | 99.6 | 99.6 |

[1] Source: Kowal and Zirkes (1983).

The data in the Table 6 indicate the geometric mean of CDU levels observed among the general population is 0.52 µ/g Cd/l urine (unadjusted), with a geometric standard deviation of 3.0. Normalized for creatinine, the geometric mean for the population is 0.66 µ/g CRTU, with a geometric standard deviation of 2.7. Table 7 provides the distributions of CDU concentrations for the general population studied by Kowal and Zirkes. The data in this table indicate that 95% of the CDU levels observed among those not occupationally exposed to cadmium are below 3 µ/g CRTU.

### 5.2.7.2 Range of CDU concentrations observed among exposed workers

Table 8 is a summary of results from available studies of CDU concentrations observed among cadmium-exposed workers. In this table, arithmetic and/or geometric means and standard deviations are provided if reported in these studies. The absolute range for the data in each study, or the 95% confidence interval around the mean of each study, also are provided when reported. The lower and upper 95th percentile of the distribution are presented for each study in which a mean and corresponding standard deviation were reported. Table 8 also provides estimates of the years of exposure, and the levels of exposure, to cadmium in the work place if reported in these studies. Concentrations reported in this table are in µ/g CRTU, unless otherwise stated.

### TABLE 8—URINE CADMIUM CONCENTRATIONS IN WORKERS EXPOSED TO CADMIUM IN THE WORKPLACE

| Study No. | Work environment (worker population monitored) | No. in study (n) | Employment in years (mean) | Mean concentration of cadmium in air (µg/m³) | Arithmetic mean (±S.D.)[b] | Absolute range or (95% C.I.)[c] | Geometric mean (GSD)[d] | Lower 95th percentile of range[e] ( )[f] | Upper 95th percentile of range[e] ( )[f] | Reference |
|---|---|---|---|---|---|---|---|---|---|---|
| 1 | Ni-Cd battery plant and Cd production plant. | ...... | 3-40 | ≤90 | ...... | ...... | ...... | ...... | ...... | Lauwerys et al. 1976. |
|  | (Workers without kidney lesions). | 96 | .......... | ........ | 16.3±16.7 | ......... | ......... | (0) | (44) |  |
|  | (Workers with kidney lesions). | 25 | .......... | ........ | 48.2±42.6 | ......... | ......... | (0) | (120) |  |
| 2 | Ni-Cd battery plant | ...... | ...... | ...... | ...... | ...... | ...... | ...... | ...... | Adamsson et al. (1979). |
|  | (Smokers) | 7 | (5) | 10.1 | 5.5 | 1.0-14.7 | ...... | ...... | ...... |  |
|  | (Nonsmokers) | 8 | (9) | 7.0 | 3.6 | 0.5-9.3 | ...... | ...... | ...... |  |
| 3 | Cadmium salts production facility. | 148 | (15.4) | ........ | 15.8 | 2-150 | ...... | ...... | ...... | Butchet et al. 1980. |
| 4 | Retrospective study of workers with renal problems. | 19 | 15-41 | ........ | ...... | ...... | ...... | ...... | ...... | Roels et al. 1982. |
|  | (Before removal) | ...... | (27.2) | ........ | 39.4±28.1 | 10.8-117 | ...... | (0) | (88) |  |
|  | (After removal) | ...... | (4.2)[g] | ........ | 16.4±9.0 | 80-42.3 | ...... | (1.0) | (32) |  |
| 5 | Cadmium production plant | ...... | .......... | ........ | ...... | ...... | ...... | ...... | ...... | Ellis et al. 1983. |
|  | (Workers without renal dysfunction). | 33 | 1-34 | ........ | 9.4±6.9 | 2-27 | ...... | (0) | (21) |  |
|  | (Workers with renal dysfunction). | 18 | 10-34 | ........ | 22.8±12.7 | 8-55 | ...... | (1) | (45) |  |
| 6 | Cd-Cu alloy plant | 75 | Up to 39 | Note[h] | 6.9±9.4 | ...... | ...... | (0) | (23) | Mason et al. 1988. |
| 7 | Cadmium recovery operation. | 45 | (19) | 87 | 9.3±6.9 | ...... | ...... | (0) | (21) | Thun et al. 1989. |
| 8 | Pigment manufacturing plant. | 29 | (12.8) | 0.18-3.0 | ...... | 0.2-9.5 | 1.1 | ...... | ...... | Mueller et al. 1989. |
| 9 | Pigment manufacturing plant. | 26 | (12.1) | ≤3.0 | ...... | ...... | 1.25±2.45 | 0.3 | 6 | Kawada et al. 1990. |

[a] Concentrations reported in µg/g Cr.
[b] S.D.—Standard Deviation.
[c] C.I.—Confidence Interval.
[d] GSD—Geometric Standard Deviation.
[e] Based on an assumed lognormal distribution.
[f] Based on an assumed normal distribution.
[g] Years following removal.
[h] Equivalent to 50 for 20-22 yrs

Data in Table 8 from Lauwerys et al. (1976) and Ellis et al. (1983) indicate that CDU concentrations are higher among those exhibiting kidney lesions or dysfunction than among those lacking these symptoms. Data from the study by Roels et al. (1982) indicate that CDU levels decrease among workers removed from occupational exposure to cadmium in comparison to workers experiencing ongoing exposure. In both cases, however, the distinction between the 2 groups is not as clear as with CDB; there is more overlap in CDU levels observed among each of the paired populations than is true for corresponding CDB levels. As with CDB levels, the data in Table 8 suggest increased CDU concentrations among workers who experienced increased overall exposure.

Although a few occupationally-exposed workers in the studies presented in Table 8 exhibit CDU levels below 3 µg/g CRTU, most of those workers exposed to cadmium levels in excess of the PEL defined in the final cadmium rule exhibit CDU levels above 3 µg/g CRTU; this level represents the upper 95th percentile of the CDU distribution observed among those who are not occupationally exposed to cadmium (Table 7).

The mean CDU levels reported in Table 8 among occupationally exposed groups studied (except 2) exceed 3 µg/g CRTU. Correspondingly, the level of exposure reported in these studies (with 1 exception) are significantly higher than what workers will experience under the final cadmium rule. The 2 exceptions are from the studies by Mueller et al. (1989) and Kawada et al. (1990); these studies indicate that workers exposed to cadmium during pigment manufacture do not exhibit CDU levels as high as those levels observed among workers exposed to cadmium in other occupations. Exposure levels, however, were lower in the pigment manufacturing plants studied. Significantly, workers removed from occupational cadmium exposure for an average of 4 years still exhibited CDU levels in excess of 3 µg/g CRTU (Roels et al. 1982). In the single-exception study with a reported level of cadmium exposure lower than levels proposed in the final rule (i.e., the study of a pigment manufacturing plant by Kawada et al. 1990), most of the workers exhibited CDU levels less than 3 µg/g CRTU (i.e., the mean value was only 1.3 µg/g CRTU). CDU levels among workers with such limited cadmium exposure are expected to be significantly lower than levels of other studies reported in Table 8.

Based on the above data, a CDU level of 3 µg/g CRTU appear to represent a threshold above which significant work place exposure to cadmium occurs over the work span of those being monitored. Note that this threshold is not as distinct as the corresponding threshold described for CDB. In general, the variability associated with CDU measurements among exposed workers appears to be higher than the variability associated with CDB measurements among similar workers.

### 5.2.8 Conclusions and Recommendations for CDU

The above evaluation supports the following recommendations for a CDU proficiency program. These recommendations address only sampling and analysis procedures for CDU determinations specifically, which are to be reported as an unadjusted µg Cd/l urine. Normalizing this result to creatinine requires a second analysis for CRTU so that the ratio of the 2 measurements can be obtained. Creatinine analysis is addressed in Section 5.4. Formal procedures for combining the 2 measurements to derive a value and a confidence limit for CDU in µg/g CRTU are provided in Section 3.3.3.

#### 5.2.8.1 *Recommended method*

The method of Pruszkowska et al. (1983) should be adopted for CDU analysis. This method is recommended because it is simple, straightforward and reliable (i.e., small variations in experimental conditions do not affect the analytical results).

A synopsis of the methods used by laboratories to determine CDU under the interlaboratory program administered by the CTQ (1991) indicates that more than 78% (24 of 31) of the participating laboratories use a dilution method to prepare urine samples for CDU analysis. Laboratories may adopt alternate methods, but it is the responsibility of the laboratory to demonstrate that the alternate methods provide results of comparable quality to the Pruszkowska method.

#### 5.2.8.2 *Data quality objectives*

The following data quality objectives should facilitate interpretation of analytical results, and are achievable based on the above evaluation.

*Limit of Detection.* A level of 0.5 µg/l (i.e., corresponding to a detection limit of 0.5 µg/g CRTU, assuming 1 g CRT/l urine) should be achievable. Pruszkowska et al. (1983) achieved a limit of detection of 0.04 g/l for CDU based on the slope of the curve for their working standards (0.35 pg Cd/0.0044, A signal=1% absorbance using GF-AAS).

The CDC reports a minimum detection limit for CDU of 0.07 µg/l using a modified Pruszkowska method. This limit of detection was defined as 3 times the standard deviation calculated from 10 repeated measurements of a "low level" CDU test sample (Attachment 8 of exhibit 106 of OSHA docket H057A).

Stoeppler and Brandt (1980) report a limit of detection for CDU of 0.2 µg/l using an aqueous dilution (1:2) of the urine samples.

*Accuracy.* A recent report from the CTQ (Weber, private communication) indicates that 36% of the laboratories in the program achieve the target of ±1 µg/l or 15% for more than 75% of the samples analyzed over the last 5 years, while 45% of participating laboratories achieve a target of ±2 µg/l or 15% for more than 75% of the samples analyzed over the same period. With time and a strong incentive for improvement, it is expected that the proportion of laboratories successfully achieving the stricter level of accuracy should increase. It should be noted, however, these indices of performance do not include variations resulting from the ancillary measurement of CRTU (which is recommended for the proper recording of results). The low cadmium levels expected to be measured indicate that the analysis of creatinine will contribute relatively little to the overall variability observed among creatinine-normalized CDU levels (see Section 5.4). The initial target value for reporting CDU under this program, therefore, is set at ±1 µg/g CRTU or 15% (whichever is greater).

*Precision.* For internal QC samples (which are recommended as part of an internal QA/QC program, Section 3.3.1), laboratories should attain an overall precision of 25%. For CDB samples with concentrations less than 2 µg/l, a target precision of 40% is acceptable, while precisions of 20% should be achievable for CDU concentrations greater than 2 µg/l. Although these values are more stringent than those observed in the CTQ interlaboratory program reported by Webber (1988), they are well within limits expected to be achievable for the method as reported by Stoeppler and Brandt (1980).

#### 5.2.8.3 *Quality assurance/quality control*

Commercial laboratories providing CDU determinations should adopt an internal QA/QC program that incorporates the following components: Strict adherence to the selected method, including calibration requirements; regular incorporation of QC samples during actual runs; a protocol for corrective actions, and documentation of such actions; and, participation in an interlaboratory proficiency program. Note that the nonmandatory program presented in Attachment 1 as an example of an acceptable QA/QC program, is based on using the Pruszkowska method for CDU analysis. Should an alternate method be adopted by a laboratory, the laboratory should develop a QA/QC program equivalent to the nonmandatory program, and which satisfies the provisions of Section 3.3.1.

### 5.3 Monitoring β-2-Microglobulin in Urine (B2MU)

As indicated in Section 4.3, B2MU appears to be the best of several small proteins that may be monitored as early indicators of cadmium-induced renal damage. Several analytic techniques are available for measuring B2M.

#### 5.3.1 Units of B2MU Measurement

Procedures adopted for reporting B2MU levels are not uniform. In these guidelines, OSHA recommends that B2MU levels be reported as µg/g CRTU, similar to reporting CDU concentrations. Reporting B2MU normalized to the concentration of CRTU requires an additional analytical process beyond the analysis of B2M: Independent analysis for creatinine so that results may be reported as a ratio of the B2M and creatinine concentrations found in the urine sample. Consequently, the overall quality of the analysis depends on the combined performance on these 2 analyses. The analysis used for B2MU determinations is described in terms of µg B2M/l urine, with analysis of creatinine addressed separately. Techniques used to measure creatinine are provided in Section 5.4. Note that Section 3.3.3 provides techniques for deriving the value of B2M as function of CRTU, and the confidence limits for independent measurements of B2M and CRTU.

#### 5.3.2 Analytical Techniques Used to Monitor B2MU

One of the earliest tests used to measure B2MU was the radial immunodiffusion technique. This technique is a simple and specific method for identification and quantitation of a number of proteins found in human serum and other body fluids when the protein is not readily differentiated by standard electrophoretic procedures. A quantitative relationship exists between the concentration of a protein deposited in a well that is cut into a thin agarose layer containing the corresponding monospecific antiserum, and the distance that the resultant complex diffuses. The wells are filled with an unknown serum and the standard (or control), and incubated in a moist environment at room temperature. After the optimal point of diffusion has been reached, the diameters of the resulting precipitation rings are measured. The diameter of a ring is related to the concentration of the constituent substance. For B2MU determinations required in the medical monitoring program, this method requires a process that may be insufficient to concentrate the protein to levels that are required for detection.

Radioimmunoassay (RIA) techniques are used widely in immunologic assays to measure the concentration of antigen or antibody in body-fluid samples. RIA procedures are based on competitive-binding techniques. If antigen concentration is being measured, the principle underlying the procedure is that radioactive-labeled antigen competes with the sample's unlabeled antigen for binding sites on a known amount of immobile antibody. When these 3 components are present in the system, an equilibrium exists. This equilibrium is followed by a separation of the free and bound forms of the antigen. Either free or bound radioactivelabeled antigen can be assessed to determine the amount of antigen in the sample. The analysis is performed by measuring the level of radiation emitted either by the bound complex following removal of the solution containing the free antigen, or by the isolated solution containing the residual-free antigen. The main advantage of the RIA method is the extreme sensitivity of detection for emitted radiation and the corresponding ability to detect trace amounts of antigen. Additionally, large numbers of tests can be performed rapidly.

The enzyme-linked immunosorbent assay (ELISA) techniques are similar to RIA techniques except that nonradioactive labels are employed. This technique is safe, specific and rapid, and is nearly as sensitive as RIA techniques. An enzyme-labeled antigen is used in the immunologic assay; the labeled antigen detects the presence and quantity of unlabeled antigen in the sample. In a representative ELISA test, a plastic plate is coated with antibody (e.g., antibody to B2M). The antibody reacts with antigen (B2M) in the urine and forms an antigen-antibody complex on the plate. A second anti-B2M antibody (i.e., labeled with an enzyme) is added to the mixture and forms an antibody-antigen-antibody complex. Enzyme activity is measured spectrophotometrically after the addition of a specific chromogenic substrate which is activated by the bound enzyme. The results of a typical test are calculated by comparing the spectrophotometric reading of a serum sample to that of a control or reference serum. In general, these procedures are faster and require less laboratory work than other methods.

In a fluorescent ELISA technique (such as the one employed in the Pharmacia Delphia test for B2M), the labeled enzyme is bound to a strong fluorescent dye. In the Pharmacia Delphia test, an antigen bound to a fluorescent dye competes with unlabeled antigen in the sample for a predetermined amount of specific, immobile antibody. Once equilibrium is reached, the immobile phase is removed from the labeled antigen in the sample solution and washed; an enhancement solution then is added that liberates the fluorescent dye from the bound antigen-antibody complex. The enhancement solution also contains a chelate that complexes with the fluorescent dye in solution; this complex increases the fluorescent properties of the dye so that it is easier to detect.

To determine the quantity of B2M in a sample using the Pharmacia Delphia test, the intensity of the fluorescence of the enhancement solution is measured. This intensity is proportional to the concentration of labeled antigen that bound to the immobile antibody phase during the initial competition with unlabeled antigen from the sample. Consequently, the intensity of the fluorescence is an inverse function of the concentration of antigen (B2M) in the original sample. The relationship between the fluorescence level and the B2M concentration in the sample is determined using a series of graded standards, and extrapolating these standards to find the concentration of the unknown sample.

### 5.3.3 Methods Developed for B2MU Determinations

B2MU usually is measured by radioimmunoassay (RIA) or enzymelinked immunosorbent assay (ELISA); however, other methods (including gel electrophoresis, radial immunodiffusion, and nephelometric assays) also have been described (Schardun and van Epps 1987). RIA and ELISA methods are preferred because they are sensitive at concentrations as low as micrograms per liter, require no concentration processes, are highly reliable and use only a small sample volume.

Based on a survey of the literature, the ELISA technique is recommended for monitoring B2MU. While RIAs provide greater sensitivity (typically about 1μg/l, Evrin et al. 1971), they depend on the use of radioisotopes; use of radioisotopes requires adherence to rules and regulations established by the Atomic Energy Commission, and necessitates an expensive radioactivity counter for testing. Radioisotopes also have a relatively short half-life, which corresponds to a reduced shelf life, thereby increasing the cost and complexity of testing. In contrast, ELISA testing can be performed on routine laboratory spectrophotometers, do *not* necessitate adherence to additional rules and regulations governing the handling of radioactive substances, and the test kits have long shelf lives. Further, the range of sensitivity commonly achieved by the recommended ELISA test (i.e., the Pharmacia Delphia test) is approximately 100 μg/l (Pharmacia 1990), which is sufficient for monitoring B2MU levels resulting from cadmium exposure. Based on the studies listed in Table 9 (Section 5.3.7), the average range of B2M concentrations among the general, nonexposed population falls between 60 and 300 μg/g CRTU. The upper 95th percentile of distributions, derived from studies in Table 9 which reported standard deviations, range between 180 and 1,140 μg/g CRTU. Also, the Pharmacia Delphia test currently is the most widely used test for assessing B2MU.

### 5.3.4 Sample Collection and Handling

As with CDB or CDU, sample collection procedures are addressed primarily to identify ways to minimize the degree of variability introduced by sample collection during medical monitoring. It is unclear the extent to which sample collection contributes to B2MU variability. Sources of variation include time-of-day effects, the interval since consuming liquids and the quantity of liquids consumed, and the introduction of external contamination during the collection process. A special problem unique to B2M sampling is the sensitivity of this protein to degradation under acid conditions commonly found in the bladder. To minimize this problem, strict adherence to a sampling protocol is recommended. The protocol should include provisions for normalizing the conditions under which the urine is collected. Clearly, it is important to minimize the interval urine spends in the bladder. It also is recommended that every effort be made to collect samples during the same time of day.

Collection of urine samples for biological monitoring usually is performed using "spot" (i.e., single-void) urine. Logistics and sample integrity become problems when efforts are made to collect urine over extended periods (e.g., 24 hrs). Unless single-void urines are used, numerous opportunities exist for measurement error cause of poor control over sample collection, storage and environmental contamination.

To minimize the interval that sample urine resides in the bladder, the following adaption to the "spot" collection procedure is recommended: The bladder should be emptied and then a large glass of water should be consumed; the sample then should be collected within an hour after the water is consumed.

### 5.3.5 Best Achievable Performance

The best achievable performance is assumed to be equivalent to the performance reported by the manufacturers of the Pharmacia Delphia test kits (Pharmacia 1990). According to the insert that comes with these kits, QC results should be within ±2 SDs of the mean for each control sample tested; a CV of less than or equal to 5.2% should be maintained. The total CV reported for test kits is less than or equal to 7.2%.

### 5.3.6 General Method Performance

Unlike analyses for CDB and CDU, the Pharmacia Delphia test is standardized in a commercial kit that controls for many sources of variation. In the absence of data to the contrary, it is assumed that the achievable performance reported by the manufacturer of this test kit will serve as an achievable performance objective. The CTQ proficiency testing program for B2MU analysis is expected to use the performance parameters defined by the test kit manufacturer as the basis of the B2MU proficiency testing program.

Note that results reported for the test kit are expressed in terms of μg B2M/l of urine, and have not been adjusted for creatinine. The indicated performance, therefore, is a measure of the performance of the B2M portion of the analyses only, and does not include variation that may have been introduced during the analysis of creatinine.

### 5.3.7 Observed B2MU Concentrations

As indicated in Section 4.3, the concentration of B2MU may serve as an early indicator of the onset of kidney damage associated with cadmium exposure.

#### 5.3.7.1 Range of B2MU Concentrations Among Unexposed Samples

Most of the studies listed in Table 9 report B2MU levels for those who were not occupationally exposed to cadmium. Studies noted in the second column of this table (which contain the footnote "d") reported B2MU concentrations among cadmium-exposed workers who, nonetheless, showed no signs of proteinuria. These latter studies are included in this table because, as indicated in Section 4.3, monitoring B2MU is intended to provide advanced warning of the onset of kidney dysfunction associated with cadmium exposure, rather than to distinguish relative exposure. This table, therefore, indicates the range of B2MU levels observed among those who had no symptoms of renal dysfunction (including cadmium-exposed workers with none of these symptoms).

## TABLE 9—B-2—MICROGLOBULIN CONCENTRATIONS OBSERVED IN URINE AMONG THOSE NOT OCCUPATIONALLY EXPOSED TO CADMIUM

| Study No. | No. in study | Geometric mean | Geometric standard deviation | Lower 95th percentile of distribution[a] | Upper 95th percentile of distribution[a] | Reference |
|---|---|---|---|---|---|---|
| 1 | 133 m[b] | 115 µg/g[c] | 4.03 | 12 | 1,140 µg/g[c] | Ishizaki et al. 1989. |
| 2 | 161 f[b] | 146 µg/g[c] | 3.11 | 23 | 940 µg/g[c] | Ishizaki et al. 1989. |
| 3 | 10 | 84 µg/g | .......... | .......... | .......... | Ellis et al. 1983. |
| 4 | 203 | 76 µg/l | .......... | .......... | .......... | Stewart and Hughes 1981. |
| 5 | 9 | 103 µg/g | .......... | .......... | .......... | Chia et al. 1989. |
| 6 | 47[d] | 86 µg/L | 1.9 | 30 µg/l | 250 µg/L | Kjellstrom et al. 1977. |
| 7 | 1,000[e] | 68.1 µg/gr Cr[f] | 3.1 m & f | <10 µg/gr Cr[h] | 320 µg/gr Cr[h] | Kowal 1983. |
| 8 | 87 | 71 µg/g[l] | .......... | 7[h] | 200[h] | Buchet et al. 1980. |
| 9 | 10 | 0.073 mg/24h | .......... | .......... | .......... | Evrin et al. 1971. |
| 10 | 59 | 156 µg/g | 1.1[j] | 130 | 180 | Mason et al. 1988. |
| 11 | 8 | 118 µg/g | .......... | .......... | .......... | Iwao et al. 1980. |
| 12 | 34 | 79 µg/g | .......... | .......... | .......... | Wibowo et al. 1982. |
| 13 | 41 m | .......... | .......... | .......... | 400 µg/gr Cr[k] | Falck et al. 1983. |
| 14 | 35[n] | 67 | .......... | .......... | .......... | Roels et al. 1991. |
| 15 | 31[d] | 63 | .......... | .......... | .......... | Roels et al. 1991. |
| 16 | 36[d] | 77[i] | .......... | .......... | .......... | Miksche et al. 1981. |
| 17 | 18[n] | 130 | .......... | .......... | .......... | Kawada et al. 1989. |
| 18 | 32[p] | 122 | .......... | .......... | .......... | Kawada et al. 1989. |
| 19 | 18[d] | 295 | 1.4 | 170 | 510 | Thun et al. 1989. |

[a] Based on an assumed lognormal distribution.
[b] m = males, f = females.
[c] Aged general population from non-polluted area; 47.9% population aged 50-69; 52.1% ≥70 years of age; values reported in study.
[d] Exposed workers without proteinuria.
[e] 492 females, 484 male.
[f] Creatinine adjusted; males = 68.1 µg/g Cr, females = 64.3 mg/g Cr.
[h] Reported in the study.
[i] Arithmetic mean.
[j] Geometric standard error.
[k] Upper 95% tolerance limits: for Falck this is based on the 24 hour urine sample.
[n] Controls.
[p] Exposed synthetic resin and pigment workers without proteinuria; Cadmium in urine levels up to 10 µg/g Cr.

To the extent possible, the studies listed in Table 9 provide geometric means and geometric standard deviations for measurements among the groups defined in each study. For studies reporting a geometric standard deviation along with a mean, the lower and upper 95th percentile for these distributions were derived and reported in the table.

The data provided from 15 of the 19 studies listed in Table 9 indicate that the geometric mean concentration of B2M observed among those who were not occupationally exposed to cadmium is 70-170 µg/g CRTU. Data from the 4 remaining studies indicate that exposed workers who exhibit no signs of proteinuria show mean B2MU levels of 60-300 µg/g CRTU. B2MU values in the study by Thun et al. (1989), however, appear high in comparison to the other 3 studies. If this study is removed, B2MU levels for those who are not occupationally exposed to cadmium are similar to B2MU levels found among cadmium-exposed workers who exhibit no signs of kidney dysfunction. Although the mean is high in the study by Thun et al., the range of measurements reported in this study is within the ranges reported for the other studies.

Determining a reasonable upper limit from the range of B2M concentrations observed among those who do not exhibit signs of proteinuria is problematic. Elevated B2MU levels are among the signs used to define the onset of kidney dysfunction. Without access to the raw data from the studies listed in Table 9, it is necessary to rely on reported standard deviations to estimate an upper limit for normal B2MU concentrations (i.e., the upper 95th percentile for the distributions measured). For the 8 studies reporting a geometric standard deviation, the upper 95th percentiles for the distributions are 180-1140 µg/g CRTU. These values are in general agreement with the upper 95th percentile for the distribution (i.e., 631 µg/g CRTU) reported by Buchet et al. (1980). These upper limits also appear to be in general agreement with B2MU values (i.e., 100-690 µg/g CRTU) reported as the normal upper limit by Iwao et al. (1980), Kawada et al. (1989), Wibowo et al. (1982), and Schardun and van Epps (1987). These values must be compared to levels reported among those exhibiting kidney dysfunction to define a threshold level for kidney dysfunction related to cadmium exposure.

### 5.3.7.2 Range of B2MU Concentrations Among Exposed Workers

Table 10 presents results from studies reporting B2MU determinations among those occupationally exposed to cadmium in the work place; in some of these studies, kidney dysfunction was observed among exposed workers, while other studies did not make an effort to distinguish among exposed workers based on kidney dysfunction. As with Table 9, this table provides geometric means and geometric standard deviations for the groups defined in each study if available. For studies reporting a geometric standard deviation along with a mean, the lower and upper 95th percentiles for the distributions are derived and reported in the table.

## TABLE 10—B-2-MICROGLOBULIN CONCENTRATIONS OBSERVED IN URINE AMONG OCCUPATIONALLY-EXPOSED WORKERS

| Study No. | N | Concentration of B-2 Microglobulin in urine ||||  Reference |
|---|---|---|---|---|---|---|
| | | Geometric mean (µg/g)[a] | Geom std dev | L 95% of range[b] | U 95% of range[b] | |
| 1 | 1,424 | 160 | 6.19 | 8.1 | 3,300 | Ishizaki et al., 1989. |
| 2 | 1,754 | 260 | 6.50 | 12 | 5,600 | Ishizaki et al., 1989. |
| 3 | 33 | 210 | ............ | ............ | ............ | Ellis et al., 1983. |
| 4 | 65 | 210 | ............ | ............ | ............ | Chia et al., 1989. |
| 5 | [c]44 | 5,700 | 6.49 | [d]300 | [d]98,000 | Kjellstrom et al., 1977. |
| 6 | 148 | [e]180 | ............ | [f]110 | [f]280 | Buchet et al., 1980. |
| 7 | 37 | 160 | 3.90 | 17 | 1,500 | Kenzaburo et al., 1979. |
| 8 | [c]45 | 3,300 | 8.7 | [d]310 | [d]89,000 | Mason et al., 1988. |
| 9 | [c]10 | 6,100 | 5.99 | [f]650 | [f]57,000 | Falck et al., 1983. |
| 10 | [c]11 | 3,900 | 2.96 | [d]710 | [d]15,000 | Elinder et al., 1985. |
| 11 | [c]12 | 300 | ............ | ............ | ............ | Roels et al., 1991. |
| 12 | [g]8 | 7,400 | ............ | ............ | ............ | Roels et al., 1991. |
| 13 | [c]23 | [h]1,800 | ............ | ............ | ............ | Roels et al., 1989. |
| 14 | 10 | 690 | ............ | ............ | ............ | Iwao et al., 1980. |
| 15 | 34 | 71 | ............ | ............ | ............ | Wibowo et al., 1982. |
| 16 | [c]15 | 4,700 | 6.49 | [d]590 | [d]93,000 | Thun et al., 1989. |

[a] Unless otherwise stated.
[b] Based on an assumed lognormal distribution.
[c] Among workers diagnosed as having renal dysfunction; for Elinder this means $\beta_2$ levels greater than 300 micrograms per gram creatinine (µg/gr Cr); for Roels, 1991, range = 31 – 35, 170 µg$\beta_2$/gr Cr and geometric mean = 63 among healthy workers; for Mason $\beta_2$>300 µg/gr Cr.
[d] Based on a detailed review of the data by OSHA.
[e] Arthmetic mean.
[f] Reported in the study.
[g] Retired workers.
[h] 1,800 mg$\beta_2$/gr Cr for first survey; second survey = 1,600; third survey = 2,600; fourth survey = 2,600; fifth survey = 2,600.

The data provided in Table 10 indicate that the mean B2MU concentration observed among workers experiencing occupational exposure to cadmium (but with undefined levels of proteinuria) is 160-7400 µg/g CRTU. One of these studies reports geometric means lower than this range (i.e., as low as 71 µg/g CRTU); an explanation for this wide spread in average concentrations is not available.

Seven of the studies listed in Table 10 report a range of B2MU levels among those diagnosed as having renal dysfunction. As indicated in this table, renal dysfunction (proteinuria) is defined in several of these studies by B2MU levels in excess of 300 µg/g CRTU (see footnote "c" of Table 10); therefore, the range of B2MU levels observed in these studies is a function of the operational definition used to identify those with renal dysfunction. Nevertheless, a B2MU level of 300 µg/g CRTU appears to be a meaningful threshold for identifying those having early signs of kidney damage. While levels much higher than 300 µg/g CRTU have been observed among those with renal dysfunction, the vast majority of those not occupationally exposed to cadmium exhibit much lower B2MU concentrations (see Table 9). Similarly, the vast majority of workers *not* exhibiting renal dysfunction are found to have levels below 300 µg/g CRTU (Table 9).

The 300 µg/g CRTU level for B2MU proposed in the above paragraph has support among researchers as the threshold level that distinguishes between cadmium-exposed workers with and without kidney dysfunction. For example, in the guide for physicians who must evaluate cadmium-exposed workers written for the Cadmium Council by Dr. Lauwerys, levels of B2M greater than 200-300 µg/g CRTU are considered to require additional medical evaluation for kidney dysfunction (exhibit 8-447, OSHA docket H057A). The most widely used test for measuring B2M (i.e., the Pharmacia Delphia test) defines B2MU levels above 300 µg/l as abnormal (exhibit L-140-1, OSHA docket H057A).

Dr. Elinder, chairman of the Department of Nephrology at the Karolinska Institute, testified at the hearings on the proposed cadmium rule. According to Dr. Elinder (exhibit L-140-45, OSHA docket H057A), the normal concentration of B2MU has been well documented (Evrin and Wibell 1972; Kjellstrom et al. 1977a; Elinder et al. 1978, 1983; Buchet et al. 1980; Jawaid et al. 1983; Kowal and Zirkes, 1983). Elinder stated that the upper 95 or 97.5 percentiles for B2MU among those without tubular dysfunction is below 300 µg/g CRTU (Kjellstrom et al. 1977a; Buchet et al. 1980; Kowal and Zirkes, 1983). Elinder defined levels of B2M above 300 µg/g CRTU as "slight" proteinuria.

### 5.3.8 Conclusions and Recommendations for B2MU

Based on the above evaluation, the following recommendations are made for a B2MU proficiency testing program. Note that the following discussion addresses only sampling and analysis for B2MU determinations (i.e., to be reported as an unadjusted µg B2M/l urine). Normalizing this result to creatinine requires a second analysis for CRTU (see Section 5.4) so that the ratio of the 2 measurements can be obtained.

#### 5.3.8.1 *Recommended method*

The Pharmacia Delphia method (Pharmacia 1990) should be adopted as the standard method for B2MU determinations. Laboratories may adopt alternate methods, but it is the responsibility of the laboratory to demonstrate that alternate methods provide results of comparable quality to the Pharmacia Delphia method.

#### 5.3.8.2 *Data quality objectives*

The following data quality objectives should facilitate interpretation of analytical results, and should be achievable based on the above evaluation.

*Limit of Detection.* A limit of 100 µg/l urine should be achievable, although the insert to the test kit (Pharmacia 1990) cites a detection limit of 150 µg/l; private conversations with representatives of Pharmacia, however, indicate that the lower limit of 100 µg/l should be achievable provided an additional standard of 100 µg/l B2M is run with the other standards to derive the calibration curve (Section 3.3.1.1). The lower detection limit is desirable due to the proximity of this detection limit to B2MU values defined for the cadmium medical monitoring program.

*Accuracy.* Because results from an interlaboratory proficiency testing program are not available currently, it is difficult to define an achievable level of accuracy. Given the general performance parameters defined by the insert to the test kits, however, an accuracy of ±15% of the target value appears achievable.

Due to the low levels of B2MU to be measured generally, it is anticipated that the analysis of creatinine will contribute relatively little to the overall variability observed among creatinine-normalized B2MU levels (see Section 5.4). The initial level of accuracy for reporting B2MU levels under this program should be set at ±15%.

*Precision.* Based on precision data reported by Pharmacia (1990), a precision value (i.e., CV) of 5% should be achievable over the defined range of the analyte. For internal QC samples (i.e., recommended as part of an internal QA/QC program, Section 3.3.1), laboratories should attain precision near 5% over the range of concentrations measured.

### 5.3.8.3 Quality assurance/quality control

Commercial laboratories providing measurement of B2MU should adopt an internal QA/QC program that incorporates the following components: Strict adherence to the Pharmacia Delphia method, including calibration requirements; regular use of QC samples during routine runs; a protocol for corrective actions, and documentation of these actions; and, participation in an interlaboratory proficiency program. Procedures that may be used to address internal QC requirements are presented in Attachment 1. Due to differences between analyses for B2MU and CDB/CDU, specific values presented in Attachment 1 may have to be modified. Other components of the program (including characterization runs), however, can be adapted to a program for B2MU.

## 5.4 Monitoring Creatinine in Urine (CRTU)

Because CDU and B2MU should be reported relative to concentrations of CRTU, these concentrations should be determined in addition CDU and B2MU determinations.

### 5.4.1 Units of CRTU Measurement

CDU should be reported as µg Cd/g CRTU, while B2MU should be reported as µg B2M/g CRTU. To derive the ratio of cadmium or B2M to creatinine, CRTU should be reported in units of g crtn/l of urine. Depending on the analytical method, it may be necessary to convert results of creatinine determinations accordingly.

### 5.4.2 Analytical Techniques Used to Monitor CRTU

Of the techniques available for CRTU determinations, an absorbance spectrophotometric technique and a high-performance liquid chromatography (HPLC) technique are identified as acceptable in this protocol.

### 5.4.3 Methods Developed for CRTU Determinations

CRTU analysise performed in support of either CDU or B2MU determinations should be performed using either of the following 2 methods:

1. The Du Pont method (i.e., Jaffe method), in which creatinine in a sample reacts with picrate under alkaline conditions, and the resulting red chromophore is monitored (at 510 nm) for a fixed interval to determine the rate of the reaction; this reaction rate is proportional to the concentration of creatinine present in the sample (a copy of this method is provided in Attachment 2 of this protocol); or,

2. The OSHA SLC Technical Center (OSLTC) method, in which creatinine in an aliquot of sample is separated using an HPLC column equipped with a UV detector; the resulting peak is quantified using an electrical integrator (a copy of this method is provided in Attachment 3 of this protocol).

### 5.4.4 Sample Collection and Handling

CRTU samples should be segregated from samples collected for CDU or B2MU analysis. Sample-collection techniques have been described under Section 5.2.4. Samples should be preserved either to stabilize CDU (with $HNO_3$) or B2MU (with NaOH). Neither of these procedures should adversely affect CRTU analysis (see Attachment 3).

### 5.4.5 General Method Performance

Data from the OSLTC indicate that a CV of 5% should be achievable using the OSLTC method (Septon, L private communication). The achievable accuracy of this method has not been determined.

Results reported in surveys conducted by the CAP (CAP 1991a, 1991b and 1992) indicate that a CV of 5% is achievable. The accuracy achievable for CRTU determinations has not been reported.

Laboratories performing creatinine analysis under this protocol should be CAP accredited and should be active participants in the CAP surveys.

### 5.4.6 Observed CRTU Concentrations

Published data suggest the range of CRTU concentrations is 1.0-1.6 g in 24-hour urine samples (Harrison 1987). These values are equivalent to about 1 g/l urine.

### 5.4.7 Conclusions and Recommendations for CRTU

#### 5.4.7.1 Recommended method

Use either the Jaffe method (Attachment 2) or the OSLTC method (Attachment 3). Alternate methods may be acceptable provided adequate performance is demonstrated in the CAP program.

#### 5.4.7.2 Data quality objectives

*Limit of Detection.* This value has not been formally defined; however, a value of 0.1 g/l urine should be readily achievable.

*Accuracy.* This value has not been defined formally; accuracy should be sufficient to retain accreditation from the CAP.

*Precision.* A CV of 5% should be achievable using the recommended methods.

## 6.0 References

Adamsson E, Piscator M, and Nogawa K. (1979). Pulmonary and gastrointestinal exposure to cadmium oxide dust in a battery factory. *Environmental Health Perspectives,* 28, 219-222.

American Conference of Governmental Industrial Hygienists (ACGIH). (1986). *Documentation of the Threshold Limit Values and Biological Exposure Indices.* 5th edition. p. BEI-55.

Bernard A, Buchet J, Roels H, Masson P, and Lauwerys R. (1979). Renal excretion of proteins and enzymes in workers exposed to cadmium. *European Journal of Clinical Investigation,* 9, 11-22.

Bernard A and Lauwerys R. (1990). Early markers of cadmium nephrotoxicity: Biological significance and predictive value. *Toxocological and Environmental Chemistry,* 27, 65-72.

Braunwald E, Isselbacher K, Petersdorf R, Wilson J, Martin J, and Fauci A (Eds.). (1987). *Harrison's Principles of Internal Medicine.* New York: Mc-Graw-Hill Book Company.

Buchet J, Roels H, Bernard I, and Lauwerys R. (1980). Assessment of renal funcion of workers exposed to inorganic lead, cadmium, or mercury vapor. *Journal of Occupational Medicine,* 22, 741-750.

CAP. (1991). Urine Chemistry, Series 1: Survey (Set U-B). College of American Pathologists.

CAP. (1991). Urine Chemistry, Series 1: Survey (Set U-C). College of American Pathologists.

CAP. (1992). Urine Chemistry, Series 1: Survey (Set U-A). College of American Pathologists.

CDC. (1986). Centers for Disease Control, Division of Environmental Health Laboratory Sciences, Center for Environmental Health, Atlanta, Georgia. Docket No. 106A. Lake Couer d'Alene, Idaho cadmium and lead study: 86-0030, Specimen collection and shipping protocol.

CDC. (1990). Centers for Disease Control, Nutritional Biochemistry Branch. 4/27/90 Draft SOP for Method 0360A "Determination of cadmium in urine by graphite furnace atomic absorption spectrometry with Zeeman background correction.

Centre de Toxicologie du Quebec. (1991). Interlaboratory comparison program report for run #2. Shipping date 3/11/91. Addition BLR 9/19.

Chia K, Ong C, Ong H, and Endo G. (1989). Renal tubular function of workers exposed to low levels of cadmium. *British Journal of Industrial Medicine,* 46, 165-170.

Claeys-Thoreau F. (1982). Determination of low levels of cadmium and lead in biological fluids with simple dilution by atomic absorption spectrophotometry using Zeeman effect background absorption and the L'Vov platform. *Atomic Spectroscopy,* 3, 188-191.

DeBenzo Z, Fraile R, and Carrion N. (1990). Electrothermal atomization atomic absorption spectrometry with stabilized aqueous standards for the determination of cadmium in whole blood. *Analytica Chimica Acta,* 231, 283-288.

Elinder C, Edling C, Lindberg E, Kagedal B, and Vesterberg O. (1985). Assessment of renal function in workers previously exposed to cadmium. *British Journal of Internal Medicine,* 42, 754.

Ellis K, Cohn S, and Smith T. (1985). Cadmium inhalation exposure estimates: Their significance with respect to kidney and liver cadmium burden. *Journal of Toxicology and Environmental Health,* 15, 173-187.

Ellis K, Yasumura S, Vartsky D, and Cohn S. (1983). Evaluation of biological indicators of body burden of cadmium in humans. *Fundamentals and Applied Toxicology,* 3, 169-174.

Ellis K, Yeun K, Yasumura S, and Cohn S. (1984). Doseresponse analysis of cadmium in man: Body burden vs kidney function. *Environmental Research,* 33, 216-226.

Evrin P, Peterson A, Wide I, and Berggard I. (1971). Radioimmunoassay of B-2-microglobulin in human biological fluids. *Scandanavian Journal of Clinical Laboratory Investigation,* 28, 439-443.

Falck F, Fine L, Smith R, Garvey J, Schork A, England B, McClatchey K, and Linton J. (1983). Metallothionein and occupational exposure to cadmium. *British Journal of Industrial Medicine, 40*, 305-313.

FEDERAL REGISTER. (1990). Occupational exposure to cadmium: Proposed rule. 55/22/4052-4147, February 6.

Friberg, Exhibit 29, (1990). Exhibit No. 29 of the OSHA Federal Docket H057A. Washington, DC.

Friberg L. (1988). Quality assurance. In T. Clarkson (Ed.), *Biological Monitoring of Toxic Metals* (pp. 103-105). New York: Plenum Press.

Friberg L, and Elinder C. (1988). Cadmium toxicity in humans. In *Essential and Trace Elements in Human Health and Disease* (pp. 559-587). Docket Number 8-660.

Friberg L, Elinder F, et al. (1986). *Cadmium and Health: A Toxicological and Epidemiological Appraisal. Volume II, Effects and Response.* Boca Raton, FL: CRC Press.

Friberg L, Piscator M, Nordberg G, and Kjellstrom T. (1974). *Cadmium in the Environment* (2nd ed.). Cleveland:CRC.

Friberg L and Vahter M. (1983). Assessment of exposure to lead and cadmium through biological monitoring: Results of a UNEP/WHO global study. *Environmental Research, 30*, 95-128.

Gunter E, and Miller D. (1986). Laboratory procedures used by the division of environmental health laboratory sciences center for environmental health, Centers for Disease Control for the hispanic health and nutrition examination survey (HHANES). Atlanta, GA: Centers for Disease Control.

Harrison. (1987). Harrison's Principles of Internal Medicine. Braunwald, E; Isselbacher, KJ; Petersdorf, RG; Wilson, JD; Martin, JB; and Fauci, AS Eds. Eleventh Ed. McGraw Hill Book Company. San Francisco.

Henry J. (1991). *Clinical Diagnosis and Management by Laboratory Methods* (18th edition). Philadelphia: WB Saunders Company.

IARC (1987). *IRAC Monographs on the Evaluation of Carcinogenic Risks to Humans. Overall Evaluation of Carcinogenicity: Update of Volume 1-42.* Supplemental 7, 1987.

Ishizaki M, Kido T, Honda R, Tsuritani I, Yamada Y, Nakagawa H, and Nogawa K. (1989). Dose-response relationship between urinary cadmium and B-2-microglobulin in a Japanese environmentally cadmium exposed population. *Toxicology, 58*, 121-131.

Iwao S, Tsuchiya K, and Sakurai H. (1980). Serum and urinary B-2-microglobulin among cadmium-exposed workers. *Journal of Occupational Medicine, 22*, 399-402.

Iwata K, Katoh T, Morikawa Y, Aoshima K, Nishijo M, Teranishi H, and Kasuya M. (1988). Urinary trehalase activity as an indicator of kidney injury due to environmental cadmium exposure. *Archives of Toxicology, 62*, 435-439.

Kawada T, Koyama H, and Suzuki S. (1989). Cadmium, NAG activity, and B-2-microglobulin in the urine of cadmium pigment workers. *British Journal of Industrial Medicine, 46*, 52-55.

Kawada T, Tohyama C, and Suzuki S. (1990). Significance of the excretion of urinary indicator proteins for a low level of occupational exposure to cadmium. *International Archives of Occupational Environmental Health, 62*, 95-100.

Kjellstrom T. (1979). Exposure and accumulation of cadmium in populations from Japan, the United States, and Sweden. *Environmental Health Perspectives, 28*, 169-197.

Kjellstrom T, Evrin P, and Rahnster B. (1977). Dose-response analysis of cadmium-induced tubular proteinuria. *Environmental Research, 13*, 303-317.

Kjellstrom T, Shiroishi K, and Evrin P. (1977). Urinary B-2microglobulin excretion among people exposed to cadmium in the general environment. *Environmental Research, 13*, 318-344.

Kneip T, & Crable J (Eds.). (1988). Method 107. Cadmium in blood. *Methods for biological monitoring* (pp.161-164). Washington, DC: American Public Health Association.

Kowal N. (1988). Urinary cadmium and B-2-microglobulin: Correlation with nutrition and smoking history. *Journal of Toxicology and Environmental Health, 25*, 179-183.

Kowal N, Johnson D, Kraemer D, and Pahren H. (1979). Normal levels of cadmium in diet, urine, blood, and tissues of inhabitants of the United States. *Journal of Toxicology and Environmental Health, 5*, 995-1014.

Kowal N and Zirkes M. (1983). Urinary cadmium and B-2microglobulin: Normal values and concentration adjustment. *Journal of Toxicology and Environmental Health, 11*, 607-624.

Lauwerys R, Buchet J, and Roels H. (1976). The relationship between cadmium exposure or body burden and the concentration of cadmium in blood and urine in man. *International Archives of Occupational and Environmental Health, 36*, 275-285

Lauwerys R, Roels H, Regniers, Buchet J, and Bernard A. (1979). Significance of cadmium concentration in blood and in urine in workers exposed to cadmium. *Environmental Research, 20*, 375-391.

Lind B, Elinder C, Friberg L, Nilsson B, Svartengren M, and Vahter M. (1987). Quality control in the analysis of lead and cadmium in blood. *Fresenius' Zeitschrift fur Analytical Chemistry, 326*, 647-655.

Mason H, Davison A, Wright A, Guthrie C, Fayers P, Venables K, Smith N, Chettle D, Franklin D, Scott M, Holden H, Gompertz D, and Newman-Taylor A. (1988). Relations between liver cadmium, cumulative exposure, and renal function in cadmium alloy workers. *British Journal of Industrial Medicine, 45*, 793-802.

Meridian Research, Inc. (1989). *Quantitative Assessment of Cancer Risks Associated with Occupational Exposure to Cd.* Prepared by Meridian Research, Inc. and Roth Associates, Inc. for the Occupational Safety & Health Administration. June 12, 1989.

Meridian Research, Inc and Roth Associates, Inc. (1989). *Quantitative Assessment of the Risk of Kidney Dysfunction Associated with Occupational Exposure to Cd.* Prepared by Meridian Research, Inc. and Roth Associates, Inc. for the Occupational Safety & Health Administration. July 31 1989.

Micheils E and DeBievre P. (1986). Method 25-Determination of cadmium in whole blood by isotope dilution mass spectrometry. O'Neill I, Schuller P, and Fishbein L (Eds.), *Environmental Carcinogens Selected Methods of Analysis* (Vol. 8). Lyon, France: International Agency for Research on Cancer.

Mueller P, Smith S, Steinberg K, and Thun M. (1989). Chronic renal tubular effects in relation to urine cadmium levels. *Nephron, 52*, 45-54.

NIOSH. (1984a). Elements in blood or tissues. Method 8005 issued 5/15/85 and Metals in urine. Method 8310 issued 2/15/84 In P. Eller (Ed.), *NIOSH Manual of Analytical Methods* (Vol. 1, Ed. 3). Cincinnati, Ohio: US-DHHS.

NIOSH. (1984b). Lowry L. Section F: Special considerations for biological samples in *NIOSH Manual of Analytical Methods* (Vol. 1, 3rd ed) P. Eller (Ed.). Cincinnati, Ohio: US-DHHS.

Nordberg G and Nordberg M. (1988). Biological monitoring of cadmium. In T. Clarkson, L. Friberg, G. Nordberg, and P. Sager (Eds.), *Biological Monitoring of Toxic Metals*, New York: Plenum Press.

Nogawa K. (1984). Biologic indicators of cadmium nephrotoxicity in persons with low-level cadmium exposure. *Environmental Health Perspectives, 54*, 163-169.

OSLTC (no date). Analysis of Creatinine for the Normalization of Cadmium and Beta-2-Microglobulin Concentrations in Urine. OSHA Salt Lake Technical Center. Salt Lake City, UT. Paschal. (1990). Attachment 8 of exhibit 106 of the OSHA docket H057A.

Perkin-Elmer Corporation. (1982). *Analytical Methods for Atomic Absorption Spectroscopy.*

Perkin-Elmer Corporation. (1977). *Analytical Methods Using the HGA Graphite Furnace.*

Pharmacia Diagnostics. (1990). Pharmacia DELFIA system B-2microglobulin kit insert. Uppsala, Sweden: Pharmacia Diagnostics.

Piscator M. (1962). Proteinuria in chronic cadmium poisoning. *Archives of Environmental Health,5*, 55-62.

Potts, C.L. (1965). Cadmium Proteinuria—The Health Battery Workers Exposed to Cadmium Oxide dust. Ann Occup Hyg, 3:55-61, 1965.

Princi F. (1947). A study of industrial exposures to cadmium. *Journal of Industrial Hygiene and Toxicology, 29*, 315-320.

Pruszkowska E, Carnick G, and Slavin W. (1983). Direct determination of cadmium in urine with use of a stabilized temperature platform furnace and Zeeman background correction. *Clinical Chemistry, 29*, 477-480.

Roberts C and Clark J. (1986). Improved determination of cadmium in blood and plasma by flameless atomic absorption spectroscopy. *Bulletin of Environmental Contamination and Toxicology, 36*, 496-499.

Roelandts I. (1989). Biological reference materials. *Soectrochimica Acta, 44B*, 281-290.

Roels H, Buchet R, Lauwerys R, Bruaux P, Clays-Thoreau F, Laafontaine A, Overschelde J, and Verduyn J. (1978). Lead and cadmium absorption among children near a nonferrous metal plant. *Environmental Research, 15*, 290-308.

Roels H, Djubgang J, Buchet J, Bernard A, and Lauwerys R. (1982). Evolution of cadmium-induced renal dysfunction in workers removed from exposure. *Scandanavian Journal of Work and Environmental Health, 8*, 191-200.

Roels H, Lauwerys R, and Buchet J. (1989). Health significance of cadmium induced renal dysfunction: A five year follow-up. *British Journal of Industrial Medicine, 46*, 755-764.

Roels J, Lauwerys R, Buchet J, Bernard A, Chettle D, Harvey T, and Al-Haddad I. (1981). In vivo measurements of liver and kidney cadmium in workers exposed to this metal: Its significance with respect to cadmium in blood and urine. *Environmental Research, 26*, 217-240.

Roels H, Lauwerys R, Buchet J, Bernard A, Lijnen P, and Houte G. (1990). Urinary kallikrein activity in workers exposed to cadmium, lead, or mercury vapor. *British Journal of Industrial Medicine, 47*, 331-337.

Sakurai H, Omae K, Toyama T, Higashi T, and Nakadate T. (1982). Cross-sectional study of pulmonary function in cadmium alloy workers. *Scandanavian Journal of Work and Environmental Health, 8*, 122-130.

Schardun G and van Epps L. (1987). B-2-microglobulin: Its significance in the evaluation of renal function. *Kidney International, 32*, 635-641.

Shaikh Z, and Smith L. (1984). Biological indicators of cadmium exposure and toxicity. *Experentia, 40*, 36-43.

Smith J and Kench J. (1957). Observations on urinary cadmium and protein excretion in men exposed to cadmium oxide dust and fume. *British Journal of Industrial Medicine, 14*, 240-245.

Smith J, Kench J, and Lane R. (1955). Determination of Cadmium in urine and observations on urinary cadmium and protein excretion in men exposed to cadmium oxide dust. *British Journal of Industrial Medicine, 12*, 698-701.

SWRI (Southwest Research Institute). (1978). The distribution of cadmium and other metals in human tissues. Health Effects Research Lab, Research Triangle Park, NC, Population Studies Division. NTIS No. PB-285-200.

Stewart M and Hughes E. (1981). Urinary B-2-microglobulin in the biological monitoring of cadmium workers. *British Journal of Industrial Medicine, 38*, 170-174.

Stoeppler K and Brandt M. (1980). Contributions to automated trace analysis. Part V. Determination of cadmium in whole blood and urine by electrothermal atomic absorption spectrophotometry. *Fresenius' Zeitschrift fur Analytical Chemistry, 300*, 372-380.

Takenaka et al. (1983). *Carcinogencity of Cd Chloride Aerosols in White Rates.* INCI 70: 367-373, 1983.

Thun M, Osorio A, Schober S, Hannon W, Lewis B, and Halperin W. (1989). Nephropathy in cadmium workers: Assessment of risk from airborne occupational exposure to cadmium. *British Journal of Industrial Medicine, 46*, 689-697.

Thun M, Schnorr T, Smith A, Halperin W, and Lemen R. (1985). Mortality among a cohort of US cadmium production workers—an update. *Journal of the National Cancer Institute, 74*, 325-333.

Travis D and Haddock A. (1980). Interpretation of the observed age-dependency of cadmium body burdens in man. *Environmental Research, 22*, 46-60.

Tsuchiya K. (1967). Proteinuria of workers exposed to cadmium fume. *Archives of Environmental Health, 14*, 875-880.

Tsuchiya K. (1976). Proteinuria of cadmium workers. *Journal of Occupational Medicine, 18*, 463-470.

Tsuchiya K, Iwao S, Sugita M, Sakurai H. (1979). Increased urinary B-2-microglobulin in cadmium exposure: Dose-effect relationship and biological significance of B-2-microglobulin. *Environmental Health Perspectives, 28*, 147-153.

USEPA. (1985). Updated Mutagenicity and Carcinogenicity Assessments of Cd: Addendum to the Health Assessment Document for Cd (May 1981). Final Report. June 1985.

Vahter M and Friberg L. (1988). Quality control in integrated human exposure monitoring of lead and cadmium. *Fresenius' Zeitschrift fur Analytical Chemistry, 332*, 726-731.

Weber J. (1988). An interlaboratory comparison programme for several toxic substances in blood and urine. *The Science of the Total Environment, 71*, 111-123.

Weber J. (1991a). Accuracy and precision of trace metal determinations in biological fluids. In K. Subramanian, G. Iyengar, and K. Okamot (Eds.), *Biological Trace Element Research*-Multidisciplinary Perspectives, ACS Symposium Series 445. Washington, DC: American Chemical Society.

Weber J. (1991b). Personal communication about interlaboratory program and shipping biological media samples for cadmium analyses.

Wibowo A, Herber R, van Deyck W, and Zielhuis R. (1982). Biological assessment of exposure in factories with second degree usage of cadmium compounds. *International Archives of Occupational Environmental Health, 49*, 265-273.

## ATTACHMENT 1:
### NONMANDATORY PROTOCOL FOR AN INTERNAL QUALITY ASSURANCE/QUALITY CONTROL PROGRAM

The following is an example of the type of internal quality assurance/quality control program that assures adequate control to satisfy OSHA requirements under this protocol. However, other approaches may also be acceptable.

As indicated in Section 3.3.1 of the protocol, the QA/QC program for CDB and CDU should address, at a minimum, the following:

- calibration;
- establishment of control limits;
- internal QC analyses and maintaining control; and
- corrective action protocols.

This illustrative program includes both initial characterization runs to establish the performance of the method and ongoing analysis of quality control samples intermixed with compliance samples to maintain control.

### Calibration

Before any analytical runs are conducted, the analytic instrument must be calibrated. This is to be done at the beginning of each day on which quality control samples and/or compliance samples are run. Once calibration is established, quality control samples or compliance samples may be run. Regardless of the type of samples run, every fifth sample must be a standard to assure that the calibration is holding.

Calibration is defined as holding if every standard is within plus or minus (±) 15% of its theoretical value. If a standard is more than plus or minus 15% of its theoretical value, then the run is out of control due to calibration error and the entire set of samples must either be reanalyzed after recalibrating or results should be recalculated based on a statistical curve derived from the measurement of all standards.

It is essential that the highest standard run is higher than the highest sample run. To assure that this is the case, it may be necessary to run a high standard at the end of the run, which is selected based on the results obtained over the course of the run.

All standards should be kept fresh, and as they get old, they should be compared with new standards and replaced if they exceed the new standards by ±15%.

### INITIAL CHARACTERIZATION RUNS AND ESTABLISHING CONTROL

A participating laboratory should establish four pools of quality control samples for each of the analytes for which determinations will be made. The concentrations of quality control samples within each pool are to be centered around each of the four target levels for the particular analyte identified in Section 4.4 of the protocol.

Within each pool, at least 4 quality control samples need to be established with varying concentrations ranging between plus or minus 50% of the target value of that pool. Thus for the medium-high cadmium in blood pool, the theoretical values of the quality control samples may range from 5 to 15 µg/l, (the target value is 10 µg/l). At least 4 unique theoretical values must be represented in this pool.

The range of theoretical values of plus or minus 50% of the target value of a pool means that there will be overlap of the pools. For example, the range of values for the medium-low pool for cadmium in blood is 3.5 to 10.5 µg/l while the range of values for the medium-high pool is 5 to 15 µg/l. Therefore, it is possible for a quality control sample from the medium-low pool to have a higher concentration of cadmium than a quality control sample from the medium-high pool.

Quality control samples may be obtained as commercially available reference materials, internally prepared, or both. Internally prepared samples should be well characterized and traced or compared to a reference material for which a consensus value for concentration is available. Levels of analyte in the quality control samples must be concealed from the analyst prior to the reporting of analytical results. Potential sources of materials that may be used to construct quality control samples are listed in Section 3.3.1 of the protocol.

Before any compliance samples are analyzed, control limits must be established. Control limits should be calculated for every pool of each analyte for which determinations will be made and control charts should be kept for each pool of each analyte. A separate set of control charts and control limits should be established for each analytical instrument in a laboratory that will be used for analysis of compliance samples.

At the beginning of this QA/QC program, control limits should be based on the results of the analysis of 20 quality control samples from each pool of each analyte. For any given pool, the 20 quality control samples should be run on 20 different days. Although no more than one sample should be run from any single pool on a particular day, a laboratory may run quality control samples from different pools on the same day. This constitutes a set of initial characterization runs.

For each quality control sample analyzed, the value F/T (defined in the glossary) should be calculated. To calculate the control limits for a pool of an analyte, it is first necessary to calculate the mean, $\bar{x}$, of the F/T values for each quality control sample in a pool and then to calculate its standard deviation. Thus, for the control limit for a pool, $\bar{x}$ is calculated as:

$$\frac{\left(\sum \frac{F}{T}\right)}{N}$$

and σ is calculated as

$$\left[\frac{\sum \left(\frac{F}{T} - \bar{X}\right)^2}{N-1}\right]^{1/2}$$

Where N is the number of quality control samples run for a pool.

The control limit for a particular pool is then given by the mean plus or minus 2 standard deviations (X ±3σ).

The control limits may be no greater than 40% of the mean F/T value. If three standard deviations are greater than 40% of the mean F/T value, then analysis of compliance samples may not begin.[1] Instead, an investigation into the causes of the large standard deviation should begin, and the inadequacies must be remedied. Then, control limits must be reestablished which will mean repeating the running 20 quality control samples from each pool over 20 days.

### Internal Quality Control Analyses and Maintaining Control

Once control limits have been established for each pool of an analyte, analysis of compliance samples may begin. During any run of compliance samples, quality control samples are to be interspersed at a rate of no less than 5% of the compliance sample workload. When quality control samples are run, however, they should be run in sets consisting of one quality control sample from each pool. Therefore, it may be necessary, at times, to intersperse quality control samples at a rate greater than 5%.

There should be at least one set of quality control samples run with any analysis of compliance samples. At a minimum, for example, 4 quality control samples should be run even if only 1 compliance sample is run. Generally, the number of quality control samples that should be run are a multiple of four with the minimum equal to the smallest multiple of four that is greater than 5% of the total number of samples to be run. For example, if 300 compliance samples of an analyte are run, then at least 16 quality control samples should be run (16 is the smallest multiple of four that is greater than 15, which is 5% of 300).

Control charts for each pool of an analyte (and for each instrument in the laboratory to be used for analysis of compliance samples) should be established by plotting F/T versus date as the quality control sample results are reported. On the graph there should be lines representing the control limits for the pool, the mean F/T limits for the pool, and the theoretical F/T of 1.000. Lines representing plus or minus (±)$\hat{\sigma}$ should also be represented on the charts. A theoretical example of a control chart is presented in Figure 1.

**FIGURE 1—THEORETICAL EXAMPLE OF A CONTROL CHART FOR A POOL OF AN ANALYTE**

1.162 (Upper Control Line)
1.096 (Upper $\hat{\sigma}$ Line)
1.000 (Theoretical Mean)
0.964 (Mean)
0.832 (Lower $\hat{\sigma}$ Line)
0.766 (Lower Control Limit)

March 2 2 3 5 6 9 10 13 16 17

All quality control samples should be plotted on the chart, and the charts should be checked for visual trends. If a quality control sample falls above or below the control limits for its pool, then corrective steps must be taken (see the section on corrective actions below). Once a laboratory's program has been established, control limits should be updated every 2 months.

The updated control limits should be calculated from the results of the last 100 quality control samples run for each pool. If 100 quality control samples from a pool have not been run at the time of the update, then the limits should be based on as many as have been run provided at least 20 quality control samples from each pool have been run over 20 different days.

The trends that should be looked for on the control charts are:

1. 10 consecutive quality control samples falling above or below the mean;

2. 3 consecutive quality control samples falling more than 2σ from the mean (above or below the 2σ lines of the chart); or

3. the mean calculated to update the control limits falls more than 10% above or below the theoretical mean of 1.000.

If any of these trends is observed, then all analysis must be stopped, and an investigation into the causes of the errors must begin. Before the analysis of compliance samples may resume, the inadequacies must be remedied and the control limits must be reestablished for that pool of an analyte. Reestablishment of control limits will entail running 20 sets of quality control samples over 20 days.

Note that alternative procedures for defining internal quality control limits may also be acceptable. Limits may be based, for example, on proficiency testing, such as ±1 µg or 15% of the mean (whichever is greater). These should be clearly defined.

### Corrective actions

Corrective action is the term used to describe the identification and remediation of errors occurring within an analysis. Corrective action is necessary whenever the result of the analysis of any quality control sample falls outside of the established control limits. The steps involved may include simple things like checking calculations of basic instrument maintenance, or it may involve more complicated actions like major instrument repair. Whatever the source of error, it must be identified and corrected (and a Corrective Action Report (CAR) must be completed. CARs should be kept on file by the laboratory.

---

[1] Note that the value, "40%" may change over time as experience is gained with the program.

## ATTACHMENT 2—
## CREATININE IN URINE (JAFFE PROCEDURE)

Intended use: The CREA pack is used in the Du Pont ACA® discrete clinical analyzer to quantitatively measure creatinine in serum and urine.

Summary: The CREA method employs a modification of the kinetic Jaffe reaction reported by Larsen. This method has been reported to be less susceptible than conventional methods to interference from non-creatinine, Jaffe-positive compounds.[1]

A split sample comparison between the CREA method and a conventional Jaffe procedure on Autoanalyzer® showed a good correlation. (See Specific Performance Characteristics).

*Note: Numbered subscripts refer to the bibliography and lettered subscripts refer to footnotes.

Autoanalyzer,® is a registered trademark of Technicon Corp., Tarrytown, NY.

Principles of Procedure: In the presence of a strong base such as NaOH, picrate reacts with creatinine to form a red chromophore. The rate of increasing absorbance at 510 nm due to the formation of this chromophore during a 17.07-second measurement period is directly proportional to the creatinine concentration in the sample.

$$\text{Creatinine} + \text{Picrate} \xrightarrow{\text{NaOH}} \text{Red chromophore (absorbs at 510 nm)}$$

Reagents:

| Compartment[a] | Form | Ingredient | Quantity[b] |
|---|---|---|---|
| No. 2, 3, & 4 | Liquid | Picrate | 0.11 mmol. |
| No. 6 | Liquid | NaOH (for pH adjustment)[c] | |

[a] Compartments are numbered 1-7, with compartment #7 located closest to pack fill position #2.
[b] Nominal value at manufacture.
[c] See Precautions.

Precautions: Compartment #6 contains 75 µL of 10 N NaOH; avoid contact; skin irritant; rinse contacted area with water. Comply with OSHA'S Bloodborne Pathogens Standard while handling biological samples (29 CFR 1910.1039).

Used packs contain human body fluids; handle with appropriate care.

### FOR IN VITRO DIAGNOSTIC USE
### Mixing and Diluting:

Mixing and diluting are automatically performed by the ACA® discrete clinical analyzer. The sample cup must contain sufficient quantity to accommodate the sample volume plus the "dead volume"; precise cup filling is not required.

### SAMPLE CUP VOLUMES (µL)

| Analyzer | Standard Dead | Standard Total | Microsystem Dead | Microsystem Total |
|---|---|---|---|---|
| II, III | 120 | 3000 | 10 | 500 |
| IV, SX | 120 | 3000 | 30 | 500 |
| V | 90 | 3000 | 10 | 500 |

Storage of Unprocessed Packs: Store at 2-8 °C. Do not freeze. Do not expose to temperatures above 35 °C or to direct sunlight.

Expiration: Refer to EXPIRATION DATE on the tray label.

Specimen Collection: Serum or urine can be collected and stored by normal procedures.[2]

### Known Interfering Substances[3]

- Serum Protein Influence—Serum protein levels exert a direct influence on the CREA assay. The following should be taken into account when this method is used for urine samples and when it is calibrated:

    Aqueous creatinine standards or urine specimens will give CREA results depressed by approximately 0.7 mg/dL [62 µmol/L][d] and will be less precise than samples containing more than 3 g/dL [30 g/L] protein.

    All urine specimens should be diluted with an albumin solution to give a final protein concentration of at least 3 g/dL [30 g/L]. Du Pont Enzyme Diluent (Cat. #790035-901) may be used for this purpose.

- High concentration of endrogenous bilirubin (.20 mg/dL [.342 µmol/L]) will give depressed CREA results (average depression 0.8 mg/dL [71 µmol/L]).[4]

- Grossly hemolyzed (hemoglobin .100 mg/dL [.62 µmol/L]) or visibly lipemic specimens may cause falsely elevated CREA results.[5,6]

- The following cephalosporin antibiotics do not interfere with the CREA method when present at the concentrations indicated. Systematic inaccuracies (bias) due to these substances are less than or equal to 0.1 mg/dL [8.84 µmol/L] at CREA concentrations of approximately 1 mg/dL [88 µmol/L].

| Antibiotic | Peak serum level[7,8,9] mg/dL | [mmol/L] | Drug concentration mg/dL | [mmol/L] |
|---|---|---|---|---|
| Cephaloridine | 1.4 | 0.3 | 25 | 6.0 |
| Cephalexin | 0.6-2.0 | 0.2-0.6 | 25 | 7.2 |
| Cephamandole | 1.3-2.5 | 0.3-0.5 | 25 | 4.9 |
| Cephapirin | 2.0 | D0.4 | 25 | 5.6 |
| Cephradine | 1.5-2.0 | 0.4-0.6 | 25 | 7.1 |
| Cefazolin | 2.5-5.0 | 0.55-1.1 | 50 | 11.0 |

- The following cephalosporin antibiotics have been shown to affect CREA results when present at the indicated concentrations. System inaccuracies (bias) due to these substances are greater that 0.1 mg/dL [8.84 µmol/L] at CREA concentrations of:

| Antibiotic | Peak serum level[7,8,9] mg/dL | [mmol/L] | Drug concentration mg/dL | [mmol/L] | Effect |
|---|---|---|---|---|---|
| Cephalothin | 1-6 | 0.2-1.5 | 100 | 25.2 | ↓20-25% |
| Cephoxitin | 2.0 | 0.5 | 5.0 | 1.2 | ↑35-40% |

- The single wavelength measurement used in this method eliminates interference from chromophores whose 510 nm absorbance is constant throughout the measurement period.

- Each laboratory should determine the acceptability of its own blood collection tubes and serum separation products. Variations in these products may exist between manufacturers and, at times, from lot to lot.

Procedure:

---

[d] Systeme International d'unites (S.I. Units) are in brackets.

## TEST MATERIALS

| Item | II, III Du Pont Cat. No. | IV, SX Du Pont Cat. No. | V Du Pont Cat. No. |
|---|---|---|---|
| ACA®CREA Analytical Test Pack | 701976901 | 701976901 | 701976901 |
| Sample System Kit or | 710642901 | 710642901 | 713697901 |
| Micro Sample System Kit and | 702694901 | 710356901 | NA |
| Micro Sample System Holders | 702785000 | NA | NA |
| DYLUX® Photosensitive | ............ | ............ | ............ |
| Printer Paper | 700036000 | NA | NA |
| Thermal Printer Paper | NA | 710639901 | 713645901 |
| Du Pont Purified Water | 704209901 | 710615901 | 710815901 |
| Cell Wash Solution | 701864901 | 710664901 | 710864901 |

Test Steps: The operator need only load the sample kit and appropriate test pack(s) into a properly prepared ACA® discrete clinical analyzer. It automatically advances the pack(s) through the test steps and prints a result(s). See the Instrument Manual of the ACA® analyzer for details of mechanical travel of the test pack(s).

### Preset Creatinine (CREA)—Test Conditions

- Sample Volume: 200 µL
- Diluent: Purified Water
- Temperature: 37.0 ±0.1 °C
- Reaction Period: 29 seconds
- Type of Measurement: Rate
- Measurement Period: 17.07 seconds
- Wavelength: 510 nm
- Units: mg/dL [µmol/L]

CALIBRATION: The general calibration procedure is described in the Calibration/Verification chapter of the Manuals.

The following information should be considered when calibrating the CREA method.

- Assay Range: 0-20 mg/mL [0-1768 µmol/L][e].
- Reference Material: Protein containing primary standards[f] or secondary calibrators such as Du Pont Elevated Chemistry Control (Cat. #790035903) and Normal Chemistry Control (Cat. •#790035905)[g].
- Suggested Calibration Levels: 1,5,20, mg/mL [88, 442, 1768 µmol/L].
- Calibration Scheme: 3 levels, 3 packs per level.
- Frequency: Each new pack lot. Every 3 months for any one pack lot.

### PRESET CREATININE (CREA) TEST CONDITIONS

| Item | ACA®, II, analyzer | ACA®, III, IV, SX, V, analyzer |
|---|---|---|
| Count by | One (1) | NA. |
| Decimal point location. | [Five (5)] 0.0 mg/dl [000. µmo/l] | 000.0 mg/dl [000 µmol/l]. |
| Assigned starting point or offset $C_o$. | 999.8 [9823.] | - 1.000 E1 [-8.840 E2]. |
| Scale factor or assigned. | 0.2000 mg/dl/count[h] | 2.004 E-1[h]. |
| Linear term $C_1$[h] | [0.3536 µmol/l/count] | [1.772 E1]. |

[h] The preset scale factor (linear term) was derived from the molar absorptivity of the indicator and is based on an absorbance to activity relationship (sensitivity) of 0.596 (mA/min)/(U/L). Due to small differences in filters and electronic components between instruments, the actual scale factor (linear term) may differ slightly from that given above.

Quality Control: Two types of quality control procedures are recommended:

- General Instrument Check. Refer to the Filter Balance Procedure and the Absorbance Test Method described in the ACA Analyzer Instrument Manual. Refer also to the ABS Test Methodology literature.
- Creatinine Method Check. At least once daily run a CREA test on a solution of known creatinine activity such as an assayed control or calibration standard other than that used to calibrate the CREA method. For further details review the Quality Assurance Section of the Chemistry Manual. The result obtained should fall within acceptable limits defined by the day-to-day variability of the system as measured in the user's laboratory. (See SPECIFIC PERFORMANCE CHARACTERISTICS for guidance.) If the result falls outside the laboratory's acceptable limits, follow the procedure outlined in the Chemistry Troubleshooting Section of the Chemistry Manual.

A possible system malfunction is indicated when analysis of a sample with five consecutive test packs gives the following results:

| Level | SD |
|---|---|
| 1 mg/dL | >0.15 mg/dL. |
| [88 µmol/L] | [>13 µmol/L]. |
| 20 mg/dL | >0.68 mg/dL. |
| [1768 µmol/L] | [>60 µmol/L]. |

Refer to the procedure outlined in the Trouble Shooting Section of the Manual.

Results: The ACA® analyzer automatically calculates and prints the CREA result in mg/dL [µmol/L].

### Limitation of Procedure: Results >20 mg/dL [1768 µmol/L]:

- Dilute with suitable protein base diluent. Reassay. Correct for diluting before reporting.

The reporting system contains error messages to warn the operator of specific malfunctions. Any report slip containing a letter code or word immediately following the numerical value should not be reported. Refer to the Manual for the definition of error codes.

### Reference Interval

Serum: [11,i]

| | |
|---|---|
| Males | 0.8-1.3 md/dL [71-115 µmol/L] |
| Females | 0.6-1.0 md/dL [53-88 µmol/L] |

Urine: [12]

| | |
|---|---|
| Males | 0.6-2.5 g/24 hr [53-221 mmol/24 hr] |
| Females | 0.6-1.5 g/24 hr [53-133 mmol/24 hr] |

[i] Reference interval data obtained from 200 apparently healthy individuals (71 males, 129 females) between the ages of 19 and 72.

---

[e] For the results in S.I. units [µmol/L] the conversion factory is 88.4.

[f] Refer to the Creatinine Standard Preparation and Calibration Procedure available on request from a Du Pont Representative.

[g] If the Du Pont Chemistry Controls are being used, prepare them according to the instructions on the product insert sheets.

Each laboratory should establish its own reference intervals for CREA as performed on the analyzer.

## Specific Performance Characteristics[j]
### REPRODUCIBILITY[k]

| Material | Mean | Standard deviation (% CV) Within-run | Between-day |
|---|---|---|---|
| Lyophilized Control | 1.3 [115] | 0.05 (3.7) [4.4] | 0.05 (3.7) [4.4] |
| Lyophilized Control | 20.6 1821] | 0.12 (0.6) [10.6] | 0.37 (1.8) [32.7] |

[j] All specific performance characteristics tests were run after normal recommended equipment quality control checks were performed (see Instrument Manual).

[k] Specimens at each level were analyzed in duplicate for twenty days. The within-run and between-day standard deviations were calculated by the analysis of variance method.

### CORRELATION—REGRESSION STATISTICS[l]

| Comparative method | Slope | Intercept | Correlation coefficient | n |
|---|---|---|---|---|
| Autoanalyzer® | 1.03 | 0.03[2.7] | 0.997 | 260 |

[l] Model equation for regression statistics is:

Result of ACA® Analyzer = Slope (Comparative method result) + intercept

### Assay Range[m]

0.0-20.0 mg/dl

[0-1768 μmol]

[m] See REPRODUCIBILITY for method performance within the assay range.

### Analytical Specificity

See KNOWN INTERFERING SUBSTANCES section for details.

### BIBLIOGRAPHY

[1] Larsen, K, Clin Chem Acta 41, 209 (1972).

[2] Tietz, NW, Fundamentals of Clinical Chemistry, W. B. Saunders Co., Philadelphia, PA, 1976, pp 47-52, 1211.

[3] Supplementary information pertaining to the effects of various drugs and patient conditions on in vivo or in vitro diagnostic levels can be found in "Drug Interferences with Clinical Laboratory Tests," Clin. Chem 21(5) (1975), and "Effects of Disease on Clinical Laboratory Tests," Clin Chem, 26 (4) 1D-476D (1980).

[4] Watkins, R. Fieldkamp, SC, Thibert, RJ, and Zak, B, Clin Chem, 21, 1002 (1975).

[5] Kawas, EE, Richards, AH, and Bigger, R, An Evaluation of a Kinetic Creatinine Test for the Du Pont ACA, Du Pont Company, Wilmington, DE (February 1973). (Reprints available from DuPont Company, Diagnostic Systems)

[6] Westgard, JO, Effects of Hemolysis and Lipemia on ACA Creatinine Method, 0.200 L, Sample Size, Du Pont Company, Wilmington, DE (October 1972).

[7] Physicians' Desk Reference, Medical Economics Company, 33 Edition, 1979.

[8] Henry, JB, Clinical Diagnosis and Management by Laboratory Methods, W.B. Saunders Co., Philadelphia, PA 1979, Vol. III.

[9] Krupp, MA, Tierney, LM Jr., Jawetz, E, Roe, RI, Camargo, CA, Physicians Handbook, Lange Medical Publications, Los Altos, CA, 1982 pp 635-636.

[10] Sarah, AJ, Koch, TR, Drusano, GL, Celoxitin Falsely Elevates Creatinine Levels, JAMA 247, 205-206 (1982).

[11] Gadsden, RH, and Phelps, CA, A Normal Range Study of Amylase in Urine and Serum on the Du Pont ACA, Du Pont Company, Wilmington, DE (March 1978). (Reprints available from DuPont Company, Diagnostic Systems)

[12] Dicht, JJ, Reference Intervals for Serum Amylase and Urinary Creatinine on the Du Pont ACA® Discrete Clinical Analyzer, Du Pont Company, Wilmington, DE (November 1984).

## ATTACHMENT 3—ANALYSIS OF CREATININE FOR THE NORMALIZATION OF CADMIUM AND BETA-2-MICROGLOBULIN CONCENTRATIONS IN URINE (OSLTC PROCEDURE)

Matrix: Urine.

Target concentration: 1.1 g/L (this amount is representative of creatinine concentrations found in urine).

Procedure: A 1.0 mL aliquot of urine is passed through a C18 SEP-PAK® (Waters Associates). Approximately 30 mL of HPLC (high performance liquid chromatography) grade water is then run through the SEP-PAK. The resulting solution is diluted to volume in a 100-mL volumetric flask and analyzed by HPLC using an ultraviolet (UV) detector.

Special requirements: After collection, samples should be appropriately stabilized for cadmium (Cd) analysis by using 10% high purity (with low Cd background levels) nitric acid (exactly 1.0 mL of 10% nitric acid per 10 mL of urine) or stabilized for Beta-2-Microglobulin (B2M) by taking to pH 7 with dilute NaOH (exactly 1.0 mL of 0.11 N NaOH per 10 mL of urine). If not immediately analyzed, the samples should be frozen and shipped by overnight mail in an insulated container.

Dated: January 1992.

David B. Armitage,

Duane Lee,

Chemists.

Organic Service Branch II, OSHA Technical Center, Salt Lake City, Utah

### 1. General Discussion
#### 1.1 Background

1.1.1. History of procedure

Creatinine has been analyzed by several methods in the past. The earliest methods were of the wet chemical type. As an example, creatinine reacts with sodium picrate in basic solution to form a red complex, which is then analyzed colorimetrically (Refs. 5.1. and 5.2.).

Since industrial hygiene laboratories will be analyzing for Cd and B2M in urine, they will be normalizing those concentrations to the concentration of creatinine in urine. A literature search revealed several HPLC methods (Refs. 5.3., 5.4., 5.5. and 5.6.) for creatinine in urine and because many industrial hygiene laboratories have HPLC equipment, it was desirable to develop an industrial hygiene HPLC method for creatinine in urine. The method of Hausen, Fuchs, and Wachter was chosen as the starting point for method development. SEP-PAKs were used for sample clarification and cleanup in this method to protect the analytical column. The urine aliquot which has been passed through the SEP-PAK is then analyzed by reverse-phase HPLC using ion-pair techniques.

This method is very similar to that of Ogata and Taguchi (Ref. 5.6.), except they used centrifugation for sample clean-up. It is also of note that they did a comparison of their HPLC results to those of the Jaffe method (a picric acid method commonly used in the health care industry) and found a linear relationship of close to 1:1. This indicates that either HPLC or colorimetric methods may be used to measure creatinine concentrations in urine.

1.1.2. Physical properties (Ref. 5.7.)

Molecular weight: 113.12

Molecular formula: $C_4H_7N_3O$

Chemical name: 2-amino-1,5-dihydro-1-methyl-4H-imidazol-4-one

CAS No.: 60-27-5

Melting point: 300 °C (decomposes)

Appearance: white powder

Solubility: soluble in water; slightly soluble in alcohol; practically insoluble in acetone, ether, and chloroform

Synonyms: 1-methylglycocyamidine, 1-methylhydantoin-2-imide

Structure: see Figure #1

# OSHA Standards for General Industry

**FIGURE #1**

### 1.2. Advantages

1.2.1. This method offers a simple, straightforward, and specific alternative method to the Jaffe method.

1.2.2. HPLC instrumentation is commonly found in many industrial hygiene laboratories.

## 2. Sample stabilization procedure

### 2.1. Apparatus

Metal-free plastic container for urine sample.

### 2.2. Reagents

2.2.1. Stabilizing Solution—

(1) Nitric acid (10%, high purity with low Cd background levels) for stabilizing urine for Cd analysis or

(2) NaOH, 0.11 N, for stabilizing urine for B2M analysis.

2.2.2. HPLC grade water

### 2.3. Technique

2.3.1. Stabilizing solution is added to the urine sample (see section 2.2.1.). The stabilizing solution should be such that for each 10 mL of urine, add exactly 1.0 mL of stabilizer solution. (Never add water or urine to acid or base. Always add acid or base to water or urine.) Exactly 1.0 mL of 0.11 N NaOH added to 10 mL of urine should result in a pH of 7. Or add 1.0 mL of 10% nitric acid to 10 mL of urine.

2.3.2. After sample collection seal the plastic bottle securely and wrap it with an appropriate seal. Urine samples should be frozen and then shipped by overnight mail (if shipping is necessary) in an insulated container. (Do not fill plastic bottle too full. This will allow for expansion of contents during the freezing process.)

### 2.4. The Effect of Preparation and Stabilization Techniques on Creatinine Concentrations

Three urine samples were prepared by making one sample acidic, not treating a second sample, and adjusting a third sample to pH 7. The samples were analyzed in duplicate by two different procedures. For the first procedure a 1.0 mL aliquot of urine was put in a 100-mL volumetric flask, diluted to volume with HPLC grade water, and then analyzed directly on an HPLC. The other procedure used SEP-PAKs. The SEP-PAK was rinsed with approximately 5 mL of methanol followed by approximately 10 mL of HPLC grade water and both rinses were discarded. Then, 1.0 mL of the urine sample was put through the SEP-PAK, followed by 30 mL of HPLC grade water. The urine and water were transferred to a 100-mL volumetric flask, diluted to volume with HPLC grade water, and analyzed by HPLC. These three urine samples were analyzed on the day they were obtained and then frozen. The results show that whether the urine is acidic, untreated or adjusted to pH 7, the resulting answer for creatinine is essentially unchanged. The purpose of stabilizing the urine by making it acidic or neutral is for the analysis of Cd or B2M respectively.

### COMPARISON OF PREPARATION AND STABILIZATION TECHNIQUES

| Sample | w/o SEP-PAK g/L creatinine | with SEP-PAK g/L creatinine |
|---|---|---|
| Acid | 1.10 | 1.10 |
| Acid | 1.11 | 1.10 |
| Untreated | 1.12 | 1.11 |
| Untreated | 1.11 | 1.12 |
| pH 7 | 1.08 | 1.02 |
| pH 7 | 1.11 | 1.08 |

### 2.5. Storage

After 4 days and 54 days of storage in a freezer, the samples were thawed, brought to room temperature and analyzed using the same procedures as in section 2.4. The results of several days of storage show that the resulting answer of creatinine is essentially unchanged.

### STORAGE DATA

| Sample | 4 days w/o SEP-Pak g/L creatinine | 4 days with SEP-PAK g/L creatinine | 54 days w/o SEP-PAK g/L creatinine | 54 days with SEP-PAK g/L creatinine |
|---|---|---|---|---|
| Acid | 1.09 | 1.09 | 1.08 | 1.09 |
| Acid | 1.10 | 1.10 | 1.09 | 1.10 |
| Acid | .......... | .......... | 1.09 | 1.09 |
| Untreated | 1.13 | 1.14 | 1.09 | 1.11 |
| Untreated | 1.15 | 1.14 | 1.10 | 1.10 |
| Untreated | .......... | .......... | 1.09 | 1.10 |
| pH 7 | 1.14 | 1.13 | 1.12 | 1.12 |
| pH 7 | 1.14 | 1.13 | 1.12 | 1.12 |
| pH 7 | .......... | .......... | 1.12 | 1.12 |

### 2.6. Interferences

None.

### 2.7. Safety precautions

2.7.1. Make sure samples are properly sealed and frozen before shipment to avoid leakage.

2.7.2. Follow the appropriate shipping procedures.

The following modified special safety precautions are based on those recommended by the Centers for Disease Control (CDC) (Ref. 5.8.). and OSHA's Bloodborne Pathogens standard (29 CFR 1910.1039).

2.7.3. Wear gloves, lab coat, and safety glasses while handling all human urine products. Disposable plastic, glass, and paper (pipet tips, gloves, etc.) that contact urine should be placed in a biohazard autoclave bag. These bags should be kept in appropriate containers until sealed and autoclaved. Wipe down all work surfaces with 10% sodium hypochlorite solution when work is finished.

2.7.4. Dispose of all biological samples and diluted specimens in a biohazard autoclave bag at the end of the analytical run.

2.7.5. Special care should be taken when handling and dispensing nitric acid. Always remember to add acid to water (or urine). Nitric acid is a corrosive chemical capable of severe eye and skin damage. Wear metal-free gloves, a lab coat, and safety glasses. If the nitric acid comes in contact with any part of the body, quickly wash with copious quantities of water for at least 15 minutes.

2.7.6. Special care should be taken when handling and dispensing NaOH. Always remember to add base to water (or urine). NaOH can cause severe eye and skin damage. Always wear the appropriate gloves, a lab coat, and safety glasses. If the NaOH comes in contact with any part of the body, quickly wash with copious quantities of water for at least 15 minutes.

### 3. Analytical procedure

#### 3.1. Apparatus

3.1.1. A high performance liquid chromatograph equipped with pump, sample injector and UV detector.

3.1.2. A C18 HPLC column; 25 cm x 4.6 mm I.D.

3.1.3. An electronic integrator, or some other suitable means of determining analyte response.

3.1.4. Stripchart recorder.

3.1.5. C18 SEP-PAKs (Waters Associates) or equivalent.

3.1.6. Luer-lock syringe for sample preparation (5 mL or 10 mL).

3.1.7. Volumetric pipettes and flasks for standard and sample preparation.

3.1.8. Vacuum system to aid sample preparation (optional).

#### 3.2. Reagents

3.2.1. Water, HPLC grade.

3.2.2. Methanol, HPLC grade.

3.2.3. PIC B-7® (Waters Associates) in small vials.

3.2.4. Creatinine, anhydrous, Sigma hemical Corp., purity not listed.

3.2.5. 1-Heptanesulfonic acid, sodium salt monohydrate.

3.2.6. Phosphoric acid.

3.2.7. Mobile phase. It can be prepared by mixing one vial of PIC B-7 into a 1 L solution of 50% methanol and 50% water. The mobile phase can also be made by preparing a solution that is 50% methanol and 50% water with 0.005M heptanesulfonic acid and adjusting the pH of the solution to 3.5 with phosphoric acid.

#### 3.3. Standard preparation

3.3.1. Stock standards are prepared by weighing 10 to 15 mg of creatinine. This is transferred to a 25-mL volumetric flask and diluted to volume with HPLC grade water.

3.3.2. Dilutions to a working range of 3 to 35 µg/mL are made in either HPLC grade water or HPLC mobile phase (standards give the same detector response in either solution).

#### 3.4. Sample preparation

3.4.1. The C18 SEP-PAK is connected to a Luer-lock syringe. It is rinsed with 5 mL HPLC grade methanol and then 10 mL of HPLC grade water. These rinses are discarded.

3.4.2. Exactly 1.0 mL of urine is pipetted into the syringe. The urine is put through the SEP-PAK into a suitable container using a vacuum system.

3.4.3. The walls of the syringe are rinsed in several stages with a total of approximately 30 mL of HPLC grade water. These rinses are put through the SEP-PAK into the same container. The resulting solution is transferred to a 100-mL volumetric flask and then brought to volume with HPLC grade water.

#### 3.5. Analysis (conditions and hardware are those used in this evaluation.)

3.5.1. Instrument conditions

Column: Zorbax® ODS, 5-6 µm particle size; 25 cm × 4.6 mm I.D.

Mobile phase: See Section 3.2.7.

Detector: Dual wavelength UV; 229 nm (primary) 254 nm (secondary)

Flow rate: 0.7 mL/minute

Retention time: 7.2 minutes

Sensitivity: 0.05 AUFS

Injection volume: 20 µl

3.5.2. Chromatogram (see Figure #2)

**CHROMATOGRAM OF A CREATININE STANDARD**
**FIGURE #2**

#### 3.6. Interferences

3.6.1. Any compound that has the same retention time as creatinine and absorbs at 229 nm is an interference.

3.6.2. HPLC conditions may be varied to circumvent interferences. In addition, analysis at another UV wavelength (i.e., 254 nm) would allow a comparison of the ratio of response of a standard to that of a sample. Any deviations would indicate an interference.

#### 3.7. Calculations

3.7.1. A calibration curve is constructed by plotting detector response versus standard concentration (See Figure #3).

3.7.2. The concentration of creatinine in a sample is determined by finding the concentration corresponding to its detector response. (See Figure #3).

**CALIBRATION CURVE FOR CREATININE**
**FIGURE #3**

3.7.3. The µg/mL creatinine from section 3.7.2. is then multiplied by 100 (the dilution factor). This value is equivalent to the micrograms of creatinine in the 1.0 mL stabilized urine aliquot or the milligrams of creatinine per liter of urine. The desired units, g/L, is determined by the following relationship:

$$g/L = \frac{\mu g/mL}{1000} = \frac{mg/L}{1000}$$

3.7.4. The resulting value for creatinine is used to normalize the urinary concentration of the desired analyte (A) (Cd or B2M) by using the following formula.

$$\frac{\mu g\ A/g}{creatinine} = \frac{\mu g\ A/L\ (experimental)}{g/L\ creatinine}$$

Where A is the desired analyte. The protocol of reporting such normalized results is µg A/g creatinine.

### 3.8. Safety precautions See section 2.7.

### 4. Conclusions

The determination of creatinine in urine by HPLC is a good alternative to the Jaffe method for industrial hygiene laboratories. Sample clarification with SEP-PAKs did not change the amount of creatinine found in urine samples. However, it does protect the analytical column. The results of this creatinine in urine procedure are unaffected by the pH of the urine sample under the conditions tested by this procedure. Therefore, no special measures are required for creatinine analysis whether the urine sample has been stabilized with 10% nitric acid for the Cd analysis or brought to a pH of 7 with 0.11 N NaOH for the B2M analysis.

### 5. References

5.1. Clark, L.C.; Thompson, H.L.; *Anal. Chem.* 1949, 21, 1218.

5.2. Peters, J.H.; *J. Biol. Chem.* 1942, 146, 176.

5.3. Hausen, V.A.; Fuchs, D.; Wachter, H.; *J. Clin. Chem. Clin. Biochem.* 1981, 19, 373-378.

5.4. Clark, P.M.S.; Kricka L.J.; Patel, A.; *J. Liq. Chrom.* 1980, 3(7), 1031-1046.

5.5. Ballerini, R.; Chinol, M.; Cambi, A.; *J. Chrom.* 1979, 179, 365-369.

5.6. Ogata, M.; Taguchi, T.; *Industrial Health* 1987, 25, 225-228.

5.7. "Merck Index", 11th ed.; Windholz, Martha Ed.; Merck: Rahway, N.J., 1989; p 403.

5.8. Kimberly, M.; *"Determination of Cadmium in Urine by Graphite Furnace Atomic Absorption Spectrometry with Zeeman Background Correction."*, Centers for Disease Control, Atlanta, Georgia, unpublished, update 1990.

[57 FR 42389, Sept. 14, 1992, as amended at 57 FR 49272, Oct. 30, 1992; 58 FR 21781, Apr. 23, 1993; 61 FR 5508, Feb. 13, 1996; 63 FR 1288, Jan. 8, 1998]

### §1910.1028 Benzene.

**1910.1028(a)** *Scope and application.* **(1)** This section applies to all occupational exposures to benzene, Chemical Abstracts Service Registry No. 71-43-2, except as provided in paragraphs (a)(2) and (a)(3) of this section.

**1910.1028(a)(2)** This section does not apply to:

**1910.1028(a)(2)(i)** The storage, transportation, distribution, dispensing, sale or use of gasoline, motor fuels, or other fuels containing benzene subsequent to its final discharge from bulk wholesale storage facilities, except that operations where gasoline or motor fuels are dispensed for more than 4 hours per day in an indoor location are covered by this section.

**1910.1028(a)(2)(ii)** Loading and unloading operations at bulk wholesale storage facilities which use vapor control systems for all loading and unloading operations, except for the provisions of 29 CFR 1910.1200 as incorporated into this section and the emergency provisions of paragraphs (g) and (i)(4) of this section.

**1910.1028(a)(2)(iii)** The storage, transportation, distribution or sale of benzene or liquid mixtures containing more than 0.1 percent benzene in intact containers or in transportation pipelines while sealed in such a manner as to contain benzene vapors or liquid, except for the provisions of 29 CFR 1910.1200 as incorporated into this section and the emergency provisions of paragraphs (g) and (i)(4) of this section.

**1910.1028(a)(2)(iv)** Containers and pipelines carrying mixtures with less than 0.1 percent benzene and natural gas processing plants processing gas with less than 0.1 percent benzene.

**1910.1028(a)(2)(v)** Work operations where the only exposure to benzene is from liquid mixtures containing 0.5 percent or less of benzene by volume, or the vapors released from such liquids until September 12, 1988; work operations where the only exposure to benzene is from liquid mixtures containing 0.3 percent or less of benzene by volume or the vapors released from such liquids from September 12, 1988, to September 12, 1989; and work operations where the only exposure to benzene is from liquid mixtures containing 0.1 percent or less of benzene by volume or the vapors released from such liquids after September 12, 1989; except that tire building machine operators using solvents with more than 0.1 percent benzene are covered by paragraph (i) of this section.

**1910.1028(a)(2)(vi)** Oil and gas drilling, production and servicing operations.

**1910.1028(a)(2)(vii)** Coke oven batteries.

**1910.1028(a)(3)** The cleaning and repair of barges and tankers which have contained benzene are excluded from paragraph (f) methods of compliance, paragraph (e)(1) exposure monitoring-general, and paragraph (e)(6) accuracy of monitoring. Engineering and work practice controls shall be used to keep exposures below 10 ppm unless it is proven to be not feasible.

**1910.1028(b)** *Definitions.*

*Action level* means an airborne concentration of benzene of 0.5 ppm calculated as an 8-hour time-weighted average.

*Assistant Secretary* means the Assistant Secretary of Labor for Occupational Safety and Health, U.S. Department of Labor, or designee.

*Authorized person* means any person specifically authorized by the employer whose duties require the person to enter a regulated area, or any person entering such an area as a designated representative of employees for the purpose of exercising the right to observe monitoring and measuring procedures under paragraph (l) of this section, or any other person authorized by the Act or regulations issued under the Act.

*Benzene* ($C_6H_6$) (CAS Registry No. 71-43-2) means liquefied or gaseous benzene. It includes benzene contained in liquid mixtures and the benzene vapors released by these liquids. It does not include trace amounts of unreacted benzene contained in solid materials.

*Bulk wholesale storage facility* means a bulk terminal or bulk plant where fuel is stored prior to its delivery to wholesale customers.

*Container* means any barrel, bottle, can, cylinder, drum, reaction vessel, storage tank, or the like, but does not include piping systems.

*Day* means any part of a calendar day.

*Director* means the Director of the National Institute for Occupational Safety and Health, U.S. Department of Health and Human Services, or designee.

*Emergency* means any occurrence such as, but not limited to, equipment failure, rupture of containers, or failure of control equipment which may or does result in an unexpected significant release of benzene.

*Employee exposure* means exposure to airborne benzene which would occur if the employee were not using respiratory protective equipment.

*Regulated area* means any area where airborne concentrations of benzene exceed or can reasonably be expected to exceed, the permissible exposure limits, either the 8-hour time weighted average exposure of 1 ppm or the short-term exposure limit of 5 ppm for 15 minutes.

*Vapor control system* means any equipment used for containing the total vapors displaced during the loading of gasoline, motor fuel or other fuel tank trucks and the displacing of these vapors through a vapor processing system or balancing the vapor with the storage tank. This equipment also includes systems containing the vapors displaced from the storage tank during the unloading of the tank truck which balance the vapors back to the tank truck.

**1910.1028(c)** *Permissible exposure limits (PELs)*—**(1)** *Time-weighted average limit (TWA).* The employer shall assure that no employee is exposed to an airborne concentration of benzene in excess of one part of benzene per million parts of air (1 ppm) as an 8-hour time-weighted average.

**1910.1028(c)(2)** *Short-term exposure limit (STEL).* The employer shall assure that no employee is exposed to an airborne concentration of benzene in excess of five (5) ppm as averaged over any 15 minute period.

**1910.1028(d)** *Regulated areas.* **(1)** The employer shall establish a regulated area wherever the airborne concentration of benzene exceeds or can reasonably be expected to exceed the permissible exposure limits, either the 8-hour time weighted average exposure of 1 ppm or the short-term exposure limit of 5 ppm for 15 minutes.

**1910.1028(d)(2)** Access to regulated areas shall be limited to authorized persons.

**1910.1028(d)(3)** Regulated areas shall be determined from the rest of the workplace in any manner that minimizes the number of employees exposed to benzene within the regulated area.

**1910.1028(e)** *Exposure monitoring*—**(1)** *General.* **(i)** Determinations of employee exposure shall be made from breathing zone air samples that are representative of each employee's average exposure to airborne benzene.

**1910.1028(e)(1)(ii)** Representative 8-hour TWA employee exposures shall be determined on the basis of one sample or samples representing the full shift exposure for each job classification in each work area.

**1910.1028(e)(1)(iii)** Determinations of compliance with the STEL shall be made from 15 minute employee breathing zone samples measured at operations where there is reason to believe exposures are high, such as where tanks are opened, filled, unloaded or gauged; where containers or process equipment are opened and where benzene is used for cleaning or as a solvent in an uncontrolled situation. The employer may use objective data, such as measurements from brief period measuring devices, to determine where STEL monitoring is needed.

**1910.1028(e)(1)(iv)** Except for initial monitoring as required under paragraph (e)(2) of this section, where the employer can document that one shift will consistently have higher employee exposures for an operation, the employer shall only be required to determine representative employee exposure for that operation during the shift on which the highest exposure is expected.

**1910.1028(e)(2)** *Initial monitoring.* **(i)** Each employer who has a place of employment covered under paragraph (a)(1) of this section shall monitor each of these workplaces and work operations to determine accurately the airborne concentrations of benzene to which employees may be exposed.

**1910.1028(e)(2)(ii)** The initial monitoring required under paragraph (e)(2)(i) of this section shall be completed by 60 days after the effective date of this standard or within 30 days of the introduction of benzene into the workplace. Where the employer has monitored within one year prior to the effective date of this standard and the monitoring satisfies all other requirements of this section, the employer may rely on such earlier monitoring results to satisfy the requirements of paragraph (e)(2)(i) of this section.

**1910.1028(e)(3)** *Periodic monitoring and monitoring frequency.* **(i)** If the monitoring required by paragraph (e)(2)(i) of this section reveals employee exposure at or above the action level but at or below the TWA, the employer shall repeat such monitoring for each such employee at least every year.

**1910.1028(e)(3)(ii)** If the monitoring required by paragraph (e)(2)(i) of this section reveals employee exposure above the TWA, the employer shall repeat such monitoring for each such employee at least every six (6) months.

**1910.1028(e)(3)(iii)** The employer may alter the monitoring schedule from every six months to annually for any employee for whom two consecutive measurements taken at least 7 days apart indicate that the employee exposure has decreased to the TWA or below, but is at or above the action level.

**1910.1028(e)(3)(iv)** Monitoring for the STEL shall be repeated as necessary to evaluate exposures of employees subject to short term exposures.

**1910.1028(e)(4)** *Termination of monitoring.* **(i)** If the initial monitoring required by paragraph (e)(2)(i) of this section reveals employee exposure to be below the action level the employer may discontinue the monitoring for that employee, except as otherwise required by paragraph (e)(5) of this section.

**1910.1028(e)(4)(ii)** If the periodic monitoring required by paragraph (e)(3) of this section reveals that employee exposures, as indicated by at least two consecutive measurements taken at least 7 days apart, are below the action level the employer may discontinue the monitoring for that employee, except as otherwise required by paragraph (e)(5).

**1910.1028(e)(5)** *Additional monitoring.* **(i)** The employer shall institute the exposure monitoring required under paragraphs (e)(2) and (e)(3) of this section when there has been a change in the production, process, control equipment, personnel or work practices which may result in new or additional exposures to benzene, or when the employer has any reason to suspect a change which may result in new or additional exposures.

**1910.1028(e)(5)(ii)** Whenever spills, leaks, ruptures or other breakdowns occur that may lead to employee exposure, the employer shall monitor (using area or personal sampling) after the cleanup of the spill or repair of the leak, rupture or other breakdown to ensure that exposures have returned to the level that existed prior to the incident.

**1910.1028(e)(6)** *Accuracy of monitoring.* Monitoring shall be accurate, to a confidence level of 95 percent, to within plus or minus 25 percent for airborne concentrations of benzene.

**1910.1028(e)(7)** *Employee notification of monitoring results.* **(i)** The employer must, within 15 working days after the receipt of the results of any monitoring performed under this section, notify each affected employee of these results either individually in writing or by posting the results in an appropriate location that is accessible to employees.

**1910.1028(e)(7)(ii)** Whenever the PELs are exceeded, the written notification required by paragraph (e)(7)(i) of this section shall contain the corrective action being taken by the employer to reduce the employee exposure to or below the PEL, or shall refer to a document available to the employee which states the corrective actions to be taken.

**1910.1028(f)** *Methods of compliance*—**(1)** *Engineering controls and work practices.* **(i)** The employer shall institute engineering controls and work practices to reduce and maintain employee exposure to benzene at or below the permissible exposure limits, except to the extent that the employer can establish that these controls are not feasible or where the provisions of paragraph (f)(1)(iii) or (g)(1) of this section apply.

**1910.1028(f)(1)(ii)** Wherever the feasible engineering controls and work practices which can be instituted are not sufficient to reduce employee exposure to or below the PELs, the employer shall use them to reduce employee exposure to the lowest levels achievable by these controls and shall supplement them by the use of respiratory protection which complies with the requirements of paragraph (g) of this section.

**1910.1028(f)(1)(iii)** Where the employer can document that benzene is used in a workplace less than a total of 30 days per year, the employer shall use engineering controls, work practice controls or respiratory protection or any combination of these controls to reduce employee exposure to benzene to or below the PELs, except that employers shall use engineering and work practice controls, if feasible, to reduce exposure to or below 10 ppm as an 8-hour TWA.

**1910.1028(f)(2)** *Compliance program.* **(i)** When any exposures are over the PEL, the employer shall establish and implement a written program to reduce employee exposure to or below the PEL primarily by means of engineering and work practice controls, as required by paragraph (f)(1) of this section.

**1910.1028(f)(2)(ii)** The written program shall include a schedule for development and implementation of the engineering and work practice controls. These plans shall be reviewed and revised as appropriate based on the most recent exposure monitoring data, to reflect the current status of the program.

**1910.1028(f)(2)(iii)** Written compliance programs shall be furnished upon request for examination and copying to the Assistant Secretary, the Director, affected employees and designated employee representatives.

**1910.1028(g)** *Respiratory protection*—**(1)** *General.* For employees who use respirators required by this section, the employer must provide respirators that comply with the requirements of this paragraph. Respirators must be used during:

**1910.1028(g)(1)(i)** Periods necessary to install or implement feasible engineering and work-practice controls.

**1910.1028(g)(1)(ii)** Work operations for which the employer establishes that compliance with either the TWA or STEL through the use of engineering and work-practice controls is not feasible; for example, some maintenance and repair activities, vessel cleaning, or other operations for which engineering and work-practice controls are infeasible because exposures are intermittent and limited in duration.

**1910.1028(g)(1)(iii)** Work operations for which feasible engineering and work-practice controls are not yet sufficient, or are not required under paragraph (f)(1)(iii) of this section, to reduce employee exposure to or below the PELs.

**1910.1028(g)(1)(iv)** Emergencies.

**1910.1028(g)(2)** *Respirator program.* **(i)** The employer must implement a respiratory protection program in accordance with 29 CFR 1910.134(b) through (d) (except (d)(1)(iii), (d)(3)(iii)(B)(1), and (2)), and (f) through (m).

**1910.1028(g)(2)(ii)** For air-purifying respirators, the employer must replace the air-purifying element at the expiration of its service life or at the beginning of each shift in which such elements are used, whichever comes first.

**1910.1028(g)(2)(iii)** If NIOSH approves an air-purifying element with an endof-service-life indicator for benzene, such an element may be used until the indicator shows no further useful life.

**1910.1028(g)(3)** *Respirator selection.* **(i)** The employer must select the appropriate respirator from Table 1 of this section.

**1910.1028(g)(3)(ii)** Any employee who cannot use a negative-pressure respirator must be allowed to use a respirator with less breathing resistance, such as a powered air-purifying respirator or supplied-air respirator.

### TABLE 1—RESPIRATORY PROTECTION FOR BENZENE

| Airborne concentration of benzene or condition of use | Respirator type |
| --- | --- |
| (a) Less than or equal to 10 ppm. | (1) Half-mask air-purifying respirator with organic vapor cartridge. |
| (b) Less than or equal to 50 ppm. | (1) Full facepiece respirator with organic vapor cartridges.<br>(1) Full facepiece gas mask with chin style canister.[1] |
| (c) Less than or equal to 100 ppm. | (1) Full facepiece powered air-purifying respirator with organic vapor canister.[1] |
| (d) Less than or equal to 1,000 ppm. | (1) Supplied air respirator with full face piece in positive-pressure mode. |
| (e) Greater than 1,000 ppm or unknown concentration. | (1) Self-contained breathing apparatus with full facepiece in positive pressure mode.<br>(2) Full facepiece positive-pressure supplied-air respirator with auxiliary self-contained air supply. |
| (f) Escape | (1) Any organic vapor gas mask; or<br>(2) Any self-contained breathing apparatus with full facepiece. |
| (g) Firefighting | (1) Full facepiece self-contained breathing apparatus in positive pressure mode. |

[1] Canisters must have a minimum service life of four (4) hours when tested at 150 ppm benzene, at a flow rate of 64 LPM, 25 °C, and 85% relative humidity for non-powered air purifying respirators. The flow rate shall be 115 LPM and 170 LPM respectively for tight fitting and loose fitting powered air-purifying respirators.

**1910.1028(h)** *Protective clothing and equipment.* Personal protective clothing and equipment shall be worn where appropriate to prevent eye contact and limit dermal exposure to liquid benzene. Protective clothing and equipment shall be provided by the employer at no cost to the employee and the employer shall assure its use where appropriate. Eye and face protection shall meet the requirements of 29 CFR 1910.133.

**1910.1028(i)** *Medical surveillance*—**(1)** *General.* **(i)** The employer shall make available a medical surveillance program for employees who are or may be exposed to benzene at or above the action level 30 or more days per year; for employees who are or may be exposed to benzene at or above the PELs 10 or more days per year; for employees who have been exposed to more than 10 ppm of benzene for 30 or more days in a year prior to the effective date of the standard when employed by their current employer; and for employees involved in the tire building operations called tire building machine operators, who use solvents containing greater than 0.1 percent benzene.

**1910.1028(i)(1)(ii)** The employer shall assure that all medical examinations and procedures are performed by or under the supervision of a licensed physician and that all laboratory tests are conducted by an accredited laboratory.

**1910.1028(i)(1)(iii)** The employer shall assure that persons other than licensed physicians who administer the pulmonary function testing required by this section shall complete a training course in spirometry sponsored by an appropriate governmental, academic or professional institution.

**1910.1028(i)(1)(iv)** The employer shall assure that all examinations and procedures are provided without cost to the employee and at a reasonable time and place.

**1910.1028(i)(2)** *Initial examination.* **(i)** Within 60 days of the effective date of this standard, or before the time of initial assignment, the employer shall provide each employee covered by paragraph (i)(1)(i) of this section with a medical examination including the following elements:

**1910.1028(i)(2)(i)(A)** A detailed occupational history which includes:

**1910.1028(i)(2)(i)(A)***(1)* Past work exposure to benzene or any other hematological toxins,

**1910.1028(i)(2)(i)(A)***(2)* A family history of blood dyscrasias including hematological neoplasms;

**1910.1028(i)(2)(i)(A)***(3)* A history of blood dyscrasias including genetic hemoglobin abnormalities, bleeding abnormalities, abnormal function of formed blood elements;

**1910.1028(i)(2)(i)(A)***(4)* A history of renal or liver dysfunction;

**1910.1028(i)(2)(i)(A)***(5)* A history of medicinal drugs routinely taken;

**1910.1028(i)(2)(i)(A)***(6)* A history of previous exposure to ionizing radiation and

**1910.1028(i)(2)(i)(A)***(7)* Exposure to marrow toxins outside of the current work situation.

**1910.1028(i)(2)(i)(B)** A complete physical examination.

**1910.1028(i)(2)(i)(C)** *Laboratory tests.* A complete blood count including a leukocyte count with differential, a quantitative thrombocyte count, hematocrit, hemoglobin, erythrocyte count and erythrocyte indices (MCV, MCH, MCHC). The results of these tests shall be reviewed by the examining physician.

**1910.1028(i)(2)(i)(D)** Additional tests as necessary in the opinion of the examining physician, based on alterations to the components of the blood or other signs which may be related to benzene exposure; and

**1910.1028(i)(2)(i)(E)** For all workers required to wear respirators for at least 30 days a year, the physical examination shall pay special attention to the cardiopulmonary system and shall include a pulmonary function test.

**1910.1028(i)(2)(ii)** No initial medical examination is required to satisfy the requirements of paragraph (i)(2)(i) of this section if adequate records show that the employee has been examined in accordance with the procedures of paragraph (i)(2)(i) of this section within the twelve months prior to the effective date of this standard.

**1910.1028(i)(3)** *Periodic examinations.* **(i)** The employer shall provide each employee covered under paragraph (i)(1)(i) of this section with a medical examination annually following the previous examination. These periodic examinations shall incude at least the following elements:

**1910.1028(i)(3)(i)(A)** A brief history regarding any new exposure to potential marrow toxins, changes in medicinal drug use, and the appearance of physical signs relating to blood disorders:

**1910.1028(i)(3)(i)(B)** A complete blood count including a leukocyte count with differential, quantitative thrombocyte count, hemoglobin, hematocrit, erythrocyte count and erythrocyte indices (MCV, MCH, MCHC); and

**1910.1028(i)(3)(i)(C)** Appropriate additional tests as necessary, in the opinion of the examining physician, in consequence of alterations in the components of the blood or other signs which may be related to benzene exposure.

**1910.1028(i)(3)(ii)** Where the employee develops signs and symptoms commonly associated with toxic exposure to benzene, the employer shall provide the employee with an additional medical examination which shall include those elements considered appropriate by the examining physician.

**1910.1028(i)(3)(iii)** For persons required to use respirators for at least 30 days a year, a pulmonary function test shall be performed every three (3) years. A specific evaluation of the cardiopulmonary system shall be made at the time of the pulmonary function test.

**1910.1028(i)(4)** *Emergency examinations.* **(i)** In addition to the surveillance required by (i)(1)(i), if an employee is exposed to benzene in an emergency situation, the employer shall have the employee provide a urine sample at the end of the employee's shift and have a urinary phenol test performed on the sample within 72 hours. The urine specific gravity shall be corrected to 1.024.

**1910.1028(i)(4)(ii)** If the result of the urinary phenol test is below 75 mg phenol/L of urine, no further testing is required.

**1910.1028(i)(4)(iii)** If the result of the urinary phenol test is equal to or greater than 75 mg phenol/L of urine, the employer shall provide the employee with a complete blood count including an erythrocyte count, leukocyte count with differential and thrombocyte count at monthly intervals for a duration of three (3) months following the emergency exposure.

**1910.1028(i)(4)(iv)** If any of the conditions specified in paragraph (i)(5)(i) of this section exists, then the further requirements of paragraph (i)(5) of this section shall be met and the employer shall, in addition, provide the employees with periodic examinations if directed by the physician.

**1910.1028(i)(5)** *Additional examinations and referrals.* **(i)** Where the results of the complete blood count required for the initial and periodic examinations indicate any of the following abnormal conditions exist, then the blood count shall be repeated within 2 weeks.

**1910.1028(i)(5)(i)(A)** The hemoglobin level or the hematocrit falls below the normal limit [outside the 95% confidence interval (C.I.)] as determined by the laboratory for the particular geographic area and/or these indices show a persistent downward trend from the individual's pre-exposure norms; provided these findings cannot be explained by other medical reasons.

**1910.1028(i)(5)(i)(B)** The thrombocyte (platelet) count varies more than 20 percent below the employee's most recent values or falls outside the normal limit (95% C.I.) as determined by the laboratory.

**1910.1028(i)(5)(i)(C)** The leukocyte count is below 4,000 per mm$^3$ or there is an abnormal differential count.

**1910.1028(i)(5)(ii)** If the abnormality persists, the examining physician shall refer the employee to a hematologist or an internist for further evaluation unless the physician has good reason to believe such referral is unnecessary. (See Appendix C for examples of conditions where a referral may be unnecessary.)

**1910.1028(i)(5)(iii)** The employer shall provide the hematologist or internist with the information required to be provided to the physician under paragraph (i)(6) of this section and the medical record required to be maintained by paragraph (k)(2)(ii) of this section.

**1910.1028(i)(5)(iv)** The hematologist's or internist's evaluation shall include a determination as to the need for additional tests, and the employer shall assure that these tests are provided.

**1910.1028(i)(6)** *Information provided to the physician.* The employer shall provide the following information to the examining physician:

**1910.1028(i)(6)(i)** A copy of this regulation and its appendices;

**1910.1028(i)(6)(ii)** A description of the affected employee's duties as they relate to the employee's exposure;

**1910.1028(i)(6)(iii)** The employee's actual or representative exposure level;

**1910.1028(i)(6)(iv)** A description of any personal protective equipment used or to be used; and

**1910.1028(i)(6)(v)** Information from previous employment-related medical examinations of the affected employee which is not otherwise available to the examining physician.

**1910.1028(i)(7)** *Physician's written opinions.* **(i)** For each examination under this section, the employer shall obtain and provide the employee with a copy of the examining physician's written opinion within 15 days of the examination. The written opinion shall be limited to the following information:

**1910.1028(i)(7)(i)(A)** The occupationally pertinent results of the medical examination and tests;

**1910.1028(i)(7)(i)(B)** The physician's opinion concerning whether the employee has any detected medical conditions which would place the employee's health at greater than normal risk of material impairment from exposure to benzene;

**1910.1028(i)(7)(i)(C)** The physician's recommended limitations upon the employee's exposure to benzene or upon the employee's use of protective clothing or equipment and respirators.

**1910.1028(i)(7)(i)(D)** A statement that the employee has been informed by the physician of the results of the medical examination and any medical conditions resulting from benzene exposure which require further explanation or treatment.

**1910.1028(i)(7)(ii)** The written opinion obtained by the employer shall not reveal specific records, findings and diagnoses that have no bearing on the employee's ability to work in a benzene-exposed workplace.

**1910.1028(i)(8)** *Medical removal plan.* **(i)** When a physician makes a referral to a hematologist/internist as required under paragraph (i)(5)(ii) of this section, the employee shall be removed from areas where exposures may exceed the action level until such time as the physician makes a determination under paragraph (i)(8)(ii) of this section.

**1910.1028(i)(8)(ii)** Following the examination and evaluation by the hematologist/internist, a decision to remove an employee from areas where benzene exposure is above the action level or to allow the employee to return to areas where benzene exposure is above the action level shall be made by the physician in consultation with the hematologist/internist. This decision shall be communicated in writing to the employer and employee. In the case of removal, the physician shall state the required probable duration of removal from occupational exposure to benzene above the action level and the requirements for future medical examinations to review the decision.

**1910.1028(i)(8)(iii)** For any employee who is removed pursuant to paragraph (i)(8)(ii) of this section, the employer shall provide a followup examination. The physician, in consultation with the hematologist/internist, shall make a decision within 6 months of the date the employee was removed as to whether the employee shall be returned to the usual job or whether the employee should be removed permanently.

**1910.1028(i)(8)(iv)** Whenever an employee is temporarily removed from benzene exposure pursuant to paragraph (i)(8)(i) or (i)(8)(ii) of this section, the employer shall transfer the employee to a comparable job for which the employee is qualified (or can be trained for in a short period) and where benzene exposures are as low as possible, but in no event higher than the action level. The employer shall maintain the employee's current wage rate, seniority and other benefits. If there is no such job available, the employer shall provide medical removal protection benefits until such a job becomes available or for 6 months, whichever comes first.

**1910.1028(i)(8)(v)** Whenever an employee is removed permanently from benzene exposure based on a physician's recommendation pursuant to paragraph (i)(8)(iii) of this section, the employee shall be given the opportunity to transfer to another position which is available or later becomes available for which the employee is qualified (or can be trained for in a short period) and where benzene exposures are as low as possible but in no event higher than the action level. The employer shall assure that such employee suffers no reduction in current wage rate, seniority or other benefits as a result of the transfer.

**1910.1028(i)(9)** *Medical removal protection benefits.* **(i)** The employer shall provide to an employee 6 months of medical removal protection benefits immediately following each occasion an employee is removed from exposure to benzene because of hematological findings pursuant to paragraphs (i)(8)(i) and (ii) of this section, unless the employee has been transferred to a comparable job where benzene exposures are below the action level.

**1910.1028(i)(9)(ii)** For the purposes of this section, the requirement that an employer provide medical removal protection benefits means that the employer shall maintain the current wage rate, seniority and other benefits of an employee as though the employee had not been removed.

**1910.1028(i)(9)(iii)** The employer's obligation to provide medical removal protection benefits to a removed employee shall be reduced to the extent that the employee receives compensation for earnings lost during the period of removal either from a publicly or employerfunded compensation program, or from employment with another employer made possible by virtue of the employee's removal.

**1910.1028(j)** *Communication of benzene hazards to employees—*(1) *Signs and labels.* **(i)** The employer shall post signs at entrances to regulated areas. The signs shall bear the following legend:

**DANGER**

**BENZENE**

**CANCER HAZARD**

**FLAMMABLE—NO SMOKING**

**AUTHORIZED PERSONNEL ONLY**

**RESPIRATOR REQUIRED**

**1910.1028(j)(1)(ii)** The employer shall ensure that lables or other appropriate forms of warning are provided for containers of benzene within the workplace. There is no requirement to label pipes. The labels shall comply with the requirements of 29 CFR 1910.1200(f) and in addition shall include the following legend:

**DANGER**

**CONTAINS BENZENE**

**CANCER HAZARD**

# OSHA Standards for General Industry

**1910.1028(j)(2)** *Material safety data sheets.* **(i)** Employers shall obtain or develop, and shall provide access to their employees, to a material safety data sheet (MSDS) which addresses benzene and complies with 29 CFR 1910.1200.

**1910.1028(j)(2)(ii)** Employers who are manufacturers or importers shall:

**1910.1028(j)(2)(ii)(A)** Comply with paragraph (a) of this section, and

**1910.1028(j)(2)(ii)(B)** Comply with the requirement in OSHA's Hazard Communication Standard, 29 CFR 1910.1200, that they deliver to downstream employers an MSDS which addresses benzene.

**1910.1028(j)(3)** *Information and training.* **(i)** The employer shall provide employees with information and training at the time of their initial assignment to a work area where benzene is present. If exposures are above the action level, employees shall be provided with information and training at least annually thereafter.

**1910.1028(j)(3)(ii)** The training program shall be in accordance with the requirements of 29 CFR 1910.1200(h)(1) and (2), and shall include specific information on benzene for each category of information included in that section.

**1910.1028(j)(3)(iii)** In addition to the information required under 29 CFR 1910.1200, the employer shall:

**1910.1028(j)(3)(iii)(A)** Provide employees with an explanation of the contents of this section, including Appendices A and B, and indicate to them where the standard is available; and

**1910.1028(j)(3)(iii)(B)** Describe the medical surveillance program required under paragraph (i) of this section, and explain the information contained in Appendix C.

**1910.1028(k)** *Recordkeeping—***(1)** *Exposure measurements.* **(i)** The employer shall establish and maintain an accurate record of all measurements required by paragraph (e) of this section, in accordance with 29 CFR 1910.20.

**1910.1028(k)(1)(ii)** This record shall include:

**1910.1028(k)(1)(ii)(A)** The dates, number, duration, and results of each of the samples taken, including a description of the procedure used to determine representative employee exposures;

**1910.1028(k)(1)(ii)(B)** A description of the sampling and analytical methods used;

**1910.1028(k)(1)(ii)(C)** A description of the type of respiratory protective devices worn, if any; and

**1910.1028(k)(1)(ii)(D)** The name, social security number, job classification and exposure levels of the employee monitored and all other employees whose exposure the measurement is intended to represent.

**1910.1028(k)(1)(iii)** The employer shall maintain this record for at least 30 years, in accordance with 29 CFR 1910.20.

**1910.1028(k)(2)** *Medical surveillance.* **(i)** The employer shall establish and maintain an accurate record for each employee subject to medical surveillance required by paragraph (i) of this section, in accordance with 29 CFR 1910.20.

**1910.1028(k)(2)(ii)** This record shall include:

**1910.1028(k)(2)(ii)(A)** The name and social security number of the employee;

**1910.1028(k)(2)(ii)(B)** The employer's copy of the physician's written opinion on the initial, periodic and special examinations, including results of medical examinations and all tests, opinions and recommendations;

**1910.1028(k)(2)(ii)(C)** Any employee medical complaints related to exposure to benzene;

**1910.1028(k)(2)(ii)(D)** A copy of the information provided to the physician as required by paragraphs (i)(6) (ii) through (v) of this section; and

**1910.1028(k)(2)(ii)(E)** A copy of the employee's medical and work history related to exposure to benzene or any other hematologic toxins.

**1910.1028(k)(2)(iii)** The employer shall maintain this record for at least the duration of employment plus 30 years, in accordance with 29 CFR 1910.20.

**1910.1028(k)(3)** *Availability.* **(i)** The employer shall assure that all records required to be maintained by this section shall be made available upon request to the Assistant Secretary and the Director for examination and copying.

**1910.1028(k)(3)(ii)** Employee exposure monitoring records required by this paragraph shall be provided upon request for examination and copying to employees, employee representatives, and the Assistant Secretary in accordance with 29 CFR 1910.20(a) through (e) and (g) through (i).

**1910.1028(k)(3)(iii)** Employee medical records required by this paragraph shall be provided upon request for examination and copying, to the subject employee, to anyone having the specific written consent of the subject employee, and to the Assistant Secretary in accordance with 29 CFR 1910.20.

**1910.1028(k)(4)** *Transfer of records.* **(i)** The employer shall comply with the requirements involving transfer of records set forth in 29 CFR 1019.20(h).

**1910.1028(k)(4)(ii)** If the employer ceases to do business and there is no successor employer to receive and retain the records for the prescribed period, the employer shall notify the Director, at least three (3) months prior to disposal, and transmit them to the Director if required by the Director within that period.

**1910.1028(l)** *Observation of monitoring—***(1)** *Employee observation.* The employer shall provide affected employees, or their designated representatives, an opportunity to observe the measuring or monitoring of employee exposure to benzene conducted pursuant to paragraph (e) of this section.

**1910.1028(l)(2)** *Observation procedures.* When observation of the measuring or monitoring of employee exposure to benzene requires entry into areas where the use of protective clothing and equipment or respirators is required, the employer shall provide the observer with personal protective clothing and equipment or respirators required to be worn by employees working in the area, assure the use of such clothing and equipment or respirators, and require the observer to comply with all other applicable safety and health procedures.

**1910.1028(m)** [Reserved.]

**1910.1028(n)** *Appendices.* The information contained in Appendices A, B, C, and D is not intended, by itself, to create any additional obligations not otherwise imposed or to detract from any existing obligations.

## APPENDIX A TO §1910.1028— SUBSTANCE SAFETY DATA SHEET, BENZENE

### I. SUBSTANCE IDENTIFICATION

A. Substance: Benzene.

B. Permissible Exposure: Except as to the use of gasoline, motor fuels and other fuels subsequent to discharge from bulk terminals and other exemptions specified in §1910.1028(a)(2):

1. Airborne: The maximum time-weighted average (TWA) exposure limit is 1 part of benzene vapor per million parts of air (1 ppm) for an 8-hour workday and the maximum short-term exposure limit (STEL) is 5 ppm for any 15-minute period.

2. Dermal: Eye contact shall be prevented and skin contact with liquid benzene shall be limited.

C. Appearance and odor: Benzene is a clear, colorless liquid with a pleasant, sweet odor. The odor of benzene does not provide adequate warning of its hazard.

### II. HEALTH HAZARD DATA

A. *Ways in which benzene affects your health.* Benzene can affect your health if you inhale it, or if it comes in contact with your skin or eyes. Benzene is also harmful if you happen to swallow it.

B. *Effects of overexposure.* 1. Short-term (acute) overexposure: If you are overexposed to high concentrations of benzene, well above the levels where its odor is first recognizable, you may feel breathless, irritable, euphoric, or giddy; you may experience irritation in eyes, nose, and respiratory tract. You may develop a headache, feel dizzy, nauseated, or intoxicated. Severe exposures may lead to convulsions and loss of consciousness.

2. Long-term (chronic) exposure. Repeated or prolonged exposure to benzene, even at relatively low concentrations, may result in various blood disorders, ranging from anemia to leukemia, an irreversible, fatal disease. Many blood disorders associated with benzene exposure may occur without symptoms.

### III. PROTECTIVE CLOTHING AND EQUIPMENT

A. *Respirators.* Respirators are required for those operations in which engineering controls or work practice controls are not feasible to reduce exposure to the permissible level. However, where employers can document that benzene is present in the workplace less than 30 days a year, respirators may be used in lieu of engineering controls. If respirators are worn, they must have joint Mine Safety and Health Administration and the National Institute for Occupational Safety and Health (NIOSH) seal of approval, and cartridge or canisters must be replaced before the end of their service life, or the end of the shift, whichever occurs first. If you experience difficulty breathing while wearing a respirator, you may request a positive pressure respirator from your employer. You must be thoroughly trained to use the assigned respirator, and the training will be provided by your employer.

B. *Protective Clothing.* You must wear appropriate protective clothing (such as boots, gloves, sleeves, aprons, etc.) over any parts of your body that could be exposed to liquid benzene.

C. *Eye and Face Protection.* You must wear splashproof safety goggles if it is possible that benzene may get into your eyes. In addition, you must wear a face shield if your face could be splashed with benzene liquid.

### IV. EMERGENCY AND FIRST AID PROCEDURES

A. *Eye and face exposure.* If benzene is splashed in your eyes, wash it out immediately with large amounts of water. If irritation persists or vision appears to be affected see a doctor as soon as possible.

B. *Skin exposure.* If benzene is spilled on your clothing or skin, remove the contaminated clothing and wash the exposed skin with large amounts of water and soap immediately. Wash contaminated clothing before you wear it again.

C. *Breathing.* If you or any other person breathes in large amounts of benzene, get the exposed person to fresh air at once. Apply artificial respiration if breathing has stopped. Call for medical assistance or a doctor as soon as possible. Never enter any vessel or confined space where the benzene concentration might be high without proper safety equipment and at least one other person present who will stay outside. A life line should be used.

D. *Swallowing.* If benzene has been swallowed and the patient is conscious, do not induce vomiting. Call for medical assistance or a doctor immediately.

### V. MEDICAL REQUIREMENTS

If you are exposed to benzene at a concentration at or above 0.5 ppm as an 8-hour time-weighted average, or have been exposed at or above 10 ppm in the past while employed by your current employer, your employer is required to provide a medical examination and history and laboratory tests within 60 days of the effective date of this standard and annually thereafter. These tests shall be provided without cost to you. In addition, if you are accidentally exposed to benzene (either by ingestion, inhalation, or skin/eye contact) under emergency conditions known or suspected to constitute toxic exposure to benzene, your employer is required to make special laboratory tests available to you.

### VI. OBSERVATION OF MONITORING

Your employer is required to perform measurements that are representative of your exposure to benzene and you or your designated representative are entitled to observe the monitoring procedure. You are entitled to observe the steps taken in the measurement procedure, and to record the results obtained. When the monitoring procedure is taking place in an area where respirators or personal protective clothing and equipment are required to be worn, you or your representative must also be provided with, and must wear the protective clothing and equipment.

### VII. ACCESS TO RECORDS

You or your representative are entitled to see the records of measurements of your exposure to benzene upon written request to your employer. Your medical examination records can be furnished to yourself, your physician or designated representative upon request by you to your employer.

### VIII. PRECAUTIONS FOR SAFE USE, HANDLING AND STORAGE

Benzene liquid is highly flammable. It should be stored in tightly closed containers in a cool, well ventilated area. Benzene vapor may form explosive mixtures in air. All sources of ignition must be controlled. Use nonsparking tools when opening or closing benzene containers. Fire extinguishers, where provided, must be readily available. Know where they are located and how to operate them. Smoking is prohibited in areas where benzene is used or stored. Ask your supervisor where benzene is used in your area and for additional plant safety rules.

## APPENDIX B TO §1910.1028—SUBSTANCE TECHNICAL GUIDLINES [sic], BENZENE

### I. PHYSICAL AND CHEMICAL DATA

A. Substance identification.

1. *Synonyms:* Benzol, benzole, coal naphtha, cyclohexatriene, phene, phenyl hydride, pyrobenzol. (Benzin, petroleum benzin and Benzine do not contain benzene).

2. *Formula:* $C_6H_6$ (CAS Registry Number: 71-43-2)

B. Physical data.

1. Boiling Point (760 mm Hg); 80.1 °C (176 °F)

2. Specific Gravity (water = 1): 0.879

3. Vapor Density (air = 1): 2.7

4. Melting Point: 5.5 °C (42 °F)

5. Vapor Pressure at 20 °C (68 °F): 75 mm Hg

6. Solubility in Water: .06%

7. Evaporation Rate (ether = 1): 2.8

8. Appearance and Odor: Clear, colorless liquid with a distinctive sweet odor.

### II. FIRE, EXPLOSION, AND REACTIVITY HAZARD DATA

A. Fire.

1. Flash Point (closed cup): -11 °C (12 °F)

2. Autoignition Temperature: 580 °C (1076 °F)

3. Flammable limits in Air. % by Volume: Lower: 1.3%, Upper: 7.5%

4. Extinguishing Media: Carbon dioxide, dry chemical, or foam.

5. Special Fire-Fighting procedures: Do not use solid stream of water, since stream will scatter and spread fire. Fine water spray can be used to keep fire-exposed containers cool.

6. Unusual fire and explosion hazards: Benzene is a flammable liquid. Its vapors can form explosive mixtures. All ignition sources must be controlled when benzene is used, handled, or stored. Where liquid or vapor may be released, such areas shall be considered as hazardous locations. Benzene vapors are heavier than air; thus the vapors may travel along the ground and be ignited by open flames or sparks at locations remote from the site at which benzene is handled.

7. Benzene is classified as a 1 B flammable liquid for the purpose of conforming to the requirements of 29 CFR 1910.106. A concentration exceeding 3,250 ppm is considered a potential fire explosion hazard. Locations where benzene may be present in quantities sufficient to produce explosive or ignitable mixtures are considered Class I Group D for the purposes of conforming to the requirements of 29 CFR 1910.309.

B. Reactivity.

1. Conditions contributing to instability: Heat.

2. Incompatibility: Heat and oxidizing materials.

3. Hazardous decomposition products: Toxic gases and vapors (such as carbon monoxide).

### III. SPILL AND LEAK PROCEDURES

A. Steps to be taken if the material is released or spilled. As much benzene as possible should be absorbed with suitable materials, such as dry sand or earth. That remaining must be flushed with large amounts of water. Do not flush benzene into a confined space, such as a sewer, because of explosion danger. Remove all ignition sources. Ventilate enclosed places.

B. Waste disposal method. Disposal methods must conform to other jurisdictional regulations. If allowed, benzene may be disposed of: (a) By absorbing it in dry sand or earth and disposing in a sanitary landfill; (b) if small quantities, by removing it to a safe location from buildings or other combustible sources, pouring it in dry sand or earth and cautiously igniting it; and (c) if large quantities, by atomizing it in a suitable combustion chamber.

### IV. MISCELLANEOUS PRECAUTIONS

A. High exposure to benzene can occur when transferring the liquid from one container to another. Such operations should be well ventilated and good work practices must be established to avoid spills.

B. Use non-sparking tools to open benzene containers which are effectively grounded and bonded prior to opening and pouring.

# OSHA Standards for General Industry

C. Employers must advise employees of all plant areas and operations where exposure to benzene could occur. Common operations in which high exposures to benzene may be encountered are: the primary production and utilization of benzene, and transfer of benzene.

## APPENDIX C TO §1910.1028—MEDICAL SURVEILLANCE GUIDELINES FOR BENZENE

### I. ROUTE OF ENTRY

Inhalation; skin absorption.

### II. TOXICOLOGY

Benzene is primarily an inhalation hazard. Systemic absorption may cause depression of the hematopoietic system, pancytopenia, aplastic anemia, and leukemia. Inhalation of high concentrations can affect central nervous system function. Aspiration of small amounts of liquid benzene immediately causes pulmonary edema and hemorrhage of pulmonary tissue. There is some absorption through the skin. Absorption may be more rapid in the case of abraded skin, and benzene may be more readily absorbed if it is present in a mixture or as a contaminant in solvents which are readily absorbed. The defatting action of benzene may produce primary irritation due to repeated or prolonged contact with the skin. High concentration are irritating to the eyes and the mucuous membranes of the nose, and respiratory tract.

### III. SIGNS AND SYMPTOMS

Direct skin contact with benzene may cause erythema. Repeated or prolonged contact may result in drying, scaling dermatitis, or development of secondary skin infections. In addition, there is benzene absorption through the skin. Local effects of benzene vapor or liquid on the eye are slight. Only at very high concentrations is there any smarting sensation in the eye. Inhalation of high concentrations of benzene may have an initial stimulatory effect on the central nervous system characterized by exhilaration, nervous excitation, and/or giddiness, followed by a period of depression, drowsiness, or fatigue. A sensation of tightness in the chest accompanied by breathlessness may occur and ultimately the victim may lose consciousness. Tremors, convulsions and death may follow from respiratory paralysis or circulatory collapse in a few minutes to several hours following severe exposures.

The detrimental effect on the blood-forming system of prolonged exposure to small quantities of benzene vapor is of extreme importance. The hematopoietic system is the chief target for benzene's toxic effects which are manifested by alterations in the levels of formed elements in the peripheral blood. These effects have occurred at concentrations of benzene which may not cause irritation of mucous membranes, or any unpleasant sensory effects. Early signs and symptoms of benzene morbidity are varied, often not readily noticed and non-specific. Subjective complaints of headache, dizziness, and loss of appetite may precede or follow clinical signs. Rapid pulse and low blood pressure, in addition to a physical appearance of anemia, may accompany a subjective complaint of shortness of breath and excessive tiredness. Bleeding from the nose, gums, or mucous membranes, and the development of purpuric spots (small bruises) may occur as the condition progresses. Clinical evidence of leukopenia, anemia, and thrombocytopenia, singly or in combination, has been frequently reported among the first signs.

Bone marrow may appear normal, aplastic, or hyperplastic, and may not, in all situations, correlate with peripheral blood forming tissues. Because of variations in the susceptibility to benzene morbidity, there is no "typical" blood picture. The onset of effects of prolonged benzene exposure may be delayed for many months or years after the actual exposure has ceased and identification or correlation with benzene exposure must be sought out in the occupational history.

### IV. TREATMENT OF ACUTE TOXIC EFFECTS

Remove from exposure immediately. Make sure you are adequately protected and do not risk being overcome by fumes. Give oxygen or artificial resuscitation if indicated. Flush eyes, wash skin if contaminated and remove all contaminated clothing. Symptoms of intoxication may persist following severe exposures. Recovery from mild exposures is usually rapid and complete.

### V. SURVEILLANCE AND PREVENTIVE CONSIDERATIONS

#### A. General

The principal effects of benzene exposure which form the basis for this regulation are pathological changes in the hematopoietic system, reflected by changes in the peripheral blood and manifesting clinically as pancytopenia, aplastic anemia, and leukemia. Consequently, the medical surveillance program is designed to observe, on a regular basis, blood indices for early signs of these effects, and although early signs of leukemia are not usually available, emerging diagnostic technology and innovative regimes make consistent surveillance for leukemia, as well as other hematopoietic effects, essential.

Initial examinations are to be provided within 60 days of the effective date of this standard, or at the time of initial assignment, and periodic examinations annually thereafter. There are special provisions for medical tests in the event of hematologic abnormalities or for emergency situations.

The blood values which require referral to a hematologist or internist are noted in the standard in paragraph (i)(5). The standard specifies that blood abnormalities that persist must be referred "unless the physician has good reason to believe such referral is unnecessary" (paragraph (i)(5)). Examples of conditions that could make a referral unnecessary despite abnormal blood limits are iron or folate deficiency, menorrhagia, or blood loss due to some unrelated medical abnormality.

Symptoms and signs of benzene toxicity can be non-specific. Only a detailed history and appropriate investigative procedures will enable a physician to rule out or confirm conditions that place the employee at increased risk. To assist the examining physician with regard to which laboratory tests are necessary and when to refer an employee to the specialist, OSHA has established the following guidelines.

#### B. Hematology Guidelines

A minimum battery of tests is to be performed by strictly standardized methods.

1. Red cell, white cell, platelet counts, white blood cell differential, hematacrit and red cell indices must be performed by an accredited laboratory. The normal ranges for the red cell and white cell counts are influenced by altitude, race, and sex, and therefore should be determined by the accredited laboratory in the specific area where the tests are performed.

Either a decline from an absolute normal or an individual's base line to a subnormal value or a rise to a supra-normal value, are indicative of potential toxicity, particularly if all blood parameters decline. The normal total white blood count is approximately 7,200/mm$^3$ plus or minus 3,000. For cigarette smokers the white count may be higher and the upper range may be 2,000 cells higher than normal for the laboratory. In addition, infection, allergies and some drugs may raise the white cell count. The normal platelet count is approximately 250,000 with a range of 140,000 to 400,000. Counts outside this range should be regarded as possible evidence of benzene toxicity.

Certain abnormalities found through routine screening are of greater significance in the benzene-exposed worker and require prompt consultation with a specialist, namely:

a. Thrombocytopenia.

b. A trend of decreasing white cell, red cell, or platelet indices in an individual over time is more worrisome than an isolated abnormal finding at one test time. The importance of trend highlights the need to compare an individual's test results to baseline and/or previous periodic tests.

c. A constellation or pattern of abnormalities in the different blood indices is of more significance than a single abnormality. A low white count not associated with any abnormalities in other cell indices may be a normal statistical variation, whereas if the low white count is accompanied by decreases in the platelet and/or red cell indices, such a pattern is more likely to be associated with benzene toxicity and merits thorough investigation.

Anemia, leukopenia, macrocytosis or an abnormal differential white blood cell count should alert the physician to further investigate and/or refer the patient if repeat tests confirm the abnormalities. If routine screening detects an abnormality, follow-up tests which may be helpful in establishing the etiology of the abnormality are the peripheral blood smear and the reticulocyte count.

The extreme range of normal for reticulocytes is 0.4 to 2.5 percent of the red cells, the usual range being 0.5 to 1.2 percent of the red cells, but the typical value is in the range of 0.8 to 1.0 percent. A decline in reticulocytes to levels of less than 0.4 percent is to be regarded as possible evidence (unless another specific cause is found) of benzene toxicity requiring accelerated surveillance. An increase in reticulocyte levels to about 2.5 percent may also be consistent with (but is not as characteristic of) benzene toxicity.

2. An important diagnostic test is a careful examination of the peripheral blood smear. As with reticulocyte count the smear should be with fresh uncoagulated blood obtained from a needle tip following venipuncture or from a drop of earlobe blood (capillary blood). If necessary, the smear may, under certain limited conditions, be made from a blood sample anticoagulated with EDTA (but never with oxalate or heparin). When the smear is to be prepared from a specimen of venous blood which has been collected by a commercial Vacutainer# type tube containing neutral EDTA, the smear should be made as soon as possible after the venesection. A delay of up to 12 hours is permissible between the drawing of the blood specimen into EDTA and the preparation of the smear if the blood is stored at refrigerator (not freezing) temperature.

3. The minimum mandatory observations to be made from the smear are:

a. The differential white blood cell count.

b. Description of abnormalities in the appearance of red cells.

c. Description of any abnormalities in the platelets.

d. A careful search must be made throughout of every blood smear for immature white cells such as band forms (in more than normal proportion, i.e., over 10 percent of the total differential count), any number of metamyelocytes, myelocytes or myeloblasts. Any nucleate or multinucleated red blood cells should be reported. Large "giant" platelets or fragments of megakaryocytes must be recognized.

An increase in the proportion of band forms among the neutrophilic granulocytes is an abnormality deserving special mention, for it may represent a change which should be considered as an early warning of benzene toxicity in the absence of other causative factors (most commonly infection). Likewise, the appearance of metamyelocytes, in the absence of another probable cause, is to be considered a possible indication of benzeneinduced toxicity.

An upward trend in the number of basophils, which normally do not exceed about 2.0 percent of the total white cells, is to be regarded as possible evidence of benzene toxicity. A rise in the eosinophil count is less specific but also may be suspicious of toxicity if the rises above 6.0 percent of the total white count.

The normal range of monocytes is from 2.0 to 8.0 percent of the total white count with an average of about 5.0 percent. About 20 percent of individuals reported to have mild but persisting abnormalities caused by exposure to benzene show a persistent monocytosis. The findings of a monocyte count which persists at more than 10 to 12 percent of the normal white cell count (when the total count is normal) or persistence of an absolute monocyte count in excess of 800/mm$^3$ should be regarded as a possible sign of benzene-induced toxicity.

A less frequent but more serious indication of benzene toxicity is the finding in the peripheral blood of the so-called "pseudo" (or acquired) Pelger-Huet anomaly. In this anomaly many, or sometimes the majority, of the neutrophilic granulocytes possess two round nuclear segments—less often one or three round segments—rather than three normally elongated segments. When this anomaly is not hereditary, it is often but not invariably predictive of subsequent leukemia. However, only about two percent of patients who ultimately develop acute myelogenous leukemia show the acquired Pelger-Huet anomaly. Other tests that can be administered to investigate blood abnormalities are discussed below; however, such procedures should be undertaken by the hematologist.

An uncommon sign, which cannot be detected from the smear, but can be elicited by a "sucrose water test" of peripheral blood, is transient paroxysmal nocturnal hemoglobinuria (PNH), which may first occur insidiously during a period of established aplastic anemia, and may be followed within one to a few years by the appearance of rapidly fatal acute myelogenous leukemia. Clinical detection of PNH, which occurs in only one or two percent of those destined to have acute myelogenous leukemia, may be difficult; if the "sucrose water test" is positive, the somewhat more definitive Ham test, also known as the acid-serum hemolysis test, may provide confirmation.

e. Individuals documented to have developed acute myelogenous leukemia years after initial exposure to benzene may have progressed through a preliminary phase of hematologic abnormality. In some instances pancytopenia (i.e., a lowering in the counts of all circulating blood cells of bone marrow origin, but not to the extent implied by the term "aplastic anemia") preceded leukemia for many years. Depression of a single blood cell type or platelets may represent a harbinger of aplasia or leukemia. The finding of two or more cytopenias, or pancytopenia in a benzene-exposed individual, must be regarded as highly suspicious of more advanced although still reversible, toxicity. "Pancytopenia" coupled with the appearance of immature cells (myelocytes, myeloblasts, erythroblasts, etc.), with abnormal cells (pseudo Pelger-Huet anomaly, atypical nuclear heterochromatin, etc.), or unexplained elevations of white blood cells must be regarded as evidence of benzene overexposure unless proved otherwise. Many severely aplastic patients manifested the ominous finding of 5-10 percent myeloblasts in the marrow, occasional myeloblasts and myelocytes in the blood and 20-30% monocytes. It is evident that isolated cytopenias, pancytopenias, and even aplastic anemias induced by benzene may be reversible and complete recovery has been reported on cessation of exposure. However, since any of these abnormalities is serious, the employee must immediately be removed from any possible exposure to benzene vapor. Certain tests may substantiate the employee's prospects for progression or regression. One such test would be an examination of the bone marrow, but the decision to perform a bone marrow aspiration or needle biopsy is made by the hematologist.

The findings of basophilic stippling in circulating red blood cells (usually found in 1 to 5% of red cells following marrow injury), and detection in the bone marrow of what are termed "ringed sideroblasts" must be taken seriously, as they have been noted in recent years to be premonitory signs of subsequent leukemia.

Recently peroxidase-staining of circulating or marrow neutrophil granulocytes, employing benzidine dihydrochloride, have revealed the disappearance of, or diminution in, peroxidase in a sizable proportion of the granulocytes, and this has been reported as an early sign of leukemia. However, relatively few patients have been studied to date. Granulocyte granules are normally strongly peroxidase positive. A steady decline in leukocyte alkaline phosphatase has also been reported as suggestive of early acute leukemia. Exposure to benzene may cause an early rise in serum iron, often but not always associated with a fall in the reticulocyte count. Thus, serial measurements of serum iron levels may provide a means of determining whether or not there is a trend representing sustained suppression of erythropoiesis.

Measurement of serum iron, determination of peroxidase and of alkaline phosphatase activity in peripheral granulocytes can be performed in most pathology laboratories. Peroxidase and alkaline phosphatase staining are usually undertaken when the index of suspecion for leukemia is high.

## APPENDIX D TO §1910.1028—SAMPLING AND ANALYTICAL METHODS FOR BENZENE MONITORING AND MEASUREMENT PROCEDURES

Measurements taken for the purpose of determining employee exposure to benzene are best taken so that the representative average 8-hour exposure may be determined from a single 8-hour sample or two (2) 4-hour samples. Short-time interval samples (or grab samples) may also be used to determine average exposure level if a minimum of five measurements are taken in a random manner over the 8-hour work shift. Random sampling means that any portion of the work shift has the same change of being sampled as any other. The arithmetic average of all such random samples taken on one work shift is an estimate of an employee's average level of exposure for that work shift. Air samples should be taken in the employee's breathing zone (air that would most nearly represent that inhaled by the employee). Sampling and analysis must be performed with procedures meeting the requirements of the standard.

There are a number of methods available for monitoring employee exposures to benzene. The sampling and analysis may be performed by collection of the benzene vapor or charcoal absorption tubes, with subsequent chemical analysis by gas chromatography. Sampling and analysis may also be performed by portable direct reading instruments, real-time continuous monitoring systems, passive dosimeters or other suitable methods. The employer has the obligation of selecting a monitoring method which meets the accuracy and precision requirements of the standard under his unique field conditions. The standard requires that the method of monitoring must have an accuracy, to a 95 percent confidence level, of not less than plus or minus 25 percent for concentrations of benzene greater than or equal to 0.5 ppm.

The OSHA Laboratory modified NIOSH Method S311 and evaluated it at a benzene air concentration of 1 ppm. A procedure for determining the benzene concentration in bulk material samples was also evaluated. This work, reported in OSHA Laboratory Method No. 12, includes the following two analytical procedures:

# OSHA Standards for General Industry

## I. OSHA Method 12 for Air Samples

Analyte: Benzene

Matrix: Air

Procedure: Adsorption on charcoal, desorption with carbon disulfide, analysis by GC.

Detection limit: 0.04 ppm

Recommended air volume and sampling rate: 10 L to 0.2 L/min.

1. Principle of the Method.

1.1 A known volume of air is drawn through a charcoal tube to trap the organic vapors present.

1.2. The charcoal in the tube is transferred to a small, stoppered vial, and the anlyte is desorbed with carbon disulfide.

1.3. An aliquot of the desorbed sample is injected into a gas chromatograph.

1.4 The area of the resulting peak is determined and compared with areas obtained from standards.

2. Advantages and disadvantages of the method.

2.1 The sampling device is small, portable, and involved no liquids. Interferences are minimal, and most of those which do occur can be eliminated by altering chromatographic conditions. The samples are analyzed by means of a quick, instrumental method.

2.2 The amount of sample which can be taken is limited by the number of milligrams that the tube will hold before overloading. When the sample value obtained for the backup section of the charcoal tube exceeds 25 percent of that found on the front section, the possibility of sample loss exists.

3. Apparatus.

3.1 A calibrated personal sampling pump whose flow can be determined within ±5 percent at the recommended flow rate.

3.2. Charcoal tubes: Glass with both ends flame sealed, 7 cm long with a 6-mm O.D. and a 4-mm I.D., containing 2 sections of 20/40 mesh activated charcoal separated by a 2-mm portion of urethane foam. The activated charcoal is prepared from coconut shells and is fired at 600 °C prior to packing. The adsorbing section contains 100 mg of charcoal, the back-up section 50 mg. A 3-mm portion of urethane foam is placed between the outlet end of the tube and the back-up section. A plug of silanized glass wool is placed in front of the adsorbing section. The pressure drop across the tube must be less than one inch of mercury at a flow rate of 1 liter per minute.

3.3. Gas chromatograph equipped with a flame ionization detector.

3.4. Column (10-ft × 1/8-in stainless steel) packed with 80/100 Supelcoport coated with 20 percent SP 2100, 0.1 percent CW 1500.

3.5. An electronic integrator or some other suitable method for measuring peak area.

3.6. Two-milliliter sample vials with Teflon-lined caps.

3.7. Microliter syringes: 10-microliter (10-μL syringe, and other convenient sizes for making standards, 1-μL syringe for sample injections.

3.8. Pipets: 1.0 mL delivery pipets

3.9. Volumetric flasks: convenient sizes for making standard solutions.

4. Reagents.

4.1. Chromatographic quality carbon disulfide ($CS_2$). Most commercially available carbon disulfide contains a trace of benzene which must be removed. It can be removed with the following procedure:

Heat under reflux for 2 to 3 hours, 500 mL of carbon disulfide, 10 mL concentrated sulfuric acid, and 5 drops of concentrated nitric acid. The benzene is converted to nitrobenzene. The carbon disulfide layer is removed, dried with anhydrous sodium sulfate, and distilled. The recovered carbon disulfide should be benzene free. (It has recently been determined that benzene can also be removed by passing the carbon disulfide through 13× molecular sieve).

4.2. Benzene, reagent grade.

4.3. p-Cymene, reagent grade, (internal standard).

4.4. Desorbing reagent. The desorbing reagent is prepared by adding 0.05 mL of p-cymene per milliliter of carbon disulfide. (The internal standard offers a convenient means correcting analytical response for slight inconsistencies in the size of sample injections. If the external standard technique is preferred, the internal standard can be eliminated).

4.5. Purified GC grade helium, hydrogen and air.

5. Procedure.

5.1. Cleaning of equipment. All glassware used for the laboratory analysis should be properly cleaned and free of organics which could interfere in the analysis.

5.2. Calibration of personal pumps. Each pump must be calibrated with a representative charcoal tube in the line.

5.3. Collection and shipping of samples.

5.3.1. Immediately before sampling, break the ends of the tube to provide an opening at least one-half the internal diameter of the tube (2 mm).

5.3.2. The smaller section of the charcoal is used as the backup and should be placed nearest the sampling pump.

5.3.3. The charcoal tube should be placed in a vertical position during sampling to minimize channeling through the charcoal.

5.3.4 Air being sampled should not be passed through any hose or tubing before entering the charcoal tube.

5.3.5. A sample size of 10 liters is recommended. Sample at a flow rate of approximately 0.2 liters per minute. The flow rate should be known with an accuracy of at least ±5 percent.

5.3.6. The charcoal tubes should be capped with the supplied plastic caps immediately after sampling.

5.3.7. Submit at least one blank tube (a charcoal tube subjected to the same handling procedures, without having any air drawn through it) with each set of samples.

5.3.8. Take necessary shipping and packing precautions to minimize breakage of samples.

5.4. Analysis of samples.

5.4.1. Preparation of samples. In preparation for analysis, each charcoal tube is scored with a file in front of the first section of charcoal and broken open. The glass wool is removed and discarded. The charcoal in the first (larger) section is transferred to a 2-ml vial. The separating section of foam is removed and discarded; the second section is transferred to another capped vial. These two sections are analyzed separately.

5.4.2. Desorption of samples. Prior to analysis, 1.0 mL of desorbing solution is pipetted into each sample container. The desorbing solution consists of 0.05 μL internal standard per mL of carbon disulfide. The sample vials are capped as soon as the solvent is added. Desorption should be done for 30 minutes with occasional shaking.

5.4.3. GC conditions. Typical operating conditions for the gas chromatograph are:

1. 30 mL/min (60 psig) helium carrier gas flow.

2. 30 mL/min (40 psig) hydrogen gas flow to detector.

3. 240 mL/min (40 psig) air flow to detector.

4. 150 °C injector temperature.

5. 250 °C detector temperature.

6. 100 °C column temperature.

5.4.4. Injection size. 1 μL.

5.4.5. Measurement of area. The peak areas are measured by an electronic integrator or some other suitable form of area measurement.

5.4.6. An internal standard procedure is used. The integrator is calibrated to report results in ppm for a 10 liter air sample after correction for desorption efficiency.

5.5. Determination of desorption efficiency.

5.5.1. Importance of determination. The desorption efficiency of a particular compound can vary from one laboratory to another and from one lot of chemical to another. Thus, it is necessary to determine, at least once, the percentage of the specific compound that is removed in the desorption process, provided the same batch of charcoal is used.

5.5.2. Procedure for determining desorption efficiency. The reference portion of the charcoal tube is removed. To the remaining portion, amounts representing 0.5X, 1X, and 2X and (X represents target concentration) based on a 10 L air sample are injected into several tubes at each level. Dilutions of benzene with carbon disulfide are made to allow injection of measurable quantities. These tubes are then allowed to equilibrate at least overnight. Following equilibration they are analyzed following the same procedure as the samples. Desorption efficiency is determined by dividing the amount of benzene found by amount spiked on the tube.

6. Calibration and standards. A series of standards varying in concentration over the range of interest is prepared and analyzed under the same GC conditions that will be used on the samples. A calibration curve is prepared by plotting concentration (µg/mL) versus peak area.

7. Calculations. Benzene air concentration can be calculated from the following equation:

$$mg/m^3 = (A)(B)/(C)(D)$$

Where: A = µg/mL benzene, obtained from the calibration curve

B = desorption volume (1 mL)

C = Liters of air sampled

D = desorption efficiency

The concentration in $mg/m^3$ can be converted to ppm (at 25 °C and 760 mm) with following equation:

$$ppm = (mg/m^3)(24.46)/(78.11)$$

Where: 24.46 = molar volume of an ideal gas

25 °C and 760 mm

78.11 = molecular weight of benzene

8. Backup Data.

8.1 Detection limit—Air Samples.

The detection limit for the analytical procedure is 1.28 ng with a coefficient of variation of 0.023 at this level. This would be equivalent to an air concentration of 0.04 ppm for a 10 L air sample. This amount provided a chromatographic peak that could be identifiable in the presence of possible interferences. The detection limit data were obtained by making 1 µL injections of a 1.283 µg/mL standard.

| Injection | Area Count | |
|---|---|---|
| 1 | 655.4 | |
| 2 | 617.5 | |
| 3 | 662.0 | $\bar{X}$ = 640.2 |
| 4 | 641.1 | SD = 14.9 |
| 5 | 636.4 | CV = 0.023 |
| 6 | 629.2 | |

8.2. Pooled coefficient of variation—Air Samples. The pooled coefficient of variation for the analytical procedure was determined by 1 µL replicate injections of analytical standards. The standards were 16.04, 32.08, and 64.16 µg/mL, which are equivalent to 0.5, 1.0, and 2.0 ppm for a 10 L air sample respectively.

| | Area Counts | | |
|---|---|---|---|
| Injection | 0.5 ppm | 1.0 ppm | 2.0 ppm |
| 1 | 3996.5 | 8130.2 | 16481 |
| 2 | 4059.4 | 8235.6 | 16493 |
| 3 | 4052.0 | 8307.9 | 16535 |
| 4 | 4027.2 | 8263.2 | 16609 |
| 5 | 4046.8 | 8291.1 | 16552 |
| 6 | 4137.9 | 8288.8 | 16618 |
| $\bar{X}$ = | 4053.3 | 8254.0 | 16548.3 |
| SD = | 47.2 | 62.5 | 57.1 |
| CV = | 0.0116 | 0.0076 | 0.0034 |
| $\bar{\bar{CV}}$ = 0.008 | | | |

8.3. Storage data—Air Samples

Samples were generated at 1.03 ppm benzene at 80% relative humidity, 22 °C, and 643 mm. All samples were taken for 50 minutes at 0.2 L/min. Six samples were analyzed immediately and the rest of the samples were divided into two groups by fifteen samples each. One group was stored at refrigerated temperature of -25 °C, and the other group was stored at ambient temperature (approximately 23 °C). These samples were analyzed over a period of fifteen days. The results are tabulated below.

### PERCENT RECOVERY

| Day analyzed | Refrigerated | | | Ambient | | |
|---|---|---|---|---|---|---|
| 0 | 97.4 | 98.7 | 98.9 | 97.4 | 98.7 | 98.9 |
| 0 | 97.1 | 100.6 | 100.9 | 97.1 | 100.6 | 100.9 |
| 2 | 95.8 | 96.4 | 95.4 | 95.4 | 96.6 | 96.9 |
| 5 | 93.9 | 93.7 | 92.4 | 92.4 | 94.3 | 94.1 |
| 9 | 93.6 | 95.5 | 94.6 | 95.2 | 95.6 | 96.6 |
| 13 | 94.3 | 95.3 | 93.7 | 91.0 | 95.0 | 94.6 |
| 15 | 96.8 | 95.8 | 94.2 | 92.9 | 96.3 | 95.9 |

8.4. Desorption data.

Samples were prepared by injecting liquid benzene onto the A section of charcoal tubes. Samples were prepared that would be equivalent to 0.5, 1.0, and 2.0 ppm for a 10 L air sample.

### PERCENT RECOVERY

| Sample | 0.5 ppm | 1.0 ppm | 2.0 ppm |
|---|---|---|---|
| 1 | 99.4 | 98.8 | 99.5 |
| 2 | 99.5 | 98.7 | 99.7 |
| 3 | 99.2 | 98.6 | 99.8 |
| 4 | 99.4 | 99.1 | 100.0 |
| 5 | 99.2 | 99.0 | 99.7 |
| 6 | 99.8 | 99.1 | 99.9 |
| $\bar{X}$ = | 99.4 | 98.9 | 99.8 |
| SD = | 0.22 | 0.21 | 0.18 |
| CV = | 0.0022 | 0.0021 | 0.0018 |

$\bar{X}$ = 99.4

8.5. Carbon disulfide.

Carbon disulfide from a number of sources was analyzed for benzene contamination. The results are given in the following table. The benzene contaminant can be removed with the procedures given in section 4.1.

| Sample | µg Benzene/ml | ppm equivalent (for 10 L air sample) |
|---|---|---|
| Aldrich Lot 83017 | 4.20 | 0.13 |
| Baker Lot 720364 | 1.01 | 0.03 |
| Baker Lot 822351 | 1.01 | 0.03 |
| Malinkrodt Lot WEMP | 1.74 | 0.05 |
| Malinkrodt Lot WDSJ | 5.65 | 0.18 |
| Malinkrodt Lot WHGA | 2.90 | 0.09 |
| Treated $CS_2$ | .......... | .......... |

## II. OSHA LABORATORY METHOD NO. 12 FOR BULK SAMPLES

Analyte: Benzene.

Matrix: Bulk Samples.

Procedure: Bulk Samples are analyzed directly by high performance liquid chromatography (HPLC).

Detection limits: 0.01% by volume.

1. Principle of the method.

1.1. An aliquot of the bulk sample to be analyzed is injected into a liquid chromatograph.

1.2. The peak area for benzene is determined and compared to areas obtained from standards.

2. Advantages and disadvantages of the method.

2.1. The analytical procedure is quick, sensitive, and reproducible.

2.2. Reanalysis of samples is possible.

2.3. Interferences can be circumvented by proper selection of HPLC parameters.

2.4. Samples must be free of any particulates that may clog the capillary tubing in the liquid chromatograph. This may require distilling the sample or clarifying with a clarification kit.

3. Apparatus.

3.1. Liquid chromatograph equipped with a UV detector.

3.2. HPLC Column that will separate benzene from other components in the bulk sample being analyzed. The column used for validation studies was a Waters uBondapack C18, 30 cm x 3.9 mm.

3.3. A clarification kit to remove any particulates in the bulk if necessary.

3.4. A micro-distillation apparatus to distill any samples if necessary.

3.5. An electronic integrator or some other suitable method of measuring peak areas.

3.6. Microliter syringes—10 µL syringe and other convenient sizes for making standards. 10 µL syringe for sample injections.

3.7. Volumetric flasks, 5 mL and other convenient sizes for preparing standards and making dilutions.

4. Reagents.

4.1. Benzene, reagent grade.

4.2. HPLC grade water, methyl alcohol, and isopropyl alcohol.

5. Collection and shipment of samples.

5.1. Samples should be transported in glass containers with Teflon-lined caps.

5.2. Samples should not be put in the same container used for air samples.

6. Analysis of samples.

6.1. Sample preparation.

If necessary, the samples are distilled or clarified. Samples are analyzed undiluted. If the benzene concentration is out of the working range, suitable dilutions are made with isopropyl alcohol.

6.2. HPLC conditions.

The typical operating conditions for the high performance liquid chromatograph are:

1. Mobile phase—Methyl alcohol/water, 50/50
1. Analytical wavelength—254 nm
3. Injection size—10 µL

6.3. Measurement of peak area and calibration.

Peak areas are measured by an integrator or other suitable means. The integrator is calibrated to report results % in benzene by volume.

7. Calculations.

Since the integrator is programmed to report results in % benzene by volume in an undiluted sample, the following equation is used:

$$\% \text{ Benzene by Volume} = A \times B$$

Where: A = % by volume on report

B = Dilution Factor

(B = 1 for undiluted sample)

8. Backup Data.

8.1. Detection limit—Bulk Samples.

The detection limit for the analytical procedure for bulk samples is 0.88 µg, with a coefficient of variation of 0.019 at this level. This amount provided a chromatographic peak that could be identifiable in the presence of possible interferences. The detection limit date were obtained by making 10 µL injections of a 0.10% by volume standard.

| Injection | | Area Count | |
|---|---|---|---|
| 1 | | 45386 | ......... |
| 2 | | 44214 | ......... |
| 3 | | 43822 | $\tilde{X}$ = 44040.1 |
| 4 | | 44062 | SD = 852.5 |
| 6 | | 42724 | CV = 0.019 |

8.2. Pooled coefficient of variation—Bulk Samples.

The pooled coefficient of variation for analytical procedure was determined by 50 µL replicate injections of analytical standards. The standards were 0.01, 0.02, 0.04, 0.10, 1.0, and 2.0% benzene by volume.

### AREA COUNT (PERCENT)

| Injection No. | 0.01 | 0.02 | 0.04 | 0.10 | 1.0 | 2.0 |
|---|---|---|---|---|---|---|
| 1 | 45386 | 84737 | 166097 | 448497 | 4395380 | 9339150 |
| 2 | 44241 | 84300 | 170832 | 441299 | 4590800 | 9484900 |
| 3 | 43822 | 83835 | 164160 | 443719 | 4593200 | 9557580 |
| 4 | 44062 | 84381 | 164445 | 444842 | 4642350 | 9677060 |
| 5 | 44006 | 83012 | 168398 | 442564 | 4646430 | 9766240 |
| 6 | 42724 | 81957 | 173002 | 443975 | 4646260 | ......... |
| $\tilde{X}$ = | 44040.1 | 83703.6 | 167872 | 444149 | 4585767 | 9564986 |
| SD = | 852.5 | 1042.2 | 3589.8 | 2459.1 | 96839.3 | 166233 |
| CV = | 0.0194 | 0.0125 | 0.0213 | 0.0055 | 0.0211 | 0.0174 |
| $\bar{CV}$ = | 0.017 | | | | | |

[52 FR 34562, Sept. 11, 1987, as amended at 54 FR 24334, June 7, 1989; 61 FR 5508, Feb. 13, 1996; 63 FR 1289, Jan. 8, 1998; 63 FR 20099, Apr. 23, 1998]

### §1910.1029 Coke oven emissions.

**1910.1029(a)** *Scope and application.* This section applies to the control of employee exposure to coke oven emissions, except that this section shall not apply to working conditions with regard to which other Federal agencies exercise statutory authority to prescribe or enforce standards affecting occupational safety and health.

**1910.1029(b)** *Definitions.* For the purpose of this section:

*Authorized person* means any person specifically authorized by the employer whose duties require the person to enter a regulated area, or any person entering such an area as a designated representative of employees for the purpose of exercising the opportunity to observe monitoring and measuring procedures under paragraph (n) of this section.

*Beehive oven* means a coke oven in which the products of carbonization other than coke are not recovered, but are released into the ambient air.

*Coke oven* means a retort in which coke is produced by the destructive distillation or carbonization of coal.

*Coke oven battery* means a structure containing a number of slot-type coke ovens.

*Coke oven emissions* means the benzene-soluble fraction of total particulate matter present during the destructive distillation or carbonization of coal for the production of coke.

*Director* means the Director, National Institute for Occupational Safety and Health, U.S. Department of Health, Education, and Welfare, or his or her designee.

*Emergency* means any occurence such as, but not limited to, equipment failure which is likely to, or does, result in any massive release of coke oven emissions.

*Existing coke oven battery* means a battery in operation or under construction on January 20, 1977, and which is not a rehabilitated coke oven battery.

*Rehabilitated coke oven battery* means a battery which is rebuilt, overhauled, renovated, or restored such as from the pad up, after January 20, 1977.

*Secretary* means the Secretary of Labor, U.S. Department of Labor, or his or her designee.

*Stage charging* means a procedure by which a predetermined volume of coal in each larry car hopper is introduced into an oven such that no more than two hoppers are discharging simultaneously.

*Sequential charging* means a procedure, usually automatically timed, by which a predetermined volume of coal in each larry car hopper is introduced into an oven such that no more than two hoppers commence or finish discharging simultaneously although, at some point, all hoppers are discharging simultaneously.

*Pipeline charging* means any apparatus used to introduce coal into an oven which uses a pipe or duct permanently mounted onto an oven and through which coal is charged.

*Green plush* means coke which when removed from the oven results in emissions due to the presence of unvolatilized coal.

**1910.1029(c)** *Permissible exposure limit.* The employer shall assure that no employee in the regulated area is exposed to coke oven emissions at concentrations greater than 150 micrograms per cubic meter of air (150 µg/m³), averaged over any 8-hour period.

**1910.1029(d)** *Regulated areas.* **(1)** The employer shall establish regulated areas and shall limit access to them to authorized persons.

**1910.1029(d)(2)** The employer shall establish the following as regulated areas:

**1910.1029(d)(2)(i)** The coke oven battery including topside and its machinery, pushside and its machinery, coke side and its machinery, and the battery ends; the wharf; and the screening station;

**1910.1029(d)(2)(ii)** The beehive oven and its machinery.

**1910.1029(e)** *Exposure monitoring and measurement*—**(1)** *Monitoring program.* **(i)** Each employer who has a place of employment where coke oven emissions are present shall monitor employees employed in the regulated area to measure their exposure to coke oven emissions.

**1910.1029(e)(1)(ii)** The employer shall obtain measurements which are representative of each employee's exposure to coke oven emissions over an eight-hour period. All measurements shall determine exposure without regard to the use of respiratory protection.

**1910.1029(e)(1)(iii)** The employer shall collect fullshift (for at least seven continuous hours) personal samples, including at least one sample during each shift for each battery and each job classification within the regulated areas including at least the following job classifications:

**1910.1029(e)(1)(iii)(a)** Lidman;

**1910.1029(e)(1)(iii)(b)** Tar chaser;

**1910.1029(e)(1)(iii)(c)** Larry car operator;

**1910.1029(e)(1)(iii)(d)** Luterman;

**1910.1029(e)(1)(iii)(e)** Machine operator, coke side;

**1910.1029(e)(1)(iii)(f)** Benchman, coke side;

**1910.1029(e)(1)(iii)(g)** Benchman, pusher side;

**1910.1029(e)(1)(iii)(h)** Heater;

**1910.1029(e)(1)(iii)(i)** Quenching car operator;

**1910.1029(e)(1)(iii)(j)** Pusher machine operator;

**1910.1029(e)(1)(iii)(k)** Screening station operator;

**1910.1029(e)(1)(iii)(l)** Wharfman;

**1910.1029(e)(1)(iii)(m)** Oven patcher;

**1910.1029(e)(1)(iii)(n)** Oven repairman;

**1910.1029(e)(1)(iii)(o)** Spellman; and

**1910.1029(e)(1)(iii)(p)** Maintenance personnel.

**1910.1029(e)(1)(iv)** The employer shall repeat the monitoring and measurements required by this paragraph (e)(1) at least every three months.

**1910.1029(e)(2)** *Redetermination.* Whenever there has been a production, process, or control change which may result in new or additional exposure to coke oven emissions, or whenever the employer has any other reason to suspect an increase in employee exposure, the employer shall repeat the monitoring and measurements required by paragraph (e)(1) of this section for those employees affected by such change or increase.

**1910.1029(e)(3)** *Employee notification.* **(i)** The employer must, within 15 working days after the receipt of the results of any monitoring performed under this section, notify each affected employee of these results either individually in writing or by posting the results in an appropriate location that is accessible to employees.

**1910.1029(e)(3)(ii)** Whenever such results indicate that the representative employee exposure exceeds the permissible exposure limit, the employer shall, in such notification, inform each employee of that fact and of the corrective action being taken to reduce exposure to or below the permissible exposure limit.

**1910.1029(e)(4)** *Accuracy of measurement.* The employer shall use a method of monitoring and measurement which has an accuracy (with a confidence level of 95%) of not less than plus or minus 35% for concentrations of coke oven emissions greater than or equal to 150 µg/m³.

**1910.1029(f)** *Methods of compliance.* The employer shall control employee exposure to coke oven emmissions by the use of engineering controls, work practices and respiratory protection as follows:

**1910.1029(f)(1)** *Priority of compliance methods*—**(i)** *Existing coke oven batteries.* **(a)** The employer shall institute the engineering and work practice controls listed in paragraphs (f)(2), (f)(3) and (f)(4) of this section in existing coke oven batteries at the earliest possible time, but not later than January 20, 1980, except to the extent that the employer can establish that such controls are not feasible. In determining the earliest possible time for institution of engineering and work practice controls, the requirement, effective August 27, 1971, to implement feasible administrative or engineering controls to reduce exposures to coal tar pitch volatiles, shall be considered. Wherever the engineering and work practice controls which can be instituted are not sufficient to reduce employee exposures to or below the permissible exposure limit, the employer shall nonetheless use them to reduce exposures to the lowest level achievable by these controls and shall supplement them by the use of respiratory protection which complies with the requirements of paragraph (g) of this section.

**1910.1029(f)(1)(i)(b)** The engineering and work practice controls required under paragraphs (f)(2), (f)(3) and (f)(4) of this section are minimum requirements generally applicable to all existing coke oven batteries. If, after implementing all controls required by paragraphs (f)(2), (f)(3) and (f)(4) of this section, or after January 20, 1980, whichever is sooner, employee exposures still exceed the permissible exposure limit, employers shall implement any other engineering and work practice controls necessary to reduce exposure to or below the permissible exposure limit except to the extent that the employer can establish that such controls are not feasible. Whenever the engineering and work practice controls which can be instituted are not sufficient to reduce employee exposures to or below the permissible exposure limit, the employer shall nonetheless use them to reduce exposures to the lowest level achievable by these controls and shall supplement them by the use of respiratory protection which complies with the requirements of paragraph (g) of this section.

**1910.1029(f)(1)(ii)** *New or rehabilitated coke oven batteries.* **(a)** The employer shall institute the best available engineering and work practice controls on all new or rehabilitated coke oven batteries to reduce and maintain employee exposures at or below the permissible exposure limit, except to the extent that the employer can establish that such controls are not feasible. Wherever the engineering and work practice controls which can be instituted are not sufficient to reduce employee exposures to or below the permissible exposure limit, the employer shall nonetheless use them to reduce exposures to the lowest level achievable by these controls and shall

**1910.1029(f)(1)(ii)(b)** If, after implementing all the engineering and work practice controls required by paragraph (f)(1)(ii)(a) of this section, employee exposures still exceed the permissible exposure limit, the employer shall implement any other engineering and work practice controls necessary to reduce exposure to or below the permissible exposure limit except to the extent that the employer can establish that such controls are not feasible. Wherever the engineering and work practice controls which can be instituted are not sufficient to reduce employee exposures to or below the permissible exposure limit, the employer shall nonetheless use them to reduce exposures to the lowest level achievable by these controls and shall supplement them by the use of respiratory protection which complies with the requirements of paragraph (g) of this section.

**1910.1029(f)(1)(iii)** *Beehive ovens.* **(a)** The employer shall institute engineering and work practice controls on all beehive ovens at the earliest possible time to reduce and maintain employee exposures at or below the permissible exposure limit, except to the extent that the employer can establish that such controls are not feasible. In determining the earliest possible time for institution of engineering and work practice controls, the requirement, effective August 27, 1971, to implement feasible administrative or engineering controls to reduce exposures to coal tar pitch volatiles, shall be considered. Wherever the engineering and work practice controls which can be instituted are not sufficient to reduce employee exposures to or below the permissible exposure limit, the employer shall nonetheless use them to reduce exposures to the lowest level achievable by these controls and shall supplement them by the use of respiratory protection which complies with the requirements of paragraph (g) of this section.

**1910.1029(f)(1)(iii)(b)** If, after implementing all engineering and work practice controls required by paragraph (f)(1)(iii)(a) of this section, employee exposures still exceed the permissible exposure limit, the employer shall implement any other engineering and work practice controls necessary to reduce exposures to or below the permissible exposure limit except to the extent that the employer can establish that such controls are not feasible. Whenever the engineering and work practice controls which can be instituted are not sufficient to reduce employee exposures to or below the permissible exposure limit, the employer shall nonetheless use them to reduce exposures to the lowest level achievable by these controls and shall supplement them by the use of respiratory protection which complies with the requirements of paragraph (g) of this section.

**1910.1029(f)(2)** *Engineering controls—(i) Charging.* The employer shall equip and operate existing coke oven batteries with all of the following engineering controls to control coke oven emissions during charging operations:

**1910.1029(f)(2)(i)(a)** One of the following methods of charging:

**1910.1029(f)(2)(i)(a)(1)** Stage charging as described in paragraph (f)(3)(i)(b) of this section; or

**1910.1029(f)(2)(i)(a)(2)** Sequential charging as described in paragraph (f)(3)(i)(b) of this section except that paragraph (f)(3)(i)(b)(3)(iv) of this section does not apply to sequential charging; or

**1910.1029(f)(2)(i)(a)(3)** Pipeline charging or other forms of enclosed charging in accordance with paragraph (f)(2)(i) of this section, except that paragraphs (f)(2)(i)(b), (d), (e), (f) and (h) of this section do not apply;

**1910.1029(f)(2)(i)(b)** Drafting from two or more points in the oven being charged, through the use of double collector mains, or a fixed or moveable jumper pipe system to another oven, to effectively remove the gases from the oven to the collector mains;

**1910.1029(f)(2)(i)(c)** Aspiration systems designed and operated to provide sufficient negative pressure and flow volume to effectively move the gases evolved during charging into the collector mains, including sufficient steam pressure, and steam jets of sufficient diameter;

**1910.1029(f)(2)(i)(d)** Mechanical volumetric controls on each larry car hopper to provide the proper amount of coal to be charged through each charging hole so that the tunnel head will be sufficient to permit the gases to move from the oven into the collector mains;

**1910.1029(f)(2)(i)(e)** Devices to facilitate the rapid and continuous flow of coal into the oven being charged, such as stainless steel liners, coal vibrators or pneumatic shells;

**1910.1029(f)(2)(i)(f)** Individually operated larry car drop sleeves and slide gates designed and maintained so that the gases are effectively removed from the oven into the collector mains;

**1910.1029(f)(2)(i)(g)** Mechanized gooseneck and standpipe cleaners;

**1910.1029(f)(2)(i)(h)** Air seals on the pusher machine leveler bars to control air infiltration during charging; and

**1910.1029(f)(2)(i)(i)** Roof carbon cutters or a compressed air system or both on the pusher machine rams to remove roof carbon.

**1910.1029(f)(2)(ii)** *Coking.* The employer shall equip and operate existing coke oven batteries with all of the following engineering controls to control coke oven emissions during coking operations;

**1910.1029(f)(2)(ii)(a)** A pressure control system on each battery to obtain uniform collector main pressure;

**1910.1029(f)(2)(ii)(b)** Ready access to door repair facilities capable of prompt and efficient repair of doors, door sealing edges and all door parts;

**1910.1029(f)(2)(ii)(c)** An adequate number of spare doors available for replacement purposes;

**1910.1029(f)(2)(ii)(d)** Chuck door gaskets to control chuck door emissions until such door is repaired, or replaced; and

**1910.1029(f)(2)(ii)(e)** Heat shields on door machines.

**1910.1029(f)(3)** *Work practice controls—(i) Charging.* The employer shall operate existing coke oven batteries with all of the following work practices to control coke oven emissions during the charging operation:

**1910.1029(f)(3)(i)(a)** Establishment and implementation of a detailed, written inspection and cleaning procedure for each battery consisting of at least the following elements:

**1910.1029(f)(3)(i)(a)(1)** Prompt and effective repair or replacement of all engineering controls;

**1910.1029(f)(3)(i)(a)(2)** Inspection and cleaning of goosenecks and standpipes prior to each charge to a specified minimum diameter sufficient to effectively move the evolved gases from the oven to the collector mains;

**1910.1029(f)(3)(i)(a)(3)** Inspection for roof carbon build-up prior to each charge and removal of roof carbon as necessary to provide an adequate gas channel so that the gases are effectively moved from the oven into the collector mains;

**1910.1029(f)(3)(i)(a)(4)** Inspection of the steam aspiration system prior to each charge so that sufficient pressure and volume is maintained to effectively move the gases from the oven to the collector mains;

**1910.1029(f)(3)(i)(a)(5)** Inspection of steam nozzles and liquor sprays prior to each charge and cleaning as necessary so that the steam nozzles and liquor sprays are clean;

**1910.1029(f)(3)(i)(a)(6)** Inspection of standpipe caps prior to each charge and cleaning and luting or both as necessary so that the gases are effectively moved from the oven to the collector mains; and

**1910.1029(f)(3)(i)(a)(7)** Inspection of charging holes and lids for cracks, warpage and other defects prior to each charge and removal of carbon to prevent emissions, and application of luting material to standpipe and charging hole lids where necessary to obtain a proper seal.

**1910.1029(f)(3)(i)(b)** Establishment and implementation of a detailed written charging procedure, designed and operated to eliminate emissions during charging for each battery, consisting of at least the following elements:

**1910.1029(f)(3)(i)(b)(1)** Larry car hoppers filled with coal to a predetermined level in accordance with the mechanical volumetric controls required under paragraph (f)(2)(i)(d) of this section so as to maintain a sufficient gas passage in the oven to be charged;

**1910.1029(f)(3)(i)(b)(2)** The larry car aligned over the oven to be charged, so that the drop sleeves fit tightly over the charging holes; and

**1910.1029(f)(3)(i)(b)(3)** The oven charged in accordance with the following sequence of requirements:

**1910.1029(f)(3)(i)(b)(3)(i)** The aspiration system turned on;

**1910.1029(f)(3)(i)(b)(3)(ii)** Coal charged through the outermost hoppers, either individually or together depending on the capacity of the aspiration system to collect the gases involved;

**1910.1029(f)(3)(i)(b)(3)(iii)** The charging holes used under paragraph (f)(3)(i)(b)(3)(ii) of this section relidded or otherwise sealed off to prevent leakage of coke oven emissions;

**1910.1029(f)(3)(i)(b)(3)(iv)** If four hoppers are used, the third hopper discharged and relidded or otherwise sealed off to prevent leakage of coke oven emissions;

**1910.1029(f)(3)(i)(b)(3)(v)** The final hopper discharged until the gas channel at the top of the oven is blocked and then the chuck door opened and the coal leveled;

**1910.1029(f)(3)(i)(b)(3)(vi)** When the coal from the final hopper is discharged and the leveling operation complete, the charging hole relidded or otherwise sealed off to prevent leakage of coke oven emissions; and

**1910.1029(f)(3)(i)(b)(3)(vii)** The aspiration system turned off only after the charging holes have been closed.

**1910.1029(f)(3)(i)(c)** Establishment and implementation of a detailed written charging procedure, designed and operated to eliminate emissions during charging of each pipeline or enclosed charged battery.

**1910.1029(f)(3)(ii)** *Coking.* The employer shall operate existing coke oven batteries pursuant to a detailed written procedure established and implemented for the control of coke oven emissions during coking, consisting of at least the following elements:

**1910.1029(f)(3)(ii)(a)** Checking oven back pressure controls to maintain uniform pressure conditions in the collecting main;

**1910.1029(f)(3)(ii)(b)** Repair, replacement and adjustment of oven doors and chuck doors and replacement of door jambs so as to provide a continuous metal-to-metal fit;

**1910.1029(f)(3)(ii)(c)** Cleaning of oven doors, chuck doors and door jambs each coking cycle so as to provide an effective seal;

**1910.1029(f)(3)(ii)(d)** An inspection system and corrective action program to control door emissions to the maximum extent possible; and

**1910.1029(f)(3)(ii)(e)** Luting of doors that are sealed by luting each coking cycle and reluting, replacing or adjusting as necessary to control leakage.

**1910.1029(f)(3)(iii)** *Pushing.* The employer shall operate existing coke oven batteries with the following work practices to control coke oven emissions during pushing operations:

**1910.1029(f)(3)(iii)(a)** Coke and coal spillage quenched as soon as practicable and not shoveled into a heated oven; and

**1910.1029(f)(3)(iii)(b)** A detailed written procedure for each battery established and implemented for the control of emissions during pushing consisting of the following elements:

**1910.1029(f)(3)(iii)(b)(1)** Dampering off the ovens and removal of charging hole lids to effectively control coke oven emissions during the push;

**1910.1029(f)(3)(iii)(b)(2)** Heating of the coal charge uniformly for a sufficient period so as to obtain proper coking including preventing green pushes;

**1910.1029(f)(3)(iii)(b)(3)** Prevention of green pushes to the maximum extent possible;

**1910.1029(f)(3)(iii)(b)(4)** Inspection, adjustment and correction of heating flue temperatures and defective flues at least weekly and after any green push, so as to prevent green pushes;

**1910.1029(f)(3)(iii)(b)(5)** Cleaning of heating flues and related equipment to prevent green pushes, at least weekly and after any green push.

**1910.1029(f)(3)(iv)** *Maintenance and repair.* The employer shall operate existing coke oven batteries pursuant to a detailed written procedure of maintenance and repair established and implemented for the effective control of coke oven emissions consisting of the following elements:

**1910.1029(f)(3)(iv)(a)** Regular inspection of all controls, including goosenecks, standpipes, standpipe caps, charging hold lids and castings, jumper pipes and air seals for cracks, misalignment or other defects and prompt implementation of the necessary repairs as soon as possible;

**1910.1029(f)(3)(iv)(b)** Maintaining the regulated area in a neat, orderly condition free of coal and coke spillage and debris;

**1910.1029(f)(3)(iv)(c)** Regular inspection of the damper system, aspiration system and collector main for cracks or leakage, and prompt implementation of the necessary repairs;

**1910.1029(f)(3)(iv)(d)** Regular inspection of the heating system and prompt implementation of the necessary repairs;

**1910.1029(f)(3)(iv)(e)** Prevention of miscellaneous fugitive topside emissions;

**1910.1029(f)(3)(iv)(f)** Regular inspection and patching of oven brickwork;

**1910.1029(f)(3)(iv)(g)** Maintenance of battery equipment and controls in good working order;

**1910.1029(f)(3)(iv)(h)** Maintenance and repair of coke oven doors, chuck doors, door jambs and seals; and

**1910.1029(f)(3)(iv)(i)** Repairs instituted and completed as soon as possible, including temporary repair measures instituted and completed where necessary, including but not limited to:

**1910.1029(f)(3)(iv)(i)(1)** Prevention of miscellaneous fugitive topside emissions; and

**1910.1029(f)(3)(iv)(i)(2)** Chuck door gaskets, which shall be installed prior to the start of the next coking cycle.

**1910.1029(f)(4)** *Filtered air.* **(i)** The employer shall provided positive-pressure, temperature controlled filtered air for larry car, pusher machine, door machine, and quench car cabs.

**1910.1029(f)(4)(ii)** The employer shall provide standby pulpits on the battery topside, at the wharf, and at ther screening station, equipped with positive-pressure, temperature controlled filtered air.

**1910.1029(f)(5)** *Emergencies.* Whenever an emergency occurs, the next coking cycle may not begin until the cause of the emergency is determined and corrected, unless the employer can establish that it is necessary to initiate the next coking cycle in order to determine the cause of the emergency.

**1910.1029(f)(6)** *Compliance program.* **(i)** Each employer shall establish and implement a written program to reduce exposures solely by means of the engineering and work practice controls required in paragraph (f) of this section.

**1910.1029(f)(6)(ii)** The written program shall include at least the following:

**1910.1029(f)(6)(ii)(a)** A description of each coke oven operation by battery, including work force and operating crew, coking time, operating procedures and maintenance practices;

**1910.1029(f)(6)(ii)(b)** Engineering plans and other studies used to determine the controls for the coke battery;

**1910.1029(f)(6)(ii)(c)** A report of the technology considered in meeting the permissible exposure limit;

**1910.1029(f)(6)(ii)(d)** Monitoring data obtained in accordance with paragraph (e) of this section;

**1910.1029(f)(6)(ii)(e)** A detailed schedule for the implementation of the engineering and work practice controls required in paragraph (f) of this section; and

**1910.1029(f)(6)(ii)(f)** Other relevant information.

**1910.1029(f)(6)(iii)** If, after implementing all controls required by paragraph (f)(2)-(f)(4) of this section, or after January 20, 1980, whichever is sooner, or after completion of a new or rehabilitated battery the permissible exposure limit is still exceeded, the employer shall develop a detailed written program and schedule for the implementation of any additional engineering controls and work practices necessary to reduce exposure to or below the permissible exposure limit.

**1910.1029(f)(6)(iv)** Written plans for such programs shall be submitted, upon request, to the Secretary and the Director, and shall be available at the worksite for examination and copying by the Secretary, the Director, and the authorized employee representative. The plans required under paragraph (f)(6) of this section shall be revised and updated at least annually to reflect the current status of the program.

**1910.1029(f)(7)** *Training in compliance procedures.* The employer shall incorporate all written procedures and schedules required under this paragraph (f) in the information and training program required under paragraph (k) of this section and, where appropriate, post in the regulated area.

**1910.1029(g)** **Respiratory protection—(1)** *General.* For employees who use respirators required by this section, the employer must provide respirators that comply with the requirements of this paragraph. Compliance with the permissible exposure limit may not be achieved by the use of respirators except during:

**1910.1029(g)(1)(i)** Periods necessary to install or implement feasible engineering and work-practice controls.

**1910.1029(g)(1)(ii)** Work operations, such as maintenance and repair activity, for which engineering and work-practice controls are technologically not feasible.

**1910.1029(g)(1)(iii)** Work operations for which feasible engineering and work-practice controls are not yet sufficient to reduce employee exposure to or below the permissible exposure limit.

**1910.1029(g)(1)(iv)** Emergencies.

**1910.1029(g)(2)** *Respirator program.* The employer must implement a respiratory protection program in accordance with 29 CFR 1910.134 (b) through (d) (except (d)(1)(iii)), and (f) through (m).

**1910.1029(g)(3)** *Respirator selection.* The employer must select appropriate respirators or combination of respirators from Table I of this section.

### TABLE I—RESPIRATORY PROTECTION FOR COKE OVEN EMISSIONS

| Airborne concentration of coke oven emissions | Required respirator |
|---|---|
| (a) Any concentration | (1) A Type C supplied air respirator operated in pressure demand or other positive pressure or continuous flow mode; or |
|  | (2) A powered air-purifying particulate filter respirator for dust and mist or |
|  | (3) A powered air-purifying particulate filter respirator or combination chemical cartridge and particulate filter respirator for coke oven emissions. |
| (b) Concentrations not greater than 1500 ug/m$_3$. | (1) Any particulate filter respirator for dust and mist except single-use respirator; or |
|  | (2) Any particulate filter respirator or combination chemical cartridge and particulate filter respirator for coke oven emissions; or |
|  | (3) Any respirator listed in paragraph (g)(3)(a) of this section. |

**1910.1029(h)** *Protective clothing and equipment*—**(1)** Provision and use. The employer shall provide and assure the use of appropriate protective clothing and equipment, such as but not limited to:

**1910.1029(h)(1)(i)** Flame resistant jacket and pants;

**1910.1029(h)(1)(ii)** Flame resistant gloves;

**1910.1029(h)(1)(iii)** Face shields or vented goggles which comply with §1910.133(a)(2) of this part;

**1910.1029(h)(1)(iv)** Footwear providing insulation from hot surfaces for footwear;

**1910.1029(h)(1)(v)** Safety shoes which comply with §1910.136 of this part; and

**1910.1029(h)(1)(vi)** Protective helmets which comply with §1910.135 of this part.

**1910.1029(h)(2)** *Cleaning and replacement.* **(i)** The employer shall provide the protective clothing required by paragraphs (h)(1) (i) and (ii) of this section in a clean and dry condition at least weekly.

**1910.1029(h)(2)(ii)** The employer shall clean, launder, or dispose of protective clothing required by paragraphs (h)(1) (i) and (ii) of this section.

**1910.1029(h)(2)(iii)** The employer shall repair or replace the protective clothing and equipment as needed to maintain their effectiveness.

**1910.1029(h)(2)(iv)** The employer shall assure that all protective clothing is removed at the completion of a work shift only in change rooms prescribed in paragraph (i)(1) of this section.

**1910.1029(h)(2)(v)** The employer shall assure that contaminated protective clothing which is to be cleaned, laundered, or disposed of, is placed in a closable container in the change room.

**1910.1029(h)(2)(vi)** The employer shall inform any person who cleans or launders protective clothing required by this section, of the potentially harmful effects of exposure to coke oven emissions.

**1910.1029(i)** *Hygiene facilities and practices*—**(1)** *Change rooms.* The employer shall provide clean change rooms equipped with storage facilities for street clothes and separate storage facilities for protective clothing and equipment whenever employees are required to wear protective clothing and equipment in accordance with paragraph (h)(1) of this section.

**1910.1029(i)(2)** *Showers.* **(i)** The employer shall assure that employees working in the regulated area shower at the end of the work shift.

**1910.1029(i)(2)(ii)** The employer shall provide shower facilities in accordance with §1910.141(d)(3) of this part.

**1910.1029(i)(3)** *Lunchrooms.* The employer shall provide lunchroom facilities which have a temperature controlled, positive pressure, filtered air supply, and which are readily accessible to employees working in the regulated area.

**1910.1029(i)(4)** *Lavatories.* **(i)** The employer shall assure that employees working in the regulated area wash their hands and face prior to eating.

**1910.1029(i)(4)(ii)** The employer shall provide lavatory facilities in accordance with §1910.141(d)(1) and (2) of this part.

**1910.1029(i)(5)** *Prohibition of activities in the regulated area.* **(i)** The employer shall assure that in the regulated area, food or beverages are not present or consumed, smoking products are not present or used, and cosmetics are not applied, except that these activities may be conducted in the lunchrooms, change rooms and showers required under paragraphs (i)(1)-(i)(3) of this section.

**1910.1029(i)(5)(ii)** Drinking water may be consumed in the regulated area.

**1910.1029(j)** *Medical surveillance*—**(1)** *General requirements.* **(i)** Each employer shall institute a medical surveillance program for all employees who are employed in a regulated area at least 30 days per year.

**1910.1029(j)(1)(ii)** This program shall provide each employee covered under paragraph (j)(1)(i) of this section with an opportunity for medical examinations in accordance with this paragraph (j).

**1910.1029(j)(1)(iii)** The employer shall inform any employee who refuses any required medical examination of the possible health consequences of such refusal and shall obtain a signed statement from the employee indicating that the employee understands the risk involved in the refusal to be examined.

**1910.1029(j)(1)(iv)** The employer shall assure that all medical examinations and procedures are performed by or under the supervision of a licensed physician, and are provided without cost to the employee.

**1910.1029(j)(2)** *Initial examinations.* At the time of initial assignment to a regulated area or upon the institution of the medical surveillance program, the employer shall provide a medical examination for employees covered under paragraph (j)(1)(i) of this section including at least the following elements:

**1910.1029(j)(2)(i)** A work history and medical history which shall include smoking history and the presence and degree of respiratory symptoms, such as breathlessness, cough, sputum production, and wheezing;

**1910.1029(j)(2)(ii)** A standard posterior-anterior chest x-ray;

**1910.1029(j)(2)(iii)** Pulmonary function tests including forced vital capacity (FVC) and forced expiratory volume at one second (FEV 1.0) with recording of type of equipment used;

**1910.1029(j)(2)(iv)** Weight;

**1910.1029(j)(2)(v)** A skin examination;

**1910.1029(j)(2)(vi)** Urinalysis for sugar, albumin, and hematuria; and

**1910.1029(j)(2)(vii)** A urinary cytology examination.

**1910.1029(j)(3)** *Periodic examinations.* **(i)** The employer shall provide the examinations specified in paragraphs (j)(2) (i)-(vi) of this section at least annually for employees covered under paragraph (j)(1)(i) of this section.

**1910.1029(j)(3)(ii)** The employer must provide the examinations specified in paragraphs (j)(2)(i) through (j)(2)(vii) of this section at least annually for employees 45 years of age or older or with five (5) or more years employment in the regulated area.

**1910.1029(j)(3)(iii)** Whenever an employee who is 45 years of age or older or with five (5) or more years employment in a regulated area transfers or is transferred from employment in a regulated area, the employer must continue to provide the examinations specified in paragraphs (j)(2)(i)

through (j)(2)(vii) of this section at least annually as long as that employee is employed by the same employer or a successor employer.

**1910.1029(j)(3)(iv)** Whenever an employee has not taken the examinations specified in paragraphs (j)(3) (i)-(iii) of this section with the six (6) months preceding the termination of employment the employer shall provide such examinations to the employee upon termination of employment.

**1910.1029(j)(4)** *Information provided to the physician.* The employer shall provide the following information to the examining physician:

**1910.1029(j)(4)(i)** A copy of this regulation and its Appendixes;

**1910.1029(j)(4)(ii)** A description of the affected employee's duties as they relate to the employee's exposure;

**1910.1029(j)(4)(iii)** The employee's exposure level or estimated exposure level;

**1910.1029(j)(4)(iv)** A description of any personal protective equipment used or to be used; and

**1910.1029(j)(4)(v)** Information from previous medical examinations of the affected employee which is not readily available to the examining physician.

**1910.1029(j)(5)** *Physician's written opinion.* (i) The employer shall obtain a written opinion from the examining physician which shall include:

**1910.1029(j)(5)(i)***(a)* The results of the medical examinations;

**1910.1029(j)(5)(i)***(b)* The physician's opinion as to whether the employee has any detected medical conditions which would place the employee at increased risk of material impairment of the employee's health from exposure to coke oven emissions;

**1910.1029(j)(5)(i)***(c)* Any recommended limitations upon the employee's exposure to coke oven emissions or upon the use of protective clothing or equipment such as respirators; and

**1910.1029(j)(5)(i)***(d)* A statement that the employee has been informed by the physician of the results of the medical examination and any medical conditions which require further explanation or treatment.

**1910.1029(j)(5)(ii)** The employer shall instruct the physician not to reveal in the written opinion specific findings or diagnoses unrelated to occupational exposure.

**1910.1029(j)(5)(iii)** The employer shall provide a copy of the written opinion to the affected employee.

**1910.1029(k)** *Employee information and training*—(1) *Training program.* (i) The employer shall institute a training program for employees who are employed in the regulated area and shall assure their participation.

**1910.1029(k)(1)(ii)** The training program shall be provided as of January 27, 1977 for employees who are employed in the regulated area at that time or at the time of initial assignment to a regulated area.

**1910.1029(k)(1)(iii)** The training program shall be provided at least annually for all employees who are employed in the regulated area, except that training regarding the occupational safety and health hazards associated with exposure to coke oven emissions and the purpose, proper use, and limitations of respiratory protective devices shall be provided at least quarterly until January 20, 1978.

**1910.1029(k)(1)(iv)** The training program shall include informing each employee of:

**1910.1029(k)(1)(iv)***(a)* The information contained in the substance information sheet for coke oven emissions (Appendix A);

**1910.1029(k)(1)(iv)***(b)* The purpose, proper use, and limitations of respiratory protective devices required in accordance with paragraph (g) of this section;

**1910.1029(k)(1)(iv)***(c)* The purpose for and a description of the medical surveillance program required by paragraph (j) of this section including information on the occupational safety and health hazards associated with exposure to coke oven emissions;

**1910.1029(k)(1)(iv)***(d)* A review of all written procedures and schedules required under paragraph (f) of this section; and

**1910.1029(k)(1)(iv)***(e)* A review of this standard.

**1910.1029(k)(2)** *Access to training materials.* (i) The employer shall make a copy of this standard and its appendixes readily available to all employees who are employed in the regulated area.

**1910.1029(k)(2)(ii)** The employer shall provide upon request all materials relating to the employee information and training program to the Secretary and the Director.

**1910.1029(l)** *Precautionary signs and labels*—(1) *General.* (i) The employer may use labels or signs required by other statutes, regulations or ordinances in addition to, or in combination with, signs and labels required by this paragraph.

**1910.1029(l)(1)(ii)** The employer shall assure that no statement appears on or near any sign required by this paragraph which contradicts or detracts from the effects of the required sign.

**1910.1029(l)(1)(iii)** The employer shall assure that signs required by this paragraph are illuminated and cleaned as necessary so that the legend is readily visible.

**1910.1029(l)(2)** *Signs.* (i) The employer shall post signs in the regulated area bearing the legends:

**DANGER**

**CANCER HAZARD**

**AUTHORIZED PERSONNEL ONLY**

**NO SMOKING OR EATING**

**1910.1029(l)(2)(ii)** In addition, not later than January 20, 1978, the employer shall post signs in the areas where the permissible exposure limit is exceeded bearing the legend:

**DANGER**

**RESPIRATOR REQUIRED**

**1910.1029(l)(3)** *Labels.* The employer shall apply precautionary labels to all containers of protective clothing contaminated with coke oven emissions bearing the legend:

**CAUTION**

**CLOTHING CONTAMINATED WITH COKE EMISSIONS**

**DO NOT REMOVE DUST BY BLOWING OR SHAKING**

**1910.1029(m)** *Recordkeeping*—(1) *Exposure measurements.* The employer shall establish and maintain an accurate record of all measurements taken to monitor employee exposure to coke oven emissions required in paragraph (e) of this section.

**1910.1029(m)(1)(i)** This record shall include:

**1910.1029(m)(1)(i)***(a)* Name, social security number, and job classification of the employees monitored;

**1910.1029(m)(1)(i)***(b)* The date(s), number, duration and results of each of the samples taken, including a description of the sampling procedure used to determine representative employee exposure where applicable;

**1910.1029(m)(1)(i)***(c)* The type of respiratory protective devices worn, if any;

**1910.1029(m)(1)(i)***(d)* A description of the sampling and analytical methods used and evidence of their accuracy; and

**1910.1029(m)(1)(i)***(e)* The environmental variables that could affect the measurement of employee exposure.

**1910.1029(m)(1)(ii)** The employer shall maintain this record for at lest 40 years or for the duration of employment plus 20 years, whichever is longer.

**1910.1029(m)(2)** *Medical surveillance.* The employer shall establish and maintain an accurate record for each employee subject to medical surveillance as required by paragraph (j) of this section.

**1910.1029(m)(2)(i)** The record shall include:

**1910.1029(m)(2)(i)***(a)* The name, social security number, and description of duties of the employee;

**1910.1029(m)(2)(i)***(b)* A copy of the physician's written opinion;

**1910.1029(m)(2)(i)***(c)* The signed statement of any refusal to take a medical examination under paragraph (j)(1)(ii) of this section; and

**1910.1029(m)(2)(i)***(d)* Any employee medical complaints related to exposure to coke oven emissions.

**1910.1029(m)(2)(ii)** The employer shall keep, or assure that the examining physician keeps, the following medical records:

**1910.1029(m)(2)(ii)***(a)* A copy of the medical examination results including medical and work history required under paragraph (j)(2) of this section;

**1910.1029(m)(2)(ii)***(b)* A description of the laboratory procedures used and a copy of any standards or guidelines used to interpret the test results;

**1910.1029(m)(2)(ii)(c)** The initial x-ray;

**1910.1029(m)(2)(ii)(d)** The x-rays for the most recent five (5) years;

**1910.1029(m)(2)(ii)(e)** Any x-ray with a demonstrated abnormality and all subsequent x-rays;

**1910.1029(m)(2)(ii)(f)** The initial cytologic examination slide and written description;

**1910.1029(m)(2)(ii)(g)** The cytologic examination slide and written description for the most recent 10 years; and

**1910.1029(m)(2)(ii)(h)** Any cytologic examination slides with demonstrated atypia, if such atypia persists for 3 years, and all subsequent slides and written descriptions.

**1910.1029(m)(2)(iii)** The employer shall maintain medical records required under paragraph (m)(2) of this section for at least 40 years, or for the duration of employment plus 20 years, whichever is longer.

**1910.1029(m)(3)** *Availability.* **(i)** The employer shall make available upon request all records required to be maintained by paragraph (m) of this section to the Secretary and the Director for examination and copying.

**1910.1029(m)(3)(ii)** Employee exposure measurement records and employee medical records required by this paragraph shall be provided upon request to employees, designated representatives, and the Assistant Secretary in accordance with 29 CFR 1910.1020(a)-(e) and (g)-(i).

**1910.1029(m)(4)** *Transfer of records.* **(i)** Whenever the employer ceases to do business, the successor employer shall receive and retain all records required to be maintained by paragraph (m) of this section.

**1910.1029(m)(4)(ii)** Whenever the employer ceases to do business and there is no successor employer to receive and retain the records for the prescribed period, these records shall be transmitted by registered mail to the Director.

**1910.1029(m)(4)(iii)** At the expiration of the retention period for the records required to be maintained under paragraphs (m)(1) and (m)(2) of this section, the employer shall transmit these records by registered mail to the Director or shall continue to retain such records.

**1910.1029(m)(4)(iv)** The employer shall also comply with any additional requirements involving transfer of records set forth in 29 CFR 1910.1020(h).

**1910.1029(n)** *Observation of monitoring—***(1)** *Employee observation.* The employer shall provide affected employees or their representatives an opportunity to observe any measuring or monitoring of employee exposure to coke oven emissions conducted pursuant to paragraph (e) of this section.

**1910.1029(n)(2)** *Observation procedures.* **(i)** Whenever observation of the measuring or monitoring of employee exposure to coke oven emissions requires entry into an area where the ues of protective clothing or equipment is required, the employer shall provide the observer with and assure the use of such equipment and shall require the observer to comply with all other applicable safety and health procedures.

**1910.1029(n)(2)(ii)** Without interfering with the measurement, observers shall be entitled to:

**1910.1029(n)(2)(ii)(a)** An Explanation of the measurement procedures;

**1910.1029(n)(2)(ii)(b)** Observe all steps related to the measurement of coke oven emissions performed at the place of exposure; and

**1910.1029(n)(2)(ii)(c)** Record the results obtained.

**1910.1029(o)** [Reserved.]

**1910.1029(p)** *Appendices.* The information contained in the appendixes to this section is not intended, by itself, to create any additional obligations not otherwise imposed or to detract from any existing obligation.

[39 FR 23502, June 27, 1974, as amended at 63 FR 33468, June 18, 1998]

## APPENDIX A TO §1910.1029—COKE OVEN EMISSIONS SUBSTANCE INFORMATION SHEET

### I. SUBSTANCE IDENTIFICATION

A. *Substance:* Coke Oven Emissions

B. *Definition:* The benzene-soluble fraction of total particulate matter present during the destructive distillation or carbonization of coal for the production of coke.

C. *Permissible Exposure Limit:* 150 micrograms per cubic meter of air determined as an average over an 8-hour period.

D. *Regulated areas:* Only employees authorized by your employer should enter a regulated area. The employer is required to designate the following areas as regulated areas: the coke oven battery, including topside and its machinery, pushside and its machinery, cokeside and its machinery, and the battery ends; the screening station; and the wharf; and the beehive ovens and their machinery.

### II. HEALTH HAZARD DATA

Exposure to coke oven emissions is a cause of lung cancer, and kidney cancer, in humans. Although there have not been an excess number of skin cancer cases in humans, repeated skin contact with coke oven emissions should be avoided.

### III. PROTECTIVE CLOTHING AND EQUIPMENT

A. *Respirators:* Respirators will be provided by your employer for routine use if your employer is in the process of implementing engineering and work practice controls or where engineering and work practice controls are not feasible or insufficient to reduce exposure to or below the PEL. You must wear respirators for non-routine activities or in emergency situations where you are likely to be exposed to levels of coke oven emissions in excess of the permissible exposure limit. Until January 20, 1978, the routine wearing of respirators is voluntary. Until that date, if you choose not to wear a respirator you do not have to do so. You must still have your respirator with you and you must still wear it if you are near visible emissions. Since how well your respirator fits your face is very important, your employer is required to conduct fit tests to make sure the respirator seals properly when you wear it. These tests are simple and rapid and will be explained to you during your training sessions.

B. *Protective clothing:* Your employer is required to provide, and you must wear, appropriate, clean, protective clothing and equipment to protect your body from repeated skin contact with coke oven emissions and from the heat generated during the coking process. This clothing should include such items as jacket and pants and flame resistant gloves. Protective equipment should include face shield or vented goggles, protective helmets and safety shoes, insulated from hot surfaces where appropriate.

### IV. HYGIENE FACILITIES AND PRACTICES

You must not eat, drink, smoke, chew gum or tobacco, or apply cosmetics in the regulated area, except that drinking water is permitted. Your employer is required to provide lunchrooms and other areas for these purposes.

Your employer is required to provide showers, washing facilities, and change rooms. If you work in a regulated area, you must wash your face, and hands before eating. You must shower at the end of the work shift. Do not take used protective clothing out of the change rooms without your employer's permission. Your employer is required to provide for laundering or cleaning of your protective clothing.

### V. SIGNS AND LABELS

Your employer is required to post warning signs and labels for your protection. Signs must be posted in regulated areas. The signs must warn that a cancer hazard is present, that only authorized employees may enter the area, and that no smoking or eating is allowed. In regulated areas where coke oven emissions are above the permissible exposure limit, the signs should also warn that respirators must be worn.

### VI. MEDICAL EXAMINATIONS

If you work in a regulated area at least 30 days per year, your employer is required to provide you with a medical examination every year. The medical examination must include a medical history, a chest x-ray, pulmonary function test, weight comparison, skin examination, a urinalysis, and a urine cytology exam for early detection of urinary cancer. The urine cytology exam is only included in the initial exam until you are either 45 years or older, or have 5 or more years employment in the regulated areas when the medical exams including this test, but excepting the x-ray exam, are to be given every six months; under these conditions, you are to be given an x-ray exam at least once a year. The examining physician will provide a written opinion to your employer containing the results of the medical exams. You should also receive a copy of this opinion.

### VII. OBSERVATION OF MONITORING

Your employer is required to monitor your exposure to coke oven emissions and you are entitled to observe the monitoring procedure. You are entitled to receive an explanation of the measurement procedure, observe the steps taken in the measurement procedure, and to record the results obtained. When the monitoring procedure is taking place in an area where respirators or personal protective clothing and equipment are required to be worn, you must also be provided with and must wear the protective clothing and equipment.

### VIII. ACCESS TO RECORDS

You or your representative are entitled to records of your exposure to coke oven emissions upon request to your employer. Your medical examination records can be furnished to your physician upon request to your employer.

### IX. TRAINING AND EDUCATION

Additional information on all of these items plus training as to hazards of coke oven emissions and the engineering and work practice controls associated with your job will also be provided by your employer.

[39 FR 23502, June 27, 1974, as amended at 63 FR 33468, June 18, 1998]

## APPENDIX B TO §1910.1029—INDUSTRIAL HYGIENE AND MEDICAL SURVEILLANCE GUIDELINES

### I. INDUSTRIAL HYGIENE GUIDELINES

A. *Sampling* (Benzene-Soluble Fraction Total Particulate Matter).

Samples collected should be full shift (at least 7-hour) samples. Sampling should be done using a personal sampling pump with pulsation damper at a flow rate of 2 liters per minute. Samples should be collected on 0.8 micrometer pore size silver membrane filters (37 mm diameter) preceded by Gelman glass fiber type A-E filters encased in three-piece plastic (polystyrene) field monitor cassettes. The cassette face cap should be on and the plug removed. The rotameter should be checked every hour to ensure that proper flow rates are maintained.

A minimum of three full-shift samples should be collected for each job classification on each battery, at least one from each shift. If disparate results are obtained for particular job classification, sampling should be repeated. It is advisable to sample each shift on more than one day to account for environmental variables (wind, precipitation, etc.) which may affect sampling. Differences in exposures among different work shifts may indicate a need to improve work practices on a particular shift. Sampling results from different shifts for each job classification should not be averaged. Multiple samples from same shift on each battery may be used to calculate an average exposure for a particular job classification.

B. *Analysis.*

1. All extraction glassware is cleaned with dichromic acid cleaning solution, rinsed with tap water, then dionized water, acetone, and allowed to dry completely. The glassware is rinsed with nanograde benzene before use. The Teflon cups are cleaned with benzene then with acetone.

2. Pre-weigh the 2 ml Teflon cups to one hundredth of a milligram (0.01 mg) on a autobalance AD 2 Tare weight of the cups is about 50 mg.

3. Place the silver membrane filter and glass fiber filter into a 15 ml test tube.

4. Extract with 5 ml of benzene for five minutes in an ultrasonic cleaner.

5. Filter the extract in 15 ml medium glass fritted funnels.

6. Rinse test tube and filters with two 1.5 ml aliquots of benzene and filter through the fritted glass funnel.

7. Collect the extract and two rinses in a 10 ml Kontes graduated evaporative concentrator.

8. Evaporate down to 1 ml while rinsing the sides with benzene.

9. Pipet 0.5 ml into the Teflon cup and evaporate to dryness in a vacuum oven at 40 °C for 3 hours.

10. Weigh the Teflon cup and the weight gain is due to the benzene soluble residue in half the Sample.

### II. MEDICAL SURVEILLANCE GUIDELINES

A. *General.* The minimum requirements for the medical examination for coke oven workers are given in paragraph (j) of the standard. The initial examination is to be provided to all coke oven workers who work at least 30 days in the regulated area. The examination includes a posterior-anterior chest x-ray reading, pulmonary function tests (FVC and FEV 1.0), weight, urinalysis, skin examination, and a urinary cytologic examination. These tests are needed to serve as the baseline for comparing the employee's future test results. Periodic exams include all the elements of the initial exam, except that the urine cytologic test is to be performed only on those employees who are 45 years or older or who have worked for 5 or more years in the regulated area; periodic exams, with the exception of x-rays, are to be performed semiannually for this group instead of annually; for this group, x-rays will continue to be given at least annually. The examination contents are minimum requirements; additional tests such as lateral and oblique x-rays or additional pulmonary function tests may be performed if deemed necessary.

B. *Pulmonary function tests.*

Pulmonary function tests should be performed in a manner which minimizes subject and operator bias. There has been shown to be learning effects with regard to the results obtained from certain tests, such as FEV 1.0. Best results can be obtained by multiple trials for each subject. The best of three trials or the average of the last three of five trials may be used in obtaining reliable results. The type of equipment used (manufacturer, model, etc.) should be recorded with the results as reliability and accuracy varies and such information may be important in the evaluation of test results. Care should be exercised to obtain the best possible testing equipment.

[41 FR 46784, Oct. 22, 1976, as amended at 42 FR 3304, Jan. 18, 1977; 45 FR 35283, May 23, 1980; 50 FR 37353, 37354, Sept. 13, 1985; 54 FR 24334, June 7, 1989; 61 FR 5508, Feb. 13, 1996; 63 FR 1290, Jan. 8, 1998; 63 FR 33468, June 18, 1998]

## §1910.1030 Bloodborne pathogens.

**1910.1030(a)** *Scope and Application.* This section applies to all occupational exposure to blood or other potentially infectious materials as defined by paragraph (b) of this section.

**1910.1030(b)** *Definitions.* For purposes of this section, the following shall apply:

*Assistant Secretary* means the Assistant Secretary of Labor for Occupational Safety and Health, or designated representative.

*Blood* means human blood, human blood components, and products made from human blood.

*Bloodborne Pathogens* means pathogenic microorganisms that are present in human blood and can cause disease in humans. These pathogens include, but are not limited to, hepatitis B virus (HBV) and human immunodeficiency virus (HIV).

*Clinical Laboratory* means a workplace where diagnostic or other screening procedures are performed on blood or other potentially infectious materials.

*Contaminated* means the presence or the reasonably anticipated presence of blood or other potentially infectious materials on an item or surface.

*Contaminated Laundry* means laundry which has been soiled with blood or other potentially infectious materials or may contain sharps.

*Contaminated Sharps* means any contaminated object that can penetrate the skin including, but not limited to, needles, scalpels, broken glass, broken capillary tubes, and exposed ends of dental wires.

*Decontamination* means the use of physical or chemical means to remove, inactivate, or destroy bloodborne pathogens on a surface or item to the point where they are no longer capable of transmitting infectious particles and the surface or item is rendered safe for handling, use, or disposal.

*Director* means the Director of the National Institute for Occupational Safety and Health, U.S. Department of Health and Human Services, or designated representative.

*Engineering controls* means controls (e.g., sharps disposal containers, self-sheathing needles, safer medical devices, such as sharps with engineered sharps injury protections and needleless systems) that isolate or remove the bloodborne pathogens hazard from the workplace.

*Exposure Incident* means a specific eye, mouth, other mucous membrane, non-intact skin, or parenteral contact with blood or other potentially infectious materials that results from the performance of an employee's duties.

*Handwashing Facilities* means a facility providing an adequate supply of running potable water, soap and single use towels or hot air drying machines.

*Licensed Healthcare Professional* is a person whose legally permitted scope of practice allows him or her to independently perform the activities required by paragraph (f) Hepatitis B Vaccination and Post-exposure Evaluation and Followup.

*HBV* means hepatitis B virus.

*HIV* means human immunodeficiency virus.

*Needleless systems* means a device that does not use needles for:

(1) The collection of bodily fluids or withdrawal of body fluids after initial venous or arterial access is established;

(2) The administration of medication or fluids; or

(3) Any other procedure involving the potential for occupational exposure to bloodborne pathogens due to percutaneous injuries from contaminated sharps.

*Occupational Exposure* means reasonably anticipated skin, eye, mucous membrane, or parenteral contact with blood or other potentially infectious materials that may result from the performance of an employee's duties.

*Other Potentially Infectious Materials* means

(1) The following human body fluids: semen, vaginal secretions, cerebrospinal fluid, synovial fluid, pleural fluid, pericardial fluid, peritoneal fluid, amniotic fluid, saliva in dental procedures, any body fluid that is visibly contaminated with blood, and all body fluids in situations where it is difficult or impossible to differentiate between body fluids;

(2) Any unfixed tissue or organ (other than intact skin) from a human (living or dead); and

(3) HIV-containing cell or tissue cultures, organ cultures, and HIV- or HBV-containing culture medium or other solutions; and blood, organs, or other tissues from experimental animals infected with HIV or HBV.

*Parenteral* means piercing mucous membranes or the skin barrier through such events as needlesticks, human bites, cuts, and abrasions.

*Personal Protective Equipment* is specialized clothing or equipment worn by an employee for protection against a hazard. General work clothes (e.g., uniforms, pants, shirts or blouses) not intended to function as protection against a hazard are not considered to be personal protective equipment.

*Production Facility* means a facility engaged in industrial-scale, large-volume or high concentration production of HIV or HBV.

*Regulated Waste* means liquid or semi-liquid blood or other potentially infectious materials; contaminated items that would release blood or other potentially infectious materials in a liquid or semi-liquid state if compressed; items that are caked with dried blood or other potentially infectious materials and are capable of releasing these materials during handling; contaminated sharps; and pathological and microbiological wastes containing blood or other potentially infectious materials.

*Research Laboratory* means a laboratory producing or using research-laboratory-scale amounts of HIV or HBV. Research laboratories may produce high concentrations of HIV or HBV but not in the volume found in production facilities.

*Sharps with engineered sharps injury protections* means a nonneedle sharp or a needle device used for withdrawing body fluids, accessing a vein or artery, or administering medications or other fluids, with a built-in safety feature or mechanism that effectively reduces the risk of an exposure incident.

*Source Individual* means any individual, living or dead, whose blood or other potentially infectious materials may be a source of occupational exposure to the employee. Examples include, but are not limited to, hospital and clinic patients; clients in institutions for the developmentally disabled; trauma victims; clients of drug and alcohol treatment facilities; residents of hospices and nursing homes; human remains; and individuals who donate or sell blood or blood components.

*Sterilize* means the use of a physical or chemical procedure to destroy all microbial life including highly resistant bacterial endospores.

*Universal Precautions* is an approach to infection control. According to the concept of Universal Precautions, all human blood and certain human body fluids are treated as if known to be infectious for HIV, HBV, and other bloodborne pathogens.

*Work Practice Controls* means controls that reduce the likelihood of exposure by altering the manner in which a task is performed (e.g., prohibiting recapping of needles by a two-handed technique).

**1910.1030(c)** *Exposure control*—**(1)** *Exposure Control Plan.* **(i)** Each employer having an employee(s) with occupational exposure as defined by paragraph (b) of this section shall establish a written Exposure Control Plan designed to eliminate or minimize employee exposure.

**1910.1030(c)(1)(ii)** The Exposure Control Plan shall contain at least the following elements:

**1910.1030(c)(1)(ii)(A)** The exposure determination required by paragraph(c)(2),

**1910.1030(c)(1)(ii)(B)** The schedule and method of implementation for paragraphs (d) Methods of Compliance, (e) HIV and HBV Research Laboratories and Production Facilities, (f) Hepatitis B Vaccination and PostExposure Evaluation and Follow-up, (g) Communication of Hazards to Employees, and (h) Recordkeeping, of this standard, and

**1910.1030(c)(1)(ii)(C)** The procedure for the evaluation of circumstances surrounding exposure incidents as required by paragraph (f)(3)(i) of this standard.

**1910.1030(c)(1)(iii)** Each employer shall ensure that a copy of the Exposure Control Plan is accessible to employees in accordance with 29 CFR 1910.1020(e).

**1910.10301020(e)(c)(1)(iv)** The Exposure Control Plan shall be reviewed and updated at least annually and whenever necessary to reflect new or modified tasks and procedures which affect occupational exposure and to reflect new or revised employee positions with occupational exposure. The review and update of such plans shall also:

**1910.1030(c)(1)(iv)(A)** Reflect changes in technology that eliminate or reduce exposure to bloodborne pathogens; and

**1910.1030(c)(1)(iv)(B)** Document annually consideration and implementation of appropriate commercially available and effective safer medical devices designed to eliminate or minimize occupational exposure.

**1910.1030(c)(1)(v)** An employer, who is required to establish an Exposure Control Plan shall solicit input from non-managerial employees responsible for direct patient care who are potentially exposed to injuries from contaminated sharps in the identification, evaluation, and selection of effective engineering and work practice controls and shall document the solicitation in the Exposure Control Plan.

**1910.1030(c)(1)(vi)** The Exposure Control Plan shall be made available to the Assistant Secretary and the Director upon request for examination and copying.

**1910.1030(c)(2)** *Exposure determination.* **(i)** Each employer who has an employee(s) with occupational exposure as defined by paragraph (b) of this section shall prepare an exposure determination. This exposure determination shall contain the following:

**1910.1030(c)(2)(i)(A)** A list of all job classifications in which all employees in those job classifications have occupational exposure;

**1910.1030(c)(2)(i)(B)** A list of job classifications in which some employees have occupational exposure, and

**1910.1030(c)(2)(i)(C)** A list of all tasks and procedures or groups of closely related task and procedures in which occupational exposure occurs and that are performed by employees in job classifications listed in accordance with the provisions of paragraph (c)(2)(i)(B) of this standard.

**1910.1030(c)(2)(ii)** This exposure determination shall be made without regard to the use of personal protective equipment.

**1910.1030(d)** *Methods of compliance*—**(1)** *General.* Universal precautions shall be observed to prevent contact with blood or other potentially infectious materials. Under circumstances in which differentiation between body fluid types is difficult or impossible, all body fluids shall be considered potentially infectious materials.

**1910.1030(d)(2)** *Engineering and work practice controls.* **(i)** Engineering and work practice controls shall be used to eliminate or minimize employee exposure. Where occupational exposure remains after institution of these controls, personal protective equipment shall also be used.

**1910.1030(d)(2)(ii)** Engineering controls shall be examined and maintained or replaced on a regular schedule to ensure their effectiveness.

**1910.1030(d)(2)(iii)** Employers shall provide handwashing facilities which are readily accessible to employees.

**1910.1030(d)(2)(iv)** When provision of handwashing facilities is not feasible, the employer shall provide either an appropriate antiseptic hand cleanser in conjunction with clean cloth/paper towels or antiseptic towelettes. When antiseptic hand cleansers or towelettes are used, hands shall be washed with soap and running water as soon as feasible.

**1910.1030(d)(2)(v)** Employers shall ensure that employees wash their hands immediately or as soon as feasible after removal of gloves or other personal protective equipment.

**1910.1030(d)(2)(vi)** Employers shall ensure that employees wash hands and any other skin with soap and water, or flush mucous membranes with water immediately or as soon as feasible following contact of such body areas with blood or other potentially infectious materials.

**1910.1030(d)(2)(vii)** Contaminated needles and other contaminated sharps shall not be bent, recapped, or removed except as noted in paragraphs (d)(2)(vii)(A) and (d)(2)(vii)(B) below. Shearing or breaking of contaminated needles is prohibited.

**1910.1030(d)(2)(vii)(A)** Contaminated needles and other contaminated sharps shall not be bent, recapped or removed unless the employer can demonstrate that no alternative is feasible or that such action is required by a specific medical or dental procedure.

**1910.1030(d)(2)(vii)(B)** Such bending, recapping or needle removal must be accomplished through the use of a mechanical device or a onehanded technique.

**1910.1030(d)(2)(viii)** Immediately or as soon as possible after use, contaminated reusable sharps shall be placed in appropriate containers until properly reprocessed. These containers shall be:

**1910.1030(d)(2)(viii)(A)** Puncture resistant;

**1910.1030(d)(2)(viii)(B)** Labeled or color-coded in accordance with this standard;

**1910.1030(d)(2)(viii)(C)** Leakproof on the sides and bottom; and

**1910.1030(d)(2)(viii)(D)** In accordance with the requirements set forth in paragraph (d)(4)(ii)(E) for reusable sharps.

**1910.1030(d)(2)(ix)** Eating, drinking, smoking, applying cosmetics or lip balm, and handling contact lenses are prohibited in work areas where there is a reasonable likelihood of occupational exposure.

**1910.1030(d)(2)(x)** Food and drink shall not be kept in refrigerators, freezers, shelves, cabinets or on countertops or benchtops where blood or other potentially infectious materials are present.

**1910.1030(d)(2)(xi)** All procedures involving blood or other potentially infectious materials shall be performed in such a manner as to minimize splashing, spraying, spattering, and generation of droplets of these substances.

**1910.1030(d)(2)(xii)** Mouth pipetting/suctioning of blood or other potentially infectious materials is prohibited.

**1910.1030(d)(2)(xiii)** Specimens of blood or other potentially infectious materials shall be placed in a container which prevents leakage during collection, handling, processing, storage, transport, or shipping.

**1910.1030(d)(2)(xiii)(A)** The container for storage, transport, or shipping shall be labeled or color-coded according to paragraph (g)(1)(i) and closed prior to being stored, transported, or shipped. When a facility utilizes Universal Precautions in the handling of all specimens, the labeling/color-coding of specimens is not necessary provided containers are recognizable as containing specimens. This exemption only applies while such specimens/containers remain within the facility. Labeling or color-coding in accordance with paragraph (g)(1)(i) is required when such specimens/containers leave the facility.

**1910.1030(d)(2)(xiii)(B)** If outside contamination of the primary container occurs, the primary container shall be placed within a second container which prevents leakage during handling, processing, storage, transport, or shipping and is labeled or color-coded according to the requirements of this standard.

**1910.1030(d)(2)(xiii)(C)** If the specimen could puncture the primary container, the primary container shall be placed within a secondary container which is puncture-resistant in addition to the above characteristics.

**1910.1030(d)(2)(xiv)** Equipment which may become contaminated with blood or other potentially infectious materials shall be examined prior to servicing or shipping and shall be decontaminated as necessary, unless the employer can demonstrate that decontamination of such equipment or portions of such equipment is not feasible.

**1910.1030(d)(2)(xiv)(A)** A readily observable label in accordance with paragraph (g)(1)(i)(H) shall be attached to the equipment stating which portions remain contaminated.

**1910.1030(d)(2)(xiv)(B)** The employer shall ensure that this information is conveyed to all affected employees, the servicing representative, and/or the manufacturer, as appropriate, prior to handling, servicing, or shipping so that appropriate precautions will be taken.

**1910.1030(d)(3)** *Personal protective equipment—(i) Provision.* When there is occupational exposure, the employer shall provide, at no cost to the employee, appropriate personal protective equipment such as, but not limited to, gloves, gowns, laboratory coats, face shields or masks and eye protection, and mouthpieces, resuscitation bags, pocket masks, or other ventilation devices. Personal protective equipment will be considered "appropriate" only if it does not permit blood or other potentially infectious materials to pass through to or reach the employee's work clothes, street clothes, undergarments, skin, eyes, mouth, or other mucous membranes under normal conditions of use and for the duration of time which the protective equipment will be used.

**1910.1030(d)(3)(ii)** *Use.* The employer shall ensure that the employee uses appropriate personal protective equipment unless the employer shows that the employee temporarily and briefly declined to use personal protective equipment when, under rare and extraordinary circumstances, it was the employee's professional judgment that in the specific instance its use would have prevented the delivery of health care or public safety services or would have posed an increased hazard to the safety of the worker or co-worker. When the employee makes this judgement, the circumstances shall be investigated and documented in order to determine whether changes can be instituted to prevent such occurences in the future.

**1910.1030(d)(3)(iii)** *Accessibility.* The employer shall ensure that appropriate personal protective equipment in the appropriate sizes is readily accessible at the worksite or is issued to employees. Hypoallergenic gloves, glove liners, powderless gloves, or other similar alternatives shall be readily accessible to those employees who are allergic to the gloves normally provided.

**1910.1030(d)(3)(iv)** *Cleaning, Laundering, and Disposal.* The employer shall clean, launder, and dispose of personal protective equipment required by paragraphs (d) and (e) of this standard, at no cost to the employee.

**1910.1030(d)(3)(v)** *Repair and Replacement.* The employer shall repair or replace personal protective equipment as needed to maintain its effectiveness, at no cost to the employee.

**1910.1030(d)(3)(vi)** If a garment(s) is penetrated by blood or other potentially infectious materials, the garment(s) shall be removed immediately or as soon as feasible.

**1910.1030(d)(3)(vii)** All personal protective equipment shall be removed prior to leaving the work area.

**1910.1030(d)(3)(viii)** When personal protective equipment is removed it shall be placed in an appropriately designated area or container for storage, washing, decontamination or disposal.

**1910.1030(d)(3)(ix)** *Gloves.* Gloves shall be worn when it can be reasonably anticipated that the employee may have hand contact with blood, other potentially infectious materials, mucous membranes, and non-intact skin; when performing vascular access procedures except as specified in paragraph (d)(3)(ix)(D); and when handling or touching contaminated items or surfaces.

**1910.1030(d)(3)(ix)(A)** Disposable (single use) gloves such as surgical or examination gloves, shall be replaced as soon as practical when contaminated or as soon as feasible if they are torn, punctured, or when their ability to function as a barrier is compromised.

**1910.1030(d)(3)(ix)(B)** Disposable (single use) gloves shall not be washed or decontaminated for re-use.

**1910.1030(d)(3)(ix)(C)** Utility gloves may be decontaminated for re-use if the integrity of the glove is not compromised. However, they must be discarded if they are cracked, peeling, torn, punctured, or exhibit other signs of deterioration or when their ability to function as a barrier is compromised.

**1910.1030(d)(3)(ix)(D)** If an employer in a volunteer blood donation center judges that routine gloving for all phlebotomies is not necessary then the employer shall:

**1910.1030(d)(3)(ix)(D)*(1)*** Periodically reevaluate this policy;

**1910.1030(d)(3)(ix)(D)*(2)*** Make gloves available to all employees who wish to use them for phlebotomy;

**1910.1030(d)(3)(ix)(D)*(3)*** Not discourage the use of gloves for phlebotomy; and

**1910.1030(d)(3)(ix)(D)(4)** Require that gloves be used for phlebotomy in the following circumstances:

**1910.1030(d)(3)(ix)(D)(4)(i)** When the employee has cuts, scratches, or other breaks in his or her skin;

**1910.1030(d)(3)(ix)(D)(4)(ii)** When the employee judges that hand contamination with blood may occur, for example, when performing phlebotomy on an uncooperative source individual; and

**1910.1030(d)(3)(ix)(D)(4)(iii)** When the employee is receiving training in phlebotomy.

**1910.1030(d)(3)(x)** *Masks, Eye Protection, and Face Shields.* Masks in combination with eye protection devices, such as goggles or glasses with solid side shields, or chin-length face shields, shall be worn whenever splashes, spray, spatter, or droplets of blood or other potentially infectious materials may be generated and eye, nose, or mouth contamination can be reasonably anticipated.

**1910.1030(d)(3)(xi)** *Gowns, Aprons, and Other Protective Body Clothing.* Appropriate protective clothing such as, but not limited to, gowns, aprons, lab coats, clinic jackets, or similar outer garments shall be worn in occupational exposure situations. The type and characteristics will depend upon the task and degree of exposure anticipated.

**1910.1030(d)(3)(xii)** Surgical caps or hoods and/or shoe covers or boots shall be worn in instances when gross contamination can reasonably be anticipated (e.g., autopsies, orthopaedic surgery).

**1910.1030(d)(4)** *Housekeeping—(i) General.* Employers shall ensure that the worksite is maintained in a clean and sanitary condition. The employer shall determine and implement an appropriate written schedule for cleaning and method of decontamination based upon the location within the facility, type of surface to be cleaned, type of soil present, and tasks or procedures being performed in the area.

**1910.1030(d)(4)(ii)** All equipment and environmental and working surfaces shall be cleaned and decontaminated after contact with blood or other potentially infectious materials.

**1910.1030(d)(4)(ii)(A)** Contaminated work surfaces shall be decontaminated with an appropriate disinfectant after completion of procedures; immediately or as soon as feasible when surfaces are overtly contaminated or after any spill of blood or other potentially infectious materials; and at the end of the work shift if the surface may have become contaminated since the last cleaning.

**1910.1030(d)(4)(ii)(B)** Protective coverings, such as plastic wrap, aluminum foil, or imperviously-backed absorbent paper used to cover equipment and environmental surfaces, shall be removed and replaced as soon as feasible when they become overtly contaminated or at the end of the workshift if they may have become contaminated during the shift.

**1910.1030(d)(4)(ii)(C)** All bins, pails, cans, and similar receptacles intended for reuse which have a reasonable likelihood for becoming contaminated with blood or other potentially infectious materials shall be inspected and decontaminated on a regularly scheduled basis and cleaned and decontaminated immediately or as soon as feasible upon visible contamination.

**1910.1030(d)(4)(ii)(D)** Broken glassware which may be contaminated shall not be picked up directly with the hands. It shall be cleaned up using mechanical means, such as a brush and dust pan, tongs, or forceps.

**1910.1030(d)(4)(ii)(E)** Reusable sharps that are contaminated with blood or other potentially infectious materials shall not be stored or processed in a manner that requires employees to reach by hand into the containers where these sharps have been placed.

**1910.1030(d)(4)(iii)** *Regulated Waste—(A) Contaminated Sharps Discarding and Containment. (1)* Contaminated sharps shall be discarded immediately or as soon as feasible in containers that are:

**1910.1030(d)(4)(iii)(A)(1)(i)** Closable;

**1910.1030(d)(4)(iii)(A)(1)(ii)** Puncture resistant;

**1910.1030(d)(4)(iii)(A)(1)(iii)** Leakproof on sides and bottom; and

**1910.1030(d)(4)(iii)(A)(1)(iv)** Labeled or color-coded in accordance with paragraph (g)(1)(i) of this standard.

**1910.1030(d)(4)(iii)(A)(2)** During use, containers for contaminated sharps shall be:

**1910.1030(d)(4)(iii)(A)(2)(i)** Easily accessible to personnel and located as close as is feasible to the immediate area where sharps are used or can be reasonably anticipated to be found (e.g., laundries);

**1910.1030(d)(4)(iii)(A)(2)(ii)** Maintained upright throughout use; and

**1910.1030(d)(4)(iii)(A)(2)(iii)** Replaced routinely and not be allowed to overfill.

**1910.1030(d)(4)(iii)(A)(3)** When moving containers of contaminated sharps from the area of use, the containers shall be:

**1910.1030(d)(4)(iii)(A)(3)(i)** Closed immediately prior to removal or replacement to prevent spillage or protrusion of contents during handling, storage, transport, or shipping;

**1910.1030(d)(4)(iii)(A)(3)(ii)** Placed in a secondary container if leakage is possible. The second container shall be:

**1910.1030(d)(4)(iii)(A)(3)(ii)(A)** Closable;

**1910.1030(d)(4)(iii)(A)(3)(ii)(B)** Constructed to contain all contents and prevent leakage during handling, storage, transport, or shipping; and

**1910.1030(d)(4)(iii)(A)(3)(ii)(C)** Labeled or color-coded according to paragraph (g)(1)(i) of this standard.

**1910.1030(d)(4)(iii)(A)(4)** Reusable containers shall not be opened, emptied, or cleaned manually or in any other manner which would expose employees to the risk of percutaneous injury.

**1910.1030(d)(4)(iii)(B)** *Other Regulated Waste Containment—(1)* Regulated waste shall be placed in containers which are:

**1910.1030(d)(4)(iii)(B)(1)(i)** Closable;

**1910.1030(d)(4)(iii)(B)(1)(ii)** Constructed to contain all contents and prevent leakage of fluids during handling, storage, transport or shipping;

**1910.1030(d)(4)(iii)(B)(1)(iii)** Labeled or color-coded in accordance with paragraph (g)(1)(i) this standard; and

**1910.1030(d)(4)(iii)(B)(1)(iv)** Closed prior to removal to prevent spillage or protrusion of contents during handling, storage, transport, or shipping.

**1910.1030(d)(4)(iii)(B)(2)** If outside contamination of the regulated waste container occurs, it shall be placed in a second container. The second container shall be:

**1910.1030(d)(4)(iii)(B)(2)(i)** Closable;

**1910.1030(d)(4)(iii)(B)(2)(ii)** Constructed to contain all contents and prevent leakage of fluids during handling, storage, transport or shipping;

**1910.1030(d)(4)(iii)(B)(2)(iii)** Labeled or color-coded in accordance with paragraph (g)(1)(i) of this standard; and

**1910.1030(d)(4)(iii)(B)(2)(iv)** Closed prior to removal to prevent spillage or protrusion of contents during handling, storage, transport, or shipping.

**1910.1030(d)(4)(iii)(C)** Disposal of all regulated waste shall be in accordance with applicable regulations of the United States, States and Territories, and political subdivisions of States and Territories.

**1910.1030(d)(4)(iv)** *Laundry.* **(A)** Contaminated laundry shall be handled as little as possible with a minimum of agitation. *(1)* Contaminated laundry shall be bagged or containerized at the location where it was used and shall not be sorted or rinsed in the location of use.

**1910.1030(d)(4)(iv)(A)(2)** Contaminated laundry shall be placed and transported in bags or containers labeled or color-coded in accordance with paragraph (g)(1)(i) of this standard. When a facility utilizes Universal Precautions in the handling of all soiled laundry, alternative labeling or color-coding is sufficient if it permits all employees to recognize the containers as requiring compliance with Universal Precautions.

**1910.1030(d)(4)(iv)(A)(3)** Whenever contaminated laundry is wet and presents a reasonable likelihood of soak-through of or leakage from the bag or container, the laundry shall be placed and transported in bags or containers which prevent soak-through and/or leakage of fluids to the exterior.

**1910.1030(d)(4)(iv)(B)** The employer shall ensure that employees who have contact with contaminated laundry wear protective gloves and other appropriate personal protective equipment.

**1910.1030(d)(4)(iv)(C)** When a facility ships contaminated laundry off-site to a second facility which does not utilize Universal Precautions in the handling of all laundry, the facility generating the contaminated laundry must place such laundry in bags or containers which are labeled or color-coded in accordance with paragraph (g)(1)(i).

**1910.1030(e)** *HIV and HBV Research Laboratories and Production Facilities.* **(1)** This paragraph applies to research laboratories and production facilities engaged in the culture, production, concentration, experimentation, and manipulation of HIV and HBV. It does not apply to clinical or diagnostic laboratories engaged solely in the analysis of blood, tissues, or organs. These requirements apply in addition to the other requirements of the standard.

**1910.1030(e)(2)** Research laboratories and production facilities shall meet the following criteria:

**1910.1030(e)(2)(i)** *Standard microbiological practices.* All regulated waste shall either be incinerated or decontaminated by a method such as autoclaving known to effectively destroy bloodborne pathogens.

**1910.1030(e)(2)(ii)** *Special practices.* **(A)** Laboratory doors shall be kept closed when work involving HIV or HBV is in progress.

**1910.1030(e)(2)(ii)(B)** Contaminated materials that are to be decontaminated at a site away from the work area shall be placed in a durable, leakproof, labeled or color-coded container that is closed before being removed from the work area.

**1910.1030(e)(2)(ii)(C)** Access to the work area shall be limited to authorized persons. Written policies and procedures shall be established whereby only persons who have been advised of the potential biohazard, who meet any specific entry requirements, and who comply with all entry and exit procedures shall be allowed to enter the work areas and animal rooms.

**1910.1030(e)(2)(ii)(D)** When other potentially infectious materials or infected animals are present in the work area or containment module, a hazard warning sign incorporating the universal biohazard symbol shall be posted on all access doors. The hazard warning sign shall comply with paragraph (g)(1)(ii) of this standard.

**1910.1030(e)(2)(ii)(E)** All activities involving other potentially infectious materials shall be conducted in biological safety cabinets or other physical-containment devices within the containment module. No work with these other potentially infectious materials shall be conducted on the open bench.

**1910.1030(e)(2)(ii)(F)** Laboratory coats, gowns, smocks, uniforms, or other appropriate protective clothing shall be used in the work area and animal rooms. Protective clothing shall not be worn outside of the work area and shall be decontaminated before being laundered.

**1910.1030(e)(2)(ii)(G)** Special care shall be taken to avoid skin contact with other potentially infectious materials. Gloves shall be worn when handling infected animals and when making hand contact with other potentially infectious materials is unavoidable.

**1910.1030(e)(2)(ii)(H)** Before disposal all waste from work areas and from animal rooms shall either be incinerated or decontaminated by a method such as autoclaving known to effectively destroy bloodborne pathogens.

**1910.1030(e)(2)(ii)(I)** Vacuum lines shall be protected with liquid disinfectant traps and high-efficiency particulate air (HEPA) filters or filters of equivalent or superior efficiency and which are checked routinely and maintained or replaced as necessary.

**1910.1030(e)(2)(ii)(J)** Hypodermic needles and syringes shall be used only for parenteral injection and aspiration of fluids from laboratory animals and diaphragm bottles. Only needle-locking syringes or disposable syringe-needle units (i.e., the needle is integral to the syringe) shall be used for the injection or aspiration of other potentially infectious materials. Extreme caution shall be used when handling needles and syringes. A needle shall not be bent, sheared, replaced in the sheath or guard, or removed from the syringe following use. The needle and syringe shall be promptly placed in a puncture-resistant container and autoclaved or decontaminated before reuse or disposal.

**1910.1030(e)(2)(ii)(K)** All spills shall be immediately contained and cleaned up by appropriate professional staff or others properly trained and equipped to work with potentially concentrated infectious materials.

**1910.1030(e)(2)(ii)(L)** A spill or accident that results in an exposure incident shall be immediately reported to the laboratory director or other responsible person.

**1910.1030(e)(2)(ii)(M)** A biosafety manual shall be prepared or adopted and periodically reviewed and updated at least annually or more often if necessary. Personnel shall be advised of potential hazards, shall be required to read instructions on practices and procedures, and shall be required to follow them.

**1910.1030(e)(2)(iii)** *Containment equipment.* **(A)** Certified biological safety cabinets (Class I, II, or III) or other appropriate combinations of personal protection or physical containment devices, such as special protective clothing, respirators, centrifuge safety cups, sealed centrifuge rotors, and containment caging for animals, shall be used for all activities with other potentially infectious materials that pose a threat of exposure to droplets, splashes, spills, or aerosols.

**1910.1030(e)(2)(iii)(B)** Biological safety cabinets shall be certified when installed, whenever they are moved and at least annually.

**1910.1030(e)(3)** HIV and HBV research laboratories shall meet the following criteria:

**1910.1030(e)(3)(i)** Each laboratory shall contain a facility for hand washing and an eye wash facility which is readily available within the work area.

**1910.1030(e)(3)(ii)** An autoclave for decontamination of regulated waste shall be available.

**1910.1030(e)(4)** HIV and HBV production facilities shall meet the following criteria:

**1910.1030(e)(4)(i)** The work areas shall be separated from areas that are open to unrestricted traffic flow within the building. Passage through two sets of doors shall be the basic requirement for entry into the work area from access corridors or other contiguous areas. Physical separation of the high-containment work area from access corridors or other areas or activities may also be provided by a double-doored clothes-change room (showers may be included), airlock, or other access facility that requires passing through two sets of doors before entering the work area.

**1910.1030(e)(4)(ii)** The surfaces of doors, walls, floors and ceilings in the work area shall be water resistant so that they can be easily cleaned. Penetrations in these surfaces shall be sealed or capable of being sealed to facilitate decontamination.

**1910.1030(e)(4)(iii)** Each work area shall contain a sink for washing hands and a readily available eye wash facility. The sink shall be foot, elbow, or automatically operated and shall be located near the exit door of the work area.

**1910.1030(e)(4)(iv)** Access doors to the work area or containment module shall be self-closing.

**1910.1030(e)(4)(v)** An autoclave for decontamination of regulated waste shall be available within or as near as possible to the work area.

**1910.1030(e)(4)(vi)** A ducted exhaust-air ventilation system shall be provided. This system shall create directional airflow that draws air into the work area through the entry area. The exhaust air shall not be recirculated to any other area of the building, shall be discharged to the outside, and shall be dispersed away from occupied areas and air intakes. The proper direction of the airflow shall be verified (i.e., into the work area).

**1910.1030(e)(5)** *Training Requirements.* Additional training requirements for employees in HIV and HBV research laboratories and HIV and HBV production facilities are specified in paragraph (g)(2)(ix).

**1910.1030(f)** *Hepatitis B vaccination and post-exposure evaluation and follow-up*—**(1)** *General.* **(i)** The employer shall make available the hepatitis B vaccine and vaccination series to all employees who have occupational exposure, and post-exposure evaluation and follow-up to all employees who have had an exposure incident.

**1910.1030(f)(1)(ii)** The employer shall ensure that all medical evaluations and procedures including the hepatitis B vaccine and vaccination series and post-exposure evaluation and follow-up, including prophylaxis, are:

**1910.1030(f)(1)(ii)(A)** Made available at no cost to the employee;

**1910.1030(f)(1)(ii)(B)** Made available to the employee at a reasonable time and place;

**1910.1030(f)(1)(ii)(C)** Performed by or under the supervision of a licensed physician or by or under the supervision of another licensed healthcare professional; and

**1910.1030(f)(1)(ii)(D)** Provided according to recommendations of the U.S. Public Health Service current at the time these evaluations and procedures take place, except as specified by this paragraph (f).

**1910.1030(f)(1)(iii)** The employer shall ensure that all laboratory tests are conducted by an accredited laboratory at no cost to the employee.

**1910.1030(f)(2)** *Hepatitis B Vaccination.* **(i)** Hepatitis B vaccination shall be made available after the employee has received the training required in paragraph (g)(2)(vii)(I) and within 10 working days of initial assignment to all employees who have occupational exposure unless the employee has previously received the complete hepatitis B vaccination series, antibody testing has revealed that the employee is immune, or the vaccine is contraindicated for medical reasons.

**1910.1030(f)(2)(ii)** The employer shall not make participation in a pre-screening program a prerequisite for receiving hepatitis B vaccination.

**1910.1030(f)(2)(iii)** If the employee initially declines hepatitis B vaccination but at a later date while still covered under the standard decides to accept the vaccination, the employer shall make available hepatitis B vaccination at that time.

**1910.1030(f)(2)(iv)** The employer shall assure that employees who decline to accept hepatitis B vaccination offered by the employer sign the statement in appendix A.

**1910.1030(f)(2)(v)** If a routine booster dose(s) of hepatitis B vaccine is recommended by the U.S. Public Health Service at a future date, such booster dose(s) shall be made available in accordance with section (f)(1)(ii).

**1910.1030(f)(3)** *Post-exposure Evaluation and Follow-up.* Following a report of an exposure incident, the employer shall make immediately available to the exposed employee a confidential medical evaluation and follow-up, including at least the following elements:

**1910.1030(f)(3)(i)** Documentation of the route(s) of exposure, and the circumstances under which the exposure incident occurred;

**1910.1030(f)(3)(ii)** Identification and documentation of the source individual, unless the employer can establish that identification is infeasible or prohibited by state or local law;

**1910.1030(f)(3)(ii)(A)** The source individual's blood shall be tested as soon as feasible and after consent is obtained in order to determine HBV and HIV infectivity. If consent is not obtained, the employer shall establish that legally required consent cannot be obtained. When the source individual's consent is not required by law, the source individual's blood, if available, shall be tested and the results documented.

**1910.1030(f)(3)(ii)(B)** When the source individual is already known to be infected with HBV or HIV, testing for the source individual's known HBV or HIV status need not be repeated.

**1910.1030(f)(3)(ii)(C)** Results of the source individual's testing shall be made available to the exposed employee, and the employee shall be informed of applicable laws and regulations concerning disclosure of the identity and infectious status of the source individual.

**1910.1030(f)(3)(iii)** Collection and testing of blood for HBV and HIV serological status;

**1910.1030(f)(3)(iii)(A)** The exposed employee's blood shall be collected as soon as feasible and tested after consent is obtained.

**1910.1030(f)(3)(iii)(B)** If the employee consents to baseline blood collection, but does not give consent at that time for HIV serologic testing, the sample shall be preserved for at least 90 days. If, within 90 days of the exposure incident, the employee elects to have the baseline sample tested, such testing shall be done as soon as feasible.

**1910.1030(f)(3)(iv)** Post-exposure prophylaxis, when medically indicated, as recommended by the U.S. Public Health Service;

**1910.1030(f)(3)(v)** Counseling; and

**1910.1030(f)(3)(vi)** Evaluation of reported illnesses.

**1910.1030(f)(4)** *Information Provided to the Healthcare Professional.* **(i)** The employer shall ensure that the healthcare professional responsible for the employee's Hepatitis B vaccination is provided a copy of this regulation.

**1910.1030(f)(4)(ii)** The employer shall ensure that the healthcare professional evaluating an employee after an exposure incident is provided the following information:

**1910.1030(f)(4)(ii)(A)** A copy of this regulation;

**1910.1030(f)(4)(ii)(B)** A description of the exposed employee's duties as they relate to the exposure incident;

**1910.1030(f)(4)(ii)(C)** Documentation of the route(s) of exposure and circumstances under which exposure occurred;

**1910.1030(f)(4)(ii)(D)** Results of the source individual's blood testing, if available; and

**1910.1030(f)(4)(ii)(E)** All medical records relevant to the appropriate treatment of the employee including vaccination status which are the employer's responsibility to maintain.

**1910.1030(f)(5)** *Healthcare Professional's Written Opinion.* The employer shall obtain and provide the employee with a copy of the evaluating healthcare professional's written opinion within 15 days of the completion of the evaluation.

**1910.1030(f)(5)(i)** The healthcare professional's written opinion for Hepatitis B vaccination shall be limited to whether Hepatitis B vaccination is indicated for an employee, and if the employee has received such vaccination.

**1910.1030(f)(5)(ii)** The healthcare professional's written opinion for postexposure evaluation and follow-up shall be limited to the following information:

**1910.1030(f)(5)(ii)(A)** That the employee has been informed of the results of the evaluation; and

**1910.1030(f)(5)(ii)(B)** That the employee has been told about any medical conditions resulting from exposure to blood or other potentially infectious materials which require further evaluation or treatment. (iii) All other findings or diagnoses shall remain confidential and shall not be included in the written report.

**1910.1030(f)(6)** *Medical recordkeeping.* Medical records required by this standard shall be maintained in accordance with paragraph (h)(1) of this section.

**1910.1030(g)** *Communication of hazards to employees—***(1)** *Labels and signs—***(i)** *Labels.* **(A)** Warning labels shall be affixed to containers of regulated waste, refrigerators and freezers containing blood or other potentially infectious material; and other containers used to store, transport or ship blood or other potentially infectious materials, except as provided in paragraph (g)(1)(i)(E), (F) and (G).

**1910.1030(g)(1)(i)(B)** Labels required by this section shall include the following legend:

**1910.1030(g)(1)(i)(C)** These labels shall be fluorescent orange or orange-red or predominantly so, with lettering and symbols in a contrasting color.

**1910.1030(g)(1)(i)(D)** Labels shall be affixed as close as feasible to the container by string, wire, adhesive, or other method that prevents their loss or unintentional removal.

**1910.1030(g)(1)(i)(E)** Red bags or red containers may be substituted for labels.

**1910.1030(g)(1)(i)(F)** Containers of blood, blood components, or blood products that are labeled as to their contents and have been released for transfusion or other clinical use are exempted from the labeling requirements of paragraph (g).

**1910.1030(g)(1)(i)(G)** Individual containers of blood or other potentially infectious materials that are placed in a labeled container during storage, transport, shipment or disposal are exempted from the labeling requirement.

**1910.1030(g)(1)(i)(H)** Labels required for contaminated equipment shall be in accordance with this paragraph and shall also state which portions of the equipment remain contaminated.

**1910.1030(g)(1)(i)(I)** Regulated waste that has been decontaminated need not be labeled or color-coded.

**1910.1030(g)(1)(ii)** *Signs.* **(A)** The employer shall post signs at the entrance to work areas specified in paragraph (e), HIV and HBV Research Laboratory and Production Facilities, which shall bear the following legend:

> BIOHAZARD
> (Name of the Infectious Agent)
> (Special requirements for entering the area)
> (Name, telephone number of the laboratory director or other responsible person.)

**1910.1030(g)(1)(ii)(B)** These signs shall be fluorescent orange-red or predominantly so, with lettering and symbols in a contrasting color.

**1910.1030(g)(2)** *Information and Training.* **(i)** Employers shall ensure that all employees with occupational exposure participate in a training program which must be provided at no cost to the employee and during working hours.

**1910.1030(g)(2)(ii)** Training shall be provided as follows:

**1910.1030(g)(2)(ii)(A)** At the time of initial assignment to tasks where occupational exposure may take place;

**1910.1030(g)(2)(ii)(B)** At least annually thereafter.

**1910.1030(g)(2)(iii)** [Reserved.]

**1910.1030(g)(2)(iv)** Annual training for all employees shall be provided within one year of their previous training.

**1910.1030(g)(2)(v)** Employers shall provide additional training when changes such as modification of tasks or procedures or institution of new tasks or procedures affect the employee's occupational exposure. The additional training may be limited to addressing the new exposures created.

**1910.1030(g)(2)(vi)** Material appropriate in content and vocabulary to educational level, literacy, and language of employees shall be used.

**1910.1030(g)(2)(vii)** The training program shall contain at a minimum the following elements:

**1910.1030(g)(2)(vii)(A)** An accessible copy of the regulatory text of this standard and an explanation of its contents;

**1910.1030(g)(2)(vii)(B)** A general explanation of the epidemiology and symptoms of bloodborne diseases;

**1910.1030(g)(2)(vii)(C)** An explanation of the modes of transmission of bloodborne pathogens;

**1910.1030(g)(2)(vii)(D)** An explanation of the employer's exposure control plan and the means by which the employee can obtain a copy of the written plan;

**1910.1030(g)(2)(vii)(E)** An explanation of the appropriate methods for recognizing tasks and other activities that may involve exposure to blood and other potentially infectious materials;

**1910.1030(g)(2)(vii)(F)** An explanation of the use and limitations of methods that will prevent or reduce exposure including appropriate engineering controls, work practices, and personal protective equipment;

**1910.1030(g)(2)(vii)(G)** Information on the types, proper use, location, removal, handling, decontamination and disposal of personal protective equipment;

**1910.1030(g)(2)(vii)(H)** An explanation of the basis for selection of personal protective equipment;

**1910.1030(g)(2)(vii)(I)** Information on the hepatitis B vaccine, including information on its efficacy, safety, method of administration, the benefits of being vaccinated, and that the vaccine and vaccination will be offered free of charge;

**1910.1030(g)(2)(vii)(J)** Information on the appropriate actions to take and persons to contact in an emergency involving blood or other potentially infectious materials;

**1910.1030(g)(2)(vii)(K)** An explanation of the procedure to follow if an exposure incident occurs, including the method of reporting the incident and the medical follow-up that will be made available;

**1910.1030(g)(2)(vii)(L)** Information on the post-exposure evaluation and followup that the employer is required to provide for the employee following an exposure incident;

**1910.1030(g)(2)(vii)(M)** An explanation of the signs and labels and/or color coding required by paragraph (g)(1); and

**1910.1030(g)(2)(vii)(N)** An opportunity for interactive questions and answers with the person conducting the training session.

**1910.1030(g)(2)(viii)** The person conducting the training shall be knowledgeable in the subject matter covered by the elements contained in the training program as it relates to the workplace that the training will address.

**1910.1030(g)(2)(ix)** Additional Initial Training for Employees in HIV and HBV Laboratories and Production Facilities. Employees in HIV or HBV research laboratories and HIV or HBV production facilities shall receive the following initial training in addition to the above training requirements.

**1910.1030(g)(2)(ix)(A)** The employer shall assure that employees demonstrate proficiency in standard microbiological practices and techniques and in the practices and operations specific to the facility before being allowed to work with HIV or HBV.

**1910.1030(g)(2)(ix)(B)** The employer shall assure that employees have prior experience in the handling of human pathogens or tissue cultures before working with HIV or HBV.

**1910.1030(g)(2)(ix)(C)** The employer shall provide a training program to employees who have no prior experience in handling human pathogens. Initial work activities shall not include the handling of infectious agents. A progression of work activities shall be assigned as techniques are learned and proficiency is developed. The employer shall assure that employees participate in work activities involving infectious agents only after proficiency has been demonstrated.

**1910.1030(h)** *Recordkeeping*—**(1)** *Medical Records.* **(i)** The employer shall establish and maintain an accurate record for each employee with occupational exposure, in accordance with 29 CFR 1910.1020.

**1910.1030(h)(1)(ii)** This record shall include:

**1910.1030(h)(1)(ii)(A)** The name and social security number of the employee;

**1910.1030(h)(1)(ii)(B)** A copy of the employee's hepatitis B vaccination status including the dates of all the hepatitis B vaccinations and any medical records relative to the employee's ability to receive vaccination as required by paragraph (f)(2);

**1910.1030(h)(1)(ii)(C)** A copy of all results of examinations, medical testing, and follow-up procedures as required by paragraph (f)(3);

**1910.1030(h)(1)(ii)(D)** The employer's copy of the healthcare professional's written opinion as required by paragraph (f)(5); and

**1910.1030(h)(1)(ii)(E)** A copy of the information provided to the healthcare professional as required by paragraphs (f)(4)(ii)(B)(C) and (D).

**1910.1030(h)(1)(iii)** Confidentiality. The employer shall ensure that employee medical records required by paragraph (h)(1) are:

**1910.1030(h)(1)(iii)(A)** Kept confidential; and

**1910.1030(h)(1)(iii)(B)** Not disclosed or reported without the employee's express written consent to any person within or outside the workplace except as required by this section or as may be required by law.

**1910.1030(h)(1)(iv)** The employer shall maintain the records required by paragraph (h) for at least the duration of employment plus 30 years in accordance with 29 CFR 1910.1020.

**1910.1030(h)(2)** *Training Records.* **(i)** Training records shall include the following information:

**1910.1030(h)(2)(i)(A)** The dates of the training sessions;

**1910.1030(h)(2)(i)(B)** The contents or a summary of the training sessions;

**1910.1030(h)(2)(i)(C)** The names and qualifications of persons conducting the training; and

**1910.1030(h)(2)(i)(D)** The names and job titles of all persons attending the training sessions.

**1910.1030(h)(2)(ii)** Training records shall be maintained for 3 years from the date on which the training occurred.

**1910.1030(h)(3)** *Availability.* **(i)** The employer shall ensure that all records required to be maintained by this section shall be made available upon request to the Assistant Secretary and the Director for examination and copying.

**1910.1030(h)(3)(ii)** Employee training records required by this paragraph shall be provided upon request for examination and copying to employees, to employee representatives, to the Director, and to the Assistant Secretary.

**1910.1030(h)(3)(iii)** Employee medical records required by this paragraph shall be provided upon request for examination and copying to the subject employee, to anyone having written consent of the subject employee, to the Director, and to the Assistant Secretary in accordance with 29 CFR 1910.1020.

**1910.1030(h)(4)** *Transfer of Records.* **(i)** The employer shall comply with the requirements involving transfer of records set forth in 29 CFR 1910.1020(h).

**1910.1030(h)(4)(ii)** If the employer ceases to do business and there is no successor employer to receive and retain the records for the prescribed period, the employer shall notify the Director, at least three months prior to their disposal and transmit them to the Director, if required by the Director to do so, within that three month period.

**1910.1030(h)(5)** *Sharps injury log.* **(i)** The employer shall establish and maintain a sharps injury log for the recording of percutaneous injuries from contaminated sharps. The information in the sharps injury log shall be recorded and maintained in such manner as to protect the confidentiality of the injured employee. The sharps injury log shall contain, at a minimum:

**1910.1030(h)(5)(i)(A)** The type and brand of device involved in the incident,

**1910.1030(h)(5)(i)(B)** The department or work area where the exposure incident occurred, and

**1910.1030(h)(5)(i)(C)** An explanation of how the incident occurred.

**1910.1030(h)(5)(ii)** The requirement to establish and maintain a sharps injury log shall apply to any employer who is required to maintain a log of occupational injuries and illnesses under 29 CFR 1904.

**1910.1030(h)(5)(iii)** The sharps injury log shall be maintained for the period required by 29 CFR 1904.6.

**1910.1030(i)** *Dates*—**(1)** *Effective Date.* The standard shall become effective on March 6, 1992.

**1910.1030(i)(2)** The Exposure Control Plan required by paragraph (c) of this section shall be completed on or before May 5, 1992.

**1910.1030(i)(3)** Paragraph (g)(2) Information and Training and (h) Recordkeeping shall take effect on or before June 4, 1992.

**1910.1030(i)(4)** Paragraphs (d)(2) Engineering and Work Practice Controls, (d)(3) Personal Protective Equipment, (d)(4) Housekeeping, (e) HIV and HBV Research Laboratories and Production Facilities, (f) Hepatitis B Vaccination and PostExposure Evaluation and Follow-up, and (g) (1) Labels and Signs, shall take effect July 6, 1992.

[56 FR64004, Dec. 6, 1991, as amended at 57 FR 12717, April 13, 1992; 57 FR 29206, July 1, 1992; 61 FR 5507, Feb. 13, 1996]

## APPENDIX A TO SECTION 1910.1030— HEPATITIS B VACCINE DECLINATION (MANDATORY)

I understand that due to my occupational exposure to blood or other potentially infectious materials I may be at risk of acquiring hepatitis B virus (HBV) infection. I have been given the opportunity to be vaccinated with hepatitis B vaccine, at no charge to myself. However, I decline hepatitis B vaccination at this time. I understand that by declining this vaccine, I continue to be at risk of acquiring hepatitis B, a serious disease. If in the future I continue to have occupational exposure to blood or other potentially infectious materials and I want to be vaccinated with hepatitis B vaccine, I can receive the vaccination series at no charge to me.

[56 FR 64175, Dec. 6, 1991, as amended at 57 FR 12717, Apr. 13, 1992; 57 FR 29206, July 1, 1992; 61 FR 5508, Feb. 13, 1996]

## §1910.1043 Cotton dust.

**1910.1043(a)** *Scope and application.* **(1)** This section, in its entirety, applies to the control of employee exposure to cotton dust in all workplaces where employees engage in yarn manufacturing, engage in slashing and weaving operations, or work in waste houses for textile operations.

**1910.1043(a)(2)** This section does not apply to the handling or processing of woven or knitted materials; to maritime operations covered by 29 CFR Parts 1915 and 1918; to harvesting or ginning of cotton; or to the construction industry.

**1910.1043(a)(3)** Only paragraphs (h) Medical surveillance, (k)(2) through (4) Recordkeeping—Medical Records, and Appendices B, C and D of this section apply in all work places where employees exposed to cotton dust engage in cottonseed processing or waste processing operations.

**1910.1043(a)(4)** This section applies to yarn manufacturing and slashing and weaving operations exclusively using washed cotton (as defined by paragraph (n) of this section) only to the extent specified by paragraph (n) of this section.

**1910.1043(a)(5)** This section, in its entirety, applies to the control of all employees exposure to the cotton dust generated in the preparation of washed cotton from opening until the cotton is thoroughly wetted.

**1910.1043(a)(6)** This section does not apply to knitting, classing or warehousing operations except that employers with these operations, if requested by NIOSH, shall grant NIOSH access to their employees and workplaces for exposure monitoring and medical examinations for purposes of a health study to be performed by NIOSH on a sampling basis.

**1910.1043(b)** *Definitions.* For the purpose of this section:

*Assistant Secretary* means the Assistant Secretary of Labor for Occupational Safety and Health, U.S. Department of Labor, or designee;

*Blow down* means the general cleaning of a room or a part of a room by the use of compressed air.

*Blow off* means the use of compressed air for cleaning of short duration and usually for a specific machine or any portion of a machine.

*Cotton dust* means dust present in the air during the handling or processing of cotton, which may contain a mixture of many substances including ground up plant matter, fiber, bacteria, fungi, soil, pesticides, non-cotton plant matter and other contaminants which may have accumulated with the cotton during the growing, harvesting and subsequent processing or storage periods. Any dust present during the handling and processing of cotton through the weaving or knitting of fabrics, and dust present in other operations or manufacturing processes using raw or waste cotton fibers or cotton fiber byproducts from textile mills are considered cotton dust within this definition. Lubricating oil mist associated with weaving operations is not considered cotton dust.

*Director* means the Director of the National Institute for Occupational Safety and Health (NIOSH), U.S. Department of Health and Human Services, or designee.

*Equivalent Instrument* means a cotton dust sampling device that meets the vertical elutriator equivalency requirements as described in paragraph (d)(1)(iii) of this section.

*Lint-free respirable cotton dust* means particles of cotton dust of approximately 15 micrometers or less aerodynamic equivalent diameter;

*Vertical elutriator cotton dust sampler* or *vertical elutriator* means a dust sampler which has a particle size cut-off at approximately 15 micrometers aerodynamic equivalent diameter when operating at the flow rate of 7.4 ±0.2 liters of air per minute;

*Waste processing* means waste recycling (sorting, blending, cleaning and willowing) and garnetting.

*Yarn manufacturing* means all textile mill operations from opening to, but not including, slashing and weaving.

**1910.1043(c)** *Permissible exposure limits and action levels*—**(1)** *Permissible exposure limits (PEL).* **(i)** The employer shall assure that no employee who is exposed to cotton dust in yarn manufacturing and cotton washing operations is exposed to airborne concentrations of lint-free respirable cotton dust greater than 200 µg/m³ mean concentration, averaged over an eight-hour period, as measured be a vertical elutriator or an equivalent instrument.

**1910.1043(c)(1)(ii)** The employer shall assure that no employee who is exposed to cotton dust in textile mill waste house operations or is exposed in yarn manufacturing to dust from "lower grade washed cotton" as defined in paragraph (n)(5) of this section is exposed to airborne concentrations of lint-free respirable cotton dust greater than 500 µg/m³ mean concentration, averaged over an eight-hour period, as measured by a vertical elutriator or an equivalent instrument.

**1910.1043(c)(1)(iii)** The employer shall assure that no employee who is exposed to cotton dust in the textile processes known as slashing and weaving is exposed to airborne concentrations of lint-free respirable cotton dust greater than 750 µg/m³ mean concentration, averaged over an eight hour period, as measured by a vertical elutriator or an equivalent instrument.

**1910.1043(c)(2)** *Action levels.* **(i)** The action level for yarn manufacturing and cotton washing operations is an airborne concentration of lint-free respirable cotton dust of 100 µg/m³ mean concentration, averaged over an eight-hour period, as measured by a vertical elutriator or an equivalent instrument.

**1910.1043(c)(2)(ii)** The action level for waste houses for textile operations is an airborne concentration of lint-free respirable cotton dust of 250 µg/m³ mean concentration, averaged over an eight-hour period, as measured by a vertical elutriator or an equivalent instrument.

**1910.1043(c)(2)(iii)** The action level for the textile processes known as slashing and weaving is an airborne concentration of lint-free respirable cotton dust of 375 µg/m³ mean concentration, averaged over an eight-hour period, as measured by a vertical elutriator or an equivalent instrument.

**1910.1043(d)** *Exposure monitoring and measurement*—**(1)** *General.* **(i)** For the purposes of this section, employee exposure is that exposure which would occur if the employee were not using a respirator.

**1910.1043(d)(1)(ii)** The sampling device to be used shall be either the vertical elutriator cotton dust sampler or an equivalent instrument.

**1910.1043(d)(1)(iii)** If an alternative to the vertical elutriator cotton dust sampler is used, the employer shall establish equivalency by reference to an OSHA opinion or by documenting, based on data developed by the employer or supplied by the manufacturer, that the alternative sampling devices meets the following criteria:

**1910.1043(d)(1)(iii)(A)** It collects respirable particulates in the same range as the vertical elutriator (approximately 15 microns);

**1910.1043(d)(1)(iii)(B)** Replicate exposure data used to establish equivalency are collected in side-by-side field and laboratory comparisons; and

**1910.1043(d)(1)(iii)(C)** A minimum of 100 samples over the range of 0.5 to 2 times the permissible exposure limit are collected, and 90% of these samples have an accuracy range of plus or minus 25 per cent of the vertical elutriator reading with a 95% confidence level as demonstrated by a statistically valid protocol. (An acceptable protocol for demonstrating equivalency is described in Appendix E of this section.)

**1910.1043(d)(1)(iv)** OSHA will issue a written opinion stating that an instrument is equivalent to a vertical elutriator cotton dust sampler if

**1910.1043(d)(1)(iv)(A)** A manufacturer or employer requests an opinion in writing and supplies the following information:

**1910.1043(d)(1)(iv)(A)***(1)* Sufficient test data to demonstrate that the instrument meets the requirements specified in this paragraph and the protocol specified in Appendix E of this section;

**1910.1043(d)(1)(iv)(A)***(2)* Any other relevant information about the instrument and its testing requested by OSHA; and

**1910.1043(d)(1)(iv)(A)***(3)* A certification by the manufacturer or employer that the information supplied is accurate, and

**1910.1043(d)(1)(iv)(B)** if OSHA finds, based on information submitted about the instrument, that the instrument meets the requirements for equivalency specified by paragraph (d) of this section.

**1910.1043(d)(2)** *Initial monitoring.* Each employer who has a place of employment within the scope of paragraph (a)(1), (a)(4), or (a)(5) of this section shall conduct monitoring by obtaining measurements which are representative of the exposure of all employees to airborne concentrations of lint-free respirable cotton dust over an eight-hour period. The sampling program shall include at least one determination during each shift for each work area.

**1910.1043(d)(3)** *Periodic monitoring.* **(i)** If the initial monitoring required by paragraph (d)(2) of this section or any subsequent monitoring reveals employee exposure to be at or below the permissible exposure limit, the employer shall repeat the monitoring for those employees at least annually.

**1910.1043(d)(3)(ii)** If the initial monitoring required by paragraph (d)(2) of this section or any subsequent monitoring reveals employee exposure to be above the PEL, the employer shall repeat the monitoring for those employees at least every six months.

**1910.1043(d)(3)(iii)** Whenever there has been a production, process, or control change which may result in new or additional exposure to cotton dust, or whenever the employer has any other reason to suspect an increase in employee exposure, the employer shall repeat the monitoring and measurements for those employees affected by the change or increase.

**1910.1043(d)(4)** *Employee notification.* **(i)** The employer must, within 15 working days after the receipt of the results of any monitoring performed under this section, notify each affected employee of these results either individually in writing or by posting the results in an appropriate location that is accessible to employees.

**1910.1043(d)(4)(ii)** Whenever the results indicate that the employee's exposure exceeds the applicable permissible exposure limit specified in paragraph (c) of this section, the employer shall include in the written notice a statement that the permissible exposure limit was exceeded and a description of the corrective action taken to reduce exposure below the permissible exposure limit.

**1910.1043(e)** *Methods of compliance*—**(1)** *Engineering and work practice controls.* The employer shall institute engineering and work practice controls to reduce and maintain employee exposure to cotton dust at or below the permissible exposure limit specified in paragraph (c) of this section, except to the extent that the employer can establish that such controls are not feasible.

**1910.1043(e)(2)** Whenever feasible engineering and work practice controls are not sufficient to reduce employee exposure to or below the permissible exposure limit, the employer shall nonetheless institute these controls to reduce exposure to the lowest feasible level, and shall supplement these controls with the use of respirators which shall comply with the provisions of paragraph (f) of this section.

**1910.1043(e)(3)** *Compliance program.* **(i)** Where the most recent exposure monitoring data indicates that any employee is exposed to cotton dust levels greater than the permissible exposure limit, the employer shall establish and implement a written program sufficient to reduce exposures to or below the permissible exposure limit solely by means of engineering controls and work practices as required by paragraph (e)(1) of this section.

**1910.1043(e)(3)(ii)** The written program shall include at least the following:

**1910.1043(e)(3)(ii)(A)** A description of each operation or process resulting in employee exposure to cotton dust at levels greater than the PEL;

**1910.1043(e)(3)(ii)(B)** Engineering plans and other studies used to determine the controls for each process;

**1910.1043(e)(3)(ii)(C)** A report of the technology considered in meeting the permissible exposure limit;

**1910.1043(e)(3)(ii)(D)** Monitoring data obtained in accordance with paragraph (d) of this section;

**1910.1043(e)(3)(ii)(E)** A detailed schedule for development and implementation of engineering and work practice controls, including exposure levels projected to be achieved by such controls;

**1910.1043(e)(3)(ii)(F)** Work practice program; and

**1910.1043(e)(3)(ii)(G)** Other relevant information.

**1910.1043(e)(3)(iii)** The employer's schedule as set forth in the compliance program, shall project completion of the implementation of the compliance program no later than March 27, 1984 or as soon as possible if monitoring after March 27, 1984 reveals exposures over the PEL, except as provided in paragraph (m)(2)(ii)(B) of this section.

**1910.1043(e)(3)(iv)** The employer shall complete the steps set forth in his program by the dates in the schedule.

**1910.1043(e)(3)(v)** Written programs shall be submitted, upon request, to the Assistant Secretary and the Director, and shall be available at the worksite for examination and copying by the Assistant Secretary, the Director, and any affected employee or their designated representatives.

**1910.1043(e)(3)(vi)** The written program required under paragraph (e)(3) of this section shall be revised and updated when necessary to reflect the current status of the program and current exposure levels.

**1910.1043(e)(4)** *Mechanical ventilation.* When mechanical ventilation is used to control exposure, measurements which demonstrate the effectiveness of the system to control exposure, such as capture velocity, duct velocity, or static pressure shall be made at reasonable intervals.

**1910.1043(f)** *Respiratory protection*—**(1)** *General.* For employees who are required to use respirators by this section, the employer must provide respirators that comply with the requirements of this paragraph. Respirators must be used during:

**1910.1043(f)(1)(i)** Periods necessary to install or implement feasible engineering and work-practice controls.

**1910.1043(f)(1)(ii)** Maintenance and repair activities for which engineering and work-practice controls are not feasible.

**1910.1043(f)(1)(iii)** Work operations for which feasible engineering and work-practice controls are not yet sufficient to reduce employee exposure to or below the permissible exposure limits.

**1910.1043(f)(1)(iv)** Work operations specified under paragraph (g)(1) of this section.

**1910.1043(f)(1)(v)** Periods for which an employee requests a respirator.

**1910.1043(f)(2)** *Respirator program.* **(i)** The employer must implement a respiratory protection program in accordance with 29 CFR 1910.134(b) through (d) (except (d)(1)(iii)), and (f) through (m).

**1910.1043(f)(2)(ii)** Whenever a physician determines that an employee who works in an area in which the cotton-dust concentration exceeds the PEL is unable to use a respirator, including a powered airpurifying respirator, the employee must be given the opportunity to transfer to an available position, or to a position that becomes available later, that has a cotton-dust concentration at or below the PEL. The employer must ensure that such employees retain their current wage rate or other benefits as a result of the transfer.

**1910.1043(f)(3)** *Respirator selection.* **(i)** The employer must select the appropriate respirator from Table I of this section.

### TABLE I

| Cotton dust concentration | Required respirator |
|---|---|
| Not greater than: | |
| (a) 5 × the applicable permissible exposure limit (PEL). | A disposable respirator with a particulate filter. |
| (b) 10 × the applicable PEL | A quarter or half-mask respirator, other than a disposable respirator, equipped with particulate filters. |
| (c) 100 × the applicable PEL. | A full facepiece respirator equipped with high-efficiency particulate filters |
| (d) Greater than 100 × the applicable PEL. | A powered air-purifying respirator equipped with high-efficiency particulate filters. |

Notes:
1. A disposable respirator means the filter element is an inseparable part of the respirator.
2. Any respirators permitted at higher environmental concentrations can be used at lower concentrations.
3. Self-contained breathing apparatus are not required respirators but are permitted respirators.
4. Supplied air respirators are not required but are permitted under the following conditions: Cotton dust concentration not greater than 10 × the PEL—Any supplied air respirator; not greater than 100 × the PEL—Any supplied air respirator with full facepiece, helmet or hood; greater than 100 × the PEL—A supplied air respirator operated in positive pressure mode.

**1910.1043(f)(3)(ii)** Whenever respirators are required by this section for cotton-dust concentrations that do not exceed the applicable permissible exposure limit by a multiple of 100 (100 X), the employer must, when requested by an employee, provide a powered air-purifying respirator with a high-efficiency particulate filter instead of the respirator specified in paragraphs (a), (b), or (c) of Table I of this section.

**1910.1043(g)** *Work practices.* Each employer shall, regardless of the level of employee exposure, immediately establish and implement a written program of work practices which shall minimize cotton dust exposure. The following shall be included were applicable:

**1910.1043(g)(1)** Compressed air "blow down" cleaning shall be prohibited where alternative means are feasible. Where compressed air is used for cleaning, the employees performing the "blow down" or "blow off" shall wear suitable respirators. Employees whose presence is not required to perform "blow down" or "blow of" shall be required to leave the area affected by the "blow down" or "blow off" during this cleaning operation.

**1910.1043(g)(2)** Cleaning of clothing or floors with compressed air shall be prohibited.

**1910.1043(g)(3)** Floor sweeping shall be performed with a vacuum or with methods designed to minimize dispersal of dust.

**1910.1043(g)(4)** In areas where employees are exposed to concentrations of cotton dust greater than the permissible exposure limit, cotton and cotton waste shall be stacked, sorted, baled, dumped, removed or otherwise handled by mechanical means, except where the employer can show that it is infeasible to do so. Where infeasible, the method used for handling cotton and cotton waste shall be the method which reduces exposure to the lowest level feasible.

**1910.1043(h)** *Medical surveillance*—**(1)** *General.* **(i)** Each employer covered by the standard shall institute a program of medical surveillance for all employees exposed to cotton dust.

**1910.1043(h)(1)(ii)** The employer shall assure that all medical examinations and procedures are performed by or under the supervision of a licensed physician and are provided without cost to the employee.

**1910.1043(h)(1)(iii)** Persons other than licensed physicians, who administer the pulmonary function testing required by this section shall have completed a NIOSH-approved training course in spirometry.

**1910.1043(h)(2)** *Initial examinations.* The employer shall provide medical surveillance to each employee who is or may be exposed to cotton dust. For new employees, this examination shall be provided prior to initial assignment. The medical surveillance shall include at least the following:

**1910.1043(h)(2)(i)** A medical history;

**1910.1043(h)(2)(ii)** The standardized questionnaire contained in Appendix B; and

**1910.1043(h)(2)(iii)** A pulmonary function measurement, including a determination of forced vital capacity (FVC) and forced expiratory volume in one second ($FEV_1$), the $FEV_1$/FVC ratio, and the percentage that the measured values of $FEV_1$ and FVC differ from the predicted values, using the standard tables in Appendix C. These determinations shall be made for each employee before the employee enters the workplace on the first day of the work week, preceded by at least 35 hours of no exposure to cotton dust. The tests shall be repeated during the shift, no less than 4 and no more than 10 hours after the beginning of the work shift; and, in any event, no more than one hour after cessation of exposure. Such exposure shall be typical of the employee's usual workplace exposure. The predicted $FEV_1$ and FVC for blacks shall be multiplied by 0.85 to adjust for ethnic differences.

**1910.1043(h)(2)(iv)** Based upon the questionnaire results, each employee shall be graded according to Schilling's byssinosis classification system.

**1910.1043(h)(3)** *Periodic examinations.* **(i)** The employer shall provide at least annual medical surveillance for all employees exposed to cotton dust above the action level in yarn manufacturing, slashing and weaving, cotton washing and waste house operations. The employer shall provide medical surveillance at least every two years for all employees exposed to cotton dust at or below the action level, for all employees exposed to cotton dust from washed cotton (except from washed cotton defined in paragraph (n)(3) of this section), and for all employees exposed to cotton dust in cottonseed processing and waste processing operations. Periodic medical surveillance shall include at least an update of the medical history, standardized questionnaire (App. B-111), Schilling byssinosis grade, and the pulmonary function measurements in paragraph (h)(2)(iii) of this section.

**1910.1043(h)(3)(ii)** Medical surveillance as required in paragraph (h)(3)(i) of this section shall be provided every six months for all employees in the following categories:

**1910.1043(h)(3)(ii)(A)** An FEV$_1$ of greater than 80 percent of the predicted value, but with an FEV$_1$ decrement of 5 percent or 200 ml. on a first working day;

**1910.1043(h)(3)(ii)(B)** An FEV$_1$ of less than 80 percent of the predicted value; or

**1910.1043(h)(3)(ii)(C)** Where, in the opinion of the physician, any significant change in questionnaire findings, pulmonary function results, or other diagnostic tests have occurred.

**1910.1043(h)(3)(iii)** An employee whose FEV$_1$ is less than 60 percent of the predicted value shall be referred to a physician for a detailed pulmonary examination.

**1910.1043(h)(3)(iv)** A comparison shall be made between the current examination results and those of previous examinations and a determination made by the physician as to whether there has been a significant change.

**1910.1043(h)(4)** *Information provided to the physician.* The employer shall provide the following information to the examination physician:

**1910.1043(h)(4)(i)** A copy of this regulation and its Appendices;

**1910.1043(h)(4)(ii)** A description of the affected employee's duties as they relate to the employee's exposure;

**1910.1043(h)(4)(iii)** The employee's exposure level or anticipated exposure level;

**1910.1043(h)(4)(iv)** A description of any personal protective equipment used or to be used; and

**1910.1043(h)(4)(v)** Information from previous medical examinations of the affected employee which is not readily available to the examining physician.

**1910.1043(h)(5)** *Physician's written opinion.* (i) The employer shall obtain and furnish the employee with a copy of a written opinion from the examining physician containing the following:

**1910.1043(h)(5)(i)(A)** The results of the medical examination and tests including the FEV$_1$, FVC, AND FEV$_1$/FVC ratio;

**1910.1043(h)(5)(i)(B)** The physician's opinion as to whether the employee has any detected medical conditions which would place the employee at increased risk of material impairment of the employee's health from exposure to cotton dust;

**1910.1043(h)(5)(i)(C)** The physician's recommended limitations upon the employee's exposure to cotton dust or upon the employee's use of respirators including a determination of whether an employee can wear a negative pressure respirator, and where the employee cannot, a determination of the employee's ability to wear a powered air purifying respirator; and,

**1910.1043(h)(5)(i)(D)** A statement that the employee has been informed by the physician of the results of the medical examination and any medical conditions which require further examination or treatment.

**1910.1043(h)(5)(ii)** The written opinion obtained by the employer shall not reveal specific findings or diagnoses unrelated to occupational exposure.

**1910.1043(i)** *Employee education and training*—(1) *Training program.* (i) The employer shall provide a training program for all employees exposed to cotton dust and shall assure that each employee is informed of the following:

**1910.1043(i)(1)(i)(A)** The acute and long term health hazards associated with exposure to cotton dust;

**1910.1043(i)(1)(i)(B)** The names and descriptions of jobs and processes which could result in exposure to cotton dust at or above the PEL.

**1910.1043(i)(1)(i)(C)** The measures, including work practices required by paragraph (g) of this section, necessary to protect the employee from exposures in excess of the permissible exposure limit;

**1910.1043(i)(1)(i)(D)** The purpose, proper use and limitations of respirators required by paragraph (f) of this section;

**1910.1043(i)(1)(i)(E)** The purpose for and a description of the medical surveillance program required by paragraph (h) of this section and other information which will aid exposed employees in understanding the hazards of cotton dust exposure; and

**1910.1043(i)(1)(i)(F)** The contents of this standard and its appendices.

**1910.1043(i)(1)(ii)** The training program shall be provided prior to initial assignment and shall be repeated annually for each employee exposed to cotton dust, when job assignments or work processes change and when employee performance indicates a need for retraining.

**1910.1043(i)(2)** *Access to training materials.* (i) Each employer shall post a copy of this section with its appendices in a public location at the workplace, and shall, upon request, make copies available to employees.

**1910.1043(i)(2)(ii)** The employer shall provide all materials relating to the employee training and information program to the Assistant Secretary and the Director upon request.

**1910.1043(j)** *Signs.* The employer shall post the following warning sign in each work area where the permissible exposure limit for cotton dust is exceeded:

**WARNING**

**COTTON DUST WORK AREA**

**MAY CAUSE ACUTE OR DELAYED**

**LUNG INJURY**

**(BYSSINOSIS)**

**RESPIRATORS**

**REQUIRED IN THIS AREA**

**1910.1043(k)** *Recordkeeping*—(1) *Exposure measurements.* (i) The employer shall establish and maintain an accurate record of all measurements required by paragraph (d) of this section.

**1910.1043(k)(1)(ii)** The record shall include:

**1910.1043(k)(1)(ii)(A)** A log containing the items listed in paragraph IV (a) of Appendix A, and the dates, number, duration, and results of each of the samples taken, including a description of the procedure used to determine representative employee exposure;

**1910.1043(k)(1)(ii)(B)** The type of protective devices worn, if any, and length of time worn; and

**1910.1043(k)(1)(ii)(C)** The names, social security numbers, job classifications, and exposure levels of employees whose exposure the measurement is intended to represent.

**1910.1043(k)(1)(iii)** The employer shall maintain this record for at least 20 years.

**1910.1043(k)(2)** *Medical surveillance.* (i) The employer shall establish and maintain an accurate medical record for each employee subject to medical surveillance required by paragraph (h) of this section.

**1910.1043(k)(2)(ii)** The record shall include:

**1910.1043(k)(2)(ii)(A)** The name and social security number and description of the duties of the employee;

**1910.1043(k)(2)(ii)(B)** A copy of the medical examination results including the medical history, questionnaire response, results of all tests, and the physician's recommendation;

**1910.1043(k)(2)(ii)(C)** A copy of the physician's written opinion;

**1910.1043(k)(2)(ii)(D)** Any employee medical complaints related to exposure to cotton dust;

**1910.1043(k)(2)(ii)(E)** A copy of this standard and its appendices, except that the employer may keep one copy of the standard and the appendices for all employees, provided that he references the standard and appendices in the medical surveillance record of each employee; and

**1910.1043(k)(2)(ii)(F)** A copy of the information provided to the physician as required by paragraph (h)(4) of this section.

**1910.1043(k)(2)(iii)** The employer shall maintain this record for at least 20 years.

**1910.1043(k)(3)** *Availability.* (i) The employer shall make all records required to be maintained by paragraph (k) of this section available to the Assistant Secretary and the Director for examination and copying.

**1910.1043(k)(3)(ii)** Employee exposure measurement records and employee medical records required by this paragraph shall be provided upon request to employees, designated representatives, and the Assistant Secretary in accordance with 29 CFR 1910.1020(a) through (e) and (g) through (i).

**1910.1043(k)(4)** *Transfer of records.* (i) Whenever the employer ceases to do business, the successor employer shall receive and retain all records required to be maintained by paragraph (k) of this section.

**1910.1043(k)(4)(ii)** Whenever the employer ceases to do business, and there is no successor employer to receive and retain the records for the prescribed period, these records shall be transmitted to the Director.

# OSHA Standards for General Industry

**1910.1043(k)(4)(iii)** At the expiration of the retention period for the records required to be maintained by this section, the employer shall notify the Director at least 3 months prior to the disposal of such records and shall transmit those records to the Director if the Director requests them within that period.

**1910.1043(k)(4)(iv)** The employer shall also comply with any additional requirements involving transfer of records set forth in 29 CFR 1910.1020(h).

**1910.1043(l)** *Observation of monitoring.* **(1)** The employer shall provide affected employees or their designated representatives an opportunity to observe any measuring or monitoring of employee exposure to cotton dust conducted pursuant to paragraph (d) of this section.

**1910.1043(l)(2)** Whenever observation of the measuring or monitoring of employee exposure to cotton dust requires entry into an area where the use of personal protective equipment is required, the employer shall provide the observer with and assure the use of such equipment and shall require the observer to comply with all other applicable safety and health procedures.

**1910.1043(l)(3)** Without interfering with the measurement, observers shall be entitled to:

**1910.1043(l)(3)(i)** An explanation of the measurement procedures;

**1910.1043(l)(3)(ii)** An opportunity to observe all steps related to the measurement of airborne concentrations of cotton dust performed at the place of exposure; and

**1910.1043(l)(3)(iii)** An opportunity to record the results obtained.

**1910.1043(m)** *Washed Cotton*—**(1)** *Exemptions.* Cotton, after it has been washed by the processes described in this paragraph, is exempt from all or parts of this section as specified if the requirements of this paragraph are met.

**1910.1043(m)(2)** *Initial requirements.* **(i)** In order for an employer to qualify as exempt or partially exempt from this standard for operations using washed cotton, the employer must demonstrate that the cotton was washed in a facility which is open to inspection by the Assistant Secretary and the employer must provide sufficient accurate documentary evidence to demonstrate that the washing methods utilized meet the requirements of this paragraph.

**1910.1043(m)(2)(ii)** An employer who handles or processes cotton which has been washed in a facility not under the employer's control and claims an exemption or partial exemption under this paragraph, must obtain from the cotton washer and make available at the worksite, to the Assistant Secretary, to any affected employee, or to their designated representative the following:

**1910.1043(m)(2)(ii)(A)** A certification by the washer of the cotton of the grade of cotton, the type of washing process, and that the batch meets the requirements of this paragraph;

**1910.1043(m)(2)(ii)(B)** Sufficient accurate documentation by the washer of the cotton grades and washing process; and

**1910.1043(m)(2)(ii)(C)** An authorization by the washer that the Assistant Secretary or the Director may inspect the washer's washing facilities and documentation of the process.

**1910.1043(m)(3)** *Medical and dyed cotton.* Medical grade (USP) cotton, cotton that has been scoured, bleached and dyed, and mercerized yarn shall be exempt from all provisions of this standard.

**1910.1043(m)(4)** *Higher grade washed cotton.* The handling or processing of cotton classed as "low middling light spotted or better" (color grade 52 or better and leaf grade code 5 or better according to the 1993 USDA classification system) shall be exempt from all provisions of the standard except the requirements of paragraphs (h) medical surveillance, (k)(2) through (4) recordkeeping—medical records, and Appendices B, C, and D of this section, if they have been washed on one of the following systems:

**1910.1043(m)(4)(i)** On a continuous batt system or a rayon rinse system including the following conditions:

**1910.1043(m)(4)(i)(A)** With water,

**1910.1043(m)(4)(i)(B)** At a temperature of no less than 60 °C,

**1910.1043(m)(4)(i)(C)** With a water-to-fiber ratio of no less than 40:1; and

**1910.1043(m)(4)(i)(D)** With bacterial levels in the wash water controlled to limit bacterial contamination of the cotton.

**1910.1043(m)(4)(ii)** On a batch kier washing system including the following conditions:

**1910.1043(m)(4)(ii)(A)** With water;

**1910.1043(m)(4)(ii)(B)** With cotton fiber mechanically opened and thoroughly prewetted before forming the cake;

**1910.1043(m)(4)(ii)(C)** For low-temperature processing, at a temperature of no less than 60 °C with a water-to-fiber ratio of no less than 40:1; or, for high-temperature processing, at a temperature of no less than 93 °C with a water-to-fiber ratio of no less than 15:1;

**1910.1043(m)(4)(ii)(D)** With a minimum of one wash cycle followed by two rinse cycles for each batch, using fresh water in each cycle, and

**1910.1043(m)(4)(ii)(E)** With bacterial levels in the wash water controlled to limit bacterial contamination of the cotton.

**1910.1043(m)(5)** *Lower grade washed cotton.* The handling and processing of cotton of grades lower than "low middling light spotted," that has been washed as specified in paragraph (n)(4) of this section and has also been bleached, shall be exempt from all provisions of the standard except the requirements of paragraphs (c)(1)(ii) Permissible Exposure Limit, (d) Exposure Monitoring, (h) Medical Surveillance, (k) Recordkeeping, and Appendices B, C and D of this section.

**1910.1043(m)(6)** *Mixed grades of washed cotton.* If more than one grade of washed cotton is being handled or processed together, the requirements of the grade with the most stringent exposure limit, medical and monitoring requirements shall be followed.

**1910.1043(n)** *Appendices.* **(1)** Appendices B, C, and D of this section are incorporated as part of this section and the contents of these appendices are mandatory.

**1910.1043(n)(2)** Appendix A of this section contains information which is not intended to create any additional obligations not otherwise imposed or to detract from any existing obligations.

**1910.1043(n)(3)** Appendix E of this section is a protocol which may be followed in the validation of alternative measuring devices as equivalent to the vertical elutriator cotton dust sampler. Other protocols may be used if it is demonstrated that they are statistically valid, meet the requirements in paragraph (d)(l)(iii) of this section, and are appropriate for demonstrating equivalency.

## APPENDIX A TO §1910.1043—AIR SAMPLING AND ANALYTICAL PROCEDURES FOR DETERMINING CONCENTRATIONS OF COTTON DUST

### I. SAMPLING LOCATIONS

The sampling procedures must be designed so that samples of the actual dust concentrations are collected accurately and consistently and reflect the concentrations of dust at the place and time of sampling. Sufficient number of 6-hour area samples in each distinct work area of the plant should be collected at locations which provide representative samples of air to which the worker is exposed. In order to avoid filter overloading, sampling time may be shortened when sampling in dusty areas. Samples in each work area should be gathered simultaneously or sequentially during a normal operating period. The daily timeweighted average (TWA) exposure of each worker can then be determined by using the following formula:

**Summation of hours spent in each location and the dust concentration in that location.**

**Total hours exposed**

A time-weighted average concentration should be computed for each worker and properly logged and maintained on file for review.

### II. SAMPLING EQUIPMENT

(a) Sampler. The instrument selected for monitoring is the Lumsden-Lynch vertical elutriator. It should operate at a flow rate of 7.4 ±0.2 liters/minute.

The samplers should be cleaned prior to sampling. The pumps should be monitored during sampling.

(b) Filter Holder. A three-piece cassette constructed of polystyrene designed to hold a 37-mm diameter filter should be used. Care must be exercised to insure that an adequate seal exists between elements of the cassette.

(c) Filers and Support Pads. The membrane filters used should be polyvinyl chloride with a 5-um pore size and 37-mm diameter. A support pad, commonly called a backup pad, should be used under the filter membrane in the field monitor cassette.

(d) Balance. A balance sensitive to 10 micrograms should be used.

(e) Monitoring equipment for use in Class III hazardous locations must be approved for use in such locations, in accordance with the requirements of the OSHA electrical standards in Subpart S of Part 1910.

## III. INSTRUMENT CALIBRATION PROCEDURE

Samplers shall be calibrated when first received from the factory, after repair, and after receiving any abuse. The samplers should be calibrated in the laboratory both before they are used in the field and after they have been used to collect a large number of field samples. The primary standard, such as a spirometer or other standard calibrating instruments such as a wet test meter or a large bubble meter or dry gas meter, should be used. Instructions for calibration with the wet test meter follow. If another calibration device is selected, equivalent procedures should be used:

(a) Level wet test meter. Check the water level which should just touch the calibration point at the left side of the meter. If water level is low, add water 1-2 °F. warmer than room temperature of till point. Run the meter for 30 minutes before calibration;

(b) Place the polyvinyl chloride membrane filter in the filter cassette;

(c) Assemble the calibration sampling train;

(d) Connect the wet test meter to the train.

The pointer on the meter should run clockwise and a pressure drop of not more than 1.0 inch of water indicated. If the pressure drop is greater than 1.0, disconnect and check the system;

(e) Operate the system for ten minutes before starting the calibration;

(f) Check the vacuum gauge on the pump to insure that the pressure drop across the orifice exceeds 17 inches of mercury;

(g) Record the following on calibration data sheets:

(1) Wet test meter reading, start and finish;

(2) Elapsed time, start and finish (at least two minutes);

(3) Pressure drop at manometer;

(4) Air temperature;

(5) Barometric pressure; and

(6) Limiting orifice number;

(h) Calculate the flow rate and compare against the flow of 7.4 ±0.2 liters/minute. If flow is between these limits, perform calibration again, average results, and record orifice number and flow rate. If flow is not within these limits, discard or modify orifice and repeat procedure;

(i) Record the name of the person performing the calibration, the date, serial number of the wet test meter, and the number of the critical orifices being calibrated.

## IV. SAMPLING PROCEDURE

(a) Sampling data sheets should include a log of:

(1) The date of the sample collection;

(2) The time of sampling;

(3) The location of the sampler;

(4) The sampler serial number;

(5) The cassette number;

(6) The time of starting and stopping the sampling and the duration of sampling;

(7) The weight of the filter before and after sampling;

(8) The weight of dust collected (corrected for controls);

(9) The dust concentration measured;

(10) Other pertinent information; and

(11) Name of person taking sample

(b) Assembly of filter cassette should be as follows:

(1) Loosely assemble 3-piece cassette;

(2) Number cassette;

(3) Place absorbent pad in cassette;

(4) Weigh filter to an accuracy of 10 µg;

(5) Place filter in cassette;

(6) Record weight of filter in log, using cassette number for identification;

(7) Fully assemble cassette, using pressure to force parts tightly together;

(8) Install plugs top and bottom;

(9) Put shrink band on cassette, covering joint between center and bottom parts of cassette; and

(10) Set cassette aside until shrink band dries thoroughly.

(c) Sampling collection should be performed as follows:

(1) Clean lint out of the motor and elutriator;

(2) Install vertical elutriator in sampling locations specified above with inlet 4 1/2 to 5 1/2 feet from floor (breathing zone height);

(3) Remove top section of cassette;

(4) Install cassette in ferrule of elutriator;

(5) Tape cassette to ferrule with masking tape or similar material for airtight seal;

(6) Remove bottom plug of cassette and attach hose containing critical orifice;

(7) Start elutriator pump and check to see if gauge reads above 17 in. of Hg vacuum;

(8) Record starting time, cassette number, and sampler number;

(9) At end of sampling period stop pump and record time; and

(10) Controls with each batch of samples collected, two additional filter cassettes should be subjected to exactly the same handling as the samples, except that they are not opened. These control filters should be weighed in the same manner as the sample filters.

Any difference in weight in the control filters would indicate that the procedure for handling sample filters may not be adequate and should be evaluated to ascertain the cause of the difference, whether and what necessary corrections must be made, and whether additional samples must be collected.

(d) Shipping. The cassette with samples should be collected, along with the appropriate number of blanks, and shipped to the analytical laboratory in a suitable container to prevent damage in transit.

(e) Weighing of the sample should be achieved as follows:

(1) Remove shrink band;

(2) Remove top and middle sections of cassette and bottom plug;

(3) Remove filter from cassette and weigh to an accuracy of 10 µg; and

(4) Record weight in log against original weight

(f) Calculation of volume of air sampled should be determined as follows:

(1) From starting and stopping times of sampling period, determine length of time in minutes of sampling period; and

(2) Multiply sampling time in minutes by flow rate of critical orifice in liters per minute and divide by 1000 to find air quantity in cubic meters.

(g) Calculation of Dust Concentrations should be made as follows:

(1) Subtract weight of clean filter from dirty filter and apply control correction to find actual weight of sample. Record this weight (in µg) in log; and

(2) Divide mass of sample in µg by air volume in cubic meters to find dust concentration in µg/m. Record in log.

## APPENDIX B-1
## RESPIRATORY QUESTIONNAIRE

# RESPIRATORY QUESTIONNAIRE

## A. IDENTIFICATION DATA

PLANT _____

SOCIAL SECURITY NO. _____

DATE OF INTERVIEW ___ DAY ___ MONTH (figures) ___ YEAR (last 2 digits)

NAME _____
(Surname)

_____
(First Names)

DATE OF BIRTH _____

ADDRESS _____ AGE ___ (8, 9) SEX M___ F___ (10)

RACE W___ N___ IND. ___ OTHER ___ (11)

INTERVIEWER: 1 2 3 4 5 6 8 (12)

WORK SHIFT: 1st___ 2nd___ 3rd___ STANDING HEIGHT ___ (14, 15)

WEIGHT ___ (16, 18)

PRESENT WORK AREA

If working in more than one specified work area, X area where most of the work shift is spent. If "other", but spending 25% of the work shift in one of the specified work areas, classify in that work area. If carding department employee, check area within that department where most of the work shift is spent (if in doubt, check "throughout"). For work areas such as spinning and weaving where many work rooms may be involved, be sure to check the specific work room to which the employee is assigned— if he works in more than one work room within a department classify as 7 (all) for that department.

| | (19) | (20) | (21) Card #1 | (22) Card #2 | (23) | (24) | (25) | (26) | (27) | (28) | (29) | (30) |
|---|---|---|---|---|---|---|---|---|---|---|---|---|
| Workroom Number | Open | Pick | Area | | Spin | Wind | Twist | Spool | Warp | Slash | Weave | Other |
| AT RISK (cotton & cotton blend) 1 | | | Cards | | | | | | | | | |
| 2 | | | Draw | | | | | | | | | |
| 3 | | | Comb | | | | | | | | | |
| 4 | | | Rove | | | | | | | | | |
| 5 | | | Thru Out | | | | | | | | | |
| 6 | | | | | | | | | | | | |
| 7 (all) | | | | | | | | | | | | |
| Control (synthetic & wood) 8 | | | | | | | | | | | | |
| Ex-Worker (cotton) 9 | | | | | | | | | | | | |

Use actual wording of each question. Put X in appropriate square after each question. When in doubt record 'No'. When no square, circle appropriate answer.

## B. COUGH

Do you usually cough first thing in the morning? (on getting up)†
(Count a cough with first smoke or on "first going out of doors."
Exclude clearing throat or a single cough.) Yes ___ No ___ (31)

Do you usually cough during the day or at night?
(Ignore an occasional cough.) Yes ___ No ___ (32)

If 'Yes' to either question (31-32):
Do you cough like this on most days for as much
as three months a year? Yes ___ No ___ (33)

Do you cough on any particular day of the week? Yes ___ No ___ (34)

If 'Yes': which day?
(1) Mon. (2) Tues. (3) Wed. (4) Thur. (5) Fri. (6) Sat (7) Sun. (35)

## C. PHLEGM or alternative word to suit local custom.

Do you usually bring up any phlegm from your chest first thing in the morning? (on getting up)†
(Count phlegm with the first smoke or on "first going out of doors.
" Exclude phlegm from the nose. Count swallowed phlegm.) Yes ___ No ___ (36)

†These words are for subjects who work at night

Do you usually bring up any phlegm from your chest during the day or at night?
(Accept twice or more.) Yes ___ No ___ (37)

If 'Yes' to either question (36) or (37):
Do you bring up phlegm like this on most days for as much
as three months each year? Yes ___ No ___ (38)

If 'Yes' to question (33) or (38):
How long have you had this phlegm (cough)? (39)
(Write in number of years)
(1) ☐ 2 years or less
(2) ☐ More than 2 years-9 years
(3) ☐ 10-19 years
(4) ☐ 20+ years

## D. CHEST ILLNESSES

In the past three years, have you had a period
of (increased) †cough and phlegm lasting for
3 weeks or more? Yes ___ No ___ (40)
(1) ☐ No
(2) ☐ Yes, only one period
(3) ☐ Yes, two or more periods

†For subjects who usually have phlegm

During the past 3 years have you had any chest illness which has
kept you off work, indoors at home or in bed?
(For as long as one week, flu?) Yes ___ No ___ (41)

If 'Yes' to (41):
Did you bring up (more) phlegm than usual in any of these illnesses? Yes ___ No ___ (42)

If 'Yes' to (42):
During the past three years have you had: (43)
Only one such illness with increased phlegm? (44)
More than one such illness:
(1) ☐
(2) ☐
Br. Grade ___

## E. TIGHTNESS

Does your chest ever feel tight or your breathing become difficult? Yes ___ No ___ (45)
Is your chest tight or your breathing difficult on any particular day
of the week? (after a week or 10 days away from the mill) Yes ___ No ___ (46)

If 'Yes': Which day?
(1) Mon. (2) Tues. (3) Wed. (4) Thur. (5) Fri. (6) Sat. (7) Sun. (47)
Sometimes Always

If 'Yes' Monday: At what time on Monday does your chest feel tight or your breathing difficult? (48)
1 ☐ Before entering the mill
2 ☐ After entering the mill
(Ask only if NO to Question (45))

In the past, has your chest ever been tight or your breathing difficult on any particular day of the week? Yes ___ No ___ (49)

If 'Yes': Which day?
(1) Mon. (2) Tues. (3) Wed. (4) Thur. (5) Fri. (6) Sat. (7) Sun. (8) (50)
Sometimes Always

## F. BREATHLESSNESS

If disabled from walking by any condition other than heart or lung disease put "X" here
and leave questions (52-60) unasked. ☐ (51)
Are you ever troubled by shortness of breath, when hurrying on the level or walking up a slight hill? Yes ___ No ___ (52)

If 'No', grade is 1. If 'Yes', proceed to next question
Do you get short of breath walking with other people at an ordinary pace on the level? Yes ___ No ___ (53)

If 'No', grade is 2. If 'Yes', proceed to next question
Do you have to stop for breath when walking at your own pace on the level? Yes ___ No ___ (54)

If 'No', grade is 3. If 'Yes', proceed to next question
Are you short of breath on washing or dressing? Yes ___ No ___ (55)

If 'No', grade is 4. If 'Yes', grade is 5. Dyspnea Grd. ___ (56)

ON MONDAYS
Are you ever troubled by shortness of breath, when hurrying on the level or walking up a slight hill? Yes ___ No ___ (57)

# Subpart Z—Toxic and Hazardous Substances

If 'No', grade is 1. If 'Yes', proceed to next question
Do you get short of breath walking with other people at an ordinary pace on the level?
Yes _____ No _____ (58)

If 'No', grade is 2. If 'Yes', proceed to next question
Do you have to stop for breath when walking at your own pace on the level?
Yes _____ No _____ (59)

If 'No', grade is 3. If 'Yes', proceed to next question
Are you short of breath on washing or dressing?
Yes _____ No _____ (60)

If 'No', grade is 4. If 'Yes', grade is 5          B. Grd. _____ (61)

G. OTHER ILLNESSES AND ALLERGY HISTORY
Do you have a heart condition for which you are under a doctor's care? Yes _____ No _____ (62)
Have you ever had asthma? Yes _____ No _____ (63)
If 'Yes', did it begin          (1) ☐ Before age 30
                                (2) ☐ After age 30      (64)

If 'Yes' before 30 did you have asthma before ever going to work in a textile mill?
Yes _____ No _____ (65)

Have you ever had hay fever or other allergies (other than above)? Yes _____ No _____ (66)

H. TOBACCO SMOKING:*
Do you smoke?
Record 'Yes' if regular smoker up to one month ago (Cigarettes, cigar or pipe)
Yes _____ No _____ (67)

If 'No' to (63)
Have you ever smoked? (Cigarettes, cigars, pipe. Record 'No' if subject has never smoked as much as one cigarette a day, or 1 oz of tobacco a month, for as long as one year.) Yes _____ No _____ (<40)
If 'Yes' to (63) or (64), what have you smoked and for how many years?
(Write in specific number of years in the appropriate square)

|  | (1) (<5) | (2) (5-9) | (3) (10-14) | (4) (15-19) | (5) (20-24) | (6) (25-29) | (7) (30-34) | (8) (35-39) | (9) (<40) |
|---|---|---|---|---|---|---|---|---|---|
| Years |  |  |  |  |  |  |  |  |  |
| Cigarettes |  |  |  |  |  |  |  |  |  |
| Pipe |  |  |  |  |  |  |  |  |  |
| Cigars |  |  |  |  |  |  |  |  |  |

If cigarettes, how many packs per day?
(Write in number of cigarettes)
(1) _____ less than 1/2 pack
(2) _____ 1/2 pack, but less than 1 pack
(3) _____ 1 pack, but less than 1 1/2 packs
(4) _____ 1-1/2 packs or more

Number of pack years: _____ (72, 73)

If an ex smoker (cigarettes, cigar or pipe), how long since you stopped?
(Write in number of years)
(1) _____ 0-1 year
(2) _____ 1-4 years
(3) _____ 5-9 years
(4) _____ 10+ years       (74)

*Have you changed your smoking habits since last interview? If yes, specify what changes.

I. OCCUPATIONAL HISTORY**
Have you ever worked in. A foundry? (As long as one year) Yes _____ No _____ (75)
Stone or mineral mining, quarrying or processing?
(As long as one year) Yes _____ No _____ (76)
Asbestos milling or processing? (Ever) Yes _____ No _____ (77)
Other dusts, fumes or smoke? If yes, specify. Yes _____ No _____ (78)
Types of exposure _____
Length of exposure _____

**Ask only on first interview.

At what age did you first go to work in a textile mill?
(Write in specific age in appropriate square)

| (1) | (2) | (3) | (4) | (5) | (6) |
|---|---|---|---|---|---|
| <20 | 20-24 | 25-29 | 30-34 | 35-39 | 40+ |

When you first worked in a textile mill, did you work with
(1) ☐ Cotton or cotton blend  (79)
(2) ☐ Synthetic or wool  (80)

## APPENDIX B-II
## RESPIRATORY QUESTIONNAIRE FOR NON-TEXTILE WORKERS FOR THE COTTON INDUSTRY

**Respiratory Questionnaire for Non-Textile Workers for the Cotton Industry**

Identification No. _____         Interviewer Code _____
Location _____                    Date of Interview _____

### A. IDENTIFICATION

1. NAME _____ _____ _____           6. AGE LAST BIRTHDAY _____
      (Last)  (First)  (Middle Initial)
2. CURRENT ADDRESS _____                7. SEX 1 ☐ Male  2 ☐ Female
   (Number, Street, or Rural Route)
   _____ _____ _____ _____        8. ETHNIC GROUP OR ANCESTRY
   (City or Town) (County) (State) (Zip Code)   1. ☐ White, not of Hispanic Origin
3. PHONE NUMBER                                  2. ☐ Black, not of Hispanic Origin
   AREA CODE ( ___ ) NO. _____                 3. ☐ Hispanic
4. SOCIAL SECURITY (optional see below)          4. ☐ American Indian or Alaskan Native
   ___-___-___                                   5. ☐ Asian or Pacific Islander
5. BIRTHDATE _____                             6. ☐ Other: _____
   (Mo., Day, Yr.)
                                          9. STANDING HEIGHT _____ (cm)
                                          10. WEIGHT _____
                                          11. WORK SHIFT  1st ☐  2nd ☐  3rd ☐

12. PRESENT WORK AREA
Please indicate primary assigned work area and percent of time spent at that site.
If at other locations, please indicate and note percent of time for each.
PRIMARY WORK AREA _____
SPECIFIC JOB _____

13. APPROPRIATE INDUSTRY
1 ☐ Garnetting          2 ☐ Cottonseed Oil Mill        3 ☐ Cotton Warehouse
4 ☐ Utilization         5 ☐ Cotton Classification      6 ☐ Cotton Ginning

(Furnishing your Social Security number is voluntary. Your refusal to provide this number will not affect any right, benefit, or privilege to which you would be entitled if you did provide your Social Security number. Your Social Security number is being requested since it will permit use in future determinations in, statistical research studies.)

### B. OCCUPATIONAL HISTORY TABLE

Complete the following table showing the entire work history of the individual from present to initial employment. Sporadic, part-time periods of employment, each of no significant duration, should be grouped if possible.

| INDUSTRY AND LOCATION | TENURE OF EMPLOYMENT FROM 19__ TO 19__ | SPECIFIC OCCUPATION | AVERAGE NO. DAYS WORKED PER WEEK | HAZARDOUS HEALTH EXPOSURE ASSOCIATED WITH WORK |  |
|---|---|---|---|---|---|
|  |  |  |  | YES NO | IF YES, DESCRIBE |
|  |  |  |  |  |  |
|  |  |  |  |  |  |
|  |  |  |  |  |  |
|  |  |  |  |  |  |
|  |  |  |  |  |  |
|  |  |  |  |  |  |

# OSHA Standards for General Industry

## C. SYMPTOMS

Use actual wording of each question. Put X in appropriate square after each question. When in doubt record "No".

**COUGH**

1. Do you usually cough first thing in the morning (on getting up)?*    1 ☐ Yes   2 ☐ No
   (Count a cough with first smoke or on "first going out of doors". Exclude clearing throat or a single cough.)
2. Do you usually cough during the day or at night?    1 ☐ Yes   2 ☐ No
   (Ignore an occasional cough.)
   If YES to either question 1 or 2:
3. Do you cough like this on most days for as much as three months a year?
      1 ☐ Yes   2 ☐ No   9 ☐ NA
4. Do you cough on any particular day of the week?    1 ☐ Yes   2 ☐ No
   If YES:
5. Which day?   Mon.   Tue.   Wed.   Thur.   Fri.   Sat.   Sun.

**PHLEGM**

6. Do you usually bring up any phlegm from your chest first thing in the morning (on getting up)?* (Count phlegm with the first smoke or on "first going out of doors." Exclude phlegm from the nose. Count swallowed phlegm.)
      1 ☐ Yes   2 ☐ No
7. Do you usually bring up any phlegm from your chest during the day or at night? (Accept twice or more.)
      1 ☐ Yes   2 ☐ No
   If YES to either question 6 or 7:
8. Do you bring up phlegm like this on most days for as much as three months each year?    1 ☐ Yes   2 ☐ No
   If YES to question 3 or 8:
9. How long have you had this phlegm? (cough) (Write in number of years)
   (1) ☐ 2 years or less
   (2) ☐ More than 2 years - 9 years
   (3) ☐ 10-19 years
   (4) ☐ 20☐plus; years

**CHEST ILLNESS**

10. In the past three years, have you had a period of (increased) cough and phlegm lasting for 3 weeks or more?
    (1) ☐ No
    (2) ☐ Yes, only one period
    (3) ☐ Yes, two or more periods
    For subjects who usually have phlegm:
11. During the past 3 years have you had any chest illness which has kept you off work, indoors at home or in bed?
    (For as long as one week, flu?)    1 ☐ Yes   2 ☐ No
    If YES to 11:
12. Did you bring up (more) phlegm than usual in any of these illnesses?    1 ☐ Yes   2 ☐ No
    If YES to 12: During the past three years have you had:
13. Only one such illness with increased phlegm:    1 ☐ Yes   2 ☐ No
14. More than one such illness.    Br. Brade _____

**TIGHTNESS**

15. Does your chest ever feel tight or your breathing become difficult?    1 ☐ Yes   2 ☐ No
16. Is your chest tight or your breathing difficult on any particular day of the week? (after a week or 10 days away from the mill)    1 ☐ Yes   2 ☐ No
    (3)   (4)   (5)   (6)   (7)   (8)
17. If 'Yes': Which day?
    (1) Mon.   (2) Tues.   (3) Wed.   (4) Thur.   (5) Fri.   (6) Sat.   (7) Sun.
    Sometimes    Always
18. If YES Monday:   At what time on Monday does your chest feel tight or your breathing difficult?
    ☐ Before entering mill    ☐ After entering mill
19. In the past, has your chest ever been tight or your breathing difficult on any particular day of the week?
       1 ☐ Yes   2 ☐ No
    (ASK ONLY IF NO TO QUESTION 15)
20. If 'Yes': Which day?
    (1) Mon.   (2) Tues.   (3) Wed.   (4) Thur.   (5) Fri.   (6) Sat.   (7) Sun.
    Sometimes    Always

\* These words are for subjects who work at night

## APPENDIX B-III
## ABBREVIATED RESPIRATORY QUESTIONNAIRE

**ABBREVIATED RESPIRATORY QUESTIONNAIRE**

**A. IDENTIFICATION DATA**

PLANT _____    SOCIAL SECURITY NO. _____

NAME _____ (Surname)    DATE OF INTERVIEW _____ DAY __ MONTH __ YEAR __ (figures) (last 2 digits)

_____ (First Names)    DATE OF BIRTH _____

ADDRESS _____    AGE _____ (8, 9) SEX M ☐ F ☐ (10)

RACE W ___ N ___ IND. ___ OTHER ___ (11)

INTERVIEWER: 1 2 3 4 5 6 7 8 (12)

WORK SHIFT: 1st ___ 2nd ___ 3rd ___ (13) STANDING HEIGHT ___ (14, 15)    WEIGHT ___ (16, 18)

**PRESENT WORK AREA**

If working in more than one specified work area, X area where most of the work shift is spent. If "other," but spending 25% of the work shift in one of the specified work areas, classify in that work area. If carding department employee, check area within that department where most of the work shift is spent (if in doubt, check "throughout"). For work areas such as spinning and weaving where many work rooms may be involved, be sure to check the specific work room to which the employee is assigned. If he works in more than one work room within a department classify as 7 (all) for that department.

| Workroom Number | Area | (19) Open | (20) Pick | (21) Card #1 | (22) Card #2 | (23) Spin | (24) Wind | (25) Twist | (26) Spool | (27) Warp | (28) Slash | (29) Weave | (30) Other |
|---|---|---|---|---|---|---|---|---|---|---|---|---|---|
| 1 | Cards | | | | | | | | | | | | |
| 2 | Draw | | | | | | | | | | | | |
| 3 | Comb | | | | | | | | | | | | |
| 4 | Rove | | | | | | | | | | | | |
| 5 | Thru Out | | | | | | | | | | | | |
| 6 | | | | | | | | | | | | | |
| 7 (all) | | | | | | | | | | | | | |

AT RISK (cotton & cotton blend)

Control (synthetic & wood)    8

Ex-Worker (cotton)    9

Use actual wording of each question. Put X in appropriate square after each question. When in doubt record 'No'. When no square, circle appropriate answer.

**B. COUGH**

Do you usually cough first thing in the morning? (on getting up)†    Yes ___ No ___ (31)
(Count a cough with first smoke or on "first going out of doors."
Exclude clearing throat or a single cough.)

Do you usually cough during the day or at night?    Yes ___ No ___ (32)
(Ignore an occasional cough.)

If 'Yes' to either question (31-32):
Do you cough like this on most days for as much as three months a year?    Yes ___ No ___ (33)

†These words are for subjects who work at night

614

**Subpart Z—Toxic and Hazardous Substances**

BREATHLESSNESS

21. If disabled from walking by any condition other than heart or lung disease put "X" in the space and leave questions (22-30) unasked. ☐
  1 ☐ Yes  2 ☐ No
22. Are you ever troubled by shortness of breath, when hurrying on the level or walking up a slight hill?
  1 ☐ Yes  2 ☐ No
If NO, grade is 1. If YES, proceed to next question
23. Do you get short of breath walking with other people at an ordinary pace on the level?  1 ☐ Yes  2 ☐ No
If NO, grade is 2. If YES, proceed to next question
24. Do you have to stop for breath when walking at your own pace on the level?  1 ☐ Yes  2 ☐ No
If NO, grade is 3. If YES, proceed to next question
25. Are you short of breath on washing or dressing?  1 ☐ Yes  2 ☐ No
If NO, grade is 4. If YES, grade is 5.                           Dyspnea Grd. _____

ON MONDAYS:
26.
27. Are you ever troubled by shortness of breath, when hurrying on the level or walking up a slight hill?  1 ☐ Yes  2 ☐ No
If NO, grade is 1. If YES, proceed to next question
28. Do you get short of breath walking with other people at an ordinary pace on the level?  1 ☐ Yes  2 ☐ No
If NO, grade is 2. If YES, proceed to next question
29. Do you have to stop for breath when walking at your own pace on the level?  1 ☐ Yes  2 ☐ No
If NO, grade is 3. If YES, proceed to next question
30. Are you short of breath on washing or dressing?  1 ☐ Yes  2 ☐ No
If NO, grade is 4. If YES, grade is 5
31.                                                              B. Grd. _____

OTHER ILLNESSES AND ALLERGY HISTORY
32. Do you have a heart condition for which you are under a doctor's care?  1 ☐ Yes  2 ☐ No
OTHER ILLNESSES AND ALLERGY HISTORY CONTINUED:
33. Have you ever had asthma?  1 ☐ Yes  2 ☐ No
  If yes, did it begin:  (1) Before age 30 ☐  (2) After age 30 ☐
34. If yes before 30: did you have asthma before ever going to work in a textile mill?  1 ☐ Yes  2 ☐ No
35. Have you ever had hay fever or other allergies (other than above)?  1 ☐ Yes  2 ☐ No

TOBACCO SMOKING
36. Do you smoke?  1 ☐ Yes  2 ☐ No
  Record Yes if regular smoker up to one month ago. (Cigarettes, cigar or pipe)
  If NO to (33).
37. Have you ever smoked? (Cigarettes, cigars, pipe. Record NO if subject has never smoked as much as one cigarette a day, or 1 oz. of tobacco a month, for as long as one year.)  1 ☐ Yes  2 ☐ No
If Yes to (33) or (34): what have you smoked for how many years?
(Write in specific number of years in the appropriate square)

| Years | (1) <5 | (2) 5-9 | (3) 10-14 | (4) 15-19 | (5) 20-24 | (6) 25-29 | (7) 30-34 | (8) 35-39 | (9) (<40) |
|---|---|---|---|---|---|---|---|---|---|
| 38. Cigarettes | | | | | | | | | |
| 39. Pipe | | | | | | | | | |
| 40. Cigars | | | | | | | | | |

41. If cigarettes, how many packs per day?
  Write in number of cigarettes _____
  ☐ Less than 1/2 pack
  ☐ 1/2 pack, but less than 1 pack
  ☐ 1 pack, but less than 1 1/2 packs
  ☐ 1-1/2 packs or more
42. Number of pack years: _____
43. If an ex-smoker (cigarettes, cigar or pipe), how long since you stopped? (Write in number of years.)
  ☐ 0-1 year  ☐ 1-4 years  ☐ 5-9 years  ☐ 10+ years

OCCUPATIONAL HISTORY
Have you ever worked in:
44. A foundry? (As long as one year)  1 ☐ Yes  2 ☐ No
45. Stone or mineral mining, quarrying or processing? (As long as one year)  1 ☐ Yes  2 ☐ No
46. Asbestos milling or processing? (Ever)  1 ☐ Yes  2 ☐ No
47. Cotton or cotton blend mill? (For controls only)  1 ☐ Yes  2 ☐ No
48. Other dusts, fumes or smoke? If yes, specify.  1 ☐ Yes  2 ☐ No
  Type of exposure _____
  Length of exposure _____

---

Do you cough on any particular day of the week?                Yes _____ No _____ (34)
  (1) Mon.  (2) Tues.  (3) Wed.  (4) Thur.  (5) Fri.  (6) Sat.  (7) Sun.
If 'Yes': which day?                                                                    (35)

C. PHLEGM or alternative word to suit local custom.
Do you usually bring up any phlegm from your chest first thing in the morning? (on getting up)†
(Count phlegm with the first smoke or on "first going out of doors. "Exclude phlegm from the nose. Count swallowed phlegm.)  Yes _____ No _____ (36)

Do you usually bring up any phlegm from your chest during the day or at night? (Accept twice or more.)  Yes _____ No _____ (37)

If 'Yes' to either question (36) or (37):
Do you bring up phlegm like this on most days for as much as three months each year?  Yes _____ No _____ (38)

If 'Yes' to question (33) or (38):  (cough)
How long have you had this phlegm?
(Write in number of years)
  (1) ☐ 2 years or less
  (2) ☐ More than 2 years-9 years
  (3) ☐ 10-19 years
  (4) ☐ 20+ years

D. TIGHTNESS
Does your chest ever feel tight or your breathing become difficult?  Yes _____ No _____ (39)
Is your chest tight or your breathing difficult on any particular day of the week? (after a week or 10 days away from the mill)  Yes _____ No _____ (40)
If 'Yes': Which day?
  (3) Mon.  (4) Tues.  (5) Wed.  (6) Thur.  (7) Fri.  (8) Sat.  Sun.   (41)
  (1) Sometimes  (2) Always

If 'Yes' Monday: At what time on Monday does your chest feel tight or your breathing difficult?
  1 ☐ Before entering the mill    2 ☐ After entering the mill  (42)
(Ask only if NO to Question (39)
In the past, has your chest ever been tight or your breathing difficult on any particular day of the week?  Yes _____ No _____ (43)
If 'Yes': Which day?
  (3) Mon.  (4) Tues.  (5) Wed.  (6) Thur.  (7) Fri.  (8) Sat.  Sun.   (44)
  (1) Sometimes  (2) Always

E. TOBACCO SMOKING
*Have you changed your smoking habits since last interview?
If yes, specify what changes.
_____
_____
_____
_____

# APPENDIX C—SPIROMETRY PREDICTION TABLES FOR NORMAL MALES AND FEMALES

## TABLE 1. PREDICTED FVC FOR MALES (KNUDSON, ET AL.: AM. REV. RESPIR. DIS. 1976, 113, 587.)

| HT | AGE 17 | 19 | 21 | 23 | 25 | 27 | 29 | 31 | 33 | 35 | 37 | 39 | 41 |
|---|---|---|---|---|---|---|---|---|---|---|---|---|---|
| 60.0 | 3.44 | 3.59 | 3.75 | 3.91 | 3.72 | 3.66 | 3.61 | 3.55 | 3.49 | 3.43 | 3.37 | 3.32 | 3.26 |
| 60.5 | 3.50 | 3.66 | 3.81 | 3.97 | 3.80 | 3.75 | 3.69 | 3.63 | 3.57 | 3.51 | 3.46 | 3.40 | 3.34 |
| 61.0 | 3.56 | 3.72 | 3.88 | 4.03 | 3.89 | 3.83 | 3.77 | 3.71 | 3.66 | 3.60 | 3.54 | 3.48 | 3.42 |
| 61.5 | 3.63 | 3.78 | 3.94 | 4.10 | 3.97 | 3.91 | 3.85 | 3.80 | 3.74 | 3.68 | 3.62 | 3.56 | 3.51 |
| 62.0 | 3.69 | 3.85 | 4.00 | 4.16 | 4.05 | 3.99 | 3.94 | 3.88 | 3.82 | 3.76 | 3.70 | 3.65 | 3.59 |
| 62.5 | 3.76 | 3.91 | 4.07 | 4.22 | 4.13 | 4.08 | 4.02 | 3.96 | 3.90 | 3.84 | 3.79 | 3.73 | 3.67 |
| 63.0 | 3.82 | 3.97 | 4.13 | 4.29 | 4.22 | 4.16 | 4.10 | 4.04 | 3.99 | 3.93 | 3.87 | 3.81 | 3.75 |
| 63.5 | 3.88 | 4.04 | 4.19 | 4.35 | 4.30 | 4.25 | 4.18 | 4.13 | 4.07 | 4.01 | 3.95 | 3.89 | 3.84 |
| 64.0 | 3.95 | 4.10 | 4.26 | 4.41 | 4.38 | 4.32 | 4.27 | 4.21 | 4.15 | 4.09 | 4.03 | 3.98 | 3.92 |
| 64.5 | 4.01 | 4.17 | 4.32 | 4.48 | 4.46 | 4.41 | 4.35 | 4.29 | 4.23 | 4.17 | 4.12 | 4.06 | 4.00 |
| 65.0 | 4.07 | 4.23 | 4.39 | 4.54 | 4.55 | 4.49 | 4.43 | 4.37 | 4.32 | 4.26 | 4.20 | 4.14 | 4.08 |
| 65.5 | 4.14 | 4.29 | 4.45 | 4.60 | 4.63 | 4.57 | 4.51 | 4.46 | 4.40 | 4.34 | 4.28 | 4.22 | 4.17 |
| 66.0 | 4.20 | 4.36 | 4.51 | 4.67 | 4.71 | 4.65 | 4.60 | 4.54 | 4.48 | 4.42 | 4.36 | 4.31 | 4.25 |
| 66.5 | 4.26 | 4.42 | 4.58 | 4.73 | 4.80 | 4.74 | 4.68 | 4.62 | 4.56 | 4.51 | 4.45 | 4.39 | 4.33 |
| 67.0 | 4.33 | 4.48 | 4.64 | 4.80 | 4.88 | 4.82 | 4.76 | 4.70 | 4.65 | 4.59 | 4.53 | 4.47 | 4.41 |
| 67.5 | 4.39 | 4.55 | 4.70 | 4.86 | 4.96 | 4.90 | 4.84 | 4.79 | 4.73 | 4.67 | 4.61 | 4.55 | 4.50 |
| 68.0 | 4.45 | 4.61 | 4.77 | 4.92 | 5.04 | 4.98 | 4.93 | 4.87 | 4.81 | 4.75 | 4.69 | 4.64 | 4.58 |
| 68.5 | 4.52 | 4.67 | 4.83 | 4.99 | 5.13 | 5.07 | 5.01 | 4.95 | 4.89 | 4.84 | 4.78 | 4.72 | 4.66 |
| 69.0 | 4.58 | 4.74 | 4.89 | 5.05 | 5.21 | 5.15 | 5.09 | 5.03 | 4.98 | 4.92 | 4.86 | 4.80 | 4.74 |
| 69.5 | 4.64 | 4.80 | 4.96 | 5.11 | 5.29 | 5.23 | 5.17 | 5.12 | 5.06 | 5.00 | 4.94 | 4.88 | 4.83 |
| 70.0 | 4.71 | 4.86 | 5.02 | 5.18 | 5.37 | 5.32 | 5.26 | 5.20 | 5.14 | 5.08 | 5.02 | 4.97 | 4.91 |
| 70.5 | 4.77 | 4.93 | 5.08 | 5.24 | 5.46 | 5.40 | 5.34 | 5.28 | 5.22 | 5.17 | 5.11 | 5.05 | 4.99 |
| 71.0 | 4.83 | 4.99 | 5.15 | 5.30 | 5.54 | 5.48 | 5.42 | 5.36 | 5.31 | 5.25 | 5.19 | 5.13 | 5.07 |
| 71.5 | 4.90 | 5.05 | 5.21 | 5.37 | 5.62 | 5.56 | 5.50 | 5.45 | 5.39 | 5.33 | 5.27 | 5.21 | 5.16 |
| 72.0 | 4.96 | 5.12 | 5.27 | 5.43 | 5.70 | 5.65 | 5.59 | 5.53 | 5.47 | 5.41 | 5.36 | 5.30 | 5.24 |
| 72.5 | 5.03 | 5.18 | 5.34 | 5.49 | 5.79 | 5.73 | 5.67 | 5.61 | 5.55 | 5.50 | 5.44 | 5.38 | 5.32 |
| 73.0 | 5.09 | 5.24 | 5.40 | 5.56 | 5.87 | 5.81 | 5.75 | 5.69 | 5.64 | 5.58 | 5.52 | 5.46 | 5.40 |
| 73.5 | 5.15 | 5.31 | 5.46 | 5.62 | 5.95 | 5.89 | 5.83 | 5.78 | 5.70 | 5.66 | 5.60 | 5.54 | 5.49 |
| 74.0 | 5.22 | 5.37 | 5.53 | 5.68 | 6.03 | 5.98 | 5.92 | 5.86 | 5.80 | 5.74 | 5.69 | 5.63 | 5.57 |
| 74.5 | 5.28 | 5.44 | 5.59 | 5.75 | 6.12 | 6.06 | 6.00 | 5.94 | 5.88 | 5.83 | 5.77 | 5.71 | 5.65 |
| 75.0 | 5.34 | 5.50 | 5.65 | 5.81 | 6.20 | 6.14 | 6.08 | 6.02 | 5.97 | 5.91 | 5.85 | 5.79 | 5.73 |
| 75.5 | 5.41 | 5.56 | 5.72 | 5.87 | 6.28 | 6.22 | 6.17 | 6.11 | 6.05 | 5.99 | 5.93 | 5.88 | 5.82 |
| 76.0 | 5.47 | 5.63 | 5.78 | 5.94 | 6.36 | 6.31 | 6.25 | 6.19 | 6.13 | 6.07 | 6.02 | 5.96 | 5.90 |
| 76.5 | 5.53 | 5.69 | 5.85 | 6.00 | 6.45 | 6.39 | 6.33 | 6.27 | 6.21 | 6.16 | 6.10 | 6.04 | 5.98 |
| 77.0 | 5.60 | 5.75 | 5.91 | 6.06 | 6.53 | 6.47 | 6.41 | 6.35 | 6.30 | 6.24 | 6.18 | 6.12 | 6.06 |
| 77.5 | 5.66 | 5.82 | 5.97 | 6.13 | 6.61 | 6.55 | 6.50 | 6.44 | 6.38 | 6.32 | 6.26 | 6.21 | 6.15 |
| 78.0 | 5.72 | 5.88 | 6.04 | 6.19 | 6.69 | 6.64 | 6.58 | 6.52 | 6.46 | 6.40 | 6.35 | 6.29 | 6.23 |
| 78.5 | 5.79 | 5.94 | 6.10 | 6.26 | 6.78 | 6.72 | 6.66 | 6.60 | 6.54 | 6.49 | 6.43 | 6.37 | 6.31 |
| 79.0 | 5.85 | 6.01 | 6.16 | 6.32 | 6.86 | 6.80 | 6.74 | 6.68 | 6.63 | 6.57 | 6.51 | 6.45 | 6.39 |
| 79.5 | 5.91 | 6.07 | 6.23 | 6.38 | 6.94 | 6.88 | 6.83 | 6.77 | 6.71 | 6.65 | 6.59 | 6.54 | 6.48 |
| 80.0 | 5.98 | 6.13 | 6.29 | 6.45 | 7.02 | 6.97 | 6.91 | 6.85 | 6.79 | 6.73 | 6.68 | 6.62 | 6.56 |
| 80.5 | 6.04 | 6.20 | 6.35 | 6.51 | 7.11 | 7.05 | 6.99 | 6.93 | 6.87 | 6.82 | 6.76 | 6.70 | 6.64 |
| 81.0 | 6.10 | 6.26 | 6.42 | 6.57 | 7.19 | 7.13 | 7.07 | 7.02 | 6.96 | 6.90 | 6.84 | 6.78 | 6.73 |
| 81.5 | 6.17 | 6.32 | 6.48 | 6.64 | 7.27 | 7.21 | 7.16 | 7.10 | 7.04 | 6.98 | 6.92 | 6.87 | 6.81 |
| 82.0 | 6.23 | 6.39 | 6.54 | 6.70 | 7.35 | 7.30 | 7.24 | 7.18 | 7.12 | 7.06 | 7.01 | 6.95 | 6.89 |
| 82.5 | 6.30 | 6.45 | 6.61 | 6.76 | 7.44 | 7.38 | 7.32 | 7.26 | 7.20 | 7.15 | 7.09 | 7.03 | 6.97 |
| 83.0 | 6.36 | 6.51 | 6.67 | 6.83 | 7.52 | 7.46 | 7.40 | 7.35 | 7.29 | 7.23 | 7.17 | 7.11 | 7.06 |
| 83.5 | 6.42 | 6.58 | 6.73 | 6.89 | 7.60 | 7.54 | 7.49 | 7.43 | 7.37 | 7.31 | 7.25 | 7.20 | 7.14 |
| 84.0 | 6.49 | 6.64 | 6.80 | 6.95 | 7.68 | 7.63 | 7.57 | 7.51 | 7.45 | 7.39 | 7.34 | 7.28 | 7.22 |
| 84.5 | 6.55 | 6.71 | 6.86 | 7.02 | 7.77 | 7.71 | 7.65 | 7.59 | 7.53 | 7.48 | 7.42 | 7.36 | 7.30 |
| 85.0 | 6.61 | 6.77 | 6.92 | 7.08 | 7.85 | 7.79 | 7.73 | 7.68 | 7.62 | 7.56 | 7.50 | 7.44 | 7.39 |

## TABLE 1. PREDICTED FVC FOR MALES (KNUDSON, ET AL.: AM. REV. RESPIR. DIS. 1976, 113, 587.) (CONTINUED)

| HT | AGE 43 | 45 | 47 | 49 | 51 | 53 | 55 | 57 | 59 | 61 | 63 | 65 |
|---|---|---|---|---|---|---|---|---|---|---|---|---|
| 60.0 | 3.20 | 3.14 | 3.08 | 3.03 | 2.97 | 2.91 | 2.85 | 2.79 | 2.74 | 2.68 | 2.62 | 2.56 |
| 60.5 | 3.28 | 3.22 | 3.17 | 3.11 | 3.05 | 2.99 | 2.93 | 2.88 | 2.82 | 2.76 | 2.70 | 2.64 |
| 61.0 | 3.37 | 3.31 | 3.25 | 3.19 | 3.13 | 3.08 | 3.02 | 2.96 | 2.90 | 2.84 | 2.79 | 2.73 |
| 61.5 | 3.45 | 3.39 | 3.33 | 3.27 | 3.22 | 3.16 | 3.10 | 3.04 | 2.98 | 2.93 | 2.87 | 2.81 |
| 62.0 | 3.53 | 3.47 | 3.41 | 3.36 | 3.30 | 3.24 | 3.18 | 3.12 | 3.07 | 3.01 | 2.95 | 2.89 |
| 62.5 | 3.61 | 3.55 | 3.50 | 3.44 | 3.38 | 3.32 | 3.26 | 3.21 | 3.15 | 3.09 | 3.03 | 2.97 |
| 63.0 | 3.70 | 3.64 | 3.58 | 3.52 | 3.46 | 3.41 | 3.35 | 3.29 | 3.23 | 3.17 | 3.12 | 3.06 |
| 63.5 | 3.78 | 3.72 | 3.66 | 3.60 | 3.55 | 3.49 | 3.43 | 3.37 | 3.31 | 3.26 | 3.20 | 3.14 |
| 64.0 | 3.86 | 3.80 | 3.74 | 3.69 | 3.63 | 3.57 | 3.51 | 3.45 | 3.40 | 3.34 | 3.28 | 3.22 |
| 64.5 | 3.94 | 3.88 | 3.83 | 3.77 | 3.71 | 3.65 | 3.59 | 3.54 | 3.48 | 3.42 | 3.36 | 3.30 |
| 65.0 | 4.03 | 3.97 | 3.91 | 3.85 | 3.79 | 3.74 | 3.68 | 3.62 | 3.56 | 3.50 | 3.45 | 3.39 |
| 65.5 | 4.11 | 4.05 | 3.99 | 3.93 | 3.88 | 3.82 | 3.76 | 3.70 | 3.64 | 3.59 | 3.53 | 3.47 |
| 66.0 | 4.19 | 4.13 | 4.07 | 4.02 | 3.96 | 3.90 | 3.84 | 3.78 | 3.73 | 3.67 | 3.61 | 3.55 |
| 66.5 | 4.27 | 4.22 | 4.16 | 4.10 | 4.04 | 3.98 | 3.93 | 3.87 | 3.81 | 3.75 | 3.69 | 3.64 |
| 67.0 | 4.36 | 4.30 | 4.24 | 4.18 | 4.12 | 4.07 | 4.01 | 3.95 | 3.89 | 3.83 | 3.78 | 3.72 |
| 67.5 | 4.44 | 4.38 | 4.32 | 4.26 | 4.21 | 4.15 | 4.09 | 4.03 | 3.97 | 3.92 | 3.86 | 3.80 |
| 68.0 | 4.52 | 4.46 | 4.40 | 4.35 | 4.29 | 4.23 | 4.17 | 4.11 | 4.06 | 4.00 | 3.94 | 3.88 |
| 68.5 | 4.60 | 4.55 | 4.49 | 4.43 | 4.37 | 4.31 | 4.26 | 4.20 | 4.14 | 4.08 | 4.02 | 3.97 |
| 69.0 | 4.69 | 4.63 | 4.57 | 4.51 | 4.45 | 4.40 | 4.34 | 4.28 | 4.22 | 4.16 | 4.11 | 4.05 |
| 69.5 | 4.77 | 4.71 | 4.65 | 4.59 | 4.54 | 4.48 | 4.42 | 4.36 | 4.30 | 4.25 | 4.19 | 4.13 |
| 70.0 | 4.85 | 4.79 | 4.74 | 4.68 | 4.62 | 4.56 | 4.50 | 4.44 | 4.39 | 4.33 | 4.27 | 4.21 |
| 70.5 | 4.93 | 4.88 | 4.82 | 4.76 | 4.70 | 4.64 | 4.59 | 4.53 | 4.47 | 4.41 | 4.35 | 4.30 |
| 71.0 | 5.02 | 4.96 | 4.90 | 4.84 | 4.78 | 4.73 | 4.67 | 4.61 | 4.55 | 4.49 | 4.44 | 4.38 |
| 71.5 | 5.10 | 5.04 | 4.98 | 4.92 | 4.87 | 4.81 | 4.75 | 4.69 | 4.63 | 4.58 | 4.52 | 4.46 |
| 72.0 | 5.18 | 5.12 | 5.07 | 5.01 | 4.95 | 4.89 | 4.83 | 4.78 | 4.72 | 4.66 | 4.60 | 4.54 |
| 72.5 | 5.26 | 5.21 | 5.15 | 5.09 | 5.03 | 4.97 | 4.92 | 4.86 | 4.80 | 4.74 | 4.68 | 4.63 |
| 73.0 | 5.35 | 5.29 | 5.23 | 5.17 | 5.11 | 5.06 | 5.00 | 4.94 | 4.88 | 4.82 | 4.77 | 4.71 |
| 73.5 | 5.43 | 5.37 | 5.31 | 5.25 | 5.20 | 5.14 | 5.08 | 5.02 | 4.96 | 4.91 | 4.85 | 4.79 |
| 74.0 | 5.51 | 5.45 | 5.40 | 5.34 | 5.28 | 5.22 | 5.16 | 5.11 | 5.05 | 4.99 | 4.93 | 4.87 |
| 74.5 | 5.59 | 5.54 | 5.48 | 5.42 | 5.36 | 5.30 | 5.25 | 5.19 | 5.13 | 5.07 | 5.01 | 4.96 |
| 75.0 | 5.68 | 5.62 | 5.56 | 5.50 | 5.44 | 5.39 | 5.33 | 5.27 | 5.21 | 5.15 | 5.10 | 5.04 |
| 75.5 | 5.76 | 5.70 | 5.64 | 5.59 | 5.53 | 5.47 | 5.41 | 5.35 | 5.30 | 5.24 | 5.18 | 5.12 |
| 76.0 | 5.84 | 5.78 | 5.73 | 5.67 | 5.61 | 5.55 | 5.49 | 5.44 | 5.38 | 5.32 | 5.26 | 5.20 |
| 76.5 | 5.92 | 5.87 | 5.81 | 5.75 | 5.69 | 5.63 | 5.58 | 5.52 | 5.46 | 5.40 | 5.34 | 5.29 |
| 77.0 | 6.01 | 5.95 | 5.89 | 5.83 | 5.77 | 5.72 | 5.66 | 5.60 | 5.54 | 5.48 | 5.43 | 5.37 |
| 77.5 | 6.09 | 6.03 | 5.97 | 5.92 | 5.86 | 5.80 | 5.74 | 5.68 | 5.63 | 5.57 | 5.51 | 5.45 |
| 78.0 | 6.17 | 6.11 | 6.06 | 6.00 | 5.94 | 5.88 | 5.82 | 5.77 | 5.71 | 5.65 | 5.59 | 5.53 |
| 78.5 | 6.25 | 6.20 | 6.14 | 6.08 | 6.02 | 5.96 | 5.91 | 5.85 | 5.79 | 5.73 | 5.67 | 5.62 |
| 79.0 | 6.34 | 6.28 | 6.22 | 6.16 | 6.10 | 6.05 | 5.99 | 5.93 | 5.87 | 5.81 | 5.76 | 5.70 |
| 79.5 | 6.42 | 6.36 | 6.30 | 6.25 | 6.19 | 6.13 | 6.07 | 6.01 | 5.96 | 5.90 | 5.84 | 5.78 |
| 80.0 | 6.50 | 6.44 | 6.39 | 6.33 | 6.27 | 6.21 | 6.15 | 6.10 | 6.04 | 5.98 | 5.92 | 5.86 |
| 80.5 | 6.58 | 6.53 | 6.47 | 6.41 | 6.35 | 6.29 | 6.24 | 6.18 | 6.12 | 6.06 | 6.00 | 5.95 |
| 81.0 | 6.67 | 6.61 | 6.55 | 6.49 | 6.44 | 6.38 | 6.32 | 6.26 | 6.20 | 6.15 | 6.09 | 6.03 |
| 81.5 | 6.75 | 6.69 | 6.63 | 6.58 | 6.52 | 6.46 | 6.40 | 6.34 | 6.29 | 6.23 | 6.17 | 6.11 |
| 82.0 | 6.83 | 6.77 | 6.72 | 6.66 | 6.60 | 6.54 | 6.48 | 6.43 | 6.37 | 6.31 | 6.25 | 6.19 |
| 82.5 | 6.91 | 6.86 | 6.80 | 6.74 | 6.68 | 6.62 | 6.57 | 6.51 | 6.45 | 6.39 | 6.33 | 6.28 |
| 83.0 | 7.00 | 6.94 | 6.88 | 6.82 | 6.77 | 6.71 | 6.65 | 6.59 | 6.53 | 6.48 | 6.42 | 6.36 |
| 83.5 | 7.08 | 7.02 | 6.96 | 6.91 | 6.85 | 6.79 | 6.73 | 6.67 | 6.62 | 6.56 | 6.50 | 6.44 |
| 84.0 | 7.16 | 7.10 | 7.05 | 6.99 | 6.93 | 6.87 | 6.81 | 6.76 | 6.70 | 6.64 | 6.58 | 6.52 |
| 84.5 | 7.24 | 7.19 | 7.13 | 7.07 | 7.01 | 6.95 | 6.90 | 6.84 | 6.78 | 6.72 | 6.66 | 6.61 |
| 85.0 | 7.33 | 7.27 | 7.21 | 7.15 | 7.10 | 7.04 | 6.98 | 6.92 | 6.86 | 6.81 | 6.75 | 6.69 |

## TABLE 2. PREDICTED FEV$^1$ FOR MALES (KNUDSON, ET AL.: AM. REV. RESPIR. DIS. 1976, 113, 587.)

| HT | AGE 17 | 19 | 21 | 23 | 25 | 27 | 29 | 31 | 33 | 35 | 37 | 39 | 41 |
|---|---|---|---|---|---|---|---|---|---|---|---|---|---|
| 60.0 | 2.97 | 3.06 | 3.15 | 3.24 | 3.05 | 2.99 | 2.94 | 2.88 | 2.83 | 2.78 | 2.72 | 2.67 | 2.61 |
| 60.5 | 3.03 | 3.12 | 3.21 | 3.30 | 3.11 | 3.06 | 3.00 | 2.95 | 2.90 | 2.84 | 2.79 | 2.73 | 2.68 |
| 61.0 | 3.08 | 3.17 | 3.26 | 3.35 | 3.18 | 3.12 | 3.07 | 3.02 | 2.96 | 2.91 | 2.85 | 2.80 | 2.75 |
| 61.5 | 3.14 | 3.23 | 3.32 | 3.41 | 3.24 | 3.19 | 3.14 | 3.08 | 3.03 | 2.97 | 2.92 | 2.87 | 2.81 |
| 62.0 | 3.20 | 3.29 | 3.38 | 3.47 | 3.31 | 3.26 | 3.20 | 3.15 | 3.09 | 3.04 | 2.99 | 2.93 | 2.88 |
| 62.5 | 3.26 | 3.35 | 3.44 | 3.53 | 3.38 | 3.32 | 3.27 | 3.22 | 3.16 | 3.11 | 3.05 | 3.00 | 2.95 |
| 63.0 | 3.32 | 3.41 | 3.50 | 3.59 | 3.44 | 3.39 | 3.34 | 3.28 | 3.23 | 3.17 | 3.12 | 3.07 | 3.01 |
| 63.5 | 3.38 | 3.47 | 3.56 | 3.65 | 3.51 | 3.46 | 3.40 | 3.35 | 3.29 | 3.24 | 3.19 | 3.13 | 3.08 |
| 64.0 | 3.43 | 3.52 | 3.61 | 3.70 | 3.58 | 3.52 | 3.47 | 3.41 | 3.36 | 3.31 | 3.25 | 3.20 | 3.14 |
| 64.5 | 3.49 | 3.58 | 3.67 | 3.76 | 3.64 | 3.59 | 3.53 | 3.48 | 3.43 | 3.37 | 3.32 | 3.26 | 3.21 |
| 65.0 | 3.55 | 3.64 | 3.73 | 3.82 | 3.71 | 3.65 | 3.60 | 3.55 | 3.49 | 3.44 | 3.38 | 3.33 | 3.28 |
| 65.5 | 3.61 | 3.70 | 3.79 | 3.88 | 3.77 | 3.72 | 3.67 | 3.61 | 3.56 | 3.50 | 3.45 | 3.40 | 3.34 |
| 66.0 | 3.67 | 3.76 | 3.85 | 3.94 | 3.84 | 3.79 | 3.73 | 3.68 | 3.62 | 3.57 | 3.52 | 3.46 | 3.41 |
| 66.5 | 3.73 | 3.82 | 3.91 | 4.00 | 3.91 | 3.85 | 3.80 | 3.74 | 3.69 | 3.64 | 3.58 | 3.53 | 3.47 |
| 67.0 | 3.79 | 3.88 | 3.97 | 4.06 | 3.97 | 3.92 | 3.86 | 3.81 | 3.76 | 3.70 | 3.65 | 3.59 | 3.54 |
| 67.5 | 3.84 | 3.93 | 4.02 | 4.11 | 4.04 | 3.98 | 3.93 | 3.88 | 3.82 | 3.77 | 3.71 | 3.66 | 3.61 |
| 68.0 | 3.90 | 3.99 | 4.08 | 4.17 | 4.10 | 4.05 | 4.00 | 3.94 | 3.89 | 3.83 | 3.78 | 3.73 | 3.67 |
| 68.5 | 3.96 | 4.05 | 4.14 | 4.23 | 4.17 | 4.12 | 4.06 | 4.01 | 3.95 | 3.90 | 3.85 | 3.79 | 3.74 |
| 69.0 | 4.02 | 4.11 | 4.20 | 4.29 | 4.24 | 4.18 | 4.13 | 4.07 | 4.02 | 3.97 | 3.91 | 3.86 | 3.80 |
| 69.5 | 4.08 | 4.17 | 4.26 | 4.35 | 4.30 | 4.25 | 4.19 | 4.14 | 4.09 | 4.03 | 3.98 | 3.92 | 3.87 |
| 70.0 | 4.14 | 4.23 | 4.32 | 4.41 | 4.37 | 4.31 | 4.26 | 4.21 | 4.15 | 4.10 | 4.04 | 3.99 | 3.94 |
| 70.5 | 4.19 | 4.28 | 4.37 | 4.46 | 4.43 | 4.38 | 4.33 | 4.27 | 4.22 | 4.16 | 4.11 | 4.06 | 4.00 |
| 71.0 | 4.25 | 4.34 | 4.43 | 4.52 | 4.50 | 4.45 | 4.39 | 4.34 | 4.28 | 4.23 | 4.18 | 4.12 | 4.07 |
| 71.5 | 4.31 | 4.40 | 4.49 | 4.58 | 4.57 | 4.51 | 4.46 | 4.40 | 4.35 | 4.30 | 4.24 | 4.19 | 4.13 |
| 72.0 | 4.37 | 4.46 | 4.55 | 4.64 | 4.63 | 4.58 | 4.52 | 4.47 | 4.42 | 4.36 | 4.31 | 4.25 | 4.20 |
| 72.5 | 4.43 | 4.52 | 4.61 | 4.70 | 4.70 | 4.64 | 4.59 | 4.54 | 4.48 | 4.43 | 4.37 | 4.32 | 4.27 |
| 73.0 | 4.49 | 4.58 | 4.67 | 4.76 | 4.76 | 4.71 | 4.66 | 4.60 | 4.55 | 4.49 | 4.44 | 4.39 | 4.33 |
| 73.5 | 4.54 | 4.63 | 4.72 | 4.81 | 4.83 | 4.78 | 4.72 | 4.67 | 4.61 | 4.56 | 4.51 | 4.45 | 4.40 |
| 74.0 | 4.60 | 4.69 | 4.78 | 4.87 | 4.90 | 4.84 | 4.79 | 4.73 | 4.68 | 4.63 | 4.57 | 4.52 | 4.46 |
| 74.5 | 4.66 | 4.75 | 4.84 | 4.93 | 4.96 | 4.91 | 4.85 | 4.80 | 4.75 | 4.69 | 4.64 | 4.58 | 4.53 |
| 75.0 | 4.72 | 4.81 | 4.90 | 4.99 | 5.03 | 4.97 | 4.92 | 4.87 | 4.81 | 4.76 | 4.70 | 4.65 | 4.60 |
| 75.5 | 4.78 | 4.87 | 4.96 | 5.05 | 5.09 | 5.04 | 4.99 | 4.93 | 4.88 | 4.82 | 4.77 | 4.72 | 4.66 |
| 76.0 | 4.84 | 4.93 | 5.02 | 5.11 | 5.16 | 5.11 | 5.05 | 5.00 | 4.94 | 4.89 | 4.84 | 4.78 | 4.73 |
| 76.5 | 4.90 | 4.99 | 5.08 | 5.17 | 5.23 | 5.17 | 5.12 | 5.06 | 5.01 | 4.96 | 4.90 | 4.85 | 4.79 |
| 77.0 | 4.95 | 5.04 | 5.13 | 5.22 | 5.29 | 5.24 | 5.18 | 5.13 | 5.08 | 5.02 | 4.97 | 4.91 | 4.86 |
| 77.5 | 5.01 | 5.10 | 5.19 | 5.28 | 5.36 | 5.30 | 5.25 | 5.20 | 5.14 | 5.09 | 5.03 | 4.98 | 4.93 |
| 78.0 | 5.07 | 5.16 | 5.25 | 5.34 | 5.42 | 5.37 | 5.32 | 5.26 | 5.21 | 5.15 | 5.10 | 5.05 | 4.99 |
| 78.5 | 5.13 | 5.22 | 5.31 | 5.40 | 5.49 | 5.44 | 5.38 | 5.33 | 5.27 | 5.22 | 5.17 | 5.11 | 5.06 |
| 79.0 | 5.19 | 5.28 | 5.37 | 5.46 | 5.56 | 5.50 | 5.45 | 5.39 | 5.34 | 5.29 | 5.23 | 5.18 | 5.12 |
| 79.5 | 5.25 | 5.34 | 5.43 | 5.52 | 5.62 | 5.57 | 5.51 | 5.46 | 5.41 | 5.35 | 5.30 | 5.24 | 5.19 |
| 80.0 | 5.30 | 5.39 | 5.48 | 5.57 | 5.69 | 5.63 | 5.58 | 5.53 | 5.47 | 5.42 | 5.36 | 5.31 | 5.26 |
| 80.5 | 5.36 | 5.45 | 5.54 | 5.63 | 5.75 | 5.70 | 5.65 | 5.59 | 5.54 | 5.48 | 5.43 | 5.38 | 5.32 |
| 81.0 | 5.42 | 5.51 | 5.60 | 5.69 | 5.82 | 5.77 | 5.71 | 5.66 | 5.60 | 5.55 | 5.50 | 5.44 | 5.39 |
| 81.5 | 5.48 | 5.57 | 5.66 | 5.75 | 5.89 | 5.83 | 5.78 | 5.72 | 5.67 | 5.62 | 5.56 | 5.51 | 5.45 |
| 82.0 | 5.54 | 5.63 | 5.72 | 5.81 | 5.95 | 5.90 | 5.84 | 5.79 | 5.74 | 5.68 | 5.63 | 5.57 | 5.52 |
| 82.5 | 5.60 | 5.69 | 5.78 | 5.87 | 6.02 | 5.96 | 5.91 | 5.86 | 5.80 | 5.75 | 5.69 | 6.64 | 5.59 |
| 83.0 | 5.65 | 5.74 | 5.83 | 5.92 | 6.08 | 6.03 | 5.98 | 5.92 | 5.87 | 5.81 | 5.76 | 5.71 | 5.65 |
| 83.5 | 5.71 | 5.80 | 5.90 | 5.98 | 6.15 | 6.10 | 6.04 | 5.99 | 5.93 | 5.88 | 5.83 | 5.77 | 5.72 |
| 84.0 | 5.77 | 5.86 | 5.95 | 6.04 | 6.22 | 6.16 | 6.11 | 6.05 | 6.00 | 5.95 | 5.89 | 5.84 | 5.78 |
| 84.5 | 5.83 | 5.92 | 6.01 | 6.10 | 6.28 | 6.23 | 6.12 | 6.17 | 6.07 | 6.01 | 5.96 | 5.90 | 5.85 |
| 85.0 | 5.89 | 5.98 | 6.07 | 6.16 | 6.36 | 6.29 | 6.24 | 6.19 | 6.13 | 6.06 | 6.02 | 5.97 | 5.92 |

## TABLE 2. PREDICTED FEV$^1$ FOR MALES (KNUDSON, ET AL.: AM. REV. RESPIR. DIS. 1976, 113, 587.) (CONTINUED)

| HT | AGE 43 | 45 | 47 | 49 | 51 | 53 | 55 | 57 | 59 | 61 | 63 | 65 |
|---|---|---|---|---|---|---|---|---|---|---|---|---|
| 60.0 | 2.56 | 2.51 | 2.45 | 2.40 | 2.34 | 2.29 | 2.24 | 2.18 | 2.13 | 2.07 | 2.02 | 1.97 |
| 60.5 | 2.63 | 2.57 | 2.52 | 2.46 | 2.41 | 2.36 | 2.30 | 2.25 | 2.19 | 2.14 | 2.09 | 2.03 |
| 61.0 | 2.69 | 2.64 | 2.58 | 2.53 | 2.48 | 2.42 | 2.37 | 2.31 | 2.26 | 2.21 | 2.15 | 2.10 |
| 61.5 | 2.76 | 2.70 | 2.65 | 2.60 | 2.54 | 2.49 | 2.43 | 2.38 | 2.33 | 2.27 | 2.22 | 2.16 |
| 62.0 | 2.82 | 2.77 | 2.72 | 2.66 | 2.61 | 2.55 | 2.50 | 2.45 | 2.39 | 2.34 | 2.28 | 2.23 |
| 62.5 | 2.89 | 2.84 | 2.78 | 2.73 | 2.68 | 2.62 | 2.57 | 2.51 | 2.46 | 2.41 | 2.35 | 2.30 |
| 63.0 | 2.96 | 2.90 | 2.85 | 2.80 | 2.74 | 2.69 | 2.63 | 2.58 | 2.53 | 2.47 | 2.42 | 2.36 |
| 63.5 | 3.02 | 2.97 | 2.92 | 2.86 | 2.81 | 2.75 | 2.70 | 2.65 | 2.59 | 2.54 | 2.48 | 2.43 |
| 64.0 | 3.09 | 3.04 | 2.98 | 2.93 | 2.87 | 2.82 | 2.77 | 2.71 | 2.66 | 2.60 | 2.55 | 2.50 |
| 64.5 | 3.16 | 3.10 | 3.05 | 2.99 | 2.94 | 2.89 | 2.83 | 2.78 | 2.72 | 2.67 | 2.62 | 2.56 |
| 65.0 | 3.22 | 3.17 | 3.11 | 3.06 | 3.01 | 2.95 | 2.90 | 2.84 | 2.79 | 2.74 | 2.68 | 2.63 |
| 65.5 | 3.29 | 3.23 | 3.18 | 3.13 | 3.07 | 3.02 | 2.96 | 2.91 | 2.86 | 2.80 | 2.75 | 2.69 |
| 66.0 | 3.35 | 3.30 | 3.25 | 3.19 | 3.14 | 3.08 | 3.03 | 2.98 | 2.92 | 2.87 | 2.81 | 2.76 |
| 66.5 | 3.42 | 3.37 | 3.31 | 3.26 | 3.20 | 3.15 | 3.10 | 3.04 | 2.99 | 2.93 | 2.88 | 2.83 |
| 67.0 | 3.49 | 3.43 | 3.38 | 3.32 | 3.27 | 3.22 | 3.16 | 3.11 | 3.05 | 3.00 | 2.95 | 2.89 |
| 67.5 | 3.55 | 3.50 | 3.44 | 3.39 | 3.34 | 3.28 | 3.23 | 3.17 | 3.12 | 3.07 | 3.01 | 2.96 |
| 68.0 | 3.62 | 3.56 | 3.51 | 3.46 | 3.40 | 3.35 | 3.29 | 3.24 | 3.19 | 3.13 | 3.08 | 3.02 |
| 68.5 | 3.68 | 3.63 | 3.58 | 3.52 | 3.47 | 3.41 | 3.36 | 3.31 | 3.25 | 3.20 | 3.14 | 3.09 |
| 69.0 | 3.75 | 3.70 | 3.64 | 3.59 | 3.53 | 3.48 | 3.43 | 3.37 | 3.32 | 3.26 | 3.21 | 3.16 |
| 69.5 | 3.82 | 3.76 | 3.71 | 3.65 | 3.60 | 3.55 | 3.49 | 3.44 | 3.38 | 3.33 | 3.28 | 3.22 |
| 70.0 | 3.88 | 3.83 | 3.77 | 3.72 | 3.67 | 3.61 | 3.56 | 3.50 | 3.45 | 3.40 | 3.34 | 3.29 |
| 70.5 | 3.95 | 3.89 | 3.84 | 3.79 | 3.73 | 3.68 | 3.62 | 3.57 | 3.52 | 3.46 | 3.41 | 3.35 |
| 71.0 | 4.01 | 3.96 | 3.91 | 3.85 | 3.80 | 3.74 | 3.69 | 3.64 | 3.58 | 3.53 | 3.47 | 3.42 |
| 71.5 | 4.08 | 4.03 | 3.97 | 3.92 | 3.86 | 3.81 | 3.76 | 3.70 | 3.65 | 3.59 | 3.54 | 3.49 |
| 72.0 | 4.15 | 4.09 | 4.04 | 3.98 | 3.93 | 3.88 | 3.82 | 3.77 | 3.71 | 3.66 | 3.61 | 3.55 |
| 72.5 | 4.21 | 4.16 | 4.10 | 4.05 | 4.00 | 3.94 | 3.89 | 3.83 | 3.78 | 3.73 | 3.67 | 3.62 |
| 73.0 | 4.28 | 4.22 | 4.17 | 4.12 | 4.06 | 4.01 | 3.95 | 3.90 | 3.85 | 3.79 | 3.74 | 3.68 |
| 73.5 | 4.34 | 4.29 | 4.24 | 4.18 | 4.13 | 4.07 | 4.02 | 3.97 | 3.91 | 3.86 | 3.80 | 3.75 |
| 74.0 | 4.41 | 4.36 | 4.30 | 4.25 | 4.19 | 4.14 | 4.09 | 4.03 | 3.98 | 3.92 | 3.87 | 3.82 |
| 74.5 | 4.48 | 4.42 | 4.37 | 4.31 | 4.26 | 4.21 | 4.15 | 4.10 | 4.04 | 3.99 | 3.94 | 3.88 |
| 75.0 | 4.54 | 4.49 | 4.43 | 4.38 | 4.33 | 4.27 | 4.22 | 4.16 | 4.11 | 4.06 | 4.00 | 3.95 |
| 75.5 | 4.61 | 4.55 | 4.50 | 4.45 | 4.39 | 4.34 | 4.28 | 4.23 | 4.18 | 4.12 | 4.07 | 4.01 |
| 76.0 | 4.67 | 4.62 | 4.57 | 4.51 | 4.46 | 4.40 | 4.35 | 4.30 | 4.24 | 4.19 | 4.13 | 4.08 |
| 76.5 | 4.74 | 4.69 | 4.63 | 4.58 | 4.52 | 4.47 | 4.42 | 4.36 | 4.31 | 4.25 | 4.20 | 4.15 |
| 77.0 | 4.81 | 4.75 | 4.70 | 4.64 | 4.59 | 4.54 | 4.48 | 4.43 | 4.37 | 4.32 | 4.27 | 4.21 |
| 77.5 | 4.87 | 4.82 | 4.76 | 4.71 | 4.66 | 4.60 | 4.55 | 4.49 | 4.44 | 4.39 | 4.33 | 4.28 |
| 78.0 | 4.94 | 4.88 | 4.83 | 4.78 | 4.72 | 4.67 | 4.61 | 4.56 | 4.51 | 4.45 | 4.40 | 4.34 |
| 78.5 | 5.00 | 4.95 | 4.90 | 4.84 | 4.79 | 4.73 | 4.68 | 4.63 | 4.57 | 4.52 | 4.46 | 4.41 |
| 79.0 | 5.07 | 5.02 | 4.96 | 4.91 | 4.85 | 4.80 | 4.75 | 4.69 | 4.64 | 4.58 | 4.53 | 4.48 |
| 79.5 | 5.14 | 5.08 | 5.03 | 4.97 | 4.92 | 4.87 | 4.81 | 4.76 | 4.70 | 4.65 | 4.60 | 4.54 |
| 80.0 | 5.20 | 5.15 | 5.09 | 5.04 | 4.99 | 4.93 | 4.88 | 4.82 | 4.77 | 4.72 | 4.66 | 4.61 |
| 80.5 | 5.27 | 5.21 | 5.16 | 5.11 | 5.05 | 5.00 | 4.94 | 4.89 | 4.84 | 4.78 | 4.73 | 4.67 |
| 81.0 | 5.33 | 5.28 | 5.23 | 5.17 | 5.12 | 5.06 | 5.01 | 4.96 | 4.90 | 4.85 | 4.79 | 4.74 |
| 81.5 | 5.40 | 5.35 | 5.29 | 5.24 | 5.18 | 5.13 | 5.08 | 5.02 | 4.97 | 4.91 | 4.86 | 4.81 |
| 82.0 | 5.47 | 5.41 | 5.36 | 5.30 | 5.25 | 5.20 | 5.14 | 5.09 | 5.03 | 4.98 | 4.93 | 4.87 |
| 82.5 | 5.53 | 5.48 | 5.42 | 5.37 | 5.32 | 5.26 | 5.21 | 5.15 | 5.10 | 5.05 | 4.99 | 4.94 |
| 83.0 | 5.60 | 5.54 | 5.49 | 5.44 | 5.38 | 5.33 | 5.27 | 5.22 | 5.17 | 5.11 | 5.06 | 5.00 |
| 83.5 | 5.66 | 5.61 | 5.56 | 5.50 | 5.45 | 5.39 | 5.34 | 5.29 | 5.23 | 5.18 | 5.12 | 5.07 |
| 84.0 | 5.73 | 5.68 | 5.62 | 5.57 | 5.51 | 5.46 | 5.41 | 5.35 | 5.30 | 5.24 | 5.19 | 5.14 |
| 84.5 | 5.80 | 5.74 | 5.69 | 5.63 | 5.58 | 5.53 | 5.47 | 5.42 | 5.36 | 5.31 | 5.26 | 5.20 |
| 85.0 | 5.86 | 5.81 | 5.75 | 5.70 | 5.65 | 5.59 | 5.54 | 5.58 | 5.43 | 5.38 | 5.32 | 5.27 |

## TABLE 3. PREDICTED FVC FOR FEMALES (KNUDSON, ET AL.: AM. REV. RESPIR. DIS. 1976, 113, 587.)

| HT | AGE 17 | 19 | 21 | 23 | 25 | 27 | 29 | 31 | 33 | 35 | 37 | 39 | 41 |
|---|---|---|---|---|---|---|---|---|---|---|---|---|---|
| 52.0 | 2.45 | 2.64 | 2.65 | 2.61 | 2.56 | 2.52 | 2.47 | 2.43 | 2.39 | 2.34 | 2.30 | 2.25 | 2.21 |
| 52.5 | 2.50 | 2.68 | 2.70 | 2.65 | 2.61 | 2.57 | 2.52 | 2.48 | 2.43 | 2.39 | 2.35 | 2.30 | 2.26 |
| 53.0 | 2.54 | 2.72 | 2.74 | 2.70 | 2.66 | 2.61 | 2.57 | 2.52 | 2.48 | 2.44 | 2.39 | 2.35 | 2.30 |
| 53.5 | 2.58 | 2.76 | 2.79 | 2.75 | 2.70 | 2.66 | 2.62 | 2.57 | 2.53 | 2.48 | 2.44 | 2.40 | 2.35 |
| 54.0 | 2.62 | 2.81 | 2.84 | 2.79 | 2.75 | 2.71 | 2.66 | 2.62 | 2.57 | 2.53 | 2.49 | 2.44 | 2.40 |
| 54.5 | 2.66 | 2.85 | 2.89 | 2.84 | 2.80 | 2.75 | 2.71 | 2.67 | 2.62 | 2.58 | 2.53 | 2.49 | 2.45 |
| 55.0 | 2.71 | 2.89 | 2.93 | 2.89 | 2.84 | 2.80 | 2.76 | 2.71 | 2.67 | 2.62 | 2.58 | 2.54 | 2.49 |
| 55.5 | 2.75 | 2.93 | 2.98 | 2.94 | 2.89 | 2.85 | 2.80 | 2.76 | 2.72 | 2.67 | 2.63 | 2.58 | 2.54 |
| 56.0 | 2.79 | 2.97 | 3.03 | 2.98 | 2.94 | 2.89 | 2.85 | 2.81 | 2.76 | 2.72 | 2.67 | 2.63 | 2.59 |
| 56.5 | 2.83 | 3.01 | 3.07 | 3.03 | 2.99 | 2.94 | 2.90 | 2.85 | 2.81 | 2.77 | 2.72 | 2.68 | 2.63 |
| 57.0 | 2.87 | 3.06 | 3.12 | 3.08 | 3.03 | 2.99 | 2.94 | 2.90 | 2.86 | 2.81 | 2.77 | 2.72 | 2.68 |
| 57.5 | 2.91 | 3.10 | 3.17 | 3.12 | 3.08 | 3.04 | 2.99 | 2.95 | 2.90 | 2.86 | 2.82 | 2.77 | 2.73 |
| 58.0 | 2.96 | 3.14 | 3.21 | 3.17 | 3.13 | 3.08 | 3.04 | 2.99 | 2.95 | 2.91 | 2.86 | 2.82 | 2.77 |
| 58.5 | 3.00 | 3.18 | 3.26 | 3.22 | 3.17 | 3.13 | 3.09 | 3.04 | 3.00 | 2.95 | 2.91 | 2.87 | 2.82 |
| 59.0 | 3.04 | 3.22 | 3.31 | 3.26 | 3.22 | 3.18 | 3.13 | 3.09 | 3.04 | 3.00 | 2.96 | 2.91 | 2.87 |
| 59.5 | 3.08 | 3.27 | 3.36 | 3.31 | 3.27 | 3.22 | 3.18 | 3.14 | 3.09 | 3.05 | 3.00 | 2.96 | 2.92 |
| 60.0 | 3.12 | 3.31 | 3.40 | 3.36 | 3.31 | 3.27 | 3.23 | 3.18 | 3.14 | 3.09 | 3.05 | 3.01 | 2.96 |
| 60.5 | 3.17 | 3.35 | 3.45 | 3.41 | 3.36 | 3.32 | 3.27 | 3.23 | 3.19 | 3.14 | 3.10 | 3.05 | 3.01 |
| 61.0 | 3.21 | 3.39 | 3.50 | 3.45 | 3.41 | 3.36 | 3.32 | 3.28 | 3.23 | 3.19 | 3.14 | 3.10 | 3.06 |
| 61.5 | 3.25 | 3.43 | 3.54 | 3.50 | 3.46 | 3.41 | 3.37 | 3.32 | 3.28 | 3.24 | 3.19 | 3.15 | 3.10 |
| 62.0 | 3.29 | 3.48 | 3.59 | 3.55 | 3.50 | 3.46 | 3.41 | 3.37 | 3.33 | 3.28 | 3.24 | 3.19 | 3.15 |
| 62.5 | 3.33 | 3.52 | 3.64 | 3.59 | 3.55 | 3.51 | 3.46 | 3.42 | 3.37 | 3.33 | 3.29 | 3.24 | 3.20 |
| 63.0 | 3.38 | 3.56 | 3.68 | 3.64 | 3.60 | 3.55 | 3.51 | 3.46 | 3.42 | 3.38 | 3.33 | 3.29 | 3.24 |
| 63.5 | 3.42 | 3.60 | 3.73 | 3.69 | 3.64 | 3.60 | 3.56 | 3.51 | 3.47 | 3.42 | 3.38 | 3.34 | 3.29 |
| 64.0 | 3.46 | 3.64 | 3.78 | 3.73 | 3.69 | 3.65 | 3.60 | 3.56 | 3.51 | 3.47 | 3.43 | 3.38 | 3.34 |
| 64.5 | 3.50 | 3.69 | 3.83 | 3.78 | 3.74 | 3.69 | 3.65 | 3.61 | 3.56 | 3.52 | 3.47 | 3.43 | 3.39 |
| 65.0 | 3.54 | 3.73 | 3.87 | 3.83 | 3.78 | 3.74 | 3.70 | 3.65 | 3.61 | 3.56 | 3.52 | 3.48 | 3.43 |
| 65.5 | 3.59 | 3.77 | 3.92 | 3.88 | 3.83 | 3.79 | 3.74 | 3.70 | 3.66 | 3.61 | 3.57 | 3.52 | 3.48 |
| 66.0 | 3.63 | 3.81 | 3.97 | 3.92 | 3.88 | 3.83 | 3.79 | 3.75 | 3.70 | 3.66 | 3.61 | 3.57 | 3.53 |
| 66.5 | 3.67 | 3.85 | 4.01 | 3.97 | 3.93 | 3.88 | 3.84 | 3.79 | 3.75 | 3.71 | 3.66 | 3.62 | 3.57 |
| 67.0 | 3.71 | 3.89 | 4.06 | 4.02 | 3.97 | 3.93 | 3.88 | 3.84 | 3.80 | 3.75 | 3.71 | 3.66 | 3.62 |
| 67.5 | 3.75 | 3.94 | 4.11 | 4.06 | 4.02 | 3.98 | 3.93 | 3.89 | 3.84 | 3.80 | 3.76 | 3.71 | 3.67 |
| 68.0 | 3.79 | 3.98 | 4.15 | 4.11 | 4.07 | 4.02 | 3.98 | 3.93 | 3.89 | 3.85 | 3.80 | 3.76 | 3.71 |
| 68.5 | 3.84 | 4.02 | 4.20 | 4.16 | 4.11 | 4.07 | 4.03 | 3.98 | 3.94 | 3.89 | 3.85 | 3.81 | 3.76 |
| 69.0 | 3.88 | 4.06 | 4.25 | 4.20 | 4.16 | 4.12 | 4.07 | 4.03 | 3.98 | 3.94 | 3.90 | 3.85 | 3.81 |
| 69.5 | 3.92 | 4.10 | 4.30 | 4.29 | 4.21 | 4.16 | 4.12 | 4.08 | 4.03 | 3.99 | 3.94 | 3.90 | 3.86 |
| 70.0 | 3.96 | 4.15 | 4.34 | 4.30 | 4.25 | 4.21 | 4.17 | 4.12 | 4.08 | 4.03 | 3.99 | 3.95 | 3.90 |
| 70.5 | 4.00 | 4.19 | 4.39 | 4.35 | 4.30 | 4.26 | 4.21 | 4.17 | 4.13 | 4.08 | 4.04 | 3.99 | 3.95 |
| 71.0 | 4.05 | 4.23 | 4.44 | 4.39 | 4.35 | 4.30 | 4.26 | 4.22 | 4.17 | 4.13 | 4.08 | 4.04 | 4.00 |
| 71.5 | 4.09 | 4.27 | 4.48 | 4.44 | 4.40 | 4.35 | 4.31 | 4.26 | 4.22 | 4.18 | 4.13 | 4.09 | 4.04 |
| 72.0 | 4.13 | 4.31 | 4.53 | 4.49 | 4.44 | 4.40 | 4.35 | 4.31 | 4.27 | 4.22 | 4.18 | 4.13 | 4.09 |
| 72.5 | 4.17 | 4.36 | 4.58 | 4.53 | 4.49 | 4.45 | 4.40 | 4.36 | 4.31 | 4.27 | 4.23 | 4.18 | 4.14 |
| 73.0 | 4.21 | 4.40 | 4.62 | 4.58 | 4.54 | 4.49 | 4.45 | 4.40 | 4.36 | 4.32 | 4.27 | 4.23 | 4.18 |
| 73.5 | 4.26 | 4.44 | 4.67 | 4.63 | 4.50 | 4.54 | 4.50 | 4.45 | 4.41 | 4.36 | 4.32 | 4.28 | 4.23 |
| 74.0 | 4.30 | 4.48 | 4.72 | 4.67 | 4.63 | 4.59 | 4.54 | 4.50 | 4.45 | 4.41 | 4.37 | 4.32 | 4.28 |
| 74.5 | 4.34 | 4.52 | 4.77 | 4.72 | 4.68 | 4.63 | 4.59 | 4.55 | 4.50 | 4.46 | 4.41 | 4.37 | 4.33 |
| 75.0 | 4.38 | 4.57 | 4.81 | 4.77 | 4.72 | 4.68 | 4.64 | 4.59 | 4.55 | 4.50 | 4.46 | 4.42 | 4.37 |
| 75.5 | 4.42 | 4.61 | 4.86 | 4.82 | 4.77 | 4.73 | 4.68 | 4.64 | 4.60 | 4.55 | 4.51 | 4.46 | 4.42 |
| 76.0 | 4.47 | 4.65 | 4.91 | 4.86 | 4.82 | 4.77 | 4.73 | 4.69 | 4.64 | 4.60 | 4.55 | 4.51 | 4.47 |
| 76.5 | 4.51 | 4.69 | 4.95 | 4.91 | 4.87 | 4.82 | 4.78 | 4.73 | 4.69 | 4.65 | 4.60 | 4.56 | 4.51 |
| 77.0 | 4.55 | 4.73 | 5.00 | 4.96 | 4.91 | 4.87 | 4.82 | 4.78 | 4.74 | 4.69 | 4.65 | 4.60 | 4.56 |

## TABLE 3. PREDICTED FVC FOR FEMALES (KNUDSON, ET AL.: AM. REV. RESPIR. DIS. 1976, 113, 587.) (CONTINUED)

| HT | AGE 43 | 45 | 47 | 49 | 51 | 53 | 55 | 57 | 59 | 61 | 63 | 65 |
|---|---|---|---|---|---|---|---|---|---|---|---|---|
| 52.0 | 2.17 | 2.12 | 2.08 | 2.03 | 1.99 | 1.95 | 1.90 | 1.86 | 1.81 | 1.77 | 1.73 | 1.68 |
| 52.5 | 2.21 | 2.17 | 2.13 | 2.08 | 2.04 | 1.99 | 1.95 | 1.91 | 1.86 | 1.82 | 1.77 | 1.73 |
| 53.0 | 2.26 | 2.22 | 2.17 | 2.13 | 2.08 | 2.04 | 2.00 | 1.95 | 1.91 | 1.86 | 1.82 | 1.78 |
| 53.5 | 2.31 | 2.26 | 2.22 | 2.18 | 2.13 | 2.09 | 2.04 | 2.00 | 1.96 | 1.91 | 1.87 | 1.82 |
| 54.0 | 2.35 | 2.31 | 2.27 | 2.22 | 2.18 | 2.13 | 2.09 | 2.05 | 2.00 | 1.96 | 1.91 | 1.87 |
| 54.5 | 2.40 | 2.36 | 2.31 | 2.27 | 2.23 | 2.18 | 2.14 | 2.09 | 2.05 | 2.01 | 1.96 | 1.92 |
| 55.0 | 2.45 | 2.40 | 2.36 | 2.32 | 2.27 | 2.23 | 2.18 | 2.14 | 2.10 | 2.05 | 2.01 | 1.96 |
| 55.5 | 2.50 | 2.45 | 2.41 | 2.36 | 2.32 | 2.28 | 2.23 | 2.19 | 2.14 | 2.10 | 2.06 | 2.01 |
| 56.0 | 2.54 | 2.50 | 2.45 | 2.41 | 2.37 | 2.32 | 2.28 | 2.23 | 2.19 | 2.15 | 2.10 | 2.06 |
| 56.5 | 2.59 | 2.55 | 2.50 | 2.46 | 2.41 | 2.37 | 2.33 | 2.28 | 2.24 | 2.19 | 2.15 | 2.11 |
| 57.0 | 2.64 | 2.59 | 2.55 | 2.50 | 2.46 | 2.42 | 2.37 | 2.33 | 2.28 | 2.24 | 2.20 | 2.15 |
| 57.5 | 2.68 | 2.64 | 2.60 | 2.55 | 2.51 | 2.46 | 2.42 | 2.38 | 2.33 | 2.29 | 2.24 | 2.20 |
| 58.0 | 2.73 | 2.69 | 2.64 | 2.60 | 2.55 | 2.51 | 2.47 | 2.42 | 2.38 | 2.33 | 2.29 | 2.25 |
| 58.5 | 2.78 | 2.73 | 2.69 | 2.65 | 2.60 | 2.56 | 2.51 | 2.47 | 2.43 | 2.38 | 2.34 | 2.29 |
| 59.0 | 2.82 | 2.78 | 2.74 | 2.69 | 2.65 | 2.60 | 2.56 | 2.52 | 2.47 | 2.43 | 2.38 | 2.34 |
| 59.5 | 2.87 | 2.83 | 2.78 | 2.74 | 2.70 | 2.65 | 2.61 | 2.56 | 2.52 | 2.48 | 2.43 | 2.39 |
| 60.0 | 2.92 | 2.87 | 2.83 | 2.79 | 2.74 | 2.70 | 2.65 | 2.61 | 2.57 | 2.52 | 2.48 | 2.43 |
| 60.5 | 2.97 | 2.92 | 2.88 | 2.83 | 2.79 | 2.75 | 2.70 | 2.66 | 2.61 | 2.57 | 2.53 | 2.48 |
| 61.0 | 3.01 | 2.97 | 2.92 | 2.88 | 2.84 | 2.79 | 2.75 | 2.70 | 2.66 | 2.62 | 2.57 | 2.53 |
| 61.5 | 3.06 | 3.02 | 2.97 | 2.93 | 2.88 | 2.84 | 2.80 | 2.75 | 2.71 | 2.66 | 2.62 | 2.58 |
| 62.0 | 3.11 | 3.06 | 3.02 | 2.97 | 2.93 | 2.89 | 2.84 | 2.80 | 2.75 | 2.71 | 2.67 | 2.62 |
| 62.5 | 3.15 | 3.11 | 3.07 | 3.02 | 2.98 | 2.93 | 2.89 | 2.85 | 2.80 | 2.76 | 2.71 | 2.67 |
| 63.0 | 3.20 | 3.16 | 3.11 | 3.07 | 3.02 | 2.98 | 2.94 | 2.89 | 2.85 | 2.80 | 2.76 | 2.72 |
| 63.5 | 3.25 | 3.20 | 3.16 | 3.12 | 3.07 | 3.03 | 2.98 | 2.94 | 2.90 | 2.85 | 2.81 | 2.76 |
| 64.0 | 3.29 | 3.25 | 3.21 | 3.16 | 3.12 | 3.07 | 3.03 | 2.99 | 2.94 | 2.90 | 2.85 | 2.81 |
| 64.5 | 3.34 | 3.30 | 3.25 | 3.21 | 3.17 | 3.12 | 3.08 | 3.03 | 2.99 | 2.95 | 2.90 | 2.86 |
| 65.0 | 3.39 | 3.34 | 3.30 | 3.26 | 3.21 | 3.17 | 3.12 | 3.08 | 3.04 | 2.99 | 2.95 | 2.90 |
| 65.5 | 3.44 | 3.39 | 3.35 | 3.30 | 3.26 | 3.22 | 3.17 | 3.13 | 3.08 | 3.04 | 3.00 | 2.95 |
| 66.0 | 3.48 | 3.44 | 3.39 | 3.35 | 3.31 | 3.26 | 3.22 | 3.17 | 3.13 | 3.09 | 3.04 | 3.00 |
| 66.5 | 3.53 | 3.49 | 3.44 | 3.40 | 3.35 | 3.31 | 3.27 | 3.22 | 3.18 | 3.13 | 3.09 | 3.05 |
| 67.0 | 3.58 | 3.53 | 3.49 | 3.44 | 3.40 | 3.36 | 3.31 | 3.27 | 3.22 | 3.18 | 3.14 | 3.09 |
| 67.5 | 3.62 | 3.58 | 3.54 | 3.49 | 3.45 | 3.40 | 3.36 | 3.32 | 3.27 | 3.23 | 3.18 | 3.14 |
| 68.0 | 3.67 | 3.63 | 3.58 | 3.54 | 3.49 | 3.45 | 3.41 | 3.36 | 3.32 | 3.27 | 3.23 | 3.19 |
| 68.5 | 3.72 | 3.67 | 3.63 | 3.59 | 3.54 | 3.50 | 3.45 | 3.41 | 3.37 | 3.32 | 3.28 | 3.23 |
| 69.0 | 3.76 | 3.72 | 3.68 | 3.63 | 3.59 | 3.54 | 3.50 | 3.46 | 3.41 | 3.37 | 3.32 | 3.28 |
| 69.5 | 3.81 | 3.77 | 3.72 | 3.68 | 3.64 | 3.59 | 3.55 | 3.50 | 3.46 | 3.42 | 3.37 | 3.33 |
| 70.0 | 3.86 | 3.81 | 3.77 | 3.73 | 3.68 | 3.64 | 3.59 | 3.55 | 3.51 | 3.46 | 3.42 | 3.37 |
| 70.5 | 3.91 | 3.86 | 3.82 | 3.77 | 3.73 | 3.69 | 3.64 | 3.60 | 3.55 | 3.51 | 3.47 | 3.42 |
| 71.0 | 3.95 | 3.91 | 3.86 | 3.82 | 3.78 | 3.73 | 3.69 | 3.64 | 3.60 | 3.56 | 3.51 | 3.47 |
| 71.5 | 4.00 | 3.96 | 3.91 | 3.87 | 3.82 | 3.78 | 3.74 | 3.69 | 3.65 | 3.60 | 3.56 | 3.52 |
| 72.0 | 4.05 | 4.00 | 3.96 | 3.91 | 3.87 | 3.83 | 3.78 | 3.74 | 3.69 | 3.65 | 3.61 | 3.56 |
| 72.5 | 4.09 | 4.05 | 4.01 | 3.96 | 3.92 | 3.87 | 3.83 | 3.79 | 3.74 | 3.70 | 3.65 | 3.61 |
| 73.0 | 4.14 | 4.10 | 4.05 | 4.01 | 3.96 | 3.92 | 3.88 | 3.83 | 3.79 | 3.74 | 3.70 | 3.66 |
| 73.5 | 4.19 | 4.14 | 4.10 | 4.06 | 4.01 | 3.97 | 3.92 | 3.88 | 3.84 | 3.79 | 3.75 | 3.70 |
| 74.0 | 4.23 | 4.19 | 4.15 | 4.10 | 4.06 | 4.01 | 3.97 | 3.93 | 3.88 | 3.84 | 3.79 | 3.75 |
| 74.5 | 4.28 | 4.24 | 4.19 | 4.15 | 4.11 | 4.06 | 4.02 | 3.97 | 3.93 | 3.89 | 3.84 | 3.80 |
| 75.0 | 4.33 | 4.28 | 4.24 | 4.20 | 4.15 | 4.11 | 4.06 | 4.02 | 3.98 | 3.93 | 3.89 | 3.84 |
| 75.5 | 4.38 | 4.33 | 4.29 | 4.24 | 4.20 | 4.16 | 4.11 | 4.07 | 4.02 | 3.98 | 3.94 | 3.89 |
| 76.0 | 4.42 | 4.38 | 4.33 | 4.29 | 4.25 | 4.20 | 4.16 | 4.11 | 4.07 | 4.03 | 3.98 | 3.94 |
| 76.5 | 4.47 | 4.43 | 4.38 | 4.34 | 4.29 | 4.25 | 4.21 | 4.16 | 4.12 | 4.07 | 4.03 | 3.99 |
| 77.0 | 4.52 | 4.47 | 4.43 | 4.38 | 4.34 | 4.30 | 4.25 | 4.21 | 4.16 | 4.12 | 4.08 | 4.03 |

## TABLE 4. PREDICTED FEV¹ FOR FEMALES (KNUDSON, ET AL.: AM. REV. RESPIR. DIS. 1976, 113, 587.)

| HT | AGE 17 | 19 | 21 | 23 | 25 | 27 | 29 | 31 | 33 | 35 | 37 | 39 | 41 |
|---|---|---|---|---|---|---|---|---|---|---|---|---|---|
| 52.0 | 2.31 | 2.48 | 2.33 | 2.29 | 2.25 | 2.21 | 2.16 | 2.12 | 2.08 | 2.04 | 2.00 | 1.95 | 1.91 |
| 52.5 | 2.34 | 2.51 | 2.37 | 2.32 | 2.28 | 2.24 | 2.20 | 2.16 | 2.11 | 2.07 | 2.03 | 1.99 | 1.95 |
| 53.0 | 2.38 | 2.55 | 2.40 | 2.36 | 2.32 | 2.27 | 2.23 | 2.19 | 2.15 | 2.11 | 2.06 | 2.02 | 1.98 |
| 53.5 | 2.41 | 2.58 | 2.43 | 2.39 | 2.35 | 2.31 | 2.27 | 2.22 | 2.18 | 2.14 | 2.10 | 2.06 | 2.01 |
| 54.0 | 2.45 | 2.62 | 2.47 | 2.43 | 2.38 | 2.34 | 2.30 | 2.26 | 2.22 | 2.17 | 2.13 | 2.09 | 2.05 |
| 54.5 | 2.48 | 2.65 | 2.50 | 2.46 | 2.42 | 2.38 | 2.33 | 2.29 | 2.25 | 2.21 | 2.17 | 2.12 | 2.08 |
| 55.0 | 2.51 | 2.68 | 2.54 | 2.49 | 2.45 | 2.41 | 2.37 | 2.33 | 2.28 | 2.24 | 2.20 | 2.16 | 2.12 |
| 55.5 | 2.55 | 2.72 | 2.57 | 2.53 | 2.49 | 2.45 | 2.40 | 2.36 | 2.32 | 2.28 | 2.24 | 2.19 | 2.15 |
| 56.0 | 2.58 | 2.75 | 2.61 | 2.56 | 2.52 | 2.48 | 2.44 | 2.40 | 2.35 | 2.31 | 2.27 | 2.23 | 2.19 |
| 56.5 | 2.62 | 2.79 | 2.64 | 2.60 | 2.56 | 2.51 | 2.47 | 2.43 | 2.39 | 2.35 | 2.30 | 2.26 | 2.22 |
| 57.0 | 2.65 | 2.82 | 2.67 | 2.63 | 2.59 | 2.55 | 2.51 | 2.46 | 2.42 | 2.38 | 2.34 | 2.30 | 2.25 |
| 57.5 | 2.69 | 2.86 | 2.71 | 2.67 | 2.62 | 2.58 | 2.54 | 2.50 | 2.46 | 2.41 | 2.37 | 2.33 | 2.29 |
| 58.0 | 2.72 | 2.89 | 2.74 | 2.70 | 2.66 | 2.62 | 2.57 | 2.53 | 2.49 | 2.45 | 2.41 | 2.36 | 2.32 |
| 58.5 | 2.75 | 2.92 | 2.78 | 2.73 | 2.69 | 2.65 | 2.61 | 2.57 | 2.52 | 2.48 | 2.44 | 2.40 | 2.36 |
| 59.0 | 2.79 | 2.96 | 2.81 | 2.77 | 2.73 | 2.69 | 2.64 | 2.60 | 2.56 | 2.52 | 2.48 | 2.43 | 2.39 |
| 59.5 | 2.82 | 2.99 | 2.85 | 2.80 | 2.76 | 2.72 | 2.68 | 2.64 | 2.59 | 2.55 | 2.51 | 2.47 | 2.43 |
| 60.0 | 2.86 | 3.03 | 2.88 | 2.84 | 2.80 | 2.75 | 2.71 | 2.67 | 2.63 | 2.59 | 2.54 | 2.50 | 2.46 |
| 60.5 | 2.89 | 3.06 | 2.91 | 2.87 | 2.83 | 2.79 | 2.75 | 2.70 | 2.66 | 2.62 | 2.58 | 2.54 | 2.49 |
| 61.0 | 2.93 | 3.10 | 2.95 | 2.91 | 2.86 | 2.82 | 2.78 | 2.74 | 2.70 | 2.65 | 2.61 | 2.57 | 2.53 |
| 61.5 | 2.96 | 3.13 | 2.98 | 2.94 | 2.90 | 2.86 | 2.81 | 2.77 | 2.73 | 2.69 | 2.65 | 2.60 | 2.56 |
| 62.0 | 2.99 | 3.16 | 3.02 | 2.97 | 2.93 | 2.89 | 2.85 | 2.81 | 2.76 | 2.72 | 2.68 | 2.64 | 2.60 |
| 62.5 | 3.03 | 3.20 | 3.05 | 3.01 | 2.97 | 2.93 | 2.88 | 2.84 | 2.80 | 2.76 | 2.72 | 2.67 | 2.63 |
| 63.0 | 3.06 | 3.23 | 3.09 | 3.04 | 3.00 | 2.96 | 2.92 | 2.88 | 2.83 | 2.79 | 2.75 | 2.71 | 2.67 |
| 63.5 | 3.10 | 3.27 | 3.12 | 3.08 | 3.04 | 2.99 | 2.95 | 2.91 | 2.87 | 2.83 | 2.78 | 2.74 | 2.70 |
| 64.0 | 3.13 | 3.30 | 3.15 | 3.11 | 3.07 | 3.03 | 2.99 | 2.94 | 2.90 | 2.86 | 2.82 | 2.78 | 2.73 |
| 64.5 | 3.17 | 3.34 | 3.19 | 3.15 | 3.10 | 3.06 | 3.02 | 2.98 | 2.94 | 2.89 | 2.85 | 2.81 | 2.77 |
| 65.0 | 3.20 | 3.37 | 3.22 | 3.18 | 3.14 | 3.10 | 3.05 | 3.01 | 2.97 | 2.93 | 2.89 | 2.84 | 2.80 |
| 65.5 | 3.23 | 3.40 | 3.26 | 3.21 | 3.17 | 3.13 | 3.09 | 3.05 | 3.00 | 2.96 | 2.92 | 2.88 | 2.84 |
| 66.0 | 3.27 | 3.44 | 3.29 | 3.25 | 3.21 | 3.17 | 3.12 | 3.08 | 3.04 | 3.00 | 2.96 | 2.91 | 2.87 |
| 66.5 | 3.30 | 3.47 | 3.33 | 3.28 | 3.24 | 3.20 | 3.16 | 3.12 | 3.07 | 3.03 | 2.99 | 2.95 | 2.91 |
| 67.0 | 3.34 | 3.51 | 3.36 | 3.32 | 3.28 | 3.23 | 3.19 | 3.15 | 3.11 | 3.07 | 3.02 | 2.98 | 2.94 |
| 67.5 | 3.37 | 3.54 | 3.39 | 3.35 | 3.31 | 3.27 | 3.23 | 3.18 | 3.14 | 3.10 | 3.06 | 3.02 | 2.97 |
| 68.0 | 3.41 | 3.58 | 3.43 | 3.39 | 3.34 | 3.30 | 3.26 | 3.22 | 3.18 | 3.13 | 3.09 | 3.05 | 3.01 |
| 68.5 | 3.44 | 3.61 | 3.46 | 3.42 | 3.38 | 3.34 | 3.29 | 3.25 | 3.21 | 3.17 | 3.13 | 3.08 | 3.04 |
| 69.0 | 3.47 | 3.64 | 3.50 | 3.46 | 3.41 | 3.37 | 3.33 | 3.29 | 3.25 | 3.20 | 3.16 | 3.12 | 3.08 |
| 69.5 | 3.51 | 3.68 | 3.53 | 3.49 | 3.45 | 3.41 | 3.36 | 3.32 | 3.28 | 3.24 | 3.20 | 3.15 | 3.11 |
| 70.0 | 3.54 | 3.71 | 3.57 | 3.52 | 3.48 | 3.44 | 3.40 | 3.36 | 3.31 | 3.27 | 3.23 | 3.19 | 3.15 |
| 70.5 | 3.58 | 3.75 | 3.60 | 3.56 | 3.52 | 3.47 | 3.43 | 3.39 | 3.35 | 3.31 | 3.26 | 3.22 | 3.18 |
| 71.0 | 3.61 | 3.78 | 3.63 | 3.59 | 3.55 | 3.51 | 3.47 | 3.42 | 3.38 | 3.34 | 3.30 | 3.26 | 3.21 |
| 71.5 | 3.65 | 3.82 | 3.67 | 3.63 | 3.58 | 3.54 | 3.50 | 3.46 | 3.42 | 3.37 | 3.33 | 3.29 | 3.25 |
| 72.0 | 3.68 | 3.85 | 3.70 | 3.66 | 3.62 | 3.58 | 3.53 | 3.49 | 3.45 | 3.41 | 3.37 | 3.32 | 3.28 |
| 72.5 | 3.71 | 3.88 | 3.74 | 3.70 | 3.65 | 3.61 | 3.57 | 3.53 | 3.49 | 3.44 | 3.40 | 3.36 | 3.32 |
| 73.0 | 3.75 | 3.92 | 3.77 | 3.73 | 3.69 | 3.65 | 3.60 | 3.56 | 3.52 | 3.48 | 3.44 | 3.39 | 3.35 |
| 73.5 | 3.78 | 3.95 | 3.81 | 3.76 | 3.72 | 3.68 | 3.64 | 3.60 | 3.55 | 3.51 | 3.47 | 3.43 | 3.39 |
| 74.0 | 3.82 | 3.99 | 3.84 | 3.80 | 3.76 | 3.71 | 3.67 | 3.63 | 3.59 | 3.55 | 3.50 | 3.46 | 3.42 |
| 74.5 | 3.85 | 4.02 | 3.87 | 3.83 | 3.79 | 3.75 | 3.71 | 3.66 | 3.62 | 3.58 | 3.54 | 3.50 | 3.45 |
| 75.0 | 3.89 | 4.06 | 3.91 | 3.87 | 3.82 | 3.78 | 3.74 | 3.70 | 3.66 | 3.61 | 3.57 | 3.53 | 3.49 |
| 75.5 | 3.92 | 4.09 | 3.94 | 3.90 | 3.86 | 3.82 | 3.77 | 3.73 | 3.69 | 3.65 | 3.61 | 3.56 | 3.52 |
| 76.0 | 3.95 | 4.12 | 3.98 | 3.94 | 3.89 | 3.85 | 3.81 | 3.77 | 3.73 | 3.68 | 3.64 | 3.60 | 3.56 |
| 76.5 | 3.99 | 4.16 | 4.01 | 3.97 | 3.93 | 3.89 | 3.84 | 3.80 | 3.76 | 3.72 | 3.68 | 3.63 | 3.59 |
| 77.0 | 4.02 | 4.19 | 4.05 | 4.00 | 3.96 | 3.92 | 3.88 | 3.84 | 3.79 | 3.75 | 3.71 | 3.67 | 3.63 |

**TABLE 4. PREDICTED FEV¹ FOR FEMALES (KNUDSON, ET AL.: AM. REV. RESPIR. DIS. 1976, 113, 587.) (CONTINUED)**

| HT | AGE 43 | 45 | 47 | 49 | 51 | 53 | 55 | 57 | 59 | 61 | 63 | 65 |
|---|---|---|---|---|---|---|---|---|---|---|---|---|
| 52.0 | 1.87 | 1.83 | 1.79 | 1.74 | 1.70 | 1.66 | 1.62 | 1.58 | 1.53 | 1.49 | 1.45 | 1.41 |
| 52.5 | 1.90 | 1.86 | 1.82 | 1.78 | 1.74 | 1.69 | 1.65 | 1.61 | 1.57 | 1.53 | 1.48 | 1.44 |
| 53.0 | 1.94 | 1.90 | 1.85 | 1.81 | 1.77 | 1.73 | 1.69 | 1.64 | 1.60 | 1.56 | 1.52 | 1.48 |
| 53.5 | 1.97 | 1.93 | 1.89 | 1.85 | 1.80 | 1.76 | 1.72 | 1.68 | 1.64 | 1.59 | 1.55 | 1.51 |
| 54.0 | 2.01 | 1.96 | 1.92 | 1.88 | 1.84 | 1.80 | 1.75 | 1.71 | 1.67 | 1.63 | 1.59 | 1.54 |
| 54.5 | 2.04 | 2.00 | 1.96 | 1.91 | 1.87 | 1.83 | 1.79 | 1.75 | 1.70 | 1.66 | 1.62 | 1.58 |
| 55.0 | 2.07 | 2.03 | 1.99 | 1.95 | 1.91 | 1.86 | 1.82 | 1.78 | 1.74 | 1.70 | 1.65 | 1.61 |
| 55.5 | 2.11 | 2.07 | 2.03 | 1.98 | 1.94 | 1.90 | 1.86 | 1.82 | 1.77 | 1.73 | 1.69 | 1.65 |
| 56.0 | 2.14 | 2.10 | 2.06 | 2.02 | 1.98 | 1.93 | 1.89 | 1.85 | 1.81 | 1.77 | 1.72 | 1.68 |
| 56.5 | 2.18 | 2.14 | 2.09 | 2.05 | 2.01 | 1.97 | 1.93 | 1.88 | 1.84 | 1.80 | 1.76 | 1.72 |
| 57.0 | 2.21 | 2.17 | 2.13 | 2.09 | 2.04 | 2.00 | 1.96 | 1.92 | 1.88 | 1.83 | 1.79 | 1.75 |
| 57.5 | 2.25 | 2.20 | 2.16 | 2.12 | 2.08 | 2.04 | 1.99 | 1.95 | 1.91 | 1.87 | 1.83 | 1.78 |
| 58.0 | 2.28 | 2.24 | 2.20 | 2.15 | 2.11 | 2.07 | 2.03 | 1.99 | 1.94 | 1.90 | 1.86 | 1.82 |
| 58.5 | 2.31 | 2.27 | 2.23 | 2.19 | 2.15 | 2.10 | 2.06 | 2.02 | 1.98 | 1.94 | 1.89 | 1.85 |
| 59.0 | 2.35 | 2.31 | 2.27 | 2.22 | 2.18 | 2.14 | 2.10 | 2.06 | 2.01 | 1.97 | 1.93 | 1.89 |
| 59.5 | 2.38 | 2.34 | 2.30 | 2.26 | 2.22 | 2.17 | 2.13 | 2.09 | 2.05 | 2.01 | 1.96 | 1.92 |
| 60.0 | 2.42 | 2.38 | 2.33 | 2.29 | 2.25 | 2.21 | 2.17 | 2.12 | 2.08 | 2.04 | 2.00 | 1.96 |
| 60.5 | 2.45 | 2.41 | 2.37 | 2.33 | 2.28 | 2.24 | 2.20 | 2.16 | 2.12 | 2.07 | 2.03 | 1.99 |
| 61.0 | 2.49 | 2.44 | 2.40 | 2.36 | 2.32 | 2.28 | 2.23 | 2.19 | 2.15 | 2.11 | 2.07 | 2.02 |
| 61.5 | 2.52 | 2.48 | 2.44 | 2.39 | 2.35 | 2.31 | 2.27 | 2.23 | 2.18 | 2.14 | 2.10 | 2.06 |
| 62.0 | 2.55 | 2.51 | 2.47 | 2.43 | 2.39 | 2.34 | 2.30 | 2.26 | 2.22 | 2.18 | 2.13 | 2.09 |
| 62.5 | 2.59 | 2.55 | 2.51 | 2.46 | 2.42 | 2.38 | 2.34 | 2.30 | 2.25 | 2.21 | 2.17 | 2.13 |
| 63.0 | 2.62 | 2.58 | 2.54 | 2.50 | 2.46 | 2.41 | 2.37 | 2.33 | 2.29 | 2.25 | 2.20 | 2.16 |
| 63.5 | 2.66 | 2.62 | 2.57 | 2.53 | 2.49 | 2.45 | 2.41 | 2.36 | 2.32 | 2.28 | 2.24 | 2.20 |
| 64.0 | 2.69 | 2.65 | 2.61 | 2.57 | 2.52 | 2.48 | 2.44 | 2.40 | 2.36 | 2.31 | 2.27 | 2.23 |
| 64.5 | 2.73 | 2.68 | 2.64 | 2.60 | 2.56 | 2.52 | 2.47 | 2.43 | 2.39 | 2.35 | 2.31 | 2.26 |
| 65.0 | 2.76 | 2.72 | 2.68 | 2.63 | 2.59 | 2.55 | 2.51 | 2.47 | 2.42 | 2.38 | 2.34 | 2.30 |
| 65.5 | 2.79 | 2.75 | 2.71 | 2.67 | 2.63 | 2.58 | 2.54 | 2.50 | 2.46 | 2.42 | 2.37 | 2.33 |
| 66.0 | 2.83 | 2.79 | 2.75 | 2.70 | 2.66 | 2.62 | 2.58 | 2.54 | 2.49 | 2.45 | 2.41 | 2.37 |
| 66.5 | 2.86 | 2.82 | 2.78 | 2.74 | 2.70 | 2.65 | 2.61 | 2.57 | 2.53 | 2.49 | 2.44 | 2.40 |
| 67.0 | 2.90 | 2.86 | 2.81 | 2.77 | 2.73 | 2.69 | 2.65 | 2.60 | 2.56 | 2.52 | 2.48 | 2.44 |
| 67.5 | 2.94 | 2.89 | 2.85 | 2.81 | 2.76 | 2.72 | 2.68 | 2.64 | 2.60 | 2.55 | 2.51 | 2.47 |
| 68.0 | 2.97 | 2.92 | 2.88 | 2.84 | 2.80 | 2.76 | 2.71 | 2.67 | 2.63 | 2.59 | 2.55 | 2.50 |
| 68.5 | 3.00 | 2.96 | 2.92 | 2.87 | 2.83 | 2.79 | 2.75 | 2.71 | 2.66 | 2.62 | 2.58 | 2.54 |
| 69.0 | 3.04 | 2.99 | 2.95 | 2.91 | 2.87 | 2.83 | 2.78 | 2.74 | 2.70 | 2.66 | 2.62 | 2.57 |
| 69.5 | 3.07 | 3.03 | 2.99 | 2.94 | 2.90 | 2.86 | 2.82 | 2.78 | 2.73 | 2.69 | 2.65 | 2.61 |
| 70.0 | 3.10 | 3.06 | 3.02 | 2.98 | 2.94 | 2.89 | 2.85 | 2.81 | 2.77 | 2.73 | 2.68 | 2.64 |
| 70.5 | 3.14 | 3.10 | 3.05 | 3.01 | 2.97 | 2.93 | 2.89 | 2.84 | 2.80 | 2.76 | 2.72 | 2.68 |
| 71.0 | 3.17 | 3.13 | 3.09 | 3.05 | 3.00 | 2.96 | 2.92 | 2.88 | 2.84 | 2.79 | 2.75 | 2.71 |
| 71.5 | 3.21 | 3.16 | 3.12 | 3.08 | 3.04 | 3.00 | 2.95 | 2.91 | 2.87 | 2.83 | 2.79 | 2.74 |
| 72.0 | 3.24 | 3.20 | 3.16 | 3.11 | 3.07 | 3.03 | 2.99 | 2.95 | 2.90 | 2.86 | 2.82 | 2.78 |
| 72.5 | 3.28 | 3.23 | 3.19 | 3.15 | 3.11 | 3.07 | 3.02 | 2.98 | 2.94 | 2.90 | 2.86 | 2.81 |
| 73.0 | 3.31 | 3.27 | 3.23 | 3.18 | 3.14 | 3.10 | 3.06 | 3.02 | 2.97 | 2.93 | 2.89 | 2.85 |
| 73.5 | 3.34 | 3.30 | 3.26 | 3.22 | 3.18 | 3.13 | 3.09 | 3.05 | 3.01 | 2.97 | 2.92 | 2.88 |
| 74.0 | 3.38 | 3.34 | 3.29 | 3.25 | 3.21 | 3.17 | 3.13 | 3.08 | 3.04 | 3.00 | 2.96 | 2.92 |
| 74.5 | 3.41 | 3.37 | 3.33 | 3.29 | 3.24 | 3.20 | 3.16 | 3.12 | 3.08 | 3.03 | 2.99 | 2.95 |
| 75.0 | 3.45 | 3.40 | 3.36 | 3.32 | 3.28 | 3.24 | 3.19 | 3.15 | 3.11 | 3.07 | 3.03 | 2.98 |
| 75.5 | 3.48 | 3.44 | 3.40 | 3.35 | 3.31 | 3.27 | 3.23 | 3.19 | 3.14 | 3.10 | 3.06 | 3.02 |
| 76.0 | 3.52 | 3.47 | 3.43 | 3.39 | 3.35 | 3.31 | 3.26 | 3.22 | 3.18 | 3.14 | 3.10 | 3.05 |
| 76.5 | 3.55 | 3.51 | 3.47 | 3.42 | 3.38 | 3.34 | 3.30 | 3.26 | 3.21 | 3.17 | 3.13 | 3.09 |
| 77.0 | 3.58 | 3.54 | 3.50 | 3.46 | 3.42 | 3.37 | 3.33 | 3.29 | 3.25 | 3.21 | 3.16 | 3.12 |

**APPENDIX D TO §1910.1043—PULMONARY FUNCTION STANDARDS FOR COTTON DUST STANDARD**

The spirometric measurements of pulmonary function shall conform to the following minimum standards, and these standards are not intended to preclude additional testing or alternate methods which can be determined to be superior.

### I. APPARATUS

a. The instrument shall be accurate to within ±50 milliliters or within ±3 percent of reading, whichever is greater.

b. The instrument should be capable of measuring vital capacity from 0 to 7 liters BTPS.

c. The instrument shall have a low inertia and offer low resistance to airflow such that the resistance to airflow at 12 liters per second must be less than 1.5 cm $H_2O$/(liter/sec).

# OSHA Standards for General Industry

d. The zero time point for the purpose of timing the $FEV_1$ shall be determined by extrapolating the steepest portion of the volume time curve back to the maximal inspiration volume (1, 2, 3, 4) or by an equivalent method.

e. Instruments incorporating measurements of airflow to determine volume shall conform to the same volume accuracy stated in (a) of this section when presented with flow rates from at least 0 to 12 liters per second.

f. The instrument or user of the instrument must have a means of correcting volumes to body temperature saturated with water vapor (BTPS) under conditions of varying ambient spirometer temperatures and barometric pressures.

g. The instrument used shall provide a tracing or display of either flow versus volume or volume versus time during the entire forced expiration. A tracing or display is necessary to determine whether the patient has performed the test properly. The tracing must be stored and available for recall and must be of sufficient size that hand measurements may be made within requirement of paragraph (a) of this section. If a paper record is made it must have a paper speed of at least 2 cm/sec and a volume sensitivity of at least 10.0 mm of chart per liter of volume.

h. The instrument shall be capable of accumulating volume for a minimum of 10 seconds and shall not stop accumulating volume before (1) the volume change for a 0.5 second interval is less than 25 milliliters, or (2) the flow is less than 50 milliliters per second for a 0.5 second interval.

i. The forced vital capacity (FVC) and forced expiratory volume in 1 second ($FEV_{1.0}$) measurements shall comply with the accuracy requirements stated in paragraph (a) of this section. That is, they should be accurately measured to within ±50 ml or within ±3 percent of reading, whichever is greater.

j. The instrument must be capable of being calibrated in the field with respect to the $FEV_1$ and FVC. This calibration of the $FEV_1$ and FVC may be either directly or indirectly through volume and time base measurements. The volume calibration source should provide a volume displacement of at least 2 liters and should be accurate to within ±30 milliliters.

## II. TECHNIQUE FOR MEASUREMENT OF FORCED VITAL CAPACITY MANEUVER

a. Use of a nose clip is recommended but not required. The procedures shall be explained in simple terms to the patient who shall be instructed to loosen any tight clothing and stand in front of the apparatus. The subject may sit, but care should be taken on repeat testing that the same position be used and, if possible, the same spirometer. Particular attention shall be given to insure that the chin is slightly elevated with the neck slightly extended. The patient shall be instructed to make a full inspiration from a normal breathing pattern and then blow into the apparatus, without interruption, as hard, fast, and completely as possible. At least three forced expirations shall be carried out. During the maneuvers, the patient shall be observed for compliance with instruction. The expirations shall be checked visually for reproducibility from flow-volume or volume-time tracings or displays. The following efforts shall be judged unacceptable when the patient:

1. Has not reached full inspiration preceding the forced expiration,

2. Has not used maximal effort during the entire forced expiration,

3. Has not continued the expiration for at least 5 seconds or until an obvious plateau in the volume time curve has occurred,

4. Has coughed or closed his glottis,

5. Has an obstructed mouthpiece or a leak around the mouthpiece (obstruction due to tongue being placed in front of mouthpiece, false teeth falling in front of mouthpiece, etc.)

6. Has an unsatisfactory start of expiration, one characterized by excessive hesitation (or false starts), and therefore not allowing back extrapolation of time 0 (extrapolated volume on the volume time tracing must be less than 10 percent of the FVC.)

7. Has an excessive variability between the three acceptable curves. The variation between the two largest FVC's and $FEV_1$'s of the three satisfactory tracings should not exceed 10 percent or ±100 milliliters, whichever is greater.

b. Periodic and routine recalibration of the instrument or method for recording FVC and $FEV_{1.0}$ should be performed using a syringe or other volume source of at least 2 liters.

## III. INTERPRETATION OF SPIROGRAM

a. The first step in evaluating a spirogram should be to determine whether or not the patient has performed the test properly or as described in II above. From the three satisfactory tracings, the forced vital capacity (FVC) and forced expiratory volume in 1 second ($FEV_{1.0}$) shall be measured and recorded. The largest observed FVC and largest observed $FEV_1$ shall be used in the analysis regardless of the curve(s) on which they occur.

b. The following guidelines are recommended by NIOSH for the evaluation and management of workers exposed to cotton dust. It is important to note that employees who show reductions in $FEV_1$/FVC ratio below .75 or drops in Monday $FEV_1$ of 5 percent or greater on their initial screening exam, should be re-evaluated within a month of the first exam. Those who show consistent decrease in lung function, as shown on the following table, should be managed as recommended.

## IV. QUALIFICATIONS OF PERSONNEL ADMINISTERING THE TEST

Technicians who perform pulmonary function testing should have the basic knowledge required to produce meaningful results. Training consisting of approximately 16 hours of formal instruction should cover the following areas.

a. Basic physiology of the forced vital capacity maneuver and the determinants of airflow limitation with emphasis on the relation to reproducibility of results.

b. Instrumentation requirements including calibration procedures, sources of error and their correction.

c. Performance of the testing including subject coaching, recognition of improperly performed maneuvers and corrective actions.

d. Data quality with emphasis on reproducibility.

e. Actual use of the equipment under supervised conditions.

f. Measurement of tracings and calculations of results.

## APPENDIX E TO §1910.1043— VERTICAL ELUTRIATOR EQUIVALENCY PROTOCOL

a. *Samples to be taken*—In order to ascertain equivalency, it is necessary to collect a total of 100 samples from at least 10 sites in a mill. That is, there should be 10 replicate readings at each of 10 sites. The sites should represent dust levels which vary over the allowable range of 0.5 to 2 times the permissible exposure limit. Each sample requires the use of two vertical elutriators (VE's) and at least one but not more than two alternative devices (AD's). Thus, the end result is 200 VE readings and either 100 or 200 AD readings. The 2 VE readings and the 1 or 2 AD readings at each time and site must be made simultaneously. That is, the two VE's and one or two AD's must be arranged together in such a way that they are measuring essentially the same dust levels.

b. *Data averaging*—The two VE readings taken at each site are then averaged. These averages are to be used as the 100 VE readings. If two alternate devices were used, their test results are also averaged. Thus, after this step is accomplished, there will be 100 VE readings and 100 AD readings.

c. *Differences*—For each of the 100 sets of measurements (VE and AD) the difference is obtained as the average VE reading minus the AD reading. Call these differences $D_i$. Thus, we have.

$$D_i = VE_i - AD_i, \quad i = 1, 2, \ldots, 100 \quad (1)$$

Next we compute the arithmetic mean and standard deviations of the differences, using equations (2) and (3), respectively.

$$\overline{X_D} = \frac{1}{N} \sum_{i=1}^{N} D_i \quad (2)$$

$$S_D = \sqrt{\frac{\sum D_i^2 - \frac{(\sum D_i)^2}{N}}{N-1}} \quad (3)$$

where N equals the number of differences (100 in this case), $X_D$ is the arithmetic mean and $S_D$ is the standard deviation.

We next calculate the critical value as $T = KS_D + |\overline{X}_D|$ where K = 1.87, based on 100 samples.

d. *Equivalency test.* The next step is to obtain the average of the 100 VE readings. This is obtained by equation (4)

$$\overline{X}_{VE} = \frac{1}{N}\left(\sum_{i=1}^{N} VE_i\right) \quad (4)$$

We next multiply 0.25 by $\overline{X}_{VE}$. If $T \leq 0.25\,\overline{X}_{VE}$, we can say that the alternate device has passed the equivalency test.

[43 FR 27394, June 23, 1978; 43 FR 35035, Aug. 8, 1978, as amended at 45 FR 67340, Oct. 10, 1980; 50 FR 51173, Dec. 13, 1985; 51 FR 24325, July 3, 1986; 54 FR 24334, June 7, 1989; 61 FR 5508, Feb. 13, 1996; 63 FR 1290, Jan. 8, 1998]

### §1910.1044 1,2-dibromo-3-chloropropane.

**1910.1044(a)** *Scope and application.* **(1)** This section applies to occupational exposure to 1,2-dibromo-3-chloropropane (DBCP).

**1910.1044(a)(2)** This section does not apply to:

**1910.1044(a)(2)(i)** Exposure to DBCP which results solely from the application and use of DBCP as a pesticide; or

**1910.1044(a)(2)(ii)** The storage, transportation, distribution or sale of DBCP in intact containers sealed in such a manner as to prevent exposure to DBCP vapors or liquid, except for the requirements of paragraphs (i), (n) and (o) of this section.

**1910.1044(b)** *Definitions. Authorized person* means any person required by his duties to be present in regulated areas and authorized to do so by his employer, by this section, or by the Act. *Authorized person* also includes any person entering such areas as a designated representative of employees exercising an opportunity to observe employee exposure monitoring.

*DBCP* means 1,2-dibromo-3-chloropropane, Chemical Abstracts Service Registry Number 96-12-8, and includes all forms of DBCP.

*Director* means the Director, National Institute for Occupational Safety and Health, U.S. Department of Health and Human Services, or designee.

*Emergency* means any occurrence such as, but not limited to equipment failure, rupture of containers, or failure of control equipment which may, or does, result in an unexpected release of DBCP.

*OSHA Area Office* means the Area Office of the Occupational Safety and Health Administration having jurisdiction over the geographic area where the affected workplace is located.

*Assistant Secretary* means the Assistant Secretary of Labor for Occupational Safety and Health, U.S. Department of Labor, or designee.

**1910.1044(c)** *Permissible exposure limit*—**(1)** *Inhalation.* The employer shall assure that no employee is exposed to an airborne concentration of DBCP in excess of 1 part DBCP per billion parts of air (ppb) as an 8-hour time-weighted average.

**1910.1044(c)(2)** *Dermal and eye exposure.* The employer shall assure that no employee is exposed to eye or skin contact with DBCP.

**1910.1044(d)** Reserved.

**1910.1044(e)** *Regulated areas.* **(1)** The employer shall establish, within each place of employment, regulated areas wherever DBCP concentrations are in excess of the permissible exposure limit.

**1910.1044(e)(2)** The employer shall limit access to regulated areas to authorized persons.

**1910.1044(f)** *Exposure monitoring*—**(1)** *General.* **(i)** Determinations of airborne exposure levels shall be made from air samples that are representative of each employee's exposure to DBCP over an 8-hour period.

**1910.1044(f)(1)(ii)** For the purposes of this paragraph, employee exposure is that exposure which would occur if the employee were not using a respirator.

**1910.1044(f)(2)** *Initial.* Each employer who has a place of employment in which DBCP is present, shall monitor each workplace and work operation to accurately determine the airborne concentrations of DBCP to which employees may be exposed.

**1910.1044(f)(3)** *Frequency.* **(i)** If the monitoring required by this section reveals employee exposures to be at or below the permissible exposure limit, the employer must repeat these measurements at least every 6 months.

**1910.1044(f)(3)(ii)** If the monitoring required by this section reveals employee exposures to be in excess of the permissible exposure limit, the employer must repeat these measurements for each such employee at least quarterly. The employer must continue quarterly monitoring until at least two consecutive measurements, taken at least seven (7) days apart, are at or below the permissible exposure limit. Thereafter the employer must monitor at least every 6 months.

**1910.1044(f)(4)** *Additional.* Whenever there has been a production, process, control, or personnel change which may result in any new or additional exposure to DBCP, or whenever the employer has any reason to suspect new or additional exposures to DBCP, the employer shall monitor the employees potentially affected by such change for the purpose of redetermining their exposure.

**1910.1044(f)(5)** *Employee notification.* **(i)** The employer must, within 15 working days after the receipt of the results of any monitoring performed under this section, notify each employee of these results either individually in writing or by posting the results in an appropriate location that is accessible to employees.

**1910.1044(f)(5)(ii)** Whenever the results indicate that employee exposure exceeds the permissible exposure limit, the employer shall include in the written notice a statement that the permissible exposure limit was exceeded and a description of the corrective action being taken to reduce exposure to or below the permissible exposure limit.

**1910.1044(f)(6)** *Accuracy of measurement.* The employer shall use a method of measurement which has an accuracy, to a confidence level of 95 percent, of not less than plus or minus 25 percent for concentrations of DBCP at or above the permissible exposure limit.

**1910.1044(g)** *Methods of compliance*—**(1)** *Priority of compliance methods.* The employer shall institute engineering and work practice controls to reduce and maintain employee exposures to DBCP at or below the permissible exposure limit, except to the extent that the employer establishes that such controls are not feasible. Where feasible engineering and work practice controls are not sufficient to reduce employee exposures to within the permissible exposure limit, the employer shall nonetheless use them to reduce exposures to the lowest level achievable by these controls, and shall supplement them by use of respiratory protection.

**1910.1044(g)(2)** *Compliance program.* **(i)** The employer shall establish and implement a written program to reduce employee exposures to DBCP to or below the permissible exposure limit solely by means of engineering and work practice controls as required by paragraph (g)(1) of this section.

**1910.1044(g)(2)(ii)** The written program shall include a detailed schedule for development and implementation of the engineering and work practice controls. These plans must be revised at least annually to reflect the current status of the program.

**1910.1044(g)(2)(iii)** Written plans for these compliance programs shall be submitted upon request to the Assistant Secretary and the Director, and shall be available at the worksite for examination and copying by the Assistant Secretary, the Director, and any affected employee or designated representative of employees.

**1910.1044(g)(2)(iv)** The employer shall institute and maintain at least the controls described in his most recent written compliance program.

**1910.1044(h)** *Respiratory protection*—**(1)** *General.* For employees who are required to use respirators by this section, the employer must provide respirators that comply with the requirements of this paragraph. Respirators must be used during:

**1910.1044(h)(1)(i)** Periods necessary to install or implement feasible engineering and work-practice controls.

**1910.1044(h)(1)(ii)** Maintenance and repair activities for which engineering and work-practice controls are not feasible.

**1910.1044(h)(1)(iii)** Work operations for which feasible engineering and work-practice controls are not yet sufficient to reduce employee exposure to or below the permissible exposure limit.

**1910.1044(h)(1)(iv)** Emergencies.

**1910.1044(h)(2)** *Respirator program.* The employer must implement a respiratory protection program in accordance with 29 CFR 1910.134 (b) through (d) (except (d)(1)(iii)), and (f) through (m).

**1910.1044(h)(3)** *Respirator selection.* The employer must select the appropriate respirator from Table 1 of this section.

## TABLE 1—RESPIRATORY PROTECTION FOR DBCP

| Airborne concentration of DBCP or condition of use | Respirator type |
|---|---|
| (a) Less than or equal to 10 ppb | (1) Any supplied-air respirator; or (2) any self-contained breathing apparatus. |
| (b) Less than or equal to 50 ppb | (1) Any supplied-air respirator with full facepiece, helmet, or hood; or (2) any self-contained breathing apparatus with full facepiece. |
| (c) Less than or equal to 1,000 ppb | (1) A Type C supplied-air respirator operated in pressure-demand or other positive pressure or continuous flow mode. |
| (d) Less than or equal to 2,000 ppb | (1) A Type C supplied-air respirator with full facepiece operated in pressure-demand or other positive pressure mode, or with full facepiece, helmet, or hood operated in continuous flow mode. |
| (e) Greater than 2,000 ppb or entry and escape from unknown concentrations | (1) A combination respirator which includes a Type C supplied-air respirator with full facepiece operated in pressure-demand or other positive pressure or continuous flow mode and an auxiliary self-contained breathing apparatus operated in pressure-demand or positive pressure mode; or (2) a self-contained breathing apparatus with full facepiece operated in pressure-demand or other positive pressure mode. |
| (f) Firefighting | (1) A self-contained breathing apparatus with full facepiece operated in pressure-demand or other positive pressure mode. |

**1910.1044(i)** *Emergency situations*—**(1)** *Written plans.* **(i)** A written plan for emergency situations shall be developed for each workplace in which DBCP is present.

**1910.1044(i)(1)(ii)** Appropriate portions of the plan shall be implemented in the event of an emergency.

**1910.1044(i)(2)** Employees engaged in correcting emergency conditions shall be equipped as required in paragraphs (h) and (j) of this section until the emergency is abated.

**1910.1044(i)(3)** *Evacuation.* Employees not engaged in correcting the emergency shall be removed and restricted from the area and normal operations in the affected area shall not be resumed until the emergency is abated.

**1910.1044(i)(4)** *Alerting employees.* Where there is a possibility of employee exposure to DBCP due to the occurrence of an emergency, a general alarm shall be installed and maintained to promptly alert employees of such occurrences.

**1910.1044(i)(5)** *Medical surveillance.* For any employee exposed to DBCP in an emergency situation, the employer shall provide medical surveillance in accordance with paragraph (m)(6) of this section.

**1910.1044(i)(6)** *Exposure monitoring.* **(i)** Following an emergency, the employer shall conduct monitoring which complies with paragraph (f) of this section.

**1910.1044(i)(6)(ii)** In workplaces not normally subject to periodic monitoring, the employer may terminate monitoring when two consecutive measurements indicate exposures below the permissible exposure limit.

**1910.1044(j)** *Protective clothing and equipments*—**(1)** *Provision and use.* Where there is any possibility of eye or dermal contact with liquid or solid DBCP, the employer shall provide, at no cost to the employee, and assure that the employee wears impermeable protective clothing and equipment to protect the area of the body which may come in contact with DBCP. Eye and face protection shall meet the requirements of §1910.133 of this part.

**1910.1044(j)(2)** *Removal and storage.* **(i)** The employer shall assure that employees remove DBCP contaminated work clothing only in change rooms provided in accordance with paragraph (l)(1) of this section.

**1910.1044(j)(2)(ii)** The employer shall assure that employees promptly remove any protective clothing and equipment which becomes contaminated with DBCP-containing liquids and solids. This clothing shall not be reworn until the DBCP has been removed from the clothing or equipment.

**1910.1044(j)(2)(iii)** The employer shall assure that no employee takes DBCP contaminated protective devices and work clothing out of the change room, except those employees authorized to do so for the purpose of laundering, maintenance, of disposal.

**1910.1044(j)(2)(iv)** DBCP-contaminated protective devices and work clothing shall be placed and stored in closed containers which prevent dispersion of the DBCP outside the container.

**1910.1044(j)(2)(v)** Containers of DBCP contaminated protective devices or work clothing which are to be taken out of change rooms or the workplace for cleaning, maintenance or disposal, shall bear labels in accordance with paragraph (o)(3) of this section.

**1910.1044(j)(3)** *Cleaning and replacement.* **(i)** The employer shall clean, launder, repair, or replace protective clothing and equipment required by this paragraph to maintain their effectiveness. The employer shall provide clean protective clothing and equipment at least daily to each affected employee.

**1910.1044(j)(3)(ii)** The employer shall inform any person who launders or clean DBCP-contaminated protective clothing or equipment of the potentially harmful effects of exposure to DBCP.

**1910.1044(j)(3)(iii)** The employer shall prohibit the removal of DBCP from protective clothing and equipment by blowing or shaking.

**1910.1044(k)** *Housekeeping*—**(1)** *Surfaces.* **(i)** All workplace surfaces shall be maintained free of visible accumulations of DBCP.

**1910.1044(k)(1)(ii)** Dry sweeping and the use of compressed air for the cleaning of floors and other surfaces is prohibited where DBCP dusts or liquids are present.

**1910.1044(k)(1)(iii)** Where vacuuming methods are selected to clean floors and other surfaces, either portable units or a permanent system may be used.

**1910.1044(k)(1)(iii)(a)** If a portable unit is selected, the exhaust shall be attached to the general workplace exhaust ventilation system or collected within the vacuum unit, equipped with high efficiency filters or other appropriate means of contaminant removal, so that DBCP is not reintroduced into the workplace air; and

**1910.1044(k)(1)(iii)(b)** Portable vacuum units used to collect DBCP may not be used for other cleaning purposes and shall be labeled as prescribed by paragraph (o)(3) of this section.

**1910.1044(k)(1)(iv)** Cleaning of floors and other surfaces contaminated with DBCP-containing dusts shall not be performed by washing down with a hose, unless a fine spray has first been laid down.

**1910.1044(k)(2)** *Liquids.* Where DBCP is present in a liquid form, or as a resultant vapor, all containers or vessels containing DBCP shall be enclosed to the maximum extent feasible and tightly covered when not in use.

**1910.1044(k)(3)** *Waste disposal.* DBCP waste scrap, debris, containers or equipment, shall be disposed of in sealed bags or other closed containers which prevent dispersion of DBCP outside the container.

**1910.1044(l)** *Hygiene facilities and practices*—**(1)** *Change rooms.* The employer shall provide clean change rooms equipped with storage facilities for street clothes and separate storage facilities for protective clothing and equipment whenever employees are required to wear protective clothing and equipment in accordance with paragraphs (h) and (j) of this section.

**1910.1044(l)(2)** *Showers.* **(i)** The employer shall assure that employees working in the regulated area shower at the end of the work shift.

**1910.1044(l)(2)(ii)** The employer shall assure that employees whose skin becomes contaminated with DBCP-containing liquids or solids immediately wash or shower to remove any DBCP from the skin.

**1910.1044(l)(2)(iii)** The employer shall provide shower facilities in accordance with 29 CFR 1910.141(d)(3).

**1910.1044(l)(3)** *Lunchrooms.* The employer shall provide lunchroom facilities which have a temperature controlled, positive pressure, filtered air supply, and which are readily accessible to employees working in regulated areas.

**1910.1044(l)(4)** *Lavatories.* **(i)** The employer shall assure that employees working in the regulated area remove protective clothing and wash their hands and face prior to eating.

**1910.1044(l)(4)(ii)** The employer shall provide a sufficient number of lavatory facilities which comply with 29 CFR 1910.141(d)(1) and (2).

**1910.1044(l)(5)** *Prohibition of activities in regulated areas.* The employer shall assure that, in regulated areas, food or beverages are not present or consumed, smoking products and implements are not present or used, and cosmetics are not present or applied.

**1910.1044(m)** *Medical surveillance*—**(1)** *General.* **(i)** The employer shall make available a medical surveillance program for employees who work in regulated areas and employees who are subjected to DBCP exposures in an emergency situation.

**1910.1044(m)(1)(ii)** All medical examinations and procedures shall be performed by or under the supervision of a licensed physician, and shall be provided without cost to the employee.

**1910.1044(m)(2)** *Frequency and content.* At the time of initial assignment, and annually thereafter, the employer shall provide a medical examination for employees who work in regulated areas, which includes at least the following:

**1910.1044(m)(2)(i)** A medical and occupational history including reproductive history.

**1910.1044(m)(2)(ii)** A physical examination, including examination of the genito-urinary tract, testicle size and body habitus, including a determination of sperm count.

**1910.1044(m)(2)(iii)** A serum specimen shall be obtained and the following determinations made by radioimmunoassay techniques utilizing National Institutes of Health (NIH) specific antigen or one of equivalent sensitivity:

**1910.1044(m)(2)(iii)***(a)* Serum follicle stimulating hormone (FSH);

**1910.1044(m)(2)(iii)***(b)* Serum luteinizing hormone (LH); and

**1910.1044(m)(2)(iii)***(c)* Serum total estrogen (females).

**1910.1044(m)(2)(iv)** Any other tests deemed appropriate by the examining physician.

**1910.1044(m)(3)** *Additional examinations.* If the employee for any reason develops signs or symptoms commonly associated with exposure to DBCP, the employer shall provide the employee with a medical examination which shall include those elements considered appropriate by the examining physician.

**1910.1044(m)(4)** *Information provided to the physician.* The employer shall provide the following information to the examining physician:

**1910.1044(m)(4)(i)** A copy of this regulation and its appendices;

**1910.1044(m)(4)(ii)** A description of the affected employee's duties as they relate to the employee's exposure;

**1910.1044(m)(4)(iii)** The level of DBCP to which the employee is exposed; and

**1910.1044(m)(4)(iv)** A description of any personal protective equipment used or to be used.

**1910.1044(m)(5)** *Physician's written opinion.* **(i)** For each examination under this section, the employer shall obtain and provide the employee with a written opinion from the examining physician which shall include:

**1910.1044(m)(5)(i)***(a)* The results of the medical tests performed;

**1910.1044(m)(5)(i)***(b)* The physician's opinion as to whether the employee has any detected medical condition which would place the employee at an increased risk of material impairment of health from exposure to DBCP; and

**1910.1044(m)(5)(i)***(c)* Any recommended limitations upon the employee's exposure to DBCP or upon the use of protective clothing and equipment such as respirators.

**1910.1044(m)(5)(ii)** The employer shall instruct the physician not to reveal in the written opinion specific findings or diagnoses unrelated to occupational exposure.

**1910.1044(m)(6)** *Emergency situations.* If the employee is exposed to DBCP in an emergency situation, the employer shall provide the employee with a sperm count test as soon as practicable, or, if the employee has been vasectionized or is unable to produce a semen specimen, the hormone tests contained in paragraph (m)(2)(iii) of this section. The employer shall provide these same tests three months later.

**1910.1044(n)** *Employee information and training*—**(1)** *Training program.* **(i)** The employer shall institute a training program for all employees who may be exposed to DBCP and shall assure their participation in such training program.

**1910.1044(n)(1)(ii)** The employer shall assure that each employee is informed of the following:

**1910.1044(n)(1)(ii)***(a)* The information contained in Appendix A;

**1910.1044(n)(1)(ii)***(b)* The quantity, location, manner of use, release or storage of DBCP and the specific nature of operations which could result in exposure to DBCP as well as any necessary protective steps;

**1910.1044(n)(1)(ii)***(c)* The purpose, proper use, and limitations of respirators;

**1910.1044(n)(1)(ii)***(d)* The purpose and description of the medical surveillance program required by paragraph (m) of this section; and

**1910.1044(n)(1)(ii)***(e)* A review of this standard, including appendices.

**1910.1044(n)(2)** *Access to training materials.* **(i)** The employer shall make a copy of this standard and its appendices readily available to all affected employees.

**1910.1044(n)(2)(ii)** The employer shall provide, upon request, all materials relating to the employee information and training program to the Assistant Secretary and the Director.

**1910.1044(o)** *Signs and labels*—**(1)** *General.* **(i)** The employer may use labels or signs required by other statutes, regulations, or ordinances in addition to or in combination with, signs and labels required by this paragraph.

**1910.1044(o)(1)(ii)** The employer shall assure that no statement appears on or near any sign or label required by this paragraph which contradicts or detracts from the required sign or label.

**1910.1044(o)(2)** *Signs.* **(i)** The employer shall post signs to clearly indicate all regulated areas. These signs shall bear the legend:

**DANGER**

**1,2-Dibromo-3-chloropropane**

(Insert Appropriate Trade or Common Names)

**CANCER HAZARD**

**AUTHORIZED PERSONNEL ONLY**

**RESPIRATOR REQUIRED**

**1910.1044(o)(3)** *Labels.* **(i)** The employer shall assure that precautionary labels are affixed to all containers of DBCP and of products containing DBCP in the workplace, and that the labels remain affixed when the DBCP or products containing DBCP are sold, distributed, or otherwise leave the employer's workplace. Where DBCP or products containing DBCP are sold, distributed or otherwise leave the employer's workplace bearing appropriate labels required by EPA under the regulations in 40 CFR Part 162, the labels required by this paragraph need not be affixed.

**1910.1044(o)(3)(ii)** The employer shall assure that the precautionary labels required by this paragraph are readily visible and legible. The labels shall bear the following legend:

**DANGER**

**1,2-Dibromo-3-chloropropane**

**CANCER HAZARD**

**1910.1044(p)** *Recordkeeping*—**(1)** *Exposure monitoring.* **(i)** The employer shall establish and maintain an accurate record of all monitoring required by paragraph (f) of this section.

# OSHA Standards for General Industry

**1910.1044(p)(1)(ii)** This record shall include:

**1910.1044(p)(1)(ii)(a)** The dates, number, duration and results of each of the samples taken, including a description of the sampling procedure used to determine representative employee exposure;

**1910.1044(p)(1)(ii)(b)** A description of the sampling and analytical methods used;

**1910.1044(p)(1)(ii)(c)** Type of respiratory protective devices worn, if any; and

**1910.1044(p)(1)(ii)(d)** Name, social security number, and job classification of the employee monitored and of all other employees whose exposure the measurement is intended to represent.

**1910.1044(p)(1)(iii)** The employer shall maintain this record for at least 40 years or the duration of employment plus 20 years, whichever is longer.

**1910.1044(p)(2)** *Medical surveillance.* **(i)** The employer shall establish and maintain an accurate record for each employee subject to medical surveillance required by paragraph (m) of this section.

**1910.1044(p)(2)(ii)** This record shall include:

**1910.1044(p)(2)(ii)(a)** The name and social security number of the employee;

**1910.1044(p)(2)(ii)(b)** A copy of the physician's written opinion;

**1910.1044(p)(2)(ii)(c)** Any employee medical complaints related to exposure to DBCP;

**1910.1044(p)(2)(ii)(d)** A copy of the information provided the physician as required by paragraphs (m)(4)(ii) through (m)(4)(iv) of this section; and

**1910.1044(p)(2)(ii)(e)** A copy of the employee's medical and work history.

**1910.1044(p)(2)(iii)** The employer shall maintain this record for at least 40 years or the duration of employment plus 20 years, whichever is longer.

**1910.1044(p)(3)** *Availability.* **(i)** The employer shall assure that all records required to be maintained by this section be made available upon request to the Assistant Secretary and the Director for examination and copying.

**1910.1044(p)(3)(ii)** Employee exposure monitoring records and employee medical records required by this paragraph shall be provided upon request to employees, designated representatives, and the Assistant Secretary in accordance with 29 CFR 1910.1020(a) through (e) and (g) through (i).

**1910.1044(p)(4)** *Transfer of records.* **(i)** If the employer ceases to do business, the successor employer shall receive and retain all records required to be maintained by paragraph (p) of this section for the prescribed period.

**1910.1044(p)(4)(ii)** If the employer ceases to do business and there is no successor employer to receive and retain the records for the prescribed period, the employer shall transmit these records by mail to the Director.

**1910.1044(p)(4)(iii)** At the expiration of the retention period for the records required to be maintained under paragraph (p) of this section, the employer shall transmit these records by mail to the Director.

**1910.1044(p)(4)(iv)** The employer shall also comply with any additional requirements involving transfer of records set forth in 29 CFR 1910.1020(h).

**1910.1044(q)** *Observation of monitoring*—**(1)** *Employee observation.* The employer shall provide affected employees, or their designated representatives, with an opportunity to observe any monitoring of employee exposure to DBCP required by this section.

**1910.1044(q)(2)** *Observation procedures.* **(i)** Whenever observation of the measuring or monitoring of employee exposure to DBCP requires entry into an area where the use of protective clothing or equipment is required, the employer shall provide the observer with personal protective clothing or equipment required to be worn by employees working in the area, assure the use of such clothing and equipment, and require the observer to comply with all other applicable safety and health procedures.

**1910.1044(q)(2)(ii)** Without interfering with the monitoring or measurement, observers shall be entitled to:

**1910.1044(q)(2)(ii)(a)** Receive an explanation of the measurement procedures;

**1910.1044(q)(2)(ii)(b)** Observe all steps related to the measurement of airborne concentrations of DBCP performed at the place of exposure; and

**1910.1044(q)(2)(ii)(c)** Record the results obtained.

**1910.1044(r)** *Appendices.* The information contained in the appendices is not intended, by itself, to create any additional obligations not otherwise imposed or to detract from any existing obligation.

## APPENDIX A TO §1910.1044—SUBSTANCE SAFETY DATA SHEET FOR DBCP

### I. SUBSTANCE IDENTIFICATION

A. Synonyms and trades names: DBCP; Dibromochloropropane; Fumazone (Dow Chemical Company TM); Nemafume; Nemagon (Shell Chemical Co. TM); Nemaset; BBC 12; and OS 1879.

B. Permissible exposure:

1. *Airborne.* 1 part DBCP vapor per billion parts of air (1 ppb); time-weighted average (TWA) for an 8-hour workday.

2. *Dermal.* Eye contact and skin contact with DBCP are prohibited.

C. Appearance and odor: Technical grade DBCP is a dense yellow or amber liquid with a pungent odor. It may also appear in granular form, or blended in varying concentrations with other liquids.

D. Uses: DBCP is used to control nematodes, very small wormlike plant parasites, on crops including cotton, soybeans, fruits, nuts, vegetables and ornamentals.

### II. HEALTH HAZARD DATA

A. Routes of entry: Employees may be exposed:

1. Through inhalation (breathing);

2. Through ingestion (swallowing);

3. Skin contact; and

4. Eye contact.

B. Effects of exposure:

1. *Acute exposure.* DBCP may cause drowsiness, irritation of the eyes, nose, throat and skin, nausea and vomiting. In addition, overexposure may cause damage to the lungs, liver or kidneys.

2. *Chronic exposure.* Prolonged or repeated exposure to DBCP has been shown to cause sterility in humans. It also has been shown to produce cancer and sterility in laboratory animals and has been determined to constitute an increased risk of cancer in man.

3. *Reporting Signs and Symptoms.* If you develop any of the above signs or symptoms that you think are caused by exposure to DBCP, you should inform your employer.

### III. EMERGENCY FIRST AID PROCEDURES

A. *Eye exposure.* If DBCP liquid or dust containing DBCP gets into your eyes, wash your eyes immediately with large amounts of water, lifting the lower and upper lids occasionally. Get medical attention immediately. Contact lenses should not be worn when working with DBCP.

B. *Skin exposure.* If DBCP liquids or dusts containing DBCP get on your skin, immediately wash using soap or mild detergent and water. If DBCP liquids or dusts containing DBCP penetrate through your clothing, remove the clothing immediately and wash. If irritation is present after washing get medical attention.

C. *Breathing.* If you or any person breathe in large amounts of DBCP, move the exposed person to fresh air at once. If breathing has stopped, perform artificial respiration. Do not use mouth-to-mouth. Keep the affected person warm and at rest. Get medical attention as soon as possible.

D. *Swallowing.* When DBCP has been swallowed and the person is conscious, give the person large amounts of water immediately. After the water has been swallowed, try to get the person to vomit by having him touch the back of his throat with his finger. Do not make an unconscious person vomit. Get medical attention immediately.

E. *Rescue.* Notify someone. Put into effect the established emergency rescue procedures. Know the locations of the emergency rescue equipment before the need arises.

### IV. RESPIRATORS AND PROTECTIVE CLOTHING

A. *Respirators.* You may be required to wear a respirator in emergencies and while your employer is in the process of reducing DBCP exposures through engineering controls. If respirators are worn, they must have a National Institute for Occupational Safety and Health (NIOSH) approval label (Older respirators may have a Bureau of Mines Approval label). For effective protection, a respirator must fit your face and head snugly. The respirator should not be loosened or removed in work situations where its use is required. DBCP does not have a detectable odor except at 1,000 times or more above the permissible exposure limit. If you can smell DBCP while wearing a respirator, the respirator is not working correctly; go immediately to fresh air. If you experience difficulty breathing while wearing a respirator, tell your employer.

B. *Protective clothing.* When working with DBCP you must wear for your protection impermeable work clothing provided by your employer. (Standard rubber and neoprene protective clothing do not offer adequate protection).

DBCP must never be allowed to remain on the skin. Clothing and shoes must not be allowed to become contaminated with DBCP, and if they do, they must be promptly removed and not worn again until completely free of DBCP. Turn in impermeable clothing that has developed leaks for repair or replacement.

C. *Eye protection.* You must wear splash-proof safety goggles where there is any possibility of DBCP liquid or dust contacting your eyes.

### V. PRECAUTIONS FOR SAFE USE, HANDLING, AND STORAGE

A. DBCP must be stored in tightly closed containers in a cool, well-ventilated area.

B. If your work clothing may have become contaminated with DBCP, or liquids or dusts containing DBCP, you must change into uncontaminated clothing before leaving the work premises.

C. You must promptly remove any protective clothing that becomes contaminated with DBCP. This clothing must not be reworn until the DBCP is removed from the clothing.

D. If your skin becomes contaminated with DBCP, you must immediately and thoroughly wash or shower with soap or mild detergent and water to remove any DBCP from your skin.

E. You must not keep food, beverages, cosmetics, or smoking materials, nor eat or smoke, in regulated areas.

F. If you work in a regulated area, you must wash your hands thoroughly with soap or mild detergent and water, before eating, smoking or using toilet facilities.

G. If you work in a regulated area, you must remove any protective equipment or clothing before leaving the regulated area.

H. Ask your supervisor where DBCP is used in your work area and for any additional safety and health rules.

### VI. ACCESS TO INFORMATION

A. Each year, your employer is required to inform you of the information contained in this Substance Safety Data Sheet for DBCP. In addition, your employer must instruct you in the safe use of DBCP, emergency procedures, and the correct use of protective equipment.

B. Your employer is required to determine whether you are being exposed to DBCP. You or your representative have the right to observe employee exposure measurements and to record the result obtained. Your employer is required to inform you of your exposure. If your employer determines that you are being overexposed, he is required to inform you of the actions which are being taken to reduce your exposure.

C. Your employer is required to keep records of your exposure and medical examinations. Your employer is required to keep exposure and medical data for at least 40 years or the duration of your employment plus 20 years, whichever is longer.

D. Your employer is required to release exposure and medical records to you, your physician, or other individual designated by you upon your written request.

### APPENDIX B TO §1910.1044— SUBSTANCE TECHNICAL GUIDELINES FOR DBCP

#### I. PHYSICAL AND CHEMICAL DATA

A. Substance Identification

1. Synonyms: 1,2-dibromo-3-chloropropane; DBCP, Fumazone; Nemafume; Nemagon; Nemaset; BBC 12; OS 1879. DBCP is also included in agricultural pesticides and fumigants which include the phrase "Nema—" in their name.

2. Formula: $C_3H_5Br_2$ Cl.

3. Molecular Weight: 236.

B. Physical Data:

1. Boiling point (760 mm HG): 195 °C (383 °F)

2. Specific gravity (water = 1): 2.093.

3. Vapor density (air = 1 at boiling point of DBCP): Data not available.

4. Melting point: 6 °C (43 °F).

5. Vapor pressure at 20 °C (68 °F): 0.8 mm Hg

6. Solubility in water: 1000 ppm.

7. Evaporation rate (Butyl Acetate = 1): very much less than 1.

8. Appearance and odor: Dense yellow or amber liquid with a pungent odor at high concentrations. Any detectable odor of DBCP indicates overexposure.

### II. FIRE EXPLOSION AND REACTIVITY HAZARD DATA

A. Fire

1. Flash point: 170 °F (77 °C)

2. Autoignition temperature: Data not available.

3. Flammable limits in air, percent by volume: Data not available.

4. Extinguishing media: Carbon dioxide, dry chemical.

5. Special fire-fighting procedures: Do not use a solid stream of water since a stream will scatter and spread the fire. Use water spray to cool containers exposed to a fire.

6. Unusual fire and explosion hazards: None known.

7. For purposes of complying with the requirements of §1910.106, liquid DBCP is classified as a Class III A combustible liquid.

8. For the purpose of complying with §1910.309, the classification of hazardous locations as described in article 500 of the National Electrical Code for DBCP shall be Class I, Group D.

9. For the purpose of compliance with §1910.157, DBCP is classified as a Class B fire hazard.

10. For the purpose of compliance with §1910.178, locations classified as hazardous locations due to the presence of DBCP shall be Class I, Group D.

11. Sources of ignition are prohibited where DBCP presents a fire or explosion hazard.

B. Reactivity

1. Conditions contributing to instability: None known.

2. Incompatibilities: Reacts with chemically active metals, such as aluminum, magnesium and tin alloys.

3. Hazardous decomposition products: Toxic gases and vapors (such as HBr, HCl and carbon monoxide) may be released in a fire involving DBCP.

4. Special precautions: DBCP will attack some rubber materials and coatings.

### III. SPILL, LEAK AND DISPOSAL PROCEDURES

A. If DBCP is spilled or leaked, the following steps should be taken:

1. The area should be evacuated at once and re-entered only after thorough ventilation.

2. Ventilate area of spill or leak.

3. If in liquid form, collect for reclamation or absorb in paper, vermiculite, dry sand, earth or similar material.

4. If in solid form, collect spilled material in the most convenient and safe manner for reclamation or for disposal.

B. Persons not wearing protective equipment must be restricted from areas of spills or leaks until cleanup has been completed.

C. Waste Disposal Methods:

1. For small quantities of liquid DBCP, absorb on paper towels, remove to a safe place (such as a fume hood) and burn the paper. Large quantities can be reclaimed or collected and atomized in a suitable combustion chamber equipped with an appropriate effluent gas cleaning device. If liquid

# OSHA Standards for General Industry

DBCP is absorbed in vermiculite, dry sand, earth or similar material and placed in sealed containers it may be disposed of in a Stateapproved sanitary landfill.

2. If in solid form, for small quantities, place on paper towels, remove to a safe place (such as a fume hood) and burn. Large quantities may be reclaimed. However, if this is not practical, dissolve in a flammable solvent (such as alcohol) and atomize in a suitable combustion chamber equipped with an appropriate effluent gas cleaning device. DBCP in solid form may also be disposed in a state-approved sanitary landfill.

## IV. MONITORING AND MEASUREMENT PROCEDURES

A. Exposure above the permissible exposure limit.

1. *Eight Hour Exposure Evaluation:* Measurements taken for the purpose of determining employee exposure under this section are best taken so that the average 8-hour exposure may be determined from a single 8-hour sample or two (2) 4-hour samples. Air samples should be taken in the employee's breathing zone (air that would most nearly represent that inhaled by the employee).

2. *Monitoring Techniques:* The sampling and analysis under this section may be performed by collecting the DBCP vapor on petroleum based charcoal absorption tubes with subsequent chemical analyses. The method of measurement chosen should determine the concentration of airborne DBCP at the permissible exposure limit to an accuracy of plus or minus 25 percent. If charcoal tubes are used, a total volume of 10 liters should be collected at a flow rate of 50 cc. per minute for each tube. Analyze the resultant samples as you would samples of halogenated solvent.

B. Since many of the duties relating to employee protection are dependent on the results of monitoring and measuring procedures, employers should assure that the evaluation of employee exposures is performed by a competent industrial hygienist or other technically qualified person.

## V. PROTECTIVE CLOTHING

Employees should be required to wear appropriate protective clothing to prevent any possibility of skin contact with DBCP. Because DBCP is absorbed through the skin, it is important to prevent skin contact with both liquid and solid forms of DBCP. Protective clothing should include impermeable coveralls or similar fullbody work clothing, gloves, headcoverings, and workshoes or shoe coverings. Standard rubber and neoprene gloves do not offer adequate protection and should not be relied upon to keep DBCP off the skin. DBCP should never be allowed to remain on the skin. Clothing and shoes should not be allowed to become contaminated with the material, and if they do, they should be promptly removed and not worn again until completely free of the material. Any protective clothing which has developed leaks or is otherwise found to be defective should be repaired or replaced. Employees should also be required to wear splash-proof safety goggles where there is any possibility of DBCP contacting the eyes.

## VI. HOUSEKEEPING AND HYGIENE FACILITIES

1. The workplace must be kept clean, orderly and in a sanitary condition;

2. Dry sweeping and the use of compressed air is unsafe for the cleaning of floors and other surfaces where DBCP dust or liquids are found. To minimize the contamination of air with dust, vacuuming with either portable or permanent systems must be used. If a portable unit is selected, the exhaust must be attached to the general workplace exhaust ventilation system, or collected within the vacuum unit equipped with high efficiency filters or other appropriate means of contamination removal and not used for other purposes. Units used to collect DBCP must be labeled.

3. Adequate washing facilities with hot and cold water must be provided, and maintained in a sanitary condition. Suitable cleansing agents should also be provided to assure the effective removal of DBCP from the skin.

4. Change or dressing rooms with individual clothes storage facilities must be provided to prevent the contamination of street clothes with DBCP. Because of the hazardous nature of DBCP, contaminated protective clothing must be stored in closed containers for cleaning or disposal.

## VII. MISCELLANEOUS PRECAUTIONS

A. Store DBCP in tightly closed containers in a cool, well ventilated area.

B. Use of supplied-air suits or other impervious clothing (such as acid suits) may be necessary to prevent skin contact with DBCP. Supplied-air suits should be selected, used, and maintained under the supervision of persons knowledgeable in the limitations and potential life-endangering characteristics of supplied-air suits.

C. The use of air-conditioned suits may be necessary in warmer climates.

D. Advise employees of all areas and operations where exposure to DBCP could occur.

## VIII. COMMON OPERATIONS

Common operations in which exposure to DBCP is likely to occur are: during its production; and during its formulation into pesticides and fumigants.

### APPENDIX C TO §1910.1044—MEDICAL SURVEILLANCE GUIDELINES FOR DBCP

#### I. ROUTE OF ENTRY

Inhalation; skin absorption

#### II. TOXICOLOGY

Recent data collected on workers involved in the manufacture and formulation of DBCP has shown that DBCP can cause sterility at very low levels of exposure. This finding is supported by studies showing that DBCP causes sterility in animals. Chronic exposure to DBCP resulted in pronounced necrotic action on the parenchymatous organs (i.e., liver, kidney, spleen) and on the testicles of rats at concentrations as low as 5 ppm. Rats that were chronically exposed to DBCP also showed changes in the composition of the blood, showing low RBC, hemoglobin, and WBC, and high reticulocyte levels as well as functional hepatic disturbance, manifesting itself in a long prothrombin time. Reznik et al. noted a single dose of 100 mg produced profound depression of the nervous system of rats. Their condition gradually improved. Acute exposure also resulted in the destruction of the sex gland activity of male rats as well as causing changes in the estrous cycle in female rats. Animal studies have also associated DBCP with an increased incidence of carcinoma. Olson, et al. orally administered DBCP to rats and mice 5 times per week at experimentally predetermined maximally tolerated doses and at half those doses. As early as ten weeks after initiation of treatment, DBCP induced a high incidence of squamous cell carcinomas of the stomach with metastases in both species. DBCP also induced mammary adenocarcinomas in the female rats at both dose levels.

#### III. SIGNS AND SYMPTOMS

A. Inhalation: Nausea, eye irritation, conjunctivitis, respiratory irritation, pulmonary congestion or edema, CNS depression with apathy, sluggishness, and ataxia.

B. Dermal: Erythema or inflammation and dermatitis on repeated exposure.

#### IV. SPECIAL TESTS

A. *Semen analysis:* The following information excerpted from the document "Evaluation of Testicular Function", submitted by the Corporate Medical Department of the Shell Oil Company (exhibit 39-3), may be useful to physicians conducting the medical surveillance program;

In performing semen analyses certain minimal but specific criteria should be met:

1. It is recommended that a minimum of three valid semen analyses be obtained in order to make a determination of an individual's average sperm count.

2. A period of sexual abstinence is necessary prior to the collection of each masturbatory sample. It is recommended that intercourse or masturbation be performed 48 hours before the actual specimen collection. A period of 48 hours of abstinence would follow; then the masturbatory sample would be collected.

3. Each semen specimen should be collected in a clean, widemouthed, glass jar (not necessarily pre-sterilized) in a manner designated by the examining physician. Any part of the seminal fluid exam should be initialed *only after liquifaction* is complete, i.e., 30 to 45 minutes after collection.

4. Semen volume should be measured to the nearest 1/10 of a cubic centimeter.

5. Sperm density should be determined using routine techniques involving the use of a white cell pipette and a hemocytometer chamber. The immobilizing fluid most effective and most easily obtained for this process is distilled water.

6. Thin, dry smears of the semen should be made for a morphologic classification of the sperm forms and should be stained with either he-

matoxalin or the more difficult, yet more precise, Papanicolaou technique. Also of importance to record is obvious sperm agglutination, pyospermia, delayed liquifaction (greater than 30 minutes), and hyperviscosity. In addition, pH, using nitrazine paper, should be determined.

7. A total morphology evaluation should include percentages of the following:

a. Normal (oval) forms,

b. Tapered forms,

c. Amorphous forms (include large and small sperm shapes),

d. Duplicated (either heads or tails) forms, and

e. Immature forms.

8. Each sample should be evaluated for sperm *viability* (percent viable sperm moving at the time of examination) as well as sperm *motility* (subjective characterization of "purposeful forward sperm progression" of the majority of those viable sperm analyzed) within two hours after collection, ideally by the same or equally qualified examiner.

B. *Serum determinations:* The following serum determinations should be performed by radioimmuno-assay techniques using National Institutes of Health (NIH) specific antigen or antigen preparations of equivalent sensitivity:

1. Serum follicle stimulating hormone (FSH);

2. Serum luteinizing hormone (LH); and

3. Serum total estrogen (females only).

### V. TREATMENT

Remove from exposure immediately, give oxygen or artificial resuscitation if indicated. Contaminated clothing and shoes should be removed immediately. Flush eyes and wash contaminated skin. If swallowed and the person is conscious, induce vomiting. Recovery from mild exposures is usually rapid and complete.

### VI. SURVEILLANCE AND PREVENTIVE CONSIDERATIONS

A. *Other considerations.* DBCP can cause both acute and chronic effects. It is important that the physician become familiar with the operating conditions in which exposure to DBCP occurs. Those with respiratory disorders may not tolerate the wearing of negative pressure respirators.

B. *Surveillance and screening.* Medical histories and laboratory examinations are required for each employee subject to exposure to DBCP. The employer should screen employees for history of certain medical conditions (listed below) which might place the employee at increased risk from exposure.

1. *Liver disease.* The primary site of biotransformation and detoxification of DBCP is the liver. Liver dysfunctions likely to inhibit the conjugation reactions will tend to promote the toxic actions of DBCP. These precautions should be considered before exposing persons with impaired liver function to DBCP.

2. *Renal disease.* Because DBCP has been associated with injury to the kidney it is important that special consideration be given to those with possible impairment of renal function.

3. *Skin desease.* DBCP can penetrate the skin and can cause erythema on prolonged exposure. Persons with pre-existing skin disorders may be more susceptible to the effects of DBCP.

4. *Blood dyscrasias.* DBCP has been shown to decrease the content of erythrocytes, hemoglobin, and leukocytes in the blood, as well as increase the prothrombin time. Persons with existing blood disorders may be more susceptible to the effects of DBCP.

5. *Reproductive disorders.* Animal studies have associated DBCP with various effects on the reproductive organs. Among these effects are atrophy of the testicles and changes in the estrous cycle. Persons with pre-existing reproductive disorders may be at increased risk to these effects of DBCP.

### REFERENCES

1. Reznik, Ya. B. and Sprinchan, G. K.: Experimental Data on the Gonadotoxic effect of Nemagon, *Gig. Sanit.*, (6), 1975, pp. 101-102, (translated from Russian).

2. Faydysh, E. V., Rakhmatullaev, N. N. and Varshavskii, V. A.: The Cytotoxic Action of Nemagon in a Subacute Experiment, *Med. Zh. Uzbekistana*, (No. 1), 1970, pp. 64-65, (translated from Russian).

3. Rakhmatullaev, N. N.: Hygienic Characteristics of the Nematocide Nemagon in Relation to Water Pollution Control, *Hyg. Sanit.*, 36(3), 1971, pp. 344-348, (translated from Russian).

4. Olson, W. A. *et al.*: Induction of Stomach Cancer in Rats and Mice by Halogenated Aliphatic Fumigants, *Journal of the National Cancer Institute*, (51), 1973, pp. 1993-1995.

5. Torkelson, T. R. *et al.*: Toxicologic Investigations of 1,2-Dibromo-3-chloropropane, *Toxicology and Applied Pharmacology*, 3, 1961 pp. 545-559.

[43 FR 11527, Mar. 17, 1978, as amended at 45 FR 35283, May 23, 1980; 49 FR 18295, Apr. 30, 1984; 54 FR 24334, June 7, 1989; 58 FR 35310, June 30, 1993; 61 FR 5508, Feb. 13, 1996; 63 FR 1291, Jan. 8, 1998]

### §1910.1045 Acrylonitrile.

**1910.1045(a)** *Scope and application.* **(1)** This section applies to all occupational exposures to acrylonitrile (AN), Chemical Abstracts Service Registry No. 000107131, except as provided in paragraphs (a)(2) and (a)(3) of this section.

**1910.1045(a)(2)** This section does not apply to exposures which result solely from the processing, use, and handling of the following materials:

**1910.1045(a)(2)(i)** ABS resins, SAN resins, nitrile barrier resins, solid nitrile elastomers, and acrylic and modacrylic fibers, when these listed materials are in the form of finished polymers, and products fabricated from such finished polymers;

**1910.1045(a)(2)(ii)** Materials made from and/or containing AN for which objective data is reasonably relied upon to demonstrate that the material is not capable of releasing AN in airborne concentrations in excess of 1 ppm as an eight (8)-hour time-weighted average, under the expected conditions of processing, use, and handling which will cause the greatest possible release; and

**1910.1045(a)(2)(iii)** Solid materials made from and/or containing AN which will not be heated above 170 °F during handling, use, or processing.

**1910.1045(a)(3)** An employer relying upon exemption under paragraph (a)(2)(ii) shall maintain records of the objective data supporting that exemption, and of the basis of the employer's reliance on the data, as provided in paragraph (q) of this section.

**1910.1045(b)** *Definitions. Acrylonitrile* or *AN* means acrylonitrile monomer, chemical formula $CH_2 = CHCN$.

*Action level* means a concentration of AN of 1 ppm as an eight (8)-hour time-weighted average.

*Assistant Secretary* means the Assistant Secretary of Labor for Occupational Safety and Health, U.S. Department of Labor, or designee.

*Authorized person* means any person specifically authorized by the employer whose duties require the person to enter a regulated area, or any person entering such an area as a designated representative of employees for the purpose of exercising the opportunity to observe monitoring procedures under paragraph (r) of this section.

*Decontamination* means treatment of materials and surfaces by water washdown, ventilation, or other means, to assure that the materials will not expose employees to airborne concentrations of AN above 1 means the Director, National Institute for Occupational Safety and Health, U.S. Department of Health and Human Services, or designee.

*Emergency* means any occurrence such as, but not limited to, equipment failure, rupture of containers, or failure of control equipment, which results in an unexpected massive release of AN.

*Liquid AN* means AN monomer in liquid form, and liquid or semiliquid polymer intermediates, including slurries, suspensions, emulsions, and solutions, produced during the polymerization of AN.

*OSHA Area Office* means the Area Office of the Occupational Safety and Health Administration having jurisdiction over the geographic area where the affected workplace is located.

**1910.1045(c)** *Permissible exposure limits*—**(1)** *Inhalation.* (i) *Time weighted average limit (TWA).* The employer shall assure that no employee is exposed to an airborne concentration of acrylonitrile in excess of two (2) parts acrylonitrile per million parts of air (2 ppm) as an eight (8)-hour time-weighted average.

**1910.1045(c)(1)(ii)** *Ceiling limit.* The employer shall assure that no employee is exposed to an airborne concentration of acrylonitrile in excess of ten (10) ppm as averaged over any fifteen (15)-minute period during the work day.

# OSHA Standards for General Industry

**1910.1045(c)(2)** *Dermal and eye exposure.* The employer shall assure that no employee is exposed to skin contact or eye contact with liquid AN.

**1910.1045(d)** Reserved.

**1910.1045(e)** *Exposure monitoring*—**(1)** *General.* **(i)** Determinations of airborne exposure levels shall be made from air samples that are representative of each employee's exposure to AN over an eight (8)-hour period.

**1910.1045(e)(1)(ii)** For the purposes of this section, employee exposure is that exposure which would occur if the employee were not using a respirator.

**1910.1045(e)(2)** *Initial monitoring.* Each employer who has a place of employment in which AN is present shall monitor each such workplace and work operation to accurately determine the airborne concentrations of AN to which employees may be exposed.

**1910.1045(e)(3)** *Frequency.* **(i)** If the monitoring required by this section reveals employee exposure to be below the action level, the employer may discontinue monitoring for that employee.

**1910.1045(e)(3)(ii)** (ii) If the monitoring required by this section reveals employee exposure to be at or above the action level but at or below the permissible exposure limits, the employer must repeat such monitoring for each such employee at least every 6 months. The employer must continue these measurements every 6 months until at least two consecutive measurements taken at least seven (7) days a part, are below the action level, and thereafter the employer may discontinue monitoring for that employee.

**1910.1045(e)(3)(iii)** If the monitoring required by this section reveals employee exposure to be in excess of the permissible exposure limits, the employer must repeat these determinations for each such employee at least quarterly. The employer must continue these quarterly measurements until at least two consecutive measurements, taken at least seven (7) days apart, are at or below the permissible exposure limits, and thereafter the employer must monitor at least every 6 months.

**1910.1045(e)(4)** *Additional monitoring.* Whenever there has been a production, process, control, or personnel change which may result in new or additional exposures to AN, or whenever the employer has any other reason to suspect a change which may result in new or additional exposures to AN, additional monitoring which complies with this paragraph shall be conducted.

**1910.1045(e)(5)** *Employee notification.* **(i)** The employer must, within 15 working days after the receipt of the results of any monitoring performed under this section, notify each affected employee of these results either individually in writing or by posting the results in an appropriate location that is accessible to employees.

**1910.1045(e)(5)(ii)** Whenever the results indicate that the representative employee exposure exceeds the permissible exposure limits, the employer shall include in the written notice a statement that the permissible exposure limits were exceeded and a description of the corrective action being taken to reduce exposure to or below the permissible exposure limits.

**1910.1045(e)(6)** *Accuracy of measurement.* The method of measurement of employee exposures shall be accurate to a confidence level of 95 percent, to within plus or minus 35 percent for concentrations of AN at or above the permissible exposure limits, and plus or minus 50 percent for concentrations of AN below the permissible exposure limits.

**1910.1045(f)** *Regulated areas.* **(1)** The employer shall establish regulated areas where AN concentrations are in excess of the permissible exposure limits.

**1910.1045(f)(2)** Regulated areas shall be demarcated and segregated from the rest of the workplace, in any manner that minimizes the number of persons who will be exposed to AN.

**1910.1045(f)(3)** Access to regulated areas shall be limited to authorized persons or to persons otherwise authorized by the act or regulations issued pursuant thereto.

**1910.1045(f)(4)** The employer shall assure that food or beverages are not present or consumed, tobacco products are not present or used, and cosmetics are not applied in the regulated area.

**1910.1045(g)** *Methods of compliance*—**(1)** *Engineering and work practice controls.* **(i)** By November 2, 1980, the employer shall institute engineering and work practice controls to reduce and maintain employee exposures to AN, to or below the permissible exposure limits, except to the extent that the employer establishes that such controls are not feasible.

**1910.1045(g)(1)(ii)** Wherever the engineering and work practice controls which can be instituted are not sufficient to reduce employee exposures to or below the permissible exposure limits, the employer shall nonetheless use them to reduce exposures to the lowest levels achievable by these controls, and shall supplement them by the use of respiratory protection which complies with the requirements of paragraph (h) of this section.

**1910.1045(g)(2)** *Compliance program.* **(i)** The employer shall establish and implement a written program to reduce employee exposures to or below the permissible exposure limits solely by means of engineering and work practice controls, as required by paragraph (g)(1) of this section.

**1910.1045(g)(2)(ii)** Written plans for these compliance programs shall include at least the following:

**1910.1045(g)(2)(ii)(A)** A description of each operation or process resulting in employee exposure to AN above the permissible exposure limits;

**1910.1045(g)(2)(ii)(B)** An outline of the nature of the engineering controls and work practices to be applied to the operation or process in question;

**1910.1045(g)(2)(ii)(C)** A report of the technology considered in meeting the permissible exposure limits;

**1910.1045(g)(2)(ii)(D)** A schedule for implementation of engineering and work practice controls for the operation or process, which shall project completion no later than November 2, 1980; and

**1910.1045(g)(2)(ii)(E)** Other relevant information.

**1910.1045(g)(2)(iii)** The employer shall complete the steps set forth in the compliance program by the dates in the schedule.

**1910.1045(g)(2)(iv)** Written plans shall be submitted upon request to the Assistant Secretary and the Director, and shall be available at the worksite for examination and copying by the Assistant Secretary, the Director, or any affected employee or representative.

**1910.1045(g)(2)(v)** The plans required by this paragraph must be revised and updated at least annually to reflect the current status of the program.

**1910.1045(h)** *Respiratory protection*—**(1)** *General.* For employees who use respirators required by this section, the employer must provide respirators that comply with the requirements of this paragraph. Respirators must be used during:

**1910.1045(h)(1)(i)** Periods necessary to install or implement feasible engineering and work-practice controls.

**1910.1045(h)(1)(ii)** Work operations, such as maintenance and repair activities or reactor cleaning, for which the employer establishes that engineering and work-practice controls are not feasible.

**1910.1045(h)(1)(iii)** Work operations for which feasible engineering and work-practice controls are not yet sufficient to reduce employee exposure to or below the permissible exposure limits.

**1910.1045(h)(1)(iv)** Emergencies.

**1910.1045(h)(2)** *Respirator program.* **(i)** The employer must implement a respiratory protection program in accordance with 29 CFR 1910.134(b) through (d) (except (d)(1)(iii), (d)(3)(iii)(B)(1), and (2)), and (f) through (m).

**1910.1045(h)(2)(ii)** If air-purifying respirators (chemical-cartridge or chemical-canister types) are used:

**1910.1045(h)(2)(ii)(A)** The air-purifying canister or cartridge must be replaced prior to the expiration of its service life or at the completion of each shift, whichever occurs first.

**1910.1045(h)(2)(ii)(B)** A label must be attached to the cartridge or canister to indicate the date and time at which it is first installed on the respirator.

**1910.1045(h)(3)** *Respirator selection.* The employer must select the appropriate respirator from Table I of this section.

## TABLE I—RESPIRATORY PROTECTION FOR ACRYLONITRILE (AN)

| Concentration of AN or condition of use | Respirator type |
|---|---|
| (a) Less than or equal to 20 ppm. | (1) Chemical cartridge respirator with organic vapor cartridge(s) and half-mask facepiece; or<br>(2) Supplied air respirator with half-mask facepiece. |
| (b) Less than or equal to 100 ppm or maximum use concentration (MUC) of cartridges or canisters, whichever is lower. | (1) Full facepiece respirator with (A) organic vapor cartridges, (B) organic vapor gas mask chin-style, or (C) organic vapor gas mask canister, front-or back-mounted;<br>(2) Supplied air respirator with full facepiece; or<br>(3) Self-contained breathing apparatus with full facepiece. |
| (c) Less than or equal to 4,000 ppm. | (1) Supplied air respirator operated in the positive pressure mode with full facepiece, helmet, suit, or hood. |
| (d) Greater than 4,000 ppm or unknown concentration | (1) Supplied air and auxiliary self-contained breathing apparatus with full facepiece in positive pressure mode; or<br>(2) Self-contained breathing apparatus with full facepiece in positive pressure mode. |
| (e) Firefighting | Self-contained breathing apparatus with full facepiece in positive pressure mode. |
| (f) Escape | (1) Any organic vapor respirator, or<br>(2) Any self-contained breathing apparaus. |

**1910.1045(i) Emergency situations—(1)** *Written plans.* **(i)** A written plan for emergency situations shall be developed for each workplace where liquid AN is present. Appropriate portions of the plan shall be implemented in the event of an emergency.

**1910.1045(i)(1)(ii)** The plan shall specifically provide that employees engaged in correcting emergency conditions shall be equipped as required in paragraph (h) of this section until the emergency is abated.

**1910.1045(i)(1)(iii)** Employees not engaged in correcting the emergency shall be evacuated from the area and shall not be permitted to return until the emergency is abated.

**1910.1045(i)(2)** *Alerting employees.* Where there is the possibility of employee exposure to AN in excess of the ceiling limit, a general alarm shall be installed and used to promptly alert employees of such occurrences.

**1910.1045(j)** *Protective clothing and equipment—***(1)** *Provision and use.* Where eye or skin contact with liquid AN may occur, the employer shall provide at no cost to the employee, and assure that employees wear, impermeable protective clothing or other equipment to protect any area of the body which may come in contact with liquid AN. The provision of §§1910.132 and 1910.133 shall be complied with.

**1910.1045(j)(2)** *Cleaning and replacement.* **(i)** The employer shall clean, launder, maintain, or replace protective clothing and equipment required by this section as needed to maintain their effectiveness.

**1910.1045(j)(2)(ii)** The employer shall assure that impermeable protective clothing which contacts or is likely to have contacted liquid AN shall be decontaminated before being removed by the employee.

**1910.1045(j)(2)(iii)** The employer shall assure that an employee whose nonimpermeable clothing becomes wetted with liquid AN shall immediately remove that clothing and proceed to shower. The clothing shall be decontaminated before it is removed from the regulated area.

**1910.1045(j)(2)(iv)** The employer shall assure that no employee removes protective clothing or equipment from the change room, except for those employees authorized to do so for the purpose of laundering, maintenance, or disposal.

**1910.1045(j)(2)(v)** The employer shall inform any person who launders or cleans protective clothing or equipment of the potentially harmful effects of exposure to AN.

**1910.1045(k)** *Housekeeping.* **(1)** All surfaces shall be maintained free of visible accumulations of liquid AN.

**1910.1045(k)(2)** For operations involving liquid AN, the employer shall institute a program for detecting leaks and spills of liquid AN, including regular visual inspections.

**1910.1045(k)(3)** Where spills of liquid AN are detected, the employer shall assure that surfaces contacted by the liquid AN are decontaminated. Employees not engaged in decontamination activities shall leave the area of the spill, and shall not be permitted in the area until decontamination is completed.

**1910.1045(l)** *Waste disposal.* AN waste, scrap, debris, bags, containers, or equipment shall be decontaminated before being incorporated in the general waste disposal system.

**1910.1045(m)** *Hygiene facilities and practices.* **(1)** Where employees are exposed to airborne concentrations of AN above the permissible exposure limits, or where employees are required to wear protective clothing or equipment pursuant to paragraph (j) of this section, the facilities required by 29 CFR 1910.141, including clean change rooms and shower facilities, shall be provided by the employer for the use of those employees, and the employer shall assure that the employees use the facilities provided.

**1910.1045(m)(2)** The employer shall assure that employees wearing protective clothing or equipment for protection from skin contact with liquid AN shall shower at the end of the work shift.

**1910.1045(m)(3)** The employer shall assure that, in the event of skin or eye exposure to liquid AN, the affected employee shall shower immediately to minimize the danger of skin absorption.

**1910.1045(m)(4)** The employer shall assure that employees working in the regulated area wash their hands and faces prior to eating.

**1910.1045(n)** *Medical surveillance—***(1)** *General.* **(i)** The employer shall institute a program of medical surveillance for each employee who is or will be exposed to AN at or above the action level, without regard to the use of respirators. The employer shall provide each such employee with an opportunity for medical examinations and tests in accordance with this paragraph.

**1910.1045(n)(1)(ii)** The employer shall assure that all medical examinations and procedures are performed by or under the supervision of a licensed physician, and that they shall be provided without cost to the employee.

**1910.1045(n)(2)** *Initial examinations.* At the time of initial assignment, or upon institution of the medical surveillance program, the employer shall provide each affected employee an opportunity for a medical examination, including at least the following elements:

**1910.1045(n)(2)(i)** A work history and medical history with special attention to skin, respiratory, and gastrointestinal systems, and those nonspecific symptoms, such as headache, nausea, vomiting, dizziness, weakness, or other central nervous system dysfunctions that may be associated with acute or with chronic exposure to AN;

**1910.1045(n)(2)(ii)** A complete physical examination giving particular attention to the peripheral and central nervous system, gastrointestinal system, respiratory system, skin, and thyroid;

**1910.1045(n)(2)(iii)** A 14- by 17-inch posteroanterior chest X-ray; and

**1910.1045(n)(2)(iv)** Further tests of the intestinal tract, including fecal occult blood screening, for all workers 40 years of age or older, and for any other affected employees for whom, in the opinion of the physician, such testing is appropriate.

**1910.1045(n)(3)** *Periodic examinations.* **(i)** The employer shall provide the examinations specified in paragraph (n)(2) of this section at least annually for all employees specified in paragraph (n)(1) of this section.

**1910.1045(n)(3)(ii)** If an employee has not had the examination specified in paragraph (n)(2) of this section within 6 months preceding termination of employment, the employer shall make such examination available to the employee prior to such termination.

## OSHA Standards for General Industry

**1910.1045(n)(4)** *Additional examinations.* If the employee for any reason develops signs or symptoms which may be associated with exposure to AN, the employer shall provide an appropriate examination and emergency medical treatment.

**1910.1045(n)(5)** *Information provided to the physician.* The employer shall provide the following information to the examining physician:

**1910.1045(n)(5)(i)** A copy of this standard and its appendixes;

**1910.1045(n)(5)(ii)** A description of the affected employee's duties as they relate to the employee's exposure;

**1910.1045(n)(5)(iii)** The employee's representative exposure level;

**1910.1045(n)(5)(iv)** The employee's anticipated or estimated exposure level (for preplacement examinations or in cases of exposure due to an emergency);

**1910.1045(n)(5)(v)** A description of any personal protective equipment used or to be used; and

**1910.1045(n)(5)(vi)** Information from previous medical examinations of the affected employee, which is not otherwise available to the examining physician.

**1910.1045(n)(6)** *Physician's written opinion.* **(i)** The employer shall obtain a written opinion from the examining physician which shall include:

**1910.1045(n)(6)(i)(A)** The results of the medical examination and test performed;

**1910.1045(n)(6)(i)(B)** The physician's opinion as to whether the employee has any detected medical condition(s) which would place the employee at an increased risk of material impairment of the employee's health from exposure to AN;

**1910.1045(n)(6)(i)(C)** Any recommended limitations upon the employee's exposure to AN or upon the use of protective clothing and equipment such as respirators; and

**1910.1045(n)(6)(i)(D)** A statement that the employee has been informed by the physician of the results of the medical examination and any medical conditions which require further examination or treatment.

**1910.1045(n)(6)(ii)** The employer shall instruct the physician not to reveal in the written opinion specific findings or diagnoses unrelated to occupational exposure to AN.

**1910.1045(n)(6)(iii)** The employer shall provide a copy of the written opinion to the affected employee.

**1910.1045(o)** *Employee information and training*—**(1)** *Training program.* **(i)** By January 2, 1979, the employer shall institute a training program for and assure the participation of all employees exposed to AN above the action level, all employees whose exposures are maintained below the action level by engineering and work practice controls, and all employees subject to potential skin or eye contact with liquid AN.

**1910.1045(o)(1)(ii)** Training shall be provided at the time of initial assignment, or upon institution of the training program, and at least annually thereafter, and the employer shall assure that each employee is informed of the following:

**1910.1045(o)(1)(ii)(A)** The information contained in appendixes A and B;

**1910.1045(o)(1)(ii)(B)** The quantity, location, manner of use, release, or storage of AN, and the specific nature of operations which could result in exposure to AN, as well as any necessary protective steps;

**1910.1045(o)(1)(ii)(C)** The purpose, proper use, and limitations of respirators and protective clothing;

**1910.1045(o)(1)(ii)(D)** The purpose and a description of the medical surveillance program required by paragraph (n) of this section;

**1910.1045(o)(1)(ii)(E)** The emergency procedures developed, as required by paragraph (i) of this section;

**1910.1045(o)(1)(ii)(F)** Engineering and work practice controls, their function, and the employee's relationship to these controls; and

**1910.1045(o)(1)(ii)(G)** A review of this standard.

**1910.1045(o)(2)** *Access to training materials.* **(i)** The employer shall make a copy of this standard and its appendixes readily available to all affected employees.

**1910.1045(o)(2)(ii)** The employer shall provide, upon request, all materials relating to the employee information and training program to the Assistant Secretary and the Director.

**1910.1045(p)** *Signs and labels*—**(1)** *General.* **(i)** The employer may use labels or signs required by other statutes, regulations, or ordinances in addition to, or in combination with, signs and labels required by this paragraph.

**1910.1045(p)(1)(ii)** The employer shall assure that no statement appears on or near any sign or label required by this paragraph which contradicts or detracts from the required sign or label.

**1910.1045(p)(2)** *Signs.* **(i)** The employer shall post signs to clearly indicate all workplaces where AN concentrations exceed the permissible exposure limits. The signs shall bear the following legend:

**DANGER**

**ACRYLONITRILE (AN)**

**CANCER HAZARD**

**AUTHORIZED PERSONNEL ONLY**

**RESPIRATORS MAY BE**

**REQUIRED**

**1910.1045(p)(2)(ii)** The employer shall assure that signs required by this paragraph are illuminated and cleaned as necessary so that the legend is readily visible.

**1910.1045(p)(3)** *Labels.* **(i)** The employer shall assure that precautionary labels are affixed to all containers of liquid AN and AN-based materials not exempted under paragraph (a)(2) of this standard. The employer shall assure that the labels remain affixed when the materials are sold, distributed, or otherwise leave the employer's workplace.

**1910.1045(p)(3)(ii)** The employer shall assure that the precautionary labels required by this paragraph are readily visible and legible. The labels shall bear the following legend:

**DANGER**

**CONTAINS ACRYLONITRILE (AN)**

**CANCER HAZARD**

**1910.1045(q)** *Recordkeeping*—**(1)** *Objective data for exempted operations.* **(i)** Where the processing, use, and handling of materials made from or containing AN are exempted pursuant to paragraph (a)(2)(ii) of this section, the employer shall establish and maintain an accurate record of objective data reasonably relied upon in support of the exemption.

**1910.1045(q)(1)(ii)** This record shall include at least the following information:

**1910.1045(q)(1)(ii)(A)** The material qualifying for exemption;

**1910.1045(q)(1)(ii)(B)** The source of the objective data;

**1910.1045(q)(1)(ii)(C)** The testing protocol, results of testing, and/or analysis of the material for the release of AN;

**1910.1045(q)(1)(ii)(D)** A description of the operation exempted and how the data supports the exemption; and

**1910.1045(q)(1)(ii)(E)** Other data relevant to the operations, materials, and processing covered by the exemption.

**1910.1045(q)(1)(iii)** The employer shall maintain this record for the duration of the employer's reliance upon such objective data.

**1910.1045(q)(2)** *Exposure monitoring.* **(i)** The employer shall establish and maintain an accurate record of all monitoring required by paragraph (e) of this section.

**1910.1045(q)(2)(ii)** This record shall include:

**1910.1045(q)(2)(ii)(A)** The dates, number, duration, and results of each of the samples taken, including a description of the sampling procedure used to determine representative employee exposure;

**1910.1045(q)(2)(ii)(B)** A description of the sampling and analytical methods used and the data relied upon to establish that the methods used meet the accuracy and precision requirements of paragraph (e)(6) of this section;

**1910.1045(q)(2)(ii)(C)** Type of respiratory protective devices worn, if any; and

**1910.1045(q)(2)(ii)(D)** Name, social security number, and job classification of the employee monitored and of all other employees whose exposure the measurement is intended to represent.

**1910.1045(q)(2)(iii)** The employer shall maintain this record for at least forty (40) years, or for the duration of employment plus twenty (20) years, whichever is longer.

**1910.1045(q)(3)** *Medical surveillance.* **(i)** The employer shall establish and maintain an accurate record for each employee subject to medical surveillance as required by paragraph (n) of this section.

**1910.1045(q)(3)(ii)** This record shall include:

**1910.1045(q)(3)(ii)(A)** A copy of the physician's written opinions;

**1910.1045(q)(3)(ii)(B)** Any employee medical complaints related to exposure to AN;

**1910.1045(q)(3)(ii)(C)** A copy of the information provided to the physician as required by paragraph (n)(5) of this section; and

**1910.1045(q)(3)(ii)(D)** A copy of the employee's medical and work history.

**1910.1045(q)(3)(iii)** The employer shall assure that this record be maintained for at least forty (40) years, or for the duration of employment plus twenty (20) years, whichever is longer.

**1910.1045(q)(4)** *Availability.* **(i)** The employer shall make all records required to be maintained by this section available, upon request, to the Assistant Secretary and the Director for examination and copying.

**1910.1045(q)(4)(ii)** Records required by paragraphs (q)(1) through (q)(3) of this section shall be provided upon request to employees, designated representatives, and the Assistant Secretary in accordance with 29 CFR 1910.1020(a) through (e) and (g) through (i). Records required by paragraph (q)(1) shall be provided in the same manner as exposure monitoring records.

**1910.1045(q)(5)** *Transfer of records.* **(i)** Whenever the employer ceases to do business, the successor employer shall receive and retain all records required to be maintained by this section for the prescribed period.

**1910.1045(q)(5)(ii)** Whenever the employer ceases to do business and there is no successor employer to receive and retain the records for the prescribed period, these records shall be transmitted to the Director.

**1910.1045(q)(5)(iii)** At the expiration of the retention period for the records required to be maintained pursuant to this section, the employer shall notify the Director at least 3 months prior to the disposal of the records, and shall transmit them to the Director upon request.

**1910.1045(q)(5)(iv)** The employer shall also comply with any additional requirements involving transfer of records set forth in 29 CFR 1910.1020(h).

**1910.1045(r)** *Observation of monitoring*—**(1)** *Employee observation.* The employer shall provide affected employees, or their designated representatives, an opportunity to observe any monitoring of employee exposure to AN conducted pursuant to paragraph (e) of this section.

**1910.1045(r)(2)** *Observation procedures.* **(i)** Whenever observation of the monitoring of employee exposure to AN requires entry into an area where the use of protective clothing or equipment is required, the employer shall provide the observer with personal protective clothing and equipment required to be worn by employees working in the area, assure the use of such clothing and equipment, and require the observer to comply with all other applicable safety and health procedures.

**1910.1045(r)(2)(ii)** Without interfering with the monitoring, observers shall be entitled:

**1910.1045(r)(2)(ii)(A)** To receive an explanation of the measurement procedures;

**1910.1045(r)(2)(ii)(B)** To observe all steps related to the measurement of airborne concentrations of AN performed at the place of exposure; and

**1910.1045(r)(2)(ii)(C)** To record the results obtained.

**1910.1045(s)** [Reserved.]

**1910.1045(t)** *Appendixes.* The information contained in the appendixes is not intended, by itself, to create any additional obligation not otherwise imposed, or to detract from any obligation.

## APPENDIX A TO §1910.1045—SUBSTANCE SAFETY DATA SHEET FOR ACRYLONITRILE

### I. SUBSTANCE IDENTIFICATION

A. Substance: Acrylonitrile ($CH_2$ CHCN).

B. Synonyms: Propenenitrile; vinyl cyanide; cyanoethylene; AN; VCN; acylon; carbacryl; fumigrian; ventox.

C. Acrylonitrile can be found as a liquid or vapor, and can also be found in polymer resins, rubbers, plastics, polyols, and other polymers having acrylonitrile as a raw or intermediate material.

D. AN is used in the manufacture of acrylic and modiacrylic fibers, acrylic plastics and resins, speciality polymers, nitrile rubbers, and other organic chemicals. It has also been used as a fumigant.

E. Appearance and odor: Colorless to pale yellow liquid with a pungent odor which can only be detected at concentrations above the permissible exposure level, in a range of 13-19 parts AN per million parts of air (13-19 ppm).

F. Permissible exposure: Exposure may not exceed either:

1. Two parts AN per million parts of air (2 ppm) averaged over the 8-hour workday; or

2. Ten parts AN per million parts of air (10 ppm) averaged over any 15-minute period in the workday.

3. In addition, skin and eye contact with liquid AN is prohibited.

### II. HEALTH HAZARD DATA

A. Acrylonitrile can affect your body if you inhale the vapor (breathing), if it comes in contact with your eyes or skin, or if you swallow it. It may enter your body through your skin.

B. Effects of overexposure: 1. Short-term exposure: Acrylonitrile can cause eye irritation, nausea, vomiting, headache, sneezing, weakness, and light-headedness. At high concentrations, the effects of exposure may go on to loss of consciousness and death. When acrylonitrile is held in contact with the skin after being absorbed into shoe leather or clothing, it may produce blisters following several hours of no apparent effect. Unless the shoes or clothing are removed immediately and the area washed, blistering will occur. Usually there is no pain or inflammation associated with blister formation.

2. Long-term exposure: Acrylonitrile has been shown to cause cancer in laboratory animals and has been associated with higher incidences of cancer in humans. Repeated or prolonged exposure of the skin to acrylonitrile may produce irritation and dermatitis.

3. Reporting signs and symptoms: You should inform your employer if you develop any signs or symptoms and suspect they are caused by exposure to acrylonitrile.

### III. EMERGENCY FIRST AID PROCEDURES

A. Eye exposure: If acrylonitrile gets into your eyes, wash your eyes immediately with large amounts of water, lifting the lower and upper lids occasionally. Get medical attention immediately. Contact lenses should not be worn when working with this chemical.

B. Skin exposure: If acrylonitrile gets on your skin, immediately wash the contaminated skin with water. If acrylonitrile soaks through your clothing, especially your shoes, remove the clothing immediately and wash the skin with water. If symptoms occur after washing, get medical attention immediately. Thoroughly wash the clothing before reusing. Contaminated leather shoes or other leather articles should be discarded.

C. Inhalation: If you or any other person breathes in large amounts of acrylonitrile, move the exposed person to fresh air at once. If breathing has stopped, perform artificial respiration. Keep the affected person warm and at rest. Get medical attention as soon as possible.

D. Swallowing: When acrylonitrile has been swallowed, give the person large quantities of water immediately. After the water has been swallowed, try to get the person to vomit by having him touch the back of his throat with his finger. Do not make an unconscious person vomit. Get medical attention immediately.

E. Rescue: Move the affected person from the hazardous exposure. If the exposed person has been overcome, notify someone else and put into effect the established emergency procedures. Do not become a casualty yourself.

# OSHA Standards for General Industry

Understand your emergency rescue procedures and know the location of the emergency equipment before the need arises.

F. Special first aid procedures: First aid kits containing an adequate supply (at least two dozen) of amyl nitrite pearls, each containing 0.3 ml, should be maintained at each site where acrylonitrile is used. When a person is suspected of receiving an overexposure to acrylonitrile, immediately remove that person from the contaminated area using established rescue procedures. Contaminated clothing must be removed and the acrylonitrile washed from the skin immediately. Artificial respiration should be started at once if breathing has stopped. If the person is unconscious, amyl nitrite may be used as an antidote by a properly trained individual in accordance with established emergency procedures. Medical aid should be obtained immediately.

## IV. RESPIRATORS AND PROTECTIVE CLOTHING

A. Respirators. You may be required to wear a respirator for nonroutine activities, in emergencies, while your employer is in the process of reducing acrylonitrile exposures through engineering controls, and in areas where engineering controls are not feasible. If respirators are worn, they must have a label issued by the National Institute for Occupational Safety and Health under the provisions of 42 CFR part 84 stating that the respirators have been approved for use with organic vapors. For effective protection, respirators must fit your face and head snugly. Respirators must not be loosened or removed in work situations where their use is required.

Acrylonitrile does not have a detectable odor except at levels above the permissible exposure limits. Do not depend on odor to warn you when a respirator cartridge or canister is exhausted. Cartridges or canisters must be changed daily or before the end-of-service-life, whichever comes first. Reuse of these may allow acrylonitrille to gradually filter through the cartridge and cause exposures which you cannot detect by odor. If you can smell acrylonitrile while wearing a respirator, proceed immediately to fresh air. If you experience difficulty breathing while wearing a respirator, tell your employer.

B. Supplied-air suits: In some work situations, the wearing of supplied-air suits may be necessary. Your employer must instruct you in their proper use and operation.

C. Protective clothing: You must wear impervious clothing, gloves, face shield, or other appropriate protective clothing to prevent skin contact with liquid acrylonitrile. Where protective clothing is required, your employer is required to provide clean garments to you as necessary to assume that the clothing protects you adequately.

Replace or repair impervious clothing that has developed leaks.

Acrylonitrile should never be allowed to remain on the skin. Clothing and shoes which are not impervious to acrylonitrile should not be allowed to become contaminated with acrylonitrile, and if they do the clothing and shoes should be promptly removed and decontaminated. The clothing should be laundered or discarded after the AN is removed. Once acrylonitrile penetrates shoes or other leather articles, they should not be worn again.

D. Eye protection: You must wear splashproof safety goggles in areas where liquid acrylonitrile may contact your eyes. In addition, contact lenses should not be worn in areas where eye contact with acrylonitrile can occur.

## V. PRECAUTIONS FOR SAFE USE, HANDLING, AND STORAGE

A. Acrylonitrile is a flammable liquid, and its vapors can easily form explosive mixtures in air.

B. Acrylonitrile must be stored in tightly closed containers in a cool, well-ventilated area, away from heat, sparks, flames, strong oxidizers (especially bromine), strong bases, copper, copper alloys, ammonia, and amines.

C. Sources of ignition such as smoking and open flames are prohibited wherever acrylonitrile is handled, used, or stored in a manner that could create a potential fire or explosion hazard.

D. You should use non-sparking tools when opening or closing metal containers of acrylonitrile, and containers must be bonded and grounded when pouring or transferring liquid acrylonitrile.

E. You must immediately remove any non-impervious clothing that becomes wetted with acrylonitrile, and this clothing must not be reworn until the acrylonitrile is removed from the clothing.

F. Impervious clothing wet with liquid acrylonitrile can be easily ignited. This clothing must be washed down with water before you remove it.

G. If your skin becomes wet with liquid acrylonitrile, you must promptly and thoroughly wash or shower with soap or mild detergent to remove any acrylonitrile from your skin.

H. You must not keep food, beverages, or smoking materials, nor are you permitted to eat or smoke in regulated areas where acrylonitrile concentrations are above the permissible exposure limits.

I. If you contact liquid acrylonitrile, you must wash your hands thoroughly with soap or mild detergent and water before eating, smoking, or using toilet facilities.

J. Fire extinguishers and quick drenching facilities must be readily available, and you should know where they are and how to operate them.

K. Ask your supervisor where acrylonitrile is used in your work area and for any additional plant safety and health rules.

## VI. ACCESS TO INFORMATION

A. Each year, your employer is required to inform you of the information contained in this Substance Safety Data Sheet for acrylonitrile. In addition, you employer must instruct you in the proper work practices for using acrylonitrile, emergency procedures, and the correct use of protective equipment.

B. Your employer is required to determine whether you are being exposed to acrylonitrile. You or your representative has the right to observe employee measurements and to record the results obtained. Your employer is required to inform you of your exposure. If your employer determines that you are being overexposed, he or she is required to inform you of the actions which are being taken to reduce your exposure to within permissible exposure limits.

C. Your employer is required to keep records of your exposures and medical examinations. These records must be kept by the employer for at least forty (40) years or for the period of your employment plus twenty (20) years, whichever is longer.

D. Your employer is required to release your exposure and medical records to you or your representative upon your request.

## APPENDIX B TO §1910.1045—SUBSTANCE TECHNICAL GUIDELINES FOR ACRYLONITRILE

### I. PHYSICAL AND CHEMICAL DATA

A. Substance identification: 1. Synonyms: AN; VCN; vinyl cyanide; propenenitrile; cyanoethylene; Acrylon; Carbacryl; Fumigrain; Ventox.

2. Formula: $CH_2 = CHCN$.

3. Molecular weight: 53.1.

B. Physical data: 1. Boiling point (760 mm Hg): 77.3 °C (171 °F);

2. Specific gravity (water = 1): 0.81 (at 20 °C or 68 °F);

3. Vapor density (air = 1 at boiling point of acrylonitrile): 1.83;

4. Melting point: -83 °C (-117 °F);

5. Vapor pressure (@20 °F): 83 mm Hg;

6. Solubility in water, percent by weight @20 °C (68 °F): 7.35;

7. Evaporation rate (Butyl Acetate = 1): 4.54; and

8. Appearance and odor: Colorless to pale yellow liquid with a pungent odor at concentrations above the permissible exposure level. Any detectable odor of acrylonitrile may indicate overexposure.

### II. FIRE, EXPLOSION, AND REACTIVITY HAZARD DATA

A. Fire: 1. Flash point: -1 °C (30 °F) (closed cup).

2. Autoignition temperature: 481 °C (898 °F).

3. Flammable limits air, percent by volume: Lower: 3, Upper: 17.

4. Extinguishing media: Alcohol foam, carbon dioxide, and dry chemical.

5. Special fire-fighting procedures: Do not use a solid stream of water, since the stream will scatter and spread the fire. Use water to cool containers exposed to a fire.

6. Unusual fire and explosion hazards: Acrylonitrile is a flammable liquid. Its vapors can easily form explosive mixtures with air. All ignition sources must be controlled where acrylonitrile is handled, used, or stored in a

manner that could create a potential fire or explosion hazard. Acrylonitrile vapors are heavier than air and may travel along the ground and be ignited by open flames or sparks at locations remote from the site at which acrylonitrile is being handled.

7. For purposes of compliance with the requirements of 29 CFR 1910.106, acrylonitrile is classified as a class IB flammable liquid. For example, 7,500 ppm, approximately one-fourth of the lower flammable limit, would be considered to pose a potential fire and explosion hazard.

8. For purposes of compliance with 29 CFR 1910.157, acrylonitrile is classified as a Class B fire hazard.

9. For purpose of compliance with 29 CFR 1919.309, locations classified as hazardous due to the presence of acrylonitrile shall be Class I, Group D.

B. Reactivity:

1. Conditions contributing to instability: Acrylonitrile will polymerize when hot, and the additional heat liberated by the polymerization may cause containers to explode. Pure AN may self-polymerize, with a rapid build-up of pressure, resulting in an explosion hazard. Inhibitors are added to the commercial product to prevent self-polymerization.

2. Incompatibilities: Contact with strong oxidizers (especially bromine) and strong bases may cause fires and explosions. Contact with copper, copper alloys, ammonia, and amines may start serious decomposition.

3. Hazardous decompostion products: Toxic gases and vapors (such as hydrogen cyanide, oxides of nitrogen, and carbon monoxide) may be released in a fire involving acrylonitrile and certain polymers made from acrylonitrile.

4. Special precautions: Liquid acrylonitrile will attack some forms of plastics, rubbers, and coatings.

### III. SPILL, LEAK, AND DISPOSAL PROCEDURES

A. If acrylonitrile is spilled or leaked, the following steps should be taken:

1. Remove all ignition sources.

2. The area should be evacuated at once and re-entered only after the area has been thoroughly ventilated and washed down with water.

3. If liquid acrylonitrile or polymer intermediate, collect for reclamation or absorb in paper, vermiculite, dry sand, earth, or similar material, or wash down with water into process sewer system.

B. Persons not wearing protective equipment should be restricted from areas of spills or leaks until clean-up has been completed.

C. Waste disposal methods: Waste material shall be disposed of in a manner that is not hazardous to employees or to the general population. Spills of acrylonitrile and flushing of such spills shall be channeled for appropriate treatment or collection for disposal. They shall not be channeled directly into the sanitary sewer system. In selecting the method of waste disposal, applicable local, State, and Federal regulations should be consulted.

### IV. MONITORING AND MEASUREMENT PROCEDURES

A. Exposure above the Permissible Exposure Limit:

1. Eight-hour exposure evaluation: Measurements taken for the purpose of determining employee exposure under this section are best taken so that the average 8-hour exposure may be determined from a single 8-hour sample or two (2) 4-hour samples. Air samples should be taken in the employee's breathing zone (air that would most nearly represent that inhaled by the employee.)

2. Ceiling evaluation: Measurements taken for the purpose of determining employee exposure under this section must be taken during periods of maximum expected airborne concentrations of acrylonitrile in the employee's breathing zone. A minimum of three (3) measurements should be taken on one work shift. The average of all measurements taken is an estimate of the employee's ceiling exposure.

3. Monitoring techniques: The sampling and analysis under this section may be performed by collecting the acrylonitrile vapor on charcoal adsorption tubes or other composition adsorption tubes, with subsequent chemical analysis. Sampling and analysis may also be performed by instruments such as real-time continuous monitoring systems, portable direct-reading instruments, or passive dosimeters. Analysis of resultant samples should be by gas chromatograph.

Appendix D lists methods of sampling and analysis which have been tested by NIOSH and OSHA for use with acrylonitrile. NIOSH and OSHA have validated modifications of NIOSH Method S-156 (See Appendix D) under laboratory conditions for concentrations below 1 ppm. The employer has the obligation of selecting a monitoring method which meets the accuracy and precision requirements of the standard under his unique field conditions. The standard requires that methods of monitoring must be accurate, to a 95-percent confidence level, to ±35-percent for concentrations of AN at or above 2 ppm, and to ±50-percent for concentrations below 2 ppm. In addition to the methods described in Appendix D, there are numerous other methods available for monitoring for AN in the workplace. Details on these other methods have been submitted by various companies to the rulemaking record, and are available at the OSHA Docket Office.

B. Since many of the duties relating to employee exposure are dependent on the results of monitoring and measuring procedures, employers shall assure that the evaluation of employee exposures is performed by a competent industrial hygienist or other technically qualified person.

### V. PROTECTIVE CLOTHING

Employees shall be provided with and required to wear appropriate protective clothing to prevent any possibility of skin contact with liquid AN. Because acrylonitrile is absorbed through the skin, it is important to prevent skin contact with liquid AN. Protective clothing shall include impermeable coveralls or similar full-body work clothing, gloves, head-coverings, as appropriate to protect areas of the body which may come in contact with liquid AN.

Employers should ascertain that the protective garmets are impermeable to acrylonitrile. Non-impermeable clothing and shoes should not be allowed to become contaminated with liquid AN. If permeable clothing does become contaminated, it should be promptly removed, placed in a regulated area for removal of the AN, and not worn again until the AN is removed. If leather footwear or other leather garments become wet from acrylonitrile, they should be replaced and not worn again, due to the ability of leather to absorb acrylonitrile and hold it against the skin. Since there is no pain associated with the blistering which may result from skin contact with liquid AN, it is essential that the employee be informed of this hazard so that he or she can be protected.

Any protective clothing which has developed leaks or is otherwise found to be defective shall be repaired or replaced. Clean protective clothing shall be provided to the employee as necessary to assure its protectiveness. Whenever impervious clothing becomes wet with liquid AN, it shall be washed down with water before being removed by the employee. Employees are also required to wear splash-proof safety goggles where there is any possibility of acrylonitrile contacting the eyes.

### VI. HOUSEKEEPING AND HYGIENE FACILITIES

For purposes of complying with 29 CFR 1910.141, the following items should be emphasized:

A. The workplace should be kept clean, orderly, and in a sanitary condition. The employer is required to institute a leak and spill detection program for operations involving liquid AN in order to detect sources of fugitive AN emissions.

B. Dry sweeping and the use of compressed air is unsafe for the cleaning of floors and other surfaces where liquid AN may be found.

C. Adequate washing facilities with hot and cold water are to be provided, and maintained in a sanitary condition. Suitable cleansing agents are also to be provided to assure the effective removal of acrylonitrile from the skin.

D. Change or dressing rooms with individual clothes storage facilities must be provided to prevent the contamination of street clothes with acrylonitrile. Because of the hazardous nature of acrylonitrile, contaminated protective clothing should be placed in a regulated area designated by the employer for removal of the AN before the clothing is laundered or disposed of.

### VII. MISCELLANEOUS PRECAUTIONS

A. Store acrylonitrile in tightly-closed containers in a cool, well-ventilated area and take necessary precautions to avoid any explosion hazard.

B. High exposures to acrylonitrile can occur when transferring the liquid from one container to another.

# OSHA Standards for General Industry

C. Non-sparking tools must be used to open and close metal acrylonitrile containers. These containers must be effectively grounded and bonded prior to pouring.

D. Never store uninhibited acrylonitrile.

E. Acrylonitrile vapors are not inhibited. They may form polymers and clog vents of storage tanks.

F. Use of supplied-air suits or other impervious coverings may be necessary to prevent skin contact with and provide respiratory protection from acrylonitrile where the concentration of acrylonitrile is unknown or is above the ceiling limit. Supplied-air suits should be selected, used, and maintained under the immediate supervision of persons knowledgeable in the limitations and potential life-endangering characteristics of supplied-air suits.

G. Employers shall advise employees of all areas and operations where exposure to acrylonitrile could occur.

## VIII. COMMON OPERATIONS

Common operations in which exposure to acrylonitrile is likely to occur include the following: Manufacture of the acrylonitrile monomer; synthesis of acrylic fibers, ABS, SAN, and nitrile barrier plastics and resins, nitrile rubber, surface coatings, specialty chemicals, use as a chemical intermediate, use as a fumigant and in the cyanoethylation of cotton.

## APPENDIX C TO §1910.1045—MEDICAL SURVEILLANCE GUIDELINES FOR ACRYLONITRILE

### I. ROUTE OF ENTRY

Inhalation; skin absorption; ingestion.

### II. TOXICOLOGY

Acrylonitrile vapor is an asphyxiant due to inhibitory action on metabolic enzyme systems. Animals exposed to 75 or 100 ppm for 7 hours have shown signs of anoxia; in some animals which died at the higher level, cyanomethemoglobin was found in the blood. Two human fatalities from accidental poisoning have been reported; one was caused by inhalation of an unknown concentration of the vapor, and the other was thought to be caused by skin absorption or inhalation. Most cases of intoxication from industrial exposure have been mild, with rapid onset of eye irritation, headache, sneezing, and nausea. Weakness, lightheadedness, and vomiting may also occur. Exposure to high concentrations may produce profound weakness, asphyxia, and death. The vapor is a severe eye irritant. Prolonged skin contract with the liquid may result in absorption with systemic effects, and in the formation of large blisters after a latent period of several hours. Although there is usually little or no pain or inflammation, the affected skin resembles a second-degree thermal burn. Solutions spilled on exposed skin, or on areas covered only by a light layer of clothing, evaporate rapidly, leaving no irritation, or, at the most, mild transient redness. Repeated spills on exposed skin may result in dermatitis due to solvent effects.

Results after 1 year of a planned 2-year animal study on the effects of exposure to acrylonitrile have indicated that rats ingesting as little as 35 ppm in their drinking water develop tumors of the central nervous system. The interim results of this study have been supported by a similar study being conducted by the same laboratory, involving exposure of rats by inhalation of acrylonitrile vapor, which has shown similar types of tumors in animals exposed to 80 ppm.

In addition, the preliminary results of an epidemiological study being performed by duPont on a cohort of workers in their Camden, S.C. acrylic fiber plant indicate a statistically significant increase in the incidence of colon and lung cancers among employees exposed to acrylonitrile.

### III. SIGNS AND SYMPTOMS OF ACUTE OVEREXPOSURE

Asphyxia and death can occur from exposure to high concentrations of acrylonitrile. Symptoms of overexposure include eye irritation, headache, sneezing, nausea and vomiting, weakness, and light-headedness. Prolonged skin contact can cause blisters on the skin with appearance of a second-degree burn, but with little or no pain. Repeated skin contact may produce scaling dermatits.

### IV. TREATMENT OF ACUTE OVEREXPOSURE

Remove employee from exposure. Immediately flush eyes with water and wash skin with soap or mild detergent and water. If AN has been swallowed, and person is conscious, induce vomiting. Give artificial resuscitation if indicated. More severe cases, such as those associated with loss of consciousness, may be treated by the intravenous administration of sodium nitrite, followed by sodium thiosulfate, although this is not as effective for acrylonitrile poisoning as for inorganic cyanide poisoning.

### V. SURVEILLANCE AND PREVENTIVE CONSIDERATIONS

A. As noted above, exposure to acrylonitrile has been linked to increased incidence of cancers of the colon and lung in employees of the duPont acrylic fiber plant in Camden, S.C. In addition, the animal testing of acrylonitrile has resulted in the development of cancers of the central nervous system in rats exposed by either inhalation or ingestion. The physician should be aware of the findings of these studies in evaluating the health of employees exposed to acrylonitrile.

Most reported acute effects of occupational exposure to acrylonitrile are due to its ability to cause tissue anoxia and asphyxia. The effects are similar to those caused by hydrogen cyanide. Liquid acrylonitrile can be absorbed through the skin upon prolonged contact. The liquid readily penetrates leather, and will produce burns of the feet if footwear contaminated with acrylonitrile is not removed.

It is important for the physician to become familiar with the operating conditions in which exposure to acrylonitrile may occur. Those employees with skin diseases may not tolerate the wearing of whatever protective clothing may be necessary to protect them from exposure. In addition, those with chronic respiratory disease may not tolerate the wearing of negativepressure respirators.

B. Surveillance and screening. Medical histories and laboratory examinations are required for each employee subject to exposure to acrylonitrile above the action level. The employer must screen employees for history of certain medical conditions which might place the employee at increased risk from exposure.

1. *Central nervous system dysfunction.* Acute effects of exposure to acrylonitrile generally involve the central nervous system. Symptoms of acrylonitrile exposure include headache, nausea, dizziness, and general weakness. The animal studies cited above suggest possible carcinogenic effects of acrylonitrile on the central nervous system, since rats exposed by either inhalation or ingestion have developed similar CNS tumors.

2. *Respiratory disease.* The du Pont data indicate an increased risk of lung cancer among employees exposed to acrylonitrile.

3. *Gastrointestinal disease.* The du Pont data indicate an increased risk of cancer of the colon among employees exposed to acrylonitrile. In addition, the animal studies show possible tumor production in the stomachs of the rats in the ingestion study.

4. *Skin disease.* Acrylonitrile can cause skin burns when prolonged skin contact with the liquid occurs. In addition, repeated skin contact with the liquid can cause dermatitis.

5. *General.* The purpose of the medical procedures outlined in the standard is to establish a baseline for future health monitoring. Persons unusually susceptible to the effects of anoxia or those with anemia would be expected to be at increased risk. In addition to emphasis on the CNS, respiratory and gastro-intestinal systems, the cardiovascular system, liver, and kidney function should also be stressed.

## APPENDIX D TO §1910.1045—SAMPLING AND ANALYTICAL METHODS FOR ACRYLONITRILE

There are many methods available for monitoring employee exposures to acrylonitrile. Most of these involve the use of charcoal tubes and sampling pumps, with analysis by gas chromatograph. The essential differences between the charcoal tube methods include, among others, the use of different desorbing solvents, the use of different lots of charcoal, and the use of different equipment for analysis of the samples.

Besides charcoal, considerable work has been performed on methods using porous polymer sampling tubes and passive dosimeters. In addition, there are several portable gas analyzers and monitoring units available on the open market.

This appendix contains details for the methods which have been tested at OSHA Analytical Laboratory in Salt Lake City, and NIOSH in Cincinnati. Each is a variation on NIOSH Method S-156, which is also included for reference. This does not indicate that these methods are the only ones which will be satisfactory. There also may be workplace situations in which these methods are not adequate, due to such factors as high humidity. Copies of the other methods available to OSHA are available in the rulemaking record, and may be obtained from the OSHA Docket Office. These include, the Union Carbide, Monsanto, Dow Chemical and Dow Badische methods, as well as NISOH Method P & CAM 127.

Employers who note problems with sample breakthrough should try larger charcoal tubes. Tubes of larger capacity are available, and are often used for sampling vinyl chloride. In addition, lower flow rates and shorter sampling times should be beneficial in minimizing breakthrough problems.

Whatever method the employer chooses, he must assure himself of the method's accuracy and precision under the unique conditions present in his workplace.

### NIOSH METHOD S-156 (UNMODIFIED)

Analyte: Acrylonitrile.

Matrix: Air.

Procedure: Absorption on charcoal, desorption with methanol, GC.

1. *Principle of the method* (Reference 11.1).

1.1 A known volume of air is drawn through a charcoal tube to trap the organic vapors present.

1.2 The charcoal in the tube is transferred to a small, stoppered sample container, and the analyte is desorbed with methanol.

1.3 An aliquot of the desorbed sample is injected into a gas chromatograph.

1.4 The area of the resulting peak is determined and compared with areas obtained for standards.

2. *Range and sensitivity.*

2.1 This method was validated over the range of 17.5-70.0 mg/cu m at an atmospheric temperature and pressure of 22 °C and 760 MM Hg, using a 20-liter sample. Under the conditions of sample size (20-liters) the probable useful range of this method is 4.5-135 mg-cu m. The method is capable of measuring much smaller amounts if the desorption efficiency is adequate. Desorption efficiency must be determined over the range used.

2.2 The upper limit of the range of the method is dependent on the adsorptive capacity of the charcoal tube. This capacity varies with the concentrations of acrylonitrile and other substances in the air. The first section of the charcoal tube was found to hold at least 3.97 mg of acrylonitrile when a test atmosphere containing 92.0 mg/cu m of acrylonitrile in air was sampled 0.18 liter per minute for 240 minutes; at that time the concentration of acrylonitrile in the effluent was less than 5 percent of that in the influent. (The charcoal tube consists of two sections of activated charcoal separated by a section of urethane foam. See section 6.2.) If a particular atmosphere is suspected of containing a large amount of contaminant, a smaller sampling volume should be taken.

3. *Interference.*

3.1 When the amount of water in the air is so great that condensation actually occurs in the tube, organic vapors will not be trapped efficiently. Preliminary experiments using toluene indicate that high humidity severely decreases the breakthrough volume.

3.2 When interfering compounds are known or suspected to be present in the air, such information, including their suspected identities, should be transmitted with the sample.

3.3 It must be emphasized that any compound which has the same retention time as the analyte at the operating conditions described in this method is an interference. Retention time data on a single column cannot be considered proof of chemical identity.

3.4 If the possibility of interference exists, separation conditions (column packing, temperature, etc.) must be changed to circumvent the problem.

4. *Precision and accuracy.*

4.1 The Coefficient of Variation (CVT) for the total analytical and sampling method in the range of 17.5-70.0 mg/cu m was 0.073. This value corresponds to a 3.3 mg/cu m standard deviation at the (previous) OSHA standard level (20 ppm). Statistical information and details of the validation and experimental test procedures can be found in Reference 11.2.

4.2 On the average the concentrations obtained at the 20 ppm level using the overall sampling and analytical method were 6.0 percent lower than the "true" concentrations for a limited number of laboratory experiments. Any difference between the "found" and "true" concentrations may not represent a bias in the sampling and analytical method, but rather a random variation from the experimentally determined "true" concentration. Therefore, no recovery correction should be applied to the final result in section 10.5.

5. *Advantages and disadvantages of the method.*

5.1 The sampling device is small, portable, and involves no liquids. Interferences are minimal, and most of those which do occur can be eliminated by altering chromatographic conditions. The tubes are analyzed by means of a quick, instrumental method.

The method can also be used for the simultaneous analysis of two or more substances suspected to be present in the same sample by simply changing gas chromatographic conditions.

5.2 One disadvantage of the method is that the amount of sample which can be taken is limited by the number of milligrams that the tube will hold before overloading. When the sample value obtained for the backup section of the charcoal tube exceeds 25 percent of that found on the front section, the possibility of sample loss exists.

5.3 Furthermore, the precision of the method is limited by the reproducibility of the pressure drop across the tubes. This drop will affect the flow rate and cause the volume to be imprecise, because the pump is usually calibrated for one tube only.

6. *Apparatus.*

6.1 A calibrated personal sampling pump whose flow can be determined within ±5 percent at the recommended flow rate. (Reference 11.3).

6.2 Charcoal tubes: Glass tubes with both ends flame sealed, 7 cm long with a 6-mm O.D. and a 4-mm I.D., containing 2 sections of 20/40 mesh activated charcoal separated by a 2-mm portion of urethane foam. The activated charcoals prepared from coconut shells and is fired at 600 °C prior to packing. The adsorbing section contains 100 mg of charcoal, the backup section 50 mg. A 3-mm portion of urethane foam is placed between the outlet end of the tube and the backup section. A plug of silicated glass wool is placed in front of the adsorbing section. The pressure drop across the tube must be less than 1 inch of mercury at a flow rate of 1 liter per minute.

6.3 Gas chromatograph equipped with a flame ionization detector.

6.4 Column (4-ft × 1/4-in stainless steel) packed with 50/80 mesh Poropak, type Q.

6.5 An electronic integrator or some other suitable method for measuring peak areas.

6.6 Two-milliliter sample containers with glass stoppers or Teflon-lined caps. If an automatic sample injector is used, the associated vials may be used.

6.7 Microliter syringes: 10-microliter and other convenient sizes for making standards.

6.8 Pipets: 1.0-ml delivery pipets.

6.9 Volumetric flask: 10-ml or convenient sizes for making standard solutions.

7. *Reagents.*

7.1 Chromatographic quality methanol.

7.2 Acrylonitrile, reagent grade.

7.3 Hexane, reagent grade.

7.4 Purified nitrogen.

7.5 Prepurified hydrogen.

7.6 Filtered compressed air.

8. *Procedure.*

8.1 Cleaning of equipment. All glassware used for the laboratory analysis should be detergent washed and thoroughly rinsed with tap water and distilled water.

8.2 Calibration of personal pumps. Each personal pump must be calibrated with a representative charcoal tube in the line. This will minimize errors associated with uncertainties in the sample volume collected.

8.3 Collection and shipping of samples.

8.3.1 Immediately before sampling, break the ends of the tube to provide an opening at least one-half the internal diameter of the tube (2 mm).

8.3.2 The smaller section of charcoal is used as a backup and should be positioned nearest the sampling pump.

8.3.3 The charcoal tube should be placed in a vertical direction during sampling to minimize channeling through the charcoal.

8.3.4 Air being sampled should not be passed through any hose or tubing before entering the charcoal tube.

8.3.5 A maximum sample size of 20 liters is recommended. Sample at a flow of 0.20 liter per minute or less. The flow rate should be known with an accuracy of at least ±5 percent.

8.3.6 The temperature and pressure of the atmosphere being sampled should be recorded. If pressure reading is not available, record the elevation.

8.3.7 The charcoal tubes should be capped with the supplied plastic caps immediately after sampling. Under no circumstances should rubber caps be used.

8.3.8 With each batch of 10 samples submit one tube from the same lot of tubes which was used for sample collection and which is subjected to exactly the same handling as the samples except that no air is drawn through it. Label this as a blank.

8.3.9 Capped tubes should be packed tightly and padded before they are shipped to minimize tube breakage during shipping.

8.3.10 A sample of the bulk material should be submitted to the laboratory in a glass container with a Teflon-lined cap. This sample should not be transported in the same container as the charcoal tubes.

8.4 *Analysis of samples.*

8.4.1 Preparation of samples. In preparation for analysis, each charcoal tube is scored with a file in front of the first section of charcoal and broken open. The glass wool is removed and discarded. The charcoal in the first (larger) section is transferred to a 2-ml stoppered sample container. The separating section of foam is removed and discarded; the second section is transferred to another stoppered container. These two sections are analyzed separately.

8.4.2 Desorption of samples. Prior to analysis, 1.0 ml of methanol is pipetted into each sample container. Desorption should be done for 30 minutes. Tests indicate that this is adequate if the sample is agitated occasionally during this period. If an automatic sample injector is used, the sample vials should be capped as soon as the solvent is added to minimize volatilization.

8.4.3 GC conditions. The typical operating conditions for the gas chromatograph are:

1. 50 ml/min (60 psig) nitrogen carrier gas flow.
2. 65 ml/min (24 psig) hydrogen gas flow to detector.
3. 500 ml/min (50 psig) air flow to detector.
4. 235 °C injector temperature.
5. 255 °C manifold temperature (detector).
6. 155 °C column temperature.

8.4.4 Injection. The first step in the analysis is the injection of the sample into the gas chromatograph. To eliminate difficulties arising from blowback or distillation within the syringe needle, one should employ the solvent flush injection technique. The 10-microliter syringe is first flushed with solvent several times to wet the barrel and plunger. Three microliters of solvent are drawn into the syringe to increase the accuracy and reproducibility of the injected sample volume. The needle is removed from the solvent, and the plunger is pulled back about 0.2 microliter to separate the solvent flush from the sample with a pocket of air to be used as a marker. The needle is then immersed in the sample, and a 5-microliter aliquot is withdrawn, taking into consideration the volume of the needle, since the sample in the needle will be completely injected. After the needle is removed from the sample and prior to injection, the plunger is pulled back 1.2 microliters to minimize evaporation of the sample from the tip of the needle. Observe that the sample occupies 4.9-5.0 microliters in the barrel of the syringe. Duplicate injections of each sample and standard should be made. No more than a 3 percent difference in area is to be expected. An automatic sample injector can be used if it is shown to give reproducibility at least as good as the solvent flush method.

8.4.5 Measurement of area. The area of the sample peak is measured by an electronic integrator or some other suitable form of area measurement, and preliminary results are read from a standard curve prepared as discussed below.

8.5 *Determination of desorption efficiency.*

8.5.1 Importance of determination. The desorption efficiency of a particular compound can vary from one laboratory to another and also from one batch of charcoal to another. Thus, it is necessary to determine at least once the percentage of the specific compound that is removed in the desorption process, provided the same batch of charcoal is used.

8.5.2 Procedure for determining desorption efficiency. Activated charcoal equivalent to the amount in the first section of the sampling tube (100 mg) is measured into a 2.5 in, 4-mm I.D. glass tube, flame sealed at one end. This charcoal must be from the same batch as that used in obtaining the samples and can be obtained from unused charcoal tubes. The open end is capped with Parafilm. A known amount of hexane solution of acrylonitrile containing 0.239 g/ml is injected directly into the activated charcoal with a microliter syringe, and tube is capped with more Parafilm. When using an automatic sample injector, the sample injector vials, capped with Teflon-faced septa, may be used in place of the glass tube.

The amount injected is equivalent to that present in a 20 liter air sample at the selected level.

Six tubes at each of three levels (0.5X, 1X, and 2X of the standard) are prepared in this manner and allowed to stand for at least overnight to assure complete adsorption of the analyte onto the charcoal. These tubes are referred to as the sample. A parallel blank tube should be treated in the same manner except that no sample is added to it. The sample and blank tubes are desorbed and analyzed in exactly the same manner as the sampling tube described in section 8.4.

Two or three standards are prepared by injecting the same volume of compound into 1.0 ml of methanol with the same syringe used in the preparation of the samples. These are analyzed with the samples.

The desorption efficiency (D.E.) equals the average weight in mg recovered from the tube divided by the weight in mg added to the tube, or

$$\text{D.E.} = \frac{\text{Average weight recovered (mg)}}{\text{weight added (mg)}}$$

The desorption efficiency is dependent on the amount of analyte collected on the charcoal. Plot the desorption efficiency versus weight of analyte found. This curve is used in section 10.4 to correct for adsorption losses.

9. *Calibration and standards.*

It is convenient to express concentration of standards in terms of mg/1.0 ml methanol, because samples are desorbed in this amount of methanol. The density of the analyte is used to convert mg into microliters for easy measurement with a microliter syringe. A series of standards, varying in concentration over the range of interest, is prepared and analyzed under the same GC conditions and during the same time period as the unknown samples. Curves are established by plotting concentration in mg/1.0 ml versus peak area.

**NOTE:** Since no internal standard is used in the method, standard solutions must be analyzed at the same time that the sample analysis is done. This will minimize the effect of known day-to-day variations and variations during the same day of the FID response.

10. *Calculations.*

10.1 Read the weight, in mg, corresponding to each peak area from the standard curve. No volume corrections are needed, because the standard curve is based on mg/1.0 ml methanol and the volume of sample injected is identical to the volume of the standards injected.

10.2 Corrections for the blank must be made for each sample.

$$\text{mg} = \text{mg sample} - \text{mg blank}$$

Where:

mg sample = mg found in front section of sample tube.

mg sample = mg found in front section of blank tube.

A similar procedure is followed for the backup sections.

10.3 Add the weights found in the front and backup sections to get the total weight in the sample.

10.4 Read the desorption efficiency from the curve (see sec. 8.5.2) for the amount found in the front section. Divide the total weight by this desorption efficiency to obtain the corrected mg/sample.

**Corrected mg/sample = Total weight/D.E.**

10.5 The concentration of the analyte in the air sampled can be expressed in mg/cu m.

**mg/cu m = Corrected mg (section 10.4) × 1,000 (liter/cu m)/air volume sampled (liter)**

10.6 Another method of expressing concentration is ppm.

**ppm = m mg/cu × 24.45/M.W. × 760/P × T. + 273/298**

Where:

$P$ = Pressure (mm Hg) of air sampled.

$T$ = Temperature (°C) of air sampled.

24.45 = Molar volume (liter/mole) at 25 °C and 760 mm Hg.

$M.W.$ = Molecular weight (g/mole) of analyte.

760 = Standard pressure (mm Hg).

298 = Standard temperature (°K).

11. *References.*

11.1 White, L. D. et al., "A Convenient Optimized Method for the Analysis of Selected Solvent Vapors in the Industrial Atmosphere," *Amer. Ind. Hyg. Assoc. J.*, 31: 225 (1970).

11.2 Documentation of NIOSH Validation Tests, NIOSH Contract No. CDC-99-74-45.

11.3 Final Report, NIOSH Contract HSM-99-71-31, "Personal Sampler Pump for Charcoal Tubes," September 15, 1972.

### NIOSH Modification of NIOSH Method S-156

The NIOSH recommended method for low levels for acrylonitrile is a modification of method S-156. It differs in the following respects:

(1) Samples are desorbed using 1 ml of 1 percent acetone in CS2 rather than methanol.

(2) The analytical column and conditions are:

Column: 20 percent SP-1000 on 80/100 Supelcoport 10 feet × 1/8 inch S.S.

Conditions:

Injector temperature: 200 °C.

Detector temperature: 100 °C.

Column temperature: 85 °C.

Helium flow: 25 ml/min.

Air flow: 450 ml/min.

Hydrogen flow: 55 ml/min.

(3) A 2 µl injection of the desorbed analyte is used.

(4) A sampling rate of 100 ml/min is recommended.

### OSHA Laboratory Modification of NIOSH Method S-156

Analyte: Acrylonitrile.

Matrix: Air.

Procedure: Adsorption on charcoal, desorption with methanol, GC.

1. *Principle of the Method* (Reference 1).

1.1 A known volume of air is drawn through a charcoal tube to trap the organic vapors present.

1.2 The charcoal in the tube is transferred to a small, stoppered sample vial, and the analyte is desorbed with methanol.

1.3 An aliquot of the desorbed sample is injected into a gas chromatograph.

1.4 The area of the resulting peak is determined and compared with areas obtained for standards.

2. *Advantages and disadvantages of the method.*

2.1 The sampling device is small, portable, and involves no liquids. Interferences are minimal, and most of those which do occur can be eliminated by altering chromatographic conditions. The tubes are analyzed by means of a quick, instrumental method.

2.2 This method may not be adequate for the simultaneous analysis of two or more substances.

2.3 The amount of sample which can be taken is limited by the number of milligrams that the tube will hold before overloading. When the sample value obtained for the backup section of the charcoal tube exceeds 25 percent of that found on the front section, the possibility of sample loss exists.

2.4 The precision of the method is limited by the reproducibility of the pressure drop across the tubes. This drop will affect the flow rate and cause the volume to be imprecise, because the pump is usually calibrated for one tube only.

3. *Apparatus.*

3.1 A calibrated personal sampling pump whose flow can be determined within ±5 percent at the recommended flow rate.

3.2 Charcoal tubes: Glass tube with both ends flame sealed, 7 cm long with a 6-mm O.D. and a 4-mm I.D., containing 2 sections of 20/40 mesh activated charcoal separated by a 2-mm portion of urethane foam. The activated charcoal is prepared from coconut shells and is fired at 600 °C prior to packing. The adsorbing section contains 100 mg of charcoal, the back-up section 50 mg. A 3-mm portion of urethane foam is placed between the outlet end of the tube and the back-up section. A plug of sililated glass wool is placed in front of the adsorbing section. The pressure drop across the tube must be less than one inch of mercury at a flow rate of 1 liter per minute.

3.3 Gas chromatograph equipped with a nitrogen phosphorus detector.

3.4 Column (10-ft × 1/8-in stainless steel) packed with 100/120 Supelcoport coated with 10 percent SP 1000.

3.5 An electronic integrator or some other suitable method for measuring peak area.

3.6 Two-milliliter sample vials with Teflon-lined caps

3.7 Microliter syringes: 10-microliter, and other convenient sizes for making standards.

3.8 Pipets: 1.0-ml delivery pipets.

3.9 Volumetric flasks: convenient sizes for making standard solutions.

4. *Reagents.*

4.1 Chromatographic quality methanol.

4.2 Acrylonitrile, reagent grade.

4.3 Filtered compressed air.

4.4 Purified hydrogen.

4.5 Purified helium.

5. *Procedure.*

5.1 Cleaning of equipment. All glassware used for the laboratory analysis should be properly cleaned and free of organics which could interfere in the analysis.

5.2 Calibration of personal pumps. Each pump must be calibrated with a representative charcoal tube in the line.

5.3 Collection and shipping of samples.

5.3.1 Immediately before sampling, break the ends of the tube to provide an opening at least one-half the internal diameter of the tube (2 mm).

5.3.2 The smaller section of the charcoal is used as the backup and should be placed nearest the sampling pump.

5.3.3 The charcoal should be placed in a vertical position during sampling to minimize channeling through the charcoal.

5.3.4 Air being sampled should not be passed through any hose or tubing before entering the charcoal tube.

5.3.5 A sample size of 20 liters is recommended. Sample at a flow rate of approximately 0.2 liters per minute. The flow rate should be known with an accuracy of at least ±5 percent.

5.3.6 The temperature and pressure of the atmosphere being sampled should be recorded.

5.3.7 The charcoal tubes should be capped with the supplied plastic caps immediately after sampling. Rubber caps should not be used.

5.3.8 Submit at least one blank tube (a charcoal tube subjected to the same handling procedures, without having any air drawn through it) with each set of samples.

# OSHA Standards for General Industry

5.3.9. Take necessary shipping and packing precautions to minimize breakage of samples.

5.4 Analysis of samples.

5.4.1 Preparation of samples. In preparation for analysis, each charcoal tube is scored with a file in front of the first section of charcoal and broken open. The glass wool is removed and discarded. The charcoal in the first (larger) section is transferred to a 2-ml vial. The separating section of foam is removed and discarded; the section is transferred to another capped vial. These two sections are analyzed separately.

5.4.2 Desorption of samples. Prior to analysis, 1.0 ml of methanol is pipetted into each sample container. Desorption should be done for 30 minutes in an ultrasonic bath. The sample vials are recapped as soon as the solvent is added.

5.4.3 GC conditions. The typical operating conditions for the gas chromatograph are:

1. 30 ml/min (60 psig) helium carrier gas flow.
2. 3.0 ml/min (30 psig) hydrogen gas flow to detector.
3. 50 ml/min (60 psig) air flow to detector.
4. 200 °C injector temperature.
5. 200 °C dejector temperature.
6. 100 °C column temperature.

5.4.4 Injection. Solvent flush technique or equivalent.

5.4.5 Measurement of area. The area of the sample peak is measured by an electronic integrator or some other suitable form of area measurement, and preliminary results are read from a standard curve prepared as discussed below.

5.5 Determination of desorption efficiency.

5.5.1 Importance of determination. The desorption efficiency of a particular compound can vary from one laboratory to another and also from one batch of charcoal to another. Thus, it is necessary to determine, at least once, the percentage of the specific compound that is removed in the desorption process, provided the same batch of charcoal is used.

5.5.2 Procedure for determining desorption efficiency. The reference portion of the charcoal tube is removed. To the remaining portion, amounts representing 0.5X, 1X, and 2X (X represents TLV) based on a 20 l air sample are injected onto several tubes at each level. Dilutions of acrylonitrile with methanol are made to allow injection of measurable quantities. These tubes are then allowed to equilibrate at least overnight. Following equilibration they are analyzed following the same procedure as the samples A curve of the desorption efficiency amt recovered/amt added is plotted versus amount of analyte found. This curve is used to correct for adsorption losses.

6. *Calibration and standards.*

A series of standards, varying in concentration over the range of interest, is prepared and analyzed under the same GC conditions and during the same time period as the unknown samples. Curves are prepared by plotting concentration versus peak area.

**NOTE:** Since no internal standard is used in the method, standard solutions must be analyzed at the same time that the sample analysis is done. This will minimize the effect of known day-to-day variations and variations during the same day of the NPD response. Multiple injections are necessary.

7. *Calculations.*

Read the weight, corresponding to each peak area from the standard curve, correct for the blank, correct for the desorption efficiency, and make necessary air volume corrections.

8. *Reference.* NIOSH Method S-156.

[43 FR 45809, Oct. 3, 1978, as amended at 45 FR 35283, May 23, 1980; 54 FR 24334, June 7, 1989; 58 FR 35310, June 30, 1993; 61 FR 5508, Feb. 13, 1996; 63 FR 1291, Jan. 8, 1998; 63 FR 20099, Apr. 23, 1998]

## §1910.1047 Ethylene oxide.

**1910.1047(a)** *Scope and application.* **(1)** This section applies to all occupational exposures to ethylene oxide (EtO), Chemical Abstracts Service Registry No. 75-21-8, except as provided in paragraph (a)(2) of this section.

**1910.1047(a)(2)** This section does not apply to the processing, use, or handling of products containing EtO where objective data are reasonably relied upon that demonstrate that the product is not capable of releasing EtO in airborne concentrations at or above the action level, and may not reasonably be foreseen to release EtO in excess of the excursion limit, under the expected conditions of processing, use, or handling that will cause the greatest possible release.

**1910.1047(a)(3)** Where products containing EtO are exempted under paragraph (a)(2) of this section, the employer shall maintain records of the objective data supporting that exemption and the basis for the employer's reliance on the data, as provided in paragraph (k)(1) of this section.

**1910.1047(b)** *Definitions:* For the purpose of this section, the following definitions shall apply:

*Action level* means a concentration of airborne EtO of 0.5 ppm calculated as an eight (8)-hour time-weighted average.

*Assistant Secretary* means the Assistant Secretary of Labor for Occupational Safety and Health, U.S. Department of Labor, or designee.

*Authorized person* means any person specifically authorized by the employer whose duties require the person to enter a regulated area, or any person entering such an area as a designated representative of employees for the purpose of exercising the right to observe monitoring and measuring procedures under paragraph (l) of this section, or any other person authorized by the Act or regulations issued under the Act.

*Director* means the Director of the National Institute for Occupational Safety and Health, U.S. Department of Health and Human Services, or designee.

*Emergency* means any occurrence such as, but not limited to, equipment failure, rupture of containers, or failure of control equipment that is likely to or does result in an unexpected significant release of EtO.

*Employee exposure* means exposure to airborne EtO which would occur if the employee were not using respiratory protective equipment.

*Ethylene oxide* or *EtO* means the three-membered ring organic compound with chemical formula $C_2H_4O$.

**1910.1047(c)** *Permissible exposure limits*—**(1)** *8-hour time weighted average (TWA).* The employer shall ensure that no employee is exposed to an airborne concentration of EtO in excess of one (1) part EtO per million parts of air (1 ppm) as an 8-hour time-weighted average (8-hour TWA).

**1910.1047(c)(2)** *Excursion limit.* The employer shall ensure that no employee is exposed to an airborne concentration of EtO in excess of 5 parts of EtO per million parts of air (5 ppm) as averaged over a sampling period of fifteen (15) minutes.

**1910.1047(d)** *Exposure monitoring*—**(1)** *General.* **(i)** Determinations of employee exposure shall be made from breathing zone air samples that are representative of the 8-hour TWA and 15-minute short-term exposures of each employee.

**1910.1047(d)(1)(ii)** Representative 8-hour TWA employee exposure shall be determined on the basis of one or more samples representing fullshift exposure for each shift for each job classification in each work area. Representative 15-minute short-term employee exposures shall be determined on the basis of one or more samples representing 15-minute exposures associated with operations that are most likely to produce exposures above the excursion limit for each shift for each job classification in each work area.

**1910.1047(d)(1)(iii)** Where the employer can document that exposure levels are equivalent for similar operations in different work shifts, the employer need only determine representative employee exposure for that operation during one shift.

**1910.1047(d)(2)** *Initial monitoring.* **(i)** Each employer who has a workplace or work operation covered by this standard, except as provided for in paragraph (a)(2) or (d)(2)(ii) of this section, shall perform initial monitoring to determine accurately the airborne concentrations of EtO to which employees may be exposed.

**1910.1047(d)(2)(ii)** Where the employer has monitored after June 15, 1983 and the monitoring satisfies all other requirements of this section, the employer may rely on such earlier monitoring results to satisfy the requirements of paragraph (d)(2)(i) of this section.

**1910.1047(d)(2)(iii)** Where the employer has previously monitored for the excursion limit and the monitoring satisfies all other requirements of this sections, the employer may rely on such earlier monitoring results to satisfy the requirements of paragraph (d)(2)(i) of this section.

**1910.1047(d)(3)** *Monitoring frequency (periodic monitoring).* **(i)** If the monitoring required by paragraph (d)(2) of this section reveals employee exposure at or above the action level but at or below the 8-hour TWA, the employer shall repeat such monitoring for each such employee at least every 6 months.

**1910.1047(d)(3)(ii)** If the monitoring required by paragraph (d)(2)(i) of this section reveals employee exposure above the 8-hour TWA, the employer shall repeat such monitoring for each such employee at least every 3 months.

**1910.1047(d)(3)(iii)** The employer may alter the monitoring schedule from quarterly to semiannually for any employee for whom two consecutive measurements taken at least 7 days apart indicate that the employee's exposure has decreased to or below the 8-hour TWA.

**1910.1047(d)(3)(iv)** If the monitoring required by paragraph (d)(2)(i) of this section reveals employee exposure above the 15 minute excursion limit, the employer shall repeat such monitoring for each such employee at least every 3 months, and more often as necessary to evaluate exposure the employee's short-term exposures.

**1910.1047(d)(4)** *Termination of monitoring.* **(i)** If the initial monitoring required by paragraph (d)(2)(i) of this section reveals employee exposure to be below the action level, the employer may discontinue TWA monitoring for those employees whose exposures are represented by the initial monitoring.

**1910.1047(d)(4)(ii)** If the periodic monitoring required by paragraph (d)(3) of this section reveals that employee exposures, as indicated by at least two consecutive measurements taken at least 7 days apart, are below the action level, the employer may discontinue TWA monitoring for those employees whose exposures are represented by such monitoring.

**1910.1047(d)(4)(iii)** If the initial monitoring required by paragraph (d)(2)(1) of this section reveals employee exposure to be at or below the excursion limit, the employer may discontinue excursion limit monitoring for those employees whose exposures are represented by the initial monitoring.

**1910.1047(d)(4)(iv)** If the periodic monitoring required by paragraph (d)(3) of this section reveals that employee exposures, as indicated by at least two consecutive measurements taken at least 7 days apart, are at or below the excursion limit, the employer may discontinue excursion limit monitoring for those employees whose exposures are represented by such monitoring.

**1910.1047(d)(5)** *Additional monitoring.* Notwithstanding the provisions of paragraph (d)(4) of this section, the employer shall institute the exposure monitoring required under paragraphs (d)(2)(i) and (d)(3) of this section whenever there has been a change in the production, process, control equipment, personnel or work practices that may result in new or additional exposures to EtO or when the employer has any reason to suspect that a change may result in new or additional exposures.

**1910.1047(d)(6)** *Accuracy of monitoring.* **(i)** Monitoring shall be accurate, to a confidence level of 95 percent, to within plus or minus 25 percent for airborne concentrations of EtO at the 1 ppm TWA and to within plus or minus 35 percent for airborne concentrations of EtO at the action level of 0.5 ppm.

**1910.1047(d)(6)(ii)** Monitoring shall be accurate, to a confidence level of 95 percent, to within plus or minus 35 percent for airborne concentrations of EtO at the excursion limit.

**1910.1047(d)(7)** *Employee notification of monitoring results.* **(i)** The employer must, within 15 working days after the receipt of the results of any monitoring performed under this section, notify each affected employee of these results either individually in writing or by posting the results in an appropriate location that is accessible to employees.

**1910.1047(d)(7)(ii)** The written notification required by paragraph (d)(7)(i) of this section shall contain the corrective action being taken by the employer to reduce employee exposure to or below the TWA and/or excursion limit, wherever monitoring results indicated that the TWA and/or excursion limit has been exceeded.

**1910.1047(e)** *Regulated areas.* **(1)** The employer shall establish a regulated area wherever occupational exposure to airborne concentrations of EtO may exceed the TWA or wherever the EtO concentration exceeds or can reasonably be expected to exceed the excursion limit.

**1910.1047(e)(2)** Access to regulated areas shall be limited to authorized persons.

**1910.1047(e)(3)** Regulated areas shall be demarcated in any manner that minimizes the number of employees within the regulated area.

**1910.1047(f)** *Methods of compliance.* **(1)** *Engineering controls and work practices.* **(i)** The employer shall institute engineering controls and work practices to reduce and maintain employee exposure to or below the TWA and to or below the excursion limit, except to the extent that such controls are not feasible.

**1910.1047(f)(1)(ii)** Wherever the feasible engineering controls and work practices that can be instituted are not sufficient to reduce employee exposure to or below the TWA and to or below the excursion limit, the employer shall use them to reduce employee exposure to the lowest levels achievable by these controls and shall supplement them by the use of respiratory protection that complies with the requirements of paragraph (g) of this section.

**1910.1047(f)(1)(iii)** Engineering controls are generally infeasible for the following operations: collection of quality assurance sampling from sterilized materials removal of biological indicators from sterilized materials: loading and unloading of tank cars; changing of ethylene oxide tanks on sterilizers; and vessel cleaning. For these operations, engineering controls are required only where the Assistant Secretary demonstrates that such controls are feasible.

**1910.1047(f)(2)** *Compliance program.* **(i)** Where the TWA or excursion limit is exceeded, the employer shall establish and implement a written program to reduce exposure to or below the TWA and to or below the excursion limit by means of engineering and work practice controls, as required by paragraph (f)(1) of this section, and by the use of respiratory protection where required or permitted under this section.

**1910.1047(f)(2)(ii)** The compliance program shall include a schedule for periodic leak detection surveys and a written plan for emergency situations, as specified in paragraph (h)(i) of this section.

**1910.1047(f)(2)(iii)** Written plans for a program required in paragraph (f)(2) shall be developed and furnished upon request for examination and copying to the Assistant Secretary, the Director, affected employees and designated employee representatives. Such plans shall be reviewed at least every 12 months, and shall be updated as necessary to reflect significant changes in the status of the employer's compliance program.

**1910.1047(f)(2)(iv)** The employer shall not implement a schedule of employee rotation as a means of compliance with the TWA or excursion limit.

**1910.1047(g)** *Respiratory protection and personal protective equipment*—**(1)** *General.* For employees who use respirators required by this section, the employer must provide respirators that comply with the requirements of this paragraph. Respirators must be used during:

**1910.1047(g)(1)(i)** Periods necessary to install or implement feasible engineering and work-practice controls.

**1910.1047(g)(1)(ii)** Work operations, such as maintenance and repair activities and vessel cleaning, for which engineering and workpractice controls are not feasible.

**1910.1047(g)(1)(iii)** Work operations for which feasible engineering and work-practice controls are not yet sufficient to reduce employee exposure to or below the TWA.

**1910.1047(g)(1)(iv)** Emergencies.

**1910.1047(g)(2)** *Respirator program.* The employer must implement a respiratory protection program in accordance with 29 CFR 1910.134 (b) through (d) (except (d)(1)(iii)), and (f) through (m).

**1910.1047(g)(3)** *Respirator selection.* The employer must select the appropriate respirator from Table 1 of this section.

### TABLE 1—MINIMUM REQUIREMENTS FOR RESPIRATORY PROTECTION FOR AIRBORNE EtO

| Condition of use or concentration of airborne EtO (ppm) | Minimum required respirator |
|---|---|
| Equal to or less than 50 | (a) Full facepiece respirator with EtO approved canister, front-or back-mounted. |
| Equal to or less than 2,000 | (a) Positive-pressure supplied air respirator, equipped with full facepiece, hood, or helmet, or<br>(b) Continuous-flow supplied air respirator (positive pressure) equipped with hood, helmet or suit. |
| Concentration above 2,000 or unknown concentration (such as in emergencies). | (a) Positive-pressure self-contained breathing apparatus (SCBA), equipped with full facepiece, or<br>(b) Positive-pressure full facepiece supplied air respirator equipped with an auxiliary positive-pressure self-contained breathing apparatus. |
| Firefighting | (a) Positive pressure self-contained breathing apparatus equipped with full facepiece. |
| Escape | (a) Any respirator described above. |

Note. Respirators approved for use in higher concentrations are permitted to be used in lower concentrations.

**1910.1047(g)(4)** *Protective clothing and equipment.* When employees could have eye or skin contact with EtO or EtO solutions, the employer must select and provide, at no cost to the employee, appropriate protective clothing or other equipment in accordance with 29 CFR 1910.132 and 1910.133 to protect any area of the employee's body that may come in contact with the EtO or EtO solution, and must ensure that the employee wears the protective clothing and equipment provided.

**1910.1047(h)** *Emergency situations—*(1) *Written plan.* (i) A written plan for emergency situations shall be developed for each workplace where there is a possibility of an emergency. Appropriate portions of the plan shall be implemented in the event of an emergency.

**1910.1047(h)(1)(ii)** The plan shall specifically provide that employees engaged in correcting emergency conditions shall be equipped with respiratory protection as required by paragraph (g) of this section until the emergency is abated.

**1910.1047(h)(1)(iii)** The plan shall include the elements prescribed in 29 CFR 1910.38 and 29 CFR 1910.39 "Emergency action plans" and "Fire prevention plans."

**1910.1047(h)(2)** *Alerting employees.* Where there is the possibility of employee exposure to EtO due to an emergency, means shall be developed to alert potentially affected employees of such occurrences promptly. Affected employees shall be immediately evacuated from the area in the event that an emergency occurs.

**1910.1047(i)** *Medical Surveillance—*(1) *General—*(i) *Employees covered.* (A) The employer shall institute a medical surveillance program for all employees who are or may be exposed to EtO at or above the action level, without regard to the use of respirators, for at least 30 days a year.

**1910.1047(i)(1)(i)(B)** The employer shall make available medical examinations and consultations to all employees who have been exposed to EtO in an emergency situation.

**1910.1047(i)(1)(ii)** *Examination by a physician.* The employer shall ensure that all medical examinations and procedures are performed by or under the supervision of a licensed physician, and are provided without cost to the employee, without loss of pay, and at a reasonable time and place.

**1910.1047(i)(2)** *Medical examinations and consultations—*(i) *Frequency.* The employer shall make available medical examinations and consultations to each employee covered under paragraph (i)(1)(i) of this section on the following schedules:

**1910.1047(i)(2)(i)(A)** Prior to assignment of the employee to an area where exposure may be at or above the action level for at least 30 days a year.

**1910.1047(i)(2)(i)(B)** At least annually each employee exposed at or above the action level for at least 30 days in the past year.

**1910.1047(i)(2)(i)(C)** At termination of employment or reassignment to an area where exposure to EtO is not at or above the action level for at least 30 days a year.

**1910.1047(i)(2)(i)(D)** As medically appropriate for any employee exposed during an emergency.

**1910.1047(i)(2)(i)(E)** As soon as possible, upon notification by an employee either (1) that the employee has developed signs or symptoms indicating possible overexposure to EtO, or (2) that the employee desires medical advice concerning the effects of current or past exposure to EtO on the employee's ability to produce a healthy child.

**1910.1047(i)(2)(i)(F)** If the examining physician determines that any of the examinations should be provided more frequently than specified, the employer shall provide such examinations to affected employees at the frequencies recommended by the physician.

**1910.1047(i)(2)(ii)** *Content.* **(A)** Medical examinations made available pursuant to paragraphs (i)(2)(i)(A)-(D) of this section shall include:

**1910.1047(i)(2)(ii)(A)***1)* A medical and work history with special emphasis directed to symptoms related to the pulmonary, hematologic, neurologic, and reproductive systems and to the eyes and skin.

**1910.1047(i)(2)(ii)(A)***2)* A physical examination with particular emphasis given to the pulmonary, hematologic, neurologic, and reproductive systems and to the eyes and skin.

**1910.1047(i)(2)(ii)(A)***3)* A complete blood count to include at least a white cell count (including differential cell count), red cell count, hematocrit, and hemoglobin.

**1910.1047(i)(2)(ii)(A)***4)* Any laboratory or other test which the examining physician deems necessary by sound medical practice.

**1910.1047(i)(2)(ii)(B)** The content of medical examinations or consultation made available pursuant to paragraph (i)(2)(i)(E) of this section shall be determined by the examining physician, and shall include pregnancy testing or laboratory evaluation of fertility, if requested by the employee and deemed appropriate by the physician.

**1910.1047(i)(3)** *Information provided to the physician.* The employer shall provide the following information to the examining physician:

**1910.1047(i)(3)(i)** A copy of this standard and Appendices A, B, and C.

**1910.1047(i)(3)(ii)** A description of the affected employee's duties as they relate to the employee's exposure.

**1910.1047(i)(3)(iii)** The employee's representative exposure level or anticipated exposure level.

**1910.1047(i)(3)(iv)** A description of any personal protective and respiratory equipment used or to be used.

**1910.1047(i)(3)(v)** Information from previous medical examinations of the affected employee that is not otherwise available to the examining physician.

**1910.1047(i)(4)** *Physician's written opinion.* (i) The employer shall obtain a written opinion from the examining physician. This written opinion shall contain the results of the medical examination and shall include:

**1910.1047(i)(4)(i)(A)** The physician's opinion as to whether the employee has any detected medical conditions that would place the employee at an increased risk of material health impairment from exposure to EtO;

**1910.1047(i)(4)(i)(B)** Any recommended limitations on the employee or upon the use of personal protective equipment such as clothing or respirators; and

**1910.1047(i)(4)(i)(C)** A statement that the employee has been informed by the physician of the results of the medical examination and of any medical conditions resulting from EtO exposure that require further explanation or treatment.

**1910.1047(i)(4)(ii)** The employer shall instruct the physician not to reveal in the written opinion given to the employer specific findings or diagnoses unrelated to occupational exposure to EtO.

**1910.1047(i)(4)(iii)** The employer shall provide a copy of the physician's written opinion to the affected employee within 15 days from its receipt.

**1910.1047(j)** *Communication of EtO hazards to employees*—**(1)** *Signs and labels.* **(i)** The employer shall post and maintain legible signs demarcating regulated areas and entrances or accessways to regulated areas that bear the following legend:

DANGER

ETHYLENE OXIDE

CANCER HAZARD AND REPRODUCTIVE HAZARD

AUTHORIZED PERSONNEL ONLY

RESPIRATORS AND PROTECTIVE CLOTHING MAY BE REQUIRED

TO BE WORN IN THIS AREA

**1910.1047(j)(1)(ii)** The employer shall ensure that precautionary labels are affixed to all containers of EtO whose contents are capable of causing employee exposure at or above the action level or whose contents may reasonably be foreseen to cause employee exposure above the excursion limit, and that the labels remain affixed when the containers of EtO leave the workplace. For the purpose of this paragraph, reaction vessels, storage tanks, and pipes or piping systems are not considered to be containers. The labels shall comply with the requirements of 29 CFR 1910.1200(f) of OSHA's Hazard Communication standard, and shall include the following legend:

(A) DANGER

CONTAINS ETHYLENE OXIDE

CANCER HAZARD AND REPRODUCTIVE HAZARD;

and

**1910.1047(j)(1)(ii)(B)** A warning statement against breathing airborne concentrations of EtO.

**1910.1047(j)(1)(iii)** The labeling requirements under this section do not apply where EtO is used as a pesticide, as such term is defined in the Federal Insecticide, Fungicide, and Rodenticide Act (7 U.S.C. 136 et seq.), when it is labeled pursuant to that Act and regulations issued under that Act by the Environmental Protection Agency.

**1910.1047(j)(2)** *Material safety data sheets.* Employers who are manufacturers or importers of EtO shall comply with the requirements regarding development of material safety data sheets as specified in 29 CFR 1910.1200(g) of OSHA's Hazard Communication standard.

**1910.1047(j)(3)** *Information and training.* **(i)** The employer shall provide employees who are potentially exposed to EtO at or above the action level or above the excursion limit with information and training on EtO at the time of initial assignment and at least annually thereafter.

**1910.1047(j)(3)(ii)** Employees shall be informed of the following:

**1910.1047(j)(3)(ii)(A)** The requirements of this section with an explanation of its contents, including Appendices A and B;

**1910.1047(j)(3)(ii)(B)** Any operations in their work area where EtO is present;

**1910.1047(j)(3)(ii)(C)** The location and availability of the written EtO final rule; and

**1910.1047(j)(3)(ii)(D)** The medical surveillance program required by paragraph (i) of this section with an explanation of the information in Appendix C.

**1910.1047(j)(3)(iii)** Employee training shall include at least:

**1910.1047(j)(3)(iii)(A)** Methods and observations that may be used to detect the presence or release of EtO in the work area (such as monitoring conducted by the employer, continuous monitoring devices, etc.);

**1910.1047(j)(3)(iii)(B)** The physical and health hazards of EtO;

**1910.1047(j)(3)(iii)(C)** The measures employees can take to protect themselves from hazards associated with EtO exposure, including specific procedures the employer has implemented to protect employees from exposure to EtO, such as work practices, emergency procedures, and personal protective equipment to be used; and

**1910.1047(j)(3)(iii)(D)** The details of the hazard communication program developed by the employer, including an explanation of the labeling system and how employees can obtain and use the appropriate hazard information.

**1910.1047(k)** *Recordkeeping*—**(1)** *Objective data for exempted operations.* **(i)** Where the processing, use, or handling of products made from or containing EtO are exempted from other requirements of this section under paragraph (a)(2) of this section, or where objective data have been relied on in lieu of initial monitoring under paragraph (d)(2)(ii) of this section, the employer shall establish and maintain an accurate record of objective data reasonably relied upon in support of the exemption.

**1910.1047(k)(1)(ii)** This record shall include at least the following information:

**1910.1047(k)(1)(ii)(A)** The product qualifying for exemption;

**1910.1047(k)(1)(ii)(B)** The source of the objective data;

**1910.1047(k)(1)(ii)(C)** The testing protocol, results of testing, and/or analysis of the material for the release of EtO;

**1910.1047(k)(1)(ii)(D)** A description of the operation exempted and how the data support the exemption; and

**1910.1047(k)(1)(ii)(E)** Other data relevant to the operations, materials, processing, or employee exposures covered by the exemption.

**1910.1047(k)(1)(iii)** The employer shall maintain this record for the duration of the employer's reliance upon such objective data.

**1910.1047(k)(2)** *Exposure measurements.* **(i)** The employer shall keep an accurate record of all measurements taken to monitor employee exposure to EtO as prescribed in paragraph (d) of this section.

**1910.1047(k)(2)(ii)** This record shall include at least the following information:

**1910.1047(k)(2)(ii)(A)** The date of measurement;

**1910.1047(k)(2)(ii)(B)** The operation involving exposure to EtO which is being monitored;

**1910.1047(k)(2)(ii)(C)** Sampling and analytical methods used and evidence of their accuracy;

**1910.1047(k)(2)(ii)(D)** Number, duration, and results of samples taken;

**1910.1047(k)(2)(ii)(E)** Type of protective devices worn, if any; and

**1910.1047(k)(2)(ii)(F)** Name, social security number and exposure of the employees whose exposures are represented.

**1910.1047(k)(2)(iii)** The employer shall maintain this record for at least thirty (30) years, in accordance with 29 CFR 1910.1020.

**1910.1047(k)(3)** *Medical surveillance.* **(i)** The employer shall establish and maintain an accurate record for each employee subject to medical surveillance by paragraph (i)(1)(i) of this section, in accordance with 29 CFR 1910.1020.

**1910.1047(k)(3)(ii)** The record shall include at least the following information:

**1910.1047(k)(3)(ii)(A)** The name and social security number of the employee;

**1910.1047(k)(3)(ii)(B)** Physicians' written opinions;

**1910.1047(k)(3)(ii)(C)** Any employee medical complaints related to exposure to EtO; and

**1910.1047(k)(3)(ii)(D)** A copy of the information provided to the physician as required by paragraph (i)(3) of this section.

**1910.1047(k)(3)(iii)** The employer shall ensure that this record is maintained for the duration of employment plus thirty (30) years, in accordance with 29 CFR 1910.1020.

**1910.1047(k)(4)** *Availability.* **(i)** The employer, upon written request, shall make all records required to be maintained by this section available to the Assistant Secretary and the Director for examination and copying.

**1910.1047(k)(4)(ii)** The employer, upon request, shall make any exemption and exposure records required by paragraphs (k)(1) and (2) of this section available for examination and copying to affected employees, former employees, designated representatives and the Assistant Secretary, in accordance with 29 CFR 1910.1020(a) through (e) and (g) through (i).

**1910.1047(k)(4)(iii)** The employer, upon request, shall make employee medical records required by paragraph (k)(3) of this section available for examination and copying to the subject employee, anyone having the specific written consent of the subject employee, and the Assistant Secretary, in accordance with 29 CFR 1910.1020.

# OSHA Standards for General Industry

**1910.1047(k)(5)** *Transfer of records.* **(i)** The employer shall comply with the requirements concerning transfer of records set forth in 29 CFR 1910.1020(h).

**1910.1047(k)(5)(ii)** Whenever the employer ceases to do business and there is no successor employer to receive and retain the records for the prescribed period, the employer shall notify the Director at least 90 days prior to disposal and transmit them to the Director.

**1910.1047(l)** *Observation of monitoring*—**(1)** *Employee observation.* The employer shall provide affected employees or their designated representatives an opportunity to observe any monitoring of employee exposure to EtO conducted in accordance with paragraph (d) of this section.

**1910.1047(l)(2)** *Observation procedures.* When observation of the monitoring of employee exposure to EtO requires entry into an area where the use of protective clothing or equipment is required, the observer shall be provided with and be required to use such clothing and equipment and shall comply with all other applicable safety and health procedures.

**1910.1047(m)** [Reserved.]

**1910.1047(n)** *Appendices.* The information contained in the appendices is not intended by itself to create any additional obligations not otherwise imposed or to detract from any existing obligation.

## APPENDIX A TO §1910.1047— SUBSTANCE SAFETY DATA SHEET FOR ETHYLENE OXIDE (NON-MANDATORY)

### I. SUBSTANCE IDENTIFICATION

A. Substance: Ethylene oxide ($C_2H_4O$).

B. Synonyms: dihydrooxirene, dimethylene oxide, EO, 1,2-epoxyethane, EtO, ETO, oxacyclopropane, oxane, oxidoethane, alpha/beta-oxidoethane, oxiran, oxirane.

C. Ethylene oxide can be found as a liquid or vapor.

D. EtO is used in the manufacture of ethylene glycol, surfactants, ethanolamines, glycol ethers, and other organic chemicals. EtO is also used as a sterilant and fumigant.

E. Appearance and odor: Colorless liquid below 10.7 °C (51.3 °F) or colorless gas with ether-like odor detected at approximately 700 parts EtO per million parts of air (700 ppm).

F. Permissible Exposure: Exposure may not exceed 1 part EtO per million parts of air averaged over the 8-hour workday.

### II. HEALTH HAZARD DATA

A. Ethylene oxide can cause bodily harm if you inhale the vapor, if it comes into contact with your eyes or skin, or if you swallow it.

B. Effects of overexposure:

1. Ethylene oxide in liquid form can cause eye irritation and injury to the cornea, frostbite, and severe irritation and blistering of the skin upon prolonged or confined contact. Ingestion of EtO can cause gastric irritation and liver injury. Acute effects from inhalation of EtO vapors include respiratory irritation and lung injury, headache, nausea, vomiting, diarrhea, shortness of breath, and cyaonosis (blue or purple coloring of skin). Exposure has also been associated with the occurrence of cancer, reproductive effects, mutagenic changes, neurotoxicity, and sensitization.

1. EtO has been shown to cause cancer in laboratory animals and has been associated with higher incidences of cancer in humans. Adverse reproductive effects and chromosome damage may also occur from EtO exposure.

a. Reporting signs and symptoms: You should inform your employer if you develop any signs or symptoms and suspect that they are caused by exposure to EtO.

### III. EMERGENCY FIRST AID PROCEDURES

A. Eye exposure: If EtO gets into your eyes, wash your eyes immediately with large amounts of water, lifting the lower and upper eyelids. Get medical attention immediately. Contact lenses should not be worn when working with this chemical.

B. Skin exposure: If EtO gets on your skin, immediately wash the contaminated skin with water. If EtO soaks through your clothing, especially your shoes, remove the clothing immediately and wash the skin with water using an emergency deluge shower. Get medical attention immediately. Thoroughly wash contaminated clothing before reusing. Contaminated leather shoes or other leather articles should not be reused and should be discarded.

C. Inhalation: If large amounts of EtO are inhaled, the exposed person must be moved to fresh air at once. If breathing has stopped, perform cardiopulmonary resuscitation. Keep the affected person warm and at rest. Get medical attention immediately.

D. Swallowing: When EtO has been swallowed, give the person large quantities of water immediately. After the water has been swallowed, try to get the person to vomit by having him or her touch the back of the throat with his or her finger. Do not make an unconscious person vomit. Get medical attention immediately.

E. Rescue: Move the affected person from the hazardous exposure. If the exposed person has been overcome, attempt rescue only after notifying at least one other person of the emergency and putting into effect established emergency procedures. Do not become a casualty yourself. Understand your emergency rescue procedures and know the location of the emergency equipment before the need arises.

### IV. RESPIRATORS AND PROTECTIVE CLOTHING

A. Respirators. You may be required to wear a respirator for nonroutine activities, in emergencies, while your employer is in the process of reducing EtO exposures through engineering controls, and in areas where engineering controls are not feasible. As of the effective date of this standard, only air-supplied, positive-pressure, full-facepiece respirators are approved for protection against EtO. If air-purifying respirators are worn in the future, they must have a label issued by the National Institute for Occupational Safety and Health under the provisions of 42 CFR part 84 stating that the respirators have been approved for use with ethylene oxide. For effective protection, respirators must fit your face and head snugly. Respirators must not be loosened or removed in work situations where their use is required.

EtO does not have a detectable odor except at levels well above the permissible exposure limits. If you can smell EtO while wearing a respirator, proceed immediately to fresh air. If you experience difficulty breathing while wearing a respirator, tell your employer.

B. Protective clothing: You may be required to wear impermeable clothing, gloves, a face shield, or other appropriate protective clothing to prevent skin contact with liquid EtO or EtO-containing solutions. Where protective clothing is required, your employer must provide clean garments to you as necessary to assure that the clothing protects you adequately.

Replace or repair protective clothing that has become torn or otherwise damaged.

EtO must never be allowed to remain on the skin. Clothing and shoes which are not impermeable to EtO should not be allowed to become contaminated with EtO, and if they do, the clothing should be promptly removed and decontaminated. Contaminated leather shoes should be discarded. Once EtO penetrates shoes or other leather articles, they should not be worn again.

C. Eye protection: You must wear splashproof safety goggles in areas where liquid EtO or EtO-containing solutions may contact your eyes. In addition, contact lenses should not be worn in areas where eye contact with EtO can occur.

### V. PRECAUTIONS FOR SAFE USE, HANDLING, AND STORAGE

A. EtO is a flammable liquid, and its vapors can easily form explosive mixtures in air.

B. EtO must be stored in tightly closed containers in a cool, well-ventilated area, away from heat, sparks, flames, strong oxidizers, alkalines, and acids, strong bases, acetylide-forming metals such as cooper, silver, mercury and their alloys.

C. Sources of ignition such as smoking material, open flames and some electrical devices are prohibited wherever EtO is handled, used, or stored in a manner that could create a potential fire or explosion hazard.

D. You should use non-sparking tools when opening or closing metal containers of EtO, and containers must be bonded and grounded in the rare instances in which liquid EtO is poured or transferred.

E. Impermeable clothing wet with liquid EtO or EtO-containing solutions may be easily ignited. If your are wearing impermeable clothing and are splashed with liquid EtO or EtO-containing solution, you should immediately remove the clothing while under an emergency deluge shower.

F. If your skin comes into contact with liquid EtO or EtO-containing solutions, you should immediately remove the EtO using an emergency deluge shower.

G. You should not keep food, beverages, or smoking materials in regulated areas where employee exposures are above the permissible exposure limits.

H. Fire extinguishers and emergency deluge showers for quick drenching should be readily available, and you should know where they are and how to operate them.

I. Ask your supervisor where EtO is used in your work area and for any additional plant safety and health rules.

### VI. ACCESS TO INFORMATION

A. Each year, your employer is required to inform you of the information contained in this standard and appendices for EtO. In addition, your employer must instruct you in the proper work practices for using EtO emergency procedures, and the correct use of protective equipment.

B. Your employer is required to determine whether you are being exposed to EtO. You or your representative has the right to observe employee measurements and to record the results obtained. Your employer is required to inform you of your exposure. If your employer determine that you are being overexposed, he or she is required to inform you of the actions which are being taken to reduce your exposure to within permissible exposure limits.

C. Your employer is required to keep records of your exposures and medical examinations. These exposure records must be kept by the employer for at least thirty (30) years. Medical records must be kept for the period of your employment plus thirty (30) years.

D. Your employer is required to release your exposure and medical records to your physician or designated representative upon your written request.

### VII. STERILANT USE OF ETO IN HOSPITALS AND HEALTH CARE FACILITIES

This section of Appendix A, for informational purposes, sets forth EPA's recommendations for modifications in workplace design and practice in hospitals and health care facilities for which the Environmental Protection Agency has registered EtO for uses as a sterilant or fumigant under the Federal Insecticide, Funigicide, and Rodenticide Act, 7 U.S.C. 136 *et seq.* These new recommendations, published in the **FEDERAL REGISTER** by EPA at 49 FR 15268, as modified in today's **REGISTER**, are intended to help reduce the exposure of hospital and health care workers to EtO to 1 ppm. EPA's recommended workplace design and workplace practice are as follows:

#### 1. Workplace Design

a. *Installation of gas line hand valves.* Hand valves must be installed on the gas supply line at the connection to the supply cylinders to minimize leakage during cylinder change.

b. *Installation of capture boxes.* Sterilizer operations result in a gas/water discharge at the completion of the process. This discharge is routinely piped to a floor drain which is generally located in an equipment or an adjacent room. When the floor drain is not in the same room as the sterilizer and workers are not normally present, all that is necessary is that the room be well ventilated.

The installation of a "capture box" will be required for those work place layouts where the floor drain is located in the same room as the sterilizer or in a room where workers are normally present. A "capture box" is a piece of equipment that totally encloses the floor drain where the discharge from the sterilizer is pumped. The "capture box" is to be vented directly to a non-recirculating or dedicated ventilation system. Sufficient air intake should be allowed at the bottom of the box to handle the volume of air that is ventilated from the top of the box. The "capture box" can be made of metal, plastic, wood or other equivalent material. The box is intended to reduce levels of EtO discharged into the work room atmosphere. The use of a "capture box" is not required if: (1) The vacuum pump discharge floor drain is located in a well ventilated equipment or other room where workers are not normally present or (2) the water sealed vacuum pump discharges directly to a closed sealed sewer line (check local plumbing codes).

If it is impractical to install a vented "capture box" and a well ventilated equipment or other room is not feasible, a box that can be sealed over the floor drain may be used if: (1) The floor drain is located in a room where workers are not normally present and EtO cannot leak into an occupied area, and (2) the sterilizer in use is less than 12 cubic feet in capacity (check local plumbing codes).

c. *Ventilation of aeration units* i. *Existing aeration units.* Existing units must be vented to a non-recirculating or dedicated system or vented to an equipment or other room where workers are not normally present and which is well ventilated. Aerator units must be positioned as close as possible to the sterilizer to minimize the exposure from the off-gassing of sterilized items.

ii. *Installation of new aerator units (where none exist).* New aerator units must be vented as described above for existing aerators. Aerators must be in place by July 1, 1986.

d. *Ventilation during cylinder change.* Workers may be exposed to short but relatively high levels of EtO during the change of gas cylinders. To reduce exposure from this route, users must select one of three alternatives designed to draw off gas that may be released when the line from the sterilizer to the cylinder is disconnected:

i. Location of cylinders in a well ventilated equipment room or other room where workers are not normally present.

ii. Installation of a flexible hose (at least 4" in diameter) to a non-recirculating or dedicated ventilation system and located in the area of cylinder change in such a way that the hose can be positioned at the point where the sterilizer gas line is disconnected from the cylinder.

iii. Installation of a hood that is part of a nonrecirculating or dedicated system and positioned no more than one foot above the point where the change of cylinders takes place.

e. *Ventilation of sterilizer door area.* One of the major sources of exposure to EtO occurs when the sterilizer door is opened following the completion of the sterilization process. In order to reduce this avenue of exposure, a hood or metal canopy closed on each end must be installed over the sterilizer door. The hood or metal canopy must be connected to a nonrecirculating or dedicated ventilation system or one that exhausts gases to a well ventilated equipment or other room where workers are not normally present. A hood or canopy over the sterilizer door is required for use even with those sterilizers that have a purge cycle and must be in place by July 1, 1986.

f. *Ventilation of sterilizer relief valve.* Sterilizers are typically equipped with a safety relief device to release gas in case of increased pressure in the sterilizer. Generally, such relief devices are used on pressure vessels. Although these pressure relief devices are rarely opened for hospital and health care sterilizers, it is suggested that they be designed to exhaust vapor from the sterilizer by one of the following methods:

i. Through a pipe connected to the outlet of the relief valve ventilated directly outdoors at a point high enough to be away from passers by, and not near any windows that open, or near any air conditioning or ventilation air intakes.

ii. Through a connection to an existing or new nonrecirculating or dedicated ventilation system.

iii. Through a connection to a well ventilated equipment or other room where workers are not normally present.

g. *Ventilation systems.* Each hospital and health care facility affected by this notice that uses EtO for the sterilization of equipment and supplies must have a ventilation system which enables compliance with the requirements of section (b) through (f) in the manner described in these sections and within the timeframes allowed. Thus, each affected hospital and health care facility must have or install a non-recirculating or dedicated ventilation equipment or other room where workers are not normally present in which to vent EtO.

h. *Installation of alarm systems.* An audible and visual indicator alarm system must be installed to alert personnel of ventilation system failures, i.e., when the ventilation fan motor is not working.

#### 2. Workplace Practices

All the workplace practices discussed in this unit must be permanently posted near the door of each sterilizer prior to use by any operator.

a. *Changing of supply line filters.* Filters in the sterilizer liquid line must be changed when necessary, by the following procedure:

i. Close the cylinder valve and the hose valve.

ii. Disconnect the cylinder hose (piping) from the cylinder.

iii. Open the hose valve and bleed slowly into a proper ventilating system at or near the in-use supply cylinders.

iv. Vacate the area until the line is empty.

v. Change the filter.

vi. Reconnect the lines and reverse the value position.

vii. Check hoses, filters, and valves for leaks with a fluorocarbon leak detector (for those sterilizers using the 88 percent chlorofluorocarbon, 12 percent ethylene oxide mixture (12/88)).

b. *Restricted access area.* i. Areas involving use of EtO must be designated as restricted access areas. They must be identified with signs or floor marks near the sterilizer door, aerator, vacuum pump floor drain discharge, and in-use cylinder storage.

ii. All personnel must be excluded from the restricted area when certain operations are in progress, such as discharging a vacuum pump, emptying a sterilizer liquid line, or venting a nonpurge sterilizer with the door ajar or other operations where EtO might be released directly into the face of workers.

c. *Door opening procedures.* i. *Sterilizers with purge cycles.* A load treated in a sterilizer equipped with a purge cycle should be removed immediately upon completion of the cycle (provided no time is lost opening the door after cycle is completed). If this is not done, the purge cycle should be repeated before opening door.

ii. *Sterilizers without purge cycles.* For a load treated in a sterilizer not equipped with a purge cycle, the sterilizer door must be ajar 6" for 15 minutes, and then fully opened for at least another 15 minutes before removing the treated load. The length of time of the second period should be established by peak monitoring for one hour after the two 15-minute periods suggested. If the level is above 10 ppm time-weighted average for 8 hours, more time should be added to the second waiting period (door wide open). However, in no case may the second period be shortened to less than 15 minutes.

d. *Chamber unloading procedures.* i. Procedures for unloading the chamber must include the use of baskets or rolling carts, or baskets and rolling tables to transfer treated loads quickly, thus avoiding excessive contact with treated articles, and reducing the duration of exposures.

ii. If rolling carts are used, they should be pulled not pushed by the sterilizer operators to avoid offgassing exposure.

e. *Maintenance.* A written log should be instituted and maintained documenting the date of each leak detection and any maintenance procedures undertaken. This is a suggested use practice and is not required.

i. *Leak detection.* Sterilizer door gaskets, cylinder and vacuum piping, hoses, filters, and valves must be checked for leaks under full pressure with a Fluorocarbon leak detector (for 12/88 systems only) every two weeks by maintenance personnel. Also, the cylinder piping connections must be checked after changing cylinders. Particular attention in leak detection should be given to the automatic solenoid valves that control the flow of EtO to the sterilizer. Specifically, a check should be made at the EtO gasline entrance port to the sterilizer, while the sterilizer door is open and the solenoid valves are in a closed position.

ii. *Maintenance procedures.* Sterilizer/areator door gaskets, valves, and fittings must be replaced when necessary as determined by maintenance personnel in their bi-weekly checks; in addition, visual inspection of the door gaskets for cracks, debris, and other foreign substances should be conducted daily by the operator.

### APPENDIX B TO §1910.1047— SUBSTANCE TECHNICAL GUIDELINES FOR ETHYLENE OXIDE (NON-MANDATORY)

#### I. PHYSICAL AND CHEMICAL DATA

A. Substance identification:

1. Synonyms: dihydrooxirene, dimethylene oxide, EO, 1,2epoxyethane, EtO ETO oxacyclopropane, oxane, oxidoethane, alpha/beta-oxidoethane, oxiran, oxirane.

2. Formula: ($C_2H_4O$).

3. Molecular weight: 44.06

B. Physical data:

1. Boiling point (760 mm Hg): 10.70 °C (51.3 °F);

2. Specific gravity (water = 1): 0.87 (at 20 °C or 68 °F)

3. Vapor density (air = 1): 1.49;

4. Vapor pressure (at 20 °C); 1,095 mm Hg;

5. Solubility in water: complete;

6. Appearance and odor: colorless liquid; gas at temperature above 10.7 °F or 51.3 °C with ether-like odor above 700 ppm.

#### II. FIRE, EXPLOSION, AND REACTIVITY HAZARD DATA

A. Fire:

1. Flash point: less than 0 °F (open cup);

2. Stability: decomposes violently at temperatures above 800 °F;

3. Flammable limits in air, percent by volume: Lower: 3, Upper: 100;

4. Extinguishing media: Carbon dioxide for small fires, polymer or alcohol foams for large fires;

5. Special fire fighting procedures: Dilution of ethylene oxide with 23 volumes of water renders it non-flammable;

6. Unusual fire and explosion hazards: Vapors of EtO will burn without the presence of air or other oxidizers. EtO vapors are heavier than air and may travel along the ground and be ignited by open flames or sparks at locations remote from the site at which EtO is being used.

7. For purposes of compliance with the requirements of 29 CFR 1910.106, EtO is classified as a flammable gas. For example, 7,500 ppm, approximately one-fourth of the lower flammable limit, would be considered to pose a potential fire and explosion hazard.

8. For purposes of compliance with 29 CFR 1910.155, EtO is classified as a Class B fire hazard.

9. For purpose of compliance with 29 CFR 1919.307, locations classified as hazardous due to the presence of EtO shall be Class I.

B. Reactivity:

1. Conditions contributing to instability: EtO will polymerize violently if contaminated with aqueous alkalies, amines, mineral acids, metal chlorides, or metal oxides. Violent decomposition will also occur at temperatures above 800 °F;

2. Incompatabilities: Alkalines and acids;

3. Hazardous decomposition products: Carbon monoxide and carbon dioxide.

#### III. SPILL, LEAK, AND DISPOSAL PROCEDURES

A. If EtO is spilled or leaked, the following steps should be taken:

1. Remove all ignition sources.

2. The area should be evacuated at once and re-entered only after the area has been thoroughly ventilated and washed down with water.

B. Persons not wearing appropriate protective equipment should be restricted from areas of spills or leaks until cleanup has been completed.

C. Waste disposal methods: Waste material should be disposed of in a manner that is not hazardous to employees or to the general population. In selecting the method of waste disposal, applicable local, State, and Federal regulations should be consulted.

#### IV. MONITORING AND MEASUREMENT PROCEDURES

A. Exposure above the Permissible Exposure Limit:

1. Eight-hour exposure evaluation: Measurements taken for the purpose of determining employee exposure under this section are best taken with consecutive samples covering the full shift. Air samples should be taken in the employee's breathing zone (air that would most nearly represent that inhaled by the employee.)

2. Monitoring techniques: The sampling and analysis under this section may be performed by collection of the EtO vapor on charcoal adsorption tubes or other composition adsorption tubes, with subsequent chemical analysis. Sampling and analysis may also be performed by instruments such as real-time continuous monitoring systems, portable direct reading instruments, or passive dosimeters as long as measurements taken using these methods accurately evaluate the concentration of EtO in employees' breathing zones.

Appendix D describes the validated method of sampling and analysis which has been tested by OSHA for use with EtO. Other available methods are also described in Appendix D. The employer has the obligation of selecting a monitoring method which meets the accuracy and precision requirements of the standard under his unique field conditions. The standard requires that the method of monitoring should be accurate, to a 95 percent confidence level, to plus or minus 25 percent for concentrations of EtO at 1 ppm, and to plus or minus 35 percent for concentrations at 0.5 ppm. In addition to the method described in Appendix D, there are numerous other methods available for monitoring for EtO in the workplace. Details on these other methods have been submitted by various companies to the rulemaking record, and are available at the OSHA Docket Office.

B. Since many of the duties relating to employee exposure are dependent on the results of measurement procedures, employers should assure that the evaluation of employee exposures is performed by a technically qualified person.

### V. PROTECTIVE CLOTHING AND EQUIPMENT

Employees should be provided with and be required to wear appropriate protective clothing wherever there is significant potential for skin contact with liquid EtO or EtO-containing solutions. Protective clothing shall include impermeable coveralls or similar full-body work clothing, gloves, and head coverings, as appropriate to protect areas of the body which may come in contact with liquid EtO or EtO-containing solutions.

Employers should ascertain that the protective garments are impermeable to EtO. Permeable clothing, including items made of rubber, and leather shoes should not be allowed to become contaminated with liquid EtO. If permeable clothing does become contaminated, it should be immediately removed, while the employer is under an emergency deluge shower. If leather footwear or other leather garments become wet from EtO they should be discarded and not be worn again, because leather absorbs EtO and holds it against the skin.

Any protective clothing that has been damaged or is otherwise found to be defective should be repaired or replaced. Clean protective clothing should be provided to the employee as necessary to assure employee protection. Whenever impermeable clothing becomes wet with liquid EtO, it should be washed down with water before being removed by the employee. Employees are also required to wear splash-proof safety goggles where there is any possibility of EtO contacting the eyes.

### VI. MISCELLANEOUS PRECAUTIONS

A. Store EtO in tightly closed containers in a cool, wellventilated area and take all necessary precautions to avoid any explosion hazard.

B. Non-sparking tools must be used to open and close metal containers. These containers must be effectively grounded and bonded.

C. Do not incinerate EtO cartridges, tanks or other containers.

D. Employers should advise employees of all areas and operations where exposure to EtO occur.

### VII. COMMON OPERATIONS

Common operations in which exposure to EtO is likely to occur include the following: Manufacture of EtO, surfactants, ethanolamines, glycol ethers, and specialty chemicals, and use as a sterilant in the hospital, health product and spice industries.

## APPENDIX C TO §1910.1047—MEDICAL SURVEILLANCE GUIDELINES FOR ETHYLENE OXIDE (NON-MANDATORY)

### I. ROUTE OF ENTRY

Inhalation.

### II. TOXICOLOGY

Clinical evidence of adverse effects associated with the exposure to EtO is present in the form of increased incidence of cancer in laboratory animals (leukemia, stomach, brain), mutation in offspring in animals, and resorptions and spontaneous abortions in animals and human populations respectively. Findings in humans and experimental animals exposed to airborne concentrations of EtO also indicate damage to the genetic material (DNA). These include hemoglobin alkylation, unsecheduled DNA synthesis, sister chromatid exchange chromosomal aberration, and functional sperm abnormalities.

Ethylene oxide in liquid form can cause eye irritation and injury to the cornea, frostbite, severe irritation, and blistering of the skin upon prolonged or confined contact. Ingestion of EtO can cause gastric irritation and liver injury. Other effects from inhalation of EtO vapors include respiratory irritation and lung injury, headache, nausea, vomiting, diarrhea, dyspnea and cyanosis.

### III. SIGNS AND SYMPTOMS OF ACUTE OVEREXPOSURE

The early effects of acute overexposure to EtO are nausea and vomiting, headache, and irritation of the eyes and respiratory passages. The patient may notice a "peculiar taste" in the mouth. Delayed effects can include pulmonary edema, drowsiness, weakness, and incoordination. Studies suggest that blood cell changes, an increase in chromosomal aberrations, and spontaneous abortion may also be causally related to acute overexposure to EtO.

Skin contact with liquid or gaseous EtO causes characteristic burns and possibly even an allergic-type sensitization. The edema and erythema occurring from skin contact with EtO progress to vesiculation with a tendency to coalesce into blebs with desquamation. Healing occurs within three weeks, but there may be a residual brown pigmentation. A 40-80% solution is extremely dangerous, causing extensive blistering after only brief contact. Pure liquid EtO causes frostbite because of rapid evaporation. In contrast, the eye is relatively insensitive to EtO, but there may be some irritation of the cornea.

Most reported acute effects of occupational exposure to EtO are due to contact with EtO in liquid phase. The liquid readily penetrates rubber and leather, and will produce blistering if clothing or footwear contaminated with EtO are not removed.

### IV. SURVEILLANCE AND PREVENTIVE CONSIDERATIONS

As noted above, exposure to EtO has been linked to an increased risk of cancer and reproductive effects including decreased male fertility, fetotoxicity, and spontaneous abortion. EtO workers are more likely to have chromosomal damage than similar groups not exposed to EtO. At the present, limited studies of chronic effects in humans resulting from exposure to EtO suggest a causal association with leukemia. Animal studies indicate leukemia and cancers at other sites (brain, stomach) as well. The physician should be aware of the findings of these studies in evaluating the health of employees exposed to EtO.

Adequate screening tests to determine an employee's potential for developing serious chronic diseases, such as cancer, from exposure to EtO do not presently exist. Laboratory tests may, however, give evidence to suggest that an employee is potentially overexposed to EtO. It is important for the physician to become familiar with the operating conditions in which exposure to EtO is likely to occur. The physician also must become familiar with the signs and symptoms that indicate a worker is receiving otherwise unrecognized and unacceptable exposure to EtO. These elements are especially important in evaluating the medical and work histories and in conducting the physical exam. When an unacceptable exposure in an active employee is identified by the physician, measures taken by the employer to lower exposure should also lower the risk of serious long-term consequences.

The employer is required to institute a medical surveillance program for all employees who are or will be exposed to EtO at or above the action level (0.5 ppm) for at least 30 days per year, without regard to respirator use. All examinations and procedures must be performed by or under the supervision of a licensed physician at a reasonable time and place for the employee and at no cost to the employee.

Although broad latitude in prescribing specific tests to be included in the medical surveillance program is extended to the examining physician, OSHA requires inclusion of the following elements in the routine examination:

**1910.1047(m)(3)(i)** Medical and work histories with special emphasis directed to symptoms related to the pulmonary, hematologic, neurologic, and reproductive systems and to the eyes and skin.

**1910.1047(m)(3)(ii)** Physical examination with particular emphasis given to the pulmonary, hematologic, neurologic, and reproductive systems and to the eyes and skin.

**1910.1047(m)(3)(iii)** Complete blood count to include at least a white cell count (including differential cell count), red cell count, hematocrit, and hemoglobin.

**1910.1047(m)(3)(iv)** Any laboratory or other test which the examining physician deems necessary by sound medical practice.

If requested by the employee, the medical examinations shall include pregnancy testing or laboratory evaluation of fertility as deemed appropriate by the physician.

In certain cases, to provide sound medical advice to the employer and the employee, the physician must evaluate situations not directly related to EtO. For example, employees with skin diseases may be unable to tolerate wearing protective clothing. In addition those with chronic respiratory diseases may not tolerate the wearing of negative pressure (air purifying) respirators. Additional tests and procedures that will help the physician determine which employees are medically unable to wear such respirators should include: An evaluation of cardiovascular function, a baseline chest x-ray to be repeated at five year intervals, and a pulmonary function test to be repeated every three years. The pulmonary function test should include measurement of the employee's forced vital capacity (FVC), forced expiratory volume at one second (FEV1), as well as calculation of the ratios of FEV1 to FVC, and measured FVC and measured FEV1 to expected values corrected for variation due to age, sex, race, and height.

The employer is required to make the prescribed tests available at least annually to employees who are or will be exposed at or above the action level, for 30 or more days per year; more often than specified if recommended by the examining physician; and upon the employee's termination of employment or reassignment to another work area. While little is known about the long term consequences of high short-term exposures, it appears prudent to monitor such affected employees closely in light of existing health data. The employer shall provide physician recommended examinations to any employee exposed to EtO in emergency conditions. Likewise, the employer shall make available medical consultations including physician recommended exams to employees who believe they are suffering signs or symptoms of exposure to EtO.

The employer is required to provide the physician with the following informatin: a copy of this standard and its appendices; a description of the affected employee's duties as they relate to the employee exposure level; and information from the employee's previous medical examinations which is not readily available to the examining physician. Making this information available to the physician will aid in the evaluation of the employee's health in relation to assigned duties and fitness to wear personal protective equipment, when required.

The employer is required to obtain a written opinion from the examining physician containing the results of the medical examinations; the physician's opinion as to whether the employee has any detected medical conditions which would place the employee at increased risk of material impairment of his or her health from exposure to EtO; any recommended restrictions upon the employee's exposure to EtO, or upon the use of protective clothing or equipment such as respirators; and a statement that the employee has been informed by the physician of the results of the medical examination and of any medical conditions which require further explanation or treatment. This written opinion must not reveal specific findings or diagnoses unrelated to occupational exposure to EtO, and a copy of the opinion must be provided to the affected employee.

The purpose in requiring the examining physician to supply the employer with a written opinion is to provide the employer with a medical basis to aid in the determination of initial placement of employees and to assess the employee's ability to use protective clothing and equipment.

### APPENDIX D TO §1910.1047— SAMPLING AND ANALYTICAL METHODS FOR ETHYLENE OXIDE (NON-MANDATORY)

A number of methods are available for monitoring employee exposures to EtO. Most of these involve the use of charcoal tubes and sampling pumps, followed by analysis of the samples by gas chromatograph. The essential differences between the charcoal tube methods include, among others, the use of different desorbing solvents, the use of different lots of charcoal, and the use of different equipment for analysis of the samples.

Besides charcoal, methods using passive dosimeters, gas sampling bags, impingers, and detector tubes have been utilized for determination of EtO exposure. In addition, there are several commercially available portable gas analyzers and monitoring units.

This appendix contains details for the method which has been tested at the OSHA Analytical Laboratory in Salt Lake City. Inclusion of this method in the appendix does not mean that this method is the only one which will be satisfactory. Copies of descriptions of other methods available are available in the rulemaking record, and may be obtained from the OSHA Docket Office. These include the Union Carbide, Dow Chemical, 3M, and DuPont methods, as well as NIOSH Method S-286. These methods are briefly described at the end of this appendix.

Employers who note problems with sample breakthrough using the OSHA or other charcoal methods should try larger charcoal tubes. Tubes of larger capacity are available. In addition, lower flow rates and shorter sampling times should be beneficial in minimizing breakthrough problems. Whatever method the employer chooses, he must assure himself of the method's accuracy and precision under the unique conditions present in his workplace.

### ETHYLENE OXIDE

Method No.: 30.

Matrix: Air.

Target Concentration: 1.0 ppm (1.8 mg/m$^3$).

Procedure: Samples are collected on two charcoal tubes in series and desorbed with 1% $CS_2$ in benzene. The samples are derivatized with HBr and treated with sodium carbonate. Analysis is done by gas chromatography with an electron capture detector.

Recommended Air Volume and Sampling Rate: 1 liter and 0.05 Lpm.

Detection Limit of the Overall Procedure: 13.3 ppb (0.024 mg/m$^3$) (Based on 1.0 liter air sample).

Reliable Quantitation Limit: 52.2 ppb (0.094 mg/m$^3$) (Based on 1.0 liter air sample).

Standard Error of Estimate: 6.59% (See Backup Section 4.6).

Special Requirements: Samples must be analyzed within 15 days of sampling date.

Status of Method: The sampling and analytical method has been subjected to the established evaluation procedures of the Organic Method Evaluations Branch.

Date: August 1981.

Chemist: Wayne D. Potter.

### ORGANIC SOLVENTS BRANCH, OSHA ANALYTICAL LABORATORY, SALT LAKE CITY, UTAH

1. General Discussion.

1.1 Background.

1.1.1 History of Procedure.

Ethylene oxide samples analyzed at the OSHA Laboratory have normally been collected on activated charcoal and desorbed with carbon disulfide. The analysis is performed with a gas chromatograph equipped with a FID (Flame ionization detector) as described in NIOSH Method S286 (Ref. 5.1). This method is based on a PEL of 50 ppm and has a detection limit of about 1 ppm.

Recent studies have prompted the need for a method to analyze and detect ethylene oxide at very low concentrations.

Several attempts were made to form an ultraviolet (UV) sensitive derivative with ethylene oxide for analysis with HPLC. Among those tested that gave no detectable product were: panisidine, methylimidazole, aniline, and 2,3,6-trichlorobenzoic acid. Each was tested with catalysts such as triethylamine, aluminum chloride, methylene chloride and sulfuric acid but no detectable derivative was produced.

The next derivatization attempt was to react ethylene oxide with HBr to form 2-bromoethanol. This reaction was successful. An ECD (electron capture detector) gave a very good response for 2bromoethanol due to the presence of bromine. The use of carbon disulfide as the desorbing solvent gave too large a response and masked the 2-bromoethanol. Several other solvents were tested for both their response on the ECD and their ability to desorb ethylene oxide from the charcoal. Among those tested were toluene, xylene, ethyl benzene, hexane, cyclohexane and benzene. Benzene was the only solvent tested that gave a suitable response on the ECD and a high desorption. It was found that the desorption efficiency was improved by using 1% $CS_2$ with the benzene. The carbon disulfide did not significantly improve the recovery with the other solvents. SKC Lot 120 was used in all tests done with activated charcoal.

1.1.2 Physical Properties (Ref. 5.2-5.4).

Synonyms: Oxirane; dimethylene oxide, 1,2-epoxy-ethane; oxane; $C_2H_4O$; ETO;

Molecular Weight: 44.06

Boiling Point: 10.7 °C (51.3 °F)

Melting Point: -111 °C

Description: Colorless, flammable gas

Vapor Pressure: 1095 mm. at 20 °C

Odor: Ether-like odor

Lower Explosive Limits: 3.0% (by volume)

Flash Point (TOC): Below 0 °F

Molecular Structure: $CH_2-CH_2$

1.2 Limit Defining Parameters.

1.2.1 Detection Limit of the Analytical Procedure.

The detection limit of the analytical procedure is 12.0 picograms of ethylene oxide per injection. This is the amount of analyte which will give a peak whose height is five times the height of the baseline noise. (See Backup Data Section 4.1).

1.2.2 Detection Limit of the Overall Procedure.

The detection limit of the overall procedure is 24.0 ng of ethylene oxide per sample.

This is the amount of analyte spiked on the sampling device which allows recovery of an amount of analyte equivalent to the detection limit of the analytical procedure. (See Backup Data Section 4.2).

1.2.3 Reliable Quantitation Limit.

The reliable quantitation limit is 94.0 nanograms of ethylene oxide per sample. This is the smallest amount of analyte which can be quantitated within the requirements of 75% recovery and 95% confidence limits. (See Backup Data Section 4.2).

It must be recognized that the reliable quantitation limit and detection limits reported in the method are based upon optimization of the instrument for the smallest possible amount of analyte. When the target concentration of an analyte is exceptionally higher than these limits, they may not be attainable at the routine operating parameters. In this case, the limits reported on analysis reports will be based on the operating parameters used during the analysis of the samples.

1.2.4 Sensitivity.

The sensitivity of the analytical procedure over a concentration range representing 0.5 to 2 times the target concentration based on the recommended air volume is 34105 area units per µg/ml. The sensitivity is determined by the slope of the calibration curve (See Backup Data Section 4.3).

The sensitivity will vary somewhat with the particular instrument used in the analysis.

1.2.5 Recovery.

The recovery of analyte from the collection medium must be 75% or greater. The average recovery from spiked samples over the range of 0.5 to 2 times the target concentration is 88.0% (See Backup Section 4.4). At lower concentrations the recovery appears to be non-linear.

1.2.6 Precision (Analytical Method Only).

The pooled coefficient of variation obtained from replicate determination of analytical standards at 0.5X, 1X and 2X the target concentration is 0.036 (See Backup Data Section 4.5).

1.2.7 Precision (Overall Procedure).

The overall procedure must provide results at the target concentration that are 25% of better at the 95% confidence level. The precision at the 95% confidence level for the 15 day storage test is plus or minus 12.9% (See Backup Data Section 4.6).

This includes an additional plus or minus 5% for sampling error.

1.3 Advantages.

1.3.1 The sampling procedure is convenient.

1.3.2 The analytical procedure is very sensitive and reproducible.

1.3.3 Reanalysis of samples is possible.

1.3.4 Samples are stable for at least 15 days at room temperature.

1.3.5 Interferences are reduced by the longer GC retention time of the new derivative.

1.4 Disadvantages.

1.4.1 Two tubes in series must be used because of possible breakthrough and migration.

1.4.2 The precision of the sampling rate may be limited by the reproducibility of the pressure drop across the tubes. The pumps are usually calibrated for one tube only.

1.4.3 The use of benzene as the desorption solvent increases the hazards of analysis because of the potential carcinogenic effects of benzene.

1.4.4 After repeated injections there can be a buildup of residue formed on the electron capture detector which decreases sensitivity.

1.4.5 Recovery from the charcoal tubes appears to be nonlinear at low concentrations.

2. Sampling Procedure.

2.1 Apparatus.

2.1.1 A calibrated personal sampling pump whose flow can be determined within plus or minus 5% of the recommended flow.

2.1.2 SKC Lot 120 Charcoal tubes: glass tube with both ends flame sealed, 7 cm long with a 6 mm O.D. and a 4-mm I.D., containing 2 sections of coconut shell charcoal separated by a 2-mm portion of urethane foam. The adsorbing section contains 100 mg of charcoal, the backup section 50 mg. A 3-mm portion of urethane foam is placed between the outlet end of the tube and the backup section. A plug of silylated glass wool is placed in front of the adsorbing section.

2.2 Reagents.

2.2.1 None required.

2.3 Sampling Technique.

2.3.1 Immediately before sampling, break the ends of the charcoal tubes. All tubes must be from the same lot.

2.3.2 Connect two tubes in series to the sampling pump with a short section of flexible tubing. A minimum amount of tubing is used to connect the two sampling tubes together. The tube closer to the pump is used as a backup. This tube should be identified as the backup tube.

2.3.3 The tubes should be placed in a vertical position during sampling to minimize channeling.

2.3.4 Air being sampled should not pass through any hose or tubing before entering the charcoal tubes.

2.3.5 Seal the charcoal tubes with plastic caps immediately after sampling. Also, seal each sample with OSHA seals lengthwise.

2.3.6 With each batch of samples, submit at least one blank tube from the same lot used for samples. This tube should be subjected to exactly the same handling as the samples (break, seal, transport) except that no air is drawn through it.

2.3.7 Transport the samples (and corresponding paperwork) to the lab for analysis.

2.3.8 If bulk samples are submitted for analysis, they shoud [sic] be transported in glass containers with Teflon-lined caps. These samples must be mailed separately from the container used for the charcoal tubes.

2.4 Breakthrough.

2.4.1 The breakthrough (5% breakthrough) volume for a 3.0 mg/m ethylene oxide sample stream at approximately 85% relative humidity, 22 °C and 633 mm is 2.6 liters sampled at 0.05 liters per minute. This is equivalent to 7.8 µg of ethylene oxide. Upon saturation of the tube it appeared that the water may be displacing ethylene oxide during sampling.

2.5 Desorption Efficiency.

2.5.1 The desorption efficiency, from liquid injection onto charcoal tubes, averaged 88.0% from 0.5 to 2.0 x the target concentration for a 1.0 liter air sample. At lower ranges it appears that the desorption efficiency is non-linear (See Backup Data Section 4.2).

## OSHA Standards for General Industry

2.5.2 The desorption efficiency may vary from one laboratory to another and also from one lot of charcoal to another. Thus, it is necessary to determine the desorption efficiency for a particular lot of charcoal.

2.6 Recommended Air Volume and Sampling Rate.

2.6.1 The recommended air volume is 1.0 liter.

2.6.2 The recommended maximum sampling rate is 0.05 Lpm.

2.7 Interferences.

2.7.1 Ethylene glycol and Freon 12 at target concentration levels did not interfere with the collection of ethylene oxide.

2.7.2 Suspected interferences should be listed on the sample data sheets.

2.7.3 The relative humidity may affect the sampling procedure.

2.8 Safety Precautions.

2.8.1 Attach the sampling equipment to the employee so that it does not interfere with work performance.

2.8.2 Wear safety glasses when breaking the ends of the sampling tubes.

2.8.3 If possible, place the sampling tubes in a holder so the sharp end is not exposed while sampling.

3. Analytical Method.

3.1 Apparatus.

3.1.1 Gas chromatograph equipped with a linearized electron capture detector.

3.1.2 GC column capable of separating the derivative of ethylene oxide (2-bromoethanol) from any interferences and the 1% $CS_2$ in benzene solvent. The column used for validation studies was: 10 ft x 1/8 inch stainless steel 20% SP-2100, .1% Carbowax 1500 on 100/120 Supelcoport.

3.1.3 An electronic integrator or some other suitable method of measuring peak areas.

3.1.4 Two milliliter vials with Teflon-lined caps.

3.1.5 Gas tight syringe—500 µl or other convenient sizes for preparing standards.

3.1.6 Microliter syringes—10 µl or other convenient sizes for diluting standards and 1 µL for sample injections.

3.1.7 Pipets for dispensing the 1% $CS_2$ in benzene solvent. The Glenco 1 ml dispenser is adequate and convenient.

3.1.8 Volumetric flasks—5 ml and other convenient sizes for preparing standards.

3.1.9 Disposable Pasteur pipets.

3.2 Reagents.

3.2.1 Benzene, reagent grade.

3.2.2 Carbon Disulfide, reagent grade.

3.2.3 Ethylene oxide, 99.7% pure.

3.2.4 Hydrobromic Acid, 48% reagent grade.

3.2.5 Sodium Carbonate, anhydrous, reagent grade.

3.2.6 Desorbing reagent, 99% Benzene/1% $CS_2$.

3.3 Sample Preparation.

3.3.1 The front and back sections of each sample are transferred to separate 2-ml vials.

3.3.2 Each sample is desorbed with 1.0 ml of desorbing reagent.

3.3.3 The vials are sealed immediately and allowed to desorb for one hour with occasional shaking.

3.3.4 Desorbing reagent is drawn off the charcoal with a disposable pipet and put into clean 2-ml vials.

3.3.5 One drop of HBr is added to each vial. Vials are resealed and HBr is mixed well with the desorbing reagent.

3.3.6 About 0.15 grams of sodium carbonate is carefully added to each vial. Vials are again resealed and mixed well.

3.4 Standard Preparation.

3.4.1 Standards are prepared by injecting the pure ethylene oxide gas into the desorbing reagent.

3.4.2 A range of standards are prepared to make a calibration curve. A concentration of 1.0 µL of ethylene oxide gas per 1 ml desorbing reagent is equivalent to 1.0 ppm air concentration (all gas volumes at 25 °C and 760 mm) for the recommended 1 liter air sample. This amount is uncorrected for desorption efficiency (See Backup Data Section 4.2. for desorption efficiency corrections).

3.4.3 One drop of HBr per ml of standard is added and mixed well.

3.4.4 About 0.15 grams of sodium carbonate is carefully added for each drop of HBr (A small reaction will occur).

3.5 Analysis.

3.5.1 GC Conditions.

Nitrogen flow rate—10ml/min.

Injector Temperature—250 °C

Detector Temperature—300 °C

Column Temperature—100 °C

Injection size—0.8 µL

Elution time—3.9 minutes

3.5.2 Peak areas are measured by an integrator or other suitable means.

3.5.3 The integrator results are in area units and a calibration curve is set up with concentration vs. area units.

3.6 Interferences.

3.6.1 Any compound having the same retention time of 2-bromoethanol is a potential interference. Possible interferences should be listed on the sample data sheets.

3.6.2 GC parameters may be changed to circumvent interferences.

3.6.3 There are usually trace contaminants in benzene. These contaminants, however, posed no problem of interference.

3.6.4 Retention time data on a single column is not considered proof of chemical identity. Samples over the 1.0 ppm target level should be confirmed by GC/Mass Spec or other suitable means.

3.7 Calculations

3.7.1 The concentration in µg/ml for a sample is determined by comparing the area of a particular sample to the calibration curve, which has been prepared from analytical standards.

3.7.2 The amount of analyte in each sample is corrected for desorption efficiency by use of a desorption curve.

3.7.3 Analytical results (A) from the two tubes that compose a particular air sample are added together.

3.7.4 The concentration for a sample is calculated by the following equation:

$$\text{ETO, mg/m}^3 = \frac{A \times B}{C}$$

where:

A = µg/ml

B = desorption volume in milliliters

C = air volume in liters.

3.7.5 To convert mg/m³ to parts per million (ppm) the following relationship is used:

$$\text{ETO, ppm} = \frac{\text{mg/m}^3 \times 24.45}{44.05}$$

where:

mg/m³ = results from 3.7.4

24.45 = molar volume at 25 °C and 760 mm Hg

44.05 = molecular weight of ETO.

3.8 Safety Precautions

3.8.1 Ethylene oxide and benzene are potential carcinogens and care must be exercised when working with these compounds.

3.8.2 All work done with the solvents (preparation of standards, desorption of samples, etc.) should be done in a hood.

3.8.3 Avoid any skin contact with all of the solvents.

3.8.4 Wear safety glasses at all times.

3.8.5 Avoid skin contact with HBr because it is highly toxic and a strong irritant to eyes and skin.

4. Backup Data.

4.1 Detection Limit Data.

The detection limit was determined by injecting 0.8 μl of a 0.015 μg/ml standard of ethylene oxide into 1% $CS_2$ in benzene. The detection limit of the analytical procedure is taken to be $1.20 \times 10^{-5}$ μg per injection. This is equivalent to 8.3 ppb (0.015 mg/m³) for the recommended air volume.

4.2 Desorption Efficiency.

Ethylene oxide was spiked onto charcoal tubes and the following recovery data was obtained.

| Amount spiked (μg) | Amount recovered (μg) | Percent recovery |
|---|---|---|
| 4.5 | 4.32 | 96.0 |
| 3.0 | 2.61 | 87.0 |
| 2.25 | 2.025 | 90.0 |
| 1.5 | 1.365 | 91.0 |
| 1.5 | 1.38 | 92.0 |
| .75 | .6525 | 87.0 |
| .375 | .315 | 84.0 |
| .375 | .312 | 83.2 |
| .1875 | .151 | 80.5 |
| .094 | .070 | 74.5 |

At lower amounts the recovery appears to be non-linear.

4.3 Sensitivity Data.

The following data was used to determine the calibration curve.

| Injection | 0.5 × .75 μg/mL | 1 × 1.5 μg/mL | 2 × 3.0 μg/mL |
|---|---|---|---|
| 1 | 30904 | 59567 | 111778 |
| 2 | 30987 | 62914 | 106016 |
| 3 | 32555 | 58578 | 106122 |
| 4 | 32242 | 57173 | 109716 |
| X | 31672 | 59558 | 108408 |

Slope = 34.105.

4.4 Recovery.

The recovery was determined by spiking ethylene oxide onto lot 120 charcoal tubes and desorbing with 1% $CS_2$ in Benzene. Recoveries were done at 0.5, 1.0, and 2.0 X the target concentration (1 ppm) for the recommended air volume.

### PERCENT RECOVERY

| Sample | 0.5× | 1.0× | 2.0× |
|---|---|---|---|
| 1 | 88.7 | 95.0 | 91.7 |
| 2 | 83.8 | 95.0 | 87.3 |
| 3 | 84.2 | 91.0 | 86.0 |
| 4 | 88.0 | 91.0 | 83.0 |
| 5 | 88.0 | 86.0 | 85.0 |
| X | 86.5 | 90.5 | 87.0 |

Weighted Average = 88.2.

4.5 Precision of the Analytical Procedure.

The following data was used to determine the precision of the analytical method:

| Concentration | 0.5×.75 μg/mL | 1×1.5 μg/mL | 2×3.0 μg/mL |
|---|---|---|---|
| Injection | .7421 | 1.4899 | 3.1184 |
| | .7441 | 1.5826 | 3.0447 |
| | .7831 | 1.4628 | 2.9149 |
| | .7753 | 1.4244 | 2.9185 |
| Average | .7612 | 1.4899 | 2.9991 |
| Standard Deviation | .0211 | .0674 | .0998 |
| CV | .0277 | .0452 | .0333 |

$$CV = \frac{3(.0277)^2 + 3(.0452)^2 + 3(.0333)^2}{3 + 3 + 3}$$

CV + 0.036

4.6 Storage Data.

Samples were generated at 1.5 mg/m3 ethylene oxide at 85% relative humidity, 22 °C and 633 mm. All samples were taken for 20 minutes at 0.05 Lpm. Six samples were analyzed as soon as possible and fifteen samples were stored at refrigerated temperature (5 °C) and fifteen samples were stored at ambient temperature (23 °C). These stored samples were analyzed over a period of nineteen days.

### PERCENT RECOVERY

| Day analyzed | Refrigerated | Ambient |
|---|---|---|
| 1 | 87.0 | 87.0 |
| 1 | 93.0 | 93.0 |
| 1 | 94.0 | 94.0 |
| 1 | 92.0 | 92.0 |
| 4 | 92.0 | 91.0 |
| 4 | 93.0 | 88.0 |
| 4 | 91.0 | 89.0 |
| 6 | 92.0 | .......... |
| 6 | 92.0 | .......... |
| 8 | .......... | 92.0 |
| 8 | .......... | 86.0 |
| 10 | 91.7 | .......... |
| 10 | 95.5 | .......... |
| 10 | 95.7 | .......... |
| 11 | | 90.0 |
| 11 | | 82.0 |
| 13 | 78.0 | .......... |
| 13 | 81.4 | .......... |
| 13 | 82.4 | .......... |
| 14 | | 78.5 |
| 14 | | 72.1 |
| 18 | 66.0 | .......... |
| 18 | 68.0 | .......... |
| 19 | | 64.0 |
| 19 | | 77.0 |

4.7 Breakthrough Data.

Breakthrough studies were done at 2 ppm (3.6 mg/m³) at approximately 85% relative humidity at 22 °C (ambient temperature). Two charcoal tubes were used in series. The backup tube was changed every 10 minutes and analyzed for breakthrough. The flow rate was 0.050 Lpm.

| Tube No. | Time (minutes) | Percent breakthrough |
|---|---|---|
| 1 | 10 | (¹) |
| 2 | 20 | (¹) |
| 3 | 30 | (¹) |
| 4 | 40 | 1.23 |
| 5 | 50 | 3.46 |
| 6 | 60 | 18.71 |
| 7 | 70 | 39.2 |
| 8 | 80 | 53.3 |
| 9 | 90 | 72.0 |
| 10 | 100 | 96.0 |
| 11 | 110 | 113.0 |
| 12 | 120 | 133.9 |

¹ None.

The 5% breakthrough volume was reached when 2.6 liters of test atmosphere were drawn through the charcoal tubes.

5. References.

5.1 "NIOSH Manual of Analytical Methods," 2nd ed. NIOSH: Cincinnati, 1977; Method S286.

5.2 "IARC Monographs on the Evaluation of Carcinogenic Risk of Chemicals to Man," International Agency for Research on Cancer: Lyon, 1976; Vol. II, p. 157.

5.3 Sax., N.I. "Dangerous Properties of Industrial Materials," 4th ed.; Van Nostrand Reinhold Company. New York, 1975; p. 741.

5.4 "The Condensed Chemical Dictionary", 9th ed.; Hawley, G.G., ed.; Van Nostrand Reinhold Company, New York, 1977; p. 361.

### Summary of Other Sampling Procedures

OSHA believes that served other types of monitoring equipment and techniques exist for monitoring time-weighted averages. Considerable research and method development is currently being performed, which will lead to improvements and a wider variety of monitoring techniques. A combination of monitoring procedures can be used. There probably is no one best method for monitoring personal exposure to ethylene oxide in all cases. There are advantages, disadvantages, and limitations to each method. The method of choice will depend on the need and requirements. Some commonly used methods include the use of charcoal tubes, passive dosimeters, Tedler gas sampling bags, detector tubes, photoionization detection units, infrared detection units and gas chromatographs. A number of these methods are described below.

### A. CHARCOAL TUBE SAMPLING PROCEDURES

*Qazi-Ketcham method* (Ex. 11-133)—This method consists of collecting EtO on Columbia JXC activated carbon, desorbing the EtO with carbon disulfide and analyzing by gas chromatography with flame ionization detection. Union Carbide has recently updated and revalidated this monitoring procedures. This method is capable of determining both eight-hour time-weighted average exposures and short-term exposures. The method was validated to 0.5 ppm. Like other charcoal collecting procedures, the method requires considerable analytical expertise.

*ASTM-proposed method*—The Ethylene Oxide Industry Council (EOIC) has contracted with Clayton Environmental Consultants, Inc. to conduct a collaborative study for the proposed method. The ASTM-Proposed method is similar to the method published by Qazi and Ketcham is the November 1977 American Industrial Hygiene Association Journal, and to the method of Pilney and Coyne, presented at the 1979 American Industrial Hygiene Conference. After the air to be sampled is drawn through an activated charcoal tube, the ethylene oxide is desorbed from the tube using carbon disulfide and is quantitated by gas chromatography utilizing a flame ionization detector. The ASTM-proposed method specifies a large two-section charcoal tube, shipment in dry ice, storage at less than -5 °C, and analysis within three weeks to prevent migration and sample loss. Two types of charcoal tubes are being tested—Pittsburgh Coconut-Based (PCB) and Columbia JXC charcoal. This collaborative study will give an indication of the inter- and intralaboratory precision and accuracy of the ASTM-proposed method. Several laboratories have considerable expertise using the Qazi-Ketcham and Dow methods.

B. Passive Monitors—Ethylene oxide diffuses into the monitor and is collected in the sampling media. The DuPont Pro-Tek badge collects EtO in an absorbing solution, which is analyzed colorimetrically to determine the amount of EtO present. The 3M 350 badge collects the EtO on chemically treated charcoal. Other passive monitors are currently being developed and tested. Both 3M and DuPont have submitted data indicating their dosimeters meet the precision and accuracy requirements of the proposed ethylene oxide standard. Both presented laboratory validation data to 0.2 ppm (Exs. 11-65, 4-20, 108, 109, 130).

C. Tedlar Gas Sampling Bags-Samples are collected by drawing a known volume of air into a Tedlar gas sampling bag. The ethylene oxide concentration is often determined on-site using a portable gas chromatograph or portable infrared spectometer.

D. Detector tubes—A known volume of air is drawn through a detector tube using a small hand pump. The concentration of EtO is related to the length of stain developed in the tube. Detector tubes are economical, easy to use, and give an immediate readout. Unfortunately, partly because they are nonspecific, their accuracy is often questionable. Since the sample is taken over a short period of time, they may be useful for determining the source of leaks.

E. Direct Reading Instruments—There are numerous types of direct reading instruments, each having its own strengths and weaknesses (Exs. 135B, 135C, 107, 11-78, 11-153). Many are relatively new, offering greater sensitivity and specificity. Popular ethylene oxide direct reading instruments include infrared detection units, photoionization detection units, and gas chromatographs.

Portable infrared analyzers provide an immediate, continuous indication of a concentration value; making them particularly useful for locating high concentration pockets, in leak detection and in ambient air monitoring. In infrared detection units, the amount of infrared light absorbed by the gas being analyzed at selected infrared wavelengths is related to the concentration of a particular component. Various models have either fixed or variable infrared filters, differing cell pathlengths, and microcomputer controls for greater sensitivity, automation, and interference elimination.

A fairly recent detection system is photoionization detection. The molecules are ionized by high energy ultraviolet light. The resulting current is measured. Since different substances have different ionization potentials, other organic compounds may be ionized. The lower the lamp energy, the better the selectivity. As a continuous monitor, photoionization detection can be useful for locating high concentration pockets, in leak detection, and continuous ambient air monitoring. Both portable and stationary gas chromatographs are available with various types of detectors, including photoionization detectors. A gas chromatograph with a photoionization detector retains the photionization sensitivity, but minimizes or eliminates interferences. For several GC/PID units, the sensitivity is in the 0.1-0.2 ppm EtO range. The GC/PID with microprocessors can sample up to 20 sample points sequentially, calculate and record data, and activate alarms or ventilation systems. Many are quite flexible and can be configured to meet the specific analysis needs for the workplace.

DuPont presented their laboratory validation data of the accuracy of the Qazi-Ketcham charcoal tube, the PCB charcoal tube, Miran 103 IR analyzer, 3M #3550 monitor and the Du Pont C-70 badge. Quoting Elbert V. Kring:

We also believe that OSHA's proposed accuracy in this standard is appropriate. At plus or minus 25 percent at one part per million, and plus or minus 35 percent below that. And, our data indicates there's only one monitoring method, right now, that we've tested thoroughly, that meets that accuracy requirements. That is the Du Pont Pro-Tek badge***. We also believe that this kind of data should be confirmed by another independent laboratory, using the same type dynamic chamber testing (Tr. 1470)

Additional data by an independent laboratory following their exact protocol was not submitted. However, information was submitted on comparisons and precision and accuracy of those monitoring procedures which indicate far better precision and accuracy of those monitoring procedures than that obtained by Du Pont (Ex. 4-20, 130, 11-68, 11-133, 130, 135A).

The accuracy of any method depends to a large degree upon the skills and experience of those who not only collect the samples but also those who analyze the samples. Even for methods that are collaboratively tested, some laboratories are closer to the true values than others. Some laboratories may meet the precision and accuracy requirements of the method; others may consistently far exceed them for the same method.

[49 FR 25796, June 22, 1984, as amended at 50 FR 9801, Mar. 12, 1985; 50 FR 41494, Oct. 11, 1985; 51 FR 25053, July 10, 1986; 53 FR 11436, 11437, Apr. 6, 1988; 53 FR 27960, July 26, 1988; 54 FR 24334, June 7, 1989; 61 FR 5508, Feb. 13, 1996; 63 FR 1292, Jan. 8, 1998]

### §1910.1048 Formaldehyde.

**1910.1048(a)** *Scope and application.* This standard applies to all occupational exposures to formaldehyde, i.e., from formaldehyde gas, its solutions, and materials that release formaldehyde.

**1910.1048(b)** *Definitions.* For purposes of this standard, the following definitions shall apply:

*Action level* means a concentration of 0.5 part formaldehyde per million parts of air (0.5 ppm) calculated as an eight (8)-hour time-weighted average (TWA) concentration.

*Assistant Secretary* means the Assistant Secretary of Labor for the Occupational Safety and Health Administration, U.S. Department of Labor, or designee.

*Authorized person* means any person required by work duties to be present in regulated areas, or authorized to do so by the employer, by this section, or by the OSH Act of 1970.

*Director* means the Director of the National Institute for Occupational Safety and Health, U.S. Department of Health and Human Services, or designee.

*Emergency* is any occurrence, such as but not limited to equipment failure, rupture of containers, or failure of control equipment that results in an uncontrolled release of a significant amount of formaldehyde.

*Employee exposure* means the exposure to airborne formaldehyde which would occur without corrections for protection provided by any respirator that is in use.

*Formaldehyde* means the chemical substance, HCHO, Chemical Abstracts Service Registry No. 50-00-0.

**1910.1048(c)** *Permissible Exposure Limit (PEL)*—**(1)** *TWA:* The employer shall assure that no employee is exposed to an airborne concentration of formaldehyde which exceeds 0.75 parts formaldehyde per million parts of air (0.75 ppm) as an 8-hour TWA.

**1910.1048(c)(2)** *Short Term Exposure Limit (STEL):* The employer shall assure that no employee is exposed to an airborne concentration of formaldehyde which exceeds two parts formaldehyde per million parts of air (2 ppm) as a 15-minute STEL.

**1910.1048(d)** *Exposure monitoring*—**(1)** *General.* **(i)** Each employer who has a workplace covered by this standard shall monitor employees to determine their exposure to formaldehyde.

**1910.1048(d)(1)(ii)** *Exception.* Where the employer documents, using objective data, that the presence of formaldehyde or formaldehyde-releasing products in the workplace cannot result in airborne concentrations of formaldehyde that would cause any employee to be exposed at or above the action level or the STEL under foreseeable conditions of use, the employer will not be required to measure employee exposure to formaldehyde.

**1910.1048(d)(1)(iii)** When an employee's exposure is determined from representative sampling, the measurements used shall be representative of the employee's full shift or short-term exposure to formaldehyde, as appropriate.

**1910.1048(d)(1)(iv)** Representative samples for each job classification in each work area shall be taken for each shift unless the employer can document with objective data that exposure levels for a given job classification are equivalent for different work shifts.

**1910.1048(d)(2)** *Initial monitoring.* The employer shall identify all employees who may be exposed at or above the action level or at or above the STEL and accurately determine the exposure of each employee so identified.

**1910.1048(d)(2)(i)** Unless the employer chooses to measure the exposure of each employee potentially exposed to formaldehyde, the employer shall develop a representative sampling strategy and measure sufficient exposures within each job classification for each worksheet to correctly characterize and not underestimate the exposure of any employee within each exposure group.

**1910.1048(d)(2)(ii)** The initial monitoring process shall be repeated each time there is a change in production, equipment, process, personnel, or control measures which may result in new or additional exposure to formaldehyde.

**1910.1048(d)(2)(iii)** If the employer receives reports of signs or symptoms of respiratory or dermal conditions associated with formaldehyde exposure, the employer shall promptly monitor the affected employee's exposure.

**1910.1048(d)(3)** *Periodic monitoring.* **(i)** The employer shall periodically measure and accurately determine exposure to formaldehyde for employees shown by the initial monitoring to be exposed at or above the action level or at or above the STEL.

**1910.1048(d)(3)(ii)** If the last monitoring results reveal employee exposure at or above the action level, the employer shall repeat monitoring of the employees at least every 6 months.

**1910.1048(d)(3)(iii)** If the last monitoring results reveal employee exposure at or above the STEL, the employer shall repeat monitoring of the employees at least once a year under worst conditions.

**1910.1048(d)(4)** *Termination of monitoring.* The employer may discontinue periodic monitoring for employees if results from two consecutive sampling periods taken at least 7 days apart show that employee exposure is below the action level and the STEL. The results must be statistically representative and consistent with the employer's knowledge of the job and work operation.

**1910.1048(d)(5)** *Accuracy of monitoring.* Monitoring shall be accurate, at the 95 percent confidence level, to within plus or minus 25 percent for airborne concentrations of formaldehyde at the TWA and the STEL and to within plus or minus 35 percent for airborne concentrations of formaldehyde at the action level.

**1910.1048(d)(6)** *Employee notification of monitoring results.* The employer must, within 15 working days after the receipt of the results of any monitoring performed under this section, notify each affected employee of these results either individually in writing or by posting the results in an appropriate location that is accessible to employees. If employee exposure is above the PEL, affected employees shall be provided with a description of the corrective actions being taken by the employer to decrease exposure.

**1910.1048(d)(7)** *Observation of monitoring.* **(i)** The employer shall provide affected employees or their designated representatives an opportunity to observe any monitoring of employee exposure to formaldehyde required by this standard.

**1910.1048(d)(7)(ii)** When observation of the monitoring of employee exposure to formaldehyde requires entry into an area where the use of protective clothing or equipment is required, the employer shall provide the clothing and equipment to the observer, require the observer to use such clothing and equipment, and assure that the observer complies with all other applicable safety and health procedures.

**1910.1048(e)** *Regulated areas.* **(1)** The employer shall establish regulated areas where the concentration of airborne formaldehyde exceeds either the TWA or the STEL and post all entrances and accessways with signs bearing the following information:

**DANGER**

**FORMALDEHYDE**

**IRRITANT AND POTENTIAL CANCER HAZARD**

**AUTHORIZED PERSONNEL ONLY**

**1910.1048(e)(2)** The employer shall limit access to regulated areas to authorized persons who have been trained to recognize the hazards of formaldehyde.

**1910.1048(e)(3)** An employer at a multiemployer worksite who establishes a regulated area shall communicate the access restrictions and locations of these areas to other employers with work operations at that worksite.

**1910.1048(f)** *Methods of compliance*—**(1)** *Engineering controls and work practices.* The employer shall institute engineering and work practice controls to reduce and maintain employee exposures to formaldehyde at or below the TWA and the STEL.

**1910.1048(f)(2)** *Exception.* Whenever the employer has established that feasible engineering and work practice controls cannot reduce employee exposure to or below either of the PELs, the employer shall apply these controls to reduce employee exposures to the extent feasible and shall supplement them with respirators which satisfy this standard.

**1910.1048(g)** *Respiratory protection*—**(1)** *General.* For employees who use respirators required by this section, the employer must provide respirators that comply with the requirements of this paragraph. Respirators must be used during:

**1910.1048(g)(1)(i)** Periods necessary to install or implement feasible engineering and work-practice controls.

**1910.1048(g)(1)(ii)** Work operations, such as maintenance and repair activities or vessel cleaning, for which the employer establishes that engineering and work-practice controls are not feasible.

**1910.1048(g)(1)(iii)** Work operations for which feasible engineering and work-practice controls are not yet sufficient to reduce employee exposure to or below the PELs.

**1910.1048(g)(1)(iv)** Emergencies.

**1910.1048(g)(2)** *Respirator program.* **(i)** The employer must implement a respiratory protection program in accordance with 29 CFR 1910.134(b) through (d) (except (d)(1)(iii), (d)(3)(iii)(B)(1), and (2)), and (f) through (m).

**1910.1048(g)(2)(ii)** If air-purifying chemical-cartridge respirators are used, the employer must:

**1910.1048(g)(2)(ii)(A)** Replace the cartridge after three (3) hours of use or at the end of the workshift, whichever occurs first, unless the cartridge contains a NIOSH-approved end-of-service-life indicator (ESLI) to show when breakthrough occurs.

**1910.1048(g)(2)(ii)(B)** Unless the canister contains a NIOSH-approved ESLI to show when breakthrough occurs, replace canisters used in atmospheres up to 7.5 ppm (10xPEL) every four (4) hours and industrial-sized canisters used in atmospheres up to 75 ppm (100xPEL) every two (2) hours, or at the end of the workshift, whichever occurs first.

**1910.1048(g)(3)** *Respirator selection.* **(i)** The employer must select appropriate respirators from Table 1 in this section.

### TABLE 1—MINIMUM REQUIREMENTS FOR RESPIRATORY PROTECTION AGAINST FORMALDEHYDE

| Condition of use or formaldehyde concentration (ppm) | Minimum required respirator[1] |
|---|---|
| Up to 7.5 ppm. (10 × PEL) | Full facepiece with cartridges or canisters specifically approved for protection against formaldehyde.[2] |
| Up to 75 ppm. (100 × PEL) | Full-face mask with chin style or chest or back mounted type, with industrial size canister specifically approved for protection against formaldehyde. Type C supplied air respirator, demand type, or continuous flow type, with full facepiece, hood, or helmet. |
| Above 75 ppm or unknown. (emergencies). (100 × PEL). | Self-contained breathing apparatus (SCBA) with positive pressure full facepiece. Combination supplied-air, full facepiece positive pressure respirator with auxiliary self-contained air supply. |
| Firefighting | SCBA with positive pressure in full facepiece. |
| Escape | SCBA in demand or pressure demand mode. Full-face mask with chin style or front or back mounted type industrial size canister specifically approved for protection against formaldehyde. |

[1] Respirators specified for use at higher concentrations may be used at lower concentrations.
[2] A half-mask respirator with cartridges specifically approved for protection against formaldehyde can be substituted for the full facepiece respirator providing that effective gas-proof goggles are provided and used in combination witht he half-mask repirator.

**1910.1048(g)(3)(ii)** The employer must provide a powered air-purifying respirator adequate to protect against formaldehyde exposure to any employee who has difficulty using a negative-pressure respirator.

**1910.1048(h)** *Protective equipment and clothing.* Employers shall comply with the provisions of 29 CFR 1910.132 and 29 CFR 1910.133. When protective equipment or clothing is provided under these provisions, the employer shall provide these protective devices at no cost to the employee and assure that the employee wears them.

**1910.1048(h)(1)** *Selection.* The employer shall select protective clothing and equipment based upon the form of formaldehyde to be encountered, the conditions of use, and the hazard to be prevented.

**1910.1048(h)(1)(i)** All contact of the eyes and skin with liquids containing 1 percent or more formaldehyde shall be prevented by the use of chemical protective clothing made of material impervious to formaldehyde and the use of other personal protective equipment, such as goggles and face shields, as appropriate to the operation.

**1910.1048(h)(1)(ii)** Contact with irritating or sensitizing materials shall be prevented to the extent necessary to eliminate the hazard.

**1910.1048(h)(1)(iii)** Where a face shield is worn, chemical safety goggles are also required if there is a danger of formaldehyde reaching the area of the eye.

**1910.1048(h)(1)(iv)** Full body protection shall be worn for entry into areas where concentrations exceed 100 ppm and for emergency reentry into areas of unknown concentration.

**1910.1048(h)(2)** *Maintenance of protective equipment and clothing.* **(i)** The employer shall assure that protective equipment and clothing that has become contaminated with formaldehyde is cleaned or laundered before its reuse.

**1910.1048(h)(2)(ii)** When ventilating formaldehyde-contaminated clothing and equipment, the employer shall establish a storage area so that employee exposure is minimized. Containers for contaminated clothing and equipment and storage areas shall have labels and signs containing the following information:

**DANGER**

**FORMALDEHYDE-CONTAMINATED [CLOTHING] EQUIPMENT**

**AVOID INHALATION AND SKIN CONTACT**

**1910.1048(h)(2)(iii)** The employer shall assure that only persons trained to recognize the hazards of formaldehyde remove the contaminated material from the storage area for purposes of cleaning, laundering, or disposal.

**1910.1048(h)(2)(iv)** The employer shall assure that no employee takes home equipment or clothing that is contaminated with formaldehyde.

**1910.1048(h)(2)(v)** The employer shall repair or replace all required protective clothing and equipment for each affected employee as necessary to assure its effectiveness.

**1910.1048(h)(2)(vi)** The employer shall inform any person who launders, cleans, or repairs such clothing or equipment of formaldehyde's potentially harmful effects and of procedures to safely handle the clothing and equipment.

**1910.1048(i)** *Hygiene protection.* **(1)** The employer shall provide change rooms, as described in 29 CFR 1910.141 for employees who are required to change from work clothing into protective clothing to prevent skin contact with formaldehyde.

**1910.1048(i)(2)** If employees' skin may become spashed with solutions containing 1 percent or greater formaldehyde, for example, because of equipment failure or improper work practices, the employer shall provide conveniently located quick drench showers and assure that affected employees use these facilities immediately.

**1910.1048(i)(3)** If there is any possibility that an employee's eyes may be splashed with solutions containing 0.1 percent or greater formaldehyde, the employer shall provide acceptable eyewash facilities within the immediate work area for emergency use.

**1910.1048(j)** *Housekeeping.* For operations involving formaldehyde liquids or gas, the employer shall conduct a program to detect leaks and spills, including regular visual inspections.

**1910.1048(j)(1)** Preventative maintenance of equipment, including surveys for leaks, shall be undertaken at regular intervals.

**1910.1048(j)(2)** In work areas where spillage may occur, the employer shall make provisions to contain the spill, to decontaminate the work area, and to dispose of the waste.

**1910.1048(j)(3)** The employer shall assure that all leaks are repaired and spills are cleaned promptly by employees wearing suitable protective equipment and trained in proper methods for cleanup and decontamination.

**1910.1048(j)(4)** Formaldehyde-contaminated waste and debris resulting from leaks or spills shall be placed for disposal in sealed containers bearing a label warning of formaldehyde's presence and of the hazards associated with formaldehyde.

**1910.1048(k)** *Emergencies.* For each workplace where there is the possibility of an emergency involving formaldehyde, the employer shall assure appropriate procedures are adopted to minimize injury and loss of life. Appropriate procedures shall be implemented in the event of an emergency.

**1910.1048(l)** *Medical surveillance—(1)* *Employees covered.* **(i)** The employer shall institute medical surveillance programs for all employees exposed to formaldehyde at concentrations at or exceeding the action level or exceeding the STEL.

**1910.1048(l)(1)(ii)** The employer shall make medical surveillance available for employees who develop signs and symptoms of overexposure to formaldehyde and for all employees exposed to formaldehyde in emergencies. When determining whether an employee may be experiencing signs and symptoms of possible overexposure to formaldehyde, the employer may rely on the evidence that signs and symptoms associated with formaldehyde exposure will occur only in exceptional circumstances when airborne exposure is less than 0.1 ppm and when formaldehyde is present in material in concentrations less than 0.1 percent.

**1910.1048(l)(2)** *Examination by a physician.* All medical procedures, including administration of medical disease questionnaires, shall be performed by or under the supervision of a licensed physician and shall be provided without cost to the employee, without loss of pay, and at a reasonable time and place.

**1910.1048(l)(3)** *Medical disease questionnaire.* The employer shall make the following medical surveillance available to employees prior to assignment to a job where formaldehyde exposure is at or above the action level or above the STEL and annually thereafter. The employer shall also make the following medical surveillance available promptly upon determining that an employee is experiencing signs and symptoms indicative of possible overexposure to formaldehyde.

**1910.1048(l)(3)(i)** Administration of a medical disease questionnaire, such as in appendix D, which is designed to elicit information on work history, smoking history, any evidence of eye, nose, or throat irritation; chronic airway problems or hyperreactive airway disease; allergic skin conditions or dermatitis; and upper or lower respiratory problems.

**1910.1048(l)(3)(ii)** A determination by the physician, based on evaluation of the medical disease questionnaire, of whether a medical examination is necessary for employees not required to wear respirators to reduce exposure to formaldehyde.

**1910.1048(l)(4)** *Medical examinations.* Medical examinations shall be given to any employee who the physician feels, based on information in the medical disease questionnaire, may be at increased risk from exposure to formaldehyde and at the time of initial assignment and at least annually thereafter to all employees required to wear a respirator to reduce exposure to formaldehyde. The medical examination shall include:

**1910.1048(l)(4)(i)** A physical examination with emphasis on evidence of irritation or sensitization of the skin and respiratory system, shortness of breath, or irritation of the eyes.

**1910.1048(l)(4)(ii)** Laboratory examinations for respirator wearers consisting of baseline and annual pulmonary function tests. As a minimum, these tests shall consist of forced vital capacity (FVC), forced expiratory volume in one second ($FEV_1$), and forced expiratory flow (FEF).

**1910.1048(l)(4)(iii)** Any other test which the examining physician deems necessary to complete the written opinion.

**1910.1048(l)(4)(iv)** Counseling of employees having medical conditions that would be directly or indirectly aggravated by exposure to formaldehyde on the increased risk of impairment of their health.

**1910.1048(l)(5)** *Examinations for employees exposed in an emergency.* The employer shall make medical examinations available as soon as possible to all employees who have been exposed to formaldehyde in an emergency.

**1910.1048(l)(5)(i)** The examination shall include a medical and work history with emphasis on any evidence of upper or lower respiratory problems, allergic conditions, skin reaction or hypersensitivity, and any evidence of eye, nose, or throat irritation.

**1910.1048(l)(5)(ii)** Other examinations shall consist of those elements considered appropriate by the examining physician.

**1910.1048(l)(6)** *Information provided to the physician.* The employer shall provide the following information to the examining physician:

**1910.1048(l)(6)(i)** A copy of this standard and appendix A, C, D, and E;

**1910.1048(l)(6)(ii)** A description of the affected employee's job duties as they relate to the employee's exposure to formaldehyde;

**1910.1048(l)(6)(iii)** The representative exposure level for the employee's job assignment;

**1910.1048(l)(6)(iv)** Information concerning any personal protective equipment and respiratory protection used or to be used by the employee; and

**1910.1048(l)(6)(v)** Information from previous medical examinations of the affected employee within the control of the employer.

**1910.1048(l)(6)(vi)** In the event of a nonroutine examination because of an emergency, the employer shall provide to the physician as soon as possible: A description of how the emergency occurred and the exposure the victim may have received.

**1910.1048(l)(7)** *Physician's written opinion.* **(i)** For each examination required under this standard, the employer shall obtain a written opinion from the examining physician. This written opinion shall contain the results of the medical examination except that it shall not reveal specific findings or diagnoses unrelated to occupational exposure to formaldehyde. The written opinion shall include:

**1910.1048(l)(7)(i)(A)** The physician's opinion as to whether the employee has any medical condition that would place the employee at an increased risk of material impairment of health from exposure to formaldehyde;

**1910.1048(l)(7)(i)(B)** Any recommended limitations on the employee's exposure or changes in the use of personal protective equipment, including respirators;

**1910.1048(l)(7)(i)(C)** A statement that the employee has been informed by the physician of any medical conditions which would be aggravated by exposure to formaldehyde, whether these conditions may have resulted from past formaldehyde exposure or from exposure in an emergency, and whether there is a need for further examination or treatment.

**1910.1048(l)(7)(ii)** The employer shall provide for retention of the results of the medical examination and tests conducted by the physician.

**1910.1048(l)(7)(iii)** The employer shall provide a copy of the physician's written opinion to the affected employee within 15 days of its receipt.

**1910.1048(l)(8)** *Medical removal.* **(i)** The provisions of paragraph (l)(8) apply when an employee reports significant irritation of the mucosa of the eyes or the upper airways, respiratory sensitization, dermal irritation, or dermal sensitization attributed to workplace formaldehyde exposure. Medical removal provisions do not apply in the case of dermal irritation or dermal sensitization when the product suspected of causing the dermal condition contains less than 0.05% formaldehyde.

**1910.1048(l)(8)(ii)** An employee's report of signs or symptoms of possible overexposure to formaldehyde shall be evaluated by a physician selected by the employer pursuant to paragraph (l)(3). If the physician determines that a medical examination is not necessary under paragraph (l)(3)(ii), there shall be a two-week evaluation and remediation period to permit the employer to ascertain whether the signs or symptoms subside untreated or with the use of creams, gloves, first aid treatment or personal protective equipment. Industrial hygiene measures that limit the employee's exposure to formaldehyde may also be implemented during this period. The employee shall be referred immediately to a physician prior to expiration of the two-week period if the signs or symptoms worsen. Earnings, seniority and benefits may not be altered during the two-week period by virtue of the report.

**1910.1048(l)(8)(iii)** If the signs or symptoms have not subsided or been remedied by the end of the two-week period, or earlier if signs or symptoms warrant, the employee shall be examined by a physician selected by the employer. The physician shall presume, absent contrary evidence, that observed dermal irritation or dermal sensitization are not attributable to formaldehyde when products to which the affected employee is exposed contain less than 0.1% formaldehyde.

**1910.1048(l)(8)(iv)** Medical examinations shall be conducted in compliance with the requirements of paragraph (l)(5) (i) and (ii). Additional guidelines for conducting medical exams are contained in appendix C.

**1910.1048(l)(8)(v)** If the physician finds that significant irritation of the mucosa of the eyes or of the upper airways, respiratory sensitization, dermal irritation, or dermal sensitization result from workplace formaldehyde exposure and recommends restrictions or removal, the employer shall promptly comply with the restrictions or recommendation of removal. In the event of a recommendation of removal, the employer shall remove the effected employee from the current formaldehyde exposure and if possible, transfer the employee to work having no or significantly less exposure to formaldehyde.

**1910.1048(l)(8)(vi)** When an employee is removed pursuant to paragraph (l)(8)(v), the employer shall transfer the employee to comparable work for which the employee is qualified or can be trained in a short period (up to 6 months), where the formaldehyde exposures are as low as possible, but not higher than the action level. The employer shall maintain the employee's current earnings, seniority, and other benefits. If there is no such work available, the employer shall maintain the employee's current earnings, seniority and other benefits until such work becomes available, until the employee is determined to be unable to return to workplace formaldehyde exposure, until the employee is determined to be able to return to the original job status, or for six months, whichever comes first.

**1910.1048(l)(8)(vii)** The employer shall arrange for a follow-up medical examination to take place within six months after the employee is removed pursuant to this paragraph. This examination shall determine if the employee can return to the original job status, or if the removal is to be permanent. The physician shall make a decision within six months of the date the employee was removed as to whether the employee can be returned to the original job status, or if the removal is to be permanent.

**1910.1048(l)(8)(viii)** An employer's obligation to provide earnings, seniority and other benefits to a removed employee may be reduced to the extent that the employee receives compensation for earnings lost during the period of removal either from a publicly or employer-funded compensation program or from employment with another employer made possible by virtue of the employee's removal.

**1910.1048(l)(8)(ix)** In making determinations of the formaldehyde content of materials under this paragraph the employer may rely on objective data.

**1910.1048(l)(9)** *Multiple physician review.* **(i)** After the employer selects the initial physician who conducts any medical examination or consultation to determine whether medical removal or restriction is appropriate, the employee may designate a second physician to review any findings, determinations or recommendations of the initial physician and to conduct such examinations, consultations, and laboratory tests as the second physician deems necessary and appropriate to evaluate the effects of formaldehyde exposure and to facilitate this review.

**1910.1048(l)(9)(ii)** The employer shall promptly notify an employee of the right to seek a second medical opinion after each occasion that an initial physician conducts a medical examination or consultation for the purpose of medical removal or restriction.

**1910.1048(l)(9)(iii)** The employer may condition its participation in, and payment for, the multiple physician review mechanism upon the employee doing the following within fifteen (15) days after receipt of the notification of the right to seek a second medical opinion, or receipt of the initial physician's written opinion, whichever is later;

**1910.1048(l)(9)(iii)(A)** The employee informs the employer of the intention to seek a second medical opinion, and

**1910.1048(l)(9)(iii)(B)** The employee initiates steps to make an appointment with a second physician.

**1910.1048(l)(9)(iv)** If the findings, determinations or recommendations of the second physician differ from those of the initial physician, then the employer and the employee shall assure that efforts are made for the two physicians to resolve the disagreement. If the two physicians are unable to quickly resolve their disagreement, then the employer and the employee through their respective physicians shall designate a third physician who shall be a specialist in the field at issue;

**1910.1048(l)(9)(iv)(A)** To review the findings, determinations or recommendations of the prior physicians; and

**1910.1048(l)(9)(iv)(B)** To conduct such examinations, consultations, laboratory tests and discussions with the prior physicians as the third physician deems necessary to resolve the disagreement of the prior physicians.

**1910.1048(l)(9)(v)** In the alternative, the employer and the employee or authorized employee representative may jointly designate such third physician.

**1910.1048(l)(9)(vi)** The employer shall act consistent with the findings, determinations and recommendations of the third physician, unless the employer and the employee reach an agreement which is otherwise consistent with the recommendations of at least one of the three physicians.

**1910.1048(m)** *Hazard communication*—**(1)** *General.* Communication of the hazards associated with formaldehyde in the workplace shall be governed by the requirements of paragraph (m). The definitions of 29 CFR 1910.1200(c) shall apply under this paragraph.

**1910.1048(m)(1)(i)** The following shall be subject to the hazard communication requirements of this paragraph: Formaldehyde gas, all mixtures or solutions composed of greater than 0.1 percent formaldehyde, and materials capable of releasing formaldehyde into the air, under reasonably foreseeable conditions of use, at concentrations reaching or exceeding 0.1 ppm.

**1910.1048(m)(1)(ii)** As a minimum, specific health hazards that the employer shall address are: Cancer, irritation and sensitization of the skin and respiratory system, eye and throat irritation, and acute toxicity.

**1910.1048(m)(2)** Manufacturers and importers who produce or import formaldehyde or formaldehyde-containing products shall provide downstream employers using or handling these products with an objective determination through the required labels and MSDSs if these items may constitute a health hazard within the meaning of 29 CFR 1910.1200(d) under normal conditions of use.

**1910.1048(m)(3)** *Labels.* **(i)** The employer shall assure that hazard warning labels complying with the requirements of 29 CFR 1910.1200(f) are affixed to all containers of materials listed in paragraph (m)(1)(i), except to the extent that 29 CFR 1910.1200(f) is inconsistent with this paragraph.

**1910.1048(m)(3)(ii)** *Information on labels.* As a minimum, for all materials listed in paragraph (m)(1)(i) capable of releasing formaldehyde at levels of 0.1 ppm to 0.5 ppm, labels shall identify that the product contains formaldehyde; list the name and address of the responsible party; and state that physical and health hazard information is readily available from the employer and from material safety data sheets.

**1910.1048(m)(3)(iii)** For materials listed in paragraph (m)(1)(i) capable of releasing formaldehyde at levels above 0.5 ppm, labels shall appropriately address all hazards as defined in 29 CFR 1910.1200(d) and 29 CFR 1910.1200 appendices A and B, including respiratory sensitization, and shall contain the words "Potential Cancer Hazard."

**1910.1048(m)(3)(iv)** In making the determinations of anticipated levels of formaldehyde release, the employer may rely on objective data indicating the extent of potential formaldehyde release under reasonably foreseeable conditions of use.

**1910.1048(m)(3)(v)** *Substitute warning labels.* The employer may use warning labels required by other statutes, regulations, or ordinances which impart the same information as the warning statements required by this paragraph.

**1910.1048(m)(4)** *Material safety data sheets.* **(i)** Any employer who uses formaldehyde-containing materials listed in paragraph (m)(1)(i) shall comply with the requirements of 29 CFR 1910.1200(g) with regard to the development and updating of material safety data sheets.

**1910.1048(m)(4)(ii)** Manufacturers, importers, and distributors of formaldehyde-containing materials listed in paragraph (m)(1)(i) shall assure that material safety data sheets and updated information are provided to all employers purchasing such materials at the time of the initial shipment and at the time of the first shipment after a material safety data sheet is updated.

**1910.1048(m)(5)** *Written hazard communication program.* The employer shall develop, implement, and maintain at the workplace, a written hazard communication program for formaldehyde exposures in the workplace, which at a minimum describes how the requirements specified in this paragraph for labels and other forms of warning and material safety data sheets, and paragraph (n) for employee information and training, will be met. Employers in multi-employer workplaces shall comply with the requirements of 29 CFR 1910.1200(e)(2).

**1910.1048(n)** *Employee information and training*—**(1)** *Participation.* The employer shall assure that all employees who are assigned to workplaces where there is exposure to formaldehyde participate in a training program, except that where the employer can show, using objective data, that employees are not exposed to formaldehyde at or above 0.1 ppm, the employer is not required to provide training.

**1910.1048(n)(2)** *Frequency.* Employers shall provide such information and training to employees at the time of initial assignment, and whenever a new exposure to formaldehyde is introduced into the work area. The training shall be repeated at least annually.

**1910.1048(n)(3)** *Training program.* The training program shall be conducted in a manner which the employee is able to understand and shall include:

**1910.1048(n)(3)(i)** A discussion of the contents of this regulation and the contents of the Material Safety Data Sheet.

**1910.1048(n)(3)(ii)** The purpose for and a description of the medical surveillance program required by this standard, including:

**1910.1048(n)(3)(ii)(A)** A description of the potential health hazards associated with exposure to formaldehyde and a description of the signs and symptoms of exposure to formaldehyde.

**1910.1048(n)(3)(ii)(B)** Instructions to immediately report to the employer the development of any adverse signs or symptoms that the employee suspects is attributable to formaldehyde exposure.

**1910.1048(n)(3)(iii)** Description of operations in the work area where formaldehyde is present and an explanation of the safe work practices appropriate for limiting exposure to formaldehyde in each job;

**1910.1048(n)(3)(iv)** The purpose for, proper use of, and limitations of personal protective clothing and equipment;

**1910.1048(n)(3)(v)** Instructions for the handling of spills, emergencies, and clean-up procedures;

**1910.1048(n)(3)(vi)** An explanation of the importance of engineering and work practice controls for employee protection and any necessary instruction in the use of these controls; and

**1910.1048(n)(3)(vii)** A review of emergency procedures including the specific duties or assignments of each employee in the event of an emergency.

**1910.1048(n)(4)** *Access to training materials.* **(i)** The employer shall inform all affected employees of the location of written training materials and shall make these materials readily available, without cost, to the affected employees.

**1910.1048(n)(4)(ii)** The employer shall provide, upon request, all training materials relating to the employee training program to the Assistant Secretary and the Director.

**1910.1048(o)** *Recordkeeping*—**(1)** *Exposure measurements.* The employer shall establish and maintain an accurate record of all measurements taken to monitor employee exposure to formaldehyde. This record shall include:

**1910.1048(o)(1)(i)** The date of measurement;

**1910.1048(o)(1)(ii)** The operation being monitored;

**1910.1048(o)(1)(iii)** The methods of sampling and analysis and evidence of their accuracy and precision;

**1910.1048(o)(1)(iv)** The number, durations, time, and results of samples taken;

**1910.1048(o)(1)(v)** The types of protective devices worn; and

**1910.1048(o)(1)(vi)** The names, job classifications, social security numbers, and exposure estimates of the employees whose exposures are represented by the actual monitoring results.

**1910.1048(o)(2)** *Exposure determinations.* Where the employer has determined that no monitoring is required under this standard, the employer shall maintain a record of the objective data relied upon to support the determination that no employee is exposed to formaldehyde at or above the action level.

**1910.1048(o)(3)** *Medical surveillance.* The employer shall establish and maintain an accurate record for each employee subject to medical surveillance under this standard. This record shall include:

**1910.1048(o)(3)(i)** The name and social security number of the employee;

**1910.1048(o)(3)(ii)** The physician's written opinion;

**1910.1048(o)(3)(iii)** A list of any employee health complaints that may be related to exposure to formaldehyde; and

**1910.1048(o)(3)(iv)** A copy of the medical examination results, including medical disease questionnaires and results of any medical tests required by the standard or mandated by the examining physician.

**1910.1048(o)(4)** *Respirator fit testing.* **(i)** The employer shall establish and maintain accurate records for employees subject to negative pressure respirator fit testing required by this standard.

**1910.1048(o)(4)(ii)** This record shall include:

**1910.1048(o)(4)(ii)(A)** A copy of the protocol selected for respirator fit testing.

**1910.1048(o)(4)(ii)(B)** A copy of the results of any fit testing performed.

**1910.1048(o)(4)(ii)(C)** The size and manufacturer of the types of respirators available for selection.

**1910.1048(o)(4)(ii)(D)** The date of the most recent fit testing, the name and social security number of each tested employee, and the respirator type and facepiece selected.

**1910.1048(o)(5)** *Record retention.* The employer shall retain records required by this standard for at least the following periods:

**1910.1048(o)(5)(i)** Exposure records and determinations shall be kept for at least 30 years.

**1910.1048(o)(5)(ii)** Medical records shall be kept for the duration of employment plus 30 years.

**1910.1048(o)(5)(iii)** Respirator fit testing records shall be kept until replaced by a more recent record.

**1910.1048(o)(6)** *Availability of records.* **(i)** Upon request, the employer shall make all records maintained as a requirement of this standard available for examination and copying to the Assistant Secretary and the Director.

**1910.1048(o)(6)(ii)** The employer shall make employee exposure records, including estimates made from representative monitoring and available upon request for examination, and copying to the subject employee, or former employee, and employee representatives in accordance with 29 CFR 1910.1020(a)-(e) and (g)(i).

**1910.1048(o)(6)(iii)** Employee medical records required by this standard shall be provided upon request for examination and coying, to the subject employee or former employee or to anyone having the specific written consent of the subject employee or former employee in accordance with 29 CFR 1910.1020(a)-(e) and (g)-(i).

## APPENDIX A TO §1910.1048— SUBSTANCE TECHNICAL GUIDELINES FOR FORMALIN

The following Substance Technical Guideline for Formalin provides information on uninhibited formalin solution (37% formaldehyde, no methanol stabilizer). It is designed to inform employees at the production level of their rights and duties under the formaldehyde standard whether their job title defines them as workers or supervisors. Much of the information provided is general; however, some information is specific for formalin. When employee exposure to formaldehyde is from resins capable of releasing formaldehyde, the resin itself and other impurities or decomposition products may also be toxic, and employers should include this information as well when informing employees of the hazards associated with the materials they handle. The precise hazards associated with exposure to formaldehyde depend both on the form (solid, liquid, or gas) of the material and the concentration of formaldehyde present. For example, 37-50 percent solutions of formaldehyde present a much greater hazard to the skin and eyes from spills or splashes than solutions containing less than 1 percent formaldehyde. Individual Substance Technical Guidelines used by the employer for training employees should be modified to properly give information on the material actually being used.

# OSHA Standards for General Industry

## Substance Identification

*Chemical Name:* Formaldehyde

*Chemical Family:* Aldehyde

*Chemical Formula:* HCHO

*Molecular Weight:* 30.03

*Chemical Abstracts Service Number (CAS Number):* 50-00-0

*Synonyms:* Formalin; Formic Aldehyde; Paraform; Formol; Formalin (Methanol-free); Fyde; Formalith; Methanal; Methyl Aldehyde; Methylene Glycol; Methylene Oxide; Tetraoxymethalene; Oxomethane; Oxymethylene

## Components and Contaminants

*Percent:* 37.0 Formaldehyde

*Percent:* 63.0 Water

**(NOTE—Inhibited solutions contain methanol.)**

*Other Contaminants:* Formic acid (alcohol free)

*Exposure Limits:*

OSHA TWA—0.75 ppm

OSHA STEL—2 ppm

## Physical Data

*Description:* Colorless liquid, pungent odor

*Boiling point:* 214 °F (101 °C)

*Specific Gravity:* 1.08 (H$_2$O = 1 @ 20 °C)

*pH:* 2.8-4.0

*Solubility in Water:* Miscible

*Solvent Solubility:* Soluble in alcohol and acetone

*Vapor Density:* 1.04 (Air = 1 @ 20 °C)

*Odor Threshold:* 0.8-1 ppm

## Fire and Explosion Hazard

Moderate fire and explosion hazard when exposed to heat or flame.

The flash point of 37% formaldehyde solutions is above normal room temperature, but the explosion range is very wide, from 7 to 73% by volume in air.

Reaction of formaldehyde with nitrogen dioxide, nitromethane, perchloric acid and aniline, or peroxyformic acid yields explosive compounds.

*Flash Point:* 185 °F (85 °C) closed cup

*Lower Explosion Limit:* 7%

*Upper Explosion Limit:* 73%

*Autoignition Temperature:* 806 °F (430 °C)

*Flammability Class (OSHA):* III A

*Extinguishing Media:* Use dry chemical, "alcohol foam", carbon dioxide, or water in flooding amounts as fog. Solid streams may not be effective. Cool fire-exposed containers with water from side until well after fire is out.

Use of water spray to flush spills can also dilute the spill to produce nonflammable mixtures. Water runoff, however, should be contained for treatment.

*National Fire Protection Association Section 325M Designation:*

*Health:* 2—Materials hazardous to health, but areas may be entered with full-faced mask self-contained breathing apparatus which provides eye protection.

*Flammability:* 2—Materials which must be moderately heated before ignition will occur. Water spray may be used to extinguish the fire because the material can be cooled below its flash point.

*Reactivity:* D—Materials which (in themselves) are normally stable even under fire exposure conditions and which are not reactive with water. Normal fire fighting procedures may be used.

## Reactivity

*Stability:* Formaldehyde solutions may self-polymerize to form paraformaldehyde which precipitates.

*Incompatibility (Materials to Avoid):* Strong oxidizing agents, caustics, strong alkalies, isocyanates, anhydrides, oxides, and inorganic acids. Formaldehyde reacts with hydrochloric acid to form the potent carcinogen, bis-chloromethyl ether. Formaldehyde reacts with nitrogen dioxide, nitromethane, perchloric acid and aniline, or peroxyformic acid to yield explosive compounds. A violent reaction occurs when formaldehyde is mixed with strong oxidizers.

*Hazardous Combustion or Decomposition Products:* Oxygen from the air can oxidize formaldehyde to formic acid, especially when heated. Formic acid is corrosive.

## Health Hazard Data

### Acute Effects of Exposure

*Ingestion (Swallowing):* Liquids containing 10 to 40% formaldehyde cause severe irritation and inflammation of the mouth, throat, and stomach. Severe stomach pains will follow ingestion with possible loss of consciousness and death. Ingestion of dilute formaldehyde solutions (0.03-0.04%) may cause discomfort in the stomach and pharynx.

*Inhalation (Breathing):* Formaldehyde is highly irritating to the upper respiratory tract and eyes. Concentrations of 0.5 to 2.0 ppm may irritate the eyes, nose, and throat of some individuals. Concentrations of 3 to 5 ppm also cause tearing of the eyes and are intolerable to some persons. Concentrations of 10 to 20 ppm cause difficulty in breathing, burning of the nose and throat, cough, and heavy tearing of the eyes, and 25 to 30 ppm causes severe respiratory tract injury leading to pulmonary edema and pneumonitis. A concentration of 100 ppm is immediately dangerous to life and health. Deaths from accidental exposure to high concentrations of formaldehyde have been reported.

*Skin (Dermal):* Formalin is a severe skin irritant and a sensitizer. Contact with formalin causes white discoloration, smarting, drying, cracking, and scaling. Prolonged and repeated contact can cause numbness and a hardening or tanning of the skin. Previously exposed persons may react to future exposure with an allergic eczematous dermatitis or hives.

*Eye Contact:* Formaldehyde solutions splashed in the eye can cause injuries ranging from transient discomfort to severe, permanent corneal clouding and loss of vision. The severity of the effect depends on the concentration of formaldehyde in the solution and whether or not the eyes are flushed with water immediately after the accident.

**NOTE.—**The perception of formaldehyde by odor and eye irritation becomes less sensitive with time as one adapts to formaldehyde. This can lead to overexposure if a worker is relying on formaldehyde's warning properties to alert him or her to the potential for exposure.

*Acute Animal Toxicity:*

*Oral,* rats: LD50 = 800 mg/kg

*Oral,* mouse: LD50 = 42 mg/kg

*Inhalation,* rats: LCLo = 250 mg/kg

*Inhalation,* mouse: LCLo = 900 mg/kg

*Inhalation,* rats: LC50 = 590 mg/kg

### Chronic Effects of Exposure

*Carcinogenicity:* Formaldehyde has the potential to cause cancer in humans. Repeated and prolonged exposure increases the risk. Various animal experiments have conclusively shown formaldehyde to be a carcinogen in rats. In humans, formaldehyde exposure has been associated with cancers of the lung, nasopharynx and oropharynx, and nasal passages.

*Mutagenicity:* Formaldehyde is genotoxic in several *in vitro* test systems showing properties of both an initiator and a promoter.

*Toxicity:* Prolonged or repeated exposure to formaldehyde may result in respiratory impairment. Rats exposed to formaldehyde at 2 ppm developed benign nasal tumors and changes of the cell structure in the nose as well as inflamed mucous membranes of the nose. Structural changes in the epithelial cells in the human nose have also been observed. Some persons have developed asthma or bronchitis following exposure to formaldehyde, most often as the result of an accidental spill involving a single exposure to a high concentration of formaldehyde.

### Emergency and First Aid Procedures

*Ingestion (Swallowing):* If the victim is conscious, dilute, inactivate, or absorb the ingested formaldehyde by giving milk, activated charcoal, or water. Any organic material will inactivate formaldehyde. Keep affected person warm and at rest. Get medical attention immediately. If vomiting occurs, keep head lower than hips.

*Inhalation (Breathing):* Remove the victim from the exposure area to fresh air immediately. Where the formaldehyde concentration may be very high, each rescuer must put on a selfcontained breathing apparatus before attempting to remove the victim, and medical personnel should be informed of the formal dehyde exposure immediately. If breathing has stopped, give artificial respiration. Keep the affected person warm and at rest. Qualified first-aid or medical personnel should administer oxygen, if available, and maintain the patient's airways and blood pressure until the victim can be transported to a medical facility. If exposure results in a highly irritated upper respiratory tract and coughing continues for more than 10 minutes, the worker should be hospitalized for observation and treatment.

*Skin Contact:* Remove contaminated clothing (including shoes) immediately. Wash the affected area of your body with soap or mild detergent and large amounts of water until no evidence of the chemical remains (at least 15 to 20 minutes). If there are chemical burns, get first aid to cover the area with sterile, dry dressing, and bandages. Get medical attention if you experience appreciable eye or respiratory irritation.

*Eye Contact:* Wash the eyes immediately with large amounts of water occasionally lifting lower and upper lids, until no evidence of chemical remains (at least 15 to 20 minutes). In case of burns, apply sterile bandages loosely without medication. Get medical attention immediately. If you have experienced appreciable eye irritation from a splash or excessive exposure, you should be referred promptly to an opthamologist for evaluation.

### Emergency Procedures

*Emergencies:* If you work in an area where a large amount of formaldehyde could be released in an accident or from equipment failure, your employer must develop procedures to be followed in event of an emergency. You should be trained in your specific duties in the event of an emergency, and it is important that you clearly understand these duties. Emergency equipment must be accessible and you should be trained to use any equipment that you might need. Formaldehyde contaminated equipment must be cleaned before reuse.

If a spill of appreciable quantity occurs, leave the area quickly unless you have specific emergency duties. Do not touch spilled material. Designated persons may stop the leak and shut off ignition sources if these procedures can be done without risk. Designated persons should isolate the hazard area and deny entry except for necessary people protected by suitable protective clothing and respirators adequate for the exposure. Use water spray to reduce vapors. Do not smoke, and prohibit all flames or flares in the hazard area.

*Special Firefighting Procedures:* Learn procedures and responsibilities in the event of a fire in your workplace. Become familiar with the appropriate equipment and supplies and their location. In firefighting, withdraw immediately in case of rising sound from venting safety device or any discoloration of storage tank due to fire.

### Spill, Leak, and Disposal Procedures

*Occupational Spill:* For small containers, place the leaking container in a well ventilated area. Take up small spills with absorbent material and place the waste into properly labeled containers for later disposal. For larger spills, dike the spill to minimize contamination and facilitate salvage or disposal. You may be able to neutralize the spill with sodium hydroxide or sodium sulfite. Your employer must comply with EPA rules regarding the clean-up of toxic waste and notify state and local authorities, if required. If the spill is greater than 1,000 lb/day, it is reportable under EPA's Superfund legislation.

*Waste Disposal:* Your employer must dispose of waste containing formaldehyde in accordance with applicable local, state, and Federal law and in a manner that minimizes exposure of employees at the site and of the clean-up crew.

### Monitoring and Measurement Procedures

*Monitoring Requirements:* If your exposure to formaldehyde exceeds the 0.5 ppm action level or the 2 ppm STEL, your employer must monitor your exposure. Your employer need not measure every exposure if a "high exposure" employee can be identified. This person usually spends the greatest amount of time nearest the process equipment. If you are a "representative employee", you will be asked to wear a sampling device to collect formaldehyde. This device may be a passive badge, a sorbent tube attached to a pump, or an impinger containing liquid. You should perform your work as usual, but inform the person who is conducting the monitoring of any difficulties you are having wearing the device.

*Evaluation of 8-hour Exposure:* Measurements taken for the purpose of determining time-weighted average (TWA) exposures are best taken with samples covering the full shift. Samples collected must be taken from the employee's breathing zone air.

*Short-term Exposure Evaluation:* If there are tasks that involve brief but intense exposure to formaldehyde, employee exposure must be measured to assure compliance with the STEL. Sample collections are for brief periods, only 15 minutes, but several samples may be needed to identify the peak exposure.

*Monitoring Techniques:* OSHA's only requirement for selecting a method for sampling and analysis is that the methods used accurately evaluate the concentration of formaldehyde in employees' breathing zones. Sampling and analysis may be performed by collection of formaldehyde on liquid or solid sorbents with subsequent chemical analysis. Sampling and analysis may also be performed by passive diffusion monitors and shortterm exposure may be measured by instruments such as real-time continuous monitoring systems and portable direct reading instruments.

*Notification of Results:* Your employer must inform you of the results of exposure monitoring representative of your job. You may be informed in writing, but posting the results where you have ready access to them constitutes compliance with the standard.

### Protective Equipment and Clothing

[Material impervious to formaldehyde is needed if the employee handles formaldehyde solutions of 1% or more. Other employees may also require protective clothing or equipment to prevent dermatitis.]

*Respiratory Protection:* Use NIOSH-approved full facepiece negative pressure respirators equipped with approved cartridges or canisters within the use limitations of these devices. (Present restrictions on cartridges and canisters do not permit them to be used for a full workshift.) In all other situations, use positive pressure respirators such as the positive-pressure air purifying respirator or the self-contained breathing apparatus (SCBA). If you use a negative pressure respirator, your employer must provide you with fit testing of the respirator at least once a year.

*Protective Gloves:* Wear protective (impervious) gloves provided by your employer, at no cost, to prevent contact with formalin. Your employer should select these gloves based on the results of permeation testing and in accordance with the ACGIH Guidelines for Selection of Chemical Protective Clothing.

*Eye Protection:* If you might be splashed in the eyes with formalin, it is essential that you wear goggles or some other type of complete protection for the eye. You may also need a face shield if your face is likely to be splashed with formalin, but you must not substitute face shields for eye protection. (This section pertains to formaldehyde solutions of 1% or more.)

*Other Protective Equipment:* You must wear protective (impervious) clothing and equipment provided by your employer at no cost to prevent repeated or prolonged contact with formaldehyde liquids. If you are required to change into wholebody chemical protective clothing, your employer must provide a change room for your privacy and for storage of your normal clothing.

If you are splashed with formaldehyde, use the emergency showers and eyewash fountains provided by your employer immediately to prevent serious injury. Report the incident to your supervisor and obtain necessary medical support.

### ENTRY INTO AN IDLH ATMOSHPERE

Enter areas where the formaldehyde concentration might be 100 ppm or more only with complete body protection including a self-contained breathing apparatus with a full facepiece operated in a positive pressure mode or a supplied air respirator with full facepiece and operated in a positive pressure mode. This equipment is essential to protect your life and health under such extreme conditions.

### Engineering Controls

Ventilation is the most widely applied engineering control method for reducing the concentration of airborne substances in the breathing zones of workers. There are two distinct types of ventilation.

## OSHA Standards for General Industry

*Local Exhaust:* Local exhaust ventilation is designed to capture airborne contaminants as near to the point of generation as possible. To protect you, the direction of contaminant flow must always be toward the local exhaust system inlet and away from you.

*General (Mechanical):* General dilution ventilation involves continuous introduction of fresh air into the workroom to mix with the contaminated air and lower your breathing zone concentration of formaldehyde. Effectiveness depends on the number of air changes per hour. Where devices emitting formaldehyde are spread out over a large area, general dilution ventilation may be the only practical method of control.

*Work Practices:* Work practices and administrative procedures are an important part of a control system. If you are asked to perform a task in a certain manner to limit your exposure to formaldehyde, it is extremely important that you follow these procedures.

### Medical Surveillance

Medical surveillance helps to protect employees' health. You are encouraged strongly to participate in the medical surveillance program.

Your employer must make a medical surveillance program available at no expense to you and at a reasonable time and place if you are exposed to formaldehyde at concentrations above 0.5 ppm as an 8-hour average or 2 ppm over any 15-minute period. You will be offered medical surveillance at the time of your initial assignment and once a year afterward as long as your exposure is at least 0.5 ppm (TWA) or 2 ppm (STEL). Even if your exposure is below these levels, you should inform your employer if you have signs and symptoms that you suspect, through your training, are related to your formaldehyde exposure because you may need medical surveillance to determine if your health is being impaired by your exposure.

The surveillance plan includes:

(a) A medical disease questionnaire.

(b) A physical examination if the physician determines this is necessary.

If you are required to wear a respirator, your employer must offer you a physical examination and a pulmonary function test every year.

The physician must collect all information needed to determine if you are at increased risk from your exposure to formaldehyde. At the physician's discretion, the medical examination may include other tests, such as a chest x-ray, to make this determination.

After a medical examination the physician will provide your employer with a written opinion which includes any special protective measures recommended and any restrictions on your exposure. The physician must inform you of any medical conditions you have which would be aggravated by exposure to formaldehyde.

All records from your medical examinations, including disease surveys, must be retained at your employer's expense.

### EMERGENCIES

If you are exposed to formaldehyde in an emergency and develop signs or symptoms associated with acute toxicity from formaldehyde exposure, your employer must provide you with a medical examination as soon as possible. This medical examination will include all steps necessary to stabilize your health. You may be kept in the hospital for observation if your symptoms are severe to ensure that any delayed effects are recognized and treated.

### APPENDIX B TO §1910.1048—SAMPLING STRATEGY AND ANALYTICAL METHODS FOR FORMALDEHYDE

To protect the health of employees, exposure measurements must be unbiased and representative of employee exposure. The proper measurement of employee exposure requires more than a token commitment on the part of the employer. OSHA's mandatory requirements establish a baseline; under the best of circumstances all questions regarding employee exposure will be answered. Many employers, however, will wish to conduct more extensive monitoring before undertaking expensive commitments, such as engineering controls, to assure that the modifications are truly necessary. The following sampling strategy, which was developed at NIOSH by Nelson A. Leidel, Kenneth A. Busch, and Jeremiah R. Lynch and described in NIOSH publication No. 77-173 (Occupational Exposure Sampling Strategy Manual) will assist the employer in developing a strategy for determining the exposure of his or her employees.

There is no one correct way to determine employee exposure. Obviously, measuring the exposure of every employee exposed to formaldehyde will provide the most information on any given day. Where few employees are exposed, this may be a practical solution. For most employers, however, use of the following strategy will give just as much information at less cost.

Exposure data collected on a single day will not automatically guarantee the employer that his or her workplace is always in compliance with the formaldehyde standard. This does not imply, however, that it is impossible for an employer to be sure that his or her worksite is in compliance with the standard. Indeed, a properly designed sampling strategy showing that all employees are exposed below the PELs, at least with a 95 percent certainty, is compelling evidence that the exposure limits are being achieved provided that measurements are conducted using valid sampling strategy and approved analytical methods.

There are two PELs, the TWA concentration and the STEL. Most employers will find that one of these two limits is more critical in the control of their operations, and OSHA expects that the employer will concentrate monitoring efforts on the critical component. If the more difficult exposure is controlled, this information, along with calculations to support the assumptions, should be adequate to show that the other exposure limit is also being achieved.

### Sampling Strategy
#### Determination of the Need for Exposure Measurements

The employer must determine whether employees may be exposed to concentrations in excess of the action level. This determination becomes the first step in an employee exposure monitoring program that minimizes employer sampling burdens while providing adequate employee protection. If employees may be exposed above the action level, the employer must measure exposure. Otherwise, an objective determination that employee exposure is low provides adequate evidence that exposure potential has been examined.

The employer should examine all available relevant information, *eg.* insurance company and trade association data and information from suppliers or exposure data collected from similar operations. The employer may also use previouslyconducted sampling including area monitoring. The employer must make a determination relevant to each operation although this need not be on a separate piece of paper. If the employer can demonstrate conclusively that no employee is exposed above the action level or the STEL through the use of objective data, the employer need proceed no further on employee exposure monitoring until such time that conditions have changed and the determination is no longer valid.

If the employer cannot determine that employee exposure is less than the action level and the STEL, employee exposure monitoring will have to be conducted.

#### Workplace Material Survey

The primary purpose of a survey of raw material is to determine if formaldehyde is being used in the work environment and if so, the conditions under which formaldehyde is being used.

The first step is to tabulate all situations where formaldehyde is used in a manner such that it may be released into the workplace atmosphere or contaminate the skin. This information should be available through analysis of company records and information on the MSDSs available through provisions of this standard and the Hazard Communication standard.

If there is an indication from materials handling records and accompanying MSDSs that formaldehyde is being used in the following types of processes or work operations, there may be a potential for releasing formaldehyde into the workplace atmosphere:

(1) Any operation that involves grinding, sanding, sawing, cutting, crushing, screening, sieving, or any other manipulation of material that generates formaldehyde-bearing dust

(2) Any processes where there have been employee complaints or symptoms indicative of exposure to formaldehyde

(3) Any liquid or spray process involving formaldehyde

(4) Any process that uses formaldehyde in preserved tissue

(5) Any process that involves the heating of a formaldehyde-bearing resin

Processes and work operations that use formaldehyde in these manners will probably require further investigation at the worksite to determine the extent of employee monitoring that should be conducted.

### Workplace Observations

To this point, the only intention has been to provide an indication as to the existence of potentially exposed employees. With this information, a visit to the workplace is needed to observe work operations, to identify potential health hazards, and to determine whether any employees may be exposed to hazardous concentrations of formaldehyde.

In many circumstances, sources of formaldehyde can be identified through the sense of smell. However, this method of detection should be used with caution because of olfactory fatigue.

Employee location in relation to source of formaldehyde is important in determining if an employee may be significantly exposed to formaldehyde. In most instances, the closer a worker is to the source, the higher the probability that a significant exposure will occur.

Other characteristics should be considered. Certain high temperature operations give rise to higher evaporation rates. Locations of open doors and windows provide natural ventilation that tend to dilute formaldehyde emissions. General room ventilation also provides a measure of control.

### Calculation of Potential Exposure Concentrations

By knowing the ventilation rate in a workplace and the quantity of formaldehyde generated, the employer may be able to determine by calculation if the PELs might be exceeded. To account for poor mixing of formaldehyde into the entire room, locations of fans and proximity of employees to the work operation, the employer must include a safety factor. If an employee is relatively close to a source, particularly if he or she is located downwind, a safety factor of 100 may be necessary. For other situations, a factor of 10 may be acceptable. If the employer can demonstrate through such calculations that employee exposure does not exceed the action level or the STEL, the employer may use this information as objective data to demonstrate compliance with the standard.

### Sampling Strategy

Once the employer determines that there is a possibility of substantial employee exposure to formaldehyde, the employer is obligated to measure employee exposure.

The next step is selection of a maximum risk employee. When there are different processes where employees may be exposed to formaldehyde, a maximum risk employee should be selected for each work operation.

Selection of the maximum risk employee requires professional judgment. The best procedure for selecting the maximum risk employee is to observe employees and select the person closest to the source of formaldehyde. Employee mobility may affect this selection; *e.g.*, if the closest employee is mobile in his tasks, he may not be the maximum risk employee. Air movement patterns and differences in work habits will also affect selection of the maximum risk employee.

When many employees perform essentially the same task, a maximum risk employee cannot be selected. In this circumstance, it is necessary to resort to random sampling of the group of workers. The objective is to select a subgroup of adequate size so that there is a high probability that the random sample will contain at least one worker with high exposure if one exists. The number of persons in the group influences the number that need to be sampled to ensure that at least one individual from the highest 10 percent exposure group is contained in the sample. For example, to have 90 percent confidence in the results, if the group size is 10, nine should be sampled; for 50, only 18 need to be sampled.

If measurement shows exposure to formaldehyde at or above the action level or the STEL, the employer needs to identify all other employees who may be exposed at or above the action level or STEL and measure or otherwise accurately characterize the exposure of these employees.

Whether representative monitoring or random sampling are conducted, the purpose remains the same—to determine if the exposure of any employee is above the action level. If the exposure of the most exposed employee is less than the action level and the STEL, regardless of how the employee is identified, then it is reasonable to assume that measurements of exposure of the other employees in that operation would be below the action level and the STEL.

### Exposure Measurements

There is no "best" measurement strategy for all situations. Some elements to consider in developing a strategy are:

(1) Availability and cost of sampling equipment

(2) Availability and cost of analytic facilities

(3) Availability and cost of personnel to take samples

(4) Location of employees and work operations

(5) Intraday and interday variations in the process

(6) Precision and accuracy of sampling and analytic methods, and

(7) Number of samples needed.

Samples taken for determining compliance with the STEL differ from those that measure the TWA concentration in important ways. STEL samples are best taken in a nonrandom fashion using all available knowledge relating to the area, the individual, and the process to obtain samples during periods of maximum expected concentrations. At least three measurements on a shift are generally needed to spot gross errors or mistakes; however, only the highest value represents the STEL.

If an operation remains constant throughout the workshift, a much greater number of samples would need to be taken over the 32 discrete nonoverlapping periods in an 8-hour workshift to verify compliance with a STEL. If employee exposure is truly uniform throughout the workshift, however, an employer in compliance with the 1 ppm TWA would be in compliance with the 2 ppm STEL, and this determination can probably be made using objective data.

### Need to Repeat the Monitoring Strategy

Interday and intraday fluctuations in employee exposure are mostly influenced by the physical processes that generate formaldehyde and the work habits of the employee. Hence, in-plant process variations influence the employer's determination of whether or not additional controls need to be imposed. Measurements that employee exposure is low on a day that is not representative of worst conditions may not provide sufficient information to determine whether or not additional engineering controls should be installed to achieve the PELs.

The person responsible for conducting sampling must be aware of systematic changes which will negate the validity of the sampling results. Systematic changes in formaldehyde exposure concentration for an employee can occur due to:

(1) The employee changing patterns of movement in the workplace

(2) Closing of plant doors and windows

(3) Changes in ventilation from season to season

(4) Decreases in ventilation efficiency or abrupt failure of engineering control equipment

(5) Changes in the production process or work habits of the employee.

Any of these changes, if they may result in additional exposure that reaches the next level of action (i.e., 0.5 or 1.0 ppm as an 8-hr average or 2 ppm over 15 minutes) require the employer to perform additional monitoring to reassess employee exposure.

A number of methods are suitable for measuring employee exposure to formaldehyde or for characterizing emissions within the worksite. The preamble to this standard describes some methods that have been widely used or subjected to validation testing. A detailed analytical procedure derived from the OSHA Method 52 for acrolein and formaldehyde is presented below for informational purposes.

Inclusion of OSHA's method in this appendix in no way implies that it is the only acceptable way to measure employee exposure to formaldehyde. Other methods that are free from significant interferences and that can determine formaldehyde at the permissible exposure limits within ±25 percent of the "true" value at the 95 percent confidence level are also acceptable. Where applicable, the method should also be capable of measuring formaldehyde at the action level to ±35 percent of the "true" value with a 95 percent confidence level. OSHA encourages employers to choose methods that will be best for their individual needs. The employer must exercise caution, however, in choosing an appropriate method since some techniques suffer from interferences that are likely to be present in workplaces of certain industry sectors where formaldehyde is used.

### OSHA's Analytical Laboratory Method

*Method No:* 52

*Matrix:* Air

*Target Concentration:* 1 ppm (1.2 mg/m$^3$)

# OSHA Standards for General Industry

*Procedures:* Air samples are collected by drawing known volumes of air through sampling tubes containing XAD-2 adsorbent which have been coated with 2-(hydroxymethyl) piperidine. The samples are desorbed with toluene and then analyzed by gas chromatography using a nitrogen selective detector.

*Recommended Sampling Rate and Air Volumes:* 0.1 L/min and 24 L

*Reliable Quantitation Limit:* 16 ppb (20 µg/m$^3$)

*Standard Error of Estimate at the Target Concentration:* 7.3%

*Status of the Method:* A sampling and analytical method that has been subjected to the established evaluation procedures of the Organic Methods Evaluation Branch.

*Date:* March 1985

## 1. General Discussion

1.1 *Background:* The current OSHA method for collecting acrolein vapor recommends the use of activated 13X molecular sieves. The samples must be stored in an ice bath during and after sampling and also they must be analyzed within 48 hours of collection. The current OSHA method for collecting formaldehyde vapor recommends the use of bubblers containing 10% methanol in water as the trapping solution.

This work was undertaken to resolve the sample stability problems associated with acrolein and also to eliminate the need to use bubblers to sample formaldehyde. A goal of this work was to develop and/or to evaluate a common sampling and analytical procedure for acrolein and formaldehyde.

NIOSH has developed independent methodologies for acrolein and formaldehyde which recommend the use of reagent-coated adsorbent tubes to collect the aldehydes as stable derivatives. The formaldehyde sampling tubes contain Chromosorb 102 adsorbent coated with N-benzylethanolamine (BEA) which reacts with formaldehyde vapor to form a stable oxazolidine compound. The acrolein sampling tubes contain XAD-2 adsorbent coated with 2(hydroxymethyl)piperidine (2-HMP) which reacts with acrolein vapor to form a different, stable oxazolidine derivative. Acrolein does not appear to react with BEA to give a suitable reaction product. Therefore, the formaldehyde procedure cannot provide a common method for both aldehydes. However, formaldehyde does react with 2-HMP to form a very suitable reaction product. It is the quantitative reaction of acrolein and formaldehyde with 2-HMP that provides the basis for this evaluation.

This sampling and analytical procedure is very similar to the method recommended by NIOSH for acrolein. Some changes in the NIOSH methodology were necessary to permit the simultaneous determination of both aldehydes and also to accommodate OSHA laboratory equipment and analytical techniques.

1.2 *Limit-defining parameters:* The analyte air concentrations reported in this method are based on the recommended air volume for each analyte collected separately and a desorption volume of 1 ml. The amounts are presented as acrolein and/or formaldehyde, even though the derivatives are the actual species analyzed.

1.2.1 *Detection limits of the analytical procedure:* The detection limit of the analytical procedure was 386 pg per injection for formaldehyde. This was the amount of analyte which gave a peak whose height was about five times the height of the peak given by the residual formaldehyde derivative in a typical blank front section of the recommended sampling tube.

1.2.2 *Detection limits of the overall procedure:* The detection limits of the overall procedure were 482 ng per sample (16 ppb or 20 µg/m$^3$ for formaldehyde). This was the amount of analyte spiked on the sampling device which allowed recoveries approximately equal to the detection limit of the analytical procedure.

1.2.3 *Reliable quantitation limits:* The reliable quantitation limit was 482 ng per sample (16 ppb or 20 µg/m$^3$) for formaldehyde. These were the smallest amounts of analyte which could be quantitated within the limits of a recovery of at least 75% and a precision (±1.96 SD) of ±25% or better.

The reliable quantitation limit and detection limits reported in the method are based upon optimization of the instrument for the smallest possible amount of analyte. When the target concentration of an analyte is exceptionally higher than these limits, they may not be attainable at the routine operating parameters.

1.2.4 *Sensitivity:* The sensitivity of the analytical procedure over concentration ranges representing 0.4 to 2 times the target concentration, based on the recommended air volumes, was 7,589 area units per µg/ml for formaldehyde. This value was determined from the slope of the calibration curve. The sensitivity may vary with the particular instrument used in the analysis.

1.2.5 *Recovery:* The recovery of formaldehyde from samples used in an 18-day storage test remained above 92% when the samples were stored at ambient temperature. These values were determined from regression lines which were calculated from the storage data. The recovery of the analyte from the collection device must be at least 75% following storage.

1.2.6 *Precision (analytical method only):* The pooled coefficient of variation obtained from replicate determinations of analytical standards over the range of 0.4 to 2 times the target concentration was 0.0052 for formaldehyde (Section 4.3).

1.2.7 *Precision (overall procedure):* The precision at the 95% confidence level for the ambient temperature storage tests was ±14.3% for formaldehyde. These values each include an additional ±5% for sampling error. The overall procedure must provide results at the target concentrations that are ±25% at the 95% confidence level.

1.2.8 *Reproducibility:* Samples collected from controlled test atmospheres and a draft copy of this procedure were given to a chemist unassociated with this evaluation. The formaldehyde samples were analyzed following 15 days storage. The average recovery was 96.3% and the standard deviation was 1.7%.

1.3 *Advantages:*

1.3.1 The sampling and analytical procedures permit the simultaneous determination of acrolein and formaldehyde.

1.3.2 Samples are stable following storage at ambient temperature for at least 18 days.

1.4 *Disadvantages:* None.

## 2. Sampling Procedure

2.1 *Apparatus:*

2.1.1 Samples are collected by use of a personal sampling pump that can be calibrated to within ±5% of the recommended 0.1 L/min sampling rate with the sampling tube in line.

2.1.2 Samples are collected with laboratory prepared sampling tubes. The sampling tube is constructed of silane treated glass and is about 8-cm long. The ID is 4 mm and the OD is 6 mm. One end of the tube is tapered so that a glass wool end plug will hold the contents of the tube in place during sampling. The other end of the sampling tube is open to its full 4-mm ID to facilitate packing of the tube. Both ends of the tube are firepolished for safety. The tube is packed with a 75-mg backup section, located nearest the tapered end and a 150-mg sampling section of pretreated XAD-2 adsorbent which has been coated with 2-HMP. The two sections of coated adsorbent are separated and retained with small plugs of silanized glass wool. Following packing, the sampling tubes are sealed with two 7/32 inch OD plastic end caps. Instructions for the pretreatment and the coating of XAD-2 adsorbent are presented in Section 4 of this method.

2.1.3 Sampling tubes, similar to those recommended in this method, are marketed by Supelco, Inc. These tubes were not available when this work was initiated; therefore, they were not evaluated.

2.2 *Reagents:* None required.

2.3 *Technique:*

2.3.1 Properly label the sampling tube before sampling and then remove the plastic end caps.

2.3.2 Attach the sampling tube to the pump using a section of flexible plastic tubing such that the large, front section of the sampling tube is exposed directly to the atmosphere. Do not place any tubing ahead of the sampling tube. The sampling tube should be attached in the worker's breathing zone in a vertical manner such that it does not impede work performance.

2.3.3 After sampling for the appropriate time, remove the sampling tube from the pump and then seal the tube with plastic end caps.

2.3.4 Include at least one blank for each sampling set. The blank should be handled in the same manner as the samples with the exception that air is not drawn through it.

2.3.5 List any potential interferences on the sample data sheet.

## 2.4 Breakthrough:

2.4.1 Breakthrough was defined as the relative amount of analyte found on a backup sample in relation to the total amount of analyte collected on the sampling train.

2.4.2 For formaldehyde collected from test atmospheres containing 6 times the PEL, the average 5% breakthrough air volume was 41 L. The sampling rate was 0.1 L/min and the average mass of formaldehyde collected was 250 µg.

2.5 *Desorption Efficiency:* No desorption efficiency corrections are necessary to compute air sample results because analytical standards are prepared using coated adsorbent. Desorption efficiencies were determined, however, to investigate the recoveries of the analytes from the sampling device. The average recovery over the range of 0.4 to 2 times the target concentration, based on the recommended air volumes, was 96.2% for formaldehyde. Desorption efficiencies were essentially constant over the ranges studied.

2.6 *Recommended Air Volume and Sampling Rate:*

2.6.1 The recommended air volume for formaldehyde is 24 L.

2.6.2 The recommended sampling rate is 0.1 L/min.

2.7 *Interferences:*

2.7.1 Any collected substance that is capable of reacting 2-HMP and thereby depleting the derivatizing agent is a potential interference. Chemicals which contain a carbonyl group, such as acetone, may be capable or reacting with 2-HMP.

2.7.2 There are no other known interferences to the sampling method.

2.8 *Safety Precautions:*

2.8.1 Attach the sampling equipment to the worker in such a manner that it well not interfere with work performance or safety.

2.8.2 Follow all safety practices that apply to the work area being sampled.

### 3. Analytical Procedure

3.1 *Apparatus:*

3.1.1 A gas chromatograph (GC), equipped with a nitrogen selective detector. A Hewlett-Packard Model 5840A GC fitted with a nitrogen-phosphorus flame ionization detector (NPD) was used for this evaluation. Injections were performed using a Hewlett-Packard Model 7671A automatic sampler.

3.1.2 A GC column capable of resolving the analytes from any interference. A 6 ft × 1/4 in OD (2mm ID) glass GC column containing 10% UCON 50-HB-5100 + 2% KOH on 80/100 mesh Chromosorb W-AW was used for the evaluation. Injections were performed on-column.

3.1.3 Vials, glass 2-ml with Teflon-lined caps.

3.1.4 Volumetric flasks, pipets, and syringes for preparing standards, making dilutions, and performing injections.

3.2 *Reagents:*

3.2.1 Toluene and dimethylformamide. Burdick and Jackson solvents were used in this evaluation.

3.2.2 Helium, hydrogen, and air, GC grade.

3.2.3 Formaldehyde, 37%, by weight, in water. Aldrich Chemical, ACS Reagent Grade formaldehyde was used in this evaluation.

3.2.4 Amberlite XAD-2 adsorbent coated with 2-(hydroxymethyl—piperidine (2-HMP), 10% by weight (Section 4).

3.2.5 Desorbing solution with internal standard. This solution was prepared by adding 20 µl of dimethylformamide to 100 ml of toluene.

3.3 *Standard preparation:*

3.3.1 *Formaldehyde:* Prepare stock standards by diluting known volumes of 37% formaldehyde solution with methanol. A procedure to determine the formaldehyde content of these standards is presented in Section 4. A standard containing 7.7 mg/ml formaldehyde was prepared by diluting 1 ml of the 37% reagent to 50 ml with methanol.

3.3.2 It is recommended that analytical standards be prepared about 16 hours before the air samples are to be analyzed in order to ensure the complete reaction of the analytes with 2-HMP. However, rate studies have shown the reaction to be greater than 95% complete after 4 hours. Therefore, one or two standards can be analyzed after this reduced time if sample results are outside the concentration range of the prepared standards.

3.3.3 Place 150-mg portions of coated XAD-2 adsorbent, from the same lot number as used to collect the air samples, into each of several glass 2-ml vials. Seal each vial with a Teflon-lined cap.

3.3.4 Prepare fresh analytical standards each day by injecting appropriate amounts of the diluted analyte directly onto 150-mg portions of coated adsorbent. It is permissible to inject both acrolein and formaldehyde on the same adsorbent portion. Allow the standards to stand at room temperature. A standard, approximately the target levels, was prepared by injecting 11 µl of the acrolein and 12 µl of the formaldehyde stock standards onto a single coated XAD-2 adsorbent portion.

3.3.5 Prepare a sufficient number of standards to generate the calibration curves. Analytical standard concentrations should bracket sample concentrations. Thus, if samples are not in the concentration range of the prepared standards, additional standards must be prepared to determine detector response.

3.3.7 Desorb the standards in the same manner as the samples following the 16-hour reaction time.

3.4 *Sample preparation:*

3.4.1 Transfer the 150-mg section of the sampling tube to a 2-ml vial. Place the 75-mg section in a separate vial. If the glass wool plugs contain a significant number of adsorbent beads, place them with the appropriate sampling tube section. Discard the glass wool plugs if they do not contain a significant number of adsorbent beads.

3.4.2 Add 1 ml of desorbing solution to each vial.

3.4.3 Seal the vials with Teflon-lined caps and then allow them to desorb for one hour. Shake the vials by hand with vigorous force several times during the desorption time.

3.4.4 Save the used sampling tubes to be cleaned and recycled.

3.5 *Analysis:*

3.5.1 *GC Conditions*

*Column Temperature:*

Bi-level temperature program—First level: 100 to 140 °C at 4 °C/min following completion of the first level.

Second level: 140 to 180 °C at 20 °C/min following completion of the first level.

Isothermal period: Hold column at 180 °C until the recorder pen returns to baseline (usually about 25 min after injection).

*Injector temperature:* 180 °C

*Helium flow rate:* 30 ml/min (detector response will be reduced if nitrogen is substituted for helium carrier gas).

*Injection volume:* 0.8 µl

*GC column:* Six-ft × 1/4-in OD (2 mm ID) glass GC column containing 10% UCON 50-HB-5100 + 2% KOH on 80/100 Chromosorb W-AW.

*NPD conditions:*

Hydrogen flow rate: 3 ml/min

Air flow rate: 50 ml/min

Detector temperature: 275 °C

3.5.2 *Chromatogram:* For an example of a typical chromatogram, see Figure 4.11 in OSHA Method 52.

3.5.3 Use a suitable method, such as electronic integration, to measure detector response.

3.5.4 Use an internal standard method to prepare the calibration curve with several standard solutions of different concentrations. Prepare the calibration curve daily. Program the integrator to report results in g/ml.

3.5.5 Bracket sample concentrations with standards.

3.6 *Interferences (Analytical)*

3.6.1 Any compound with the same general retention time as the analytes and which also gives a detector response is a potential interference. Possible interferences should be reported to the laboratory with submitted samples by the industrial hygienist.

3.6.2 GC parameters (temperature, column, etc.) may be changed to circumvent interferences.

# OSHA Standards for General Industry

3.6.3 A useful means of structure designation is GC/MS. It is recommended this procedure be used to confirm samples whenever possible.

3.6.4 The coated adsorbent usually contains a very small amount of residual formaldehyde derivative (Section 4.8).

3.7 *Calculations:*

3.7.1 Results are obtained by use of calibration curves. Calibration curves are prepared by plotting detector response against concentration for each standard. The best line through the data points is determined by curve fitting.

3.7.2 The concentration, in µg/ml, for a particular sample is determined by comparing its detector response to the calibration curve. If either of the analytes is found on the backup section, it is added to the amount found on the front section. Blank corrections should be performed before adding the results together.

3.7.3 The acrolein and/or formaldehyde air concentration can be expressed using the following equation:

$$mg/m^3 = (A)(B)/C$$

where A = µg/ml from 3.7.2, B = desorption volume, and C = L of air sampled.

No desorption efficiency corrections are required.

3.7.4 The following equation can be used to convert results in mg/m³ to ppm.

$$ppm = (mg/m^3)(24.45)/MW$$

where mg/m³ = result from 3.7.3, 24.45 = molar volume of an ideal gas at 760 mm Hg and 25 °C, MW = molecular weight (30.0).

## 4. Backup Data

4.1 Backup data on detection limits, reliable quantitation limits, sensitivity and precision of the analytical method, breakthrough, desorption efficiency, storage, reproducibility, and generation of test atmospheres are available in OSHA Method 52, developed by the Organics Methods Evaluation Branch, OSHA Analytical Laboratory, Salt Lake City, Utah.

4.2 *Procedure to Coat XAD-2 Adsorbent with 2-HMP:*

4.2.1 *Apparatus:* Soxhlet extraction apparatus, rotary evaporation apparatus, vacuum dessicator, 1-L vacuum flask, 1-L round-bottomed evaporative flask, 1-L Erlenmeyer flask, 250-ml Buchner funnel with a coarse fritted disc, etc.

4.2.2 *Reagents:*

4.2.2.1 Methanol, isooctane, and toluene.

4.2.2.2 2-(Hydroxymethyl)piperidine.

4.2.2.3 Amberlite XAD-2 non-ionic polymeric adsorbent, 20 to 60 mesh, Aldrich Chemical XAD-2 was used in this evaluation.

4.2.3 *Procedure:* Weigh 125 g of crude XAD-2 adsorbent into a 1-L Erlenmeyer flask. Add about 200 ml of water to the flask and then swirl the mixture to wash the adsorbent. Discard any adsorbent that floats to the top of the water and then filter the mixture using a fritted Buchner funnel. Air dry the adsorbent for 2 minutes. Transfer the adsorbent back to the Erlenmeyer flask and then add about 200 ml of methanol to the flask. Swirl and then filter the mixture as before. Transfer the washed adsorbent back to the Erlenmeyer flask and then add about 200 ml of methanol to the flask. Swirl and then filter the mixture as before. Transfer the washed adsorbent to a 1-L roundbottomed evaporative flask, add 13 g of 2-HMP and then 200 ml of methanol, swirl the mixture and then allow it to stand for one hour. Remove the methanol at about 40 °C and reduced pressure using a rotary evaporation apparatus. Transfer the coated adsorbent to a suitable container and store it in a vacuum desiccator at room temperature overnight. Transfer the coated adsorbent to a Soxhlet extractor and then extract the material with toluene for about 24 hours. Discard the contaminated toluene, add methanol in its place and then continue the Soxhlet extraction for an additional 4 hours. Transfer the adsorbent to a weighted 1-L round-bottom evaporative flask and remove the methanol using the rotary evaporation apparatus. Determine the weight of the adsorbent and then add an amount of 2HMP, which is 10% by weight of the adsorbent. Add 200 ml of methanol and then swirl the mixture. Allow the mixture to stand for one hour. Remove the methanol by rotary evaporation. Transfer the coated adsorbent to a suitable container and store it in a vacuum desiccator until all traces of solvents are gone. Typically, this will take 2-3 days. The coated adsorbent should be protected from contamination. XAD-2 adsorbent treated in this manner will probably not contain residual acrolein derivative. However, this adsorbent will often contain residual formaldehyde derivative levels of about 0.1 µg per 150 mg of adsorbent. If the blank values for a batch of coated adsorbent are too high, then the batch should be returned to the Soxhlet extractor, extracted with toluene again and then recoated. This process can be repeated until the desired blank levels are attained.

The coated adsorbent is now ready to be packed into sampling tubes. The sampling tubes should be stored in a sealed container to prevent contamination. Sampling tubes should be stored in the dark at room temperature. The sampling tubes should be segregated by coated adsorbent lot number. A sufficient amount of each lot number of coated adsorbent should be retained to prepare analytical standards for use with air samples from that lot number.

4.3 *A Procedure to Determine Formaldehyde by Acid Titration:* Standardize the 0.1 N HCl solution using sodium carbonate and methyl orange indicator.

Place 50 ml of 0.1 M sodium sulfite and three drops of thymophthalein indicator into a 250-ml Erlenmeyer flask. Titrate the contents of the flask to a colorless endpoint with 0.1 N HCl (usually one or two drops is sufficient). Transfer 10 ml of the formaldehyde/methanol solution (prepared in 3.3.1) into the same flask and titrate the mixture with 0.1 N HCl, again, to a colorless endpoint. The formaldehyde concentration of the standard may be calculated by the following equation:

$$\text{Formaldehyde, mg/mL} = \frac{\text{acid titer} \times \text{acid normality} \times 30.0}{\text{mL of sample}}$$

This method is based on the quantitative liberation of sodium hydroxide when formaldehyde reacts with sodium sulfite to form the formaldehyde-bisulfite addition product. The volume of sample may be varied depending on the formaldehyde content but the solution to be titrated must contain excess sodium sulfite. Formaldehyde solutions containing substantial amounts of acid or base must be neutralized before analysis.

## APPENDIX C TO §1910.1048—MEDICAL SURVEILLANCE—FORMALDEHYDE

### I. Health Hazards

The occupational health hazards of formaldehyde are primarily due to its toxic effects after inhalation, after direct contact with the skin or eyes by formaldehyde in liquid or vapor form, and after ingestion.

### II. Toxicology

#### A. Acute Effects of Exposure

1. *Inhalation (breathing):* Formaldehyde is highly irritating to the upper airways. The concentration of formaldehyde that is immediately dangerous to life and health is 100 ppm. Concentrations above 50 ppm can cause severe pulmonary reactions within minutes. These include pulmonary edema, pneumonia, and bronchial irritation which can result in death. Concentrations above 5 ppm readily cause lower airway irritation characterized by cough, chest tightness and wheezing. There is some controversy regarding whether formaldehyde gas is a pulmonary sensitizer which can cause occupational asthma in a previously normal individual. Formaldehyde can produce symptoms of bronchial asthma in humans. The mechanism may be either sensitization of the individual by exposure to formaldehyde or direct irritation by formaldehyde in persons with pre-existing asthma. Upper airway irritation is the most common respiratory effect reported by workers and can occur over a wide range of concentrations, most frequently above 1 ppm. However, airway irritation has occurred in some workers with exposures to formaldehyde as low as 0.1 ppm. Symptoms of upper airway irritation include dry or sore throat, itching and burning sensations of the nose, and nasal congestion. Tolerance to this level of exposure may develop within 1-2 hours. This tolerance can permit workers remaining in an environment of gradually increasing formaldehyde concentrations to be unaware of their increasingly hazardous exposure.

2. *Eye contact:* Concentrations of formaldehyde between 0.05 ppm and 0.5 ppm produce a sensation of irritation in the eyes with burning, itching, redness, and tearing. Increased rate of blinking and eye closure generally protects the eye from damage at these low levels, but these protective mechanisms may interfere with some workers' work abilities. Tolerance can occur in workers continuously exposed to concentrations of formaldehyde in this range. Accidental splash injuries of human eyes to aqueous solutions of formaldehyde (formalin) have resulted in a wide

range of ocular injuries including corneal opacities and blindness. The severity of the reactions have been directly dependent on the concentration of formaldehyde in solution and the amount of time lapsed before emergency and medical intervention.

3. *Skin contact:* Exposure to formaldehyde solutions can cause irritation of the skin and allergic contact dermatitis. These skin diseases and disorders can occur at levels well below those encountered by many formaldehyde workers. Symptoms include erythema, edema, and vesiculation or hives. Exposure to liquid formalin or formaldehyde vapor can provoke skin reactions in sensitized individuals even when airborne concentrations of formaldehyde are well below 1 ppm.

4. *Ingestion:* Ingestion of as little as 30 ml of a 37 percent solution of formaldehyde (formalin) can result in death. Gastrointestinal toxicity after ingestion is most severe in the stomach and results in symptoms which can include nausea, vomiting, and servere abdominal pain. Diverse damage to other organ systems including the liver, kidney, spleen, pancreas, brain, and central nervous systems can occur from the acute response to ingestion of formaldehyde.

### B. Chronic Effects of Exposure

Long term exposure to formaldehyde has been shown to be associated with an increased risk of cancer of the nose and accessory sinuses, nasopharyngeal and oropharyngeal cancer, and lung cancer in humans. Animal experiments provide conclusive evidence of a causal relationship between nasal cancer in rats and formaldehyde exposure. Concordant evidence of carcinogenicity includes DNA binding, genotoxicity in short-term tests, and cytotoxic changes in the cells of the target organ suggesting both preneoplastic changes and a dose-rate effect. Formaldehyde is a complete carcinogen and appears to exert an effect on at least two stages of the carcinogenic process.

## III. Surveillance considerations

### A. History

1. *Medical and occupational history:* Along with its acute irritative effects, formaldehyde can cause allergic sensitization and cancer. One of the goals of the work history should be to elicit information on any prior or additional exposure to formaldehyde in either the occupational or the nonoccupational setting.

2. *Respiratory history:* As noted above, formaldehyde has recognized properties as an airway irritant and has been reported by some authors as a cause of occupational asthma. In addition, formaldehyde has been associated with cancer of the entire respiratory system of humans. For these reasons, it is appropriate to include a comprehensive review of the respiratory system in the medical history. Components of this history might include questions regarding dyspnea on exertion, shortness of breath, chronic airway complaints, hyperreactive airway disease, rhinitis, bronchitis, bronchiolitis, asthma, emphysema, respiratory allergic reaction, or other preexisting pulmonary disease.

In addition, generalized airway hypersensitivity can result from exposures to a single sensitizing agent. The examiner should, therefore, elicit any prior history of exposure to pulmonary irritants, and any short- or long-term effects of that exposure.

Smoking is known to decrease mucociliary clearance of materials deposited during respiration in the nose and upper airways. This may increase a worker's exposure to inhaled materials such as formaldehyde vapor. In addition, smoking is a potential confounding factor in the investigation of any chronic respiratory disease, including cancer. For these reasons, a complete smoking history should be obtained.

3. *Skin Disorders:* Because of the dermal irritant and sensitizing effects of formaldehyde, a history of skin disorders should be obtained. Such a history might include the existence of skin irritation, previously documented skin sensitivity, and other dermatologic disorders. Previous exposure to formaldehyde and other dermal sensitizers should be recorded.

4. *History of atopic or allergic diseases:* Since formaldehyde can cause allergic sensitization of the skin and airways, it might be useful to identify individuals with prior allergen sensitization. A history of atopic disease and allergies to formaldehyde or any other substances should also be obtained. It is not definitely known at this time whether atopic diseases and allergies to formaldehyde or any other substances should also be obtained. Also it is not definitely known at this time whether atopic individuals have a greater propensity to develop formaldehyde sensitivity than the general population, but identification of these individuals may be useful for ongoing surveillance.

5. *Use of disease questionnaires:* Comparison of the results from previous years with present results provides the best method for detecting a general deterioration in health when toxic signs and symptoms are measured subjectively. In this way recall bias does not affect the results of the analysis. Consequently, OSHA has determined that the findings of the medical and work histories should be kept in a standardized form for comparison of the year-to-year results.

### B. Physical Examination

1. *Mucosa of eyes and airways:* Because of the irritant effects of formaldehyde, the examining physician should be alert to evidence of this irritation. A speculum examination of the nasal mucosa may be helpful in assessing possible irritation and cytotoxic changes, as may be indirect inspection of the posterior pharynx by mirror.

2. *Pulmonary system:* A conventional respiratory examination, including inspection of the thorax and auscultation and percussion of the lung fields should be performed as part of the periodic medical examination. Although routine pulmonary function testing is only required by the standard once every year for persons who are exposed over the TWA concentration limit, these tests have an obvious value in investigating possible respiratory dysfunction and should be used wherever deemed appropriate by the physician. In cases of alleged formaldehydeinduced airway disease, other possible causes of pulmonary disfunction (including exposures to other substances) should be ruled out. A chest radiograph may be useful in these circumstances. In cases of suspected airway hypersensitivity or allergy, it may be appropriate to use bronchial challenge testing with formaldehyde or methacholine to determine the nature of the disorder. Such testing should be performed by or under the supervision of a physician experienced in the procedures involved.

3. *Skin:* The physician should be alert to evidence of dermal irritation of sensitization, including reddening and inflammation, urticaria, blistering, scaling, formation of skin fissures, or other symptoms. Since the integrity of the skin barrier is compromised by other dermal diseases, the presence of such disease should be noted. Skin sensitivity testing carries with it some risk of inducing sensitivity, and therefore, skin testing for formaldehyde sensitivity should not be used as a routine screening test. Sensitivity testing may be indicated in the investigation of a suspected existing sensitivity. Guidelines for such testing have been prepared by the North American Contact Dermatitis Group.

### C. Additional Examinations or Tests

The physician may deem it necessary to perform other medical examinations or tests as indicated. The standard provides a mechanism whereby these additional investigations are covered under the standard for occupational exposure to formaldehyde.

### D. Emergencies

The examination of workers exposed in an emergency should be directed at the organ systems most likely to be affected. Much of the content of the examination will be similar to the periodic examination unless the patient has received a severe acute exposure requiring immediate attention to prevent serious consequences. If a severe overexposure requiring medical intervention or hospitalization has occurred, the physician must be alert to the possibility of delayed symptoms. Followup nonroutine examinations may be necessary to assure the patient's well-being.

### E. Employer Obligations

The employer is required to provide the physician with the following information: A copy of this standard and appendices A, C, D, and E; a description of the affected employee's duties as they relate to his or her exposure concentration; an estimate of the employee's exposure including duration (e.g., 15 hr/wk, three 8-hour shifts, full-time); a description of any personal protective equipment, including respirators, used by the employee; and the results of any previous medical determinations for the affected employee related to formaldehyde exposure to the extent that this information is within the employer's control.

### F. Physician's Obligations

The standard requires the employer to obtain a written statement from the physician. This statement must contain the physician's opinion as to whether the employee has any medical condition which would place him or her at increased risk of impaired health from exposure to formaldehyde or use of respirators, as appropriate. The physician must also state his opinion regarding any restrictions that should be placed on the employee's exposure to formaldehyde or upon the use of protective clothing or equipment such as respirators. If the employee wears a respirator as a result of his or her exposure to formaldehyde, the physician's opinion must also contain a statement regarding the suitability of the employee to wear the type of respirator assigned. Finally, the physician must inform the employer that the employee has been told the results of the medical examination and of any medical conditions which require further explanation or treatment.

This written opinion is not to contain any information on specific findings or diagnoses unrelated to occupational exposure to formaldehyde.

The purpose in requiring the examining physician to supply the employer with a written opinion is to provide the employer with a medical basis to assist the employer in placing employees initially, in assuring that their health is not being impaired by formaldehyde, and to assess the employee's ability to use any required protective equipment.

## APPENDIX D TO §1910.1048—NONMANDATORY MEDICAL DISEASE QUESTIONNAIRE

**NONMANDATORY MEDICAL DISEASE QUESTIONNAIRE**

### A. Identification

Plant Name _____ Date _____
Employee Name _____ S.S. # _____
Job Title _____ Birthdate: _____
Age: _____ Sex: _____
Height: _____ Weight: _____

### B. Medical History

1. Have you ever been in the hospital as a patient? Yes ☐ No ☐
   If yes, what kind of problem were you having? _____
2. Have you ever had any kind of operation? Yes ☐ No ☐
   If yes, what kind? _____
3. Do you take any kind of medicine regularly? Yes ☐ No ☐
   If yes, what kind?
4. Are you allergic to any drugs, foods, or chemicals? Yes ☐ No ☐
   If yes, what kind of allergy is it?
   What causes the allergy?
5. Have you ever been told that you have asthma, hayfever, or sinusitis? Yes ☐ No ☐
6. Have you ever been told that you have emphysema, bronchitis, or any other respiratory problems? Yes ☐ No ☐
7. Have you ever been told you had hepatitis? Yes ☐ No ☐
8. Have you ever been told that you had cirrhosis? Yes ☐ No ☐
9. Have you ever been told that you had cancer? Yes ☐ No ☐
10. Have you ever had arthritis or joint pain? Yes ☐ No ☐
11. Have you ever been told that you had high blood pressure? Yes ☐ No ☐
12. Have you ever had a heart attack or heart trouble? Yes ☐ No ☐

### B-1. Medical History Update

1. Have you been in the hospital as a patient any time within the past year? Yes ☐ No ☐
   If so, for what condition? _____
2. Have you been under the care of a physician during the past year? Yes ☐ No ☐
   If so, for what condition? _____
3. Is there any change in your breathing since last year? Yes ☐ No ☐ No change? _____
   Better? _____ Worse? _____
   If change, do you know why?
4. Is your general health different this year from last year? Yes ☐ No ☐
   If different, in what way?
5. Have you in the past year or are you now taking any medication on a regular basis? Yes ☐ No ☐
   Name Rx
   Condition being treated

### C. Occupational History

1. How long have you worked for your present employer?
2. What jobs have you held with this employer? Include job title and length of time in each job.
3. In each of these jobs, how many hours a day were you exposed to chemicals?
4. What chemicals have you worked with most of the time?
5. Have you ever noticed any type of skin rash you feel was related to your work? Yes ☐ No ☐
6. Have you ever noticed that any kind of chemical makes you cough? Yes ☐ No ☐;
   Become short of breath or cause your chest to become tight? Yes ☐ No ☐
   Wheeze? Yes ☐ No ☐
7. Are you exposed to any dust or chemicals at home? Yes ☐ No ☐
   If yes, explain:
8. In other jobs, have you ever had exposure to: Wood dust? Yes ☐ No ☐
   Nickel or chromium? Yes ☐ No ☐ Silica (foundry, sand blasting)? Yes ☐ No ☐
   Arsenic or asbestos? Yes ☐ No ☐ Organic solvents? Yes ☐ No ☐ Urethane foams? Yes ☐ No ☐

### C-1. Occupational History Update

1. Are you working on the same job this year as you were last year? Yes ☐ No ☐
   If not, how has your job changed?
2. What chemicals are you exposed to on your job?
3. How many hours a day are you exposed to chemicals?
4. Have you noticed any skin rash within the past year you feel was related to your work? Yes ☐ No ☐
   If so, explain circumstances:

[Questionnaire content - rotated sideways on page:]

5. Have you noticed that any chemical makes you cough, be short of breath, or wheeze? Yes ☐ No ☐
If so, can you identify it? _____

**D. Miscellaneous**

1. Do you smoke? Yes ☐ No ☐
If so, how much and for how long? Pipe ☐ Cigarettes ☐ Cigars ☐
2. Do you drink alcohol in any form? Yes ☐ No ☐
If so, how much, how long, and how often?
3. Do you wear glasses or contact lenses? Yes ☐ No ☐
4. Do you get any physical exercise other than that required to do your job? Yes ☐ No ☐
If so, explain: _____
5. Do you have any hobbies or "side jobs" that require you to use chemicals, such as furniture stripping, sand blasting, insulation or manufacture of urethane foam, furniture, etc? Yes ☐ No ☐
If so, please describe, giving type of business or hobby, chemicals used and length of exposures.

**E. Symptoms Questionnaire**

1. Do you ever have any shortness of breath? Yes ☐ No ☐
If yes, do you have to rest after climbing several flights of stairs? Yes ☐ No ☐
If yes, if you walk on the level with people your own age, do you walk slower than they do? Yes ☐ No ☐
If yes, if you walk slower than a normal pace, do you have to limit the distance that you walk? Yes ☐ No ☐
If yes, do you have to stop and rest while bathing or dressing? Yes ☐ No ☐
2. Do you cough as much as three months out of the year? Yes ☐ No ☐
If yes, have you had this cough for more than two years? Yes ☐ No ☐
If yes, do you ever cough anything up from chest? Yes ☐ No ☐
3. Do you ever have a feeling of smothering, unable to take a deep breath, or tightness in your chest? Yes ☐ No ☐
If yes, do you notice that this on any particular day of the week? Yes ☐ No ☐
If yes, what day or the week? _____
If yes, do you notice that this occurs at any particular place? Yes ☐ No ☐
If yes, do you notice that this is worse after you have returned to work after being off for several days? Yes ☐ No ☐
Is this caused by exposure to any kind of dust or other material? Yes ☐ No ☐
If yes, what kind?
4. Have you ever noticed any wheezing in your chest? Yes ☐ No ☐
If yes, explain circumstances:
5. Have you noticed any burning, tearing, or redness of your eyes when you are at work? Yes ☐ No ☐
6. Have you noticed any sore or burning throat or itchy or burning nose when you are at work? Yes ☐ No ☐
If so, explain circumstances:
7. Have you noticed any stuffiness or dryness of your nose? Yes ☐ No ☐
8. Do you ever have swelling of the eyelids or face? Yes ☐ No ☐
9. Have you ever been jaundiced? Yes ☐ No ☐
If yes, was this accompanied by any pain? Yes ☐ No ☐
10. Have you ever had a tendency to bruise easily or bleed excessively? Yes ☐ No ☐
11. Do you have frequent headaches that are not relieved by aspirin or tylenol? Yes ☐ No ☐
If yes, do they occur at any particular time of the day or week? Yes ☐ No ☐
If yes, when do they occur?
12. Do you have frequent episodes of nervousness or irritability? Yes ☐ No ☐
13. Do you tend to have trouble concentrating or remembering? Yes ☐ No ☐
14. Do you ever feel dizzy, light-headed, excessively drowsy or like you have been drugged? Yes ☐ No ☐
15. Does your vision ever become blurred? Yes ☐ No ☐
16. Do you have numbness or tingling of the hands or feet or other parts of your body? Yes ☐ No ☐
17. Have you ever had chronic weakness or fatigue? Yes ☐ No ☐
18. Have you ever had any swelling of your feet or ankles to the point where you could not wear your shoes? Yes ☐ No ☐
19. Are you bothered by heartburn or indigestion? Yes ☐ No ☐
20. Do you ever have itching, dryness, or peeling and scaling of the hands? Yes ☐ No ☐
21. Do you ever have a burning sensation in the hands, or reddening of the skin? Yes ☐ No ☐
22. Do you ever have cracking or bleeding of the skin on your hands? Yes ☐ No ☐
23. Are you under a physician's care? Yes ☐ No ☐
If yes, for what are you being treated?
24. Do you have any physical complaints today? Yes ☐ No ☐
If yes, explain?
25. Do you have other health conditions not covered by these questions? Yes ☐ No ☐
If yes, explain: _____

## APPENDIX E TO §1910.1048—QUALITATIVE AND QUANTITATIVE FIT TESTING PROCEDURES [Removed.]

[§1910.1048, App. E, removed at 63 FR 1293, Jan. 8, 1998, eff. Apr. 8, 1998.]

[57 FR 22310, May 27, 1992; 57 FR 27161, June 18, 1992; 61 FR 5508, Feb. 13, 1996; 63 FR 1292, Jan. 8, 1998; 63 FR 20099, Apr. 23, 1998]

## §1910.1050 Methylenedianiline.

**1910.1050(a)** *Scope and application.* **(1)** This section applies to all occupational exposures to MDA, Chemical Abstracts Service Registry No. 101-77-9, except as provided in paragraphs (a)(2) through (a)(7) of this section.

**1910.1050(a)(2)** Except as provided in paragraphs (a)(8) and (e)(5) of this section, this section does not apply to the processing, use, and handling of products containing MDA where initial monitoring indicates that the product is not capable of releasing MDA in excess of the action level under the expected conditions of processing, use, and handling which will cause the greatest possible release; and where no "dermal exposure to MDA" can occur.

**1910.1050(a)(3)** Except as provided in paragraph (a)(8) of this section, this section does not apply to the processing, use, and handling of products containing MDA where objective data are reasonably relied upon which demonstrate the product is not capable of releasing MDA under the expected conditions of processing, use, and handling which will cause the greatest possible release; and where no "dermal exposure to MDA" can occur.

**1910.1050(a)(4)** This section does not apply to the storage, transportation, distribution or sale of MDA in intact containers sealed in such a manner as to contain the MDA dusts, vapors, or liquids, except for the provisions of 29 CFR 1910.1200 and paragraph (d) of this section.

**1910.1050(a)(5)** This section does not apply to the construction industry as defined in 29 CFR 1910.12(b). (Exposure to MDA in the construction industry is covered by 29 CFR 1926.60).

**1910.1050(a)(6)** Except as provided in paragraph (a)(8) of this secton, this section does not apply to materials in any form which contain less than 0.1% MDA by weight or volume.

**1910.1050(a)(7)** Except as provided in paragraph (a)(8) of this section, this section does not apply to "finished articles containing MDA."

**1910.1050(a)(8)** Where products containing MDA are exempted under paragraphs (a)(2) through (a)(7) of this section, the employer shall maintain records of the initial monitoring results or objective data supporting that exemption and the basis for the employer's reliance on the data, as provided in the recordkeeping provision of paragraph (n) of this section.

**1910.1050(b)** *Definitions.* For the purpose of this section, the following definitions shall apply:

*Action level* means a concentration of airborne MDA of 5 ppb as an eight (8)-hour time-weighted average.

*Assistant Secretary* means the Assistant Secretary of Labor for Occupational Safety and Health, U.S. Department of Labor, or designee.

*Authorized person* means any person specifically authorized by the employer whose duties require the person to enter a regulated area, or any person entering such an area as a designated representative of employees, for the purpose of exercising the right to observe monitoring and measuring procedures under paragraph (o) of this section, or any other person authorized by the Act or regulations issued under the Act.

*Container* means any barrel, bottle, can, cylinder, drum, reaction vessel, storage tank, commercial packaging or the like, but does not include piping systems.

*Dermal exposure to MDA* occurs where employees are engaged in the handling, application or use of mixtures or materials containing MDA, with any of the following non-airborne forms of MDA:

(i) Liquid, powdered, granular, or flaked mixtures containing MDA in concentrations greater than 0.1% by weight or volume; and

(ii) Materials other than "finished articles" containing MDA in concentrations greater than 0.1% by weight or volume.

*Director* means the Director of the National Institute for Occupational Safety and Health, U.S. Department of Health and Human Services, or designee.

*Emergency* means any occurrence such as, but not limited to, equipment failure, rupture of containers, or failure of control equipment which results in an unexpected and potentially hazardous release of MDA.

*Employee exposure* means exposure to MDA which would occur if the employee were not using respirators or protective work clothing and equipment.

*Finished article containing MDA* is defined as a manufactured item:

(i) Which is formed to a specific shape or design during manufacture;

(ii) Which has end use function(s) dependent in whole or part upon its shape or design during end use; and

(iii) Where applicable, is an item which is fully cured by virtue of having been subjected to the conditions (temperature, time) necessary to complete the desired chemical reaction.

*4,4' Methylenedianiline or MDA* means the chemical, 4,4'-diaminodiphenylmethane, Chemical Abstract Service Registry number 101-77-9, in the form of a vapor, liquid, or solid. The definition also includes the salts of MDA.

*Regulated areas* means areas where airborne concentrations of MDA exceed or can reasonably be expected to exceed, the permissible exposure limits, or where dermal exposure to MDA can occur.

*STEL* means short term exposure limit as determined by any 15 minute sample period.

**1910.1050(c)** *Permissible exposure limits (PEL).* The employer shall assure that no employee is exposed to an airborne concentration of MDA in excess of ten parts per billion (10 ppb) as an 8-hour time-weighted average or a STEL of 100 ppb.

**1910.1050(d)** *Emergency situations*—**(1)** *Written plan.* **(i)** A written plan for emergency situations shall be developed for each workplace where there is a possibility of an emergency. Appropriate portions of the plan shall be implemented in the event of an emergency.

**1910.1050(d)(1)(ii)** The plan shall specifically provide that employees engaged in correcting emergency conditions shall be equipped with the appropriate personal protective equipment and clothing as required in paragraphs (h) and (i) of this section until the emergency is abated.

**1910.1050(d)(1)(iii)** The plan shall specifically include provisions for alerting and evacuating affected employees as well as the elements prescribed in 29 CFR 1910.38 and 29 CFR 1910.39, "Emergency action plans" and "Fire prevention plans," respectively.

**1910.1050(d)(2)** *Alerting employees.* Where there is the possibility of employee exposure to MDA due to an emergency, means shall be developed to alert promptly those employees who have the potential to be directly exposed. Affected employees not engaged in correcting emergency conditions shall be evacuated immediately in the event that an emergency occurs. Means shall also be developed and implemented for alerting other employees who may be exposed as a result of the emergency.

**1910.1050(e)** *Exposure monitoring*—**(1)** *General.* **(i)** Determinations of employee exposure shall be made from breathing zone air samples that are representative of each employee's exposure to airborne MDA over an eight (8) hour period. Determination of employee exposure to the STEL shall be made from breathing zone air samples collected over a 15 minute sampling period.

**1910.1050(e)(1)(ii)** Representative employee exposure shall be determined on the basis of one or more samples representing full shift exposure for each shift for each job classification in each work area where exposure to MDA may occur.

**1910.1050(e)(1)(iii)** Where the employer can document that exposure levels are equivalent for similar operations in different work shifts, the employer shall only be required to determine representative employee exposure for that operation during one shift.

**1910.1050(e)(2)** *Initial monitoring.* Each employer who has a workplace or work operation covered by this standard shall perform initial monitoring to determine accurately the airborne concentrations of MDA to which employees may be exposed.

**1910.1050(e)(3)** *Periodic monitoring and monitoring frequency.* **(i)** If the monitoring required by paragraph (e)(2) of this section reveals employee exposure at or above the action level, but at or below the PELs, the employer shall repeat such representative monitoring for each such employee at least every six (6) months.

**1910.1050(e)(3)(ii)** If the monitoring required by paragraph (e)(2) of this section reveals employee exposure above the PELs, the employer shall repeat such monitoring for each such employee at least every three (3) months.

**1910.1050(e)(3)(iii)** The employer may alter the monitoring schedule from every three months to every six months for any employee for whom two consecutive measurements taken at least 7 days apart indicate that the employee exposure has decreased to below the TWA but above the action level.

**1910.1050(e)(4)** *Termination of monitoring.* **(i)** If the initial monitoring required by paragraph (e)(2) of this section reveals employee exposure to be below the action level, the employer may discontinue the monitoring for that employee, except as otherwise required by paragraph (e)(5) of this section.

**1910.1050(e)(4)(ii)** If the periodic monitoring required by paragraph (e)(3) of this section reveals that employee exposures, as indicated by at least two consecutive measurements taken at least 7 days apart, are below the action level the employer may discontinue the monitoring for that employee, except as otherwise required by paragraph (e)(5) of this section.

**1910.1050(e)(5)** *Additional monitoring.* The employer shall institute the exposure monitoring required under paragraphs (e)(2) and (e)(3) of this section when there has been a change in production process, chemicals present, control equipment, personnel, or work practices which may result in new or additional exposures to MDA, or when the employer has any reason to suspect a change which may result in new or additional exposures.

**1910.1050(e)(6)** *Accuracy of monitoring.* Monitoring shall be accurate, to a confidence level of 95 percent, to within plus or minus 25 percent for airborne concentrations of MDA.

**1910.1050(e)(7)** *Employee notification of monitoring results.* **(i)** The employer shall, within 15 working days after the receipt of the results of any monitoring performed under this standard, notify each employee of these results, in writing, either individually or by posting of results in an appropriate location that is accessible to affected employees.

**1910.1050(e)(7)(ii)** The written notification required by paragraph (e)(7)(i) of this section shall contain the corrective action being taken by the employer to reduce the employee exposure to or below the PELs, wherever the PELs are exceeded.

**1910.1050(e)(8)** *Visual monitoring.* The employer shall make routine inspections of employee hands, face and forearms potentially exposed to MDA. Other potential dermal exposures reported by the employee must be referred to the appropriate medical personnel for observation. If the employer determines that the employee has been exposed to MDA the employer shall:

**1910.1050(e)(8)(i)** Determine the source of exposure;

**1910.1050(e)(8)(ii)** Implement protective measures to correct the hazard; and

**1910.1050(e)(8)(iii)** Maintain records of the corrective actions in accordance with paragraph (n) of this section.

**1910.1050(f)** *Regulated areas*—**(1)** *Establishment*—**(i)** *Airborne exposures.* The employer shall establish regulated areas where airborne concentrations of MDA exceed or can reasonably be expected to exceed, the permissible exposure limits.

**1910.1050(f)(1)(ii)** *Dermal exposures.* Where employees are subject to dermal exposure to MDA the employer shall establish those work areas as regulated areas.

**1910.1050(f)(2)** *Demarcation.* Regulated areas shall be demarcated from the rest of the workplace in a manner that minimizes the number of persons potentially exposed.

**1910.1050(f)(3)** *Access.* Access to regulated areas shall be limited to authorized persons.

**1910.1050(f)(4)** *Personal protective equipment and clothing.* Each person entering a regulated area shall be supplied with, and required to use, the appropriate personal protective clothing and equipment in accordance with paragraphs (h) and (i) of this section.

**1910.1050(f)(5)** *Prohibited activities.* The employer shall ensure that employees do not eat, drink, smoke, chew tobacco or gum, or apply cosmetics in regulated areas.

**1910.1050(g)** *Methods of compliance*—**(1)** *Engineering controls and work practices.* **(i)** The employer shall institute engineering controls and work practices to reduce and maintain employee exposure to MDA at or below the PELs except to the extent that the employer can establish that these controls are not feasible or where the provisions of paragraph (g)(1)(ii) or (h)(1) (i) through (iv) of this section apply.

**1910.1050(g)(1)(ii)** Wherever the feasible engineering controls and work practices which can be instituted are not sufficient to reduce employee exposure to or below the PELs, the employer shall use them to reduce employee exposure to the lowest levels achievable by these controls and shall supplement them by the use of respiratory protective devices which comply with the requirements of paragraph (h) of this section.

**1910.1050(g)(2)** *Compliance program.* **(i)** The employer shall establish and implement a written program to reduce employee exposure to or below the PELs by means of engineering and work practice controls, as required by paragraph (g)(1) of this section, and by use of respiratory protection where permitted under this section. The program shall include a schedule for periodic maintenance (e.g., leak detection) and shall include the written plan for emergency situations as specified in paragraph (d) of this section.

**1910.1050(g)(2)(ii)** Upon request this written program shall be furnished for examination and copying to the Assistant Secretary, the Director, affected employees, and designated employee representatives. The employer shall review and, as necessary, update such plans at least once every 12 months to make certain they reflect the current status of the program.

**1910.1050(g)(3)** *Employee rotation.* Employee rotation shall not be permitted as a means of reducing exposure.

**1910.1050(h)** *Respiratory protection—***(1)** *General.* For employees who use respirators required by this section, the employer must provide respirators that comply with the requirements of this paragraph. Respirators must be used during:

**1910.1050(h)(1)(i)** Periods necessary to install or implement feasible engineering and work-practice controls.

**1910.1050(h)(1)(ii)** Work operations for which the employer establishes that engineering and work-practice controls are not feasible.

**1910.1050(h)(1)(iii)** Work operations for which feasible engineering and work-practice controls are not yet sufficient to reduce employee exposure to or below the PEL.

**1910.1050(h)(1)(iv)** Emergencies.

**1910.1050(h)(2)** *Respirator program.* The employer must implement a respiratory protection program in accordance with 29 CFR 1910.134 (b) through (d) (except (d)(1)(iii)), and (f) through (m).

**1910.1050(h)(3)** *Respirator selection.* **(i)** The employer must select, and ensure that employees use, the appropriate respirator from Table 1 in this section.

### TABLE 1—RESPIRATORY PROTECTION FOR MDA

| Airborne concentration of MDA or condition of use | Respirator type |
| --- | --- |
| a. Less than or equal to 10 × PEL | (1) Half-Mask Respirator with HEPA[1] Cartridge.[2] |
| b. Less than or equal to 50 × PEL | (1) Full facepiece Respirator with HEPA[1] Cartridge or Canister.[2] |
| c. Less than or equal to 1000 × PEL | (1) Full facepiece powered air-purifying respirator with HEPA[1] cartridges.[2] |
| d. Greater than 1000 × PEL or unknown concentrations. | (1) Self-contained breathing apparatus with full facepiece in positive pressure mode. (2) Full facepiece positive pressure demand supplied-air respirator with auxiliary self-contained air supply. |
| e. Escape | (1) Any full facepiece air-purifying respirator with HEPA[1] cartridges;[2] (2) Any positive pressure or continuous flow self-contained breathing apparatus with full facepiece or hood. |
| f. Firefighting | (1) Full facepiece self-contained breathing apparatus in positive pressure demand mode. |

NOTE: Respirators assigned for higher environmental concentrations may be used at lower concentrations.

[1] High Efficiency Particulate in Air filter (HEPA) means a filter that is at least 99.97 percent efficient against mono-dispersed particles of 0.3 micrometers or larger.

[2] Combination HEPA/Organic Vapor Cartridges shall be used whenever MDA in liquid form or a process requiring heat is used.

**1910.1050(h)(3)(ii)** Any employee who cannot use a negative-pressure respirator must be given the option of using a positive-pressure respirator, or a supplied-air respirator operated in the continuous-flow or pressure-demand mode.

**1910.1050(i)** *Protective work clothing and equipment—***(1)** *Provision and use.* Where employees are subject to dermal exposure to MDA, where liquids containing MDA can be splashed into the eyes, or where airborne concentrations of MDA are in excess of the PEL, the employer shall provide, at no cost to the employee, and ensure that the employee uses, appropriate protective work clothing and equipment which prevent contact with MDA such as, but not limited to:

**1910.1050(i)(1)(i)** Aprons, coveralls or other full-body work clothing;

**1910.1050(i)(1)(ii)** Gloves, head coverings, and foot coverings; and

**1910.1050(i)(1)(iii)** Face shields, chemical goggles; or

**1910.1050(i)(1)(iv)** Other appropriate protective equipment which comply with §1910.133.

**1910.1050(i)(2)** *Removal and storage.* **(i)** The employer shall ensure that, at the end of their work shift, employees remove MDA-contaminated protective work clothing and equipment that is not routinely removed throughout the day in change rooms provided in accordance with the provisions established for change rooms.

**1910.1050(i)(2)(ii)** The employer shall ensure that, during their work shift, employees remove all other MDA-contaminated protective work clothing or equipment before leaving a regulated area.

**1910.1050(i)(2)(iii)** The employer shall ensure that no employee takes MDA-contaminated work clothing or equipment out of the change room, except those employees authorized to do so for the purpose of laundering, maintenance, or disposal.

**1910.1050(i)(2)(iv)** MDA-contaminated work clothing or equipment shall be placed and stored in closed containers which prevent dispersion of the MDA outside the container.

**1910.1050(i)(2)(v)** Containers of MDA-contaminated protective work clothing or equipment which are to be taken out of change rooms or the workplace for cleaning, maintenance, or disposal, shall bear labels warning of the hazards of MDA.

**1910.1050(i)(3)** *Cleaning and replacement.* **(i)** The employer shall provide the employee with clean protective clothing and equipment. The employer shall ensure that protective work clothing or equipment required by this paragraph is cleaned, laundered, repaired, or replaced at intervals appropriate to maintain its effectiveness.

**1910.1050(i)(3)(ii)** The employer shall prohibit the removal of MDA from protective work clothing or equipment by blowing, shaking, or any methods which allow MDA to re-enter the workplace.

**1910.1050(i)(3)(iii)** The employer shall ensure that laundering of MDA-contaminated clothing shall be done so as to prevent the release of MDA in the workplace.

**1910.1050(i)(3)(iv)** Any employer who gives MDA-contaminated clothing to another person for laundering shall inform such person of the requirement to prevent the release of MDA.

**1910.1050(i)(3)(v)** The employer shall inform any person who launders or cleans protective clothing or equipment contaminated with MDA of the potentially harmful effects of exposure.

# OSHA Standards for General Industry

**1910.1050(i)(3)(vi)** MDA-contaminated clothing shall be transported in properly labeled, sealed, impermeable bags or containers.

**1910.1050(j)** *Hygiene facilities and practices*—**(1)** *Change rooms.* **(i)** The employer shall provide clean change rooms for employees, who must wear protective clothing, or who must use protective equipment because of their exposure to MDA.

**1910.1050(j)(1)(ii)** Change rooms must be equipped with separate storage for protective clothing and equipment and for street clothes which prevents MDA contamination of street clothes.

**1910.1050(j)(2)** *Showers.* **(i)** The employer shall ensure that employees, who work in areas where there is the potential for exposure resulting from airborne MDA (e.g., particulates or vapors) above the action level, shower at the end of the work shift.

**1910.1050(j)(2)(i)(A)** Shower facilities required by this paragraph shall comply with §1910.141(d)(3).

**1910.1050(j)(2)(i)(B)** The employer shall ensure that employees who are required to shower pursuant to the provisions contained herein do not leave the workplace wearing any protective clothing or equipment worn during the work shift.

**1910.1050(j)(2)(ii)** Where dermal exposure to MDA occurs, the employer shall ensure that materials spilled or deposited on the skin are removed as soon as possible by methods which do not facilitate the dermal absorption of MDA.

**1910.1050(j)(3)** *Lunch facilities*—**(i)** *Availability and construction.* **(A)** Whenever food or beverages are consumed at the worksite and employees are exposed to MDA at or above the PEL or are subject to dermal exposure to MDA the employer shall provide readily accessible lunch areas.

**1910.1050(j)(3)(i)(B)** Lunch areas located within the workplace and in areas where there is the potential for airborne exposure to MDA at or above the PEL shall have a positive pressure, temperature controlled, filtered air supply.

**1910.1050(j)(3)(i)(C)** Lunch areas may not be located in areas within the workplace where the potential for dermal exposure to MDA exists.

**1910.1050(j)(3)(ii)** The employer shall ensure that employees who have been subjected to dermal exposure to MDA or who have been exposed to MDA above the PEL wash their hands and faces with soap and water prior to eating, drinking, smoking, or applying cosmetics.

**1910.1050(j)(3)(iii)** The employer shall ensure that employees exposed to MDA do not enter lunch facilities with MDA-contaminated protective work clothing or equipment.

**1910.1050(k)** *Communication of hazards to employees*—**(1)** *Signs and labels.* **(i)** The employer shall post and maintain legible signs demarcating regulated areas and entrances or accessways to regulated areas that bear the following legend:

**DANGER**

**MDA MAY CAUSE CANCER LIVER TOXIN**

**AUTHORIZED PERSONNEL ONLY**

**RESPIRATORS AND PROTECTIVE CLOTHING**

**MAY BE REQUIRED TO BE WORN IN THIS AREA**

**1910.1050(k)(1)(ii)** The employer shall ensure that labels or other appropriate forms of warning are provided for containers of MDA within the workplace. The labels shall comply with the requirements of 29 CFR 1910.1200(f) and shall include the following legend:

**1910.1050(k)(1)(ii)(A)** For Pure MDA

**DANGER**

**CONTAINS MDA**

**MAY CAUSE CANCER LIVER TOXIN**

**1910.1050(k)(1)(ii)(B)** For mixtures containing MDA

**DANGER**

**CONTAINS MDA**

**CONTAINS MATERIALS WHICH MAY CAUSE**

**CANCER LIVER TOXIN**

**1910.1050(k)(2)** *Material safety data sheets* (MSDS). **(i)** Employers shall obtain or develop, and shall provide access to their employees, to a material safety data sheet (MSDS) for MDA. In meeting this obligation, employers shall make appropriate use of the information found in Appendices A and B.

**1910.1050(k)(2)(ii)** Employers who are manufacturers or importers shall:

**1910.1050(k)(2)(ii)(A)** Comply with paragraph (k) (1) (ii) of this section as appropriate, and

**1910.1050(k)(2)(ii)(B)** Comply with the requirement in OSHA's Hazard Communication standard, 29 CFR 1910.1200, that they deliver to downstream employers an MSDS for MDA.

**1910.1050(k)(3)** *Information and training.* **(i)** The employer shall provide employees with information and training on MDA, in accordance with 29 CFR 1910.1200(h), at the time of initial assignment and at least annually thereafter.

**1910.1050(k)(3)(ii)** In addition to the information required under 29 CFR 1910.1200, the employer shall:

**1910.1050(k)(3)(ii)(A)** Provide an explanation of the contents of this section, including appendices A and B, and indicate to employees where a copy of the standard is available;

**1910.1050(k)(3)(ii)(B)** Describe the medical surveillance program required under paragraph (m) of this section, and explain the information contained in Appendix C; and

**1910.1050(k)(3)(ii)(C)** Describe the medical removal provision required under paragraph (m) of this section.

**1910.1050(k)(4)** *Access to training materials.* **(i)** The employer shall make readily available to all affected employees, without cost, all written materials relating to the employee training program, including a copy of this regulation.

**1910.1050(k)(4)(ii)** The employer shall provide to the Assistant Secretary and the Director, upon request, all information and training materials relating to the employee information and training program.

**1910.1050(l)** *Housekeeping.* **(1)** All surfaces shall be maintained as free as practicable of visible accumulations of MDA.

**1910.1050(l)(2)** The employer shall institute a program for detecting MDA leaks, spills, and discharges, including regular visual inspections of operations involving liquid or solid MDA.

**1910.1050(l)(3)** All leaks shall be repaired and liquid or dust spills cleaned up promptly.

**1910.1050(l)(4)** Surfaces contaminated with MDA may not be cleaned by the use of compressed air.

**1910.1050(l)(5)** Shoveling, dry sweeping, and other methods of dry cleanup of MDA may be used where HEPA-filtered vacuuming and/or wet cleaning are not feasible or practical.

**1910.1050(l)(6)** Waste, scrap, debris, bags, containers, equipment, and clothing contaminated with MDA shall be collected and disposed of in a manner to prevent the re-entry of MDA into the workplace.

**1910.1050(m)** *Medical surveillance*—**(1)** *General.* **(i)** The employer shall make available a medical surveillance program for employees exposed to MDA:

**1910.1050(m)(1)(i)(A)** Employees exposed at or above the action level for 30 or more days per year;

**1910.1050(m)(1)(i)(B)** Employees who are subject to dermal exposure to MDA for 15 or more days per year;

**1910.1050(m)(1)(i)(C)** Employees who have been exposed in an emergency situation;

**1910.1050(m)(1)(i)(D)** Employees whom the employer, based on results from compliance with paragraph (e)(8) of this section, has reason to believe are being dermally exposed; and

**1910.1050(m)(1)(i)(E)** Employees who show signs or symptoms of MDA exposure.

**1910.1050(m)(1)(ii)** The employer shall ensure that all medical examinations and procedures are performed by, or under the supervision of, a licensed physician, at a reasonable time and place, and provided without cost to the employee.

**1910.1050(m)(2)** *Initial examinations.* **(i)** Within 150 days of the effective date of this standard, or before the time of initial assignment, the employer shall provide each employee covered by paragraph (m)(1)(i) of this section with a medical examination including the following elements:

**1910.1050(m)(2)(i)(A)** A detailed history which includes:

**1910.1050(m)(2)(i)(A)***1)* Past work exposure to MDA or any other toxic substances;

**1910.1050(m)(2)(i)(A)***2)* A history of drugs, alcohol, tobacco, and medication routinely taken (duration and quantity); and

**1910.1050(m)(2)(i)(A)***3)* A history of dermatitis, chemical skin sensitization, or previous hepatic disease.

**1910.1050(m)(2)(i)(B)** A physical examination which includes all routine physical examination parameters, skin examination, and signs of liver disease.

**1910.1050(m)(2)(i)(C)** Laboratory tests including:

**1910.1050(m)(2)(i)(C)***1)* Liver function tests and

**1910.1050(m)(2)(i)(C)***2)* Urinalysis.

**1910.1050(m)(2)(i)(D)** Additional tests as necessary in the opinion of the physician.

**1910.1050(m)(2)(ii)** No initial medical examination is required if adequate records show that the employee has been examined in accordance with the requirements of this section within the previous six months prior to the effective date of this standard or prior to the date of initial assignment.

**1910.1050(m)(3)** *Periodic examinations.* **(i)** The employer shall provide each employee covered by this section with a medical examination at least annually following the initial examination. These periodic examinations shall include at least the following elements:

**1910.1050(m)(3)(i)(A)** A brief history regarding any new exposure to potential liver toxins, changes in drug, tobacco, and alcohol intake, and the appearance of physical signs relating to the liver, and the skin;

**1910.1050(m)(3)(i)(B)** The appropriate tests and examinations including liver function tests and skin examinations; and

**1910.1050(m)(3)(i)(C)** Appropriate additional tests or examinations as deemed necessary by the physician.

**1910.1050(m)(3)(ii)** If in the physicians' opinion the results of liver function tests indicate an abnormality, the employee shall be removed from further MDA exposure in accordance with paragraph (m)(9) of this section. Repeat liver function tests shall be conducted on advice of the physician.

**1910.1050(m)(4)** *Emergency examinations.* If the employer determines that the employee has been exposed to a potentially hazardous amount of MDA in an emergency situation as addressed in paragraph (d) of this section, the employer shall provide medical examinations in accordance with paragraphs (m)(3)(i) and (ii) of this section. If the results of liver function testing indicate an abnormality, the employee shall be removed in accordance with paragraph (m)(9) of this section. Repeat liver function tests shall be conducted on the advice of the physician. If the results of the tests are normal, tests must be repeated two to three weeks from the initial testing. If the results of the second set of tests are normal and, on the advice of the physician, no additional testing is required.

**1910.1050(m)(5)** *Additional examinations.* Where the employee develops signs and symptoms associated with exposure to MDA, the employer shall provide the employee with an additional medical examination including a liver function test. Repeat liver function tests shall be conducted on the advice of the physician. If the results of the tests are normal, tests must be repeated two to three weeks from the initial testing. If the results of the second set of tests are normal and, on the advice of the physician, no additional testing is required.

**1910.1050(m)(6)** *Multiple physician review mechanism.* **(i)** If the employer selects the initial physician who conducts any medical examination or consultation provided to an employee under this section, and the employee has signs or symptoms of occupational exposure to MDA (which could include an abnormal liver function test), and the employee disagrees with the opinion of the examining physician, and this opinion could affect the employee's job status, the employee may designate an appropriate, mutually acceptable second physician:

**1910.1050(m)(6)(i)(A)** To review any findings, determinations, or recommendations of the initial physician; and

**1910.1050(m)(6)(i)(B)** To conduct such examinations, consultations, and laboratory tests as the second physician deems necessary to facilitate this review.

**1910.1050(m)(6)(ii)** The employer shall promptly notify an employee of the right to seek a second medical opinion after each occasion that an initial physician conducts a medical examination or consultation pursuant to this section. The employer may condition its participation in, and payment for, the multiple physician review mechanism upon the employee doing the following within fifteen (15) days after receipt of the foregoing notification, or receipt of the initial physician's written opinion, whichever is later:

**1910.1050(m)(6)(ii)(A)** The employee informing the employer that he or she intends to seek a second medical opinion, and

**1910.1050(m)(6)(ii)(B)** The employee initiating steps to make an appointment with a second physician.

**1910.1050(m)(6)(iii)** If the findings, determinations, or recommendations of the second physician differ from those of the initial physician, then the employer and the employee shall assure that efforts are made for the two physicians to resolve any disagreement.

**1910.1050(m)(6)(iv)** If the two physicians have been unable to resolve quickly their disagreement, then the employer and the employee through their respective physicians shall designate a third physician;

**1910.1050(m)(6)(iv)(A)** To review any findings, determinations, or recommendations of the prior physicians; and

**1910.1050(m)(6)(iv)(B)** To conduct such examinations, consultations, laboratory tests, and discussions with the prior physicians as the third physician deems necessary to resolve the disagreement of the prior physicians.

**1910.1050(m)(6)(v)** The employer shall act consistent with the findings, determinations, and recommendations of the third physician, unless the employer and the employee reach an agreement which is otherwise consistent with the recommendations of at least one of the three physicians.

**1910.1050(m)(7)** *Information provided to the examining and consulting physicians.* **(i)** The employer shall provide the following information to the examining physician:

**1910.1050(m)(7)(i)(A)** A copy of this regulation and its appendices;

**1910.1050(m)(7)(i)(B)** A description of the affected employee's duties as they relate to the employee's potential exposure to MDA;

**1910.1050(m)(7)(i)(C)** The employee's current actual or representative MDA exposure level;

**1910.1050(m)(7)(i)(D)** A description of any personal protective equipment used or to be used; and

**1910.1050(m)(7)(i)(E)** Information from previous employment-related medical examinations of the affected employee.

**1910.1050(m)(7)(ii)** The employer shall provide the foregoing information to a second physician under this section upon request either by the second physician, or by the employee.

**1910.1050(m)(8)** *Physician's written opinion.* **(i)** For each examination under this section, the employer shall obtain, and provide the employee with a copy of, the examining physician's written opinion within 15 days of its receipt. The written opinion shall include the following:

**1910.1050(m)(8)(i)(A)** The occupationally-pertinent results of the medical examination and tests;

**1910.1050(m)(8)(i)(B)** The physician's opinion concerning whether the employee has any detected medical conditions which would place the employee at increased risk of material impairment of health from exposure to MDA;

**1910.1050(m)(8)(i)(C)** The physician's recommended limitations upon the employee's exposure to MDA or upon the employee's use of protective clothing or equipment and respirators; and

**1910.1050(m)(8)(i)(D)** A statement that the employee has been informed by the physician of the results of the medical examination and any medical conditions resulting from MDA exposure which require further explanation or treatment.

**1910.1050(m)(8)(ii)** The written opinion obtained by the employer shall not reveal specific findings or diagnoses unrelated to occupational exposures.

**1910.1050(m)(9)** *Medical removal*—**(i)** *Temporary medical removal of an employee*—**(A)** *Temporary removal resulting from occupational exposure.* The employee shall be removed from work environments in which exposure to MDA is at or above the action level or where dermal exposure to MDA may occur, following an initial examination (paragraph (m)(2) of this section), periodic examinations (paragraph (m)(3) of this section), an emergency situation paragraph (m)(4) of this section, or an additional examination (paragraph (m)(5) of this section) in the following circumstances:

**1910.1050(m)(9)(i)(A)**(1) When the employee exhibits signs and/or symptoms indicative of acute exposure to MDA; or

**1910.1050(m)(9)(i)(A)**(2) When the examining physician determines that an employee's abnormal liver function tests are not associated with MDA exposure but that the abnormalities may be exacerbated as a result of occupational exposure to MDA.

**1910.1050(m)(9)(i)(B)** *Temporary removal due to a final medical determination.* **(1)** The employer shall remove an employee from work environments in which exposure to MDA is at or above the action level or where dermal exposure to MDA may occur, on each occasion that there is a final medical determination or opinion that the employee has a detected medical condition which places the employee at increased risk of material impairment to health from exposure to MDA.

**1910.1050(m)(9)(i)(B)**(2) For the purposes of this section, the phrase "final medical determination" shall mean the outcome of the physician review mechanism used pursuant to the medical surveillance provisions of this section.

**1910.1050(m)(9)(i)(B)**(3) Where a final medical determination results in any recommended special protective measures for an employee, or limitations on an employee's exposure to MDA, the employer shall implement and act consistent with the recommendation.

**1910.1050(m)(9)(ii)** *Return of the employee to former job status.* **(A)** The employer shall return an employee to his or her former job status:

**1910.1050(m)(9)(ii)(A)**(1) When the employee no longer shows signs or symptoms of exposure to MDA, or upon the advice of the physician.

**1910.1050(m)(9)(ii)(A)**(2) When a subsequent final medical determination results in a medical finding, determination, or opinion that the employee no longer has a detected medical condition which places the employee at increased risk of material impairment to health from exposure to MDA.

**1910.1050(m)(9)(ii)(B)** For the purposes of this section, the requirement that an employer return an employee to his or her former job status is not intended to expand upon or restrict any rights an employee has or would have had, absent temporary medical removal, to a specific job classification or position under the terms of a collective bargaining agreement.

**1910.1050(m)(9)(iii)** *Removal of other employee special protective measure or limitations.* The employer shall remove any limitations placed on an employee, or end any special protective measures provided to an employee, pursuant to a final medical determination, when a subsequent final medical determination indicates that the limitations or special protective measures are no longer necessary.

**1910.1050(m)(9)(iv)** *Employer options pending a final medical determination.* Where the physician review mechanism used pursuant to the medical surveillance provisions of this section, has not yet resulted in a final medical determination with respect to an employee, the employer shall act as follows:

**1910.1050(m)(9)(iv)(A)** *Removal.* The employer may remove the employee from exposure to MDA, provide special protective measures to the employee, or place limitations upon the employee, consistent with the medical findings, determinations, or recommendations of any of the physicians who have reviewed the employee's health status.

**1910.1050(m)(9)(iv)(B)** *Return.* The employer may return the employee to his or her former job status, and end any special protective measures provided to the employee, consistent with the medical findings, determinations, or recommendations of any of the physicians who have reviewed the employee's health status, with two exceptions.

**1910.1050(m)(9)(iv)(B)**(1) If the initial removal, special protection, or limitation of the employee resulted from a final medical determination which differed from the findings, determinations, or recommendations of the initial physician; or

**1910.1050(m)(9)(iv)(B)**(2) If the employee has been on removal status for the preceding six months as a result of exposure to MDA, then the employer shall await a final medical determination.

**1910.1050(m)(9)(v)** *Medical removal protection benefits—***(A)** *Provisions of medical removal protection benefits.* The employer shall provide to an employee up to six (6) months of medical removal protection benefits on each occasion that an employee is removed from exposure to MDA or otherwise limited pursuant to this section.

**1910.1050(m)((9)(v)(B)** *Definition of medical removal protection benefits.* For the purposes of this section, the requirement that an employer provide medical removal protection benefits means that the employer shall maintain the earnings, seniority, and other employment rights and benefits of an employee as though the employee had not been removed from normal exposure to MDA or otherwise limited.

**1910.1050(m)(9)(v)(C)** *Follow-up medical surveillance during the period of employee removal or limitations.* During the period of time that an employee is removed from normal exposure to MDA or otherwise limited, the employer may condition the provision of medical removal protection benefits upon the employee's participation in follow-up medical surveillance made available pursuant to this section.

**1910.1050(m)(9)(v)(D)** *Workers' compensation claims.* If a removed employee files a claim for workers' compensation payments for a MDA-related disability, then the employer shall continue to provide medical removal protection benefits pending disposition of the claim. To the extent that an award is made to the employee for earnings lost during the period of removal, the employer's medical removal protection obligation shall be reduced by such amount. The employer shall receive no credit for workers' compensation payments received by the employee for treatment-related expenses.

**1910.1050(m)(9)(v)(E)** *Other credits.* The employer's obligation to provide medical removal protection benefits to a removed employee shall be reduced to the extent that the employee receives compensation for earnings lost during the period of removal either from a publicly or employer-funded compensation program, or receives income from non-MDA-related employment with any employer made possible by virtue of the employee's removal.

**1910.1050(m)(9)(v)(F)** *Employees who do not recover within the 6 months of removal.* The employer shall take the following measures with respect to any employee removed from exposure to MDA:

**1910.1050(m)(9)(v)(F)**(1) The employer shall make available to the employee a medical examination pursuant to this section to obtain a final medical determination with respect to the employee;

**1910.1050(m)(9)(v)(F)**(2) The employer shall assure that the final medical determination obtained indicates whether or not the employee may be returned to his or her former job status, and, if not, what steps should be taken to protect the employee's health;

**1910.1050(m)(9)(v)(F)**(3) Where the final medical determination has not yet been obtained, or, once obtained indicates that the employee may not yet be returned to his or her former job status, the employer shall continue to provide medical removal protection benefits to the employee until either the employee is returned to former job status, or a final medical determination is made that the employee is incapable of ever safely returning to his or her former job status; and

**1910.1050(m)(9)(v)(F)**(4) Where the employer acts pursuant to a final medical determination which permits the return of the employee to his or her former job status, despite what would otherwise be an abnormal liver function test, later questions concerning removing the employee again shall be decided by a final medical determination. The employer need not automatically remove such an employee pursuant to the MDA removal criteria provided by this section.

**1910.1050(m)(9)(vi)** *Voluntary removal or restriction of an employee.* Where an employer, although not required by this section to do so, removes an employee from exposure to MDA or otherwise places limitations on an employee due to the effects of MDA exposure on the employee's medical condition, the employer shall provide medical removal protection benefits to the employee equal to that required by paragraph (m)(9)(v) of this section.

**1910.1050(n)** *Recordkeeping—***(1)** *Monitoring data for exempted employers.* **(i)** Where as a result of the initial monitoring the processing, use, or handling of products made from or containing MDA are exempted from other requirements of this section under paragraph (a)(2) of this section, the employer shall establish and maintain an accurate record of monitoring relied on in support of the exemption.

**1910.1050(n)(1)(ii)** This record shall include at least the following information:

**1910.1050(n)(1)(ii)(A)** The product qualifying for exemption;

**1910.1050(n)(1)(ii)(B)** The source of the monitoring data (e.g., was monitoring performed by the employer or a private contractor);

**1910.1050(n)(1)(ii)(C)** The testing protocol, results of testing, and/or analysis of the material for the release of MDA;

**1910.1050(n)(1)(ii)(D)** A description of the operation exempted and how the data support the exemption (e.g., are the monitoring data representative of the conditions at the affected facility); and

**1910.1050(n)(1)(ii)(E)** Other data relevant to the operations, materials, processing, or employee exposures covered by the exemption.

**1910.1050(n)(1)(iii)** The employer shall maintain this record for the duration of the employer's reliance upon such objective data.

**1910.1050(n)(2)** *Objective data for exempted employers.* **(i)** Where the processing, use, or handling of products made from or containing MDA are exempted from other requirements of this section under paragraph (a) of this section, the employer shall establish and maintain an accurate record of objective data relied upon in support of the exemption.

**1910.1050(n)(2)(ii)** This record shall include at least the following information:

**1910.1050(n)(2)(ii)(A)** The product qualifying for exemption;

**1910.1050(n)(2)(ii)(B)** The source of the objective data;

**1910.1050(n)(2)(ii)(C)** The testing protocol, results of testing, and/or analysis of the material for the release of MDA;

**1910.1050(n)(2)(ii)(D)** A description of the operation exempted and how the data support the exemption; and

**1910.1050(n)(2)(ii)(E)** Other data relevant to the operations, materials, processing, or employee exposures covered by the exemption.

**1910.1050(n)(2)(iii)** The employer shall maintain this record for the duration of the employer's reliance upon such objective data.

**1910.1050(n)(3)** *Exposure measurements.* **(i)** The employer shall establish and maintain an accurate record of all measurements required by paragraph (e) of this section, in accordance with 29 CFR 1910.1020.

**1910.1050(n)(3)(ii)** This record shall include:

**1910.1050(n)(3)(ii)(A)** The dates, number, duration, and results of each of the samples taken, including a description of the procedure used to determine representative employee exposures;

**1910.1050(n)(3)(ii)(B)** Identification of the sampling and analytical methods used;

**1910.1050(n)(3)(ii)(C)** A description of the type of respiratory protective devices worn, if any; and

**1910.1050(n)(3)(ii)(D)** The name, social security number, job classification and exposure levels of the employee monitored and all other employees whose exposure the measurement is intended to represent.

**1910.1050(n)(3)(iii)** The employer shall maintain this record for at least 30 years, in accordance with 29 CFR 1910.1020.

**1910.1050(n)(4)** *Medical surveillance.* **(i)** The employer shall establish and maintain an accurate record for each employee subject to medical surveillance required by paragraph (m) of this section, in accordance with 29 CFR 1910.1020.

**1910.1050(n)(4)(ii)** This record shall include:

**1910.1050(n)(4)(ii)(A)** The name, social security number and description of the duties of the employee;

**1910.1050(n)(4)(ii)(B)** The employer's copy of the physician's written opinion on the initial, periodic, and any special examinations, including results of medical examination and all tests, opinions, and recommendations;

**1910.1050(n)(4)(ii)(C)** Results of any airborne exposure monitoring done for that employee and the representative exposure levels supplied to the physician; and

**1910.1050(n)(4)(ii)(D)** Any employee medical complaints related to exposure to MDA;

**1910.1050(n)(4)(iii)** The employer shall keep, or assure that the examining physician keeps, the following medical records:

**1910.1050(n)(4)(iii)(A)** A copy of this standard and its appendices, except that the employer may keep one copy of the standard and its appendices for all employees provided the employer references the standard and its appendices in the medical surveillance record of each employee;

**1910.1050(n)(4)(iii)(B)** A copy of the information provided to the physician as required by any paragraphs in the regulatory text;

**1910.1050(n)(4)(iii)(C)** A description of the laboratory procedures and a copy of any standards or guidelines used to interpret the test results or references to the information;

**1910.1050(n)(4)(iii)(D)** A copy of the employee's medical and work history related to exposure to MDA; and

**1910.1050(n)(4)(iv)** The employer shall maintain this record for at least the duration of employment plus 30 years, in accordance with 29 CFR 1910.1020.

**1910.1050(n)(5)** *Medical removals.* **(i)** The employer shall establish and maintain an accurate record for each employee removed from current exposure to MDA pursuant to paragraph (m) of this section.

**1910.1050(n)(5)(ii)** Each record shall include:

**1910.1050(n)(5)(ii)(A)** The name and social security number of the employee;

**1910.1050(n)(5)(ii)(B)** The date of each occasion that the employee was removed from current exposure to MDA as well as the corresponding date on which the employee was returned to his or her former job status;

**1910.1050(n)(5)(ii)(C)** A brief explanation of how each removal was or is being accomplished; and

**1910.1050(n)(5)(ii)(D)** A statement with respect to each removal indicating the reason for the removal.

**1910.1050(n)(5)(iii)** The employer shall maintain each medical removal record for at least the duration of an employee's employment plus 30 years.

**1910.1050(n)(6)** *Availability.* **(i)** The employer shall assure that records required to be maintained by this section shall be made available, upon request, to the Assistant Secretary and the Director for examination and copying.

**1910.1050(n)(6)(ii)** Employee exposure monitoring records required by this section shall be provided upon request for examination and copying to employees, employee representatives, and the Assistant Secretary in accordance with 29 CFR 1910.1020(a)-(e) and (g)-(i).

**1910.1050(n)(6)(iii)** Employee medical records required by this section shall be provided upon request for examination and copying, to the subject employee, to anyone having the specific written consent of the subject employee, and to the Assistant Secretary in accordance with 29 CFR 1910.1020.

**1910.1050(n)(7)** *Transfer of records.* **(i)** The employer shall comply with the requirements involving transfer of records set forth in 29 CFR 1910.1020(h).

**1910.1050(n)(7)(ii)** If the employer ceases to do business and there is no successor employer to receive and retain the records for the prescribed period, the employer shall notify the Director, at least 90 days prior to disposal, and transmit the records to the Director if so requested by the Director within that period.

**1910.1050(o)** *Observation of monitoring*—**(1)** *Employee observation.* The employer shall provide affected employees, or their designated representatives, an opportunity to observe the measuring or monitoring of employee exposure to MDA conducted pursuant to paragraph (e) of this section.

**1910.1050(o)(2)** *Observation procedures.* When observation of the measuring or monitoring of employee exposure to MDA requires entry into areas where the use of protective clothing and equipment or respirators is required, the employer shall provide the observer with personal protective clothing and equipment or respirators required to be worn by employees working in the area, assure the use of such clothing and equipment or respirators, and require the observer to comply with all other applicable safety and health procedures.

**1910.1050(p)** [Reserved.]

**1910.1050(q)** *Appendices.* The information contained in appendices A, B, C and D of this section is not intended by itself, to create any additional obligations not otherwise imposed by this standard nor detract from any existing obligation.

## APPENDIX A TO §1910.1050—SUBSTANCE DATA SHEET, FOR 4,4'-METHYLENEDIANILINE

### I. Substance Identification

A. Substance: Methylenedianiline (MDA)

B. Permissible Exposure:

1. Airborne: Ten parts per billion parts of air (10 ppb), time-weighted average (TWA) for an 8-hour workday and an action level of five parts per billion parts of air (5 ppb).

2. *Dermal:* Eye contact and skin contact with MDA are not permitted.

C. *Appearance and odor:* White to tan solid; amine odor

### II. Health Hazard Data

A. *Ways in which MDA affects your health.* MDA can affect your health if you inhale it, or if it comes in contact with your skin or eyes. MDA is also harmful if you happen to swallow it. Do not get MDA in eyes, on skin, or on clothing.

B. *Effects of overexposure.* 1. Short-term (acute) overexposure: Overexposure to MDA may produce fever, chills, loss of appetite, vomiting, jaundice. Contact may irritate skin, eyes and mucous membranes. Sensitization may occur.

2. *Long-term (chronic) exposure.* Repeated or prolonged exposure to MDA, even at relatively low concentrations, may cause cancer. In addition, damage to the liver, kidneys, blood, and spleen may occur with long term exposure.

3. *Reporting signs and symptoms.* You should inform your employer if you develop any signs or symptoms which you suspect are caused by exposure to MDA including yellow staining of the skin.

### III. Protective Clothing and Equipment

A. *Respirators.* Respirators are required for those operations in which engineering controls or work-practice controls are not adequate or feasible to reduce exposure to the permissible limit. If respirators are worn, they must have a label issued by the National Institute for Occupational Safety and Health under the provisions of 42 CFR part 84 stating that the respirators have been approved for this purpose, and cartridges and canisters must be replaced in accordance with the requirements of 29 CFR 1910.134. If you experience difficulty breathing while wearing a respirator, you can request a positivepressure respirator from your employer. You must be thoroughly trained to use the assigned respirator, and the training must be provided by your employer.

MDA does not have a detectable odor except at levels well above the permissible exposure limits. Do not depend on odor to warn you when a respirator canister is exhausted. If you can smell MDA while wearing a respirator, proceed immediately to fresh air. If you experience difficulty breathing while wearing a respirator, tell your employer.

B. *Protective Clothing.* You may be required to wear coveralls, aprons, gloves, face shields, or other appropriate protective clothing to prevent skin contact with MDA. Where protective clothing is required, your employer is required to provide clean garments to you, as necessary, to assure that the clothing protects you adequately. Replace or repair impervious clothing that has developed leaks.

MDA should never be allowed to remain on the skin. Clothing and shoes which are not impervious to MDA should not be allowed to become contaminated with MDA, and if they do, the clothing and shoes should be promptly removed and decontaminated. The clothing should be laundered to remove MDA or discarded. Once MDA penetrates shoes or other leather articles, they should not be worn again.

C. *Eye protection.* You must wear splashproof safety goggles in areas where liquid MDA may contact your eyes. Contact lenses should not be worn in areas where eye contact with MDA can occur. In addition, you must wear a face shield if your face could be splashed with MDA liquid.

### IV. Emergency and First Aid Procedures

A. *Eye and face exposure.* If MDA is splashed into the eyes, wash the eyes for at least 15 minutes. See a doctor as soon as possible.

B. *Skin exposure.* If MDA is spilled on your clothing or skin, remove the contaminated clothing and wash the exposed skin with large amounts of soap and water immediately. Wash contaminated clothing before you wear it again.

C. *Breathing.* If you or any other person breathes in large amounts of MDA, get the exposed person to fresh air at once. Apply artificial respiration if breathing has stopped. Call for medical assistance or a doctor as soon as possible. Never enter any vessel or confined space where the MDA concentration might be high without proper safety equipment and at least one other person present who will stay outside. A life line should be used.

D. *Swallowing.* If MDA has been swallowed and the patient is conscious, do not induce vomiting. Call for medical assistance or a doctor immediately.

### V. Medical Requirements

If you are exposed to MDA at a concentration at or above the action level for more than 30 days per year, or exposed to liquid mixtures more than 15 days per year, your employer is required to provide a medical examination, including a medical history and laboratory tests, within 60 days of the effective date of this standard and annually thereafter. These tests shall be provided without cost to you. In addition, if you are accidentally exposed to MDA (either by ingestion, inhalation, or skin/eye contact) under conditions known or suspected to constitute toxic exposure to MDA, your employer is required to make special examinations and tests available to you.

### VI. Observation of Monitoring

Your employer is required to perform measurements that are representative of your exposure to MDA and you or your designated representative are entitled to observe the monitoring procedure. You are entitled to observe the steps taken in the measurement procedure and to record the results obtained. When the monitoring procedure is taking place in an area where respirators or personal protective clothing and equipment are required to be worn, you and your representative must also be provided with, and must wear, the protective clothing and equipment.

### VII. Access to Records

You or your representative are entitled to see the records of measurements of your exposure to MDA upon written request to your employer. Your medical examination records can be furnished to your physician or designated representative upon request by you to your employer.

### VIII. Precautions for Safe Use, Handling and Storage

A. *Material is combustible.* Avoid strong acids and their anhydrides. Avoid strong oxidants. Consult supervisor for disposal requirements.

B. *Emergency clean-up.* Wear self-contained breathing apparatus and fully clothe the body in the appropriate personal protective clothing and equipment.

## APPENDIX B TO §1910.1050— SUBSTANCE TECHNICAL GUIDELINES, MDA

### I. Identification

A. Substance identification.

1. Synonyms: CAS No. 101-77-9. 4,4'-methylenedianiline; 4,4'methylenebisaniline; methylenedianiline; dianilinomethane.

2. Formula: $C_{13}H_{14}N_2$

### II. Physical Data

1. Appearance and Odor: White to tan solid; amine odor

2. Molecular Weight: 198.26

3. Boiling Point: 398-399 °C at 760 mm Hg

4. Melting Point: 88-93 °C (190-100 °F)

5. Vapor Pressure: 9 mm Hg at 232 °C

6. Evaporation Rate (n-butyl acetate = 1): Negligible

7. Vapor Density (Air = 1): Not Applicable

8. Volatile Fraction by Weight: Negligible

9. Specific Gravity (Water = 1): Slight

10. Heat of Combustion: -8.40 kcal/g

11. Solubility in Water: Slightly soluble in cold water, very soluble in alcohol, benzene, ether, and many organic solvents.

### III. Fire, Explosion, and Reactivity Hazard Data

1. Flash Point: 190 °C (374 °F) Setaflash closed cup

2. Flash Point: 226 °C (439 °F) Cleveland open cup

3. Extinguishing Media: Water spray; Dry Chemical; Carbon dioxide.

4. Special Fire Fighting Procedures: Wear self-contained breathing apparatus and protective clothing to prevent contact with skin and eyes.

5. Unusual Fire and Explosion Hazards: Fire or excessive heat may cause production of hazardous decomposition products.

### IV. Reactivity Data

1. Stability: Stable

2. Incompatibility: Strong oxidizers

3. Hazardous Decomposition Products: As with any other organic material, combustion may produce carbon monoxide. Oxides of nitrogen may also be present.

4. Hazardous Polymerization: Will not occur.

### V. Spill and Leak Procedures

1. Sweep material onto paper and place in fiber carton.

2. Package appropriately for safe feed to an incinerator or dissolve in compatible waste solvents prior to incineration.

3. Dispose of in an approved incinerator equipped with afterburner and scrubber or contract with licensed chemical waste disposal service.

4. Discharge treatment or disposal may be subject to federal, state, or local laws.

5. Wear appropriate personal protective equipment.

### VI. Special Storage and Handling Precautions

A. High exposure to MDA can occur when transferring the substance from one container to another. Such operations should be well ventilated and good work practices must be established to avoid spills.

B. Pure MDA is a solid with a low vapor pressure. Grinding or heating operations increase the potential for exposure.

C. Store away from oxidizing materials.

D. Employers shall advise employees of all areas and operations where exposure to MDA could occur.

### VII. Housekeeping and Hygiene Facilities

A. The workplace should be kept clean, orderly, and in a sanitary condition.

The employer should institute a leak and spill detection program for operations involving MDA in order to detect sources of fugitive MDA emissions.

B. Adequate washing facilities with hot and cold water are to be provided and maintained in a sanitary condition. Suitable cleansing agents should also be provided to assure the effective removal of MDA from the skin.

### VIII. Common Operations

Common operations in which exposure to MDA is likely to occur include the following: Manufacture of MDA; Manufacture of Methylene diisocyanate; Curing agent for epoxy resin structures; Wire coating operations; and filament winding.

## APPENDIX C TO §1910.1050—
## MEDICAL SURVEILLANCE GUIDELINES FOR MDA

### I. Route of Entry

Inhalation; skin absorption; ingestion. MDA can be inhaled, absorbed through the skin, or ingested.

### II. Toxicology

MDA is a suspect carcinogen in humans. There are several reports of liver disease in humans and animals resulting from acute exposure to MDA. A well documented case of an acute cardiomyopathy secondary to exposure to MDA is on record. Numerous human cases of hepatitis secondary to MDA are known. Upon direct contact MDA may also cause damage to the eyes. Dermatitis and skin sensitization have been observed. Almost all forms of acute environmental hepatic injury in humans involve the hepatic parenchyma and produce hepatocellular jaundice. This agent produces intrahepatic cholestasis. The clinical picture consists of cholestatic jaundice, preceded or accompanied by abdominal pain, fever, and chills. Onset in about 60% of all observed cases is abrupt with severe abdominal pain. In about 30% of observed cases, the illness presented and evolved more slowly and less dramatically, with only slight abdominal pain. In about 10% of the cases only jaundice was evident. The cholestatic nature of the jaundice is evident in the prominence of itching, the histologic predominance of bile stasis, and portal inflammatory infiltration, accompanied by only slight parenchymal injury in most cases, and by the moderately elevated transaminase values. Acute, high doses, however, have been known to cause hepatocellular damage resulting in elevated SGPT, SGOT, alkaline phosphatase and bilirubin.

Absorption through the skin is rapid. MDA is metabolized and excreted over a 48-hour period. Direct contact may be irritating to the skin, causing dermatitis. Also MDA which is deposited on the skin is not thoroughly removed through washing.

MDA may cause bladder cancer in humans. Animal data supporting this assumption is not available nor is conclusive human data. However, human data collected on workers at a helicopter manufacturing facility where MDA is used suggests a higher incidence of bladder cancer among exposed workers.

### III. Signs and Symptoms

Skin may become yellow from contact with MDA.

Repeated or prolonged contact with MDA may result in recurring dermatitis (red-itchy, cracked skin) and eye irritation. Inhalation, ingestion or absorption through the skin at high concentrations may result in hepatitis, causing symptoms such as fever and chills, nausea and vomiting, dark urine, anorexia, rash, right upper quadrant pain and jaundice. Corneal burns may occur when MDA is splashed in the eyes.

### IV. Treatment of Acute Toxic Effects/Emergency Situation

If MDA gets into the eyes, immediately wash eyes with large amounts of water. If MDA is splashed on the skin, immediately wash contaminated skin with mild soap or detergent. Employee should be removed from exposure and given proper medical treatment. Medical tests required under the emergency section of the medical surveillance section (M)(4) must be conducted.

If the chemical is swallowed do not induce vomiting but remove by gastric lavage.

## APPENDIX D TO §1910.1050—
## SAMPLING AND ANALYTICAL METHODS FOR MDA MONITORING AND MEASUREMENT PROCEDURES

Measurements taken for the purpose of determining employee exposure to MDA are best taken so that the representative average 8-hour exposure may be determined from a single 8-hour sample or two (2) 4-hour samples. Short-time interval samples (or grab samples) may also be used to determine average exposure level if a minimum of five measurements are taken in a random manner over the 8-hour work shift. Random sampling means that any portion of the work shift has the same chance of being sampled as any other. The arithmetic average of all such random samples taken on one work shift is an estimate of an employee's average level of exposure for that work shift. Air samples should be taken in the employee's breathing zone (air that would most nearly represent that inhaled by the employee).

There are a number of methods available for monitoring employee exposures to MDA. The method OSHA currently uses is included below.

The employer, however, has the obligation of selecting any monitoring method which meets the accuracy and precision requirements of the standard under his unique field conditions. The standard requires that the method of monitoring must have an accuracy, to a 95 percent confidence level, of not less than plus or minus 25 percent for the select PEL.

### OSHA Methodology

### Sampling Procedure

#### Apparatus

Samples are collected by use of a personal sampling pump that can be calibrated within ±5% of the recommended flow rate with the sampling filter in line.

Samples are collected on 37 mm Gelman type A/E glass fiber filters treated with sulfuric acid. The filters are prepared by soaking each filter with 0.5 ml of 0.26N $H_2SO_4$. (0.26 N $H_2SO_4$ can be prepared by diluting 1.5 ml of 36N $H_2SO_4$ to 200 ml with deionized water.) The filters are dried in an oven at 100 °C for one hour and then assembled into two-piece 37 mm polystyrene cassettes with backup pads. The cassettes are sealed with shrink bands and the ends are plugged with plastic plugs.

After sampling, the filters are carefully removed from the cassettes and individually transferred to small vials containing approximately 2 ml deionized water. The vials must be tightly sealed. The water can be added before or after the filters are transferred. The vials must be sealable and capable of holding at least 7 ml of liquid. Small glass scintillation vials with caps containing Teflon liners are recommended.

#### Reagents

Deionized water is needed for addition to the vials.

#### Sampling Technique

Immediately before sampling, remove the plastic plugs from the filter cassettes.

Attach the cassette to the sampling pump with flexible tubing and place the cassette in the employee's breathing zone.

After sampling, seal the cassettes with plastic plugs until the filters are transferred to the vials containing deionized water.

At some convenient time within 10 hours of sampling, transfer the sample filters to vials.

Seal the small vials lengthwise.

Submit at least one blank filter with each sample set. Blanks should be handled in the same manner as samples, but no air is drawn through them.

Record sample volumes (in L of air) for each sample, along with any potential interferences.

### Retention Efficiency

A retention efficiency study was performed by drawing 100 L of air (80% relative humidity) at 1 L/min through sample filters that had been spiked with 0.814 µg MDA. Instead of using backup pads, blank acid-treated filters were used as backups in each cassette. Upon analysis, the top filters were found to have an average of 91.8% of the spiked amount. There was no MDA found on the bottom filters, so the amount lost was probably due to the slight instability of the MDA salt.

### Extraction Efficiency

The average extraction efficiency for six filters spiked at the target concentration is 99.6%.

The stability of extracted and derivatized samples was verified by reanalyzing the above six samples the next day using fresh standards. The average extraction efficiency for the reanalyzed samples is 98.7%.

### Recommended Air Volume and Sampling Rate

The recommended air volume is 100 L.

The recommended sampling rate is 1 L/min.

### Interferences (Sampling)

MDI appears to be a positive interference. It was found that when MDI was spiked onto an acid-treated filter, the MDI converted to MDA after air was drawn through it.

Suspected interferences should be reported to the laboratory with submitted samples.

### Safety Precautions (Sampling)

Attach the sampling equipment to the employees so that it will not interfere with work performance or safety.

Follow all safety procedures that apply to the work area being sampled.

### Analytical Procedure

Apparatus: The following are required for analysis.

A GC equipped with an electron capture detector. For this evaluation a Tracor 222 Gas Chromatograph equipped with a Nickel 63 High Temperature Electron Capture Detector and a Linearizer was used.

A GC column capable of separating the MDA derivative from the solvent and interferences. A 6 ft X 2 mm ID glass column packed with 3% OV-101 coated on 100/120 Gas Chrom Q was used in this evaluation.

A electronic integrator or some other suitable means of measuring peak areas or heights.

Small resealable vials with Teflon-lined caps capable of holding 4 ml.

A dispenser or pipet for toluene capable of delivering 2.0 ml.

Pipets (or repipets with plastic or Teflon tips) capable of delivering 1 ml for the sodium hydroxide and buffer solutions.

A repipet capable of delivering 25 µl HFAA.

Syringes for preparation of standards and injection of standards and samples into a GC.

Volumetric flasks and pipets to dilute the pure MDA in preparation of standards.

Disposable pipets to transfer the toluene layers after the samples are extracted.

### Reagents

0.5 NaOH prepared from reagent grade NaOH.

Toluene, pesticide grade. Burdick and Jackson distilled in glass toluene was used.

Heptafluorobutyric acid anhydride (HFAA). HFAA from Pierce Chemical Company was used.

pH 7.0 phosphate buffer, prepared from 136 g potassium dihydrogen phosphate and 1 L deionized water. The pH is adjusted to 7.0 with saturated sodium hydroxide solution.

4,4'-Methylenedianiline (MDA), reagent grade.

### Standard Preparation

Concentrated stock standards are prepared by diluting pure MDA with toluene. Analytical standards are prepared by injecting uL amounts of diluted stock standards into vials that contain 2.0 ml toluene.

25 uL HFAA are added to each vial and the vials are capped and shaken for 10 seconds.

After 10 min, 1 ml of buffer is added to each vial.

The vials are recapped and shaken for 10 seconds.

After allowing the layers to separate, aliquots of the toluene (upper) layers are removed with a syringe and analyzed by GC.

Analytical standard concentrations should bracket sample concentrations. Thus, if samples fall out of the range of prepared standards, additional standards must be prepared to ascertain detector response.

### Sample Preparation

The sample filters are received in vials containing deionized water.

1 ml of 0.5N NaOH and 2.0 ml toluene are added to each vial.

The vials are recapped and shaken for 10 min.

After allowing the layers to separate, approximately 1 ml aliquots of the toluene (upper) layers are transferred to separate vials with clean disposable pipets.

The toluene layers are treated and analyzed.

### Analysis
### GC conditions

Zone temperatures:

Column—220 °C

Injector—235 °C

Detector—335 °C

Gas flows, Ar/CH4 Column—28 ml/min

(95/5) Purge—40 ml/min

Injection volume: 5.0 uL

Column: 6 ft × 1/8 in ID glass, 3% OV-101 on 100/120 Gas Chrom Q

Retention time of MDA derivative: 3.5 min

### Chromatogram

Peak areas or heights are measured by an integrator or other suitable means.

A calibration curve is constructed by plotting response (peak areas or heights) of standard injections versus ug of MDA per sample. Sample concentrations must be bracketed by standards.

### Interferences (Analytical)

Any compound that gives an electron capture detector response and has the same general retention time as the HFAA derivative of MDA is a potential interference. Suspected interferences reported to the laboratory with submitted samples by the industrial hygienist must be considered before samples are derivatized.

GC parameters may be changed to possibly circumvent interferences.

Retention time on a single column is not considered proof of chemical identity. Analyte identity should be confirmed by GC/MS if possible.

### Calculations

The analyte concentration for samples is obtained from the calibration curve in terms of ug MDA per sample. The extraction efficiency is 100%. If any MDA is found on the blank, that amount is subtracted from the sample amounts. The air concentrations are calculated using the following formulae.

$$\mu g/m^3 = (\mu g \text{ MDA per sample}) (1000)/(L \text{ of air sampled})$$

$$ppb = (\mu g/m^3) (24.46)/(198.3) = (\mu g/m^3) (0.1233)$$

where 24.46 is the molar volume at 25 °C and 760 mm Hg

### Safety Precautions (Analytical)

Avoid skin contact and inhalation of all chemicals.

Restrict the use of all chemicals to a fume hood if possible.

Wear safety glasses and a lab coat at all times while in the lab area.

### APPENDIX E TO §1910.1050— QUALITATIVE AND QUANTITATIVE FIT TESTING PROCEDURES [Removed.]

[57 FR 35666, Aug. 10, 1992, as amended at 57 FR 49649, Nov. 3, 1992; 61 FR 5508, Feb. 13, 1996; 63 FR 1293, Jan. 8, 1998]

### §1910.1051 1,3-Butadiene.

**1910.1051(a)** *Scope and application.* **(1)** This section applies to all occupational exposures to 1,3-Butadiene (BD), Chemical Abstracts Service Registry No. 106-99-0, except as provided in paragraph (a)(2) of this section.

**1910.1051(a)(2)(i)** Except for the recordkeeping provisions in paragraph (m)(1) of this section, this section does not apply to the processing, use, or handling of products containing BD or to other work operations and streams in which BD is present where objective data are reasonably relied upon that demonstrate the work operation or the product or the group of products or operations to which it belongs may not reasonably be foreseen to release BD in airborne concentrations at or above the action level or in excess of the STEL under the expected conditions of processing, use, or handling that will cause the greatest possible release or in any plausible accident.

**1910.1051(a)(2)(ii)** This section also does not apply to work operations, products or streams where the only exposure to BD is from liquid mixtures containing 0.1% or less of BD by volume or the vapors released from such liquids, unless objective data become available that show that airborne concentrations generated by such mixtures can exceed the action level or STEL under reasonably predictable conditions of processing, use or handling that will cause the greatest possible release.

**1910.1051(a)(2)(iii)** Except for labeling requirements and requirements for emergency response, this section does not apply to the storage, transportation, distribution or sale of BD or liquid mixtures in intact containers or in transportation pipelines sealed in such a manner as to fully contain BD vapors or liquid.

**1910.1051(a)(3)** Where products or processes containing BD are exempted under paragraph (a)(2) of this section, the employer shall maintain records of the objective data supporting that exemption and the basis for the employer's reliance on the data, as provided in paragraph (m)(1) of this section.

**1910.1051(b)** *Definitions:* For the purpose of this section, the following definitions shall apply:

*Action level* means a concentration of airborne BD of 0.5 ppm calculated as an eight (8)-hour time-weighted average.

*Assistant Secretary* means the Assistant Secretary of Labor for Occupational Safety and Health, U.S. Department of Labor, or designee.

*Authorized person* means any person specifically designated by the employer, whose duties require entrance into a regulated area, or a person entering such an area as a designated representative of employees to exercise the right to observe monitoring and measuring procedures under paragraph (d)(8) of this section, or a person designated under the Act or regulations issued under the Act to enter a regulated area.

*1,3-Butadiene* means an organic compound with chemical formula $CH_2=CH\text{-}CH=CH_2$ that has a molecular weight of approximately 54.15 gm/mole.

*Business day* means any Monday through Friday, except those days designated as federal, state, local or company specific holidays.

*Complete Blood Count (CBC)* means laboratory tests performed on whole blood specimens and includes the following: White blood cell count (WBC), hematocrit (Hct), red blood cell count (RBC), hemoglobin (Hgb), differential count of white blood cells, red blood cell morphology, red blood cell indices, and platelet count.

*Day* means any part of a calendar day.

*Director* means the Director of the National Institute for Occupational Safety and Health (NIOSH), U.S. Department of Health and Human Services, or designee.

*Emergency situation* means any occurrence such as, but not limited to, equipment failure, rupture of containers, or failure of control equipment that may or does result in an uncontrolled significant release of BD.

*Employee exposure* means exposure of a worker to airborne concentrations of BD which would occur if the employee were not using respiratory protective equipment.

*Objective data* means monitoring data, or mathematical modelling or calculations based on composition, chemical and physical properties of a material, stream or product.

*Permissible Exposure Limits, PELs* means either the 8 hour Time Weighted Average (8-hr TWA) exposure or the Short-Term Exposure Limit (STEL).

*Physician or other licensed health care professional* is an individual whose legally permitted scope of practice (i.e., license, registration, or certification) allows him or her to independently provide or be delegated the responsibility to provide one or more of the specific health care services required by paragraph (k) of this section.

*Regulated area* means any area where airborne concentrations of BD exceed or can reasonably be expected to exceed the 8-hour time weighted average (8-hr TWA) exposure of 1 ppm or the short-term exposure limit (STEL) of 5 ppm for 15 minutes.

*This section* means this 1,3-butadiene standard.

**1910.1051(c)** *Permissible exposure limits (PELs)*—**(1)** *Time-weighted average (TWA) limit.* The employer shall ensure that no employee is exposed to an airborne concentration of BD in excess of one (1) part BD per million parts of air (ppm) measured as an eight (8)-hour time-weighted average.

**1910.1051(c)(2)** *Short-term exposure limit (STEL).* The employer shall ensure that no employee is exposed to an airborne concentration of BD in excess of five parts of BD per million parts of air (5 ppm) as determined over a sampling period of fifteen (15) minutes.

**1910.1051(d)** *Exposure monitoring*—**(1)** *General.* **(i)** Determinations of employee exposure shall be made from breathing zone air samples that are representative of the 8-hour TWA and 15-minute short-term exposures of each employee.

**1910.1051(d)(1)(ii)** Representative 8-hour TWA employee exposure shall be determined on the basis of one or more samples representing fullshift exposure for each shift and for each job classification in each work area.

**1910.1051(d)(1)(iii)** Representative 15-minute short-term employee exposures shall be determined on the basis of one or more samples representing 15-minute exposures associated with operations that are most likely to produce exposures above the STEL for each shift and for each job classification in each work area.

**1910.1051(d)(1)(iv)** Except for the initial monitoring required under paragraph (d)(2) of this section, where the employer can document that exposure levels are equivalent for similar operations on different work shifts, the employer need only determine representative employee exposure for that operation from the shift during which the highest exposure is expected.

**1910.1051(d)(2)** *Initial monitoring.* **(i)** Each employer who has a workplace or work operation covered by this section, shall perform initial monitoring to determine accurately the airborne concentrations of BD to which employees may be exposed, or shall rely on objective data pursuant to paragraph (a)(2)(i) of this section to fulfill this requirement. The initial monitoring shall be completed within 60 days of the introduction of BD into the workplace.

**1910.1051(d)(2)(ii)** Where the employer has monitored within two years prior to the effective date of this section and the monitoring satisfies all other requirements of this section, the employer may rely on such earlier monitoring results to satisfy the requirements of paragraph (d)(2)(i) of this section, provided that the conditions under which the initial monitoring was conducted have not changed in a manner that may result in new or additional exposures.

**1910.1051(d)(3)** *Periodic monitoring and its frequency.* **(i)** If the initial monitoring required by paragraph (d)(2) of this section reveals employee exposure to be at or above the action level but at or below both the 8-hour TWA limit and the STEL, the employer shall repeat the representative monitoring required by paragraph (d)(1) of this section every twelve months.

**1910.1051(d)(3)(ii)** If the initial monitoring required by paragraph (d)(2) of this section reveals employee exposure to be above the 8-hour TWA limit, the employer shall repeat the representative monitoring required by paragraph (d)(1)(ii) of this section at least every three months until the employer has collected two samples per quarter (each at least 7 days apart) within a twoyear period, after which such monitoring must occur at least every six months.

**1910.1051(d)(3)(iii)** If the initial monitoring required by paragraph (d)(2) of this section reveals employee exposure to be above the STEL, the employer shall repeat the representative monitoring required by paragraph (d)(1)(iii) of this section at least every three months until the employer has collected two samples per quarter (each at least 7 days apart) within a two-year period, after which such monitoring must occur at least every six months.

**1910.1051(d)(3)(iv)** The employer may alter the monitoring schedule from every six months to annually for any required representative monitoring for which two consecutive measurements taken at least 7 days apart indicate that employee exposure has decreased to or below the 8-hour TWA, but is at or above the action level.

**1910.1051(d)(4)** *Termination of monitoring.* **(i)** If the initial monitoring required by paragraph (d)(2) of this section reveals employee exposure to be below the action level and at or below the STEL, the employer may discontinue the monitoring for employees whose exposures are represented by the initial monitoring.

**1910.1051(d)(4)(ii)** If the periodic monitoring required by paragraph (d)(3) of this section reveals that employee exposures, as indicated by at least two consecutive measurements taken at least 7 days apart, are below the action level and at or below the STEL, the employer may discontinue the monitoring for those employees who are represented by such monitoring.

**1910.1051(d)(5)** *Additional monitoring.* **(i)** The employer shall institute the exposure monitoring required under paragraph (d) of this section whenever there has been a change in the production, process, control equipment, personnel or work practices that may result in new or additional exposures to BD or when the employer has any reason to suspect that a change may result in new or additional exposures.

**1910.1051(d)(5)(ii)** Whenever spills, leaks, ruptures or other breakdowns occur that may lead to employee exposure above the 8-hr TWA limit or above the STEL, the employer shall monitor [using leak source, such as direct reading instruments, area or personal monitoring], after the cleanup of the spill or repair of the leak, rupture or other breakdown, to ensure that exposures have returned to the level that existed prior to the incident.

**1910.1051(d)(6)** *Accuracy of monitoring.* Monitoring shall be accurate, at a confidence level of 95 percent, to within plus or minus 25 percent for airborne concentrations of BD at or above the 1 ppm TWA limit and to within plus or minus 35 percent for airborne concentrations of BD at or above the action level of 0.5 ppm and below the 1 ppm TWA limit.

**1910.1051(d)(7)** *Employee notification of monitoring results.* **(i)** The employer must, within 15 working days after the receipt of the results of any monitoring performed under this section, notify each affected employee of these results either individually in writing or by posting the results in an appropriate location that is accessible to employees.

**1910.1051(d)(7)(ii)** The employer shall, within 15 business days after receipt of any monitoring performed under this section indicating the 8-hour TWA or STEL has been exceeded, provide the affected employees, in writing, with information on the corrective action being taken by the employer to reduce employee exposure to or below the 8-hour TWA or STEL and the schedule for completion of this action.

**1910.1051(d)(8)** *Observation of monitoring*—**(i)** *Employee observation.* The employer shall provide affected employees or their designated representatives an opportunity to observe any monitoring of employee exposure to BD conducted in accordance with paragraph (d) of this section.

**1910.1051(d)(8)(ii)** *Observation procedures.* When observation of the monitoring of employee exposure to BD requires entry into an area where the use of protective clothing or equipment is required, the employer shall provide the observer at no cost with protective clothing and equipment, and shall ensure that the observer uses this equipment and complies with all other applicable safety and health procedures.

**1910.1051(e)** *Regulated areas.* **(1)** The employer shall establish a regulated area wherever occupational exposures to airborne concentrations of BD exceed or can reasonably be expected to exceed the permissible exposure limits, either the 8-hr TWA or the STEL.

**1910.1051(e)(2)** Access to regulated areas shall be limited to authorized persons.

**1910.1051(e)(3)** Regulated areas shall be demarcated from the rest of the workplace in any manner that minimizes the number of employees exposed to BD within the regulated area.

**1910.1051(e)(4)** An employer at a multi-employer worksite who establishes a regulated area shall communicate the access restrictions and locations of these areas to other employers with work operations at that worksite whose employees may have access to these areas.

**1910.1051(f)** *Methods of compliance*—**(1)** *Engineering controls and work practices.* **(i)** The employer shall institute engineering controls and work practices to reduce and maintain employee exposure to or below the PELs, except to the extent that the employer can establish that these controls are not feasible or where paragraph (h)(1)(i) of this section applies.

**1910.1051(f)(1)(ii)** Wherever the feasible engineering controls and work practices which can be instituted are not sufficient to reduce employee exposure to or below the 8-hour TWA or STEL, the employer shall use them to reduce employee exposure to the lowest levels achievable by these controls and shall supplement them by the use of respiratory protection that complies with the requirements of paragraph (h) of this section.

**1910.1051(f)(2)** *Compliance plan.* **(i)** Where any exposures are over the PELs, the employer shall establish and implement a written plan to reduce employee exposure to or below the PELs primarily by means of engineering and work practice controls, as required by paragraph (f)(1) of this section, and by the use of respiratory protection where required or permitted under this section. No compliance plan is required if all exposures are under the PELs.

**1910.1051(f)(2)(ii)** The written compliance plan shall include a schedule for the development and implementation of the engineering controls and work practice controls including periodic leak detection surveys.

**1910.1051(f)(2)(iii)** Copies of the compliance plan required in paragraph (f)(2) of this section shall be furnished upon request for examination and copying to the Assistant Secretary, the Director, affected employees and designated employee representatives. Such plans shall be reviewed at least every 12 months, and shall be updated as necessary to reflect significant changes in the status of the employer's compliance program.

**1910.1051(f)(2)(iv)** The employer shall not implement a schedule of employee rotation as a means of compliance with the PELs.

**1910.1051(g)** *Exposure Goal Program.* **(1)** For those operations and job classifications where employee exposures are greater than the action level, in addition to compliance with the PELs, the employer shall have an exposure goal program that is intended to limit employee exposures to below the action level during normal operations.

**1910.1051(g)(2)** Written plans for the exposure goal program shall be furnished upon request for examination and copying to the Assistant Secretary, the Director, affected employees and designated employee representatives.

**1910.1051(g)(3)** Such plans shall be updated as necessary to reflect significant changes in the status of the exposure goal program.

**1910.1051(g)(4)** Respirator use is not required in the exposure goal program.

**1910.1051(g)(5)** The exposure goal program shall include the following items unless the employer can demonstrate that the item is not feasible, will have no significant effect in reducing employee exposures, or is not necessary to achieve exposures below the action level:

**1910.1051(g)(5)(i)** A leak prevention, detection, and repair program.

**1910.1051(g)(5)(ii)** A program for maintaining the effectiveness of local exhaust ventilation systems.

**1910.1051(g)(5)(iii)** The use of pump exposure control technology such as, but not limited to, mechanical double-sealed or seal-less pumps.

**1910.1051(g)(5)(iv)** Gauging devices designed to limit employee exposure, such as magnetic gauges on rail cars.

**1910.1051(g)(5)(v)** Unloading devices designed to limit employee exposure, such as a vapor return system.

**1910.1051(g)(5)(vi)** A program to maintain BD concentration below the action level in control rooms by use of engineering controls.

**1910.1051(h)** *Respiratory protection*—**(1)** *General.* For employees who use respirators required by this section, the employer must provide respirators that comply with the requirements of this paragraph. Respirators must be used during:

**1910.1051(h)(1)(i)** Periods necessary to install or implement feasible engineering and work-practice controls.

**1910.1051(h)(1)(ii)** Non-routine work operations that are performed infrequently and for which employee exposures are limited in duration.

**1910.1051(h)(1)(iii)** Work operations for which feasible engineering and work-practice controls are not yet sufficient to reduce employee exposures to or below the PELs.

**1910.1051(h)(1)(iv)** Emergencies.

**1910.1051(h)(2)** *Respirator program.* **(i)** The employer must implement a respiratory protection program in accordance with 29 CFR 1910.134(b) through (d) (except (d)(1)(iii), (d)(3)(iii)(B)(1), and (2)), and (f) through (m).

**1910.1051(h)(2)(ii)** If air-purifying respirators are used, the employer must replace the air-purifying filter elements according to the replacement schedule set for the class of respirators listed in Table 1 of this section, and at the beginning of each work shift.

**1910.1051(h)(2)(iii)** Instead of using the replacement schedule listed in Table 1 of this section, the employer may replace cartridges or canisters at 90% of their expiration service life, provided the employer:

**1910.1051(h)(2)(iii)(A)** Demonstrates that employees will be adequately protected by this procedure.

**1910.1051(h)(2)(iii)(B)** Uses BD breakthrough data for this purpose that have been derived from tests conducted under worst-case conditions of humidity, temperature, and air-flow rate through the filter element, and the employer also describes the data supporting the cartridge-or canister-change schedule, as well as the basis for using the data in the employer's respirator program.

**1910.1051(h)(2)(iv)** A label must be attached to each filter element to indicate the date and time it is first installed on the respirator.

**1910.1051(h)(2)(v)** If NIOSH approves an end-of-service-life indicator (ESLI) for an air-purifying filter element, the element may be used until the ESLI shows no further useful service life or until the element is replaced at the beginning of the next work shift, whichever occurs first.

**1910.1051(h)(2)(vi)** Regardless of the air-purifying element used, if an employee detects the odor of BD, the employer must replace the air-purifying element immediately.

**1910.1051(h)(3)** *Respirator selection.* **(i)** The employer must select appropriate respirators from Table 1 of this section.

### TABLE 1—MINIMUM REQUIREMENTS FOR RESPIRATORY PROTECTION FOR AIRBORNE BD

| Concentration of airborne BD (ppm) or condition of use | Minimum required respirator |
|---|---|
| Less than or equal to 5 ppm (5 times PEL) | (a) Air-purifying half mask or full facepiece respirator equipped with approved BD or organic vapor cartridges or canisters. Cartridges or canisters shall be replaced every 4 hours. |
| Less than or equal to 10 ppm (10 times PEL) | (a) Air-purifying half mask or full facepiece respirator equipped with approved BD or organic vapor cartridges or canisters. Cartridges or canisters shall be replaced every 3 hours. |
| Less than or equal to 25 ppm (25 times PEL) | (a) Air-purifying full facepiece respirator equipped with approved BD or organic vapor cartridges or canisters. Cartridges or canisters shall be replaced every 2 hours.<br>(b) Any powered air-purifying respirator equipped with approved BD or organic vapor cartridges. PAPR cartridges shall be replaced every 2 hours.<br>(c) Continuous flow supplied air respirator equipped with a hood or helmet. |
| Less than or equal to 50 ppm (50 times PEL) | (a) Air-purifying full facepiece respirator equipped with approved BD or organic vapor cartridges or canisters. Cartridges or canisters shall be replaced every (1) hour.<br>(b) Powered air-purifying respirator equipped with a tight-fitting facepiece and an approved BD or organic vapor cartridges. PAPR cartridges shall be replaced every (1) hour. |
| Less than or equal to 1,000 ppm (1,000 times PEL) | (a) Supplied air respirator equipped with a half mask of full facepiece and operated in a pressure demand or other positive pressure mode. |
| Greater than 1000 ppm unknown concentration, or firefighting. | (a) Self-contained breathing apparatus equipped with a full facepiece and operated in a pressure demand or other positive pressure mode.<br>(b) Any supplied air respirator equipped with a full facepiece and operated in a pressure demand or other positive pressure mode in combination with an auxiliary self-contained breathing apparatus operated in a pressure demand or other positive pressure mode. |
| Escape from IDLH conditions | (a) Any positive pressure self-contained breathing apparatus with an appropriate service life.<br>(b) A air-purifying full facepiece respirator equipped with a front or back mounted BD or organic vapor canister. |

Notes: Respirators approved for use in higher concentrations are permitted to be used in lower concentrations. Full facepiece is required when eye irritation is anticipated.

**1910.1051(h)(3)(ii)** Air-purifying respirators must have filter elements approved by NIOSH for organic vapors or BD.

**1910.1051(h)(3)(iii)** When an employee whose job requires the use of a respirator cannot use a negative-pressure respirator, the employer must provide the employee with a respirator that has less breathing resistance than the negative-pressure respirator, such as a powered air-purifying respirator or supplied-air respirator, when the employee is able to use it and if it provides the employee adequate protection.

**1910.1051(i)** *Protective clothing and equipment.* Where appropriate to prevent eye contact and limit dermal exposure to BD, the employer shall provide protective clothing and equipment at no cost to the employee and shall ensure its use. Eye and face protection shall meet the requirements of 29 CFR 1910.133.

**1910.1051(j)** *Emergency situations. Written plan.* A written plan for emergency situations shall be developed, or an existing plan shall be modified, to contain the applicable elements specified in 29 CFR 1910.38 and 29 CFR 1910.39, "Emergency action plans" and "Fire prevention plans," respectively and in 29 CFR 1910.120 "Hazardous Waste Operations and Emergency Responses," for each workplace where there is a possibility of an emergency.

**1910.1051(k)** *Medical screening and surveillance*—**(1)** *Employees covered.* The employer shall institute a medical screening and surveillance program as specified in this paragraph for:

**1910.1051(k)(1)(i)** Each employee with exposure to BD at concentrations at or above the action level on 30 or more days or for employees who have or may have exposure to BD at or above the PELs on 10 or more days a year;

**1910.1051(k)(1)(ii)** Employers (including successor owners) shall continue to provide medical screening and surveillance for employees, even after transfer to a non-BD exposed job and regardless of when the employee is transferred, whose work histories suggest exposure to BD:

**1910.1051(k)(1)(ii)(A)** At or above the PELs on 30 or more days a year for 10 or more years;

**1910.1051(k)(1)(ii)(B)** At or above the action level on 60 or more days a year for 10 or more years; or

**1910.1051(k)(1)(ii)(C)** Above 10 ppm on 30 or more days in any past year; and

**1910.1051(k)(1)(iii)** Each employee exposed to BD following an emergency situation.

**1910.1051(k)(2)** *Program administration.* **(i)** The employer shall ensure that the health questionnaire, physical examination and medical procedures are provided without cost to the employee, without loss of pay, and at a reasonable time and place.

**1910.1051(k)(2)(ii)** Physical examinations, health questionnaires, and medical procedures shall be performed or administered by a physician or other licensed health care professional.

**1910.1051(k)(2)(iii)** Laboratory tests shall be conducted by an accredited laboratory.

**1910.1051(k)(3)** *Frequency of medical screening activities.* The employer shall make medical screening available on the following schedule:

**1910.1051(k)(3)(i)** For each employee covered under paragraphs (j)(1) (i)(ii) of this section, a health questionnaire and complete blood count with differential and platelet count (CBC) every year, and a physical examination as specified below:

**1910.1051(k)(3)(i)(A)** An initial physical examination that meets the requirements of this rule, if twelve months or more have elapsed since the last physical examination conducted as part of a medical screening program for BD exposure;

**1910.1051(k)(3)(i)(B)** Before assumption of duties by the employee in a job with BD exposure;

**1910.1051(k)(3)(i)(C)** Every 3 years after the initial physical examination;

**1910.1051(k)(3)(i)(D)** At the discretion of the physician or other licensed health care professional reviewing the annual health questionnaire and CBC;

**1910.1051(k)(3)(i)(E)** At the time of employee reassignment to an area where exposure to BD is below the action level, if the employee's past exposure history does not meet the criteria of paragraph (j)(1)(ii) of this section for continued coverage in the screening and surveillance program, and if twelve months or more have elapsed since the last physical examination; and

**1910.1051(k)(3)(i)(F)** At termination of employment if twelve months or more have elapsed since the last physical examination.

**1910.1051(k)(3)(ii)** Following an emergency situation, medical screening shall be conducted as quickly as possible, but not later than 48 hours after the exposure.

**1910.1051(k)(3)(iii)** For each employee who must wear a respirator, physical ability to perform the work and use the respirator must be determined as required by 29 CFR 1910.134.

**1910.1051(k)(4)** *Content of medical screening.* **(i)** Medical screening for employees covered by paragraphs (j)(1) (i)-(ii) of this section shall include:

**1910.1051(k)(4)(i)(A)** A baseline health questionnaire that includes a comprehensive occupational and health history and is updated annually. Particular emphasis shall be placed on the hematopoietic and reticulo-endothelial systems, including exposure to chemicals, in addition to BD, that may have an adverse effect on these systems, the presence of signs and symptoms that might be related to disorders of these systems, and any other information determined by the examining physician or other licensed health care professional to be necessary to evaluate whether the employee is at increased risk of material impairment of health from BD exposure. Health questionnaires shall consist of the sample forms in Appendix C to this section, or be equivalent to those samples;

**1910.1051(k)(4)(i)(B)** A complete physical examination, with special emphasis on the liver, spleen, lymph nodes, and skin;

**1910.1051(k)(4)(i)(C)** A CBC; and

**1910.1051(k)(4)(i)(D)** Any other test which the examining physician or other licensed health care professional deems necessary to evaluate whether the employee may be at increased risk from exposure to BD.

**1910.1051(k)(4)(ii)** Medical screening for employees exposed to BD in an emergency situation shall focus on the acute effects of BD exposure and at a minimum include: A CBC within 48 hours of the exposure and then monthly for three months; and a physical examination if the employee reports irritation of the eyes, nose throat, lungs, or skin, blurred vision, coughing, drowsiness, nausea, or headache. Continued employee participation in the medical screening and surveillance program, beyond these minimum requirements, shall be at the discretion of the physician or other licensed health care professional.

**1910.1051(k)(5)** *Additional medical evaluations and referrals.* **(i)** Where the results of medical screening indicate abnormalities of the hematopoietic or reticuloendothelial systems, for which a nonoccupational cause is not readily apparent, the examining physician or other licensed health care professional shall refer the employee to an appropriate specialist for further evaluation and shall make available to the specialist the results of the medical screening.

**1910.1051(k)(5)(ii)** The specialist to whom the employee is referred under this paragraph shall determine the appropriate content for the medical evaluation, e.g., examinations, diagnostic tests and procedures, etc.

**1910.1051(k)(6)** *Information provided to the physician or other licensed health care professional.* The employer shall provide the following information to the examining physician or other licensed health care professional involved in the evaluation:

**1910.1051(k)(6)(i)** A copy of this section including its appendices;

**1910.1051(k)(6)(ii)** A description of the affected employee's duties as they relate to the employee's BD exposure;

**1910.1051(k)(6)(iii)** The employee's actual or representative BD exposure level during employment tenure, including exposure incurred in an emergency situation;

**1910.1051(k)(6)(iv)** A description of pertinent personal protective equipment used or to be used; and

**1910.1051(k)(6)(v)** Information, when available, from previous employment-related medical evaluations of the affected employee which is not otherwise available to the physician or other licensed health care professional or the specialist.

**1910.1051(k)(7)** *The written medical opinion.* **(i)** For each medical evaluation required by this section, the employer shall ensure that the physician or other licensed health care professional produces a written opinion and provides a copy to the employer and the employee within 15 business days of the evaluation. The written opinion shall be limited to the following information:

**1910.1051(k)(7)(i)(A)** The occupationally pertinent results of the medical evaluation;

**1910.1051(k)(7)(i)(B)** A medical opinion concerning whether the employee has any detected medical conditions which would place the employee's health at increased risk of material impairment from exposure to BD;

**1910.1051(k)(7)(i)(C)** Any recommended limitations upon the employee's exposure to BD; and

**1910.1051(k)(7)(i)(D)** A statement that the employee has been informed of the results of the medical evaluation and any medical conditions resulting from BD exposure that require further explanation or treatment.

**1910.1051(k)(7)(ii)** The written medical opinion provided to the employer shall not reveal specific records, findings, and diagnoses that have no bearing on the employee's ability to work with BD.

**NOTE:** However, this provision does not negate the ethical obligation of the physician or other licensed health care professional to transmit any other adverse findings directly to the employee.

**1910.1051(k)(8)** *Medical surveillance.* **(i)** The employer shall ensure that information obtained from the medical screening program activities is aggregated (with all personal identifiers removed) and periodically reviewed, to ascertain whether the health of the employee population of that employer is adversely affected by exposure to BD.

**1910.1051(k)(8)(ii)** Information learned from medical surveillance activities must be disseminated to covered employees, as defined in paragraph (k)(1) of this section, in a manner that ensures the confidentiality of individual medical information.

**1910.1051(l)** *Communication of BD hazards to employees*—**(1)** *Hazard communication.* The employer shall communicate the hazards associated with BD exposure in accordance with the requirements of the Hazard Communication Standard, 29 CFR 1910.1200, 29 CFR 1915.1200, and 29 CFR 1926.59.

**1910.1051(l)(2)** *Employee information and training.* **(i)** The employer shall provide all employees exposed to BD with information and training in accordance with the requirements of the Hazard Communication Standard, 29 CFR 1910.1200, 29 CFR 1915.1200, and 29 CFR 1926.59.

**1910.1051(l)(2)(ii)** The employer shall institute a training program for all employees who are potentially exposed to BD at or above the action level or the STEL, ensure employee participation in the program and maintain a record of the contents of such program.

**1910.1051(l)(2)(iii)** Training shall be provided prior to or at the time of initial assignment to a job potentially involving exposure to BD at or above the action level or STEL and at least annually thereafter.

**1910.1051(l)(2)(iv)** The training program shall be conducted in a manner that the employee is able to understand. The employee shall ensure that each employee exposed to BD over the action level or STEL is informed of the following:

**1910.1051(l)(2)(iv)(A)** The health hazards associated with BD exposure, and the purpose and a description of the medical screening and surveillance program required by this section;

**1910.1051(l)(2)(iv)(B)** The quantity, location, manner of use, release, and storage of BD and the specific operations that could result in exposure to BD, especially exposures above the PEL or STEL;

**1910.1051(l)(2)(iv)(C)** The engineering controls and work practices associated with the employee's job assignment, and emergency procedures and personal protective equipment;

**1910.1051(l)(2)(iv)(D)** The measures employees can take to protect themselves from exposure to BD.

**1910.1051(l)(2)(iv)(E)** The contents of this standard and its appendices, and

**1910.1051(l)(2)(iv)(F)** The right of each employee exposed to BD at or above the action level or STEL to obtain:

**1910.1051(l)(2)(iv)(F)***(1)* medical examinations as required by paragraph (j) of this section at no cost to the employee;

**1910.1051(l)(2)(iv)(F)***(2)* the employee's medical records required to be maintained by paragraph (m)(4) of this section; and

**1910.1051(l)(2)(iv)(F)***(3)* all air monitoring results representing the employee's exposure to BD and required to be kept by paragraph (m)(2) of this section.

**1910.1051(l)(3)** *Access to information and training materials.* **(i)** The employer shall make a copy of this standard and its appendices readily available without cost to all affected employees and their designated representatives and shall provide a copy if requested.

**1910.1051(l)(3)(ii)** The employer shall provide to the Assistant Secretary or the Director, or the designated employee representatives, upon request, all materials relating to the employee information and the training program.

**1910.1051(m)** *Recordkeeping*—**(1)** *Objective data for exemption from initial monitoring.* **(i)** Where the processing, use, or handling of products or streams made from or containing BD are exempted from other requirements of this section under paragraph (a)(2) of this section, or where objective data have been relied on in lieu of initial monitoring under paragraph (d)(2)(ii) of this section, the employer shall establish and maintain a record of the objective data reasonably relied upon in support of the exemption.

**1910.1051(m)(1)(ii)** This record shall include at least the following information:

**1910.1051(m)(1)(ii)(A)** The product or activity qualifying for exemption;

**1910.1051(m)(1)(ii)(B)** The source of the objective data;

**1910.1051(m)(1)(ii)(C)** The testing protocol, results of testing, and analysis of the material for the release of BD;

**1910.1051(m)(1)(ii)(D)** A description of the operation exempted and how the data support the exemption; and

**1910.1051(m)(1)(ii)(E)** Other data relevant to the operations, materials, processing, or employee exposures covered by the exemption.

**1910.1051(m)(1)(iii)** The employer shall maintain this record for the duration of the employer's reliance upon such objective data.

**1910.1051(m)(2)** *Exposure measurements.* **(i)** The employer shall establish and maintain an accurate record of all measurements taken to monitor employee exposure to BD as prescribed in paragraph (d) of this section.

**1910.1051(m)(2)(ii)** The record shall include at least the following information:

**1910.1051(m)(2)(ii)(A)** The date of measurement;

**1910.1051(m)(2)(ii)(B)** The operation involving exposure to BD which is being monitored;

**1910.1051(m)(2)(ii)(C)** Sampling and analytical methods used and evidence of their accuracy;

**1910.1051(m)(2)(ii)(D)** Number, duration, and results of samples taken;

**1910.1051(m)(2)(ii)(E)** Type of protective devices worn, if any; and

**1910.1051(m)(2)(ii)(F)** Name, social security number and exposure of the employees whose exposures are represented.

**1910.1051(m)(2)(ii)(G)** The written corrective action and the schedule for completion of this action required by paragraph (d)(7)(ii) of this section.

**1910.1051(m)(2)(iii)** The employer shall maintain this record for at least 30 years in accordance with 29 CFR 1910.1020.

**1910.1051(m)(3)** *Respirator Fit-test.* **(i)** The employer shall establish a record of the fit tests administered to an employee including:

**1910.1051(m)(3)(i)(A)** The name of the employee,

**1910.1051(m)(3)(i)(B)** Type of respirator,

**1910.1051(m)(3)(i)(C)** Brand and size of respirator,

**1910.1051(m)(3)(i)(D)** Date of test, and

**1910.1051(m)(3)(i)(E)** Where QNFT is used, the fit factor, strip chart recording or other recording of the results of the test.

**1910.1051(m)(3)(ii)** Fit test records shall be maintained for respirator users until the next fit test is administered.

**1910.1051(m)(4)** *Medical screening and surveillance.* **(i)** The employer shall establish and maintain an accurate record for each employee subject to medical screening and surveillance under this section.

**1910.1051(m)(4)(ii)** The record shall include at least the following information:

**1910.1051(m)(4)(ii)(A)** The name and social security number of the employee;

**1910.1051(m)(4)(ii)(B)** Physician's or other licensed health care professional's written opinions as described in paragraph (k)(7) of this section;

**1910.1051(m)(4)(ii)(C)** A copy of the information provided to the physician or other licensed health care professional as required by paragraphs (k)(7)(ii)-(iv) of this section.

**1910.1051(m)(4)(iii)** Medical screening and surveillance records shall be maintained for each employee for the duration of employment plus 30 years, in accordance with 29 CFR 1910.1020.

**1910.1051(m)(5)** *Availability.* **(i)** The employer, upon written request, shall make all records required to be maintained by this section available for examination and copying to the Assistant Secretary and the Director.

**1910.1051(m)(5)(ii)** Access to records required to be maintained by paragraphs (l)(1)-(3) of this section shall be granted in accordance with 29 CFR 1910.1020(e).

**1910.1051(m)(6)** *Transfer of records.* **(i)** Whenever the employer ceases to do business, the employer shall transfer records required by this section to the successor employer. The successor employer shall receive and maintain these records. If there is no successor employer, the employer shall notify the Director, at least three (3) months prior to disposal, and transmit them to the Director if requested by the Director within that period.

**1910.1051(m)(6)(ii)** The employer shall transfer medical and exposure records as set forth in 29 CFR 1910.1020(h).

# OSHA Standards for General Industry

**1910.1051(n)** [Reserved.]

**1910.1051(o)** **Appendices.** **(1)** Appendix E to this section is mandatory.

**1910.1051(o)(2)** Appendices A, B, C, D, and F to this section are informational and are not intended to create any additional obligations not otherwise imposed or to detract from any existing obligations.

## APPENDIX A TO §1910.1051—SUBSTANCE SAFETY DATA SHEET FOR 1,3-BUTADIENE (NON-MANDATORY)

### I. SUBSTANCE IDENTIFICATION

A. Substance: 1,3-Butadiene ($CH_2$=CH-CH=$CH_2$).

B. Synonyms: 1,3-Butadiene (BD); butadiene; biethylene; bi-vinyl; divinyl; butadiene-1,3; buta-1,3-diene; erythrene; NCI-C50602; CAS-106-99-0.

C. BD can be found as a gas or liquid.

D. BD is used in production of styrene-butadiene rubber and polybutadiene rubber for the tire industry. Other uses include copolymer latexes for carpet backing and paper coating, as well as resins and polymers for pipes and automobile and appliance parts. It is also used as an intermediate in the production of such chemicals as fungicides.

E. Appearance and odor: BD is a colorless, non-corrosive, flammable gas with a mild aromatic odor at standard ambient temperature and pressure.

F. Permissible exposure: Exposure may not exceed 1 part BD per million parts of air averaged over the 8-hour workday, nor may short-term exposure exceed 5 parts of BD per million parts of air averaged over any 15-minute period in the 8-hour workday.

### II. HEALTH HAZARD DATA

A. BD can affect the body if the gas is inhaled or if the liquid form, which is very cold (cryogenic), comes in contact with the eyes or skin.

B. Effects of overexposure: Breathing very high levels of BD for a short time can cause central nervous system effects, blurred vision, nausea, fatigue, headache, decreased blood pressure and pulse rate, and unconsciousness. There are no recorded cases of accidental exposures at high levels that have caused death in humans, but this could occur. Breathing lower levels of BD may cause irritation of the eyes, nose, and throat. Skin contact with liquefied BD can cause irritation and frostbite.

C. Long-term (chronic) exposure: BD has been found to be a potent carcinogen in rodents, inducing neoplastic lesions at multiple target sites in mice and rats. A recent study of BDexposed workers showed that exposed workers have an increased risk of developing leukemia. The risk of leukemia increases with increased exposure to BD. OSHA has concluded that there is strong evidence that workplace exposure to BD poses an increased risk of death from cancers of the lymphohematopoietic system.

D. Reporting signs and symptoms: You should inform your supervisor if you develop any of these signs or symptoms and suspect that they are caused by exposure to BD.

### III. EMERGENCY FIRST AID PROCEDURES

In the event of an emergency, follow the emergency plan and procedures designated for your work area. If you have been trained in first aid procedures, provide the necessary first aid measures. If necessary, call for additional assistance from coworkers and emergency medical personnel.

A. Eye and Skin Exposures: If there is a potential that liquefied BD can come in contact with eye or skin, face shields and skin protective equipment must be provided and used. If liquefied BD comes in contact with the eye, immediately flush the eyes with large amounts of water, occasionally lifting the lower and the upper lids. Flush repeatedly. Get medical attention immediately. Contact lenses should not be worn when working with this chemical. In the event of skin contact, which can cause frostbite, remove any contaminated clothing and flush the affected area repeatedly with large amounts of tepid water.

B. Breathing: If a person breathes in large amounts of BD, move the exposed person to fresh air at once. If breathing has stopped, begin cardiopulmonary resuscitation (CPR) if you have been trained in this procedure. Keep the affected person warm and at rest. Get medical attention immediately.

C. Rescue: Move the affected person from the hazardous exposure. If the exposed person has been overcome, call for help and begin emergency rescue procedures. Use extreme caution so that you do not become a casualty. Understand the plant's emergency rescue procedures and know the locations of rescue equipment before the need arises.

### IV. RESPIRATORS AND PROTECTIVE CLOTHING

A. Respirators: Good industrial hygiene practices recommend that engineering and work practice controls be used to reduce environmental concentrations to the permissible exposure level. However, there are some exceptions where respirators may be used to control exposure. Respirators may be used when engineering and work practice controls are not technically feasible, when such controls are in the process of being installed, or when these controls fail and need to be supplemented or during brief, nonroutine, intermittent exposure. Respirators may also be used in situations involving non-routine work operations which are performed infrequently and in which exposures are limited in duration, and in emergency situations. In some instances cartridge respirator use is allowed, but only with strict time constraints. For example, at exposure below 5 ppm BD, a cartridge (or canister) respirator, either full or half face, may be used, but the cartridge must be replaced at least every 4 hours, and it must be replaced every 3 hours when the exposure is between 5 and 10 ppm. If the use of respirators is necessary, the only respirators permitted are those that have been approved by the National Institute for Occupational Safety and Health (NIOSH). In addition to respirator selection, a complete respiratory protection program must be instituted which includes regular training, maintenance, fit testing, inspection, cleaning, and evaluation of respirators. If you can smell BD while wearing a respirator, proceed immediately to fresh air, and change cartridge (or canister) before re-entering an area where there is BD exposure. If you experience difficulty in breathing while wearing a respirator, tell your supervisor.

B. Protective Clothing: Employees should be provided with and required to use impervious clothing, gloves, face shields (eightinch minimum), and other appropriate protective clothing necessary to prevent the skin from becoming frozen by contact with liquefied BD (or a vessel containing liquid BD).

Employees should be provided with and required to use splashproof safety goggles where liquefied BD may contact the eyes.

### V. PRECAUTIONS FOR SAFE USE, HANDLING, AND STORAGE

A. Fire and Explosion Hazards: BD is a flammable gas and can easily form explosive mixtures in air. It has a lower explosive limit of 2%, and an upper explosive limit of 11.5%. It has an autoignition temperature of 420 °C (788 °F). Its vapor is heavier than air (vapor density, 1.9) and may travel a considerable distance to a source of ignition and flash back. Usually it contains inhibitors to prevent self-polymerization (which is accompanied by evolution of heat) and to prevent formation of explosive peroxides. At elevated temperatures, such as in fire conditions, polymerization may take place. If the polymerization takes place in a container, there is a possibility of violent rupture of the container.

B. Hazard: Slightly toxic. Slight respiratory irritant. Direct contact of liquefied BD on skin may cause freeze burns and frostbite.

C. Storage: Protect against physical damage to BD containers. Outside or detached storage of BD containers is preferred. Inside storage should be in a cool, dry, well-ventilated, noncombustible location, away from all possible sources of ignition. Store cylinders vertically and do not stack. Do not store with oxidizing material.

D. Usual Shipping Containers: Liquefied BD is contained in steel pressure apparatus.

E. Electrical Equipment: Electrical installations in Class I hazardous locations, as defined in Article 500 of the National Electrical Code, should be in accordance with Article 501 of the Code. If explosion-proof electrical equipment is necessary, it shall be suitable for use in Group B. Group D equipment may be used if such equipment is isolated in accordance with Section 5015(a) by sealing all conduit 1/2- inch size or larger. See Venting of Deflagrations (NFPA No. 68, 1994), National Electrical Code (NFPA No. 70, 1996 ), Static Electricity (NFPA No. 77, 1993), Lightning Protection Systems (NFPA No. 780, 1995), and Fire Hazard Properties of Flammable Liquids, Gases and Volatile Solids (NFPA No. 325, 1994).

F. Fire Fighting: Stop flow of gas. Use water to keep fire-exposed containers cool. Fire extinguishers and quick drenching facilities must be readily available, and you should know where they are and how to operate them.

G. Spill and Leak: Persons not wearing protective equipment and clothing should be restricted from areas of spills or leaks until clean-up has been completed. If BD is spilled or leaked, the following steps should be taken:

1. Eliminate all ignition sources.

2. Ventilate area of spill or leak.

3. If in liquid form, for small quantities, allow to evaporate in a safe manner.

4. Stop or control the leak if this can be done without risk. If source of leak is a cylinder and the leak cannot be stopped in place, remove the leaking cylinder to a safe place and repair the leak or allow the cylinder to empty.

H. Disposal: This substance, when discarded or disposed of, is a hazardous waste according to Federal regulations (40 CFR part 261). It is listed as hazardous waste number D001 due to its ignitability. The transportation, storage, treatment, and disposal of this waste material must be conducted in compliance with 40 CFR parts 262, 263, 264, 268 and 270. Disposal can occur only in properly permitted facilities. Check state and local regulation of any additional requirements as these may be more restrictive than federal laws and regulation.

I. You should not keep food, beverages, or smoking materials in areas where there is BD exposure, nor should you eat or drink in such areas.

J. Ask your supervisor where BD is used in your work area and ask for any additional plant safety and health rules.

## VI. MEDICAL REQUIREMENTS

Your employer is required to offer you the opportunity to participate in a medical screening and surveillance program if you are exposed to BD at concentrations exceeding the action level (0.5 ppm BD as an 8-hour TWA) on 30 days or more a year, or at or above the 8 hr TWA (1 ppm) or STEL (5 ppm for 15 minutes) on 10 days or more a year. Exposure for any part of a day counts. If you have had exposure to BD in the past, but have been transferred to another job, you may still be eligible to participate in the medical screening and surveillance program. The OSHA rule specifies the past exposures that would qualify you for participation in the program. These past exposure are work histories that suggest the following: (1) That you have been exposed at or above the PELs on 30 days a year for 10 or more years; (2) that you have been exposed at or above the action level on 60 days a year for 10 or more years; or (3) that you have been exposed above 10 ppm on 30 days in any past year. Additionally, if you are exposed to BD in an emergency situation, you are eligible for a medical examination within 48 hours. The basic medical screening program includes a health questionnaire, physical examination, and blood test. These medical evaluations must be offered to you at a reasonable time and place, and without cost or loss of pay.

## VII. OBSERVATION OF MONITORING

Your employer is required to perform measurements that are representative of your exposure to BD and you or your designated representative are entitled to observe the monitoring procedure. You are entitled to observe the steps taken in the measurement procedure, and to record the results obtained. When the monitoring procedure is taking place in an area where respirators or personal protective clothing and equipment are required to be worn, you or your representative must also be provided with, and must wear, the protective clothing and equipment.

## VIII. ACCESS TO INFORMATION

A. Each year, your employer is required to inform you of the information contained in this appendix. In addition, your employer must instruct you in the proper work practices for using BD, emergency procedures, and the correct use of protective equipment.

B. Your employer is required to determine whether you are being exposed to BD. You or your representative has the right to observe employee measurements and to record the results obtained. Your employer is required to inform you of your exposure. If your employer determines that you are being overexposed, he or she is required to inform you of the actions which are being taken to reduce your exposure to within permissible exposure limits and of the schedule to implement these actions.

C. Your employer is required to keep records of your exposures and medical examinations. These records must be kept by the employer for at least thirty (30) years.

D. Your employer is required to release your exposure and medical records to you or your representative upon your request.

## APPENDIX B TO §1910.1051—SUBSTANCE TECHNICAL GUIDELINES FOR 1,3-BUTADIENE (NON-MANDATORY)

### I. Physical and Chemical Data

A. Substance identification:

1. Synonyms: 1,3-Butadiene (BD); butadiene; biethylene; bivinyl; divinyl; butadiene-1,3; buta-1,3-diene; erythrene; NCI-C50620; CAS-106-99-0.

2. Formula: $CH_2=CH-CH=CH_2$.

3. Molecular weight: 54.1.

B. Physical data:

1. Boiling point (760 mm Hg): -4.7 °C (23.5 °F).

2. Specific gravity (water = 1): 0.62 at 20 °C (68 °F).

3. Vapor density (air = 1 at boiling point of BD): 1.87.

4. Vapor pressure at 20 °C (68 °F): 910 mm Hg.

5. Solubility in water, g/100 g water at 20 °C (68 °F): 0.05.

6. Appearance and odor: Colorless, flammable gas with a mildly aromatic odor. Liquefied BD is a colorless liquid with a mildly aromatic odor.

### II. Fire, Explosion, and Reactivity Hazard Data

A. Fire:

1. Flash point: -76 °C (-105 °F) for take out; liquefied BD; Not applicable to BD gas.

2. Stability: A stabilizer is added to the monomer to inhibit formation of polymer during storage. Forms explosive peroxides in air in absence of inhibitor.

3. Flammable limits in air, percent by volume: Lower: 2.0; Upper: 11.5.

4. Extinguishing media: Carbon dioxide for small fires, polymer or alcohol foams for large fires.

5. Special fire fighting procedures: Fight fire from protected location or maximum possible distance. Stop flow of gas before extinguishing fire. Use water spray to keep fire-exposed cylinders cool.

6. Unusual fire and explosion hazards: BD vapors are heavier than air and may travel to a source of ignition and flash back. Closed containers may rupture violently when heated.

7. For purposes of compliance with the requirements of 29 CFR 1910.106, BD is classified as a flammable gas. For example, 7,500 ppm, approximately one-fourth of the lower flammable limit, would be considered to pose a potential fire and explosion hazard.

8. For purposes of compliance with 29 CFR 1910.155, BD is classified as a Class B fire hazard.

9. For purposes of compliance with 29 CFR 1910.307, locations classified as hazardous due to the presence of BD shall be Class I.

B. Reactivity:

1. Conditions contributing to instability: Heat. Peroxides are formed when inhibitor concentration is not maintained at proper level. At elevated temperatures, such as in fire conditions, polymerization may take place.

2. Incompatibilities: Contact with strong oxidizing agents may cause fires and explosions. The contacting of crude BD (not BD monomer) with copper and copper alloys may cause formations of explosive copper compounds.

3. Hazardous decomposition products: Toxic gases (such as carbon monoxide) may be released in a fire involving BD.

4. Special precautions: BD will attack some forms of plastics, rubber, and coatings. BD in storage should be checked for proper inhibitor content, for self-polymerization, and for formation of peroxides when in contact with air and iron. Piping carrying BD may become plugged by formation of rubbery polymer.

C. Warning Properties:

1. Odor Threshold: An odor threshold of 0.45 ppm has been reported in The American Industrial Hygiene Association (AIHA) Report, *Odor Thresholds for Chemicals with Established Occupational Health Standards*. (Ex. 32-28C)

2. Eye Irritation Level: Workers exposed to vapors of BD (concentration or purity unspecified) have complained of irritation of eyes, nasal passages, throat, and lungs. Dogs and rabbits exposed experimentally to as much as

6700 ppm for 7 1/2 hours a day for 8 months have developed no histologically demonstrable abnormality of the eyes.

3. Evaluation of Warning Properties: Since the mean odor threshold is about half of the 1 ppm PEL, and more than 10-fold below the 5 ppm STEL, most wearers of air purifying respirators should still be able to detect breakthrough before a significant overexposure to BD occurs.

### III. Spill, Leak, and Disposal Procedures

A. Persons not wearing protective equipment and clothing should be restricted from areas of spills or leaks until cleanup has been completed. If BD is spilled or leaked, the following steps should be taken:

1. Eliminate all ignition sources.

2. Ventilate areas of spill or leak.

3. If in liquid form, for small quantities, allow to evaporate in a safe manner.

4. Stop or control the leak if this can be done without risk. If source of leak is a cylinder and the leak cannot be stopped in place, remove the leaking cylinder to a safe place and repair the leak or allow the cylinder to empty.

B. Disposal: This substance, when discarded or disposed of, is a hazardous waste according to Federal regulations (40 CFR part 261). It is listed by the EPA as hazardous waste number D001 due to its ignitability. The transportation, storage, treatment, and disposal of this waste material must be conducted in compliance with 40 CFR parts 262, 263, 264, 268 and 270. Disposal can occur only in properly permitted facilities. Check state and local regulations for any additional requirements because these may be more restrictive than federal laws and regulations.

### IV. Monitoring and Measurement Procedures

A. Exposure above the Permissible Exposure Limit (8-hr TWA) or Short-Term Exposure Limit (STEL):

1. 8-hr TWA exposure evaluation: Measurements taken for the purpose of determining employee exposure under this standard are best taken with consecutive samples covering the full shift. Air samples must be taken in the employee's breathing zone (air that would most nearly represent that inhaled by the employee).

2. STEL exposure evaluation: Measurements must represent 15 minute exposures associated with operations most likely to exceed the STEL in each job and on each shift.

3. Monitoring frequencies: Table 1 gives various exposure scenarios and their required monitoring frequencies, as required by the final standard for occupational exposure to butadiene.

#### TABLE 1—FIVE EXPOSURE SCENARIOS AND THEIR ASSOCIATED MONITORING FREQUENCIES

| Action level | 8-hr TWA | STEL | Required monitoring activity |
|---|---|---|---|
| - * | - | - | No 8-hr TWA or STEL monitoring required. |
| + * | - | - | No STEL monitoring required. Monitor 8-hr TWA annually. |
| + | + | - | No STEL monitoring required. Periodic monitoring 8-hr TWA, in accordance with (d)(3)(ii).** |
| + | + | + | Periodic monitoring 8-hr TWA, in accordance with (d)(3)(ii)**. Periodic monitoring STEL, in accordance with (d)(3)(iii). |
| + | | + | Periodic monitoring STEL, in accordance with (d)(3)(iii). Monitor 8-hr TWA, annually. |

\* Exposure Scenario, Limit Exceeded: + = Yes, - = No.

\*\* The employer may decrease the frequency of exposure monitoring to annually when at least 2 consecutive measurements taken at least 7 days apart show exposures to be below the 8 hr TWA, but at or above the action level.

4. Monitoring techniques: Appendix D describes the validated method of sampling and analysis which has been tested by OSHA for use with BD.

The employer has the obligation of selecting a monitoring method which meets the accuracy and precision requirements of the standard under his or her unique field conditions. The standard requires that the method of monitoring must be accurate, to a 95 percent confidence level, to plus or minus 25 percent for concentrations of BD at or above 1 ppm, and to plus or minus 35 percent for concentrations below 1 ppm.

### V. Personal Protective Equipment

A. Employees should be provided with and required to use impervious clothing, gloves, face shields (eight-inch minimum), and other appropriate protective clothing necessary to prevent the skin from becoming frozen from contact with liquid BD.

B. Any clothing which becomes wet with liquid BD should be removed immediately and not re-worn until the butadiene has evaporated.

C. Employees should be provided with and required to use splash proof safety goggles where liquid BD may contact the eyes.

### VI. Housekeeping and Hygiene Facilities

For purposes of complying with 29 CFR 1910.141, the following items should be emphasized:

A. The workplace should be kept clean, orderly, and in a sanitary condition.

B. Adequate washing facilities with hot and cold water are to be provided and maintained in a sanitary condition.

### VII. Additional Precautions

A. Store BD in tightly closed containers in a cool, wellventilated area and take all necessary precautions to avoid any explosion hazard.

B. Non-sparking tools must be used to open and close metal containers. These containers must be effectively grounded.

C. Do not incinerate BD cartridges, tanks or other containers.

D. Employers must advise employees of all areas and operations where exposure to BD might occur.

## APPENDIX C TO §1910.1051— MEDICAL SCREENING AND SURVEILLANCE FOR 1,3-BUTADIENE (NON-MANDATORY)

### I. Basis for Medical Screening and Surveillance Requirements

#### A. Route of Entry Inhalation

#### B. Toxicology

Inhalation of BD has been linked to an increased risk of cancer, damage to the reproductive organs, and fetotoxicity. Butadiene can be converted via oxidation to epoxybutene and diepoxybutane, two genotoxic metabolites that may play a role in the expression of BD's toxic effects.

BD has been tested for carcinogenicity in mice and rats. Both species responded to BD exposure by developing cancer at multiple primary organ sites. Early deaths in mice were caused by malignant lymphomas, primarily lymphocytic type, originating in the thymus.

Mice exposed to BD have developed ovarian or testicular atrophy. Sperm head morphology tests also revealed abnormal sperm in mice exposed to BD; lethal mutations were found in a dominant lethal test. In light of these results in animals, the possibility that BD may adversely affect the reproductive systems of male and female workers must be considered.

Additionally, anemia has been observed in animals exposed to butadiene. In some cases, this anemia appeared to be a primary response to exposure; in other cases, it may have been secondary to a neoplastic response.

#### C. Epidemiology

Epidemiologic evidence demonstrates that BD exposure poses an increased risk of leukemia. Mild alterations of hematologic parameters have also been observed in synthetic rubber workers exposed to BD.

### II. Potential Adverse Health Effects

#### A. Acute

Skin contact with liquid BD causes characteristic burns or frostbite. BD is gaseous form can irritate the eyes, nasal passages, throat, and lungs. Blurred vision, coughing, and drowsiness may also occur. Effects are mild at 2,000 ppm and pronounced at 8,000 ppm for exposures occurring over the full workshift.

At very high concentrations in air, BD is an anesthetic, causing narcosis, respiratory paralysis, unconsciousness, and death. Such concentrations are unlikely, however, except in an extreme emergency because BD poses an explosion hazard at these levels.

### B. Chronic

The principal adverse health effects of concern are BD-induced lymphoma, leukemia and potential reproductive toxicity. Anemia and other changes in the peripheral blood cells may be indicators of excessive exposure to BD.

### C. Reproductive

Workers may be concerned about the possibility that their BD exposure may be affecting their ability to procreate a healthy child. For workers with high exposures to BD, especially those who have experienced difficulties in conceiving, miscarriages, or stillbirths, appropriate medical and laboratory evaluation of fertility may be necessary to determine if BD is having any adverse effect on the reproductive system or on the health of the fetus.

## III. Medical Screening Components At-A-Glance

### A. Health Questionnaire

The most important goal of the health questionnaire is to elicit information from the worker regarding potential signs or symptoms generally related to leukemia or other blood abnormalities. Therefore, physicians or other licensed health care professionals should be aware of the presenting symptoms and signs of lymphohematopoietic disorders and cancers, as well as the procedures necessary to confirm or exclude such diagnoses. Additionally, the health questionnaire will assist with the identification of workers at greatest risk of developing leukemia or adverse reproductive effects from their exposures to BD.

Workers with a history of reproductive difficulties or a personal or family history of immune deficiency syndromes, blood dyscrasias, lymphoma, or leukemia, and those who are or have been exposed to medicinal drugs or chemicals known to affect the hematopoietic or lymphatic systems may be at higher risk from their exposure to BD. After the initial administration, the health questionnaire must be updated annually.

### B. Complete Blood Count (CBC)

The medical screening and surveillance program requires an annual CBC, with differential and platelet count, to be provided for each employee with BD exposure. This test is to be performed on a blood sample obtained by phlebotomy of the venous system or, if technically feasible, from a fingerstick sample of capillary blood. The sample is to be analyzed by an accredited laboratory.

Abnormalities in a CBC may be due to a number of different etiologies. The concern for workers exposed to BD includes, but is not limited to, timely identification of lymphohematopoietic cancers, such as leukemia and non-Hodgkin's lymphoma. Abnormalities of portions of the CBC are identified by comparing an individual's results to those of an established range of normal values for males and females. A substantial change in any individual employee's CBC may also be viewed as "abnormal" for that individual even if all measurements fall within the population-based range of normal values. It is suggested that a flowsheet for laboratory values be included in each employee's medical record so that comparisons and trends in annual CBCs can be easily made.

A determination of the clinical significance of an abnormal CBC shall be the responsibility of the examining physician, other licensed health care professional, or medical specialist to whom the employee is referred. Ideally, an abnormal CBC should be compared to previous CBC measurements for the same employee, when available. Clinical common sense may dictate that a CBC value that is very slightly outside the normal range does not warrant medical concern. A CBC abnormality may also be the result of a temporary physical stressor, such as a transient viral illness, blood donation, or menorrhagia, or laboratory error. In these cases, the CBC should be repeated in a timely fashion, i.e., within 6 weeks, to verify that return to the normal range has occurred. A clinically significant abnormal CBC should result in removal of the employee from further exposure to BD. Transfer of the employee to other work duties in a BD-free environment would be the preferred recommendation.

### C. Physical Examination

The medical screening and surveillance program requires an initial physical examination for workers exposed to BD; this examination is repeated once every three years. The initial physical examination should assess each worker's baseline general health and rule out clinical signs of medical conditions that may be caused by or aggravated by occupational BD exposure. The physical examination should be directed at identification of signs of lymphohematopoietic disorders, including lymph node enlargement, splenomegaly, and hepatomegaly.

Repeated physical examinations should update objective clinical findings that could be indicative of interim development of a lymphohematopoietic disorder, such as lymphoma, leukemia, or other blood abnormality. Physical examinations may also be provided on an as needed basis in order to follow up on a positive answer on the health questionnaire, or in response to an abnormal CBC. Physical examination of workers who will no longer be working in jobs with BD exposure are intended to rule out lymphohematopoietic disorders.

The need for physical examinations for workers concerned about adverse reproductive effects from their exposure to BD should be identified by the physician or other licensed health care professional and provided accordingly. For these workers, such consultations and examinations may relate to developmental toxicity and reproductive capacity.

Physical examination of workers acutely exposed to significant levels of BD should be especially directed at the respiratory system, eyes, sinuses, skin, nervous system, and any region associated with particular complaints. If the worker has received a severe acute exposure, hospitalization may be required to assure proper medical management. Since this type of exposure may place workers at greater risk of blood abnormalities, a CBC must be obtained within 48 hours and repeated at one, two, and three months.

## APPENDIX D TO §1910.1051— SAMPLING AND ANALYTICAL METHOD FOR 1,3-BUTADIENE (NON-MANDATORY)

*OSHA Method No.:* 56.

*Matrix:* Air.

*Target concentration:* 1 ppm (2.21 mg/m$^3$)

*Procedure:* Air samples are collected by drawing known volumes of air through sampling tubes containing charcoal adsorbent which has been coated with 4-tert-butylcatechol. The samples are desorbed with carbon disulfide and then analyzed by gas chromatography using a flame ionization detector.

*Recommended sampling rate and air volume:* 0.05 L/min and 3 L.

*Detection limit of the overall procedure:* 90 ppb (200 ug/m$^3$) (based on 3 L air volume).

*Reliable quantitation limit:* 155 ppb (343 ug/m$^3$) (based on 3 L air volume).

*Standard error of estimate at the target concentration:* 6.5%.

*Special requirements:* The sampling tubes must be coated with 4-tert-butylcatechol. Collected samples should be stored in a freezer.

*Status of method:* A sampling and analytical method has been subjected to the established evaluation procedures of the Organic Methods Evaluation Branch, OSHA Analytical Laboratory, Salt Lake City, Utah 84165.

### 1. Background

This work was undertaken to develop a sampling and analytical procedure for BD at 1 ppm. The current method recommended by OSHA for collecting BD uses activated coconut shell charcoal as the sampling medium (Ref. 5.2). This method was found to be inadequate for use at low BD levels because of sample instability.

The stability of samples has been significantly improved through the use of a specially cleaned charcoal which is coated with 4-tert-butylcatechol (TBC). TBC is a polymerization inhibitor for BD (Ref. 5.3).

#### 1.1.1 Toxic effects

Symptoms of human exposure to BD include irritation of the eyes, nose and throat. It can also cause coughing, drowsiness and fatigue. Dermatitis and frostbite can result from skin exposure to liquid BD. (Ref. 5.1)

NIOSH recommends that BD be handled in the workplace as a potential occupational carcinogen. This recommendation is based on two inhalation studies that resulted in cancers at multiple sites in rats and in mice. BD has also demonstrated mutagenic activity in the presence of a liver microsomal activating system. It has also been reported to have adverse reproductive effects. (Ref. 5.1)

#### 1.1.2. Potential workplace exposure

About 90% of the annual production of BD is used to manufacture styrene-butadiene rubber and Polybutadiene rubber. Other uses include:

Polychloroprene rubber, acrylonitrile butadiene-stryene resins, nylon intermediates, styrene-butadiene latexes, butadiene polymers, thermoplastic elastomers, nitrile resins, methyl methacrylate-butadiene styrene resins and chemical intermediates. (Ref. 5.1)

### 1.1.3. Physical properties (Ref. 5.1)

CAS No.: 106-99-0

Molecular weight: 54.1

Appearance: Colorless gas

Boiling point: -4.41 °C (760 mm Hg)

Freezing point: -108.9 °C

Vapor pressure: 2 atm @ 15.3 °C; 5 atm @ 47 °C

Explosive limits: 2 to 11.5% (by volume in air)

Odor threshold: 0.45 ppm

Structural formula: $H_2C{:}CHCH{:}CH_2$

Synonyms: BD; biethylene; bivinyl; butadiene; divinyl; buta-1,3diene; alpha-gamma-butadiene; erythrene; NCI-C50602; pyrrolylene; vinylethylene.

### 1.2. Limit defining parameters

The analyte air concentrations listed throughout this method are based on an air volume of 3 L and a desorption volume of 1 ml. Air concentrations listed in ppm are referenced to 25 °C and 760 mm Hg.

#### 1.2.1. Detection limit of the analytical procedure

The detection limit of the analytical procedure was 304 pg per injection. This was the amount of BD which gave a response relative to the interferences present in a standard.

#### 1.2.2. Detection limit of the overall procedure

The detection limit of the overall procedure was 0.60 µg per sample (90 ppb or 200 µg/m³). This amount was determined graphically. It was the amount of analyte which, when spiked on the sampling device, would allow recovery approximately equal to the detection limit of the analytical procedure.

#### 1.2.3. Reliable quantitation limit

The reliable quantitation limit was 1.03 µg per sample (155 ppb or 343 µg/m₃). This was the smallest amount of analyte which could be quantitated within the limits of a recovery of at least 75% and a precision (±1.96 SD) of ±25% or better.

#### 1.2.4. Sensitivity[1]

The sensitivity of the analytical procedure over a concentration range representing 0.6 to 2 times the target concentration, based on the recommended air volume, was 387 area units per g/ml. This value was determined from the slope of the calibration curve. The sensitivity may vary with the particular instrument used in the analysis.

#### 1.2.5. Recovery

The recovery of BD from samples used in storage tests remained above 77% when the samples were stored at ambient temperature and above 94% when the samples were stored at refrigerated temperature. These values were determined from regression lines which were calculated from the storage data. The recovery of the analyte from the collection device must be at least 75% following storage.

#### 1.2.6. Precision (analytical method only)

The pooled coefficient of variation obtained from replicate determinations of analytical standards over the range of 0.6 to 2 times the target concentration was 0.011.

#### 1.2.7. Precision (overall procedure)

The precision at the 95% confidence level for the refrigerated temperature storage test was ±12.7%. This value includes an additional ±5% for sampling error. The overall procedure must provide results at the target concentrations that are ±25% at the 95% confidence level.

#### 1.2.8. Reproducibility

Samples collected from a controlled test atmosphere and a draft copy of this procedure were given to a chemist unassociated with this evaluation. The average recovery was 97.2% and the standard deviation was 6.2%.

## 2. Sampling procedure

### 2.1. Apparatus

*2.1.1.* Samples are collected by use of a personal sampling pump that can be calibrated to within ±5% of the recommended 0.05 L/min sampling rate with the sampling tube in line.

*2.1.2.* Samples are collected with laboratory prepared sampling tubes. The sampling tube is constructed of silane-treated glass and is about 5-cm long. The ID is 4 mm and the OD is 6 mm. One end of the tube is tapered so that a glass wool end plug will hold the contents of the tube in place during sampling. The opening in the tapered end of the sampling tube is at least one-half the ID of the tube (2 mm). The other end of the sampling tube is open to its full 4-mm ID to facilitate packing of the tube. Both ends of the tube are fire-polished for safety. The tube is packed with 2 sections of pretreated charcoal which has been coated with TBC. The tube is packed with a 50-mg backup section, located nearest the tapered end, and with a 100-mg sampling section of charcoal. The two sections of coated adsorbent are separated and retained with small plugs of silanized glass wool. Following packing, the sampling tubes are sealed with two 7/32 inch OD plastic end caps. Instructions for the pretreatment and coating of the charcoal are presented in Section 4.1 of this method.

### 2.2. Reagents

None required.

### 2.3. Technique

*2.3.1.* Properly label the sampling tube before sampling and then remove the plastic end caps.

*2.3.2.* Attach the sampling tube to the pump using a section of flexible plastic tubing such that the larger front section of the sampling tube is exposed directly to the atmosphere. Do not place any tubing ahead of the sampling tube. The sampling tube should be attached in the worker's breathing zone in a vertical manner such that it does not impede work performance.

*2.3.3.* After sampling for the appropriate time, remove the sampling tube from the pump and then seal the tube with plastic end caps. Wrap the tube lengthwise.

*2.3.4.* Include at least one blank for each sampling set. The blank should be handled in the same manner as the samples with the exception that air is not drawn through it.

*2.3.5.* List any potential interferences on the sample data sheet.

*2.3.6.* The samples require no special shipping precautions under normal conditions. The samples should be refrigerated if they are to be exposed to higher than normal ambient temperatures. If the samples are to be stored before they are shipped to the laboratory, they should be kept in a freezer. The samples should be placed in a freezer upon receipt at the laboratory.

### 2.4. Breakthrough

(Breakthrough was defined as the relative amount of analyte found on the backup section of the tube in relation to the total amount of analyte collected on the sampling tube. Five-percent breakthrough occurred after sampling a test atmosphere containing 2.0 ppm BD for 90 min at 0.05 L/min. At the end of this time 4.5 L of air had been sampled and 20.1 µg of the analyte was collected. The relative humidity of the sampled air was 80% at 23 °C.)

Breakthrough studies have shown that the recommended sampling procedure can be used at air concentrations higher than the target concentration. The sampling time, however, should be reduced to 45 min if both the expected BD level and the relative humidity of the sampled air are high.

---

[1] The reliable quantitation limit and detection limits reported in the method are based upon optimization of the instrument for the smallest possible amount of analyte. When the target concentration of an analyte is exceptionally higher than these limits, they may not be attainable at the routine operation parameters.

### 2.5. Desorption efficiency

The average desorption efficiency for BD from TBC coated charcoal over the range from 0.6 to 2 times the target concentration was 96.4%. The efficiency was essentially constant over the range studied.

### 2.6. Recommended air volume and sampling rate

2.6.1. The recommended air volume is 3L.

2.6.2. The recommended sampling rate is 0.05 L/min for 1 hour.

### 2.7. Interferences

There are no known interferences to the sampling method.

### 2.8. Safety precautions

2.8.1. Attach the sampling equipment to the worker in such a manner that it will not interfere with work performance or safety.

2.8.2. Follow all safety practices that apply to the work area being sampled.

## 3. Analytical procedure

### 3.1. Apparatus

3.1.1. A gas chromatograph (GC), equipped with a flame ionization detector (FID).[2]

3.1.2. A GC column capable of resolving the analytes from any interference.[3]

3.1.3. Vials, glass 2-ml with Teflon-lined caps.

3.1.4. Disposable Pasteur-type pipets, volumetric flasks, pipets and syringes for preparing samples and standards, making dilutions and performing injections.

### 3.2. Reagents

3.2.1. Carbon disulfide.[4]

The benzene contaminant that was present in the carbon disulfide was used as an internal standard (ISTD) in this evaluation.

3.2.2. Nitrogen, hydrogen and air, GC grade.

3.2.3. BD of known high purity.[5]

### 3.3. Standard preparation

3.3.1. Prepare standards by diluting known volumes of BD gas with carbon disulfide. This can be accomplished by injecting the appropriate volume of BD into the headspace above the 1-ml of carbon disulfide contained in sealed 2-ml vial. Shake the vial after the needle is removed from the septum.[6]

3.3.2. The mass of BD gas used to prepare standards can be determined by use of the following equations:

$$MV = (760/BP)(273+t)/(273)(22.41)$$

Where:

MV = ambient molar volume

BP = ambient barometric pressure

T = ambient temperature

$$\mu g/\mu l = 54.09/MV$$

µg/standard = (µg/µl)(µl) BD used to prepare the standard

### 3.4. Sample preparation

3.4.1. Transfer the 100-mg section of the sampling tube to a 2-ml vial. Place the 50-mg section in a separate vial. If the glass wool plugs contain a significant amount of charcoal, place them with the appropriate sampling tube section.

3.4.2. Add 1-ml of carbon disulfide to each vial.

3.4.3. Seal the vials with Teflon-lined caps and then allow them to desorb for one hour. Shake the vials by hand vigorously several times during the desorption period.

3.4.4. If it is not possible to analyze the samples within 4 hours, separate the carbon disulfide from the charcoal, using a disposable Pasteur-type pipet, following the one hour. This separation will improve the stability of desorbed samples.

3.4.5. Save the used sampling tubes to be cleaned and repacked with fresh adsorbent.

### 3.5. Analysis

3.5.1. GC Conditions

Column temperature: 95 °C

Injector temperature: 180 °C

Detector temperature: 275 °C

Carrier gas flow rate: 30 ml/min

Injection volume: 0.80 µl

GC column: 20-ft x 1/8-in OD stainless steel GC column containing 20% FFAP on 80/100 Chromabsorb W-AW-DMCS.

3.5.2. Chromatogram. See Section 4.2.

3.5.3. Use a suitable method, such as electronic or peak heights, to measure detector response.

3.5.4. Prepare a calibration curve using several standard solutions of different concentrations. Prepare the calibration curve daily. Program the integrator to report the results in µg/ml.

### 3.5.5. Bracket sample concentrations with standards.

### 3.6. Interferences (analytical)

3.6.1. Any compound with the same general retention time as the analyte and which also gives a detector response is a potential interference. Possible interferences should be reported by the industrial hygienist to the laboratory with submitted samples.

3.6.2. GC parameters (temperature, column, etc.) may be changed to circumvent interferences.

3.6.3. A useful means of structure designation is GC/MS. It is recommended that this procedure be used to confirm samples whenever possible.

### 3.7. Calculations

3.7.1. Results are obtained by use of calibration curves. Calibration curves are prepared by plotting detector response against concentration for each standard. The best line through the data points is determined by curve fitting.

3.7.2. The concentration, in ug/ml, for a particular sample is determined by comparing its detector response to the calibration curve. If any analyte is found on the backup section, this amount is added to the amount found on the front section. Blank corrections should be performed before adding the results together.

3.7.3. The BD air concentration can be expressed using the following equation:

$$mg/m^3 = (A)(B)/(C)(D)$$

Where:

A = µg/ml from Section 3.7.2

B = volume

C = L of air sampled

D = efficiency

3.7.4. The following equation can be used to convert results in $mg/m^3$ to ppm:

$$ppm = (mg/m^3)(24.46)/54.09$$

Where:

$mg/m^3$ = result from Section 3.7.3.

24.46 = molar volume of an ideal gas at 760 mm Hg and 25 °C.

---

[2] A Hewlett-Packard Model 5840A GC was used for this evaluation. Injections were performed using a Hewlett-Packard Model 7671A automatic sampler.

[3] A 20-ft x 1/8-inch OD stainless steel GC column containing 20% FFAP on 80/100 mesh Chromabsorb W-AW-DMCS was used for this evaluation.

[4] Fisher Scientific Company A.C.S. Reagent Grade solvent was used in this evaluation.

[5] Matheson Gas Products, CP Grade 1,3-butadiene was used in this study.

[6] A standard containing 7.71 µg/ml (at ambient temperature and pressure) was prepared by diluting 4 µl of the gas with 1-ml of carbon disulfide.

## 3.8. Safety precautions (analytical)

*3.8.1.* Avoid skin contact and inhalation of all chemicals.

*3.8.2.* Restrict the use of all chemicals to a fume hood whenever possible.

*3.8.3.* Wear safety glasses and a lab coat in all laboratory areas.

## 4. Additional Information

*4.1.* A procedure to prepare specially cleaned charcoal coated with TBC

*4.1.1.* Apparatus.

*4.1.1.1.* Magnetic stirrer and stir bar.

*4.1.1.2.* Tube furnace capable of maintaining a temperature of 700 °C and equipped with a quartz tube that can hold 30 g of charcoal.[8]

*4.1.1.3.* A means to purge nitrogen gas through the charcoal inside the quartz tube.

*4.1.1.4.* Water bath capable of maintaining a temperature of 60 °C.

*4.1.1.5.* Miscellaneous laboratory equipment: Oneliter vacuum flask, 1-L Erlenmeyer flask, 350-M1 Buchner funnel with a coarse fitted disc, 4-oz brown bottle, rubber stopper, Teflon tape etc.

### 4.1.2. Reagents

*4.1.2.1.* Phosphoric acid, 10% by weight, in water.[9]

*4.1.2.2.* 4-tert-Butylcatechol (TBC).[10]

*4.1.2.3.* Specially cleaned coconut shell charcoal, 20/40 mesh.[11]

*4.1.2.4.* Nitrogen gas, GC grade.

*4.1.3.* Procedure.

Weigh 30g of charcoal into a 500-ml Erlenmeyer flask. Add about 250 ml of 10% phosphoric acid to the flask and then swirl the mixture. Stir the mixture for 1 hour using a magnetic stirrer. Filter the mixture using a fitted Buchner funnel. Wash the charcoal several times with 250-ml portions of deionized water to remove all traces of the acid. Transfer the washed charcoal to the tube furnace quartz tube. Place the quartz tube in the furnace and then connect the nitrogen gas purge to the tube. Fire the charcoal to 700 °C. Maintain that temperature for at least 1 hour. After the charcoal has cooled to room temperature, transfer it to a tared beaker. Determine the weight of the charcoal and then add an amount of TBC which is 10% of the charcoal, by weight.

CAUTION-TBC is toxic and should only be handled in a fume hood while wearing gloves.

Carefully mix the contents of the beaker and then transfer the mixture to a 4-oz bottle. Stopper the bottle with a clean rubber stopper which has been wrapped with Teflon tape. Clamp the bottle in a water bath so that the water level is above the charcoal level. Gently heat the bath to 60 °C and then maintain that temperature for 1 hour. Cool the charcoal to room temperature and then transfer the coated charcoal to a suitable container.

The coated charcoal is now ready to be packed into sampling tubes. The sampling tubes should be stored in a sealed container to prevent contamination. Sampling tubes should be stored in the dark at room temperature. The sampling tubes should be segregated by coated adsorbent lot number.

### 4.2 Chromatograms

The chromatograms were obtained using the recommended analytical method. The chart speed was set at 1 cm/min for the first three min and then at 0.2 cm/min for the time remaining in the analysis.

The peak which elutes just before BD is a reaction product between an impurity on the charcoal and TBC. This peak is always present, but it is easily resolved from the analyte. The peak which elutes immediately before benzene is an oxidation product of TBC.

## 5. References

*5.1.* "Current Intelligence Bulletin 41, 1,3-Butadiene", U.S. Dept. of Health and Human Services, Public Health Service, Center for Disease Control, NIOSH.

*5.2.* "NIOSH Manual of Analytical Methods", 2nd ed; U.S. Dept. of Health Education and Welfare, National Institute for Occupational Safety and Health: Cincinnati, OH. 1977, Vol. 2, Method No. S91 DHEW (NIOSH) Publ. (US), No. 77-157-B.

*5.3.* Hawley, G.C., Ed. "The Condensed Chemical Dictionary", 8th ed.; Van Nostrand Rienhold Company: New York, 1971; 139.5.4. *Chem. Eng. News* (June 10, 1985), (63), 22-66.

### APPENDIX E—RESPIRATOR FIT TESTING PROCEDURES (MANDATORY) [Reserved]

[§1910.1051, App. E. reserved at 63 FR 1294, Jan. 8, 1998, eff. Apr. 8, 1998.]

### APPENDIX F TO §1910.1051— MEDICAL QUESTIONNAIRES (NON-MANDATORY)

[61 FR 56831, Nov. 4, 1996, as amended at 63 FR 1294, Jan. 8, 1998]

### §1910.1052 Methylene Chloride.

This occupational health standard establishes requirements for employers to control occupational exposure to methylene chloride (MC). Employees exposed to MC are at increased risk of developing cancer, adverse effects on the heart, central nervous system and liver, and skin or eye irritation. Exposure may occur through inhalation, by absorption through the skin, or through contact with the skin. MC is a solvent which is used in many different types of work activities, such as paint stripping, polyurethane foam manufacturing, and cleaning and degreasing. Under the requirements of paragraph (d) of this section, each covered employer must make an initial determination of each employee's exposure to MC. If the employer determines that employees are exposed below the action level, the only other provisions of this section that apply are that a record must be made of the determination, the employees must receive information and training under paragraph (l) of this section and, where appropriate, employees must be protected from contact with liquid MC under paragraph (h) of this section. The provisions of the MC standard are as follows:

**1910.1052(a)** *Scope and application.* This section applies to all occupational exposures to methylene chloride (MC), Chemical Abstracts Service Registry Number 75-09-2, in general industry, construction and shipyard employment.

**1910.1052(b)** *Definitions.* For the purposes of this section, the following definitions shall apply:

*Action level* means a concentration of airborne MC of 12.5 parts per million (ppm) calculated as an eight (8)-hour time-weighted average (TWA).

*Assistant Secretary* means the Assistant Secretary of Labor for Occupational Safety and Health, U.S. Department of Labor, or designee.

*Authorized person* means any person specifically authorized by the employer and required by work duties to be present in regulated areas, or any person entering such an area as a designated representative of employees for the purpose of exercising the right to observe monitoring and measuring procedures under paragraph (d) of this section, or any other person authorized by the OSH Act or regulations issued under the Act.

*Director* means the Director of the National Institute for Occupational Safety and Health, U.S. Department of Health and Human Services, or designee.

*Emergency* means any occurrence, such as, but not limited to, equipment failure, rupture of containers, or failure of control equipment, which results, or is likely to result in an uncontrolled release of MC. If an incidental release of MC can be controlled by employees such as maintenance personnel at the time of release and in accordance with the leak/spill provisions required by paragraph (f) of this section, it is not considered an emergency as defined by this standard.

*Employee exposure* means exposure to airborne MC which occurs or would occur if the employee were not using respiratory protection.

*Methylene chloride* (MC) means an organic compound with chemical formula, $CH_2Cl_2$. Its Chemical Abstracts Service Registry Number is 75-09-2. Its molecular weight is 84.9 g/mole.

---

[8] A Lindberg Type 55035 Tube furnace was used in this evaluation.

[9] Baker Analyzed" Reagent grade was diluted with water for use in this evaluation.

[10] The Aldrich Chemical Company 99% grade was used in this evaluation.

[11] Specially cleaned charcoal was obtained from Supelco, Inc. for use in this evaluation. The cleaning process used by Supelco is proprietary.

Subpart Z—Toxic and Hazardous Substances

12. Were you in the military? yes no
If yes, what did you do in the military? _____

### Family Health History

1. In the FAMILY MEMBER column, across from the disease name, write which family member, if any, had the disease.

| DISEASE | FAMILY MEMBER |
|---|---|
| Cancer | _____ |
| Lymphoma | _____ |
| Sickle Cell Disease or Trait | _____ |
| Immune Disease | _____ |
| Leukemia | _____ |
| Anemia | _____ |

2. Please fill in the following information about family health:

| Relative | Alive? | Age at death? | Cause of death? |
|---|---|---|---|
| Father | _____ | _____ | _____ |
| Mother | _____ | _____ | _____ |
| Brother/Sister | _____ | _____ | _____ |
| Brother/Sister | _____ | _____ | _____ |
| Brother/Sister | _____ | _____ | _____ |

### Personal Health History

Birth Date ___/___/___   Age _____   Sex _____   Height _____   Weight _____

1. Do you smoke any tobacco products? yes no
Please circle your answer.
2. Have you ever had any kind of surgery or operation? yes no
If yes, what type of surgery: _____
3. Have you ever been in the hospital for any other reasons? yes no
If yes, please describe the reason: _____
4. Do you have any on-going or current medical problems or conditions? yes no
If yes, please describe: _____
5. Do you now have or have you ever had any of the following? Please check all that apply to you.

unexplained fever __   bruising easily __   still birth __
anemia ("low blood") __   lupus __   eye redness __
HIV/AIDS __   weight loss __   lumps you can feel __
weakness __   kidney problems __   child with birth defect __
sickle cell __   enlarged lymph nodes __   autoimmune disease __
miscarriage __   liver disease __   overly tired __
skin rash __   cancer __   lung problems __
bloody stools __   infertility __   rheumatoid arthritis __
leukemia/lymphoma __   drinking problems __   mononucleosis ("mono") __
neck mass/swelling __   thyroid problems __   nagging cough __
wheezing __   night sweats __
yellowing of skin __   chest pain __

Please circle your answer.
6. Do you have any symptoms or health problems that you think may be related to your work with BD? yes no don't know
If yes, please describe: _____
7. Have any of your co-workers had similar symptoms or problems? yes no
If yes, please describe: _____

---

## MEDICAL QUESTIONNAIRES (NON-MANDATORY)

### 1,3 -Butadiene (BD) Initial Health Questionnaire

**DIRECTIONS:**
You have been asked to answer the questions on this form because you work with BD (butadiene). These questions are about your work, medical history, and health concerns. Please do your best to answer all of the questions. If you need help, please tell the doctor or health care professional who reviews this form.

This form is a confidential medical record. Only information directly related to your health and safety on the job may be given to your employer. Personal health information will not be given to anyone without your consent.

Date: _____
Name: _____ _____ _____   SSN ___-___-___
       Last             First              MI
Job Title: _____   Company's Name: _____
Supervisor's Name: _____   Supervisor's Phone No.: (___) ___-____

### Work History

1. Please list all jobs you have had in the past, starting with the job you have now and moving back in time to your first job. (For more space, write on the back of this page.)

| Main Job Duty | Years | Company Name City, State | Chemicals |
|---|---|---|---|
| 1. | | | |
| 2. | | | |
| 3. | | | |
| 4. | | | |
| 5. | | | |
| 6. | | | |
| 7. | | | |
| 8. | | | |

2. Please describe what you do during a typical work day. Be sure to tell about your work with BD.
_____

3. Please check any of these chemicals that you work with now or have worked with in the past:

benzene __   carbon tetrachloride ("carbon tet") __
glues __   arsine __
toluene __   carbon disulfide __
inks, dyes __   lead __
other solvents, grease cutters __   cement __
insecticides (like DDT, lindane, etc.) __   petroleum products __
paints, varnishes, thinners, strippers __   nitrites __
dusts __

4. Please check the protective clothing or equipment you use at the job you have now:
gloves __   coveralls __
respirator __   dust mask __
safety glasses, goggles __

Please circle your answer of yes or no.
5. Does your protective clothing or equipment fit you properly? yes no
6. Have you ever made changes in your protective clothing or equipment to make it fit better? yes no
7. Have you ever been exposed to BD when you were not wearing protective clothing or equipment? yes no
8. Where do you eat, drink and/or smoke when you are at work? (Please check all that apply.)
Cafeteria/restaurant/snack bar __   Break room/employee lounge __
Smoking lounge __   At my work station __

Please circle your answer.
9. Have you been exposed to radiation (like x-rays or nuclear material) at the job you have now or at past jobs? yes no
10. Do you have any hobbies that expose you to dusts or chemicals (including paints, glues, etc.)? yes no
11. Do you have any second or side jobs? yes no
If yes, what are your duties there? _____

4. Do you have any of the following? Please place a check for all that apply to you.

| | | |
|---|---|---|
| unexplained fever | bruising easily | still birth |
| anemia ("low blood") | lupus | eye redness |
| HIV/AIDS | weight loss | lumps you can feel |
| weakness | kidney problems | child with birth defect |
| sickle cell | enlarged lymph nodes | autoimmune disease |
| miscarriage | liver disease | overly tired |
| skin rash | cancer | lung problems |
| bloody stools | infertility | rheumatoid arthritis |
| leukemia/lymphoma | drinking problems | mononucleosis ("mono") |
| neck mass/swelling | thyroid problems | nagging cough |
| wheezing | night sweats | yellowing of skin |
| chest pain | | |

Please circle your answer.

5. Do you have any symptoms or health problems that you think may be related to your work with BD? yes no don't know
   If yes, please describe: _____
6. Have any of your co-workers had similar symptoms or problems? yes no
   If yes, please describe: _____
7. Do you notice any irritation of your eyes, nose, throat, lungs, or skin when working with BD? yes no
   If yes, please describe: _____
8. Do you notice any blurred vision, coughing, drowsiness, nausea, or headache when working with BD? yes no
   Have you been taking any NEW medications (including birth control or over-the-counter)? yes no
   If yes, please list: _____
10. Have you developed any NEW allergies to medications, foods, or chemicals? yes no
    If yes, please list: _____
11. Do you have any health conditions not covered by this questionnaire that you think are affected by your work with BD? yes no
    If yes, please explain: _____
12. Did you understand all the questions? yes no
    Signature _____

[61 FR 56831, Nov. 4, 1996, as amended at 63 FR 1294, Jan. 8, 1998]

---

8. Do you notice any irritation of your eyes, nose, throat, lungs, or skin when working with BD? yes no
9. Do you notice any blurred vision, coughing, drowsiness, nausea or headache when working with BD? yes no
10. Do you take any medications (including birth control or over-the-counter)? yes no
    If yes, please list: _____
11. Are you allergic to any medication, food, or chemicals? yes no
    If yes, please list: _____
12. Do you have any health conditions not covered by this questionnaire that you think are affected by your work with BD? yes no
    If yes, please explain: _____
13. Did you understand all the questions? yes no
    Signature _____

1,3-Butadiene (BD) Update Health Questionnaire

DIRECTIONS: You have been asked to answer the questions on this form because you work with BD (butadiene). These questions ask about changes in your work, medical history, and health concerns since the last time you were evaluated. Please do your best to answer all of the questions. If you need help, please tell the doctor or health care professional who reviews this form.

This form is a confidential record. Only information directly related to your health and safety on the job may be given to your employer. Personal health information will not be given to anyone without your consent.

Date: _____
Name: _____ _____ _____ SSN _____ / _____ / _____
       Last     First    MI
Job Title: _____ Company's Name: _____
Supervisor's Name: _____ Supervisor's Phone No.: ( ___ ) ___ - ____

### Present Work History

1. Please describe any NEW duties that you have at your job: _____

2. Please list any additional job titles you have: _____

Please circle your answer.

3. Are you exposed to any other chemicals in your work since the last time you were evaluated for exposure to BD? yes no
   If yes, please list what they are: _____
4. Does your personal protective equipment and clothing fit you properly? yes no
5. Have you made changes in this equipment or clothing to make it fit better? yes no
6. Have you been exposed to BD when you were not wearing protective equipment or clothing? yes no
7. Are you exposed to any NEW chemicals at home or while working on hobbies? yes no
   If yes, please list what they are: _____
8. Since your last BD health evaluation, have you started working any new second or side jobs? yes no
   If yes, what are your duties there? _____

### Personal Health History

1. What is your current weight? _____ pounds
2. Have you been diagnosed with any new medical conditions or illness since your last evaluation? yes no
   If yes, please tell what they are: _____
3. Since your last evaluation, have you been in the hospital for any illnesses, injuries, or surgery? yes no
   If yes, please describe: _____

*Physician or other licensed health care professional* is an individual whose legally permitted scope of practice (i.e., license, registration, or certification) allows him or her to independently provide or be delegated the responsibility to provide some or all of the health care services required by paragraph (j) of this section.

*Regulated area* means an area, demarcated by the employer, where an employee's exposure to airborne concentrations of MC exceeds or can reasonably be expected to exceed either the 8-hour TWA PEL or the STEL.

*Symptom* means central nervous system effects such as headaches, disorientation, dizziness, fatigue, and decreased attention span; skin effects such as chapping, erythema, cracked skin, or skin burns; and cardiac effects such as chest pain or shortness of breath.

*This section* means this methylene chloride standard.

**1910.1052(c)** *Permissible exposure limits (PELs)*—**(1)** *Eight-hour time-weighted average (TWA) PEL.* The employer shall ensure that no employee is exposed to an airborne concentration of MC in excess of twenty-five parts of MC per million parts of air (25 ppm) as an 8-hour TWA.

**1910.1052(c)(2)** *Short-term exposure limit (STEL).* The employer shall ensure that no employee is exposed to an airborne concentration of MC in excess of one hundred and twenty-five parts of MC per million parts of air (125 ppm) as determined over a sampling period of fifteen minutes.

**1910.1052(d)** *Exposure monitoring*—**(1)** *Characterization of employee exposure.* **(i)** Where MC is present in the workplace, the employer shall determine each employee's exposure by either:

**1910.1052(d)(1)(i)(A)** Taking a personal breathing zone air sample of each employee's exposure; or

**1910.1052(d)(1)(i)(B)** Taking personal breathing zone air samples that are representative of each employee's exposure.

**1910.1052(d)(1)(ii)** *Representative samples.* The employer may consider personal breathing zone air samples to be representative of employee exposures when they are taken as follows:

**1910.1052(d)(1)(ii)(A)** *8-hour TWA PEL.* The employer has taken one or more personal breathing zone air samples for at least one employee in each job classification in a work area during every work shift, and the employee sampled is expected to have the highest MC exposure.

**1910.1052(d)(1)(ii)(B)** *Short-term exposure limits.* The employer has taken one or more personal breathing zone air samples which indicate the highest likely 15-minute exposures during such operations for at least one employee in each job classification in the work area during every work shift, and the employee sampled is expected to have the highest MC exposure.

**1910.1052(d)(1)(ii)(C)** *Exception.* Personal breathing zone air samples taken during one work shift may be used to represent employee exposures on other work shifts where the employer can document that the tasks performed and conditions in the workplace are similar across shifts.

**1910.1052(d)(1)(iii)** *Accuracy of monitoring.* The employer shall ensure that the methods used to perform exposure monitoring produce results that are accurate to a confidence level of 95 percent, and are:

**1910.1052(d)(1)(iii)(A)** Within plus or minus 25 percent for airborne concentrations of MC above the 8-hour TWA PEL or the STEL; or

**1910.1052(d)(1)(iii)(B)** Within plus or minus 35 percent for airborne concentrations of MC at or above the action level but at or below the 8-hour TWA PEL.

**1910.1052(d)(2)** *Initial determination.* Each employer whose employees are exposed to MC shall perform initial exposure monitoring to determine each affected employee's exposure, except under the following conditions:

**1910.1052(d)(2)(i)** Where objective data demonstrate that MC cannot be released in the workplace in airborne concentrations at or above the action level or above the STEL. The objective data shall represent the highest MC exposures likely to occur under reasonably foreseeable conditions of processing, use, or handling. The employer shall document the objective data exemption as specified in paragraph (m) of this section;

**1910.1052(d)(2)(ii)** Where the employer has performed exposure monitoring within 12 months prior to April 10, 1997 and that exposure monitoring meets all other requirements of this section, and was conducted under conditions substantially equivalent to existing conditions; or

**1910.1052(d)(2)(iii)** Where employees are exposed to MC on fewer than 30 days per year (e.g., on a construction site), and the employer has measurements by direct-reading instruments which give immediate results (such as a detector tube) and which provide sufficient information regarding employee exposures to determine what control measures are necessary to reduce exposures to acceptable levels.

**1910.1052(d)(3)** *Periodic monitoring.* Where the initial determination shows employee exposures at or above the action level or above the STEL, the employer shall establish an exposure monitoring program for periodic monitoring of employee exposure to MC in accordance with Table 1:

### TABLE 1—INITIAL DETERMINATION EXPOSURE SCENARIOS AND THEIR ASSOCIATED MONITORING FREQUENCIES

| Exposure scenario | Required monitoring activity |
|---|---|
| Below the action level and at or below the STEL. | No 8-hour TWA or STEL monitoring required. |
| Below the action level and above the STEL. | No 8-hour TWA monitoring required; monitor STEL exposures every three months. |
| At or above the action level, at or below the TWA, and at or below the STEL. | Monitor 8-hour TWA exposures every six months. |
| At or above the action level, at or below the TWA, and above the STEL. | Monitor 8-hour TWA exposures every six months and monitor STEL exposures every three months. |
| Above the TWA and at or below the STEL. | Monitor 8-hour TWA exposures every three months. In addition, without regard to the last sentence of the note to paragraph (d)(3), the following employers must monitor STEL exposures paragraph (n) of this section or the date by which they in fact achieve the 8-hour TWA PEL, whichever comes first: employers engaged in polyurethane foam manufacturing; foam fabrication; furniture refinishing; general aviation aircraft stripping; product formulation; use of MC-based adhesives for boat building and repair, recreational vehicle manufacture, van conversion, or upholstery; and use of MC in construction work for restoration and preservation of buildings, painting and paint removal, cabinet making, or floor refinishing and resurfacing. |
| Above the TWA and above the STEL. | Monitor 8-hour TWA exposures and STEL exposures every three months. |

**[Note to paragraph (d)(3):** The employer may decrease the frequency of 8-hour TWA exposure monitoring to every six months when at least two consecutive measurements taken at least seven days apart show exposures to be at or below the 8-hour TWA PEL. The employer may discontinue the periodic 8-hour TWA monitoring for employees where at least two consecutive measurements taken at least seven days apart are below the action level. The employer may discontinue the periodic STEL monitoring for employees where at least two consecutive measurements taken at least 7 days apart are at or below the STEL.]

**1910.1052(d)(4)** *Additional monitoring.* **(i)** The employer shall perform exposure monitoring when a change in workplace conditions indicates that employee exposure may have increased. Examples of situations that may require additional monitoring include changes in production, process, control equipment, or work practices, or a leak, rupture, or other breakdown.

**1910.1052(d)(4)(ii)** Where exposure monitoring is performed due to a spill, leak, rupture or equipment breakdown, the employer shall clean-up the MC and perform the appropriate repairs before monitoring.

**1910.1052(d)(5)** *Employee notification of monitoring results.* **(i)** The employer shall, within 15 working days after the receipt of the results of any monitoring performed under this section, notify each affected employee of these results in writing, either individually or by posting of results in an appropriate location that is accessible to affected employees.

**1910.1052(d)(5)(ii)** Whenever monitoring results indicate that employee exposure is above the 8-hour TWA PEL or the STEL, the employer shall describe in the written notification the corrective action being taken to reduce employee exposure to or below the 8-hour TWA PEL or STEL and the schedule for completion of this action.

**1910.1052(d)(6)** *Observation of monitoring*—**(i)** *Employee observation.* The employer shall provide affected employees or their designated representatives an opportunity to observe any monitoring of employee exposure to MC conducted in accordance with this section.

**1910.1052(d)(6)(ii)** *Observation procedures.* When observation of the monitoring of employee exposure to MC requires entry into an area where the use of protective clothing or equipment is required, the employer shall provide, at no cost to the observer(s), and the observer(s) shall be required to use such clothing and equipment and shall comply with all other applicable safety and health procedures.

**1910.1052(e)** *Regulated areas.* **(1)** The employer shall establish a regulated area wherever an employee's exposure to airborne concentrations of MC exceeds or can reasonably be expected to exceed either the 8-hour TWA PEL or the STEL.

**1910.1052(e)(2)** The employer shall limit access to regulated areas to authorized persons.

**1910.1052(e)(3)** The employer shall supply a respirator, selected in accordance with paragraph (h)(3) of this section, to each person who enters a regulated area and shall require each affected employee to use that respirator whenever MC exposures are likely to exceed the 8-hour TWA PEL or STEL.

[Note to paragraph (e)(3): An employer who has implemented all feasible engineering, work practice and administrative controls (as required in paragraph (f) of this section), and who has established a regulated area (as required by paragraph (e)(1) of this section) where MC exposure can be reliably predicted to exceed the 8-hour TWA PEL or the STEL only on certain days (for example, because of work or process schedule) would need to have affected employees use respirators in that regulated area only on those days.]

**1910.1052(e)(4)** The employer shall ensure that, within a regulated area, employees do not engage in non-work activities which may increase dermal or oral MC exposure.

**1910.1052(e)(5)** The employer shall ensure that while employees are wearing respirators, they do not engage in activities (such as taking medication or chewing gum or tobacco) which interfere with respirator seal or performance.

**1910.1052(e)(6)** The employer shall demarcate regulated areas from the rest of the workplace in any manner that adequately establishes and alerts employees to the boundaries of the area and minimizes the number of authorized employees exposed to MC within the regulated area.

**1910.1052(e)(7)** An employer at a multi-employer worksite who establishes a regulated area shall communicate the access restrictions and locations of these areas to all other employers with work operations at that worksite.

**1910.1052(f)** *Methods of compliance*—**(1)** *Engineering and work practice controls.* The employer shall institute and maintain the effectiveness of engineering controls and work practices to reduce employee exposure to or below the PELs except to the extent that the employer can demonstrate that such controls are not feasible. Wherever the feasible engineering controls and work practices which can be instituted are not sufficient to reduce employee exposure to or below the 8-TWA PEL or STEL, the employer shall use them to reduce employee exposure to the lowest levels achievable by these controls and shall supplement them by the use of respiratory protection that complies with the requirements of paragraph (g) of this section.

**1910.1052(f)(2)** *Prohibition of rotation.* The employer shall not implement a schedule of employee rotation as a means of compliance with the PELs.

**1910.1052(f)(3)** *Leak and spill detection.* **(i)** The employer shall implement procedures to detect leaks of MC in the workplace. In work areas where spills may occur, the employer shall make provisions to contain any spills and to safely dispose of any MC-contaminated waste materials.

**1910.1052(f)(3)(ii)** The employer shall ensure that all incidental leaks are repaired and that incidental spills are cleaned promptly by employees who use the appropriate personal protective equipment and are trained in proper methods of cleanup.

[Note to paragraph (f)(3)(ii): See Appendix A of this section for examples of procedures that satisfy this requirement. Employers covered by this standard may also be subject to the hazardous waste and emergency response provisions contained in 29 CFR 1910.120(q).]

**1910.1052(g)** *Respiratory protection*—**(1)** *General.* For employees who use respirators required by this section, the employer must provide respirators that comply with the requirements of this paragraph. Respirators must be used during:

**1910.1052(g)(1)(i)** Periods when an employee's exposure to MC exceeds the 8-hour TWA PEL, or STEL (for example, when an employee is using MC in a regulated area).

**1910.1052(g)(1)(ii)** Periods necessary to install or implement feasible engineering and work-practice controls.

**1910.1052(g)(1)(iii)** A few work operations, such as some maintenance operations and repair activities, for which the employer demonstrates that engineering and work-practice controls are infeasible.

**1910.1052(g)(1)(iv)** Work operations for which feasible engineering and work-practice controls are not sufficient to reduce employee exposures to or below the PELs.

**1910.1052(g)(1)(v)** Emergencies.

**1910.1052(g)(2)** *Respirator program.* **(i)** The employer must implement a respiratory protection program in accordance with 29 CFR 1910.134 (b) through (m) (except (d)(1)(iii) and (d)(3)(iii)(B) (1) and (2)).

**1910.1052(g)(2)(ii)** Employers who provide employees with gas masks with organic-vapor canisters for the purpose of emergency escape must replace the canisters after any emergency use and before the gas masks are returned to service.

**1910.1052(g)(3)** *Respirator selection.* The employer must select appropriate atmosphere-supplying respirators from Table 2 of this section.

## TABLE 2—MINIMUM REQUIREMENTS FOR RESPIRATORY PROTECTION FOR AIRBORNE METHYLENE CHLORIDE

| Methylene chloride airborne concentration (ppm) or condition of use | Minimum respirator required [1] |
|---|---|
| Up to 625 ppm (25 × PEL) | (1) Continuous flow supplied-air respirator, hood or helmet. |
| Up to 1250 ppm (50 × 8-TWA PEL) | (1) Full facepiece supplied-air respirator operated in negative pressure (demand) mode.<br>(2) Full facepiece self-contained breathing apparatus (SCBA) operated in negative pressure (demand) mode. |
| Up to 5000 ppm (200 × 8-TWA PEL) | (1) Continuous flow supplied-air respirator, full facepiece.<br>(2) Pressure demand supplied-air respirator, full facepiece.<br>(3) Positive pressure full facepiece SCBA. |
| Unknown concentration, or above 5000 ppm (Greater than 200 × 8-TWA PEL) | (1) Positive pressure full facepiece SCBA.<br>(2) Full facepiece pressure demand supplied-air respirator with an auxiliary self-contained air supply. |
| Fire fighting | Positive pressure full facepiece SCBA. |
| Emergency escape | (1) Any continuous flow or pressure demand SCBA.<br>(2) Gas mask with organic vapor canister. |

[1] Respirators assigned for higher airborne concentrations may be used at lower concentrations.

**1910.1052(g)(4)** *Medical evaluation.* Before having an employee use a supplied-air respirator in the negative-pressure mode, or a gas mask with an organic-vapor canister for emergency escape, the employer must:

**1910.1052(g)(4)(i)** Have a physician or other licensed health-care professional (PLHCP) evaluate the employee's ability to use such respiratory protection.

**1910.1052(g)(4)(ii)** Ensure that the PLHCP provides their findings in a written opinion to the employee and the employer.

**1910.1052(h)** *Protective Work Clothing and Equipment.* **(1)** Where needed to prevent MC-induced skin or eye irritation, the employer shall provide clean protective clothing and equipment which is resistant to MC, at no cost to the employee, and shall ensure that each affected employee uses it. Eye and face protection shall meet the requirements of 29 CFR 1910.133 or 29 CFR 1915.153, as applicable.

**1910.1052(h)(2)** The employer shall clean, launder, repair and replace all protective clothing and equipment required by this paragraph as needed to maintain their effectiveness.

**1910.1052(h)(3)** The employer shall be responsible for the safe disposal of such clothing and equipment.

[Note to paragraph (h)(4): See Appendix A for examples of disposal procedures that will satisfy this requirement.]

**1910.1052(i)** *Hygiene facilities.* **(1)** If it is reasonably foreseeable that employees' skin may contact solutions containing 0.1 percent or greater MC (for example, through splashes, spills or improper work practices), the employer shall provide conveniently located washing facilities capable of removing the MC, and shall ensure that affected employees use these facilities as needed.

**1910.1052(i)(2)** If it is reasonably foreseeable that an employee's eyes may contact solutions containing 0.1 percent or greater MC (for example through splashes, spills or improper work practices), the employer shall provide appropriate eyewash facilities within the immediate work area for emergency use, and shall ensure that affected employees use those facilities when necessary.

**1910.1052(j)** *Medical surveillance*—**(1)** *Affected employees.* The employer shall make medical surveillance available for employees who are or may be exposed to MC as follows:

**1910.1052(j)(1)(i)** At or above the action level on 30 or more days per year, or above the 8-hour TWA PEL or the STEL on 10 or more days per year;

**1910.1052(j)(1)(ii)** Above the 8-TWA PEL or STEL for any time period where an employee has been identified by a physician or other licensed health care professional as being at risk from cardiac disease or from some other serious MC-related health condition and such employee requests inclusion in the medical surveillance program;

**1910.1052(j)(1)(iii)** During an emergency.

**1910.1052(j)(2)** *Costs.* The employer shall provide all required medical surveillance at no cost to affected employees, without loss of pay and at a reasonable time and place.

**1910.1052(j)(3)** *Medical personnel.* The employer shall ensure that all medical surveillance procedures are performed by a physician or other licensed health care professional, as defined in paragraph (b) of this section.

**1910.1052(j)(4)** *Frequency of medical surveillance.* The employer shall make medical surveillance available to each affected employee as follows:

**1910.1052(j)(4)(i)** *Initial surveillance.* The employer shall provide initial medical surveillance under the schedule provided by paragraph (n)(2)(iii) of this section, or before the time of initial assignment of the employee, whichever is later. The employer need not provide the initial surveillance if medical records show that an affected employee has been provided with medical surveillance that complies with this section within 12 months before April 10, 1997.

**1910.1052(j)(4)(ii)** *Periodic medical surveillance.* The employer shall update the medical and work history for each affected employee annually. The employer shall provide periodic physical examinations, including appropriate laboratory surveillance, as follows:

**1910.1052(j)(4)(ii)(A)** For employees 45 years of age or older, within 12 months of the initial surveillance or any subsequent medical surveillance; and

**1910.1052(j)(4)(ii)(B)** For employees younger than 45 years of age, within 36 months of the initial surveillance or any subsequent medical surveillance.

**1910.1052(j)(4)(iii)** *Termination of employment or reassignment.* When an employee leaves the employer's workplace, or is reassigned to an area where exposure to MC is consistently at or below the action level and STEL, medical surveillance shall be made available if six months or more have elapsed since the last medical surveillance.

**1910.1052(j)(4)(iv)** *Additional surveillance.* The employer shall provide additional medical surveillance at frequencies other than those listed above when recommended in the written medical opinion. (For example, the physician or other licensed health care professional may determine an examination is warranted in less than 36 months for employees younger than 45 years of age based upon evaluation of the results of the annual medical and work history.)

**1910.1052(j)(5)** *Content of medical surveillance*—**(i)** *Medical and work history.* The comprehensive medical and work history shall emphasize neurological symptoms, skin conditions, history of hematologic or liver disease, signs or symptoms suggestive of heart disease (angina, coronary artery disease), risk factors for cardiac disease, MC exposures, and work practices and personal protective equipment used during such exposures.

[Note to paragraph (j)(5)(i): See Appendix B of this section for an example of a medical and work history format that would satisfy this requirement.]

**1910.1052(j)(5)(ii)** *Physical examination.* Where physical examinations are provided as required above, the physician or other licensed health care professional shall accord particular attention to the lungs, cardiovascular system (including blood pressure and pulse), liver, nervous system, and skin. The physician or other licensed health care professional shall determine the extent and nature of the physical examination based on the health status of the employee and analysis of the medical and work history.

**1910.1052(j)(5)(iii)** *Laboratory surveillance.* The physician or other licensed health care professional shall determine the extent of any required laboratory surveillance based on the employee's observed health status and the medical and work history.

[**Note to paragraph (j)(5)(iii):** See Appendix B of this section for information regarding medical tests. Laboratory surveillance may include before- and after-shift carboxyhemoglobin determinations, resting ECG, hematocrit, liver function tests and cholesterol levels.]

**1910.1052(j)(5)(iv)** *Other information or reports.* The medical surveillance shall also include any other information or reports the physician or other licensed health care professional determines are necessary to assess the employee's health in relation to MC exposure.

**1910.1052(j)(6)** *Content of emergency medical surveillance.* The employer shall ensure that medical surveillance made available when an employee has been exposed to MC in emergency situations includes, at a minimum:

**1910.1052(j)(6)(i)** Appropriate emergency treatment and decontamination of the exposed employee;

**1910.1052(j)(6)(ii)** Comprehensive physical examination with special emphasis on the nervous system, cardiovascular system, lungs, liver and skin, including blood pressure and pulse;

**1910.1052(j)(6)(iii)** Updated medical and work history, as appropriate for the medical condition of the employee; and

**1910.1052(j)(6)(iv)** Laboratory surveillance, as indicated by the employee's health status.

[**Note to paragraph (j)(6)(iv):** See Appendix B for examples of tests which may be appropriate.]

**1910.1052(j)(7)** *Additional examinations and referrals.* Where the physician or other licensed health care professional determines it is necessary, the scope of the medical examination shall be expanded and the appropriate additional medical surveillance, such as referrals for consultation or examination, shall be provided.

**1910.1052(j)(8)** *Information provided to the physician or other licensed health care professional.* The employer shall provide the following information to a physician or other licensed health care professional who is involved in the diagnosis of MC-induced health effects:

**1910.1052(j)(8)(i)** A copy of this section including its applicable appendices;

**1910.1052(j)(8)(ii)** A description of the affected employee's past, current and anticipated future duties as they relate to the employee's MC exposure;

**1910.1052(j)(8)(iii)** The employee's former or current exposure levels or, for employees not yet occupationally exposed to MC, the employee's anticipated exposure levels and the frequency and exposure levels anticipated to be associated with emergencies;

**1910.1052(j)(8)(iv)** A description of any personal protective equipment, such as respirators, used or to be used; and

**1910.1052(j)(8)(v)** Information from previous employment-related medical surveillance of the affected employee which is not otherwise available to the physician or other licensed health care professional.

**1910.1052(j)(9)** *Written medical opinions.* **(i)** For each physical examination required by this section, the employer shall ensure that the physician or other licensed health care professional provides to the employer and to the affected employee a written opinion regarding the results of that examination within 15 days of completion of the evaluation of medical and laboratory findings, but not more than 30 days after the examination. The written medical opinion shall be limited to the following information:

**1910.1052(j)(9)(i)(A)** The physician or other licensed health care professional's opinion concerning whether exposure to MC may contribute to or aggravate the employee's existing cardiac, hepatic, neurological (including stroke) or dermal disease or whether the employee has any other medical condition(s) that would place the employee's health at increased risk of material impairment from exposure to MC.

**1910.1052(j)(9)(i)(B)** Any recommended limitations upon the employee's exposure to MC, including removal from MC exposure, or upon the employee's use of respirators, protective clothing, or other protective equipment.

**1910.1052(j)(9)(i)(C)** A statement that the employee has been informed by the physician or other licensed health care professional that MC is a potential occupational carcinogen, of risk factors for heart disease, and the potential for exacerbation of underlying heart disease by exposure to MC through its metabolism to carbon monoxide; and

**1910.1052(j)(9)(i)(D)** A statement that the employee has been informed by the physician or other licensed health care professional of the results of the medical examination and any medical conditions resulting from MC exposure which require further explanation or treatment.

**1910.1052(j)(9)(ii)** The employer shall instruct the physician or other licensed health care professional not to reveal to the employer, orally or in the written opinion, any specific records, findings, and diagnoses that have no bearing on occupational exposure to MC.

[**Note to paragraph (j)(9)(ii):** The written medical opinion may also include information and opinions generated to comply with other OSHA health standards.]

**1910.1052(j)(10)** *Medical presumption.* For purposes of this paragraph (j) of this section, the physician or other licensed health care professional shall presume, unless medical evidence indicates to the contrary, that a medical condition is unlikely to require medical removal from MC exposure if the employee is not exposed to MC above the 8-hour TWA PEL. If the physician or other licensed health care professional recommends removal for an employee exposed below the 8-hour TWA PEL, the physician or other licensed health care professional shall cite specific medical evidence, sufficient to rebut the presumption that exposure below the 8-hour TWA PEL is unlikely to require removal, to support the recommendation. If such evidence is cited by the physician or other licensed health care professional, the employer must remove the employee. If such evidence is not cited by the physician or other licensed health care professional, the employer is not required to remove the employee.

**1910.1052(j)(11)** *Medical Removal Protection (MRP).* **(i)** Temporary medical removal and return of an employee.

**1910.1052(j)(11)(i)(A)** Except as provided in paragraph (j)(10) of this section, when a medical determination recommends removal because the employee's exposure to MC may contribute to or aggravate the employee's existing cardiac, hepatic, neurological (including stroke), or skin disease, the employer must provide medical removal protection benefits to the employee and either:

**1910.1052(j)(11)(i)(A)(1)** Transfer the employee to comparable work where methylene chloride exposure is below the action level; or

**1910.1052(j)(11)(i)(A)(2)** Remove the employee from MC exposure.

**1910.1052(j)(11)(i)(B)** If comparable work is not available and the employer is able to demonstrate that removal and the costs of extending MRP benefits to an additional employee, considering feasibility in relation to the size of the employer's business and the other requirements of this standard, make further reliance on MRP an inappropriate remedy, the employer may retain the additional employee in the existing job until transfer or removal becomes appropriate, provided:

**1910.1052(j)(11)(i)(B)(1)** The employer ensures that the employee receives additional medical surveillance, including a physical examination at least every 60 days until transfer or removal occurs; and

**1910.1052(j)(11)(i)(B)(2)** The employer or PLHCP informs the employee of the risk to the employee's health from continued MC exposure.

**1910.1052(j)(11)(i)(C)** The employer shall maintain in effect any job-related protective measures or limitations, other than removal, for as long as a medical determination recommends them to be necessary.

**1910.1052(j)(11)(ii)** End of MRP benefits and return of the employee to former job status.

**1910.1052(j)(11)(ii)(A)** The employer may cease providing MRP benefits at the earliest of the following:

**1910.1052(j)(11)(ii)(A)(1)** Six months;

**1910.1052(j)(11)(ii)(A)(2)** Return of the employee to the employee's former job status following receipt of a medical determination concluding that the employee's exposure to MC no longer will aggravate any cardiac, hepatic, neurological (including stroke), or dermal disease;

**1910.1052(j)(11)(ii)(A)(3)** Receipt of a medical determination concluding that the employee can never return to MC exposure.

**1910.1052(j)(11)(ii)(B)** For the purposes of this paragraph (j), the requirement that an employer return an employee to the employee's former job status is not intended to expand upon or restrict any rights an employee has or would have had, absent temporary medical removal, to a specific job classification or position under the terms of a collective bargaining agreement.

**1910.1052(j)(12)** *Medical removal protection benefits.* **(i)** For purposes of this paragraph (j), the term medical removal protection benefits means that, for each removal, an employer must maintain for up to six months the earnings, seniority, and other employment rights and benefits of the employee as though the employee had not been removed from MC exposure or transferred to a comparable job.

**1910.1052(j)(12)(ii)** During the period of time that an employee is removed from exposure to MC, the employer may condition the provision of medical removal protection benefits upon the employee's participation in follow-up medical surveillance made available pursuant to this section.

**1910.1052(j)(12)(iii)** If a removed employee files a workers' compensation claim for a MC-related disability, the employer shall continue the MRP benefits required by this paragraph until either the claim is resolved or the 6-month period for payment f MRP benefits has passed, whichever occurs first. To the extent the employee is entitled to indemnity payments for earnings lost during the period of removal, the employer's obligation to provide medical removal protection benefits to the employee shall be reduced by the amount of such indemnity payments.

**1910.1052(j)(12)(iv)** The employer's obligation to provide medical removal protection benefits to a removed employee shall be reduced to the extent that the employee receives compensation for earnings lost during the period of removal from either a publicly or an employer-funded compensation program, or receives income from employment with another employer made possible by virtue of the employee's removal.

**1910.1052(j)(13)** *Voluntary removal or restriction of an employee.* Where an employer, although not required by this section to do so, removes an employee from exposure to MC or otherwise places any limitation on an employee due to the effects of MC exposure on the employee's medical condition, the employer shall provide medical removal protection benefits to the employee equal to those required by paragraph (j)(12) of this section.

**1910.1052(j)(14)** *Multiple health care professional review mechanism.* **(i)** If the employer selects the initial physician or licensed health care professional (PLHCP) to conduct any medical examination or consultation provided to an employee under this paragraph (j)(11), the employer shall notify the employee of the right to seek a second medical opinion each time the employer provides the employee with a copy of the written opinion of that PLHCP.

**1910.1052(j)(14)(ii)** If the employee does not agree with the opinion of the employer-selected PLHCP, notifies the employer of that fact, and takes steps to make an appointment with a second PLHCP within 15 days of receiving a copy of the written opinion of the initial PLHCP, the employer shall pay for the PLHCP chosen by the employee to perform at least the following:

**1910.1052(j)(14)(ii)(A)** Review any findings, determinations or recommendations of the initial PLHCP; and

**1910.1052(j)(14)(ii)(B)** Conduct such examinations, consultations, and laboratory tests as the PLHCP deems necessary to facilitate this review.

**1910.1052(j)(14)(iii)** If the findings, determinations or recommendations of the second PLHCP differ from those of the initial PLHCP, then the employer and the employee shall instruct the two health care professionals to resolve the disagreement.

**1910.1052(j)(14)(iv)** If the two health care professionals are unable to resolve their disagreement within 15 days, then those two health care professionals shall jointly designate a PLHCP who is a specialist in the field at issue. The employer shall pay for the specialist to perform at least the following:

**1910.1052(j)(14)(iv)(A)** Review the findings, determinations, and recommendations of the first two PLHCPs; and

**1910.1052(j)(14)(iv)(B)** Conduct such examinations, consultations, laboratory tests and discussions with the prior PLHCPs as the specialist deems necessary to resolve the disagreements of the prior health care professionals.

**1910.1052(j)(14)(v)** The written opinion of the specialist shall be the definitive medical determination. The employer shall act consistent with the definitive medical determination, unless the employer and employee agree that the written opinion of one of the other two PLHCPs shall be the definitive medical determination.

**1910.1052(j)(14)(vi)** The employer and the employee or authorized employee representative may agree upon the use of any expeditious alternate health care professional determination mechanism in lieu of the multiple health care professional review mechanism provided by this paragraph so long as the alternate mechanism otherwise satisfies the requirements contained in this paragraph.

**1910.1052(k)** *Hazard communication.* The employer shall communicate the following hazards associated with MC on labels and in material safety data sheets in accordance with the requirements of the Hazard Communication Standard, 29 CFR 1910.1200, 29 CFR 1915.1200, or 29 CFR 1926.59, as appropiate: cancer, cardiac effects (including elevation of carboxyhemoglobin), central nervous system effects, liver effects, and skin and eye irritation.

**1910.1052(l)** *Employee information and training.* **(1)** The employer shall provide information and training for each affected employee prior to or at the time of initial assignment to a job involving potential exposure to MC.

**1910.1052(l)(2)** The employer shall ensure that information and training is presented in a manner that is understandable to the employees.

**1910.1052(l)(3)** In addition to the information required under the Hazard Communication Standard at 29 CFR 1910.1200, 29 CFR 1915.1200, or 29 CFR 1926.59, as appropriate:

**1910.1052(l)(3)(i)** The employer shall inform each affected employee of the requirements of this section and information available in its appendices, as well as how to access or obtain a copy of it in the workplace;

**1910.1052(l)(3)(ii)** Wherever an employee's exposure to airborne concentrations of MC exceeds or can reasonably be expected to exceed the action level, the employer shall inform each affected employee of the quantity, location, manner of use, release, and storage of MC and the specific operations in the workplace that could result in exposure to MC, particularly noting where exposures may be above the 8-hour TWA PEL or STEL;

**1910.1052(l)(4)** The employer shall train each affected employee as required under the Hazard Communication standard at 29 CFR 1910.1200, 29 CFR 1915.1200, or 29 CFR 1926.59, as appropriate.

**1910.1052(l)(5)** The employer shall re-train each affected employee as necessary to ensure that each employee exposed above the action level or the STEL maintains the requisite understanding of the principles of safe use and handling of MC in the workplace.

**1910.1052(l)(6)** Whenever there are workplace changes, such as modifications of tasks or procedures or the institution of new tasks or procedures, which increase employee exposure, and where those exposures exceed or can reasonably be expected to exceed the action level, the employer shall update the training as necessary to ensure that each affected employee has the requisite proficiency.

**1910.1052(l)(7)** An employer whose employees are exposed to MC at a multi-employer worksite shall notify the other employers with work operations at that site in accordance with the requirements of the Hazard Communication Standard, 29 CFR 1910.1200, 29 CFR 1915.1200, or 29 CFR 1926.59, as appropiate.

**1910.1052(l)(8)** The employer shall provide to the Assistant Secretary or the Director, upon request, all available materials relating to employee information and training.

**1910.1052(m)** *Recordkeeping*—**(1)** *Objective data.* **(i)** Where an employer seeks to demonstrate that initial monitoring is unnecessary through reasonable reliance on objective data showing that any materials in the workplace containing MC will not release MC at levels which exceed the action level or the STEL under foreseeable conditions of exposure, the employer shall establish and maintain an accurate record of the objective data relied upon in support of the exemption.

**1910.1052(m)(1)(ii)** This record shall include at least the following information:

**1910.1052(m)(1)(ii)(A)** The MC-containing material in question;

**1910.1052(m)(1)(ii)(B)** The source of the objective data;

**1910.1052(m)(1)(ii)(C)** The testing protocol, results of testing, and/or analysis of the material for the release of MC;

**1910.1052(m)(1)(ii)(D)** A description of the operation exempted under paragraph (d)(2)(i) of this section and how the data support the exemption; and

**1910.1052(m)(1)(ii)(E)** Other data relevant to the operations, materials, processing, or employee exposures covered by the exemption.

**1910.1052(m)(1)(iii)** The employer shall maintain this record for the duration of the employer's reliance upon such objective data.

**1910.1052(m)(2)** *Exposure measurements.* **(i)** The employer shall establish and keep an accurate record of all measurements taken to monitor employee exposure to MC as prescribed in paragraph (d) of this section.

**1910.1052(m)(2)(ii)** Where the employer has 20 or more employees, this record shall include at least the following information:

**1910.1052(m)(2)(ii)(A)** The date of measurement for each sample taken;

**1910.1052(m)(2)(ii)(B)** The operation involving exposure to MC which is being monitored;

**1910.1052(m)(2)(ii)(C)** Sampling and analytical methods used and evidence of their accuracy;

**1910.1052(m)(2)(ii)(D)** Number, duration, and results of samples taken;

**1910.1052(m)(2)(ii)(E)** Type of personal protective equipment, such as respiratory protective devices, worn, if any; and

**1910.1052(m)(2)(ii)(F)** Name, social security number, job classification and exposure of all of the employees represented by monitoring, indicating which employees were actually monitored.

**1910.1052(m)(2)(iii)** Where the employer has fewer than 20 employees, the record shall include at least the following information:

**1910.1052(m)(2)(iii)(A)** The date of measurement for each sample taken;

**1910.1052(m)(2)(iii)(B)** Number, duration, and results of samples taken; and

**1910.1052(m)(2)(iii)(C)** Name, social security number, job classification and exposure of all of the employees represented by monitoring, indicating which employees were actually monitored.

**1910.1052(m)(2)(iv)** The employer shall maintain this record for at least thirty (30) years, in accordance with 29 CFR 1910.1020.

**1910.1052(m)(3)** *Medical surveillance.* **(i)** The employer shall establish and maintain an accurate record for each employee subject to medical surveillance under paragraph (j) of this section.

**1910.1052(m)(3)(ii)** The record shall include at least the following information:

**1910.1052(m)(3)(ii)(A)** The name, social security number and description of the duties of the employee;

**1910.1052(m)(3)(ii)(B)** Written medical opinions; and

**1910.1052(m)(3)(ii)(C)** Any employee medical conditions related to exposure to MC.

**1910.1052(m)(3)(iii)** The employer shall ensure that this record is maintained for the duration of employment plus thirty (30) years, in accordance with 29 CFR 1910.1020.

**1910.1052(m)(4)** *Availability.* **(i)** The employer, upon written request, shall make all records required to be maintained by this section available to the Assistant Secretary and the Director for examination and copying in accordance with 29 CFR 1910.1020.

**[Note to paragraph (m)(4)(i):** All records required to be maintained by this section may be kept in the most administratively convenient form (for example, electronic or computer records would satisfy this requirement).]

**1910.1052(m)(4)(ii)** The employer, upon request, shall make any employee exposure and objective data records required by this section available for examination and copying by affected employees, former employees, and designated representatives in accordance with 29 CFR 1910.1020.

**1910.1052(m)(4)(iii)** The employer, upon request, shall make employee medical records required to be kept by this section available for examination and copying by the subject employee and by anyone having the specific written consent of the subject employee in accordance with 29 CFR 1910.1020.

**1910.1052(m)(5)** *Transfer of records.* The employer shall comply with the requirements concerning transfer of records set forth in 29 CFR 1910.1020(h).

**1910.1052(n)** [Reserved.]

**1910.1052(o)** *Appendices.* The information contained in the appendices does not, by itself, create any additional obligations not otherwise imposed or detract from any existing obligation.

**[Note to paragraph (o):** The requirement of 29 CFR 1910.1052(g)(1) to use respiratory protection whenever an employee's exposure to methylene chloride exceeds or can reasonably be expected to exceed the 8-hour TWA PEL is hereby stayed until August 31, 1998 for employers engaged in polyurethane foam manufacturing; foam fabrication; furniture refinishing; general aviation aircraft stripping; formulation of products containing methylene chloride; boat building and repair; recreational vehicle manufacture; van conversion; upholstery; and use of methylene chloride in construction work for restoration and preservation of buildings, painting and paint removal, cabinet making and/or floor refinishing and resurfacing.

The requirement of 29 CFR 1910.1052(f)(1) to implement engineering controls to achieve the 8-hour TWA PEL and STEL is hereby stayed until December 10, 1998 for employers with more than 100 employees engaged in polyurethane foam manufacturing and for employers with more than 20 employees engaged in foam fabrication; furniture refinishing; general aviation aircraft stripping; formulation of products containing methylene chloride; boat building and repair; recreational vehicle manufacture; van conversion; upholstery; and use of methylene chloride in construction work for restoration and preservation of buildings, painting and paint removal, cabinet making and/or floor refinishing and resurfacing.]

## APPENDIX A TO SECTION 1910.1052—SUBSTANCE SAFETY DATA SHEET AND TECHNICAL GUIDELINES FOR METHYLENE CHLORIDE

### I. SUBSTANCE IDENTIFICATION

A. Substance: Methylene chloride ($CH_2Cl_2$).

B. Synonyms: MC, Dichloromethane (DCM); Methylene dichloride; Methylene bichloride; Methane dichloride; CAS: 75-09-2; NCI-C50102.

C. Physical data:

1. Molecular weight: 84.9.

2. Boiling point (760 mm Hg): 39.8 °C (104 °F).

3. Specific gravity (water = 1): 1.3.

4. Vapor density (air = 1 at boiling point): 2.9.

5. Vapor pressure at 20 °C (68 °F): 350 mm Hg.

6. Solubility in water, g/100 g water at 20 °C (68 °F) = 1.32.

7. Appearance and odor: colorless liquid with a chloroform-like odor.

D. Uses:

MC is used as a solvent, especially where high volatility is required. It is a good solvent for oils, fats, waxes, resins, bitumen, rubber and cellulose acetate and is a useful paint stripper and degreaser. It is used in paint removers, in propellant mixtures for aerosol containers, as a solvent for plastics, as a degreasing agent, as an extracting agent in the pharmaceutical industry and as a blowing agent in polyurethane foams. Its solvent property is sometimes increased by mixing with methanol, petroleum naphtha or tetrachloroethylene.

E. Appearance and odor:

MC is a clear colorless liquid with a chloroform-like odor. It is slightly soluble in water and completely miscible with most organic solvents.

F. Permissible exposure:

Exposure may not exceed 25 parts MC per million parts of air (25 ppm) as an eight-hour time-weighted average (8-hour TWA PEL) or 125 parts of MC per million parts of air (125 ppm) averaged over a 15-minute period (STEL).

### II. HEALTH HAZARD DATA

A. MC can affect the body if it is inhaled or if the liquid comes in contact with the eyes or skin. It can also affect the body if it is swallowed.

B. Effects of overexposure:

1. Short-term Exposure:

MC is an anesthetic. Inhaling the vapor may cause mental confusion, light-headedness, nausea, vomiting, and headache. Continued exposure may cause increased light-headedness, staggering, unconsciousness, and even death. High vapor concentrations may also cause irritation of the eyes and respiratory tract. Exposure to MC may make the symptoms of angina (chest pains) worse. Skin exposure to liquid MC may cause irritation. If liquid MC remains on the skin, it may cause skin burns. Splashes of the liquid into the eyes may cause irritation.

### 2. Long-term (chronic) exposure:

The best evidence that MC causes cancer is from laboratory studies in which rats, mice and hamsters inhaled MC 6 hours per day, 5 days per week for 2 years. MC exposure produced lung and liver tumors in mice and mammary tumors in rats. No carcinogenic effects of MC were found in hamsters.

There are also some human epidemiological studies which show an association between occupational exposure to MC and increases in biliary (bile duct) cancer and a type of brain cancer. Other epidemiological studies have not observed a relationship between MC exposure and cancer. OSHA interprets these results to mean that there is suggestive (but not absolute) evidence that MC is a human carcinogen.

### C. Reporting signs and symptoms:

You should inform your employer if you develop any signs or symptoms and suspect that they are caused by exposure to MC.

### D. Warning Properties:

#### 1. Odor Threshold:

Different authors have reported varying odor thresholds for MC. Kirk-Othmer and Sax both reported 25 to 50 ppm; Summer and May both reported 150 ppm; Spector reports 320 ppm. Patty, however, states that since one can become adapted to the odor, MC should not be considered to have adequate warning properties.

#### 2. Eye Irritation Level:

Kirk-Othmer reports that "MC vapor is seriously damaging to the eyes." Sax agrees with Kirk-Othmer's statement. The ACGIH Documentation of TLVs states that irritation of the eyes has been observed in workers exposed to concentrations up to 5000 ppm.

#### 3. Evaluation of Warning Properties:

Since a wide range of MC odor thresholds are reported (25-320 ppm), and human adaptation to the odor occurs, MC is considered to be a material with poor warning properties.

## III. EMERGENCY FIRST AID PROCEDURES

In the event of emergency, institute first aid procedures and send for first aid or medical assistance.

### A. Eye and Skin Exposures:

If there is a potential for liquid MC to come in contact with eye or skin, face shields and skin protective equipment must be provided and used. If liquid MC comes in contact with the eye, get medical attention. Contact lenses should not be worn when working with this chemical.

### B. Breathing:

If a person breathes in large amounts of MC, move the exposed person to fresh air at once. If breathing has stopped, perform cardiopulmonary resuscitation. Keep the affected person warm and at rest. Get medical attention as soon as possible.

### C. Rescue:

Move the affected person from the hazardous exposure immediately. If the exposed person has been overcome, notify someone else and put into effect the established emergency rescue procedures. Understand the facility's emergency rescue procedures and know the locations of rescue equipment before the need arises. Do not become a casualty yourself.

## IV. RESPIRATORS, PROTECTIVE CLOTHING, AND EYE PROTECTION

### A. Respirators:

Good industrial hygiene practices recommend that engineering controls be used to reduce environmental concentrations to the permissible exposure level. However, there are some exceptions where respirators may be used to control exposure. Respirators may be used when engineering and work practice controls are not feasible, when such controls are in the process of being installed, or when these controls fail and need to be supplemented. Respirators may also be used for operations which require entry into tanks or closed vessels, and in emergency situations.

If the use of respirators is necessary, the only respirators permitted are those that have been approved by the Mine Safety and Health Administration (MSHA) or the National Institute for Occupational Safety and Health (NIOSH). Supplied-air respirators are *required* because air-purifying respirators do not provide adequate respiratory protection against MC.

In addition to respirator selection, a complete written respiratory protection program should be instituted which includes regular training, maintenance, inspection, cleaning, and evaluation. If you can smell MC while wearing a respirator, proceed immediately to fresh air. If you experience difficulty in breathing while wearing a respirator, tell your employer.

### B. Protective Clothing:

Employees must be provided with and required to use impervious clothing, gloves, face shields (eight-inch minimum), and other appropriate protective clothing necessary to prevent repeated or prolonged skin contact with liquid MC or contact with vessels containing liquid MC. Any clothing which becomes wet with liquid MC should be removed immediately and not reworn until the employer has ensured that the protective clothing is fit for reuse. Contaminated protective clothing should be placed in a regulated area designated by the employer for removal of MC before the clothing is laundered or disposed of. Clothing and equipment should remain in the regulated area until all of the MC contamination has evaporated; clothing and equipment should then be laundered or disposed of as appropriate.

### C. Eye Protection:

Employees should be provided with and required to use splash-proof safety goggles where liquid MC may contact the eyes.

## V. HOUSEKEEPING AND HYGIENE FACILITIES

For purposes of complying with 29 CFR 1910.141, the following items should be emphasized:

A. The workplace should be kept clean, orderly, and in a sanitary condition. The employer should institute a leak and spill detection program for operations involving liquid MC in order to detect sources of fugitive MC emissions.

B. Emergency drench showers and eyewash facilities are recommended. These should be maintained in a sanitary condition. Suitable cleansing agents should also be provided to assure the effective removal of MC from the skin.

C. Because of the hazardous nature of MC, contaminated protective clothing should be placed in a regulated area designated by the employer for removal of MC before the clothing is laundered or disposed of.

## VI. PRECAUTIONS FOR SAFE USE, HANDLING, AND STORAGE

### A. Fire and Explosion Hazards:

MC has no flash point in a conventional closed tester, but it forms flammable vapor-air mixtures at approximately 100 °C (212 °F), or higher. It has a lower explosion limit of 12%, and an upper explosion limit of 19% in air. It has an autoignition temperature of 556.1 °C (1033 °F), and a boiling point of 39.8 °C (104 °F). It is heavier than water with a specific gravity of 1.3. It is slightly soluble in water.

### B. Reactivity Hazards:

Conditions contributing to the instability of MC are heat and moisture. Contact with strong oxidizers, caustics, and chemically active metals such as aluminum or magnesium powder, sodium and potassium may cause fires and explosions. Special precautions: Liquid MC will attack some forms of plastics, rubber, and coatings.

### C. Toxicity:

Liquid MC is painful and irritating if splashed in the eyes or if confined on the skin by gloves, clothing, or shoes. Vapors in high concentrations may cause narcosis and death. Prolonged exposure to vapors may cause cancer or exacerbate cardiac disease.

### D. Storage:

Protect against physical damage. Because of its corrosive properties, and its high vapor pressure, MC should be stored in plain, galvanized or lead lined, mild steel containers in a cool, dry, well ventilated area away from direct sunlight, heat source and acute fire hazards.

### E. Piping Material:

All piping and valves at the loading or unloading station should be of material that is resistant to MC and should be carefully inspected prior to connection to the transport vehicle and periodically during the operation.

### F. Usual Shipping Containers:

Glass bottles, 5- and 55-gallon steel drums, tank cars, and tank trucks.

**NOTE:** This section addresses MC exposure in marine terminal and longshore employment only where leaking or broken packages allow MC exposure

# OSHA Standards for General Industry

that is not addressed through compliance with 29 CFR parts 1917 and 1918, respectively.

G. Electrical Equipment:

Electrical installations in Class I hazardous locations as defined in Article 500 of the National Electrical Code, should be installed according to Article 501 of the code; and electrical equipment should be suitable for use in atmospheres containing MC vapors. See Flammable and Combustible Liquids Code (NFPA No. 325M), Chemical Safety Data Sheet SD-86 (Manufacturing Chemists' Association, Inc.).

H. Fire Fighting:

When involved in fire, MC emits highly toxic and irritating fumes such as phosgene, hydrogen chloride and carbon monoxide. Wear breathing apparatus and use water spray to keep fire-exposed containers cool. Water spray may be used to flush spills away from exposures. Extinguishing media are dry chemical, carbon dioxide, foam. For purposes of compliance with 29 CFR 1910.307, locations classified as hazardous due to the presence of MC shall be Class I.

I. Spills and Leaks:

Persons not wearing protective equipment and clothing should be restricted from areas of spills or leaks until cleanup has been completed. If MC has spilled or leaked, the following steps should be taken:

1. Remove all ignition sources.

2. Ventilate area of spill or leak.

3. Collect for reclamation or absorb in vermiculite, dry sand, earth, or a similar material.

J. Methods of Waste Disposal:

Small spills should be absorbed onto sand and taken to a safe area for atmospheric evaporation. Incineration is the preferred method for disposal of large quantities by mixing with a combustible solvent and spraying into an incinerator equipped with acid scrubbers to remove hydrogen chloride gases formed. Complete combustion will convert carbon monoxide to carbon dioxide. Care should be taken for the presence of phosgene.

K. You should not keep food, beverage, or smoking materials, or eat or smoke in regulated areas where MC concentrations are above the permissible exposure limits.

L. Portable heating units should not be used in confined areas where MC is used.

M. Ask your supervisor where MC is used in your work area and for any additional plant safety and health rules.

## VII. MEDICAL REQUIREMENTS

Your employer is required to offer you the opportunity to participate in a medical surveillance program if you are exposed to MC at concentrations at or above the action level (12.5 ppm 8-hour TWA) for more than 30 days a year or at concentrations exceeding the PELs (25 ppm 8-hour TWA or 125 ppm 15-minute STEL) for more than 10 days a year. If you are exposed to MC at concentrations over either of the PELs, your employer will also be required to have a physician or other licensed health care professional ensure that you are able to wear the respirator that you are assigned. Your employer must provide all medical examinations relating to your MC exposure at a reasonable time and place and at no cost to you.

## VIII. MONITORING AND MEASUREMENT PROCEDURES

A. Exposure above the Permissible Exposure Limit:

1. Eight-hour exposure evaluation: Measurements taken for the purpose of determining employee exposure under this section are best taken with consecutive samples covering the full shift. Air samples must be taken in the employee's breathing zone.

2. Monitoring techniques: The sampling and analysis under this section may be performed by collection of the MC vapor on two charcoal adsorption tubes in series or other composition adsorption tubes, with subsequent chemical analysis. Sampling and analysis may also be performed by instruments such as real-time continuous monitoring systems, portable direct reading instruments, or passive dosimeters as long as measurements taken using these methods accurately evaluate the concentration of MC in employees" breathing zones.

OSHA method 80 is an example of a validated method of sampling and analysis of MC. Copies of this method are available from OSHA or can be downloaded from the Internet at http://www.osha.gov. The employer has the obligation of selecting a monitoring method which meets the accuracy and precision requirements of the standard under his or her unique field conditions. The standard requires that the method of monitoring must be accurate, to a 95 percent confidence level, to plus or minus 25 percent for concentrations of MC at or above 25 ppm, and to plus or minus 35 percent for concentrations at or below 25 ppm. In addition to OSHA method 80, there are numerous other methods available for monitoring for MC in the workplace.

B. Since many of the duties relating to employee exposure are dependent on the results of measurement procedures, employers must assure that the evaluation of employee exposure is performed by a technically qualified person.

## IX. OBSERVATION OF MONITORING

Your employer is required to perform measurements that are representative of your exposure to MC and you or your designated representative are entitled to observe the monitoring procedure. You are entitled to observe the steps taken in the measurement procedure, and to record the results obtained. When the monitoring procedure is taking place in an area where respirators or personal protective clothing and equipment are required to be worn, you or your representative must also be provided with, and must wear, protective clothing and equipment.

## X. ACCESS TO INFORMATION

A. Your employer is required to inform you of the information contained in this Appendix. In addition, your employer must instruct you in the proper work practices for using MC, emergency procedures, and the correct use of protective equipment.

B. Your employer is required to determine whether you are being exposed to MC. You or your representative has the right to observe employee measurements and to record the results obtained. Your employer is required to inform you of your exposure. If your employer determines that you are being over exposed, he or she is required to inform you of the actions which are being taken to reduce your exposure to within permissible exposure limits.

C. Your employer is required to keep records of your exposures and medical examinations. These records must be kept by the employer for at least thirty (30) years.

D. Your employer is required to release your exposure and medical records to you or your representative upon your request.

E. Your employee is required to provide labels and material safety data sheets (MSDS) for all materials, mixtures or solutions composed of greater than 0.1 percent MC. An example of a label that would satisfy these requirements would be:

**DANGER**
**CONTAINS METHYLENE CHLORIDE**
**POTENTIAL CANCER HAZARD**

May worsen heart disease because methylene chloride is converted to carbon monoxide in the body.

May cause dizziness, headache, irritation of the throat and lungs, loss of consciousness and death at high concentrations (for example, if used in a poorly ventilated room).

*Avoid Skin Contact.* Contact with liquid causes skin and eye irritation.

## XI. COMMON OPERATIONS AND CONTROLS

The following list includes some common operations in which exposure to MC may occur and control methods which may be effective in each case:

| Operations | Controls |
| --- | --- |
| Use as solvent in paint and varnish removers; manufacture of aerosols; cold cleaning and ultrasonic cleaning; and as a solvent in furniture stripping. | General dilution ventilation; local exhaust ventilation; personal protective equipment; substitution. |
| Use as solvent in vapor degreasing. | Process enclosure; local exhaust ventilation; chilling coils; substitution. |
| Use as a secondary refrigerant in air conditioning and scientific testing. | General dilution ventilation; local exhaust ventilation; personal protective equipment. |

# APPENDIX B TO SECTION 1910.105 [sic 1910.1052]— MEDICAL SURVEILLANCE FOR METHYLENE CHLORIDE

## I. PRIMARY ROUTE OF ENTRY

Inhalation.

## II. TOXICOLOGY

Methylene Chloride (MC) is primarily an inhalation hazard. The principal acute hazardous effects are the depressant action on the central nervous system, possible cardiac toxicity and possible liver toxicity. The range of CNS effects are from decreased eye/hand coordination and decreased performance in vigilance tasks to narcosis and even death of individuals exposed at very high doses. Cardiac toxicity is due to the metabolism of MC to carbon monoxide, and the effects of carbon monoxide on heart tissue. Carbon monoxide displaces oxygen in the blood, decreases the oxygen available to heart tissue, increasing the risk of damage to the heart, which may result in heart attacks in susceptible individuals. Susceptible individuals include persons with heart disease and those with risk factors for heart disease.

Elevated liver enzymes and irritation to the respiratory passages and eyes have also been reported for both humans and experimental animals exposed to MC vapors.

MC is metabolized to carbon monoxide and carbon dioxide via two separate pathways. Through the first pathway, MC is metabolized to carbon monoxide as an end-product via the P-450 mixed function oxidase pathway located in the microsomal fraction of the cell. This biotransformation of MC to carbon monoxide occurs through the process of microsomal oxidative dechlorination which takes place primarily in the liver. The amount of conversion to carbon monoxide is significant as measured by the concentration of carboxyhemoglobin, up to 12% measured in the blood following occupational exposure of up to 610 ppm. Through the second pathway, MC is metabolized to carbon dioxide as an end product (with formaldehyde and formic acid as metabolic intermediates) via the glutathione dependent enzyme found in the cytosolic fraction of the liver cell. Metabolites along this pathway are believed to be associated with the carcinogenic activity of MC.

MC has been tested for carcinogenicity in several laboratory rodents. These rodent studies indicate that there is clear evidence that MC is carcinogenic to male and female mice and female rats. Based on epidemiologic studies, OSHA has concluded that there is suggestive evidence of increased cancer risk in MC-related worker populations. The epidemiological evidence is consistent with the finding of excess cancer in the experimental animal studies. NIOSH regards MC as a potential occupational carcinogen and the International Agency for Research Cancer (IARC) classifies MC as an animal carcinogen. OSHA considers MC as a suspected human carcinogen.

## III. MEDICAL SIGNS AND SYMPTOMS OF ACUTE EXPOSURE

Skin exposure to liquid MC may cause irritation or skin burns. Liquid MC can also be irritating to the eyes. MC is also absorbed through the skin and may contribute to the MC exposure by inhalation.

At high concentrations in air, MC may cause nausea, vomiting, light-headedness, numbness of the extremities, changes in blood enzyme levels, and breathing problems, leading to bronchitis and pulmonary edema, unconsciousness and even death.

At lower concentrations in air, MC may cause irritation to the skin, eye, and respiratory tract and occasionally headache and nausea. Perhaps the greatest problem from exposure to low concentrations of MC is the CNS effects on coordination and alertness that may cause unsafe operations of machinery and equipment, leading to self-injury or accidents.

Low levels and short duration exposures do not seem to produce permanent disability, but chronic exposures to MC have been demonstrated to produce liver toxicity in animals, and therefore, the evidence is suggestive for liver toxicity in humans after chronic exposure.

Chronic exposure to MC may also cause cancer.

## IV. SURVEILLANCE AND PREVENTIVE CONSIDERATIONS

As discussed above, MC is classified as a suspect or potential human carcinogen. It is a central nervous system (CNS) depressant and a skin, eye and respiratory tract irritant. At extremely high concentrations, MC has caused liver damage in animals.

MC principally affects the CNS, where it acts as a narcotic. The observation of the symptoms characteristic of CNS depression, along with a physical examination, provides the best detection of early neurological disorders.

Since exposure to MC also increases the carboxyhemoglobin level in the blood, ambient carbon monoxide levels would have an additive effect on that carboxyhemoglobin level. Based on such information, a periodic post-shift carboxyhemoglobin test as an index of the presence of carbon monoxide in the blood is recommended, but not required, for medical surveillance.

Based on the animal evidence and three epidemiologic studies previously mentioned, OSHA concludes that MC is a suspect human carcinogen. The medical surveillance program is designed to observe exposed workers on a regular basis. While the medical surveillance program cannot detect MC-induced cancer at a preneoplastic stage, OSHA anticipates that, as in the past, early detection and treatments of cancers leading to enhanced survival rates will continue to evolve.

### A. Medical and Occupational History:

The medical and occupational work history plays an important role in the initial evaluation of workers exposed to MC. It is therefore extremely important for the examining physician or other licensed health care professional to evaluate the MC-exposed worker carefully and completely and to focus the examination on MC's potentially associated health hazards. The medical evaluation must include an annual detailed work and medical history with special emphasis on cardiac history and neurological symptoms.

An important goal of the medical history is to elicit information from the worker regarding potential signs or symptoms associated with increased levels of carboxyhemoglobin due to the presence of carbon monoxide in the blood. Physicians or other licensed health care professionals should ensure that the smoking history of all MC exposed employees is known. Exposure to MC may cause a significant increase in carboxyhemoglobin level in all exposed persons. However, smokers as well as workers with anemia or heart disease and those concurrently exposed to carbon monoxide are at especially high risk of toxic effects because of an already reduced oxygen carrying capacity of the blood.

A comprehensive or interim medical and work history should also include occurrence of headache, dizziness, fatigue, chest pain, shortness of breath, pain in the limbs, and irritation of the skin and eyes.

In addition, it is important for the physician or other licensed health care professional to become familiar with the operating conditions in which exposure to MC is likely to occur. The physician or other licensed health care professional also must become familiar with the signs and symptoms that may indicate that a worker is receiving otherwise unrecognized and exceptionally high exposure levels of MC.

An example of a medical and work history that would satisfy the requirement for a comprehensive or interim work history is represented by the following:

The following is a list of recommended questions and issues for the self-administered questionnaire for methylene chloride exposure.

### QUESTIONNAIRE FOR METHYLENE CHLORIDE EXPOSURE

#### I. Demographic Information

1. Name
2. Social Security Number
3. Date
4. Date of Birth
5. Age
6. Present occupation
7. Sex
8. Race

#### II. Occupational History

1. Have you ever worked with methylene chloride, dichloromethane, methylene dichloride, or $CH_2 Cl_2$ (all are different names for the same chemical)? Please list which on the occupational history form if you have not already.

2. If you have worked in any of the following industries and have not listed them on the occupational history form, please do so.

Furniture stripping

Polyurethane foam manufacturing

Chemical manufacturing or formulation

Pharmaceutical manufacturing

Any industry in which you used solvents to clean and degrease equipment or parts

# OSHA Standards for General Industry

Construction, especially painting and refinishing

Aerosol manufacturing

Any industry in which you used aerosol adhesives

3. If you have not listed hobbies or household projects on the occupational history form, especially furniture refinishing, spray painting, or paint stripping, please do so.

## III. Medical History

### A. General

1. Do you consider yourself to be in good health? If no, state reason(s).

2. Do you or have you ever had:

a. Persistent thirst

b. Frequent urination (three times or more at night)

c. Dermatitis or irritated skin

d. Non-healing wounds

3. What prescription or non-prescription medications do you take, and for what reasons?

4. Are you allergic to any medications, and what type of reaction do you have?

### B. Respiratory

1. Do you have or have you ever had any chest illnesses or diseases? Explain.

2. Do you have or have you ever had any of the following:

a. Asthma

b. Wheezing

c. Shortness of breath

3. Have you ever had an abnormal chest X-ray? If so, when, where, and what were the findings?

4. Have you ever had difficulty using a respirator or breathing apparatus? Explain.

5. Do any chest or lung diseases run in your family? Explain.

6. Have you ever smoked cigarettes, cigars, or a pipe? Age started:

7. Do you now smoke?

8. If you have stopped smoking completely, how old were you when you stopped?

9. On the average of the entire time you smoked, how many packs of cigarettes, cigars, or bowls of tobacco did you smoke per day?

### C. Cardiovascular

1. Have you ever been diagnosed with any of the following: Which of the following apply to you now or did apply to you at some time in the past, even if the problem is controlled by medication? Please explain any yes answers (i.e., when problem was diagnosed, length of time on medication).

a. High cholesterol or triglyceride level

b. Hypertension (high blood pressure)

c. Diabetes

d. Family history of heart attack, stroke, or blocked arteries

2. Have you ever had chest pain? If so, answer the next five questions.

a. What was the quality of the pain (i.e., crushing, stabbing, squeezing)?

b. Did the pain go anywhere (i.e., into jaw, left arm)?

c. What brought the pain out?

d. How long did it last?

e. What made the pain go away?

3. Have you ever had heart disease, a heart attack, stroke, aneurysm, or blocked arteries anywhere in you body? Explain (when, treatment).

4. Have you ever had bypass surgery for blocked arteries in your heart or anywhere else? Explain.

5. Have you ever had any other procedures done to open up a blocked artery (balloon angioplasty, carotid endarterectomy, clot-dissolving drug)?

6. Do you have or have you ever had (explain each):

a. Heart murmur

b. Irregular heartbeat

c. Shortness of breath while lying flat

d. Congestive heart failure

e. Ankle swelling

f. Recurrent pain anywhere below the waist while walking

7. Have you ever had an electrocardiogram (EKG)? When?

8. Have you ever had an abnormal EKG? If so, when, where, and what were the findings?

9. Do any heart diseases, high blood pressure, diabetes, high cholesterol, or high triglycerides run in your family? Explain.

### D. Hepatobiliary and Pancreas

1. Do you now or have you ever drunk alcoholic beverages? Age started:____Age stopped: _____ .

2. Average numbers per week:

a. Beers: _____, ounces in usual container:

b. Glasses of wine: _____ , ounces per glass:

c. Drinks: _____ , ounces in usual container:

3. Do you have or have you ever had (explain each):

a. Hepatitis (infectious, autoimmune, drug-induced, or chemical)

b. Jaundice

c. Elevated liver enzymes or elevated bilirubin

d. Liver disease or cancer

### E. Central Nervous System

1. Do you or have you ever had (explain each):

a. Headache

b. Dizziness

c. Fainting

d. Loss of consciousness

e. Garbled speech

f. Lack of balance

g. Mental/psychiatric illness

h. Forgetfulness

### F. Hematologic

1. Do you have, or have you ever had (explain each):

a. Anemia

b. Sickle cell disease or trait

c. Glucose-6-phosphate dehydrogenase deficiency

d. Bleeding tendency disorder

2. If not already mentioned previously, have you ever had a reaction to sulfa drugs or to drugs used to prevent or treat malaria? What was the drug? Describe the reaction.

### B. Physical Examination

The complete physical examination, when coupled with the medical and occupational history, assists the physician or other licensed health care professional in detecting pre-existing conditions that might place the employee at increased risk, and establishes a baseline for future health monitoring. These examinations should include:

1. Clinical impressions of the nervous system, cardiovascular function and pulmonary function, with additional tests conducted where indicated or determined by the examining physician or other licensed health care professional to be necessary.

2. An evaluation of the advisability of the worker using a respirator, because the use of certain respirators places an additional burden on the cardiopulmonary system. It is necessary for the attending physician or other licensed health care professional to evaluate the cardiopulmonary function of these workers, in order to inform the employer in a written

medical opinion of the worker's ability or fitness to work in an area requiring the use of certain types of respiratory protective equipment. The presence of facial hair or scars that might interfere with the worker's ability to wear certain types of respirators should also be noted during the examination and in the written medical opinion.

Because of the importance of lung function to workers required to wear certain types of respirators to protect themselves from MC exposure, these workers must receive an assessment of pulmonary function before they begin to wear a negative pressure respirator and at least annually thereafter. The recommended pulmonary function tests include measurement of the employee's forced vital capacity (FVC), forced expiratory volume at one second (FEV1), as well as calculation of the ratios of FEV1 to FVC, and the ratios of measured FVC and measured FEV1 to expected respective values corrected for variation due to age, sex, race, and height. Pulmonary function evaluation must be conducted by a physician or other licensed health care professional experienced in pulmonary function tests.

The following is a summary of the elements of a physical exam which would fulfill the requirements under the MC standard:

### PHYSICAL EXAM

#### I. Skin and appendages
1. Irritated or broken skin
2. Jaundice
3. Clubbing cyanosis, edema
4. Capillary refill time
5. Pallor

#### II. Head
1. Facial deformities
2. Scars
3. Hair growth

#### III. Eyes
1. Scleral icterus
2. Corneal arcus
3. Pupillary size and response
4. Fundoscopic exam

#### IV. Chest
1. Standard exam

#### V. Heart
1. Standard exam
2. Jugular vein distension
3. Peripheral pulses

#### VI. Abdomen
1. Liver span

#### VII. Nervous System
1. Complete standard neurologic exam

#### VIII. Laboratory
1. Hemoglobin and hematocrit
2. Alanine aminotransferase (ALT, SGPT)
3. Post-shift carboxyhemoglobin

#### IX. Studies
1. Pulmonary function testing
2. Electrocardiogram

An evaluation of the oxygen carrying capacity of the blood of employees (for example by measured red blood cell volume) is considered useful, especially for workers acutely exposed to MC.

It is also recommended, but not required, that end of shift carboxyhemoglobin levels be determined periodically, and any level above 3% for non-smokers and above 10% for smokers should prompt an investigation of the worker and his workplace. This test is recommended because MC is metabolized to CO, which combines strongly with hemoglobin, resulting in a reduced capacity of the blood to transport oxygen in the body. This is of particular concern for cigarette smokers because they already have a diminished hemoglobin capacity due to the presence of CO in cigarette smoke.

### C. Additional Examinations and Referrals

1. Examination by a Specialist

When a worker examination reveals unexplained symptoms or signs (i.e., in the physical examination or in the laboratory tests), follow-up medical examinations are necessary to assure that MC exposure is not adversely affecting the worker's health. When the examining physician or other licensed health care professional finds it necessary, additional tests should be included to determine the nature of the medical problem and the underlying cause. Where relevant, the worker should be sent to a specialist for further testing and treatment as deemed necessary.

The final rule requires additional investigations to be covered and it also permits physicians or other licensed health care professionals to add appropriate or necessary tests to improve the diagnosis of disease should such tests become available in the future.

2. Emergencies

The examination of workers exposed to MC in an emergency should be directed at the organ systems most likely to be affected. If the worker has received a severe acute exposure, hospitalization may be required to assure proper medical intervention. It is not possible to precisely define "severe," but the physician or other licensed health care professional's judgement should not merely rest on hospitalization. If the worker has suffered significant conjunctival, oral, or nasal irritation, respiratory distress, or discomfort, the physician or other licensed health care professional should instigate appropriate follow-up procedures. These include attention to the eyes, lungs and the neurological system. The frequency of follow-up examinations should be determined by the attending physician or other licensed health care professional. This testing permits the early identification essential to proper medical management of such workers.

### D. Employer Obligations

The employer is required to provide the responsible physician or other licensed health care professional and any specialists involved in a diagnosis with the following information: a copy of the MC standard including relevant appendices, a description of the affected employee's duties as they relate to his or her exposure to MC; an estimate of the employee's exposure including duration (e.g., 15hr/wk, three 8-hour shifts/wk, full time); a description of any personal protective equipment used by the employee, including respirators; and the results of any previous medical determinations for the affected employee related to MC exposure to the extent that this information is within the employer's control.

### E. Physicians' or Other Licensed Health Care Professionals' Obligations

The standard requires the employer to ensure that the physician or other licensed health care professional provides a written statement to the employee and the employer. This statement should contain the physician's or licensed health care professional's opinion as to whether the employee has any medical condition placing him or her at increased risk of impaired health from exposure to MC or use of respirators, as appropriate. The physician or other licensed health care professional should also state his or her opinion regarding any restrictions that should be placed on the employee's exposure to MC or upon the use of protective clothing or equipment such as respirators. If the employee wears a respirator as a result of his or her exposure to MC, the physician or other licensed health care professional's opinion should also contain a statement regarding the suitability of the employee to wear the type of respirator assigned. Furthermore, the employee should be informed by the physician or other licensed health care professional about the cancer risk of MC and about risk factors for heart disease, and the potential for exacerbation of underlying heart disease by exposure to MC through its metabolism to carbon monoxide. Finally, the physician or other licensed health care professional should inform the employer that the employee has been told the results of the medical examination and of any medical conditions which require further explanation or treatment. This written opinion must not contain any information on specific findings or diagnosis unrelated to employee's occupational exposures.

The purpose in requiring the examining physician or other licensed health care professional to supply the employer with a written opinion is to provide the employer with a medical basis to assist the employer in placing employees initially, in assuring that their health is not being impaired by exposure to MC, and to assess the employee's ability to use any required protective equipment.

# APPENDIX C TO SECTION 1910.1052—QUESTIONS AND ANSWERS—METHYLENE CHLORIDE CONTROL IN FURNITURE STRIPPING

## DISCLAIMER

Mention of any company or product does not constitute endorsement by the Centers for Disease Control and Prevention, National Institute for Occupational Safety and Health.

This document is in the public domain and may be freely copied or reprinted.

Copies of this and other NIOSH documents are available from

Publications Dissemination, DSDTT
National Institute for Occupational Safety and Health
4676 Columbia Parkway
Cincinnati, OH 45226
FAX (513) 533-8573

For information about other occupational safety and health problems, call 1-800-35-NIOSH

DHHS (NIOSH) Publication NO. 93-133

---

# Questions and Answers— Methylene Chloride Control in Furniture Stripping

U.S. DEPARTMENT OF HEALTH AND HUMAN SERVICES
Public Health Service
Centers for Disease Control and Prevention
National Institute for Occupational Safety and Health
September 1993

## Q's & A's

### Introduction

This Pamphlet answers commonly asked questions about the hazards from exposure to methylene chloride. It also describes approaches to controlling methylene chloride exposure during the most common furniture stripping processes. Although these approaches were developed and field tested by NIOSH, each setting requires custom installation because of the different air flow interferences at each site.

### What is the Stripping Solution Base?

The most common active ingredient in paint removers is a chemical called methylene chloride. Methylene chloride is present in the paint remover to penetrate, blister, and finally lift the old finish. Other chemicals in paint removers work to accelerate the stripping process, to retard evaporation, and to act as thickening agents. These other ingredients may include: methanol, toluene, acetone, or paraffin.[1]

### Is Methylene Chloride Bad for Me?

Exposure to methylene chloride may cause short-term health effects or long-term health effects.

### Short-Term (acute) Health Effects

Exposure to high levels of paint removers over short periods of time can cause irritation to the skin, eyes, mucous membranes, and respiratory tract. Other symptoms of high exposure are dizziness headache, and lack of coordination. The occurrence of any of these symptoms indicates that you are being exposed to high levels of the methylene chloride. At the onset of any of these symptoms, you should leave the work area, get some fresh air, and determine why the levels were high.

A portion of inhaled methylene chloride is converted by the body to carbon monoxide, which can lower the blood's ability to carry oxygen. When the solvent is used properly, however, the levels of carbon monoxide should not be hazardous. Individuals with cardiovascular or pulmonary health problems should check with their physician before using the paint stripper. Individuals experiencing severe symptoms such as shortness of breath or chest pains should obtain proper medical care immediately.[2]

### Long-term (Chronic) Health Effects

Methylene chloride has been shown to cause cancer in certain laboratory animal tests. The available human studies do not provide the necessary information to determine whether methylene chloride causes cancer in humans. However, as a result of the animal studies, methylene chloride is

Figure 1 — Slot Hood

## Q's & A's

### How Can I Be Exposed to Methylene Chloride while Stripping Furniture?

Methylene chloride can be inhaled when vapors are in the air. Inhalation of the methylene chloride vapors is generally the most important source of exposure. Methylene chloride evaporates quicker than most chemicals. The odor threshold of methylene chloride is 300 ppm.[6] Therefore, once you smell methylene chloride, you are being over-exposed. Pouring, moving, or stirring the chemical will increase the rate of evaporation.

Methylene chloride can be absorbed through the skin either by directly touching the chemical or through your gloves. Methylene chloride can be swallowed if it gets on your hands, clothes, or beard, or if food or drinks become contaminated.

### How Can Breathing Exposures be Reduced?

### Install a Local Exhaust Ventilation System

Local exhaust ventilation can be used to control exposures. Local exhaust ventilation systems

Figure 2 — Downdraft Hood

considered a potential occupational carcinogen. There is also considerable indirect evidence to suggest that workers exposed to methylene chloride may be at increased risk of developing ischemic heart disease. Therefore, it is prudent to minimize exposures to solvent vapors.[3]

### What Do Federal Agencies Say About Methylene Chloride?

In 1991, the Occupational Safety and Health Administration published a Notice of Proposed Rulemaking for methylene chloride. The proposed standard would establish an eight-hour time-weighted average exposure limit of 25 parts per million (ppm), as well as a short-term exposure limit of 125 ppm determined from a 15 minute sampling period. That is a sharp reduction from the current limit of 500 ppm. The proposed standard would also set a 12.5 ppm action level (a level that would trigger periodic exposure monitoring and medical surveillance provisions.[4]

The National Institute for Occupational Safety and Health recommends that methylene chloride be regarded as a "potential occupational carcinogen." NIOSH further recommends that occupational exposure to methylene chloride be controlled to the lowest feasible limit. This recommendation was based on the observation of cancers and tumors in both rats and mice exposed to methylene chloride in air.[5]

capture contaminated air from the source before it spreads into the workers' breathing zone.[7] If engineering controls are not effective, only a self-contained breathing apparatus equipped with a full facepiece and operated in a positive-pressure mode or a supplied-air respirator affords the necessary level of protection. Air-purifying respirators such as organic vapor cartridges can only be used for escape situations.[8]

A local exhaust system consists of the following: a hood, a fan, ductwork, and a replacement air system.[9,10,11] Two processes are commonly used in furniture stripping: flow-over and dip tanks. For flow-over systems there are two common local exhaust controls for methylene chloride — a slot hood and a downdraft hood. A slot hood of different design is most often used for dip tanks. (See Figures 1, 2, and 3)

The hood is made of sheet metal and connected to the tank. All designs require a centrifugal fan to exhaust the fumes, ductwork connecting the hood and the fan, and a replacement air system to bring conditioned air into the building to replace the air exhausted.

In constructing or designing a slot or downdraft hood, use the following data:

### Slot hood (Figure 1)
- At least 2200 cfm per 8' X 4' tank
- 1 - 2 inch slots
- Slot velocity - 1000 fpm
- 3 - 5 slots
- Plenum at least 1 foot deep

### Downdraft hood (Figure 2)
- At least 1600 cfm per 8' X 4' tank
- Plenum at least 9" deep

### Slot hood for Dip Tank (Figure 3)
- At least 2900 cfm per 8' X 4' tank
- 3/4" slot that runs the length of the front and back of the tank
- Slot velocity — 3200 fpm
- Plenum on the sides of the tank should be 6" deep by 36" long.
- 12" duct leads from the center of the front plenum to the fan

### Safe work practices

Workers can lower exposures by decreasing their access to the methylene chloride.[12]

1) Turn on dip tank control system several minutes before entering the stripping area.
2) Avoid unnecessary transferring or moving of stripping solution.
3) Keep face out of the air stream between the solution-covered furniture and the exhaust system.
4) Keep face out of vapor zone above the stripping solution and dip tank.
5) Retrieve dropped items with a long handled tool.
6) Keep the solution-recycling system off when not in use. Cover reservoir for recycling system.
7) Cover dip tank when not in use.
8) Provide adequate ventilation for rinse area.

**Figure 3 — Slot Hood for Dip Tank**

---

## Q's & A's

### How Can Skin Exposures be Reduced?

Skin exposures can be reduced by wearing gloves whenever you are in contact with the stripping solution.[13]

1) Two gloves should be worn. The inner glove should be made from polyethylene/ethylene vinyl alcohol (e.g. Silver Shield® or 4H®). This material, however, does not provide good physical resistance against tears, so an outer glove made from nitrile or neoprene should be worn.
2) Shoulder-length gloves will be more protective.
3) Change gloves before the break-through time occurs. Rotate several pairs of gloves throughout the day. Let the gloves dry in a warm well ventilated area at least over night before reuse.
4) Keep gloves clean by rinsing often. Keep gloves in good condition. Inspect the gloves before use for pin-holes, cracks, thin spots, and stiffer than normal or sticky surfaces.
5) Wear a face shield or goggles to protect face and eyes.

### What Other Problems Occur?

#### Stripping Solution Temperature

Most manufacturers of stripping solution recommend controlling the solution to a temperature of 70°F. This temperature is required for the wax in the solution to form a vapor barrier on top of the solution to keep the solution from evaporating too quickly. If the temperature is too high, the wax will not form the vapor barrier. If it is too cold, the wax will solidify and separate from the solvent causing increased evaporation. Use a belt heater to heat the solution to the correct temperature. Call your solution manufacturer for the correct temperature for your solution.[14]

#### Make-Up Air

Air will enter a building in an amount to equal the amount of air exhausted whether or not provision is made for this replacement. If a local exhaust system is added a make-up or replacement air system must be added to replace the air removed. Without a replacement air system, air will enter the building through cracks causing uncontrollable eddy currents. If the building perimeter is tightly sealed, it will prevent the air from entering and severely decrease the amount exhausted from the ventilation system. This will cause the building to be under negative pressure and decrease the performance of the exhaust system.[15]

#### Dilution Ventilation

With general or dilution ventilation, uncontaminated air is moved through the workroom by means of fans or open windows, which dilutes the pollutants in the air. Dilution ventilation does not provide effective protection to other workers and does not confine the methylene chloride vapors to one area.[16]

#### Phosgene Poisoning from Use of Kerosene Heaters

Do not use kerosene heaters or other open flame heaters while stripping furniture. Use of kerosene heaters in connection with methylene chloride can create lethal or dangerous concentrations of phosgene. Methylene chloride vapor is mixed with the air used for the combustion of kerosene in kerosene stoves. The vapor thus passes through the flames, coming into close contact with carbon monoxide at high temperatures. Any chlorine formed by decomposition may, under these conditions, react with carbon monoxide and form phosgene.[17]

## REFERENCES

[1] Halogenated Solvents Industry Alliance and Consumer Product Safety Commission [1990]. Stripping Paint from Wood (Pamphlet for consumers on how to strip furniture and precautions to take). Washington DC: Consumer Product Safety Commission.

[2] Ibid.

[3] NIOSH [1992]. NIOSH Testimony on Occupational Safety and Health Administration's proposed rule on occupational exposure to methylene chloride, September 21, 1992, OSHA Docket No. H-71. NIOSH policy statements. Cincinnati, OH: U.S. Department of Health and Human Services, Public Health Service, Centers for Disease Control, National Institute for Occupational Safety and Health.

[4] 56 Fed. Reg. 57036 [1991]. Occupational Safety and Health Administration: Proposed rule on occupational exposure to methylene chloride.

[5] NIOSH [1992].

[6] Kirk, R.E. and P.F. Othmer, Eds. [1978]. Encyclopedia of Chemical Technology. 3rd Ed., Vol. 5:690. New York: John Wiley & Sons, Inc.

[7] ACGIH [1988]. Industrial Ventilation: A Manual of Recommended Practice, 20th Ed. Cincinnati, OH: American Conference of Governmental Industrial Hygienists.

[8] NIOSH [1992].

[9] Fairfield, C.L. and A.A. Beasley [1991]. In-depth Survey Report at the Association for Retarded Citizens, Meadowlands, PA. The Control of Methylene Chloride During Furniture Stripping. Cincinnati, OH: U.S. Department of Health and Human Services, Public Health Service, Centers for Disease Control, National Institute for Occupational Safety and Health.

[10] Fairfield, C.L. [1991]. In-depth Survey Report at the J.M. Murray Center, Cortland, NY. The Control of Methylene Chloride During Furniture Stripping. Cincinnati, OH: U.S. Department of Health and Human Services, Public Health Service, Centers for Disease Control, National Institute for Occupational Safety and Health.

[11] Hall, R.M., K.F. Martinez, and P.A. Jensen [1992]. In-depth Survey Report at Tri-County Furniture Stripping and Refinishing, Cincinnati, OH. The Control of Methylene Chloride During Furniture Stripping. Cincinnati, OH: U.S. Department of Health and Human Services, Public Health Service, Centers for Disease Control, National Institute for Occupational Safety and Health.

[12] Fairfield, C.L. and A.A. Beasley [1991]. In-depth Survey Report at the Association for Retarded Citizens, Meadowlands, PA. The Control of Methylene Chloride During Furniture Stripping. Cincinnati, OH: U.S. Department of Health and Human Services, Public Health Service, Centers for Disease Control, National Institute for Occupational Safety and Health.

[13] Roder, M. [1991]. Memorandum of March 11, 1991 from Michael Roder of the Division of Safety Research to Cheryl L. Fairfield of the Division of Physical Sciences and Engineering, National Institute for Occupational Safety and Health, Centers for Disease Control, Public Health Service, U.S. Department of Health and Human Services.

[14] Kwick Kleen Industrial Solvents, Inc. [1981]. Operations Manual, Kwick Kleen Industrial Solvents, Inc., Vincennes, IN.

[15] ACGIH [1988].

[16] Ibid.

[17] Gerritsen, W.B. and C.H. Buschmann [1960]. Phosgene Poisoning Caused by the Use of Chemical Paint Removers containing Methylene Chloride in Ill-Ventilated Rooms Heated by Kerosene Stoves. British Journal of Industrial Medicine 17:187.

## Q's & A's

**Where Should I go for More Information?**

The NIOSH 800-number is a toll-free technical information service that provides convenient public access to NIOSH and its information resources. Callers may request information about any aspect of occupational safety and health.

**1-800-35-NIOSH**
(1-800-356-4674)

U.S. GOVERNMENT PRINTING OFFICE: 1993, 350-147/80026

---

[62 FR 1601, Jan. 10, 1997, as amended at 62 FR 42667, Aug. 8, 1997; 62 FR 54383, Oct. 20, 1997; 62 FR 66277, Dec. 18, 1997; 63 FR 1295, Jan. 8, 1998; 63 FR 20099, Apr. 23, 1998; 63 FR 50729, Sept. 22, 1998]

### §1910.1096 Ionizing radiation.

**1910.1096(a)** *Definitions applicable to this section.* **(1)** *Radiation* includes alpha rays, beta rays, gamma rays, X-rays, neutrons, high-speed electrons, high-speed protons, and other atomic particles; but such term does not include sound or radio waves, or visible light, or infrared or ultraviolet light.

**1910.1096(a)(2)** *Radioactive material* means any material which emits, by spontaneous nuclear disintegration, corpuscular or electromagnetic emanations.

**1910.1096(a)(3)** *Restricted area* means any area access to which is controlled by the employer for purposes of protection of individuals from exposure to radiation or radioactive materials.

**1910.1096(a)(4)** *Unrestricted area* means any area access to which is not controlled by the employer for purposes of protection of individuals from exposure to radiation or radioactive materials.

**1910.1096(a)(5)** *Dose* means the quantity of ionizing radiation absorbed, per unit of mass, by the body or by any portion of the body. When the provisions in this section specify a dose during a period of time, the dose is the total quantity of radiation absorbed, per unit of mass, by the body or by any portion of the body during such period of time. Several different units of dose are in current use. Definitions of units used in this section are set forth in paragraphs (a) (6) and (7) of this section.

**1910.1096(a)(6)** *Rad* means a measure of the dose of any ionizing radiation to body tissues in terms of the energy absorbed per unit of mass of the tissue. One rad is the dose corresponding to the absorption of 100 ergs per gram of tissue (1 millirad (mrad)=0.001 rad).

**1910.1096(a)(7)** *Rem* means a measure of the dose of any ionizing radiation to body tissue in terms of its estimated biological effect relative to a dose of 1 roentgen (r) of X-rays (1 millirem (mrem)=0.001 rem). The relation of the rem to other dose units depends upon the biological effect under consideration and upon the conditions for irradiation. Each of the following is considered to be equivalent to a dose of 1 rem:

**1910.1096(a)(7)(i)** A dose of 1 roentgen due to X- or gamma radiation;

**1910.1096(a)(7)(ii)** A dose of 1 rad due to X-, gamma, or beta radiation;

**1910.1096(a)(7)(iii)** A dose of 0.1 rad due to neutrons or high energy protons;

**1910.1096(a)(7)(iv)** A dose of 0.05 rad due to particles heavier than protons and with sufficient energy to reach the lens of the eye;

**1910.1096(a)(7)(v)** If it is more convenient to measure the neutron flux, or equivalent, than to determine the neutron dose in rads, as provided in paragraph (a)(7)(iii) of this section, 1 rem of neutron radiation may, for purposes of the provisions in this section be assumed to be equivalent to 14 million neutrons per square centimeter incident upon the body; or, if there is sufficient information to estimate with reasonable accuracy the approximate distribution in energy of the neutrons, the incident number of neutrons per square centimeter equivalent to 1 rem may be estimated from Table G-17:

### TABLE G-17—NEUTRON FLUX DOSE EQUIVALENTS

| Neutron energy (million electron volts (Mev)) | Number of neutrons per square centimeter equivalent to a dose of 1 rem (neutrons/cm²) | Average flux to deliver 100 millirem in 40 hours (neutrons/cm² per sec.) |
|---|---|---|
| Thermal | 970×10⁶ | 670 |
| 0.0001 | 720×10⁶ | 500 |
| 0.005 | 820×10⁶ | 570 |
| 0.02 | 400×10⁶ | 280 |
| 0.1 | 120×10⁶ | 80 |
| 0.5 | 43×10⁶ | 30 |
| 1.0 | 26×10⁶ | 18 |
| 2.5 | 29×10⁶ | 20 |
| 5.0 | 26×10⁶ | 18 |
| 7.5 | 24×10⁶ | 17 |
| 10 | 24×10⁶ | 17 |
| 10 to 30 | 14×10⁶ | 10 |

**1910.1096(a)(8)** For determining exposures to X- or gamma rays up to 3 Mev., the dose limits specified in this section may be assumed to be equivalent to the "air dose". For the purpose of this section *air dose* means that the dose is measured by a properly calibrated appropriate instrument in air at or near the body surface in the region of the highest dosage rate.

**1910.1096(b)** *Exposure of individuals to radiation in restricted areas.* **(1)** Except as provided in paragraph (b)(2) of this section, no employer shall possess, use, or transfer sources of ionizing radiation in such a manner as to cause any individual in a restricted area to receive in any period of one calendar quarter from sources in the employer's possession or control a dose in excess of the limits specified in Table G-18:

### TABLE G-18

|  | Rems per calendar quarter |
|---|---|
| Whole body: Head and trunk; active blood-forming organs; lens of eyes; or gonads | 1 1/4 |
| Hands and forearms; feet and ankles | 18 3/4 |
| Skin of whole body | 7 1/2 |

**1910.1096(b)(2)** An employer may permit an individual in a restricted area to receive doses to the whole body greater than those permitted under subparagraph (1) of this paragraph, so long as:

**1910.1096(b)(2)(i)** During any calendar quarter the dose to the whole body shall not exceed 3 rems; and

**1910.1096(b)(2)(ii)** The dose to the whole body, when added to the accumulated occupational dose to the whole body, shall not exceed 5 (N-18) rems, where "N" equals the individual's age in years at his last birthday; and

**1910.1096(b)(2)(iii)** The employer maintains adequate past and current exposure records which show that the addition of such a dose will not cause the individual to exceed the amount authorized in this subparagraph. As used in this subparagraph *Dose to the whole body* shall be deemed to include any dose to the whole body, gonad, active bloodforming organs, head and trunk, or lens of the eye.

**1910.1096(b)(3)** No employer shall permit any employee who is under 18 years of age to receive in any period of one calendar quarter a dose in excess of 10 percent of the limits specified in Table G-18.

**1910.1096(b)(4)** *Calendar quarter* means any 3-month period determined as follows:

**1910.1096(b)(4)(i)** The first period of any year may begin on any date in January: *Provided,* That the second, third, and fourth periods accordingly begin on the same date in April, July, and October, respectively, and that the fourth period extends into January of the succeeding year, if necessary to complete a 3-month quarter. During the first year of use of this method of determination, the first period for that year shall also include any additional days in January preceding the starting date for the first period; or

**1910.1096(b)(4)(ii)** The first period in a calendar year of 13 complete, consecutive calendar weeks; the second period in a calendar year of 13 complete, consecutive calendar weeks; the third period in a calendar year of 13 complete, consecutive calendar weeks; the fourth period in a calendar year of 13 complete, consecutive calendar weeks. If at the end of a calendar year there are any days not falling within a complete calendar week of that year, such days shall be included within the last complete calendar week of that year. If at the beginning of any calendar year there are days not falling within a complete calendar week of that year, such days shall be included within the last complete calendar week of the previous year; or

**1910.1096(b)(4)(iii)** The four periods in a calendar year may consist of the first 14 complete, consecutive calendar weeks; the next 12 complete, consecutive calendar weeks, the next 14 complete, consecutive calendar weeks, and the last 12 complete, consecutive calendar weeks. If at the end of a calendar year there are any days not falling within a complete calendar week of that year, such days shall be included (for purposes of this section) within the last complete calendar week of the year. If at the beginning of any calendar year there are days not falling within a complete calendar week of that year, such days shall be included (for purposes of this section) within the last complete week of the previous year.

**1910.1096(c)** *Exposure to airborne radioactive material.* **(1)** No employer shall possess, use or transport radioactive material in such a manner as to cause any employee, within a restricted area, to be exposed to airborne radioactive material in an average concentration in excess of the limits specified in Table 1 of appendix B to 10 CFR part 20. The limits given in Table 1 are for exposure to the concentrations specified for 40 hours in any workweek of 7 consecutive days. In any such period where the number of hours of exposure is less than 40, the limits specified in the table may be increased proportionately. In any such period where the number of hours of exposure is greater than 40, the limits specified in the table shall be decreased proportionately.

**1910.1096(c)(2)** No employer shall possess, use, or transfer radioactive material in such a manner as to cause any individual within a restricted area, who is under 18 years of age, to be exposed to airborne radioactive material in an average concentration in excess of the limits specified in Table II of appendix B to 10 CFR part 20. For purposes of this paragraph, concentrations may be averaged over periods not greater than 1 week.

**1910.1096(c)(3)** *Exposed* as used in this paragraph means that the individual is present in an airborne concentration. No allowance shall be made for the use of protective clothing or equipment, or particle size.

**1910.1096(d)** *Precautionary procedures and personal monitoring.* **(1)** Every employer shall make such surveys as may be necessary for him to comply with the provisions in this section. *Survey* means an evaluation of the radiation hazards incident to the production, use, release, disposal, or presence of radioactive materials or other sources of radiation under a specific set of conditions. When appropriate, such evaluation includes a physical survey of the location of materials and equipment, and measurements of levels of radiation or concentrations of radioactive material present.

**1910.1096(d)(2)** Every employer shall supply appropriate personnel monitoring equipment, such as film badges, pocket chambers, pocket dosimeters, or film rings, and shall require the use of such equipment by:

**1910.1096(d)(2)(i)** Each employee who enters a restricted area under such circumstances that he receives, or is likely to receive, a dose in any calendar quarter in excess of 25 percent of the applicable value specified in paragraph (b)(1) of this section; and

**1910.1096(d)(2)(ii)** Each employee under 18 years of age who enters a restricted area under such circumstances that he receives, or is likely to receive, a dose in any calendar quarter in excess of 5 percent of the applicable value specified in paragraph (b)(1) of this section; and

**1910.1096(d)(2)(iii)** Each employee who enters a high radiation area.

**1910.1096(d)(3)** As used in this section:

**1910.1096(d)(3)(i)** *Personnel monitoring equipment* means devices designed to be worn or carried by an individual for the purpose of measuring the dose received (e.g., film badges, pocket chambers, pocket dosimeters, film rings, etc.);

**1910.1096(d)(3)(ii)** *Radiation area* means any area, accessible to personnel, in which there exists radiation at such levels that a major portion of the body could receive in any 1 hour a dose in excess of 5 millirem, or in any 5 consecutive days a dose in excess of 100 millirem; and

**1910.1096(d)(3)(iii)** *High radiation area* means any area, accessible to personnel, in which there exists radiation at such levels that a major portion of the body could receive in any one hour a dose in excess of 100 millirem.

**1910.1096(e)** *Caution signs, labels, and signals*—**(1)** *General.* **(i)** Symbols prescribed by this paragraph shall use the conventional radiation caution colors (magenta or purple on yellow background). The symbol prescribed by this paragraph is the conventional three-bladed design:

**RADIATION SYMBOL**
1. Cross-hatched area is to be magenta or purple.   2. Background is to be yellow.

**FIGURE G-10**

**1910.1096(e)(1)(ii)** [Reserved]

**1910.1096(e)(2)** *Radiation area.* Each radiation area shall be conspicuously posted with a sign or signs bearing the radiation caution symbol described in subparagraph (1) of this paragraph and the words:

CAUTION
RADIATION AREA

**1910.1096(e)(3)** *High radiation area.* (i) Each high radiation area shall be conspicuously posted with a sign or signs bearing the radiation caution symbol and the words:

CAUTION
HIGH RADIATION AREA

**1910.1096(e)(3)(ii)** Each high radiation area shall be equipped with a control device which shall either cause the level of radiation to be reduced below that at which an individual might receive a dose of 100 millirems in 1 hour upon entry into the area or shall energize a conspicuous visible or audible alarm signal in such a manner that the individual entering and the employer or a supervisor of the activity are made aware of the entry. In the case of a high radiation area established for a period of 30 days or less, such control device is not required.

**1910.1096(e)(4)** *Airborne radioactivity area.* (i) As used in the provisions of this section, *airborne radioactivity area* means:

**1910.1096(e)(4)(i)***(a)* Any room, enclosure, or operating area in which airborne radioactive materials, composed wholly or partly of radioactive material, exist in concentrations in excess of the amounts specified in column 1 of Table 1 of appendix B to 10 CFR part 20 or

**1910.1096(e)(4)(i)***(b)* Any room, enclosure, or operating area in which airborne radioactive materials exist in concentrations which, averaged over the number of hours in any week during which individuals are in the area, exceed 25 percent of the amounts specified in column 1 of Table 1 of appendix B to 10 CFR part 20.

**1910.1096(e)(4)(ii)** Each airborne radioactivity area shall be conspicuously posted with a sign or signs bearing the radiation caution symbol described in paragraph (e)(1) of this section and the words:

CAUTION
AIRBORNE RADIOACTIVITY AREA

**1910.1096(e)(5)** *Additional requirements.* (i) Each area or room in which radioactive material is used or stored and which contains any radioactive material (other than natural uranium or thorium) in any amount exceeding 10 times the quantity of such material specified in appendix C to 10 CFR part 20 shall be conspicuously posted with a sign or signs bearing the radiation caution symbol described in paragraph (e)(1) of this section and the words:

CAUTION
RADIOACTIVE MATERIALS

**1910.1096(e)(5)(ii)** Each area or room in which natural uranium or thorium is used or stored in an amount exceeding 100 times the quantity of such material specified in 10 CFR part 20 shall be conspicuously posted with a sign or signs bearing the radiation caution symbol described in paragraph (e)(1) of this section and the words:

CAUTION
RADIOACTIVE MATERIALS

**1910.1096(e)(6)** *Containers.* (i) Each container in which is transported, stored, or used a quantity of any radioactive material (other than natural uranium or thorium) greater than the quantity of such material specified in appendix C to 10 CFR part 20 shall bear a durable, clearly visible label bearing the radiation caution symbol described in paragraph (e)(1) of this section and the words:

CAUTION
RADIOACTIVE MATERIALS

**1910.1096(e)(6)(ii)** Each container in which natural uranium or thorium is transported, stored, or used in a quantity greater than 10 times the quantity specified in appendix C to 10 CFR part 20 shall bear a durable, clearly visible label bearing the radiation caution symbol described in paragraph (e)(1) of this section and the words:

CAUTION
RADIOACTIVE MATERIALS

**1910.1096(e)(6)(iii)** Notwithstanding the provisions of paragraphs (e)(6) (i) and (ii) of this section a label shall not be required:

**1910.1096(e)(6)(iii)***(a)* If the concentration of the material in the container does not exceed that specified in column 2 of Table 1 of appendix B to 10 CFR part 20, or

**1910.1096(e)(6)(iii)***(b)* For laboratory containers, such as beakers, flasks, and test tubes, used transiently in laboratory procedures, when the user is present.

**1910.1096(e)(6)(iv)** Where containers are used for storage, the labels required in this subparagraph shall state also the quantities and kinds of radioactive materials in the containers and the date of measurement of the quantities.

**1910.1096(f)** *Immediate evacuation warning signal*—(1) *Signal characteristics.* (i) The signal shall be a midfrequency complex sound wave amplitude modulated at a subsonic frequency. The complex sound wave in free space shall have a fundamental frequency (f1) between 450 and 500 hertz (Hz) modulated at a subsonic rate between 4 and 5 hertz.

**1910.1096(f)(1)(ii)** The signal generator shall not be less than 75 decibels at every location where an individual may be present whose immediate, rapid, and complete evacuation is essential.

**1910.1096(f)(1)(iii)** A sufficient number of signal units shall be installed such that the requirements of paragraph (f)(1)(ii) of this section are met at every location where an individual may be present whose immediate, rapid, and complete evacuation is essential.

**1910.1096(f)(1)(iv)** The signal shall be unique in the plant or facility in which it is installed.

**1910.1096(f)(1)(v)** The minimum duration of the signal shall be sufficient to insure that all affected persons hear the signal.

**1910.1096(f)(1)(vi)** The signal-generating system shall respond automatically to an initiating event without requiring any human action to sound the signal.

**1910.1096(f)(2)** *Design objectives.* (i) The signal-generating system shall be designed to incorporate components which enable the system to produce the desired signal each time it is activated within one-half second of activation.

**1910.1096(f)(2)(ii)** The signal-generating system shall be provided with an automatically activated secondary power supply which is adequate to simultaneously power all emergency equipment to which it is connected, if operation during power failure is necessary, except in those systems using batteries as the primary source of power.

**1910.1096(f)(2)(iii)** All components of the signal-generating system shall be located to provide maximum practicable protection against damage in case of fire, explosion, corrosive atmosphere, or other environmental extremes consistent with adequate system performance.

**1910.1096(f)(2)(iv)** The signal-generating system shall be designed with the minimum number of components necessary to make it function as intended, and should utilize components which do not require frequent servicing such as lubrication or cleaning.

**1910.1096(f)(2)(v)** Where several activating devices feed activating information to a central signal generator, failure of any activating device shall not render the signal-generator system inoperable to activating information from the remaining devices.

**1910.1096(f)(2)(vi)** The signal-generating system shall be designed to enhance the probability that alarm occurs only when immediate evacuation is warranted.

The number of false alarms shall not be so great that the signal will come to be disregarded and shall be low enough to minimize personal injuries or excessive property damage that might result from such evacuation.

**1910.1096(f)(3)** *Testing.* **(i)** Initial tests, inspections, and checks of the signal-generating system shall be made to verify that the fabrication and installation were made in accordance with design plans and specifications and to develop a thorough knowledge of the performance of the system and all components under normal and hostile conditions.

**1910.1096(f)(3)(ii)** Once the system has been placed in service, periodic tests, inspections, and checks shall be made to minimize the possibility of malfunction.

**1910.1096(f)(3)(iii)** Following significant alterations or revisions to the system, tests and checks similar to the initial installation tests shall be made.

**1910.1096(f)(3)(iv)** Tests shall be designed to minimize hazards while conducting the tests.

**1910.1096(f)(3)(v)** Prior to normal operation the signal-generating system shall be checked physically and functionally to assure reliability and to demonstrate accuracy and performance. Specific tests shall include:

**1910.1096(f)(3)(v)***(a)* All power sources.

**1910.1096(f)(3)(v)***(b)* Calibration and calibration stability.

**1910.1096(f)(3)(v)***(c)* Trip levels and stability.

**1910.1096(f)(3)(v)***(d)* Continuity of function with loss and return of required services such as AC or DC power, air pressure, etc.

**1910.1096(f)(3)(v)***(e)* All indicators.

**1910.1096(f)(3)(v)***(f)* Trouble indicator circuits and signals, where used.

**1910.1096(f)(3)(v)***(g)* Air pressure (if used)

**1910.1096(f)(3)(v)***(h)* Determine that sound level of the signal is within the limit of paragraph (f)(1)(ii) of this section at all points that require immediate evacuation.

**1910.1096(f)(3)(vi)** In addition to the initial startup and operating tests, periodic scheduled performance tests and status checks must be made to insure that the system is at all times operating within design limits and capable of the required response. Specific periodic tests or checks or both shall include:

**1910.1096(f)(3)(vi)***(a)* Adequacy of signal activation device.

**1910.1096(f)(3)(vi)***(b)* All power sources.

**1910.1096(f)(3)(vi)***(c)* Function of all alarm circuits and trouble indicator circuits including trip levels.

**1910.1096(f)(3)(vi)***(d)* Air pressure (if used).

**1910.1096(f)(3)(vi)***(e)* Function of entire system including operation without power where required.

**1910.1096(f)(3)(vi)***(f)* Complete operational tests including sounding of the signal and determination that sound levels are adequate.

**1910.1096(f)(3)(vii)** Periodic tests shall be scheduled on the basis of need, experience, difficulty, and disruption of operations. The entire system should be operationally tested at least quarterly.

**1910.1096(f)(3)(viii)** All employees whose work may necessitate their presence in an area covered by the signal shall be made familiar with the actual sound of the signal—preferably as it sounds at their work location. Before placing the system into operation, all employees normally working in the area shall be made acquainted with the signal by actual demonstration at their work locations.

**1910.1096(g)** *Exceptions from posting requirements.* Notwithstanding the provisions of paragraph (e) of this section:

**1910.1096(g)(1)** A room or area is not required to be posted with a caution sign because of the presence of a sealed source, provided the radiation level 12 inches from the surface of the source container or housing does not exceed 5 millirem per hour.

**1910.1096(g)(2)** Rooms or other areas in onsite medical facilities are not required to be posted with caution signs because of the presence of patients containing radioactive material, provided that there are personnel in attendance who shall take the precautions necessary to prevent the exposure of any individual to radiation or radioactive material in excess of the limits established in the provisions of this section.

**1910.1096(g)(3)** Caution signs are not required to be posted at areas or rooms containing radioactive materials for periods of less than 8 hours: *Provided,* That

**1910.1096(g)(3)(i)** The materials are constantly attended during such periods by an individual who shall take the precautions necessary to prevent the exposure of any individual to radiation or radioactive materials in excess of the limits established in the provisions of this section; and

**1910.1096(g)(3)(ii)** Such area or room is subject to the employer's control.

**1910.1096(h)** *Exemptions for radioactive materials packaged for shipment.* Radioactive materials packaged and labeled in accordance with regulations of the Department of Transportation published in 49 CFR Chapter I, are exempt from the labeling and posting requirements of this subpart during shipment, provided that the inside containers are labeled in accordance with the provisions of paragraph (e) of this section.

**1910.1096(i)** *Instruction of personnel, posting.* **(1)** Employers regulated by the Nuclear Regulatory Commission shall be governed by 10 CFR part 20 standards. Employers in a State named in paragraph (p)(3) of this section shall be governed by the requirements of the laws and regulations of that State. All other employers shall be regulated by the following:

**1910.1096(i)(2)** All individuals working in or frequenting any portion of a radiation area shall be informed of the occurrence of radioactive materials or of radiation in such portions of the radiation area; shall be instructed in the safety problems associated with exposure to such materials or radiation and in precautions or devices to minimize exposure; shall be instructed in the applicable provisions of this section for the protection of employees from exposure to radiation or radioactive materials; and shall be advised of reports of radiation exposure which employees may request pursuant to the regulations in this section.

**1910.1096(i)(3)** Each employer to whom this section applies shall post a current copy of its provisions and a copy of the operating procedures applicable to the work conspicuously in such locations as to insure that employees working in or frequenting radiation areas will observe these documents on the way to and from their place of employment, or shall keep such documents available for examination of employees upon request.

**1910.1096(j)** *Storage of radioactive materials.* Radioactive materials stored in a nonradiation area shall be secured against unauthorized removal from the place of storage.

**1910.1096(k)** *Waste disposal.* No employer shall dispose of radioactive material except by transfer to an authorized recipient, or in a manner approved by the Nuclear Regulatory Commission or a State named in paragraph (p)(3) of this section.

**1910.1096(l)** *Notification of incidents*—**(1)** *Immediate notification.* Each employer shall immediately notify the Assistant Secretary of Labor or his duly authorized representative, for employees not protected by the Nuclear Regulatory Commission by means of 10 CFR part 20; paragraph (p)(2) of this section, or the requirements of the laws and regulations of States named in paragraph (p)(3) of this section, by telephone or telegraph of any incident involving radiation which may have caused or threatens to cause:

**1910.1096(l)(1)(i)** Exposure of the whole body of any individual to 25 rems or more of radiation; exposure of the skin of the whole body of any individual to 150 rems or more of radiation; or exposure of the feet, ankles, hands, or forearms of any individual to 375 rems or more of radiation; or

**1910.1096(l)(1)(ii)** The release of radioactive material in concentrations which, if averaged over a period of 24 hours, would exceed 5,000 times the limit specified for such materials in Table II of appendix B to 10 CFR part 20.

**1910.1096(l)(2)** *Twenty-four hour notification.* Each employer shall within 24 hours following its occurrence notify the Assistant Secretary of Labor or his duly authorized representative for employees not protected by the Nuclear Regulatory Commission by means of 10 CFR part 20; paragraph (p)(2) of this section, or the requirements of the laws and applicable regulations of States named in paragraph (p)(3) of this section, by telephone or telegraph of any incident involving radiation which may have caused or threatens to cause:

**1910.1096(l)(2)(i)** Exposure of the whole body of any individual to 5 rems or more of radiation; exposure of the skin of the whole body of any individual to 30 rems or more of radiation; or exposure of the feet, ankles, hands, or forearms to 75 rems or more of radiation; or

**1910.1096(l)(2)(ii)** [Reserved]

**1910.1096(m)** *Reports of overexposure and excessive levels and concentrations.* **(1)** In addition to any notification required by paragraph (l) of this section each employer shall make a report in writing within 30 days to the Assistant Secretary of Labor or his duly authorized representative, for employees not protected by the Nuclear Regulatory Commission by means of 10 CFR part 20; or under paragraph (p)(2) of this section, or the requirements of the laws and regulations of States named in paragraph (p)(3) of this section, of each exposure of an individual to radiation or concentrations of radioactive material in excess of any applicable limit in this section. Each report required under this paragraph shall describe the extent of exposure of persons to radiation or to radioactive material; levels of radiation and concentration of radioactive material involved, the cause of the exposure, levels of concentrations; and corrective steps taken or planned to assure against a recurrence.

**1910.1096(m)(2)** In any case where an employer is required pursuant to the provisions of this paragraph to report to the U.S. Department of Labor any exposure of an individual to radiation or to concentrations of radioactive material, the employer shall also notify such individual of the nature and extent of exposure. Such notice shall be in writing and shall contain the following statement: "You should preserve this report for future reference."

**1910.1096(n)** *Records.* **(1)** Every employer shall maintain records of the radiation exposure of all employees for whom personnel monitoring is required under paragraph (d) of this section and advise each of his employees of his individual exposure on at least an annual basis.

**1910.1096(n)(2)** Every employer shall maintain records in the same units used in tables in paragraph (b) of this section and appendix B to 10 CFR part 20.

**1910.1096(o)** *Disclosure to former employee of individual employee's record.* **(1)** At the request of a former employee an employer shall furnish to the employee a report of the employee's exposure to radiation as shown in records maintained by the employer pursuant to paragraph (n)(1) of this section. Such report shall be furnished within 30 days from the time the request is made, and shall cover each calendar quarter of the individual's employment involving exposure to radiation or such lesser period as may be requested by the employee. The report shall also include the results of any calculations and analysis of radioactive material deposited in the body of the employee. The report shall be in writing and contain the following statement: "You should preserve this report for future reference."

**1910.1096(o)(2)** [Reserved]

**1910.1096(p)** *Nuclear Regulatory Commission licensees—NRC contractors operating NRC plants and facilities—NRC Agreement State licensees or registrants.* **(1)** Any employer who possesses or uses source material, byproduct material, or special nuclear material, as defined in the Atomic Energy Act of 1954, as amended, under a license issued by the Nuclear Regulatory Commission and in accordance with the requirements of 10 CFR part 20 shall be deemed to be in compliance with the requirements of this section with respect to such possession and use.

**1910.1096(p)(2)** NRC contractors operating NRC plants and facilities: Any employer who possesses or uses source material, byproduct material, special nuclear material, or other radiation sources under a contract with the Nuclear Regulatory Commission for the operation of NRC plants and facilities and in accordance with the standards, procedures, and other requirements for radiation protection established by the Commission for such contract pursuant to the Atomic Energy Act of 1954 as amended (42 U.S.C. 2011 et seq.), shall be deemed to be in compliance with the requirements of this section with respect to such possession and use.

**1910.1096(p)(3)** NRC-agreement State licensees or registrants:

**1910.1096(p)(3)(i)** *Atomic Energy Act sources.* Any employer who possesses or uses source material, byproduct material, or special nuclear material, as defined in the Atomic Energy Act of 1954, as amended (42 U.S.C. 2011 et seq.), and has either registered such sources with, or is operating under a license issued by, a State which has an agreement in effect with the Nuclear Regulatory Commission pursuant to section 274(b) (42 U.S.C. 2021(b)) of the Atomic Energy Act of 1954, as amended, and in accordance with the requirements of that State's laws and regulations shall be deemed to be in compliance with the radiation requirements of this section, insofar as his possession and use of such material is concerned, unless the Secretary of Labor, after conference with the Nuclear Regulatory Commission, shall determine that the State's program for control of these radiation sources is incompatible with the requirements of this section. Such agreements currently are in effect only in the States of Alabama, Arkansas, California, Kansas, Kentucky, Florida, Mississippi, New Hampshire, New York, North Carolina, Texas, Tennessee, Oregon, Idaho, Arizona, Colorado, Louisiana, Nebraska, Washington, Maryland, North Dakota, South Carolina, and Georgia.

**1910.1096(p)(3)(ii)** *Other sources.* Any employer who possesses or uses radiation sources other than source material, byproduct material, or special nuclear material, as defined in the Atomic Energy Act of 1954, as amended (42 U.S.C. 2011 et seq.), and has either registered such sources with, or is operating under a license issued by a State which has an agreement in effect with the Nuclear Regulatory Commission pursuant to section 274(b) (42 U.S.C. 2021(b)) of the Atomic Energy Act of 1954, as amended, and in accordance with the requirements of that State's laws and regulations shall be deemed to be in compliance with the radiation requirements of this section, insofar as his possession and use of such material is concerned, provided the State's program for control of these radiation sources is the subject of a currently effective determination by the Assistant Secretary of Labor that such program is compatible with the requirements of this section. Such determinations currently are in effect only in the States of Alabama, Arkansas, California, Kansas, Kentucky, Florida, Mississippi, New Hampshire, New York, North Carolina, Texas, Tennessee, Oregon, Idaho, Arizona, Colorado, Louisiana, Nebraska, Washington, Maryland, North Dakota, South Carolina, and Georgia.

[39 FR 23502, June 27, 1974, as amended at 43 FR 49746, Oct. 24, 1978; 43 FR 51759, Nov. 7, 1978; 49 FR 18295, Apr. 30, 1984; 58 FR 35309, June 30, 1993. Redesignated at 61 FR 31430, June 20, 1996]

### §1910.1200 Hazard communication.

**1910.1200(a)** *Purpose.* **(1)** The purpose of this section is to ensure that the hazards of all chemicals produced or imported are evaluated, and that information concerning their hazards is transmitted to employers and employees. This transmittal of information is to be accomplished by means of comprehensive hazard communication programs, which are to include container labeling and other forms of warning, material safety data sheets and employee training.

**1910.1200(a)(2)** This occupational safety and health standard is intended to address comprehensively the issue of evaluating the potential hazards of chemicals, and communicating information concerning hazards and appropriate protective measures to employees, and to preempt any legal requirements of a state, or political subdivision of a state, pertaining to this subject. Evaluating the potential hazards of chemicals, and communicating information concerning hazards and appropriate protective measures to employees, may include, for example, but is not limited to, provisions for: developing and maintaining a written hazard communication program for the workplace, including lists of hazardous chemicals present; labeling of containers of chemicals in the workplace, as well as of containers of chemicals being shipped to other workplaces; preparation and distribution of material safety data sheets to employees and downstream employers; and development and implementation of employee training programs regarding hazards of chemicals and protective measures. Under section 18 of the Act, no state or political subdivision of a state may adopt or enforce, through any court or agency, any requirement relating to the issue addressed by this Federal standard, except pursuant to a Federally-approved state plan.

**1910.1200(b)** *Scope and application.* **(1)** This section requires chemical manufacturers or importers to assess the hazards of chemicals which they produce or import, and all employers to provide information to their employees about the hazardous chemicals to which they are exposed, by means of a hazard communication program, labels and other forms of warning, material safety data sheets, and information and training. In addition, this section requires distributors to transmit the required information to employers. (Employers who do not produce or import chemicals need only focus on those parts of this rule that deal with establishing a workplace program and communicating information to their workers. Appendix E of this section is a general guide for such employers to help them determine their compliance obligations under the rule.)

**1910.1200(b)(2)** This section applies to any chemical which is known to be present in the workplace in such a manner that employees may be exposed under normal conditions of use or in a foreseeable emergency.

**1910.1200(b)(3)** This section applies to laboratories only as follows:

**1910.1200(b)(3)(i)** Employers shall ensure that labels on incoming containers of hazardous chemicals are not removed or defaced;

**1910.1200(b)(3)(ii)** Employers shall maintain any material safety data sheets that are received with incoming shipments of hazardous chemicals, and ensure that they are readily accessible during each workshift to laboratory employees when they are in their work areas;

**1910.1200(b)(3)(iii)** Employers shall ensure that laboratory employees are provided information and training in accordance with paragraph (h) of this section, except for the location and availability of the written hazard communication program under paragraph (h)(2)(iii) of this section; and,

**1910.1200(b)(3)(iv)** Laboratory employers that ship hazardous chemicals are considered to be either a chemical manufacturer or a distributor under this rule, and thus must ensure that any containers of hazardous chemicals leaving the laboratory are labeled in accordance with paragraph (f)(1) of this section, and that a material safety data sheet is provided to distributors and other employers in accordance with paragraphs (g)(6) and (g)(7) of this section.

**1910.1200(b)(4)** In work operations where employees only handle chemicals in sealed containers which are not opened under normal conditions of use (such as are found in marine cargo handling, warehousing, or retail sales), this section applies to these operations only as follows:

**1910.1200(b)(4)(i)** Employers shall ensure that labels on incoming containers of hazardous chemicals are not removed or defaced;

**1910.1200(b)(4)(ii)** Employers shall maintain copies of any material safety data sheets that are received with incoming shipments of the sealed containers of hazardous chemicals, shall obtain a material safety data sheet as soon as possible for sealed containers of hazardous chemicals received without a material safety data sheet if an employee requests the material safety data sheet, and shall ensure that the material safety data sheets are readily accessible during each work shift to employees when they are in their work area(s); and,

**1910.1200(b)(4)(iii)** Employers shall ensure that employees are provided with information and training in accordance with paragraph (h) of this section (except for the location and availability of the written hazard communication program under paragraph (h)(2)(iii) of this section), to the extent necessary to protect them in the event of a spill or leak of a hazardous chemical from a sealed container.

**1910.1200(b)(5)** This section does not require labeling of the following chemicals:

**1910.1200(b)(5)(i)** Any pesticide as such term is defined in the Federal Insecticide, Fungicide, and Rodenticide Act (7 U.S.C. 136 et seq.), when subject to the labeling requirements of that Act and labeling regulations issued under that Act by the Environmental Protection Agency;

**1910.1200(b)(5)(ii)** Any chemical substance or mixture as such terms are defined in the Toxic Substances Control Act (15 U.S.C. 2601 et seq.), when subject to the labeling requirements of that Act and labeling regulations issued under that Act by the Environmental Protection Agency.

**1910.1200(b)(5)(iii)** Any food, food additive, color additive, drug, cosmetic, or medical or veterinary device or product, including materials intended for use as ingredients in such products (e.g., flavors and fragrances), as such terms are defined in the Federal Food, Drug, and Cosmetic Act (21 U.S.C. 301 et seq.) or the Virus-Serum-Toxin Act of 1913 (21 U.S.C. 151 et seq.), and regulations issued under those Acts, when they are subject to the labeling requirements under those Acts by either the Food and Drug Administration or the Department of Agriculture;

**1910.1200(b)(5)(iv)** Any distilled spirits (beverage alcohols), wine, or malt beverage intended for nonindustrial use, as such terms are defined in the Federal Alcohol Administration Act (27 U.S.C. 201 et seq.) and regulations issued under that Act, when subject to the labeling requirements of that Act and labeling regulations issued under that Act by the Bureau of Alcohol, Tobacco, and Firearms;

**1910.1200(b)(5)(v)** Any consumer product or hazardous substance as those terms are defined in the Consumer Product Safety Act (15 U.S.C. 2051 et seq.) and Federal Hazardous Substances Act (15 U.S.C. 1261 et seq.) respectively, when subject to a consumer product safety standard or labeling requirement of those Acts, or regulations issued under those Acts by the Consumer Product Safety Commission; and,

**1910.1200(b)(5)(vi)** Agricultural or vegetable seed treated with pesticides and labeled in accordance with the Federal Seed Act (7 U.S.C. 1551 et seq.) and the labeling regulations issued under that Act by the Department of Agriculture.

**1910.1200(b)(6)** This section does not apply to: **(i)** Any hazardous waste as such term is defined by the Solid Waste Disposal Act, as amended by the Resource Conservation and Recovery Act of 1976, as amended (42 U.S.C. 6901 et seq.), when subject to regulations issued under that Act by the Environmental Protection Agency;

**1910.1200(b)(6)(ii)** Any hazardous substance as such term is defined by the Comprehensive Environmental Response, Compensation and Liability ACT (CERCLA) (42 U.S.C. 9601 et seq.) when the hazardous substance is the focus of remedial or removal action being conducted under CERCLA in accordance with Environmental Protection Agency regulations;

**1910.1200(b)(6)(iii)** Tobacco or tobacco products;

**1910.1200(b)(6)(iv)** Wood or wood products, including lumber which will not be processed, where the chemical manufacturer or importer can establish that the only hazard they pose to employees is the potential for flammability or combustibility (wood or wood products which have been treated with a hazardous chemical covered by this standard, and wood which may be subsequently sawed or cut, generating dust, are not exempted);

**1910.1200(b)(6)(v)** Articles (as that term is defined in paragraph (c) of this section);

**1910.1200(b)(6)(vi)** Food or alcoholic beverages which are sold, used, or prepared in a retail establishment (such as a grocery store, restaurant, or drinking place), and foods intended for personal consumption by employees while in the workplace;

**1910.1200(b)(6)(vii)** Any drug, as that term is defined in the Federal Food, Drug, and Cosmetic Act (21 U.S.C. 301 et seq.), when it is in solid, final form for direct administration to the patient (e.g., tablets or pills); drugs which are packaged by the chemical manufacturer for sale to consumers in a retail establishment (e.g., over-the-counter drugs); and drugs intended for personal consumption by employees while in the workplace (e.g., first aid supplies);

**1910.1200(b)(6)(viii)** Cosmetics which are packaged for sale to consumers in a retail establishment, and cosmetics intended for personal consumption by employees while in the workplace;

**1910.1200(b)(6)(ix)** Any consumer product or hazardous substance, as those terms are defined in the Consumer Product Safety Act (15 U.S.C. 2051 et seq.) and Federal Hazardous Substances Act (15 U.S.C. 1261 et seq.) respectively, where the employer can show that it is used in the workplace for the purpose intended by the chemical manufacturer or importer of the product, and the use results in a duration and frequency of exposure which is not greater than the range of exposures that could reasonably be experienced by consumers when used for the purpose intended;

**1910.1200(b)(6)(x)** Nuisance particulates where the chemical manufacturer or importer can establish that they do not pose any physical or health hazard covered under this section;

**1910.1200(b)(6)(xi)** Ionizing and nonionizing radiation; and,

**1910.1200(b)(6)(xii)** Biological hazards.

**1910.1200(c)** *Definitions.*

*Article* means a manufactured item other than a fluid or particle: (i) which is formed to a specific shape or design during manufacture; (ii) which has end use function(s) dependent in whole or in part upon its shape or design during end use; and (iii) which under normal conditions of use does not release more than very small quantities, e.g., minute or trace amounts of a hazardous chemical (as determined under paragraph (d) of this section), and does not pose a physical hazard or health risk to employees.

*Assistant Secretary* means the Assistant Secretary of Labor for Occupational Safety and Health, U.S. Department of Labor, or designee.

*Chemical* means any element, chemical compound or mixture of elements and/or compounds.

*Chemical manufacturer* means an employer with a workplace where chemical(s) are produced for use or distribution.

*Chemical name* means the scientific designation of a chemical in accordance with the nomenclature system developed by the International Union of Pure and Applied Chemistry (IUPAC) or the Chemical Abstracts Service (CAS) rules of nomenclature, or a name which will clearly identify the chemical for the purpose of conducting a hazard evaluation.

*Combustible liquid* means any liquid having a flashpoint at or above 100 °F (37.8 °C), but below 200 °F (93.3 °C), except any mixture having components with flashpoints of 200 °F (93.3 °C), or higher, the total volume of which make up 99 percent or more of the total volume of the mixture.

*Commercial account* means an arrangement whereby a retail distributor sells hazardous chemicals to an employer, generally in large quantities over time and/or at costs that are below the regular retail price.

*Common name* means any designation or identification such as code name, code number, trade name, brand name or generic name used to identify a chemical other than by its chemical name.

*Compressed gas* means:

(i) A gas or mixture of gases having, in a container, an absolute pressure exceeding 40 psi at 70 °F (21.1 °C); or

(ii) A gas or mixture of gases having, in a container, an absolute pressure exceeding 104 psi at 130 °F (54.4 °C) regardless of the pressure at 70 °F (21.1 °C); or

(iii) A liquid having a vapor pressure exceeding 40 psi at 100 °F (37.8 °C) as determined by ASTM D-323-72.

*Container* means any bag, barrel, bottle, box, can, cylinder, drum, reaction vessel, storage tank, or the like that contains a hazardous chemical. For purposes of this section, pipes or piping systems, and engines, fuel tanks, or other operating systems in a vehicle, are not considered to be containers.

*Designated representative* means any individual or organization to whom an employee gives written authorization to exercise such employee's rights under this section. A recognized or certified collective bargaining agent shall be treated automatically as a designated representative without regard to written employee authorization.

*Director* means the Director, National Institute for Occupational Safety and Health, U.S. Department of Health and Human Services, or designee.

*Distributor* means a business, other than a chemical manufacturer or importer, which supplies hazardous chemicals to other distributors or to employers.

*Employee* means a worker who may be exposed to hazardous chemicals under normal operating conditions or in foreseeable emergencies. Workers such as office workers or bank tellers who encounter hazardous chemicals only in non-routine, isolated instances are not covered.

*Employer* means a person engaged in a business where chemicals are either used, distributed, or are produced for use or distribution, including a contractor or subcontractor.

*Explosive* means a chemical that causes a sudden, almost instantaneous release of pressure, gas, and heat when subjected to sudden shock, pressure, or high temperature.

*Exposure* or *exposed* means that an employee is subjected in the course of employment to a chemical that is a physical or health hazard, and includes potential (e.g., accidental or possible) exposure. "Subjected" in terms of health hazards includes any route of entry (e.g., inhalation, ingestion, skin contact or absorption.)

*Flammable* means a chemical that falls into one of the following categories:

(i) *Aerosol, flammable* means an aerosol that, when tested by the method described in 16 CFR 1500.45, yields a flame projection exceeding 18 inches at full valve opening, or a flashback (a flame extending back to the valve) at any degree of valve opening;

(ii) *Gas, flammable* means: (A) A gas that, at ambient temperature and pressure, forms a flammable mixture with air at a concentration of thirteen (13) percent by volume or less; or

(ii)(B) A gas that, at ambient temperature and pressure, forms a range of flammable mixtures with air wider than twelve (12) percent by volume, regardless of the lower limit;

(iii) *Liquid, flammable* means any liquid having a flashpoint below 100 °F (37.8 °C), except any mixture having components with flashpoints of 100 °F (37.8 °C) or higher, the total of which make up 99 percent or more of the total volume of the mixture.

(iv) *Solid, flammable* means a solid, other than a blasting agent or explosive as defined in §1910.109(a), that is liable to cause fire through friction, absorption of moisture, spontaneous chemical change, or retained heat from manufacturing or processing, or which can be ignited readily and when ignited burns so vigorously and persistently as to create a serious hazard. A chemical shall be considered to be a flammable solid if, when tested by the method described in 16 CFR 1500.44, it ignites and burns with a self-sustained flame at a rate greater than one-tenth of an inch per second along its major axis.

*Flashpoint* means the minimum temperature at which a liquid gives off a vapor in sufficient concentration to ignite when tested as follows:

(i) Tagliabue Closed Tester (See American National Standard Method of Test for Flash Point by Tag Closed Tester, Z11.24-1979 (ASTM D 56-79)) for liquids with a viscosity of less than 45 Saybolt Universal Seconds (SUS) at 100 °F (37.8 °C), that do not contain suspended solids and do not have a tendency to form a surface film under test; or

(ii) Pensky-Martens Closed Tester (see American National Standard Method of Test for Flash Point by Pensky-Martens Closed Tester, Z11.7-1979 (ASTM D 93-79)) for liquids with a viscosity equal to or greater than 45 SUS at 100 °F (37.8 °C), or that contain suspended solids, or that have a tendency to form a surface film under test; or

(iii) Setaflash Closed Tester (see American National Standard Method of Test for Flash Point by Setaflash Closed Tester (ASTM D 3278-78)).

Organic peroxides, which undergo autoaccelerating thermal decomposition, are excluded from any of the flashpoint determination methods specified above.

*Foreseeable emergency* means any potential occurrence such as, but not limited to, equipment failure, rupture of containers, or failure of control equipment which could result in an uncontrolled release of a hazardous chemical into the workplace.

*Hazardous chemical* means any chemical which is a physical hazard or a health hazard.

*Hazard warning* means any words, pictures, symbols, or combination thereof appearing on a label or other appropriate form of warning which convey the specific physical and health hazard(s), including target organ effects, of the chemical(s) in the container(s). (See the definitions for "physical hazard" and "health hazard" to determine the hazards which must be covered.)

*Health hazard* means a chemical for which there is statistically significant evidence based on at least one study conducted in accordance with established scientific principles that acute or chronic health effects may occur in exposed employees. The term "health hazard" includes chemicals which are carcinogens, toxic or highly toxic agents, reproductive toxins, irritants, corrosives, sensitizers, hepatotoxins, nephrotoxins, neurotoxins, agents which act on the hematopoietic system, and agents which damage the lungs, skin, eyes, or mucous membranes. Appendix A provides further definitions and explanations of the scope of health hazards covered by this section, and Appendix B describes the criteria to be used to determine whether or not a chemical is to be considered hazardous for purposes of this standard.

*Identity* means any chemical or common name which is indicated on the material safety data sheet (MSDS) for the chemical. The identity used shall permit cross-references to be made among the required list of hazardous chemicals, the label and the MSDS.

*Immediate use* means that the hazardous chemical will be under the control of and used only by the person who transfers it from a labeled container and only within the work shift in which it is transferred.

*Importer* means the first business with employees within the Customs Territory of the United States which receives hazardous chemicals produced in other countries for the purpose of supplying them to distributors or employers within the United States.

*Label* means any written, printed, or graphic material displayed on or affixed to containers of hazardous chemicals.

*Material safety data sheet (MSDS)* means written or printed material concerning a hazardous chemical which is prepared in accordance with paragraph (g) of this section.

*Mixture* means any combination of two or more chemicals if the combination is not, in whole or in part, the result of a chemical reaction.

*Organic peroxide* means an organic compound that contains the bivalent -O-O-structure and which may be considered to be a structural derivative of hydrogen peroxide where one or both of the hydrogen atoms has been replaced by an organic radical.

*Oxidizer* means a chemical other than a blasting agent or explosive as defined in §1910.109(a), that initiates or promotes combustion in other materials, thereby causing fire either of itself or through the release of oxygen or other gases.

*Physical hazard* means a chemical for which there is scientifically valid evidence that it is a combustible liquid, a compressed gas, explosive, flammable, an organic peroxide, an oxidizer, pyrophoric, unstable (reactive) or water-reactive.

# OSHA Standards for General Industry

*Produce* means to manufacture, process, formulate, blend, extract, generate, emit, or repackage.

*Pyrophoric* means a chemical that will ignite spontaneously in air at a temperature of 130 °F (54.4 °C) or below.

*Responsible party* means someone who can provide additional information on the hazardous chemical and appropriate emergency procedures, if necessary.

*Specific chemical identity* means the chemical name, Chemical Abstracts Service (CAS) Registry Number, or any other information that reveals the precise chemical designation of the substance.

*Trade secret* means any confidential formula, pattern, process, device, information or compilation of information that is used in an employer's business, and that gives the employer an opportunity to obtain an advantage over competitors who do not know or use it. Appendix D sets out the criteria to be used in evaluating trade secrets.

*Unstable (reactive)* means a chemical which in the pure state, or as produced or transported, will vigorously polymerize, decompose, condense, or will become self-reactive under conditions of shocks, pressure or temperature.

*Use* means to package, handle, react, emit, extract, generate as a by-product, or transfer.

*Water-reactive* means a chemical that reacts with water to release a gas that is either flammable or presents a health hazard.

*Work area* means a room or defined space in a workplace where hazardous chemicals are produced or used, and where employees are present.

*Workplace* means an establishment, job site, or project, at one geographical location containing one or more work areas.

**1910.1200(d)** *Hazard determination.* **(1)** Chemical manufacturers and importers shall evaluate chemicals produced in their workplaces or imported by them to determine if they are hazardous. Employers are not required to evaluate chemicals unless they choose not to rely on the evaluation performed by the chemical manufacturer or importer for the chemical to satisfy this requirement.

**1910.1200(d)(2)** Chemical manufacturers, importers or employers evaluating chemicals shall identify and consider the available scientific evidence concerning such hazards. For health hazards, evidence which is statistically significant and which is based on at least one positive study conducted in accordance with established scientific principles is considered to be sufficient to establish a hazardous effect if the results of the study meet the definitions of health hazards in this section. Appendix A shall be consulted for the scope of health hazards covered, and Appendix B shall be consulted for the criteria to be followed with respect to the completeness of the evaluation, and the data to be reported.

**1910.1200(d)(3)** The chemical manufacturer, importer or employer evaluating chemicals shall treat the following sources as establishing that the chemicals listed in them are hazardous:

**1910.1200(d)(3)(i)** 29 CFR part 1910, subpart Z, Toxic and Hazardous Substances, Occupational Safety and Health Administration (OSHA); or,

**1910.1200(d)(3)(ii)** *Threshold Limit Values for Chemical Substances and Physical Agents in the Work Environment,* American Conference of Governmental Industrial Hygienists (ACGIH) (latest edition). The chemical manufacturer, importer, or employer is still responsible for evaluating the hazards associated with the chemicals in these source lists in accordance with the requirements of this standard.

**1910.1200(d)(4)** Chemical manufacturers, importers and employers evaluating chemicals shall treat the following sources as establishing that a chemical is a carcinogen or potential carcinogen for hazard communication purposes:

**1910.1200(d)(4)(i)** National Toxicology Program (NTP), *Annual Report on Carcinogens* (latest edition);

**1910.1200(d)(4)(ii)** International Agency for Research on Cancer (IARC) *Monographs* (latest editions); or

**1910.1200(d)(4)(iii)** 29 CFR part 1910, subpart Z, Toxic and Hazardous Substances, Occupational Safety and Health Administration.

**NOTE:** The *Registry of Toxic Effects of Chemical Substances* published by the National Institute for Occupational Safety and Health indicates whether a chemical has been found by NTP or IARC to be a potential carcinogen.

**1910.1200(d)(5)** The chemical manufacturer, importer or employer shall determine the hazards of mixtures of chemicals as follows:

**1910.1200(d)(5)(i)** If a mixture has been tested as a whole to determine its hazards, the results of such testing shall be used to determine whether the mixture is hazardous;

**1910.1200(d)(5)(ii)** If a mixture has not been tested as a whole to determine whether the mixture is a health hazard, the mixture shall be assumed to present the same health hazards as do the components which comprise one percent (by weight or volume) or greater of the mixture, except that the mixture shall be assumed to present a carcinogenic hazard if it contains a component in concentrations of 0.1 percent or greater which is considered to be a carcinogen under paragraph (d)(4) of this section;

**1910.1200(d)(5)(iii)** If a mixture has not been tested as a whole to determine whether the mixture is a physical hazard, the chemical manufacturer, importer, or employer may use whatever scientifically valid data is available to evaluate the physical hazard potential of the mixture; and,

**1910.1200(d)(5)(iv)** If the chemical manufacturer, importer, or employer has evidence to indicate that a component present in the mixture in concentrations of less than one percent (or in the case of carcinogens, less than 0.1 percent) could be released in concentrations which would exceed an established OSHA permissible exposure limit or ACGIH Threshold Limit Value, or could present a health risk to employees in those concentrations, the mixture shall be assumed to present the same hazard.

**1910.1200(d)(6)** Chemical manufacturers, importers, or employers evaluating chemicals shall describe in writing the procedures they use to determine the hazards of the chemical they evaluate. The written procedures are to be made available, upon request, to employees, their designated representatives, the Assistant Secretary and the Director. The written description may be incorporated into the written hazard communication program required under paragraph (e) of this section.

**1910.1200(e)** *Written hazard communication program.* **(1)** Employers shall develop, implement, and maintain at each workplace, a written hazard communication program which at least describes how the criteria specified in paragraphs (f), (g), and (h) of this section for labels and other forms of warning, material safety data sheets, and employee information and training will be met, and which also includes the following:

**1910.1200(e)(1)(i)** A list of the hazardous chemicals known to be present using an identity that is referenced on the appropriate material safety data sheet (the list may be compiled for the workplace as a whole or for individual work areas); and,

**1910.1200(e)(1)(ii)** The methods the employer will use to inform employees of the hazards of non-routine tasks (for example, the cleaning of reactor vessels), and the hazards associated with chemicals contained in unlabeled pipes in their work areas.

**1910.1200(e)(2)** *Multi-employer workplaces.* Employers who produce, use, or store hazardous chemicals at a workplace in such a way that the employees of other employer(s) may be exposed (for example, employees of a construction contractor working on-site) shall additionally ensure that the hazard communication programs developed and implemented under this paragraph (e) include the following:

**1910.1200(e)(2)(i)** The methods the employer will use to provide the other employer(s) on-site access to material safety data sheets for each hazardous chemical the other employer(s)' employees may be exposed to while working;

**1910.1200(e)(2)(ii)** The methods the employer will use to inform the other employer(s) of any precautionary measures that need to be taken to protect employees during the workplace's normal operating conditions and in foreseeable emergencies; and,

**1910.1200(e)(2)(iii)** The methods the employer will use to inform the other employer(s) of the labeling system used in the workplace.

**1910.1200(e)(3)** The employer may rely on an existing hazard communication program to comply with these requirements, provided that it meets the criteria established in this paragraph (e).

**1910.1200(e)(4)** The employer shall make the written hazard communication program available, upon request, to employees, their designated representatives, the Assistant Secretary and the Director, in accordance with the requirements of 29 CFR 1910.20(e).

**1910.1200(e)(5)** Where employees must travel between workplaces during a workshift, i.e., their work is carried out at more than one geographical location, the written hazard communication program may be kept at the primary workplace facility.

**1910.1200(f)** *Labels and other forms of warning.* **(1)** The chemical manufacturer, importer, or distributor shall ensure that each container of hazardous chemicals leaving the workplace is labeled, tagged or marked with the following information:

**1910.1200(f)(1)(i)** Identity of the hazardous chemical(s);

**1910.1200(f)(1)(ii)** Appropriate hazard warnings; and

**1910.1200(f)(1)(iii)** Name and address of the chemical manufacturer, importer, or other responsible party.

**1910.1200(f)(2)(i)** For solid metal (such as a steel beam or a metal casting), solid wood, or plastic items that are not exempted as articles due to their downstream use, or shipments of whole grain, the required label may be transmitted to the customer at the time of the initial shipment, and need not be included with subsequent shipments to the same employer unless the information on the label changes;

**1910.1200(f)(2)(ii)** The label may be transmitted with the initial shipment itself, or with the material safety data sheet that is to be provided prior to or at the time of the first shipment; and,

**1910.1200(f)(2)(iii)** This exception to requiring labels on every container of hazardous chemicals is only for the solid material itself, and does not apply to hazardous chemicals used in conjunction with, or known to be present with, the material and to which employees handling the items in transit may be exposed (for example, cutting fluids or pesticides in grains).

**1910.1200(f)(3)** Chemical manufacturers, importers, or distributors shall ensure that each container of hazardous chemicals leaving the workplace is labeled, tagged, or marked in accordance with this section in a manner which does not conflict with the requirements of the Hazardous Materials Transportation Act (49 U.S.C. 1801 *et seq.*) and regulations issued under that Act by the Department of Transportation.

**1910.1200(f)(4)** If the hazardous chemical is regulated by OSHA in a substance-specific health standard, the chemical manufacturer, importer, distributor or employer shall ensure that the labels or other forms of warning used are in accordance with the requirements of that standard.

**1910.1200(f)(5)** Except as provided in paragraphs (f)(6) and (f)(7) of this section, the employer shall ensure that each container of hazardous chemicals in the workplace is labeled, tagged or marked with the following information:

**1910.1200(f)(5)(i)** Identity of the hazardous chemical(s) contained therein; and,

**1910.1200(f)(5)(ii)** Appropriate hazard warnings, or alternatively, words, pictures, symbols, or combination thereof, which provide at least general information regarding the hazards of the chemicals, and which, in conjunction with the other information immediately available to employees under the hazard communication program, will provide employees with the specific information regarding the physical and health hazards of the hazardous chemical.

**1910.1200(f)(6)** The employer may use signs, placards, process sheets, batch tickets, operating procedures, or other such written materials in lieu of affixing labels to individual stationary process containers, as long as the alternative method identifies the containers to which it is applicable and conveys the information required by paragraph (f)(5) of this section to be on a label. The written materials shall be readily accessible to the employees in their work area throughout each work shift.

**1910.1200(f)(7)** The employer is not required to label portable containers into which hazardous chemicals are transferred from labeled containers, and which are intended only for the immediate use of the employee who performs the transfer. For purposes of this section, drugs which are dispensed by a pharmacy to a health care provider for direct administration to a patient are exempted from labeling.

**1910.1200(f)(8)** The employer shall not remove or deface existing labels on incoming containers of hazardous chemicals, unless the container is immediately marked with the required information.

**1910.1200(f)(9)** The employer shall ensure that labels or other forms of warning are legible, in English, and prominently displayed on the container, or readily available in the work area throughout each work shift. Employers having employees who speak other languages may add the information in their language to the material presented, as long as the information is presented in English as well.

**1910.1200(f)(10)** The chemical manufacturer, importer, distributor or employer need not affix new labels to comply with this section if existing labels already convey the required information.

**1910.1200(f)(11)** Chemical manufacturers, importers, distributors, or employers who become newly aware of any significant information regarding the hazards of a chemical shall revise the labels for the chemical within three months of becoming aware of the new information. Labels on containers of hazardous chemicals shipped after that time shall contain the new information. If the chemical is not currently produced or imported, the chemical manufacturer, importers, distributor, or employer shall add the information to the label before the chemical is shipped or introduced into the workplace again.

**1910.1200(g)** *Material safety data sheets.* **(1)** Chemical manufacturers and importers shall obtain or develop a material safety data sheet for each hazardous chemical they produce or import. Employers shall have a material safety data sheet in the workplace for each hazardous chemical which they use.

**1910.1200(g)(2)** Each material safety data sheet shall be in English (although the employer may maintain copies in other languages as well), and shall contain at least the following information:

**1910.1200(g)(2)(i)** The identity used on the label, and, except as provided for in paragraph (i) of this section on trade secrets:

**1910.1200(g)(2)(i)(A)** If the hazardous chemical is a single substance, its chemical and common name(s);

**1910.1200(g)(2)(i)(B)** If the hazardous chemical is a mixture which has been tested as a whole to determine its hazards, the chemical and common name(s) of the ingredients which contribute to these known hazards, and the common name(s) of the mixture itself; or,

**1910.1200(g)(2)(i)(C)** If the hazardous chemical is a mixture which has not been tested as a whole:

**1910.1200(g)(2)(i)(C)***(1)* The chemical and common name(s) of all ingredients which have been determined to be health hazards, and which comprise 1% or greater of the composition, except that chemicals identified as carcinogens under paragraph (d) of this section shall be listed if the concentrations are 0.1% or greater; and,

**1910.1200(g)(2)(i)(C)***(2)* The chemical and common name(s) of all ingredients which have been determined to be health hazards, and which comprise less than 1% (0.1% for carcinogens) of the mixture, if there is evidence that the ingredient(s) could be released from the mixture in concentrations which would exceed an established OSHA permissible exposure limit or ACGIH Threshold Limit Value, or could present a health risk to employees; and,

**1910.1200(g)(2)(i)(C)***(3)* The chemical and common name(s) of all ingredients which have been determined to present a physical hazard when present in the mixture;

**1910.1200(g)(2)(ii)** Physical and chemical characteristics of the hazardous chemical (such as vapor pressure, flash point);

**1910.1200(g)(2)(iii)** The physical hazards of the hazardous chemical, including the potential for fire, explosion, and reactivity;

**1910.1200(g)(2)(iv)** The health hazards of the hazardous chemical, including signs and symptoms of exposure, and any medical conditions which are generally recognized as being aggravated by exposure to the chemical;

**1910.1200(g)(2)(v)** The primary route(s) of entry;

**1910.1200(g)(2)(vi)** The OSHA permissible exposure limit, ACGIH Threshold Limit Value, and any other exposure limit used or recommended by the chemical manufacturer, importer, or employer preparing the material safety data sheet, where available;

**1910.1200(g)(2)(vii)** Whether the hazardous chemical is listed in the National Toxicology Program (NTP) Annual Report on Carcinogens (latest edition) or has been found to be a potential carcinogen in the International Agency for Research on Cancer (IARC) Monographs (latest editions), or by OSHA;

**1910.1200(g)(2)(viii)** Any generally applicable precautions for safe handling and use which are known to the chemical manufacturer, importer or employer preparing the material safety data sheet, including appropriate hygienic practices, protective measures during repair and maintenance of contaminated equipment, and procedures for clean-up of spills and leaks;

**1910.1200(g)(2)(ix)** Any generally applicable control measures which are known to the chemical manufacturer, importer or employer preparing the material safety data sheet, such as appropriate engineering controls, work practices, or personal protective equipment;

**1910.1200(g)(2)(x)** Emergency and first aid procedures;

**1910.1200(g)(2)(xi)** The date of preparation of the material safety data sheet or the last change to it; and,

**1910.1200(g)(2)(xii)** The name, address and telephone number of the chemical manufacturer, importer, employer or other responsible party preparing or distributing the material safety data sheet, who can provide additional information on the hazardous chemical and appropriate emergency procedures, if necessary.

**1910.1200(g)(3)** If no relevant information is found for any given category on the material safety data sheet, the chemical manufacturer, importer or employer preparing the material safety data sheet shall mark it to indicate that no applicable information was found.

**1910.1200(g)(4)** Where complex mixtures have similar hazards and contents (i.e., the chemical ingredients are essentially the same, but the specific composition varies from mixture to mixture), the chemical manufacturer, importer or employer may prepare one material safety data sheet to apply to all of these similar mixtures.

**1910.1200(g)(5)** The chemical manufacturer, importer or employer preparing the material safety data sheet shall ensure that the information recorded accurately reflects the scientific evidence used in making the hazard determination. If the chemical manufacturer, importer or employer preparing the material safety data sheet becomes newly aware of any significant information regarding the hazards of a chemical, or ways to protect against the hazards, this new information shall be added to the material safety data sheet within three months. If the chemical is not currently being produced or imported the chemical manufacturer or importer shall add the information to the material safety data sheet before the chemical is introduced into the workplace again.

**1910.1200(g)(6)(i)** Chemical manufacturers or importers shall ensure that distributors and employers are provided an appropriate material safety data sheet with their initial shipment, and with the first shipment after a material safety data sheet is updated;

**1910.1200(g)(6)(ii)** The chemical manufacturer or importer shall either provide material safety data sheets with the shipped containers or send them to the distributor or employer prior to or at the time of the shipment;

**1910.1200(g)(6)(iii)** If the material safety data sheet is not provided with a shipment that has been labeled as a hazardous chemical, the distributor or employer shall obtain one from the chemical manufacturer or importer as soon as possible; and,

**1910.1200(g)(6)(iv)** The chemical manufacturer or importer shall also provide distributors or employers with a material safety data sheet upon request.

**1910.1200(g)(7)(i)** Distributors shall ensure that material safety data sheets, and updated information, are provided to other distributors and employers with their initial shipment and with the first shipment after a material safety data sheet is updated;

**1910.1200(g)(7)(ii)** The distributor shall either provide material safety data sheets with the shipped containers, or send them to the other distributor or employer prior to or at the time of the shipment;

**1910.1200(g)(7)(iii)** Retail distributors selling hazardous chemicals to employers having a commercial account shall provide a material safety data sheet to such employers upon request, and shall post a sign or otherwise inform them that a material safety data sheet is available;

**1910.1200(g)(7)(iv)** Wholesale distributors selling hazardous chemicals to employers over-the-counter may also provide material safety data sheets upon the request of the employer at the time of the over-the-counter purchase, and shall post a sign or otherwise inform such employers that a material safety data sheet is available;

**1910.1200(g)(7)(v)** If an employer without a commercial account purchases a hazardous chemical from a retail distributor not required to have material safety data sheets on file (i.e., the retail distributor does not have commercial accounts and does not use the materials), the retail distributor shall provide the employer, upon request, with the name, address, and telephone number of the chemical manufacturer, importer, or distributor from which a material safety data sheet can be obtained;

**1910.1200(g)(7)(vi)** Wholesale distributors shall also provide material safety data sheets to employers or other distributors upon request; and,

**1910.1200(g)(7)(vii)** Chemical manufacturers, importers, and distributors need not provide material safety data sheets to retail distributors that have informed them that the retail distributor does not sell the product to commercial accounts or open the sealed container to use it in their own workplaces.

**1910.1200(g)(8)** The employer shall maintain in the workplace copies of the required material safety data sheets for each hazardous chemical, and shall ensure that they are readily accessible during each work shift to employees when they are in their work area(s). (Electronic access, microfiche, and other alternatives to maintaining paper copies of the material safety data sheets are permitted as long as no barriers to immediate employee access in each workplace are created by such options.)

**1910.1200(g)(9)** Where employees must travel between workplaces during a workshift, i.e., their work is carried out at more than one geographical location, the material safety data sheets may be kept at the primary workplace facility. In this situation, the employer shall ensure that employees can immediately obtain the required information in an emergency.

**1910.1200(g)(10)** Material safety data sheets may be kept in any form, including operating procedures, and may be designed to cover groups of hazardous chemicals in a work area where it may be more appropriate to address the hazards of a process rather than individual hazardous chemicals. However, the employer shall ensure that in all cases the required information is provided for each hazardous chemical, and is readily accessible during each work shift to employees when they are in in their work area(s).

**1910.1200(g)(11)** Material safety data sheets shall also be made readily available, upon request, to designated representatives and to the Assistant Secretary, in accordance with the requirements of 29 CFR 1910.20(e). The Director shall also be given access to material safety data sheets in the same manner.

**1910.1200(h)** *Employee information and training.* **(1)** Employers shall provide employees with effective information and training on hazardous chemicals in their work area at the time of their initial assignment, and whenever a new physical or health hazard the employees have not previously been trained about is introduced into their work area. Information and training may be designed to cover categories of hazards (e.g., flammability, carcinogenicity) or specific chemicals. Chemical-specific information must always be available through labels and material safety data sheets.

**1910.1200(h)(2)** *Information.* Employees shall be informed of:

**1910.1200(h)(2)(i)** The requirements of this section;

**1910.1200(h)(2)(ii)** Any operations in their work area where hazardous chemicals are present; and,

**1910.1200(h)(2)(iii)** The location and availability of the written hazard communication program, including the required list(s) of hazardous chemicals, and material safety data sheets required by this section.

**1910.1200(h)(3)** *Training.* Employee training shall include at least:

**1910.1200(h)(3)(i)** Methods and observations that may be used to detect the presence or release of a hazardous chemical in the work area (such as monitoring conducted by the employer, continuous monitoring devices, visual appearance or odor of hazardous chemicals when being released, etc.);

**1910.1200(h)(3)(ii)** The physical and health hazards of the chemicals in the work area;

**1910.1200(h)(3)(iii)** The measures employees can take to protect themselves from these hazards, including specific procedures the employer has implemented to protect employees from exposure to hazardous chemicals, such as appropriate work practices, emergency procedures, and personal protective equipment to be used; and,

**1910.1200(h)(3)(iv)** The details of the hazard communication program developed by the employer, including an explanation of the labeling system and the material safety data sheet, and how employees can obtain and use the appropriate hazard information.

**1910.1200(i)** *Trade secrets.* **(1)** The chemical manufacturer, importer, or employer may withhold the specific chemical identity, including the chemical name and other specific identification of a hazardous chemical, from the material safety data sheet, provided that:

**1910.1200(i)(1)(i)** The claim that the information withheld is a trade secret can be supported;

**1910.1200(i)(1)(ii)** Information contained in the material safety data sheet concerning the properties and effects of the hazardous chemical is disclosed;

**1910.1200(i)(1)(iii)** The material safety data sheet indicates that the specific chemical identity is being withheld as a trade secret; and,

**1910.1200(i)(1)(iv)** The specific chemical identity is made available to health professionals, employees, and designated representatives in accordance with the applicable provisions of this paragraph.

**1910.1200(i)(2)** Where a treating physician or nurse determines that a medical emergency exists and the specific chemical identity of a hazardous chemical is necessary for emergency or first-aid treatment, the chemical manufacturer, importer, or employer shall immediately disclose the specific chemical identity of a trade secret chemical to that treating physician or nurse, regardless of the existence of a written statement of need or a confidentiality agreement. The chemical manufacturer, importer, or employer may require a written statement of need and confidentiality agreement, in accordance with the provisions of paragraphs (i) (3) and (4) of this section, as soon as circumstances permit.

**1910.1200(i)(3)** In non-emergency situations, a chemical manufacturer, importer, or employer shall, upon request, disclose a specific chemical identity, otherwise permitted to be withheld under paragraph (i)(1) of this section, to a health professional (i.e., physician, industrial hygienist, toxicologist, epidemiologist, or occupational health nurse) providing medical or other occupational health services to exposed employee(s), and to employees or designated representatives, if:

**1910.1200(i)(3)(i)** The request is in writing;

**1910.1200(i)(3)(ii)** The request describes with reasonable detail one or more of the following occupational health needs for the information:

**1910.1200(i)(3)(ii)(A)** To assess the hazards of the chemicals to which employees will be exposed;

**1910.1200(i)(3)(ii)(B)** To conduct or assess sampling of the workplace atmosphere to determine employee exposure levels;

**1910.1200(i)(3)(ii)(C)** To conduct pre-assignment or periodic medical surveillance of exposed employees;

**1910.1200(i)(3)(ii)(D)** To provide medical treatment to exposed employees;

**1910.1200(i)(3)(ii)(E)** To select or assess appropriate personal protective equipment for exposed employees;

**1910.1200(i)(3)(ii)(F)** To design or assess engineering controls or other protective measures for exposed employees; and,

**1910.1200(i)(3)(ii)(G)** To conduct studies to determine the health effects of exposure.

**1910.1200(i)(3)(iii)** The request explains in detail why the disclosure of the specific chemical identity is essential and that, in lieu thereof, the disclosure of the following information to the health professional, employee, or designated representative, would not satisfy the purposes described in paragraph (i)(3)(ii) of this section:

**1910.1200(i)(3)(iii)(A)** The properties and effects of the chemical;

**1910.1200(i)(3)(iii)(B)** Measures for controlling workers' exposure to the chemical;

**1910.1200(i)(3)(iii)(C)** Methods of monitoring and analyzing worker exposure to the chemical; and,

**1910.1200(i)(3)(iii)(D)** Methods of diagnosing and treating harmful exposures to the chemical;

**1910.1200(i)(3)(iv)** The request includes a description of the procedures to be used to maintain the confidentiality of the disclosed information; and,

**1910.1200(i)(3)(v)** The health professional, and the employer or contractor of the services of the health professional (i.e., downstream employer, labor organization, or individual employee), employee, or designated representative, agree in a written confidentiality agreement that the health professional, employee, or designated representative, will not use the trade secret information for any purpose other than the health need(s) asserted and agree not to release the information under any circumstances other than to OSHA, as provided in paragraph (i)(6) of this section, except as authorized by the terms of the agreement or by the chemical manufacturer, importer, or employer.

**1910.1200(i)(4)** The confidentiality agreement authorized by paragraph (i)(3)(iv) of this section:

**1910.1200(i)(4)(i)** May restrict the use of the information to the health purposes indicated in the written statement of need;

**1910.1200(i)(4)(ii)** May provide for appropriate legal remedies in the event of a breach of the agreement, including stipulation of a reasonable pre-estimate of likely damages; and,

**1910.1200(i)(4)(iii)** May not include requirements for the posting of a penalty bond.

**1910.1200(i)(5)** Nothing in this standard is meant to preclude the parties from pursuing non-contractual remedies to the extent permitted by law.

**1910.1200(i)(6)** If the health professional, employee, or designated representative receiving the trade secret information decides that there is a need to disclose it to OSHA, the chemical manufacturer, importer, or employer who provided the information shall be informed by the health professional, employee, or designated representative prior to, or at the same time as, such disclosure.

**1910.1200(i)(7)** If the chemical manufacturer, importer, or employer denies a written request for disclosure of a specific chemical identity, the denial must:

**1910.1200(i)(7)(i)** Be provided to the health professional, employee, or designated representative, within thirty days of the request;

**1910.1200(i)(7)(ii)** Be in writing;

**1910.1200(i)(7)(iii)** Include evidence to support the claim that the specific chemical identity is a trade secret;

**1910.1200(i)(7)(iv)** State the specific reasons why the request is being denied; and,

**1910.1200(i)(7)(v)** Explain in detail how alternative information may satisfy the specific medical or occupational health need without revealing the specific chemical identity.

**1910.1200(i)(8)** The health professional, employee, or designated representative whose request for information is denied under paragraph (i)(3) of this section may refer the request and the written denial of the request to OSHA for consideration.

**1910.1200(i)(9)** When a health professional, employee, or designated representative refers the denial to OSHA under paragraph (i)(8) of this section, OSHA shall consider the evidence to determine if:

**1910.1200(i)(9)(i)** The chemical manufacturer, importer, or employer has supported the claim that the specific chemical identity is a trade secret;

**1910.1200(i)(9)(ii)** The health professional, employee, or designated representative has supported the claim that there is a medical or occupational health need for the information; and,

**1910.1200(i)(9)(iii)** The health professional, employee or designated representative has demonstrated adequate means to protect the confidentiality.

**1910.1200(i)(10)(i)** If OSHA determines that the specific chemical identity requested under paragraph (i)(3) of this section is not a *bona fide* trade secret, or that it is a trade secret, but the requesting health professional, employee, or designated representative has a legitimate medical or occupational health need for the information, has executed a written confidentiality agreement, and has shown adequate means to protect the confidentiality of the information, the chemical manufacturer, importer, or employer will be subject to citation by OSHA.

**1910.1200(i)(10)(ii)** If a chemical manufacturer, importer, or employer demonstrates to OSHA that the execution of a confidentiality agreement would not provide sufficient protection against the potential harm from the unauthorized disclosure of a trade secret specific chemical identity, the Assistant Secretary may issue such orders or impose such additional limitations or conditions upon the disclosure of the requested chemical information as may be appropriate to assure that the occupational health services are provided without an undue risk of harm to the chemical manufacturer, importer, or employer.

**1910.1200(i)(11)** If a citation for a failure to release specific chemical identity information is contested by the chemical manufacturer, importer, or employer, the matter will be adjudicated before the Occupational Safety and Health Review Commission in accordance with the Act's enforcement scheme and the applicable Commission rules of procedure. In accordance with the Commission rules, when a chemical manufacturer, importer, or employer continues to withhold the information during the contest, the Administrative Law Judge may review the citation and supporting documentation *in camera* or issue appropriate orders to protect the confidentiality of such matters.

**1910.1200(i)(12)** Notwithstanding the existence of a trade secret claim, a chemical manufacturer, importer, or employer shall, upon request, disclose to the Assistant Secretary any information which this section requires the chemical manufacturer, importer, or employer to make available. Where there is a trade secret claim, such claim shall be made no later than at the time the information is provided to the Assistant Secretary so that suitable determinations of trade secret status can be made and the necessary protections can be implemented.

**1910.1200(i)(13)** Nothing in this paragraph shall be construed as requiring the disclosure under any circumstances of process or percentage of mixture information which is a trade secret.

**1910.1200(j)** *Effective dates.* Chemical manufacturers, importers, distributors, and employers shall be in compliance with all provisions of this section by March 11, 1994.

**NOTE:** The effective date of the clarification that the exemption of wood and wood products from the Hazard Communication standard in paragraph (b)(6)(iv) only applies to wood and wood products including lumber which will not be processed, where the manufacturer or importer can establish that the only hazard they pose to employees is the potential for flammability or combustibility, and that the exemption does not apply to wood or wood products which have been treated with a hazardous chemical covered by this standard, and wood which may be subsequently sawed or cut generating dust has been stayed from March 11, 1994 to August 11, 1994.

## APPENDIX A TO §1910.1200— HEALTH HAZARD DEFINITIONS (MANDATORY)

Although safety hazards related to the physical characteristics of a chemical can be objectively defined in terms of testing requirements (e.g., flammability), health hazard definitions are less precise and more subjective. Health hazards may cause measurable changes in the body—such as decreased pulmonary function. These changes are generally indicated by the occurrence of signs and symptoms in the exposed employees—such as shortness of breath, a non-measurable, subjective feeling. Employees exposed to such hazards must be apprised of both the change in body function and the signs and symptoms that may occur to signal that change.

The determination of occupational health hazards is complicated by the fact that many of the effects or signs and symptoms occur commonly in non-occupationally exposed populations, so that effects of exposure are difficult to separate from normally occurring illnesses. Occasionally, a substance causes an effect that is rarely seen in the population at large, such as angiosarcomas caused by vinyl chloride exposure, thus making it easier to ascertain that the occupational exposure was the primary causative factor. More often, however, the effects are common, such as lung cancer. The situation is further complicated by the fact that most chemicals have not been adequately tested to determine their health hazard potential, and data do not exist to substantiate these effects.

There have been many attempts to categorize effects and to define them in various ways. Generally, the terms "acute" and "chronic" are used to delineate between effects on the basis of severity or duration. "Acute" effects usually occur rapidly as a result of short-term exposures, and are of short duration. "Chronic" effects generally occur as a result of long-term exposure, and are of long duration.

The acute effects referred to most frequently are those defined by the American National Standards Institute (ANSI) standard for Precautionary Labeling of Hazardous Industrial Chemicals (Z129.1-1988)—irritation, corrosivity, sensitization and lethal dose. Although these are important health effects, they do not adequately cover the considerable range of acute effects which may occur as a result of occupational exposure, such as, for example, narcosis.

Similarly, the term chronic effect is often used to cover only carcinogenicity, teratogenicity, and mutagenicity. These effects are obviously a concern in the workplace, but again, do not adequately cover the area of chronic effects, excluding, for example, blood dyscrasias (such as anemia), chronic bronchitis and liver atrophy.

The goal of defining precisely, in measurable terms, every possible health effect that may occur in the workplace as a result of chemical exposures cannot realistically be accomplished. This does not negate the need for employees to be informed of such effects and protected from them. Appendix B, which is also mandatory, outlines the principles and procedures of hazard assessment.

For purposes of this section, any chemicals which meet any of the following definitions, as determined by the criteria set forth in Appendix B are health hazards. However, this is not intended to be an exclusive categorization scheme. If there are available scientific data that involve other animal species or test methods, they must also be evaluated to determine the applicability of the HCS.

1. *Carcinogen:* A chemical is considered to be a carcinogen if:

(a) It has been evaluated by the International Agency for Research on Cancer (IARC), and found to be a carcinogen or potential carcinogen; or

(b) It is listed as a carcinogen or potential carcinogen in the Annual Report on Carcinogens published by the National Toxicology Program (NTP) (latest edition); or,

(c) It is regulated by OSHA as a carcinogen.

2. *Corrosive:* A chemical that causes visible destruction of, or irreversible alterations in, living tissue by chemical action at the site of contact. For example, a chemical is considered to be corrosive if, when tested on the intact skin of albino rabbits by the method described by the U.S. Department of Transportation in appendix A to 49 CFR part 173, it destroys or changes irreversibly the structure of the tissue at the site of contact following an exposure period of four hours. This term shall not refer to action on inanimate surfaces.

3. *Highly toxic:* A chemical falling within any of the following categories:

(a) A chemical that has a median lethal dose ($LD_{50}$) of 50 milligrams or less per kilogram of body weight when administered orally to albino rats weighing between 200 and 300 grams each.

(b) A chemical that has a median lethal dose ($LD_{50}$) of 200 milligrams or less per kilogram of body weight when administered by continuous contact for 24 hours (or less if death occurs within 24 hours) with the bare skin of albino rabbits weighing between two and three kilograms each.

(c) A chemical that has a median lethal concentration ($LC_{50}$) in air of 200 parts per million by volume or less of gas or vapor, or 2 milligrams per liter or less of mist, fume, or dust, when administered by continuous inhalation for one hour (or less if death occurs within one hour) to albino rats weighing between 200 and 300 grams each.

4. *Irritant:* A chemical, which is not corrosive, but which causes a reversible inflammatory effect on living tissue by chemical action at the site of contact. A chemical is a skin irritant if, when tested on the intact skin of albino rabbits by the methods of 16 CFR 1500.41 for four hours exposure or by other appropriate techniques, it results in an empirical score of five or more. A chemical is an eye irritant if so determined under the procedure listed in 16 CFR 1500.42 or other appropriate techniques.

5. *Sensitizer:* A chemical that causes a substantial proportion of exposed people or animals to develop an allergic reaction in normal tissue after repeated exposure to the chemical.

6. *Toxic.* A chemical falling within any of the following categories:

(a) A chemical that has a median lethal dose ($LD_{50}$) of more than 50 milligrams per kilogram but not more than 500 milligrams per kilogram of body weight when administered orally to albino rats weighing between 200 and 300 grams each.

(b) A chemical that has a median lethal dose ($LD_{50}$) of more than 200 milligrams per kilogram but not more than 1,000 milligrams per kilogram of body weight when administered by continuous contact for 24 hours (or less if death occurs within 24 hours) with the bare skin of albino rabbits weighing between two and three kilograms each.

(c) A chemical that has a median lethal concentration ($LC_{50}$) in air of more than 200 parts per million but not more than 2,000 parts per million by volume of gas or vapor, or more than two milligrams per liter but not more than 20 milligrams per liter of mist, fume, or dust, when administered by continuous inhalation for one hour (or less if death occurs within one hour) to albino rats weighing between 200 and 300 grams each.

7. *Target organ effects.*

The following is a target organ categorization of effects which may occur, including examples of signs and symptoms and chemicals which have been found to cause such effects. These examples are presented to illustrate the range and diversity of effects and hazards found in the workplace, and the broad scope employers must consider in this area, but are not intended to be all-inclusive.

a. Hepatotoxins: Chemicals which produce liver damage3

Signs & Symptoms: Jaundice; liver enlargement

Chemicals: Carbon tetrachloride; nitrosamines

b. Nephrotoxins: Chemicals which produce kidney damage

Signs & Symptoms: Edema; proteinuria

Chemicals: Halogenated hydrocarbons; uranium

c. Neurotoxins: Chemicals which produce their primary toxic effects on the nervous system

Signs & Symptoms: Narcosis; behavioral changes; decrease in motor functions

Chemicals: Mercury; carbon disulfide

d. Agents which act on the blood or hemato-poietic system: Decrease hemoglobin function; deprive the body tissues of oxygen

Signs & Symptoms: Cyanosis; loss of consciousness

Chemicals: Carbon monoxide; cyanides

e. Agents which damage the lung: Chemicals which irritate or damage pulmonary tissue

Signs & Symptoms: Cough; tightness in chest; shortness of breath

Chemicals: Silica; asbestos

f. Reproductive toxins: Chemicals which affect the reproductive capabilities including chromosomal damage (mutations) and effects on fetuses (teratogenesis)

Signs & Symptoms: Birth defects; sterility

Chemicals: Lead; DBCP

g. Cutaneous hazards: Chemicals which affect the dermal layer of the body

Signs & Symptoms: Defatting of the skin; rashes; irritation

Chemicals: Ketones; chlorinated compounds

h. Eye hazards: Chemicals which affect the eye or visual capacity

Signs & Symptoms: Conjunctivitis; corneal damage

Chemicals: Organic solvents; acids

## APPENDIX B TO §1910.1200— HAZARD DETERMINATION (MANDATORY)

The quality of a hazard communication program is largely dependent upon the adequacy and accuracy of the hazard determination. The hazard determination requirement of this standard is performance-oriented. Chemical manufacturers, importers, and employers evaluating chemicals are not required to follow any specific methods for determining hazards, but they must be able to demonstrate that they have adequately ascertained the hazards of the chemicals produced or imported in accordance with the criteria set forth in this Appendix.

Hazard evaluation is a process which relies heavily on the professional judgment of the evaluator, particularly in the area of chronic hazards. The performance-orientation of the hazard determination does not diminish the duty of the chemical manufacturer, importer or employer to conduct a thorough evaluation, examining all relevant data and producing a scientifically defensible evaluation. For purposes of this standard, the following criteria shall be used in making hazard determinations that meet the requirements of this standard.

1. *Carcinogenicity:* As described in paragraph (d)(4) of this section and Appendix A of this section, a determination by the National Toxicology Program, the International Agency for Research on Cancer, or OSHA that a chemical is a carcinogen or potential carcinogen will be considered conclusive evidence for purposes of this section. In addition, however, all available scientific data on carcinogenicity must be evaluated in accordance with the provisions of this Appendix and the requirements of the rule.

2. *Human data:* Where available, epidemiological studies and case reports of adverse health effects shall be considered in the evaluation.

3. *Animal data:* Human evidence of health effects in exposed populations is generally not available for the majority of chemicals produced or used in the workplace. Therefore, the available results of toxicological testing in animal populations shall be used to predict the health effects that may be experienced by exposed workers. In particular, the definitions of certain acute hazards refer to specific animal testing results (see Appendix A).

4. *Adequacy and reporting of data.* The results of any studies which are designed and conducted according to established scientific principles, and which report statistically significant conclusions regarding the health effects of a chemical, shall be a sufficient basis for a hazard determination and reported on any material safety data sheet. *In vitro* studies alone generally do not form the basis for a definitive finding of hazard under the HCS since they have a positive or negative result rather than a statistically significant finding.

The chemical manufacturer, importer, or employer may also report the results of other scientifically valid studies which tend to refute the findings of hazard.

## APPENDIX C TO §1910.1200— INFORMATION SOURCES (ADVISORY) [Reserved]

Appendix C to §1910.1200 removed at 61 FR 9245, Mar. 7, 1996, effective May 6, 1996.]

## APPENDIX D TO §1910.1200— DEFINITION OF "TRADE SECRET" (MANDATORY)

The following is a reprint of the *Restatement of Torts* section 757, comment *b* (1939):

*b. Definition of trade secret.* A trade secret may consist of any formula, pattern, device or compilation of information which is used in one's business, and which gives him an opportunity to obtain an advantage over competitors who do not know or use it. It may be a formula for a chemical compound, a process of manufacturing, treating or preserving materials, a pattern for a machine or other device, or a list of customers. It differs from other secret information in a business (see s759 of the *Restatement of Torts* which is not included in this Appendix) in that it is not simply information as to single or ephemeral events in the conduct of the business, as, for example, the amount or other terms of a secret bid for a contract or the salary of certain employees, or the security investments made or contemplated, or the date fixed for the announcement of a new policy or for bringing out a new model or the like. A trade secret is a process or device for continuous use in the operations of the business. Generally it relates to the production of goods, as, for example, a machine or formula for the production of an article. It may, however, relate to the sale of goods or to other operations in the business, such as a code for determining discounts, rebates or other concessions in a price list or catalogue, or a list of specialized customers, or a method of bookkeeping or other office management.

*Secrecy.* The subject matter of a trade secret must be secret. Matters of public knowledge or of general knowledge in an industry cannot be appropriated by one as his secret. Matters which are completely disclosed by the goods which one markets cannot be his secret. Substantially, a trade secret is known only in the particular business in which it is used. It is not requisite that only the proprietor of the business know it. He may, without losing his protection, communicate it to employees involved in its use. He may likewise communicate it to others pledged to secrecy. Others may also know of it independently, as, for example, when they have discovered the process or formula by independent invention and are keeping it secret. Nevertheless, a substantial element of secrecy must exist, so that, except by the use of improper means, there would be difficulty in acquiring the information. An exact definition of a trade secret is not possible. Some factors to be considered in determining whether given information is one's trade secret are: (1) The extent to which the information is known outside of his business; (2) the extent to which it is known by employees and others involved in his business; (3) the extent of measures taken by him to guard the secrecy of the information; (4) the value of the information to him and his competitors; (5) the amount of effort or money expended by him in developing the information; (6) the ease or difficulty with which the information could be properly acquired or duplicated by others.

*Novelty and prior art.* A trade secret may be a device or process which is patentable; but it need not be that. It may be a device or process which is clearly anticipated in the prior art or one which is merely a mechanical improvement that a good mechanic can make. Novelty and invention are not requisite for a trade secret as they are for patentability. These requirements are essential to patentability because a patent protects against unlicensed use of the patented device or process even by one who discovers it properly through independent research. The patent monopoly is a reward to the inventor. But such is not the case with a trade secret. Its protection is not based on a policy of rewarding or otherwise encouraging the development of secret processes or devices. The protection is merely against breach of faith and reprehensible means of learning another's secret. For this limited protection it is not appropriate to require also the kind of novelty and invention which is a requisite of patentability. The nature of the secret is, however, an important factor in determining the kind of relief that is appropriate against one who is subject to liability under the rule stated in this Section. Thus, if the secret consists of a device or process which is a novel invention, one who acquires the secret wrongfully is ordinarily enjoined from further use of it and is required to account for the profits derived from his past use. If, on the other hand, the secret consists of mechanical improvements that a good mechanic can make without resort to the secret, the wrongdoer's liability may be limited to damages, and an injunction against future use of the improvements made with the aid of the secret may be inappropriate.

## APPENDIX E TO §1910.1200—(ADVISORY)—GUIDELINES FOR EMPLOYER COMPLIANCE

The Hazard Communication Standard (HCS) is based on a simple concept—that employees have both a need and a right to know the hazards and identities of the chemicals they are exposed to when working. They also need to know what protective measures are available to prevent adverse effects from occurring. The HCS is designed to provide employees with the information they need.

Knowledge acquired under the HCS will help employers provide safer workplaces for their employees. When employers have information about the chemicals being used, they can take steps to reduce exposures, substitute less hazardous materials, and establish proper work practices. These efforts will help prevent the occurrence of work-related illnesses and injuries caused by chemicals.

The HCS addresses the issues of evaluating and communicating hazards to workers. Evaluation of chemical hazards involves a number of technical concepts, and is a process that requires the professional judgment of experienced experts. That's why the HCS is designed so that employers who simply use chemicals, rather than produce or import them, are not required to evaluate the hazards of those chemicals. Hazard determination is the responsibility of the producers and importers of the materials. Producers and importers of chemicals are then required to provide the hazard information to employers that purchase their products.

Employers that don't produce or import chemicals need only focus on those parts of the rule that deal with establishing a workplace program and communicating information to their workers. This appendix is a general guide for such employers to help them determine what's required under the rule. It does not supplant or substitute for the regulatory provisions, but rather provides a simplified outline of the steps an average employer would follow to meet those requirements.

### 1. Becoming Familiar With The Rule.

OSHA has provided a simple summary of the HCS in a pamphlet entitled "Chemical Hazard Communication," OSHA Publication Number 3084. Some employers prefer to begin to become familiar with the rule's requirements by reading this pamphlet. A copy may be obtained from your local OSHA Area Office, or by contacting the OSHA Publications Office at (202) 523-9667.

The standard is long, and some parts of it are technical, but the basic concepts are simple. In fact, the requirements reflect what many employers have been doing for years. You may find that you are already largely in compliance with many of the provisions, and will simply have to modify your existing programs somewhat. If you are operating in an OSHA-approved State Plan State, you must comply with the State's requirements, which may be different than those of the Federal rule. Many of the State Plan States had hazard communication or "right-to-know" laws prior to promulgation of the Federal rule. Employers in State Plan States should contact their State OSHA offices for more information regarding applicable requirements.

The HCS requires information to be prepared and transmitted regarding all hazardous chemicals. The HCS covers both physical hazards (such as flammability), and health hazards (such as irritation, lung damage, and cancer). Most chemicals used in the workplace have some hazard potential, and thus will be covered by the rule.

One difference between this rule and many others adopted by OSHA is that this one is performance-oriented. That means that you have the flexibility to adapt the rule to the needs of your workplace, rather than having to follow specific, rigid requirements. It also means that you have to exercise more judgment to implement an appropriate and effective program.

The standard's design is simple. Chemical manufacturers and importers must evaluate the hazards of the chemicals they produce or import. Using that information, they must then prepare labels for containers, and more detailed technical bulletins called material safety data sheets (MSDS).

Chemical manufacturers, importers, and distributors of hazardous chemicals are all required to provide the appropriate labels and material safety data sheets to the employers to which they ship the chemicals. The information is to be provided automatically. Every container of hazardous chemicals you receive must be labeled, tagged, or marked with the required information. Your suppliers must also send you a properly completed material safety data sheet (MSDS) at the time of the first shipment of the chemical, and with the next shipment after the MSDS is updated with new and significant information about the hazards.

You can rely on the information received from your suppliers. You have no independent duty to analyze the chemical or evaluate the hazards of it.

Employers that "use" hazardous chemicals must have a program to ensure the information is provided to exposed employees. "Use" means to package, handle, react, or transfer. This is an intentionally broad scope, and includes any situation where a chemical is present in such a way that employees may be exposed under normal conditions of use or in a foreseeable emergency.

The requirements of the rule that deal specifically with the hazard communication program are found in this section in paragraphs (e), written hazard communication program; (f), labels and other forms of warning; (g), material safety data sheets; and (h), employee information and training. The requirements of these paragraphs should be the focus of your attention. Concentrate on becoming familiar with them, using paragraphs (b), scope and application, and (c), definitions, as references when needed to help explain the provisions.

There are two types of work operations where the coverage of the rule is limited. These are laboratories and operations where chemicals are only handled in sealed containers (e.g., a warehouse). The limited provisions for these workplaces can be found in paragraph (b) of this section, scope and application. Basically, employers having these types of work operations need only keep labels on containers as they are received; maintain material safety data sheets that are received, and give employees access to them; and provide information and training for employees. Employers do not have to have written hazard communication programs and lists of chemicals for these types of operations.

The limited coverage of laboratories and sealed container operations addresses the obligation of an employer to the workers in the operations involved, and does not affect the employer's duties as a distributor of chemicals. For example, a distributor may have warehouse operations where employees would be protected under the limited sealed container provisions. In this situation, requirements for obtaining and maintaining MSDSs are limited to providing access to those received with containers while the substance is in the workplace, and requesting MSDSs when employees request access for those not received with the containers. However, as a distributor of hazardous chemicals, that employer will still have responsibilities for providing MSDSs to downstream customers at the time of the first shipment and when the MSDS is updated. Therefore, although they may not be required for the employees in the work operation, the distributor may, nevertheless, have to have MSDSs to satisfy other requirements of the rule.

### 2. Identify Responsible Staff

Hazard communication is going to be a continuing program in your facility. Compliance with the HCS is not a "one shot deal." In order to have a successful program, it will be necessary to assign responsibility for both the initial and ongoing activities that have to be undertaken to comply with the rule. In some cases, these activities may already be part of current job assignments. For example, site supervisors are frequently responsible for on-the-job training sessions. Early identification of the responsible employees, and involvement of them in the development of your plan of action, will result in a more effective program design. Evaluation of the effectiveness of your program will also be enhanced by involvement of affected employees.

For any safety and health program, success depends on commitment at every level of the organization. This is particularly true for hazard communication, where success requires a change in behavior. This will only occur if employers understand the program, and are committed to its success, and if employees are motivated by the people presenting the information to them.

### 3. Identify Hazardous Chemicals in the Workplace.

The standard requires a list of hazardous chemicals in the workplace as part of the written hazard communication program. The list will eventually serve as an inventory of everything for which an MSDS must be maintained. At this point, however, preparing the list will help you complete the rest of the program since it will give you some idea of the scope of the program required for compliance in your facility.

The best way to prepare a comprehensive list is to survey the workplace. Purchasing records may also help, and certainly employers should establish procedures to ensure that in the future purchasing procedures result in MSDSs being received before a material is used in the workplace.

The broadest possible perspective should be taken when doing the survey. Sometimes people think of "chemicals" as being only liquids in containers. The HCS covers chemicals in all physical forms—liquids, solids, gases, vapors, fumes, and mists—whether they are "contained" or not. The hazardous nature of the chemical and the potential for exposure are the factors which determine whether a chemical is covered. If it's not hazardous, it's not covered. If there is no potential for exposure (e.g., the chemical is inextricably bound and cannot be released), the rule does not cover the chemical.

Look around. Identify chemicals in containers, including pipes, but also think about chemicals generated in the work operations. For example, welding fumes, dusts, and exhaust fumes are all sources of chemical exposures. Read labels provided by suppliers for hazard information. Make a list of all chemicals in the workplace that are potentially hazardous. For your own information and planning, you may also want to note on the list the location(s) of the products within the workplace, and an indication of the hazards as found on the label. This will help you as you prepare the rest of your program.

Paragraph (b) of this section, scope and application, includes exemptions for various chemicals or workplace situations. After compiling the complete list of chemicals, you should review paragraph (b) of this section to determine if any of the items can be eliminated from the list because they are exempted materials. For example, food, drugs, and cosmetics brought into the workplace for employee consumption are exempt. So rubbing alcohol in the first aid kit would not be covered.

Once you have compiled as complete a list as possible of the potentially hazardous chemicals in the workplace, the next step is to determine if you have received material safety data sheets for all of them. Check your files against the inventory you have just compiled. If any are missing, contact your supplier and request one. It is a good idea to document these requests, either by copy of a letter or a note regarding telephone conversations. If you have MSDSs for chemicals that are not on your list, figure out why. Maybe you don't use the chemical anymore. Or maybe you missed it in your survey. Some suppliers do provide MSDSs for products that are not hazardous. These do not have to be maintained by you.

You should not allow employees to use any chemicals for which you have not received an MSDS. The MSDS provides information you need to ensure proper protective measures are implemented prior to exposure.

### 4. Preparing and Implementing a Hazard Communication Program

All workplaces where employees are exposed to hazardous chemicals must have a written plan which describes how the standard will be implemented in that facility. Preparation of a plan is not just a paper exercise—all of the elements must be implemented in the workplace in order to be in compliance with the rule. See paragraph (e) of this section for the specific requirements regarding written hazard communication programs. The only work operations which do not have to comply with the written plan requirements are laboratories and work operations where employees only handle chemicals in sealed containers. See paragraph (b) of this section, scope and application, for the specific requirements for these two types of workplaces.

The plan does not have to be lengthy or complicated. It is intended to be a blueprint for implementation of your program—an assurance that all aspects of the requirements have been addressed.

Many trade associations and other professional groups have provided sample programs and other assistance materials to affected employers. These have been very helpful to many employers since they tend to be tailored to the particular industry involved. You may wish to investigate whether your industry trade groups have developed such materials.

Although such general guidance may be helpful, you must remember that the written program has to reflect what you are doing in your workplace. Therefore, if you use a generic program it must be adapted to address the facility it covers. For example, the written plan must list the chemicals present at the site, indicate who is to be responsible for the various aspects of the program in your facility, and indicate where written materials will be made available to employees.

If OSHA inspects your workplace for compliance with the HCS, the OSHA compliance officer will ask to see your written plan at the outset of the inspection. In general, the following items will be considered in evaluating your program.

The written program must describe how the requirements for labels and other forms of warning, material safety data sheets, and employee information and training, are going to be met in your facility. The following discussion provides the type of information compliance officers will be looking for to decide whether these elements of the hazard communication program have been properly addressed:

### A. Labels and Other Forms of Warning

In-plant containers of hazardous chemicals must be labeled, tagged, or marked with the identity of the material and appropriate hazard warnings. Chemical manufacturers, importers, and distributors are required to ensure that every container of hazardous chemicals they ship is appropriately labeled with such information and with the name and address of the producer or other responsible party. Employers purchasing chemicals can rely on the labels provided by their suppliers. If the material is subsequently transferred by the employer from a labeled container to another container, the employer will have to label that container unless it is subject to the portable container exemption. See paragraph (f) of this section for specific labeling requirements.

The primary information to be obtained from an OSHA-required label is an identity for the material, and appropriate hazard warnings. The identity is any term which appears on the label, the MSDS, and the list of chemicals, and thus links these three sources of information. The identity used by the supplier may be a common or trade name ("Black Magic Formula"), or a chemical name (1,1,1,-trichloroethane). The hazard warning is a brief statement of the hazardous effects of the chemical ("flammable," "causes lung damage"). Labels frequently contain other information, such as precautionary measures ("do not use near open flame"), but this information is provided voluntarily and is not required by the rule. Labels must be legible, and prominently displayed. There are no specific requirements for size or color, or any specified text.

With these requirements in mind, the compliance officer will be looking for the following types of information to ensure that labeling will be properly implemented in your facility:

1. Designation of person(s) responsible for ensuring labeling of in-plant containers;

2. Designation of person(s) responsible for ensuring labeling of any shipped containers;

3. Description of labeling system(s) used;

4. Description of written alternatives to labeling of in-plant containers (if used); and,

5. Procedures to review and update label information when necessary.

Employers that are purchasing and using hazardous chemicals—rather than producing or distributing them—will primarily be concerned with ensuring that every purchased container is labeled. If materials are transferred into other containers, the employer must ensure that these are labeled as well, unless they fall under the portable container exemption (paragraph (f)(7) of this section). In terms of labeling systems, you can simply choose to use the labels provided by your suppliers on the containers. These will generally be verbal text labels, and do not usually include numerical rating systems or symbols that require special training. The most important thing to remember is that this is a continuing duty—all in-plant containers of hazardous chemicals must always be labeled. Therefore, it is important to designate someone to be responsible for ensuring that the labels are maintained as required on the containers in your facility, and that newly purchased materials are checked for labels prior to use.

### B. Material Safety Data Sheets

Chemical manufacturers and importers are required to obtain or develop a material safety data sheet for each hazardous chemical they produce or import. Distributors are responsible for ensuring that their customers are provided a copy of these MSDSs. Employers must have an MSDS for each hazardous chemical which they use. Employers may rely on the information received from their suppliers. The specific requirements for material safety data sheets are in paragraph (g) of this section.

There is no specified format for the MSDS under the rule, although there are specific information requirements. OSHA has developed a non-mandatory format, OSHA Form 174, which may be used by chemical manufacturers and importers to comply with the rule. The MSDS must be in English. You are entitled to receive from your supplier a data sheet which includes all of the information required under the rule. If you do not receive one automatically, you should request one. If you receive one that is obviously inadequate, with, for example, blank spaces that are not completed, you should request an appropriately completed one. If your request for a data sheet or for a corrected data sheet does not produce the information needed, you should contact your local OSHA Area Office for assistance in obtaining the MSDS.

The role of MSDSs under the rule is to provide detailed information on each hazardous chemical, including its potential hazardous effects, its physical and chemical characteristics, and recommendations for appropriate protective measures. This information should be useful to you as the employer responsible for designing protective programs, as well as to the workers. If you are not familiar with material safety data sheets and with chemical terminology, you may need to learn to use them yourself. A glossary of MSDS terms may be helpful in this regard. Generally speaking, most employers using hazardous chemicals will primarily be concerned with MSDS information regarding hazardous effects and recommended protective measures. Focus on the sections of the MSDS that are applicable to your situation.

MSDSs must be readily accessible to employees when they are in their work areas during their workshifts. This may be accomplished in many different ways. You must decide what is appropriate for your particular workplace. Some employers keep the MSDSs in a binder in a central location (e.g., in the pick-up truck on a construction site). Others, particularly in workplaces with large numbers of chemicals, computerize the information and provide access through terminals. As long as employees can get the information when they need it, any approach may be used. The employees must have access to the MSDSs themselves—simply having a system where the information can be read to them over the phone is only permitted under the mobile worksite provision, paragraph (g)(9) of this section, when employees must travel between workplaces during the shift. In this situation, they have access to the MSDSs prior to leaving the primary worksite, and when they return, so the telephone system is simply an emergency arrangement.

In order to ensure that you have a current MSDS for each chemical in the plant as required, and that employee access is provided, the compliance officers will be looking for the following types of information in your written program:

1. Designation of person(s) responsible for obtaining and maintaining the MSDSs;

2. How such sheets are to be maintained in the workplace (e.g., in notebooks in the work area(s) or in a computer with terminal access), and how employees can obtain access to them when they are in their work area during the work shift;

3. Procedures to follow when the MSDS is not received at the time of the first shipment;

4. For producers, procedures to update the MSDS when new and significant health information is found; and,

5. Description of alternatives to actual data sheets in the workplace, if used.

For employers using hazardous chemicals, the most important aspect of the written program in terms of MSDSs is to ensure that someone is responsible for obtaining and maintaining the MSDSs for every hazardous chemical in the workplace. The list of hazardous chemicals required to be maintained as part of the written program will serve as an inventory. As new chemicals are purchased, the list should be updated. Many companies have found it convenient to include on their purchase orders the name and address of the person designated in their company to receive MSDSs.

### C. Employee Information and Training

Each employee who may be "exposed" to hazardous chemicals when working must be provided information and trained prior to initial assignment to work with a hazardous chemical, and whenever the hazard changes. "Exposure" or "exposed" under the rule means that "an employee is subjected to a hazardous chemical in the course of employment through any route of entry (inhalation, ingestion, skin contact or absorption, etc.) and includes potential (e.g., accidental or possible) exposure." See paragraph (h) of this section for specific requirements. Information and training may be done either by individual chemical, or by categories of hazards (such as flammability or carcinogenicity). If there are only a few chemicals in the workplace, then you may want to discuss each one individually. Where there are large numbers of chemicals, or the chemicals change frequently, you will probably want to train generally based on the hazard categories (e.g., flammable liquids, corrosive materials, carcinogens). Employees will have access to the substance-specific information on the labels and MSDSs.

Information and training is a critical part of the hazard communication program. Information regarding hazards and protective measures are provided to workers through written labels and material safety data sheets. However, through effective information and training, workers will learn to read and understand such information, determine how it can be obtained and used in their own workplaces, and understand the risks of exposure to the chemicals in their workplaces as well as the ways to protect themselves. A properly conducted training program will ensure comprehension and understanding. It is not sufficient to either just read material to the workers, or simply hand them material to read. You want to create a climate where workers feel free to ask questions. This will help you to ensure that the information is understood. You must always remember that the underlying purpose of the HCS is to reduce the incidence of chemical source illnesses and injuries. This will be accomplished by modifying behavior through the provision of hazard information and information about protective measures. If your program works, you and your workers will better understand the chemical hazards within the workplace. The procedures you establish regarding, for example, purchasing, storage, and handling of these chemicals will improve, and thereby reduce the risks posed to employees exposed to the chemical hazards involved. Furthermore, your workers' comprehension will also be increased, and proper work practices will be followed in your workplace.

If you are going to do the training yourself, you will have to understand the material and be prepared to motivate the workers to learn. This is not always an easy task, but the benefits are worth the effort. More information regarding appropriate training can be found in OSHA Publication No. 2254 which contains voluntary training guidelines prepared by OSHA's Training Institute. A copy of this document is available from OSHA's Publications Office at (202) 219-4667.

In reviewing your written program with regard to information and training, the following items need to be considered:

1. Designation of person(s) responsible for conducting training;

2. Format of the program to be used (audiovisuals, classroom instruction, etc.);

3. Elements of the training program (should be consistent with the elements in paragraph (h) of this section); and,

4. Procedure to train new employees at the time of their initial assignment to work with a hazardous chemical, and to train employees when a new hazard is introduced into the workplace.

The written program should provide enough details about the employer's plans in this area to assess whether or not a good faith effort is being made to train employees. OSHA does not expect that every worker will be able to recite all of the information about each chemical in the workplace. In general, the most important aspects of training under the HCS are to ensure that employees are aware that they are exposed to hazardous chemicals, that they know how to read and use labels and material safety data sheets, and that, as a consequence of learning this information, they are following the appropriate protective measures established by the employer. OSHA compliance officers will be talking to employees to determine if they have received training, if they know they are exposed to hazardous chemicals, and if they know where to obtain substance-specific information on labels and MSDSs.

The rule does not require employers to maintain records of employee training, but many employers choose to do so. This may help you monitor your own program to ensure that all employees are appropriately trained. If you already have a training program, you may simply have to supplement it with whatever additional information is required under the

HCS. For example, construction employers that are already in compliance with the construction training standard (29 CFR 1926.21) will have little extra training to do.

An employer can provide employees information and training through whatever means are found appropriate and protective. Although there would always have to be some training on-site (such as informing employees of the location and availability of the written program and MSDSs), employee training may be satisfied in part by general training about the requirements of the HCS and about chemical hazards on the job which is provided by, for example, trade associations, unions, colleges, and professional schools. In addition, previous training, education and experience of a worker may relieve the employer of some of the burdens of informing and training that worker. Regardless of the method relied upon, however, the employer is always ultimately responsible for ensuring that employees are adequately trained. If the compliance officer finds that the training is deficient, the employer will be cited for the deficiency regardless of who actually provided the training on behalf of the employer.

### D. Other Requirements

In addition to these specific items, compliance officers will also be asking the following questions in assessing the adequacy of the program:

Does a list of the hazardous chemicals exist in each work area or at a central location?

Are methods the employer will use to inform employees of the hazards of non-routine tasks outlined?

Are employees informed of the hazards associated with chemicals contained in unlabeled pipes in their work areas?

On multi-employer worksites, has the employer provided other employers with information about labeling systems and precautionary measures where the other employers have employees exposed to the initial employer's chemicals?

Is the written program made available to employees and their designated representatives?

If your program adequately addresses the means of communicating information to employees in your workplace, and provides answers to the basic questions outlined above, it will be found to be in compliance with the rule.

### 5. Checklist for Compliance

The following checklist will help to ensure you are in compliance with the rule:

Obtained a copy of the rule. _____

Read and understood the requirements. _____

Assigned responsibility for tasks. _____

Prepared an inventory of chemicals. _____

Ensured containers are labeled. _____

Obtained MSDS for each chemical. _____

Prepared written program. _____

Made MSDSs available to workers. _____

Conducted training of workers. _____

Established procedures to maintain current program. _____

Established procedures to evaluate effectiveness. _____

### 6. Further Assistance

If you have a question regarding compliance with the HCS, you should contact your local OSHA Area Office for assistance. In addition, each OSHA Regional Office has a Hazard Communication Coordinator who can answer your questions. Free consultation services are also available to assist employers, and information regarding these services can be obtained through the Area and Regional offices as well.

The telephone number for the OSHA office closest to you should be listed in your local telephone directory. If you are not able to obtain this information, you may contact OSHA's Office of Information and Consumer Affairs at (202) 219-8151 for further assistance in identifying the appropriate contacts.

[59 FR 6170, Feb. 9, 1994, as amended at 59 FR 17479, Apr. 13, 1994; 59 FR 65948, Dec. 22, 1994; 61 FR 9245, Mar. 7, 1996]

## §1910.1201 Retention of DOT markings, placards and labels.

**1910.1201(a)** Any employer who receives a package of hazardous material which is required to be marked, labeled or placarded in accordance with the U. S. Department of Transportation's Hazardous Materials Regulations (49 CFR Parts 171 through 180) shall retain those markings, labels and placards on the package until the packaging is sufficiently cleaned of residue and purged of vapors to remove any potential hazards.

**1910.1201(b)** Any employer who receives a freight container, rail freight car, motor vehicle, or transport vehicle that is required to be marked or placarded in accordance with the Hazardous Materials Regulations shall retain those markings and placards on the freight container, rail freight car, motor vehicle or transport vehicle until the hazardous materials which require the marking or placarding are sufficiently removed to prevent any potential hazards.

**1910.1201(c)** Markings, placards and labels shall be maintained in a manner that ensures that they are readily visible.

**1910.1201(d)** For non-bulk packages which will not be reshipped, the provisions of this section are met if a label or other acceptable marking is affixed in accordance with the Hazard Communication Standard (29 CFR 1910.1200).

**1910.1201(e)** For the purposes of this section, the term "hazardous material" and any other terms not defined in this section have the same definition as in the Hazardous Materials Regulations (49 CFR Parts 171 through 180).

[59 FR 36700, July 19, 1994]

## §1910.1450 Occupational exposure to hazardous chemicals in laboratories.

**1910.1450(a)** *Scope and application.* **(1)** This section shall apply to all employers engaged in the laboratory use of hazardous chemicals as defined below.

**1910.1450(a)(2)** Where this section applies, it shall supersede, for laboratories, the requirements of all other OSHA health standards in 29 CFR part 1910, subpart Z, except as follows:

**1910.1450(a)(2)(i)** For any OSHA health standard, only the requirement to limit employee exposure to the specific permissible exposure limit shall apply for laboratories, unless that particular standard states otherwise or unless the conditions of paragraph (a)(2)(iii) of this section apply.

**1910.1450(a)(2)(ii)** Prohibition of eye and skin contact where specified by any OSHA health standard shall be observed.

**1910.1450(a)(2)(iii)** Where the action level (or in the absence of an action level, the permissible exposure limit) is routinely exceeded for an OSHA regulated substance with exposure monitoring and medical surveillance requirements, paragraphs (d) and (g)(1)(ii) of this section shall apply.

**1910.1450(a)(3)** This section shall not apply to:

**1910.1450(a)(3)(i)** Uses of hazardous chemicals which do not meet the definition of laboratory use, and in such cases, the employer shall comply with the relevant standard in 29 CFR part 1910, subpart Z, even if such use occurs in a laboratory.

**1910.1450(a)(3)(ii)** Laboratory uses of hazardous chemicals which provide no potential for employee exposure. Examples of such conditions might include:

**1910.1450(a)(3)(ii)(A)** Procedures using chemically-impregnated test media such as Dip-and-Read tests where a reagent strip is dipped into the specimen to be tested and the results are interpreted by comparing the color reaction to a color chart supplied by the manufacturer of the test strip; and

**1910.1450(a)(3)(ii)(B)** Commercially prepared kits such as those used in performing pregnancy tests in which all of the reagents needed to conduct the test are contained in the kit.

**1910.1450(b)** *Definitions*—

*Action level* means a concentration designated in 29 CFR part 1910 for a specific substance, calculated as an eight (8)-hour time-weighted average, which initiates certain required activities such as exposure monitoring and medical surveillance.

*Assistant Secretary* means the Assistant Secretary of Labor for Occupational Safety and Health, U.S. Department of Labor, or designee.

*Carcinogen* (see *select carcinogen*).

# OSHA Standards for General Industry

*Chemical Hygiene Officer* means an employee who is designated by the employer, and who is qualified by training or experience, to provide technical guidance in the development and implementation of the provisions of the Chemical Hygiene Plan. This definition is not intended to place limitations on the position description or job classification that the designated individual shall hold within the employer's organizational structure.

*Chemical Hygiene Plan* means a written program developed and implemented by the employer which sets forth procedures, equipment, personal protective equipment and work practices that (i) are capable of protecting employees from the health hazards presented by hazardous chemicals used in that particular workplace and (ii) meets the requirements of paragraph (e) of this section.

*Combustible liquid* means any liquid having a flashpoint at or above 100 °F (37.8 °C), but below 200 °F (93.3 °C), except any mixture having components with flashpoints of 200 °F (93.3 °C), or higher, the total volume of which make up 99 percent or more of the total volume of the mixture.

*Compressed gas* means:

(i) A gas or mixture of gases having, in a container, an absolute pressure exceeding 40 psi at 70 °F (21.1 °C); or

(ii) A gas or mixture of gases having, in a container, an absolute pressure exceeding 104 psi at 130 °F (54.4 °C) regardless of the pressure at 70 °F (21.1 °C); or

(iii) A liquid having a vapor pressure exceeding 40 psi at 100 °F (37.8 °C) as determined by ASTM D-323-72.

*Designated area* means an area which may be used for work with "select carcinogens," reproductive toxins or substances which have a high degree of acute toxicity. A designated area may be the entire laboratory, an area of a laboratory or a device such as a laboratory hood.

*Emergency* means any occurrence such as, but not limited to, equipment failure, rupture of containers or failure of control equipment which results in an uncontrolled release of a hazardous chemical into the workplace.

*Employee* means an individual employed in a laboratory workplace who may be exposed to hazardous chemicals in the course of his or her assignments.

*Explosive* means a chemical that causes a sudden, almost instantaneous release of pressure, gas, and heat when subjected to sudden shock, pressure, or high temperature.

*Flammable* means a chemical that falls into one of the following categories:

(i) *Aerosol, flammable* means an aerosol that, when tested by the method described in 16 CFR 1500.45, yields a flame protection exceeding 18 inches at full valve opening, or a flashback (a flame extending back to the valve) at any degree of valve opening;

(ii) *Gas, flammable* means:

(ii)(A) A gas that, at ambient temperature and pressure, forms a flammable mixture with air at a concentration of 13 percent by volume or less; or

(ii)(B) A gas that, at ambient temperature and pressure, forms a range of flammable mixtures with air wider than 12 percent by volume, regardless of the lower limit.

(iii) *Liquid, flammable* means any liquid having a flashpoint below 100 °F (37.8 °C), except any mixture having components with flashpoints of 100 °F (37.8 °C) or higher, the total of which make up 99 percent or more of the total volume of the mixture.

(iv) *Solid, flammable* means a solid, other than a blasting agent or explosive as defined in §1910.109(a), that is liable to cause fire through friction, absorption of moisture, spontaneous chemical change, or retained heat from manufacturing or processing, or which can be ignited readily and when ignited burns so vigorously and persistently as to create a serious hazard. A chemical shall be considered to be a flammable solid if, when tested by the method described in 16 CFR 1500.44, it ignites and burns with a self-sustained flame at a rate greater than one-tenth of an inch per second along its major axis.

*Flashpoint* means the minimum temperature at which a liquid gives off a vapor in sufficient concentration to ignite when tested as follows:

(i) Tagliabue Closed Tester (See American National Standard Method of Test for Flash Point by Tag Closed Tester, Z11.24-1979 (ASTM D 56-79))-for liquids with a viscosity of less than 45 Saybolt Universal Seconds (SUS) at 100 °F (37.8 °C), that do not contain suspended solids and do not have a tendency to form a surface film under test; or

(ii) Pensky-Martens Closed Tester (see American National Standard Method of Test for Flash Point by Pensky-Martens Closed Tester, Z11.7-1979 (ASTM D 93-79))-for liquids with a viscosity equal to or greater than 45 SUS at 100 °F (37.8 °C), or that contain suspended solids, or that have a tendency to form a surface film under test; or

(iii) Setaflash Closed Tester (see American National Standard Method of Test for Flash Point by Setaflash Closed Tester (ASTM D 3278-78)).

Organic peroxides, which undergo autoaccelerating thermal decomposition, are excluded from any of the flashpoint determination methods specified above.

*Hazardous chemical* means a chemical for which there is statistically significant evidence based on at least one study conducted in accordance with established scientific principles that acute or chronic health effects may occur in exposed employees. The term *health hazard* includes chemicals which are carcinogens, toxic or highly toxic agents, reproductive toxins, irritants, corrosives, sensitizers, hepatotoxins, nephrotoxins, neurotoxins, agents which act on the hematopoietic systems, and agents which damage the lungs, skin, eyes, or mucous membranes.

Appendices A and B of the Hazard Communication Standard (29 CFR 1910.1200) provide further guidance in defining the scope of health hazards and determining whether or not a chemical is to be considered hazardous for purposes of this standard.

*Laboratory* means a facility where the "laboratory use of hazardous chemicals" occurs. It is a workplace where relatively small quantities of hazardous chemicals are used on a non-production basis.

*Laboratory scale* means work with substances in which the containers used for reactions, transfers, and other handling of substances are designed to be easily and safely manipulated by one person. "Laboratory scale" excludes those workplaces whose function is to produce commercial quantities of materials.

*Laboratory-type hood* means a device located in a laboratory, enclosure on five sides with a moveable sash or fixed partial enclosed on the remaining side; constructed and maintained to draw air from the laboratory and to prevent or minimize the escape of air contaminants into the laboratory; and allows chemical manipulations to be conducted in the enclosure without insertion of any portion of the employee's body other than hands and arms.

Walk-in hoods with adjustable sashes meet the above definition provided that the sashes are adjusted during use so that the airflow and the exhaust of air contaminants are not compromised and employees do not work inside the enclosure during the release of airborne hazardous chemicals.

*Laboratory use of hazardous chemicals* means handling or use of such chemicals in which all of the following conditions are met:

(i) Chemical manipulations are carried out on a "laboratory scale;"

(ii) Multiple chemical procedures or chemicals are used;

(iii) The procedures involved are not part of a production process, nor in any way simulate a production process; and

(iv) "Protective laboratory practices and equipment" are available and in common use to minimize the potential for employee exposure to hazardous chemicals.

*Medical consultation* means a consultation which takes place between an employee and a licensed physician for the purpose of determining what medical examinations or procedures, if any, are appropriate in cases where a significant exposure to a hazardous chemical may have taken place.

*Organic peroxide* means an organic compound that contains the bivalent –O–O– structure and which may be considered to be a structural derivative of hydrogen peroxide where one or both of the hydrogen atoms has been replaced by an organic radical.

*Oxidizer* means a chemical other than a blasting agent or explosive as defined in §1910.109(a), that initiates or promotes combustion in other materials, thereby causing fire either of itself or through the release of oxygen or other gases.

*Physical hazard* means a chemical for which there is scientifically valid evidence that it is a combustible liquid, a compressed gas, explosive, flammable, an organic peroxide, an oxidizer, pyrophoric, unstable (reactive) or water-reactive.

*Protective laboratory practices and equipment* means those laboratory procedures, practices and equipment accepted by laboratory health and safety experts as effective, or that the employer can show to be effective, in minimizing the potential for employee exposure to hazardous chemicals.

*Reproductive toxins* means chemicals which affect the reproductive capabilities including chromosomal damage (mutations) and effects on fetuses (teratogenesis)

*Select carcinogen* means any substance which meets one of the following criteria:

(i) It is regulated by OSHA as a carcinogen; or

(ii) It is listed under the category, "known to be carcinogens," in the Annual Report on Carcinogens published by the National Toxicology Program (NTP) (latest edition); or

(iii) It is listed under Group 1 ("carcinogenic to humans") by the International Agency for Research on Cancer Monographs (IARC) (latest editions); or

(iv) It is listed in either Group 2A or 2B by IARC or under the category, "reasonably anticipated to be carcinogens" by NTP, and causes statistically significant tumor incidence in experimental animals in accordance with any of the following criteria:

(iv)(A) After inhalation exposure of 6-7 hours per day, 5 days per week, for a significant portion of a lifetime to dosages of less than 10 mg/m$^3$;

(iv)(B) After repeated skin application of less than 300 (mg/kg of body weight) per week; or

(iv)(C) After oral dosages of less than 50 mg/kg of body weight per day.

*Unstable (reactive)* means a chemical which is the pure state, or as produced or transported, will vigorously polymerize, decompose, condense, or will become self-reactive under conditions of shocks, pressure or temperature.

*Water-reactive* means a chemical that reacts with water to release a gas that is either flammable or presents a health hazard.

**1910.1450(c)** *Permissible exposure limits.* For laboratory uses of OSHA regulated substances, the employer shall assure that laboratory employees' exposures to such substances do not exceed the permissible exposure limits specified in 29 CFR part 1910, subpart Z.

**1910.1450(d)** *Employee exposure determination*—**(1)** *Initial monitoring.* The employer shall measure the employee's exposure to any substance regulated by a standard which requires monitoring if there is reason to believe that exposure levels for that substance routinely exceed the action level (or in the absence of an action level, the PEL).

**1910.1450(d)(2)** *Periodic monitoring.* If the initial monitoring prescribed by paragraph (d)(1) of this section discloses employee exposure over the action level (or in the absence of an action level, the PEL), the employer shall immediately comply with the exposure monitoring provisions of the relevant standard.

**1910.1450(d)(3)** *Termination of monitoring.* Monitoring may be terminated in accordance with the relevant standard.

**1910.1450(d)(4)** *Employee notification of monitoring results.* The employer shall, within 15 working days after the receipt of any monitoring results, notify the employee of these results in writing either individually or by posting results in an appropriate location that is accessible to employees.

**1910.1450(e)** *Chemical hygiene plan*—*General.* (Appendix A of this section is non-mandatory but provides guidance to assist employers in the development of the Chemical Hygiene Plan.)

**1910.1450(e)(1)** Where hazardous chemicals as defined by this standard are used in the workplace, the employer shall develop and carry out the provisions of a written Chemical Hygiene Plan which is:

**1910.1450(e)(1)(i)** Capable of protecting employees from health hazards associated with hazardous chemicals in that laboratory and

**1910.1450(e)(1)(ii)** Capable of keeping exposures below the limits specified in paragraph (c) of this section.

**1910.1450(e)(2)** The Chemical Hygiene Plan shall be readily available to employees, employee representatives and, upon request, to the Assistant Secretary.

**1910.1450(e)(3)** The Chemical Hygiene Plan shall include each of the following elements and shall indicate specific measures that the employer will take to ensure laboratory employee protection:

**1910.1450(e)(3)(i)** Standard operating procedures relevant to safety and health considerations to be followed when laboratory work involves the use of hazardous chemicals;

**1910.1450(e)(3)(ii)** Criteria that the employer will use to determine and implement control measures to reduce employee exposure to hazardous chemicals including engineering controls, the use of personal protective equipment and hygiene practices; particular attention shall be given to the selection of control measures for chemicals that are known to be extremely hazardous;

**1910.1450(e)(3)(iii)** A requirement that fume hoods and other protective equipment are functioning properly and specific measures that shall be taken to ensure proper and adequate performance of such equipment;

**1910.1450(e)(3)(iv)** Provisions for employee information and training as prescribed in paragraph (f) of this section;

**1910.1450(e)(3)(v)** The circumstances under which a particular laboratory operation, procedure or activity shall require prior approval from the employer or the employer's designee before implementation;

**1910.1450(e)(3)(vi)** Provisions for medical consultation and medical examinations in accordance with paragraph (g) of this section;

**1910.1450(e)(3)(vii)** Designation of personnel responsible for implementation of the Chemical Hygiene Plan including the assignment of a Chemical Hygiene Officer and, if appropriate, establishment of a Chemical Hygiene Committee; and

**1910.1450(e)(3)(viii)** Provisions for additional employee protection for work with particularly hazardous substances. These include "select carcinogens," reproductive toxins and substances which have a high degree of acute toxicity. Specific consideration shall be given to the following provisions which shall be included where appropriate:

**1910.1450(e)(3)(viii)(A)** Establishment of a designated area;

**1910.1450(e)(3)(viii)(B)** Use of containment devices such as fume hoods or glove boxes;

**1910.1450(e)(3)(viii)(C)** Procedures for safe removal of contaminated waste; and

**1910.1450(e)(3)(viii)(D)** Decontamination procedures.

**1910.1450(e)(4)** The employer shall review and evaluate the effectiveness of the Chemical Hygiene Plan at least annually and update it as necessary.

**1910.1450(f)** *Employee information and training.* **(1)** The employer shall provide employees with information and training to ensure that they are apprised of the hazards of chemicals present in their work area.

**1910.1450(f)(2)** Such information shall be provided at the time of an employee's initial assignment to a work area where hazardous chemicals are present and prior to assignments involving new exposure situations. The frequency of refresher information and training shall be determined by the employer.

**1910.1450(f)(3)** *Information.* Employees shall be informed of:

**1910.1450(f)(3)(i)** The contents of this standard and its appendices which shall be made available to employees;

**1910.1450(f)(3)(ii)** The location and availability of the employer's Chemical Hygiene Plan;

**1910.1450(f)(3)(iii)** The permissible exposure limits for OSHA regulated substances or recommended exposure limits for other hazardous chemicals where there is no applicable OSHA standard;

**1910.1450(f)(3)(iv)** Signs and symptoms associated with exposures to hazardous chemicals used in the laboratory; and

**1910.1450(f)(3)(v)** The location and availability of known reference material on the hazards, safe handling, storage and disposal of hazardous chemicals found in the laboratory including, but not limited to, Material Safety Data Sheets received from the chemical supplier.

**1910.1450(f)(4)** *Training.* (i) Employee training shall include:

**1910.1450(f)(4)(i)(A)** Methods and observations that may be used to detect the presence or release of a hazardous chemical (such as monitoring conducted by the employer, continuous monitoring devices, visual appearance or odor of hazardous chemicals when being released, etc.);

**1910.1450(f)(4)(i)(B)** The physical and health hazards of chemicals in the work area; and

**1910.1450(f)(4)(i)(C)** The measures employees can take to protect themselves from these hazards, including specific procedures the employer has implemented to protect employees from exposure to hazardous chemicals, such as appropriate work practices, emergency procedures, and personal protective equipment to be used.

**1910.1450(f)(4)(ii)** The employee shall be trained on the applicable details of the employer's written Chemical Hygiene Plan.

**1910.1450(g)** *Medical consultation and medical examinations.* **(1)** The employer shall provide all employees who work with hazardous chemicals an opportunity to receive medical attention, including any follow-up examinations which the examining physician determines to be necessary, under the following circumstances:

**1910.1450(g)(1)(i)** Whenever an employee develops signs or symptoms associated with a hazardous chemical to which the employee may have been exposed in the laboratory, the employee shall be provided an opportunity to receive an appropriate medical examination.

**1910.1450(g)(1)(ii)** Where exposure monitoring reveals an exposure level routinely above the action level (or in the absence of an action level, the PEL) for an OSHA regulated substance for which there are exposure monitoring and medical surveillance requirements, medical surveillance shall be established for the affected employee as prescribed by the particular standard.

**1910.1450(g)(1)(iii)** Whenever an event takes place in the work area such as a spill, leak, explosion or other occurrence resulting in the likelihood of a hazardous exposure, the affected employee shall be provided an opportunity for a medical consultation. Such consultation shall be for the purpose of determining the need for a medical examination.

**1910.1450(g)(2)** All medical examinations and consultations shall be performed by or under the direct supervision of a licensed physician and shall be provided without cost to the employee, without loss of pay and at a reasonable time and place.

**1910.1450(g)(3)** *Information provided to the physician.* The employer shall provide the following information to the physician:

**1910.1450(g)(3)(i)** The identity of the hazardous chemical(s) to which the employee may have been exposed;

**1910.1450(g)(3)(ii)** A description of the conditions under which the exposure occurred including quantitative exposure data, if available; and

**1910.1450(g)(3)(iii)** A description of the signs and symptoms of exposure that the employee is experiencing, if any.

**1910.1450(g)(4)** *Physician's written opinion.* **(i)** For examination or consultation required under this standard, the employer shall obtain a written opinion from the examining physician which shall include the following:

**1910.1450(g)(4)(i)(A)** Any recommendation for further medical follow-up;

**1910.1450(g)(4)(i)(B)** The results of the medical examination and any associated tests;

**1910.1450(g)(4)(i)(C)** Any medical condition which may be revealed in the course of the examination which may place the employee at increased risk as a result of exposure to a hazardous chemical found in the workplace; and

**1910.1450(g)(4)(i)(D)** A statement that the employee has been informed by the physician of the results of the consultation or medical examination and any medical condition that may require further examination or treatment.

**1910.1450(g)(4)(ii)** The written opinion shall not reveal specific findings of diagnoses unrelated to occupational exposure.

**1910.1450(h)** *Hazard identification.* **(1)** With respect to labels and material safety data sheets:

**1910.1450(h)(1)(i)** Employers shall ensure that labels on incoming containers of hazardous chemicals are not removed or defaced.

**1910.1450(h)(1)(ii)** Employers shall maintain any material safety data sheets that are received with incoming shipments of hazardous chemicals, and ensure that they are readily accessible to laboratory employees.

**1910.1450(h)(2)** The following provisions shall apply to chemical substances developed in the laboratory:

**1910.1450(h)(2)(i)** If the composition of the chemical substance which is produced exclusively for the laboratory's use is known, the employer shall determine if it is a hazardous chemical as defined in paragraph (b) of this section. If the chemical is determined to be hazardous, the employer shall provide appropriate training as required under paragraph (f) of this section.

**1910.1450(h)(2)(ii)** If the chemical produced is a byproduct whose composition is not known, the employer shall assume that the substance is hazardous and shall implement paragraph (e) of this section.

**1910.1450(h)(2)(iii)** If the chemical substance is produced for another user outside of the laboratory, the employer shall comply with the Hazard Communication Standard (29 CFR 1910.1200) including the requirements for preparation of material safety data sheets and labeling.

**1910.1450(i)** *Use of respirators.* Where the use of respirators is necessary to maintain exposure below permissible exposure limits, the employer shall provide, at no cost to the employee, the proper respiratory equipment. Respirators shall be selected and used in accordance with the requirements of 29 CFR 1910.134.

**1910.1450(j)** *Recordkeeping.* **(1)** The employer shall establish and maintain for each employee an accurate record of any measurements taken to monitor employee exposures and any medical consultation and examinations including tests or written opinions required by this standard.

**1910.1450(j)(2)** The employer shall assure that such records are kept, transferred, and made available in accordance with 29 CFR 1910.20.

**1910.1450(k)** [Reserved.]

**1910.1450(l)** *Appendices.* The information contained in the appendices is not intended, by itself, to create any additional obligations not otherwise imposed or to detract from any existing obligation.

[55 FR 3327, Jan. 31, 1990, 55 FR 7967, Mar. 6, 1990, 55 FR 12111, Mar. 30, 1990]

## APPENDIX A TO §1910.1450— NATIONAL RESEARCH COUNCIL RECOMMENDATIONS CONCERNING CHEMICAL HYGIENE IN LABORATORIES (NON-MANDATORY)

### TABLE OF CONTENTS

#### Foreword
Corresponding Sections of the Standard and This Appendix

#### A. General Principles
1. Minimize all Chemical Exposures
2. Avoid Underestimation of Risk
3. Provide Adequate Ventilation
4. Institute a Chemical Hygiene Program
5. Observe the PELs and TLVs

#### B. Responsibilities
1. Chief Executive Officer
2. Supervisor of Administrative Unit
3. Chemical Hygiene Officer
4. Laboratory Supervisor
5. Project Director
6. Laboratory Worker

#### C. The Laboratory Facility
1. Design
2. Maintenance
3. Usage
4. Ventilation

### D. Components of the Chemical Hygiene Plan

1. Basic Rules and Procedures
2. Chemical Procurement, Distribution, and Storage
3. Environmental Monitoring
4. Housekeeping, Maintenance and Inspections
5. Medical Program
6. Personal Protective Apparel and Equipment
7. Records
8. Signs and Labels
9. Spills and Accidents
10. Training and Information
11. Waste Disposal

### E. General Procedures for Working With Chemicals

1. General Rules for all Laboratory Work with Chemicals
2. Allergens and Embryotoxins
3. Chemicals of Moderate Chronic or High Acute Toxicity
4. Chemicals of High Chronic Toxicity
5. Animal Work with Chemicals of High Chronic Toxicity

### F. Safety Recommendations

### G. Material Safety Data Sheets

### Foreword

As guidance for each employer's development of an appropriate laboratory Chemical Hygiene Plan, the following non-mandatory recommendations are provided. They were extracted from "Prudent Practices for Handling Hazardous Chemicals in Laboratories" (referred to below as "Prudent Practices"), which was published in 1981 by the National Research Council and is available from the National Academy Press, 2101 Constitution Ave., NW., Washington DC 20418.

"Prudent Practices" is cited because of its wide distribution and acceptance and because of its preparation by members of the laboratory community through the sponsorship of the National Research Council. However, none of the recommendations given here will modify any requirements of the laboratory standard. This Appendix merely presents pertinent recommendations from "Prudent Practices", organized into a form convenient for quick reference during operation of a laboratory facility and during development and application of a Chemical Hygiene Plan. Users of this appendix should consult "Prudent Practices" for a more extended presentation and justification for each recommendation.

"Prudent Practices" deals with both safety and chemical hazards while the laboratory standard is concerned primarily with chemical hazards. Therefore, only those recommendations directed primarily toward control of toxic exposures are cited in this appendix, with the term "chemical hygiene" being substituted for the word "safety". However, since conditions producing or threatening physical injury often pose toxic risks as well, page references concerning major categories of safety hazards in the laboratory are given in section F.

The recommendations from "Prudent Practices" have been paraphrased, combined, or otherwise reorganized, and headings have been added. However, their sense has not been changed.

### Corresponding Sections of the Standard and this Appendix

The following table is given for the convenience of those who are developing a Chemical Hygiene Plan which will satisfy the requirements of paragraph (e) of the standard. It indicates those sections of this appendix which are most pertinent to each of the sections of paragraph (e) and related paragraphs.

| Paragraph and topic in laboratory standard | Relevant appendix section |
|---|---|
| (e)(3)(i) Standard operating procedures for handling toxic chemicals | C, D, E |
| (e)(3)(ii) Criteria to be used for implementation of measures to reduce exposures | D |
| (e)(3)(iii) Fume hood performance | C4b |
| (e)(3)(iv) Employee information and training (including emergency procedures) | D10, D9 |
| (e)(3)(v) Requirements for prior approval of laboratory activities | E2b, E4b |
| (e)(3)(vi) Medical consultation and medical examinations | D5, E4f |
| (e)(3)(vii) Chemical hygiene responsibilities | B |
| (e)(3)(viii) Special precautions for work with particularly hazardous substances | E2, E3, E4 |

In this appendix, those recommendations directed primarily at administrators and supervisors are given in sections A-D. Those recommendations of primary concern to employees who are actually handling laboratory chemicals are given in section E. (Reference to page numbers in "Prudent Practices" are given in parentheses.)

### A. General Principles for Work with Laboratory Chemicals

In addition to the more detailed recommendations listed below in sections B-E, "Prudent Practices" expresses certain general principles, including the following:

1. *It is prudent to minimize all chemical exposures.* Because few laboratory chemicals are without hazards, general precautions for handling all laboratory chemicals should be adopted, rather than specific guidelines for particular chemicals (2, 10). Skin contact with chemicals should be avoided as a cardinal rule (198).

2. *Avoid underestimation of risk.* Even for substances of no known significant hazard, exposure should be minimized; for work with substances which present special hazards, special precautions should be taken (10, 37, 38). One should assume that any mixture will be more toxic than its most toxic component (30, 103) and that all substances of unknown toxicity are toxic (3, 34).

3. *Provide adequate ventilation.* The best way to prevent exposure to airborne substances is to prevent their escape into the working atmosphere by use of hoods and other ventilation devices (32, 198).

4. *Institute a chemical hygiene program.* A mandatory chemical hygiene program designed to minimize exposures is needed; it should be a regular, continuing effort, not merely a standby or short-term activity (6, 11). Its recommendations should be followed in academic teaching laboratories as well as by full-time laboratory workers (13).

5. *Observe the PELs, TLVs.* The Permissible Exposure Limits of OSHA and the Threshold Limit Values of the American Conference of Governmental Industrial Hygienists should not be exceeded (13).

### B. Chemical Hygiene Responsibilities

Responsibility for chemical hygiene rests at all levels (6, 11, 21) including the:

1. *Chief executive officer,* who has ultimate responsibility for chemical hygiene within the institution and must, with other administrators, provide continuing support for institutional chemical hygiene (7, 11).

2. *Supervisor of the department or other administrative unit,* who is responsible for chemical hygiene in that unit (7).

3. *Chemical hygiene officer(s),* whose appointment is essential (7) and who must:

(a) Work with administrators and other employees to develop and implement appropriate chemical hygiene policies and practices (7);

(b) Monitor procurement, use, and disposal of chemicals used in the lab (8);

(c) See that appropriate audits are maintained (8);

(d) Help project directors develop precautions and adequate facilities (10);

(e) Know the current legal requirements concerning regulated substances (50); and

(f) Seek ways to improve the chemical hygiene program (8, 11).

4. *Laboratory supervisor,* who has overall responsibility for chemical hygiene in the laboratory (21) including responsibility to:

(a) Ensure that workers know and follow the chemical hygiene rules, that protective equipment is available and in working order, and that appropriate training has been provided (21, 22);

(b) Provide regular, formal chemical hygiene and housekeeping inspections including routine inspections of emergency equipment (21, 171);

(c) Know the current legal requirements concerning regulated substances (50, 231);

(d) Determine the required levels of protective apparel and equipment (156, 160, 162); and

(e) Ensure that facilities and training for use of any material being ordered are adequate (215).

5. *Project director or director of other specific operation,* who has primary responsibility for chemical hygiene procedures for that operation (7).

6. *Laboratory worker,* who is responsible for:

(a) Planning and conducting each operation in accordance with the institutional chemical hygiene procedures (7, 21, 22, 230); and

(b) Developing good personal chemical hygiene habits (22).

### C. The Laboratory Facility

1. *Design.* The laboratory facility should have:

(a) An appropriate general ventilation system (see C4 below) with air intakes and exhausts located so as to avoid intake of contaminated air (194);

(b) Adequate, well-ventilated stockrooms/storerooms (218, 219);

(c) Laboratory hoods and sinks (12, 162);

(d) Other safety equipment including eyewash fountains and drench showers (162, 169); and

(e) Arrangements for waste disposal (12, 240).

2. *Maintenance.* Chemical-hygiene-related equipment (hoods, incinerator, etc.) should undergo continuing appraisal and be modified if inadequate (11, 12).

3. *Usage.* The work conducted (10) and its scale (12) must be appropriate to the physical facilities available and, especially, to the quality of ventilation (13).

4. *Ventilation*—(a) *General laboratory ventilation.* This system should: Provide a source of air for breathing and for input to local ventilation devices (199); it should not be relied on for protection from toxic substances released into the laboratory (198); ensure that laboratory air is continually replaced, preventing increase of air concentrations of toxic substances during the working day (194); direct air flow into the laboratory from non-laboratory areas and out to the exterior of the building (194).

(b) *Hoods.* A laboratory hood with 2.5 linear feet of hood space per person should be provided for every 2 workers if they spend most of their time working with chemicals (199); each hood should have a continuous monitoring device to allow convenient confirmation of adequate hood performance before use (200, 209). If this is not possible, work with substances of unknown toxicity should be avoided (13) or other types of local ventilation devices should be provided (199). See pp. 201-206 for a discussion of hood design, construction, and evaluation.

(c) *Other local ventilation devices.* Ventilated storage cabinets, canopy hoods, snorkels, etc. should be provided as needed (199). Each canopy hood and snorkel should have a separate exhaust duct (207).

(d) *Special ventilation areas.* Exhaust air from glove boxes and isolation rooms should be passed through scrubbers or other treatment before release into the regular exhaust system (208). Cold rooms and warm rooms should have provisions for rapid escape and for escape in the event of electrical failure (209).

(e) *Modifications.* Any alteration of the ventilation system should be made only if thorough testing indicates that worker protection from airborne toxic substances will continue to be adequate (12, 193, 204).

(f) *Performance.* Rate: 4-12 room air changes/hour is normally adequate general ventilation if local exhaust systems such as hoods are used as the primary method of control (194).

(g) *Quality.* General air flow should not be turbulent and should be relatively uniform throughout the laboratory, with no high velocity or static areas (194, 195); airflow into and within the hood should not be excessively turbulent (200); hood face velocity should be adequate (typically 60-100 lfm) (200, 204).

(h) *Evaluation.* Quality and quantity of ventilation should be evaluated on installation (202), regularly monitored (at least every 3 months) (6, 12, 14, 195), and reevaluated whenever a change in local ventilation devices is made (12, 195, 207). See pp. 195-198 for methods of evaluation and for calculation of estimated airborne contaminant concentrations.

### D. Components of the Chemical Hygiene Plan

#### 1. Basic Rules and Procedures (Recommendations for these are given in section E, below)

#### 2. Chemical Procurement, Distribution, and Storage

(a) *Procurement.* Before a substance is received, information on proper handling, storage, and disposal should be known to those who will be involved (215, 216). No container should be accepted without an adequate identifying label (216). Preferably, all substances should be received in a central location (216).

(b) *Stockrooms/storerooms.* Toxic substances should be segregated in a well-identified area with local exhaust ventilation (221). Chemicals which are highly toxic (227) or other chemicals whose containers have been opened should be in unbreakable secondary containers (219). Stored chemicals should be examined periodically (at least annually) for replacement, deterioration, and container integrity (218-19).

Stockrooms/storerooms should not be used as preparation or repackaging areas, should be open during normal working hours, and should be controlled by one person (219).

(c) *Distribution.* When chemicals are hand carried, the container should be placed in an outside container or bucket. Freight-only elevators should be used if possible (223).

(d) *Laboratory storage.* Amounts permitted should be as small as practical. Storage on bench tops and in hoods is inadvisable. Exposure to heat or direct sunlight should be avoided. Periodic inventories should be conducted, with unneeded items being discarded or returned to the storeroom/stockroom (225-6, 229).

#### 3. Environmental Monitoring

Regular instrumental monitoring of airborne concentrations is not usually justified or practical in laboratories but may be appropriate when testing or redesigning hoods or other ventilation devices (12) or when a highly toxic substance is stored or used regularly (e.g., 3 times/week) (13).

#### 4. Housekeeping, Maintenance, and Inspections

(a) *Cleaning.* Floors should be cleaned regularly (24).

(b) *Inspections.* Formal housekeeping and chemical hygiene inspections should be held at least quarterly (6, 21) for units which have frequent pesonnel changes and semiannually for others; informal inspections should be continual (21).

(c) *Maintenance.* Eye wash fountains should be inspected at intervals of not less than 3 months (6). Respirators for routine use should be inspected periodically by the laboratory supervisor (169). Safety showers should be tested routinely (169). Other safety equipment should be inspected regularly. (e.g., every 3-6 months) (6, 24, 171). Procedures to prevent restarting of out-of-service equipment should be established (25).

(d) *Passageways.* Stairways and hallways should not be used as storage areas (24). Access to exits, emergency equipment, and utility controls should never be blocked (24).

### 5. Medical Program

(a) *Compliance with regulations.* Regular medical surveillance should be established to the extent required by regulations (12).

(b) *Routine surveillance.* Anyone whose work involves regular and frequent handling of toxicologically significant quantities of a chemical should consult a qualified physician to determine on an individual basis whether a regular schedule of medical surveillance is desirable (11, 50).

(c) *First aid.* Personnel trained in first aid should be available during working hours and an emergency room with medical personnel should be nearby (173). See pp. 176-178 for description of some emergency first aid procedures.

### 6. Protective Apparel and Equipment

These should include for each laboratory:

(a) Protective apparel compatible with the required degree of protection for substances being handled (158-161);

(b) An easily accessible drench-type safety shower (162, 169);

(c) An eyewash fountain (162);

(d) A fire extinguisher (162-164);

(e) Respiratory protection (164-9), fire alarm and telephone for emergency use (162) should be available nearby; and

(f) Other items designated by the laboratory supervisor (156, 160).

### 7. Records

(a) Accident records should be written and retained (174).

(b) Chemical Hygiene Plan records should document that the facilities and precautions were compatible with current knowledge and regulations (7).

(c) Inventory and usage records for high-risk substances should be kept as specified in sections E3e below.

(d) Medical records should be retained by the institution in accordance with the requirements of state and federal regulations (12).

### 8. Signs and Labels

Prominent signs and labels of the following types should be posted:

(a) Emergency telephone numbers of emergency personnel/facilities, supervisors, and laboratory workers (28);

(b) Identity labels, showing contents of containers (including waste receptacles) and associated hazards (27, 48);

(c) Location signs for safety showers, eyewash stations, other safety and first aid equipment, exits (27) and areas where food and beverage consumption and storage are permitted (24); and

(d) Warnings at areas or equipment where special or unusual hazards exist (27).

### 9. Spills and Accidents

(a) A written emergency plan should be established and communicated to all personnel; it should include procedures for ventilation failure (200), evacuation, medical care, reporting, and drills (172).

(b) There should be an alarm system to alert people in all parts of the facility including isolation areas such as cold rooms (172).

(c) A spill control policy should be developed and should include consideration of prevention, containment, cleanup, and reporting (175).

(d) All accidents or near accidents should be carefully analyzed with the results distributed to all who might benefit (8, 28).

### 10. Information and Training Program

(a) Aim: To assure that all individuals at risk are adequately informed about the work in the laboratory, its risks, and what to do if an accident occurs (5, 15).

(b) Emergency and Personal Protection Training: Every laboratory worker should know the location and proper use of available protective apparel and equipment (154, 169).

Some of the full-time personnel of the laboratory should be trained in the proper use of emergency equipment and procedures (6).

Such training as well as first aid instruction should be available to (154) and encouraged for (176) everyone who might need it.

(c) Receiving and stockroom/storeroom personnel should know about hazards, handling equipment, protective apparel, and relevant regulations (217).

(d) Frequency of Training: The training and education program should be a regular, continuing activity—not simply an annual presentation (15).

(e) Literature/Consultation: Literature and consulting advice concerning chemical hygiene should be readily available to laboratory personnel, who should be encouraged to use these information resources (14).

### 11. Waste Disposal Program.

(a) Aim: To assure that minimal harm to people, other organisms, and the environment will result from the disposal of waste laboratory chemicals (5).

(b) Content (14, 232, 233, 240): The waste disposal program should specify how waste is to be collected, segregated, stored, and transported and include consideration of what materials can be incinerated. Transport from the institution must be in accordance with DOT regulations (244).

(c) Discarding Chemical Stocks: Unlabeled containers of chemicals and solutions should undergo prompt disposal; if partially used, they should not be opened (24, 27). Before a worker's employment in the laboratory ends, chemicals for which that person was responsible should be discarded or returned to storage (226).

(d) Frequency of Disposal: Waste should be removed from laboratories to a central waste storage area at least once per week and from the central waste storage area at regular intervals (14).

(e) Method of Disposal: Incineration in an environmentally acceptable manner is the most practical disposal method for combustible laboratory waste (14, 238, 241).

Indiscriminate disposal by pouring waste chemicals down the drain (14, 231, 242) or adding them to mixed refuse for landfill burial is unacceptable (14).

Hoods should not be used as a means of disposal for volatile chemicals (40, 200).

Disposal by recycling (233, 243) or chemical decontamination (40, 230) should be used when possible.

## E. Basic Rules and Procedures for Working with Chemicals

The Chemical Hygiene Plan should require that laboratory workers know and follow its rules and procedures. In addition to the procedures of the sub programs mentioned above, these should include the rules listed below.

### 1. General Rules

The following should be used for essentially all laboratory work with chemicals:

(a) *Accidents and spills*—Eye Contact: Promptly flush eyes with water for a prolonged period (15 minutes) and seek medical attention (33, 172).

Ingestion: Encourage the victim to drink large amounts of water (178).

Skin Contact: Promptly flush the affected area with water (33, 172, 178) and remove any contaminated clothing (172, 178). If symptoms persist after washing, seek medical attention (33).

Clean-up. Promptly clean up spills, using appropriate protective apparel and equipment and proper disposal (24 33). See pp. 233-237 for specific clean-up recommendations.

(b) *Avoidance of "routine" exposure:* Develop and encourage safe habits (23); avoid unnecessary exposure to chemicals by any route (23);

Do not smell or taste chemicals (32). Vent apparatus which may discharge toxic chemicals (vacuum pumps, distillation columns, etc.) into local exhaust devices (199).

Inspect gloves (157) and test glove boxes (208) before use.

Do not allow release of toxic substances in cold rooms and warm rooms, since these have contained recirculated atmospheres (209).

(c) *Choice of chemicals:* Use only those chemicals for which the quality of the available ventilation system is appropriate (13).

(d) *Eating, smoking, etc.:* Avoid eating, drinking, smoking, gum chewing, or application of cosmetics in areas where laboratory chemicals are present (22, 24, 32, 40); wash hands before conducting these activities (23, 24).

Avoid storage, handling or consumption of food or beverages in storage areas, refrigerators, glassware or utensils which are also used for laboratory operations (23, 24, 226).

(e) *Equipment and glassware:* Handle and store laboratory glassware with care to avoid damage; do not use damaged glassware (25). Use extra care with Dewar flasks and other evacuated glass apparatus; shield or wrap them to contain chemicals and fragments should implosion occur (25). Use equipment only for its designed purpose (23, 26).

(f) *Exiting:* Wash areas of exposed skin well before leaving the laboratory (23).

(g) *Horseplay:* Avoid practical jokes or other behavior which might confuse, startle or distract another worker (23).

(h) *Mouth suction:* Do not use mouth suction for pipeting or starting a siphon (23, 32).

(i) *Personal apparel:* Confine long hair and loose clothing (23, 158). Wear shoes at all times in the laboratory but do not wear sandals, perforated shoes, or sneakers (158).

(j) *Personal housekeeping:* Keep the work area clean and uncluttered, with chemicals and equipment being properly labeled and stored; clean up the work area on completion of an operation or at the end of each day (24).

(k) *Personal protection:* Assure that appropriate eye protection (154-156) is worn by all persons, including visitors, where chemicals are stored or handled (22, 23, 33, 154).

Wear appropriate gloves when the potential for contact with toxic materials exists (157); inspect the gloves before each use, wash them before removal, and replace them periodically (157). (A table of resistance to chemicals of common glove materials is given p. 159).

Use appropriate (164-168) respiratory equipment when air contaminant concentrations are not sufficiently restricted by engineering controls (164-5), inspecting the respirator before use (169).

Use any other protective and emergency apparel and equipment as appropriate (22, 157-162).

Avoid use of contact lenses in the laboratory unless necessary; if they are used, inform supervisor so special precautions can be taken (155).

Remove laboratory coats immediately on significant contamination (161).

(l) *Planning:* Seek information and advice about hazards (7), plan appropriate protective procedures, and plan positioning of equipment before beginning any new operation (22, 23).

(m) *Unattended operations:* Leave lights on, place an appropriate sign on the door, and provide for containment of toxic substances in the event of failure of a utility service (such as cooling water) to an unattended operation (27, 128).

(n) *Use of hood:* Use the hood for operations which might result in release of toxic chemical vapors or dust (198-9).

As a rule of thumb, use a hood or other local ventilation device when working with any appreciably volatile substance with a TLV of less than 50 ppm (13).

Confirm adequate hood performance before use; keep hood closed at all times except when adjustments within the hood are being made (200); keep materials stored in hoods to a minimum and do not allow them to block vents or air flow (200).

Leave the hood "on" when it is not in active use if toxic substances are stored in it or if it is uncertain whether adequate general laboratory ventilation will be maintained when it is "off" (200).

(o) *Vigilance:* Be alert to unsafe conditions and see that they are corrected when detected (22).

(p) *Waste disposal:* Assure that the plan for each laboratory operation includes plans and training for waste disposal (230).

Deposit chemical waste in appropriately labeled receptacles and follow all other waste disposal procedures of the Chemical Hygiene Plan (22, 24).

Do not discharge to the sewer concentrated acids or bases (231); highly toxic, malodorous, or lachrymatory substances (231); or any substances which might interfere with the biological activity of waste water treatment plants, create fire or explosion hazards, cause structural damage or obstruct flow (242).

(q) *Working alone:* Avoid working alone in a building; do not work alone in a laboratory if the procedures being conducted are hazardous (28).

## 2. Working with Allergens and Embryotoxins

(a) *Allergens* (examples: diazomethane, isocyanates, bichromates): Wear suitable gloves to prevent hand contact with allergens or substances of unknown allergenic activity (35).

(b) *Embryotoxins* (34-5) (examples: organomercurials, lead compounds, formamide): If you are a woman of childbearing age, handle these substances only in a hood whose satisfactory performance has been confirmed, using appropriate protective apparel (especially gloves) to prevent skin contact.

Review each use of these materials with the research supervisor and review continuing uses annually or whenever a procedural change is made.

Store these substances, properly labeled, in an adequately ventilated area in an unbreakable secondary container.

Notify supervisors of all incidents of exposure or spills; consult a qualified physician when appropriate.

## 3. Work with Chemicals of Moderate Chronic or High Acute Toxicity

EXAMPLES: diisopropylflurophosphate (41), hydrofluoric acid (43), hydrogen cyanide (45).

Supplemental rules to be followed in addition to those mentioned above (Procedure B of "Prudent Practices", pp. 39-41):

(a) *Aim:* To minimize exposure to these toxic substances by any route using all reasonable precautions (39).

(b) *Applicability:* These precautions are appropriate for substances with moderate chronic or high acute toxicity used in significant quantities (39).

(c) *Location:* Use and store these substances only in areas of restricted access with special warning signs (40, 229).

Always use a hood (previously evaluated to confirm adequate performance with a face velocity of at least 60 linear feet per minute) (40) or other containment device for procedures which may result in the generation of aerosols or vapors containing the substance (39); trap released vapors to prevent their discharge with the hood exhaust (40).

(d) *Personal protection:* Always avoid skin contact by use of gloves and long sleeves (and other protective apparel as appropriate) (39). Always wash hands and arms immediately after working with these materials (40).

(e) *Records:* Maintain records of the amounts of these materials on hand, amounts used, and the names of the workers involved (40, 229).

(f) *Prevention of spills and accidents:* Be prepared for accidents and spills (41).

Assure that at least 2 people are present at all times if a compound in use is highly toxic or of unknown toxicity (39).

Store breakable containers of these substances in chemically resistant trays; also work and mount apparatus above such trays or cover work and storage surfaces with removable, absorbent, plastic backed paper (40).

If a major spill occurs outside the hood, evacuate the area; assure that cleanup personnel wear suitable protective apparel and equipment (41).

(g) *Waste:* Thoroughly decontaminate or incinerate contaminated clothing or shoes (41). If possible, chemically decontaminate by chemical conversion (40).

Store contaminated waste in closed, suitably labeled, impervious containers (for liquids, in glass or plastic bottles half-filled with vermiculite) (40).

## 4. Work with Chemicals of High Chronic Toxicity

(Examples: dimethylmercury and nickel carbonyl (48), benzo-a-pyrene (51), N-nitrosodiethylamine (54), other human carcinogens or substances with high carcinogenic potency in animals (38).)

Further supplemental rules to be followed, in addition to all these mentioned above, for work with substances of known high chronic toxicity (in quantities above a few milligrams to a few grams, depending on the substance) (47). (Procedure A of "Prudent Practices" pp. 47-50).

(a) *Access:* Conduct all transfers and work with these substances in a "controlled area": a restricted access hood, glove box, or portion of a lab, designated for use of highly toxic substances, for which all people with access are aware of the substances being used and necessary precautions (48).

(b) *Approvals:* Prepare a plan for use and disposal of these materials and obtain the approval of the laboratory supervisor (48).

(c) *Non-contamination/Decontamination:* Protect vacuum pumps against contamination by scrubbers or HEPA filters and vent them into the hood (49). Decontaminate vacuum pumps or other contaminated equipment, including glassware, in the hood before removing them from the controlled area (49, 50).

Decontaminate the controlled area before normal work is resumed there (50).

(d) *Exiting:* On leaving a controlled area, remove any protective apparel (placing it in an appropriate, labeled container) and thoroughly wash hands, forearms, face, and neck (49).

(e) *Housekeeping:* Use a wet mop or a vacuum cleaner equipped with a HEPA filter instead of dry sweeping if the toxic substance was a dry powder (50).

(f) *Medical surveillance:* If using toxicologically significant quantities of such a substance on a regular basis (e.g., 3 times per week), consult a qualified physician concerning desirability of regular medical surveillance (50).

(g) *Records:* Keep accurate records of the amounts of these substances stored (229) and used, the dates of use, and names of users (48).

(h) *Signs and labels:* Assure that the controlled area is conspicuously marked with warning and restricted access signs (49) and that all containers of these substances are appropriately labeled with identity and warning labels (48).

(i) *Spills:* Assure that contingency plans, equipment, and materials to minimize exposures of people and property in case of accident are available (233-4).

(j) *Storage:* Store containers of these chemicals only in a ventilated, limited access (48, 227, 229) area in appropriately labeled, unbreakable, chemically resistant, secondary containers (48, 229).

(k) *Glove boxes:* For a negative pressure glove box, ventilation rate must be at least 2 volume changes/hour and pressure at least 0.5 inches of water (48). For a positive pressure glove box, thoroughly check for leaks before each use (49). In either case, trap the exit gases or filter them through a HEPA filter and then release them into the hood (49).

(l) *Waste:* Use chemical decontamination whenever possible; ensure that containers of contaminated waste (including washings from contaminated flasks) are transferred from the controlled area in a secondary container under the supervision of authorized personnel (49, 50, 233).

### 5. Animal Work with Chemicals of High Chronic Toxicity

(a) *Access:* For large scale studies, special facilities with restricted access are preferable (56).

(b) *Administration of the toxic substance:* When possible, administer the substance by injection or gavage instead of in the diet. If administration is in the diet, use a caging system under negative pressure or under laminar air flow directed toward HEPA filters (56).

(c) *Aerosol suppression:* Devise procedures which minimize formation and dispersal of contaminated aerosols, including those from food, urine, and feces (e.g., use HEPA filtered vacuum equipment for cleaning, moisten contaminated bedding before removal from the cage, mix diets in closed containers in a hood) (55, 56).

(d) *Personal protection:* When working in the animal room, wear plastic or rubber gloves, fully buttoned laboratory coat or jumpsuit and, if needed because of incomplete suppression of aerosols, other apparel and equipment (shoe and head coverings, respirator) (56).

(e) *Waste disposal:* Dispose of contaminated animal tissues and excreta by incineration if the available incinerator can convert the contaminant to non-toxic products (238); otherwise, package the waste appropriately for burial in an EPA-approved site (239).

### F. Safety Recommendations

The above recommendations from "Prudent Practices" do not include those which are directed primarily toward prevention of physical injury rather than toxic exposure. However, failure of precautions against injury will often have the secondary effect of causing toxic exposures. Therefore, we list below page references for recommendations concerning some of the major categories of safety hazards which also have implications for chemical hygiene:

1. Corrosive agents: (35-6)

2. Electrically powered laboratory apparatus: (179-92)

3. Fires, explosions: (26, 57-74, 162-4, 174-5, 219-20, 226-7)

4. Low temperature procedures: (26, 88)

5. Pressurized and vacuum operations (including use of compressed gas cylinders): (27, 75-101)

### G. Material Safety Data Sheets

Material safety data sheets are presented in "Prudent Practices" for the chemicals listed below. (Asterisks denote that comprehensive material safety data sheets are provided).

*Acetyl peroxide (105)
*Acrolein (106)
*Acrylonilrile (107)
Ammonia (anhydrous) (91)
*Aniline (109)
*Benzene (110)
*Benzo[a]pyrene (112)
*Bis(chloromethyl) ether (113)
Boron trichloride (91)
Boron trifluoride (92)
Bromine (114)
*Tert-butyl hydroperoxide (148)
*Carbon disulfide (116)
Carbon monoxide (92)
*Carbon tetrachloride (118)
*Chlorine (119)
Chlorine trifluoride (94)
*Chloroform (121)
Chloromethane (93)
*Diethyl ether (122)
Diisopropyl fluorophosphate (41)
*Dimethylformamide (123)
*Dimethyl sulfate (125)
*Dioxane (126)

*Ethylene dibromide (128)
*Fluorine (95)
*Formaldehyde (130)
*Hydrazine and salts (132)
Hydrofluoric acid (43)
Hydrogen bromide (98)
Hydrogen chloride (98)
*Hydrogen cyanide (133)
*Hydrogen sulfide (135)
Mercury and compounds (52)
*Methanol (137)
*Morpholine (138)
*Nickel carbonyl (99)
*Nitrobenzene (139)
Nitrogen dioxide (100)
N-nitrosodiethylamine (54)
*Peracetic acid (141)
*Phenol (142)
*Phosgene (143)
*Pyridine (144)
*Sodium azide (145)
*Sodium cyanide (147)
Sulfur dioxide (101)
*Trichloroethylene (149)
*Vinyl chloride (150)

## APPENDIX B TO §1910.1450—REFERENCES (NON-MANDATORY)

The following references are provided to assist the employer in the development of a Chemical Hygiene Plan. The materials listed below are offered as non-mandatory guidance. References listed here do not imply specific endorsement of a book, opinion, technique, policy or a specific solution for a safety or health problem. Other references not listed here may better meet the needs of a specific laboratory. (a) Materials for the development of the Chemical Hygiene Plan:

1. American Chemical Society, Safety in Academic Chemistry Laboratories, 4th edition, 1985.

2. Fawcett, H.H. and W. S. Wood, Safety and Accident Prevention in Chemical Operations, 2nd edition, Wiley-Interscience, New York, 1982.

3. Flury, Patricia A., Environmental Health and Safety in the Hospital Laboratory, Charles C. Thomas Publisher, Springfield IL, 1978.

4. Green, Michael E. and Turk, Amos, Safety in Working with Chemicals, Macmillan Publishing Co., NY, 1978.

5. Kaufman, James A., Laboratory Safety Guidelines, Dow Chemical Co., Box 1713, Midland, MI 48640, 1977.

6. National Institutes of Health, NIH Guidelines for the Laboratory use of Chemical Carcinogens, NIH Pub. No. 81-2385, GPO, Washington, DC 20402, 1981.

7. National Research Council, Prudent Practices for Disposal of Chemicals from Laboratories, National Academy Press, Washington, DC, 1983.

8. National Research Council, Prudent Practices for Handling Hazardous Chemicals in Laboratories, National Academy Press, Washington, DC, 1981.

9. Renfrew, Malcolm, Ed., Safety in the Chemical Laboratory, Vol. IV, *J. Chem. Ed.*, American Chemical Society, Easlon, PA, 1981.

10. Steere, Norman V., Ed., Safety in the Chemical Laboratory, *J. Chem. Ed.* American Chemical Society, Easlon, PA, 18042, Vol. I, 1967, Vol. II, 1971, Vol. III 1974.

11. Steere, Norman V., Handbook of Laboratory Safety, the Chemical Rubber Company Cleveland, OH, 1971.

12. Young, Jay A., Ed., Improving Safety in the Chemical Laboratory, John Wiley & Sons, Inc. New York, 1987.

(b) Hazardous Substances Information:

1. American Conference of Governmental Industrial Hygienists, Threshold Limit Values for Chemical Substances and Physical Agents in the Workroom Environment with Intended Changes, 6500 Glenway Avenue, Bldg. D-7 Cincinnati, OH 45211-4438 (latest edition).

2. Annual Report on Carcinogens, National Toxicology Program U.S. Department of Health and Human Services, Public Health Service, U.S. Government Printing Office, Washington, DC, (latest edition).

3. Best Company, Best Safety Directory, Vols. I and II, Oldwick, N.J., 1981.

4. Bretherick, L., Handbook of Reactive Chemical Hazards, 2nd edition, Butterworths, London, 1979.

5. Bretherick, L., Hazards in the Chemical Laboratory, 3rd edition, Royal Society of Chemistry, London, 1986.

6. Code of Federal Regulations, 29 CFR part 1910 subpart Z. U.S. Govt. Printing Office, Washington, DC 20402 (latest edition).

7. IARC Monographs on the Evaluation of the Carcinogenic Risk of Chemicals to Man, World Health Organization Publications Center, 49 Sheridan Avenue, Albany, New York 12210 (latest editions).

8. NIOSH/OSHA Pocket Guide to Chemical Hazards. NIOSH Pub. No. 85-114, U.S. Government Printing Office, Washington, DC, 1985 (or latest edition).

9. Occupational Health Guidelines, NIOSH/OSHA NIOSH Pub. No. 81-123 U.S. Government Printing Office, Washington, DC, 1981.

10. Patty, F.A., Industrial Hygiene and Toxicology, John Wiley & Sons, Inc., New York, NY (Five Volumes).

11. Registry of Toxic Effects of Chemical Substances, U.S. Department of Health and Human Services, Public Health Service, Centers for Disease Control, National Institute for Occupational Safety and Health, Revised Annually, for sale from Superintendent of Documents U.S. Govt. Printing Office, Washington, DC 20402.

12. The Merck Index: An Encyclopedia of Chemicals and Drugs. Merck and Company Inc. Rahway, N.J., 1976 (or latest edition).

13. Sax, N.I. Dangerous Properties of Industrial Materials, 5th edition, Van Nostrand Reinhold, NY., 1979.

14. Sittig, Marshall, Handbook of Toxic and Hazardous Chemicals, Noyes Publications, Park Ridge, NJ, 1981.

(c) Information on Ventilation:

1. American Conference of Governmental Industrial Hygienists Industrial Ventilation (latest edition), 6500 Glenway Avenue, Bldg. D-7, Cincinnati, Ohio 45211-4438.

2. American National Standards Institute, Inc. American National Standards Fundamentals Governing the Design and Operation of Local Exhaust Systems ANSI Z 9.2-1979 American National Standards Institute, N.Y. 1979.

3. Imad, A.P. and Watson, C.L. Ventilation Index: An Easy Way to Decide about Hazardous Liquids, Professional Safety pp 15-18, April 1980.

4. National Fire Protection Association, Fire Protection for Laboratories Using Chemicals NFPA-45, 1982.

Safety Standard for Laboratories in Health Related Institutions, NFPA, 56c, 1980.

Fire Protection Guide on Hazardous Materials, 7th edition, 1978.

National Fire Protection Association, Batterymarch Park, Quincy, MA 02269.

5. Scientific Apparatus Makers Association (SAMA), Standard for Laboratory Fume Hoods, SAMA LF7-1980, 1101 16th Street, NW., Washington, DC 20036.

(d) Information on Availability of Referenced Material:

1. American National Standards Institute (ANSI), 1430 Broadway, New York, NY 10018.

2. American Society for Testing and Materials (ASTM), 1916 Race Street, Philadelphia, PA 19103.

[55 FR 3327, Jan. 31, 1990; 55 FR 7967, Mar. 6, 1990; 57 FR 29204, July 1, 1992; 61 FR 5508, Feb. 13, 1996]

### §1910.1499 Source of standards [Removed].

[§1910.1499 removed at 61 FR 9245, Mar. 7, 1996, effective May 6, 1996.]

### §1910.1500 Standards organizations [Removed].

[§1910.1500 added at 49 FR 23073, May 28, 1975; amended at 43 FR 57603, December 8, 1978, corrected at 57 FR 29204, July 1, 1992; removed at 61 FR 9245, Mar. 7, 1996, effective May 6, 1996.]

# Occupational Safety and Health Act, Sec. 5. Duties "General Duty Clause"

[The following is commonly referred to as the "General Duty Clause"—CCH.]

Sec. 5. (a) Each employer—

(1) shall furnish to each of his employees employment and a place of employment which are free from recognized hazards that are causing or are likely to cause death or serious physical harm to his employees;

(2) shall comply with occupational safety and health standards promulgated under this Act.

# 29 CFR Part 1903— Inspections, Citations, and Proposed Penalties

> **AUTHORITY:** Sections 8 and 9 of the Occupational Safety and Health Act of 1970 (29 U.S.C. 657, 658); 5 U.S.C. 553; Secretary of Labor's Order No. 1-90 (55 FR 9033) or 6-96 (62 FR 111), as applicable.
>
> **SOURCE:** 1903.7 also issued under 5 U.S.C. 553.
>
> **SOURCE:** 36 FR 17850, Sept. 4, 1971, unless otherwise noted.

### §1903.1 Purpose and scope.

The Williams-Steiger Occupational Safety and Health Act of 1970 (84 Stat. 1590 et seq., 29 U.S.C. 651 et seq.) requires, in part, that every employer covered under the Act furnish to his employees employment and a place of employment which are free from recognized hazards that are causing or are likely to cause death or serious physical harm to his employees. The Act also requires that employers comply with occupational safety and health standards promulgated under the Act, and that employees comply with standards, rules, regulations and orders issued under the Act which are applicable to their own actions and conduct. The Act authorizes the Department of Labor to conduct inspections, and to issue citations and proposed penalties for alleged violations. The Act, under section 20(b), also authorizes the Secretary of Health, Education, and Welfare to conduct inspections and to question employers and employees in connection with research and other related activities. The Act contains provisions for adjudication of violations, periods prescribed for the abatement of violations, and proposed penalties by the Occupational Safety and Health Review Commission, if contested by an employer or by an employee or authorized representative of employees, and for judicial review. The purpose of this Part 1903 is to prescribe rules and to set forth general policies for enforcement of the inspection, citation, and proposed penalty provisions of the Act. In situations where this Part 1903 sets forth general enforcement policies rather than substantive or procedural rules, such policies may be modified in specific circumstances where the Secretary or his designee determines that an alternative course of action would better serve the objectives of the Act.

### §1903.2 Posting of notice; availability of Act, regulations and applicable standards.

**1903.2(a)(1)** Each employer shall post and keep posted a notice or notices, to be furnished by the Occupational Safety and Health Administration, U.S. Department of Labor, informing employees of the protections and obligations provided for in the Act, and that for assistance and information, including copies of the Act and of specific safety and health standards, employees should contact the employer or the nearest office of the Department of Labor. Such notice or notices shall be posted by the employer in each establishment in a conspicuous place or places where notices to employees are customarily posted. Each employer shall take steps to insure that such notices are not altered, defaced, or covered by other material.

**1903.2(a)(2)** Where a State has an approved poster informing employees of their protections and obligations as defined in §1952.10 of this chapter, such poster, when posted by employers covered by the State plan, shall constitute compliance with the posting requirements of section 8(c)(1) of the Act. Employers whose operations are not within the issues covered by the State plan must comply with paragraph (a)(1) of this section.

**1903.2(a)(3)** Reproductions or facsimiles of such Federal or State posters shall constitute compliance with the posting requirements of section 8(c)(1) of the Act where such reproductions or facsimiles are at least 8 1/2 inches by 14 inches, and the printing size is at least 10 pt. Whenever the size of the poster increases, the size of the print shall also increase accordingly. The caption or heading on the poster shall be in large type, generally not less than 36 pt.

**1903.2(b)** *Establishment* means a single physical location where business is conducted or where services or industrial operations are performed. (For example: A factory, mill, store, hotel, restaurant, movie theatre, farm, ranch, bank, sales office, warehouse, or central administrative office.) Where distinctly separate activities are performed at a single physical location (such as contract construction activities from the same physical location as a lumber yard), each activity shall be treated as a separate physical establishment, and a separate notice or notices shall be posted in each such establishment, to the extent that such notices have been furnished by the Occupational Safety and Health Administration, U.S. Department of Labor. Where employers are engaged in activities which are physically dispersed, such as agriculture, construction, transportation, communications, and electric, gas and sanitary services, the notice or notices required by this section shall be posted at the location to which employees report each day. Where employees do not usually work at, or report to, a single establishment, such as longshoremen, traveling salesmen, technicians, engineers, etc., such notice or notices shall be posted at the location from which the employees operate to carry out their activities. In all cases, such notice or notices shall be posted in accordance with the requirements of paragraph (a) of this section.

**1903.2(c)** Copies of the Act, all regulations published in this chapter and all applicable standards will be available at all Area Offices of the Occupational Safety and Health Administration, U.S. Department of Labor. If an employer has obtained copies of these materials, he shall make them available upon request to any employee or his authorized representative for review in the establishment where the employee is employed on the same day the request is made or at the earliest time mutually convenient to the employee or his authorized representative and the employer.

**1903.2(d)** Any employer failing to comply with the provisions of this section shall be subject to citation and penalty in accordance with the provisions of section 17 of the Act.

[36 FR 17850, Sept. 4, 1971, as amended at 39 FR 39036, Nov. 5, 1974]

### §1903.3 Authority for inspection.

**1903.3(a)** Compliance Safety and Health Officers of the Department of Labor are authorized to enter without delay and at reasonable times any factory, plant, establishment, construction site, or other area, workplace or environment where work is performed by an employee of an employer; to inspect and investigate during regular working hours and at other reasonable times, and within reasonable limits and in a reasonable manner, any such place of employment, and all pertinent conditions, structures, machines, apparatus, devices, equipment and materials therein; to question privately any employer, owner, operator, agent or employee; and to review records required by the Act and regulations published in this chapter, and other records which are directly related to the purpose of the inspection. Representatives of the Secretary of Health, Education, and Welfare are authorized to make inspections and to question employers and employees in order to carry out the functions of the Secretary of Health, Education, and Welfare under the Act. Inspections conducted by Department of Labor Compliance Safety and Health Officers and representatives of the Secretary of Health, Education, and Welfare under section 8 of the Act and pursuant to this Part 1903 shall not affect the authority of any State to conduct inspections in accordance with agreements and plans under section 18 of the Act.

**1903.3(b)** Prior to inspecting areas containing information which is classified by an agency of the United States Government in the interest of national security, Compliance Safety and Health Officers shall have obtained the appropriate security clearance.

### §1903.4 Objection to inspection.

**1903.4(a)** Upon a refusal to permit the Compliance Safety and Health Officer, in exercise of his official duties, to enter without delay and at reasonable times any place of employment or any place therein, to inspect, to review records, or to question any employer, owner, operator, agent, or employee, in accordance with §1903.3 or to permit a representative of employees to accompany the Compliance Safety and Health Officer during the physical inspection of any workplace in accordance with §1903.8, the Safety and Health Officer shall terminate the inspection or confine the inspection to other areas, conditions, structures, machines, apparatus, devices, equipment, materials, records, or interviews concerning which no objection is raised. The Compliance Safety and Health Officer shall endeavor to ascertain the reason for such refusal, and shall immediately report the refusal and the reason therefor to the Area Director. The Area Director shall consult with the Regional Solicitor, who shall take appropriate action, including compulsory process, if necessary.

**1903.4(b)** Compulsory process shall be sought in advance of an attempted inspection or investigation if, in the judgment of the Area Director and the Regional Solicitor, circumstances exist which make such preinspection process desirable or necessary. Some examples of circumstances in which it may be desirable or necessary to seek compulsory process in advance of an attempt to inspect or investigate include (but are not limited to):

**1903.4(b)(1)** When the employer's past practice either implicitly or explicitly puts the Secretary on notice that a warrantless inspection will not be allowed;

**1903.4(b)(2)** When an inspection is scheduled far from the local office and procuring a warrant prior to leaving to conduct the inspection would avoid, in case of refusal of entry, the expenditure of significant time and resources to return to the office, obtain a warrant and return to the worksite;

**1903.4(b)(3)** When an inspection includes the use of special equipment or when the presence of an expert or experts is needed in order to properly conduct the inspection, and procuring a warrant prior to an attempt to inspect would alleviate the difficulties or costs encountered in coordinating the availability of such equipment or expert.

**1903.4(c)** With the approval of the Regional Administrator and the Regional Solicitor, compulsory process may also be obtained by the Area Director or his designee.

**1903.4(d)** For purposes of this section, the term compulsory process shall mean the institution of any appropriate action, including *ex parte* application for an inspection warrant or its equivalent. *Ex parte* inspection warrants shall be in the preferred form of compulsory process in all circumstances where compulsory process is relied upon to seek entry to a workplace under this section.

[45 FR 65923, Oct. 3, 1980]

### §1903.5 Entry not a waiver.

Any permission to enter, inspect, review records, or question any person, shall not imply or be conditioned upon a waiver of any cause of action, citation, or penalty under the Act. Compliance Safety and Health Officers are not authorized to grant any such waiver.

### §1903.6 Advance notice of inspections.

**1903.6(a)** Advance notice of inspections may not be given, except in the following situations:

**1903.6(a)(1)** In cases of apparent imminent danger, to enable the employer to abate the danger as quickly as possible;

**1903.6(a)(2)** In circumstances where the inspection can most effectively be conducted after regular business hours or where special preparations are necessary for an inspection;

**1903.6(a)(3)** Where necessary to assure the presence of representatives of the employer and employees or the appropriate personnel needed to aid in the inspection; and

**1903.6(a)(4)** In other circumstances where the Area Director determines that the giving of advance notice would enhance the probability of an effective and thorough inspection.

**1903.6(b)** In the situations described in paragraph (a) of this section, advance notice of inspections may be given only if authorized by the Area Director, except that in cases of apparent imminent danger, advance notice may be given by the Compliance Safety and Health Officer without such authorization if the Area Director is not immediately available. When advance notice is given, it shall be the employer's responsibility promptly to notify the authorized representative of employees of the inspection, if the identity of such representative is known to the employer. (See §1903.8(b) as to situations where there is no authorized representative of employees.) Upon the request of the employer, the Compliance Safety and Health Officer will inform the authorized representative of employees of the inspection, provided that the employer furnishes the Compliance Safety and Health Officer with the identity of such representative and with such other information as is necessary to enable him promptly to inform such representative of the inspection. An employer who fails to comply with his obligation under this paragraph promptly to inform the authorized representative of employees of the inspection or to furnish such information as is necessary to enable the Compliance Safety and Health Officer promptly to inform such representative of the inspection, may be subject to citation and penalty under section 17(c) of the Act. Advance notice in any of the situations described in paragraph (a) of this section shall not be given more than 24 hours before the inspection is scheduled to be conducted, except in apparent imminent danger situations and in other unusual circumstances.

**1903.6(c)** The Act provides in section 17(f) that any person who gives advance notice of any inspection to be conducted under the Act, without authority from the Secretary or his designees, shall, upon conviction, be punished by fine of not more than $1,000 or by imprisonment for not more than 6 months, or by both.

### §1903.7 Conduct of inspections.

**1903.7(a)** Subject to the provisions of §1903.3, inspections shall take place at such times and in such places of employment as the Area Director or the Compliance Safety and Health Officer may direct. At the beginning of an inspection, Compliance Safety and Health Officers shall present their credentials to the owner, operator, or agent in charge at the establishment; explain the nature and purpose of the inspection; and indicate generally the scope of the inspection and the records specified in §1903.3 which they wish to review. However, such designation of records shall not preclude access to additional records specified in §1903.3.

**1903.7(b)** Compliance Safety and Health Officers shall have authority to take environmental samples and to take or obtain photographs related to the purpose of the inspection, employ other reasonable investigative techniques, and question privately any employer, owner, operator, agent or employee of an establishment. (See §1903.9 on trade secrets.) As used herein, the term "employ other reasonable investigative techniques" includes, but is not limited to, the use of devices to measure employee exposures and the attachment of personal sampling equipment such as dosimeters, pumps, badges and other similar devices to employees in order to monitor their exposures.

[§1903.7(b) amended at 47 FR 55478 December 10, 1982, effective January 10, 1983.]

**1903.7(c)** In taking photographs and samples, Compliance Safety and Health Officers shall take reasonable precautions to insure that such actions with flash, spark-producing, or other equipment would not be hazardous. Compliance Safety and Health Officers shall comply with all employer safety and health rules and practices at the establishment being inspected, and they shall wear and use appropriate protective clothing and equipment.

**1903.7(d)** The conduct of inspections shall be such as to preclude unreasonable disruption of the operations of the employer's establishment.

**1903.7(e)** At the conclusion of an inspection, the Compliance Safety and Health Officer shall confer with the employer or his representative and informally advise him of any apparent safety or health violations disclosed by the inspection. During such conference, the employer shall be afforded an opportunity to bring to the attention of the Compliance Safety and Health Officer any pertinent information regarding conditions in the workplace.

**1903.7(f)** Inspections shall be conducted in accordance with the requirements of this part.

[36 FR 17850, Sept. 14, 1971, as amended at 47 FR 6533, Feb. 12, 1982; 47 FR 55481, Dec. 10, 1982]

### §1903.8 Representatives of employers and employees.

**1903.8(a)** Compliance Safety and Health Officers shall be in charge of inspections and questioning of persons. A representative of the employer and a representative authorized by his employees shall be given an opportunity to accompany the Compliance Safety and Health Officer during the physical inspection of any workplace for the purpose of aiding such inspection. A Compliance Safety and Health Officer may permit additional employer representatives and additional representatives authorized by employees to accompany him where he determines that such additional representatives will further aid the inspection. A different employer and employee representative may accompany the Compliance Safety and Health Officer during each different phase of an inspection if this will not interfere with the conduct of the inspection.

**1903.8(b)** Compliance Safety and Health Officers shall have authority to resolve all disputes as to who is the representative authorized by the employer and employees for the purpose of this section. If there is no authorized representative of employees, or if the Compliance Safety and Health Officer is unable to determine with reasonable certainty who is such representative, he shall consult with a reasonable number of employees concerning matters of safety and health in the workplace.

**1903.8(c)** The representative(s) authorized by employees shall be an employee(s) of the employer. However, if in the judgment of the Compliance Safety and Health Officer, good cause has been shown why accompaniment by a third party who is not an employee of the employer (such as an industrial hygienist or a safety engineer) is reasonably necessary to the conduct of an effective and thorough physical inspection of the workplace, such third party may accompany the Compliance Safety and Health Officer during the inspection.

**1903.8(d)** Compliance Safety and Health Officers are authorized to deny the right of accompaniment under this section to any person whose conduct interferes with a fair and orderly inspection. The right of accompaniment

# OSHA Standards for General Industry

in areas containing trade secrets shall be subject to the provisions of §1903.9(d). With regard to information classified by an agency of the U.S. Government in the interest of national security, only persons authorized to have access to such information may accompany a Compliance Safety and Health Officer in areas containing such information.

### §1903.9 Trade secrets.

**1903.9(a)** Section 15 of the Act provides: "All information reported to or otherwise obtained by the Secretary or his representative in connection with any inspection or proceeding under this Act which contains or which might reveal a trade secret referred to in section 1905 of title 18 of the United States Code shall be considered confidential for the purpose of that section, except that such information may be disclosed to other officers or employees concerned with carrying out this Act or when relevant in any proceeding under this Act. In any such proceeding the Secretary, the Commission, or the court shall issue such orders as may be appropriate to protect the confidentiality of trade secrets." Section 15 of the Act is considered a statute within the meaning of section 552(b)(3) of title 5 of the United States Code, which exempts from the disclosure requirements matters that are "specifically exempted from disclosure by statute."

**1903.9(b)** Section 1905 of title 18 of the United States Code provides: "Whoever, being an officer or employee of the United States or of any department or agency thereof, publishes, divulges, discloses, or makes known in any manner or to any extent not authorized by law any information coming to him in the course of his employment or official duties or by reason of any examination or investigation made by, or return, report or record made to or filed with, such department or agency or officer or employee thereof, which information concerns or relates to the trade secrets, processes, operations, style of work, or apparatus, or to the identity, confidential statistical data, amount or source of any income, profits, losses, or expenditures of any person, firm, partnership, corporation, or association; or permits any income return or copy thereof or any book containing any abstract or particulars thereof to be seen or examined by any person except as provided by law; shall be fined not more than $1,000, or imprisoned not more than 1 year, or both; and shall be removed from office or employment."

**1903.9(c)** At the commencement of an inspection, the employer may identify areas in the establishment which contain or which might reveal a trade secret. If the Compliance Safety and Health Officer has no clear reason to question such identification, information obtained in such areas, including all negatives and prints of photographs, and environmental samples, shall be labeled "confidential—trade secret" and shall not be disclosed except in accordance with the provisions of section 15 of the Act.

**1903.9(d)** Upon the request of an employer, any authorized representative of employees under §1903.8 in an area containing trade secrets shall be an employee in that area or an employee authorized by the employer to enter that area. Where there is no such representative or employee, the Compliance Safety and Health Officer shall consult with a reasonable number of employees who work in that area concerning matters of safety and health.

### §1903.10 Consultation with employees.

Compliance Safety and Health Officers may consult with employees concerning matters of occupational safety and health to the extent they deem necessary for the conduct of an effective and thorough inspection. During the course of an inspection, any employee shall be afforded an opportunity to bring any violation of the Act which he has reason to believe exists in the workplace to the attention of the Compliance Safety and Health Officer.

### §1903.11 Complaints by employees.

**1903.11(a)** Any employee or representative of employees who believe that a violation of the Act exists in any workplace where such employee is employed may request an inspection of such workplace by giving notice of the alleged violation to the Area Director or to a Compliance Safety and Health Officer. Any such notice shall be reduced to writing, shall set forth with reasonable particularity the grounds for the notice, and shall be signed by the employee or representative of employees. A copy shall be provided the employer or his agent by the Area Director or Compliance Safety Health Officer no later than at the time of inspection, except that, upon the request of the person giving such notice, his name and the names of individual employees referred to therein shall not appear in such copy or on any record published, released, or made available by the Department of Labor.

**1903.11(b)** If upon receipt of such notification the Area Director determines that the complaint meets the requirements set forth in paragraph (a) of this section, and that there are reasonable grounds to believe that the alleged violation exists, he shall cause an inspection to be made as soon as practicable, to determine if such alleged violation exists. Inspections under this section shall not be limited to matters referred to in the complaint.

**1903.11(c)** Prior to or during any inspection of a workplace, any employee or representative of employees employed in such workplace may notify the Compliance Safety and Health Officer, in writing, of any violation of the Act which they have reason to believe exists in such workplace. Any such notice shall comply with the requirements of paragraph (a) of this section.

**1903.11(d)** Section 11(c)(1) of the Act provides: "No person shall discharge or in any manner discriminate against any employee because such employee has filed any complaint or instituted or caused to be instituted any proceeding under or related to this Act or has testified or is about to testify in any such proceeding or because of the exercise by such employee on behalf of himself or others of any right afforded by this Act."

(Approved by the Office of Management and Budget under control number 1218-0064)

[36 FR 17850, Sept. 4, 1973, as amended at 54 FR 24333, June 7, 1989]

### §1903.12 Inspection not warranted; informal review.

**1903.12(a)** If the Area Director determines that an inspection is not warranted because there are no reasonable grounds to believe that a violation or danger exists with respect to a complaint under §1903.11, he shall notify the complaining party in writing of such determination. The complaining party may obtain review of such determination by submitting a written statement of position with the Assistant Regional Director and, at the same time, providing the employer with a copy of such statement by certified mail. The employer may submit an opposing written statement of position with the Assistant Regional Director and, at the same time, provide the complaining party with a copy of such statement by certified mail. Upon the request of the complaining party or the employer, the Assistant Regional Director at his discretion, may hold an informal conference in which the complaining party and the employer may orally present their views. After considering all written and oral views presented, the Assistant Regional Director shall affirm, modify, or reverse the determination of the Area Director and furnish the complaining party and the employer and written notification of his decision and the reasons therefor. The decision of the Assistant Regional Director shall be final and not subject to further review.

**1903.12(b)** If the Area Director determines that an inspection is not warranted because the requirements of §1903.11(a) have not been met, he shall notify the complaining party in writing of such determination. Such determination shall be without prejudice to the filing of a new complaint meeting the requirements of §1903.11(a).

### §1903.13 Imminent danger.

Whenever and as soon as a Compliance Safety and Health Officer concludes on the basis of an inspection that conditions or practices exist in any place of employment which could reasonably be expected to cause death or serious physical harm immediately or before the imminence of such danger can be eliminated through the enforcement procedures otherwise provided by the Act, he shall inform the affected employees and employers of the danger and that he is recommending a civil action to restrain such conditions or practices and for other appropriate relief in accordance with the provisions of section 13(a) of the Act. Appropriate citations and notices of proposed penalties may be issued with respect to an imminent danger even though, after being informed of such danger by the Compliance Safety and Health Officer, the employer immediately eliminates the imminence of the danger and initiates steps to abate such danger.

### §1903.14 Citations; notices of de minimis violations; policy regarding employee rescue activities.

**1903.14(a)** The Area Director shall review the inspection report of the Compliance Safety and Health Officer. If, on the basis of the report the Area Director believes that the employer has violated a requirement of section 5 of the Act, of any standard, rule or order promulgated pursuant to section 6 of the Act, or of any substantive rule published in this chapter, he shall, if appropriate, consult with the Regional Solicitor, and he shall issue to the employer either a citation or a notice of de minimis violations which have no direct or immediate relationship to safety or health. An appropriate citation or notice of de minimis violations shall be issued even though after being informed of an alleged violation by the Compliance Safety and Health Officer, the employer immediately abates, or initiates steps to abate, such alleged violation. Any citation or notice of de minimis violations shall be issued with reasonable promptness after termination of the inspection. No citation may be issued under this section after the expiration of 6 months following the occurrence of any alleged violation.

**1903.14(b)** Any citation shall describe with particularity the nature of the alleged violation, including a reference to the provision(s) of the Act, standard, rule, regulation, or order alleged to have been violated. Any citation shall also fix a reasonable time or times for the abatement of the alleged violation.

**1903.14(c)** If a citation or notice of de minimis violations is issued for a violation alleged in a request for inspection under §1903.11(a) or a notification of violation under §1903.11(c), a copy of the citation or notice of de minimis violations shall also be sent to the employee or representative of employees who made such request or notification.

**1903.14(d)** After an inspection, if the Area Director determines that a citation is not warranted with respect to a danger or violation alleged to exist in a request for inspection under §1903.11(a) or a notification of violation under §1903.11(c), the informal review procedures prescribed in §1903.12(a) shall be applicable. After considering all views presented, the Assistant Regional Director shall affirm the determination of the Area Director, order a reinspection, or issue a citation if he believes that the inspection disclosed a violation. The Assistant Regional Director shall furnish the complaining party and the employer with written notification of his determination and the reasons therefor. The determination of the Assistant Regional Director shall be final and not subject to review.

**1903.14(e)** Every citation shall state that the issuance of a citation does not constitute a finding that a violation of the Act has occurred unless there is a failure to contest as provided for in the Act or, if contested, unless the citation is affirmed by the Review Commission.

**1903.14(f)** No citation may be issued to an employer because of a rescue activity undertaken by an employee of that employer with respect to an individual in imminent danger unless:

**1903.14(f)(1)(i)** Such employee is designated or assigned by the employer to have responsibility to perform or assist in rescue operations, and

**1903.14(f)(1)(ii)** The employer fails to provide protection of the safety and health of such employee, including failing to provide appropriate training and rescue equipment; or

**1903.14(f)(2)(i)** Such employee is directed by the employer to perform rescue activities in the course of carrying out the employee's job duties, and

**1903.14(f)(2)(ii)** The employer fails to provide protection of the safety and health of such employee, including failing to provide appropriate training and rescue equipment; or

**1903.14(f)(3)(i)** Such employee is employed in a workplace that requires the employee to carry out duties that are directly related to a workplace operation where the likelihood of life-threatening accidents is foreseeable, such as a workplace operation where employees are located in confined spaces or trenches, handle hazardous waste, respond to emergency situations, perform excavations, or perform construction over water; and

**1903.14(f)(3)(ii)** Such employee has not been designated or assigned to perform or assist in rescue operations and voluntarily elects to rescue such an individual; and

**1903.14(f)(3)(iii)** The employer has failed to instruct employees not designated or assigned to perform or assist in rescue operations of the arrangements for rescue, not to attempt rescue, and of the hazards of attempting rescue without adequate training or equipment.

**1903.14(f)(4)** For purposes of this policy, the term "imminent danger" means the existence of any condition or practice that could reasonably be expected to cause death or serious physical harm before such condition or practice can be abated.

[36 FR 17850, Sept. 4, 1971, as amended at 59 FR 66613, Dec. 27, 1994]

### §1903.14a Petitions for modification of abatement date.

**1903.14a(a)** An employer may file a petition for modification of abatement date when he has made a good faith effort to comply with the abatement requirements of a citation, but such abatement has not been completed because of factors beyond his reasonable control.

**1903.14a(b)** A petition for modification of abatement date shall be in writing and shall include the following information:

**1903.14a(b)(1)** All steps taken by the employer, and the dates of such action, in an effort to achieve compliance during the prescribed abatement period.

**1903.14a(b)(2)** The specific additional abatement time necessary in order to achieve compliance.

**1903.14a(b)(3)** The reasons such additional time is necessary, including the unavailability of professional or technical personnel or of materials and equipment, or because necessary construction or alteration of facilities cannot be completed by the original abatement date.

**1903.14a(b)(4)** All available interim steps being taken to safeguard the employees against the cited hazard during the abatement period.

**1903.14a(b)(5)** A certification that a copy of the petition has been posted and, if appropriate, served on the authorized representative of affected employees, in accordance with subsection (c)(1) of this section and a certification of the date upon which such posting and service was made.

**1903.14a(c)** A petition for modification of abatement date shall be filed with the Area Director of the United States Department of Labor who issued the citation no later than the close of the next working day following the date on which abatement was originally required. A later-filed petition shall be accompanied by the employer's statement of exceptional circumstances explaining the delay.

**1903.14a(c)(1)** A copy of such petition shall be posted in a conspicuous place where all affected employees will have notice thereof or near such location where the violation occurred. The petition shall remain posted for a period of ten (10) working days. Where affected employees are represented by an authorized representative, said representative shall be served with a copy of such petition.

**1903.14a(c)(2)** Affected employees or their representatives may file an objection in writing to such petition with the aforesaid Area Director. Failure to file such objection within ten (10) working days of the date of posting of such petition or of service upon an authorized representative shall constitute a waiver of any further right to object to said petition.

**1903.14a(c)(3)** The Secretary or his duly authorized agent shall have the authority to approve any petition for modification of abatement date filed pursuant to paragraphs (b) and (c) of this section. Such uncontested petitions shall become final orders pursuant to sections 10(a) and (c) of the Act.

**1903.14a(c)(4)** The Secretary or his authorized representative shall not exercise his approval power until the expiration of fifteen (15) working days from the date the petition was posted or served pursuant to paragraphs (c)(1) and (2) of this section by the employer.

**1903.14a(d)** Where any petition is objected to by the Secretary or affected employees, the petition, citation, and any objections shall be forwarded to the Commission within three (3) working days after the expiration of the fifteen (15) day period set out in paragraph (c)(4) of this section.

[40 FR 6334, Feb. 11, 1975; 40 FR 11351, Mar. 11, 1975]

### §1903.15 Proposed penalties.

**1903.15(a)** After, or concurrent with, the issuance of a citation, and within a reasonable time after the termination of the inspection, the Area Director shall notify the employer by certified mail or by personal service by the Compliance Safety and Health Officer of the proposed penalty under section 17 of the Act, or that no penalty is being proposed. Any notice of proposed penalty shall state that the proposed penalty shall be deemed to be the final order of the Review Commission and not subject to review by any court or agency unless, within 15 working days from the date of receipt of such notice, the employer notifies the Area Director in writing that he intends to contest the citation or the notification of proposed penalty before the Review Commission.

**1903.15(b)** The Area Director shall determine the amount of any proposed penalty, giving due consideration to the appropriateness of the penalty with respect to the size of the business of the employer being charged, the gravity of the violation, the good faith of the employer, and the history of previous violations, in accordance with the provisions of section 17 of the Act.

**1903.15(c)** Appropriate penalties may be proposed with respect to an alleged violation even though after being informed of such alleged violation by the Compliance Safety and Health Officer, the employer immediately abates, or initiates steps to abate, such alleged violation. Penalties shall not be proposed for de minimis violations which have no direct or immediate relationship to safety or health.

# OSHA Standards for General Industry

### §1903.16 Posting of citations.

**1903.16(a)** Upon receipt of any citation under the Act, the employer shall immediately post such citation, or a copy thereof, unedited, at or near each place an alleged violation referred to in the citation occurred, except as provided below. Where, because of the nature of the employer's operations, it is not practicable to post the citation at or near each place of alleged violation, such citation shall be posted, unedited, in a prominent place where it will be readily observable by all affected employees. For example, where employers are engaged in activities which are physically dispersed (see §1903.2(b)), the citation may be posted at the location to which employees report each day. Where employees do not primarily work at or report to a single location (see §1903.2(b)), the citation may be posted at the location from which the employees operate to carry out their activities. The employer shall take steps to ensure that the citation is not altered, defaced, or covered by other material. Notices of de minimis violations need not be posted.

**1903.16(b)** Each citation, or a copy thereof, shall remain posted until the violation has been abated, or for 3 working days, whichever is later. The filing by the employer of a notice of intention to contest under §1903.17 shall not affect his posting responsibility under this section unless and until the Review Commission issues a final order vacating the citation.

**1903.16(c)** An employer to whom a citation has been issued may post a notice in the same location where such citation is posted indicating that the citation is being contested before the Review Commission, and such notice may explain the reasons for such contest. The employer may also indicate that specified steps have been taken to abate the violation.

**1903.16(d)** Any employer failing to comply with the provisions of paragraphs (a) and (b) of this section shall be subject to citation and penalty in accordance with the provisions of section 17 of the Act.

### §1903.17 Employer and employee contests before the Review Commission.

**1903.17(a)** Any employer to whom a citation or notice of proposed penalty has been issued may, under section 10(a) of the Act, notify the Area Director in writing that he intends to contest such citation or proposed penalty before the Review Commission. Such notice of intention to contest shall be postmarked within 15 working days of the receipt by the employer of the notice of proposed penalty. Every notice of intention to contest shall specify whether it is directed to the citation or to the proposed penalty, or both. The Area Director shall immediately transmit such notice to the Review Commission in accordance with the rules of procedure prescribed by the Commission.

**1903.17(b)** Any employee or representative of employees of an employer to whom a citation has been issued may, under section 10(c) of the Act, file a written notice with the Area Director alleging that the period of time fixed in the citation for the abatement of the violation is unreasonable. Such notice shall be postmarked within 15 working days of the receipt by the employer of the notice of proposed penalty or notice that no penalty is proposed. The Area Director shall immediately transmit such notice to the Review Commission in accordance with the rules of procedure prescribed by the Commission.

### §1903.18 Failure to correct a violation for which a citation has been issued.

**1903.18(a)** If an inspection discloses that an employer has failed to correct an alleged violation for which a citation has been issued within the period permitted for its correction, the Area Director shall, if appropriate, consult with the Regional Solicitor, and he shall notify the employer by certified mail or by personal service by the Compliance Safety and Health Officer of such failure and of the additional penalty proposed under section 17(d) of the Act by reason of such failure. The period for the correction of a violation for which a citation has been issued shall not begin to run until the entry of a final order of the Review Commission in the case of any review proceedings initiated by the employer in good faith and not solely for delay or avoidance of penalties.

**1903.18(b)** Any employer receiving a notification of failure to correct a violation and of proposed additional penalty may, under section 10(b) of the Act, notify the Area Director in writing that he intends to contest such notification or proposed additional penalty before the Review Commission. Such notice of intention to contest shall be postmarked within 15 working days of the receipt by the employer of the notification of failure to correct a violation and of proposed additional penalty. The Area Director shall immediately transmit such notice to the Review Commission in accordance with the rules of procedure prescribed by the Commission.

**1903.18(c)** Each notification of failure to correct a violation and of proposed additional penalty shall state that it shall be deemed to be the final order of the Review Commission and not subject to review by any court or agency unless, within 15 working days from the date of receipt of such notification, the employer notifies the Area Director in writing that he intends to contest the notification or the proposed additional penalty before the Review Commission.

### §1903.19 Abatement verification.

***Purpose.*** OSHA's inspections are intended to result in the abatement of violations of the Occupational Safety and Health Act of 1970 (the OSH Act). This section sets forth the procedures OSHA will use to ensure abatement. These procedures are tailored to the nature of the violation and the employer's abatement actions.

**1903.19(a)** ***Scope and application.*** This section applies to employers who receive a citation for a violation of the Occupational Safety and Health Act.

**1903.19(b)** ***Definitions.*** **(1)** *Abatement* means action by an employer to comply with a cited standard or regulation or to eliminate a recognized hazard identified by OSHA during an inspection.

**1903.19(b)(2)** *Abatement date* means:

**1903.19(b)(2)(i)** For an uncontested citation item, the later of:

**1903.19(b)(2)(i)(A)** The date in the citation for abatement of the violation;

**1903.19(b)(2)(i)(B)** The date approved by OSHA or established in litigation as a result of a petition for modification of the abatement date (PMA); or

**1903.19(b)(2)(i)(C)** The date established in a citation by an informal settlement agreement.

**1903.19(b)(2)(ii)** For a contested citation item for which the Occupational Safety and Health Review Commission (OSHRC) has issued a final order affirming the violation, the later of:

**1903.19(b)(2)(ii)(A)** The date identified in the final order for abatement; or

**1903.19(b)(2)(ii)(B)** The date computed by adding the period allowed in the citation for abatement to the final order date;

**1903.19(b)(2)(ii)(C)** The date established by a formal settlement agreement.

**1903.19(b)(3)** *Affected employees* means those employees who are exposed to the hazard(s) identified as violation(s) in a citation.

**1903.19(b)(4)** *Final order date* means:

**1903.19(b)(4)(i)** For an uncontested citation item, the fifteenth working day after the employer's receipt of the citation;

**1903.19(b)(4)(ii)** For a contested citation item:

**1903.19(b)(4)(ii)(A)** The thirtieth day after the date on which a decision or order of a commission administrative law judge has been docketed with the commission, unless a member of the commission has directed review; or

**1903.19(b)(4)(ii)(B)** Where review has been directed, the thirtieth day after the date on which the Commission issues its decision or order disposing of all or pertinent part of a case; or

**1903.19(b)(4)(ii)(C)** The date on which a federal appeals court issues a decision affirming the violation in a case in which a final order of OSHRC has been stayed.

**1903.19(b)(5)** *Movable equipment* means a hand-held or non-hand-held machine or device, powered or unpowered, that is used to do work and is moved within or between worksites.

**1903.19(c)** ***Abatement certification.*** **(1)** Within 10 calendar days after the abatement date, the employer must certify to OSHA (the Agency) that each cited violation has been abated, except as provided in paragraph (c)(2) of this section.

**1903.19(c)(2)** The employer is not required to certify abatement if the OSHA Compliance Officer, during the on-site portion of the inspection:

**1903.19(c)(2)(i)** Observes, within 24 hours after a violation is identified, that abatement has occurred; and

**1903.19(c)(2)(ii)** Notes in the citation that abatement has occurred.

**1903.19(c)(3)** The employer's certification that abatement is complete must include, for each cited violation, in addition to the information required by paragraph (h) of this section, the date and method of abatement and a statement that affected employees and their representatives have been informed of the abatement.

**NOTE TO PARAGRAPH (c):** Appendix A contains a sample Abatement Certification Letter.

**1903.19(d)** *Abatement documentation.* **(1)** The employer must submit to the Agency, along with the information on abatement certification required by paragraph (c)(3) of this section, documents demonstrating that abatement is complete for each willful or repeat violation and for any serious violation for which the Agency indicates in the citation that such abatement documentation is required.

**1903.19(d)(2)** Documents demonstrating that abatement is complete may include, but are not limited to, evidence of the purchase or repair of equipment, photographic or video evidence of abatement, or other written records.

**1903.19(e)** *Abatement plans.* **(1)** The Agency may require an employer to submit an abatement plan for each cited violation (except an other-than-serious violation) when the time permitted for abatement is more than 90 calendar days. If an abatement plan is required, the citation must so indicate.

**1903.19(e)(2)** The employer must submit an abatement plan for each cited violation within 25 calendar days from the final order date when the citation indicates that such a plan is required. The abatement plan must identify the violation and the steps to be taken to achieve abatement, including a schedule for completing abatement and, where necessary, how employees will be protected from exposure to the violative condition in the interim until abatement is complete.

**NOTE TO PARAGRAPH (e):** Appendix B contains a Sample Abatement Plan form.

**1903.19(f)** *Progress reports.* **(1)** An employer who is required to submit an abatement plan may also be required to submit periodic progress reports for each cited violation. The citation must indicate:

**1903.19(f)(1)(i)** That periodic progress reports are required and the citation items for which they are required;

**1903.19(f)(1)(ii)** The date on which an initial progress report must be submitted, which may be no sooner than 30 calendar days after submission of an abatement plan;

**1903.19(f)(1)(iii)** Whether additional progress reports are required; and

**1903.19(f)(1)(iv)** The date(s) on which additional progress reports must be submitted.

**1903.19(f)(2)** For each violation, the progress report must identify, in a single sentence if possible, the action taken to achieve abatement and the date the action was taken.

**NOTE TO PARAGRAPH (f):** Appendix B contains a Sample Progress Report Form.

**1903.19(g)** *Employee notification.* **(1)** The employer must inform affected employees and their representative(s) about abatement activities covered by this section by posting a copy of each document submitted to the Agency or a summary of the document near the place where the violation occurred.

**1903.19(g)(2)** Where such posting does not effectively inform employees and their representatives about abatement activities (for example, for employers who have mobile work operations), the employer must:

**1903.19(g)(2)(i)** Post each document or a summary of the document in a location where it will be readily observable by affected employees and their representatives; or

**1903.19(g)(2)(ii)** Take other steps to communicate fully to affected employees and their representatives about abatement activities.

**1903.19(g)(3)** The employer must inform employees and their representatives of their right to examine and copy all abatement documents submitted to the Agency.

**1903.19(g)(3)(i)** An employee or an employee representative must submit a request to examine and copy abatement documents within 3 working days of receiving notice that the documents have been submitted.

**1903.19(g)(3)(ii)** The employer must comply with an employee's or employee representative's request to examine and copy abatement documents within 5 working days of receiving the request.

**1903.19(g)(4)** The employer must ensure that notice to employees and employee representatives is provided at the same time or before the information is provided to the Agency and that abatement documents are:

**1903.19(g)(4)(i)** Not altered, defaced, or covered by other material; and

**1903.19(g)(4)(ii)** Remain posted for three working days after submission to the Agency.

**1903.19(h)** *Transmitting abatement documents.* **(1)** The employer must include, in each submission required by this section, the following information:

**1903.19(h)(1)(i)** The employer's name and address;

**1903.19(h)(1)(ii)** The inspection number to which the submission relates;

**1903.19(h)(1)(iii)** The citation and item numbers to which the submission relates;

**1903.19(h)(1)(iv)** A statement that the information submitted is accurate; and

**1903.19(h)(1)(v)** The signature of the employer or the employer's authorized representative.

**1903.19(h)(2)** The date of postmark is the date of submission for mailed documents. For documents transmitted by other means, the date the Agency receives the document is the date of submission.

**1903.19(i)** *Movable equipment.* **(1)** For serious, repeat, and willful violations involving movable equipment, the employer must attach a warning tag or a copy of the citation to the operating controls or to the cited component of equipment that is moved within the worksite or between worksites.

**NOTE TO PARAGRAPH (i)(1):** Attaching a copy of the citation to the equipment is deemed by OSHA to meet the tagging requirement of paragraph (i)(1) of this section as well as the posting requirement of 29 CFR 1903.16.

**1903.19(i)(2)** The employer must use a warning tag that properly warns employees about the nature of the violation involving the equipment and identifies the location of the citation issued.

**NOTE TO PARAGRAPH (i)(2):** Non-Mandatory Appendix C contains a sample tag that employers may use to meet this requirement.

**1903.19(i)(3)** If the violation has not already been abated, a warning tag or copy of the citation must be attached to the equipment:

**1903.19(i)(3)(i)** For hand-held equipment, immediately after the employer receives the citation; or

**1903.19(i)(3)(ii)** For non-hand-held equipment, prior to moving the equipment within or between worksites.

**1903.19(i)(4)** For the construction industry, a tag that is designed and used in accordance with 29 CFR 1926.20(b)(3) and 29 CFR 1926.200(h) is deemed by OSHA to meet the requirements of this section when the information required by paragraph (i)(2) is included on the tag.

**1903.19(i)(5)** The employer must assure that the tag or copy of the citation attached to movable equipment is not altered, defaced, or covered by other material.

**1903.19(i)(6)** The employer must assure that the tag or copy of the citation attached to movable equipment remains attached until:

**1903.19(i)(6)(i)** The violation has been abated and all abatement verification documents required by this regulation have been submitted to the Agency;

**1903.19(i)(6)(ii)** The cited equipment has been permanently removed from service or is no longer within the employer's control; or

**1903.19(i)(6)(iii)** The Commission issues a final order vacating the citation.

## APPENDICES TO §1903.19— ABATEMENT VERIFICATION

**NOTE:** Appendices A through C provide information and nonmandatory guidelines to assist employers and employees in complying with the appropriate requirements of this section.

## APPENDIX A TO SECTION 1903.19—SAMPLE ABATEMENT-CERTIFICATION LETTER (NONMANDATORY)

(Name), Area Director

U.S. Department of Labor—OSHA

Address of the Area Office (on the citation)

[Company's Name]

[Company's Address]

The hazard referenced in Inspection Number [insert 9-digit #] for violation identified as:

Citation [insert #] and item [insert #] was corrected on [insert date] by:
_____

Citation [insert #] and item [insert #] was corrected on [insert date] by:
_____

Citation [insert #] and item [insert #] was corrected on [insert date] by:
_____

Citation [insert #] and item [insert #] was corrected on [insert date] by:
_____

Citation [insert #] and item [insert #] was corrected on [insert date] by:
_____

Citation [insert #] and item [insert #] was corrected on [insert date] by:
_____

Citation [insert #] and item [insert #] was corrected on insert date by:
_____

Citation [insert #] and item [insert #] was corrected on [insert date] by:
_____

I attest that the information contained in this document is accurate.

Signature

Typed or Printed Name

## APPENDIX B TO SECTION 1903.19—SAMPLE ABATEMENT PLAN OR PROGRESS REPORT (NONMANDATORY)

(Name), Area Director

U. S. Department of Labor—OSHA

Address of Area Office (on the citation)

[Company's Name]

[Company's Address]

Check one:

Abatement Plan [ ]

Progress Report [ ]

Inspection Number _____

Page _____ of _____

Citation Number(s)* _____

Item Number(s)* _____

| Action | Proposed Completion Date (for abatement plans only) | Completion Date (for progress reports only) |
|---|---|---|
| 1. | ........................ | ........................ |
| 2. | ........................ | ........................ |
| 3. | ........................ | ........................ |
| 4. | ........................ | ........................ |
| 5. | ........................ | ........................ |
| 6. | ........................ | ........................ |
| 7. | ........................ | ........................ |

Date required for final abatement: _____

I attest that the information contained in this document is accurate.

Signature

Typed or Printed Name

Name of primary point of contact for questions: [optional]

Telephone number: _____

*Abatement plans or progress reports for more than one citation item may be combined in a single abatement plan or progress report if the abatement actions, proposed completion dates, and actual completion dates (for progress reports only) are the same for each of the citation items.

## APPENDIX C TO SECTION 1903.19—Sample Warning Tag (Nonmandatory)

**WARNING:**

EQUIPMENT HAZARD

CITED BY OSHA

EQUIPMENT CITED:
_____
_____
_____

HAZARD CITED:
_____
_____
_____

FOR DETAILED INFORMATION

SEE OSHA CITATION POSTED AT:
_____

BACKGROUND COLOR—ORANGE
MESSAGE COLOR—BLACK

[62 FR 15337, Mar. 31, 1997]

### §1903.20 Informal conferences.

At the request of an affected employer, employee, or representative of employees, the Assistant Regional Director may hold an informal conference for the purpose of discussing any issues raised by an inspection, citation, notice of proposed penalty, or notice of intention to contest. The settlement of any issue at such conference shall be subject to the rules of procedure prescribed by the Review Commission. If the conference is requested by the employer, an affected employee or his representative shall be afforded an opportunity to participate, at the discretion of the Regional Administrator. If the conference is requested by an employee or representative of employees, the employer shall be afforded an opportunity to participate, at the discretion of the Regional Administrator. Any party may be represented by counsel at such conference. No such conference or request for such conference shall operate as a stay of any 15-working-day period for filing a notice of intention to contest as prescribed in §1903.17.

[36 FR 17850, Sept. 4, 1971. Redesignated at 62 FR 15337, Mar. 31, 1997]

### §1903.21 State administration.

Nothing in this Part 1903 shall preempt the authority of any State to conduct inspections, to initiate enforcement proceedings or otherwise to implement the applicable provisions of State law with respect to State occupational safety and health standards in accordance with agreements and plans under section 18 of the Act and Parts 1901 and 1902 of this chapter.

[36 FR 17850, Sept. 4, 1971. Redesignated at 62 FR 15337, Mar. 31, 1997]

### §1903.22 Definitions.

1903.22(a) *Act* means the Williams-Steiger Occupational Safety and Health Act of 1970. (84 Stat. 1590 *et seq.*, 29 U.S.C. 651 *et seq.*)

**1903.22(b)** The definitions and interpretations contained in section 3 of the Act shall be applicable to such terms when used in this Part 1903.

**1903.22(c)** *Working days* means Mondays through Fridays but shall not include Saturdays, Sundays, or Federal holidays. In computing 15 working days, the day of receipt of any notice shall not be included, and the last day of the 15 working days shall be included.

**1903.22(d)** *Compliance Safety and Health Officer* means a person authorized by the Occupational Safety and Health Administration, U.S. Department of Labor, to conduct inspections.

**1903.22(e)** *Area Director* means the employee or officer regularly or temporarily in charge of an Area Office of the Occupational Safety and Health Administration, U.S. Department of Labor, or any other person or persons who are authorized to act for such employee or officer. The latter authorizations may include general delegations of the authority of an Area Director under this part to a Compliance Safety and Health Officer or delegations to such an officer for more limited purposes, such as the exercise of the Area Director's duties under §1903.14(a). The term also includes any employee or officer exercising supervisory responsibilities over an Area Director. A supervisory employee or officer is considered to exercise concurrent authority with the Area Director.

**1903.22(f)** *Assistant Regional Director* means the employee or officer regularly or temporarily in charge of a Region of the Occupational Safety and Health Administration, U.S. Department of Labor, or any other person or persons who are specifically designated to act for such employee or officer in his absence. The term also includes any employee or officer in the Occupational Safety and Health Administration exercising supervisory responsibilities over the Assistant Regional Director. Such supervisory employee or officer is considered to exercise concurrent authority with the Assistant Regional Director. No delegation of authority under this paragraph shall adversely affect the procedures for independent informal review of investigative determinations prescribed under §1903.12 of this part.

**1903.22(g)** *Inspection* means any inspection of an employer's factory, plant, establishment, construction site, or other area, workplace or environment where work is performed by an employee of an employer, and includes any inspection conducted pursuant to a complaint filed under §1903.11(a) and (c), any reinspection, followup inspection, accident investigation or other inspection conducted under section 8(a) of the Act.

[36 FR 17850, Sept. 4, 1971, as amended at 38 FR 22624, Aug. 23, 1973. Redesignated at 62 FR 15337, Mar. 31, 1997]

# 29 CFR Part 1904—Recording and Reporting Occupational Injuries and Illnesses

**Authority:** : 29 U.S.C. 657, 658, 660, 666, 669, 673, Secretary of Labor's Order No. 3-2000 (65 FR 50017), and 5 U.S.C. 533.

## Subpart A—Purpose

### §1904.0 Purpose.

The purpose of this rule (Part 1904) is to require employers to record and report work-related fatalities, injuries and illnesses.

**Note to §1904.0:** Recording or reporting a work-related injury, illness, or fatality does not mean that the employer or employee was at fault, that an OSHA rule has been violated, or that the employee is eligible for workers' compensation or other benefits.

[66 FR 5915, Jan. 19, 2001]

## Subpart B—Scope

**Note to Subpart B:** All employers covered by the Occupational Safety and Health Act (OSH Act) are covered by these Part 1904 regulations. However, most employers do not have to keep OSHA injury and illness records unless OSHA or the Bureau of Labor Statistics (BLS) informs them in writing that they must keep records. For example, employers with 10 or fewer employees and business establishments in certain industry classifications are partially exempt from keeping OSHA injury and illness records.

### §1904.1 Partial exemption for employers with 10 or fewer employees.

**1904.1(a)** *Basic requirement.* **(1)** If your company had ten (10) or fewer employees at all times during the last calendar year, you do not need to keep OSHA injury and illness records unless OSHA or the BLS informs you in writing that you must keep records under §1904.41 or §1904.42. However, as required by §1904.39, all employers covered by the OSH Act must report to OSHA any workplace incident that results in a fatality or the hospitalization of three or more employees.

**1904.1(a)(2)** If your company had more than ten (10) employees at any time during the last calendar year, you must keep OSHA injury and illness records unless your establishment is classified as a partially exempt industry under §1904.2.

**1904.1(b)** *Implementation.* **(1)** *Is the partial exemption for size based on the size of my entire company or on the size of an individual business establishment?* The partial exemption for size is based on the number of employees in the entire company.

**1904.1(b)(2)** *How do I determine the size of my company to find out if I qualify for the partial exemption for size?* To determine if you are exempt because of size, you need to determine your company's peak employment during the last calendar year. If you had no more than 10 employees at any time in the last calendar year, your company qualifies for the partial exemption for size.

[66 FR 5915, Jan. 19, 2001]

### §1904.2 Partial exemption for establishments in certain industries.

**1904.2(a)** *Basic requirement.* **(1)** If your business establishment is classified in a specific low hazard retail, service, finance, insurance or real estate industry listed in Appendix A to this Subpart B, you do not need to keep OSHA injury and illness records unless the government asks you to keep the records under §1904.41 or §1904.42. However, all employers must report to OSHA any workplace incident that results in a fatality or the hospitalization of three or more employees (see §1904.39).

**1904.2(a)(2)** If one or more of your company's establishments are classified in a non-exempt industry, you must keep OSHA injury and illness records for all of such establishments unless your company is partially exempted because of size under §1904.1.

**1904.2(b)** *Implementation.* **(1)** *Does the partial industry classification exemption apply only to business establishments in the retail, services, finance, insurance or real estate industries (SICs 52-89)?* Yes, business establishments classified in agriculture; mining; construction; manufacturing; transportation; communication, electric, gas and sanitary services; or wholesale trade are not eligible for the partial industry classification exemption.

**1904.2(b)(2)** *Is the partial industry classification exemption based on the industry classification of my entire company or on the classification of individual business establishments operated by my company?* The partial industry classification exemption applies to individual business establishments. If a company has several business establishments engaged in different classes of business activities, some of the company's establishments may be required to keep records, while others may be exempt.

**1904.2(b)(3)** *How do I determine the Standard Industrial Classification code for my company or for individual establishments?* You determine your Standard Industrial Classification (SIC) code by using the Standard Industrial Classification Manual, Executive Office of the President, Office of Management and Budget. You may contact your nearest OSHA office or State agency for help in determining your SIC.

[66 FR 5915, Jan. 19, 2001]

### §1904.3 Keeping records for more than one agency.

If you create records to comply with another government agency's injury and illness recordkeeping requirements, OSHA will consider those records as meeting OSHA's Part 1904 recordkeeping requirements if OSHA accepts the other agency's records under a memorandum of understanding with that agency, or if the other agency's records contain the same information as this Part 1904 requires you to record. You may contact your nearest OSHA office or State agency for help in determining whether your records meet OSHA's requirements.

[66 FR 5915, Jan. 19, 2001]

### Non-Mandatory Appendix A to Subpart B— Partially Exempt Industries

Employers are not required to keep OSHA injury and illness records for any establishment classified in the following Standard Industrial Classification (SIC) codes, unless they are asked in writing to do so by OSHA, the Bureau of Labor Statistics ( BLS), or a state agency operating under the authority of OSHA or the BLS. All employers, including those partially exempted by reason of company size or industry classification, must report to OSHA any workplace incident that results in a fatality or the hospitalization of three or more employees (see §1904.39).

| SIC code | Industry description |
| --- | --- |
| 525 | Hardware Stores |
| 542 | Meat and Fish Markets |
| 544 | Candy, Nut, and Confectionery Stores |
| 545 | Dairy Products Stores |
| 546 | Retail Bakeries |
| 549 | Miscellaneous Food Stores |
| 551 | New and Used Car Dealers |
| 552 | Used Car Dealers |
| 554 | Gasoline Service Stations |
| 557 | Motorcycle Dealers |
| 56 | Apparel and Accessory Stores |
| 573 | Radio, Television, & Computer Stores |
| 58 | Eating and Drinking Places |
| 591 | Drug Stores and Proprietary Stores |
| 592 | Liquor Stores |
| 594 | Miscellaneous Shopping Goods Stores |
| 599 | Retail Stores, Not Elsewhere Classified |
| 60 | Depository Institutions (banks & savings institutions) |
| 61 | Nondepository |
| 62 | Security and Commodity Brokers |
| 63 | Insurance Carriers |
| 64 | Insurance Agents, Brokers & Services |
| 653 | Real Estate Agents and Managers |
| 654 | Title Abstract Offices |
| 67 | Holding and Other Investment Offices |
| 722 | Photographic Studios, Portrait |
| 723 | Beauty Shops |
| 724 | Barber Shops |
| 725 | Shoe Repair and Shoeshine Parlors |

| SIC code | Industry description |
|---|---|
| 726......... | Funeral Service and Crematories. |
| 729......... | Miscellaneous Personal Services. |
| 731......... | Advertising Services. |
| 732......... | Credit Reporting and Collection Services. |
| 733......... | Mailing, Reproduction, & Stenographic Services. |
| 737......... | Computer and Data Processing Services. |
| 738......... | Miscellaneous Business Services. |
| 764......... | Reupholstery and Furniture Repair. |
| 78........... | Motion Picture. |
| 791......... | Dance Studios, Schools, and Halls. |
| 792......... | Producers, Orchestras, Entertainers. |
| 793......... | Bowling Centers. |
| 801......... | Offices & Clinics Of Medical Doctors. |
| 802......... | Offices and Clinics Of Dentists. |
| 803......... | Offices Of Osteopathic. |
| 804......... | Offices Of Other Health Practitioners. |
| 807......... | Medical and Dental Laboratories. |
| 809......... | Health and Allied Services, Not Elsewhere Classified. |
| 81........... | Legal Services. |
| 82........... | Educational Services (schools, colleges, universities and libraries). |
| 832......... | Individual and Family Services. |
| 835......... | Child Day Care Services. |
| 839......... | Social Services, Not Elsewhere Classified. |
| 841......... | Museums and Art Galleries. |
| 86........... | Membership Organizations. |
| 87........... | Engineering, Accounting, Research, Management, and Related Services. |
| 899......... | Services, not elsewhere classified. |

### Subpart C—Recordkeeping Forms and Recording Criteria

**Note to Subpart C:** This Subpart describes the work-related injuries and illnesses that an employer must enter into the OSHA records and explains the OSHA forms that employers must use to record work-related fatalities, injuries, and illnesses.

### §1904.4 Recording criteria.

**1904.4(a)** *Basic requirement.* Each employer required by this Part to keep records of fatalities, injuries, and illnesses must record each fatality, injury and illness that:

**1904.4(a)(1)** Is work-related; and

**1904.4(a)(2)** Is a new case; and

**1904.4(a)(3)** Meets one or more of the general recording criteria of §1904.7 or the application to specific cases of §1904.8 through §1904.12.

**1904.4(b)** *Implementation.* **(1)** *What sections of this rule describe recording criteria for recording work-related injuries and illnesses?* The table below indicates which sections of the rule address each topic.

**1904.4(b)(1)(i)** Determination of work-relatedness. See §1904.5.

**1904.4(b)(1)(ii)** Determination of a new case. See §1904.6.

**1904.4(b)(1)(iii)** General recording criteria. See §1904.7.

**1904.4(b)(1)(iv)** Additional criteria. (Needlestick and sharps injury cases, tuberculosis cases, hearing loss cases, medical removal cases, and musculoskeletal disorder cases). See §1904.8 through §1904.12.

**1904.4(b)(2)** *How do I decide whether a particular injury or illness is recordable?* The decision tree for recording work-related injuries and illnesses below shows the steps involved in making this determination.

[66 FR 5915, Jan. 19, 2001]

### §1904.5 Determination of work-relatedness.

**1904.5(a)** *Basic requirement.* You must consider an injury or illness to be work-related if an event or exposure in the work environment either caused or contributed to the resulting condition or significantly aggravated a pre-existing injury or illness. Work-relatedness is presumed for injuries and illnesses resulting from events or exposures occurring in the work environment, unless an exception in §1904.5(b)(2) specifically applies.

**1904.5(b)** *Implementation.* **(1)** *What is the "work environment"?* OSHA defines the work environment as "the establishment and other locations where one or more employees are working or are present as a condition of their employment. The work environment includes not only physical locations, but also the equipment or materials used by the employee during the course of his or her work."

**1904.5(b)(2)** *Are there situations where an injury or illness occurs in the work environment and is not considered work-related?* Yes, an injury or illness occurring in the work environment that falls under one of the following exceptions is not work-related, and therefore is not recordable.

| 1904.5(b)(2) | You are not required to record injuries and illnesses if . . . |
|---|---|
| (i) | At the time of the injury or illness, the employee was present in the work environment as a member of the general public rather than as an employee. |
| (ii) | The injury or illness involves signs or symptoms that surface at work but result solely from a non-work-related event or exposure that occurs outside the work environment. |
| (iii) | The injury or illness results solely from voluntary participation in a wellness program or in a medical, fitness, or recreational activity such as blood donation, physical examination, flu shot, exercise class, racquetball, or baseball. |
| (iv) | The injury or illness is solely the result of an employee eating, drinking, or preparing food or drink for personal consumption (whether bought on the employer's premises or brought in). For example, if the employee is injured by choking on a sandwich while in the employer's establishment, the case would not be considered work-related. **NOTE:** If the employee is made ill by ingesting food contaminated by workplace contaminants (such as lead), or gets food poisoning from food supplied by the employer, the case would be considered work-related. |
| (v) | The injury or illness is solely the result of an employee doing personal tasks (unrelated to their employment) at the establishment outside of the employee's assigned working hours. |
| (vi) | The injury or illness is solely the result of personal grooming, self medication for a non work-related condition, or is intentionally self inflicted. |
| (vii) | The injury or illness is caused by a motor vehicle accident and occurs on a company parking lot or company access road while the employee is commuting to or from work. |
| (viii) | The illness is the common cold or flu (**NOTE**: contagious diseases such as tuberculosis, brucellosis, hepatitis A, or plague are considered work-related if the employee is infected at work). |
| (ix) | The illness is a mental illness. Mental illness will not be considered work-related unless the employee voluntarily provides the employer with an opinion from a physician or other licensed health care professional with appropriate training and experience (psychiatrist, psychologist, psychiatric nurse practitioner, etc.) stating that the employee has a mental illness that is work-related. |

**1904.5(b)(3)** *How do I handle a case if it is not obvious whether the precipitating event or exposure occurred in the work environment or occurred away from work?* In these situations, you must evaluate the employee's work duties and environment to decide whether or not one or more events or exposures in the work environment either caused or contributed to the resulting condition or significantly aggravated a pre-existing condition.

**1904.5(b)(4)** *How do I know if an event or exposure in the work environment "significantly aggravated" a preexisting injury or illness?* A preexisting injury or illness has been significantly aggravated, for purposes of OSHA injury and illness recordkeeping, when an event or exposure in the work environment results in any of the following:

**1904.5(b)(4)(i)** Death, provided that the preexisting injury or illness would likely not have resulted in death but for the occupational event or exposure.

**1904.5(b)(4)(ii)** Loss of consciousness, provided that the preexisting injury or illness would likely not have resulted in loss of consciousness but for the occupational event or exposure.

**1904.5(b)(4)(iii)** One or more days away from work, or days of restricted work, or days of job transfer that otherwise would not have occurred but for the occupational event or exposure.

**1904.5(b)(4)(iv)** Medical treatment in a case where no medical treatment was needed for the injury or illness before the workplace event or exposure, or a change in medical treatment was necessitated by the workplace event or exposure.

**1904.5(b)(5)** *Which injuries and illnesses are considered pre-existing conditions?* An injury or illness is a preexisting condition if it resulted solely from a non-work-related event or exposure that occurred outside the work environment.

**1904.5(b)(6)** *How do I decide whether an injury or illness is work-related if the employee is on travel status at the time the injury or illness occurs?* Injuries and illnesses that occur while an employee is on travel status are work-related if, at the time of the injury or illness, the employee was engaged in work activities "in the interest of the employer." Examples of such activities include travel to and from customer contacts, conducting job tasks, and entertaining or being entertained to transact, discuss, or promote business (work-related entertainment includes only entertainment activities being engaged in at the direction of the employer).

Injuries or illnesses that occur when the employee is on travel status do not have to be recorded if they meet one of the exceptions listed below.

| 1904.5(b)(6) | If the employee has . . . | You may use the following to determine if an injury or illness is work related |
|---|---|---|
| (i) | checked into a hotel or motel for one or more days. | When a traveling employee checks into a hotel, motel, or into a other temporary residence, he or she establishes a "home away from home." You must evaluate the employee's activities after he or she checks into the hotel, motel, or other temporary residence for their work-relatedness in the same manner as you evaluate the activities of a non traveling employee. When the employee checks into the temporary residence, he or she is considered to have left the work environment. When the employee begins work each day, he or she re-enters the work environment. If the employee has established a "home away from home" and is reporting to a fixed worksite each day, you also do not consider injuries or illnesses work-related if they occur while the employee is commuting between the temporary residence and the job location. |
| (ii) | taken a detour for personal reasons. | Injuries or illnesses are not considered work-related if they occur while the employee is on a personal detour from a reasonably direct route of travel (e.g., has taken a side trip for personal reasons). |

**1904.5(b)(7)** *How do I decide if a case is work-related when the employee is working at home?* Injuries and illnesses that occur while an employee is working at home, including work in a home office, will be considered work-related if the injury or illness occurs while the employee is performing work for pay or compensation in the home, and the injury or illness is directly related to the performance of work rather than to the general home environment or setting. For example, if an employee drops a box of work documents and injures his or her foot, the case is considered work-related. If an employee's fingernail is punctured by a needle from a sewing machine used to perform garment work at home, becomes infected and requires medical treatment, the injury is considered work-related. If an employee is injured because he or she trips on the family dog while rushing to answer a work phone call, the case is not considered work-related. If an employee working at home is electrocuted because of faulty home wiring, the injury is not considered work-related.

[66 FR 5915, Jan. 19, 2001]

## §1904.6 Determination of new cases.

**1904.6(a)** *Basic requirement.* You must consider an injury or illness to be a "new case" if:

**1904.6(a)(1)** The employee has not previously experienced a recorded injury or illness of the same type that affects the same part of the body, or

**1904.6(a)(2)** The employee previously experienced a recorded injury or illness of the same type that affected the same part of the body but had recovered completely (all signs and symptoms had disappeared) from the previous injury or illness and an event or exposure in the work environment caused the signs or symptoms to reappear.

**1904.6(b)** *Implementation.* **(1)** *When an employee experiences the signs or symptoms of a chronic work-related illness, do I need to consider each recurrence of signs or symptoms to be a new case?* No, for occupational illnesses where the signs or symptoms may recur or continue in the absence of an exposure in the workplace, the case must only be recorded once. Examples may include occupational cancer, asbestosis, byssinosis and silicosis.

**1904.6(b)(2)** *When an employee experiences the signs or symptoms of an injury or illness as a result of an event or exposure in the workplace, such*

as an episode of occupational asthma, must I treat the episode as a new case? Yes, because the episode or recurrence was caused by an event or exposure in the workplace, the incident must be treated as a new case.

**1904.6(b)(3)** *May I rely on a physician or other licensed health care professional to determine whether a case is a new case or a recurrence of an old case?* You are not required to seek the advice of a physician or other licensed health care professional. However, if you do seek such advice, you must follow the physician or other licensed health care professional's recommendation about whether the case is a new case or a recurrence. If you receive recommendations from two or more physicians or other licensed health care professionals, you must make a decision as to which recommendation is the most authoritative (best documented, best reasoned, or most authoritative), and record the case based upon that recommendation.

[66 FR 5915, Jan. 19, 2001]

### §1904.7 General recording criteria.

**1904.7(a)** *Basic requirement.* You must consider an injury or illness to meet the general recording criteria, and therefore to be recordable, if it results in any of the following: death, days away from work, restricted work or transfer to another job, medical treatment beyond first aid, or loss of consciousness. You must also consider a case to meet the general recording criteria if it involves a significant injury or illness diagnosed by a physician or other licensed health care professional, even if it does not result in death, days away from work, restricted work or job transfer, medical treatment beyond first aid, or loss of consciousness.

**1904.7(b)** *Implementation.* (1) *How do I decide if a case meets one or more of the general recording criteria?* A work-related injury or illness must be recorded if it results in one or more of the following:

**1904.7(b)(1)(i)** Death. See §1904.7(b)(2).

**1904.7(b)(1)(ii)** Days away from work. See §1904.7(b)(3).

**1904.7(b)(1)(iii)** Restricted work or transfer to another job. See §1904.7(b)(4).

**1904.7(b)(1)(iv)** Medical treatment beyond first aid. See §1904.7(b)(5).

**1904.7(b)(1)(v)** Loss of consciousness. See §1904.7(b)(6).

**1904.7(b)(1)(vi)** A significant injury or illness diagnosed by a physician or other licensed health care professional. See §1904.7(b)(7).

**1904.7(b)(2)** *How do I record a work-related injury or illness that results in the employee's death?* You must record an injury or illness that results in death by entering a check mark on the OSHA 300 Log in the space for cases resulting in death. You must also report any work-related fatality to OSHA within eight (8) hours, as required by §1904.39.

**1904.7(b)(3)** *How do I record a work-related injury or illness that results in days away from work?* When an injury or illness involves one or more days away from work, you must record the injury or illness on the OSHA 300 Log with a check mark in the space for cases involving days away and an entry of the number of calendar days away from work in the number of days column. If the employee is out for an extended period of time, you must enter an estimate of the days that the employee will be away, and update the day count when the actual number of days is known.

**1904.7(b)(3)(i)** *Do I count the day on which the injury occurred or the illness began?* No, you begin counting days away on the day after the injury occurred or the illness began.

**1904.7(b)(3)(ii)** *How do I record an injury or illness when a physician or other licensed health care professional recommends that the worker stay at home but the employee comes to work anyway?* You must record these injuries and illnesses on the OSHA 300 Log using the check box for cases with days away from work and enter the number of calendar days away recommended by the physician or other licensed health care professional. If a physician or other licensed health care professional recommends days away, you should encourage your employee to follow that recommendation. However, the days away must be recorded whether the injured or ill employee follows the physician or licensed health care professional's recommendation or not. If you receive recommendations from two or more physicians or other licensed health care professionals, you may make a decision as to which recommendation is the most authoritative, and record the case based upon that recommendation.

**1904.7(b)(3)(iii)** *How do I handle a case when a physician or other licensed health care professional recommends that the worker return to work but the employee stays at home anyway?* In this situation, you must end the count of days away from work on the date the physician or other licensed health care professional recommends that the employee return to work.

**1904.7(b)(3)(iv)** *How do I count weekends, holidays, or other days the employee would not have worked anyway?* You must count the number of calendar days the employee was unable to work as a result of the injury or illness, regardless of whether or not the employee was scheduled to work on those day(s). Weekend days, holidays, vacation days or other days off are included in the total number of days recorded if the employee would not have been able to work on those days because of a work-related injury or illness.

**1904.7(b)(3)(v)** *How do I record a case in which a worker is injured or becomes ill on a Friday and reports to work on a Monday, and was not scheduled to work on the weekend?* You need to record this case only if you receive information from a physician or other licensed health care professional indicating that the employee should not have worked, or should have performed only restricted work, during the weekend. If so, you must record the injury or illness as a case with days away from work or restricted work, and enter the day counts, as appropriate.

**1904.7(b)(3)(vi)** *How do I record a case in which a worker is injured or becomes ill on the day before scheduled time off such as a holiday, a planned vacation, or a temporary plant closing?* You need to record a case of this type only if you receive information from a physician or other licensed health care professional indicating that the employee should not have worked, or should have performed only restricted work, during the scheduled time off. If so, you must record the injury or illness as a case with days away from work or restricted work, and enter the day counts, as appropriate.

**1904.7(b)(3)(vii)** *Is there a limit to the number of days away from work I must count?* Yes, you may "cap" the total days away at 180 calendar days. You are not required to keep track of the number of calendar days away from work if the injury or illness resulted in more than 180 calendar days away from work and/or days of job transfer or restriction. In such a case, entering 180 in the total days away column will be considered adequate.

**1904.7(b)(3)(viii)** *May I stop counting days if an employee who is away from work because of an injury or illness retires or leaves my company?* Yes, if the employee leaves your company for some reason unrelated to the injury or illness, such as retirement, a plant closing, or to take another job, you may stop counting days away from work or days of restriction/job transfer. If the employee leaves your company because of the injury or illness, you must estimate the total number of days away or days of restriction/job transfer and enter the day count on the 300 Log.

**1904.7(b)(3)(ix)** *If a case occurs in one year but results in days away during the next calendar year, do I record the case in both years?* No, you only record the injury or illness once. You must enter the number of calendar days away for the injury or illness on the OSHA 300 Log for the year in which the injury or illness occurred. If the employee is still away from work because of the injury or illness when you prepare the annual summary, estimate the total number of calendar days you expect the employee to be away from work, use this number to calculate the total for the annual summary, and then update the initial log entry later when the day count is known or reaches the 180-day cap.

**1904.7(b)(4)** *How do I record a work-related injury or illness that results in restricted work or job transfer?* When an injury or illness involves restricted work or job transfer but does not involve death or days away from work, you must record the injury or illness on the OSHA 300 Log by placing a check mark in the space for job transfer or restriction and an entry of the number of restricted or transferred days in the restricted workdays column.

**1904.7(b)(4)(i)** *How do I decide if the injury or illness resulted in restricted work?* Restricted work occurs when, as the result of a work-related injury or illness:

**1904.7(b)(4)(i)(A)** You keep the employee from performing one or more of the routine functions of his or her job, or from working the full workday that he or she would otherwise have been scheduled to work; or

**1904.7(b)(4)(i)(B)** A physician or other licensed health care professional recommends that the employee not perform one or more of the routine functions of his or her job, or not work the full workday that he or she would otherwise have been scheduled to work.

**1904.7(b)(4)(ii)** *What is meant by "routine functions"?* For recordkeeping purposes, an employee's routine functions are those work activities the employee regularly performs at least once per week.

**1904.7(b)(4)(iii)** *Do I have to record restricted work or job transfer if it applies only to the day on which the injury occurred or the illness began?* No, you do not have to record restricted work or job transfers if you, or the physician or other licensed health care professional, impose the restriction or transfer only for the day on which the injury occurred or the illness began.

**1904.7(b)(4)(iv)** *If you or a physician or other licensed health care professional recommends a work restriction, is the injury or illness automatically recordable as a "restricted work" case?* No, a recommended work restriction is recordable only if it affects one or more of the employee's routine job functions. To determine whether this is the case, you must evaluate the restriction in light of the routine functions of the injured or ill employee's job. If the restriction from you or the physician or other licensed health care professional keeps the employee from performing one or more of his or her routine job functions, or from working the full workday the injured or ill employee would otherwise have worked, the employee's work has been restricted and you must record the case.

**1904.7(b)(4)(v)** *How do I record a case where the worker works only for a partial work shift because of a work-related injury or illness?* A partial day of work is recorded as a day of job transfer or restriction for record-keeping purposes, except for the day on which the injury occurred or the illness began.

**1904.7(b)(4)(vi)** *If the injured or ill worker produces fewer goods or services than he or she would have produced prior to the injury or illness but otherwise performs all of the routine functions of his or her work, is the case considered a restricted work case?* No, the case is considered restricted work only if the worker does not perform all of the routine functions of his or her job or does not work the full shift that he or she would otherwise have worked.

**1904.7(b)(4)(vii)** *How do I handle vague restrictions from a physician or other licensed health care professional, such as that the employee engage only in "light duty" or "take it easy for a week"?* If you are not clear about the physician or other licensed health care professional's recommendation, you may ask that person whether the employee can do all of his or her routine job functions and work all of his or her normally assigned work shift. If the answer to both of these questions is "Yes," then the case does not involve a work restriction and does not have to be recorded as such. If the answer to one or both of these questions is "No," the case involves restricted work and must be recorded as a restricted work case. If you are unable to obtain this additional information from the physician or other licensed health care professional who recommended the restriction, record the injury or illness as a case involving restricted work.

**1904.7(b)(4)(viii)** *What do I do if a physician or other licensed health care professional recommends a job restriction meeting OSHA's definition, but the employee does all of his or her routine job functions anyway?* You must record the injury or illness on the OSHA 300 Log as a restricted work case. If a physician or other licensed health care professional recommends a job restriction, you should ensure that the employee complies with that restriction. If you receive recommendations from two or more physicians or other licensed health care professionals, you may make a decision as to which recommendation is the most authoritative, and record the case based upon that recommendation.

**1904.7(b)(4)(ix)** *How do I decide if an injury or illness involved a transfer to another job?* If you assign an injured or ill employee to a job other than his or her regular job for part of the day, the case involves transfer to another job. Note: This does not include the day on which the injury or illness occurred.

**1904.7(b)(4)(x)** *Are transfers to another job recorded in the same way as restricted work cases?* Yes, both job transfer and restricted work cases are recorded in the same box on the OSHA 300 Log. For example, if you assign, or a physician or other licensed health care professional recommends that you assign, an injured or ill worker to his or her routine job duties for part of the day and to another job for the rest of the day, the injury or illness involves a job transfer. You must record an injury or illness that involves a job transfer by placing a check in the box for job transfer.

**1904.7(b)(4)(xi)** *How do I count days of job transfer or restriction?* You count days of job transfer or restriction in the same way you count days away from work, using §1904.7(b)(3)(i) to (viii), above. The only difference is that, if you permanently assign the injured or ill employee to a job that has been modified or permanently changed in a manner that eliminates the routine functions the employee was restricted from performing, you may stop the day count when the modification or change is made permanent. You must count at least one day of restricted work or job transfer for such cases.

**1904.7(b)(5)** *How do I record an injury or illness that involves medical treatment beyond first aid?* If a work-related injury or illness results in medical treatment beyond first aid, you must record it on the OSHA 300 Log. If the injury or illness did not involve death, one or more days away from work, one or more days of restricted work, or one or more days of job transfer, you enter a check mark in the box for cases where the employee received medical treatment but remained at work and was not transferred or restricted.

**1904.7(b)(5)(i)** *What is the definition of medical treatment?* "Medical treatment" means the management and care of a patient to combat disease or disorder. For the purposes of Part 1904, medical treatment does not include:

**1904.7(b)(5)(i)(A)** Visits to a physician or other licensed health care professional solely for observation or counseling;

**1904.7(b)(5)(i)(B)** The conduct of diagnostic procedures, such as x-rays and blood tests, including the administration of prescription medications used solely for diagnostic purposes (e.g., eye drops to dilate pupils); or

**1904.7(b)(5)(i)(C)** "First aid" as defined in paragraph (b)(5)(ii) of this section.

**1904.7(b)(5)(ii)** *What is "first aid"?* For the purposes of Part 1904, "first aid" means the following:

**1904.7(b)(5)(ii)(A)** Using a non-prescription medication at nonprescription strength (for medications available in both prescription and non-prescription form, a recommendation by a physician or other licensed health care professional to use a non-prescription medication at prescription strength is considered medical treatment for recordkeeping purposes);

**1904.7(b)(5)(ii)(B)** Administering tetanus immunizations (other immunizations, such as Hepatitis B vaccine or rabies vaccine, are considered medical treatment);

**1904.7(b)(5)(ii)(C)** Cleaning, flushing or soaking wounds on the surface of the skin;

**1904.7(b)(5)(ii)(D)** Using wound coverings such as bandages, Band-Aids™, gauze pads, etc.; or using butterfly bandages or Steri-Strips™ (other wound closing devices such as sutures, staples, etc., are considered medical treatment);

**1904.7(b)(5)(ii)(E)** Using hot or cold therapy;

**1904.7(b)(5)(ii)(F)** Using any non-rigid means of support, such as elastic bandages, wraps, non-rigid back belts, etc. (devices with rigid stays or other systems designed to immobilize parts of the body are considered medical treatment for recordkeeping purposes);

**1904.7(b)(5)(ii)(G)** Using temporary immobilization devices while transporting an accident victim (e.g., splints, slings, neck collars, back boards, etc.).

**1904.7(b)(5)(ii)(H)** Drilling of a fingernail or toenail to relieve pressure, or draining fluid from a blister;

**1904.7(b)(5)(ii)(I)** Using eye patches;

**1904.7(b)(5)(ii)(J)** Removing foreign bodies from the eye using only irrigation or a cotton swab;

**1904.7(b)(5)(ii)(K)** Removing splinters or foreign material from areas other than the eye by irrigation, tweezers, cotton swabs or other simple means;

**1904.7(b)(5)(ii)(L)** Using finger guards;

**1904.7(b)(5)(ii)(M)** Using massages (physical therapy or chiropractic treatment are considered medical treatment for recordkeeping purposes); or

**1904.7(b)(5)(ii)(N)** Drinking fluids for relief of heat stress.

**1904.7(b)(5)(iii)** *Are any other procedures included in first aid?* No, this is a complete list of all treatments considered first aid for Part 1904 purposes.

**1904.7(b)(5)(iv)** *Does the professional status of the person providing the treatment have any effect on what is considered first aid or medical treatment?* No, OSHA considers the treatments listed in §1904.7(b)(5)(ii) of this Part to be first aid regardless of the professional status of the person providing the treatment. Even when these treatments are provided by a physician or other licensed health care professional, they are considered first aid for the purposes of Part 1904. Similarly, OSHA considers treatment beyond first aid to be medical treatment even when it is provided by someone other than a physician or other licensed health care professional.

**1904.7(b)(5)(v)** *What if a physician or other licensed health care professional recommends medical treatment but the employee does not follow the recommendation?* If a physician or other licensed health care professional recommends medical treatment, you should encourage the injured or ill employee to follow that recommendation. However, you must record the case even if the injured or ill employee does not follow the physician or other licensed health care professional's recommendation.

**1904.7(b)(6)** *Is every work-related injury or illness case involving a loss of consciousness recordable?* Yes, you must record a work-related injury or illness if the worker becomes unconscious, regardless of the length of time the employee remains unconscious.

**1904.7(b)(7)** *What is a "significant" diagnosed injury or illness that is recordable under the general criteria even if it does not result in death, days away from work, restricted work or job transfer, medical treatment beyond first aid, or loss of consciousness?* Work-related cases involving cancer, chronic irreversible disease, a fractured or cracked bone, or a punctured eardrum must always be recorded under the general criteria at the time of diagnosis by a physician or other licensed health care professional.

**Note to §1904.7:** OSHA believes that most significant injuries and illnesses will result in one of the criteria listed in §1904.7(a): death, days away from work, restricted work or job transfer, medical treatment beyond first aid, or loss of consciousness. However, there are some significant injuries, such as a punctured eardrum or a fractured toe or rib, for which neither medical treatment nor work restrictions may be recommended. In addition, there are some significant progressive diseases, such as byssinosis, silicosis, and some types of cancer, for which medical treatment or work restrictions may not be recommended at the time of diagnosis but are likely to be recommended as the disease progresses. OSHA believes that cancer, chronic irreversible diseases, fractured or cracked bones, and punctured eardrums are generally considered significant injuries and illnesses, and must be recorded at the initial diagnosis even if medical treatment or work restrictions are not recommended, or are postponed, in a particular case.

[66 FR 5915, Jan. 19, 2001]

## §1904.8 Recording criteria for needlestick and sharps injuries.

**1904.8(a)** *Basic requirement.* You must record all work-related needlestick injuries and cuts from sharp objects that are contaminated with another person's blood or other potentially infectious material (as defined by 29 CFR 1910.1030). You must enter the case on the OSHA 300 Log as an injury. To protect the employee's privacy, you may not enter the employee's name on the OSHA 300 Log (see the requirements for privacy cases in paragraphs 1904.29(b)(6) through 1904.29(b)(9)).

**1904.8(b)** *Implementation.* **(1)** *What does "other potentially infectious material" mean?* The term "other potentially infectious materials" is defined in the OSHA Bloodborne Pathogens standard at §1910.1030(b). These materials include:

**1904.8(b)(1)(i)** Human bodily fluids, tissues and organs, and

**1904.8(b)(1)(ii)** Other materials infected with the HIV or hepatitis B (HBV) virus such as laboratory cultures or tissues from experimental animals.

**1904.8(b)(2)** *Does this mean that I must record all cuts, lacerations, punctures, and scratches?* No, you need to record cuts, lacerations, punctures, and scratches only if they are work-related and involve contamination with another person's blood or other potentially infectious material. If the cut, laceration, or scratch involves a clean object, or a contaminant other than blood or other potentially infectious material, you need to record the case only if it meets one or more of the recording criteria in §1904.7.

**1904.8(b)(3)** *If I record an injury and the employee is later diagnosed with an infectious bloodborne disease, do I need to update the OSHA 300 Log?* Yes, you must update the classification of the case on the OSHA 300 Log if the case results in death, days away from work, restricted work, or job transfer. You must also update the description to identify the infectious disease and change the classification of the case from an injury to an illness.

**1904.8(b)(4)** *What if one of my employees is splashed or exposed to blood or other potentially infectious material without being cut or scratched?* Do I need to record this incident? You need to record such an incident on the OSHA 300 Log as an illness if:

**1904.8(b)(4)(i)** It results in the diagnosis of a bloodborne illness, such as HIV, hepatitis B, or hepatitis C; or

**1904.8(b)(4)(ii)** It meets one or more of the recording criteria in §1904.7.

[66 FR 5915, Jan. 19, 2001]

## §1904.9 Recording criteria for cases involving medical removal under OSHA standards.

**1904.9(a)** *Basic requirement.* If an employee is medically removed under the medical surveillance requirements of an OSHA standard, you must record the case on the OSHA 300 Log.

**1904.9(b)** *Implementation.* **(1)** *How do I classify medical removal cases on the OSHA 300 Log?* You must enter each medical removal case on the OSHA 300 Log as either a case involving days away from work or a case involving restricted work activity, depending on how you decide to comply with the medical removal requirement. If the medical removal is the result of a chemical exposure, you must enter the case on the OSHA 300 Log by checking the "poisoning" column.

**1904.9(b)(2)** *Do all of OSHA's standards have medical removal provisions?* No, some OSHA standards, such as the standards covering bloodborne pathogens and noise, do not have medical removal provisions. Many OSHA standards that cover specific chemical substances have medical removal provisions. These standards include, but are not limited to, lead, cadmium, methylene chloride, formaldehyde, and benzene.

**1904.9(b)(3)** *Do I have to record a case where I voluntarily removed the employee from exposure before the medical removal criteria in an OSHA standard are met?* No, if the case involves voluntary medical removal before the medical removal levels required by an OSHA standard, you do not need to record the case on the OSHA 300 Log.

[66 FR 5915, Jan. 19, 2001]

## §1904.10 Recording criteria for cases involving occupational hearing loss.

**1904.10(a)** *Basic requirement.* If an employee's hearing test (audiogram) reveals that the employee has experienced a work-related Standard Threshold Shift (STS) in hearing in one or both ears, and the employee's total hearing level is 25 decibels (dB) or more above audiometric zero (averaged at 2000, 3000, and 4000 Hz) in the same ear(s) as the STS, you must record the case on the OSHA 300 Log.

**1904.10(b)** *Implementation.* **(1)** *What is a Standard Threshold Shift?* A Standard Threshold Shift, or STS, is defined in the occupational noise exposure standard at 29 CFR 1910.95(g)(10)(i) as a change in hearing threshold, relative to the baseline audiogram for that employee, of an average of 10 decibels (dB) or more at 2000, 3000, and 4000 hertz (Hz) in one or both ears.

**1904.10(b)(2)** *How do I evaluate the current audiogram to determine whether an employee has an STS and a 25-dB hearing level?* If the employee has never previously experienced a recordable hearing loss, you must compare the employee's current audiogram with that employee's baseline audiogram. If the employee has previously experienced a recordable hearing loss, you must compare the employee's current audiogram with the employee's revised baseline audiogram (the audiogram reflecting the employee's previous recordable hearing loss case).

**1904.10(b)(2)(i)** *STS.* If the employee has never previously experienced a recordable hearing loss, you must compare the employee's current audiogram with that employee's baseline audiogram. If the employee has previously experienced a recordable hearing loss, you must compare the employee's current audiogram with the employee's revised baseline audiogram (the audiogram reflecting the employee's previous recordable hearing loss case).

**1904.10(b)(2)(ii)** *25-dB loss.* Audiometric test results reflect the employee's overall hearing ability in comparison to audiometric zero. Therefore, using the employee's current audiogram, you must use the average hearing level at 2000, 3000, and 4000 Hz to determine whether or not the employee's total hearing level is 25 dB or more.

**1904.10(b)(3)** *May I adjust the current audiogram to reflect the effects of aging on hearing?* Yes. When you are determining whether an STS has occurred, you may age adjust the employee's current audiogram results by using Tables F-1 or F-2, as appropriate, in Appendix F of 29 CFR 1910.95. You may not use an age adjustment when determining whether the employee's total hearing level is 25 dB or more above audiometric zero.

**1904.10(b)(4)** *Do I have to record the hearing loss if I am going to retest the employee's hearing?* No, if you retest the employee's hearing within 30 days of the first test, and the retest does not confirm the recordable STS, you are not required to record the hearing loss case on the OSHA 300 Log. If the retest confirms the recordable STS, you must record the hearing loss illness within seven (7) calendar days of the retest. If subsequent audiometric testing performed under the testing requirements of the §1910.95 noise standard indicates that an STS is not persistent, you may erase or line-out the recorded entry.

**1904.10(b)(5)** *Are there any special rules for determining whether a hearing loss case is work-related?* No. You must use the rules in §1904.5 to determine if the hearing loss is work-related. If an event or exposure in the work environment either caused or contributed to the hearing loss, or significantly aggravated a pre-existing hearing loss, you must consider the case to be work related.

**1904.10(b)(6)** *If a physician or other licensed health care professional determines the hearing loss is not work-related, do I still need to record the case?* If a physician or other licensed health care professional determines that the hearing loss is not work-related or has not been significantly aggravated by occupational noise exposure, you are not required to consider the case work-related or to record the case on the OSHA 300 Log.

**1904.10(b)(7)** *How do I complete the 300 Log for a hearing loss case?* When you enter a recordable hearing loss case on the OSHA 300 Log, you must check the 300 Log column for hearing loss.

**Note to §1904.10(b)(7):** The applicability of paragraph (b)(7) is delayed until further notice.

**1904.10(c)** *Recording criteria for calendar year 2002.* From January 1, 2002 until December 31, 2002, you are required to record a work-related hearing loss averaging 25dB or more at 2000, 3000, and 4000 hertz in either ear on the OSHA 300 Log. You may use the employee's original baseline audiogram for comparison. You may make a correction for presbycusis (aging) by using the tables in Appendix F of 29 CFR 1910.95. The requirement of §1904.37(b)(1) that States with OSHA-approved state plans must have the same requirements for determining which injuries and illnesses are recordable and how they are recorded shall not preclude the states from retaining their existing criteria with regard to this section during calendar year 2002.

**Note to §1904.10:** Paragraphs (a) and (b) of this section are effective on January 1, 2003. Paragraph (c) of this section applies from January 1, 2002 until December 31, 2002.

[66 FR 5915, Jan. 19, 2001]

## §1904.11 Recording criteria for work-related tuberculosis cases.

**1904.11(a)** *Basic requirement.* If any of your employees has been occupationally exposed to anyone with a known case of active tuberculosis (TB), and that employee subsequently develops a tuberculosis infection, as evidenced by a positive skin test or diagnosis by a physician or other licensed health care professional, you must record the case on the OSHA 300 Log by checking the "respiratory condition" column.

**1904.11(b)** *Implementation.* **(1)** *Do I have to record, on the Log, a positive TB skin test result obtained at a pre-employment physical?* No, you do not have to record it because the employee was not occupationally exposed to a known case of active tuberculosis in your workplace.

**1904.11(b)(2)** *May I line-out or erase a recorded TB case if I obtain evidence that the case was not caused by occupational exposure?* Yes, you may line-out or erase the case from the Log under the following circumstances:

**1904.11(b)(2)(i)** The worker is living in a household with a person who has been diagnosed with active TB;

**1904.11(b)(2)(ii)** The Public Health Department has identified the worker as a contact of an individual with a case of active TB unrelated to the workplace; or

**1904.11(b)(2)(iii)** A medical investigation shows that the employee's infection was caused by exposure to TB away from work, or proves that the case was not related to the workplace TB exposure.

[66 FR 5915, Jan. 19, 2001]

## §1904.12 [Removed]

## §1904.13—1904.28 [Reserved]

[66 FR 5915, Jan. 19, 2001]

## §1904.29 Forms

**1904.29(a)** *Basic requirement.* You must use OSHA 300, 300-A, and 301 forms, or equivalent forms, for recordable injuries and illnesses. The OSHA 300 form is called the Log of Work-Related Injuries and Illnesses, the 300-A is the Summary of Work-Related Injuries and Illnesses, and the OSHA 301 form is called the Injury and Illness Incident Report.

**1904.29(b)** *Implementation.* **(1)** *What do I need to do to complete the OSHA 300 Log?* You must enter information about your business at the top of the OSHA 300 Log, enter a one or two line description for each recordable injury or illness, and summarize this information on the OSHA 300-A at the end of the year.

**1904.29(b)(2)** *What do I need to do to complete the OSHA 301 Incident Report?* You must complete an OSHA 301 Incident Report form, or an equivalent form, for each recordable injury or illness entered on the OSHA 300 Log.

**1904.29(b)(3)** *How quickly must each injury or illness be recorded?* You must enter each recordable injury or illness on the OSHA 300 Log and 301 Incident Report within seven (7) calendar days of receiving information that a recordable injury or illness has occurred.

**1904.29(b)(4)** *What is an equivalent form?* An equivalent form is one that has the same information, is as readable and understandable, and is completed using the same instructions as the OSHA form it replaces. Many employers use an insurance form instead of the OSHA 301 Incident Report, or supplement an insurance form by adding any additional information required by OSHA.

**1904.29(b)(5)** *May I keep my records on a computer?* Yes, if the computer can produce equivalent forms when they are needed, as described under §§1904.35 and 1904.40, you may keep your records using the computer system.

**1904.29(b)(6)** *Are there situations where I do not put the employee's name on the forms for privacy reasons?* Yes, if you have a "privacy concern case," you may not enter the employee's name on the OSHA 300 Log. Instead, enter "privacy case" in the space normally used for the employee's name. This will protect the privacy of the injured or ill employee when another employee, a former employee, or an authorized employee representative is provided access to the OSHA 300 Log under §1904.35(b)(2). You must keep a separate, confidential list of the case numbers and employee names for your privacy concern cases so you can update the cases and provide the information to the government if asked to do so.

**1904.29(b)(7)** *How do I determine if an injury or illness is a privacy concern case?* You must consider the following injuries or illnesses to be privacy concern cases:

**1904.29(b)(7)(i)** An injury or illness to an intimate body part or the reproductive system;

**1904.29(b)(7)(ii)** An injury or illness resulting from a sexual assault;

**1904.29(b)(7)(iii)** Mental illnesses;

**1904.29(b)(7)(iv)** HIV infection, hepatitis, or tuberculosis; (**Note:** The first sentence of this §1904.29(b)(7)(iv) is effective on January 1, 2002. The second sentence is effective beginning on January 1, 2003.)

**1904.29(b)(7)(v)** Needlestick injuries and cuts from sharp objects that are contaminated with another person's blood or other potentially infectious material (see §1904.8 for definitions); and

**1904.29(b)(7)(vi)** Other illnesses, if the employee voluntarily requests that his or her name not be entered on the log.

**1904.29(b)(8)** *May I classify any other types of injuries and illnesses as privacy concern cases?* No, this is a complete list of all injuries and illnesses considered privacy concern cases for Part 1904 purposes.

**1904.29(b)(9)** *If I have removed the employee's name, but still believe that the employee may be identified from the information on the forms, is there anything else that I can do to further protect the employee's privacy?* Yes, if you have a reasonable basis to believe that information describing the privacy concern case may be personally identifiable even though the employee's name has been omitted, you may use discretion in describing the injury or illness on both the OSHA 300 and 301 forms. You must enter enough information to identify the cause of the incident and the general severity of the injury or illness, but you do not need to include details of an intimate or private nature. For example, a sexual assault case could be described as "injury from assault," or an injury to a reproductive organ could be described as "lower abdominal injury."

**1904.29(b)(10)** *What must I do to protect employee privacy if I wish to provide access to the OSHA Forms 300 and 301 to persons other than government representatives, employees, former employees or authorized representatives?* If you decide to voluntarily disclose the Forms to persons other than government representatives, employees, former employees or authorized representatives (as required by §§1904.35 and 1904.40), you must remove or hide the employees' names and other personally identifying information, except for the following cases. You may disclose the Forms with personally identifying information only:

**1904.29(b)(10)(i)** to an auditor or consultant hired by the employer to evaluate the safety and health program;

**1904.29(b)(10)(ii)** to the extent necessary for processing a claim for workers' compensation or other insurance benefits; or

**1904.29(b)(10)(iii)** to a public health authority or law enforcement agency for uses and disclosures for which consent, an authorization, or opportunity to agree or object is not required under Department of Health and Human Services Standards for Privacy of Individually Identifiable Health Information, 45 CFR 164.512.

[66 FR 5915, Jan. 19, 2001]

## Subpart D—Other OSHA Injury and Illness Recordkeeping Requirements

### §1904.30 Multiple business establishments.

**1904.30(a)** *Basic requirement.* You must keep a separate OSHA 300 Log for each establishment that is expected to be in operation for one year or longer.

**1904.30(b)** *Implementation.* (1) *Do I need to keep OSHA injury and illness records for short-term establishments (i.e., establishments that will exist for less than a year)?* Yes, however, you do not have to keep a separate OSHA 300 Log for each such establishment. You may keep one OSHA 300 Log that covers all of your short-term establishments. You may also include the short-term establishments' recordable injuries and illnesses on an OSHA 300 Log that covers short-term establishments for individual company divisions or geographic regions.

**1904.30(b)(2)** *May I keep the records for all of my establishments at my headquarters location or at some other central location?* Yes, you may keep the records for an establishment at your headquarters or other central location if you can:

**1904.30(b)(2)(i)** Transmit information about the injuries and illnesses from the establishment to the central location within seven (7) calendar days of receiving information that a recordable injury or illness has occurred; and

**1904.30(b)(2)(ii)** Produce and send the records from the central location to the establishment within the time frames required by §1904.35 and §1904.40 when you are required to provide records to a government representative, employees, former employees or employee representatives.

**1904.30(b)(3)** *Some of my employees work at several different locations or do not work at any of my establishments at all. How do I record cases for these employees?* You must link each of your employees with one of your establishments, for recordkeeping purposes. You must record the injury and illness on the OSHA 300 Log of the injured or ill employee's establishment, or on an OSHA 300 Log that covers that employee's short-term establishment.

**1904.30(b)(4)** *How do I record an injury or illness when an employee of one of my establishments is injured or becomes ill while visiting or working at another of my establishments, or while working away from any of my establishments?* If the injury or illness occurs at one of your establishments, you must record the injury or illness on the OSHA 300 Log of the establishment at which the injury or illness occurred. If the employee is injured or becomes ill and is not at one of your establishments, you must record the case on the OSHA 300 Log at the establishment at which the employee normally works.

[66 FR 5915, Jan. 19, 2001]

### §1904.31 Covered employees.

**1904.31(a)** *Basic requirement.* You must record on the OSHA 300 Log the recordable injuries and illnesses of all employees on your payroll, whether they are labor, executive, hourly, salary, part-time, seasonal, or migrant workers. You also must record the recordable injuries and illnesses that occur to employees who are not on your payroll if you supervise these employees on a day-to-day basis. If your business is organized as a sole proprietorship or partnership, the owner or partners are not considered employees for recordkeeping purposes.

**1904.31(b)** *Implementation.* (1) *If a self-employed person is injured or becomes ill while doing work at my business, do I need to record the injury or illness?* No, self-employed individuals are not covered by the OSH Act or this regulation.

**1904.31(b)(2)** *If I obtain employees from a temporary help service, employee leasing service, or personnel supply service, do I have to record an injury or illness occurring to one of those employees?* You must record these injuries and illnesses if you supervise these employees on a day-to-day basis.

**1904.31(b)(3)** *If an employee in my establishment is a contractor's employee, must I record an injury or illness occurring to that employee?* If the contractor's employee is under the day-to-day supervision of the contractor, the contractor is responsible for recording the injury or illness. If you supervise the contractor employee's work on a day-to-day basis, you must record the injury or illness.

**1904.31(b)(4)** *Must the personnel supply service, temporary help service, employee leasing service, or contractor also record the injuries or illnesses occurring to temporary, leased or contract employees that I supervise on a day-to-day basis?* No, you and the temporary help service, employee leasing service, personnel supply service, or contractor should coordinate your efforts to make sure that each injury and illness is recorded only once: either on your OSHA 300 Log (if you provide day-to-day supervision) or on the other employer's OSHA 300 Log (if that company provides day-to-day supervision).

[66 FR 5915, Jan. 19, 2001]

### §1904.32 Annual summary.

**1904.32(a)** *Basic requirement.* At the end of each calendar year, you must:

**1904.32(a)(1)** Review the OSHA 300 Log to verify that the entries are complete and accurate, and correct any deficiencies identified;

**1904.32(a)(2)** Create an annual summary of injuries and illnesses recorded on the OSHA 300 Log;

**1904.32(a)(3)** Certify the summary; and

**1904.32(a)(4)** Post the annual summary.

**1904.32(b)** *Implementation.* (1) *How extensively do I have to review the OSHA 300 Log entries at the end of the year?* You must review the entries as extensively as necessary to make sure that they are complete and correct.

**1904.32(b)(2)** *How do I complete the annual summary?* You must:

**1904.32(b)(2)(i)** Total the columns on the OSHA 300 Log (if you had no recordable cases, enter zeros for each column total); and

**1904.32(b)(2)(ii)** Enter the calendar year covered, the company's name, establishment name, establishment address, annual average number of employees covered by the OSHA 300 Log, and the total hours worked by all employees covered by the OSHA 300 Log.

**1904.32(b)(2)(iii)** If you are using an equivalent form other than the OSHA 300-A summary form, as permitted under §1904.6(b)(4), the summary you use must also include the employee access and employer penalty statements found on the OSHA 300-A Summary form.

**1904.32(b)(3)** *How do I certify the annual summary?* A company executive must certify that he or she has examined the OSHA 300 Log and that he or she reasonably believes, based on his or her knowledge of the process by which the information was recorded, that the annual summary is correct and complete.

**1904.32(b)(4)** *Who is considered a company executive?* The company executive who certifies the log must be one of the following persons:

**1904.32(b)(4)(i)** An owner of the company (only if the company is a sole proprietorship or partnership);

**1904.32(b)(4)(ii)** An officer of the corporation;

**1904.32(b)(4)(iii)** The highest ranking company official working at the establishment; or

**1904.32(b)(4)(iv)** The immediate supervisor of the highest ranking company official working at the establishment.

**1904.32(b)(5)** *How do I post the annual summary?* You must post a copy of the annual summary in each establishment in a conspicuous place or places where notices to employees are customarily posted. You must ensure that the posted annual summary is not altered, defaced or covered by other material.

**1904.32(b)(6)** *When do I have to post the annual summary?* You must post the summary no later than February 1 of the year following the year covered by the records and keep the posting in place until April 30.

[66 FR 5915, Jan. 19, 2001]

## §1904.33 Retention and updating.

**1904.33(a)** *Basic requirement.* You must save the OSHA 300 Log, the privacy case list (if one exists), the annual summary, and the OSHA 301 Incident Report forms for five (5) years following the end of the calendar year that these records cover.

**1904.33(b)** *Implementation.* **(1)** *Do I have to update the OSHA 300 Log during the five-year storage period?* Yes, during the storage period, you must update your stored OSHA 300 Logs to include newly discovered recordable injuries or illnesses and to show any changes that have occurred in the classification of previously recorded injuries and illnesses. If the description or outcome of a case changes, you must remove or line out the original entry and enter the new information.

**1904.33(b)(2)** *Do I have to update the annual summary?* No, you are not required to update the annual summary, but you may do so if you wish.

**1904.33(b)(3)** *Do I have to update the OSHA 301 Incident Reports?* No, you are n t required to update the OSHA 301 Incident Reports, but you may do so if you wish.

[66 FR 5915, Jan. 19, 2001]

## §1904.34 Change in business ownership.

If your business changes ownership, you are responsible for recording and reporting work-related injuries and illnesses only for that period of the year during which you owned the establishment. You must transfer the Part 1904 records to the new owner. The new owner must save all records of the establishment kept by the prior owner, as required by §1904.33 of this Part, but need not update or correct the records of the prior owner.

[66 FR 5915, Jan. 19, 2001]

## §1904.35 Employee involvement.

**1904.35(a)** *Basic requirement.* Your employees and their representatives must be involved in the recordkeeping system in several ways.

**1904.35(a)(1)** You must inform each employee of how he or she is to report an injury or illness to you.

**1904.35(a)(2)** You must provide limited access to your injury and illness records for your employees and their representatives.

**1904.35(b)** *Implementation.* **(1)** *What must I do to make sure that employees report work-related injuries and illnesses to me?*

**1904.35(b)(1)(i)** You must set up a way for employees to report work-related injuries and illnesses promptly; and

**1904.35(b)(1)(ii)** You must tell each employee how to report work-related injuries and illnesses to you.

**1904.35(b)(2)** *Do I have to give my employees and their representatives access to the OSHA injury and illness records?* Yes, your employees, former employees, their personal representatives, and their authorized employee representatives have the right to access the OSHA injury and illness records, with some limitations, as discussed below.

**1904.35(b)(2)(i)** *Who is an authorized employee representative?* An authorized employee representative is an authorized collective bargaining agent of employees.

**1904.35(b)(2)(ii)** *Who is a "personal representative" of an employee or former employee?* A personal representative is:

**1904.35(b)(2)(ii)(A)** Any person that the employee or former employee designates as such, in writing; or

**1904.35(b)(2)(ii)(B)** The legal representative of a deceased or legally incapacitated employee or former employee.

**1904.35(b)(2)(iii)** *If an employee or representative asks for access to the OSHA 300 Log, when do I have to provide it?* When an employee, former employee, personal representative, or authorized employee representative asks for copies of your current or stored OSHA 300 Log(s) for an establishment the employee or former employee has worked in, you must give the requester a copy of the relevant OSHA 300 Log(s) by the end of the next business day.

**1904.35(b)(2)(iv)** *May I remove the names of the employees or any other information from the OSHA 300 Log before I give copies to an employee, former employee, or employee representative?* No, you must leave the names on the 300 Log. However, to protect the privacy of injured and ill employees, you may not record the employee's name on the OSHA 300 Log for certain "privacy concern cases," as specified in paragraphs 1904.29(b)(6) through 1904.29(b)(9).

**1904.35(b)(2)(v)** *If an employee or representative asks for access to the OSHA 301 Incident Report, when do I have to provide it?*

**1904.35(b)(2)(v)(A)** When an employee, former employee, or personal representative asks for a copy of the OSHA 301 Incident Report describing an injury or illness to that employee or former employee, you must give the requester a copy of the OSHA 301 Incident Report containing that information by the end of the next business day.

**1904.35(b)(2)(v)(B)** When an authorized employee representative asks for a copies of the OSHA 301 Incident Reports for an establishment where the agent represents employees under a collective bargaining agreement, you must give copies of those forms to the authorized employee representative within 7 calendar days. You are only required to give the authorized employee representative information from the OSHA 301 Incident Report section titled "Tell us about the case." You must remove all other information from the copy of the OSHA 301 Incident Report or the equivalent substitute form that you give to the authorized employee representative.

**1904.35(b)(2)(vi)** *May I charge for the copies?* No, you may not charge for these copies the first time they are provided. However, if one of the designated persons asks for additional copies, you may assess a reasonable charge for retrieving and copying the records.

[66 FR 5915, Jan. 19, 2001]

## §1904.36 Prohibition against discrimination.

Section 11(c) of the Act prohibits you from discriminating against an employee for reporting a work-related fatality, injury or illness. That provision of the Act also protects the employee who files a safety and health complaint, asks for access to the Part 1904 records, or otherwise exercises any rights afforded by the OSH Act.

[66 FR 5915, Jan. 19, 2001]

## §1904.37 State recordkeeping regulations.

**1904.37(a)** *Basic requirement.* Some States operate their own OSHA programs, under the authority of a State Plan approved by OSHA. States operating OSHA-approved State Plans must have occupational injury and illness recording and reporting requirements that are substantially identical to the requirements in this Part (see 29 CFR 1902.3(k), 29 CFR 1952.4 and 29 CFR 1956.10(i)).

**1904.37(b)** *Implementation.* **(1)** State-Plan States must have the same requirements as Federal OSHA for determining which injuries and illnesses are recordable and how they are recorded.

**1904.37(b)(2)** For other Part 1904 provisions (for example, industry exemptions, reporting of fatalities and hospitalizations, record retention, or employee involvement), State-Plan State requirements may be more stringent than or supplemental to the Federal requirements, but because of the unique nature of the national recordkeeping program, States must consult with and obtain approval of any such requirements.

**1904.37(b)(3)** Although State and local government employees are not covered Federally, all State-Plan States must provide coverage, and must develop injury and illness statistics, for these workers. State Plan recording and reporting requirements for State and local government entities may differ from those for the private sector but must meet the requirements of paragraphs 1904.37(b)(1) and (b)(2).

**1904.37(b)(4)** A State-Plan State may not issue a variance to a private sector employer and must recognize all variances issued by Federal OSHA.

**1904.37(b)(5)** A State Plan State may only grant an injury and illness recording and reporting variance to a State or local government employer within the State after obtaining approval to grant the variance from Federal OSHA.

[66 FR 5915, Jan. 19, 2001]

## §1904.38 Variances from the recordkeeping rule.

**1904.38(a)** *Basic requirement.* If you wish to keep records in a different manner from the manner prescribed by the Part 1904 regulations, you may submit a variance petition to the Assistant Secretary of Labor for Occupational Safety and Health, U.S. Department of Labor, Washington, DC 20210. You can obtain a variance only if you can show that your alternative recordkeeping system:

**1904.38(a)(1)** Collects the same information as this Part requires;

**1904.38(a)(2)** Meets the purposes of the Act; and

**1904.38(a)(3)** Does not interfere with the administration of the Act.

**1904.38(b)** *Implementation.* **(1)** *What do I need to include in my variance petition?* You must include the following items in your petition:

**1904.38(b)(1)(i)** Your name and address;

**1904.38(b)(1)(ii)** A list of the State(s) where the variance would be used;

**1904.38(b)(1)(iii)** The address(es) of the business establishment(s) involved;

**1904.38(b)(1)(iv)** A description of why you are seeking a variance;

**1904.38(b)(1)(v)** A description of the different recordkeeping procedures you propose to use;

**1904.38(b)(1)(vi)** A description of how your proposed procedures will collect the same information as would be collected by this Part and achieve the purpose of the Act; and

**1904.38(b)(1)(vii)** A statement that you have informed your employees of the petition by giving them or their authorized representative a copy of the petition and by posting a statement summarizing the petition in the same way as notices are posted under §1903.2(a).

**1904.38(b)(2)** *How will the Assistant Secretary handle my variance petition?* The Assistant Secretary will take the following steps to process your variance petition.

**1904.38(b)(2)(i)** The Assistant Secretary will offer your employees and their authorized representatives an opportunity to submit written data, views, and arguments about your variance petition.

**1904.38(b)(2)(ii)** The Assistant Secretary may allow the public to comment on your variance petition by publishing the petition in the **Federal Register**. If the petition is published, the notice will establish a public comment period and may include a schedule for a public meeting on the petition.

**1904.38(b)(2)(iii)** After reviewing your variance petition and any comments from your employees and the public, the Assistant Secretary will decide whether or not your proposed recordkeeping procedures will meet the purposes of the Act, will not otherwise interfere with the Act, and will provide the same information as the Part 1904 regulations provide. If your procedures meet these criteria, the Assistant Secretary may grant the variance subject to such conditions as he or she finds appropriate.

**1904.38(b)(2)(iv)** If the Assistant Secretary grants your variance petition, OSHA will publish a notice in the **Federal Register** to announce the variance. The notice will include the practices the variance allows you to use, any conditions that apply, and the reasons for allowing the variance.

**1904.38(b)(3)** *If I apply for a variance, may I use my proposed recordkeeping procedures while the Assistant Secretary is processing the variance petition?* No, alternative recordkeeping practices are only allowed after the variance is approved. You must comply with the Part 1904 regulations while the Assistant Secretary is reviewing your variance petition.

**1904.38(b)(4)** *If I have already been cited by OSHA for not following the Part 1904 regulations, will my variance petition have any effect on the citation and penalty?* No, in addition, the Assistant Secretary may elect not to review your variance petition if it includes an element for which you have been cited and the citation is still under review by a court, an Administrative Law Judge (ALJ), or the OSH Review Commission.

**1904.38(b)(5)** *If I receive a variance, may the Assistant Secretary revoke the variance at a later date?* Yes, the Assistant Secretary may revoke your variance if he or she has good cause. The procedures revoking a variance will follow the same process as OSHA uses for reviewing variance petitions, as outlined in paragraph 1904.38(b)(2). Except in cases of willfulness or where necessary for public safety, the Assistant Secretary will:

**1904.38(b)(5)(i)** Notify you in writing of the facts or conduct that may warrant revocation of your variance; and

**1904.38(b)(5)(ii)** Provide you, your employees, and authorized employee representatives with an opportunity to participate in the revocation procedures.

[66 FR 5915, Jan. 19, 2001]

## Subpart E—Reporting Fatality, Injury and Illness Information to the Government

### §1904.39 Reporting fatalities and multiple hospitalization incidents to OSHA.

**1904.39(a)** *Basic requirement.* Within eight (8) hours after the death of any employee from a work-related incident or the in-patient hospitalization of three or more employees as a result of a work-related incident, you must orally report the fatality/multiple hospitalization by telephone or in person to the Area Office of the Occupational Safety and Health Administration (OSHA), U.S. Department of Labor, that is nearest to the site of the incident. You may also use the OSHA toll-free central telephone number, 1-800-321-OSHA (1-800-321-6742).

**1904.39(b)** *Implementation.* **(1)** *If the Area Office is closed, may I report the incident by leaving a message on OSHA's answering machine, faxing the area office, or sending an e-mail?* No, if you can't talk to a person at the Area Office, you must report the fatality or multiple hospitalization incident using the 800 number.

**1904.39(b)(2)** *What information do I need to give to OSHA about the incident?* You must give OSHA the following information for each fatality or multiple hospitalization incident:

**1904.39(b)(2)(i)** The establishment name;

**1904.39(b)(2)(ii)** The location of the incident;

**1904.39(b)(2)(iii)** The time of the incident;

**1904.39(b)(2)(iv)** The number of fatalities or hospitalized employees;

**1904.39(b)(2)(v)** The names of any injured employees;

**1904.39(b)(2)(vi)** Your contact person and his or her phone number; and

**1904.39(b)(2)(vii)** A brief description of the incident.

**1904.39(b)(3)** *Do I have to report every fatality or multiple hospitalization incident resulting from a motor vehicle accident?* No, you do not have to report all of these incidents. If the motor vehicle accident occurs on a public street or highway, and does not occur in a construction work zone, you do not have to report the incident to OSHA. However, these injuries must be recorded on your OSHA injury and illness records, if you are required to keep such records.

**1904.39(b)(4)** *Do I have to report a fatality or multiple hospitalization incident that occurs on a commercial or public transportation system?* No, you do not have to call OSHA to report a fatality or multiple hospitalization incident if it involves a commercial airplane, train, subway or bus accident. However, these injuries must be recorded on your OSHA injury and illness records, if you are required to keep such records.

**1904.39(b)(5)** *Do I have to report a fatality caused by a heart attack at work?* Yes, your local OSHA Area Office director will decide whether to investigate the incident, depending on the circumstances of the heart attack.

**1904.39(b)(6)** *Do I have to report a fatality or hospitalization that occurs long after the incident?* No, you must only report each fatality or multiple hospitalization incident that occurs within thirty (30) days of an incident.

**1904.39(b)(7)** *What if I don't learn about an incident right away?* If you do not learn of a reportable incident at the time it occurs and the incident would otherwise be reportable under paragraphs (a) and (b) of this section, you must make the report within eight (8) hours of the time the incident is reported to you or to any of your agent(s) or employee(s).

[66 FR 5915, Jan. 19, 2001]

### §1904.40 Providing records to government representatives.

**1904.40(a)** *Basic requirement.* When an authorized government representative asks for the records you keep under Part 1904, you must provide copies of the records within four (4) business hours.

**1904.40(b)** *Implementation.* **(1)** *What government representatives have the right to get copies of my Part 1904 records?* The government representatives authorized to receive the records are:

**1904.40(b)(1)(i)** A representative of the Secretary of Labor conducting an inspection or investigation under the Act;

**1904.40(b)(1)(ii)** A representative of the Secretary of Health and Human Services (including the National Institute for Occupational Safety and Health—NIOSH) conducting an investigation under section 20(b) of the Act, or

**1904.40(b)(1)(iii)** A representative of a State agency responsible for administering a State plan approved under section 18 of the Act.

**1904.40(b)(2)** *Do I have to produce the records within four (4) hours if my records are kept at a location in a different time zone?* OSHA will consider your response to be timely if you give the records to the government representative within four (4) business hours of the request. If you maintain the records at a location in a different time zone, you may use the business hours of the establishment at which the records are located when calculating the deadline.

[66 FR 5915, Jan. 19, 2001]

### §1904.41 Annual OSHA injury and illness survey of ten or more employers.

**1904.41(a)** *Basic requirement.* If you receive OSHA's annual survey form, you must fill it out and send it to OSHA or OSHA's designee, as stated on the survey form. You must report the following information for the year described on the form:

**1904.41(a)(1)** the number of workers you employed;

**1904.41(a)(2)** the number of hours worked by your employees; and

**1904.41(a)(3)** the requested information from the records that you keep under Part 1904.

**1904.41(b)** *Implementation.* **(1)** *Does every employer have to send data to OSHA?* No, each year, OSHA sends injury and illness survey forms to employers in certain industries. In any year, some employers will receive an OSHA survey form and others will not. You do not have to send injury and illness data to OSHA unless you receive a survey form.

**1904.41(b)(2)** *How quickly do I need to respond to an OSHA survey form?* You must send the survey reports to OSHA, or OSHA's designee, by mail or other means described in the survey form, within 30 calendar days, or by the date stated in the survey form, whichever is later.

**1904.41(b)(3)** *Do I have to respond to an OSHA survey form if I am normally exempt from keeping OSHA injury and illness records?* Yes, even if you are exempt from keeping injury and illness records under §1904.1 to §1904.3, OSHA may inform you in writing that it will be collecting injury and illness information from you in the following year. If you receive such a letter, you must keep the injury and illness records required by §1904.5 to §1904.15 and make a survey report for the year covered by the survey.

**1904.41(b)(4)** *Do I have to answer the OSHA survey form if I am located in a State-Plan State?* Yes, all employers who receive survey forms must respond to the survey, even those in State-Plan States.

**1904.41(b)(5)** *Does this section affect OSHA's authority to inspect my workplace?* No, nothing in this section affects OSHA's statutory authority to investigate conditions related to occupational safety and health.

[66 FR 5915, Jan. 19, 2001]

### §1904.42 Requests from the Bureau of Labor Statistics for data.

**1904.42(a)** *Basic requirement.* If you receive a Survey of Occupational Injuries and Illnesses Form from the Bureau of Labor Statistics (BLS), or a BLS designee, you must promptly complete the form and return it following the instructions contained on the survey form.

**1904.42(b)** *Implementation.* **(1)** *Does every employer have to send data to the BLS?* No, each year, the BLS sends injury and illness survey forms to randomly selected employers and uses the information to create the Nation's occupational injury and illness statistics. In any year, some employers will receive a BLS survey form and others will not. You do not have to send injury and illness data to the BLS unless you receive a survey form.

**1904.42(b)(2)** *If I get a survey form from the BLS, what do I have to do?* If you receive a Survey of Occupational Injuries and Illnesses Form from the Bureau of Labor Statistics (BLS), or a BLS designee, you must promptly complete the form and return it, following the instructions contained on the survey form.

**1904.42(b)(3)** *Do I have to respond to a BLS survey form if I am normally exempt from keeping OSHA injury and illness records?* Yes, even if you are exempt from keeping injury and illness records under §1904.1 to §1904.3, the BLS may inform you in writing that it will be collecting injury and illness information from you in the coming year. If you receive such a letter, you must keep the injury and illness records required by §1904.5 to §1904.15 and make a survey report for the year covered by the survey.

**1904.42(b)(4)** *Do I have to answer the BLS survey form if I am located in a State-Plan State?* Yes, all employers who receive a survey form must respond to the survey, even those in State-Plan States.

[66 FR 5915, Jan. 19, 2001]

### Subpart F—Transition From the Former Rule

### §1904.43 Summary and posting of the 2001 data.

**1904.43(a)** *Basic requirement.* If you were required to keep OSHA 200 Logs in 2001, you must post a 2000 annual summary from the OSHA 200 Log of occupational injuries and illnesses for each establishment.

**1904.43(b)** *Implementation.* **(1)** *What do I have to include in the summary?*

**1904.43(b)(1)(i)** You must include a copy of the totals from the 2001 OSHA 200 Log and the following information from that form:

**1904.43(b)(1)(i)(A)** The calendar year covered;

**1904.43(b)(1)(i)(B)** Your company name;

**1904.43(b)(1)(i)(C)** The name and address of the establishment; and

**1904.43(b)(1)(i)(D)** The certification signature, title and date.

**1904.43(b)(1)(ii)** If no injuries or illnesses occurred at your establishment in 2001, you must enter zeros on the totals line and post the 2001 summary.

**1904.43(b)(2)** *When am I required to summarize and post the 2001 information?*

**1904.43(b)(2)(i)** You must complete the summary by February 1, 2002; and

**1904.43(b)(2)(ii)** You must post a copy of the summary in each establishment in a conspicuous place or places where notices to employees are customarily posted. You must ensure that the summary is not altered, defaced or covered by other material.

**1904.43(b)(3)** You must post the 2001 summary from February 1, 2002 to March 1, 2002.

[66 FR 5915, Jan. 19, 2001]

### §1904.44 Retention and updating of old forms.

You must save your copies of the OSHA 200 and 101 forms for five years following the year to which they relate and continue to provide access to the data as though these forms were the OSHA 300 and 301 forms. You are not required to update your old 200 and 101 forms.

[66 FR 5915, Jan. 19, 2001]

### §1904.45 OMB control numbers under the Paperwork Reduction Act

The following sections each contain a collection of information requirement which has been approved by the Office of Management and Budget under the control number listed

| 29 CFR citation | OMB Control No. |
| --- | --- |
| 1904.4-35 | 1218-0176 |
| 1904.39-41 | 1218-0176 |
| 1904.42 | 1220-0045 |
| 1904.43-44 | 1218-0176 |

[66 FR 5915, Jan. 19, 2001]

### Subpart G—Definitions

### §1904.46 Definitions

*The Act.* The Act means the Occupational Safety and Health Act of 1970 (29 U.S.C. 651 *et seq.*). The definitions contained in section 3 of the Act (29 U.S.C. 652) and related interpretations apply to such terms when used in this Part 1904.

*Establishment.* An establishment is a single physical location where business is conducted or where services or industrial operations are performed. For activities where employees do not work at a single physical location, such as construction; transportation; communications, electric, gas and sanitary services; and similar operations, the establishment is represented by main or branch offices, terminals, stations, etc. that either supervise such activities or are the base from which personnel carry out these activities.

**1904.46(1)** *Can one business location include two or more establishments?* Normally, one business location has only one establishment. Under limited conditions, the employer may consider two or more separate businesses that share a single location to be separate establishments. An employer may divide one location into two or more establishments only when:

**1904.46(1)(i)** Each of the establishments represents a distinctly separate business;

**1904.46(1)(ii)** Each business is engaged in a different economic activity;

**1904.46(1)(iii)** No one industry description in the Standard Industrial Classification Manual (1987) applies to the joint activities of the establishments; and

**1904.46(1)(iv)** Separate reports are routinely prepared for each establishment on the number of employees, their wages and salaries, sales or receipts, and other business information. For example, if an employer operates a construction company at the same location as a lumber yard, the employer may consider each business to be a separate establishment.

**1904.46(2)** *Can an establishment include more than one physical location?* Yes, but only under certain conditions. An employer may combine two or more physical locations into a single establishment only when:

**1904.46(2)(i)** The employer operates the locations as a single business operation under common management;

**1904.46(2)(ii)** The locations are all located in close proximity to each other; and

**1904.46(2)(iii)** The employer keeps one set of business records for the locations, such as records on the number of employees, their wages and salaries, sales or receipts, and other kinds of business information. For example, one manufacturing establishment might include the main plant, a warehouse a few blocks away, and an administrative services building across the street.

**1904.46(3)** *If an employee telecommutes from home, is his or her home considered a separate establishment?* No, for employees who telecommute from home, the employee's home is not a business establishment and a separate 300 Log is not required. Employees who telecommute must be linked to one of your establishments under §1904.30(b)(3).

*Injury or illness.* An injury or illness is an abnormal condition or disorder. Injuries include cases such as, but not limited to, a cut, fracture, sprain, or amputation. Illnesses include both acute and chronic illnesses, such as, but not limited to, a skin disease, respiratory disorder, or poisoning. (Note: Injuries and illnesses are recordable only if they are new, work-related cases that meet one or more of the Part 1904 recording criteria.)

*Physician or Other Licensed Health Care Professional.* A physician or other licensed health care professional is an individual whose legally permitted scope of practice (i.e., license, registration, or certification) allows him or her to independently perform, or be delegated the responsibility to perform, the activities described by this regulation.

*You.* "You" means an employer as defined in Section 3 of the Occupational Safety and Health Act of 1970 (29 U.S.C. 652).

[66 FR 5915, Jan. 19, 2001]

# 29 CFR Part 1990—Identification, Classification, and Regulation of Potential Occupational Carcinogens

**AUTHORITY**: Secs. 4, 6, 8, Occupational Safety and Health Act of 1970 (29 U.S.C. 653, 655, 657); Secretary of Labor's Order 8-76 (41 FR 25059); and 29 CFR Part 1911.

**SOURCE**: 45 FR 5282, Jan. 22, 1980, unless otherwise noted.

## GENERAL

### §1990.101 Scope.

This part establishes criteria and procedures for the identification, classification, and regulation of potential occupational carcinogens found in each workplace in the United States regulated by the Occupational Safety and Health Act of 1970 (the Act). The procedures contained in this part supplement the procedural regulations in other parts of this chapter. In the event of a conflict, the procedures contained in this part shall govern the identification, classification, and regulation of potential occupational carcinogens. This part may be referred to as "The OSHA Cancer Policy."

### §1990.102 Purpose.

The Act provides, among other things, that the Secretary, in promulgating standards dealing with toxic materials or harmful physical agents under this subsection, shall set the standard which most adequately assures, to the extent feasible, on the basis of the best available evidence, that no employee will suffer material impairment of health or functional capacity even if such employee has regular exposure to the hazard dealt with by such standard for the period of his or her working life. Development of standards under this section shall be based upon research, demonstrations, experiments, and such other information as may be appropriate. In addition to the attainment of the highest degree of health and safety protection for the employee, other considerations shall be the latest available scientific data in the field, the feasibility of the standards, and experience gained under this and other health and safety laws. Whenever practicable, the standard promulgated shall be expressed in terms of objective criteria and of the performance desired (section 6(b)(5)).

It is the purpose of the regulations of this part to carry out the intent of the Act with respect to the identification, classification, and regulation of potential occupational carcinogens.

### §1990.103 Definitions.

Terms used in this part shall have the meanings set forth in the Act. In addition, as used in this part, the following terms shall have the meanings set forth below:

*Act* means the Occupational Safety and Health Act of 1970 (Pub. L. 91-596, 84 Stat. 1590 et seq., 29 U.S.C. 551 et seq. ).

*Administrator of EPA* means the Administrator of the United States Environmental Protection Agency, or designee.

*Chairperson of CPSC* means the Chairman of the United States Consumer Product Safety Commission, or designee.

*Commissioner of FDA* means the Commissioner of the Food and Drug Administration, United States Department of Health and Human Services, or designee.

*Director of NCI* means the Director of the National Cancer Institute, United States Department of Health and Human Services, or designee.

*Director of NIEHS* means the Director of the National Institute of Environmental Health Sciences, United States Department of Health and Human Services, or designee.

*Director of NIOSH* means the Director of the National Institute for Occupational Safety and Health, United States Department of Health and Human Services, or designee.

*Mutagenesis* means the induction of heritable changes in the genetic material of either somatic or germinal cells.

*Positive results in short-term tests* means positive results in assays for two or more of the following types of effect:

(1) The induction of DNA damage and/or repair;

(2) Mutagenesis in bacteria, yeast, *Neurospora* or *Drosophila melanogaster*;

(3) Mutagenesis in mammalian somatic cells;

(4) Mutagenesis in mammalian germinal cells; or

(5) Neoplastic transformation of mammalian cells in culture.

*Potential occupational carcinogen* means any substance, or combination or mixture of substances, which causes an increased incidence of benign and/or malignant neoplasms, or a substantial decrease in the latency period between exposure and onset of neoplasms in humans or in one or more experimental mammalian species as the result of any oral, respiratory or dermal exposure, or any other exposure which results in the induction of tumors at a site other than the site of administration. This definition also includes any substance which is metabolized into one or more potential occupational carcinogens by mammals.

*Secretary of HHS* means the Secretary of the United States Department of Health and Human Services, or designee.

### §1990.104 Scientific review panel.

**1990.104(a)** *General.* At any time, the Secretary may request the Director of NCI, the Director of NIEHS and/or the Director of NIOSH to convene a scientific review panel ("the panel") to provide recommendations to the Secretary in the identification, classification, or regulation of any potential occupational carcinogen.

**1990.104(b)** *Membership.* The panel will consist of individuals chosen by the respective Director(s). The panel will consist of individuals who are appropriately qualified in the disciplines relevant to the issues to be considered, and who are employed by the United States. The panel does not constitute an advisory committee within the meaning of section 6(b) or 7(b) of the Act, or the Federal Advisory Committee Act (Pub. L. 92-463, 86 Stat. 770).

1990.104(c) Report. The Secretary shall request that the panel submit a report of its evaluation within ninety (90) days after the appointment of the members of the panel. The Secretary shall place a copy of the report in the record of any relevant rulemaking undertaken pursuant to this part and allow an appropriate time for public review and comment. If a panel is not established or fails to file a timely report, or if the Secretary determines that it is necessary to proceed without waiting for the panel's report, the Secretary may proceed in making any determination without such report.

**1990.104(d)** *Other aid and assistance.* Nothing herein precludes the Secretary from obtaining advice or other aid from any person or organization including NCI, NIEHS, and NIOSH.

### §1990.105 Advisory committees.

The Secretary may appoint an Advisory Committee, pursuant to sections 6(b) and 7 of the Act, and 29 CFR part 1912, concerning any potential occupational carcinogen. The Secretary shall require the Advisory Committee to submit its recommendations to assist the Secretary in standard setting no later than ninety (90) days from the date of the Advisory Committee's appointment, unless extended by the Secretary for exceptional circumstances. If an Advisory Committee fails to file a timely report, the Secretary may proceed in standard setting activities without such a report.

### §1990.106 Amendments to this policy.

**1990.106(a)** *Initiation of review of this policy.*—**(1)** *Secretary's request.* No later than every three (3) years from the effective date of this part, or from the last general review, the Secretary shall request the Director of NCI, the Director of NIEHS and/or the Director of NIOSH, to review this part and render their opinions on whether significant scientific or technical advances made since the effective date of this part warrant any amendment to this part. The request shall ask that the answer be provided to the Secretary within one hundred twenty (120) days.

**1990.106(a)(2)** *Recommendations by the institutes.* At any time, the Director of NCI, the Director of NIEHS and/or the Director of NIOSH may submit recommendations to the Secretary for amendments to this part whenever any of them believes that scientific or technical advances justify such amendments.

**1990.106(a)(3)** *Petitions from the public.* **(i)** Any interested person may petition the Secretary concerning amendments to this part based upon substantial new issues or substantial new evidence.

**1990.106(a)(3)(ii)** For the purposes of this part, substantial new evidence is evidence which differs significantly from that presented in establishing this part, including amendments.

**1990.106(a)(3)(iii)** For the purposes of this part, substantial new issues are issues which differ significantly from those upon which the Secretary has reached a conclusion in the rulemaking establishing this part (including the conclusions reached in the preamble).

**1990.106(a)(3)(iv)** Each petition to amend this part shall contain at least the following information:

**1990.106(a)(3)(iv)(A)** Name and address of petitioner;

**1990.106(a)(3)(iv)(B)** The provisions which the petitioner believes are inappropriate;

**1990.106(a)(3)(iv)(C)** All data, views and arguments relied upon by the petitioner; and

**1990.106(a)(3)(iv)(D)** A detailed statement and analysis as to why the petitioner believes that the data, views and arguments presented by petitioner:

**1990.106(a)(3)(iv)(D)(i)** Constitute substantial new issues or substantial new evidence; and

**1990.106(a)(3)(iv)(D)(ii)** Are so significant as to warrant amendment of this part.

**1990.106(b)** *Response to recommendations and petitions.*—(1) *By the institutes.* Whenever any Director recommends an amendment to this part, the Secretary shall, within one hundred twenty (120) days after receipt of the recommendation, publish in the **FEDERAL REGISTER**, a notice which:

**1990.106(b)(1)(i)** States the reasons why the Secretary has determined not to commence a rulemaking proceeding to amend this part, in whole or in part, at that time; or

**1990.106(b)(1)(ii)** Commences a rulemaking proceeding to consider amending this part accordingly; or

**1990.106(b)(1)(iii)** Appoints an Advisory Committee as provided for by §1990.105 of this part and sections 6(b) and 7 of the Act.

**1990.106(b)(2)** *By the public.* Within ninety (90) days, or as soon thereafter as possible, after receipt of a petition pursuant to section 1990.106(a)(3), the Secretary shall:

**1990.106(b)(2)(i)** Refer the petition to the Director of NCI, the Director of NIEHS and/or the Director of NIOSH, in which case the provisions of §1990.106(a)(1) and (b)(1) are applicable; or

**1990.106(b)(2)(ii)** Appoint an advisory committee;

**1990.106(b)(2)(iii)** Deny the petition, briefly giving the reasons therefor; or

**1990.106(b)(2)(iv)** Commence a rulemaking proceeding to consider amending this part accordingly.

**1990.106(b)(3)** *On the Secretary's motion.* At any time, the Secretary may, on his own motion, commence a rulemaking proceeding to amend this part.

[45 FR 5282, Jan. 22, 1980; 45 FR 43405, June 27, 1980]

## THE OSHA CANCER POLICY

### §1990.111 General statement of regulatory policy.

**1990.111(a)** This part establishes the criteria and procedures under which substances will be regulated by OSHA as potential occupational carcinogens. Although the conclusive identification of "carcinogens" is a complex matter "on the frontiers of science," (*IUD v. Hodgson* 499 F.2d 467, 474 (D.C. Cir. 1974)), responsible health regulatory policy requires that criteria should be specified for the identification of substances which should be regulated as posing potential cancer risks to workers.

**1990.111(b)** The criteria established by this part are based on an extensive review of scientific data and opinions. The part provides for amending these criteria in light of new scientific developments. Decisions as to whether any particular substance meets the criteria or not will be consistent with the policies and procedures established by this part and will be based upon scientific evaluation of the evidence on that substance.

**1990.111(c)** This part applies to individual substances, groups of substances, or combinations or mixtures of substances which may be found in workplaces in the United States. In individual rulemaking proceedings under this part, the identity and range of substances and mixtures to be covered by the standard will be specified and the appropriateness of applying the available evidence to the range of substances and mixtures proposed for regulation will be subject to scientific and policy review.

**1990.111(d)** Potential occupational carcinogens will be identified and classified on the basis of human epidemiological studies and/or experimental carcinogenesis bioassays in mammals. Positive results in short term tests will also be used as concordant evidence.

**1990.111(e)** Potential occupational carcinogens will be classified and regulated in accordance with the policy. The scientific evidence as to whether individual substances meet these criteria will be considered in individual rulemakings. The issues which may be considered in these rulemakings will be limited as specified herein.

**1990.111(f)** This policy provides for the classification of potential occupational carcinogens into two categories depending on the nature and extent of the available scientific evidence. The two categories of potential occupational carcinogens may be regulated differently.

**1990.111(g)** The policy establishes a procedure for setting priorities and making them public.

**1990.111(h)** Worker exposure to Category I Potential Carcinogens will be reduced primarily through the use of engineering and work practice controls.

**1990.111(i)** Worker exposure to Category II Potential Carcinogens will be reduced as appropriate and consistent with the statutory requirements on a case-by-case basis in the rulemaking proceedings on individual substances. Any permissible exposure level so established shall be met primarily through engineering and work practice controls.

**1990.111(j)** The assessment of cancer risk to workers resulting from exposure to a potential occupational carcinogen will be made on the basis of available data. Because of the uncertainties and serious consequences to workers if the estimated risk is understated, cautious and prudent assumptions will be utilized to perform risk assessments.

**1990.111(k)** Where the Secretary determines that one or more suitable substitutes exist for certain uses of Category I Potential Carcinogens that are less hazardous to humans, a no occupational exposure level shall be set for those uses, to be achieved solely through the use of engineering and work practice controls to encourage substitution. In determining whether a substitute is suitable, the Secretary will consider the technological and economic feasibility of the introduction of the substitute, including its relative effectiveness and other relevant factors, such as regulatory requirements and the time needed for an orderly transition to the substitute.

[45 FR 5282, Jan. 22, 1980, as amended at 46 FR 5881, Jan. 21, 1981]

### §1990.112 Classification of potential carcinogens.

The following criteria for identification, classification and regulation of potential occupational carcinogens will be applied, unless the Secretary considers evidence under the provisions of §§1990.143, 1990.144 and 1990.145 and determines that such evidence warrants an exception to these criteria.

**1990.112(a)** *Category I Potential Carcinogens.* A substance shall be identified, classified, and regulated as a Category I Potential Carcinogen if, upon scientific evaluation, the Secretary determines that the substance meets the definition of a potential occupational carcinogen in (1) humans, or (2) in a single mammalian species in a long-term bioassay where the results are in concordance with some other scientifically evaluated evidence of a potential carcinogenic hazard, or (3) in a single mammalian species in an adequately conducted long-term bioassay, in appropriate circumstances where the Secretary determines the requirement for concordance is not necessary. Evidence of concordance is any of the following: positive results from independent testing in the same or other species, positive results in short-term tests, or induction of tumors at injection or implantation sites.

**1990.112(b)** *Category II Potential Carcinogens.* A substance shall be identified, classified, and regulated as a Category II Potential Carcinogen if, upon scientific evaluation, the Secretary determines that:

**1990.112(b)(i)** The substance meets the criteria set forth in §1990.112(a), but the evidence is found by the Secretary to be only "suggestive"; or

**1990.112(b)(ii)** The substance meets the criteria set forth in §1990.112(a) in a single mammalian species without evidence of concordance.

## PRIORITY SETTING

### §1990.121 Candidate list of potential occupational carcinogens.

**1990.121(a)** *Contents.* The Secretary shall prepare a list of substances (the "Candidate List") which are reported to be present in any American workplace and which, on the basis of a brief scientific review of available data, may be considered candidates for further scientific review and possible regulation

# OSHA Standards for General Industry

as Category I Potential Carcinogens or Category II Potential Carcinogens. For the purposes of this paragraph, "available data" means:

**1990.121(a)(1)** The data submitted by any person;

**1990.121(a)(2)** Any data referred to by the Secretary of HHS or by the Director of NIOSH, either in the latest list entitled "Suspected Carcinogens" or any other communication;

**1990.121(a)(3)** Literature referred to in U.S. Public Health Service, Publication No. 149;

**1990.121(a)(4)** Data summarized and reviewed in Monographs of the International Agency for Research on Cancer (IARC) of the World Health Organization;

**1990.121(a)(5)** The Toxic Substances Control Act Inventory of Chemical Substances, published by the Administrator of EPA;

**1990.121(a)(6)** The Secretary of HHS's Annual Report to the President and the Congress as required by the Community Mental Health Centers Extension Act of 1978, sec. 404(a)(9), 42 U.S.C. sec. 285.

**1990.121(a)(7)** Any other relevant data of which the Secretary has actual knowledge.

**1990.121(b)** *Tentative classification.* The Secretary may tentatively designate substances on the Candidate List as candidates for classification as Category I Potential Carcinogens or as Category II Potential Carcinogens, or may list substances without a tentative designation, based on the brief scientific review of available data for the purpose of initiating a more extensive scientific review.

**1990.121(c)** *No legal rights established.* The inclusion or exclusion of any substance from the Candidate List shall not be subject to judicial review nor be the basis of any legal action, nor shall the exclusion of any substance from the list prevent the regulation of that substance as a potential occupational carcinogen. The inclusion of a substance on the Candidate List and its possible tentative designation as a Category I Potential Carcinogen or a Category II Potential Carcinogen therein do not reflect a final scientific determination that the substance is, in fact, a Category I Potential Carcinogen or a Category II Potential Carcinogen. It is a policy determination based on the brief scientific review that the Secretary should conduct a thorough review of all relevant scientific data concerning the substance.

EFFECTIVE DATE NOTE: Paragraphs (a) and (b) of §1990.121 were stayed at 48 FR 243, Jan. 4, 1983, in order to evaluate the impact of publishing the Candidate Lists and Priority List and to reconsider the criteria used in establishing the lists (see also 47 FR 187, Jan. 5, 1982).

## §1990.122 Response to petitions.

Whenever the Secretary receives any information submitted in writing by any interested person concerning the inclusion or omission of any substance from the Candidate List, the Secretary shall briefly review the information and any other available data, as defined in §1990.121(a). The results of the Secretary's review shall be transmitted to the petitioner, together with a short statement of the Secretary's reasons therefor, and made public upon request.

EFFECTIVE DATE NOTE: Section 1990.122 was stayed at 48 FR 243, Jan. 4, 1983, in order to evaluate the impact of publishing the Candidate List and Priority Lists and to reconsider the criteria used in establishing the lists (see also 47 FR 187, Jan. 5, 1982).

## §1990.131 Priority lists for regulating potential occupational carcinogens.

The Secretary shall establish two priority lists for regulating potential occupational carcinogens. One list should include approximately ten (10) candidates for rulemaking as Category I Potential Carcinogens; the other approximately ten (10) candidates for rulemaking as Category II Potential Carcinogens. The order of placement of substances on these lists will not reflect the Secretary's determination of the exact order in which these substances should be regulated in rulemaking proceedings but rather a policy determination that the Secretary plans to address some or all of these substances prior to proceeding with a thorough scientific review of data concerning other substances on the Candidate List. The inclusion or exclusion of any substance on these lists shall not be subject to judicial review or be the basis for any legal action. The Secretary may regulate a potential occupational carcinogen which has not been placed on these lists. The inclusion of a substance on either of these lists does not reflect a final scientific determination that the substance is, in fact, a Category I Potential Carcinogen or a Category II Potential Carcinogen.

EFFECTIVE DATE NOTE: Section 1990.131 was stayed at 48 FR 243, Jan. 4, 1983, in order to evaluate the impact of publishing the Candidate List and Priority Lists and to reconsider the criteria used in establishing the lists (see also 47 FR 187, Jan. 5, 1982).

## §1990.132 Factors to be considered.

**1990.132(a)** The setting of priorities is a complex matter which requires subjective and policy judgments. It is not appropriate to establish a rigid formula or to assign predetermined weight to each factor. The identification of some of the elements is to guide the OSHA staff and inform the public on the development of priorities. It is not intended to create any legal rights with respect to the setting of priorities.

**1990.132(b)** Some factors which may be taken into account in setting priorities for regulating potential occupational carcinogens, when such data are available, are:

**1990.132(b)(1)** The estimated number of workers exposed;

**1990.132(b)(2)** The estimated levels of human exposure;

**1990.132(b)(3)** The levels of exposure to the substance which have been reported to cause an increased incidence of neoplasms in exposed humans, animals or both;

**1990.132(b)(4)** The extent to which regulatory action could reduce not only risks of contracting cancer but also other occupational and environmental health hazards;

**1990.132(b)(5)** Whether the molecular structure of the substance is similar to the molecular structure of another substance which meets the definition of a potential occupational carcinogen;

**1990.132(b)(6)** Whether there are substitutes that pose a lower risk of cancer or other serious human health problems, or available evidence otherwise suggests that the social and economic costs of regulation would be small; and

**1990.132(b)(7)** OSHA will also consider its responsibilities for dealing with other health and safety hazards and will consider the actions being taken or planned by other governmental agencies in dealing with the same or similar health and safety hazards.

## §1990.133 Publication.

**1990.133(a)** The Secretary shall publish the Candidate List in the **FEDERAL REGISTER** at least annually.

**1990.133(b)** The Secretary shall publish the Priority Lists in the **FEDERAL REGISTER** at least every six months and may seek public comment thereon.

**1990.133(c)** The Secretary may periodically publish in the **FEDERAL REGISTER** a notice requesting information concerning the classification and establishment of priorities for substances on the Candidate List together with a brief statement describing the type of information being sought.

EFFECTIVE DATE NOTE: Section 1990.133 was stayed at 48 FR 243, Jan. 4, 1983, in order to evaluate the impact of publishing the Candidate List and Priority Lists and to reconsider the criteria used in establishing the lists (see also 47 FR 187, Jan. 5, 1982).

## REGULATION OF POTENTIAL OCCUPATIONAL CARCINOGENS

## §1990.141 Advance notice of proposed rulemaking.

**1990.141(a)** Within thirty (30) days after OSHA initiates a study concerning the economic and/or technological feasibility of specific standards that might be applied in the regulation of a potential occupational carcinogen, the Secretary will normally publish, in the **FEDERAL REGISTER**, a notice which includes at least the following:

**1990.141(a)(1)** The name of the substance(s),

**1990.141(a)(2)** The scope of the study, including where possible,

**1990.141(a)(2)(i)** Affected industries,

**1990.141(a)(2)(ii)** Levels of exposure being studied,

**1990.141(a)(2)(iii)** The anticipated completion date of the study;

**1990.141(a)(3)** A brief summary of the available data on health effects;

**1990.141(a)(4)** An estimate of when the Secretary anticipates the issuance of a proposal;

**1990.141(a)(5)** An invitation to interested parties to provide relevant information;

**1990.141(a)(6)** A statement that persons wishing to provide OSHA with their own study should complete it within 30 days after the anticipated proposal date; and

**1990.141(a)(7)** A statement of the procedural requirements that must be met before substantial new issues or substantial new evidence will be considered in the proceeding pursuant to §1990.145.

**1990.141(b)** Where the Secretary determines to discontinue a feasibility study, the Secretary should publish, within 30 days, a notice in the **FEDERAL REGISTER** so indicating.

### §1990.142 Initiation of a rulemaking.

Where the Secretary decides to regulate a potential occupational carcinogen, the Secretary shall initiate a rulemaking proceeding in accordance with one of the following procedures, as appropriate.

**1990.142(a)** *Notice of proposed rulemakings (section 6(b) of the Act)*—**(1)** *General.* The Secretary may issue a notice of proposed rulemaking in the **FEDERAL REGISTER**, pursuant to section 6(b) of the Act and part 1911 of this chapter. The notice shall provide for no more than a sixty (60) day comment period, and may provide for a hearing, which shall be scheduled for no later than one hundred (100) days after publication of the Notice of Proposed Rulemaking. The commencement of the hearing may be postponed once, for no more than thirty (30) days, for good cause shown.

**1990.142(a)(2)** *Provisions of the proposed standard for Category I Potential Carcinogens.* Whenever the Secretary issues a notice of proposed rulemaking to regulate a substance as a Category I Potential Carcinogen:

**1990.142(a)(2)(i)** The proposed standard shall contain at least provisions for scope and application, definitions, notification of use, a permissible exposure limit, monitoring, regulated areas, methods of compliance including the development of a compliance plan, respiratory protection, protective clothing and equipment, housekeeping, waste disposal, hygiene facilities, medical surveillance, employee information and training, signs and labels, recordkeeping, and employee observation of monitoring as set forth in §1990.151, unless the Secretary explains why any or all such provisions are not appropriate;

**1990.142(a)(2)(ii)** The model standard set forth in §1990.151 shall be used as a guideline, and

**1990.142(a)(2)(iii)** The permissible exposure limit shall be achieved primarily through engineering and work practice controls except that if a suitable substitute is available for one or more uses no occupational exposure shall be permitted for those uses.

**1990.142(a)(3)** *Provisions of the proposed standard for Category II Potential Carcinogens.* Whenever the Secretary issues a Notice of Proposed Rulemaking to regulate a substance as a Category II Potential Carcinogen:

**1990.142(a)(3)(i)** The proposed standard shall contain at least provisions for scope and application, definitions, notification of use, monitoring, respiratory protection, protective clothing and equipment, housekeeping, waste disposal, medical surveillance, employee information and training, recordkeeping and employee observation of monitoring as set forth in §1990.151, unless the Secretary explains why any or all such provisions are not appropriate; and

**1990.142(a)(3)(ii)** The model standard set forth in §1990.151 shall be used as a guideline; and

**1990.142(a)(3)(iii)** Worker exposure to Category II Potential Carcinogens will be reduced as appropriate and consistent with the statutory requirements on a case-by-case basis in the individual rulemaking proceedings. Any permissible exposure level so established shall be met primarily through engineering and work practice controls.

**1990.142(b)** *Emergency temporary standards* (Section 6(c) of the Act)—**(1)** *General.* The Secretary may issue an Emergency Temporary Standard (ETS) for a Category I Potential Carcinogen in accordance with section 6(c) of the Act.

**1990.142(b)(2)** *Provisions of the ETS.* **(i)** The ETS shall contain at least provisions for scope and application, definitions, notification of use, a permissible exposure limit, monitoring, methods of compliance including the development of a compliance plan, respiratory protection, protective clothing and equipment, housekeeping, waste disposal, medical surveillance, employee information and training, signs and labels, recordkeeping and employee observation of monitoring, unless the Secretary explains why any or all such provisions are not appropriate.

**1990.142(b)(2)(ii)** The model standard set forth in §1990.152 shall be used as a guideline.

**1990.142(b)(2)(iii)** The permissible exposure limit shall be achieved through any practicable combination of engineering controls, work practice controls and respiratory protection.

[45 FR 5282, Jan. 22, 1980, as amended at 46 FR 5881, Jan. 21, 1981]

### §1990.143 General provisions for the use of human and animal data.

Human and animal data which are scientifically evaluated to be positive evidence for carcinogenicity including the following policies shall be uniformly relied upon for the identification of potential occupational carcinogens. Arguments challenging the following provisions or their application to specific substances will be considered in individual rulemaking proceedings only if the evidence presented in support of the arguments meets the criteria for consideration specified in §1990.144 or §1990.145.

**1990.143(a)** *Positive human studies.* Positive results obtained in one or more human epidemiologic studies will be used to establish the qualitative inference of carcinogenic hazards to workers.

**1990.143(b)** *Positive animal studies.* Positive results obtained in one or more experimental studies conducted in one or more mammalian species will be used to establish the qualitative inference of carcinogenic hazard to workers. Arguments that positive results obtained in mammalian species should not be relied upon will be considered only if evidence is presented which meets the criteria for consideration specified in §1990.144(c) or 1990.144(f).

**1990.143(c)** *Non-positive human studies.* Positive results in human or mammalian studies generally will be used for the qualitative identification of potential occupational carcinogens, even where non-positive results from human studies exist. Such non-positive results will be considered by the Secretary only if the studies or results meet the criteria set forth in §1990.144(a).

**1990.143(d)** *Non-positive animal studies.* Positive results in one or more mammalian studies will be used for the qualitative identification of potential occupational carcinogens, even where non-positive studies exist in other mammalian species. Where non-positive and positive results exist in studies in the same species, the non-positive results will be evaluated.

**1990.143(e)** *Spontaneous tumors.* Positive results in human or mammalian studies for the induction or acceleration of induction of tumors of a type which occurs "spontaneously" in unexposed individuals will be used for the qualitatative identification of potential occupational carcinogens.

**1990.143(f)** *Routes of exposure.* **(1)** Positive results in studies in which mammals are exposed via the oral, respiratory or dermal routes will be used for the qualitative identification of potential occupational carcinogens, whether tumors are induced at the site of application or distant sites.

**1990.143(f)(2)** Positive results in studies in which mammals are exposed via any route of exposure and in which tumors are induced at sites distant from the site of administration will be used for the qualitative identification of potential occupational carcinogens.

**1990.143(f)(3)(i)** Positive results in mammalian studies in which tumors are induced only at the site of administration, in which a substance or mixture of substances is administered by routes other than oral, respiratory or dermal, will be used as "concordant" evidence that a substance is a potential occupational carcinogen.

**1990.143(f)(3)(ii)** Arguments that such studies should not be relied upon will be considered only if evidence which meets the criteria set forth in §1990.144(b) is provided.

**1990.143(g)** *Use of high doses in animal testing.* Positive results for carcinogenicity obtained in mammals exposed to high doses of a substance will be used to establish the qualitative inference of carcinogenic hazard to workers. Arguments that such studies should not be relied upon will be considered only if evidence which meets the criteria set forth in §1990.144(d) is provided.

**1990.143(h)** *"Threshold" or "No-effect" Levels.* No determination will be made that a "threshold" or "no-effect" level of exposure can be established for a human population exposed to carcinogens in general, or to any specific substance.

**1990.143(i)** *Benign tumors.* Results based on the induction of benign or malignant tumors, or both, will be used to establish a qualitative inference of carcinogenic hazard to workers. Arguments that substances that induce benign tumors do not present a carcinogenic risk to workers will be considered only if evidence that meets the criteria set forth in §1990.144(e) is provided.

**1990.143(j)** *Statistical evaluation.* Statistical evaluation will be used in the determination of whether results in human, animal or short-term studies provide positive evidence for carcinogenicity, but will not be the exclusive means for such evaluation.

**1990.143(k)** *Carcinogenicity of metabolites.* A substance which is metabolized by mammals to yield one or more potential occupational carcinogens will itself be identified and classified as a potential occupational carcinogen, whether or not there is direct evidence that it induces tumors in humans or experimental animals. Evidence for such metabolism will normally be derived from *in vivo* studies in mammals. In appropriate circumstances, evidence may be derived from *in vitro* studies of mammalian tissues or fractions thereof. Arguments that evidence from *in vitro* metabolic studies in mammals is not relevant to the inference of carcinogenic hazard to humans will be considered only if such evidence meets the criteria set forth in §1990.144(c).

[45 FR 5282, Jan. 22, 1980; 45 FR 43405, June 27, 1980]

## §1990.144 Criteria for consideration of arguments on certain issues.

Arguments on the following issues will be considered by the Secretary in identifying or classifying any substance pursuant to this part, if evidence for the specific substance subject to the rulemaking conforms to the following criteria. Such arguments and evidence will be evaluated based upon scientific and policy judgments.

**1990.144(a)** *Non-positive results obtained in human epidemiologic studies.* Non-positive results obtained in human epidemiologic studies regarding the substance subject to the rulemaking or to a similar or closely related substance will be considered by the Secretary only if they meet the following criteria:

*Criteria.* (i) The epidemiologic study involved at least 20 years' exposure of a group of subjects to the substance and at least 30 years' observation of the subjects after initial exposure;

(ii) Documented reasons are provided for predicting the site(s) at which the substance would induce cancer if it were carcinogenic in humans; and

(iii) The group of exposed subjects was large enough for an increase in cancer incidence of 50% above that in unexposed controls to have been detected at any of the predicted sites.

Arguments that non-positive results obtained in human epidemiologic studies should be used to establish numerical upper limits on potential risks to humans exposed to specific levels of a substance will be considered only if criteria (i) and (ii) are met and, in addition:

(iv) Specific data on the level of exposure of the group of workers are provided, based either on direct measurements made periodically throughout the period of exposure, or upon other data which provide reliable evidence of the magnitude of exposure.

**1990.144(b)** *Tumors induced at site of administration.* Arguments that tumors at the site of administration should not be considered will be considered only if:

**1990.144(b)(i)** The route of administration is not oral, respiratory or dermal; and

**1990.144(b)(ii)** Evidence is provided which establishes that induction of local tumors is related to the physical configuration or formulation of the material administered (*e.g.*, crystalline form or dimensions of a solid material, or matrix of an impregnated implant) and that tumors are not induced when the same material is administered in a different configuration or formula.

**1990.144(c)** *Metabolic differences.* Arguments that differences in metabolic profiles can be used to demonstrate that a chemical found positive in an experimental study in a mammalian species would pose no potential carcinogenic risk to exposed workers will be considered by the Secretary only if the evidence presented for the specific substance subject to the rulemaking meets the following criteria:

*Criteria* (i) A complete metabolic profile, including identities of trace metabolites, is presented for the experimental animal species;

(ii) A complete metabolic profile, including identities of trace metabolites, is available for a human population group representative of those who are occupationally exposed;

(iii) Documented evidence is provided for ascribing the carcinogenic activity of the substance in the test animal species to metabolite(s) produced only in that species and not in humans; and

(iv) Documented evidence is provided to show that other metabolites produced also in humans have been adequately tested and have not been shown to be carcinogenic.

**1990.144(d)** *Use of high doses in animal testing.* Arguments that positive results obtained in carcinogenesis bioassays with experimental animals subjected to high doses of a substance are not relevant to potential carcinogenic risks to exposed workers will be considered by the Secretary only if the evidence for the specific substance subject to the rulemaking meets the following criteria:

*Criteria.* (i) Documented evidence is presented to show that the substance in question is metabolized by the experimental animal species exposed at the dose levels used in the bioassay(s) to metabolic products which include one or more that are not produced in the same species at lower doses.

(ii) Documented evidence is presented to show that the metabolite(s) produced only at high doses in the experimental animal species are the ultimate carcinogen(s) and that the metabolites produced at low doses are not also carcinogenic; and

(iii) Documented evidence is presented to show that the metabolite(s) produced only at high doses in the experimental animal species are not produced in humans exposed to low doses.

**1990.144(e)** *Benign tumors.* The Secretary will consider evidence that the substance subject to the rulemaking proceeding is capable only of inducing benign tumors in humans or experimental animals provided that the evidence for the specific substance meets the following criteria:

*Criteria* (i) Data are available from at least two well-conducted biosassays in each of two species of mammals (or from equivalent evidence in more than two species);

(ii) Each of the bioassays to be considered has been conducted for the full lifetime of the experimental animals;

(iii) The relevant tissue slides are made available to OSHA or its designee and the diagnoses of the tumors as benign are made by at least one qualified pathologist who has personally examined each of the slides and who provides specific diagnostic criteria and descriptions and

(iv) All of the induced tumors must be shown to belong to a type which is known not to progress to malignancy or to be at a benign stage when observed. In the latter case, data must be presented to show that multiple sections of the affected organ(s) were adequately examined to search for invasion of the tumor cells into adjacent tissue, and that multiple sections of other organs were adequately examined to search for tumor metastases.

**1990.144(f)** *Indirect mechanisms.* The Secretary will consider evidence that positive results obtained in a carcinogenesis bioassay with experimental animals are not relevant to a determination of a carcinogenic risk to exposed workers, if the evidence demonstrates that the mechanism by which the observed tumor incidence is effected is indirect and would not occur if humans were exposed. As examples, evidence will be considered that a substance causes a carcinogenic effect by augmenting caloric intake or that the carcinogenic effect from exposure to a substance is demonstrated to be the result of the presence of a carcinogenic virus and it is demonstrated that, in either case, the effect would not take place in the absence of the particular carcinogenic virus or the augmented caloric intake.

[45 FR 5282, Jan. 22, 1980, as amended at 46 FR 5881, Jan. 21, 1981]

## §1990.145 Consideration of substantial new issues or substantial new evidence.

**1990.145(a)** *Substantial new issues.* Notwithstanding any other provision of this part, the Secretary will consider in a rulemaking proceeding on a specific substance any substantial new issues upon which the Secretary did not reach a conclusion in the rulemaking proceeding(s) underlying this part including conclusions presented in the preamble.

**1990.145(b)** *Substantial new evidence.* Notwithstanding any other provision of this part, the Secretary will consider in a rulemaking proceeding on a specific substance any arguments, data or views which he determines are based upon substantial new evidence which may warrant the amendment of one or more provisions of this part. For the purposes of this part, "substantial new evidence" is evidence directly relevant to any provision of this part and is based upon data, views or arguments which differ significantly from those presented in establishing this part, including amendments thereto.

**1990.145(c)** *Petitions for consideration of substantial new evidence.*—**(1)** *Petition.* Any interested person may file a written petition with the Secretary to consider "substantial new evidence" or one or more "substantial new issues" which contains the information specified in paragraph (c)(2) of this section. The Secretary shall treat such a petition as a request to amend this part, as well as a petition to consider "substantial new evidence".

**1990.145(c)(2)** *Contents.* Each petition for consideration of "substantial new evidence" or one or more "substantial new issues" shall contain at least the following information:

**1990.145(c)(2)(i)** Name and address of the petitioner;

**1990.145(c)(2)(ii)** All of the data, views and arguments that the petitioner would like the Secretary to consider;

**1990.145(c)(2)(iii)** The provision or provisions that petitioner believes are inappropriate or should be added to this part in light of the new data, views, and arguments;

**1990.145(c)(2)(iv)** A statement which demonstrates that the data, views, and arguments relied upon by petitioners are directly relevant to the substance or class of substances that is the subject of a rulemaking or an Advance Notice of Proposed Rulemaking;

**1990.145(c)(2)(v)** A detailed statement and analysis as to why the petitioner believes that the data, views, and arguments presented by the petitioner:

**1990.145(c)(2)(v)(A)** Differ significantly from those presented in the proceeding(s) which establish this part;

**1990.145(c)(2)(v)(B)** Are so substantial as to warrant amendment of this part; and

**1990.145(c)(2)(v)(C)** Constitute a new issue or new evidence within the meaning of paragraphs (a) and (b) of this section.

**1990.145(c)(3)** *Deadline for petitions.* **(i)** Petitions which comply with paragraph (c) of this section, shall be filed in accordance with the schedule set forth in the Advance Notice of Proposed Rulemaking.

**1990.145(c)(3)(ii)** In extraordinary cases the Secretary may consider evidence submitted after the deadline if the petitioner establishes that the evidence relied upon was not available and could not have reasonably been available in whole or substantial part by the deadline and that it is being submitted at the earliest possible time.

**1990.145(d)** *Secretary's response.* **(1)** The Secretary shall respond to petitions under this paragraph in accordance with §1990.106.

**1990.145(d)(2)** Whenever the Secretary determines that the "substantial new issue" or the "substantial new evidence" submitted under this paragraph is sufficient to initiate a proceeding to amend this part, the Secretary shall:

**1990.145(d)(2)(i)** Issue a notice to consider amendment to this part and not proceed on the rulemaking concerning the individual substance until completion of the amendment proceeding; or

**1990.145(d)(2)(ii)** Issue a notice to consider amendment to this part and consolidate it with the proceeding on the individual substance.

### §1990.146 Issues to be considered in the rulemaking.

Except as provided in §1990.145, after issuance of the advance notice of rulemaking, the proceedings for individual substances under this part shall be limited to consideration of the following issues:

**1990.146(a)** Whether the substance, group of substances or combination of substances subject to the proposed rulemaking is appropriately considered in a single proceeding;

**1990.146(b)** Whether the substance or group of substances subject to the rulemaking meets the definition of a potential occupational carcinogen set forth in §1990.103, including whether the scientific studies are reliable;

**1990.146(c)** Whether the available data can appropriately be applied to the substance, group of substances or combination of substances covered by the rulemaking;

**1990.146(d)** Whether information, data, and views that are submitted in accordance with §1990.144 are sufficient to warrant an exception to this part;

**1990.146(e)** Whether the data, views and arguments that are submitted in accordance with §1990.145 are sufficient to warrant amendment of this part;

**1990.146(f)** Whether the potential occupational carcinogen meets the criteria for a Category I Potential Carcinogen or a Category II Potential Carcinogen;

**1990.146(g)** The environmental impact arising from regulation of the substance;

**1990.146(h)** Any issues required by statute or executive order;

**1990.146(i)** The determination of the level to control exposures to Category I Potential Carcinogens primarily through the use of engineering and work practice controls including technological and economic considerations.

**1990.146(j)** The determination of the appropriate employee exposure level, consistent with the Act's requirements, for Category II Potential Carcinogens;

**1990.146(k)** Whether suitable substitutes are available for one or more uses of Category I Potential Carcinogens and; if so, the no occupational exposure level to be achieved solely with engineering and work practice controls and other issues relevant to substitution; and

**1990.146(l)** Whether the provisions of the proposal and of §§1990.151 and 1990.152 (model standards) are appropriate, except as limited by §1990.142 and whether additional regulatory provisions may be appropriate.

[45 FR 5282, Jan. 22, 1980, as amended at 46 FR 5881, Jan. 21, 1981]

### §1990.147 Final action.

**1990.147(a)** Within one hundred twenty (120) days from the last day of any hearing or ninety (90) days from the close of any post hearing comment period, whichever occurs first, the Secretary shall publish in the **FEDERAL REGISTER:**

**1990.147(a)(1)** A final standard based upon the record in the proceeding; or

**1990.147(a)(2)** A statement that no final standard will be issued, and the reasons therefor, or

**1990.147(a)(3)** A statement that the Secretary intends to issue a final rule, but that he is unable to do so at the present time, including:

**1990.147(a)(3)(i)** The reasons therefor; and

**1990.147(a)(3)(ii)** The date by which the standard will be published, which may not exceed one hundred twenty (120) days thereafter.

**1990.147(a)(3)(iii)** The Secretary may issue no more than one such notice, unless the Secretary determines that (A) new evidence which was unavailable during the rulemaking proceeding has just become available; (B) the evidence is so important that a final rule could not reasonably be issued without this evidence, and; (C) the record is reopened for receipt of comments and/or a hearing on this evidence. This paragraph does not require the Secretary to consider any evidence which is submitted after the dates established for the submission of evidence.

**1990.147(b)** The failure of the Secretary to comply with the required timeframes shall not be a basis to set aside any standard or to require the issuance of a new proposal on any individual substance.

**1990.147(c)** The final standard shall state whether the substance or group of substances subject to the rulemaking is classified as a Category I Potential Carcinogen or as a Category II Potential Carcinogen. If the classification differs from that in the notice of proposed rulemaking, the Secretary shall explain the reasons for the change in classification in the preamble to the final standard.

**1990.147(d)** If the substance is classified as a Category I Potential Carcinogen, the final standard shall conform to the provisions of §1990.142(a)(2)(iii). If the final standard contains other provisions that substantially differ from the proposed provisions, the Secretary shall explain the reasons for the changes in the preamble to the final standard.

**1990.147(e)** If the substance is classified as a Category II potential carcinogen, the final standard shall conform to the provisions of §1990.142(a)(3)(iii). If the final standard contains other provisions that substantially differ from the proposed provisions, the Secretary shall explain the reasons for the changes in the preamble to the final standard.

**1990.147(f)** If the substance is classified as a Category II potential carcinogen, the Secretary shall notify the applicable federal and state agencies, including the Administrator of EPA, the Director of NCI, the Director of NIEHS, the Director of NIOSH, the Commissioner of FDA and the Chair-

# OSHA Standards for General Industry

person of CPSC of such determination and request that the applicable agencies engage in, or stimulate, further research pursuant to their legislative authority, to develop new and additional scientific data.

**1990.147(g)** If, after a rulemaking, the Secretary determines that the substance under consideration should not be classified as a Category I potential carcinogen or a Category II potential carcinogen, the Secretary shall publish a notice of this determination in the **FEDERAL REGISTER**, together with the reasons therefor.

## MODEL STANDARDS

### §1990.151 Model standard pursuant to section 6(b) of the Act.

Occupational Exposure to _____
Permanent Standard (insert section number of standard)

**1990.151(a)** *Scope and application*—**(1)** *General.* This section applies to all occupational exposures to _____ or to (specify those uses or classes of uses of _____ [Chemical Abstracts Service Registry Number 0000] which are covered by the standard, including, where appropriate, the type of exposure to be regulated by the standard) except as provided in paragraph (a)(2).

**1990.151(a)(2)** *Exemptions.* This section does not apply to (insert those uses or classes of uses of _____ which are exempted from compliance with the standard, including, where appropriate,

**1990.151(a)(2)(i)** Workplaces where exposure to _____ results from solid or liquid mixtures containing a specified percentage of _____ or less;

**1990.151(a)(2)(ii)** Workplaces where another Federal agency is exercising statutory authority to prescribe or enforce standards or regulations affecting occupational exposure to _____; or

**1990.151(a)(2)(iii)** Workplaces which are appropriately addressed in a separate standard).

**1990.151(b)** *Definitions.*

_____ means (definition of the substance, group of substances, *or combination of substances, to be regulated* ).

*Action level* means an airborne concentration of _____ of (insert appropriate level of exposure).

**NOTE:** Where appropriate, consider an action level as a limitation on requirements for periodic monitoring (para. (e)(3)), medical surveillance (para. (n)), training (para. (o)), labels (para. (p)(3)), and other provisions.

*Assistant Secretary* means the Assistant Secretary of Labor for Occupational Safety and Health, U.S. Department of Labor, or designee.

*Authorized person* means any person specifically authorized by the employer whose duties require the person to enter regulated areas or any person entering such an area as a designated representative of employees for the purpose of exercising the opportunity to observe monitoring procedures under paragraph (r) of this section.

*Director* means the Director, National Institute for Occupational Safety and Health, U.S. Department of Health, and Health Services, or designee.

*Emergency* means in any occurrence such as, but not limited to, equipment failure, rupture of containers, or failure of control equipment which may result in a release of _____ which is (insert appropriate quantitative or qualitative level of release which constitutes an emergency).

*OSHA Area Office* means the Area Office of the Occupational Safety and Health Administration having jurisdiction over the geographic area where the affected workplace is located.

**1990.151(c)** *Permissible exposure limits provisions*—**(1)** *Inhalation*—**(i)** *Time weighted average limit (TWA).* Within (insert appropriate time period) of the effective date of this section, the employer shall assure that no employee is exposed to an airborne concentration of _____ in excess of: (insert appropriate exposure limit or when it is determined by the Secretary that there are available suitable substitutes for uses or classes of uses that are less hazardous to humans, the proposal shall permit no occupational exposure) as an eight (8)-hour-time-weighted average.

(Where the Secretary finds that suitable substitutes for _____ may exist, the determination of the _____ level shall include consideration of the availability, practicability, relative degree of hazard, and economic consequences of the substitutes.)

**1990.151(c)(1)(ii)** *Ceiling limit (If appropriate).* Within (insert appropriate time period) of the effective date of this section, the employer shall assure that no employee is exposed to an airborne concentration of _____ in excess of: (insert exposure limit) as averaged over any: (insert appropriate time period) during the working day.

**1990.151(c)(2)** *Dermal and eye exposure.* (As appropriate.) (i) Within (insert appropriate time period) of the effective date of this section, the employer shall (If eye exposure to _____ does not create a risk of cancer, insert exposure level or criteria which will prevent other adverse health effects of eye exposure to _____ if any. If eye exposure creates a risk of cancer, insert exposure level or criteria which represents the level of eye exposure to _____).

**1990.151(c)(2)(ii)** Within (insert appropriate time period) of the effective date of this section, the employer shall (If skin exposure to _____ does not create a risk of cancer, insert exposure level or criteria which will prevent other adverse health effects of skin exposure to _____ if any. If skin exposure creates a risk of cancer, insert exposure level or criteria which represents the level of skin exposure to _____).

**1990.151(d)** *Notification of use and emergencies*—**(1)** *Use.* Within (insert appropriate time period and additional information requirements if appropriate), of the effective date of this standard or within thirty days of the introduction of _____ into the workplace, every employer who has a place of employment in which _____, is present shall report the address and location of each place of employment to the OSHA Area Office and an estimate of the number of employees exposed.

**1990.151(d)(2)** *Emergencies.* Emergencies, and the facts obtainable at that time, shall be reported within (insert appropriate number) hours of, or during the first federal working day after, the time the employer becomes aware of the emergency to the OSHA Area Office, whichever is longer. Upon request of the OSHA Area Office, the employer shall submit additional information in writing relevant to the nature and extent of employee exposures and measures taken to prevent future emergencies of a similar nature.

**1990.151(e)** *Exposure monitoring*—**(1)** *General.* **(i)** Determinations of airborne exposure levels shall be made from air samples that are representative of each employee's exposure to _____, over an eight (8) hour period. (Modify the time period as appropriate to be practical in the relevant industries yet reasonably representative of full shift exposures.) Monitoring of exposure levels required under this paragraph shall be made as follows: [insert method or alternative methods to be used to meet the requirements of this paragraph].

**1990.151(e)(1)(ii)** For the purpose of this section, employee exposure is that exposure which would occur if the employee were not using a respirator.

**1990.151(e)(2)** *Initial monitoring.* Each employer who has one or more workplaces where (specify the types of workplaces subject to the monitoring requirement) shall, within (insert appropriate period) of the effective date of this section (insert requirements for initial monitoring, as appropriate).

**1990.151(e)(3)** *Frequency.* (Insert, if appropriate, provisions prescribing the minimum frequency at which monitoring must be repeated, the conditions under which such frequency must be increased or may be reduced, and conditions under which such routine monitoring may be discontinued (for example, where the action level is not exceeded). Where appropriate, specify different frequency requirements for certain types of workplaces where, for example, exposure levels are subject to greater or less variability.)

**1990.151(e)(4)** *Additional monitoring.* (Insert, if appropriate, provisions for monitoring, in addition to the requirements (if any) of paragraph (e)(3). This may include a production, process, control or personnel change which might result in new or additional exposure to _____, or whenever the employer has any other reason to suspect a change which might result in new or additional exposures to _____.)

**1990.151(e)(5)** *Employee notification.* **(i)** Within (insert appropriate period) after the receipt of monitoring results, the employer shall notify each employee in writing of the results which represent that employee's exposure.

**1990.151(e)(5)(ii)** Whenever the results indicate that the representative employee exposure exceeds the permissible exposure limits, the employer shall include in the written notice a statement that the permissible exposure limits were exceeded and a description of the corrective action being taken to reduce exposure to or below the permissible exposure limits.

**1990.151(e)(6)** *Accuracy of measurement.* (Insert requirements for accuracy of methods of measurement or detection used to comply with the paragraph).

**1990.151(f)** *Regulated areas*—**(1)** Within (insert appropriate time period) of the effective date of this section, the employer shall, where practicable, establish regulated areas where _____, concentrations are in excess of the permissible exposure limits.

**1990.151(f)(2)** Regulated areas shall be demarcated and segregated from the rest of the workplace, in any manner that minimizes the number of persons who will be exposed to _____.

**1990.151(f)(3)** Access to regulated areas shall be limited to authorized persons or to persons otherwise authorized by the Act or regulations issued pursuant thereto.

**1990.151(f)(4)** The employer shall assure that in the regulated area, food or beverages are not present or consumed, smoking products are not present or used, and cosmetics are not applied (except that these activities may be conducted in the lunchroom, change rooms and showers required under paragraphs (m)(1) through (m)(3) of this section).

**1990.151(g)** *Methods of compliance*—**(1)** *Engineering and work practice controls.* **(i)** The employer shall institute engineering or work practice controls to reduce and maintain employee exposures to _____ to or below the permissible exposure limits, except to the extent that the employer establishes that such controls are not feasible.

**1990.151(g)(1)(ii)** Engineering and work practice controls shall be implemented to reduce exposures even if they will not be sufficient to reduce exposures to or below the permissible exposure limits.

**1990.151(g)(2)** *Compliance program.* **(i)** Within (insert appropriate period) of the effective date of this section, the employer shall establish and implement a written program to reduce exposures to or below the permissible exposure limits by means of engineering and work practice controls, as required by paragraph (g)(1) of this section.

**1990.151(g)(2)(ii)** Written plans for these compliance programs shall include at least the following:

**1990.151(g)(2)(ii)(A)** A description of each operation or process resulting in employee exposure to _____;

**1990.151(g)(2)(ii)(B)** Engineering plans and other studies contemplated or used to determine the controls for each process;

**1990.151(g)(2)(ii)(C)** A report of the technology considered or to be considered in meeting the permissible exposure limits;

**1990.151(g)(2)(ii)(D)** A detailed schedule for the implementation of engineering or work practice controls; and

**1990.151(g)(2)(ii)(E)** Other relevant information reasonably requested by OSHA.

**1990.151(g)(2)(iii)** Written plans for such a program shall be submitted, upon request, to the Assistant Secretary and the Director, and shall be available at the worksite for examination and copying by the Assistant Secretary, the Director, or any affected employee or designated representative.

**1990.151(g)(2)(iv)** The plans required by this paragraph shall be revised and updated periodically to reflect the current status of the program.

**1990.151(h)** *Respiratory protection*—**(1)** *General.* The employer shall assure that respirators are used where required pursuant to this section to reduce employee exposures to or below the permissible exposure limits and in emergencies. Compliance with the permissible exposure limits may not be achieved by the use of respirators except:

**1990.151(h)(1)(i)** During the time period necessary to install or implement feasible engineering and work practice controls; or

**1990.151(h)(1)(ii)** In work operations in which the employer establishes that engineering and work practice controls are not feasible; or

**1990.151(h)(1)(iii)** In work situations where feasible engineering and work practice controls are not yet sufficient to reduce exposure to or below the permissible exposure limits; or

**1990.151(h)(1)(iv)** In emergencies.

**1990.151(h)(2)** *Respirator selection.* **(i)** Where respiratory protection is required under this section, the employer shall select and provide at no cost to the employee, the appropriate type of respirator from Table 1 below and shall assure that the employee wears the respirator provided.

**TABLE 1—RESPIRATORY PROTECTION FOR** _____

(The table will contain a listing of the appropriate type of respirator for various conditions of exposure to _____).

**1990.151(h)(2)(ii)** The employer shall select respirators from those approved by the National Institute for Occupational Safety and Health under the provisions of 30 CFR part 11.

**1990.151(h)(3)** *Respirator program.* **(i)** The employer shall institute a respiratory protection program in accordance with 29 CFR 1910.134(b), (d), (e), and (f).

**1990.151(h)(3)(ii)** Employees who wear respirators shall be allowed to wash their face and respirator facepiece to prevent potential skin irritation associated with respirator use.

**1990.151(h)(3)(iii)** The employer shall assure that the respirator issued to each employee is properly fitted (as appropriate, indicate the requirement for a qualitative or quantitative respirator fit testing program).

**1990.151(i)** *Emergency situations*—**(1)** *Written plans.* **(i)** A written plan for emergency situations shall be developed for each workplace where _____ is present. Appropriate portions of the plan shall be implemented in the event of an emergency.

**1990.151(i)(1)(ii)** The plan shall specifically provide that employees engaged in correcting emergency conditions shall be equipped with respirators as required in paragraph (h) of this section and other necessary personal protective equipment as required in paragraph (j) until the emergency is abated.

**1990.151(i)(2)** *Alerting employees*—**(i)** *Alarms.* Where there is the possibility of employee exposure to _____, due to the occurrence of an emergency, a general alarm shall be installed and maintained to promptly alert employees of such occurrences.

**1990.151(i)(2)(ii)** *Evacuation.* Employees not engaged in correcting the emergency shall be restricted from the area and shall not be permitted to return until the emergency is abated.

**1990.151(j)** *Protective clothing and equipment*—**(1)** *Provision and use.* Where employees are exposed to eye or skin contact with _____ (insert criteria which trigger this requirement as appropriate), the employer shall, within (insert appropriate time period) of the effective date of this section provide at no cost to such employees, and assure that such employees wear, appropriate protective clothing or other equipment in accordance with 29 CFR 1910.132 and 1910.133 to protect the area of the body which may come in contact with _____.

**1990.151(j)(2)** *Cleaning and replacement.* **(i)** The employer shall clean, launder, maintain, or replace protective clothing and equipment required to maintain their effectiveness.

**1990.151(k)** *Housekeeping*—**(1)** *General.* The employer shall, within appropriate time period of the effective date of this section, implement a housekeeping program to minimize accumulation of _____.

**1990.151(k)(2)** *Specific provisions.* The program shall include (insert appropriate elements).

**1990.151(k)(2)(i)** Periodic scheduling of routine housekeeping.

**1990.151(k)(2)(ii)** Provision for periodic cleaning of dust collection systems.

**1990.151(k)(2)(iii)** Provision for maintaining clean surfaces.

**1990.151(k)(2)(iv)** Provision for assigning personnel to housekeeping procedures; and the

**1990.151(k)(2)(v)** Provision for informing employees about the housekeeping program.

**1990.151(l)** *Waste disposal*—**(1)** *General.* The employer shall assure that no waste material containing _____ is dispersed into the workplace, to the extent practicable.

**1990.151(l)(2)** The employer shall label, or otherwise inform employees who may contact waste material containing _____, the contents of such waste material.

**1990.151(l)(3)** (Insert specific disposal methods, as appropriate.)

**1990.151(m)** *Hygiene facilities and practices.* Where employees are exposed to airborne concentrations of _____ in excess of the permissible exposure limits specified in paragraph (c)(1), or where employees are required to wear protective clothing or equipment pursuant to

# OSHA Standards for General Industry

paragraph (j) of this section, or where otherwise found to be appropriate, the following facilities shall be provided by the employer for the use of those employees and the employer shall assure that the employees use the facilities provided.

[Specify appropriate hygiene facilities and practices such as]:

**1990.151(m)(1)** *Change rooms.* The employer shall provide clean change rooms in accordance with 29 CFR 1910.141(e).

**1990.151(m)(2)** *Showers.* (i) The employer shall provide shower facilities in accordance with 29 CFR 1910.141(d)(3).

**1990.151(m)(2)(ii)** The employer shall assure that employees exposed to _____ shower at the end of the work shift.

**1990.151(m)(3)** Lunchrooms. (if appropriate or other suitable requirements depending on the circumstances). Whenever food or beverages are consumed in the workplace, the employer shall provide lunchroom facilities which have a temperature controlled, positive pressure, filtered air supply, and which are readily accessible to employees exposed to _____.

**1990.151(n)** *Medical surveillance*—**(1)** *General.* **(i)** The employer shall institute a program of medical surveillance for (specify the types of employees subject to the medical surveillance requirement, for example, by specifying the level, duration, and frequency of exposure to _____ which make medical surveillance appropriate for individual employees). The employer shall provide each such employee with an opportunity for medical examinations and tests in accordance with this paragraph.

**1990.151(n)(1)(ii)** The employer shall assure that all medical examinations and procedures are performed by or under the supervision of a licensed physician, and shall be provided without cost to the employee.

**1990.151(n)(2)** *Initial examinations.* Within (insert appropriate time period) of the effective date of this section or thereafter at the time of initial assignment, the employer shall provide each employee specified in paragraph (n)(1) of this section an opportunity for a medical examination, including at least the following elements:

**1990.151(n)(2)(i)** A work history and a medical history which shall include: (insert specific areas to be covered pertinent to the health hazards posed by _____).

**1990.151(n)(2)(ii)** A physical examination which shall include: (insert specific tests, procedures, etc., pertinent to the health hazards posed by _____). Where appropriate, provide that the examining physician shall conduct such additional examinations and tests as are needed according to his professional judgment).

**NOTE:** Where appropriate, require or permit different medical protocols, or different frequencies of medical examinations, for separate sub-populations of employees covered under paragraph (n)(1).

**1990.151(n)(3)** *Periodic examinations.* (i) The employer shall provide the examinations specified below in this subparagraph at least (insert appropriate time) for all employees specified in paragraph (n)(1) of this section: (insert appropriate medical protocol for periodic examinations).

**1990.151(n)(3)(ii)** If an employee has not had the examinations prescribed in paragraph (n)(3)(i) of this section within (insert appropriate time period) prior to termination of employment, the employer shall make such examination available to the employee upon such termination.

**1990.151(n)(4)** *Additional examinations.* If the employee for any reason develops signs or symptoms commonly associated with exposure to _____, the employer shall provide appropriate examination and emergency medical treatment.

**1990.151(n)(5)** *Information provided to the physician.* The employer shall provide the following information to the examining physician:

**1990.151(n)(5)(i)** A copy of this standard and its appendices;

**1990.151(n)(5)(ii)** A description of the affected employee's duties as they relate to the employee's exposure;

**1990.151(n)(5)(iii)** The employee's actual or representative exposure level;

**1990.151(n)(5)(iv)** The employee's anticipated or estimated exposure level (for preplacement examinations or in cases of exposure due to an emergency);

**1990.151(n)(5)(v)** A description of any personal protective equipment used or to be used; and

**1990.151(n)(5)(vi)** The names and addresses of physicians who, under the sponsorship of the employer, provided previous medical examinations of the affected employee, if such records are not otherwise available to the examining physician.

**1990.151(n)(6)** *Physician's written opinion.* (i) The employer shall obtain a written opinion from the examining physician which shall include:

**1990.151(n)(6)(i)(A)** The physician's certification that he has received the information from the employer required under the paragraph (n)(5) and has performed all medical examinations and tests which are in his opinion appropriate under this standard;

**1990.151(n)(6)(i)(B)** The physician's opinion as to whether the employee has any detected medical condition which would place the employee at an increased risk of material impairment of the employee's health from exposure to _____;

**1990.151(n)(6)(i)(C)** Any recommended limitations upon the employee's exposure to _____ or upon the use of protective clothing and equipment such as respirators; and

**1990.151(n)(6)(i)(D)** A statement that the employee has been informed by the physician of the results of the medical examination and any medical conditions which require further examination or treatment.

**1990.151(n)(6)(ii)** The employer shall instruct the physician not to reveal in the written opinion specific findings or diagnoses unrelated to occupational exposure to _____;

**1990.151(n)(6)(iii)** The employer shall provide a copy of the written opinion to the affected employee.

**1990.151(o)** *Employee Information and Training*—**(1)** *Training Program.* (i) Within (insert appropriate time period) from the effective date of this section, the employer shall institute a training program for all employees who (specify the employees subject to the training requirement), and shall assure their participation in the training program.

**1990.151(o)(1)(ii)** The training program shall be provided at the time of initial assignment, or upon institution of the training program, and at least (insert appropriate time period) thereafter, and the employer shall assure that each employee is informed of the following:

**NOTE:** Specify, as appropriate, some or all of the following information, or any other appropriate information. Where appropriate, require training programs with different contents, or different frequencies, for separate subpopulations of the employees specified in paragraph (o)(1).

**1990.151(o)(1)(ii)(A)** The information contained in the Appendices;

**1990.151(o)(1)(ii)(B)** The quantity, location, manner of use, release or storage of _____ and the specific nature of operations which could result in exposure to _____, as well as any necessary protective steps;

**1990.151(o)(1)(ii)(C)** The purpose, proper use, and limitations of respirators;

**1990.151(o)(1)(ii)(D)** The purpose and a description of the medical surveillance program required by paragraph (n) of this section;

**1990.151(o)(1)(ii)(E)** The emergency procedures developed, as required by paragraph (i) of of this section;

**1990.151(o)(1)(ii)(F)** The engineering and work practice controls, their function and the employee's relationship thereto; and

**1990.151(o)(1)(ii)(G)** A review of this standard.

**1990.151(o)(2)** *Access to training materials.* (i) The employer shall make a copy of this standard and its appendices readily available to all affected employees.

**1990.151(o)(2)(ii)** The employer shall provide, upon request, all materials relating to the employee information and training program to the Assistant Secretary and the Director.

**1990.151(p)** *Signs and labels*—**(1)** *General.* (i) The employer may use labels or signs required by other statutes, regulations, or ordinances in addition to, or in combination with, signs and labels required by this paragraph.

**1990.151(p)(1)(ii)** The employer shall assure that no statement appears on or near any sign or label, required by this paragraph, which contradicts or detracts from the meaning of the required sign or label.

**1990.151(p)(2)** *Signs.* (i) The employer shall post signs to clearly indicate all workplaces (Specify as appropriate the description of the area to be signposted such as "where employees are exposed to _____," or "where exposures exceed the action level," or "where exposures exceed the PEL," or "which are regulated areas"). The signs shall bear the following legend:

**DANGAR**

_____
(insert appropriate trade or common names)
**CANCER HAZARD**
**AUTHORIZED PERSONNEL ONLY**

**1990.151(p)(2)(ii)** The employer shall assure that signs required by this paragraph are illuminated and cleaned as necessary so that the legend is readily visible.

**1990.151(p)(2)(iii)** Where airborne concentrations of _____ exceed the permissible exposure limits, the signs shall bear the additional legend: "Respirator Required" or "Respirator May Be Required" as appropriate.

**1990.151(p)(3)** *Labels.* **(i)** The employer shall assure that precautionary labels are affixed to all containers of _____ and of products containing _____ (specify if appropriate suitable modifications), and that the labels remain affixed when the _____ or products containing _____ are sold, distributed or otherwise leave the employer's workplace.

**1990.151(p)(3)(ii)** The employer shall assure that the precautionary labels required by this paragraph are readily visible and legible. The labels shall bear the following legend:

**DANGER**
**CONTAINS** _____
**CANCER HAZARD**

**NOTE:** Utilize the clause "POTENTIAL CANCER HAZARD" if it is appropriate to include a signs and labels provision for a Category II potential carcinogen.

**1990.151(q)** *Recordkeeping*—**(1)** *Exposure monitoring.* **(i)** The employer shall establish and maintain an accurate record of all monitoring required by paragraph (e) of this section.

**1990.151(q)(1)(ii)** This record shall include:

**1990.151(q)(1)(ii)(A)** The dates, number, duration, and results of each of the samples taken, including a description of the sampling procedure used to determine representative employees exposure;

**1990.151(q)(1)(ii)(B)** A description of the sampling and analytical methods used;

**1990.151(q)(1)(ii)(C)** Type of respiratory protective devices worn, if any; and

**1990.151(q)(1)(ii)(D)** Name, social security number and job classification of the employees monitored and of all other employees whose exposure the measurement is intended to represent.

**1990.151(q)(1)(iii)** The employer shall maintain this record for (insert appropriate period) or for the duration of employment plus (insert appropriate period) whichever is longer.

**1990.151(q)(2)** *Medical surveillance.* **(i)** The employer shall establish and maintain an accurate record of each employee subject to medical surveillance as required by paragraph (n) of this section.

**1990.151(q)(2)(ii)** This record shall include:

**1990.151(q)(2)(ii)(A)** A copy of the physicians' written opinions or a written explanation of the absence of any such opinion or employee refusal to take the medical examination;

**1990.151(q)(2)(ii)(B)** Any employees medical complaints related to exposure to _____;

**1990.151(q)(2)(ii)(C)** A copy of the information provided to the physician as required by paragraphs (n)(5)(ii) through (v) of this section unless it is systematically retained elsewhere by the employer for the period of time specified in paragraph (q)(2)(ii); and

**1990.151(q)(2)(ii)(D)** A copy of the employee's work history.

**1990.151(q)(2)(iii)** The employer shall assure that this record be maintained for (insert appropriate period) or for the duration of employment plus (insert appropriate period) whichever is longer.

**1990.151(q)(3)** *Availability.* **(i)** The employer shall assure that all records required to be maintained by this section be made available upon request to the Assistant Secretary and the Director for examination and copying.

**1990.151(q)(3)(ii)** Employee exposure measurement records and employee medical records required by this section shall be provided upon request to employees, designated representatives, and the Assistant Secretary in accordance with 29 CFR 1910.20(a) through (e) and (g) through (i).

**1990.151(q)(4)** *Transfer of records.* **(i)** Whenever the employer ceases to do business, the successor employer shall receive and retain all records required to be maintained by this section.

**1990.151(q)(4)(ii)** Whenever the employer ceases to do business and there is no successor employer to receive and retain the records for the prescribed period, these records shall be transmitted to the Director.

**1990.151(q)(4)(iii)** At the expiration of the retention period for the records required to be maintained pursuant to this section, the employer shall transmit these records to the Director.

**1990.151(q)(4)(iv)** The employer shall also comply with any additional requirements involving transfer of records set forth in 29 CFR 1910.20(h).

**NOTE:** Include other recordkeeping requirements if appropriate.

**1990.151(r)** *Observation of monitoring*—**(1)** *Employee observation.* The employer shall provide affected employees, or their designated representatives, an opportunity to observe any monitoring of employee exposure to _____ conducted pursuant to paragraph (e) of this section.

**1990.151(r)(2)** *Observation procedures.* **(i)** Whenever observation of the monitoring of employee exposure to _____ requires entry into an area where the use of protective clothing or equipment is required, the employer shall provide the observer with personal protective clothing or equipment required to be worn by employees working in the area, assure the use of such clothing and equipment, and require the observer to comply with all other applicable safety and health procedures.

**1990.151(r)(2)(ii)** Without interfering with the monitoring, observers shall be entitled to:

**1990.151(r)(2)(ii)(A)** Receive an explanation of the measurement procedures;

**1990.151(r)(2)(ii)(B)** Observe all steps related to the measurement of airborne concentrations of _____ performed at the place of exposure; and

**1990.151(r)(2)(ii)(C)** Record the results obtained and receive results supplied by the laboratory.

**1990.151(s)** *Effective date.* This section shall become effective (insert effective date).

**1990.151(t)** *Appendices.* The information contained in the appendices is not intended, by itself, to create any additional obligations not otherwise imposed or to detract from any existing obligation. (In normal circumstances three appendices will be included in each standard, an "Appendix A—Substance Safety Data Sheet," an "Appendix B—Substance Technical Guidelines," and an "Appendix C—Medical Surveillance Guidelines." Insert additional appendices or delete any of the suggested appendices as appropriate).

[45 FR 5282, Jan. 22, 1980; 45 FR 43405-43406, June 27, 1980, as amended at 46 FR 5881, Jan. 21, 1981]

### §1990.152 Model emergency temporary standard pursuant to section 6(c) of the Act.

Occupational Exposure to _____

Emergency Temporary Standard
(insert section number of standard.)

**1990.152(a)** *Scope and Application*—**(1)** *General.* This section applies to all occupational exposures to _____, or to (specify the uses of classes of uses of _____ [Chemical Abstracts Service Registry Number 00000], which are covered by the standard, including, where appropriate, the type of exposure to be regulated by the standard) except as provided in paragraph (a)(2).

**1990.152(a)(2)** *Exemption.* This section does not apply to (insert those uses or classes of uses of _____ which are exempted from compliance with the standard, including, where appropriate,

**1990.152(a)(2)(i)** Workplaces where exposure to _____ results from solid or liquid mixtures containing a specified percentage of _____ or less;

**1990.152(a)(2)(ii)** Workplaces where another Federal agency is exercising statutory authority to prescribe or enforce standards or regulations affecting occupational exposure to _____ or

# OSHA Standards for General Industry

**1990.152(a)(2)(iii)** Workplaces which are appropriately addressed in a separate standard.

**1990.152(b)** *Definitions.*

_____ means (definition of the substance, group of substances, or combination of substances, to be regulated).

*Action level* means an airborne concentration of _____ of (insert appropriate level of exposure).

**NOTE:** Where appropriate, consider an action level as a limitation on requirements for periodic monitoring (para. (e)(3)), medical surveillance (para. (n)), training (para. (o)), and other provisions.

*Assistant Secretary* means the Assistant Secretary of Labor for Occupational Safety and Health, U.S. Department of Labor, or designee.

*Authorized person* means any person specifically authorized by the employer whose duties require the person to enter a regulated area or any person entering such an area as a designated representative of employees exercising the opportunity to observe monitoring procedures under paragraph (r) of this section.

*Director* means the Director, National Institute for Occupational Safety and Health, U.S. Department of Health, Education and Welfare, or designee.

*Emergency* means any occurrence such as, but not limited to, equipment failure, rupture of containers, or failure of control equipment which may result in a release of _____ which is (insert appropriate quantitative or qualitative level of release which constitutes an emergency).

*OSHA Area Office* means the Area Office of the Occupational Safety and Health Administration having jurisdiction over the geographic area where the affected workplace is located.

**1990.152(c)** *Permissible exposure limits*—**(1)** *Inhalation*—**(i)** *Time-weighted average limit (TWA).* Within (insert appropriate time) from the effective date of this emergency temporary standard, the employer shall assure that no employee is exposed to an airborne concentration of _____ in excess of: (insert appropriate exposure limit representing a level that can be complied with immediately) as an eight (8)-hour-time-weighted average.

**1990.152(c)(1)(ii)** *Ceiling limit (If appropriate.)* The employer shall assure that no employee is exposed to an airborne concentration of _____ in excess of: (insert appropriate exposure limit representing a level that can be complied with immediately) as averaged over any: (insert appropriate time period) during the working day.

**1990.152(c)(2)** *Dermal and eye exposure.* (As appropriate.) **(i)** Within (insert appropriate time period) of the effective date of this section, the employer shall (If eye exposure to _____ does not create a risk of cancer, insert exposure level or criteria which will prevent other adverse effects of eye exposure to _____, if any. If eye exposure creates a risk of cancer, insert exposure level or criteria which represent the level of eye exposure to _____.)

**1990.152(c)(2)(ii)** Within (insert appropriate time period) of the effective date of this section, the employer shall (If skin exposure to _____ does not create a risk of cancer, insert exposure level or criteria which will prevent other adverse health affects of skin exposure to _____ if any. If skin exposure creates a risk of cancer, insert exposure level or criteria which represents the level of skin exposure to _____).

**1990.152(d)** *Notification of Use.* Within (insert appropriate time and omit specific categories of information if appropriate) of the effective date of this section, or within fifteen (15) days following the introduction of _____ into the workplace, every employer shall report the following information to the nearest OSHA Area Office for each such workplace:

**1990.152(d)(1)** The address and location of each workplace in which _____ is present;

**1990.152(d)(2)** A brief description of each process or operation which may result in employee exposure to _____;

**1990.152(d)(3)** The number of employees engaged in each process or operation who may be exposed to _____ and an estimate of the frequency and degree of exposure that occurs; and

**1990.152(d)(4)** A brief description of the employer's safety and health program as it relates to limitation of employee exposure to _____;

**1990.152(e)** *Exposure monitoring*—**(1)** *General.* **(i)** Determinations of airborne exposure levels shall be made from air samples that are representative of each employee's exposure to _____ over an eight (8) hour period. (Modify the time period as appropriate to be practical in the relevant industries yet reasonably representative of full shift exposures.) Monitoring of exposure levels required under this paragraph shall be made as follows: [insert method or alternative methods to be used to meet the requirements of this paragraph].

**1990.152(e)(1)(ii)** For the purposes of this section, employee exposure is the exposure which would occur if the employee were not using a respirator.

**1990.152(e)(2)** *Initial monitoring.* Each employer who has one or more workplaces where (specify the types of workplaces subject to the monitoring requirement), shall within (insert appropriate period) of the effective date of this section (insert requirements for initial monitoring, as appropriate).

**1990.152(e)(3)** *Frequency.* (Insert, if appropriate, provisions prescribing the minimum frequency at which monitoring must be repeated, the conditions under which such frequency must be increased, or may be reduced, and conditions under which such routine monitoring may be discontinued (for example where the action level is not exceeded). Where appropriate, specify different frequency requirements for certain types of workplaces where, for example, exposure levels are subject to greater or less variability.)

**1990.152(e)(4)** *Additional monitoring.* (Insert, if appropriate, provisions for monitoring, in addition to the requirements (if any) of paragraph (e)(3). This may include a production, process, control or personnel change which might result in new or additional exposure to _____ or whenever the employer has any other reason to suspect a change which might result in new or additional exposures to _____.)

**1990.152(e)(5)** *Employee notification.* **(i)** Within (insert appropriate period) after the receipt of monitoring results, the employer shall notify each employee in writing of the results which represent that employee's exposure.

**1990.152(e)(5)(ii)** Whenever the results indicate that the representative employee exposure exceeds the permissible exposure limits, the employer shall include in the written notice a statement that permissible exposure limits were exceeded and a description of the corrective action being taken to reduce exposure to or below the permissible exposure limits.

**1990.152(e)(6)** *Accuracy of measurement.* (Insert requirements for accuracy of methods of measurement or detection used to comply with the paragraph.)

**1990.152(f)** [Reserved]

**1990.152(g)** *Methods of compliance*—**(1)** *General.* **(i)** Employee exposures to _____ shall be controlled to or below the permissible exposure limits by any practicable combination of engineering controls, work practices and personal protective devices and equipment, during the effective period of this emergency temporary standard.

**NOTE:** Where engineering controls or work practices can reduce employee exposures to _____ it is recommended that they be implemented where practicable, even where they do not themselves reduce exposures to, or below the permissible exposure limits. Work practices which can be implemented by the employer to help reduce employee exposures to _____ include limiting access to work areas to authorized personnel, prohibiting smoking and consumption of food and beverages in work areas, and establishing good maintenance and housekeeping practices, including the prompt clean-up of spills and repair of leaks.)

**1990.152(g)(2)** *Engineering and work practice control plan.* **(i)** Within (insert appropriate time period) of the effective date of this emergency temporary standard, the employer shall develop a written plan describing proposed means to reduce employee exposures to the lowest feasible level by means of engineering and work practice controls (which will be eventually required by a permanent standard for occupational exposure to _____, as provided for by §1990.151(g) of this subpart).

**1990.152(g)(2)(ii)** Written plans required by this paragraph shall be submitted, upon request, to the Assistant Secretary and the Director and shall be available at the worksite for examination and copying by the Assistant Secretary, the Director, and any affected employee or designated representative.

**1990.152(h)** *Respiratory protection*—**(1)** *Required use.* The employer shall assure that respirators are used where required pursuant to this section to reduce employee exposures to within the permissible exposure limits and in emergencies.

**1990.152(h)(2)** *Respirator selection.* (i) Where respiratory protection is required under this section, the employer shall select and provide at no cost to the employee, the appropriate respirator from Table 1 below and shall assure that the employee wears the respirator provided.

TABLE 1.—RESPIRATORY PROTECTION FOR _____
_____

(The table will contain a listing of the appropriate type

of respirator for various conditions of exposure to _____.)

**1990.152(h)(2)(ii)** The employer shall select respirators from those approved by the National Institute for Occupational Safety and Health under the provisions of 30 CFR part 11.

**1990.152(h)(3)** *Respirator program.* (i) The employer shall institute a respirator protection program in accordance with 29 CFR 1910.134(b), (d), (e) and (f).

**1990.152(h)(3)(ii)** Employees who wear respirators shall be allowed to wash their face and respirator face piece to prevent potential skin irritation associated with respirator use.

**1990.152(h)(3)(iii)** The employer shall assure that the respirator issued to each employee is properly fitted (as appropriate, indicate the requirement for a qualitative or quantitative respirator fit testing program.)

**1990.152(i)** [Reserved]

**1990.152(j)** *Protective clothing and equipment*—**(1)** *Provision and use.* Where employees are exposed to eye or skin contact with _____ (insert criteria which trigger this requirement as appropriate), the employer shall within (insert appropriate time period) of the effective date of this standard provide, at no cost to the employees, and assure that employees wear, appropriate protective clothing or other equipment in accordance with 29 CFR 1910.132 and 1910.133 to protect the area of the body which may come in contact with _____.

**1990.152(j)(2)** *Cleaning and replacement.* (i) The employer shall clean, launder, maintain, or replace protective clothing and equipment required by this paragraph, as needed to maintain their effectiveness.

**1990.152(k)** *Housekeeping*—**(1)** *General.* The employer shall, within (insert appropriate time period) of the effective date of this section, implement a housekeeping program to minimize accumulations of _____.

**1990.152(k)(2)** *Specific provisions.* The program shall include (insert appropriate elements):

**1990.152(k)(2)(i)** Periodic scheduling of routine housekeeping procedures;

**1990.152(k)(2)(ii)** Provision for periodic cleaning of dust collection systems;

**1990.152(k)(2)(iii)** Provision for maintaining clean surfaces;

**1990.152(k)(2)(iv)** Provision for assigning personnel to housekeeping procedures; and

**1990.152(k)(2)(v)** Provision for informing employees about the housekeeping program.

**1990.152(l)** *Waste disposal*—**(1)** *General.* The employer shall assure that no waste material containing _____ is dispersed into the workplace, to the extent practicable.

**1990.152(l)(2)** The employer shall label, or otherwise inform employees who may contact waste material containing _____ of the contents of such waste material.

**1990.152(l)(3)** (Insert specific disposal methods, as appropriate.)

**1990.152(m)** [Reserved]

**1990.152(n)** *Medical surveillance*—**(1)** *General.* (i) The employer shall institute a program of medical surveillance for (specify the types of employees subject to the medical surveillance requirement, for example, by specifying the level, duration, and frequency of exposure to _____ which make medical surveillance appropriate for individual employees). The employer shall provide each such employee with an opportunity for medical examinations and tests in accordance with this paragraph.

**1990.152(n)(1)(ii)** The employer shall assure that all medical examinations and procedures are performed by or under the supervision of a licensed physician, and shall be provided without cost to the employee.

**1990.152(n)(2)** *Initial examinations.* Within (insert appropriate time period) of the effective date of this section, or thereafter at the time of initial assignment, the employer shall provide each employee specified in paragraph (n)(1) of this section an opportunity for a medical examination, including at least the following elements:

**1990.152(n)(2)(i)** A work history and a medical history which shall include (insert specific areas to be covered pertinent to the health hazards posed by _____).

**1990.152(n)(2)(ii)** A physical examination which shall include: (insert specific tests, procedures, etc., pertinent to the health hazards posed by _____. Where appropriate, provide that the examining physician shall conduct such additional examinations and tests as are needed according to his professional judgement).

NOTE: Where appropriate, require or permit different medical protocols, or different frequencies of medical examinations, for separate sub-populations of employees covered under paragraph (n)(1).

**1990.152(n)(3)** *Periodic examinations.* (If appropriate insert appropriate medical protocol and time.)

**1990.152(n)(4)** *Additional examinations.* If the employee for any reason develops signs or symptoms commonly associated with exposure to _____, the employer shall provide an appropriate examination and emergency medical treatment.

**1990.152(n)(5)** *Information provided to the physician.* The employer shall provide the following information to the examining physician:

**1990.152(n)(5)(i)** A copy of this emergency temporary standard and its appendices;

**1990.152(n)(5)(ii)** A description of the affected employee's duties as they relate to the employee's exposure;

**1990.152(n)(5)(iii)** The employee's actual or representative exposure level;

**1990.152(n)(5)(iv)** The employee's anticipated or estimated exposure level (for preplacement examinations or in cases of exposures due to an emergency);

**1990.152(n)(5)(v)** A description of any personal protective equipment used or to be used; and

**1990.152(n)(5)(vi)** The names and addresses of physicians who, under the sponsorship of the employer, provided previous medical examinations of the affected employee, if such records are not otherwise available to the examining physician.

**1990.152(n)(6)** *Physician's written opinion.* (i) The employer shall obtain a written opinion from the examining physician which shall include:

**1990.152(n)(6)(i)(A)** The results of the medical tests performed;

**1990.152(n)(6)(i)(B)** The physician's opinion as to whether the employee has any detected medical condition which would place the employee at an increased risk of material impairment of the employee's health from exposure to _____;

**1990.152(n)(6)(i)(C)** Any recommended limitations upon the employee's exposure to _____ or upon the use of protective clothing and equipment such as respirators; and

**1990.152(n)(6)(i)(D)** A statement that the employee has been informed by the physician of the results of the medical examination and any medical conditions which require further examination or treatment.

**1990.152(n)(6)(ii)** The employer shall instruct the physician not to reveal in the written opinion specific findings or diagnoses unrelated to occupational exposure to _____;

**1990.152(n)(6)(iii)** The employer shall provide a copy of the written opinion to the affected employee.

**1990.152(o)** *Employee information and training*—**(1)** *Training program.* (i) Within (insert appropriate time period) from the effective date of this standard, the employer shall institute a training program for all employees who (specify the employees subject to the training requirement), and shall assure their participation in the training program.

**1990.152(o)(1)(ii)** The employer shall assure that each employee is informed of the following:

**1990.152(o)(1)(ii)(A)** The information contained in the Appendices;

**1990.152(o)(1)(ii)(B)** The quantity, location, manner of use, release, or storage of _____ and the specific nature of operations

# OSHA Standards for General Industry

which could result in exposure to _____, as well as any necessary protective steps;

**1990.152(o)(1)(ii)(C)** The purpose, proper use, and limitations of respirators;

**1990.152(o)(1)(ii)(D)** The purpose and description of the medical surveillance program required by paragraph (n) of this section; and

**1990.152(o)(1)(ii)(E)** A review of this standard.

**1990.152(o)(2)** *Access to training materials.* **(i)** The employer shall make a copy of this standard and its appendices readily available to all affected employees.

**1990.152(o)(2)(ii)** The employer shall provide, upon request, all materials relating to the employee information and training program to the Assistant Secretary and the Director.

**1990.152(p)** *Signs and labels.* (include a signs or a signs and labels provision if it is appropriate for the duration of the ETS)—**(1)** *General.* **(i)** The employer may use labels or signs required by other statutes, regulations, or ordinances in addition to, or in combination with, signs and labels required by this paragraph.

**1990.152(p)(1)(ii)** The employer shall assure that no statement appears on or near any sign or label, required by this paragraph, which contradicts or detracts from the meaning of the required sign or label.

**1990.152(p)(2)** *Signs.* **(i)** The employer shall post signs to clearly indicate all workplaces (specify as appropriate the description of the area to be signposted such as "where employees are exposed to _____," or "where exposures exceed the PEL," or "which are regulated areas"). The signs shall bear the following legend:

### DANGER

_____

(insert appropriate trade or common names)

### CANCER HAZARD

### AUTHORIZED PERSONNEL ONLY

**1990.152(p)(2)(ii)** The employer shall assure that signs required by this paragraph are illuminated and cleaned as necessary so that the legend is readily visible.

**1990.152(p)(2)(iii)** Where airborne concentrations of _____ exceed the permissible exposure limits, the signs shall bear the additional legend: ("Respirator Required" or "Respirator may be Required" as appropriate).

**1990.152(p)(3)** *Labels.* **(i)** The employer shall assure that precautionary labels are affixed to all containers of _____ and of products containing _____ (specify if appropriate suitable modifications), and that the labels remain affixed when _____ or products containing _____ are sold, distributed or otherwise leave the employer's workplace.

**1990.152(p)(3)(ii)** The employer shall assure that the precautionary labels required by this paragraph are readily visible and legible. The labels shall bear the following legend:

### DANGER
### CONTAINS _____
### CANCER HAZARD

**1990.152(q)** *Recordkeeping*—**(1)** *Exposure monitoring.* **(i)** The employer shall establish and maintain an accurate record of all monitoring required by paragraph (e) of this section.

**1990.152(q)(1)(ii)** This record shall include:

**1990.152(q)(1)(ii)(A)** The dates, number, duration, and results of each of the samples taken, including a description of the sampling procedures used to determine representative employee exposure;

**1990.152(q)(1)(ii)(B)** A description of the sampling and analytical methods used;

**1990.152(q)(1)(ii)(C)** Type of respiratory protective devices worn, if any; and

**1990.152(q)(1)(ii)(D)** Name, social security number, and job classification of the employee monitored and of all other employees whose exposure the measurement is intended to represent.

**1990.152(q)(1)(iii)** The employer shall maintain this record for the effective period of this emergency temporary standard, and for any additional period required by the permanent standard.

**1990.152(q)(2)** *Medical surveillance.* **(i)** The employer shall establish and maintain an accurate record for each employee subject to medical surveillance as required by paragraph (n) of this section.

**1990.152(q)(2)(ii)** This record shall include:

**1990.152(q)(2)(ii)(A)** A copy of the physicians' written opinions or a written explanation of the absence of any such opinion or employee refusal to take the medical examination;

**1990.152(q)(2)(ii)(B)** Any employee medical complaints related to exposure to _____;

**1990.152(q)(2)(ii)(C)** A copy of the information provided to the physician as required by paragraphs (n)(5)(ii)—(iv) of this section unless it is systematically retained elsewhere by the employer for the period of time specified in paragraph (q)(2)(iii); and

**1990.152(q)(2)(ii)(D)** A copy of the employee's work history. **(1)** The employer shall assure that employee exposure measurement records, as required by this section, be made available upon request to the Assistant Secretary and the Director for examination and copying.

**1990.152(q)(2)(iii)** The employer shall assure that this record be maintained for the effective period of this emergency temporary standard, and for any additional period required by the permanent standard.

**1990.152(q)(3)** *Availability.* **(i)** The employer shall assure that all records required to be maintained by this section be made available upon request, to the Assistant Secretary and the Director for examination and copying.

**1990.152(q)(3)(ii)** Employee exposure measurement records and employee medical records required by this section shall be provided upon request to employees, designated representatives, and the Assistant Secretary in accordance with 29 CFR 1910.20 (a) through (e) and (g) through (i).

**1990.152(r)** *Observation of monitoring.* **(1)** Employee observation. The employer shall provide affected employees, or their designated representatives, an opportunity to observe any monitoring of employee exposure to _____ conducted pursuant to paragraph (e) of this section.

**1990.152(r)(2)** *Observation procedures.* **(i)** Whenever observation of the monitoring of employee exposure to _____ requires entry into an area where the use of protective clothing or equipment is required, the employer shall provide the observer with personal protective clothing or equipment required to be worn by employees working in the area, assure the use of such clothing and equipment, and require the observer to comply with all other applicable safety and health procedures.

**1990.152(r)(2)(ii)** Without interfering with the monitoring, observers shall be entitled to:

**1990.152(r)(2)(ii)(A)** Receive an explanation of measurement procedures;

**1990.152(r)(2)(ii)(B)** Observe all steps related to the measurement of airborne concentrations of _____ performed at the place of exposure; and

**1990.152(r)(2)(ii)(C)** Record the results obtained and receive results supplied by the laboratory.

**1990.152(s)** *Effective date.* This section shall become effective (insert effective date).

**1990.152(t)** *Appendices.* The information contained in the appis not intended, itself, to create any additional obligations not otherwise imposed or to detract from any existing obligation. (In normal circumstances three appendices will be included in each standard, an "Appendix A—Substance Safety Data Sheet," an "Appendix B—Substance Technical Guidelines," and an "Appendix C—Medical Surveillance Guidelines." Insert additional appendices or delete any of the suggested appendices as appropriate.)

[45 FR 5282, Jan. 22, 1980; 45 FR 43406-43407, June 27, 1980, as amended at 46 FR 5882, Jan. 21, 1981]

## The Williams-Steiger Occupational Safety and Health Act of 1970

To assure safe and healthful working conditions for working men and women; by authorizing enforcement of the standards developed under the Act; by assisting and encouraging the States in their efforts to assure safe and healthful working conditions; by providing for research, information, education, and training in the field of occupational safety and health; and for other purposes.

*Be it enacted by the Senate and House of Representatives of the United States of America in Congress assembled,* That this Act may be cited as the "Occupational Safety and Health Act of 1970."

[The following is commonly referred to as the "General Duty Clause"—CCH.]

**Sec. 5. (a)** Each employer—

**(1)** shall furnish to each of his employees employment and a place of employment which are free from recognized hazards that are causing or are likely to cause death or serious physical harm to his employees;

**(2)** shall comply with occupational safety and health standards promulgated under this Act.

[The following is not part of the "General Duty Clause" but is provided for convenience—CCH.]

**(b)** Each employee shall comply with occupational safety and health standards and all rules, regulations, and orders issued pursuant to this Act which are applicable to his own actions and conduct.

# NOTES

# Part 1977 — Discrimination Against Employees

**Authority:** Sec. 8, 11, Occupational Safety and Health Act of 1970 (29 U.S.C. 657, 660); Secretary of Labor's Order No. 12-71 (36 FR 8754). SOURCE: 38 FR 2681, Jan. 29, 1973 unless otherwise noted.

## Part 1977 — Discrimination Against Employees Exercising Rights Under The Williams-steiger Occupational Safety And Health Act Of 1970

### GENERAL

| | |
|---|---|
| 1977.1 | Introductory statement. |
| 1977.2 | Purpose of this part. |
| 1977.3 | General requirements of section 11(c) of the Act. |
| 1977.4 | Persons prohibited from discriminating. |
| 1977.5 | Persons protected by section 11(c). |
| 1977.6 | Unprotected activities distinguished. |

### SPECIFIC PROTECTIONS

| | |
|---|---|
| 1977.9 | Complaints under or related to the Act. |
| 1977.10 | Proceedings under or related to the Act. |
| 1977.11 | Testimony. |
| 1977.12 | Exercise of any right afforded by the Act. |

### PROCEDURES

| | |
|---|---|
| 1977.15 | Filing of complaint for discrimination. |
| 1977.16 | Notification of Secretary of Labor's determination. |
| 1977.17 | Withdrawal of complaint. |
| 1977.18 | Arbitration or other agency proceedings. |

### SOME SPECIFIC SUBJECTS

| | |
|---|---|
| 1977.22 | Employee refusal to comply with safety rules. |
| 1977.23 | State plans. |

### §1977.1 Introductory Statement.

**1977.1(a)** The Occupational Safety and Health Act of 1970 (29 U.S.C. 651, et seq.), hereinafter referred to as the Act, is a Federal statute of general application designed to regulate employment conditions relating to occupational safety and health and to achieve safer and healthier workplaces throughout the Nation. By terms of the Act, every person engaged in a business affecting commerce who has employees is required to furnish each of his employees employment and a place of employment free from recognized hazards that are causing or likely to cause death or serious physical harm, and, further, to comply with occupational safety and health standards promulgated under the Act. See Part 1975 of this chapter concerning coverage of the Act.

**1977.1(b)** The Act provides, among other things, for the adoption of occupational safety and health standards, research and development activities, inspections, and investigations of workplaces, and recordkeeping requirements. Enforcement procedures initiated by the Department of Labor, review proceedings before an independent quasi-judicial agency (the Occupational Safety and Health Review Commission), and express judicial review are provided by the Act. In addition, States which desire to assume responsibility for development and enforcement of standards which are at least as effective as the Federal standards published in this chapter may submit plans for such development and enforcement of the Secretary of Labor.

**1977.1(c)** Employees and representatives of employees are afforded a wide range of substantive and procedural rights under the Act. moreover, effective implementation of the Act and achievement of its goals depend in large part upon the active but orderly participation of employees, individually and through their representatives, at every level of safety and health activity.

**1977.1(d)** This part deals essentially with the rights of employees afforded under section 11(c) of the Act. Section 11(c) of the Act prohibits reprisals, in any form, against employees who exercise rights under the Act.

### §1977.2 Purpose of This Part.

The purpose of this part is to make available in one place interpretations of the various provisions of section 11(c) of the Act which will guide the Secretary of Labor in the performance of his duties thereunder unless and until otherwise directed by authoritative decisions of the courts, or concluding, upon re-examination of an interpretation, that it is incorrect.

### §1977.3 General Requirements of Section 11(c) of the Act.

Section 11(c) provides in general that no person shall discharge or in any manner discriminate against any employee because the employee has: (a) Filed any complaint under or related to the Act; (b) Instituted or caused to be instituted any proceeding under or related to the Act; (c) Testified or is about to testify in any proceeding under the Act or related to the Act; or (d) Exercised on his own behalf or on behalf of others any right afforded by the Act. Any employee who believes that he has been discriminated against in violation of section 11(c) of the Act may, within 30 days after such violation occurs, lodge a complaint with the Secretary of Labor alleging such violation. The Secretary shall then cause appropriate investigation to be made. If, as a result of such investigation, the Secretary determines that the provisions of section 11(c) have been violated civil action may be instituted in any appropriate United States district court, to restrain violations of section 11(c)(1) and to obtain other appropriate relief, including rehiring or reinstatement of the employee to his former position with back pay. Section 11(c) further provides for notification of complainants by the Secretary of determinations made pursuant to their complaints.

### 1977.4 Persons Prohibited from Discriminating.

Section 11(c) specifically states that "no person shall discharge or in any manner discriminate against any employee" because the employee has exercised rights under the Act. Section 3(4) of the Act defines "person" as "one or more individuals, partnerships, associations, corporations, business trusts, legal representatives, or any group of persons." Consequently, the prohibitions of section 11(c) are not limited to actions taken by employers against their own employees. A person may be chargeable with discriminatory action against an employee of another person. Section 11(c) would extend to such entities as organizations representing employees for collective bargaining purposes, employment agencies, or any other person in a position to discriminate against an employee. *See, Meek v. United States*, 136 F. 2d 679 (6th Cir., 1943); *Bowe v. Judson C. Burns*, 137 F 2d 37 (3rd Cir., 1943).

### §1977.5 Persons Protected by Section 11(c).

**1977.5(a)** All employees are afforded the full protection of section 11(c). For purposes of the Act, an employee is defined as "an employee of an employer who is employed in a business of his employer which affects commerce." The Act does not define the term "employ." However, the broad remedial nature of this legislation demonstrates a clear congressional intent that the existence of an employment relationship, for purposes of section 11(c), is to be based upon economic realities rather than upon common law doctrines and concepts. *See, U.S. v. Silk*, 331 U.S. 704 (1947); *Rutherford Food Corporation v. McComb*, 331 U.S. 722 (1947).

**1977.5(b)** For purposes of section 11(c), even an applicant for employment could be considered an employee. See, NLRB v Lamar Creamery, 246 F. 2d 8 (5th Cir., 1957). Further, because section 11(c) speaks in terms of any employee, it is also clear that the employee need not be an employee of the discriminator. The principal consideration would be whether the person alleging discrimination was an "employee" at the time of engaging in protected activity.

**1977.5(c)** In view of the definitions of "employee" contained in the Act, employees of a State or political subdivision thereof would not ordinarily be within the contemplated coverage of section 11(c).

### §1977.6 Unprotected Activities Distinguished.

**1977.6(a)** Actions taken by an employer, or others, which adversely affect an employee may be predicated upon nondiscriminatory grounds. the proscriptions of section 11(c) apply when the adverse action occurs because the employee has engaged in protected activities. An employee's engagement in activities protected by the Act does not automatically render him immune from discharge or discipline for legitimate reasons, or from adverse action dictated by non-prohibited considerations. *See, NLRB v Dixie Motor Coach Corp.*, 128 F. 2d 201 (5th Cir., 1942).

**1977.6(b)** At the same time, to establish a violation of section 11(c), the employee's engagement in protected activity need not be the sole consideration behind discharge or other adverse action. If protected activity was a substantial reason for the action, or if the discharge or other adverse action would not have taken place "but for" engagement in protected activity, section 11(c) has been violated. See, Mitchell v. Goodyear Tire & Rubber Co., 278 F. 2d 562 (8th Cir., 1960); Goldberg v. Bama Manufacturing, 302 F. 2d 152 (5th Cir., 1962). Ultimately, the issue as to whether a discharge was because of protected activity will have to be determined on the basis of the facts in the particular case.

### §1977.9 Complaints Under or Related to the Act.

**1977.9(a)** Discharge of, or discrimination against, an employee because the employee has filed "any complaint ... under or related to this Act ..." is prohibited by section 11(c). An example of a complaint made "under" the Act would be an employee request for inspection pursuant to section 8(f). However, this would not be the only type of complaint protected by section 11(c). The range of complaints "related to" the Act is commensurate with the broad remedial purposes of this legislation and the sweeping scope of its application, which entails the full extent of the commerce power. (See Cong. Rec., vol. 116 p. P. 42206 Dec. 17, 1970).

**1977.9(b)** Complaints registered with other federal agencies which have the authority to regulate or investigate occupational safety and health conditions are complaints "related to" this Act. Likewise, complaints made to State or local agencies regarding occupational safety and health conditions would be "related to" the Act. Such complaints, however, must relate to conditions at the workplace, as distinguished from complaints touching only upon general public safety and health.

**1977.9(c)** Further, the salutary principles of the Act would be seriously undermined if employees were discouraged from lodging complaints about occupational safety and health matters with their employers. (Section 2(1), (2), and (3).) Such complaints to employers, if made in good faith, therefore would be related to the Act, and an employee would be protected against discharge or discrimination caused by a complaint to the employer.

### §1977.10 Proceedings Under or Related to the Act.

**1977.10(a)** Discharge of, or discrimination against, any employee because the employee has "instituted or caused to be instituted any proceeding under or related to this Act" is also prohibited by section 11(c). Examples of proceedings which could arise specifically under the Act would be inspections of worksites under section 8 of the Act, employee contest of abatement date under section 10(c) of the Act, employee initiation of proceedings for promulgation of an occupational safety and health standard under section 6(b) of the Act and part 1911 of this chapter, employee application for modification of revocation of a variance under section 6(d) of the Act and part 1905 of this chapter, employee judicial challenge to a standard under section 6(f) of the Act and employee appeal of an Occupational Safety and Health review commission order under section 11(a) of the Act. In determining whether a "proceeding" is "related to" the Act, the considerations discussed in 1977.9 would also be applicable.

**1977.10(b)** An employee need not himself directly institute the proceedings. It is sufficient if he sets into motion activities of others which result in proceedings under or related to the Act.

### §1977.11 Testimony.

Discharge of, or discrimination against, any employee because the employee has testified or is about to testify" in proceedings under or related to the Act is also prohibited by section 11(c). This protection would of course not be limited to testimony in proceedings instituted or caused to be instituted by the employee, but would extend to any statements given in the course of judicial, quasi-judicial, and administrative proceedings, including inspections, investigations, and administrative rule making or adjudicative functions. If the employee is giving or is about to give testimony in any proceeding under or related to the Act, he would be protected against discrimination resulting from such testimony.

### §1977.12 Exercise of any Right Afforded by the Act.

**1977.12(a)** In addition to protecting employees who file complaints, institute proceedings, or testify in proceedings under or related to the Act, section 11(c) also protects employees from discrimination occurring because of the exercise "of any right afforded by this Act." Certain rights are explicitly provided in the Act; for example, there is a right to participate as a party in enforcement proceedings (sec. 10). Certain other rights exist by necessary implication. For example, employees may request information from the Occupational Safety and Health administration; such requests would constitute the exercise of a right afforded by the Act. Likewise, employees interviewed by agents of the Secretary in the course of inspections or investigations could not subsequently be discriminated against because of their cooperation.

**1977.12(b) 1977.12(b)(1)** On the other hand, review of the Act and examination of the legislative history discloses that, as a general matter, there is no right afforded by the Act which would entitle employees to walk off the job because of potential unsafe conditions at the workplace. Hazardous conditions which may be violative of the Act will ordinarily be corrected by the employer, once brought to his attention. If corrections are not accomplished, or if there is dispute about the existence of a hazard, the employee will normally have opportunity to request inspection of the workplace pursuant to section 8(f) of the Act, or to seek the assistance of other public agencies which have responsibility in the field of safety and health. Under such circumstances, therefore, an employer would not ordinarily be in violation of section 11(c) by taking action to discipline an employee for refusing to perform normal job activities because of alleged safety or health hazards.

**1977.12(b)(2)** However, occasions might arise when an employee is confronted with a choice between not performing assigned tasks or subjecting himself to serious injury or death arising from a hazardous condition at the workplace. If the employee, with no reasonable alternative, refuses in good faith to expose himself to the dangerous condition, he would be protected against subsequent discrimination. The condition causing the employee's apprehension of death or injury must be of such a nature that a reasonable person, under the circumstances then confronting the employee, would conclude that there is a real danger of death or serious injury and that there is insufficient time, due to the urgency of the situation, to eliminate the danger through resort to regular statutory enforcement channels. In addition, in such circumstances, the employee, where possible, must also have sought from his employer, and been unable to obtain, a correction of the dangerous condition.

[38 FR 2681, Jan. 29, 1973 as amended at 38 FR 4577, Feb. 16, 1973]

### §1977.15 Filing of Complaint for Discrimination.

**1977.15(a)** Who may file. A complaint of section 11(c) discrimination may be filed by the employee himself, or by a representative authorized to do so on his behalf.

**1977.15(b)** Nature of filing. No particular form of complaint is required.

**1977.15(c)** Place of filing. Complaint should be filed with the Area Director (Occupational Safety and Health Administration) responsible for enforcement activities in the geographical area where the employee resides or was employed.

**1977.15(d)** Time for filing.

**1977.15(d)(1)** Section 11(c)(2) provides that an employee who believes that he has been discriminated against in violation of section 11(c)(1) "may, within 30 days after such violation occurs," file a complaint with the Secretary of Labor.

**1977.15(d)(2)** A major purpose of the 30-day period in this provision is to allow the Secretary to decline to entertain complaints which have become stale. Accordingly, complaints not filed within 30 days of an alleged violation will ordinarily be presumed to be untimely.

**1977.15(d)(3)** However, there may be circumstances which would justify tolling of the 30-day period on recognized equitable principles or because of strongly extenuating circumstances, e.g., where the employer has concealed, or misled the employee regarding the grounds for discharge or other adverse action; or where the discrimination is in the nature of a continuing violation. The pendency of grievance-arbitration proceedings or filing with another agency, among others, are circumstances which do not justify tolling the 30-day period. In the absence of circumstances justifying a tolling of the 30-day period, untimely complaints will not be processed.

[38 FR 2681, Jan. 29, 1973, as amended at 50 FR 32846, Aug. 15, 1985]

### §1977.16 Notification of Secretary of Labor's Determination.

Section 11(c)(3) provides that the Secretary is to notify a complainant within 90 days of the complaint of his determination whether prohibited discrimination has occurred. This 90-day provision is considered directory in nature. While every effort will be made to notify complainants of the Secretary's determination within 90 days, there may be instances when it is not possible to meet the directory period set forth in section 11(c)(3).

### §1977.17 Withdrawal of Complaint.

Enforcement of the provisions of section 11(c) is not only a matter of protecting rights of individual employees, but also of public interest. Attempts by an employee to withdraw a previously filed complaint will not necessarily result in termination of the Secretary's investigation. The Secretary's jurisdiction cannot be foreclosed as a matter of law by unilateral action of the employee. However, a voluntary and uncoerced request from a complainant to withdraw his complaint will be given careful consideration and substantial weight as a matter of policy and sound enforcement procedure.

### §1977.18 Arbitration or Other Agency Proceedings

**1977.18(a)** General.

**1977.18(a)(1)** An employee who files a complaint under section 11(c) of the Act may also pursue remedies under grievance arbitration proceedings in collective bargaining agreements. In addition, the complainant may concurrently resort to other agencies for relief, such as the National Labor Relations Board. The Secretary's jurisdiction to entertain section 11(c) complaints, to investigate, and to determine whether discrimination has occurred, is independent of the jurisdiction of other agencies or bodies. The secretary may file action in U.S. district court regardless of the pendency of other proceedings.

**1977.18(a)(2)** However, the Secretary also recognizes the national policy favoring voluntary resolution of disputes under procedures in collective bargaining agreements. See, e.g., Boy's Markets, Inc. v retail Clerks, 398 U.S. 235 (1970); Republic Steel Corp. c. Maddox, 379 U.S. 650 (1965); Carey v. Westinghouse Electric Co., 375 U.S. 261 (1964); Collier Insulated Wire, 192 NLRB No. 150 (1971). By the same token, due deference should be paid to the jurisdiction of other forums established to resolve disputes which may also be related to section 11(c) complaints.

**1977.18(a)(3)** Where a complainant is in fact pursuing remedies other than those provided by section 11(c), postponement of the Secretary's determination and deferral to the results of such proceedings may be in order. See, Burlington Truck Lines, Inc., v. U.S., 371 U.S. 156 (1962).

**1977.18(b)** Postponement of determination. Postponement of determination would be justified where the rights asserted in other proceedings are substantially the same as rights under section 11(c), and those proceedings are not likely to violate the rights guaranteed by section 11(c). The factual issues in such proceedings must be substantially the same as those raised by section 11(c) complaint, and the forum hearing the matter must have the power to determine the ultimate issue of discrimination. See, Rios v. Reynolds Metals Co., F. 2d (5th Cir., 1972), 41 U.S.L.W. 1049 (Oct. 10, 1972); Newman b. Avco Corp., 451 F. 2d 743 (6th Cir., 1971).

**1977.18(c)** Deferral to outcome of other proceedings. A determination to defer to the outcome of other proceedings initiated by a complainant must necessarily be made on a case-to-case basis, after careful scrutiny of all available information. Before deferring to the results of other proceedings, it must be clear that those proceedings dealt adequately with all factual issues, that the proceedings were fair, regular, and free of procedural infirmities, and that the outcome of the proceedings was not repugnant to the purpose and policy of the Act. In this regard, if such other actions initiated by a complainant are dismissed without adjudicatory hearing thereof, such dismissal will not ordinarily be regarded as determinative of the section 11(c) complaint.

### §1977.22 Employee refusal to Comply with Safety Rules.

Employees who refuse to comply with occupational safety and health standards or valid safety rules implemented by the employer in furtherance of the Act are not exercising any rights afforded by the Act. Disciplinary measures taken by employers solely in response to employee refusal to comply with appropriate safety rules and regulations, will not ordinarily be regarded as discriminatory action prohibited by section 11(c). This situation should be distinguished from refusals to work, as discussed in 1977.12.

### §1977.23 State Plans.

A State which is implementing its own occupational safety and health enforcement program pursuant to section 18 of the Act and Parts 1902 and 1952 of this chapter must have provisions as effective as those of section 11(c) to protect employees from discharge or discrimination. Such provisions do not divest either the Secretary of Labor or federal district courts of jurisdiction over employee complaints of discrimination. However, the Secretary of Labor may refer complaints of employees adequately protected by State Plans' provisions to the appropriate State Agency. The basic principles outlined in 1977.18, supra will be observed as to deferrals to findings of State agencies.

## NOTES

# Part 1980 — Handling of Discrimination Complaints

**Authority:** 18 U.S.C. 1514A; Secretary of Labor's Order No. 5-2002, 67 FR 65008 (October 22, 2002). [69 FR 52113, August 24, 2004]

## Part 1980 — Procedures for the Handling of Discrimination Complaints Under Section 806 of the Corporate and Criminal Fraud Accountability Act of 2002, Title VIII of the Sarbanes-Oxley Act of 2002

### SUBPART A — COMPLAINTS, INVESTIGATIONS, FINDINGS AND PRELIMINARY ORDERS

| | |
|---|---|
| 1980.100 | Purpose and scope. |
| 1980.101 | Definitions. |
| 1980.102 | Obligations and prohibited acts. |
| 1980.103 | Filing of discrimination complaint. |
| 1980.104 | Investigation. |
| 1980.105 | Issuance of findings and preliminary orders. |

### SUBPART B — LITIGATION

| | |
|---|---|
| 1980.106 | Objections to the findings and the preliminary order and request for a hearing. |
| 1980.107 | Hearings. |
| 1980.108 | Role of Federal agencies. |
| 1980.109 | Decision and orders of the administrative law judge. |
| 1980.110 | Decision and orders of the Administrative Review Board. |

### SUBPART C — MISCELLANEOUS PROVISIONS

| | |
|---|---|
| 1980.111 | Withdrawal of complaints, objections, and findings; settlement. |
| 1980.112 | Judicial review. |
| 1980.113 | Judicial enforcement. |
| 1980.114 | District Court jurisdiction of discrimination complaints. |
| 1980.115 | Special circumstances; waiver of rules. |

### §1980.100 Purpose and Scope.

**1980.100(a)** This part implements procedures under section 806 of the Corporate and Criminal Fraud Accountability Act of 2002, Title VIII of the Sarbanes-Oxley Act of 2002 ("Sarbanes-Oxley" or "Act"), enacted into law July 30, 2002. Sarbanes-Oxley provides for employee protection from discrimination by companies and representatives of companies because the employee has engaged in protected activity pertaining to a violation or alleged violation of 18 U.S.C. 1341, 1343, 1344, or 1348, or any rule or regulation of the Securities and Exchange Commission, or any provision of Federal law relating to fraud against shareholders.

**1980.100(b)** This part establishes procedures pursuant to Sarbanes-Oxley for the expeditious handling of discrimination complaints made by employees, or by persons acting on their behalf. These rules, together with those rules codified at 29 CFR part 18, set forth the procedures for submission of complaints under Sarbanes-Oxley, investigations, issuance of findings and preliminary orders, objections to findings and orders, litigation before administrative law judges, post-hearing administrative review, and withdrawals and settlements.

### §1980.101 Definitions.

**Act** means section 806 of the Corporate and Criminal Fraud Accountability Act of 2002, Title VIII of the Sarbanes-Oxley Act of 2002, Public Law No.107-204, July 30, 2002, codified at 18 U.S.C. 1514A.

**Assistant Secretary** means the Assistant Secretary of Labor for Occupational Safety and Health or the person or persons to whom he or she delegates authority under the Act.

**Company** means any company with a class of securities registered under section 12 of the Securities Exchange Act of 1934 (15 U.S.C. 78l) and any company required to file reports under section 15(d) of the Securities Exchange Act of 1934 (15 U.S.C. 78o(d)).

**Company representative** means any officer, employee, contractor, subcontractor, or agent of a company.

**Complainant** means the employee who filed a complaint under the Act or on whose behalf a complaint was filed.

**Employee** means an individual presently or formerly working for a company or company representative, an individual applying to work for a company or company representative, or an individual whose employment could be affected by a company or company representative.

**Named person** means the employer and/or the company or company representative named in the complaint who is alleged to have violated the Act.

**OSHA** means the Occupational Safety and Health Administration of the United States Department of Labor.

**Person** means one or more individuals, partnerships, associations, corporations, business trusts, legal representatives or any group of persons.

**Secretary** means the Secretary of Labor or persons to whom authority under the Act has been delegated.

### §1980.102 Obligations and Prohibited Acts.

**1980.102(a)** No company or company representative may discharge, demote, suspend, threaten, harass or in any other manner discriminate against any employee with respect to the employee's compensation, terms, conditions, or privileges of employment because the employee, or any person acting pursuant to the employee's request, has engaged in any of the activities specified in paragraphs (b)(1) and (2) of this section.

**1980.102(b)** An employee is protected against discrimination (as described in paragraph (a) of this section) by a company or company representative for any lawful act:

**1980.102(b)(1)** To provide information, cause information to be provided, or otherwise assist in an investigation regarding any conduct which the employee reasonably believes constitutes a violation of 18 U.S.C. 1341, 1343, 1344, or 1348, any rule or regulation of the Securities and Exchange Commission, or any provision of Federal law relating to fraud against shareholders, when the information or assistance is provided to or the investigation is conducted by --

**1980.102(b)(1)(i)** A Federal regulatory or law enforcement agency;

**1980.102(b)(1)(ii)** Any Member of Congress or any committee of Congress; or

**1980.102(b)(1)(iii)** A person with supervisory authority over the employee (or such other person working for the employer who has the authority to investigate, discover, or terminate misconduct); or

**1980.102(b)(2)** To file, cause to be filed, testify, participate in, or otherwise assist in a proceeding filed or about to be filed (with any knowledge of the employer) relating to an alleged violation of 18 U.S.C. 1341, 1343, 1344, or 1348, any rule or regulation of the Securities and Exchange Commission, or any provision of Federal law relating to fraud against shareholders.

### §1980.103 Filing of Discrimination Complaint.

**1980.103(a)** *Who may file.* An employee who believes that he or she has been discriminated against by a company or company representative in violation of the Act may file, or have filed by any person on the employee's behalf, a complaint alleging such discrimination.

**1980.103(b)** *Nature of filing.* No particular form of complaint is required, except that a complaint must be in writing and should include a full statement of the acts and omissions, with pertinent dates, which are believed to constitute the violations.

**1980.103(c)** *Place of filing.* The complaint should be filed with the OSHA Area Director responsible for enforcement activities in the geographical area where the employee resides or was employed, but may be filed with any OSHA officer or employee. Addresses and telephone numbers for these officials are set forth in local directories and at the following Internet address: http://www.osha.gov/index.html.

**1980.103(d)** *Time for filing.* Within 90 days after an alleged violation of the Act occurs (*i.e.*, when the discriminatory decision has been both made and communicated to the complainant), an employee who believes that he or she has been discriminated against in violation of the Act may file, or have filed by any person on the employee's behalf, a complaint alleging such discrimination. The date of the postmark, facsimile transmittal, or e-mail communication will be considered to be the date of filing; if the complaint is filed in person, by hand-delivery or other means, the complaint is filed upon receipt.

### §1980.104 Investigation.

**1980.104(a)** Upon receipt of a complaint in the investigating office, the Assistant Secretary will notify the named person (or named persons) of the filing of the complaint, of the allegations contained in the complaint, and of the substance of the evidence supporting the complaint (redacted to protect the identity of any confidential informants). The Assistant Secretary also will notify the named person of its right under paragraphs (b) and (c) of this section and paragraph (e) of ? 1980.110. A copy of the notice to the named person will also be provided to the Securities and Exchange Commission.

**1980.104(b)** A complaint of alleged violation shall be dismissed unless the complainant has made a prima facie showing that protected behavior or conduct was a contributing factor in the unfavorable personnel action alleged in the complaint.

**1980.104(b)(1)** The complaint, supplemented as appropriate by interviews of the complainant, must allege the existence of facts and evidence to make a prima facie showing as follows:

**1980.104(b)(1)(i)** The employee engaged in a protected activity or conduct;

**1980.104(b)(1)(ii)** The named person knew or suspected, actually or constructively, that the employee engaged in the protected activity;

**1980.104(b)(1)(iii)** The employee suffered an unfavorable personnel action; and

**1980.104(b)(1)(iv)** The circumstances were sufficient to raise the inference that the protected activity was a contributing factor in the unfavorable action.

**1980.104(b)(2)** For purposes of determining whether to investigate, the complainant will be considered to have met the required burden if the complaint on its face, supplemented as appropriate through interviews of the complainant, alleges the existence of facts and either direct or circumstantial evidence to meet the required showing, *i.e.*, to give rise to an inference that the named person knew or suspected that the employee engaged in protected activity and that the protected activity was a contributing factor in the unfavorable personnel action. Normally the burden is satisfied, for example, if the complaint shows that the adverse personnel action took place shortly after the protected activity, giving rise to the inference that it was a factor in the adverse action. If the required showing has not been made, the complainant will be so advised and the investigation will not commence.

**1980.104(c)** Notwithstanding a finding that a complainant has made a prima facie showing, as required by this section, an investigation of the complaint shall not be conducted if the named person, pursuant to the procedures provided in this paragraph, demonstrates by clear and convincing evidence that it would have taken the same unfavorable personnel action in the absence of the complainant's protected behavior or conduct. Within 20 days of receipt of the notice of the filing of the complaint, the named person may submit to the Assistant Secretary a written statement and any affidavits or documents substantiating its position. Within the same 20 days, the named person may request a meeting with the Assistant Secretary to present its position.

**1980.104(d)** If the named person fails to demonstrate by clear and convincing evidence that it would have taken the same unfavorable personnel action in the absence of the behavior protected by the Act, the Assistant Secretary will conduct an investigation. Investigations will be conducted in a manner that protects the confidentiality of any person who provides information on a confidential basis, other than the complainant, in accordance with part 70 of this title.

**1980.104(e)** Prior to the issuance of findings and a preliminary order as provided for in ? 1980.105, if the Assistant Secretary has reasonable cause, on the basis of information gathered under the procedures of this part, to believe that the named person has violated the Act and that preliminary reinstatement is warranted, the Assistant Secretary will again contact the named person to give notice of the substance of the relevant evidence supporting the complainant's allegations as developed during the course of the investigation. This evidence includes any witness statements, which will be redacted to protect the identity of confidential informants where statements were given in confidence; if the statements cannot be redacted without revealing the identity of confidential informants, summaries of their contents will be provided. The named person will be given the opportunity to submit a written response, to meet with the investigators to present statements from witnesses in support of its position, and to present legal and factual arguments. The named person will present this evidence within 10 business days of the Assistant Secretary's notification pursuant to this paragraph, or as soon afterwards as the Assistant Secretary and the named person can agree, if the interests of justice so require.

### §1980.105 Issuance of Findings and Preliminary Orders.

**1980.105(a)** After considering all the relevant information collected during the investigation, the Assistant Secretary shall issue, within 60 days of filing of the complaint, written findings as to whether or not there is reasonable cause to believe that the named person has discriminated against the complainant in violation of the Act.

**1980.105(a)(1)** If the Assistant Secretary concludes that there is reasonable cause to believe that a violation has occurred, he or she shall accompany the findings with a preliminary order providing relief to the complainant. The preliminary order shall include all relief necessary to make the employee whole, including, where appropriate: reinstatement with the same seniority status that the employee would have had but for the discrimination; back pay with interest; and compensation for any special damages sustained as a result of the discrimination, including litigation costs, expert witness fees, and reasonable attorney's fees. Where the named person establishes that the complainant is a security risk (whether or not the information is obtained after the complainant's discharge), a preliminary order of reinstatement would not be appropriate.

**1980.105(a)(2)** If the Assistant Secretary concludes that a violation has not occurred, the Assistant Secretary will notify the parties of that finding.

**1980.105(b)** The findings and the preliminary order will be sent by certified mail, return receipt requested, to all parties of record. The letter accompanying the findings and order will inform the parties of their right to file objections and to request a hearing, and of the right of the named person to request attorney's fees from the ALJ, regardless of whether the named person has filed objections, if the named person alleges that the complaint was frivolous or brought in bad faith. The letter also will give the address of the Chief Administrative Law Judge. At the same time, the Assistant Secretary will file with the Chief Administrative Law Judge, U.S. Department of Labor, a copy of the original complaint and a copy of the findings and order.

**1980.105(c)** The findings and preliminary order will be effective 30 days after receipt by the named person pursuant to paragraph (b) of this section, unless an objection and a request for a hearing has been filed as provided at ? 1980.106. However, the portion of any preliminary order requiring reinstatement will be effective immediately upon receipt of the findings and preliminary order.

### §1980.106 Objections to the Findings and the Preliminary Order and Request for a Hearing.

**1980.106(a)** Any party who desires review, including judicial review, of the findings and preliminary order, or a named person alleging that the complaint was frivolous or brought in bad faith who seeks an award of attorney's fees, must file any objections and/or a request for a hearing on the record within 30 days of receipt of the findings and preliminary order pursuant to paragraph (b) of § 1980.105. The objection or request for attorney's fees and request for a hearing must be in writing and state whether the objection is to the findings, the preliminary order, and/or whether there should be an award of attorney's fees. The date of the postmark, facsimile transmittal, or e-mail communication will be considered to be the date of filing; if the objection is filed in person, by hand-delivery or other means, the objection is filed upon receipt. Objections must be filed with the Chief Administrative Law Judge, U.S. Department of Labor, Washington, DC 20001, and copies of the objections must be mailed at the same time to the other parties of record, the OSHA official who issued the findings and order, and the Associate Solicitor, Division of Fair Labor Standards, U.S. Department of Labor, Washington, DC 20210.

**1980.106(b)(1)** If a timely objection is filed, all provisions of the preliminary order will be stayed, except for the portion requiring preliminary reinstatement, which shall not be automatically stayed. The portion of the preliminary order requiring reinstatement will be effective immediately upon the named person's receipt of the findings and preliminary order, regardless of any objections to the order. The named person may file a motion with the Office of Administrative Law Judges for a stay of the Assistant Secretary's preliminary order of reinstatement.

**1980.106(b)(2)** If no timely objection is filed with respect to either the findings or the preliminary order, the findings or preliminary order, as the case may be, shall become the final decision of the Secretary, not subject to judicial review.

### §1980.107 Hearings.

**1980.107(a)** Except as provided in this part, proceedings will be conducted in accordance with the rules of practice and procedure for administrative hearings before the Office of Administrative Law Judges, codified at subpart A, part 18 of title 29 of the Code of Federal Regulations.

**1980.107(b)** Upon receipt of an objection and request for hearing, the Chief Administrative Law Judge will promptly assign the case to a judge who will notify the parties, by certified mail, of the day, time, and place of hearing. The hearing is to commence expeditiously, except upon a showing of good cause or unless otherwise agreed to by the parties. Hearings will be conducted ***de novo***, on the record. Administrative law judges have broad discretion to limit discovery in order to expedite the hearing.

**1980.107(c)** If both the complainant and the named person object to the findings and/or order, the objections will be consolidated and a single hearing will be conducted.

**1980.107(d)** Formal rules of evidence will not apply, but rules or principles designed to assure production of the most probative evidence will be applied. The administrative law judge may exclude evidence that is immaterial, irrelevant, or unduly repetitious.

### § 1980.108 Role of Federal Agencies.

**1980.108(a)(1)** The complainant and the named person will be parties in every proceeding. At the Assistant Secretary's discretion, the Assistant Secretary may participate as a party or as ***amicus curiae*** at any time at any stage of the proceedings. This right to participate includes, but is not limited to, the right to petition for review of a decision of an administrative law judge, including a decision approving or rejecting a settlement agreement between the complainant and the named person.

**1980.108(a)(2)** Copies of pleadings in all cases, whether or not the Assistant Secretary is participating in the proceeding, must be sent to the Assistant Secretary, Occupational Safety and Health Administration, and to the Associate Solicitor, Division of Fair Labor Standards, U.S. Department of Labor, Washington, DC 20210.

**1980.108(b)** The Securities and Exchange Commission may participate as ***amicus curiae*** at any time in the proceedings, at the Commission's discretion. At the request of the Securities and Exchange Commission, copies of all pleadings in a case must be sent to the Commission, whether or not the Commission is participating in the proceeding.

### §1980.109 Decision and Orders of the Administrative Law Judge.

**1980.109(a)** The decision of the administrative law judge will contain appropriate findings, conclusions, and an order pertaining to the remedies provided in paragraph (b) of this section, as appropriate. A determination that a violation has occurred may only be made if the complainant has demonstrated that protected behavior or conduct was a contributing factor in the unfavorable personnel action alleged in the complaint. Relief may not be ordered if the named person demonstrates by clear and convincing evidence that it would have taken the same unfavorable personnel action in the absence of any protected behavior. Neither the Assistant Secretary's determination to dismiss a complaint without completing an investigation pursuant to ? 1980.104(b) nor the Assistant Secretary's determination to proceed with an investigation is subject to review by the administrative law judge, and a complaint may not be remanded for the completion of an investigation or for additional findings on the basis that a determination to dismiss was made in error. Rather, if there otherwise is jurisdiction, the administrative law judge will hear the case on the merits.

**1980.109(b)** If the administrative law judge concludes that the party charged has violated the law, the order will provide all relief necessary to make the employee whole, including reinstatement of the complainant to that person's former position with the seniority status that the complainant would have had but for the discrimination, back pay with interest, and compensation for any special damages sustained as a result of the discrimination, including litigationcosts, expert witness fees, and reasonable attorney's fees. If, upon the request of the named person, the administrative law judge determines that a complaint was frivolous or was brought in bad faith, the judge may award to the named person a reasonable attorney's fee, not exceeding $1,000.

**1980.109(c)** The decision will be served upon all parties to the proceeding. Any administrative law judge's decision requiring reinstatement or lifting an order of reinstatement by the Assistant Secretary will be effective immediately upon receipt of the decision by the named person, and will not be stayed. All other portions of the judge's order will be effective 10 business days after the date of the decision unless a timely petition for review has been filed with the Administrative Review Board.

### §1980.110 Decision and Orders of the Administrative Review Board.

**1980.110(a)** Any party desiring to seek review, including judicial review, of a decision of the administrative law judge, or a named person alleging that the complaint was frivolous or brought in bad faith who seeks an award of attorney's fees, must file a written petition for review with the Administrative Review Board ("the Board"), which has been delegated the authority to act for the Secretary and issue final decisions under this part. The decision of the administrative law judge will become the final order of the Secretary unless, pursuant to this section, a petition for review is timely filed with the Board. The petition for review must specifically identify the findings, conclusions or orders to which exception is taken. Any exception not specifically urged ordinarily will be deemed to have been waived by the parties. To be effective, a petition must be filed within 10 business days of the date of the decision of the administrative law judge. The date of the postmark, facsimile transmittal, or e-mail communication will be considered to be the date of filing; if the petition is filed in person, by hand-delivery or other means, the petition is considered filed upon receipt. The petition must be served on all parties and on the Chief Administrative Law Judge at the time it is filed with the Board. Copies of the petition for review and all briefs must be served on the Assistant Secretary, Occupational Safety and Health Administration, and on the Associate Solicitor, Division of Fair Labor Standards, U.S. Department of Labor, Washington, DC 20210.

**1980.110(b)** If a timely petition for review is filed pursuant to paragraph (a) of this section, the decision of the administrative law judge will become the final order of the Secretary unless the Board, within 30 days of the filing of the petition, issues an order notifying the parties that the case has been accepted for review. If a case is accepted for review, the decision of the administrative law judge will be inoperative unless and until the Board issues an order adopting the decision, except that a preliminary order of reinstatement will be effective while review is conducted by the Board, unless the Board grants a motion to stay the order. The Board will specify the terms under which any briefs are to be filed. The Board will review the factual determinations of the administrative law judge under the substantial evidence standard.

**1980.110(c)** The final decision of the Board shall be issued within 120 days of the conclusion of the hearing, which will be deemed to be the conclusion of all proceedings before the administrative law judge — ***i.e.***, 10 business days after the date of the decision of the administrative law judge unless a motion for reconsideration has been filed with the administrative law judge in the interim. The decision will be served upon all parties and the Chief Administrative Law Judge by mail to the last known address. The final decision will also be served on the Assistant Secretary, Occupational Safety and Health Administration, and on the Associate Solicitor, Division of Fair Labor Standards, U.S. Department of Labor, Washington, DC 20210, even if the Assistant Secretary is not a party.

**1980.110(d)** If the Board concludes that the party charged has violated the law, the final order will order the party charged to provide all relief necessary to make the employee whole, including reinstatement of the complainant to that person's former position with the seniority status that the complainant would have had but for the discrimination, back pay with interest, and compensation for any special damages sustained as a result of the discrimination, including litigation costs, expert witness fees, and reasonable attorney's fees.

**1980.110(e)** If the Board determines that the named person has not violated the law, an order will be issued denying the complaint. If, upon the request of the named person, the Board determines that a complaint was frivolous or was brought in bad faith, the Board may award to the named person a reasonable attorney's fee, not exceeding $1,000.

## §1980.111 Withdrawal of Complaints, Objections, and Findings; Settlement.

**1980.111(a)** At any time prior to the filing of objections to the findings or preliminary order, a complainant may withdraw his or her complaint under the Act by filing a written withdrawal with the Assistant Secretary. The Assistant Secretary will then determine whether to approve the withdrawal. The Assistant Secretary will notify the named person of the approval of any withdrawal. If the complaint is withdrawn because of settlement, the settlement will be approved in accordance with paragraph (d) of this section.

**1980.111(b)** The Assistant Secretary may withdraw his or her findings or a preliminary order at any time before the expiration of the 30-day objection period described in ? 1980.106, provided that no objection has yet been filed, and substitute new findings or preliminary order. The date of the receipt of the substituted findings or order will begin a new 30-day objection period.

**1980.111(c)** At any time before the findings or order become final, a party may withdraw his or her objections to the findings or order by filing a written withdrawal with the administrative law judge or, if the case is on review, with the Board. The judge or the Board, as the case may be, will determine whether to approve the withdrawal. If the objections are withdrawn because of settlement, the settlement will be approved in accordance with paragraph (d) of this section.

**1980.111(d)(1) *Investigative settlements*.** At any time after the filing of a complaint, and before the findings and/or order are objected to or become a final order by operation of law, the case may be settled if the Assistant Secretary, the complainant and the named person agree to a settlement.

**1980.111(d)(2) *Adjudicatory settlements*.** At any time after the filing of objections to the Assistant Secretary's findings and/or order, the case may be settled if the participating parties agree to a settlement and the settlement is approved by the administrative law judge if the case is before the judge, or by the Board if a timely petition for review has been filed with the Board. A copy of the settlement will be filed with the administrative law judge or the Board, as the case may be.

**1980.111(e)** Any settlement approved by the Assistant Secretary, the administrative law judge, or the Board, will constitute the final order of the Secretary and may be enforced pursuant to § 1980.113.

## §1980.112 Judicial Review.

**1980.112(a)** Within 60 days after the issuance of a final order by the Board (Secretary) under ? 1980.110, any person adversely affected or aggrieved by the order may file a petition for review of the order in the United States Court of Appeals for the circuit in which the violation allegedly occurred or the circuit in which the complainant resided on the date of the violation. A final order of the Board is not subject to judicial review in any criminal or other civil proceeding.

**1980.112(b)** If a timely petition for review is filed, the record of a case, including the record of proceedings before the administrative law judge, will be transmitted by the Board to the appropriate court pursuant to the rules of the court.

## §1980.113 Judicial Enforcement.

**1980.113** Whenever any person has failed to comply with a preliminary order of reinstatement or a final order or the terms of a settlement agreement, the Secretary or a person on whose behalf the order was issued may file a civil action seeking enforcement of the order in the United States district court for the district in which the violation was found to have occurred.

## §1980.114 District Court Jurisdiction of Discrimination Complaints.

**1980.114(a)** If the Board has not issued a final decision within 180 days of the filing of the complaint, and there is no showing that there has been delay due to the bad faith of the complainant, the complainant may bring an action at law or equity for ***de novo*** review in the appropriate district court of the United States, which will have jurisdiction over such an action without regard to the amount in controversy.

**1980.114(b)** Fifteen days in advance of filing a complaint in federal court, a complainant must file with the administrative law judge or the Board, depending upon where the proceeding is pending, a notice of his or her intention to file such a complaint. The notice must be served upon all parties to the proceeding. If the Assistant Secretary is not a party, a copy of the notice must be served on the Assistant Secretary, Occupational Safety and Health Administration, and on the Associate Solicitor, Division of Fair Labor Standards, U.S. Department of Labor, Washington, DC 20210.

## §1980.115 Special Circumstances; Waiver of Rules.

In special circumstances not contemplated by the provisions of this part, or for good cause shown, the administrative law judge or the Board on review may, upon application, after three days notice to all parties and interveners, waive any rule or issue any orders that justice or the administration of the Act requires.

NOTES

# Regional Office Locations

### REGION 1
Regional Office
JFK Federal Building, Room E340
Boston, Massachusetts 02203
(617) 565-9860
(617) 565-9827 FAX

### REGION 2
Regional Office
201 Varick Street, Room 670
New York, New York 10014
(212) 337-2378
(212) 337-2371 FAX

### REGION 3
Regional Office
U.S. Department of Labor/OSHA
The Curtis Center-Suite 740 West
170 S. Independence Mall West
Philadelphia, PA 19106-3309
TELE: (215) 861-4900
FAX: (215) 861-4904

### REGION 4
Regional Office
61 Forsyth Street, SW
Atlanta, Georgia 30303
(404) 562-2300
(404) 562-2295 FAX

### REGION 5
Regional Office
230 South Dearborn Street, Room 3244
Chicago, Illinois 60604
(312) 353-2220
(312) 353-7774 FAX

### REGION 6
Regional Office
525 Griffin Street, Room 602
Dallas, Texas 75202
(214) 767-4731
(214) 767-4693 FAX

### REGION 7
Regional Office
City Center Square
1100 Main Street, Suite 800
Kansas City, Missouri 64105
(816) 426-5861
(816) 426-2750 FAX

### REGION 8
Regional Office
1999 Broadway, Suite 1690
P.O. Box 46550
Denver, Colorado 80201-6550
(303) 844-1600
(303) 844-1616 FAX

### REGION 9
Region IX Federal Contact Numbers
71 Stevenson Street, Room 420
San Francisco, California 94105
(415) 975-4310 (Main Public -
    8:00 AM - 4:30 PM Pacific)
(800) 475-4019 (For Technical Assistance)
(800) 475-4020 (For Complaints -
    Accidents/Fatalities)
(800) 475-4022 (For Publication Requests)
(415) 975-4319 FAX

### REGION 10
Regional Office
1111 Third Avenue, Suite 715
Seattle, Washington 98101-3212
(206) 553-5930
(206) 553-6499 FAX

### OSHA Office Directory
National Office
U.S. Department of Labor
Occupational Safety and Health
    Administration (OSHA)
200 Constitution Avenue, N.W.
Washington, D.C. 20210

## State Office Locations

### Alaska
Alaska Department of Labor
(Mailing address)
P.O. Box 21149
Juneau, Alaska 99801-1149
(Physical location)
1111 W. 8th Street, Room 304
Juneau, Alaska 99801-1149
(907) 465-2700
(907) 465-2784 FAX
http://www.labor.state.ak.us/lss/lss.htm

### Arizona
Industrial Commission of Arizona
800 W. Washington
Phoenix, Arizona 85007-2922
(602) 542-4411
(602) 542-1614 FAX
http://www.ica.state.az.us/ADOSH/oshatop.htm

### California
California Department of Industrial Relations
455 Golden Gate Avenue - 10th Floor
San Francisco, California 94102
(415) 703-5050
(415) 703-5114 FAX
http://www.dir.ca.gov/occupational_safety.html

### Connecticut
Connecticut Department of Labor
200 Folly Brook Boulevard
Wethersfield, Connecticut 06109
(860) 566-5123
(860) 566-1520 FAX
http://www.ctdol.state.ct.us/osha/osha.htm

### Hawaii
Hawaii Department of Labor and Industrial Relations
830 Punchbowl Street
Honolulu, Hawaii 96813
(808) 586-8844
(808) 586-9099 FAX
http://www.state.hi.us/dlir/hiosh/

### Indiana
Indiana Department of Labor
State Office Building
402 West Washington Street, Room W195
Indianapolis, Indiana 46204-2751
(317) 232-2378
(317) 233-3790 FAX
http://www.in.gov/labor/iosha/

### Iowa
Iowa Division of Labor Services
1000 E. Grand Avenue
Des Moines, Iowa 50319-0209
(515) 281-6432
(515) 281-4698 FAX
http://www.state.ia.us/iwd/labor/index.html

### Kentucky
Kentucky Labor Cabinet
1047 U.S. Highway 127 South, Suite 4
Frankfort, Kentucky 40601
(502) 564-3070
(502) 564-5387 FAX
http://www.kylabor.net/kyosh/index.htm

### Maryland
Maryland Department of Labor, Licensing and Regulation
1100 North Eutaw Street, Room 613
Baltimore, Maryland 21201-2206
(410) 767-2241
(410) 767-2986 FAX
http://www.dllr.state.md.us/labor/mosh.html

### Michigan
Michigan Department of Consumer and Industry Services
Kathleen M. Wilbur, Director
Bureau of Safety and Regulation
P.O. Box 30643
Lansing, Michigan 48909-8143
(517) 322-1814
(517) 322-1775 FAX
http://www.cis.state.mi.us/bsr/

### Minnesota
Minnesota Department of Labor and Industry
443 Lafayette Road North
St. Paul, Minnesota 55155-4307
(651) 284-5050
(651) 282-5405 FAX
http://www.doli.state.mn.us/mnosha.html

### Nevada
Nevada Division of Industrial Relations
400 West King Street
Carson City, Nevada 89710
(775) 687-3032
(775) 687-6305 FAX
http://dirweb.state.nv.us

### New Jersey
New Jersey Department of Labor
John Fitch Plaza - Labor Building
Market and Warren Streets
P.O. Box 110
Trenton, New Jersey 0826-0010
(609) 292-2975
(609) 633-9271 FAX
http://www.state.nj.us/labor/lsse/lspeosh.html

### New Mexico
New Mexico Environment Department
1190 St. Francis Drive
P.O. Box 26110
Santa Fe, New Mexico 87502
(505) 827-2850
(505) 827-2836 FAX
http://www.nmenv.state.nm.us/OHSB_website/ohsb_home.htm

### New York
New York Department of Labor
W. Averell Harriman State Office Building - 12, Room 500
Albany, New York 12240
(518) 457-2746
(518) 457-6908 FAX
http://www.labor.state.ny.us/working_ny/worker_rights/safety_health.html

### North Carolina
North Carolina Department of Labor (NCDOL)
4 West Edenton Street
Raleigh, North Carolina 27601-1092
(919) 807-2900
(919) 807-2855 FAX
http://www.nclabor.com/osha/osh.htm

### Oregon
Oregon Occupational Safety and Health Division
Department of Consumer & Business Services
350 Winter Street, NE, Room 430
Salem, Oregon 97301
(503) 378-3272
(800) 922-2689 TOLL FREE in Oregon
http://www.orosha.org/

### Puerto Rico
Puerto Rico Department of Labor and Human Resources
Prudencio Rivera Martínez Building, 21st Floor
505 Muñoz Rivera Avenue
Hato Rey, Puerto Rico 00918
(787) 754-2119
(787) 753-9550 FAX
http://dtrh.prstar.net

### South Carolina
South Carolina Department of Labor, Licensing, and Regulation
Koger Office Park, Kingstree Building
110 Centerview Drive
PO Box 11329
Columbia, South Carolina 29210
(803) 896-4300
(803) 896-4393 FAX
http://www.llr.state.sc.us/osha.asp

### Tennessee
Tennessee Department of Labor and Workforce Development
710 James Robertson Parkway
Nashville, Tennessee 37243-0659
(615) 741-2582
(615) 741-5078 FAX
http://www.state.tn.us/labor-wfd

### Utah
Utah Labor Commission
160 East 300 South, 3rd Floor
PO Box 146650
Salt Lake City, Utah 84114-6650
(801) 530-6901
(801) 536-7906 FAX
http://www.labor.state.ut.us/Utah_Occupational_Safety___Hea/utah_occupational_safety___hea.html

### Vermont
Vermont Department of Labor and Industry
National Life Building - Drawer 20
National Life Drive Montpelier, Vermont 05620-3401
(802) 828-5098
(802) 828-2195 FAX
http://www.state.vt.us/labind/vosha.htm

### Virgin Islands
Virgin Islands Department of Labor
2203 Church Street
Christiansted
St. Croix, Virgin Islands 00820-4660
(340) 773-1994
(340) 773-1858 FAX

### Virginia
Virginia Department of Labor and Industry
Powers-Taylor Building
13 South 13th Street
Richmond, Virginia 23219
(804) 786-2377
(804) 371-6524 FAX
http://www.doli.state.va.us/

### Washington
Washington Department of Labor and Industries
PO Box 44001
Olympia, Washington 98504-4001
(360) 902-4200
(360) 902-4202 FAX
http://www.lni.wa.gov/wisha/

### Wyoming
Wyoming Department of Employment
Worker's Safety and Compensation Division
Cheyenne Business Center
1510 East Pershing Boulevard
Cheyenne, Wyoming 82002
(307) 777-7786
(307) 777-3646 FAX
http://wydoe.state.wy.us/doe.asp?!D=7

# You Have a Right to a Safe and Healthful Workplace.
# IT'S THE LAW!

- You have the right to notify your employer or OSHA about workplace hazards. You may ask OSHA to keep your name confidential.

- You have the right to request an OSHA inspection if you believe that there are unsafe and unhealthful conditions in your workplace. You or your representative may participate in the inspection.

- You can file a complaint with OSHA within 30 days of discrimination by your employer for making safety and health complaints or for exercising your rights under the *OSH Act*.

- You have a right to see OSHA citations issued to your employer. Your employer must post the citations at or near the place of the alleged violation.

- Your employer must correct workplace hazards by the date indicated on the citation and must certify that these hazards have been reduced or eliminated.

- You have the right to copies of your medical records or records of your exposure to toxic and harmful substances or conditions.

- Your employer must post this notice in your workplace.

The *Occupational Safety and Health Act of 1970 (OSH Act)*, P.L. 91-596, assures safe and healthful working conditions for working men and women throughout the Nation. The Occupational Safety and Health Administration, in the U.S. Department of Labor, has the primary responsibility for administering the *OSH Act*. The rights listed here may vary depending on the particular circumstances. To file a complaint, report an emergency, or seek OSHA advice, assistance, or products, call 1-800-321-OSHA or your nearest OSHA office: • Atlanta (404) 562-2300 • Boston (617) 565-9860 • Chicago (312) 353-2220 • Dallas (214) 767-4731 • Denver (303) 844-1600 • Kansas City (816) 426-5861 • New York (212) 337-2378 • Philadelphia (215) 861-4900 • San Francisco (415) 975-4310 • Seattle (206) 553-5930. Teletypewriter (TTY) number is 1-877-889-5627. To file a complaint online or obtain more information on OSHA federal and state programs, visit OSHA's website at www.osha.gov. If your workplace is in a state operating under an OSHA-approved plan, your employer must post the required state equivalent of this poster.

## 1-800-321-OSHA
## www.osha.gov

U.S. Department of Labor • Occupational Safety and Health Administration • OSHA 3165

# Usted Tiene el Derecho a un Lugar de Trabajo Seguro y Saludable.

# ¡LO ESTABLECE LA LEY!

- Tiene el derecho de notificar a su empleador o a la OSHA sobre cualquier peligro en su lugar de trabajo. Puede pedir a la OSHA que mantenga su nombre en reserva.
- Tiene el derecho de solicitar una inspección de la OSHA si considera que existen condiciones peligrosas y poco saludables en su lugar de trabajo. Usted o su representante puede participar en la inspección.
- Puede presentar un reclamo a OSHA durante un plazo de 30 días si su empleador lo discrimina por presentar reclamos de seguridad y sanidad o por ejercer sus derechos de acuerdo con la Ley.
- Tiene el derecho de ver las citaciones de la OSHA enviadas a su empleador. Su empleador debe colocar las citaciones en un lugar visible en el sitio de la supuesta infracción o cerca de él.
- Su empleador debe corregir los peligros en el lugar de trabajo dentro del plazo indicado en la citación y debe certificar que dichos peligros se hayan reducido o eliminado.
- Tiene el derecho de recibir copias de su historial médico o de los registros de su exposición a sustancias o condiciones tóxicas y peligrosas.
- Su empleador debe colocar este aviso en un lugar visible de su lugar de trabajo.

La Ley de Seguridad y Salud Ocupacionales de 1970 (la Ley), P.L. 91-596, garantiza condiciones ocupacionales seguras y saludables para los hombres y las mujeres que desempeñen algún trabajo en toda la Nación. La Administración de Seguridad y Salud Ocupacionales (OSHA), dependiente del Departamento del Trabajo de los Estados Unidos, es la responsable principal de supervisar la Ley. Los derechos que se indican en este documento pueden variar según las circunstancias particulares. Para presentar un reclamo, informar sobre una emergencia o pedir consejo, asistencia o productos de la OSHA, llame al 1-800-321-OSHA o a la oficina de la OSHA más cercana a usted: • Atlanta (404) 562-2300 • Boston (617) 565-9860 • Chicago (312) 353-2220 • Dallas (214) 767-4731 • Denver (303) 844-1600 • Ciudad de Kansas (816) 426-5861 • Nueva York (212) 337-2378 • Filadelfia (215) 861-4900 • San Francisco (415) 975-4310 • Seattle (206) 553-5930. El número TTY es 1-877-889-5627. Para presentar un reclamo en línea u obtener más información sobre los programas federales y estatales de la OSHA, visite el sitio Web de la OSHA en www.osha.gov. Si su lugar de trabajo se encuentra en un estado que funciona según un plan aprobado por la OSHA, su empleador debe colocar en un sitio visible el equivalente estatal de este afiche.

# 1-800-321-OSHA
# www.osha.gov

Departamento del Trabajo de los E.E. UU. • Administración de Seguridad y Salud Ocupacionales • OSHA 3167

# OSHA
# Forms for Recording
# Work-Related Injuries and Illnesses

## Dear Employer:

This booklet includes the forms needed for maintaining occupational injury and illness records for 2004. These new forms have changed in several important ways from the 2003 recordkeeping forms.

In the December 17, 2002 Federal Register (67 FR 77165-77170), OSHA announced its decision to add an occupational hearing loss column to OSHA's Form 300, Log of Work-Related Injuries and Illnesses. This forms package contains modified Forms 300 and 300A which incorporate the additional column M(5) Hearing Loss. Employers required to complete the injury and illness forms must begin to use these forms on January 1, 2004.

In response to public suggestions, OSHA also has made several changes to the forms package to make the recordkeeping materials clearer and easier to use:

- On Form 300, we've switched the positions of the day count columns. The days "away from work" column now comes before the days "on job transfer or restriction."
- We've clarified the formulas for calculating incidence rates.
- We've added new recording criteria for occupational hearing loss to the "Overview" section.
- On Form 300, we've made the column heading "Classify the Case" more prominent to make it clear that employers should mark only one selection among the four columns offered.

The Occupational Safety and Health Administration shares with you the goal of preventing injuries and illnesses in our nation's workplaces. Accurate injury and illness records will help us achieve that goal.

*Occupational Safety and Health Administration*
*U.S. Department of Labor*

## What's Inside...

In this package, you'll find everything you need to complete OSHA's *Log* and the *Summary of Work-Related Injuries and Illnesses* for the next several years. On the following pages, you'll find:

▼ **An Overview: Recording Work-Related Injuries and Illnesses** — General instructions for filling out the forms in this package and definitions of terms you should use when you classify your cases as injuries or illnesses.

▼ **How to Fill Out the Log** — An example to guide you in filling out the *Log* properly.

▼ **Log of Work-Related Injuries and Illnesses** — Several pages of the *Log* (but you may make as many copies of the *Log* as you need.) Notice that the *Log* is separate from the *Summary*.

▼ **Summary of Work-Related Injuries and Illnesses** — Removable *Summary* pages for easy posting at the end of the year. Note that you post the *Summary* only, not the *Log*.

▼ **Worksheet to Help You Fill Out the Summary** — A worksheet for figuring the average number of employees who worked for your establishment and the total number of hours worked.

▼ **OSHA's 301: Injury and Illness Incident Report** — A copy of the OSHA 301 to provide details about the incident. You may make as many copies as you need or use an equivalent form.

Take a few minutes to review this package. If you have any questions, *visit us online at www.osha.gov* **or** *call your local OSHA office.* We'll be happy to help you.

**U.S. Department of Labor**
Occupational Safety and Health Administration

# An Overview:
# Recording Work-Related Injuries and Illnesses

The Occupational Safety and Health (OSH) Act of 1970 requires certain employers to prepare and maintain records of work-related injuries and illnesses. Use these definitions when you classify cases on the Log. OSHA's recordkeeping regulation (see 29 CFR Part 1904) provides more information about the definitions below.

The *Log of Work-Related Injuries and Illnesses* (Form 300) is used to classify work-related injuries and illnesses and to note the extent and severity of each case. When an incident occurs, use the *Log* to record specific details about what happened and how it happened. The *Summary* — a separate form (Form 300A) — shows the totals for the year in each category. At the end of the year, post the *Summary* in a visible location so that your employees are aware of the injuries and illnesses occurring in their workplace.

Employers must keep a *Log* for each establishment or site. If you have more than one establishment, you must keep a separate *Log* and *Summary* for each physical location that is expected to be in operation for one year or longer.

Note that your employees have the right to review your injury and illness records. For more information, see 29 Code of Federal Regulations Part 1904.35, *Employee Involvement*.

Cases listed on the *Log of Work-Related Injuries and Illnesses* are not necessarily eligible for workers' compensation or other insurance benefits. Listing a case on the *Log* does not mean that the employer or worker was at fault or that an OSHA standard was violated.

## When is an injury or illness considered work-related?

An injury or illness is considered work-related if an event or exposure in the work environment caused or contributed to the condition or significantly aggravated a preexisting condition. Work-relatedness is presumed for injuries and illnesses resulting from events or exposures occurring in the workplace, unless an exception specifically applies. See 29 CFR Part 1904.5(b)(2) for the exceptions. The work environment includes the establishment and other locations where one or more employees are working or are present as a condition of their employment. See 29 CFR Part 1904.5(b)(1).

## Which work-related injuries and illnesses should you record?

Record those work-related injuries and illnesses that result in:

- death,
- loss of consciousness,
- days away from work,
- restricted work activity or job transfer, or
- medical treatment beyond first aid.

You must also record work-related injuries and illnesses that are significant (as defined below) or meet any of the additional criteria listed below.

You must record any significant work-related injury or illness that is diagnosed by a physician or other licensed health care professional. You must record any work-related case involving cancer, chronic irreversible disease, a fractured or cracked bone, or a punctured eardrum. See 29 CFR 1904.7.

## What are the additional criteria?

You must record the following conditions when they are work-related:

- any needlestick injury or cut from a sharp object that is contaminated with another person's blood or other potentially infectious material;
- any case requiring an employee to be medically removed under the requirements of an OSHA health standard;
- tuberculosis infection as evidenced by a positive skin test or diagnosis by a physician or other licensed health care professional after exposure to a known case of active tuberculosis.
- an employee's hearing test (audiogram) reveals 1) that the employee has experienced a Standard Threshold Shift (STS) in hearing in one or both ears (averaged at 2000, 3000, and 4000 Hz) and 2) the employee's total hearing level is 25 decibels (dB) or more above audiometric zero (also averaged at 2000, 3000, and 4000 Hz) in the same ear(s) as the STS.

## What is medical treatment?

Medical treatment includes managing and caring for a patient for the purpose of combating disease or disorder. The following are not considered medical treatments and are NOT recordable:

- visits to a doctor or health care professional solely for observation or counseling;

## What do you need to do?

1. Within 7 calendar days after you receive information about a case, decide if the case is recordable under the OSHA recordkeeping requirements.
2. Determine whether the incident is a new case or a recurrence of an existing one.
3. Establish whether the case was work-related.
4. If the case is recordable, decide which form you will fill out as the injury and illness incident report.

   You may use *OSHA's 301: Injury and Illness Incident Report* or an equivalent form. Some state workers compensation, insurance, or other reports may be acceptable substitutes, as long as they provide the same information as the OSHA 301.

## How to work with the Log

1. Identify the employee involved unless it is a privacy concern case as described below.
2. Identify when and where the case occurred.
3. Describe the case, as specifically as you can.
4. Classify the seriousness of the case by recording the most serious outcome associated with the case, with column G (Death) being the most serious and column J (Other recordable cases) being the least serious.
5. Identify whether the case is an injury or illness. If the case is an injury, check the injury category. If the case is an illness, check the appropriate illness category.

**U.S. Department of Labor**
Occupational Safety and Health Administration

diagnostic procedures, including administering prescription medications that are used solely for diagnostic purposes; and

▶ any procedure that can be labeled first aid. (*See below for more information about first aid.*)

## What is first aid?

If the incident required only the following types of treatment, consider it first aid. Do NOT record the case if it involves only:

▶ using non-prescription medications at non-prescription strength;

▶ administering tetanus immunizations;

▶ cleaning, flushing, or soaking wounds on the skin surface;

▶ using wound coverings, such as bandages, BandAids™, gauze pads, etc., or using SteriStrips™ or butterfly bandages.

▶ using hot or cold therapy;

▶ using any totally non-rigid means of support, such as elastic bandages, wraps, non-rigid back belts, etc.;

▶ using temporary immobilization devices while transporting an accident victim (splints, slings, neck collars, or back boards).

▶ drilling a fingernail or toenail to relieve pressure, or draining fluids from blisters;

▶ using eye patches;

▶ using simple irrigation or a cotton swab to remove foreign bodies not embedded in or adhered to the eye;

▶ using irrigation, tweezers, cotton swab or other simple means to remove splinters or foreign material from areas other than the eye;

▶ using finger guards;

▶ using massages;

▶ drinking fluids to relieve heat stress

## How do you decide if the case involved restricted work?

Restricted work activity occurs when, as the result of a work-related injury or illness, an employer or health care professional keeps, or recommends keeping, an employee from doing the routine functions of his or her job or from working the full workday that the employee would have been scheduled to work before the injury or illness occurred.

## How do you count the number of days of restricted work activity or the number of days away from work?

Count the number of calendar days the employee was on restricted work activity or was away from work as a result of the recordable injury or illness. Do not count the day on which the injury or illness occurred in this number. Begin counting days from the day after the incident occurs. If a single injury or illness involved both days away from work and days of restricted work activity, enter the total number of days for each. You may stop counting days of restricted work activity or days away from work once the total of either or the combination of both reaches 180 days.

## Under what circumstances should you NOT enter the employee's name on the OSHA Form 300?

You must consider the following types of injuries or illnesses to be privacy concern cases:

▶ an injury or illness to an intimate body part or to the reproductive system,

▶ an injury or illness resulting from a sexual assault,

▶ a mental illness,

▶ a case of HIV infection, hepatitis, or tuberculosis,

▶ a needlestick injury or cut from a sharp object that is contaminated with blood or other potentially infectious material (see 29 CFR Part 1904.8 for definition), and

▶ other illnesses, if the employee independently and voluntarily requests that his or her name not be entered on the log.

You must not enter the employee's name on the OSHA 300 *Log* for these cases. Instead, enter "privacy case" in the space normally used for the employee's name. You must keep a separate, confidential list of the case numbers and employee names for the establishment's privacy concern cases so that you can update the cases and provide information to the government if asked to do so.

If you have a reasonable basis to believe that information describing the privacy concern case may be personally identifiable even though the employee's name has been omitted, you may use discretion in describing the injury or illness on both the OSHA 300 and 301 forms. You must enter enough information to identify the cause of the incident and the general severity of the injury or illness, but you do not need to include details of an intimate or private nature.

## What if the outcome changes after you record the case?

If the outcome or extent of an injury or illness changes after you have recorded the case, simply draw a line through the original entry or, if you wish, delete or white-out the original entry. Then write the new entry where it belongs. Remember, you need to record the most serious outcome for each case.

## Classifying injuries

An injury is any wound or damage to the body resulting from an event in the work environment.

*Examples:* Cut, puncture, laceration, abrasion, fracture, bruise, contusion, chipped tooth, amputation, insect bite, electrocution, or a thermal, chemical, electrical, or radiation burn. Sprain and strain injuries to muscles, joints, and connective tissues are classified as injuries when they result from a slip, trip, fall or other similar accidents.

## Classifying illnesses

### Skin diseases or disorders

Skin diseases or disorders are illnesses involving the worker's skin that are caused by work exposure to chemicals, plants, or other substances.

*Examples:* Contact dermatitis, eczema, or rash caused by primary irritants and sensitizers or poisonous plants; oil acne; friction blisters, chrome ulcers; inflammation of the skin.

### Respiratory conditions

Respiratory conditions are illnesses associated with breathing hazardous biological agents, chemicals, dust, gases, vapors, or fumes at work.

*Examples:* Silicosis, asbestosis, pneumonitis, pharyngitis, rhinitis or acute congestion; farmer's lung, beryllium disease, tuberculosis, occupational asthma, reactive airways dysfunction syndrome (RADS), chronic obstructive pulmonary disease (COPD), hypersensitivity pneumonitis, toxic inhalation injury, such as metal fume fever, chronic obstructive bronchitis, and other pneumoconioses.

### Poisoning

Poisoning includes disorders evidenced by abnormal concentrations of toxic substances in blood, other tissues, other bodily fluids, or the breath that are caused by the ingestion or absorption of toxic substances into the body.

*Examples:* Poisoning by lead, mercury, cadmium, arsenic, or other metals; poisoning by carbon monoxide, hydrogen sulfide, or other gases; poisoning by benzene, benzol, carbon tetrachloride, or other organic solvents; poisoning by insecticide sprays, such as parathion or lead arsenate; poisoning by other chemicals, such as formaldehyde.

### Hearing Loss

Noise-induced hearing loss is defined for recordkeeping purposes as a change in hearing threshold relative to the baseline audiogram of an average of 10 dB or more in either ear at 2000, 3000 and 4000 hertz, and the employee's total hearing level is 25 decibels (dB) or more above audiometric zero (also averaged at 2000, 3000, and 4000 hertz) in the same ear(s).

### All other illnesses

All other occupational illnesses.

*Examples:* Heatstroke, sunstroke, heat exhaustion, heat stress and other effects of environmental heat; freezing, frostbite, and other effects of exposure to low temperatures; decompression sickness; effects of ionizing radiation (isotopes, x-rays, radium); effects of nonionizing radiation (welding flash, ultra-violet rays, lasers); anthrax; bloodborne pathogenic diseases, such as AIDS, HIV, hepatitis B or hepatitis C; brucellosis; malignant or benign tumors; histoplasmosis; coccidioidomycosis.

---

### When must you post the Summary?

You must post the *Summary* only — not the *Log* — by February 1 of the year following the year covered by the form and keep it posted until April 30 of that year.

### How long must you keep the Log and Summary on file?

You must keep the *Log* and *Summary* for 5 years following the year to which they pertain.

### Do you have to send these forms to OSHA at the end of the year?

No. You do not have to send the completed forms to OSHA unless specifically asked to do so.

### How can we help you?

If you have a question about how to fill out the *Log*,

- ☐ visit us online at www.osha.gov or
- ☐ call your local OSHA office.

---

**U.S. Department of Labor**
Occupational Safety and Health Administration

## Optional
# Calculating Injury and Illness Incidence Rates

### What is an incidence rate?
An incidence rate is the number of recordable injuries and illnesses occurring among a given number of full-time workers (usually 100 full-time workers) over a given period of time (usually one year). To evaluate your firm's injury and illness experience over time or to compare your firm's experience with that of your industry as a whole, you need to compute your incidence rate. Because a specific number of workers and a specific period of time are involved, these rates can help you identify problems in your workplace and/or progress you may have made in preventing work-related injuries and illnesses.

### How do you calculate an incidence rate?
You can compute an occupational injury and illness incidence rate for all recordable cases or for cases that involved days away from work for your firm quickly and easily. The formula requires that you follow instructions in paragraph (a) below for the total recordable cases or those in paragraph (b) for cases that involved days away from work, and for both rates the instructions in paragraph (c).

(a) *To find out the total number of recordable injuries and illnesses that occurred during the year*, count the number of line entries on your OSHA Form 300, or refer to the OSHA Form 300A and sum the entries for columns (G), (H), (I), and (J).

(b) *To find out the number of injuries and illnesses that involved days away from work*, count the number of line entries on your OSHA Form 300 that received a check mark in column (H), or refer to the entry for column (H) on the OSHA Form 300A.

(c) *The number of hours all employees actually worked during the year.* Refer to OSHA Form 300A and optional worksheet to calculate this number.

You can compute the incidence rate for all recordable cases of injuries and illnesses using the following formula:

*Total number of injuries and illnesses x 200,000 ÷ Number of hours worked by all employees = Total recordable case rate*

(The 200,000 figure in the formula represents the number of hours 100 employees working 40 hours per week, 50 weeks per year would work, and provides the standard base for calculating incidence rates.)

You can compute the incidence rate for recordable cases involving days away from work, days of restricted work activity or job transfer (DART) using the following formula:

*(Number of entries in column H + Number of entries in column I) x 200,000 ÷ Number of hours worked by all employees = DART incidence rate*

You can use the same formula to calculate incidence rates for other variables such as cases involving restricted work activity (column (I) on Form 300A), cases involving skin disorders (column (M-2) on Form 300A), etc. Just substitute the appropriate total for these cases, from Form 300A, into the formula in place of the total number of injuries and illnesses.

### What can I compare my incidence rate to?
The Bureau of Labor Statistics (BLS) conducts a survey of occupational injuries and illnesses each year and publishes incidence rate data by various classifications (e.g., by industry, by employer size, etc.). You can obtain these published data at www.bls.gov/iif or by calling a BLS Regional Office.

### Worksheet

Total number of injuries and illnesses [ ] × 200,000 ÷ Number of hours worked by all employees [ ] = Total recordable case rate [ ]

Number of entries in Column H + Column I [ ] × 200,000 ÷ Number of hours worked by all employees [ ] = DART incidence rate [ ]

**U.S. Department of Labor**
Occupational Safety and Health Administration

# How to Fill Out the Log

The *Log of Work-Related Injuries and Illnesses* is used to classify work-related injuries and illnesses and to note the extent and severity of each case. When an incident occurs, use the *Log* to record specific details about what happened and how it happened.

If your company has more than one establishment or site, you must keep separate records for each physical location that is expected to remain in operation for one year or longer.

We have given you several copies of the *Log* in this package. If you need more than we provided, you may photocopy and use as many as you need.

The *Summary* — a separate form — shows the work-related injury and illness totals for the year in each category. At the end of the year, count the number of incidents in each category and transfer the totals from the *Log* to the *Summary*. Then post the *Summary* in a visible location so that your employees are aware of injuries and illnesses occurring in their workplace.

**You don't post the *Log*. You post only the *Summary* at the end of the year.**

## OSHA's Form 300 (Rev 01/2004)
### Log of Work-Related Injuries and Illnesses

**Attention:** This form contains information relating to employee health and must be used in a manner that protects the confidentiality of employees to the extent possible while the information is being used for occupational safety and health purposes.

Year 20___
U.S. Department of Labor
Occupational Safety and Health Administration
Form approved OMB no. 1218-0176

Establishment name _XYZ Company_
City _Anywhere_   State _MA_

| (A) Case no. | (B) Employee's name | (C) Job title | (D) Date of injury or onset of illness | (E) Where the event occurred | (F) Describe injury or illness, parts of body affected, and object/substance that directly injured or made person ill |
|---|---|---|---|---|---|
| 1 | Mark Bagin | Welder | 5/25 | basement | fracture, left arm and left leg, fell from ladder |
| 2 | Shana Alexander | Foundry man | 7/2 | pouring deck | poisoning from lead fumes |
| 3 | Sam Sander | Electrician | 8/5 | 2nd floor storeroom | broken left foot, fell over box |
| 4 | Ralph Baccella | Laborer | 9/17 | packaging dept | Back strain lifting boxes |
| 5 | Jarrod Daniels | Machine opr. | 10/23 | production floor | dust in eye |

**Classify the case** — CHECK ONLY ONE box for each case based on the most serious outcome for that case:

Columns: (G) Death, (H) Days away from work, (I) Job transfer or restriction, (J) Other recordable cases

**Enter the number of days the injured or ill worker was:**
(K) Away from work / (L) On job transfer or restriction
- 12 days / 15 days
- ___ days / 30 days
- 7 days / 30 days
- 3 days / ___ days

**Check the "injury" column or choose one type of illness:**
(M) (1) Injury (2) Skin disorder (3) Respiratory condition (4) Poisoning (5) Hearing loss (6) All other illnesses

- **Be as specific as possible.** You can use two lines if you need more room.
- **Revise the log** if the injury or illness progresses and the outcome is more serious than you originally recorded for the case. Cross out, erase, or white-out the original entry.
- **Choose ONLY ONE of these categories.** Classify the case by recording the most serious outcome of the case, with column G (Death) being the most serious and column J (Other recordable cases) being the least serious.
- **Note whether the case involves an injury or an illness.**

## OSHA's Form 300 (Rev. 01/2004)

# Log of Work-Related Injuries and Illnesses

**Attention:** This form contains information relating to employee health and must be used in a manner that protects the confidentiality of employees to the extent possible while the information is being used for occupational safety and health purposes.

**U.S. Department of Labor**
Occupational Safety and Health Administration

Form approved OMB no. 1218-0176

Year 20 ___

Establishment name _____
City _____ State _____

You must record information about every work-related death and about every work-related injury or illness that involves loss of consciousness, restricted work activity or job transfer, days away from work, or medical treatment beyond first aid. You must also record significant work-related injuries and illnesses that are diagnosed by a physician or licensed health care professional. You must also record work-related injuries and illnesses that meet any of the specific recording criteria listed in 29 CFR Part 1904.8 through 1904.12. Feel free to use two lines for a single case if you need to. You must complete an Injury and Illness Incident Report (OSHA Form 301) or equivalent form for each injury or illness recorded on this form. If you're not sure whether a case is recordable, call your local OSHA office for help.

### Identify the person

| (A) Case no. | (B) Employee's name | (C) Job title (e.g., Welder) |
|---|---|---|

### Describe the case

| (D) Date of injury or onset of illness | (E) Where the event occurred (e.g., Loading dock north end) | (F) Describe injury or illness, parts of body affected, and object/substance that directly injured or made person ill (e.g., Second degree burns on right forearm from acetylene torch) |
|---|---|---|
| month/day | | |

### Classify the case

CHECK ONLY ONE box for each case based on the most serious outcome for that case:

- Death (G)
- Days away from work (H)
- **Remained at Work** — Job transfer or restriction (I)
- Other recordable cases (J)

Enter the number of days the injured or ill worker was:

- Away from work (K) _____ days
- On job transfer or restriction (L) _____ days

Check the "injury" column or choose one type of illness:

(M)
(1) Injury
(2) Skin disorder
(3) Respiratory condition
(4) Poisoning
(5) Hearing loss
(6) All other illnesses

**Page totals** _____

Be sure to transfer these totals to the Summary page (Form 300A) before you post it.

Public reporting burden for this collection of information is estimated to average 14 minutes per response, including time to review the instructions, search and gather the data needed, and complete and review the collection of information. Persons are not required to respond to the collection of information unless it displays a currently valid OMB control number. If you have any comments about these estimates or any other aspects of this data collection, contact: US Department of Labor, OSHA Office of Statistical Analysis, Room N-3644, 200 Constitution Avenue, NW, Washington, DC 20210. Do not send the completed forms to this office.

Page _____ of _____

# OSHA's Form 300A (Rev. 01/2004)
## Summary of Work-Related Injuries and Illnesses

**U.S. Department of Labor**
Occupational Safety and Health Administration

Year 20____

Form approved OMB no. 1218-0176

*All establishments covered by Part 1904 must complete this Summary page, even if no work-related injuries or illnesses occurred during the year. Remember to review the Log to verify that the entries are complete and accurate before completing this summary.*

*Using the Log, count the individual entries you made for each category. Then write the totals below, making sure you've added the entries from every page of the Log. If you had no cases, write "0."*

*Employees, former employees, and their representatives have the right to review the OSHA Form 300 in its entirety. They also have limited access to the OSHA Form 301 or its equivalent. See 29 CFR Part 1904.35, in OSHA's recordkeeping rule, for further details on the access provisions for these forms.*

### Number of Cases

Total number of deaths

_____ (G)

Total number of cases with days away from work

_____ (H)

Total number of cases with job transfer or restriction

_____ (I)

Total number of other recordable cases

_____ (J)

### Number of Days

Total number of days away from work

_____ (K)

Total number of days of job transfer or restriction

_____ (L)

### Injury and Illness Types

Total number of . . .
(M)

(1) Injuries _____
(2) Skin disorders _____
(3) Respiratory conditions _____

(4) Poisonings _____
(5) Hearing loss _____
(6) All other illnesses _____

**Post this Summary page from February 1 to April 30 of the year following the year covered by the form.**

### Establishment information

**Your establishment name** _____

Street _____
City _____ State ____ ZIP ____

Industry description (e.g., Manufacture of motor truck trailers) _____

Standard Industrial Classification (SIC), if known (e.g., 3715) ___ ___ ___ ___

OR

North American Industrial Classification (NAICS), if known (e.g., 336212) ___ ___ ___ ___ ___ ___

### Employment information *(If you don't have these figures, see the Worksheet on the back of this page to estimate.)*

Annual average number of employees _____

Total hours worked by all employees last year _____

### Sign here

Knowingly falsifying this document may result in a fine.

I certify that I have examined this document and that to the best of my knowledge the entries are true, accurate, and complete.

_____ _____
Company executive Title

( ) ___ - ____ ___/___/___
Phone Date

Public reporting burden for this collection of information is estimated to average 50 minutes per response, including time to review the instructions, search and gather the data needed, and complete and review the collection of information. Persons are not required to respond to the collection of information unless it displays a currently valid OMB control number. If you have any comments about these estimates or any other aspects of this data collection, contact: US Department of Labor, OSHA Office of Statistical Analysis, Room N-3644, 200 Constitution Avenue, NW, Washington, DC 20210. Do not send the completed forms to this office.

**Optional**

# Worksheet to Help You Fill Out the Summary

At the end of the year, OSHA requires you to enter the average number of employees and the total hours worked by your employees on the summary. If you don't have these figures, you can use the information on this page to estimate the numbers you will need to enter on the Summary page at the end of the year.

## How to figure the average number of employees who worked for your establishment during the year:

**❶ Add** the total number of employees your establishment paid in all pay periods during the year. Include all employees: full-time, part-time, temporary, seasonal, salaried, and hourly.

The number of employees paid in all pay periods = _____ ❶

**❷ Count** the number of pay periods your establishment had during the year. Be sure to include any pay periods when you had no employees.

The number of pay periods during the year = _____ ❷

**❸ Divide** the number of employees by the number of pay periods.

❶ _____
❷ _____ = _____ ❸

**❹ Round the answer** to the next highest whole number. Write the rounded number in the blank marked *Annual average number of employees.*

The number rounded = _____ ❹

---

For example, Acme Construction figured its average employment this way:

| For pay period... | Acme paid this number of employees... |
|---|---|
| 1 | 10 |
| 2 | 0 |
| 3 | 15 |
| 4 | 30 |
| 5 | 40 |
| ▶ | |
| 24 | 20 |
| 25 | 15 |
| 26 | +10 |
|   | 830 |

Number of employees paid = 830     ❶

Number of pay periods = 26           ❷

830 ÷ 26 = 31.92                         ❸

31.92 rounds to 32                       ❹

32 is the annual average number of employees

## How to figure the total hours worked by all employees:

Include hours worked by salaried, hourly, part-time and seasonal workers, as well as hours worked by other workers subject to day to day supervision by your establishment (e.g., temporary help services workers).

Do not include vacation, sick leave, holidays, or any other non-work time, even if employees were paid for it. If your establishment keeps records of only the hours paid or if you have employees who are not paid by the hour, please estimate the hours that the employees actually worked.

If this number isn't available, you can use this optional worksheet to estimate it.

### Optional Worksheet

**Find** the number of full-time employees in your establishment for the year.

_____

**Multiply** by the number of work hours for a full-time employee in a year.

× _____

This is the number of full-time hours worked.

_____

**Add** the number of any overtime hours as well as the hours worked by other employees (part-time, temporary, seasonal)

+ _____

**Round** the answer to the next highest whole number. Write the rounded number in the blank marked *Total hours worked by all employees last year.*

_____

**U.S. Department of Labor**
Occupational Safety and Health Administration

## OSHA's Form 301

# Injury and Illness Incident Report

**U.S. Department of Labor**
Occupational Safety and Health Administration

Form approved OMB no. 1218-0176

**Attention:** This form contains information relating to employee health and must be used in a manner that protects the confidentiality of employees to the extent possible while the information is being used for occupational safety and health purposes.

This *Injury and Illness Incident Report* is one of the first forms you must fill out when a recordable work-related injury or illness has occurred. Together with the *Log of Work-Related Injuries and Illnesses* and the accompanying *Summary*, these forms help the employer and OSHA develop a picture of the extent and severity of work-related incidents.

Within 7 calendar days after you receive information that a recordable work-related injury or illness has occurred, you must fill out this form or an equivalent. Some state workers' compensation, insurance, or other reports may be acceptable substitutes. To be considered an equivalent form, any substitute must contain all the information asked for on this form.

According to Public Law 91-596 and 29 CFR 1904, OSHA's recordkeeping rule, you must keep this form on file for 5 years following the year to which it pertains.

If you need additional copies of this form, you may photocopy and use as many as you need.

Completed by _____

Title _____

Phone ( ___ ) ___ - ____   Date ___ / ___ / ___

### Information about the employee

1) Full name _____

2) Street _____
   City _____ State _____ ZIP _____

3) Date of birth ___ / ___ / ___
4) Date hired ___ / ___ / ___
5) ☐ Male
   ☐ Female

### Information about the physician or other health care professional

6) Name of physician or other health care professional _____

7) If treatment was given away from the worksite, where was it given?
   Facility _____
   Street _____
   City _____ State _____ ZIP _____

8) Was employee treated in an emergency room?
   ☐ Yes
   ☐ No

9) Was employee hospitalized overnight as an in-patient?
   ☐ Yes
   ☐ No

### Information about the case

10) Case number from the Log _____ *(Transfer the case number from the Log after you record the case.)*

11) Date of injury or illness ___ / ___ / ___

12) Time employee began work _____ AM / PM

13) Time of event _____ AM / PM  ☐ Check if time cannot be determined

14) **What was the employee doing just before the incident occurred?** Describe the activity, as well as the tools, equipment, or material the employee was using. Be specific. *Examples:* "climbing a ladder while carrying roofing materials"; "spraying chlorine from hand sprayer"; "daily computer key-entry."

15) **What happened?** Tell us how the injury occurred. *Examples:* "When ladder slipped on wet floor, worker fell 20 feet"; "Worker was sprayed with chlorine when gasket broke during replacement"; "Worker developed soreness in wrist over time."

16) **What was the injury or illness?** Tell us the part of the body that was affected and how it was affected; be more specific than "hurt," "pain," or sore." *Examples:* "strained back"; "chemical burn, hand"; "carpal tunnel syndrome."

17) **What object or substance directly harmed the employee?** *Examples:* "concrete floor"; "chlorine"; "radial arm saw." *If this question does not apply to the incident, leave it blank.*

18) **If the employee died, when did death occur?** Date of death ___ / ___ / ___

---

Public reporting burden for this collection of information is estimated to average 22 minutes per response, including time for reviewing instructions, searching existing data sources, gathering and maintaining the data needed, and completing and reviewing the collection of information. Persons are not required to respond to the collection of information unless it displays a current valid OMB control number. If you have any comments about this estimate or any other aspects of this data collection, including suggestions for reducing this burden, contact: US Department of Labor, OSHA Office of Statistical Analysis, Room N-3644, 200 Constitution Avenue, NW, Washington, DC 20210. Do not send the completed forms to this office.

# If You Need Help...

*If you need help deciding whether a case is recordable, or if you have questions about the information in this package, feel free to contact us. We'll gladly answer any questions you have.*

▼ **Visit us online at www.osha.gov**

▼ **Call your OSHA Regional office and ask for the recordkeeping coordinator**

or

▼ **Call your State Plan office**

### Federal Jurisdiction

Region 1 - 617 / 565-9860
**Connecticut; Massachusetts; Maine; New Hampshire; Rhode Island**

Region 2 - 212 / 337-2378
**New York; New Jersey**

Region 3 - 215 / 861-4900
**DC; Delaware; Pennsylvania; West Virginia**

Region 4 - 404 / 562-2300
**Alabama; Florida; Georgia; Mississippi**

Region 5 - 312 / 353-2220
**Illinois; Ohio; Wisconsin**

Region 6 - 214 / 767-4731
**Arkansas; Louisiana; Oklahoma; Texas**

Region 7 - 816 / 426-5861
**Kansas; Missouri; Nebraska**

Region 8 - 303 / 844-1600
**Colorado; Montana; North Dakota; South Dakota**

Region 9 - 415 / 975-4310

Region 10 - 206 / 553-5930
**Idaho**

### State Plan States

Alaska - 907 / 269-4957
Arizona - 602 / 542-5795
California - 415 / 703-5100
*Connecticut - 860 / 566-4380
Hawaii - 808 / 586-9100
Indiana - 317 / 232-2688
Iowa - 515 / 281-3661
Kentucky - 502 / 564-3070
Maryland - 410 / 767-2371
Michigan - 517 / 322-1848
Minnesota - 651 / 284-5050
Nevada - 702 / 486-9020
*New Jersey - 609 / 984-1389
New Mexico - 505 / 827-4230
*New York - 518 / 457-2574
North Carolina - 919 / 807-2875
Oregon - 503 / 378-3272
Puerto Rico - 787 / 754-2172
South Carolina - 803 / 734-9669
Tennessee - 615 / 741-2793
Utah - 801 / 530-6901
Vermont - 802 / 828-2765
Virginia - 804 / 786-6613
Virgin Islands - 340 / 772-1315
Washington - 360 / 902-5554
Wyoming - 307 / 777-7786

*Public Sector only

**U.S. Department of Labor**
Occupational Safety and Health Administration

# OSHA Training Institute Education Centers

## REGION I

Keene State College
OSHA Education Center
175 Ammon Drive
Manchester, NH 03103-3308
Phone: (800) 449-6742
Fax: (603) 358-2569

## REGION II

Rochester Institute of Technology
31 Lomb Memorial Dr.
Rochester, NY 14623-5603
Phone: (866) 385-7470 x-2919
Fax: (585) 475-6292

Atlantic OSHA Training Center
Univ. of Medicine & Dentistry NJ
683 Has Lane West
Piscataway, NJ 68854
Phone: (732) 235-9450
Fax: (732) 235-9460

Atlantic OSHA Training Center
University at Buffalo
3435 Main Street Room 134
Buffalo, NY 14214-3000
Phone: (716) 829-2125
Fax: (716) 829-2806

Atlantic OSHA Training Center
Universidad Metropolitana
PO Box 21150
San Juan, PR 00928-1150
Phone: (787) 766-1717 x-6553
Fax: (787) 751-5540

## REGION III

George Meany Center for Labor Studies, Inc.
10000 New Hampshire Avenue
Silver Spring, MD 20903-1706
Phone: (800) 367-6724
Fax: (301) 431-5411

Bldg. Const. Trades Dept. AFL-CIO/
Center to Protect Workers' Rights
815 16th St. NW Ste 600
Washington, D.C. 20006-4101
Phone: (202) 756-4636
Fax: (202) 756-4675

West Virginia University
Safety and Health Extension
130 Tower Lane
Morgantown, WV 26506-6615
Phone: (800) 626-4748
Fax: (304) 293-5905

Keystone Occupational Safety and Health Center (KOSH)
Science Center
3701 Market St. 3rd Floor
Philadelphia, PA 19104
Phone: (800) 318-4846
Fax: (215) 387-6321

Keystone Occupational Safety and Health Center (KOSH)
Indiana University of Pennsylvania
1010 Oakland Ave
Indiana, PA 15705-1087
Phone: (724) 357-3019
Fax: (724) 357-3992

## REGION IV

Eastern Kentucky University
521 Lancaster Ave. Room 202
Richmond, KY 40475-3100
Phone: (888) 401-1956
Fax: (859) 622-6205

Georgia Institute of Technology
151 Sixth Street
Atlanta, GA 30332-0837
Phone: (800) 653-3629
Fax: (404) 894-8275

University of South Florida
13201 Bruce B. Downs Blvd.
MDC 56
Tampa, FL 33612-3805
Phone: (866) 697-0975
Fax: (813) 974-9972

## REGION V

Great Lakes Regional OTI Education Center
University of Cincinnati
P.O. Box 670567
Cincinnati, OH 45267-0567
Phone: (800) 207-9399
Fax: (513) 558-1756

Great Lakes Region
OSHA Training Consortium
MWC for Occup. Safety & Health
2221 Univ. Ave. SE Ste. 350
Minneapolis, MN 55414-3078
Phone: (800) 493-2060
Fax: (612) 626-4525

Great Lakes Region
OTI Training Center
UAW Health and Safety Dept.
8000 East Jefferson Ave
Detroit, MI 48214-3963
Phone: (800) 605-2046
Fax: (313) 824-4473

Great Lakes Regional OTI Education Center
Eastern Michigan University
2000 Huron River Drive, Ste. 101
Ypsilanti, MI 48197-1699
Phone: (800) 932-8689
Fax: (734) 481-0509

National Safety Education Center
Northern Illinois University
590 Garden Rd. RM 318
DeKalb, IL 60115-2854
Phone: (800) 656-5317
Fax: (815) 753-4203

National Safety Education Center
Construction Safety Council
4100 Madison Street
Hillside, IL 60162-1768
Phone: (800) 552-7744
Fax: (708) 544-2371

National Safety Education Center
National Safety Council
1121 Spring Lake Drive
Itasca, IL 60143-3201
Phone: (800) 621-7615
Fax: (630) 285-1613

Mid-America OSHA Trng. Inst.
Sinclair Community College
444 W. 3rd St.
Dayton, OH 45402-1460
Phone: (937) 512-3242
Fax: (937) 512-2279

Mid-America OSHA Trng. Inst.
Ohio Valley Const. Ed. Foundation
33 Greenwood Lane
Springboro, OH 45066-3034
Phone: (866) 444-4412
Fax: (937) 704-9394

Minnesota Safety Council
Continuing Education
474 Concordia Avenue
St. Paul, Minnesota 55103
Phone: (800) 444-9150
Fax: (651) 291-7584

## REGION VI

Southwest Education Center
Texas Engineering Ext. Service
15515 IH-20 at Lumley
Mesquite, TX 75181-3710
Phone: (800) 723-3811
Fax: (972) 222-2978

## REGION VII

Metropolitan Community Colleges/
Business & Technology College
1775 Universal Avenue
Kansas City, MO 64120-1313
Phone: (800) 841-7158
Fax: (816) 482-5408

Midwest OSHA Education Centers
Saint Louis University
3545 Lafayette Ste. 300
St. Louis, MO 63104-8150
Phone: (888) 382-3756
Fax: (314) 977-8150

Midwest OSHA Education Center
Kirkwood Community College
6301 Kirkwood Blvd. SW
Cedar Rapids, IA 52404-5260
Phone: (800) 464-6874
Fax: (319) 398-1250

Midwest OSHA Education Center
National Safety Council
11620 M Circle
Omaha, NE 68137-2231
Phone: (800) 592-9004
Fax: (402) 896-6331

## REGION VIII

Rocky Mountain Education Center
Red Rocks Community College
13300 West Sixth Avenue
Lakewood, CO 80228-1255
Phone: (800) 933-8394
Fax: (303) 980-8339

# OSHA Training Institute Education Centers

Mountain West
OSHA Training
Consortium of Salt Lake C.C./
University of Utah
75 South 2000 East
Salt Lake City, UT 84112-5120
Phone: (801) 581-4055
Fax: (801) 585-5275

## REGION IX

OSHA Training Center University of California, San Diego
15373 Innovation Drive, Ste. 105
San Diego, CA 92128-3424
Phone: (800) 358-9206
Fax: (858) 485-7390

WESTEC
Westside Energy Services
210 East Center Street
Taft, CA 93268-3605
Phone: (866) 493-7832
Fax: (661) 763-5162

## REGION X

Region X OSHA Training Center
University of Washington
4225 Roosevelt Way NE #100
Seattle, WA 98105-6099
Phone: (800) 326-7568
Fax: (206) 685-3872

# Multi-Employer Citation Policy

**DIRECTIVE NUMBER:** CPL 2-0.124
**EFFECTIVE DATE:** December 10, 1999
**SUBJECT:** Multi-Employer Citation Policy

## ABSTRACT

| | |
|---|---|
| **Purpose:** | To Clarify the Agency's multi-employer citation policy |
| **Scope:** | OSHA-wide |
| **References:** | OSHA Instruction CPL 2.103 (the FIRM) |
| **Suspensions:** | Chapter III, Paragraph C. 6. of the FIRM is suspended and replaced by this directive |
| **State Impact:** | This Instruction describes a Federal Program Change. Notification of State intent is required, but adoption is not. |
| **Action Offices:** | National, Regional, and Area Offices |
| **Originating Office:** | Directorate of Compliance Programs |
| **Contact:** | Carl Sall (202) 693-2345
Directorate of Construction
N3468 FPB
200 Constitution Ave., NW
Washington, DC 20210 |

By and Under the Authority of R. Davis Layne Deputy Assistant Secretary, OSHA

## TABLE OF CONTENTS

I. Purpose
II. Scope
III. Suspension
IV. References
V. Action Information
VI. Federal Program Change
VII. Force and Effect of Revised Policy
VIII. Changes in Web Version of FIRM
IX. Background
    A. Continuation of Basic Policy
    B. No Changes in Employer Duties
X. Multi-employer Worksite Policy
    A. Multi-employer Worksites
    B. The Creating Employer
    C. The Exposing Employer
    D. The Correcting Employer
    E. The Controlling Employer
    F. Multiple Roles

I. **Purpose.** This Directive clarifies the Agency's multi-employer citation policy and suspends Chapter III. C. 6. of OSHA's Field Inspection Reference Manual (FIRM).

II. **Scope.** OSHA-Wide

III. **Suspension.** Chapter III. Paragraph C. 6. of the FIRM (CPL 2.103) is suspended and replaced by this Directive.

IV. **References.** OSHA Instructions:
- CPL 02-00.103; OSHA Field Inspection Reference Manual (FIRM), September 26, 1994.
- ADM 08-0.1C, OSHA Electronic Directive System, December 19, 1997.

V. **Action Information**
    A. **Responsible Office.** Directorate of Construction.
    B. **Action Offices.** National, Regional and Area Offices
    C. **Information Offices.** State Plan Offices, Consultation Project Offices

VI. **Federal Program Change.** This Directive describes a Federal Program Change for which State adoption is not required. However, the States shall respond via the two-way memorandum to the Regional Office as soon as the State's intent regarding the multi-employer citation policy is known, but no later than 60 calendar days after the date of transmittal from the Directorate of Federal-State Operations.

VII. **Force and Effect of Revised Policy.** The revised policy provided in this Directive is in full force and effect from the date of its issuance. It is an official Agency policy to be implemented OSHA-wide.

VIII. **Changes in Web Version of FIRM.** A note will be included at appropriate places in the FIRM as it appears on the Web indicating the suspension of Chapter III paragraph 6. C. and its replacement by this Directive, and a hypertext link will be provided connecting viewers with this Directive.

XI. **Background.** OSHA's Field Inspection Reference Manual (FIRM) of September 26, 1994 (CPL 2.103), states at Chapter III, paragraph 6. C., the Agency's citation policy for multi-employer worksites. The Agency has determined that this policy needs clarification. This directive describes the revised policy.

    A. **Continuation of Basic Policy.** This revision continues OSHA's existing policy for issuing citations on multi-employer worksites. However, it gives clearer and more detailed guidance than did the earlier description of the policy in the FIRM, including new examples explaining when citations should and should not be issued to exposing, creating, correcting, and controlling employers. These examples, which address common situations and provide general policy guidance, are not intended to be exclusive. In all cases, the decision on whether to issue citations should be based on all of the relevant facts revealed by the inspection or investigation.

    B. **No Changes in Employer Duties.** This revision neither imposes new duties on employers nor detracts from their existing duties under the OSH Act. Those duties continue to arise from the employers' statutory duty to comply with OSHA standards and their duty to exercise reasonable diligence to determine whether violations of those standards exist.

X. **Multi-employer Worksite Policy.** The following is the multi-employer citation policy:

    A. **Multi-employer Worksites.** On multi-employer worksites (in all industry sectors), more than one employer may be citable for a hazardous condition that violates an OSHA standard. A two-step process must be followed in determining whether more than one employer is to be cited.

        1. **Step One.** The first step is to determine whether the employer is a creating, exposing, correcting, or controlling employer. The definitions in paragraphs (B) - (E) below explain and give examples of each. Remember that an employer may have multiple roles (see paragraph H). Once you determine the role of the employer, go to Step Two to determine if a citation is appropriate (NOTE: only exposing employers can be cited for General Duty Clause violations).

        2. **Step Two.** If the employer falls into one of these categories, it has obligations with respect to OSHA requirements. Step Two is to determine if the employer's actions were sufficient to meet those obligations. The extent of the actions required of employers varies based on which category applies. Note that the extent of the measures that a controlling employer must take to satisfy its duty to exercise reasonable care to prevent and detect violations is less than what is required of an employer with respect to protecting its own employees.

## B. The Creating Employer

1. **Step 1: Definition**: The employer that caused a hazardous condition that violates an OSHA standard.

2. **Step 2: Actions Taken**: Employers must not create violative conditions. An employer that does so is citable even if the only employees exposed are those of other employers at the site.

   a. Example 1: Employer Host operates a factory. It contracts with Company S to service machinery. Host fails to cover drums of a chemical despite S's repeated requests that it do so. This results in airborne levels of the chemical that exceed the Permissible Exposure Limit.

      **Analysis: Step 1**: Host is a creating employer because it caused employees of S to be exposed to the air contaminant above the PEL. **Step 2**: Host failed to implement measures to prevent the accumulation of the air contaminant. It could have met its OSHA obligation by implementing the simple engineering control of covering the drums. Having failed to implement a feasible engineering control to meet the PEL, Host is citable for the hazard.

   b. **Example 2**: Employer M hoists materials onto Floor 8, damaging perimeter guardrails. Neither its own employees nor employees of other employers are exposed to the hazard. It takes effective steps to keep all employees, including those of other employers, away from the unprotected edge and informs the controlling employer of the problem. Employer M lacks authority to fix the guardrails itself.

      **Analysis: Step 1**: Employer M is a creating employer because it caused a hazardous condition by damaging the guardrails. **Step 2**: While it lacked the authority to fix the guardrails, it took immediate and effective steps to keep all employees away from the hazard and notified the controlling employer of the hazard. Employer M is not citable since it took effective measures to prevent employee exposure to the fall hazard.

## C. The Exposing Employer

1. **Step 1: Definition:** An employer whose own employees are exposed to the hazard. See Chapter III, section (C)(1)(b) for a discussion of what constitutes exposure.

2. **Step 2: Actions taken**: If the exposing employer created the violation, it is citable for the violation as a creating employer. If the violation was created by another employer, the exposing employer is citable if it (1) knew of the hazardous condition or failed to exercise reasonable diligence to discover the condition, and (2) failed to take steps consistent with its authority to protect is employees. If the exposing employer has authority to correct the hazard, it must do so. If the exposing employer lacks the authority to correct the hazard, it is citable if it fails to do each of the following: (1) ask the creating and/or controlling employer to correct the hazard; (2) inform its employees of the hazard; and (3) take reasonable alternative protective measures. In extreme circumstances (e.g., imminent danger situations), the exposing employer is citable for failing to remove its employees from the job to avoid the hazard.

3. **Example 3**: Employer Sub S is responsible for inspecting and cleaning a work area in Plant P around a large, permanent hole at the end of each day. An OSHA standard requires guardrails. There are no guardrails around the hole and Sub S employees do not use personal fall protection, although it would be feasible to do so. Sub S has no authority to install guardrails. However, it did ask Employer P, which operates the plant, to install them. P refused to install guardrails.

   **Analysis: Step 1**: Sub S is an exposing employer because its employees are exposed to the fall hazard. **Step 2**: While Sub S has no authority to install guardrails, it is required to comply with OSHA requirements to the extent feasible. It must take steps to protect its employees and ask the employer that controls the hazard - Employer P - to correct it. Although Sub S asked for guardrails, since the hazard was not corrected, Sub S was responsible for taking reasonable alternative protective steps, such as providing personal fall protection. Because that was not done, Sub S is citable for the violation.

   b. **Example 4:** Unprotected rebar on either side of an access ramp presents an impalement hazard. Sub E, an electrical subcontractor, does not have the authority to cover the rebar. However, several times Sub E asked the general contractor, Employer GC, to cover the rebar. In the meantime, Sub E instructed its employees to use a different access route that avoided most of the uncovered rebar and required them to keep as far from the rebar as possible.

      **Analysis: Step 1:** Since Sub E employees were still exposed to some unprotected rebar, Sub E is an exposing employer. **Step 2:** Sub E made a good faith effort to get the general contractor to correct the hazard and took feasible measures within its control to protect its employees. Sub E is not citable for the rebar hazard.

## D. The Correcting Employer

1. **Step 1: Definition:** An employer who is engaged in a common undertaking, on the same worksite, as the exposing employer and is responsible for correcting a hazard. This usually occurs where an employer is given the responsibility of installing and/or maintaining particular safety/health equipment or devices.

2. **Step 2: Actions taken**: The correcting employer must exercise reasonable care in preventing and discovering violations and meet its obligations of correcting the hazard.

   a. **Example 5:** Employer C, a carpentry contractor, is hired to erect and maintain guardrails throughout a large, 15-story project. Work is proceeding on all floors. C inspects all floors in the morning and again in the afternoon each day. It also inspects areas where material is delivered to the perimeter once the material vendor is finished delivering material to that area. Other subcontractors are required to report damaged/missing guardrails to the general contractor, who forwards those reports to C. C repairs damaged guardrails immediately after finding them and immediately after they are reported. On this project few instances of damaged guardrails have occurred other than where material has been delivered. Shortly after the afternoon inspection of Floor 6, workers moving equipment accidentally damage a guardrail in one area. No one tells C of the damage and C has not seen it. An OSHA inspection occurs at the beginning of the next day, prior to the morning inspection of Floor 6. None of C's own employees are exposed to the hazard, but other employees are exposed.

      **Analysis: Step 1:** C is a correcting employer since it is responsible for erecting and maintaining fall protection equipment. **Step 2:** The steps C implemented to discover and correct damaged guardrails were reasonable in light of the amount of activity and size of the project. It exercised reasonable care in preventing and discovering violations; it is not citable for the damaged guardrail since it could not reasonably have known of the violation.

## E. The Controlling Employer

1. **Step 1: Definition**: An employer who has general supervisory authority over the worksite, including the power to correct safety and health violations itself or require others to correct them. Control can be established by contract or, in the absence of explicit contractual provisions, by the exercise of control in practice. Descriptions and examples of different kinds of controlling employers are given below.

2. **Step 2: Actions Taken:** A controlling employer must exercise reasonable care to prevent and detect violations on the site. The extent of the measures that a controlling employer must implement to satisfy this duty of reasonable care is less than what is required of an employer with respect to protecting its own employees. This means that the controlling employer is not normally required to inspect for hazards

as frequently or to have the same level of knowledge of the applicable standards or of trade expertise as the employer it has hired.

3. **Factors Relating to Reasonable Care Standard**. Factors that affect how frequently and closely a controlling employer must inspect to meet its standard of reasonable care include:

   a. The scale of the project;
   b. The nature and pace of the work, including the frequency with which the number or types of hazards change as the work progresses;
   c. How much the controlling employer knows both about the safety history and safety practices of the employer it controls and about that employer's level of expertise.
   d. More frequent inspections are normally needed if the controlling employer knows that the other employer has a history of non-compliance. Greater inspection frequency may also be needed, especially at the beginning of the project, if the controlling employer had never before worked with this other employer and does not know its compliance history.
   e. Less frequent inspections may be appropriate where the controlling employer sees strong indications that the other employer has implemented effective safety and health efforts. The most important indicator of an effective safety and health effort by the other employer is a consistently high level of compliance. Other indicators include the use of an effective, graduated system of enforcement for non-compliance with safety and health requirements coupled with regular jobsite safety meetings and safety training.

4. **Evaluating Reasonable Care**. In evaluating whether a controlling employer has exercised reasonable care in preventing and discovering violations, consider questions such as whether the controlling employer:

   a. Conducted periodic inspections of appropriate frequency (frequency should be based on the factors listed in G.3.);
   b. Implemented an effective system for promptly correcting hazards;
   c. Enforces the other employer's compliance with safety and health requirements with an effective, graduated system of enforcement and follow-up inspections.

5. **Types of Controlling Employers**

   a. **Control Established by Contract**. In this case, **the Employer Has a Specific Contract Right to Control Safety:** To be a controlling employer, the employer must itself be able to prevent or correct a violation or to require another employer to prevent or correct the violation. One source of this ability is explicit contract authority. This can take the form of a specific contract right to require another employer to adhere to safety and health requirements and to correct violations the controlling employer discovers.

      **(1) Example 6:** Employer GH contracts with Employer S to do sandblasting at GH's plant. Some of the work is regularly scheduled maintenance and so is general industry work; other parts of the project involve new work and are considered construction. Respiratory protection is required. Further, the contract explicitly requires S to comply with safety and health requirements. Under the contract GH has the right to take various actions against S for failing to meet contract requirements, including the right to have non-compliance corrected by using other workers and back-charging for that work. S is one of two employers under contract with GH at the work site, where a total of five employees work. All work is done within an existing building. The number and types of hazards involved in S's work do not significantly change as the work progresses. Further, GH has worked with S over the course of several years. S provides periodic and other safety and health training and uses a graduated system of enforcement of safety and health rules. S has consistently had a high level of compliance at its previous jobs and at this site. GH monitors S by a combination of weekly inspections, telephone discussions and a weekly review of S's own inspection reports. GH has a system of graduated enforcement that it has applied to S for the few safety and health violations that had been committed by S in the past few years. Further, due to respirator equipment problems S violates respiratory protection requirements two days before GH's next scheduled inspection of S. The next day there is an OSHA inspection. There is no notation of the equipment problems in S's inspection reports to GH and S made no mention of it in its telephone discussions.

      **Analysis: Step 1:** GH is a controlling employer because it has general supervisory authority over the worksite, including contractual authority to correct safety and health violations. **Step 2:** GH has taken reasonable steps to try to make sure that S meets safety and health requirements. Its inspection frequency is appropriate in light of the low number of workers at the site, lack of significant changes in the nature of the work and types of hazards involved, GH's knowledge of S's history of compliance and its effective safety and health efforts on this job. GH has exercised reasonable care and is not citable for this condition.

      **(2) Example 7:** Employer GC contracts with Employer P to do painting work. GC has the same contract authority over P as Employer GH had in Example 6. GC has never before worked with P. GC conducts inspections that are sufficiently frequent in light of the factors listed above in (G)(3). Further, during a number of its inspections, GC finds that P has violated fall protection requirements. It points the violations out to P during each inspection but takes no further actions.

      **Analysis: Step 1:** GC is a controlling employer since it has general supervisory authority over the site, including a contractual right of control over P. **Step 2:** GC took adequate steps to meet its obligation to discover violations. However, it failed to take reasonable steps to require P to correct hazards since it lacked a graduated system of enforcement. A citation to GC for the fall protection violations is appropriate.

      **(3) Example 8:** Employer GC contracts with Sub E, an electrical subcontractor. GC has full contract authority over Sub E, as in Example 6. Sub E installs an electric panel box exposed to the weather and implements an assured equipment grounding conductor program, as required under the contract. It fails to connect a grounding wire inside the box to one of the outlets. This incomplete ground is not apparent from a visual inspection. Further, GC inspects the site with a frequency appropriate for the site in light of the factors discussed above in (G)(3). It saw the panel box but did not test the outlets to determine if they were all grounded because Sub E represents that it is doing all of the required tests on all receptacles. GC knows that Sub E has implemented an effective safety and health program. From previous experience it also knows Sub E is familiar with the applicable safety requirements and is technically competent. GC had asked Sub E if the electrical equipment is OK for use and was assured that it is.

**Analysis: Step 1:** GC is a controlling employer since it has general supervisory authority over the site, including a contractual right of control over Sub E. **Step 2:** GC exercised reasonable care. It had determined that Sub E had technical expertise, safety knowledge and had implemented safe work practices. It conducted inspections with appropriate frequency. It also made some basic inquiries into the safety of the electrical equipment. Under these circumstances GC was not obligated to test the outlets itself to determine if they were all grounded. It is not citable for the grounding violation.

b. **Control Established by a Combination of Other Contract Rights:** Where there is no explicit contract provision granting the right to control safety, or where the contract says the employer does not have such a right, an employer may still be a controlling employer. The ability of an employer to control safety in this circumstance can result from a combination of contractual rights that, together, give it broad responsibility at the site involving almost all aspects of the job. Its responsibility is broad enough so that its contractual authority necessarily involves safety. The authority to resolve disputes between subcontractors, set schedules and determine construction sequencing are particularly significant because they are likely to affect safety. (NOTE: citations should only be issued in this type of case after consulting with the Regional Solicitor's office).

(1) **Example 9:** Construction manager M is contractually obligated to: set schedules and construction sequencing, require subcontractors to meet contract specifications, negotiate with trades, resolve disputes between subcontractors, direct work and make purchasing decisions, which affect safety. However, the contract states that M does not have a right to require compliance with safety and health requirements. Further, Subcontractor S asks M to alter the schedule so that S would not have to start work until Subcontractor G has completed installing guardrails. M is contractually responsible for deciding whether to approve S's request.

**Analysis: Step 1:** Even though its contract states that M does not have authority over safety, the combination of rights actually given in the contract provides broad responsibility over the site and results in the ability of M to direct actions that necessarily affect safety. For example, M's contractual obligation to determine whether to approve S's request to alter the schedule has direct safety implications. M's decision relates directly to whether S's employees will be protected from a fall hazard. M is a controlling employer. **Step 2:** In this example, if M refused to alter the schedule, it would be citable for the fall hazard violation.

(2) **Example 10:** Employer ML's contractual authority is limited to reporting on subcontractors' contract compliance to owner/developer O and making contract payments. Although it reports on the extent to which the subcontractors are complying with safety and health infractions to O, ML does not exercise any control over safety at the site.

**Analysis: Step 1:** ML is not a controlling employer because these contractual rights are insufficient to confer control over the subcontractors and ML did not exercise control over safety. Reporting safety and health infractions to another entity does not, by itself (or in combination with these very limited contract rights), constitute an exercise of control over safety. **Step 2:** Since it is not a controlling employer it had no duty under the OSH Act to exercise reasonable care with respect to enforcing the subcontractors' compliance with safety; there is therefore no need to go to Step 2.

c. **Architects and Engineers:** Architects, engineers, and other entities are controlling employers only if the breadth of their involvement in a construction project is sufficient to bring them within the parameters discussed above.

(1) **Example 11:** Architect A contracts with owner O to prepare contract drawings and specifications, inspect the work, report to O on contract compliance, and to certify completion of work. A has no authority or means to enforce compliance, no authority to approve/reject work and does not exercise any other authority at the site, although it does call the general contractor's attention to observed hazards noted during its inspections.

**Analysis: Step 1:** A's responsibilities are very limited in light of the numerous other administrative responsibilities necessary to complete the project. It is little more than a supplier of architectural services and conduit of information to O. Its responsibilities are insufficient to confer control over the subcontractors and it did not exercise control over safety. The responsibilities it does have are insufficient to make it a controlling employer. Merely pointing out safety violations did not make it a controlling employer. NOTE: In a circumstance such as this it is likely that broad control over the project rests with another entity. **Step 2:** Since A is not a controlling employer it had no duty under the OSH Act to exercise reasonable care with respect to enforcing the subcontractors' compliance with safety; there is therefore no need to go to Step 2.

(2) **Example 12:** Engineering firm E has the same contract authority and functions as in Example 9.

**Analysis: Step 1:** Under the facts in Example 9, E would be considered a controlling employer. **Step 2:** The same type of analysis described in Example 9 for Step 2 would apply here to determine if E should be cited.

d. **Control Without Explicit Contractual Authority.** Even where an employer has no explicit contract rights with respect to safety, an employer can still be a controlling employer if, in actual practice, it exercises broad control over subcontractors at the site (see Example 9). NOTE: Citations should only be issued in this type of case after consulting with the Regional Solicitor's office.

(1) **Example 13:** Construction manager MM does not have explicit contractual authority to require subcontractors to comply with safety requirements, nor does it explicitly have broad contractual authority at the site. However, it exercises control over most aspects of the subcontractors' work anyway, including aspects that relate to safety.

**Analysis: Step 1:** MM would be considered a controlling employer since it exercises control over most aspects of the subcontractor's work, including safety aspects. **Step 2:** The same type of analysis on reasonable care described in the examples in (G)(5)(a) would apply to determine if a citation should be issued to this type of controlling employer.

F. **Multiple Roles**

1. A creating, correcting or controlling employer will often also be an exposing employer. Consider whether the employer is an exposing employer before evaluating its status with respect to these other roles.

2. Exposing, creating and controlling employers can also be correcting employers if they are authorized to correct the hazard.

# Emergency Preparedness and Response Guidelines

## INTRODUCTION

The importance of an effective workplace safety and health program cannot be overemphasized. There are many benefits from such a program including increased productivity, improved employee morale, reduced absenteeism and illness, and reduced workers' compensation rates; however, incidents still occur in spite of efforts to prevent them. Therefore, proper planning for emergencies is necessary to minimize employee injury and property damage.

## PURPOSE

This discussion details the basic steps to handle emergencies in the workplace. These emergencies include accidental releases of toxic gases, chemical spills, fires, explosions, and bodily harm and trauma caused by workplace violence. This discussion is intended to assist small businesses that do not have safety and health professionals. It is not intended as an all inclusive safety program but rather to provide guidelines for planning for emergencies.

## PLANNING

The effectiveness of response during emergencies depends on the amount of planning and training performed. Management must show its support for plant safety programs and the importance of emergency planning. If management is not interested in employee protection and in minimizing property loss, little can be done to promote a safe workplace. It is therefore management's responsibility to see that a program is instituted and that it is frequently reviewed and updated. The input and support of all employees must be obtained to ensure an effective program. The emergency response plan should be developed locally and should be comprehensive enough to deal with all types of emergencies specific to that site. When emergency action plans are required by a particular OSHA standard, the plan must be in writing; except for firms with 10 or fewer employees, the plan may be communicated orally to employees. The plan must include, as a minimum, the following elements:

(1) Emergency escape procedures and emergency escape route assignments,

(2) Procedures to be followed by employees who remain to perform (or shut down) critical plant operations before the plant is evacuated,

(3) Procedures to account for all employees after emergency evacuation has been completed,

(4) Rescue and medical duties for those employees who are to perform them,

(5) The preferred means for reporting fires and other emergencies, and

(6) Names or regular job titles of persons or departments to be contacted for further information or explanation of duties under the plan.

The emergency action plan should address all potential emergencies that can be expected in the workplace. Therefore, it will be necessary to perform a hazard assessment to determine toxic materials in the workplace, hazards, and potentially dangerous conditions. For information on chemicals, the manufacturer or supplier can be contacted to obtain Material Safety Data Sheets. These forms describe the hazards that a chemical may present, list precautions to take when handling, storing, or using the substance, and outline emergency and first-aid procedures.

The employer must list in detail the procedures to be taken by those employees who must remain behind to care for essential plant operations until their evacuation becomes absolutely necessary. This may include monitoring plant power supplies, water supplies, and other essential services that cannot be shut down for every emergency alarm, and use of fire extinguishers.

For emergency evacuation, the use of floor plans or workplace maps that clearly show the emergency escape routes and safe or refuge areas should be included in the plan. All employees must be told what actions they are to take in emergency situations that may occur in the workplace, such as a designated meeting location after evacuation.

This plan must be reviewed with employees initially when the plan is developed, whenever the employees' responsibilities under the plan change, and whenever the plan is changed. A copy should be kept where employees can refer to it at convenient times. In fact, to go a step further, the employer could provide the employees with a copy of the plan, particularly all new employees.

## CHAIN OF COMMAND

A chain of command should be established to minimize confusion so that employees will have no doubt about who has authority for making decisions. Responsible individuals should be selected to coordinate the work of the emergency response team. In larger organizations, there may be a plant coordinator in charge of plant-wide operations, public relations, and ensuring that outside aid is called in. Because of the importance of these functions, adequate backup must be arranged so that trained personnel are always available. The duties of the Emergency Response Team Coordinator should include the following:

(1) Assessing the situation and determining whether an emergency exists that requires activating the emergency procedures,

(2) Directing all efforts in the area including evacuating personnel,

(3) Ensuring that outside emergency services such as medical aid and local fire departments are called in when necessary, and

(4) Directing the shutdown of plant operations when necessary.

## COMMUNICATIONS

During a major emergency involving a fire or explosion it may be necessary to evacuate offices in addition to manufacturing areas. Also, normal services, such as electricity, water, and telephones, may be nonexistent. Under these conditions, it may be necessary to have an alternate area to which employees can report or that can act as a focal point for incoming and outgoing calls. Since time is an essential element for adequate response, the person designated as being in charge should make this the alternate headquarters so that he/she can be easily reached.

Emergency communications equipment such as amateur radio systems, public address systems, or portable radio units should be present for notifying employees of the emergency and for contacting local authorities, such as law enforcement officials, private sector charitable groups, and the fire department.

A method of communication also is needed to alert employees to the evacuation or to take other action as required in the plan. Alarms must be audible or seen by all people in the plant and have an auxiliary power supply in the event electricity is affected. The alarm must be distinctive and recognizable as a signal to evacuate the work area or perform actions designated under the emergency action plan. The employer must explain to each employee the means for reporting emergencies, such as manual pull box alarms, public address systems, or telephones. Emergency phone numbers should be posted on or near telephones, on employees' notice boards, or in other conspicuous locations. The warning plan should be in writing and management must be sure each employee knows what it means and what action is to be taken.

It may be necessary to notify other key personnel such as the plant manager or physician during off-duty hours. An updated written list of key personnel should be kept listed in order of priority.

## ACCOUNTING FOR PERSONNEL

Management will need to know when all personnel have been accounted for. This can be difficult during shift changes or if contractors are on site. A responsible person in the control center must be appointed to account for personnel and to inform police or Emergency Response Team members of those persons believed missing.

## EMERGENCY RESPONSE TEAMS

Emergency Response Teams are the first line of defense in emergencies. Before assigning personnel to these teams, the employer must assure that employees are physically capable of performing the duties that may be assigned to them. Depending on the size of the plant there may be one or several teams trained in the following areas:

(1) Use of various types of fire extinguishers,

(2) First aid, including cardiopulmonary resuscitation (CPR),

(3) Shutdown procedures,

(4) Evacuation procedures,

(5) Chemical spill control procedures,

(6) Use of self-contained breathing apparatus (SCBA),

(7) Search and emergency rescue procedures,

(8) Incipient and advanced stage firefighting, and

(9) Trauma counseling.

The type and extent of the emergency will depend on the plant operations and the response will vary according to the type of process, the material handled, the number of employees, and the availability of outside resources. OSHA's Hazard Communication Standard (29 CFR part 1910.1200) is designed to ensure that the hazards of all chemicals produced or imported are evaluated and that information concerning their hazards is transmitted to employers and employees. This is done by means of comprehensive hazard communication programs including container labeling and other forms of warnings, material safety data sheets, and employee training.

Emergency Response Teams should be trained in the types of possible emergencies and the emergency actions to be performed. They are to be informed about special hazards - such as storage and use of flammable materials, toxic chemicals, radioactive sources, and water-reactive substances-to which they may be exposed during fire and other emergencies. It is important to determine when not to intervene. For example, team members must be able to determine if the fire is too large for them to handle or whether search and emergency rescue procedures should be performed. If there is the possibility of members of the Emergency Response Team receiving fatal or incapacitating injuries, they should wait for professional firefighters or emergency response groups.

## TRAINING

Training is important to the effectiveness of an emergency plan. Before implementing an emergency action plan, a sufficient number of persons must be trained to assist in the safe and orderly evacuation of employees. Training for each type of disaster response is necessary so that employees know what actions are required.

In addition to the specialized training for Emergency Response Team members, all employees should be trained in the following:

(1) Evacuation plans,

(2) Alarm systems,

(3) Reporting procedures for personnel,

(4) Shutdown procedures, and

(5) Types of potential emergencies.

These training programs must be provided as follows:

(1) Initially when the plan is developed,

(2) For all new employees,

(3) When new equipment, materials, or processes are introduced,

(4) When procedures have been updated or revised,

(5) When exercises show that employee performance must be improved, and

(6) At least annually.

The emergency control procedures should be written in concise terms and be made available to all personnel. A drill should be held for all personnel, at random intervals at least annually, and an evaluation of performance made immediately by management and employees. When possible, drills should include groups supplying outside services such as fire and police departments. In buildings with several places of employment, the emergency plans should be coordinated with other companies and employees in the building. Finally, the emergency plan should be reviewed periodically and updated to maintain adequate response personnel and program efficiency.

## PERSONAL PROTECTION

Effective personal protection is essential for any person who may be exposed to potentially hazardous substances. In emergency situations employees may be exposed to a wide variety of hazardous circumstances, including:

(1) Chemical splashes or contact with toxic materials,

(2) Falling objects and flying particles,

(3) Unknown atmospheres that may contain toxic gases, vapors or mists, or inadequate oxygen to sustain life,

(4) Fires and electrical hazards, and

(5) Violence in the workplace.

It is extremely important that employees be adequately protected in these situations. Some of the safety equipment that may be used includes:

(1) Safety glasses, goggles, or face shields for eye protection,

(2) Hard hats and safety shoes for head and foot protection,

(3) Proper respirators for breathing protection,

(4) Whole body coverings chemical suits, gloves, hoods, and boots for body protection from chemicals, and

(5) Body protection for abnormal environmental conditions such as extreme temperatures.

The equipment selected must meet the criteria contained in the OSHA standards or described by a nationally recognized standards producing organization. The choice of proper equipment is not a simple matter and consultation should be made with health and safety professionals before making any purchases. Manufacturers and distributors of health and safety products may be able to answer questions if they have enough information about the potential hazards involved.

Professional consultation will most likely be needed in providing adequate respiratory protection. Respiratory protection is necessary for toxic atmospheres of dust, mists, gases, or vapors and for oxygen-deficient atmospheres. There are four basic categories of respirators:

(1) Air-purifying devices (filters, gas masks, and chemical cartridges), which remove contaminants from the air but cannot be used in oxygen-deficient atmospheres.

(2) Air-supplied respirators (hose masks, air line respirators), which should not be used in atmospheres that are immediately dangerous to life or health.

(3) Positive-pressure self-contained breathing apparatus (SCBA), which are required for unknown atmospheres, oxygen-deficient atmospheres, or atmospheres immediately dangerous to life or health.

(4) Escape masks.

Before assigning or using respiratory equipment the following conditions must be met:

(1) A medical evaluation should be made to determine if the employees are physically able to use the respirator.

(2) Written procedures must be prepared covering safe use and proper care of the equipment, and employees must be trained in these procedures and in the use and maintenance of respirators.

(3) A fit test must be made to determine a proper match between the facepiece of the respirator and the face of the wearer. This testing must be repeated periodically. Training must provide the employee an opportunity to handle the respirator, have it fitted properly, test its facepiece-to-face seal, wear it in normal air for a familiarity period, and wear it in a test atmosphere.

(4) A regular maintenance program must be instituted including cleaning, inspecting, and testing of all respiratory equipment. Respirators used for emergency response must be inspected after each use and at least monthly to assure that they are in satisfactory working condition. A written record of inspection must be maintained.

(5) Distribution areas for equipment used in emergencies must be readily accessible to employees.

A positive-pressure self-contained breathing apparatus (SCBA) offers the best protection to employees involved in controlling emergency situations. It must have a minimum service life rating of at least 30 minutes. Conditions that require a positive-pressure SCBA include the following:

(1) Leaking cylinders or containers, smoke from chemical fires, or chemical spills that indicate high potential for exposure to toxic substances.

(2) Atmospheres with unknown contaminants or unknown contaminant concentrations, confined spaces that may contain toxic substances, or oxygen-deficient atmospheres.

Emergency situations may involve entering confined spaces to rescue employees who are overcome by toxic compounds or who lack oxygen. These permit-required confined spaces include tanks, vaults, pits, sewers, pipelines, and vessels. Entry into permit-required confined spaces can expose the employee to a variety of hazards, including toxic gases, explosive atmospheres, oxygen deficiency, electrical hazards, and hazards created by mixers and impellers that have not been deactivated and locked out. Personnel must never enter a permit-required confined space unless the atmosphere has been tested for adequate oxygen, combustibility, and toxic substances. Conditions in a permit-required confined space must be considered immediately dangerous to life and health unless shown otherwise. If a permit-required confined space must be entered in an emergency, the following precautions must be adhered to:

(1) All lines containing inert, toxic, flammable, or corrosive materials must be disconnected or blocked off before entry.

(2) All impellers, agitators, or other moving equipment inside the vessel must be locked out.

(3) Appropriate personal protective equipment must be worn by employees before entering the vessel. Mandatory use of harnesses must be stressed.

(4) Rescue procedures must be specifically designed for each entry. A trained stand-by person must be present. This person should be assigned a fully charged, positive-pressure, self-contained breathing apparatus with a full facepiece. The stand-by person must maintain unobstructed lifelines and communications to all workers within the permit-required confined space and be prepared to summon rescue personnel if necessary. The stand-by person should not enter the confined space until adequate assistance is present. While awaiting rescue personnel, the stand-by person may make a rescue attempt utilizing lifelines from outside the permit-required confined space.

A more complete description of procedures to follow while working in confined spaces may be found in the OSHA standard for permit-required confined spaces, 29 CFR 1910.145 and the National Institute for Occupational Safety and Health (NIOSH) Publication Number 80-106, Criteria for a Recommended Standard...Working in Confined Spaces.

## MEDICAL ASSISTANCE

In a major emergency, time is critical factor in minimizing injuries. Most small businesses do not have a formal medical program, but they are required to have the following medical and first-aid services:

(1) In the absence of an infirmary, clinic, or hospital in close proximity to the workplace that can be used for treatment of all injured employees, the employer must ensure that a person or persons are adequately trained to render first aid. The first aid is to begin within 3 to 4 minutes of the incident if the injury is of a serious nature.

(2) Where the eyes or body of any employee may be exposed to injurious corrosive materials, eye washes or suitable equipment for quick drenching or flushing must be provided in the work area for immediate emergency use. Employees must be trained to use the equipment.

(3) The employer must ensure the ready availability of medical personnel for advice and consultation on matters of employees' health. This does not mean that health care must be provided, but rather that, if health problems develop in the workplace, medical help will be available to resolve them.

To fulfill the above requirements, the following actions should be considered:

(1) Survey the medical facilities near the place of business and make arrangements to handle routine and emergency cases. A written emergency medical procedure should then be prepared for handling accidents with minimum confusion.

(2) If the business is located far from medical facilities, at least one and preferably more employees on each shift must be adequately trained to render first aid. The American Red Cross, some insurance carriers, local safety councils, fire departments, and others may be contacted for this training.

(3) First-aid supplies should be provided for emergency use. This equipment should be ordered through consultation with a physician.

(4) Emergency phone numbers should be posted in conspicuous places near or on telephones.

(5) Sufficient ambulance service should be available to handle any emergency. This requires advance contact with ambulance services to ensure they become familiar with plant location, access routes, and hospital locations.

## SECURITY

During an emergency, it is often necessary to secure the area to prevent unauthorized access and to protect vital records and equipment. An off-limits area must be established by cordoning off the area with ropes and signs. It may be necessary to notify local law enforcement personnel or to employ private security personnel to secure the area and prevent the entry of unauthorized personnel.

Certain records also may need to be protected, such as essential accounting files, legal documents, and lists of employees' relatives to be notified in case of emergency. These records may be stored in duplicate outside the plant or in protected secure locations within the plant.

## SOME OSHA REQUIREMENTS

The following is a list of some of the OSHA requirements pertaining to emergency response. These references refer to appropriate sections of the Occupational Safety and Health Standards (Title 29, Code of Federal Regulations, Part 1910, which are the OSHA General Industry Standards).

### Subpart E—Means of Egress

1910.37 Means of egress

1910.38 Employee emergency plans and fire prevention plans

Appendix to Subpart E—Means of Egress

### Subpart H—Hazardous Materials

1910.119 Process safety management of highly hazardous chemicals

1910.120 Hazardous waste operations and emergency response.

### Subpart I—Personal Protective Equipment

1910.132 General requirements - personnel protection

1910.133 Eye and face protection

1910.134 Respiratory protection

1910.135 Occupational head protection

1910.136 Occupational foot protection

1910.138 Hand protection

### Subpart J—General Environmental Controls

1910.146 Permit-required confined spaces

1910.147 Control of hazardous energy sources

### Subpart K—Medical and First Aid

1910.151 Medical services and first aid

### Subpart L—Fire Protection

1910.155-156 Fire protection and fire brigades

1910.157- 163 Fire suppression equipment

1910.164 Fire detection systems

1910.165 Employee alarm systems

Appendix A-E of Subpart L

### Subpart R—Special Industries, Electrical Power

Generation, Transmission, and Distribution

### Subpart Z—Toxic and Hazardous Substances

1910.1030 Bloodborne pathogens

1910.1200 Hazard communication

### INFORMATION AND CONSULTATION SERVICES

Much of the planning and program development for responding to occupational emergencies will require professional assistance. Many public and private agencies provide information and services free or at minimal cost (e.g., Federal, State, and local health and labor departments, insurance carriers, and local universities). After having exhausted these sources, consider using a private consultant selected by matching his/her specialty with your specific needs.

If there is a carrier for workers' compensation insurance, that company probably has safety and health specialists on staff who are familiar with minimum standards and technical information currently available and may be quite helpful in advising about accident and illness prevention and control.

Trade associations often have technical materials, programs, and industry data available for specific needs.

The Department of Labor through the Occupational Safety and Health Administration (OSHA) provides information in interpreting the law and on meeting the applicable standards. This information is available free of charge or obligation. The OSHA Area Office or State Plan Office nearest to the plant may be contacted for this information.

The Department of Health and Human Services through the National Institute for Occupational Safety and Health (NIOSH) provides printed material relating to employee safety and health in the workplace. Staff from this agency will perform industrial hygiene surveys of plants upon request of employers or employees.

Machine or product manufacturers can be helpful in providing additional information on precautions to take in using their products. Any special problems should be referred to them first. Professional societies in the safety, industrial hygiene, and medical fields issue publications in the form of journals, pamphlets, and books that may be quite useful (e.g., American Society of Safety Engineers or the Occupational Health Institute). They can also recommend individuals from their societies to serve as consultants.

Local colleges and universities sometimes have industrial hygiene, public health, medical, or other relevant departments with faculty and libraries to assist.

### OTHER SOURCES OF OSHA ASSISTANCE

Effective management of worker safety and health protection is a decisive factor in reducing the extent and severity of work-related injuries and their related costs. To assist employers and employees in developing effective safety and health programs, OSHA published recommended *Safety and Health Management Guidelines* [Federal Register 54(18): 3908-3916, January 26, 1988]. These voluntary guidelines apply to all places of employment covered by OSHA.

The guidelines identify four general elements that are critical to the development of a successful safety and health management program:

(1) Management commitment and employee involvement;

(2) Worksite analysis;

(3) Hazard prevention and control; and

(4) Safety and health training.

The guidelines recommend specific actions, under each of these general elements, to achieve an effective safety and health program. A single free copy of the guidelines can be obtained from the OSHA Publications Office, U.S. Department of Labor, 200 Constitution Avenue, N.W., Room N3101, Washington DC 20210, by sending a self-addressed mail label with your request.

### STATE OCCUPATIONAL SAFETY AND HEALTH PLANS

The *Occupational Safety and Health Act of 1970*, under Section 18(b), encourages States to develop and operate their own State job safety and health plans under the approval and monitoring of OSHA. Twenty-five states and territories operate such plans. They are required to set standards that are at least as effective as the federal, conduct inspections to enforce those standards (including inspections in response to workplace complaints), cover State and local government employees, and operate occupational safety and health training and education programs. In addition, all States provide on-site consultation to help employers to identify and correct workplace hazards. Such consultation may be provided either under the plan or through a special agreement under section 7(c)(1) of the Act. Federal OSHA does not conduct enforcement activities in the state plan States, except in very limited circumstances.

A listing of those States that operate approved State plans can be obtained from your local OSHA Area Office.

A comprehensive customer service poster listing OSHA services and how to contact agency Regional, Area, and District offices is available from OSHA's Publications Office, 200 Constitution Avenue, N.W. Washington D.C. 20210, Rm N3101. Telephone (202) 219-4667.

### FREE ON-SITE CONSULTATION

Free on-site safety and health consultation services are available to employers in all states who want help in establishing and maintaining a safe and healthful workplace. This service is largely funded by OSHA. Primarily developed for smaller employers with more hazardous operations, the consultation service is delivered by state governments employing professional safety consultants and health consultants. Comprehensive assistance includes an appraisal of all mechanical systems, physical work practices, and environmental hazards of the workplace and all aspects of the employer's present job safety and health program.

This program is completely separate from OSHA's inspection efforts. No penalties are proposed or citations issued for any safety or health problems identified by the consultant. The service is confidential.

For more information concerning consultation services, contact your local OSHA Area Office.

### VOLUNTARY PROTECTION PROGRAMS

Voluntary Protection programs (VPPs) and onsite consultation services, when coupled with an effective enforcement program, expand worker protector to help meet the goals of the OSH Act. The three VPPs - Star, Merit, and Demonstration - are designed to recognize outstanding achievement by companies that have successfully incorporated comprehensive safety and health programs into their total management system. They motivate others to achieve excellent safety and health results in the same outstanding way, and they establish a cooperative relationship among employers, employees, and OSHA.

For additional information on VPPs and how to apply, contact your local OSHA Area Office.

### TRAINING AND EDUCATION

OSHA's area offices offer a variety of informational services, such as publications, audiovisual aids, technical advice, and speakers for special events. OSHA's Training Institute in Des Plaines, IL, provides basic and advanced courses in safety and health for federal and state compliance officers, state consultants, federal agency personnel, and private sector employers, employees, and their representatives.

OSHA also provides funds to nonprofit organizations, through grants, to conduct workplace training and education in subjects where OSHA believes there is a lack of workplace training. Grants are awarded annually. Grant recipients are expected to contribute 20 percent of the total grant cost.

For more information on grants, training, and education, contact the OSHA Training Institute, Office of Training and Education, 1555 Times Drive, Des Plaines, IL 60018, (847) 297-4810.

For further information on any OSHA program, contact your nearest OSHA area or regional office.

## OSHA AREA OFFICE SERVICES

OSHA Area Offices are prime sources of information, publications, and assistance in understanding the requirements of standards.

They can furnish:

1. Job Safety and Health Protection (the OSHA workplace poster),
2. The necessary forms for OSHA recordkeeping requirements,
3. Information on applying for variances,
4. Off-site advice on controlling various hazards,
5. Copies of various publications and fact sheets,
6. Safety and health complaint investigations,
7. Investigations of complaints alleging discrimination for exercising safety and health rights,
8. Speakers at public events on safety and health topics, and
9. Advice and consultation on maintaining and calibrating some monitoring measuring equipment.

In addition they can provide referral services regarding:

1. Free on-site consultation,
2. Grant recipients with projects, products, or services related to hazards,
3. Training and education delivery resources,
4. Other Federal agencies and their areas of jurisdiction,
5. Voluntary protection programs under which employers with exemplary programs and safety records can be exempted from routine OSHA inspections (not all States have implemented this program), and
6. The National Institute for Occupational Safety and Health for health hazard evaluations.

These offices may be contacted by phone, by mail, by Fax, or in person, without fear of initiating an inspection.

## ADDITIONAL SOURCES OF INFORMATION

### Safety Data Sheets, Guides and Manuals

*AIHA Hygienic Guide Series.* American Industrial Hygiene Association, 2700 Prosperity Ave., Fairfax, VA 22031. Separate data sheets on specific substances giving hygienic standards, properties, industrial hygiene practices, specific procedures, and references.

*ANSI Standards, Z37 Series, Acceptable Concentrations of Toxic Dusts and Gases.* American National Standards Institute, 11 West 42nd Street, New York, NY 10036. These guides represent a consensus of interested parties concerning minimum safety requirements for the storage, transportation, and handling of toxic substances; they are intended to aid the manufacturers, the consumer, and the general public.

*ASTM Standards with Related Material.* American Society for Testing and Materials, 1916 Race Street, Philadelphia, PA 19103.

### Standards and Specification Groups

American National Standards Institute, 11 West 42nd Street, New York, NY 10036, coordinates and administers the federated voluntary standardization system in the United States.

American Society for Testing and Materials, 1916 Race Street, Philadelphia, PA 10103. World's largest source of voluntary consensus standards for materials, products, systems, and services.

### Fire Protection Organizations

Factory Insurance Association, 85 Woodland Street, Hartford, CT 06105. Composed of capital stock insurance companies to provide engineering, inspections, and loss adjustment service to industry.

Factory Mutual System, 1151 Boston-Providence Turnpike, Norwood, MA 02062. An industrial fire protection, engineering, and inspection bureau established and maintained by mutual fire insurance companies.

National Fire Protection Association, 470 Batterymarch Park, Quincy, MA 02269. The clearinghouse for information on fire protection and fire prevention also writes NFPA standards. Nonprofit technical and educational organization.

Underwriter Laboratories, Inc., 207 East Ohio Street, Chicago, IL 60611. Not-for-profit organization whose laboratories publish annual lists of manufacturers whose products proved acceptable under appropriate standards.

### Medical Consultation

Arrange for a local doctor to advise on workplace medical matters. Contact the local Red Cross chapter for assistance in first-aid training. If a local chapter cannot be located write:

American National Red Cross
National Headquarters Safety Programs
18th and E Streets, N.W.
Washington, D.C. 20006

### References

Chemical Industries Association, Chemical Industry Safety and Health Council. *Recommended Procedures for Handling Major Emergencies.* Alembic House. 93 Albert Embarkment. London, SEIO 7TU, July 1976.

Krikorian, Michael. "Advanced Planning is the Key to Controlling Emergencies and Disasters in the Workplace." *Prof Safety:* 39-42, December 1977.

Lee, W.R. Sources of Consultation and Reference Aids. Section XI, in H.M. Key, A.F. Henschel, J.Butler, R.N. Ligo, I.R. Tabershaw, and L. Ede (Eds): *Occupational Diseases: A Guide to Their Recognition.* NIOSH Publication No. 77-181. Cincinnati, 1977. Pp. 523-556. Also available as Lee, W.R. Consultation and Reference Sources for Occupational Health. J Occu Med 17(7): 446-456, July 1975.

National Safety Council. *Accident Prevention Manual for Industrial Operations Administration and Programs.* 8th ed. Chicago, 1981. Pp. 439-471.

U.S. Department of Labor. Occupational Safety and Health Administration. *OSHA Handbook for Small Businesses.* OSHA 2209. Washington, DC, 1996.

Occupational Safety and Health Administration. *Principal Emergency Response and Preparedness Requirements in OSHA Standards and Guidance for Safety and Health Programs.* OSHA 3122. Washington, DC, 1990.

U.S. Department of Health and Human Services. National Institute for Occupational Safety and Health, *Safety and Health Alert: Request for Assistance In Preventing Homicide In the Workplace.* U.S. Department of Health and Human Services, Cincinnati, Ohio, September 1993, Number 93-109.

Public Health Service. National Institute for Occupational Safety and Health. *A Guide to Industrial Respiratory Protection.* NIOSH Publication No. 76-189. Cincinnati, 1976.

*Criteria for a Recommended Standard...Working in Confined Spaces.* NIOSH Publication. No.80-106. Cincinnati, 1980.

*Respiratory Protection...An Employer's Manual.* NIOSH Publication No. 78-198A. Cincinnati, October 1978.

*Self-Evaluation of Occupational Safety and Health Programs.* NIOSH Publication No. 78-187. Cincinnati, 1978.

**U.S. Department of Labor
Program Highlights**

### Fact Sheet No. OSHA 92-19
### RESPONDING TO WORKPLACE EMERGENCIES

Employers should establish effective safety and health programs and prepare their workers to handle emergencies before they arise.

### Planning

Where required by the Occupational Safety and Health Administration (OSHA), firms with more than 10 employees must have a written emergency action plan; smaller companies may communicate their plans orally. [See 29 Code of Federal Regulation (CFR) Part 1910.38(a) for further information.] Essential to an effective emergency action plan are top management support and commitment and the involvement of all employees.

Management should review plans with employees initially and whenever the plan itself, or employees responsibilities under it, change. Plans should be re-evaluated and updated periodically. Emergency procedures, including the handling of any toxic chemicals, should include:

- Escape procedures and escape route assignments.
- Special procedures for employees who perform or shut down critical plant operations.
- A system to account for all employees after evacuation.
- Rescue and medical duties for employees who perform them.
- Means for reporting fires and other emergencies.
- Contacts for information about the plan.
- Chain of Command

An emergency response coordinator and a back-up coordinator must be designated. The coordinator may be responsible for plant-wide operations, public information and ensuring that outside aid is called in. A back-up coordinator ensures that a trained person is always available. Duties of the coordinator include:

- Determining what emergencies may occur and seeing that emergency procedures are developed to address them.
- Directing all emergency activities including evacuation of personnel.
- Ensuring that outside emergency services such as medical aid and local fire departments are called when necessary.
- Directing the shutdown of plant operations when necessary.

### Emergency Response Teams

Members of emergency response teams should be thoroughly trained for potential emergencies and physically capable of carrying out their duties; know about toxic hazards in the workplace and be able to judge when to evacuate personnel or depend on outside help (e.g. when a fire is too large for them to handle). One or more teams must be trained in:

- Use of various types of fire extinguishers.
- First aid, including cardiopulmonary resuscitation (CPR).
- The requirements of the OSHA bloodborne pathogens standard.
- Shutdown procedures.
- Chemical spill control procedures.
- Use of self-contained breathing apparatus (SCBA).
- Search and emergency rescue procedures.
- Hazardous materials emergency response in accordance with 29 CFR 1910.120.

### Response Activities

Effective emergency communication is vital. An alternate area for a communications center other than management offices should be established in the plans and the emergency response coordinator should operate from this center. Management should provide emergency alarms and ensure that employees know how to report emergencies. An updated list of key personnel and off-duty telephone numbers should be maintained.

A system should be established for accounting for personnel once workers have been evacuated with a person in the control center responsible for notifying police or emergency response team members of persons believed missing.

Effective security procedures, such as cordoned off areas, can prevent unauthorized access and protect vital records and equipment. Duplicate records can be kept in off-site locations for essential accounting files, legal documents and lists of employees relatives to be notified in case of emergency.

### Training

Every employee needs to know details of the emergency action plan, including evacuation plans, alarm systems, reporting procedures for personnel, shutdown procedures, and types of potential emergencies. Drills should be held at random intervals, at least annually, and include, if possible, outside police and fire authorities.

Training must be conducted initially, when new employees are hired, and at least annually. Additional training is needed when new equipment, materials, or processes are introduced, when procedures have been updated or revised, or when exercises show that employee performance is inadequate.

### Personal Protection

Employees exposed to accidental chemical splashes, falling objects, flying particles, unknown atmospheres with inadequate oxygen or toxic gases, fires, live electrical wiring, or similar emergencies need personal protective equipment, including:

- Safety glasses, goggles, or face shields for eye protection.
- Hard hats and safety shoes.
- Properly selected and fitted respirators.
- Whole body coverings, gloves, hoods, and boots.
- Body protection for abnormal environmental conditions such as extreme temperatures.

### Medical Assistance

Employers not near an infirmary, clinic, or hospital should have someone on-site trained in first aid, have medical personnel readily available for advice and consultation, and develop written emergency medical procedures.

It is essential that first aid supplies are available to the trained medical personnel, that emergency phone numbers are placed in conspicuous places near or on telephones, and prearranged ambulance services for any emergency are available.

### Further Information

More detailed information on workplace emergencies is provided in "How to Prepare for Workplace Emergencies" (OSHA 3088) available free from OSHA Publications, Room N3101, 200 Constitution Ave., N.W., Washington, D.C. 20210, telephone 202-219-4667, or local OSHA offices.

---

This is one of a series of fact sheets highlighting U.S. Department of Labor programs. It is intended as a general description only and does not carry the force of legal opinion. This information will be made available to sensory impaired individuals upon request. Voice phone: 202-219-8151. TDD message referral phone: 1-800-326-2577.

# Safety and Health Management Voluntary Guidelines

## Safety and Health Management Guidelines

Scope and Application. (1) This guideline applies to all places of employment which are covered by OSHA standards in 29 CFR Parts 1910, 1915, 1917 and 1918.

(2) This guideline does not apply to places of employment which are covered by OSHA standards found in 29 CFR Part 1926.

Introduction. The Occupational Safety and Health Administration (OSHA) has concluded that effective management of worker safety and health protection is a decisive factor in reducing the extent and the severity of work-related injuries and illnesses. Effective management addresses all work-related hazards, including those potential hazards which could result from a change in worksite conditions or practices. It addresses hazards whether or not they are regulated by government standards.

OSHA has reached this conclusion in the course of its evaluation of worksites in its enforcement program, its State-operated consultation program, and its Voluntary Protection Programs. These evaluations have revealed a basic relationship between effective management of worker safety and health protection and a low incidence and severity of employee injuries. Such management also correlates with the elimination or adequate control of employee exposure to toxic substances and other unhealthful conditions.

OSHA's experience in the Voluntary Protection Programs has also indicated that effective management of safety and health protection improves employee moral and productivity, as well as significantly reducing workers' compensation costs and other less obvious costs of work-related injuries and illnesses.

Through an analysis of public comment received in response to its request and through an earlier review of literature. OSHA has found that the conclusions it has reached from its own experience are supported by a substantial body of expert and practitioner opinion.

Based on this cumulative evidence that systematic management policies, procedures and practices are fundamental to the reduction of work-related injuries an dillnesses an dtheir attendant economic costs. OSHA offers the following guidelines for effective management of worker safety and health protection. OSHA urges all employers to establish and to maintain programs which meet these guidelines in a manner which addresses the specific operations and conditions of their worksites.

## The Guidelines

(a) *General.* (1) Employers are advised and encouraged to institute and maintain in their establishments a program which provides systematic policies, procedures, and practices that are adequate to recognize and protect their employees from occupational safety and health hazards.

(2) An effective program includes provisions for the systematic identification, evaluation, and prevention or control of general workplace hazards, specific job hazards, and potential hazards which may arise from foreseeable conditions.

(3) Although compliance with the law, including specific OSHA standards, is an important objective, and effective program looks beyond specific requirements of law to address all hazards. It will seek to prevent injuries and illnesses, whether or not compliance is at issue.

(4) The extent to which the program is described in writing is less important then how effective it is in practice. As the size of a worksite or the complexity of a hazardous operation increases, however, the need for written guidance increases to ensure clear communications of policies and priorities and consistent and fair application of rules.

(b) *Major Elements.* An effective occupational safety and health program will include the following four elements. To implement these elements, it will include the actions described in paragraph (c).

(1) **Management commitment and employee involvement** are complementary. Management commitment provides the motivating force and the resources for organizing and controlling activities within an organization. In an effective program, management regards workers safety and health as a fundamental value of the organization and applies its commitment to safety and health protection with as much vigor as to other organizational purposes. Employee involvement provides the means through which workers develop and/or express their own commitment to safety and health protection, for themselves and for their fellow workers.

(2) **Worksite analysis** involves a variety of worksite examinations, to identify not only existing hazards but also conditions and operations in which changes might occur to create hazards. Unawareness of a hazard which stems from failure to examine the worksite is a sure sign that safety and health policies and/or practices are ineffective. Effective management actively analyzes the work and worksite, to anticipate and prevent harmful occurrences.

(3) **Hazard prevention and controls** are triggered by a determination that a hazard or potential hazard exists. Where feasible, hazards are prevented by effective design of the jobsite or job. Where it is not feasible to eliminate them, they are controlled to prevent unsafe and unhealthful exposure. Elimination or controls is accomplished in a timely manner, once a hazard or potential hazard is recognized.

(4) **Safety and health training** addresses the safety and health responsibilities of all personnel concerned with the site, whether salaried or hourly. If is often most effective when incorporated into other training about performance requirements and job practices. Its complexity depends on the size and complexity of the worksite, and the nature of the hazards and potential hazards at the site.

(c) *Recommended Actions* (i) **Management Commitment and Employee Involvement**. (i) State clearly a worksite policy on safe and healthful work and working conditions, so that all personnel with responsibility at the site and personnel at other locations with responsibility for the site understand the priority of safety and health protection in relation to other organizational values.

(ii) Establish and communicate a clear goal for the safety and health program and objectives for meeting that goal, so that all members of the organization understand the results desired and the measures planned for achieving them.

(iii) Provide visible top management involvement in implementing the program, so that all will understand that management's commitments is serious.

(iv) Provides for the encourage employee involvement in the structure and operation of the program and in decisions that affect their safety and health, so that they will commit their insight and energy to achieving the safety and health program's goal and objectives.

(v) Assign and communicate responsibility for all aspects of the program so that manages, supervisors, and employees in all parts of the organization know what performance is expected of them.

(vi) Provide adequate authority and resources to responsible parties, so that assigned responsibilities can be met.

(vii) Hold managers, supervisors, and employees accountable for meeting their responsibilities, so that essential tasks will be performed.

(viii) Review program operations at least annually to evaluate their success in meeting the goal and objectives, so that deficiencies can be identified and the program and/or the objectives can be revised when they do not meet the goal of effective safety and health protection.

(2) **Worksite Analysis.** (i) So that all hazards are identified:

(A) Conduct comprehensive baseline worksite surveys for safety and health and periodic comprehensive update surveys:

(B) Analyze planned and new facilities, processes, materials, and equipment; and

(C) Perform routine job hazard analyses.

(ii) Provide for regular site safety and health inspection, so that new or previously missed hazards and failures in hazard controls are identified.

(iii) So that employee insight and experience in safety and health protection may be utilized and employee concerns may be addressed, provide a reliable system for employees, without fear of reprisal, to notify management personnel about conditions that appear hazardous and to receive timely and appropriate responses; and encourage employees to use the system. (iv) Provide for investigation of accidents and "near miss" incidents, so that their causes and means for their prevention are identified.

(v) Analyze injury and illness trends over time, so that patterns with common causes can be identified and prevented.

(3) **Hazard Prevention and Control.** (i) So that all current and potential hazards, however detected, are corrected or controlled in a timely manner, established procedures for that purpose, using the following measures:

(A) Engineering techniques where feasible and appropriate:

(B) Procedures for safe work which are understood and followed by all affected parties, as a result of training, positive reinforcement, correction of unsafe performance, and, if necessary, enforcement through a clearly communicated disciplinary system:

(C) Provision of personal protective equipment; and

(D) Administrative controls, such as reducing the duration of exposure.

(ii) Provide for facility and equipment maintenance, so that hazardous breakdown is prevented.

(iii) Plan and prepare for emergencies, and conduct training and drills as needed, so that the response of all parties to emergencies will be "second nature."

(iv) Establish a medical program which includes availability of first aid on site and of physician and emergency medical care nearby, so that harm will be minimized if any injury or illness does occur.

(4) **Safety and Health Training.** (i) Ensure that all employees understand the hazards to which they may be exposed and how to prevent harm to themselves and others from exposure to these hazards, so that employees accept and follow established safety and health protections.

(ii) So that supervisors will carry out their safety and health responsibilities effectively, ensure that they understand those responsibilities and the reasons for them, including:

(A) Analyzing the work under their supervision to identify unrecognized potential hazards:

(B) Maintaining physical protections in their work areas; and

(C) Reinforcing employee training on the nature of potential hazards in their work and on needed protective measures, through continual performance feedback and, if necessary, through enforcement of safe work practices.

(iii) Ensure that managers understand their safety and health responsibilities, as described under (c)(1). "Management Commitment and Employee Involvement," so that the managers will effectively carry out those responsibilities.

### The Commentary
(Paragraph by Paragraph)

This Commentary indicates the background and rationale for each part of the guidelines. To facilitate its use, each segment of the guidelines except the introduction is repented just before it is discussed. The background of the introduction immediately follows this paragraph.

*Introduction*

Comment on Introduction. Over the years, OSHA and State enforcement and consultation staff have seen many examples of exemplary workplaces where safety and health programs were well managed and where injury raters were exceptionally low. The common characteristics observed at these sites were the use of organized and systematic methods to assign appropriate responsibility to all managers, supervisors, and employees, to inspect regularly for and control existing and potential hazards, and to orient and train all employees in the ways and means to eliminate or avoid those hazards.

The fundamental importance of such methods has been reflected in decisions of the Occupational Safety and Health Review Commission and the U.S. Courts of Appeal, especially in cases involving an employer claim that a violative workplace condition or action resulted from unpreventable employee misconduct. Such misconduct has been recognized as a defense against citation only when an employer had a work rule prohibiting the conduct, had provided training to ensure that the rule was understood, and had supplied adequate supervision (including regular inspections and work rule enforcement) to ensure that the work rule was followed. These criteria have been applied by the courts in cases involving the citation of OSHA standards as well as the general duty clause. The implication of these cases is that an employer has the duty to establish and maintain such management practices, to the extent that they are necessary to ensure that safe and healthful working conditions are maintained and that safe and healthful work practices are followed.

OSHA has reflected it increasing recognition of the importance of effective safety and health program management by including program management requirements in standards; by recommending safety and health program improvements in conjunction with inspections; by issuing citations under the general duty clause of the Occupational Safety and Health Act of 1970 (Sec. 5(a)(1), 29 U.S.C. 654) which include safety and health management factors; by revising its State-operated consultation program to focus on the promotion of effective safety and health management; and by a range of other promotional efforts.

To further encourage employers and employees to adopt and improve existing safety and health programs, OSHA established on July 2, 1982 (47 FR 29025), the Voluntary Protection Programs (VPP) to recognize worksites with exemplary safety and health management. The participation requirements embodied in the VPP are a distillation of the means, methods, and processes already in use at worksites where safety and health conditions are exceptionally good.

Because VPP participating worksites are officially recognized and are excluded form routine programmed OSHA inspections, the quality of the safety and health programs at these sites must be maintained as models of effectiveness. In 1988, 62 sites were participating in the VPP, and several had been in the program for five or more years. Collectively, during their participation in the VPP, these sites experienced lost-time injuries that were approximately one-fifth to one-third of the average for their industrial classifications. (Unpublished statistics, U.S. Department of Labor, OSHA, 1988).

The fact the VPP participants have injury rates which are so much lower than their industry averages demonstrates that significant reduction is possible. It also strongly indicates that the requirements of the VPP, distilled in the management policies, procedures, and practices described in these recommended guidelines, are major means to achieve the reduction.

In addition, employers at these sites reported improved morale and productivity benefits, as well as significantly reduced workers' compensation and other costs. One plant manager found that the implementation of a single safe work practice at his 44-employee plant during the first three years of participation in the VPP resulted in a greater volume of product and a reduction in rejected project. This change alone saved $265,000 a year. (Proceedings of Public Information Gathering Meeting on Suggested Guidelines for General Safety and Health Programs. U.S. Department of Labor, OSHA, Docket No. C-02. P.77 (October 6, 1988).)

The reduction in workers' compensation and other costs and the improvements in worksite morale and productivity reported by VPP participants reflect significant economic benefits which complement the substantial safety and health benefits of improvement management of worker protection. A Business Roundtable report (Improving Construction Safety Performance (New York, The Business Roundtable. Report A-3. January, 1982). p. 16) concludes that, for construction, the savings from effective administration of safety and health protection is 3.2 times the cost. OSHA has no independent confirmation of this ratio nor of its relevance to industries other than construction. Based on its experience with VPP sites and the conclusions of experienced safety and health professionals, however, OSHA believes that the long-term benefits of effective safety and health management consistently exceed its costs.

To understand this conclusion, it is essential to understand the indirect as well as the direct costs of occupational injuries and illnesses. According to commonly accepted safety management concepts as outlined by Frank E. Bird, Jr. in his Management Guide to Loss Control (Loganville, GA: Institute Press, 1978), for every $1 in medical or insurance compensation costs ("direct costs") for a worker injury, $5-50 more are likely to be spent on "indirect costs" to repair building, tool or equipment damage; to replace damage products or materials; and to make up for losses from production delays and interruptions. An additional $1-3 in indirect costs will be spent for hiring and training replacements and for time to investigate the incident. Mr. Bird's figures do not consider the impact of reduced commitment to work when employees operate in a situation in which injuries are common. Because they frequently involve longer absences, the impact of job- related illnesses can be even greater.

Although economic incentives are secondary to human health and safety as motives for safety and health protection, an employer may find it useful to calculate the total (direct and indirect) costs of injuries and illnesses as a means of determining the economic benefits which might be achieved by preventing the injuries and illnesses. By determining the average cost of an injury and of an illness, the employer can estimate the incremental impact of reducing the rate of injuries and illnesses at the site and therefore the potential economic benefit of such reduction.

Some employers may wish to compare their savings or costs in relation to the nation average for their industries. A method which can be used for that comparison with respect to occupational injuries is described by David R. Bell, a former OSHA employee, in his article, "Gauging Safety Outlays and Objectives," in Occupational Hazards. June, 1987. If the lost workday case rate (LWCR) for a site is below the national average, a formula provided by Bell can be used to calculate how many fewer injuries occurred than would have occurred if the site rate had equalled the national average. (Lost workdays case rates are published annually by the Bureau of Labor Statistics in "Occupational Injuries and Illnesses in the United States by Industry", available from the U.S. Government Printing Office, Washington, DC 20402. The rate for each industry represents the average number of lost workday cases that occurred per 100 employees in the industry).

The number of cases which would have occurred if the site rate had been average Bell calls "expected cases" and the actual cases he calls "injuries avoided." His formula, in which "employment at the site" means the number of equivalent work-years at the site during the year, is as follows:

Industry LWCR = Employment at the site

100  = Expected LW Cases --

Actual LW Cases =

Number of Injuries

Avoided

If the site lost workday case rate is above, the national average, the number of cases by which the site exceeds the national average can be determined by subtracting "expected cases" from "actual cases," once the former number has been calculated.

By multiplying the number of "injuries avoided" or the number of injuries above the average by the average cost of an injury at the site, the employer can estimate the savings or losses which resulted from the quality of its management of safety protection relative to national performance. (Because national data on the incidence of occupational illnesses is incomplete, the formula is less useful in relation to occupational health protection.)

(a) *General*

"(a) General. (1) Employers are advised and encouraged to institute and maintain in their establishments a program which provides systematic policies, procedures, and practices that are adequate to recognize and protect their employees from occupational safety and health hazards."

Comment: In essence, this paragraph states that the end (protection of employees from occupational safety and health hazards) determines the means. The criterion for determining what is needed in a safety and health program at a particular site is: whatever feasible action it takes to protect the workers from the safety and health hazards at that specific site. The form of the safety and health program elements and implementing actions will vary at each site according to the nature of site organization and the nature of the hazards and potential hazards at the site.

"(2) An effective program includes provisions for the systematic identification, evaluation, and prevention or control of general workplace hazards, specific job hazards and potential hazards, which may arise from foreseeable conditions."

Comment: Provisions for identifying and preventing hazards are systematic. If not, hazards or potential hazards will be missed and/or preventive controls will break down, and the chance of injury or illness will significantly increase.

General workplace hazards include such conditions as tripping hazards in walking areas and poor illumination. Specific job hazards may relate to the specific conditions in a job, such as exposure to a saw blade, or to the inherent hazardousness of an operation required in the job, such as the removal of jammed material from a point of operation. Potential hazards include such situations as the possibility of exposure to toxic chemicals as a result of a rupture of piping from the impact of a forklift.

"(3) Although compliance with the law, including specific OSHA standards, is an important objective, an effective program looks beyond specific requirements of law to address all hazards. It will seek to prevent injuries and illnesses, whether or not compliance is at issue."

Comment: OSHA and other government standards provide important guidance on the identification and control of hazards, but they are not always enough. Although compliance with the law is an important objective of and motive for an effective program. OSHA has found that the most successful programs look beyond government standards and legal requirements. They look for other sources of information about hazards, such as the National Electrical Code (NEC), the American Conference of Government Industrial Hygienists (ACGIH), and the American National Standards Institute (ANSI): and they use their own seasoned analytical abilities to look for and address hazards not covered by government or other standards. Their motive is to prevent injuries and illnesses and the attendant human and economic costs, whether or not compliance with the law is at issue.

This approach is essential in view of the difficulty that regulatory agencies have in moving quickly to set standards for every possible hazard in the workplace and to revise them when new information becomes available.

"(4) The extent to which the program is described in writing is less important than how effective it is in practice. As the size of a worksite or the complexity of a hazardous operation increases, however, the need for written guidance increases to ensure clear communication of policies and priorities and consistent and fair application of rules."

Comment: OSHA recognizes that relatively simple, unwritten policies, practices, and procedures are adequate to address the hazards in many smaller or less hazardous establishments. The more complex and hazardous and operation is, the more formal (written) and complex the program will probably need to be. A written program which is revised regularly can clarify policy, create consistency and continuity in its interpretation, serve as a checkpoint whenever there is a question of priority between safety and production, and support fair and equitable enforcement of safe work rules and practices.

(b) *Major Elements*

"(b) Major Elements. An effect occupational safety and health program will include the following four elements. To implement these elements, it will include the actions described in paragraph (c).

(1) **Management commitment and employee involvement** are complementary. Management commitment provides the motivation force and the resources for organizing and controlling activities within an organization. In an effective program, management regards worker safety and health as a fundamental value of the organization and applies its commitment to safety and health protection with as much vigor as to other organizational purposes. Employee involvement provides the means through which workers develop and/or express their own commitment to safety and health protection, for themselves and for their fellow workers.

(2) **Worksite analysis** involves a variety of worksite examinations, to identify not only existing hazards but also conditions and operations in which changes might occur to create hazards. Unawareness of a

hazard which stems from failure to examine the worksite is a sure sign that safety and health policies and/or practices are ineffective. Effective management actively analyzes the work and worksite, to anticipate and prevent harmful occurrences.

(3) **Hazard prevention and control** are triggered by a determination that a hazard or potential hazard exists. Where feasible, hazards are prevented by effective design of the job site or job. Where it is not feasible to eliminate them, they are controlled to prevent unsafe or unhealthful exposure. Elimination or control is accomplished in a timely manner, once a hazard or potential hazard is recognized.

(4) **Safety and health training** addresses the safety and health responsibilities of all personnel concerned with the site, whether salaried or hourly. It is often most effective when incorporated into other training about performance requirements and job practices. Its complexity depends on the size and complexity of the worksite, and the nature of the hazards and potential hazards at the site."

Comment: These paragraphs set forth the areas of managerial practice which are essential to effective safety and health protection. These practices, means, and methods are consistent with those used by employers to achieve other organizational objectives, such as cost control, quality, and productivity. Giving safety and health equal organizational priority in relation to these other objectives is fundamental to the protection of individual employees and to the effectiveness of the organization itself.

These elements consist of methods historically used to accomplish organizational objectives. They are generic in that they are generally applicable regardless of unique operations or conditions of particular firms. Only the form which they take varies. Though at points they are expressed in the terms of the "hierarchical" organizations most common in American industry (i.e., by reference to "managers," "supervisors," "employees"), they can easily be adapted to other organizational forms or styles of operation. They relate to essential concerns and activities of any organization. It is on this basis that OSHA considers them applicable in shipyard employment, marine terminals, and longshoring as well as general industry.

(c) *Recommended Actions.*

(c)(1) **Management Commitment and Employee Involvement**

Comment: Each action listed in this section represents the application to occupational safety and health of a key means for organizing, motivating and controlling activities within an organization.

"(c)(1)(i) State clearly a worksite policy on safe and healthful work and working conditions, so that all personnel with responsibility at the site and personnel at other locations with responsibility for the site understand the priority of safety and health protection in relation to other organizational values."

Comment: A statement of policy is the foundation of safety and health management. It communicates the value in which safety and health protection is held in the business organization. If it is absorbed by all in the organization, it becomes the basic point of reference for all decisions affecting safety and health. It also becomes the criterion by which the adequacy of protective actions is measured.

"(c)(1)(ii) Establish and communicate a clear goal for the safety and health program and objectives for meeting that goal, so that all members of the organization understand the results desired and the measures planned for achieving them."

Comment: A goal, and implementing objectives, make the safety and health policy more specific. Communicating them ensures that all in the organization understand the direction it is taking.

"(c)(1)(iii) Provide visible top management involvement in implementing the program so that all will understand that management's commitment is serious."

Comment: Actions speak louder than words. If top management gives high priority to safety and health protection in practice, other will see and follow. If not, a written or spoken policy of high priority for safety and health will have little credibility, and others will not follow it. Plant managers who wear required personal protective equipment in work areas, perform periodic "housekeeping" inspections, and personally track performance in safety and health protection demonstrate such involvement.

"(c)(1)(iv) Provide for and encourage employee involvement in the structure and operation of the program and in decisions that affect their safety and health, so that they will commit their insight and energy to achieving the safety and health program's goal and objectives."

Comment: Since an effective program depends on commitment by employees as well as managers, it is important for their concerns to be reflected in it. An effective program includes all personnel in the organization--managers, supervisors, and other--in policy development, planning, and operations.

This does not mean transfer of responsibility to employees. the Occupational Safety and Health Act of 1970 clearly places responsibility for safety and health protection on the employer. However, employees intimate knowledge of the jobs they perform and the special concerns they bring to the job give them a unique perspective which can be used to make the program more effective.

Employee participation may take any or all of a number of forms. For instance, the system for notifying management personnel about conditions that appear hazardous serves as a major means of worksite analysis to identify hazards and is therefore included as paragraph (c)(2)(iii). Such a system is, however, by itself not sufficient to provide for effective employee involvement. Forms of participation which engage employees more fully in systematic prevention include (1) inspecting for hazards and recommending corrections or controls; (2) analyzing jobs to locate potential hazards and develop safe work procedures; (3) developing or revising general rules for safe work; (4) training newly hired employees in safe work procedures and rules, and/or training their co-workers in newly revised safe work procedures; (5) providing programs and presentations for safety meeting; and (6) assisting in accident investigations.

Such functions can be carried out in a number of organizational contexts. Joint labor-management committees are most common. Other means include labor safety committees, safety circle teams, rotational assignment of employees to such functions, and acceptance of employee volunteers for the functions.

Employee involvement is effective only when the employer welcomes it and provides protection from any discrimination, including unofficial harassment, to the employees involved. However, inclusion of employees in one or more of the suggested activities, or in any way that fits the individual worksite and provides an employee role that has impact on decisions about safety and health protection, will strengthen the employer's overall program of safety and health protection.

"(c)(1)(v) Assign and communicate responsibility for all aspects of the program, so that managers, supervisors, and employees in all parts of the organization know that performance is expected of them."

Comment: Assignment of responsibility for safety and health protection to a single staff member, or even a small group, will leave other members feeling that someone else is taking care of safety and health problems. Everyone in an organization has some responsibility for safety and health.

A clear statement of that responsibility, as it relates both to organizational goals and objectives and to the specific functions of individuals, is essential. If all persons in an organization do not know what is expected of them, they are unlikely to perform as desired.

"(c)(1)(vi) Provide adequate authority and resources to responsible parties, so that assigned responsibilities can be met."

Comment: It is unreasonable to assign responsibility without providing adequate authority and resources to get the job done. For example, a person with responsibility for the safety of a piece of machinery need the authority to shut it down and get it repaired. Needed resources may include adequately trained and equipped personnel and adequate operational and capital expenditure funds.

"(c)(1)(vii) Hold managers, supervisors, and employees accountable for meeting their responsibilities, so that essential tasks will be performed."

Comment: Stating expectations of managers, supervisors, and other employees means little if management is not serious enough to track performance, to reward it when it is competent and to correct it when it is not. Holding everyone accountable for meeting their responsibilities is at the heart of effective workers safety and health protection. If management states high expectations for such protection but pays greater attention to productivity or other values, safety and health protection will be neglected.

To be effective, a system of accountability must be applied to everyone, from senior management to hourly employees. If some are held firmly to expected performance and other are not, the system will lose its credibility. Those held to expectations will be resentful; those allowed to neglect expectations may increase their neglect. Consequently, the chance of injury and illness will increase.

"(c)(1)(viii) Review program operations at least annually to evaluate their success in meeting the goal and objectives, so that deficiencies can be identified and the program and/or the objectives can be revised when they do not meet the goal of effective safety and health protection."

Comment: A Comprehensive program audit is essential periodically to evaluate the whole set of safety and health management means, methods, and processes, to ensure that they are adequate to protect against the potential hazards at the specific worksite. The audit determines whether policies and procedures are implemented as planned and whether in practice they have met the objectives set for the program. It also determines whether the objectives provide sufficient challenge to lead the organization to meet the program goal of effective safety and health protection. When either performance or the objectives themselves are found inadequate, revisions are made. Without such a comprehensive review, program flaws and their interrelationship may not be caught and corrected.

### (c)(2) Worksite Analysis

Comment: The identification of hazards and potential hazards at a worksite requires an active, on-going examination and analysis of work processes and working conditions. Because many hazards are by nature difficult to recognize, effective examination and analysis will approach the work and working conditions from several perspectives. Each of the activities recommended in this paragraph represents a different perspective.

The recognition of hazards which could result from changes in work practices or conditions requires especially thorough observation and thought, both from those who perform the work and those who are specially trained for that purpose. Since such divergence from the routine and familiar is often the occasion for injuries and health hazard exposures to occur, the anticipation of such changes is critical.

Identification at a worksite of those safety and health hazards which are recognized in its industry is a critical foundation for safety and health protection. It is the general duty of the employer under the Occupational Safety and Health Act of 1970. Successful employers will actively seek the benefit of the experience of others in their industry, through trade associations, equipment manufacturers, and other sources.

An effective program does not stop at this point, however. It continually reviews working conditions and operations to identify hazards which have not previously been recognized in the industry.

Implicit in the provision for the survey, reviews, and analyses recommended in this section is the need for employers to seek competent advice and assistance when they lack needed expertise and to use appropriate means and methods to examine and assess all existing and foreseeable hazards. Personnel who perform comprehensive baseline and update surveys, analysis of new facilities, processes, procedures, and equipment, and job hazard analyses may require greater expertise than those who conduct routine inspections, since the former are conducting a boarder and/or deeper review.

Personnel performing regular inspections should, however, possess a degree of experience and competence adequate to recognize hazards in the areas they review and to identify reasonable means for their correction or control. Such competence should normally be expected of ordinary employees who are capable of safely supervising or performing the operations of the specific workplace. Smaller businesses which need assistance in the development of such competence can receive free assistance from a number of sources, including OSHA and a nationwide network of OSHA-funded, State- operated consultation projects.

"(c)(2)(i) So that all hazards and potential hazards are identified:

(A) conduct comprehensive baseline worksite survey for safety and health and periodic comprehensive update surveys;

(B) analyze planned and new facilities, processes, materials, and equipment; and

(C) perform routine job hazard analyses."

Comment: A comprehensive baseline survey of the work and working conditions at a site permits a systematic recording of those hazards and potential hazards which can be recognized without intensive analysis. This baseline record provides a checklist for the more frequent routine inspections, recommended in paragraph (c)(2)(ii). With those hazards under control, attention can be given to the intensive analysis required to recognized less obvious hazards.

Subsequent comprehensive surveys provide an opportunity to step back from the routine check on control of previously recognized hazards and look for others. With the baseline established, these subsequent reviews are one occasion for focusing more intensive analysis in areas with the highest potential for new or less obvious hazards. The frequency with which comprehensive examinations are needed depends on the complexity, hazardousness, and changeability of the worksite. Many successful worksites conduct such reviews on an annual or biannual basis.

Analysis of new facilities, processes, materials, and equipment in the course of their design and early use (sometimes called "change analysis") provides a check against the introduction of new hazards with them. Effective management ensures the conduct of such analyses during the planning phase, just before their first use, and during the early phases of their use. Numerous specific OSHA standards require inspection of particular equipment, conditions, and activities as a safety precaution prior to operation or use. This guideline makes clear that, in effective safety and health programs, this generally recognized inspection practice is applied more broadly to all conditions and activities.

Job hazard analysis is an important tool for more intensive analysis to identify hazards and potential hazards not previously recognized, and to determine protective measures. Through more careful attention to the work processes in a particular job, analysis can recognize new points at which exposure to hazards may occur or at which foreseeable changes in practice or conditions could result in new hazards.

"(c)(2)(ii) Provide for regular site safety and health inspections, so that new or previously missed hazards and failures in hazard controls are identified."

Comment: Once a comprehensive examination of the workplace has been conducted and hazard controls have been established, routine site safety and health inspections are necessary to ensure that changes in conditions and activities do not create new hazards and that hazard controls remain in place and are effective. Routine industrial hygiene monitoring and sampling are essential components of such inspections in many workplaces.

Personnel conducting these inspections also look out for new or previously unrecognized hazards, but not as thoroughly as those conducting comprehensive surveys.

The frequency and scope of these "routine" inspection depends on the nature and severity of the hazards which could be present and the relative stability and complexity of worksite operations.

"(c)(2)(iii) So that employee insight and experience is safety and health protection may be utilized and employee concerns may be addressed, provide a reliable system for employees, without fear of reprisal, to notify management personnel about conditions that appear hazardous and to receive timely and appropriate responses; and encourage employees to use the system."

Comment: A reliable system for employees to notify management of conditions or practices that appear hazardous and to receive a timely and appropriate response serves a dual purpose. It gives management the benefit of many more points of observations and more experienced insight in recognizing hazards or other symptoms of breakdown in safety and health protection systems. It also gives employees assurance that their investment in safety and health is worthwhile.

A system is reliable only if it ensures employees a credible and timely response. The response will include both timely action to address any problems identified and a timely explanation of why particular actions were or were not taken.

Since the employer benefits from employee notices, effective management will not only guard against reprisals to avoid discouraging them but will take positive steps to encourage their submission.

"(c)(2)(iv) Provide for investigation of accidents and `near miss' incidents, so that their causes and means for preventing repetitions are identified."

Comments: Accidents, and incidents in which employees narrowly escape injury, clearly expose hazards. Analysis to identify their causes permits development of measures to prevent future injury or illness. Although a first look may suggest that "employee error" is a major factor, it is rarely sufficient to stop there. Even when an employee has disobeyed a required work practice, it is critical to ask, "Why?" A thorough analysis will generally reveal a number of deeper factors, which permitted or even encouraged an employee's action. Such factors may include a supervisor's allowing or pressuring the employee to take short cuts in the interest of production, inadequate equipment, or a work practice which is difficult for the employee to carry out safely. An effective analysis will identify actions to address each of the causal factors in an accident or "near miss" incident.

"(c)(2)(v) Analyze injury and illness trends over time, so that patterns will common causes can be identified and prevented."

Comment: A review of injury experience over a period of time may reveal patterns of injury with common causes which can be addressed. Correlation of changes in injury experience with changes in safety and health program operations, personnel, and production processes may help to identify causes.

### (c)(3) Hazard Prevention and Control

Comment: Effective management prevents or controls identified hazards and prepares to minimize the harm form job-related injuries and illnesses when they do occur.

"(c)(3)(i) So that all current and potential hazards, however detected, are corrected or controlled in a timely manner, establish procedures for that purpose, using the following measures:

(A) engineering techniques where feasible and appropriate;

(B) procedures for safe work which are understood and followed by all affected parties, as a result of training, positive reinforcement, and, if necessary, endorsement through a clearly communicated disciplinary system;

(C) provision of personal protective equipment; and

(D) administrative controls, such as reducing the duration of exposure."

Comment: Hazards, once recognized, are promptly prevented or controlled. Management action in this respect determines the credibility of its safety and health management policy and the usefulness of it entire program.

An effective program relies on the means for prevention or control which provides the best feasible protection of employee safety and health.

It regards legal requirements as a minimum. When there are alternative ways to address a hazard, effective managers have found that involving employees in discussions of methods can identify useful prevention and control measures, serve as a means for communicating the rational for decisions made, and encourage employee acceptance of the decisions.

When safe work procedures are the means of protection, ensuring that they are followed becomes critical. Ensuring safe work practices involves discipline in both a positive sense and a corrective sense. Every component of effective safety and health management is designed to create a disciplined environment in which all personnel act on the basic that worker safety and health protection is a fundamental value of the organization. Such an environment depends on the credibility of management's involvement in safety and health matters, inclusion of employees in decisions which affect their safety and health, rigorous worksite analysis to identify hazards and potential hazards, stringent prevention and control measures, and thorough training. In such an environment, all personnel will understand the hazards to which they are exposed, why the hazards pose a threat, and how to protect themselves and others form the hazards. Training for the purpose is reinforced by encouragement of attempt to work safely and by positive recognition of safe behavior.

If, in such a context, an employee, supervisor, or manager fails to follow a safe procedure, it is advisable not only to stop the unsafe action but also to determine whether some condition of the work has made it difficult to follow the procedure or whether some management system has failed to communicate the danger of the action and the means for avoiding it. If the unsafe action was not based on an external condition or a lack of understanding, or if, after such external condition or lack of understanding has been corrected, the person repeats the action, it is essential that corrective discipline be applied. To allow an unsafe action to continue not only continues to endanger the actor and perhaps others; it also undermines the positive discipline of the entire safety and health program. To be effective, corrective discipline must be applied consistently to all, regardless of role or rank; but it must be applied.

Factors which may affect the time required for correction of hazards include: (1) The complexity abatement technology; (2) the degree of risk; and (3) the availability of necessary equipment, materials, and staff qualified to complete the correction. Because conditions affecting hazard correction and control vary widely, it is impractical of OSHA to recommend specific time limits for all situations. An effective program corrects hazards in the shortest time permitted by the technology required and the availability of needed personnel and materials. It also provides for interim protection when immediate correction is not possible.

"(c)(3)(ii) Provide for facility and equipment maintenance, so that hazardous breakdown is prevented."

Comment: Maintenance of equipment of facilities in an especially important means of anticipating potential hazards and preventing their development. Planning, scheduling, and tracking preventive maintenance activities provides a systematic way of ensuring that they are not neglected.

"(c)(3)(iii) Plan and prepares for emergencies, and conduct training and drills as needed, so that the response of all parties to emergencies will be "second nature."

Comment: Planning and training for emergencies is essential in minimizing the harmful consequences of an accident or other threat if it does occur. If personnel are not so thoroughly trained to react to emergencies that their responses are immediate and precise, they may expose themselves and others to greater danger rather than reduce their exposure. The nature of potential emergencies depends on the nature of site operations and its geographical location. The extent to which training and drills are needed depends on the severity and complexity of the emergencies which may arise.

"(c)(2)(iv) Establish a medical program which includes availability of first aid on site and of physician and emergency medical care nearby, so that harm will be minimized if an injury or illness does occur."

Comment: The availability of first aid and emergency medial care are essential in minimizing the harmful consequences of injuries and illnesses if they do occur. The nature of services needed will depend on the seriousness of injuries or health hazard exposures which may occur. Minimum requirements are addressed in OSHA standards.

### (c)(4) Safety and Health Training

Comment: Education and training are essential means for communicating practical understanding of the requirements of effective safety and health protection to all personnel. Without such understanding, managers, supervisors, and other employees will not perform their responsibilities for safety and health protection effectively.

It is not suggested that elaborate or formal training programs solely related to safety and health are always needed. Integrating consideration of safety and health protection into all organizational activities is the key to its effectiveness. Safety and health information and instruction is, therefore, often most effective when incorporated into other training about performance requirements and job practices, such as management training on performance evaluation, problem solving, or managing change; supervisors' training on the reinforcement of good work practices and the correction of poor ones; and employee training on the operation of a particular machine or the conduct of a specific task.

Each paragraph in this section recommends that the employer ensure understanding of safety and health information by employees, supervisors, and managers. The act of training itself is not sufficient to endure practical comprehension. Some means of verifying comprehension is essential. Formal testing, oral questioning, observation, and other means can be useful. In its Voluntary Protection Programs. OSHA has found that observing and interviewing employees, supervisors, and managers are the most effective measures for determining their understanding of what is expected of them in practice. Although there is no fully reliable means for ensuring understanding, effective safety and health management will apply the same diligence with respect to safety and health protection as is applied to ensuring an understanding of other operational requirements, such as time and attendance, production schedules, and jog skills.

"(c)(4)(i) Ensure that all employees understand the hazards to which they may be exposed and how to prevent harm to themselves and others from exposure to these hazards, so that employees accept and follow established safety and health protections."

Comment: The commitment and cooperation of employees in preventing and controlling exposure to hazards is critical, not only for their own safety and health but for that of others as well. That commitment and cooperation depends on their understanding what hazards they may be exposed to, why the hazards pose a threat, and how they can protect themselves and others from the hazards. The means of protection which they need to understand include not only the immediate protections from hazards in their work processes and locations, but also the management systems which commit the organization to safety and health protection and provide for employee involvement in hazard identification and prevention.

OSHA's Hazard Communication Standard specifies, for chemical hazards, an employer duty to inform employees about workplace hazards and to provide training that will enable them to avoid work-related injuries or illnesses. Other standards set forth training requirements, as summarized in OSHA Publication 2254. "Training Requirements in OSHA Standards and Training Guidelines." The rational for these standards requirements is, however, applicable in relation to all hazards.

Education and training in safety and health protection is especially critical for employees who are assuming new duties. This fact is reflected by the disproportionately high injury rates among workers newly assigned to work tasks. Although some of these injuries may be attributable to other causes, a substantial number are directly related to inadequate knowledge of job hazards and safe work practices. The Bureau of Labor Statistics reports that in 1979. 48 percent of workers injured had been on the job less than one year. ("The New Worker Factor Associated with Occupational Injuries and Illnesses," U.S. Department of Labor, Bureau of Labor Statistics, 1982.) These figures make clear the importance of training employees on job hazards and safe work practices before they assume new duties.

The extent of hazard information which is needed by employees will vary, but includes at least; (1) The general hazards and safety rules of the worksite; (2) specific hazards, safety rules, and practices related to particular work assignments; and (3) the employee's role in emergency situations. Such information and training is particularly relevant to hazards that may not be readily apparent to, to within the ordinary experience and knowledge of, the employee.

"(c)(4)(ii) So that supervisors will carry out their safety and health responsibilities effectively, ensure that they understand those responsibilities and the reasons for them, including;

(A) analyzing the work under their supervision to identify unrecognized potential hazards;

(B) maintaining physical protections in their work areas; and

(C) reinforcing employee training on the nature of potential hazards in their work and on needed protective measures, through continual performance feedback and, if necessary, through enforcement of safe work practices."

Comment: First-line supervisors have an especially critical role in safety and health protection because of their immediate responsibility for workers and for the work being performed. Effective training of supervisors will address their safety and health management responsibilities as well as information on hazards, hazard prevention, and response to emergencies. Although they may have other safety and health responsibilities, those listed in these guidelines merit particular attention.

"(c)(4)(iii) Ensure that managers understand their safety and health responsibilities" and described under (c)(1). "Management Commitment and Employee Involvement," so that the managers will effectively carry out those responsibilities."

Comment: Because there is a tendency in some businesses to consider safety and health a staff function and to neglect the training of managers in safety and health responsibilities, the importance of managerial training is noted separately. Managers who understand both the way and the extend to which effective safety and health protection impacts on the overall effectiveness of the business itself are far more likely to ensure that the necessary safety and health management systems operates as needed.

[FR Doc. 89-1594 Filed 1-25-89; 8:45 am]

# Local Emphasis Programs

Local Emphasis Programs (LEPs) are enforcement stratgies designed and implemented at the Regional Office and/or Area Office levels. These programs are intended to address hazards or industries that pose a particular risk to workers in the Office's jurisdiction. The emphasis programs may be implemented by a single Area Office, or at the Regional level (Regional Emphasis Programs) and applied to all of the Area Offices within the Region. Often times, these LEPs will be accompanied by outreach intended to make employers in the area aware of the program as well as the hazards that the programs are designed to reduce or eliminate. This outreach may be in the form of informational mailings, training at local tradeshows, or speeches at meetings of industry groups or labor organizations.

The following is a list of Local (Area Office) and Regional Emphasis Programs, separated by Region and noting the office where appropriate. This list is current as of July 12, 2005, but may change without notice.

### Region I

- Autobody Shops - Maine Area Offices
- Bridge and Tunnel Construction (SIC 1622, NAICS 234120) - Providence
- Construction Repair, Shutdown and Maintenance Activities (not on Dodge reports) - Augusta
- EISA Pilot - Augusta
- Expedited Informal Settlement Agreement - Regionwide
- Fall Hazards - Regionwide
- Food and Kindred Products - Maine Area Offices
- Methylene Chloride - Regionwide
- Methylene Diphenyl Isocyanate - Regionwide
- Mobile Crane Operations in Construction - Braintree & Methuen
- Powered Industrial Trucks - Regionwide
- Residential Construction - Hartford
- Tunneling and Underground Construction - Braintree

### Region II

- Amputations - Regionwide
- Bridge, Highway and Street Construction - New Jersey Area Offices
- Concrete and Plaster Products - Regionwide
- Construction Worksites - Local Targeting - Manhattan
- Fall Hazards - Regionwide
- Food Processing - Syracuse & Buffalo
- Fuel Oil Tank/Boiler Cleaning, Repair, Maintenance, and Manufacturing - Tarrytown
- Gut Rehabilitation and Demolition - Manhattan
- Hotels - Puerto Rico
- Landscape and Horticultural Service Worksites - Regionwide
- Logging - Syracuse
- Marina: Marinas, Ship and Boat Building and Repair - Puerto Rico & Albany
- Methylene Chloride - Albany
- Processing and Printing of Paper Products - Hasbrouck Heights
- Refuse Handlers and Haulers (SICs 4212 and 4953) - Parsippany
- Silica in Road, Highway, and Utility Work - New Jersey Area Offices
- Small Employers - Warehousing, Storage, and Distribution Industries - Regionwide
- Warehousing, Storage and Distribution Industries - Regionwide
- Woodworking Industries - Syracuse

### Region III

- Blast Furnaces - Charleston
- Bloodborne Pathogens in the Health Care Industry - Pittsburgh, Philadelphia, & Wilkes-Barre
- Chemical Process Safety Management - Charleston
- Concrete Block/Brick, Concrete Products - Pittsburgh
- Diving - Norfolk
- Dust Explosions - Harrisburg
- Dust Explosions - Philadelphia
- Fall Hazards in Construction - Regionwide
- Hazards Associated with the Spray-on Bedliner Industry - Regionwide
- High Injury LWDII - Harrisburg
- Landscape and Horticultural Services on Military Reservations - Norfolk
- Lockout/Tagout - Erie
- Logging - Charleston
- Never Before Inspected High Hazard Manufacturing - Allentown & Wilkes-Barre
- Oil and Gas - Erie
- Oil and Gas Services - Charleston
- PSM in Facilities Using Anhydrous Ammonia - Philadelphia
- Public Warehousing and Storage - Allentown
- Residential Construction - Philadelphia
- Residential Construction - Wilmington
- Ship/Boat Building and Repair - Norfolk
- Silica in Concrete Manufacturing and Ready-Mix Facilities - Charleston
- Silica in Dental Labs - Erie
- Silica in Masonry Restoration Operations - Allentown
- Warehouses - Wilkes-Barre
- Water Transportation Services - Norfolk

### Region IV

- Amputation Hazards in Industries Using Wood Saws - Atlanta West
- Canned, Frozen, and Preserved Fruits, Vegetables, and Food Specialties Industry - Savannah
- Electrical Hazards - Regionwide
- Falls in Construction - Regionwide
- Hazards in the Scrap Waste Industry - Jackson
- Isocyanate Exposures in the Bedliner Industry - Jacksonville
- Landscaping and Horticultural Services - Regionwide
- Lead Hazards - Atlanta West
- Lead Hazards in Radiator Repair Shops - Jackson
- Maritime Cranes and Related Equipment Hazards - Nashville
- Reducing Formaldehyde Exposure - Atlanta West
- Safety and Health Hazards at Establishments Engaged in Fruit and Vegetable Processing - Fort Lauderdale
- Sanitation and Cleanup Operations in Meat Packing, Poultry and Fish Processing Industries - Regionwide
- Scaffolding - Mobile
- Ship & Boat Building and Repair - Columbia, Frankfort, Nashville, & Raleigh

- Ship & Boat Building and Repair - Tampa
- Silica and Other Safety Hazards in Concrete Block, Concrete Products, and Ready-Mixed Concrete Manufacturing - Tampa
- Work Zone Safety - Jackson

### Region V
- Amputations - Regionwide
- Brake Repair Industry - Cleveland
- Building Renovation/Rehabilitation - Calumet City, Madison & Milwaukee
- Canned, Frozen, and Preserved Fruits and Vegetables Industry - Illinois & Wisconsin Area Offices
- Concrete Industry - Wisconsin Area Offices
- Construction of Telecommunication Towers - Regionwide
- Electroplating - Calumet City
- Ergonomics - Regionwide
- Fall Hazards in Construction - Regionwide
- High Rise Building Construction for Inspections in Downtown Chicago-
- Isocyanates - Regionwide
- Logging - Wisconsin Area Offices
- Powered Industrial Trucks in Construction and General Industry - Illinois Area Offices
- Process Safety Management - Illinois Area Offices
- Residential Construction Hazards - Cleveland & Toledo
- Road Construction Work Zone Activities - Regionwide
- Scheduling Construction Inspections - Columbus
- Silica in the Cut Stone Industry - Aurora & Calumet City
- Working in the Vicinity of Power Lines - Illinois Area Offices

### Region VI
- Bloodborne Pathogens - Little Rock
- Concrete, Gypsum, and Plaster - Baton Rouge
- Destruction and Renovation of Fixed Structures - Regionwide
- Fabricated Metal Products - Corpus Christi
- Fall Protection in Construction - Regionwide
- Fruits and Vegetables - Corpus Christi
- Landscaping and Horticultural Services - Regionwide
- Logging - Baton Rouge
- MDI in the Spray-On Bed Liner Industry - Regionwide
- Methylene Chloride - Regionwide
- Oil and Gas - Regionwide
- Oil and Gas Drilling Industry - Fort Worth, Corpus Christi Power Line Safety - Regionwide
- Residential Construction - Regionwide
- Shipbuilding and Repair - Baton Rouge
- Silica in Construction - Houston South
- Warehousing - Lubbock
- Warehousing Ergo - Regionwide
- Work zone Safety and Health - Regionwide

### Region VII
- Amputations - Regionwide
- Autobody Shops - St. Louis
- Communication Towers - St. Louis
- Demolition Activities that Occur During the Destruction, Restoration, and Renovation of Buildings and Structures - St. Louis
- Electrical Hazards - St. Louis
- Employers in the State of Nebraska With Potentially High Injury Rates as Identified by the OSHA Strategic Plan - Omaha
- Falls, Scaffolds, and Electrocutions from Overhead Power Lines PSI - Regionwide
- Federal Agency Establishments - Des Moines
- Maritime Activities - Des Moines
- Meat Processing - Omaha
- Powered Industrial Trucks in Construction and Gen. Industry - Omaha & St. Louis
- Residential Construction - Kansas City & St. Louis
- Residential Construction - Wichita
- Road Construction Work Zone Activities - St. Louis & Kansas City
- Trenching and Excavation in Construction - Omaha

### Region VIII
- Amputations - Englewood, Denver, Billings & Bismarck
- Concrete and Gypsum Products - Regionwide
- Construction Fall Hazards - Denver
- Construction Fall Hazards and for Landscaping Operations - Billings & Bismarck
- Grain Handling Facilities - Bismarck
- Lead in Radiator Shops - Billings
- Logging Operations - Englewood, Billings & Denver
- Lumber and Wood Products - Billings
- Methylene Bisphenyl Isocyanate - Bismarck, Denver & Englewood
- Methylene Chloride - Billings
- Occupational Asthma Associated With Exposure to Isocyanates in Aircraft Painting (SIC 4581) - Billings
- Oil and Gas Well Drilling - Denver & Englewood
- Residential Scaffold Fall Hazards and Residential Landscaping Operations - Englewood
- Silica and Beryllium in Dental Offices - Englewood
- Silica and Slab Handling - Englewood
- Silica in Construction - Billings
- Silica in SICS 3281 & 5999 - Denver
- Silica in the Cut Stone and Stone Products Industry - Billings
- Tree-trimming - Regionwide
- Workzone Safety (Englewood)

### Region IX
- Amputations - Regionwide
- Forklifts/Warehousing - Regionwide
- Garment Industry - Saipan
- Labor Barracks - Commonwealth of the Northern Marianna Islands
- Programmed Construction Inspections - Regionwide
- Shipbuilding, Breaking, and Boat Repair - Regionwide

### Region X
- Commercial Fish Processing Operations in Alaska - Alaska
- Construction Fall Hazards - Idaho
- Construction Inspections Under Federal Jurisdiction in State Plan State - Alaska, Oregon & Washington
- Inspections in Federal Agencies Using Workers' Compensation Data - - Washington, Oregon, Alaska & Idaho
- Logging Operations in Idaho - Idaho

- Logging Operations in Oregon and Washington - Oregon & Washington
- Military Bases, National Parks, and National Cemeteries - Washington
- Native Health Care Facilities - Alaska
- Offshore Oil and Gas Drilling Platforms - Alaska
- Residential Construction - Boise
- Shipbuilding and Ship Repair - Alaska, Oregon & Washington

# Subject Index for 29 CFR 1910, Sec. 5, 1903, 1904, and 1990— Occupational Safety And Health Standards

| Subject term, Section No. | Page |
|---|---|
| A-Frame Derricks: | |
| (see also Derricks), .181 | 272 |
| AIDS | |
| (see Bloodborne pathogens), .1030 | 594 |
| Aboveground storage tanks, flammable | |
| and combustible liquid, .106(b)(2) | 91 |
| Spill control, .106(b)(2) (viii) | 93 |
| Spacing, .106(b)(2)(ii) | 91 |
| Venting, .106(b)(2) (iv), (v), (vi) | 92, 93 |
| Abrasive Blasting: | |
| (see also Ventilation), .94 | 63 |
| Air Compressors, Breathing Air, .94(a)(6) | 64 |
| Air Supply, Breathing, .94(a)(6) | 64 |
| Blast Cleaning Enclosures, .94(a)(3) | 63 |
| Cleaning Nozzles, .244(b) | 333 |
| Dust Hazards, .94(a)(2) | 63 |
| Abrasive Wheel Machinery: | |
| Blotters, .215(c)(6) | 303 |
| Definitions, .211(b) | 293 |
| Effective Dates, .220 | 328 |
| Excluded Machinery, .215(a)(5) | 300 |
| Flanges, .215(a)(3), (c) | 300, 302 |
| Guard Design, .215(a)(2) | 300 |
| Specifications, .215(b)(12) | 302 |
| Guard Exposure Angles, .215(b)(2) | 300 |
| Band Type, .215(b)(11) | 302 |
| Bench and Floor Stands, .215(b)(3) | 300 |
| Cup Wheels, .215(b)(1) | 300 |
| Cylindrical Grinders, .215(b)(4) | 301 |
| Dimensions, .215(b)(10) | 301 |
| Material Requirements, .215(b)(10) | 301 |
| Snagging Machines, .215(b)(7) | 301 |
| Surface Grinding, .215(b)(5) | 301 |
| Swing Frame, .215(b)(6) | 301 |
| Guarding, .215(a)(1), (b) | 300 |
| Mounting, .215(d) | 303 |
| Arbor Size, .215(d)(2) | 304 |
| Blotters, .215(d)(5) | 304 |
| Bushings, .215(d)(4) | 304 |
| Inspections, .215(d)(1) | 303 |
| Multiple Wheel, .215(d)(6) | 304 |
| Ring Test, .215(d)(1) | 303 |
| Surface Conditions, .215(d)(3) | 304 |
| Standards Sources, .221 | 328 |
| Work Rests, .215(a)(4) | 300 |
| Abrasive Wheel Machinery, Portable: | |
| Definitions, .241(b) | 329 |
| Guarding, .243(c) | 330 |
| Cup Wheels, .243(c)(2) | 330 |
| General Requirements, .243(c)(1) | 330 |
| Other Type Grinders, .243(c)(4) | 331 |
| Vertical Grinders, .243(c)(3) | 331 |
| Inspection, .243(c)(5) | 331 |
| Mounting, .243(c)(5) | 331 |
| Abrasive Wheels: | |
| (see Abrasive Wheel Machinery) | |
| Access: | |
| Bulk Oxygen Systems, .104(b)(2)(ii) | 88 |
| Cranes, .179(c)(2) | 265 |
| Exposure and medical records, .1020 | 500 |
| Industrial Plants, .106(e)(9)(ii) | 100 |
| Powered Platforms, .66 | 41 |
| Processing Plants, .106(h)(8)(ii) | 106 |
| Spraying Operations, Vents, .107(d)(10) | 108 |
| Sprinkler valve, .107(f)(2) | 109 |
| Accident Prevention Signs and Tags: | |
| (see also Signs and Tags), .145 | 214 |
| Standards Sources, .150 | 230 |
| 2-Acetylaminofluorene, .1003 | 488 |
| Area requirements, .1003(c) | 488 |
| Closed system operation, .1003(c)(2) | 488 |
| Isolated systems, .1003(c)(1) | 488 |
| Maintenance and decontamination activities, .1003(c)(5) | 489 |
| Open-vessel system operations, .1003(c)(3) | 488 |
| Transfer from a closed operation, .1003(c)(4) | 489 |
| Medical surveillance, .1003(g) | 490 |
| Examinations, .1003(g)(1) | 490 |
| Records, .1003(g)(2) | 490 |
| Regulated area requirements, .1003(d) | 489 |
| Contamination control, .1003(d)(4) | 489 |
| Emergencies, .1003(d)(2) | 489 |
| Hygiene facilities and practices, .1003(d)(3) | 489 |
| Signs, information, and training, .1003(e) | 489 |
| Container contents identification, .1003(e)(2) | 489 |
| Lettering, .1003(e)(3) | 489 |
| Prohibited statements, .1003(e)(4) | 489 |
| Signs, .1003(e)(1) | 489 |
| Training and indoctrination, .1003(e)(5) | 489 |
| Acetylene, .102 | 83 |
| Cylinders, .102(a), (c) | 83 |
| Generators, .102(c) | 83 |
| Pipe Systems, .102(b) | 83 |
| Acetylene Generators, .253(f) | 343 |
| Approval, .253(f)(1) | 343 |
| Location, .253(f)(3) | 343 |
| Maintenance, .253(f)(7) | 344 |
| Marking, .253(f)(1) | 343 |
| Operation, .253(f)(7) | 344 |
| Portable, .253(f)(5) | 344 |
| Pressure Limits, .253(f)(2) | 343 |
| Rating, .253(f)(2) | 343 |
| Stationary, .253(f)(4) | 343 |
| Houses and Rooms, .253(f)(6) | 344 |
| Acid Carboys, .262(nn) | 358 |
| Acrylonitrile, .1045 | 630 |
| Emergency situations, .1045(i) | 632 |
| Employee information and training, .1045(o) | 633 |
| Exposure monitoring, .1045(e) | 631 |
| Housekeeping, .1045(k) | 632 |
| Hygiene facilities and practices, .1045(m) | 632 |
| Medical surveillance, .1045(n) | 632 |
| Methods of compliance, .1045(g) | 631 |
| Observation and monitoring, .1045(r) | 634 |
| Permissible exposure limit, .1045(c) | 630 |
| Protective clothing and equipment, .1045(j) | 632 |
| Recordkeeping, .1045(q) | 633 |
| Regulated areas, .1045(f) | 631 |
| Respiratory protection, .1045(h) | 631 |
| Signs and labels, .1045(p) | 633 |
| Waste disposal, .1045(l) | 632 |
| Adjustments: | |
| Cranes, .179(l)(3) | 268 |
| Derricks, .181(f)(2), (3) | 275 |
| Air Compressors, Abrasive Blasting, .94(a)(6), .134(d)(2)(ii) | 64, 190 |
| Air Contaminants, .1000, .1001 | 453, 462 |
| Effective Dates, .98, .1000 | 82 |
| Exposure Limits, .1000, .1001 | 454, 463 |
| Permissible exposure limits, .1000 | 453 |
| Air Controlling Equipment, Power Presses, .217(b)(10) | 311 |
| Air Lift Hammers, Forging, .218(e)(1) | 325 |
| Air Quality, .134(d) | 190 |
| Air Receivers: | |
| Application, .169(a)(1) | 253 |
| Compressed Air, .169 | 253 |
| Equipment: | |
| Drains, .169(b)(2) | 253 |
| Installation, .169(b)(1) | 253 |
| Pressure Gages, .169(b)(3) | 253 |
| Traps, .169(b)(2) | 253 |
| Valves, .169(b)(3) | 253 |
| Standards Sources, .169(a)(2), .170 | 253 |
| Air Supply, .94(a)(6), .134(d) | 64, 190 |
| Airhoses, .243(b)(2) | 330 |

## Subject Index For 29 CFR 1910, Sec. 5, 1903, 1904, and 1990

| Subject term, Section No. | Page |
|---|---|
| Aisles: | |
|    Working Surfaces, .22(b) | 20 |
| Alarms: | |
|    (see also Fire Alarms, Sprinklers, Warning Devices) | |
|    Employee alarm systems, .165 | 240 |
| 4-Aminodiphenyl, .1003 | 488 |
|    Area requirements, .1003(c) | 488 |
|       Closed system operation, .1003(c)(2) | 488 |
|       Isolated systems, .1003(c)(1) | 488 |
|       Maintenance and decontamination activities, .1003(c)(5) | 489 |
|       Open-vessel system operations, .1003(c)(3) | 488 |
|       Transfer from a closed operation, .1003(c)(4) | 488 |
|    Medical surveillance, .1003(g) | 490 |
|       Examinations, .1003(g)(1) | 490 |
|       Records, .1003(g)(2) | 490 |
|    Regulated area requirements, .1003(d) | 489 |
|       Contamination control, .1003(d)(4) | 489 |
|       Emergencies, .1003(d)(2) | 489 |
|       Hygiene facilities and practices, .1003(d)(3) | 489 |
|    Signs, information, and training, .1003(e) | 48 |
|    Container contents identification, .1003(e)(2) | 485 |
|       Lettering, .1003(e)(3) | 489 |
|       Prohibited statements, .1003(e)(4) | 489 |
|       Signs, .1003(e)(1) | 489 |
|       Training and indoctrination, .1003(e)(5) | 489 |
| Ammonia, Anhydrous: | 140 |
|    (see also Anhydrous Ammonia), .111 | |
| Ammonium Nitrate, .109(i) | 119 |
|    Bulk Storage, .109(i)(4) | 120 |
|    Containers, .109(i)(3) | 119 |
|    Contaminants, .109(i)(5) | 120 |
|    Electrical Installations, .109(i)(6) | 120 |
|    Fire Protection, .109(i)(7) | 120 |
|    Separation Walls, .109(i)(5) | 120 |
|    Warehouses, .109(i)(4) | 120 |
| Anchoring Fixed Machinery, .212(a)(5), (b) | 297 |
| Anhydrous Ammonia: | |
|    Containers: | |
|       Appurtenances, .111(b)(6) | 142 |
|       Charging, .111(b)(11) | 144 |
|       DOT, .111(e) | 146 |
|       Farm Vehicles, .111(g), (h) | 147, 148 |
|       Location, .111(b)(5) | 142 |
|       Markings, .111(b)(3) | 141 |
|       Motor Vehicle, .111(f) | 147 |
|       Non-Refrigerated, .111(b)(2), (c) | 141, 145 |
|       Refrigerated, .111(d) | 145 |
|          Markings, .111(b)(4) | 142 |
|       Safety Relief Devices, .111(b)(9), (c)(3), (d)(4), (f)(5) | 143, 145, 146, 147 |
|    Electrical Systems, .111(b)(16) | 145 |
|    Fittings, .111(b)(7) | 142 |
|    Handling, .111 | 140 |
|    Hoses, .111(b)(8) | 142 |
|    Liquid Level Gaging Devices, .111(b)(14) | 144 |
|    Liquid Transfer, .111(b)(12), (f)(6) | 144, 147 |
|    Piping, .111(b)(7) | 142 |
|    Standards Sources, .115 | 148 |
|    Storage, .111 | 140 |
|    Tank Car Unloading, .111(b)(13) | 144 |
|    Tubing, .111(b)(7) | 142 |
| Appliances: | |
|    Electric, .306(d)(1) | 427 |
|    Liquefied Petroleum Gases, .110(b)(20) | 131 |
| Arbor Grinding Wheels, .215(d)(2) | 304 |
| Arc Welding, .254 | 345 |
|    Environmental Conditions, .254(b)(2) | 345 |
|    Equipment: | |
|       Design, .254(b)(4) | 345 |
|       Disconnecting Means, .305(j)(3) | 426 |
|       Grounding, .254(c)(2) | 346 |
|       Installation, .254(c) | 346 |
|       Maintenance, .254(d)(9) | 346 |
|       Operation, .254(d) | 346 |
|       Personnel Protection, .252(b) | 336 |
|       Protection from Rays, .252(b)(2)(iii) | 336 |

| Subject term, Section No. | Page |
|---|---|
|       Supply Connections, .254(c)(3), (d)(3) | 346 |
|    Health Protection, .252(c) | 337 |
|    Ventilation, .252(b)(4)(ii), (c) | 337 |
|    Voltage, .254(b)(3) | 345 |
| Arsenic, Inorganic, .1018 | 493 |
| Asbestos: | |
|    Airborne Concentration, .1001(c) | 463 |
|    Caution Signs and Labels, .1001(j) | 465 |
|    Change Rooms, .1001(h)(2) | 465 |
|    Compliance, .1001(f) | 464 |
|    Definitions, .1001(b) | 463 |
|    Exposure, Permissible, .1001(c) | 463 |
|    Hazard Communication, .1001(j) | 465 |
|    Housekeeping, .1001(k) | 467 |
|    Hygiene Facilities and Practices, .1001(i) | 465 |
|    Medical Examinations, .1001(l) | 467 |
|    Medical Surveillance, .1001(l) | 467 |
|    Monitoring, .1001(d) | 463 |
|    Personal Protective Equipment, .1001(h) | 465 |
|    Recordkeeping, .1001(m) | 468 |
|    Regulated Areas, .1001(e) | 464 |
|    Respiratory protection, .1001(g) | 464 |
|    Special Clothing, .1001(d)(3) | 463 |
| Atmospheric Contaminants: | |
|    (see Air Contaminants) | |
| Atmospheric Tanks, .106(b)(1)(iii) | 91 |
| Attendants: | |
|    Liquefied Hydrogen Systems, .103(c)(4)(ii) | 88 |
|    Liquefied Petroleum Gases, .110(b)(14) | 128 |
| Automatic Sprinkler Systems: | |
|    (see also Sprinkler Systems, Automatic), .159 | 238 |
| Automobile Undercoatings, .107(k) | 110 |
| Baffle Plates: | |
|    Spray Booths, .107(b)(4) | 107 |
| Bakery Equipment: | |
|    Air Conditioning, .263(i)(14) | 360 |
|    Bag Chutes and Lifts, .263(d)(2) | 358 |
|    Biscuit Equipment, .263(k) | 360 |
|    Blenders, .263(d)(3) | 359 |
|    Conveyors, .263(d)(7), (i)(7) | 359, 360 |
|    Cracker Equipment, .263(k) | 360 |
|    Dividers, .263(f) | 359 |
|    Dough Brakes, .263(h) | 359 |
|    Dumpbins, .263(d)(3) | 359 |
|    Flour Handling Equipment, .263(d) | 358 |
|    Machine Guarding, .263(c) | 358 |
|    Miscellaneous Equipment, .263(i) | 360 |
|    Mixers, .263(e) | 359 |
|    Moulders, .263(g) | 359 |
|    Ovens, .263(l) | 361 |
|    Pulverizers, .263(k)(2) | 360 |
|    Scales, Flour, .263(d)(9) | 359 |
|    Sifters, .263(d)(8) | 359 |
|    Slicers, .263(j) | 360 |
|    Storage Bins, .263(d)(6) | 359 |
|    Wrappers, .263(j) | 360 |
| Ballast, Cranes, .180(i)(2) | 272 |
| Band Saws and Resaws, .213(i) | 298 |
| Barking Devices: | |
|    Hydraulic, .261(e)(14) | 351 |
|    Pulp Wood and Chips, .261(c), (e)(8) | 349, 350 |
|    Sawmills, .265(d)(4) | 368 |
| Barrels: | |
|    Guarding, .212(a)(4) | 297 |
| Basket Derricks: | |
|    (see Derricks), .181 | 273 |
| Bathing Facilities: | |
|    Labor Camps, .142(f) | 213 |
| Battery Changing and Charging, .178(g), .305(j)(7) | 260, 426 |
| Bearings, .219(j), (p)(3) | 327, 328 |
| Belts: | |
|    Definitions, .211(f)(1)-(3) | 296 |
|    Manlifts, .68(c)(1) | 60 |
|    Power Transmission Apparatus, .219(e)(1), (o)(3), (p)(6) | 326, 327, 328 |
| Bench and Floor Stands Guarding, .215(b)(3) | 300 |
| Benzene, .1028 | 581 |
|    Communication of benzene hazards to employees, .1028(j) | 584 |

# OSHA Standards for the General Industry

| Subject term, Section No. | Page |
|---|---|
| Exposure monitoring and measurement, .1028(e) | 582 |
| Medical surveillance, .1028(i) | 583 |
| Methods of compliance, .1028(f) | 582 |
| Observation of monitoring, .1028(l) | 585 |
| Permissible exposure limit, .1028(c) | 581 |
| Protective clothing and equipment, .1028(h) | 583 |
| Recordkeeping, .1028(k) | 585 |
| Regulated areas, .1028(d) | 581 |
| Respiratory protection, .1028(g) | 582 |
| Benzidine, .1003 | 488 |
| Area requirements, .1003(c) | 488 |
| Closed system operation, .1003(c)(2) | 488 |
| Isolated systems, .1003(c)(1) | 488 |
| Maintenance and decontamination activities, .1003(c)(5) | 489 |
| Open-vessel system operations, .1003(c)(3) | 488 |
| Transfer from a closed operation, .1003(c)(4) | 488 |
| Medical surveillance, .1003(g) | 490 |
| Examinations, .1003(g)(1) | 490 |
| Records, .1003(g)(2) | 490 |
| Regulated area requirements, .1003(d) | 489 |
| Contamination control, .1003(d)(4) | 489 |
| Emergencies, .1003(d)(2) | 489 |
| Hygiene facilities and practices, .1003(d)(3) | 489 |
| Signs, information, and training, .1003(e) | 489 |
| Container contents identification, .1003(e)(2) | 489 |
| Lettering, .1003(e)(3) | 489 |
| Prohibited statements, .1003(e)(4) | 489 |
| Signs, .1003(e)(1) | 489 |
| Training and indoctrination, .1003(e)(5) | 489 |
| Beryllium, .1000, Table Z-2 | 462 |
| Bins, Bulk Storage of Explosives, .109(g)(4) | 117 |
| Biological Hazards Signs and Tags, .145(e)(4), (f)(8) | 214, 215 |
| Blades Exposure, .212(a)(5) | 297 |
| Blankets, Rubber Insulating, .137 | 204 |
| Blasting Agents (see also Explosives and Blasting Agents), .109(g), (k) (1), (2), .119 | 115, 121, 148 |
| Bulk Delivery, .109(g)(3), (h)(4) | 116, 119 |
| Bulk Storage Bins, .109(g)(4) | 117 |
| Effective Dates, .114 | 148 |
| Mixing, Fixed Location, .109(g)(2), (h)(3) | 115, 118 |
| Mixing Vehicles, .109(g)(3), (h)(4) | 116, 119 |
| Slurries, .109(h) | 118 |
| Standards Sources, .115 | 148 |
| Storage, .109(g)(5) | 117 |
| Transportation, .109(g)(6) | 118 |
| Use, .109(g)(7) | 118 |
| Water Gels, .109(h) | 118 |
| Bleaching: | |
| Pulp and Paper Mills, .261(h) | 352 |
| Textiles, .262(p) | 357 |
| Bloodborne pathogens, .1030 | 594 |
| Effective dates, .1030(i) | 605 |
| Engineering and work-practice controls, .1030(d)(2) | 599 |
| Housekeeping, .1030(d)(4) | 601 |
| Laboratories and production facilities, HIV and HBVresearch, .1030(e) | 602 |
| Personal protective equipment, .1030(c)(2)(ii), (d)(2)(i), (3) | 599 |
| Recordkeeping, .1030(f)(6), (h) | 603, 604 |
| Training, .1030(e)(5), (g)(2) | 602, 604 |
| Vaccinations, HBV, .1030(f) | 602 |
| Warning labels and signs, .1030(g)(1) | 603 |
| Blotters, .215(c)(1)(v), (c)(6), (d)(5) | 302, 303, 304 |
| Board Drop Hammers, .218(e)(2) | 325 |
| Boatswain's Chair Scaffolds, .28(j) | 33 |
| Employee Protection, .28(j)(4) | 33 |
| Fiber Ropes, .28(j)(2) | 33 |
| Life Belts, .28(j)(4) | 33 |
| Roof Irons, Hooks, .28(j)(6) | 33 |
| Seat Slings, .28(j)(3) | 33 |
| Size, .28(j)(1) | 33 |
| Tackle, .28(j)(5) | 33 |
| Boom Guards: | |
| Cranes, .179(e) | 265 |

| Subject term, Section No. | Page |
|---|---|
| Derricks, .181(j)(1) | 276 |
| Booms, Derricks, .181(i)(6) | 276 |
| Boring Machines, .213(l) | 299 |
| Brakes: | |
| Bandsaws, .213(j)(1) | 298 |
| Bridges, .179(f)(4), (6) | 266 |
| Control, .179(f)(3) | 266 |
| Cranes, .179(f) | 266 |
| Friction, Power Presses, .217(b)(2) | 309 |
| Hoists, .179(f)(1) | 266 |
| Holding, .179(f)(2) | 266 |
| Industrial Trucks, .178(g), (h), (m)(5) | 260, 261 |
| Manlifts, .68(c)(1)(i) | 60 |
| Power Control, .179(f)(3) | 266 |
| Trolleys, .179(f)(4), (5) | 266 |
| Brazing: | |
| (see also Welding), .252 | 335 |
| Definitions, .251 | 335 |
| Standards Sources, .256 | 347 |
| Breast Derricks: | |
| (see also Derricks), .181(a)(4) | 273 |
| Bricklayers' Square Scaffolds, .28(l) | 33 |
| Bridge Bumpers, Cranes, .179(e)(2) | 265 |
| Bridge Plates: | |
| (see also Dockboards), .30(a) | 36 |
| Buffing: | |
| (see Grinding, Polishing and Buffing) | |
| Building Maintenance Powered Platforms, .66 | 41 |
| Buildings, Sawmills, .265(c) | 363 |
| Bulk Delivery: | |
| Blasting Agents, .109(g)(3), (h)(4) | 116, 119 |
| Explosives, .109(h)(4) | 119 |
| Bulk Oxygen Systems, .104 | 88 |
| Accessibility, .104(b)(2)(ii) | 88 |
| Cleaning, .104(b)(8)(i) | 89 |
| Clear Zone, .104(b)(10) | 89 |
| Containers, .104(b)(4), (6) | 88, 89 |
| Gaseous, .104(b)(4)(iii) | 89 |
| Liquid, .104(b)(4)(ii) | 89 |
| Dikes, .104(b)(2)(v) | 88 |
| Distances from Hazards, .104(b)(3) | 88 |
| Combustible Liquids, .104(b)(3)(vii), (viii) | 88 |
| Combustible Materials, .104(b)(3)(x) | 88 |
| Combustible Structures, .104(b)(3)(ii) | 88 |
| Congested Areas, .104(b)(3)(xiii) | 88 |
| Fire Resistant Structures, .104(b)(3)(iii) | 88 |
| Flammable Gases, .104(b)(3)(ix) | 88 |
| Flammable Liquids, .104(b)(3)(v), (vi) | 88 |
| Openings, .104(b)(3)(iv) | 88 |
| Slow-Burning Materials, .104(b)(3)(xi) | 88 |
| Electrical Wiring, .104(b)(3)(ix) | 89 |
| Firewalls, .104(b)(3)(viii) | 88 |
| Fittings, .104(b)(5) | 89 |
| Installation, .104(b)(8)(iv) | 89 |
| Joints, .104(b)(8)(ii) | 89 |
| Leakage, .104(b)(2)(iii) | 88 |
| Liquid Oxygen Vaporizers, .104(b)(7) | 89 |
| Grounding, .104(b)(7)(iv) | 89 |
| Location, .104(b)(2) | 88 |
| Maintenance, .104(b)(10) | 89 |
| Marking, .104(b)(8)(viii) | 89 |
| Operating Instructions, .104(b)(9) | 89 |
| Piping, .104(b)(5) | 89 |
| Placarding, .104(b)(8)(viii) | 89 |
| Safety Relief Devices, .104(b)(6), (7)(ii) | 89 |
| All Containers, .104(b)(6)(i) | 89 |
| ASME Containers, .104(b)(6)(iii) | 89 |
| DOT Containers, .104(b)(6)(ii) | 89 |
| Security, .104(b)(8)(vi) | 89 |
| Storage Containers, .104(b)(4), (6) | 88, 89 |
| Testing, .104(b)(8)(v) | 89 |
| Tubing, .104(b)(5) | 89 |
| Vaporizers, .104(b)(7) | 89 |
| Ventilation, .104(b)(3)(xii) | 88 |
| Venting, .104(b)(8)(vii) | 89 |
| Bulk Plants, Flammable and Combustible Liquids, .106(f) | 100 |
| Buildings, .106(f)(2) | 100 |

| Subject term, Section No. | Page |
|---|---|
| Drainage, .106(f)(7) | 101 |
| Electrical Equipment, .106(f)(5) | 101 |
| Fire Protection, .106(f)(8) | 101 |
| Ignition Sources, .106(f)(6) | 101 |
| Liquid Storage, .106(f)(1) | 100 |
| Loading, .106(f)(3) | 100 |
| Waste Disposal, .106(f)(7) | 101 |
| Wharves, .106(f)(4) | 100 |
| Bumpers: | |
| Bridge, .179(e)(2) | 265 |
| Trolley, .179(e)(3) | 265 |
| 1,3 Butadiene:. | |
| Communication of BD Hazards to Employees, .1051(l) | 682 |
| Emergency Situations, .1051(j) | 680 |
| Exposure Goal Program, .1051(g) | 679 |
| Exposure Monitoring, .1051(d) | 678 |
| Medical Screening and Surveillance, .1051(k) | 680 |
| Methods of Compliance, .1051(f) | 679 |
| Permissible Exposure Limits, .1051(c) | 678 |
| Protective Clothing and Equipment, .1051(i) | 680 |
| Recordkeeping, .1051(m) | 682 |
| Regulated Areas, .1051(e) | 679 |
| Respiratory Protection, .1051(h) | 680 |
| Cabinets, Flammable and Combustible Liquid storage, .106(d)(3) | 96 |
| Fire resistance, .106(d)(3)(ii) | 96 |
| Size, .106(d)(3)(i) | 96 |
| Cabs: | |
| Cranes, .179(c), (o)(2); 180(i)(3) | 265, 269, 272 |
| Derricks, .181(j)(6) | 276 |
| Cadmium, .252(c)(9) | 338 |
| Airborne Concentration, .1027(c) | 526 |
| Compliance, .1027(f) | 526 |
| Confined Spaces, .252(c)(9)(ii) | 338 |
| Emergency Situations, .1027(h) | 528 |
| Exposure, Permissible, .1027(c) | 526 |
| Hazard Communication, .1027(m) | 533 |
| Housekeeping, .1027(k) | 529 |
| Hygiene Areas and Practices, .1027(j) | 528 |
| Indoors, .252(c)(9)(i) | 338 |
| Medical Surveillance, .1027(l) | 529 |
| Monitoring, .1027(d) | 526 |
| Personal Protective Equipment, .1027(i) | 528 |
| Clothing, .1027(i) | 528 |
| Recordkeeping, .1027(n) | 533 |
| Regulated Areas, .1027(e) | 526 |
| Respiratory protection, .1027(g) | 527 |
| Warning Labels, .1027(m) | 533 |
| Calcium Carbide: | |
| Indoors, .253(g)(2) | 345 |
| Packaging, .253(g)(1) | 344 |
| Storage, .253(g)(2), (3) | 345 |
| Calenders, .262(ee) | 358 |
| Rubber and Plastics Industry: | |
| Location Protection, .216(d)(2) | 309 |
| Safety Controls, .216(c) | 309 |
| Stopping Limits, .216(f)(1), (3) | 309 |
| Switches, Trip and Emergency, .216(e) | 309 |
| Textiles, .262(ee) | 358 |
| Canisters, Gas Mask: | |
| (see Gas Mask Canisters, Respirators) | |
| Cantilever Gantry Cranes: | |
| (see Gantry Cranes) | |
| Carcinogens: | |
| Advisory committees, 1990.105 | 755 |
| Amendments to part 1990, 1990.106 | 755 |
| Initiation of review part 1990, 1990.106(a) | 755 |
| Response to recommendations and petitions, 1990.106(b) | 756 |
| Arguments, criteria for consideration, 1990.144 | 759 |
| Benign tumors, 1990.144(e) | 759 |
| Indirect mechanisms, 1990.144(f) | 759 |
| Metabolic differences, 1990.144(c) | 759 |
| Non-positive results obtained in human epidemiologic studies, 1990.144(a), | 759 |

| Subject term, Section No. | Page |
|---|---|
| Tumors induced at site of administration, 1990.144(b). | 759 |
| Use of high doses in animal testing, 1990.144(d), | 759 |
| Candidate list of potential occupational carcinogens, 1990.121 | 756 |
| Contents, 1990.121(a) | 756 |
| No legal rights established, 1990.121(c) | 757 |
| Response to petitions, 1990.122 | 757 |
| Tentative classification, 1990.121(b), | 757 |
| Classification of potential carcinogens, 1990.112 | 756 |
| Category I potential carcinogens, 1990.112(a) | 756 |
| Category II potential carcinogens, 1990.112(b) | 756 |
| Definitions, 1990.103 | 755 |
| Human and animal data, 1990.143 | 758 |
| Benign tumors, 1990.143(i) | 759 |
| Carinogenicity of metabolites, 1990.143(k), | 759 |
| Non-positive animal studies, 1990.143(d), | 758 |
| Non-positive human studies, 1990.143(c), | 758 |
| Positive animal studies, 1990.143(b) | 758 |
| Positive human studies, 1990.143(a) | 758 |
| Routes of exposure, 1990.143(f) | 758 |
| Spontaneous tumors, 1990.143(e) | 758 |
| Statistical evaluation, 1990.143(j) | 759 |
| "Threshold" or "no-effect" levels, 1990.143(h) | 758 |
| Use of high doses in animal testing, 1990.143(g) | 758 |
| Metabolic profiles, 1990.144(c) | 759 |
| Model standards | |
| Model emergency temporary standard pursuant to section 6(c) of the Act, 1990.152 | 764 |
| Definitions, 1990.152(b) | 765 |
| Effective date, 1990.152(e) | 765 |
| Employee information and training, 1990.152(o) | 766 |
| Exposure monitoring, 1990.152(e) | 765 |
| Housekeeping, 1990.152(k) | 766 |
| Medical surveillance, 1990.152(n) | 766 |
| Methods of compliance, 1990.152(g) | 765 |
| Notification of use, 1990.152(d) | 765 |
| Observation of monitoring 1990.152(r), | 767 |
| Permissible exposure limits, 1990.152(c), | 765 |
| Protective clothing and equipment, 1990.152(j), | 766 |
| Recordkeeping, 1990.152(q) | 767 |
| Respiratory protection, 1990.152(h) | 765 |
| Scope and application, 1990.152(a) | 764 |
| Signs and labels, 1990.152(p) | 767 |
| Waste disposal, 1990.152(l) | 766 |
| Model standard pursuant to section 6(b) of the Act, 1990.151 | 761 |
| Appendices, 1990.151(t) | 764 |
| Definitions, 1990.151(b) | 761 |
| Effective date, 1990.151(s) | 764 |
| Emergency situations, 1990.151(i) | 762 |
| Employee information and training, 1990.151(o) | 763 |
| Exposure monitoring, 1990.151(e), | 761 |
| Housekeeping, 1990.151(k) | 762 |
| Hygiene facilities and practices, 1990.151(m) | 762 |
| Medical surveillance, 1990.151(n) | 763 |
| Methods of compliance, 1990.151(g) | 762 |
| Notification of use and emergencies, 1990.151(d) | 761 |
| Observation of monitoring, 1990.151(r) | 764 |
| Permissible exposure limits provisions, 1990.151(c) | 761 |
| Protective clothing and equipment, 1990.151(j) | 762 |
| Recordkeeping, 1990.151(q) | 764 |
| Regulated areas, 1990.151(f) | 762 |
| Respiratory protection, 1990.151(h) | 762 |
| Scope and application, 1990.151(a) | 761 |
| Signs and labels, 1990.151(p) | 763 |
| Waste disposal, 1990.151(l) | 762 |
| Priority lists for regulating potential occupational carcinogens, 1990.131 | 757 |
| Factors to be considered, 1990.132 | 757 |
| Publication, 1990.133 | 757 |
| Purpose of 1990, 1990.102 | 755 |
| Regulatory policy on carcinogens, 1990.111 | 756 |
| Rulemaking | |
| Advance notice of proposed rulemaking, 1990.141 | 757 |
| Emergency temporary standards, 1990.142(b) | 758 |
| Final action, 1990.147 | 760 |
| Initiation of a rulemaking, 1990.142 | 758 |
| Emergency temporary standards, 1990.142(b) | 758 |

# OSHA Standards for the General Industry

**Subject term,** *Section No.* — Page

| | |
|---|---|
| Notice of proposed rulemakings (section 6(b) of the Act), *1990.142(a)* | 758 |
| Issues to be considered in the rulemaking, *1990.146* | 760 |
| Notice of proposed rulemakings, *1990.142(a)* | 758 |
| Scientific review panel, *1990.104* | 755 |
| General, *1990.104(a)* | 755 |
| Membership, *1990.104(b)* | 755 |
| Other aid and assistance, *1990.104(d)* | 755 |
| Report, *1990.104(c)* | 755 |
| Scope of Part 1990, *1990.101* | 755 |
| Substantial new evidence, *1990.145(b)* | 759 |
| Petitions for consideration, *1990.145(c)* | 760 |
| Secretary's response, *1990.145(d)* | 760 |
| Substantial new issues, *1990.145(a)* | 759 |
| Secretary's response, *1990.145(d)* | 760 |
| Carpenters' Bracket Scaffolds, *.28(k)* | 33 |
| Bracket Attachment, *.28(k)(2)* | 33 |
| Bracket Dimensions, *.28(k)(1)* | 33 |
| Employee Protection, *.28(k)(3)* | 33 |
| Guardrails, *.28(k)(5)* | 33 |
| Platform Size, *.28(k)(4)* | 33 |
| Caustics, *.262(oo)* | 358 |
| Emergency Showers, *.261(g)(18)(i)* | 352 |
| Pipeline Identification, *.261(h)(3)(vi)* | 353 |
| Caution Signs and Labels, *.145(c)(2), (d)(4)* | 214 |
| Asbestos, *.1001(j)* | 465 |
| Fluorides, *.252(c)(1)(iv)* | 337 |
| Ionizing Radiation, *.96(e), (g), (h)* | 82 |
| Welding, *.252(c)(1)(iv)* | 337 |
| Ceiling Workers' Scaffolds: (see also *Plasterers' Scaffolds*), *.28(o)* | 34 |
| Chain Guarding, *.219(f)* | 327 |
| Change Rooms: | |
| Asbestos, *.1001(i)(1)* | 465 |
| Drying Facilities, *.141(f)* | 212 |
| Separate Facilities, *.141(e)* | 212 |
| Charge Initiation, *.109(e)(4)* | 115 |
| Chemical Plants: (see also *Refineries, Chemical Plants and Distilleries*), *.106(i)* | 106 |
| Chemicals, hazard communication, *.1200* | 710 |
| Chemicals, hazardous; occupational exposure in laboratories: | |
| Chemical Hygiene Plan, *.1450(e)* | 724 |
| Exposure determination, monitoring, *.1450(d)* | 724 |
| Hazard identification, *.1450(h)* | 725 |
| Hygiene recommendations, *.1450, App. A* | 725 |
| Medical consultations and examinations, *.1450(g)* | 725 |
| Permissible exposure limits, *.1450(c)* | 724 |
| Recordkeeping, *.1450(j)* | 725 |
| Respirator use, *.1450(i)* | 725 |
| Training, *.1450(f)* | 724 |
| Chemicals, highly hazardous; process safety management, *.119* | 148 |
| Chemicals, toxic and reactive, threshold list, *.119, App. A* | 152 |
| Compliance guidelines, *.119, App. A* | 152 |
| Contractor, employer responsibilities, *.119(h)* | 150 |
| Emergency planning and response, *.119(h)* | 150 |
| Hot work (welding) permits, *.119(n)* | 151 |
| Trade secrets, *.119(p)* | 151 |
| Training, *.119(g)* | 150 |
| Chicago Boom Derricks: (see also *Derricks*), *.181* | 273 |
| Chicken Ladders: (see also *Crawling Boards*), *.28(t)* | 35 |
| bis-Chloromethyl ether, *.1003* | 488 |
| Area requirements, *.1003(c)* | 488 |
| Closed system operation, *.1003(c)(2)* | 488 |
| Isolated systems, *.1003(c)(1)* | 488 |
| Maintenance and decontamination activities, *.1003(c)(5)* | 489 |
| Open-vessel system operations, *.1003(c)(3)* | 488 |
| Transfer from a closed operation, *.1003(c)(4)* | 488 |
| Medical surveillance, *.1003(g)* | 490 |
| Examinations, *.1003(g)(1)* | 490 |
| Records, *.1003(g)(2)* | 490 |
| Regulated area requirements, *.1003(d)* | 489 |
| Contamination control, *.1003(d)(4)* | 489 |

**Subject term,** *Section No.* — Page

| | |
|---|---|
| Emergencies, *.1003(d)(2)* | 489 |
| Hygiene facilities and practices, *.1003(d)(3)* | 489 |
| Signs, information, and training, *.1003(e)* | 489 |
| Container contents identification, *.1003(e)(2)* | 489 |
| Lettering, *.1003(e)(3)* | 489 |
| Prohibited statements, *.1003(e)(4)* | 489 |
| Signs, *.1003(e)(1)* | 489 |
| Training and indoctrination, *.1003(e)(5)* | 489 |
| Chromium (VI) | 522 |
| Communication of chromium (VI) hazard to employees, *1026(i)* | 525 |
| Dates, *1026(n)* | 525 |
| Exposure determination, *.1026(d)* | 523 |
| Housekeeping, *1026(h)* | 524 |
| Hygiene areas and practices, *1026(i)* | 524 |
| Medical surveillance, *1026(k)* | 524 |
| Methods of compliance, *1026(f)* | 523 |
| Permissible exposure limits, *1026(c)* | 523 |
| Protective work clothing and equipment, *1026(h)* | 523 |
| Record keeping, *1026(m)* | 525 |
| Regulated areas, *1026(e)* | 523 |
| Respiratory protection, *1026(g)* | 523 |
| Chute Openings, *.23(a)(2)* | 20 |
| Circular Resaws, *.213(e)* | 298 |
| Circular Saws, *.213(f)* | 298 |
| Arbors, *.213(s)(4)* | 300 |
| Portable, *.243(a)(1)* | 330 |
| Clean Air, Spray Finishing, *.94(c)(7)* | 73 |
| Cleaning: | |
| Air Receivers, *.169* | 253 |
| Bulk Oxygen Systems, *.104(b)(8)(i)* | 89 |
| Compressed Air, *.242(b)* | 330 |
| Powder Coatings, *.107(l)(4)(i)* | 110 |
| Powered Platforms, *.66(e)(7)* | 43 |
| Respirators, *.134(f)(3)* | 191 |
| Solvents, *.107(g)(5)* | 109 |
| Spray Booths, *.107(b)(9)* | 107 |
| Spraying Operations, *.107(g)(2)* | 109 |
| Cleaning Compounds, *.252(c)(11)* | 338 |
| Degreasing, *.252(c)(11)(ii)* | 338 |
| Manufacturer's Instructions, *.252(c)(11)(i)* | 338 |
| Cleaning Solvents: | |
| Spraying, *.107(g)(5)* | 109 |
| Clear Zones: | |
| Bulk Oxygen Systems, *.104(b)(10)* | 89 |
| Industrial Plants, *.106(e)(9)(iv)* | 100 |
| Liquefied Hydrogen Systems, *.103(c)(5)* | 88 |
| Processing Plants, *.106(h)(8)(iv)* | 106 |
| Clearances: | |
| Cranes, *.179(b)(6)* | 265 |
| Derricks, *.181(j)(5)* | 276 |
| Fixed Ladders, *.27(c)* | 25 |
| Back, *.27(c)(4)* | 25 |
| Climbing Side, *.27(c)(1)* | 25 |
| Grab Bars, *.27(c)(5)* | 25 |
| Hatch Covers, *.27(c)(7)* | 26 |
| Step-Across Distance, *.27(c)(6)* | 25 |
| With Cages or Baskets, *.27(c)(3)* | 25 |
| Without Cages or Wells, *.27(c)(2)* | 25 |
| Manlifts, *.68(b)(11)* | 60 |
| Spraying Discharges, *.107(d)(8)* | 108 |
| Stairs, *.24(i)* | 22 |
| Clothing, Protective: (see also *Personal Protective Equipment*), *.132, .252(b)(3)* | 187, 336 |
| Asbestos, *.1001(d)(3)* | 463 |
| Body, *.156(e)(3)* | 235 |
| Electrical, *.137* | 204 |
| Eye, *.133, .156(e)(5)* | 187, 235 |
| Face, *.133, .156(e)(5)* | 187, 235 |
| Fire brigade, *.156* | 234 |
| Footwear, *.136, .156(e)(2)* | 204, 235 |
| Goggles, *.133* | 187 |
| Hand, *.156(e)(4)* | 235 |
| Head, *.135, .156(e)(5)* | 203, 235 |
| Helmets, *.135* | 203 |
| Leg, *.156(e)(3)* | 235 |

## Subject Index For 29 CFR 1910, Sec. 5, 1903, 1904, and 1990

| Subject term, Section No. | Page |
|---|---|
| Rubber, .137 | 204 |
| Storage, .107(g)(4) | 109 |
| Welders, .252(b)(3) | 336 |
| Clutches, .217(b)(3), (7) | 309, 310 |
| Definition, .180(a)(19) | 269 |
| Power Transmission Apparatus, .219(k) | 327 |
| Coatings, Spray: | |
| Dual Component, .107(m) | 110 |
| Organic Peroxide, .107(m) | 110 |
| Powder, .107(l) | 110 |
| Undercoatings, .107(k) | 110 |
| Coke oven emissions, .1029 | 591 |
| Employee information and training, .1029(k) | 596 |
| Exposure monitoring and measurement, .1029(e) | 592 |
| Hygiene facilities and practices, .1029(i) | 595 |
| Medical surveillance, .1029(j) | 595 |
| Methods of compliance, .1029(f) | 592 |
| Observation of monitoring, .1029(n) | 597 |
| Permissible exposure limit, .1029(c) | 592 |
| Protective clothing and equipment, .1029(h) | 595 |
| Precautionary signs and labels, .1029(l) | 596 |
| Recordkeeping, .1029(m) | 596 |
| Regulated areas, .1029(d) | 592 |
| Respiratory protection, .1029(g) | 594 |
| Collars, .219(i) | 327 |
| Color Codes: | |
| Danger, .144(a)(1)(ii), .145(d)(2) | 213, 214 |
| Identification of filters, cartridges, and canisters, .134(j) | 193 |
| Physical Hazards, Colors, .144(a), .145(d)(2) | 213, 214 |
| Standards Sources, .150 | 230 |
| Stop, .144(a)(1)(iii) | 214 |
| Combustible Dust Atmospheres, | |
| Use of Trucks, .178(c)(2)(vi), (vii) | 259 |
| Combustible Liquids: | |
| (see Flammable and Combustible Liquids), .106 | 89 |
| Effective Dates, .114 | 148 |
| Standards Sources, .115 | 148 |
| Combustible Materials: | |
| Welding, .252(a)(2) | 335 |
| Communicable Diseases Reporting: | |
| Labor Camps, .142(l) | 213 |
| Communications, Powered Platforms, .66(e)(11)(vi) | 43 |
| Compressed Air, Cleaning, .242(b) | 330 |
| Compressed Air Equipment: | |
| (see also Compressed Gas Equipment) | |
| Air Receivers, .169 | 253 |
| Compressed Gas Cylinders: | |
| (see also Compressed Gases), .253(a)(2) | 339 |
| Approval, .252(b)(1) | 336 |
| Inspection, .101(a) | 83 |
| Manifolding, .253(c) | 340 |
| Markings, .253(b)(1) | 339 |
| Operating Procedures, .253(b)(5) | 339 |
| Oxygen Manifolds, .253(c)(2), (3) | 340 |
| Public Protection, .252(d)(1)(iv) | 338 |
| Safety Relief Valves, .101(c) | 83 |
| Standards Sources, .170 | 253 |
| Storage, .253(b)(2)-(4) | 339 |
| Compressed Gas Equipment: | |
| (see Compressed Gas Cylinders) | |
| Compressed Gases: | |
| (see also Compressed Gas Cylinders), .101 | 83 |
| Effective Dates, .114 | 148 |
| Handling, .101(b) | 83 |
| Safety Relief Devices, .101(c) | 83 |
| Standards Sources, .115 | 148 |
| Storage, .101(b) | 83 |
| Conductors: | |
| (see also Electric Wiring) | |
| Cranes, .179(g)(1)(iv), (6) | 266, 267 |
| General Wiring, .305(f) | 425 |
| Confined Spaces: | |
| Atmospheric testing flow charts, .146, App. B | 221 |
| Attendant duties, .146(i) | 219 |
| Authorized entrant duties, .146(h) | 219 |
| Entry supervisor duties, .146(j) | 220 |
| Flow chart, decisions, .146, App. A | 221 |
| Instruction of employees relating to, .21(b)(6) | 17 |

| Subject term, Section No. | Page |
|---|---|
| Permit samples, .146, App. D | 224 |
| Program examples, .146, App. C | 221 |
| Rescue and emergency services, .146(k) | 220 |
| Sewer system entry, .146, App. E | 225 |
| Training, .146(g) | 219 |
| Confined Spaces, Hazardous work, .120(b)(4)(ii)(I), (c)(3) | 162 |
| Electrical safety-related work practices, .333(c)(5) | 434 |
| Containers: | |
| (see also Tank Storage, Portable) | |
| Ammonium Nitrate, .109(i)(3) | 119 |
| Bulk Oxygen Systems, .104(b)(4), (6) | 88, 89 |
| Gaseous, .104(b)(4)(iii) | 89 |
| Liquid, .104(b)(4)(ii) | 89 |
| Flammable and Combustible Liquids, .106(d) | 96 |
| Bulk plants, storage, .106(f) | 100 |
| Design, .106(d)(2) | 96 |
| Industrial plants, storage, .106(e)(2)(ii) | 98 |
| Service stations, storage, .106(g)(1) | 101 |
| Processing plants, .106(h)(4) | 105 |
| Gaseous Hydrogen Systems, .103(b)(1)(i) | 83 |
| Guarding, .212(a)(4) | 297 |
| Liquefied Hydrogen Systems, .103(c)(1)(i) | 85 |
| Liquefied Petroleum Gases, .110 | 121 |
| Spraying, .107(e)(3), (5) | 108 |
| Welding, Gas, .253(a), (b) | 339 |
| Containers, Liquefied Petroleum Gases, .110 | 121 |
| Accessories, .110(b)(7); (c)(6); (d)(3), (8); (e)(5) | 122, 132, 133, 135, 136 |
| Awaiting Use or Resale, .110(f) | 137 |
| Capacity, .110(d)(6), (h)(5) | 134, 139 |
| Charging Plants, .110(d)(13) | 135 |
| Construction, .110(b)(3) | 121 |
| Cylinder Systems, .110(c) | 131 |
| Accessories, .110(c)(6) | 132 |
| Indoor, .110(c)(5) | 131 |
| Markings, .110(c)(2) | 131 |
| Outdoor, .110(c)(4) | 131 |
| Valves, .110(c)(6) | 132 |
| Filling Densities, .110(b)(12) | 127 |
| Fire Protection, .110(d)(14) | 135 |
| Fittings, .110(b)(8); (e)(6); (h)(7), (9) | 123, 136, 139 |
| Hoses, .110(b)(9) | 124 |
| Industrial Plants, .110(d)(12) | 135 |
| Installation, .110(e)(4), (h)(6) | 136, 139 |
| Lighting, .110(d)(16) | 135 |
| Location, .110(b)(6), (f)(5) | 123, 138 |
| Markings, .110(b)(5), (c)(2) | 122, 131 |
| Non-DOT Containers, .110(d) | 133 |
| Accessories, .110(d)(3) | 133 |
| Capacity, .110(d)(6) | 134 |
| Installation, .110(d)(7) | 134 |
| Pipes, .110(d)(3) | 133 |
| Pressure, Design, .110(d)(2) | 133 |
| Reinstallation, .110(d)(5) | 134 |
| Safety Relief Devices, .110(d)(4) | 133 |
| Valves, .110(d)(3) | 133 |
| Original Testing, .110(b)(4) | 121 |
| Piping, .110(b)(8), (d)(3), (e)(6), (h)(9), (h)(10) | 123, 133, 136, 139, 140 |
| Pressure Design, .110(d)(2), (e)(3) | 133, 136 |
| Safety Relief Devices, .110(b)(10), (c)(7), (d)(4), (e)(7), (h)(4) | 124, 132, 133, 137, 138 |
| Tubing, .110(b)(8), (e)(6) | 123, 136 |
| Valves, .110(b)(7), (c)(6), (d)(3), (e)(5), (h)(9) | 122, 132, 133, 136, 139 |
| Vaporizers, .110(b)(11), (d)(17), (e)(8) | 126, 135, 137 |
| Welding, .110(b)(4) | 121 |
| Controllers: | |
| Cranes, .179(g)(3) | 266 |
| Conveyors: | |
| Bakeries, .263(d)(7), (i)(7) | 359, 360 |

# OSHA Standards for the General Industry

| Subject term, Section No. | Page |
|---|---|
| Electrostatic Spraying, .107(h)(7) | 109 |
| Forging Machines, .218(j)(3) | 325 |
| Sawmills, .265(c)(18) | 364 |
| Spray Booths, .107(b)(7) | 107 |
| Corrosion Protection: | |
| Piping, Valves and Fittings, .106(c)(5) | 96 |
| Storage Tanks, .106(b)(1)(vi) | 91 |
| Underground Tanks, .106(b)(3)(iii) | 93 |
| Cotton Dust, .1043 | 597 |
| Counterbalances, .217(b)(9) | 310 |
| Counterweights: | |
| Cranes, .180(i)(2) | 272 |
| Covers, Openings: | |
| Working Surfaces, .22(c); 23(a)(1), (3)(i), (5), (6), (8)(ii), (9) | 20 |
| Cranes: | |
| Crawler, .180 | 269 |
| Definitions, .179(a) | 264 |
| Effective Dates, .179(b)(2), .180(b)(2), .182 | 265, 270, 276 |
| Electric, .306(b) | 427 |
| Gantry, .179 | 264 |
| Locomotive, .180 | 269 |
| Overhead, .179 | 264 |
| Pulp and Paper Mills, .261(c)(8) | 349 |
| Standards Sources, .183 | 276 |
| Truck, .180 | 269 |
| Crawler Cranes: | |
| (see also Crawler, Locomotive and Truck Cranes), .180 | 269 |
| Crawler, Locomotive and Truck Cranes, .180 | 269 |
| Cabs, .180(i)(3) | 272 |
| Electric Power Lines, .180(j) | 272 |
| Fire Extinguishers, .180(i)(5) | 272 |
| Inspections, .180(d) | 270 |
| Frequent, .180(d)(3) | 271 |
| Idle (Irregular), .180(d)(5) | 271 |
| Initial, .180(d)(1) | 270 |
| Periodic, .180(d)(4) | 271 |
| Records, .180(d)(6) | 271 |
| Ropes, .180(g) | 271 |
| Load Handling, .180(h) | 272 |
| Load Ratings, .180(c) | 270 |
| Maintenance, .180(f) | 271 |
| Refueling, .180(i)(4) | 272 |
| Rope Inspection, .180(g) | 271 |
| Standards Sources, .183 | 276 |
| Swinging Locomotives, .180(i)(6) | 272 |
| Testing, .180(e) | 271 |
| Crawling Boards, .28(t) | 35 |
| Crosscut Table Saws, .213(d) | 298 |
| Cup Wheels, .243(c)(2) | 330 |
| Flaring-Cup, Type 11, .241(b)(8) | 329 |
| Straight-Cup, Type 6, .241(b)(9) | 329 |
| Straight, Type 1, .241(b)(10) | 330 |
| Curing Apparatus: | |
| (see Drying, Curing and Fusion Apparatus) | |
| Cutting: | |
| (see also Welding), .252 | 335 |
| Containers, .252(a)(3) | 336 |
| Definitions, .251 | 335 |
| Ventilation, .252(c) | 337 |
| Cutting-Off Machines, .215(b)(5) | 301 |
| Cutoff Couplings, .219(k)(1) | 327 |
| Cutoff Saws, Swing, .213(g) | 298 |
| Cylinders, Welding Gas, .253(b) | 339 |
| Manifolding, .253(c) | 340 |
| Operating Procedures, .253(b)(5) | 339 |
| Storage, .253(b)(2)-(4) | 339 |
| Cylindrical Grinders, .215(b)(4) | 301 |
| Danger: | |
| Color Codes, .144(a)(1)(ii) | 213 |
| Signs, .145(c)(1), (d)(2) | 214 |
| Tag, .145(f)(5) | 215 |
| Dates, Effective: | |
| (see Effective Dates) | |
| DBCP (1,2-Dibromo 3-Chloropropane), .1044 | 624 |
| Emergency situations, .1044(i) | 625 |
| Employee information and training, .1044(n) | 626 |
| Exposure monitoring, .1044(f) | 624 |

| Subject term, Section No. | Page |
|---|---|
| Housekeeping, .1044(k) | 625 |
| Hygiene facilities and practices, .1044(l) | 625 |
| Medical surveillance, .1044(m) | 626 |
| Methods of compliance, .1044(g) | 624 |
| Notification of use, .1044(d) | 624 |
| Observation of monitoring, .1044(q) | 627 |
| Permissible exposure limit, .1044(c) | 624 |
| Protective clothing and equipment, .1044(j) | 625 |
| Recordkeeping, .1044(p) | 626 |
| Regulated areas, .1044(e) | 624 |
| Respirators, .1044(h) | 624 |
| Signs and labels, .1044(o) | 626 |
| Dead-Man Controls, .243(a)(2) | 330 |
| Decorators' Scaffolds: | |
| (see also Plasterers' Scaffolds), .28(o) | 34 |
| Degreasing: | |
| Cleaning Compounds, .252(c)(11)(ii) | 338 |
| Derricks: | |
| Adjustments, .181(f)(3) | 275 |
| Cabs, .181(j)(6) | 276 |
| Electric Power Lines, .181(j)(5) | 276 |
| Fire Extinguishers, .181(j)(3) | 276 |
| Guards, .181(j)(1) | 276 |
| Hooks, .181(j)(2) | 276 |
| Inspections, .181(d), (g) | 275 |
| Load Handling, .181(i) | 276 |
| Load Ratings, .181(c) | 274 |
| Maintenance, .181(f) | 275 |
| Operations, .181(h) | 275 |
| Refueling, .181(j)(4) | 276 |
| Repairs, .181(f)(3) | 275 |
| Rope Inspections, .181(g) | 275 |
| Standards Sources, .183 | 276 |
| Testing, .181(e) | 275 |
| 3,3'-Dichlorobenzidine (and its salts), .1003 | 488 |
| Area requirements, .1003(c) | 488 |
| Closed system operation, .1003(c)(2) | 488 |
| Isolated systems, .1003(c)(1) | 488 |
| Maintenance and decontamination activities, .1003(c)(5) | 489 |
| Open-vessel system operations, .1003(c)(3) | 488 |
| Transfer from a closed operation, .1003(c)(4) | 488 |
| Medical surveillance, .1003(g) | 490 |
| Examinations, .1003(g)(1) | 490 |
| Records, .1003(g)(2) | 490 |
| Regulated area requirements, .1003(d) | 489 |
| Contamination control, .1003(d)(4) | 489 |
| Emergencies, .1003(d)(2) | 489 |
| Hygiene facilities and practices, .1003(d)(3) | 489 |
| Signs, information, and training, .1003(e) | 489 |
| Container contents identification, .1003(e)(2) | 489 |
| Lettering, .1003(e)(3) | 489 |
| Prohibited statements, .1003(e)(4) | 489 |
| Signs, .1003(e)(1) | 489 |
| Training and Indoctrination, .1003(e)(5) | 489 |
| Dies, .217(d) | 312 |
| Changing, .218(h)(5) | 325 |
| Fastening, .217(d)(7) | 312 |
| Guide Post Hazards, .217(d)(4) | 312 |
| Handling, .217(d)(3), (8) | 312, 313 |
| Requirements, .217(d)(1) | 312 |
| Scrap: | |
| Ejecting, .217(d)(2) | 312 |
| Handling, .217(d)(3) | 312 |
| Selling, .217(d)(9) | 313 |
| Stroke, .217(d)(6) | 312 |
| Tonnage, .217(d)(6) | 312 |
| Unitized Tooling, .217(d)(5) | 312 |
| Weight, .217(d)(6) | 312 |
| Diesel Powered Trucks, .178(b)(1)-(3) | 257 |
| Dikes: | |
| Bulk Oxygen Systems, .104(b)(2)(v) | 88 |
| Storage Tanks, .106(b)(2)(vii), (c) | 93, 95 |
| 4-Dimethylaminoazobenzene, .1003 | 488 |
| Area Requirements, .1003(c) | 488 |
| Closed System Operation, .1003(c)(2) | 488 |

# Subject Index For 29 CFR 1910, Sec. 5, 1903, 1904, and 1990

| Subject term, Section No. | Page |
|---|---|
| Isolated Systems, .1003(c)(1) | 488 |
| Maintenance and decontamination activities, .1003(c)(5) | 489 |
| Open-vessel System Operations, .1003(c)(3) | 488 |
| Transfer from a closed operation, .1003(c)(4) | 488 |
| Medical surveillance, .1003(g) | 490 |
| Examinations, .1003(g)(1) | 490 |
| Records, .1003(g)(2) | 490 |
| Regulated Area Requirements, .1003(d) | 489 |
| Contamination Control, .1003(d)(4) | 489 |
| Emergencies, .1003(d)(2) | 489 |
| Hygiene facilities and practices, .1003(d)(3) | 489 |
| Signs, Information, and Training, .1003(e) | 489 |
| Container Contents Identification, .1003(e)(2) | 489 |
| Lettering, .1003(e)(3) | 489 |
| Prohibited Statements, .1003(e)(4) | 489 |
| Signs, .1003(e)(1) | 489 |
| Training and Indoctrination, .1003(e)(5) | 489 |
| **Dining Facilities:** | |
| *(see also Lunchrooms)* | |
| Labor Camps, .142(i) | 213 |
| **Dip Tanks:, .123-.126** | 184-186 |
| Application, .123(a) | 184 |
| Bottom Drains, .125(c) | 185 |
| Construction, .124(a), .125(a) | 184, 185 |
| Conveyors, .125(d), .126(g)(2) | 185, 186 |
| Covers, .125(f)(3) | 185 |
| Electrical Ignition Sources, .125(e)(1) | 185 |
| Electrostatic Apparatus, .126(g) | 186 |
| Fire Extinguishers, .125(f)(2)(i) | 185 |
| Fire Protection, .125(f) | 185 |
| Flow Coating, .126(b) | 186 |
| Hardening, .126(a)(1)(i),(ii) | 186 |
| Heating, .125(g) | 185 |
| Ignition Sources, .125(e) | 185 |
| Inspections, .124(j)(1),(3) | 185 |
| Liquid Storage, .125(e)(2) | 185 |
| Maintenance, .125(e)(4) | 185 |
| Overflow Pipes, .125(b) | 185 |
| Sprinklers, .125(f) | 185 |
| Tempering, .126(a) | 186 |
| Ventilation, .124(b), .125(d)(2) | 184, 185 |
| Warning Signs, .125(e)(2) | 185 |
| Waste Cans, .125(e)(4)(ii),(iii) | 185 |
| **Disposal Systems:** | |
| *(see Waste Disposal Systems)* | |
| **Distances From Hazards:** | |
| Ammonium Nitrate, .109(i)(5) | 120 |
| Bulk Oxygen Systems, .104(b)(3) | 88 |
| Electrostatic Spraying, .107(h)(6) | 109 |
| Explosives Storage, .109(c) | 111 |
| Ignition Sources, Separation, .107(c)(2) | 107 |
| Spray Booths, Separations, .107(b)(8) | 107 |
| **Distilleries:** | |
| *(see also Refineries, Chemical Plants and Distilleries), .106(i)* | 106 |
| **Distribution Plates:** | |
| Spray Booths, .107(b)(4) | 107 |
| Dividers, Bakery Equipment, .263(f) | 359 |
| **Diving, Commercial, .401 – 441** | 441, 442-445, 446 |
| Air-Compressor System, .430(b) | 445 |
| Breathing hoses, .430(c) | 445 |
| Communications, During Dive, .422(c) | 443 |
| Decompression Chambers, .430(f) | 445 |
| Decompression Tables, .422(d) | 443 |
| Definitions, .402 | 441 |
| Designated Person in Charge, .410(c) | 442 |
| Dive Profile, .422(e) | 443 |
| Effective Dates, .441 | 445 |
| Emergencies, .401(b) | 441 |
| Employee Briefings, .421(f) | 443 |
| Equipment Inspection, .421(g) | 443 |
| Explosives, .422(h) | 443 |
| First-Aid Supplies, .421(c) | 442 |
| Gauges and Timekeeping Devices, .430(g) | 445 |
| Hand-Held Power Tools and Equipment, .422(f) | 443 |
| Hazardous Activities, .421(e) | 442 |
| Liveboating, .427 | 445 |
| Masks and Helmets, .430(h) | 446 |
| Mixed-Gas Diving, .426 | 444 |
| Oxygen Safety, .430(i) | 446 |
| Qualifications of Dive Team, .410 | 442 |
| Planning and assessment, .421(d) | 442 |
| Pre-dive Procedures, .421 | 442 |
| Post-Dive Procedures, .423 | 443 |
| Procedures During Dive, .422 | 443 |
| Recompression, .423(c) | 443 |
| Recordkeeping, .423(d), .440 | 444, 446 |
| Safe Practices Manual, .420 | 442 |
| SCUBA Diving, .424 | 444 |
| Scope and application, .401 | 441 |
| Surface-Supplied Air Diving, .425 | 444 |
| Termination of Dive, .422(i) | 443 |
| Warning Signals, .421(h) | 443 |
| Water Entry and Exit, .422(b) | 443 |
| Welding and Burning, .422(g) | 443 |
| Scientific, .401(a)(2)(iv), 402, Appendix B | 441 |
| Dockboards, .30(a) | 36 |
| Dough Brakes, Manually Fed, .263(h) | 359 |
| Drag Saws, .213(r) | 300 |
| **Drainage:** | |
| Bulk Plants, .106(f)(7) | 101 |
| Industrial Plants, .106(e)(3)(iii) | 99 |
| Labor Camps, .142(a) | 212 |
| Processing Plants, .106(h)(3)(ii) | 105 |
| Service Stations, .106(g)(7) | 104 |
| Sprinkler Systems, .159(c)(7) | 238 |
| Storage Tanks, .106(b)(2)(vii)(c) | 93 |
| **Drains:** | |
| Air Receivers, .169(b)(2) | 253 |
| Dressing Rooms, Personnel, .141(e) | 212 |
| Drips, Condensed Gas, .110(d)(11) | 135 |
| Drives—Belt, Rope and Chain, .219(e), (g), (o)(3) | 326, 327 |
| Belt Tighteners, .219(e)(6) | 327 |
| Cone-Pulley Belts, .219(e)(5) | 326 |
| Horizontal Belts and Ropes, .219(e)(1)(i) | 326 |
| Inclined Belts, .219(e)(3) | 326 |
| Overhead Horizontal Belts, .219(e)(2) | 326 |
| Vertical Belts, .219(e)(3), (4) | 332 |
| Drums, .212(a)(4) | 297 |
| Dry Chemical Extinguishing Systems, Fixed, .161 | 239 |
| Scope and Application, .161(a) | 239 |
| Specific Requirements, .161(b) | 239 |
| **Drying:** | |
| Spraying Operations, .107(d)(12) | 108 |
| Drying, Curing, and Fusion Apparatus, .107(j) | 110 |
| Adjacent System, .107(j)(3) | 110 |
| Alternate Use: | |
| Permitted, .107(j)(4) | 110 |
| Prohibited, .107(j)(2) | 110 |
| Conformance, .107(j)(1) | 110 |
| Powder Coatings, .107(l)(3) | 110 |
| Spraying Rooms, .107(j)(2) | 110 |
| Dual Component Coatings, .107(m) | 110 |
| **Dust Hazards:** | |
| Abrasive Blasting, .94(a)(2) | 63 |
| Asbestos, .19(a), .1001 | 12, 462 |
| Employee Exposure, .1000(a) | 453 |
| Grain handling facilities, .272 | 412 |
| **Effective Dates:** | |
| Accident Prevention Signs and Tags, .149 | 230 |
| Air Contaminants, .98 | 82 |
| Asbestos, .98, .1001(o) | 82, 468 |
| Color Codes, .149 | 230 |
| Crawler, Locomotive, and Truck Cranes, .180(b)(2), .182 | 270, 276 |
| Derricks, .181(b)(2), .182 | 274, 276 |
| Dip Tanks, .114 | 148 |
| Environmental Controls, .98, .149 | 82, 230 |
| Federal Standards, .17 | 12 |
| Indoor Storage, .182 | 276 |
| Ionizing Radiation, .98 | 82 |
| Labor Camps, .142(d)(7), | 213 |

# OSHA Standards for the General Industry

| Subject term, Section No. | Page |
|---|---|
| Liquefied Petroleum Gases, *.110(b)(19)(i), (i)(3);, .11* | 130, 140 |
| Materials Handling, *.182* | 276 |
| Mechanical Power Presses, *.220* | 328 |
| Mechanical Power Transmission Apparatus, *.221* | 328 |
| Noise Exposure, *.98* | 82 |
| Nonionizing Radiation, *.98* | 82 |
| Nonwater Disposal Systems, *.149* | 230 |
| Occupational Health, *.98* | 82 |
| Overhead and Gantry Cranes, *.179(b)(2), .182* | 265, 276 |
| Physical Hazards Markings, *.149* | 230 |
| Pulpwood Logging, *.266(j)* | 375 |
| Safety Color Codes, *.149* | 230 |
| Sanitation, *.149* | 230 |
| Ventilation, *.98* | 82 |
| Egress Means: | |
| (see also Exit Routes) | |
| Electric controls, Mechanical Power Presses, *.217(b)(8)* | 310 |
| Electric energy, hazardous; control of | |
| (see Lockout/tagout) | |
| Electric Equipment: | |
| (see Electrical Wiring) | |
| Electric Ignition Sources: | |
| (see Ignition Sources), *.107(c), (d)(5)* | 107, 108 |
| Electric Motor Ignition Sources, *.107(d)(5)* | 108 |
| Electric power generation, transmission, and distribution, *.269* | 384 |
|   Definitions, *.269(x)* | 401 |
|   Enclosed spaces, *.269(e)* | 388 |
|   Fall protection, *.269(g)(1)* | 389 |
|   Hazardous energy control (Lockout/tagout), *.269(d)* | 386 |
|   Ladders, platforms, steps, etc., *.269(h)* | 389 |
|   Live-line tools, *.269(j)* | 390 |
|   Materials handling and storage, *.269(k)* | 390 |
|   Mechanical equipment, *.269(p)* | 395 |
|   Medical services and first aid, *.269(b)* | 385 |
|   Overhead lines, *.269(q)* | 395 |
|   Personal protective equipment, *.269(g), (n)(4), (r)(2)(v), (r)(4)(ii)* | 389, 394, 397 |
|   Power tools, hand and portable, *.269(i)* | 389 |
|   Telecommunications facilities, *.269(s)* | 398 |
|   Testing and test facilities, *.269(o)* | 394 |
|   Training, *.269(a)(2), (b)(1), (d)(2), (e)(2), (q)(3)(i), (r)(1)(vi)* | 385, 386, 388, 396, 397 |
|   Tree trimming, line-clearance, *.269 (a)(1)(i)(E), (r)* | 385, 397 |
|   Water, work near, *.269(w)(5)* | 401 |
| Electric Power Lines: | |
|   Crane Operations, *.180(j), .333(c)(3)* | 272, 433 |
|   Derrick Operations, *.181(j)(5), .268(j)(4), .333(c)(3)* | 276, 379, 433 |
| Electric Powered Trucks, *.120(g)(3), .178(b)(4)-(7)* | 164, 257 |
| Electric wiring: | |
|   Ammonium Nitrate, *.109(i)(6)* | 120 |
|   Appliances, *.305(j)(3)* | 426 |
|   Approval, *.303(a)* | 420 |
|   Arcing Parts, *.303(d)* | 420 |
|   Attachment Plugs (Caps), *.305(j)(2)* | 425 |
|   Boxes, *.305(b)* | 424 |
|   Branch Circuits, *.304(b)* | 421 |
|   Bulk Oxygen Systems, *.104(b)(8)(ix)* | 89 |
|   Bulk Plants, *.106(f)(5)* | 101 |
|   Cabinets, *.305(b)* | 424 |
|   Capacitors, *.305(j)(6)* | 426 |
|   Communications Systems, *.308(e)* | 430 |
|   Conductors, *.305(f)* | 425 |
|   Cranes, *.179(g), .306(b)* | 266, 427 |
|   Data Processing Systems, *.306(e)* | 427 |
|   Electrolytic Cells, *.306(h)* | 427 |
|   Elevators, *.306(c)* | 427 |
|   Emergency Systems, *.308(b)* | 430 |
|   Enclosures for Damp or Wet Locations, *.305(e)* | 425 |
|   Escalators, *.306(c)* | 427 |
|   Examination of Equipment, *.303(b)(1)* | 420 |
|   Fire Protective Signaling Circuits, *.308(d)* | 430 |
|   Fittings, *.305(b)* | 424 |
|   Fixture Wires, *.305(i)* | 425 |
|   Flexible Cords and Cables, *.305(g)* | 425 |
|   General requirements, *.303* | 420 |
|   Grounded and Grounding Conductors, Installation and Use, *.304(a)* | 421 |
|   Grounding, *.304(f)* | 422 |
|   Guarding Live Parts, *.303(g)(2), (h)(2)* | 420 |
|   Hand Spraying, *.107(i)(5)* | 109 |
|   Hazardous (Classified) Locations, *.307* | 429 |
|   Heating Equipment, *.306(g)* | 427 |
|   High Voltage (Over 600 Volts): | |
|     General, *.308(a)* | 429 |
|     Grounding, *.304(f)(7)* | 423 |
|     Guarding, *.303(h)(2)* | 420 |
|     Workspace, *.303(h)(3), (h)(4)* | 421 |
|   Hoists, *.306(b)* | 427 |
|   Identification of Disconnecting Means and Circuits, *.303(f)* | 420 |
|   Ignition Sources, *.107(c)(4), (6)* | 107 |
|   Industrial Plants, *.106(e)(7)* | 99 |
|   Installation and Use of Equipment, *.303(b)(2)* | 420 |
|   Irrigation Machines, *.306(i)* | 428 |
|   Lamps, *.305(j)(1)* | 425 |
|   Liquefied Hydrogen Systems, *.103(c)(1)(ix)* | 86 |
|   Liquefied Petroleum Systems, *.110(b)(17), (18), (h)(13)* | 128, 130, 140 |
|   Marking, *.303(e)* | 420 |
|   Motors, *.305(j)(4)* | 426 |
|   Moving Walks, *.306(c)* | 427 |
|   Outline Lighting, *.306(a)* | 426 |
|   Outside Conductors, *.304(c)* | 421 |
|   Overcurrent Protection, *.304(e)* | 422 |
|   Panelboards, *.305(d)* | 425 |
|   Portable Cables, *.305(h)* | 425 |
|   Powder Coatings, *.107(l)(1)* | 110 |
|   Power-Limited Circuits, *.308(c)* | 430 |
|   Processing Plants, *.106(h)(7)(iii)* | 106 |
|   Receptacles, *.305(j)(2)* | 425 |
|   Remote Control Circuits, *.308(c)* | 430 |
|   Service Stations, *.106(g)(5)* | 104 |
|   Services, *.304(d)* | 422 |
|   Signaling Circuits, *.308(c)* | 430 |
|   Signs, *.306(a)* | 426 |
|   Splices, *.303(c)* | 420 |
|   Spraying Operations, *.107(c)(4), (6), (i)(1)-(5)* | 107, 109 |
|   Storage Batteries, *.305(j)(7)* | 426 |
|   Storage Rooms, *.106(d)(4)(iii)* | 97 |
|   Swimming Pools, *.306(j)* | 428 |
|   Switchboards, *.305(d)* | 425 |
|   Switches, *.305(c)* | 425 |
|   Transformers, *.305(j)(5)* | 426 |
|   Welders, *.306(d)* | 427 |
|   Wiring Design and Protection, *.304* | 421 |
|   Wiring Methods, *.305(a)* | 424 |
|   Work practices, safety-related, *.331-.335* | 431-435 |
|   Working Space about Electric Equipment, *.303(g)(1), (h)(3), (h)(4)* | 420, 421 |
|   X-Ray Equipment, *.306(f)* | 427 |
| Electrical Installations, *.301-.399* | 419-435 |
| Electrical Protective Equipment, *.137, .268(f)* | 204, 377 |
|   Care and use, in-service, *.137(b)* | 204 |
|   Design, *.137(a)* | 204 |
| Electrical safety-related work practices, *.331-.335* | 431-435 |
|   Confined spaces, *.333(c)(5)* | 434 |
|   Illumination, *.333(c)(4)* | 434 |
|   Ladders, portable, *.333(c)(7)* | 434 |
|   Lockout and tags, *.333(b), .335(b)(1)* | 432, 435 |
|   Personal protective equipment, *.333(c)(2), .335(a)* | 433, 435 |
|   Portable electric equipment, *.334(a)* | 434 |
|   Power lines, overhead, *.333(c)(3)* | 433 |
|   Training, *.332* | 431 |
| Electromagnetic Radiation: | |
|   Definitions, *.97(a)(1)* | 82 |
|   Nonionizing Radiation, *.97(a)* | 82 |
|   Protection Guide, *.97(a)(2)* | 82 |
|   Warning Symbol, *.97(a)(3)* | 82 |

Index Dim—Ele

# Subject Index For 29 CFR 1910, Sec. 5, 1903, 1904, and 1990

| Subject term, Section No. | Page |
|---|---|
| Electrostatic Apparatus: | |
| *(see also Electrostatic Apparatus, Fixed; Electrostatic Hand Spraying Equipment)* | |
| Powder Coatings, .107(l)(5)-(7) | 109, 110 |
| Electrostatic Apparatus, Fixed, .107(h) | 109 |
| Powder Coatings, .107(l)(5) | 109 |
| Spraying: | |
| Conformance, .107(h)(1) | 109 |
| Conveyors, .107(h)(7) | 109 |
| Fail-Safe Controls, .107(h)(9) | 109 |
| Fire Protection, .107(h)(12) | 109 |
| Grounding, .107(h)(5) | 109 |
| Guarding, .107(h)(10) | 109 |
| Insulators, .107(h)(5) | 109 |
| Location, .107(h)(3) | 109 |
| Safe Distances, .107(h)(6) | 109 |
| Supports, .107(h)(4) | 109 |
| Ventilation, .107(h)(11) | 109 |
| Electrostatic Hand Spraying Equipment, .107(i) | 109 |
| Application, .107(i)(1) | 109 |
| Approval, .107(i)(3) | 109 |
| Conformance, .107(i)(2) | 109 |
| Electrical Support Equipment, .107(i)(4) | 109 |
| Grounding, .107(i)(5)-(7) | 109, 110 |
| Interlocks, .107(i)(8) | 110 |
| Powder Coatings, .107(i)(6) | 110 |
| Specifications, .107(i)(3), (4) | 109 |
| Spray Gun Grounding, .107(i)(5) | 109 |
| Ventilation, .107(i)(9) | 110 |
| Elevating Work Platforms: | |
| *(see Vehicle-Mounted Work Platforms), .67* | 58 |
| Hazardous waste operations and emergency response, exemption, .120(l)(1)(ii) | 166 |
| Highly hazardous chemicals, process safety management, .119(n) | 151 |
| Emergency action plans | |
| Alarm system, .38(d) | 39 |
| Application, .38(a) | 38 |
| Elements of, .38(c) | 39 |
| Oral, .38(b) | 39 |
| Review, .38(f) | 39 |
| Training, .38(e) | 39 |
| Written, .38(b) | 39 |
| Emergency Lighting, pulp and paper mills, .261(b)(2) | 349 |
| Emergency Preparedness and Response Guidelines | 801 |
| Emergency Response, Hazardous Waste, .120 | 159 |
| Employee Alarm Systems, .165 | 240 |
| Installation and Restoration, .165(c) | 240 |
| Maintenance and Testing, .165(d) | 240 |
| Manual Operation, .165(e) | 241 |
| Employee-Owned Protective Equipment, .132(b) | 187 |
| Employee Protection: | |
| *(see also Personal Protective Equipment)* | |
| Engine Room Guardrails, .219(k)(2) | 327 |
| Engineering Controls: | |
| Air Contaminants, .1000(e) | 453 |
| Asbestos, .1001(f) | 464 |
| Effective Dates, .98, .149 | 82, 230 |
| Labor Camps, .142 | 212 |
| Marking Physical Hazards, .144 | 213 |
| Noise Exposure, .95 | 73 |
| Physical Hazards Markings, .144 | 213 |
| Radiation: | |
| Ionizing, .96 | 82 |
| Nonionizing, .97 | 82 |
| Safety Color Codes, .144 | 213 |
| Sanitation, .141 | 211 |
| Signs and Tags, .145 | 214 |
| Standards Sources, .99, .150 | 82, 230 |
| Ventilation, .94 | 63 |
| Equalizers, Crane Hoists, .179(h)(3) | 267 |
| Ethylene Oxide, .1047 | 641 |
| Ethyleneimine, .1003 | 488 |
| Area requirements, .1003(c) | 488 |
| Closed system operation, .1003(c)(2) | 488 |
| Isolated systems, .1003(c)(1) | 488 |

| Subject term, Section No. | Page |
|---|---|
| Maintenance and decontamination activities, .1003(c)(5) | 489 |
| Open-vessel system operations, .1003(c)(3) | 488 |
| Transfer from a closed operation, .1003(c)(4) | 488 |
| Medical surveillance, .1003(g) | 490 |
| Examinations, .1003(g)(1) | 490 |
| Records, .1003(g)(2) | 490 |
| Regulated area requirements, .1003(d) | 489 |
| Contamination control, .1003(d)(4) | 489 |
| Emergencies, .1003(d)(2) | 489 |
| Hygiene facilities and practices, .1003(d)(3) | 489 |
| Signs, information, and training, .1003(e) | 489 |
| Container contents identification, .1003(e)(2) | 489 |
| Lettering, .1003(e)(3) | 489 |
| Prohibited statements, .1003(e)(4) | 489 |
| Signs, .1003(e)(1) | 489 |
| Training and indoctrination, .1003(e)(5) | 489 |
| Evacuation, Ionizing Radiation, .1096(f) | 708 |
| Exhaust Air Filters, Spray Booths, .107(b)(5) | 108 |
| Exhaust Systems: *(see also Ventilation)* | |
| Abrasive Blasting, .94(a)(4) | 63 |
| Grinding, Polishing, Buffing, .94(b)(4) | 65 |
| Sawmills, .265(c)(20) | 364 |
| Exhausts, Spraying Operations, .107(d) (3), (7), (9) | 108 |
| Exit Routes | |
| Alarm system, .37(e) | 38 |
| Basic Requirements, .36(a) | 37 |
| Capacity, .36(f) | 38 |
| Compliance, .35 | 37 |
| Coverage, .34(a), .34(b) | 37 |
| Danger, .37(a) | 38 |
| Design and construction, 36 | 37 |
| During construction, .37(d) | 38 |
| Electroluminescent, .34(c) | 37 |
| Exit access, .34(c) | 37 |
| Exit discharge, .36(c) | 38 |
| Exit discharge, .34(c) | 37 |
| Exit route, .34(c) | 37 |
| Exit, .34(c) | 37 |
| Fire retardant, .37(c) | 38 |
| Height, .36(g) | 38 |
| High hazard area, .34(c) | 37 |
| Life Safety Code, .35 | 37 |
| Lighting, .37(b) | 38 |
| Locking, .36(d) | 38 |
| Maintenance, 37 | 38 |
| Manlifts, .68(b)(8) | 59 |
| Number of exit routes, .36(b) | 37 |
| Obstruction, .37(a)(3) | 38 |
| Occupant load, .34(c) | 37 |
| Outdoor, .36(h) | 38 |
| Refuge area, .34(c) | 37 |
| Safeguards, .37(a)(4) | 38 |
| Self-luminous, .34(c) | 37 |
| Side hinged door, .36(e) | 38 |
| Width, .36(g) | 38 |
| Explosive-Actuated Fastening Tools, .243(d) | 331 |
| Definitions, .241(a) | 329 |
| Fasteners, .243(d)(3) | 332 |
| High-Velocity Tools, .243(d)(2)(i) | 331 |
| Inspection, .243(d)(2) | 331 |
| Loads, .243(d)(3) | 332 |
| Low-Velocity Tools, .243(d)(2)(ii) | 331 |
| Maintenance, .243(d)(2) | 331 |
| Explosives and Blasting Agents, .109 | 111 |
| Blast Holes, .109(e)(3) | 115 |
| Bulk Delivery, .109(h)(4) | 119 |
| Charge Initiation, .109(e)(4) | 115 |
| Warning, .109(e)(5) | 115 |
| Chemicals, highly hazardous, process safety management, .119 | 148 |
| Hours of Transfer, .109(f)(5) | 115 |
| Loading, .109(e)(3) | 115 |
| Logging , .266(d)(10) | 371 |
| Magazines, .109(c)(2) | 112 |
| Mixing Vehicles, .109(h)(4) | 119 |
| Piers, .109(f) | 115 |

# OSHA Standards for the General Industry

| Subject term, Section No. | Page |
|---|---|
| Pulpwood Logging, .261(a)(4)(iii), (c)(14)(ii) | 349, 350 |
| Railroad Cars and Stations, .109(f) | 115 |
| Slurries, .109(h) | 118 |
| Small Arms Ammunition, .109(j) | 120 |
| Smoking, .109(e)(1) | 115 |
| Storage, .109(c), (f)(4) | 111, 115 |
| Transportation, .109(d) | 114 |
| Use, .109(e) | 115 |
| Vessels, .109(f) | 115 |
| Water Gels, .109(h) | 118 |
| Exposure: | |
| Air Contaminants, .1000 | 453 |
| Airborne Radioactive Material, .1096(c) | 707 |
| Asbestos, .1001(c) | 463 |
| Asbestos Fibers, .1001(b) | 463 |
| Limits (Tables Z-1 to Z-3), .1000 | 453 |
| Mineral Dusts, .1000 | 453 |
| Minors, .1096(b)(3), (c)(2), (d)(2)(ii) | 707 |
| Noise, .95 | 73 |
| Extension Ladders, Portable: | |
| Metal, .26(a)(2) | 23 |
| Metal, Trestle, .26(a)(4) | 24 |
| Wood, .25 | 22 |
| Wood, Trestle, .25(c)(3)(v) | 23 |
| Extension Lamps, Cranes, .179(g)(7) | 267 |
| Extractors, .262(y) | 357 |
| Eye and Face Protection, .133 | 187 |
| Markings, .133(a)(4) | 187 |
| Optical Corrections, .133(a)(3) | 187 |
| Protectors, .133(a)(2) | 187 |
| Welding, .252(b)(2) | 336 |
| Face Protection: | |
| (see also Eye and Face Protection; Personal Protective Equipment), .133 | 187 |
| Facilities, Labor Camps: | |
| (see Labor Camps, Temporary) | |
| Fail-Safe Controls, Spraying, .107(h)(9) | 109 |
| Overhead and Gantry Cranes, .179(a)(40), (g)(3)(viii) | 264, 266 |
| Fan-Rotating Element, .107(d)(4) | 108 |
| Farm Vehicles, Anhydrous Ammonia, .111(g), (h) | 147, 148 |
| Fasteners, .243(d)(3) | 332 |
| Fastening Tools, .243(d) | 331 |
| Fibers, Asbestos, .1001(b) | 463 |
| Filling Densities, Liquefied Petroleum Gases, .110(b)(12) | 127 |
| Filters, Spraying, .107(b)(5) | 107 |
| Fire Alarms: | |
| (see also Fire Protection) | |
| Fire Brigades, .156 | 234 |
| Fire Fighting Equipment, .156(d) | 234 |
| Organization, .156(b) | 234 |
| Protective Clothing, .156(e)(1) | 235 |
| Body protection, .156(e)(3) | 235 |
| Foot and Leg Protection, .156(e)(2) | 235 |
| Hand Protection, .156(e)(4) | 235 |
| Head, Eye, Face Protection, .156(e)(5) | 235 |
| Respiratory Protection, .156(f) | 235 |
| General Requirements, .156(f)(1) | 235 |
| Positive-Pressure Breathing Apparatus, .156(f)(2) | 235 |
| Selection and Distribution, .157(d) | 236 |
| Training and Education, .156(c) | 234 |
| Fire Detection Systems, .164 | 240 |
| Installation and Restoration, .164(b) | 240 |
| Maintenance and Testing, .164(c) | 240 |
| Number, Location, Spacing, .164(f) | 240 |
| Protection of Detectors, .164(d) | 240 |
| Response Time, .164(e) | 240 |
| Fire Extinguishers | |
| Cranes, .179(c)(3), (o)(3), .180(i)(5) | 265, 269, 272 |
| Dip Tanks, .125(f)(2)(i) | 185 |
| Explosives Transportation, .109(d)(2)(ii) | 114 |
| Powered working platforms, .66(f)(5)(ii)(L) | 45 |
| Welding, .252(a)(2)(ii) | 335 |
| Fire Extinguishers, Portable, .157 | 236 |
| Exemptions, .157(b) | 236 |
| General Requirements, .157(c) | 236 |
| Hydrostatic Testing, .157(f) | 236 |

| Subject term, Section No. | Page |
|---|---|
| Inspection, Maintenance Testing, .157(e) | 236 |
| Selection and Distribution, .157(d) | 236 |
| Training and Education, .157(g) | 237 |
| Fire prevention plans | |
| Application, .39(a) | 39 |
| Elements, .39(c) | 39 |
| Flammable waste, .39(c)(2) | 39 |
| Hazards, .39(c)(1) | 39 |
| Responsibilities, .39(c)(4), .39(c)(5) | 39 |
| Safeguards, .39(c)(3) | 39 |
| Employee information, .39(d) | 39 |
| Oral, .39(b) | 39 |
| Written, .39(b) | 39 |
| Fire Protection: | |
| Ammonium Nitrate, .109(i)(7) | 120 |
| Blasting Agents, .109(i)(7) | 120 |
| Bulk Plants, .106(f)(4)(ix), (8) | 101 |
| Chemical Plants, .106(i)(5) | 106 |
| Definitions, .155 | 233 |
| Distilleries, .106(i)(5) | 106 |
| Electrostatic Apparatus, .107(h)(12) | 109 |
| Explosives, .109(i)(7) | 120 |
| Flammable Liquids, .106(d)(7), (e)(5), (f)(8), (g)(9), (h)(6), (i)(5) | 98, 99, 101, 104, 105, 106 |
| Industrial Plants, .106(e)(5) | 99 |
| Liquified Petroleum Gases, .110(d)(14), (f)(7), (h)(14) | 135, 138, 140 |
| Local Fire Alarms, .163 | 240 |
| Processing Plants, .106(h)(6) | 105 |
| Refineries, .106(i)(5) | 106 |
| Service Stations, .106(g)(9) | 104 |
| Spray Booths, .107(f) | 109 |
| Cleaning, .107(f)(3) | 109 |
| Conformance, .107(f)(1) | 109 |
| Extinguishers, Portable, .107(f)(4) | 109 |
| Valve Access, .107(f)(2) | 109 |
| Storage Tanks, .106(d)(7) | 109 |
| Trucks, .178 | 257 |
| Fire Protection Equipment: | |
| Color Identification, .144(a)(1) | 213 |
| Fire Brigades, .156(d) | 234 |
| Fire Resistance (Rating): | |
| Inside Storage Rooms, .106(d)(4)(ii) | 97 |
| Storage Cabinets, .106(d)(3)(ii) | 96 |
| Tank Supports, .106(b)(5)(ii) | 94 |
| Fire Watch, Welding, .252(a)(2)(iii) | 335 |
| Fireworks: | |
| (see Pyrotechnics) | |
| First Aid, .262(pp) | 358 |
| Eye Flushing, .151(c) | 231 |
| Labor Camps, .142(k) | 213 |
| Logging, .266(d)(2) | 370 |
| Standards Sources, .153 | 231 |
| Welding, .252(c)(13) | 332 |
| Fittings: | |
| (see Piping, Fittings and Tubing; Piping, Valves and Fittings) | |
| Fixed Extinguishing Systems, .160, .161, .162, .163 | 238, 239, 240 |
| Dry Chemical Agent Systems, .161 | 239 |
| Gaseous Agent Systems, .162 | 239 |
| General Requirements, .160(b) | 238 |
| Total Flooding Systems, .160(c) | 239 |
| Water Spray and Foam Systems, .163 | 240 |
| Fixed Industrial Stairs: | |
| (see Stairs, Fixed Industrial), .24 | 22 |
| Flammable and Combustible Liquids: | |
| Bulk Plants, .106(f) | 100 |
| Chemical Plants, .106(i) | 106 |
| Container Marking, color codes, .144(a)(1)(ii) | 213 |
| Containers, .106(d) | 96 |
| Dip Tanks, .123-.126 | 184-186 |
| Distilleries, .106(i) | 106 |
| Effective Dates, .114 | 148 |
| Hazardous communication, .1200 | 708 |

| Subject term, Section No. | Page |
|---|---|
| Ignition Sources, .106(b)(6), (e)(6), (f)(6), (g)(8), (h)(7) | 95, 99, 101, 104, 105 |
|     Industrial Plants, .106(e) | 98 |
|     Piping, Valves and Fittings, .106(c) | 95 |
|     Pressure Vessels, .106(b)(1)(v) | 91 |
|     Process safety management of highly hazardous chemicals, .119 | 148 |
|     Processing Plants, .106(h) | 105 |
|     Refineries, .106(i) | 106 |
|     Service Stations, .106(g) | 101 |
|     Spray Finishing, .107 | 106 |
|         Storage and Handling, .107(e) | 108 |
|             Conformance, .107(e)(1) | 108 |
|             Containers, .107(e)(3), (5) | 108 |
|             Hoses, .107(e)(6) | 108 |
|             Grounding, .107(c)(9), (e)(9) | 107, 109 |
|             Liquid Heaters, .107(e)(7) | 108 |
|             Liquid Transfer, .107(e)(4) | 108 |
|             Pipes, .107(e)(6) | 108 |
|             Pump Relief, .107(e)(8) | 108 |
|             Quantity, .107(e)(2) | 108 |
|             Safety Relief Devices, .107(e)(8) | 108 |
|             Spraying Containers, .107(e)(5) | 108 |
|     Standards Sources, .115 | 148 |
|     Storage Containers, .106(d) | 96 |
|     Storage Tanks, .106(b), (d) | 91, 96 |
|     Tanks, .106(b), (d) | 91, 96 |
| Flammable Materials, Trucks Used, .178(c)(2) | 258 |
| Flanges, Abrasive Wheel Machinery, .215(a)(3), (c) | 300, 302 |
|     Balance, .215(c)(3) | 303 |
|     Blotters, .215(c)(1)(v), (c)(6), (d)(5) | 302, 303, 304 |
|     Diameter Uniformity, .215(c)(4) | 303 |
|     Dimensions, .215(c)(8) | 303 |
|     Driving, .215(c)(7) | 303 |
|     Finish, .215(c)(3) | 303 |
|     General Requirements, .215(c)(1) | 302 |
|     Maintenance, .215(c)(9) | 303 |
|     Recess, .215(c)(5) | 303 |
|     Repairs, .215(c)(9) | 303 |
|     Types, .215(c)(1)(iv) | 302 |
|     Undercut, .215(c)(5) | 303 |
| Flash Welding Equipment, .255(d) | 347 |
|     Fire Curtains, .255(d)(2) | 347 |
|     Ventilation, .255(d)(1) | 347 |
| Float Scaffolds, .28(u) | 35 |
| Flooding, Tank Areas, .106(b)(5)(vi) | 94 |
| Floor Loading, .22(d) | 20 |
| Floor Openings (Holes), .23(a) | 20 |
|     Manlifts, .68(b) (5), (7) | 59 |
| Floors: | |
|     Covers, Hinged, .23(a)(3)(i) | 20 |
|     Open-Sided, .23(c) | 20 |
|     Spray Booths, .107(b)(3) | 107 |
| Flow Coatings, .126(b) | 186 |
| Fluidized Beds, .107(l)(7) | 110 |
| Fluorine Compounds, Welding: | |
|     (see also Air Contaminants), .252(c)(5) | 338 |
| Foam Extinguishing Systems, Fixed, .163 | 240 |
| Food Handling, .120(n)(4), .141(h) | 167, 212 |
| Foot Pedals, Power Presses, .217(b)(4) | 310 |
| Foot Protection, .136 | 204 |
| Footwalks: | |
|     Cranes, .179(d) | 265 |
| Forging Hammers, .218(a)(3) | 324 |
|     Foot-Operated Devices, .218(b)(2) | 325 |
|     Gravity, .218(e) | 325 |
|         Air Lifts, .218(e)(1) | 325 |
|         Board Drop Hammers, .218(e)(2) | 325 |
|     Keys, .218(b)(1) | 325 |
|     Power-Driven, .218(d) | 325 |
|         Cylinder Draining, .218(d)(3) | 325 |
|         Pressure Pipes, .218(d)(4) | 325 |
|         Safety Cylinder Heads, .218(d)(1) | 325 |
|         Shutoff Valves, .218(d)(2) | 325 |
| Forging Machine Area, .30(b) | 36 |
| Forging Machines: | |

| Subject term, Section No. | Page |
|---|---|
|     Billet Shears, .218(j)(1) | 325 |
|     Boltheading, .218(i)(1) | 325 |
|     Conveyors, .218(j)(3) | 325 |
|     Definitions, .211(e) | 296 |
|     Effective Dates, .220 | 328 |
|     Grinding, .218(j)(5) | 325 |
|     Hammers, .218(a)(3), (b), (d), (e) | 324, 325 |
|     Inspections, .218(a)(2) | 324 |
|     Lead Use, .218(a)(1) | 324 |
|     Maintenance, .218(a)(2) | 324 |
|     Presses, .218(a)(3), (c), (f), (g) | 324, 325 |
|     Rivet Making, .218(i)(2) | 325 |
|     Saws, .218(j)(2) | 325 |
|     Shot Blast, .218(j)(4) | 325 |
|     Standards Sources, .221 | 328 |
|     Upsetters, .218(h) | 325 |
| Forging Presses, .218(f) | 325 |
| Fork Trucks: | |
|     (see also Powered Industrial Trucks), .178 | 257 |
| Formaldehyde, .1048, App. A, App. B, App. D | 654, 658, 661, 667 |
|     Airborne Concentration, .1048(c) | 654 |
|     Compliance, .1048(f) | 654 |
|     Emergencies, .1048(k) | 656 |
|     Exposure, Permissible, .1048(c) | 654 |
|     Hazard Communication, .1048(m) | 657 |
|     Housekeeping, .1048(j) | 656 |
|     Hygiene Protection, .1048(i) | 655 |
|     Medical Surveillance, .1048(l), App. C | 656, 665 |
|     Monitoring, .1048(d) | 654 |
|     Personal Protective Clothing, .1048(h) | 655 |
|     Recordkeeping, .1048(o) | 658 |
|     Regulated Areas, .1048(e) | 654 |
|     Respiratory protection, .1048(g) | 655 |
|     Training, Employee, .1048(n) | 658 |
| Fuel-Gas Systems: | |
|     (see also Oxygen-Fuel Gas Systems), .253 | 339 |
| Fuels: | |
|     (see also Refueling) | |
|     Handling and Storage, .178(f) | 259 |
| Fumishings: | |
| Fusion Apparatus: | |
|     (see Drying, Curing and Fusion Apparatus) | |
| Gaging Devices, .110(b)(19) | 130 |
| Gantry Cranes: | |
|     (see also Overhead and Gantry Cranes), .179 | 263 |
| Garages, Undercoating Operations, .107(k) | 110 |
| Garnett Machines, .262(f) | 356 |
| Gas Cylinder Inspection, .101(a) | 83 |
| Gaseous agent extinguishing systems; fixed, .162 | 239 |
|     Scope and Application, .162(a) | 239 |
|     Specific Requirements, .162(b) | 239 |
| Gaseous Hydrogen Systems: | |
|     (see Hydrogen) | |
| Gasoline Powered Trucks, .178(b)(8), (9) | 257 |
| Gears, .219(f) | 327 |
| General Duty Clause 5(a) | 733 |
| Gill Boxes, .262(k) | 357 |
| Gin Pole Derricks: | |
|     (see also Derricks), .181(a)(6) | 273 |
| Gloves, Rubber Insulating, .137 | 203 |
| Glue Spreaders, .213(r) | 300 |
| Goggles: | |
|     (see also Eye Protection; Eye and Face Protection), .133 | 187 |
| Grain Handling, .272 | 412 |
|     Application, .272(b) | 412 |
|     Continuous flow bulk raw grain dryers, .272(o) | 414 |
|     Contractors, .272(h) | 414 |
|     Emergency action plan, .272(d) | 413 |
|     Emergency escape, .272(n) | 414 |
|     Entry into bins, silos, tanks, .272(g) | 413 |
|     Filter collectors, .272(k) | 414 |
|     Grain stream processing equipment, .272(m) | 414 |
|     Grate openings, .272(j) | 414 |
|     Hot work permit, .272(f) | 413 |
|     Housekeeping, .272(i) | 414 |
|     Inside bucket elevators, .272(p) | 414 |
|     Preventive maintenance, .272(l) | 414 |

# OSHA Standards for the General Industry

| Subject term, Section No. | Page |
|---|---|
| Training, .272(e) | 413 |
| Gravity Hammers, .218(e) | 325 |
| **Grinders:** | |
| (see also Abrasive Wheel) | |
|     Machinery; Cutoff Wheels, .243(c)(3), (4) | 331 |
| Grinding, Forging Equipment, .218(j)(5) | 325 |
| **Grinding Machines:** | |
|     Cylindrical, .215(b)(4) | 301 |
|     Flanges, .215(a)(3) | 300 |
|     Surface Grinders, .215(b)(5) | 301 |
|     Swing Frame Grinders, .215(b)(6) | 301 |
|     Top Grinding, .215(b)(8) | 301 |
|     Work Rest, .215(a)(4) | 300 |
| **Grinding, Polishing, and Buffing:** | |
|     Branch Pipes, .94(b)(3) | 64 |
|     Enclosure Design, .94(b)(5) | 65 |
|     Exhaust Systems, .94(b)(4) | 65 |
|     Hoods, .94(b)(3), (5) | 64, 65 |
| Grinding, Top, .215(b)(8) | 301 |
| **Grounding:** | |
|     Bulk Oxygen Systems, .104(b)(7)(iv) | 89 |
|     Circuits, .334(b) | 431 |
|     Electrostatic Spraying, .107(h)(5), (i)(5)-(7) | 109, 110 |
|     Flammable and Combustible Liquids, .106(e)(6)(ii), (f)(3)(iv) | 99, 100 |
|     General, .304(f) | 422 |
|     Hand Spraying, .107(i)(5)-(7) | 109, 110 |
|     Ignition Sources, .107(c)(9) | 107 |
|     Liquefied Hydrogen Systems, .103(c)(4)(iv) | 88 |
|     Liquid Transfer, .107(e)(9) | 109 |
|     Methods, .304(f)(6), (f)(7) | 423 |
|     Spray Booths, .107(h)(10) | 109 |
|     Spraying Operations, .107(c)(9), (e)(9), (i)(5)-(7) | 107, 109, 110 |
|     Welding, .254(c)(2), (d)(3);, .255(b)(9), (c)(6) | 346, 347 |
|     Woodworking Tools, .243(a)(5) | 330 |
| **Guarding:** | |
| (see also Term To Which It Applies), .211-.222 | 293-328 |
|     Abrasive Wheels, Portable, .243(c) | 330 |
|     Floor Openings (Holes), .23(a) | 20 |
|     Hatchways, .23(a)(3) | 20 |
|     Ladderways, .23(a)(2) | 20 |
|     Live Parts, .303(g)(2), (h)(2) | 420 |
|     Machinery, .211-.219, .220-222 | 293-328 |
|     Mechanical Power Transmission Apparatus, .219 | 325 |
|         Clutches, .219(k) | 327 |
|         Friction Drives, .219(g) | 327 |
|         Prime Movers, .219(b) | 325 |
|         Pulleys, .219(d) | 326 |
|     Open-Sided Floors, .23(c) | 20 |
|     Platforms, .23(c) | 20 |
|     Powered Tools, Portable, .243 | 330 |
|     Railings, .23(e) | 21 |
|     Runways, .23(c) | 20 |
|     Skylight, .23(a)(4) | 20 |
|     Spraying Equipment, .107(h)(10) | 109 |
|     Stairways, .23(a)(1), (d) | 20, 21 |
|     Wall Openings (Holes), .23(b) | 20 |
| **Guardrails:** | |
|     Definitions, .21(f)(10) | 19 |
|     Manlifts, .68(b)(8)(i), (10)(iv) | 59, 60 |
|     Power Transmission Apparatus, .219(o)(5) | 328 |
|     Powered and Working Platforms, .66(e)(3), (f)(3)(i)(K), (f)(5)(i)(G), (f)(5)(ii)(K) | 43, 44, 45 |
|     Removable, .23(a)(3) | 20 |
|     Working Surfaces, .22(c) | 20 |
| **Guardrails, Scaffolds:** | |
| (see Listing Under Specific Type Scaffold) | |
| **Guards:** | |
| (see also Guardrails) | |
|     Derricks, .181(j)(1) | 276 |
|     Hoisting Ropes, .179(e)(5) | 266 |
|     Manlifts, .68(b)(7)-(9) | 59, 60 |
|     Moving Parts, .179(e)(6) | 266 |
|     Trucks, .178(e) | 259 |
| Guide Posts, .217(d)(4) | 307 |

| Subject term, Section No. | Page |
|---|---|
| **Guidelines** | |
|     Emergency Preparedness and Response | 801 |
|     Safety and Health Management Voluntary | 807 |
| Gudgeon Pin, .181(a)(20) | 274 |
| Guy Derricks, .181(a)(7) | 273 |
| **Hammers, Forging:** | |
| (see Forging Hammers) | |
| Hand protection, .138 | 206 |
| **Hand Spraying Equipment:** | |
| (see Electrostatic Hand Spraying Equipment) | |
| Hand Tools, .242 | 330 |
|     Dead-Man Controls, .243(a)(2) | 330 |
|     Pulp and Paper Mills, .261(c)(13) | 350 |
| Handholds, Manlifts, .68(c)(4) | 60 |
| **Handling:** | |
| (see also Materials Handling and Storage) | |
|     Anhydrous Ammonia, .111 | 140 |
|     Compressed Gases, .101(b) | 83 |
|     Liquefied Hydrogen Systems, .103(c)(2)(iii) | 87 |
|     Liquefied Petroleum Gases, .110 | 121 |
|     Liquids, .106(h)(4) | 105 |
|     Service Stations, .106(g)(1) | 101 |
| Handrails, .24(h) | 22 |
|     Cranes, .179(d)(3), (4)(ii) | 265 |
|     Mobile Ladder Stands, .29(f)(4) | 36 |
| Hangers, .219(p)(4) | 328 |
| Hardening Tanks, .126(a)(1)(i),(ii) | 186 |
| Hatchways Guarding, .23(a)(3) | 20 |
| **Hazard Communication,** | |
| chemical information, transmittal, .1200, App. A, App. E | 710, 717, 719 |
|     Hazard Communication Program, .1200(e) | 713 |
|     Hazard Determination, .1200(d), App. B | 713, 718 |
|     Information and Training, .1200(h) | 715 |
|     Labels and Warnings, .1200(f) | 714 |
|     Material Safety Data Sheets, .1200(g) | 714 |
|     Trade Secrets, .1200(i), App. D | 715, 718 |
| Hazardous chemicals, highly, process safety management (see Chemicals, etc.), .119 | 148 |
| Hazardous chemicals, occupational exposure to in laboratories (see Chemicals, hazardous), .1450 | 722 |
| Hazardous energy; control of (see Lockout/tagout) | |
| **Hazardous Materials:** | |
|     Acetylene, .102 | 83 |
|     Anhydrous Ammonia, .111 | 140 |
|     Blasting Agents, .109 | 111 |
|     Bulk Oxygen Systems, .104 | 88 |
|     Chemicals (see entries under Chemicals, etc.), .119 | 148 |
|     Combustible Liquids, .106 | 89 |
|     Compressed Gases, .101 | 83 |
|     Dip Tanks, .122-.126 | 183-186 |
|     Effective Dates, .114 | 148 |
|     Explosives, .109 | 111 |
|     Flammable Liquids, .106 | 89 |
|     Hazardous Wastes, .120 | 159 |
|     Hydrogen, .103 | 83 |
|     Liquefied Petroleum Gases, .110 | 121 |
|     Nitrous Oxide, .105 | 89 |
|     Oxygen, .104 | 88 |
|     Packages, transport vehicles, etc, retention of DOT markings, .1201 | 722 |
|     Spray Finishing, .107 | 106 |
|     Storage and Handling: | |
|         Anhydrous Ammonia, .111 | 140 |
|         DOT markings, retention, .1201 | 718 |
|         Liquefied Petroleum Gases, .110 | 121 |
|     Trucks Used, .178(c)(2) | 258 |
| **Hazardous Waste Operations:** | |
|     Contractors and Subcontractors, .120(b)(1)(iv) | 161 |
|     Decontamination, .120(k), (p)(4) | 166, 168 |
|     Drums and Containers, .120(j) | 165 |
|     Emergency Response, .120(e)(7), (l), (p)(8), (q) | 163, 166, 168, 169 |
|     Engineering Controls and Personal Protective Equipment, .120(g) | 164 |

## Subject Index For 29 CFR 1910, Sec. 5, 1903, 1904, and 1990

| Subject term, Section No. | Page |
|---|---|
| Illumination, .120(m) | 167 |
| Information Program, .120(b), (i) | 161, 165 |
| Laboratory Waste Packs, .120(j)(6) | 166 |
| Material Handling, .120(j) | 165 |
| Medical Surveillance, .120(f), (p)(3), (q)(9) | 163, 168, 171 |
| Monitoring, .120(c)(6), (h) | 162, 165 |
| Post-emergency Response, .120(a)(3)(D), (q)(11) | 160, 171 |
| Radioactive Wastes, .120(j)(4) | 166 |
| Recordkeeping, .120(f)(7) | 164 |
| RCRA Facilities, .120(p) | 168 |
| Safety and Health Program, .120(b) | 161 |
| Sanitation, .120(n) | 167 |
| Shock-sensitive Wastes, .120(j)(5) | 166 |
| Site Characterization and Analysis, .120(c) | 162 |
| Site Control, .120(d) | 162 |
| Site Safety and Health Plan, .120(b)(4) | 161 |
| Tank and Vault Procedures, .120(j)(9) | 166 |
| Totally-encapsulating Chemical Protective Suits, .120(g)(4) | 165 |
| Training, .120(e), (p)(8)(iii), (q)(6) | 162, 168, 170 |
| Uncontrolled sites, emergency responses, .120(l) | 166 |
| Head Protection, .135 | 204 |
| Health, Safety and Health Management Voluntary Guidelines | 807 |
| Healthcare professions and related industries, exposures to bloodborne pathogens, .1030 | 594 |
| Hearing Conservation Program, .95(c) | 74 |
| Heating: | |
| Dip Tanks, .125(g) | 185 |
| Bulk Plants, .106(f)(2)(ii) | 100 |
| Service Stations, .106(g)(6) | 104 |
| Helicopters, .183 | 276 |
| Helmets, .135, .252(b)(2), .266(d)(1)(vi) | 204, 336, 370 |
| Hepatitis B | |
| (see also Bloodborne pathogens), .1030 | 594 |
| Hexavalent Chromium | XXX |
| Hinged Floor Covers, .23(a)(3)(i) | 20 |
| Hoist Limit Switches, .179(n)(4) | 269 |
| Hoisting Equipment: | |
| Cranes, .179(e)(5), (h) | 266, 267 |
| Powered Platforms, .66(f)(4), (g)(6) | 44, 48 |
| Rope Guards, .179(e)(5) | 266 |
| Holding Brakes, .179(f)(2) | 266 |
| Holes: | |
| (see also Floor Openings (Holes); Wall Openings (Holes)), .23 | 20 |
| Hooks: | |
| Cranes, .179(h)(4) | 267 |
| Derricks, .181(j)(2) | 276 |
| Horse Scaffolds, .28(m) | 34 |
| Hoses: | |
| Flammable Liquids, .107(e)(6) | 108 |
| Liquefied Petroleum Gases, .110(b)(9) | 124 |
| Semiconductors, .109(a)(12) | 111 |
| Sprinkler Systems, .159(c)(5) | 238 |
| Standpipe and hose systems, .158(c)(3) | 237 |
| Welding and Cutting, .253(e) | 342 |
| Hot Sources, .107(c)(3) | 107 |
| Hot-work permits, process safety management of highly hazardous chemicals, .119(k) | 151 |
| Hours of Transfer, Explosives, .109(f)(5) | 115 |
| Household Stepladders, Type III, .25(c)(2) | 22 |
| Housekeeping, .141(a)(3) | 211 |
| Asbestos, .1001(d) | 458 |
| Flammable Liquids, .106(e)(9) | 100 |
| Walking-Working Surfaces, .22(a) | 20 |
| Hydraulic Barkers, .261(e)(14) | 351 |
| Hydraulic Equipment, .217(b)(11) | 311 |
| Hydraulically designed sprinkler systems, .159(c)(11) | 238 |
| Hydrogen, .103 | 83 |
| Gaseous Hydrogen Systems, .103(a)(2)(i), (b) | 83 |
| Clear Zone, .103(b)(5) | 85 |
| Containers, .103(b)(1)(i) | 83 |
| Design, .103(b)(1) | 83 |
| Equipment Assembly, .103(b)(1)(iv) | 83 |
| Fittings, .103(b)(1)(iii) | 83 |
| Inspection, .103(b)(5) | 85 |

| Subject term, Section No. | Page |
|---|---|
| Location, .103(b)(2) | 84 |
| Outdoor, .103(b)(3)(i) | 85 |
| Separate Buildings, .103(b)(3)(ii) | 85 |
| Operating Instructions, .103(b)(4) | 85 |
| Piping, .103(b)(1)(iii) | 83 |
| Safety Relief Devices, .103(b)(1)(ii) | 83 |
| Testing, .103(b)(1)(vi) | 84 |
| Tubing, .103(b)(1)(iii) | 83 |
| Liquefied Hydrogen Systems, .103(a)(2)(ii), .103(c) | 83, 85 |
| Clear Zone, .103(c)(5) | 88 |
| Containers, .103(c)(1)(i) | 85 |
| Design, .103(c)(1) | 85 |
| Electrical Systems, .103(c)(1)(ix) | 86 |
| Equipment Assembly, .103(c)(1)(vi) | 86 |
| Fittings, .103(c)(1)(v) | 85 |
| Grounding, .103(c)(4)(iv) | 88 |
| Location, .103(c)(2) | 86 |
| Outdoor, .103(c)(3)(i) | 87 |
| Separate Buildings, .103(c)(3)(ii) | 87 |
| Special Rooms, .103(c)(3)(iii) | 87 |
| Maintenance, .103(c)(5) | 88 |
| Markings, .103(c)(1)(iii) | 85 |
| Operating Instructions, .103(c)(4) | 87 |
| Attendants, .103(c)(4)(ii) | 88 |
| Security, .103(c)(4)(iii) | 88 |
| Piping, .103(c)(1)(v) | 85 |
| Safety Relief Devices, .103(c)(1)(iv) | 85 |
| Supports, .103(c)(1)(ii) | 85 |
| Testing, .103(c)(1)(vii) | 86 |
| Tubing, .103(c)(1)(v) | 85 |
| Vaporizers, .103(c)(1)(viii) | 86 |
| Hydrostatic Tests: | |
| (see also Testing) | |
| Fire Extinguishers, .157(f) | 236 |
| Piping, .106(c)(7) | 96 |
| Ignition Sources: | |
| Bulk Plants, .106(f)(6) | 101 |
| Dip Tanks, .125(e) | 185 |
| Industrial Plants, .106(e)(6) | 99 |
| Powder Coatings, .107(l)(1) | 110 |
| Processing Plants, .106(h)(7) | 105 |
| Service Stations, .106(g)(8) | 104 |
| Spraying Operations, .107(c) | 107 |
| Combustible Residues, .107(c)(5) | 107 |
| Conformance, .107(c)(1) | 107 |
| Electrical Wiring, .107(c)(4), (6) | 107 |
| Grounding, .107(c)(9) | 107 |
| Hot Sources, .107(c)(3) | 107 |
| Lamps, .107(c)(7), (8) | 107 |
| Separation Minimum, .107(c)(2) | 107 |
| Storage Tanks, .106(b)(6) | 95 |
| Illumination: | |
| (see Lighting) | |
| Indoor Storage: | |
| Flammable and Combustible Liquids, .106(b)(4), (d)(4), (d)(5), (e)(5), (g)(1)(iii), (h)(4)(i) | 94, 97, 99, 102, 105 |
| Rooms, .106(d)(4) | 97 |
| Standards Sources, .183 | 276 |
| Industrial Plants: | |
| Flammable and Combustible Liquids, .106(e) | 98 |
| Electrical Systems, .106(e)(7) | 99 |
| Fire Protection, .106(e)(5) | 99 |
| Housekeeping, .106(e)(9) | 100 |
| Incidental Storage, .106(e)(2) | 98 |
| Ignition Sources, .106(e)(6) | 99 |
| Maintenance, .106(e)(9) | 100 |
| Repairs, Equipment, .106(e)(8) | 100 |
| Tank Loading, .106(e)(4) | 99 |
| Unit Physical Operations, .106(e)(3) | 99 |
| Liquefied Petroleum, .110(d)(12), (f)(4) | 135, 137 |
| Industrial Stepladders, Type I, .25(c)(2) | 22 |
| Insect Control, .141(a)(5) | 211 |
| Labor Camps, .142(j) | 213 |
| Inspection: | |
| (see also Term To Which It Applies) | |
| Abatement date, petitions for modification, 1903.14a | 738 |

# OSHA Standards for the General Industry

| Subject term, Section No. | Page |
|---|---|
| Abatement verification: *1903.19* | **739** |
|   Abatement certification, *1903.19(c)* | **739** |
|     Sample letter, *Appendix A to 1903.19* | **741** |
|   Abatement documentation, *1903.19(d)* | **740** |
|   Abatement plans, *1903.19(e)* | **740** |
|     Sample plan, *Appendix B to 1903.19* | **739** |
|   Definitions, *1903.19(b)* | **739** |
|   Employee notification, *1903.19(g)* | **740** |
|   Movable equipment, *1903.19(i)* | **740** |
|   Progress reports, *1903.19(f)* | **740** |
|   Purpose, *1903.19* | **739** |
|   Sample warning tag, *Appendix C to 1903.19* | **741** |
|   Scope and application, *1903.19(a)* | **739** |
|   Transmitting abatement documents, *1903.19(h)* | **740** |
| Advance notice of inspection, *1903.6* | **736** |
| Authority for inspection, *1903.3* | **735** |
| Availability of Act, *1903.2* | **735** |
| Citations, *1903.14* | **737** |
| Complaints by employees, *1903.11* | **737** |
| Compressed Gas Cylinders, *.101(a)* | **83** |
| Conduct of inspections, *1903.7* | **736** |
| Consultation with employees, *1903.10* | **737** |
| Contests before the Review Commission, *1903.17* | **739** |
| Cranes, *.179(j), .180(d)* | **267, 270** |
|   Crawler, *.180(d)* | **270** |
|   Gantry, *.179(j)* | **267** |
|     Ropes, *.179(m)* | **268** |
|   Locomotive, *.180(d)* | **270** |
|   Overhead, *.179(j)* | **267** |
|     Ropes, *.179(m)* | **268** |
|   Truck, *.180(d)* | **270** |
| Cylinders, *.101(a)* | **83** |
| De minimis violations, *1903.14* | **737** |
| Definitions, *1903.22* | **741** |
| Derricks, *.181(d)* | **274** |
| Employee rescue activities, *1903.14* | **737** |
| Entry not a waiver, *1903.5* | **736** |
| Failure to correct a violation for which a citation has been issued, *1903.18* | **739** |
| Fire Extinguishers, *.157(e)* | **236** |
| Flooding, Tank Areas, *.106(b)(5)(v), (vi)* | **94** |
| Gas Cylinders, *.101(a)* | **83** |
| Gaseous Hydrogen, *.103(b)(5)* | **85** |
| Imminent danger, *1903.13* | **737** |
| Informal conferences, *1903.20* | **741** |
| Informal review, *1903.12* | **737** |
| Inspection not warranted, *1903.12* | **737** |
| Manlifts, *.68(e)* | **61** |
| Objection to inspection, *1903.4* | **735** |
| Posting of citations, *1903.16* | **739** |
| Posting of notice; availability of Act, regulations and applicable standards, *1903.2* | **735** |
| Power Presses, *.217(e)* | **313** |
| Powered Platforms, *.66(g)* | **47** |
| Proposed penalties, *1903.15* | **738** |
| Purpose and scope, *1903.1* | **735** |
| Regulations and applicable standards, *1903.2* | **735** |
| Representatives of employers and employees, *1903.8* | **736** |
| Respirators, *.134(f)* | **191** |
| Ropes, Cranes, *.179(m)* | **268** |
| Scope, *1903.1* | **735** |
| State administration, *1903.21* | **741** |
| Trade secrets, *1903.9* | **737** |
| Woodworking Machines, *.213(s)* | **300** |
| Instruction Signs, Manlifts, *.68(c)(7)* | **61** |
| Insulators, *.107(h)(5)* | **109** |
| Interior Hung Scaffolds, *.28(p)* | **34** |
| Ionizing Radiation, *.1096* | **706** |
|   AEC Licensees, *.1096(p)* | **710** |
|   Airborne Radioactive Materials, *.1096(c)* | **707** |
|   Caution Signs and Labels, *.1096(e)* | **707** |
|   Employees: | |
|     Disclosure, *.1096(o)* | **710** |
|     Exposure Records, *.1096(m), (n)* | **710** |
|     Incident Reporting, *.1096(l)* | **709** |
|     Instruction Posting, *.1096(i)* | **709** |

| Subject term, Section No. | Page |
|---|---|
|   Evacuation, *.1096(f)* | **708** |
|   Exemptions, *.1096(g), (h)* | **709** |
|   Exposure, *.1096(b)* | **707** |
|     Airborne Radioactive Materials, *.1096(c)* | **707** |
|     Minors, *.1096(b)(3), (c)(2), (d)(2)(ii)* | **707** |
|   Exposure Records, *.1096(m)-(o)* | **710** |
|   Incident Reporting, *.1096(l)* | **709** |
|   Monitoring, *.1096(d)* | **707** |
|   Overexposure Reports, *.1096(m)* | **710** |
|   Personnel Instructions, Posting, *.1096(i)* | **709** |
|   Radioactive Materials: | |
|     Packaged, *.1096(h)* | **709** |
|     Storage, *.1096(j)* | **709** |
|   Warning Signals, *.1096(f)* | **708** |
|   Waste Disposal, *.1096(k)* | **709** |
| Jacks: | |
|   Definitions, *.241(d)* | **330** |
|   Fixed Truck, *.178(k)(3)* | **259** |
|   Loading, *.244(a)(1)* | **333** |
|   Maintenance, *.244(a)(2)* | **333** |
|   Marking, *.244(a)(1)* | **333** |
|   Truck, *.178(k)(3)* | **260** |
| Jointers, *.213(j)* | **298** |
|   Blades, *.213(s)(12)* | **300** |
| Keys, Projecting, *.219(h)* | **327** |
| Kiers, *.262(q)* | **357** |
| Kilns, *.265(f)* | **369** |
| Kitchens, Labor Camps, *.142(i)* | **213** |
| Labeling, Hazardous Chemicals, *.1200* | **710** |
| Labor Camps, Temporary, *.142* | **212** |
|   Bathing Facilities, *.142(f)* | **213** |
|   Bedding, *.142(b)(3)* | **212** |
|   Communicable Diseases Reportings, *.142(l)* | **213** |
|   Dining Facilities, *.142(j)* | **213** |
|   Facilities, *.142(b)* | **212** |
|   First Aid, *.142(k)* | **213** |
|   Floors, *.142(b)(4), (5)* | **212** |
|   Furnishings, *.142(b)* | **212** |
|   Grounds, *.142(a)(3)* | **212** |
|   Heating Equipment, *.142(b)(11)* | **212** |
|   Insect Control, *.142(j)* | **213** |
|   Kitchens, *.142(i)* | **213** |
|   Laundry Facilities, *.142(f)* | **213** |
|   Lighting, *.142(g)* | **213** |
|   Refuse Disposal, *.142(h)* | **213** |
|   Rodent Control, *.142(j)* | **213** |
|   Screening, *.142(b)(8)* | **212** |
|   Sewage Disposal, *.142(e)* | **213** |
|   Shelters, *.142(b)* | **212** |
|   Site, *.142(a)* | **212** |
|   Size, *.142(a)(2)* | **212** |
|   Sleeping, *.142(b)(2), (3)* | **212** |
|   Space, *.142(b)(2), (9)* | **212** |
|   Stoves, *.142(b)(10)* | **212** |
|   Toilet Facilities, *.142(d)* | **213** |
|   Washing, *.142(f)* | **213** |
|   Waste Disposal, *.142(h)* | **213** |
|   Water Supply, *.142(c)* | **212** |
|   Windows, *.142(b)(7), (8)* | **212** |
| Laboratories, occuptional exposures to hazardous chemicals in (see *Chemicals, hazardous*), *.1450* | **722** |
| Laboratories and production facilities, HIV and HBV research, *.1030(e)* | **598** |
| Ladder-Jack Scaffolds, *.28(q)* | **34** |
| Ladder Stands, Manual Mobile; (see *Work Platforms, Mobile*), *.29(f)* | **36** |
| Ladders: | |
|   Cranes, *.179(d)(4), (o)(1)* | **265, 269** |
|   Fixed, *.27* | **24** |
|   Manlifts, *.68(b)(12)* | **60** |
|   Portable Metal, *.26* | **23** |
|   Portable Wood, *.25* | **22** |
| Ladders, Fixed, *.27* | **24** |
|   Cages, *.27(c)(3), (d)(1)* | **25, 26** |
|   Clearances, *.27(c)* | **25** |
|   Cleats, *.27(b)(1)* | **24** |
|   Design, *.27(a)* | **24** |

# Subject Index For 29 CFR 1910, Sec. 5, 1903, 1904, and 1990

| Subject term, Section No. | Page |
|---|---|
| Stresses, .27(a)(2) | 24 |
| Deterioration, .27(b)(7) | 25 |
| Electrolytic Action, .27(b)(5) | 25 |
| Extensions, .27(d)(3) | 27 |
| Fastenings, .27(b)(3) | 25 |
| Grab Bars, .27(c)(5), (d)(4) | 25, 27 |
| Ladder Extensions, .27(d)(3) | 27 |
| Landing Platforms, .27(d)(2) | 27 |
| Maintenance, .27(f) | 27 |
| Pitch, .27(e) | 27 |
| Rungs, .27(b)(1) | 24 |
| Safety Devices, .27(d)(5) | 27 |
| Side Rails, .27(b)(2) | 25 |
| Splices, .27(b)(4) | 25 |
| Welding, .27(b)(6) | 25 |
| Wells, .27(d)(1) | 26 |
| Ladders, Portable Metal, .26 | 23 |
| Care, .26(c)(2) | 24 |
| Electrical safety-related work practices, .333(c)(7) | 434 |
| Extension Ladders, .26(a)(2), (4) | 23, 24 |
| General Requirements, .26(a)(1) | 23 |
| Platform Ladders, .26(a)(5) | 24 |
| Stepladders, .26(a)(3) | 24 |
| Straight Ladders, .26(a)(2), (4) | 23, 24 |
| Use, .26(c)(3) | 24 |
| Ladders, Portable Wood, .25 | 22 |
| Care, .25(d)(1) | 23 |
| Rung Ladders, .25(c)(3) | 22 |
| Sectional, .25(c)(3)(iv) | 22 |
| Single, .25(c)(3)(ii) | 22 |
| Trestle, .25(c)(3)(v) | 23 |
| Two-Section, .25(c)(3)(iii) | 22 |
| Side-Rolling Ladders, .25(c)(5) | 23 |
| Special Purpose Ladders, .25(c)(4) | 23 |
| Masons', .25(c)(4)(iii) | 23 |
| Painters', .25(c)(4)(ii) | 23 |
| Stepladders, .25(c)(2) | 22 |
| Trolley Ladders, .25(c)(5) | 23 |
| Materials, .25(b) | 22 |
| Use, .25(d)(2) | 23 |
| Ladderway Guarding, .23(a)(2) | 20 |
| Lamps: | |
| (see also Lighting), .107(c)(7)(8);, .305(j)(1) | 107, 425 |
| Landings, Manlifts, .68(b)(6), (10) | 59, 60 |
| Lathers' Scaffolds: | |
| (see also Plasterers' Scaffolds), .28(o) | 34 |
| Lathes, .213(o) | 299 |
| Laundry Facilities, Labor Camps, .142(f) | 213 |
| Laundry Operations, .264 | 362 |
| Miscellaneous Equipment, .264(c)(4) | 362 |
| Operating Rules, .264(d) | 362 |
| Markers, .264(d)(1)(iii) | 362 |
| Mechanical Safeguards, .264(d)(2) | 362 |
| Point-of-Operation Guards, .264(c) | 362 |
| Washroom Machines, .264(c)(1) | 362 |
| Lavatories, .141(d)(2) | 213 |
| Lawn Mowers, Power, .243(e) | 332 |
| Forging Machines, .218(a)(1) | 324 |
| General Requirements, .243(e)(1) | 332 |
| Riding Rotary, .243(e)(2), (4) | 332, 333 |
| Walk-Behind, .243(e)(2), (3) | 332, 333 |
| Lead, .252(c)(7), .1025 | 505 |
| Compliance, .1025(e) | 506 |
| Confined Spaces, .252(c)(7)(i), (iii) | 338 |
| Housekeeping, .1025(h) | 507 |
| Hygiene Facilities and Practices, .1025(i) | 507 |
| Indoors, .252(c)(7)(ii), (iii) | 338 |
| Medical Removal, .1025(k) | 509 |
| Medical Surveillance, .1025(j) | 508 |
| Monitoring, .1025(d) | 505 |
| Monitoring, Observation of, .1025(o) | 511 |
| Protective Equipment and, .1025(g) | 507 |
| Clothing, .1025(g) | 507 |
| Recordkeeping, .1025(n) | 511 |
| Respiratory Protection, .1025(f) | 507 |
| Signs, .1025(m) | 511 |
| Training, Employee, .1025(l) | 510 |
| Ventilation, .252(c)(7)(iii) | 338 |

| Subject term, Section No. | Page |
|---|---|
| Leakage, Bulk Oxygen Systems, .104(b)(2)(iii) | 88 |
| Levers, Hand-Operated, .217(b)(5) | 310 |
| Lifelines: | |
| (see also Safety Belts) | |
| Chicken Ladders, .28(t)(2) | 35 |
| Confined Spaces, .252(b)(4)(iv) | 337 |
| Crawling Boards, .28(t)(2) | 35 |
| Powered Platforms, .66(e), App. C | 42, 51 |
| Welding, .252(b)(4)(iv) | 337 |
| Lighting: | |
| (see also Lamps) | |
| Container Areas, .110(d)(16) | 135 |
| Cranes, .179(c)(4), (g)(7) | 265, 267 |
| Electric Equipment, | |
| Workspace About, .303(g)(1)(v), (h)(3)(ii) | 420, 421 |
| Electrical safety-related work practices, .333(c)(4) | 434 |
| Hazardous Waste Operations, .120(m) | 167 |
| Labor Camps, .142(g) | 213 |
| Machinery, Basement Areas, .219(c)(5) | 326 |
| Manlifts, .68(b)(6)(iii), (14) | 59, 60 |
| Operating Areas, Industrial Trucks, .178(h) | 260 |
| Pulp and Paper Mills, .261(b)(2), (c)(10), (k)(21) | 349, 350, 354 |
| Sawmills, .265(c)(5)(iii), (9), (23)(iii) | 364, 365 |
| Spray Booths, .107(b)(10) | 107 |
| Storage Areas, .177(f)(1), .178(h) | 256, 260 |
| Lighting Receptacles: | |
| Cranes, .179(g)(7) | 267 |
| Liquefied Hydrogen Systems: | |
| (see Hydrogen) | |
| Liquefied, Petroleum Gases: | |
| (see also Containers, Liquefied Petroleum Gases), .110 | 121 |
| Appliances, .110(b)(20) | 131 |
| Attendant, .110(b)(14) | 128 |
| Buildings: | |
| Engines Use, .110(e)(11), (12) | 137 |
| Industrial Trucks, .110(e)(13) | 137 |
| Inside Storage, .110(f) | 137 |
| Piping Into, .110(b)(13) | 127 |
| Condensed Gas Drips, .110(d)(9) | 135 |
| Definitions, .110(a) | 121 |
| Effective Dates, .110(b)(19)(i) | 130 |
| Electrical Equipment, .110(b)(17), (18); (h)(13) | 128, 130, 140 |
| Engines in Buildings, .110(e)(11), (12) | 137 |
| Equipment Approval, .110(b)(2) | 121 |
| Fire Protection, .110(d)(14), (f)(7), (h)(14) | 135, 138, 140 |
| Fuel Handling and Storage, .178(f) | 259 |
| Gaging Devices, .110(b)(19) | 130 |
| Garaging Vehicles, .110(e)(14) | 137 |
| Handling, .110 | 121 |
| Liquid Level Gaging Device, .110(b)(19) | 130 |
| Liquid Transfer, .110(b)(14) | 128 |
| Loading, .110(b)(15) | 128 |
| Motor Fuel, .110(e) | 135 |
| Odorizing Gases, .110(b)(1) | 121 |
| Pits and Drains, .110(d)(11) | 135 |
| Regulating Equipment, .110(b)(6); (c)(5); (d)(9); (e)(9) | 122, 131, 135, 137 |
| Indoor, .110(c)(5) | 131 |
| Location, .110(b)(6) | 122 |
| Outdoor, .110(c)(4) | 131 |
| Service Stations, .110(h) | 138 |
| Storage, .110 | 121 |
| Tank Car Loading, .110(b)(15) | 128 |
| Transport Trucks, .110(b)(15) | 128 |
| Trucks, .178(b)(10), (11) | 257 |
| Trucks Conversion, .178(d), (q)(12) | 259, 262 |
| Liquid Fuels: | |
| Handling and Storage, .178(f) | 259 |
| Service Stations, .106(g) | 101 |
| Liquid Heaters, Spray, .107(e)(7) | 108 |
| Liquid Transfer: | |
| Anhydrous Ammonia, .111(b)(12), (f)(6) | 144, 147 |
| Flammable Liquids, .106(e)(2)(iv), (e)(3)(vi), (f)(3)(vi), (g), (h)(4), .107(e)(4), (9) | 99, 100, 101, |

# OSHA Standards for the General Industry

| Subject term, Section No. | Page |
|---|---|
| | 105, 108, 109 |
| Liquefied Petroleum Gases, .110(b)(14) | 128 |
| **Load Handling:** | |
| Crawler, Locomotive and Truck Cranes, .180(h) | 272 |
| Attaching, .180(h)(2) | 272 |
| Holding, .180(h)(4) | 272 |
| Moving, .180(h)(3) | 272 |
| Size, .180(h)(1) | 272 |
| Derricks, .181 | 272 |
| Attaching, .181(i)(2) | 276 |
| Boom Securing, .181(i)(6) | 276 |
| Holding, .181(i)(4) | 276 |
| Moving, .181(i)(3) | 276 |
| Size, .181(i)(1) | 276 |
| Winch Heads, .181(i)(5) | 276 |
| Overhead and Gantry Cranes, .179(n) | 268 |
| Attaching, .179(n)(2) | 268 |
| Hoist Limit Switches, .179(n)(4) | 269 |
| Moving, .179(n)(3) | 269 |
| Size, .179(n)(1) | 268 |
| **Load Ratings:** | |
| Cranes, .180(c) | 270 |
| Derricks, .181(c) | 274 |
| Overhead and Gantry Cranes, .179(b)(5) | 265 |
| Powered Platforms, .66(f)(5)(i)(C) | 45 |
| **Loading:** | |
| Bulk Plants, .106(f)(3) | 100 |
| Explosives, .109(e)(3) | 115 |
| Industrial Plants, .106(e)(4) | 99 |
| Liquefied Petroleum Gases, .110(b)(15) | 128 |
| Processing Plants, .106(h)(5) | 105 |
| Scaffolds, .29(a)(2) | 35 |
| Lockout/tagout of hazardous energy, .147 | 226 |
| Control sequence, .147(d) | 229 |
| Electrical safety-related work practices, .333(b) | 432 |
| Inspection, .147(c)(6) | 228 |
| Powered platforms, .66(f)(3)(i)(J) | 44 |
| Release procedures, .147(e) | 229 |
| Testing, .147(f)(1) | 229 |
| Training, .147(c)(7) | 228 |
| **Locomotive Cranes:** | |
| (see also Crawler, Locomotive and Truck Cranes), .180 | 269 |
| **Log Handling:** | |
| (see also Sawmills), .265(d) | 366 |
| Longshoring, .16(a) | 11 |
| Looms, .262(n) | 357 |
| Low Pressure Tanks, .106(b)(1)(iv) | 91 |
| **LP-Gases:** | |
| (see Liquefied Petroleum Gases) | |
| Lumber Handling, .265(c)(27), (28) | 366 |
| Lunchrooms, .141(g) | 212 |
| Location, .141(g)(1), (2) | 212 |
| Waste Disposal Containers, .141(g)(3) | 212 |
| **Machine Guarding:** | |
| (see Machine(ry) Guarding) | |
| Machine(ry) Guarding, .211-.222 | 293-328 |
| Abrasive Wheel Machinery, .215 | 300 |
| Anchoring Fixed Machinery, .212(a), (b) | 297 |
| Bakeries, .263(c) | 358 |
| Barrels, .212(a)(4) | 297 |
| Blades Exposure, .212(a)(5) | 297 |
| Calendars, .216 | 309 |
| Containers, .212(a)(4) | 297 |
| Definitions, .211 | 293 |
| Drums, .212(a)(4) | 297 |
| Forging Machines, .218 | 324 |
| Mills, .216 | 309 |
| Point of Operation, .212(a)(3) | 297 |
| Power Presses, .217 | 309 |
| Power Transmission Equipment, .219 | 325 |
| Types, .212(a)(1) | 297 |
| Woodworking Machinery, .213 | 297 |
| **Machines:** | |
| Abrasive Wheels, .215 | 300 |
| Definitions, .211 | 293 |
| Forging, .218 | 324 |

| Subject term, Section No. | Page |
|---|---|
| Laundry, .264 | 362 |
| Mills and Calenders, .216 | 309 |
| Power Transmission, Mechanical, .219 | 325 |
| Presses, Mechanical, .217 | 309 |
| Textiles, .262 | 355 |
| Woodworking, .213 | 297 |
| Magazines, Explosives, .109(c)(2) | 112 |
| Class I, .109(c)(3) | 113 |
| Class II, .109(c)(4) | 113 |
| Storage, .109(c)(5) | 113 |
| **Maintenance:** | |
| (see also Term To Which It Applies) | |
| Bulk Oxygen Systems, .104(b)(10) | 89 |
| Cranes, .179(l), .180(f) | 268, 271 |
| Derricks, .181(f) | 275 |
| Fire Extinguishers, .157(e) | 236 |
| Gaseous Hydrogen Systems, .103(b)(5) | 85 |
| Industrial Plants, .106(e)(9) | 100 |
| Liquefied Hydrogen Systems, .103(c)(5) | 88 |
| Powder Coatings, .107(l)(4) | 110 |
| Powered Industrial Trucks, .178(q) | 262 |
| Powered Platforms, .66(e)(6) | 43 |
| Processing Plants, .106(h)(8) | 106 |
| Respirators, .134(f) | 191 |
| Sprinkler Systems, .159(c)(2) | 238 |
| Standpipe and Hose System, .158(e) | 238 |
| Management, Safety and Health Management Voluntary Guidelines | 807 |
| Manifolding Gas Cylinders, .253(c) | 340 |
| Fuel-Gas, .253(c)(1) | 340 |
| Operating Procedures, .253(c)(5) | 341 |
| Oxygen, .253(c)(2), (3) | 340 |
| Portable Outlet Headers, .253(c)(4) | 341 |
| Manholes, .23(a)(6), .268(o) | 20, 381 |
| Manlifts, .68 | 59 |
| Belts, .68(c)(1)(ii) | 60 |
| Brakes, .68(c)(1)(i) | 60 |
| Clearances, .68(b)(11) | 60 |
| Design, .68(b)(3) | 59 |
| Exit Protection, .68(b)(8) | 59 |
| Floor Openings, .68(b)(5), (7) | 59 |
| Guardrails, .68(b)(8)(i), (10)(iv) | 59, 60 |
| Guards, .68(b)(7), (9) | 59, 60 |
| Handholds, .68(c)(4) | 60 |
| Inspections, .68(e) | 61 |
| Instruction Signs, .68(c)(7) | 61 |
| Ladders, .68(b)(12) | 60 |
| Landings, .68(b)(6) | 59 |
| Lighting, .68(b)(6)(iii), (14) | 59, 60 |
| Machinery, .68(c) | 60 |
| Mechanical Requirements, .68(c) | 60 |
| Operating Rules, .68(d) | 61 |
| Platforms, .68(c)(3) | 60 |
| Speed, .68(c)(2) | 60 |
| Steps, .68(c)(3) | 60 |
| Stops, .68(c)(5), (6) | 60, 61 |
| Warning Signs, .68(c)(7) | 61 |
| Weather Protection, .68(b)(15) | 60 |
| Marine Service Stations, .106(g)(4) | 103 |
| Marine Terminals, .16(b) | 12 |
| Marking Physical Hazards, .144 | 213 |
| Sawmills, .265(c)(11) | 364 |
| **Markings:** | |
| (see also Signs and Tags) | |
| Bulk Oxygen Systems, .104(b)(8)(viii) | 89 |
| Compressed Gas Cylinders, .253(b)(1) | 339 |
| Electric equipment: | |
| General, .303(e) | 420 |
| Hazardous locations, .307(b)(2)(ii) | 429 |
| Explosive Actuated Tools, .243(d)(3) | 332 |
| Explosives, .109(d)(2)(ii) | 114 |
| Eye and Face Protection, .133(a)(4) | 187 |
| Gaseous Hydrogen Systems, .103(b)(1)(v) | 84 |
| Hazardous materials, retention of DOT markings, .1201 | 722 |
| Liquefied Hydrogen Systems, .103(c)(1)(iii) | 85 |
| Liquefied Petroleum Gases, .110(b)(5), (c) | 122, 131 |

Index  Lad—Mar

| Subject term, Section No. | Page |
|---|---|
| Load Ratings: | |
|     Cranes, .180(c)(2) | 270 |
|     Derricks, .181(c) | 274 |
|     Powered Industrial Trucks, .178(a)(3) | 257 |
|     Powered Platforms, .66(f)(7) | 46 |
|   Physical Hazards, .144 | 213 |
|   Respirators, .134(g) | 191 |
|   Sawmills, .265(c)(11) | 364 |
| Mason's Adjustable Multiple Point Suspension Scaffolds, .28(f) | 31 |
| Masons' Ladders, .25(c)(4)(iii) | 23 |
| Matching Machines, .213(n) | 299 |
| Material Safety Data Sheets, chemical hazards information, .1200 | 710 |
| Materials Handling and Storage: | |
|   Aisles and Passageways, .176(a) | 255 |
|   Clearance Signs, .176(e) | 255 |
|   Cranes—Crawler, Locomotive and Truck, .180 | 269 |
|   Cranes, Overhead and Gantry, .179 | 264 |
|   Derricks, .181 | 272 |
|   Guarding Openings, .176(g) | 255 |
|   Hazardous materials, retention of DOT markings, .1201 | 722 |
|   Hazardous Waste Operations, .120(j), (p)(6) | 165, 168 |
|   Helicopters, .183 | 276 |
|   Housekeeping, .176(c) | 255 |
|   Mechanical Equipment, .176(a) | 255 |
|   Powered Industrial Trucks, .178 | 257 |
|   Pulp and Paper Mills, .261(c), (d), (m) | 349, 350, 354 |
|   Railroad Car Blocks, .176(f) | 255 |
|   Securing, .176(b) | 255 |
| Maximum Allowable Concentration: | |
|   Fluorine, .252(c)(5)(ii) | 338 |
|   Welding Contamination, .252(c)(1)(iii) | 337 |
| Means of Egress: | |
|   (see Egress Means) | |
| Mechanical Handling Equipment: | |
|   Clearances, .176(a) | 255 |
|   Powered Industrial Trucks, .177(e), .178 | 256, 257 |
| Mechanical Power Presses: | |
|   Air Controlling Equipment, .217(b)(10) | 311 |
|   Brakes, Friction, .217(b)(2) | 309 |
|   Clearances, Work Area, .217(f)(3) | 313 |
|   Clutches: | |
|     Full Revolution, .217(b)(3) | 309 |
|     Part Revolution, .217(b)(7) | 310 |
|   Definitions, .211(d) | 294 |
|   Dies, .217(d) | 312 |
|   Electrical Controls, .217(b)(8) | 310 |
|   Excluded Machines, .217(a)(5) | 309 |
|   Foot Pedals, .217(b)(4) | 310 |
|   Guarding, .217(b) | 309 |
|   Guide Posts, .217(d)(4) | 312 |
|   Hand Feeding Tools, .217(c)(4) | 312 |
|   Hazards: | |
|     Guide Posts, .217(d)(4) | 312 |
|     Personnel, .217(b)(1) | 309 |
|   Hydraulic Equipment, .217(b)(11) | 311 |
|   Inspection, Records, .217(e)(1) | 313 |
|   Instructions, .217(f)(2) | 313 |
|   Lever, Hand-Operated, .217(b)(5) | 310 |
|   Maintenance: | |
|     Records, .217(e)(1) | 313 |
|     Training Personnel, .217(e)(3) | 313 |
|   Modifications, .217(a)(4), (e)(2) | 309, 313 |
|   Operating Instructions, .217(f)(2) | 313 |
|     Overloading, .217(f)(4) | 313 |
|     Point of Operation, .217(c) | 311 |
|   Pressure Vessels, .217(b)(12) | 311 |
|   Slide Counterbalances, .217(b)(9) | 310 |
|     Air, .217(b)(9)(iii)-(v) | 311 |
|     Spring, .217(b)(9)(i), (ii) | 310 |
|   Training Maintenance Personnel, .217(e)(3) | 313 |
|   Treadles, .217(b)(4) | 310 |
|   Trips, Two-Hand, .217(b)(6) | 310 |
|   Unitized Tooling, .217(d)(5) | 312 |
| Mechanical Power Transmission Apparatus, .219 | 325 |

| Subject term, Section No. | Page |
|---|---|
|   Bearings, .219(j), (p)(3) | 327, 328 |
|   Belts: | |
|     Care, .219(p)(6) | 328 |
|     Fasteners, .219(l)(4) | 327 |
|     Perches, .219(l)(3) | 327 |
|     Shifters, .219(l)(1) | 327 |
|     Shippers, .219(l)(2) | 327 |
|   Chains, .219(f) | 327 |
|   Clutches, .219(k), (l) | 327 |
|   Collars, .219(i) | 327 |
|   Couplings, .219(i), (k)(1) | 327 |
|   Cutoff Couplings, .219(k)(1) | 327 |
|   Definitions, .211(f) | 296 |
|   Drives: | |
|     Belt, Rope, and Chain, .219(e) | 326 |
|     Friction, .219(g) | 327 |
|   Engine Rooms, .219(k)(2) | 327 |
|   Equipment Care, .219(p) | 328 |
|   Excluded Apparatus, .219(a)(1) | 325 |
|   Gears, .219(f) | 327 |
|   Guarding, .219 | 325 |
|   Guards: | |
|     Disks, .219(m)(1) | 327 |
|     Horizontal Overhead: | |
|       Belts, .219(o)(3) | 327 |
|       Rope and Chain Drives, .219(o)(4) | 328 |
|     Materials, .219(m)(1), (o) | 327 |
|     Prime Mover, .219(b) | 325 |
|     Shields, .219(m)(2) | 327 |
|     Standard, .219(m) | 327 |
|       Manufacturing Methods, .219(m)(2) | 327 |
|       Materials, .219(m)(1) | 327 |
|     Toeboards, .219(o)(5) | 328 |
|     Wooden, .219(o)(2) | 327 |
|   Hangers, .219(p)(4) | 328 |
|   Keys, .219(h) | 327 |
|   Located in Basements, Towers, and Rooms, .219(c)(5) | 326 |
|   Personnel Protection, .219(p)(7) | 328 |
|   Prime Mover Guards: | |
|     Connecting Rods, .219(b)(2) | 326 |
|     Cranks, .219(b)(2) | 326 |
|     Extension Piston Rods, .219(b)(3) | 326 |
|     Flywheels, .219(b)(1) | 325 |
|     Tail Rods, .219(b)(3) | 326 |
|   Projections, .219(h) | 327 |
|   Pulleys, .219(d), (k), (p)(5) | 326, 327, 328 |
|   Setscrews, .219(h) | 327 |
|   Shafting, .219(c) | 326 |
|     Care, .219(p)(2) | 328 |
|     Guarding, .219(c)(2), (3) | 326 |
|     Installation, .219(c)(1) | 326 |
|     Projecting Shafts, .219(c)(4) | 326 |
|   Sprockets, .219(f) | 327 |
|   Standards Sources, .221 | 328 |
|   Textile Industry, .219(a)(3) | 325 |
| Medical Services: | |
|   (see also First Aid Personnel Protective Equipment), .151 | 231 |
|   Asbestos, .1001(j) | 465 |
|   First Aid, .151 | 231 |
|     Labor Camps, .142(k) | 213 |
|     Logging, .266(d)(2) | 370 |
|     Textiles, .262(pp) | 358 |
|     Welding, .252(c)(13) | 338 |
|   Labor Camps, .142(k) | 213 |
|   Radiation Exposure Records, .1096(n) | 710 |
|   Standards Sources, .153 | 231 |
| Medical Surveillance, .120(f) | 163 |
| Mercantile Occupancies, .106(d)(5)(iv) | 97 |
| Mercury, .252(c)(10) | 338 |
|   Exposure Limit, .95(b) | 73 |
| Metal Cutting: | |
|   (see Cutting and Welding) | |
| Metal Ladders, Portable: | |
|   (see also Ladders, Portable Metal), .26 | 23 |
| Methyl chloromethyl ether, .1003 | 488 |
|   Area requirements, .1003(c) | 488 |

# OSHA Standards for the General Industry

| Subject term, Section No. | Page |
|---|---|
| Closed system operation, .1003(c)(2) | 488 |
| Isolated systems, .1003(c)(1) | 488 |
| Maintenance and decontamination activities, .1003(c)(5) | 489 |
| Open-vessel system operations, .1003(c)(3) | 488 |
| Transfer from a closed operation, .1003(c)(4) | 488 |
| Medical surveillance, .1003(g) | 490 |
| Examinations, .1003(g)(1) | 490 |
| Records, .1003(g)(2) | 490 |
| Regulated area requirements, .1003(d) | 489 |
| Contamination control, .1003(d)(4) | 489 |
| Emergencies, .1003(d)(2) | 489 |
| Hygiene facilities and practices, .1003(d)(3) | 489 |
| Signs, information, and training, .1003(e) | 489 |
| Container contents identification, .1003(e)(2) | 489 |
| Lettering, .1003(e)(3) | 489 |
| Prohibited statements, .1003(e)(4) | 489 |
| Signs, .1003(e)(1) | 489 |
| Training and indoctrination, .1003(e)(5) | 489 |
| Methylene Chloride: | |
| App. A, App. C, .1052 | 697, 703 |
| Employee Information and Training, .1052(l) | 696 |
| Exposure Monitoring, .1052(d) | 662 |
| Hazard Communications, .1052(k) | 696 |
| Hygiene Facilities, .1052(i) | 694 |
| Medical Surveillance, .1052(j), App. B | 694, 700 |
| Methods of Compliance, .1052(f) | 693 |
| Permissible Exposure Limits, .1052(c) | 692 |
| Protective Work Clothing and Equipment, .1052(h) | 694 |
| Recordkeeping, .1052(m) | 696 |
| Regulated Areas, .1052(e) | 693 |
| Respiratory Protection, .1052(g) | 693 |
| 4,4-Methylenedianiline: | |
| Airborne Concentration, .1050(c) | 668 |
| App. A, App. B, App. E, .1050 | 675, 678 |
| Compliance, .1050(g) | 669 |
| Emergency Situations, .1050(d) | 669 |
| Exposure, Permissible, .1050(c) | 669 |
| Hazard Communication, .1050(k) | 671 |
| Housekeeping, .1050(l) | 671 |
| Hygiene Facilities and Practices, .1050(j) | 671 |
| Medical Surveillance, .1050(m), App. C | 671, 676 |
| Monitoring, .1050(e), App. D | 669, 676 |
| Personal Protective Equipment, .1050(i) | 670 |
| Clothing, .1050(i) | 670 |
| Recordkeeping, .1050(n) | 673 |
| Regulated Areas, .1050(f) | 669 |
| Respiratory protection, .1050(h) | 670 |
| Mill Roll Heights, .216(a)(4) | 309 |
| Mills, Pulp, Paper and Paperboard (see also Pulp, Paper and Paperboard Mills), .261 | 349 |
| Mills, Rubber and Plastics Industry: | |
| Definitions, .211(c) | 294 |
| Location Protection, .216(d)(1) | 309 |
| Roll Heights, .216(a)(4) | 309 |
| Safety Controls, .216(b) | 309 |
| Auxiliary Equipment, .216(b)(3) | 309 |
| Safety Trip Control, .216(b)(1) | 309 |
| Stopping Limits, .216(f)(1), (2) | 309 |
| Switches, Trip and Emergency, .216(e) | 309 |
| Minors: | |
| Ionizing Radiation Exposure, .1096(b)(3), (c)(2), (d)(2)(ii) | 707 |
| Mixing: | |
| Blasting Agents, .109(g)(2), (3); (h)(3), (4) | 115, 116, 118, 119 |
| Explosives, .109(h)(3), (4) | 118, 119 |
| Molding Machines, .213(n) | 299 |
| Monitoring: | |
| Asbestos, .1001(f) | 464 |
| Ionizing Radiation, .1096(d) | 707 |
| Mortising Machines, .213(e) | 298 |
| Motor Fuels, .110(e) | 135 |
| Motor Vehicles: | |
| Anhydrous Ammonia, .111(f) | 147 |
| Motorized Hand Trucks: | |
| (see also Powered Industrial Trucks), .178 | 257 |
| Multi-piece Rim Wheels, .177 | 255 |

| Subject term, Section No. | Page |
|---|---|
| alpha-Napthylamine, .1003 | 488 |
| Area requirements, .1003(c) | 488 |
| Closed system operation, .1003(c)(2) | 488 |
| Contamination control, .1003(d)(4) | 489 |
| Emergencies, .1003(d)(2) | 489 |
| General regulated area requirements, .1003(d) | 489 |
| Hygiene facilities and practices, .1003(d)(3) | 489 |
| Isolated systems, .1003(c)(1) | 488 |
| Maintenance and decontamination activities, .1003(c)(5) | 489 |
| Open-vessel system operations, .1003(c)(3) | 488 |
| Transfer from a closed operation, .1003(c)(4) | 488 |
| Medical surveillance, .1003(g) | 490 |
| Examinations, .1003(g)(1) | 490 |
| Records, .1003(g)(2) | 490 |
| Regulated area requirements, .1003(d) | 489 |
| Contamination control, .1003(d)(4) | 489 |
| Emergencies, .1003(d)(2) | 489 |
| Hygiene facilities and practices, .1003(d)(3) | 489 |
| Signs, information, and training, .1003(e) | 489 |
| Container contents identification, .1003(e)(2) | 489 |
| Lettering, .1003(e)(3) | 489 |
| Prohibited statements, .1003(e)(4) | 488 |
| Signs, .1003(e)(1) | 489 |
| Training and indoctrination, .1003(e)(5) | 489 |
| beta-Napthylamine, .1003 | 488 |
| Area requirements, .1003(c) | 488 |
| Closed system operation, .1003(c)(2) | 488 |
| Contamination control, .1003(d)(4) | 489 |
| Emergencies, .1003(d)(2) | 489 |
| General regulated area requirements, .1003(d) | 489 |
| Hygiene facilities and practices, .1003(d)(3) | 489 |
| Isolated systems, .1003(c)(1) | 488 |
| Maintenance and decontamination activities, .1003(c)(5) | 489 |
| Open-vessel system operations, .1003(c)(3) | 488 |
| Transfer from a closed operation, .1003(c)(4) | 488 |
| Medical surveillance, .1003(g) | 490 |
| Examinations, .1003(g)(1) | 490 |
| Records, .1003(g)(2) | 490 |
| Regulated area requirements, .1003(d) | 489 |
| Contamination control, .1003(d)(4) | 489 |
| Emergencies, .1003(d)(2) | 489 |
| Hygiene facilities and practices, .1003(d)(3) | 489 |
| Signs, information, and training, .1003(e) | 489 |
| Container contents identification, .1003(e)(2) | 489 |
| Lettering, .1003(e)(3) | 489 |
| Prohibited statements, .1003(e)(4) | 489 |
| Signs, .1003(e)(1) | 489 |
| Training and indoctrination, .1003(e)(5) | 489 |
| Needle Beam Scaffolds, .28(n) | 34 |
| 4-Nitrobiphenyl, .1003 | 488 |
| Area requirements, .1003(c) | 488 |
| Closed system operation, .1003(c)(2) | 488 |
| Isolated systems, .1003(c)(1) | 488 |
| Maintenance and decontamination activities, .1003(c)(5) | 489 |
| Open-vessel system operations, .1003(c)(3) | 488 |
| Transfer from a closed operation, .1003(c)(4) | 488 |
| Medical surveillance, .1003(g) | 490 |
| Examinations, .1003(g)(1) | 490 |
| Records, .1003(g)(2) | 490 |
| Regulated area requirements, .1003(d) | 489 |
| Contamination control, .1003(d)(4) | 489 |
| Emergencies, .1003(d)(2) | 489 |
| Hygiene facilities and practices, .1003(d)(3) | 489 |
| Signs, information, and training, .1003(e) | 489 |
| Container contents identification, .1003(e)(2) | 489 |
| Lettering, .1003(e)(3) | 489 |
| Prohibited statements, .1003(e)(4) | 489 |
| Signs, .1003(e)(1) | 489 |
| Training and indoctrination, .1003(e)(5) | 489 |
| N-Nitrosodimethylamine, .1003 | 488 |
| Area requirements, .1003(c) | 488 |
| Closed system operation, .1003(c)(2) | 488 |
| Isolated systems, .1003(c)(1) | 488 |

| Subject term, Section No. | Page |
|---|---|
| Maintenance and decontamination activities, .1003(c)(5) | 489 |
| Open-vessel system operations, .1003(c)(3) | 488 |
| Transfer from a closed operation, .1003(c)(4) | 488 |
| Medical surveillance, .1003(g) | 490 |
| Examinations, .1003(g)(1) | 490 |
| Records, .1003(g)(2) | 490 |
| Regulated area requirements, .1003(d) | 489 |
| Contamination control, .1003(d)(4) | 489 |
| Emergencies, .1003(d)(2) | 489 |
| Hygiene facilities and practices, .1003(d)(3) | 489 |
| Signs, information, and training, .1003(e) | 489 |
| Container contents identification, .1003(e)(2) | 489 |
| Lettering, .1003(e)(3) | 489 |
| Prohibited statements, .1003(e)(4) | 489 |
| Signs, .1003(e)(1) | 489 |
| Training and indoctrination, .1003(e)(5) | 489 |
| Nitrous Oxide, .105 | 89 |
| Noise Exposure, .95, App. A-H | 76, 78, 80, 81 |
| Effective Dates, .98 | 82 |
| Personal Protective Equipment, .95(b)(1), (c), (i), (j) | 73, 74, 75 |
| Nonionizing Radiation, .97 | 82 |
| Effective Dates, .98 | 82 |
| Electromagnetic Radiation, .97(a) | 82 |
| Nonpotable Water, .120(n)(2) | 167 |
| Noxious Gases, Storage Areas, .178(i) | 260 |
| Nozzles: | |
| Abrasive Blasting, .94(a)(2)(iii), .244(b) | 63, 333 |
| Gasoline, .106(g)(3)(vi) | 103 |
| Standpipe, .158(c)(4) | 238 |
| Occupational Noise Exposure: | |
| (see Noise Exposure) | |
| Odorizing Gases, .110(b)(1) | 121 |
| Open-Sided Floors, .23(c) | 20 |
| Open Surface Tanks: | |
| (see Tanks: Open Surface) | |
| Openings: | |
| (see also Floor Openings (Holes), Wall Openings (Holes)), .23 | 20 |
| Tanks: | |
| Inside, .106(b)(4)(iv) | 94 |
| Organic Peroxide Coatings: | |
| (see also Dual Component Coatings), .107(m) | 110 |
| Outdoor Storage: | |
| Flammable Liquids, .106(d)(6) | 98 |
| Outlet Headers, welding, .253(c)(4) | 341 |
| Protective Equipment, .253(e)(4) | 341 |
| Outrigger Scaffolds, .28(e) | 31 |
| Outside Storage Trucks, .178(c)(2)(ix), (xi) | 259 |
| Ovens, .263(l) | 361 |
| Direct-Fire, .263(l)(10) | 362 |
| Direct Recirculating, .263(l)(11) | 362 |
| Electrical Heating Equipment, .263(l)(8) | 361 |
| General Requirements, .263(l)(9) | 361 |
| Indirect Recirculating, .263(l)(15) | 362 |
| Location, .263(l)(1) | 361 |
| Mechanical Parts, .263(l)(3) | 361 |
| Overflow Pipes, Dip Tanks, .125(b) | 185 |
| Overhead Cranes: | |
| (see also Overhead and Gantry Cranes), .179 | 263 |
| Overhead and Gantry Cranes: | |
| Access, .179(c)(2) | 265 |
| Adjustments, .179(l)(3) | 268 |
| Brakes, .179(f) | 266 |
| Bridge Bumpers, .179(e)(2) | 265 |
| Cabs, .179(c) | 265 |
| Clearances, .179(b)(6) | 265 |
| Electric Equipment, .179(g) | 266 |
| Fire Extinguishers, .179(c)(3), (o)(3) | 265, 269 |
| Footwalks, .179(d) | 265 |
| Guards, .179(e)(5), (6) | 266 |
| Handrails, .179(d)(3), (4)(ii) | 265 |
| Hoisting Equipment, .179(h) | 267 |
| Hoisting Rope Guards, .179(e)(5) | 266 |
| Inspections, .179(j), (m) | 267, 268 |
| Ladders, .179(d)(4) | 265 |

| Subject term, Section No. | Page |
|---|---|
| Lighting, .179(c)(4) | 265 |
| Load Handling, .179(n) | 268 |
| Maintenance, .179(l) | 268 |
| Modifications, .179(b)(3) | 265 |
| Moving Part Guards, .179(e)(6) | 266 |
| Rail Clamps, .179(b)(4) | 265 |
| Rail Sweeps, .179(e)(4) | 266 |
| Rated Load: | |
| Markings, .179(b)(5) | 265 |
| Tests, .179(k)(2) | 268 |
| Repairs, .179(l)(3) | 268 |
| Rope inspection, .179(m) | 268 |
| Stairways, .179(d)(3) | 265 |
| Testing, .179(k) | 268 |
| Toeboards, .179(d)(3) | 265 |
| Trolley Bumpers, .179(e)(3) | 265 |
| Trolley Stops, .179(e)(1) | 265 |
| Warning Devices, .179(i) | 267 |
| Wind Indicators, .179(b)(4) | 265 |
| Overhead Wires: | |
| Cranes, .180(j) | 272 |
| Derricks, .181(j)(5) | 276 |
| Overspray Collectors, .107(b)(6) | 107 |
| Oxygen: | |
| (see also Bulk Oxygen Systems), .104 | 88 |
| Effective Dates, .114 | 148 |
| Standards Sources, .115 | 148 |
| Storage, .104 (b) | 88 |
| Oxygen-Fuel Gas Systems, .253 | 339 |
| Outlet Headers, .253(c)(4) | 341 |
| Piping Systems, .253(d) | 341 |
| Protective Equipment, .253(e) | 342 |
| Oxygen Manifolds: | |
| High Pressure, .253(c)(2) | 340 |
| Low Pressure, .253(c)(3) | 340 |
| Painters' Stepladders, .25(c)(4)(ii) | 23 |
| Paints: | |
| Color Code, .144 | 213 |
| Paper and Paperboard Mills: | |
| (see also Pulp, Paper and Paperboard Mills), .261 | 349 |
| Passageways, Working Surfaces, .22(b) | 20 |
| Penalties, Proposed, 1903.15 | 738 |
| Permissible Exposure Limits, .1000 (c) | 463 |
| Personal Protection: | |
| (see also Personal Protective Equipment), .219(p)(7) | 328 |
| Personal Protective Equipment: | |
| (see also Lifelines, and Other Terms Listed Below) | |
| Abrasive Blasting, .94(a)(5) | 63 |
| Asbestos Exposure, .1001(d) | 463 |
| Clothing, .1001(h) | 465 |
| Bloodborne pathogens, exposure to, use of ppe, .1030(c)(2)(ii), (d)(2)(i) and (3) | 595 |
| Electrical Protective Equipment, .137, .268(f) | 204, 377 |
| Electrical safety-related work practices, use of ppe, .333(c)(2), .335(a) | 433, 435 |
| Emergency Showers and Fountains: | |
| Pulp, Paper and Paperboard Mills, .261(g)(5), (18) | 351, 352 |
| Eye Protection, .133 | 187 |
| Face Protection, .133 | 187 |
| Fire Brigades, .156 | 234 |
| Foot Protection, .136 | 204 |
| General Requirements, .132 | 187 |
| Hand protection, .138 | 206 |
| Hazardous Waste Operations, .120(g) | 164 |
| Head Protection, .135 | 204 |
| Noise Exposure, .95(b)(1) | 73 |
| Pulp and Paper Mills, .261(g)(2), (i)(4), (k)(3) | 351, 353 |
| Respiratory Protection, .134 | 188 |
| Welding, .252-.257 | 335-347 |
| Booths, .252(b)(2)(iii) | 336 |
| Cable, .252(b)(1)(ii) | 336 |
| Clothing, .252(b)(3) | 336 |
| Eye Protection, .252(b)(2) | 336 |
| Helmets, .252(b)(2) | 336 |
| Railing, .252(b)(1)(i) | 336 |
| Shade Numbers, Lenses, .252(b)(2)(ii)(H) | 336 |

# OSHA Standards for the General Industry

| Subject term, Section No. | Page |
|---|---|
| Petitions, Modification of abatement, *1903.14a*, | 738 |
| Physical Hazards Markings: | |
| *(see also Color Codes, Physical Hazards: Markings)*, *.144* | 213 |
|     Effective Dates, *.149* | 230 |
|     Standards Sources, *.150* | 230 |
| Piers and Wharves: | |
| *(see also Wharves)* | |
|     Trucks Used, *.178(c)(2)(x)* | 259 |
| Pipes: | |
|     Dip Tanks, *.125(b)* | 185 |
|     Flammable Liquids, *.107(e)(6)* | 108 |
|     Overflow, *.125(b)* | 185 |
| Piping: | |
| *(see Piping, Fittings and Tubing; Piping, Valves and Tubing)* | |
| Piping, Fittings and Tubing: | |
|     Anhydrous Ammonia, *.111(b)(7)* | 142 |
|     Bulk Oxygen Systems, *.104(b)(5)* | 89 |
|     Gaseous Hydrogen Systems, *.103(b)(1)(ii), (iii)* | 83 |
|     Liquefied Hydrogen Systems, *.103(c)(1)(iv), (v)* | 85 |
|     Liquefied Petroleum Gases, *.110(b)(8)* | 123 |
|     Safety Relief Devices, *.103(b)(1)(ii), (c)(1)(iv)* | 83, 85 |
| Piping Systems, Oxygen-Fuel, *.253(d)* | 341 |
|     Fittings, *.253(d)(1)* | 341 |
|     Installation, *.253(d)(3)* | 341 |
|     Painting, *.253(d)(4)* | 342 |
|     Piping, *.253(d)(1)* | 341 |
|     Piping Joints, *.253(d)(2)* | 341 |
|     Pressure Relief Devices, *.253(e)(2)* | 342 |
|     Protective Equipment, *.253(e)(3), (4)* | 342 |
|     Signs, *.253(d)(4)* | 342 |
|     Station Outlets, *.253(e)(4)* | 342 |
|     Testing, *.253(d)(5)* | 342 |
|     X-ray Inspections, *.252(d)(1)(vii)* | 338 |
| Piping, Valves, and Fittings: | |
|     Flammable and Combustible Liquids, *.106(c)* | 95 |
|         Corrosion Protection, *.106(c)(5)* | 96 |
|         Design, *.106(c)(1)(i)* | 95 |
|         Joints, *.106(c)(3)* | 96 |
|         Materials, *.106(c)(2)* | 95 |
|         Supports, *.106(c)(4)* | 96 |
|         Testing, *.106(c)(7)* | 96 |
|         Valves, *.106(c)(6)* | 96 |
|     Liquefied Petroleum Gases, *.110(h)(7)* | 139 |
|     Processing Plants, *.106(h)(4)(ii)* | 105 |
| Pits, *.23(a)(5)* | 20 |
|     Drains, *.110(d)(11)* | 135 |
| Planing Machines, *.213(n)* | 299 |
| Plasterers' Scaffolds, *.28(o)* | 34 |
| Plastics Industry: | |
| *(see also Mills, Rubber and Plastics Industry)* | |
|     Auxiliary Equipment, *.216(a)(3)* | 309 |
|     Installations, *.216(a)* | 309 |
|     Mills and Calenders, *.216* | 309 |
|     Standards Sources, *.221* | 328 |
| Platform Lift Trucks: | |
| *(see also Powered Industrial Trucks)*, *.178* | 257 |
| Platforms, Scaffolds: | |
| *(see also Listings Under Specific Type Scaffold)* | |
|     Guarding, *.23(c)* | 20 |
|     Manlifts, *.68(c)(3)* | 60 |
| Pneumatic Powered Tools, *.243(b)* | 330 |
|     Airhoses, *.243(b)(2)* | 330 |
|     ToolRetainer, *.243(b)(1)* | 330 |
| Point of Operation Guarding, *.217(c)* | 297, 311 |
| Polishing: | |
| *(see Grinding, Polishing and Buffing)* | |
| Portable Fire Extinguishers: | |
| *(see also Fire Extinguishers, Portable)*, *.157* | 236 |
| Portable Metal Ladders: | |
| *(see also Ladders, Portable Metal)*, *.26* | 23 |
| Portable Stepladders: | |
| *(see Stepladders, Portable)* | |
| Portable Tank Storage: | |
| *(see Tank Storage, Portable)* | |
| Portable Tanks: | |
| *(see Tanks, Portable)* | |
| Portable Tools: | |
| *(see also Powered Tools,* | |
|   *Hand and Portable)*, *.244* | 333 |

| Subject term, Section No. | Page |
|---|---|
| Portable Welding Machines: | |
| *(see Welding Machines, Portable)* | |
| Portable Wood Ladders: | |
| *(see also Ladders, Portable Wood)*, *.25* | 22 |
| Posting, Citations, *1903.16*, | 739 |
| Powder Coatings, *.107(l)* | 110 |
| Power Presses, Mechanical: | |
| *(see Mechanical Power Presses)* | |
| Powered Industrial Trucks: | |
|     Approval Labels, *.178(a)(3), (7)* | 257 |
|     Batteries, *.178(g)* | 260 |
|     Combustible Dusts, *.178(c)(2)(vi)* | 259 |
|     Conversion, *.178(d), (q)(12)* | 259, 262 |
|     Design and Construction, *.178(a)(2)* | 257 |
|     Designated Locations, *.178(c)(1)* | 258 |
|     Designations, Trucks: | |
|         D, *.178(b)(1)* | 257 |
|         DS, *.178(b)(2)* | 257 |
|         DY, *.178(b)(3)* | 257 |
|         E, *.178(b)(4)* | 257 |
|         EE, *.178(b)(6)* | 257 |
|         ES, *.178(b)(5)* | 257 |
|         EX, *.178(b)(7)* | 257 |
|         G, *.178(b)(8)* | 257 |
|         GS, *.178(b)(9)* | 257 |
|         LP, *.178(b)(10)* | 257 |
|         LPS, *.178(b)(11)* | 257 |
|     Effective Dates, *.182* | 276 |
|     Fire Protection, *.178(a)(1)* | 257 |
|     Front End Attachments, *.178(a)(5)* | 257 |
|     Fuel Handling, *.178(f)* | 259 |
|     Gases and Fumes, *.178(i)* | 260 |
|     Grain Handling, *.178(c)(2)(vi)* | 257, 259 |
|     Hazardous Materials, *.178(c)(2)* | 258 |
|     Lighting, *.178(h)* | 260 |
|     Loading, *.178(o)* | 261 |
|     Maintenance, *.178(q)* | 262 |
|     Markings, *.178(a)(6)* | 257 |
|     Modifications, *.178(a)(4)* | 257 |
|     Operations, *.178(p)* | 262 |
|     Repairs, *.178(q)* | 262 |
|     Safety Guards, *.178(e)* | 259 |
|     Training Operators, *.178(l)* | 260 |
|     Traveling, *.178(n)* | 261 |
|     Truck Operations, *.178(m)* | 261 |
| Powered Platforms, *.66-.70* | 41-61 |
|     Access, *.66(f)(3)(i)(K), (ii)(D), (iii)(C)(2), (f)(5)(ii)(J)* | 44, 45 |
|     Application, *.66(b)* | 41 |
|     Buildings, affected parts, *.66(e)* | 42 |
|     Definitions, *.66(d)* | 41 |
|     Electrical, *.66(e)(11), (f)(8)* | 43, 47 |
|     Equipment, *.66(f)* | 43 |
|     Fall Arrest systems *66(f)(5)(ii)(L), (M), (j), App. C* | 45, 48, 51 |
|     Hoisting Equipment, *.66(f)(4), (g)(6)* | 44, 48 |
|     Inspections, *.66(g)* | 47 |
|     Lockout, *.66(f)(3)(i)(J)* | 44 |
|     Maintenance, *.66(e)(5), (10), (g), (h)* | 43, 47, 48 |
|     Manlifts, *.68* | 59 |
|     Reshackling Hoists, *.66(h)(4)* | 48 |
|     Ropes, *.66(f)(7), (g)(5), (h)(3), (4)* | 46, 47, 48 |
|     Standards Sources, *.69* | 61 |
|     Tests, *.66(g)* | 47 |
|     Vehicle-Mounted, *.67* | 58 |
| Powered Tools, Hand and Portable: | |
|     Abrasive Wheels, *.243(c)* | 330 |
|     Compressed Air Cleaning, *.242(b)* | 330 |
|     Definitions, *.241* | 329 |
|     Effective Dates, *.245* | 333 |
|     Explosive Actuated Fastening, *.243(d)* | 331 |
|     Guarding, *.243* | 330 |
|     Lawn Mowers, Power, *.243(e)* | 332 |
|     Pneumatic Powered, *.243(b)* | 330 |
|     Standards Sources, *.246* | 333 |
|     Woodworking, *.243(a)* | 330 |
| Presses: | |
| *(see also Mechanical Power Presses)* | |
|     Cold Trimming, *.218(g)(2)* | 325 |

# Subject Index For 29 CFR 1910, Sec. 5, 1903, 1904, and 1990

| Subject term, Section No. | Page |
|---|---|
| Forging, .218(f) | 325 |
| Hot Trimming, .218(g)(1) | 325 |
| Hydraulic Forging, .218(f)(2) | 325 |
| Trimming, .218(g) | 325 |
| Pressure Gages, Air Receivers, .169(b)(3) | 253 |
| Pressure Vessels, .106(b)(1)(v), .217(b)(12) | 91, 311 |
| Chemical Plants, .106(i)(3) | 106 |
| Distilleries, .106(i)(3) | 106 |
| Pulp and Paper Mills, .261(g)(16), (17) | 352 |
| Refineries, .106(i)(3) | 106 |
| Pressures: | |
| (see Safety Relief Devices) | |
| Prime Mover Guards, .219(b) | 325 |
| Primers, Ammunition, .109(j)(4) | 120 |
| Process safety management of highly hazardous chemicals (see Chemicals, etc.), .119 | 148 |
| Processing Plants, Flammable and Combustible Liquids, .106(h) | 105 |
| Application, .106(h)(1) | 105 |
| Buildings, .106(h)(3) | 105 |
| Fire Protection, .106(h)(6) | 105 |
| Housekeeping, .106(h)(8) | 106 |
| Ignition Sources, .106(h)(7) | 105 |
| Liquid Handling, .106(h)(4) | 105 |
| Loading, .106(h)(5) | 105 |
| Location, .106(h)(2) | 105 |
| Maintenance, .106(h)(8) | 106 |
| Profile Lathes, .213(o) | 299 |
| Projections, .219(h) | 327 |
| beta-Propiolactone, .1003 | 488 |
| Area requirements, .1003(c) | 488 |
| Closed system operation, .1003(c)(2) | 488 |
| Isolated systems, .1003(c)(1) | 488 |
| Maintenance and decontamination activities, .1003(c)(5) | 489 |
| Open-vessel system operations, .1003(c)(3) | 488 |
| Transfer from a closed operation, .1003(c)(4) | 488 |
| Medical surveillance, .1003(g) | 490 |
| Examinations, .1003(g)(1) | 490 |
| Records, .1003(g)(2) | 4906 |
| Regulated area requirements, .1003(d) | 489 |
| Contamination control, .1003(d)(4) | 489 |
| Emergencies, .1003(d)(2) | 489 |
| Hygiene facilities and practices, .1003(d)(3) | 489 |
| Signs, information, and training, .1003(e) | 489 |
| Container contents identification, .1003(e)(2) | 489 |
| Lettering, .1003(e)(3) | 489 |
| Prohibited statements, .1003(e)(4) | 489 |
| Signs, .1003(e)(1) | 489 |
| Training and indoctrination, .1003(e)(5) | 489 |
| Protective Clothing: | |
| (see Clothing, Protective and Personal Protective Equipment) | |
| Protective Equipment, Piping: | |
| (see also Personal Protective Equipment), .253(e) | 342 |
| Hoses and Connections, .253(e)(5) | 342 |
| Pressure-Reducing Regulations, .253(e)(6) | 343 |
| Stations Outlet, .253(e)(4) | 342 |
| Pulleys, .219(d), (k), (p)(5) | 326, 327, 328 |
| Pulp, Paper and Paperboard Mills: | |
| Barking Devices, .261(c)(12) | 350 |
| Belt Conveyors, .261(c)(15) | 350 |
| Bleaching, .261(h) | 352 |
| Chemical Processes, .261(g) | 351 |
| Cranes, .261(c)(8) | 349 |
| Finishing Rooms, .261(l) | 354 |
| Hand Tools, .261(c)(13) | 350 |
| Handling, .261(c), (d) | 349, 350 |
| Lighting, .261(b)(2), (c)(10), (k)(21) | 349, 350, 354 |
| Lockouts, .261(b)(1) | 349 |
| Machine Rooms, .261(k) | 353 |
| Materials Handling, .261(m) | 354 |
| Mechanical Pulp Processes, .261(i) | 353 |
| Personal Protective Equipment, .261(d)(1) | 350 |
| Pulpwood: | |
| Preparation, .261(e) | 350 |
| Removal, .261(c)(14) | 350 |

| Subject term, Section No. | Page |
|---|---|
| Rags and Old Paper, .261(f) | 351 |
| Safe Practices, .261(b) | 349 |
| Signs: | |
| Conveyors, .261(c)(16) | 350 |
| Traffic, .261(c)(9) | 350 |
| Standards Sources, .261(a)(3), (4);, .268 | 349, 376 |
| Stock Preparation, .261(j) | 353 |
| Storage, .261(c), (d) | 349, 350 |
| Chocking Rolls, .261(d)(4)(ii), (iii) | 350 |
| Clearances, .261(d)(2) | 350 |
| Piling, .261(d)(3) | 350 |
| Traffic Warning Signs, .261(c)(9) | 350 |
| Pulpwood Logging, .266 | 369 |
| Chain Saw Operations, .266(e)(2) | 371 |
| Chipping, .266(h)(4) | 374 |
| Environmental Conditions, .266(d)(5) | 371 |
| Explosives, .266(d)(10) | 371 |
| First Aid, .266(d)(2), (i)(7), App. A, App. B | 370, 375 |
| Hand and Portable Powered Tools, .266(e) | 371 |
| Harvesting, .266(h) | 373 |
| Bucking, .266(h)(3) | 374 |
| Felling, .266(h)(2) | 373 |
| Limbing, .266(h)(3) | 374 |
| Loading, .266(h)(6) | 374 |
| Machines for Moving Materials, .266(f) | 372 |
| brakes, .266(f)(7) | 373 |
| designated operator, .266(f)(2) | 372 |
| exhaust system, .266(f)(6) | 373 |
| FOPS/ROPS, .266(f)(3) | 372 |
| guarding, .266(f)(8) | 373 |
| machine access, .266(f)(5) | 373 |
| overhead guard, .266(f)(4) | 373 |
| Personal Protective Equipment, .266(d)(1) | 370 |
| Seat Belts, .266(d)(3) | 370 |
| Storage, .266(h)(8) | 374 |
| Training, .266(i) | 374 |
| certification, .266(i)(10) | 375 |
| content, .266(i)(3) | 374 |
| designated trainer, .266(i)(8) | 375 |
| first-aid, .266(i)(7) | 375 |
| frequency, .266(i)(2) | 374 |
| meetings, .266(i)(11) | 375 |
| Vehicles, .266(g) | 373 |
| inspection, .266(g)(2) | 373 |
| instructions, .266(g)(3) | 373 |
| maintenance, .266(g)(1) | 373 |
| Work Areas, .266(d)(6) | 371 |
| Pumps, Gasoline: | |
| (see also Service Stations), .106(g)(3), (4) | 103 |
| Pyrotechnics, .109(k), .119 | 121, 148 |
| Radial Saws, .213(h) | 298 |
| Radiation: | |
| Ionizing, .96 | 82 |
| Nonionizing, .97 | 82 |
| Radioactive Materials: | |
| Packaged, .1096(h) | 709 |
| Storage, .1096(j) | 709 |
| Rail Clamps, .179(b)(4), .180(i)(1) | 265, 272 |
| Rail Sweeps, .179(e)(4) | 266 |
| Railroad Cars, .176, .178(k)(2)-(4) | 255, 260 |
| Explosives, .109(f) | 115 |
| Rated Load Markings: | |
| Cranes, .179(b)(5) | 265 |
| Derricks, .181(c) | 274 |
| Rated Load Test: | |
| Crawler, Locomotive, and Truck Cranes, .180(e)(2) | 271 |
| Overhead and Gantry Cranes, .179(k)(2) | 268 |
| Recordkeeping: | |
| 2001 Data, summary and posting 1904.43 | 753 |
| Annual OSHA Survey 1904.41 | 753 |
| Annual summary 1904.32; 1904.33 | 750, 751 |
| Asbestos, .1001(m) | 468 |
| Bloodborne pathogens, exposure to, .1030(f)(6), (h) | 603, 604 |
| Building Inspection, assurance for powered platform use, .66(c) | 41 |
| Bureau of Labor Statistics (BLS) Survey 1904.42 | 753 |
| Change in business ownership 1904.34 | 751 |

# OSHA Standards for the General Industry

| Subject term, Section No. | Page |
|---|---|
| Communicable Diseases, .142(l)(1) and (2) | 213 |
| Covered employees 1904.31 | 750 |
| Cranes: | |
|     Crawler, Locomotive and Truck, .180(d)(2) and (6), (e)(2), (g)(1) and (2) | 270, 271 |
|     Overhead and Gantry, .179(k)(2), (m)(1) and (2) | 268 |
| Definitions 1904.46 | 753 |
| Derricks, .181(g)(1) and (3) | 275 |
| Employee involvement 1904.35 | 751 |
| Establishment, defined 1904.46 | 753 |
| Exemptions: | |
|     Partial exemption for small employers, 1904.1 | 743 |
|     Partially exempt industries, 1904.3 Appendix A to Subpart B | 743 |
|     Standard Industrial Classification (SIC) Codes, 1904.3; Appendix A to Subpart B | 743 |
| Fatalities, reporting to OSHA, 1904.39 | 752 |
| Forging Equipment Inspection, .218(a)(2) | 324 |
| Forms: | |
|     OSHA 300 Log, 1904.29 | 749 |
|     OSHA 300-A Summary, 1904.29 | 749 |
|     OSHA 301 Incident Report, 1904.29 | 749 |
|     Retention and updating: | |
|         New forms (OSHA 300, OSHA 301), 1904.33 | 751 |
|         Old forms (OSHA 101; OSHA 200), 1904.44 | 753 |
| Hazardous Waste Operations, .120(f)(7) | 164 |
| Hearing loss, 1904.10 | 748 |
| Injury Reporting, Welding, .252(c)(13) | 338 |
| Ionizing Radiation Exposure, .1096(m), (o) | 710 |
| Keeping records for more than one government agency, 1904.3 | 743 |
| Labor Camps, .142(l)(1) and (2) | 213 |
| Liquid Storage Tanks, Class I, .106(g)(1) | 101 |
| Manlifts, .68(e)(3) | 61 |
| Mechanical Power Presses, .217(e)(1) | 313 |
| Multiple business establishments, 1904.30 | 750 |
| Multiple hospitalization incidents, reporting to OSHA, 1904.39 | 752 |
| Needlestick and sharps injuries 1904.8 | 748 |
| Office of Management and Budget (OMB) control numbers 1904.45 | 753 |
| Personal Monitoring: | |
|     Asbestos, .1001(i)( d),(1), (n) | 463, 468 |
| Power Presses Inspection, .217(e)(1) | 313 |
| Powered Platforms Inspection, .66(g) | 47 |
| Preexisting conditions, 1904.5 | 744 |
| Privacy concern cases, 1904.29(b)(6)-(10); 1904.33(a) | 749, 751 |
| Prohibition against discrimination, 1904.36 | 751 |
| Providing records to government representatives, 1904.40 | 752 |
| Purpose of Part 1904, 1904.0 | 743 |
| Radiation Exposure, .1096(b)(2)(iii), (m)(1), (n), (o)(1) | 707, 710 |
| Records, Disclosure, Ionizing Radiation, .1096(o)(1) | 708 |
| Respirators, .134 (m) | 194 |
| Records: | |
|     Asbestos, .1001 (m) | 468 |
|     Ionizing Radiation, .1096(m), (n) | 710 |
|     Mechanical Power Presses, .217(e)(1) | 313 |
| Refineries, Chemical Plants and Distilleries, .106(i) | 106 |
|     Application, .106(j) | 106 |
|     Fire Protection, .106(i)(5) | 106 |
|     Pressure Vessels, .106(i)(3) | 106 |
|     Process Unit Location, .106(i)(4) | 106 |
|     Storage Tanks, .106(i)(1) | 106 |
|     Wharves, .106(i)(2) | 106 |
| Refrigerated Containers: | |
|     Anhydrous Ammonia, .111(d) | 145 |
| Refueling: | |
|     Cranes, .180(i)(4) | 272 |
|     Derricks, .181(j)(4) | 276 |
|     Trucks, .178(p)(2) | 262 |
| Refuse: | |
|     Disposal, .142(h) | 213 |
|     Receptacles, .141(a)(4) | 211 |
| Relief Devices: | |
|     (see Safety Relief Devices) | |
| Remote Gas Pumping Systems, .106(g)(3)(v) | 103 |

| Subject term, Section No. | Page |
|---|---|
| Residue Disposal: | |
|     (see Waste Disposal) | |
| Resistance Welding Equipment, .255(c) | 347 |
|     Capacitor Discharge Welding, .255(b)(2) | 346 |
|     Disconnecting Means, .306(d)(2) | 427 |
|     Foot Switches, .255(b)(6) | 347 |
|     Grounding, .255(b)(9) | 347 |
|     Guarding, .255(a)(4), (b)(4) | 346 |
|     Installation, .255(a)(1) | 346 |
|     Interlocks, .255(b)(3) | 346 |
|     Safety Pins, .255(b)(8) | 347 |
|     Shields, .255(b)(5) | 346 |
|     Spot and Seam Welding, .255(b) | 346 |
|     Stop Buttons, .255(b)(7) | 347 |
|     Thermal Protection, .255(a)(2) | 346 |
| Resistors: | |
|     Cranes, .179(g)(4) | 267 |
| Respirators: | |
|     (see also Gas Mask Canisters), .134 | 188 |
|     Abrasive Blasting, .94 (a)(1)(ii), (a)(5) | 63 |
|     Air Supply, .94(a)(6), .134(d) | 64, 190 |
|     Asbestos, .1001(g)(1), (2) | 464 |
|     Cleaning, .134(c)(1)(v) | 189 |
|     Color Codes, .134(j) | 193 |
|     Employer Provided, .134(a)(2) | 188 |
|     Fire brigades, .156(f) | 235 |
|     Identification, .134(g) | 191 |
|     Inspection, .134(h) | 192 |
|     Labeling, .134(g) | 191 |
|     Maintenance, .134(f), (h) | 192 |
|     Medical evaluation, .134(e) | 190 |
|     Positive pressure, .156(f)(2) | 235 |
|     Pulp, and Paper Mills, .261(g)(2), (6), (10), (15)(ii) | 351, 352 |
|     Repairs, .134(f)(4) | 191 |
|     Selection, .134(c) | 189 |
|     Storage, .134(h) | 192 |
|     Training, .134(c)(1)(vii), (viii) | 189 |
|     Welding, .252(c)(4)(ii), (iii); (5)(ii); (7)(ii); (8); (9); (10) | 338 |
| Respiratory Protection: | |
|     (see also Respirators), .134 | 188 |
|     Air Quality, .94(a)(6), .134(d) | 64, 190 |
|     Air Supply, .94(a)(6), .134(d) | 64, 190 |
|     Fire brigades, .156(f) | 235 |
|     Fit testing, .134(c)(1)(iii) App. C, | 189, 200, 664 |
|     Gas Mask Canister Identification, .134(g) | 191 |
|     Minimum Acceptable Program, .134(b) | 188 |
|     Permissible Practices, .134(a)(1) | 188 |
|     Respirators, .134(a)(2), (b), (c), (e) | 188, 189, 190 |
|     Use, .134(e)(5) | 191 |
| Review Commission, 1903.17, | 739 |
| Right to know, .1200 | 710 |
| Rim wheels, multi-piece, .177 | 255 |
| Ring Test, .215(d)(1) | 303 |
| Ripsaws, .213(c) | 298 |
| Rodent Control, .141(a)(5) | 211 |
|     Labor Camps, .142(j) | 213 |
| Rolling Scaffolds: | |
|     (see Work Platforms, Mobile). | |
| Roofing Brackets, .28(s) | 34 |
|     Catch Platforms, .28(s)(3) | 34 |
|     Construction, .28(s)(1) | 34 |
|     Supports, .28(s)(2) | 34 |
| Rope Inspections: | |
|     Cranes, .179(m), .180(g) | 268, 271 |
|     Derricks, .181(g) | 275 |
| Ropes: | |
|     Cranes, .179(m), .180(g) | 268, 271 |
|         Hoists, .179(h)(2) | 267 |
|         Inspections, .179(m), .180(g) | 268, 271 |
|         Running, .179(m)(1) | 268 |
|     Derricks, .181(g) | 275 |
|         Idle Ropes, .181(g)(3) | 275 |
|         Limited Travel, .181(g)(2) | 275 |
|         Nonrotating Ropes, .181(g)(4) | 275 |
|         Running, .181(g)(1) | 275 |

| Subject term, Section No. | Page |
|---|---|
| Powered Platforms, .66(f)(7), (g)(5), (h)(3), (4) | 46, 47, 48 |
| Rotary Lawn Mowers, .243(e)(1), (4) | 332, 333 |
| Rotating Work Platforms: | |
| (see also Vehicle-Mounted WorkPlatforms), .67 | 58 |
| Rubber Industry: (see also Mills, Rubber and Plastics Industry) | |
|     Auxiliary Equipment, .216(a)(3), (b)(3) | 309 |
|     Effective Dates, .220 | 328 |
|     Mills and Calenders, .216 | 309 |
|     Standards Sources, .221 | 328 |
| Rung Ladders, Portable, .25(c)(3) | 22 |
| Running Ropes: | |
|     Cranes, .179(m)(1), .180(g)(1) | 268, 271 |
|     Derricks, .181(g)(1) | 275 |
| Runway Conductors: | |
|     Cranes, .179(g)(6) | 267 |
| Runway Protection, .23(c) | 20 |
| Safety, Safety and Health Management Voluntary Guidelines | 805 |
| Safety Belts: | |
| (see also Lifelines) | |
|     Pulp, Paper, and Paperboard Mills, .261(g)(4), (15) | 351, 352 |
|     Scaffolding, .28(j)(4), (n)(8), (s)(3), (t)(2), (u)(6) | 33, 34, 35 |
|     Welding, .252(b)(4)(iv) | 337 |
| Safety Color Codes: | |
|     Effective Dates, .149 | 230 |
| Safety Devices: | |
|     Ladders, .27(d)(5) | 27 |
| Safety Guard Design, Abrasive Wheel Machinery, .215(a)(2), (b)(10)-(12) | 300, 301-302 |
| Safety Instruction Signs, .145(c)(3), (d)(6) | 214 |
| Safety Relief Devices: | |
|     Bulk Oxygen Systems, .104(b)(6), (7)(ii) | 89 |
|     Flammable Liquids, .107(e)(8) | 108 |
|     Gaseous Hydrogen Systems, .103(b)(1)(iii) | 83 |
|     Liquefied Hydrogen Systems, .103(c)(1)(iv) | 85 |
|     Liquefied Petroleum Gases, .110(b)(10), (c)(7), (d)(4), (e)(7), (h)(4) | 124, 132, 133, 137, 138 |
|     Non-DOT Containers, .110(d)(4) | 133 |
|     Spraying, .107(e)(8) | 108 |
| Safety-Toe Footwear: | |
| (see Foot Protection) | |
| Sanding Machines, .213(p), .243(a)(3) | 299, 330 |
| Sanitation, .141 | 211 |
|     Application, .141(a)(1) | 211 |
|     Change Rooms, .141(e) | 212 |
|     Effective Dates, .149 | 230 |
|     Food Handling, .141(h) | 212 |
|     Hazardous Waste Operations, .120 (n) | 167 |
|     Housekeeping, .141(a)(3) | 211 |
|     Insect Control, .141(a)(5) | 211 |
|     Lunchrooms, .141(g) | 212 |
|     Rodent Control, .141(a)(5) | 211 |
|     Standards Sources, .150 | 230 |
|     Toilet Facilities, .141(c) | 211 |
|     Vermin Control, .141(a)(5) | 211 |
|     Washing Facilities, .141(d) | 212 |
|     Waste Disposal, .141(a)(4) | 211 |
|     Water Supply, .141(b) | 211 |
| Sawmills: | |
|     Bins, Bunkers, Hoppers, and Fuel Houses, .265(c)(23) | 365 |
|         Lighting, .265(c)(23)(iii) | 365 |
|         Loading Bins, .265(c)(23)(ii) | 365 |
|     Blower Systems, .265(c)(20) | 364 |
|     Building Facilities, .265(c) | 363 |
|         Docks, .265(c)(4) | 364 |
|         Emergency Exits, .265(c)(6) | 364 |
|         Fire Escapes, .265(c)(6) | 364 |
|         Floors, .265(c)(3) | 363 |
|         Lighting, .265(c)(9) | 364 |
|         Platforms, .265(c)(4) | 364 |
|         Stairways, .265(c)(5) | 364 |
|             Handrails, .265(c)(5)(ii) | 364 |
|             Lighting, .265(c)(5)(iii) | 364 |
|         Tanks, .265(c)(8) | 364 |
|         Vats, .265(c)(8) | 364 |
|         Walkways, .265(c)(4) | 364 |

| Subject term, Section No. | Page |
|---|---|
|         Work Areas, .265(c)(2) | 363 |
|     Burners, .265(c)(29) | 366 |
|     Chippers, .265(c)(21) | 364 |
|     Conveyors, .265(c)(18) | 364 |
|     Definitions, .265(b) | 362 |
|     Exhaust Systems, .265(c)(20) | 364 |
|     Gas Piping and Appliances, .265(c)(15) | 364 |
|     General Requirements, .265(a) | 362 |
|     Hydraulic Systems, .265(c)(13) | 364 |
|     Kilns, Dry, .265(f) | 369 |
|     Log Breakdown, .265(e) | 368 |
|     Log Handling, Sorting, and Storage, .265(d) | 366 |
|         Barking Devices, .265(d)(4) | 368 |
|         Log Decks, .265(d)(3) | 367 |
|         Storage Areas, .265(d)(2) | 367 |
|         Unloading, .265(d)(1) | 366 |
|         Unloading Areas, .265(d)(2) | 367 |
|     Lumber: | |
|         Loading, .265(c)(28) | 366 |
|         Piling, .265(c)(27) | 366 |
|         Storage, .265(c)(27) | 366 |
|     Marking Physical Hazards, .265(c)(11) | 364 |
|     Refuse Removal, .265(c)(20)(vi) | 364 |
|     Ropes, Cables, Slings, and Chains, .265(c)(24) | 365 |
|     Stackers and Unstackers, .265(c)(26) | 365 |
|     Traffic Control, .265(c)(31) | 366 |
|     Tramways, .265(c)(19) | 364 |
|     Trestles, .265(c)(19) | 364 |
|     Vehicles, .265(c)(30) | 366 |
| Saws: | |
|     Band, .213(i) | 298 |
|     Band Resaws, .213(i) | 298 |
|     Circular, .213(f);, .243(a)(1) | 298, 330 |
|     Circular Resaws, .213(e) | 298 |
|     Cracked, .243(a)(4) | 330 |
|     Drag, .213(r) | 300 |
|     Forging Machines, .218(j)(2) | 325 |
|     Inspection, .213(s) | 300 |
|     Radial, .213(h) | 298 |
|     Ripsaws, .213(c) | 298 |
|     Swing Cutoff, .213(g) | 298 |
|     Table, .213(d) | 298 |
| Scaffolding: | |
| (see also Scaffolds) | |
|     Safety Requirements, .28 | 27 |
| Scaffolds: | |
| (see also Ladder Stands Listings by Names of Scaffolds) | |
|     Boatswain's Chair, .28(j) | 33 |
|     Bricklayers' Square, .28(l) | 33 |
|     Carpenters' Bracket, .28(k) | 33 |
|     Chicken Ladders, .28(t) | 35 |
|     Coupler, Mobile, .29(d) | 36 |
|     Crawling Boards, .28(t) | 35 |
|     Decorators', .28(o) | 34 |
|     Float, .28(u) | 35 |
|     Horse, .28(m) | 34 |
|     Interior Hung, .28(p) | 34 |
|     Ladder-Jack, .28(q) | 34 |
|     Masons' Adjustable Multiple-Point Suspension, .28(f) | 31 |
|     Needle Beam, .28(n) | 34 |
|     Outrigger, .28(e) | 31 |
|     Plasterers', .28(o) | 34 |
|     Powered platforms, .66 | 41 |
|     Roofing Brackets, .28(s) | 34 |
|     Ship, .28(u) | 35 |
|     Single-Point Adjustable Suspension, .28(i) | 33 |
|     Stone Setters' Adjustable Multiple Point Suspension, .28(h) | 32 |
|     Suspension, .28(f), (g), (h), (i) | 31, 32, 33 |
|     Swinging, .28(g) | 32 |
|     Tube and Coupler, .28(c) | 30 |
|     Tube and Coupler, Mobile, .29(d) | 36 |
|     Tubular Welded Frame, .28(d), .29(b) | 31, 35 |
|     Tubular Welded Sectional Folding, .29(c) | 36 |
|     Two-Point Suspension, .28(g) | 32 |
|     Window-Jack, .28(r) | 34 |

# OSHA Standards for the General Industry

| Subject term, Section No. | Page |
|---|---|
| Wood Pole, .28(b) | 28 |
| **Scaffolds, Manual Mobile:** | |
| *(see also Work Platforms, Mobile Scaffolds), .29* | 35 |
| **Scientific Diving** | |
| *(see Diving, Scientific)* | |
| **Semigantry Cranes:** | |
| *(see Gantry Cranes)* | |
| **Separation Walls:** | |
| *(see also Distances From Hazards)* | |
| Ammonium Nitrate, .109(i)(5) | 120 |
| **Service Stations:** | |
| Dispensing Systems, .106(g)(3) | 103 |
| Drainage, .106(g)(7) | 104 |
| Electrical Equipment, .106(g)(5) | 104 |
| Fire Protection, .106(g)(9) | 104 |
| Flammable and Combustible Liquids, .106(g) | 101 |
| Handling, .106(g)(1) | 101 |
| Heating Equipment, .106(g)(6) | 104 |
| Ignition Sources, .106(g)(8) | 104 |
| Marine Stations, .106(g)(4) | 103 |
| Multi-piece rim wheels, servicing, .177 | 255 |
| Storage, .106(g)(1) | 101 |
| Waste Disposal, .106(g)(7) | 104 |
| Liquefied Petroleum Gases, .110(h) | 138 |
| Accessories, .110(h)(3) | 138 |
| Capacity, .110(h)(5) | 139 |
| Containers, .110(h)(2) | 138 |
| Installation, .110(h)(6) | 139 |
| Protecting Fittings, .110(h)(7), (9) | 139 |
| Valves, .110(h)(3) | 138 |
| Dispensing Devices, .110(h)(11) | 140 |
| Electrical Systems, .110(h)(13) | 140 |
| Fire Protection, .110(h)(14) | 140 |
| Fittings, .110(h)(7) | 139 |
| Piping, .110(h)(7) | 139 |
| Pumps, .110(h)(10) | 140 |
| Safety Relief Valves, .110(h)(4) | 138 |
| Truck Unloading, .110(h)(8) | 139 |
| Valves, .110(h)(7) | 139 |
| Setscrews, .219(h) | 327 |
| Sewage Disposal, .142(e) | 213 |
| **Shafting Guarding:** | |
| Horizontal, .219(c)(2) | 326 |
| Inclined, .219(c)(3) | 326 |
| Vertical, .219(c)(3) | 326 |
| **Sheaves:** | |
| Crane Hoists, .179(h)(1) | 266 |
| **Shelters, Labor Camps:** | |
| *(see also Facilities, Labor Camps), .142(b)* | 212 |
| **Ship Scaffolds:** | |
| *(see also Float Scaffolds), .28(u)* | 35 |
| Side-Rolling Ladders, .25(c)(5) | 23 |
| **Signs and Tags:** | |
| *(see also Markings)* | |
| Accident Prevention, .145 | 214 |
| Classification, .145(c) | 214 |
| Definitions, .145(b) | 214 |
| Use Classification, .145(c) | 214 |
| Biological Hazards, .145(e)(4), (f)(8) | 214, 215 |
| Caution, .145(c)(2), (d)(4), (f)(6); .1001(g), | 214, 215, 460 |
| Colors, .145(d) | 214 |
| Danger, .145(c)(1), (d)(2), (e)(3), (f)(5), .1001(j)(4)(ii) | 214, 215, 466 |
| Design, .145(d), (f)(4) | 214, 215 |
| Gas Mask Canisters, .134(g) | 191 |
| Hazardous materials, retention of DOT markings, .1201... | 722 |
| Powered Platforms, .66(f)(7)(vi) | 46 |
| Pulp and Paper Mills, .261(c)(9), (16) | 350 |
| Radiation Warning, .97(a)(3) | 82 |
| Respirators, .134(g) | 191 |
| Safety Instruction, .145(c)(3), (d)(6) | 214 |
| Slow-Moving Vehicles, .145(d)(10) | 214 |
| Wordings, .145(e) | 214 |
| Single-Point Adjustable Suspension Scaffolds, .28(i) | 33 |
| Single-Rung Ladders, .25(c)(3)(ii) | 22 |
| Mason's, .25(c)(4)(iii) | 23 |

| Subject term, Section No. | Page |
|---|---|
| Skylight Floor Openings, .23(a)(4) | 20 |
| Sleeping Facilities, temporary, .120(n)(5) | 167 |
| Sleeves, Rubber Insulating, .137 | 204 |
| Slings, .184 | 277 |
| Slurries, Explosive, .109(h) | 118 |
| Small Arms Ammunition, .109(j) | 120 |
| Primers, .109(j)(4) | 120 |
| Smokeless Propellants, .109(j)(3) | 120 |
| Storage, .109(j) | 120 |
| Smokeless Propellants, .109(j)(3) | 120 |
| **Smoking:** | |
| Dual Component Coatings, .107(m)(2) | 111 |
| Explosives, .109(e)(1) | 115 |
| Flammable Liquids, .106(d)(7)(iii) | 98 |
| Organic Peroxide, .107(m)(2) | 111 |
| Powder Coatings, .107(l)(4)(iii) | 110 |
| Spraying, .107(g)(7), (l)(4)(iii), (m)(2) | 109, 110, 111 |
| Snagging Machines, .215(b)(7) | 301 |
| **Sources of Standards:** | |
| *(see Standards Sources)* | |
| **Special Industries:** | |
| Bakeries, .263 | 358 |
| Forging, .218 | 324 |
| Hazardous Waste Operations, .120 | 159 |
| Laundries, .264 | 362 |
| Paper and Paperboard Mills, .261 | 349 |
| Plastics Industry, .216 | 309 |
| Pulp Mills, .261 | 349 |
| Pulpwood Logging, .266 | 369 |
| Rubber Industry, .216 | 309 |
| Sawmills, .265 | 362 |
| Standards Sources, Telecommunications, .268 | 376 |
| Textiles, .219(a)(3), .262 | 325, 355 |
| Woodworking, .213 | 297 |
| Spill Containment, .106(d)(6)(iii) | 98 |
| Spot and Seam Welding Machines, .255(b) | 346 |
| Spray Booths, .107(b) | 107 |
| **Spray Finishing, .107** | 106 |
| Air Flow, .94(c)(6) | 72 |
| Application, .107(n) | 111 |
| Automobile Undercoatings, .107(k) | 110 |
| Clean Air, .94(c)(7) | 73 |
| Combustible Liquids Storage, .107(e) | 108 |
| Curing Apparatus, .107(j) | 110 |
| Drying Apparatus, .107(j) | 110 |
| Dual Component Coatings, .107(m) | 110 |
| Electrical Systems, .107(c) | 107 |
| Electrostatic Apparatus, .107(h), (1) | 109 |
| Fire Protection, .107(f) | 109 |
| Flammable Liquids Storage, .107(e) | 108 |
| Fusion Apparatus, .107(j) | 110 |
| Ignition Sources, .107(c) | 107 |
| Location, .94(c)(2) | 71 |
| Maintenance, .106(g) | 101 |
| Make-Up Air, .94(c)(7) | 73 |
| Organic Peroxide Coatings, .107(m) | 110 |
| Powder Coatings, .107(l) | 110 |
| Spray Booths, .94(c)(3), .107(b) | 71, 107 |
| Spray Rooms, .94(c)(4) | 71 |
| Undercoatings, .107(k) | 110 |
| Velocity, .94(c)(6) | 72 |
| Ventilation, .94(c)(5), .107(d) | 71, 108 |
| Spray Liquid Heaters, .107(e)(7) | 108 |
| Spraying Operations, .107(g) | 109 |
| **Sprinkler Systems:** | |
| Sprinkler Systems, Automatic, .159 | 238 |
| Acceptance tests, .159(c)(3) | 238 |
| Design, .159(c)(1) | 238 |
| Drainage, .159(c)(7) | 238 |
| Exemptions, .159(b) | 238 |
| Hose Connections, .159(c)(5) | 238 |
| Hydraulically Designed, .159(c)(11) | 238 |
| Maintenance, .159(c)(2) | 238 |
| Protection of Piping, .159(c)(6) | 238 |
| Sprinkler Alarms, .159(c)(9) | 238 |
| Sprinkler Spacing, .159(c)(10) | 238 |

Index Rop—Spr

## Subject Index For 29 CFR 1910, Sec. 5, 1903, 1904, and 1990

| Subject term, Section No. | Page |
|---|---|
| Water supply, .159(c)(4) | 238 |
| Sprinklers | |
|     Dip Tanks, .125(f) | 185 |
|     Sprockets, .219(f) | 327 |
| Stability Margin: | |
|     Crane Loads, .180(c)(1)(i)-(iv) | 270 |
| Stainless Steel Cutting, .252(c)(12) | 338 |
| Stairs, Fixed Industrial, .24 | 22 |
|     Handrails, .24(h) | 22 |
|     Length of Stairways, .24(g) | 22 |
|     Railings, .24(h) | 22 |
|     Rise Angle, .24(e) | 22 |
|     Strength, .24(c) | 22 |
|     Treads, .24(f) | 22 |
|     Vertical Clearance, .24(i) | 22 |
|     Width, .24(d) | 22 |
| Standpipe and Hose Systems, .158 | 237 |
|     Equipment, .158(c) | 237 |
|         Hose, .158(c)(3) | 237 |
|         Hose Outlets and Connections, .158(c)(2) | 237 |
|         Nozzles, .158(c)(4) | 238 |
|         Reels and Equipment, .158(c)(1) | 237 |
|     Exceptions, .158(a)(2) | 237 |
|     Protection, .158(b) | 237 |
|     Scope and Application, .158(a)(1) | 237 |
|     Tests and Maintenance, .158(e) | 238 |
|         Acceptance Tests, .158(e)(1) | 238 |
|         Maintenance, .158(e)(2) | 238 |
|     Water Supply, .158(d) | 238 |
| Stands, Ladder: | |
|     (see also Scaffolds; Work Platforms, Mobile) | |
| Stationary Derricks: | |
|     (see also Derricks), .181 | 271 |
| Static Sparks, .219(p)(2)(ii) | 328 |
| Stepladders: | |
|     Portable Metal, .26(a)(3) | 24 |
| Stepladders, Portable, .25(c)(2) | 22 |
| Steps: | |
|     (see Stairs) | |
| Sticking Machines, .213(n) | 299 |
| Stiffleg Derricks: | |
|     (see also Derricks), .181 | 271 |
| Stone Setters' Adjustable Multiple-Point: | |
|     Suspension Scaffolds, .28(h) | 32 |
| Stopping Limits, Mills and Calenders, .216(f) (1)-(3) | 309 |
| Stops: | |
|     (see also Safety Devices) | |
|     Manlifts, .68(c)(5), (6) | 60, 61 |
| Storage: | |
|     (see also Materials Storage: Storage Areas; Tank Storage: Tank Storage, Portable) | |
|     Ammonium Nitrate, .109(i) | 119 |
|     Anhydrous Ammonia, .111 | 140 |
|     Blasting Agents, .109(g)(5) | 117 |
|     Buildings, .106(d)(5) | 97 |
|         Mercantile Occupancies, .106(d)(5)(iv) | 97 |
|         Office Occupancies, .106(d)(5)(iii) | 97 |
|         Warehouses, .106(d)(5)(v) | 97 |
|     Clothing, Spray Finishing,..107(g)(4) | 109 |
|     Compressed Gases, .101(b) | 83 |
|     Containers, Bulk Oxygen, .104(b)(4), (6) | 88, 89 |
|     Explosives, .109(b)(1), (c), (e)(2) | 111, 115 |
|     Flammable and Combustible Liquids, .106(b), (d) | 91, 96 |
|         Inside Storage Rooms, .106(d)(4) | 97 |
|         Storage Inside Buildings, .106(d)(5) | 97 |
|         Storage Outside Buildings, .106(d)(6) | 98 |
|     Indoor Rooms, .106(d)(5) | 97 |
|     Liquefied Petroleum Gases, .110 | 121 |
|     Logs, .265(d) | 366 |
|     Lumber, .265(c)(27) | 366 |
|     Pulp and Paper Mills, .261(c), (d) | 349, 350 |
|     Service Stations, .106(g)(1) | 101 |
| Storage Areas: | |
|     Aisles and Passageways, .176(a) | 255 |
|     Bridge Plates, .178(j), (k)(4) | 260 |
|     Clearance Signs, .176(e) | 255 |
|     Clearances, .176(a) | 255 |
|     Dockboards, .178(j), (k)(4) | 260 |

| Subject term, Section No. | Page |
|---|---|
|     Housekeeping, .176(c) | 255 |
|     Lighting, .178(h) | 260 |
|     Noxious Gases, .178(i) | 260 |
|     Railroad Cars, .178(k)(2)-(4) | 260 |
|     Securing, .176(b) | 255 |
|     Trucks, Highway, .178(k)(1), (3); (m) | 260, 261 |
| Storage Batteries: | |
|     (see Battery Changing and Charging) | |
| Storage Bridge Cranes: | |
|     (see Gantry Cranes) | |
| Storage, Tanks: | |
|     (see Tank Storage; Tank Storage, Portable) | |
| Straight Ladders, Portable Metal, .26(a)(2) | 23 |
| Surface Grinders, .215(b)(5) | 301 |
| Swing Frame Grinders, .215(b)(6) | 301 |
| Swing-Head Lathes, .213(o) | 299 |
| Swinging Locomotive Cranes, .180(i)(6) | 272 |
| Swinging Scaffolds: | |
|     (see also Two-Point Suspension Scaffolds), .28(g) | 32 |
| Switches: | |
|     Electric, .305(c) | 425 |
|     Cranes, .179(g)(5) | 267 |
|     Trip and Emergency, .216(e) | 309 |
| Table Saws, .213(d) | 298 |
| Tags: | |
|     (see Signs and Tags) | |
| Tanks: | |
|     (see also Cargo Tanks—Portable Tanks) | |
|     Hardening, .126(a)(1)(i),(ii) | 186 |
|     Tempering, .126(a) | 186 |
| Tanks, Dip: | |
|     (see also Dip Tanks), .123-.126 | 184-186 |
| Tanks, Storage: | |
|     Flammable and Combustible Liquids, .106(b) | 91 |
|         Atmospheric Tanks, .106(b)(1)(iii) | 91 |
|         Construction, .106(b)(1) | 91 |
|         Corrosion, .106(b)(1)(vi) | 91 |
|         Diking, .106(b)(2)(vii) | 93 |
|         Ignition Sources, .106(b)(6) | 95 |
|         Installation: | |
|             Above Ground, Outside, .106(b)(2) | 91 |
|             Inside Buildings, .106(b)(4) | 94 |
|             Underground, .106(b)(3) | 93 |
|         Low Pressure Tanks, .106(b)(1)(iv) | 91 |
|         Materials, .106(b)(1)(i) | 91 |
|         Pressure Vessels, .106(b)(1)(v) | 91 |
|         Supports, .106(b)(5) | 94 |
|         Testing, .106(b)(7) | 95 |
|         Venting, .106(b)(2)(iv), (v), (vi), (3)(iv), (4)(ii), (iii) | 92 93, 94 |
| Tanks, Storage, Portable: | |
|     Flammable and Combustible Liquids, .106(d) | 96 |
|         Application, .106(d)(1)(i) | 96 |
|         Capacity, .106(d)(2) | 96 |
|         Design, .106(d)(2) | 96 |
|         Exceptions, .106(d)(1), (2) | 96 |
|         Fire Protection, .106(d)(7) | 98 |
|         Indoor Storage, .106(d)(4), (5) | 97 |
|         Outdoor Storage, .106(d)(6) | 98 |
|         Storage Cabinets, .106(d)(3) | 96 |
| Telecommunications, .268 | 376 |
| Tempering Tanks, .126(a) | 186 |
| Temporary Floor Openings, .23(a)(7) | 20 |
| Temporary Labor Camps: | |
|     (see also Labor Camps, Temporary), .142 | 212 |
| Tenoning Machines, .213(k) | 299 |
| Testing: | |
|     Bulk Oxygen, .104(b)(8)(v) | 89 |
|     Cranes, .179(k), .180(e) | 268, 271 |
|     Derricks, .181(e) | 275 |
|     Fire Extinguishers:, .157(e) | 236 |
|     Gaseous Hydrogen Systems, .103(b) | 83 |
|     Liquefied Hydrogen Systems, .103(c)(1)(vii) | 86 |
|     Piping, .106(c)(7) | 96 |
|     Powered Platforms, .66(g) | 47 |
|     Radiation Alarm, .1096(f)(3) | 709 |
|     Sprinkler Systems, .159(c)(3) | 238 |
|     Standpipe and hose systems, .158(e) | 238 |
|     Storage Tanks, .106(b)(7) | 95 |

# OSHA Standards for the General Industry

| Subject term, Section No. | Page |
|---|---|
| Textiles: | |
|    Acid Carboys, .262(nn) | 358 |
|    Bleaching, .262(p) | 357 |
|    Calenders, .262(ee) | 358 |
|    Caustics, .262(oo) | 358 |
|    Color-Mixing Room, .262(kk) | 358 |
|    Cotton Cards, .262(e) | 356 |
|    Cotton Combers, .262(j) | 357 |
|    Drawing Frames, .262(j) | 357 |
|    Drying Cans, .262(w) | 357 |
|    Drying Tumblers, .262(cc) | 358 |
|    Dye Vats, .262(mm) | 358 |
|    Dyeing Jigs, .262(u) | 357 |
|    Extractors, .262(y) | 357 |
|    First Aid, .262(pp) | 358 |
|    Flat Work Ironers, .262(x) | 357 |
|    Folders, Overhead, .262(jj) | 358 |
|    Garnet Machines, .262(f) | 356 |
|    Gill Boxes, .262(k) | 357 |
|    Hand Boiling Machines, .262(hh) | 358 |
|    Kiers, .262(q) | 357 |
|    Lappers, .262(m) | 357 |
|    Looms, .262(n) | 357 |
|    Mercerizing Ranges, .262(s) | 357 |
|    Nip Guards, .262(v), (z), (dd)(1) | 357, 358 |
|    Openers, .262(d) | 356 |
|    Padders, .262(v) | 357 |
|    Pickers, .262(d) | 356 |
|    Power Transmission, .219(a)(3) | 325 |
|    Printing Machines, .262(dd) | 358 |
|    Rings Frames, .262(j) | 357 |
|    Roll Bench, .262(ii) | 358 |
|    Rope Washers, .262(bb) | 358 |
|    Sanforizing and Palmer Machines, .262(aa) | 358 |
|    Shearing Machines, .262(o) | 357 |
|    Slashers, .262(h) | 356 |
|    Slubbers, .262(j) | 357 |
|    Spinning Mules, .262(g) | 356 |
|    Standards Incorporated, .262(a)(2) | 355 |
|    Staple Cullers, .262(ff) | 358 |
|    Tanks, Open, .262(ll) | 358 |
|    Tenter Frames, .262(t) | 357 |
|    Tumblers, .262(cc) | 358 |
|    Warpers, .262(i) | 357 |
|    Worsted Drawing, .262(l) | 357 |
| Toe Protection: | |
|    (see Foot Protection) | |
| Toeboards: | |
|    Cranes, .179(d)(3) | 265 |
|    Definition, .21(a)(9) | 17 |
|    Power Transmission Apparatus, .219(o)(5) | 328 |
|    Powered Platforms, .66(f)(5)(i)(G) | 45 |
|    Walking-Working Surfaces, .23(a)(2), (3)(ii), (e) | 20, 21 |
| Toilet Facilities: | |
|    (see also Toilets), .141(c) | 211 |
|    Construction, .141(c)(2) | 212 |
|    Hazardous Waste, .120(n)(3) | 167 |
|    Labor Camps, .142(d) | 213 |
|    Lavatories, .141(d)(2) | 212 |
|    Minimum Numbers, .141(c)(1), (d)(2) | 211, 212 |
|    Towels, .141(d)(3)(v) | 212 |
|    Washing Facilities, .141(d) | 212 |
| Tongs, Upsetters, .218(h)(4) | 325 |
| Tooling, .217(d)(5) | 312 |
| Torch Valves, Welding, .252(a)(4)(ii) | 336 |
| Towels, .141(d)(3)(v) | 212 |
| Towers, Scaffolds: | |
|    (see Ladder Stands and Scaffolds; Scaffolds; Work Platforms, Mobile), .29 | 35 |
| Tractors: | |
|    (see also Powered Industrial Trucks), .178 | 257 |
| Trailers, .111(d)(7) | 146 |
| Training Personnel, .217(e)(3), .1096(i), | 313, 703 |
|    Bloodborne pathogens, exposure to, .1030(e)(5), (g)(2) | 602, 603 |
|    Electrical safety-related work practices, .332 | 431 |
|    Fire brigades, .156 | 234 |
|    Fire extinguishers, .157(g) | 237 |

| Subject term, Section No. | Page |
|---|---|
|    Hazardous chemicals, .1200 | 710 |
|    Hazardous chemicals, highly, process safety management, .119(g) | 150 |
|    Hazardous waste operations, .120(e), (p)(7), (q)(6) | 162, 168, 170 |
|    Respirators, .134(c)(1)(vii), (viii)) | 189 |
|    Telecommunications, .268(c) | 370 |
|    Truck Operators, .178(l) | 260 |
|    Working platform operations, .66(i)(1) | 48 |
| Transmission Pipeline Welding, .252(d)(1) | 338 |
|    Construction Standards, .252(d)(1)(v) | 338 |
|    Electric Shock, .252(d)(1)(iii) | 338 |
|    Field Shop Operations, .252(d)(1)(ii) | 338 |
|    Flammable Substances, .252(d)(1)(vi) | 338 |
|    Pressure Testing, .252(d)(1)(iv) | 338 |
|    X-ray Inspection, .252(d)(1)(vii) | 338 |
| Transportation: | |
|    Blasting Agents, .109(g)(6) | 118 |
|    Explosives, .109(d) | 114 |
|       Fire Extinguishers, .109(d)(2)(iii) | 114 |
|       Markings, .109(d)(2)(ii) | 114 |
|       Vehicles, .109(d)(2), (3) | 114 |
| Trapdoors, .23(a)(5) | 20 |
| Traps, Air Receivers, .169(b)(2) | 253 |
| Treadles, .217(b)(4) | 310 |
| Treads, Stairs, .24(f) | 22 |
| Trestle Ladders, Portable: | |
|    Metal, .26(a)(4) | 24 |
|    Wood, .25(c)(3)(v) | 23 |
| Trimming Presses, .218(g) | 325 |
| Trips, Two-Hand, .217(b)(6) | 310 |
| Trolley Bumpers, Cranes, .179(e)(3) | 265 |
| Trolley Ladders, Portable, .25(c)(5) | 23 |
| Trolley Stops, Cranes, .179(e)(1) | 265 |
| Truck Cranes: | |
|    (see Crawler, Locomotive and Truck Cranes), .180 | 269 |
| Trucks, .178(k), (m) | 260, 261 |
|    Forklift, .261(c)(1) | 349 |
|    Hand, .261(m)(1) | 354 |
|    Highway, .178(k), (m) | 260, 261 |
|    Powered Industrial, .178 | 257 |
| Trucks, Powered Industrial: | |
|    (see also Powered Industrial Trucks), .178 | 257 |
| Tube and Coupler Scaffolds, .28(c) | 30 |
| Tube and Coupler Scaffolds, Mobile, .29(d) | 36 |
| Tubing: | |
|    (see Piping, Fittings, and Tubing) | |
| Tubular Welded Frame Scaffolds, .28(d) | 31 |
| Tubular Welded Frame Scaffolds, Mobile, .29(b) | 35 |
| Tubular Welded Sectional Folding Scaffolds, .29(c) | 36 |
| Tumors: | |
|    Benign, 1990.143(i); 1990.144(e) | 758, 759 |
|    Indirect mechanisms, 1990.144(f) | 759 |
|    Induced at site of administration, 1990.144(b) | 759 |
|    Spontaneous, 1990.143(e) | 758 |
| Turning Machines, .213(o) | 299 |
| Two-Point Suspension Scaffolds, .28(g) | 32 |
| Two-Section Rung Ladders, .25(c)(3)(iii) | 22 |
| Underground Storage Tanks, Flammable and CombustibleLiquids, .106(b)(3) | 93 |
|    Location, .106(b)(3)(i) | 93 |
|    Depth and Cover, .106(b)(3)(ii) | 93 |
|    Corrosion Protection, .106(b)(3)(iii) | 93 |
|    Vents, .106(b)(3)(iv) | 93 |
| Unit Physical Operations, Ventilation, .106(e)(3)(v) | 99 |
| Upsetters, .218(h) | 325 |
|    Dies Changing, .218(h)(5) | 325 |
|    Lockouts, .218(h)(2) | 325 |
|    Manual Controls, .218(h)(3) | 325 |
|    Supporting Foundations, .218(h)(1) | 325 |
|    Tongs, .218(h)(4) | 325 |
| Valves: | |
|    (see also Piping, Valves, and Fittings) | |
|    Air Receivers, .169(b)(3) | 253 |
|    Liquefied Petroleum Gases, .110(b)(7) | 122 |
|    Non-DOT Containers, .110(d)(3) | 133 |
| Vaporizers: | |
|    Liquefied Petroleum Gases, .110(b)(11) | 126 |

# Subject Index For 29 CFR 1910, Sec. 5, 1903, 1904, and 1990

| Subject term, Section No. | Page |
|---|---|
| Liquid Hydrogen, .103(c)(1)(viii) | 86 |
| Liquid Oxygen, .104(b)(7) | 89 |
| Vehicles, slow-moving, signs, .145(d)(10) | 214 |
| Veneer Machinery, .30(c) | 36 |
|     Cutters, .213(q), (s)(13) | 299, 300 |
| Ventilation, .94, .107(d) | 63, 108 |
|     Abrasive Blasting, .94(a) | 63 |
|     Asbestos, .1001(f)(1)(iv),(v),(ix) | 464 |
|     Bulk Oxygen Systems, .104(b)(3)(xii) | 88 |
|     Bulk Plants, .106(f)(2)(iii) | 100 |
|     Confined Spaces, .146(c) | 216 |
|     Dip Tanks, .124(b), .125(d)(2) | 184, 185 |
|     Effective Dates, .98 | 82 |
|     Electrostatic Spraying, .107(i), (h)(11) | 109 |
|     Exhaust Duct System, .107(d)(3), (7) | 108 |
|         Fan-Rotating Element, .107(d)(4) | 108 |
|         Independent Exhaust, .107(d)(3) | 108 |
|         Room Intakes, .107(d)(11) | 108 |
|     Grinding, Polishing, and Buffing, .94(b) | 64 |
|     Inside Storage Rooms, .106(d)(4)(iv) | 97 |
|     Powder Coatings, .107(l)(2) | 110 |
|     Processing Buildings, .106(h)(3)(iii) | 105 |
|     Sawmills, .265(c)(7) | 364 |
|     Spray Finishing, .94(c)(5) | 71 |
|     Spraying Operations, .94(c), .107(d) | 71, 108 |
| Venting, Tanks: | |
|     Aboveground, .106(b)(2)(iv)-(vi) | 92-93 |
|     Inside, .106(b)(4)(ii) | 94 |
|     Portable, .106(d)(2)(ii) | 96 |
|     Underground, .106(b)(3)(iv) | 93 |
| Vents: | |
|     (see Venting) | |
| Vermin Control, .141(a)(5) | 211 |
| Vinyl Chloride, .1017 | 490 |
|     Emergency situations, .1017(i) | 492 |
|     Hazardous operations, .1017(h) | 492 |
|     Medical surveillance, .1017(k) | 492 |
|     Methods of compliance, .1017(f) | 491 |
|     Monitoring, .1017(d) | 491 |
|     Permissible exposure limit, .1017(c) | 491 |
|     Regulated area, .1017(e) | 491 |
|     Respiratory protection, .1017(g) | 491 |
|     Signs and labels, .1017(l) | 493 |
|     Training, .1017(j) | 492 |
| Walking-Working Surfaces: | |
|     Aisles, .22(b) | 20 |
|     Covers, .22(c) | 20 |
|     Definitions, .21 | 17 |
|     Fixed Industrial Stairs, .24 | 22 |
|     Floor Loading, .22(d) | 20 |
|     Floor Openings Guard, .23 | 20 |
|     General Requirements, .22 | 20 |
|     Guardrails, .22(c) | 20 |
|     Housekeeping, .22(a) | 20 |
|     Ladders: | |
|         Fixed, .27 | 24 |
|         Portable: | |
|             Metal, .26 | 23 |
|             Wood, .25 | 22 |
|         Stands, .29 | 35 |
|     Passageways, .22(b) | 20 |
|     Scaffolding Safety, .28 | 27 |
|     Scaffolds (Towers), .29 | 35 |
|     Stairs, Fixed Industrial, .24 | 22 |
|     Wall Openings Guarding, .23 | 20 |
|     Working Surfaces, .30 | 36 |
| Wall Cranes: | |
|     (see Gantry Cranes) | |
| Wall Openings (Holes), .23(b) | 20 |
| Warehouses: | |
|     Ammonium Nitrate, .109(i)(4) | 120 |
|     Flammable Liquids, .106(d)(5)(v) | 97 |
| Warning Devices and Signs: (see also Signs and Tags) | |
|     Bloodborne pathogens, .1030(g)(1) | 603 |
|     Cranes, .179(i) | 267 |
|     Ionizing Radiation, .1096(f) | 708 |
|     Manlifts, .68(c)(7) | 61 |

| Subject term, Section No. | Page |
|---|---|
| Nonionizing Radiation, .97(a)(3) | 82 |
| Washing Facilities, .120(n)(6), .141(d), .142(f) | 167, 212, 213 |
| Waste Disposal, .141(a)(4) | 211 |
|     Asbestos, .1001(k) | 467 |
|     Bulk Plants, .106(f)(7) | 101 |
|     Containers, .141(g)(3) | 212 |
|     Dip Tanks, .125(e)(4)(ii),(iii) | 185 |
|     Ionizing Radiation, .1096(k) | 709 |
|     Labor Camps, .142(e), (h) | 213 |
|     Processing Plants, .106(h)(8)(iii) | 106 |
|     Radiation, .1096(k) | 709 |
|     Service Stations, .106(g)(7) | 104 |
|     Spraying, .107(g)(3) | 109 |
| Water Gels, .109(h) | 118 |
| Water Spray Extinguishing Systems, Fixed, .163 | 240 |
| Water Supply: | |
|     Hazardous waste operations, .120(n) | 167 |
|     Labor Camps, .142(c) | 212 |
|     Nonpotable Water, .141(b)(2) | 211 |
|     Potable Water, .141(b)(1) | 211 |
|     Sprinkler Systems, .159(c)(4) | 238 |
|     Standpipe and Hose Systems, .158(d) | 238 |
| Weather Protection Manlifts, .68(b)(15) | 60 |
| Welding: | |
|     (see also Acetylene Generators; Arc Welding; Flash Welding Equipment; Resistance Welding Equipment; Welding Machines, Portable), .251-.255 | 335-346 |
|     Beryllium, .252(c)(8) | 338 |
|     Cadmium, .252(c)(1)(iv)(B), (c)(9) | 337, 338 |
|     Chemicals, highly hazardous, process safety management; hot-work permits, .119(k) | 151 |
|     Cleaning Compounds, .252(c)(11) | 338 |
|     Concentrations, Maximum Allowable, .252(c)(1)(iii) | 337 |
|     Containers, .252(a)(3) | 336 |
|     Contamination, .252(c)(1)(i) | 337 |
|     Definitions, .251 | 335 |
|     Exhaust Hoods, .252(c)(3) | 337 |
|     Fire Protection, .252(a)(2)(i), (ii), (xv) | 335, 336 |
|     First Aid Equipment, .252(c)(13) | 338 |
|     Fluorine Compounds, .252(c)(1)(iv)(C), (c)(5) | 337, 338 |
|     Labels, .252(c)(1)(iv) | 337 |
|     Ladders, Fixed Welding, .27(b)(6) | 25 |
|     Lead, .252(c)(7) | 338 |
|     Liquefied Petroleum Gases, .110(b)(4) | 121 |
|     Mercury, .252(c)(10) | 338 |
|     Piping Systems, Mechanical, .252(d)(2) | 338 |
|     Personnel Protection, .252(b) | 336 |
|     Precautions, .252(a)(2), .255(e) | 335, 347 |
|     Prohibited Areas, .252(a)(2)(vi) | 335 |
|     Screens, .252(c)(1)(ii) | 337 |
|     Spot and Seam, .255(b) | 346 |
|     Stainless Steels, .252(c)(12) | 338 |
|     Supervisory Responsibility, .252(a)(2)(xiv) | 335 |
|     Transmission Pipelines, .252(d)(1) | 338 |
|     Ventilation, .252(c)(1)(ii), (c)(2)-(4) | 337, 338 |
|     Zinc, .252(c)(6) | 338 |
| Welding Machines, Portable, .255(c) | 347 |
|     Clevis, .255(c)(3) | 347 |
|     Counterbalance, .255(c)(1) | 347 |
|     Grounding, .255(c)(6) | 347 |
|     Holder, Movable, .255(c)(5) | 347 |
|     Safety Chains, .255(c)(2) | 347 |
|     Switch Guards, .255(c)(4) | 347 |
| Wharves: | |
|     Bulk Plants, .106(f)(4) | 100 |
|     Chemical Plants, .106(i)(2) | 106 |
|     Distilleries, .106(i)(2) | 106 |
|     Explosives, .109(f) | 115 |
|     Marine Service Stations, .106(g)(4) | 103 |
|     Refineries, .106(i)(2) | 106 |
| Wheels, Multi-Piece Rim: Servicing, .177 | 255 |
| Winch Heads, Derricks, .181(i)(5) | 276 |
| Wind Indicators, .179(b)(4) | 265 |
| Window-Jack Scaffolds, .28(r) | 34 |
|     Guardrails, .28(r)(3) | 34 |
|     Use, .28(r)(1), (2) | 34 |

# OSHA Standards for the General Industry

| Subject term, Section No. | Page |
|---|---|
| Wood Heel Turning Machines, .213(o) | 299 |
| Wood Ladders, Portable: | |
| (see also Ladders, Portable Wood), .25 | 22 |
| Wood Pole Scaffolds, .28(b) | 28 |
| Wood Shapers, .213(m) | 299 |
| Wooden Guards, .219(o)(2) | 327 |
| Woodworking Machinery, .213 | 297 |
|     Band Saws and Resaws, .213(i) | 298 |
|     Boring Machines, .213(l) | 299 |
|     Circular Resaws, .213(e) | 298 |
|     Construction, .213(a) | 297 |
|     Controls, .213(b) | 297 |
|     Crosscut Table Saws, .213(d) | 298 |
|     Definitions, .211(a) | 293 |
|     Drag Saws, .213(r) | 300 |
|     Glue Spreaders, Roll-Type, .213(r) | 300 |
|     Hand-Fed Crosscut Table Saws, .213(d) | 298 |
|     Hand-Fed Ripsaws, .213(c) | 298 |
|     Inspection, .213(s) | 300 |
|     Jointers, .213(j) | 298 |
|     Maintenance, .213(s) | 300 |
|     Matching Machines, .213(n) | 299 |
|     Molding Machines, .213(n) | 299 |
|     Mortising Machines, .213(l) | 299 |
|     Planing Machines, .213(n) | 299 |
|     Profile Lathes, .213(o) | 299 |
|     Radial Saws, .213(h) | 298 |
|     Ripsaws, .213(c) | 298 |
|     Sanding Machines, .213(p) | 299 |
|     Self-Fed Circular Saws, .213(f) | 298 |
|     Sticking, .213(n) | 299 |
|     Swing Cutoff Saws, .213(g) | 298 |
|     Swing Head Lathes, .213(o) | 299 |
|     Table Saws, .213(d) | 298 |
|     Tenoning Machines, .213(k) | 299 |
|     Turning Machines, .213(o) | 299 |
|     Veneer Cutters, .213(q) | 299 |
|     Wood Heel Turning Machines, .213(o) | 299 |
|     Wood Shapers, .213(m) | 299 |
| Woodworking Tools, Portable Powered, .243(a) | 330 |
|     Belt Sanding Machines, .243(a)(3) | 330 |
|     Circular Saws, .243(a)(1) | 330 |
|     Cracked Saws, .243(a)(4) | 330 |
|     Dead-Man Controls, .243(a)(2) | 330 |
|     Grounding, .243(a)(5) | 330 |
|     Sanding Machines, .243(a)(3) | 330 |
| Work Platforms, .66, .67 | 41, 58 |
|     Elevating and Rotating, .67 | 58 |
|     Powered Platforms, .66 | 41 |
|     Vehicle-Mounted, .67 | 58 |
|         Application, .67(b)(1) | 58 |
|         Design, .67(b)(2) | 58 |
| Work Platforms, Mobile: | |
| (see also Scaffolds), .29(e) | 36 |
| Working Surfaces, Other: | |
| (see also Walking-Working Surfaces), .30 | 36 |
| X-ray Inspections, Mechanical Piping Systems, .252(d)(2)(ii) | 338 |
| Zinc, .252(c)(6) | 338 |
|     Confined Spaces, .252(c)(6)(i) | 338 |
|     Indoors, .252(c)(6)(ii) | 338 |

## NOTES

# NOTES

# Interpretation Letters

### §§1910.219(c)(2)(i); 1910.219(c)(5)

**01/09/2006 - OSHA's machine guarding requirements provide no exemption from guarding based on shaft size or speed.**

January 9, 2006

Mr. John Tignor
Safety & Ergonomics Director
Cargill, Inc.
East Highway 154
Dodge City, KS 67801

Dear Mr. Tignor:

Thank you for your February 9, 2005, letter to the Occupational Safety and Health Administration's (OSHA's) Directorate of Enforcement Programs (DEP). This letter constitutes OSHA's interpretation only of the requirements discussed and may not be applicable to any questions not delineated within your original correspondence. You are requesting an interpretation of the OSHA standard 29 CFR 1910.219, and whether or not rotating round shafts must be guarded. Your paraphrased scenario, question, and our response are provided below.

**Scenario:** The exposed round shafts are completely smooth and have no bolts, holes, or keyways. The shaft's diameters that can be used are 3-inch, 4-inch and 4 5/16-inches. The length of the exposed shaft is approximately 7 inches. The shafts revolutions range from 17 to 54 per minute (rpm). The measurement from the outside of the motor mounts to the outside of the shaft varies from 8 inches to 13 inches depending on the equipment.

**Question:** Would OSHA provide an interpretation of 29 CFR 1910.219 as to whether these shafts must be guarded, because the manufacturer believes that the American National Standards Institute's (ANSI's) guarding requirements refer to rectangular shafts only?

**Response:** From your incoming letter, we were not able to determine the installation heights of the shafts. As you may know, §1910.219, *Mechanical Power-Transmission Apparatus*, does not reference square or rectangular shafts, nor does it exempt round shafts from its guarding requirements. Also, OSHA's standard provides no exemption for shaft size or speed (rpm).

OSHA standard, §1910.219(c)(2)(i) states, "All exposed parts of horizontal shafting seven (7) feet or less from floor or working platform, excepting runways used exclusively for oiling, or running adjustments, shall be protected by a stationary casing enclosing shafting completely or by a trough enclosing sides and top or sides and bottom of shafting as location requires."

However, 29 CFR 1910.219(c)(5), provides guarding exemptions for all mechanical power transmission apparatus located in basements, towers, and rooms used exclusively for power transmission equipment.

In addition, if the housing/mount is designed or constructed in such a manner as to prevent employees from having any part of their body, clothing, jewelry, or the like (such as neckties, scarves, and necklaces) contact the shaft, then it would be considered as guarded by location and would be compliant with the provisions of §1910.219(c)(2)(i). Otherwise, all rotating shafts seven feet or less from a floor or work platform must be completely enclosed as required by the standard.

Please note that ANSI's standards are promulgated for voluntary use, whereas OSHA's standards have enforceable mandatory provisions. Further, OSHA does not interpret consensus standards. For an official interpretation of American Society of Mechanical Engineers' standard, you may contact ANSI/ASME at:

The American Society of Mechanical Engineers
Secretary of the B15 Committee
Three Park Avenue
New York, NY 10016-5990
Phone: (212) 642-4900

Thank you for your interest in occupational safety and health. OSHA requirements are set by statute, standards, and regulations. Our interpretation letters explain these requirements and how they apply to particular circumstances, but they cannot create additional employer obligations. This letter constitutes OSHA's interpretation of the requirements discussed. Note that our enforcement guidance may be affected by changes to OSHA rules. Also, from time to time we update our guidance in response to new information. To keep apprised of such developments, you can consult OSHA's website at http://www.osha.gov. If you have any further questions, please contact the Office of General Industry Enforcement at 202-693-1850.

Sincerely,

Richard E. Fairfax, Director
Directorate of Enforcement Programs

### §§1910.1030; 1910.1030(c)(1)(iv); 1910.1030(f)(3); 1910.1030(h)(5)

**01/12/2006 - Employer's obligation to assure the accuracy of the sharps injury log.**

January 12, 2006

Mr. William A. Hyman
Professor
Texas A&M University
Department of Environmental Engineering
233 Zachry Engineering Center
3120 TAMU
College Station, TX 77843-3120

Dear Mr. Hyman:

Thank you for your letter to the Occupational Safety and Health Administration's (OSHA's) Directorate of Enforcement Programs (DEP). Your letter requests OSHA's interpretation of several provisions under 29 CFR 1910.1030, the Bloodborne Pathogens standard. This letter constitutes OSHA's interpretation only of the requirements discussed and may not be applicable to any question not delineated within your original correspondence. For clarification, your specific questions are paraphrased below, followed by OSHA's response.

**Question 1:** Is there an obligation for employers to assure the accuracy and meaningfulness of the information that is reported on a facility's sharps injury log?

**Response 1:** The purpose of the log, which is required by paragraph (h)(5) of the standard, is to aid in the evaluation of devices being used in the workplace and to quickly identify problem areas in the facility. The standard implicitly requires that the information be accurate. Furthermore, any person who makes any false statement in any document required to be maintained under the Occupational Safety and Health Act of 1970 may be punished by fine or imprisonment [29 U.S.C. §666(g)].

Employers must include the type and brand of device involved in the incident, the department or work area where the exposure incident occurred and an explanation of how the incident occurred so that the intended evaluation of risk and device effectiveness can be accomplished. Complete information regarding the type and brand of the device is crucial; this is especially true for the proper documentation of device evaluations required by paragraph (c)(1)(iv) of the standard. Entries on the sharps log need to be complete enough for evaluators to determine accurately which particular product was actually being used when the incidents occurred. Because manufacturers frequently have several product lines, the log entries for "brand" must include, in addition to the manufacturer's name, the specific product in order to accurately describe which particular device was being used. In fact, in many cases, if only the manufacturer's name were entered on the log, it would be difficult to know exactly which device was being used. As stated in OSHA's compliance directive, *Enforcement Procedures for the Occupational Exposure to Bloodborne Pathogens*, CPL 02-02-069, XIII.H.3, employers are encouraged to write as much information as is necessary to aid in the evaluation of problem areas; however, the confidentiality of the injured employee must be maintained throughout the process. Employers may include information in addition to that which is required by the standard, but may not include information that would identify the employee involved in the incident. However, for incidents, such as a housekeeper being stuck through a trash bag, where determining the type and brand of the device would increase the potential for additional exposure, the type/brand may be recorded as "Unknown" [CPL 02-02-069, XIII.H.3].

**Question 2:** Is the employer required to include an analysis of the data contained on the "sharps injury log" as part of the annual review of the facility's Exposure Control Plan? Should this information be presented so as to indicate whether one or more specific area(s) or issue(s) require action in order to reduce the incidence of sharps injury?

**Response 2:** Although an analysis of the log required by (h)(5) is not mandated, a review of the sharps log can be very instrumental in identifying the causes of exposure incidents and serve as a valuable tool for reducing future sharps injuries. It may also indicate where further training is needed and identify ineffective devices which may need replacement. As stated in the response to Question #1, the purpose of the sharps injury log is to aid in the evaluation of devices and in the identification of problem areas in the facility. Therefore, OSHA recommends that the log be reviewed periodically and as part of the annual review and update of the exposure control plan. As you know, employers are required to review and update the exposure control plan at least annually **and** whenever necessary to reflect new or modified tasks and procedures which affect occupational exposure [29 CFR 1910.1030(c)(1)(iv)]. Employers are additionally required to annually document their consideration and implementation of appropriate commercially available and effective safer medical devices [29 CFR 1910.1030(c)(1)(iv)(B)]. They are also required to document exposure incidents [29 CFR 1910.1030(f)(3)(i) and (ii)].

**Question 3:** Employers are required in their annual review of the exposure control plan to reflect changes in technology and document annual consideration and implementation of commercially available and effective safer medical devices. Should this review include reconsideration of technologies not currently selected even if they are not new, especially if there are indications from the analysis of the sharps injury log that currently used products may not be sufficiently effective?

**Response 3:** There is no requirement under (c)(1)(iv) to reconsider already considered medical devices. Please keep in mind that it is common for facilities to observe a slight initial increase in injuries after the selection and implementation of a new device. Additional training is often needed when staff attempts to become familiar with a newly selected device. Any analysis of the sharps injury log should take these factors into consideration in arriving at a conclusion about the effectiveness of a newer device. However, if an analysis of the sharps injury log identifies a particular selected engineering device as the root cause of injuries (i.e., as opposed to work practices, employee training, or other issues), then the employer may choose to include one or more of the previously evaluated devices in the annual re-evaluation to determine if a better selection can be made.

**Question 4:** Employers are required to solicit input from non-managerial employees responsible for direct patient care. Is a simple open request for input adequate, even if none is received or does this require a serious study of user observations, opinions and testing of currently used and alternative devices?

**Response 4:** A simple open request for input is adequate. The employer must solicit employee input in a manner appropriate to the circumstances in the workplace. Methods for soliciting employee input may include joint labor-management safety committees; involvement in informal problem-solving groups; participation in safety meetings and audits, employee surveys, worksite inspections, or exposure incident investigations; using a suggestion box or other effective methods for obtaining written employee comments; and participation in the evaluation of devices through pilot testing. The opportunities for employee input shall be effectively communicated to employees [CPL 02-02-069, XIII.C.6]. The failure of employees to provide input is not in itself a violation of the standard. No studies are required under this provision.

**Question 5:** Does the maintenance of the sharps injury log relieve a hospital of the requirement to report sharps injuries under the FDA's Medical Device Reporting requirements, and in accordance with the Food and Drug Administration's (FDA's) *Needlesticks: Medical Device Reporting Guidance for User Facilities, Manufacturers, and Importers?*

**Response 5:** No, the requirement under OSHA's bloodborne pathogens standard, 29 CFR 1910.1030(h)(5)(i), does not exempt healthcare employers from having to report deaths and serious injuries related to medical devices to the FDA. The FDA's Medical Device Reporting (MDR) regulation, 21 CFR Part 803, is a separate and independent set of requirements under which facilities using medical devices must report such information to the FDA and/or the device manufacturers. The FDA has specific forms that must be completed and submitted for incidental and annual reporting. You may wish to contact the FDA directly for specific guidance on the information that must be included on the MDR forms.

Thank you for your interest in occupational safety and health. We hope you find this information helpful. OSHA requirements are set by statute, standards, and regulations. Our interpretation letters explain these requirements and how they apply to particular circumstances, but they cannot create additional employer obligations. This letter constitutes OSHA's interpretation of the requirements discussed. Note that our enforcement guidance may be affected by changes to OSHA rules. Also, from time to time we update our guidance in response to new information. To keep apprised of such developments, you can consult OSHA's website at www.osha.gov. If you have any further questions, please feel free to contact the Office of Health Enforcement at 202-693-2190.

Sincerely,

Richard E. Fairfax, Director
Directorate of Enforcement Programs

## §§1910.147; 1910.147(a)(2)(ii); 1910.212

**01/18/2006 - Interlock requirements for access door(s) of trash compactor.**

January 18, 2006
Mr. Dennis O'Hagan
Columbia University
B-230 East Campus
Mail Code 3405
410 West 118th Street
New York, NY 10027

Dear Mr. O'Hagan:

Thank you for your January 21, 2005 letter to the Occupational Safety and Health Administration's (OSHA's) Directorate of Enforcement Programs. This letter constitutes OSHA's interpretation only of the requirements discussed and may not be applicable to any questions not delineated within your original correspondence. You had specific questions regarding the use of interlocks on a trash compactor. We apologize for the delay in responding to your letter.

We have restated your questions below.

**Question #1:** Is an interlock required on the side door of a trash compactor, if the side access door is located 55 inches from ground level, and the door is closed and latched while the compactor is cycling?

**Response:** If an employee may be exposed to machine hazards while the machine operates, the hazardous areas of the machine must be guarded. The Occupational Safety and Health Review Commission has found employee exposure where it is reasonably predictable, either by operational necessity or otherwise (including inadvertence), that employees have been, are, or will be in the machine's zone of danger. While it is not possible to conclusively determine the answer to your question without observing the trash compactor and considering the manner in which it is situated and used, it is likely that employees would be exposed to moving parts associated with the trash compactor during the course of their employment. As such, OSHA's machine guarding standard (29 CFR §1910.212) likely would require guarding to prevent employee exposure to the hazards.

While properly designed, constructed, and maintained interlocks generally provide effective machine guarding, §1910.212 does not require the use of interlocks or any other specific guarding methods. The standard describes several acceptable methods for guarding machine hazards, but the standard is performance-oriented, and an employer can comply with the standard by using any method(s) of guarding that provides effective employee protection against the hazard. In the scenario that you described, access to the trash compactor's point-of-operation is covered by a latched, solid door. If the door effectively prevents employee access to the point-of-operation or other machine hazards, the door would be considered effective guarding. **However**, since barrier guards must be affixed such that they cannot be easily defeated, simply latching the door in a closed position would not be sufficient. Thus, the door would need to be secured such that it could not be easily circumvented (e.g., by use of means such as fasteners not readily removable by the operator, a lock, etc.) or, if access is desired using only a latch, an interlock or other guarding method must be used to prevent employee exposure to the trash compactor's point-of-operation.

Please be aware that, should employees be exposed to hazardous energy during servicing or maintenance activities associated with the trash compactor, which require the side access door to be opened, employees can access the danger zone only pursuant to a lockout/tagout program that is compliant with §1910.147. Pursuant to that standard, certain "minor servicing activities" may be performed using alternative guarding measures, including barrier guards with properly designed interlocks, which provide effective employee protection. See 29 CFR §1910.147(a)(2)(ii) (Note).

**Question #2:** Does federal OSHA have an interpretation of ANSI Z245.2?

**Response:** OSHA does not issue citations for violations of ANSI standards that have not been adopted as (or incorporated into) OSHA standards, but ANSI or other industry standards are sometimes referred to in citations and used as evidence of a recognized hazard or a feasible means of abatement. Employers should contact the appropriate ANSI committee for official interpretations of ANSI standards.

While not interpreting ANSI Z245.2 (2004), *American National Standard for Equipment Technology and Operations for Wastes and Recyclable Materials — Stationary Compactors — Safety Requirements for Installation, Maintenance and Operation*, we note that this ANSI standard requires that "[a]ccess door(s) of loading chambers [on trash compactors] shall have an interlock system that prevents cycling motion while the access door(s) is open." Section 5.8.1. We are aware of no provision in ANSI Z245.2 (2004) that would permit a side access door to be accessible to employees without the incorporation of an interlock system merely because the side access door was 55 inches above ground level. Rather, the ANSI standard provides that the access door must be interlocked unless "[s]ustained-manual-pressure controls [are] located in such a way that the operator cannot reach the point-of-operation and [are] within an area fully visible to the operator of the complete point-of-operation area when the guard is not used." Section 5.9.1(c). Your letter provides no indication that the compactor at your facility has such controls. We also recognize that the ANSI standard permits an opening into a compactor if the length of a fixed guard (or loading hopper) in conjunction with the height of the opening from the ground exceeds 84 inches, so as to "prevent any persons from contacting moving parts at the point-of-operation." Section 5.9.1(a). However, this recognized guarding method is not applicable to the present situation, in which the compactor has a side access door that is only 55 inches from the ground and lacks a fixed guard or loading hopper designed to prevent access to the compactor's moving parts.

**Question #3:** Does federal OSHA recognize a state Occupational Safety and Health Division interpretation letter as proper guidance and interpretation for regions outside of that state?

**Response:** Federal OSHA enforces federal standards in states not covered by state OSHA plans. When a state official interprets a state standard promulgated pursuant to a state occupational safety and health statute, the interpretation definitively addresses only the application of the relevant state standard.[1] Employers subject to federal OSHA authority should seek an interpretation from federal OSHA regarding the application of federal OSHA standards promulgated pursuant to the Occupational Safety and Health Act.

Thank you for your interest in occupational safety and health. We hope you find this information helpful. OSHA requirements are set by statute, standards, and regulations. Our interpretation letters explain these requirements and how they apply to particular circumstances, but they cannot create additional employer obligations. This letter constitutes OSHA's interpretation of the requirements discussed. Note that our enforcement guidance may be affected by changes to OSHA rules. Also, from time to time we update our guidance in response to new information. To keep apprised of such developments, you can consult OSHA's website at http://www.osha.gov. If you have any further questions, please feel free to contact the Office of General Industry Enforcement at (202) 693-1850.

Sincerely,

Richard E. Fairfax, Director
Directorate of Enforcement Programs

[1] We note, however, that the opinion offered in the Oregon Occupational Safety and Health Division (OSHD) letter that you referenced (July 8, 1997 from Marilyn K. Schuster to Dale Anderson) is consistent with the provisions of the ANSI Z245.2 standard, which are discussed in response to your second question (above). The Oregon OSHD letter addresses a loading chute that is 42 inches above the floor and is further guarded by a 4-foot, fixed loading hopper that extends out from the compactor at that height. Thus, this would meet the 84-inch total protective distance that is contemplated in Section 5.9.1 of the ANSI standard.

### §§1910.120; 1910.120(a)(2)(iii); 1910.120(p)(8)
**02/01/2006 - Applicability of HAZWOPER standard to hazardous waste generators that are not conditionally exempt as small quantity generators.**

February 1, 2006

Mr. Eric Clark
Manager, Government Relations
Synthetic Organic Chemical Manufacturers Association (SOCMA)
1850 M Street NW, Suite 700
Washington DC, 20036-5810

Dear Mr. Clark:

Thank you for your October 27, 2005 letter to the Occupational Safety and Health Administration's (OSHA's) Directorate of Enforcement Programs. You have a question regarding the applicability of the Hazardous Waste Operations and Emergency Response (HAZWOPER) standard to hazardous waste generators. This letter constitutes OSHA's interpretation only of the requirements discussed and may not be applicable to any question not delineated within your original correspondence. We apologize for the delay in our response. The scenario, your paraphrased question, and OSHA's response are below.

**Scenario:** In your letter to OSHA, you raise several issues that lead to a question regarding the applicability of paragraph (p)(8) of HAZWOPER to hazardous waste generators that are not conditionally exempt as small quantity generators. Specifically, your concern involves the following statement that comes from the *Notes and Exceptions* under 1910.120(a)(2)(iii): "Excepted employers who are required by the EPA or state agency to have their employees engage in emergency response or who direct their employees to engage in emergency response are covered by paragraph (p)(8) of this section, and cannot be exempted by (p)(8)(i) of this section." You state that the language of the rule, available guidance, and interpretation letters make it clear that HAZWOPER does not apply when (1) employees are not exposed or potentially exposed to health or safety hazards, or (2) the only releases of hazardous substances are incidental in nature. You point out, however, that two interpretation letters (January 9, 1990, addressed to Mr. John D'Aloia, Jr., and September 22, 1992, addressed to Mr. Thomas A. Wehrenberg) appear to blur these exemptions because they are silent on these two conditions and therefore make it appear that all generators, other than conditionally exempt small quantity generators, are covered by (p)(8) where the EPA requires some level of involvement in emergency response. Finally, you state that the confusion stems from the fact that the EPA does not specifically define "emergency response" and the performance of some EPA-required duties (40 CFR 265.55 & 56) would not result in employee exposure or the spill could be classified as an incidental release.

**Question:** Does (p)(8) of HAZWOPER apply to hazardous waste generators, other than conditionally exempt small quantity generators, if an assessment of reasonably anticipated emergency scenarios indicates there are no hazards to employees or the only potential spill scenarios are incidental in nature?

**Response:** OSHA agrees with the premise that (p)(8) of HAZWOPER would not apply to hazardous waste generators if the employer can demonstrate that the operation does not involve any potential employee exposures or the reasonable possibility of employee exposure to safety or health hazards, as is stated in 1910.120(a)(1). Similarly, HAZWOPER would not apply to small quantity incidental releases, as is discussed in Appendix E to OSHA Directive CPL 02-02-059. Employers making such a determination must use careful judgment and consider all the applicable factors involved such as the inherent properties of the hazardous substances themselves (e.g., toxicity, volatility, flammability, explosiveness, corrosiveness, etc.), the particular circumstances of the release itself (e.g., quantity spilled, permit-required confined spaces, etc.), employee knowledge of the hazardous substance in the immediate release area, personal protective equipment at hand, pre-established standard operating procedures, and other relative factors. If HAZWOPER does not apply, employers must still protect employees through other applicable OSHA standards such as 29 CFR 1910.1200 (Hazard Communication) and 29 CFR 1910.132 (Personal Protective Equipment).

In your letter to OSHA, you stated that you feel the confusion can be resolved by annotating or amending the January 9, 1990, and September 22, 1992, interpretation letters with the qualification that when an assessment of reasonably anticipated emergency scenarios indicates there are no hazards to employees or when the only potential spill scenarios are incidental releases, employers who operate hazardous waste generators exempted under 40 CFR 262.34 can be excluded from coverage under HAZWOPER. Although we generally concur with your understanding of the applicability of HAZWOPER to hazardous waste generators, OSHA cannot alter text in existing interpretation letters. To provide additional clarity, however, a reference to this response will be added to the two letters indicated above.

Thank you for your interest in occupational safety and health. We hope this provides the clarification you were seeking. OSHA requirements are set by statute, standards, and regulations. Our interpretation letters explain these requirements and how they apply to particular circumstances, but they cannot create additional employer obligations. This letter constitutes OSHA's interpretation of the requirements discussed. Note that our enforcement guidance may be affected by changes to OSHA rules. Also, from time to time we update our guidance in response to new information. To keep apprised of such developments, you can consult OSHA's website at www.osha.gov. If you have any further questions, please feel free to contact the Office of Health Enforcement at 202-693-2190.

Sincerely,

Richard E. Fairfax, Director
Directorate of Enforcement Programs

## §§1910.156; 1910.156(a)

### 02/06/2006 - Applicability of the Fire brigade standard to health care facilities.

February 6, 2006

Linda Chaff
Chaff & Co.
The Concierge Level
600 Republic Centre
Chattanooga, TN 37450

Dear Ms. Chaff:

Thank you for your December 15, 2005 letter to the Occupational Safety and Health Administration's (OSHA's) Directorate of Enforcement Programs (DEP). This letter constitutes OSHA's interpretation only of the requirements discussed and may not be applicable to any questions not delineated within your original correspondence. You have a question regarding OSHA's interpretation of 29 CFR 1910.156.

**Question:** Must health care facilities with groups designated as *Fire Brigades or Fire Response Teams*, whose functions may require the use of fire extinguishers and standpipe hose systems to control and/or extinguish small fires, evacuation of patients, and actions involving flammable liquids and gases, toxic chemicals, and radioactive sources comply with 29 CFR 1910.156?

**Reply:** In accordance with 29 CR 1910.156(a), the Fire brigades standard contains requirements for the organization, training, and personal protective equipment (PPE) of fire brigades whenever they are established by an employer. Therefore, if an employer of a health care facility designates a group of employees to perform the tasks that you have described in your question, the employer must comply with all requirements set forth in 29 CFR 1910.156.

Thank you for your interest in occupational safety and health. We hope you find this information helpful. OSHA requirements are set by statute, standards and regulations. Our interpretation letters explain these requirements and how they apply to particular circumstances, but they cannot create additional employer obligations. This letter constitutes OSHA's interpretation of the requirements discussed. Note that our enforcement guidance may be affected by changes to OSHA rules. Also, from time to time we update our guidance in response to new information. To keep apprised of such developments, you can consult OSHA's website at www.osha.gov.

If you have any further questions, please feel free to contact the Office of General Industry Enforcement at (202) 693-1850.

Sincerely,

Richard E. Fairfax, Director
Directorate of Enforcement Programs

## §§1910.95; 1910.95(g)(9)

### 02/10/2006 - Clarification on revising a baseline audiogram if a standard threshold shift occurs in only one ear.

February 10, 2006

Dr. Robert Williams
Director Audiology
TK Group Incorporated
Stewart Square, Suite 250
308 West State Street
Rockford, IL 61101

Dear Dr. Williams:

Thank you for your August 3, 2005 letter to the Occupational Safety and Health Administration (OSHA) regarding the requirements for revising baseline audiograms under the Occupational Noise standard, 29 CFR 1910.95. In particular, you are requesting further clarification of our May 8, 2003 letter to Ms. Linda Ballas. This letter constitutes OSHA's interpretation only of the requirements discussed and may not be applicable to any question not delineated within your original correspondence.

In the May 8th 2003 letter, we stated a baseline may only be revised if there is a standard threshold shift of greater than 10 dB. If a shift occurs in only one ear, then only the baseline for that ear would be permitted to be revised under the standard. In your letter you state that many professionals had previously adopted a methodology of revising the baseline for both ears, even when there was only a shift in one ear. This confusion may have been partially due to the available commercial software that was available to track hearing loss which offered that methodology as a configurable option.

Since the May 8th 2003 letter, many of the computer programs that track hearing loss have been corrected to properly record shifts that occur in only one ear. Professionals who are correcting their databases may find it useful to completely rebuild them in order to reflect accurate historical tracking of standard threshold shift events. Some computer programs may even do this automatically, when they are updated to the single ear criteria. However, OSHA will usually look at whether the employer is currently out of compliance and whether the employer is evaluating what noise sources or work procedures may be currently contributing to employee hearing loss. Therefore, the employer would not need to go back further than the previous year's audiogram and ensure all future audiograms are revised in the same manner.

Thank you for your interest in occupational safety and health. We hope you find this information helpful. OSHA requirements are set by statute, standards and regulations. Our interpretation letters explain these requirements and how they apply to particular circumstances, but they cannot create additional employer obligations. This letter constitutes OSHA's interpretation of the requirements discussed. Note that our enforcement guidance may be affected by changes to OSHA rules. Also, from time to time we update our guidance in response to new information. To keep apprised of such developments, you can consult OSHA's website at www.osha.gov. If you have any further questions, please feel free to contact the Office of Health Enforcement at (202) 693-2190.

Sincerely,

Richard E. Fairfax, Director
Directorate of Enforcement Programs

### §§1910.23; 1910.23(c)

**02/27/2006 - Alternative fall protection program when guardrails are infeasible to install.**

February 14, 2006
Mr. Todd B. Logsdon
Greenebaum, Doll, & McDonald PLLC
3500 National City Tower
101 South Fifth Street
Louisville, KY 40202-3197

Dear Mr. Logsdon:

Thank you for your March 31, 2005, letter to the Occupational Safety and Health Administration's (OSHA's) Directorate of Enforcement Programs (DEP). This letter constitutes OSHA's interpretation only of the requirements discussed and may not be applicable to any questions not delineated within your original correspondence. You are requesting clarification, for your client, on whether the proposed procedures would meet the intent for an alternative fall protection program when guardrails are infeasible to install. Your paraphrased scenario, question, and our responses are provided below.

**Scenario:** Employees gain access to rooftops via roof hatches, interior stairwells that are located in the center of rooftops, and at the rooftops' edges via exterior fixed ladders, staircases, and portable extension ladders to perform post-construction. The rooftop's surface ranges from 300,000 to 1,000,000 square feet with different configurations and has a 4:12 slope. Employees are never closer than 15 feet to the roof's edge.

**Question:** Would a painted, large red stripe 15 feet away from the roof's edge designating a work zone, plus the following conditions be an acceptable alternative for guardrails?

- Personnel working inside of the marked area (i.e., more than 15 feet away from the roof's edge) would not be required to utilize personal fall protection.
- The work rules would require personnel working less than 15 feet from the roof's edge to employ a full fall protection program.

**Response:** The use of a painted line is not permitted under the standard at 29 CFR 1910.23(c), *Protection of open-sided floors, platforms, and runways*, which requires the use of a guardrail system to protect workers from falls of four feet or more to lower levels. Under OSHA's policy for *de minimis* violations, however, in place of complying with an existing standard an employer may comply with a proposed standard or amendment that clearly provides equal or greater employee protection. OSHA has proposed to revise the existing rules for Working and Walking Surfaces, including the requirement at 1910.23(c). See 55 FR 13369, April 10, 1990; republished, 68 FR 23528, May 2, 2003.

Proposed 1910.28(d)(1) would address "work...of a temporary nature, such as maintenance on roof top equipment." For such work, "[e]mployers may establish designated areas which comply with the provisions of this paragraph as an alternative to installing guardrails, where employers demonstrate that employees within the designated areas are not exposed to fall hazards." Under the proposed standard, however, a painted line cannot be used to demarcate the designated area. A designated area must be surrounded by a rope, wire or chain and supporting stanchions. For your convenience in reviewing the proposed standard in its entirely, we are enclosing a copy of the 2003 *Federal Register* reprint.

If an employer follows the requirements for designated areas in lieu of conventional guardrails, OSHA would consider it to be a *de minimis* violation. *De minimis* violations are violations of standards which have no direct or immediate relationship to safety or health. Under the *de minimis* policy, an employer may comply with a proposed standard or amendment or a consensus standard, rather than with the standard in effect at the time of the inspection, only if the employer's action clearly provides equal or greater employee protection or the employer complies with a written interpretation issued by the OSHA Regional or National Office. With *de minimis* violations, OSHA does not issue citations, or require abatement.

As you may know, the State of Kentucky administers its own occupational safety and health program under a plan approved and monitored by Federal OSHA. States are required to have regulations that are, at least as effective as the federal standards, although they may be more stringent. In order to obtain Kentucky's policy on this issue, you may direct your inquiry to:

Philip J. Anderson, Commissioner
Kentucky Department of Labor
1047 U. S. Highway 127 South, Suite 4
Frankfort, KY 40601

Telephone: (502) 564-3070

Thank you for your interest in occupational safety and health. We hope you find this information helpful. OSHA requirements are set by statute, standards and regulations. Our interpretation letters explain these requirements and how they apply to particular circumstances, but they cannot create additional employer obligations. This letter constitutes OSHA's interpretation of the requirements discussed. Note that our enforcement guidance may be affected by changes to OSHA rules. Also, from time to time we update our guidance in response to new information. To keep apprised of such developments, you can consult OSHA's website at www.osha.gov. If you have any further questions, please feel free to contact the Office of General Industry Enforcement at (202) 693-1850.

Sincerely,

Richard E. Fairfax, Director
Directorate of Enforcement Programs

Enclosures